ISBN 978-1-5277-7945-7
PIBN 10893682

FB &c Ltd, Dalton House, 60 Windsor Avenue, London, SW19 2RR.
Company number 08720141. Registered in England and Wales.

For support please visit www.forgottenbooks.com

INDEX

A

B

C

D

E

F

Page

G

H

I

J

K

L

M

N

O

P

U

V

W

EDITORIAL

T. A. RICKARD, Editor

ATTENTION is directed to the sub-heading on our cover-page. Some readers may not have noticed it. We are glad to be able to assert that this is "an independent paper, owned, edited, and managed by engineers."

OUR British friend *The Mining Magazine* informs us that the Board of Inland Revenue has been induced to regard a mine as a wasting asset, and in accordance with this decision some allowance for amortization of capital will be made in connection with the assessment for the excess-profit tax. Thus the British tax-collector has learned a fundamental truth. War is an effective teacher of economics.

NITRIC ACID to the extent of 7000 tons per annum will be manufactured at a new plant being erected at Trollhättan, Sweden. The Birkeland-Eyde process will be employed. The new company is being financed in part by Germans, and the output, naturally, will be utilized in the manufacture of explosives, some of which will likely be directed to the destruction of Americans. This brings home to us anew the impossible position of a nation pretending neutrality in a world-racking war.

FROM New York it is reported that the Government has placed an order for 60,000,000 pounds of copper at 25 cents per pound. Meanwhile arrangements are being made for a much larger contract at a price not yet determined. There is talk of higher market quotations owing to the curtailment of output at the mines due to labor troubles, but any rise at this time would only give aid to the labor agitator and draw attention to the artificial methods by which the high price of copper is maintained.

NEWS comes from London that the Giant mine, in Rhodesia, has been closed because "there is no hope of running the property at a profit." This sepulchral announcement may be compared with the recklessly optimistic statements that sent the shares of the Giant to $32.50 in 1909, giving the mine a valuation at that time of $8,500,000. Only $5 per share has been paid in dividends since then. This was one of the mines that brought discredit not so much on Rhodesia as upon Rhodesian finance, and more particularly on the stock-jobbing methods of the Consolidated Gold Fields of South Africa.

TUNNEL-DRIVING at Pine Mountain is the subject of a highly informing article in this issue, written by Mr. [illegible] Devereux. It will interest every engineer to follow the close analysis of detail given to show in what manner, during this period of high costs, Mr. M. M. O'Shaughnessy, the engineer in charge, succeeded in driving and lining an 8700-ft. tunnel, with a section of 8 by 8 feet inside the timbers, at the average rate of 19 feet per diem and at a total cost per foot of $16.12. Interesting comparisons with adits driven by the Alaska-Gastineau Mining Company, the Arizona Copper Company, and others form a measure of the successful work accomplished at Pine Mountain.

FERRO-MANGANESE requirements in this country for the remainder of the current year have been subjected to critical analysis by *The Iron Age,* which shows, first, that a steel output in 1917 of 45,000,000 tons almost certainly will be realized. This would be an increase of 45% over the production in 1913. Of that total, 40,320,000 tons would require ferro-manganese at the average rate of 17 pounds per ton of steel, indicating 306,000 long tons of ferro-manganese required. To meet that demand the calculated output of our own furnaces is 240,000 tons, and the anticipated imports will add 70,000, making a total of 310,000 tons. Any decrease in the productivity of foreign mines would easily produce a shortage. It is important to investigate further the possibilities of developing domestic deposits, in conjunction with concentration to meet the market standard.

CIRCULARS have been issued from the head office of the American Institute of Mining Engineers for the purpose of eliciting opinions in regard to a new definition of eligibility for membership. We note, with pleasure, that no attempt is to be made to disturb the status of those already members. Some years ago sundry reactionary members suggested the idea of creating a special class of superior persons to be called 'Fellows', who were to be placed on a plane above the run-of-mine 'Members'. That would, we hope, have provoked a riot, for once a man is a full member of a society he has a right to object to any later sorting and stratification. The plan now proposed is something quite different; it is to render the election of members more selective and so automatically to improve the average of the whole membership, making the Institute more truly representative of the profession.

RECENTLY one of our readers wrote to protest, vigorously, against the use of pounds, shillings, and pence, instead of American currency, in an article on mining in South Africa. It annoyed him. We replied that the rate of exchange between pounds and dollars had ranged from $7 to $4.60 during the last three years and that it might vary considerably between the date of writing an article and the date of publication. The same holds true for Mexican economics; who shall state the equivalent—in cents—of the peso at a given moment? Uniformity is admirable, but it may be the cause of confusion. The article to which our protestant referred was written by an American mining engineer, Mr. H. Foster Bain, and it was intended for the reading, among others, of American engineers on the Rand. They would have been annoyed, at least as much as our kicker, if we had given them our criticisms, through Mr. Bain, in terms of dollars and cents. Gentlemen, think of the other fellow sometimes.

WE publish a short statement issued by the National Research Council concerning submarines and their capabilities. This summary of facts is intended to guide inventors and engineers now at work in devising a means to overcome the assassin of the seas. Most of the ideas transmitted to Washington emanate from fanatics. Mr. Lawrence Addicks confessed recently that the Naval Consulting Board, of which he is a member, is "overwhelmed with people who have fantastic ideas of hanging nets and magnets on ships to stop torpedoes." We feel confident that an invitation for suggestions from our engineering friends will not add to the burdens of those at Washington, but that any ideas presented by them will be sane and scientific. The submarine is still a menace. We must find some way of circumventing it. Vice-Admiral Sims, now in command of our destroyer flotilla on the other side, has laid emphasis on the fact that the submarine must come to the surface at intervals to breathe, as it were, or, more technically, to recharge its batteries. That requires from three to five hours. It is then that the patrol-boats have a chance to hunt their quarry. The nets used to entrap a submarine when moving under water are made of steel wire a quarter of an inch thick in meshes 12 feet wide. Already the Germans have invented a tool to cut such nets; therefore it remains for us to devise a kind of net that they cannot cut in the limited time available. Mr. Arthur H. Pollen, a British naval expert of high standing, advocates an active search for submarine bases and the destroying of them. Unfortunately the Germans use neutral coasts, such as those of Spain and Norway, as a rendezvous, and until these neutrals join us it will be impossible to carry out such a policy with any degree of completeness. Rear-Admiral Goodrich, writing in *The Nation,* advocates this same idea of tracking the enemy to his lair. The stretching of a steel net from Scotland to Norway, across 240 miles of shallow water, has been suggested as a costly but feasible means of defence, the idea being based on the successful use of similar netting for protecting traffic from England to France. Such fine and large schemes are less likely to prove effective than the research now being made into means for detecting the submarine by sound or exposing it to attack by light. Once the secrecy of the submarine's movements has been pierced its power for harm will be enormously decreased. The only idea that ever came to us on the subject was to use oil to cover the surface of the sea and thereby obscure vision through the periscope, but we decided that we would not bother our friends on the Naval Consulting Board with such a proposal, because the ocean is wide and oil is costly. At one time the aeroplane, more particularly the hydroplane, was expected to ferret the marine vermin, but apparently this method of hunting has proved disappointing. It is too much like looking for a needle in a haystack. A vigilant patrol by small boats supplemented by destroyers appears to be the one sure method. It is a great and honorable sport in which thousands of gallant men are now engaged. Meanwhile it is incumbent upon American ingenuity to devise something more directly effective.

Labor Unrest

Work is being stopped at the copper mines of Montana and Arizona. At Butte, a strike, started by the Metal Miners Union, has spread among the electrical workers, crippling the operation of the mines to a point where the entire local industry is paralyzed. In Arizona the strike at Jerome had no sooner been adjusted than demands were made upon the companies operating in the Clifton-Morenci district, where a bitter strike was ended only a year ago. Serious trouble has developed at Globe, Miami, and Bisbee. Apparently the labor population of a number of important mining districts is seething with unrest, and unless some measure of patriotic spirit is shown on both sides the supply of copper will be seriously diminished. We have received copies of the bulletins issued by the strikers at Butte. They are written intelligently and forcibly. We have also received a copy of the *Arizona Labor Journal* containing a statement by Mr. J. L. Donnelly, president of the Arizona State Federation of Labor. This also is a clever brief for the workers. The demand for higher wages, despite the drop in the price of copper, is based upon the increased cost of living. For instance, Mr. Donnelly claims that "$3.50 per day eighteen months ago would purchase the equivalent of $5.50 today." Even if this be a slight exaggeration, it is more than likely that the taste for luxury excited by the sudden rise in wages has caused the miner to be so reckless in his purchasing that today he is saving less than he did eighteen months ago. Our own recent observations of life in several mining districts would tend to confirm this inference. The copper-mining companies made a great blunder, of course, in not paying the bonus separately from the wages; they should have given the regular pay in one check and the bonus, based on the rise in the price of copper, in another check. That would have reminded the men of the temporary nature of the big advance in their income. But, say the leaders of organized labor, the increased pay, even with the bonus, is not enough, now that the cost of living has advanced faster that the bonus dependent upon the price of copper. Copper may have fallen in price, they say, but it is still selling at a figure more than twice the cost of producing it; the companies are making enormous profits; we are not profiting at all, owing to the high cost of living; so we demand another increase of wages. Not content with that the Arizonan unions insist that the contract system shall be abandoned, and they even murmur something about a 6-hour day. Here bad faith becomes manifest; for these further demands have nothing to do with the high cost of living and are merely clubs shaken at the managers with the idea of intimidating them. At Butte the demand for higher wages, based on vital conditions, is tied to a protest against the 'rustling card,' meaning the exercise of discrimination by the companies in giving employment. Protest is also made against the unsafe conditions underground, this last being prompted by the recent Speculator disaster. If the loss of life was due, as is asserted, to the existence of concrete bulkheads that blocked the necessary exits, then the men have a legitimate grievance not only against the companies but against the State Mine Inspector; but the charge of black-listing means merely that the employer tries to keep out men that are known to be trouble-makers. Of the secondary issues, the contract system is much the most important because it goes to the root of the question whether the employer is to have a fair deal; whether he is to pay for what service he gets, or whether he must distribute uniformly high wages for a varying return in work, giving equal pay to the incompetent and the competent, the lazy and the energetic among his employees. We believe that the contract system is fair to all concerned and the only system that upholds the self-respect of the working-man; we recognize, however, that it is not always fairly applied and that it then becomes vicious. Here we come to the root of the whole matter: fair dealing as between men. We sympathize sincerely with the managers of mines, for they are often placed between the devil of the greedy capitalist and the deep sea of the workman's ignorance; it needs—always—a man of strong character and high principle to safeguard the interests of a company and yet protect the welfare of the men on the pay-roll. The mining profession as a whole is more than willing to be fair to the toilers underground, and its keen interest in the welfare movement is an indication of that fact, so that the use of the high cost of living as an argument for higher pay would be received sympathetically if it did not come on top of a period of extravagance in the mining communities and if it were not made at such a time and in such a manner as to suggest a 'hold-up.' The strike bulletin at Butte exclaims joyously that the companies cannot furnish metal to fulfill their contracts now that so many men are idle. Who will suffer? The companies? No! The Nation; the Great Cause to which we are committed; these are the ones to suffer, and when the public awakens to this fact we expect to see a popular demand for a speedy settlement of this sinister attempt to clog the wheels of industry at a period of great crisis. Thus we started with unconcealed sympathy for the predicament in which economic conditions, due to the War, had placed the miner and his comrades, but we arrive inevitably at the conclusion that sympathy must be withheld if advantage is to be taken of the Nation's necessity in order not only to obtain a reasonable concession but to enforce an unreasonable demand. One word more. We have referred to the propaganda of union labor. It is well prepared and likely to be effective. The labor element is organized and articulate. The employers, namely, the mining companies and the resident managers for the companies, are deficient in these two respects. Only too often petty jealousies interfere with united action and the differences of policy prevent an outspoken statement of purpose. We believe that the companies should take the public into their confidence, for, in the end, public opinion is the arbiter, just as public opinion is the power behind the law. The propaganda of the labor agitators should be met with a frank explanation of the case as it seems to the mine managers, and the explanation ought to be put in printed form so that everybody in the mining community can read it. In some localities the companies control one or more local papers; they should acknowledge the fact frankly so that the control may lose any sinister suggestiveness. Nobody thinks it improper for the union to run a paper. Let both sides come into the open and argue their contentions at the bar of public opinion. This is a democracy. The methods of the Star Chamber and those of the Black Hand are equally repugnant to good citizens. Publicity is the anti-toxin of wrong.

Misgoverned Mexico

Mexico is testing our mettle. The wholesale confiscation of Mexican mines owned by citizens of this country may be regarded as a deliberate challenge. Abandonment of the copper mines and smelters at Cananea has been forced by executive orders with which it was impossible for the owners to comply. At the same time the Department of Finance in Mexico City is proceeding to confiscate 7702 mining claims belonging to the Cananea company on the ground of its refusal to re-pay taxes that had been paid once to other officials, but which the Carranza administration insists must be forthcoming again, with extravagant penalties accrued under arbitrary decrees skilfully contrived to work the ruin of alien investors. The plain logic of the situation is ignored. For years no constitutional government has existed in Mexico. Carranza was not in line of succession to the presidency after the forced resignation of Madero. He had no legal status as a Federal officer under the constitution of 1857. He took up arms ostensibly to re-instate a government that would adhere to the terms of that constitution, and he was entitled to the credit of good intentions while he continued to pursue that purpose. Nevertheless, this did not entitle him to revenue from such Mexican territory as he did not effectively hold and administer. Administration involves protection, and the Cananea company was not protected by Carranza while it was being robbed by Maytorena's looters. Moreover, Maytorena had a sounder title to executive authority in

Sonora than Carranza, because he was actually there, and he was likewise in arms professedly for the re-establishment of a stable government under the old constitution. He had declined to co-operate with the First Chief for reasons that were honorable. In the meantime, the question of the right to rule and to collect duties and taxes was being rudely discussed on the battle-field. It is preposterous for Carranza to claim that he represents a government that has suffered by loss of the revenues wrested by officers in control before he had established his authority. The United States has treaties with the government created under the constitution of 1857, and those treaties concede no right to penalize American property-owners for having paid moneys under duress to revolutionary upstarts. Carranza does not represent the Mexico of the old constitution with which we entered into treaty-relations, for he did not re-establish that constitution; he evaded it while acting as dictator, and he ended by discarding it altogether and framing a new constitution, written with intent to win the adherence of the rabble by means of labor-clauses that express the principles of the I. W. W., while it reserves powers to the executive that make him in effect an autocrat. Such an international situation has been created by these confiscatory acts of the Carranza government as to call for more astute diplomacy on our part than has characterized our previous dealings with revolutionary Mexico.

Not only has Cananea suffered, but the El Tigre silver mine, belonging to the Lucky Tiger-Combination Company, has been obliged to close, as well as the rich Pedrazzini silver property near Arizpe, at which point the manager, an Englishman, has been placed in confinement. It is not the mines alone that have suffered. The important agricultural project in the Yaqui valley, belonging to the Richardson Construction Company, in which Messrs. John Hays Hammond and Harry Payne Whitney are the chief shareholders, has been seized by the Mexican officials for refusal to pay confiscatory taxes and to comply with impossible regulations. This form of Turkish justice is being administered to corporations belonging exclusively to non-Teutonic Americans whose properties lie comparatively near the border. While arbitrary proceedings have hampered and oppressed foreign corporations operating elsewhere in the republic, the most conspicuous violations of alien rights have taken place where they would attract the most attention in the United States. This implies an ulterior motive. It is clearly meant to ascertain the strength and decision of the United States in the first instance; to see whether in the face of the German peril we dare risk hostilities with Mexico. We know, and Carranza knows, that a war with him would ultimately absorb more soldiers than we now have trained and ready. If we hesitate, if we fail to send an ultimatum that would involve armed intervention in case of a diplomatic rebuff, he will be emboldened to increase the embarrassments and financial burdens of the oil-producers. Evidently he is feeling us out under the direction of his German advisors. Meanwhile he is preparing for aggressive action by rushing the construction of new munition-plants furnished by the Japanese. The latest of these is being erected near the City of Mexico. If we continue our time-worn policy of reasoning and coaxing he may be indiscreet enough to undertake to prohibit the exportation of oil. That is what his new-found friends desire. It might wreck Mexico, but Mexico could also wreck the oilfields before the Allied fleets could take possession. It is even possible that the marines could not cope with the situation, and the interruption of oil-deliveries for a few weeks might imperil the operation of the British fleet. We ought to take drastic action. We have on hand the business of crushing the Enemy, and he who is not with us is against us. Carranza is not only not with us, but by every act and word he has shown himself hostile; nevertheless, he understands plain English when spoken in no uncertain tone. He does not wish to be dealt with as we and his own people will deal with him if he falls from power in a clash with us. He is less loved even than Santa Ana, and would less easily resist the general opprobrium that would follow an unnecessary and calamitous conflict with the United States. Had he played fair with Villa he might have ridden into power with the support of all classes remaining in Mexico, and he would not at the present time be defied by Zapata in Morelos, by Felix Diaz in Puebla, by Meixueiro in Oaxaca, by Villa and Salazar in Chihuahua, and by the brothers Cedillo in San Luis Potosí. Had he played fair with the United States and respected the treaty-rights of our citizens such a flood of American capital would have poured across the Rio Grande to develop the resources of that marvelous store-house of mineral riches that he would not now be floundering in financial quagmires, nor would he have had to resort to the looting of the Mexican banks in a last desperate effort to retain the loyalty of his soldiers by paying them with specie that should have been jealously guarded in the interest of the banking-credit of the country. Evidently the question now before him is whether he can borrow from the United States to keep himself going, or whether, failing in this, he can so cripple us and our Allies as to promote the triumph of Prussianism. Germany cannot finance him at the moment; we can if we will; and, being a shrewd man, he is naturally weighing the chances of ready American coin against Teutonic promises to pay him if he turns the trick. If we finance him once we shall have to administer the same medicine repeatedly. It is for this country to decide whether to pay tribute to Carranza or, if he prove recalcitrant, whether the hour be propitious for putting some worthy Mexican in power charged to maintain a respectable and self-respecting government. We do not want Mexico; we merely desire a decent administration in Mexico, representative of the finer intelligence and spirit of the republic and true to its constitution. We believe there are men in Mexico as capable of bringing this about as were Mitre and Sarmiento in Argentina, whose courage and wisdom started that country on its triumphant career of genuinely democratic progress.

DISCUSSION

Our readers are invited to use this department for the discussion of technical and other matters pertaining to mining and metallurgy. The Editor welcomes expressions of views contrary to his own, believing that careful criticism is more valuable than casual compliment.

Sampling Large Low-Grade Orebodies

The Editor:

Sir—This matter seems to have reduced itself to the question as to whether such bodies can be better sampled by mill-tests or by ordinary moil-sampling, and anent this you ask several pertinent questions in your issue of May 26.

I believe Mr. L. A. Parsons has struck the keynote when he says: "representative—that is the crucial point." If mill-tests could be made representative of the ore-blocks or of the mine from which they are taken, I believe the results arrived at would in many cases be nearer the correct value than that obtained by moil-sampling, but this necessitates a large number of mill-tests from any given block of ore, in a wide deposit, and a great number of such tests from the various cross-cuts, drifts, and raises, that so frequently expose but two of three sides of a block.

In order therefore to make a mill test 'representative' there must be at least several hundred, perhaps a thousand, of them, each a separate run with a separate clean-up. Has any one ever heard of such a thing having been done? Yet in the examination of large ore deposits by moil-sampling it is a common thing to take anywhere from 1500 to 3000 samples, while during the exploration and development of the low-grade copper deposits the number of small samples derived from both drilling and moiling runs into the tens of thousands.

It is evident then that the small-sample method has and always will prevail over the mill-test, chiefly because it is far more 'representative' of the ore-mass. In either case when the deposit is a large one an enormous amount of development work is required, involving heavy cost. If to this be added the far greater expense, as compared with moil-sampling, of cutting hundreds of small stopes and making an individual test on each lot of ore the preliminary outlay is too great to be practicable.

In some of the large deposits of low-grade gold ore even the development work necessary to expose enough faces to make moil-sampling fairly 'representative' has not been done, not only because of the great expense involved but because the resultant work cannot be used in the subsequent ore extraction on account of the mine having to be worked in huge stopes of unusual height, in order to reduce the cost of mining.

In the case you refer to where 375,000 tons of ore were milled before the value of the ore was considered to be established, most of us, I think, would have considered the mining and milling of such a huge amount to have been a 'representative' test of the mine. But subsequent results have shown that it was not, and the natural conclusion is that driving, cross-cutting, etc., on several levels and for long distances would have paid in the end. It is easy to be wise after the event, but I think this experience should teach one the almost insuperable difficulties in sampling a large and very low-grade mass of gold ore. As Mr. Jackling has pointed out, the expense involved makes it impracticable financially.

Again, we have all met the mine-owner who says, "This mine cannot be sampled; only a mill-run will give you accurate and dependable values." Such a mine is almost sure to be a good mine—for the owner to keep.

Moil-sampling, done with care and judgment, can and should be more 'representative' of the ore behind the sample, because of the greater number of samples that can be taken, because of the greater elasticity of the method where the metal contents occur in streaks and patches, because the method is readily applicable to remote places and workings more difficult of access, and because it is far cheaper than mill-tests made in sufficient numbers to make them 'representative.'

After all, what is the sampling of a mine other than a means of obtaining "the best possible approximation" of its value. In the nature of the case it can only be an approximation due to several causes into which it is not necessary to enter here. Enough to say that no perfect sampling method has yet been advised. That moil-sampling comes closest to this "best possible approximation" has, I think, been definitely proved by practical experience.

New York, June 20. THOS. H. LEGGETT.

Forest Reserve Again

The Editor:

Sir—I send you an open letter addressed to H. G. Merrill, supervisor of the Monterey National Forest Reserve, King City, California, revealing a situation that is detrimental to operators of mines on Forest Reserves. It is a subject that deserves attention. The letter follows:

Mr. H. G. Merrill, King City, California.

Dear Sir: I would respectfully call your attention to the following facts: Since 1906 annually the officers of the reserve have caused to be issued grazing permits for the section of country known as 'Gold Ridge', which is private property, being located as mineral land. The stock is turned loose without herders, fences, or corrals, with the result that they destroy the trails, roads, ditches, and reservoirs every year, that have been built at great expense. It is impossible to make headway without protection against the above acts. We have protested in vain to the officers; and the Government is practising the same method, in that they sued Messrs. Shannon and Little for the State of Montana and Colorado. The county of Monterey passed the 'no fence' law to check the very acts that the officers of the reserve are committing. You leave us no feed upon our own ground for our stock to work our mine. The hue and cry now is, where has the prospector gone? He is not in evidence upon the public domain any more. Get the records in the case of U. S. v. Copper Mountain Mining Co., and study them well, and you won't ask the reason why. Never a forest ranger comes around to ascertain what damage is being done. You say in your USE BOOKS "we foster the small home-builder, mining not interfered with", and the like. The records here and in the Pinnacles and National Monument Reserve do not show it; it is the reverse. See the protest filed at Washington in the Interior Department against the unlawful acts of the above officers;; it will show how business has been done in the above reserves, and so far there has not been any redress. The above protest was filed by Wm. D. MacPhie, of Soledad, on behalf of the above copper company some 3½ years ago. For full particulars apply to the undersigned, at Gold Ridge.

HENRY F. MELVILLE.

Jolon, California, March 31.

The Editor:

Sir—Upon examining the records of this office I find that a similar protest to the one transmitted by Mr. Melville was filed on December 18, 1909, and that nearly each year since that date a similar protest has been submitted. Upon receipt of each one of these protests the local Forest officers have endeavored to make an adjustment between the mining claimants and the stockmen who grazed their stock under permit in that particular locality of the Monterey National Forest. From the records it appears that there are between 25 and 30 head of stock under permit in this locality, the owners of which own ranch property and grazing-land near the Forest. These permittees have occupied the Forest ever since its establishment and have gained what we term a preference in the use of National Forest range; however, our regulations provide specifically that subsisting mining locations are not taken into consideration when grazing permits are granted for National Forest land. I quote from our instructions to local Forest officers the following: "Mining claims—Persons holding unpatented mining claims within a National Forest have the right to the grass or other forage upon such claim needed for stock used in connection with the development of the claims, but they have no right to dispose of the forage to any other person or to collect rental for the use of the claims for grazing purposes. Such unperfected mining claims, therefore, cannot be accepted as the basis for a permit under this regulation." In view of the fact that we have no jurisdiction over these claims, and, further, from the fact that the permit issued to each stock-man states specifically that it does not grant the right to the use of any other than National Forest lands, you will see, I believe, that we have no jurisdiction in this matter, and if the damage done by the grazing of this stock actually occurs it is a matter that must be handled under the State law between the mining claimants and the owners of the stock. In the past we have endeavored to have all parties interested come to some agreement so that the proper drift-fences could be constructed, or other means devised by which this complaint would not be continually coming up. Our efforts along this line have been unsuccessful, and the only course left open for us to pursue is to withdraw from the controversy entirely and let the mining claimants and the owners of the stock adjust their differences.

C. E. RACHFORD,
Assistant District Forester.

San Francisco, California, June 13.

Canadian Mining Regulations

The Editor:

Sir—on March 5, 1917, under the provisions of the War Measures Act, 1914, the Canadian Government enacted, by order-in-council, certain regulations restricting the rights of aliens as to holding or acquiring, directly or indirectly, mining and other property in Canada; where such rights, lands, and so on, were vested in or administered by the Federal Government. As aliens who were not enemies of Canada were also included, mining men and others deemed the regulations too severe; therefore they induced the Government to rescind the objectionable clauses and substitute others bearing only on alien enemies, companies and corporations.

The revised order-in-council, No. 1268, is dated May 8, 1917, and a copy of the same is herewith enclosed.

As you published a summary of the regulations dated March 5, 1917, I hope you will be agreeable to give the same publicity to the revised copy now enclosed.

WM. THOMLINSON.

New Denver, B. C., June 20.

Clauses 3, 4, and 5 of the Regulations established by Order in Council of the 5th of March, 1917, (P.C. 572), are hereby rescinded and the following Clauses are hereby made and enacted in lieu thereof:

"3. No company shall acquire or hold any of the rights, powers or benefits hereinbefore referred to if such company be an alien enemy company, or registered in an alien enemy country, or having its principal place of business within such country, or if the chairman of such company or any of the directors are subjects of an alien enemy country, or if such company is controlled, either directly or indirectly, by an alien enemy or alien enemies, or by an alien enemy corporation or alien enemy corporations.

"4. Any alteration in the Memorandum of Articles of Association, or in the constitution, or in the laws of any company holding any rights, powers or benefits hereinafter referred to shall be reported by the proper officer of the company to the Minister of the Interior, and two months previous notice in writing shall be given to the Minister of the Interior of the intention to make any alteration which might conceivably, either directly or indirectly, affect the character or control of any such company, and if, in the opinion of the Minister of the Interior, the said alteration shall be contrary to the cardinal principal that the said company shall be and remain a company not of alien enemy origin or control, the Minister of the Interior may refuse his consent to such alteration, and if his refusal is not obeyed, may declare such company to be an alien enemy company and may cancel the said rights, powers and benefits under the provisions of the next following regulation.

"5. If any company which has acquired any right, power or benefit hereinbefore referred to shall, at any time, become subject to the control of an alien enemy, or alien enemies, or an alien enemy corporation or corporations, or shall assign any of the rights, powers or benefits aforesaid, without the consent in writing of the Minister of the Interior being first had and obtained, or if the said right, power and benefit has been acquired through error, misrepresentation or fraud, the Minister of the Interior may cancel the grant of such right, power or benefit and thereupon the same shall *ipso facto* be cancelled and any moneys or fees paid to or deposited with His Majesty shall be *ipso facto* forfeited to His Majesty.

Magmatic Ore Segregation

The Editor:

Sir—In view of a suggestive article by Dr. J. T. Singewald in the MINING & SCIENTIFIC PRESS for May 26, where the mode of occurrence of chromite is referred to, a review of the conditions under which chromite occurs in southern Quebec may be of interest. The examination of the Quebec chromite deposits has not yet been completed, and further investigation may throw new light on their origin. A preliminary report (Memoir 22, Geological Survey of Canada) covering this area was prepared by myself in 1911 and published in 1913. Since that date a topographic map of the district has been drawn and Dr. R. Harvie of the Geological Survey staff is preparing to make a final report. It is therefore suitable to confine this communication to an interim description, mainly of the features bearing on the subject of Singewald's article. The chromite-producing district of Quebec lies south-east of the St. Lawrence river, and is distant by rail about 80 miles from the City of Quebec and 160 miles from Montreal. The main production has been from a small area about 6 by 12 miles in extent, but other occurrences have been found outside of this area which, in the present favorable market, are being developed with good prospects of success. The production since 1886 has been somewhat more than 100,000 tons.

The ore occurs in a complex of basic igneous rocks, intrusive into folded and altered sediments consisting of slates and quartzites of Cambrian and Ordovician age. The principal types of igneous rock are peridotite partly altered to serpentine, pyroxenite, and diabase. Other allied rock-varieties appear in places as gabbro and porphyrite. Granite, also, is present in the form of dikes and small bosses. The basic rocks are plainly differentiates from a single intrusion, while the granite has been intruded later, though probably while the wall-rocks were still heated. The basic rocks occur in the form of stocks and of sills that in places show laccolithic thickening. The different rocks are arranged in the sills in the order of decreasing basicity and density, that is, peridotite, pyroxenite, and diabase, with the peridotite on the bottom, while in the stocks the order of basicity is from the centre outwards.

Chromite occurs widely, almost universally, disseminated throughout the peridotite and the pyroxenite as an accessory constituent. Microscopic evidence shows that it is a primary mineral and has been one of the earliest to crystallize in these rocks. It also occurs in masses, often forming single orebodies containing several thousand tons. These occur usually, if not only, in the transition-rock between the peridotite and pyroxenite. Without recounting the evidences at length, it may safely be said that the field-relations indicate that the orebodies are primary members of the igneous complex. The walls are frequently ill defined, irregular, and not bounded by structural features. There is no evidence adequate to prove either solution and re-deposition in masses of this difficultly soluble mineral, nor its injection in its present position after the solidification of the country-rock. Certain minerals denoting pneumatolytic action, such as garnet, vesuvianite, and molybdenite are found at some of the deposits, but they are associated with later granite dikes. In brief, the ore deposits seem to be phases of the rock in which they occur.

The place in the igneous complex in which the chromite occurs in mass, between peridotite and pyroxenite, is the principal point of interest. It neither agrees with the arrangement of the rocks of the complex in order of density, nor with its own place where disseminated as an accessory mineral in these rocks, in which it was one of the first minerals to crystallize. While chromite shows a strong tendency toward early crystallization, its maximum development seems to have taken place near the close of the period of crystallization of the olivine. This seems to imply a retarded crystallization of the chromite in mass, such as Singewald has noted in ilmenite and magnetite deposits, and which, with apparent good reason, he ascribes to the physical action of the mineralization agents of the French petrologists. Broadly speaking, the chromite-concentrations in Quebec seem to be best described as products of magmatic segregation occupying a position in the series of differentiates that has been influenced by the action of the mineralizers.

Montreal, June 9. JOHN A. DRESSER.

MAGNETIC-IRON ores often contain large amounts of manganese; also many manganese deposits are mixtures of iron sequi-oxide, or of limonite, with manganese dioxide. The iron in these can be removed, when not present in a pulverulent form, by various types of magnetic separators, such as the Wetherill. The residue containing the manganese can then be further concentrated, if necessary, to eliminate silicious minerals so as to bring the manganese product within the limits set by the makers of ferro-manganese, which are a minimum of 40% metallic manganese, and a maximum of 12% silica. Also the maximum permissible phosphorus content is 0.225%.

Phosphate Rock

By R. W. STONE

*Prior to 1914 the United States was producing annually close to 3,000,000 tons of phosphate rock, of which over 99% came from Florida, Tennessee, and South Carolina. With the beginning of the war the facilities for shipping phosphate rock to Europe were greatly decreased. Many Florida plants were shut down, and they have not resumed operations. In 1916 the industry was in some areas practically demoralized, but there was nevertheless a gain over 1915. The total output in 1916 was 1,980,000 tons, valued at $5,897,000. Tennessee, in spite of decreased production, has yielded a larger proportion of the country's output since the War began. It would seem that the Tennessee industry, not having been so related to the export trade, and being equipped in part with modern machinery, should develop while the European trade is restricted and while the industry in Florida and South Carolina is more dormant. The deposits of phosphate rock in the United States are confined very definitely to the southeastern part of the country and to the Rocky mountain region from the latitude of Salt Lake City, Utah, to that of Helena, Montana. Although by far the largest deposits are in the Western States, the production from that region is less than 1% of the whole, owing to the lack of a nearby large market at present and to high freight rates on the crude rock. The western rock-phosphate deposits are so extensive that, even if the entire world depended on them for its supply, they would not be exhausted in many generations. The Florida phosphate deposits comprise three classes—hard rock, land pebble, and river pebble. The hard rock is the highest grade, the land pebble is produced in the largest quantity, and the river pebble is not mined at present. The area of hard-rock deposits forms a narrow strip along the western part of the Florida Peninsula from Suwannee county to Pasco county, a distance of approximately 100 miles. The land-pebble phosphate area lies east of Tampa and is about 30 miles long and 10 miles wide. The South Carolina output consists of land-rock phosphate mined in the vicinity of Charleston. River-pebble phosphate occurs in the same area but is not mined. The Tennessee deposits of rock phosphate are in the west-central part and extreme northeast corner of the State. The latter have not been mined. Three types are recognized and known by their colors as brown, blue, and white rock. The brown rock comes from Maury, Giles, Hickman, Lewis, and Sumner counties and is sold under a guaranty of 70 to 80% tricalcium phosphate. The blue rock is mined in Lewis and Maury counties and varies considerably in its phosphatic content. The phosphate deposits of Kentucky lie between Frankfort and Lexington, and considerable quantities of rock have been mined near Wallace. Deposits occur interruptedly for a distance of 80 miles in the north-central part of Arkansas, and a small quantity is produced at Anderson, Independence county.

Four of the Western States possess vast deposits of high-grade rock phosphate, but the western production amounts to less than 5000 tons per year. Idaho, Utah, and Wyoming are the producers. Montana is not yet a producer, although at Elliston, Garrison, Philipsburg, and Melrose are extensive deposits easy of access and close to rail transportation. In the southeastern part of Idaho an extensive supply of high-grade phosphate occurs along both sides of Blackfoot river, in Fort Hall Indian Reservation, near Montpelier, and north of Bear lake. The Utah deposits are east of Great Salt lake, in the Wasatch and Uinta ranges, and east of Bear lake. These deposits are extensive, but the material averages only about 60% tricalcium phosphate. Western Wyoming also is rich in rock phosphate, the deposits being mostly in the Owl creek, Wind river, Gros Ventre, and Salt river ranges. Some of them are thick beds carrying 80% tricalcium phosphate and extending for many miles, and they constitute a reserve supply that is almost inexhaustible.

Estimated quantity of phosphate rock in the United States:

	Long tons
Eastern States:	
Florida	227,000,000
Tennessee	88,000,000
South Carolina	9,000,000
Kentucky	1,000,000
Arkansas	20,000,000
	345,000,000
Western States (Montana, Idaho, Utah, and Wyoming)	5,367,000,000
	5,712,000,000

Although the total reserves as shown by this estimate are extremely large, the supply of high-grade rock is much less and should not be considered inexhaustible.

While the War continues phosphate rock cannot be sent to the largest consumer, Germany, and high ocean-freight rates practically stop shipments to other European countries. The demand for sulphuric acid for making munitions has raised the prices of acid so that manufacturers of acid-phosphate have been obliged to curtail production. It seems reasonable to believe that at the end of the War European nations will need increased quantities of phosphate, as their stores of foodstuffs will be low, and intensive cultivation of the soil will be necessary.

*U. S. Geol. Surv. Bull. 666-J.

Principles of Flotation—I

By T. A. RICKARD

*INTRODUCTION. The understanding of the principles governing flotation has been delayed mainly because the explanation of the phenomena—or appearances— characteristic of the process is to be found in physics rather than in chemistry. Modern metallurgy has been in the hands of men primarily chemists, rather than physicists. Cyanidation and chlorination, for example, may be explained by chemical formulas, even if they cannot be expressed in their entirety by the language of elemental symbols; but flotation is not to be interpreted in that way; it is controlled by physical laws that are obscure and that hardly came within the cognizance of the metallurgist until the need for study was felt by him within a period so recent that the full results of scientific research are not yet available.

To understand the rationale of the flotation process we must return to the amusements of our boyhood; in the soap-bubble and in the greased needle we shall find an inkling of the forces at play in the flotation machine. Everybody knows the trick of the greased needle. If a needle be greased and then placed carefully on the surface of tap-water in a bowl it will float, despite the fact that steel is eight times heavier than water. Even the natural oil on the fingers, or that obtainable by passing the fingers through the hair, will suffice for the purpose of assisting the needle to float.

The first idea is that the buoyant effect of the oil adhering to the needle prevents it from being drowned. However, the quantity of oil thus attached to the needle is not enough to buoy it; the specific gravity of the oil is, say, 0.9 as compared with water, which is the unit of specific gravity; therefore the flotative margin is only one-tenth, and for the oil to float a piece of steel, having a specific gravity of 8, its volume would have to be more than 80 times that of the steel. So the buoyancy of the oil does not do it. Moreover, an ungreased needle also will float. This experiment must be conducted carefully. To be certain that the needle was free from grease[1] I held it in metallic pincers, dipped it in a solution of washing-soda (sodium carbonate, which is a solvent for grease), and then dried it, taking care to use a clean cloth and not to touch it with my fingers. Then I placed a piece of tissue-paper on the water in a cup and laid the needle, held in the pincers, upon the paper, which was depressed gently into the water by the point of a wooden match, until the paper became soggy and finally sank, leaving the needle floating. It lay in a depression of the water-surface, which appeared to be bent under it.

The needle that will float after being greased is larger than the one that floats without being greased,[2] so the oil seems to aid flotation; but when the needle is too large it cannot be made to float, greased or not. It is too

Fig. 1

heavy; that is, the force of gravity multiplied by mass is sufficient to overcome the peculiar resistance offered by the surface of the water. What causes that resistance?

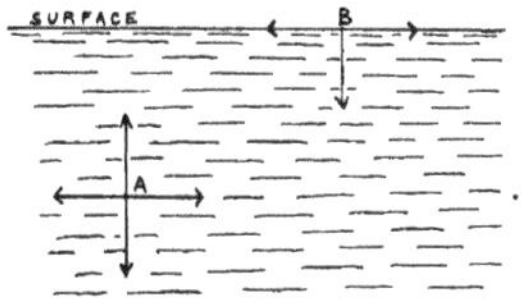

SURFACE-TENSION. The force responsible for the floating of the needle is called 'surface-tension.' It is a man-

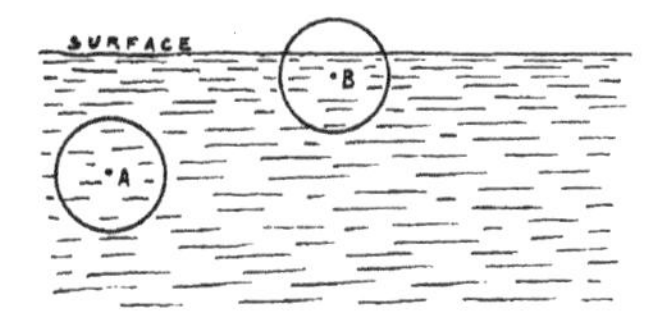

Fig. 2

ifestation of cohesion, which is the attraction that binds molecules of like kind to each other. Each molecule within the interior of the liquid is imagined as surrounded by molecules like itself to which it is attracted and which it attracts equally in every direction, whereas the molecules at the free surface of the liquid are attracted only by those internal to themselves, the result being to constrict the free surface of the liquid. In consequence, the surface acts as if it were a stretched membrance or an elastic film. These molecular conditions may be represented graphically. See Fig. 1. The attractive forces acting on a molecule (A) in the body of the liquid may be represented by four resultant axial components, which are equal, so that the molecule is perfectly free to move,

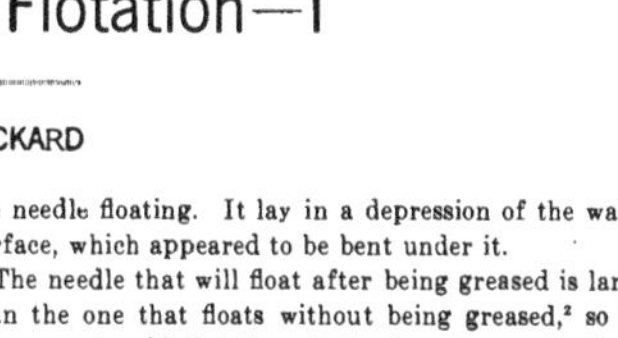

*This is an attempt to re-state the fundamental principles of flotation. The author will welcome corrections or criticisms.

[1]New needles are slightly greasy, as I ascertained by means of the camphor test, described later. The grease protects the needles from rusting.

[2]I tried five large greased needles, all of which floated; then I tried the same needles after they had been washed in the soda solution and wiped dry on a clean cloth. One time all five sank; the other time four sank.

except for viscous resistance. At the surface itself the upward component disappears and the pull downward on the molecule (B) is uncompensated, any extension of the surface being opposed by a force the horizontal component of which is 'surface-tension'.

This can be illustrated in another way. Each particle of water is attracted by all the particles that lie within its range, which is definitely small, about 0.00000015 cm., therefore the scope of molecular attraction may be considered as a sphere of influence. Thus A in Fig. 2 is attracted, and attracts, within a definite sphere, while B, which is close to the surface, is more attracted inward than outward, since a part of its sphere of attraction lies outside the water.

Such a hypothesis is largely an abstraction; a concrete idea of the nature of surface-tension can be obtained by noting some of its various manifestations.

1. The drawing, or 'soaking up', of water by a sponge.
2. The penetration of wood by varnish.
3. The rising of oil in a lamp-wick.
4. The clinging of ink to a pen.
5. The running of the ink from the pen to the paper.
6. The absorption of the excess of ink by blotting-paper.
7. The cohesion between two plates that have been wetted.
8. Dip a camel's hair brush in water, remove it from the water, and observe how the hairs cling together. Immerse the brush in the water and note how the hairs separate.
9. Watch the water-spiders running over a pool, like boys skating on thin ice. H. H. Dixon actually measured the pressure exerted by the spider's feet on the water. He photographed the shadow of the dimple, then mounted one of the spider's feet on a delicate balance, and made it press on the water until it made a dimple of the same depth as that previously recorded.
10. Pour colored water in a thin layer over the bottom of a white dish; then touch a part of its surface with a glass rod that has been dipped in alcohol. The colored water shrinks from the part touched, leaving an irregular patch of white bottom dry. This is due to the tension of the pure water being greater than that of the alcoholized water, so that the liquid is pulled away from the place where the tension is weak to the place where it is strong.[3] The lively movements of the particles of dye in the water indicate the conflict between the forces of diffusion and surface-tension.
11. The formation of a drop at the end of a tube or from the small mouth of a bottle is another example of surface-tension. Note how the drop grows slowly until it has attained a definite size, and then breaks away suddenly. The size of the drop is always the same for the same liquid coming through the same orifice. It hangs as if suspended in an elastic bag that ruptures when the weight becomes excessive. The contractile character of surface-tension is manifested in the formation of the drop, the force tending to draw the fragment of liquid into the most compact form, that presenting the least surface in relation to volume, namely, a sphere.

Similarly, if we admit air through a glass tube of given size into various liquids, we shall obtain the biggest bubble in the liquid with the highest surface-tension. If various liquids in succession are allowed to run out of an opening of given size, the largest drop will be that of the liquid having the highest surface-tension.

12. When an iron ring is dipped into a solution of soap

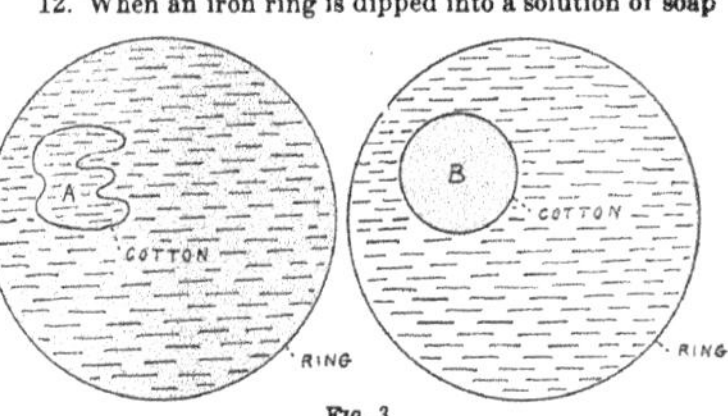

FIG. 3

and then taken out, it will be seen that a film of solution stretches across the ring, covering the whole interior circular space. If a small loop of cotton, previously moistened in the soapy solution, is placed on the film stretched across the circle of the ring, this loop can be made to

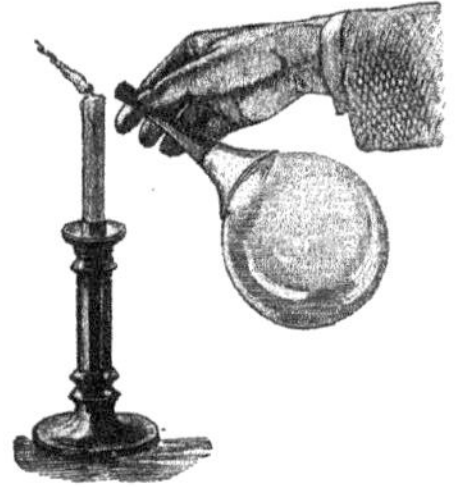

FIG. 4

assume, and to retain, any form, such as is shown at A in Fig. 3. If, however, this film within the loop is broken, the loop immediately assumes the form of a perfect circle, as shown at B; and if it is now deformed in any way, it springs back at once to a circle as soon as it is released. Evidently the surface of the solution assumes the shape covering the smallest area. The surface-tension of the liquid acts equally on both sides of the cotton so long as it is wholly immersed, but when the film of liquid inside the loop is broken, the tension acts on one side only—on the open side, where it is in contact with air—and hence draws the loop into a circle, which involves the minimum of extension.

13. The contractile force of surface-tension is shown

[3]This simple experiment is a fascinating exhibition of surface-tension and it should be made by every student of flotation.

in a simple way by blowing a soap-bubble on the large end of a pipe and then holding the other end of the pipe to a candle, whereupon the air escaping from the shrinking bag of the bubble will extinguish the flame, as in Fig. 4.[4]

14. When water is sprinkled on a dusty floor, the dust prevents the wetting of the floor by obstructing the coalescence of the drops, that is, the spreading of the water over the floor. The water draws itself into rolling spherules that become armored by particles of dust. They are nearly round, the larger ones showing a flattening, because the gravitational stress overcomes the contractibility or sphericity of the film. This flattening is shown by a drop of mercury on glass or by the beads of gold on an assayer's cupel.

15. The globular form assumed by water when spilled on a hot stove is another manifestation of these forces. The water is protected from the hot iron by a film of steam, which, as it is formed, decreases the size of the globule until it disappears. If the iron is not sufficiently hot, it becomes cooler and therefore wetted, by spreading of the water, which is instantly converted into steam.

16. Some of the physics of flotation can be illustrated at the dinner-table.

A. Fill a tumbler little over full of water and note the convex surface, indicating the play of a force that prevents the liquid from spilling. It is a contractile force.

B. Fill a wine-glass half-full with port and observe how the wine climbs up the side of the glass, forming a meniscus around the circumference of the surface. This liquid consists of alcohol and water, both of which evaporate, the alcohol faster than the water, so that the surficial layer becomes watery. In the middle of the glass the surficial layer recovers its strength by diffusion from below, but the film adhering to the glass, being more exposed to the air, loses its alcohol by evaporation more quickly and therefore acquires a surface-tension higher than that of the undiluted wine. It creeps up the side of the glass dragging the strong wine after it, and this continues until the quantity of fluid pulled upward collects into drops—called the 'tears of wine'—that run back into the glass.

C. Fill a glass two-thirds full from a 'siphon' containing water that is effervescent because it contains gas in solution. Take three or four small grapes, preferably of the Californian seedless variety. The grapes will sink to the bottom of the glass, but soon they become restless and rise to the surface, one after the other. They do not remain there long; first one and then the other sinks. They will continue the performance for half an hour, bobbing up and down; their activities slowly diminish, and eventually they are left inert at the bottom of the glass. What happens is simple enough. The siphon has come from the refrigerator; the warmth of the room and the lowering of pressure release the carbonic-acid gas, which, in the form of minute bubbles, attaches itself to the grapes, buoying them to the surface as mineral particles are raised to the surface of a pulp in the Potter process. There the bubbles burst, causing the grapes to fall back. If a couple of grapes collide, the bubbles become detached, dropping their freight, and themselves rising to the surface. At first the grapes rise rapidly and rebound from the surface of the water as if it were an elastic membrane. This is a remarkable effect and should be noted carefully. After the evolution of gas has diminished the bubbles become too few to buoy the grapes, and the performance ends.

[4]C. V. Boys in 'Soap Bubbles'.

Surface-tension is identified with 'capillarity', because it is so marked in a tube the bore of which is only large enough to admit a *capillus*, or hair. When the lower end of a wide tube is held in water, the water inside rises to about the same level as that outside the tube, in accordance with the law of hydrostatic pressure; but when the lower end of a glass tube of small bore, say, 1 mm., open at both ends, is inserted into water, the water rises within the tube and stands at a level higher than the water outside. If, again, the tube be held vertically with its lower end immersed in mercury, the liquid metal inside the tube sinks to a level below that of the mercury outside. See Fig. 5. This is explained by saying that the molecu-

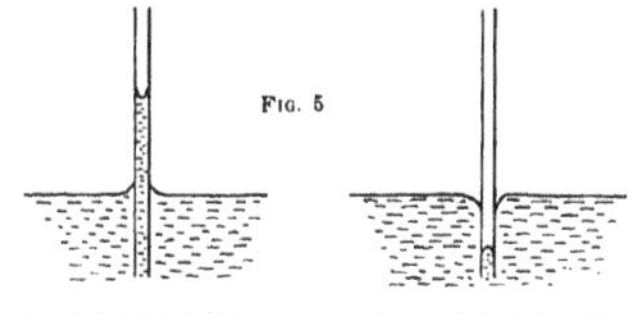

Fig. 5

lar attraction of water to glass is greater than that of water to water; whereas the attraction of mercury to glass is less than that of mercury to mercury. The forces of cohesion in a substance and of adhesion between various substances have been measured. Quincke and others have ascertained by experiment that the effect is sensible within a range of one thousandth and one twenty-thousandth of a millimetre. Such is the scope of molecular attraction. The liquid rises in a capillary tube until the weight of the vertical column between the free surface and the level of the liquid in the tube balances the resultant of the surface-tension.

The surface-tension of liquids can be modified. It is decreased by a rise of temperature. For example, place two matches an inch apart on the surface of pure water in a bowl and then touch the water between them with a hot wire. They draw apart promptly, because the surface-tension of the water between them has been lowered relatively to that of the rest of the liquid in the bowl, so that the pull of the water-surface under normal tension is stronger than that of the surface of the warm water between the matches.

The addition of an impurity or contaminant will lower the surface-tension of water. We have seen how this effect is caused both by alcohol and soap. Distilled water has a maximum surface-tension, which is lowered by almost any substance that is soluble or miscible in it. The soluble substance, or solute, modifies the tension directly, whereas the minutely divisible substance, forming an emulsion, creates a great number of interfaces, or surfaces of contact, each having a lower tension. The particular contaminant, or modifying agent, associated with the early history of flotation was oil, which is partly soluble and readily dispersible. The oil generally used at first was a heavy oil, like oleic acid.[6] By the addition of sufficient oil the surface-tension of water is lowered from 73 to 14 dynes per linear centimetre. The following experiment illustrates this fact. If a wooden match be laid on the surface of tap-water in a pan, so that it remains at rest, and if then a drop of olive-oil be placed on the surface of the water near the match, the match will draw away smartly, because the oil has reduced the tension of part of the water-surface and caused the uncontaminated water to pull away. This modification of the surface-tension of water by a contaminant is one of the fundamental factors in flotation, as we shall see.

Let us now go back to the floating needle. If it is greased, does the grease lower the surface-tension of the water? That can be ascertained by a pretty experiment. If camphor is whittled with a knife above a bowl of water the shavings, dropping on the water, will dance on the surface in a life-like manner suggesting insects in a fit. This phenomenon, as shown by Marangoni, is due to the dissolving of the camphor—a crystalline vegetal distillate—preferably at the pointed end, where the largest area per unit of volume is presented for solution. The dissolving of the camphor lowers the surface-tension of the water in contact and thereby causes the uncontaminated water, with its stronger tension, to pull away from the spot affected by the camphor—as in the colored water and alcohol experiment, No. 10. This causes the chips of camphor to turn and move spasmodically. In order to incite such activity the surface-tension of the water must be greater than that of the camphor solution. As soon as enough camphor has dissolved to modify the whole surface of the water in the bowl or cup, the chips become inert. Likewise if the surface-tension be lowered by the addition of grease the camphor remains quiet. For example, if, while the chips of camphor are lively, the water be touched by a greasy finger—all fingers are slightly greasy—the camphor is quieted immediately. No ordinary 'clean' cooking-utensil is sufficiently free from grease to allow an exhibition of the camphor dance.

Here we have a simple means of detecting the presence of grease or oil in the water upon which the needle is floating. I introduced some camphor chips into the water on which the ungreased needle was floating and they became lively. Then I repeated the experiment with a needle that was slightly greased, by rubbing it with the fingers that had touched my hair, and the camphor appeared unaffected thereby; it was lively. Finally, I smeared the needle with olive-oil; an iridescence on the surface of the water indicated diffusion of the oil. This time the chips of camphor fell dead as a door-nail and remained wholly inert on the water. Apparently, therefore, the needle will hold to itself a limited proportion of oil, which adheres so selectively as not to contaminate the water; but an excess of oil, more than the needle can hold, will be set free at once to modify the water and lower its surface-tension.

This is a classic experiment, as I ascertained afterward. Raleigh showed that the decrease of surface-tension begins as soon as the quantity of oil is about half that required to stop the camphor movements, and he suggested that this stage may synchronize with a complete coating of the surface with a single layer of molecules.[6]

A reference has been made already to the measuring of surface-tension. It can be done in several ways. For example, a framework, such as is shown in Fig. 6, is constructed[7] out of a transverse bar AB and two grooved slips CD and EF, so as to allow a piece of wire $GHIJ$ to slip freely up and down. The wire HI is pushed against AB and some of the liquid is applied between them. The little pan X is loaded with sand so that the wire HI is pulled gently from AB. The minimum force required to do this is mg, the weight of M grammes. This weight suspended on the film of liquid between AB and HI equals the tension of the film on the wire. If the film stretches until the wire HI is at p, then the film has an area CE, CP. The total weight mg is distributed over the breadth CE, whence if T represents the surficial tension across the unit of length CE, then

$$mg = T.CE, \text{ or } T = \frac{mg}{CE}$$

Another simple way of measuring surface-tension is to make a wire-frame of which one side is movable; thus (Fig. 7) let ABC represent a bent wire and DE a straight piece. If a film of liquid is spread over the space DBE then the surface-tension acting on DE will support not only the weight of the wire DE but also a small weight X. If W be the mass of the cross-wire DE and its attached weight, then the surface-tension of the film supports W and exerts a force Wg. The surface-tension acts all along that part (l) of the wire DE that lies in contact with the film, and it acts at right angles to DE. Since the film has two surfaces, if the force exerted on a unit length of DE and on one side of the film be T, then the upward force on DE due to surface-tension is $2Tl$. Hence[8] if there is equilibrium $2Tl = Wg$, or $T = \frac{Wg}{2l}$

This method was suggested by Clerk Maxwell. An ingenious mechanical model for illustrating the definition of surface-tension has been devised by Frank B. Kenrick, of the University of Toronto. He gives the definition as "the maximum quantity of work that can be gained when a surface is decreased in area by one square centimetre", and describes his device as follows: "A pro-

[6]No wonder the judges were puzzled by the technical terms used in flotation lawsuits. 'Oleic acid' is called an oil, whereas 'oil of vitriol' is an acid.

[6]The Encyclopædia Britannica. 11th Edition. Page 267.

[7]Alfred Danniell. 'A Text Book of the Principles of Physics'.

[8]W. Watson. 'A Text-book of Physics', p. 182.

jection cell 40 mm. by 10 mm. and 60 mm. high, the upper edges of which have been coated with a film of paraffine-wax, is filled almost to overflowing with water. On the surface is floated a thin shaving of cork 30 mm. by 5 mm. by 1 mm., to which is attached a fine cotton

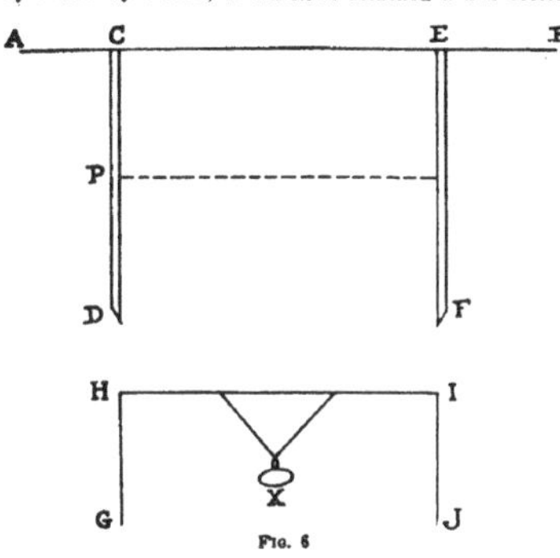

FIG. 6

thread about 40 mm. long terminating in a little glass hook. The thread passes over a small pulley made from a pill-box and a pin resting in a double Y-shaped glass bearing. Three weights of glass or bent wire weighing

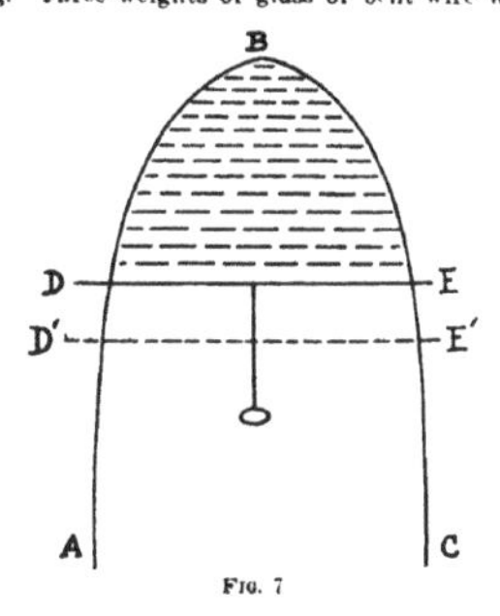

FIG. 7

about 0.1 gramme, 0.07 gm., and 0.04 gm. may be hung on the hook. The middle weight approximately balances the surface-tension, while the lighter one on being pulled down with a pair of tweezers is lifted again by the surface-tension. A fall of 1 cc. produces one square centimetre of surface, namely, 0.5 cm^2 on the forward under side of the cork that is wet with water and 0.5 cm^2 on the upper surface of the liquid in the cell."[9] For the accompanying sketch (Fig. 8) I am indebted to Professor Kenrick, who sent it to me on request. The waxing of the upper edge of the glass cell allows the water, which does not wet paraffine, to rise slightly higher than the level of the glass without overflowing.

[9]*Jour. of Phys. Chem.*, Vol. XVI, page 513.

By such experiments the force of surface-tension between water and air has been determined to be 3.14 grammes per linear inch or 72.62 dynes per centimetre at 20°C.[10] Many disturbing factors enter into the meas-

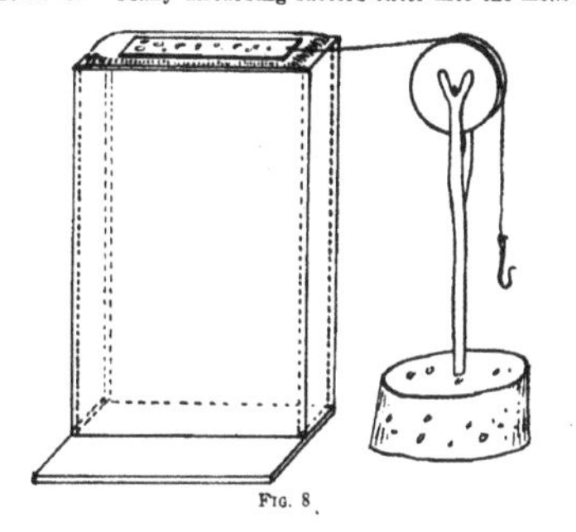

FIG. 8

urement of this force, so that divers figures, ranging from 70.6 to 81 have been announced at various times.[11]

This force may seem small, yet the actual tensile strength per unit-area of cross-section of the film is about one-fourth that of the iron or mild steel used in the shells of steam-boilers, although its density is not much more than one-eighth as great as that of the iron.[12]

The surface-tension of a liquid must be stated with reference to the fluid—gas or liquid—in contact, for it is modified by the nature of the substance on either side of the interface. An interfacial tension exists at any surface separating two substances and it has a particular value for each pair of substances. For example, the tension separating mercury from water is 418 dynes per centimetre whereas that separating olive-oil from air is only 36.9 dynes. A drop of water will not spread over

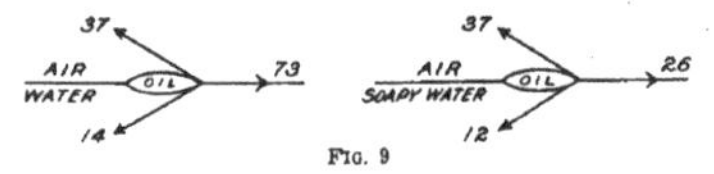

FIG. 9

the surface of mercury but oil will spread over water. The balance of forces is different in the two cases. When a globule of oil is placed on water, the tension of the water-

[10]Theodore W. Richards and Leslie B. Coombs. 'The Surface-Tension of Water, Alcohols, etc.' Jour. Amer. Chem. Soc., July 1915. One dyne is equal to 1.02 milligrammes.

[11]T. J. Hoover in his valuable book 'Concentrating Ores by Flotation' quotes from Clerk Maxwell's article on 'Capillarity' in the Encyclopædia Britannica and gives the figure as 81, but he makes the mistake of saying that it is 81 dynes "per *square* centimetre." It is a tension, not a pressure.

[12]M. M. Garver. *Jour. Phys. Chem.*, Vol. XVI, page 243.

air surface exerts a pull of 73 dynes as against the joint pull (37 plus 14) of the air-oil and oil-water surfaces. Thus $14 + 37 < 73$. (Fig. 9) The oil spreads. If soap, in the form of ¼% sodium oleate, be added to the water its surface-tension will be lowered to 26 and the oil-water tension will also be decreased, how much I do not know, but certainly decreased, say, to 12; therefore $37 + 12 > 26$, and the oil will not spread over the water. On the other hand, the tension of the mercury-air surface has been given as 436 dynes and that of the mercury-water surface as 418. If this be so, then a drop of water will not spread, because $418 + 73 > 456$. But Quincke showed long ago that pure water will spread on pure mercury, although the presence of an impurity, such as a slight greasiness, on the surface of the mercury will prevent spreading. According to later determinations of the interfacial tensions, by Freundlich, that of mercury-air is 445 dynes and that of mercury-water 370, so that $73 + 370 < 445$, and the pure water ought to spread on the pure mercury, as Quincke stated. If the water be contaminated, so as to lower its surface-tension, it will spread readily even on ordinary mercury, which is not chemically pure and on which pure water will not spread.

WETTING. A steel needle floats on water, but a glass rod of the same size sinks immediately; yet the specific gravity of steel is to that of glass as 8 to 2.75. The surface of the water resists rupture by the steel but it is readily broken by the glass; in other words, the glass is readily 'wetted,' while the steel is not. Again, if the glass rod be greased it will float; it ceases to be easily wetted. Here we face one of the underlying phenomena of flotation. The understanding of what constitutes 'wetting' is essential to the subject.

If a drop of pure water be placed on a clean piece of glass, it will flatten itself out so as to increase the space it first touched. If a similar drop of water be placed on a cabbage-leaf, it will not spread, but will retain its spherical form. We say that water 'wets' a glassy surface and does not 'wet' a waxy vegetal surface. A drop of mercury spreads eagerly over gold, but does not spread on glass; mercury wets gold but not glass. The statement is not absolute; it is a question of degree.

If I press the surface of water with a piece of glass the water rises to meet the glass, forming a mound, whereas if I make the same test with a piece of steel the water shrinks away from it, forming a depression. The tendency is for the water to lap the glass but to avoid the steel; the one substance is easily 'wetted,' the other not. The glass and the steel typify the gangue and the sulphide respectively in an ore treated by flotation. If we look carefully at the steel and glass, at the instant of touching the water, we see the conditions sketched in Fig. 10.

Note how ink from a pen will not run on paper that is at all greasy. The paper refuses to be wetted where it is greased. That is why new pens are refractory: the steel has been greased to prevent rusting, like the needles. I used to burn the point of a new pen by aid of a match in order to cause it to deliver the ink to the paper comfortably. That burned the grease, but spoiled the temper of the pen-point.

The free surface of a liquid is horizontal, but at the contact with a solid the surface is curved, the direction and amount of curvature varying as between different liquids and solids. The water curves upward against glass, whereas it curves downward against steel; it tends to drown the one, but to float the other until gravity

FIG. 10

overmasters surface-tension. The way in which a liquid impinges on a solid is called the 'angle of contact.' For example, in Fig. 11 water is shown in contact with glass. Consider the conditions at the point O. The gravitational pull on a minute quantity of the water is negligible in comparison with its own cohesive force; so we can disregard the effect of gravity. The force of adhesion exerted by the surface of the glass is represented by $O\ A$, the force of cohesion in the water is represented by $O\ B$, and the resultant of these two forces is $O\ C$. If the adhesive force of the liquid to the solid exceeds the cohesive force of the liquid, the resultant will lie to the left of the vertical, $E\ D$, that is, within the solid; and

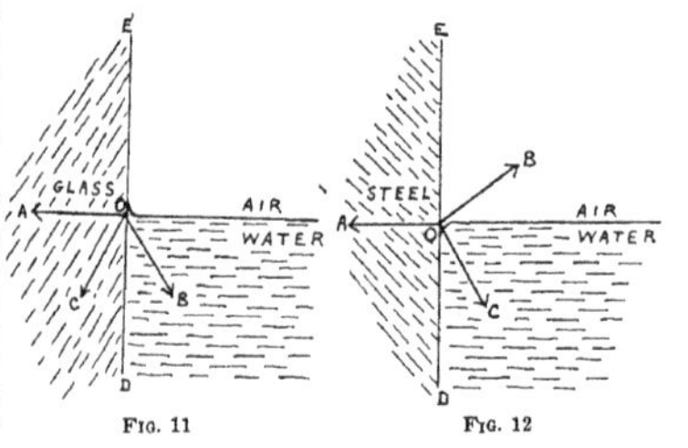

FIG. 11 FIG. 12

since the surface of a liquid assumes a position at right angles to this resultant force, the water rises on the face of the glass. If, on the other hand, as in Fig. 12, where steel is shown in water, the cohesion of the liquid is greater than the adhesion of the liquid to the solid, then the resultant force lies to the right of the vertical, or within the liquid, which accordingly is depressed at the face of the solid.

In Fig. 11 and 12 the contact-angle is DOB. Since the surface of the liquid always assumes a position at right angles to the resultant force, the water will tend to rise on the glass and to sink on the steel. This angle of contact between a liquid surface and a solid is usually the same for the same pair of substances, but there is a subtle variation, which is called 'hysteresis' and it is said to play an important part in flotation. The variation is connected with the ability of a solid to condense a film of gas upon its surface. This gas-condensing power, or adsorption, can be modified, by acidulation, for example. Sulman has stated that "whereas the angular hysteresis of silica in plain water may exceed 30°, thus indicating that substance to have a definite power to occlude gas and to float, it drops from 4° to nil in water acidulated with sulphuric acid. Galena, on the other hand, retains its full measure of angular variation, or is but slightly affected."[13] This effect of the surface-energy of solids is apparently an important factor in flotation, and it is a pity that the exigencies of patent litigation have prevented Mr. Sulman from contributing more to the technology of the subject.

The angle of contact between water and glass is so acute as to be more nearly zero the purer the water and the cleaner the glass; between turpentine and glass it is 17°; between mercury and glass it is 148°. In a general way, subject to the variation already noted, the size of the contact-angle measures the capacity for 'wetting.' This angle can be changed by modifying the surface-tension of the water by means of a contaminant, such as oil, or the angle can be altered by modifying the surface of the solid, also by oiling. The oiling of the steel needle increased the angle of contact with the water so that it did not impinge as directly on the needle, and it did the same to the glass rod, but the effect was relatively less on the steel than on the glass because of the higher specific gravity of the former. The force tending to prevent sinking depends upon the radius of the needle, its density relative to that of the water, the surface-tension of the water, and the cosine of the contact-angle.[14] In metallurgical practice the pull of gravity is decisive in so far as it limits the size of particle that can be floated in water. If our needle is too large, it sinks, no matter how favorable the other conditions may be. So the flotation of a particle of mineral is conditioned on the size to which it has been reduced by crushing in the mill. The oiling of the needle increased the upward component of the surface-tension by enlarging the angle of contact, but the use of an excess of oil, that is, more than the needle could hold of itself, served to lower the surface-tension of the water and therefore to diminish the resultant force operating against wetting and in favor of flotation. Thus the oil used in flotation has two possible functions, and they may interfere with each other.

If to the water in which a needle is floating I add a drop of pine-oil, the needle sinks at once because the lowering of the surface-tension enables the water to wet the needle, that is, to diminish the angle of contact so that the water envelopes the steel. Let us make some other simple experiments. Take a piece of chalcocite that presents a smooth surface. A drop of water will not spread over it as it will on glass; the globule of water flattens itself on the glass but tends to retain its spherical form on the chalcocite. The glass may typify quartz or some other gangue-mineral. A drop of flotation-oil, such as coal-tar creosote, flattens on the chalcocite, whereas water maintains its sphericity. Coal-tar spreads less on glass than on water, but water spreads more on glass than on chalcocite. Thus water wets mineral less easily than gangue, whereas oil coats mineral more readily than gangue. So we say that gangue has a greater affinity for water than mineral, which, on the contrary, has a greater affinity for oil.

Water drips off oiled copper more quickly than off the unoiled; there is more adhesion between the water and the unoiled metal; the oil prevents wetting by the water. The effect of the density of the surrounding medium is shown by placing a piece of glass under water, dropping a globule of coal-tar upon the glass, and then raising it out of the water. The globule of oil spreads when lifted out of the denser medium and shrinks when returned to the water, although not quite to its first shape, on account of the adhesive surface. The oil on the galena replaces the water on its surface, but the oil on the quartz is unable to prevent the water from pushing itself underneath and over the surface of the quartz. Thus we have "an instance of the selective action of oil on a metallic sulphide in the presence of water, and the selective action of water on a gangue-mineral in the presence of oil."[16] On this phenomenon largely depends the process for separating valuable mineral from worthless gangue by flotation.

If a piece of galena and a piece of quartz are placed under water on the bottom of a beaker and if a few drops of oil, such as wood-creosote, are dropped upon the water, they will descend through the water owing "to their momentum and the releasing of the surface tension of the water"[17] until one may fall on the galena, on which the oil will spread, while another falls on the quartz, on which it tends to draw into globular form, instead of spreading. Flotation is essentially a selective process. If I throw powdered ore on water, the particles of gangue sink and the particles of mineral float, in accord with our expectation, based on the foregoing experiments and the deductions therefrom, but some of the small particles of gangue will float and some of the larger particles of mineral will sink, because the play of forces is so complex that any single one of them is not uniformly decisive. Flotation is preferential, not absolute.

(To be continued)

FLOTATION depends upon the presence of substances that will lower the surface-tension of water and are adsorbed by the mineral particles that it is desired to float.

SULPHURIC ACID and other electrolytes increase the surface-tension of water, but this increase is negligible unless acid is added in strong proportions.

[13]H. L. Sulman. Presidential address. Trans. I. M. & M., Vol. XX, p. XLVII.

[14]Joel H. Hildebrand. 'Principles Underlying Flotation.' M. & S. P., July 29, 1916.

[16]A. F. Taggart, as witness in the recent trial, at Butte.

[17]Taggart. *Op. cit.*

Methods of Driving Pine Mountain Tunnel

By H. DEVEREUX

The Pine Mountain tunnel is being driven in Marin county, California, at a point about three miles west of Fairfax. The tunnel is part of the proposed system for the Marin Municipal Water District, connecting the Lagunitas water-shed with the water supply system for the eastern part of Marin county. It is of horseshoe section, 8 by 8 ft. net, inside the concrete lining. The total length is 8700 ft., of which about 600 ft. near the two portals will have a 12-in. concrete lining, and the remaining 8100 ft., a 6-in. lining. The quantities per lineal foot are as follows: excavation, 3 cu. yd., theoretical amount of concrete in 12-in. lining, 1.2 cu. yd.; concrete in 6-in. lining, 0.6 cu. yd. The actual quantities of concrete will be about 35% greater, on account of overbreakage. The work is being done under the direction of M. M. O'Shaughnessy, consulting engineer for the district, A. R. Baker, engineer for the district, and C. T. Broughton, resident engineer on the work. The contract was awarded early in December 1916 to McLeran & Peterson, of San Francisco, at $257,400 for the entire work, or $29.70 per lin. ft. for driving and lining the tunnel. Work was commenced in the middle of December. Hand-drills were used until about February 1, machine-drills since that time. The actual cost of driving has been about the same for both methods, but the machine work is much more rapid. Up to May 15 about 2000 ft. of progress was made on the east end, and 1000 ft. on the west end. Concreting commenced about June 1. The maximum progress in driving for any one month has been 568 ft. in the east heading, or 19 ft. per day. The average progress since the machine drills were installed has been 13 ft. per day in the east end, and 10 ft. in the west end.

It will prove of interest to compare records of other American tunnels having nearly equal sections. On the Sheep Creek tunnel in greenstone and slate the maximum progress was 661 ft. per month and the average for a period of six months, 596 ft. per month. At the Alaska-Gastineau mine, an 8 by 10-ft. tunnel was driven 8800 ft. at a rate of 544 ft. per month. The maximum monthly progress in the Roosevelt tunnel, which had a 6 by 10-ft. section and was driven in Pikes Peak granite, was 435 ft., and the average was 292 ft. The Elizabeth Lake tunnel on the Los Angeles Aqueduct was driven 604 ft. in one month through black shale. The section was 12 by 13 ft. The Red Rock tunnel, also on the Aqueduct was driven 1061 ft. in one month through cemented sandstone. An 8 by 8-ft. tunnel for the Arizona Copper Co. was driven 799 ft. in one month through porphyry. The average monthly progress was 669 ft. The Gunnison tunnel, 6 by 10½ ft., was driven 824 ft. in one month through soft limestone. The Mt. Royal tunnel, 8 by 12 ft., was driven 810 ft. in one month through limestone. An 8½ by 9½-ft. tunnel at Mammoth, California, was driven 395 ft. in one month. The average progress was 316 ft. The Laramie-Poudre tunnel, 7½ by 9½ ft., was driven 653 ft. in one month through granite. The average was 525 feet.

EAST PORTAL OF TUNNEL.

The power-plant for the Pine Mountain tunnel is on the Bolinas road about three miles west of Fairfax. The east portal is about 1000 ft. from the power-house. The west portal is nearly two miles southwest from the power-plant. Until recently the west portal could be reached only by pack-train, but the trail has recently been improved so that light loads can be hauled over the mountain.

The compressor-plant consists of three 25-hp. Fairbanks-Morse Y-type semi-diesel engines, and three 8 by 8-in. Sullivan compressors. The pressure maintained at the compressor is 100 lb. per sq. in. A study of records on 25 other long tunnels shows that this is the average pressure maintained. The lowest was 85 lb., on the Strawberry tunnel, and the highest, 120 lb. on the Laramie-Poudre. Six tunnels used 100 pounds.

Angeles Aqueduct recommends a 25-lb. rail where concrete is to be hauled. Where heavy cars are used, a mechanical dumping-system is required.

Considerable trouble has been experienced at the west end of the Pine Mountain tunnel on account of water. There are 'pockets' of water that give a large flow for a few days and then run dry, making pumping a difficult matter. This has been remedied by cutting a ditch to drain the water back from the face of the tunnel. This matter of drainage is frequently a considerable item of expense in long tunnels. The cost of pumping at the Mile Rock tunnel was $1.30 per lin. ft. At the east end of the Strawberry tunnel, the wet end, the cost of pumping was $1.36 per lin. ft. At the Roosevelt tunnel a drainage ditch 4 by 6 ft. was excavated at a cost of $1.10 per ft. If an 8 by 8-in. ditch is carried under the track in the middle of the heading, lined on the sides and covered with a 2-in. plank at an elevation of 18-in. below grade at the same time as the rest of the drilling is done, the water-problem at the lower end of a tunnel will be easily and economically handled. A similar arrangement is advisable at the upper end of the tunnel, whenever the grade is such that the ditch need not be over 4 or 5 ft. deep at the portal.

WEST PORTAL OF TUNNEL

So far, but little timbering has been required. About 10% of the tunnel has been side and top-lagged, while for about 5%, timber sets in the arch only have been used with the top-segments lagged. In the first 2000 ft. of tunnel 20,000 ft. B.M. of timber has been used, or 10 ft. B.M. per foot of tunnel. The cost of placing timber full-lagged is about $15 per M.B.M. Where arch-sets and crown-bars are used the cost will be greater, about $25 per M.B.M. Where timber is cut in the woods nearby a man can cut and frame 1000 ft. B.M. in 4½ days.

The overbreakage thus far has been small, not exceeding 35% of the net area of the concrete section. On the east end, the ground is being drilled to the full section, on the west end to about 80% of the full section, requiring trimming along the entire length in the latter case. On the Los Angeles Aqueduct some tunnels were driven and trimmed so closely that the excess yardage did not exceed 15 or 20% of the theoretical yardage of concrete, but the cost of trimming amounted to as much as $2 per lin. ft. of tunnel. The conclusion with regard to hard-rock tunnels was that the excess yardage of concrete lining should not be over 30 to 40%. When a cubic yard of concrete in the net section was required per lin. ft. of tunnel, a 100% excess was valued at $6 to $7 and a 30% excess at $1.80 per lin. ft. It was found to be the best practice to excavate the sub-grade at the start so that the top of ties is at the bottom of the theoretical sub-grade, so as to avoid expensive trimming and delays when the concrete lining is placed. In the Mile Rock tunnel the theoretical quantity of concrete per lin. ft. of tunnel was 1.6 cu. yd., and the actual 2 cu. yd., or a 25% excess.

The cost of driving the Mile Rock tunnel was $4.75 per cu. yd. or $27 per lin. ft. of tunnel. The cost of lining was $9 per cu. yd. of concrete, or $18 per lin. ft. of tunnel. In general the cost of lining small-section tunnels where compressed air is used to make and place the concrete, and wooden forms are employed, is $1.60 per lin. ft. for forms and $1.40 per cu. yd. for labor and royalty in placing concrete, plus the cost of materials, aggregate, and power, and the distributed general expense and liability insurance. On the Mile Rock tunnel the overhead for general expense and insurance was 15% of the total cost to the contractor. On the Pine Mountain tunnel the liability insurance and bond are carried by the water district and do not appear in the contract price.

The following is the scale of wages paid on the Pine Mountain tunnel:

Drillers	$3.75	Dump-men	$3.00
Helpers	3.00	Foremen	6.00
Shovelers	3.25	Compressor-engineers	3.50
Teamsters	3.00	Pipe-men	3.00
Blacksmiths	4.00	Packers	3.00
Helpers	3.00	Common labor	2.50

The maximum monthly progress has been 568 ft. or 19 ft. per day. For this rate of progress, which was at the east end, the cost per lin. ft. to the contractor is as follows:

Drilling	$1.67	Teaming	$0.30
Shoveling	1.90	Packing	0.25
Hauling	0.67	Miscellaneous plant, pipe-line, track	2.00
Dumping	0.48	Heavy plant, engines and compressors	1.00
Blacksmith	0.37	Drill steel, repairs	0.30
Pipe-men	0.16	Timbering, material	0.20
Powder	1.80	Timbering, labor	0.12
Power	0.65		
Bonus	2.00		
Constr'n and maint'nce roads and trails	0.25	Total cost per foot	$16.12

A bonus of $0.25 per man per ft. is paid for each ad-

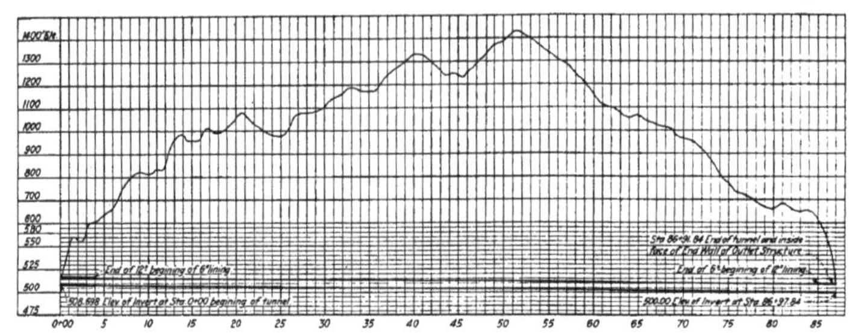

PROFILE OF TUNNEL

Air is conveyed through a 3½-in pipe, 1000 ft. to the east portal, and 10,000 ft. to the west portal. Three hundred cubic feet of free air per minute is supplied, which is sufficient to run three drills and also the forges. The loss of pressure at the west end does not exceed 2 lb. Records on nine long tunnels show capacities of plant ranging from 247 to 868 cu. ft. of free air per minute, with an average of 550 pounds.

The D'Arcy-Cox formula for the conveyance of compressed air in pipes is

$$D = \frac{c\sqrt{d^5}}{\sqrt{L}} \times \frac{\sqrt{P_1 - P_2}}{W_1}$$

Where

D = volume of compressed air in cubic feet per minute discharged at final pressure.

c = a coefficient, ranging from 45 for a 1-in. to 60 for a 6-in. pipe.

d = diameter of pipe in inches.

L = length of pipe in feet.

P_1 = initial gauge-pressure in pounds per square inch.

P_2 = final gauge-pressure in pounds per square inch.

W_1 = density of the air or its weight in pounds per cubic foot.

TABLE I

Diameter of pipe, inches	Value of $c\sqrt{d^5}$
1	45
1¼	105
1½	155
2	300
2½	530
3	875
3½	1300
4	1860
5	3300
6	5270

TABLE II

Values of $\frac{\sqrt{P_1 - P_2}}{W_1}$

Final pressure, lb. per sq. in.	Losses of pressure $P_1 - P_2$ in pounds 1	2	3	4	5	6	8	10
70	1.5	2.1	2.6	3.0	3.3	3.6	4.1	4.5
80	1.4	2.0	2.4	2.8	3.1	4.0	3.9	4.3
90	1.3	1.9	2.3	2.7	3.0	3.2	3.7	4.1
100	..	1.8	2.2	2.6	2.8	3.1	3.6	3.9

Example: Given a 3½-in. pipe, 10,000 ft. long, how many cubic feet of air per minute at an initial pressure of 90 lb. can be transmitted, with a loss of pressure of not more than 2 pounds?

From Table I, opposite a 3½-in. pipe, find 1300.

Square root of 10,000 = 100.

1300/100 = 13

From Table II, for drop of 2 lb. at 90 lb. final pressure, find 1.9.

$$13 \times 1.9 = 23.7$$

$$90 \text{ lb.} + 14.7 \text{ (atmospheric)} = 104.7$$

$$\frac{23.7 \times 194.7}{14.7} = 170 \text{ cu. ft. (approx.) of free air per minute.}$$

In using Table II, it will be noted that the initial and final pressures are taken as the same. Should greater

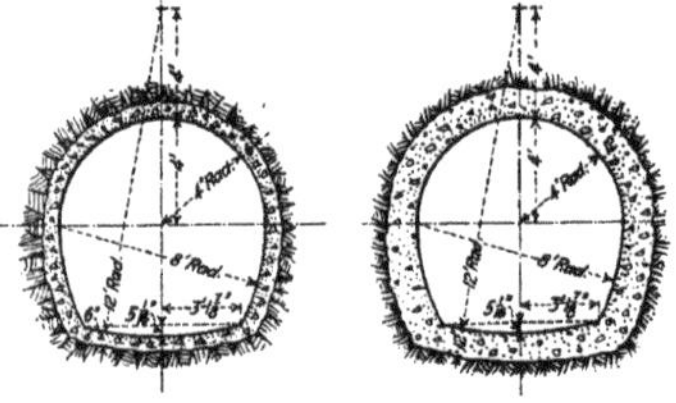

STANDARD SECTIONS FOR TUNNEL

refinement be required, interpolate for the difference, remembering, however, that a slight leak in the line may change the results.

Three No. 18 Leyner drills with a 1¼-in. chuck, 24-in. feed, and six sets of 1¼-in. hollow steel from 24 to 96 in. long, are used. There is an 18-gal. water-tank and an air-line manifold. The drills are mounted on horizontal bars. Jackhamers are used for trimming. On other long tunels, preference was about equally divided between vertical columns and horizontal bars.

The duty of a No. 18 Leyner drill in this tunnel is 5 ft. of hole per hour, using 1000 cu. ft. of free air at 100 lb. pressure per lin. ft. of hole drilled. The cost of drilling, including bonus is $0.20 per ft. This estimate allows for delays.

The cost of fuel and lubricating oil for a 135-day run was as follows: 11,000 gal. fuel-oil, 24 Baumé gravity, at $0.032 per gallon = $352; 250 gal. valvoline, strained and used twice, at $1.35 = $324; hauling, three miles from railroad, at $0.005 per gallon, = $55; total, $731.

The plant was run continuously and 62½ hp. was developed. This gives a cost of 0.36c. per hp. hour for fuel and lubricating-oil. This amount of power is sufficient for three drills and two forges. The cost of labor at the power-house is $320 per month. During March the total progress in both headings was 854 ft., and the cost for power was $0.65 per ft., $0.29 for fuel and $0.36 for labor. A 25-hp. Fairbanks-Morse Y-type engine has just been installed to operate a 220-volt, 25 kilowatt generator, which will be used to run a plant for lighting the tunnel and a rock-crusher for crushing aggregate for the concrete lining.

Purchased electric power on four other long tunnels cost from $1.65 to $2.15 per lin. ft. or an average of $1.90. Electric power generated on the work for the Elizabeth Lake tunnel on the Los Angeles Aqueduct cost $5.25 per ft., and on the Strawberry tunnel, where it was transmitted a distance of 23 miles, $5.50 per ft. On another tunnel, where steam was used, with wood for fuel, the cost for power was $2.50 per ft. Where steam was used with crude-oil for fuel, the cost was $2.28 for fuel, and $0.80 for labor.

INTERIOR OF COMPRESSOR PLANT

For ventilation, 1000 ft. of 10-in. pipe has been laid at each heading, reducing to 8-in. for the remainder of the distance. This pipe has been found to be too small, and is to be replaced with a 12-in. pipe, using a blower working at a pressure of 4.5 lb. per sq. in. To clear a tunnel of foul air in 15 minutes, which is the maximum time that should be allowed for delays after blasting, requires a capacity of 4000 cu. ft. of air per minute. For respiration allow 75 cu. ft. per man per minute, and 150 cu. ft. per animal. The average rated capacity of ventilating apparatus used on 16 long tunnels was 3400 cu. ft. per minute. The size of the ventilating-pipe ranged from 10 in. for the Carter and Mission tunnels, to 18 in. for the Elizabeth Lake and 19 in. for the Central tunnel. Where light-gauge sheet-metal pipes are used for ventilation, it is advisable to build a small bulkhead of track-ties in front of the pipe before blasting in order to prevent collapse of the pipe.

Table III gives the diameter of pipe in inches required to deliver 4000 cu. ft. of free air per minute for lengths of pipe from 1000 to 14,000 ft., and for pressúres from 1 to 6 pounds.

TABLE III

Length of pipe, ft.	Pressure, lb. 1	2	3	4	6
1,000	12	10	..	..	..
2,000	14	12	11	10	..
3,000	15	13	12	11	10
4,000	16	14	13	12	11
5,000	17	15	13½	12½	11½
6,000	..	15½	14	13	12
8,000	..	16	15	14	12½
10,000	..	17	15½	14½	13
12,000	..	..	16	15	13½
14,000	..	..	..	..	14

For illumination, acetylene lamps on the men's caps and candles have been used, but as already noted, an electric-lighting plant has now been installed. Acetylene lamps have been used on several long tunnels, small lamps being used on the men's caps and larger stationary lights being placed 150 ft. apart along the tunnel. This is an economical method of lighting. Electric lights are usually employed where electric hauling is done. The cost of lighting the Mile Rock tunnel with electricity was $0.50 per ft. of tunnel. In wet tunnels, electric lights are uncertain.

The rock penetrated so far in the east end of the tunnel is sedimentary with intrusive igneous rock. On the west end there is Franciscan sandstone, black serpentine, and hard boulders, with some diabase. The rock at the west end is harder than at the east end. The number of holes per round at the east end is 9 to 14, and 14 to 16 at the west end. Six feet of hole is drilled per round, the wedge-cut system of arrangement being used. Forty per cent L. F. gelatine powder is used, supplied by the Hercules Powder Co. In the east end, 9 lb. of powder, 19 ft. of fuse, and 4 caps are used per lin. ft. of progress. In the west end, 15 lb. of powder, 26 ft. of fuse and 4 caps are required.

Records of other long tunnels show that the wedge-cut was used in 19 instances, the pyramid-cut in 4, and the bottom cut in 7. An analysis of the depth of holes generally used in American practice, would lead one to infer that the most successful driving was secured when the average depth of the holes was from 60 to 80% of the

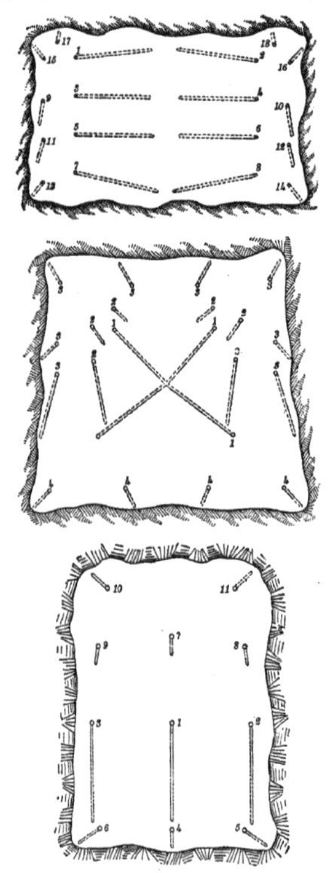

WEDGE-CUT, PYRAMID-CUT, AND BOTTOM-CUT SYSTEMS (From Brunton and Davis, Modern Tunneling)

width of the heading for wedge and pyramid cuts, and 60 to 80% of the height for the bottom cut. In cases where deeper holes had been used, and the depth of holes was later reduced according to the above rule, there was an increase of speed of as much as 20% and a decrease of powder used of as much as 25%. The number of holes per round is dependent upon the character of the rock. An approximate rule for sedimentary formations is one hole to 5 or 6 sq. ft. of face of heading, and in igneous formations, one hole to 2.5 to 4 sq. ft. of face.

The amount of explosive used in small tunnels ranges from 3.5 to 10 lb. per cu. yd. of material removed. Bottom-cut holes appear to require more powder than pyramid or wedge-cut holes. As regards percentage of gelatine powder used, practice has ranged from 40% in six tunnels to 100% in the Roosevelt tunnel, which was driven through Pikes Peak granite. Loading the bottom of the hole with 80 to 100% powder and the rest with 40 to 60% has given good results in a number of cases. On the Grapevine division of the Los Angeles Aqueduct 40 and 60% ammonia powder was used. There was comparatively little difference in the effect of the two grades. Ammonia powders are affected by moisture and are not suited to wet tunnels. The gases from the ammonia are disliked by the workmen.

It is best to place the cap near the top of the charge and tamped with powder, although many powder-men prefer placing the cap at the bottom of the hole. A cap has recently been placed on the market that acts like a time-fuse, enabling the cut-holes, relievers, back-holes, and lifters to be fired at intervals and in rotation by an electric battery.

On the east end there are two drillers and one helper per shift, and one driller and one helper per shift on the west end. Three shifts are worked and as many as five rounds or 30 ft. of progress has been made in 24 hours, requiring 150 ft. of drill-hole per drill for each 24 hours. Four shovelers are employed on a shift in each end, two working at any one time and two resting. They handle from 10 to 15 cu. yd. per man per shift. The shovelers use square-pointed shovels, and shovel from steel floor-plates, ¼ in. by 4 by 6 feet.

The material is transported by mules in turn-table end-dump cars of 25 cu. ft. capacity. When the work has progressed so far that mules cannot handle the material economically, it is proposed to remodel White or Stanley steam-automobiles to do the hauling. Such remodeled machines are now being used successfully on the Twin Peaks tunned. At the east end the material is dumped close to the portal. On the west end it has to be hauled for some distance. The track has an 18-in. gauge and is laid with 27 lb.-rails. The original intention was to use 25-lb. rail, but it happened that a quantity of 27-lb. rail was available.

There seems to be no established practice regarding types of cars or track for this class of work. On 7 tunnels, turn-table end-dump cars were used. on 4 common end-dump cars, on 7 side-dump cars, and on 4 rocker-dump cars. On the Los Angeles Aqueduct, rocker-dump cars with a capacity of 32 cu. ft. were used, while on the Catskill Aqueduct side-dump cars with a capacity of 40 cu. ft. were employed. Sizes of cars varied from 1 ton up to 40 cu. ft. The smaller cars are more convenient since in narrow tunnels the empty cars can be made to pass the loaded ones by simply tipping the former off the track. Weights of rails have ranged from 12 to 36 lb., and gauges from 18 to 36 in. The final report of the Los

ditional foot over 14 ft. per day, to all men who remain during the month. The cost at the west end is about the same as at the east end. The rate of progress is less, but on the other hand there is no bonus. Assuming that the excavation runs 3 cu. yd. per lin. ft. of tunnel, the cost per cu. yd. is approximately $5.40.

Other miscellaneous items are as follows: Cost of camp-buildings, etc., $25 per man for a 70-man camp. The cost of laying the 3½-in. pipe for the compressed-air line over the mountain was $0.08 per ft. for distribution, and $0.04 for laying, a total of $0.12. The cook-house, caring for 40 men, uses 1 bbl. fuel-oil in 8 days.

Memoranda on Submarines

The Engineering Committee of the National Research Council issues the following data to guide those desirous of helping to circumvent the Enemy's submarine campaign by means of invention and suggestion. Any communication on the subject should be addressed to W. F. Durand, Vice-Chairman of the Committee, at Washington.

Submarines operate singly or in groups as may seem best suited to local or special conditions.

They are supposed, where circumstances favor, to lie on the bottom at rest and with listening devices attempt to detect the approach of vessels. On receipt of evidence that a vessel is approaching, they rise to a level permitting observation with periscope, and then manœuver accordingly. When in water too deep to permit lying on bottom, the submarine must maintain steerage way in order to hold its level of submergence. The minimum speed at which this can be done will range with circumstances from 2 to 4 knots. The maximum depth of submergence is about 200 ft. The usual depth of running is from 50 to 100 feet.

They have been supposed to return to the home base at intervals of 30 to 35 days. The total radius of action will presumably range from 5000 to 8000 miles at a moderate cruising speed of 10 or 11 knots. The high speed emerged will range from 14 to 18 knots, or possibly more in latest designs. The maximum submerged speed is about 10 knots.

Hidden bases have been presumably used off the Irish and other coasts. There have also been suspicions of bases on the coasts of Greenland and Iceland. Submerged bases for oil and supplies have also been employed.

The time required from emergence to submergence will range from 1 to 3 or 4 minutes, according to circumstances. When submerged near the surface, the time required to raise the periscope, take a quick observation and lower it again, may range from 15 to 30 seconds. If desired, the submarine can follow an undulating path, rising and submerging alternately, at frequent intervals, at will. Or otherwise it may run fully submerged but near the surface and take frequent observation through the periscope. Modern submarines are provided with two or three periscopes. The loss or destruction of one, therefore, will not necessarily disable the boat.

Torpedoes fired from submarines are presumably aimed by changing the direction of the boat. This, however, is not assured in all cases. The torpedo, in order to run true, must travel at an immersion of about 10 ft. In smooth water it may be run at a shallower depth than in rough water.

Submarines may operate at night with less liability of detection, but with, of course, greater difficulty in picking up their target.

Submarines use the gyroscopic compass.

Sounds produced by the movement of a submarine through the water, including those traceable to the propeller, to movements of the rudder, etc., should permit of detection by the use of modern refined sound detecting devices.

The distance at which a protecting net, plate, or shield or other means of exploding the torpedo before reaching the side of the ship must be located in order that such distance will render the effect of the torpedo harmless, will depend primarily upon (1) Weight of explosive charge, (2) Depth of torpedo when exploded, (3) Strength of the ship's structure. With modern torpedoes and a depth of 10 or 12 ft. and with the structure of modern merchant ships, distances of 20 or 30 ft. would perhaps be required in order to give good assurance against injury. With rough water and possibly much less submergence at the time of explosion, reduced distances of 15 or 20 ft. might prove sufficient. Experimental investigations on this subject show a very wide divergence among the results and no precise rule can be given. It may be added, however, that naval constructors generally are satisfied that the distance at which protecting plates or shields would have to be placed in order to secure immunity is so great as to render their use of very doubtful practicability.

British export prohibitions indicate the relative importance of many of the minerals and metals in the conduct of war as revealed by the experience of our Allies. The prohibited articles are classified under three heads, in the order of their importance. Class A are the most necessary, and their exportation is absolutely forbidden; class B may be exported to other parts of the British Empire; while class C may be sent to the Allies of Great Britain but not to neutral countries. Under class A are the following: lead compounds; manganese compounds; mercury nitrate; nickel nitrate; sulphur and preparations containing that element; compounds of titanium and zirconium; all articles manufactured wholly or in part of copper; galvanized sheets; iron and iron alloys containing chrome, cobalt, molybdenum, nickel, tungsten, or vanadium; magnesite; magnesium and its alloys; mercury, platinum and alloys of platinum; railway materials; silicon-manganese; special steels containing tungsten, vanadium, or molybdenum; uranium in any form; iron wire; zinc ashes, zinc and its alloys, aluminum in any form; and zirconium and its alloys.

Recent Vulcanism in Salvador

By C. ERB WUENSCH

San Salvador, the capital of El Salvador, the diminutive republic of Central America, is situated in a region of unusual vulcanism. It was the volcano San Salvador, five miles west of the capital, that was responsible for the recent disaster. This volcano has been quiet since its terrible eruption late in the 16th century. In the year 1625, Thomas Sage, the celebrated English traveler, mentions the appearance of a new volcanic vent in the place now occupied by the volcano Izalco, 35 miles west of the capital and 15 miles north-east from the Pacific seaport of Acajutla. It was, however, not until 1770 that this new vent assumed the status of a volcano. It has, until a year ago, been continuously active. Where the volcano now stands was formerly a large level plain, upon which was situated one of the richest and largest cattle estates of the old Spanish days. During this short period of activity it has built its cone from practically sea-level to its present altitude, which, according to Sonnensternst, is 4973 ft. above tide. Ever since its beginning, great columns of smoke, accompanied by frequent irregular eruptions of small magnitude, continually rose from its crater. It was a marvelous spectacle for the passengers on Pacific steamers as they passed the port of Acajutla, especially at night, when the heavens were illuminated periodically with weird red reflections accompanied by deep heavy rumbling sounds. Often the molten lava could be seen pouring out of the crater and running down the slopes of the volcano. Izalco was well named 'the safety valve of Salvador.' During its period of activity no great eruptions have taken place in Salvador, but slight earthquakes were frequently felt and small amounts of smoke were occasionally seen arising from some of the older volcanoes in the Republic.

About eight months ago, shortly after Izalco ceased its activity, the old San Miguel volcano 70 miles east of San Salvador, situated near the city of San Miguel, commenced to emit unusual volumes of smoke from its huge crater. This renewed activity gave the inhabitants considerable fear, but as the appearance of the smoke waned they ceased to be uneasy. In this connection it might be of interest to relate a strange bit of prophesy on the part of a geologic friend of mine in San Francisco. On June 7 I called upon him, and in the course of conversation he inquired particularly about the volcanoes of Salvador. When informed that Izalco had become quiescent he asked "Is there any new volcanic activity?" Upon being told no, he shook his head and said, "That means trouble." The very next morning cablegrams were received telling of the volcanic eruption and earthquake at San Salvador. This eruption took place through its vent on the western side of the volcano, on the side distant from the city. It was because of the position of the vent that the capital was not more severely damaged. Quezaltepeque, a town nine miles north-west of the city of San Salvador, suffered the most damage.

In this region calcareous and argillaceous formations, derived from the weathering of the volcanic tuffs and ashes, predominate. They possess features characteristic of sedimentary rocks for which they might easily be mistaken. In the immediate vicinity of San Salvador volcano is a large granite porphyry intrusion through the basaltic lava of which the cone is composed. In the lava-beds of Izalco is found a mineral salt, chloride of ammonia associated with sulphur. The lava is stained a variety of colors; yellow, green, red, and purple, due to the oxidation of small amounts of various metallic sulphides.

A possible explanation of the transference of the activity from volcano Izalco to that of San Salvador may be found in the sealing up of the vent of Izalco, the molten magma then stoping its way laterally until it made a connection with the older vent in the volcano of San Salvador, which had been occupied by a crater-lake. The heat from the magma coming in contact with the water seepage from the lake generated sufficient steam to shatter the rocks, relieve the pressure of the superincumbent column of rock, and start the eruption.

It might be of interest to recall the unusual volcanic activity that occurred in 1880 in Lake Illopango, a crater-lake eight miles long and five miles wide and from 500 to 1500 ft. deep, situated six miles east of the capital. The surface of the water is 1200 ft. below the mean level of the surrounding plain. This suggests that here was once a volcano of great size. In that year two volcanic cones rose from the depths of the lake and extended about 200 ft. above the surface, and ejected smoke and ashes. The water subsided 40 ft. and found an outlet into the Jiboa river and flooded the surrounding country. Goodyear estimated that 635,000,000 cubic metres of water was released from the volcano. Other famous volcanoes in Salvador are Santa Ana, with an elevation of 6615 ft., very slightly active at the present time; San Miguel, elevation 6500 ft., also slightly active; and San Vicente, elevation 7793 ft., long extinct. Along the slopes of many of the volcanoes are situated the richest coffee plantations in the country. The volcanic ashes make an exceedingly rich soil. A cablegram received from President Melendez of Salvador stated that the ashes from the present eruption would prove beneficial to the soil and offset some of the damage done by the earthquake.

CO-OPERATION of American and French and English physicists has been sought by the National Research Council in an effort to find means to combat the submarine. A conference for this purpose has been held in Washington, at which Charles Fabry, Henri Abraham, M. le Duc de Guiche, Sir Ernest Rutherford, and Commander Cyprian Bridge were present, along with a number of American physicists. It is hoped to obtain the co-operation of experimenters throughout the country having laboratory facilities at command. It is pointed out that the best laboratory equipment now available for work of this kind is found at the Universities in the United States.

Mining in Utah

By L. O. HOWARD

Several problems are being faced by local operators. At a time when metal prices stimulate intensified production, numerous strikes have served to interfere seriously with operations. Tintic and Park City have both been handicapped by labor troubles that were ultimately settled through the agency of a member of the new State Industrial Commission. Some annoyance was also caused at the plants of the Utah Copper Co. through small strikes of men engaged not directly in operation but in construction, repair, and maintenance. Finally came the strike at the International smelter that resulted in closing the works. Although the Federal Department of Labor has a representative on the ground, a settlement has not been effected at the time of this writing. The principal point of difference is an increase in wages of 50 cents per day. There has been no violence, but it is to be feared that when the furnaces are again blown-in the scarcity of labor will be felt severely, inasmuch as many of the men have left for other parts.

There is considerable inquiry for lead mines. Bingham, Park City, and Tintic are pushing production to the limit. Transportation in Big Cottonwood is still difficult, and maximum shipments have not yet been attained. The Cardiff has several hundred tons of ore scattered up and down the canyon during the winter, but is adding to its fleet of motor-trucks and expects to clean-up the accumulation soon. The Maxfield, which has been the only other persistent shipper, has been closed pending an investigation of its affairs. A new faction in the directorate has obtained control and has stopped all work. The ore has been coming from points several hundred feet below the adit-level, and incidentally below the level of the creek; therefore the excessive cost of pumping has prevented a profit. While the ore was rich, it occurred in small lenses so irregularly distributed as to cause development to bear an undue share of expense. It is probable that work will be continued on the [illegible]-ft. or adit-level in an effort to find other ore-shoots above that horizon, similar to those that proved profitable in the earlier history of the mine.

The Kennebec Mining Co., owning property adjoining the Cardiff, proposes to begin development with funds loaned to it *pro rata* by the stockholders, at 8%. This method of financing has been adopted owing to the non-assessable character of the stock. Development in many other properties in the canyon is said to be encouraging, notably on the Big Cottonwood Coalition and the American Consolidated Copper.

Little Cottonwood is active as never before in recent years. In the first three weeks of June the Michigan-Utah Consolidated Mines Co. made settlements on 36 [illegible], averaging for the most part $25 per ton. [illegible] cars netted over $70 per ton and two cars [illegible]. The company is able to ship $20 ore at a profit. It reports the beginning of shipments of copper-silver ore from the Copper Prince tunnel. The ore is said to be of greater extent than elsewhere in this locality. While the limits of the deposit have not been determined, it has been found that over a width of 30 ft. there is a uniform high iron content and that the copper varies from traces to as high as 20%, and much shipping ore has been blocked out. This development is daily gaining greater importance, and it is distinctly possible that a fair-sized copper deposit is to be opened in the Cottonwoods at last. Many engineers have been confident that excellent copper deposits would be found at the east end of both the Big and Little Cottonwood districts. The Michigan-Utah development tends to confirm this opinion as to Little Cottonwood, and showings recently made in the Big Cottonwood Coalition are likewise favorable.

The Copper Prince tunnel has been connected to the aerial tramway of the Michigan-Utah company. This conveys much Alta ore to Tanner's Flat, about four miles down the canyon, where it is loaded into narrow-gauge cars and dropped down to Wasatch, another four miles, on the tracks of the Little Cottonwood Transportation Co. At Wasatch it is again transferred into cars of the Salt Lake & Alta railroad, a branch of the D. & R. G., which carries it to the smelters at Midvale and Murray. Eighty-five men are now on the company's pay-roll, and it is intended to keep a force of 100 men at work during the year.

During the first quarter the Emma Consolidated shipped 33 cars of ore from Alta that netted nearly $30 per ton. This company is shipping 60 to 70 tons per day at present, and expects to go on a 100-ton basis soon. There has been some dissatisfaction with the facilities afforded by the Salt Lake & Alta railroad, and some ore is being hauled to the smelters from Wasatch in auto-trucks. The railroad company is preparing to lay heavier rail, after which better service is anticipated. The Alta Consolidated has started the shipment of some rich ore taken out during the winter. The Sells is at the point of shipping and is expected to make a steady output during the shipping season. The South Hecla maintains a regular production.

At Bingham mining is on a larger scale than ever. The Utah Copper Co. is milling 38,000 tons per day. The completion of the leaching-plant has been again delayed by poor deliveries of structural material and is now expected about the first of September. The work is 75% finished. The initial capacity is set at 4000 tons, although it is planned to increase to 10,000 tons per day as soon as possible. There is about 40,000,000 tons of oxidized material that should yield close to 13 lb. copper per ton. A short time ago the County Assessor announced his intention of taxing the tailing-dumps of the company, estimating that there is $68,000,000 worth of available copper in 57,000,000 tons of tailing, representing 35 to 40% of the original content of the ore, most of which could be recovered by flotation. If this policy of taxation is carried out it will affect many other companies in the State that have treated sulphide ores and accumulated tailing-piles containing recoverable metals.

If the County officials seek to enforce their ruling it will doubtless mean another legal battle. If the tax-gatherer is successful, what is to prevent his applying the same logic to oxidized tailing, or practically any kind of tailing that may contain economic minerals?

Further troubles are imminent in connection with the insurance rate, which the State Industrial Commission is seeking to establish under the provisions of the Workmen's Compensation Act. This act has been declared obligatory on workmen and on employers of more than four men. Some of the premiums per $100 of pay-roll are: assaying, $1.68; ore concentration, $4.04; smelting, $6.11; metalliferous mines, $5.59; and coal mines, $9. These rates are said to have been adopted from figures furnished by the Workmen's Compensation Service Bureau of New York, which, incidentally, is also engaged in an attempt to boost the Colorado rates another 25%. Colorado rates furnish an interesting commentary. They are $3.85 per $100 of pay-roll for metalliferous mines and mill operators. These rates enabled the State Insurance Fund to provide for all expenses, losses, compensation, and surplus, besides paying a dividend of 23% of the amount paid in. The local chapter of the American Mining Congress, in collaboration with the mine operators, has been conducting an exhaustive investigation into the matter and is making an earnest attempt to obtain a fair rate. The casualty expert engaged by these interests has recommended that the rates for metalliferous mines be reduced to $4.25, coal mines to $6.04, and that mill-rates be also reduced.

Reports of last month's ore shipments show that Tintic mines shipped 889 cars, bringing the total for five months to 4337 cars, and that Park City mines produced 8447 tons. The Judge Mining & Smelting Co. held second place with 2376 tons, leading the Silver King Consolidated, which reported 1595 tons, and trailing the Silver King Coalition, which shipped 2506 tons. It is expected that the large increase in shipments by the Judge company in June will place it at the head of the list. The Tintic 'car' seems to be as precise a measure as the Joplin 'can.' No close estimate of tonnage is possible from Tintic reports.

Among the Bingham mines, Ohio Copper reports a monthly profit of $70,000. A break in a reservoir on the Price river has resulted in the destruction of several miles of the main line of the Denver & Rio Grande track, and has completely isolated the Carbon County coal mines, aggravating the usual coal shortage in Utah. Several thousand men are thrown out of work. All through service on the railroad between Denver and Salt Lake City has been suspended. Several steel bridges went out and it is estimated that temporary repairs cannot be made in less than two weeks. According to later news the wash-out on the Denver & Rio Grande railroad will cause suspension of traffic for probably three weeks, and it is feared that this may result in the closing of the smelters for want of coke, the supply on hand being small. The district is completely isolated, so that normal conditions will not be restored for six weeks.

Government's Lead Purchase

The following communication has been sent by the Lead Committee to the mine operators, smelters, and refiners of lead throughout the country: The Committee on Lead appointed by the Advisory Commission of the Council of National Defense has agreed on behalf of the domestic producers of lead to furnish 8000 tons of pig lead between now and August 1, to meet the requirements of the United States Government. The price set for this tonnage is 8c. per lb. East St. Louis. As the probable production for the month of July will be about 48,000 tons of pig lead, this represents one-sixth of the total production for that month. If every producer agrees to furnish his share of this sale, it means that each will sell one-sixth of his July production on the basis of 8c. per lb. Are you willing to participate in this sale? If so, please advise the chairman of the Committee on Lead promptly. If you own a smelter you will receive shipping instructions from the committee for the amount which you agree to furnish. If you do not own a smelter, may we ask that you instruct the smelter which smelts your ores to furnish one-sixth of the lead-content of the ore which it accepts from you in July on this government order and notify them that you will accept in settlement for that amount of lead in your ore the price that the Government is paying. If you give such notice to your smelter, please inform the chairman of the Committee on Lead so that the Committee can make their plans accordingly and can see that your smelter is required to furnish that amount of lead to the Government at the 8-cent price. An early reply will be greatly appreciated.

CLINTON H. CRANE,
Chairman, Committee on Lead.

JAPANESE buyers are paying any price demanded in a scramble to purchase ship-plates. Recently 9.9 cents per pound was paid in order to switch 1000 tons of such material to the new ship-building Empire that evidently has dreams of becoming the world's carrier after the War. It is plain that control of supplies required in our preparations for effective warfare must come speedily if we are to do our part in resisting the common enemy. There is a serious defect in our organization when we cannot build ships to meet our own urgent needs, yet allow Japan to ship out of the country 1000 tons of ready-made ship-plate. It is important to recall in this connection, the suggestion of Great Britain that Japanese merchant vessels might relieve the pressure in the trans-Atlantic movement of supplies, which brought the soft reply that Japan would be pleased to supplant the British ships plying in Eastern waters so as to relieve them for service in the submarine zone.

BARYTES is being mined on a large scale near Pulantien, in the Kwangtung leased territory in Manchuria, by a company of Japanese resident at Pulantien. The corporation is called the Manchuria Barium Co., and is capitalized at $25,000.

REVIEW OF MINING

As seen at the world's great mining centres by our own correspondents.

CRIPPLE CREEK, COLORADO

LESSEES THROUGHOUT THE DISTRICT CONTINUE TO MAKE A LARGE AND PROFITABLE PRODUCTION.—NEW STRIKES REPORTED FROM VARIOUS PARTS OF THE DISTRICT.—IMPORTANT DEVELOPMENT FROM THE ROOSEVELT TUNNEL.

Block 8 of School Section 16, on the north-east slope of Bull Cliffs, is again in the producing list. Charles Eaton and associates are operating a sub-lease from the Co-operative Mining & Development Co., that holds the original lease on the school lands from the State. Eaton & Co. loaded out their initial shipment of milling ore last week. This property, the only one in the Cripple Creek district, located on school lands that is paying royalties to the State, was operated for 15 years consecutively by the late Alfred La Montaigne, a French Canadian. La Montaigne, it is estimated, took out about $250,000 gross from above the 600-ft. level.

The Millasier Mines Corporation, operating and owning the Clyde property on the north-east slope of Battle mountain, has unwatered the Clyde shaft 800 ft. deep, with about 3000 ft. of laterals at the bottom level. The shaft is to be sunk an additional 1500 ft. G. F. Lasier of Detroit, Michigan, one of the owners, is at the mine.

The Dante mine on Bull hill, owned by the Dante Gold Mining Co. is to resume activity. The lease held by the Consolidated Mines & Development Co. is to be transferred to a new company being organized to operate the mine.

The Hahnewald brothers, Olsen & Co., operating the property of the Gold Sovereign Mining & Tunnel Co. on the south-west slope of Bull hill, are cutting a station at the 1500-ft. level. Extensive development both north and south of the shaft has been planned. The workings on the property are 1350 ft. deep.

A shipment of milling-grade ore was loaded out last week from the Coriolanus mine on Battle and Squaw mountains, by Matt Edr and Aitken of Victor, lessees, operating under lease from the Aloha Gold Mining Co., T. B. Burbridge, of Denver, president. The ore shipped was of milling grade.

The Catherine Gold Mining Co., Charles Walden, of Victor, general manager, holding a lease and option on the properties of the Last Dollar Gold Mining Co. on Bull hill, is extending a drift south-east at a depth of 1500-ft. toward the Modoc mine and on the extension of the Modoc-Last Dollar vein. The ground under development is virgin and with the value of the ore improving it is expected an orebody is near. The breast of the drift is about 600 ft. from the line so that there is ample ground ahead.

The Acacia Gold Mining Co. is cutting a station at 1350 ft. preparatory to beginning lateral work at this depth. In the meantime the company continues production from the 1250-ft. level, and lessees are operating in the levels above. About 250 tons of milling ore has been loaded out from the mine this month.

Last week a strike was reported from the 400-ft. level of the Jerry Johnson mine and during the past few days, the Cripple Creek Deep Leasing Co., operating below the 650-ft. level of the mine, has entered the downward extension of the Catey shoot at the 650-ft. level, by a raise from the 950-ft. level. Two feet of the 4-ft. vein is sampling 5 oz. gold per ton and the ore broken 3½ to 4 ft. wide will ship at 2 oz. gold per ton.

A carload settlement, under date of June 25, on 63,000 lb. net of ore from the Caley lease, was at a rate of $46.90 per ton. The check to the lessee after deduction of freight, treatment, and royalty amounted to $1233.19.

The Beacon & Raven Hill Gold Mining Co., owning 17 acres patented, at the Arequa townsite on the southern slope of Raven hill, has contracted 250-ft. of driving from the line of the Roosevelt tunnel, of the Cripple Creek Deep Drainage & Tunnel Co., to get under the orebody developed to a depth of 700 ft. by the Elkton Mining & Milling Co. The Beacon & Raven Hill property adjoins the Elkton mine, and that company mined a good grade of milling-ore to that depth. The work from the drainage-tunnel level, at an elevation of approximately 8100 ft. above sea-level, is being followed with interest by mining men. The lateral will cut under the ore-shoot 900 ft. deeper than any previous development in this part of the district. Low assays are already obtained from the drift, and there remains about 30 ft. to drive before the objective point under the shoot developed above is reached.

A reported strike on the 20th level of the Golden Cycle mine of the Vindicator Consolidated Gold Mining Co. was practically confirmed last week by a visit of Guildford S. Wood, president; Adolph Zank, the treasurer; George Stahl, the secretary; and Mr. Sigel, a director of the mine. From a local source it is learned that a strong orebody has been opened at a depth slightly exceeding 2000 ft. in the main shaft of the Golden Cycle mine, on one of the main veins, and the ore exposed as broken from 6 to 8 ft. wide, is all of a good milling grade, with a central strip several inches wide that may be classed as of smelting grade.

TREADWELL, ALASKA

MINERS AT KENNECOTT STRIKE ON BEING REFUSED AN INCREASE IN WAGES.—THE TREADWELL COMPANY TAKES AN OPTION ON THE RED DIAMOND GROUP OF MINES.—EXTENSIVE PREPARATIONS BEING MADE TO EQUIP AND OPERATE THIS PROPERTY.

Two hundred miners working in the Bonanza and Jumbo mines of the Kennecott Corporation have walked out. They demanded an increase of wages of 15 to 50%, according to the price of copper. An offer of arbitration was made in behalf of the company, but it was refused and the strike followed. About half of the men have refused to strike, however, and these include the men in the shops, mills, and leaching-plant, and the construction men.

The threat to strike came several days ago and at that time the United States marshal took the situation in hand and as a precautionary measure closed the saloons at McCarthy.

The Treadwell company has taken an option on the Red Diamond group of claims adjoining its holdings on the west and will begin prospecting at once. This enterprise will create a greater demand for labor than now exists and just how that demand will be satisfied is a problem. Any kind of labor is very scarce in this part of the country, hundreds of men having left for the West and the interior within the past few weeks. Employers of labor, however, expect better conditions in the near future.

The Red Diamond property is the first to be bonded on Douglas island by the Treadwell companies with over $1,000,000 development fund available. The property is situated south by west of the Nevada Creek group fronting toward Stephens passage. Engineers have surveyed an outlet from back of the Ready Bullion property, which would give a passage for the ore by tram or surface railway to the company's mills at Treadwell. The Red Diamond group comprises 12 claims. Work has been carried on at various times for the past ten years, development consisting of a shaft and drifts on the orebody at several levels. A summer-camp is to be established next week at the property. In the meantime, provisions, tools, and other necessities are being delivered to the property.

P. R. Bradley, general manager, has announced that his engineers are ready and willing to investigate any property presented.

A new compressor has recently been installed at the Mineral King Mining Co.'s property at Bettles bay.

Articles of incorporation have been filed at Valdez for the Q & Q Gold Mining Co., the capital stock being $50,000. The incorporators are Frank Cockrell, Dr. H. Cockrell, and Robert L. Hawkins. The company is organized for the purpose of owning and operating property near Port Wells.

TORONTO, ONTARIO

HEAVY SHIPMENTS OF SILVER ORE AND BULLION FROM COBALT.—OLD MINES RE-OPENED PRODUCING HIGH-GRADE ORE.—HOLLINGER PASSES A DIVIDEND.—SERIOUS COAL SHORTAGE FEARED IN CANADA AND OFFICIAL STEPS TAKEN TO PROVIDE AGAINST IT.

The silver mining industry of Cobalt was never more active than at present, operators being desirous of maintaining production at the highest point in order to take advantage of a favorable market. Some 2000 men and about 200 machines are steadily employed, and despite the talk of strikes the majority of the workers are well satisfied with existing conditions. They are receiving in addition to wages a bonus of 50c. per day so long as the price of silver remains above 70c. per oz., and should it rise above 80c. the bonus will be increased to 75c. per day.

Shipments of both ore and bullion have recently been heavy. During the week ended June 16 approximately 956,404 lb. of ore was sent out with 517,666 oz. of bullion, being the largest aggregate for several months. The Nipissing maintains its lead as the largest producer. During May the company mined ore of an estimated value of $261,668 and shipped bullion from its own and custom-ores of an estimated net value of $405,000. Driving on the Cobalt Lake fault has been nearly completed and about 1500 ft. of the vein developed at the 425 and 520-ft. levels. With the payment of its regular 5% dividend, due in July, the total returns to shareholders made by the company will amount to $16,240,000, being 264% on the issued capital. The production of the Mining Corporation of Canada for the present year, up to April 22, aggregated 1,400,123 oz. of silver, the output showing a steady increase.

Development work at the Ophir has resulted in the discovery of a promising 6-in. vein, carrying a small silver content, south of shaft No. 2. It will be cross-cut and developed on the keewatin-diabase contact.

The report of the Beaver Consolidated for the quarter ended May 31 shows silver in bullion, due from smelters, and in ore bagged aggregating 252,948 oz. and $49,915 cash in hand. High-grade ore and mill-rock are being recovered from orebodies on the 400 and 600-ft. levels. A raise has been driven for 100 ft. on the vein at the 1600-ft. level. The demand for mining machinery for new prospects in Northern Ontario, which cannot always be filled promptly by the manufacturers, has created a market for the disused equipment of many of the closed-down properties of Cobalt and some of these small plants, which were considered as of little value, have been sold at good prices.

The Gowganda silver area has been attracting a good deal of attention, largely on account of the size and richness of the vein found on the Miller Lake-O Brien. The Reeves-Dobie property has been re-opened with encouraging results, and high-grade ore is being sacked. The T. C. 177 company is developing a property adjacent to the Miller Lake-O'Brien and will sink to the 300-ft. level. At the Silverado machinery is being installed. Power for this district will be obtained from Hanging Stone Falls, but construction has been considerably retarded on account of labor shortage.

Though the labor situation in Porcupine and the outlying gold district is not yet altogether satisfactory, some improvement has been effected by bringing in laborers from other points. The Dome Mines is continuing operations as well as its reduced forces will permit, attention being principally centred upon the cross-cut at the 700-ft. level to open up a large high-grade orebody discovered at that depth by diamond-drilling. It is stated to be 120 ft. wide with an average of $17 per ton, and is expected to be reached early in July. Hitherto the average grade of the mine has been below $6 per ton.

The Hollinger Consolidated has passed the dividend due this month. President Timmins states that this action was taken owing to the uncertainty of labor conditions and the difficulty of securing enough men for underground work to mine sufficient ore to keep the mill in operation. The directors are considering the advisability of devoting all efforts to development so long as present conditions continue and to put the mine into a position to considerably increase the output as soon as an adequate supply of labor becomes available. The suspension of the dividend will enable the company to wipe out the deficit of $174,184 and establish a cash surplus. Construction work on the new mill has been suspended except so far as contracts already let are concerned.

During May the McIntyre milled 15,064 tons of ore of an average value of $9.83 with a total production of $142,476. The West Dome Consolidated has completed 1200 ft. of workings on the 300-ft. level and it is estimated that between 60,000 and 70,000 tons of ore of an average grade of $9 per ton has been blocked out. The policy of the management is to confine operations to this level and create a large ore reserve for the proposed mill.

At the 200-ft. level of the Kirkland Lake mine a 4-ft. vein carrying high-grade ore has been discovered. It is believed to be the vein of the Wright-Hargreaves and the Teck-Hughes. The Lake Shore mine lying between these two will also benefit by the find. Construction work on the new Lake Shore mill is being pushed.

The question of providing against a serious coal shortage during the coming winter is receiving much attention from the Canadian government and the municipal authorities. The situation has become much more threatening since the United States declared war, as, owing to the increased home demand with a diminished supply of labor and a lack of adequate transportation facilities, a great shortage of the importations from the American coal mines, on which Ontario and a large part of the West are dependent for food, apears inevitable. Many municipal bodies are endeavoring to arrange for supplies to be stored as a provision against a winter coal-famine, but so far with little success. The Canadian government has appointed C. A. Magrath, of Lethbridge, Alberta, a prominent Western man, as fuel controller with full powers to regulate the price and distribution of coal, encourage increased home production, and arrange for importations. It is hoped that at least he will be able to prevent undue accumulations of coal stocks, and to check extortionate prices. Dr. Ruttan, of the Advisory Council of Industrial and Scientific Research, announces that their investigations have resulted in a discovery which will solve the problem as regards Western Canada. The extensive lignite deposits of Saskatchewan, he claims, may be made commercially valuable by a process of briquetting, which will produce a fuel nearly equal in heating power to anthracite at about two-thirds the cost of coal in the West. The Advisory Council has asked the Government to devote $400,000 for the construction of a plant in which the briquetting process can be carried on.

PORCUPINE, ONTARIO

Labor Troubles Are Being Gradually but Quietly Adjusted Though All of the Operating Mines Are in Need of More Men.

Within the past week much of the uncertainty that for many weeks has characterized the labor situation in the Porcupine district has been cleared up. It is rumored that the strike vote taken Sunday, June 14, resulted in a majority of more than 300 against going out on strike. However, the officials of the union are keeping the returns to themselves. As yet, no definite arrangement has been reached at any of the producing mines, except the McIntyre-Porcupine and Dome Lake. The former company has granted a 50-cent increase to meet the high cost of living, and appears to have clear sailing. Dome Lake has granted an increase almost equal to that of the McIntyre, but it is understood the increase does not cover every branch of labor at the mine. At practically every mine but the Hollinger Consolidated, the employees have taken action, independent of the Miner's Union, and by forming themselves into committees representative of the particular mine at which they are employed, have approached the management, that in turn has in every instance given consideration to these requests.

It is the opinion that within a few days most of the producing mines will have taken action similar to that of the McIntyre-Porcupine.

The most peculiar fact about the situation is that at Timmins and South Porcupine reports are current that certain companies have granted an increase. These reports sometimes are convincing, but when investigated appear to have as yet no foundation in fact.

Hollinger Consolidated, although with only about half enough men available, is operating as aggressively as circumstances will permit. Outwardly there would appear to be no difference about the mine from a few months ago, when operations were at the maximum, but the tonnage coming from below speaks plainly, and convinces one that inwardly present working forces are not anywhere near the requirements. However, despite the fact that the regular disbursement of dividends at the Hollinger have been discontinued, it does not by any means signify that this mine has taken second place to any other gold mine in the Dominion. In fact, with present [illegible] forces, the net profit-earning power of this company is [illegible] greater than any other gold or silver mine in the country. [illegible] figures showed net profits of $194,000 for the four-week period.

McIntyre-Porcupine is perhaps in the most enviable position of all the Porcupine producers, in that the company has very little construction work pending, and at present has a full complement of men. Operations at this property are going with greater speed than ever before. The tonnage treated sometimes rises to 550 tons per day, and the monthly average is well above 500 tons per day. Mill-heads range from $7 to $15. This is due to the fact that the management is not endeavoring to establish an average grade, but is taking run-of-mine ore, and anything containing $2 per ton and upward is being sent to the mill. The average grade of ore treated within the past 12 months has approximated $10, while the average throughout the mine is about $12.50. Preparations to carry development work first to the 1300 and then to the 1600-ft. level are now under way. The most interesting development at the mine recently is that the main orebody at the 1000-ft. level has now widened out to 52 ft. and is high-grade milling ore. The face of the main drive is now within 20 ft. of the Jupiter line.

Mill-construction at the Schumacher is proceeding. About 80 men are now on the pay-roll as compared with a desired force of 120. The old mill is running at a capacity of about 100 tons daily and mill-heads are now up to around $10 per ton.

Porcupine V. N. T. is working about 90 men and here also the mill is treating about 100 tons of ore per day. Developments underground are proceeding satisfactorily. At the 600-ft. level the main vein is 22 ft. wide of a good milling grade of ore.

Porcupine-Crown now has about 90 men employed, and is maintaining production satisfactorily.

COBALT, ONTARIO

Production Being Forced Under the Stimulus of the High Price of Silver.—Labor Agitators Continue Their Efforts to Bring About a Strike, Though the Majority of the Workers Are Satisfied.

Operations at the producing mines of Cobalt were never conducted more vigorously than at present. Over 2000 men and about 200 machines are employed. The high quotation for bar-silver is undoubtedly the main reason for the keen desire to force production to the maximum. The mine-workers are all receiving a bonus of 50c. per day when the price of silver remains above 70c. an ounce. This, in itself, has, to a certain extent, won the good-will of the workers. However, should silver rise to over 80c. an ounce the bonus will be increased to 75c. per day. With a 75c. bonus the workers would be receiving considerably more per day than that outlined under the wage-scale now demanded. Under the present system the workers are able to share the prosperity of the mine operators, and the married men, together with the efficient workers, would apparently not welcome any change in the present form of pay. The agitators in the union are largely unmarried men with little responsibility.

During the first week of June a total of nine cars of ore containing approximately 682,210 lb. of ore was sent from Cobalt. Bullion shipments for the corresponding period totalled $114,063.

The report of the Mining Corporation of Canada for the first 16 weeks of the current year shows that a total of 1,400,123.96 oz. of silver was produced having a value of $1,000,000. Total dividends paid to date by this company are $1,711,875.

The usual quarterly dividend of 5% has been declared payable July 20 by Nipissing Mining Company, a disbursement of $300,000, and the third dividend of this amount to be paid during the current year. The total paid by Nipissing to date is $16,240,000, or equal to 264% on the issued capital of the company. The financial statement of June 2 shows cash in bank $1,255,034; ore and bullion in transit, $351,860; ore and bullion on hand, at the mine, $853,614; making a total of $2,460,508.64.

The Temiskaming has also declared a dividend of 3%, payable July 16 to shareholders of record June 30. This disbursement amounts to $75,000 and is the second dividend during the current year. Temiskaming has now paid a total of $1,834,156.25, which is equal to 74% of the issued capital stock of the company. Generally speaking, the labor supply at Cobalt is comparatively satisfactory.

SUDBURY, ONTARIO

Experiments Being Made at Iron Mountain Said to Be Satisfactory.—A New Mill Completed and Running.—Large Mass of Iron Ore Available.

Iron Mountain, in the Sudbury district, is receiving a thorough test, and results so far obtained are said to be satisfactory. About 150 men are employed, and should the experimental work now going on justify it, the number probably will be increased to 1000 men. A mill with a capacity for 300 tons of ore per day has been installed and preparations for the concentration of ore are about completed. Mining is done by glory-hole method, and the ore is loaded direct into a train of cars and conveyed to the mill. The mill concentrate is shipped to smelters.

The Iron Mountain iron deposit is understood to contain at least 7,000,000 tons of ore. The deposit is situated four miles from Milnet Junction on the main line of the Canadian Northern railway. A spur-line has been built to the mine. Later on the facilities for iron ore production will probably be greatly increased.

ELDORA, COLORADO

Large Output of Tungsten from Numerous Leases on Prominent Mines.—Mills Running at Capacity and Extensions Being Made.

The Vasco company at Tungsten, which has been steadily operating its mines and mills through the winter, sees a continuation or even an increase in its prosperity this summer. The company's large mill is working at full capacity to handle the continuous shipments from over 12 different workings on the Vasco property. Shipments range from 3 to 50% tungsten, with an average of about 10%. Fred Barrett, one of the oldest mill-men in Boulder county, is superintendent of the mill. Fred B. Copeland, of the Copeland Sampling Co., in Cripple Creek, is in charge of the sampling for the company.

H. S. O'Neil and Harry O'Day are leasing on the Vasco No. 3. Bradford Black has leased No. 4, and is now clearing out one of the old stopes. No. 5 has been taken by John McKenna, who is doing exceptionally well. Cox and McKenzie have leased No. 6, and Thomas McGrath No. 7. John Walsh is leasing No. 8, and has been making regular shipments for some time. George L. Holland and associates have made the No. 10 tunnel the largest producer on the Vasco at the present time. J. B. Newham on 12, and Petro & O'Day on 13, are doing fine.

The change of this district from the Stevens Camp, of a little more than a year to the prosperous little town of Tungsten can hardly be realized. Great credit is due the Vasco company in opening up this district.

Tom McGrath has a position with the Morgan-Tungsten Mining Co., where he is superintendent. He was until recently with the Vasco people in the same capacity. This mill was recently taken over from Diggs and Clark. Verne Collins has a position with the Boulder Products Company.

The Clark tunnel, at Tungsten, has cut two veins and has leased them both. The Clark mill is running nearly to full capacity, and is receiving a good supply from nearly all the workings on the Clark property. Ex-sheriff Baxter is leasing on the Clark No. 1 and has been making steady shipments. Dan Gillett and Goddard are working the Clark 3, and taking out a good grade of ore regularly. Jacob Wade reports he has made a big strike on the Clark No. 5.

The Caribou mill at Caribou, Colorado, has been shut-down and will suspend operation as soon as it runs the ore on hand. It is reported that this is due to some dispute about the water-rights. Walter and Humphrey are preparing to wash the dump at the Vasco 6, on the Clark property. This is one of the richest Vasco dumps and is sure to pay well.

President Howe, of the Keystone Mining & Milling Co., at Magnolia, Colorado, visited the company's property this week. Manager Clifford Staley has been repairing the mill and adding to it. A. Ganvey, of Farmersville, Ohio, reports success from the Doss mine at Wallstreet that he recently leased.

C. A. DeWitt, superintendent of the Wolf Tongue mill at Nederland, reports the mill running to its fullest capacity.

The biennial report of the Colorado State Bureau of Mines in district No. 1, Boulder county, contains a résumé of the mining situation and conditions here. The Colorado State School of Mines will re-open in the fall as usual.

Woodring and Dupont, on the Huron, at Eldora, are steadily pushing development work. The Dixie mine above Lake Eldora has let a contract and work will be started at once on this property.

KIRKLAND LAKE, ONTARIO

The Important Gold Mines of This District.—Large Amount of High-Grade Ore in Sight in the Various Mines, and Further Development Planned.

Second in importance of the quartz-gold mining districts of Canada is Kirkland Lake that already is yielding upward of $1,000,000 in gold annually.

The Tough-Oakes is the pioneer and the premier mine of the district and is equipped with the largest mill. The ore at this mine is considerably higher in value than at any other gold mine in any of the Northern Ontario gold districts. It is true that costs are comparatively high, due chiefly to the hardness of the rock, but the high average gold content of the ore allows a large margin of profit and a handsome surplus for dividends is piling up. The regular dividend paid by the Tough-Oakes company is 2½% quarterly. During 1916 the company paid a total of $260,750. The ore reserves of this mine are estimated at upward of $1,000,000.

The second gold producer of the district is the Teck-Hughes, which only recently installed a ball-mill having a capacity of 75 tons of ore daily. The result of the first month's run, which is understood to be satisfactory, will probably soon be made known. The ore reserves of this property are said to be sufficiently large to warrant a further addition to the milling equipment, that is, provided the amount of labor available were to become normal. The ore throughout the mine is of comparatively high grade.

The third producer will be the Lake Shore mine. Already on this property there are upward of $500,000 worth of ore blocked out, and a new 75-ton mill is being transported to the property. Recently a new vein 4 ft. wide and carrying remarkably rich ore was discovered in a cross-cut underneath the lake at a depth of 200 feet.

The property of the Kirkland Lake Gold Mines Ltd. may be considered as the fourth in importance. This property has been opened to a depth of 600 ft. and approximately 5000 tons of $10 ore has been blocked out. About 5000 tons of good milling ore has been accumulated on the dumps. The lower workings of this property are the deepest in the Kirkland Lake camp. The president of the company has intimated that the property is about ready for a 100-ton mill.

The Wright-Hargreaves is another valuable property, although occupying fifth place in point of development work done and ore reserves blocked out. It is, nevertheless, considered as one of the leaders of this district. The main vein has been traced for something like 3000 ft. and at all points where opened up contains a high average gold content. At the 100-ft. level of No. 2 and 3 shafts, the grade of ore is said to range around $30 per ton across a width of 12 ft. The treasury of the company is in excellent shape, upward of $100,000 being on hand, and development will be continued to the 300-ft. level of both shafts, where a 90-ft. drift will be driven to connect the two workings.

For the time being, work on the La Belle Kirkland has been suspended. It has been developed to a depth of 350 ft., and a large tonnage of good-grade mill-ore has been blocked out. Also, as a result of 5000 ft. of diamond-drilling, done during the past winter, ore has been indicated to a depth of about 700 feet.

All the properties mentioned are either producers, or considered as probable producers.

THE MINING SUMMARY

The news of the week as told by our special correspondents and compiled from the local press.

ALASKA

(Special Correspondence.)—About 200 men at the Jumbo and Bonanza mines of the Kennecott Mining Co. walked out on a strike on the morning of June 16. This is about half the number of men on the pay-roll, and includes all the miners and shovelers, the tramway-men at the upper stations, and the Japanese cooks at the mines. The strikers are demanding an increase in wages of from 18 to 50%. The walk-out is confined to the mines. The employees at the lower camp, machine-shops, mill, leaching-plant, and those engaged in construction work refused to join the strikers. All the strikers left peaceably after being paid off and have gone to Blackburn, four miles from Kennecott.

A demand was made by the men for higher wages and but three days was given to the management to accede to the demand. E. T. Stannard, the manager, is on his way to New York, and H. D. Smith, assistant manager, lacking authority to grant such a demand, requested of the men a reasonable time in which to lay the matter properly before the New York officials of the company.

Any additional time was refused by the men, as well as an offer to arbitrate and also a proposal to have Mr. Stannard return immediately for conference with the men.

On two previous occasions during the last year the employees at the Kennecott mines have made demands for increase in wages upon short notice, but in each instance sufficient time was given in which to refer the matter to the New York office and the demands were either met in full or otherwise amicably settled.

It is not claimed by the men that the company has not fairly lived up to its part of these former agreements.

The demand now made is for $5.75 per day with $1.25 off for board, for miners and shovelers with even a greater per cent increase for cooks.

The scale at the mines at the time of the walk-out this morning was as follows:

Base-rate for miners $4.25 per day, and for shovelers $3.75 per day, with $1.25 off for board, plus the following bonuses:

Copper between 18 and 20c., 25c. per day; between 22 and 26c., 50c. per day; between 26 and 30c., 75c. per day; over 30c., $1 per day. The local officials of the company insist that the men have not lived up to their former agreements, and that they are not fair in refusing to allow a reasonable time for the company to consider their demands.

United States Judge Fred M. Brown, of the Third Division of the District of Alaska, has ordered all saloons closed at McCarthy, which is five miles from Kennecott, and is the nearest town to the mines in which saloons are allowed.

United States Marshal Brenneman is on his way to McCarthy to take personal charge in case of any violence or disorders.

Cordova, June 16.

(Special Correspondence.)—The Alaska Treadwell Gold Mining Co. has taken a working-option on the Red Diamond group of claims, situated near Bullion creek, on Douglas island. Development work on the group will start at once. The claims are owned by the Winn interests, of Juneau.

Juneau, June 17.

The Shelekoff Mining Co. has shipped machinery to its mines on Kodiak island. The property consists of seven copper claims situated near the head of Kuliak bay.

ARIZONA

Gila County

The Inspiration Consolidated Copper Co. at Miami has issued the following circular letter:

Miami, June 29, 1917.

To the employees of the Inspiration Company: The following communication is being presented to the various operating Companies of this district by a Committee of the Local Miners' Union:

June 28, 1917.

"Dear Sir:

"Resolved that we request a conference with the representatives of the mining companies of the Globe-Miami District for the purpose of discussing the following demands of the local unions of the International Union of Mine, Mill and Smelter Workers of America.

"No. 1. Recognition of the Grievance Committee of the Local Unions of the Mine, Mill and Smelter Workers International Union, and those of the other organized trades now represented in the mining industry in the Globe-Miami district.

"No. 2. Representatives of the unions to be allowed on company property at any time for the purpose of organization, it being understood that such representatives will in no way interfere with men in the discharge of their duties.

"No. 3. Reinstatement of men discharged for cause other than incompetency, competency or incompetency to be determined jointy by the Grievance Committee of the unions and representatives of the companies. The spirit and intent of this clause is that no man shall be discharged or refused employment on account of personal prejudice or on account of activities in union affairs.

"No. 4. Equal representation on the Board of Control of the Hospital.

"We request an answer by 10 A.M. Friday morning, failing to receive an answer accepting or rejecting these demands, a strike vote will be taken on Saturday, and if carried, a strike will be called on Monday morning at 7 A.M."

The Inspiration Company deems it proper, at this time, to advise its employes, labor organizations, and the public at large, of its position on the subject of meeting with delegations from labor organizations or recognizing committees from those bodies in its operations.

First and foremost, we reserve the right and privilege to conduct our own affairs. We have always respected the rights and principles of Union organizations and their members, and have never attempted to disrupt these organizations or to discriminate against any of our employes because of their affiliation with the Union. On the other hand, we have always recognized the rights of those who did not care to join these organizations. This policy will be maintained in the future.

There appears to be no need of any conference with outside labor organizations. If the matters which they wish to bring to our attention concern our employes, the men have a representative Committee elected by themselves, through which such matters, whether they be complaints or suggestions for improved conditions or service in any department, can be brought to the management. If the matters which they wish to discuss concern recognition of the Unions, we say, frankly and with full knowledge of gravity of the situation, that we will recognize no such labor organizations or delegations from them, and that we will grant no demands for recognizing Union Committees.

In short, if we are to continue to operate it can only be possible if we are allowed to conduct our own affairs with the help and suggestions of our own employes.

INSPIRATION CONSOLIDATED COPPER CO.
By C. E. MILLS, Gen. Mgr.

Metal Mine Workers' Industrial Union, No. 800, affiliated with the Industrial Workers of the World, formulated its demands on the evening of June 29 for presentation to the managers of the three big copper companies in the district, says the *Arizona Record* of June 30. The demands are identical to those which have been made by striking miners at Butte and Bisbee.

According to a representative of the metal workers, this organization will hold a mass-meeting tonight at which time, unless the demands are met by the companies, a strike vote will be taken. If they are refused, the members will possibly quit work Sunday night or Monday morning.

Following is the official statement of demands which was issued:

"Miami, Arizona, June 29, 1917.

"Mine Managers of the Globe and Miami District:

"We, the joint committees of the Metal Mine Workers' Industrial Union, No. 800, present the following demands:

"1. Two men to work on all piston and Leyner machines.
"2. Two men to work together in all raises.
"3. No blasting in stopes, drifts, or raises during shift.
"4. Abolition of the 'rustling-card' system.
"5. Abolition of the contract-bonus system.
"6. Abolition of the sliding-scale.
"7. Water-sprays shall be used on all machines.
"8. No discrimination against any member of any union.
"9. Representation in the control of hospital.
"10. Minimum wage of $6 for all men working underground.
"11. Minimum wage of $5.50 for all men working on surface.

"Respectfully submitted,

"METAL MINE WORKERS' INDUSTRIAL UNION No. 800."

The local authorities at Miami claim that the strike situation has gone beyond their power of control, and the War Department at Washington late on July 3 instructed the Southern Department to take the steps necessary to handle the situation, and troops will be sent into the mining districts where needed if the request is officially made by the authorities in Arizona.

CALAVERAS COUNTY

(Special Correspondence.)—The Sheep Ranch mine, after a long idleness, is to be reopened by New York capital under the direction of H. R. Plate, of New York. The mine has been worked to a depth of about 1200 ft. and for many years was a large producer. The ore was free-milling and often was filled with visible gold. Ore of this character was generally of dark bluish color and was in demand by jewelers who converted the rock into various forms of jewelry. The vein was about 18 in. wide and the average of the ore for years was about $15 per ton.

Sheep Ranch, June 24.

(Special Correspondence.)—The Maypole mine, formerly known as the Shear mine, has been re-opened by L. G. Blakemore. The mine is half a mile north-west of Mokelumne Hill, at what is known as the Italian gardens. The claim, which is 2700 ft. long, is located on a north-east south-west vein that dips north-west in greenstone schist. The quartz is 6 ft. wide at the face of the old tunnel, which has been cleaned out by Mr. Blakemore. It was caved throughout its entire length, having been idle and forgotten for more than 30 years. On the dump was 300 tons of rock that sampled $5 per ton, which was considered good enough to justify an examination. By driving 300 ft. farther there will be available from 350 to 400 ft. of backs. The outlook for this old property, one of the first, if not the first, quartz claim located in this county, is considered promising.

The old Boston (Esperanza) mine is to be re-opened by a Mr. Baker, who it is reported has made a payment of $17,500 on the purchase price. This mine has been extensively worked in years gone by and has had not less than three mills, and at one time a chlorination works. The vein is large, 40 to 60 ft. wide, and consists of a zone of silicified amphibolite schist. It carries 2 to 10% auriferous sulphide, together with free gold, but is low grade and requires careful and economical handling.

The Gardella brothers have closed the sale of their Garibaldi mine in the Mill Valley district to W. M. Stiver, who represents San Francisco capitalists. The mine is seven miles from Mokelumne Hill. A company has been organized to operate this property, which will be known as the Garibaldi Mining Co. There is a shaft 100 ft. deep in which some good ore has been found. Machinery will be placed on this property by the new owners.

Mokelumne Hill, June 24.

(Special Correspondence.)—There is much prospecting being done about Railroad Flat, and some good discoveries are announced. The district is one of many possibilities. It was only a few years ago that the Comet mine was discovered, almost by accident, and within a short time produced $125,000 at large profit. There are a number of old mines, at one time substantial producers, that are again being investigated.

Railroad Flat, June 24.

(Special Correspondence.)—The Utica mill has been closed-down for a week because of an unusual flow of salt water, but is about to resume operations. The Angels Quartz mine has been thoroughly cleaned up and is now ready for a long period of activity.

Angels, June 24.

(Special Correspondence.)—The proposal to re-open the Blazing Star has been abandoned, it is stated, owing to the War. The Sawyer mine, controlled by William Foltz of Seattle, is about to resume operations, San Francisco capital having been interested, it is reported.

West Point, June 25.

DEL NORTE COUNTY

(Special Correspondence.)—Tom Galvin, who is representing the American Exploration Co., which is operating chrome properties on French hill and Low divide, expects to spend many months here. There now is 8000 tons of high-grade chrome ore ready for shipment. This week the company expects five auto-trucks with which to haul the ore to this place. It will be shipped by steamer to San Francisco.

The Duley and Lauff copper property on Patricks creek is developed by many hundred feet of adits. Work has been suspended for some time and an Eastern company that has examined the property now contemplates its purchase. O. B. Lauff and West Duley are hotel owners of this place.

Large deposits of chrome ore have been found and located on Diamond creek by G. W. Gravlin, and other large deposits have been found and located on Cedar creek by John Taggert, in this county.

Crescent City, June 23.

SISKIYOU COUNTY

(Special Correspondence.)—John Hays of Gold Hill, Oregon, and Charles Moon, of Hornbrook, left for their gold quartz mine, 8 miles from Hornbrook. They discovered the vein several years ago and from time to time have done considerable development work, and now are preparing to erect a small mill on the property. The ore is rich and free-milling, but is only 8 to 10 in. wide, where uncovered.

Mike G. Womack, of Medford, Oregon, and M. A. Carter, and

L. D. Corbitt of Ashland, Oregon, have a promising gold prospect, the Golden Gem, which is situated 16 miles west of Hilts on Hungry creek. They discovered the vein last year and have done enough work to justify the further development of the property. On account of the inaccessability of the wagon-road to the mine they contemplate erecting a small mill on the property. The ore assays $70 per ton of gold and is free-milling. The vein has a width of 6 to 22 in. There is several hundred feet of work on the property, though the greatest depth attained on the vein is 70 feet.

Mr. Womack and his associates will resume operation on a vein of gold, silver, and galena situated 16 miles west of Gazelle, in the Etna Mills district. The vein is in limestone and averages 10 ft. wide. On account of the distance to a shipping-point the owners are contemplating erecting a mill on the property this season and reducing the ore to concentrate for shipping.——Work has been resumed on a number of old quartz properties in the Hornbrook district.

Hornbrook, June 25.

COLORADO

Boulder County

(Special Correspondence.)—C. E. Brandenburg, on Left Hand, has made a shipment of gold ore from his new discovery that seems destined to prove a bonanza. This discovery is one of the best that has been made here in years. The vein is from 3 to 4 ft. wide with a rich streak from 10 to 12 in. of solid copper-iron.

Cowdry & Co. has leased the Sunday mine at Rollinsville and will start operations on a large scale.

Frank Arondel, of Nederland, reports that the Last Chance has 3 ft. of good tungsten ore.

Eldora, June 26.

Teller County

(Special Correspondence.)—The directors of the Granite Gold Mining Co. declared a dividend of 1c. per share on the issued stock of the company, payable July 5 to stockholders of record June 26. The amount of the dividend is $16,500. The company now will go on a bi-monthly dividend basis. Previous to this new dividend the last disbursement was that of November 1912.

Cripple Creek, June 26.

MONTANA

Granite County

(Special Correspondence.)—The manganese mines of the Flint Creek district, which is just east of Philipsburg, are producing from 200 to 400 tons of ore per day; the Philipsburg Mining Co. is shipping from 60 to 100 tons per day; the Mullins-Hynes property, from 60 to 100 tons; the Courtenay lease, from 50 to 60 tons; the Cape property, 25 to 50 tons; and other mines varying amounts. Philipsburg, which his been a silver-producing camp for 50 years, has awakened to the possibilities of her manganese deposits, and it is believed this district will produce more manganese this year than any other district in the United States.

W. C. Phalen, of the U. S. Bureau of Mines, has just paid a visit to the mines here and states that he has actually seen more manganese here than at any other place in the United States.

Most of the ore is being shipped East and a 45% ore will net at the mines at present $30 per long ton.

Philipsburg, June 25.

Silver Bow County

Miners at Butte are interested in the discovery of a large body of manganese ore in the Hibernia mine of the Davis Daly company. The orebody is 100 ft. wide, and between 700,000 and 1,000,000 tons of ore is in sight. It is close to the surface and can be handled by open-cut method.

W. L. Creden, general manager for the Davis Daly company, has gone East to close contracts for the disposal of this ore. Two of the largest steel companies are negotiating for the entire output. Arrangements have been made by the Davis Daly company with the Butte Detroit company by which the latter will handle the ores at Butte. In addition to the orebody on the Hibernia claim, a body of manganese has been found in the New Republic claim, which also belongs to the Davis Daly company. This body of ore is said to exceed in size that in the Hibernia. The opening of these two bodies of manganese has started a hunt in scores of claims in the West Butte district.

NEVADA

Clark County

(Special Correspondence.)—The Duplex mill is running on ore from the J. E. Griffith lease on the Duplex mine, and there is no indication of a shut-down, as other lessees are getting out good ore.

George Colton, one of the owners of the Duplex mine, and who is now leasing, says that Jack Ellison, one of his employees, has opened a fine body of rich ore. Several carloads recently shipped ran well in gold, copper, and lead.

Wells, Lind, and Ray, who are operating the Searchlight Mining & Milling Co.'s property, have shipped a large tonnage of good lead ore. Two cars shipped recently to the smelter returned $18,000, and during this period several hundred tons of good milling ore was also produced by them. They found a new vein of rich ore a few days ago.

The mill of the Chief of the Hills company is nearly completed, a test run having been made to ascertain what changes, if any, would be advisable. Ben Stevens is superintendent.

Searchlight, June 23.

Eureka County

(Special Correspondence.)—Thomas Brown and associates, of San Francisco, have taken a bond on the Holly mine of this place and will commence work July 1. The Holly is fully equipped with a 60-hp. gasoline-hoist and arrangements have been made to get distillate from a carload just shipped in for the Connelly mine.——A body of shipping ore has been opened up and two and a half tons per day is being taken out of the Will Heubner property, recently purchased by A. G. Burritt, of Salt Lake City. This property adjoins the California. It is expected that as soon as the development work now being done is completed large orebodies will have been found, as the ore becomes larger, richer, and better as work progresses.

Eureka, June 24.

Humboldt County

(Special Correspondence.)—The Adams vein on the 800-ft. level of the Rochester mine, at Rochester, has widened to 14 ft. Much of the ore is high-grade with free gold and native silver showing plentifully. Drifts are being extended from the 900-ft. level in expectation of intersecting the Adams vein. The East vein is yielding good ore on the 700, 800, 900, and upper levels. The mill clean-up for the last half of May yielded $26,943.

Superintendent Wilkey, of the Rochester Combined Co., expects to have mill-grading completed by July 10 and construction of the plant started. Over 100,000 ft. of lumber has arrived and considerable machinery is daily expected. A 160-ft. raise has been started from the main working-tunnel in the Shepherd claim to connect with No. 1 level. The main tunnel is in 480 ft. and has entered good ore. The Maynaugh tunnel, on the Happy Jack, has also entered excellent ore about 380 ft. from the portal. Developments are proceeding rapidly on the Bacchus and Happy Jack claims.

New work has begun on the 1600-ft. level of the Seven Troughs Coalition. The main drift is advancing toward the shoot, which yielded specimen ore in the upper workings. Good ore is being mined above this point.

An orebody ranging from 4 to 12 ft. wide has been opened in the Cheefoo mine, at National. All the ore is stated to be of good milling grade with streaks showing silver sulphide and native silver. A little gold occurs. Construction of mine buildings is proceeding. N. P. R. Hatch is general manager. J. C. Sullivan and H. L. Schreck, of San Francisco, are heavily interested.

A new mill is in operation at the Buckaroo mine, in the Pine Forest district. Developments have been proceeding for several years and a large tonnage of good ore is stated to be in sight. The mine is 12 miles north-west of Quinn River crossing, and is operated by the Oklahoma Gold Mining Co. Thomas Ewing is general manager.

Ore averaging $250 per ton in gold and silver is being shipped from the 700-ft. level of the Seven Troughs mine, at Vernon, and on the dumps a large tonnage of $25 ore is being placed with a view to treating it at the mine. From the winze on the 800-ft. level a cross-cut is advancing in promising territory.

Lovelocks, June 28.

NYE COUNTY

(Special Correspondence.)—Foundations are being placed for the auxiliary power-station of the Tonopah Extension, and by August 1 it is believed the plant will be ready for service. Its completion will be the signal for resumption of work at the Great Western and Tonopah Bonanza mines with three shifts, and extension of the deep cross-cut of the Western Tonopah Co., controlled by Boston and Butte capitalists. This work is designed to determine the possibilities for ore in the extreme western part of the district.

The mill clean-up of the Tonopah Extension for the first half of June yielded 31 bars of bullion valued at $58,000. Most of the ore is coming from No. 2 shaft workings, but as soon as the auxiliary power-plant is in operation a heavy output from the deep levels of the Victor shaft is planned.

The Tonopah Belmont Co. expects to have the mill at its Surf Inlet gold mine, on Princess Royal island, above Vancouver, B. C., in operation about August 15. Mine conditions are reported highly satisfactory. Work on the adjoining Pugsley property is proceeding steadily. It is stated the Eagle-Shawmut mine, near Sonora, California, is also developing satisfactorily. The company reports net earnings for May at $98,525, exceeding the April net earnings by over 15%. The Jim Butler reports May profits of $44,136, and a total of $191,-956 for the year. The weekly output is about 700 tons.

Tonopah, June 28.

(Special Correspondence.)—Shaft sinking in the White Caps has reached a depth of 110 ft. below the fourth level. The fifth station will be cut after a lift of 127 ft. has been made, and as the fourth level is 436 ft. below the surface, the next level will have a vertical depth of 563 ft. The water from the several faces in the workings has steadily diminished during the week. The ground cut in the shaft-sinking is silicious shale and is the hardest known in the camp, and for this reason the footage made during the week has not broken any records. In the west drift on the third level 15 ft. has been made during the week, with no water encountered. The bulk-heads and flood-gate are in place in the east drift and as the water now only approximates 75 gal. per minute the drift will be extended. It is the intention to commence the filling of the ore-bins during the first week in July, preparatory to the mills starting. In the mill five hearths of the Wedge roaster have been completed and the bricklayers are at work on the sixth. The crusher-house has been completed and machinery installed. The tightening of the mill-tanks after being filled with water is in progress. The landing-platform and trestle from the head-frame to No. 1 ore-bin is nearing completion. A large-size Steele Harvey tilting-furnace for reducing the gold precipitate has been received and is on its foundations in the refinery. The cut-off wagon road from the White Caps to Pipe Springs has been completed and is in use, both to the White Caps and the White Caps Extension.

At the shaft of the White Caps Extension five car-loads of lumber are coming in for the mine buildings, employees' bunk-house, and superintendent's house. The shaft will be the largest in the district, having three compartments. Grading for the buildings is in progress. An air-line has been laid from the White Caps to furnish air for the rock-breaker, until the Extension's compressors arrive. Three 50-kw. transformers have arrived for the mine. It is expected to purchase the necessary hoisting machinery at Goldfield, and a double-drum hoist will be brought in immediately.

At the Union Amalgamated the face continues in hard blue lime. The east drift from the north cross-cut has exposed one foot of ore averaging $100 per ton in gold. This probably is the continuation of the Bath ore-shoot, that in the upper part of the mine yielded many thousand dollars to lessees. No. 2 cross-cut raise has been started to reach the ore left in the hanging wall of the west drift, 110 ft. west from the shaft. It was found impracticable to carry all of the vein in the drift at this point. In the raise is 4 ft. of ore of milling grade, which makes the vein 8 ft. wide at this point. Since the above was written, further development shows the ore in the new orebody mentioned as the probable continuation of the Bath orebody to assay $176.

Rapid progress is being made in sinking the shaft of the Extension company below the 100-ft. level. An average of four feet per day has been made in the sinking, with one shift of miners. The vein has opened out as the shaft is being sunk from 16 in. to over 3 ft. It is the intention to continue the shaft-sinking to a depth of at least 100 ft. farther before exploring the medium-grade ore-shoot developed on the 100-ft. level.

At a depth of 110 ft. water was struck in the Red Top shaft. For a few days an endeavor was made to keep the flow down by bailing, but this has been found impracticable, and to prospect the lime belt, a drift on the lime-shale contact has been started to the north-east. Gold was found and the drift is being extended. The management has ordered a three-drill compressor.

The installation of new machinery has occupied most of the time at the Morning Glory recently. The new hoist has been connected up, and the buildings have been erected. Installation of the compressor is proceeding. In the No. 3 shaft-sinking all of the rock taken out is pay ore. The average value for the last 10 ft. of sinking has been $10 in gold.

The drift on the 300-ft. level of the Manhattan Consolidated is out 200 ft. from the shaft, with 20 ft. to go to reach a point beneath the rich shoot worked on the 200-ft. level. The 'mud-fault' has been reached, and important developments are anticipated.

Manhattan, June 21.

It is reported that the drift on the 300-ft. level of the Manhattan Consolidated has been extended under the ore-shoot developed on the 200-ft. level, and has broken into high-grade ore.

OREGON

JOSEPHINE COUNTY

(Special Correspondence.)—Operations have been resumed on 'The Diamond Creek' cinnabar property, 16 miles south-west of here. This property is owned and operated by W. Ehrman, John Taggert, and L. C. Cole. Preparations are being made to install machinery.

The Preston Peak copper mine is under option to J. F. Reddy of Grants Pass. It is reported that an Eastern company is ready to take over the option and begin operations. This mine is opened to the depth of 1300 ft. There are many thousand tons of ore in sight showing from 3½ to 20% copper and from $4 to $8 per ton gold.

A. Justin Townsend, of Lynn, Massachusetts, owner of the Pacific placer mine, is planning to put in a dredge capable of handling 2000 yards of gravel daily. This will be used in addition to the hydraulic equipment.

The Collard-Moore and Collard chrome mine is one of the largest shippers in this district. The owners have installed a concentrator, and are shipping some high-grade massive chrome ore that requires no concentration.

A body of chrome ore was found 6 miles south-east of Waldo by W. Bunch and son and Walter Smith. It is only a few hundred feet from the Kerby-Holland stage-road, near the old Sly ranch. Ore is being mined and shipped by way of Grants Pass.

The Osgood placer mine, located in Fry gulch, is a dependable gold producer. It is owned by F. H. Osgood, of Seattle, and has been leased by James Logan for the past four years. This property comprises about 640 acres. Three giants are in operation. Water is taken from the east fork of the Illinois river.

Waldo, June 23.

TEXAS

Llano County

Ceylon E. Lyman, president of the Wakefield Iron & Coal Land Improvement Co., of Minneapolis, Minnesota, says in the *Manufacturers Record*:

We are continuing the development of a large graphite deposit which we were working last year, and it is proving to extend over an area two miles long and embracing several veins, some of which are 50 to 150 ft. wide. Assays from these veins run from 12 to 25% in graphite.

We are not doing anything in manganese at present, owing to the distance from a railroad, although we may do some work to determine the extent of one deposit in view of the possible wants of the Government. In this course we may be influenced by a report recently made by a representative of the Government, who pronounced it a large and remarkable deposit.

CANADA

Yukon Territory

It is reported from Dawson that when transportation on the Yukon opens, $9,000,000 in gold will be shipped out from the Yukon Basin districts as the result of the season's clean-up. Most of this is from gravel mined during the winter months and washed during the early summer when water was available.

MEXICO

Vera Cruz

(Special Correspondence.)—According to advices received here from Teziutlan, State of Vera Cruz, the large smelter of the Compañía Metalúrgica at that place is now in full operation. This company has been operating large copper mines in that district for many years and its output of metal was large before the revolutionary period. The conditions of banditry interfered with railroad transportaton to and from the company's property, but these conditions in that particular locality are said to have shown improvement of late. There has been a project on foot for a long time to construct a railroad from Teziutlan to the prospective deep-water port of Nautla, situated about midway between Vera Cruz and Tuxpam, and the Federal Government has signified its intention of carrying out this work at the earliest possible moment. Connected with the construction of the railroad will be the building of extensive harbor and port works at Nautla, at an estimated cost of about $2,000,000.

Monterrey, June 23.

A recent report from Cananea states that everything there is quiet at present. The authorities have taken charge of the mines, mills, and smelters of the Greene-Cananea company and has placed guards of soldiers at all the properties. It is also reported that the authorities contemplate starting up the mines and works, but nothing definite is yet known concerning this.

Personal

Note: The Editor invites members of the profession to send particulars of their work and appointments. This information is interesting to our readers.

Rush M. Hess is at Bouse, Arizona.

William Truvan is in Plumas county.

C. Erb Wuensch is in Calaveras county, California.

E. A. S. Whittard was in San Francisco from Tuolumne.

P. T. McGrath is in San Francisco from Phoenix, Arizona.

Augustus Locke, of Boston, is visiting California mining districts.

Frederick Laist has returned to Anaconda from South America.

Waldemar Lindgren has returned to Boston from South America.

Charles Janin, on his return from Russia, has gone to Washington.

H. R. Plate, of New York, is at the Sheep Ranch mine, in Calaveras county.

Andrew C. Lawson has completed an examination of the coalfields of northern Arizona.

Frank H. Probert has been making a geologic examination of the Rochester mines in Nevada.

S. F. Shaw is re-opening the Panaco mine, in Coahuila, Mexico, for the A. S. & R. Company.

L. P. Pressler has returned from Mexico and is now with the Tonopah Mining Company at Tonopah, Nevada.

D'Arcy Weatherbe and Ross B. Hoffmann sailed by the 'Tenyo Maru' for China and Siberia on June 30.

Sumner S. Smith has been named resident engineer for the Alaskan Engineering Commission at Anchorage.

G. H. Wolhaupter has left the Magma mill of the Utah Copper Co. to join the Michigan College of Mines Batallion.

Will D. Coghill has been appointed metallurgist at the Seattle experiment station of the U. S. Bureau of Mines.

H. J. Wendler has been at Mokelumne Hill from Sonora, Mexico, making an examination of the Mokelumne group of mines.

E. S. Boalich has returned from Washington, and has taken the position as mining engineer with the California State Mining Bureau.

J. S. Diller will make San Francisco his headquarters for about three weeks while studying chrome deposits for the U. S. Geological Survey.

Charles S. Galbraith has resigned from his connection with the Callow flotation business and has opened an office at Webb City, Missouri, as a metallurgical and civil engineer.

Howland Bancroft, having finished examinations in Colorado, Wyoming, and New Mexico, has gone to New York, where he will be at the Engineer's Club until the middle of July.

Douglas A. Mutch, manager for the Hudson Bay Mines, Ltd., at Cobalt, Ontario, has been appointed consulting engineer for the Dome Lake Mining & Milling Co., at South Porcupine.

Frank R. Corwin has resigned his position as assistant superintendent of the International smelter to take charge of the Consolidated Arizona Smelter Co.'s smelter at Humboldt, Arizona, in the capacity of superintendent.

A. M. Swartley is acting as director of the Oregon Bureau of Mines and Geology during the absence of Henry M. Parks, who is a captain of engineers in the Officers Reserve Corps, and is now at Vancouver Barracks, Washington.

Alfred H. Brooks, formerly in charge of the Division of Alaskan Mineral Resources of the U. S. Geological Survey, has been appointed a captain in the Engineer Officers Reserve Corps and ordered to report for training. During Mr. Brooks' absence on military duty, Mr. George C. Martin will be geologist, acting in charge of Alaskan work.

THE METAL MARKET

METAL PRICES

San Francisco, July 3.

Antimony, cents per pound	18.50—22.00
Electrolytic copper, cents per pound	34.50
Pig lead, cents per pound	12.25—12.50
Platinum, soft and hard metal, per ounce	$105—111
Quicksilver, per flask of 75 lb	$85
Spelter, cents per pound	11.50
Tin, cents per pound	59
Zinc-dust, cents per pound	20

ORE PRICES

San Francisco, July 3.

Aluminum-dust (100-lb. lots), per lb	$1.00
Aluminum-dust (ton lots), per lb	$0.95
Antimony, 50% metal, per unit	$1.35
Chrome, 40% and over, f.o.b. cars California, cents per unit	50—55
Magnesite, crude, per ton	$8.00—12.00
Tin, cents per pound	60
Tungsten, 60% WO_3, per unit	$25.00—30.00
Molybdenite, per unit for MoS_2 contained	40.00
Manganese, 45% (under 35% metal not desired), cents, unit	33—37

Manganese prices and specifications, as per the quotations of the Carnegie Steel Co. schedule of prices per ton of 2240 lb. for domestic manganese ore delivered, freight prepaid, at Pittsburg, Pa., or Chicago, Ill. For ore containing

	Per unit
Above 49% metallic manganese	$1.00
46 to 49% metallic manganese	0.98
43 to 46% metallic manganese	0.95
40 to 43% metallic manganese	0.90

Prices are based on ore containing not more than 8% silica nor more than 0.2% phosphorus, and are subject to deductions as follows: (1) for each 1% in excess of 8% silica, a deduction of 15c. per ton, fractions in proportion; (2) for each 0.02% in excess of 0.2% phosphorus, a deduction of 2c. per unit of manganese per ton; fractions in proportion; (3) ore containing less than 40% manganese, or more than 12% silica, or 0.225% phosphorus, subject to acceptance or refusal at buyer's option; settlements based on analysis of sample dried at 212° F., the percentage of moisture in the sample as taken to be deducted from the weight. Prices are subject to change without notice unless specially agreed upon.

Tungsten has taken a sharp advance, owing to continued and increasing demand, and scarcity of the supply to meet it. From a nominal price of $20 to $22 per unit for 60% ore, the price has risen to $25 to $30 per unit, and unless indications are misleading it will go still higher. Some ore of low grade has been shipped from the mines at Atolia to Boulder county, Colorado, for concentration.

EASTERN METAL MARKET

(By wire from New York)

July 3.—Copper dull and nominal at 32.25c. Lead is quiet at 11.70 to 11.50c. Zinc is dead and lower at 9.37 to 9.25c. Platinum shows no change, being $105 for soft and $111 for hard metal. The average price of tin in the month of June was 61.93c. per pound.

COPPER

Prices of electrolytic in New York, in cents per pound.

Date		Average week ending	
June 27	32.25	May 22	32.25
" 28	32.25	" 29	32.50
" 29	32.25	June 5	32.62
" 30	32.25	" 12	32.75
July 1 Sunday		" 19	32.58
" 2	32.25	" 26	32.42
" 3	32.25	July 3	32.25

Monthly Averages

	1915	1916	1917		1915	1916	1917
Jan.	13.60	24.30	29.53	July	19.09	25.66	....
Feb.	14.38	26.62	34.57	Aug.	17.27	27.03	....
Mch.	14.80	26.65	36.00	Sept.	17.69	28.28	....
Apr.	16.64	28.02	33.16	Oct.	17.90	28.50	....
May	18.71	29.02	31.69	Nov.	18.88	31.95	....
June	19.75	27.47	32.57	Dec.	20.67	32.89	....

SILVER

Below are given the average New York quotations, in cents per ounce, of fine silver.

Date		Average week ending	
June 27	78.25	May 22	74.78
" 28	78.50	" 29	74.62
" 29	77.87	June 5	74.80
" 30	77.87	" 12	75.83
July 1 Sunday		" 19	77.00
" 2	77.87	" 26	78.12
" 3	77.87	July 3	77.98

Monthly Averages

	1915	1916	1917		1915	1916	1917
Jan.	48.85	56.76	75.14	July	47.52	63.06	....
Feb.	48.45	56.74	77.54	Aug.	47.11	66.07	....
Mch.	50.61	57.89	74.13	Sept.	48.77	68.51	....
Apr.	50.25	64.37	72.51	Oct.	49.40	67.86	....
May	49.87	74.27	74.61	Nov.	51.88	71.60	....
June	49.03	65.04	76.44	Dec.	55.34	75.70	....

High-record levels have been reached by silver sales at 80⅝c. per oz. at Vancouver. This new top price, indicating a premium of about 2c. per oz. over the New York quotation, was made by Nipissing Mines Co. for shipment to the Far East. Producers talk of even higher prices for this metal, basing their contention on the big demand both at home and abroad and the fact that production shows no material increase. Curtailment of operations at Butte, Cananea, and other mining centres where silver figures prominently as a by-product, will cut down the yield of this metal, rather than provide the increase that had been looked for at this time. The Federal Government has been buying silver heavily for some weeks past, the purchases aggregating 400,000 oz. weekly. The last purchase, however, was for 800,000 ounces.

Much of the silver for the Far East now goes direct from the United States or Canada from Pacific ports rather than through the medium of the London market which necessitates crossing the Atlantic. This is because of the high shipping-costs with Atlantic war-risk insurance to London. Until recently it has been next to impossible to break the hold which English brokers have always held upon the Far Eastern markets.

With an advance of 6⅝c. from the low point of the year, March 27, silver prices have reached the highest level since 1892. They are now actually higher than at any time during the sensational advance of May 1916, when 77¼c. was twice touched. Every indication points to 80c. silver in the near future.

LEAD

Lead is quoted in cents per pound. New York delivery.

Date		Average week ending	
June 27	11.70	May 22	10.50
" 28	11.70	" 29	10.93
" 29	11.50	June 5	11.46
" 30	11.50	" 12	11.83
July 1 Sunday		" 19	12.00
" 2	11.50	" 26	11.75
" 3	11.50	July 3	11.57

Monthly Averages

	1915	1916	1917		1915	1916	1917
Jan.	3.73	5.95	7.64	July	5.59	6.40	....
Feb.	3.83	6.23	9.01	Aug.	4.67	6.28	....
Mch.	4.04	7.26	10.07	Sept.	4.62	6.86	....
Apr.	4.21	7.70	9.38	Oct.	4.62	7.02	....
May	4.24	7.38	10.29	Nov.	5.15	7.07	....
June	5.75	6.88	11.74	Dec.	5.34	7.55	....

The Bunker Hill & Sullivan Mining & Concentrating Co., of Kellogg, Idaho, paid dividend No. 254, of $81,750, on June 4, and a dividend, No. 255, of $81,750, on July 3. An extra dividend, No. 256, of $81,750, was also paid July 3. These dividends bring the grand total paid by this company to date to $19,716,000.

The Hecla Mining Co. of Wallace, Idaho, has declared dividend No. 169 of 15c. per share, being $150,000. Total for 1917, $900,000. Total paid to date, $6,205,000.

It is reported that the leading lead producers of the Coeur d'Alene mines have agreed to sell 2000 tons of lead to the Government at 8c. per pound.

ZINC

Zinc is quoted as spelter, standard Western brands, New York delivery, in cents per pound

Date		Average week ending	
June 27	9.37	May 22	9.37
" 28	9.37	" 29	9.50
" 29	9.37	June 5	9.66
" 30	9.25	" 12	9.75
July 1 Sunday		" 19	9.72
" 2	9.25	" 26	9.43
" 3	9.25	July 3	9.32

Monthly Averages

	1915	1916	1917		1915	1916	1917
Jan.	6.30	18.21	9.75	July	20.54	9.90	....
Feb.	9.05	19.90	10.45	Aug.	14.17	9.03	....
Mch.	8.40	18.40	10.78	Sept.	14.14	9.18	....
Apr.	9.78	18.62	10.20	Oct.	14.05	9.92	....
May	17.03	16.01	9.41	Nov.	17.20	11.81	....
June	22.20	12.85	9.63	Dec.	16.75	11.26	....

QUICKSILVER

The primary market for quicksilver is San Francisco, California being the largest producer. The price is fixed in the open market, according to quantity. Prices, in dollars per flask of 75 pounds:

Week ending

Date			
June 5	90.00	June 19	82.00
" 12	90.00	" 26	80.00
		July 3	85.00

Monthly Averages

	1915	1916	1917		1915	1916	1917
Jan.	51.90	222.00	81.00	July	95.00	81.20	....
Feb.	60.00	295.00	126.25	Aug.	93.75	74.50	....
Mch.	78.00	219.00	113.75	Sept.	91.00	75.00	....
Apr.	77.50	141.60	114.50	Oct.	92.90	78.20	....
May	75.00	90.00	104.00	Nov.	101.50	79.50	....
June	90.00	74.70	85.50	Dec.	123.00	80.00	....

TIN

Prices in New York, in cents per pound.

Monthly Averages

	1915	1916	1917		1915	1916	1917
Jan.	34.40	41.76	44.10	July	37.38	38.37	....
Feb.	37.23	42.60	51.47	Aug.	34.37	38.88	....
Mch.	48.76	50.50	54.27	Sept.	33.12	36.66	....
Apr.	48.25	51.49	55.63	Oct.	33.00	41.10	....
May	39.28	49.10	63.21	Nov.	39.50	44.12	....
June	40.26	42.07	61.93	Dec.	38.71	42.55	....

MOLYBDENUM

Such small quantities as were offered on the New York market changed hands at $2.10 per pound MoS_2 for 90% concentrate.

ANTIMONY

The market is very dull and little business has been reported. Nominally, prompt antimony is quoted from 19 to 19½c., with future quotations ranging from 14 to 16c., according to position. Needle-antimony, however, remains firm at 12c. for spot and 9 to 9½c. for future delivery.

MANGANESE

Manganese is unchanged with the schedule price of $1 per unit for high-grade ore.

Eastern Metal Market

New York, June 27.

All the metals grow more inactive rather than otherwise as the weeks go by—due to the continued uncertainty regarding Government decisions as to the quantity and price of probable purchases. The net effect is general stagnation.

Copper quotations are practically nominal with actual business of small volume.

Lead is in small demand and lower.

Tin is dull and has declined.

Zinc is dead but a little business has been done on lower prices.

Antimony is inactive and unchanged.

In the steel market a new and serious uncertainty has been injected due to the agitation of Government control of prices, including coal and coke, thus increasing the perplexity of buyers. Pig-iron buyers and purchasers of finished steel are much disturbed over the proposed fixing of maximum prices in private transactions. Government buying continues to grow in volume, coming from various sources. Japanese buyers of ship-plates are grabbing every bit of tonnage they can lay hands on, paying as high as 9.90c. for 1000 tons for early shipment. The whole market continues to advance in the scramble for the small quantity of steel available for early shipment.

COPPER

All sorts of rumors are emanating from Washington as to the negotiations in reference to copper purchases by this and the other belligerent governments. Estimates of the amount our own army and navy will need for the remainder of 1917 approximates 225,000,000 lb., while others state that these needs for 12 months will exceed 380,000,000 lb. As to price, common talk is that to the Government 25c. will be the settled figure with 28c. per lb. to the Allies. It is even stated that deliveries are now being made in certain urgent cases on Government account at 25c. per lb. Other reports are disquieting and are to the effect that there is a decided controversy regarding the price-question as it relates to the cost of copper production and the whole situation is disquieting. Business has come to a halt and the hesitancy is expected to continue until the problem is cleared up. This is not the only unsettling factor. It is stated that certain large brass interests have been compelled to shut down indefinitely. Consumers will not come into the market under present conditions, and sales have been light. The market is a stale and drifting one. Price quotations are almost distinctly nominal. Early delivery Lake and electrolytic is held at about 32.25c., New York, with third quarter at 30.50c., and last quarter 29c. to 29.50c. Copper exports for May are returned as 45,241 gross tons, bringing the total to June 1 to 225,967 tons. The contrast is revealed by the fact that exports to May 1 this year are 180,726 tons, against only 92,286 tons in the same four months of 1916. The London market is unchanged at £142 for spot and £138 for future electrolytic copper.

TIN

Announcement has been made of the personnel of the sub-committee on tin of the Council of National Defense as appointed by the American Iron and Steel Institute. It is made up of John Hughes, chairman, assistant to the president, United States Steel Corporation; E. R. Crawford, president, McKeesport Tin Plate Co.; A. B. Hall, manager, metal department National Lead Co.; Theo. Pratt, assistant manager, manufacturing department Standard Oil Co., and John A. Fry, purchasing agent, American Can Co. This committee has been approved and confirmed by the Government but its personnel does not cause satisfaction among importers and dealers, made up as it is almost entirely of consumers of tin. The market late last week was featureless and characterized by lack of business. Considerable complaint was voiced by dealers and importers. The appearance and offering of Chinese tin was also an unsettling influence. Late in the week sales were made of both Banca and Chinese tin but not in large volume and mostly for future delivery. Early this week more activity was experienced, sales on Monday amounting to 200 to 300 tons, nearly all futures, with a little spot, and mostly by two sellers. On Tuesday about 150 tons of futures was sold, and the arrivals of about 300 tons from London caused the spot market to fall to 62c., New York, a decline of 1¾c. per lb. since June 20, when sales were made at 63.75c. Arrivals to June 26 inclusive were 1225 tons, with 3081 tons of Straits tin reported afloat. The London quotation on the 25th, the last cable received, was £243 15s. for spot Straits and £241 for futures.

LEAD

Many believe that a reaction downward is now due, with a probability of its continuance. At any rate the market is easier and slightly lower, with demand light. This is due to the easing of the situation occasioned by the announcement, reported in this market last week almost exclusively, that the Government had made arrangements to purchase its July requirements of about 8000 tons at 8c. per lb., St. Louis. The quantity was considerably less than anticipated, and as a result some producers and dealers, who had been holding stocks against Government needs, had some lead as a surplus which they at once offered for sale. Lead has changed hands recently at 11.50c., St. Louis, or 11.65c. to 11.75c., New York, but at present the market is quiet and has come to a halt. Some dealers held out for 12c. but were left out of the running. The London lead market is £30 10s. for spot and £29 10s. for future delivery lead, unchanged for some time.

ZINC

With production at a record rate but with demand for the metal at a low ebb, it is rather remarkable that the price holds as firmly as it does. Expectation of large Government needs and the firmness of ore prices are the explanation, but in the last few days there has been a slight weakening. Producers, large and small, are getting weary of their 'watchful waiting' attitude as to the Government's needs and the prices it will pay. Actual buying is not large and there is talk of attempts to find other fields for zinc as substitutes for other metals in which the Germans are said to have achieved marked success. Future deliveries continue to command higher prices than nearby deliveries, but there is more tendency to shade the earlier positions than the other. Quotations for early delivery range around 9.25c., St. Louis, with the forward position held at 9.50c., St. Louis. Sales have been made of fair tonnages for early delivery at a shade under 9.25c., St. Louis, or at 9.12½c., St. Louis, and 9.37½c., New York, while for the future position 9.50c., St. Louis, or 9.75c., New York, has been done with resistance on the part of sellers to shade this. Reports are coming from Washington of extended controversies as to the amount and price of the high-grade and other kinds of zinc soon to be needed by the Government, but nothing definite is forthcoming. By a week from now a clearer outlook is hoped for.

ANTIMONY

The market continues dull and uninteresting, with demand very light. It is reported that Cookson's antimony is being offered for shipment from England at about 22c., New York, duty paid. Chinese and Japanese grades are practically unchanged at 19c. to 19.50c., New York, duty paid, for early delivery. Futures are in more demand than early deliveries.

INDUSTRIAL PROGRESS

Hydraulic Pressure Tests of Oxy-Acetylene Welded and Screwed Pipe Connections

*The interest created by an article that appeared about a year ago giving a report of tests of oxy-acetylene welded pipe connections and screwed pipe connections tested under tension and compression, and the desire of readers for further informa-

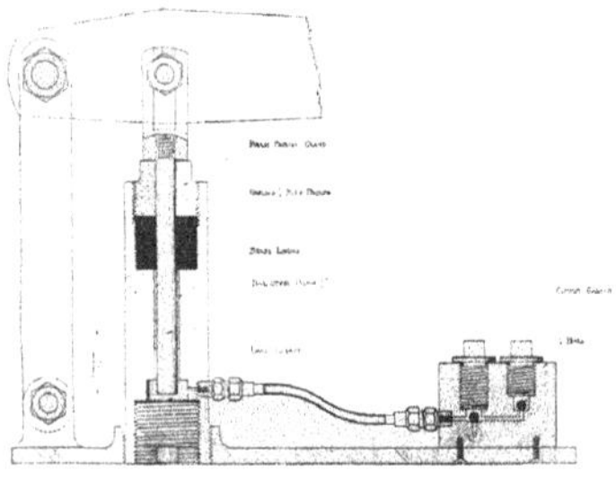

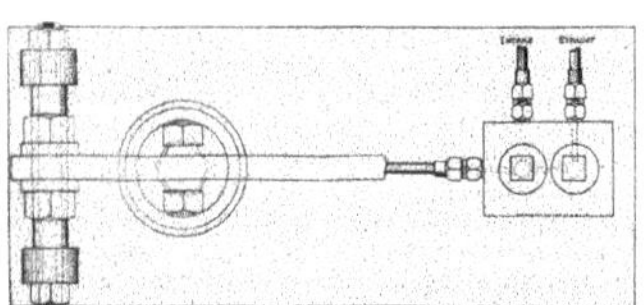

Fig. 1

tion regarding the relative strength of welded and screwed pipe connections, encouraged the experimenters to carry on a second series of tests.

These tests were conducted in the machine-construction laboratory of the University of Kansas and had for their purpose the determination of the relative strength of welded and screwed connections in steel pipe of various sizes when subjected to internal hydraulic pressure.

The pipe samples, which were cut from standard black steel pipe, were from the same stock and hence probably of uniform quality. The welded specimens were made by operators of the Oxweld Acetylene Co., Chicago. The screwed connections were made up with malleable-iron couplings and tees by expert pipe fitters. The pieces for the butt welds were cut at an angle of about 60° in a pipe-cutting machine to get the necessary 'V' groove for welding. The 'T' welds were made by cutting a hole in the run and butting the outlet against the run. The ends of all the specimens were sealed by welding in plugs or discs made from boiler plate punchings. Two of these discs are shown at the bottom of Fig. 3.

*F. H. Sibley, Professor of Mechanical Engineering, University of Kansas, assisted by Messrs. Maris, Ruth, Schooley, Jesperson, and Dryden.

The specimens were subjected to pressure by means of a small hydraulic pressure pump, Fig. 1, which was made especially for this work. The specimen under test was placed about 25 ft. from the pump and connected to it by means of a ¼-in. copper tube. A pressure-gage with a check-valve opening toward the gage was placed between the pump and the specimen. The check-valve was necessary to steady the pressure in order that satisfactory readings could be obtained because some of the samples carried pressures greater than 5000 lb. per sq. in. before failing. The illustrations show several of the specimens tested and their modes of failure.

Welded 2 and 3-in. specimens are shown in Fig. 2. These

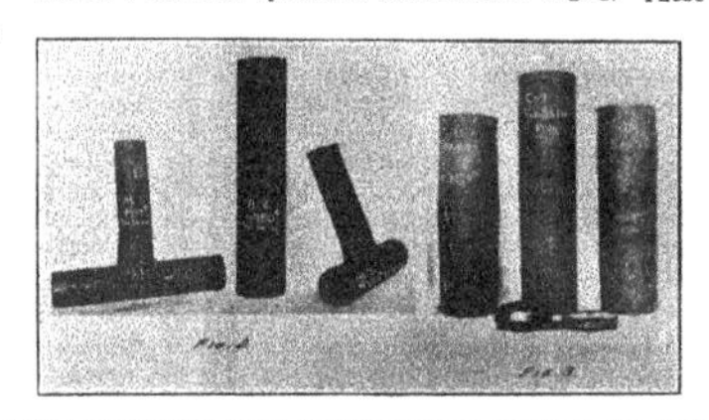

failed by splitting along the longitudinal seams of the pipe, the split stopping at the welded section. The 4-in. welded specimens in Fig. 3 bulged under the high pressure but did not fail in either the weld or the pipe-seam.

The mode of failure of two of the screwed connections are shown in Fig. 4. The bursting pressures for the screwed

connections were far below that of the welded specimens and all failed in the fitting. Great difficulty was experienced in testing the specimens made up with screwed fittings because sand-holes developed and the water leaked through the castings to such an extent that it was almost impossible to reach the point of rupture. Fig. 5 shows one of the screwed 'T' specimens under test and illustrates clearly the leaky condition just mentioned.

Examination of the data given in the table shows that in only one case was there failure in the weld and that was merely a leak which did not develop until 3860 lb. per sq. in. pressure was applied. This brings out the point that while leaks are much less likely to occur in welded than in screwed connections they are the principal cause of difficulty. Therefore, pipe-lines that are to be subjected to high pressure, if properly tested for leaks when installed, should give no trouble under service.

The results of these tests bear out the conclusions given in the previous series, namely:

a. The strength of a welded pipe connection is practically the same as that of unwelded pipe. By building up the weld slightly it can be made stronger than the rest of the pipe.

b. The strength of the welded pipe connection is very much greater than that of the malleable-iron screwed fittings.

c. Although a careless or inexperienced operator might produce a leaky joint, nevertheless, if the pipe-line is tested for leaks when installed it should give no difficulty in service.

HYDRAULIC-PRESSURE TEST OF OXY-ACETYLENE AND SCREWED PIPE CONNECTIONS

Size pipe in.	Type joint	Pressure at failure, lb. per sq. in.	Maximum pressure, lb. per sq. in.	Nature of failure	Condition of weld
2	Welded 'T'	4400	4400	Tube seam split	O.K.
2	" "	2200	2200	Leak in tube seam	O.K.
2	" "	4750	4750	Tube seam split	O.K.
2	Screwed 'T'	2350	2750	Sand holes in fitting	....
2	" "	500	2000	Sand holes in fitting	....
3	Butt weld	....	5300		O.K.
3	" "	4950	4950	Tube seam split	O.K.
3	" "	....	4250		O.K.
3	Coupling	3950	3950	Coupling split	....
3	"	3400	4400	Leak in coupling	....
3	Welded 'T'	....	3500		O.K.
3	" "	....	4250		O.K.
3	" "	....	3505		O.K.
3	Screwed 'T'	350	2700	Sand holes in fitting	....
3	" "	300	3100	Sand holes in fitting	....
4	Butt weld	....	5100	Pipe bulged	O.K.
4	" "	....	3250		O.K.
4	Coupling	300	3000	Leak at threads	....
4	"	750	2600	Leak at threads	....
4	Welded 'T'	3850	5100	Leak in weld	Leaked
4	Screwed 'T'	1000	1950	Sand holes in fitting	....

Saves $19,000 Annually Hauliug Manganese Ore

Fifty years ago manganese ore was in such demand that the owners of the Ladd mine, a manganese property in San Joaquin county, California, found it profitable to transport it on mule-back to Stockton, a distance of 32 miles. It was then loaded on river boats and carried 90 miles to San Francisco. At this point it was placed aboard sailing ships for a 15,000 mile journey to England, via Cape Horn. As new deposits of this ore were discovered in more accessible places the Ladd mine could not meet the prices of its competitors and the property was closed down.

Manganese ore, such as is found in the Ladd mine, is used for making ferro-manganese steel, and the oxide is particularly sought by glass-works on account of the low percentage of iron it contains. It lies in a formation of sandstone, chert, and serpentine. Jasper lies on the hanging wall, sandstone on the foot-wall. There are three adits tapping the vein at a depth of 300 ft. The pay-streak is from 1 to 6 ft. wide, averaging 3 ft. The ore runs from 43 to 55%. A few months after the beginning of the War the price of manganese went up, and M. C. Seagrave, of San Francisco, realizing that the Ladd mine was a 'War Bride' worth cultivating, purchased the property and reopened the mine.

Mr. Seagrave invested $2750 in transportation-equipment, purchasing eight mules at $225 each, two 3-ton bottom-dump wagons at $250 each, and six sets of double-harness at $75 each. Operating records show that the two mule-teams could make three trips daily and haul 18 tons from the mine of the crushing-plant situated in Corral Hollow on a switch of the Western Pacific railroad, a distance of three miles. The cost of this hauling, including the wages of two drivers and one stable-man, feed for the animals, repairing, shoeing, depreciation, interest on investment, taxes, and insurance, totalled $17.07 per day. This was an expense of 32c. per ton-mile of ore hauled, 95c. per ton of ore, or $6.95 cost per trip.

The road over which the mules hauled their loads is black adobe. From the upper ore-chute at the mine there is a 20% grade to the main road, which then skirts the mountain, twisting and turning down a 10% grade for one and a half miles. Here a broad wash is reached. It is the natural drainage of the surrounding mountains. In the summer the road is dusty, full of chuck-holes, and deep adobe ruts, baked to the hardness of cement by the burning rays of the sun.

In the winter the ore-wagons would often sink axle-deep in the black quagmire, and these conditions usually forced the abandonment of all activities during wet weather. For weeks after a heavy rain four mules could barely haul one ton of ore over these roads. Two big mule-teams could not haul enough ore to keep the crusher busy.

Eastern steel-mills began demanding regular tonnage and later went so far as to offer attractive premiums to miners who could meet their requirements. Mr. Seagrave was unable to fill his contracts. He summoned a mining expert, who, after a careful analysis of conditions, recommended the purchase of a White Good Roads Truck and a 5-ton trailer to supplant the mules and 3-ton wagons.

The new motor-equipment cost the mine-owner $7825, almost three times the original outlay for the mules. But the truck and trailer were able to haul every nine hours a total of 80 tons of ore to the crusher and 80 tons of gravel for repairing

THE LADD MANGANESE MINE, SAN JOAQUIN COUNTY, CALIFORNIA

the road. This equipment reduced the cost to 4c. per ton-mile, 12.3c. per ton of ore, or $2.47 per trip. The truck made eight trips per day, registering 48 miles. The total cost for operating the truck and trailer, including salaries of driver and helper, distillate, oil, grease, depreciation, interest, taxes, insurance, maintenance and repairs on truck, and maintenance, repairs, and tires for the trailer, amounted to $19.77 per day.

For 300 working days the motor equipment saved the owner $66.40 per day or a total of $19,671.05 per year, over the cost of performing the same work with the old facilities.

When the truck and trailer were delivered at the mine the rainy season was but 30 days off, and it required quick work on the part of the owner to place the roads in condition before the wet weather set in. Every ton of ore mined and not delivered to the crusher meant a loss of $25. To stop hauling ore and repair the road with the mule equipment was not practicable. Building a road by ordinary means would have made necessary the hireing of additional men, wagons, and mules.

The mule-teams were taken off the job, a gravel-elevator was erected in the creek-bed near the ore-crusher, and the truck began its work of hauling both ore and gravel. The ore was loaded at the chute, transported to the crusher, and dumped in less than 30 seconds. The truck was then driven to the gravel-elevator where it was loaded with 10 tons of gravel, and on the return trip to the mine spread it along the road where it was required. The broad steel wheels of the truck rolled the surface hard and smooth.

The truck and trailer with one driver and one helper completed the road and delivered all the ore that was mined to the crusher without interruption, in 26 days. Dangerous curves were straightened out, culverts were built, the road widened in many places, turnouts made, and in many other ways it was improved.

When the road was nearly completed the truck and trailer delivered 10 tons of ore three miles, picked up 10 tons of gravel on the return trip, spread and rolled it, and returned to the chute at the mine in one hour and ten minutes. On the completed road a four-mule team was able to deliver but three tons of ore to the crusher and return empty to the chute in three hours.

LOADING CHUTE FOR LADD MANGANESE MINE

The BOOTH-HALL Co. has been formed to conduct an electric-furnace building, engineering, and metallurgical business, with offices at 565 West Washington Blvd., Chicago. The new company is composed of five former officers and employees of the Snyder Electric Furnace Company.

The Absorption Method of Extracting Gasoline From Mineral Oils

The annual report of the State Oil and Gas Supervisor for California contains a description of the absorption method by B. E. Lindsly, engineer for the Honolulu Consolidated Oil Co. The method consists of subjecting gas to intimate contact with oil that is completely devoid of gasoline. The gasoline vapors in the gas are deposited in the oil and subsequently recovered by distillation. The gravity of the oil should be about 34° Baumé, and the boiling point sufficiently high to permit an easy separation from the absorbed gasoline.

The cycle of the absorbing-oil is as follows. The oil enters the heat-exchanger, where it transmits heat to the incoming oil. Then it passes to a cooler, to a horizontal absorber, and to a vertical spray-absorber. The gas-pressure forces it to a separator where the pressure is released. The excess gas given off in the separator is collected. This oil passes to the exchanger and to the still, where the gas is extracted and the cycle is repeated.

It is claimed that the absorption method has the following advantages over the compressor method: greater recovery, lower first cost, and lower operating-cost. A recovery of at least 0.1 gal. per 1000 ft. of gas can usually be obtained from even so-called 'dry' gas, or gas that has already been treated by the compressor method. Engineers connected with the U. S. Bureau of Mines have estimated that the first cost of an absorption plant with a capacity of 60,000,000 cu. ft. of gas per day would be $1 to $1.50 per 1000 cu. ft. capacity. Doubling this maximum estimate for a small plant, the cost of a 2,000,000 cu. ft. plant would be only $6000. The output of such a plant, assuming a recovery of 0.25 gal. per 1000 cu. ft. would be 500 gal. of gasoline per day. The cost for labor would not be over $15 per day. Since the method is patented and controlled by the Hope Natural Gas Co. of Pittsburg, Pennsylvania, there would also be a royalty charge.

The gravity of the gasoline recovered is usually higher than the commercial product, and therefore requires blending with a lower grade refinery-naphtha to fit it for the retail market.

EDITORIAL

T. A. RICKARD, Editor

INDEX for Vol. 114, January to June 1917, is now ready and may be obtained by writing to this office.

SEVENTEEN copper-mining companies distributed $91,669,281 in dividends for the first half of the current year, as compared with $65,046,051 during the corresponding period of last year.

WHAT kind of postal facilities exist in parts of Mexico, such as Chihuahua, is illustrated by the fact that a letter sent from this office on March 5, 1912, has just come back marked, in French and Spanish, "Returned owing to interruption of communications."

REFERRING to the use of alpha-naphthylamine in flotation, as patented recently by Mr. Harry P. Corliss, we are able to state that since the use of it was introduced at the Magma mill, in Arizona, the tailing has been reduced to 0.3% copper, as against 0.6% formerly, on a 4 to 4½% feed. Mr. Corliss has another patent for the use of nitro-naphthalene.

ANNOUNCEMENT is made from Petrograd that it is intended to place many Russian mines in charge of Americans. This is said to have been decided by the Mining Commission of the Ministry of Trade. It is also reported that American capitalists will be offered the island of Saghalien, or the Russian half of it, for exploiting petroleum and coal deposits. All this is important, if confirmed.

ST. JOHN DEL REY, the deepest metal mine in the world, is, as our readers know, a gold mine in Brazil. According to the latest report the shaft is now 6800 feet deep, and yet the manager is able to state that "a change has undoubtedly taken place in the lower horizons [levels] which seems favorable both as regards the quality of the mineral [ore] and also as regards the size of the lode, as shown by its sectional area." We congratulate Mr. George Chalmers and his excellent staff.

ZINC-PRODUCTION is far below the present capacities of the mines and reduction-works. This has led to a consideration of possible new uses. It is pointed out that the excessive demand for iron and steel has resulted in a shortage of galvanized sheet-metal; likewise tin-plate is harder to get, and sheet-copper is selling at almost prohibitive prices. Considering the applicability of sheet-zinc to many purposes for which galvanized and tinned sheet and sheet-copper have ordinarily been employed, the New Jersey Zinc Company has decided to erect a sheet-zinc mill at Palmerton, Pennsylvania. This may encourage a permanent increase in the utilization of zinc in that form. The price of sheet-zinc averages today about 7½ cents per pound more than the spelter from which it is made. The cost of rolling is small compared with this difference, thus indicating an opportunity to enlarge the demand for the metal in that direction.

ESTIMATES of cost have been hard hit by the War, especially if they were such as might be considered sanguine even in time of peace. Thus the Chile, or Chuquicamata, forecast of producing copper for 4 cents per pound may be compared with the 9 cents that is declared as the average of producing copper "at the plant" during the last quarter. To this cost "at the plant" must be added 3.75 cents per pound for freight and insurance—both abnormal just now—and marketing, so that the real cost is 12.75, which is in violent contrast to the cheerful 4 cents of the prospectus.

IF we have failed on occasion to be impressed by the doings of the Mining and Metallurgical Society of America, we are all the more glad to express respect for the manner in which the leading members of the Society assembled to consider ways and means of doing patriotic service immediately after the declaration of war. We have read the account of that meeting as recorded in Bulletin No. 108 and confess gladly that it does honor not only to the group of men participating but to the profession that they represented so worthily. Mr. W. R. Ingalls, as president and chairman, steered the meeting most happily and made a number of pertinent suggestions. He said, for example: "In this national crisis we professional men are anxious to know what we can do and to do what we can. Nobody yet knows. We must find out what we can do." Mr. Lawrence Addicks, with his experience as a member of the Naval Consulting Board, described the difficulty of making use of the tremendous volume of offers to serve. He quoted, from a recent proclamation of the President, the statement that "if the metal industries fail, the work of the statesman and the soldier is absolutely useless." A telegram was read from Colonel Robert M. Thompson in which that excellent citizen said: "My advice would be, pull together to increase production and urge every individual to decrease his consumption." That is to the point; it combines the purposes alike of the organization identified with Mr. Bernard Baruch and of Mr. Herbert Hoover, of raw materials and food administration. Capt. Stuart Godfrey, of the Engineer Corps of the Army, made a fine speech crowned by the quotation of Henley's poem on 'Peace and War.' Incidentally he stated that the average normal size of the gallery made by the tunnelling engineers at the front, as at Messines, is 4 feet high, by 2½ feet wide, and that the average rate of progress is 12 feet per day. Three men work at the face, one excavating, one filling the sand-bags, and one resting. He said: "This is a day of great opportunity, not only for the individual, but for the Nation as a whole. We shall suffer, if the War lasts; we shall feel it, but we shall grow." Yes, indeed, grow great and fine instead of big and fat. After Capt. Godfrey came Major Dwight, whom most of us know better as Arthur S. Dwight. He was one of the first to join the Engineer Officers' Reserve Corps and has taken an honorably prominent part in organizing the First Reserve Engineers, a volunteer regiment recruited in New York. Others contributed to the discussion, notably Mr. P. E. Barbour, who also holds a military commission. Then came an interesting suggestion, made by Mr. S. C. Thomson, well known in South Africa, that American coal-miners should be recruited, and sent under the direction of American mining engineers, to re-open the coal mines around Lens as soon as that district has been re-conquered by the British. Mr. Thomson had written to the Director of the U. S. Bureau of Mines and had received a sympathetic reply. Reference was also made to a resolution forwarded by the San Francisco section of the Society recommending that one or more representatives or commissioners be sent to Europe with a view to ascertaining what was the most useful service that could be rendered by the Society as a whole or by its members.

THINGS are not always what they seem. Sometimes we wish they were. Among the items of news that would be labeled 'important, if true' are two appearing in the daily press. The first states that the new 'grenade' employed by our destroyers has proved most effective against the Enemy's submarines, this grenade being sufficiently violent when exploded even within 50 feet of a U-boat to send it to the bottom. Therefore less accuracy of fire, as compared with an ordinary shell, suffices to destroy the assassin of the seas. This report is given some measure of dignity by being tagged as a 'Special Dispatch' from Washington. Next we are told, from El Paso, that General Gonzales, commanding in Northern Mexico, has expressed sentiments friendly to the United States and to our Allies. "A well defined movement favoring an open break with Germany" has developed, and it is predicted that Mexico will declare war against the Enemy within 30 days. Furthermore, the Mexican government intends to make the Tampico oil-fields safe as a source of supply for the allied fleets by declaring a zone within which traffic will be restricted. This sounds cheerful, but the best is yet to come. It is announced, from the same not wholly reliable source that the mines, mills, and smelters of Mexico are to be placed under the protection (or confiscation?) of the Government of Señor Carranza in order that they may produce metals and munitions for the Allies. The introduction of a device to destroy submarines and the protection of American mining enterprise in Mexico would be two developments so gratifying as to border on the marvelous. For such a turn of events we hope devoutly.

The Striking Miners

Since we wrote on this subject, last week, the trend of events has been all the wrong way. Strikes have multiplied and spread in the principal copper-mining districts, particularly in the South-West. At Butte only a tenth of the pay-roll is at work. Circumstantial evidence would appear to indicate that the four recent underground fires in this district were of incendiary origin. The testimony of a survivor from the Speculator disaster shows that the fire in that mine was started by a sub-foreman who ignited the insulation of an electric wire, the lead-pipe covering of which had been broken, in the shaft. The managers have been slow to ventilate these facts because they feared that men would be intimidated from coming to work. However, the suspicion of incendiarism was sufficient to cause many men to stop work several weeks before the strike. In the Globe-Miami district the three big mines are idle, and regular troops have arrived to prevent disorder. The mines of the Clifton-Morenci district are practically shut-down. Bisbee is much in the same plight. Austrians, who constitute a considerable part of the labor element in Arizona, have shown themselves truculent; they have attacked American miners and threatened violence to county officials. On July 5 the President telegraphed to ex-Governor Hunt asking him to use his influence as a mediator. Meanwhile a favorable sign is the evidence of jealousy and antagonism between the I. W. W.—the Industrial Workers of the World, known also as 'I Won't Work'—and the International Union of Mine, Mill, and Smelter Workers, which is headed by Charles H. Moyer, formerly identified with that anarchistic organization the Western Federation of Miners. Mr. Moyer has charged the rival organization, now the principal agent in fomenting these unpatriotic proceedings, with being financed by German money and stimulated by German propaganda. We are slow to believe anything so damnable, but Senator Charles S. Thomas, of Colorado, has come out with a positive statement that domestic war is being made behind our backs "by individuals in the employ of our enemies." The sheriff at Clifton has stated in print that it is difficult for him "to believe, especially at this time, that any loyal American would contemplate anything which might in the slightest degree cripple the Government of our country. The closing down or hindrance of the great copper industry in this section would prove a direct blow, and a heavy one, to the Government of the United States." It would indeed; and yet, Mr. Sheriff, you must know that nearly 50% of the men on the pay-rolls of the copper mines in Arizona are Americans only from the fact of domicile. They came to the United States to obtain higher wages; most of them do not speak our language; and only a few of them have become citizens in spirit as well as in form.

The problem of labor-control in these copper-mining districts is largely that of disciplining a mob of uneducated aliens, for whom the fact of the United States being at war is only an opportunity to make unreasonable demands upon their employers. In our last issue we published, under 'Mining News,' the demands made by the union at Miami and the straightforward reply made by Mr. C. E. Mills, the manager for the Inspiration Consolidated Copper Company. The demands indicate a 'hold-up' during time of war. That is all that need be said of them; they include the abolition of the contract-bonus and of the sliding-scale, besides a minimum wage of $6 for all men working underground. We are informed by an engineer, not now in the employ of the local companies, that two organizers of trouble at Miami were formerly members of the Western Federation and one of them was a colleague of Moyer during the bitter strike at Houghton. Our correspondent listened to the open-air speech-making, and informs us that there was very little said about the scale of wages, but a great deal about the 'rustler's card,' which is simply the black-list of those known to be fomenters of trouble, sometimes legitimate labor-leaders, but usually lawless agitators. Our correspondent has talked with a great many American miners, carpenters, and engineers and has ascertained that these had no wish to strike. The miners are "disgruntled over the cost of living and some blame it on the mining companies," he says. At Miami breakfast costs 30 cents; lunch (put up at a restaurant), 40 cents; supper, 50 cents; $10 per month is paid for a room, 25c. for the round-trip fare in an automobile, and $1.35 for the hospital fee, making in all $56.30 per month for necessaries as against wages at $5.40 per day, or $151.20 per month. Our informant himself lived in this way; he asserts that a man and wife can live at Miami for nearly the same cost as the single man, so that he is of the opinion that the strikers are "unreasonable." It is not the cost of living but the spending for luxury that has impoverished the mine-worker. The statement has been made that the miner can save no more at $5.50 per day than he could formerly when his wages were $3.50. This is untrue, if any idea of thrift be implicit. When wages were $3.50, the miner could save $1; now, granting an increase of 50% in the cost of living, he should save $1.75. He does not do so, but that is not the fault of his employer. The idea that the copper company must increase its scale of wages in proportion to the dividends it pays is neither just nor practicable. It would work a hardship on other forms of industry, notably gold mining. The Ford system of paying extravagant wages may prove beneficial to a particular business while it is prospering, but it means the diversion of labor from other sources and the continuous fomenting of unrest. We agree that it seems inequitable that the copper companies should gain so greatly from the abnormal market created by the War while other businesses are suffering from the same cause, but we do not see why any further part of their dividends should go to laborers that are already receiving 60% more than before the War. If the profits of the copper companies are to be taxed, they should be taxed by the Government for the benefit of the nation and for the success of the struggle in which we are engaged.

The New Metallurgy

Hydro-metallurgy of the common metals had a brief vogue about 70 years ago, and aroused the enthusiastic hopes of the mining world. There are fashions in technology as in other things, and back of them lies a compelling principle. It is a part of the spirit of progress, which the biologist would call variation. An invariable metallurgy would be possible only in a world of shrunken intelligence. It is associated with the empiricisms of mastercraftsmanship, exaggerating the importance of the 'cunning' workman hedged about by secrecy. The earlier development of hydro-metallurgy coincided with the beginnings of conscious strength in the field of chemistry when men realized that they were exchanging the juvenilities of an art for scientific understanding. Scarcely more than a hundred years ago the phlogiston theory of combustion still inflamed the imagination of scientific enquirers. Although overthrown by Lavoisier near the end of the 18th century, the notion lingered far into the 19th. The first half of the last century was essentially a period of investigation into chemical laws and of weeding out the absurd medieval conceptions of pseudo-science. It is not surprising that the chemists, in the flush of epoch-making discoveries, should then have challenged the supremacy of Vulcan. It was at this time that Ziervogel worked out the delicate balance between iron and copper in an argentiferous matte, enabling him, by a complicated and equally delicate method of roasting, to extract the silver as a water-soluble sulphate. Augustin had preceded him with a chloridizing roast of copper matte to admit of leaching the silver chloride in hot brine. Following on the heels of these came the hyposulphite system, which endured until replaced by the cyanide process. For the treatment of copper the Longmaid-Henderson method was brought forward about 1842 in an effort to solve the problem of cheap and high extraction from cupriferous cinder coming from the pyrite-burners at sulphuric-acid works; this consisted of chloridizing the copper by a final roast with salt, condensing the gases and vapors, and leaching the residual 'cinder' with water and tower-acid, whereby the copper, gold, and silver were extracted. These are but examples of many hydro-metallurgic processes that aroused interest before the impetus of mechanical improvement, accompanying the reduction in cost of power-generation, gave to pyro-metallurgy a decided lead. The principle of mere bigness also had much to do with this development; enlargement of shaft-furnaces, deeper blast-penetration, water-jacketing of the smelting zone, and finally the bessemerizing of copper matte, coming to its consummation in capacity and economy through the Great Falls type of huge basic-lined converter. Perhaps pyro-metallurgy may not have reached the limit of its evolution, but a new science has come to maturity as a lusty younger brother of chemistry. Electro-chemistry has renewed the assault on smelting. Mr. A. E. Drucker says this week in our Discussion department that "present-day smelting methods of extracting copper . . . will be replaced gradually by a combined roasting and hydro-metallurgical treatment on the spot." A similar prophesy was made boldly several years ago by Mr. Pope Yeatman. These opinions point the direction of metallurgic advance. Other conditions must soon give way to practices more scientific and savoring less of the muscular metallurgy of Tubal Cain, such as the forging of steel that will ultimately be set aside through control of the physico-chemical relations of its constituents to produce the qualities desired. Clearly the future development of metallurgy will be in the utilization of wet recovery and electrolysis on the one hand, and the direct use of energy in the form of high current-density in the electric furnace on the other. This seems to be indicated further by the growing possibilities not only of high recoveries from ores by flotation but by the possibility of producing cleaner concentrate than ever before, and also by the making of suitably refined salts for reduction, all of which will be demanded by the newer metallurgy.

Opportunity for Small Ore-Producers

Small producers of lead will be interested in the circular issued by Mr. Clinton H. Crane, the chairman of the Committee on Lead, a sub-committee of the Council of National Defense, calling for offers of that metal at the rate of 8 cents per pound. We reproduced the circular last week, and we direct special attention to it as the embodiment of a principle that we hope will be expanded in the administration of the work coming under the purview of other committees for supplying raw materials to the Government. Complaint has been general because the organization of the committees on metals seemed to exclude participation of the independent producer in the prevailing high prices. As a result the agitation for State smelters, and for other means of control to ensure a fair deal, has been intensified recently. Now comes Mr. Crane, saying in his circular, "If you do not own a smelter, may we ask you to instruct the smelter which smelts your ores to furnish one-sixth of the lead-content of the ore which it accepts from you in July on this Government order and notify them that you will accept in settlement for that amount of lead in your ore the price that the Government is paying." That price is 8 cents per pound, a material improvement over the 4½ cents that the smelters have been paying on acceptances. The small shipper will point sarcastically to the "one-sixth" of the metal-content to be paid at the higher rate, and to the joker in the phrase that limits this enjoyment to those whose ore the smelter "accepts." The imputation of favoritism apparently has a peg to hang on; that must be admitted. Even Mr. Crane must concede it. Nevertheless, it is a symptom that the Government recognizes its duty to treat equally the offerings of the great and the small. It would be interesting to see what would happen if shippers familiar with accurate sampling, but having no contracts with the smelters, were to offer ore of favorable composition in response to the Government's call. It is doubtful if it would be rejected. The difficulty is that the owners of small mines, who are being crowded out of the boom-market today, lack the financial nerve to maintain their rights. A man may have moral backbone enough to fight for his democratic privileges, but he must also have a large financial cord in his spine to carry the war into the smelting company's preserves so that it may become aware of the invasion. It is for this reason that we have urged organization as the surest and most rational means of gaining recognition from the smelters. In his inner consciousness the small shipper must realize that he rarely can deliver a large tonnage of ore on a long-time contract. He may talk about three carloads a week when making an offer, but when the time of delivery comes it too often happens that he may deliver three carloads in the first week and then stop until he has developed more ore. Likewise, having neither adequate means for sampling, nor proper training for doing it correctly, he over-guesses the metal-content of his ore. To the smelter, who must look ahead in calculating furnace-charges that will maintain a nearly uniform metal-burden and invariable slag-type, without which economical working-conditions are impossible, it is of the utmost importance to have dependable supplies of ore at command. Until the small producers can render themselves dependable by effective organization they will suffer as a class because of the general shortcomings of the individual. On the other hand, if 40 small operators with mines tributary to a single shipping-point were to form an association, maintaining a sampling-works, experienced samplers, and an assay-office, with a good business man as general manager, and this organization could guarantee 20 carloads of ore per week, of known composition, we believe the smelters would be ready to make contracts. Whenever dozens of such organizations in a State are associated for doing business on a larger scale, the moral force of numbers and their consequent responsibility will soon make them a power in the metal world. We have sympathy with the hardships of the small worker; he is indeed crowded out of the banquet-hall, but effectiveness comes from momentum, and momentum is a product of factors in which mass is more easily attained than velocity. The power of democracies lies in the association of groups rather than in the possession of a potential. The ability of the small ore-producer to reach the market and to receive respectful treatment lies in association. Through joining an organization he will be forced to take his own measure more correctly, and the development of such groups of independent producers, enjoying both technical and financial credit, will lead to a recognition that men standing alone, and having but a driblet to offer, cannot expect to command.

DISCUSSION

Our readers are invited to use this department for the discussion of technical and other matters pertaining to mining and metallurgy. The Editor welcomes expressions of views contrary to his own, believing that careful criticism is more valuable than casual compliment.

The Development of Flotation

The Editor:

Sir—On this late date I find in your publication an article written by Rudolf Gahl on the 'Future Development of the Flotation Process' in your issue of December 30, 1916. Dr. Gahl's interesting observations still hold good, although they are at this time nearly one year old and that means a lot in the flotation process. He discusses in this article my patent No. 807,501, and I would like to add a little to the cold legal language in which such patents are naturally written.

In my work in the flotation process, which dates back to 1903, I never have been and I never am, up to this present date, a believer in what is known as the 'adhesion' theory of the flotation process. One of the detriments to the flotation process is our distinction between organic and inorganic chemistry. Of course, there is no such thing, and it is only a term of convenience. Students of the flotation process are mostly accustomed to either organic or inorganic work, and will often overlook the possible reactions between substances belonging to either class, yet the formation of compounds between organic and inorganic substances is very well known and I believe plays an important part in flotation. The formation of insoluble soaps, such as those of magnesia, alumina, lime, and barium, will undoubtedly influence the process very materially.

Dr. Gahl, in the latter part of his article, makes the observation that he finds a great deal of support for the assumption that froth is caused by water-soluble substances. We have, like a red thread going all through the process, a combination of sulphide minerals and hydrocarbons or their derivatives. We can also take almost any flotation agent and extract something water-soluble out of it. We can take that same agent after the extraction is made and its usefulness is materially decreased. This shows that the soluble substances are of great importance, and it is my theory that a chemical reaction between the mineral and these soluble substances takes place. When I described in my patent a sulphide-coating I had in mind not only a sulphide mineral but a sulphide compound which I believe formed between the so-called organic oils and the so-called inorganic minerals.

Since this patent was applied for I have spent close to 15 years more or less on experimentation on the flotation process and have practically verified this theory, that is, that the flotation process depends for its success upon the presence of water-soluble substances which actually react with the mineral and also form certain insoluble soaps. Insoluble oils may be present and may help the process, but solely through the action of modifying the surface-tension.

The successful float means to obtain the proper reagent (mostly of organic character) to promote the proper chemical reaction between the mineral and this reagent, and to select modifying agents to obtain the right surface-tension, without, however, interfering with the chemical reactions of the water-soluble substances; and both reagents must be used in quantity insufficient to cause chemical enclosure of gangue-particles by excessive frothing.

A. SCHWARZ.

Webb City, Missouri, June 18.

Fire-Protection in Shafts

The Editor:

Sir—I am interested in learning the best methods of preventing shaft-fires in vertical and inclined shafts. I have seen no recent articles on this interesting phase of mine-safety, and I would appreciate a discussion of fire-prevention in shafts.

Take the case of an inclined shaft, say 68° dip, of three compartments, timbered with hemlock laths, dry, containing steam-pipes, pump-discharge pipe, and electric cable; the shaft is up-cast seven months in the year and down-cast the remainder of the time. In case of a fire the current of air would undoubtedly reverse from down-cast to up-cast. On the assumption that the shaft will be up-cast when fire breaks out, how shall we guard this shaft to control the flames? Two solutions present themselves. The first is the usual type of protection. The discharge-line from the pump is tapped at every level, a valve inserted, and a rubber hose, 15 to 50 ft. long, connected. Assume that fire has broken out on the 15th level, and must be fought from the 14th. The head, if the valve were opened wide, would be over 1400 ft. Provided that two or three men could hold this hose with that terrific pressure and that the hose did not burst, could they play down on the fire in an up-cast shaft? The heat would be great and I believe that their efforts would be fruitless. We could only seal the shaft and let the men in the mine take their chances through the other exits. The other solution is a modification of the Grinnell automatic sprinkler system. Three sprinkler-heads would be placed every 25 ft. or 50 ft. in the shaft, one in each compartment. The sprinkler-lines could be connected to the discharge-pipe from the pump and thus have the advantage of the full head on the water-column.

The heads could be so placed that when the fusible plugs melted the full play of the water would be on the hanging side and from there drop to the foot. The result would be a curtain of water. There is one strong objection to this system. The fire could crawl up the cedar-blocking behind the laths before the laths would catch fire, and since the play of water would be on the laths the crawl of the fire in the blocking would render the scheme valueless unless blocking and laths were omitted from each set where the sprinkler-head is installed. I am desirous of knowing what other methods of fire-protection in shafts are known.

FRANK A. MADSON.

Bessemer, Michigan, June 18.

The Extra-Lateral Right

The Editor:

Sir—I am greatly interested in V. G. Hills' comment on my extra-lateral right article in your issue of June 23, for it was the desire to stimulate intelligent criticism and discussion of this subject that prompted me to write the series. I might suggest that if any of your readers are interested in contributing to this discussion I will gladly send each one addressing me in care of the MINING AND SCIENTIFIC PRESS, a re-print of a series of the four complete articles. I make this suggestion as it will save unnecessary discussion on points which are fully covered in the re-print, and which were necessarily eliminated to save space in the abstract of two of these articles which appeared in your columns. Referring to Mr. Hills' comment I thorouhgly appreciate that there are many extra-lateral complaints that cause trouble but which are never actually filed in court, and that many suits are filed and then compromised before determination by the courts. This, however, is an incident of all litigation no matter what its character. Intimate knowledge of these problems in California and in other Western States during the past 15 years or more leads me to state that the extra-lateral situation has no more than its proportion of this fringe or penumbra of threatened and compromised litigation which inevitably accompanies other equally important property-right problems. The law yet remains to be devised that will be free from such difficulties.

I would take exception to Mr. Hills' conclusion that the litigated cases are only a small proportion of "the host of cases which are not appealed;" it is certainly not the experience here in California and neighboring States; nor do I agree with him when he states that the cause of the multitude of surface-contests "is directly traceable to the apex-law." An examination of the reported cases, and the experience with such cases covering a long period of years, does not support this conclusion. Disputes as to priority of location, exact position of surface-boundaries, performance of the various acts of location and of annual labor, are the major causes underlying such litigation, and the fact that such litigation is common wherever placer claims are numerous is absolute proof that the extra-lateral right is not responsible for most of these surface cases. The moment the ownership of the surface controls ownership of the vein vertically beneath, it is inevitable that the right to the surface is going to be more frequently and more bitterly assailed and contested than in the past. To argue that dimunition of surface-litigation will result from abolishing the extra-lateral right is to ignore the logic of the situation. Every foot of surface-ground will, in such event, have an added value, and there will be just as many 'jumpers' and 'blackmailers' left in the world. I am glad that Mr. Hills and I can agree so thoroughly on the importance and necessity of the complete severance of surface and mineral titles in the event that the extra-lateral right is abolished. In fact I gave Mr. Hills credit for and quoted his excellent statement of the advantage to be derived from this policy, in a foot-note to my original article which was necessarily eliminated from the MINING AND SCIENTIFIC PRESS abstract because of lack of space.

WM. E. COLBY.

San Francisco, June 23.

The Editor:

Sir—I have read with much interest Mr. Colby's dissertation on the extra-lateral right in your issue of June 2. My acquaintance with the mining law is chiefly from the standpoint of the engineer, and the mere engineer who ventures to discuss this subject with a lawyer of Mr. Colby's standing, may seem like one who rushes in where angels fear to tread, but I confess that I have found myself in much the same situation as that of the cynical Persian poet who evermore came out the same door wherein he went, a situation in which I felt that Mr. Colby's article also leaves us. Much of Mr. Colby's argument against the abolition of the apex law seems to be predicated on a fear that Congress will remove that part of the mining code as an offending appendix is removed by cutting it out and putting nothing else in its place. Many representatives of the mining industry and of mining organizations, however, are watching the situation, so it is hardly conceivable that such an atrocity would be perpetrated, and an atrocity it unquestionably would be.

I was surprised at the small proportion of cases due to apex litigation, but just what are considered as mining cases in this summary? I am writing where I have no opportunity to refer to Morrison's or any other reports, but I have seen many cases, such as trespass, master and servant, leasehold, and the like, reported and discussed as mining cases because they arose in connection with mining, but if we are to consider only such questions as are peculiar to mining, would not the proportion of apex-cases exceed the 1.9% quoted from Shamel? Mr. Colby quotes from Charles S. Thomas of Colorado: "Now the vast amount of mining controversy—and I am speaking of numbers of actions—has not been apex-litigation. They have been the most expensive and the most far-reaching. They have perhaps resulted in the greater proportion of injustice; but the conflicting (surface) locations have produced the multitude of cases, a small percentage of which perhaps reach the Court of Appeals, but whose aggregate has burdened the prospector and locator with an expense almost unbearable." Note that these cases "have been the most expensive and the most far-reaching. They have perhaps resulted in the greater proportion of injustice." Is not the fact that a law works an injustice a potent argument in favor of providing means for its elimination? Instances of such injustice are numerous and well-known. The Butte litigation had its origin in the extra-lateral right. It has been stated that a certain wealthy Californian bought a rundown mine, simply because he realized the possibilities of an apex-suit against his rich neighbor. After the expenditure of thousands of dollars in attorneys fees, geologists, and engineers, the defendant won the case, the judge deciding on plain common-sense grounds, although he stated that the plaintiff's contentions as to geological conditions were entitled to consideration. A more technical judge might have decided differently. This case probably is not reported, as I think it was not appealed, the plaintiff having died before the decision. In another case a company spent many months and thousands of dollars developing a supposed blind-lead. After obtaining a gratifying showing of ore, ordering new equipment, completing tests and making plans for a mill, they were served with notice of suit by their neighbor, on the pretense that their so-called blind-lead was merely the faulted portion of a vein having its apex on the neighbor's ground. When any system opens the way to practices that are little short of blackmail, and gives the strong a chance to oppress the weak by ruinous litigation, we are justified in taking steps to correct it, even though it appear drastic at the outset.

It is a notable fact that many neighboring companies in different mining districts have voluntarily agreed to disregard the extra-lateral rights they might have against each other; they have accepted the vertical planes as their boundaries. Such agreements have been entered into by some of the large mining companies of the country, not because they were afraid to stand up for what was rightfully theirs, but, as a straight business proposition, because they recognized the futility and the ruinous nature of such litigation as apex-suits usually entail.

Note the last sentence of the quotation from Senator Thomas. Why do we have conflicting surface locations? Aside from the man who goes out and deliberately jumps a claim, the one who makes a conflicting location does so for the purpose of obtaining an extra-lateral right on a vein which exists in the senior location and is so situated that it will cross one of the end-lines of the latter on its dip. A potent argument in favor of the abolition of the extra-lateral right is that it will do away with conflicting locations, excepting, of course, those made for the avowed purpose of claim-jumping. There can be no doubt that such a change would be radical and would call for other changes as well, but I do not see wherein the difficulties will be insuperable nor do I think that the laws need be as "profoundly amended" as Mr. Colby suggests. He himself states, in his first conclusion, that in nearly every case where the extra-lateral feature has been incorporated in the laws of other countries it has been abolished eventually. My own experience has been confined to this country, but I have read and heard many statements, particularly with reference to British Columbia, that such abolition has been accomplished without confusion and that the general effect has been beneficial.

I have been told that in the boom-days of Leadville it was not uncommon to see several groups of men within the area of one claim, each sinking madly in an effort to be the first to reach a vein and claim a discovery. I do not think any one would wish to impose such a requirement under present-day conditions, but Mr. Colby seems to fear that such might be a necessity if the extra-lateral right were abolished. Every one, I think, must recognize the fact that the abolition of the extra-lateral right will entail a modification of the requirement of discovery; but need this modification be so radical, or do away with such rights as a discovery would naturally confer? In other words, the fact that a discovery might not be required in every case need not do away with certain indisputable rights that it would confer if actually made. An actual discovery in a controversy over the character of land would have just as much weight as ever, and perhaps more. The discovery of a vein should entitle the discoverer to locate one or more additional claims on the dip. I believe that it might be practicable to carry the policy still farther and allow the location of mining claims on geologic evidence. It has been held before the courts and the Land Department as well that such indications as would put a man of ordinary prudence on his guard as to the mineral character of a piece of land would preclude its entry under the agricultural law. This does not actually require the visible presence of mineral of value, and there have been numbers of coal-land cases so decided where there was no outcrop within miles. If such evidence is sufficient to preclude an agricultural location certainly it should be sufficient to sustain a mineral location.

I do not appreciate the weight of Mr. Colby's argument that "such elimination of discovery would destroy the simplest test whereby mineral lands are now practically and easily classified under existing laws so that mineral locators are able readily to obtain the same lands." The actual discovery would always be held to be the most conclusive evidence as to the mineral character of the land. The modification would be that where such an actual discovery were not feasible it would not be insisted upon; but other competent evidence tending to prove the character of the land might be accepted.

In discussing the feasibility of a classification of mineral lands by the Government, Mr. Colby says, "It would mean aggravating delays where mines were discovered in rugged or desert regions remote from centres of travel." Is it his idea that under a law providing for classification a man who might make a discovery in a remote region would be required to wait for Government classification of the land before he could perfect a location? Such a law would be wholly inconsistent with past and present practice. Our laws provide for the classification of coal-lands, and land once classified as containing coal can only be disposed of by taking the coal into consideration, that is, by purchasing at the appraisal price under the coal-land laws or by making a filing under the non-mineral law which reserves the coal to the Government; but this does not for a minute prevent a man from filing on coal-land that has not yet been withdrawn or classified as such. Anyone who makes a discovery of coal on land not withdrawn or classified may purchase it under the coal-land laws, provided he furnish evidence as to the likely existence of coal. He is not required to furnish evidence of outcrop or actual discovery on the tract that he seeks, provided he can show reasonable geologic evidence as to its character. This is submitted in the form of affidavits with the filing so that it readily can be seen that if we follow out this same principle in our mineral laws there need be no barrier to immediate location of a mining claim wherever an actual discovery is made.

Nevertheless, any classification of land which depends on an opinion, geologic or otherwise, is bound to provoke litigation, and I am coming to agree with Mr. Colby that the proper solution is a separation of surface and underground rights. This should be done by incorporating into every agricultural patent a clause reserving to the Government all mineral existing in the ground conveyed, and making it available for prospecting, provided the claimant to the surface is properly secured against damage. This is only a step farther than the coal-land act of June 17, 1910. It provides that anyone making an agricultural entry on lands withdrawn or classified as valuable for coal must reserve to the Government or its agents all coal therein, and the right to prospect and mine for the same. The act also fixes the means by which the agricultural claimant shall be secured against damage to the surface. The purchaser of the coal under an agricultural claim acquires such surface rights as are absolutely necessary to mining, but one who purchases a tract of coal-land on which no agricultural claim has been located obtains surface-title as well.

My conclusion is that the extra-lateral right is an anachronism. It is bound to go sooner or later, and should be disposed of, as suggested by Mr. Colby, by segregation of surface and underground rights, and also, as he likewise suggests, revision must be general and only after careful consideration by most competent men. To this I might add: if eventually, why not now?

LEROY A. PALMER.

San Francisco, June 25.

Hydro-Metallurgy v. Smelting

The Editor:

Sir—The present-day smelting methods of extracting copper from table and flotation concentrates will be gradually replaced by a combined roasting and hydro-metallurgical treatment on the spot. Such is proving to be the case with zinc-blende and gold and silver-bearing concentrate. During the next few years we shall see some remarkable advances in the treatment of copper sulphides by hydro-metallurgical methods.

Why go to the extra expense of handling and transporting copper concentrate to a smelter when it can be treated at the mine at a greater profit by combined roasting and leaching, producing refined electrolytic copper direct? Let us have some discussion on this important subject.

A. E. DRUCKER.

New York, June 28.

Misfires

The Editor:

Sir—Every miner, like myself, is interested in the matter of safety, and as misfires constitute one of the most dangerous elements of the miner's daily work this particular matter has interested me greatly. I have read with unfailing interest the several contributions on this subject appearing under 'Discussion' in your paper, and it seems to me that another contribution analyzing all that has been said in the way of experience and suggestion would be timely and appreciated by all who realize the great importance of this subject.

MINER.

Angels, California, June 20.

Seale-Shellshear 'Cascade' Process

*This flotation process was invented by Seale and Shellshear, previously of the Junction North mine, Broken Hill. The patents have been acquired by the Minerals Separation and De Bavay's Process Co. Fleury James Lyster, of the Zinc Corporation, and James Hebbard, of the Sulphide Corporation, have modified and improved the design of the boxes. The latter company has installed the process on the lead mill of the Central mine, Broken Hill, and is making experiments with a view to adapting it also to the zinc mill. The plant consists of a series of five boxes arranged above one another, the total height being about 24 ft. The pulp is elevated to the first box and in cascading from one box to another the necessary agitation is provided without any revolving impellors. The only power absorbed is that involved in elevating the pulp to the top box. On the Central mine considerable savings have been made possible, not the least of which is a reduction of fully 150 hp. in treating the same amount of material, while repair-costs are reduced to a minimum. This plant is displacing a large number of tables. The introduction of the process has widely extended the scope of flotation separation. The Broken Hill South mine is also experimenting with the 'cascades' in the lead-section of their slimes plant, where it is proposed to install six boxes in series, each containing three 11/16th in. water-jets. Lieut. H. V. Seale—one of the inventors—is at present on active service with the A. I. F., while Wilton Shellshear, the other inventor, is with the Burma Mines, Ltd.

**Mining & Engineering Review.*

Principles of Flotation—II

By T. A. RICKARD

BUBBLES. We saw how the floating of the needle was aided by bubbles of air attached to it. That suggests, but does not explain, the latest and most successful phase of flotation. To understand it we must go back to the small boy's soap-bubble. The man that understands the physics of a soap-bubble has mastered the chief mystery of flotation. The boy, who, as pictured by Millais, watches the birth, ascent, and disappearance of the iridescent sphere of his own making, is the type of our modern metallurgist, who makes the multitudinous bubbles constituting a froth and then wonders to what natural laws his filmy product owes its existence.

To put it briefly, the boy, having dissolved soap in water, holds a little of the liquid in the bowl of his clay pipe while he blows through the stem. The soapy water forms a film that is distended by the boy's warm breath into a lovely sphere, which is lighter than the surrounding air, and therefore rises, while the sunlight falling upon it undergoes refraction into the colors of the spectrum. When the boy blows through his pipe into pure water, he makes bubbles likewise, but they burst instantly. The high tension shatters them. They do not burst explosively by expansion of the air within their envelope, but by lateral displacement of the substance composing their incompletely elastic films. To prevent such immediate collapse it is necessary to lessen the tension, that is, to diminish the contractile force at work in the watery substance constituting the exterior of the bubble. This can be done by introducing an impurity or contaminant. Water has the highest surface-tension of any common liquid, so that the addition of almost any other liquid—such as oil, alcohol, or acid—will lower the tension. The boy rubs the soap between his wet hands and dissolves it in the water. The soluble soaps contain an alkaline base, such as potash or soda, combined with a fatty acid, such as oleic or palmitic, extracted from tallow or oil. The boy uses oleate of soda, a compound of soda and oleic acid. The flotationist uses oleic acid, and much of the early work was done with this thick oil. In both cases, boy or man, playing at bubbles or working at metallurgy, the oil serves to lower the surface-tension of the water and to prolong the life of the bubbles that are made out of this modified water.

Two phases of the subject may be compared: The needle that floats on tap-water will sink in distilled water, because the latter lacks the air-bubbles that assist flotation. Although the tap-water has a lower surface-tension on account of its slight impurity, that effect is less decisive than the aeration. The bubble blown in pure water will break almost as soon as it comes into existence, but the solution of a little soap in the water will enable a boy to blow bubbles that sail away beautifully. The lowering of the surface-tension by the contaminant lessens the tendency of the bubbles to collapse. We have seen, in the camphor experiment, how the oil would lower the surface-tension not only of the bubble-film but also of the water in which it might be generated; that lowering of the surface-tension promotes wetting, which is antithetic to floating. If, to water on which mineral particles are floating, an addition of alcohol or caustic soda be made, or even the vapor of alcohol be allowed to play over the surface of the water, the mineral particles sink. The intense local contamination of the water has decreased its surface-tension so much as to increase the relative effect of gravity. Instant wetting ensues. It is evident therefore that oil can be used effectively in flotation in two ways: Either in such large quantity as to raise the mineral by sheer buoyancy or in such small quantity as to coat the particles of mineral, in preference to the gangue, and also decrease the surface-tension of the water in such a way as to promote the formation of a stable froth. Luckily the increased wetting power of the water due to the solution or emulsification of the oil is rendered largely ineffective by the oiling of the mineral particles themselves, on the surfaces of which the oil displaces the water and thus prevents wetting, while the lack of adhesion between oil and gangue serves differentially to aid the wetting of the latter by the water.

The changing colors of the bubble indicate that the thickness of the film is not constant; on the contrary, it may vary within wide limits without noteworthy variation of the surface-tension. That makes an important difference between a liquid film and any ordinary elastic membrane. "The tension in a liquid film is independent of the stretching, provided that it is not so great as to reduce the thickness of the film below about five millionths of a centimetre."[1] This result is promoted by the use of a solute that will be strongly adsorbed at the surface of the solution.[2] As the film is being stretched, the new surface formed at the thinner portion will contain less solute, owing to the time needed for adsorption, so that the new surface will be stronger than the old. Likewise, when water has been modified by a relatively insoluble contaminant, the components of the film can so dispose themselves that the surficial forces will be the same everywhere, that is, they tend to remain in equilibrium, including the force of gravity, which otherwise would pull them apart. Thus the tension at the surface

[1] Poynting & Thomson, *op. cit.*, page 137.

[2] Hildebrand. Fig. 2, page 169, M. & S. P., July 29, 1916. Also Willard Gibbs' 'Thermodynamics,' page 313.

of a contaminated liquid is able to adjust itself within fairly wide limits, and a film made of such a liquid can remain in equilibrium, whereas a film of pure liquid breaks at once. A soap-bubble will last for hours, a pure-water bubble persists for a fraction of a second. Moreover, the presence of a contaminant in water may also affect its viscosity, or internal friction, whereby it offers resistance to change of shape. This strengthens the film of a bubble generated in modified water. It has been asserted[5] that a concentration of the contaminant occurs at the surface of such a liquid, causing the viscosity to be magnified as compared with the body of the liquid. This statement is well founded.

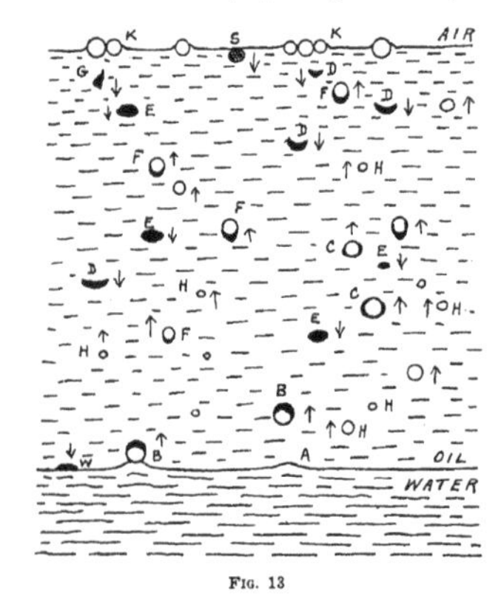

FIG. 13

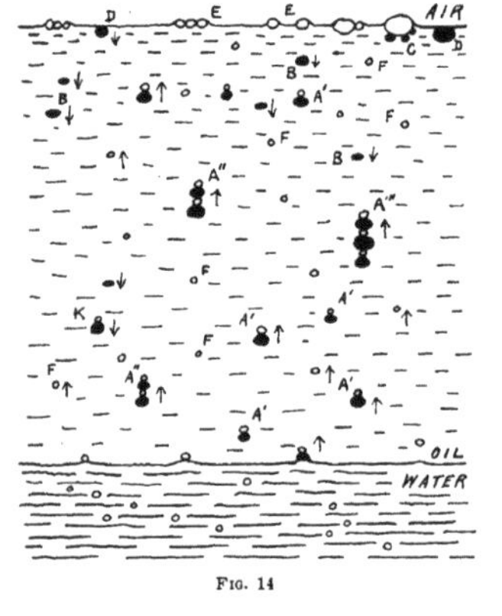

FIG. 14

An interesting experiment[6] to illustrate this phase of the subject can be made by floating kerosene over blue-colored water and then passing air into the lower liquid. When bubbles are formed in the oil, they are short-lived, but they last long enough to indicate that the oil is not a pure and perfectly homogeneous liquid. In such a liquid, the bubble would break on arrival at the surface. The fact that two bubbles touch without coalescing (*K*, *K*, Fig. 13) proves that there is a film of variable composition between them. When I blow air gently into the colored water,[7] the bubbles that rise into the oil are colorless. They accumulate at the upper surface of the oil, where they show an attraction for each other and also for the sides of the glass vessel. They last longer than the bubbles blown in oil because they are made out of a liquid containing a decided contaminant, the dye. Next, I blow air more energetically, and I note that when the bubble is about to escape from the blue water it raises the surface into a mound (A in Fig. 13), emerging at the point of it (as at *B*) as if the air had dragged the water in an effort to overcome a viscous layer. This indeed is the fact. I caught one bubble in the act; it came slowly through the little heap of water and remained poised at the top of the mound, finally breaking away, while the water subsided sluggishly to its level. Finally, I introduced air more rapidly into the water. The bubbles broke through the viscous water-oil interface and carried portions of water with them. These portions slipped from the north (*B*, *B*) to the south pole (F, *F*) of the bubbles and fell away, sometimes not until the bubbles had reached the upper surface of the oil. An intermediate stage is shown by C, C. This water that detached itself from the air-bubble was not a stable film but a viscous coating. It assumed various forms, crescent, hemispherical (D, D), lenticular, flatly globular (*E*, *E*), or even shapeless (*G*). The retention of a form that is not spherical is proof that the force of surface tension is overcome by the high viscosity of the film at the water-oil interface.[8] Occasionally some of the blue water remains as a globule attached to the surface of the oil, as at *S*. On reaching the oil-water interface the globule (as at *W*) will merge itself slowly with the liquid from which it originated.

If a similar experiment is made with carbonated water, in which minute bubbles of nearly equal size are generated quickly, one can see the little bubbles, like bright colorless beads, leading a much bigger globule of

[5] Samuel S. Sadtler, in Minerals Separation *v.* Miami suit, 1915. Emphasized recently in the Butte & Superior case.

[6] How variously it can be seen and interpreted is shown by the descriptions given by Messrs. Durell, Norris, and Rickard, in 'The Flotation Process,' pp. 137, 315, 358; also by Messrs. Taggart and Beach in Trans. A. I. M. E., September 1916.

[7] Some of these experiments may seem almost childish to the supercilious, but I can commend them not only as giving insight into fundamental principles but as likely to stimulate thoughtful discussion.

[8] As elucidated recently by A. F. Taggart in the Butte case.

blue water upward (as at A′, A′ in Fig. 14) through the oil to the surface, where the bubble breaks and the globule of water falls back through the oil in oblately spheroidal shape (*B*, *B*). Sometimes two, or even three, couples rise tandem (as at *A″* and *A‴*). At the surface of the oil the coalescence of several bubbles may leave one large bubble to which several small globules of water are attached (as at C), or globules of blue water (*D*) may remain floating in the oil, as if hanging from the surface of it. Sometimes the bubble may be overweighted and, after rising a little way, it descends (*K*). If the couples collide, the bubbles are released and leave their freight of water, which drops back. The interesting feature is the air-bubble's ability to lift a water-globule so much larger than itself. This is due to the fact that the water comes from the water-oil interface and includes oil.

The amount of the contaminant in the froth of a flotation-cell can be measured by analysis. The concentration in a film may proceed so far as to form a solid, as when using hard water. The use of oil as a modifying agent is advantageous because it is not prone to enter into chemical reactions with impurities in the mill-water even when thus concentrated in the bubble-films; otherwise some other contaminant might be used. Indeed, it is likely that oil will be replaced by some contaminant that is cheaper and that may also induce some desirable chemical reaction. Several such substitutes are now being tried in flotation plants.

The question has been asked, when a bubble is formed in a liquid, is it a spherical hole filled with gas or is it a sac; in short, has it a skin or not? The reply to this question involves the whole theory of surface-tension and bubble-making. When a pure gas is blown into a pure liquid, the bubbles rise rapidly to the surface, where they burst instantly. The gas injected into the liquid is subject to the gas-liquid tension, therefore the surface of the liquid enclosing the portion of gas assumes a spherical shape in obedience to that tension, because a sphere occupies the least space. The liquid in contact with the gas will have a different orientation of its molecules and it will be slightly denser than the internal liquid. These conditions will accompany the globule of gas in its passage upward. The form of the liquid periphery persists but the substance of the liquid in contact with the gas is changing as the bubble rises. An analogy is furnished by the motionless cloud on a mountain. The cloud retains its shape, although its substance is fleeting. Ascend the mountain and you find yourself surrounded by a mist that is traveling at the rate of 20 or 30 miles per hour, or even faster; yet as seen from the valley the cloud seems fixed. The explanation is that the moisture-laden air sweeps into the cold area on one side, either the snowy or shady side of the peak, and there the moisture is condensed to globules of water constituting a fog or mist; these are visibly driven forward, to be expanded suddenly and dissipated into clear air as soon as they pass beyond the cold area, but their place is taken by others coming on behind, so the shape of the cloud persists although the substance of it is rushing forward at the speed of a railway-train.

Now the important question arises: What is the substance of the film of the bubble as it passes from one liquid into another? The attachment of blue water to the bubble in the water-oil experiment is confusing, because it obscures the fact that, as the coating of water slips away, the bubble acquires an oily film and when temporarily at rest on the surface it is enveloped in an oily film. No blue tinge can be detected, if the effect of reflection from below be avoided. On the other hand, if the experiment be repeated with heavy oil (colored by 'oil orange') and alcohol, it will be found that the bubbles that come to roost at the upper surface of the alcohol are orange-colored. Thus, as scientific theory would suggest, the bubbles take a film of the liquid having the lower surface-tension or less molecular cohesion. In passing from water to oil or from oil to alcohol the bubble has an oily film at the end of its journey. If a bubble were generated in water and passed successively through oil and alcohol, it would have a water, oil, and alcohol film in

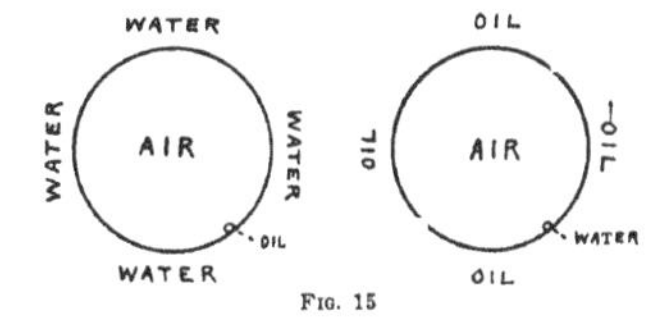

FIG. 15

sequence. If the bubble passed in the reverse direction it would have an alcoholic film in the alcohol, the oil, and the water alike, because alcohol spreads over oil and oil spreads over water, the liquid having the less cohesion or surface-tension being pulled by the molecular attraction of the liquid having the stronger cohesion or surface-tension. There is this to be added, however, that the bubble generated in water would have some water in its oily film when in the oil, and some oil in its alcoholic film when in the alcohol. Each liquid in turn serves slightly to contaminate. On the return journey, the alcoholic film, contaminated slightly by the air and by any impurity in the alcohol-air interface, would resist modification by the oil and by the water (forming the lower layers of liquid) because the alcohol would spread over to the oil-air interface and over the water-air interface. Imagine a globule of oil in an air-bubble enclosed by water (Fig. 15): the oil spreads and forms a film to enclose the air. Now imagine a globule of water in an air-bubble surrounded by oil; the water does not spread, because the pull of the air-water and water-oil surfaces is greater than that of the oil-air surface; therefore a water-filmed bubble will acquire an oil film when passing into oil; on the other hand an oil-filmed bubble will retain its film in making the same entry through water.

We have seen that mineral has a selective adsorption for oil rather than for water and that in this respect it differs from gangue. Metallic particles adsorb air, but this fact is relatively unimportant in flotation because the air approaches them when it is enclosed within a liquid envelope that is contaminated by oil. Therefore the adhesion of oil for the metallic surface becomes the dominant factor. The older notion that the affinity of air for metallic surfaces played an important part in flotation has been set aside, because of the absence in the flotation-cell of any direct contact between air and mineral. Metallic surfaces, such as those of minerals, are supposed to adsorb air and that is why they are not readily wetted. It may be due to molecular density, coupled with reduction of inter-molecular distance, which is practically the same thing as a reduction of sub-capillary porosity. Adsorption of air would also bear a relation to the higher density of the mineral. Such adsorption plays its part in the older surface-tension processes, such as those of Wood and Macquisten, but in the later flotation processes there is present insoluble oil or a soluble frothing agent, and this renders it impossible for the globule of air to come into direct contact with the mineral. It is not the air, but the film around it, that provokes the attachment of the bubble to the mineral.

Now let us consider the air-bubble made in water containing an impurity that decreases its surface-tension. In the language of flotation we would say that this impurity is a contaminant modifying the water. As soon as the air enters the water it assumes a globular form as before, but when the bubble reaches the surface it persists; it does not burst at once. The bubble in the water is a spherical hole occupied by air; the air has displaced the water and is enclosed by it; the water-surface in contact with the air is in a state of tension as compared with the interior body of water, and that causes contraction into spherical shape. The surface-tension has been lowered by the contaminant so that the bubble-film is in a state of less strain than a similar film of pure liquid, hence a diminution in the tendency to contract and to collapse. Moreover there is a tendency for the contaminant, whatever it be, to concentrate at the air-water surface; there is a differentiation of the constituents of the liquid, causing the surface to differ slightly in composition from the bulk of the solution and so to accentuate the modification due to the presence of the impurity. The bubble-film or air-liquid contact adsorbs the contaminant until equilibrium is established, and the contaminated liquid of the film carries some of the contaminant all the way to the surface, despite the interchange between molecules or particles of the contaminant on the way up. This differentiation and concentration of the contaminant at the surface of the water in contact with the air-bubble may indeed be likened to a film or membrane, so that the bubble may be regarded as a sac, but it is a sac the substance of which is not fixed while the bubble is moving upward through the water. It cannot be regarded as enclosed within a definite film until it reaches the end of its journey, and even then the film is co-terminous with the surface at which it rests, and the play of light upon it shows that the re-arrangement of its substance is still in progress, as the excess of liquid drains to the south pole. The variability in the surface-tension due to the shifting of the contaminating particles is essential to the longevity of the bubble-film. That brings us to a recognition of an important factor: viscosity.

VISCOSITY. This is defined as the internal friction of a liquid or its resistance to a change of shape. Two years ago the part played by viscosity in establishing a bubble-film was subordinated to emphasis on the lowering of the surface-tension of the water in the ore-pulp.[4] Since then this branch of the theory has been elucidated by Messrs. Taggart, Beach, and Bancroft.[5]

The addition of alcohol increases the viscosity of water up to about 47%, after which the further addition decreases the viscosity. Alcohol, of course, lowers the surface tension of water, but an experiment[9] will prove that the change of viscosity is the dominant factor in making a froth. If alcohol, to which 5% water has been added, be stirred violently in the glass-jar machine familiar to flotationists there will be no formation of froth, but if the experiment be repeated with tap-water, to which 1% of alcohol is added, then a froth is produced at once.

Such an alcohol-water froth is non-persistent, because the absolute viscosity is low. To increase it we must have a colloidal suspension; for example, the foam on beer. The colloidal protein of beer yields a froth that lasts longer than the bubbles on champagne, which are short-lived, like the alcohol-water foam of the experiment just described. To obtain a froth sufficiently persistent to serve a metallurgic purpose it is necessary to increase the viscosity of the bubble-films. This is one of the functions of the oil, and it is one that follows upon its affinity for metallic surfaces. It adsorbs or concentrates (at the surface of the bubbles) the mineral particles in the pulp so as to form an interface that is more viscous than either the oil or the water or the mixture of the two.[10] It is the presence of solid matter that contributes to the viscosity of the bubble-films in the froth.

If a needle be floated on water by means of a raft made of wooden matches and if a chip of wood be floated to one side of it, one can use a magnet to turn the raft and needle on the surface of the water without moving the chip. This shows that the surface, or water-air interface, has no noticeable viscosity.[11] If, however, the surface be dusted with finely pulverized ore, then the magnet will cause the chip to move with the rafted needle.

[4]However, I pointed to the probability of viscosity contributing to the tenacity of the film, even in the needle experiment on tap-water, and quoted Boys to show that increase of viscosity was involved in the lowering of surface-tension in enabling a bubble to persist. M. & S. P., Sept. 11, 1915, p. 385.

[5]More particularly in their expert testimony at Butte, from which I have quoted already.

[9]Described by Wilder D. Bancroft in his testimony at Butte.

[10]Taggart.

[11]Taggart. He pointed to the fact that the addition of the oil increased the viscosity of the surface so as to cause it to act as a solid within small distances, close to the raft, but considerably less than when the powdered ore was sprinkled upon the oiled surface.

THE FROTH IN A CALLOW CELL

The viscosity has been so greatly increased by the addition of solid matter to the interfacial film that the surface behaves as if it were solid. Next, if a drop of oil, sufficient to lower its surface-tension, be added to the water, the chip will not turn when the rafted needle is moved by the attraction of the magnet. Such increase of viscosity as has been caused by the oil is insufficient to form a resisting medium. Finally, if powdered ore is dusted upon the oil-contaminated surface, again the chip does not move with the raft, because "the surface has been stabilized and made highly viscous."[12]

If water and kerosene be poured successively into a glass bottle, and if then finely-divided copper, called 'bronze powder', be introduced and the contents of the bottle be subjected to vigorous shaking, and then allowed to remain quiescent, the copper powder collects at the oil-water interface and from it slowly a bronze film will separate itself and become pendant. This, when viewed

[12] I am quoting from Mr. Taggart's testimony, from which the description of the experiment also is taken.

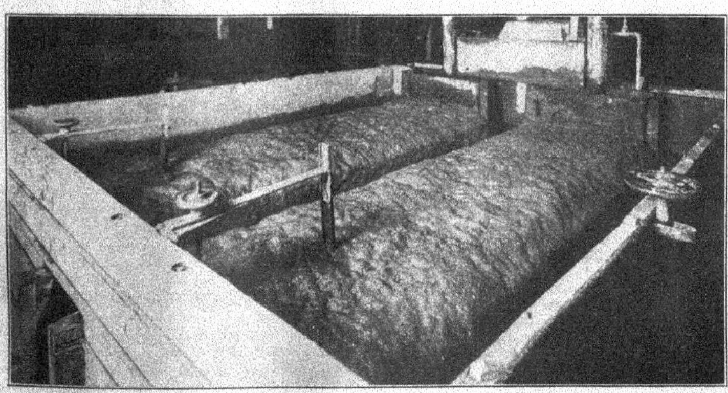

ANOTHER PHOTOGRAPH, SHOWING THE FROTH OF TWO ADJACENT CELLS

by transmitted light, is seen to be a lace-like fabric, like a cobweb that has been long exposed to dust.[13] It is a film of particles of kerosene and water so viscous, owing to the inclusion of the powdered copper, that it hangs like a curtain; it is an adsorption layer of bubble-film matter hanging from the oil-water interface. The presence of the powdered copper has stabilized the film.

It is important to note that such increase of viscosity as prolongs the life of the bubble-film need not be metallic. When pine-oil is added to water, and the mixture is agitated, the froth that comes to the surface of the water is thin and evanescent. When to this there is added lycopodium powder, which is of vegetal origin, being the spores of club-moss, the froth becomes thick and lasting.[14] If the lycopodium be used without the pine-oil, no persistent froth is made. In this case, as with the bronze powder, the effect of the solid is to stabilize the froth by making the bubble-films more viscous. The gangue would serve for this purpose if the particles of gangue could pass into the oil-water interface, but it happens, as we have seen, that the oil exerts a preference for the particles of mineral, so that they are adsorbed preferentially.

Another experiment:[15] When a needle was floated on water in a beaker and a drop of caster-oil was added, the needle did not sink. When another drop of the same oil was added, the globule moved to the needle and adhered to it. But it continued to float. When a drop of pine-oil was allowed to run down the side of the beaker, the needle sank as soon as the pine-oil touched the water, while the globule of oil remained afloat. Apparently the increase of viscosity due to the thick oil counteracted the lowering of the water's surface-tension.

The effect of saponine, noted in Hoover's book as being so detrimental to flotation, can now be explained. Although it does not increase the surface-tension of water, but tends rather to decrease it very slightly, according to Freundlich, it causes a marked increase of viscosity. The result is a good froth; but it exhibits no essential adhesion, that is, the saponine solution is not adsorbed by the mineral. Therefore the froth does not persist and the mineral is not floated.

Any substance that is adsorbed into the oil, or the oil-water interface, of the bubble will pass into the film. If it does that the substance will be floated. Mineral goes into oil in preference to gangue. If a particle of sulphide is in the vicinity of oil and water, the oil-surface of the sulphide grows larger and the water-surface grows smaller, until the sulphide at the last takes a position within the oil. Reversely, a particle of quartz takes a position within the water. The greatest possible area of sulphide that can be covered by the oil is when the sulphide is within the oil; therefore the particles of sulphide tend to encase themselves within the oily substance of the bubble-film and so not only stabilize it but give themselves the opportunity of being floated to the surface in the froth.

[13] F. E. Beach, who performed the experiment in the court-room at Butte. R. B. Yerxa repeated it for me at Miami.

[14] Bancroft, who performed the experiment in the court-room at Butte.

[15] Made for me by Mr. Yerxa in the laboratory at Miami.

Oil-Films. In the course of the first trial of the Miami lawsuit, at Wilmington, a series of demonstrations was made in court for the purpose of argument. These experiments were photographed and placed in the record. Some of them are of scientific interest. Fig. 16 shows the curved pipette employed to pass an air-bubble to the bubble-holder, which is a bell-mouthed glass tube. Fig. 17 shows the play of a bubble on the oil placed upon a particle of galena lying at the bottom of a vessel containing water. In A the particle of galena and the bubble-holder are shown. In B a globule of oil rests on the galena. The oil is 1½ times the volume of the galena particle. In C the air-bubble is adhering to the oil on the galena and drawing it up, forming a neck of oil between the bubble and the galena. The photographs exhibit the affinity of the oil for the air-bubble. If the bubble failed to raise the particle of galena, this should not occasion surprise, as it was much too large—several thousand times bigger than the average pulp treated in flotation. In Fig. 18 similar experiments on particles of unoiled galena of a reasonable size—about 20 mesh—are recorded photographically. In the first of this series the bubble-holder is approaching one of three particles, in the second it is moving away with one of them, and in the third with another. In Fig. 19 another series of experiments is shown, but with oiled particles of galena, of plus 20-mesh size. In the third member of this group it will be noted that all of the galena particles are being carried away by the bubble. Two of the particles are adhering to the third particle, which is attached directly to the bubble. Ordinary tap-water was used. These experiments, and others like them, showed that particles of galena will adhere to an air-bubble, whether they are oiled or not. The adhesion takes place even when the mineral carries an excess of oil. Particles of chalcocite do not adhere so readily to the air-bubble when they are unoiled as when they are oiled, but the evidence given in this suit was incomplete; moreover it was not shown whether a bubble made out of water suitably modified will, or will not, adhere to an unoiled particle of chalcocite. The motion-pictures of these demonstrations cost a great deal of money, but it will be acknowledged now, I believe, that they threw but little light on the theory of flotation.

The adhesion of air, as a bubble in water, to mineral particles is easy enough to prove, but such bubbles, as far as I have been able to ascertain by experiments, will adhere to almost anything that happens to be near-by. Trying some of these experiments recently with Mr. Yerxa, at Miami, I found that a large air-bubble would not lift an 8-mesh particle of chalcocite without a good deal of coaxing, but when a minute (accidental) air-bubble became poised on the chalcocite then the big bubble attached itself to the small one and thereby raised the mineral particle. When the chalcocite was oiled the bubble was lifted without hesitation. Examining the bubble-film, it will be seen (Fig. 20) that the particle of chalcocite hangs from it when in the water, but as soon as the bubble is taken out of the water into the air, the chalcocite is enclosed between an inner and

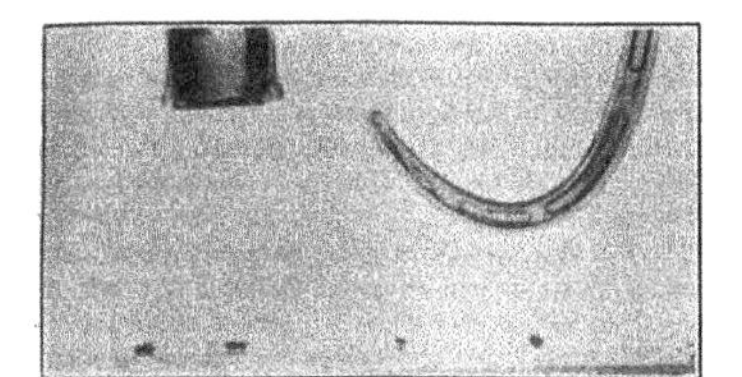

Fig. 16

A

B

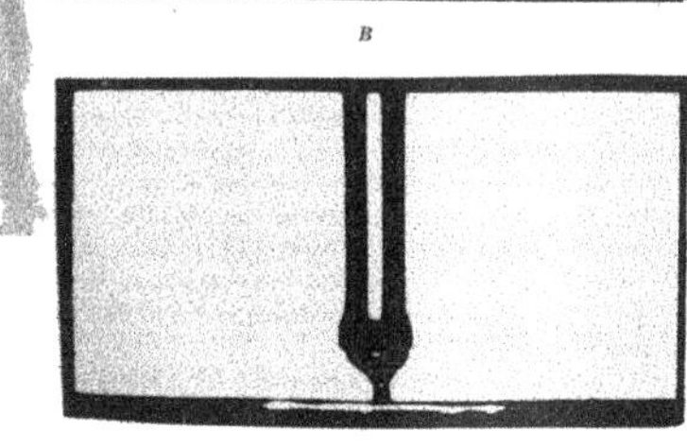

C

Fig. 17

an outer **surface,*** in both of which the oily contaminant is so concentrated as to form an adsorption layer.

*As elucidated by Taggart at Butte.

The nature of this oily-water interface is shown by another experiment. If water and pine-oil are poured successively into a test-tube and a particle of chalcocite is dropped into it, we shall find (Fig. 21) the particle floating at the oil-water interface in such a way that

A

B

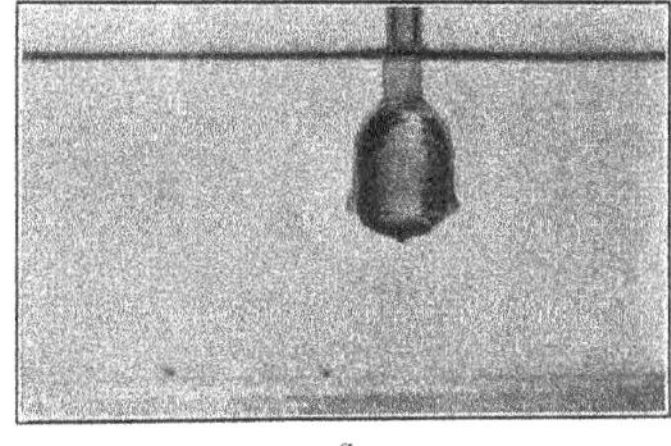

C

Fig. 18

the mineral seems to be in the water, when it is really enclosed within a downward protrusion of the oil.

When a bubble is in oily water it has only one contaminated surface, or adsorption layer, but when it emerges it has two. See Fig. 22. The oil is concentrated at the surfaces in contact with the air, outside and inside, leaving the less modified water between.

Again, when a globule of pine-oil was placed on the

smooth surface of a lump of chalcocite under water, the pine-oil was held by the chalcocite as against a bubble brought in contact with it, but when the globule of oil lay on a piece of quartz the pine-oil was adsorbed by the bubble. A particle of mineral and a bubble show mutual attraction and if the mineral particle is minute

A

B

C

Fig. 19

it becomes drawn into the interface of the bubble-film. That may be why larger particles are not floated easily; they are too big to be enveloped in this way. The mineral particles are carried within the bubble-film; they are not attached to it outside. That may explain why fine pulverization is essential to the success of flotation. Thus we arrive at the idea that it is not the air in the bubble only, but the nature of the film, that affects the floatability of the metallic particles.

The addition of oil to water—in a beaker, for example—causes an oily film to appear at the interface between water and air. When an air-bubble meets an oil-globule they will be mutually attracted and some of the oil will pass into the interface between water and air. When air occupies a hole in water, forming what is called a bubble, the periphery of this hole presents a surface—exposed to the air within—like the surface of the water in the beaker. In each case the oil tends to concentrate at that air-surface.

The old idea that the mineral particle attached itself directly to air is now relegated to one side; while this mutual attraction may exist, it plays a minor part because the air when it approaches the mineral in a pulp is always enclosed within a watery film contaminated by oil or some similar substance.

It has been disclosed by microscopic examination* that the mineral particle is not in direct contact with air, but so enclosed within the film as not to be in touch with air either inside or outside the bubble in a mass of froth. The film raises itself over the particle and wraps itself under the particle, so that the mineral is enclosed within a watery interspace. The film itself consists of an exterior surface in which the oil is concentrated, and of an interior surface in which oil also is concentrated, both of these oily concentrations grading toward the water that lies between them. The oil is concentrated at each gas-liquid interface, just as oil concentrates at the surface of water in contact with the atmosphere.

The various experiments described in the foregoing pages have shown that the oil in a pulp, consisting of crushed ore and water, performs three distinct functions:

1. It lowers the surface-tension of the water.
2. It assists in the selection of the mineral particles.
3. It promotes the formation of a stable froth.

Water is a convenient liquid for flotation work because it has a surface-tension so high that the addition of almost any other liquid will lower it. The lowering of the surface-tension diminishes the contractile force in water and lengthens the life of the bubbles that are formed by the injection of air; but this lowering of the surface-tension has another important consequence: it creates such a variable concentration of oil in the watery film of the bubble as to enable the film to adjust its strength to external forces. This variability of tension is even more important than the lowering of the surface-tension, because it serves to strengthen the film where necessary by lessening the proportion of contaminant at any weak spot. The contaminant will concentrate at the surface of the liquid because by doing so it will decrease the potential energy.

Next comes the selective adsorption of mineral particles by the oily film. The oil wets mineral in preference to gangue; it envelops the mineral, by which it is 'adsorbed' or attracted. This causes the particles of mineral to be drawn into the oily film of the bubbles, which in turn are strengthened by reason of the increase of viscosity imparted to their films by the inclusion of

*Taggart.

mineral particles. The electro-static hypothesis has been discarded in the latest investigations.

Any substance that will lower the surface-tension of water and be adsorbed by mineral particles would appear to promote flotation. The value of a flotation agent

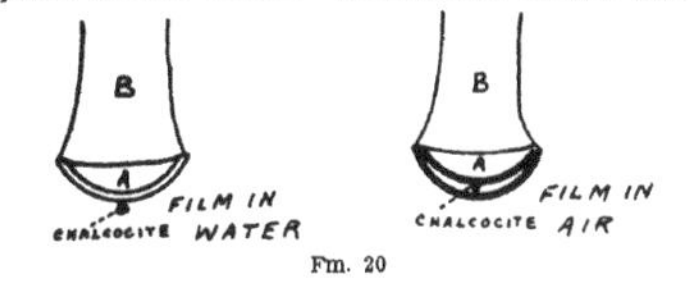

FIG. 20

depends upon its ability to 'adsorb' mineral. Most 'frothers' or bubble-makers by themselves are not satisfactory because they lack this ability, and, in order to correct the deficiency, it is customary to add a 'non-frothing' oil, which is adsorbed strongly by the mineral, thereby promoting successful flotation.* A froth made with a relatively soluble oil, like pine-oil, can be stabilized by adding a relatively insoluble viscous oil, like fuel-oil. The idea of agitation, whether of the violent and mechanical kind or of the gentle and pneumatic

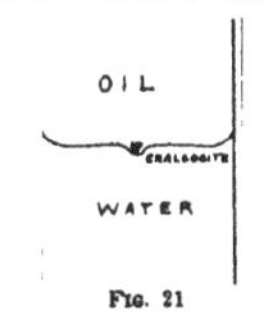

FIG. 21

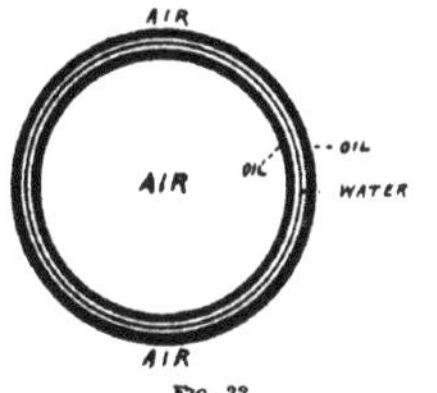

FIG. 22

kind, is to bring the particles of mineral in contact with the oily films of the air-bubbles. Whether the oil is emulsified before or after it is added to the pulp does not matter at this stage, but the oil must have been presented to the bubbles in a minutely subdivided condition, so that they may acquire oily films and so that those films may come in touch with the mineral particles. In doing so the globules of oil and the bubbles that they contaminate beneficially come in contact with particles of gangue as well as particles of mineral, but owing to the tendency of oil to replace water at the surface of the mineral particles these will be coated with oil and adsorbed into the oily film of the bubbles and rise, whereas, by reason of the tendency of water to displace oil on the surface of gangue-particles, those will become wetted and sink.

'Mineral,' 'metallic,' even 'ore' are used interchangeably in the technology of flotation. The misuse of 'ore' has caused great confusion, for the object of the process is not to recover the 'ore', but only the valuable mineral in the 'ore', rejecting the valueless portion, called 'gangue'. As between 'metallic' and 'mineral', the reference is not so much to substances containing metals, for that would include much of the gangue, such as rhodonite and feldspar, but particularly to minerals having a metallic lustre, which feature appears to be favorable to the adhesion alike of air and oil. 'Sulphide' is another synonym, because the sulphur compounds with the base metals are particularly the object of flotation, but 'sulphide' would exclude the tellurides. At least one sulphide without metallic lustre is amenable to flotation, namely, cinnabar. So is graphite, which is neither sulphidic nor metallic, except in lustre. Likewise certain forms of scheelite respond to flotation, and it has been shown by experiment that a stable froth can be made with lycopodium powder, which is of vegetal origin. So we must be careful in our use of terms. The use of 'metallic' and 'mineral' as adjectives to designate floatable substances is based on a concept of flotation that may soon be discarded. No classification of floatable minerals can be made yet and when it is made it must be based on a better understanding of the physical conditions governing flotation.

The amount of oil required in froth-flotation depends upon three factors: the proportion of mineral to be concentrated, the amount of water, and the degree of aeration. Air and water are needed to make bubbles; these bubbles must be oiled in order that they may engage the mineral in the pulp. The more numerous the mineral particles the greater the number of oily bubbles needed to arrest them. If the amount of water is doubled, there will be only half the number of mineral particles in a unit of space; therefore more oily bubbles will have to be sent in search of them than if they were herded within the smaller volume of water. The idea that a 'critical' proportion of oil—somewhere under 1%—is required to perform successful froth-flotation has no basis of evidence outside the imaginings of a group of patentees and it has been stultified by the operations of 1000-ton plants using 22 or 23 pounds of oil per ton of ore, in Utah and Montana. As Wilder D. Bancroft has said: "The hypothesis of a 'critical point' rests on unverified and unverifiable statements."

THE HYPOTHESIS. Let us recall the principal points in the evidence before venturing upon a summary of our conclusions. I write in the plural advisedly, for the evidence has come from many sources and the suggestions explaining it have been borrowed from many writers; the theory, like the practice, of flotation is the joint work of a large number of investigators.

*Bancroft, in his testimony at Butte.

(1) The needle that floats on tap-water will sink in distilled water. Although contaminants have lowered the surface-tension† of the tap-water, it has more sustaining power on account of its aeration.

(2) The bubble blown in distilled water will break as soon as it emerges, but the solution of an oily substance will enable a boy to blow bubbles that sail away beautifully.

(3) The addition of oil lowers the surface-tension and thereby promotes wetting, but the adhesion of the oil to the surface of the mineral particles causes the water to be displaced, so that the gangue preferably, not the mineral, is wetted, and drowned.

(4) Emulsification of the oil provides a means, through the subsequent breaking of the emulsion, for imparting oil in a minutely subdivided state, as needed, for oiling the bubble-films and the mineral particles.

(5) The contaminant, such as oil, in water concentrates at the air-surface and by doing so affords a surface-tension sufficiently variable to be adjustable to shock.

(6) The oil-water interface is more viscous than the body of either liquid.

(7) Oil is attracted and adsorbed by mineral particles, which therefore are pulled into the oily film of the bubbles.

(8) Bubbles will break when they collide unless there is a stable film between them, preventing coalescence. Such stability is furnished by a dissolved substance that adjusts the surface-tension and also increases the viscosity of the film.

(9) A multiplicity of bubbles, or 'froth,' will serve a metallurgic purpose if it floats valuable mineral matter long enough to facilitate a separation from the valueless components of the pulp.

The recent trend of hypothesis—it has hardly the status of a theory—is to subordinate sundry ideas prominent a year ago.* The direct 'adhesion' of air to mineral particles is not so vital as was supposed, because air and mineral rarely come in direct contact in the flotation process; usually either the air-bubble has an oily film or the mineral itself has undergone oil-filming. The lowering of the surface-tension of water is still a fundamental factor, but this modification of the water is recognized as chiefly important not for the first consequence, which promotes the wetting of the mineral, but for its secondary result, which is to create a variable tension on the surface of a bubble-film, and thereby strengthen it greatly. The addition of acid has ceased to be essential, it having been found that alkaline water is better for the treatment of many ores. The acid, like the oil, is supposed to serve more than one purpose:

(1) To adsorb on the gangue and aid the wetting of it.

(2) To promote the flocculation of gangue-particles and the separation of them from the valuable mineral.

†The layer of liquid subject to surface-tension has a thickness less than the radius of molecular action. R. S. Willows and E. Hatschek. 'Surface Energy,' page. 8.

*'The Flotation Process,' 1916.

Fine grinding of the ore is recognized as necessary, not only to separate the mineral from the gangue, but to assist the making of a froth rich in mineral. No longer is the mineral supposed to be buoyed by the bubbles, as if tied to a cork, but the minute particles of mineral are believed to be drawn into the bubble-film, so that, to pursue the simile, the life-preserver of cork surrounds and encases the thing to be floated. The idea that a fixed proportion of oil to ore is necessary has gone with the supposition that oil only will perform the absorptive function necessary to a stable froth. Colloidal sulphur, sulphur di-oxide, and salt-cake have been proved effective agents in froth-flotation; and we may expect a steady increase in the discovery of such substances until oil, which is expensive, is discarded. The part played by emulsification and the formation of colloid hydrates are becoming recognized as possibly important factors. The violent type of agitation has been found unnecessary, and, thanks to recent litigation, it is likely that the use of compressed air under low pressure will supplant the power-consuming devices of an earlier period. The trend is toward simplicity both of treatment and apparatus. When air and a cheap modifying agent are found adequate for the making of a mineral-bearing froth then the flotation process may be deemed fully developed.

THE GEOLOGY of the Telkwa district in British Columbia has recently been reported upon by Victor Dolmage who finds occurrences of interest. Among these is a characteristic type of copper deposit, assumed to be new, although similar occurrences are not uncommon in the basin region of the south-western United States and north-eastern Mexico. This consists in magnetite segregations accompanied with chalcopyrite, bornite, and tetrahedrite often associated with native silver, the latter being disseminated through the magnetite in the form of rounded grains. The ore is found in veins varying from a width of a few inches to four feet, cutting the Hazleton group of andesites, quartz porphyries, and tuffs. They are mineralized as a result of two distinct epochs of vein-formation, one following the intrusion of the Coast-range batholith of quartz diorite, which was supposedly injected during Jurassic time, and the other following the Bulkley eruptives occurring in the Tertiary age and bringing in diabases, lamprophyres, and soda-syenite porphyry. The earlier solutions deposited in the veins, in the order named, quartz, epidote, hematite, pyrite, zinc-blende, chalcopyrite, bornite, chalcocite, silver-bearing tetrahedrite, and galena. The later solutions deposited quartz, much hematite, epidote and calcite with pyrite and chalcopyrite. The district, as might be supposed from the description given of the characteristics of the veins, is one capable of yielding only moderate amounts of high-grade ore from pockets of superior enrichment and from the careful cobbing of ore derived from the wider veins.

CHINA-CLAY is in sharp demand at the present time, the domestic grades being quoted as high as $20 per ton at Eastern points. The best imported fetches $35.

What is a Metalliferous Mineral?

By L. O. HOWARD

Most persons would probably reply off-hand to the above query that a metalliferous mineral is a mineral containing a metal or metals. I might query further, "What is a metal?" I confess that I have always believed that I knew the answers to these questions and that they were axiomatic. However, I have been disabused by a document recently sent out to one of my legal friends by the Commissioner of the General Land Office at Washington.

For the past six years there have been mined in eastern and south-eastern Utah ores containing vanadium, uranium, and radium, present in the mineral carnotite in certain Jura-Trias sandstone beds. Lode locations have been the invariable method of entry. The shallow depth of most of the ore and the lack of deeper exploration have made the locators content to proceed without patent. This has been shown to be an unsafe procedure in certain cases. Lately, due perhaps to unsettled conditions, many claims have been surveyed for patent. In the Green River area in Emery county a petroleum reserve was established on March 14, 1912, as Petroleum Reserve No. 25, Utah No. 2. This covered most of the carnotite area and antedates many of the locations. Application for a patent to a certain carnotite claim was filed on September 24, 1915, consequent upon a location made on January 4, 1914. On May 25, of this year, the Commissioner of the General Land Office states that the question "now arises of the metalliferous or non-metalliferous character of the mineral sought." The entry was based on a location claiming a portion of a lode, vein, or deposit "bearing uranium and other valuable minerals," and it was stated in the application for patent that "the mineral found is carnotite ore."

The record appears clear up to this point. The claim was located and worked in good faith, assessment and patent work done, survey for patent made and accepted, and all requirements of the law fulfilled. The ground was undoubtedly open to entry and patent if containing metalliferous minerals, according to Section 2 of the Act of June 25, 1910, as amended by the Act of August 24, 1912 (35 Stat., 697), which provides that "all lands withdrawn under the provision of this act shall at all times be open to exploration, discovery, . . ., and purchase under the mining laws of the United States as far as the same apply to metalliferous minerals." Now comes the answer to the query, "What is a metalliferous mineral?" or perhaps one had better say, "What is not a metalliferous mineral?" Let me quote from a letter of the Director of the U. S. Geological Survey, of March 7, 1917, to the Commissioner of the General Land Office in relation to this case. The Director says: "Deposits of carnotite or other radium-bearing ores are mined primarily for the production, not of the radium itself, which is chemically a metal, but of radium salts, which are non-metallic. Metallic radium is seldom, if ever, produced, and the non-metallic salts, chiefly the chlorides and bromides, constitute the article of commerce. The radium ores may be considered in the same category as potash, limestone, or common salt, which, though the salts respectively of the elements potassium, calcium, and sodium, classified chemically as metals, have uniformly been considered by the courts, the Department, and in mining law as non-metalliferous minerals, as are borax, . . . and similar substances. The question is somewhat further complicated by the fact that the radium occurs only in extremely minute quantities in carnotite-bearing ore, carnotite being a mineral which contains uranium and vanadium. These ores, however, are earthy and non-metallic in chemical character.[1] Vanadium is not even chemically a metal,[2] and although uranium is chemically a metal, . . . as are calcium, the basic element in limestone, and aluminum, the basic element in clay . . ., such limited use as it has in the arts is almost exclusively as an oxide or other salt. The Survey believes that carnotite is *not* a *metalliferous* mineral in the sense in which the term is used in this act." The Commissioner quotes, in further support of his ruling, Dr. George P. Merrill, head curator of geology, U. S. National Museum, who, in his treatise on 'The Non-Metallic Minerals, Their Occurrence and Uses,' says, "Uranium is never used in the metallic state, but in the form of oxides or uranates of soda and potash, and finds a limited application in the arts."

On the basis of the above opinion the Commissioner has allowed the claimant 30 days in which to show cause why the entry should not be cancelled. The Commissioner apparently bases his ruling on the following premises:

1. Radium is not used in the metallic state but in the form of certain salts, such as bromides and chlorides.
2. Vanadium is not chemically a metal.
3. Uranium, although chemically a metal, is principally used as an oxide or salt.
4. Therefore carnotite, which is the economic mineral containing radium, vanadium, and uranium, is a non-metalliferous mineral, analogous to borax, salt, limestone, and potash.
5. Land containing only non-metalliferous minerals is not subject to entry under the terms of the Act with-

[1] Equally true as to some forms of hematite, as for instance, certain Mesabi iron ores.

[2] What is a metal?

drawing petroleum lands from entry as mining claims.

With No. 5 we have no quarrel, since the law is explicit in this respect. Considering first No. 1, although it is true that at present, for reasons well known to those familiar with radium and its uses, it is deemed expedient to market radium in the form of its salts, it becomes pertinent to enquire into certain phases of its use, considered in a broader spirit than is shown in the opinions quoted. What is the function of the salt and to what does it owe its uses? For what is it valuable? We may say without question that all its uses are due solely to the properties of the radium itself, without regard to the form in which it may be. It may be contained in barium chloride or common sand. The entire value depends wholly on the content in metallic radium. The salt, then, is not valuable owing to any particular property of its own. Convenience merely requires that, for easy and efficient use, the radium be carried in some container or packing. What this container may be is immaterial, so that it be harmless. The barium chloride or other salt performs the function of a carrier or container, and this is its sole function. One might as well consider that nails were sold in kegs, because a keg of nails *per se* was especially valuable, rather than buy the keg of nails for the sake of the nails themselves. A keg happens to be a convenient container. An evidence of the importance of the metallic radium is the present insistence that all radium salts be rated in terms of the metallic radium, for it is the metallic radium that is important, not the salt. As to No. 2, it is obvious that a strict interpretation places the metallurgists, manufacturers, and users of steel in the category of those who use meaningless and incorrect terms to describe their products. Does No. 2 imply that ferro-vanadium is not a ferro-alloy? If it does, our definition of alloy needs revision. Webster gives it as a mixture of metals. Also, in what category are we to place tungsten, molybdenum, nickel, aluminum, chromium, and other components of ferro-alloys? Would not the ruling apply to ores of these 'metals' as well?

A comparison of No. 1, 2, and 3 shows some interesting anomalies. No. 1 defeats the entry because, though radium is chemically a metal, its use is not as a metal; while vanadium is used as a metal, it is not chemically a metal; and uranium, while chemically a metal is not used as a metal; so that No. 2 and 3 are also reasons for cancelling the entry. Furthermore, how long is uranium to be used principally as a salt? How long was tungsten used only otherwise than as a metal, and how recently was it found possible to make tungsten wire? Does the Survey intend to say that vanadium and uranium are to have, or do have, but slight use in the metallic arts? If we adhere to the time-honored definition of an alloy, we must admit that the use of vanadium and uranium in the form of ferro-alloys is strictly as metallic substances. Minerals carrying either of these metals are classed correctly as metalliferous minerals. If not, pray, what is a metalliferous mineral? Wherein is the analogy to borax, salt, limestone, and clay? These are all valuable solely as such, and not because of the content in boron, sodium, calcium, and aluminum, whereas the salts af vanadium and radium are valuable solely because of their content in these metals. Ferro-uranium is being advertised extensively today, and it is surely possible that uranium may find its greatest use in this form.

Further, is it true that carnotite is, or has been, mined principally for its content in radium? It would seem that if it could be proved that at any time the carnotite ores had been diligently sought, and were exploited, as a source of vanadium and uranium for use in the metallic arts, without regard to, or thought of, the radium contained, this statement would lose its force. Early in 1912 I had charge of certain carnotite mines in the area in question. What my clients sought, mined, and shipped was a material to be utilized as a source of uranium and vanadium, and not of radium; in fact, radium was not even considered. Such a competent authority as Madame Curie had scouted the idea of these ores being a commercial source of radium. Work done by pioneers in Colorado was ridiculed and the term 'radium king' was facetiously applied to the chief exponent of the theory that these ores could be made to yield radium commercially. However, his persistence won success. It was only when the Utah ore was found to be slightly too low-grade for the extraction of vanadium and uranium at a profit that attention was turned to the possibilities lying in its radium-content; and it may be remarked that this particular ore was also rather too low-grade for profitable radium extraction.

For many years previous to the attempt to extract radium, countless attempts were made in Colorado to treat these ores for their uranium and vanadium contents. That complete commercial success did not ensue was due to the process employed and lack of experience on the part of the exploiters, not to say unsuitability of the ore itself as a profitable source of these metals. Previous to the discovery of the large Peruvian deposit, a mill was in successful operation in Colorado, furnishing vanadium salts for the manufacture of ferro-alloys, utilizing an ore as low in vanadium as many of the carnotites and containing no uranium or radium whatever; and much of the market was supplied from this source. This mill is still in operation.

Finally, is there any doubt that the leading producer of radium is also utilizing the vanadium and uranium contents of his ore on a fairly large scale in the metal industry? In fact, the Standard Alloys Co., a subsidiary, markets both ferro-vanadium and ferro-uranium, for the latter of which claims are made that may lead to extended use. Is not, then, a mineral carrying uranium and vanadium a metalliferous mineral in the sense in which the term is used in the Act, as well as chemically? Is earthiness peculiar to non-metalliferous minerals? Is it in any sense a criterion? Are not many hematites earthy, many oxidized ores of copper, of lead, and of many other metals? Are sulphide ores usually composed of metalliferous minerals and oxide ores not?

And what is a metalliferous mineral?

Oruro Tin-Silver District, Bolivia

By FRANCIS CHURCH LINCOLN

The mines of Oruro were worked by the subjects of the Incas before the Conquest. Francisco Medraño, a Spanish curate, learned of them from an Indian, and in 1595 opened the Socavón de la Virgen and Atocha mines. The mines were developed until their silver production rivalled that of Potosí, by the year 1678. With alternating periods of bonanza and borrasca, the mines of Oruro continued to flourish until the outbreak of the Bolivian war of independence. In the three years preceding this war, the silver mines of Oruro paid no less than $40,000,000 in taxes to the Spanish Crown. As these taxes were one-fifth of the production, the output must have been at least $200,000,000. The Bolivian revolution resulted in a complete paralysis of the mining industry. It was not until 1885 that revival occurred, when the Chilean Compañia Minera de Oruro purchased the ancient Socavón de la Virgen mine and began work. More recently, this company acquired the Itos property which is separated from the Socavón by the San José mine. The latter mine was purchased by the Compañia Minera San José de Oruro, also a Chilean corporation. These companies control the output of the district, which at the present time is of greater value for its tin than for its silver. According to M. G. F. Söhnlein, the silver production of the Oruro district in 1915 was 780,000 oz., while the normal monthly production of tin concentrate by the Compañia Minera de Oruro is from 2500 to 3500 quintales and that of the Cia. Minera San José from 800 to 1000 quintales.

Oruro, with its 25,000 inhabitants, is second only to La Paz in importance. It is on the Antofagasta-Bolivia Ry., 147 miles south-easterly from La Paz, and 575 miles north-easterly from the Chilean port of Antofagasta. Railroad connection is possible also with the Pacific coast by way of Viacha over the Arica-La Paz Ry. to Arica, Chile, or over the Southern Railway of Peru to Mollendo, Peru; while by way of Uyuni it will soon be possible to go by rail to the Atlantic coast at Buenos Ayres, Argentina. Oruro lies on the Bolivian plateau, at an elevation of 12,120 ft. above sea-level. To the westward, beginning at the outskirts of the town, the Oruro hills rise to heights of from 1000 to 1700 ft. above the plateau.

The Oruro ore deposit is noteworthy as presenting the best-known example of tin-silver veins. Special interest attaches to this type of deposit both because of its rarity outside the Republic of Bolivia, and because it forms a connecting link between pneumatolytic and hydrothermal deposits, tin being characteristic of the former and silver of the latter. The country-rock of the Oruro district is quartz porphyry intrusive through Paleozoic shales. The porphyry is probably of Tertiary age, and may be more closely classified as a dellenite porphyry. It contains large phenocrysts of quartz and orthoclase and has been highly altered in the vicinity of the veins, as shown by micro-photographs made by Romaña,* who has exhaustively studied the district. The shales are dark in color and frequently occur as included fragments and blocks in the porphyry. The veins worked on all

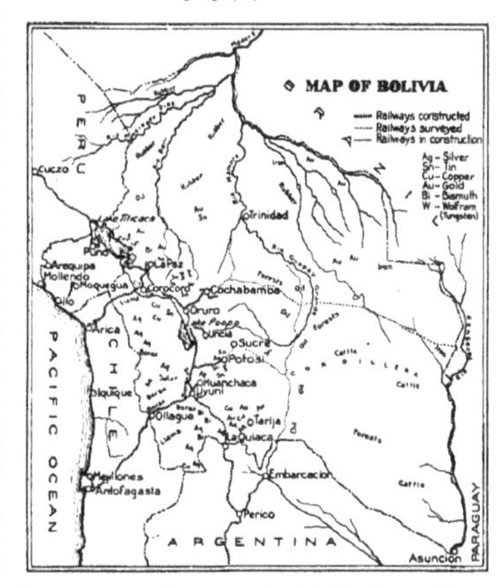

three properties form a single linked system which extends in a north-westerly direction diagonally across the northern part of the Oruro hills. The main vein, known as the Purísima, has a strike of about N. 15° W. on the Socavón ground, but has many bends and angles, and develops a westerly strike at the Itos mine. Its dip is likewise irregular, being at times as flat as 50°. The width varies from 3 to 8 ft. The vein splits into numerous narrow stringers around blocks of slate in the Itos ground which re-unite when the vein passes into the porphyry on the further side. A depth of 1250 ft. has been reached on the San José property. The veins are generally filled fissures with the filling 'frozen' to the walls, but occasionally a fault separates the vein from the country-rock and in places the filling penetrates the wall-rock without a distinct line of demarcation. As a

*Boletin del Cuerpo de Ingenieros de Minas del Perú, No. 57 (1908).

rule, when the veins enter the slates not only do they break up into stringers but the tin disappears, though the silver persists. The primary ore minerals are pyrite, cassiterite, argentiferous tetrahedrite, argentiferous jamesonite, and a little galena. The gangue-minerals are those of the country-rock with a little vein-quartz. The tetrahedrite contains about 5% silver and the jamesonite 0.2%. Cassiterite and tetrahedrite both occur as irregular patches in the pyrite, at times separate and at times intergrown, but the tetrahedrite displays a tendency to occur as stringers, while the jamesonite is found mainly in the form of tufts of fine radiating needles in druses in the pyrite. In an interesting variety known as 'ring ore,' pyrite cores are seen encrusted with cassiterite, while the remaining spaces are filled with tetrahedrite and jamesonite. The order of the ore-minerals is therefore as follows: pyrite, cassiterite, tetrahedrite, jamesonite. The pyrite stage, however, overlaps both the cassiterite and the tetrahedrite stages, while the cassiterite stage overlaps that of the tetrahedrite. Thus, while the deposition of tin was in part contemporaneous with that of the silver, the silver precipitation continued for a longer period. Rich silver chloride ores formerly occurred in the oxidized zone extending downward for several hundred feet, but these were exhausted in the early days. Along with the hornsilver ores were present tin ores known as 'pacos' containing from 4 to 5% tin, and from 6 to 9 oz. silver per ton. These were cast aside, but recently have become the object of exploitation. They are soft and porous, and contain cassiterite, clay, and iron oxides. At times the cassiterite in the 'pacos' is white and pulverulent, and it is then called 'white tin'. Below the oxidized zone there was found a zone of secondary enrichment from which high-grade silver-sulphide ores were extracted in the Colonial period. Underneath the secondary sulphide enrichments is the primary ore which has shown no sign of impoverishment in depth.

From the offices of the Cia. Minera de Oruro, in the city of Oruro, an ancient cross-cut tunnel, the Socavón de la Virgen, whence the mine takes its name, extends to the Purisima vein. The main workings of the Socavón mine extend to a depth of 725 ft. below this tunnel. The mine is worked through an incline-shaft which cuts the Socavón de la Virgen and extends to the third level, 400 ft. below. From the third level a winze has been sunk on one of the branch veins, near its junction with the Purisima, to a depth of 325 ft. Ore from the Itos mine owned by the same company is brought across the Oruro hills by means of an aerial tramway. The San José mine of the Cia. Minera San José de Oruro has a depth of 1250 ft., making it the deepest tin mine in Bolivia. The Socavón mine uses electric power, developed by oil engines. Jackhammer drills are employed in the mine. A considerable amount of the stoping is from ancient fillings and pillars. The mine timbering consists mainly of dry-walls. About 25,000 gal. of water per diem is bailed from the mine. Miners are paid three bolivianos per day, which, at the present rate of exchange, is equivalent to about $1.00. H. F. Grondijs is consulting engineer for the Cia. Minera de Oruro as well as for Abelli & Co. in the Pazña district.

All the ore is hand-sorted on the surface and the picked ore shipped by rail to the mill.

Owing to scarcity of water at Oruro, the mills of the Cia. Minera de Oruro and the Minera San José are situated on the Antofagasta-Bolivia Ry. at Machacamarca and Poopó. The ore is given a chloridizing roast and the silver is extracted by hyposulphite lixiviation, at the same time leaching and precipitating what little copper is present. The lixiviation tailing is then concentrated to recover the tin, the tin concentrate, locally known as 'barrilla', containing from 65 to 70% of that metal. The Machacamarca mill is under the superintendence of M. G. F. Söhulein. The extraction at this plant is about 80% of the tin and 85% of the silver in the ore.

ZINC ORE for retorting is roasted to a low-sulphur content. 'Faulty' sulphur, as it is known in the technology of the art, is the sulphur remaining in the calcine in the form of sulphide and as soluble zinc sulphate. This is distinguished by determining the sulphur combined with lead and calcium, and deducting that from the total sulphur found. The metallurgist assumes that the 'faulty' sulphur exerts a deleterious effect on the zinc recovery, while the sulphur present as calcium sulphate and as lead sulphate is not released from combination in retorting. Edward M. Johnson, superintendent of the Eagle Picher Lead Co., says that it is still an unsettled question whether the detrimental effect of the sulphur held by zinc in the calcine is due to chemical reactions taking place in the retort, thereby retaining zinc as a sulphate, or whether it is because of poor condensation of zinc-vapor. He states as his experience that no means has been found for overcoming the difficulty, and the old rule remains true that "one per cent of 'faulty' sulphur retains two per cent of zinc". Sulphide sulphur in the calcine has a bad effect in forming matte, which exerts a highly corrosive effect on the retorts. "The fireman, in order to save his retorts from 'butchering' does not carry the distillation as far as he might otherwise do, with a consequent loss of zinc in the residue. On the other hand, if the fireman insists upon working off the furnace a high retort-loss ensues through the absorption of zinc."

CHEMICAL industries will figure at the third National Exposition under the auspices of the American Chemical Society at the Grand Central Palace in New York during the week beginning September 24. The advisory committee consists of Charles H. Herty, chairman, Raymond F. Bacon, L. H. Baekeland, Henry B. Faber, Colin G. Fink, Bernhard C. Hesse, A. D. Little, Utley Wedge, and others. This will be the most notable representation of American progress in chemical manufacture that has ever been made. There will be meetings of the Chemical and other societies, with addresses by the foremost scientists of the country. The United States government is also taking a hand in the exposition, and will contribute in many ways to the interest of the occasion.

REVIEW OF MINING

As seen at the world's great mining centres by our own correspondents.

ALASKA

NEW DEVELOPMENT ON KNIGHTS ISLAND AND FIDALGO BAY.—THE STRIKE AT THE KENNECOTT COPPER MINE IS STILL IN FORCE BUT A SETTLEMENT IS ANTICIPATED.—ACTIVITY NEAR VALDEZ.

W. A. Dickey, of Landlock, is developing what is known as the Rua property near Rua cove on the east shore of Knights island. This property was bought by W. A. Dickey and Fred B. Snyder, the latter of Minneapolis, last fall. It is a large pyrrhotite vein carrying copper. The development to date consists of a 500-ft. adit and about 300 ft. of cross-cuts from this adit. The orebody is exposed on the surface at intervals by open-cuts for about 400 ft. and the outcrop can be traced for 2000 ft. Under ground 300 ft. of development has been done this spring under the supervision of Thomas Blakney, the engineer in charge. Fifteen men are regularly employed. A water-driven Ingersoll compressor furnishes power for two jack-hammers, which work effectively. Development will be continued during the summer, and it is expected that a large body of low-grade copper ore will be blocked out.

At Fidalgo bay the Alaska Mines Corporation, under the management of Byron Wilson, is working what is known as the Schlosser property under a bond and lease. The company is shipping to Tacoma from 400 to 500 tons of copper ore monthly that will average about 13% copper. Twenty-five men are employed, and all mining is done by hand. The ore is conveyed from the mine to the bunkers on the dock by a Trenton aerial tram. The ore occurs in irregular lenses in a shear-zone and is a very good grade of chalcopyrite. In prospecting for the high-grade lenses considerable low-grade ore is being found that may be milled at some time in the future. The mine is opened on three levels by adits. Recently a lens 80 ft. long by from 5 to 10 ft. wide was found on the lowest level. This is the largest body of shipping-ore so far found during the intermittent working of the mine in the last seven or eight years. This discovery makes the future of the property look good.

At Cordova the strike at the Kennecott Company properties is still on. The men are camped at Blackburn, four miles from Kennecott. Saloons at McCarthy nearby are closed tight and everything is peaceable and quiet. E. T. Stannard, general manager for the Kennecott company, is to return by June 25 and it seems to be the general opinion that the strike will be settled within a few days. A few of the American and Scandinavian miners employed at the mines refused to affiliate with the strikers on the ground that the men did not keep faith with the company. These men are leaving, or have left the camp. The mill-men and all the men at the lower camp took the same view and are still working. The strike will not affect the mill-output for some time as there is a large stock-pile upon which to draw.

W. H. Seagrave, formerly general manager for the Kennecott company, who now has an office in Seattle, is at McCarthy directing the preliminary work for the Tjosevig-Kennecott company, for whom he is consulting engineer. This company is building a horse-trail from McCarthy to the Tjosevig property, which is across the glacier from the Bonanza mine, preparatory to doing considerable development work during the coming summer.

The Ramsey-Rutherford mine, a small gold mine near the Valdez glacier, nine miles from Valdez, shut-down on June 1. Some prospecting is being carried on and it is expected that the mill will start up again in a couple of months. This property has been a steady producer for the last three years. It is owned and operated by local people who installed a larger compressor last winter, at considerable expense on account of location, with which they had hoped to keep ore developed far enough ahead to operate the mill continuously.

CRIPPLE CREEK, COLORADO

OUTPUT OF THE DISTRICT INCREASING.—LOW-GRADE ORE FORMERLY CONSIDERED AS WASTE NOW BEING MILLED AT A PROFIT.—CRESSON CONSOLIDATED HAS A NEW AND RICH VEIN.—NEW DEVELOPMENT WORK.—ROOSEVELT TUNNEL DRAINAGE.

The output of gold ores from the mines of the Cripple Creek district during the month of June, as compiled from the reports of mill managers and district smelter representatives, totaled 89,740 tons, with a bullion value of $1,064,465.50.

As shown in the accompanying table, local mills of the Portland Gold Mining Company treated the heaviest tonnage of record, 43,650 tons, with a low average value of $2.016 per ton, and a gross bullion value of $88,008. This tonnage of low-grade necessarily brought down the average, and it is the lowest on record, $11.86 per ton. Ore treated at the Independence mill of this company carried a gold content of only $1.80 per ton. The treatment figures of the several companies and plants follow:

Plant and location.	Tons treated	Average value per ton	Gross value
Golden Cycle M. & R. Co., Colorado Springs	32,000	$20.00	$640,000.00
Portland G. M. Co., Colo. Springs	10,190	19.25	196,157.50
Smelters, Denver and Pueblo	3,500	55.00	137,500.00
Portland G. M. Co., Victor mill	18,150	2.32	42,108.00
Portland G. M. Co., Independence mill	25,500	1.80	45,900.00
Rex mill, Kavanaugh lease	1,400	2.00	2,800.00
	89,740	$11.86	$1,064,465.50

Including the June output the production for the six months of 1917 has totalled 431,738 tons with the gross bullion value of $6,201,668. The Roosevelt tunnel of the Cripple Creek Deep Drainage & Tunnel Company, according to the measurements taken by T. R. Countryman, consulting engineer for the tunnel company, was advanced 153 ft. in June. The flow of water from the tunnel, passing through the weir at the portal, and flowing thence into Cripple creek, measured 5152 cu. ft. per minute. This is the lowest flow recorded since the tapping of the C. K. & N. water-course in 1912. Work was commenced the last week in June, on a lateral to extend from the Roosevelt tunnel to a point under the Cresson Consolidated Mining & Milling Company's main shaft on Raven hill. A drift has been started from the west side-line of the Old Ironsides claim of the United Gold Mines Company, on Battle mountain, and is headed north-east for the objective point. Low-grade ore is found in the dike in which the drift is carried.

Stock-transfer books of the Cresson Consolidated Gold Mining & Milling Company, and Golden Cycle Mining & Reduction Company closed on Saturday, June 30, preparatory to the payment of the usual monthly dividends by these corporations on Tuesday, July 10. The Cresson dividend at the regular rate of 10c. per share will amount to $122,000; the Golden Cycle dividend at 3c. per share to $45,000. The directors of the Portland and Vindicator Gold Mining companies will meet this week, when the regular quarterly dividends of 3c. each are expected to be declared.

The Granite Gold Mining Company has gone upon a bi-monthly dividend basis and will pay its first dividend of 1c. per share on July 5, to all stockholders of record on June 30. The amount is $16,500. It is the opinion of men in the confidence of the management that the Granite company will shortly be placed on a monthly dividend basis. The Granite mine is at Victor.

An important new discovery in the Cresson mine was authenticated by A. E. Carlton, president of the Cresson company, the last week in June. A new vein has been opened up at the 14th level of the main shaft, at a distance of 29 ft. from the main Cresson vein. According to the president's statement, the vein is 7 ft. wide and he further states that the ore is of very good grade and will certainly average $35 per ton.

At the annual stockholders' meeting of the Modoc Consolidated Mines Company, held in Denver the last week of June, the former officers and directors were re-elected as follows: Frank Cannon, president; A. H. Frankenberg, vice-president; Richard Roelofs, Thomas Arneal, Mark A. Skinner, E. D. Avery.

The operations of the company, as shown by the company report, are being steadily enlarged and the upper levels of the old incline-shaft have been leased and are producing. The company is sinking a new vertical shaft on the Battle mountain end of its property on Bull hill and Battle mountain, and has raised from the old workings at the 1100-ft. point to surface. The shaft is being timbered with Oregon-pine square-sets. The timber was held up by a Government embargo for several weeks but was recently released, and is now at the mine in quantity sufficient to square-set to the 1500-ft. level. The Excelsior Mining, Milling & Electric Company, holding a long-time lease on the Longfellow Gold Mining Company's Bull Hill mine, through the Stratton Estate control of that company, has sunk the new shaft to a depth of 300 ft. and has timbered it down to 280 ft. Laterals will be run out to connect with the workings of the Golden Cycle mine, extended into the Longfellow at the 500 and 600-ft. levels, when these depths have been attained. The leasing company was operating through the Golden Cycle and has a large tonnage of ore in sight, but ceased operating through that mine and commenced sinking, conditional on the lease extension. The Stratton Estate will save the low-grade ore rejected by the company operating through the Longfellow shaft, and dumping this low-grade and waste ore on the ground.

COBALT, ONTARIO

Encouraging Prospect for a Settlement of the Pending Labor Dispute.—Aggressive Work Going on in the Mines.

During the fourth week of June ore and bullion shipments from the Cobalt continued comparatively heavy. A total of 10 cars weighing approximately 763,190 lb. was sent out. Six companies contributed to the ore shipments, Nipissing with five cars leading the list. Bullion shipments for the week totalled 313 bars, weighing 363,250.97 oz. and valued at $288,356.64. Bullion shipments so far during the current year aggregate upward of 5,250,000 oz. valued at over $4,000,000.

The labor situation is still in a state of plasticity, and, although the strike-vote taken Sunday last resulted in a large majority in favor of striking, as a means of getting the 50c. increase in the present base-wage, the situation is, nevertheless, viewed with more or less optimism. It is generally believed the men will finally, and before calling a general strike, decide upon treating with their employers apart from their union affiliations either individually or by committee. Should the men follow this course, there would appear to be a possibility of getting a 'high cost of living allowance' instead of the present 'high price of silver bonus'. It was along these lines that a settlement was effected at the Dome and McIntyre mines at Porcupine.

The Kerr Lake Mining Company has declared a special dividend of 15c. per share, payable August 10 to shareholders of record July 5. Half of this dividend will be paid the shareholders and half will be devoted to patriotic purposes. The camp is comparatively well supplied with labor, and in a general way, developments are going forward aggressively.

PORCUPINE, ONTARIO

The Labor Situation Much Improved.—Wages Are Increased and a Better Class of Miners Will Now Be Available.—The Dome and McIntyre Have High-grade Ore and Will Increase Output.

The situation at Porcupine has improved during the past week or so, and it is now certain that there will be no general tie-up here. The Dome Mines company has decided to grant its employees approximately 50c. per day above the regular base-wage, which is along similar lines to the action taken over a week ago by the management of the McIntyre-Porcupine. Labor leaders have stated that although the McIntyre management did not deal with and did not recognize the union, it nevertheless has dealt squarely with its employees and will from now on benefit by any favors within the power of the union to confer, in the way of sending the best available men to work at the mine. The fact of the Dome Mines company having followed the lead set by the McIntyre is expected to have a beneficial effect, not only at Porcupine, but at Cobalt as well. At present the Dome Mines employs only 350 men and the increase of 50c. per day will amount to $175 daily, or about $5250 per month. During 1916 the company paid $800,000 in dividends, and at present there is upward of $700,000 in the treasury, so that the $63,000 payable yearly under the recent increase in wages will not greatly affect the earning power of the company. In fact, the probable higher efficiency that will result will, it is anticipated, more than make up for the added outlay in wages. Early next week the cross-cut at the 700-ft. level of the Dome will probably enter the orebody 119 ft. wide, that was indicated by the diamond-drill core and mentioned but not included in the estimate of the annual report of the company issued a few weeks ago. This ore is officially stated to carry an average gold content of $17 per ton. By early August the cross-cut will probably have crossed the entire width of the body, at which time driving, winzing, and stoping will be commenced. With ore once going to the mill from the stopes in this high-grade ore the grade of the mill-feed will probably immediately rise to new high-record for this mine. Hitherto the average grade at Dome has been below $5 per ton. When it is considered that the milling capacity of the Dome is about 1500 tons per day, it can at once be seen that every $1 added to the grade of ore treated would increase the daily output approximately $1500, which would be a large net profit.

On June 27 mining and milling operations at the Schumacher mine were suspended owing to the decision of the directorate and management not to grant an increase in wages to their men, and the impossibility of securing sufficient labor at the old rate of pay. Mill construction, however, is being continued at the Schumacher and the new mill will increase the milling capacity to about 280 tons daily.

The four-weekly report of Hollinger for the period ending May 20 was somewhat disappointing, in that gross production had fallen off to $92,000 as compared with around $194,000 during the preceding period. The management, it is understood, is now devoting more energy toward centralizing the underground work pending the return to more normal labor conditions and cost of supplies. This policy will ultimately result in added net profit to those interested in the mine, and with a return to pre-war conditions a new high record in the rate of dividends will probably be established.

The McIntyre is now employing about 350 men and producing approximately $150,000 per month. Net profits amount to nearly $90,000 monthly, which is almost on a par with the Hollinger. Recently, the main orebody was cut by diamond-drills at a depth of 1550 ft. and determined to be over 30 ft. wide of high-grade milling ore. The main body, at the 1000-ft. level, averages around $12.50 per ton.

It is the general opinion that the period of prolonged strain through which the Porcupine district is passing is at an end, and that increasing prosperity is at hand.

ELDORA, COLORADO

CHARACTER OF THE TUNGSTEN ORES OF BOULDER COUNTY, COLORADO.—AN IMPORTANT CHANGE WITH DEPTH.—OPERATING COMPANIES.—MILL METHODS COMPARED.—DUTY ON TUNGSTEN ORES SOUGHT.

Tungsten ores follow the same law as the ores of gold, silver, and copper—free-milling at the surface, from which much high-grade ore can be cobbed. In the early days milling of tungsten was an easy problem, but as the ore-shoots increased in depth the ores became more complicated and more tightly bound to the rock in which they occurred and so more difficult to handle; and at the same time they did not offer the amount of free ore that might be cobbed from it as at the surface.

Ores that have been sent to mills for milling have, up to about two years ago, averaged over 10% tungsten tri-oxide, where as now they are of an average of about 5% tungsten tri-oxide. This with the fact that they are of a more complicated nature has caused great changes in the method of milling.

The depth to which tungsten goes is a problem. Some mines give promise of good ore to considerable depth and others do not. The Condor of the Primos company was one that did well to the 900-ft. level. The Clyde mine has great bodies of low-grade ore at 300 ft. and all indications are that they will continue to greater depth, at least the Wolf Tongue company, which owns it, will continue to sink the shaft for investigation.

The Primos company is running at full capacity.

The Vasco Mining Co., associated with the Vanadium Alloys Steel Co., of Pennsylvania, is running steadily. It purchased the upper Rodgers tract and erected a 50-ton mill last year. It also purchased the Boyd mill at Boulder and remodeled it for tungsten ores, and it is running.

The Rare Metals Co., at Rollinsville, has erected a 50-ton mill and is operating it, purchasing ores from the Beaver district. This company has leased mines on Beaver creek and will operate them.

Mr. Caudray, of Denver, has leased the Lone Chance mine on Beaver creek, and also the Smith-Ardouel mill and will mine and mill the ore from this mine.

Jack Clark, who has been in the tungsten business for some time, is mining and milling ores at Stevens camp.

The old Colborn mill, near Boulder, has been remodeled for tungsten ores and will handle it for lessees and from its own

The Wolf Tongue Mining Co., at Nederland, which has been in the tungsten business for 13 years and owns considerable tungsten land, is producing heavily. It mines its own ores and also leases ground. It has ten mines on the company payroll and 20 sets of lessees taking out ore. The Clyde mine is at present the heaviest producer. The Cross No. 1, operated by lease, gives promise of becoming a great producer.

The Bonanza, Star, Orange Blossom, Hoosier, Tenderfoot, and Town Lot are all on the shipping-list. Several sets of lessees are running the float material over jigs and tables and shipping concentrate. The same plan of operation is used by the Primos company, the Vasco company, and others in the district.

The mills of the district vary in the methods of handling the ores. Some believe, and no doubt have sufficient reason for it, that out one grade of concentrate should be made, and that of a 60% grade. The Wolf Tongue company from years of experience has arrived at the conclusion that two grades are necessary; one of 60% and one of about 20%, called second-grade; and, if necessary, to run the second grade through a mill in the same way as crude ore and thus get the first-grade from this second-grade. Some mills in one operation return all second-grade ore to the re-grinding machinery as soon as made and get the first-grade from it while the crude ore is also going through the mill. This difference of opinion has never been compared to see just which is the best, but the Wolf Tongue mill, from careful work in both ways, has sufficient proof to say that to mix the re-ground sand from the tables or the second jig-product from the jigs with the crude ore and mill both at the same time does not produce the best results. The above statement is based on very careful work and several trials. The best results con be secured by re-running the second-grade by itself and not by treating it with the crude ore. This, however, is a matter of difference of opinion between mill-men here, and each no doubt has secured results to prove his side of the question.

The price of tungsten due to the War has varied. California with its large deposits of sheelite is ready to meet all excessive demands, as well as Bolivia and one or two other South American countries. The talked of tariff on tungsten seems to be coming along well. The opinion of all the old tungsten operators is that a tariff of $10 per ton will be sufficient to meet all demands and insure a permanent business for some years, and will enable the United States to meet foreign competition. The tariff is a broader question than is generally believed, for it involves the products in which tungsten is used.

Two or three companies are erecting refining laboratories at Boulder for the making of tungstic acid and the Rare Metals Co. at Rollinsville is going to make the acid also.

The future of the tungsten business looks good for many years to come, or until the field here is worked out. As the Wolf Tongue and Vasco companies are directly and indirectly connected with steel companies in the East that will take their products they will be doing business for some time. The situation for the other companies is also good, from all that can be learned.

All the dumps of former years of mining are being jigged by lessees and screened and sent to the various mills. Greater depth has increased the cost of mining, the necessity of pumping water being one reason. The Wolf Tongue company has been furnishing to lessees all equipments, pumps, pipe, and motors, and charging a royalty of 25% on ore taken out. At present it is not offering any more leases to anyone.

MEXICO

EFFECT OF THE NEW CONSTITUTION ON THE MINING INDUSTRY.—A SHORT-LIVED STRIKE.—INCREASE IN WAGES ANTICIPATES TROUBLE.

Mining conditions at Pachuca are fairly stable, with the Real del Monte Co. milling 1800 tons per day, the Santa Gertrudis Co. 1000 tons per day, and La Blanca Co. 500 tons per day. Owing to the crazy labor provisions of the new constitution, and to the increasing independence of the Mexican workmen, the mining companies are anticipating a rough and stormy voyage. At a meeting of the laborers recently a strike was called in which all the mechanics, hoist-men, pump-men, electricians, and the rest participated. In sympathy with them most of the miners and mill-men walked out. The strike lasted about a week, at the end of which hunger and thirst for pulque forced them back to work. This is the usual outcome of these strikes. The Santa Gertrudis Company, however, granted an increase of pay to the workmen. This was probably a far-sighted move, as all the companies will undoubtedly be forced by the Government to accede to the demands of the workmen in the near future. Santa Gertrudis has thus gained the goodwill of the powers that be.

CLIFTON, ARIZONA

THE ARIZONA COPPER COMPANY ADDRESSES AN OPEN LETTER TO ITS EMPLOYEES IN REPLY TO THAT ISSUED BY THE MINERS.—TEXT OF THE LETTER.

The Arizona Copper Co. has replied to the letter from the union miners, the text of which appeared in the issue of the MINING AND SCIENTIFIC PRESS of July 7. Folowing is the letter of the company in full:

TO THE EMPLOYEES OF
THE ARIZONA COPPER COMPANY, LTD.:

In order that you may be fully informed of the answer returned by this Company to the demands recently presented by your Employees' Committee, I desire to put a copy of it in each of your hands. The answer given was as follows:

To the Members of the Employees' Conference Committee, of the Arizona Copper Company, Ltd.

Gentlemen: The conference committee, representing the employees of The Arizona Copper Company, Phelps-Dodge Corporation, Morenci Branch, and Shannon Copper Company, have presented for the consideration of the managers a number of demands.

With regard to these demands The Arizona Copper Company has the following answer to give:

No. 1. That any grievance arising among men working on contracts shall be taken up in regular form by the grievance committee.

The agreement at present in effect between the employers and employees in this district provides that any employee, believing himself to be the subject of unfair or unjust treatment, has the right of appeal through the duly appointed grievance committee of the department or company in which he is employed, and that every employee shall have the right of ultimate appeal to the manager of such company concerning any conditions or treatment to which he may be subjected and which he may deem unfair; under the above provision this company recognizes the right of any employee working either by the day or under contract to use such means in seeking redress of any grievance which he may have.

No. 2. That any employee refusing to accept a contract shall not be discriminated against or discharged for refusing the same.

No men who are working in places where a contract is let shall be discriminated against or discharged, on account of same.

The present agreement fully covers such cases and no further rule seems necessary.

No. 3. That seniority rule must prevail both in increasing and decreasing the force.

This company cannot adopt such a rule in the operation of its mines, as it would serve neither the interest of the employer nor employees generally, and would be in direct conflict with the spirit of the agreement which provides that the right to hire and discharge, the management of the property, and the direction of the working force, shall be vested exclusively in the company. The practice of this company in the past when forces were increased or reduced has had regard both for the efficient prosecution of the work and for the personal claims of the workmen, as for instance: When preference has been given to married employees when it has been necessary to curtail the working forces. The adoption of any rule that would limit the company's rights in this respect cannot be considered.

No. 4. That time and one-half be paid for all overtime, and that time and one-half be paid to all craftsmen and their helpers for all Sunday work, the 4th of July and Christmas.

Inquiry amongst our employees developed the fact that a large proportion of those who would be affected were such a rule adopted is not in favor of this demand. We will, therefore, defer consideration of it until we can be shown that substantially the majority of those who would be directly affected by it are in favor of such a rule.

No. 5. That the living conditions in this district are such that we are compelled to ask for the Miami scale of wages.

This company recognizes the present conditions with respect to the cost of living, as given by the committee as a reason for the demand for an increase in wages at this time, such conditions being due to the European war, and it is prepared to offer as an offset to the present increased cost of living an increase of 50c. per day to those employees who are receiving on the present scale a base rate of wages of from 25c. to 46½c., inclusive, per hour, and a raise of 25c. per day to employees receiving a base rate higher than 46½c. per hour. This offer to take effect July 1, 1917, and to continue in effect until after 30 days' notice has been given its employees by this company. This offer is made upon the following conditions, viz:

That its employees do not go out on strike, and with the further understanding that such offer will be withdrawn in the event of a strike occurring which shall have the effect of suspending the operations of the company.

The company also desires to notify its employees that in event of a strike it will refuse all guarantees with respect to reinstatement of any employee, upon resumption of operations thereafter.

THE ARIZONA COPPER COMPANY, LTD.
By NORMAN CARMICHAEL, General Manager.

Clifton, Arizona, June 30, 1917.

When demands upon a company are made on behalf of employees through their committee and an answer is rendered, it is customary for such answer to be communicated to the employees generally, especially when offers of compromise are contained, in order to ascertain the wish of the majority as to whether such answer is satisfactory or not. Such a course indicates a proper conception of responsibility on the part of the leaders and shows their respect for the intelligence of the employees they represent.

A strike, involving loss of work to several thousand men and the suspension of operations which sustain a large community, is a very serious matter, and the hasty manner in which the present strike was called without giving you an opportunity to even read the reply which we made to your demands and without giving you an opportunity to show, by your vote whether you preferred to accept the answer given or to go out on strike, indicates that your interests were not consulted in the matter, and that your faith in your leaders is not well placed.

The offer which this Company made of an increase of 50 cents per day to those at present receiving from 25 cents to 46½ cents per hour, base rate, meant that the largest proportion of you would receive this increase to offset the present high cost of living, and which you would enjoy during the continuance of such conditions. Do you realize that this offer was made upon condition there would be no strike, and that, in calling you out on strike without putting these facts before you, your leaders have deprived you of that offer?

THE ARIZONA COPPER COMPANY, LTD.
By NORMAN CARMICHAEL, General Manager.

Clifton, Arizona, July 2, 1917.

THE MINING SUMMARY

The news of the week as told by our special correspondents and compiled from the local press.

ALASKA

(**Special Correspondence.**)—Secretary Lane has directed the Alaskan Engineering Commission to start mining operations in the Matanuska coalfields immediately. Federal mine inspector, Sumner S. Smith, has been named as resident engineer and will take charge of the operations of the mines. The purpose of the work is to furnish an immediate supply of coal to the Alaskan Engineering Commission for its use in the operation of its trains and construction work. It is expected that the mines will be opened up on a comprehensive plan and the work will be carried on under the supervision of Mr. Smith with the co-operation of George W. Evans, district engineer for the Bureau of Mines, stationed at Seattle, and who already has had experience in the field. The reason for this action is that the private operators not having commenced operations on a scale that would insure the Commission an adequate supply of coal, it was thought best for the Commission to operate its own mines for the time being at least. Leasing Unit No. 7, with the equipment on the property, has been purchased by the Commission to obtain an immediate supply of coal and work will be started on Unit No. 12, at Chickaloon, to supplement this. With these two mines in operation the Commission will be assured of an ample supply of coal, not only for its use locally but to supply the Government boats visiting this harbor.

Reports received at Anchorage from Lewis river are encouraging and operators are satisfied that the camp will be prosperous. Breunerman and Hamilton reached their holdings last week and will start operating on June 20. Peterson & Co. started work on June 11. Sam Wagner and partner are sluicing with good results.

The Alaska Copper company has filed location notices in the office of the United States commissioner covering the Silver Dollar, Phoenix, and Black Bear lode mining-claims situated in the Harris mining district.

The officers of the steamer 'Alaska,' which reached Juneau June 21, report that the people of Cordova have been unable to gain any news of the Kennecott strike. It is known that 200 men walked out and that the ore is not arriving as rapidly as formerly, but otherwise there is no further news of the strike. E. T. Standard, manager of the mine, is returning north on the 'Northwestern,' having left Seattle June 19. The Alaska Treadwell company is contemplating the erection of a trestle around the 'cave in' as the most feasible means of getting out the heavy machinery. This will restore the transportation system to its former status and will do away with tramming over the 'high line' through the central crushing plant.

Treadwell, June 30.

ARIZONA

Cochise County

(**Special Correspondence.**)—The Mascot Copper Co. of Willcox, controlling 50 claims and operating a railroad 16 miles long between Willcox and the mines, has been taken over by the American Smelting & Refining Co. under a long-time lease. The Mascot company has been shipping between 500 and 600 tons of ore per week. This production will probably be in-

Willcox, June 20.

On June 27 about 2500 of the 5000 miners employed in the mines at Bisbee went on strike. They demand an increase in pay and better working conditions. The companies most seriously affected are the Copper Queen, the Shattuck-Arizona, and the Calumet & Arizona. The managers of these companies are reported as saying they will close down every mine under their control in the district rather than submit.

The Bisbee *Review* of July 3 says: Raising their sights, by some action, the I. W. W. at Bisbee, have decided that, in the words of one of last night's speakers at the City Park, "we will not sign peace terms until every other company which now has labor trouble on its hands acceeds to the demands of the strikers, in every part of the country." This action followed a similar announcement made by the Butte, Montana, organization.

But for two mass meetings, held by the I. W. W. yesterday, one in the afternoon and one last evening, the day was quiet. No arrests were reported from any part of the district. One Slavonian resident of Bisbee, however, reports having been threatened with personal violence if he continued at work. He was standing in front of the Busy Bee restaurant when accosted by several men. He did not know any of them.

More miners and other underground workers reported for work at the various mines in the district yesterday. The exact increase in the working forces was not obtainable, but it is said to be a substantial one. It is felt and expressed by many men, thoroughly conversant with the situation, that nearly complete forces will be on hand at the mines after July 4, on which day the two companies have declared a holiday.

Many men continue to leave the district for other parts of the country. The railroad ticket offices are crowded before every outgoing train departs. It is estimated that upward of 500 men have left Bisbee in four days. There is another movement of the I. W. W. This seems to be in the direction of the Globe-Miami and Clifton-Morenci districts. During the first day or two of the trouble the influx of I. W. W.'s was considerable. For some reason, attributed by many to a feeling that the strike has been lost, the tide has set in an another direction.

Gila County

Telegrams received at I. W. W. headquarters at Miami late on the afternoon of July 2 assert that the metal-mine workers at Jerome will strike in sympathy with the branches that have walked out in the Globe-Miami district. The messages were from the secretary of the Jerome branch of the I. W. W. organization. Local metal-mine workers say that many other camps throughout the country will shortly follow with strike orders in their effort to cripple the copper industry until their demands are granted.

Mohave County

(Special Correspondence.)—Three Dorr agitators are being installed at the United Eastern mill, which will increase the total daily capacity of the plant to 300 tons or more per day. The crushing capacity of the mill is 400 tons daily, but tank capacity was originally planned on a basis of 200 tons. Six months operation of the plant convinced the management that daily capacity should be increased to at least 300 tons. Development is progressing on both the 565 and the 665-ft. levels, opening up new ore reserves. The main shaft is being sunk 300 ft. deeper, a supplementary plant having been installed for that purpose.

While a large amount of ore is being opened up at the Gold Road property, it is not probable that milling operations will be resumed there for some time. The increased cost of milling and mining is given as a reason.

As high as 340 tons per day has been milled at the new Tom Reed plant. In a supplementary report to the stockholders, the board of directors stated that the company has ore reserves of over 100,000 tons, sufficient to supply the new mill for one year. Development is progressing on the Aztec vein, but shipping of ore to the mill is confined to motor-truck haulage pending the installation of the aerial-tramway. At the Adams Mining Company's property in the Black Range, two parallel veins or fault-fissures have been explored by driving about 350 ft. from where it was first cross-cut from the shaft at the 400-ft. level. On one of them valuable ore has been opened up for 50 ft. In anticipation of a heavy flow of water the main cross-cut from the shaft is provided with a concrete bulk-head and a steel door which can be made water tight, thus protecting the pumping-plant on this level.

Though gold is heavier than tailing dust, it has been proved at the Tom Reed plant that the heat of the sun will cause it to come to the surface. Assays made of the tailing-dump last year by Mr. Rabb, the superintendent, showed that the top half inch of the pond carried about $15 per ton in gold, while that below was practically of no value. Accordingly the pond was systematically scraped and about $10,000 in gold recovered. This process is being repeated this year.

Oatman, July 3.

(Special Correspondence.)—Ou the 400-ft. level of the Adams mine a body of payable gold quartz has been opened up in a vein 350 ft. from the working-shaft. The vein is 14 ft. wide and driving is in progress along the foot-wall. C. H. Palmer, Jr., is engineer for the Adams company and N. A. D'Arcy is manager. The vein parallels one that has been driven for 350 ft. without satisfactory results.

Diamond-drilling will be done to explore the Telluride property adjoining the Tom Reed and the Sunnyside, by J. L. McIver, one of the discoverers of the United Eastern. The same method will be tried at the Mohawk Central by M. J. Monnette. This is the first time this method of prospecting has been tried here, except on the old Moss mine, where the Santa Gertrudis Corporation ran two diamond-drill holes.

J. P. Loftus and J. K. Turner have started operation of a new 30-ton ball and amalgamating-mill on the Oatman Gold Top property in the Secret Pass district.

The Gold Ore Mining Co., adjoining the Gold Road, announces that it has commissioned Otto Wartenweiler, of Los Angeles, to design a mill with a capacity of 200 tons per day. A unit of 50 tons will be built first. Mr. Wartenweiler designed the United Eastern mill.

Oatman, June 24.

Pima County

(Special Correspondence.)—The Growler mines, 16 miles south-west of Ajo, have been bonded for $250,000 by a group of Ajo capitalists. The Growler was owned by George H. Morrill, of Boston. It was formerly known as the Colonial Copper Co. There are 26 claims in the group. The mines are worked through a shaft 325 ft. deep and a 200-ft. incline.

Tuscon, June 20.

Yavapai County

Ninety per cent of the day-shift at the United Verde copper mine at Jerome reported for work on Sunday, following the rejection Saturday by members of the Jerome local of the International Union of Mine, Mill and Smelter Workers of a proposal to submit to the membership a strike vote on the question of joining the Metal Mine Workers' Industrial Union No. 800, of the Industrial Workers of the World, in a strike declared Saturday by the latter organization in the Jerome copper district.

All of the smaller mines, which closed down on July 7 pending announcement of the result of the vote, were at work with practically full forces today.

Yuma County

(Special Correspondence.)—Mining is active in the hills adjacent to Wenden and Salome. At Salome the Navajo Mines, Cobrita Verdi, Glory-Hole, and several others are coming into prominence.

At the Navajo Mines there is an encouraging showing, and an adit on the main vein is being driven. Charles Redall, of the Cobrita Verdi, has recently shipped in machinery from Goldfield, Nevada, and is installing it preparatory to starting work. At the Glory-Hole, with a complete equipment of machinery, including jack-hammers, Ernest Hall, in charge, has men developing that property. The Harqua Hala mine, under direction of John Martin, is developing and is expected to become as prominent again as in the bonanza days when over $2,000,000 was taken from one level. It is said that a considerable tonnage of ore has been opened up there, but it is of a different character from the ore on the upper levels. It is more base, and carries copper as well as gold. Formerly all ore taken from here was free-milling. The old 40-stamp mill and a complete equipment is on the property. There has been talk of putting in either a concentrating-plant or flotation, as too much of the value is lost by simple amalgamation.

George Easton and E. A. Stent, lessees of the Critic mine, which is owned by George B. Layton, of New York, are steadily operating their property in Cunningham pass and have discontinued leasing to miners and will work the mine themselves. The last car shipped from there returned 1.05 oz. in gold and 19.84% in copper per ton. From two to three cars is shipped from there monthly, and the former lessees were making big money.

The superintendent of the Black Reef mine in Cunningham pass, has been getting chalcopyrite from the bottom of his shaft, which is now down 300 ft. Mr. Scott, one of those interested in the Black Reef, and who is also interested in the Superstition and known among his friends as 'Lucky Scott,' has been called to Canada to join his regiment.

Mr. Ormsby, of the Wenden Copper Co., is steadily sinking a shaft to reach the permanent water-level. He is backed by Globe people who are interested in the Old Dominion.

Wenden, June 28.

CALIFORNIA

The amount of oil available in the various areas of California has been summarized in a report made to the State Council of Defense by R. P. McLaughlin of the State Mining Bureau. The visible supply of oil has been rapidly decreasing during the past year. A special committee, headed by Max Thelen, of the Railroad Commission, has, for several weeks, been engaged in investigating the oil-supply of the State, which furnishes power for railroads and many other industries of vital importance at the present time. The report dealing with the oil remaining underground has been compiled from production records and well-logs filed with the Mining Bureau during the past two years incidental to the work of protecting the fields from damage by improper well-drilling. Future use of the information will be determined entirely by the State Council of Defense.

El Dorado County

(Special Correspondence.)—The Cincinnati gold quartz mine, situated 10 miles by auto-road north-west of Placerville, in the Kelsey mining district, now has 10,000 tons of friable free-milling quartz practically blocked-out, and the management is proceeding to have installed a 5-stamp mill of 1000-lb. stamps, to be operated by an oil-engine. Burr Evans, the local mining engineer, states that such a mill will readily crush 25 tons per day of the kind of material to be reduced. No rock-breaker or concentrator will be required. The free-gold will all be recovered on the battery-plates.

The ore consists of a yellow ochre-like material, stratified with numerous small stringers of crumbly quartz, with an occasional egg-shaped lens of high-grade ore about the size of a foot-ball. The pay-shoot averages 16 ft. wide, and will run not less than $3.50 free-gold per ton, without including the high-grade lenses. The ore can be mined by auger-boring, light blasting, and picking, and conveyed by gravity through a 250-ft. adit to the mill at a cost of 26c. per ton.

The cost of milling the ore is estimated not to exceed 54c. per ton; thus making a total cost of 80c. per ton for mining and milling. Then, after allowing 70c. per ton for continuous new development work and overhead expenses, it is figured that it will net not less than $2 per ton. The mine is being operated by Berkeley and San Francisco capital. In this district in the 80's the Kelsey mine was operated for some time at a cost of less than 50c. per ton for mining and milling.

Placerville, July 3.

Fresno County

(Special Correspondence.)—Dan Yokovich, of Shawmut, Tuolumne county, has discovered what may prove to be a valuable vein of molybdenite-bearing rock in the granite at or near the Inyo-Fresno county line in the Sierra Nevada. Analyses of samples have returned 3.4% of molybdenum sulphide.

Shawmut, July 2.

Modoc County

At Copper Peak, the Valley View Mining Co. is reported as ready to start a new adit at the foot of the mountain. This will cut the shaft and the orebody from which ore was extracted last year at about 1200 ft. depth. There is from 5 to 7 ft. of azurite and native copper. The adit will also cut the white metal nickel-cobalt veins before it reaches the copper veins. The copper ore averages 45% copper and $4.50 gold. The nickel veins give 14% copper and $8.40 gold on assay. No assay was made for nickel. Owners of the copper group on the east of the Valley View prospect have given an option on their property to R. Kemp Welch.

Parties in the East are negotiating with Mr. Welch, owner of the Copper Gold group, 16 claims, with from 12 to 14 ft. of oxidized copper ore at surface. In the group are several veins which pan copper. Two veins pan gold.

The mines are within three-quarters of a mile of the foot of the mountain, and are reached from Alturas, 34 miles, on the N. C. O. R. R., by auto-stage, or 14 miles distant from Miller's Ranch station on the same railroad.

Colorado people came in last week and are at work, having secured options and paid some cash on a claim seven miles south. There will be great activity here this summer.

Shasta County

(Special Correspondence.)—The Noble Electric Steel Co. is constructing a blast-furnace at its Heroult iron-smelter, with a capacity of 30 tons per day. It will be operated in conjunction with the electric-furnaces and is scheduled to go into service within 30 days. Difficulty in securing electrodes is interfering with electric-smelting and plans have been prepared for manufacture of electrodes at San Francisco. It is planned to keep five electric-furnaces in constant operation for the production of pig-iron, ferro-chrome, terro-manganese, and similar products.

The Afterthought Copper Co. has arranged for the immediate building of a large roaster at its Ingot plant. The copper-zinc ore will be treated by leaching and flotation, according to recent advices. Several buildings are under construction and the mine has been placed in condition for a heavy output. Besides copper and zinc, the developed ore is said to average better than $1.50 per ton in gold and silver.

The Mammoth Copper Co. has acquired the Keystone copper property on Flat creek, from Robert Strenson, George A. Grotofend, and William Stennon, of Redding. Four thousand dollars has been paid and the balance will be met in installments. The deal calls for extensive development pending consummation of the purchase.

Prospecting of dredging-ground is active around Redding. The El Oro company is vigorously exploring a wide area north of town, and the Gardella interests are busy along Clear creek. Construction of four dredges by this company will be pressed as fast as lumber and other material can be secured. Several other companies are testing gravel with promising results. It is reported another effort will be made to work the extensive gravel deposits in the vicinity of Igo.

Redding, June 28.

COLORADO

San Juan County

The Lackawana mine near Silverton has been cleaned out and promises to become once more a substantial producer. The J. B. Smith tunnel, which is in 1200 ft., is once more in good condition. A cross-cut was run at the face of the old tunnel and a vein was found in 15 ft., which has now been followed for 200 ft. The vein is from 3½ to 5 ft. wide, and a stope has been carried up 90 ft. Net returns on a carload of ore recently sent out were $108 in excess of any carload that has been shipped from there. There is also in another place 3½ ft. of galena.

San Miguel County

Manager Barnhart of the Mountain Top Mining Co. says that the new mill, recently completed 400 ft. underground in the Mountain Top mine, is doing good work since it was finished several months ago, says the Ouray *Plaindealer*. He is working three shifts each 24 hours and getting about 80% of the possible capacity of the mill. He has ordered steel-chrome balls for the mill and as soon as they arrive he expects to be able to put through approximately 50 tons of ore daily. Edward Treweek, former mine superintendent of the Wanakah and Vernon mines at Ouray took charge of the mine and mill about one month ago, and the results of his experience and executive ability are already apparent in increased production of both the mine and mill.

IDAHO

Shoshone County

(Special Correspondence.)—As a result of the sale of the Keystone mine, also known as the Blacktail, on Pend Oreille lake, nearly $250,000 is being distributed to the former stockholders on an issue of 1,746,440 shares at the rate of 13½c. per share. There are upward of 50 beneficiaries, nearly all of whom are residents of Spokane, Washington, and of British Columbia. Volney D. Williamson sold the property to Henry H. Armstead for $250,000 several months ago. The Keystone mine was acquired and developed to the point of important production by Mr. Williamson and associates, and is now in the possession of the Armstead mines, of which Henry H. Armstead is the controlling owner. Associated with Mr. Armstead is a group of the foremost tobacco manufacturers of the East and South.

The Tamarack & Custer Consolidated Mining Co. will add $900,000 to $1,250,000 to the production of the Coeur d'Alene region in the last half of this year, according to estimates. It has been estimated that the equipment is capable of handling 300 to 400 tons per day and of producing 60 to 80 tons of concentrate containing 40 to 48% lead.

The average content of the ore is 8 to 10% lead, 6 to 8 oz. silver, and a small quantity of zinc. The zinc will be saved by flotation. About 250 men will be employed in the mine and mill when operations are under way. The labor and other costs will be $1500 per day. The cost of production has been running close to that of other large properties of the district, although hauling has been a large item of expense. It is believed this item will be reduced now that a tramway has been installed, especially if a means is found of handling timber on the tramway.

The mill of the Hercules Mining Co. at Wallace is receiving

ore at the rate of 700 tons per day, according to a report recieved this week. This is 50 tons in excess of its expected capacity. If the report is correct the company is producing in excess of its production for 1916. The official figures for that period show a shipping product of 87,179 tons having a gross value of $7,278,258. This is at the rate of nearly 250 tons per day and $90 per ton. The Hercules paid $1,501,129 for extraction in 1916. This cost now will be greater as a result of higher charges for labor and supplies. It is believed the net profits will be greater than those of 1916, notwithstanding an exceptionally low grade of ore. Much ore from the Hercules mine has gone to the Consolidated smelter at Trail, B. C., and the Day smelter at Northport, Washington. Some of it will now go to the Bunker Hill & Sullivan smelter at Kellogg.

Spokane, July 1.

The Snowstorm Consolidated produced 7400 tons of ore in May, the first month of its operation. The rate of production was a little better in June—probably 7600 tons. The mill has been receiving 240 to 250 tons daily since the start. The extraction was good from the beginning.

The Dreadnaught Mining Co. has cut an orebody and the conditions are favorable for the development of an important tonnage. The first orebody opened in the Dreadnaught, 3½ ft. wide, was cut at a depth of 730 ft. The average content is 10 oz. of silver, 12% lead, and 4 to 6% zinc. In a better part of the shoot it runs 36% lead, 19% zinc, and a larger quantity of silver.

NEVADA

Humboldt County

(Special Correspondence.)—In the spring of 1916 Lovelocks was much excited over the discovery that tungsten minerals occurred at many places in the vicinity, and many claims were located, followed by the construction of two large concentrating mills. Now Lovelocks is again excited over a new discovery—this time potash. J. C. Smith, Herman Markes, E. F. Hunter, and C. Offers, who had located several claims at the mouth of Cole canyon, 6 miles north-east of Lovelocks, at the south end of the Humboldt range, discovered potash on their claims. L. B. Snipes who saw the rock thought it contained potash, and an analysis proved his guess to be correct. The owners told their friends, and Lovelocks was promptly practically deserted, all hands hurrying to the scene of the new discovery, at the place that has been called Kopatka, where many claims have been located. There is a little soft earthy incrustation, but the best ore, which occurs in large amount, is hard and compact, resembling rhyolite.

Lovelocks, June 26.

It is reported that a discovery of importance has been made in the Seven Troughs Coalition mines at Vernon, where 2 ft. of ore running $600 per ton in gold has been found.

Nye County

(Special Correspondence.)—The shaft-sinking at the White Caps has progressed steadily during the past week. The bottom of the shaft is 120 ft. below the fourth level. One more set of timbers will be placed and then the fifth-station set will go in. The ground has been so hard that piston-drills have been used in the shaft. The east drift on the 300-ft. level has been advanced 15 ft. during the week, through marbleized limestone, and this characteristic has been observed in the vicinity of the orebodies both east and west from the shaft. It seems probable that a new orebody will shortly be reached. In the west drift 300 ft. has been made with no material change in the face. The water has decreased to some extent during the week, and is now easily handled by the pumps. The total flow is 115 gal. per minute. Seven hearths of the roaster have been completed and the drying-hearth will be finished shortly. The crusher-house is complete and ready for operation. The mill will be ready to crush ore by July 1. The cyanide department of the mill also is ready for operation. The roasting-flue and stacks are not yet completed. Dahl oil-burners will be used for heating the roaster and the material for the burners is at the mill ready for installation. In mill-construction little remains to be done except as noted above.

The manager of the White Caps Extension, O. McCraney, has returned from Goldfield, where he purchased a Hendrie & Bolthoff No. 4 electric-hoist. A 40-ft. head-frame was also obtained. Negotiations are on for a large air-compressor. The first round in the Extension shaft has been shot, and as soon as the machinery is installed, shaft-sinking will be urged as fast as possible.

The installation of the 40-hp. electric-hoist at the Morning Glory No. 1 shaft has been completed and shaft-sinking is proceeding. The shaft has been sampled each day. The last assay, taken from the shaft bottom, shows an average of $5.65 in gold. The 4-drill compressor is about complete and as soon as the necessary air-pipe reaches the mine machine-drills will be used in the shaft. The development in the No. 3 shaft of the Morning Glory has reached a depth of 40 ft. This shaft is close to the White Caps west side-line, and about 100 ft. north-east from the No. 1 shaft of that company, which for the past 15 ft. has been in pay-ore, with considerable calcite in the hanging wall. The past three days work in the shaft has been in ore with assays of $40. Two samples assayed $200 and $140 respectively. This shaft is being sunk on a lime and shale contact and is developing ore underlying the various orebodies developed in the White Caps property west of the old White Caps Leasing Co. shaft.

In the Amalgamated property the east drift on the 600-ft. level has been extended 16 ft. and is out 364 ft. from the shaft. Work was retarded during the early part of the week by a heavy flow of water from a longitudinal fault carrying good gold ore, evidently the edge of an orebody.

At the Manhattan Consolidated a drift was started along the ore on the foot-wall to determine the length of the shoot, which on the second level was 85 ft. long and 25 to 40 ft. wide. One round was fired in this drift, breaking 6 ft., and the greatest high-grade ore-pocket ever developed in Manhattan was exposed. To a depth of 6 ft. and for 8 ft. in the vein from the foot-wall, manganese mixed with yellow oxide, a soft gouge-like material shows full of free gold. Some of the gold is in crystals, but it is mostly wire and flake-gold. One panning made from the soft material, without mortaring produced $20 worth of gold with strings of wire-gold welded together. Eliminating the pockets where the gold shows, a sample of the ore was taken, two ore sacks being filled, to get an average of the 8 ft. This sample assayed $531 in gold.

White Pine County

(Special Correspondence.)—The Nevada Consolidated Copper Co. has started up the north half of its new crushing-plant and it is running satisfactorily, handling between 600 and 700 tons per hour, reducing the ore to about ½-in. size. Formerly the ore went to the rolls, in lumps the size of one's two fists. The ore carries normally 3 to 4% moisture.

The Consolidated Copper Mines Co. is shipping two cars of 8% ore to the McGill smelter per day from the old Alpha workings. It is an oxidized silicious ore. The mill is handling between 500 and 600 tons per day of 2% ore from the Morris workings. The other unit is expected to be in commission this month. Two drills are in operation.

The old Ward mine is shipping 50 to 75 tons daily. It is being hauled with the Knox auto-trucks.

Several lessees are working on lead-silver ores throughout the county; there are five separate leases on the Hunter mine, 18 miles south-east of Cherry creek. The man in charge representing the Eastern owners charges each shipper $15 per car shipped. This is paid him by the smelting company as an 'inspection charge,' which the shippers consider exorbitant.

The ores from Hamilton will be hauled to Kimberly and go

out over the Nevada Northern railroad because Mr. Sexton, the manager of the Eureka-Palisades railroad, refuses to handle it. The State Railroad Commission will probably make an investigation.

Steve Pappas, who made the first discovery of gold ore in Willow creek, 100 miles south-west of Ely, a few years since, recently made another discovery on the extension of one of the known veins. On the surface he took out a pocket of 1100 lb. for which the smelter paid him $591.

Some Salt Lake people have made him a small cash payment and taken a bond on it. Some of the same people did the same on the original discovery, did a little work and quit. There have been several small pockets of rich ore found there during the past three or four years, but they have all been dug out quickly.

The Consolidated Copper Mines Co. has closed contracts with the Nevada Consolidated Copper Co. to mine by August 15, 1918, from its Oro claim, at the entrance of the Liberty pit, 425,000 tons of ore and to mill and smelt the same. At the end of this period the company will continue to mine and treat at least 75,000 tons per annum from the same place for a period of five years.

The Nevada Consolidated also will treat all of the concentrate made by the Consolidated Copper Mines Co. for a period of five years from May 1, 1917. This latter includes sufficient high-grade smelting-ores (equal to about two carloads per day) to make the total 150 tons per day.

The Copper Mines Co. (old Giroux) is treating in its new mill upward of 15,000 tons per month of its ores that average around 2% copper, from the Morris workings. The second unit, of equal size, is expected to be ready for operation some time during July. Delay has been caused by non-delivery of material. On the basis of 25c. copper, with costs at 12½c., when the other unit is in operation, this company should make over $2,500,000 net during the next year.

The direct-smelting ore, now going to the smelter, runs about 12% copper. The development from the Giroux shaft, on the west side, in and beyond the old Alpha workings, is showing up some rich oxide in bodies up to 50 ft. wide.

The management appears to be in good hands, where the stockholders will get a square deal. Humphreys, Burgess, and Merritt, who have the operating control, are good business men, who desire to develop the property as an investment.

Ely, July 1.

NEW MEXICO

Dona Ana County

El Paso men have taken one year's bond and lease on the Atlas Apex group in the Quartzite mining district. The Atlas and the Apex are the principal mines in the group. Twenty tons of ore was shipped this week from the Willow Creek mine on the upper Pecos to the smelter at Salt Lake City for a test run. D. C. Jackling and other owners of the Salt Lake smelter are reported to have interested themselves in the upper Pecos and Dalton districts. The extensive development work now going on within 30 miles of Santa Fe is said also to be at their behest.

Sierra County

The famous Bridal Chamber mine, at Lake Valley, is again being worked, but for high-grade manganese ore of which three carloads per day are being shipped to the steel mills at Joliet, Ill. One hundred men are employed and two new hoists have been installed.

The Empire Zinc Co. is making a topographical survey of its 42 claims in the Kingston district in the Black range. The Kangaroo Mining Co. will install a gasoline-compressor and machine-drills for the driving of a tunnel 800 ft. from Saw Pit

Socorro County

(Special Correspondence.)—The Socorro Mining & Milling Co. cleaned up 1800 lb. of gold and silver bullion for the first half of June. New ore-bins are being added to take care of the increasing custom business.

The Oaks Co. has increased ore shipments from the Maud S. property. Another lot of burros has been added to the pack-train. The new wagon-road to the Central shaft has been completed and lumber is being delivered at the shaft collar. The head-frame is nearing completion.

Complete surface and underground surveys are being made on the Confidence property and the indications point toward an early resumption of operations at this old producer.

Mogollon, June 28.

TEXAS

Burnet County

(Special Correspondence.)—It is announced that the Texas Graphite Company soon will be re-organized and taken out of the hands of the receiver. The company owns a large deposit of graphite, situated near here, upon which it invested more than $200,000 in improvements. It is planned to resume the development of the property about August 1. McCarty Moore has installed a graphite-concentrating mill and other equipment upon a large deposit of graphite which he is developing near Llano. He is making shipments of the graphite concentrate to Cincinnati and Cleveland, Ohio.

R. H. Downman, of New Orleans, who is developing a deposit of graphite 30 miles north of Burnet will construct a mill for the purpose of treating the raw material. This property consists of 2450 acres of perpetual mineral rights.

Burnet, July 1.

CANADA

British Columbia

Ore valued at over $350 per ton in gold, after payment of all freight and treatment charges, has been discovered by the operators of the Emancipation group of claims near Hope. The group from which this ore was obtained is situated on the Coquahalla river, 16 miles from Hope, and close to Jessica station, on the Kettle Valley railway. The property is owned by Michael Merrick, Herbert Beech, and William Thomson, who are the discoverers. It is bonded to C. H. Lighthall at $125,000. Mr. Lighthall represents New York capitalists.

Recently a shipment of ore was sent to the Tacoma smelter, and after freight and treatment charges were deducted, $18,295 was netted from 53 tons. The vein is at the contact of slate and diorite and can be traced for almost 2000 ft. It pans free gold for almost the entire distance and in places is 12 ft. wide.

At 23-mile camp, on the Princeton road, large bodies of copper and silver-lead ore are awaiting development. At Jones Lake the Foley, Welch & Stewart interests are diamond-drilling their property recently bonded for $100,000.

Ontario

(Special Correspondence.)—The La Rose directors have voted $30,000 for development work on their Violet claim adjoining the O'Brien mine at Cobalt.

The Buffalo Mining Co., of Cobalt, is this year helping its employees to defeat the high cost of living. The mine cleared up the Watash claim for farming purposes, supplying teams, explosives, and labor free for the work. The married employees organized into squads under shift-bosses and each individual in the organization is required to do at least 50 hours work on the farm. The mine furnished the seed and the produce is to be divided among the men in proportion to the size of their families. Sixty miners are in the association and about 20 acres of land has been cleared. J. G. Dickenson of the O'Brien mine is in Nova Scotia; Stanley Graham of the Technical College, at Halifax, is in the Cobalt district; the Huronia Gold Mines are being re-opened by a syndicate composed of Quebec and Cobalt men. J. Young, underground superintendent of the Hollinger mines, has returned from Montana.

The flotation-process is becoming increasingly important at Cobalt. The present users of the process are the Buffalo mines, McKinley-Darragh-Savage mines, the Nipissing mines, the Coulagas mines, the Dominion Reduction Co., the Northern Customs Concentrators, the National mines, and the Trethewey mines. The Buffalo mines, with a 600-ton per day plant, has the largest installation.

Cobalt, June 30.

MEXICO

SONORA

An official of Green-Cananea Copper Co., at New York, says he doubts correctness of dispatches from Mexico attributing certain accusations to Secretary Nieto of Mexican department of finance against the Cananea Consolidated, the operating company of Greene-Cananea, as the company is acting and has always acted entirely within mining laws of Mexico. Statements purporting to have been given out by Secretary Nieto state that the Cananea Consolidated Copper Co. had refused to pay taxes on 7702 mining claims, which were overdue and owing to the Mexican government, had moved machinery across the border, closed up its hospital and ejected its patients.

It is true the company has abandoned these 7702 mining claims. This was done in accordance with laws of Mexico and followed the inauguration of the high-rate mining tax put into effect more than a year ago. Failure to pay taxes on mining claims constitutes an abandonment of all rights, and there is no provision in the law calling for any specific notification to the government. The company has paid taxes on all claims retained.

Last April a representative of the stamp-tax office and the principal administrator of the tax at Nogales conferred with officials of the company in Nogales. Nothing further was heard until June 12, when Secretary Nieto telegraphed that if the taxes (which would have been due if the claims had not been abandoned) were not paid within 15 days the right of export of bullion would be withdrawn. As it was manifestly impossible to operate if the company was not permitted to export metals, and as it was an illegal and arbitrary action, it was decided to cease operations before the date set. As to the other charges the official says: "There is no truth in the charge that the company secretly sent out some of the machinery nor in the charge that the hospital was closed and patients ejected. The company left its own physician and staff in charge of hospital patients, and the local Mexican officials arrested them and put a Mexican physician in charge, but the following day released the American physician and attendants and they are now in charge of the hospital under direction of a Mexican agent."

Edward Steidle, engineer in charge of Rescue Car No. 1 of the U. S. Bureau of Mines, announces that Car No. 1 will be at the several places named below, at the times specified: Salt Lake City, Utah, July 1 to July 10; Park City, Utah, July 11 to July 25; Bingham, Utah, July 26 to Aug. 11; Milford, Utah, Aug. 12 to Aug. 18; Eureka, Utah, Aug. 19 to Sep. 1; Sandy, Utah, Sep. 2 to Sep. 15; Scofield, Utah, Sep. 16 to Sep. 22; Castlegate, Utah, Sep. 23 to Sep. 29; Hiawatha, Utah, Sep. 30 to Oct. 6; Sunnyside, Utah, Oct. 7 to Oct. 13.

In case of mine disaster Car 1 can be reached indirectly through the University of Nevada, Reno, Nevada, (headquarters); A. J. Stinson, Nevada State Mine Inspector, Carson City, Nevada; Henry M. Rives, Secretary Nevada Mine Operators' Association, Reno, Nevada; P. A. Thatcher, Utah Industrial Accident Commission, State Capitol, Salt Lake City, Utah; A. G. Mackenzie, Secretary Utah Chapter American Mining Congress, also representative of Coal & Metal Producers Association of Utah, Boston Bdg., Salt Lake City, Utah; H. M. Wolflin, California Industrial Accident Commission, Underwood Bdg., San Francisco, California; and Robert I. Kerr, Secretary, California Metal Producers Association, Merchants National Bank Bdg., San Francisco, California.

Personal

Note: The Editor invites members of the profession to send particulars of their work and appointments. This information is interesting to our readers.

D. M. RIORDAN is in New York.

D. V. KEEDY is in French Guiana.

H. FOSTER BAIN writes from Shanghai.

MORTON WEBBER has gone to southern Arizona.

JOHN ROSS, JR., is in San Francisco from Nevada.

W. S. NOYES has returned from Ashland, Oregon.

ALBERT BURCH has returned from the Coeur d'Alene.

H. G. CANNON is mining in Mono county, California.

A. CHESTER BEATTY is traveling in China and Japan.

C. S. GALBRAITH has opened an office at Webb City, Missouri.

ERNEST A. HERSAM has returned from New York to Berkeley.

E. B. REESE has returned to Los Angeles from San Franc'sco.

GEORGE J. BANCROFT, of Denver, passed through San Francisco on his way to Idaho.

C. YADE, engineer to the Sumitomo Besshi copper mine, is visiting our mining districts.

C. W. PURINGTON lectured recently at King's College, London, on 'Pacific Routes to Siberia.'

C. E. VAN BARNEVELD passed through San Francisco on his way from Salt Lake City to Tucson.

C. T. GRISWOLD, geologist of The Associated Geological Engineers, of New York, has gone to Wyoming.

J. B. TYRRELL has been appointed Canadian representative of the Consolidated Mines Selection Co., London.

FRANCIS A. THOMSON has been appointed dean of the School of Mines of Washington University at Pullman, Washington.

HERBERT G. THOMSON has been appointed superintendent for the Nevada Packard Mines Co., at Lower Rochester, Nevada.

JAY A. CARPENTER has resigned as general superintendent for the Nevada Packard Mines Co. and is now at Wonder, Nevada.

ROBERT C. STICHT has had to go to Butte, so his address before the local section of the A. I. M. E. is postponed until further notice.

E. S. KING has been appointed manager for the Waihi Grand Junction mine, in New Zealand. succeeding WILLIAM F. GRACE, who has retired owing to ill health.

H. C. HOOVER has been awarded the Cross of a Commander of the Legion of Honor by the French government in recognition of his services in provisioning Belgium and northern France.

A REGIMENT OF ARTILLERY is being organized in San Francisco as a part of the National Guard of California, but with the intention of mustering it into the Federal service on August 5. Mining engineers are invited to enlist. A few commissions are available. Further particulars will be given in our next issue. Communications may be addressed to W. G. Devereux, care of this office.—EDITOR.

Obituary

W. GUY SCOTT died at his home near Soulsbyville, Tuolumne county, California, June 24, Mr. Scott with his brother, Proctor Scott, for many years operated the Black Oak mine, near Soulsbyville. He was a native of California, having been born at Diamond Springs, in El Dorado county. For some years past Mr. Scott has been employed by the Government as forest ranger.

JAMES E. DYE, well known as a mine manager in Amador county, California, died at the home of his son at Vancouver, B. C., June 25, at the age of 62 years. He was manager for the Exploration Company, of London, in Amador county for several years, where he had charge of the Amador Queen mines. More recently he had been identified with the Bank of Amador, at Jackson.

THE METAL MARKET

METAL PRICES

San Francisco, July 10

Antimony, cents per pound	18.50—22.00
Electrolytic copper, cents per pound	34.50
Pig lead, cents per pound	12.25—12.50
Platinum, soft and hard metal, per ounce	$105—111
Quicksilver, per flask of 75 lb.	$100
Spelter, cents per pound	11.50
Tin, cents per pound	59
Zinc-dust, cents per pound	20

ORE PRICES

San Francisco, July 10

Aluminum-dust (100-lb. lots), per lb.	$1.00
Aluminum-dust (ton lots), per lb.	$0.95
Antimony, 50% metal, per unit	$1.35
Chrome, 40% and over, f.o.b. cars California, cents per unit	50—55
Magnesite, crude, per ton	$8.00—10.00
Tin, cents per pound	60
Tungsten, 60% WO_3, per unit	$25.00—30.00
Molybdenite, per unit for MoS_2 contained	40.00
Manganese, 45% (under 35% metal not desired), cents, unit	33—37

Manganese prices and specifications, as per the quotations of the Carnegie Steel Co. schedule of prices per ton of 2240 lb. for domestic manganese ore delivered, freight prepaid, at Pittsburg, Pa., or Chicago, Ill. For ore containing

	Per Unit
Above 49% metallic manganese	$1.00
46 to 49% metallic manganese	0.98
43 to 46% metallic manganese	0.95
40 to 43% metallic manganese	0.90

Prices are based on ore containing not more than 8% silica nor more than 0.2% phosphorus, and are subject to deductions as follows: (1) for each 1% in excess of 8% silica, a deduction of 15c. per ton, fractions in proportion; (2) for each 0.02% in excess of 0.2% phosphorus, a deduction of 2c. per unit of manganese per ton, fractions in proportion; (3) ore containing less than 40% manganese, or more than 12% silica, or 0.225% phosphorus, subject to acceptance or refusal at buyer's option; settlements based on analysis of sample dried at 212° F., the percentage of moisture in the sample as taken to be deducted from the weight. Prices are subject to change without notice unless specially agreed upon.

Tungsten has taken a sharp advance, owing to continued and increasing demand, and scarcity of the supply to meet it. From a nominal price of $20 to $22 per unit for 60% ore, the price has risen to $25 to $30 per unit, and unless indications are misleading it will go still higher. Some ore of low grade has been shipped from the mines at Atolia to Boulder county, Colorado, for concentration.

EASTERN METAL MARKET

(By wire from New York)

July 10.—Copper is dull and weaker at 31c. Lead is quiet and lower at 11.12c. Zinc is dead and lower at 9.12c. Platinum remains unchanged at $105 for soft and $111 for the hard metal.

COPPER

Prices of electrolytic in New York, in cents per pound.

Date		Average week ending	
July 4	Holiday	May 29	32.50
" 5	32.00	June 5	32.62
" 6	31.75	" 12	32.75
" 7	31.50	" 19	32.58
" 8	Sunday	" 26	32.42
" 9	31.25	July 3	32.25
" 10	31.00	" 10	31.50

Monthly Averages

	1915	1916	1917		1915	1916	1917
Jan.	13.60	24.30	29.53	July	19.09	25.66	....
Feb.	14.38	26.62	34.57	Aug.	17.27	27.03	....
Mch.	14.80	26.65	36.00	Sept	17.69	28.28	....
Apr.	16.64	28.02	33.16	Oct.	17.90	28.50	....
May	18.71	29.02	31.69	Nov.	18.88	31.95	....
June	19.75	27.47	32.57	Dec.	20.67	32.89	....

June copper dividends approximated $29,000,000. For the half year ended June 30 a new record was established in the payment of copper mining company dividends; the total disbursement having been $91,669,281. The amount paid in the first six months of 1916 was $65,046,051.

SILVER

Below are given the average New York quotations, in cents per ounce, of fine silver.

Date		Average week ending	
July 4	Holiday	May 29	74.62
" 5	78.50	June 5	74.80
" 6	78.37	" 12	75.83
" 7	78.37	" 19	77.00
" 8	Sunday	" 26	78.12
" 9	78.75	July 3	77.98
" 10	79.50	" 10	78.70

Monthly Averages

	1915	1916	1917		1915	1916	1917
Jan	48.85	56.76	75.14	July	47.52	63.06	....
Feb	48.45	56.74	77.54	Aug.	47.11	66.07	....
Mch	50.61	57.89	74.13	Sept.	48.77	68.51	....
Apr	50.25	64.37	73.51	Oct.	49.40	67.86	....
May	49.87	74.27	74.61	Nov.	51.88	71.60	....
June	49.03	65.04	76.44	Dec.	55.34	75.70	....

The weekly letter of Samuel Montagu & Co. of London, dated June 14, contains the following regarding silver: The market has at last left the doldrums in which it has remained for more than a month past. The change was accompanied by an abrupt movement of the price upward.

Disquieting news from China, where the political horizon is overcast, has aggravated the firmness of the market. Whether the Chinese position will clear without civil strife or not, the evident unrest must affect the movements of trade, and, also to a certain extent, the means of communication, particularly the railways now occupied by military exigencies. It must not be assumed that the whole of the heavy transfers of silver in the form of sycee from China to India and elsewhere, that have taken place during the period of the War, will necessarily, have to be replaced at a subsequent date. Much of the silver was derived from hoards in the interior, when it has been drawn by the tempting rise in its exchange value. Moreover, substitutes have been adopted, one of which, by no means the least important, is thus indicated by the 'North China Herald' under date of April 21. "There is another factor contributing to the depletion of silver in China, and that is, the imports of gold-bars and gold coins to China by exporters to pay for their purchases, as this way of settling bills has been found cheaper than sending the white metal here."

On the other hand, quantities of copper cash are being smelted under private auspices in Shanghai. Such an operation was a serious crime under the Ching dynasty, and is also a punishable offence under the Republican regime. Of course a scarcity of copper cash would probably create a local demand for silver currency.

LEAD

Lead is quoted in cents per pound, New York delivery.

Date		Average week ending	
July 4	Holiday	May 29	10.93
" 5	11.37	June 5	11.46
" 6	11.37	" 12	11.83
" 7	11.25	" 19	12.00
" 8	Sunday	" 26	11.75
" 9	11.12	July 3	11.57
" 10	11.12	" 10	11.25

Monthly Averages

	1915	1916	1917		1915	1916	1917
Jan.	3.73	5.95	7.64	July	5.59	6.40	....
Feb.	3.83	6.23	9.01	Aug.	4.67	6.28	....
Mch.	4.04	7.26	10.07	Sept.	4.62	6.86	....
Apr.	4.21	7.70	9.38	Oct.	4.62	7.02	....
May	4.24	7.38	10.29	Nov.	5.15	7.07	....
June	5.75	6.88	11.74	Dec.	5.34	7.55	....

The Standard Silver-Lead Mining Co. declared a quarterly dividend of $100,000, payable July 15 to stockholders of record July 1. This disbursement will be the second of $100,000 made this year, the first having been made on April 15. The rate of disbursement is 5% per quarter on a capitalization of $2,000,000. The forthcoming dividend will raise the total to $2,000,000, a greater part of which was paid at the rate of $600,000 per year in monthly disbursements of $50,000. The report of the company for April showed earnings of $34,911 in that month and a surplus of $230,420. The property is at Silverton, B. C.

ZINC

Zinc is quoted as spelter, standard Western brands, New York delivery, in cents per pound

Date		Average week ending	
July 4	Holiday	May 29	9.50
" 5	9.25	June 5	9.66
" 6	9.25	" 12	9.75
" 7	9.25	" 19	9.72
" 8	Sunday	" 26	9.43
" 9	9.12	July 3	9.32
" 10	9.12	" 10	9.20

Monthly Averages

	1915	1916	1917		1915	1916	1917
Jan.	6.30	18.21	9.75	July	20.54	9.90	....
Feb.	9.05	19.90	10.45	Aug.	14.17	9.03	....
Mch.	8.40	18.40	10.78	Sept.	14.14	9.18	....
Apr.	9.78	18.62	10.20	Oct.	14.05	9.92	....
May	17.03	16.01	9.41	Nov.	17.20	11.81	....
June	22.20	12.85	9.63	Dec.	16.75	11.26	....

The United States government has closed negotiations for 23,000,000 lb. of high-grade spelter at 13½c. per lb. The transaction was conducted through the spelter committee, representing zinc producers, and was the second effected in the past few weeks. When the zinc committee was formed a few weeks ago a small amount of spelter was purchased apportioned among the various grades, known in this connection as A, B, C, and D. The total approximated 4500 tons and the prices ranged according to grade.

QUICKSILVER

The primary market for quicksilver is San Francisco, California being the largest producer. The price is fixed in the open market, according to quantity. Prices, in dollars per flask of 75 pounds:

Week ending

Date			
June 12	90.00	June 26	80.00
" 19	82.00	July 3	85.00
		" 10	100.00

Monthly Averages

	1915	1916	1917		1915	1916	1917
Jan.	51.90	222.00	81.00	July	95.00	81.20	....
Feb.	60.00	295.00	126.25	Aug.	93.75	74.50	....
Mch.	78.00	219.00	113.75	Sept.	91.00	75.00	....
Apr.	77.50	141.60	114.50	Oct.	92.90	78.20	....
May	75.00	90.00	104.00	Nov.	101.50	79.50	....
June	90.00	74.70	85.50	Dec.	123.00	80.00	....

TIN

Prices in New York, in cents per pound.

Monthly Averages

	1915	1916	1917		1915	1916	1917
Jan.	34.40	41.76	44.10	July	37.38	38.37	....
Feb.	37.23	42.60	51.47	Aug.	34.37	38.88	....
Mch.	48.76	50.50	54.27	Sept.	33.12	36.66	....
Apr.	48.25	51.49	55.63	Oct.	33.00	41.10	....
May	39.28	49.10	63.21	Nov.	39.50	44.12	....
June	40.26	42.07	61.93	Dec.	38.71	42.55	....

Eastern Metal Market

New York, July 3.

Almost complete stagnation describes the market of practically all of the metals. The continued uncertainty of any decisive Government action regarding purchases and prices, that will give some idea to the trade as to what to expect for some definite and relatively lengthy period, instead of spasmodic buying for a month's needs or less, is acting as a decided drag on the market.

Copper manifests a weaker tendency and prices are nominal.

Tin is steady but inactive and dull.

Lead is unsettled and lower, with the tendency soft.

Spelter is absolutely dead and weaker.

Antimony is quiet and unchanged.

The steel market is also more or less at sea awaiting something definite as to what measures are to be taken regarding its regulation both as to prices for materials and as to taxes on excess profits. The coal fiasco has not tended to lessen anxiety. An end to the chaos is earnestly desired and absolutely necessary. Pig-iron output for June was lower than that of May because of coke-troubles. The total for June was 3,270,055 tons, or 109,002 tons per day, against 3,417,340 tons in May, or 110,238 tons per day. The output for the first half of 1917 is less than that of the first half of 1916, or 19,069,892 tons against 19,410,453 tons.

COPPER

Daily-press reports are persistent that the Government has purchased 60,000,000 lb. of copper at 25c. per lb., the price being a tentative one subject to change later if it is decided that it is necessary as a result of Government investigations. This scale is credited by some in the trade and discredited by others. It is also rumored that a purchase of about 15,000,000 lb. is to be made at any moment. The whole matter seems to be shrouded in mystery and secrecy. It would be better for the trade in this and other metal markets if some definite conclusion could be arrived at. The entire copper market is quiet and easier, with the general tendency downward. No business is reported worth talking about outside of the Government order referred to, and the market is a drifting and nominal one, with early-delivery metal quoted at 31.75 to 32c., New York, third quarter at 30c. and fourth quarter at 29c., New York. Small lots have changed hands but quietness rules. Reports of strikes and excessive wage-demands among copper miners in the West are attracting more attention than anything else, but in some quarters these are regarded as exaggerated for political and economic effect. The London market is unchanged at £142 for spot electrolytic and £138 for futures.

TIN

The absence of regular receipts of cables from London until late in the day, or more often until the next day, is exerting an unsettling influence. Sellers are in the dark each day as to what quotations to make, and buyers are in doubt as to what action to take. As a result business is held in check. The entire market is slow and permeated by extreme cautiousness on the part of everyone. Yesterday, July 2, no spot business was reported, but about 100 tons of futures were sold in the shape of September-October shipment from the East at 56.25c. Inquiry for June shipment from the East developed a scarcity for this position and none was to be had. The spot market came to a standstill. Last week business for any position was meager. A little spot-business was done on June 27 at 62c., New York, but on the 28th no sales were reported, though inquiry on that day and the day following was fairly good, amounting to probably 100 tons in all. There are no developments as to what the sub-committee on tin has done or is doing, and the entire market is quiet and dull with quotations at 62c., New York, on every day since our last letter. Deliveries of tin for June, according to the New York Metal Exchange, were 6398 tons, of which 2798 tons arrived at Pacific ports. The quantity afloat July 2 was 3081 tons, with the arrivals on that day 280 tons. The London quotation for spot Straits was £244 on July 2, a decline of £1 from the previous quotation.

LEAD

The reaction, which was forecasted last week, has developed in a mild form, and today the market is quoted at 11.25c., St. Louis, or 11.37½c., New York. It is bare of features, however, and the tendency is to lower levels if anything. More metal is being offered than the demand seems to be able to absorb, and business generally is dull and unsatisfactory. Late last week a little business was done as low as 11 to 11.25c., St. Louis, but at the close of the week better inquiry developed, and a fairly good business was reported. The sale to the Government of 8000 tons at 8c. per lb. as its July requirements, with nothing settled as to future needs and prices, has not been a stabilizing influence, and the continued uncertainty has acted as an unsettling factor. Further reaction is looked for by some.

ZINC

Unconfirmed reports are to the effect that the Government has purchased 11,000 tons of high-grade spelter at 13.50c. per lb., St. Louis, or 2c. per lb. above the price paid for the 6700 tons bought early last May. Definite information regarding the whole question of Government needs and buying seems unobtainable, and in the meantime the market is growing weaker almost daily. The continued suspense and uncertainty is acting as a great drag on initiative and enterprise, and soon will become serious unless something is settled. One daily trade paper is responsible for the above report but no one in the trade has definite knowledge regarding it. Also it is stated that the Government has bought sheet-zinc at 16c. per lb., as compared with the present quotation of 19c. This has not been confirmed. Prices have declined recently. Early delivery of prime Western is quoted at 9c., St. Louis, with August and September metal at 9.12½c., St. Louis, but sales have been very few. It is acknowledged that many producers cannot operate at a profit at these prices and it will not be surprising if numerous small ones are obliged to shut-down before many weeks unless the situation clears decidedly. Exports of spelter in May were large—18,533 tons.

ANTIMONY

Conditions are unchanged, and demand is light, with Chinese and Japanese grades quoted at 19 to 19.50c., duty paid, New York. A consignment of Cookson's antimony from England, the first to be released to this market in a long time, has been sunk by a submarine. It was quoted at about 22c., New York.

ALUMINUM

Demand is not active and the market is a little easier. No. 1 virgin aluminum, 98 to 99% pure, is quoted at 58 to 60c., New York, for early delivery.

ORES

Tungsten: The ore market is quiet, with demand reported as only of moderate volume. Quotations are unchanged at $20 to $22 per unit for 60% concentrate. The ferro-tungsten market is also dull. Offers to sell at $2.20 per lb. of contained tungsten have been made for export, with quotations generally varying from that to $2.50 per pound.

Antimony and Molybdenum: There have been no changes reported, and conditions and quotations are unchanged from those prevailing in last week's letter.

Company Reports

NORTH BUTTE MINING COMPANY

The annual report of the North Butte Mining Company for the year ended December 31, 1916, shows that during the year there was shipped 560,673 wet tons of ore and 120 wet tons of precipitate, and there was treated at the smelter 544,365 dry tons of ore and 89 dry tons of precipitate. Of this ore 49,252 dry tons, or 9.1%, was first-class, 423,118 dry tons, or 77.7%, was second-class, and 71,995 dry tons, or 13.2%, was third-class. This ore produced 24,498,181 lb. of fine copper, 1,047,063.56 oz. of silver, and 1,712,004 oz. of gold. During the months of November and December there was also mined 1652 wet tons of zinc ore and there was treated 1625 dry tons, which produced 412,953 lb. of zinc and 2510.24 oz. of silver.

Deliveries of copper, silver, gold, and zinc made during the year, in amounts and at average prices received, were as follows:

Copper, pounds	21,505,584	23.295c.
Silver, ounces	960,246.62	66.371c.
Gold, ounces	1,712.004	$20.00
Zinc, pounds	412,953	10.674c.

Four dividends were paid during the year, as follows:

No. 37, January 26, 1916	$ 215,000
No. 38, April 26, 1916	215,000
No. 39, July 29, 1916	322,500
No. 40, October 23, 1916	322,500
Total	$1,075,000

The mine was in operation 349½ days, the average number of men employed was 1160, and the average tonnage hoisted daily was 1605.

The net cost of producing copper was 15.51c. per lb. as shown in the following statement:

Classification	Cost per lb.
Mining and development work	$0.108784
Freight on ore	0.002749
Concentrating, smelting, freight on bullion, refining and selling expenses	0.075352
General and miscellaneous expense, personal and federal taxes	0.002539
Total	$0.189424
Less value of silver, gold, and zinc	0.033724
Net cost	$0.155700

THE BROKEN HILL PROPRIETARY COMPANY, LIMITED

The half yearly report of the Broken Hill Proprietary Company, Limited, for the year ended November 30, 1916, shows the following:

At the mine operations were carried on continuously, except on several occasions of 'stop-work' resolutions carried by the miners. Costs show a material increase, due to this cause, and to the higher cost of supplies. The quantity of ore mined was 110,276 tons. Exploration was carried on during the half-year period, but nothing of value was disclosed.

The zinc concentration plant was again operated during the last month of the half-year, and produced 2805 tons of zinc concentrate of a slightly higher grade than had been previously obtained, the average being 47.41% zinc. The installation of a slime-flotation plant was completed and operations started in the middle of the half-year. The results obtained were eminently satisfactory, a lead concentrate of about 56% lead and 80 oz. silver; and a zinc concentrate of about 46% zinc being produced.

At the Newcastle steel works a succession of strikes of workmen greatly hampered production and increased expenses, while the shipment of finished products available for delivery was seriously hampered. Since the issue of the last report the Imperial Government has completed the purchase of 30,000 tons of shell-steel, steel rails, and fish-plates. The successful issue of £400,000 of 6% debentures enabled the management to decide upon the duplication of the blast-furnace plant; and everything is being urged forward as rapidly as possible.

The gross profit for the half-year amounted to £171,598 15s. 3d., which after deducting £15,169 4s. for depreciation, leaves a net profit of £156,429 11s.3d. During the term the sum of £325,969 4s. was expended in construction, the principal item being in connection with the Newcastle steel works, which amounted to £318,732 6s.8d.

After providing for all outstanding liabilities there remain liquid assets in cash, bullion, and other convertible stocks, representing a total value of £326,322 16s.6d.

CANADA COPPER CORPORATION, LIMITED

The annual report of the Canada Copper Corporation for the year ended December 31, 1916, shows the following:

Operations at the Greenwood smelter were continued during the year. After writing off to depreciation $235,238.37, a profit of $215,304.85 remained. The total amount of ore smelted during the year was 306,450 dry tons, of which 23,243 tons was custom-ore. From this was produced 5,196,239 lb. of fine copper; 49,928.71 oz. of silver, and 12,366.24 oz. of gold. It was only possible to operate the smelter profitably because of the high price of copper. High costs resulted due to the ore being taken from pillars and caved areas remaining in the mine and also to the high cost of labor and supplies. The smelting operations were interfered with, due to continuous shortage of coke, which condition still exists.

In the last annual report, reference was made to the decision to proceed with underground work at the Copper Mountain property, the purpose of which was to confirm the results previously secured from diamond-drilling. It was also planned to lay out the work as part of the permanent programme, looking to the underground development of the mine, for extraction of ore on a large scale.

In order to provide for the rapid completion of the amount of underground work planned, a power transmission-line 13.6 miles long was brought in from Princeton, B. C., where a lease upon a power-plant had been secured. Ample compressor facilities, machine, blacksmith, and carpenter-shops were installed, also warehouses, bunk-houses, and additional dwellings. Heat, light, water, sewer, and telephone-systems were also provided.

A permanent pumping plant has been installed at the river and a Gould triplex pump is operating under a head of 1700 ft. in one lift.

Since the first of the present year, a 50-ton experimental flotation mill has been placed in operation. The purpose of this plant is to outline definitely in advance the metallurgical procedure which is to be adopted in the large mill.

A tunnel 9 ft. by 9 ft. in the clear was driven 2100 ft. on the 3950-ft. level. In addition to this, numerous lateral drifts and raises were made, the total amount of driving and raising up to the end of the year amounting to 5206 ft. As soon as it became apparent that the results secured from diamond-drilling were reliable, drilling from the surface was resumed, and 8007 ft. of diamond-drilling was accomplished during the year. In addition to this 2364 ft. of surface-trenches was sunk on newly located claims.

The net cash expenditure during the year on the development of the Copper Mountain property was $396,000. This includes payments on account of the purchase price on claims under bond as well as administrative and engineering expense.

There remains a total of $18,000 not yet due, to complete payment for all the mining area at present desired on Copper mountain.

The development work at Copper Mountain during the year demonstrated the accuracy of estimates previously made and upon which tonnage estimates had been established.

No material increase in ore reserves can be reported because the underground work was directed along the line of diamond-drilling previously performed. However, since the beginning of the year, underground diamond-drilling was started, which work could not be undertaken until underground work had been advanced sufficiently for the purpose. New ore is being found.

Prior to the underground development campaign, it was deemed expedient to class the ore as reasonably assured and probable. It is now estimated that there are 10,000,000 tons of definitely assured ore and 2,000,000 tons of probable ore. The average grade of this tonnage is 1.74% copper and 20c. per ton recoverable in gold and silver.

The possibilities for still further increases are considered excellent, in view of the results being secured at the present time, and in view of the existing geological conditions.

Allen H. Rogers made an independent report on the properties, and his report confirms our estimates with regard to tonnage and value of ore. His conclusion is that the property is sufficiently developed to warrant the erection of a mill having a capacity of 3000 tons per day. He estimates the cost of producing copper at 9.57c. per lb., based on existing smelting contracts for the treatment of similar product elsewhere in British Columbia. A conservative figure for the cost of transportation is assumed.

ST. JOHN DEL REY MINING COMPANY

The 86th annual report of the St. John Del Rey Mining Co. for the year ended February 28, 1917, was submitted by the directors at a meeting held in London, June 21, and shows the following: Tons of ore hoisted, 198,586. Tons of ore crushed, in 130-stamp mill, 187,400. Recovery in gold and silver, 110,-552 oz. Value realized in London, £471,247. The proportion of mineral rejected was but 2.89%. The yield per ton was 50s. 3½d. The profit for the year was £155,593 and a balance of £7574 was brought forward.

The superintendent's report states that a favorable change has taken place in the lower levels, both as regards the size of the lode and the value of the ore.

During the year the amount of new ore blocked out greatly exceeds the amount of ore extracted.

Nevada Section A. I. M. E.

The second annual field-meeting of the Nevada section of the American Institute of Mining Engineers was held at Ely, McGill, and Ruth, Nevada, on June 22 and 23. The meeting was highly successful and the various visiting engineers and their ladies were loud in their praises of the hospitality of the engineers of the Ely district. The first day's session was held at McGill, where 45 members from various parts of the State were present. This was a technical session; it opened with an address of welcome by C. V. Jenkins, business manager for the Nevada Consolidated Copper Co. After a response by the chairman, J. W. Hutchinson, five papers prepared by members of the Nevada Consolidated staff were read and discussed. The titles of these are:

'Present operation of steam-shovel mines'. By Robert Marsh, Jr.

'Branch-raise system at the Ruth mine'. By Walter S. Larsh.

'Ball-mill practice'. By Geo. C. Riser.

'Coarse-crushing practice'. By Curtis H. Lindley, Jr.

'Handling and roasting fine-slime concentrate at the Steptoe plant': By R. E. H. Pomeroy, and J. C. Kinnear.

'Water supply'. By Lindsay Duncan.

An adjournment was then taken after which lunch was served at the company mess. During the afternoon an inspection was made of the smelter, concentrator, and crushing plant of the company. In the evening a banquet was tendered to members and their ladies at the Steptoe hotel at East Ely, after which the 54 guests participated in a dance. On the following morning the party proceeded to Ruth, where many went underground in the Nevada Consolidated mine, while others inspected the operation of the churn-drills and the steam-shovels in the great pit. After a lunch served by the company, a paper was presented by Edward Steidle, engineer in charge of the U. S. Bureau of Mines Rescue-Car No. 1. An inspection was then made of the rescue-car stationed temporarily at Ruth, after which a demonstration by rescue-teams of the Nevada Consolidated Copper Co. and Giroux Consolidated Mines Co. was presented. The ladies, who on the previous day had been entertained at luncheon by Mrs. C. V. Jenkins of McGill, attended a luncheon given by Mrs. Walter Larsh of Ruth.

The officers elected for the ensuing year were R. E. H. Pomeroy of McGill, chairman; J. C. Jones of Reno, vice-chairman; Henry M. Rives of Reno, secretary-treasurer. The new executive committee, which includes the chairman and vice-chairman, consists of the following: Emmet D. Boyle, Governor of Nevada, John G. Kirchen of Reno, J. W. Hutchinson of Goldfield, C. B. Lakenan of McGill, Whitman Symmes of Virginia City, Frederick Bradshaw of Tonopah, and W. H. Blackburn of Tonopah.

In the evening, after the return from Ruth, members attended the Red Cross dance at Ely in lieu of one that had been planned for the visiting engineers and their guests.

Mining Decisions

SEVERANCE OF MINERAL RIGHTS—ADVERSE POSSESSION

One who purchases surface rights to land and then forms the intention of holding the mineral interests therein, does not establish adverse possession unless his claim is open and notorious. The mere use and possession of the surface is not enough to constitute adverse possession of the mineral rights, even if he actually mines coal for domestic purposes only, where he is entitled to mine for domestic purposes under his deed. Paying taxes on the surface raises no adverse claim to the minerals beneath it in such a case.

Pond Creek Coal Co. *v.* Hatfield (Kentucky), 239 Federal, 622. February 6, 1917.

OIL AND GAS LEASE—ONE DOLLAR CONSIDERATION SUFFICIENT

An oil and gas lease was granted upon $1 consideration, together with a covenant on the part of the lessee to drill a test well within a year and pay royalties thereafter based on minimum rentals. Five months later, nothing having been done under the lease, the land owners declared the same void and without consideration and made new leases which were assigned to defendants. Held, on suit by the first lessee to enjoin any assertion of rights by the assignees of the second lease, that the one dollar consideration paid for the first lease was sufficient to support the same during the period in which test wells were to be bored and an injunction was awarded.

Lindlay *v.* Raydure (Kentucky), 239 Federal, 928. February 3, 1917.

ADVERSE CLAIM—ABANDONMENT

In an action brought on an adverse claim in patent proceedings, the defendant abandoned his application for patent to the area alleged to be in conflict after the suit was commenced and the plaintiff waived his right to secure patent to that area also. Held, that as the action then became merely one for possession and there being no proof of a conflict, it was an error for the court to direct a verdict for plaintiff.

Lucky Four Gold Mining Co. *v.* Bacon (Colorado), 163 Pacific, 862. March 5, 1917.

EDITORIAL

T. A. RICKARD, Editor

CONGRESS as yet has not taken final action to relieve locators from doing annual labor on their mining claims during the War. The Senate has agreed to the House resolution exempting those persons who may be mustered into service in the Army or Navy from the obligation of doing the statutory $100 worth of labor, and the Senate has adopted another resolution, introduced by Mr. John F. Shafroth of Colorado, which provides for acceptance, in lieu of the required assessment work, of an equal expenditure in producing supplies needed for the support of the Army or Navy or of the people of the United States.

REACTIONS between sulphur and oil are discussed by Mr. G. Sherburne Rogers in an article appearing in this issue. The source of the sulphur is of interest to the student of ore deposits as well as to the specialist in petroleum. For the most part it is traceable to waters carrying sulphates in solution, coming under the influence of reducing substances in the sediments and in the oil itself. A point of particular significance is the statement that oil will also reduce sulphur from pyrite. This has been suggested also by investigations in connection with the flotation process, it having been claimed that this reaction may exert an influence upon the surface of sulphide mineral particles, making them more readily floatable.

ALUMINUM has been selling during the War at approximately 60 cents per pound. In June, 1914, it sold for 17.5 cents, and the upward trend, beginning in April, 1915, was like a balloon ascension. This was occasioned by its demonstrated value in the making of explosives. It is significant of the general inflation of prices, in no wise corresponding to increases in the cost of production, that Mr. Arthur V. Davis, president of the Aluminum Company of America, has offered to supply the Government with all the aluminum it needs at 27.5 cents per pound. He is said to have arrived at this figure by adding 2 cents to the average market-price of the metal during the past decade. Mr. Davis deserves commendation for the merit of frankness in revealing one more truth regarding costs of production.

OUR Toronto correspondent, in last week's issue, records a judicial decision giving damages to residents in the Sudbury district for injury done by smelter-fume, but refusing to grant an injunction against the two smelting companies in that famous nickel-mining district because the Court was of the opinion that "individual rights could not be maintained against the interests of the whole community." The Court refused to destroy the local mining industry even to save a few farms. This seems just and for the public good. The smelting companies have expressed willingness to buy at a liberal price the lands of those that claim to be suffering from the effect of the smelter-smoke. It is time that the public showed resentment against the blackmailing of a basic industry, the industry that in many localities was the chief cause in giving a value to the farm-products by affording employment, and therefore buying-power, to the local population.

SILVER is selling for 81 cents, a price not quoted since 1892. The reasons for the rise have been stated in these columns on several occasions: first, they are the increased purchasing power of China and India, both of which produce commodities that have been in steadily growing demand and for which, according to their custom, they ask silver in exchange; second, the coinage of silver for the use of the armies in Europe, particularly those of Great Britain, France, and Russia. To these now must be added the United States; for it is the purchase of silver by our Government that is the proximate cause of the recent rise. San Francisco has become the principal point of export for silver, instead of New York. The Mexican, Canadian, and American production goes this way to the Far East, in order to avoid the submarine menace in the Mediterranean. London still fixes the price of silver, because business there opens five hours earlier than at New York. The improved prospect for the metal should not only benefit Tonopah and Cobalt, the two chief silver-mining districts in northern America, but it should swell the profits of the copper-mining companies, many of which, particularly in Montana, recover silver as a by-product. Another consequence is to help Mexican finance, for the peso is appreciating rapidly. The rise may also put it into the head of the Mexicans to assist and stimulate the mining of silver ore instead of looting right and left. A revival of silver mining in Mexico would help greatly to extricate that country from its difficulties.

MEETINGS of scientific societies have been discouraged by the leaders of a number of these organizations on account of the War. The American Electrochemical Society, however, has vigorously opposed such a policy, and we agree that the helpfulness of personal contact, the inspiration gained through an exchange of thought between the men who are doing the world's work, and the dissemination of practical knowledge that is elicited in discussion, should not be sacrificed at a time when the country needs more abundantly than ever the advantage of these quickening influences that develop proficiency and stimulate inventive ability. Under the stress of these abnormal times, we are learning the high patriotism that resides in work for the common welfare, and gatherings of scientific men will assume a new importance. The banquet and the oratory will be subordinated to more serious things. Never were conferences of technical men more justifiable than at the present moment, when exchange of information is urgently needed. This is the basis of intelligent co-operative effort, and we may add that the spirit of solidarity should reach further. The tendency to erect a medieval wall of secrecy around many metallurgical and chemical works, that has grown more pronounced in recent years, is distinctly opposed to progress. The open door to men of appreciation and understanding brings its reward through enlightening criticism and suggestion. The example set by Dr. James Douglas, who extended a cordial welcome to every intelligent visitor at a time when a contrary custom almost universally prevailed, is worthy of emulation; his policy of free-trade in technical ideas did not hinder the economic development of the Phelps-Dodge industries. A similar policy will help all America in the winning of this War.

LAST week our Leadville correspondent sent a most interesting account of the hearing given by the State Industrial Commission in Colorado to an argument made by representatives of the local union when demanding higher wages. The Commission found that the cost of living had not risen as much as claimed, that the operators do not receive anything like the full benefit of the increase in metal-prices, and that the margin of profit is so narrow that any considerable increase of cost would jeopardize local industry. This reminds us that the legislature of Colorado at its last session appointed a Smelter and Ore Sales Investigation Committee. This committee was ordered to investigate the smelting business and report to the Public Utilities Commission before January 1, 1918. The committee of investigation is described by mining engineers in Colorado as "pretty good," which means above the average. The members of the committee are all men well informed in mining affairs and likely to make an intelligent enquiry. They have been authorized to investigate all custom-mills, smelters, and sampling-works; also to ascertain whether the customers receive weights, moisture-deductions, samples of ore, and assays; also to find out whether proper prices are paid for copper, lead, and zinc. For instance, they are expected to be curious concerning the payment by the American Smelting & Refining Company of 4½ cents per pound for lead when the market-price stands at 10 cents. They have been requested to make recommendations for correcting any wrongs they may detect, and if they do this, it is expected that they will suggest the placing of custom-mills and smelters under the Public Utilities Commission of Colorado, after submitting the proposal to public vote on a constitutional amendment. The investigation is backed by the Metal Miners Association of the State. We understand that the dominant smelting company has already made sundry concessions, and we are informed that the most unscrupulous of the Denver papers—it is not necessary to specify further—has already attempted to blackmail Mr. Simon Guggenheim. The report of the Committee is awaited with keen interest.

DURING the past week the I. W. W. movement has gone through a serio-comic phase. It may seem comical for the outraged citizens of a mining community to take the law into their hands, herd a lot of anarchists and loafers into an enclosure, and then ship them on a train to some distant point, 'passing the buck' to another community, preferably a neighboring State. It is a serious matter when citizens, in order to assert law and order, at a time when the naval and military forces of our country are fighting to "make the world safe for democracy," feel compelled to stultify government of the people by arming themselves and using force to eject an undesirable element from their midst. Viewed from any standpoint it is deeply regrettable that matters should have come to such a pass, and somebody is to blame for it. The only good that has resulted is the opening of the eyes of the public to the seriousness of the crisis and possibly to some realization of the forces of misrule set in motion by this organization of men unwilling to work and unwilling to let others work. The argument based on the high cost of living and the rest of it has been exposed as a mere pretense. They want the three sixes, namely, $6 wages, six hours of work, six days per week, and when they have obtained that they will demand $7 wages, five hours work, and a five-day week. Even this will not be the limit. No thoughtful observer of the conditions obtaining in the mining communities of the West at the moment can any longer regard the excuse for strikes as founded on the reasonable aspirations of honest labor. Whether financed by German money or stimulated by Enemy propaganda, or not, the whole campaign of these I. W. W. agitators is calculated to give aid and comfort to the Enemy, and therefore it is treasonable, not to be endured, and calling for prompt action by the executive. Much of the trouble, as usual, is due to politics. During the week the disturbance has spread beyond the immediate vicinity of the mines. Even around our own academic environs, at Berkeley, we read that "to prevent the threatened I. W. W. invasion of this county, the sheriff has stationed duputies upon all the roads leading from Contra Costa county," because "the I. W. Ws. have threatened to burn the county's grain supply." Things have come to a pretty pass when we are menaced by a Boxer rebellion or a Villa raid, as it were, in the heart of an orderly community. We are engaged in a great war; it is time to put a summary end to these antics and assert the dignity of this democracy.

The Mexican Menace Again

Our Mexican correspondent this week calls attention to new regulations affecting the producers of oil that would be confiscatory, unless we and our Allies oppose the taking of their property by administrative process. The present situation is not by any means new; we have been pointing out for months what was coming, and it is possible that the Department of State has not been idle. Not long ago Senator Frank B. Kellogg, at a hearing before the Public Lands Committee on sources of oil, asked if we might depend upon Mexico for this important necessary, and Commander James Richardson responded in the way of the man whose business it is to win a fight when he has gone into it, "Don't you believe that if it depended on getting oil from Mexico we'd get it, even if we had to take Mexico to do it"? It is this sort of straightforward talk that has helped to bring clearly to the mind of Carranza the expediency of trying to make a virtue of necessity. The peculiarity of the Carranza mind is that it is essentially subtle, and, just as he has done so often in his revolutionary career, he is now resorting to clever diplomacy and to what might be called the strategy of statecraft to wring success out of failure. We confess to no little admiration for the astuteness of this Mexican fox; if we cannot comprehend his moves and circumvent them we deserve to be beaten at the game, and to accept the fate of the geese.

The operators in the Mexican oilfields fully understand what they are contending against; they have perceived it ever since the new constitution made at Querétaro declared the nationalization of the oil-lands. To put oil on the same footing as gold, silver, copper, and lead was to recognize the ancient right of the sovereign state in mineral resources, and to give to anyone the privilege of entering upon land, wherever held, to explore for oil, and, when found, to 'denounce' it according to the prescriptions of the mining law. In due course a leasehold title would issue. The plain intent of the constitution was to make this retroactive, the civil code to the contrary notwithstanding, since this was in line with uniformity of procedure in departmental control of oil-production, and theoretically it should work no legal hardship upon the present holders of oil-land leases; it would merely necessitate their repairing to the local mining agent to file their applications for so-called mining titles. Of course, it would mean that enormous sums would have to be paid immediately into the Government coffers in the form of fees accompanying the denouncements; Mexican surveyors would be given abundant employment in delimiting the claims in accord with the mining statute, and later in proving the monumenting of these areas. Even in normal times this always has been a heavy expense, and one can imagine what it would entail in times such as those brought about in Mexico through the turmoil of revolution, with the starvation produced through cessation of industrial activity, ending in the supremacy of a clique that knows it cannot last long and [illegible] fatten while it has the chance. Moreover, the new system subjects the oil-claims to taxation under mining laws that have been seriously impaired by decrees issued by the First Chief before he made himself president. These not only have increased the taxes but have made them cumulative proportionately with an increase in area. It would be well for anyone having business in Mexico, whether in oil or in other enterprises, to acquaint himself with these dictatorial decrees. They may be had in Spanish and English in bound form, indexed with reference to the subjects treated. In no other way is it possible to ascertain what Mexican law is at the present moment. The decrees will show at least the situation at the time this publication was issued; by acts of the Mexican Congress the Department of Hacienda is empowered with extraordinary functions, which are legislative as well as administrative, so that decrees are still coming forward and the status of any industry may be seriously altered over-night.

The foregoing is an example of the utter disregard of titles to real property that exists in Mexico; the whole foundation of industry is unsettled by it. In the case of the owners of oil property, it is clear that there would be many a slip in the routine of putting their applications for leaseholds through to what the Mexican calls a *titulo;* something is sure to go wrong in the process of *tramitacion*, that is, in the wearisome bureaucratic red-tape leading at last to the signature, alongside the seal and blue ribbon, by the departmental head. In the discovery of technical error in legal proceedings the Mexican is a past-master. The door is thereby opened for infinite graft, for burdensome fines, and for wholesale confiscations accomplished through officially declaring applications invalid. It is hardly to be supposed that the original applicant would have time to correct his denouncement to the satisfaction of the highly discriminating officer before an application by favored competitors, under effective protection of scheming bureaucrats, would be filed with the mining agent. In our comments upon the new constitution prepared by Carranza we showed months ago what the effect would be upon the oil industry. The nationalization of the oil-lands was arranged in order to carry out the game that is now being played openly. The interference with mining companies, as detailed in our editorial pages last week, was merely a part of the same programme, intended to bring Mexican affairs prominently before the people of this country so as to increase our anxiety in preparation of the public for the next move. What this is now appears through the fact that Carranza, who has long been hinting at financial assistance from American bankers, has recently secured from his congress authorization for bond issues amounting to ₱300,000,000. With this authorization he is ready to offer exemptions from interference with mining and oil companies, and to assume the rôle of a defender of foreign interests in his country. Dispatches from Washington since the arrival there of Mr. Henry P. Fletcher show that our diplomatic representative has not been asleep. Mr. Fletcher has sounded a warning against the intentions of the Carranza government, definitely affirming that its purpose is the confiscation of the oil-lands. If we concede the demand for a loan we shall be obliged to pay the bills for running the Mexican government as a private monopoly of Carranza and his followers throughout the War. We may be forced to undertake the operation of the Mexican republic on the side, but, if we do, it will not be in the way that Carranza has conceived. We shall run it as a real democracy for the welfare of the Mexican people, not for a clique of grafters.

The Physics of Flotation

During the recent trial at Butte, Mr. Wilder D. Bancroft was asked, "Is that a book to which you would refer as an authority?" "Oh, dear no," he retorted; "I would not refer to any book as authority." It is not clear whether Dr. Bancroft assumed, quite properly, that he was better informed on the matter than any book extant or whether he meant that no book on a subject so obscure could be authoritative. We have read the evidence given by him and his talented colleagues in the latest litigation over flotation patents, and we acknowledge the value of their contribution to current knowledge; and yet it must be confessed that the physics of flotation still lacks scientific elucidation. We are nearer the truth, undoubtedly, for several truths have been elucidated by experiment and induction, but we have not arrived at a coherent hypothesis. By 'we', of course, we mean the whole body of earnest enquirers. Some fallacies have been exposed. That is so much to the good. The 'critical point' has been thrown into the limbo of false assumptions; the electro-static theory has slunk into the dark; the adhesion of air to metallic surfaces has been retired into the background; but who can tell us what is 'oil', that is, what is the characteristic of the oil that causes it to function favorably in flotation? The answer is coming, for metallurgic froth is being made by substances that are not 'oil'. Who can define 'metallic' lustre, not in the terms of mineralogy, but in the language of the froth-maker? What is the molecular arrangement in a surface capable of absorbing oil or in the substance that functions as oil? What is 'emulsification'? We confess to being among those that only put faith in scientific men that can define in simple terms; we do not believe that an archdeacon is defined when he is said to be a man that performs archidiaconal functions. That leaves us as we were—or slightly worse. One good sign, however, we detect, and that is the growing interest of engineers and metallurgists in the physics of flotation. Some there are that assume a supercilious attitude toward simple experiments and a top-lofty pose toward pure science in general, but they are unimportant. Mining departments still undertake to teach flotation without the aid of the department of physics, and metallurgists are half-ashamed to ask elementary questions at the door of the professor of physics, but even some of this shamefacedness is becoming changed by the spirit of co-operation, without which no man can accomplish anything worth while in this world. Much remains to be done. We urge teachers in mining-schools not to waste time in trying to study flotation without the guidance of those versed in modern physics. More particularly we advise those interested in the technique of the process to make the simple experiments that illustrate fundamental principles. If two or more men will make an experiment, and then discuss it, they will find how often they fail to see similarly, and how different may be the ideas suggested by their observations. From such experiments may come a clarifying discussion and the stimulation of scientific curiosity. In a branch of metallurgy in which theory is so hazy it is of the utmost importance to lay firm hands on a few facts, and that can be done best by making experiments. Consider, for example, how long Messrs. Sulman, Picard, and Nutter were able to bluff the profession with their statement that there resided a mysterious quality in a given ratio of oil and, what is more absurd, in a ratio that was proportioned to the 'ore', whereas the essence of the process was the differentiation of the constituents of the ore, the separating effect upon the mineral and the gangue. For 12 years this unscientific assertion remained unmasked; it was not until the early months of this year, after the Supreme Court had been misled into endorsing the absurd dictum, that several plants were operated under conditions that finally smothered the 'critical point' in oil, and ridicule. Remember how we were told that a certain patent, or the method supposed to be described in that patent, produced a thick, coherent, and persistent froth, so sustaining that a shovel could rest upon it, whereas other froths were so thin, flimsy, and evanescent that a match-stick would sink in them. We were told, by the Minerals Separation people, that their miraculous kind of froth was the only one that would float mineral successfully and that it could only be made by aid of a special kind of particularly violent agitation. Many believed it because few tested the assertion by experiment. Look at the fable of the greased needle. It was one of the amusing contradictions at which our childish imaginations boggled. The experiment was held to typify the flotation of mineral by aid of oil, until somebody showed that the ungreased needle likewise floated. It only remained for some iconoclast to prove that the making of a soap-bubble by a boy was due, not to the lowering of the surface-tension of the water, but to viscosity caused by the insoluble matter of the soap, or by something equally alien to drawing-room philosophy! Indeed, something of the kind did happen. Among the papers read at the Arizona meeting of the American Institute of Mining Engineers, a meeting that was in effect a symposium in flotation, there was one by Messrs. A. F. Taggart and F. E. Beach, called 'An Explanation of the Flotation Process'. This paper was writen by two professors in Yale University, it discussed the principles of flotation, and it stated, among other things, that "water displaces air more readily on an oiled solid surface than on a clean surface of the same solid." The subversive statement was fortified by the description of an experiment with an aluminum ring, which sank when oiled, but floated when unoiled. This, of course, was rank heresy. Given forth at that meeting of enthusiastic votaries of the new process of flotation, it was like rushing into a tea-party and shouting that muffins predisposed one to measles. Nevertheless, the thoughtful statement of two undoubtedly clever scientific men was allowed to pass without comment or criticism. In vain do we search in the record of the discussion at Globe for any reference to this paper. It could not be for lack of respect for the authors of it nor for want of interest in their opinions. We venture to say that the failure to refute what seemed an error was due to the insecure foundation on which the theory of flotation rested and the unwillingness of mill-men and metallurgists to venture a rebuttal. The failure to confirm or to disprove so momentous an assertion as this—that a metallic surface is more readily wetted when oiled and that presumably a metallic mineral is less floatable when oiled than when unoiled—is a reflection upon the Institute. It illustrates how easy it is to fill bulletins and volumes with reading-matter, and how difficult it is to obtain real criticism, without which we only swamp ourselves with a undigested mass of material—as if a haystack fell upon us. We have made experiments to test the extremely interesting statement of Messrs. Taggart and Beach, and we think these gentlemen wrong, if we interpret them correctly, but the point is that every student of the subject that has read their paper should be able by this time to express his own opinion upon their statement regarding the effect of oil. If we do not know what that effect is, we are astray on one of the elementary, although possibly not fundamental, explanations of flotation. So we say that the understanding of the process, and the further understanding of the physics of it, must depend upon intelligent experimentation and the scientific discussion that will ensue, and, above all, frank criticism.

Misfires

The proper manner of loading a hole for blasting would seem to be a simple matter, but the numerous articles and letters that have been published in the MINING AND SCIENTIFIC PRESS during the last six months indicate a wide variety of opinion. The object universally sought is efficiency with safety, and that is attainable through practice conforming to the lessons of experience. It is important to note that valuable experience in blasting comes not alone from the daily work of the miner, but is derived equally from the elaborate tests made by manufacturers of explosive materials. The sum of all these observations, critically examined and sifted by trained technologists, has resulted in the development of a body of rules that may be accepted as authoritative. The wrong way of using explosives is more general than the right way, and faulty custom is difficult to correct because of prejudices held by the miners. Accustomed as so many of them are to wrong methods of charging holes, they resent attempts by their foremen to shake them from familiar habit. The manager of a mine, however, is entrusted with responsibility for securing an economic result. He cannot concede free rein to his workmen in using or misusing materials according to their whim. A misfire represents not only a waste of explosive, but loss of time on the part of employees in searching for the unexploded charge or in digging it out; the cost of drilling the missed hole has been thrown away; and the ground broken per pound of powder on that shift is less. That is not the whole of the indictment. The majority of misfires are the result of methods in themselves opposed to the most economical use of explosives. Errors culminating in a misfire involve a loss in efficiency distributed through the entire round of shots. Under the best management scrupulous attention is given to the proper use of explosives, and the tonnage of ore and waste-rock broken per pound of powder and per foot of fuse are subjects of daily record.

Mr. E. F. Brooks rightly insisted on the use of high-force caps. He recommends 5X and 6X. His method of inserting the cap into the primer is similar to one suggested by the Bureau of Mines, punching a diagonal hole with a stick near the end of the cartridge; but he advises loading another stick of powder on top of the primer. Mr. W. S. Weeks criticizes this practice, as does Mr. Edward Higgins, adhering to another recommendation by the Bureau of Mines, with which apparently all powder manufacturers are in agreement, that is, to place the cap centrally in the end of the last stick in the hole, tying the paper of the cartridge-end around the fuse. As expressed by Mr. Weeks, the cap should point in the direction in which the wave of detonation is to be propagated. That accords with the principles governing the detonation of high explosives, as demonstrated experimentally by Bertholet. The explosion of a charge of dynamite or blasting gelatine, though it may seem to be instantaneous, is not so in reality. A measurable time-interval elapses in the transmission of the wave from its initial point to the end of the charge; moreover, the wave progresses in the direction of the initial impulse; it does not expand uniformly in all directions from the place of origin. Unlike the spherical waves of compression generated by an earthquake shock, it is rectilinear in its motion. The fulminate is exploded by the fire spitting from the end of the fuse, that is, by an incandescent spark. The heat dissociates that part of the fulminate with which the fire comes into contact, generating still more intense heat by the highly exothermic reaction; the explosion proceeds through the train of sensitive molecules with increasing velocity to the bottom of the charge, from which a blow of enormous intensity is delivered in the direction of propagation. This is easily demonstrated by the familiar experiment of placing a cap, with fuse attached, in a hole bored through a block of one-inch plank, set upon a sheet of mild steel. On exploding, the cap will punch a hole through the steel without doing serious injury to the wooden block.

The velocity of the initial impulse determines the

velocity of the detonating wave transmitted through the charge of powder. The heat generated, and the expansive effort of the confined gases that develop the explosive energy, merely react upon any part of the powder lying out of the path of progression of the detonating wave. The result is an explosion of a lower order for that residual portion of the charge. The degree of explosion depends on the velocity of the wave, and the velocity of the wave that causes detonation of the explosive lying in the path of propagation is not transmitted to the powder behind the cap. The expansive effort, on which the explosive effect depends, is directly due to the velocity of the chemical reaction, and any portion of a charge that explodes with less rapidity than the rest fails to deliver its full quota of potential energy in the form of useful work. It is apparent, therefore, that the proper place for the cap is at the end of the last stick of powder, otherwise called the primer. It is also certain that to place it elsewhere is to waste the powder.

Our correspondents display a variety of opinion regarding the likelihood of 'side-spitting' from fuse when bent. The fact that it does not happen in the majority of cases when fuse is subjected to severe handling merely argues for the excellence of its manufacture. If all articles in common use were as honestly made as explosive materials the world might draw moral inspiration from commerce. Although "the function of an explosive is to explode," as was laconically stated by Lieut. Walker of the United States Navy, the extreme care taken by the manufacturers to afford an ample margin of safety against the vicissitudes of handling such delicately balanced agents of destruction, permits rougher usage than intelligent caution might commend. Mr. Higgins points out that the practice of inserting the cap in the wrong end of the cartridge, and doubling back the fuse, is the fertile cause of misfires, and misfires are responsible for one-fourth of the fatalities overtaking those engaged in mining within the State of California, which State does not stand alone in this respect. Fuse not infrequently does spit fire through the walls; seldom will a coil cut into 3-ft. lengths, and tested, fail to yield one example of this defect. That should be sufficient warrant for taking pains to avoid any chance of accident arising from this source. The advice of manufacturers is so to load a hole that the fuse may not come into contact with the powder. The fuse should be dry, as our correspondents insist; it should be cut with a sharp clean knife, never with the scissors-type of cutter often combined with a crimper, because the fuse becomes pinched, shutting off the powder-train and causing a side-spit near the end that may fail to explode the cap; the cut should be square across the fuse; the cap should be placed upon the fuse, preferably holding the latter upward and drawing the inverted cap down upon it. Just as the housewife will explain the difference in efficiency between threading a needle and 'needling' the thread, so is there a difference, tending to security and efficiency, between 'fusing' a cap, and 'capping' a fuse. The double crimp is preferable; it insures holding the fuse in firm contact with the fulminate in the cap. The man who bites the cap upon the fuse deserves to lose his lower jaw, as he frequently does.

We venture to say that any mine breaking as much as 100 tons per diem of ore and rock will find it economical to employ a man to make primers, and distribute all explosives to the miners. In no other way can the proper making and use of primers be insured. The 'powder monkey,' as this employee is generally called, can be trained to observe all the proprieties in making reliable primers, and in tracing the inefficient use of blasting materials if he be given responsibilty as powder-foreman. It then becomes possible also to make primers of powder containing a larger proportion of nitro-glycerine, which insures a higher order of detonation, developing increased useful effort in the lower-grade powder used in regular blasting. Primers of 60% dynamite, three inches long, with 6X caps, will be found to effect a saving in powder per unit of rock broken. The wooden borer with a shoulder to limit the depth of hole for the cap, not only protects against premature ignition of the charge by side-spitting from the fuse, but it prevents an air-space being left beneath the cap. Insignificant as it may appear, the tiny air-cell left under a cap when a pointed object is used to make the hole reduces the initial force of the explosion by reason of the compression of that air before it can transmit the detonating wave. It reduces the velocity and the order of detonation, which means loss of power, waste of powder, less ore produced per pound of dynamite.

Tamping opens another interesting question, where room for argument may exist. The suddenness of detonation of a high explosive develops almost instantaneously an enormous expansive effort. On account of its suddenness the maximum resistance of the air due to its inertia is supposed to be realized. With unabsorbed nitro-glycerine this result is approximately obtained, especially when the superficial area of the explosive exposed to the air is considerable. The case is different with practical blasting-powders; their velocity of detonation is highly modified by the absorbent. Moreover the projection of a slender column of air into the surrounding atmosphere introduces new physical conditions; the power required to impel a jet is different from that required to overcome the inertia of the air surrounding a 'sand-blast' in bulldozing. Plastic tamping is first compressed and tightened in the hole; then the friction developed between the tamping and the wall, under the enormous suddenly applied pressure due to the detonation, makes the stemming almost as resistant as the solid rock.

The proper application of explosives deserves consideration by all thoughtful engineers. It is a fruitful subject for discussion, having to do with the protection of human life against carelessness and ignorance as well as touching, at so many points, the commercial result of the complex operations of mining that finally are centred upon the relation between the cost of the ore broken and the price obtained for it.

THE BUTTE & SUPERIOR MINE AND MILL

Butte Re-Visited

By ROBERT E. BRINSMADE

When I landed in Butte recently, there had elapsed just 16 years since my last sight of the city. The train that brought me, the Columbian express of the Chicago, Milwaukee & St. Paul system, with its steel cars and its electric locomotives, was an indication of the engineering progress that the interval had produced, a record of achievement in transportation that the mining industry of the locality might find difficulty in matching.

In the business district around Main street and Broadway, skyscrapers have arisen to keep company with the solitary Hennessy block of former days. The luxurious post-office, the regal county courthouse, occupying nearly a whole square, the Silverbow Club with its tasteful façade, as well as new hotels and apartment-houses, indicate that builders have become hopeful of the future of Butte and are staking their fortunes with the belief that its stability as a centre of population is assured, whatever may be the ultimate fate of its mineral resources. Indeed, the change from the Helena to the Butte branch railway for the passage of the main-line traffic of the Northern Pacific system, and its selection, in spite of the difficult topography, as the route of the new Chicago, Milwaukee & St. Paul extension to the coast, demonstrate that the city has finally shed the swaddling clothes of a transient mining settlement.

On entering Butte from the south-west, no vestiges are visible of the old Colorado smelting works, while the adjoining plant, the Butte Reduction Works, remains only as a cold and lifeless reminder of its erstwhile productivity. Farther south, the Flat, formerly so bare and forsaken, is becoming populous as a residence section, and the same may be said of Silverbow valley on the east, where, in place of the three smelters, the Montana Ore Purchasing, the Butte & Boston, and that famous belcher of pungent sulphur-smoke, the Parrot, there functions only the infant plant of the East Butte company. Not only have the buildings and machinery of the famous old smelters been removed, but what still remains of their tailing and slag-piles is being shipped away so rapidly for re-treatment during the present bonanza era that any future delvers in the 'ruins of ancient Butte' will find few clues to the activities of former inhabitants.

The Anaconda and St. Lawrence mines, as well as adjoining shafts of the 'richest hill in the world', covering the Anaconda lode-system, seem to have changed little in superficial appearance. The existing steel head-frames had all been erected in 1900, and many of the adjoining long lines of iron boiler-stacks are still standing. Though the mines have drawn electric current from Great Falls and other Montana water-powers, the old boilers, when hardly worth moving elsewhere, form a valuable reserve in case of line-troubles. To the north, at the High Ore shaft, have been erected the great buildings that house the mighty electric compressors, which supply air to the re-constructed steam-hoists of the deep shafts of the Anaconda company. On Syndicate hill, farther west, the old land-marks, the Bell, the Diamond, the Green Mountain, the Mountain Con., and the Coulin mines, are still active, and have been supplemented by the great new Beaver State mine. This last is a visible evidence of the extension of the copper zone, which formerly was thought to be as contracted as the throat of a sperm-whale. It has now been followed to the North Butte, the Butte-Ballaklava, the Tuolumne, the Tropic and the East Butte on the north and east, and to the Butte-Duluth, the Bullwhacker, and the Davis-Daly on the south. In 1900 the Silver Bow No. 3 vein was believed to be the southern, and the Continental fault the eastern boundary, of the copper-zone, even though Heinze had already begun to work east of Columbia Garden in what later became the productive Receiver mine. Thus the work of the past decade has demonstrated the fallacy of such a belief. Boring through the sediments of the Flat into bedrock has demonstrated the presence of copper veins far to the south, and such productive mines as the Bullwhacker and the Butte-Duluth have proved the presence of abundant copper to the east of the Continental fault. These discoveries caused a boom in the corresponding mineral rights, so that owners of town lots around the Race Track on the Flat found themselves in clover, and the claim-owners near Columbia Garden unloaded their undeveloped holdings upon the North Butte company for over a million dollars.

Even as late as 1906, the Butte report of the U. S. Geological Survey affirmed the non-existence of commercial oxidized copper ore in Butte, and the dictum was then well founded, for the outcrops of the marvelous veins of Anaconda hill, with their hungry iron-stained quartz, show less signs of the red metal than a German kitchen in war-time, and even the copper-stained Syndicate outcrops farther north have scarcely any ore of value till the sulphide-zone is reached. Yet the opening of the great granite stockworks of 5% oxidized copper ore in the Bullwhacker and Butte-Duluth have discredited the former dogma and have brought to light the belt of oxidized ore that lies east of the Continental fault, and extends south from Park canyon for more than a mile. The oxidized veins are sub-vertical and mainly contain chrysocolla and malachite, both in the veins proper and in the disseminations of their granite walls. They have been mined open-cast, by Lake Superior methods, and treated by leaching with sulphuric acid in the Butte-Duluth mill of Captain Wolvin, at the rate of 500 tons daily. Their oxidized filling extends to water-level, which is at 300 ft. in the Butte-Duluth, and then merges into the typical Butte copper sulphides; in fact, these distant veins have the easterly strike of the great veins of Anaconda hill. It appears plausible to believe that the oxidized-copper zone depends on the Continental fault, which, being 200 to 1000 ft. wide, and with a vertical throw of 1500 ft., has generated quite different conditions for surficial leaching between its opposite sides.

The saying that 'Fools rush in where angels fear to tread' has perhaps never had a more unexpected and happier outcome than in the recent history of Butte; the two 'fools' in this case being Ralph Baggaley and Captain Wolvin. The first, a Pittsburg steel-man, and consequently a novice in copper, was so ignorant as to insist that copper matte could be blown in basic-lined converters. Backing his belief with his large fortune, he conducted costly eperiments, and finally had the satisfaction of accomplishing the impossible and of proving himself less a fool than his critics. Captain Wolvin had gained title and fortune as a ship-owner at Duluth; so, a decade ago, when he entered the race for mining honors at Butte, he was booked by the wise-acres far down on the list of 'rank outsiders'; but the Captain, undaunted by scoffers, took hold of the old Black Rock silver mine, organized the Butte & Superior Copper Co., and proceeded to sink for another copper bonanza. Though a fool's luck did not give Wolvin a copper mine, it soon handed him a body of zinc that has out-classed in value half the copper bonanzas of the world. After the completion of a concentration mill and the flooding of the Kansas smelters with its output, he found himself able to sell his interest for a million profit, and he then embarked as boldly, but with less financial success, in the opening of the Butte-Duluth mine. The new owners of the Butte & Superior omitted last April the word 'copper' from their company's name; evidently they were too profitably united to Dame Zinc to desire further flirtations with Mistress Copper. Yet Captain Wolvin was not 'going it blind' at the Black Rock. The astute Heinze had, as early as 1902, found rich zinc ore beneath the silver-manganese ore of the old Lexington mine, and this discovery established the probable existence of zinc at depth in all the silver mines of the Rainbow lode-system extending eastward from the Amy and Alice mines to the Continental fault. The huge chimney of zinc ore, 300 ft. wide, that is now supplying the 2000-ton mill at the Black Rock mine seems to have been formed by the step-faulting of a very wide vein, and is thus not essentially different in structure from the step-fault chimneys of the Anaconda lode-system, as seen in the West Colusa or Minnie Healy mines.

The Black Rock zinc ore is a mixture of sphalerite, galena, pyrite, and quartz. As sent to the mill, it assays about 16% zinc, 2% lead, and 12 oz. silver, and produces 25% of its weight as a concentrate with 56% zinc, besides yielding a little high-grade galena. Since the adaptation to this ore of the Hyde flotation process, and the consequent saving of over 90% of the zinc, the Butte zinc-belt has attracted more attention than the copper, because less explored and therefore of more romantic possibilities. When former Senator Clark, with his usual luck, found himself owning the western extension of the Black Rock vein, in his Elm Orlu ground, he proceeded to explore his bonanza, and later built the Timber Butte 500-ton flotation-mill that is now in operation on the first hill-slope south of the Flat.

The greatest ultimate beneficiary of the pioneer work of Heinze and Wolvin may prove to be the Anaconda Copper Mining Co., which seems destined to be as important in the world of zinc as it has been for three decades in that of copper. Not only has Anaconda long controlled such famous mines of the Rainbow lode-system as the Alice and the Lexington, but it has lately acquired the Nettie silver mine, west of the Big Butte intrusion of rhyolite, and the Emma, south-west of town. In all of these Anaconda has developed zinc orebodies, and soon this company's new electrolytic zinc works at Great Falls, producing 100 tons of spelter daily, will depend on their output of ore; in fact, Butte bids fair to possess enough natural resources to make her for an indefinite time a rival, as a zinc-producer, of the famous Broken·Hill lode in Australia.

Though development has shown that the silver veins change in depth to zinc, the hypothesis that the zinc ore changes in turn to copper is not yet fully demonstrated. The fact noted by Sales* that the great copper veins of Anaconda hill contain less copper and more zinc as one recedes from the central copper belt does not nullify this hypothesis. The Gagnon, the Beaver State, and the Speculator furnish good examples of zinc changing to copper in depth, and there are said to be similar indications in some of the new zinc mines. It may be that the copper belt merely comprises those veins that have been the most corroded and that have thus lost their upper silver zone and their middle zinc zone. This corrosion would result in their existing apices still persisting in the copper zones. If such be the fact, a prolonged span of life for Butte as the world's leading copper district is well assured.

AT THE COLLAR OF THE SHAFT

Underground, Butte adhered steadfastly to square-set stoping until 1916. The original excuses for adopting this expensive practice were the cheap timber of western Montana and the ready adaptation of this system to the huge soft irregular orebodies with their resemblance to those of the Comstock lode, whence came many of the early Butte miners. Timber is now growing scarcer, and even the general introduction of the old Gagnon scheme of using round, instead of square, timbers for the sets, has given little relief, in view of the doubled output of ore; but, owing to the presence of many valuable surface-structures, any timberless system of mining that involves caving the surface is not permissible. At present the

*Trans. A. I. M. E., Vol. XLVI, p. 3.

Anaconda company is employing a sort of 'rill system' in some of the veins that are not over three sets wide, such as the Beaver State. Levels are spaced 200 ft. apart and then connected by three-compartment raises, 50 to 100 ft. apart, timbered with square sets. The central compartment of each raise is a man-way, one side-compartment being used for sending out broken ore and the other for admitting waste intended for filling. Stoping starts at the first floor of a waste-compartment by slicing the back on an incline of 30° along the vein, that being about the angle of repose for broken rock. The back is drilled by Rand-stopers, and the ore is broken down and dropped through the plank-lagging into cars on the drift below. The excavated space is then filled with waste dropped down the raise from the level above, and is then covered with plank, spread on its angle of repose, and with enough space between plank and back to permit the miners to drill the vein for another slice. After dropping the ore of the second slice into the cars below, another fill of waste is run in above the uncovered waste, and the planks are then put back as a cover to separate the waste from the broken ore of the third slice. When, after two or three slices, the waste-floor extends as far as the next raise, the broken ore can henceforth be run into the ore-compartment, and dropping it direct into the drift can cease. The slicing process may then proceed upward, ejecting the broken ore from the chute at one end of the 50-ft. stope, and drawing down the waste for filling from the other end, until the level 200-ft. above is reached. By this system sorting can be done within the stope, leaving the reject there as filling. The back, of course, must be kept well arched and carefully scaled down after each blast before starting to drill.

The earlier custom of drilling either by piston-drill or by hand is now obsolete. The old 265-lb., or even heavier, piston-drill has been replaced by the 165-lb. water-Leyner for driving horizontally or sinking. For cutting out square-set stopes with piston-drills, breast-stoping was most convenient; but now that Rand-stopers with solid bits are used, back-stoping must be employed. For breaking dry holes the cheaper ammonia-powder has been found to be as good as the nitro-glycerine type, and at the Leonard mine about half of each kind is consumed, the average service being 1.8 tons of broken rock per pound of explosive. For a daily output of 1400 to 1500 tons of ore and 300 tons of waste two shifts of one foreman, two assistant-formen, 26 shift-boses, and 568 miners and shovelers are employed, with a complement of 45 mechanical and other surface-men, or a total of about 700 men in the 24 hours. This gives an average output of over two tons per man.

In 1900 mules were used underground by the Boston & Montana company, and compressed-air locomotives by the Anaconda. Since the use of the hydro-electric current, the electric motor has become the favorite tractor; not having to deal with explosive gases as in coal mines it is possible to use a 440-volt pressure. When self-dumping skips were first introduced in 1898, they were filled by dumping the cars directly into them at each level. Now bins are excavated in the floor of each shaft-station, and the 4½-ton skips are filled with one run of the bin instead of with three dumps of the 1500-lb. cars.

When hydro-electric power was introduced a decade ago, it gradually replaced steam as a prime mover in the mines. However, each of the Anaconda's deep shafts was then equipped with steam-hoists, and to discard a dozen serviceable engines that had cost nearly $50,000 apiece meant a huge loss. The dilemma was avoided by continuing to use the hoists and moving them by air, compressed electrically, instead of by steam. This required only a special pipe-system to feed air to each hoist from a central compressor-plant, and the replacement of the steam-cylinders at the hoists by others that would act as air-compressors to force air back into the system whenever the load was negative. Each hoist is further equipped with a pre-heater for raising the air to 330° F. For all machines save the great hoists, such as drills, portable pumps, and timber-hoists, air is supplied from an individual electric-compressor at each mine. The ownership of numerous adjacent mines by the Anaconda company will permit a common system of levels to be established in all the new lower workings of a group. This will mean easier ventilation; also cheaper transport and drainage. Soon the water from all the mines of the Boston & Montana group will flow to the Leonard and be handled at its 1400 and 2800-ft. levels. At the higher level there are now installed five vertical electric pumps, each with five single-acting plungers and a capacity of 600 gal. per min. against a 1400-ft. head. The lead-lined wrought-iron water-pipes, formerly used in the Anaconda shaft to resist the acid water, are superseded in the later installations by the same pipes lined with creosoted wood. The pump-pistons and adjacent parts are still made of bronze, but a successful experiment was made last year in substituting porcelain for bronze in acid-water pumps.

In 1900 Butte was still in the throes of the consolidation of companies and of the famous lawsuits between Heinze and the Amalgamated company. The latter had been born in 1899 and within 18 months it had absorbed the Anaconda, the Washoe, the Boston & Montana, the Butte & Boston, the Colorado, and the Parrot companies. The lawsuits had been begun against Heinze in 1898 for alleged ore-stealing by the Boston & Montana company, and were a part of the 'assets' inherited by the Amalgamated when it bought out the latter company. Heinze had retaliated by 'carrying the war into Africa' and locating every fraction of unclaimed ground he could find within the central copper zone. He then brought suit against the Amalgamated for the ore removed from the veins, of which his new fractions were the 'apex'. Such are the vagaries of the famous apex law that Heinze's lawsuits, with their rich pickings for lawyers and experts, might have continued to this day had he not grown tired of the fight and sold his copper holdings to his opponent for a big sum. He then went to New York to beard the Standard Oil in its den. This last adventure was more rash than wise, and Heinze's brief plunge into Wall Street nearly wrecked him financially

GRANITE MOUNTAIN SHAFT, WHERE 171 MINERS PERISHED RECENTLY

MAIN STREET, BUTTE, LOOKING NORTH

during the panic of 1907, seven years before his death.

With the absorption, a few years later, of the Original, the Colusa Parrot, and the smaller copper mines of W. A. Clark, the Amalgamated had achieved the ownership of practically all the copper companies that were independent in 1898. Nevertheless the district is still far from being consolidated in the hands of a trust, since many new companies hold ground in the newly developed extensions of the old copper zone. Among these are the Davis-Daly, the East Butte, the North Butte, the Butte Ballaklava, the Tuolumne, the Butte & London, the Butte Duluth, and the Butte & Bacorn, without mentioning such zinc giants as the Elm Orlu and the Butte & Superior. The absorption by the Amalgamated, now the Anaconda, of so many Butte mines has not produced the dire results for labor that were freely predicted as certain to follow the entrance of Standard Oil capital. In 1900 the miners worked 10 hours by day and 9 by night, while smelter-men worked 11 hr. by day and 13 hr. by night. The State 8-hour law was secured by the unions as a by-product of their support of the Heinze-Clark combination against Marcus Daly in the election of 1900. This law is now well observed, and in addition much of the former unnecessary Sunday work has been abolished. A liberal State workmen's compensation law was recently introduced, and has proved a great stimulus to the safety-first movement in Butte. For rescue-work the Anaconda company alone has two stations and 60 Draeger helmets, while the U. S. Bureau of Mines has given valuable instruction in first-aid to resident miners.

It is curious to see Butte an open shop where once was the impregnable stronghold of unionism. There were two reasons for the prolonged prosperity of the Butte Miners' Union. The first was the favor of Marcus Daly, who used to say, "We may pay better here than anywhere else, but then we can afford to, and the bonus will keep them from getting jealous of us mine-owners who got here first and located the bonanza claims." The second was because this union was free from such vicious union practices as lead to restricted membership and output, and never tried to dictate who should be hired or who discharged. Provided all new mine-workers were required by the big companies to pay their union dues of $1 per month, and the union wages and hours were observed, the union kept the peace between man and master; but like all political democracies, the labor democracies called unions are apt to fall into the hands of a clique. This happened to the Western Federation of Miners, with which the Butte union was always affiliated. The Western Federation clique, justly or unjustly, incurred the enmity of a large number of Butte miners, and the ill feeling culminated on Labor Day in 1913, when the two factions came to blows, and during the battle the local union-hall was blown up by dynamite. Later the State militia had to be called to suppress the disorder, which had undoubtedly been augmented by members of the I. W. W., the American branch of syndicalism or revolutionary anarchy. Since that date the mine-owners have refused to make labor-contracts with either faction, and the miners have suffered nothing from the change in wages or hours. In fact, the sliding scale of payment, which grants a minimum daily wage to everyone underground of $3.50 and advances it 25c. for every 3c. advance in copper above 15c. per lb., has resulted in equalizing the income of the operatives with the advancing prices of commodities.

The cloud of expensive lawsuits that hung over Butte for a decade proved to have a silver lining. The studies made of the old workings, and the new pits and trenches driven for the elucidation of theories advanced by the mining 'experts', revealed so much of practical value in the finding of orebodies that the geological departments of the companies, founded originally for litigation only, have since continued in action as an indispensable part of their operations. Before 1900 the college graduate was viewed with disfavor in Butte; in fact, to get a job with many managers the less said about one's technical education the better. The practical Cousin Jack, or the 'Paddy-practical,' provided he came from Daly's home county, had his pick of jobs in the mines, mills, and smelters. The Anaconda was the happy haven for the Green Islanders, while the Parrot and the others with Cornish foremen favored men from Cymric England; but today all this is changed, and the technician receives the recognition to which his education entitles him. The Anaconda Consolidated Co. now employs as many in its strictly technical department as did all of the parent companies; while in its geologic bureau, it has over 18 engineers. Moreover it is the policy of this company to introduce into all executive positions, such as superintendent, foreman, or shift-boss, as many technicians as are found suitable. The Butte copper zone is sprinkled so thickly with big rich orebodies that almost every graduate from the end of a shovel could give satisfaction as a boss in the old days merely by getting out plenty of ore, but now 'book-larnin' is having its innings and the 'scientific dudes' are recovering from some of the 'exhausted' upper levels almost as much ore as was ever taken from them in their virgin days. The vast old tailing-piles and slag-dumps, the output of muscular milling and smelting, are also being largely re-treated and they often show a considerably larger metallic content than the old assay-reports indicate. It seems to be proving true that "Assays don't lie, but liars will sample." Besides the above recovery of unmined ore in the upper levels, it has been the custom since 1900 to extract, during all periods of high-priced copper, the filling of low-grade ore in the old stopes. Those dating from the 'eighties may run as high as 10% copper, and those from the 'nineties up to 5%. An old ore-fill is replaced by fresh waste, which is dumped from the cars on the level above, taking waste from cross-cuts in barren rock, and also from old dumps on the surface. Even though it is the custom to fill abandoned drifts as well as stopes with waste, the whole formation around Anaconda hill seems to be moving, and survey-plugs placed in the 'nineties are often found many feet away from their correct position in space.

Sources of Sulphur in Oils

By G. SHERBURNE ROGERS

*There are several possible sources of sulphur, though in the Californian fields sulphide waters are probably the most important. The surface and shallow ground-waters in the Californian fields carry large amounts of sodium, calcium, and magnesium sulphate, and outside of the oil-fields the deeper waters are also strongly sulphatic in character. The waters in and near the oil-measures, however, are almost or quite sulphate-free, and are usually solutions of carbonates and chlorides. Between the sulphate surface-waters and the sulphate-free waters associated with the oil, every gradation may be found; and near the horizon at which sulphate begins to decrease and carbonate to increase the waters usually contain hydrogen sulphide. As sulphate is abundant in the shallower waters everywhere along the Californian coast ranges, whereas sulphide is found only in the oilfields and near the oil and gas, it is reasonable to suppose that the sulphide has been formed through the reduction of sulphate by the hydrocarbons. The reaction supposed to be involved is usually written:

$$CaSO_4 + CH_4 = CaS + CO_2 + 2H_2O = CaCO_3 + H_2S + H_2O$$

Although the field evidence in favor of some such reaction is strong it must be admitted that it has apparently never been experimentally proved. In any event the reaction as written can be considered only as a condensed representation of the type of change that takes place, the intermediate stages in the decomposition of the hydrocarbons on the one hand and of the sulphate on the other being as yet unknown.

In some regions gypsum may be disseminated through the strata near the oil-measures and if taken into solution and carried to the oil may be reduced to sulphide. This is, of course, essentially the same as the reduction of sulphate surface-waters. Gypsum in the anhydrous condition, however, is a very stable compound, and even with an active reducing agent, such as carbon monoxide, a temperature of about 700° C. is required for its reduction. Whether the reduction of sulphate by hydrocarbons takes place or not, it is certain that many of the oilfield waters carry hydrogen sulphide or alkaline sulphide in amounts ranging up to more than 300 parts per million. The tendency of alkaline sulphide to become free H_2S, and the tendency of this gas to oxidize to sulphur, are well known. In this connection the following personal communication from Clifford Richardson is of interest: "Some years ago I collected in a sealed tube 200 or 300 cc. of a natural gas in Trinidad which contained hydrogen sulphide. This was allowed to stand for about ten years without observation, but at the end of that time it was found that the sulphur of the H_2S was deposited on the walls of the tube in colorless crystals." The oxidation of hydrogen sulphide proceeds even under very feebly oxidizing conditions, as on the floor of the ocean, and it doubtless takes place even in deeply buried strata. In the light of other corroborative evidence, it seems probable that considerable amounts of hydrogen sulphide are oxidized to sulphur, which is precipitated. As the strata directly above the oil-measures have not been tested for sulphur, this supposition cannot be definitely proved, but it is significant that small deposits of disseminated sulphur are not uncommon along the western edges of the Coalinga and Midway-Sunset fields. Moreover, a commercial deposit of sulphur has been found near the southern end of the Sunset field, in the same formation that contains the oil-measures in the field nearby. An interesting feature of this sulphur, to which my attention was first directed by E. A. Starke of the Standard Oil Co., is its intimate mixture with material containing hydrocarbon, which seems to constitute 20% or more of the amorphous substance.

It may be added that the waters associated with the oil in many regions are known to be free from sulphate. In many fields there are strong chloride waters which doubtless represent the sea-water entrapped in the sediments when they were laid down, and which therefore never contained a large concentration of sulphate; but in some fields the low chloride and the high carbonate indicate that the waters are in part altered meteoric waters from which considerable quantities of sulphate have been removed. In some Tertiary and Cretaceous fields in which the normal surface-waters are strongly sulphate in character the reduction of the sulphate by the stages outlined above may afford abundant supplies of sulphur to react with the oil. Hence, as meteoric waters carry oxygen and also salts that eventually yield sulphur, it is probable that in many regions waters are the chief agents in the alteration of the oil. The apparent increase in the gravity of oil that has been associated with certain types of water is recognized by many practical oil men.

As pyrite is said to react with and to yield sulphur to petroleum it is probable that in some localities the action of both pyrite and its less stable isomer, marcasite, have been important. These minerals have been found in many wells in the Californian fields. They probably formed in part during the deposition of the sediments, through the reducing action of organic matter on iron-sulphate solutions; but they may also have originated later through the direct action of hydrogen sulphide on

*Abstract: Trans. A. I. M. E., St. Louis meeting, 1917.

chalybeate waters. J. J. Hern, a Californian oil man of wide experience, told me that in wells in which large quantities of iron sulphide are found the oil below is likely to be abnormally warm. If this observation is well founded, it may be significant as indicating chemical reaction between the sulphide and the oil. In some regions sulphur is doubtless derived from still other sources. Hydrogen sulphide and sulphur di-oxide are common components of volcanic emanations, and the former is found in many thermal springs supposed to represent the last stages of igneous activity. Much of the Mexican oil, which is heavy, asphaltic, and high in sulphur, is found near igneous intrusions and may well have been affected by the sulphurous gases that doubtless accompanied them. The oil in the salt-domes of the Gulf Coast is somewhat similar in character and has evidently been altered by sulphur, but the origin of the sulphur in this case is related to that of the salt-domes themselves, and has never been satisfactorily explained. Again, it has long been known that some varieties of bacteria have the property of generating hydrogen sulphide through the reduction of sulphate solutions. Some of these bacteria are anærobic, being able to exist in the absence of air, and their action has been repeatedly observed in ocean water, but whether they can exist and function in deeply buried strata is open to question. Finally, there are oils, like those of the Appalachian fields, that have apparently never been subjected to the action of sulphur; and others, like the Trenton limestone oil of Ohio, that are generally supposed to owe their sulphur to the character of the organic remains from which they were formed. If the organic origin of petroleum is accepted the old idea that the contained sulphur indicates derivation from animal remains is not necessarily valid. It is generally recognized, however, that the character of the original organic material has a bearing on the composition of the oil derived from it.

Mine Found by Thieves

How a Brazilian physician owes a prospective fortune to the cupidity of negro thieves is revealed in a little story which comes from Rio Janeiro. According to this tale, Dr. Marques da Silva rented a house in a suburb of Rio Janeiro to a family of negroes. After remaining in the house long enough to run up a good-sized unpaid rent bill, the negroes suddenly decamped, taking with them all the electric wiring and plumbing fixtures in the house. They even tore up a lead pipe leading underground to a water-main.

Dr. da Silva went through the looted house, sadly noting his losses and the damage done to the premises, and sat down on the veranda to think over the iniquity of his missing tenants. Suddenly he noticed a peculiar metallic gleam in the trench where the pipe had been torn out. The gleam was caused by mercury oozing from the clay. The mercury mine probably will make the doctor a millionaire.—*Daily Metal Reporter.*

A Blast-Furnace Record

Ten tons per square foot of hearth-area is the record made by a 14-ft. blast-furnace at the smelting plant of the Consolidated Arizona Smelting Co.'s smelter at Humboldt, Arizona. It is a rectangular water-jacketed furnace, 14 ft. long and 52 in. wide at the tuyeres. It taps from the side through a water-cooled copper breast-jacket with a water-jacketed spout, and a water-cooled copper lip into a pear-shaped settler 27 ft. long and 13 ft. wide and 55 in. deep. It has practically no crucible, as it is bricked up to within 3 in. of the tuyeres on each end and the bottom slopes toward the middle so that it is just below the connection hole at the centre. The furnace is fed mechanically with six charge-cars of the old Anaconda type, having a capacity of 60 cu. ft. each, and dumped by means of an air-lift.

Details of Furnace

Width at tuyeres	52 in.
Length of furnace	168 in.
Square feet hearth-area	60.66
Height of furnace	11 ft. 6 in.
Distance from tuyeres to top	8 ft. 6 in.
Distance from tuyeres to sole plate	3 ft. 6 in.
Diameter of tuyeres	4 in.
Centre to centre between tuyeres	15 in.
Number of tuyeres	22
Tuyere-area	276.46 sq. in.
Tuyere-area per square foot of hearth-area	4.56 in.
Cubic feet of air per minute	20,000
Air-pressure	26 to 30 oz.
Tons smelted per 24 hr	609
Tons per square foot of hearth-area per 24 hr.	10.04
Per cent coke	9.3%

The charge is put in as follows: two cars of coke holding 800 lb. each are dumped in, and then two ore-cars holding 8000 lb. each are dumped on top of it from the same side of the furnace, and this is repeated on the other side of the furnace for the next charge. The charge consists of ore, converter-slag, and limestone, and has the following composition:

	%
SiO_2	33.5
Fe	26.8
CaO	8.6
S	10.4
Cu	3.12

With an 80% sulphur-elimination this gives a 38% matte and a slag assaying:

	%
SiO_2	40
Fe	30
CaO	10.2
Cu	0.25

The smelter superintendent in charge is F. K. Brunton.

PETROLEUM development in Louisiana is approaching near to the city of New Orleans. Wells are being drilled within 40 miles of that city by the Concordia Gas & Oil Co. The oil and gas zone has now been shown to extend from Natchez to a point south of New Orleans.

Survey of Inclines Without Auxiliaries

By A. J. SALE

In an out of the way district, an engineer may sometimes be called upon to make a survey down a steep incline when he has no auxiliary telescope available. If this should happen, he will find that the following method readily solves the difficulty. Necessity having forced me to evolve this procedure, I became so attached to it that I have since ceased to use auxiliaries where accuracy is an important factor. At best, either a top or side telescope has many disadvantages. They are hard to keep in adjustment, they greatly detract from the stability of the 'set up,' and introduce awkward corrections into the calculations. The essence of the problem is to generate a vertical plane down the incline; the remainder is merely detail. For this type of work, it stands to reason that the instrument must be in perfect adjustment, but, in addition to the usual transit adjustments, including both the one for leveling and the one of the vertical vernier, there is another condition that must be investigated. By the 'peg adjustment,' or otherwise, one can apparently make the bubble-tube of the telescope parallel to the line-of-sight; but, in reality, it only causes them to lie in parallel planes, and, in an inclined position of the plate, they would show false conditions. This special adjustment is made in the following manner: Select two parallel walls, about 100 ft. apart, and mark an approximate centre line between them. Have all three of the legs of the transit at equal length, and place them as indicated in Fig. 1. Loosen the leveling screws under the plate and turn the head of the machine until a pair of opposite screws lies in a line approximately parallel to the marked centre line. This forces the other pair into a position approximately perpendicular to this line. Tighten all of the leveling screws, and level the plate. With the vertical motion, bring the bubble of the telescope into a central position, causing the vertical vernier to read 0. Set the plate-vernier to read 0, and, with the lower motion, turn until the telescope points forward along the marked centre line. This causes one of the plate bubble-tubes to be approximately perpendicular to the centre line. With the upper motion, turn to 90°. (As the upper motion is used throughout the remainder of this adjustment, I will not repeat the mention of it.) Mark a point where the line-of-sight intersects the wall. Then, leaving the vertical motion fixed, mark another point at the same level, but at about 2 ft. in advance of the last. With a straight-edge, draw a line connecting these points. In like manner, after turning to 270°, mark another line on the opposite wall. By this method there will have been drawn two horizontal lines at the same elevation. Now, by shortening the forward leg, cause the plate to become inclined to any desired angle, which may be called θ. It is best to make this inclination to some special angle such as 15° or 30°, depending upon the steepness of the incline to be surveyed. For very steep inclines, 30° or more may be necessary.

It is a simple matter to set this angle of inclination. Suppose 20° is desired: Set the vertical vernier to read + 20°; lower the forward leg of the instrument until the telescope-bubble approaches a level position, then bring it exactly level by means of the leveling-screws which lie parallel to the centre line. After having set θ, level the bubble-tube which lies perpendicular to the centre line. (This operation is not absolutely necessary, but saves time by bringing the instrument into approximate position.) Set the vertical vernier to read 0, and turn the plate to 90°. (As the instrument has been brought into perfect adjustment, the line of sight will be parallel to the plate when the vertical vernier reads 0.) Mark the point where the line of sight cuts the wall. (Since the general position of the machine has been lowered, this point will usually be a few inches below the horizontal line already marked.) Now, leaving the vertical motion fixed, turn to 270° and mark the point where the line of sight cuts that wall. Measure the distance from each of these points to its corresponding horizontal line. From each point mark off one-half of the difference between these measurements in the direction in which it occurs.

As an example: Suppose, at the 90° position the point is 3 in. below its line, while at the 270° position it is 3½ in. below. Then, at the 90° position, mark a new point ¼ in. below the original; while at the 270° position mark a new point ¼ in. above the original. Through each of these new points draw a short horizontal line. (These lines may not be necessary; but, as there is a tendency toward a horizontal movement in the next change of the telescope, they will probably save time.) Leaving the vertical motion fixed, and by means of the pair of leveling-screws which lie perpendicular to the centre line, raise or lower the line of sight until it cuts the new short horizontal line just marked. Then turn back to the 90° position, where the line of sight should exactly cut the new line which has just been marked for that position. If there is a slight discrepancy, correct half at both positions as has been described. Now turn back to the position and re-set θ, if it shows any change. If any correction is made in θ, it will be necessary to turn back to both the 90° and 270° positions and correct any small errors that may have occurred, after which it will not be necessary to re-set θ.

By the above operations a horizontal line is fixed into

which the line of sight falls at both the 90° and 270° positions. Notice the position of the telescope-bubble at each of them. (The inclination of the plate causes the bubble to be near the upper side of its tube, but that does not prevent a satisfactory determination of its position). If the bubble-tube happen to be exactly parallel to the line of sight, the bubble will be in the central point at both the 90° and 270° positions. But if it is warped from a parallel by a small angle, which may be called μ, the bubble will be off centre at one position, and an equal amount in the opposite direction at the other. If one did not account for the effect of the angle μ, and assumed that the line of sight was horizontal when the bubble was in a central point at either 90° or 270°, the result would be that the line-of-sight would make a small angle with the horizontal at the 90° position, and an equal angle with opposite sign at 270°. In later operations, either turn the telescope from 270° to 360°, or from 90° to 0, which will bring the horizontal axis of the telescope to either the 270° or 90° position of the line-of-sight, and the final effect is the same as if the plate were in normal position, and the horizontal axis of the telescope inclined by a small angle, which call ν; this angle being the vertical projection of the angle μ.

Now it might be possible to shift the bubble-tube so that it did lie parallel to the line of sight, and make the regular bubble-adjustment afterward; but this amount of work is entirely unnecessary. If it so happen that the bubble-tube is exactly parallel to the line of sight, trouble of inclining the plate to a fixed angle will be saved, but that is the only advantage. The vital feature of the proposition is to mark the exact positions of the bubble which force a horizontal condition of the line of sight at both the 90° and 270° positions. This marking can be done by taking narrow strips of paper and pasting them on the bubble-tube at the ends of the bubble. Considerable error is allowable in setting θ, without seriously affecting the carrying of true bearings.

By descriptive geometry, it can be demonstrated that $tg\,\nu = tg\,\mu \sin\theta$. (In Fig. 2, I indicate how this demonstration is made.) To study the relative errors, apply a simple principle of calculus and differentiate relative to ν and θ. Differentiating:

$$tg\,\nu = tg\,\mu \sin\theta$$

$$\frac{1}{\cos^2\nu} \,.\, d,\nu = tg\,\mu \,.\, \cos\theta \,.\, d,\theta$$

But $\dfrac{1}{\cos^2\nu} = 1 + tg^2\,\nu = 1 + tg^2\,\mu \,.\, \sin^2\theta$

Therefore

$$d,\nu = \frac{tg\,\mu \,.\, \cos\theta}{1 + tg^2\,\mu \,.\, \sin^2\theta} \,.\, d,\theta$$

Now, since μ is a very small angle, $tg^2\,\mu \,.\, \sin^2\theta$ is negligible as compared with unity.

Therefore $d,\ \nu = tg\,\mu \,.\, \cos\theta \,.\, d,\theta$.

Now suppose that it is required to look down a steep angle, which call ϕ, with the result that the inclination of the horizontal axis of the telescope (angle ν), causes an error in the bearing, which call ψ.

It can be demonstrated that

$$tg\,\psi = tg\,\phi \,.\, \sin\nu$$

(This demonstration is indicated in Fig. 3.)

Differentiating relative to ψ and ν, it appears that

$$\frac{1}{\cos^2\psi} \,.\, d,\psi = tg\,\phi \,.\, \cos\nu \,.\, d,\nu$$

But $\dfrac{1}{\cos^2\psi} = 1 + tg^2\,\psi = 1 + tg^2\,\phi \,.\, \sin^2\nu$.

Therefore $d,\psi = \dfrac{tg\,\phi \,.\, \cos\nu}{1 + tg^2\,\phi \,.\, \sin^2\nu} \,.\, d,\nu$

But, since ν is even less than μ, $tg^2\,\phi \,.\, \sin^2\nu$ is negligible as compared to unity; and, for practical results, $\cos\nu$ can be taken as unity, making $d,\ \psi = tg\,\phi \,.\, d,\nu$.

Substituting for $d,\ \nu$ its previously derived value, there results

$$d,\psi = tg\,\phi \,.\, d,\nu = tg\,\phi \,.\, tg\,\mu \,.\, \cos\theta \,.\, d,\ \theta.$$

To apply this formula practically, assume:

$\theta = 75°$

$\mu = 30'$. (It is extremely improbable that μ will ever be as much as 10'.)

$\theta = 30°$

Also assume that, in attempting to set θ at 30°, an error of 30' is made.

Applying the above values in the formula:

$d,\psi = tg\ 75° \,.\, tg\ 30' \,.\, \cos 30° \,.\, d,\theta = 0.0282 \,.\, d,\theta$.

And, since d,θ was taken at 30', $d,\ \psi = 0.846' = 0°$, 00', 50''.

In other words, an error of half a degree in setting the plate only amounts to 50 seconds in future bearings.

The method of operation of the adjusted machine is as follows: Place the transit in a firm position at the top or the incline, with two of its legs close to the collar, and the third in a symmetrical position. Loosen the leveling screws and turn the head of the machine until a pair of opposite screws is approximately in the line of the incline. Tighten the leveling screws, and incline the plate to the required angle θ as previously described. Set the upper motion to read 0, and, with the lower motion, turn until the telescope points along the general direction of the incline. This will bring one of the plate-bubbles in line with and perpendicular to this line. Level this bubble, using the pair of leveling screws which lie perpendicular to the line of sight. (The position of the machine is shown in Fig. 4.)

Now, at some convenient point close to the incline in the underground workings, place an over-head plug containing a perforated horse-shoe nail or other suitable eye-point (X, Fig. 4). Drop a plumb-line to the floor and mark the point X' directly under X. Place an illumination behind X'; and, using the lower motion, focus upon the plumb-line at any convenient point. (The vertical cross-hair will probably be at a small angle to the plumb-line.) The plate bubble-tube will probably get out of level; but that is of no importance, as it was only levelled to bring the machine into approximate position. With the upper motion, turn an angle of 90°; and, by means of the vertical motion, bring the telescope-bubble to the point marked for a horizontal condition of the line-of-sight at the 90° position. Continuing with the upper motion, turn to 270°. The telescope-bubble will deviate by a considerable amount from the point which has been marked to enforce a horizontal line of sight at the 270° position. Correct half of this amount by means of the vertical motion, and the other half with the pair of leveling screws which lie at right angles to the line-of-sight. Continuing with the upper motion, turn back to 0. The vertical cross-hair should now be very nearly parallel with the plumb-line. Using the lower motion, focus exactly upon the plumb-line. Recheck θ, and, if it is in error, re-set it. (This is the last time that it will be necessary to bother with this angle, as it has been demonstrated that a small change in it will not materially affect the accuracy of the work.)

If any change has been made in θ, re-focus upon the plumb-line, using the lower motion. Repeat the operations at the 90° and 270° positions. This time the amount of bubble-correction will probably be very small; and, on the final turn to 0, the vertical cross-hair should exactly coincide with the plumb-line; it not, repeat until it does. (After the final leveling, the vertical vernier should read exactly 0 at both the 90° and 270° positions.)

The problem is now practically solved, except for a few working details which will be described later, for there has been developed such a condition that, when the telescope is focused on the plumb-line, and rotated around its horizontal axis, it will generate a vertical plane through the plumb-line. The demonstration of this is as follows: In Fig. 5 the operator is looking along the line of the incline and parallel to the plate. *M-N* represents a horizontal line and *A-B* the position of the line-of-sight, after leveling at the 90° position. *O-P* represents the axis of the socket which, since the instrument is in perfect adjustment, passes through the line-of-sight, and is also perpendicular to the plane of the plate. Assume a plane passing through the line-of-sight and the axis of the socket and let it intersect the plane of the plate at *C-D*. Now, $\angle\, OPC = 90° = \angle\, OPD$; because the axis of the socket, being perpendicular to the plane of the plate, must be perpendicular to an intersecting line in that plane. Also assume that *C-D* makes an angle ($\angle\, OMP$, Fig. 5) with *M-N*. Now the telescope is rotated 180° (that is, from 90° to 270°) around the axis *O-P*. Let *A′-B′* represent the rotated position of *A-B*. The total deviation is $\angle\, AOB'$, or $\angle\, A'OB$. Half of this is to be corrected by the vertical motion, and half by the leveling screws which lie in a line perpendicular to the line of the incline. Assume that the first correction brings *A′-B′* to *A″-B″*, in which case $\angle\, A'OA'' = \frac{1}{2}\angle\, A'OB$ or $\angle\, AOB'' = \frac{1}{2}\angle\, AOB'$, which means $\angle\, AOB' = 2\angle\, AOB''$.

Next consider $\triangle MPO$: $\angle\, MPO = \angle\, OPC = 90°$. Therefore $\angle\, MOP = 90° - \angle\, OMP$. But, since *A-B* is rotated around *O-P*, $\angle\, A'OP = \angle\, AOP = \angle\, MOP = 90° - \angle\, OMP$.

Therefore

$$\angle\, AOA' = \angle\, AOP + \angle\, A'OP = 2\angle\, AOP = 2\,(90° - \angle\, OMP) = 180° - 2\angle\, OMP.$$

But, *A-B* being a straight line, $\angle\, A'OB = 180° -$

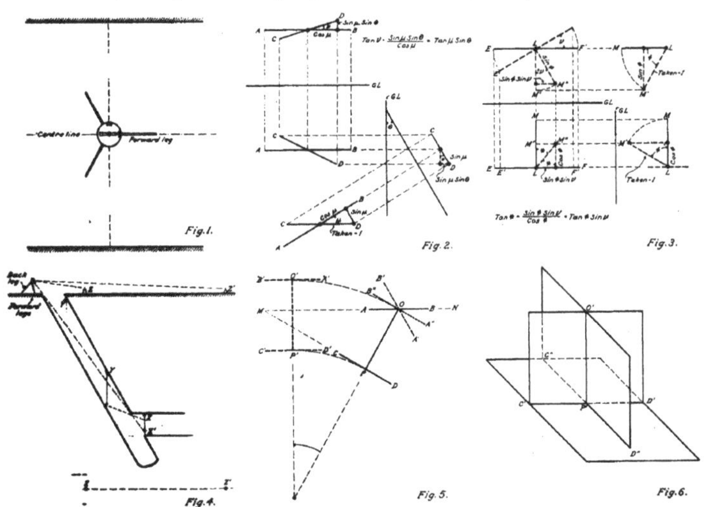

$\angle A O A' = 180° - (180° - 2 \angle O M P) = 2 \angle O M P$.

But $\angle A' O B = \angle A O B'$. Therefore $\angle A O B' = 2 \angle O M P$.

But $\angle A O B' = 2 \angle A O B''$.

Therefore $\angle A O B'' = \angle O M P$, or A''-B'' is parallel to C-D.

Now the second correction causes a rotation around the centre of the ball and socket joint below the plate to such an extent that the line-of-sight becomes horizontal. O goes to O'; A''-B'' to 'A'-'B'; P to P'; and C-D to C'-D'. But, as this second correction does not change the relative position of these lines as regards each other, C'-D' will be parallel to 'A'-'B', and $\angle O' P' D' = \angle O' P' C' = \angle O P C = 90°$. But this rotation caused 'A'-'B' to be a horizontal line, therefore, C'-D' is a horizontal line.

Now consider the plane which would be generated by the rotation of the line-of-sight around the horizontal axis of the telescope in the final (or 360°), position: Referring to Fig. 6, this plane is shown as $O'C'' P' D''$. (C'' and D'' being the respective positions of C' and D', after the rotation from the 270° to the 360° position.) The $\angle O'P'D' = 90°$, from the previous demonstration. Also $\angle D'P'D'' = 90°$, due to the turn from 270° to 360°. Therefore the line C'-D', being perpendicular to the line O'-P', also perpendicular to the intersecting line P'-D'', must be perpendicular to the plane $O'C''P'D'$. Therefore the plane $O'C''P'D''$ is perpendicular to the line C'-D'. But the line C'-D' has been demonstrated to be a horizontal line. Therefore the plane of rotation will be a vertical plane.

This has prepared the way for the final details: At some convenient position (Y, Fig. 4) place a survey-plug in the line-of-sight and drive in a perforated horse-shoe nail. Drop a plumb-line from this point, and drive over the horse-shoe nail until it comes exactly into line. The point Y should be so selected that when the plumb-line is let down as far as possible the lower end will be visible from the 'set up' under X by raising the telescope to an angle of about 45°. As this system is only required for inclines whose dip exceeds 45°, it is evident that the plumb-line will reach a considerable way down the shaft before touching the floor. It is best to set at least two of these points to make sure of obtaining a back-sight with as long a base as possible. After setting Y, rotate the telescope around its horizontal axis, until it looks over the collar of the incline, and place a hub (Z, Fig. 4) in line and close to the incline. Drive a tack exactly in the line of sight. Also place another hub and tack in the same line but several hundred feet away (Z', Fig. 4). Since it has been demonstrated that the line-of-sight rotates in a vertical plane, these hubs can be either forward of back-sights. Having set the hubs, the next operation is to read the exact value of θ, after leveling the telescope-bubble at the 0 position. It was previously demonstrated that, without seriously affecting the generating of a vertical plane, it is allowable to take considerable margin in setting θ to a fixed angle; but the exact value of θ must be used to obtain true vertical angles. To obtain any true vertical angle, algebraically subtract the value of θ from the vertical angle as read. Only angles lying in the generated vertical plane are to be used.

Suppose $\theta = 30°$:

Then, for an apparent vertical angle of $+40°$, the actual vertical angle will be $+40 - 30° = +10°$, while for an apparent vertical angle of $-40°$ it will be $-40 - 30 = -70°$.

If it become necessary to use back-sight positions, involving the use of a reversed telescope, the angle θ must be algebraically added. In the above case an apparent back-angle of $-10°$ will be a true angle of $-10° + 30° = +20°$. An apparent back angle of $-30°$ will be a true angle of $-30° + 30° = 0°$, or horizontal, while an apparent back angle of $-60''$ will be a true angle of $-60° + 30° = -30°$.

The remaining operation from this 'set-up' is to tape the distances to the nearest hub (Z, Fig. 4), and to the point X' (Fig. 4), and read the apparent vertical angles. Both the horizontal and vertical distances are obtained by this operation.

Suppose: $\theta = 30°, 30'$; and the tape reading to Z is 15.15 ft., with an apparent vertical angle of $+ (25°, 20')$; then the actual vertical angle is $+ (25°, 20') - (30°, 30') = - (5°, 10')$, and the horizontal distance between the instrument and the hub is 15.15 ft. times cos (5°, 10'); while the 'H. I.,' relative to the hub, will be + [15.15 ft. times sin (5°, 10')].

Also suppose the tape reading to X' to be 225.20 ft. with an apparent vertical angle of $-(38°, 10')$; this makes the actual vertical angle $-(38°, 10')-(30°, 30') = (68°, 40')$, and the horizontal distance between the instrument and point X will be 225.20 ft. times cos (68°, 40'); while the floor of the level (point X') will be 225.20 ft. times sin (68°, 40') below the 'H. I.'

The final operation to complete the surface-work is to tie in Z from some other station; set on Z and get the bearing of Z-Z'; and to obtain the elevation of Z relative to a known bench mark.

There is now a complete survey down the incline, covering bearings, horizonal distances, and elevations. It only remains to set under X (that is, over X'), back-sight to the plumb-line from Y, and continue as desired. In making the back-sight from the 'set-up' under X, it is best to raise the telescope as much as convenient in order to look as far back up the incline as possible. If several points Y have been set, there is a chance that a much longer back-base can be obtained than was even suspected at the start.

If more than one level is to be tied in, it is best to make the first point (X) at the bottom level, and then place points on the other levels in the manner described for setting Y. If it is necessary to carry a survey up an incline, it is best to work downward from the top, and tie in the line below afterward. A triangular eye-piece is a convenient help to save kinks in the neck of the operator, but its use is not absolutely necessary.

The method described will rapidly and accurately

solve the problem of carrying a survey down any steep incline of a reasonable depth, but, for a very deep incline, some special considerations are advisable. Suppose the incline to be very steep, and a 1000 ft. or more deep. Under these conditions, as it is not probable that the point Y can be placed at more than 20 or 30 ft. from point X, there will be involved all the errors connected with a long sight in setting both the points X and Y; while one will have to depend upon a very short back-sight under ground. Even under these conditions, if due care be used, the errors will be comparatively small, and this method will prove sufficiently accurate for any ordinary work, but, for work where unusual accuracy is required, I would recommend the following procedure: Set point Y where it can be easily seen from an ordinary 'set-up' under X. Also set an additional point, which call Y', in line and up near the collar of the incline. Everything else is done as previously described. When prepared to work below, mark a line on the floor, through X', and approximately in the general direction of the incline. Have the three legs of the instrument at equal length and set over X' (that is, under X), in such a position that the back leg of the machine, from a position looking up the incline, lies in the line just drawn, while the forward ones are placed symmetrically. The reverse of the position shown in Fig. 1 will produce this condition. Make an accurate 'set-up' over X' with as little relative change in the legs as possible. Loosen the leveling screws under the plate and turn the head of the instrument until a pair of opposite leveling screws is in an approximate line with the incline. Tighten the leveling screws, level the plate, set the upper-motion to read 0, and, with the lower-motion, turn until the telescope points in a line directly reversed to that up the incline. Set the required angle θ, tilt the plate, and do the leveling at the 90° position and the half and half correction at 270°, as previously described. Loosen the vertical motion, set the upper-motion at 180° and, by means of the lower-motion, focus upon a plumb-line dropped from Y'.

If proper precaution has been taken in regard to symmetry, a plumb-line dropped from Y will come within the range of vision by changing the focus; and a lateral shifting of the plate should bring the line-of-sight into the line Y'-Y, without altering the position of the legs. Should the amount of required shifting exceed the limits of the plate, move each leg by the same amount in the same direction so as to keep the general position uniform. After the line-of-sight has been brought into the line Y'-Y, turn back to 0; re-set θ, and repeat the operations. It is now probable that, when the line-of-sight has been focused upon Y', Y will only deviate by a very small amount from being in line, and, after making the required shift of the plate, and repeating the complete operation, the line-of-sight should be exactly in the line Y'-Y; if not, repeat until it does. If proper preliminary judgment as to symmetry has been used this whole amount of maneuvering should not take more than twice the time usually required to set in line with two plumb-lines, and its accuracy can be more fully depended upon.

After these operations have been completed, if the line-of-sight is brought into position and rotated around the horizontal axis of the telescope, it will generate a vertical plane up the incline through the points Y' and Y. The only remaining operation is to fix this plane by setting two points, one close to the machine, and the other as far away as convenient. In continuing the underground work it is advisable to use only this new line for meridian, as the points X and X' may vary by a small amount from it, but, as this variation will be very small, it will be safe enough to carry distances and elevations from the old points.

While it has taken considerable space to describe this modus operandi and to make clear its mathematical demonstration, a practical trial by any engineer should convince him that in its field application the method is rapid, the resultant office calculations simple, and the final result the attainment of a degree of accuracy which cannot be reached by any auxiliary attachments.

BARYTES, or barium sulphate, is used chiefly in making mixed paints, in which white, ground, and water-floated barite is employed as a pigment. Ground barite is also used in the rubber industry and to some extent by the makers of heavy glazed paper and ink. Lithopone, a chemically prepared white pigment consisting of about 70% barium sulphate and 30% zinc sulphate, is one of the chief constituents of the 'flat' wall-paints so extensively used in office-buildings and hospitals, replacing the less desirable paper and calcimine wall-finishes. Its larger use is in the manufacture of linoleum, and as an adulterant in making rubber tires. Since the beginning of the War a barium chemical industry has been established in the United States to supply barium carbonate, nitrate, chloride, chlorate, hydrate, and dioxide, formerly imported largely from Germany. In 1915 this consumed 10% of the domestic barite, but the consumption in 1916 was somewhat larger. The barium chemicals have a wide variety of applications, perhaps the most important being the use of barium dioxide in the preparation of hydrogen peroxide, that of barium chloride as a water softener; other various salts are used in the manufacture of optical glass. Barytes is mined principally in Missouri, northwestern Georgia, east-central Tennessee, central and western Kentucky, north-eastern Alabama, south-western North Carolina, north-western South Carolina, and south-western Virginia. The price now ranges from $28 to $32 per ton for prime white or floated material.

ARGENTINA is developing oil-fields in Patagonia and in other parts of the republic. A recent report by the Argentine Bureau of Mines affirms that the petroleum deposits of Rivadavia on the Patagonian coast promise to rank among the most important of the world. The oil is found at a depth of about 1600 ft. below sea-level. By distillation it yields 5% benzene, 16.27% of illuminating oil, and 67% heavy oil.

Chromite

By J. S. DILLER

*The importance of chromite in the manufacture of armor plate, armor-piercing projectiles, stellite for high-speed tools, and automobile and other special steels can scarcely be overestimated. The chief sources of supply for the United States during the last few years have been Rhodesia, New Caledonia, Turkey, and Greece. Embargoes were placed on the shipment of chrome ore from some of the principal sources, and it was feared that the supply for the United States would be cut off, but after the producers received a guaranty that the ore would not be re-shipped to enemy-belligerents the imports, as shown in the following table, greatly increased, especially those from Rhodesia, New Caledonia, and Canada, though those from Greece have declined slightly and those from Turkey have entirely ceased.

CHROMIC IRON IMPORTED INTO THE UNITED STATES, 1913-1916, IN LONG TONS

	1913	1916
Cuba	....	34
Canada	....	10,930
England	....	5
Greece	....	7,900
Japan	322	.....
French Oceania	6,620	30,950
Australia	....	2,986
British South Africa	....	23,000
Portuguese Africa	29,000	38,850
Turkey in Asia	13,830	.....
	49,772	114,655

Chromic iron produced and sold in the United States, 1913 and 1916, was respectively 255 and 40,000 long tons. The greatly increased trade, especially in steel, and the consequently larger demand for chromite, have stimulated the search for it in the United States, as shown by the large increase in production. On the Atlantic coast and in Wyoming there has been only a small production, but in the Pacific Coast States, especially California, the advance in the output has been remarkable. It is evident that, for some time to come, California will furnish the chief domestic supply. The production from some deposits in 1917 is expected to exceed that of 1916. It is possible, however, that some counties, Del Norte, for instance, which produced no chromite in 1916, will produce much in 1917 on account of better transportation facilities, both by land and sea. There are two main belts of production in California, one in the Klamath mountains and the Coast Range from Siskiyou county to San Luis Obispo county, and the other in the Sierra Nevada from Plumas to Tulare county. The larger output has come from the Klamath mountains, because the orebodies there are larger and railroad transportation is more convenient.

The production in Oregon is increasing in both the Klamath and Blue mountains. The ores west of Riddle are the richest yet mined in the State, in some places assaying as high as 55% chromic oxide, and much of the ore contains about 50%. Most of the Oregon ore, however, like that of California, averages about 40% chromic oxide, and ore of that grade is commonly the basis of sale. It generally contains 38 to 45% chromic oxide, 6 to 8% silica, and 17 to 25% alumina. The largest orebody and producing mine thus far developed in Oregon is owned and operated by Collard & Moore near Holland, about 20 miles southeast of Kerby, in Josephine county. Much of the ore may be improved by concentration, and a plant of 90-ton capacity for that purpose is nearly completed. It is claimed that the ore can be concentrated to a content of 55% chromic oxide. The concentration of the lower-grade ore would give it a wider market and increase its value and the demand for it. Without concentration the Pacific Coast deposits cannot furnish a dependable supply of high-grade chrome ore, but with successful concentration industries based on high-grade ore may be attracted to the Coast. The Sawyer Tanning Co., established on tidewater at Napa, California, has had great difficulty in obtaining sufficient high-grade ore for its use. T. W. Gruetter has recently established at Kerby, Oregon, a custom-plant for concentrating black sand to win its gold and platinum. The black sand of the Klamath mountains usually contains a considerable amount of chromite, and it is believed that by adding magnetic separators to Gruetter's plant to remove the other minerals from the tailing sufficient chromite may be obtained from the black sand in chromiferous serpentine areas to make the operation financially successful.

The relation of these experiments in concentration to the problem of obtaining high-grade chrome ore on the Pacific Coast will be better understood when attention is called to the fact that by the disintegration and washing away of the weathered serpentine, in which practically all the chromite deposits occur, the heavy grains of chromite are left behind, and, consequently, the soil or surface-wash in the water-courses of serpentine areas becomes enriched by the accumulation of residual chromite. Chromite boulders and sand are therefore, as a rule, more abundant in the soil than in the solid serpentine beneath. Many prospectors finding boulders of chromite on the surface feel confident that there is a large body beneath, but a few shallow prospect holes usually give disappointing results.

*Abstract: U. S. Geol. Surv., Bull. 666-A.

In the Atlantic States, where most of the chromite produced in this country is used, the only production is in the vicinity of Baltimore, where the chrome industry of the United States was started by the Tysons many years ago. At first bodies of chromite were quarried from the serpentine areas about Baltimore and northeastward into Pennsylvania, but as the supply became exhausted the chromiferous residual deposits in the soil and the stream-gravel within the serpentine areas were washed for chrome-sand. A small output of chrome-sand is now obtained at Soldiers Delight, near Baltimore. This enterprise suggests the possibility of considerable expansion in the utilization of chromite sand in Maryland and Pennsylvania.

Ferro-chrome, the alloy used in making chrome-steel, is now manufactured in the United States by electro-metallurgic methods, almost wholly in the East, at the plants of the Electro Metallurgical Co. at Niagara Falls and elsewhere. The Noble Electric Steel Co. has three furnaces at Heroult, California, however, operating to full capacity producing manganese, chrome, and silica-steels. The metallurgy of chromium has apparently been so developed in the hydro-electric process as to utilize relatively low-grade ores such as are most abundant in the United States, and the further development of that process on the Pacific Coast, where water-power abounds, would greatly diminish the handicap of long transportation.

Missouri Zinc and Lead 1915-16

The following tables give at a glance an interesting analysis of the lead and zinc ores of Missouri and their beneficiation, comparing the years 1915 and 1916, the statistics discriminating between the so-called 'soft ground' and the 'sheet ground' in the southwestern part of the State. The tables were compiled by the U. S. Geological Survey.

ZINC AND LEAD OUTPUT, SOUTHWEST MISSOURI, 1915-16

	1915	1916
SOFT GROUND		
Total crude ore ... short tons	4,004,900	4,711,700
Total concentrates in crude ore ... per cent	3.88	3.48
Lead ... do	.32	.38
Zinc ... do	3.56	3.10
Metal content of crude ore ... do	2.26	1.92
Lead ... do	.25	.22
Zinc ... do	2.01	1.70
Average lead content of galena concentrates ... do	79.2	75.0
Average lead content of lead carbonate concentrates ... do	56.1	58.5
Average zinc content of sphalerite concentrates ... do	57.7	58.1
Average zinc content of silicates and carbonates ... do	30.3	30.4
Average value per ton:		
Galena concentrates	$48.19	$82.00
Lead carbonate concentrates	$32.56	$51.48
Sphalerite concentrates	$72.68	$77.94
Zinc silicates and carbonates	$45.00	$50.21
SHEET GROUND		
Total crude ore ... short tons	8,301,000	8,484,700
Total concentrates in crude ore ... per cent	2.25	2.19
Lead ... do	.34	.33
Zinc ... do	1.91	1.86
Metal content of crude ore ... do	1.38	1.35
Lead ... do	.25	.25
Zinc ... do	1.13	1.10
Average lead content of galena concentrates ... do	78.4	79.5
Average content of sphalerite concentrates ... do	59.1	59.2
Average value per ton:		
Galena concentrates	$54.63	$85.62
Sphalerite concentrates	$78.79	$86.02

LEAD OUTPUT, SOUTHEAST MISSOURI, 1915-16

	1915	1916
Total crude lead ore ... short tons	5,067,800	5,467,800
Galena concentrates in crude ore ... per cent	5.47	5.90
Lead content in crude ore ... do	3.62	3.86
Average lead content of galena concentrates ... do	66.5	65.66
Average value per ton of galena concentrates	$45.90	$67.47

Quicksilver Industry of Texas

By WM. B. PHILLIPS

*The total value of the quicksilver produced in Texas since the beginning of the industry, in 1899, exceeds $3,500,000. The industry has been confined to a comparatively small area in the southern part of Brewster county, about 100 miles south of Marfa, on the Southern Pacific railway. Hauling to the railroad is done by wagons, the ordinary rate being from 50 to 60c. per 100 lb. While some native quicksilver has been found, by far the greater production has been from cinnabar occurring in limestone, in a bituminous shale, and in an acid igneous rock. The larger production was formerly from the cinnabar in the limestone, but in recent years the shales have yielded the greater part. When the D-shaped retort was used in distilling the metal it was not uncommon to obtain quicksilver, a combustible gas, and oil, from the same charge, but this practice did not long survive, as the retorts required much richer ore than the Scott furnace.

The ore is crushed and charged into a brick stack provided with staggered shelves from top to bottom, heated by a wood fire, and discharging into a series of brick chambers into which all the smoke and fume is conducted. These brick chambers are the condensers. The condensing-chambers have to be made with great care, for quicksilver, although it is nearly 14 times as heavy as water, will go wherever air will go, and there are considerable losses due to the escape of the metal as a fine mist through joints and cracks. The chambers have sloping floors, and the metal drains to a small plugged opening in an iron door. It flows through this opening, at intervals, and into a cement trough communicating with a cement tank in the collecting-room. Here it is ladled into wrought-iron bottles or flasks holding 75 lb. net.

The quicksilver ores in Brewster county occur in two principal horizons, the Eagle Ford shales of the Upper Cretaceous and the limestones of the uppermost members of the Lower, or Comanchean, Cretaceous, especially the Washita limestone. In a general way the Terlingua quicksilver district is divided into two great groups by Vogel's draw, which runs from north to south. On the west side of this draw are the limestones of the Lower Cretaceous and on the east are the shales of the Upper Cretaceous. This draw marks the approximate course of a fault that has brought the Lower Cretaceous up from its normal position. There seems to be no doubt that igneous intrusions had a good deal to do with the occurrence of quicksilver ores in this district, especially at California hill and at Study butte and Maverick mountain. The richer ores have not been obtained directly from the igneous rocks nor from the limestones and shales in immediate association with them. The average content of quicksilver in the Texas ores has been higher than in the California deposits mined in recent years. The content of quicksilver in the Texas ores is from 1 to 1.10% as against 0.50 to 0.60% in California.

*Abstract: *Manufacturers Record.*

Cyanidation v. Flotation at Pachuca

In the latest issue of the 'Boletin Minero' published by the Department of Industry and Commerce, of Mexico, appears an article on experiments that are being made with flotation upon the silver minerals in the ores of the Pachuca y Real del Monte and the Santa Gertrudis properties, written by Simeón Ramirez, inspector of mines for the Government. He says that at the Guerrero plant belonging to the Real del Monte y Pachuca company, situated at Real del Monte near the village of Omitlán, there has been installed a flotation testing-plant employing the Minerals Separation system with a capacity of 50 tons of ore per diem. The average assay-value of the ore treated is 11 oz. silver and 0.055 oz. gold per ton,.and an average recovery of more than 60% has been realized, but this percentage will be increased when the apparatus has been brought under better adjustment and when a better understanding of the proper mixtures of oils has been reached. There has been variation in the quantity of oil employed as well as variation in the speed of the agitators, air-pressure, and dilution of pulp, depending on the class of minerals contained in the slime. The average cost of concentration by this method has proved to be about $1.57 U. S. gold per ton of ore. However, the duration of the experiments has not been sufficient to admit of safe deductions as to the cost when working on a large scale. The ores treated in the Guerrero mill come from the San Ignacio, Dolores, Cabrera, and Escobar mines, all situated in Real del Monte, and having a total output of 850 tons daily. A part of this is treated by flotation, and the rest by cyanidation when there is sufficient cyanide to be had. The following statement shows the results obtained in this plant by flotation: Recovery, by difference, silver 63.9%, gold 60%; tonnage, dry pulp sent to the machine, 1.141 tons per hr.; dilution, 4.91 parts of water to 1 part of pulp; solution shows traces of CaO.

Oils employed:	Lb. per ton
Tar-oil, Barret No. 1	1.319
Creosote oil, Barret No. 606	1.319
Pensacola pine-tar No. 80	0.200
Total oil used	2.838

Assay:	Silver, oz. per ton	Gold, oz. per ton
Heading	11.96	0.04
Tailing	4.32	0.016
Extraction	7.64	0.024

ASSAYS OF FROTH FROM DIFFERENT COMPARTMENTS

Compartments	Silver, oz. per ton	Gold, oz. per ton
1, 2, and 3	160.5	0.707
4	115.7	0.458
5 and 6	51.9	0.192
7 and 8	37.3	0.128
9 and 10	57.9	0.192
11 and 12	75.6	0.041

The froth in the Dorr thickener assays 422.9 oz. silver and 1.608 oz. gold per ton. This test covered a period of 5½ hours.

The Santa Gertrudis company has three flotation machines installed close to the Santa Gertrudis mine on the slope of the hill upon which is situated the new mill and the cyanide annex. In this plant they are using Callow cells for the cleaning of the concentrate produced by 'K & K' machines. Although Minerals Separation machines were also used in the beginning, the 'K & K' is now preferred, and has given recoveries as high as 78%, with an average of 72%. These machines are simple in construction and easy to adjust and maintain.

The principal mining companies of Pachuca and El Chico are installing flotation machines and others have ordered them from the United States with the object of working out the problem of flotation for modifying their metallurgical plants. Meanwhile, they are using cyanide, the price of which is beginning to decrease. Among these new attempts may be noted the treatment of the ore sent from the Arévalo mine which has given very good results in a small Callow apparatus. Here the effect of varying quantities of oil was clearly seen in the modification of the consistence of the froth, these variations being for the purpose of obtaining the requisite persistence of froth to discharge with its load of concentrate into the concentrate-launder. This ore contains 3% lead, with an average content of 800 grammes silver (2.57 oz. per ton) of which, according to Max Kraut, 98% is readily recovered. In the ore from the Arévalo vein, which contains no lead, the recovery has been from 92.6% to 94% of the silver.

It has been observed that the ores which resist treatment by flotation in these tests have been those that contain the oxides, corresponding to the upper zone of the mine, which had formerly been abandoned as not yielding sufficient metalliferous content to pay the cost until cyanidation was introduced, but, in accordance with investigations made in the United States, it now appears possible to float these carbonates and oxides after sulphidization by soluble sulphides. On account of the difficulty in cyaniding the flotation-concentrate, due either to the fact that the oil behaves as a cyanicide or contributes in some other manner to produce a similar effect, experiments are being carried out with the object of overcoming these disadvantages. Eliminating the oil by treating the concentrate with alkaline or acid solutions or washing with water has made it appear that it is not the acid which is prejudicial but the ferrous salts formed in the concentrate itself. This indicates the necessity of further investigation in order to reach an understanding of the causes which lead to the difficulty mentioned. For the present, on account of the cost of cyanide, it appears to be profitable to extract the silver from these ores in the form of concentrate for export or for smelting locally. In order to admit of beneficiating these ores, either by cyaniding, or by floating and smelting the concentrate on the spot, it is necessary to regulate railroad traffic in order to admit of the introduction of fuels at a favorable rate.

Tungsten Mines of Inyo County, California

Mining of tungsten near Bishop, Inyo county, California, began actively about one year ago. The ore is a low-grade scheelite associated with a garnet gangue, originating by replacement of limestone which supplied the calcium needed for the development of the garnet and epidote in the process of contact-metamorphism. No limestone, however, remains in connection with the larger bodies of ore. The granitic rocks adjoin some of the tungsten deposits, and have been profoundly altered. According to Adolph Knopf, of the U. S. Geological Survey, the metasomatic alterations of the adjoining granitic rocks occurred some time after the consolidation of the granite. The tungsten mineralization was subsequent to the intrusion both of the granite and of the monzonite of the district. The minerals included with the ore indicate that the mineralizing solutions probably entered at so high a temperature as to be in a gaseous state. These solutions were rich in silicon, aluminum, and ferric iron, were poor in sulphur, and carried tungsten, which was fixed by the calcite in the limestones as calcium tungstate (scheelite).

Owing to the garnet in the gangue the ore is rather difficult to concentrate. The average grade being treated assays close to 0.75% tungstic trioxide. The deposits occur usually in large masses of sedimentary rocks associated with the surrounding granite country-rock. The greatest depth so far reached in any of the mines is about 120 ft. below the surface and the tungsten-content seems to be increasing at this depth. Two mills are now in operation, namely, that of the Standard Tungsten Co., having a capacity of 70 tons daily, and the Tungsten Mines Co., having a capacity of about 250 tons daily. The production of scheelite concentrate is about 6000 lb. per day. At the property of the Round Valley Tungsten Co., of which Cooper Shapley is manager, a mill is being completed which will have a daily capacity of about 100 tons. A new departure from the usual milling practice is being tried at this plant, consisting in the use of a Marcy ball-mill for the fine crushing. In order to avoid sliming of the scheelite, a grate opening of ¼ in. is used and the ore after leaving the ball-mill is screened to 10 mesh on Colorado Iron Works impact screens. By employing an excess of water in the Marcy mill, it is expected that the ore, as soon as it is crushed to pass a ⅛-in. screen, will be flushed out and sliming avoided. The oversize from the impact screen is returned to the Marcy mill for further grinding. Mining and milling can be done cheaply because of the low-priced electric power and the accessibility of the properties to the railroad and to sources of supplies.

Besides the above-mentioned properties, another large tungsten property is now being examined with the object of installing a 1000-ton mill. Diamond-drills are to be used in the development and examination of this property, and it is probable that next year the mill will be in operation.

Zinc Situation in Australia

*When William Morris, the Premier of the Commonwealth, was in England he arranged a 10-year contract for the treatment of a certain tonnage of our zinc concentrate in Great Britain, and he also obtained an advance of £500,000 toward the establishment of spelter-works in Australia with a promise that the output would be purchased by the British government. What Mr. Hughes has not told us is how this half million pounds is to be allocated. The works of the Electrolytic Zinc Co. Proprietary Ltd. at Hobart, in course of erection, will call for the expenditure of £1,000,000; the Mount Lyell Co. is also considering the erection of an electrolytic plant for the treatment of the Mount Read-Rosebery ores; Gilbert Rigg, metallurgist, is investigating the subject of zinc reduction both by the retort and electrolytic methods; while a scheme for utilizing the brown coal deposits of Victoria in connection with the industry has been mooted. The need for such works is urgent, as, under present circumstances, it is imposible to find a market for the current production, notwithstanding the shipments going to the United States and to another of our allies. As a consequence large tonnages are being stacked for future treatment. It will therefore be realized that the allocation of the money obtained from the British government for this purpose is a matter of great public interest and importance, and an announcement of the government's policy in this respect should not be longer delayed. With respect to the Electrolytic Zinc Co. it may be said that the erection of the first unit capable of producing 10 to 11 tons of zinc per day is proceeding satisfactorily, and, provided no delays are experienced due to non-delivery of certain equipment under order in Great Britain, it should be in commission in August next. The experimental test-unit capable of producing 250 to 300 lb. of electrolytic zinc per day is now running, and the results confirm those obtained in America. This unit is supplying useful data, and will facilitate the earlier successful operation of the first unit. The Zinc Corporation, Ltd., has acquired a one-fifth interest in this company. David Meredith has been appointed general manager of the Amalgamated Zinc Co., in place of H. W. Gepp, who has been appointed general manager of the Electrolytic Zinc Co.

FERRO-URANIUM produced from the uranium oxide obtained as a by-product in the extraction of radium from its ores is being investigated by the Bureau of Mines. Ferro-uranium is used in making uranium steel, employed in Germany for the linings of big guns, which, it is claimed, stand up at a rate of fire so rapid that other steels fail. Work will be started on the production of sample lots of uranium steel and other special steels, for test by the Bureau of Ordnance of the War Department as to their suitability for use in guns. The work on gun-steel will also require the use of electric furnaces.

*The Min. & Eng. Review.

Recent Patents

1,228,183. FLOTATION OF MINERALS. Harry P. Corliss, Pittsburgh, Pa., assignor to Metals Recovery Company, New York, N. Y., a Corporation of Maine. Filed Mar. 21, 1917. Serial No. 156,491.

1. The method of effecting the concentration of minerals by flotation, which comprises adding to the mineral pulp a small amount of alpha-naphthylamin and subjecting the resulting mixture to a flotation operation; substantially as described.

2. The method of effecting the concentration of minerals by flotation, which comprises adding to the mineral pulp a small amount of alpha-naphthylamin and of oil and subjecting the resulting mixture to a flotation operation; substantially as described.

1,228,184. FLOTATION OF MINERALS. Harry P. Corliss, Pittsburgh, Pa., assignor to Metals Recovery Company, New York, N. Y., a Corporation of Maine. Filed Mar. 21, 1917. Serial No. 156,492.

1. The method of effecting the concentration of minerals by flotation, which comprises adding to the mineral pulp a small amount of nitro-naphthalene and subjecting the resulting mixture to a flotation operation; substantially as described.

2. The method of effecting the concentration of minerals by flotation, which comprises adding to the mineral pulp a small amount of nitro-naphthalene and of oil and subjecting the resulting mixture to a flotation operation; substantially as described.

1,212,334. TREATMENT OF NICKEL ORES. Frederick A. Eustis, Milton, Mass.

1. The method of recovering nickel from oxidized or silicate nickel-bearing ores, which comprises mixing the ore with a small proportion of sulfur-bearing material such as pyrite, and roasting the mixture in its raw state in a suitable furnace at such temperature and for such a time that a relatively large amount of the nickel is made soluble while a relatively small amount of the gangue is made soluble.

2. The method of recovering nickel from an oxidized or silicate nickel-iron ore, which comprises mixing the ore with a small proportion of sulfur-bearing material such as pyrite, and roasting the mixture in its raw state in a suitable furnace at such temperature and for such time that a relatively small amount of the iron is made soluble.

1,227,615. TREATMENT OF ORES. Arthur Howard Higgins, London, England, assignor to Minerals Separation Limited, London, England. Filed Nov. 14, 1913. Serial No. 800,966.

1. A process for the concentration of ores which consists in treating an ore pulp conjointly with a mineral frothing agent for the separation of metalliferous constituents by flotation and with a chemical agent that facilitates the separation of some metalliferous constituents by flotation and dissolves other constituents.

5. A process for the concentration of ores which consists in first roasting the ore and thereafter agitating it with water containing a mineral frothing agent and also a chemical agent that facilitates the separation of some metalliferous constituents by flotation and dissolves certain other constituents and thereafter precipitating the dissolved metal in such a manner as to regenerate the chemical agent.

1,228,609. PROCESS FOR MAKING INSULATING MATERIAL FROM BASIC MAGNESIUM CARBONATE AND FIBROUS SUBSTANCES. Karl Schmid, Alt-Mugeln, near Mugeln, Leipzig, Germany, assignor to the Firm of "Lipsia" Chemische Fabrik, Actien-Gessellschaft, Mugeln, Leipzig, Germany. Filed Nov. 9, 1916.

1. The process of making insulating material from basic magnesium carbonate and fibrous substances, consisting in depositing basic magnesium carbonate on fibrous substances *in statu nascendi*.

5. The process of making insulating material consisting in depositing normal magnesium carbonate on a mass of fibrous substances suspended in a solution containing magnesium hydroxid by the reaction of a carbonate compound of ammonia therewith, thereupon filtering off the liquid and drying the remaining mass, the said normal magnesium carbonate being thereby converted into the basic magnesium carbonate.

1,228,078. RECORDING-SAMPLER. Arthur E. Truesdell, Adams, Mass. Filed June 15, 1912. Serial No. 703,803.

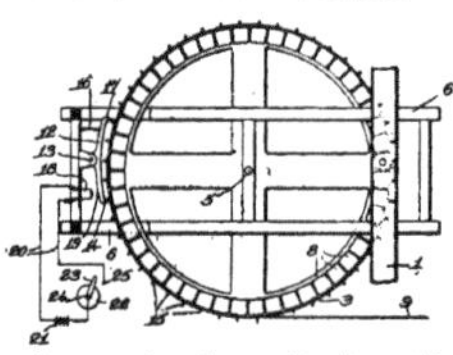

In a recording sampler, the combination with a chute or passageway through which the material to be sampled passes and which is provided with an opening, of a permanently-open spout communicating with said opening and adapted to receive material therefrom, a sample-receiving element rotatable in a horizontal plane beneath the delivery end of said spout and closely adjacent thereto, said element having projections extending from its periphery, automatically, operative means acting on said element and tending to rotate it forwardly, a pivoted escapement level coöperating with said projections and normally restraining said element from movement, and time-controlled electric means for periodically actuating the escapement lever to release the sample-receiving element and permit it to rotate one step forward.

1,227,867. CRUDE-OIL BURNER. John Young, New Westminster, British Columbia, Canada. Filed Jan 6, 1917.

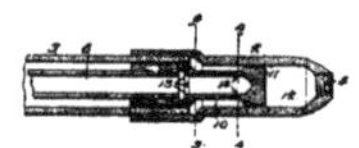

5. A crude-oil burner, comprising in combination, an outer burner member axially bored and threaded at one end to fit an oil service pipe and therebeyond conically reduced to a cylindrical bore which at the farther end is abruptly reduced to a circular delivery outlet threaded with a screw thread, an inner burner member the outside diameter of which is at one end larger than the cylindrical bore of the outer member and at the other end has a diameter slightly less than the said cylindrical bore and intermediate of its ends is reduced below the diameter of the outer end, the reduction of the larger end being conical to correspond with the conical reduction in the bore of the outer member, the outer end of the inner member terminating a short distance from the outlet reduction of the outer member so as to leave a chambered interspace, the inner member being axially bored to a short distance from its outer end and having a series of small apertures through it at the intermediate reduced portion, and means for delivering a steam service to the bore of the inner member.

1,215,517. PROCESS FOR OBTAINING POTASH FROM POTASH-ROCKS. Frederick C. Gillen, Milwaukee, Wis., assignor to William A. Krasselt, Milwaukee, Wis.

1. The process of decomposing potash rock which consists in

mixing ground potash rock with an excess of fixed alkali hydrate, adding water to this mixture and heating it under pressure to form a solution of fixed alkali silicate and potassium aluminate, adding alkaline borate to this solution to temporarily prevent the formation of the double salt of alkali aluminum silicate, and then adding a reagent to this solution while it is under the temporary action of the alkaline borate to separate the silica therefrom and to form a solution of potassium salt and fixed alkali carbonate from which the potassium salt may be separated.

2. The process of decomposing potash rock which consists in mixing ground potash rock with an excess of fixed alkali carbonate, adding water to this mixture and heating it to form a solution of fixed alkali silicate and potassium aluminate, adding alkaline borate to prevent the formation of a double salt of alkali aluminum silicate, and then adding carbon dioxid to this solution to form a solution containing potassum carbonate, fixed alkali carbonate and alkaline borate from which the potassium carbonate may be separated.

1,228,338. Centrifugal Impact Pulverizing Apparatus. Donn O. Marks, San Francisco, Cal., assignor of one-half to Lynn S. Atkinson, Los Angeles, Cal. Filed Dec. 11, 1913.

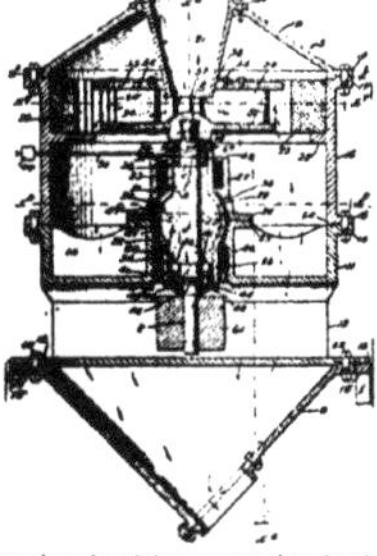

1. In an impact pulverizing apparatus having an annular impact wall; a rotor provided with impellers pivotally connected to the rotor body at one side of their centers of gravity and extending only part of the way from their pivots to said wall, so as to leave at all times an unobstructed space between the impeller and the wall, and so that the impellers are free to swing their longitudinal axes to and fro across those radii of the rotor in which the impeller pivots are located, respectively, without touching the impeller wall; for the purpose of yieldingly taking on the material to be impelled outward for disintegrating impact with the surrounding wall.

5. In a centrifugal impact pulverizer; a rotor comprising a shaft, a head on said shaft, said head being constructed with two disks spaced apart, one above the other; centrifugally cushioned impellers pivoted between the disks; a universal joint; a bearing for the shaft; said bearing located near the lower end of the shaft and supported by the universal joint, and power applying means connected below the bearing to drive the shaft.

1,214,991. Production of Alumina and Potassium Sulfate from Alunite. Earl Blough and Thomas McIntosh, Pittsburgh, Pa., assignor to Aluminum Company of America, Pittsburgh, Pa.

1. The process of treating alunite, comprising mixing the alunite with common salt, heating the mixture whereby the salt is decomposed, dissolving out sodium sulfate and potassium sulfate from the reaction product separating the latter sulfate from the solution, heating together sodium sulfate thus obtained and the residue from the solution, whereby the sodium sulfate is broken up, and from the residue of the latter heating dissolving out alumina as sodium aluminate.

1,227,816. Air-Lift-Pump Booster. Frank S. Miller, deceased, Indianapolis, Ind., by Donald S. Morris, administrator *de bonis non*, Indianapolis, Ind., assignor to William Langsenkamp, Jr., trustee. Filed May 20, 1916.

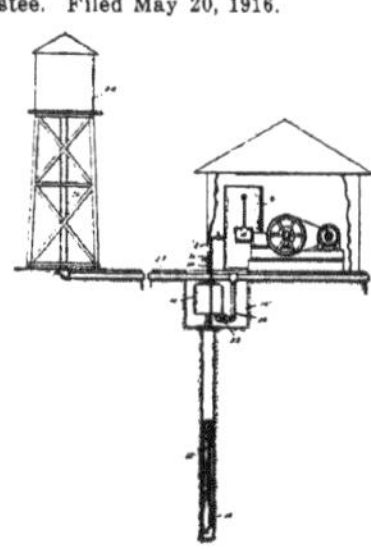

5. In combination, a closed chamber, a pipe leading from a well into such chamber, an air nozzle in such pipe below such chamber, a pipe leading from such chamber and opening into such chamber below the liquid level therein, said chamber having a restricted air outlet opening therefrom which air outlet opening communicates with the chamber at a point higher than the water level therein so as to permit a restricted escape of air while maintaining the pressure of the retained air above atmospheric pressure, and a safety valve responsive to an excessive pressure within said chamber for permitting a greater escape of air from the chamber.

1,227,831. Briquetting-Machine. Grant W. Rigby, Pittsburgh, Pa. Filed Aug. 8, 1913.

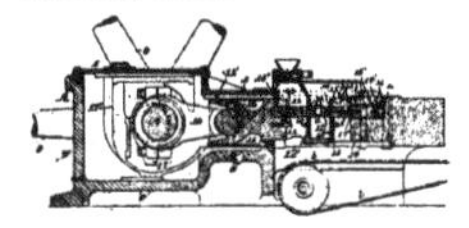

1. A plunger mechanism for a molding machine having a reciprocating feed box for the material, comprising reciprocating plungers coöperating with said feed box and having material-receiving cavities in the end thereof, movable resisting plungers in line with said first-named plungers and having material-receiving cavities in the ends thereof, said resisting plungers being held normally stationary against pressure and having fluid cylinders therein, ejector bars within said resisting plungers for expelling the molded material from the cavities thereof, and differential piston heads connected with said bars and contained in said cylinder.

4. In a molding machine, a receptacle for containing a quantity of material considerably greater than the quantity molded at each operation, a pair of opposed molding plungers having mold-cavities at their opposing ends and arranged, when brought together, to form a mold complete in itself, means for effecting a relative movement between said plungers to effect the molding operation and the discharging operation, and means for effecting a relative movement between said receptacle and said plungers such that the molding operation takes place within said receptacle and the discharging operation takes place outside said receptacle.

REVIEW OF MINING

As seen at the world's great mining centres by our own correspondents.

LORDSBURG, NEW MEXICO

AN OLD DISTRICT RESUMES OPERATIONS UNDER THE STIMULUS OF HIGH PRICE FOR METALS.—LARGE PRODUCTION IN 1917.—CHARACTER OF THE ORES AND GEOLOGY OF THE DISTRICT.

Among the copper producing camps of the South-west that have lately come into prominence owing to the entrance of new companies and the enlarging of operations by the established companies, Lordsburg, situated in southern Grant county, stands out prominently.

The ore of the district is in most cases of shipping-grade. There are no treatment plants in the district. Owing to the highly silicious character of the ore it is in demand at both the El Paso and Douglas smelters. Occasionally it is shipped elsewhere. The ores contain copper, gold, and silver, the first named predominating. I would estimate the average assay at 0.11 oz. gold, 4.07 oz. silver, and 3% copper.

During the years past the district was hindered, as many meritorious districts are in the South-west, by the lack of transportation. Up until 1915 all ore had to be hauled from 3 to 5 miles by wagon, truck, caterpillars, and other traction methods. Wet weather, bad roads, and breakdowns caused the usual delays. In 1915 the Arizona & New Mexico railroad built a spur from their main line to the 85 mine. From that time on the production of the camp was continuous and has increased monthly. The demand for the local ore has also been steady. The camp is now in a position of permanency and many new operators are taking hold of the dormant and half-dormant properties.

The largest producer in the Lordsburg district is the 85 Mining Company. The Lawrence Mining Co., operating the Bonney mine is second in output. These two companies are at present producing practically all the ore shipped from the camp. Lessees ship occasional lots but to no great extent. At the present time the Atwood mine, near the 85 property, is being unwatered and cleaned out for producing by the 85 Extension Copper Mining Co. The Valedon Mining Co., Octo Mining Co., Monte Rico Mining & Milling Co., Hecla Mining Co., and individual property holders have valuable developed and undeveloped claims in the Virginia district; none of which, however, are shipping ore. In the Pyramid district the Nellie Bly, Robert E. Lee, Last Chance, Viola, and other copper, gold, and silver claims are operated occasionally. The production from the Pyramid district, however, will be light this year.

The Lordsburg district is favored with an excellent geographical location. It is near the smelters and larger supply points. Mexican and skilled American labor may be readily obtained. The climate is mild, high, and dry. Fuel-oil is comparatively cheap, being on the main line of the Southern Pacific railroad. No labor troubles have ever arisen in the district. The people of Lordsburg stand ready to encourage and aid new mining enterprises. The proverbial camp-knocker is dying out in this section.

The production of the Lordsburg district in 1916 was about $2,100,700 from 186,000 tons of ore shipped. The forecast of ore shipments for 1917 probably will total $2,800,000 at the present rate of shipment.

The ore-bearing lodes of the district occur as fissures in diorite changing in places to andesite and andesite-porphyry. There are at irregular intervals intrusions of the diorite into the andesite. Where these intrusions occur the ore is generally in large quantities and of high value. The fracturing of the rocks, resulting in the fissures, was followed by ascending solutions, which deposited the ores. Later, descending solutions have percolated through the silicious vein-matter depositing the mineral. This is more fully verified by the fact that the richer gold and silver ores are nearer the surface. There has been no intricate faulting where development work has opened up the lodes although a slight shearing has taken place. The dip of the veins on the 85 property, in the Virginia district, is 76° south-west and the strike is north 40° east. The width of the veins is from 5 to 50 ft. and the value ranges from $7 to $25 per ton. The character of the ore is oxide, chloride, and sulphide, with occasional high-grade chalcocite pockets occurring as ore-shoots in the vein.

LORDSBURG, NEW MEXICO, AND THE 85 MINE

The entire district is traversed by prominent dikes, which caused miners to first turn their attention to the district. These dikes or wall-like veins have an east-northeast trend intersecting at about Lee's peak. The dikes are highly silicious and brecciated, occurring in the porphyry and withstanding erosion.

In the western part of the field diorite-porphyry is common and in the eastern exposure of andesite they are noticeable.

The Pyramid range of mountains are of Tertiary age, the characteristic country rock being andesite-porphyry.

No attempts have been made in later days to treat the ore locally, although a number of experiments in flotation have been made by the 85 Mining Company. A test-plant was erected at El Paso, Texas, by J. W. Crowdus, the smelter representative, and a number of experiments made. At the present time the company is experimenting at the mine with roasting and precipitation.

A dry-mill was erected in 1915 for the treating of some low-grade lead ore found on the east side of the mineralized zone. The mill did not prove successful and was soon closed down after several months' testing was done. Save for an abandoned stamp-mill in the Pyramid camp, there are no further treatment-plants in the district.

Almost as well known as the district itself is the 85 mine and adjoining developed property owned by the 85 Mining Co., which is employing about 600 men and shipping on an average 10,000 tons of ore per month to the smelters. The 85 mine is situated three miles south-west of Lordsburg.

In 1915 the 85 Mining Co. turned in to the State Tax Commission of New Mexico a gross production value of $762,921.78, and in 1916, $1,456,587.29. The net value determined by the State Tax Commission for the 85 Mining Co. was $566,613 for 1916.

The company has 12 patented claims in the Virginia district. The most extensive work, however, is being done on the '85,' '99,' Superior, Emerald, and Mohawk claims.

The ore of the 85 mine is mined by the shrinkage-stope method, with chutes every 20 ft. There are six levels, the lowest being 550 ft. from the adit-level and about 820 ft. from the surface. A main adit has been driven from the east side of the 85 mountain, a distance of 1500 ft. to the winze, from which all hoisting is done. The adit connects with the ore-bins and power-plant. Haulage is done with gasoline-locomotives, being a recent innovation here.

Mining is in progress on practically all the levels and especially the 450 and 550-ft. levels. On the 450-ft. level lateral work extends about 2000 ft. The property is connected with the Superior and other mines on the west.

The shaft has three compartments with double-deck cages. The hoist is electrically driven, Ottumwa type. Driving is done under contract, with Leyner drills. There are four electrical pumps in the mine lifting about 550 gal. of water per minute.

The company intends to continue shaft-sinking until the 950-ft. level is reached. This work is now under way.

A handsome new store building has been constructed; also a large number of houses for employees. A hotel for the single men, and a hospital, which will be used jointly between the 85 Mining Co. and the Lawrence Mining Co., are to be erected this year. Strict attention is being given the sanitation of the camp.

A. J. Interrieden, manager for the 85 Mining Co., is constantly looking out for the welfare of the employees. He has found this a paramount factor in the success of mining. In every way he has bettered the living conditions of the men and is using every means to make it an ideal camp. As a reward for his efforts there has never been any labor trouble in the camp and an efficient working force has been organized.

The 85 Mining Co. is composed of Wisconsin capitalists and business men. Arthur P. Warner and Charles H. Warner, inventors of the auto-meter, are president and secretary, respectively, of the company. B. P. Yates, president of the Berlin Machine Co., manufacturers of wood-working machinery, is vice-president, and a large interest in the company is held by James Barclay, of Hot Springs, Arkansas. J. A. Leahy, W. F. Ritter, and John Gleeson are other large owners of stock in the company.

The power-house of the 85 Mining Co. is one of the most complete and modern of any in the South-west. Power is generated by two 450-hp. Lyons Atlas Deisel-type crude-oil engines which are installed in a new steel power-house 200 by 100 ft. A 250-kva. Allis-Chalmers generator is installed to generate electricity for the mines, and air-pressure is secured from a late type Ingersoll-Rand compressor with an air displacement of 1600 cu. ft. Leyner machine-drills are used in the mine, also Aldrich pumps. There is an auxiliary power-plant equipped with an 80-hp. Fairbanks-Morse crude-oil engine, a 6600-volt generator, and a compressor. J. H. Clark is master mechanic and G. A. Biersach is his assistant.

PLATTEVILLE, WISCONSIN

A Falling Market Disappoints Holders of Zinc and Lead Ore.—Output of the Mines in June.—New Strikes and Diamond-Drill Development.

Metals in general, with the exception of spelter, prospered during June. Zinc was neglected. The situation in the spelter industry could be described in few words—smelters carried unsold stocks. The softening of the spelter market toward the end of June met with a sympathetic response in the offerings for blende.

Operators viewed the situation at the close of the month with misgivings. Such as were protected in their deliveries by long-term contracts are resting quietly. Others affiliated with smelter interests, or magnetic-separating plants, operating in the field, were enabled to produce consistently. The independent joint-stock corporation, compelled to solicit bids, fared badly except in the Linden district, where nearness to reduction-plants at Mineral Point obviated excessive transportation costs. The independent producers that had facilities for raising ore to top grade have been successful in disposing of the finished product but have not always had a free field. Only when competition becomes accentuated between the large buyers, and there is a keen demand for high-grade ore for ready metal, have they been permitted to enjoy a period of real prosperity. The high cost of living leaves the large employers of mining help fearful to even hint at reductions in wages.

Lead, the last metal to respond to the War's demands, has shown increased activity, but with a price movement comparatively slow, that has given way to a market full of interest. In the Wisconsin field prices of lead ore advanced steadily during the first half of June until offerings at one time stood at $132 per ton, base of 80% metal. Competition developed between buyers and for a time some ore came out of unexpected places, and it looked as if miners would clean-up but the nature of the pig-lead market encouraged many to hold in the belief that still better prices would be obtainable. With pig-lead almost three times as high in price as before the War, and ready metal scarce for prompt and nearby shipments, producers considered that they were safe in holding. In the midst of the hour of greatest hope came a drop to $120 per ton.

Deliveries of ore from mines direct to reduction-plants in the field and to smelters from May 28 to June 30 were: Zinc, 48,605,000 lb.; lead, 1,190,000 lb.; and pyrite, 6,230,000 pounds. High-grade blende from refining-plants to smelters was sent out as follows: Benton, 6,968,000 lb.; Mineral Point, 5,706,000 lb.; Cuba City, 5,066,000 lb.; Linden, 342,000 lb. Total, 18,082,000 pounds.

The total recovery of mine-run product for the field from May 28 to June 30 aggregated 24,510 tons; total net deliveries out of the field to smelter, 15,430 tons. The reserve in the field at the close of June exceeded 2000 tons of raw ore at all mines; no high-grade ore was carried over; 1000 tons of lead ore and 2500 tons of pyrite.

Sales and distribution of zinc ore made during June was as follows: to Mineral Point Zinc Co., 7106 tons; Wisconsin Zinc Co., 4989 tons; Grasselli Chemical Co., 4689 tons; National Separating Co., 4691 tons; American Zinc Co., Hillsboro, Illinois, 4009 tons; Linden Zinc Co., 1465 tons; American Metal Co., 927 tons; Illinois Zinc Co., 869 tons; Matthiesen & Hegeler Zinc Co., LaSalle, Illinois, 666 tons; Lanyon Zinc Co., Joplin, Missouri, 602 tons; Benton Roasters, 551 tons; Edgar Zinc Co., 175 tons; total, 30,739 tons; the period being considered one of the best showings yet made in the Wisconsin field and far below normal outputting capacity, there being a dearth of mining labor at all points in the field.

During June local producers of carbonate zinc ore at Highland had few offerings for this class of ore, and shipments were negligible, but the New Jersey Zinc Co., operating with new producers, increased its deliveries. The Linden district found a ready demand for its low-grade sphalerite with the Mineral Point Zinc Co. The Optimo Mining Co. had an unusually successful month. Mine No. 3 produced at the rate of 125 tons of raw concentrate per week. Mine No. 4, just opened up on the James J. Rule lease of 120 acres, discovered an extensive deposit of rich zinc ore. Drill-squads were employed on the lease all month and will continue exploration until the range has been fully prospected. The Spring-Hill Mining Co. struck new deposits on the William Ross land, and drills were kept at work proving the ground. Kletszch Bros. drilled a vein of zinc ore on the John Wasley farm and found sufficient to warrant both mine development and surface-rig. A new local mining enterprise, organized as the Super-Six Mining Co., developed a lead-ore producer on the Frank Bottoms farm. The Weigle mine, recently developed, reached, through underground workings, a new deposit of zinc ore after months of effort and recoveries are running from 10 to 12 tons of 40% zinc concentrate per shift of nine hours daily. The new producer is owned by the Milwaukee-Linden Development Co. The local magnetic-separating plant, controlled by the Zinc Concentrating Co. in New York, finished blende assaying as high as 62% zinc, calling for top prices. Official announcement was made that this plant, and that operated by the same company at Cuba City, will be doubled in capacity. Additional Dings separating-machines are to be installed. New devices for consuming gases will be a feature of the plants. The O. P. David Mining Co., in the Montfort district, drove west toward the zinc deposits found by drilling during May, and increased output was the result. Several new mining companies were organized to explore and develop new leases that are presumed to be on the trend of the main range.

In the Dodgeville district the North Survey Mining Co., in 50 working days from the time ground was first broken for foundations, built and set in operation a new 150-ton concentrator. Shipment of the first car of blende was made June 28, ore assaying over 50% zinc. The deposit occurs at a depth of 50 ft. Electric power is used throughout the mill.

Mineral Point and its vicinity, which has shown little activity in mining for many years, is being rejuvenated through the efforts of the Utt-Thorne Mining Co., which has a re-built mill in operation, a heavy pump installed, and three contiguous leases being explored. The mill makes a concentrate assaying 55% zinc. Magnetic-separation is rarely undertaken here where such a quantity of zinc is shown. Sometimes operators mix high-grade mill-feed with the low-grade to insure a better recovery.

MEXICO

New Law Regulating Oil Industry.—Petroleum Committees' Recommendations.—Operation of Law Will be Almost Equivalent to Confiscation.—Productive Capacity of the Oil District.

Printed copies of the first draft of the proposed law for the control and regulation of practically all phases of the oil industry in Mexico have been distributed among the interests that are to be affected. Provisions of this measure which will soon be laid before Congress are of such a drastic nature that their enforcement would mean the virtual confiscation of the landed holdings of foreign as well as native investors throughout the oil-producing region. The bill in its present form contains more than 150 printed pages. Most of these are devoted to new regulations which, if put into law, would make the oil industry of that country so complicated as to almost prohibit it being continued, it is claimed. The proposed law is entirely separate and distinct from the taxation decree which went into effect on July 1 and which imposes a heavy burden upon oil producers. The new bill as now drawn is largely devoted to the land feature of the oil industry. As one illustration of its possible far-reaching effects it may be stated that it places the oil of the country in the same category as the metalliferous minerals, and it applies to oil development many of the regulations now contained in the mining law of that country. All land that has been leased or is owned in fee simple for the purpose of exploiting oil is required, under this proposed new law, to pay to the Government an annual tax of five pesos per hectare, which is equivalent to a tax of $1.01 gold per acre on the land. Several of the larger oil companies own 1,000,000 to 4,000,000 acres of land and the imposition of this tax would mean an enormous annual outlay. It is estimated that approximately 25,000,000 acres of land are situated in what is known as the proved oil region bordering that part of the gulf coast around Tampico. At a tax of $1.01 gold per acre it would mean the bringing to the Government of an annual revenue of about $25,000,000 from this source alone; but this is not the worst and most drastic provision of the bill. Placing oil under the mining law makes all privately and publicly owned lands subject to denouncement for its possible oil wealth. In other words, if this pending measure is enacted into law, any person can go upon the land of another, whether the oil-rights upon it be already leased or not, and locate a claim of a certain size upon which to bore prospect wells, and if he happen to strike oil, which is a posible result in nearly all parts of the Tampico territory, he will have a clear title to it. It is provided, however, that the producer of the oil shall pay to the Federal government one-half of whatever royalty the original lessee may have agreed to pay the owner of the land. The new law gives to the Government a royalty on all oil that may be produced.

It is claimed that if the oil lands are made subject to denouncement by anyone who may want to go upon them and put down a well, it will mean the financial ruin of every large American and British oil-producing company now operating in that country. It is reported to be the announced purpose of President Carranza and those that occupy high positions under his administration, to cause the dividing of the oil-land holdings of the foreign investors in the Tampico region. The application of the mining law to oil development will cause a great rush of prospectors into the different fields, and instead of this enormous underground wealth being in the possession of a comparatively few concerns, it would be divided among many, and in this way the Mexicans themselves, who have up to this time neglected their opportunities for engaging in this industry, will be placed in a position to acquire, with but little cost to themselves, valuable oil holdings.

The preparation of this drastic bill was begun by a specially appointed congressional commission about six months ago and it is said to be now ready for final action of Congress. Although there has been comparatively little done in the matter of boring new wells during the past two years, a careful survey which has just been taken of the completed wells shows that they have at this time a total available output of about 365,000,000 barrels per annum. The exportations aggregate about 4,500,000 barrels per month or at the rate of about 54,000,000 barrels per annum, which leaves an underground developed supply of approximately 311,000,000 barrels. It is stated by leading oil operators here that it would be an easy matter to more than double the oil production of the Tampico territory and that this could probably be brought about within a period of six months. In fact it might be done in that many weeks as it has already been proved that it is possible to bring in wells with a daily flow of upward of 200,000 barrels each. It is not a problem of available or possible production there, but it is one of transportation and marketing facilities which has to be met before the present export figures can be materially increased.

LEADVILLE, COLORADO

A GENERAL STRIKE OF UNION MEN THREATENED, AND STATE TROOPS ARE QUIETLY BEING SENT INTO THE DISTRICT TO PROTECT PROPERTY.—THE PRINCIPAL MINES CONTINUE TO WORK, THOUGH MANY MEN HAVE LEFT THE DISTRICT.

Cloud City Miners' Union No. 33 of the International Union of Mine, Mill, and Smelter Workers, by a referendum vote taken July 3, declared in favor of a strike in the mines of the Leadville district, by 641 to 72. The strike has not yet been called, as the local organization is waiting for the arrival of a member of the executive committee of the International Union. President Carpenter and Secretary Follett, of the Cloud City Union, also stated after the result of the voting was announced that a strike would not be called until every effort to bring about a settlement with the operators had failed. This decision has tended to release the tension of the situation, although it is the general opinion that a strike will be declared, as the operators persistently refuse to make any concessions or to offer a compromise.

President Charles H. Moyer of the International Union of Mine, Mill, and Smelter Workers, predicts that a settlement will be reached before it becomes necessary to call a strike. After having been informed of the situation here showing that a majority of the union members favor a strike, he wired the following statement to the union headquarters.

"I advise and earnestly request that you rturn to your employment Thursday morning. Also that you elect a committee with full power to act with representatives of your international with regard to the wage demands that you have made to your employers."

This caused many men to return to work who had previously decided to draw their time-checks, or to remain away from the mines until a settlement had been reached. Nearly all of the mines continue to operate with nearly full crews. A number of Austrian miners failed to return to work, many of them having drawn their time. A few non-union men have left the district. The general opinion among the operators is that the Austrian element in the union is controling its action, and that they have determined upon trouble that will cripple the mines. The union leaders claim that the organization has over 1000 members. If this is true, the vote taken on the strike was light, and there is no doubt that the members who did not vote were the men not favoring a strike. It was noticed that those who went to the polls were mostly foreigners, which bears out the contention of the operators that the Austrian element is in control.

A strike at this time is unpopular here among business men, mechanics, pump-men, and engineers in the mines, non-union miners, and miners generally. It is estimated that there are 2300 miners in the district, and the union leaders only claim to have 1035, many of them being smelter-men. Sentiment is not against the demands of the union, as it is conceded that under existing living costs the men are running to the limit of their wages for general expenses, and that they must have relief before long; but it is believed that action which would permanently cripple the mines of the district, throw 3000 men out of work, bring hardship to the families of as many more, possibly destroy millions of dollars worth of property, and stop the revival of mining in the Leadville district, at a time when it promises so much for the future.

On the morning of July 4, following the day of voting, one of the guards patroling at the power-station of the Colorado Power Co., near the Yak tunnel, was fired on by a sniper hiding behind one of the buildings. The bullet pierced the hat of the guard but did him no injury. He returned the fire, but as it was dark, he could see nothing to shoot at. A search for the sniper followed but no clue was found. On the night of July 4, a heavy charge of dynamite was exploded in an old shaft on the Sixth Street property, one of the Down Town mines that is idle. The shock broke windows in surrounding residences and store-buildings and caved the shaft.

State troops are being quietly brought into the district and stationed to protect the city light and water-plants, the smelter, and other important points.

The operators held a meeting July 6 to discuss the situation, but nothing was done that would tend to bring about a settlement. They are awaiting the action of the union, believing that should a strike be called it will prove ineffective, as many of the miners will continue to work.

New development in the district is retarded because of the pending labor trouble. A number of large enterprises that had planned to start operations about July 1 have made no attempts to proceed. On the borders of the district, however, a few companies are taking up development regardless of the situation, as it is necessary to pay higher wages to men in these out-of-the-way places than at the properties near town.

At a meeting of the union held July 9, it was decided to call the strike at noon on the 14th unless a settlement is reached in the meantime. It has also been announced that former Judge Musser, of the State Supreme Court, and Verner Z. Reed, both of Denver, have been appointed by Secretary of Labor Wilson to act as mediators in the situation. The union officials have stated their willingness to call a special meeting and further postpone the strike if the government mediators find that they need more time.

A. S. Sharp, cashier for the Leadville Water Co., and associates, have secured a long-term lease on the Houston claim in Iowa gulch, lying immediately south-east of the Ella Beeler property; and have started work cleaning out the adit.

KENNECOTT, ALASKA

STRIKING MINERS AT KENNECOTT MAKE ARBITRARY DEMANDS ON THE MANAGER OF THE KENNECOTT COPPER CO., WHICH DEMANDS ARE REFUSED.—FEDERAL TROOPS GUARD THE COMPANY'S PROPERTY.

E. T. Stannard, manager for the Kennecott Copper Corporation, returned from New York on June 25. The miners, about 200 in number, that had walked out about two weeks ago, in his absence, renewed their demand for $5.75 per day, flat wage, and gave notice through their committee that they would not even agree that that scale would obtain for any definite period. Under such an uncertain demand the management refused further to deal with the committee and immediately posted a new scale of wages effective June 16, 1917, as follows: The standard wage-scale will continue as the base-rate—that is $4.25 per day for miners and $3.75 for shovelers when copper is under 15c. per lb.; copper between 15 and 18c. per lb., bonus 25c. per day; copper between 18 and 21c. per lb., bonus 50c. per day; copper between 21 and 24c. per lb., bonus 75c. per day; copper between 24 and 27½c. per lb., bonus $1 per day; copper over 27½c. per lb., bonus $1.25 per day; settlement to be made on the average price of copper for the preceding month as given in the *Engineering and Mining Journal* quotations. Mess employees on a monthly basis will receive a bonus of $4 per month for each 25c. change in bonus for employees on daily basis. The men agreed to return to work on this scale if the manager would take them all back. This the manager refused to do, reserving the right to reject any individual. None of the strikers have returned to work yet. The management is preparing to resume work at the mines as soon as the necessary men can be secured.

A squad of 24 United States soldiers was sent from Fort Liscum, near Valdez, to Kennecott on June 27, and are now guarding the company's property. It seems to be the general impression that plenty of men can be had within a week or two to resume operations at the mines.

THE MINING SUMMARY

The news of the week as told by our special correspondents and compiled from the local press.

ALASKA

(Special Correspondence.)—Charles Davis, of the Alaska Petroleum & Coal Co., is authority for the statement that his company will be prepared to deliver coal to the towns along the Alaska coast early this fall. A railroad from the mines to a point on Controller bay, where the coal can be picked up by barges, is now under construction. The length of the rail-haul is 25 miles.

Clark Davis, vice-president and general manager of the company, is expected at Katalla at an early date.

Katalla, July 1.

In view of the disappointing results at the Alaska Gold property, the ore running 50 to 75c. per ton below expectations, the starting of the mill on the Alaska-Juneau property adjoining has been awaited with much interest. The results to date show a very low yield of gold and compare with an average assay at Alaska Gold of $1.19 in 1916 and $1.15 in 1915. F. W. Bradley, president of the Alaska-Juneau company, says to the *Boston News Bureau:* "One unit of the new mill started up March 31, and in April a few additional units were started, resulting in a total crushing for the month of April of 7046 tons, having an average gold assay-value of 72.6c. per ton. In the month of May, 37,000 tons was crushed, having an average gold assay-value of 77c. per ton. For the first 10 days of June, 17,229 tons were crushed, and during the second 10 days 16,000 tons were crushed. All starting-up troubles should be gradually overcome during this summer and the mill will be in full operation by the end of the year. The milling and auxiliary surface-equipment has cost to date a total of $2,532,027. The ore, as hauled from the mine, is first delivered for coarse crushing to a rock-house, following which is the mill proper composed of 12 units with a primary-crushing ball-mill at the head of each unit. These ball-mills were originally designed to have a capacity of 700 tons each; but it was afterward planned to push this capacity up to 1000 tons each per day, if possible; so the whole plant should have an eventual capacity of anywhere from 8400 to 12,000 tons per day. At an average daily capacity of 11,000 tons, we expect to recover 80c. per ton at a cost of between 40 and 45c. per ton."

ARIZONA

There has been a great deal of excitement in the copper districts of Arizona the past week, every important copper producing district except Ray being affected. At Jerome an attempt to call a strike was promptly defeated and one of the I. W. W. contingent killed Orson P. McRae, shift boss in the Copper Queen mine at Bisbee, by shooting through a door. The murderer was promptly killed by deputy sheriffs.

A citizens committee and deputy sheriffs loaded over 1100 I. W. Ws. in box and cattle-cars and took them into New Mexico, where they now are guarded by Federal troops near Hermanas.

At Globe, John McBride and G. W. P. Hunt, ex-governor of Arizona, Federal mediators, appealed to President Wilson to take action to stop further deportations of strikers from Bisbee and other copper districts of Arizona. Governor Campbell wired President Wilson to send Federal troops into Arizona mining regions where strikes are in progress. These are at Clifton, Morenci, Bisbee, Jerome, and Ajo copper districts; Humboldt smelter in Yavapai county, and two districts in Mohave county.

President Wilson directed General Parker, commander of the Southern Department, to take whatever steps were necessary to protect life and property in Arizona.

Drastic action was urged in Washington against the lawless element that seems bent on destruction of industries and crops in the West.

It is reported that the Western Union Telegraph Co. will investigate the alleged censorship of its lines leading out of Bisbee by Robert Rae, general auditor for the Copper Queen Co., and H. H. Stout, superintendent of the smelter.

MOHAVE COUNTY

(Special Correspondence.)—A strike described by J. A. Burgess, J. L. McIver, and other competent authorities, as

MAP OF ARIZONA

second only in importance to that of the United Eastern was made last week on the Telluride property, which lies southwest of and adjoining the Tom Reed. The Telluride is controlled by J. L. McIver, one of the discoverers of the United Eastern. The vein is presumed to be an extension of the Aztec-Tom Reed vein, which is faulted near the end-line of the two properties. A drift had been advanced for 300 ft. from the shaft on the 535-ft. level along the hanging wall of a well-defined vein carrying ore intermittently. At the point of the discovery referred to the vein widened and became more solid, being composed of quartz and spar. High assays are reported. In order to explore the vein at greater depth a winze is now being sunk.

Oatman, July 10.

CALIFORNIA

The scarcity of oil-well casing is resulting in the abandonment of various old wells for the purpose of salvaging the casing. The State Mining Bureau insists on strict compliance with its recommendations as to methods of abandoning wells so as to prevent future damage to oil lands. Each well requires special treatment and no general rules can be issued. Owners of land upon which such wells may be situated are particularly cautioned to see that the work complies with regulations, as the land is made liable under the law for the cost of remedying improper work.

Reports filed with the State Mining Bureau for the week ended July 7, show 20 new wells started, making a total of 597 since the first of the year; 16 wells were reported ready for test of water shut-off, 12 deepening or re-drilling, and two abandoned.

CALAVERAS COUNTY

(Special Correspondence.)—The Lockwood mine, situated two and one-half miles north-east of West Point, is controlled by W. O. Pray, who is also superintendent. He is putting in a cyanide-plant. There is in the mine thousands of tons of ore in sight that averages $7 to $9 per ton. The Rehfus Bros. are cyaniding their tailing at the Gilbertson mine, near West Point.——The Extension of the Comet mine near Railroad Flat has started a small crew of men working two shifts.——The Black Wonder mine, situated in the Blue Mountain mining district, has started operations. It is a contact vein of slate and granite and assays well. From the Eureka Gold Mining Co. at Virginia City, Nevada, the owners bought a 60-hp. hoisting-engine and a compressor. The ore though base is high-grade.——Some Eastern capitalists, reported to be Kansas people, are sampling the Sawyer claims, which are owned by Wm. Foltz of Seattle.——The Comet mine is still working and prospects are getting better every day.——The owners of the Mason & Phillips mine have just taken out a large quantity of ore, which was hauled to the Porteous custom-mill and the bullion shipped to the San Francisco mint. Both walls of this mine are granite. It is situated in the East Belt.

The Secretary mine has just cleaned up and the mill returns went $50 per ton.——W. O. Pray, of the Lockwood mine, has an option on the Lone Star mine two and a half miles from West Point on the Mokelumne river.——The difficulties at the Blazing Star mine, which has been in litigation for many years, have been finally adjusted and in the transaction which followed, W. O. Pray & Co. have acquired the controlling interest and now virtually own it outright. It will soon start up.——A new strike has been made by a miner, doing assessment work at the Camille, or 'Corn-Meal' mine, who uncovered a vein 14 ft. wide in slate. A moderate excitement took place locally over this development.——Wendell Phillips, of Lodi, has cleaned-up 20 tons of free-milling ore, which averaged $30 per ton.——Judge Condon is doing assessment work at the Enterprise mine for Fred Plagemann, and has found rich sulphide ore.——At the North Star mine, situated midway between the Lockwood and the Blazing Star, the shaft has reached a depth of 57 ft. The vein is 22 in. wide with a greenstone hanging wall and a granite foot-wall. The vein consists of blue quartz with sulphide, assaying all the way from $2 to $110 per ton. The pay-shoot goes $100 per ton. The present owner has had the claim for 33 years. It would seem a good proposition for some company to develop it, as the owner is unable to do the work that the mine justifies.——At the Deerfoot mine, controlled by the Rehfus Bros., they are cyaniding 500 tons of ore that runs from $5 to $9 per ton.

West Point, July 10.

DEL NORTE COUNTY

(Special Correspondence.)—Chrome ore mining is becoming increasingly important in the hills back of Adams station on the Crescent City stage-road. This station is on Smith river, just west of Gasquet. A large number of claims are leased by R. J. Rowen, M. E. Young, and Geo. S. Barton, all of Grants Pass, Oregon. About 40 openings have been started, in most of which a good showing of chrome has been made. At one place 5000 tons of ore has been developed. The claims are on what is locally called French hill. Shipments will be made by trucks to Crescent City and from there by water to San Francisco. Development work has been in progress only two months, but the ore developed warrants the commencement of shipments. The ore runs from 50 to 62% chromium oxide.

Near the Old Altaville mine, in the Low Divide district, John L. Childs and Mr. McMurray have located what is proving to be a large body of chrome. This property adjoins the Tysen Chrome and produces the highest grade of ore in the district.

PART OF CHROME REGION OF CALIFORNIA

Several auto-trucks are now hauling ore to Crescent City from this district.

W. J. Ehrman has been in Grants Pass for the past week making arrangements to get machinery hauled in to his Diamond Creek cinnabar property. There is a good road to the mine from the main stage-road, and the machinery is to be at the mine this week. This is the only developed cinnabar property in this district. Most of the ore is high-grade and it is thought that it will soon be a profitable producer.

Crescent City, July 6.

EL DORADO COUNTY

The Noble and Farmer chrome mines are employing more than 100 men in mining and transporting chromite to the railroad, and the industry is growing as other deposits are being opened in various places.

(Special Correspondence.)—The Teddy Bear gold-quartz mine, situated on the main Mother Lode, between the old Church-Union and the Laus Padre mines, about three miles south of El Dorado railroad station, will be extensively developed by the El Dorado Exploration Co. of Seattle, under the management of John W. Cover. Five men have been working in the mine for some time, and Mr. Cover recently returned from Seattle for the purpose of pushing the development of the property. June 29 a 5-ft. vein of ore, similar in appearance to the high-grade ore of the old Church-Union mine, was found in sinking a winze, on the Teddy Bear claim, near the face of the 150-ft. cross-cut adit, at a vertical depth of about 160 ft. below the surface. The average assay-value of the ore is $60 gold per ton. Burr Evans, consulting engineer, of Placerville, who recently visited the mine, states that the management will sink the winze to a depth of 500 ft. and open up the pay-shoot by driving north and south. After the orebody is developed a raise over the winze will be cut, thus making a shaft to the surface, and a mill of 50 tons daily capacity will be installed. A mill-test run of a dozen tons or more of the rich ore from the vein in the winze will be made in a 5-stamp mill on Grant Busick's mine.

William Vaughn, of Georgetown, arrived at Placerville on July 9 with his six-mule team loaded with two and a half tons of high-grade chrome ore in sacks, from the Bald Mountain chrome deposit, 23 miles north-east of Placerville. Over 250 tons of ore has been shipped from this deposit, and there is a large amount still to be mined. The ore averages 63% chrome oxide and less than 3% silica.

Placerville, July 10.

INYO COUNTY

(Special Correspondence.)—The Darwin Development Co. is steadily increasing the development of its group of mines near Darwin and has blocked out an extensive tonnage of gold-bearing ore. A 25-ton mill, using flotation and a magnetic system of separation, is in operation and its capacity is to be increased to 75 tons daily. Late mine developments have been reported as satisfactory.

The lower tunnel of the Santa Rosa, in the White mountains, 26 miles from Keeler, has cut the orebody at a vertical depth of 400 ft. Approximately 200 veins and shoots have been cut, ranging from a few inches to 8 ft. wide. The ore contains silver and lead, and heavy shipments are being made to the Midvale smelter. Ten leases are also active. In all 55 men are employed.

Shipments from the Cerro Gordo are averaging around 50 tons daily, with much of the ore running high in silver, lead, and zinc. Shipments of slag from the old smelter-dumps continues to be profitably made. The large zinc deposit recently uncovered is stated to be developing well.

Keeler, July 10.

NEVADA COUNTY

(Special Correspondence.)—The Cherokee and Columbia Hill districts, near the famous old hydraulic mining camp of North Bloomfield, are active. Fully 100 men are employed in prospecting gravel deposits and tailing from pioneer hydraulic operations, and several drills are in operation. It is reported the area will be drilled to Lake City. Many of the properties in this field have been idle since the anti-hydraulic mining laws went into effect. It is understood the Guggenheims are interested in the new work, and that it is purposed to work the lower gravels with dredges.

The Roger group of quartz claims at Beauty Flat, near Washington, is being developed by the North Star Mines Co. of Grass Valley. The tunnel is advancing in promising ground and equipment will be installed. The company is understood to be considering the purchase of several placer properties in this district. The passing of the June dividend by the North Star Co., due to increased mining costs and lower gold content of the ore, has caused alarm in Grass Valley, where the regular quarterly dividends have come to be taken as a matter of course.

Dewatering of the Allison Ranch mine is proceeding and considerable work has been commenced above the 400-ft. level. Rich shoots have been exposed and the management expects to have the mill in operation shortly. The company has acquired the Benoit tract of 168 acres, adjoining the Allison Ranch on the east, and plans a thorough exploration of the area. The company, the Grass Valley Consolidated Mines, now owns 300 acres of mineral land, all to be worked from the Allison Ranch shaft. C. K. Brockington is manager.

Fifteen men are working the Goodwin placer property at You Bet, under management of Louis Gidette. It is said Austrian capitalists have purchased the property and that an effort may be made to work the gravel by hydraulicking, as other methods have not proved satisfactory.

Grass Valley, July 11.

The Golden Centre Mining Co. at Grass Valley has decided to discontinue all stoping and milling operations and to devote all energies to development of orebodies. The board of directors has authorized the expenditure of $200,000 to carry on this campaign of development, which includes the sinking of the shaft an additional 1000 ft. This will give the mine a total depth of 2000 feet.

SHASTA COUNTY

(Special Correspondence.)—At the electric smelter at Heroult, the blast-furnace will be completed by the end of the month. This furnace will have a capacity of 30 tons per day, and smelting will be done by blast as well as by electricity. Great inconvenience is caused by the inability to get electrodes for the electric process, as they are needed. The electric furnace is turning out ferro-manganese at the rate of 8 to 10 tons per day, the ore coming from Livermore and Mendocino county. Ferro-manganese is selling in New York around $450 per ton.

The Afterthought Copper Co. at Ingot is employing 150 men. The flotation-plant will be completed by the close of the month. When that comes about the mining of ore will begin. The mine and smelter had been idle since January 1908 until recently. George L. Porter is superintendent.

At the Bell Cow mine, twelve miles west of Ono, on Arbuckle mountain, C. L. Wilson, who has the property under bond, has completed an assay-office, which with equipment has cost $1000. The 5-stamp mill is being overhauled and will be in operating condition by the end of the month. The mill has a daily capacity of 25 tons. It will be used in making a working-test of the ore taken out in development work.

The Mountain Copper Co. keeps 20 men at work on its New Year's claim in the heart of Balaklala ground. Ore shipments are made regularly by way of Coram to the smelter at Martinez.

At the Silver King mine, four miles west of Redding, L. C. Parker has put in heavier machinery and will soon begin to sink 125 ft. deeper to open up a new level.

The Star shaft at Bully Hill has been unwatered to the 900-ft. level.——J. Barnes, operating the Summit mine west of French Gulch, has leased the Black Tom mine in the same neighborhood.

Redding, July 9.

TRINITY COUNTY

(Special Correspondence.)—The Trinity Star Co. is hauling lumber to the site of its new dredge to be built at Lewiston, on the Paulsen ranch. The pit for the dredge has been dug. It is 100 ft. square and 9 ft. deep. The dredge will have a wooden hull. Most of the machinery is in Redding awaiting transportation.

Redding, July 9.

TULARE COUNTY

(Special Correspondence.)—Approximately 72,000 tons of magnesite has been taken from mines in the district around Porterville, in Tulare county, during the six months ended June 30. The total value of this magnesite is nearly $1,000,000, about $160,000 being the average amount paid out monthly for magnesite both crude and calcined during the past six months. This is an increase of 200% over the output of ore for the same period last year.

From 600 to 700 men are employed at present in the various phases of the magnesite industry in this district, mining and trucking the ore, and in and around the calcining-plants and offices of the several companies operating in the Porterville district. This number of employees represents an average monthly pay-roll of about $85,000 during the past six months.

The American Refractories Company, of Pittsburg, Pennsylvania, is the biggest buyer of magnesite in this district at present, handling approximately two-thirds of the output. Daily shipments of magnesite from Porterville now average about 400 tons. This district is at present producing from one-third to one-half the total output of magnesite in the United States.

Porterville, July 10.

SISKIYOU COUNTY

(Special Correspondence.)—W. R. Beal, of Happy Camp, the

owner of the Know-Nothing gold quartz mine on Know-Nothing creek, in the Salmon River district, adjacent to Happy Camp, recently took the mine under lease. It had been idle for 20 years. He is now having the 8-stamp mill overhauled and remodeled, and has bought a compressor and will use machine-drills. The machinery is operated by water-power, which is available throughout the year. He is purchasing a large quantity of pipe and auxiliary machinery and has contracted with the Ladd Bros., of Happy Camp, to pack the machinery 7 miles to the mine from Forks of Salmon. He contemplates having the mine and mill running within six weeks. There is enough good ore in sight to operate for three years.

The Fledderman Bros., of Etna, have purchased a 3-ton Packard truck and are hauling chrome ore from their mine near Etna to the railroad. They have shipped two car-loads of ore and are loading the third car.

J. B. Nesbit and Wm. Magill, of Happy Camp, have made a road and established a camp at their copper mines on Deer Lick gulch on Indian creek. This property adjoins the Gray Eagle mines.

Harry C. Dannehower, of Philadelphia, who represents capital from that city, is investigating the copper deposits in Happy Camp district.

Dan. L. Harrington, manager for the Sullivan Machinery Co. of Chicago, has been superintending the drilling for his company on the Gray Eagle mine on Indian creek near Happy Camp. He reports that the contract for 6000 ft. of drilling on the property is about completed, and that his company has entered into a new contract for several thousand additional feet.

Harry Wilson, of Happy Camp, who has been employed at the Clear Creek copper mines, which were closed down, has left for his holding on Elk creek, which he will further develop.

While freighting a massive boiler up the Indian Creek road one of the Reichman freight teams crashed through the bridge at the George Crumpton ranch, completely demolishing the bridge, but with little damage to the freighting equipment or machinery. A new and safe bridge was constructed without delay.

Hornbrook, July 8.

Tuolumne County

O. A. Ellis, manager of the Chaparral mine, which adjoins the Buchanan mine on the south, has placed a Chile mill of novel design on his property for the purpose of testing the ore, and later to regularly operate the property if results are satisfactory. The mill weighs 5500 lb. and instead of the usual upright grinding mullers that roll around the basin, iron balls weighing 400 lb. each have been substituted. The upper part of the mill, which revolves, rests on the balls and causes them to roll around the groove at the edge of the basin of the mill, crushing the ore as it falls beneath them.

COLORADO

Park County

(Special Correspondence.)—Because of the lateness of the opening of summer, work here has been held back more than a month. However, the good weather during June has made the resumption of mining possible and several properties are working, and many more will receive attention during the summer. C. R. Welsh and associates, who have been working the Wheeler mine since August last, are planning the installation of a compressor and air-drills. The power will be furnished by a gas-engine. A new adit is to be driven to cut the [illegible] 140 ft. below that now opened up in No. 2 level. The ore contains from 25 to 48% lead, 2 oz. gold, and 2% copper per ton, and occurs as a fissure vein in the granite. Besides the shipping ore the mine contains a fair tonnage of milling ore. Mr. Welsh has purchased a burro train for his own use, but will be able to do considerable packing for other miners.

W. J. H. Miller is removing the ice from the adits of the Atlantic and Pacific properties which are reported to contain a large tonnage of low-grade gold ore. Any milling ore found in these properties will be treated by the Commonwealth Mining Co. The Hock Hocking mill was started June 1 and is exceeding the expectations of the management. The new oil-engines are satisfactory. Development work at the mine is being carried well in advance of the daily mill capacity. The Commonwealth Mining Co. has again started work on the tramway, which will be completed about August 1. The mill soon will be ready and will handle custom ore in addition to that from the company mine.

Alma, July 10.

San Juan County

The United States Smelting, Refining & Mining Exploration Co. has acquired controlling interest in the Sunnyside and Gold Prince groups of mines on Hanson's peak near Silverton. The properties are extensively developed and are reported to have blocked-out 850,000 tons of ore. The Sunnyside and No Name are the important veins, and these have been explored for a horizontal distance of 4590 feet. There are eight levels in the Sunnyside that aggregate 11,580 ft. of work. The veins run from 6 to 7 ft. wide. The ore is uniform in character and averages approximately 0.1 oz. gold, 6 oz. silver, 5½% lead, ½% copper, and 9% zinc per ton. At Eureka, the town nearest the mines, a 500-ton flotation mill is being built. Much of the steel frame work of the old Gold Prince mill is being used in the construction of the new mill.

IDAHO

Shoshone County

(Special Correspondence.)—The million-dollar smelter of the Bunker Hill & Sullivan M. & C. Co., at Kellogg, was blown-in a few days ago. There was not a hitch of any kind, everything working smoothly from the start. Ingots bearing the initials of the company, 'B. H. S.,' were run from the first metal handled, and distributed among the visitors as souvenirs. The smelter has been so constructed that it may be enlarged from time to time. The next improvement will be a zinc electrolytic plant.

Spokane, July 11.

(Special Correspondence.)—On the 400-ft. level of the Giant Ledge Mining Co.'s property, on the North Fork of the Coeur d'Alene river, 25 ft. of good milling ore has been developed and the company has decided to build a mill of 150 tons capacity, to be enlarged later if a greater output can be made. Charles G. Taylor, of Murray, Idaho, is manager of the property. This mill and the development of the property was made possible by the Washington Water Power Co., of Spokane, which extended its power-line into the Murray district to supply the Guggenheim dredging operations. The power-line will pass over the ground of the Giant Ledge company and is expected to be completed by early fall.

Stockholders of the Northern Light Mining & Milling company at their annual meeting at Wallace elected directors and decided to build a 150-ton concentrator at the property on Pine creek. Plans already are being prepared and the plant will be built this summer. This is a lead-zinc property which has been well developed. It has three veins, on one of which work has been done to the 400-ft. level, where an ore-shoot 250 ft. long with an average width reported of 57 ft. The stockholders are mostly Eastern men.

The flotation-plant in course of construction for the Consolidated Interstate-Callahan company will increase the production 25%, according to an announcement by John A. Percival, the president, at Wallace. The shipment of 8000 tons monthly is expected when the flotation-plant is completed. The lead shipments for June aggregated 1000 tons, the largest in the history of the mine. It is believed this output can be maintained. When the flotation-plant is in service the zinc output is expected to be 7000 tons monthly.

Lead producers of the Coeur d'Alene have agreed to divert one-sixth of their July output to the Government. A meeting was held in response to a call issued by Harry L. Day, member of the lead committee, at the request of Chairman Crane. Nearly all big producers of the district were present.

Spokane, July 2.

MONTANA

SILVER BOW COUNTY

Press dispatches indicate that the mining industry at Butte is almost at a standstill, thousands of men being idle as a result of the strike of electricians, who are demanding of the Montana Power Co. a higher scale of wages. The power company has made concessions and there is hope that the difficulties soon will be settled.

The Butte *Miner* states that letters have been received from a high official of the American Federation of Labor urging a speedy termination of the strike and that the men return peaceably to work, for there is good reason to believe that if this is not done the Federal authorities at Washington will conscript the strikers, take control of the copper mines, and require the miners to work for a much lower rate of wages than they have ever received at Butte. It is argued that the Government would not be willing to pay men that were thus drafted to serve at their occupation as miners a much higher rate of wages than is paid men in the trenches, where the risks are infinitely greater. Foreseeing that the liberal wages now paid in the Butte district are in jeopardy, the high officials of organized labor are urging the men to take a common-sense view of the situation, to cease agitation, and hold on to their jobs.

As a result of a disagreement between the striking miners and electricians at Butte, it was proposed to form a new union. At a meeting of the Metal Mine Workers' Union, held July 12, a resolution was passed without a dissenting vote, to the effect that one delegate for every 500 men in the union be sent to Denver not later than August 1 for the purpose of meeting to form an international union.

NEVADA

ESMERALDA COUNTY

(Special Correspondence.)—The east cross-cut from the 320-ft. level of the Cracker Jack mine has entered what appears to be the Columbia Mountain fault-vein 700 ft. from the shaft. The rock is badly crushed and appears to be a widely shattered zone. It is stated to assay around $7 gold per ton. Cross-cutting continues in expectation of intersecting the Rabbit Trail vein.

The main north drift from the 880-ft. level of the Jumbo Junior is being extended through the leased section of the Kewanas mine to connect with the main workings of the latter. The orebody has been exposed for about 150 ft., ranging from 1 to 3 ft. of good ore. The latest assays average $56 per ton. It is planned to make a small shipment.

The Red Hill Florence Co. is developing the Florence vein on the 500-ft. level and reports encouraging results. Work is also proceeding from several points higher. Negotiations for control of the Florence mine continue and most of the large stockholders and creditors are said to have signified their approval of the project.

The Goldfield Consolidated Co. is using the aerial-tramway between the tailing-pond and mill and is treating some of the old mill material in its cyanide-plant. Developments are now largely confined to the deeper levels for the purpose of augmenting the reserves of copper-gold ore. Recent work in the deep levels of the Laguna and Mohawk are reported to be particularly satisfactory. The company has not yet begun to treat ore from the Atlanta, although arrangements to this end were made several months ago.

Goldfield, July 12.

Personal

Note: The Editor invites members of the profession to send particulars of their work and appointments. This information is interesting to our readers.

H. R. WAGNER has returned from Chile to New York.

F. LE ROI THURMOND has opened an assay-office at Anchorage, Alaska.

L. A. PARSONS has returned from Cobalt to Copper Cliff, Ontario.

JOHN H. BANKS has returned to New York from Jerome, Arizona.

MILTON A. ALLEN is now with the Arizona Bureau of Mines at Tucson.

HOMER L. CARR, of New York, is visiting the mining districts of California.

A. O. GATES has obtained a commission in the U. S. Naval Reserve Forces.

J. B. TYRRELL now represents the Consolidated Mines Selection Co. in Canada.

E. B. HOPKINS, geologist, has gone to Mexico for the Associated Geological Engineers.

S. E. WOODWORTH has become an ensign in the Navy and is undergoing training at Annapolis.

GEORGE J. YOUNG, professor of metallurgy in the Colorado School of Mines, is in Tuolumne county.

E. C. GAMBLE and W. S. STEWART of Oakland, California, have been examining mines in Mariposa county.

S. C. DICKINSON is now safety engineer with the Arizona State Bureau of Mines, with headquarters at Tucson.

F. C. FREY, manager of the Redjang Lebong mines, in Sumatra, has returned to Reno, Nevada, on a holiday.

RICHARD B. MOORE, of the U. S. Bureau of Mines, passed through San Francisco on his way from Denver to Tucson.

ARTHUR C. TERRILL has returned to Lawrence, Kansas, from inspection of Osage City coal mines and Blue Rapids gypsum mines.

WALTER R. VIDLER, of Cripple Creek, has been in the Llano-Burnet district of Texas, and at Bowie, Arizona, on professional business.

F. J. HOENIGMANN and W. G. FARNLACHER have returned to San Francisco from Nevada and Utah, respectively, to do military service.

H. J. SHEAFE, superintendent of the Globe Mines, California, has obtained a commission as captain in the Engineer Officers Reserve Corps.

FOREST RUSHERFORD has resigned as general superintendent of reduction works for the Copper Queen Co. at Douglas, Arizona, and will open an office as consulting metallurgical engineer after a summer's rest. His present address is Pueblo, Colorado.

Obituary

FRANK M. MURPHY, a prominent mine and railroad operator, of Arizona, died at Prescott, on June 24, at the age of 62 years. For many years Mr. Murphy had been identified with the development of northern Arizona. In 1887 he succeeded in interesting "Diamond Jo" Reynolds in the Congress mine, near Wickenburg, and for years Mr. Murphy was the successful manager of that famous gold mine. It was largely through his efforts that the Santa Fe built its connecting line between Ash Fork and Phoenix, and was for a long time its manager, and secured the building of a number of important branches of the road. He also established the Prescott National Bank, and was prominent in many other important business enterprises, including the extensive operation of several groups of mines in the Bradshaw mountains and elsewhere in Arizona. He was a man of recognized business ability and integrity. His loss will be sincerely regretted by all who knew him.

THE METAL MARKET

METAL PRICES

San Francisco, July 17

Antimony, cents per pound	20
Electrolytic copper, cents per pound	33
Pig lead, cents per pound	12.25—12.50
Platinum, soft and hard metal, per ounce	$105—111
Quicksilver, per flask of 75 lb	$105
Spelter, cents per pound	11
Tin, cents per pound	60
Zinc-dust, cents per pound	20

ORE PRICES

San Francisco, July 17

Aluminum-dust (100-lb. lots), per lb	$1.00
Aluminum-dust (ton lots), per lb	$0.95
Antimony, 50% metal, per unit	$1.35
Chrome, 40% and over, f.o.b. cars California, cents per unit	50—55
Magnesite, crude, per ton	$8.00—10.00
Tin, cents per pound	60
Tungsten, 60% WO_3, per unit	$25.00—30.00
Molybdenite, per unit for MoS_2 contained	40.00
Manganese, 45% (under 35% metal not desired), cents, unit	33—37

Manganese prices and specifications, as per the quotations of the Carnegie Steel Co. schedule of prices per ton of 2240 lb. for domestic manganese ore delivered, freight prepaid, at Pittsburg, Pa., or Chicago, Ill. For ore containing

	Per unit
Above 49% metallic manganese	$1.00
46 to 49% metallic manganese	0.98
43 to 46% metallic manganese	0.95
40 to 43% metallic manganese	0.90

Prices are based on ore containing not more than 8% silica nor more than 0.2% phosphorus, and are subject to deductions as follows: (1) for each 1% in excess of 8% silica, a deduction of 15c. per ton, fractions in proportion; (2) for each 0.02% in excess of 0.2% phosphorus, a deduction of 2c. per unit of manganese per ton, fractions in proportion; (3) ore containing less than 40% manganese, or more than 12% silica, or 0.225% phosphorus, subject to acceptance or refusal at buyer's option; settlements based on analysis of sample dried at 212° F., the percentage of moisture in the sample as taken to be deducted from the weight. Prices are subject to change without notice unless specially agreed upon.

EASTERN METAL MARKET

(By wire from New York)

July 17.—Copper is weak and nominal at 30 to 29.75c. Lead is dull and easy at 11 to 10.87c. Zinc is inactive and lower at 9c. Platinum remains unchanged at $105 for soft and $111 for hard metal.

COPPER

Prices of electrolytic in New York, in cents per pound.

Date		Average week ending	
July 11	30.75	June 5	32.62
" 12	30.50	" 12	32.75
" 13	30.50	" 19	32.58
" 14	30.25	" 26	32.42
" 15 Sunday		July 3	32.25
" 16	30.00	" 10	31.50
" 17	29.75	" 17	30.29

Monthly Averages

	1915	1916	1917		1915	1916	1917
Jan.	13.60	24.30	29.53	July	19.09	25.66	....
Feb.	14.38	26.62	34.57	Aug.	17.27	27.03	....
Mch.	14.80	26.65	36.00	Sept.	17.69	28.28	....
Apr.	16.64	28.02	33.16	Oct.	17.90	28.50	....
May	18.71	29.02	31.69	Nov.	18.88	31.95	....
June	19.75	27.47	32.57	Dec.	20.67	32.89	....

Secretary of the Navy Daniels has agreed to pay for copper 75% of 25c. per lb. for 60,000,000 lb. of copper and leave 25% of 25c. per lb. for adjustment when cost of producing copper shall have been determined by the Federal Trade Commission. It is not known whether the copper producers will accept, without further parleys, the offer of Secretary Daniels to purchase 60,000,000 pounds of copper at what is the equivalent of 18¾c. (75% of 25c.), with adjustment later on the 6¼c. (25% of 25c.), which is the balance of the 25c. figure named by the producers. Any price less than 25c. would involve serious labor controversies and just now labor is demanding more than it has already agreed to accept on the sliding-scale basis, and has tied up the copper producing industry of Arizona, the biggest producing section of the country, in order to force its demands.

SILVER

Below are given the average New York quotations, in cents per ounce, of fine silver.

Date		Average week ending	
July 11	80.00	June 5	74.80
" 12	80.25	" 12	75.83
" 13	80.75	" 19	77.00
" 14	80.75	" 26	78.12
" 15 Sunday		July 3	77.98
" 16	81.26	" 10	78.70
" 17	80.67	" 17	80.62

Monthly Averages

	1915	1916	1917		1915	1916	1917
Jan.	48.85	56.76	75.14	July	47.52	63.06	....
Feb.	48.45	56.74	77.54	Aug.	47.11	66.07	....
Mch.	50.61	57.89	74.13	Sept.	48.77	68.51	....
Apr.	50.25	64.37	72.81	Oct.	49.40	67.86	....
May	49.27	74.27	74.61	Nov.	51.88	71.60	....
June	49.63	65.04	76.44	Dec.	55.34	75.70	....

'As high as 83c. per oz. has been paid for silver in San Francisco, according to the Boston 'News Bureau,' against an 80c. market in New York, the cost of transportation from the Pacific Coast to New York being but one-half cent per ounce. The premium paid for the metal over New York was 2½c. per oz. over the New York quotation.

The absence of stiff war risk premiums, prevalent between Atlantic ports and England, on the Pacific route to the Far East, has been the dominant factor in switching considerable activity in the silver market from London to this country. The copper market, once centred in that city, notwithstanding the fact that the United States produced about 60% of the world's output, has shifted here. Such a large proportion cannot be claimed for this country's yield of silver but in conjunction with Canada and Mexico the North American silver output under normal conditions ranks high.

At the present time the silver market has no real central location. Quotations still come out of London and they have much effect on the world's price. The actual metal itself, however, does not go through that point as formerly and spread out to India, China, and other consuming centres. New York now occupies to a greater extent than ever before the predominant position.

A world wide demand for silver for coinage purposes taken in conjunction with the curtailment in output in Mexico and elsewhere must be regarded as the chief cause for the present strength in silver which has forced it to price levels higher than at any other time in the past quarter century.

The president of a prominent silver-producing company says. "The world is clamoring for silver, and production is not up to requirements. As a result prices are climbing and are going higher. Gold is tight and the people want a hard metal money instead of paper currency, with the result that silver is much in demand."

LEAD

Lead is quoted in cents per pound, New York delivery.

Date		Average week ending.	
July 11	11.12	June 5	11.46
" 12	11.00	" 12	11.83
" 13	11.00	" 19	12.00
" 14	11.00	" 26	11.75
" 15 Sunday		July 3	11.57
" 16	11.00	" 10	11.25
" 17	10.87	" 17	10.98

Monthly Averages

	1915	1916	1917		1915	1916	1917
Jan.	3.73	5.95	7.64	July	5.59	6.40	....
Feb.	3.83	6.23	9.01	Aug.	4.67	6.28	....
Mch.	4.04	7.26	10.07	Sept.	4.62	6.86	....
Apr.	4.21	7.70	9.38	Oct.	4.62	7.02	....
May	4.24	7.38	10.29	Nov.	5.15	7.07	....
June	5.75	6.88	11.74	Dec.	5.34	7.55	....

ZINC

Zinc is quoted as spelter, standard Western brands, New York delivery, in cents per pound

Date		Average week ending	
July 11	9.12	June 5	9.66
" 12	9.12	" 12	9.75
" 13	9.12	" 19	9.72
" 14	9.00	" 26	9.43
" 15 Sunday		July 3	9.32
" 16	9.00	" 10	9.20
" 17	9.00	" 17	9.06

Monthly Averages

	1915	1916	1917		1915	1916	1917
Jan.	6.30	18.21	9.75	July	20.54	9.90	....
Feb.	9.05	19.99	10.45	Aug.	14.17	9.03	....
Mch.	8.40	18.40	10.78	Sept.	14.14	9.18	....
Apr.	9.78	18.62	10.20	Oct.	14.05	9.92	....
May	17.03	16.01	9.41	Nov.	17.20	11.81	....
June	22.20	12.85	9.63	Dec.	16.75	11.26	....

QUICKSILVER

The primary market for quicksilver is San Francisco, California being the largest producer. The price is fixed in the open market, according to quantity. Prices, in dollars per flask of 75 pounds:

Week ending

Date			
June 19	82.00	July 3	85.00
" 26	80.00	" 10	100.00
		" 17	105.00

Monthly Averages

	1915	1916	1917		1915	1916	1917
Jan.	51.90	222.00	81.00	July	95.00	81.20	....
Feb.	60.00	295.00	126.25	Aug.	93.75	74.50	....
Mch.	78.00	219.00	113.75	Sept.	91.00	75.00	....
Apr.	77.50	141.60	114.50	Oct.	92.90	78.20	....
May	75.00	90.00	104.00	Nov.	101.50	79.50	....
June	90.00	74.70	85.50	Dec.	123.00	80.00	....

Spain's production of cinnabar in 1915 was 20,717 tons, an increase of 3003 tons over 1914. In the Province of Ciudad Real, containing the famous mine of Almaden, 10,094 tons was mined, 1062 tons less than in 1914; but the decrease was offset by the increased output of Cranada's two mines and Oviedo's 14, all much smaller. At Almaden 297 excavations were made in the mineral deposits consuming 125 days and costing about $70,000 for labor.

The mines of Oviedo yielded 8153 tons of ore, which also contained arsenic, and those of Granada 2407 tons in 1915. The output of refined quicksilver at these works was 22 tons, 20.6 tons coming from Ovedo. The Oviedo works—La Pena, El Terronal, La Margarita—also produced 83 tons of arsenic. The mones of Granada are the Ella and Resurreccion. These and the mines of Oveida are the property of private companies.

TIN

Prices in New York, in cents per pound.

Monthly Averages

	1915	1916	1917		1915	1916	1917
Jan.	34.40	41.76	44.10	July	37.38	38.37	....
Feb.	37.23	42.60	51.47	Aug.	34.37	38.88	....
Mch.	48.76	50.50	54.27	Sept.	33.12	36.66	....
Apr.	48.25	51.49	55.63	Oct.	33.00	41.10	....
May	39.28	49.10	63.21	Nov.	39.50	44.12	....
June	40.26	42.07	61.93	Dec.	38.71	42.55	....

Eastern Metal Market

New York, July 11.

Decided weakness is manifested in nearly every metal, except tin which is higher. Continued uncertainty as to purchases by the Government and the Allies is the main cause.

Copper is stagnant and weaker.

Tin is higher but not specially active.

Lead is dull, uncertain, and lower.

Zinc is almost paralyzed and is again lower.

Antimony has declined and the demand is very poor.

Aluminum is a little lower, but still inactive.

The chief market influence in the steel-world is the expectation that some form of price-regulation, either by the Government or by producers under Government sanction, will be effected. A dictatorship of steel manufacture and distribution seems less a possibility today than last week. Estimates, by steel-makers, of the Government and Allies' total buying of steel-products, expressed in terms of ingots, approximate 12,000,000 tons for the coming year, or about 30% of the country's present steel-production. Government use of plates and shapes is expected to reach 40 or 50%. New business is less than in two years and export dealings are being held up by Washington. The actual effect of the embargo, set for July 15, in the domestic market, is not expected to be important, since exports of the principal steel-products to neutrals have been small.

COPPER

The market is stagnant and weaker. Yesterday, after a week or so of easy tendencies, decided softness was discernible, and both Lake and electrolytic were again lower at 30.75c., New York. One explanation of the fall to lower levels is that, in the absence of business, some sellers may have become tired of sitting still and may be offering at concessions. Whether this is coming from first or second-hands is difficult to decide. One reason for this change in prices and sentiment is attributed to the possible conviction on the part of some that future prices for copper, purchased by the Government and its Allies, will be lower than 25c. per pound, perhaps as low as cost-price plus a reasonable profit. The continued haggling at Washington on this subject may result in a dead-lock, in which case the Government would take a hand and have its own way and the copper producers would be deprived of the expected cinch. Added to this consideration is the reported rejection by the Government of the proposal of the aluminum producers to furnish that metal at 27½c. per lb. While the market is without features, there is at the same time much less gossip. Strikes among metal miners and producers continue disturbing. Quotations for later positions have also eased off about ½c. per lb. to 29.50c. for the third quarter, and to 28.50c. for the fourth quarter. There are no changes in the London quotations as reported last week.

TIN

The tin market, while stronger, has developed some peculiar features. Demand has been varied and spotty during the past week. While inquiry for futures one day has been good, with that for spot poor, the next day the contrary has been the case, with futures neglected and nearby metal in demand. For instance, on July 5 there was a good future demand amounting to 200 to 250 tons, which ended in business, but spot delivery was dull. On the 10th, however, futures were neglected, with nearby inquiry for tin afloat resulting in fair sales, but the spot market was quiet. Prices have consistently advanced the past week to 63c., New York, yesterday for spot Straits, an increase of 1c. per lb. since July 2. This has been due largely to the fact that stocks of spot Straits are now light and it is easier to buy 5 tons than 25 tons. The cable situation has continued to be disturbing. With hardly an exception London cables have been delayed almost daily, hampering business here decidedly. The weekly complaint regarding Government indecision as to taxes and other matters is also a factor, and these two elements have almost caused a halt in general business. Tin arrivals to July 10 inclusive have been 700 tons, with the quantity afloat at 4354 tons. The London market has advanced £3 over that of July 2, spot Straits being quoted at £247 there yesterday.

LEAD

Lead is again lower and yesterday was quoted at 11c., St. Louis, or 11.12½c., New York. Demand has declined almost daily and lower prices have consequently resulted. Some lots to large buyers have been offered at under 11c., St. Louis. There is very little if any difference now between the trust and the outside market, the quotation of the former still being unchanged at 11c., New York. It is reported that some of the large producers have no metal to sell for July, while others have, and the situation is unusual. The attitude of the Government is a disturbing feature, and while more orders from this source are expected, developments may be such in the near future as to cause a decided reaction further downward.

ZINC

The entire market is extremely dull and is weakening. Demand is almost nothing and sales scarce. Attempts are being made to hold the quotation for early delivery at not much under 9c., St. Louis, but it is believed that some quiet scalping has been done and that sales have been made at 8.87½c., St. Louis, or 9.12½c., New York. At these prices, however, some producers are certain to be operating at a loss and many may have to shut-down soon entirely. Some have already done so, giving needed repairs as the excuse. Futures continue a little higher than early deliveries, perhaps ½c. higher, but demand is quiet and there is not much inclination to sell for this position. It is not unlikely that labor troubles will be an important factor in the not distant future. Labor difficulties of great seriousness are reported from the Butte and Lake Superior districts. No details are yet available as to the reported purchase by the Government of 11,000 tons of high-grade spelter at 13.50c. per lb. Some credit it; others believe it is not a fact. Continued uncertainty as to Government needs and prices exert a demoralizing influence and the prospect is not bright. According to statistics completed by W. R. Ingalls of the *Engineering and Mining Journal*, the production of spelter in the last quarter of 1916 was 189,572 net tons, the largest of any quarter since or before the War, the largest just before the War having been 92,816 tons in the second quarter of 1914. His classification of the 1916 consumption of spelter shows 207,849 tons as consumed for galvanizing, 175,435 tons for brass, 40,053 tons for sheet zinc, with the balance of a total of 450,304 tons for other purposes.

ANTIMONY

The antimony market is not only dull but lifeless. Demand is so slack that quotations have fallen to 17c. to 17.50c., New York, for Chinese and Japanese grades.

ALUMINUM

No. 1 virgin metal, 98 to 99% pure, is a litle lower because of the slack demand and for early delivery is quoted at 57c. to 59c., New York. It is reported that the Government has rejected the producers' proposal to furnish aluminum at 27.50c. per lb.—the 10-year average price plus 2c. per pound.

Company Reports

TONOPAH MINING COMPANY OF NEVADA

The fifteenth annual report of the Tonopah Mining Co. of Nevada for the year ended December 31, 1916, states that a total of 81,782 tons of ore was milled, containing 15,636 oz. of gold and 1,387,557 oz. of silver, of a gross value of $1,279,157.86. The average recovery of the value was 94.2% of the gold and 90.3% of the silver. The recovery of value based on net smelter returns was 90.1%. In October the crusher-plant was almost totally destroyed by fire. The plant was re-built upon a less elaborate plan, but sufficient to meet present requirements. In the meantime the crushing-plant of the Western Ore Company at Millers was used where all the ore was crushed and sent to the mill. By this arrangement little time was lost. Operating statistics follow:

OPERATING STATISTICS

Tons of ore mined	74,991
Tons of ore shipped from dump	5,542
Tons of ore treated at Desert mill	81,782

COSTS PER TON

The average cost to mine and mill the ore and market the products for the past year was as follows:

Mining costs and costs of handling dump ore	$ 4.61
Milling costs	3.16
Freight on ore milled	0.72
Marketing mill products	0.22
Total costs per ton	$ 8.71
Metal losses in milling and refining	1.34
Profit per ton	5.60
Average gross value of ore milled	$15.65

SUBSIDIARY COMPANIES OF THE TONOPAH MINING COMPANY OF NEVADA

TONOPAH PLACERS COMPANY

Operating at Breckenridge, Colorado

The three dredges were put in operation during the month of March, and continued in operation until the end of December, at which time the two large dredges were closed down, and the smaller dredge continued in operation until the latter part of January 1917. The dredges were closed down on account of the winter weather and for necessary repairs. An indebtedness of $56,000 to the Tonopah Mining Co. was paid off during the year 1916. No. 2 dredge was operating during the whole season upon property owned by the Farncomb Hill Gold Dredging Co., under a contract with that company. The net earnings of the dredge, while on this property, will be divided equally between the Farncomb Hill Gold Dredging Co. and the Tonopah Placers Company.

THE MANDY MINING COMPANY

Operating in Manitoba, Canada

A property located in the Province of Manitoba, Canada, was acquired during the past year, and the Mandy Mining Co. was organized to own and operate it. The Tonopah Canadian Mines Co. owns 85% of the stock of the Mandy Mining Co. About 20,000 tons of high-grade copper ore has been developed on this property, and shipments of this ore to the smelters are being made.

THE EDEN MINING COMPANY

Operating in Nicaragua

The electric-power plant of the Tunky Transportation & Power Co. was completed and put into operation during the month of May, and has been in continuous operation since that time.

The mill made its first run on ore March 1, 1917.

In the month of June, J. L. Phillips resigned as general superintendent, and Robert Hawxhurst, Jr., was appointed as his successor.

TONOPAH NICARAGUA COMPANY

Operating in Nicaragua

A property known as the Santa Rita Mines, located about thirty miles from the property of the Eden Mining Co. in Nicaragua, was purchased during the past year for $10,000, and the Tonopah Nicaragua Co. was organized to own and operate this property. A force of men is now engaged in clearing the property and in development work, and it is expected that shipment of the products from this property will be made during 1917. The Tonopah Mining Co. owns about 92% of the stock of this company.

THE TONOPAH CANADIAN MINES COMPANY AND BRUTUS MINING COMPANY

Mining claims adjoining those of the Mandy Mining Co. were acquired and located during the past year, and the Tonopah Canadian Mines Co., and the Brutus Mining Co., which is controlled by Tonopah Canadian Mines Co., were organized to own and operate these properties. The Tonopah Mining Co. owns about 92% of the stock of the Tonopah Canadian Mines Company.

The option upon the property of the Mispah Extension Co. was extended in January 1917, for one year, and work is being continued upon the property, but no decision has as yet been made as to its acquisition.

THE GREAT BOULDER PROPRIETARY GOLD MINES, LTD.

The annual report of the Great Boulder Proprietary Mines, Ltd., in Western Australia, for the year ended December 31, 1916, shows the following:

	£	s.	d.
Expenses in opening up 120,900 tons of ore	7,574	6	10
Expense stoping 175,787 tons of ore	112,271	2	8
Sulphide mill expense, 175,787 tons of ore	73,912	17	2
Cyanide mill expense, 175,787 tons	33,376	17	8
Residue re-treatment, 206,443 tons	14,958	13	1
General charges	8,917	6	3
Sundries	5,418	18	2
Total expense	256,430	1	10
Gold realized	442,629	18	6
In process of realization	81,767	7	8
	524,397	6	2
Less cost of minting	916	3	8
	523,481	2	6
Sundry receipts	101	9	7
	523,582	12	1

The total cost per ton of ore treated in 1916 was 27s.10d.

Extensive diamond-drilling operations were carried on during the year, a total of 54,921 ft., equivalent to 10.401 miles having been bored.

There is estimated to be still available in the mine 372,791 long tons, having a total gross value of 271,706 oz., equivalent to about $5,434,000.

Labor shortage was responsible for less development having been done than had been planned. With a view to acquiring another property, many mines offered were examined, but, so far, none was found sufficiently promising to warrant development. Dividends paid during the year were as follows:

June 22	£65,625
Sept. 29	65,625
Dec. 23	65,625

A fourth dividend of £65,625 was paid March 24, 1917, making a total within a year of £262,500.

Recent Publications

THE PLIOCENE CITRONELLE FORMATION OF THE GULF COASTAL PLAIN AND ITS FLORA. By George Charlton Matson and Edward Wilber Berry. Professional Paper No. 98-L. U. S. Geological Survey. Pp. 41. Ill., Index.

THE INORGANIC CONSTITUENTS OF MARINE INVERTEBRATES. By Frank Wigglesworth Clarke and Walter Calhoun Wheeler. Professional Paper No. 102. U. S. Geological Survey. Pp. 56. Washington, 1917.

THE OREGON BASIN GAS AND OIL FIELD OF PARK COUNTY, WYOMING. By Victor Ziegler. Bulletin No. 15 of the office of the State Geologist of Wyoming. Pp. 32. Ill. and maps. Cheyenne, Wyoming, 1917.

This bulletin describes the location, physiography, and geology of the oil-gas fields of Park county, together with a brief description of numerous wells.

A BIBLIOGRAPHY OF THE GEOLOGY AND MINING INTERESTS OF THE BLACK HILLS REGION. By C. C. O'Harra. Issued by the South Dakota School of Mines, as Bulletin No. 11. Pp. 216, with index and map. Rapid City, 1917.

This will prove to be a useful volume to those who are in search of information relating to the greatly varied mineral resources of that wonderful mineral province known as the Black Hills.

NOTES ON THE GEOLOGY AND IRON ORES OF THE CUYUNA DISTRICT OF MINNESOTA. By E. C. Harder and W. A. Johnston. Bulletin 660-A. U. S. Geological Survey, prepared in cooperation with the Minnesota Geological Survey. Pp. 26. Maps. Washington, 1917.

Describes the general geology of Minnesota and more particularly the rocks and geology of the iron deposits of the Cuyuna district.

GEOLOGY AND MINERAL RESOURCES OF THE REEFTON SUBDIVISION, THE WESTPORT AND NORTH WESTLAND DIVISIONS OF NEW ZEALAND. By J. Henderson. Bulletin 18 (new series) of the New Zealand Geological Survey. Pp. 232. Ill., maps, and index. Wellington, N. Z., 1917.

This publication is a general geological treatise on the ore deposits, geology, and mining industry in the several districts mentioned in the title.

TWENTY-FIFTH ANNUAL REPORT OF THE ONTARIO BUREAU OF MINES. Pp. 311. Ill., maps, index. Toronto, 1916.

This publication reviews the mineral industry of Ontario in 1916, devotes a chapter to mining accidents, describes the mines of Ontario, giving a detailed geological description of numerous deposits. One chapter is devoted to a scientific study of certain minerals of the Cobalt district. The metallurgy of the ores is also described with flow-sheets of some of the mills.

The following publications have recently been issued by the United States Bureau of Mines, Washington, D. C.:

BULLETIN 124. Sandstone quarrying in the United States, by Oliver Bowles. 1917. 143 pp., 6 pl., 19 fig.

TECHNICAL PAPER 82. Oxygen mine-rescue apparatus and physiological effects on users, by Yandell Henderson and James W. Paul. 1917. 106 pp., 5 pl., 6 fig.

TECHNICAL PAPER 135. Bibliography of recent literature on flotation of ores, January to June, 1916, compiled by D. A. Lyon, O. C. Ralston, F. B. Laney, and R. S. Lewis. 1917. 20 pp.

TECHNICAL PAPER 140. The primary volatile products of the carbonization of coal, by G. B. Taylor and H. C. Porter. 1916. 59 pp., 1 pl., 25 fig.

TECHNICAL PAPER 143. The ores of copper, lead, gold, and silver, by C. H. Fulton. 1916. 45 pp.

TECHNICAL PAPER 160. The determination of nitrogen in substances used in explosives, by W. C. Cope and G. B. Taylor. 1917. 46 pp., 1 pl., 4 fig.

TECHNICAL PAPER 166. Motor gasoline; properties, laboratory tests, and practical specifications, by E. W. Dean. 1917. 27 pp.

Book Reviews

THE EFFICIENT PURCHASE AND UTILIZATION OF MINE SUPPLIES. By Herbert N. Stronck. Pp. 97. Ill. diagrammatically. John Wiley & Sons, New York and London. For sale by the MINING AND SCIENTIFIC PRESS.

This little book will be found most useful to mine superintendents who wish to handle their mine-supplies in a methodical and economical manner. It includes chapters on the purchasing department; the receiving and testing department; the stores system with accurate accounting; the issuing system; reports on consumption of supplies, and methods of preventing waste. There are numerous styles of blank forms for all purposes connected with the warehouse and stock-rooms of a mine, and many valuable suggestions to the store-keeper and accountant.

STEAM TURBINES. By James Ambrose Moyer. Third Edition. Pp. 460 and index. Ill. John Wiley & Sons, New York and London. For sale by the MINING AND SCIENTIFIC PRESS. Price, $3.50.

This excellent work is a compendium for power users, being a practical and theoretical treatise for engineers and students. It reviews completely the recent improvements in the economy of steam-turbines, and as the author says: The low cost of power where fuel is cheap makes the large turbine-electric generating-plant almost an unrivaled competitor of water-power for metallurgical purposes. Many changes have been made in some departments of the book and it is in every sense up-to-date. All who contemplate large power installations should secure this book in order to be fully informed in the most modern practice in the generation of power for any purpose whatsoever. In an appendix is a series of questions, with answers, for the student who soon may have these very questions presented to him in actual practice.

PROPERTIES OF THE CALCIUM SILICATES AND CALCIUM ALUMINATES OCCURRING IN NORMAL PORTLAND CEMENT. Technologic Paper of the Bureau of Standards, No. 78. By P. H. Bates and A. A. Klein. Pp. 38, ill., has no index. Washington, D. C., 1917.

This pamphlet is replete with new data concerning the characteristics of cement, revealing by microscopic investigation the constitution of the clinker obtained from different mixes, and the resultant compounds after wetting, and setting in briquettes. It throws light on the conditions that give strength and weakness to cements, and should be carefully studied by engineers. The conclusions are succinctly set forth in a summary which shows among other things that tri-calcium silicate has all the important properties of portland cement, especially those of the rate of setting and strength developed. It also shows that plaster of paris when added to any of the compounds or mixtures studied generally increased the strength at all periods. The authors state that the ideal cement should apparently have an excess of di-calcium silicate, which would give a moderately dense hydrated material that will gain strength with aging of the concrete.

EDITORIAL

T. A. RICKARD, Editor

SUBMARINE "is the *last* argument of kings," said the Crown Prince. We believe that to be true, but we do not place the emphasis on 'kings.'

MARK SULLIVAN, in *Collier's*, speaks of "Thou shalt not make money out of the War" as "an ethical hallelujah." Most men would rather make their money and then have the privilege of giving it to patriotic purpose; but when they have made it, some of them are content with the 'hallelujah.'

ALTHOUGH the embargo on exports may prove annoying not only to neutrals but to our own people, it must be noted that the value of exports of the more important commodities to Denmark, Norway, Sweden, Holland, and Switzerland increased from $66,053,595 in 1913 to $177,144,085 in 1916, whereas other exports, not required by the Enemy, decreased slightly, from $201,000,000 to $192,000,000. These figures will take a lot of explaining by neutral governments.

FROM the acting Chief Inspector of Mines at Cairo we have received a letter objecting to the statement made by Mr. Ernest H. S. Sampson in our issue of March 3, saying that the Egyptian government "has wisely made laws that preclude any possibility of individual enterprise." We are informed by the Chief Inspector that "this statement is incorrect and liable to create misapprehension," so he sends us a copy of the 'Rules and Regulations as to Mining,' published by the Department of Mines of the Ministry of Finance, Egypt, for the year 1916. The text of these regulations shows a willingness on the part of the Government to facilitate individual mining enterprise in Egypt.

CHAIRMEN of English mining companies sometimes make technical statements such as are authoritative only when backed by the name of a responsible engineer. Unfortunately, shareholders are apt to swallow assertions of a technical character when made by untechnical gentlemen prominent in finance. Thus Sir Lionel Phillips, at the Central Mining meeting, made sundry positive assertions concerning deep developments on the Rand, rich zones being said to succeed poor zones in depth. Now, if in giving such information he had quoted Mr. H. F. Marriott or some other reputable engineer in the employ of the Central Mining Corporation, we would be inclined to consider the information as having scientific value. Under the circumstances, we do not; for we remember that Sir Lionel said, in March 1910, at the East Rand Proprietary meeting, that "there seemed to be no evidence whatever that the gold contents at the deepest levels are not fully as high as they were at the surface, or within 300 ft. of the actual surface." At that time he had the evidence of Mr. Frederick Hellman, the manager of the East Rand Proprietary, to the very contrary. We read in a recent copy of the *Financial Times* that the present chairman of this same company, Mr. E. A. Wallers, at the last annual meeting had to announce that owing to "the extremely poor development experienced during recent years", it was impossible to supply the mills with adequate ore, and that there was "no prospect of dividends for some time to come."

IN the evening paper that published the details of the draft for an American army, we were given part of the text of the new German Chancellor's speech before the Reichstag, in which he sneered at our military effort. A lack of the sense of humor is the mark of "the beast with the brains of an engineer," as Upton Sinclair phrased it; otherwise the recollection of the former Chancellor's unfortunate reference to another "contemptible little army" might have prevented Dr. Michaelis from repeating a stupidity. His predecessor has gone into the discard with his "scrap of paper" and it is now the task of General Pershing and his brave men to make the present exponent of Prussianism feel the force of another organized democracy. Our Allies have 'done their bit,' our men will 'do their durndest.'

WE publish a letter from Mr. W. G. Devereux, an engineer known to our readers as the manager of the Melones Mining Company, in which he makes a bid for volunteers to serve in the regiment of Field Artillery now being raised as a Californian unit in the army going to Europe. This is to be a Western regiment of Western men fighting under the sign of the Bear. We commend the service to our young mining engineers and miners. The artillery is a technical branch of the military organization and one for which the members of our profession are particularly fitted. We hope that they will become keenly interested in the opportunity offered to them and that they will get some of the younger men in their employ to enlist in the same regiment. Everybody cannot obtain a commission, but those that are capable and keen can feel assured of promotion in due course. In the fighting of today the intelligent and skilful soldier is not overlooked, as we know from the rapid promotion that has come to our mining-engineer friends already in the field. Rally to the Californian Field Artillery!

ACCORDING to press reports, the Secretary of the Navy has offered to purchase 60,000,000 pounds of copper at 18¾ cents per pound with an adjustment later by the Federal Trade Commission for the difference between this price and the 25 cents quoted by the producers of the metal. The latter are said to claim that they should receive not less than 25 cents per pound, in view of the rising cost of labor. Moreover, the Government has agreed to pay 8 cents for lead in contrast with the average of 4.59 per pound for the last 15 years. A price of 18¾ cents for copper would, it is said, entail further labor controversies, because wages would have to be reduced in accordance with the sliding scale, but this is an ingenuous argument, seeing that 60,000,000 pounds represents only about 3% of the annual production or about one-third of the monthly production of copper in this country at this time, disregarding the temporary effect of current labor disturbances. The President has said that such prices should be fixed by the Government as "will sustain the industries concerned in a high state of efficiency, provide a living for those who conduct them, enable them to pay good wages, and make possible expansion of their enterprises," all of which depends upon the definition of 'efficiency,' 'living,' 'good' wages, and 'expansion.' It seems to us that this can be effected better by allowing the market to take its course and then imposing a heavy tax on excess profits. The lowering of price may not be much more artificial than the process by which the quotation for copper has been raised so high, and kept so high, during the last three years, but in the effort to treat everybody equitably it must be conceded that a drastic lowering of price will fall unequally on the various producers. One produces copper for 10 cents and another for 20 cents per pound, so that a lowering of 5 cents in the price from 30 cents reduces the profit of the first by 25% and of the other by 50%; and if the intention be to promote intensive exploitation of our copper resources it is best to give everybody a nearly equal inducement to go ahead. By the way, we are beginning to get new estimates of the cost of production. The *Boston News Bureau*, for example, says that "the average cost of producing American copper is not far from 15 cents per pound," whereas not long ago this same paper published plentiful statistics to show how the principal copper-mining companies were winning copper for 6 to 8 cents, and even as low as 4 cents per pound. Most of these low figures were fictitious, because they represented the bare operating cost, disregarding much of the 'overhead' expenditure, most of the development and equipment, and ignoring amortization of capital. Our Boston contemporary remarks piously: "Unfortunately many producers have not heretofore made due allowance on their cost-sheets for construction and development expenditures and depreciation, to say nothing of any allowance for depletion of mineral assets," and yet the *Boston News Bureau* has publlished such essentially fictitious statements with infinite gusto. "When the Devil was sick a monk he would be." In these days of stress we shall learn real economics, just as in England the mining financier has succeeded at last in persuading the tax-collector to understand that a mine is a wasting asset and that a dividend is not necessarily income.

Steel on the Pacific Coast

One year ago pig-iron was selling at $18 per ton. The steel-makers were excited when the price touched $20 in February. The market has grown so used to the excuse of the War for boosting prices that it no longer thrills when the current price of pig-iron is over $50. These fancy prices, which are not in any respect an expression of cost plus a reasonable profit, now turn to the advantage of the Pacific Coast and may effect a permanent change in the industrial relationships between the two sides of the continent. It might have been otherwise had the Government committed itself frankly and fairly to price regulation. That is one aspect of the situation. From the standpoint of the man in the street it was hoped that a plain, sensible, democratic principle would be applied, but he is now disillusioned. Certain commodities only are to be regulated, and we are not sure that, under the law, these can be controlled effectively. For awhile Congressional committees dallied with amendments that included the iron and steel industry, and then, when the Administration undertook to hurry Congress in order to get some kind of a food bill, the steel incumbrance to legislative speed was thrown to one side, in consequence of which, whether the legislation be democratic or not, a new chapter in the industrial history of the Pacific Coast may be written. At $50 per ton for pig-iron, and on the assumption that the War will last for two years, it would seem possible for blast-furnaces established in California or in the State of Washington to produce sufficient steel in one year of active operation to gain a profit that would amortize at least the larger part of the capital required. Even though only 5% of the total steel output of the country is consumed on this Coast, this proportion represents a big quantity. Just what the cost of producing pig-iron would be is difficult to say, but it is evident that there is an opportunity to establish enterprises of this character on this Coast that would survive the War. One corporation has already broken ground for an iron furnace on the Carson river, in Nevada. This is the Merrimac Steel & Smelting Company. The coke will come from Utah and the iron ore will be derived from magnetite mines in the vicinity. Deposits of hematite also exist near-by, and these could be utilized if necessary. This plant will be in operation within about nine months. A still larger enterprise is contemplated by one of the strongest capitalistic groups in California, including Messrs. W. H. Crocker, S. F. B. Morse, and B. L. Thane. It is hinted that Mr. D. C. Jackling and his friends also are interested in the venture. For two years these gentlemen have been making a survey extending from San Diego to Alaska. They have examined every iron deposit along the Coast, and have made elaborate investigations into the available source of iron ore and have completed arrangements for making by-product coke. The smelting-plant will be established at Lake Washington on the outskirts of Seattle. The plans of the company have not been made public, and, in fact, we understand that the financing of the operation has not yet been perfected. This seems, however, to be a mere detail, for it is certain that ground will soon be broken for a plant if arrangements can be made to insure delivery of the necessary equipment within such a period of time as will enable them to take advantage of the prevailing high prices. It is this great margin that offers the opportunity for establishing an enterprise that may endure.

The increase in the manufacture of steel on the West Coast has become important. Four years ago there was not a single steel-plant fronting the Pacific Ocean. To-day 16 open-hearth furnaces are in operation, mainly on scrap, which is selling at the rate of $29 for cast-iron and $42 for steel. The Noble Electric Steel Company, which began many years ago to experiment on the manufacture of steel with the Heroult type of furnace, has passed through a costly experimental stage to one of profitable production as a manufacturer of ferro-silicon and ferro-manganese. A special type of furnace has been developed known as the Frickey, possessing features said to introduce greater economy than was possible by the original Heroult process. Each of these is equipped with four graphite electrodes manufactured at Niagara Falls. In the production of ferro-silicon they use a local supply of exceptionally pure silica and a remarkably high-grade magnetite, which, as charged into the furnace, assays 70% metallic iron. The same company is erecting three single-phase furnaces, each of 2½ tons capacity, for the manufacture of ferro-chrome. The two larger furnaces making ferro-silicon and ferro-manganese have a capacity of eight tons each per diem. All of these products are being sold to steel-makers on the Pacific Coast that ventured to engage in the industry under the stimulus of War prices. In view of the fact that the expected benefit from the Panama Canal in low freight-rates for our domestic inter-oceanic commerce is not to be realized, the new iron and steel industry developing on the West Coast will always enjoy the protective differential corresponding to the trans-continental freight-rate on pig-iron. An academic discussion of the iron and steel problem on the Pacific Coast by Mr. Ernest A. Hersam is printed elsewhere in this issue. It will be read with interest because it reveals the difficulties under which the development of such an enterprise has hitherto labored. Professor Hersam is justified in saying that the special process for producing iron without the use of blast-furnaces and high-grade coke deserve to be carefully investigated with reference to Western needs, and it would seem, furthermore, that the great abundance of electric power cheaply available in the mountains of the West presents an opportunity for a greater development of the electric iron and steel industry than is possible anywhere else in the world, with the exception of Norway. About 44% of the total available water-power in the United States lies west of the Sierra Nevada, nevertheless it is interesting to see that the starting of the iron industry on the Pacific Coast will be along old lines, sustained by a favoring margin of profit made possible because of inflated prices due to the War.

Sampling Large Low-Grade Orebodies

The editorial article published in our issue of May 26 on this subject has elicited the discussion desired. Mr. W. J. Loring lays emphasis on the fact that it is not the number or weight of the samples, but their representativeness that counts in the appraisal of a mine. The two examples he quotes, and the description of the method by which the work was done, should prove helpful to our young engineers, for Mr. Loring has 'made good' on the samplings he describes by developing profitable enterprises on both of them. The collection of evidence is only half the task: the deduction from that evidence is at least as important. The first must be accurate in order that the second may be correct; but it is easier to make a careful sampling than a true diagnosis; therefore the personal equation continues to be the basic factor in mine-appraisal. Next we have the letter from Mr. Morton Webber, who also contributes a thoughtful article to this issue. We are grateful to an engineer so busily engaged for finding time to enrich our pages so effectively. In his letter, which followed Mr. Loring's, he returns to his former friendly controversy with Mr. L. A. Parsons, whose article we re-published in our issue of May 26. Mr. Webber insists that there are mines that cannot be sampled; in such mines the distribution of rich ore is so sporadic and irregular that no rigid system of sampling is of use as a means of obtaining reliable data for valuation. The argument is that if a large cake contains a dozen raisins distributed at random, no two, nor even four, cross-sectionings of the cake by a knife would furnish a basis for estimating the number of raisins in that cake. The only way to find out would be to eat the whole cake. Skill in sampling, therefore, is wasted on a 'pockety' mine. Mr. Albert Burch, who writes in this issue, agrees with Mr. Webber, instancing the Tightner as a case in point. He also describes a similar mine in Oregon. A wise man knows the limitations of his technique. To Mr. Richard A. Parker we are indebted for the suggestion that sub-sampling is necessary for discriminating inferences; he states a truth known to experienced practitioners, like himself, that any pronounced departure from the normal average of an orebody should be investigated by re-sampling the abnormal length of lode, taking fresh samples at short intervals, in order to ascertain the importance that should be given to the abnormality. Mr. Parsons contributes a valuable article, in the writing of which, he informs us, he consulted Mr. William W. Mein. These engineers are vigorous exponents of the moil and hammer, although recognizing the proper function of the mill-test. Samples must be representative and accurate, he says rightly. The mill-test may be representative, but it represents only one or two parts of the mine; it is likely to be "broadly selective." That is a true word and right worthy of acceptation. On the other hand, the mill-test is more accurate because the bulk of it is so large as to minimize the aberrancy caused by particles of gold. Such free gold vitiates the accuracy of a 50-pound sample much more than that of a 100-ton mill-test. In the one it may be abnormal; in the other it and others like it are normal. The accuracy of the moil-sample is dependent upon the human factor; mind no less than muscle is required for taking a true sample. On the other hand, the breaking of many moil-samples distributed at uniform intervals over the length and breadth of the orebodies gives an opportunity for collecting much collateral information concerning the geological structure and other peculiarities sure to exert an important influence upon the subsequent exploitation of the mine. All these points are brought out by Mr. Parsons most lucidly. Next, our friend Mr. T. H. Leggett lays stress on the word 'representative,' and thereupon suggests that mill-tests may lack this quality because the ore mined for milling usually does not come from a sufficient number of places, or, rather, that not a sufficient number of mill-tests of ore from various parts of the mine are made. That is why moil-sampling is usually more reliable. To make mill-tests so numerous as to be representative may entail a costly scheme of development. The engineer has to cut the garment according to his cloth; for time and money are two basic factors in engineering. Mr. Leggett says, a mine that cannot be sampled is a good one—for the owner—to keep. With this opinion we are inclined to agree. It is unwise to buy a pig in a poke. The unsamplable mine is rarely appraisable; the element of risk is so large; the venture is so much of a gamble, that such mines are best left to men that have plenty of capital and their own experience to guide them. If Messrs. A. DeW. Foote and William Hague made money out of the Tightner mine, for example, it was because they were living in the district and knew not only the past history of the mine, but the idiosyncracies of the deposits in that locality. Mr. Burch encores Mr. Leggett in the emphasis upon 'representative' and closes his own contribution with the sensible advice that where reliable data are available a preliminary moil-sampling may be set aside in favor of mill-testing, although it is clear from the context that he does not expect the data furnished by the vendor to be often so reliable as to warrant abstention from the moil-sampling. He refers to the 'cleaning-up' of the mill as a source of error. Of course, it is; and one that always predisposed the present writer against mill-tests on a small scale.

The foregoing summary of the high lights in the preceding discussion of a most practical subject will serve to increase interest in the article by Mr. Morton Webber on the respective merits of moil-sampling and mill-tests. He has gone to some pains to make clear his argument by citing three examples taken out of his own recent experience. He lays stress on the fact that the terms of an option necessarily influence the choice of method; the mine-valuer must have business acumen as well as technical skill if he is to prove a safe adviser to the buyer of mines. The choice between the moil and the mill is conditioned upon the circumstances of each option and of the mine optioned. In making a representative shipment of ore, it is necessary that each stope shall contribute its representative weight, says Mr. Webber. That is a point well taken; so also is the suggestion that the comparison of the moil-sampling with the mill-test of a stope gives a sampling-factor, or ratio of sample-assay to mill-bullion, that is most valuable. He lays stress on a detail usually overlooked, namely, the uniform mixing of samples after they have been crushed, by allotting a definite period for the operation. He believes that a mill-test is useful for ascertaining the metallurgical character of the ore and for establishing the sampling-error, and in his third example he shows that moil-sampling of ore already exposed may not furnish the information most requisite for estimating the future prospects of a mine. His article is one of the most useful we have had the pleasure of publishing; it is a clear gain to the technology of the subject. It remains to state once more that sampling of any kind must be made to yield results indicative of future mining operations, those operations anticipated by the purchaser of the mine. Sampling misleads because it is done on a laboratory scale and style, as it were. That is why sometimes a mill-test furnishes a check, because in blasting ore for a mill it is broken in a manner more nearly like ordinary stoping. As Mr. Parsons suggests, the moil-sampling of a narrow width of ore is likely to disregard the stoping-width. The moil breaks the ore more cleanly than the dynamite in a drill-hole; the sample is free from the casing, wall-rock, and other diluents that create discrepancies between the report on a mine and its subsequent life-history. What we need today is a sufficient collection of post-mortems, say, two or three hundred reports by capable engineers supplemented by the later records of the mines themselves, to furnish us with the factors of error, and the reasons for them. One factor is incontestable, and that is the lowering of the grade. Most reports on most mines give a grade of ore lower than is subsequently mined. This is due not only to the cleaner breaking of ore by the samplers, but to the advancement of metallurgy, the betterment of ore-breaking, and other improvements that enable the mine-manager to reduce the grade below that of the appraising engineer's report and to prolong the life of the mine economically, that is, within limits that recognize the value of money and the need for amortizing capital. Finally, we suggest to the profession that there is an inherent discrepancy between the sampling of a mine and the actual exploitation of it. The sampler's ideal is to obtain a true average, to be non-selective, whereas the miner, the manager of a mine, inevitably selects, he extracts ore that will yield the profit on which the enterprise is predicated. That is why the sampling of a mine under option so rarely corroborates the estimate based on past production. To be a good appraiser therefore the engineer must be a man of constructive imagination, a man with foresight, able to foresee how the mine is going to be worked. He must be able, in a measure, not only to write last year's almanac but also to predict the future; and to do that he must have been, at some time, a mine-manager himself.

We shall be glad to publish further discussion.

DISCUSSION

Our readers are invited to use this department for the discussion of technical and other matters pertaining to mining and metallurgy. The Editor welcomes expressions of views contrary to his own, believing that careful criticism is more valuable than casual compliment.

Sampling Large Low-Grade Orebodies

The Editor:

Sir—I have read with considerable interest the discussion of this subject, as printed in recent issues of your paper, which, I understand, had for its inception articles from the pens of Messrs. Morton Webber and L. A. Parsons.

Mr. Webber took the ground in the beginning that there are mines (his class 5) the sampling of which is useless and on this point Mr. Parsons takes issue, but I decidedly agree with Mr. Webber, though I also think that, fortunately, mines of class 5 are rare and therefore relatively unimportant. They are "freak" mines, but they do exist and, in a later letter, Mr. Webber cites the well-known case of the Tightner as an example. From my own experience I might mention one that is even more extreme than the Tightner, being a mine in eastern Oregon which I visited several years ago.

At the time of my visit, the un-stoped portion of the vein was well blocked out by drifts, cross-cuts, and raises, which could have readily been sampled, but the mine had been well managed and an examination of the well-kept records of production extending over a period of several years showed that sampling would have been entirely useless. These records showed that the average ore had about paid expenses, and that the dividends paid coincided almost exactly with the combined values of two pockets of extremely high-grade ore totaling less than two tons in weight. I understand that a few other mines of a similar character have been found in the same district, but this is the only one of which I have personal knowledge. Such examples would seem to me to establish the existence of mines belonging to Mr. Webber's class 5, and nothing that has been said by his critics, including Mr. Parsons, has convinced me that it is not useless to sample them. The discussion, commenced in this way, has been broadened by your editorial of May 26 and applied to mines of a different and more important class, namely, those containing large bodies of rock which, taken as a whole, constitute payable ore, but within which the gold is so unevenly distributed as to render them difficult but not impossible of sampling, and the question has narrowed to one of moil *versus* mill-test sampling. On this question, I believe, Mr. Leggett has hit the nail squarely on the head by the use of the word "representative"; for if the mill-tests can be made on rock as truly representative of the average ore as the moil-samples, then it would seem that the mill-test method should be more accurate than that by moil-samples, because of the elimination of many of the minor inaccuracies incident to such sampling. The three most frequent causes of error by this method are probably (1) tendency of operator to cut a disproportionate amount of either hard or soft rock, (2) liability to mixture of sample with particles of dust and pieces of rock falling from points outside the sample cut, and (3) danger of salting from a single small piece of gold which finds its way into the very small proportion of the original sample that finally goes to the assayer's crucible.

Admitting then, for the sake of argument, that the mill-test can be made the more accurate of the two methods, if it is quite as representative, how can it be made representative? First, the openings to be sampled must be in such positions within the vein as to develop ore of the same average grade as that of the entire orebody, and this applies quite as much to openings which are to be sampled by hammer and moil as to those from which mill-tests are to be made; and in the failure to so make the preliminary openings upon the vein, I believe we find the source of future disappointments more often than in inaccurate sampling. Given properly placed openings the next step in sampling by mill-test should be to cut a sample of uniform width, but this must also be done in moil-sampling, and except in cases of soft or caving ground is quite as easy in the one case as the other. It is true that the length of a cut for a mill-test must be made several times as great as that of a single moil sample in order to avoid frequent cleaning up of the mill, which is itself a source of error, but, as the object is to determine the average value of a large orebody which previous development has shown to be probably valuable and *not* to pick out pay-streaks, for selective mining, there could be no objection to including within a single mill-run all of the samples from an entire cross-cut, or a long section of a drift.

As to the valuation of a mine upon the basis of past production, there can be no doubt that such a method is erroneous unless the past production has come from workings uniformly distributed through sections of the orebody to be mined in the future. Otherwise such past production is representative of nothing except the part of the mine already exhausted, and as you have very well pointed out, a good miner is quite likely to select the best part of his mine for his first stoping. I have frequently said that one could pick out the best part of any orebody by examining a longitudinal section of the workings and noting what portions had been carried in advance of the remainder.

Regarding the questions of cost and of time consumed,

there would seem to be no doubt that ordinarily the advantage lies with moil-sampling, though this advantage in the matter of cost need not be so very great, if, as is usually the case under mining options, the purchaser has the right to retain all of the mineral recovered, less a small royalty. Furthermore, it costs a great deal of money to build an operating plant for a large low-grade mine, and no one financially able to build such a plant should neglect to forestall the chance of losing his entire investment by failing to use the best means available for determining in advance the value of his mine, even though the best method should be more expensive than the next best.

In conclusion, I would advocate for a large low-grade mine at which no reliable data of the results of development are available, a preliminary moil-sampling followed by mill-test sampling, but, where such data are available, I believe one might safely dispense with the moil-sampling and depend solely upon mill-tests for determining the value of the property.

San Francisco, July 18. ALBERT BURCH.

California Field Artillery

The Editor:

Sir—It may be interesting to those of your readers who contemplate military service, but have not as yet determined in what arm or branch they prefer to serve, to know that a Field Artillery unit of four batteries has been authorized to be raised through the National Guard of California which will, in effect, be a volunteer organization. The recruiting of this organization has been given to the San Francisco Cavalry Troop, which is an officers' training organization that has been commanded and trained by a regular army officer for the past two years and which will be used to form the nucleus of the commissioned personnel.

The movement has the hearty endorsement of such eminent officers as Major-General John J. Pershing, and Major-General Hunter Liggett, Commander of the Western Department.

The commander of the new organization will be an expert artillery officer from the Regular Army. Some of the commissioned positions are open to be filled from the enlistments, as well as most of the non-commissioned personnel.

The Field Artillery is the branch of the service that pre-eminently appeals to the mining profession and to the technical man, and in the present war has rapidly grown to be the most important arm.

The most desirable qualifications, second to actual military experience, are a knowledge of horses, and the mental and manual training obtained from having handled machinery, but any able-bodied man is acceptable. It is expected that this organization will see active service just as soon as its training reaches the required degree of perfection, and it is hoped that the quality of the men who will join will reduce this period to a minimum.

Full information can be obtained at the Recruiting Headquarters, 210 Montgomery St., where enlistments are rapidly coming in. We have assurance that as fast as batteries are recruited to the necessary strength, they will be recognized and the men who have enrolled will be exempt from the draft.

CALIFORNIA FIELD ARTILLERY.
By WILLIAM G. DEVEREUX,
Vice-Chairman Recruiting Committee.

Relief From Annual Assessment Work

[The following resolution exempting enlisted men from the obligation to perform annual labor on mining locations has passed both houses of Congress, and is merely awaiting the signature of the President to become law. For the benefit of those whom it may affect we reproduce it verbatim.—EDITOR.]

JOINT RESOLUTION

To relieve the owners of mining claims who have been mustered into the military or naval service of the United States as officers or enlisted men from performing assessment work during the term of such service.

Resolved by the Senate and House of Representatives of the United States of America in Congress assembled, That the provisions of section twenty-three hundred and twenty-four of the Revised Statutes of the United States, which require that on each mining claim located after the tenth day of May, eighteen hundred and seventy-two, and until patent has been issued therefor, not less than $100 worth of labor shall be performed or improvements made during each year, shall not apply to claims or parts of claims owned by officers or enlisted men who have been mustered into the military or naval service of the United States, so that no mining claim or any part thereof owned by such person which has been regularly located and recorded shall be subject to forfeiture for nonperformance of the annual assessments until six months after such owner is mustered out of the service or until six months after his death in the service: *Provided*, That the claimant of any mining location, in order to obtain the benefits of this resolution, shall file, or cause to be filed, in the office where the location notice or certificate is recorded, within ninety days from and after the passage and approval of this resolution, a notice of his muster into the service of the United States and of his desire to hold said mining location under this resolution.

MANGANESE oxides in large quantities are characteristic of a number of the silver deposits at Philipsburg, Montana, notably in the Cliff and Trout mines. The mineral occurs as pyrolusite derived from the alteration of rhodocrosite, which is abundant in the gangue of the Philipsburg ores. The accumulations of pyrolusite in these deposits have again attracted attention owing to the high price of manganese, and the shipments of this ore from the district now amount to approximately 500 tons weekly.

Outlook for Iron and Steel on the Pacific Coast

By ERNEST A. HERSAM

Whether or not the Pacific Coast states can profitably produce iron and steel from the ore, or can develop later an extensive iron industry, is a question that demands an answer. If the industry is practicable, steps to promote it are important and should be encouraged. If there are qualifications or limitations to iron smelting in the West, the people ought to know of it. To consider the possibility of developing an extensive iron and steel industry through the actual smelting of iron ore, leads to some fundamental scientific and economic considerations. It is an easy matter to define the requirements of iron manufacture as represented in present practice, and to observe whether or not these can conform to given local conditions in future practice. It appears at once to be extraordinary that the iron ores of the West Coast have not contributed more abundantly toward the iron-output of the United States, yet the Pacific-Coast States are importing from the East and from abroad practically all of the iron used by them in local construction and manufacture. This backwardness is seen to limit the local expansion of manufacturing industry, dependent as it is upon pig-iron and the iron and steel products derived from pig-iron.

In the ordinary process of iron and steel manufacture the mills and foundries designed to supply the metal in the forms required in commerce are based upon supplies of pig-iron that are easily accessible in ample quantity. Without such an abundant supply of pig-iron, no well-developed and extensive manufacturing industry can exist. Foundries and rolling-mills that rely on distant sources, produce only with difficulty and at high cost, and the manufacture of small articles, machinery, and varied appliances, in which iron and steel are essential, is monopolized by districts favorably situated.

As a fundamental commodity pig-iron ranks with timber, food, and fuel; and these elements of the commercial world are the determinants of industrial growth, as are air, water, and sunlight of human life. An available supply of iron indicates that a region will become densely populated, that the harbors will be crowded with shipping and that great cities will be built. The industries on the western coast of the United States, that started from resources other than iron, and have grown despite the absence of a local iron-supply, have remained at the limit set by the local cost of iron and steel.

There are some who hope for an abundant supply of Californian pig-iron in the near future, at a price not higher than that prevailing in the iron regions of the East. Others believe that the Pacific Coast does not possess the essentials for a self-supporting iron-industry, and that no relief from the high prices would result even if local production were attempted. A potential supply of iron does not consist solely in the presence of the ore, but requires also a fuel such as can be used to smelt it. The economic tendency of the iron-industry, the world over, is toward centralization, and is opposed by the extension of iron-production to the West. Though there are ores, and oil, and electric power, and markets on the Pacific Coast, similarly there are hundreds of millions of tons of unmined ore available in the densely peopled states of the East, but remaining unused because of cheaper ore 1000 miles distant.

The ore resources of the Pacific Coast appear to warrant the development of an iron-smelting industry. The hematite deposits of Eagle mountain, in southern California, have been shown to be ample for operating a blast-furnace for 200 years. Some of this ore is pronounced to be of the finest quality. Much is judged to be capable of being enriched by sorting. Shasta county, in California, is also capable of supplying iron ore for an important industry. In Madera county, California, there is a body of ore estimated at 50 to 100 million tons of good quality. Elsewhere, as in San Bernardino county, in the Kingston mountains, and also in Oregon and Washington, there is ore said to be suitable in quality and amount for smelting operations. While the deposits of iron ore on the western coast are small compared with the great reserves in the East and in the Lake Superior region, they are adequate, so far as the ore is concerned, to meet the local demand, and are worthy of attention. In recent years the extensive deposits of black-sand, which abounds on the Pacific Coast and elsewhere, has been looked to as an attractive source of iron. These sands are known to contain, amongst other minerals, a small proportion of magnetite. While there is little possibility of using widely scattered material of this character for extensive iron-manufacture, yet the improvements in separation and the demand for the contained minerals, other than the magnetite, suggest using the material for iron or steel-manufacture in a small way. The backwardness of the iron-industry in the West is the direct outcome of the economy of production in the East; and it is the relative cheapness of the Eastern supply that renders the western condition tolerable. Moreover the demand for iron and steel in the West covers a wide variety of forms; and plants to supply the Western needs are called upon to furnish shapes difficult to produce at any single establishment. The local consumers would be limited if made dependent upon a local supply. A single blast-furnace would fail to supply the varied demand for local consumption. The limitation, however, that is most serious to production in the West is the absence of fuel of the kind commonly required in iron smelting. To obviate the cost of transportation of coke there arise the problems of novel metallurgical methods utilizing the heat-energy that locally can be had.

The sources of coke available in the West are those from China or abroad, the development of the coking coals of Alaska, the coke of Utah and Colorado, and finally there arises the question of adaptation of practice to permit of employing coke that can be made from the coals of Washington. There remains the possible substitution of other fuels for the needed coke. The production of a satisfactory coke has been claimed to be possible as a by-product from petroleum, when treated by a special process in gas manufacture, with this end in view. Commercial demonstration of the possibility of doing this, however, is as yet wanting. The use of natural oil, with regenerative appliances, is possible, but only under important limitations and with diminished heat-efficiency. California, which is one of the most productive oil fields of the world, would be in a position to profit by the use of its natural resources in this way. The use of electric energy in iron-smelting has been demonstrated as not only possible, but practicable, under certain limitations. In this particular the West, possessing available water-power in the Sierra Nevada and the Cascade ranges, offers a cost comparing favorably with such power elsewhere, and is in a position to profit by this extension of the industry.

The treatment of iron ores, by the standard processes, is an industry involving large tonnages and broad markets, while the mining, transportation, and smelting are circumscribed by the low value for a given weight. The average value of iron ore, such as is now derived from the Lake Superior region, or from Cuba, Spain, Canada, and elsewhere, is normally about $3 per ton at seaboard, or at the smelting-centers. The cost of the ore at the mine may be rated at 50c. to $1 per ton. The cost of ore, at $3, at the Eastern smelting-centers consists, therefore, largely in the cost of transportation, which commonly amounts to two or three times the cost of mining. The handling and transportation of large tonnages of iron ore of this relatively low value calls for accessibility, either through natural ways of transportation or the actual proximity of the ore to regions where the metal is consumed. The average value of coke at tidewater, for example, at Baltimore, may be taken as normally about $2.50 per ton. Coke in the State of Washington costs twice this amount. Nearly two tons of ore and one ton of coke are required to produce one ton of pig-iron. This ton of pig-iron, of average grade, and at normal times, is worth approximately $15. The price at present is three times this sum. In terms of normal operation, over a long period, three tons of ore and fuel, having a value of $8 or more, are required to produce one ton of pig-iron valued at $15. When such prices and production are contemplated for the West, the cost of transportation eclipses all other items. The increment in the ton-value of the material affected by the transportation and smelting, based upon the combined weight of fuel and ore, is less than $2 per ton. A slight increase in the cost of transportation easily consumes any allowance for the cost and profit in smelting. Water-transportation in the movement of iron-ore has been rated at 0.09c. per ton-mile, and land-transportation at 0.66c. on well developed lines. At this rate, the entire increment of operation would be consumed by a land-haul of the coke of 1000 to 1100 miles, or of the ore for 500 miles or more, or of the ore and coke for about 333 miles. In water-transportation, the distance would be seven or more times as great at a corresponding cost.

The freight-cost on pig-iron, at 60 to 75c. per hundred, from Pittsburg or New York, or 55c. from Chicago, brings war prices to the western coast in times of peace. Iron produced locally can be marketed, normally, at nearly $30 per ton. At any time, however, local smelting brought into competition only with existing prices, with the freight-costs added, brings no relief to local industries as a whole. The cost of production in the West must more nearly approach that in the East to warrant interest in the turning of industry from its established channels.

The extension to the West Coast of the standard industry of iron-smelting with coke is seen to require, not so much a gradual growth of many industrial interests as a single independent and energetic interest. By the standard methods of treatment, we could consider the iron industry to comprise five dependent steps. The ore is reduced to metal and manufactured into rails, rods, bars, plates, sheets, and structural shapes, or cast into forms for the consumer. The first, and primary operation, that of smelting the ore in the blast-furnace, is the only missing element to a complete chain of independent production in the West. This operation, carried on extensively in the Central, Eastern, and South-eastern parts of the United States, and in the West as far as Colorado, supplies pig-iron for the iron-foundry and the steel-plant, which may be regarded the second and the third branches of the industry. No large-scale substitute for the production of pig-iron in the blast-furnace has been demonstrated, as yet, to be practicable. The iron foundries are commensurate with the local need. By these the waste cast-iron, gathered from the surrounding country, is brought again to useful application. In the West, many foundries are relatively small, and are operating at a profit. Coke is the fuel commonly employed, but while a superior quality is to be preferred, grades inferior in some regards can be used. Oil fuel has been used in the foundry-cupola with the promise of economy and success. In experiments at one of the foundries in San Francisco, it was stated that about 8 gallons of crude oil per ton of iron would serve the cupola. The steel-plants on the Pacific Coast are relatively small, but for some years have been growing in importance. Plants of the open-hearth and bessemer types are represented. In the open-hearth plant, as operated in the East, pig-iron furnishes the greater part of the metal for the steel product, but with the pig-iron is used a suitable proportion of pure iron ore and also steel and iron scrap. So far as

ELECTRIC FURNACES AT HEROULT, CALIFORNIA, PRODUCING FERRO-SILICON

NOBLE ELECTRIC STEEL PLANT, HEROULT, CALIFORNIA

ore is used, this furnishes metal direct to the steel. The plant, to this extent, becomes a direct producer of steel from ore. The fuel commonly desired is producer-gas, but for the manufacture of gas a wide variety of fuels, including oil, can be used. The adaptation of the process to the West has led to the use of oil-fuel direct, with much success. Since, however, the tonnage of metal produced at the steel-mill practically equals that received in the raw condition, there is always the competition of steel made in proximity to the iron-furnace. The limitation falls more seriously upon the bessemer process than upon that of the open-hearth. The rolling-mills may be regarded as the fourth division of the industry. These supply the smaller shapes of structural steel to the building-trades and to the manufactories of machinery, including bolts, nails, rivets, and a great variety of iron and steel articles. These industries are acquiring importance in the West. The steel-foundries may be considered the fifth branch of the industry. There is an increasing demand for steel castings, which gives reason to believe that the time will come when light steel-castings will replace much of the heavy iron-structure in machinery. Waste steel has offered a supply of metal at a price not before equalled. The local supply of iron and steel is becoming depleted, however, by the heavy demand and high prices prevailing. The cost of producing steel from scrap in the West has been such that, in the past, in spite of the local supply of waste metal, manufacturers have found it less costly to purchase steel from the East than to produce the metal locally from the scrap-material available. The steel-foundry can operate with coal, coke, producer-gas, oil, or with electric energy. In these plants throughout the country the labor is skilled, and the demand for the products is largely local. The plant in many cases is an adjunct of a factory for making special machinery. The appliances are varied in which the melting of steel is performed. Much cast-steel is made by the crucible process. In the past few years the electric-arc re-melting furnace has come into application in many places.

There has been no success, so far, in making pig-iron on an extensive scale without coke. The use of fuel-oil, or of electric energy for the production of iron from ore, would be the first expedient in the development of a Western industry. Whether or not such methods can be developed to compete with existing practice is the problem to be considered. For the treatment of iron ore in the manufacture either of steel or iron, it is seen that the direct use of the ore along with a due proportion of the reduced metal is possible to a limited extent. Here the ore, or oxide, not only furnishes metal, but serves a useful purpose as an oxidizing agent. The oxidizing action, however, pre-supposes that there has been a previous reduction of some constituent of the charge. The direct use of iron ore in large foundry-furnaces for producing cast-iron directly from ore reverts back to the earliest days of blast-furnace smelting, and promises little in the way of practical economy. The direct reduction of iron, as in the American bloomary-process, is primitive and impracticable under existing labor conditions. The reduction of iron sponge, in a separate operation by the use of fuel other than coke, followed by the melting and casting of this reduced metal into steel-ingots of low carbon-content, involves a double operation which, compared with the standard processes is costly, since it divides the operation which, in the blast-furnace is conducted with the fullest economy in a single operation, into two separate steps, performed at different times, and with two separate applications of heat.

In the smelting of iron there is chemical and physical work to be done, only in part by the application of heat. There is slag to be melted, water-vapor, carbon di-oxide and other gases to be expelled, and various heat-absorbing side-reactions to be satisfied. A part of the heat is wholly lost in the liquefaction of products and in the formation and expansion of escaping gas. Notwithstanding the complex distribution of the energy, there is a part that is to be regarded as peculiarly and unalterably necessary to the process. This energy performs the service of reducing the oxide to metal. The heat consumed in fusing the metal may be partly recovered upon solidification, by improvements in construction that have been suggested from time to time. So also heat that is lost by conduction and radiation may be further decreased by improved heat-insulation or by improved furnace design.

In the case of the energy that is required to reduce the iron and form the desired iron-carbon alloy, it will be seen that the interchange of the forms of energy is more difficult. In any case, the low value of the material and the high tonnage to be treated require low cost. Hydro-electric energy comes into open competition with the energy derived from petroleum or from coal. Electric power, light, and heat, however, with the improvements in modern industrial and domestic applications, has become a refined form of energy meeting an increasing general demand. Oil, similarly, as a direct source of heat, is to be expected to find an indefinitely increasing market through the extension of known devices and methods of application and facilities for transportation. The maintenance of the high temperature required in the hearth of a blast-furnace requires consideration apart from that for other appliances. A reaction is sought in the blast-furnace by which metallic iron and a carbide of iron are formed. The reaction, which occurs only at a high temperature, absorbs heat at that temperature. Such reactions are likely to require appliances that are wasteful of energy. There is loss of heat in elevating the materials to the temperature of reaction and of obtaining a return from the heat contained in the products. Here the reduction of the oxide to metal requires a temperature ranging from 800° to 1800° F. In electric smelting, it is a simple matter to produce the heat, required for the reduction, by the arrangement of the electric arc, or by other suitable electric resistance. Reduction work, however, is not obtained by the heat alone. Carbon, in some form, is required in addition to the electric energy. The electrolytic capacity of the appliance is too small and the electrolytic action too slight to justify the maintenance of the high temperature to secure it. In the effort to use oil in iron smelting, as a source of the heat, the products of combustion become troublesome. The gas from the combustion contains moisture from the hydrogen content of the oil. The water-vapor, along with the carbon di-oxide from final combustion, produce an oxidizing effect upon the iron and carbon, and interfere with the reduction.

The energy required in iron-smelting can be expressed in simple units. The reduction of one pound of iron from the sesquioxide (hematite) requires the consumption of fuel capable of yielding 3141 B.t.u. This heat, capable of yielding 3141 B.t.u., is absorbed in the process, and is not available as a heating effect. Aside from any consideration of heat required to satisfy this consumption, a given amount of carbon or carbon monoxide is necessary to carry on the action. However the reduction be performed, whether with carbon, carbon monoxide, hydrogen, or hydrocarbons, and whatever the heat-value of the fuel used to perform the reduction, this constant amount of heat-energy is absorbed from the reaction at the high temperature necessary. The heat-balance of the reaction of oxide with the fuel shows a slight excess of heat, which results in elevating the temperature after satisfying the heat-absorption. With solid carbon fuel, producing carbon monoxide as the end-product, there is an absorption of heat throughout, and the reaction is not self-sustaining but requires supplementary heat. The carbon necessary to effect this reduction would be approximately $\frac{1}{6}$ of the weight of the iron, in a case where carbon escaped from the reaction in the state of the higher oxide. Where the product of the reaction is carbon monoxide, the weight of carbon involved would be twice that amount. Similarly, if this carbon had been previously oxidized to carbon monoxide, and the reduction of the iron were performed by the carbon monoxide, the amount of carbon represented in the reaction would be approximately $\frac{1}{3}$ the weight of the iron. The reduction of such ore requires the application of carbon ranging from $\frac{1}{6}$ to $\frac{1}{3}$ lb. per pound of iron produced; and in the use of this carbon little or no heat is produced over and above that consumed by the reduction of the iron.

NOBLE ELECTRIC STEEL PLANT: IRON MINE ON RIGHT

The fuel consumed in blast-furnace practice approximates one pound of coke per pound of iron produced. The heat-value of coke, or of a medium grade coal, may be taken as 14400 B.t.u. per pound. This pound of coal or coke is equivalent to about 5.6 hp. hr. or 4.18 kw. It is the heat that would be expected from the complete combustion of about $7\frac{1}{2}$ cu. ft. of natural gas, $2\frac{1}{2}$ lb. of dry wood, $^2/_3$ lb. or $^1/_{12}$ gal. of petroleum. Its service in reduction is to furnish 3141 B.t.u. for the conversion of the oxide into 1 lb. of the metal. The balance, or $\frac{5}{8}$ of its heat-value, is available for the liberation of heat. For the economy of coke and the substitution of as much coke as possible by fuel other than coke, or by energy derived otherwise than by the combustion of coke, it is natural to look to some process that may permit of employing the $\frac{1}{6}$ to $\frac{1}{3}$ lb. of coke as the reducing agent, and of supplying the remainder from other fuel or other agents that can be locally provided. The application of heat by means of any exterior contact is attended with serious physical losses. The application of supplementary fuel within a reducing appliance, on the other hand, contaminates the gases which are desired to exert a reducing effect. The use of hydrocarbon fuel particularly, in the blast-furnace, either in the zone of fusion or of reduction, is attended by incomplete reaction and incomplete recovery of the heat-value. The employment of electric energy is opposed by a high unit-cost. The electric furnaces thus far devised, however, are small and ill-adapted to conserve the heat. In smelting iron by the use of any fuel, the full efficiency of the process can be secured only by utilizing the chemical energy contained in the waste-gas. Such waste-gas, of necessity strongly reducing, when issuing from the smelting appliance, is a fuel superior to any that easily can be supplied from an auxiliary source, produced from other fuel. The heat-units described above represent the effect of complete combustion of the gas. In modern blast-furnace smelting, the heat, latent in the waste-gas, is utilized by burning the gas where complete combustion can be secured, and employing the heat thereby developed for pre-heating the entering air, or for the generation of steam. The recovery of the waste-heat is an essential economic factor.

Efficiency under the conditions of high-temperature reactions generally implies a need of recovering the heat latent in the products issuing from the appliance after the reaction has occurred, and the avoidance of loss by radiation and conduction. In certain metallurgical operations the desired reactions occur spontaneously, and the production of heat sustains the action, when once the initial means are provided to produce the requisite temperature.

When the magnitude of the blast-furnace is considered, with its height of 75 ft. or more, and its diameter of 15 ft. and its large content, relative to its exterior exposure, and when the long experience with the blast-furnace is considered, with the improvements that time and ingenuity have wrought in its details, there will be less expectation of substituting small appliances for it with an idea of securing comparable efficiency. No one objects to the blast-furnace as an economic appliance. The objection is to the restriction in securing practical fuel. The effort to use a volatile fuel in the blast-furnace, or in any direct modification of a blast-furnace involves ineffectual reduction, and poor economy in heat units. In the shaft-furnace there is difficulty in keeping the charge open for the passage of the necessary gas. The use of solid fuel of high fixed-carbon provides the means for this. The use of electric energy for the production of the pig-iron fails to provide the required reducing action. In addition to heat derived from this source, some $\frac{1}{10}$ lb. of fixed-carbon is required for each pound of pig-iron reduced. The gaseous product is irregularly supplied, and its utilization has not thus far been made practicable. The consumption of electric energy is found practically to amount to 2200 kw.-hr. per ton of iron produced, in addition to the coke, charcoal, or other fixed-carbon fuel consumed. Here the substitution of electric energy for coke results in the saving of approximately 0.6 tons of coke per ton of metal, at the expense of 2200 kw.-hr. In Sweden, electric power has been used for the production of pig-iron, but the economic conditions differ from our own. At Heroult, California, with iron ore and power available, the electric furnace has given excellent service, not in the production of ordinary pig-iron, but in the manufacture of special alloys. For the manufacture of steel, either from blast-furnace products, or directly from ore, the electric furnace is coming more generally into prominence. There are in use in the United States 43 furnaces of the Heroult type, the largest of which has a capacity of 20 tons, and 14 of the Snyder pattern. The industry is growing, and the present conditions offer extraordinary encouragement for its development. The manufacture of steel, however, is an industry subordinate to the manufacture of pig-iron.

The development of a complete iron industry in the West, to include the production of pig-iron, demands special encouragement. Ideal conditions are not to be found. The abundant coking coal, found elsewhere adjacent to iron ore and limestone, is denied the West; yet its absence has engendered the greater need. The requirement of the West is not to be measured by standards elsewhere. The need is to discover possibilities under conditions that have been thought of with disfavor. Herein the proposed and untried processes that have appeared from time to time merit final and critical consideration. The blast-furnace must be thought of, not for what it can do at its best, but for what it could do if operated with the fuel that is to be had. The ores of the Pacific Coast must be more thoroughly examined and made better known. Transportation, for the movement of coke, must be encouraged by industry as a whole, and the local and national benefit from it fully recognized. The use of ore must be fostered in steel manufacture and in every local process where it can be employed; and until a way is found to produce sufficient pig-iron for local industry, local ores should be developed to supply this need. Beneath the high prices of the present time there can be read, in terms of past prices, the possibility of continuous and profitable future operation. In every phase of the industry the way is now cleared for expansion by the supreme commercial inducement for industrial activity. An iron industry in the West required, perhaps, nothing less than a world-war for its conception, but in coming it will find a permanent abode.

Work of the Council of National Defense

A report was recently issued by the Council of National Defense setting forth the accomplishments of the sub-committees up to the first week of June. Among these achievements attention is directed to the mobilization of 262,000 miles of railroad for Government use, as well as the organization of the telegraphic and telephonic lines of communication, arrangement for the purchase of copper, lead and zinc for Government needs at reduced prices, thus saving large sums of money, the completion of the inventory for military purposes of 27,000 manufacturing plants, the creation of the Aircraft Production Board which has started to construct 3500 aeroplanes and to train 6000 aviators during the current year, the enlistment of reserve engineer-regiments for re-habilitating the railroads of France, and of another corps for improvement of the Russian railway service. The report is elaborate and reveals that the work of organization has been carried out so that definite results in providing the necessary supplies have been obtained. This has advanced to a point that will obviate the scramble for material that characterized the opening months of the War in France and England, and which gave an opportunity for shameless extortion and for graft through long chains of middlemen who exacted burdensome commissions. An orderly mobilization of the country's resources is in progress, with the co-operation of an army of producers of raw material and of manufacturers working in a well-unified effort toward the desired ends. In the matter of small arms and rifles arrangements have been concluded for supplying 1,000,000 men, but the work of manufacturing artillery has not proceeded so far with the exception of gun-forgings which are now being made. On the other hand sources of supply of artillery and accessories have been definitely found, and bases for contracts have been arrived at. Armored cars in large quantity are being manufactured, and an increased output of machine-guns is assured.

In coming to an agreement for the manufacture of aeroplanes the committee on aircraft has acted as mediator between the Wright-Martin Aircraft Corporation and the Curtiss Aeroplane Co. and has arranged a satisfactory basis of settlement of the patent controversy that hampered the operations of these competitors, thus leading to friendly co-operation for the welfare of the country. This has been accomplished by a legal cross-licensing agreement. The preparations for efficiency in this branch of the service give promise of realizing the expectations held by many as to the vital contribution of aeronautics toward the winning of the War. The medical service has been thoroughly organized under the general supervision of Dr. Franklin H. Martin of the Advisory Commission, with Dr. F. F. Simpson in direct charge. This Board is taking energetic measures for safe-guarding the health of the recruits to the Army and Navy.

One important step in the work of the Council of National Defense has been to create an Inter-departmental Advisory Committee to co-ordinate the efforts of the many special committees so as to obviate reduplication, and to serve as a sort of clearing house so that the different committees may without loss of time be able to secure any information and material needed for their specific purposes. A very brief statement is given of the activities of the Navy Consulting Board, which, however, has made important contributions toward the solving of some of the more pressing problems. Being the oldest of the Boards it is to be expected that it should display early efficiency. The Commercial Economy Board has attacked some important problems in conservation by undertaking to cut out many of the enormous costs in the delivery of goods to individual consumers. This has become a most wasteful practice in American retail trade, and the committee has secured the co-operation of leading merchants and Boards of Trade throughout the country to effect a reform in this particular, and also to avoid excessive costs in wrapping and packing, and also in the distribution of catalogues and circulars. The Committee on Coal Production, under the chairmanship of F. S. Peabody, of the Peabody Coal Co. of Chicago, has been doing good work in assuring supplies of fuel for the Navy, but since the date of this report the Federal Trade Commission has made a special report to Congress indicating the need of suitable action to provide for Government control of the sources of coal supply and of transportation systems so as to prevent interruption of manufacturing in the manner that so seriously embarrassed all branches of industry last winter.

Few portions of the committee-work have yielded such far-reaching results for aggressive as well as defensive purposes as the National Research Council, under the direction of George E. Hale and R. A. Millikan, who represent the highest ability of the country in physico-chemical research. This committee is working on means for detecting submarines and mines, in the perfection of range-finders, detectors for invisible aircraft and sapping parties, wireless apparatus, military photography, new explosives, and the prevention of corrosion and electrolytic action on the hulls of vessels.

Raw materials, minerals, and metals are under control of a committee headed by Bernard M. Baruch, and it has organized the field with regard to alcohol, aluminum, asbestos, magnesia and roofing, brass, cement, chemicals, coal-tar by-products, copper, lead, lumber, mica, nickel, oil, pig-iron, iron ore and lake transportation, rubber, steel and steel-products, sulphur, wood and zinc. Co-operative committees have been formed under the foregoing headings. Several of these committees maintain representatives in Washington in order to keep in close touch with the executive departments of the Goverment, and the chairmen of all of the co-operative committees visit Washington from time to time and get in touch with the Departments with which they have to deal in an endeavor to become acquainted with their requirments and specifications. While those committees were formed to mobilize the industries mentioned and to act only in an advisory capacity, experience has shown that they have been of great immediate value to the Government in perfecting early deliveries and in securing prices below market-quotations. The concrete accomplishments of the Committee on Raw Materials, Minerals and Metals, may be itemized as follows: (a) 45,000 lb. of copper has been offered to the Government by the copper producers, acting at the instance of this committee, at an approximate saving to the Government of $10,000,000; (b) The co-operative committee on zinc has contracted for about 25,000,000 lb. of zinc at about two-thirds of the market price, and is prepared to use its efforts to effect still further savings on the vast quantities that must be purchased; (c) Through the co-operative committee on steel the Navy Department contracted for several hundred thousand tons of ship-plates and other materials at great concessions. When ship-plates were selling at $160 per ton, the Navy bought them at $58 per ton; (d) The co-operative committee on aluminum purchased for the Government its requirements in aluminum at 27½c. per pound, when the regular price to large purchasers was 38c., and the price in the open market was 60c. per pound; (e) The co-operative committee on chemicals is now engaged in negotiations with the fertilizer interests of the country to stabilize and lower prices; (f) The co-operative committee on oil has closed contracts, for the delivery of oil to the Navy, at satisfactory prices. None of the foregoing committees is exercising executive functions, but each has advised with and assisted the executive departments of the Government in the purchase of their supplies.

CUPEL-ABSORPTIONS with low-copper, in tests made by Frederic P. Dewey, given in the Trans. A. I. M. E., varied from 0.23 to 0.33 mg., and on the high-copper from 0.24 to 0.34 mg., but there was a decided preponderance of low absorptions with low-copper and high ones with high-copper. The average absorptions were:

	Mg.
High-copper	0.2991
Low-copper	0.2773
Difference	0.0218

In order to keep the conditions the same, more lead and a higher temperature were used on the copper than were necessary, but the test clearly shows that copper increases the cupel-absorption. The same fact is also shown in a general way by 26 rows of cupels with copper in the ends, but without copper in the centres. The copper was alloyed with silver and two proportions of copper, 62.5 mg. and 125 mg. were each used with 4, 5, 8, and 10 gm. lead. The end-cupels yielded 9.33 mg. gold, and the centres 4.12, or 0.545 mg., less than half of the ends.

INFLUENCE OF BASE METALS in gold-bullion assaying, by F. P. Dewey, are summarized in the Trans. A. I. M. E. as follows: In small amounts, Zn alone has no effect upon the cupel Au. In large amounts the Zn-Au-Ag alloy is formed, comes to the surface, and carries Au to the cupel. A tendency to do this occurs when small amounts of Zn are cupelled with large amounts of Pb. In all the comparative tests Cd showed a distinct protection of the Au, even when Zn was also present. The presence of Cu *per se* directly increases the absorption of Au, but its effect is intensified by the higher temperature and the larger amount of Pb required. The effect of Pb varies greatly with the surrounding conditions. The presence of Ag is a good protection to the Au, but may be overbalanced by other conditions.

Need of Chemical Research

No amount of warning on the subject of the importance of pure chemistry seems to have the slightest effect upon American opinion. It has been dinned into the ear of our public that Germany owes her triumphs in the field to her pure chemistry. Our failure to profit by the warning, observes Dr. L. Charles Raiford in a recent address before the Chemical Department of Oklahoma University, is due to a typically American mental attitude toward what so many of us are pleased to call 'applied science.' Many highly intelligent educated people do not understand the meaning and the value of research, and hence they call it theoretical and impractical. Their cry is for that magic thing called 'applied science,'—for something 'practical.' They fail to recognize the fact that there can be no applied science until there is science to apply. Was Roentgen thinking of the extraction of bullets, the reduction of dislocated limbs, or the setting of broken bones, when he worked his way to the finding of the famous rays? By no means. Again, did Helmholtz have in mind the prevention of eye diseases or their cure when he worked out the principle of the ophthalmoscope? Not at all. Was Cavendish thinking of providing food and war munitions when, more than a century and a quarter ago, he read his paper to a learned society on the fixation of nitrogen. At this moment the republic is clamoring for preparedness, with no perception that the preparedness of the foe is chemical—based on the pure chemistry of the research-worker. Here in our own land cheap fertilizer must be furnished in times of peace, and nitric acid and other materials for explosives in time of war. The real problem is to furnish the means—pure chemistry—by which this programme can be carried out. Two specific requirements confront us. First there must be adequate training in the fundamentals of chemistry, and second there must be opportunity for chemical research. Now, if you ask the first person you meet, or if you get the opinion of the influential citizen, he is apt to tell you that the chemist is a man who can analyze substances. Do we stop to think of pure chemistry when we read the headlines in the newspaper bearing a wireless message flashed across the ocean? Who, as he steps into his automobile or watches the movies, thinks of the chemist in his research laboratory? Chemistry! It will win the War; and no American, befogged by 'applied science,' can get that truth into his head.

Mill-Tests v. Hand-Sampling in Valuing Mines

By MORTON WEBBER

Two methods may be chosen by the mine-valuer in the examination of a mine. Much has been written as to which is the better. In fact, so divergent have been the views expressed that there has developed what really amounts to two schools of thought. The one is the practice of basing the report of an examination on hand-sampling and the other is the process of procuring for the purpose of a mill-test a representative sample of the expected profitable ore of the mine.

During a recent discussion in the MINING AND SCIENTIFIC PRESS, I was asked by the editor to contribute my opinion as to the method I regarded as the more reliable, especially as applied to the examination of large low-grade ore deposits. There is no doubt in my mind that the choice of either method should depend primarily on the ore-structure. The size of the deposit, as a factor in the choice of method, is only of importance in relation to the cost of examination. I believe each mine is a problem unto itself. My experience has taught me that there is generally a particular method of sampling that will possess decided advantages over any other method of examination. This may be a combination of both methods or only one, or it may be the application of one method applied to the mine-workings as a whole, with a variation or combination centralized on a certain part of the mine to throw additional light on some specific feature.

The more experienced the engineer the quicker and more accurately will he decide on the method most expedient to obtain the essential data. In other words, the proper choice of the examination programme, for a particular case, as against rule of thumb, is the criterion of the valuation specialist as against the general practitioner. There is also another point. The terms of purchase may govern the method of examination. I suggest that this feature has been considerably overlooked in discussions on this subject. It was only through somewhat varied and extensive examinations on behalf of a purchaser—an important corporation—in the course of which I had to negotiate for options prior to examination, that I realized the importance of this additional burden on the valuing engineer. In recent years the mine-valuer has had to become a consultant on mining finance generally, and particularly on the method of payment in the purchase of mining property. When a very young man, I was introduced to the famous Cecil Rhodes, and he indicated that there would always be more than ample room 'at the top' for the mine-valuer who could co-ordinate natural business ability with mine-valuation.

The following description of three examinations conducted under my direction may be of interest. For reasons that will be sufficiently obvious, I cannot furnish the names of the properties, so I shall call them the Red, White, and Blue mines, and they will be dealt with in this sequence.

The Red mine represents a case in which, because of the structural nature of the orebody, I considered a mill-test was the proper method of examination. The mine was subjected to a preliminary sampling prior to the mill-test in order to determine the locus of the shoots and a more detailed sampling was undertaken subsequent to the procuring of the representative shipment. I give this example, therefore, as an actual case of a combination of the two methods. The White mine, on the other hand, represents a case in which the orebody was uniform in value. In other words, the ore was not localized in what

THE SAMPLING CREW THAT CUT 6½ TONS OF SAMPLES IN 30 DAYS AT THE WHITE MINE

are generally regarded as 'shoots,' nor did there exist the variation in value and mineral characteristics between one shoot and another, one of the principal reasons for applying the mill-test in the case of the Red mine. The Blue mine, finally, represents a case as mentioned above, where it was desirable to apply ordinary sampling over the general workings, but where a representative shipment was obtained from a certain part of the mine to throw additional light on a specific feature. This was also a case where the terms of purchase governed the method of examination.

The ores of the first case (Red mine) contained gold, silver, and lead. Briefly, the mine was composed of several irregular ore-shoots occurring along a fault-zone in granite. The thickness of the zone was about 50 ft. and considerable brecciation and post-mineral faulting were evident. The shoots were located on the hanging wall and also on the foot-wall. Ore-shoots were also encountered running diagonally from wall to wall. An important feature was the marked difference in the mineral characteristics of the ore-shoots. One shoot was clean lead-silver ore, carrying about 12% lead and approximately 10 oz. silver. Another shoot was a heavy iron pyrite (about 7 cu. ft. per ton), carrying an unimportant quantity of lead and averaging about $17 per ton in gold and silver. A third shoot was a friable fine-grained iron pyrite, carrying about 50 oz. silver and about $10 in gold.

This was a difficult type of mine to examine. Evidently the examination of the property involved a treatment problem. I believe that the specialist in valuation will assume too much if he atteempts to encroach upon the domain of the not less important specialist on metallurgy. It is, however, the duty of the valuing engineer, if doubt exists as to the best method of treatment, to procure a shipment from the mine-workings, representative of conditions as they exist at the date of examination. An important point is to submit to his client a clear opinion stating how far such a shipment is likely to represent the ore upon which the future life of the enterprise will depend. I considered it my duty therefore to obtain this representative shipment. If the shipment thus obtained differed from the product of the mine under actual operation, the error lay with me. If it did not differ, but the metallurgical process was defective, or a failure, the blame lay with the metallurgist.

In accordance with my observation that there is generally 'one best way' to conduct the examination of a particular case, it was evident that the solution did not lie in shooting down a composite shipment from each of the stopes. At the outset there was one factor outstanding in relation to the subsequent executive or administrative features. This was that the widely varying types of orebodies, could be mined separately, if desirable. Therefore it was important to get the representative shipment to the metallurgist in such a way that this administrative feature remained undisturbed. In other words, it was important that the metallurgist could deal with each stope or individual source of ore-suply separately, that he could eliminate the ore from certain stopes at his discretion, or make such combinations as he desired, in addition to dealing with the average ore of the mine, as a whole.

The method of examination to cover these features was as follows. It was necessary from the start to decide on a pre-determined weight per foot sampled. In order to obtain a representative shipment of the average product of the mine, it is important that each stope should contribute its *representative* weight. If every care is not exercised in this particular, it will throw the representative shipment out. I am convinced it is here that most mill-tests go wrong. In the majority of mill-tests that have come under my observation, little attention was paid to this point, which I regard as a most critical feature in an examination. In this case, I decided on a foot-weight of 500 lb. per foot. What the foot-weight should be depends on the individual case. The main point is that once it is chosen, the foot-weight should be rigidly adhered to.

It was impossibile to break down the ore from the stopes in definite proportion to their width. A round of shots from a stope eight feet wide would break 6000 lb., or 50% above the composite proportion, whereas another stope five feet wide would break 5000 lb., or 100% above its composite proportion. These simple figures will illustrate where the usual mill-test fails. In all cases I made certain before firing the rounds that the shots would break more rather than less than the required foot-weight. This was to avoid going back over the same area for the deficient portion, as such a step would be, in effect, tampering or unnecessarily introducing the personal factor. Once the sampling programme is laid out, this should be avoided as much as possible.

When the product was obtained from a particular stope, it was broken to a convenient size for thorough mixing. After being thoroughly mixed, the over-weight was discarded. The mixing was done on a mixing-floor arranged for the purpose. As an additional precaution against error, great care was exercised in removing the over-weight. If the sample was found to be, say, 20% over-weight, the whole sample was hand-shoveled to another part of the mixing-floor. In the transfer every fifth shovel was discarded. This was introduced to anticipate an error in event of the sample being insufficiently mixed. The samples from the various stopes were numbered according to the stope from which they came.

In the above examination, therefore, I accomplished what I set out to do, namely to obtain a shipment of the workings as a whole that would represent the average grade and also the average mineral characteristics of the mine, as a whole, if the stopes were operated simultaneously. I obtained the sample in such a way that each stope could be dealt with separately by the metallurgist in his research work, as reflecting the underground administrative policy, should it be desired to keep the product of the stopes separate. I also found, by comparing the result of the hand-sampling with the assay of the composite pulp from each stope, the sampling error for each class of ore—a most important feature in a mine of so divergent stope characteristics. By comparing the assay of the pulp of the total shipment with the subsequent hand-sampling, I obtained also a valuable index of the sampling factor of the mine as a whole.

The White mine represents an entirely different case. It contained a large low-grade gold deposit. It was known that the ore of this mine was simple in treatment. The factors of existing and expected quantity of ore and the average grade were therefore primary, and the metallurgical problem was relatively subordinate. This, coupled with the fact that there was no localizing of the gold into shoots and particularly into shoots of dissimilar mineral characteristics, made an examination by means of a mill-test inexpedient. In fact, I characterize this case as being essentially a straight hand-sampling job. The sampling was conducted as follows, as quoted from the text of my report:

"The workings were sampled throughout the supposed orebody at intervals of 10 ft. In the raises and winzes sectionalized sampling was employed wherever desirable. In view of the low tenor of the ore, it was important that sampling should be accurate. An error that may be unimportant in medium or high-grade ore becomes a large proportion of the total content when the ore is low-grade. The sampling and cutting down was performed as follows: The original cut averaged about 10 lb. per foot. This mine-sample was reduced on the surface to pass half-inch holes. The entire sample was then mixed on a mat for a minimum of 5 minutes. Great care was taken to insure each sample being mixed for at least this period. A clock was furnished for this purpose. The work was under the constant supervision of a trained engineer. His duties were confined to this department. After the sample had been mixed, as described, it was then reduced by a Jones sampler, when it was ground to pass quarter-inch holes. The sample was then mixed again on a mat for 5 minutes, when it was cut down by a Jones sampler to approximately one pound and a half. The sample was then, without further reduction, ground in its entirety to pass 80-mesh, when it was again mixed for 5 minutes, prior to being split in duplicate. Great importance was attached to mixing. By this process each sample in its reduction from the original, which averaged between 50 and 60 lb., had been mixed for at least 15 minutes.

"In order to check the accuracy of the sample as cut in the mine and also in its reduction to duplicate pulps, two systems of checking were employed. A series of cuts was re-cut and given different numbers from the original tags and were treated in every respect as separate samples in subsequent reduction and assay. The series checked to 25 cents per ton. This checked the accuracy in the cutting of the sample and its subsequent reduction. The sampling work was also checked by the assay of a series of duplicate pulps. The assays were in duplicate and they checked with the original pulps, as assayed at the mine, to 20 cents. The assaying at the mine was checked by a third system whereby a group of pulps that had been assayed were put into fresh envelopes and different numbers attached. The assayer was thereby caused to check himself, which he did to 20 cents. A flux and slag assay was run every day."

A feature of this work was the heavy sampling. I am not an advocate of cutting a light sample where it is necessary to obtain the highest possible accuracy under practical sampling conditions. It may be of interest to state that while engaged by a client to search for mining property, it was my custom to make a preliminary sampling of a mine and then to option the property before I undertook a final sampling. In the preliminary examination I always cut light samples, about 1½ lb. per foot, because I was not equipped to take heavy samples. In cases that withstood the 'preliminary,' and final sampling was undertaken, I found frequently that lower results were recorded by heavy sampling.

In the sampling of this mine it was desired to average 10 lb. per foot. In my experience this is the weight that gives the highest accuracy compatible with convenience. It was my object, if possible, to have the average of my sampling within 30 cents per ton of the mill-heads of actual operations. The photograph will indicate the size of the original mine-samples, and it will also illustrate the size of sampling crew necessary on a fair-sized mine to do a job where much heavy sampling is necessary. In this case, about 6½ tons of samples were cut in 30 days, and the samples in the photograph represent about one day's work.

THE POP-SHOT SAMPLES TAKEN IN THE BLUE MINE

The Blue mine—the third mine—represents a hand-sampling job where a mill-test was made covering a part of the workings to obtain special data. This example represents a case where the terms of purchase governed the method of examination. The property was a silver-gold mine, the silver being in the form of argentite. The vein was about 15 ft. thick and the country-rock was andesite. The vein outcropped for nearly two miles. The larger portion was controlled by the property under examination. In the early history of the operations ore had been mined and shipped from several shafts situated at irregular intervals along the enormous outcrop. That these surface shoots did not persist in depth was evidenced by the fact that one after another these small operations had been abandoned. The property had then been lying idle for a period of years until an optionee sunk a shaft through one of the old workings, passed through what, for simplicity of expression, could be called a barren zone, when a profitable ore-shoot was again encountered. A level was driven to determine the length of the shoot, which was, in round figures, about 300 ft. As the vein was very hard, the optionee drove this level along one of the walls and drove cross-cuts through the orebodies at intervals of 50 ft. These cross-cuts indicated the width of the shoot to be about 10 ft., and of profitable grade. The shaft was then extended with the intention of demonstrating the shoot at the 600-ft. horizon and at further depth. The shaft was in the foot-wall. The fact that the orebody on the 500-ft. level was only demonstrated by cross-cuts at 50-ft. intervals should be noted, because this influenced the conduct of the examination, as I shall attempt to show.

The above was roughly the situation at the date I made my examination on behalf of a new purchaser. The sketch will demonstrate the position of the workings. Under the terms of my client's option, about one-third of the purchase price fell due in about four months from the comencement of my examination. I mention this as I have pointed out that the terms of purchase may govern the conduct of examination.

The speculative attraction was briefly in finding a new and distinct ore-shoot at depth below a surface shoot that had been exhausted. I considered that if the lower shoot could be demonstrated to be of importance, and in itself was likely to amortize a fair proportion of the purchase price, it would make the development of the lower horizons of the other old surface-workings an extremely interesting development speculation. This was how I regarded the project at the date of my preliminary examination. There was still ore left in many of the old workings, but I regarded such ore as merely a bonus of some importance, in case we took the property. I regarded the standing ore as only of subordinate importance as an index of the future worth of the mine. The latter feature was the major item of interest to me as representing a purchaser with ample financial resources intending to put the enterprise on its feet. In order to ascertain the grade of the new shoot, I decided to put 'pop-shots' in both sides of the cross-cuts, for I considered there were insufficient faces to sample over the area involved on the 500-ft. level to rely sufficiently on the law of average. In other words, it was essential to have each sample represent the place from which it came as accurately as possible. Therefore I fell back on the principle I have found from experience to be the best (as employed in the White mine, and at variance with the advocates of the light sample), namely, heavy sampling. Only in this case I obtained enough ore for a mill-test, for the foot-weight of each cross-cut was about 120 lb. per foot. The ore from each cross-cut was sacked separately and reduced separately to duplicate pulps. The accompanying photograph will indicate about half the shipment on its arrival at the smelter. As in the case of the Red mine, the sacks were numbered in groups, depending from where they came.

The average of the cross-cut pulps and the assay of the pulp of the composite sample checked to exactly one cent. I mention this as indicating the care taken in mixing the sample before cutting down. This was done at the sampling-plant of a near-by smelter. This concern turned over to me some of their most experienced sampling crew. While these men were only the ordinary day's-pay employees used at smelters for sampling, they had been accustomed to this class of work for years. This and the use of the power-driven sampling-plant gave me every facility for accuracy.

I desire to call attention to a matter that may seem elementary. It is most important to mix a sample thoroughly before the 'first' cutting. Many engineers give too little care to this important stage. I cannot too strongly emphasize that the careful mixing of a sample *before* stage-cutting is commenced, is one of the most important factors in this work. For example, if a sample is cut before it is sufficiently mixed and the assay of the opposite quarters is in one case $10 per ton and that of the other opposite quarters is $12 per ton, it does not matter what care is exhibited in the subsequent crushing, mixing, and cutting, the above error at the 'fountain head' cannot be eradicated.

I shall now attempt to show wherein the terms of purchase governed the form of examination. By consulting the sketch of the horizontal projection, the probable position of the lower ore-shoot will become clear. The position of the shaft-bottom at the date of examination is shown in the sketch. I had considerable trouble in arranging for the delay of the first payment, about one-third of the purchase price, until four months from date of examination. As these were the best terms obtainable, I optioned the property, because I considered it would be possible by forcing the work to sink the shaft to the necessary depth and to cross-cut the workings at the intervening lifts and to extend drifts sufficiently to obtain a fair indication as to the advisability of exercising the option. I repeat that, if it was possible to demonstrate an important ore-shoot on this lower horizon, the enterprise would present an attractive possibility of finding recurring ore-shoots by developing the lower horizons of the other surface shoots distributed at irregular intervals along the long length of outcrop. The point, therefore, was to ascertain, if possible, the prospect of developing such a shoot. Getting an organization started on a job like this and performing the work, which had to be rushed, was a matter of considerable expense. It was important therefore to forestall a situation which might arise in four months, namely, of having insufficient data on which to advise my client as to whether the large payment that would then mature should be made or not. If such a situation arose, my client would have been in the unfortunate position of having to ask for a re-arrangement of terms. In the event of refusal, this would have meant either assuming a big risk, or dropping out and leaving improvements and the additional development to the benefit of the owner. To avoid this, a double crew under the charge of an assistant was employed to drive a cross-cut from the bottom of the shaft to penetrate the orebody. By referring to the sketch, it will be noted that the shaft is in the foot-wall and approximately in the centre of the lower shoot. The cross-cut is marked X. In this way the shoot was penetrated from wall to wall. The orebody at this intersection proved quite unprofitable. On the basis of these data, I put the position frankly before the owner. I explained that it was possible that the cross-cut X might be situated in a barren patch in the orebody. While I did not demand a modification of the total price, I insisted that the first payment be delayed to give my client time could have been cheaply sampled on day's-pay. The organization would have been on the ground in any case.

The foregoing will indicate that the method of the examination of a mine is a matter of the diagnosis of the individual case. In the absence of an actual example it is dangerous to generalize as to whether hand-sampling or mill-tests are preferable. It may be assumed, however, that in the majority of cases mill-tests should be limited to ascertaining the sampling-error. This can be obtained by comparing the average of successive sampling of the stope-faces during the removal of the ore for the mill-test with recovery plus tailing-loss. Therefore the valuation of the ore reserves of a mine and the sampling of the various points critical in the estimation of the future possibilities of the enterprise should depend in most cases on hand-sampling. This may be corrected by procuring a sampling-error for various

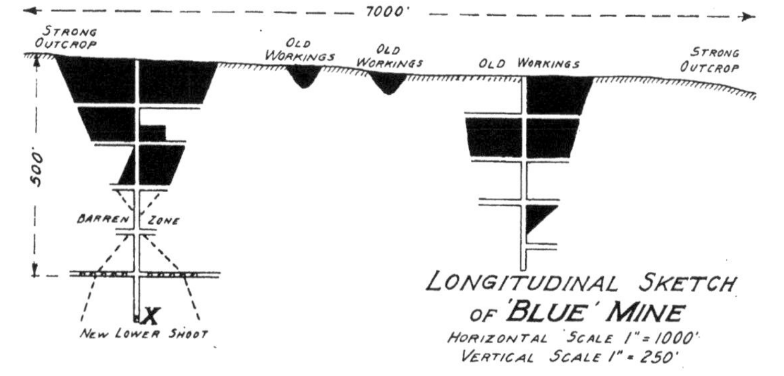

LONGITUDINAL SKETCH OF 'BLUE' MINE
HORIZONTAL SCALE 1" = 1000'
VERTICAL SCALE 1" = 250'

to determine by further development whether the adverse evidence was local or general. The owner declined to concede the required modification, and negotiations terminated.

Shortly afterward the mine was examined on behalf of a new prospective purchaser, a large corporation. All the workings were sampled at 5-ft. intervals, shafts, cross-cuts, and old workings alike. Apart from the question of the fee of the valuing engineer, the cost of sampling alone must have been three times the cost of my examination including the cost of driving the cross-cut X. I mention this as an interesting example of how two engineers may view a proposed purchase from entirely different standpoints. I do not see what was gained by sampling the old workings at that time notwithstanding the probability that they contained some important portions of ore. As an index to the future life of the enterprise, I regarded their importance as purely subordinate. Had my client taken physical possession as the outcome of the examination centralized on what I regarded as the 'life extension' of the enterprise, the old workings zones. The area of zones, or, in other words, the adjacency of mill-tests to each other will depend on the nature of the ore. In this way it may be possible to divide any ore deposit into zones depending on the engineering expediency of the particular case.

POTASH EXTRACTION from the materials entering into the manufacture of cement is increased by adding to the raw mix a calculated amount of calcium fluoride (fluor spar) which causes the potash to volatilize more readily as potassium fluoride. The flue-dust and fume are saved in the Cottrell apparatus, and the dust so collected is leached with water. The lime-salts present again react with the fluorine in the potassium fluoride, regenerating the insoluble calcium fluoride, which remains in the filter-cake and is returned to the kilns, while the soluble potassium salts are recovered by evaporation. This method of augmenting the yield of potash as a by-product in the cement industry was patented by Huber and Reith of the Riverside Portland Cement Co., Riverside, California.

Flotation—The Butte & Superior Case—I

Part of Argument by Walter A. Scott, for the Defendant

When we started out with this case we studied the patent suit and found that it sets forth no definite principle and contains no definite statement whatever, except that a new phenomenon came into existence when this young gentleman (A. H. Higgins) in London, in trying some experiment, reduced the quantity of oil to an amount less than 1%, and that this new phenomenon was a froth, a useful froth, a mineral froth, capable of wide utility. We proceeded upon that theory in the Hyde case, that the patent meant something and that it could only mean one thing. When the evidence was taken in the Hyde case we began to feel that we had come to the right conclusion and were upon sure ground, because Dr. Chandler, their expert, after quoting the statement in the patent that when the quantity of oil was reduced to less than 1% the mineral rises as a froth, added, "This is the foundation of the invention of the patent in suit". We had no refined theory there about soluble ingredients in the oils; we had nothing of frothing agents, of inert oils; we simply had 'oils'. And the foundation of that invention was said to be reducing that oil to less than 1%.

When the Hyde case was heard by this Court our presentation was simply an oral argument illustrated by one simple demonstration in a hand-manipulated test-tube. Our opponents produced an elaborate system of demonstration and experiment wherein they represented to this Court that all of the processes of the prior art, without exception, were failures. This was purely negative evidence, but we did not meet it by repeating the demonstrations before this Court that had been performed by our witnesses and by them described in their testimony; and quite naturally this Court was influenced by plaintiff's demonstrations, as I have always thought.

But the presentation of this case here in the past few weeks has served to emphasize the danger of reliance on merely negative testimony. They had learned to infallibly cause all of the suggestions of the prior art to seem useless, and, when put to practice, to end in failure. That such evidence should have no probative force is shown clearly to us by the fact that Mr. Higgins, the skilled demonstrator of this plaintiff, was obliged to try six times before he could concentrate Butte & Superior ore in this Court by the very process of the patent in suit. These demonstrators come here, they exhibit to this Court the processes of the prior art with every inducement, motive, and desire to make them fail, they desire this Court to believe that that is evidence that these things cannot be made to succeed. They then come before this Court with this elaborate setting of failure within which to cause their invention to sparkle and try the invention of the patent in suit upon the Butte & Superior ore that the Butte & Superior company has been able to treat successfully for years, and after five successive attempts to apply that invention to that ore, they meet only with failure. How do they account for that? How does that harmonize with their efforts to prove the failure of the prior art by exhibiting these ludicrous, these studied attempts at failure to the Court? Do they think, with every motive, inducement, and facility to make the process of the patent in suit successful, it does not bear at all against the process of the patent in suit to have it fail five times after elaborate rehearsal and preparation, but that it is convincing proof when studied attempts to cause failure of the prior art processes do eventuate in failure?

When we appeared before the Court of Appeals in the Hyde case we made demonstrations of the prior-art processes, made them in strict conformity with the directions of the prior art. Now, in practising flotation there is very little to follow either in the patent in suit or in any of the prior-art patents. What has made flotation a success today is not found in any of these, much less in the patent in suit. The only essentials that we find in the patent in suit are the grinding of the ore, its admixture with water, the addition of oil and agitation. And by 'oil,' as the prior-art patents say, is meant any substance which will adhere selectively to the metalliferous mineral. Therefore 'oil' is a generic term including, as stated by Cattermole and other patentees, all of the substances used in the prior art, oils proper, phenol, and creosol, which are soluble, and which this plaintiff seeks to make the subject of another patent ten years later, oleic acid, and, as Mrs. Everson says, all animal, vegetable, and mineral oils. Now, we have these four items, and the illustrations which we presented before the Court of Appeals in San Francisco embodied these four items, and these four items are stated with no more precision in the patent in suit than in the prior-art patents.

Of course this plaintiff can go to the prior-art patents and say they don't tell us how fast to agitate; that they say nothing of this violent agitation practised today. They do not in terms, and neither does the patent in suit. None of them have anything but descriptive words. And the link that I claim is lacking in this case is the link which, if the position of this plaintiff is good, is something which shall connect this violent agitation of today with their patent. There is none. Their patent says, "thorough agitation." Other and older patents say "thorough agitation" (Everson patent, p. 2, line 100; Kirby patent, 809,959, p. 1, line 73); and some of them say "violent agitation" (Kirby patent 809,959, claims

1, 2, 3)—mere descriptive language in all of them. Kirby also refers to vigorous agitation (patent 809,959, claim 9). The California Journal* refers to thorough agitation to produce a foam or froth. And it is my contention that unless in the patent itself there can be pointed out something more directly connecting modern flotation with that patent than with these other prior-art documents, this plaintiff has no right to claim that this patent has disclosed flotation to the world and that the world owes flotation to this patent.

The process has developed during all of these years, to be sure, but the patent in suit is a disclosure to the world of only what is in it, and that patent should, in all justice, stand on the same footing as all of the other prior-art patents.

Therefore, I say that when we did these demonstrations before the Court of Appeals in San Francisco we followed the prior art with precision. We derived nothing from the patent in suit that was not equally well disclosed in the prior art, namely, the grinding of the ore, the addition of water to make a pulp, the addition of oil, agitation and the resulting froth. The Court of Appeals, looking at these things, came to the conclusion that there was nothing in the disclosure in the patent in suit, and so decided.

Now we come to the Supreme Court, where the definition of the alleged invention met its first alteration at the hands of counsel for plaintiff. The patent began to take on the protective coloring demanded by its new surroundings. In the course of his argument before the Supreme Court Mr. Justice McReynolds asked Mr. Kenyon when in the process of reducing oil the invention came into existence, to which Mr. Kenyon replied, "at about one-half of 1% of oil." Mr. Justice McReynolds then asked whether before getting to one-half of 1% there was any invention, to which Mr. Kenyon replied that the invention was not reached, he should say from certain figures in the record, until about one-half of 1% of oil was reached. Mr. Justice McReynolds followed with a question as to whether there was any invention at 1%. Mr. Kenyon replied "No," but that the invention began to come remotely at one-half of 1%. The final question of this colloquy put by Mr. Justice McReynolds was an inquiry as to whether when a float has more than one-half of 1% of oil it does not infringe, and Mr. Kenyon replied that it did not infringe.

Now, before the Supreme Court we presented our demonstrations of the prior art as we have in the case at bar before this Court; and the products resulting from these prior-art processes were absolutely indistinguishable, indistinguishable to the eye and to analysis as to the amount of metal recovered, as shown by the testimony of the witnesses whose demonstrations were repeated, but we were there met by still another argument, or assertion I should say, and a false assertion. We were met by the assertion that while these froths look alike, while opposing counsel themselves said they couldn't tell them apart by looking at them, still, your Honor, they said to the Supreme Court: "These things cannot be done in a mill."

THE COURT: You told the Supreme Court it could, I suppose?

MR. SCOTT: I told the Supreme Court, your Honor, that the record did not enable me to answer the question. That was my precise answer. I would like to have told them what your Honor suggests, but the record contained no support for such a statement.

THE COURT: Well, did you suppose that the Supreme Court instead of taking the Hyde record would take the statement of counsel if they were made? Of course, this is outside the record.

MR. SCOTT: I would not have expected them to accept as a fact a statement of counsel in the absence of any testimony of any witness. It did not appear to me to ask them to accept my statement as a statement of fact on these operations.

THE COURT: Their statements, which you say you made, don't you suppose after all that this decision was rendered, not on the statements of counsel nor the claimed estoppel, but upon the record in the Hyde case? Can we go back of that?

MR. SCOTT: Well, if it was decided upon the record in the Hyde case, your Honor, the suggestion of opposing counsel that this could not be done in a mill might have been considered argumentative by the Supreme Court, and then looking at the Hyde record the Court found no evidence that it was done in a mill. They found simply the testimony regarding laboratory demonstrations.

The reason that in the Hyde case we could not show any operations in a mill with any large quantity of oil was that at the date of the trial of that case, seven years after the application for the patent in suit, the process was not being carried out in any mill in the United States excepting the Butte & Superior mill, and at the time the testimony was taken in the Hyde case the process was of doubtful utility even there, as testified to by Mr. Frank Janney, who said in this Court that it was not until the spring of 1913 that the officials of the Butte & Superior company became convinced that flotation would be a success. So, on that state of facts, it was an absolute impossibility for such evidence to be produced.

Now, in deciding the Hyde case I think we find in the opinion of the Court abundant evidence that Mr. Kenyon's statement had great influence upon the Court, this statement about no infringement below one-half of 1% of oil. I will quote eight or ten lines of it:

"Convinced as we are that the small amount of oil makes it clear that the lifting force which separates the metallic particles of the pulp from the other substances of it is not to be found principally in the buoyancy of the oil used, as was the case in prior processes, but that this force is to be found, chiefly, in the buoyancy of the air bubles introduced into the mixture by an agitation greater than and different from that which had been resorted to before, and that this advance on the prior art and the resulting froth concentrate so different from the

*Referring to the article by three students in *The California Journal of Technology*. See M. & S. P., July 31, 1915.—EDITOR.

product of other processes make it a patentable discovery as new and original as it has proved useful and economical." And in the closing passage of the opinion, the Court referred to the critical proportion of oil.

Now, analyzing these passages, we find that this plaintiff in this Court has repudiated every single one of these elements upon which the Supreme Court based its finding of validity. The plaintiff now claims the use of any amount of oil which produces a certain result. The Supreme Court in this passage which I have just read says:

"Convinced as we are that the small amount of oil used makes it clear"—

Now plaintiff discards the small amount. Any amount that produces a certain result, they claim.

Then their witnesses go further in the repudiation of this passage from the Supreme Court opinion. The Supreme Court opinion says that

"The small amount of oil used makes it clear that the lifting force which separates the metallic particles of the pulp from the other substances of it is not to be found principally in the buoyancy of the oil used."

Mr. Higgins now says of prior-art quantities of oil that they do result in this 'air-froth' of the patent in suit. Mr. Higgins says that 20 lb. of oil will only float 2 lb. of mineral. Of course, that is negligible, showing that that amount of oil cannot account for anything. Mr. Higgins says the demonstration in this model pyramid machine that was exhibited in the grand-jury room was a demonstration of the process of the patent in suit. Now, in that operation there were 42 lb. of oil used per ton or 2.1% of oil. That is a prior-art quantity of oil and that is a quantity of oil which the Supreme Court was prevailed upon to believe must result in oil-buoyancy. The Supreme Court says that the small amount of oil in this patented process makes it clear that the lifting force is not to be found principally in the buoyancy of the oil as is the case in the prior processes. The Supreme Court was prevailed upon to believe that it is the reduction of the amount of oil below 1%, or one-half of 1%, that led to the distinction between the mineral being lifted by the buoyancy of air and its being lifted by the buoyancy of oil. As a matter of fact—

THE COURT: I was just going to observe—of course, there is no impropriety in your argument—but don't you think that ought to be addressed to the Supreme Court in a motion for rehearing? Could this Court take it into consideration?

MR. SCOTT: My idea of that is this: Opposing counsel referred to the doctrine of *stare decisis*. But the state of facts is absolutely different in this case. Even counsel who told the Supreme Court that there was no infringement, with over one-half of 1%, now comes to this Court and says there may be infringement with any percentage of oil.

THE COURT: I was just trying to make it clear to myself how far your argument ought to weigh with the Court, that was the idea. You may proceed.

MR. SCOTT: The statement in the Hyde case seems to me to be plainly a judicial admission that there is no invention with over one-half of 1% and no possible infringement with over one-half of 1%, and my point is that such a variance in the presentation of the matter by counsel, to say nothing of the variance that is involved in the testimony of the witnesses, removes the decision of the Supreme Court from any possible interference with the exercise of judgment *de novo* in this case now before this Court. I fail to see how we could make the Supreme Court case a precedent, the facts and testimony and admissions of counsel being so different.

THE COURT: I understand your attitude.

Rapid Shaft-Sinking

Recently rapid progress was made in sinking at the Gadsden shaft, in the Warren district, Arizona. Through the courtesy of J. K. Hooper, superintendent of the Calumet & Arizona Mining Co., at Jerome, Arizona, some of the details of the work and the conditions under which it was accomplished are here given. The shaft has three compartments, two of which are for hoisting, being 4 ft. 6 in. by 5 ft. in the clear, one being for pumps, pipe-lines, and ladders, having a section 7 ft. 6 in. by 5 ft. in the clear. The shaft required an opening 21 ft. long and 8 ft. 6 in. wide. From the morning of May 8 to the morning of May 15, that is seven days or 21 shifts, the shaft was sunk 63 ft., and 12 sets of timbers, spaced at 5 ft. centres, with guides, ladders, and pipe-lines, were put in place. The work was done on company-account, plus a bonus for a footage exceeding 130 ft. per month. The distance usually carried between the timbers and the bottom was from 9 to 14 ft. The rock passed through was limestone, and it required from 26 to 34 holes for a round. The cut-holes were 10 ft. deep, the others averaging 7½ ft. The men worked three shifts daily, with 6 men on a shift. They shoveled the broken rock into buckets for hoisting to the surface. For the work of drilling four Ingersoll-Rand BCRW-430 jackhammers and one Sullivan DP-33 jackhammer-drills were employed, using ⅞-in. hexagon-steel with cross-bits. Hercules 1-in. 40 and 60% gelatin-powder was the explosive used exclusively, the 60% powder being placed in the bottom of the holes. Triple-tape fuse and No. 8 caps were used, all shots being fired by hand, spitting in the usual manner. There was little water to contend with, nearly all of it being taken up by the fine rock with which it was hoisted in the buckets. During the week referred to a total of about 12 hr. was lost in blowing out smoke after blasting. The hoisting-equipment consists of a direct-motion, double-drum electric hoist, and an Ingersoll-Rand 'Imperial' compressor of a capacity of 888 cu. ft. of free air per minute.

RESEARCH into the constitution and character of cement by the U. S. Bureau of Standards has shown that alumina is apparently not essential to obtain the desired setting qualities, but it is necessary as a flux in the burning of the mix to make the clinker. Otherwise the manufacture of cement as now practised would be commercially impossible.

Electric Blasting-Caps and Delay-Electric Igniters

By I. H. BARKDOLL

An electric blasting-cap is a special form of instantaneous detonator fired or exploded by an electric current. The delay-electric igniter is a cylindrical copper tube into one end of which the necessary wire, plug, and other parts are inserted, while the other end is crimped upon a piece of ordinary fuse with a blasting-cap attached. The difference in the length of the fuse designates the different delays, that is, two, four, six, and eight inches of fuse gives one, two, three, and four delays. They are made waterproof with well insulated wires and long waterproof rubber covers for the igniter, fuse, and detonator; this is further waterproofed by a heavy coating of gutta-percha over the entire length. In blasting in a drift, shaft, or raise, it is necessary to blast each round of holes in sections or in rotation. The difference in the length of the fuse used with the electric igniter, explodes the charge in the proper sequence. There is practically no limit to the number of charges that can be fired in sequence by this method. Under most conditions there is nothing to be gained by dividing the round of holes into more than three or four sections. This permits of cuts, relief, and corner-shots.

Electric blasting-caps and electric igniters have been developed for such conditions so that these four classes of shots, or more, may be fired in rotation with a single set of wiring and with but one application of the electric current. The different delays are so constructed that there is a brief lapse of time after the current is applied before the first delay explodes and a longer delay before the second delay explodes. It should not be understood that all first-delay caps explode simultaneously, but the variation in time is very little. This variation is caused by the irregularity in the burning-time of the fuse and the slight difference that may occur in the fuse-lengths. Connecting-wire used in their manufacture is of insulated copper, No. 20 gauge. The ends of the connecting-wires must be scraped bright before the connections are made, and the joints should not be permitted to lie in water or on wet ground. If this cannot be prevented, the joints should be covered with insulating tape. The wire for connecting blasting-caps and other electric detonators to the blasting-machine is known as the leading-wire. It is of insulated copper wire, may be furnished in desired lengths, and should always be long enough to keep the blaster well out of the danger-zone. Leading-wires are of two classes, single-strand and duplex. Duplex is made by bringing together two insulated copper wires with an outer insulation, thus giving a return-circuit that may be handled as a single wire. The single strand would be the same two wires, but not bound together, requiring the handling of two wires separately. There are advantages and disadvantages in the standardization of either type of leading-wire. All leading-wires 300 ft. long, or less, should be solid wire of a gauge not smaller than No. 14; from 300 to 400 ft., No. 12 gauge solid; and for lengths greater than 400 ft., No. 10 gauge solid should be used.

Electric blasting may be accomplished either with a blasting-machine or with a power or lighting circuit. Blasting-machines are small portable dynamos used to generate the current for firing blasts by electricity. The downward thrust of the rack-bar converts muscular energy into electric energy, which is short circuited through a field-magnet for the purpose of building, intensifying, and storing the current until the end of the stroke, when the accumulated current is sent through the firing line. They are rated according to the number of electric blasting-caps they will fire when connected in series. There are five different sizes, known as two-post machines, No. 2, 3, 4, 5, and 6. No. 2 has a capacity of from 1 to 10 electric blasting-caps; No. 3 from 1 to 40; No. 4 from 1 to 60; No. 5 from 1 to 100; No. 6 from 1 to 150.

When operating a blasting-machine, set it squarely on a solid level place, connect the wires to the binding-posts, and screw down the wing-nuts tight, lift the bar by the handle to its full extent, and with one quick hard stroke push it to the bottom of the box with a solid thud, using both hands. Try to knock the bottom out of the machine. As the bar approaches the bottom it becomes more difficult to operate because of the building up of the current, but the speed of the thrust should not be diminished as the finish of the operation is more important than the start. A galvanometer is a small instrument used by some blasters to determine whether a blasting circuit is open or closed. An open circuit is one that is broken or faulty, due to poor connections or to broken wires. Galvanometers are also useful to indicate the existence of leaks or short circuits. It is a magnetic device which, if the circuit is closed, that is, if there are no faulty connections nor broken wires, moves a pointer across a scale, due to the flow of electric current through the closed circuit. In making the test with a galvanometer, connect or touch the leading-wires to the two binding-posts after all connections are made. If the circuit is perfect, the needle will move along the scale. If the needle does not move, or not as far as it should, there is a break or some high resistance like a bad joint. Single electric blasting-caps can also be tested, both before and after loading them in the hole, simply by touching the end of the cap-wires to the two posts. It is excellent practice to test electric blasting-caps and delay-electric igniters before and after loading. Wiring for electric blasts must be well and carefully done to insure success, the work of wiring may be divided into three parts, connecting the detonator-wires to the leading-wires and connecting the leading-

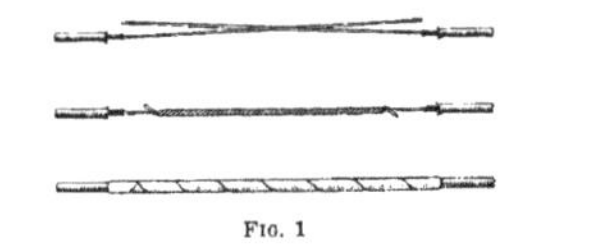

FIG. 1

wires to the blasting-machine. Connections between the detonator-wires must be well made; first scrape the bare ends of the wires with a knife-blade, and join them with a long hard twist, known as the Western Union twist (see Fig. 1). Never hook wires together as shown in

FIG. 2

Fig 2. In making connections between detonator-wires and leading-wires, the same precautions must be observed with regard to scraping the ends of the wires. Wrap the ends of the detonator-wire tightly around the end of the leading-wire about 2 in. from the end, as in Fig. 3. Then

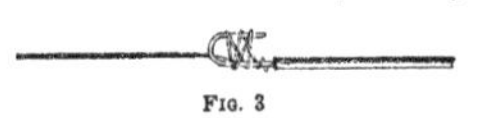

FIG. 3

bend the end of the leading-wire back sharply and take a turn or two of the detonator-wire around the hook. Connecting the leading-wires to the blasting-machine is done by loosening the wing-nuts on the two binding-posts and inserting the end of the leading-wire into the small

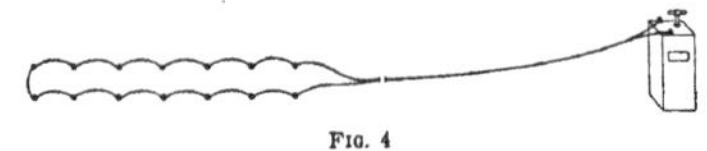

FIG. 4

hole in the post and tightening the wing-nut on the wire.

When using a blasting-machine all circuits must be in series, as shown in Fig. 4. In wiring up a blast known as series-connections, one wire from each hole should be connected to one wire in the next hole, and so

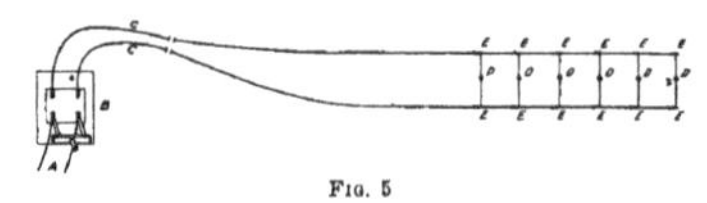

FIG. 5

on to the end, when only the two end-wires are left free. These are connected to the end of the leading-wire. For electric blasting where lighting or power-circuits are used the parallel-connections should be made as in Fig. 5. This type of connection cannot be used successfully with a blasting-machine. The parallel-connections differ materially from the series-connections and should not be confused with one another. In order to obtain satisfactory results the series must be used with the machine and the parallel with a power-circuit. A parallel-connection is made by connecting one wire from each hole with the leading-wire. In wiring a blasting-circuit all joints must be protected against short circuiting, especially in wet ground. This is done in several ways. When conections lie on moist ground or in water they must be held away by bending them on sticks or blocks so that only the insulated part of the wire touches the ground or the supports; or the joint should be insulated with tape. All joints covered with water or dirt should be well insulated with good waterproof tape and made as near damp-proof as possible.

TUNGSTEN CONCENTRATE is usually pulverized and digested with acids for extraction of the metal. Any method of decomposing wolframite with acids for precipitating the tungstic acid should leave the manganese in solution, but in practice small amounts of the manganese will be carried down mechanically in the precipitate, and the quantity will be greater with an increase of the manganese-content in the original concentrate. In the treatment of scheelite the ore or concentrate is digested with boiling hydrochloric acid, to which small quantities of sodium or potassium chlorate are added as oxidizers. This method is also adapted to any ore of tungsten. The objection to it is that the silica remains with the tungstic acid, and either has to be left in it or removed by a separate process, such as leaching with ammonia in which the tungstic acid is soluble to the exclusion of the silica. If tungsten concentrate is first fused with alkalies the tungsten is converted into a soluble tungstate, such as sodium tungstate. After dissolving the salt from the fusion the tungstic acid may be precipitated by adding an acid, leaving the manganese in solution. Part of the silica would remain in the solution, but a part would be precipitated with the tungstic acid. According to R. B. Moore of the Bureau of Mines, the best way to convert tungstic acid into metallic tungsten is by heating it with powdered charcoal in a furnace at a bright red heat. This reaction can be obtained either in graphite crucibles or in cast-steel pots, or in steel-tubes.

ASSAYS for private parties are not made by the Bureau of Mines nor by other branches of the Government service free of charge, as this would interfere with professional chemists and assayers. The Bureau, however, furnishes lists of assayers and chemists on request. The Director of the Mint, on the other hand, announces that assays will be made for private parties at the mints situated in Philadelphia; New Orleans; Carson City, Nevada; Boise, Idaho; Helena, Montana; Seattle; and Salt Lake, as follows: determination of gold and silver, $1; copper. tin, zinc, iron, lead, and tungsten, $1 for each metal.

REVIEW OF MINING

As seen at the world's great mining centres by our own correspondents.

CRIPPLE CREEK, COLORADO

Important Development of New Veins in the Cresson Consolidated Mine.—Diamond-Drilling Discovers New Orebodies.—Lessees Continue to Operate at Good Profit.—Disastrous Fire at Last Dollar.

A new orebody of great richness is under development at the 1400-ft. level of the main shaft of the Cresson Consolidated Gold Mining & Milling Co., on Raven hill, according to A. E. Carlton, president of the company.

The dividend was paid July 10 at the regular monthly rate of 10c. per share and amounted to $122,000.

During May new ore of the net value of $19,572 was added to ore reserves, which June 1 were estimated to have a net value of $3,817,000.

The 16th level has cut the contact and a drift has been turned north-east to cut the 1500-ft. level orebody partly developed by a winze from the 15th level. It will be necessary to drive about 600 ft. and the orebody should be reached about October 1. The main cross-cut has intersected a vein showing a 6-in. streak sampling $20 to $28. This did not show any value on the 15th level.

The diamond-drill has cut an orebody on the 14th level which averages $35 over a width of 8 ft. A second hole 30 ft. away showed two feet of ore assaying over $1500 per ton. This is ore which probably would not have been found in the ordinary course of development. A cross-cut is now being driven. A cross-cut from the Roosevelt tunnel to intersect the Cresson shaft is now under way.

The annual stockholders meeting has been called for September 4. The Golden Cycle M. & R. Co. paid its regular monthly dividend of 3c. per share, amounting to $45,000, on July 10. A. E. Carlton is also president of the Golden Cycle company.

The directors of the Vindicator Consolidated Gold Mining Co. declared the regular quarterly dividend of 3c. per share last week, to be paid on July 25 and will amount to $45,000.

The Granite Gold Mining Co. on July 5 paid 1c. per share, amounting to $15,600. The company will pay regular monthly dividends and it is expected that the rate either will be increased or that the company will go on a monthly dividend basis.

A call has been issued for the annual stockholders meeting of the Camp Bird Mining, Leasing & Power Co., to be held at Pitkin, Colorado, on August 8, for the election of directors. This company holds a six years' lease on the property on the Rose Nicol G. M. Co., on Battle mountain, in this district, and sinking has been started from the 650-ft. level. The shaft will connect with a lateral from the Portland No. 2 shaft at the 1000-ft. level.

Sinking has been resumed by Edwin Gaylord in the Forest Queen shaft, on Ironclad hill. The shaft, now 650 ft. deep, is to be sunk to the 800-ft. point. The station at the 650-ft. level has been enlarged and an underground-hoist installed. Production from the upper levels will continue while sinking is in progress. California parties are part owners of this mine.

The Eclipse Leasing Co., holding a 5 years' lease on the Joe Dandy mine, to the Joe Dandy Mining Co., on Raven hill, has finished sinking and cutting of stations at the 600, 700, and 800-ft. levels, and is mining a good grade of ore in both the new 6th and 7th levels. Production has been resumed. The Joe Dandy Mining Co.'s stock is not listed on Colorado exchanges.

A consolidation was effected last week, at meetings of stockholders of the Albert Beacon Gold Mining Co., the Cripple Creek General Mining & Exploitation Co., and the Gold Camp Syndicate, whereby the holdings of the companies in this district, comprising 57 acres patented on Bull, Ironclad, Newton, and Beacon hills, will be taken over and operated by the Victory Gold Mining Co., a Colorado corporation, to be incorporated this month. Harry J. Newton, of Denver, will be the president of the new company.

The July production of the mines of the United Gold Mines Co., operated under lease, totaled approximately 2500 tons of an average value of $20 per ton—a gross production of $50,000. The mines from which the ore was shipped were the Trail, on Bull hill, with 3 sets of lessees; W. P. H. mine, Ironclad hill, leased, and the Mohican, a fractional claim on Battle mountain, operated by the Granite Gold Mining Company.

The output from the main shaft of the Jerry Johnson Gold Mines Co., on Ironclad hill, operated by Frank Galey and son, and the Cripple Creek Deep Leasing Co., totaled 425 tons. The ore from the Caley lease averaged better than $30 per ton, one car bringing settlement at $47.80, and a second car $57 per ton.

The El Paso Consolidated Gold Mining Co.'s lessees, 15 in number, mined and shipped 500 tons of $20 ore during June. The company is not operating.

The mine-plant, shaft-house, ore-house, blacksmith-shop, powder-magazine, and equipment at the Last Dollar mine, on Bull hill, with contents were totalley destroyed on July 7. The mine is under lease to the Catherine Gold Mining Co., Charles Walden, of Victor, general manager, and is owned by the Last Dollar Gold Mining Co. The fire was caused by lightning setting fire to the ore-house. The men were brought to the surface in safety, the engineer sticking to his post despite the fact that the magazine was enveloped by fire. The magazine caught fire from the flying embers and with its contents, 40 boxes of explosives, burned without explosion, except the caps. The Catherine company has recently installed a powerful electric hoisting-plant, and 12-drill electrically-driven compressor, at a cost exceeding $15,000. The total loss is in excess of $25,000. Mr. Walden leaves for Chicago this week to consult with Eastern directors of his company, and it is expected that when the insurance has been adjusted the plant will be replaced and operations resumed.

TORONTO, ONTARIO

The Threatened Strike of Miners at Porcupine is Averted by the Mine Managers Raising Wages.—Hope for Similar Action at Cobalt.—The Crows Nest Pass Coal Mines to be Operated Under Direction of the Government.

All danger of a general strike in the Porcupine district has been averted by the course taken by the managers of several of the leading mines. While declining to recognize the union, on the ground of its affiliation with the Western Federation of Miners, they agreed to hold conferences with representatives of their own employees, with the result that in several cases the question of wages has been satisfactorily adjusted. The McIntyre and Dome Mines, and the Tough Oakes, at Kirkland

Lake, have granted an increase to mine and mill-workers of 50c. per day, to continue as long as the cost of living as shown by the Dominion Food Commission remains higher than in August 1914, all other bonuses being discontinued. The Hollinger Consolidated has granted a flat-rate of $4 per day to all its underground workers, also discontinuing the 'loyal service' bonus. The mill-men are not affected by the change. Other companies have held conferences with their employees, the results of which have not been announced. One effect of the amicable settlement of the difficulty has been to attract to Porcupine many of the miners of Cobalt where the situation, though less menacing, is still uncertain. A strike vote taken on June 24 by the Cobalt miners was, by a large majority, in favor of a strike if their demands were not granted, but it has not been followed by any decisive action, as it is understood the matter has been referred to the union headquarters at Denver. In the meantime the settlement effected at Porcupine leads to the hope that similar action may be taken by the Cobalt mine-owners, as the local union leaders say that they will not press for the recognition of their organization provided the demands of the men as to wages are met. As the workers are now receiving bonuses based on the high price of silver, practically equivalent to the wage increase asked for, it is unlikely that a strike will result.

At the Dome Mines the cross-cut on the 700-ft. level is expected shortly to cut the large body of ore averaging $17 per ton in gold, as indicated by diamond-drilling. It is anticipated that early in August the cross-cut will have penetrated the full width of the ore, 119 ft., and that stoping will have begun. The 4-weekly statement of the Hollinger Consolidated shows a continued decline in production. Gross profits for the period ended May 20 were $92,809, as compared with $194,688 for the previous 4 weeks, the tonnage treated being 35,337 tons of the average value of $7.49 per ton, as compared with 42,849 tons of the average value of $9.20. Working costs were $4.66 per ton. The mill ran 65% of possible running time. The deficit was reduced to $81,375. The central shaft is in operation and will be used exclusively for hoisting ore. A circular issued to the shareholders states that the directors have no intention of closing down the mine, but as long as unfavorable labor conditions continue may devote their attention mainly to development so as to get the mine in a position largely to increase the output when an adequate supply of efficient labor is to be had. The Schumacher has temporarily closed down both the mine and mill owing to labor shortage. At the McIntyre the main drift at the 1000-ft. level is over 1100 ft. long with a face of ore 52 ft. wide. The average grade throughout the mine has risen to $12 per ton. The diamond-drill working on the 1000-ft. level has passed through a vein at the 1325-ft. level 22 ft. wide, which is stated to run $20.50 per ton. New equipment is being installed preparatory to deeper sinking. At the Dome Extension diamond-drilling is in progress from a station at the 600-ft. level, 2000 ft. distant from the Dome glory-hole, and preparations are being made to sink an incline to 1600 feet.

The La Rose Consolidated is developing the Violet property to the east of the O'Brien and will sink to the 300-ft. level to prospect the veins coming in from that direction. The Adanac is opening up a 4-in. vein on the 312-ft. level, which in places assays 1000 oz. per ton. The Mining Corporation of Canada, now operating the Little Nipissing, has encountered, in a winze put down from the 200-ft. level, a 4-in. vein carrying native silver at a depth of 50 ft. At the Peterson Lake a 3-in. vein carrying high-grade ore, and believed to be a continuation of Nipissing vein No. 10, has been found and is being opened up.

The protracted strike of coal miners in the Crows Nest Pass district and the adjacent parts of Alberta and British Columbia, where 6000 men have been idle for several months, has been terminated by the action of the Government under the War Measures Act. W. H. Armstrong, of Vancouver, has been appointed commissioner with power to operate the mines, or to compel the owners to do so. He has ordered the mines to be re-opened at once, the men to receive an advance of 22½% over former wages; a commission to investigate the cost of living to be appointed every four months if requested by either party. This action was taken owing to the failure of negotiations which had resulted in a tentative agreement as to wages, but which were discontinued because the mine-owners insisted on a clause penalizing the men should they quit work. This provision will be eliminated. As the men have been widely scattered during the strike, it was not expected that work could be resumed before July 3. The mines will be worked day and night to secure as large an output as possible. As the coke supply of the smelters at Grand Forks and Trail, B. C., is dependant upon the Crows Nest Pass coal mines, the resumption of operations at the collieries is a matter of great importance to the metallurgical industries of British Columbia.

JUNEAU, ALASKA

THE READY BULLION, ALASKA GASTINEAU, AND ALASKA JUNEAU MINES ALL RUNNING AT REDUCED CAPACITY OWING TO SHORTAGE OF LABOR.—A FINNISH MINER'S IDEA OF A CONTRACT.

The shortage of labor in the Juneau district is fast becoming a serious matter. The Ready Bullion mill on Douglas island has been reduced to day-shift operation only, to allow all the available men to be sent underground, in an effort to rush development work on the 2400 and 2600-ft. levels and also to complete the filling of the old stopes with waste rock, which is being mined from the foot-wall on the 450-ft. level. This filling of old stopes will avert all possibility of a disaster in the Ready Bullion mine similar to that which occurred at the Treadwell mine.

The Alaska Gastineau Gold Mining Co. is finding it difficult to keep up the tonnage of the mine, owing to the shortage of men. The mine is running with approximately 100 men short. The Alaska Juneau is suffering similarily, though not to the same extent, as their daily output at present is averaging but 2000 tons.

W. S. Pecovitch, superintendent of the Funter Bay mines, at Funter Bay on Admiralty island, recently returned from the East to Juneau. He has made financial arrangements to develop the property and will leave for the mine as soon as possible with 15 men to commence work. The present plans include the driving of a cross-cut adit to cut the 18 veins which outcrop on the surface. The result of this cross-cut will determine the future course to be pursued in equipping and developing the property.

Very seldom is a humorous occurrence reported as happening in the operation of a mine. There are good stories of salting samples during an examination, and shrewd bits of financial finesse in mining deals, but for a clever slip-over, the following has few equals:

In one of the big mines of the district, recently, a Finlander had a contract to drive three raises from various points to the surface. The work was started late last fall and two raises were completed during the winter. The last raise being 100 ft. or so, longer than the others, took longer to drive. The rock was hard, and as the raise neared the surface, the flow of water increased until the stream in the raise was like a small waterfall—cold water too. The raise was driven from an adit, to reach which from the mine, a pack of perhaps 200 yd. was necessary. And the trail was through deep snow, generally fresh fallen every day. Labor was almost impossible to get, and to send a man to that raise, was equivalent to giving him a time-check. The Finn even put in some shifts himself, packing steel, drilling, packing powder, and blasting, alone.

One morning the contractor met the foreman and announced that "she was in—snow on the bulkhead." The measurement on the previous day showed the raise up 253 ft., which was about 10 ft. short of the calculated length, but that was accounted for by the fact that there could easily have been 6 or 8 ft. of dirt on bedrock at the surface. The foreman went up the raise, found snow on the bulkhead, and took no trouble to go to the face, which would, he judged, be loose and dangerous. The raise was reported as finished and the contractor was paid in full for the work.

Two months passed, and on the surface, no hole appeared in the 15 ft. of snow that covered the raise. This was considered so strange that the raise was investigated and the face was found to be in solid rock, with not a sign of daylight. That Finn had packed that snow up the raise. After one trip up the raise to the top, the foreman concluded that the contractor wasn't so crazy after all.

The Alaska Treadwell Gold Mining Co. has men at work on the Red Diamond claims, on which a working option was recently taken. This development work is in charge of Monte Benson, recently a foreman of the Alaska Mexican mine.

Junk dealers from Seattle are making a clean-up in the district, made possible by the high price of scrap-iron. The abandoned equipment of the old Nowell Red mill, in Silver Bow basin, and of the old Alaska Juneau 30-stamp mill, and of the old mills in Sheep Creek basin, has brought a price averaging $12.50 per ton, on the ground.

The Pacific Coast Gypsum Co.'s property at Gypsum, on Admiralty island, is making regular shipments of 3000 tons per month to Tacoma. There are 25 men working at the property. Mining is in progress on the 190-ft. level, and the 300-ft. level is being developed. D. C. Stappleton is superintendent.

At William Henry bay, on the west shore of Lynn canal, E. Aldrich is developing a copper prospect. He is working seven men, establishing a camp and installing machinery, preparatory to development.

J. E. Berg, superintendent of the Pueblo mines, at White Horse, Y. T., has returned there to re-open the mines, which were recently shut-down by order of the Canadian authorities, following a serious cave at the mine, in which seven men were killed.

TREADWELL, ALASKA

THE READY BULLION MINE AND MILL AGAIN IN OPERATION.—A SHORTAGE OF LABOR RESTRICTS OPERATIONS.—COAL OUTPUT OF ALASKA INCREASING.—MINERAL PRODUCTION.

Owing to the scarcity of men to work in the mine, the Ready Bullion mill is now being operated during the day shift only, in order that more men can be put in the mine to prepare for increasing the output which insures both continued and increased operation of that property. The shortage of labor is still being felt on the island, although several men have signed on recently. It is believed, however, that more men will be available in a few weeks.

The Treadwell company has consummated the sale of the building and structural steel of the Central crushing plant. The work of dismantling started during the week. The six crushers will be stored for the present. Development work on the Diamond group will be started by the Treadwell company during the coming week.

Two mines in the Matanuska coal district, into which the United States railroad is being constructed, are producing coal for the market. The mine of William Martin on Eska creek is employing 50 men, and up to May 29 had shipped 3000 tons of coal to Anchorage. The daily production is about 40 tons. The mine of the Doherty Coal Company, at Moose Creek, has sold 12,100 tons of coal to the Alaskan Engineering Commission since last August besides a considerable amount to the residents of Anchorage. Fifty men are employed at the Doherty mine. Other properties are being opened. All these coal areas are leased from the Federal government. A statement of the value of the exports from Alaska for the six months ended June 30, 1917, has been prepared by the collector of customs, J. F. Pugh. The following are the value of mineral exports: copper ore, $15,007,211; gypsum, $29,600; lead ore, $54,316; marble, $28,252; gold and silver, $4,136,831.

YERINGTON, NEVADA

THREE DIFFERENT COPPER-LEACHING PLANTS IN OPERATION.—EMPIRE NEVADA.—THE KENNEDY CONSOLIDATED SALE.—MONTANA YERINGTON.—SUCCESSFUL COPPER LEACHING.—THE ROCKLAND MINE HAS NEW EQUIPMENT AND MAKING GOOD RECOVERY.

There are two copper-leaching plants working in the district and a third is being built on the property of the Empire Nevada Copper Company.

The first plant was built at the Thompson smelter a few years ago and the Weidline process adopted. This is not a commercial plant, having been built for experimental purposes only. Leaching is done with sulphuric acid, and the copper-bearing solution is then brought to digestor-tanks under pressure in the presence of steam. Native copper is precipitated and the sulphuric acid regenerated. The troubles have been mostly mechanical and are being gradually overcome. It is understood that the process is a success from a metallurgical standpoint.

The Nevada Douglas Con. Copper Co. has spent a large amount of money experimenting and has the only commercial plant in the district. The daily capacity is 150 tons at present. Crushing is done dry and the ore is leached with sulphuric acid, and the copper is precipitated on scrap-iron. The sulphide ores are first roasted in a Wedge furnace.

The Walker River Copper Co., of which J. H. Banks of New York is consulting engineer, is building a 20-ton plant at the Empire Nevada mine. The ore is crushed dry and leaching takes place in tanks. The mill equipment is all of Dorr make. Ferric chloride is used for dissolving the copper, which is precipitated on scrap-iron, and the ferric chloride regenerated by electrolysis. The flow-sheet was worked out after several years of laboratory experimentation. If the present plant shows the success anticipated, a large plant will be built. The Empire Nevada is considered by many engineers to be the largest property in the district. Churn-drilling has disclosed several million tons of oxidized ore averaging about 2% copper. J. E. Gelder is the local manager of both the Walker River Copper Co. and the Empire Nevada holdings.

The old Kennedy Consolidated property, at Buckskin, has been sold to J. F. Cowan and associates of Salt Lake City. This property is considered to be valuable but former operators were unable to work it during the years when copper was below 20c. per lb. The mine makes considerable water. The orebody is about 30 ft. wide and of undetermined length. The ore is a heavy sulphide mixture of barren iron pyrite and chalcopyrite averaging about 2% copper and containing considerable gold. It is reported that flotation experiments have been successful.

The Montana Yerington is under option to J. H. Brady, of Tonopah. P. Ober is in charge and a few men are employed in development work. This and several other properties of the district would be shipping ore to the Thompson smelter but for the fact that the Mason Valley Mines Co. will not receive any custom ore. The opening of the smelter has not been of any benefit to the small shipper and several anticipated sales have not materialized because the prospective purchasers could not market their ores. The general run of ore in the district will not average over 3 or 4% copper and shipment to

Salt Lake City smelters at a profit is therefore not possible. At the present time the smelter is treating about 800 tons of ore per day. This comes from the Bluestone mine and the Mason Valley. Occasional small shipments of high-grade from Ludwig are also treated as this ore has very desirable fluxing qualities. The capacity of the smelter is about 1800 tons daily, but there has been such a serious shortage of coke that running at capacity has been impossible.

This, however, is not the reason why custom ore is refused. The general opinion in the district is that the contract between the smelter and the Bluestone mine, owned by Capt. Delamar, stipulates that Bluestone ore up to 1400 tons per day must be accepted first, and that leaves little chance for ores from other properties.

The only gold mine in the district, which is located in Mineral county south of Yerington, is the Rockland mine that has been operated by the Pittsburg-Dolores Mining Co. for several years. Recently this company has installed an electric hoist wherewith to explore the main fissure at depth. It is understood that the lowest workings show ore of good grade and that the shoots are more persistent than was the case in the old workings. For the past two years the ore has averaged about $9 and has largely come from the upper zones. A complete cyanide-plant is operated, and extraction is around 90%. The ore is difficult of treatment. E. J. Schrader is manager of the property.

LEADVILLE, COLORADO

The Miners of the District, After an Angry Session, Decide to Strike.—The Mine Operators Remain Inflexible and are Determined to Work the Mines as Well as They Can With the Available Crews.

The strike which was to have been called in the mines of the Leadville district on Saturday morning, July 14, has been postponed until Saturday July 21, pending the outcome of a conference, which Chas. H. Moyer, president of the International Union of Mine, Mill, and Smelter Workers, and John R. Lawson, a noted Colorado labor worker, are to hold with Secretary of Labor Wilson, at Washington. The decision to postpone the strike was reached Friday night, July 13, when both Moyer and Lawson addressed a closed meeting of the union men from 9 o'clock until past midnight. The proceedings of the meeting were not made public with the exception of the statement announcing that the strike had been postponed for one week.

The efforts of Verner Z. Reed and George W. Musser, federal mediators appointed by Secretary Wilson to arbitrate between the Leadville union and the operators, failed to bring about a compromise. Following these conferences between the committees from both sides and the mediators which were held at the home of Mr. Reed in Denver, it was announced that Moyer and Lawson would come to Leadville to address the union. Their mission was unknown but it is believed that Mr. Reed brought Government influence to bear on the situation, which asked that the International heads persuade the local union to postpone the strike. Their success in this is reported to have come only after a bitter debate among the union men, and it is stated that many of the foreign-born members of the union left the meeting after the vote for postponment had carried and tore up their membership cards.

The strike question is not entirely settled, but there are many more possibilities for averting it than appeared a week ago. The result of the conferences at Washington will have a great influence on the future action of the union.

The operators have not made any public statements regarding the situation although many of them have declared their willingness to settle the labor-union question in the district once and for all time, while the matter is in the air. They are determined upon ignoring the union and have taken steps to protect their properties, prepared to continue operations with such force of men as they can secure. From 200 to 300 members of the State militia are on guard duty in the district. On the other hand, it is generally believed that should the efforts of the union to call a strike fail entirely, the operators will increase wages 50c. per shift. Such an action is eagerly looked forward to by the citizens generally, as it is believed it would go a long way toward ending labor troubles in the district for many years.

The mines are all working as usual but with decreased forces in many cases. Several hundred men, anticipating the strike call, left the district last week.

A canvas of the district shows that there are 49 properties producing ore with an output ranging from 100 tons to 10,000 tons per month. The combined output during the month of June was 63,300 tons. There are 1830 men employed in the mines of the district. At the two smelters, the Arkansas Valley plant of the A. S. & R. Co., and the Western Zinc Oxide plant, 900 men are employed and a combined total of 26,500 tons of ore is treated each month.

C. J. Dold, manager for the Dold Mining Co., is going ahead with operations at the Northern shaft near the head of East Ninth street, regardless of the pending strike. Sinking has been undertaken with two shifts, while another force of men is cleaning out the old drift leading to the Newell shaft, where it is planned to install a hoisting plant.

The Northern shaft was 640 ft. deep before sinking was started, the bottom being in the gray porphyry, which is estimated to be about 50 ft. thick. Underlying this formation, the second contact, in the blue lime, will be encountered. This second contact, which has been extensively developed in the Penrose and Coronado properties—the latter adjoining the Northern on the south—was the most productive area of the Down Town section. Huge bodies of iron, manganese, lead carbonate, silver chlorides, and zinc carbonate were uncovered in these properties. Their extension into the Northern ground is taken as assured by mining authorities here who predict that the Dold Mining Co. will develop one of the richest mines in the district.

The miners affiliated with the union struck according to programme on the morning of July 21, by which 37 mines were affected. At each of the mines 8 or 10 men were allowed to remain for the purpose of operating pumping machinery to keep the mine workings free from water.

SUTTER CREEK, CALIFORNIA

Bonanza Ore in the Lower Levels of the Central Eureka Mine Reported.—Result of Unfailing Effort to Prove the Mine at Great Depth.—Development at the South Keystone.

After several years of unsatisfactory development, but unremitting effort, the management of the Central Eureka mine has found what promises to be a shoot of bonanza ore on the 3350-ft. level, where pay ore was found in both the north and south faces of drifts run from the shaft. It is reported that the average value of the ore is about $11 per ton, which, if true, means substantial dividends for the fortunate shareholders, for the Central paid dividends on $6 ore some years ago. In addition to this new and gratifying discovery on the 3350, good ore has been found and is being developed on the 3425-ft. level.

These discoveries will illustrate the need for courage as well as financial ability in persistently sinking and exploring the veins of the Mother Lode in Amador county.

At the South Keystone development is proceeding from the North Star shaft, where a level is being driven north into the South Keystone mine, the object being to reach a body of rich ore at a depth of 600 ft. Some of this ore was high-grade when it was worked many years ago.

THE MINING SUMMARY

The news of the week as told by our special correspondents and compiled from the local press.

ALASKA

(Special Correspondence.)—Monte Benson with 12 men left Treadwell on June 6 for the west side of Douglas island, where a drift will be run on the Red Diamond mining property, on which the Treadwell company has secured an option. Later it is expected a tunnel will be driven through the range separating Treadwell from the Red Diamond property, so that ore may be treated at the Treadwell mills. F. W. Ketchmark, of Windham Bay, has struck a good body of ore on his property near the mouth of Spruce creek.

According to advices received in Juneau, troops from Fort Liscum have gone to Kennecott to remain during the strike of the copper miners at that place. While no trouble has occurred and none is anticipated, it was thought wise to have troops there. The striking miners are quietly encamped at Blackburn with their camp gaily decorated with American flags and bunting, and they are maintaining good order, administering the affairs of the camp themselves. Now that the superintendent is back at the mine, it is expected that the trouble will be adjusted. The men who were in charge during his absence were unable to handle the situation for lack of authority.

R. G. Wayland, Peter Johnson, and Thos. McDonald, are preparing to leave for the Atlin district on the next trip of the 'Mariposa.' The party will be gone a month or more and will examine and report on mining properties in that district. Within a few days the Treadwell company will have cars running for the accommodation of the workmen going and coming from the Ready Bullion mine.

Treadwell, July 8.

The Alaska United Gold Mining Co. reports returns for May as follows: Ready Bullion mill ran 30 days 7 hr. 15 min.; by water power, 24 days 2 hr. 39 min.; by electric power, 6 days 4 hr. 36 min. Crushed 24,346 tons, concentrate saved 669.74 tons. (700-mill 44.7 tons.) No free gold saved. Value of Ready Bullion concentrate, $44,862.10; of 700 concentrate, $4918.50. Total realizable value, $49,282.80. Operating expense, $44,711.24. Construction expense, $5380.28. Estimated net returns Ready Bullion, $4193.13 loss. 700-mill, $1187.15 profit. Other income, $2681.88. Yield per ton of ore $1.84.

ARIZONA

At the great copper mines in the districts of Warren, Globe, and Miami, the members of the I. W. W. fraternity are being warned to keep away—they are not wanted nor needed as men from other districts and other States are coming in, and the management of the various mines have decided to take on only American miners hereafter. At the Copper Queen it is said that the working crew has been recruited to about 75% of the usual number. Every man inquiring for work is closely questioned before being accepted, if he is taken on at all. It is thought at Bisbee that the trouble is at an end, but at Globe there is still some dissatisfaction among the men, who complain that they were misled by the union officials. In the Globe district the Miami wage-scale has been adopted. A general increase of 10c. per shift was made to conform to the Miami sliding scale, which is based on the price of copper. Wages are now $5.50 per shift, the highest ever paid in the Globe-Miami district.

Press dispatches dated July 19, at Globe, state that two German miners were arrested at Miami, at the instigation of the Federal Department of Justice, and that a search of their room was made resulting in the discovery of shotguns and revolvers with ammunition. One of them carried a certificate of registration with a German consul as a German reserve officer.

Mohave County

(Special Correspondence.)—All labor difficulties have been adjusted by committees of mine operators and the International Union of Mine, Mill, and Smelter Workers. The operators raised wages and the union is working with the operators to keep the turbulent labor-element out of the union and consequently out of employment here. All mines here have resumed with a full force.

The Minnesota-Connor is sending 200 tons of dump-ore to the new Steffy custom-mill. The old workings have been unwatered.

The Payroll is shipping ore from the 200-ft. level to the custom mill. Stopes are being opened on the 200-ft. level.

The Schenectady has closed down until machinery parts arrive. A fire destroyed the shaft-house and slightly injured the machinery. Work will be resumed soon.

The Steffy custom-mill expects to be running on custom ore by July 20. Its capacity is 300 tons daily.

The Copper Age mill is being tested out and will be placed in operation in a few days.

The New Tennessee has a good ore-shoot, which has been continuous to the present bottom of shaft, at 235 ft. from the surface. The vein is from 4 to 5 ft. wide and runs well in gold and lead.

At the Schuylkill the pumps have been lowered to the 800-ft. level, a new boiler has been installed, and water can now be handled from the lower levels.

The Bullion mine, Jesse Knight's property, is going ahead with three shifts. The electrically driven hoisting-machinery and compressor are working smoothly.

The New Mohawk soon will begin cross-cutting from the bottom of No. 2 shaft. A new shaft is soon to be started near the intersection of the Silver Key vein with the Juno vein.

Chloride, July 16.

(Special Correspondence.)—The advantage of maintaining harmonious relations between the mine-owner and the miner has been shown during the recent revolutionary campaign of the I. W. W.'s in Arizona. Oatman was protected from the I. W. W.'s not by troops, peace officers, or citizens' committees, but by the miners. When the first advance guard of agitators appeared here they were met and told not to attempt to start any trouble. The men who told them this were working miners who represented the majority of the men working in the mines and mills. The leaders departed and have not been seen at Oatman since. This incident is typical of the conditions that prevail in this district. Nearly all the working miners are stockholders in Oatman companies, or are owners of mining property here, and therefore understand the difficulties of the mine managers. The operators recently granted an increase of 50c. per day in pay. In order to revive some of the companies that have been insufficiently financed, a movement is under way to change them to assessable corporations. Nearly all Arizona mining corporations are in the non-assessable class, but many of the stockholders and officers begin to doubt the wisdom of this.

Three Oatman companies have financed new development by means of assessments. One of these is the Vivian Mining Co., owner of the Leland and Vivian properties near Oatman.

Another company that has adopted the assessment method is the Gold Range, which has changed its charter and has levied an assessment on all stock.

Motor-trucks will be used to transport ore from the Aztec shaft of the Tom Reed Gold Mines Co. to the mill, as the Government has conscripted the entire output of the A. Leschen company that had contracted to deliver a cable to the company. Two additional trucks have been put on to supply the mill with 300 tons per day.

Oatman, July 16.

PINAL COUNTY

The management of the Inspiration Copper Co. has paid off employees, both the faithful and the strikers, and closed-down its plant. The mine does not have to be unwatered, owing to its natural drainage, and this makes it possible to shut-down at a moment's notice and reopen quickly.

The mill had reached capacity operations a short time before the miners went on strike July 2, in the treatment of more than 20,000 tons of ore per day. Upon resumption of operations it will take some time before it can again be brought up to the same level of production.

YUMA COUNTY

(Special Correspondence.)—At the Swansea mine two strikes have occurred in the last 30 days. The first was for a raise to $5.40 per day, which was granted. At the time of the first strike when men were given the order to appear on top, the majority of them did not know what it was about. The second walk-out was on account of the boarding-house. The superintendent, Mr. Lane, discharged them all, as he considered their alleged greivance trival and not justified. This affected 120 men. Recently the W. A. Clark interests took over the Swansea mine on a 10-year lease. This called for a three-compartment shaft 1000 ft. deep and a 50% royalty on ores, which was not to be under $5,000,000 in the 10 years. This property is owned by French and Belgium interests, only a small part of the stock being held in the United States. A great deal of stock was sold and the money was spent on the Clara Consolidated properties, of which the Swansea was one. The property became involved in debt and a receiver was appointed who placed Mr. Lane in charge. He paid about 60% of the debt while it was under his management. In the last year he has shipped to Humboldt from three to five cars of ore daily and occasionally some to Sasco. When the debt has been paid, the property will be turned back to the old stockholders.

All of the ore shipped lately has been taken from No. 2 and No. 5 shafts. The ore averages 3% in copper. It occurs with specular iron in the form of a sulphide. The ore is desirable at the Humboldt smelter on account of its fluxing properties. It is reported that the old smelter at Swansea will be moved to the Bill Williams river, and also that Clark has bought the Swansea railroad, running from Bouse to Swansea.

Mr. Wolacott, one of the lessees on the Bullard property, says they will soon have the drills running that were recently installed.

Wenden, July 18.

CALIFORNIA

The magnesite producers of the Pacific Coast organized an association at the Palace hotel, at San Francisco, July 17, the purpose of which was to take steps toward securing congressional action to place a duty on imported magnesite. The organization will be known as the Western Magnesite Association. R. Adams, of Porterville, California, is president, F. S. White, of the Tulare Mining Co., is first vice-president, F. S. Sweasey, of White Rock mine, Napa county, is second vice-president, and A. N. Sears is secretary.

Reports filed with the State Mining Bureau for the week ended June 23 show 12 new wells started, making a total of 540 since the first of the year; 25 wells were reported ready for test of water shut-off; 14 deepening or re-drilling, and 3 abandoned.

FRESNO COUNTY

(Special Correspondence.)—The property of the Fresno Magnesite Co., near Piedra station, which was formerly owned by M. F. Tarpey, has been sold to the International Magnesite Co. of Los Angeles. A plant, consisting of crushing machinery and a rotary-kiln, will be purchased, and the property worked to capacity.

Fresno, July 18.

NAPA COUNTY

(Special Correspondence.)—The Socrates quicksilver mine has been examined by H. W. Turner in the interest of Honolulu capitalists who will follow the recommendations of Mr. Turner as to equipment, development, and operation of the mine.

The Etna quicksilver mine is now under the management of E. De Golia. E. B. Frost is superintendent, and recently took charge of the property.

Calistoga, July 18.

NEVADA COUNTY

(Special Correspondence.)—The Golden Center Mining Co. will spend $200,000 for new development and additional equipment. The 1000-ft. shaft will be sunk to 2000 ft. and an extensive system of cross-cuts and drifts run. Because of the high price of materials, and the difficulty in securing good miners, it has been decided temporarily to discontinue all mine and mill production.

John P. Clark, of Truckee, has practically closed a deal with Salt Lake men for the sale of his High Grade property, in Meadow Lake district, where 25 men are employed and shipments of copper-silver-gold ore are being made. The Utah company is reported also to be negotiating for the Great Eastern property.

An examination of the Bitner copper mine, at Spenceville, has been made by J. C. Campbell, of Nevada City, for San Francisco capitalists. The Bitner contains large bodies of low-grade ore.

Hydraulic mining will be resumed at the Omega mine, three miles north-east of North Bloomfield. More pipes have been added and the restraining dams have been re-inforced. There is a large yardage of gravel and sufficient water is available to insure a long period of activity.

Grass Valley, July 17.

The Hoff-Price Co., of San Francisco, is developing an occurrence of chrysotile asbestos on the Yuba river a mile above Washington. The asbestos occurs scattered in veinlets through a large mass of serpentine. The veinlets are from ½ to 1 in. or more wide and the mineral appears to be of fine quality. Machinery will be placed on the property to prepare the asbestos for market.

SHASTA COUNTY

(Special Correspondence.)—L. Gardella, the Oroville dredge operator, is assembling material on Clear creek, three miles south of Redding, for the construction of dredge No. 2. Three of seven cars of lumber have arrived at the site and most of the machinery is on the ground.

The Accident Gold Mining Co. of French Gulch has made application for voluntary dissolution. Its properties and other claims in the neighborhood were recently taken over by the Shasta Hills Mining Co. Ed Hillman is superintendent.

The electrolytic zinc-plant of the Mammoth smelter, at Kennett, is working successfully and turning out a satisfactory amount of the metal. The structural difficulties encountered at the start have been overcome in the main.

Redding, July 14.

TUOLUMNE COUNTY

H. Huston and associates are about to reopen the Keltz mine, situated 10 miles north of Soulsbyville, on the north slope of Elizabeth mountain overlooking the canyon of the South Fork of the Stanislaus river. The mine was operated for years by the late William Sharwood. It is equipped with a 10-stamp mill and the necessary buildings. All the work has been done through adits.

COLORADO

BOULDER COUNTY

(Special Correspondence.)—Encouraged by the increased price of tungsten, and the subsidence of the surface water, many lessees have resumed work. The prevailing idea is that the market is advancing. The result is that considerable ore is being held for better prices. The Long Chance Mining Co. has finished sinking in the Tungsten shaft. This makes the second 100-ft. lift made at this shaft in the last nine months, and gives them a depth of 300 ft. As soon as the ore-pocket and pump-station are cut, driving will start. This is now the deepest shaft in the Beaver Creek district. From indications it is believed that their next lift will pass through the schist. Prospecting on the 200-ft. level disclosed two new ore-shoots, which from the 1200 tons of ore already milled will exceed in value the rich surface deposits.

Nederland, July 11.

BOULDER COUNTY

(Special Correspondence.)—The Illini Mines Co. has been organized. It is composed of Boulder and Denver men. James Ashmore, Andrew Whitehead, and Edwin K. Whitehead, of Denver, C. S. Sicklestear and Thos. S. Bailey, of Boulder, compose the directory. The company will operate on Wood mountain, at Wall Street. The Wood mountain mines have been big producers, and have large orebodies of good value. The group consists of 60 acres of proved mineral ground.

W. G. Spaulding, of Ward, this week shipped a car of high-grade gold ore, carrying a small amount of silver. The ore came from the Sullivan No. 2. John T. Bottom, a Denver attorney and owner of the St. Louis mine at Caribou, has been looking at his property.

The Boulder Tungsten Production Co. has begun operations in the big mill at the mouth of Boulder canyon. The acid-mill is about completed. This plant is said to be one of the largest in the United States. Jack Clark of the Boulder Production Co. says the tungsten outlook is first-class and he looks for a further advance in price. Feberite (Eagle Rock) is a thriving mining camp. Two mills are running on the ores of that district. The Black Prince mine has a vein from 1 to 2 ft. wide and has been supplying ore to the mill all winter.

W. A. Keeler & Son have a good lease on the Duncan property, which has been a good paying mine in the past. Bud Conger and Gordon Taylor have a paying property.

Broughton, Holmes, Shipley, and Hewitt are making regular shipments from the Wheelman property. Bow and Casten are taking out good ore from the Dora mine. A shipment of two lots of ore from the Brandenburg properties, on Left Hand, is attracting attention. It has a big, rich, and continuous ore-body. It is predicted a new mining town will spring up in this district.

Phil Worchester of the Colorado State Geological Survey, who has examined the molybdenum veins in Allen's Park, says not enough work has been done to determine the amount or quality of the ore, though several properties there have good indications and should be developed.

Eldora, July 12.

IDAHO

IDAHO COUNTY

(Special Correspondence.)—After sending a large quantity of ore from the Black Pine mine near Elk City, to the Washoe smelter, at Helena, Montana, it was found that the ore was amenable to flotation and J. Reibcon, president of the company, has purchased all the machinery and equipment necessary for installation of a flotation plant with the expectation of having it in operation by August 15. Machine-drills are in operation and ore-bins are full.

Molybdenite and uranium ores have been discovered in the Knob Hill gold mine, owned by Frank Peck, in the Oro Grande district. More than 100,000 ft. of lumber will be used in the construction of the 300-ton mill, in course of construction, and in building the flume.

Elk City, July 16.

SHOSHONE COUNTY

(Special Correspondence.)—D. W. Peoples recently obtained a 10-year lease on the dumps of the upper workings of the Bunker Hill mine up Milo gulch, above Wardner. In the gulch, at the base of the dump, he has constructed a concentrating-plant of 200 tons capacity each 24 hours. The mill has two Blake crushers, 5 jigs, 3 Deister tables, and a Standard ball-mill, from the Colorado Iron Works Co. of Denver, Colorado. The intention of Mr. Peoples is to add flotation-cells. The value in this dump, which covers 10 acres, is about one-third ounce of silver to a unit of lead. The rock of the dump averages about 3 to 5% lead. The dump came from miles of workings made in the '80s. Through lack of facilities at that time it could not be handled. The mill is run by electric power obtained from the Washington Water & Power Co., of Spokane.

Wardner, July 14.

MONTANA

BROADWATER COUNTY

(Special Correspondence.)—The Iron Mask mining claim, near Hassel, is being unwatered. The shaft is 350 ft. deep. No work has been done at the property for some time. The ore consisting of lead and zinc was formerly not profitable to mine, but with modern improvements in concentration it is believed it will pay well.

In Iron Age gulch, at Winston, two cars of silver-lead ore from the 300-ft. level is shipped weekly to the East Helena smelter.

James H. Smith is opening the Indian creek placer near the old towns of Hogem and Florence. He is putting in a bed-rock flume and is sluicing some surface ground near by. The gravel at the head of the old flume was found to be 6 to 7 ft. deeper at the upper end than at the lower, which necessitated the deepening of the bedrock drain. The work is almost completed and the giant will be turned on a 50-ft. gravel bank.

Winston, July 16.

GRANITE COUNTY

(Special Correspondence.)—The Brooklyn Mines Co.'s property in the Philipsburg mining district, 7½ miles from the railway station of Maxville, consists of 10 claims. The principal development is a 700-ft. adit and much over-head stoping. The ore contains lead, silver, copper, and zinc, the greater part being in lead and silver, although the high-grade ore has run 300 oz. silver and 17% copper per ton. Electrical power is provided by the Washington Water & Power Co. of Spokane, Washington. Two shifts get out 75 tons per day, which is treated at the concentration-plant at the mine, to which lately has been added three Wilfley tables, making six in all, and six more flotation-cells. The concentrate is shipped to the Washoe smelter, at East Helena, H. Whitworth, of Drummond, is president of the company.

Drummond, July 16.

SILVERBOW COUNTY

(Special Correspondence.)—The country rock of the Summit Valley mining district, east of Butte, at base of the Continental Divide, is the Butte granite. There are two distinct systems of fracture—the east-west forms fissure veins, and contains the ore deposits. The vein-filling consists of iron pyrite, quartz, and copper sulphide, both chalcocite and bornite. The property of the Butte Czar Copper Co. is situated in this district and its development may mean much to the future of Butte. Two east-west fissure veins course through this property. The main vein is 40 ft. wide, with an apex of 1477 ft. This vein has been opened by a 300-ft. adit cross-cutting it 100 ft. below the surface. An 80-ft. drift on the hanging wall and one of 30 ft. on the foot-wall show continuous orebodies. Following the hanging wall is 4 ft. of sulphide ore containing 1 to 6% copper, and 2 ft. of the same class of ore on the foot-wall. The remainder of the vein is crushed granite over 30 ft. wide, with veinlets of quartz, iron, and copper. The machinery furnishes power for hoisting, for running a compressor, and the pumps when necessary. A two-compartment shaft is to be sunk 300 ft. A leaching-plant will be built to save the copper in the water coming from the 40-ft. vein. The company is well financed, with C. F. Murphy, superintendent of the Butte, Anaconda & Pacific Railway, as president. Development is under the supervision of E. C. Meiklejohn and R. M. Green, of Butte.

Butte, July 16.

NEVADA

Esmeralda County

(Special Correspondence.)—Because of the unwillingness of some of the creditors to wait longer in presenting their claims, negotiations for the transfer of the control of the Florence mine to the Red Hill Florence Co. have been discontinued.

The Silvermines Co. is erecting a 250-ton mill at its Great Western mine, at Hornsilver. The incline shaft is being sunk to 800 ft. A large tonnage of profitable ore is stated to be exposed. S. H. Brady, of Reno, is manager.

Goldfield, July 16.

(Special Correspondence.)—Shaft sinking at the White Caps has been discontinued; though drilling for a sump is in progress. The last few feet in the shaft passed through the quartzite into easier ground, and progress has been more rapid. The east drift on the 300-level cut the fault east of the lime, which contains the east orebody. The mine ore-bin will now be filled and the crushers, rolls, and conveyor-belts will be tried out. The mill construction is nearly at an end. An additional cooling-hearth has been built under the heating-hearths of the roaster.

Good progress in shaft-sinking has been made at the White Caps Extension. The three-compartment vertical shaft, each compartment being 4 by 4½ ft. in the clear, was sunk from the surface 22 ft. in the first 24 hours, three shifts being employed, using jackhammer-drills and sinking in soft shale. The shaft has reached a depth of 28 feet.

At the Manhattan Consolidated mines the east orebody, being developed from the third level of the Consolidated workings, still remains strong with high-grade ore in the face. It was reported that a fault had displaced the eastern segment, as a pronounced torsion in the foot-wall caused the limestone to appear in the face of the drift; but it was demonstrated that the ore occurred back of the raise in the foot-wall, and the drift has penetrated the orebody for 35 ft., with no sign of the end. Some remarkable samples recently were taken. The soft manganese ore assayed $1294. The first 9 ft. next to the hanging wall went $75.40; 9 ft. next to the foot-wall, $10, making an average of $42.70 for the 18 ft. across the vein, with more ore in the hanging not yet uncovered. A sample of the hard rock taken from the face of the drift returned $370 per ton in gold. It is the intention to drive on the vein 50 ft., and then to cross-cut the orebody from hanging to foot-wall.

At the Red Top a pump-station is being cut, and the compressor foundations are in. The east cross-cut 25 ft. from the shaft shows an increase in the value of the ore.

A new orebody has been exposed on the Mustang, 150 ft. south from the Train-Chase lease workings. A north-south cross-fissure in the schist has been developed from the surface which assays across 2 ft. $36 in gold.

Manhattan, July 14.

(Special Correspondence.)—The Morning Glory Co., which a few months ago began sinking on the outcrop, has now

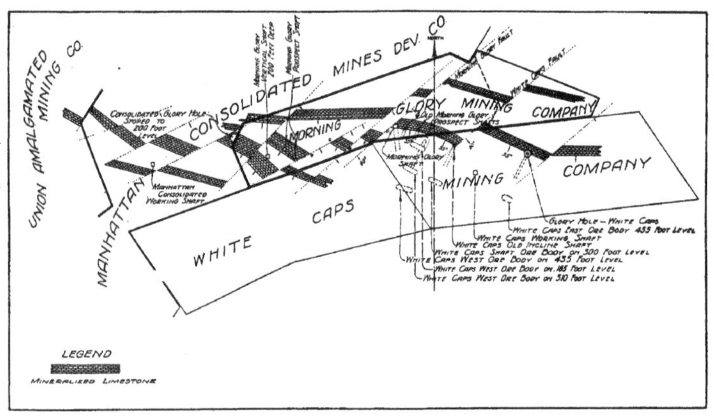

MAP SHOWING THE COMPLICATED SYSTEM OF BLOCK-FAULTING AT MANHATTAN

reached a depth of 170 ft. on the incline, in shaft No. 1, and a second shaft, No. 3, started later, is down 55 ft. Both shafts have had ore from the surface, some of it being high-grade. The shafts, which are both in the hanging-wall side of the vein, are 60 and 30 ft., respectively, from the White Caps side-line. At the bottom of the deeper shaft, known as No. 1, 5 ft. of ore runs $5 per ton in gold. In shaft No. 3 there is 2 ft. of $26 ore, and the vein is widening with depth. It is interesting to know that in the days of the early excitement at Manhattan in 1905, the prospectors, almost without exception, assumed that the strong outcropping quartzite strata were valuable veins, but experience has shown that the quartzite contains little payable ore. The valuable ore deposits are replacements of the limestone strata, and this fact accounts for some of the veins now worked being so close to the side-lines of the claims. The Morning Glory is a good example of this state of affairs. The Morning Glory is being developed under the direction of F. B. Caldwell, who outlined the plan of development after an examination of the complicated geological structure of the district several months ago, and this plan has been endorsed by W. H. Wiley, who recently made an examination of the Morning Glory and several other properties in the district.

Manhattan, July 18.

NEW MEXICO

Socorro County

(Special Correspondence.)—The district tonnage for the past week amounted to 2500 tons. Development work is being carried on in all the principal mines. The present price of silver is the highest for 20 years or more, and lower grade ore can now be handled at a profit. Daily shipments are being made from the Pacific, Little Charlie, and Johnson mines.

The Oaks Co. installed a hoist on the Central shaft during the week, the head-frame is nearing completion, and sinking for the general development of the central group will soon be under way. Driving is now being done from the first level. New ore is being found in the Maud S. mine and daily mill-runs are being made. The company also made a shipment from the Eberle mine during the week.

Mogollon, July 10.

OREGON

Jackson County

(Special Correspondence.)—The cinnabar strike recently made in the Butte Creek district is situated in the eastern part of the county 20 miles west from Mt. Pitt. This district is of recent volcanic formation, much broken up, and where mineral has never before been discovered. The first strike was made in basalt on the Terrill ranch six miles south-east of Eagle Point, the nearest railroad shipping place. The vein or dike contains cinnabar and assays high in quicksilver. Since the first discovery the dike has been traced for several miles north and south, and running from 100 to 200 ft. wide. It seems to be an extension of the cinnabar dike in Ramsey canyon, and in Meadows district. The new strike has caused many claims to be located. The new find extends in a heavily timbered region on the slopes of the Cascade mountains.

F. F. Childers, manager of the Greenback mine, says that the new equipment now at the property will be installed and ready for operation this fall. Modern machinery has displaced much of the old plant, which has been electrified. Four other properties besides the Greenback are under his management; the Jim Blaine, the Elk Basin, and the Illinois River properties. Three of these are copper, the others being gold. The new truck road to the chrome deposits in western Josephine county runs within three miles of the Illinois River properties; a connecting road will be constructed.

The war-time prices have given an impetus to the development of the sulphur deposits 45 miles north-east of Gold Hill, on the farm of W. T. Grieve, near Prospect. He is preparing to put it on the local market for fertilizer. It is being used extensively in this valley for that purpose.

Gold Hill, July 16.

Josephine County

(Special Correspondence.)—The Gold King mine, 6 miles west of Kerby, after a long idleness has been re-located by T. P. Johnson and Mrs. J. M. Finch. The mine is being unwatered. There was considerable development done a few years ago, but it was given up. With modern machinery the new owners believe it will pay.

Kerby, July 16.

(Special Correspondence.)—On July 7 the final payment of $60,000 cash was made on the Queen of Bronze copper mine by John Hampshire, representing the purchasers. Less than two years ago Hampshire saw the property and began negotiation for its purchase. An option for 90 days was granted in December, 1915, but the size of the property called for longer time, and in March, 1916, a working-bond was secured, expiring January 1, 1918. Under that bond work has been in progress up to last Saturday, when the final payment was made six months before it became due.

The property is purchased by John Towhy, R. B. Miller, John Hampshire, M. S. Boss, T. F. Ryan, and Roy H. Clarke. A corporation will be formed. Hampshire will remain general manager, with R. C. Crowell, superintendent; Roy H. Clarke, consulting engineer, and Edward Strong, foreman.

The full price paid for the mine was $150,000. There has been shipped out of the mine in little over a year $283,000 worth of ore. The shipments have averaged 9.48% copper, and gold, $3.50 per ton. The property has been shipping 700 tons per month. The monthly pay-roll is $12,000, of which $4000 goes for hauling. The expense of transportation is $4.25 per ton. This property, which had been turned down on account of the phyrrotite in the ore, has been paid for out of the mine itself. If the experiments now being conducted by the United States Bureau of Mines jointly with the State Bureau, looking to the perfection of a flotation-process of separating the phyrrotite from the chalcopyrite, is successful, a much larger percentage of the ore can be shipped profitably. These experiments are being made at this and the Waldo mine, both companies assisting in the work. The process will be given to the public as soon as experiments have been completed.

Grants Pass, July 12.

(Special Correspondence.)—Local men who have had the Waldo copper mine leased for some time have sold their leases to the owners of the mine, who are repairing the concentrator and putting in new track in adit No. 4. They will soon begin operations.

The Queen of Bronze copper mine now has 80 men employed and the force is being increased as rapidly as miners can be obtained.

K. J. Khoeery and Chas. Johnson, who sold the Lily copper mine to M. A. Delano some time ago, received another payment on the property July 1.

Takilma, July 17.

(Special Correspondence.)—At the Boswell gold mine, three miles from Holland, a gas-engine has been installed to operate the Huntington mill, recently placed on the property.

Holland, July 16.

UTAH

The mineral output of Utah in 1916, as announced in an advance sheet by V. C. Heikes, statistician for the U. S. Geological Survey, was as follows: Number of producing mines, 318; tons of ore treated, 13,920,643; gold produced, 172,938.06 oz.; silver, 13,253,037 oz.; copper, 240,275,222 lb.; lead, 201,490,075 lb.; recoverable zinc, 29,572,522 lb. Total value, $89,268,684. This output as compared with that in 1915 shows an increase in every metal except gold, which was 1652.59 oz. less than in the previous year. The total increase in value was

$34,163,614. The foremost county in the State was Salt Lake, which produced in 1916 metals valued at $68,797,428; Juab county coming next with $8,696,661.

WASHINGTON

SPOKANE COUNTY

Tailing from the mines in the Wardner-Kellogg district in Idaho are to be used in the construction of a highway in Spokane county this summer. The tailing is obtained free, the only expense being the freight.

The tailing is to be used in the paving of four miles of the Truax road leading out of Fairfield. The county will require approximately 250 carloads of tailing. This material was used in the construction of a highway out of Fairfield 10 years ago, and for the paving of the principal streets in Fairfield and Rockford, and has proved to be the best and most lasting ever placed in the county.

Spokane, July 10.

WHITMAN COUNTY

(Special Correspondence.)—The Washington School of Mines is given an important place in the reorganization of the Washington State College, at Pullman, 55 miles south of Spokane, perfected by the board of regents at its last meeting and recently announced by President E. O. Holland. Under the new plan the mining branch becomes one of the eight colleges of the university and Professor Francis A. Thompson, head of the mining department, becomes dean of the School of Mines. Facilities will be available for instruction in the treatment of ores by standard methods, including amalgamation, concentration, roasting, smelting, and leaching. A special laboratory will be devoted to the flotation-process.

Spokane, July 2.

STEVENS COUNTY

(Special Correspondence.)—The Electric Point Mining Co. has paid a dividend of $23,805 at its office at Northport. This will be at the rate of 3c. per share on an issue of 793,500 shares and will increase the total of disbursements to $47,810, all made within a year. An intention of maintaining a surplus of $100,000 was announced several months ago.

Spokane, July 2.

CANADA

BRITISH COLUMBIA

John O'Connell and Tony Angelo have a large quartz vein two miles north-west of the Emancipation group on the same contact. Their claims carry considerable free gold and will be developed. There is a good deal of ground prospected between the Emancipation and the O'Connell claims, but so far it has not been staked. There is also said to be a large area of gold-bearing country in the vicinity of the Coquahalla river that has not been prospected for quartz although some of the small streams have been worked for placer gold.

It is generally believed that the railway transportation now afforded by the Kettle Valley railroad will result in important development in the mineral region of the Hope, with the opening up of many new and valuable claims and the establishment of several permanent towns.

(Special Correspondence.)—A new compressor-plant, with a capacity of 1500 cu. ft. of air and a 300-hp. electric motor is being installed at the Granite-Poorman mine, at Nelson. The company will have power enough to treble its output of ore. In the concentrating section of the mill four new tables have been added, which gives the mill a capacity of 100 tons per day as far as concentrating is concerned. The amalgamation and stamp department of the mill will be improved later.

Thirty-five men are employed at the property. F. H. Skeels is superintendent; A. G. Larson, of Spokane, is consulting engineer for the company.

Spokane, July 10.

Personal

Note: The Editor invites members of the profession to send particulars of their work and appointments. This information is interesting to our readers.

C. ERB WUENSCH is in Nevada.

THOMAS KIDDIE is at Vancouver, B. C.

EDSON S. BASTIN is in South America.

SEELEY W. MUDD is visiting San Francisco.

R. H. TOLL is examining mines at Chloride, Arizona.

R. E. CLAPP has returned from Carson City, Nevada.

LUCIUS W. MAYER, of New York, is in San Francisco.

CLEMENT H. MACE, of Denver, has been commissioned captain in the Engineer Officers Reserve Corps.

W. M. HOLMAN of Minneapolis is in the Cripple Creek district.

HENRY M. PAYNE has been examining lignite deposits in Washington.

W. H. JANNEY is manager of the Republic mine at Hanover, New Mexico.

F. LYNWOOD GARRISON has returned from the West Indies to Philadelphia.

J. H. COLLIER has returned to San Francisco from Alaska and Montreal.

FORBES RICKARD has been examining an alunite deposit at Rico, Colorado.

G. C. CROWE is in San Francisco from eastern Oregon and western Nevada.

WAYLAND H. YOUNG has returned to San Francisco from northern Ontario.

A. CHESTER BEATTY is at the St. Francis hotel, having arrived here from India.

F. W. MACLENNAN, assistant-manager to the Miami Copper Co., is here on a visit.

A. R. WEIGALL has been appointed general manager for the Seoul Mining Co., in Korea.

A. P. COLEMAN, Professor of Geology in the University of Toronto, is in South America.

OSCAR LACHMUND, manager for the British Columbia Copper Co., has been in San Francisco.

SHIGETARO KAWASAKI, Chief Geologist of Chosen (Korea), is visiting our Western mining regions.

F. B. CALDWELL has returned to San Francisco from Manhattan, Nevada, and Sierra county, California.

ISSAKU GOTOH, metallurgist to the Mitsubishi company, in Japan, is visiting the steel plants of this country.

SIDNEY H. BALL sailed for Vladivostok on his way to Western Siberia on the 'Nippon Maru,' which left on July 18.

FRANK R. WICKS has succeeded W. H. JANNEY as manager of the mill of the Chino Copper Co. at Hurley, New Mexico.

W. C. PHALEN is in San Francisco. He will investigate the manganese deposits of California for the U. S. Bureau of Mines.

KOSAKU YAMADA, mining engineer to the Osarusawa mine, of the Mitsubishi company, is visiting our Western mining districts.

W. TOVOTE has resigned as examining engineer to Phelps, Dodge & Co. in order to engage in consulting practice at Tucson, Arizona.

P. N. MOORE, S. J. JENNINGS, B. B. LAWRENCE, J. PARKE CHANNING, and EDWIN LUDLOW are the representatives of mining on the Engineering Council, which is co-operating with the Government.

The next meeting of the San Francisco section of the A. I. M. E. will be held on the evening of July 31 at the Engineers Club, Denver, at 6:30; meeting at 8 p.m. A paper will be read by Professor D. M. FOLSOM on 'The Californian Oil Crisis.'

THE METAL MARKET

METAL PRICES

San Francisco, July 24

Aluminum-dust (100-lb. lots), per lb	$1.00
Aluminum-dust (ton lots), per lb	$0.95
Antimony, cents per pound	15.50—18.00
Electrolytic copper, cents per pound	29.50
Pig lead, cents per pound	12.25—12.50
Platinum, soft and hard metal, per ounce	$105—111
Quicksilver, per flask of 75 lb	$110
Spelter, cents per pound	10.50
Tin, cents per pound	60
Zinc-dust, cents per pound	20

ORE PRICES

San Francisco, July 24

Antimony, 50% metal, per unit	$1.20
Chrome, 40% and over, f.o.b. cars California, cents per unit	50—55
Magnesite, crude, per ton	$8.00—10.00
Tungsten, 60% WO_3, per unit	26.00
Molybdenite, per unit for MoS_2 contained	40.00
Manganese, 45% (under 35% metal not desired), cents, unit	33—37

Manganese prices and specifications, as per the quotations of the Carnegie Steel Co. schedule of prices per ton of 2240 lb. for domestic manganese ore delivered, freight prepaid, at Pittsburg, Pa., or Chicago, Ill. For ore containing

	Per unit
Above 49% metallic manganese	$1.00
46 to 49% metallic manganese	0.98
43 to 46% metallic manganese	0.95
40 to 43% metallic manganese	0.90

Prices are based on ore containing not more than 8% silica nor more than 0.2% phosphorus, and are subject to deductions as follows: (1) for each 1% in excess of 8% silica, a deduction of 15c. per ton, fractions in proportion; (2) for each 0.02% in excess of 0.2% phosphorus, a deduction of 2c. per unit of manganese per ton, fractions in proportion; (3) ore containing less than 40% manganese, or more than 12% silica, or 0.225% phosphorus, subject to acceptance or refusal at buyer's option; settlements based on analysis of sample dried at 212° F., the percentage of moisture in the sample as taken to be deducted from the weight. Prices are subject to change without notice unless specially agreed upon.

Tungsten: There is still a strong demand for tungsten ore and so far very little ore has come on the market. Such quantities as were disposed of during the past week were sold at increased prices ranging from $22 to $23 for wolframite, with scheelite selling on the basis of $26, New York. Several large users are in the market for regular supplies of tungsten ores to cover the balance of this year and the early part of 1918, but so far, none of the producers who are not already sold out has been found willing to make a price on the present basis, looking forward to a much higher market.

EASTERN METAL MARKET

(By wire from New York)

July 24.—Copper is lower and nominal at 26.75c. to 26.50c. Lead is dull and lower, at 10 to 10.25c. Zinc remains quiet at 8.37 to 8.50c. Platinum is unchanged at $105 for soft metal and $111 for hard.

SILVER

Below are given the average New York quotations, in cents per ounce, of fine silver.

Date			Average week ending	
July	18	80.25	June 12	75.83
"	19	79.50	" 19	77.00
"	20	78.87	" 26	78.12
"	21	78.87	July 3	77.98
"	22 Sunday		" 10	78.70
"	23	78.62	" 17	80.62
"	24	78.62	" 24	79.12

Monthly Averages

	1915	1916	1917		1915	1916	1917
Jan.	48.85	56.76	75.14	July	47.52	63.06	....
Feb.	48.45	56.74	77.54	Aug.	47.11	66.07	....
Mch.	50.61	57.89	74.13	Sept.	48.77	68.51	....
Apr.	50.25	64.37	72.51	Oct.	49.40	67.86	....
May	49.87	74.27	74.61	Nov.	51.88	71.60	....
June	49.03	65.04	76.44	Dec.	55.34	75.70	....

High-price records continue to be made for silver, the best figure to date being slightly better than 84c. per oz., for delivery on the Pacific Coast.

Shortage of supplies in London and the fact that quotations emanating from that city still have their effect upon the world's silver market have combined to put the New York price up to its present high level. The lower rates across the Pacific than those prevailing to London and thence to the Far East and the absence of war risk insurance have made it almost obligatory on the part of the shipping companies to route their silver for India and China from Pacific Coast ports rather than from Montreal and New York.

England and France have been persistent buyers of silver for coinage, largely to pay off their troops in the field. The United States government has bought in the past two months about 3,000,000 oz. of silver, with further purchases thought likely. This also represents coinage requirements which should grow materially from now on.

COPPER

Prices of electrolytic in New York, in cents per pound.

Date			Average week ending	
July	18	28.00	June 12	32.75
"	19	27.50	" 19	32.58
"	20	27.00	" 26	32.42
"	21	26.75	July 3	32.25
"	22 Sunday		" 10	31.50
"	23	26.50	" 17	30.29
"	24	26.50	" 24	27.04

Monthly Averages

	1915	1916	1917		1915	1916	1917
Jan.	13.60	24.30	29.53	July	19.09	25.66	....
Feb.	14.38	26.62	34.57	Aug.	17.27	27.03	....
Mch.	14.80	26.65	36.00	Sept.	17.69	28.28	....
Apr.	16.64	28.02	33.16	Oct.	17.90	28.50	....
May	18.71	29.02	31.69	Nov.	18.88	31.95	....
June	19.75	27.47	32.57	Dec.	20.67	32.89	....

The copper-ore output in Japan in 1916 amounted to 111,562 tons, as compared with 83,017 tons in 1915 and 78,700 tons in 1914; while exports amounted to 57,402 tons in 1916, as compared with 56,528 tons in 1915 and 43,305 tons in 1914. Russia now buys 60% of Japan's copper exports; the United Kingdom takes 20%, while France, the United States, and India share the remainder. The consumption of copper ore in Japan has increased considerably during the last three years, amounting in 1916 to 59,690 tons, as compared with 27,723 tons in 1915 and 32,045 tons in 1914.

The July production of metals by the Anaconda Copper Mining Co. will be extremely small, owing to the strike, says the 'Boston News Bureau.' Not only has copper output dwindled to a minimum but the electrolytic zinc plant has materially cut down its operations.

Certain overhead charges must of course be maintained regardless of operations, while pumping and maintenance of good conditions at mines and plants entail heavy expenditures.

It is estimated that Anaconda's earnings during the first half of the year were in the neighborhood of $23,000,000, based on 28c. copper and a 14c. copper cost, this item including all taxes and other extraordinary charges. Smelter production of copper was 163,000,000 pounds.

The United States Geological Survey gives the total production of smelter copper in the United States for 1916 at 1,927,850,848 pounds.

Last quarter copper has been offered at 27c. per pound, against sales for that delivery at 30½c. a short time ago. Bids have been asked for the closing three months of this year.

Copper for shipment in the first quarter of 1918, which has sold at 29¼c. and better, may now be secured at 25 cents.

LEAD

Lead is quoted in cents per pound, New York delivery.

Date			Average week ending	
July	18	10.75	June 12	11.83
"	19	10.50	" 19	12.00
"	20	10.25	" 26	11.75
"	21	10.00	July 3	11.57
"	22 Sunday		" 10	11.25
"	23	10.00	" 17	10.98
"	24	10.25	" 24	10.29

Monthly Averages

	1915	1916	1917		1915	1916	1917
Jan.	3.73	5.95	7.64	July	5.59	6.40	....
Feb.	3.83	6.23	9.01	Aug.	4.67	6.28	....
Mch.	4.04	7.26	10.07	Sept.	4.62	6.86	....
Apr.	4.21	7.70	9.38	Oct.	4.62	7.02	....
May	4.24	7.38	10.29	Nov.	5.15	7.07	....
June	5.75	6.88	11.74	Dec.	5.34	7.55	..

The Hecla Mining Co., of Wallace, Idaho, has declared dividend No. 170 of 15c. per share, making $150,000, and a total for 1917 of $1,050,000 and a total of all dividends to date of $6,355,000.

ZINC

Zinc is quoted as spelter, standard Western brands, New York delivery, in cents per pound.

Date			Average week ending	
July	18	8.87	June 12	9.75
"	19	8.87	" 19	9.72
"	20	8.50	" 26	9.43
"	21	8.37	July 3	9.32
"	22 Sunday		" 10	9.20
"	23	8.50	" 17	9.06
"	24	8.50	" 24	8.00

Monthly Averages

	1915	1916	1917		1915	1916	1917
Jan.	6.30	18.21	9.75	July	20.54	9.90	....
Feb.	9.05	19.90	10.45	Aug.	14.17	9.03	....
Mch.	8.40	18.40	10.78	Sept.	14.14	9.18	....
Apr.	9.78	18.62	10.20	Oct.	14.05	9.92	....
May	17.03	16.01	9.41	Nov.	17.20	11.81	....
June	22.20	12.85	9.63	Dec.	16.75	11.26	....

QUICKSILVER

The primary market for quicksilver is San Francisco, California being the largest producer. The price is fixed in the open market, according to quantity. Prices, in dollars per flask of 75 pounds:

Week ending

Date				
June	26	80.00	July 10	100.00
July	3	85.00	" 17	105.00
			" 24	110.00

Monthly Averages

	1915	1916	1917		1915	1916	1917
Jan.	51.90	222.00	81.00	July	95.00	81.20	....
Feb.	60.00	295.00	126.25	Aug.	93.75	74.50	....
Mch.	78.00	219.00	113.75	Sept.	91.00	75.00	....
Apr.	77.50	141.60	114.50	Oct.	92.90	78.20	....
May	75.00	90.00	104.00	Nov.	101.50	79.50	....
June	90.00	74.70	85.50	Dec.	123.00	80.00	....

TIN

Prices in New York, in cents per pound.

Monthly Averages

	1915	1916	1917		1915	1916	1917
Jan.	34.40	41.76	44.10	July	37.38	38.37	....
Feb.	37.23	42.60	51.47	Aug.	34.37	38.88	....
Mch.	48.76	50.50	54.27	Sept.	33.12	36.66	....
Apr.	48.25	51.49	55.63	Oct.	33.00	41.10	....
May	39.28	49.10	63.21	Nov.	39.50	44.12	....
June	40.26	42.07	61.93	Dec.	38.71	42.55	....

Eastern Metal Market

New York, July 18.

Stagnation and demoralization characterize the condition of practically every metal except tin. Actual prices are hard to gauge, demand and orders being of small volume. The proposed one-price-to-all policy of the Administration has caused to halt a market which before that was decidedly lame and uncertain.

Copper is growing weak each day.

Tin is fairly steady with not much business.

Lead is sagging with consumers uninterested.

Zinc is stagnant and lower.

Antimony has again declined, demand being slack.

Aluminum is a little lower and dull.

In the steel market new business has decidedly slackened. The effect, on buyers, of the President's advocacy of putting all steel and metal buyers on an equal price-basis, and a lower one, has had just the effect which was to be expected. The one policy now is postponement of buying until the Federal Commission's cost-finding inquiry is completed. There is much speculation as to the effect of the new policy, and many varied opinions as to its practicability. The embargo on exports of certain steel products has held up some shipments, but vessel-space is really a more serious factor. Mill output has been less in July than in June, which also was less than May. The entire trade awaits the outcome of the vital controversy with anxiety.

COPPER

Prices are falling and a decided weakness is looked for if not already here. The cause is not hard to determine. On Friday last week the President's statement, advocating the same price to consumers as well as to the Government for steel, metals, and other materials, came as a distinct shock. Its effect on the non-ferrous metal market has been pronounced, and prices have fallen daily. Added to the foregoing factor was the announcement that for the 60,000,000 lb. of copper recently bought by the Government at a provisional price of 25c. per lb., only 75% of it, or 18.75c. per lb., would be paid, the remainder, if any, to be determined by an investigation of costs. This was also an unsettling factor. Commenting on this phase of the subject, it is freely stated here among dealers that it is not likely that the Government will pay as high as 20c. per lb. for any of its future needs. If this is to apply to the Allies and all other consumers, then the future could hardly be more indefinite and uncertain. It is rumored that producers have refused to accede to the Government's position. Under such dominating influence it is a buyers' market. Consumers will not come into the market when lower prices are reasonably expected, and are holding back decidedly. Lower levels are generally looked for. One dealer states that he thinks any large inquiry would receive very favorable consideration, and be taken at most any price—at figures regarded as absurd only a week or 10 days ago. Prices are entirely nominal, the quotation yesterday for both Lake and electrolytic having fallen to 28c. for July delivery. For last quarter, opinions differ, offerings ranging from 24.50 to 25.50c., New York. The question of price fixing is so important and far reaching a matter that it is fraught with many disturbing factors. The fact should not be lost sight of that there are many consumers who are under contract to buy millions of tons of copper for delivery this year and into next year at almost 25c. per lb. What is to be their status if the general price is fixed at under 20c. per pound? The London market is unchanged at £142 for spot and £138 for future electrolytic metal.

TIN

This metal is the only one that exhibits any steadiness. While the market the past week has been a narrow one, it has behaved unexpectedly well in comparison with copper, lead, and zinc. The deliberations or decisions of the Tin Committee have not been made public, and have not been a factor, but the unsettlement of the other markets by the President's advocacy of a one-price-to-all policy has had its effect and has unsettled business. Many think that if a general price re-adjustment by compulsion is to result, it will be stupendous and that a business depression is not improbable. The tin market in such an event would suffer with the others. The latter part of last week a fair business was quietly done in futures, and early this week a moderate business for the same position is reported. There has been some demand for spot Straits, but it has not been large. Prices have ranged from 63 down to 62c., New York, on Monday, with the quotation yesterday at 62.50c. Stocks of Straits tin are heavier than last week. Arrivals up to July inclusive are reported at 1925 tons, with the quantity afloat at 3729 tons. The London market has declined in the week about £8 per ton, the quotation yesterday having been £239 10s. for spot Straits metal.

LEAD

What has been said about copper applies equally well to lead. The one-price-to-all policy has weakened the market and buyers are extremely cautious as to their commitments. They naturally expect, after such a public avowal from a high source, much lower prices, especially because the recent Government purchase was fixed at 8c. per lb. The tendency is decidedly lower with producers holding the market up as near 11c. per lb., New York, as possible. Early delivery metal today is quoted at 10.50c., New York, and 10.37½c., St. Louis, but the American Smelting & Refining Co. still maintains the quotation at 11c., New York. Small sales have been made, but aloofness on both sides is the attitude. One carload is reported to have gone at 10.37½c., New York. The future of the lead market is uncertain and extremely hazy. Strikes in the West have added to the difficulties of the situation.

ZINC

There may be less gloom in the zinc market than in most of the others, but if so it is because it is believed that prices have reached so low a state that a definite fixing of them by the Government would put them higher instead of lower if settled upon a cost plus a reasonable profit basis. It is generally recognized that a price below 9c. per lb. is in many cases less than cost and hence not a fair basis for the industry as a whole. Owing to recent and past edicts from Washington as to prices in general, quotations for prime Western spelter are chaotic. In the words of one dealer: "They are all over the map." One sale of 100 tons for early delivery is said to have been made at 8.62½c., St. Louis, or 8.87½c., New York, while another went at 9c., St. Louis, or 9.25c., New York. But the volume as a whole is insignificant and no guide as to real values. Consumers will not buy under present conditions nor will producers commit themselves far into the future. The little business done is all hand to mouth. A fair estimate of present prices, which are mostly nominal, is 8.62c., St. Louis, or 8.87c., New York, for early delivery, with future shipments higher if anything. The fact that the Govrnment bought sheet-zinc at 16c. per lb., or 3c. under the market, is pointed to as not at all in line with other conditions in the zinc market and as really a boon to sheet-zinc makers. They are said to be highly satisfied with their bargain.

ANTIMONY

A further decline is recorded in the face of small demand and a stagnant market. Chinese and Japanese grades are offered at 16 to 16.50c., New York, duty paid. Possibly 15.50c. could be done.

Company Reports

ALASKA UNITED GOLD MINING CO.

The annual report of the Alaska United Gold Mining Co. for the year ended December 31, 1916, shows the following: At the Ready Bullion mine 286,078 tons of ore was crushed producing in free gold $235,846.89, and base and copper bars to the value of $27,813.59. Concentrate treated, 6629.77 tons, which produced $338,533.83, making a total production of $602,194.31. Mining cost (377,547 tons stoped) was $0.9983 per ton. Milling expense (286,078 tons milled) was $0.344 per ton. Sulphide expense (6629.77 tons treated) was $0.1209 per ton milled. Total operating cost including various sundries, $1.5118 per ton, a total of $432,504.63. Construction costs were only $0.0026 per ton milled.

At the 700-Ft. claim 262,850 tons was crushed, producing $179,861.72 or at the rate of $0.6843 per ton. Base and copper bars yielded $0.0612 per ton treated, making $16,082.72. Concentrate treated, 5341.98 tons, yielded $0.9906 per ton of ore crushed, amounting to $260,375.48. Making total receipts from all sources of $456,319.92.

Mining cost $1.8128 per ton, and included 7771 ft. of development, and 284,566 tons stoped. The 700-mill crushed 262,850 tons; 5341.98 tons of concentrate was saved, the cost of treatment of which was $30,771.35 or at the rate of $0.1171 per ton of ore treated. The total operating and construction cost at the 700-Ft. mine was $623,913.36, or at the rate of $2.3737 per ton treated. The net loss for the year at this mine was $167,593.44.

MOUNT BOPPY GOLD MINING CO., LTD.

The annual report of the Mount Boppy Gold Mining Co. (Limited), Cobar district, New South Wales, for the year ended December 31, 1916, shows the following:

In 1916, 72,420 tons of ore was milled, which produced, by amalgamation, 10,172 oz. of bullion; 12,474 tons of tailing treated yielded 3253 oz. of bullion, and 55,010 tons of slime, 15,263 oz. of bullion; in addition to which 1600 oz. of fine gold was recovered from 361 tons 4 cwt. of concentrate; the total production being 30,288 oz. of bullion, or 20,512 oz. of fine gold, of a realized value of £85,837 17s.2d.

Sundry receipts, including interest and discount, and transfer fees, brought up the total income to £87,591 12s.3d. The expenditure on revenue account amounted to £95,314 19s.4d., and a balance of expenditure for the year of £7723 7s.1d.

The chief factors militating against the regular working of the mine, rendering the conditions abnormal, and accounting for the unusual occurrence of a loss being shown instead of a profit were:

(*a*) A shortage of labor and the consequent curtailment of development and stoping operations; (*b*) the lower average grade of the ore milled; (*c*) an unusual rainfall, which was almost continuous throughout the 12 months; (*d*) the temporary closing down of the mine on two occasions owing to strikes by the engine drivers and firemen for increased wages.

With regard to the shortage of labor, this was due to some 30% of the best of the employees having joined the army. These men readily responded to the appeal of the military authorities, thus showing a sense of their appreciation of their duty toward the mother-country.

NEVADA WONDER MINING CO.

The tenth annual report of the Nevada Wonder Mining Co. for the year ended December 31, 1916, shows the following: Development work during the year, 10,509 ft. The sub-shaft was completed from the 1300-ft. level to the 1600-ft. level. The average amount of ore treated in mill daily was 157 tons. Total ore treated 72,241 tons from which was recovered 10,933 oz. gold, and 1,243,753 oz. silver. The average value of the ore treated during 1915 was 0.159 oz. gold, and 18.72 oz. silver. For silver the average price received was $0.647 per ounce.

DETAILS OF COST

	1916
Total mining cost per ton of ore, including transportation	$ 5.33
Total milling cost per ton of ore	3.31
Marketing cost per ton of ore	0.26
Total cost per ton	$ 8.90
Average value per ton of ore	15.40
Loss in taling per ton of ore	1.13
Recovery per ton	$14.27
Less total cost per ton	8.90
Profit per ton of ore mined and milled	$ 5.37

The following improvements were made during this period: A new Ingersoll-Rand No. 5 drill-sharpener, with full equipment for sharpening all kinds and sizes of steel.

A new 50-hp. double-drum electric hoist was placed in operation at the sub-shaft to enable the sinking of this shaft to the 1900-ft. level.

There has been installed a Vulcan, 1000-lb. per day, ice-plant, which has proved a great benefit and comfort to the employees; an electric tractor on the 1300-ft. level for moving ore from the stopes, and waste from the sub-shaft; five piston-drills, 11 stoping-drills in the mine, 2 air-hoists, and 30 one-ton mine cars.

THE TONOPAH MINING COMPANY OF NEVADA

The fifteenth annual report announces dividends to date since April 1905 of $13,750,000, including the last paid on December 18, 1916, of $150,000. In addition the company has retired its outstanding preferred stock, amounting to $380,-557.51. During the year 1916, 74,991 tons of ore was mined, and the Desert mill treated 81,782. Also 5542 tons was shipped from the dump. The average gross value of the ore milled was $15.65 per ton. The estimated tonnage blocked at the end of the year was 72,100 tons, assaying $16.74 per ton, making reserves valued at $1,206,821. The average cost per ton of mining and milling the ore and marketing the products for the year was as follows:

Mining and handling dump-ore	$ 4.61
Milling	3.16
Freight on ore milled	0.72
Marketing product	0.22
Total cost per ton	$ 8.71
Metal losses in milling and refining	1.34
Profit per ton	5.60
Average gross value of ore milled	$15.65

The report of F. F. Heydenfeldt, superintendent of the mill, states that the ore treated carried 15,636 oz. gold, and 1,387,557 oz. silver, of which 94.2% of the gold was recovered, and 90.3% of the silver. The recoveries were less than during the previous year, owing to the delivery to mill of ore containing manganese-silver compounds that are not amenable to treatment with cyanide. The amount of silver so locked up was not large enough to repay the cost of the additional treatment that would have been required for its recovery. The increased cost of treatment was due to higher costs of supplies. The milling cost rose from $2.61 in 1915 to $2.767 in 1916. Of this difference of 15.7c. per ton 9.2c. was due to the higher cost of cyanide, and 6.5c. to other supplies. The mining cost, under W. H. Blackburn, mine superintendent, was $3.94 per ton, plus 67c. indirect cost.

Mining Decisions

COAL LEASE—ROYALTIES

Minimum royalties must be paid under a coal lease providing for them unless the lessee sustains the burden of proving that coal does not exist in merchantable quantities.

Rowland *v.* Anderson Coal Co. (Iowa), 162 North-Western, 321. April 6, 1917.

COAL LEASE—ROYALTIES LIMITED

Minimum royalty provisions which were specified in a coal lease as being applicable to certain coal veins held not applicable to veins not known to exist at the time the lease was made, although within the general limits of the ground covered by it.

Birdsall *v.* Delaware & H. Co. (Pennsylvania), 240 Federal, 618. April 10, 1917.

OIL LEASE—ROYALTIES RECOVERED

An oil lease gave the lessees three years in which to drill wells after which royalties were to be paid. Eight years elapsed before any wells were driven and the lessees then being still in possession refused to pay the royalties, claiming that the lease was forfeited at the end of the three years. Held, lessees still liable for royalties and an accounting.

Andrews *v.* Andrews (Pennsylvania), 100 Atlantic, 521. January 8, 1917.

GAS WELL—DRILLING OPERATIONS DEFINED

The driving of a stake locating a gas well and of another stake locating a place for a boiler to be used in connection with driving the same does not constitute a commencement of operations to drill such as would prevent a sale made prior to the commencement of drilling operations under a right reserved in the lease.

Henning *v.* Wichita Natural Gas Co. (Kansas), 164 Pacific, 297. April 7, 1917.

MINING DONE UNDER CITY STREET—INJUNCTION REFUSED

An injunction was denied the city of Scranton in a suit to enjoin the removal of coal underlying a street where it appeared that the street had been dedicated over land from which the mineral rights had already been severed without reservation to the owner of the surface of the right of subjacent support.

City of Scranton *v.* Scranton Coal Co. (Pennsylvania), 100 Atlantic, 812. January 22, 1917.

MINING LEASE—MINNESOTA RULE

Under the laws of Minnesota a lessee under a mining lease calling for minimum advance rentals and royalties on ore extracted cannot apply payments of such advance rentals on future royalties which become due. In Minnesota a mining lease is a rental of land, not a sale of ore. The advance payments are considered as ground rent and the royalties as payment for the ore.

Nelson *v.* Republic Iron & Steel Co. (Minnesota), 240 Federal, 285. February 22, 1917.

COAL LEASE—DAMAGES FOR FLOODING MINE

An owner of coal land leased the mineral rights and a right-of-way for removing coal mined beneath his and other ground. The lessee mined a part of the coal and left the rest for a period of years. The lessor diverted a stream of water into the workings, which carried with it and deposited a débris, for the cost of removing which the lessee brought suit. Held, damages properly awarded to the lessee. His term not having expired, he was entitled to have his mine protected whether he was actually working it or not.

Sorg *v.* Frederick (Pennsylvania), 100 Atlantic, 481. January 8, 1917.

Book Reviews

THE GOLD DEPOSITS OF THE RAND. By C. Baring Horwood. 12 mo. Pp. 393. Ill. Charles Griffin & Co., London. For sale also by the MINING AND SCIENTIFIC PRESS. Price, $5.

This is a re-print, by a well-known British publisher, of the series of articles contributed to the MINING AND SCIENTIFIC PRESS during 1913, together with the exhaustive discussion that ensued. The author, C. Baring Horwood, is a mining engineer formerly in charge of important mines on the Rand, so that to a scientific training of more than usual thoroughness he has added the qualification of systematic observation underground. The book presents a minutely detailed examination and exposition of the geologic structure, mineralogic composition, and origin of the greatest deposits of gold ore ever uncovered by the miner's pick. The genesis of the banket has been a fruitful theme of discussion during the last 15 years and in this book both the evidence and the argument are presented in scholarly fashion. The interest of the controversy has been heightened by the development of two opposing hypotheses, one imputing the gold in the conglomerate to agencies similar to those by which ordinary veins are formed, that is, fracture of the rock and hydro-thermal action in the wake of eruptive intrusion, the other supposing the gold to be an integral and original constituent of a sedimentary deposit now tilted in conformity with a regional syncline of large amplitude. Between the two antagonistic hypotheses there is also the 're-solution' theory, which supposes that the original detrital gold has been dissolved and re-deposited in place. Thus the Rand has furnished a subject involving more than one phase of the science of ore-genesis as elaborated in modern days, and for that reason the book is interesting and instructive even to those to whom the Rand is only the name of an intense expression of industrial activity. The author gives the reader photographs of microscopic sections and other painstaking records of his conscientious investigation. He argues vigorously in behalf of his own thesis and does not hesitate to break a lance with any that hold views differing from his own. The book is not one of academic dignity or disarming phraseology; on the contrary, it is alive with uncompromising argument accompanied by a barrage of closely-sifted facts, which are fired with the smokeless powder of a keen wit. Whether right or wrong—and the reviewer thinks that Mr. Horwood is right in the main—such a book is a notable contribution to the library of economic geology, for it clears the air and illuminates the essential points of a great disputation. Heat there is, as the subsequent discussion shows, but light also—light on obscure problems of vital concern to the miner. Of the discussion it can be said that it is worthy of the author's effort. The list of contributors includes distinguished British and South African geologists, such as F. H. Hatch, David Draper, E. T. Mellor, and J. W. Gregory; and among American geologists Waldemar Lindgren, James F. Kemp, H. Foster Bain, and the reviewer, besides several mining engineers familiar with the Rand, such as E. T. McCarthy, S. J. Speak, and F. P. Mennell. Mr. Horwood replies to his critics and so rounds the discussion to a satisfactory conclusion. As is customary with the house of Griffin, the book is attractive in its make-up and printing. A wealth of illustration, including photographs and maps, lightens the reading-matter and assists the exposition. We commend this volume to all students of ore deposition, and particularly to mining engineers.—T. A. R.

EDITORIAL

T. A. RICKARD, Editor

NOTHING seems to have been decided about the Government's purchase of copper, mainly because the producers of this metal do not see why they should accept so low a price as 18⅔ cents per pound at a time when the producers of other commodities are not asked to make a similar sacrifice. Meanwhile it is reported from Utah that three Federal inspectors have arrived from Washington to examine the books of the copper companies with a view to ascertaining the exact cost of producing the metal.

TYPOGRAPHICAL blunders are annoying to the reader, but we can assure him that they are far more annoying to the author and editor. In an article on 'The Principles of Flotation,' in our issue of July 7, the author was made to say that surface-tension was equal to "3.14 grammes per linear inch or 72.62 dynes per centimetre at 20°C." By that perversity which one of our Allies would call *la malice des choses* the word 'grammes' had been substituted for 'grains.' Of course, no careful writer would use 'grammes' with 'inches.' We are reminded of an equally irritating error that appeared in Mr. T. J. Hoover's book on flotation, in which, when quoting from Clerk Maxwell's article on 'Capillarity' in the Encyclopædia Britannica, the measure of surface-tension was spoiled by inserting the word 'square' before 'centimetre' in the phrase "81 dynes per centimetre." Eternal vigilance is the price of safety.

RUSSIAN affairs are of such engrossing interest at this time that we are glad to publish a record of recent observations made by Mr. Charles Janin in the course of a journey to Petrograd by way of Siberia. Excellent accounts of the revolution in Russia have been published, for example, in the *New Republic*, but to our readers an article by one of themselves, by a mining engineer, will have a particular value. Most of us are curious to know how things look to one having a mental bias like our own. Since the happenings described by Mr. Janin the Russian position has undergone several changes and the future of the new democracy has been darkened by the excesses of its friends no less than by the efforts at reaction on the part of its enemies. The national crisis has produced a strong man, A. F. Kerensky, who, if he is supported loyally, is likely to establish a stable government and organize an effective army. We confess to a feeling of optimism, despite the prevailing gloom. The Russian peasant is docile and mystical, as well as courageous and enduring; if the friends of liberty can succeed in establishing order out of chaos then Russia will play her proper part in a war that to her is no longer one of territorial expansion or political aggression. The sympathy and help of the American nation will go far to promote that result.

FURTHER protection for the public from the wiles of fraudulent promoters is promised by means of the new Blue Sky law, or Corporate Securities Act, which went into effect in California on July 27. The former law was avoided by registering wild-cat companies in other States and then peddling their stocks in California, thereby evading the discipline of the State Corporation department. The new law, drafted by Mr. H. L. Carnahan, the Commissioner of Corporations, who has done the public a great service ever since he assumed this responsibility, compels every person or company engaged in the sale of securities in this State, whether issued originally here or elsewhere, to obtain a certificate authorizing him or it to act as a broker. Such broker must bear a good reputation in order to obtain a license, which may be revoked if he engages in a fraudulent transaction. The law provides also that copies of all advertising matter relating to the sale of securities issued or published by brokers must be filed with the Commissioner, and the circulation or publication of such matter if interdicted is punishable by law. We congratulate Mr. Carnahan on the further powers thus granted to him in pursuance of his good work, for, while the last to discourage mining ventures even of a risky kind, we are as anxious as he is to stop the diversion of good money into fraudulent schemes. The more capital given to frauds the less there is for legitimate business.

THE editor of the 'Colorado School of Mines Magazine' sends us a marked copy of his July issue, containing a reply to the account of the recent trouble appearing in our issue of June 16. We have read this reply carefully and patiently. Our critic is in error in supposing that because the present writer was at Golden "a short time on Commencement day" he was unable to elicit the facts and that "he pretends to quote the students." The writer was at Golden longer than his critic supposes and he discussed the matter with a number of students, not all of them at Denver. The School has graduates all over the West and the matter was discussed with some of them, as well as with other persons thoroughly informed concerning the events that have given the School a temporarily unpleasant notoriety. The editor of the Magazine undertakes to defend the three trustees whom we criticized as unfitted for their responsibilities; we do not begrudge them a defence and we do not care to return to the attack by making further comment. Our criticism of them was, we believe, true and in the best interests of the School. One mistake we did make in referring to the other trustees: Mr. F. G. Willis is not an alumnus, but he is a mining engineer in good standing in Colorado, and this fact serves equally well for the purpose of our argument. Our humorous comparison with the strike at Telluride was borrowed from a mine-manager at Telluride and the quotation concerning the students "willingness to work in harmony with the faculty" was taken verbatim from the published statement of a student. The editor of the Magazine would have done better by using his columns in appealing to the alumni to support the faculty and warn the students that their unruliness has besmirched the reputation of the School.

MINING just across the border in Mexico is still rather mixed. Cananea remains closed, although negotiations are in progress with Governor P. Elias Calles for a settlement of the trouble. Nevertheless the bulk of the laborers have been shipped south by order of the Governor, ostensibly to prevent food-riots. This action cannot well be criticized, because the cessation of work and of importation of supplies would quickly result in starvation. On the other hand, the scattering of several thousand workmen will make the re-organization of a crew more difficult and costly, although the availability of so many men will be of immediate benefit to the farmers. It has been increasingly difficult to obtain farm-labor in Mexico, and it will be noted that the shut-down at Cananea came just at the period for planting corn in that part of the world. After this will follow the bean-planting. There is work to be done from this time throughout the summer and autumn, and the newly-favored holders of confiscated estates will be highly pleased to have so many peons thrown upon the country. This is not suggested as a motive for the pressure brought to bear upon the mining companies, but it is a feature of the situation that may not be disregarded. It is quite certain that these men will be of such use as producers of food as to overweigh the importance of the mines to the State, and those that are favored by the authorities will profit in consequence. The troubles at the Tigre mine near Nacozari are 'officially' settled, so far as gubernatorial promises go, and the technical staff has returned to re-open the property. In this case, however, it must be noted that the number of workmen is less, while the output is of exceptional value, the ore being rich in silver. The local government is said to have attempted to produce from this property on its own account, but the plant is operated throughout by electricity transmitted from the Phelps-Dodge generating station at Douglas, Arizona. It was, accordingly, impossible to turn a wheel, and it was equally impossible to install a steam-plant adequate either for mining or milling. The heavy taxes upon the silver shipped by the company are worth more than the net value of any product that the Mexicans could obtain by working the mine. This would offer a reason for special concessions to this company. Governor Calles has the reputation of being more favorable to American enterprise than his predecessor, Governor de la Huerta, who was a Carranza appointee.

The Third Anniversary

On August 2, 1914, the German army invaded Belgium. Today 20 nations are in a state of war, including all the supposed leaders of modern civilization. To thoughtful men in every country the catastrophe has seemed appalling; to them the three long years have been a nightmare of horror. To many of our own people the crowning event of participation in the struggle is so new that the decisive action of a President elected 'because he kept us out of the War' is still a bewildering event rather than a logical consequence. The thoughtless—and there are plenty of them—are asking even yet why we did it. The President has said: "It is a fearful thing to lead this great peaceful people into war, into the most terrible and disastrous of all wars." It was. His decision to do so was the result of a judgment as deliberate and painful as history records; the United States became the declared enemy of Germany only after trying every device of diplomacy and enduring the full measure of insult. We saw Belgium made a road for the gray swarm of the invader, and did nothing; we held our hand even after helpless populations were driven into slavery by the ruthless conqueror; we only protested when our citizens were murdered on the ocean; we hesitated to interfere because we loved peace and abhorred war; to us the catastrophe was already so tremendous that we hated to enlarge the scope of it; we had inherited an honorable tradition of non-interference in European troubles; we hoped to do more good to suffering humanity by helping the victims of war than by increasing the number of them through going to war ourselves; we dreamed of playing the part of peace-maker. At the end of last year the President gave the German government an opportunity to state its purpose and to justify its claim that it was waging a defensive campaign. The answer disclosed a plan of impenitent aggression, the objects of which were the subjugation of Belgium and the permanent occupation of a part of France, to destroy the freedom of the one and to seize the iron mines of the other, to extend German hegemony over the weaker nations of south-eastern Europe, and to accomplish this not only by ruthlessness on land but by the destruction of the merchant shipping of the neutral countries. That did it. We could no longer remain detached from the great struggle.

On April 6 we entered the fight "to make the world safe for democracy," as the President announced. In March one of the few surviving autocracies had been swept aside and Russia stepped through the darkness of revolution into the light of liberty. Thus the four principal Allies are united in their political ideals. But some organs of misinformation, like the San Francisco *Chronicle*, assert, "To say that we are fighting for democracy is nonsense. We already have it, and we are certainly not at war to compel others to accept it." Yes, we have democracy, but how long should we preserve it if the German purpose prevailed in Europe, if the French army were overcome and the British navy destroyed? The democracies of Europe are fighting to preserve the democratic ideal as against the insistence of autocratic absolutism; they are fighting for the freedom to live and let live against the paranoiac insanity of Junker militarism. Is that no concern of ours, seeing that we are threatened by a visitation of the same pestilent policy if England and France are beaten? There is no danger of that, says our ignoramus. The Allies will win. We believe it, but the result is too vital to be left to the slightest uncertainty. And if they win, are we to allow them to do all the fighting? No, we are going to do as Washington and Lafayette, as Lincoln and Lee, would have done. Even if we had so little sympathy with other peoples as to make a political island of ourselves; even if we attempted—under the leadership of men as ignorant as he of the *Chronicle*—to ignore the wrongs perpetrated on other democracies; even if we were willing to "pass by on the other side" while the people of Belgium were enslaved and the people of Armenia butchered wholesale, would we escape either the risk or the responsibility? Those that deny the responsibility also belittle the risk. Europe is far away, they say—7000 miles distant from San Francisco, for example; the powers of the Entente will worry through somehow they assume—without knowledge; let the European nations settle their account with Germany and let us—these fools insist—proceed on our peaceful way, not without some supercilious regret that civilized peoples should have descended to such a depth of depravity. These political Pharisees and ignorant Scribes are now the enemies of the great cause to which the United States is committed, heart and soul, pacifist and jingo alike. Here we venture to interject a quotation from an article by Mr. Vernon Kellogg, in the current *Atlantic Monthly:* "I say it dispassionately, but with conviction: if I understand the German point of view, it is one that will never allow any land or people controlled by it to exist peacefully by the side of a people governed by our point of view." Mr. Kellogg is a lover of peace, a scholar by profession, a man eager to aid suffering humanity, as shown by his work with the American Relief Commission in Belgium and France. Again he says: "I have seen that side of the horror and waste and outrage of war which is worse than the side revealed on the battlefield. How I hope for the end of all war! But I have come out believing that that cannot come until any people which has dedicated itself to the philosophy and practice of war as a means of human advancement is put into a position of impotence to indulge its belief at will. My conviction is that Germany is such a people, and that it can be put into this position only by the result of the war itself. It knows no other argument and it will accept no other decision." And to this we may add the statement made a few days ago by Mr. Robert Lansing, the Secretary of State: "If any of you have the idea that we are fighting others' battle and not our own, the sooner he gets away from that idea the better it will be for him, the better it will be for all of us." It is the intention of Our Allies—no longer The Entente—to re-assert the public law and to save civilization from red ruin. We have many reasons for drawing the sword—plenty of them. We are fighting to avenge the American men, women, and children that have been assassinated on the high seas; we are fighting to redress the wrongs of the innocent Belgians; we are fighting to prevent the domination of the world by a cult that considers an international treaty "a scrap of paper," that believes "necessity knows no law," that asserts the submarine to be "the last word of kings"; we fight against a system of government that is without conscience and a mode of warfare that is without shame; we are fighting for our own safety, for the preservation of democracy, for the peace and liberty of the world.

Oil Legislation Needed

How long would it take the Board of Directors of the United States Steel Corporation to decide on the re-opening of a closed but not exhausted iron mine if more ore were needed to supply the furnaces at Gary and if that mine were the only immediately available source of increased output? We believe that Judge Gary, with a war on hand that needed an accelerated production of steel billets, would see to it that a resolution authorizing the opening of that mine were put to vote and carried in about 15 minutes. We need oil for industry and for driving battleships; we shall be drawing down our last margin of ten million barrels in storage in the State of California in September 1918 unless the hopeless obstacle to development that exists because of the wholesale withdrawals of land may be removed by appropriate legislation. To be sure, the power to withdraw involves the power to restore, but it would mean a return to the evils of forced over-production on land that would instantly be located, and the locations pending could and would be forced to patent; in short, the intention of the Government when the withdrawals were made would be defeated, and this was mainly to regulate the output of the oil needed so as to preclude wasteful over-production and the ruination of large resources of oil by the flooding of productive strata through injudicious methods of drilling. Congress has trifled with the obligation it owes to hosts of oil-operators, actual and prospective, by failing through six years of talk to reach a subject of such vital national importance; in consequence we face a situation calling for 1,083,000 barrels of oil per month in excess of the production, with only 38,000,000 barrels in storage, and no addition to the available supply can be made. A committee of the Senate could ascertain at one sitting the best-digested thought of the country regarding an enabling act under which rational exploration and development of the oilfields could be undertaken without prejudice to public or private interests that would be one-tenth as dangerous as the injury now inflicted on both through inaction. The Committee on Petroleum of the California State Council of Defense has just issued a report on the oil crisis that every citizen should read. It insists upon the need of instant action by the Government. Without it we shall find ourselves impotent; we shall be unable to maintain our manufacturing, our transportation, our naval activity. When will Congress act?

The Nitrate Fizzle

Mobilization of the chemical industry of the country has been effected, according to a statement by Mr. W. H. Nichols, chairman of the Chemical Committee of the Council of National Defense. This is true so far as co-ordination of the chemical manufacturers is concerned. It does not mean that the chemical resources of the nation are mobilized, because we are deficient in manufacturing capacity with which to utilize our enormous supplies of raw material. Moreover, the Chemical Committee has not been given authority over special committees of the Government that are supposed to contribute toward the urgent needs of the moment. The most glaring example of failure to mobilize our chemical resources is shown in the absolute neglect to take advantage of the power conferred by an act of Congress for establishing a nitrogen-fixation plant in order that we might not remain dependent upon the Chilean nitrate deposits, which are available only by means of transportation across 7000 miles of ocean. The Act that authorized the construction of a Government nitrogen-fixation plant carried with it an appropriation of $20,000,000 for the purpose. A Nitrate Committee was appointed, the duty of which was supposed to be chiefly the selection of a site that would be so demonstrably superior to any other as to defeat the objections of the pork-barrel type of congressional representative. It is lamentable to record that the intent of this law was assailed at the very beginning by a group of men distinguished in the chemical industry. Their reputation in the world of technology was high enough to command attention. At a time when Germany was erecting nitrogen-fixation plants of various types, and when similar works were being established in Sweden, the impression was created in this country that the fixation of nitrogen was still in an experimental state. The details of the methods used abroad are known, however, to American chemists; there are men among us that are familiar with every ramification of those methods as carried out in commercial practice. Nevertheless, the Government was counseled to halt until we could put through a series of independent tests. It was bad enough to slow down our preparations at a time when wiser patriots were insisting that we take steps to meet what seemed an inevitably impending conflict. An excuse may have been urged on the ground of financial caution in the spending of the public money, but no such flimsy justification is now possible for the tactics of delay that have robbed this measure of its larger purpose. It indicates that malign influence is active at Washington. Mr. N. D. Baker, Secretary of War, issued a statement on July 16 in which he announced that a few small plants would be constructed at a cost of about $4,000,000 for the fixation of nitrogen by a modification of processes previously known, but not involving the use of water-power. He added, "the committee is engaged in the making of further engineering studies, and the subject is temporarily closed to further discussion by localities and communities desiring to be considered as possible sites for the plants." The Secretary of War appears to overlook the fact that we are actually at war! yet the three governmental departments represented on the Nitrate Committee calmly undertake research upon modifications of tried processes instead of embarking on a course charted by others that have been making fixed nitrogen successfully abroad. It is well enough to blaze new trails; it would be eminently proper to expend $4,000,000 in efforts to improve upon the older processes. The trouble is that these experiments are definitely intended to supersede other efforts to create plants to produce fixed nitrogen for an immediate emergency. The only things that will be accomplished will be to check our preparations for war and to endanger our supply of nitric acid for the manufacture of explosives. This action by the Nitrate Committee cannot be taken on the counsel of a true American. If the Government has become deaf to entreaty to hasten the construction of the nitrate plant authorized nearly a year ago, perhaps there do remain patriotic citizens with money enough to build one while Secretary Baker's experiments are being made. It might be worth while to risk the possibility that some of the foreign methods would act as well on this continent and in our special atmosphere as they do in Europe. The arc process could be quite safely relied upon to behave reasonably well, even on the Pacific Coast, where a good deal of energy is still available at the switch-boards of some of the producers of hydro-electric power, especially if advantage be taken of the valleys between the peak-loads. The cyanamide process could be expected to give analogous results in the United States to those obtained in Canada and elsewhere. Some progressive man might undertake to utilize the Serpek process for making ammonia through the treatment of bauxite with producer-gas, whereby purified alumina is obtained from the aluminum nitride produced in the furnace. This suggests the rationality of converting the relatively cheap fuel available near many of the American bauxite deposits into producer-gas, from which this valuable nitrogen by-product would be recoverable in conjunction with the refining of the bauxite for the manufacture of aluminum. Failure to secure departmental action because of faulty organization at least might have the excuse of good intentions to offer for this delay, but the deliberate shelving of so important a project as the immediate domestic production of nitrogen in suitable form for making explosives with which to wage the War before us, is not so easily explicable.

DISCUSSION

Our readers are invited to use this department for the discussion of technical and other matters pertaining to mining and metallurgy. The Editor welcomes expressions of views contrary to his own, believing that careful criticism is more valuable than casual compliment.

Patents

The Editor:

Sir—During the spring of last year you gave a good deal of space to a discussion of patents, to which I contributed the letter that you published on May 27, 1916. In your editorial of March 3, 1917, you published some statistics of the past year's work of the Patent Office that shows the fallacy of our present system. Frequent articles, letters, and editorials in recent issues have shown the impractability of our patent laws, and have indicated the dangerous monopolies that can result, and the utter lack of reward to sundry meritorious pioneers.

The whole question of the patent law is becoming critical, and as I do not see any suggestions being put forward that offer a solution of the problem, I take the liberty of asking for space enough for a rather lengthy statement. The subject is too complex to be dismissed in a short letter.

As the present law fails so completely in meeting our economic and social needs, we can best base a proposed new law on scientific reasoning; and I am convinced that, as in the case of flotation, there will be much criss-cross of thought before we understand the position. One thing I stoutly maintain, and that is that the formulating of a new patent law, a mining law, or laws affecting engineering education, will be better handled by engineers than by lawyers, patent attorneys, or politicians, to all of whom human variance and distress are meat and drink.

The real difficulty is to postulate an objective, as in such an extremely technical case, the referendum of public opinion is impossible; and the interests of the rest of the relatively small groups directly affected, and having knowledge of the subject, are almost diametrically opposed. Incidentally the discussion in the columns of the Mining and Scientific Press is most appropriate, as I believe that this paper was originally founded to register improvements in invention, particularly with relation to mining and metallurgy.

An invention is a thing of generic form; the resulting machine is a thing of concrete form; the first may have value for a time, but, commercially considered, it soon ceases to have value, generally, from two causes, (1) the expiration of the patent and the ceasing of monopoly; and (2) the fact that at any moment, and surely ultimately, it will be superseded by better inventions; but nationally it always has value, as being the basis of subsequent progress. An invention in the end is more of a public and national acquisition than a personal one; this is the real reason why it is never treated from the same legal perspective as real estate.

I believe that it will be best and most convenient not to give the ultimate and unrestricted power of monopoly to either of the interested parties, but to consider the inventions as of public utility, to be held in trust by the Government for the Nation. This then is a suggestion for an entirely new point of view, namely, that the Patent Law and the working of the Patent Office shall be based on the national ownership of patents. This is not to be confused with the national exploitation of patents. The inventor and the public would have the right to make, use, and sell, subject to license from, and the payment of royalty to, the Government. Half, or some suitable proportion of these royalties, would be distributed *pro rata* to the inventor or inventors, as also a subordinate amount of the royalties accruing from subsequent inventions or improvements developed from the parent idea. These royalties would be subject to certain general conditions to be described later.

The balance of the funds, which would be considerable, would pay for the running expenses of the Patent Office, so that registration could be made free. Funds would also be available for technical assistance and advice to the inventor, for research laboratories, publications, and other forms of special work, assistance, and reward, for meritorious workers by methods also to be described.

The Patent Office would be a much more important department of the Government than it is now, and would probably need representation by branch offices in all the important cities.

The fundamentals of this project may be summarized as follows:

(1) The act of applying for a patent would automatically assign the invention to the Nation, thus preventing personal ownership, but not interfering with commercial initiative in manufacture, sale, or use.

(2) The copying, manufacture, sale, or use of all designs or inventions, once declared novel or patentable, by the Patent Office, and for which a certificate has been issued, is to be subject to a royalty, said royalty to be payable only to the Government, and in amount and manner to be determined by the Patent Office. Such royalties should when possible be codified, and be based on a percentage at the sale-value; all questions regarding royalties to be settled by a Board of Commissioners of the Patent Office.

(3) Full protection for the inventors by a free, simple, and assisted means of registering creative ideas, in what-

ever department of art, science, mechanics, or general industry.

(4) Avoidance of all disputes and lawsuits, as to interference or infringement, by the Federal Government becoming the trustee for the national ownership of all patents, and simply adjusting the apportionment of reward to inventors by a distribution of a share of all royalties accruing to the Government. The share being in accord with the merit of the inventor's contribution, whether a basic idea, the fundamental of many others, or merely an improvement in the art.

(5) The cost of running the Patent Office to be paid out of the Government's share of royalties, instead of out of fees for registration. This would be a tax on success, whereas the present system is a tax on effort and enterprise.

(6) Practical assistance and stimulus for inventors, by the establishment of laboratories, work-shops, and test-plants, where, subject to proof of legitimate endeavor, merit, and capacity, the inventors could develop, test, and prove their ideas free of charge, and place them on exhibit, so that when meritorious they would get a chance of being brought to public use as soon as possible. See C. This work could be done on a generous scale as the amount of funds available would be so much greater than those accruing from the present system of fees.

(7) Great stimulus would be given to the inventor, by the ease of registration, by assistance, guidance, and technical suggestion from trained experts, by facilities for reference, research, and actual test, by opportunities to have their inventions demonstrated to prospective manufacturers, by freedom from the difficulties and legal expenses of protecting their ideas, and by the certainty of a moderate reward if their ideas were used and adopted. The certainty of a moderate reward would be a far greater stimulus to the average inventor than the chance of millions; and human consideration on the part of a common-sense Patent Office would be worth more yet.

(8) Protection of the manufacturing interests from undue monopoly would be obtained by the Patent Office placing the patents on the open market, so that all manufacturers could enter into competitive manufacture, subject to an equal royalty to the Government, of which the Government gives a share to the inventor or group of inventors, according to merit. When a basic idea has been developed until some time after the life of the patent, or when new discoveries, bring it within the utilitarian field, a special form of reward to be described and payable out of a special reserve fund would be provided. Evasion of royalties should be treated as a serious offence, such as the present evasion of Federal taxes. In every case the royalties should be moderate so that they would in no sense be an incubus on industry.

(9) The general public would be protected from the common practice of buying patents to prevent their competitive influence on other patents that have been already developed by large capitalistic groups. This burying of useful inventions is detrimental to the interests of the general public.

(10) If it appeared that capital required protection or stimulus, as evidenced by those cases where, after throwing the patent open for competitive manufacture, no one comes forward to place it in public use, another procedure would be followed. If, after one year of open market, due application were made, either by the inventor or by parties representing manufacturers and capital, stating that the invention had merits, and that the parties were willing to place it on the market, or use it, provided a monopoly was granted to them, claiming that, as a considerable risk in capital outlay would have to be made to test and develop the idea, they could not undertake it without a monopoly, the Patent Office could then withdraw it from the open market. The patent would be offered for sale as a close monopoly, much the same as at present, with the difference that the term of the monopoly would be only for a part of the life of the patent, the remainder being thrown open to competitive manufacture as before. Royalties would be adjusted on a fixed scale, of which a certain proportion goes to the monopoly-holder who risks his capital, a certain proportion to the Government, which the Government again divides with the patentee. Monopoly patents not exercised would be withdrawn and again thrown open to competitive manufacture, sale, and use. No possible opportunity would be given for the deliberate suppression of invention useful to the general public, and done only for the purpose of accentuating monopolies that are already an incubus on society.

(11) The whole purpose of the Patent Office would be to encourage the registration of ideas, to reject worthless ones, to bring the apparently useful ones to the notice of the manufacturer and user, to reward the inventor of those that prove useful, or that are the parent of other useful inventions, to test and develop elementary ideas on new lines, to prevent monopoly of a class that would retard further useful development.

There is probably no such thing as an absolutely new idea, it is always the outcome or growth of some prior state of the art; all that the Patent Office has to determine is that the invention put forward is not in common use, and that as put forward and assembled it is reasonably novel; minor technical infringements are immaterial, so long as the actual result is a real improvement when applied to the art. There may even be cases when a research-worker deserves reward for developing to a utilitarian point, a process or art that is already known, or we might say the clarification of an art; for in many cases the cost of such experimental work, not to speak of moral and mental energy, would not, and often could not, be commercially undertaken, without the protection of a patent and the reward accruing therefrom.

What we want is to make life materially easier and better for the whole human race, and to stimulate, protect, and reward the persons that are contributing to that desirable end.

The Patent Office should be the go-between and adjuster; hence it should be the final arbiter as to the merits, patentability, and title to inventions.

(12) The result of such a programme would be that the Patent Attorney, as such, would automatically vanish. The Patent Office would assist the inventor in elucidating his ideas, not in formulating claims put forward only to block further development. Some compensation seems to be due to the patent attorney for the abolition of his office; no doubt a great many of them might be usefully absorbed as employees of the new department and its branches. There are many technical men in the profession whose services would be invaluable, especially when the temptation was removed of encouraging palpably worthless ideas as a source of profit from fees. But we must not forget that according to our postulate we are not proposing to legislate for the benefit of any one group, but to stimulate progress in invention that will be of the greatest good to the greatest number.

(13) To admit full liberty of personal initiative on the part of the inventor, and to guard against too much power and paternalism on the part of the Patent Office in the matter of the rejection of ideas, it will be well to have some class of registration provided to meet those cases where after the technical advisors of the Patent Office have returned an idea to the inventor, stating that it is novel but worthless, the inventor could still obtain a patent if he wished to do so. This and other similar problems could be met by having several classes of certificates issued, some of which may be suggested as follows:

A. Copyright, more or less as at present.

B. Trade-mark, also as at present.

C. Registration of minor ideas, details of mechanical ideas, or improvements in arts of processes. This would be termed a Minor Patent, and would not be subject to royalties, and could be freely adopted by the manufacturing public. The Patent Office would reserve a special fund out of its general income, for the purpose of giving rewards for these Minor Patents, the rewards to take the forms of certificates of merit, medals, and small cash payments. In some cases research scholarships, and even small life-pensions, could be granted for work under this head that had proved to be meritorious, as shown by its general adoption by the public.

D. Ideas that had novelty, and therefore were patentable, but which, on technical grounds, the experts of the Patent Office considered worthless, would be returned to the inventor with a report for his consideration. If the inventor still desired a patent, a second class of Registration Patent would be issued for a short term not to exceed ten years, subject to royalty.

E. Ideas that were evidently novel, and even possibly basic, and with merit, but that were undeveloped, would not be immediately patentable, but would receive a certificate of registration, giving the inventor certain rights of experiment, at the nearest research laboratory, and possibly some financial assistance from the special fund. The nature of the invention would not be published until after it had been developed and proved, and fully patented. A time limit would be set for the inventor's experimental work, and his rights would be forfeited if he did not show due diligence in the prosecution of his research.

F. Inventions that had been developed and proved, and had been examined and passed by the expert advisor of the respective department of the Patent Office would be admissible and subject to full protection under the title of First Class Patents, which would be good for 20 years and subject to participation in royalties, and in some cases to monopoly rights; it would be unassailable in the courts, the Government guaranteeing its title to originality.

(14) The journal of the Patent Office would be different from the present publication. It would be divided under various industrial heads and ideas registered; and patents issued would be classified accordingly, so that readers could at once find the latest development in the art of any particular industry. There would also be intelligent criticisms and descriptions of the latest inventions, which would stimulate the technical press to copy and discuss. At present the inventor can get no publicity, the technical press seems to think it criminal to publish or discuss a new idea until it has been tried at some plant. How many failures at mills could be prevented by a little intelligent discussion regarding new machinery.

(15) The employees of the Patent Office should be well paid life-service appointees, subject to an examination as a branch of the Civil Service. There should be a strict rule that no employees could ever take out, or hold any interest in any patents, even after leaving the employ of the Patent Office.

The above gives at least a general scheme for a new law. If these suggestions, and those already put forward by other writers, are the means of starting a general discussion from all the interested groups, we shall soon get daylight on the subject.

JOHN NICOL.

San Francisco, June 26.

Notes on Flotation

The Editor:

Sir—I submit a few notes which may be of some assistance to the man who is undertaking practical flotation work for the first time with a machine of the Janney type. They may prove of more interest perhaps to the operator living in a locality where facilities for visiting a flotation-mill do not exist. I shall restrict myself to the consideration of a plant designed for the recovery of copper-sulphide ores. The theoretical side of flotation has been dealt with extensively in various issues of your paper, but I do not recollect having seen published any notes intended primarily for the novice.

The proportion of solid in the feed may vary between 20 and 30%. The specific gravity of the feed within these limits does not seem to have any definite influence on the recovery, as with the same oil-mixture sometimes a better, sometimes a worse, extraction is made with the same pulp-density. With a thin feed a larger quantity of oil is necessary, and the quantity required diminishes as the specific gravity of the pulp increases till a point is reached where more oil is again necessary, for the froth now becomes so heavily coated with slime that an excess of oil must be used to cause the froth to flow over the lip of the cell. A further disadvantage of a thin feed is that the passage of the pulp through the machine is too rapid to permit of a high recovery.

The feed goes to 'rougher' cells, and the first question to be considered is the degree to which emulsification should be carried. In practice where there are two emulsifying-cells, should one of the emulsifying motors be stopped the froth-level falls almost immediately and more oil has to be added to the feed. The froth now becomes of lower grade and is lighter in color owing to carrying some slime. When the feed is comparatively rich in sulphides the color of the froth in two adjacent sections, one using two emulsifying-motors and the other only one, is practically the same; but if the feed should become poorer the difference in color at once manifests itself. This indicates the importance of sufficient emulsification.

As the 'rougher' cells are followed by cells in which the tailing is re-treated, the question naturally arises as to the extent to which recovery of the sulphide content of the feed should be made in these first cells. Let us consider the two extremes. If the froth on the 'roughers' be dark, a good deal of work is thrown on the secondary cells, and the ultimate recovery may be affected adversely. If the froth be kept very light in color by the use of an additional quantity of oil the recovery is higher, but it is made at the expense of the grade of concentrate, for a larger quantity of slime is carried over and the proper cleaning of the froth becomes correspondingly difficult.

The 'water-level' in the cells is best kept at two to three inches below the lip of the cell. The level should not be kept so high as to cause breaks in the froth-layer, for at such points the mineral burden is dropped; nor should it be so low that the froth collects in masses and overflows only in a sluggish manner. It must be borne in mind that it is through the agency of froth that the sulphides are recovered and hence a sufficient and uniform overflow is imperative in good practice. As the overflow of froth would be most rapid in the first cells, especially in the froth-cleaning machine, the distance between the top of the canvas and the lip of the cell is made greater at the feed than at the discharge end, so as to ensure a more even distribution of the oil in the machine and thereby to effect a cleaner separation.

The water-level has to be raised gradually from the feed to the discharge end of the machine, as the froth is more profuse in the first cells, becoming less abundant and more fragile in the direction of pulp-flow. This is effected by means of the gateway provided for each individual cell. It is found that a somewhat higher recovery is made by 'slopping' in the end cells, that is, maintaining so high a level that the uprising air will throw over some of the pulp. This is permissible if the froth be again led to cleaning cells. The cleaned froth goes to a stock-tank from which it is distributed to the Dorr thickeners attached to the filter-plant. If the bubbles be tenacious and plentiful the froth will have to be broken down to prevent an overflow from the stock-tank. This may be done by jets of water or by the application of a vacuum; the latter method has not been found reliable in every instance. Possibly the application of heat may be of some use by causing an expansion of the enclosed air and a consequent rupture of the bubble-walls.

With respect to the volume of compressed air necessary, it is found that less air is required in the first than in the last cells. If too much air be used the bubbles burst and the froth-level sinks. To ascertain whether a low froth-level is due to the volume of air being used, the quickest method is to shut off the air for an instant and then to open the valve gradually until the best results are observed.

In summary, we may say that if the froth lies in heavily armored inert masses not enough oil or air is being employed, or the water-level is too low. If the froth consists of small bubbles that form in such profusion as to race down the sides of the cell, oil in great excess is being used. If the froth be light in color, the cause may be found in a slight excess of oil or air, a high water-level, or a pulp of too great a specific gravity. The extent to which the tailing from the roughers should be re-treated is entirely an economic question. The grade of concentrate will be determined by smelter requirements and transportation charges. The oil-mixture will depend upon the availability of the oils in the quantities demanded by the tonnage to be treated by flotation, upon the cost of the oils, and upon the recovery effected by their use.

The canvas on the air-pans should be changed when a large hole develops, for it is not possible to obtain a uniform or sufficient froth when the air rises through the pulp in strong local currents. In any plant treating large tonnages a certain amount of oversize is found at times in the flotation-feed. Where the air-pressure is between five and six pounds, this coarse material is prevented from lodging on the canvas. With air-pressures, however, of between two and three pounds the canvas has to be pounded occasionally. In any case it is best from time to time to wash the canvas, especially in the first cells, so as to remove any oversize or firmly adherent slime. Care should be taken, of course, that the stick used for pounding should neither have any sharp edges nor be wielded too lustily.

PAUL T. BRUHL.

McGill, Nevada, June 18.

WATER is absent from the petroleum-bearing Catskill sandstones of Pennsylvania and West Virginia, as shown by Frank Reeves in *Economic Geology*. These, and the Western Red-beds, are dry because of arid conditions prevailing when laid down.

THE MINE AND MILL OF THE MIAMI COPPER COMPANY

Miami, Arizona: The Discovery—I

By T. A. RICKARD

The traveler that arrives by night, reaching the end of his journey in the dark, has the satisfaction of seeing his objective without any of the detraction due to approach. I arrived at Globe on a belated train and motored to Miami, six miles, after sunset, so that there remained nothing to recall from that last lap in the journey except the hand of hospitality, a steep hillside road, and then a smooth and level pavement, groups of serried lights that betokened big reduction works, and the bracing night-air of an arid region. In the morning a faint sulphurous odor indicated the nearness of a smelter. Stepping out-doors I saw Miami—at least that better part of the settlement that surrounds the offices and dwellings of the Miami Copper Company. Behind the main buildings were smaller red-roofed white houses and behind them spread a familiar Arizonan landscape—bare hillsides dotted with chaparral and paloverde, spaced with the sparing infrequency of the desert garden. The surface of the ground along the hillside showed reddish to the west and gray to the east, a distinct line of demarcation suggesting the contact between different rocks, and, having read something about the local geology, I guessed that the gray indicated the conglomerate, which covers and limits the ore-bearing schist, itself reddened by the oxidation of its pyrite. Near-by, and south, the dump of the No. 4 shaft showed crumbly red rock that resembled the discard from a mill. This easy disintegration of the country-rock must be, said I to myself, a factor in the success of the 'slicing' system—which bore no reference to breakfast, now cheerfully imminent. From over the hill came the muffled roar of a big mill, and, turning half-round, eastward, I saw the source of the sulphurous smoke, the big black mass of the International smelter, silhouetted, like a super-dreadnaught ready for action, against the blue of the Apache mountains, dimmed by a morning mist and by the smoke of another smelter, that of the Old Do-

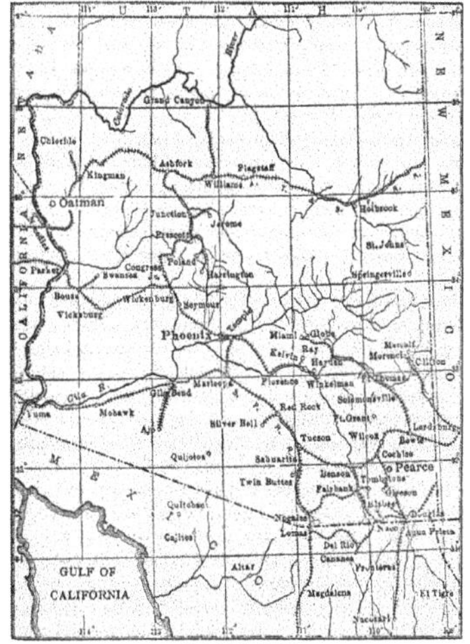

MAP OF ARIZONA

minion. The scene was eloquent of profitable activity and of human resourcefulness. Looking at the staff-quarters, the club-house, and other mitigations of life in this wilderness, I recognized how much progress had been made in the last three decades by mining companies in the effort to provide for the comfort of their employees, more particularly the technical staff. For examples of such provident forethought we owe something to the British companies formerly more numerous in the West than they are now. They may have gone too far, often building handsome houses for the management before they were assured of a persistently profitable enterprise, but they led the way in humane recognition of the vital necessity for taking proper care of the health and welfare of their people. With that kindly thought I proceeded up the hill, to breakfast.

Afterward, from the Tinkerville ridge, 200 yards west of No. 4 shaft, I saw something of the geology of the district and evidence also of the excavations being made below-ground. Underfoot were cracks that threatened an old boarding-house and that eventually must endanger the main shaft, the No. 4. These cracks come from subsidence above the stopes, such subsidence being a necessary factor in the success of the system of mining, as we shall see later. Thus the line of the subterranean orebodies is indicated by the surficial displacement due to the mining now in progress several hundred feet underground. The ore-belt follows a ridge of schist flanked on the south by granite. It is called the Schultze granite after a settler who owns a ranch on this terrain. Going north the schist loses its mineralized character. Knobs and combs of dark rock mark the remnants of comparatively recent dacitic lava-flows. On the Tinkerville ridge, where we are supposed to be standing observant, the rock underfoot is conglomerate—the Gila conglomerate—the fault-contact of which causes it to cap the ore-bearing schist at an angle of about 45°, and hinders prospecting south-eastward. This conglomerate contains fragments of schist, granite, and dacite, but the pieces are not wholly rounded; many of them are angular; the rock is the consolidation of miscellaneous detritus and might be labeled more accurately an 'agglomcrate.'

In the company of J. Parke Channing (Columbia '83), the vice-president of the Miami company, and B. Britton Gottsberger (Columbia '95), the resident manager, I was shown the signs of ore that had led to the exploratory work preceding the discovery of enormous orebodies, now yielding 17,000,000 lb. of copper per month from two adjoining mines, the Miami and the Inspiration. The story of this great mining adventure is as follows:

In November 1906 Mr. Channing was at Globe on a visit to the Old Dominion mine. On the third day of his visit, while at the Dominion hotel, he met Fred. C. Alsdorf, a graduated mining engineer (Ohio '92), whom he had know in Gilpin county, Colorado, in 1895. In the course of a friendly chat Mr. Alsdorf told Mr. Channing that he had been on a scouting expedition and that while engaged in this search he had run across a promising disseminated-copper deposit on ground about six miles west of Globe. He explained also that, having lived a number of years at Clifton, he knew the surficial signs of this type of deposit. Likewise Mr. Channing had his eyes trained to the same kind of indications, for he was familiar with the great 'porphyry-copper' mines at Bingham, he had examined the deposit of the Nevada Consolidated Copper Co. the year before, and he had brought to his examination of that mine the insight acquired by his visit to the Arizona Copper Co. at Clifton in 1897, at which time James Colquhoun was sending to the concentrator an ore that looked like burned lime with occasional specks of black chalcocite. Mr. Channing realized how Lake Superior mining methods and Montanan smelting practice could be applied to a large low-grade deposit of this character. In short, he was in a state of preparedness to appreciate the makings of a big copper enterprise based upon a large tonnage of disseminated chalcocite ore. A graduate of the Columbia School of Mines, a former assistant manager of the Calumet & Hecla, the consulting engineer and president of the Tennessee Copper Company, and the examiner of a number of famous mines in the West, Mr. Channing was ripe for the opportunity that Mr. Alsdorf put in his way.

They met on a Thursday. Mr. Alsdorf said that he had submitted a preliminary report by telegraph to some people at Colorado Springs, asking them to send their engineer to make an examination. On Saturday he told Mr. Channing that the people at Colorado Springs had telegraphed that they had no engineer available and that he was at liberty to dispose of the property as he saw fit. Thereupon he and Mr. Channing took horses and rode to what is now the Inspiration-Miami district.

Near the western edge of this mineralized area there was an adit—the Woodson tunnel—that had been driven in the course of extracting rich carbonate ore, and at its farther end this adit had penetrated low-grade material showing specks of chalcocite. Together these two engineers rode back over the ground, dismounting at intervals to examine the surface; at the eastern end of the mineralized ridge they came to the group of claims that Mr. Alsdorf had under bond and that now constitute the holdings of the Miami Copper Company. Mr. Channing examined the ground; then, recalling his experiences at Clifton, Bingham, and Ely, he told Mr. Alsdorf that he also considered the property worthy of exploration. The next day he interviewed the several claim-owners and had a talk with F. J. Elliott, an attorney at Globe, in whose name several of the options stood. These options called for cash payments such as were not justified by the condition of the property, which had only a few 10-ft. holes without a pound of ore visible. The meeting with the owners lasted until the small hours of the following morning, by which time Mr. Channing, representing the General Development Co., controlled by the Lewisohns, had taken an option to purchase the property for $250,000, of which $150,000 was to be paid in cash, and $100,000 in the stock of a

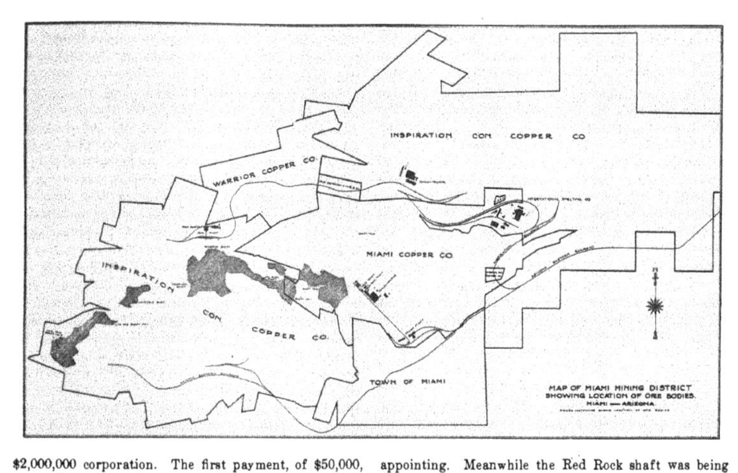

$2,000,000 corporation. The first payment, of $50,000, was to be made in six months. Mr. Channing returned to New York and sent Louis A. Wright (Michigan '04) to make an examination of the prospective mine. The report of this engineer confirmed the opinion of Messrs. Alsdorf and Channing that there existed a good chance of uncovering a large deposit of disseminated copper. In December 1906 two shafts were started under the supervision of Mr. Alsdorf: one on the Captain claim, where there was a showing of carbonate ore, and one on the Red Rock claim, where the cap was so thoroughly leached that only a rare spot of green could be detected. However, the cap on the Red Rock impressed Mr. Channing because it was well silicified and showed residual iron, which was not in such excess as on the Red Spring claim farther north. Silicification accompanies the solution and removal of the soluble portions of the rock; an excessive coloration suggests that the downward migration of the copper had not kept pace with the erosion, preventing concentration at a lower level.

The two shafts were sunk simultaneously by the aid of gasoline-hoists. At a depth of 100 ft. the Captain shaft passed out of copper-stained silicious schist into rock showing specks of chalcocite, and 70 ft. deeper it penetrated the granite, which at this point contains the primary sulphide, chalcopyrite. The granite was cut again in a drift extended north-westward at the 150-ft. level, whereas a drift in the opposite direction struck leached schist within 100 ft. The shaft was sunk to 200 ft., and at that level a drift was driven 70 ft. south-east in granite showing pyrite and chalcopyrite. Later exploration has shown that if this shaft had been placed 200 ft. farther west it would have cut the Captain ore-body at a depth of less than 75 ft. The result was disappointing. Meanwhile the Red Rock shaft was being sunk through leached cap-rock, and it was not until April, at a depth of 200 ft., that the shaft penetrated suddenly into the zone of secondary enrichment, the schist assaying 3% copper, as chalcocite. The shaft went through 50 ft. of this material; a level was started from the bottom; and at the end of May drifts had been extended 50 ft. each way to the four points of the compass. The showing was encouraging, so that in June the first payment was made. Mr. Channing went to Alaska, and on his return in October he found that the development had been so favorable that 1,000,000 tons of ore was assured. Just then came the panic of 1907, the local bank closed its doors on the day after the pay-roll checks had been issued, and Mr. Channing was under the necessity of paying the men with his personal checks upon a New York bank. Fortunately, the local bank got into good shape before long and there was no loss. By March 1908—fifteen months after the start—there was 2,000,000 tons of 3% ore assured; the shaft, then 710 ft. deep, had proved the vertical extent of the ore and most of the drifts—2518 ft. in aggregate length—were in good ground. Thereupon the Miami Copper Company was organized with a capital of 600,000 shares of $5 each, a total of $3,000,000. Inasmuch as this was a 50% increase on the capitalization specified under the option, the promoters (the General Development Co.) increased the number of shares going to the claim-owners, who then received $150,000 worth of stock. By this time about $100,000 had been spent in development, so that the total expenditure incurred in purchase and exploration was about $400,000. Of the 600,000 shares authorized, 300,000 were delivered to the General Development Co. for having found the mine and developed it to that stage. A block of 200,000 shares was sold at par in order to raise $1,000,000 of working capital. As the mine underwent development it became evident that the million would not suffice, so later the 100,000 shares remaining in the treasury were sold at $10 each, raising another million. In 1910 the capital was increased to 750,000 shares, and the extra 150,000 shares sold for $2,500,000; thus altogether there was raised for development and equipment about $4,500,000, in addition to the $400,000 paid for purchase and preliminary development. It is interesting to note how much money it took to convert a promising prospect into a productive mine. The original concentrating-mill, designed by H. Kenyon Burch, was intended to treat 2000 tons per day, but it was so well arranged that its capacity has been successively increased to 4000 tons, 5000 tons, and by the end of the present year it will be treating 6000 tons of ore and making a recovery fully up to the original expectation. At the time of my visit the Miami shares were quoted at a price that gave the mine a valuation of $30,000,000; it has paid already $11,190,000 in dividends, a sum of $1,700,000 has been put back into development and improvements, and on April 1, 1917, the company had quick assets amounting to $7,600,000, of which $2,000,000 was cash.

These facts tell a story so eloquent that no rhetorical confectionery is necessary. It shows how intelligent observation and scientific reasoning can be brought to bear even on that exploratory phase of mining which has seemed usually so haphazard. The sum of $400,000 was risked to ascertain whether there was enough ore to make a profitable mine; after that point was passed, the further development above-ground merely emphasized the bigness of the orebodies and the success of the enterprise. The story of the Miami also suggests that the successful exploitation of an orebody may involve operations on so big a scale as to require the expenditure of sums of money that make the original purchase of the bare ground seem quite cheap; it indicates that a mining claim without the intelligent use of capital is only second-rate scenery. The Miami story also shows the camaraderie that exists between engineers of the best type, and the good faith that is kept between them, as also between them and the community. It is worthy of note that Mr. Alsdorf was retained to superintend the prospecting, and that Mr. Elliott's services were engaged as legal adviser, thus emphasizing the confidence prevailing between the principal parties to the business.

A few supplementary notes may be recorded. Before Mr. Channing came to Globe in November 1906, as mentioned, he had been to the Ray district, making an inspection of that mineralized area. The prospect of developing a profitable copper mine seemed to him unattractive, so he dropped the project and went to Globe. The people at Colorado Springs that did not think it worth while to send an engineer to examine Mr. Alsdorf's option were Charles M. MacNeill and Spencer Penrose, who backed D. C. Jackling in the big Utah enterprise and became associated with him in other copper-mining ventures, among them the Ray, which Mr. Channing had 'turned down.' Thus Mr. Channing missed the Ray and Mr. Jackling missed the Miami, plus the Inspiration, but both have ample reason to be satisfied with their good luck. As to the evidence available to Mr. Channing, he showed me the Woodson tunnel, at the western or farther end of the mineralized area. This adit had penetrated into a zone of secondary enrichment, the white decomposed silicified schist showing specks and veinlets of chalcocite. In the Clipper tunnel, about 100 ft. up the slope, the oxidized ore appeared. Viewed from the standpoint of secondary enrichment such oxidation is not a good sign, for it suggests either that the leaching has not gone far enough, or that 'tertiary' action has oxidized the 'secondary' chalcocite. That is why the Red Rock was considered best; it had no green stains; on the contrary, the rock was silicified and contained not too much residual iron. The sinking of the Captain shaft close to the contact of the intrusive granite with the schist, and in rock that was stained green by malachite, was complimentary to the old ideas of copper geology, whereas the selection of the site for the second shaft, the Red Rock, was prompted by an understanding of that theory of secondary enrichment for which we are indebted to Emmons, Weed, and Winchell. The scientific prospector nowadays looks for silicification, not too much iron stain, and a mere trace of copper, just enough to indicate that originally the rock did contain copper, now leached and re-deposited below. The old-timer looked for green stains. On the dump of the Live Oak, one of the early prospects based upon oxidized ore, I saw pieces of comparatively unaltered schist exhibiting chalcocite alongside quartz stringers. Mr. Murphy[1] might have said of Miami that "the soins were moonificent; there was seriouscite for Mudd, iron for Wright, siliky for Channing, and the ox-hides of copper to indicate that it was not a tin mine, begorra." To use a gambling simile, Mr. Channing placed a chip on two numbers, and one of them was a winner. If, however, the Red Rock shaft had been 200 ft. farther west it would have penetrated some very poor stuff, so poor as to have caused an abandonment of the venture, the other shaft, the Captain, having penetrated low-grade primary mineral. To revert, the evidence of leaching at the surface, supplemented by the disclosure of chalcocite in the adit, sufficed to excite the logical expectation of finding, by sinking, orebodies of a size proportioned to the surficial area shown to have been affected by chemical solution and sub-aerial erosion. Thus advanced technical knowledge was applied to an old mining district.

Miami—the name is that of an Indian tribe formerly in Ohio and Indiana—was the scene of mining operations nearly 40 years ago. The early exploratory work done in this part of Arizona was stimulated by the richness of the Silver King mine, 19 miles south-west of Globe.

[1]See 'Mr. Murphy on Ore Deposits.' M. & S. P., August 22, 1908. Who would guess that this and other truly humorous articles were written by a person so deadly serious as the San Franciscan representative of Minerals Separation?

This discovery was made in 1874, at a time when the Apaches under Geronimo were still rampant in the region. The Silver King continued productive until 1884. Globe was selected as a distributing point on account of its good supply of water, obtained from the bed of Pinal creek. For a decade the surrounding district produced silver and gold, rather than copper. In 1883 there were 12 small mills in operation, extracting the precious metals,[2] but by 1887 the principal mines had been exhausted and were idle. In 1881 the Old Dominion company erected a small copper-furnace, about six miles nearly due west of Globe, where Miami now is, to treat oxidized copper ore found in the schist. It

THE MIAMI NO. 4 SHAFT

proved of no immediate consequence; so this pioneer smelter was moved to Globe, to treat the silicious copper ore produced by the Old Dominion mine. By 1886 there were six smelting-furnaces in the district, but they ceased work at the end of that year on account of the low price of copper, then quoted at 11 cents. In 1888 the Old Dominion company was re-organized, starting on a new and successful career, which has continued to this day. In 1895 the Black Warrior became prominent. The copper ore was found along a fault where it passed from the underlying schist into a remnant of dacite-tuff.[3] At that time the Black Copper, Keystone, and Live Oak were regarded as promising prospects, all of them working silicious copper ore in what is now the Miami belt. In the geological report by Mr. Ransome, dated 1903, it is written: "In the vicinity of Live Oak gulch the porphyritic facies of the Schultze granite has been much fissured and shattered, and is often conspicuously stained with carbonates and silicate of copper, while workable deposits of chrysocolla have been exploited on a small scale in the Live Oak and Keystone mines."

The green dumps of the Live Oak and Keystone claims, now part of the Inspiration property, serve to remind the visitor of the mining for oxidized copper that preceded the big-scale chalcocite developments of recent years, so little anticipated by the Government geologist. This oxidized ore was mainly the silicate, chrysocolla,[4] found in small veins traversing the granite.

At one time Phelps, Dodge & Co. controlled a large part of the district. In 1892 this firm purchased the United Globe mines, and shortly afterward their resident manager, Edward H. Cook, in their behalf, acquired a group of nearly a hundred claims covering the Miami area. In 1888 the Bigelow group had participated in the reorganization of the Old Dominion, but about 1903 individual members of the firm of Phelps, Dodge & Co., led by Charles Sumner Smith, obtained control of the Old Dominion also, although this property and the United Globe were operated separately until early in the current year. Phelps, Dodge & Co. held options in many parts of Arizona at the time when those claims west of Globe were cast aside and there was no reason to suspect that 2 or 3% copper ore could be exploited profitably, even if the existence of such ore under the oxidized cap had been surmised, which it was not. However, a little gossip is welcome once in a while.

Some day, when the big deposits of the Miami district have been worked out, we shall have the pleasure of reading a delicately worded obituary notice on this important district by some Government geologist. Meanwhile, the student should read Mr. Ransome's early report, in which the outlines of the local geology are laid down clearly.

The Pinal schist, in which the large orebodies are found, is of pre-Cambrian age. This crystalline basement was invaded and penetrated by the Schultze granite relatively soon after the pre-Cambrian sediments had been deposited, the result being extreme metamorphism and faulting. Emerging from the sea, these ancient rocks underwent long-continued erosion. Then followed a subsidence, during which fresh sediment was deposited on the former land-surface. The alteration of subsidence and elevation was repeated many times. Each period covered a vast space of time. To paraphrase the language of a play-bill at the theatre, the curtain is lowered to mark the passage of a million years. Marine organisms laid down the substance of limestone. Crustal unrest was shown by faulting and by intrusions of

[2]Globe Folio, U. S. G. S., 1904. F. L. Ransome.

[3]'The Geology of the Globe Copper District.' U. S. G. S. Prof. paper No. 12. F. L. Ransome. Page 156.

[4]Some of it is fine enough to be cut for jewelry, such as cuff-buttons. Locally it is called 'keystonite.'

diabase. Late in geologic time, at the beginning of the Tertiary period, there came extrusions of tuff and flows of dacitic lava, remnants of which are to be seen on the hills near Miami. Later still, in Quaternary time, the Gila conglomerate was accumulated so as to cover large parts of the district.

Turning to the economic geology: At the close of the Carboniferous period this district underwent a series of violent disturbances, culminating in fissuring, accompanied by intrusions of diabase. This was followed by cooling, settling, and then by more fracturing and fissuring; after which came further intrusions of magma. During one of the epochs of thermal activity, linked to igneous unrest, the primary copper-sulphide was precipitated in the granite and schist. Veins of copper ore cutting the diabase serve to indicate the relative age of that eruptive and the ore. The sulphide ore is older than the dacite. No sulphide ore has been found in dacite, but oxidized ore was stoped in dacitic tuff on the Montgomery claim of the Black Warrior group.[5]

Thus Mr. Ransome assigns the copper mineralization to late Mesozoic time. That probably was the period in which the process began, but it was not the period in which the orebodies were formed. It is not the original chalcopyrite distributed through the schist and granite that constitutes the mineralized rock valuable to man; the ore is a concentration caused by subsequent exchanges and migrations of metal, in operation during later geologic times. The fact that ore is found in the Tertiary tuff and in the detritus of a more recent day—a geologic yesterday—indicates that the ore-forming process has continued at work up to the very present. While the copper mineral is found in a pre-Cambrian schist, the copper concentrated to such degree as to be 'ore' is probably the product of Tertiary, or even Quaternary, activities. Hydro-thermal action, following eruptive action, caused copper-bearing solutions to rise toward the surface and precipitate their valuable contents as chalcopyrite, or copper pyrite, in the numerous fissures and shear-zones along which they found a passage. The diabase itself contains copper. The ore deposits now exploited represent enrichments made by the descent of copper solutions due to the leaching of the chalcopyrite in the overlying and eroded rock. The finding of chalcocite as a film covering the faces of fractures in chalcopyrite proves the secondary character of the chalcocite. The replacement of chalcopyrite by chalcocite, leaving a small residual core of the older mineral, is further evidence. As S. F. Emmons pointed out,[6] the occurrence of the chalcocite at the base of the oxidized zone, where carbonates and oxides give place to sulphides, is likewise significant. Indeed, the evidence obtained at Globe contributed to the formation of the theory of 'secondary enrichment' as put forward by Emmons. Mr. Ransome's report was written, under the guidance of Emmons, at the very time when this theory first saw the light. In the Old Dominion mine, the chalcocite was found first just below the 400-ft. level and it was traced as deeply as 900 ft., where it is associated with unoxidized pyrite. On the other hand, kernels of chalcocite have been found surrounded by malachite, as in the Buffalo mine, and Mr. Ransome suggests that the change has taken place "in situ." This oxidation of chalcocite would nowadays be called 'tertiary,' namely, the result of alteration on a 'secondary' mineral, itself the product of the leaching of 'primary' chalcopyrite at a higher level. Whether the bulk of the oxidized ore is "indigenous," as Mr. Ransome says, that is, occupying the place of the original chalcopyrite, is a debatable point. It is likely that he would revise his opinion in the light of later evidence, particularly at Miami, where the concentration of chalcocite has taken place over a wide area and at a level well below the surficial zone of oxidation. In his brief refcrenees to this part of the district, Mr. Ransome says:

"In the case of the chrysocolla deposits of the Black Warrior, Geneva, and Black Copper mines, the ores occur as replacements of dacitic tuff, or possibly dacite in the last-named property, and their structure is such as to indicate strongly that they have been deposited directly in the tuff as the hydrous silicate of copper, and do not represent the alteration in place of former sulphides." He points to the "lack of iron oxide" as confirmation, as also "the fact that the ore is practically a single mineral rather than a mixture such as results from the oxidation of more or less pyritic ores." He concludes therefore that the chrysocolla ore is "exotic." As a hypothesis to explain such veins of silicate copper, he suggests the lateral flow of solutions that leached the copper from beds of tuff and from the surface of shattered schist upon which this tuff was deposited, the carbonate being changed to silicate "by the abundant silica present in the glassy tuff." This explanation seems reasonable. In regard to the chrysocolla of the Keystone and other veins in granite, Mr. Ransome was unable to speak confidently, for lack of evidence. He makes a remark that is highly interesting in the light of later developments: "The occurrence of some unoxidized pyrite in a stringer near the mouth of the lower tunnel [on the Keystone claim] indicates that if the chrysocolla is to give place to sulphides in depth, the point of change should not be far off. It is hoped that the mine will be successfully exploited to a sufficient depth to thoroughly expose this interesting deposit." As the children say in the familiar game, he was "very warm" just then; he was close to a discovery of more than academic interest. An observer coming 14 years later can appreciate the scientific humor of the position. There was plenty of iron oxide to be seen in the wake of surficial oxidation in this locality and there were wide-spread signs of chemical decomposition above the big bodies of chalcocitized rock now uncovered deeply underground. However, most of us can write last year's almanac better than next year's, and it is a safe surmise that the U. S. G. S. report on Globe furnished part of the suggestion that led the mining engineer of a later day to sink shafts in search of treasure.

[5]Ransome, on whose description my geologic sketch is based.
[6]Trans. A. I. M. E., Vol. XXX, page 192.

The Russian Crisis

By CHARLES JANIN

At the Editor's request I have prepared these notes describing conditions in Russia as I saw them during my recent visit to Petrograd by way of Vladivostok and Siberia.

REVOLUTION. The active cause of this was the lack of food in Petrograd. While there was plenty of food in outlying districts, the failure of those in charge of transportation to see that sufficient supplies were sent to the capital prompted riots and fanned to a flame the smoldering fires of discontent among the people. The revolution would have occurred later in any event: it was simply advanced by a few months, and when it did take place it was a spontaneous affair entirely confined to the working people, who were without definite aims other than to obtain food. The cossacks, who had been the main strength of the Government in putting down previous riots, were in sympathy with the people, for their stomachs too were empty. Many of the soldiers had been but recently recruited from the country, and were not in sympathy with the Government; on the other hand the police were 'loyal' to the Government. They had stored large supplies of food at their stations and could look with full belly on the growing discontent of the masses. The police were unable to cope with the situation. The crowd, emboldened by the friendly attitude of the soldiers, jostled the police on all sides. The cossacks would ride carefully through the insurgents, who would greet them with cheers. Voices could be heard saying "Our little brothers won't hurt us this time." Just when the real fighting began it is hard to say. A cossack officer killed a policeman who was attacking a peasant; a group of soldiers turned on the police that had been conferring with them; workmen attacked isolated police-officers in other places. In a short time fighting became general. Then soldiers shot the officers that were leading them to put down the trouble and joined the mob; other soldiers marching to join the revolutionists were halted and fired upon by patrols from their own regiment before the latter knew of the general movement. In a short time it was all over, except for fights with the police. These held their stations and the tops of buildings, where they had posted machine-guns in what they thought advantageous positions, neglecting to make sure they were really such, so that when the trouble began they found that they could not deflect their guns sufficiently to sweep the streets. A house-to-house hunt for police and arms ensued. At first the police were killed wherever found; some were brutally murdered by the crowd. A number of regimental officers became hysterical at the threatening mobs and foolishly fired their revolvers at them. They were killed. During the first days of the revolution a party of soldiers coming down the Nevsky Prospect met other soldiers marching on a side street. Cheers for New Russia were exchanged and peace prevailed. Suddenly some enthusiast fired a gun. Intense excitement seized the whole party, who believed they were being attacked, and several hundred shots were fired in all directions at imaginary enemies. One curious American peeped

STREET-SCENE IN VLADIVOSTOK

out of his hotel-window to investigate the disturbance, but quickly withdrew when a bullet crashed through the glass above his head. Gradually order was restored by a provisional government; regimental officers and policemen were simply taken prisoners. The policing of the city was taken over by student organizations. For a few days after the revolution not an officer was to be seen on the streets. Then some of the old retired fellows appeared, and gradually the others were allowed freedom, but they had none of their former proud and arrogant bearing. Soldiers no longer saluted them when passing. The first ruling by the soldiers' deputies was, I believe, to abolish the salute. One rarely sees a soldier salute nowadays.

THE PROVISIONAL GOVERNMENT was formed hastily under Rodzianko and Miliukoff. These men were not leaders in the revolutionary outbreak and at first took no active part. They were accepted simply because the nation was without a head and the people turned to those who had offered strong opposition to the Czar's régime. No sooner was the Provisional Government appointed than the Radicals began to find fault with them and to agitate against them. Muliukoff was the first to feel the disapproval of the masses because he announced that his Government would adhere to the treaties that the Czar's ministry had made with the Allies. This did not meet with popular approval. Demonstrations occurred in the streets against the new Government and against Muliukoff, against continuing the War, against the Allies; unfriendly crowds even gathered in front of the American embassy and American consulate. These were dispersed by 'loyal' regiments. The workmen had their own leaders, Socialists, such as the man Lenine, who was subsidized by the Germans. These men exercised considerable influence and won a numerous following: they are the main cause of the later disturbances. The Provisional Government should have put a quietus on Lenine's efforts at the start, but the members of this new administration did not seem to grasp the significance of this agitation, or else the situation grew beyond them and they were powerless to cope with it when they realized what was needed. They ruled only by sufferance of the Council of Workmen's and Soldiers' Deputies. Strong measures taken at the start would have saved much trouble, not only then but in time to come. The principles of the republic, however, forbade interference with free speech. The killing of Lenine was openly advocated on the streets, but his party defended him and threatened bloody reprisals if any violence was attempted. At times it seemed as though civil war or anarchy was inevitable, and that a complete collapse of the Army and Navy could not be averted. The workmen were growing considerably more reactionary, and clashes occurred in the streets between armed bands of workmen and soldiers. When shots were fired, we could see the crowds rushing this way and that, some to get away from danger, others to see what it was all about. I did not belong to the latter.

THE FUTURE of the country depends upon the attitude taken by the soldiers and the peasants. The greatest power at present in Russia is the Council of Workmen's and Soldiers' Deputies. The Army has the actual power, however, and needs only a great leader to control the whole situation. But this leader is not forthcoming. At first it was thought Brusiloff would be the man, but he, although a great tactician, evidently lacked the will or ambition to be a dictator, or perhaps conditions at the front were not favorable for such action on his part: instead, he resigned his position, and a leader does not gain prestige by such procedure. Kerensky was a great surprise, a 'dark horse,' so to speak; in the first burst of enthusiasm he had nearly the whole country with him. Then, partly through German influence, and partly through the anarchists—and there are many of this following in Russia—he lost some of his strength, which now he appears to have regained.

LABOR. The situation before the War was becoming serious, that is, there was a shortage of labor at the mines, and the importation of Mongolians and Koreans was advocated by engineers. Now the shortage is more acute. Austrian prisoners have been used with the permission of the former government to some extent; at one mine known to me 4000 of them are employed. These men are well treated and fairly paid, so that they are much more contented with their lot than those not so occupied. On the railroad women are generally employed for section-work, and one rarely sees men in the fields, unless it be soldiers on leave or old men. The women and children plow the land. Wages throughout the country are much higher than formerly, and a further increase may be expected. Labor in general is asking for shorter hours, higher pay, and participation in profit. In places the workmen have made extravagant and outrageous demands. The 8-hour days is general now except where the workman has been strong enough to force a 6-hour day. At many properties the managers have been displaced, mostly without violence, and a committee elected from among the men to take charge. At some mining properties not only the workmen but the staff are demanding increased pay and shorter hours, and, in addition, arrears for the period since the War began. At other mines the management has deemed it wise to turn over all the profit to the workmen in order to preserve their plant, as the men have threatened to burn the buildings if their demands are not granted. In some instances it is reported that even this is not satisfactory to the workmen; they demand a portion of the invested capital. The cost of labor in general may be figured now at nearly double what it was before the Revolution and this even in districts where moderation prevails. With the increase in pay the efficiency of the individual has greatly diminished. In some cases industrial plants are closed down, as it is impossible to operate them at a profit. A complete collapse of industry was threatened shortly after the Revolution by the exorbitant demands of labor. The workmen were in command of the situation and all business had to be submitted to their committees for approval.

SOLDIERS. While the attitude of the soldiers for the most part was more favorable to the welfare of the new republic, there have been exceptions. Among the soldiers there were agitators similar to those in the workmen's parties, and some companies of soldiers paraded the streets with banners advocating peace, in opposition to the policy of the Provisional Government. There was a general lack of discipline among the different regiments; a number of desertions occurred from the armies at the front and from troops on their way to the front. The streets in Petrograd were crowded with soldiers, some marching to the front singing the Marseillaise, others returning from the trenches without leave. Many sold what clothes they could spare to the Jews. These deserters were on their way home to "get their share" of the land. Those to whom I talked said, "While we are fighting at the front, our brothers are getting all the best land." A few would add, "If it is necessary we will fight after we get the land, but why should we fight. we do not want indemnities or land from Germany." At the railway-stations these returning soldiers would crowd into the cars, occasionally putting civilian passengers off, and throwing their baggage out of the windows after them. The newspapers of April 14 contained the following 'appeal':

"The Government has issued an appeal to the soldiers asking them to stop overcrowding trains, not to expel passengers provided with tickets, and not to present demands to the railway personnel. Many passenger carriages have been damaged owing to the overcrowding that has been going on during the past four weeks."

At the front the soldiers fraternized with the opposing troops, who had been ordered to encourage an exchange of courtesies. It was said that the Russian, in his great enthusiasm over the Revolution, believed that the millennium had arrived and that peace and goodwill would be universal. Whatever the idea, this fraternizing was encouraged by the Germans and it aided in the general disorganization and demoralization of the Russian troops. Officers whom I met despaired of ever seeing discipline restored; many resigned their commissions. Some of them told me that the situation was unbearable; one, a cossack colonel, just convalescent from his wounds, said he had returned to his regiment at the front, to the men with whom he had grown up, whom he commanded and led in battle, and that they would not recognize his authority. He said, "I want to go to America." The American and English embassies were besieged by officers of like mind, some wishing to volunteer as privates in the English army.

On the other hand, regiments of soldiers under non-commissioned officers, or committees elected by the men, retained their former officers and were instrumental in keeping order in the cities. It was regiments such as these that upheld the Provisional Government during the labor riots in Petrograd and really prevented anarchy. Outside of Petrograd order was much better maintained. The soldiers elected committees to take charge of affairs, and these committees in turn selected soldiers for policing the streets. On my trip through Siberia whenever soldiers tried to force their way into the cars, members of these committees would have the courage to order them off, and as the habit to obey was still instinctive, they got off. Some of these soldiers, after being put off the train, relieved their feelings by making faces at the passengers, but as they did not go beyond such facial exercises, we did not mind.

TRAVELING in Russia at present, of course, is far from pleasant. Trains are crowded to the utmost capacity with passengers and in addition soldiers endeavor to enter at nearly every station; sometimes the corridors of the carriages are filled with them, but on other trains the soldiers simply take possession of the vestibules. This condition obtained on the train on which I traveled, though personally I suffered no inconvenience from it. Most of the soldiers and sailors with whom I talked, through my interpreter, proved to be pleasant and agreeable. Food was scanty. Anyone who is obliged to travel in Siberia should provide himself with an ample lunch-basket, which may or may not be confiscated at the border.

FOOD. Transportation improved during the first few weeks following the revolution, but food was becoming scarcer when I left Petrograd. A fair meal would cost from 7 to 10 rubles (a ruble being equal to 50 cents nominally, now 21 cents only) and this might or might not include bread; for three days or more no bread was purchasable. Those in the bread-line, after long waits, sometimes would receive a little less than a pound of black bread. This allowance was reduced to 12 ounces

A GROUP OF AUSTRIAN PRISONERS

daily per individual before I left. White bread was obtained with great difficulty at any time. The black bread is often unfit to eat; a mixed or gray bread is the kind generally furnished at the best restaurants, that is, when flour is plentiful. Those with influence or 'understanding' were able to get white flour; others purchased it through soldiers' wives. The soldiers seized freight-shipments coming into the city, and some, with

AUSTRIAN PRISONERS, A RUSSIAN GUARD ON EXTREME RIGHT

an eye to business, deputed their wives to sell the flour and sugar. In the high-class stores on the Nevsky 'luxuries' were still offered; prunes at $1 per pound; canned goods at 8 to 10 times the American prices; chocolate at $2 per pound (the further making of chocolate was prohibited in May); apples at a ruble apiece. Clothes were at fancy prices, ready-made suits cost from 250 to 300 rubles; shirts of the $1.50 kind sold for 20 rubles, undershirts, socks, handkerchiefs, and neckties in the same proportion, and shoes—well, there was a riot on the Nevsky when a dealer announced the arrival of a shipment at 80 rubles per pair, and crowds besieged the place as though it were a bargain-day. The shelves of most of the stores are bare. What a chance for a salesman in any line, and how they would be welcomed in Petrograd, if they could only guarantee deliveries! Delivery is dependent, of course, upon the transportation facilities.

RAILROADS. There is a shortage of locomotives, but the present rolling-stock of the trans-Siberian railroad is not used to one-third of its possible efficiency. Long lines of empty cars are seen on switches and on distant side-tracks. Freight-cars are sent loaded from Vladivostok, never to return. For those sent to the front some excuse may be made, but under energetic management more effective service could be obtained from the present stock. Lack of repair-shops and of mechanics is another feature. There is, strange to say, no assembly-shop for locomotives at Vladivostok, and any purchased in America must be first hauled on freight-cars to Irkutsk to be assembled. A large plant for assembling cars was built at Vladivostok. Thousands of cars were purchased in America, most of them of the small 20-ton type, but a number of large 60-ton cars were built by an American firm. I was told that this type was not popular, partly on account of the necessity for strengthening bridges and culverts, and partly because the workmen objected to filling so far behind the doors. Long before an American commission was mentioned, the Imperial Russian Government engaged an assistant traffic-manager of the Canadian Pacific Railway. This expert completed his investigations some weeks after the Revolution, and was reported to be much disgusted with the inefficiency shown; his report is in the hands of the American Railroad Commission. To illustrate the improvement possible under efficient management: An officer associated with the English transport service was engaged by the Czar's government to handle the Archangel muddle. After repeated discouragement in attempts to secure co-operation from officials, he was, through the effort of the English ambassador, given full control, and cleared the port in three months. Vladivostok has greater congestion, is much farther from terminals, and has much the same difficulties to face. The Ministry of Transportation in April expected to clear the congestion in eight months, but from personal observation I am in a position to guess that this is an optimistic estimate. At the time of my visit there were between 400,000 and 500,000 tons of freight at Vladivostok awaiting trans-Siberian shipment. Just what the American Railroad Commission can do, I do not know. It is a situation that requires infinite tact. If the workmen refuse to co-operate with their Russian managers it is a question whether they will welcome American methods and control. The Provisional Government stated that they welcomed the co-operation of America in new mining enterprises of every kind, and promised assistance in so far as it was in their power to render any. We were told that laws more favorable to mining activities were being formulated as rapidly as possible and every encouragement would be afforded to capitalists willing to develop gold, zinc, lead, copper, and iron properties, with a view to bringing them rapidly to a productive stage.

MINING. At this time I do not know just what bonus has been fixed by the Government on the baser metals, but a ruling on prices to be paid on gold was made just before I left Petrograd. This was increased to 12 rubles per zolotnik as against 7.70 rubles before the Revolution and 5.50 in normal times. A zolotnik in gold is worth about $2.83, and it can be seen therefore that the 12 rubles offsets the depreciation in Russian currency. Matters of titles, of taxation, of government ownership, and of leasing, and the treatment that will be given to existing companies holding large grants from the former government, are of vital concern to those interested in Russian mining. All the Crown lands and those belonging to members of the Czar's family have been declared confiscated for the public good. What policy will be applied to the leasing of this public land is not known; it is unreasonable to expect a new ministry struggling with matters more vital to devote much attention to details of this kind. I was told that a new director or perhaps a commission would be appointed to study the matter. Whatever the outcome of the inquiry, it will be limited to recommendations only. The policy to be adopted in the administration of public lands must be threshed out in open session of the Supreme Council; recommendations of ministers sufficiently socialistic will claim attention and be largely supported. The members of the ministry were favorably disposed toward securing foreign co-operation, and had liberal ideas in this direction. They could, however, only indulge in generalities, for they had no real power. Since then eight of the original ministers have been removed from office. There is a strong tendency to limit profits to a fixed percentage of the capital employed. This percentage varies with the individual; some, with no experience in such matters, contend that 6% is sufficient; others are more liberal. The companies now operating are much concerned over this phase, but hope that at least 15% profit will be permitted, with allowance for amortization of capital. The existing taxes, including war emergencies, now amount to 50% of the profit as shown on the books of one mining company. These are matters that must be studied in the interests of the prospective investor. Certainly, if profits are to be limited to 10% or thereabouts American capital will not be attracted to Russia.

BUSINESS. At the present time the situation is uncertain and unattractive to a conservative business man. Matters of government can only be worked out in time, and just now the vital need of the new republic is to unite against the Germans. This and the establishment of a stable government is the first consideration. If the Russians cannot do that unitedly or with sufficient force, it will not be a country to attract Americans. It will be Mexico over again.

The present financial condition is unfavorable; the ruble has dropped to 40% of its normal value. If rioting in Petrograd and other places continues, I shall expect a further decline. If, on the other hand, a stable government is formed, the ruble will rise. A number of years must elapse before Russian finances are strengthened to any great extent and care taken of external obligations, and it is only then that the ruble can go anywhere near par. There has been a vast increase in paper money during the War, for the Revolution was not allowed to interfere with the printing press. I was informed through reliable sources that 1,000,000 rubles per hour was the amount gaily clicked off in notes. Such an increase in paper should be offset by a corresponding increase in gold reserves, and Russia's gold holdings do not show any increase in this period. We shall have to finance Russia if we are to do business there.

A GROUP OF SOLDIERS

SOLDIERS CROWDING ON TRAIN

There seems to be plenty of money in Petrograd among individuals. One hears of the enormous profits made through war contracts, and I was told that the directors of one of the banks each had 1,000,000 rubles extra dividends for the year. Porters and waiters seem to have an abundance of money, and fashionable *isvoschiks* in front of the hotels won't listen to less than a 5-ruble note for a short drive, where 30 kopecks (100 kopecks = 1 ruble) would have been the ruling price before the War. Paper money is universally used in Russia, even notes of one, two, three, and five kopecks are printed, though I never saw anything that could be bought with the smaller denominations, and one often saw these little bills blowing about the streets. They were probably thrown away by Americans who felt the time wasted spent in collecting enough to buy a newspaper. Stamps were used as money of 10, 15, and 20 kopeck denominations, but the ruble bill is quite a work of art. In Eastern Siberia, copper and silver coins made in Japan are used for smaller coins, and paper from 50 kopecks and over. When conditions become more nearly normal there will be splendid opportunities for American capital, engineers, and business men. A number of American firms have had representatives in Russia and the volume of business done in some instances has reached startling figures. On account of the conditions obtaining after the Revolution many Americans left Russia feeling that all business outside of that directly undertaken with the Government was impossible. The difficulty of obtaining transportation permits and of exchange, or as they call it in Russia "getting dollars," all militate against successful and satisfactory business. I believe that this is but a temporary condition, and that when resumption of business is permitted, an influx of American goods of every description is to be expected. American firms will do well to make a study of the conditions and needs of Russia as soon as possible, so as to place orders promptly to the best advantage when order is restored.

In order to work to the best advantage in Russia one should have a working knowledge of the language, and the language is very difficult to learn. It can be learned, however, with patience and application, as is proved by the success of the American and English engineers now in the country.

GERMAN motors captured during the War have shown superiority over those made in England and France, according to official statements from Washington. Exact duplicates made in shops of the Allies have not equalled them in endurance, and the difference is ascribed to characteristics of the steel that the French and English have not yet learned to reproduce. The U. S. Bureau of Standards is also working on the problem. The superiority is thought to be due to the method of retaining the so-called combined carbon in that form by methods that are chemical rather than mechanical or purely physical.

Flotation—The Butte & Superior Case—II

Brief for Defendant. The So-Called Critical Point

The contention that in reckoning the amount of oil which is referred to in the claims, account only should be taken of the oil which is attached to the metalliferous content of the concentrate, is not based on reason, or on anything in the specifications or claims. It is contradicted by everything therein. Claim 1, for example (which is typical), describes the process as consisting in "mixing the powdered ore with water, adding a small proportion of oily liquid having a preferential affinity for metalliferous matter (amounting to a fraction of 1% on the ore), agitating the mixture until the oil-coated mineral matter forms into a froth," etc. Clearly what is described and claimed here is the use of a fraction of 1% of oil on the ore *in the mixture which is to be agitated to produce the froth.* Neither the specifications nor the claims make any reference to the amount of oil which attaches itself to the metalliferous content of the concentrate. The only reference in the specifications to the oil attached to the froth is the suggestion that the froth may be "treated with a dilute solution of acetic alkali, which removes the oleic acid in the form of soap."

Mr. Kenyon, in his oral argument, put this contention in another way; but in a way which amounts to precisely the same thing. After saying that we must consider as oil in the process only the oil which is attached to the metalliferous content of the concentrate, he said that from the total amount of oil used must be substracted all the oil which goes off with the tailing; all the oil which is absorbed in the gangue of the concentrate; all the oil which is dissolved in water, etc. This is only saying in a roundabout way that the only oil which is to be counted as the "fraction of 1%" referred to in the claims is the oil which is attached to the metalliferous content of the concentrate, because when all these things are subtracted there remains only the oil which is attached to the metalliferous content of the concentrate. We have already shown that the specification and claims directly contradict this contention, because the fraction of 1% of oil which is mentioned in them is the oil which forms part of the mixture which is to be agitated to produce the froth. It is not the oil which is attached to the metalliferous content of the concentrate when the process is completed.

If the fraction of 1% of oil in the ore referred to in the claims is limited to the oil which is attached to the concentrate, then defendant's practice is still further away from the proportions specified in the claims, for Defendant's Exhibit 158 shows that the percentage of oil in the concentrate, when it uses more than 20 lb. of oil per ton of ore, is as much as 1.86% to 2.09%.

The contention that only a fraction of 1% of the oil used by defendant is effective, the balance being inert and not to be reckoned as oil in the process is contradicted by the proofs.

Defendant has used various mixtures of oils. The mixture used by it during the joint run on April 29, 1917, which may be taken as typical, was, in round numbers, composed of 24% pine-oil, 65% fuel-oil, which is a petroleum, and 11% kerosene. Since fuel-oil and kerosene differ only in specific gravity, both being petroleum, we may simplify the formula by saying that the oil was composed of 24% pine-oil and 76% petroleum, or substantially one part vegetable oil to three parts petroleum. The amount of mixture used was 26 lb. per short ton of ore, that is, 1.3% of oil on the ore.

On behalf of complainants it is contended that petroleum is inert in the process, and should be neglected in determining the percentage of oil on the ore, within the meaning of the claims of the patent in suit. The suggestion is that petroleum was used only as a diluent to increase the bulk of oil without taking any active part in the process of the patent. This contention raises the question as to what is the "oily liquid" referred to in the claims. Does it include, or does it exclude, *mineral oils,* as the petroleums?

In the specifications the "oily liquid" of the claims is defined as "oils, fatty acids, or *other substances which have a preferential affinity for metalliferous matter over gangue.*" The specification then refers to the Cattermole patent No. 777,273 as describing the use of the same "oily substances" in larger proportions. Turning to that patent we find it states:

"The 'oil' used may be animal, vegetable, or *mineral oil* or *mixtures of these* or such coal or wood tar products or other substances which exercise, like oils, a *preferential physical affinity for metallic mineral matter as distinguished from gangue.*"

Further on the specifications of the patent in suit say:

"The proportion of mineral which floats in the form of froth varies considerably with different ores and *with different oily substances,* and before utilizing the facts above mentioned in the concentration of any particular ore a simple preliminary test is necessary to determine *which oily substance* yields the proportion of froth or scum desired."

When we come to the claims we find that they define the oil as "*an oily liquid having a preferential affinity for metalliferous matter.*" Hence we see that any liquid having, like oil, a *preferential affinity for metalliferous matter over gangue* is included within the term "oily liquid" in the claim. Since there is no dispute concerning the fact that petroleums have such *preferential affinity,* there can be no question but that they are included within the term "oily liquid" contained in the claims.

The fact is, the practice of using a mixture of vegetable oil and petroleum is not peculiar to defendant, nor is it peculiar to a process in which oil is used in quantity above 1% on the ore. On the other hand, it is a practice which is in common use by those who use quantities below one-half of 1% on the ore, and who are *operating as licensees under the patent in suit*. Thus complainants' licensee, the Braden Copper Company, uses a mixture of 1 lb. American wood-tar oil to 3 lb. of Texas oil per ton of ore. At that place, therefore, where one-fifth of 1% of oil on ore is used, the mixture of oils used is precisely like that used by defendant, to wit, one part vegetable oil to three parts petroleum. Again, complainants' licensees at the Consolidated Arizona mine use between 2 and 3 lb. of oil per ton of ore, about one-half of it being Carolina turpentine, and the other half fuel-oil and stove-oil, both of which are petroleum. In this place, therefore, where about one-tenth of 1% of oil on the ore is used, the mixture of oils used is one part vegetable oil and one part petroleum.

Furthermore, as hereinafter pointed out, the non-licensees in this country, before they adopted the use of oil in quantity above 1% on the ore, used a mixture containing petroleum as one of its components. This was true of the Utah Copper Company at its Magna plant, and also at its Arthur plant; it was true of the Chino Copper Company; and it was true of the Ray Consolidated Company.

So, as we have said, the practice of defendant in using a mixture of vegetable oil and petroleum is not peculiar to it, or to the use of quantities of oil above 1% of the weight of ore.

That petroleum is not, as contended on behalf of complainants, inert in the process, is clearly demonstrated by the mill operations at the Arthur plant of the Utah Copper Company, records of which appear in Defendant's Exhibit 31. In one run the oil used was 20.33 lb. per ton, it being a mixture composed of 89% of what complainants' witnesses call inactive oils, that is, petroleum (30% Jones fuel-oil and 59% smelter fuel-oil) and 11% of what they call active oils (10% American creosote and 1% Yaryan pine-oil). In this run the extraction was 98.4%, and the tailing carried 0.076% copper. The actual amount of so-called inactive oil used per ton was therefore (being 89% of 20.33 lb.) 18.1 lb.; and the actual amount of so-called active oil used per ton was therefore (being 11% of 20.33 lb.) 2.23 lb. In another run substantially the same amount of so-called inactive oil was used *alone* (17.84 lb. of a mixture of the same petroleums, namely, smelter-fuel and Jones fuel, in the same proportions). In this case the extraction was 95.06% and the tailing carried 0.306% copper. In another run substantially the same amount of so-called active oil was used *alone* (1.97 lb. of a mixture of the same so-called active oils, that is, American creosote and Yaryan pine, in the same proportions). In this case the extraction was 85.72%, and the tailing carried 0.81% copper. These determinations are not contradicted or questioned, and they prove that petroleum-oil used in this process is by no means inactive or inert. It is, indeed, quite as active and quite as efficient in producing the desired results as is the so-called active oil. Indeed, it will be observed that the petroleum when used alone gave higher extraction than did the so-called active oils when used alone. The highest extraction, however, was attained when they were used together in a mixture as defendant uses them.

So it is proved that it is the common practice of those licensed under the patent in suit, and others using less than one-half of 1% of oil on the ore, to use petroleum mixed with other oils in the practice of the process, and it is proved that the petroleum used is active as an oil—not inactive like milk or sawdust, as Mr. Kenyon said in argument—in effecting the concentration which is the purpose of the process.

Complainants' witnesses intimate that there is something peculiar about defendant's ore—that it contains an undefined amount of an undefined material, which Greininger called "gangue-slime" and which Chapman called "clay-gangue"—which makes it possible for defendant to use above 1% of oil—the inference sought to be deduced from this being that but for the presence of the so-called "gangue-slime" it would be impossible to practice the process with more oil than given in the examples of the patent, namely, under one-half of 1%. This testimony is mere speculation and inference, and *being adduced in rebuttal* it could not be replied to directly. It has been, however, sufficiently replied to indirectly by the proofs in the record showing that oil in excess of 1% is being regularly used at other mills than that of the defendant, where there is no suggestion that the ore contains any "gangue-slime" (whatever that may mean). The use of oil in excess of 1% on the ore has been, since the decision of the Supreme Court in the Hyde case, regularly used at the Magna mill of the Utah Copper Company, as testified to by Conrads; at the Arthur plant of the same company, as testified to by T. A. Janney; by the Chino Copper Company, as testified to by Wicks, and by the Ray Consolidated Company, as testified to by Engelman. In each case the mill-records of the plants, both before and after the use of oil above 1% on the ore was adopted as the regular mill-practice, were produced. The facts established by this testimony, in brief, are these:

At the Magna plant of the Utah Copper Company the change from below 1% to above 1% of oil on the ore was made on December 25, 1916. Before the change was made, the smallest quantity of oil used was in the month of March, 1915, when the average was 1.25 lb. per ton; and the largest quantity was used in the month of April, 1916, when the average was 5.37 lb. per ton. Before the change, a mixture of various oils, including petroleum, was used. In August, 1915, they used a mixture of Barrett creosote, Barrett No. 4, Jones oil, pine-oil, and an oil called No. 642, which is a reconstructed pine-oil. In August, 1916, they used a mixture of Jones oil, creosote, and waste oil. In December, 1916, before the change was made, they used a mixture of Jones oil and creosote. After the change was made they used a mixture of Jones oil and Yaryan pine-oil. Defendant's Exhibits 35 and 36 give a complete statement of the mill-operation before and after the adoption of the use of larger amounts of oil than 1% on the ore. Exhibit 35, which gives averages for the entire period before the adoption of 1% of oil, compared with Exhibit 36, which gives averages for the entire period after the adoption of 1% of oil, show that the extraction before the change was 97.461% and after the change was 98.161%. They show that the copper in the concentrate before the change was 30.294%, and after the change it was 28.458%.

Defendant's Exhibit 38 gives a record of experiments made with varying amounts of a given mixture of oil while other conditions were kept constant.

At the Arthur plant of the Utah Copper Company the change from below 1% to above 1% of oil on the ore was made December 21, 1916. A tabulation of the results before and after the change is contained in Defendant's Exhibit 30. Before the change, an average of 3.76 lb. of oil per ton of ore was used, and after the change an average of 21.98 lb. of oil per ton was used. Before the change, the tailings averaged 0.361% of copper; after the change, they averaged 0.238% of copper. Before the change, the recovery was 96.57%; after the change, it was 96.60%. Both before and after the change they used mixtures containing petroleum oil as one of their ingredients.

At this plant a series of 13 tests, which were full-mill operations, were made using in all the 13 tests a mixture which was made of 89% petroleum (smelter fuel-oil and Jones oil), 10% creosote, and 1% Yaryan pine-oil. In these tests the quantity of mixture used varied from 6.87 to 96.46 lb. per ton of ore. As the amount of oil was increased from the lower limits, the recovery increased until 25.50 lb. of oil per ton were used. Using oil in larger quantities than 25.50 lb. per ton of ore, and up to 96.46 lb. per ton of ore, still gave excellent results (96.39% recovery), although the tailings carried a little more copper, to wit, 2.272%.

At the mill of the Chino Copper Company the permanent change from below 1% to above 1% of oil on the ore was made December 21, 1916, although they had for three days in November, 1916, used as much as 237 lb. of oil per ton of ore. Before the change, they used a mixture of creosote (Barrett No. 4) and petroleum (Jones oil), and since the change they have been using the same mixture, but with the proportions changed from 90% Barrett and 10% Jones oil to 10% Barrett and 90% Jones oil. The tailing-loss of copper averaged, before the change, 48%; and after the change, 32%. The average recovery before the change was 95.528%; and after the change, 96.936%. After the change the average amount of oil used was 22.18 lb. per ton of ore.

In direct contradiction of the theory of complainants' witnesses that petroleum is an inactive oil and plays no part in the production of foam, Wicks describes what happened one day in the mill in the regular course of milling operations when the supply of petroleum was unintentionally shut off. He says the foam immediately disappeared, and no recoveries were obtained until the supply of petroleum was turned on again. At that time they were using 32.27 lb. of oil per ton of ore. See also testimony of Punchon as to the effect of suspending feed of petroleum at the Arthur plant.

At the mill of the Ray Consolidated Copper Company the change from below 1% to above 1% of oil on the ore was made the middle of January, 1917. A tabulation of the results before and after the change is contained in Defendant's Exhibit 44. Before the change they used a mixture of creosote (Barrett No. 4) and petroleum (fuel-oil), and since the change they have been using the same mixture, but with the proportions changed from 75% Barrett and 25% fuel-oil to 10% Barrett and 90% fuel-oil. Before the change the quantity of oil varied from 3.22 to 5.28 lb. per ton of ore. Since the change it has varied from 18.77 to 21.19 lb. per ton of ore. The average extraction before the change varied from 92.94 to 94.69% in different years; and since the change it has been 95.42%. The average copper in the tailings before the change varied from 0.617 to 0.375% in different years; and since the change it has been 4.12%. At this mill also experiments were made to determine the results of keeping the mixture of oil constant, and varying only the quantity used, which experiments showed that with the mixture now employed inferior results were obtained when a diminished quantity of oil on the ore is used.

So we see it is not anything peculiar about defendant's ore—the alleged presence of something nebulously called "clay-gangue," but not identified by any analyses, although complainants' experts had plenty of defendant's ore to analyze, and by which to prove its constituents, if they had seen fit to do so—which enables the defendant to use more than 1% of oil on the ore, because it is proved that at other mills, where the ore is not the same, amounts of oil in excess of 1% are being commercially and continuously used with satisfactory metallurgical results.

EXCESS-PROFITS TAX, as fixed by the final Senate committee-vote, increasing the maximum graduated tax on excess-profits of corporations from 40 to 50%, makes it necessary to revise previous estimates as to the extent of leading corporations to be affected. Under the new tax the old flat 8% tax is repealed and also the 12% tax on munitions exports. The following rates are substituted, based on the excess of current profits over normal pre-war profits; not in excess of 15% of the normal profits, 12%; in excess of 15 and not of 25%, 16%; in excess of 25 and not of 50%, 20%; in excess of 50 and not of 75%, 25%; in excess of 75 and not of 100%, 30%; in excess of 100 and not of 150%, 35%; in excess of 150 and not of 200%, 40%; in excess of 200 and not of 250%, 45%; in excess of 250%, 50%. This would result, for example, in collecting $30 per share of common stock from the United States Steel Corporation.

Californian Committee on Petroleum

*The extent to which industries of national importance depend on Californian oil is shown by a pamphlet just issued by the California State Council of Defense. The fuel-oil from Californian wells operates the Panama Canal; the greater part of the steam-railroads of Washington, Oregon, California, Nevada, Utah, Arizona, and New Mexico; the steamship lines along the Pacific Coast from Mexico to Alaska and across the ocean to Hawaii; the gas-plants of California, Oregon, Washington, Nevada, Arizona, and Hawaii; the mines and smelters of California, Nevada, and Arizona; the cement-works and sugar-refineries of California; and it serves a substantial portion of the manufacturing, industrial, and agricultural enterprises of the Pacific Coast States, as well as the fuel supplying the requirements of the United States Navy and Army in the West. The products of Californian petroleum, such as kerosene, gasoline, distillates, lubricants, and road-oils meet the requirements of Arizona, California, Nevada, Oregon, Washington, Alaska, and Hawaii. Californian kerosene is shipped in enormous quantities to China, Japan, India, and Australia, and to the western coast of Central and South America, while the distillates and lubricants are sold in nearly every important State in the Union, and also in England, Canada, and Australia.

The total oil in storage in California on June 1, 1917, is reported to have been slightly in excess of 38,000,000 barrels. A portion of the crude oil in storage cannot be utilized because it is situated below the outlets of tanks and reservoirs, or is being used for the operation of pipelines or for other reasons. Of the total stocks on June 1, not in excess of 32,000,000 bbl. was available for use. Of this amount 12,000,000 bbl. was refining oil, and would yield approximately 7,000,000 bbl. of residuum; furthermore there was available from crude-oil stocks approximately 27,000,000 bbl. for fuel. The present excess of consumption over production amounts to about 1,083,000 bbl. per month. If that continues, the entire available storage of Californian fuel-oil will be exhausted by June 1, 1919. If a margin of safety of 10,000,000 bbl. of fuel-oil is maintained, that margin would be reached by September 20, 1918. The field-losses of petroleum have been almost entirely eliminated and the amount of oil used in drilling and pumping has been reduced by substituting natural gas and electric power. The use of fuel-oil cannot be entirely eliminated. The refineries are working on improved processes of refining; a proportionally larger amount of gasoline and lubricants is also being secured. The surplus of kerosene is being exported to the Orient, Australia, Central and South America.

Mexican petroleum was substituted early in 1917 for Californian petroleums. The amount was approximately 2,750,000 bbl. annually, heretofore sold by the Union Oil Co. in Chile. A considerable portion of the remaining 3,000,000 bbl. of Californian fuel-oil sold in 1916 on the west coast of Central and South America, including the Panama Canal, can likewise be saved by the substitution of Mexican petroleum. By reason principally of transportation difficulties, Mexican petroleum will not be available during the War as a further substitute for more than 3% of Californian petroleum. Coal cannot be substituted for Californian fuel-oil to any important extent during the War because of difficulties in the production and transportation of coal. Approximately 1,000,000 bbl. of Californian fuel-oil will be saved in the North-West next year by the substitution of coal by the Oregon Short Line and by other industries. The Los Angeles & Salt Lake railroad and the Western Pacific railroad will use coal produced partly in the Rocky Mountain States. The Southern Pacific and the Santa Fe can also gradually turn to coal on those portions of their systems that are in proximity to the coalfields. Powdered coal has been successfully used in cement plants, stationary boilers or power-plants, and in metallurgical furnaces. One cement plant in the North-West has recently converted its plant from fuel-oil to powdered coal, saving 84,000 bbl. annually. Hydro-electric energy has already been substituted to a considerable extent in industrial and agricultural uses, but the difficulty in securing copper and other materials and the disturbance of existing industrial conditions are such that large savings of fuel-oil by the substitution of hydro-electric energy cannot be anticipated during the War. We conclude that some further saving of Californian fuel-oil is possible by elimination of losses and by substitution of other forms of fuel or power, but that no large saving can be effected without serious impairment of the efficiency of the transportation systems and industries of the Pacific Coast. The increased production should amount to more than 30,000 bbl. per day, and such increased production cannot reasonably be expected before June 1, 1918. Each difficulty standing in the way of prompt and substantial increased production must be quickly solved to forestall a serious industrial crisis.

In our opinion, the only way to meet the situation, in view of existing conditions, is to have the Federal Government call upon the manufacturers of oil-well supplies to devote a sufficient part of the capacity of their plants toward supplying the requirements of oil-producers in California and other parts of the United States to meet their needs, and to direct the railroads to transport such supplies promptly.

Of the most desirable undrilled lands, approximately 70% is involved in litigation with the Federal Government. Nearly one-half of the best undrilled proved petroleum lands of the State are claimed by the Kern Trading & Oil Co., the fuel-oil bureau of the Southern Pacific Co., under patents heretofore issued to the Southern Pacific Co. Of the remaining lands in litigation, a part of the most productive undrilled land is in posses-

*Abstract: Report of Conclusions and Recommendations by the Committee on Petroleum of the California State Council of Defense.

sion of Howard M. Payne, Federal receiver. The desirable Buena Vista Hills lands are practically embraced in Naval Reserve No. 2. A considerable area, however, of presumably productive undrilled but proved land, which in our judgment should be promptly and intensively drilled, is the Sunset field and in the east Coalinga field. The larger portion of these lands is claimed by the Kern Trading & Oil Co. A portion, particularly in the Sunset field, is in the possession of the Federal receiver. These latter should be promptly drilled. A review of the situation shows that the present emergency cannot be met without the assistance of the Federal Government. We desire, however, to emphasize the fact that a considerable portion of the proved productice territory, hitherto undrilled, lies outside of the naval reserves; that the further development of most of this land has been stopped by litigation with the Federal Government; that the policy of the Federal Government which has resulted in the creation of Naval Reserves No. 1 and 2 can have no possible application to these lands which are not claimed or needed for the Navy, and that with the help of the Federal Government substantially increased production can be secured from these lands. Unless the Federal Government can bring about radical changes in the supply of drilling material, the transportation of petroleum, and the development of the land, an increased production of over 30,000 bbl. per day, which in our judgment is necessary, cannot be secured without additional drilling in Naval Reserve No. 2.

We respectfully submit the following recommendations:

1. That every reasonable effort be made to increase the production of Californian petroleum promptly, and to this end additional drilling should be undertaken on the lands where the largest production can be developed in the least time, with the smallest expenditure of material and labor.

2. That every effort be made to conserve the supply of Californian petroleum by the diminution of field losses, the higher use of petroleum and its products, and the substitution of other forms of fuel or power.

3. That the facts with reference to the Californian petroleum situation be presented to the President, and that the Federal Government be urged to render assistance consistent with the public interest.

4. That the attention of the Federal Government be drawn to the advisability of directing the manufacturers of oil-well supplies to set aside sufficient capacity in their plants for the production of oil-well casing, drill-stems, wire-cables, and other material to supply the reasonable requirements of California.

5. That the attention of the Federal Government be drawn to the advisability of exempting from service in the armed forces of the nation all skilled workmen employed in the petroleum industry.

6. That the Federal Government be requested, in those instances in which Californian petroleum-lands now in litigation with the Federal Government are not in the hands of the Federal receiver, to consent, through the Department of Justice, to stipulations under which the claimants will be permitted to drill such lands intensively, under engagement to protect the Government in its needed supply if it win the suits.

7. That the Federal Government be requested, in those instances in which Californian petroleum-lands now in litigation with the Federal Government are in possession of the Federal receiver, to take appropriate proceedings, through the Department of Justice, so that the receiver may be directed to proceed at once to drill intensively such lands as are presumptively productive, and particularly the lands which are likely to suffer from the infiltration of water unless soon drilled. Further Congressional legislation may be needed for this purpose.

8. That the Federal Government be requested to enact promptly the necessary legislation by which such lands in the public domain as the Federal Government may consider consistent with public interest may be opened to petroleum development on terms just and reasonable both to the Federal Government and to such persons as may proceed in good faith to the exploration and development of petroleum lands. We suggest that the area of the lands in each instance be sufficiently large to permit efficient operation.

GAS-FIRES are so obstinate that wells are usually left to burn until the flow of gas ceases. A new method was adopted recently by William Guerin of the New York Fire Department, who was summoned to Monroe, Louisiana, to try to extinguish a fire from a well flowing 44,000,000 cu. ft. per diem under a pressure of 1500 lb. per sq. in. The gas being salable at 2c. per 1000 feet the loss was at the rate of $880 per day. In the course of months this would represent an enormous waste. Mr. Guerin was telegraphed for and he applied a method that promises to be generally utilized in similar cases. There was water available and pumps with which to develop pressure. Two fire streams were employed, the nozzles being brought close to the well, the firemen meanwhile protected by sheet-iron shields. The streams were played upon the gas-flow so that they met in the gas below the point at which combustion was possible, that is, where the pressure was so great as to maintain a velocity that would extinguish flame. The streams were then gradually elevated until their meeting place was at the base of the flame; the streams were then flattened out so as to make a horizontal fan; next the flames were wiped off by rapidly raising the fan formed by the meeting streams. The fire was extinguished in five minutes.

THE PRESIDIO SILVER MINE, in Texas, was in continuous operation during the first six months of 1917, and silver mining was also carried on during that period in the Van Horn and Sierra Blanca districts. Several shipments of copper ore were made from deposits in the 'Red Beds' of Foard and Knox counties. The result was a small output of copper, lead, and zinc; but a production of silver for the six months of 340,000 ounces.

REVIEW OF MINING

As seen at the world's great mining centres by our own correspondents.

TORONTO, ONTARIO

PRODUCTION OF STEEL IN CANADA BEING LARGELY INCREASED.—ELECTRIC STEEL-PLANT IN OPERATION.—COAL SHORTAGE GREATLY RELIEVED.—PORCUPINE OPERATING AT NEARLY NORMAL CONDITION.

The great scarcity and high price of steel, owing to War requirements, has given a great stimulus to the steel industry. Many new companies are being formed, and the existing plants are arranging for large extensions. President Mark Workman, of the Dominion Steel Corporation, states that all departments of the Sydney plant are operating to full capacity, and that good progress is being made with the additions and improvements now under construction at a cost of several million dollars. A new blast-furnace is expected to be in operation in the course of two or three months, which should increase the output by at least 35,000 tons yearly. A battery of 60 new coke-ovens is being built, which will be ready for operation during the winter, and a second battery of the same size will be completed in the course of a year. The Steel Company of Canada, the principal works of which are at Hamilton, Ontario, is seeking to provide its own coal and ore reserves. Acting in co-operation with large American interests, it is investing in coal and iron-ore properties situated in the Eastern States, from which its supplies will be drawn in future. Its plans for expansion include the construction of a large coke-producing plant on its Hamilton property with a capacity of 800 to 900 tons per day, work on which will be commenced promptly. Under present labor conditions, it is not expected to be completed in less than 20 months. The steel-plant of the British Forgings Ltd., of Toronto, which will be operated by electric process, using scrap-steel as raw material, is understood to be nearing completion, but as it is primarily a military enterprise, under the direction of the Imperial Munitions Board, there is a good deal of reticence as to its operations. It has 10 electrical furnaces with an aggregate capacity of from 300 to 400 tons per day at least, one of which has gone into steady operation. It is believed to be the largest electrical steel-plant in the world, and when working at full capacity will require 22,000 hp. and employ 3000 men.

The coal situation appears somewhat less threatening than earlier in the season, when fears of a serious famine prevailed. The appointment of C. A. Magrath as Canadian Fuel Controller has had a good effect. The transportation problem is being grappled with by Sir Henry Drayton, head of the Railway Commission, with a view to expediting deliveries, and the services of H. P. McCue, of Pittsburg, a leading transportation expert have been engaged to facilitate the filling of contracts, and prompt forwarding of shipments. Mr. Magrath has issued a statement urging consumers of coal to lay in their supplies as early as possible, and exercise strict economy, as, unless deliveries during the open season can be speeded up, a very heavy load will be thrown on the railways in the winter.

The Mines Branch of the Canadian Department of Mines estimates the total production of coal in Canada during the first quarter of 1917 at 3,590,991 tons, the rate of production being apparently less than the average in British Columbia and Nova Scotia, but greater in New Brunswick, Saskatchewan, and Alberta. Exports during the quarter were 501,570 tons and imports of coal 3,921,824 tons, as against exports of 501,570 tons, and imports of 4,002,892 tons during the corresponding three months of 1916. Work has been resumed at the Crows Nest Pass and other mines of Alberta and British Columbia, under the orders of W. H. Armstrong, the commissioner appointed by the Government, under the War Measure Act, but there is a great shortage of labor, many of the miners having obtained other employment during the strike. As a result of the large increase in wages secured by the men, coal prices at the mines will be increased approximately 75c. to $1 per ton.

Conditions at Porcupine show improvement, and now that the labor difficulties have been settled, many miners are coming in and finding employment. The McIntyre during June treated 14,455 tons of ore with a production of $141,208, the mill-head running $10.24 per ton. Production has been steadily maintained despite unfavorable conditions during the past six months. The drift at the 1000-ft. level is now in Jupiter ground, where the vein has widened to 55 ft. of a high milling-grade. The exploration of the Dome Extension property, undertaken by the Dome Mines, has been so delayed by insufficient working force that it is considered probable that the Dome will ask for the three months extension of their option, to which they are entitled. The option falls due on October 15. Hollinger Consolidated is pushing development, and has cut a new vein carrying free gold, and showing a high average gold content at the 425-ft. level of No. 10 shaft. The underground workings are being centralized, so as to make all the orebodies opened up readily accessible to the central shaft.

At the Kirkland Lake the shaft is being put down from the 600 to the 700-ft. level. It is estimated that the ore in sight amounts to about 50,000 tons, averaging from $9 to $10 per ton. Plans have been prepared for the installation of a ball-mill with a daily capacity of 150 tons. A water-seam has been cut at the depth of 165 ft. on the Elliott-Kirkland, which has caused the suspension of underground work pending the installation of steam equipment. Excavations for the foundation of the cyanide mill at the Lake Shore mine have been started. The recently discovered orebody at the 200-ft. level has widened to 40 ft. with ore of good milling grade over half that width. The Tough Oakes is as yet the only dividend-paying mine at Kirkland Lake, and has returned to shareholders a total of $391,125.

The price of silver having reached 80c. per oz., the Cobalt miners receive an additional bonus of 25c. per day, so long as this figure is maintained. The silver-mining industry was never more active here, though production is declining in volume as the grade of ore treated steadily grows lower. More men are employed in the mines than ever before, the total approximating 3000, as compared with 2600 a year ago. It is estimated that production for 1917 will probably be about 19,000,000 oz., as compared with 21,500,000 in 1916, though the value will be considerably higher.

The two dividends of 15c. per share recently declared by the Kerr Lake Mining Co., payable August 10, was equally divided, one-half going to the Red Cross and the remainder to the Army and Navy branch of the Y. M. C. A.

Six of the furnaces of the Grand Forks smelter of Granby Consolidated have been blown-in with the receipt of fuel from Crow's Nest Pass Coal Co. Two sections are still idle. The Grand Forks plant was closed-down in April, when the company supplying fuel was shut tight by strike. Granby has a financial interest in the Crow's Nest Pass Coal Company.

MAYER, ARIZONA

THE BINGHAMPTON COPPER MINE A NEW PRODUCER.—THE DEVELOPMENT EXTENDING AND A NEW SHAFT TO BE SUNK.—FLOTATION IN SUCCESSFUL OPERATION.

The Arizona Binghampton copper mine is five miles east of Mayer, in Yavapai county. It is an old property that has been in course of development for some time, the money for the work having been furnished by W. H. Reynolds, of New York. The company is incorporated with 340,000 shares at $5. In May 80,000 shares was advertised and much of it was taken by officers and stockholders of the United Verde and the United Verde Extension, at Jerome. Those men have made suggestions as to the method of development. There are three miles of underground workings, and judging by the reports of mining men who have examined the property, there is but a small amount of the work done that does not count. It was the similarity of the formation to that of the Jerome district that induced some of the officials and stockholders of the Jerome mines to become interested.. The ore is chalcopyrite with some iron pyrite and tetrahedrite, disseminated in the Yavapai schist. Some of the stopes are 20 ft. wide and

BINGHAMPTON MILL NEAR MAYER

from 100 to 120 ft. long. On the 400-ft. level there is a stope from which a car of ore is shipped daily, running from 13 to 14% copper.

On the 600-ft. level, in the past few days, ore has been found in four shoots that are extensions of orebodies on the 400-ft. level. One-fifth of the output of the mine is coming from the 600-ft. level from the new development work. The nearest deep mine is the Blue Bell, 9 miles south-west, which has developed orebodies 40 ft. wide at the 1000-ft. level.

On August 24, of last year, a flotation mill of 100 tons capacity was started upon the property. The first day's run showed a recovery of 95%, but this has been lowered since for economic reasons to 85 or 90%. The concentrate runs about 20% copper. The mill was built after the plans of the Inspiration mill, designed by H. Kenyon Burch, and built under the supervision of C. B. Clyne.

The mill has exceeded the estimated capacity practically every day since it started. The output for May was approximately 300,000 lb. of copper which netted the company $38,000. The mill has been doubled in capacity, and the new unit will be in operation in a few days with a daily capacity of 250 to 300 tons, which will give an output of about 500,000 lb. of copper monthly. There will be added to this a regular tonnage of high-grade ore, that at present is one car per day. This will be increased as the mine is developed.

The ore is hauled by contract from the mine to the mill, 1200 ft., in trains of 4 to 6 cars drawn by a mule. The ore is dumped from the ore-cars into the coarse-ore bin and broken to pass an 8-in. grizzly. The coarse-ore bin has a capacity of 500 tons. The ore is drawn from this bin onto a slow-moving

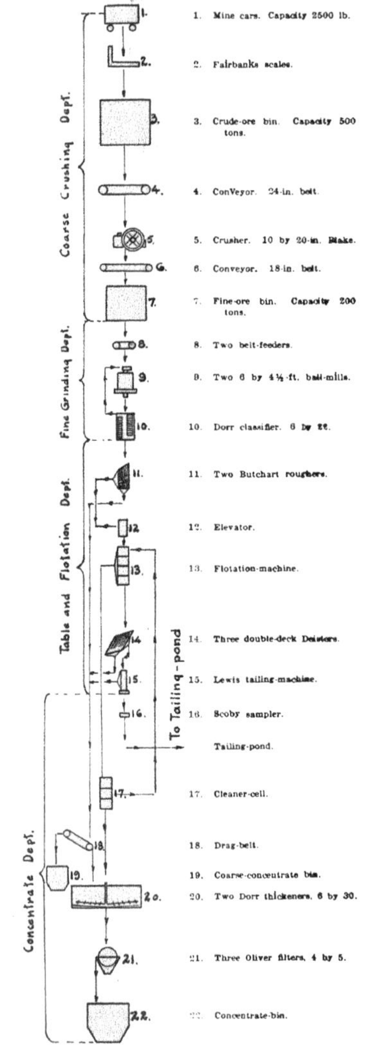

belt-conveyor which can be used as a picking-belt if desired, for sorting out either shipping-ore or waste. This conveyor dumps onto a 2-in. grizzly, the oversize going to a 10 by 20-in. Blake crusher which crushes to 3 inches. The crusher product and grizzly undersize are carried by a conveyor to the fine-ore bin of 200 tons capacity. The ore is then delivered by belt-feeders to two ball-mills which grind in closed circuit, with a 6 by 22-ft. Dorr classifier. The overflow originally went direct to the flotation machine, but later, when a marked increase in the value of the ore occurred, the much lower ratio of concentration justified the installation of tables ahead of the flotation machine, to remove as much coarse concentrate as possible, as a large tonnage which did not require fine grinding was put through the mill. The table concentrate is de-watered by a drag-belt. The coarse material then goes to the concentrate-bin and the overflow to the Dorr thickeners.

The flotation machine is of the Inspiration type. The rougher concentrate is re-treated in a cleaner, the reject from which returns to the rougher for re-treatment. Diaphragm-pumps are used for handling the cleaner reject and rougher concentrate. The reject from the flotation machine is treated by double-decked Deister tables. The table tailing is run through a Lewis tailing-machine and from that to the tailing-pond, where it is impounded and the water returned to the mill. The flotation concentrate is thickened in Dorr thickeners and dried by Oliver filters to about 12% moisture.

The concentrate is hauled by teams and motor-trucks to the railroad at Mayer, from which point it is shipped to the Consolidated Arizona smelter at Humboldt. The high-grade ore is shipped to the same place.

This mine is in the centre of a mineral district extending several miles in all directions. The Copper Queen has a mine adjoining and parallel on the east which has developed sufficient ore to warrant the erection of a 100-ton flotation mill which will be of the same design as the Binghampton mill. Work on its construction is to begin within 60 days. There are at least 20 properties in various stages of development within a radius of two miles. Half a mile south of the Binghampton mine is the Barbara property, a promising group that is under boond and lease to mining men from Bisbee.

LEADVILLE, COLORADO

All the Mines of Leadville District are Idle and Nearly 2000 Men are Out.—Pumping at the Main Drainage Plants Continues.—Text of Statement Issued by the Operators.

A complete tie-up in the mines of the Leadville district has resulted from the strike called Saturday morning, July 21, at 7 o'clock by the Cloud City Miners' Union No. 33 of the International Union of Mine, Mill, and Smelter Workers. All union and non-union men walked out leaving only a few managers, superintendents, foremen, and other salaried employees of the companies to operate the pumps. It is estimated that normally there are 1950 men employed in and around the mines. Of these the union claims 900 as members, the remaining 1050 being non-union. There were 103 operating shafts in the district prior to the strike, and of these 57 were producing. The total output for June was 65,700 tons, or approxiately 2190 tons per day. A few small lessees continue to operate, though in each case the men at work are interested in the enterprise, but these lessees are producing only a small tonnage so that the output of the district is practically stopped.

Pumping at the big draining centres of the district, the Penrose, Harvard, Wolftone, Greenback, and Yak tunnel, continues with crews of volunteers and salaried men at the pumps. The operators state that unless something unforeseen happens they will be able to continue pumping indefinitely although one or two properties are very short handed at present.

Union men picketed the roads leading to the mines on the first day of the strike, cautioning those who passed from going to work. Aside from this, they have manifested no outward interest in the situation, and their leaders have declared that they will simply remain quiet and await developments.

Both sides are apparently disappointed in the result of the strike call. The operators believed that their non-union employees would not respond to the call of the union; and the union authorities were confident that their call would so cripple the mines that the pumps would be shut-down and the mines flooded, which they believed would greatly hasten a settlement. As it is, the opposing forces have come to a deadlock, and it appears that it is now only a question of which side can hold out the longer. The operators state that rather than give in to the organization they will allow the mines to close completely.

Old-timers in the district who witnessed the strike of 1896, declare that the present situation is similar in every way to the beginning of the disastrous trouble of 20 years ago. Quiet will undoubtedly continue until the miners become fretful over their inactivity, short of money, and rapidly getting into debt. Then it is expected there will be disorder that may lead to fatalities.

Just what the outcome will be is problematical. It is generally understood that the majority of the men have very little in the way of reserve funds to go on, and that they soon will find themselves confronted by a serious alternative—go to work or starve. Already many men have left the district seeking employment in other camps, but the general unrest through the mining centres of the West makes the assurance of a job out of the question. The union will probably receive benefits from the organization as long as the funds last, but the non-union men will be thrown upon their own resources.

The failure of the government mediators, George W. Musser and Verner Z. Reed, to bring about a settlement between the union and the operators has caused some to charge the mine owners with the responsibility for the strike. At the Denver conferences, the union agreed to compromise their demands of $1 per day increase and accept 50 cents. The operators refused to consider this offer, and practically refused to give attention to anything except an offer from the Government to take over the mines.

In reply to assertions that have been made concerning their stand in the matter, the operators have issued the following statement:

"The present deplorable state of affairs in this district results from a long continued effort on the part of imported agitators to organize the miners of Lake county, and was precipitated by a demand of the Miners' Union upon the operators for a flat increase of $1 per day to the present existing wage scale.

"The union has had some success in recruiting members here but is not considered as representing the majority of the miners employed. The fact that the Leadville operators have refused to deal with the union has been deliberately misconstrued to imply that they will not meet with their men. It might be well to again state that the managers are at all times willing to confer with their own employees and are always anxious to adjust any differences which may arise.

"It will be remembered that the State Industrial Commission, acting upon the solicitation and representations made to it by the local union, recently came to the district, made a thorough investigation of the working conditions, and conducted an open hearing, at which all questions pertaining to the existing state of the mining industry here were fully presented and considered in comparison with conditions governing operations in other metal-mining districts of the State.

"The crux of the report handed down by the Commission on May 29, was that an adequate wage was being paid; and since that time there has been no material change in the conditions passed on. Despite the fact that the union originally appealed to the Commission, it promptly proceeded to repudiate the findings of the commissioners and made a demand on the operators for the $1 per day increase. Owing to the source of the demand, it was ignored, whereupon followed the decision of the union to strike.

"Attempts at arbitration have brought no good results because the arbitrators were appointed at the instigation of the labor leaders, and when granted a conference by representatives of the operators, immediately, and without first giving the operators a chance to be heard or to define their position, presented a previously arranged basis for settlement which could not, in all good faith, be accepted by the latter, having in mind the best interests of the miners as well as the owners of the properties. There can be no fair or just arbitration of any question unless both sides have some choice in the selection of the arbiters and a full and thorough investigation of all conditions effecting the issue be conducted, before a decision is arrived at.

"The operators fully realize that this strike, involving loss of work to a large number of men, and the suspension of operations which sustain an important community, is a very serious matter, and they have persistenly attempted to prevent it by all fair and reasonable means.

"The manner in which this strike was called, by discouraging pumpmen from remaining at their duties, with the evident intention of abandoning the pumps and permitting the workings to become flooded, shows clearly the pernicious influence behind this movement. Should it become impossible to keep the pumps in operation on the various large drainage enterprises now under way here, it will surely result in a disaster from which Leadville may never fully recover. Such a course is certainly not pursued in the interest of the miners themselves nor of the community at large.

"The demand for $1 per day increase is of course absolutely unreasonable. The mining industry here, in its present state, simply cannot afford to pay such wages. If the demand be persisted in, it will unquestionably result in forcing a complete shut-down of all the mines."

The Arkansas Valley plant of the American Smelting & Refining Co., employing 850 men and handling 25,000 tons of ore monthly, continues to operate as usual. It is stated that there is sufficient ore in stock at the plant to supply it for two months.

The plant of the Western Zinc Oxide Co. employing 50 men and handling 1500 tons per month is also running normally.

SILVER CITY, NEW MEXICO

LARGE SHIPMENT OF MANGANESE FROM SILVER CITY.—RICH GOLD ORE AT PINOS ALTOS.—A NEW MILL FOR PINOS ALTOS.—CHINO RAISES WAGES AND MAKING EXTENSIONS OF PLANT.—OPERATIONS IN BURRO MOUNTAINS.

Three to four cars of manganese ore is being shipped daily from the Boston Hill properties west of town. The ore averages better than 20% manganese. At Pinos Altos, rich gold-bearing quartz is being mined in the Pacific No. 2 workings of the old Skilacorn vein of the El Paso Mining Co. The company mill, which has been idle for the past two months, is about to resume operations. The shipment of lead and zinc ore from the Indian Hill shaft of the Manhattan group continues, 18 to 20 tons being hauled daily to the railroad at Silver City. The work of sinking the shaft at the end of the Manhattan adit has been resumed. The adit is in 900 ft. It is said that the shaft will be sunk 500 ft. It is reported that the Greybird claim of the Harvey property has been sold to parties from Mexico, and that they are also considering the purchase of the Mammoth mill. The U. S. Copper people suspended developments in May, but are expected to resume work August 1. Their two new double-compartment shafts are down 140 ft. and 50 ft. respectively. All machinery is in place on concrete foundations, and the new buildings are complete. The Mountain Key mine may be re-opened this fall by a Silver City company recently organized. The Key shaft is down 900 ft. on an incline and full of water. Work at the Silver Cell property was suspended when diplomatic relations were severed between United States and Germany, as two prominent Germans controled the company. The property has been kept unwatered, however. At the Calumet & New Mexico camp, development continues steadily, the ore taken out being sorted and the high-grade shipped. The average is better than 23% zinc and 25% lead. A good tonnage of low-grade ore has already been blocked out and the company contemplates the erection of a plant this fall.

W. H. Janney, president and general manager of the Re-

BLOCK OF SOLID ZINC SULPHIDE

public Mining Co., who for years was with the Chino Copper Co., at Hurley, since July 1 has devoted his entire attention to the management of the Republic property, where development is meeting with the expectation of its owners. Money has been judiciously spent and a great amount of ore is in sight. The ore carries lead, zinc, silver, and copper. The company will begin the construction of a 200-ton concentrating plant in the fall. The Hanover Copper Co. is continuously developing and shipping ore. The United States Copper Co. has cut a body of high-grade ore on its Philadalphiha holdings and is making arrangements to ship. The ore assays 18% copper, besides carrying gold, silver, and iron. The Empire Zinc Co. never misses a day, the mill being kept at its full capacity. There are reports that the Chino Del Norte Co. has developed some good ore.

At Fierro the Hanover Bessemer Steel Co.'s new mill is a success and is operating full time. The Colorado Fuel & Iron Co., that was to have ceased operations June 1, will continue until January 1, when the Hanover Bessemer Steel Co. will take over the mining of all properties under lease to the C. F. & I. Co. The McGee group of claims, 2½ miles north of Fierro, has been taken over by the Victoria Tungsten & Silver Mining Co. of Arizona, composed of Minnesota mining men. Development is to begin this summer. California men have been looking over manganese properties in this district the past month with a view to leasing.

At Santa Rita 10,000 to 11,000 tons daily is the regular output of ore made by the Chino Copper Co. Eight and nine trains per day of from 26 to 27 fifty-ton cars carry the ore to the company's plant at Hurley. Over 20,000 tons of ore and rock are handled every 24 hours. The high-grade ore is shipped to the smelter at El Paso. Recently one of the steam-shovels cut into a body of copper-glance, the extent of which has not been made public. The second big crusher installed the past year is in operation. A fine new club-house for the use of employees at Santa Rita is being erected by the company. Putting all employees on the 8-hour basis and raising the wages of all employees 25c. per day on July 1, was voluntary on the part of the company. Over 1500 men are now employed. The Federal registration last month gave over 700 men for selective draft. At Hurley the mill of the Chino Copper Co. is being crowded to the limit of its capacity. Structural work for two additional units is completed and a big crew is installing machinery. Two miles south of the mill at the tailing-dam a new plant for the re-treatment of the tailing has been erected and operation is to begin this fall. The dams at this point are over a mile wide. Experiments with new types of tables, flotation-oils, and processes are made daily. The management is looking forward to the time when the still lower-grade ore on the dumps can be treated at a profit. The 8-hour day and the 25c. per day raise is effective here also. A new club-house for the employees is also being erected at Hurley by the company. The number of registrations for war duty at Hurley was 780, the greatest in the county. Frank R. Wicks succeeds W. H. Janney in the management here.

In the Burro Mountain district the Black Hawk property, re-opened several months ago, is confirming the judgment of its owners. The Phelps-Dodge Co., at Tyrone, has created much outside interest in its development plans, not only in the mines, but in the successful operation of the new concentrator, and the recovery made by innovations. The Austin-Amazon Co. is working a small force at present, its inability to ship the ore causing the company to curtail operations at the present. The Burro Grande Co. also is waiting for an opportunity to dispose of ore. The Giant Copper Co., a new organization having acquired over 1000 acres adjoining the Phelps-Dodge holdings, has a crew of men at work on the old National diggings. Another new company, the Federal Copper Co., has acquired holdings in the Burro Mountain district, and organization is being effected.

TYPICAL SCENERY AT FIERRO, NEW MEXICO

CRIPPLE CREEK, COLORADO

A New Plant for the Catherine Company.—A New Water-Course Cut in the Roosevelt Drainage Tunnel.—Vindicator Declares a Dividend.—Old Mines Resuming Operations.

Plans for mine-buildings and plants, to replace that destroyed by the recent fire, are being prepared for the Catherine Gold Mining Co., of which Charles Walden, of Victor, is general manager. A new double-drum electric hoist, with a capacity for 2000 ft., and a 12-drill electrically driven compressor will be installed. The Catherine company holds a five-year lease and option on the Last Dollar Gold Mining Co.'s Last Dollar and Combination claims, on Bull hill.

The Roosevelt tunnel of the Cripple Creek Deep Drainage & Tunnel Co., has cut a new water-course in the Old Ironsides claim, of the United Gold Mines Co., on the west slope of Battle mountain. The flow is estimated to have reached 1000 gal. per minute when first tapped, and is now flowing steadily at 500 or 600 gal. per minute. The miners, who for many months have been working in heavy oil-skin coats and rubber boots, now work without the coats, the tapping of the water-course having an immediate effect on the drainage of the ground above. The tunnel-flow has raised again to about 6000 gal. per minute, of which about 3000 gal. is coming into the tunnel from east of the Elkton shaft, on Raven hill, and the remainder between the Elkton and Gold Dollar mine on Beacon hill, to the west.

The quarterly dividend of the Vindicator Consolidated Gold Mining Co., at the rate of 3c. per share, amounting to $45,000, will be paid July 25. This will bring the dividend total for 1917 to $135,000, and the grand total to $3,667,500. The gross production from the Vindicator properties last year was reported at $2,295,730 and $1,644,120 net.

THE MINING SUMMARY

The news of the week as told by our special correspondents and compiled from the local press.

ALASKA

(Special Correspondence.)—Cache creek operators are having a successful season. A rich strike of quartz is reported on Thunder creek in this district. Reports from the Ruby district say that returns are encouraging. Flat and Tamarack are expected to yield big returns this season. Poorman creek has been the scene of another small excitement. Stephen Capps, of the United States Geological Survey, will work in the Broad Pass region until August. He will then go to Iron creek to investigate copper deposits and then across the divide to Willow creek.

The annual meeting of the officers of the Gold Cord and Mabel Mining companies will be held at Anchorage on July 10. A rich strike has been reported in the Mabel mine. A full crew is operating at the Martin property.

Plans for a power-plant for the Willow Creek district will not be carried out this year and Colorado people are looking into the matter with a view of putting in power of some kind next year. Louis Levensaler, a mining engineer, who for years was with the Alaska Syndicate, is at Unga island making an examination of the Apollo mine. Don S. Rae of the Rae-Wallace Mining Co. and Mr. Boomer, of Wallace, Idaho, are in town and developments of an important nature will be commenced on this property in the Willow Creek district, among other things the driving of a cross-cut adit and the installation of machinery.

Willow Creek, July 10.

ARIZONA

Cochise County

Richard La Rue has discovered that the ore in his mine on White Tail creek contains molybdenite as well as lead and zinc. He had a carload of ore at the railroad station for shipment to the smelter when his attention was called to the presence of the molybdenite, a mineral with which he was unacquainted. He lost no time in reaching the railroad and stopping the shipment, until he could pick over the ore for the purpose of sorting out the molybdenite.

(Special Correspondence.)—Considerable excitement has been caused at Benson by the report of several Californian oil-men of the possibility of oil existing in the San Pedro valley. A company has been formed, the stock being issued publicly. It is only a matter of obtaining leases of sufficient area before test-drilling will commence.——The American Oil Fields Corporation has been formed with a capital of $5,000,000. Two areas have been reported as favorable for oil, one in the China valley, the other in the San Pedro valley.

Benson, July 26.

Gila County

(Special Correspondence.)—Development work is progressing favorably on the Magma Chief, according to reports of the superintendent, C. A. Kumke.

Globe, July 26.

Mohave County

(Special Correspondence.)—Within two weeks of the discovery of ore in the Telluride mine, the drift at the 400-ft. level of the neighboring property, the Sunnyside, opened up stringers of quartz of good grade. The drift was advanced 340 ft. north-west from the shaft, and to avoid timbering, the work was done in the solid ground adjoining the vein. When directly under the old 100-ft. shaft a cross-cut was run 18 ft. across the vein with no hanging wall in sight, and containing stringers of quartz and calcite, and silicified andesite.

A drift was run along the foot-wall opening up a 2-ft. vein of quartz assaying $10 per ton in gold and increasing in width and value. The drift will be continued for 100 ft. on the theory that this is the east end of an orebody.

For the first six months of operation of the United Eastern, the general superintendent, J. A. Burgess, reports a total of 37,565 tons of ore milled, having a gross value of $793,497, averaging $21.12 per ton. A net profit of $491,130 was realized from operations. The initial dividend of 5c. per share will require about $70,000, but after this and all current indebtedness incident to the construction of the mill has been paid, the company will have a considerable surplus in the treasury.

With an increase in milling capacity to 300 tons per day, the second half of the United Eastern's operation should excell the record made from January to July.

Oatman, July 26.

(Special Correspondence.)—The reported rich strike on the Telluride vein at Oatman is proving more important than at first supposed. The ore is 4 ft. wide. At the Smuggler group ore worth $300 per ton is mined from an 18-in. vein. This mine has shipped considerable high-grade ore in the past. The Golconda mine, of the Union Basin Mining Co., has adopted the Oatman-Chloride scale of wages, granting an increase of 25 to 30 cents per day in wages, and has resumed operation.——R. S. Billings, of Los Angeles, reports the discovery of large low-grade deposits of molybdenum ore. The molybdenum belt is claimed to be three miles wide and 35 miles long, covering several groups of claims, chief of which are the Copper Taylor, Acme Molybdenite, and the Great Republic.—— Financial arrangements have been made in Los Angeles for the re-opening of the old Leland property. It is expected that the new 300-ton mill of the Washington-Arizona at Mineral Park will commence operations in a few days.——J. J. Ford, representing Pittsburg financiers, has examined the property of the Grand Canyon Gold Co. Plans have been completed for the working of these placer deposits on the Colorado river.

Oatman, July 25.

Pinal County

(Special Correspondence.)—It is reported that the Silver King shaft is to be pumped out and the mine re-opened; financial arrangements are now being completed. This is a famous old silver property, but the records show that there is copper on the 800-ft. level and it is expected it will become a copper producer.

Florence, July 25.

Santa Cruz County

According to *The Oasis* of Nogales the old Mowry mine is to be re-opened and equipped with a 50-ton concentrating mill. J. C. Smith and Sidney Bennet, of western Pennsylvania, who are largely interested in the property, recently visited the mine and decided to start the work of rehabilitation.

Yavapai County

Arizona papers state that all of the important mines in the Jerome belt that were affected by the recent labor disturbances, have resumed operations. At the United Verde a 'weeding-out' process has been instituted and the undesirables are being given their time. As they will be unable to secure employment in the district they are leaving and no further trouble is anticipated.

CALIFORNIA

CALAVERAS COUNTY

The Calaveras Copper Co. at Copperopolis will sink a new vertical 3-compartment shaft to open the North Keystone claim. It is proposed to mill the old dumps, which are fairly rich in copper. The rock will be loaded into cars with a steam-shovel.

DEL NORTE COUNTY

(Special Correspondence.)—At the Diamond Creek cinnabar property the machinery is being set up by the owners, Ehrman, Cole & Taggert, and they expect to be ready to begin producing quicksilver within two weeks.

Waldo, Oregon, July 24.

EL DORADO COUNTY

The hoisting and milling machinery and buildings of the Larkin mine, near Diamond Springs, were destroyed by fire July 21. The Griffith mine property narrowly escaped similar destruction. The loss was caused by a brush fire that swept over that district.

MADERA COUNTY

It is proposed to explore the extent and character of the iron-ore deposits in the Minarets district by means of bore-holes, says *The Inyo Register*. The work is understood to be in the interest of the Noble Steel Co., and will be in charge of Henry Beck, who is now at Bishop waiting for the arrival of the drilling outfit.

MARIPOSA COUNTY

(Special Correspondence.)—E. C. Gamble and W. S. Stewart, of Oakland, have sampled the Sweetwater mine east of Mariposa and made a mill-test of the ore, and contemplate working the mine under lease and option. Thomas Doyle, of Fresno, is doing development on the Filiciana mine. The Early mine has been unwatered and re-timbered and has resumed operations. A 50-hp. gas-engine has been placed in the mill and 10 stamps are dropping. W. H. Washburn is in charge of the mill. On the 200-ft. level a shoot of rich ore has been found at the intersection of two veins. Geo. H. Hook is superintendent.

Jerseydale, July 16.

NAPA COUNTY

Napa county produces other mineral than quicksilver. Clark & McLean, of New York, have leased a number of cinnabar properties near Mt. St. Helena, and will develop them. Within the past month there has been shipped from Calistoga 500 tons of chrome, four cars of fine kaolin, 40 flasks of quicksilver, and three cars of gold and silver ore from the old long-neglected Silverado mines. Regular shipments are to be made from these mines.

SANTA CLARA COUNTY

(Special Correspondence.)—Mrs. Stock has closed a contract for a lease of her magnesite claims on Red mountain to Bodfish and McGuire, of San Francisco, who will develop the property and ship the magnesite by auto-trucks, under contract to Livermore. When the deposits have been developed, a rotary kiln will be placed on the mines. There is said to be 5000 tons of ore in sight.

Livermore, July 28.

SHASTA COUNTY

(Special Correspondence.)—F. S. Hink, of Oakland, and R. [illegible] Montgomery, of San Francisco, have taken a lease and [illegible] on the Central mine in Old Diggings. The mine will be sampled with a view to purchasing. A. A. Anthony of Old Diggings owns the mine.

The Arps Copper Co., at Copper City, has shipped its second carload of ore to the smelter at Kennett. The company contemplates providing reduction works of its own.

At the Star mine, one of the Bully Hill Copper Co.'s properties, 55 men are employed. The mine has been unwatered to the 900-ft. level and development will be continued. The mine has been idle since the Bully Hill company abandoned operations 10 years ago, on account of the smelter-fume trouble. L. C. Monahan is superintendent.

Redding, July 24.

SIERRA COUNTY

A small but rich vein of quartz has been found in the bed-rock of the Mohawk drift mine on Dock creek. It has already produced a quantity of very rich gold rock.

TUOLUMNE COUNTY

The main shaft of the Clio mine, half a mile above Jacksonville, on the Tuolumne river, is to be unwatered for the purpose of cross-cutting into the foot-wall to explore the slate vein on the 600-ft. level.

COLORADO

BOULDER COUNTY

(Special Correspondence.)—The sale of the Bradenbury group of gold properties on Left Hand has been made, and C. D. Colburn and associates, of Denver, will assume control. The annual meeting of the Degge-Clark Milling Co. was held this week. The stockholders approved the sale of the milling plant. The property is in good condition and making a large production. It is being worked by lessees.

The U. S. Gold Mining Corporation is developing its holdings. A large amount of oxidized mill ore is exposed. The noted Livingston mine, one of the properties, has a large vein. The White Raven property, one of the largest silver-lead mines in the county, is making heavy shipments. It is owned by L. A. Ewing and partners. Jamestown, situated in the centre of the Central mining district, after many years, has revived.

Eldora, July 24.

IDAHO

SHOSHONE COUNTY

(Special Correspondence.)—Free rent of company houses, free light and water, and no interest on building-loans will be the 'bonus' which employees of the Bunker Hill & Sullivan company, at Kellogg, who enlist in the army or navy will receive. A large number of men have already gone into the army or navy, and in the draft scores of men employed by the company were called.

Spokane, July 23.

MICHIGAN

HOUGHTON COUNTY

Production of copper during June by the Calumet & Hecla and subsidiary companies was as follows:

Mines	Lb.	For the year, lb.
Ahmeek	2,530,848	14,880,459
Allouez	755,399	4,806,688
C. & H.	6,346,446	40,194,517
Centennial	143,907	923,859
Isle Royale	1,240,050	7,383,234
La Salle	157,546	1,010,823
Osceola	1,327,773	8,809,579
Superior	193,050	1,211,375
Tamarack	241,703	3,004,841
White Pine	410,803	2,184,394
Total	13,347,525	84,359,769

The Calumet & Hecla Mining Co. and subsidiaries produced a total of 84,359,769 lb. of copper in the first half of 1917, an increase of 4,478,213 lb., or 5% over the same period in 1916. The subsidiaries making relatively the best showing were Ahmeek and Isle Royale, with increases of 22% and 18%, respectively. Osceola dropped off 10%. The copper product was sold at probably the highest prices ever quoted for any six-months' period in the history of the industry, but unfortunately such of it as was exported was sold f.o.b., London, the shipper being obliged to stand the heavy transportation charge, war-risk, and marine insurance. This probably jumped costs 3 to 4 cents per lb., so that combined with rising costs in other directions, the final net was probably no larger than for the last half of 1916.

MONTANA

Lewis and Clark County

(Special Correspondence.)—Ore shipments from the Scratch Gravel Gold Co. property in the Scratch Gravel hills average nearly $4000 per car at the East Helena smelter. The last car netted $3736. The ore was mined from the 300-ft. level and consists of gold-bearing quartz. The shaft is an incline, 500 ft. deep.

At the Julia mine, in Scratch Gravel hills, the shaft is down 250 ft. on the incline. Ore is coming from the 170-ft. level. It contains silver, lead, and gold. In the bottom of the shaft there is 2 ft. of low-grade ore with bunches that run well in gold and silver.

The Cruse Development Co. has paid well and is the best and most extensively developed in the Scratch Gravel hills. The incline-shaft is down 600 ft. The ore carries gold almost exclusively.

In the Grass Valley district the Helena mine is developed to a vertical depth of 300 ft. The ore is silver-lead and shipments are going out. The average smelter returns are at the rate of about $35 per ton.

The Rock Rose shaft, in the same district, is down 300 ft. vertically, in hornstone for the last 100 ft. The ground is traversed by many veins and porphyry dikes. Good ore was found at the 200-ft. level, but no ore is to be stoped or shipped before the 300-ft. level has been reached. More water is coming into the shaft, which here is regarded as a good indication. Driving on the 300-ft. level and stoping on the 200-ft. level is to begin before August 1. No ore has been mined from the property below the 100-ft. level. The ore carries silver, lead, and gold and is sent 8 miles to the East Helena smelter.

Helena, July 16.

The re-modeled Drum Lummon 25-stamp mill at Marysville is crushing 80 tons per day of stope-fillings from the Drum Lummon mine. The material is low grade, but pays well, according to the management. The property is operated by the owners, the St. Louis Mining & Milling Co. Forty men are employed. All machinery is driven by electric power.

Prospecting for new orebodies is under way at the Bald Butte mine and some good ore is reported, but as to its extent nothing has been given out. The Bald Butte mine has paid in dividends $1,500,000. Prospecting is being done in new ground.

The Helena mine, belonging to the Helena Mining Bureau (incorporated), is taking lead-silver ore from the 300-ft. level. The streak is from 1 to 3 ft. wide.

James Coffey has made a small shipment from his mine in Grass Valley to the East Helena smelter. It assayed as high as $500 in silver and lead.

The Rock Rose Co.'s vertical shaft is down 315 ft. and cross-cutting is going on at the 300-ft. level, and driving on ore at the 200-ft. level. The shaft is in hornstone from the 200-ft. level down.

The Producer mine in Dry gulch is being developed by adit and raise. A 2-stamp mill is in operation. The ore is low grade, but bunches occur that are rich in gold.

In the Katie property, in Scratch Gravel hills, an ore that is a heavy sulphide of copper and lead, and that averages about $100 per ton at the smelter, has been discovered.

There is more developing going on about here than ever before in the history of Helena. Five companies are operating in Grass Valley and three others are at work at Marysville. In the Scratch Gravel hills the Thomas Cruse Development Co., the Scratch Gravel Gold Mining Co., the Julia syndicate, the Blue Bird lessees, and the Katie are at work, besides which many small mines are being prospected.

The Snow Drift-Penobscot, at Marysville, is idle and no work of consequence has been attempted there in many years. The Whitlatch property is being worked on a small scale by lessees. They have good milling gold ore near the bottom of the 500-ft. shaft. Water, which was expected to interfere with work in the shaft, did not materialize as it percolates through the limestone to the long adit driven north to intersect other veins.

The Big Indian mine has been abandoned and the electric pole-line to the mine from East Helena has been removed as well as the 60-stamp mill and other machinery. The Big Indian ore became base with depth and was no longer profitable.

There appears to be a good chance for the Whitlatch mine to be worked again. But it may depend upon action by the owners of the Spring Hill property to the north. It was expected

MAP OF REGION ABOUT HELENA

at Helena that the Spring Hill would be purchased and worked on a large scale this summer, but nothing has been done as yet. Poor dumpage and abundant water are the principal drawbacks. The orebody is said to be extensive and is reported to average $7 gold per ton. The ore is a quartz-diorite, in which pyrrhotite prevails. The Whitlatch and Spring Hill are expected to be worked simultaneously. But when this will be is problematical.

Helena, July 24.

Silverbow County

The mining companies at Butte have announced a new sliding-scale of wages for underground miners, which grants an increase of 50c. per day, based on the present price of copper, making the daily wage $5.25, says the *Boston News Bureau*. The minimum wage remains at $3.50 per day with copper at 15c. or less, advancing to $6 per day with copper at 31 to 33c., with 25c. additional for each 2-cent advance in metal price thereafter.

All unclassified men are to get a raise of 25c. per day, and shaft-men an additional 50c. above the rate for ordinary miners. A weekly pay-day is established and the rustling-card system modified.

It is believed that the new scale will prove satisfactory to a large majority of the striking miners as they are granted practically all they asked for with the exception of absolute abolition of rustling-cards.

I. W. W. strikers, however, have not given up the fight and are threatening to begin picketing mines Wednesday. They have served warning on boarding-house keepers not to put up lunches for miners who work. They are promising miners $6 per day for four hours' work eventually if they will not return to work now. Outside of the miners, the metal trades at Butte are voting against acceptance of new rates offered to them by mining companies.

NEVADA

CLARK COUNTY

(Special Correspondence.)—The Duplex property is now expected to produce a large tonnage monthly. Last week E. F. Griffith and Andy Tomey shipped two carloads to the Selby Smelting & Lead Co., from a lease on this property. George Colten also shipped 37 tons of high-grade copper, gold, and silver ore from the property to Selby.

James Cashman, of Searchlight, and William Erwin, and William Crozier of Eldorado Canyon, shipped to Selby a carload of ore from a lease on the Rand Mining Co.'s property. The car was shipped July 8. The ore assayed over $300 per ton.

Charles Spencer, leasing on the Tex Evins property, shipped a carload of good gold and silver ore to the Garfield smelter on July 6.——Booth & Castle, lessees on the Searchlight Parallel Gold Mining Co.'s property, again shipped a brick to Selby, the result of a run at the mill on the Searchlight Spokane property.

The Edward Cebrian Cyanide Co., that has been treating the tailing at the Quartette mine, was closed down for the past few months owing to the high cost of cyanide, but expects to start up again in a few days. The handling of the water has been successfully arranged. The States Mutual Consolidated Mining Co. is making preparations for extensive development at the copper camp. Frank E. Sharp, who is interested in the Big Casino and the Chief of the Hills Mining companies, has taken over the Bonanza-Spokane. A new shaft is being sunk 300 ft. north-west of the old shaft. The new shaft is in ore from the surface. Two shifts are working and a hoist will be installed at once. The property is equipped with a 10-stamp mill and electrical equipment. The Chief of the Hills Mining Co. is overhauling its hoisting machinery, and the mill is in readiness.

Searchlight, July 25.

ELKO COUNTY

It is reported that E. R. Holden is to build a 300-ton smelter at Tuscarora to treat the ore on the dumps of the old mines of that district, according to the *Elko Independent*. There is a large amount of ore on the dumps and it is thought that under existing conditions the proposed enterprise will be profitable.

EUREKA COUNTY

(Special Correspondence.)—W. A. Barnes has taken an option on the Holly mine situated on Adams hill, two miles north of Ruby hill, of which it is a prolongation. Work is progressing for the purpose of exploration and development.

Eureka, July 24.

(Special Correspondence.)—The mill clean-up of the West End for the first half of July consisted of 29 bars valued at $49,000. The ore came from the West End and Ohio shafts, as the Halifax-Tonopah is cross-cutting on the 17th level and will not resume stoping on the 12th level until the old square-set stope has been filled.

At the West End shaft driving east is in progress in 717 D. No. 3. As the drift is in a fair grade of ore considerable territory will probably be explored. The Ohio continues to supply the mill with ore, and cross-cutting is in progress southward and raising to the vein.

The recent strike in the Monarch-Pittsburg, of a 3-ft. vein of good ore, will soon place them on the shipping list, as they are driving on the vein with 80 ft. of backs.

During June the Tonopah Mining Co. produced 8177 tons of ore of an average value of $17.66, making the total value of $136,145, of which $54,750 represented net profit. The development work of the Tonopah Mining Co. the past week has been in the Sandgrass and Silver Top shafts. The ore is also coming from the Mizpah, the value of the ore for the previous week being $35,000.

At the Tonopah Extension development consisted of 36 ft. in the Murray shaft and 79 ft. in the Victor, on the 1440-ft. level and the 1540. At the latter the completion of the 1680-ft. station, which is 140 ft. below the previous workings, will be finished in three weeks and cross-cutting south to the Murray vein will commence. The total production for the week amounted to 2400 tons.

During June the Tonopah Belmont company produced 11,416 tons of ore, making a net profit of $106,989, which is the largest June net earnings in the camp, made possible by the advanced price of silver, for the tonnage was nearly 1000 tons less than in May, while the net profits increased $8000.

Raise 90, on the 1100-ft. level of the Belmont, continues to follow the Rescue vein, which is 3 ft. wide.

The Jim Butler Tonopah Mining Co. sent 2925 tons of ore to Millers for treatment in June resulting in a net profit of $50,136.

Tonopah, July 24.

NEW MEXICO

SOCORRO COUNTY

(Special Correspondence.)—The Socorro Mining & Milling Co.'s clean-up for the last half of June amounted to 17,000 oz., making 35,000 oz. for the month. The Mogollon Mines Co. marketed during the same period about 30,000 oz., or a total of about 65,000 oz. of new gold and silver for the month. High-grade concentrate also was made and shipped direct to the smelter. Shipments are being made by this company from the Maud S. mine as well as occasional shipments from other properties being developed by it.

Mogollon, July 18.

(Special Correspondence.)—The Socorro Mining & Milling Co. cleaned up 16,000 oz. of bullion for the first half of July. The milling plant is being run to capacity.

The Oaks company has completed the head-frame for the new Central shaft. A hoist has been installed and the shaft-house is being built. This company is doing considerable development in the Maud S. mine and is maintaining daily shipments to the mill from development. The main part of the mine will be opened through the Central shaft on Deep Down ground.

The tailing-flume being built by the Mogollon Mines Co. is completed nearly to the Harry tunnel. This has been carried on steel cables through the box-canyons and is giving good satisfaction. It is the plan to run this flume to the main tailing-dam on Mineral creek, three miles below camp.

Mogollon, July 23.

OREGON

JOSEPHINE COUNTY

(Special Correspondence.)—The Turk copper mine, which is owned by the Grants Pass Hardware Co., is under option and is being prospected by M. A. Delano.

Takilma, July 25.

(Special Correspondence.)—A carload of gravel has been shipped from the Osgood placer mine, in Allen gulch, on which a test is to be made. The Preston Peak copper mine is being prospected with a diamond-drill. The mine is owned by a Chicago stock company.

Waldo, July 23.

(Special Correspondence.)—W. S. Baker, of Buffalo, New York, one of the owners of the Greenback mine and other properties in this district, which are being developed by the same interests, is here and will remain to take personal charge. It is reported that G. W. Finch is developing a copper property on Rogue river, in Curry county, at Agness.

Grants Pass, July 24.

TEXAS

Burnet County

(Special Correspondence.)—The holdings of the Texas Graphite Co., situated near Burnet, 50 miles north-west of here, were sold recently at receiver's sale to Victor Brooks, of Austin, as trustee for Albert Burrage, of New York, Ralph Arnold, of Los Angeles, California, and others. The purchase price was $50,000. The purchasers were the same men who composed the company. The property is said to represent an investment of more than $200,000. It is the purpose of the purchasers as soon as the sale is confirmed by the Federal District Court here to resume the development and operation of the property.

Austin, July 22.

Presidio County

(Special Correspondence.)—A large deposit of cinnabar has been discovered near Candelaria, a short distance from the Rio Grande, 75 miles south-west of Alpine. The original claim was taken by J. M. Ingle of Candelaria. Other promising prospects have been taken up and extensive development will follow, including the building of one or more quicksilver furnaces. This new field is situated 90 miles north-west of Terlingua where cinnabar mines have been operated for years. The Terlingua properties are 90 miles from Alpine, which is the nearest railroad shipping point. The discovery near Candelaria is 30 miles from the station of Valentine, on the Southern Pacific.

Alpine, July 23.

Reeves County

(Special Correspondence.)—Alfred Tinally, of Detroit, Michigan, and his associates have purchased large deposits of sulphur near here and are preparing to work them. Their plans include a plant for refining the crude sulphur and the construction of a railroad from Pecos 40 miles to the mines. Large quantities of crude sulphur are being mined west of here and the product is selling for $25 per ton loaded on cars at the railroad.

Toyah, July 25.

UTAH

Salt Lake County

W. J. Moore, H. A. Steinmetz, and A. S. Christian, Federal inspectors from Washington, on July 23 began a systematic investigation of the books of the Utah Copper Co., says the *Salt Lake Tribune*, presumably to arrive at the cost of producing copper at the property. It is expected that the investigation will be extended to other copper companies and to other metals than copper.

BRITISH COLUMBIA

(Special Correspondence.)—Five feet of ore containing 10% copper has been opened in the No. 1 vein on the 1000-ft. level of the Rocher de Boule mine, near New Hazelton, B. C. This is the orebody from which was shipped $1,500,000 worth of ore up to January 1, 1917, from the upper workings. The shoot is being sought at the 1200-ft. level by a cross-cut tunnel now in more than 3000 ft. This tunnel will be continued 4000 ft. where it will open on the opposite side of the mountain.

New Hazelton, July 22.

Cape Breton Island

An explosion of gas in the coal mine of the Dominion Coal Co. at New Waterford on July 25, resulted in the death of 62 miners. The work of rescue of men who were not killed was hampered by deadly gas, and by the caving of the mine workings. Again, as recently at Butte, a mere boy, Jack McKenzie, who went into the mine with the rescue crew and aided in removing a number of men finally lost his own life by asphyxiation.

Personal

Note: The Editor invites members of the profession to send particulars of their work and appointments. This information is interesting to our readers.

Robert Sticht is at Pasadena.

Ben. B. Lawrence is due here from New York.

W. A. Barnes has returned to Eureka, Nevada.

Forest Rutherford has gone to Pueblo, Colorado.

A. H. Rogers has returned to Boston from California.

H. R. Norsworthy is examining placer mines near Boise, Idaho.

Fred. B. Goetter, formerly at San Dimas, Mexico, is now at Los Angeles.

Henry E. Wood, of Denver, is here, on his return from Shasta county.

Earl B. Crane is manager of the Black Butte quicksilver mine, in Oregon.

J. S. Diller, of the U. S. Geological Survey, is in southern Oregon inspecting chrome deposits.

F. A. Beauchamp, of the firm of Beauchamp, Hamilton & Woodworth, has returned from Arizona.

Peter H. Nissen, of the Nissen stamp, is now Lieutenant-Colonel, besides having received the D. S. O.

William W. Mein and Walter Karri-Davies have gone to the Le Pas copper-mining district in Manitoba.

Charles O'Brien is on his way to Salvador mines, to serve as metallurgist with the Butters Salvador Mines.

R. A. Hardy has succeeded E. L. S. Wrampelmeier as manager for the Montana Mines Company at Ruby, Arizona.

Carl R. Davis, general manager for the Brakpan Mines, Ltd., will visit the United States in the late summer and fall.

E. Gybbon Spilsbury sailed from New York on July 26 for Central America and expects to be absent until the end of August.

Charles F. Williams, recently mining engineer at Cananea, Mexico, is with the Tonopah Belmont Development Co. at Tonopah.

W. L. Honnold, who has been serving as director, at New York, for the Commission of Relief in Belgium, is expected here next week.

Herbert W. Gepp has resigned as manager for the Amalgamated Zinc Co. to become manager for the Electrolytic Zinc Co. in Australia.

E. L. S. Wrampelmeier has resigned the management of the Montana Mines Co. at Ruby, Arizona, and now has headquarters at Tucson.

E. S. King passed through San Francisco on his way to New Zealand, having been appointed manager of the Waihi Grand Junction mine.

Howland Bancroft has gone to New York, where he will be at the Engineers Club until about the middle of August, after which he will return to Denver.

E. L. Larison has resigned from the Anaconda Copper Mining Co. to accept a position with the Garfield Chemical & Manufacturing Corporation in Utah.

John G. Kirchen will make his headquarters at Reno.

Alliene Case is superintendent of the Tonopah Extension mines, under direction of Mr. Kirchen.

Kiyoshi Shiga, physicist to the Metallurgical Research Institute, of Tokyo, has been visiting metallurgical plants in this country and is now returning to Japan.

George A. Tweedy, formerly general manager for the Minas del Tajo, Rosario, Mexico, has gone to Honduras, Central America, to examine the old long abandoned mines at Opeteca, near El Espino, in the Department of Comyagua.

THE METAL MARKET

METAL PRICES

San Francisco, July 31

Aluminum-dust (100-lb. lots), per lb	$1.00
Aluminum-dust (ton lots), per lb	$0.95
Antimony, cents per pound	15.50—18.00
Electrolytic copper, cents per pound	31.00
Pig lead, cents per pound	12.25—12.50
Platinum, soft and hard metal, per ounce	$105—111
Quicksilver, per flask of 75 lb	$110
Spelter, cents per pound	10.50
Tin, cents per pound	60
Zinc-dust, cents per pound	20

ORE PRICES

San Francisco, July 31

Antimony, 50% metal, per unit	$1.20
Chrome, 40% and over, f.o b. cars California, cents per unit	50—55
Magnesite, crude, per ton	$8.00—10.00
Tungsten, 60% WO_3, per unit	26.00
Molybdenite, per unit for MoS_2 contained	$40.00—45.00
Manganese, 45% (under 35% metal not desired), cents, unit	33—37

Manganese prices and specifications, as per the quotations of the Carnegie Steel Co. schedule of prices per ton of 2240 lb. for domestic manganese ore delivered, freight prepaid, at Pittsburg, Pa., or Chicago, Ill. For ore containing

	Per unit
Above 49% metallic manganese	$1.00
46 to 49% metallic manganese	0.98
43 to 46% metallic manganese	0.95
40 to 43% metallic manganese	0.90

Prices are based on ore containing not more than 8% silica nor more than 0.2% phosphorus, and are subject to deductions as follows: (1) for each 1% in excess of 8% silica, a deduction of 15c. per ton, fractions in proportion; (2) for each 0.02% in excess of 0.2% phosphorus, a deduction of 2c. per unit of manganese per ton, fractions in proportion; (3) ore containing less than 40% manganese, or more than 12% silica, or 0.225% phosphorus, subject to acceptance or refusal at buyer's option; settlements based on analysis of sample dried at 212° F., the percentage of moisture in the sample as taken to be deducted from the weight. Prices are subject to change without notice unless specially agreed upon.

During 1916 there was shipped from Panama mines manganese ore to the value of $328,684. Further expansion of this industry is reported as probable. Brazilian manganese mines, which are reported to be extensive, are being worked and an increasingly larger tonnage shipped. The industry there is hampered by the difficulties of cheap transportation.

EASTERN METAL MARKET

(By wire from New York)

July 31.—Copper is firmer but nominal at 28 to 28.50c. Lead is dull and higher, closing at 10.75c. Zinc is quiet and steady at 8.75c. throughout the past week. Platinum remains unchanged at $105 for soft metal and $111 for hard.

SILVER

Below are given the average New York quotations, in cents per ounce, of fine silver.

Date			Average week ending		
July	25	78.37	June	19	77.00
"	26	78.37	"	26	78.12
"	27	78.12	July	3	77.98
"	28	78.12	"	10	78.70
"	29 Sunday		"	17	80.62
"	30	78.12	"	24	79.12
"	31	78.62	"	31	78.20

Monthly Averages

	1915	1916	1917		1915	1916	1917
Jan.	48.85	56.76	75.14	July	47.52	63.06	78.92
Feb.	48.45	56.74	77.54	Aug.	47.11	66.07	....
Mch.	50.61	57.89	74.13	Sept.	48.77	68.51	....
Apr.	50.25	64.37	72.51	Oct.	49.40	67.86	....
May	49.87	74.27	74.61	Nov.	51.88	71.60	....
June	49.03	65.04	76.44	Dec.	55.34	75.70	....

A compilation by National City Bank shows that silver production of the United States is now double that of Mexico and three times that of Canada, which holds third rank among silver producing countries, says the 'Boston News Bureau.' Of 172,383,000 oz. produced in the world in 1916 the United States produced 72,883,000, or 42%, while 20 years ago, in 1896, we produced but 37% of world production, in that year 157,061,000 oz. In that 20-year period, 1896 to 1916, production of the United States has increased about 14,000,000 oz., while that of other parts of the world increased only about 1,000,000.

The chief producers are the United States, Mexico, Canada, Peru, Japan, Spain, Australia, and Chile, in order named, the United States having produced, in 1915, 74,961,000 oz., Mexico 39,570,000, Canada, 28,401,000, Peru 9,420,000, Japan 5,080,000, Spain 4,565,000, Australia 3,327,000; world total 179,574,000. In 1916, for which returns are not yet available for many smaller countries, production of the United States was 72,884,000 oz., Mexico 35,000,000, Canada 25,500,000, and the world total 172,384,000 ounces.

COPPER

Prices of electrolytic in New York, in cents per pound.

Date			Average week ending		
July	[illegible]	26.00	June	19	32.58
"	[illegible]	26.50	"	26	32.42
"	[illegible]	27.00	July	3	32.25
"	[illegible]	27.50	"	10	31.50
"	[illegible] Sunday		"	17	30.29
"	[illegible]	28.00	"	24	27.04
"	31	28.50	"	31	27.25

Monthly Averages

	1915	1916	1917		1915	1916	1917
Jan.	13.60	24.30	29.53	July	19.09	25.66	29.67
Feb.	14.38	26.62	34.57	Aug.	17.27	27.03	....
Mch.	14.80	26.65	36.00	Sept.	17.69	28.28	....
Apr.	16.64	28.02	33.16	Oct.	17.90	28.50	....
May	18.71	29.02	31.69	Nov.	18.88	31.95	....
June	19.75	27.47	32.57	Dec.	20.67	32.89	....

Copper production from the principal mines and smelters of North and South America approximated 1,085,815,413 lb. during the first six months of 1917. But for the miners' strikes which started in June, output would have been considerably greater. The Butte district in June showed a substantial falling off due principally to Anaconda's drop from 28,000,000 to 20,000,000 lb. In Arizona losses were general, all the properties with the exception of Ray Consolidated being tied up by strike. The Alaskan properties of Kennecott Copper Corporation were also shut-down with the exception of the Beatson mine.

June's total copper yield was about 167,000,000 lb., comparing with 190,000,000 lb. in May. The July production will be far less than that of June, as some of the mines will not have turned out a single pound of copper during July.

The dividend of Greene-Cananea Copper Co. which was recently declared is the first dividend of the company since the sale to it of the assets of Greene Consolidated Copper Co. The Greene-Cananea Copper Co. now receives its dividend direct from this Mexican corporation instead of through the Greene Consolidated Copper Co. as formerly. Therefore holders of Greene Consolidated Copper Co. stock who have not yet converted their shares into Greene-Cananea Co. stock will do so before the stock becomes ex-dividend on August 15 if they wish to participate in this dividend. Eighty-cent silver will be of considerable aid to copper producing companies, which also recover silver as a by-product, in offsetting to some extent their rapidly rising copper costs.

LEAD

Lead is quoted in cents per pound, New York delivery.

Date			Average week ending		
July	25	10.25	June	19	12.00
"	26	10.25	"	26	11.75
"	27	10.50	July	3	11.57
"	28	10.75	"	10	11.25
"	29 Sunday		"	17	10.98
"	30	10.75	"	24	10.29
"	31	10.75	"	31	10.55

Monthly Averages

	1915	1916	1917		1915	1916	1917
Jan.	3.73	5.95	7.64	July	5.59	6.40	10.93
Feb.	3.83	6.23	9.01	Aug.	4.67	6.28	....
Mch.	4.04	7.26	10.07	Sept.	4.62	6.86	....
Apr.	4.21	7.70	9.38	Oct.	4.62	7.02	....
May	4.24	7.38	10.29	Nov.	5.15	7.07	....
June	5.75	6.88	11.74	Dec.	5.34	7.55	....

ZINC

Zinc is quoted as spelter, standard Western brands, New York delivery, in cents per pound

Date			Average week ending		
July	25	8.75	June	19	9.72
"	26	8.75	"	26	9.43
"	27	8.75	July	3	9.32
"	28	8.75	"	10	9.20
"	29 Sunday		"	17	9.06
"	30	8.75	"	24	8.60
"	31	8.75	"	31	8.75

Monthly Averages

	1915	1916	1917		1915	1916	1917
Jan.	6.30	18.21	9.75	July	20.54	9.90	8.98
Feb.	9.05	19.90	10.45	Aug.	14.17	9.03	....
Mch.	8.40	18.40	10.78	Sept.	14.14	9.18	....
Apr.	9.78	18.62	10.20	Oct.	14.05	9.92	....
May	17.03	16.01	9.41	Nov.	17.20	11.81	....
June	22.20	12.85	9.63	Dec.	16.75	11.26	....

QUICKSILVER

The primary market for quicksilver is San Francisco, California being the largest producer. The price is fixed in the open market, according to quantity. Prices, in dollars per flask of 75 pounds:

Week ending

Date					
July	3	85.00	July	17	105.00
"	10	100.00	"	24	110.00
			"	31	110.00

Monthly Averages

	1915	1916	1917		1915	1916	1917
Jan.	51.90	222.00	81.00	July	95.00	81.20	102.00
Feb.	60.00	295.00	126.25	Aug.	93.75	74.50	....
Mch.	78.00	219.00	113.75	Sept.	91.00	75.00	....
Apr.	77.50	141.60	114.50	Oct.	92.90	78.20	....
May	75.00	90.00	104.00	Nov.	101.50	79.50	....
June	90.00	74.70	85.50	Dec.	123.00	80.00	....

TIN

Prices in New York, in cents per pound.

Monthly Averages

	1915	1916	1917		1915	1916	1917
Jan.	34.40	41.76	44.10	July	37.38	38.37	....
Feb.	37.23	42.60	51.47	Aug.	34.37	38.88	....
Mch.	48.76	50.50	54.27	Sept.	33.12	36.66	....
Apr.	48.25	51.49	55.63	Oct.	33.00	41.10	....
May	39.28	49.10	63.21	Nov.	39.50	44.12	....
June	40.26	42.07	61.93	Dec.	38.71	42.55	....

Eastern Metal Market

New York, July 25, 1917.

It is still a waiting and restricted market, almost paralyzed in some departments, held up by the indecision of the Government purchase and price-fixing policy. Quotations are largely nominal and sales are of small volume in almost every case.

Copper continues to decline and is lower.

Tin is quiet and dull but steady.

Lead has stiffened somewhat after a week of declining prices and little business.

Zinc is steadier after a period of weakness.

Antimony continues its decline, with demand small.

Aluminum has fallen on poor demand and more offerings.

Steel buyers and consumers continue to hold aloof, prompted by the fear of what will happen to prices if the Government pursues its one-price-to-all policy. Evidence is accumulating that the problem of determining a representative cost of any form of finished steel is turning out to be more difficult than was expected at the outset. It is hoped that this realization may lead to abandonment of the task. The shipbuilding programme, as now planned for the coming year, will require almost 1,000,000 tons of plates, or about 50% of what ship-plate mills can produce. In no other steel line will the Government proportion be so large—not as press reports have said: "the Government requiring the entire maximum output of the mills for some time to come."

COPPER

The influences controlling the copper situation are practically the same as those mentioned last week—Government purchases, the amount, and the price. Nor is there any definite information obtainable that will ease a situation that has nearly paralyzed not only copper but other markets. The feature of the week has been an unlooked-for drop in the London market. Spot electrolytic is now quoted at £137 with futures at £133, a decline of £5 per ton from a price which has remained stationary for a long time. Some regard this as an attempt to bear the market on this side, while others are inclined to think it will result in a better understanding as to quotations between large and small producers and perhaps have a stabilizing influence. In the past week there has been little business reported, though some sales of moderate proportions were made for September-October delivery at 24.50 to 25.50c., New York. Sales, however, for most positions were not enough to establish a market and quotations which are mostly nominal have been established by offers and not by actual transactions. Yesterday the quotation for both Lake and electrolytic copper was 26c., New York, for early delivery, or July and August, with fourth quarter generally held nominal at 24c., New York. Exports of copper are estimated at 40,000 gross tons for June, bringing the total for the half year to nearly 266,000 tons or about twice the total to July 1, 1916, of 147,943 tons. Imports for the same period have been 114,000 tons this year, against 103,500 tons last year. The June output of refinery copper is estimated at 200,000,000 lb., with the total for the first half of this year at 1,055,000,000 lb., contrasting with a total for 1916 of 2,259,000,000 lb. One commentator says that in ordinary times the requirements of public-service companies are about two-thirds of the copper consumption of the country, but at present they are probably only one-sixth to one-third of the total amount now annually consumed.

TIN

In the past week sales have been moderate in volume and importance and the market has been dull and quiet. Such sales as have been reported have been mostly early shipment from England or future shipment from the Straits Settlements. The market feature has been a pronounced scarcity of spot-tin. This has been due to unwillingness in England to grant shipping permits, and to the slowness by American buyers to commit themselves. No other attitude can be expected with the question of price regulation still dominant and holding up the market. There has been some little desire to sell tin in some quarters, but on the whole cautiousness rules. The market has been almost a nominal one at 62.50c. the entire week since July 19, the quotation on July 20 having been 62.62½c., New York. The London market is dull and unchanged from last week at £237 10s. for spot Straits.

LEAD

The conviction is firm in the trade that large quantities of lead will be wanted for war and other purposes, but until yesterday the market had not recovered from its scare of last week regarding the one-price-to-all policy as proclaimed by the President. Buyers were extremely cautious and adopted a 'play-safe' policy. Offerings were more or less liberal, coming from second-hands and from consumers, and the market declined until 10c. lead, New York and St. Louis, was the prevailing price at the end of the week. Large producers are having their labor troubles also. The latter part of last week lead was obtainable at 10c., New York, with some sales reported at 10c. for August-September shipment. Yesterday the market stiffened, due either to the withdrawal of the offerings referred to, or to their having been sold, and it was difficult to obtain lead under 11c., New York, which continues to be the quotation of the American Smelting & Refining Co. There were, however, small quantities sold here and there at 10.25c., New York and St. Louis, with also about 100 tons going at 10c. The quotation yesterday, therefore, is put at 10.25c., New York and St. Louis, though the broad market is from 10.50 to 11c., with an advance probable.

ZINC

After falling to the lowest point in some time, the market for prime Western spelter has grown somewhat stronger. There is more inquiry than for some weeks, but the volume is not large. Consumers, at least some of them, are realizing that spelter is cheap or too close to the cost of production and hence are inclined to display more interest. Added to this is the strengthening expectation, almost amounting to an assurance in some quarters, that the Government is about to negotiate for its needs for the rest of this year. A settlement of this question would be a stabilizing influence of decided benefit and importance to the whole market. Today the market is higher if anything than yesterday and is generally quoted at 8.50c., St. Louis, or 8.75c., New York, for early or future delivery. Sales of small lots of 50 or 100 tons each are reported as having been made at 8.37½ to 8.62½c., St. Louis, or 8.62½ to 8.87½c., New York. There is more interest in future delivery than in prompt or nearby, but not so much inclination to sell far ahead. A report is current today that a sale of 225,000 tons of prime Western has been made to the Government at 8.75c. per lb., St. Louis, for delivery the rest of the year, but no confirmation is obtainable.

ANTIMONY

Two lots of 25 tons each were sold in bond yesterday and Monday under the hammer on the Metal Exchange at 13.37½c. per lb., which is equal to about 14.50 to 14.75c. per lb., duty paid. The market, however, is very quiet with Chinese and Japanese grades at about 15c., New York, duty paid.

ALUMINUM

There are more sellers than buyers and the market is lower at 53 to 55c., New York, for No. 1 virgin metal, 98 to 99% pure.

EDITORIAL

T. A. RICKARD, Editor

IN our last issue we published an excerpt from the brief of the defendant in the Butte & Superior flotation suit. In this issue we give our readers a similar excerpt from the plaintiff's argument in the same case.

ON another page we reproduce a note on the cargo of copper that was lost on the 'Pewabic' in 1865. Later information states that 50 tons of copper, valued at $30,000, has actually been recovered by the divers. We note also that our correspondent at Juneau reports that a diver has contracted to retrieve the high-grade copper ore from the Kennecott mine that has been spilled into the harbor at Cordova in the course of loading the ore during the last three years. He is to be credited with half of the copper he recovers for the company.

WE are able to publish an article by Mr. Harold T. Power, who for many years directed the successful operations of the largest drift-mine in California, namely, the combined Hidden Treasure and Mountain Gate, on the Forest Hill divide, in Placer county. The reader may be surprised to learn how much timber is required in this kind of mining. When in full operation the Hidden Treasure mine consumed 2,000,000 feet annually. The tunnel that followed the course of the ancient river-channel, now buried under lava, went through a mountain, from side to side, and was four miles long.

SUNDRY smooth gentlemen in New York suggest that a market for Liberty bonds should be created by allowing the tax on excess profit to be paid in large part, say, 60 out of 78%, by purchase of such bonds. This is a lovely bit of spoofery. Mr. Otto H. Kahn has shown that the immunity from income-tax on big incomes, subject to super-tax, will give the 3½% rate of the Liberty bond a yield of 6.45% in respect of incomes of over $250,000, 8.75% on over $1,000,000, and 9.21% on over $2,000,000. In short, the plutocrats made by the War are to evade taxation and obtain a gorgeous income from their excessive profits. It is too thin.

CHEMICAL engineers are in demand. The call for men comes from a wide range of industries, many of which are intimately related to metallurgy. The development of wet processes, and the extraordinary growth of electro-chemical applications to the preparation of metallic products for the market, has rapidly [illegible] the buffer-zone between the metallurgist and the chemist. Distinction between the two is disappearing; even in ore-dressing the successful practitioner must apply recondite principles of colloid chemistry that lead also into the utilization of organic substances. It is a pity that it has seemed necessary to call into military service so many young men from the higher classes in our universities. It will leave broken ranks in the recruits of trained technologists that will be needed so urgently if the War should continue for several years. It will also leave us short of young chemico-metallurgic engineers to maintain our industrial expansion after peace has returned. In order to make good the deficiency to some extent young women might be trained for sundry branches of this work, since they have demonstrated special aptitude in laboratory practice, but the response is not likely to be great. The Government would do well to discriminate against the enlistment of young men who are receiving a university education until they have completed their training. To do otherwise is to risk setting the nation back in its provision to meet the problems of the future.

ANTIMONY is of peculiar interest, despite the fact that the market has exhibited extraordinary dullness. It is always when antimony is inactive that it pays to prepare for the sudden leaps in price that characterize this metal. When the price jumps it is too late to develop the mine to take advantage of the rise in value. Being a minor metal, with largely fluctuating demand, it is peculiarly liable to erratic quotation. It must be remembered that we are chiefly dependent upon China for our supply, and antimony is just needful enough in our domestic economy to make it a matter of concern to look curiously, not once but repeatedly, at every district on our own continent that is known to contain it in any quantity. Consequently the article on the occurrences in Arkansas, by Professor Ellsworth H. Shriver, in a recent issue, may serve a useful purpose in stimulating search and development. If done by corporations financed so as to be able to tide over depressions in the market, such exploration would appear to offer attractive opportunity. There are other districts that should be searched carefully, one of these being the Tehachapi-San Emidio zone of antimony-mineralization in California. Durango, in Mexico, also possesses notable outcrops that have not received the attention that they merit.

CARRANZA is persistently pressing the Administration at Washington to speed his plans for enjoyment of a portion of the bounty that this country is distributing so freely among the nations pledged to overthrow the common enemy. We have frequently pointed out that his interference with Americans in Mexico was inspired by the definite hope of obtaining money. It may seem anomalous to our people that anyone should undertake to badger or to injure another as a means for stimulating his generosity, but this indicates how difficult it is for the plain American to comprehend the subtle mind of a Mexican cacique. Carranza is a born psychologist; he knows that kicks constitute the humanly probable response to an appeal for sympathy, and that a war-hating people is more likely to compromise with open purse in hand than to bother about chastising him for his petty annoyances. He fully understands this nation has never taken the Mexican interests of its citizens seriously. The money invested across the Rio Grande seemed small in proportion to the total industries of the country, as in truth it is; therefore he has felt certain that his pestering would not lead to armed reprisals, especially at a time when we were busy with a war of real magnitude. The result has been what we predicted weeks ago: by his interference with our vested interests in Mexico he has called attention anew to his existence and to the possibility that if we do not sustain him he might turn to Germany, which, conceivably, might be to our disadvantage. The scheme has worked so well that the good offices of the Government have been enlisted to the extent of unofficially arranging to send a group of expert accountants, nominally representing New York financial institutions, to study the situation in Mexico with the prospect of accommodating the crafty fox with a loan of ₱300,000,000 recently authorized by his validating agents, otherwise known as the Mexican congress.

DOUBT as to the ultimate intentions of Japan has been felt by many people in this country; we have wavered between trust in her professions of friendship and suspicion of left-handed dealings with Germany. The Zimmermann episode left something to explain that has not been made perfectly clear to the public mind. Despite assurances that Japan could be depended upon to play fair as a loyal ally, her activities in aiding Mexico in preparation for an independent supply of munitions, and her quick resentment of our kind expressions to China at a moment of crisis, strongly pointed toward ulterior motives. Just what these are has been revealed in a surprisingly frank statement by Mr. Kazan Kayahara, former editor of the *Yorodzu* and of the *Third Empire* in Tokio, but now temporarily in New York as the representative of the Japanese press. The interview was given to *Las Novedades*, the leading Spanish newspaper in the United States, and an aggressive Carranza organ. Mr. Kayahara insists that Japan must have a free hand in China, and that no interference with her supremacy there will be brooked. Japan needs iron and coal from China for the further development of her steel industry, and she proposes to control the Oriental trade in cotton goods. "Only a part of the Japanese nation is disposed to fight because of exclusion," he said, "but this other question of supremacy is one regarding which the whole people thinks as one man. It is a matter of life or death for us. Isolate Japan and you crush her; but we do not intend to let her be crushed without a struggle. You know that Japan is not without friends in case of a new world-alignment of nations. If the United States, Great Britain, and France press us too hard we will not hesitate to make an alliance with Germany. There exists no hostility against Germany in Japan. In truth the Japanese admire Germany. The greater part of our scientific men have been educated in Germany. Our army has been organized on the German model. Our physicians are almost all graduates of German clinics. There do not exist any anti-German sentiments in Japan, and the events of the War have augmented our admiration for the German people and their government. In case of necessity we would turn naturally toward Germany, and this necessity would arise if the present enemies of Germany should undertake to dispute our supremacy in the Far East, that is to say, in China." After that pronouncement it would seem that we need no longer entertain any doubt as to the limit of strain that may be put upon the Japanese alliance with the warring powers whose cause has also become our own.

Concrete Ships

The announcement that a reinforced concrete ship was being built in California has attracted general attention. It appeared to be a remarkable innovation, and at first sight somewhat incongruous. As a matter of fact the construction of concrete boats is not new; as long ago as 1899 scows and barges of this material were made by Carlo Gabellini for service on the Tiber, and by 1905 his experience had resulted in the development of that method of boat-building so far as to admit of the construction of 150-ton barges for Civita Vecchia. Gabellini was, however, anticipated by a cement company in Holland that built some small boats in 1887, and gradually ventured upon more pretentious vessels of 50 to 60 tons capacity. Concrete barges are by no means uncommon in this country, the Arundel Sand & Gravel Company of Baltimore having had three such boats of 500 tons each in service for several years, and similar barges were used on the Panama Canal. A pontoon having a displacement of 783 tons was built of concrete for the Sydney Harbor Trust at Sydney, New South Wales, in 1914, and another has made a record on the Welland canal. A sea-going barge of concrete, 210 feet long, was constructed in England about four years ago and towed to South America, where it was equipped as a dredge and has seen hard usage, being subject to the excessive vibration characteristic of this class of work. Numerous patents have been issued in the United States for scows and ships of concrete, but no vessels for navigation on the high seas had so far been attempted until plans for that purpose were perfected in San Francisco. The designs were drawn by Macdonald & Kahn, under advice from the well-known marine architects, Caverley & Garden, the venture being financed by the Standard Oil Company, the Santa Cruz Portland Cement Company, and some of the strongest capitalists of California. The vessel is being built at Redwood City, on San Francisco bay, and will be ready for launching in about 100 days. It will be 320 feet long, 44 feet beam, and 32 feet molded depth, the hull and strength-deck to be made entirely of reinforced concrete without a steel frame. The net capacity will be slightly more than 5000 tons, and it is estimated to cost approximately $350,000. In view of the impossibility of obtaining the boilers and engines as soon as the hull will be completed, it is intended to tow it, when finished, to Seattle. This will provide a preliminary test. The Government has already requisitioned the vessel, and others will be constructed as rapidly as possible. The advantage offered by this type of ship is the facility for building more rapidly than when constructing in steel or wood. The limitation to launching a large number of concrete vessels this year lies in the difficulty of securing the necessary power-equipment as fast as needed.

The Lynching at Butte

On July 31 one of the I. W. W. leaders, a man named Frank Little, conspicuous in the labor agitation at Butte, was taken out of bed and hung by the neck from a trestle. The perpetrators of this crime are not known and no arrests have been made as yet. All good citizens must deplore the act. There should be none so short-sighted as to think that this is the way to get rid of irresponsible agitators and trouble-makers; otherwise our democracy is a failure. We view the event with detestation, not only because it menaces free institutions but because we believe that it does not in the least help to settle the industrial quarrel in the mining districts. Even if it succeeds, for the moment, in terrorizing the ringleaders of the mob that calls itself the I. W. W., it does nothing to check their propaganda; on the contrary, it will breed reprisals similar in kind, bringing the worst elements to the front on the opposing sides of labor and capital. Another blunder was the concession of a 50-cent increase of wages to the miners at Butte. It may cause many of the men to resume work and for a time it may frustrate the plans of the agitators that are responsible for the strike, but it settles nothing. It is a sop to Cerberus; it is a mere palliative that does nothing to cure the disease in the body politic. No principle has been recognized, save that of noise and force. The question of collective bargaining, called the 'union'; the question of the employer's right to discriminate in the selection of his employees, represented by the 'rustler's card'; the problem of participation in profits by the mine-owner and the miner; all these remain where they were, as fertile seeds for future trouble. It is easy to give 50 cents more per day with copper at twice its normal price, but what is to happen when the war-market ceases to support an abnormal quotation? Will the miners accept the logical reduction or is it to be a rule that works one way only? At Leadville also a truce has been declared on the basis of a concession of 50 cents more per shift, leaving the local food-stuff monopoly to prey on the community, leaving an unscrupulous smelting trust in control of the output, while the question of unionism remains as undecided as before. Of all the ostriches that ever buried their beaks in the sands of time, thinking thus to avoid danger, the worst is the mining industry. We shall not exorcize the demon of unrest either by violence or by bribes. To lynch an agitator or to concede a few cents per day to his followers is no way of settling these vital economic difficulties. Here is where the American Institute of Mining Engineers, daring to become both free and representative, could do a great public service. Here is where the profession, if it were independent, if it did not stand between the greed of the capitalist and the ignorance of the laborer, could guide public opinion effectively so as to stimulate both legislation and government in the right direction. It is useless to close our ears to the questions that must be answered sooner or later, and the sooner the better for all concerned. Has the worker a right to combine for collective bargaining? Has the mine-owner a similar right? Is each restricted to a local combination or may they form extra-territorial organizations, so as to bring secondary pressure to bear on a domestic disagreement? Is a strike no more and no less legal than a lock-out? Has the employer a right to select his employees, as the employees have the privilege of leaving one employer to go to another? Is working by contract beneficial to the individual and the industry? Should a mine-worker participate in the profit of the mine-owner, and, if so, should he accept lower wages when the profit decreases? Should the excessive profit of a mining company go to the workers in that mine or to the State, whose resources are being exploited, or should it all go to the shareholders? Is it fair that a copper-miner should receive $5.75, while an equally good gold-miner in an adjacent district receives only $3.50? In short, is the Devil to take the hindmost or is this policy of *laissez faire* to be corrected by a recognition of facts and an effort to make regulations that will square both with them and with the sense of fair play? What we need is a sense of social justice and a deeper interest in conditions that will imperil our social fabric unless we give heed to them. To ignore them is to be ignorant. We invite our readers to state their views.

Why We Fight

The results of the draft and the imminence of another national bond-issue suggest the necessity for a plain statement of the causes that led the United States to go to war and a clear explanation of the menace against which our armies are being mobilized. The large proportion of exemptions under the draft and the apparent desire of so many vigorous young men to evade military service is a spectacle so ignominious as to provoke anxious thought. We know that the youth of America is neither cowardly nor unpatriotic. Why then is the effort to escape active service so general, and why is it that the parents of these young men have looked around for excuses or influence to enable their sons to remain at home? The reason is not far to seek; on the contrary, it is so obvious that we marvel why it was not recognized and corrected as soon after April 6 as possible. To our mind an injustice has been done to the youth of this country. For two and a half years the young men were told by the President, and presumably by their parents, to be neutral. This meant that they were not to concern themselves with the great events in Europe, for it goes without saying that a keen interest in such events, and in the underlying causes of them, must lead to a condition of mind in which neutrality would be stultifying. So they took more interest in base-ball matches than in the invasion of Belgium, and were more excited by a Villa raid on Columbus than in the Crown Prince's attempt to conquer Verdun. After the European War—it was only that to them—had lasted two years, they had the chance to elect a new President, and the majority of them—these young men now called upon to shoulder arms—were prompted by their own ideas and by public opinion around them to vote for the President that had 'kept us out of war.' He was elected triumphantly, thereby proving that he had rightly interpreted the wishes and desires of the people of the United States. The President had not begun his second term when the dismissal of the German ambassador made war certain, but even then the peace idea was so strong that our young men still took it for granted that this country could remain detached from the European struggle. Within two months, the President, elected on a peace ticket, turned to the people of the United States with a declaration that he could no longer avoid the arbitrament of war and in the same message he called upon the manhood of America to rally to the flag for a campaign in which the Army and Navy, the capital and the industry, of the country were to be contributed wholeheartedly and without stint in an effort to "make the world safe for democracy." Is it any wonder that the young men were perplexed? Is it surprising that they did not hurry to enlist? Ever since the War began they had been told that it was none of their business, those that volunteered in the French and British armies were scolded, those that remained unperturbed were assured that their own country was not involved in the great catastrophe, that it was our duty to maintain a dignified abstention, and that it was the chief mission of the United States to act as peace-maker between the belligerents. The echoes of these teachings and preachings had hardly died away among the crash of the torpedoes that sank unarmed vessels off the coast of New York when the young men of America were told to make ready for a selective conscription. Do you wonder that they showed no enthusiasm? A large part of them would have volunteered when the call came, but, seeing that they were to be drafted under compulsion, they baulked; and when the law allowed excuses to be made, they made them—not without the prompting of parents that had deceived themselves into the notion that this was no war of ours, but one of those cat-and-dog fights for which the European nations had been preparing so long. So they made a poor response to the President's call to arms. They are not to blame. It is illogical to blame them. We had no right to expect them to make a mental right-about-face until they had been told why the policy of the Government had undergone so sudden a change. They are told that democracy must be made safe. Such abstract ideas are not convincing to everybody. To fight for democracy or against hypocrisy is an intangible issue to the young man rudely awakened from the dreams of peace and plenty. He must be told that he is to fight for his liberty, which is menaced by organized hell; for his country, which is threatened by the navies and armies of a great nation crazed by the lust to conquer; for his home, which may be devastated if things go wrong with those that are fighting the common enemy. It must be made clear to him that if the German armies overcome the Russian resistance, if the Anglo-French attack should waver, or if the German submarine campaign succeeds, then the people of the United States may be held under ransom, then a larger Belgium may be made on this continent by an inhumane and unscrupulous foe, then a German victory over the Allies would involve the surrender of the British navy, leaving the world at the mercy of a power that is without compunction and without shame, a power that bitterly resents first the neutrality and then the hostility of the United States, and that therefore would deal with this country, if it could, in no gentle manner. We fight for democracy and democratic ideas, of course, and the preciousness of this heritage from Washington and Lincoln none of us deny, but we tell every young American that a thousand flags are fluttering in the breeze to signify that we are in a fight for our safety, for the freedom to live and let live, for our hearths and our homes. It is time that all this were told in better words and at greater length to the youth of the country if we are to expect them to show eagerness to serve. Let the two ex-Presidents go on a speaking tour; let such men as Elihu Root and Henry Cabot Lodge be co-opted for the good work; let such leaders of education as President Hibben of Princeton and President Wilbur of Stanford be engaged to explain the great issue; let the diplomats that have returned recently from Europe—Ambassadors Gerard, Penfield, and Van Dyke—tell our people why and wherefore we are committed to this war, and how great is our danger if we wage it unsuccessfully. Then the claims for exemption will become few; then the next draft will have the look more of a great volunteering than a conscription; then at last our young men will go to war with an intelligent idea of the national purpose. Meanwhile, if the President and Secretary of War desire a warmer response to their call, the expeditionary force now in France should go forward into action, as soon as it is ready, not waiting for larger numbers. When American blood has been shed in pitched battle, it will be a summons more eloquent than all the speeches ever spoken. Let us wage this war with a clear understanding of the issue and a keen emotion for the justice of our cause.

Mining in Utah

By L. O. HOWARD

Dufficulties with labor appear to be ended for the present. Outside the strike at the International smelter and allied disputes, none of the troubles became serious. After tying up the plant at Tooele for several weeks, the strikers went back to work the last of June, accepting the offer of a new sliding scale of wages. This provides an increase of 10c. per day for each one cent increase in the price of lead. Eight cents is taken as the basal price, $2.85 per day being the base for unskilled labor, $3.80 for semi-skilled, and $5 for skilled. Before this settlement was made the company tried to get the producers to share part of the increased cost, but this they declined to do. Federal mediators had a large share in arranging a settlement. At the conferences the other smelters and mine operators were represented. A general increase of wages throughout the State was put into effect on July 1. The American Smelting & Refining Co. advanced wages at Murray and Garfield 25c. per day. Mines in the Tintic district posted an increase of 50c. per day. In the Bingham district employees who had received $3.25 per day will now enjoy an increase of 25c. Those receiving less will have their pay increased 20c. Common labor at the Arthur and Magna mills of the Utah Copper Co. has been advanced 20c., and skilled labor 25c. Operators from Bingham, Park City, Tintic, Cottonwoods, American Fork, Tooele county, and southern Utah, represented in the conference, adopted a uniform scale for different kinds of labor. They will pay machine-men $4.50, miners $4.25, shovelers and trammers $4, and timbermen $4.50. The Utah chapter of the American Mining Congress was unsuccessful in obtaining a reduction of the insurance rate to be paid under the terms of the Workmen's Compensation Act, which went into effect in July. The rate has been maintained at $5.59 for metal mines by the State Industrial Commission. No relief from this oppressive rate is in sight. On July 11, the D. & R. G. R. R. announced that it had restored one track to service in Carbon and Emery counties, and was ready to resume hauling coal. Some of the mines are still without transportation facilities, however, and there is a serious shortage of coal at Salt Lake; it arouses fear as to what may happen next winter. Various municipal and state authorities have been investigating the situation. An appeal was made to Washington, and an embargo sought on all shipments over the D. & R. G. except coal. Chairman Willard of the advisory commission of the Council of National Defense has had the question under consideration with the officials of the railroad, and a meeting with the authorities in Utah has been arranged that may result in good. There appears to be plenty of coal at the mines, but a lack of transportation conveniences, and it is not felt that the railroad is doing all that it might to relieve the situation.

The local United States Assay Office reports an unusual amount of work for this season of the year, an indication of the renewed activity of prospectors. Many assays are made for tin, but so far none has been found in this State.

The Garfield Chemical & Manufacturing Co., which is making sulphuric acid from Garfield smelter smoke, has increased the capacity of its plant to 150 tons of acid per day. A good market for the product is reported, and a still further increase is planned. At Bingham the principal item of interest, outside the continued prosperity of the older mines, is the development undertaken by the Montana-Bingham Consolidated Mines Co. This is a consolidation of the Montana-Bingham, with a 5000-ft. drainage and transportation tunnel; the Tiewaukee, with a record of past production and a good tonnage of milling ore broken; the Valentine, which controls the mineral rights beneath the town of Bingham, and adjoins the Utah Copper mine; the Fortuna, formerly owned by Governor Bamberger, with a record of production of $800,000, and a large tonnage of ore containing 2% copper blocked out, and a one-fifth interest in the Bingham Amalgamated. A mill employing jigs, tables, and flotation equipment, is in operation, and it is planned to increase it to an ultimate capacity of 1500 tons per day. The mill of the Ohio Copper Co. is treating 2200 tons daily. The flotation plants are handling 600 tons, and will be enlarged at the first opportunity, since the results have exceeded expectations.

Shipments from Park City in July exceeded 7000 tons, although the final figures are not available. June shipments were 6900 tons. The Three Kings mine has been equipped with a new compressor and is pushing development on the 140-ft. level, where the management expects to open the first ore-shoot. To the list of regular shippers have been added the Iowa Copper and the California-Comstock. In the week ended July 27, the former shipped 55 tons and the latter 120. The California-Comstock is equipped with electric and steam plants capable of handling 100 tons per day. There is a mill, 14 years old, but used very little and in good condition. The company owns 135 acres between the Silver King Consolidated and the Silver King Coalition. The ore is netting about $21 per ton in lead, silver, and zinc. Some of the mine-workings are in copper ore of a good grade. The mine has been opened to the 450-ft. level and is producing 30 tons per day.

Reports from Tintic are encouraging. Among the recent new producers, the Tintic Standard is making rapid progress. Its new 3-compartment shaft has reached the 1300-ft. level. Below 1180 ft. it has been in ore containing about 20% lead and 48 oz. silver. This ore is supposed to be an extension of that from which shipments have been made. The company paid its first dividend in June. Development on the 600-ft. level of the Emerald leads the management to believe that it is near the ore. The mineral so far found has been too low grade to ship, but indications that a better grade lies beneath are not wanting. To investigate this point work has been started on the 1000-ft. level. This mine is believed to be on the extension of the ore-zones of the Grand Central and the Centennial-Eureka. The Eureka Mines and the Godiva, both under the same management, are working in good ore. In June one shipment by the Eureka Mines of 45 tons netted nearly $40 per ton. This came from the new orebody on the 900-ft. level. Development on this orebody has since increased its known dimensions. Shipments are being made regularly from the 1200-ft. level, and development is going forward on the 1300 and 1400-ft. levels. The Godiva is shipping from the 500, 600, and 700-ft. levels ore that will net $1500 to $1700 per car. Work on the 1200-ft. level has cut the same vein, upon which drifts will be started to pick up the ore. The Eureka-Lily has its new electric hoist in operation. The new shaft is nearing the 1400-ft. level, and has entered ground that contains a little metal. The Copper Leaf Mining & Milling Co., adjoining the Tintic Standard, has been equipped for shaft-sinking. The two-compartment shaft was sunk 170 ft. in July to the 200-ft. level. The Iron Blossom paid a dividend of $50,000 in July, and the Chief Consolidated will pay about $88,000 early in August; 24 mines are shipping regularly.

The outstanding feature of the Cottonwoods has been the new discovery of ore in the Columbus-Rexall mine, followed by a sensational rise in the shares of that company. On June 30, 13¾c. was bid; July 31, sales were made at $1.55 to $1.60. The company is a consolidation of the Columbus Extension and the Rexall. For many years the Columbus-Extension has sought the eastward extension of the Toledo fissure, but many engineers have expected it to be found in Rexall ground. Whether it has finally been found is not yet settled. The ore just discovered lies in one of the beds of the limestone. It contains lead, silver, and copper, and early assays indicate a value of $100 per ton. It has been opened 18 ft. along the strike and is increasing in width. On July 31 it was reported to be 6 ft. wide at the face; 55 tons taken out in development has been shipped, but no returns are yet available. The mine joins the Cardiff on the east and the Kennebec on the south.

In Little Cottonwood, the Michigan-Utah continues to report favorable progress. Settlements in July are lower than usual, but the company has 28 cars of ore in the smelter yards. Ore having a net value of over $35,000 was shipped in June. One of the lessees has started shipments from an old slime-dump, which, at present high metal-prices, nets $10 to $12 per ton. The Emma Consolidated has shipped about 1800 tons this season. Development is under way to de-limit the orebody. Enough milling ore has been blocked out to induce the management to plan a 100-ton mill. The Sells is shipping 3 cars per week. Transportation facilities have been but little improved. Tramways, teams, motor-trucks, narrow-gauge railway, and standard-gauge railway, all figure in the congestion of ore, and are far from adequate to meet the conditions.

In Big Cottonwood, the Maxfield has resumed operations in a small way under new management. A few shipments of copper ore have been sent out by the Big Cottonwood Coalition, operating at the extreme easterly end of the district. The Cardiff has four trucks hauling about 80 tons per day. The city authorities of Salt Lake have forbidden the use of teams in South Fork, using the authority of a recent ordinance enacted under provisions that allow the control of the water-sheds affecting the city's supply of water. The Cardiff ore has been hauled in teams down South Fork, 2½ miles to Big Cottonwood canyon, where it was transferred to trucks for the haul to the smelter. Only one truck has been found that can make the South Fork grade, and until others can be secured the city has consented to postpone the enforcement of its order.

Production in the Deep Creek region is increasing. The Ferber Copper Co. has shipped 13 cars of ore. The Western Utah Copper Co. is shipping 150 to 250 tons per day. Lessees and smaller operators are gradually increasing the scale of their operations. The Calumet mine in Iron county, although opened to a depth of 75 ft. only, has made an enviable record. It is the first shipper of lead ore from Iron county. The mine is in the new Calumet district, 12 miles north-east of Stateline and 35 miles south-west of Frisco. The ore averages 23% lead with a little silver. The first car was sold on September 15, 1916, and the 20th car is now on its way to the smelter. The Antelope Star mine in Millard county has been equipped with a new mill using gravity concentration and K. & K. flotation machines. Expectations that it will become a big producer are high. In the Dry canyon section of the Stockton district five mines are producing. They are the Hidden Treasure, Mono Development, Brooklyn, Utah Queen, and Dry Canyon Consolidated. All are working in good ore. A scarcity of teams and labor greatly hampers the production. It has been necessary heretofore to haul the ore 9½ miles over the divide to Stockton on the main line of the Salt Lake route. The mine-owners have financed a new road down the canyon 4 miles to the Ophir & St. John railroad, which connects with the Salt Lake road at St. John. When this road is completed a larger output is expected.

AVERAGE PRICES during July were: for prime Western spelter, 8.64c.; for zinc-sheet at smelter 19c.; for pig-lead, New York, open market 10.714c., 'trust' market 11c.; electrolytic copper, New York, 28.889 cents.

Timbering in Deep Placer Mining

By HAROLD T. POWER

The Hidden Treasure gravel mines are situated on the Forest Hill divide in the eastern part of Placer county, California, about 35 miles north-east of Auburn, on the Central Pacific railway.

The deposit is known as a deep lead and consists of white quartz boulders, fine gravel, and sand, lying on and above the slate bedrock, and overlaid by a cap of andesite agglomerate and tuff varying in depth from 200 to 800 ft. This deposit occupies what was originally the bed of an ancient river, which diagonally crosses the divide that separates the North Fork from the Middle Fork of the American river, its course being nearly north-south. The discovery of this channel-system was made by a Canadian at what was subsequently known as Damascus, on the north or up-stream end. The first mining location was made at the point of discovery and was known as the Canada claim. Other locations were made subsequently, namely, the Lee, Mountain Tunnel, and Golden Gate, afterward consolidated under the name of the Mountain Gate placer mine. The operation of this property was handicapped by the fact that mining had to be carried on down-stream, necessitating the driving of successively deeper bedrock tunnels for drainage and the removal of auriferous gravel by gravity. Some members of the Mountain Gate company noted the similarity of the deposits exposed by hydraulic mining at Gas hill, and at the Big Gun mine, at Michigan Bluff, five and eight miles respectively from the point of original discovery. After surveys, and tracing of the rim-rock wherever exposed, the conclusion was reached that the channels on opposite sides of the divide were identical, and a tunnel was driven in Blacksmith ravine four miles south from Damascus. The channel was found in place at the extreme south end, where its continuation had been eroded by a more recent stream-system, although no surface indications of its existence were observable at the point of attack, owing to rock-slides and vegetation. This was the beginning of the Hidden Treasure gravel mine, the largest deep-drift mine in the world.

BREASTING GRAVEL, HIDDEN TREASURE MINE

THE MAIN GANGWAY SHOWING WIDE-SPREADING DRIFT-SETS PLACED TO RESIST PRESSURE OF SWELLING GROUND

CROSS-CUT GANGWAY IN THE HIDDEN TREASURE MINE

The bedrock of this ancient channel consists of Calaveras slate, which is friable, and when exposed to the air has a tendency to disintegrate, or, as the miners term it, to 'swell.' This gold district is an extension of the Mother Lode of California. The quartz gravel is not cemented, and therefore requires the use of little, if any, powder to break it down. It is obvious that a great quantity of timber would be required to support the roof of such a deposit and to permit the removal of that portion of it containing gold. This river-system has been mined for a distance of about four miles, over a width varying from 200 to 800 ft. The roof of the entire area had to be supported by mine-timbers, spaced from 2¼ to 5 ft. apart, which gives an idea of the enormous stumpage that went underground.

The stand of timber in the neighborhood of these mines was greater than that in the mountains of the Sierra Nevada, the average being about 25 M. ft. per acre; that on and about the Hidden Treasure and Mountain Gate properties being especially advantageous for cutting and delivery at these mines.

The principal varieties of timber indigenous to this region that were used in the mining operations, were sugar, white, and pitch-pine, Douglas fir, white and red fir, and cedar. This timber was used for two purposes underground: (1) for the maintenance of the main tunnel; (2) for temporary maintenance while the breasting operations were proceeding.

In the maintenance of the main tunnel sugar pine, Douglas fir, heart of white pine, and red cedar, especially the heart of the last, were found to resist decay to a greater extent than the other species, depending largely upon the moisture conditions in the tunnel where used, and upon the season of the year when the timber was felled—that cut in the fall and winter probably having double the life of that cut during the spring and early summer, when the sap was running. The life of white and red fir and the sap-wood of the white pine was very short where there was little or no moisture from the roof and sides; in many instances lagging 2 in. thick and timbers of 12 in. diam., of these species, had to be removed in less than one year, whereas sugar pine, heart of white pine, Douglas fir, and cedar, under variable conditions, would not have to be replaced in from three to five years. Of course, these conditions were modified by the pressure exerted upon the timbers by what has heretofore been designated as 'swelling ground.' This pressure, in many instances (especially after the ground was first opened) was so great that the timbers and lagging were crushed within a few weeks after setting, necessitating the employment of from two to four crews of timbermen directly behind the miners that were driving the main tunnel ahead. These men had to relieve the pressure on the lagging, set back the timber, and renew those sticks that had been crushed by the so-called 'swelling' of the bedrock. As an example of the action of this swelling bedrock, several instances are on record wherein this formation has been known to rise, or swell, 24 in. within 12 to 15 days, throwing out of alignment the T-rail track and the timbers, as well as destroying much lagging and many sets of timber. To overcome the effects of this swelling bedrock, the posts were set at an angle of about 30° from the vertical and 4½ ft. centre to centre, as shown in the accompanying illustration. The reason for setting the timbers at this angle was to relieve them as far as possible from the direct pressure, whereas, if they were set vertically the pressure would be irresistible. See Fig. 1.

The timber necessary in the main tunnel for each complete set was as follows:

2 posts, 12 by 12 in., 8½ ft. long	204.0 ft. B.M.
1 cap, 12 by 12 in., 5½ ft. long	66.0 " "
35 lagging, 2 by 6 in., 5 ft. long	175.0 " "
2 ft. blocks, 14 by 16 in., 4 in. thick	12.5 " "
12 cedar keys, 3 by 7 in., 8 in. long	14.0 " "
5 bridges, 2 by 6 in., 5 ft. long	25.0 "
	496.5 "
5% for wastage	24.8 " "
Total	520.3 " "

The 'keys' were made of cedar for the reason that the wood was soft and resilient, and would relieve somewhat the pressure and strain upon the posts and cap, thus preventing, in some measure, their being fractured; when these keys were crushed they could be removed and new ones driven in their place without disturbing the main sets.

After placing the first set of timbers, the lagging is started between the bridges and the main timbers, in the spaces between the keys. To prevent the caving from the top and sides, the lagging is driven ahead as fast as the material in the face is removed. When the lagging is driven about one-half its length a false set is placed to support it until it is driven home the full length. The foot-blocks, posts, and cap are then put in place, the bridges set, the keys driven home between the bridges and the main set, thus taking the weight on the lagging, heretofore supported by the false set. This false set is then removed, and the operation is repeated indefinitely. The timber-crews that follow, to relieve the timber from the swelling bedrock, place what they term centre-sets,' thus making each set 2¼ ft. centre to centre. These centre-sets are of the same size as the main sets. This gives an idea of the enormous amount of timber required for the maintenance of the main tunnel alone.

I have heretofore stated that sugar pine, Douglas fir, red cedar, and heart of white pine were most suitable for main-tunnel supports; however, there are exceptions to this. Where it is very wet and the ground is heavy, red, white, and Douglas fir have been found preferable to the other woods named. Under these conditions they are not attacked by the fungus growth common to mine timbers. They are resilient, their fibre is tough, and they do not fracture so readily as such brittle woods as sugar-pine and cedar, especially the latter. Therefore, I might say, these woods are invaluable for work of this character, their life being prolonged indefinitely where there is water. The utilization of mine-timber under the con-

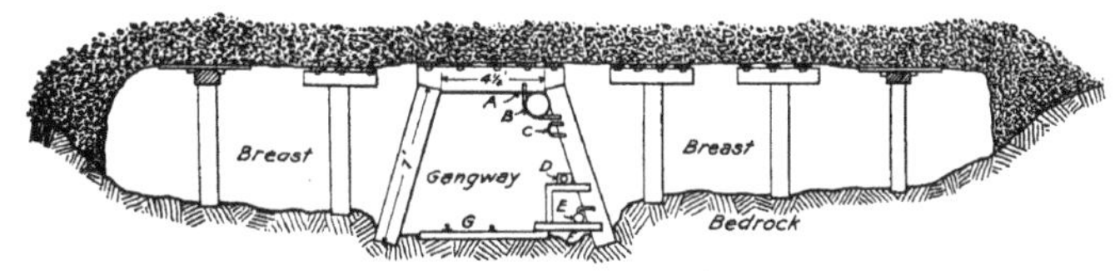

CROSS-SECTION ON WORKINGS IN HIDDEN TREASURE MINE

ditions cited is the result of over 55 years experience and close study of the life, adaptability, and best use of timber for underground work.

The removal of the gold-bearing gravel, or 'breasting,' as the miners term it, was performed in this mine as here described. Drifts, or gangways, were driven at right angles from the main tunnel, toward the rim-rock, at intervals of about 200 ft. to the outer limit of the pay-streak, which was ascertained by car-samples washed in the prospect-dump outside. When this limit was reached, what were designated as 'cross-cuts' were driven from both sides of the gangway, parallel to the main tunnel for a distance of about 100 ft., or, more correctly, 110 ft. up-stream and 90 ft. down-stream along the channel. These gangways were timbered with posts 10 by 10 in. and 6 ft. long, and caps 10 by 10 in. and 5 ft. long, the lagging being 1½ by 6 in. and 5 ft. long. The cross-cuts and breasts were timbered with posts and caps 8 by 8 in. and 5 ft. long, the lagging being of the same dimensions as that used in the gangways. When breasting commenced, the method of timbering was much the same as in the main tunnel; as the breasting proceeded toward the main tunnel all the boulders and waste were piled back into pillars in the breasted-out portion to assist in supporting the roof. If the ground was heavy, as it usually was, a post and cap of the same size as indicated above was placed under the lagging between each set, designated by the miners as 'gee-becks.' In addition to these sets, it was often necessary to place another post under the caps of each set known as a 'centre-post' further to safeguard the miners. On top of these caps lagging was driven, making a solid wooden roof. In these breasts the posts were at 2½ ft. centres over the whole area mined. In addition to this it was necessary, if the breasts were wet, to use foot-blocks to prevent the timbers from sinking into the bedrock. It was impossible to remove these timbers owing to the weight which they were compelled to support. As has been stated, breasting was toward the main tunnel, so no effort was made to keep up these breasts except insofar as the immediate safety of the miners required it. However, when water broke through the roof at the face of the breast, it became necessary for safety at that point, to blast down with dynamite the supporting timber some distance back from the face, in order to draw the water and thus relieve the weight on the timber at or near the working-face. All the species of timber mentioned were used in this part of the work, as the life or quality of the timber was not essential. See Fig. 2.

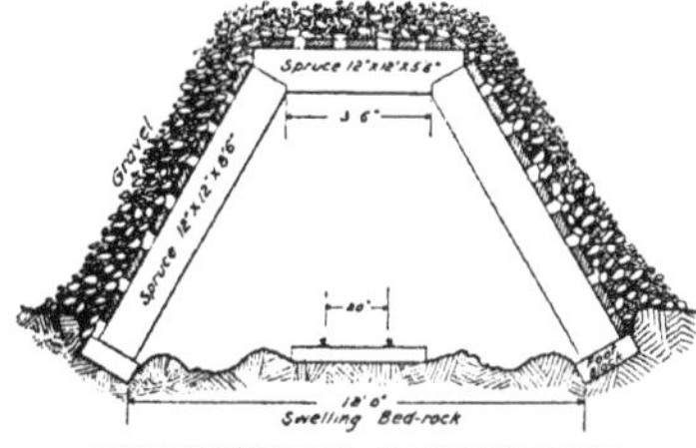

TIMBERING MAIN TUNNEL, HIDDEN TREASURE MINE

The Hidden Treasure Gravel Mining Co. owned extensive timber-land, and, prior to the creation of the Forest Reserves, miners were permitted to cut timber upon the public domain for their own use without cost or restriction. It was the custom of the company to invite bids for cutting and delivery of mine-timber for underground use for yearly periods. These contracts called for 1,500,000 to 2,000,000 ft. board measure, the dimensions being as shown in the accompanying table.

In early years, when timber was plentiful, there was great wastage in the woods, as only the straight-grained and long-bodied trees were utilized. These trees were 'bucked' into lengths as required, and then split. In fact, I am told that at one time only the finest sugar and white pines were used for making lagging. After the

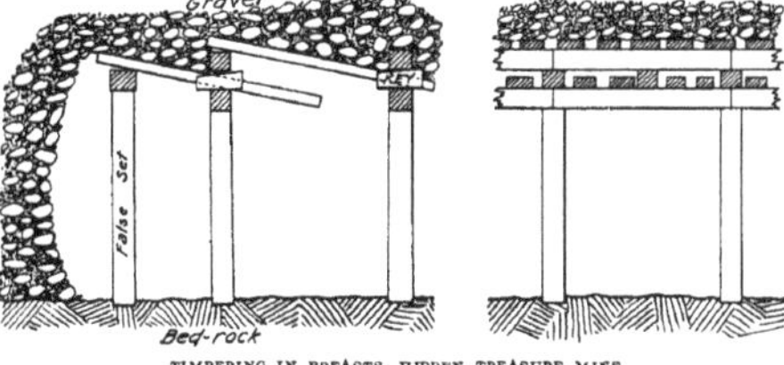

TIMBERING IN BREASTS, HIDDEN TREASURE MINE

MAIN-TUNNEL TIMBER AND LAGGING:

2 posts and 1 cap	$1.50 per set
Lagging	35.00 per 1000 pieces
Gangway-posts	0.15 each
Gangway-caps	0.12 "
Breasting-posts	0.12 "
Breasting-caps	0.12 "
Gangway and breast-lagging	30.00 per 1000 pieces
Foot-blocks	0.10 each
Bridges	0.03 "

creation of the Tahoe Forest Reserve, when suitable timber to work by this method became more difficult to obtain, the company, in 1905, for ventilation, constructed a double-compartment shaft, 850 ft. deep, to intersect the tunnel at a point more than two miles from its portal. This also greatly facilitated the transportation of mine-timber underground. A saw-mill was erected near the collar of this shaft and thereafter all timber and lumber used by the company was cut at this mill. In this manner there was little waste of standing timber, as the better qualities were sawed into lumber while the tops and slabs were made into mine-timbers and lagging. However, this method increased the cost of mine-timber at the expense of quality, I should say, 25%.

The lumber used in construction work about the mine was purchased from custom-mills in the neighborhood, the price ranging from $20 to $40 per thousand according to quality.

Sugar-pine, No. 1	$40
White pine, No. 1	40
Douglas fir, No. 1	35
Sugar-pine, No. 2	30
White pine, No. 2	30
Douglas fir, No. 2	25
Common lumber of all kinds	20

The first-class lumber was used for finishing work in and about the buildings, in car construction and repairs, and also in sluice-bottoms. The second-class lumber was used for building-construction, the common lumber for general construction work, sluice-flumes, water-flumes, and track-ties. The lumber used about the mine ranged from 50 M. ft. to 400 M. ft. per year, averaging in cost about $25 per thousand feet.

The mine-timbers were not treated for their preservation, but by regulating seasonal cutting, and by selection, the life of the timber used underground was materially lengthened. The treatment by creosote, and other methods, tends greatly to reduce its tensile strength, which, in this case, was an important factor in the use of mine-timber by this company.

EXPLOSIVES suitable for coal mining are also adapted to the breaking of ground that is loose and readily shattered by an explosion of a low order of intensity. Such explosives show a rate of detonation of 8000 to 12,000 ft. per second in cartridges of 1¼ in. diameter. An explosive capable of a high velocity of detonation develops high temperature and consequently high expansive power, and is more useful for overcoming the cohesion of dense rock. Its power is largely wasted in loose and fissured rock.

Device for Setting Wagon Tires

The operation of setting a wagon tire in the ordinary way, shrinking it upon the rim by immersion in the blacksmith's slake-tub, is unhandy and has serious disadvantages. As the cooling extends over only a portion of the tire at a time, the shrinking is not uniform, and the operation is so slow that the rim, or a portion of it at least, becomes charred. This charred portion rapidly wears away when the wheel is put into use, causing the tire to again become loose. To overcome these difficulties

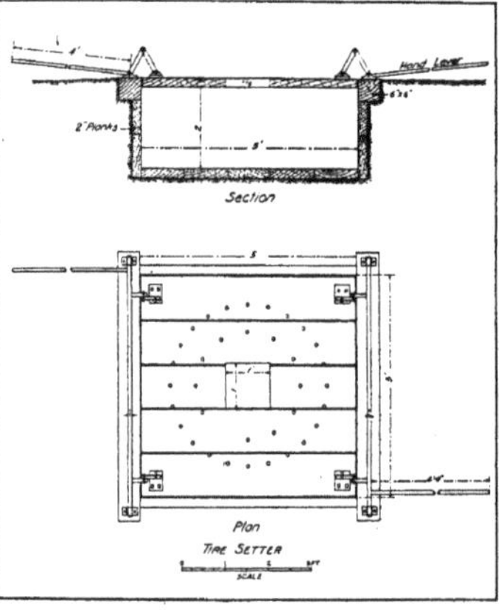

the device shown in the accompanying illustration was designed. A pit, 2 ft. deep, is lined with 3-in. plank to make it water-tight, and is fitted with a trap of 2-in. plank. This trap can be raised and lowered by the simple system of levers shown. In the centre of the trap is a hole 1 ft. square, and around it in concentric circles are bored 1-in. holes. The large hole is to accommodate the wheel-hub, permitting the rim to lie flat on the trap; the small holes and the spaces between the planks allow a rapid out-rush of water when the trap is lowered. In using this device the wheel upon which the tire is to be set is placed on the trap. The tire is heated to the required temperature and placed over the wheel. At once the two hand-levers are raised, plunging the wheel into the water. The cooling is uniform and so rapid that charring of the rim is reduced to a minimum; labor and time are also greatly reduced. This device is of great service in blacksmith shops at mines that have much wagon haulage.

PYRITE for sulphuric-acid manufacture, averaging 48 to 52% sulphur, is now selling for 16c. per unit at Atlantic Coast points.

Formation of Zinc Ferrate

By E. H. HAMILTON, G. MURRAY, and D. McINTOSH

*Early in the development of the electrolytic method of producing zinc from its ores it was noticed that a large amount of the zinc could be extracted by dilute sulphuric acid, in a few minutes, but that the remainder could be removed only by prolonged boiling. This insolubility was not due, apparently, to the particles of the zinc compounds being protected by the lead sulphate in the calcine, for on removing the lead salt by appropriate methods no increased solubility was obtained. It was noticed, further, that on treating the residue which was insoluble in cold dilute acid with boiling acid, the zinc and iron dissolved approximately in the ratio of one atom of the former to two of the latter. Finally, it was ascertained that a high temperature in roasting increased the amount of insoluble zinc, although the calcine produced might contain less sulphur as sulphide than one made at a lower temperature. The insoluble zinc must be due to the iron, since higher grade zinc ores gave greater percentages of soluble zinc. The lead sulphate has been shown to be innocuous, so that the complexes formed with iron must cause the low extraction. The compounds of zinc and iron oxides have been examined by a number of investigators. Roscoe and Schorlemmer[1] state that Ebelmen obtained black octahedra of the composition $Fe_2O_3.ZnO$ by strongly igniting the oxides with boron trioxide. Hofman (Trans. A. I. M. E., 1905) and Wells in the *Engineering and Mining Journal* showed that combination took place between these two oxides, but no attention was paid to the temperature or the time. Ingall's 'Metallurgy of Zinc' (pp. 6 and 32) mentions the formation of zinc ferrite (zinc ferrate).

The results of the present investigations corroborate the observations of Hofman and Wells, and, in addition, go more fully into the conditions governing the combination. The first part of the work consisted of experiments with pure zinc oxide and ferric oxide. Later, some roasting and leaching tests were made on various ores to find out how far the results obtained from the pure compounds applied to the minerals. Ferric oxide was prepared by treating ferric chloride with ammonia and roasting the washed hydroxide for two hours at a red heat. The zinc oxide used was Baker's 'chemically pure.' A mixture was carefully made of material ground to 120 mesh to correspond to the formula $ZnO.Fe_2O_3$. This was re-ground to bring about an intimate mixture, and the tests were made at the various temperatures in a muffle furnace heated electrically. A thermo-couple, correct at the temperature of melting zinc, indicated the temperatures, and these were maintained constant by varying a resistance in series with the muffle. After the heat-treatment the material was again ground and leached with hot ammonia and ammonium-chloride solution, and the loss in weight found. From these results the amount of zinc oxide that had entered into combination was determined. The tabulated results of tests 1 to 7 are given in table 1, and shown in the curves on plate 1.

TABLE I

Hours	%
Experiment No. 1—temperature 1600°F.	
1.15	100
3.5	100
5.00	100
5.3	100
Experiment No. 2—temperature 1500°F.	
5	100
8	100
Experiment No. 3—temperature 1400°F.	
1	88.4
3	85.5
5	89.3
7	97.1
9	100.0
Experiment No. 4—temperature 1300°F.	
1.15	67.6
2.75	76.8
5.25	82.5
7.25	85.2
Experiment No. 5—temperature 1200°F.	
1.5	35.0
3.0	73.6
4.5	70.4
6.0	70.0
Experiment No. 6—temperature 1200°F.	
2.0	76.2
4.5	84.4
6.0	71.2

Experiment No. 7—at 1100°F., no combination took place.

A few experiments to determine the formula of the compound were made by heating various mixtures so that combination was complete. In all cases where the iron oxide was in excess of the necessary amount for the formula $ZnO.Fe_2O_3$ none of the zinc went into solution. On the other hand, while the results obtained with the zinc oxide in excess did not agree as well as could be desired, it is certain that the compound does not contain more zinc than is shown by the formula $ZnO.Fe_2O_3$.

TABLE II

Ratio of ZnO to Fe_2O_3	Theoretical loss in weight if formula is $ZnO.Fe_2O_3$, %	Actual loss, %
1.5:1	14.4	17.1
2 :1	25.3	20.9
3 :1	40.3	50.9

From the curves it will be seen that the temperature has a great effect on the speed with which the reaction takes place. At 1600°F. the combination is complete at the

*Bull. Can. Min. Inst., July 1917.

[1](Page 1225.) The compound is referred to as zinc ferrite, and the name ferrate is reserved for the salt of the hypothetical acid H_2FeO_4. It would seem better to speak of the compounds of ferric oxides as ferrates, and of ferrous oxide as ferrites.

end of one hour. At 1500°F. it is complete in 5 hr. and probably sooner. At 1400°F. the reaction is 85% complete in 3 hr. and complete in 8 hr. At 1300°F. about 70% of the zinc has combined in 1 hr. and 85% in 8 hr., while at 1200°F. it is 4 hr. before the reaction is 70% completed. At 1100°F. no combination could be detected within 8 hours.

Zinc ferrate is a yellowish-brown material, magnetic, and never more basic than $ZnO.Fe_2O_3$. It is soluble in hydrochloric acid, insoluble in cold dilute H_2SO_4, H_2SO_3, and hot caustic solutions. It is, however, slowly soluble in hot dilute sulphuric acid. We were unable to find any reagent which would dissolve the zinc without bringing a considerable quantity of iron into solution. During the investigation it was found that zinc oxide also combined with freshly prepared aluminum oxide

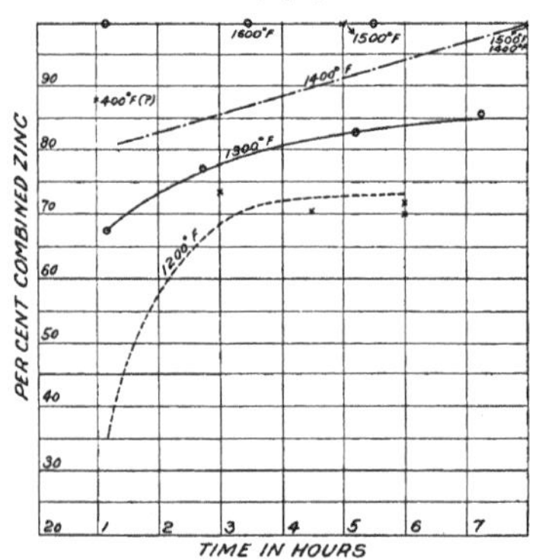

PLATE 1. COMBINATION OF ZINC AND IRON AT DIFFERENT TEMPERATURES

made by precipitating aluminum hydroxide from aluminum chloride by ammonia. This was filtered and roasted in a muffle at about 1200°F. for 2 hr. The tests were carried out exactly as with the ferrates. The results were:

TABLE III

Time, hours	Temperature, degrees F.	Zinc combined, %
1	1600	82.2
3	1600	83.3
6	1200	41.0

These results show that compounds analogous to the ferrates are obtained with alumina, but as this investigation was too wide for the time at our disposal nothing further was attempted, as an experiment made with fire-clay and zinc oxide showed practically no combination after 2, 4, and 6 hr. at 1500°F.; so that the alumina in ordinary ores would not combine with the zinc and lower the extraction. The zinc aluminate was insoluble in dilute sulphuric acid.

Ore from the Sullivan mine, Kimberly, B. C., was roasted at temperatures from 1100° to 1500°F. with hand rabbling. At 1100° and 1200°, however, once the sulphur was ignited it was not possible to keep the temperature down until a large amount of the sulphur had been burned off, so that for a time at any rate the whole body of ore was hotter than the registered temperatures, and the particles on the surface where the oxidation was rapid were at a high temperature for a short time after each rabbling. This is especially noticeable in the upper floors of the Wedge furnaces after the rabble-arm passes; for while the pyrometer in the ore registers only from 1200° to 1300°F. the ore on the surface becomes hot enough to fuse into small balls. The results are seen in

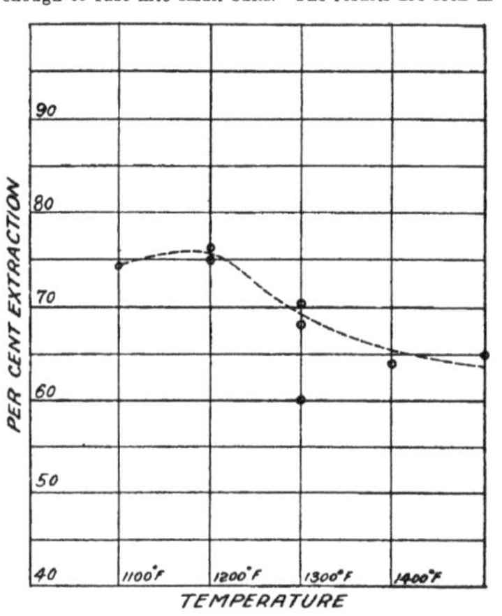

PLATE 2. EXTRACTION OF ZINC AT DIFFERENT TEMPERATURES

table IV and plate 2. The assay of the ore, of which 70% passed through 200-mesh I.M.M.. standard, was as follows:

Pb	Fe	SiO_2	Al_2O_3	CaO	S	Zn
13.2	25.3	2.4	2.9	1.4	28.5	22.6

TABLE IV.

Test	Temp.	Roast			Extraction			
		Time hr.	Total S.	S as SO_4	Zn.	Loss in Weight	Zinc in Tail	Extraction
	F.							
1	1100	7½	6.95	4.6	23.7	39.1	10.2	74.0
2	1200	4	5.45	4.3	23.0	43.4	9.6	76.1
3	1200	4	6.3	5.4	23.0	38.0	9.2	75.1
4	1300	4½	5.35	4.35	23.0	36.0	10.8	70.6
5	1300	4	4.8	4.4	22.8	29.0	10.3	68.0
6	1400	3¾	4.75	4.35	23.6	32.6	12.5	64.4
7	1500	4	5.05	4.31	23.8	27.4	11.5	65.0

In order to test the ferrate theory to see whether a greater percentage of the zinc would be soluble in an ore containing less iron than the ordinary Sullivan zinc ore, three samples were hand-picked from a carload of the mine shipment before crushing. These samples assayed as follows:

No	Pb	Fe	Insol.	CaO	S	Zn
1	3.7	11.3	1.0	0.4	31.7	46.2
2	15.2	10.0	0.4	0.3	30.1	41.6
3	10.4	10.4	0.6	0.5	30.8	44.5

These three samples were crushed until 70% would pass a 200-mesh screen and roasted for 4 hr. at a maximum temperature of 1400° + F., and leached. A sample of Sullivan concentrate from the Ding magnetic machine (- 60 mesh + 80), assaying 20% Fe and 27.2% Zn was also roasted. A good extraction was obtained, as shown in test 12 in table V.

TABLE V.

Sample	Test	Temp	Time hr.	Roast		Zn.	% Loss in Wgt. in Leaching	Zinc in Tail	Extraction.
				Total S.	S as SO_4				
1	9	1400	4	5.8	5.4	44.1	64.7	19.2	84.5
2	10	1400	4	3.8	3.6	43.9	50.0	15.9	83.2
3	11	1400	4	5.5	5.3	45.9	62.2	14.2	88.4
4	12[1]	1400	4	4.8	4.4	28.2	41.9	8.5	82.5

[1] Ding concentrate.

The size of the roasted particles has little or no influence on the extraction. This can be seen by the tests made on bucked and unbucked samples (table VI). The unbucked material was leached as it came from the roaster, while the bucked sample was ground as finely as possible in a mortar.

TABLE VI.

Test No.		Loss in Weight %	Zinc in Tail %	Extraction %
9	Bucked	64.8	19.6	84.5
	Unbucked	64.7	19.2	84.5
10	Bucked	49.9	15.5	83.5
	Unbucked	50.0	15.9	83.2
5	Bucked	29.0	10.6	66.0
	Unbucked	28.8	10.5	68.0
12	Bucked	38.7	7.3	84.4
	Unbucked	41.9	8.5	82.5

ELECTROLYTIC ZINC DEPARTMENT OF THE TRAIL SMELTER

Roasting and leaching tests were made on other ores and it is interesting to examine the results obtained from these ores in connection with the amount of ferrate formed. The ores treated were as follows:

	Pb %	Fe %	Zn %
(a)	trace	7.4	47.6
(b)	3.5	15.8	33.0
(c)	1.2	6.0	44.0
(d)	3.0	6.2	44.2
(e)	6.1	4.8	41.0
(f)	trace	9.1	45.1
(g)	trace	5.3	38.6
(h)	trace	9.9	39.9

Leaching tests on these ores are shown in table VII, the temperature being 1400° + F. on a 4-hr. roast.

In this comparison those roasts will be considered that were made for 4 hr. at a temperature of from 1400° to 1500°F. Each unit of iron will combine with 0.58 unit of zinc according to the formula $ZnO.Fe_2O_3$ and after 4 hr. at 1400° the reaction will be 90% complete, so

TABLE VII

	Roast		Zn.	On Leaching Loss in Weight %	Zinc in Tail %	Extraction %
	Total S	S as SO_4				
a	2.9	2.6	51.6	70.5	16.4	90.6
b	5.7	5.5	35.7	51.5	15.8	78.5
c	7.4	6.8	44.5	69.8	10.7	92.7
d	8.4	8.2	44.5	71.9	11.2	92.7
e	7.8	7.4	42.6	68.2	11.2	91.9
f	8.6	7.7	46.2	71.7	16.9	89.6
g	8.1	6.9	38.8	55.6	9.1	89.2
h	7.8	7.3	41.5	66.6	12.3	90.1

that we can calculate from the analysis of the ore the amount of zinc that should be insoluble after roasting. Taking 100 gm. of ore as a basis and allowing for the amount of zinc that would probably be present as sulphide and as such insoluble, the results found are tabulated in table VIII.

TABLE VIII

Sample No.	Fe. in ore	Zn. in ore	Zinc with S grams	Insoluble Zn. grams	Theoretical ferrate grams	Difference, grams	Ratio Zn:Fe.	Actual Extraction	Theoretical extraction corrected for Zn combined with S
9	11.3	46.1	0.7	6.06	5.90	+0.16	4.06:1	84.5	85.0
10	10.0	41.6	0.3	7.10	5.23	+1.87	4.16:1	83.2	87.4
11	10.4	44.5	0.3	5.06	5.30	—0.24	4.28:1	88.4	87.8
12	20.0	27.2	0.6	4.35	10.40	—6.05	1.36:1	82.5	61.0
(6 & 7)	25.3	23.0	0.5	8.05	13.20	—5.15	0.92:1	65.0	42.2
a	7.4	47.6	0.5	4.35	3.87	+0.45	6.43:1	90.6	91.5
b	15.8	33.0	0.3	7.35	8.25	—0.90	2.09:1	78.5	76.1
c	6.0	44.0	1.0	3.13	3.13	0	7.3 :1	92.2	90.7
d	6.2	44.2	0.3	2.85	3.24	—0.39	7.13:1	92.7	92.1
e	4.8	41.0	0.6	2.96	2.60	+0.36	8.55:1	92.0	92.0
f	9.1	45.1	1.8	3.98	4.75	—0.77	4.74:1	89.6	85.8
g	5.3	38.6	2.2	1.84	2.78	—0.84	7.26:1	89.2	87.2
h	9.9	39.9	0.8	3.25	5.17	—1.92	4.02:1	90.1	85.6

From an examination of these figures, it will be observed that the amount of insoluble zinc present after roasting is very close to the theoretical amount that should be formed as zinc ferrate with the exception of the regular Sullivan zinc ore (No. 6 and 7) and the magnetic concentrate (No. 12). In each of these instances the ratio of zinc to iron is close to 1 to 1; whereas, in the others, with the exception of one sample, it is at least 4 to 1. In some way, therefore, in these samples, the combination is prevented from being as complete as one should expect from the composition of the ore. One explanation may be that at this concentration of the iron and zinc another compound may be formed containing a greater ratio of iron to zinc than that represented by the formula $ZnO.Fe_2O_3$, but we have been unable to prove this and regard the existence of any combination except $ZnO.Fe_2O_3$ as unlikely. On the other hand it may be that free crystals of pyrrhotite do not combine as readily as those crystals which may be attached directly to crystals of blende or as that iron which may be isomorphous with the zinc in the blende. One test made with the addition of 20% SiO_2 to the Sullivan zinc ore before roasting indicated that the addition of some inert matter decreases the amount of insoluble zinc formed, as an extraction of 76% of the zinc was obtained in this case as against 65% without the addition of the silica. The results of the above test were as follows:

Roast Total S	S as SO_4	Zinc	Loss in weight	Zinc in tailing	Extraction
3.6	3.2	19.0	34%	6.7%	76.7

One could, of course, determine the effect of adding pyrrhotite or pyrite crystals to the ore by observing whether the extraction were decreased or not, but this has not been attempted, as the results embodied in the present paper, as far as they go, explain the reason for the failure to obtain a high extraction on the ore treated at Trail.

SUMMARY. 1. It is proved that zinc oxide and ferric oxide will combine readily at a temperature above 1200° F., the amount combining being dependent principally on the temperature and the time. In the roasting of the ore in the Wedge furnaces, however, in order to de-sulphurize the ore sufficiently, it is not possible to keep the temperature down to a point at which the formation of ferrate does not occur. In the upper part of the furnace especially, it is difficult to keep the ore from becoming too hot owing to the large amount of sulphur present, but it is possible that, by decreasing the frequency of the rabbling, the ore might be kept for a longer period in the furnace and subjected to a lower temperature, as the more rapid the stirring, while the sulphur is burning readily, the higher the temperature attained. In connection with this, in the case of the Sullivan ore as at present roasted, the maximum amount of ferrate seems to be formed, as some of the regular roast was heated to a temperature of 1500° for 5 hr. and the extraction on this material was no worse than on the original.

2. Free alumina combines with zinc oxide, but when present as a silicate the amount of combination is negligible.

3. There are four ways in which a better extraction of zinc might be obtained from an ore high in iron.

(a) By roasting at a low temperature, the remarks made at the beginning of the summary applying to this point.

(b) By eliminating the iron before roasting.

(c) By the addition of some substance to the ore which would prevent the combination of the two oxides.

(d) By finding some solvent for the zinc that would leave the iron oxide undissolved.

COLLOIDAL particles carry definite electric charges, as is brought out through a study of their precipitation or coagulation. The latter effect is produced by small quantities of electrolytes. The precipitating effect on a negatively charged colloid is dependent on, and increases with, the valency of the positive ion, and the precipitants that are most efficient are those that yield the greatest concentrations of ions at a given dilution. This property is possessed in greater degree by mineral acids than by organic acids. While the precipitation of the negatively charged colloid is dependent on the positive ion of the added electrolyte, the positively charged colloid is influenced mainly by the valency of the negative ion. Divalent negative ions are more effective than the monovalent, for example, barium chloride is more than twice as effective as NaCl added to a ferric hydrate suspension.

Flotation Tests on Gold Ores

By EDWIN JOYCE

The following tests were made at the State College of Washington on a gold ore from Eagle Creek district, Baker county, Oregon. The ore consists of quartz containing small amounts of pyrite and magnetite. The gold is carried in the quartz as well as in the pyrite, and, as shown by screen-tests, is in such condition that fine grinding is necessary to free the particles. The work was done in a Janney laboratory flotation-machine, using various oils and combinations, with both acid and neutral solutions. On account of the well-known difficulty of assaying high-grade ore, the value of the heading does not always check the value of the combined concentrate and tailing. This may be due in part also to the fact that a few grammes of pulp are almost always lost in the operation of the machine. The following table gives the results of some of the tests:

Test	Ore, gm.	Mesh	Ratio water to ore	Oil — Kind No.	Oil — gm.	Oil — %	Time, min.	Conc., %	Tail, %	Assays — Head	Assays — Conc.	Assays — Tail	Extr., %
1	500	80	4-1	5	1.0	0.2	5	12.4	87.6	$2.06	$12.36	$0.54	77.7
2	500	80	4-1	5	1.0	0.2	6	11.8	98.2	2.06	14.70	0.40	84.2
3	500	80	4-1	400	1.0	0.2	5	9.8	90.2	2.06	16.90	0.44	80.4
4	500	100	4-1	400	1.0	0.2	5	11.2	88.8	1.67	12.32	0.30	82.6
5	500	100	4-1	8	1.0	0.2	5	11.2	88.8	1.67	10.84	0.44	72.7
6	500	100	4-1	1580	0.85	0.17	5	12.4	87.6	1.67	9.76	0.38	72.4
7	500	100	4-1	22 1580	1.1 0.4	0.3	5	9.2	90.8	1.67	15.08	0.22	83.1
12	500	200	4-1	22 1580	1.6 0.4	0.4	8	18.4	81.6	2.01	9.98	0.20	91.3
13	500	200	4-1	22 1580	2.1 0.4	0.5	8	18.8	81.2	2.01	9.97	0.14	93.3
14	667	200	3-1	22 1580	2.1 0.4	0.5	8	28.0	72.0	2.01	8.54	0.18	90.9
16	667	200	3-1	....	..	..	5	4.3	95.7	0.19	2.08	0.10	47.1
17	667	200	3-1	....	..	..	5	1.8	98.2	0.19	6.20	0.08	58.7
18	400	200	5-1	....	..	..	9	37.5	62.5	10.10	24.60	1.20	91.3

In test 1, steam-distilled pine-oil No. 5 G. N. S., 0.015 gm. per drop, was used, producing a large volume of coarse lasting froth. This may be compared with the results obtained in No. 2, in which the time was increased, producing a better recovery. The oil used in test 3, crude wood-creosote, No. 400 P. T. & T. Co., 0.02 gm. per drop, produced a smaller quantity of concentrate, but the grade was improved. The froth was smaller in size of bubble and more brittle than that obtained from pine-oil. In test 4, on lower-grade heading crushed to 100 mesh, the same oil gave better recovery than in test 3, showing the effect of finer grinding. The froth obtained in test 5, using a special grade of pine-tar creosote, No. 8 G. N. S., 0.019 gm. per drop, was a dry tough froth made up of very large bubbles. On account of the more easily available supply of coal-tar creosote, tests were made to determine its adaptability to the ore, but, when used alone, the results were not satisfactory. The froth was quite fragile and broke down as fast as it was formed, prohibiting any chance of its being collected. A combination of coal-tar creosote containing carbolic acid, No. 22 G. N. S., 0.021 gm. per drop, and special pine-oil, No. 1580 P. T. & T. Co., 0.017 gm. per drop, gave results as shown in tests 7, 12, 13, and 14. A very watery fine-grained froth, in large volume, was produced. The use of 0.5% oil, made in the proportion of 1 part of pine-oil to 5 parts of creosote, was found best suited, the percentage being based on dry ore. As the object was to find the smallest amount of oil required to give good recovery, no tests were run using more than 1% of oil. Test 14 shows the result of decreasing the ratio of water to ore, the recovery falling off; this, however, may be due in part to the manipulation of the machine. In tests 16 and 17 the tailing from previous tests was re-treated and shows plainly that the complete extraction of all the floatable particles was not obtained in the first run. Twenty drops of sulphuric acid was added in each case, but subsequent tests on original ore failed to raise the percentage of extraction when acid was used. No additional oil was necessary when re-treating the tailing, the same being the case in No. 18 when the concentrate from previous tests was re-treated. In test 17, although not necessary, additional oil was introduced in small quantity. In this case when 10 drops of sulphuric acid was added, the froth appeared in small quantity, and when more creosote was added the froth was killed, appearing again when pine-oil was added, but not in such quantity as in tests 16 and 18. No oil or acid was added when re-treating the concentrate, an abundant, thick, coarse froth forming immediately.

It was found that a combination of sulphuric acid and coal-tar creosote gave an excellent froth, but did not produce as good results as were obtained from a combination of pine-oil and coal-tar creosote. In all cases where sulphuric acid was used it was introduced before the oil, giving better results than when oil was added first. The same result might be obtained by a longer period of agitation, but, as no acid seemed necessary, it was not used in succeeding tests. Examination of the concentrate and tailing revealed the fact that only a part of the magnetite was floated. In the course of the work it was found necessary to replace the worn-out packing around the shaft immediately under the agitating-chamber. As the proper material was not at hand, asbestos-wick packing, covered with hard oil and graphite, was used, and no froth could be obtained in the succeeding test. Upon examination it was found that when the gland-nut was tightened some of the hard oil and graphite had been forced into the agitation-chamber. The lubricant was apparently the source of the trouble, since it disappeared upon cleaning the machine.

In treating ore of this kind, containing only a small amount of sulphide, it is impossible to tell from the appearance of the froth when the maximum recovery has been reached. For this reason it seems best to allow all of the floatable particles to be extracted from the ore in the first operation, re-treating the 'rougher' concentrate, thus forming a high-grade concentrate and a middling that may be subjected to a further treatment.

POTASH extraction from feldspar, according to a statement by Richard K. Meade in *Metallurgical & Chemical Engineering* can be accomplished by mixing one part of ground feldspar, one part of pure limestone, and one-quarter part of calcium chloride, heating in rotary kilns and then leaching the soluble potash compound. This method will effect a recovery of about 55% of the original potash in the crude feldspar. Substituting salt for the chloride of lime gives still higher yields. The recovery varies also directly as the fineness of grinding of the charge. By comminution to a point where 98% will pass a 100-mesh screen the extraction increases to over 70%. Using glauconite-sand, containing 6% of potash, the same process yielded an extraction of 85%. The kiln is heated by direct firing with oil, producer-gas, or pulverized coal, the reaction taking place at temperatures between 800° and 1000°C. This is a lower heat than is required for burning cement-clinker; 250 lb. of pulverized coal per ton of potash-bearing mix is sufficient.

Molybdenum and Some of Its Uses

There is considerable debate as to whether molybdenum is a benefit to steel and particularly to tool-steel; in fact there is a question as to what use it may be. Some claim that it enhances the value of tool steel, and others deny it. The testimony of steel-makers as to the difficulties experienced in using molybdenum are as follows: Molybdenum tool-steels are said to be likely to crack in quenching, and not to retain their cutting edge after re-treatment, owing to the disappearance of the molybdenum from the outer skin of the steel through volatilization. In some instances tests have shown irregular cutting-speeds and a tendency of the molybdenum to render the tools brittle and weak in their bodies. Again, molybdenum-steel has developed seams and a tendency to fire-crack during treatment. These drawbacks may be overcome or avoided by special care and skill in the preparation of molybdenum-steels high in carbon. Great care is needed also in annealing after the material has been worked into bars, and before it is cut into shapes previous to the hardening. Some of the objectionable features enumerated are due to the use of impure ingredients in the manufacture of the steel or to improper heat-treatment.

In this country, and to a less extent elsewhere, the majority of the manufacturers have almost entirely discontinued the use of molybdenum as a major constituent in tool-steel, and are now using it in a minor capacity in conjunction with tungsten, cobalt, and other rare metals. Iridium-steel and 'stellite' are examples. From 2 to 3% of molybdenum is employed in the manufacture of magnet-steel, and an alloy is also employed in the manufacture of magnet-steel. An alloy containing 2 to 5% molybdenum, 10% chromium, and little or no carbon, is said to be acid-proof or stainless. Elwood Haynes, the inventor of 'stellite,' is quoted as saying that when molybdenum is added to a 15% cobalt-chromium alloy the alloy rapidly hardens as the molybdenum content increases, until, when the molybdenum reaches 40%, the material becomes exceedingly hard and brittle. It cuts keenly and deeply into glass, takes a beautiful polish, which is retained under all conditions, and its surface is not readily scratched. Mr. Haynes predicts a wide application for an alloy containing 25% molybdenum and 15% chrome for the manufacture of fine hard cutlery, but as it cannot be forged the blades would have to be cast. The melting point is not abnormally high.

The principal market for molybdenum in this country is not wide at present. The use of it as an alloy in steel for large gun-linings holds some prospect. The Germans are said to have increased the life of their guns many fold by the use of such a lining. In the steel industry itself in this country there is only a limited use of molybdenum at present. Of course, as a chemical, in the form of ammonium molybdate, it has a wide application in steel laboratories for determining phosphorus in all kinds of steel. It is indispensable for this purpose, but the amount consumed is commercially negligible.

Launder Flotation Machine

By B. M. SNYDER

The flotation machine illustrated in the accompanying drawings was designed two years ago by George Crerar, flotation metallurgist, and is now in successful use in three mills, including the Comet near Basin, Montana, and the Belmont at Belmont, Nevada. It is similar in principle to the Callow cell, making use of air under pressure forced through canvas or some other porous medium. No heating of the pulp has been found necessary, and in one plant, the pulp has a temperature of about 40° F., due to the cold stream-water used. Recently I re-designed the air-distributors to permit of their being dropped in from the top instead of being attached to the bottom of the cell as in the first machine made. This allows the removal of a distributor for repairs or for changing canvas, without interrupting the operation of the machine. The air-distributors are of cast-iron, with connections for air-pipes and water-discharge pipes. The canvas is held by iron bars which are bolted to the cast-iron pan. It will be seen by inspection of the drawings that the machine is of simple construction, being merely a vat divided into six sections, the pulp flowing through a by-pass from one cell to the other. The drop of three inches from cell to cell gives sufficient hydraulic head to keep the pulp moving freely, and the flat bottoms of the cells permit an even pressure and distribution of air in each individual cell.

The cells are narrow, only 15 in. wide, and in effect the operation of the machine is the same as if the pulp were run through an aerated launder 15 in. wide and 30 ft. long, baffles being inserted at 5-ft. intervals to check the flow. The narrow width of the cells with launders on either side permits the froth to overflow with a maximum horizontal travel of less than 8 inches. The froth-launders discharge to a main launder, the concentrate being allowed to fill the froth-launder and to make its own slope. A single small spray of water shot down the slope of each launder, under high pressure, is the most effective froth-breaker. The pulp is kept at a depth of 24 in. or less, in the last cell, the tailing-discharge valve being closed sufficiently to allow an overflow in the stand-pipe, thus maintaining a constant pulp-level. A modification of this plan consists in allowing the pulp to overflow a vertical wood-pipe 6 in. square, having a movable gate to raise or lower the pulp-level. The porous bottoms used are 4-ply, 14 to 16-oz. canvas, and they ordinarily last several months before requiring to be changed. In the first machine built, porous brick was tried for use as an air-atomizer, but the varying porosity and the difficulty of maintaining an air-tight joint around the edges of the bricks, compelled their abandonment. The air is regulated by 2-in. globe-valves conveniently placed above the cells, and once these valves are set, a single valve on the main air-pipe can be used to control the frothing.

One man can attend to a number of machines. With this, as with most flotation-machines, regularity of feed, oil, acid (if used), air, pulp-level, and attendance are necessary in order to secure the best results. The cells are of wood and can be built by any good carpenter or millwright. The iron work is largely made up of bolts and standard pipe-fittings, and the air-distributors can

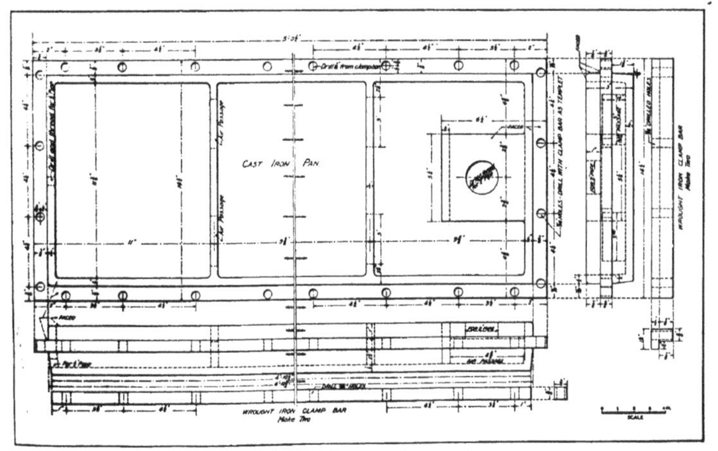

AIR-DISTRIBUTOR FOR LAUNDER FLOTATION APPARATUS

be made at any foundry. The total cost of a machine erected is about $450. In addition a positive-pressure blower, capable of supplying air under 5 lb. pressure, is required. The air-consumption will usually not exceed 10 cu. ft. of free air per square foot of frothing-area. The capacity is governed by the amount of mineral to be lifted and can be safely estimated at 250 to 300 lb. of mineral per square foot of frothing-area. This mineral usually requires cleaning in another machine to eliminate the excess of gangue and raise the grade of the concentrate.

As an indication of the results obtained with the machine, the following report of a typical 10 days' run in one plant is given. It should be remembered that the ore treated carried from 0.5% to 1.5% of zinc in the form of a carbonate:

The machine evidently has a field for usefulness and will survive competition unless all pneumatic machines are superseded by later types, such as the K. & K., that require no blower.

Date.	Flotation-head. Zinc %	Flotation-tail. Zinc %
Sept. 28	13.6	2.1
Sept. 29	13.1	1.95
Sept. 30	14.2	0.95
Oct. 1	15.2	0.7
Oct. 2	12.4	1.47
Oct. 3	11.9	1.25
Oct. 4	12.3	1.8
Oct. 5	13.6	1.2
Oct. 6	12.6	2.3
Oct. 7	13.5	1.3
Average	13.24	1.50

Recovery of zinc by flotation, 92.6%.

Screen analyses of the concentrate and tailing gave:

	%	%	%	%	%	%	%	%
Mesh	+20	+35	+48	+65	+100	+150	+200	-200
Concentrate	0	0	0	0	0.7	7.2	8.7	83.2
Tailing	0	0	0.5	1.0	6.5	17.0	11.0	64.0

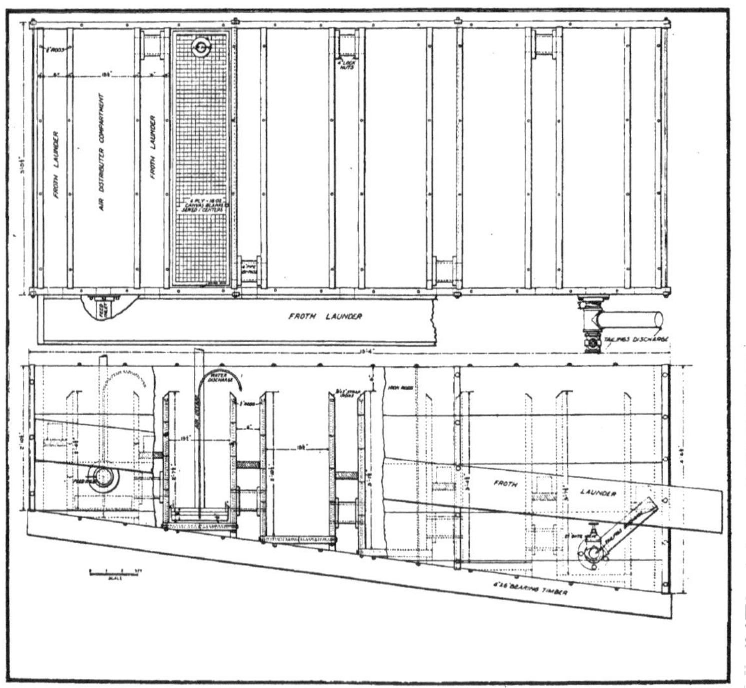

PLAN AND LONGITUDINAL VERTICAL SECTION, LAUNDER FLOTATION APPARATUS

Magnesite on the Pacific Coast

The development of the magnesite industry has been greatly stimulated by the War. Exports of magnesite from Austria and Greece ceased upon the beginning of hostilities, and investigation of our domestic supply quickly followed. As the War continued the search for magnesite became increasingly eager, until now every available deposit, no matter how remote, is given serious consideration. Roads have been constructed to magnesite deposits situated at long distances from the railroads and the motor-truck has become an important factor in the transportation of this mineral to shipping-points. The first magnesite brought to this country from Greece was imported in the early eighties for use in steel-plants, the amount taken in 1885 being about 500 tons. The yearly totals increased until in 1913 there were imported 175,000 tons of calcined product, representing fully 375,000 tons of crude mineral. Until 1914 the magnesite production of California had not exceeded 10,000 tons per annum. The Porterville district is now exceeding that tonnage every month, according to an official of the American Magnesite Company.

Magnesite is used as a plastic, large quantities being made into flooring, imitation-marble, and stucco, its non-cracking qualities and the hardness of the surface making it especially well suited for these purposes. This branch of the business is yet in its infancy, though magnesite producers anticipate its growth to large proportions when the public better appreciates its possibilities. The calcination of magnesite, through a simple process, is nevertheless interesting. The Porterville Magnesite Co., at Porterville, Tulare county, California, owns one of the most complete calcining-plants in the country, having recently installed at its plant, three miles north-east of Porterville, the largest rotary kiln in use in the world for the calcining of magnesite. It is 125 ft. long, 7 ft. diam., and weighs 60 tons. It is set in a slightly inclined position at the base of what is called Magnesite hill, which is honeycombed with mine workings. The rock is brought from the mine in small mine-cars and dumped into a chute from which it is fed to the plant below. The rock is first crushed to ½ to ¾-in. size and is then delivered to the upper end of the kiln. About 1½ hr. is required for it to pass through. When it issues from the lower end it is thoroughly calcined and has much the appearance of nuggets of gold. The temperature in the kiln is maintained at approximately 2400°F., the heat being gen-

MAGNESITE OUTCROP, SAN BENITO COUNTY, CALIFORNIA

MAGNESITE QUARRY, FITCH MINE, STEVENS COUNTY, WASHINGTON. VEIN 80 FT. WIDE

erated by the combustion of crude-oil, which is preheated in coils of pipe and is atomized by steam as it is sprayed into the kiln at the lower end. The temperature is highest at this end. In the process of calcination pure magnesite is reduced to 52% of its original weight by the elimination of the carbonic-acid. About 1¼% iron remains in the calcined material. At some works the carbonic-acid is drawn off and condensed in steel cylinders for distribution to bottling works, soda-fountains, and to chemical plants. This requires a large outlay of capital and the Porterville Magnesite Company does not save its carbonic-acid.

As the calcine drops from the kiln it is carried by conveyors to the cooling-room, and when cool enough is sacked for shipment, although some of it is shipped in bulk. Most of the calcined rock from this plant goes to Eastern makers of refractory bricks and other forms of fire-proof materials. Most of the Eastern buyers prefer that a small amount of CO_2 be left in the rock. Some of the calcine is re-ground to pass a 200-mesh screen. In this form it is used principally in the manufacture of 'sorel' cement for making tile-floors, cement-stucco, and for other purposes, this consisting of calcined magnesite wetted with a solution of magnesium chloride which sets as an extremely strong cement. The rock from the kiln is sampled every 15 minutes and these samples are combined to make a 12-hour average. Each carload shipped bears a certificate of analysis, so that the consumer knows just what grade he is getting. The big kiln of the Porterville Magnesite Co. has been in continuous operation since its installation. Practically all the machinery at the plant is operated by electricity. The time required in re-heating the ordinary kilns when once cooled off, and the increasing demand for magnesite, both crude and calcined, are the primary reasons for employing the continuous-kiln. Other magnesite companies are now adopting rotary calciners, as these have proved more satisfactory than the old upright-kiln.

The magnesite mines at Piedra, in Tulare county, California, are also being developed, and a contract has been made with the Wellman-Lewis Co. of Los Angeles to construct a calcining-plant that will cost approximately $100,000. This will be only the first unit of a plant that will eventually represent an investment of $500,000. It is announced that Eastern capitalists have become interested in the magnesite deposits at Piedra and that negotiations have been concluded for the purchase of a large part of the deposits. A plant will be constructed on land belonging to Thomas J. Curran, manager of the calcining plants near Porterville, the machinery now being in transit. This plant also will be of the rotary type and will have a daily capacity of 150 tons of crude rock.

The huge deposits of magnesite in Stevens county, Washington, recently purchased by D. C. Jackling, are proving to be of great importance. One of these masses is estimated to contain not less than 4,000,000 tons of magnesium carbonate. The accompanying illustration shows the Fitch quarry at a place where the mass is 80 ft. wide. In other places it is fully 200 ft. wide. The dump consists wholly of magnesite, and the former owners used to load it into wagons by shoveling from the front of the dump. Under the direction of Mr. Jackling a complete reform will be made in the methods of mining and transportation in order to effect economy in operation. Another mine in the same district, near Chewelah, has been taken over by a San Francisco corporation, the Northwest Magnesite Co., in which S. F. Morse, W. H. Crocker, R. S. Talbot, and B. L. Thane are interested. They are shipping crude rock, and also have a calcining plant in operation. Other operators in the same locality are the American Mineral Production Co. of Chicago, and the U. S. Magnesite Co. of Spokane. The magnesite in these Washington deposits is wholly different in appearance from that commonly mined in California. It is coarsely crystallized, gray to bluish black in color, due to the presence of microscopic scales of graphite. It resembles some varieties of marble or dolomite. It is not dolomite, however, as calcium carbonate is practically absent.

In California, along the Mother Lode, and northward through Placer, Nevada, Sierra, and Plumas counties, are large masses of rock commonly called ankerite; but not all of this rock is ankerite, for frequently no iron or manganese carbonate is present, and the rock then becomes dolomite. When iron, manganese, and calcium carbonate are absent from the rock and only magnesium carbonate remains, the rock must be classed as magnesium, even if it does not conform in physical appearance to the typical magnesite of the text-books. In California these great masses of mixed carbonate, usually, but not always, are associated with serpentine or gabbro; those in Washington are described as occurring at the contact of schist and quartzite. This fact is of no importance, however, as an analysis of the ore reveals, an average sample having been found to contain silica, 0.8%; alumina, 0.4%; iron oxide, 2.7%; lime, 3.3%; magnesium carbonate, 92.1%. In California all the deposits thus far worked, with one exception, occur in association with serpentine. The exception is that near Bissel station on the Santa Fe railroad in San Bernardino county, about 18 miles east of Mojave. There the magnesite is interbedded with shale and thin-bedded limestone. It is possible that some of the great masses of the so-called ankerite of the gold-belt of California may be found on analysis to consist principally of magnesium carbonate. In appearance it greatly resembles the deposits of Stevens county, Washington.

The conical mass shown in one of the accompanying illustrations is a typical outcrop of a large mass of magnesite in serpentine, occurring in San Benito county. It is known as the Sampson mine, and is situated 38 miles west of Mendota, to which place it is hauled for shipment. This is being worked by the John D. Hoff Asbestos Co., of San Francisco.

For use as a refractory material the specifications of the open-hearth steel-makers call for a minimum of 85% MgO, less than 6% SiO_2, less than 4% CaO, and a loss on ignition not to exceed 5%. No limit is placed on the iron oxide (Fe_2O_3) present, although calcined magnesite rarely contains more than 7%. For use in making sorel cement and as a filler for paper the silica-content may run as high as 12%. No limit is fixed for CaO, while the Fe_2O_3 must not fall below 1%.

A Lost Cargo of Copper

A cargo of copper buried beneath the waters of Lake Huron since the Civil War probably will soon be raised. Incidentally several residents of Toledo, Ohio, and Alpena, Michigan, hope to get rich through dividing the 'spoils,' which they estimate to be worth, under present war prices, at least $1,000,000. A diver has located the cargo and Detroit despatches announce that some of the copper already has been recovered. It is declared this is a record in deep water salvage. The effort to recover the valuable cargo revives again the often repeated tale of one of the numerous disasters which help make up the history of the Great Lakes. On August 9, 1865, the steamer 'Pewabic' was bound down from the Keweenaw peninsula with a cargo of copper and a crew and passenger list of more than 125 persons. When seven miles off Thunder Bay island, in Lake Huron, the 'Pewabic' was rammed by the steamer 'Meteor,' and before any of her boats could be launched she went to the bottom. Of those on board 125 persons went to the bottom with her. Like the lost treasure of Captain Kidd, the cargo of the 'Pewabic' has time after time given rise to tales that have brought dreams of wealth. Until now, however, the location of the boat remained a mystery. According to word from Alpena, Benjamin Leavitt, a diver from Toledo, found the 'Pewabic' under 180 ft. of water, and made a confidential report to his fellow adventurers. It is said to have been of such a nature as to induce his backers to lay plans for recovering the million-dollar cargo and such other property as may be worth bringing to the surface.—*The Daily Metal Reporter.*

ROTARY KILN, PORTERVILLE MAGNESITE CO., IN PROCESS OF ERECTION

THE DEEPEST American drill-hole is a diamond-drill bore in Sussex county, New Jersey, 4920 ft. deep, recently completed by the contract drilling forces of the Sullivan Machinry Co. A 2-in. diam. core was removed to a depth of 1600 ft., beyond which tools removing a $1\frac{3}{8}$-in. core were used to the completion of the hole. When a depth of 4900 ft. was reached, the long line of rods required for drilling weighed 13 tons, and it took 8 hours' work to hoist, replace, and resume drilling. It is said that the hole showed no deviation from the perpendicular. It required 20 months to complete the hole, which is 1700 ft. deeper than the deepest previous hole of which there is any record in North America.—*Eng. News-Record.*

Liberty Bell Mercury-Trap

By ALBERT G. WOLF

Mercury-traps involving many variations of the principle of conducting the pulp-flow from plates to the bottom of a receptacle and allowing the flow to discharge over the top have been described. In the accompanying illustration is shown a mercury saver built on this simple principle, which for ease and accessibility in cleaning cannot be equalled. It was used at the Liberty Bell mine, Telluride, Colorado, when amalgamation was practised there.

The receptacle of the trap is made of sheet-iron and is cylindrical in shape, 14 in. high and 10 in. diam. It is

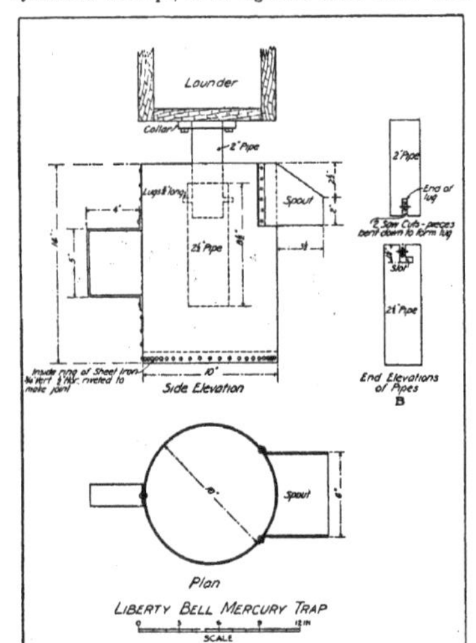

LIBERTY BELL MERCURY TRAP

riveted along one seam, and the bottom and side-pieces are held together by rivets through an angle-ring of sheet-iron. This receptacle is fitted with a handle, as shown, and a spout 6 in. wide and 4½ in. deep, tapering to 2 in. The pulp from the launder at the bottom of the plate is conducted downward through a 2-in. pipe, then through a 2½-in. pipe, 8½ in. long, to a point about 4 in. from the bottom of the trap; it then flows out through the spout to another launder. The 2-in. nipple, as it may be called, is threaded at the top and screwed into an iron collar of proper size and tapped out for the purpose; that in turn is fastened to the bottom of the launder by bolts. The nipple is of such length that its lower end will just clear the spout when the trap is pulled forward. Two lugs are formed at the bottom and on opposite sides of the nipple by making two sets of parallel hack-saw cuts ¾ in. apart and bending the two pieces outward. The piece of 2½-in. pipe is suspended from the lugs by right-angle slots, cut as shown in the diagram at B.

In cleaning the trap, the method pursued at the Liberty Bell mill was as follows: When the pulp-flow had ceased, the 2½-in. pipe was given a short turn and then pushed down from the lugs. The handle of the trap was then seized and the trap pulled forward from under the battery-plate launder. By running a hand down through the pulp just behind the spout, an opening was made to the bottom, and the quicksilver poured out. The trap was as easily returned to its place. Periodically the upper portion of the sand in the trap was skimmed by hand and the lower portion dumped into buckets to be carried to a clean-up pan for grinding. One of these traps was used below each 5-stamp battery of the 80-stamp mill, and a group of seven below the tube-mill plates, or 25 in all. That the conditions under which they worked was severe, and that they did their work satisfactorily is attested by the fact that crushing at this plant was done in cyanide solution, which necessitated the use of a moderate excess of mercury, the total amount used being 50,000 to 60,000 oz. per month in treating approximately 480 tons of ore per day, and the loss for a period of eight months averaged but 2000 oz. per month, or less than 0.15 oz. per ton ore treated. The artificial raising of the temperature of the mill-solution caused an increased loss of mercury of about 25%, undoubtedly due to an increase in the dissolving action of the cyanide-solution. These figures do not include the appreciable saving from pipes, launders, and the like, which could be cleaned only at the semi-annual Christmas and Fourth of July shutdowns. The amalgam saved in the traps varied from 2.5% to 5% by weight of the total amalgam, but was probably of lower grade.

ALASKA is something of a contributor to our much-needed tin demand. Alaska lode-tin was discovered by a geologist of the Geological Survey, while helping two prospectors by crudely smelting in a camp cup a piece of peculiar-looking ore which the prospectors rightly suspected to be tin. Specimens of stream-tin were also collected by Survey geologists in Alaska in 1900, before this metal had been known to occur in the Territory. Prospecting for stream-tin soon became active, and since 1902 nearly 1000 tons of metallic tin has been produced. The output last year was 139 tons. In the Seward Peninsula, where placer tin was first mined, the source of the stream-tin has been discovered and lode mining is now being carried on. Stream-tin is widely distributed in the Hot Springs district, but as yet few of the gold-placer miners make any pretense of saving it. Placer mining in the Hot Springs district can be carried on for about three months in summer.

MT. LASSEN, in California, shows evidence of lava having appeared in the crater within the last few months, though none has overflowed, according to a report by J. S. Diller of the U. S. Geological Survey.

Fotation—The Butte & Superior Case—III

Part of Argument by Henry D. Williams, for the Plaintiff

We stand today in this court maintaining that the Supreme Court of the United States has given us a patent for a process identifiable by the results which characterize our process, the air-froth concentrate; and has not imposed upon our patent a limitation which shall be such a limitation that a defendant who puts in 21 pounds of oil, or so-called oil, to the ton of ore, shall be permitted to escape from infringement of our patent, although he obtains the result which we first obtained in the history of metallurgy.

The evidence in the case establishes the fact that there are oils which are frothing-oils; that there are oils which produce the result of the process and they produce it when used in minute and critical proportions. The evidence in the case has established the fact that there are other oils that cannot by any possibility produce the process; they are inert; they are useless for the purpose of producing the process. And the evidence in the case establishes the fact that in the development and improvement of this new art that commenced with the patent in suit, another wonderful discovery was made; the soluble frothing-agent process. That this soluble frothing-agent process, as we demonstrated it, has extreme fineness of air-bubble formation surpassing anything that can be produced by the frothing-oils; and that when you associate together the process of the patent in suit with its critical proportion of frothing-oils, a wasteful quantity of inert oils, and a soluble frothing-agent to prevent the deleterious action or to limit it, of these inert oils, it is possible to carry out the process of the patent in suit in a procedure wherein you pour into the pulp a little more—and a very little more—than twenty pounds of oil to the ton of ore. In fact, these operations have been of exceeding interest. We have found 18 lb. and 19 lb. and 19½ lb. and 20 lb. and 20½ lb., all working the same way, because of this inert oil, and the defendant maintaining that having used more than a fraction of 1% of oil in the ore he has escaped our patent. To that we say that the Supreme Court of the United States has imposed no such limitation upon us. The patent in suit does not impose any such limitation, and the disclaimer that we have filed, following the precise language of the Supreme Court of the United States, has preserved to us—if it be necessary that it should be preserved to us by something more than the other claims—a process wherein the critical proportion of the oils that do the work of the process of the patent in suit are used and additionally an inert oil is poured in in wasteful quantities and a soluble frothing-agent used to prevent it from spoiling the process.

Now, as to the patent in suit. If ever there was a patent which described the thing that was discovered, this is such a patent. We have the history of the discovery and we have the patent in suit describing it. The first reference in the patent in suit is to two of the Cattermole patents, the principal one 777,273, which is the fundamental Cattermole patent, the other 777,274 merely because it discloses a method of generating oleic acid *in situ*, which is one of the specific things for which there are a few claims in the patent in suit. The patent says: "We have found that if the proportion of oily substance be considerably reduced"—considerably reduced from the Cattermole process, the Cattermole process then being carried out with from 2 to 3 or 4% of oil on the ore—"say to a fraction of 1% on the ore"—not such a close and limiting statement—"granulation ceases to take place, and after vigorous agitation there is a tendency for a part of the oil-coated metalliferous matter to rise to the surface of the pulp in the form of a froth or scum."

Then the factors are described as they have been developed in the work in carrying out this process—acidification preferably used—heat also of use. Fine pulverization of the ore not only a factor of the success of the process, but the tremendous advantage of the process, the thing that made it possible to rescue these slimes that had been piled in dumps all over the world and to extract the valuable metal from them.

"The proportion of mineral which floats in the form of froth varies considerably with different ores and different oily substances, and before utilizing the facts above mentioned in the concentration of any particular ore a simple preliminary test is necessary to determine which oily substance yields the proportion of froth or scum desired."

That is the direction to investigate with the means at hand in the laboratory for the purpose of determining what oily substance is best adapted, and in what proportions and under what conditions.

Then follows an example, an example which has been exactly reproduced in evidence in this case, Broken Hill ore. To this is added a mineral-acid or an acid-salt and mine or other waters containing metallic sulphate and it mentions iron sulphate. I think that mine-waters would be apt to contain copper sulphate. "To this is added a very small proportion of oleic acid, say from 0.02 to 0.05%, on the weight of the ore." "The mixture is warmed, say to 30° to 40°C., and is briskly agitated in a cone-mixer, or the like, as in the processes previously cited about 2½ to 10 minutes until the oleic acid has been brought into efficient contact with all the mineral particles in the pulp."

That procedure was exactly followed at the time of the discovery, has been reproduced in evidence in the Hyde case, and it can be reproduced at any time. The amount of oleic acid is also definitely recommended. "The minimum amount of oleic acid which can be used to effect the flotation of the mineral in the form of froth may be under 0.1% of the ore, but this proportion has been found suitable and economical."

Then: "When agitation is stopped a large proportion of the mineral present rises to the surface in the form of a froth or scum, which has derived its power of flotation mainly from the inclusion of air-bubbles introduced into the mass by the agitation, such bubbles or air-films adhering only to the mineral particles which are coated with oleic acid." We have the description there of the froth, oil-coated mineral particles, air-bubbles forming, and froth floating on the surface of the water and carrying a good portion of the metalliferous mineral.

Now, this description, which I have just read, was of the simple operation in the Gabbett, and it is a full disclosure, because the Gabbett was a well-known contrivance as we proved beyond question, although its use under such conditions as we used it in our process commenced with the Cattermole investigation. For Cattermole it was used up to 900 r.p.m. and when thus speeded up it became a really efficient aerator, although in the Cattermole process the air, if it worked at all, worked against the process. Still, it didn't seem to make any difference for the reason that metallic particles coated with oleic acid in the Cattermole proportions 2, 3, or 4% reject air-bubbles, have no attachment that is worthy of the name to air-bubbles. Putting aside all finer scientific questions as to whether it is directly attached or oil attached, they don't stick. The air rushes through in great quantities. It comes in and goes out, and the Cattermole granules maintain their integrity and their tendency to sink, and they do sink the minute you give them a chance against an up-current which carries away the sand.

Now, this description closes with the statement that the froth is removed from the pulp by spitzkast, upcast, skimming, draining, or otherwise. There is a complete disclosure of the process.

Now, the patentees might have rested upon that disclosure, but they went ahead and described an apparatus adapted to the conditions which existed in the art at that day and was a very useful apparatus, because in the art at that day 40-mesh crushing was fine crushing; because in the art at that day they feared to crush finer, because the finer they crushed the greater the proportion of slimes, and the greater amount of the metal that was wasted. Adapted to the conditions of that day, they produced a spitzkasten-classifier for the purpose of classifying their gangue and with it some particles of the mineral and especially the heavier particles of the mineral which were too heavy to be carried in the froth because, as appeared in the evidence here, it was a very serious problem in those days whether it was going to pay to crush the ore to such a degree of fineness as was necessary for the full realization of the air-froth flotation process.

A wholly new field was opened up in the metallurgical art by this process, and the inventors had not the courage to predict absolutely that such fine crushing as now characterizes the art would be feasible or possible. But with the coarser crushing, of course, the heavier particles would drop out of the froth, and they would have to be recovered by other means, and so there was the spitzkasten-classifier, such as the art knew, combined with the agitating-vessel, and the froth overflowed and the sinking products were classified, and while the particular recommendation was further treatment by agitation—in other words, doing the process over again—there was a suggestion of treatment on a shaking table, such as the art then knew, and there was also a suggestion of another method of treatment, which leads in the art in the agitation by aeration feature which now characterizes what are known as the pneumatic cells. In that method the pulp was subjected to a pressure of air or other gas of one or two atmospheres in a vessel, the result being that the air or other gas went into solution in the water, and then on release of the pressure "the gas was generated throughout the mass at once, sweeping to the surface thereof all the metalliferous matter in the form of a froth, which can be separated as before." That is the pneumatic operation, such as is carried on in the air-basket or the Callow cells, or in the Janney machine which has the air-basket in the spitzkasten.

THE COURT: Is that in the patent in suit?

MR. WILLIAMS: That is in the patent in suit. We had no occasion to refer to it in the Hyde case, because the art had not then found out how to use it.

MR. SHERIDAN: You don't mean it is claimed in the patent in suit?

MR. WILLIAMS: I say it is fully within the claims. Judge Bradford found it to be, and we are waiting for the Circuit Court of Appeals for the Third Circuit. In that language you have the sweeping to the surface of the particles of froth by air-bubbles rising upward, agitation by aeration, which was disclosed in this patent to the metallurgical art, and has now been utilized. That was the particular contention in the Wilmington suit.

Again I read from the patent, still describing this method which was called super-aeration or plus pressure; that is to say, you work at a pressure greater than the atmosphere, and the aeration is the result of relief from that high pressure.

"The whole of the mineral to which the air-bubbles are attached—say the oiled mineral—at once rises to the surface as a coherent scum or froth."

That follows the condition where you relieve the pressure and these air-bubbles come rushing up, sweeping to the surface—a sort of air upcast—or instead of relying wholly on the buoyancy of the air-bubbles to rise, you may start them up by a pressure from below, and by that pressure and buoyancy you get the agitating effect of these bubbles of air rushing through the pulp and bringing the air in contact with all the metal-bubbles, and carrying them to the surface in the form of a froth.

a rather active froth as compared with the quiescent froth which occurs when you have not this pneumatic action.

Now, briefly as to the claims.

Claim 1. "The herein described process of concentrating ores which consists in mixing the powdered ore with water, adding a small proportion of an oily liquid having a preferential affinity for metalliferous matter (amounting to a fraction of 1% on the ore), agitating the mixture until the oil-coated mineral matter forms into a froth and separating the froth from the remainder by flotation."

There is no limitation as to the means whereby that agitation is effected, as to whether it shall be effected by stirring it violently or whether it shall be effected by shooting air into it. This was the language of Judge McPherson at the argument; he said, "the question is whether you shoot it in or draw it in." The claim merely says, "Agitating the mixture until the oil-coated mineral matter forms into a froth." Agitation, whether mechanical or pneumatic, produces that froth.

Claim 2 adds acidification, which the defendant has used.

Claim 3 adds warming the mixture, which the defendant is not now using, but used up to a comparatively recent time. It uses great care now, apparently, not to warm the mixture, although the oil is warmed.

Claim 5:

THE COURT: Your claims leave it optional, do they not?

MR. WILLIAMS: Yes, claim 1 omits all reference to heating and acidification, and therefore claim 1 is infringed when they operate in the cold and without acid.

Claim 2 includes acidification and therefore claim 2 is not infringed if you omit acidification, if you omit the acid.

THE COURT: Don't you leave acid and heat optional altogether, although approved?

MR. WILLIAMS: Why, it is true as to the broad claims, claims 1, 9, and 12, of the invention, there pointed out, that those claims leave heat and acid as wholly optional, but it is true as to claims 2, 6, and 11 that the invention which is pointed out in those claims includes as an essential part of the invention the use of acid, and it is true as to claim 3 and other claims of the invention, that the invention pointed out in those claims is not infringed if the defendant does not heat the mixture, but uses the cold pulp and omits acid; that is to say, as to those claims, in the one case acid is an essential of the invention pointed out, and in the other case acid and heat are essentials of the invention pointed out.

THE COURT: As a matter of fact if they did not use the other features of your claims, merely using acid and heat, it would not be infringed.

MR. WILLIAMS: Oh, no.

THE COURT: But if they do use the other features of your invention, the mere omission of acid and heat would still leave it infringing?

MR. WILLIAMS: Yes, infringing.

THE COURT: In other words, your patent does not monopolize and limit you merely in the use of acid and heat?

MR. WILLIAMS: Yes. Your Honor gave some consideration to that in your opinion and the fundamental fact is that acid and heat are simply added on to the essential process in certain claims.

Now, that brings us to claims 9, 10, and 11. Claim 9 is the broadest claim in the patent. I will read it as it is in the patent, and then we will consider what the disclaimer has done to it to save it from the criticism of the Supreme Court of the United States.

"The process of concentrating powdered ore, which consists in separating the mineral from the gangue by coating the mineral with oil in water containing a small quantity of oil, agitating the mixture to form a froth, and separating the froth."

Now, your Honor will see that that language as it is drawn exactly describes every operation that the defendant has carried on; that the term "a small quantity of oil" cannot by any possibility be held to be definitive of a quantity of oil wherein we shall stop at 1% and say—above that you are free to use it; below that you must not use it. But of course the term "a small quantity" is indefinite, and the Supreme Court said it was indefinite.

THE COURT: Of course, we must take it as the Supreme Court says. Aren't there other patents where that term has been allowed?

MR. WILLIAMS: Oh yes; the circumstances of the case have to be considered. The circumstances considered by the Supreme Court of the United States was that here was Kirby, with 25% of oil. Now 25% of oil, as compared with Elmore, was small, but it would hardly be considered a small quantity. But the circumstances of the case, the Supreme Court said, were such that your step is not a very long one, from 25%, we will say, to a fraction of 1%. Then there was the Froment test-tube disclosure where 12½% was used; comparing that to a fraction of 1% it is not a very long step Now, the Supreme Court gave us the identifying characteristics which it would have been very difficult to write into that claim originally. I don't know just what language I would have used had I drawn the claim. The Supreme Court said "The patent must be confined to the results obtained by the use of the critical amount, a fraction of 1% of oil." Now, we have written that limitation into that claim, and wherever we find a process wherein a frothing-oil is at work, in minute quantities, producing the froth of the patent in suit, and when we see that the operation of the process as a whole produces that result which the Supreme Court says is the identifying characteristic of our process—then, under claim 9, as validated by the disclaimer, it is only necessary to consider whether or not the defendant is using a small quantity of oil.

THE COURT: I don't believe I understand you very clearly there. Would it still be infringed if it was more than 1%?

MR. WILLIAMS: If the amount of oil fed into the mixture exceeds 1%—if the amount of oil dumped into the pulp exceeds 1% is 21 lb. or 20.1 lb., or 20 lb., because 20 lb. is more than a fraction of 1%—if the amount of oil dumped into the pulp is 20 lb., and if we find upon examination that 18 lb. of that oil is an inert, useless oil —and Mr. Thomas Janney admitted that these fuel-oils and kerosene-oils—I think his was a fuel-oil—were inert oils that were not frothing-oils at all—if you dumped 18 lb. of a non-frothing oil into the pulp and 2 lb. of an oil that is a frothing-oil, and if you utilize the further discoveries of the plaintiff and put in a soluble frothing-agent so as to prevent that inert oil from spoiling the process, you have carried out the process of the patent in suit. You get the result of the process of the patent in suit, and by grace of the Supreme Court of the United States, when you obtain the results of the process of the patent in suit you use the invention of the patent in suit.

Now, your Honor asked if the words "a small quantity" did not appear in claims; they do. They appear in the claims of our solution patent, which is in evidence.

THE COURT: Of course I had reference to claims that had been upheld in the Supreme Court. I had in mind the Tilghman-Proctor case, where any degree of heat was held to be within the claim.

MR. WILLIAMS: It all depends on the circumstances of the case. Here we have a patent which has been upheld by Judge Bradford, and which we hope will be upheld by the Circuit Court of Appeals of the Third Circuit, and which we hope will not go further. Here is the claim:

"Mixing the ore with water containing in solution a small quantity of a mineral frothing agent."

There you see the characteristic is that the pulp has dissolved in it a mineral-frothing agent, and the quantity of the mineral-frothing agent is a small quantity. Now, the fact is, as has been testified here, that you can increase or diminish the amount of that frothing-agent in solution within the limits of its solubility, practically, that is, of course you have to have enough to get an action, but a very minute amount has a very wonderful action, and so that claim says "a small quantity," and Judge Bradford has found that claim to be valid and infringed, and without any question indefiniteness in claims within the permissive indefiniteness that the art permits is a perfectly proper characteristic of a claim. Of course you must not—I cannot quote any better language than the language of the Supreme Court of the United States: "The patent must be confined to the results obtained by the use of a fraction of 1% of oil." That is what the Supreme Court gave us.

THE COURT: Of course that claim itself is indefinite and allows for the use of a quantity from the smallest to the largest fraction of 1%.

MR. WILLIAMS: Yes, it does; and that range, without any doubt, will take care of the fact that the critical proportions vary with different ores and with different oils. We have not pursued that to its uttermost. For instance, we find that at the Timber Butte mill they have an ore which is very nearly the same as the defendant's ore, which is being treated with 0.7 of a pound of pine-oil. That is a frothing-agent which has a peculiarity in that it contains the oily or insoluble frothing-agent and a soluble frothing-agent, and the two processes work together there, and the total amount of the oil used is only seven-tenths of a pound. It comes pretty near to being the smallest amount that has been used.

Eucalyptus oil of the Australian variety, according to the evidence, has been used in as small an amount as half a pound to the ton of ore. That is another of those essential oils, which has a soluble and an insoluble froth-ing-agent in its make-up. The defendant uses pine-oil. Pine-oil is the basis of its whole operation. Pine-oil produces the carrying out of the process of the patent in suit with the critical proportion far below the limit of the largest fraction of 1%. Pine-oil supplies a soluble frothing-agent which gives that extremely minute emulsion which your Honor saw projected on the screen. Additional to that pine-oil they pour in great quantities of petroleum-oils that are not frothing-agents at all; they are inert and useless in the process; they do not contribute to the carrying out of the process at all. And even as to that, they have to be very, very careful. They run up a little above 20 lb. now and then, but the greater part of their operations have been below 20 lb. They get diminished recoveries, diminished grade. They are losing $100,000 a month, but by this combination they can carry on our process, they can produce our results, they can run their mine, but they are certainly using our invention.

DISSOCIATION-PRESSURES of metalliferous sulphides at comparatively high temperatures, where mercury gauges cannot be used, have been determined successfully by a new method devised and tested by E. T. Allen and Robert H. Lombard at the Geophysical Laboratory at Washington, in conjunction with the secondary enrichment investigation. The method is based upon balancing the dissociation-pressure by a known vapor-pressure of liquid sulphur. Determinations were made on the dissociation-pressures of covellite and pyrite, obtaining the figures respectively of 1 mm. to 500 mm. in terms of mercury. The method applies to other compounds where a single volatile dissociation-product is obtained which does not attack glass and which condenses at temperatures below 1200°C.

SULPHUR is readily soluble in petroleum. Some of the Texas oil contains free sulphur in solution in addition to compounds of that element. In the heavier hydro-carbon compounds the solution of sulphur takes place immediately, even at ordinary temperatures. According to H. Endemann, heavy hydro-carbons react with pyrite also, half of the sulphur being taken up, reducing the pyrite to the mono-sulphide, FeS. The sulphur in oils, as shown by G. Sherburne Rogers, is presesnt in a variety of compounds, some of which are quite complex, but free sulphur is often found in them.

REVIEW OF MINING

As seen at the world's great mining centres by our own correspondents.

McCARTHY, ALASKA

THE STRIKE AT THE KENNECOTT MINE DECLARED BROKEN.—SUCCESSFUL HYDRAULICKING ON CHITITU CREEK.—NEW COPPER PROPERTY BEING DEVELOPED ON HIDDEN CREEK.

The strikers from the Kennecott mines are dispersing and it is conceded that the strike is broken. The last offer made by the company was a raise in the bonus which would give the men $5.50 per day with the price of copper above 27½ cents. The men still demanded a raise in the base wage to $5.75 but finally voted to return to work on the company's terms, pending the arrival of a mediator who is on his way from Washington. The company refused to take them all back, reserving the right to reject any individual. On this point they deadlocked again. The company then started to bring in men from Anchorage and are now about 60% full handed. Ore shipments have been resumed and it is expected they will be normal within a week. On Chititu creek, Frank Kernan is operating his hydraulic plant on No. 11 at the confluence of White and Rex gulches. There is plenty of water this year. About 2500 cu. yd. of gravel is being handled daily with two 3-in. nozzles on the bank, and a 5-in. nozzle for stacking tailing. He has plenty of water for these lines under a 400-ft. vertical head. The gravel bank is 10 ft. thick, which, with two feet of the soft shale bedrock, is piped into the flume. The ground averages about $1 per cubic yard.

Under the direction of W. H. Seagrave, consulting engineer, the Tjosevig-Kennecott Co., at McCarthy, has commenced systematic prospecting and development on the property at the head of Hidden creek, across the Kennecott glacier from the Bonanza mine.

STRIKING MINERS AT M'CARTHY, ALASKA. NOTE THE FLAG

Work was commenced on a 13-mile wagon-road from McCarthy to the property, but has been discontinued while the more important work of developing ore progresses during the favorable summer months.

Considerable prospecting and some development work in various localities is being done along the limestone-greenstone contact for miles on either side of the Bonanza mine this summer.

MAYER, ARIZONA

DEVELOPMENT OF OLD AND NEW MINES STIMULATED BY THE IMMEDIATE PROSPECT OF A MUCH-NEEDED RAILROAD.—MUCH SHIPPING ORE AVAILABLE IN THE DISTRICT ABOUT TURKEY.

John Black, of Mayer, general manager for the Fairview Gold & Copper Mining Co., announces that extensive development of the company's property is about to begin. A gasoline hoist has been delivered at the mine and a shaft is to be sunk on a big outcrop of copper, silver, and gold ore which is traceable to Copper mountain. The mine is two miles east of Turkey. The company is fully financed. The officers are: president, F. N. Bonini, of Niles, Michigan; vice-president, W. Miller, of Kansas City; secretary and treasurer, H. M. Coffman, of South Bend, Indiana. This property has one of the largest veins of milling ore on the belt.

Negotiations are under way for the purchase of the old Parker and Great Republic mines in the Turkey district. There are a number of veins that will run high in copper and gold. A 5-stamp mill with cyanide annex is on the Parker property. The Pickaway mine, situated between the Parker and the Great Republic mines, is included in the deal.

James Cleator, who owns the Turkey Gobler mine, and several other claims on which the town of Turkey is situated, expects to close a deal with Los Angeles people for the sale of its property. The development consists of several shafts and adits in a good grade of gold ore. Turkey will be the junction point of two railroads.

The new railroad will open up the district south and east of Turkey where a number of mines have been developed up to the shipping stage, but owing to the cost of transportation heretofore, the mines have neither been worked nor equipped.

A contract is about to be let by the Big Chief Copper Co. for the sinking of a 500-ft. shaft on one of its properties on the belt between Mayer and New River. A year ago, Jordan and Grace, of Phoenix, secured a large block of mining ground on Black canyon, a mile and a half south-west of Bumblebee, and began development which determined the extent of the several veins of copper and gold ore. A shaft was sunk 50 ft. at the south end in the centre of a mineralized zone 300 ft. wide. It is estimated that 100,000 tons of milling ore can be quarried to the depth of the shaft. Some of the ore is copper glance. On another claim an adit has been run in 250 ft. which has tapped a vein lying next to an iron dike, which can be traced throughout the entire belt. Three veins have been cut which carry a milling grade of ore. In another shaft, at the north end of the property, is found gold ore running as high as $300 per ton. The officers of the company are: president, Thomas Boyd, of Mound City, Illinois; vice-president, R. E. Grace, of Phoenix; and secretary and treasurer, S. E. Jordan, of Phoenix. This will be one of the first mines to supply ore to the new railroad.

The veins which have been proved on the Big Copper Chief Co.'s ground extend north and south, and three groups of claims have been acquired by Proctor Ross and associates, of

Phoenix. They own 500 acres and plans are under way to develop them. Jasper N. Nellis, who owns an interest in all of these properties, was the original locator.

The old stage camp of Cañon will become important as the new railroad will cross the Agua Fria not far below the town. W. Jeff Martin owns several copper and gold mines in that vicinity. He recently made an important strike of copper on one of his claims, two miles north of the station.

The Kay mine is developed by several shafts along the strike of the vein for a mile and a quarter. The deepest shaft is 350 ft. The vein, small at the start, has opened out to 16 ft. at the bottom and will average 15% copper and enough gold to pay charges of mining and milling. A mile north of this shaft, another has been sunk 195 ft. on the same vein and has developed 8 ft. of ore of nearly as good a grade. The mine is owned principally by S. J. Tribolet, of Phoenix, who has spent $75,000 in its development.

The Palo Verde is one of five mines that are to be developed by E. H. Wilson, managing director of the new railway company. Recently a contract was let for sinking the old shaft at the Palo Verde another 250 ft. At the bottom of an 185-ft. shaft, the vein is 8 ft. wide and shipments of ore have run 18% copper, 11 oz. silver, and $22 in gold per ton. The other four mines are the Leland, Senia Raey, Minnie Lee, and National Amalgamated. J. M. Sweatman and C. S. Scott, of Phoenix, are also interested in these five mines, all of which have extensive development.

The Orizaba Mining Co. has been organized to take over and further develop the Orizaba copper mine which is located south-east of the Kay mine about six miles. The officers are: president, George F. Wilson, of Globe; vice-president, Parker Woodman, of Bisbee; secretary and treasurer, A. F. Muter, of Phoenix; W. W. Searles, of San Diego, J. F. Cleaveland, of Phoenix, F. M. Huddleston, of Los Angeles, and T. J. Lawrence, of Pittsburg, also are interested. It is estimated there is 15,000 tons of 7% copper ore blocked out. There are several shafts and adits, all in shipping ore.

CRIPPLE CREEK, COLORADO

IMPORTANT SURFACE DISCOVERIES CONTINUE.—LESSEES SENDING THE USUAL AMOUNT OF ORE TO THE MILLS.—OUTPUT OF THE DISTRICT FOR JULY.—THE CRESSON COMPANY SHIPS ORE UNDER ARMED GUARD.

The aerial-tram of the Granite Gold Mining Co. from the Dillon mine to the loading-station in the yard of the Midland Terminal Railroad Co., at Victor, is in operation, and two cars have been loaded each day this week. The Dillon mine is operated on company account and making a heavy production.

A surface discovery has been made on the Deerhorn, one of the Globe Hill properties of Stratton's Cripple Creek Mining & Development Co., by Mr. Reynold, a watchman for the company at the Winfield offices on Globe hill. Reynold has employed his spare time in prospecting near by, and discovered a vein near surface from which the first carload, mined at a depth of 12 ft. and shipped to the Portland mill at Colorado Springs, brought settlement at the rate of about $16 per ton.

The El Paso Extension Corporation of West Virginia, owning the Rocky Mountain and North Slope claims on the south-west slope of Beacon hill, and part of the Nellie Bly adjacent, recently commenced work. Al Campbell, a lease-operator and a former superintendent of the Gold Dollar Consolidated Mining Co., is in charge of operations. Campbell is prospecting at surface and has found rich float, and is trenching for its source.

Lessees on the Little May and Blanche fractions on the east slope of Bull cliffs, adjoining the old Victor mine, have opened up mill ore, and are getting out a shipment by windlass. A steam or electric hoist will be installed if a profit is shown by returns.

The Copper King, a patented claim on the south-east slope of Grouse mountain, idle for years, has been leased by M. Ledgerwood and B. W. Johnson, of Victor. At a depth of 41 ft. the lessees, who are old timers, after cleaning out the shaft, found exposed a narrow seam of quartz that has widened out to about eight inches and carries fair silver and low gold content. They are sinking the shaft and at 50 ft. will prospect the vein by drifts.

Lessees on the Hiawatha on the south slope of Beacon hill have opened up a new orebody and loaded out their initial shipment the first of the week. The ore is silicified granite carrying sylvanite and is returning assays as high as $45 per ton. Notices nave been mailed stockholders of the Blue Flag Gold Mining Co. of the annual stockholders meeting for the election of directors, to be held at the office of the company in the Exchange building, Denver, on Tuesday, August 14. The company owns the Blue Flag mine and mill on Raven hill in this district, now inactive, and properties near Breckenridge, in Summit county, and placer ground in California.

The Neville mill, a cyanide plant situated on the properties of the Free Coinage Gold Mining Co. on Bull hill, near the town of South Altman, is being dismantled and the milling machinery will be shipped to the Morse Bros. Machinery company at Denver. The plant was operated at loss from its inception, and the loss in purchase price, operation, and installation is estimated at fully $100,000.

The Hondo Gold Mining & Milling Co., C. W. Savery, of Denver, president, has abandoned its lease and option on the properties of the Keystone G. M. Co. on Bull hill, and has removed the mine plant.

The production of the Cripple Creek district for July totalled 87,147 tons with an average value of $12.84 per ton and a gross bullion value of $1,119,000.16.

Cripple Creek mining corporations paid dividends in July as follows: Cresson Consolidated, monthly, $122,000; Golden Cycle M. & R. Co., monthly, $45,000; Portland G. M. Co., quarterly, $45,000; Vindicator Con. G. M. Co., quarterly, $45,000; Granite G. M. Co., bi-monthly, $15,500. Dividends already have been declared by the Cresson and Golden Cycle companies, payable August 10, at the usual rates.

The work of clearing away the debris of the recent fire at the Last Dollar mine, on Bull hill, commenced July 30. The Catherine Mining Co., Charles Walden, general manager, will build a fire-proof brick and steel structure to house the machinery; an Oregon pine 50-ft. head-frame will be erected. The electric hoist and compressor that went through the fire have been shipped to Denver for repairs. The mine will be in operation inside of three months.

A Denver syndicate, headed by Frank Goudy, has leased the Dante mine, on Bull hill, owned by the Dante G. M. Co., and has commenced work.

The Reid mill, on the Dante property, has been sold to W. C. Hockenbarg, of Aurora, Nevada, and A. E. Chapman, of Denver, and is being dismantled prior to shipment to Mt. Bross, in Park county, Colorado, where it will be used on a gold-silver-lead property.

Carl Johnson, of Denver, owning the Bertha B., situated south of the Cresson mine, on Raven hill, is building an ore-house. He has opened up ore at the 300-ft. level.

W. M. Gilbert has opened up ore of shipping grade on the iron vein of the El Paso-Gold King system, in the lower Gold King adit in Poverty gulch, and is shipping to the Portland mill. He represents an Eastern syndicate.

The properties of the Christmas Gold Mining Co. are to be offered for sale. They are surrounded by the property of the Vindicator Consolidated Gold Mining Co. which owns the controlling interest in the Christmas.

The Cresson company continues to ship high-grade ore and the first of the present week three cars of high-grade ore in bulk were sent under armed guard to the mill of the Golden Cycle company at Colorado Springs. The shoot at the 1400-ft.

level is holding out, and the rich streak has widened to 15 in. of ore, rich in sylvanite, while on both sides of it are 3 to 4 ft. of ore of high grade.

The Captain vein of the Portland system is reported to have passed into the shaft at a depth of 1990 ft., and is of a higher grade than in the level above.

PORCUPINE, ONTARIO

GRADUAL ADJUSTMENT OF THE WAGE PROBLEM AT THE IMPORTANT MINES.—DOME MINES TREATING A LARGE TONNAGE OF LOW-GRADE ORE.—NEW MILLS PROPOSED FOR THE DOME AND PORCUPINE V. N. T.

Among the gold producers of the district, labor is still scarce and restless, owing to competitive bidding on the part of mine managers. The McIntyre-Porcupine, Dome Mines, and Porcupine Crown are paying a rate of wages similar to that being paid by the Cobalt companies, and consequently there is little or no friction. The Hollinger, however, is paying a straight wage of $4 to all underground workers, which is an innovation in northern Ontario. This is not viewed with general favor in mining circles here. At present it is having the effect of disorganizing to some extent the system of labor classification in vogue at all the other mines of the country.

The Hollinger forces are understood to be satisfied.

With the knowledge that in nearly every instance where deep mining has been carried on in the district, ore of higher value has been found, the Porcupine Crown management has decided to continue the shaft from the 900 to the 1100-ft. level. The 800 and 900-ft. levels have produced the highest grade ore yet taken from this mine. The 140-ton mill is running at a satisfactory rate.

Around 1500 tons of ore daily is being treated at the Dome Mines which is nearly a record in point of tonnage at this mine. The grade, however, is lower as the ore is being drawn largely from the glory-hole and other convenient workings. In some instances the ore runs only $2 per ton in gold, but being readily accessible to the mill it is being treated at a profit. A new high-grade mill will have to be erected for the economical treatment of the higher grade ores of the lower workings, and it is thought that at the next meeting of the directors it may be decided to discontinue dividends and to convert the surplus, of something like $700,000, for mill building.

The McIntyre-Porcupine, with a full force, is being developed more rapidly than any mine in the country. Ore-passes have been completed from the 7th, 8th, and 9th levels to the 1000-ft. level allowing transportation of ore along the main-haulage way of that level, thereby reducing costs. At the 1000-ft. level a 4-ft. fan, which is driven by a 15-hp. motor, has been installed. The mill is treating a record tonnage of high-grade ore. The July earnings will constitute a record at the McIntyre.

The installation of additional milling equipment at the Schumacher is proceeding and by the middle of September it is the intention to resume operations provided sufficient labor is available. The mill will then have a capacity of over 300 tons daily.

The Porcupine V. N. T. is developing at the 600-ft. level where the grade and width of the orebody has shown improvement over that in the upper levels. The mill is treating 100 tons per day. About 7000 tons of ore is in the dump, but is being held in reserve. It is the intention of the management to erect a new 400-ton mill as soon as labor conditions will permit.

At the 310-ft. level of the Adanac encouraging results were found. Good ore was found at the 200-ft. level of the Peterson Lake, on the Susquehana lease. The vein for about two rounds maintained a width of three inches, then narrowed to one and a half inches and is lower grade. Temiskaming and Beaver are still working at the 1600-ft. level with varying results.

In the gold country the labor supply shows improvement. The number of men employed now approximates 2000 as compared with 3000 necessary to run the mines of Porcupine at the maximum. The Hollinger and Dome are the principal sufferers from the labor shortage. Development at the Hollinger is adding to the ore reserves. The new 12-ft. vein at the 425-ft. level of No. 10 shaft is said to be developing satisfactorily. The underground workings are being gradually centralized, which will facilitate the handling of ore following the resumption of operations at full capacity.

The McIntyre-Porcupine with upward of 350 men is treating a higher tonnage and making a higher net profit than ever before in its history. About 530 tons of $10 ore is being treated daily. With the completion of the additional crushing equipment, the mill will be able to treat 600 tons daily. The Porcupine Crown and Porcupine V. N. T. are treating a large tonnage of ore and are maintaining a good average rate of production. Development is retarded by a shortage of miners.

COBALT, ONTARIO

UNION MINERS NOT SATISFIED WITH PROMISES AND ARE STILL THREATENING.—DISCOVERY IN THE LITTLE NIPISSING.—THE INCREASING PRICE OF SILVER GREATLY STIMULATING MINING AT COBALT.—COSTS CUT INTO PROFITS.

Despite outward appearances, the feeling of unrest so noticeable among the miner's union during recent weeks has not yet been allayed. For a time it appeared as though the assurance given by the operators would satisfy the mine workers, but the large producers have not posted notices other than those of a few months ago, and the union committee has not withdrawn its demand. Some union men believe that practically they have already won what they desire, and that the bonus now being paid is the result of the attitude of the labor organization. Should the price of silver remain at or above 80c. per oz., thereby automatically entitling the mine workers to another bonus of 25c. per day, which would be 25c. per day more than requested by the union, it would be unreasonable to call a strike. A discovery of a good grade of ore has been made in the Little Nipissing property of the Mining Corporation. The vein is from 1 to 2 ft. wide and is in the south-western continuation of the Cobalt Lake fault vein. The McKinley-Darragh mine is situated between the Little Nipissing property and Cobalt Lake and will probably be benefited by the favorable developments on Little Nipissing. During June the Nipissing Mining Co. mined ore of an estimated value of $269,469, and shipped bullion and residue from Nipissing and custom-ores of an estimated net value of $475,329. The high-grade mill treated 136 tons and shipped 567,409 fine oz. of silver. The low-grade mill treated 6052 tons. The total from this mine for the half year ended June 30 was 1,492,677 ounces.

The increasing price of silver has greatly encouraged and stimulated mining in this district. Only recently bankers in the United States bought 200,000 oz. of silver from the Mining Corporation of Canada, paying 83$\frac{7}{16}$c. per ounce, which at the time was an advance of 3$\frac{1}{4}$c. over the New York quotation. Recent additions to the Coniagas mill will permit an increase in capacity to treat the sand by flotation, up to 500 tons per day.

With mining costs so great, only the increased quotation for commercial bar-silver has permitted the continued operation of the Cobalt mines in a large way. If the metal were to go below 60c. per oz. under present conditions most of the camp's mines would be obliged to suspend or curtail their work. But in normal times, when labor and supplies can be had at reasonable figures, mining costs are so reduced that Cobalt can mine 45c. silver at a profit.

THE MINING SUMMARY

The news of the week as told by our special correspondents and compiled from the local press.

ALASKA

(Special Correspondence.)—P. R. Bradley, of the Treadwell properties, announces that the company has bonded a number of mining claims near Auk bay. The claims will be developed. They were located several years ago by the Dull brothers and two years ago options were given to Aldrich, Valentine, and Sutherland, who have been doing development work ever since. Surface showings indicate that it will make a big mine. The vein is wide and has been traced for several thousand feet.—— Over 200 tons of steel from the Treadwell mills was shipped south.——Monte Benson, who is in charge of the development being done by the Treadwell company on the recently bonded properties of the Red Diamond, reports good progress. Good reports are coming from Eagle river, where work is being done on a group of claims by M. J. O'Connor, of Douglas, and Dr. Medford, of Treadwell.——A 4-ft. seam of coal, 8 ft. below the surface and only 30 ft. from the main line of the Government railroad, has been discovered at Mile 175. The discovery was made by G. Giovaninni, who was driving a tap-tunnel to secure rock for a fill. This is the first coal discovery on the line of the railroad.——Walter McCray, a Puget Sound diver, has entered into a contract with the Kennecott Mining Co., whereby he is to receive one-half of all the copper ore he retrieves from the bottom of Cordova harbor. During the many years steamships have been loading at that port, it is believed that many tons of copper ore have been lost overboard during loading, and the high price of copper now renders its salving profitable.

Mr. Bugich, appointed by the striking miners at Kennecott to advertise the strike and to urge miners not to go there, has arrived at Juneau and says that the report that the strike has been settled is erroneous; that the miners received a request from Secretary of Labor, W. B. Wilson, to resume work until Government mediators could arrive, promising that they would endeavor to settle the dispute to the satisfaction of all concerned; that the men were ready to return to work, but the company refused to permit the strike leaders to resume work, and the miners then, by secret ballot, voted 209 to 1 against returning to work.——Reports from Cordova, on July 21, say the strike situation is improving, and that 65% of the normal force is at work, and the regular output is expected to be reached soon. The Department of Labor has sent a mediator to investigate conditions.

Three new veins of anthracite coal have been opened in the Clark Davis mine at Katalla. It is believed that the railroad from Katalla will be completed by September.

Treadwell, July 22.

(Special Correspondence.)—There is no material change in the strike situation. The men are still out. About 150 men are camped at Blackburn. They still demand a flat wage of $5.75 per day, but hinting that they will agree to a reduction when copper is below 18c. The company has offered as high as $6 per day at the present price of copper, but on the bonus system, which was formerly in effect. The saloons are all closed, and consequently order prevails. U. S. soldiers are guarding the mine and important bridges on the railroad.

The Northern Lode Copper Mines Co. property, adjoining the Bonanza mine on the McCarthy creek side of the hill, under the management of W. B. Hancock, is installing an oil-fuel power-plant at McCarthy, 15 miles from the mine, to supply 16,500 volts to drive all the machinery at the mine, and to furnish light for the town of McCarthy. A wagon road is being constructed for auto-trucks, from McCarthy to the bunkers at Lower Camp, which is connected with the mine by a 6000-ft. aerial tramway. The road is being built up McCarthy creek. There are 180 men employed. The company will commence sinking a shaft when the power is available. Regular shipments will begin as soon as the road is completed. Ore shipments heretofore have been confined to the winter sledding from the mine to the railroad at McCarthy. The ore, chalcocite, is a replacement in limestone near the lime-greenstone contact. During the last winter 2000 tons was shipped. There are two grades, the high-grade averaged 61% copper, and the low-grade, so-called, about 30% copper. This ore also carries an ounce of silver to each 4% of copper.

Cordova, July 14.

During June the Alaska United Gold Mining Co. crushed 20,559 tons of ore in the Ready Bullion mill, producing 608.68 tons of concentrate. The 700-Ft. mill was not operated, but 27.69 tons of concentrate was shipped from there. The total value of the concentrate shipped was $45,634.66. Operating expense at the Ready Bullion mine was $43,086.47, representing a loss of $634.42; and at the 700-Ft. mine the expense was $9600.20, netting a loss of $6873.08.

ARIZONA

The strike situation in south-eastern Arizona may be summarized as follows: A few of the men deported from Bisbee have returned and have been permitted to remain. These men were not miners, but engaged in business at Bisbee, one of them being a restaurant keeper. At Globe it is quiet. The mine operators have arranged to have representatives of the workman's committee meet with representatives of the mine managers twice monthly, for the purpose of discussing and adjusting personal grievances. Special meetings may be called. At Ajo three steam-shovels are operating in the pits, and 4000 tons of ore is going to the mill daily. On July 30, Fred H. Moore, who said he was the personal representative of W. D. Haywood, was escorted from Bisbee to Douglas and there placed on a train and sent to Columbus, New Mexico. The committee at Bisbee stated that they considered his presence in the district a menace.

Cochise County

(Special Correspondence.)—The New Cornelia Copper Co. has acquired the property of the Ajo Consolidated Mining Co., thus adding 24,000,000 tons of ore to its 42,000,000. This acquisition assures a large town at Ajo.

Ajo, July 31.

Gila County

Hundreds of miners are daily signing the rolls to return to work in the district, notwithstanding the threats of I. W. W. agitators and others. Those who are willing to work are promised military protection.

Mojave County

(Special Correspondence.)—J. P. Ryan has been appointed general manager for the Union Metals Co., the operating company of the Fredonia group and the Golconda Annex group of six claims. Some rich ore has been found in the bottom of the Fredonia shaft, 15 in. of which assays $125 per ton in copper, gold, and silver. Litigation has developed in the transaction whereby the holdings of the Victor Copper Co. in the Jerome district were to be sold to the Jerome-New York Copper Co. for 350,000 shares of capital stock.

Plans are under way to install a large compressor plant in a central location to supply the Banner mine, De La Fontaine mine, and Prince George mine, of the Arizona-Butte company.

It is reported that arrangements have been completed for the installation of a 200-ton dry milling plant on the McCracken Silver-Lead Mining Co.'s property.

Kingman, July 28.

PINAL COUNTY

The Magma Copper Co. has reported that its net profits for April were $195,000, the most prosperous month in its history. This is at the rate of approixmately $10 per share on the outstanding capitalization.

Kelvin, July 31.

YAVAPAI COUNTY

(Special Correspondence.)—The Henrietta mine, of the Big Ledge Copper Co., which voluntarily suspended operations in June, due to labor trouble, will soon be re-opened. Because of the closing down of this and other properties the Great Western smelter at Mayer was compelled to close down. This smelter is increasing its capacity 500 tons per day.

Prescott, July 27.

CALIFORNIA

CALAVERAS COUNTY

(Special Correspondence.)—The Keystone mine, situated a quarter of a mile from Railroad Flat, has been sold by E. C. Loftus and J. F. Buck to Eastern capitalists, represented by W. O. Duntley, president of the Pneumatic Tool Co. of Chicago and W. J. Cotton, and it will hereafter be known as the Duntley-Cotton Mines Co. The company is overhauling the stamp-mill and has installed boilers and engines, also a Pneumatic Tool compressor. The shaft is down 270 ft., with levels at 100, 200, and 250 ft. It is reported some good ore has been found. Other prospects are being examined in this district.

Railroad Flat, July 1.

(Special Correspondence.)—W. N. Mahaffey, of Stockton, has been investigating a soapstone and graphite deposit near Campo Seco. It is reported that Japanese capital is interested.

Campo Seco, July 28.

(Special Correspondence.)—The water has nearly all been removed from the Buffalo gravel mine. Sluicing will begin as soon as a dependable supply of water is available. The owners are Wisconsin people.

Mokelumne Hill, July 28.

ELDORADO COUNTY

(Special Correspondence.)—Robert G. Hart, representing an Eastern company, and Burr Evans, of Placerville, have been making an investigation of a large gold-bearing porphyry dike 15 miles north-east of Placerville. It is stated that the dike was found to be all that it was represented to be, and that the work of development will follow as a result of the examination.

Placerville, August 4.

NEVADA COUNTY

(Special Correspondence.)—At a depth of 800 ft. the cross-cut in the Pittsburg mine, two miles from Nevada City, has cut a vein of sulphide ore that is said to assay $200 per ton gold, but it will be necessary to make changes in the mill before the ore can be treated to advantage. Cross-cutting continues, to reach the main orebody, where the ore is free-milling.

Sinking is proceeding at the Ocean Star, near Washington, owned by the Columbia Consolidated Co. A vein of good-grade ore is exposed. The company has put in an electric hoist and is adding 10 stamps to the 20-stamp mill. An aerial-tramway is under construction to transport ore from the German and Ocean Star mines to the mill.

The 40-stamp mill of the Allison Ranch Co. has been tested and soon will be running at full capacity. Considerable ore of milling grade has been opened on the 400-ft. level and preparations are being made for mining in deeper workings. Dewatering of the shaft is making good progress. Most of the miners formerly employed at the Golden Center mine are now at the Allison Ranch. It is stated that the sinking of the Golden Center shaft 1000 ft. deeper soon will be commenced.

Nevada City, July 30.

SHASTA COUNTY

(Special Correspondence.)—The Mammoth Copper Co. has taken a long-term lease on the Shasta King mine of the Trinity Copper Co. The Shasta King was shut-down eight years ago. The Mammoth is employing 30 men under the superintendency of L. R. Jenkins. The force will be increased. Ore will be shipped to the Mammoth smelter at Kennett. Work is being started at the Central mine, a gold property in Old Diggings. The mine has been shut-down for 10 years. Contracts have been let to sink the shaft in the Silver King mine 120 ft. The mine is five miles west of Redding in the gold belt. The flotation plant at the Afterthought mine, at Ingot, is completed.

A flotation plant of 100 tons daily capacity will be erected at Bully Hill by W. Arnstein, who has an option on the property and has done extensive development at the Star shaft. Fifty-five men are employed. L. C. Monahan is superintendent. It is probable that the force will be increased to 250 men when the flotation-plant is in operation. The Bully Hill mines were shut-down 10 years ago owing to smoke troubles, and the difficulty in smelting ore that contained so much zinc.

Redding, August 1.

(Special Correspondence.)—The Mammoth Copper Co. is building bunkers at Coram to receive ore taken out under lease from the Trinity Copper Co.'s Shasta King mine. Three companies are shipping ore over the Balaklala's tram-line to Coram. The Balaklala ships its ore direct to Coram; the Mountain Copper Co., of Keswick, ships ore from its New Year's claim over the tram-line also. All the copper mines complain of labor shortage, partly due to the Army draft.

Redding, August 3.

SIERRA COUNTY

(Special Correspondence.)—The owners of the City of Six quartz mine have let a contract for another 100-ft. of the adit from which a raise will connect with the 40-ft. shaft. Much rich ore was formerly mined in the property. Near the mine a rich gold discovery was made recently by J. Cubit, a prospector.

Gravel mining has been resumed at the Hilo, on Chaparral hill, by the Bernhardt Bros. The workings have been placed in good condition and the water-supply system will be improved. It will be impossible to wash the gravel now being mined before the spring of 1918, because of lack of water.

The Wisconsin Mining Co. has arranged for installation of an electric hoist, compressor, and other equipment at its property in the Forest district. Besides working the gravel deposits, a quartz vein will be prospected. C. Cavagnaro is superintendent.

Additional funds have been provided for the development of the North Fork, adjoining the Wisconsin, and a mill is being talked of. The vein lately found is stated to be developing well. George F. Stone has been re-elected manager.

After prospecting an old gravel channel for five years, the owners of the Mohawk mine on Rock creek have discovered a six-inch vein of ore said to be yielding $200 per day. John N. Cubit is manager.

Downieville, July 29.

SISKIYOU COUNTY

(Special Correspondence.)—The placer season for 1916-17 in the Salmon River district has been above the average. Seven of the nine large placer mines are preparing to close-down for the season, due to the shortage of water. The principal mines being operated are the Bonally, Thomain & Teaukert, Bigelow Brothers, Joseph Frazier, John Peterson, Wukes & Mathes, J. Whitfield operating the old Ferguson mine, John Nefroney working the Bennett company's property on a lease, and the Michigan mine. There are many small properties being worked in the gulches tributary to the north fork of the Salmon.

The Bonally mine is worked throughout the year, as water is taken from the river. The other properties are dependent on water from the small streams. These mines extend for 14 miles along the river between Sawyers Bar and the forks of the Salmon.

Florian LeMay of Yreka has some high-grade manganese ore near Red Rock, north of Greenhorn, which he will develop.

It is reported that several diamond-drills are in operation in the Preston Peak district, where prospecting is active.

The Roxbury mines, near Scott bar, have shut-down for the season, due to the failure of water-supply. This company will shortly begin the construction of a 15-mile ditch to take water from Scott river several miles above the mouth of Boulder creek.

David Land, of San Francisco, who has been prospecting the Williams placer on the Klamath river near Wingate creek will resume operation in the district.

Representatives of the Spreckels interest have bought on a tonnage basis the deposits in the soda-beds near Dorris. The deposit is owned by H. E. Weed and C. U. Huff of Dunsmuir, and Fred Tebble of Weed. It is estimated that there is now in the dumps 3750 tons of the mineral.

Hornbrook, July 30.

Trinity County

(Special Correspondence.)—The Estabrook Gold Dredging Co. is putting in a saw-mill to cut lumber for building dredge No. 2, which will be three times the capacity of No. 1. The buckets will hold 18 cu. ft. The hull will be 155 by 50 ft. The company's machine shops, 75 by 155 ft., is the largest building in Trinity county.

The Pacific Gold Dredging Co. is making good progress in tearing down its all-steel dredge at the mouth of Morrison gulch on Coffee creek and moving it four miles down stream to the Graves place on Trinity river. Fifty men and 175 horses are employed. The pit, approximately 200 ft. square and 12 ft. deep, will cost $10,000. It will be more expensive to rebuild the dredge in its new position than it was to put it up at first.

Trinity Center, August 3.

Tuolumne County

(Special Correspondence.)—Work has been resumed in the shaft of the Dutch-Sweeney mine. From the 1800-ft. level it will be sunk to 2250 ft. It is reported that a large amount of ore has been blocked out. The machinery at the Yosemite mine has been repaired and sinking resumed. At the Eagle-Shawmut, since the Tonopah-Belmont Co. took the property, the shaft has been sunk 500 ft. and several hundred feet of drifts have been run, exposing a great quantity of ore. Over 100 men are employed, but the number is to be increased. It is reported, but not confirmed, that a cyanide-plant will be installed and the chlorination-plant discontinued. The company has erected a building for the recreation of its employees. It includes a pool-room and a reading-room. Other surface improvements have been made. G. F. Williamson is superintendent.

Molybdenite has been discovered on Knights creek above Columbia by W. P. Jones and E. D. Heagney.

Sonora, August 1.

The Western Exploration Co. of San Francisco has secured an option to purchase mining claims situated in the canyon of the south fork of the Stanislaus near Italian Camp. The option includes the Fortuna mine and Sirrius No. 2.

COLORADO

Lake County

A press dispatch states that the strike of metal miners in the Leadville district, which began July 21, was called off at 12:30 o'clock on the morning of August 2 at a mass-meeting of the striking miners, both union and non-union. The men voted to accept the offer of the operators of a wage increase of approximately 50c. per day.

The Mine Owners Association at Leadville addressed the following letter to Governor Gunter:

"Honorable Julius C. Gunter.

"Governor of the State of Colorado.

"Dear Sir: The undersigned, operators of mines and mining property in the Leadville district, Lake county, Colorado, respectfully submit to your Excellency the following:

"You have appointed the Honorable Platt Rogers and Harvey Riddell, Esq., to investigate and inquire into conditions affecting wages heretofore paid to miners and employees in and about the various mines in the Leadville district, we, therefore, state to you that, although the State Industrial Commission, duly authorized by law for that purpose, has, within the past two months, made such investigation and reported its findings thereon, it is satisfactory to us for the gentlemen designated by you to inquire into the conditions so existing in the Leadville district, and we will afford to these gentlemen every opportunity and assistance for such examination and for learning the conditions affecting the same and any matters which they may deem to be proper subject of inquiry in connection therewith, so as to help them in arriving at a just conclusion whether the wage scale of the Leadville district is fair and proper or otherwise.

"We take this action because we recognize the sincere efforts your Excellency is making to continue the ore production of this district in this time of national stress. We do not believe that any increase of wages is practicable in view of the low grade character of the Leadville ores, but nevertheless we will comply with the findings and conclusions which these gentlemen may, after full investigation, make and report to you.

"It must be understood, however, that no one of us is bound to, or agrees to, mine or operate any property if in his or its judgment mining operation cannot then be advantageously conducted."

The letter was signed by 23 of the more important companies, and as a result it is hoped that some sort of compromise can be effected, and that mining will be resumed. A large number of miners have left the district and others are preparing to leave. Five of the principal mines will probably be closed indefinitely, some of them having pulled the pumps and removed machinery, tools, and cars from underground.

IDAHO

(Special Correspondence.)—Ravenal Macbeth has been selected by the Idaho Mining Association to compile facts and figures on the money invested in mines of the State. Mining is carried on in 27 counties of Idaho, and, in a number of these, large expenditures are being made this year on prospects and in the further exploration of old properties. The largest expenditures, except in Shoshone county, are being made in Idaho county. Shoshone county is the wealth-producing region, but the other counties have mining resources, developed and undeveloped, that are worthy of notice.

Spokane, August 1.

NEVADA

Esmeralda County

(Special Correspondence.)—From its southern workings the Kewanas company has started an east cross-cut to open the main orebody, at a depth of 840 ft. The vein has been developed in Kewanas ground by the Jumbo Junior. A connection between the two mines will be made.

At 320 ft. depth in the Cracker Jack a mass of quartz showing streaks of high-grade copper ore and a little gold is being prospected by a drift.

At the request of Receiver H. B. Clapp, with approval of creditors, Judge Walsh has issued an order for the sale of the milling equipment of the Florence-Goldfield company. This includes the flotation-plant. The company has been receiving an average of $2000 per month as royalties from lessees, but creditors are demanding a speedy settlement of their claims. It is reported another attempt will be made to reorganize the company and resume work.

The Sunset Mining & Development Co. is preparing to work its Denver group, near Rhyolite. It is said changes will be made in the mill to make a closer saving. San Francisco people are interested.

Goldfield, July 29.

EUREKA COUNTY

(Special Correspondence.)—The Croesus Mining Co., under the management of Julius Huebner, has opened a 7-ft. vein of shipping ore on the 400-ft. level and has teams hauling 4 miles to the railroad.

Several mining engineers have been in the Eureka district looking over the different properties with a view to investing, as the large amount of what was formerly considered low-grade ore can now be shipped at a profit.

Eureka, July 28.

HUMBOLDT COUNTY

(Special Correspondence.)—The National mine and plant have been taken under lease by the National Leasing Co., which plans to drive a cross-cut from the eighth level of the National to open the extension of the vein, and expects to reach it in four months. Arthur Feust is consulting engineer, and Elmer Pfauts, superintendent.

National, August 2.

(Special Correspondence.)—Near Sulphur, a station on the Western Pacific, 50 miles west of Winnemucca, a deposit of alunite said to contain 10% potash has been bought by Salt Lake City people. It is reported that they were asked to give an option on the property for $500,000, payable in two years, and one-half the profits. The offer was accompanied by a proposition to build a complete plant to treat the alunite.

Winnemucca, July 30.

MINERAL COUNTY

(Special Correspondence.)—Ore shipments from Luning are the heaviest in the history of the district. More than 20 properties are producing with several new shippers to be added to the list. To handle custom ore as well as its own, the Pilot Copper Co., formerly known as the Nevada Champion, will erect a leaching plant. Shipments to the Hazen sampler have been increased by the Wall Street, St. Patrick, Acme Copper Hills, and several other copper companies.

Two large Fairbanks-Morse hoists have been installed at the Luning-Idaho; also a compressor and machine-drills. The Hahn adit, now in 685 ft., is being driven to open at a depth of 200 ft. a large body of ore exposed on the surface of the Sophie claim. Ore assaying 582 oz. silver was recently found there. The Erickson adit has been extended 130 ft. and is being run to reach a large orebody 200 ft. below the bottom of the McDavitt shaft. The management reports a large tonnage of 4 to 7% copper ore blocked out and shipments will soon commence. R. B. Todd, Jr., is superintendent.

The Acme Copper Hills Co. will install a 100-hp. hoist and compressor. Shipments of sulphide ore are going to the Hazen sampler. Several veins from 12 to 20 ft. wide have been cut, and the shaft is to be sunk 100 ft. deeper. The ore is stated to average 5 to 8% copper and 2 to 16 oz. silver per ton. Bernard Eastman is manager.

Following a survey and examination by S. E. Montgomery, of the Nevada Rand mine, at Rand, the company has decided on the early construction of a mill. Sufficient ore is on the dumps to defray the cost of plant, and a large tonnage of profitable ore has been exposed on the 150, 180, and deeper levels. Driving from the 450-ft. level is advancing to tap the main orebody. The principal veins carry gold and silver in about equal proportions. Charles Huber, of Tonopah, is president, W. V. Rudderow, Reno, secretary-treasurer, and Charles Koegel, of Rand, superintendent.

Sinking of shaft No. 3 is progressing rapidly at the Copper Mountain, operated on lease and option by the Jumbo Extension Co. of Goldfield. The shaft will be sunk to 250 ft. to open the orebodies exposed in No. 1 shaft. Sinking of No. 2 shaft is also proceeding. The ore is in a contact of lime and monzonite, the two main deposits occurring in parallel zones. Shipping has been discontinued pending developments.

It is reported that the second unit of the Thompson smelter will be blown-in within 30 days. The management has obtained additional supplies of coke and sufficient ore has been guaranteed by custom producers to keep a second unit operating. It is also reported the company's Mason Valley Mines is arranging for heavy shipments from its Gray Eagle mine, near Happy Camp, California.

Luning, July 28.

NYE COUNTY

(Special Correspondence.)—The No. 3 shaft of the Morning Glory is the only scene of operations at present on this property. A cross-cut from the 50-ft. point in the shaft has been started north to cut the foot-wall of the limestone. Two shifts are employed in this work. The sinking of both No. 1 main working-shaft and No. 3 shaft has ceased, due to a restraining order obtained by the White Caps Mining Co., which claims trespass into its ground by the Morning Glory workings. This means that the matter of extralateral rights will be threshed out between the two companies in the courts, unless a settlement is made out of court.

Manhattan, August 2.

(Special Correspondence.)—The Jim Butler company will begin development from the 900-ft. level of the Desert Queen into the Ophir King property, which was acquired last year. The Ophir King is undeveloped territory lying south of the Rescue Eula. The Ophir King has been recently prospected by a diamond-drill. The extension of the McNamara vein has been tapped between the 700 and 800-ft. levels, showing 4 ft. of ore. The West End vein shows in the face of the west drift of the 8th level, but it contains no ore at that point.

The Tonopah Mining Co. last week milled 2000 tons of ore valued at $30,000. Development was confined to the Sandgrass and Silver Tops shafts. At the latter a raise has been started to prospect for the extension of the West End-McNamara vein.

The recent development work in the Cash Boy has enabled the company to commence shipping ore valued at $25 per ton. The ore in sight insures weekly shipments.

The Tonopah Belmont Development Co. shipped 72 bars valued at $142,058, which represents its bi-weekly clean-up. The east drift on the Shoestring vein, at the 9th level, has increased from 2 to 4 ft. in width, while the value of the ore remains the same.

During June the Tonopah Extension Mining Co. milled 9935 tons of ore of an average value of $12 per ton, resulting in a net profit of $42,694.

The West End Consolidated Mining Co. continues to drive the 717D No. 3 toward the east in the hanging wall of the vein, and work has been started in 717D No. 2 on the east side, driving west. In the Ohio shaft cross-cuts 506 and 512 are being driven south under the vein, and the new raises 513, 516, and 517 are all in ore.

The Monarch Pittsburg made a shipment of ore from the 908 south cross-cut, where the vein was reached last week.

The McNamara Mining & Milling Co. is raising to tap the extension of the Ohio vein on the West End company.

Tonopah, August 1.

MONTANA

Lewis and Clark County

(Special Correspondence.)—The Whitlatch mine and mill are in operation at Unionville. Ore is mined from the 400-ft. level. Sherman Bros. have been working the property for several years. The ground west of the 500-ft. shaft has not been explored, but enough has been done to show that a large tonnage is available. The shaft is filled with water below the 400-ft. level.

Extensive development is planned for the Sheep Creek properties controlled by the Illinois Exploration Co. A shaft is being sunk to 1000 ft. Surface development and prospecting has disclosed copper ore in many places. A side-track on the Great Northern railway has been put in at the foot of the hill.

Shipments of ore from the Valley Forge mine, at Rimini, averages one carload per day. The ore is consigned to the New York-Montana Testing & Engineering plant at Helena, where it is concentrated four into one and the concentrate sent to the smelter. The Valley Forge mine has a large tonnage blocked out. Shipments are now made from the ore-dump. The vein is opened to a depth of 600 ft. vertically.

The 300-ft. shaft on the Rock Rose property has been completed and lead-silver ore is being stoped from the 200-ft. level for shipment.

Two carloads are going daily from the Golden Curry group, at Elkhorn. The ore is iron oxide carrying a small amount of gold and copper. The vein, which is upward of 100 ft. wide, is being worked by open-cuts and adit. The Guggenheims have the property under lease and option at approximately $160,000. Locally the property is known as the Sour Dough group. The ores are consigned to the East Helena smelter.

The Boston & Corbin Co. is mining at Corbin. The ore comes from the 1200-ft. Bertha shaft and is milled at Corbin. The concentrate is shipped to the west-side smelters for reduction. The output of copper is estimated at about 70,000 lb. per month.

A long adit into the Mt. Washington ground, at Wickes, has opened 12 ft. of silver-lead ore at a depth of 800 ft. Shipments are going to the Great Northern at Wickes.

Helena, July 28.

Silverbow County

Frank Little, an organizer and agitator of the I. W. W., who had made several inflammatory speeches to the miners of Butte, and who referred to the soldiers of the Army as "Uncle Sam's scabs in uniform," was taken from a hotel by a number of masked men at 3:30 a.m., August 1, and hanged to a railroad trestle. Federal troops are camped at Butte and are patroling the streets.

Hundreds of miners returned to work July 30, says *The Butte Miner*, in the mines of Butte, and aside from this fact there was little of significance in connection with the strike situation, except that a number of electricians, pursuant to the action of the radical element in the union, stopped work. The exact number of electricians who struck was not available, but is believed to be less than 40. The next step in the electricians' controversy will be the revocation of the charter of local No. 65 if it is ascertained that a majority stopped work and that the administration of the union sanctioned the outlaw movement.

AFRICA

The Ashanti Goldfields Corporation, Limited, has declared dividend No. 35, of 25% (1s. per share) on the issued stock of the corporation, payable August 15.

KOREA

The Seoul Mining Co., operating the Suan Concession in Whang Hai Province, Chosen, reports the following results for June: Total recovery, $125,920.

Personal

Note: The Editor invites members of the profession to send particulars of their work and appointments. This information is interesting to our readers.

Gerard Lovell has gone to New Zealand.

F. H. Mitchell, of Reno, is in San Francisco.

W. S. Noyes has gone to the Presidio mine, Texas.

Kirby Thomas has been examining mines in Ontario.

Herbert C. Enos has returned to Denver from New Mexico.

John H. Banks has returned to New York from Jerome, Arizona.

Herman Garlichs, of St. Louis, is in San Francisco on a short holiday.

G. L. Sheldon has returned to Ely, Nevada, from Madison county, Montana.

J. Volney Lewis is about to go to Alaska and northern British Columbia.

Carl R. Davis, manager of the Brakpan mine, Johannesburg, is here on a holiday.

Charles B. Croner is examining properties near Goldfield, Nevada, for Eastern capital.

F. R. Weekes has been in the vicinity of Tucson for the past month on examination work.

W. H. Emmons, of the U. S. Geological Survey, was in San Francisco on his way to Mexico.

John A. Rice has opened an office as consulting engineer at 525 Market street, San Francisco.

J. W. Boyle, of Dawson, has gone to Russia to aid in the reorganization of the railroad system.

C. Gore-Langton has gone to Chuquicamata, to serve on the staff of the Chile Copper Co., sailing on August 15.

Edwin N. Chase is in the Northern Manitoba, B. C., gold district at Herb Lake and will return to Denver on August 12.

S. W. Eccles, vice-president of the American Smelting & Refining Co., has been visiting the mining districts of northern Idaho.

B. H. Bennetts of Tacoma has been appointed consulting engineer for the Ladysmith Smelting Corporation of Ladysmith, B. C.

S. O. Harper has been appointed project manager of the Grand Valley Project, U. S. Reclamation Service, at Grand Junction, Colorado.

G. B. Marshall has resigned as assistant general superintendent of the Abangarez Gold Fields to take charge of Mina la Union, at Miramar, Costa Rica.

M. W. von Bernewitz is in New York, having joined the staff of 'The Mines Handbook and Copper Handbook,' published by Walter Harvey Weed.

Arthur C. Terrill, professor of mining in the University of Kansas, is in the new Kansas-Oklahoma zinc-fields, collecting data for the State Geological Survey of Kansas.

Frank L. Stack, superintendent of the Mina Carlotta for the Davison Sulphur & Phosphate Co., at Cumanayagua, Cuba, is returning to the United States to offer his services in the Army.

William G. Devereux has received a commission as captain in the California regiment of artillery. Charles C. Doyle, of the Mountain Copper Co., is appointed first lieutenant in the same regiment.

W. A. Whitaker, of the University of Kansas, is arranging for a metallurgical symposium to be held at Boston on September 13, in connection with the fall meeting of the American Chemical Society. Several papers dealing with flotation and metallography will be presented.

THE METAL MARKET

METAL PRICES

San Francisco, August 7

Aluminum-dust (100-lb. lots), per lb	$1.00
Aluminum-dust (ton lots), per lb	$0.95
Antimony, cents per pound	15.50—18.00
Electrolytic copper, cents per pound	31.00
Pig lead, cents per pound	12.25—12.50
Platinum, soft and hard metal, per ounce	$105—111
Quicksilver, per flask of 75 lb	$115
Spelter, cents per pound	10.50
Tin, cents per pound	60
Zinc-dust, cents per pound	17

ORE PRICES

San Francisco, August 7

Antimony, 50% metal, per unit	$1.20
Chrome, 40% and over, f.o.b. cars California, per unit	$0.70— 1.00
Magnesite, crude, per ton	$8.00—10.00
Tungsten, 60% WO_3, per unit	26.00
Molybdenite, per unit for MoS_2 contained	$40.00—45.00
Manganese, 45% (under 35% metal not desired), cents, unit	33—37

Manganese prices and specifications, as per the quotations of the Carnegie Steel Co. schedule of prices per ton of 2240 lb. for domestic manganese ore delivered, freight prepaid, at Pittsburg, Pa., or Chicago, Ill. For ore containing

	Per unit
Above 49% metallic manganese	$1.00
46 to 49% metallic manganese	0.98
43 to 46% metallic manganese	0.95
40 to 43% metallic manganese	0.90

Prices are based on ore containing not more than 8% silica nor more than 0.2% phosphorus, and are subject to deductions as follows: (1) for each 1% in excess of 8% silica, a deduction of 15c. per ton, fractions in proportion; (2) for each 0.02% in excess of 0.2% phosphorus, a deduction of 2c. per unit of manganese per ton, fractions in proportion; (3) ore containing less than 40% manganese, or more than 12% silica, or 0.225% phosphorus, subject to acceptance or refusal at buyer's option; settlements based on analysis of sample dried at 212° F., the percentage of moisture in the sample as taken to be deducted from the weight. Prices are subject to change without notice unless specially agreed upon.

EASTERN METAL MARKET

(By wire from New York)

August 7.—Copper is quiet and nominal at 28c. Lead is dull and steady at 10.87c. Zinc is quiet at 8.75c. Platinum remains unchanged at $105 for soft metal and $111 for hard. The aVerage price of tin at New York during July was 62.60c. per pound.

SILVER

Below are given the average New York quotations, in cents per ounce, of fine silver.

Date			Average week ending		
Aug.	1	79.00	June	26	78.12
"	2	80.00	July	3	77.98
"	3	80.75	"	10	78.70
"	4	80.75	"	17	80.62
"	5 Sunday		"	24	79.12
"	6	80.75	"	31	78.20
"	7	81.75	Aug.	7	80.50

Monthly Averages

	1915	1916	1917		1915	1916	1917
Jan.	48.85	56.76	75.14	July	47.52	63.06	78.92
Feb.	48.45	56.74	77.54	Aug.	47.11	66.07	....
Mch.	50.61	57.89	74.13	Sept.	48.77	68.51	....
Apr.	50.25	64.37	72.51	Oct.	49.40	67.86	....
May	49.87	74.27	74.61	Nov.	51.88	71.60	....
June	49.03	65.04	76.44	Dec.	55.34	75.70	....

Silver has again shown an advancing tendency during the week, going from 79c. on August 1 to 81.75c. at the present date.

ZINC

Zinc is quoted as spelter, standard Western brands, New York delivery, in cents per pound

Date			Average week ending		
Aug.	1	8.75	June	26	9.43
"	2	8.75	July	3	9.32
"	3	8.75	"	10	9.20
"	4	8.75	"	17	9.06
"	5 Sunday		"	24	8.60
"	6	8.75	"	31	8.75
"	7	8.75	Aug.	7	8.75

Monthly Averages

	1915	1916	1917		1915	1916	1917
Jan.	6.30	18.21	9.75	July	20.54	9.90	8.98
Feb.	9.05	19.99	10.45	Aug.	14.17	9.03	....
Mch.	8.40	18.40	10.78	Sept.	14.14	9.18	....
Apr.	9.78	18.62	10.20	Oct.	14.05	9.92	....
May	17.03	16.01	9.41	Nov.	17.20	11.81	....
June	22.20	12.85	9.63	Dec.	16.75	11.26	....

The [illegible] of 23,250,000 lb. of high-grade spelter by the United States [illegible], originally announced by the 'Boston News Bureau' early in July, [illegible] finally been consummated at these prices: 13½c. per lb. for 8,250,000 [illegible] of grade 'A' and 13c. per lb. for 15,000,000 lb. of grade 'B.' Al[illegible] the transaction was tentatively closed a month ago, a hitch subsequ[illegible] arose as to price, much the same as that developing in copper.

The last previous spelter transaction negotiated through the zinc committee of the Council of National Defense was for 20,286,000 lb., effected in May.

The New Jersey Zinc Co., which up to the outbreak of war had the high-grade spelter field to itself, now finds plenty of competition from American Zinc, Lead & Smelting Co., with its Mascot brand, Anaconda Copper Mining Co., with its Anaconda Electric brand, and other smaller concerns which re-distill a large part of their production in order to get the premium paid over the ordinary grade of spelter.

COPPER

Prices of electrolytic in New York, in cents per pound.

Date			Average week ending		
Aug.	1	29.00	June	26	32.42
"	2	28.75	July	3	32.25
"	3	28.50	"	10	31.50
"	4	28.25	"	17	30.29
"	5 Sunday		"	24	27.04
"	6	28.25	"	31	27.25
"	7	28.00	Aug.	7	28.46

Monthly Averages

	1915	1916	1917		1915	1916	1917
Jan.	13.60	24.30	29.53	July	19.09	25.66	29.67
Feb.	14.38	26.62	34.57	Aug.	17.27	27.03	....
Mch.	14.80	26.65	36.00	Sept.	17.69	28.28	....
Apr.	16.64	28.02	33.16	Oct.	17.90	28.50	....
May	18.71	29.02	31.69	Nov.	18.88	31.95	....
June	19.75	27.47	32.57	Dec.	20.67	32.89	....

LEAD

Lead is quoted in cents per pound, New York delivery.

Date			Average week ending		
Aug.	1	10.87	June	26	11.75
"	2	10.87	July	3	11.57
"	3	10.87	"	10	11.25
"	4	10.87	"	17	10.98
"	5 Sunday		"	24	10.29
"	6	10.87	"	31	10.55
"	7	10.87	Aug.	7	10.87

Monthly Averages

	1915	1916	1917		1915	1916	1917
Jan.	3.73	5.95	7.64	July	5.59	6.40	10.93
Feb.	3.83	6.23	9.01	Aug.	4.67	6.28	....
Mch.	4.04	7.26	10.07	Sept.	4.62	6.86	....
Apr.	4.21	7.70	9.38	Oct.	4.63	7.02	....
May	4.24	7.38	10.29	Nov.	5.15	7.07	....
June	5.75	6.88	11.74	Dec.	5.34	7.55	....

QUICKSILVER

The primary market for quicksilver is San Francisco, California being the largest producer. The price is fixed in the open market, according to quantity. Prices, in dollars per flask of 75 pounds:

Date			Week ending		
July	10	100.00	July	24	110.00
"	17	105.00	"	31	110.00
			Aug.	7	115.00

Monthly Averages

	1915	1916	1917		1915	1916	1917
Jan.	51.90	222.00	81.00	July	95.00	81.20	102.00
Feb.	60.00	295.00	126.25	Aug.	93.75	74.50	...
Mch.	78.00	219.00	113.75	Sept.	91.00	75.00	...
Apr.	77.50	141.60	114.50	Oct.	92.90	78.20	..
May	75.00	90.00	104.00	Nov.	101.50	79.50	
June	90.00	74.70	85.50	Dec.	123.00	80.00	.

Quicksilver recently has shown an upward tendency, rising from $85 per flask early in July to $115 at this date. The reason for this steady advance is not stated.

TIN

Prices in New York, in cents per pound.

Monthly Averages

	1915	1916	1917		1915	1916	1917
Jan.	34.40	41.76	44.10	July	37.38	38.37	62.60
Feb.	37.23	42.60	51.47	Aug.	34.37	38.88	....
Mch.	48.76	50.50	54.27	Sept.	33.12	36.66	....
Apr.	48.25	51.49	55.63	Oct.	33.00	41.10	....
May	39.28	49.10	63.21	Nov.	39.50	44.12	....
June	40.26	42.07	61.93	Dec.	38.71	42.55	...

Recent reports show steady progress in the Nigerian mining industry. The mines of this West African colony are nearly all given over to the production of tin, and, according to the latest statistics, the output of tin ore during 1915 was 6910 tons, as compared with 6144 tons in 1914, or an increase of 766 tons. The output in 1913 was 5331 tons and in 1912 only 2885 tons.

The approximate value of the 20,156,667 lb. of tin exported from Siam in the fiscal year ended March 1916 was valued at $7,807,703 gold.

The actual percentage of metallic tin in the ore cannot be ascertained, but for revenue-collecting purposes the Siamese government reckons the tin ore which is exported from the east coast of the Malay Peninsula as containing 65%, and that from the west coast 70% metallic tin.

ORES

Tungsten: There has been little change since last week. Quotations are still $22 to $25 per unit for 60% concentrate, with the demand fairly good but supplies none too large. A fair business is reported as put through for foreign and domestic account. Ferro-tungsten is in good demand at $2.25 to $2.50 per lb. of contained tungsten.

Molybdenum and Antimony: The molybdenum market continues strong at $2 to $2.25 per lb. of MoS_2 in 90% material. There is no trouble in disposing of any ore available. Antimony ore is dull and unchanged at previous quotations.

Eastern Metal Market

New York, August 1.

The metals are generally dull but steady, and most of them exhibit an upward tendency. The Government price-fixing and purchasing quandary continues retarding the forward swing of business in all metals.

Copper is nominally higher and firmer, but buying is of small volume.

Tin is stronger and steady, but dull.

Lead is higher again, but the demand is at a low ebb.

Zinc has a better tone, is steadier, but is inactive.

Antimony is stagnant and unchanged.

Aluminum is weaker and lower.

Buying of iron and steel products in the past week by the Government has been active. Prices have been stipulated in some cases, but in most instances they have been left until later. This active buying is in contrast with the abstention of private buyers from the market. Present conditions are likely to last for some time, as it is evident that inquiry into steel-making costs will take longer than expected. A more reasonable attitude at Washington on price-fixing is also not improbable. Meanwhile if some of the high prices are lowered, the trade will not suffer. The report of the Steel Corporation for the second quarter of the year shows a net income of $144,497,000 before taking out fixed charges, etc. This is greater by $15,000,000 than the total net revenue during the year 1915. The corporation set aside $87,000,000 for the half-year against expected excess-profit taxes, and $43,000,000 was reported as expended on new construction since January 1.

COPPER

The market is a narrow and drifting one with little business reported. Prices are largely nominal, and are not based on large sales, though small ones are reported for various positions. Technically the market has advanced and prices are higher than for the last two weeks. The advance has been caused not by the relations of supply and demand but has been due to sentimental rather than to other reasons. After a period of two weeks of declining prices, the market last week turned upward until the quotation yesterday for both Lake and electrolytic was 29c., New York. This bulge has been due to two causes—the withdrawal of re-sale lots and an idea that large consumers were nearly ready to buy. On the New York Metal Exchange yesterday 50 tons of spot electrolytic copper sold 'under the rule' at 27.75c. cash, New York, against refinery price. There was also a sale of 200 tons of copper rods for delivery, 50 tons each month, July, August, September, and October at 27c. cash, New York, against warehouse receipt. Fourth quarter copper is generally pegged at 27.50 to 28c., New York. Casting copper is reported scarce at 27 to 27.50 cents.

TIN

The tin market is dull and uninteresting, and the volume of business is not large. It is a halting and ragged market. Early last week there was considerable inquiry for far futures. On July 26 considerable business for these positions, with deliveries into December, was reported, as well as some sales of early shipment from London. On the same day there was a little inquiry for spot Straits, but price-cutting constituted an important factor in this business, sales being made at 62.50c., New York, with quotations as low as 62c. By Friday, July 27, interest on the part of buyers lessened and the buying died down entirely. Previous to this the sales were of a respectable volume. This week, on Monday and yesterday, the market has been uninteresting and slow, with transactions exceedinly light. In fact they were hardly enough to firmly establish price-levels. On those two days the total sales were about 150 tons. Yesterday future shipment from the East was quoted at 62c., New York. Both Banca tin and Chinese tin are reported scarce. Yesterday the quotation for spot Straits was 63.75c., New York, a progressive advance of over $1.50 per lb. since last week. The London market has experienced a decided rise in the past week, the quotation yesterday for spot Straits having been £248, an advance of nearly £9. Arrivals up to July 31 have been 2390 tons, with the quantity afloat on that date 3450 tons. The average price for tin for July was 62.60c., New York.

LEAD

It was intimated in our letter last week that a turn for the better in the lead market was to be anticipated. Events since then have corroborated the prediction. Prices have continued to advance until yesterday the quotation reached 10.87½c., New York, and 10.75c., St. Louis. One cause for this has been the buying up or withdrawal of all outside lots, diminishing the pressure on the market and resulting in a better tone. Business has been done at 10.75c., New York, but this has been the exception. While the leading interest and some of the large producers continue to quote 11c., the market can hardly yet be pegged at that figure. Demand is at a low ebb but outside lots are hard to find. The uncertainties hanging over the market are great and exercise a repressing influence. Another important factor in the situation has been the labor trouble in the West. In four localities these are serious and are limiting the output and shipments. The switchmen's strike at Chicago is delaying the movement of lead. It is no secret that the big producers in Missouri and Colorado are having their operations seriously interfered with by strikes and other labor troubles.

ZINC

The interesting feature of the week has been the official announcement of the purchase by the Government of 8,250,000 lb. of grade A spelter and 15,000,000 lb. of grade B, but the fact that no decision was made regarding grade C or prime Western was a distinct disappointment. An encouraging fact was the price paid for the two grades referred to. For grade A, 13.50c., St. Louis, was paid, and for grade B, 13c. per lb., an advance of 2c. per lb. over the prices paid for similar though smaller purchases last May. This fact has been accepted as a bullish factor on prime Western when the necessary Government purchases are finally made; nevertheless there has been no immediate effect on the market, which continues inactive and lifeless. Evidences of more interest are accumulating in the shape of inquiries, but actual selling has been of small volume and with better effect in the market. Quotations are largely nominal at 8.50c., St. Louis, or 8.75c., New York, for deliveries as far as October. One sale at this price for delivery up to January has been reported. Any material available at 8.37½c., St. Louis, is believed to have been fully absorbed. The present stagnation is expected to continue until the appearance of large foreign and Government buying. The 'watchful waiting' attitude caused by this uncertainty continues to dominate the general situation.

ANTIMONY

The market continues dull and inactive with demand at a low ebb. Chinese and Japanese grades are quoted at 15 to 15.50c., New York, duty paid.

ALUMINUM

It is reported that 100 tons of No. 1 virgin metal, 98 to 99%, was offered at 51c., New York, yesterday. The market is weak and inactive at 50 to 52c., New York, for early delivery.

EDITORIAL

T. A. RICKARD, Editor

WE shall be glad if mining engineers and metallurgists will notify us when they enlist or obtain commissions in the Army or Navy. No news is more interesting at this time.

COPPER production is being curtailed at the rate of about 50,000,000 pounds per month, a quantity equal to the annual output of one big mine. The Miami and Inspiration mines, both idle, produce 17,000,000 pounds per month.

A REVIEW of a book on another page of this issue will give keen pleasure to many of our readers, not only because it is interesting, but because it is signed by James F. Kemp and signifies that he has almost recovered from his recent illness. In behalf of Professor Kemp's many friends in the West we send hearty greetings.

BURMA'S big mine, the Bawdwin, belonging to the Burma Corporation, continues to gain in resources. The latest annual report shows a reserve of 3,644,000 tons of ore, averaging 25.2 ounces silver per ton, with 27.4% lead, 20.9% zinc, and 0.5% copper. During the first half of the current year a production of 10,387 tons of lead containing 924,000 ounces of silver has been made.

OROVILLE Dredging Company is the name of a corporation that not only exploits dredging ground at Oroville but engages in a similar kind of mining in South America. It is one of those British companies to understand which it is necessary to know something about geography. The work at Oroville is done by a subsidiary company, which reports having paid four dividends each of 12½ cents during the last fiscal year, digging 3,164,136 cubic yards averaging 7.65 cents per yard at a cost of 3.90 cents per yard. The company, through another subsidiary, also operates dredges in the Pato district, in Colombia. During 1916 the yardage excavated was 1,484,721, averaging 48.5 cents per cubic yard at a cost of 12.46 cents per cubic yard. The contrast in yield and cost is eloquent of the difference in local and technical conditions. On the Nechi, an adjoining property operated by a third subsidiary, the yardage during the year was 1,617,975, averaging 49.27 reducing his cost as much as he had expected, but he has cents, at a cost of 9 cents per cubic yard, and of this cost 5 cents is debited to repairs on the dredges. The rise in the price of materials and the increase in freight-rates have prevented Mr. W. A. Prichard, the manager, from done well and is to be congratulated on developing a highly productive dredging-area.

SOME idea of the expansive effect of the War on American commerce is afforded by the statistics of foreign trade. These show that during the first half of the current year the exports were worth $3,289,517,427, as against $1,232,942,074 in the corresponding period four years ago. The excess of exports over imports in the first six months of 1917 was $1,735,736,710, as against $326,437,957 in the first half of 1913. Another comparison is furnished by the U. S. Steel Corporation, whose net arnings were $267,619,094 in the first half of 1917 as against $38,451,381 in the half-year immediately preceding the War. Part of these gains is due to higher prices, but most of it is caused by the increased volume of business.

DEFINITIONS, if correct, serve to clarify ideas. We are glad therefore to note the definition of 'ore' given by Mr. Charles H. Fulton in Technical Paper No. 143 of the U. S. Bureau of Mines. Professor Fulton's definition is as follows: "An ore is a metalliferous mineral, or an aggregate of metalliferous minerals, more or less mixed with gangue, containing metal of commercial importance in such quantity that it may be extracted at a profit by the application of economic and skilful methods of mining and treatment." He explains that "this definition restricts the term 'ore' to natural mineral products yielding a metal, and makes only such material 'ore' from which the metal may be extracted at a profit." We commend this definition—it seems long—and length is a defect because every additional word introduces a chance for mental derailment—but it is both clear and comprehensive. Most of our readers may prefer it to the definition suggested by the present writer, namely: "Ore is metal-bearing rock that can be exploited to economic advantage."

IT is amusing to see how the *Boston News Bureau* lays stress on the high cost of producing copper and insists piously that "the real cost means the elimination of miscellaneous credits, such as 'other income' derived from sale of precious metals recovered in the ore treated, dividends from holdings in other companies, be they mining or affiliated railroads, and interest on bonds or loans." Not only these items but depletion of assets, depreciation of plant, and taxes are to be included in the 'cost.' How different from the irresponsible statements concerning 6 and 7-cent costs appearing in the same paper in the less serious boom period of an earlier day. Now the *News Bureau* states with unction that an important copper-mining company "has discovered that its 'real cost' approximates 17 cents a pound." Again we are told: "The net result is that copper costs in the United States today are nearer 20 cents a pound than 10 cents, heretofore regarded as a normal average." Even this is not enough; under the shadow of the tax-collector the *Bureau* says: "It will be found that the cost of producing copper will range from 15 to 25 cents a pound in the case of the large companies and even higher among some of the smaller ones." It is lucky that this estimation of cost stopped at this point; the editor of the *Bureau* might have wiped out the whole copper industry in his belated enthusiasm for severe statistical truth.

INTEREST continues to centre in the potash market to such an extent that we publish the current quotations elsewhere in this issue. The Government has been dilatory in legislation for the disposal of potash-lands either by locating or by leasing. In Washington we detect no sense of the necessity for haste in any matter touching our preparation to meet the exigencies of war. All known domestic sources of potash were withdrawn from location until the manner of throwing them open to the public should be determined by congressional action. A bill providing for leasing potash deposits had reached a point where agreement between the House and the Senate was in sight, and it was then delayed by an amendment limiting its application to Searles Lake in California. Meanwhile measures are being taken to produce potash as a by-product in Wyoming, and the output from cement-works and from the dust in furnace-gas at iron smelters is growing in importance. Interest in the use of raw sulphur as a fertilizer is widening, and this method of utilizing that element, when done with an understanding of the physics, chemistry, and micro-biology of soils, is remarkable for the amount of potash that it liberates in the form of assimilable potassium sulphate ready to serve as plant-food.

AT the time we wrote the editorial 'Why We Fight,' in our last issue, we had not seen the articles by Mr. Gerard appearing in the Hearst papers. Despite the disreputable medium of publicity used by the former ambassador to Germany and despite the blatant manner of their publication, it is evident that they will serve an excellent purpose in enlightening the public, particularly that part of it which has had no better source of information than the Hearst papers. Even now, ignoring the anti-German character of the Ambassador's disclosures, the *Examiner* prints a letter of flatulent praise for one of its miserable editorials, quoting a correspondent who says: "Yesterday I saw a lot of fine and nice young fellows in soldiers' clothes on the streets, and I tell you it is a crying shame to let them go to death, for a matter that don't concern us in the least, and they are thinking that they will have a 'picnic'." If this correspondent knows so little concerning the purpose for which these young men are being sent to battle, it is because the editor of the *Examiner* has been misleading him by what he chooses to call, in the headings to two congratulatory epistles, 'Editorials that Struck Home' and 'Splendid Editorial.' The lack of a sense of humor is pitiful; but the publication of such disloyal stuff is worse.

MEXICO is in need of cash, the military expenditure alone for 1917-1918 exceeding the total expenses of Madero's government in 1912-1913. So Señor Carranza is casting sheep's eyes at Washington and is beginning to talk pro-Ally instead of pro-German. It is highly probable that the Mexican government will announce itself as anti-German and propose becoming an active ally of the United States in order to obtain a loan, and to import munitions of war. Meanwhile the oil-wells and mines in Mexico are being taxed heavily in order to get some sort of revenue from a country devastated by brigandage. If the United States grants a loan, is it too much to expect that one condition for granting it will be the assurance of protection to American interests in Mexico?

Let Us Help Mr. Hoover

At last the food bill has been passed by Congress and Mr. Hoover has been appointed by the President to administer this new department of national service. The passage of this useful and necessary legislation has been so delayed as to have lost the opportunity for conserving the harvest. For this regrettable delay the country may blame Senator Reed, whose malicious and persistent attack on Mr. Hoover is yet without any sort of explanation. Let us forget it and concentrate our attention on the purpose to which Mr. Hoover has dedicated himself in a spirit of public service that makes us proud to claim him as a member of our profession. To the readers of this paper, all of whom know Mr. Hoover as an engineer of extraordinary intelligence and executive ability, it will be a matter of professional, as well as patriotic, pride to further his efforts to prevent wastage of food, so that not only we in this country but the peoples of our Allies may be adequately nourished during the times of stress incident to this world war. The day for talk has passed, each and every one of us must co-operate in the national effort. Let us 'get behind Hoover' at once. Therefore we ask our readers to sign this pledge:

I promise

1. To have at least one wheatless meal per day.
2. To eat beef, mutton, or pork not more than once per day.
3. To reduce the consumption of butter at table and in the kitchen.
4. To use less sugar in tea, coffee, and cooking.
5. To eat local products such as vegetables, fruit, and fish, so as to save the food necessary for our soldiers and allies.

6. To urge economy of food at home, and outside, especially at clubs and hotels.

Let each man sign this pledge. We signed one like it several weeks ago. In order to further the good cause we print this pledge on page 43 of our advertising supplement. Please sign it, cut it out, and send it to this office. It will be our pleasure to collect the pledges and forward them direct to Mr. Hoover as one of many proofs that the men of the metal-mining industry are supporting him whole-heartedly. Do it now!

Misuse of Terms

We have been impressed lately—as often before—by the confusion of thought produced by the incorrect application of technical terms, for example, the misleading phraseology of most of the flotation patents, in which it is stated that such and such a proportion of oil is to be used "per ton of ore." The weight of the 'ore' has nothing to do with the weight of oil required in froth-flotation. The function of the oil is to aid in separating the valuable mineral from the valueless gangue; the underlying idea is that the oil will ignore the bulk of the 'ore', namely, the gangue, and apply itself usefully to the small portion of the particular mineral it is intended to concentrate. Therefore it is foolish to talk of 1% of oil "on the ore" when the ore may contain as much as 25% or as little as 1.5% of the concentratable mineral, requiring ten or twenty times more oil in the one case than in the other. Another persistent blunder is the custom of speaking of the cost of milling, in California, for example, as, say, 35 cents per ton, when this does not include the cost of extracting the gold out of the pyritic concentrate recovered after amalgamation or alongside cyanidation. The cost of treating this concentrate is $9 to $17 per ton, equivalent to something like 30 or 35 cents per ton of mill-feed. Therefore the cost of 'milling' is nearly double the figure commonly quoted. This inaccuracy of statement arises largely from a failure to distinguish between 'recovery' and 'extraction'. A metal is 'extracted' when it is in marketable form; it is 'recovered' when it has been collected or concentrated in a form assuring successful extraction. Thus a flotation-plant 'recovers' copper in a concentrate that is sent to a smelter, where the copper is 'extracted'. The Spaniard and the educated Mexican distinguish between *recoger* and *extraer*. We read description of mills or other metallurgical plants in which we are told that some metal has been 'extracted' at a given cost; this would indicate a handsome margin of profit, if indeed the metal had been 'extracted', but the series of metallurgic operations has not been completed and considerably more cost must be incurred before the metal 'recovered' in an intermediate product is finally 'extracted' and ready for sale. The transactions of technical societies will show such blunders as the use of 'slacked' lime. 'Slacked' has no meaning; 'slaked', the correct spelling, is full of meaning. When one pours water on lime it reminds one of the Sudan thirst described by G. W. Stevens; the lime sizzles and gurgitates as if in ecstasy of absorption. The 'slaking' of lime is a pictorial phrase; the 'slacking' of it describes nothing. Others, equally careless, speak of a concentrating table for treating slime in a mill as a 'slimer', which means a maker of slime. The termination '-er' signifies agency. The proper term is 'slime-table'. Sometimes it is a 'de-sliming' table, if it happen that the object is to discard the valueless gangue-slime in order to concentrate the valuable pulverized mineral. The 'mat' of froth or the 'mat' of timber in the caving systems of mining is often spelled 'matte' incorrectly, suggesting something entirely different. 'Classify' and 'size' are used almost interchangeably. In milling, to 'classify' is to group particles of equal mass; to 'size' is to group particles of equal dimensions. One more: 'Chute' and 'raise' are employed confusedly. The opening or chimney made through rock is the 'raise', whereas the slide or inclined trough at the outlet is the 'chute'. All raises are not provided with 'chutes' nor need a chute lead from a 'raise'. The distinction is particularly important when describing the shrinkage and caving systems of mining. Such distinctions are always important, because they indicate care on the part of the writer, they favor a habit of mind that conduces to clear expression, they are the mark of the technician as against the blundering of a mere scribbler. The man that writes carelessly and expresses himself in a slovenly manner is not to be trusted with scientific research.

Flotation Litigation

Litigation over the Minerals Separation patents has reached an interesting stage. It will be useful to review the present status. On December 11, 1916, the Supreme Court of the United States reversed the decision of the Ninth Circuit Court of Appeals in the Hyde case, and found that patent No. 835,120 was valid. On May 21, 1917, the Third Circuit Court of Appeals decided that the Miami Copper Company had infringed that patent and also No. 962,678, this latter covering the use of a soluble frothing-agent. No petition for a re-hearing was filed by the Miami company. Ordinarily the mandate of the Appelate Court would have been sent down to the District Court at the expiration of 30 days, this mandate being the official notification to the lower court of the decision of the upper court, but the mandate has been stayed for a period of two months, so that the Miami company is allowed three months, or until August 21, for filing a petition in the Supreme Court praying for a writ of certiorari whereby the case can be carried to that court for final review. Evidently, therefore, it is the intention of the Miami company to make an effort to bring the case before the Supreme Court. If that court should refuse to grant the writ of certiorari, the next step ordinarily would be for the case to go back to the lower court and be referred to a Master in Chancery for the purpose of making an accounting. For the purpose of the audit, witnesses would be examined before the Master by each of the parties to the case and from their testimony the Master would decide the sum to which Minerals Separation is entitled as royalty upon the Miami operation and in the way of damages suffered by Minerals Separation.

Meanwhile the Supreme Court's decision in the Hyde case did not avail Minerals Separation against the real infringer, in that case the Butte & Superior company. After Minerals Separation people won the first round in the Hyde case before the District Court at Butte they brought suit against the Butte & Superior company and moved for a preliminary injunction. This was in the autumn of 1913. The Butte & Superior case was heard before the local court that had first tried the Hyde case and it, in the person of Judge Bourquin, ruled that no injunction would be issued if the Butte & Superior company filed a bond and also filed monthly reports of its flotation operations with the Clerk of the Court. This was done in due course. No injunction was issued and the case rested until, as a sequel to the Supreme Court's decision in the Hyde case, it was brought to trial on April 16, 1917. Whatever the decision, it is likely to be followed by an appeal, which would be heard before the Ninth Circuit Court in San Francisco, the court that gave the one decision wholly adverse to Minerals Separation, namely, the decision in the Hyde case against which they appealed successfully to the Supreme Court. Whether the Butte & Superior case will reach the court of last resort remains to be seen. The Supreme Court may refuse to grant a writ of certiorari, in which event the Butte & Superior case will not be reviewed by it. A reasonable expectation exists, having regard to the importance of the issue and the new evidence, particularly with reference to the so-called critical point in the addition of oil, that the Supreme Court will be willing to hear an appeal in the Butte & Superior case when the time comes. In that event Minerals Separation will have to explain why they have repudiated the need of an increased or different agitation, and have thrown aside their contention for the existence of a 'critical point' in the use of oil, both of these factors in controversy having been urged in the Hyde case as essential features of patent 835,120. In the Hyde case Minerals Separation persuaded the Supreme Court that the process of their patent is characterized by a different and increased agitation and by adherence to a critical quantity of oil. By reason of these contentions, denied point-blank in the Butte & Superior case, they won the most important decision given so far. Will not the Supreme Court treat any further claims of Minerals Separation with grave suspicion? The latter might say that they were mistaken in the theory presented in the Hyde case, and that subsequently they had found that no critical point exists and that the agitation of the prior art is sufficient. They might then claim that the patent is valid notwithstanding and that until the discovery set forth in patent No. 835,120 had been made nobody knew that froth-flotation could be performed with an amount of oil so small as to make the process economically practicable; that they discovered the method of conducting the process with this small amount of oil; and that thereby they took the process from the category of scientific curiosities into the class of industrial arts. Thus they might claim that the patent was valid even without the critical proportion of oil and the critical kind of agitation. The answer to such a contention, cynically regardless of all their previous testimony in the courts, would be a declaration that there was no patentable discovery in ascertaining the least amount of oil that could be used, this being merely the result of such economy as is incident to the commercial development of any metallurgical operation.

It would seem to the onlooker that if this litigation had arrived at the Supreme Court with the evidence now available, it would have proved unfavorable to the patentees. The fact is, of course, that, when Messrs. Sulman, Picard, and Ballot applied for their patent, they did not know how nor why they obtained their froth; even when the validity of the patent was first tried, eight years after it was issued, neither they nor their opponents understood the rationale of the process, and during the four years that have elapsed the various claims and assertions made in the course of litigation have been stultified by new data both from the laboratory and the mill. Not having penetrated the underlying principles of flotation, it is no wonder that their patent failed to cover the essentials of the process; nor does it seem just that they should, by an accidental decrease in the proportion of oil—a proportion now known to be non-essential—be enabled to make a blanket claim for a monopoly of a process that has proved, thanks largely to the work of others, to be of so great an importance to the mining industry. We have seen oil added to pulp in the so-called critical proportion and fail to perform any function as a flotation-agent, and we have then seen a pulp so treated suddenly develop flotative phenomena by the addition of a soluble resinous material, while another portion of the same pulp yielded its mineral more abundantly to the froth by adding oil in larger quantity. The question will be asked, has the Supreme Court been known to reverse itself? Yes, several times. We refer, for example, to a question of mining law, at issue first in Lavagnino v. Uhlig (198 U. S. 443) and then in Farrell v. Lockhart (210 U. S. 142). The decision in the second case squarely reversed the decision in the first. No man can prophesy the judgment of a court of law in a patent suit, but we, prejudiced by viewing the case from the standpoint of the welfare of the mining industry, repeat our forecast, made more than a year ago, that before this litigation is ended the use of oil will be abandoned. Shall we then have some fresh hocus-pocus to explain the real inwardness of a soluble frothing-agent? Perhaps. But that too is based on a quicksand of ignorance concerning the fundamental physics. Meanwhile the litigation is costing each side about $250,000 per annum. Is it to be a war of attrition? We hope not, for a principle is at stake, and it is well therefore that the litigation should go to a final decision.

DISCUSSION

Our readers are invited to use this department for the discussion of technical and other matters pertaining to mining and metallurgy. The Editor welcomes expressions of views contrary to his own, believing that careful criticism is more valuable than casual compliment.

Physics of Flotation

The Editor:

Sir—The many notable papers that have been published in the MINING AND SCIENTIFIC PRESS have shown that there has been rapid progress recently in applying the principles of physics to the explanation of the problems of flotation. The word 'flotation' brings to mind immediately the celebrated principle of Archimedes, that is, the principle of buoyancy, on which depends the determination of specific gravity and the sorting of bodies in water according to their relative mass. Thus we see that it is excessively easy to separate crushed quartz from sawdust, no matter how thoroughly they may be combined, merely by throwing the mixture into tranquil water, when, necessarily, the quartz will go to the bottom and the sawdust will rise to the surface. Considering that ores are a mixture of minerals which at times possess great differences in the specific gravity of their constituents, it would occur to one to form a fluid medium in which to float some of these, like sawdust, while others would submerge, as in the example cited above. Without casting thought as to the question of undertaking to prepare such a liquid practically for separation, it may be worth while to point out that since, in the usual order of things, we are accustomed to see that solid bodies determine the form and the movement of liquid bodies, the metallurgist confronts precisely the inverse relation of the solid and liquid bodies in the problem of flotation, the liquid being employed in greater mass than the solid to cause it to move, the solid previously having been reduced to exceedingly small dimensions. Thus differences in the behavior of the solid have been introduced, whereby the simple principle of Archimedes is affected by the phenomenon of capillarity. On account of the relation between the solid and liquid being distinct from that which we are accustomed to consider, and principally because of the great variety of combinations which may be formed through the changes in the several factors of which each one in the problem is susceptible, such as density, volume, and superficial area of the particle, and, on the side of the liquid, its density, fluidity or viscosity, and its ability to mix with gases or other substances to modify its action, it is probable that the professor of physics will not be able promptly to reduce the metallurgic principles of flotation to a law or to a simple formula for application in the industry. It is interesting to note, however, that in some respects the holding of the finely subdivided mineral particles in the interface between two films is analagous to the phenomenon which we call capillarity, as observed in a liquid in relation to a solid large in size as compared to the quantity of the liquid involved. While the explanation deals with energy-relations between the films, nevertheless, it may be interesting to think of the phenomenon from this standpoint.

BENJAMIN REZAS.

Denver, Colorado, July 26.

What is a Metalliferous Mineral?

The Editor:

Sir—The prevalent acceptation of terms is that generally applied in legal definitions, at least in the framing of laws. Only in the interpretation of patents and in the construction of contracts should a court undertake to penetrate the deeper significations and distinctions of science. The basis of judicial interpretation, according to the logic that the lawyers have drilled into us for generations, is the intent of the parties drawing up the agreement. In one sense legislation also partakes of the nature of contract; it is an agreement made between the representatives of the electors on behalf of the commonwealth and of the individuals of which the commonwealth is composed. The ideas held by these competent contractors on behalf of a self-governing people is what the court necessarily must endeavor to ascertain in order correctly to dispense justice. It certainly would not have occurred to the senators and congressmen that framed our land laws to undertake to distinguish between metals, non-metals, and metalloids. As a matter of fact, it would be difficult to obtain corresponding classifications by different chemists relative to such elements as hydrogen, tellurium, tin, arsenic, antimony, bismuth, vanadium, tantalum, and tungsten. These are capable of being reduced to an elemental and solid state, in which they possess the appearance of metals, but their relations in the making of chemical compounds display a divergence from the characteristics of such metallic substances as copper, silver, and lead. What does this divergence in character mean? Would you class a bat with the birds because it has wings and learned the art of aviation before that other mammal, Orville Wright, astonished the birds at Kitty Hawk, North Carolina, in 1903, by daring to invade their realm? Although the Land Office officials are not members of the judiciary, they do exercise judicial functions, and are supposed to be bound by the same rules of evidence. What evidence would the Land Office invoke to exclude vanadium from rank as a metallic element that would pass before a committee of the American Chemical Society? The same official has undoubtedly handled metallic antimony, and has been impressed with the fact that it possesses some appearances that might lead him to confound it with lead. Would he also exclude antimony? I fear that his distinctions are the outgrowth of unfamiliarity with vanadium. Many mining claims have been located according to the mining law upon antimony deposits in the West. Are these, then, illegal?

COURTENAY DE KALB.

Palo Alto, California, July 20.

The Nitrate Fizzle

The Editor:

Sir—It is, I think, unnecessary to state that I hold no brief for Secretary of War Baker nor for those working under him, when I say that I think your editorial entitled 'The Nitrate Fizzle' was a little too sweeping. Admittedly we are at war, and there should be no shilly-shallying in the assurance of a supply of so all-important a substance as nitric acid, still, does that warrant our following the charted courses of other nations when we are not navigating in their waters? While the direct combination of the nitrogen and oxygen in our atmosphere by the aid of the electric arc should theoretically result in the conversion of 42% of the gases entering the furnace into nitrous oxide, in actual practice the gases issuing from the furnace rarely contain more than 2.5% of that substance. Up to date, then, in the matter of fixing atmospheric nitrogen we seem to have 'little on' the Azotobacter, Clostridium, or the rest of the group of nitrogen-fixing organisms.

The indirect methods, of which you gave a survey, seem to offer greater possibilities. They are not nearly so wasteful of power, and are therefore not dependent on huge, cheap water-powers—or on water-power at all—for their success. If in the United States there is developed water-power to be obtained cheaply enough to make the direct fixation of nitrogen a success I am not aware of it. To develop latent water-powers of the proportions required for the process takes time. Dams have to be built, pipe-lines laid, and transmission-wires erected to a suitable shipping-point. On the other hand, many coals suitable for the manufacture of producer-gas are to be had at favorable shipping-points. United States coals are not, as a rule, high in nitrogen content, nevertheless, in converting them into producer-gas I think they may be relied on to give 70 lb. of ammonium sulphate and upward per ton of coal. This, at the present price of sulphate, which is $6.50 per 100 lb., is worth $4.55 and higher. A steam-plant fired by producer-gas can be built infinitely more quickly and cheaply than a water-power plant of the same energy, and, as the sulphate obtained should considerably more than pay for coal, sulphuric acid, and cost of making sulphate, it starts with an advantage over the water-power plant. Then, too, in war time, with representation of the enemy in our midst that is by no means negligible, is it not wiser to have our eggs in a number of baskets rather than all in one? None of us knows, of course, exactly what those in control are doing, but it seems likely that the aim will be the making of nitric acid in a number of plants by one of the indirect methods through ammonia, and if a little of the latter substance can be obtained by the way from coal, so much the better. I, for one, do not look for useless experimenting from those in charge. There is little doubt that improvements can be made on some of the indirect methods, and I believe plants will be so constructed that, if experiments fail, nitric acid will still be able to be produced in them by proved methods.

F. H. MASON.

San Diego, Cal., August 10.

[We fail to see any essential difference in opinion between Mr. Mason and ourselves, but we are glad to bring the facts once more into the open. The reasons why we applied the word 'fizzle' to the Government administration of the trust created by the passage of the nitrate law are exceedingly simple. The bill was passed a year ago; its enactment was in large part a response to petitions and communications from constituencies that regarded it is a necessary safeguard in the event of war; the sum of $20,000,000 was appropriated, and this sum, large as it is, was generally recognized as barely sufficient for a plant on a scale that might be considered commercial, and actually inadequate to supply our needs if the Chilean source of sodium nitrate were to be cut off; a commission was appointed to select a site for the plant, and this body treated its duties as if they were no more that the settlement of an economic question in piping times of peace, and the report was delayed to a point that exasperated the patriots of the country who were eager that every measure for 'preparedness' should be hastened. Finally the department responsible for action announced its intent to use but one-fifth of the sum appropriated and to apply that in the erection of several experimental plants, and no steps were to be taken to utilize the remaining $16,000,000. The result is that $4,000,000 will be expended within the coming year, and that no plant of commercial importance will be built; we will remain dependent chiefly on Chile as before. Nothing has been done to promote our independence of a foreign supply of fixed nitrogen, and, as the fund is being administered, nothing will be accomplished that will give us such independence within the next twelvemonth. If that is not a 'fizzle,' what synonym should be used to describe it? If a commission appointed by the Imperial German Government had been in charge of the matter the outcome could hardly have been more satisfactory to Kaiser Wilhelm. Private concerns have offered their plants for research and production of fixed nitrogen, among these being the American Cyanimide Co. and the Nichols Chemical Co. The latter placed its Laurel Hill works at the service of the Committee on National Defense for developing a process claimed to be superior in point of cheapness. This, however, is apart from the issue; Congress appropriated money for a plant to supplement private enterprise for national protection, and the intent of the bill has not been carried into effect.—EDITOR.]

Control of Emulsions in Flotation

By COURTENAY DE KALB

Flotation engineers have found that fine grinding of ore facilitates concentration by oils when the proper 'frother' and 'collector' have been pre-determined experimentally. On the other hand, a complaint is generally lodged against what is called 'primary slime,' derived from ore containing abundant decomposed hydrated derivatives from silicates. The presence of foreign matter, such as iron filings and undecomposed original granular constituents in the ore, will often overcome the disadvantage caused by the slime. It was, therefore, not an inherent property in the slime itself that produced the trouble; it was the absence of material that could develop favorable interfacial energy-relations that constituted the obstacle to flotation. Perhaps it might have been overcome in other ways. The use of certain salts, especially some haloid salts and hydratable sulphates, often will dispose of the slime difficulty, though for a different reason. This indicates that in these cases a condition leading to de-hydration favors the delivery of oil in flotation in such manner that it may function properly in oiling bubble-films and mineral particles.

It is important to think of the oil as being 'delivered' to the film and to the mineral particle in an effective way. The oil is not present as a miscible substance in the pulp that receives treatment in the flotation-cell; it is there as an immiscible substance, and to be effective it must first be dispersed through the pulp. A merely miscible substance, so far as we know, will not produce the effect that is observed in a foam carrying mineral to the surface of water. In order to do its work the oil must be dispersed; but that is not all; it must be locked up in a medium that will prevent the re-coalescence of the dispersed particles until wanted, and then delivered at the proper speed and in the proper amount as needed. The oil in such a system is the dispersed body, and the substance in which it is dispersed is the dispersion-medium; the two together constitute what is known as the 'dispersoid.' This is expressed in terms taken from colloid chemistry, and they apply exactly. An oil dispersed in a suitable medium is, however, a coarse dispersion compared with the more nearly molecular dimensions of the particles composing a true colloidal dispersion; therefore the old term 'emulsion' is useful to distinguish it from colloidal dispersions. The difference, nevertheless, is one of degree only. Furthermore, the difference between a persistent emulsion and one that breaks the moment the energy maintaining the dispersion is withdrawn is probably as great as that between the size of particle in a persistent emulsion and of a body brought to a colloidal degree of dispersion.

Before proceeding to a closer examination of the characteristics of emulsions it should be observed that emulsification is a result of the expenditure of energy; it represents work done in overcoming the forces of attraction which bind the molecules of the oil into masses of greater than microscopic dimensions. The practice of making emulsions commercially has shown that the application of force, in a manner similar to what is known in ore-crushing as abrasion, most readily and perfectly tears the particles of oil into lesser particles, reducing them to emulsifiable dimensions with an expenditure of power less than that required to accomplish the same result by beating the oil into the dispersion-medium. That is merely because the power, in this method of application, is concentrated more perfectly upon the work of rending the particles of oil instead of being expended largely in the useless effort of stirring a mass of material in which the rending of the oil-particles is incidental to the churning of the mixture. It will be apparent at once why the rotary mechanical agitators that have to do the work of emulsifying the oil while they also keep the pulp in suspension consume a large amount of power as compared with the effect obtained by grinding the oil with the pulp through a tube-mill and subsequently forming the foam for floating the mineral by the expansion of highly dispersed compressed air in the flotation-cell. The term 'foam' is used advisedly, because it is more accurately descriptive than 'froth,' and serves to link it, as must be done for properly understanding it, with the colloidal dispersions of which it is merely a good example of the gas-in-air type. In the Callow cell a true 'foam' is produced, although on arrival at the surface the freer expansion against the atmosphere develops a coarser mass of bubbles, which quite appropriately is called a 'froth.' There is a distinction, with a difference, between the foam produced in the body of the relatively dense pulp, and the froth with its disturbed energy-relations where the bubbles are exposed on the surface of the pulp to the light gaseous medium of the atmosphere.

Having seen that oil may be dispersed most readily in its dispersion-medium by the rending action produced between two surfaces impinging upon it and moving in opposite and parallel directions, that is, by abrasion, the nature of an emulsion may be considered. The making of emulsions and their properties have been subjected to careful investigation recently by Martin H. Fischer.[1] He defines an emulsion, in its simplest form, as a di-

[1]'Fats and Fatty Degeneration: A Physico-Chemical Study of Emulsions, and the Normal and Abnormal Distribution of Fat in Protoplasm.' Martin H. Fischer and M. O. Hooker. John Wiley & Sons, 1917.

phasic system consisting of a mixture of two immiscible liquids. Two types of emulsions are recognized, dependent upon the relative quantities of the two substances, that is to say, with much water and little oil the result would be a dispersion of oil in water, known as an oil-in-water emulsion, and, on the other hand, a water-in-oil emulsion would follow the dispersion of a given amount of water in a preponderant volume of oil. It has been claimed by some practitioners of flotation that a small amount of oil beaten into the water of the pulp makes a water-in-oil emulsion, which would signify a continuous oil-phase and a discontinuous water-phase. It is not conceivable that the oil could exist as an elastic tenuous medium under a strain that would stretch it to a point approaching its elastic limit, and hold it as a dispersion-medium involving the dispersed water-phase. If it were true that such a dispersoid could be obtained in a flotation-cell, then a contradiction to previous experience would have been found, proving that the laws of dispersed bodies, as hitherto ascertained, were in error. The dividing film, the interface, between two liquids dispersed the one in the other, is more than a wall of division; it is a surface within which energy-phenomena develop; where condensation of molecules occurs, producing what is known as 'adsorption'; where orientation of the molecules is observed in each of the surfaces of the two liquids in opposition; and where interfacial tensions are set up. These are phenomena in direct contra-distinction to the increase in tenuity of a stretched elastic medium, especially in respect of the condensation of molecules that imparts to the oil-film, for example, a density increased approximately to equality with water wherever water and oil come into contact. Molecular condensation is contraction of volume, and it is at the extreme from extension. Again, if a perfect emulsion of a mineral-oil, that is, of a stabilized emulsion of the oil-in-water type, capable of infinite dilution without breaking, be brought into contact with a mineral particle, the particle does not become oiled, but, if a water-in-oil emulsion be used, the oil, in what is then the dispersion-medium, is available, and does actually oil the surface of the particle. On the other hand, a temporary emulsion of oil in water, which endures only while energy is being applied to the dispersoid by means of agitation, will also yield oil to cover the surface of a mineral particle. Further investigation would determine the proportions at which a simple emulsion of the oil-in-water type may or may not deliver the oil out of the despersoid to form films upon solid bodies or to enter as a third phase into the water-air interface when bubbles are made. Dr. Fischer states that no spontaneous emulsion ever forms when oil and water are brought into contact; also that no persistent emulsion of the oil-in-water type can be produced in which the oil exceeds a fraction of 1%. Data as to the breaking of such an emulsion are wanting. Lewkowitsch[2] affirms that oils and fats will not emulsify in water at all, no matter how intimate the contact. Actually, as will appear on further consideration of the composition of an ordinary mill-pulp, no such simple emulsion as the oil-in-water type would be obtainable in any case, because in such a mixture conditions exist for the making of an entirely different character of emulsion. The possibility of a water-in-oil type, in the proportions used in practical flotation, is wholly contrary to experience in the formation of emulsions. It must be remembered that an oil is a neutral glyceride of a fatty acid that is not miscible with water; it is also insoluble. Mineral oil is also immiscible and insoluble. Certain so-called oils consist of mixtures of compounds, some of which are soluble, and others hydratable on reaction with some alkaline or aluminous substances. These, in part, are colloidal bodies that perform an entirely different function from oils. They may even constitute a phase facilitating the dispersion of the oil, thus serving as an assistant in emulsification. Such substances belong to the classs known as hydrated colloids.

Two kinds of colloids are recognized, namely, suspensoids and emulsoids. The prime characteristics of bodies in the colloidal state are that they do not diffuse nor dialyze, and that they display the Brownian movement. The colloidal realm, of which these two features are distinctive, is a higher state of dispersion than an emulsion, but lower than a molecular dispersion. Molecular dispersions readily dialyze. Typical molecules have diameters ranging from 0.1 $\mu\mu$ to 1 $\mu\mu$, that is, from a ten-millionth to one-millionth of a millimetre. A molecule of starch, however, has a diameter of about 5 $\mu\mu$; it is accordingly a large molecule; furthermore, it shows colloidal properties, hence it may be considered an approximate measure of the minimum colloid dimension. A substance capable of molecular dispersion in water is one that is soluble, using the word in its more popular sense; it will diffuse through a thin parchment of collodion membrane. If a test-tube be coated with collodion, as soon as the solvent has evaporated, a thin skin will be left, like a glove on a finger, that may be slipped off. If this collodion sack be filled with a solution of a salt, and the sack then be stood in a beaker of distilled water, so that the level of the water without and of the solution within coincide, the salt will gradually pass through the pores, although they are too small to be visible with the most powerful microscope, and in a short time the same quantity of the salt will be found in any unit volume of the liquid either inside or outside of the sack. This process of diffusing through such a membrane is known as dialysis. It is characteristic of molecular dispersions only. A substance with particles greater in size will not dialyze; therefore colloids, though the particles dispersed are on the limit of and beyond microscopic visibility, will not diffuse through such membranes, and they may be accurately discriminated from the molecular dispersions in this manner. The movements of molecules are persistent and extremely complicated. Colloids, though relatively large, do nevertheless approach near enough to the molecular dimensions to partake of this tendency to ceaseless mobility; the

[2]'Chemical Technology of Oils, Fats, and Waxes.' 1915. Vol. 3, p. 108.

result is a zig-zag motion of the suspended particles, particularly observed in those that do not become hydrated. This dancing of the particles, like an irregular and non-rhythmic gyration, is called the Brownian movement in honor of its discoverer, Robert Brown, an English botanist, who first noted it in 1828. It is a special characteristic of the non-hydratable colloids that constitute the class known as suspensoids. As soon as a colloid takes up water it becomes enlarged and the solution in which it is dispersed becomes more homogeneous because of this adsorption of the dispersion-medium, hence the Brownian movement ceases. Another distinction deserves mention. When substances are so dispersed as to lie within the colloid realm the solution becomes optically heterogeneous, and when a narrow pencil of light from an arc-light passes into a vessel having parallel sides, and containing the colloid dispersion in distilled water, an intense greenish-white luminous cone flashes into view. This is the famous Tyndall cone. It is characteristic of that class of colloids known as 'suspensoids.' The other class, called 'emulsoids,' fails to display this peculiar optical phenomenon because of this property of adsorbing water, which in a manner is analogous to salts that undergo hydration. Therefore, by such 'hydration,' or, more specifically, by such 'solvation,' having taken up a large quantity of the dispersion-medium, they fail to exhibit sufficient difference of refraction to produce the Tyndall cone; that is, the dispersoid containing a solvated colloid is optically more nearly homogeneous.

The significant point to the flotation metallurgist is that an emulsoid is composed to a large extent of the dispersion medium, firmly held until released by a change of state induced by a third substance that possesses the power of dehydration. Neutral salts will dehydrate many solvated colloids, thereby increasing their dispersion even to the point of diffusibility. Sulphates, citrates, phosphates, alcohol, sugar, and many other substances, cause dehydration of emulsoids, while cyanides, iodides, and chlorides increase the adsorption of water; the chlorides, however, show several maxima and minima in their effect, following the varying degrees of concentration of the salt in the solution. Most pronounced in their action in promoting hydration of the solvated colloids are acids and alkalies. Almost any hydratable colloid will swell through the adsorption of water under the influence of an acid or of an alkali. On the other hand, a colloid thus hydrated through the assistance of an acid will become dehydrated upon the introduction of an alkali, and the reverse.

Colloids of this class, because they display a desire for water, are called 'hydrophilic,' or water-loving colloids, while the suspensoids that refuse to take up water are called 'hydrophobic,' or water-hating colloids. Common hydrophilic colloids are soaps, gelatin, albumin, casein, and silicic acid. The last is produced in mill-pulp as a result of attrition on the ore. The longer and the more finely the ore is ground the larger will be the quantity of this hydrophilic gel-colloid present in the solution. A series of such gels is formed from the grinding of different silicates, and the quantity formed increases with great rapidity by prolongation of the time of grinding. If the ore be either distinctly acid or distinctly alkaline the hydration of these colloids, and their consequent swelling, is immensely increased.

Reverting to the teachings of empirical methods, it is interesting to take an old pharmaceutical description of emulsions: "Emulsions are those preparations in which oleaginous substances are suspended in water by means of gum, sugar, carrageen, yolk of egg, etc. In general it will be found that the bulk of the emulsifier must first be taken, while the oil should be added little by little, rubbing them together in a mortar, and taking care that the oil is completely absorbed or emulsified before further additions. Should too much be added, the effect is to throw out most of the oil already incorporated; it is then almost impossible to remedy the error." It should be added that soap is also commonly used as an emulsifier, as well as acacia, blood-albumin, dextrin, and other substances. All the substances mentioned are hydrophilic colloids. The peculiar and outstanding phenomenon, as pointed out by Dr. Fischer, is that a perfect emulsion is obtained only when all the water, that is, the whole of the dispersion-medium, has been absorbed by the hydrophilic colloid. A perfect emulsion, so made, possesses a high degree of persistence. It will last without breaking for days, and months, even for years. The quantity of oil that can be emulsified in this manner with certain substances, such as casein, is enormous. For example,[3] two grammes of casein, 0.8 cc. sodium carbonate (in molar dilution), and 21.2 cc. water, produce a mass that will swell enormously and acquire a ropy consistence. Into this may be incorporated gradually by mixing, with abrasion, any desired quantity of oil up to 150 cc. or more. Such an emulsion will be stiffer than a well-made mayonnaise, and will endure for an indefinite period in that condition. I have made such emulsions with ordinary flotation-oils, even using the light 'stove-oil,' and no tendency to break after prolonged keeping has been observed when the dispersion of the oil into the hydrophilic colloid had been sufficiently thorough. Want of laboratory facilities has prevented me from making similar experiments with silicic acid and other hydrophilic colloids, which result from the wet-grinding of ores. Investigations with such substances offer a field for study that promises to throw much light upon the phenomena of flotation; they may lead to more accurate control of the conditions desired for obtaining a still larger measure of economic success.

The breaking of an emulsion will take place under many conditions, depending upon the character of the hydrophilic colloid used. Soap is capable of so high a degree of dispersion as to be practically soluble. Accordingly mere dilution of an oil-in-soap emulsion will release the oil and permit of the re-coalescence of the droplets. The soap of commerce also lowers the surface-tensions of water to an extreme degree; and this is true of soaps accidentally made in flotation by combination of

[3]Fischer, *loc. cit.*, p. 36.

alkaline or aluminous bases in the mill-pulp with the fatty acids, resin, or phenol, in some oil-mixtures employed. Other emulsions, such as those with casein, may be diluted to any degree without breaking. They make, in effect, a milk, in which the minute oil-droplets of the emulsion are surrounded by protecting films of the swollen colloid. I have made such emulsions with common flotation-oils and have diluted them in the proportion of 1 of emulsion to 1000 of water. No separation of oil occurred, even after standing for weeks, but the emulsion rose to the surface of the water after the manner of a cream. An emulsion of this kind, in which an alkali has been used as a hydration-assistant, will instantly break upon adding sulphuric acid to a point beyond neutrality. A salt having the power of de-hydrating the colloid will also break the emulsion, the water then separating from it in free form, while the oil rises to the top. An emulsion made with a carbo-hydrate, such as acacia, dextrin, or starch, however, is not broken by the addition of acids, alkalies, or the ordinary dehydrating salts. The mere presence of such a hydrocarbon stabilizer in either oil-in-soap or oil-in-protein (casein, etc.) emulsions tends to restrain the breaking by acids, alkalies, or salts. In this connection it is pertinent to quote the following highly significant statement from Fischer.[4] "Following the general laws governing the distribution of a third material between two phases, an emulsion will frequently stand heavy dilution with water, or the addition of materials which dehydrate the stabilizing colloid, without breaking, if the colloid material either is from the first, or comes to be, concentrated[5] in the surface between the oil globules, and the aqueous phase. This is why oil-in-protein emulsions resist the addition of water, and will not break easily even when powerful dehydrating agents are added to them. It also explains why the addition of suspension colloids (suspensoids) or their production (as in the case of the finely divided metals) in an oil-water mixture gives the emulsion a fair stability. Through such adsorption-effects and the protecting films thereby drawn over the oil globules, these may be kept from coalescing even though every other opportunity is offered them to do so. It is only for this reason that dilution alone does not suffice to break every emulsion."

The best results in emulsification are obtained with substances that take up the most water and form fairly viscid liquids with any amount of water added to them. The viscosity of a solution containing a suspensoid, even in considerable concentration, is not greatly increased above that of the pure dispersion-medium, but emulsoids increase the viscosity with extreme rapidity as the concentration of the emulsoid in the solution becomes higher. Gelatin shows progressive augmentation of the viscosity up to the point of forming a solid jelly, when the amount in solution reaches 2%.[6] Changes in temperature, however, are of great importance, the viscosity of water decreasing 2% for every degree of rise from 0°C. to 25°C., and the decrease in viscosity of gelatin is enormously more rapid. Associated with the changes in viscosity, due to varying proportions of hydrophilic colloids added to solutions, is the phenomenon that they present of serving as a protective covering to the dispersed droplets of oil to form an emulsion. Therefore it can be comprehended why some particular degree of concentration can be found for any emulsifier in which it will most perfectly perform its function, and, referring to our old pharmaceutical recipe, we observe that the empirical maker of emulsions had discovered that the progressive dispersion of the oil was the one sure way of obtaining a stable product. This principle is not applied as yet by the flotation operator. He either incorporates the oil all at once or too rapidly, and usually under conditions admitting of no more than an exceedingly coarse dispersion. It is evident then that a large part of the oil used is unable to fulfil its purpose, except as the proper conditions are experimentally discovered in the case of each ore. The wet pulverization of ore for flotation reduces a considerable quantity of the gangue and mineral to particles so comminuted as to lie within the colloidal realm; the mineral particles that are triturated to a degree rendering them colloidal belong almost entirely to the class of suspensoids, while the gangue, for the most part, on coming within the colloid realm of dispersion, is found to belong to the class of emulsoids. Some of these are excessively hydrophilic and serve excellently for the emulsification of oil, and in such cases it is seen that there is less need of 'frothers'; others are not so hydrophilic, and assistants to the emulsification are then required. Incidentally, since suspensoids possess the faculty of stabilizing the oil-film on a bubble, while hydrophilic colloids do not possess this property, one more reason for the preference between gangue and mineral in flotation is apparent, at least as related to the highly comminuted particles. The swelling of hydrophilic colloids, moreover, does not progress rapidly in a neutral solution, but it takes place readily in a solution that is distinctly either acid or alkaline, which reveals a reason for the use of one or the other of these assistants, and it also shows that with a neutral ore it would usually not matter whether acid or alkali were added to promote the colloid solvation; but if the ore should react faintly acid then more acid will usually improve its emulsifiable quality, whereas a faintly alkaline pulp would indicate the need of more alkali. It is important, however, to discriminate between the alkaline substances present. An ore with superabundant soda or potash, or aluminates of soda or potash, will respond favorably, while an ore that liberates lime on

[4] *Loc. cit.*, p. 53.

[5] Obeying the law of colloid concentration toward a discrete phase, whether solid, liquid, or gaseous, or, as re-stated by Wolfgang Ostwald as a generalization of Willard Gibb's theorem, "adsorption will take place whenever there exists in a surface a difference in energy-potential, which can be decreased through a change in the concentration of the dispersed materials bordering upon this surface."

[6] Wolfgang Ostwald: advance proofs of a forthcoming volume, 'An Introduction to Theoretical and Applied Colloid Chemistry,' to be published in September by John Wiley & Sons.

grinding will be expected to prove stubborn in yielding its mineral to flotation. Calcium oxide requires little water for hydration; when present under conditions that yield soaps the calcium base is so strong as to make a calcium soap at the expense of the weaker alkaline bases, and this at once breaks the emulsion. A calcium soap may be used as an emulsifying agent, but will not stand dilution. The presence of lime, therefore, may prove a serious deterrent, although sometimes advantage may be taken of its properties as an emulsion-breaker to deliver oil out of an emulsion, when such an emulsion is added to a pulp, at the rate needed for supplying films to bubbles and to mineral particles. This function of lime has not been overlooked in flotation, though no systematic use of it appears to have been made; it has been applied in certain cases empirically. Colloid chemistry gives the explanation and establishes the principle for its use. In the same way the hygroscopic quality of a salt determines its function in breaking a non-diluted or a non-dilutable emulsion. A soap-emulsoid can be used as an emulsifier to make an excellent emulsion, but it cannot be diluted beyond a relatively low point without thereby breaking the emulsion; lime, or lime chloride, or a hygroscopic sulphate, will also break it without dilution. It is conceivable that some ores may produce hydrophilic colloids, which yield soap on the addition of fatty acids, resin, or phenol, and that the maximum concentration of these colloids for good emulsification may be reached. This could happen only in a practically lime-free pulp, and in such a case lime could be used as an emulsion-breaker. It must be noted that silicic acid, like soaps, is an emulsoid possessed of but little elasticity, and on dilution it behaves similarly to them. How far the presence of organic compounds in some mill-waters may serve to stabilize emulsions made with the assistance of the solvation-colloids in the pulp is an inviting subject for investigation. The function of the 'frother' is to form, with the dilute alkalies present, substances that hydrolyze and behave in a manner analogous to soaps. This occurs with compounds yielding resin and phenol. Resin, in fact, is employed to some extent as a substitute for fatty acids in the manufacture of soap. These are particularly efficacious in the emulsification of mineral-oils, although the protein emulsoids possess this power in a higher degree.

A consideration of the foregoing leads to the belief that a more perfect control of the conditions for delivering flotation-oils to the mineral in the pulp and to the foam produced in a pneumatic cell, could be accomplished by making an emulsion with the aid of a hydrophilic colloid, using either acid or alkali as an assistant, according to the character of the pulp to be treated, diluting this emulsion for convenience in adding it in proper proportion to the pulp as the pulp enters the cell, and establishing the proper conditions of acidity or alkalinity in the pulp to cause the breaking of the emulsion so as to deliver its oil at the rate needed for the flotation of the mineral. In my experiments I have done this, and have found that the oil was liberated in such manner as to oil the bubbles and attract and lift the mineral successfully, the froth being normal, and buoying its burden of mineral in orthodox fashion. This method admits of exact regulation of the oil, and of applying it effectively to do the work demanded by the quantity of mineral to be floated. The power required for the making of such an emulsion in suitable amount is extremely small compared with that needed for the types of mixers now in use, and the result is more nearly perfect. My experiments also indicate that if advantage is to be taken of the emulsoids produced by grinding the ore or those originally present in it, the nature and the quantity of these substances in any given case must be ascertained, in order that the necessary assistants to solvation may be added, and that the correct degree of dilution may not be exceeded, otherwise a good emulsion will not be formed. The organic hydrophilic colloids that could be used will rarely require as much as a half-ounce per pound of oil finally emulsified, and their cost will range from two-tenths to half of a cent per pound of oil so used, even at present exagerated prices.

DUST from cement kiln gases consists of a mixture of raw material from the 'mix,' ash from fuel-combustion, and volatilized alkalies. There is considerable potash in the dust, both in a soluble and an insoluble form, the soluble portion being potassium sulphate produced by combination of the potash in the 'mix' with the oxidized sulphur derived from the fuel; if there is a deficiency of sulphur the potash forms a carbonate. The slowly soluble potash is probably in the form of silicate resulting from combination of incandescent particles of potassium oxide with the silica from the ash. This becomes soluble by treatment for several hours with hot water, the presence of lime hastening the solution. Slowly soluble potash compounds are also formed by the interaction of potash salts in solution with silicious material, this recombination being greatly accelerated by heat. It is partly because the potash in cement kiln dust is not all given up at once on contact with water but is slowly dissolved, that it possesses peculiar value as a fertilizer. A high-grade potash-bearing dust contains about 20% K_2O, or 30% of the sulphate (K_2SO_4), and about 1.5% locked up as silicate. Cement mills now prepared for saving the potash-bearing dust are taking pains to add to the mix suitable high-potash aluminous material in order to increase the output of this valuable by-product. The average dust from cement kilns contains about 10 to 11% total K_2O, of which about 86% is readily soluble in water.—R. J. Nestell and E. Aderson in *Jour. Ind. Eng. Chem.*, July, 1917.

SILICATES containing alumina, on being ground in the presence of water, produce colloidal material of indefinite composition co-existing with a liquid containing dispersed alumina and silica.

NITRATE OF SODA is selling at New York at $4.25 per 100 lb. on the basis of material 95% pure.

An Electric Furnace

By A. W. FAHRENWALD

In the course of experimental work in determining the freezing and dripping points of certain slags, I developed a furnace suitable for high temperatures. I have described it elsewhere.* In principle this furnace is similar to all furnaces where the heat is obtained by the resistance offered to the passage of an electric current by a suitable resistor. In the furnace described, the resistor is a helix of Acheson graphite made thus by sawing around a tube of graphite of suitable inside and outside diameter, then spacing the saw-marks according to the size of furnace desired.

This is a convenient and inexpensive furnace when not larger than 6 to 8 inches high and 2 to 3 in. diameter (crucible area), but when the helical resistor is applied to turbo-furnaces of probably a couple of feet in length, the construction of them is somewhat difficult. For this reason I have devised a resistor made from Acheson graphite which is extremely simple and is suitable for tube-furnaces of any desired length.

Fig. 1 shows a longitudinal section of the furnace. Fig. 2 shows a cross-section on the line *AB* of Fig. 1. Fig. 3 illustrates the principle of the heating unit if unrolled from around the inner alundum tube.

Referring to Fig. 2 and 3, the current, for example, passes down *d* up *c*, down *b* and up *a*, and so on through the entire series of Acheson graphite rods, thus forming the resisting unit, which in this case is about 26 ft. long by $\frac{1}{4}$ in. diameter.

In making a furnace of this kind, the proper diameter and length of resistor can be found by laying out parallel to each other, on a table covered with asbestos, a certain number of graphite rods and connecting alternate ends together with copper wire and passing an electric current through the series. In this manner the proper number can be obtained beforehand. If a certain number of rods is desirable the resistance can be regulated by varying the diameter of the rods. Rods of any diameter can be obtained from the Acheson Graphite Co. at Niagara Falls.

The furnace shown requires about 30 amp. at 110 volts. I am not prepared to state how long this furnace will

*Chemical & Metallurgical Engineering, May 15, 1917.

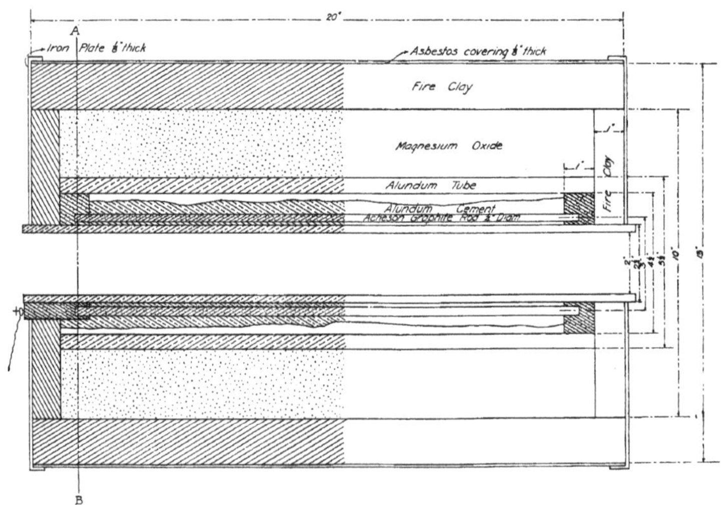

FIG. 1. LONGITUDINAL SECTION OF FURNACE

last; this, of course, depends upon the temperature of operation. The helical furnace described in the reference above cited gave excellent service, and this should do better on account of the perfect insulation of the bars, which are surrounded with alundum cement and the whole inclosed between two alundum tubes. When the bars give way they are readily and cheaply replaced

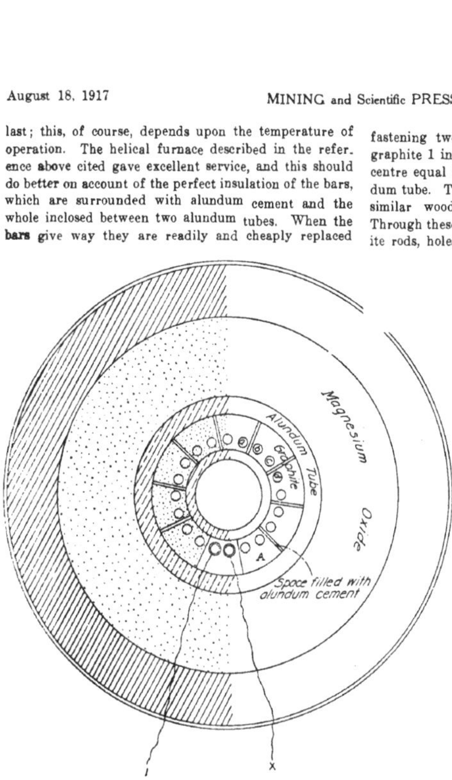

FIG. 2. CROSS-SECTION OF FURNACE ALONG A-B

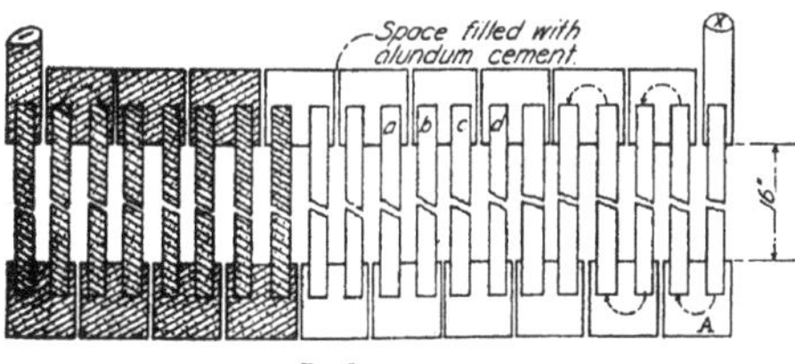

FIG. 3. RESISTOR-BARS

with new ones. The particular advantage of this construction over the helical one is the uniformity of cross-seection of the resistor-bars, thus giving a uniform heat throughout the system.

In the construction of the furnace, the required number of bars is firmly bound in a bundle and sawed the required length. The end pieces, A, in Fig. 2 and 3, fastening two bars together are made from discs of graphite 1 in. thick and 2½ in. diam. with a hole in the centre equal to the outside diameter of the inner alundum tube. These, one at a time, are fastened firmly to a similar wooden model with holes properly spaced. Through these holes, which receive the ends of the graphite rods, holes are properly distributed in the graphite disc. These are then placed over the inner alundum tube and the graphite rods, which are cemented firmly in place with alundum cement. The discs are then sawed between every other bar on the top one and likewise on the bottom one, the first cut on the bottom disc being one bar farther around than the top disc, thus giving a continuous circuit, for the passage of an electric current. The material sawed out is also replaced with alundum cement.

This furnace should be convenient for experimental heat-treatment of steels as well as for other purposes. I am also of the opinion that its use could be extended to the treatment of cyanide precipitate, by increasing the diameter of the crucible, also the number and diameter of the resistor-bars. Those who have inquired regarding making a tube-furnace from the helical resistor will find this type more convenient and easier in construction.

A SAD COMMENT upon the inability of the American people to learn the savage lessons of war is contained in a recent official statement from the War Department. Despite the persistent pleading of well-informed officers of the Army no steps have been taken in the years of menace to provide equipment for the manufacture of modern heavy artillery. In consequence we find ourselves now absolutely incapable of producing the guns that we will need, and the United States has been compelled to place orders with the already overburdened gun-factories of France in a desperate effort to meet the emergency. A number of 75-mm. guns and 150-mm. howitzers have been contracted for abroad, and it is suggested that it may prove a speedier solution of a part of the difficulty to enlarge some of the French works to accommodate our demand for artillery to use on European battlefields than immediately to attempt to provide such equipment at home.

MANGANESE deposits are being developed in the Republic of Panama, and at several points on the Pacific slope in Costa Rica and Guatemala.

Use and Testing of Oxygen Mine-Rescue Apparatus

*The value of oxygen mine-rescue apparatus is becoming more appreciated both for rescue and recovery purposes. There has been a steady growth in the establishment of rescue stations throughout the country, particularly in the Western coalfields. In training miners the Bureau of Mines teaches the use of three types of apparatus—the Draeger, the Fleuss, and the Westfalia. The bureau itself is developing a fourth type, known as the Gibbs. Each of these has a steel cylinder containing oxygen at a pressure of approximately 2000 lb. per square inch and a reducing valve that allows a definite quantity of oxygen per minute, at a pressure slightly above atmospheric, to pass from the cylinder to a reservoir from which it is breathed by the wearer. The air exhaled by the wearer flows through to a compartment containing regenerating material, usually caustic soda (sodium hydroxide), by which the carbon dioxide in the breath is removed. The regenerated air joins the stream of oxygen from the reducing valve and is breathed again. The possibilities and limitations of such apparatus are becoming more thoroughly understood. However, during the fiscal year 1916 there were two instances of men wearing apparatus to combat mine-fires with presumably little or no previous training in its use. In another instance an apparatus-party of only two men made an exploration of considerable length, and one of them lost his life. Until such practices are discontinued the dangers incident to wearing apparatus in irrespirable atmospheres will not be reduced to a minimum. Some men while wearing the apparatus fail to appreciate the fact that the added weight of the apparatus causes exhaustion more quickly than when working without it.

LUNG MOTOR AND PROTO EQUIPMENT. TWO VIEWS

During the fiscal year 1916 tests were made with the four types of mine-rescue apparatus mentioned to determine the accumulation of hydrogen and nitrogen in each type when the 'oxygen' used was similar to that furnished the various rescue-cars and stations, that is, oxygen mixed with nitrogen up to 3% and hydrogen up to 1.3%. In these tests 170 samples were taken of either the original oxygen or of the air breathed by the wearers, and the tests indicated that oxygen mixed with more than 0.2% of hydrogen, or 2.5% of nitrogen, might prove dangerous for use in mine-rescue apparatus. Because of these tests the Bureau of Mines proposes, in buying oxygen for use on its cars and stations, to specify that the percentages of hydrogen and nitrogen must not exceed the figures mentioned. These tests also served to indicate the efficiency of the regenerator furnished with the Gibbs apparatus.

*By courtesy of the Director of the U. S. Bureau of Mines.

Because of the war in Europe the German company making the Draeger apparatus has been unable to procure regenerating cartridges from Germany for use in this country, and it has been necessary to manufacture them in America. The Bureau of Mines has tested the efficiency of American-made regenerators. Two types were submitted, one containing granular sodium hydroxide and the other containing sheet sodium hydroxide. Generally speaking, all of these generators proved satisfactory, but those containing caustic soda in sheets and having absorbent blotters in the four upper trays showed the best results; they are now being used in the training work of the bureau. Two tests were made with the Fleuss apparatus to determine whether caustic soda stored for some time in a Fleuss breathing bag was still in condition to efficiently absorb the carbon di-oxide exhaled by the wearer. In one test caustic soda stored for one month in the breathing bag was tested; the caustic soda tried in a second test had been left in the bag for two months. Samples taken at the beginning and the end of the storage period in eight inhalation-air samples taken in the course of the two-hour tests of each apparatus were analyzed. The tests showed that the sodium would still act, as an inhaled air was as high as 1½%. In trials with the Fleuss apparatus, electrolytic sodium hydroxide prepared in the lump form was tested. Eight inhalation-air samples taken in the tests showed the lump sodium to be an efficient regenerating agent. In four tests of the Fleuss apparatus the wearer tried the half-mask, intended for use in fire-fighting, instead of the mouth-piece. The tests indicated that although the half-mask may seemingly have an air-tight fit on a man before he enters the smoke-room, inward leakage may occur after he has been in a noxious or poisonous atmosphere for a short time, so that the half-mask is not as safe as the mouth-piece.

An important series of tests of mine-rescue apparatus was made in Colorado to determine, if possible, whether rescue apparatus possessed defects that rendered it dangerous for use at high altitudes. The first series of tests was made at Manitou, the second at the Half Way House on Pikes Peak, and the third at the summit. The results indicated that rescue-apparatus is not more dangerous at high altitudes than at low. The oxygen supply was sufficient, and, in fact, men wearing the apparatus and getting a constant supply of oxygen from it could make physical exertions of which they were incapable when breathing the natural air because of its being rarefied and containing a less weight of oxygen per cubic foot than the air supplied by the apparatus. The conclusion reached, therefore, was that the probable cause of the excessive number of accidents with rescue-apparatus in mines at high altitudes resulted from neglect in the care of the apparatus, particularly in the renewal of rubber parts that had dried and cracked. A secondary, but, possibly, more important result of these investigations was the comparison of the Gibbs apparatus with the older forms of apparatus, particularly the Fleuss and Draeger, heretofore used by the mine-rescue crews at the bureau. The results demonstrated that the Gibbs apparatus adapts itself to the wearer's needs far better than the older forms of apparatus. In endurance tests particularly it was found that this automatic adjustment tends to insure a much longer duration of the oxygen-supply and of the power to absorb carbon di-oxide than do any of the other types.

As the Gibbs apparatus, which has been developed by W. E. Gibbs, engineer of mine-safety investigations, in co-operation with other members of the Bureau of Mines, seems in several respects to be superior to other types, a detailed description is here presented. This apparatus is a self-contained unit carried wholly on the back of the user. It is light, and its parts are well protected against injury. A special device feeds the oxygen used, and although plenty is available for the wearer when working hard none is wasted when he is resting; hence the new apparatus may be worn for a considerably longer time without re-charging than other models now in use. An unusually efficient carbon di-oxide absorber that liberates little heat is another feature. Caustic soda, which is much cheaper than the potash salt formerly thought necessary, is used as the absorbent. It is theoretically easy to supply a mine-rescue man two or three hours with an artificial atmosphere from an apparatus of no greater weight than he can readily carry on his back. In practice, however, the construction and operation of such devices offer many difficulties. The knowledge that the failure of any part to respond to the needs of the wearer means his almost certain death places a heavy responsibility on the designer and constructor.

Normal air contains roughly 20% of oxygen mixed with about 80% of nitrogen and a trace of carbon di-oxide. At each inspiration part of the oxygen breathed combines in the lungs with carbon brought by the blood, and the air expired contains about 4% of carbon di-oxide. The nitrogen of the air is unchanged by the act of respiration and takes no active part in it. For all practical purposes an artificial atmosphere of pure oxygen, or oxygen containing only a small percentage of nitrogen, is actually preferable to normal air in the conditions surrounding mine-rescue work. In spite of the general belief that pure oxygen is unsafe to breathe, no abnormal effects attend its use unless it be breathed for a much longer period than that during which rescue-apparatus is customarily worn, and even then the only symptom noted is a slight irritation of the bronchial passages. The amount of oxygen consumed in the body is precisely the same whether the gas be breathed pure or diluted with nitrogen in the form of air. Contrary to the belief held a few years ago, there is no flushing of the face, no feeling of exhilaration, no increase in the pulse-rate, nor elevation of arterial tension. If, however, the oxygen-content of the air breathed be materially reduced, unconsciousness and death are almost sure to follow without any warning symptoms, provided the carbon di-oxide content of the air remains low. For this reason it is advisable that breathing apparatus should supply an atmosphere rich in oxygen. As much of the oxygen made from liquid air contains 2 to 3% of nitrogen, which remains unchanged and accumulates in the apparatus, analyses of the atmosphere breathed by the wearer generally show a decreasing content proportionate to the length of time the apparatus is worn. If the proportion

THE DRAEGER APPARATUS

of carbon di-oxide in the artificial atmosphere rises much above 2%, deeper breathing or panting warns the wearer of danger, generally in time to let him get to safety.

Under the best conditions the exploration of mines in

THE FLEUSS-PROTO OXY-CYL EQUIPMENT

which there has been a recent fire or explosion is a hazardous undertaking, not only because of exposure to bodily injury but also because no mechanical breathing device can be made so perfect that it will never fail. Frequent inspection of the breathing apparatus and constant vigilance in keeping it in perfect repair reduce the risk of failure to a minimum.

Screen-Anaylses

*In ascertaining the efficiency of crushing-machines, as well as in the study and control of all hydro-metaliugical processes, small-scale screening tests are of the greatest value. In order that the work of one mill may be compared with that of another, it is necessary that a uniform screen-scale be used by all operators, and, furthermore, that a uniform method of making the sizing test be employed. In the past much confusion has existed, both as regards the screen-scale and the method of making the test, therefore it is often impossible to make such comparisons. It was with the end in view of establishing a standard method of performing sizing tests that this subject was taken up. In brief, the work so far done demonstrates that satisfactory sizing tests may be made either by hand or by machine. It must be emphasized, however, that certain precautions must be observed if consistent results are to be obtained, and, furthermore, it must be expected that considerable time will be consumed in making sizing tests by hand. In the sizing of crushed ores it is essential to remove the finely divided amorphous material by washing prior to screen sizing. If this is done, concordant results can be obtained, but if sizing of the original material without such washing is attempted. the time necessary for the operation is increased by reason of the clogging of the finer screens by the fine material, and, furthermore, the proportion of the coarser sizes is greater, by reason of the fact that the fine clay particles adhere to the coarse sizes. The only way to insure their removal is by preliminary washing, as previously pointed out.

PAINT may be removed from iron and steel surfaces by the following formula prescribed by the U. S. Coast Artillery Corps. The method is used in cleaning the exterior surfaces of the big guns and carriages in coast defences. One pound of concentrated powdered lye is dissolved in six pints of hot water, and enough lime is added to make the solution thick enough to spread. It is applied, freshly mixed, with a brush, and allowed to remain until almost dry. When it is then removed, it will take the paint with it. If the paint is thick and old, wash off the surface and apply a second coat of this emulsion. Before a new coat of paint is put on, the metal should be washed with a solution of one-half pound of washing-soda (sal soda) dissolved in eight quarts of water. The parts should then be thoroughly dried by wiping or warming.

SMOKE or fume consists of a dispersion of solid particles in air, in which, according to W. W. Strong, 50,000,000 may be contained in one cubic centimetre. These dispersed particles consist of small groups of molecules the relation between which, in point of size, and the molecules themselves he states in striking manner as being the difference between the island of Manhattan and the Engineers' building in New York City.

*U. S. Bureau of Mines Year-Book.

Copper-Precipitating Plant at Balaklala Mine

By S. A. HOLMAN

Construction on this plant was started on July 1, 1914, the plant being completed and the water turned in on August 7. The design is nearly the same as the larger plant of the Mammoth Copper Mining Co., but it includes several minor improvements suggested by experience. It consists of two towers approximately 4 ft. wide by 10 ft. high by 64 ft. long, there being two decks for holding the scrap-tin and a trough at the bottom for collecting the precipitate. The water flows by gravity through the Weil tunnel, at the outlet of which it enters a flume carrying it to the first precipitating-tower. These towers are filled with scrap-tin which is turned and washed clean of precipitate each day, so as to leave a clean surface for further precipitation. After passing through

COPPER-PRECIPITATING PLANT, BALAKLALA MINE

the first tower the water flows directly to the second tower, the dripping-flumes of which are on a level with the discharge of the first tower. From the bottom of the second tower the water flows into a tank filled with scrap-iron and thence to waste.

The settling-tanks used are round redwood tanks, 12 ft. in diameter and 4 ft. deep, one for each tower, and so arranged that the precipitate can be sluiced into them from the collecting troughs in the bottom of the towers. An average of about 30 gallons per minute is delivered to the first tower, the flow of mine water varying from 25 to as much as 100 gal. per minute during the heavy rains in the spring. This water contains about 0.312 grammes of copper per litre.

Samples of the water at the intake and discharge, taken at regular intervals over a period of one year, indicate a saving of about 90% of the copper-content. A clean-up of the last tank filled with scrap-iron has not been made, but undoubtedly a further saving is obtained there. Those given in the accompanying table are characteristic assays and show the work done by the two towers extracting 88.21% of the copper.

MAY 15, 1915. 60 GALLONS PER MINUTE

	Copper Gm. per litre
Water at intake	0.2205
Water at discharge first tower	0.0826
Water at discharge second tower	0.0348

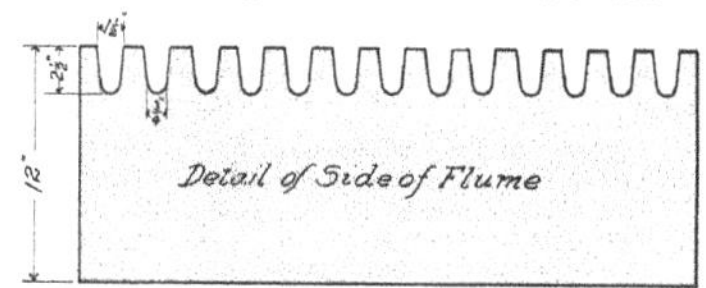

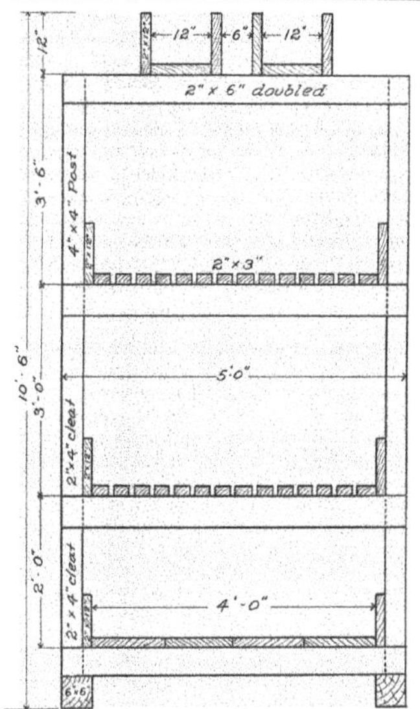

CROSS-SECTION COPPER-PRECIPITATING PLANT

The water is turned out of the towers for one five-hour period during each month when the precipitate is sluiced into the settling tanks. The supernatant water is drawn off and the cement-copper is allowed to dry by evaporation until sufficiently compact to be handled. This precipitate, still containing considerable moisture, is then sacked and shipped to the smelter for further treatment.

The cost of the installation was as follows:

Material:	
Lumber	$212.88
Nails	20.60
Oakum	1.20
Redwood tanks	62.75
Forks	1.50
Labor:	
Carpenters	132.00
Helpers	150.00
Laborers, grading	102.09
	$683.02

The results of two years' operation are summarized below:

Dry tons cement-copper produced	129.2955
Average assay, cement-copper produced	28.489%
Pounds copper produced	73785.7205
Tons scrap-tin consumed	150.694
Pounds tin consumed per lb. copper	4.08

The total costs for two years were: 150.694 tons scrap-tin, at $14, f. o. b. Coram, $2110.87; 24 months' labor at $70 per month, $1680, making a total of $3790.87. Adding the cost of installation, $683.02, the operating cost was $4473.89. The total net receipts for precipitate for this period were $14,877.77, showing a net profit of $10,403.88 or an average of $433.49 per month. The average price received for copper during this period was 24.456c. per lb. The cost of producing one pound of copper was, labor 2.276c., scrap tin and sundries 2.861c., being a total of 5.137c. It is interesting to note that if copper had been selling at 15c. per pound the profits would have been $4423.26, or an average of $184.30 per month.

PROTECTION of oil-sands by mud-laden fluid and cement was discussed recently at a hearing by the State Oil and Gas Supervisors and it was generally admitted its effectiveness was not proved. If the method can be perfected and demonstrated it will result in an enormous saving of expense for casing and drilling. Rules governing tests of the method have been formulated by the California State Mining Bureau, but only one company has volunteered to make such a test. Most operators in the immediate vicinity of Taft are using at least two strings of casing for the protection of oil-sands against water and they propose to continue the practice.

ELECTRIC STEEL is now being produced in this country at the rate of 200,000 tons per annum, and the rate of increase in the capacity of the plants for making it indicates that the output next year will exceed 300,000 tons. America is already in the lead as a producer of electric steel, notwithstanding the fact that hydro-electric power is high as compared with such countries as Norway, where it is available, under wise government control of water-powers, at prices as low as $6 per horse-power year.

Potash Market

The potash market is only fairly active, the farmers buying slowly on account of the prevailing high prices. Nevertheless there is no difficulty in disposing of potash products in car-load lots, and it would be easy to contract for steady deliveries at good prices. The market standards and the corresponding prices for immediate delivery at the middle of July were as follows:

In bags, on the basis of 80%:	Per ton
Muriate of potash, 80 to 85%	$350 to $360
Muriate of potash, minimum 95%	360
Muriate of potash, minimum 98%	nominal
Basis of 80%, in bags:	
Sulphate of potash, 90 to 95%	$275 to $300
Double manure salts, 48 to 53%	105
In bulk:	
Manure salt, minimum 20% K_2O	$50 to $60
Hardsalt, minimum 16% K_2O	40 to 50
Kainit, minimum 12.4% K_2O	40 to 50

The German 'muriate,' or commercial chloride of potassium, formerly imported, contained usually between 80 and 85% of the chloride, a typical analysis being potassium chloride 83.5%; potassium sulphate, 0.7%; magnesium sulphate, 0.4%; magnesium chloride, 0.3%; sodium chloride, 14%; calcium sulphate, 0.2%; and insoluble, 0.5%. As an indication of normal prices it may be stated that the average paid for imported kainit in 1912 was $4.66 per long ton, in which year 511,966 tons of this material was consumed in the United States.

PARAFFIN, or a paraffin-bearing oil, on being digested with sulphur at moderate temperature becomes black and heavy and finally passes to a substance resembling solid asphalt, as explained in a recent paper in the Trans. A. I. M. E. by G. Sherburne Rogers. Similarly, if a light asphaltic oil is treated with sulphur it becomes darker and more viscous, finally becoming asphalt. Under laboratory conditions the reaction is accompanied by the evolution of hydrogen sulphide, and in fact an old laboratory method of generating hydrogen sulphide consists in heating paraffin and sulphur in a retort. The sulphur atom, by extracting two atoms of hydrogen from the oil, causes a condensation or polymerization of the hydrocarbon molecule, and this change is reflected in the increase of the gravity of the oil itself as it approaches solid asphalt. A simple example of change through the action of sulphur, from the polymethylene series of the general formula C_nH_{2n}, to the heavier aromatic series of the general formula C_nH_{2n-6}, may be written:

Hexamethylene + Sulphur = Benzene + Hydrogen sulphide.

$$C_6H_{12} + 3S = C_6H_6 + 3H_2S$$

This type of reaction is taken advantage of in the Dubb process for manufacturing asphalt commercially by heating crude petroleum or petroleum-residue with sulphur. A variety of artificial asphalt known as Pittsburgh flux, made by treating with sulphur the residuum of Pennsylvania petroleum, is described as tough and sticky, and melting only at high temperature.

Labor Agitators, Mr. Roosevelt, and Others

By T. A. RICKARD

A few days ago I picked up the autobiography of Theodore Roosevelt, a book that I had bought just before the beginning of the War and had read half-way through. Mr. Roosevelt's recent altercation with Mr. Gompers, his frank and timely protest against the mob-violence at St. Louis, also his prompt offer to lead a body of volunteers to France, had re-awakened my admiration for personal qualities that, whether one likes all of them or not, are splendidly American. So I was soon in Cuba fighting the Spaniards with the Colonel of Rough Riders, reading about the Big Stick and the Square Deal, or absorbing a vigorous reformer's ideas of Social and Industrial Justice. This last is the title of Chapter XIII, in which, to my surprise, I found two pages devoted to a criticism or denunciation—it might be considered either—of an article written by me in 1903 while editor of a New York paper called the *Engineering and Mining Journal*. In the issue of November 14, 1903, I had protested mildly and politely, as I thought, against the invitation given by Mr. Roosevelt, then the President, to six representatives of the Butte labor-unions. This is what I wrote:

"It is announced that, upon the invitation of the President, six representatives of the Butte labor-unions will dine at the White House on November 18. This statement, appearing in most of the daily papers, has received but little comment from the Eastern press, although we feel assured that its significance is fully appreciated in the West, where the history and performance of the Butte Miners' Union are only too well known.

"Whether the action of Mr. Roosevelt is prompted by the impulse of hospitality in return for courtesies received by him during his recent visit to Butte, or whether it is due to political sagacity, we do not know, nor does it concern this technical journal; whether the motive be entirely creditable or distinctly blameworthy, we do know that this official recognition of the Butte labor-union is calculated to inflict grievous injury upon the mining industry of the whole country.

"Before issuing dinner invitations to these men it would have been well if the President had requested his detective service to inquire into their antecedents; of course, it is not usual to institute such search into the character of guests, but, on the other hand, it is not customary to invite people of this description to an official dinner at Washington. In default of such an inquiry we can state that what the Clan-na-gael has been to the brutalities of Irish politics, what the Mafia has been in the bloody feuds of the low-born Italians, that the Butte Miners' Union has been to the labor troubles of our Western mining camps. The unspeakable barbarities of the Coeur d'Alene strike, the anarchy which disfigured Leadville, the assassinations and murders which made Telluride a by-word, the disorders at Cripple Creek, these, every one of these, have received their mainspring and their sustenance from the gang of ruffians who have controlled the union at Butte. If they have been permitted to maintain their organization at Butte City itself, it is only because Marcus Daly used the union to control Montana politics, Clark bribed outright, while Heinze and the Amalgamated were so busy with their cat-and-dog fight that they had no time to spare for the decencies of civilization.

"If ever there was an organization which deserves to be condemned by honest men, it is the labor-union whose leaders are expected at the White House next week. Wherever a mine manager has tried to treat his men humanely and to give each that equality of opportunity to work which is the American's heritage, there this gang of Irish-Austrian-Italian anarchists has sent its emissaries to stir up trouble. Wherever a mining community has been pursuing the even tenor of its way attended by prosperity and quietness, there these agitators have incited foul murders and cowardly assassinations. It is all well known to the mining men in the West, so well known, alas, that it ceases to excite the fierce resentment which it warrants, until once in a while a friend disappears in the dark or is riddled with bullets in broad daylight. At such times, indeed, we feel like moving heaven and earth to see justice done, only to run into an impassable quagmire of muddy foulness which is called local politics! Is the chief executive, a man held by his friends to be the best type of American manhood, is he going to do an act for politics' sake which will be taken as the endorsement of the most un-American tyranny that ever existed on this continent? We hope not, most sincerely."

Shortly after this editorial appeared I had a call from Edwin Packard, vice-president of the Federal Mining company, who had drawn Mr. Roosevelt's attention to the article and had received from him the letter published in the autobiography. This follows:

November 26, 1903.

"I have your letter of the 25th instant, with enclosure. These men, not all of whom were miners, by the way, came here and were at lunch with me, in company with Mr. Carroll D. Wright, Mr. Wayne MacVeagh, and Secretary Cortelyou. They are as decent a set of men as can be. They all agreed entirely with me in my denunciation of what had been done in the Coeur d'Alene country; and it appeared that some of them were on the platform with me when I denounced this type of outrage three years ago in Butte. There is not one man who was here, who, I believe, was in any way, shape or form responsible for such outrages. I find that the ultra-Socialistic members of the unions in Butte denounced these men for coming here, in a manner as violent—and I may say as irrational—as the denunciation [by the capitalistic writer] in the article you sent me. Doubtless the gentleman of whom you speak as your general manager is an admirable man. I, of course, was not alluding to him; but I most emphatically *was* alluding to men who write such articles as that you sent me. These articles are to be paralleled by the similar articles in the Populist and Socialist papers when two years ago I had at dinner at one time Pierpont Morgan, and at another time J. J. Hill, and at another, Harriman, and at another time Schiff. Furthermore, they could be paralleled by the articles in the same type of paper which at the time of the Miller incident in the Printing Office were in a condition of nervous anxiety because I met the labor leaders to discuss it. It would have been a great misfortune if I had not met them; and it would have been an even greater misfortune if after meeting them I had yielded to their protests in the matter.

"You say in your letter that you know that I am 'on record' as opposed to violence. Pardon my saying that this seems to me not the right way to put the matter, if by 'record' you mean utterance and not action. Aside from what happened when I was Governor in connection, for instance, with the Croton dam strike riots, all you have to do is to turn back to what took place last June in Arizona—and you can find out about it from [Mr. X] of New York. The miners struck, violence followed, and the Arizona Territorial authorities notified me they could not grapple with the situation. Within twenty minutes of the receipt of the telegram, orders were issued to the nearest available troops, and twenty-four hours afterwards General Baldwin and his regulars were on the ground, and twenty-four hours later every vestige of disorder had disappeared. The Miners' Federation in their meeting, I think at Denver, a short while afterwards, passed resolutions denouncing me. I do not know whether the *Engineering and Mining Journal* paid any heed to this incident or knew of it. If the *Journal* did, I suppose it can hardly have failed to understand that to put an immedate stop to rioting by the use of the United States army is a fact of importance beside which the criticism of my having 'labor leaders' to lunch, shrinks into the same insignificance as the criticism in a different type of paper about my having 'trust magnates' to lunch. While I am President I wish the labor man to feel that he has the same right of access to me that the capitalist has; that the doors swing open as easily to the wage-worker as to the head of a big corporation—*and no easier*. Anything else seems to be not only un-American, but as symptomatic of an attitude which will cost grave trouble if persevered in. To discriminate against labor men from Butte because there is reason to believe that rioting has been excited in other districts by certain labor unions, or individuals in labor unions in Butte, would be to adopt precisely the attitude of those who desire me to discriminate against all capitalists in Wall Street because there are plenty of capitalists in Wall Street who have been guilty of bad financial practices and who have endeavored to override or evade the laws of the land. In my judgment, the only safe attitude for a private citizen, and still more for a public servant, to assume, is that he will draw the line on conduct, discriminating against neither corporation nor union as such, nor in favor of either as such; but endeavoring to make the decent member of the union and the upright capitalists alike feel that they are bound, not only by self-interest, but by every consideration of principle and duty to stand together on the matters of most moment to the nation."

Mr. Packard showed this letter to me and we discussed the subject of it, with the result that he expressed agreement with my views and asked me to write a letter addressed to him for transmission to the President. That closed the incident, and I had forgotten about it until, in my admiration for Mr. Roosevelt, I read his autobiography and found it embalmed there like a fly in the amber of a romantic story. I have no desire to retort at this late date, but the readers of the MINING AND SCIENTIFIC PRESS will, I hope, absolve me from the charge of being a "capitalistic writer" either now or in 1903. I recall several occasions on which an effort to be fair to the workers has drawn the fire of the real representatives of capital upon me. Next, the fact is worth recording that at the time of the Miller incident in the Printing Office, I had expressed hearty and respectful approval* of the President's action in his refusal to allow a union to force a foreman-printer out of the Government's service.

However, Mr. Roosevelt was not the only one to find fault with my opinion in the matter of that dinner at the White House. His guests took offence. The Butte Miners' Union officially passed a resolution of a vitriolic character condemning me to everlasting perdition. It was a three-page document written in red ink, to simulate blood, and in terms such that my natural modesty would forbid me to quote, even if the thing had not long ago gone the way of such things, into the waste-paper basket.

That is many years ago. Since then many crimes have been committed in the name of labor, and of capital too, and throughout the many outbreaks of lawlessness the two leaders of the Western Federation of Miners have been prominent—William D. Haywood and Charles H. Moyer. In 1908, after the trial of the ruffians that assassinated Governor Steunenberg of Idaho, these two men quarreled, with the result that Haywood organized the Industrial Workers of the World (I. W. W.) at Chicago, while Moyer remained at Denver in control of the Western Federation of Miners, which became a part of the American Federation of Labor. The Metal Mine Workers' Industrial Union, with headquarters at Phoenix, Arizona, is affiliated with the I. W. W., whereas the

*October 10 and October 31, 1903. *E. & M. J.*

W. F. M. is now called the International Union of Mine, Mill, and Smelter Workers, this change of name having been made last year. It is not easy to trace the relationships of the various unions, some of which are organized for a particular strike, to be supported by one or other of the larger unions, only to disappear after they have served their purpose. The strike at Butte was managed by the Metal Mine Workers' Union, organized on June 12, 1917, as is recorded on the corporate seal. They made the following demands:

"1. The Metal Mine Workers' Union to be recognized by all of the mining companies of the Butte district in its official capacity.

"2. The unconditional abolition of the rustling-card system, and to re-instate all blacklisted miners.

"3. A minimum wage of $6 per day for all men employed underground, regardless of the price of copper, as the cost of living demands more of a wage scale to the working-men. Also surface men to receive an increase in wages in proportion to raise of miners' wages.

"4. The mines to be examined at least once each month by a committee, half to be selected by this union and half by the company, the object being to avoid as far as possible fires and many other accidents.

"5. That all men starting in mines shall be shown exits to other mines, so that they shall be able to escape in case of fire or other accidents.

"6. Any member getting seven to 15 days lay-off to be given a hearing before a committee, three to be appointed by this union, and an equal number to be selected by the company. If the offense claimed is not proven, the miner or miners shall be immediately put to work and paid in full for all time lost.

"7. That all bulkheads must be guarded for the safety of miners by having manholes built therein."

These demands are not all unreasonable, even if some of them are impracticable. (1) As to the recognition of the union, I shall discuss that later. (2) The 'rustling card' is a means of identifying a miner, and, of course, of facilitating the exclusion of men known to be trouble-makers. It is irritating to organized labor, because it prevents the 'penetration' of a crew of quiet workers by agitators or union emissaries. The right or wrong of the system resolves itself into the question whether a mining company is to be allowed the right of refusing to employ those that it believes its enemies. Must an employer give work to the first man that offers or is he entitled to discriminate, choosing those that he considers likely to be docile in preference to those that incite their fellows to the making of demands for higher wages or other concessions? (3) A minimum wage of $6 per day is only 25 cents above the recent pay, which, however, is arranged by a sliding scale, itself based on the price of copper, starting at 15 cents per pound and $3.50 per day. (4) The demand for an examination of the mines, to ensure the safety of the men, by a joint committee, is prompted by the recent underground fires at Butte and by the assertion that the State Mine Inspector is not competent. That the mine-fires were of incendiary origin is believed by the managers and the men alike, and there is reason to infer that the men have their own reasons for coming to this conclusion. Whether the State Mine Inspector in Montana is competent, or not, we do not know, but it is a fact that this kind of official in our Western States may be in sympathy with the capitalist or with the worker in accordance with the leanings of the Governor that appoints him. He may be selected from the ranks of the labor-union or he may be selected on account of his antipathy to organized labor—more correctly, to the turbulent element in the organization. (5) The suggestion that the men should be shown the exits to other mines is reasonable; nor is it obvious why manholes should not be made in the concrete bulkheads separating adjoining mines, except for the possibility of their being left open at the wrong time as well as in moments of danger. (6) The lay-off clause is intended to protect a man from the arbitrary action of a foreman, sometimes due to graft.

The I. W. W. movement is altogether different in kind, as a recent circular, issued by the Metal Mine Workers, will show; it is a long statement, but I give it verbatim, so that the reader may study it at his leisure.

"The Metal Mine Workers have long represented the militant portion of labor in America. They have always given freely of their money, strength, and even of their blood to further the interests of their class, the working class.

"Though the miners have always been militants, it has been years since they have had any form of militant organization. The organization in which they have been organized, or semi-organized, in the past few years has been owned and controlled directly by the greedy mining corporations whom they were supposedly organized to combat.

"But the miner has awakened from the state of indifference and lassitude that has been his in the past. He now demands a real union, and he has it. It is Metal Mine Workers' Industrial Union No. 800 of the Industrial Workers of the World.

"The organization comes clear-cut and says to the world at large that it is after shorter hours, more wages, and better conditions today, while tomorrow we will be satisfied with no less than the complete ownership of the mines, mills, and smelters. Of course, an organization such as this is bound to meet with opposition.

"The mine-owners see in the success of this organization their final doom. The labor skates see in the growth of this organization the vanishing of their pie-cards. The apostles of law and disorder know that this organization means the end of their period of misrule. So this holy trinity combines and fights with every weapon at their command this new and growing body of men.

"But they cannot crush it. It is here. It is here to stay. Thousands are already in its ranks and applications for membership are pouring in by the hundreds. You, Mr. Digger, and you, Mr. Mill or Smelter Worker, should be in its ranks also, if you are not already there. The time is coming when you will be there. Why not now? It is the organization that you will eventually join. It voices your hopes and aspirations. It is your hope of freedom from the yoke of economic serfdom that has enthralled you in the past.

"It is the Industrial Union. Quasi-industrial unions have sprung up in the mining industry, such as the United Mine Workers of America and the Western Federation of Labor, but they never have functioned as industrial unions. In the United Mine Workers of America district organizations have killed every attempt of the miners to show their militant spirit. One district would strike, while other districts would remain at work furnishing the market with coal. Time after time the workers have been virtually sold out, and the only thing that has kept the U. M. W. A. together is the check-off system whereby the company collects the union dues. Any time the company collects the dues for the union it is dead certain there is something in it for the company. The Western Federation of Miners, while it has had its bright spots in the history of the militant working class, is now decadent. It has lost its once militant membership, and its officials today are spending more time in legislative halls and company offices than they are at the mouth of the shaft. It has lost its punch. All of the red blood that once was in that organization has come into Metal Mine Workers 800 and the headquarters of 800 has more W. F. M. cards that have been turned in in lieu of initiation fees than are being carried in the pockets of W. F. M. men throughout the country.

"The I. W. W. is the union for you as a digger to belong to. It is not an organization to restrain your rebellious spirit, but it realizes that you who work in the mines know the conditions in the mines best, and it is you who should determine when to strike and how to strike, and not some set of officials holding down an office chair. It is your union for you to use to better your condition.

"It is the union that is going to get the pay-dirt for the miner. We are starting out with real modest demands. These demands will be multiplied as soon as we have the power to enforce more demands. Among the demands that we want at once are these:

"First: A six-hour work day. Down in the hot stopes a decree has gone forth that six instead of eight hours shall be the length of a shift. That decree, backed up as it is by the organization, must be complied with.

"Second: No less than two men on a machine. This demand is exceptionally modest. Years ago we worked two men to a machine, and today, after 24 years of W. F. M. organization, one man to a machine is the rule. In the drifts and cross-cuts we have held our council and have decided to abolish that rule. Our decision, backed by real organization, must be the law.

"Third: No more speeding up. Too many of our fellow workers are rustling on the hills now. By the curtailing of production we not only will make life more bearable for ourselves, but we will also provide more jobs for the boys rustling on top of the hills.

"Fourth: The abolition of the physical examination. We do not propose to be stripped before going to work and then be stripped again when we get our pay check. One of the strippings must be abolished now. We will tend to the other one later.

"These demands are only starters. When we get to making and getting our demands in earnest we will make the boss squirm. In fact, he is squirming now with anticipation. We want no more local strikes. Our future strikes will be industrial in scope, embracing not a locality, but an industry.

"Our form of organization provides for the admittance into our ranks of every person working in the mining industry, regardless of creed, color, or nationality. Industrial Union 800 covers the entire mining districts of the West, from Alaska to Mexico. We have branches in every important mining camp. In the smaller camps we have delegates who represent the organization. There is one in your camp. See him, or write to headquarters and we will give you any and all information that you require.

"Now is the time. Arizona is teeming with agitation. Utah, Nevada and Idaho are doing wonderful work. Butte, Montana, is a tower of strength for the new union. You need its help. It needs your help. Send in your application today. The initiation fee is $2. The dues are 50c. a month. Organization is the key to success. When we are organized, then, and not until then, can we come into our own, the ownership of the earth."

This programme has been further expounded by Haywood, who stated at Chicago, on July 26: "We will go on and on until we—the roughnecks of this world—will take control of all production and work when we please and how much we please. The man who makes the wagon will ride in it himself. Hell! What's the use of talking about anything but the man who works with his hands."

This is the milk in the cocoa-nut. It is the true inwardness of the I. W. W. movement. This organization is not socialistic, of course, but anti-social. It represents a revolt not against the tyranny of capital or the prejudice of class but against any sort of orderly industry. "The man who makes the wagon is to ride in it himself;" but who is to furnish the energy to move the wagon?

The I. W. W. demand for higher wages is not based on any idea of just compensation for work, but as a step toward the confiscation of property. The insistence upon two men handling a machine-drill when one man suffices is an attack on the efficiency and economy upon which successful industry is based. So likewise is the protest against 'speeding up'; this places a premium on laziness and incapacity; it is an effort to level down, in anticipation of the time when all the "roughnecks" may sit in the wagon, doing nothing, while a donkey pulls that wagon down a steep place into the sea. The objection to a physical examination before a man is engaged for work that requires healthy vigor is against the true interest of the worker. For example, a man with a weak heart should not be allowed to work where contact with electrical machinery may prove fatal to him, although not fatal to a normal man. The second examination is made, obviously, to prevent false claims for injury during the time of employment. Both of these medical examinations are in the interest of honest labor.

The Metal Mine Workers at Butte are playing second fiddle to Haywood. In a recent bulletin they claim that the labor unrest all over the country "means that before the end of the War the United States will be operating on the German plan of State socialism, or the ownership and operation by the Government of all public utilities. It means the end of the capitalistic system of society and the inauguration of a new order." The capitalistic system has many obvious defects, but no German alternative is likely to be popular just now, and the suggestion of it argues that lack of humor which is so close to criminality. Indeed this touch is circumstantial evidence of German propaganda. In Germany Mr. Haywood and his rabble would get short shrift; it is only the good-natured tolerance of a democracy that permits such antics.

We may disregard the I. W. W. and the like of them, but we cannot ignore the reasonable insistence of the workers on the right to collective bargaining and co-operative effort in the form of a union. The mine-owner will break his teeth on that bone of contention, for he himself co-operates with his neighbors in the same district, and even in other districts. The refusal to recognize foreign unions is another matter. In that the mine-owner may be right, as he is undoubtedly justified in refusing to deal with the representatives of an organization that has been proved to be a murderous conspiracy or an anarchistic gang. Again there is the question of responsibility. Agreements made between mining companies and labor-unions are of small value because the union is not incorporated, it has no financial responsibility, and may vanish into thin air when asked to make good. It has neither property nor legal entity. On the other hand, the company is incorporated, so that it can be held subject to legal process; it has property, which can be attached on a verdict for damages. It is bound by its contract; the union is not. In consequence, the personnel of the union changes from time to time, in accordance with the spirit of the men; when everything is smooth and quiet the representatives of the law-abiding element are in control, but as soon as trouble begins they are ousted to make room for the rowdy element. If necessary, a new union is organized, usually at the instigation of a Moyer or a Haywood, who pulls the strings from Denver, Chicago, or some other distant point. What is needed is a better recognition of social justice and a more evident effort to play fair, on both sides. Most of these labor-troubles arise from injustice, plus ignorance. The mining companies would be better advised if they co-operated with the Labor Department of the Government and with any other agency that aims to do justice between the employer and the employee, and if they appealed to public opinion by publishing the facts in each case frankly. The members of the mining profession should recognize their duty, as good citizens and as men of education, not to lend themselves to the greed either of the capitalist or the worker, or to the anti-social aims of either, but to do justice as best they can. After all what is democracy except the maintenance of a recognition of the other fellow's point of view. Neither the Marcus Daly type nor anarchists like Haywood see anything beyond the tips of their noses. That is why in all this anarchistic propaganda the Butte union is "a tower of strength," as the I. W. W. circular says. I conclude by regretting that the First Citizen of the Republic should have invited the representatives of the Butte Miners' Union to dine with him at the Executive Mansion in 1903.

BORON is being experimented with in a matter that is highly suggestive of the stabilization of carbon in the hardening form in steel by those elements that perform the function of retarders of crystallization. Recently N. Tschischewsky investigated this subject, and presented interesting conclusions in the British Iron and Steel Institute. He used boron for case-hardening steel, both as pure amorphous boron and as a finely powdered ferro-alloy containing 19% boron. The iron used in the experiments had the following composition:

	Per cent
Carbon	0.12
Silicon	0.02
Manganese	0.16
Phosphorus	0.06
Sulphur	0.04

The pieces of soft iron used were cubical and were drilled to the centre, the holes then being filled with the boron or the ferro-boron. The holes were closed with iron stoppers forced home with a hydraulic press. These test-cubes were heated in a Heräeus furnace packed into silica tubes, the air having been removed from the tubes by a mercury pump. It was found that the cementation was accomplished more easily by ferro-boron than with pure boron. The hard white layer of the case-hardened portion was shown to consist of compact boric pearlite, with a twin-crystal structure, and the edges to contain a sub-eutectic alloy of ferrite-pearlite, the ferrite holding the boron in solid solution. The case-hardened part of the specimen shows dark bands representing an alloy of iron with carbon in the vicinity of the ferro-boron.

IT IS generally recognized that anticlines or arches in the rocks are the most likely places to find petroleum and natural gas, and all such places are now being tested. In consequence, the U. S. Geological Survey, after an examination by A. J. Collier, is calling attention to the Bowdoin dome, near Malta, in north-eastern Montana. It is a broadly arched part of the earth's crust from which the rocks dip away slightly on all sides. It is on the main line of the Great Northern railway east of Malta. Natural gas has already been found there in some wells drilled for water, and a deep well has been begun which should determine whether the dome contains oil. The gas already found is supposed to have been contained in the Eagle sandstone, at a depth of about 600 ft., and it is possible that oil might be found in sands near the base of the Cretaceous rocks, about 1500 ft. lower.

Recent Patents

1,227,781. Process for Recovering Potassium Salts from Organic Carbonaceous Materials. Franklin C. Grimes, Idaho Falls, Idaho. Filed Aug. 19, 1916.

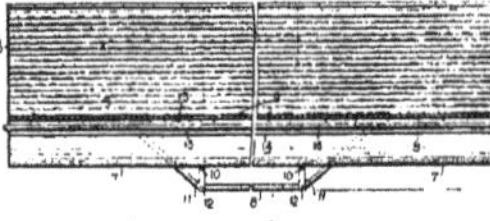

1. The process of recovering potassium salts from organic carbonaceous materials which consists in placing the material in an open chamber and incinerating the same, and adding a material which, under the action of heat, will liberate sulfur oxid gases to convert the potassium chlorid in the material into potassium sulfate.

2. The process of recovering potassium salts from organic carbonaceous materials which consist in incerating the same in an open chamber and under a forced draft, and adding sulfur to liberate sulfur oxide gases which penetrate the material during the burning and convert the potassium chlorid existing therein into potassium sulfate.

1,231,707. Apparatus for Treating Ores and the Like. Niels C. Christensen, Jr., Salt Lake City, Utah, assignor, by mesne assignments, to Holt-Christensen Process Co., Salt Lake City, Utah. Filed Mar. 30, 1915.

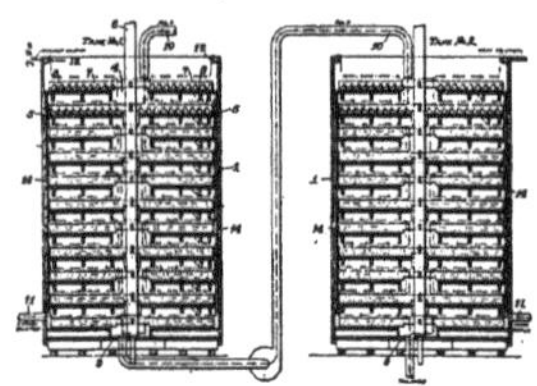

An apparatus for leaching and sedimentation comprising a vertical cylindrical tank having a central discharge opening in the bottom and an overflow rim launder at the top and with a series of stationary horizontal decks disposed one above another therein, alternates of said decks being open respectively at the centre and at the periphery, pipes connecting openings in the periphery of the upper deck with the space above one of the lower decks, a central vertical revolving shaft supplied with arms carrying rabbles above each deck and above the tank bottom, and so arranged as to move the material upon said decks to the above mentioned openings in said decks and discharge said material upon the decks below and finally discharge said material through the above mentioned discharge openings, and means for feeding ore and other substances into the upper part of the tank.

1,232,553. Process of Producing Fertilizers From Silicates. Louis L. Jackson, New York, N. Y., assignor of one-half to Odus C. Horney, New York. Filed June 16, 1916.

The process of producing a fertilizer or fertilizer base which comprises treating silicates containing alkali metals and aluminum with calcium hydrate and water and subjecting the mixture to heat and pressure; separating the aqueous solution from the insoluble and sparingly soluble constituents and treating the solution with an acid thereby forming alkali metal salts and drying; subjecting the said constituents to treatment with an acid in quantity at least sufficient to convert all the calcium present into a calcium salt and set free the aluminum as a hydrate, and separating the liquor; again treating the cake with an acid in quantity sufficient to dissolve the aluminum hydrate and separating the liquor; and mixing the cake with the alkali metal salts to form the fertilizer.

1,232,111. Rock-Drill. Michael Smith, Bisbee, Ariz. Filed Sept. 28, 1916.

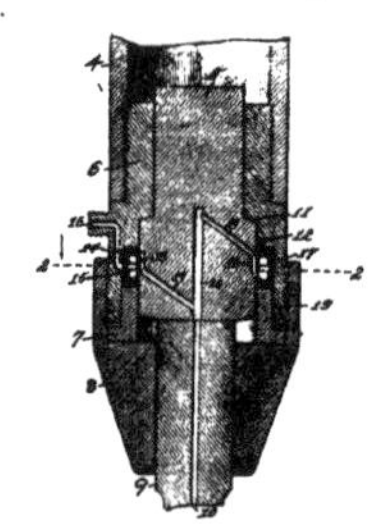

1. The combination in a rock drill adapted to a drill iron having an axial bore therethrough, a tappet having an axial and a lateral passage, and a bushing having a passage connecting the exterior with the interior thereof; of a packing consisting of a channeled ring of perforated flexible material having its open side toward said tappet lateral passage and its perforations toward said bushing passage, and a channeled ring therein of perforated metal having its open side toward said bushing passage and its perforations toward said tappet lateral passage.

2. As an article of manufacture, for use in a rock drill, a packing consisting of a channeled ring of perforated flexible material, and a channeled ring of perforated metal having its open side adjacent the perforations in said flexible ring and having its perforations adjacent the open side of said flexible ring.

1,232,977. Method of Utilizing Feldspar. John Gustaf Adolf Rhodin, Chiswick, England. Filed Dec. 4, 1915.

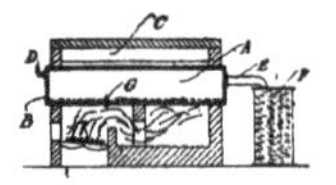

1. The method of utilizing potash feldspar which consists in mixing the ground feldspar with a small quantity of a sodium salt, causing a mixture of sulfur dioxid, steam and air to react on the heated mixture, collecting the sulfuric acid formed in suitable receivers and leaching the residue so as to obtain sulfate of potash and to leave a residue suitable for making white Roman or Portland cement.

2. The method of utilizing potash feldspar which consists in mixing the ground feldspar with a small quantity of sodium chlorid, causing a mixture of sulfur dioxid, steam and air to react on the heated mixture collecting the sulfuric acid formed in suitable receivers and leaching the residue so as to obtain sulfate of potash, and to leave a residue suitable for making white Roman or Portland cement.

1,232,611. Filter-Press. Charles Marshall Saeger, Jr., Palmerton, Pa. Filed July 11, 1916.

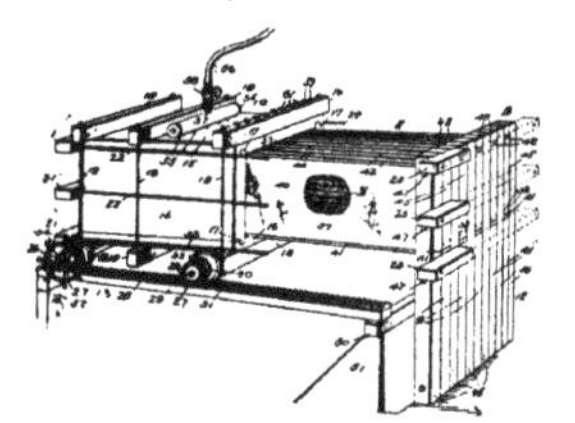

1. In a filter press, the combination of a longitudinally movable tank having an open end, a closure therefor fixed against movement, a plurality of spaced filtering elements having smooth vertical faces secured on said closure and projecting toward said tank, said filtering elements being adapted to enter and be withdrawn from said tank as the latter is moved, and a plurality of vertical wires crossing the open end of said tank and adapted to engage the vertical faces of the filtering elements and automatically clean them of collected solid matter when the tank is retracted.

1,232,754. Means for Mounting and Driving Centrifugal Machines. Melville H. Barker, Boston, and Comfort A. Adams, Cambridge, Mass.; Annie M. Barker, administratix of said Melville H. Barker, deceased; said Adams and administratix assignors to American Tool & Machine Company, Boston, Mass. Filed Apr. 22, 1910.

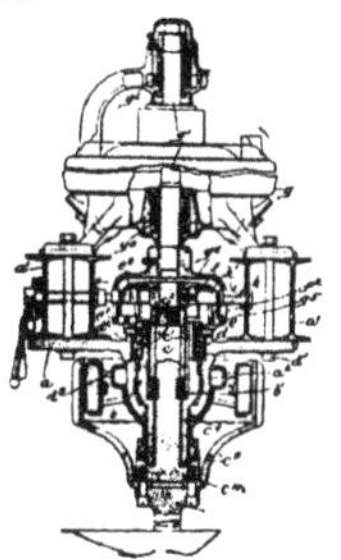

1. In a centrifugal machine the combination of a rotary gyratory basket-shaft and oscillatory bearing sleeve in which said shaft is suspended for support and rotation, a non-oscillatory motor-driven shaft arranged axially above said basket-shaft, flexible connecting means forming a driving coupling between the adjacent ends of the two shafts, said connecting means being unyielding in a circumferential direction while yielding freely for a slight distance in a radial direction, whereby the torque of the motor shaft may be transmitted to the gyratory basket-shaft without lateral strain on the motor-shaft, substantially as described.

2. In a centrifugal machine the combination of the oscillatory sleeve, a thrust bearing located therein in the plane of its centre of oscillation, a basket shaft suspended from said thrust bearing and having lateral bearing support in the lower portion of said sleeve, said shaft being divided transversely into upper and lower sections provided at their adjacent ends with opposed coupling flanges, means for detachably connecting said flanges together, an oil cup supported by the flange of the upper section by a detachable connection to extend upward outside of said sleeve beyond the plane of said thrust bearing, substantially as described.

1,233,149. Ore-Concentrating Table. Enos A. Wall, Salt Lake City, Utah. Filed Oct. 16, 1916.

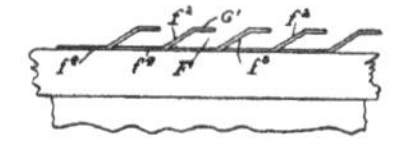

1. An ore concentrating table of the longitudinally differentially reciprocating type formed of a series of flat topped sections, each section having all its top surface in a plane common to all, and provided with parallel channels intermediate the sections extending from the head to the foot end of the table and open at the said foot end for the continuous discharge of the mineral concentrates; the said channels being of the same depth and width throughout and having their lower side walls inclined from the bottoms of the channels outward from end to end at an angle oblique to the flow of the wash water.

2. An ore concentrating table of the longitudinally differentially reciprocating type provided on its upper surface with a series of parallel channels extending from the head to the foot end of the table and open at their foot ends for the continuous discharge of the mineral concentrates, and hoods forming the upper surface of the table and projecting at one longitudinal edge over the upper sides of the channels and extending throughout the length of said channels, all portions of each hood member lying in a plane common to all the hoods of a table section.

1,231,967. Electrolytic Recovery of Metals From Their Solutions and in Apparatus Therefor. Urlyn Clifton Tainton and Malcolm Foerster Lambe Aymé Aymard, Johannesburg, Transvaal, South Africa. Filed Nov. 14, 1914.

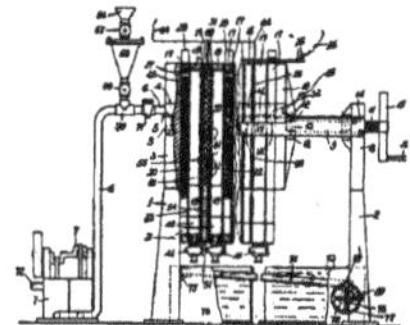

1. Apparatus for recovering metals from their solutions electrolytically comprising alternate anode and cathode elements, each cathode element comprising a porous conducting support, a coating of comminuted conducting material deposited on said support from a liquid carrying said material in suspension, means for passing the metal bearing solution through the pores or interstices of said comminuted conducting substance, and means for drawing off the comminuted conducting substance after it has been detached from said porous conducting support.

5. Apparatus for recovering metals from their solutions electrolytically, comprising alternate anode and cathode elements separated by porous diaphragms, each cathode element comprising a porous support having thereon a coating of comminuted conducting material deposited from the solution, means for adding the comminuted conducting material to the inflowing solution, means for passing the metal-bearing solution through the interstices of said coating, and means for withdrawing the comminuted conducting material after it has been detached from said porous support.

REVIEW OF MINING

As seen at the world's great mining centres by our own correspondents.

METCALF, ARIZONA

THE LABOR SITUATION IN GREENLEE COUNTY REMAINS UNCHANGED.—MINERS LEAVING THE DISTRICT.—AN OPEN LETTER.—A LACK OF INTEREST SHOWN BY WORKMEN.

There has not been much change in the strike conditions in the Clifton-Morenci-Metcalf district since the strike was declared and the men walked out, July 1. The men for the most part are loafing on the streets, apparently being supported by money furnished by the unions. However, the strike has seriously affected all classes of business; hotels report they are not making expenses; the merchants are doing very little; and freight-trains are limited to two a week. Hence, at times, there is a shortage of some staple or other. The Americans who stayed with the companies have had their salaries cut 25% and consequently are seeking other positions and are leaving as fast as they secure them.

Among the demands of the strikers is the Miami wage-scale, and Norman Carmichael, general manager for the Arizona Copper Co., has issued a pamphlet to the public as well as to the employees, which shows up the wage-question clearly. The following is taken from this pamphlet:

"In Miami in the month of May, 1917, there were mined by the Inspiration and Miami companies a total of 777,220 tons of ore, or 25,071 tons per day. The total number of men employed in their mining departments by these two companies was 1875, or a yield of 13.4 tons per man per day.

"In the Clifton-Morenci-Metcalf district, the three companies, Arizona, Phelps-Dodge, and Shannon, together mined 6710 tons per day and employ in their mining operations 3730 men, equivalent to a yield of 1.8 tons of ore per man per day.

"While it is true that the above difference in the yield per man is essentially due to the different conditions in the two districts, another factor comes into play, for it is also a fact that mining labor in this district is not equal in efficiency to the labor in other districts, nor are our miners as efficient as they were a few years ago.

"An extract from the records of one of the operating companies may be given as illustrative of this unfortunate condition.

"In the mine in question some 600 men are employed and while no change has been made in mining methods employed, which would account for any diminuation of the output per man, the following figures show the facts:

	Tons per man per shift	Average wage per shift
1914	2.52	2.30
1915	2.60	2.34
1916	2.22	3.72
1917	2.18	4.12

"Thus, while wages have steadily and rapidly advanced during the past four years, the average output per man has steadily declined.

"During 1915 many of our best miners left the district and the companies here are handicapped by the necessity of using a large number of men, some of whom are not physically able and others who are not mentally disposed to turn out an average day's work, and while there are many employees in our mines to whom this remark does not apply, inquiry among various foremen and shift-bosses in charge of the operations in the mines will corroborate this statement.

"The attitude of a certain important section of the mine employees toward their superiors has also been anything but satisfactory for some time past. The deliberate opposition shown to any method introduced for the purpose of increasing efficiency, the spirit of insubordination and the lack of co-operation and loyalty, are factors not conducive to the harmony and good feeling which should exist, and unless the companies can be assured of a better spirit and heartier co-operation with them among their employees there is little inducement to re-open the mines."

MILL AT THE MIAMI, ARIZONA

AJO, ARIZONA

LARGE FLOTATION PLANT AND SMELTER FOR AJO.—OXIDIZED ORE SUFFICIENT TO CONTINUE LEACHING FOR TEN YEARS.—THE GREAT VERDE MINE STRIKES SULPHIDE ORE.

John C. Greenway, general manager for the Calumet & Arizona and New Cornelia Co., has announced that in view of the Ajo Consolidated purchase by the New Cornelia Copper Co., a 5000-ton flotation plant and a 2000-ton smelter will be erected at Ajo. As a result of the purchase of the Ajo Consolidated by the New Cornelia, the ore reserve already developed would reach 21,000,000 tons, the greater part of which will remain untouched for 15 years. The Ajo properties are extensions of the New Cornelia and were developed by extensive mining and diamond-drilling. The moving spirit of Ajo was James Phillips, of New York, former president of the Tennessee Coal & Iron Company.

The leaching-plant of the New Cornelia produced 2,100,000 lb. of copper in July. This plant will have served its purpose in 10 years, at which time it will probably be replaced by the addition of another 5000-ton unit to the flotation-plant for the oxidized ores that are now supplying the leaching-plant at the rate of 5000 tons per day, then will have been exhausted. At the end of two years enough of the oxidized ores will be stripped from the sulphide orebodies to supply an equal tonnage of sulphides to the first unit of the flotation-plant, which, by that time will have been completed. The New Cornelia Copper Co. is shipping some oxidized ore to the smelter at Douglas. The ratio of concentration in the flotation-mill will be 15 to 1, which will provide 750 tons of concentrate per day. To ship this to Douglas would cost approximately $1,000,000 per year. The alternative, therefore, is to build a smelter at Ajo, which would pay for itself in a year and a half. In the meantime the shipments of oxide ores from both the New Cornelia and the Ajo Consolidated will continue, but the main sulphide orebodies of the latter will be held in reserve, and after stripping the oxides, will be mined to supply the flotation-mill, which eventually will have a capacity of 10,000 tons per day. A considerable tonnage of high-grade ore has been developed at depth on the Ajo Consolidated.

J. W. Hubbard, who is in charge of the operations of the Great Verde property, says that the shaft is in sandstone and limestone, in which are bunches of copper carbonate, chalcopyrite, and copper glance. These bunches of ore are in the sandstone and limestone, and not in the conglomerate. Ore was discovered during the sinking of the shaft, 36 ft. below the collar. This ore is similar to that found in the United Verde Extension on the 800-ft. level. Progress in shaft-sinking is as rapid as can be expected with hand-work, but as soon as machinery is installed, plans for which are under way, the work will be pushed.

ALASKA

MINERS AT KENNECOTT DEMAND SAME WAGES AS ARE PAID IN THE INTERIOR PLACER MINING DISTRICTS.—NEW STAMP-MILL FOR SEWARD PENINSULA.—THE KENNECOTT RAILROAD NOW OPERATING AT A PROFIT.

George E. Baldwin has gone to his copper property in the Copper River valley to do development work. W. R. Rust, of Tacoma, who is the principal owner of the Chichagoff mine, is making his annual inspection of the property. R. E. Hutchinson, secretary of the Hirst-Chichagoff Mining Co., will spend some time at the property. Extensive development work has been planned for the summer. William Conley, president of the Valdez local of the Western Federation of Miners, has sent out telegraphic information that there is no labor strike at Ellamar and that none is contemplated. It is also reported that there is no trouble at Latouche, a Kennecott property. The Kennecott miners claim that that property is an interior mine and therefore that the same wage-scale should be paid as in the placer fields of Fairbanks, Iditarod, Nome, and elsewhere. Bennett T. James, formerly of No. 17 Below Goldstream, Fairbanks, has taken in a stamp-mill which he and John Ronan will place on their property on Moss Pass near Seward. This will be the first mill in that district.

The high prices for copper, and the record shipment of ores and concentrate over its lines have now transformed the Copper River & Northwestern Railway, controlled by the Kennecott Copper Corporation, into a profitable enterprise. The capitalization against its 196 miles is $27,837,400. The road started operation in 1911 and from November of that year to the end of May, 1916, it piled up a deficit of $2,941,962. But during the period from July 1, 1916, to May 3, 1916, it showed a surplus of $487,000 after deducting interest charges. In every year but one it has shown a substantial margin above operating expenses, but the interest charges have been so heavy that the road was unable to produce a surplus before 1916. Mr. Stannard, superintendent for the Kennecott company, returned to Kennecott on July 26, intending to adjust the grievances of the miners. He offered a new sliding-scale, equivalent to 25c. per day more than the old rate. The men had demanded to be paid a flat-rate, the same as at other Alaskan mines. The highest wages they could hope for under the sliding-scale were less than the desired flat-rate and they refused to consider it. On July 30 the Kennecott officials visited the miners' camp at Blackburn and tried to reason with them, asking if the company had not kept its agreement with them; also informing them that the present Kennecott scale was 10% higher than that at either Butte or Bisbee. The men object strongly to the sliding-scale, which keeps them constantly in doubt as to their rate of pay. For instance, in May and June wages dropped 25c. at the time when they were advancing elsewhere in Alaska. As the Kennecott miners are mostly Alaskans, accustomed to wages based on the gold standard, they were opposed to such an arrangement, and demanded a fixed scale of wages last December, but were over-ruled. This is the fourth wage dispute in 18 months and the miners have appealed to the Government for a board of arbitration to settle the matter. The mine management, on the other hand, feels that it has not been fairly treated, because they entered into an agreement with the miners that was to continue throughout the year 1917. Mr. Stannard said that he had hardly left the mine before the miners, in violation of their agreement, served notice on the assistant superintendent, Mr. Smith, demanding an increase in wages, and the abandonment of the bonus-system for the flat-wage, similar to that paid to the placer-operators during the summer in the interior, where wages are $5 per day with board. Under the bonus-plan laborers were making $5.25 per day and higher, and paying $1.25 per day for board. It is said that some of the miners made as high as $5.25 per day clear. The mine management, while claiming that the miners had no right to break their agreement to adhere to the old wage-scale throughout the year 1917, conceded that there was reason for a small increase in wages, but refused to permit men who were resonsible for the agitation to return to work. Telegraphic advices from Washington (July 14) say that the miners who participated in the Kennecott strike have yielded to Secretary of Labor Wilson's request that they resume work, pending the arrival there of Labor Department mediators. Advices from Fairbanks state that Marshall L. T. Erinn has arrived from Nenana, having in custody three men charged with being ringleaders in the labor trouble at Nenana, and also charged with sedition. It is not known whether the latter charge will be pressed, but the three men will be interned in the Federal jail until the end of the War. Marshal Erinn will return to Nenana and promises to arrest other aliens unless they obey the law. The strike is over at Nenana and the men have returned to work.

PACKARD, NEVADA

RAPID MILL CONSTRUCTION AIDED BY MODERN METHODS.—ELECTRIC MOTORS FOR TRANSPORTATION.—IMPORTANT DEVELOPMENT OF SULPHIDE ORE.—PROSPECT OF ANOTHER APEX SUIT.—TUNGSTEN DISCOVERIES.

The erection of the framework of the Rochester Combined 300-ton mill is nearing completion, and sets a new record in rapid mill-construction. Compressed air is used in boring all timbers, and in driving ship-spikes. A Little Tugger hoist is used in placing timbers. Two shifts of carpenters are employed, the night-shift working by electric light. About 500,000 ft. B.M. of timber will have been used. In spite of the scarcity of supplies, the constructing engineer, K. Freitag, believes the mill will be completed by November 1.

The Combined is pushing development on several parts of

the property. Most of the work is being done on the main fault-contact between the rhyolite and rhyolite-tuff, on which the principal Packard orebodies lie. Some ore has been developed by five adits—the Shepherd No. 1 and 2, and Links No. 1, 2, and 3—for nearly a mile along this contact, the ore occurring as lenses in the rhyolite with tuff hanging wall. The Happy Jack shaft at the south-west end of the property is now down about 200 ft. on the dip of the vein. The ore from this shaft is to be hauled to the mill by electric locomotives. Tramways will deliver the ore from the Link Tunnels. A Sullivan 600-cu. ft. angle compound-compressor is now being installed, replacing the 6-drill portable Diesel-engine driven machine. The latter will be used in prospecting the outlying claims. The company is now beginning the making of a topographic map of the property, to be followed by a thorough geological examination.

The Walser adit on the Packard property is being advanced to connect with the Combined company's Shepherd No. 1 adit, and to open that part of the Shepherd orebody lying in Packard ground. A cross-cut from the 120 Tunnel recently cut a vein of sulphides nearly 25 ft. wide, showing good milling ore throughout. The same orebody was later cut on the 160-ft. level. Drifts are being run to determine its lateral extent. The finding of primary sulphide ore simultaneously in the Packard and Rochester mines was one of the most interesting developments of the past few months. Heretofore the value of the ore was almost entirely in hornsilver. It is probable that a shaft soon will be sunk to open this vein in depth. Pending the installation of a new 100-hp. motor for driving the tube-mills, the Packard mill has been shut-down for a few days.

On the Nones claims the owners have opened up a promising vein. These claims adjoin the Happy Jack group, of the Combined property.

The Packard district was recently visited by Adolph Knopf, of the U. S. Geological Survey, and a bulletin will probably be issued soon as a result of his investigation. The geology of the district was also studied by Mr. Jones, of the University of Nevada. The Nevada College of Mining Summer Camp was established on the Packard property.

The Adams vein on the 800 and 900-ft. levels of the Rochester Mines Co., at Rochester, is yielding excellent ore. This is the sulphide orebody previously referred to. No mention is being made by any of the companies regarding the threatened suit involving the Rochester Mines Co., Merger, and Nenzel Crown Point, but the recent visits to the district of Lawson, Searls, Emmons, and other geologists indicates that another big apex-law battle is about to begin.

A number of the larger stockholders of the Packard Mines Co. have bought the Nevada Austin mines at Newpass, 25 miles from Austin, and have begun the erection of a 75-ton mill and cyanide-plant. The ore is free-milling, gold predominating. Tests have shown it to be amenable to cyanidation, excellent extraction and recovery having been obtained on 80-mesh material. The flow-sheet is similar to that of the Packard mill, except that a Blake crusher will be used and rolls eliminated. The ore will go direct from the Blake crusher to a 6 by 16 Hardinge ball-mill in closed circuit with a Dorr classifier. A Dorr thickener will precede the Dorr agitator, followed by modified C. C. D. in a second set of thickeners, these to be followed by an Oliver filter. Merrill presses are to be used for precipitating, and a Monarch tilting-furnace for refining. A 150-hp. Diesel engine will furnish power. It is the intention to utilize the exhaust for heating mill-solutions. Frank Margrave, of Reno, is president and general manager of the company. Douglas Muir is constructing engineer.

A promising tungsten prospect is being developed near Oreana by E. H. Radtke. The tungsten mill at Toy is shut-down, but the plant at Toulon is in operation. Toy and Toulon are two small stations on the Southern Pacific railroad a few miles below Lovelocks.

CRIPPLE CREEK, COLORADO

CRESSON CONSOLIDATED'S HIGH-GRADE SHIPMENTS.—LESSEES WORKING OLD DUMPS AT A PROFIT.—LESSEES BREAKING HIGH-GRADE ORE.—WATER INCREASES IN THE ROOSEVELT TUNNEL.—DIVIDENDS IN AUGUST.

High-grade ore from the new east vein at the 1400-ft. level of the Cresson Consolidated mine, comprising three consignments of 90 tons each, sent to the Golden Cycle Mining & Reduction Co. at Colorado Springs, returned a gross bullion value of $153,000. One lot carried $400 per ton gold, lot No. 2, $500 per ton, and Lot No. 3, $800 per ton.

The shoot has now been proved for 125 ft., and the ore mined was taken out in driving the drift 8 ft. wide. The management is proving the vertical continuity of the shoot by raise and winze.

The July production from the Elkton mine of the Elkton Consolidated Mining & Milling Co. totaled 33 cars of mine ore, valued at $18 to $20 per ton. This production was made by 22 sets of lessees. The Tornado mine of this company under lease to Whittenberger and Collins, is again on the shipping list with a car per week. The ore mined at the 3rd and 6th levels is of about one-ounce grade.

Dump lessees who shipped heavily from the Gregory dump, have loaded out from 75 to 100 cars to the Golden Cycle mill at Colorado Springs. This ore is carrying gold not to exceed $8 per ton. The main Elkton mine dump is producing a car per week of $12 to $15 ore.

The output from the El Paso Consolidated on Beacon hill was light. There are 16 sets of lessees, who produced 550 tons, estimated at between $18 and $20 per ton. The company is still inactive, new development being dependent on the work of lessees.

The Queen Bess mine, on Tenderfoot hill, now owned and operated by Hahnewald Olsen & Company, who took up the option at $17,500, on the property, are shipping steadily, and last week 1200 lb. of sacked ore shipped to the Eagle Ore Co.'s sampler, brought settlement at a rate in excess of $1 per pound. There were seven cars of ore loaded out last week of better than 2-oz. ore. R. M. Gilber, leasing on the Iron vein of the El Paso Gold King mine in Poverty gulch, operating through the lower Gold King adit, 1200 ft. from the portal, last week loaded out his initial shipment of milling-grade ore. The vein is from 5 to 8 ft. wide, the ore shipping about one ounce gold per ton. The Iron vein strikes north, toward the Queen Bess, but has not yet been opened up on that property.

Two sets of lessees, Frank Caley of Littleton, Colorado, and the Cripple Creek Deep Leasing Co., a Denver corporation, are operating the Jerry Johnson mine on Ironclad hill, and producing steadily. Five cars of two-ounce ore were sent from the Caley lease, and the Deep Leasing Co. shipped 10 cars of $20 ore to the valley mills during July. The Caley lease extends to the 650-ft. level of the main Jerry Johnson shaft; the Cripple Creek Deep Leasing Co. is mining at the 850-ft. level.

The South Burns mine of the Acacia Gold Mining Co. has been closed down, while the plant is being overhauled and a new electric motor installed. The South Burns in July produced 9 cars of ore, 5 on company account and 4 by lessees, of an average value of $23 per ton.

The Dig Gold Mining Co., M. B. Burke, of Denver, president and general manager, has installed a steam-hoist and 80-hp. boiler at its new shaft on the line between the Alpha and Omega claims on the south slope of Gold hill, and preparatory to sinking has laid an air-line to the compressor at the Hummer shaft, of the Gold Bond Consolidated Mines Co. adjoining. The company has ore at the 100-ft. level and will mine this ore when sinking. The Rubie or Worcester mill, on the Rubie property, on Bull hill, has been purchased by Thomas Kavanaugh, Ben Morrow, and Charles Bartell, and treatment of low-grade dump-ores was resumed the first of the week. The mill is treating the dump-ore from the Rubie mine.

Dividends will be paid on August 10 by the Cresson Consolidated G. M. & M. Co. and the Golden Cycle M. & R. Co. The Cresson dividend is $122,000 and that of the Golden Cycle, $45,000. A. E. Carlton is president of both companies.

The Roosevelt Tunnel of the Cripple Creek Deep Drainage & Tunnel Co. was advanced 178 ft. with one shift in July; the lateral heading, toward the Cresson mine, was driven 100 ft. The flow from the portal raised last month to 5980 gal. per minute, due to the cutting of a water-course near to the east side-line of the Old Ironsides claim of the United Gold Mines Company.

PLATTEVILLE, WISCONSIN

OUTPUT FOR JULY SHOWS DECREASE IN QUANTITY BUT INCREASE IN PROFITS.—ENTIRE ZINC-LEAD FIELD IS ACTIVE.—FIGURES OF PRODUCTION.—SEVERAL NEW ENTERPRISES.

The month of July, regarded as the banner month in the zinc-lead districts of Wisconsin, this year failed to uphold the tradition. Investigation revealed that it was due largely to enlistments. This, combined with the great unrest among the foreigners employed, told the story. One of the leading operators of the field, employing nearly 400 men, says that his company could use an additional 500, and that only one of the seven mines under his management was being operated with a full crew. Owing to lack of sufficient labor to operate on full time no further development will be undertaken until the labor situation improves. At the Graham mine, in the Galena district, the manager said negro miners had been employed and were proving satisfactory; that they were well behaved and labored diligently with good results. While the output of zinc ore diminished during July and while mine managers deplored the high cost of operating, it was found that it had been a successful period from the point of profit-taking.

While there came in mid-month a break in the price of zinc ore there was a uniformity of prices throughout the month that promoted business, the base price at the beginning of the month standing at $75 per ton for standard and premium-grade ore, with the range down to $70. These figures held steadily for the first two weeks, when with the softening in the price of spelter came a sympathetic downward movement in zinc-ore prices. Quotations for the third week ruled at $70 base for premium and standard-grades, down to $65 for the medium and second-grade ores, which held through the month, and as the demand for high-grade ore was constant, sales and shipments held fairly well above the base. The buying lacked the spectacular features that accompany competitive bidding, and business was steady and good. The lower-grade ores, and this includes all grades from 54% down, did not fare so well.

Discrimination in offerings was marked, and independent producers who were able to market their output congratulated themselves that they had a market. During the latter half of the month 45% was bid in at $35, 40% at $29, 35% at $25, and 30% at $20. Some ore was carried over by low-grade producers, and there were a few instances of shut-downs. Some claimed that ore producers were not receiving prices compatible with the metal market, but the larger operators said that makers of ordinary spelter have been paying all they could afford for zinc ore.

Lead-ore producers complained of their treatment during the month. The price for lead ore at the beginning of the month was $120 per ton, base of 80% metal content, and some high-grade material was sold above this quotation. The trade in pig-lead was normal throughout the month and the price was less affected than was spelter. Regardless of this, and that quotations on futures were practically suspended, the price of lead ore came down to $115 per ton about the middle of the month to $110 the last week of the month. Production showed appreciable gains, good weather enabling many small companies to work old mines and shallow diggings with little expense, and the shipment of mixed lots of ore was reported frequently. Two exclusive lead-ore producers were developed, the McDermott, in the Shullsburg district being one, and the Annex, in the Cuba City district, being the other. Both reported shipments of lead ore of high grade. Shipments for the entire month were higher than for many months past and would have been greater had high prices continued. The drop in price found lead-ore producers alert, and they promptly closed up on deliveries, only two cars leaving the field the last week of July. A conservative tabulation of lead ore held in bin in the several districts showed close to 1500 tons being carried over, and yet no one district could boast more than two or three cars at the most. The lead situation is better controlled than the zinc, being in fewer hands, and competition is being restricted.

The carbonate zinc-ore producing part of the field, at one time important in output, was idle except for the operations of the New Jersey Zinc Co. in both the Centerville and Highland districts. Scores of small companies, composed of local men, have discontinued mining activity. At no time within the past five years has there been a semblance of competition in the purchase of this class of zinc ore. Occasionally outside buying-agents have dared the opposition of the New Jersey Zinc Co., which controls these districts, owning and operating the only railway line, but in no instance have they tarried long, and miners have come to recognize their dependence on this one company and have accepted the situation, there being but one alternative, that of abandoning the mines. The New Jersey Zinc Co., on the other hand, is giving employment to all miners at several newly developed and equipped properties, and the district is as prosperous as it had been under the old order of things. More men are required than can be had, and the only drawback is lack of sufficient water for concentration. This is being overcome by boring deep wells, one of which, over 700 ft. deep, is now being supplied with pumps.

Shipments of pyrite for the month were confined to separating-plants exclusively and were lighter than usual. There has been no change in price, as the ore is all sold under contract to acid makers, and sellers refuse to divulge the price agreement.

The total deliveries of zinc ore, lead ore, and iron pyrite from mines to local refineries and to smelters direct, for the month of July, up to and including the 28th, were: zinc, 36,528,000 lb.; lead, 1,591,000 lb.; pyrite, 5,946,000 pounds.

High-grade ore from refining-plants shipped to smelters was reported as follows:

	Lb.
Mineral Point Zinc Co	5,256,000
National Separators	4,038,000
Benton Roasters and Wisconsin Zinc Co	2,732,000
Linden Zinc Co	320,000
Total	12,346,000

The gross recovery from concentrators for July totaled 18,364 short tons; net deliveries out of the field both from mines to smelters and from reduction plants to smelters, 10,813 tons.

Sales and distribution of zinc ores were made as here shown to the buying companies represented in the field. Mineral Point Zinc Co., 5676 tons; Grasselli Chemical Co., 2514 tons; National Separators, 3573 tons; Wisconsin Zinc Roasters, 3024 tons; American Zinc Co., Hillsboro, Illinois, 1809 tons; Linden Zinc Co., 1360 tons; M. & H. Zinc Co., La Salle, 649 tons; Illinois Zinc Co., 630 tons; Benton Roasters, 462 tons; Lanyon Zinc Co., 355 tons; American Metal Co., 313 tons; Edgar Zinc Co., 280 tons; Sandoval Zinc Co., 35 tons. A total of 20,526 tons.

THE MINING SUMMARY

The news of the week as told by our special correspondents and compiled from the local press.

ALASKA

(Special Correspondence.)—W. A. Dickey, who has been opening up and preparing to operate the old Barrack property, owned by the Latouche Copper Mines Co., and adjoining Beatson's Bonanza, of the Kennecott company, has shipped on the steamer 'Cordova' 300 tons of ore to the Tacoma smelter. This ore is hand-sorted and it is expected to run between 5 and 6% copper. This property made two previous shipments several years ago, one of 800 tons to Tacoma smelter, and one of 250 tons to Ladysmith, the returns from neither being satisfactory to the owners at the price of copper at that time. It is now claimed by the management that a sufficient tonnage is blocked out to assure them a place among regular shippers.

Latouche, July 31.

(Special Correspondence.)—The Ellamar mine, at Virginia Bay, is now shipping from 5000 to 6000 tons per month to the Tacoma smelter. This ore is low in copper, 2% or under, but carries from $3 to $5 gold per ton, and is a much desired ore at Tacoma on account of its iron content. Eighty men are employed. The mine is under the management of L. L. Middlecamp for the Ellamar Mining Company.

Ellamar, July 30.

(Special Correspondence.)—The Midas mine, of the Granby company, across the bay from Valdez, is producing 200 tons of ore per day, which is shipped to the company's smelter at Anyox, B. C. This mine employs 75 men, and is under the management of Palmer Cook.

Valdez, July 31.

ARIZONA

Cochise County

(Special Correspondence.)—Since the taking over of the Mascot Copper property by the American Smelting & Refining Co. it is estimated that shipments for the month of July will reach 2500 tons. Report says that preparations are being made to take on a large force of miners.

Dos Cabezos, August 2.

Gila County

At Globe, after an idleness of five weeks, the smelter of the Old Dominion company was again blown-in on August 8. The concentrator also was started during the week, though both smelter and concentrator will not be operated at full capacity. The workmen, whose number is increasing daily, go to and from work under armed guards, so no disturbances occurred, though 300 union pickets lined the main street leading to the works. Of all the men accepted not a single I. W. W. was included.

Pima County

(Special Correspondence.)—A new wulfenite property has been opened up by the Vail interests on one of the claims of the old Total Wreck mining property, 9 miles south of Pantano station.

The owners are sinking and taking out about a carload of 3½% ore weekly, using a gas-hoist. The ore is sacked and shipped to Tucson, where it is concentrated.

Wulfenite was known here 30 years ago, but at that time the Total Wreck was a silver-lead property, and as the wulfenite then had no value it was thrown on the dump, which dump has lately been worked over. Work was begun six months ago, under the direction of E. L. Vail, who employs 25 men. This property has been a big producer of high-grade silver ore, and 30 years ago was equipped with a 20-stamp mill and 250-hp. Corliss engine.

The Arizona Rare Metals Co., under the management of R. O. Boykin, has erected a leaching and concentrating-plant in Tucson to recover the sodium molybdate from the slag obtained in smelting.

The slag is ground in a ball-mill and is then run into filter-tanks. The filtrate is placed in evaporating pans and the sodium molybdate recovered from the solution.

The concentration-plant was added for the treatment of wulfenite, and is handling all the output of the Total Wreck mine. The ore is fed to a crusher, elevated to shaking-screens, and the oversize passed through rolls. It then passes over a Wilfley table. The mill has a capacity of 5 tons per 8 hours.

Tucson, August 5.

(Special Correspondence.)—A strike of copper-zinc-lead-silver ore is reported by Edward G. Bush, manager on the 200-ft. level of the San Xavier Extension Copper Co.'s mine in the Twin Buttes mining district. A sample taken across the face shows silver, 32.8 oz., copper 6.34, lead 12.5, and zinc 23.5. This vein was struck on the 140-ft. level, where it assayed 50% zinc, 1% copper, 8% lead, and 4 oz. silver. The drift has been run for 25 ft. on the ore. The San Xavier Extension touches the Vulcan mine on one end and the Wakefield on the other. The vein shows also on the Wakefield claim.

It is understook that a deal is pending on the Wakefield property, which now adjoins operating companies on three sides.

Ben Heney has negotiated a deal with E. D. Sullivan on the Pandora and Fairbanks claims in the Twin Buttes district. Negotiations on three other groups controlled by Alex Rossi in this district are now pending. It is understood that Mr. Heney has interested Idaho capitalists who looked the property over in the early part of July.

Walter Harvey Weed, consulting engineer for the Magnet Copper Co., was in Tucson last week, and said that he had been asked to make a report on the conditions in Twin Buttes district with a view to the erection of a smelter.

He said that the first drill-hole on the Magnet property had verified his opinion of the existence of ore similar to that in the large poryphyry mines of this country. The bore-hole proved the ore to be of low copper content at water-level 80 ft. below the surface. At 160 ft. down, the value had gradually increased till 5 ft. of 6% copper ore was pierced. From all the assay returns of the bore-hole samples it is estimated that there is 60 ft. of ore averaging 2.08% copper. The extent of this orebody has to be proved by more drilling, and there is reason to believe that results as good as were obtained in hole No. 1 will be duplicated in the other holes which are now being bored. Development work will be continued, being directed by the results obtained by drilling.

Twin Buttes, August 3.

Pinal County

(Special Correspondence.)—The 10-hp. hoist recently bought by the Magmatic Copper Co. has been installed, and it is now expected to make better progress in shaft sinking. When the hoist was put in, the shaft was down 112 ft. The shaft is still in a grade of manganese ore, which it is believed will meet the market requirements.

Florence, August 2.

SANTA CRUZ COUNTY

(Special Correspondence.)—J. Smith and two companions, of Warren, Pennsylvania, who are said to be the principal owners of the Mowry mine, were in the district last week and have made arrangements again to work the old mine. It is announced that a concentrater will be built and a new shaft sunk, the work to be started within 30 days. It is thought that this decision of the Pennsylvania capitalists and the announcement that they intend to work the property ends the deal that has been pending for some time with the Standard Metals Co. of Los Angeles for the sale of the property.

Patagonia, August 3.

YAVAPAI COUNTY

(Special Correspondence.)—Louis Goldman, the president of the Copper Queen Gold Mining Co., has made an inspection of the company's property in this county, 6 miles east of Humboldt. The mine is now in condition to produce 100 tons of ore per day. The output will be treated at the Binghampton plant and delivery commenced in a short time.

Some interest is being caused by the revival of the report that there are possibilities for oil in the Big Chino valley. To date 100,000 acres of land has been leased, and it is reported drilling-rigs are coming in.

Prescott, August 3.

YUMA COUNTY

(Special Correspondence.)—The Billie Mack Co. is shipping a 40-ton car of ore to the A. S. & R. Co. at Hayden this week, and another car is in the course of preparation from the older workings on the property, formerly known as the Ruby. The ores are chrysocolla and malachite. Development work is being done on what is known as the Jay Bee vein. Jack-hammers are being used. W. Dowe, of Chicago, general manager for the company, will soon be on the ground when extensive plans will be made for further development of Lion hill. The advent of cooler weather will see greater activity in this district.

Parker, August 6.

CALIFORNIA

The number of new oil-wells reported to the State Mining Bureau during the week ended August 4, was six. Only once during the present year has such a low number been reported, the total to date being 673. There is as yet no definite indication of a widespread slackening of development which may be expected from a shortage of casing and other materials. The starting of a well in the Newhall field, which is one of the oldest in the State, is of almost as much interest as the fact that the Midway-Sunset fields started only three wells. The number of wells in the course of drilling and ready for test of water shut-off is 22, which is near the normal number. Wells being deepened or re-drilled, as a means to maintaining or increasing productiveness, number 15, which is also about the usual number. This week the number of wells abandoned is 6, making a total of 211 for the year.

The Bureau is engaged in formulating rules for the abandonment of wells. Before the rules are finally adopted the opinions and criticism of operators will be sought. A draft of the proposed rules may now be obtained at the local offices of the Bureau.

BUTTE COUNTY

(Special Correspondence.)—The Union Crome Co., of San Francisco, has been working several chrome deposits in this neighborhood, and shipping ore of very good grade. One deposit is near Concow, another on the Curtis ranch near Paradise, and another on Little Butte creek. The Butte creek property is working eight men. J. Davis is superintendent.

The Springer mine, near De Sabla, is working 25 men. John Kirchen, of Reno, Nevada, is the owner, and Harry Thompson is superintendent. This is a large quartz property and has been rich enough in the past to work by ground-sluice and placer methods for the pockets it contained. The present method is to work the entire vein on a large scale. The crushing, concentrating, grinding, and cyaniding plant is complete. This is the largest property working in this district.

The Mineral Slide Mining Co., controlled by Moody and Cohn, is under option to J. C. Cowan and associates of Salt Lake City, who have been working 24 men. Machine-drills have been installed, also a screen washing-plant. Some good ground has been opened up and it is likely that this property will be good for a number of years. Considerable money has been spent in development and equipment and it is thought that the option will be exercised. The property is shut-down this month due to trouble over water rights but is expected to resume operations in about two weeks. This property is on Little Butte creek, and adjoins the Lucky John mine on the west.

The Blue Hog drift mine near Magalia, is working six men. Los Angeles people are the owners.

The Kirby drift mine also is working. Seven men are employed. This property is near De Sabla. J. C. Kirby is the owner. A new shaft is being sunk.

The Paradise irrigation district is rushing work on the dam on Butte creek above Magalia, and also on the ditch system. Modern engineering methods are used and a ditching-machine is digging the main ditch. This system will supply water to an extensive district, for irrigation and domestic purposes.

It is to be regretted that here again the farming and mining interests clash, as the laws do not allow water to be sold for mining purposes. Some mining land, absolutely useless for farming, was included in this district, and the owners have the privilege of paying taxes on the improvement without being allowed to have the water.

Prospectors are active in this district, looking for chrome, manganese, cinnabar, and other ores. It is possible that this activity will result in the discovery of some deposits of value.

Paradise, August 7.

DEL NORTE COUNTY

(Special Correspondence.)—E. P. Moore, of Berkeley, who is managing chrome mines in the Low Divide district, has 3000 tons ready to ship. In one place two men are breaking 50 tons daily. The 88 tons of chrome, which was on the steamer 'Del Norte' when it was beached, was from this property.

George Hyde has located 15 chrome claims on a line between Low Divide and the north fork of Smith river.

Crescent City, August 8.

INYO COUNTY

(Special Correspondence.)—The Round Valley Tungsten Co. began the use of electric power July 31, when the mill was started up. An innovation has been introduced here in the form of a ball-mill for grinding—replacing rolls. The result of this radical change is being watched with great interest by the managers of the other tungsten mines of the district, for if the experiment should prove satisfactory others will put in ball-mills. Mr. Shapley is manager for the Round Valley company.

The Standard Tungsten company has bought a Marcy ball-mill, and a new flow-sheet is being worked out. Preparations are being made for the new machinery, which is expected to arrive within 30 days. The mill will be increased to 150 tons daily capacity. L. E. Porter is general superintendent and H. R. Potter is mill superintendent.

Bishop, August 3.

NEVADA COUNTY

(Special Correspondence.)—The 20-stamp mill at the Allison Ranch is running on good ore from the 400-ft. level. The mine has been opened to an approximate depth of 625 ft., and four stopes are being worked. Unwatering of the lower levels continues, and deep mining will begin in the early fall. C. K. Brockington is general manager.

Walter W. Bryne, representing Salt Lake capital, has completed arrangements for work at the Franklin and Ford properties, adjoining the Allison Ranch. Orders have been placed for an electric pump, a 75-hp. hoist, and a 10-drill compressor. Old records show that when last operated the mines produced ore averaging around $160 per ton.

The North Star Mines Co. has completed raising of the county road 10 to 12 ft., forming a dam around the cyanide plant, for the impounding of tailing and calculated to have a capacity for at least two years. Extensive work continues in the lower workings of the Central mine, and the Central mill is running at full capacity. The company is reported to be considering the acquisition of new holdings.

In the Rough and Ready district, controlled by King C. Gillett and associates of Los Angeles, work has been temporarily suspended at the California. New machinery has been installed and much development work done. It is stated work will be resumed when cost of materials is lower.

Work has begun at the Delhi mine, above Nevada City, recently taken under lease and option by A. A. Codd, of Reno, Nevada. The boarding-house, bunk-house, office-building, and other structures are being moved to near the portal of the lower adit.

Grass Valley, August 12.

PLUMAS COUNTY

(Special Correspondence.)—The Robinson mine, in Granite basin, has passed from the United States Exploration Co. to a new company—the Golden Feather Mines Co. The stock is now assessable and the first assessment has been levied and work will be resumed. There is stated to be a good shoot of ore on the 80-ft. level. On the 320-ft. level the ore-shoot was not found where expected and driving was done, which resulted in finding what is thought to be the north end of the shoot. The vein at that place was 10 in. wide and assays $27. Further development will be done.

Oroville, August 8.

SHASTA COUNTY

(Special Correspondence.)—Construction of a five-mile flume has been started by the Victor Power & Development Co. to supply power to the Midas mine and mill, near Knob. It will connect with a five-mile ditch from Brown's creek. Two Bessemer gas-engines have been installed to operate the generating plant during the time when water is not available. Unwatering of the shaft by bailing is proceeding and at about 600 ft. the pumps will be recovered and started. Improvements will also be made in the mill, and development of the Gold Hill mine will proceed from the Midas shaft.

It is reported that the Arnstein interests, now operating the Bully Hill group, near Copper City, have decided to build a flotation plant, of which the first 100-ton unit is to be erected before winter. The company is shipping 50 tons of ore daily to the Mammoth smelter, and has started work from the Rising Star shaft. L. C. Monahan is superintendent.

The 150-ton flotation plant of the Afterthought Copper Co. will be given a trial about September 1. Completion of the plant has been delayed by slow shipment of material from the East, but all equipment is now in place. Mine operations are being increased and the bins are filled with ore.

The Shasta Belmont Co. has arranged for the extension of the lower adit and for the driving of a cross-cut to open a small rich vein. There is some good ore in the old adit-level and the management is contemplating shipments. The mine lies near the Bully Hill and Arps properties, and the ore is noted for a high silver content in addition to good copper average, besides zinc and gold. W. E. Casson, of Carson City, Nevada, is manager.

In the French Gulch district the Sybil is being worked by the Shasta Hills company, and arrangements have been made for the development of the Black Tom. Prospecting is active.

Redding, August 7.

COLORADO

During the calendar year 1916, Colorado produced $19,148,320 in gold, 7,622,000 oz. silver, 70,358,000 lb. of lead-bullion, and leaded zinc oxide, 8,940,000 lb. of copper and 129,300,000 lb. of zinc in spelter and oxide. The total value of the production was $48,550,318.

BOULDER COUNTY

(Special Correspondence.)—Around Ward the busy mining season has commenced. The big mines of the district are free from the snow that closes many of them each winter. Ore shipments are increasing. On the Texas, work has been started. C. J. Walter is in charge.

The National Tungsten Co. will complete its mill this month. The property has been a steady producer. T. E. McKinney, manager of the Lula B. properties, has arrived from New York and will make extensive improvements. The Oronogo machinery and shaft-house has been moved to the Snowbound, on Gold hill. Fred Gushe will be in charge. The Big 5 is putting in a ball-mill and making other improvements. The Boulder Natural Gas Co. has incorporated and will operate in Boulder county. O. R. Whitaker, of Denver, has been appointed to investigate the smelter-trust on behalf of Colorado miners. The Consolidated Leasing Co., of which C. E. Kahler, of Columbus, Ohio, is president, is at work. When Mr. Kahler returns, work will be started on the mill. The Boulder Tungsten Production Co.'s refinery is said to be the largest and most complete tungsten-mill in America. Tungsten ores are coming here from other States, for treatment, and all the mines and mills in this district are preparing to ship ore here for treatment. The National Tungsten Co. has been overhauling its mill, which is now in good condition.

The Smelter and Ore-Sales Committee, of Boulder, is making good progress and will soon have its report ready for the public.

Eldora, August 4.

ROUTT COUNTY

(Special Correspondence.)—The Routt county oil-shale deposits will probably add another industry to the producti e resources of this county. Thos. J. Williams, and several others, have organized the Tow Creek Oil & Asphalt Refining Co. The company has taken an option on 2500 acres of land, upon which other minerals than oil-shales have been found, including an 8-ft. vein of paraffine. Fire-clay of high grade has also been opened.

Eldora, July 28.

TELLER COUNTY

(Special Correspondence.)—Machinery has been placed and necessary arrangements have been made to sink the shaft of the Wild Horse mine, at Midway, the purpose of which is to explore undeveloped territory. The work is being done on company account by the United Gold Mines Co., J. J. Russell, superintendent. The ore of the Cresson Consolidated mine has a value of $500 per ton without sorting.

The new Portland custom mill at Victor is about completed. It will have an initial capacity of 1500 tons.

Ben Baud of Victor has a contract to sink 400 ft. in the Rose Nichol shaft at Victor. In the Longfellow, Mr. Southerland has cut a station at 350 ft. and has a contract to sink an additional 250 feet.

Edwin Gaylord, lessee of the Forest Queen, is installing two new boilers, and expects to produce a carload of ore daily.

It is reported that Tom Kavanaugh has taken over the Worcester mill and lease on the Ruby mine and will treat the ore from the La Fayette dump.

Lessees on the Little May, on Bull Cliffs, has opened up ore on this long neglected property.

Dick Curry has secured a lease on the Star of Bethlehem mine at Independence, and is sinking a new shaft.

Victor, August 2.

MONTANA

Jefferson County

(Special Correspondence.)—A 14-ft. orebody is being mined at the Mt. Washington mine near Wickes. The ore is lead-silver and is worth $35 per ton. The strike was made in an adit about 3000 ft. from the portal and 700 ft. below the surface. The stope is three sets wide. The Mt. Washington is controlled by Duluth, Minnesota, interests.

The Gregory slag dump, which has remained undisturbed for more than 30 years, is being hauled in wagons to the railroad at Wickes and shipped to the smelter. The slag carries silver and lead. The ores were smelted with charcoal. Recoveries from the slag shipped average near $20 per ton.

Alta Mountain, at Crobin, which has produced from the Alta mines $32,000,000, is the scene of much activity. The Boston & Corbin company is mining and milling a large tonnage of copper ore. The Alta Mines Co. is developing the old Alta property and a new discovery has been made of silver-lead ore. The strike was made in a blind vein and is 6 ft. wide. The Butte & Corbin property is also producing shipping ore.

At Clancy a new corporation has been formed to work the Dan Tucker mine. New machinery has been installed and work is progressing.

Clancy, August 2.

Lewis and Clark County

(Special Correspondence.)—The Marysville district is mining and milling more ore than for many years. The Barnes-King company is working the Shannon, a new property; also the old Piegan-Gloster group. The St. Louis company is operating the Drumlummon mine. Ore is being drawn from the old stopes and is run through a 25-stamp electric-driven mill with cyanide plant having a capacity of 100 tons per day. Lessees are mining ore from a claim producing lead-silver ore at Bald Butte, and a new compressor is being sought by the Bald Butte Co. The Nile mine in Towsley gulch is shipping.

Marysville, August 3.

NEVADA

Esmeralda County

(Special Correspondence.)—The Goldfield Consolidated has purchased the Surc-ease gold mine, in the Big Bend district, near Oroville, from Phebe Bros., of Oakland, California. Work has begun under the management of Robert J. Burgess. Ore of good grade is said to be exposed. The Goldfield Consolidated is also negotiating for other mines in California. In an area recently leased from the Jumbo Extension Co., the Consolidated has opened a deposit of rich ore and is sending it to the cyanide-section of the mill. From the 450-M winze on the 600-ft. level of the Laguna, flotation-ore assaying $133 per ton is being produced. There is still difficulty in marketing of flotation concentrate, due to congestion at the custom smelters. It is reported the company is considering improvements in its refining-plant for treatment of the flotation concentrate, to be followed by placing of the second flotation-unit in operation on ore from the Atlanta, Kewanas, and other mines.

The Atlanta company has announced that the winze from the 1750-ft. level will be enlarged and sunk to 2000 ft. to open a copper-gold vein and to determine whether or not the orebodies persist beyond the shale. The company is making regular and profitable shipments to the sampler of the Western Ore Purchasing Company.

The Western Ore Purchasing Co. has filed suit for $9143 against the Florence Goldfield Co. alleging this amount was secured from plaintiff through defendant having salted an ore shipment. The complaint recites that 37 dry tons of ore was purchased by plaintiff on fraudulent samples submitted by the Florence Goldfield Co., indicating the shipment to be worth $9226, but that when reduced the total value of ore was found to be only $83.11.

The winze from the north drift on the 880-ft. level of the Jumbo Junior has exposed ore assaying around $100 per ton in the area leased from the Kewanas company. A raise at this point also shows good ore. A shipment will be made. The Kewanas company has started cross-cutting near this point and is exposing seams of rich ore, in which gold predominates, with a little copper showing.

Goldfield, August 14.

Nye County

(Special Correspondence.)—An interruption in the power-service the past week, suspended station work on the 1680-ft. level of the Victor, and dewatering the shaft has superceded the station work. The development at the Tonopah Extension consisted of 55 ft. in No. 2 shaft, and 114 ft. in the Victor, on the 1440 and 1540-ft. levels. Stoping on the 1260-ft. level of the No. 2 shaft continues on the Murray vein, which shows 8 ft. of ore; also on the O. K. and Merger veins, each of which has a 6-ft. face of ore. The past week 2380 tons of ore was sent to the mill.

The Rescue-Eula sent 188 tons to the West End mill the past week from the 950 and 1050-ft. levels.

The Tonopah Belmont sent 2744 tons of ore to the mill the past week. On the 8th level the east drift on the Rhyolite vein is in fair ore 2 ft. wide. A cross-cut has been started between the 8th and 9th levels in the hanging wall of the Belmont vein, toward the extension of the Shaft vein. On the 9th level west on the Rescue vein, there is 3 ft. of vein material. The Shoestring vein at present averages 2 ft. of fair ore.

The clean-up of the West End for the last half of July consisted of 33 bars valued at $54,000. Development in the Ohio shaft is progressing, and the West End shaft continues to supply its usual tonnage. The Halifax Tonopah has resumed stoping.

The development work of the Tonopah Mining Co. the past week consisted of 88 ft. in the Silver Top, and 35 ft. in the Sandgrass. The raise from the 540-ft. level of the Silver Top intersected the West End-McNamara vein, which shows 2 ft. of a good grade of ore. The new stope on the 1150-ft. level on the upper Sandgrass vein shows ore of fair value, and 1800 tons was sent to the mill the past week.

Tonopah, August 6.

NEW MEXICO

Grant County

(Special Correspondence.)—All of the new Phelps-Dodge buildings at Tyrone are completed. The situation which these buildings occupy has been named the New Town, and is one mile from where the Tyrone post-office was situated. The six Phelps-Dodge buildings are an up-to-date general store, and warehouse, the general office of the Burro Mountain branch, the high school, the freight warehouse, and a beautiful little station and post-office combined. All of these buildings are of stucco work, in the Spanish style, and the stucco of each building is tinted a different hue.

During June the No. 1 and 2 divisions of the mill produced 44,199 tons of ore averaging 1.9% copper. The average daily production was 1542 wet tons.

The corporation is installing two new General Electric storage-battery locomotives, Type LSB. Each motor has a rated speed of 4 miles per hour at full load. The power is furnished by a battery of 63 Edison cells at 60 volts potential. These locomotives will operate on a 20-in. gauge track. Three batteries are used in the haulage system, one of the batteries, fully charged, being kept in spare. Thus, when the battery on one of the locomotives has been run down, the locomotives will be driven to the charging-station where the engineer will replace the run-down battery with the spare and the former will be recharged and in its turn become the spare. Both locomotives will operate in the No. 3 division.

Tyrone, July 27.

SOCORRO COUNTY

(Special Correspondence.)—The tonnage for the Mogollon district during July amounted to approximately 12,000 tons, which yielded nearly 6000 lb. of gold and silver.

The first installment for a diamond-drill has arrived at the Mogollon mines.

The Oaks Co. continues to ship daily from the Maud S. mine. The property had not been worked for years, but the conditions found by the new operators seem to be satisfactory.

Mogollon, July 31.

OREGON

JACKSON COUNTY

(Special Correspondence.)—W. P. Chisholm of the Mercur and Little Jean claims, 12 miles north of Gold Hill, is installing a 12-pipe Johnson and McKay furnace on his property. The machinery is from the Joshua Hendy Iron Works, San Francisco. The claims are in a wooded district, and plenty of water is available. It will be in operation in less than 30 days. R. H. Spencer and associates, of Portland, who recently acquired the 73 group of quicksilver mines near the Chisholm group has a 3-pipe Johnson and McKay furnace, which is operating successfully. They will add 9 retorts as soon as they can be had. Other owners have ordered the same furnace, to be installed as soon as possible.

The Red Hill gold quartz mine, in the Jump-off Joe district, five miles east of the Greenback mine, and three miles from Three Pines station on the Southern Pacific railway, is being re-opened. It is owned by T. N. Anderson of Gold Hill, and Lester Lord of Danberry, Nebraska. The property was originally discovered and operated by Joseph Dysert, who mined a large amount of rich ore, which he reduced with an arrastra. The vein is at contact of red porhpyry and serpentine. The main drift, in 500 ft. on the vein, is being re-timbered, and ore shipments will be made. The power-line of the C. & O. Power Co. crosses the premises.

Samuel Carpenter and Harry Hocksworth, of Medford, are shipping antimony ore regularly to New Jersey from their Applegate mine. The ore is running 50% antimony with some gold. This property is 17 miles south-west of Jacksonville. The vein is from 18 in. to 4 ft. wide, with 400 ft. of drift.

The United Copper Co. has struck ore on its property that runs 35% copper. These mines are in the Greenback district in the north-west part of Jackson county at the head of Slate creek. The ore deposit is in a fissure in andesite. The development made last season exposed in surface cuts a vein which runs 5% copper and $2 gold. The company operates a mill and concentrates the sulphides, which are shipped. The plant is being enlarged. The company is constructing a road from the mine to Evans Creek valley leading to Rogue river, 9 miles west of Gold Hill.

E. H. Richards and A. W. Bartlett, of Grants Pass, recently leased from M. G. Womack, of Medford, and M. A. Carter, of Ashland, a vein of molybdenite near Jacksonville. The lessees are developing the prospect.

Gold Hill, August 6.

JOSEPHINE COUNTY

(Special Correspondence.)—More chrome deposits have been located in Fiddlers gulch, six miles west of Kerby, by Fred Hart and T. P. Johnson. Dave Bauer and J. W. Bigelow also are developing chrome down the Illinois river.

The Neils Success gold mine is in operation under the management of R. J. Firth, of Seattle.

Kerby, August 8.

Personal

Note: The Editor invites members of the profession to send particulars of their work and appointments. This information is interesting to our readers.

B. L. THANE is at Seattle.

W. L. HONNOLD is at Los Angeles.

F. H. MITCHELL has returned to Nevada.

WILLIAM B. FISHER is at Salt Lake City.

D. M. RIORDAN has returned from New York.

A. B. ROGERS was here, from Los Angeles, last week.

JOHN C. RHOADS is in Shasta county, at Copper City.

H. F. SCHEERER is in San Francisco from Tucson, Arizona.

M. B. YUNO sails by the 'Korea' on August 22 for Hongkong.

ROBERT E. CRANSTON has gone to Colorado, to return in two weeks.

RENSSELAER TOLL will be at Kingman, Arizona, until August 20.

DON C. BILLICK has returned to Berkeley from Latouche, Alaska.

SCOTT TURNER is at Lima, Peru, on his return from La Paz, Bolivia.

W. W. DEGGE has returned to Boulder county, Coloroda, from the East.

EUGENE C. SNEDAKER has received a commission in the Engineer Corps.

R. H. CLARKE is chemist with the Chino Copper Co., at Hurley, New Mexico.

HORACE V. WINCHELL sailed from Yokohama on August 24 on his return from China.

WALTER G. PERKINS has returned from visits to Garfield, Utah, and Hayden, Arizona.

CHESTER A. THOMAS, manager for the Yukon Gold Co., was in San Francisco last week.

A. F. HALLETT has been appointed chemist to the Khirgiz Mining Co., at Ekibastus, in Siberia.

MANUEL EISSLER, the author of many metallurgical textbooks, is here, on his return from China.

H. R. BOSTWICK is in Japan. He does not expect to return to San Francisco until early in November.

HERBERT C. ENOS is at Chihuahua, Mexico, in charge of properties for the Compania de Minerales y Metales.

C. B. CLYNE has been engaged to design and build a flotation plant for the Demming Mines Co., at Murphy, Idaho.

LIVINGSTON WERNECKE, geologist for the Treadwell company, has gone to Wrangal to examine mines in that vicinity.

TATSURO OTAGAWA has been transferred from the Kyoto Imperial University to the Northwestern Imperial University of Japan.

ORVIL R. WHITAKER has been appointed to take charge of the technical part of the official smelter and ore-sales investigation now being made in Colorado.

CHARLES R. FETTKE is engaged in making a geological survey of Clay and Macon counties, Tennessee, with reference to the possible occurrence of oil and gas.

ROBERT T. HILL has opened an office at 702 Hollingsworth Bdg., Los Angeles, having completed his report upon the geology of southern California for the U. S. Geological Survey.

GEORGE H. CLEVENGER and WILLIAM G. RAYMOND have declined the presidency of the Colorado School of Mines. Another candidate is ORVILLE HARRINGTON, editor of the School magazine.

THE METAL MARKET

METAL PRICES

San Francisco, August 14

Aluminum-dust (100-lb. lots), per lb.	$1.00
Aluminum-dust (ton lots), per lb.	$0.95
Antimony, cents per pound	15.50—18.00
Electrolytic copper, cents per pound	29.50
Pig lead, cents per pound	12.25—12.50
Platinum, soft and hard metal, per ounce	$105—111
Quicksilver, per flask of 75 lb.	$115
Spelter, cents per pound	10.75
Tin, cents per pound	60
Zinc-dust, cents per pound	20

ORE PRICES

San Francisco, August 14

Antimony, 50% metal, per unit	$1.20
Chrome, 40% and over, f.o.b. cars California, per unit	$0.60— 0.70
Magnesite, crude, per ton	$8.00—10.00
Tungsten, 60% WO_3, per unit	26.00—30.00
Molybdenite, per unit for MoS_2 contained	$40.00—45.00
Manganese, 45% (under 35% metal not desired), cents, unit	33—37

Manganese prices and specifications, as per the quotations of the Carnegie Steel Co. schedule of prices per ton of 2240 lb. for domestic manganese ore delivered, freight prepaid, at Pittsburg, Pa., or Chicago, Ill. For ore containing

	Per unit
Above 49% metallic manganese	$1.00
46 to 49% metallic manganese	0.98
43 to 46% metallic manganese	0.95
40 to 43% metallic manganese	0.90

Prices are based on ore containing not more than 8% silica nor more than 0.2% phosphorus, and are subject to deductions as follows: (1) for each 1% in excess of 8% silica, a deduction of 15c. per ton, fractions in proportion; (2) for each 0.02% in excess of 0.2% phosphorus, a deduction of 2c. per unit of manganese per ton, fractions in proportion; (3) ore containing less than 40% manganese, or more than 12% silica, or 0.225% phosphorus, subject to acceptance or refusal at buyer's option; settlements based on analysis of sample dried at 212° F., the percentage of moisture in the sample as taken to be deducted from the weight. Prices are subject to change without notice unless specially agreed upon.

Chrome prices are causing considerable comment among both buyers and consumers. A San Francisco buyer quotes today the following schedule of prices:

35 to 37%	50c. per unit
37 to 40%	55c. " "
40 to 44%	60c. " "
44 to 48%	65c. " "
48% and over	70c. " "

These quotations are f.o.b. California and Oregon main-line railroad points. A certificate of sampling and analysis by a reliable chemist must be attached to bill of lading.

EASTERN METAL MARKET

(By wire from New York)

August 14.—Copper is dull and quotations are nominal at 27.50 to 27c. Lead is lifeless, price 10.87 throughout the past week. Zinc has stood at 8.75c. all week. Platinum remains unchanged at $105 for soft metal and $111 for hard.

TIN

Prices in New York, in cents per pound.

Monthly Averages

	1915	1916	1917		1915	1916	1917
Jan.	34.40	41.76	44.10	July	37.38	38.37	62.60
Feb.	37.23	42.60	51.47	Aug.	34.37	38.88	....
Mch.	48.76	50.50	54.27	Sept.	33.12	36.66	....
Apr.	48.25	51.49	55.63	Oct.	33.00	41.10	....
May	39.28	49.10	63.21	Nov.	39.50	44.12	....
June	40.26	42.07	61.93	Dec.	38.71	42.55	....

SILVER

Below are given the average New York quotations, in cents per ounce, of fine silver.

Date			Average week ending	
Aug. 8	82.50		July 3	77.98
" 9	82.87		" 10	78.70
" 10	82.87		" 17	80.62
" 11	82.87		" 24	79.12
" 12 Sunday			" 31	78.20
" 13	82.87		Aug. 7	80.50
" 14	83.75		" 14	82.95

Monthly Averages

	1915	1916	1917		1915	1916	1917
Jan.	48.85	56.76	75.14	July	47.52	63.06	78.92
Feb.	48.45	56.74	77.54	Aug.	47.11	66.07	....
Mch.	50.61	57.89	74.13	Sept.	48.77	68.51	....
Apr.	50.25	64.37	72.51	Oct.	49.40	67.86	....
May	49.87	74.27	74.61	Nov.	51.88	71.60	....
June	49.03	65.04	76.44	Dec.	55.34	75.70	....

Silver is quoted today in New York at 83.75c. per oz.—a higher price than at any time since 1892, when it was at 87 cents.

A new ruling has been made by the Treasury Department regarding the purchase of gold-silver bullion. Heretofore the mints bought gold-silver bullion on the basis of the fineness in gold and silver, no matter what the ratio of gold to silver might be, provided the bullion contained at least 20% of the precious metals, that is, 200 fine. The Government assay offices, however, did not buy gold-silver bullion unless the bullion not only was 200 fine but that the value of the gold contained was at least ten times as great as the silver in the bullion. This restriction has now been removed, according to a press dispatch—no official notification of the order having been received at the San Francisco Mint up to August 14. The Government assay offices may, therefore, accept and pay for any bullion containing gold and silver, if the amount of the precious metals constitutes at least 20% of the weight of the bullion offered.

ZINC

Zinc is quoted as spelter, standard Western brands, New York delivery, in cents per pound.

Date			Average week ending	
Aug. 8	8.75		July 3	9.32
" 9	8.75		" 10	9.20
" 10	8.75		" 17	9.06
" 11	8.75		" 24	8.60
" 12 Sunday			" 31	8.75
" 13	8.75		Aug. 7	8.75
" 14	8.75		" 14	8.75

Monthly Averages

	1915	1916	1917		1915	1916	1917
Jan.	6.30	18.21	9.75	July	20.54	9.90	8.98
Feb.	9.05	19.99	10.45	Aug.	14.17	9.03	....
Mch.	8.40	18.40	10.78	Sept.	14.14	9.18	....
Apr.	9.78	18.62	10.20	Oct.	14.05	9.92	....
May	17.03	16.01	9.41	Nov.	17.20	11.81	....
June	22.20	12.85	9.63	Dec.	16.75	11.28	....

The United States Geological Survey, from returns representing 99% of the output, estimates that the production of spelter during the first six months of 1917 was 364,000 short tons, as compared with 351,000 short tons during the last half of 1916. Stocks on hand are estimated at 33,000 tons, as compared with 17,600 at the beginning of the year. A large number of retorts, about 35,000, including 14 complete plants, were reported idle June 30, in addition to the retorts engaged in refining prime Western metal and in re-distilling zinc ashes.

COPPER

Prices of electrolytic in New York, in cents per pound.

Date			Average week ending	
Aug. 8	28.00		July 3	32.25
" 9	28.00		" 10	31.50
" 10	27.75		" 17	30.29
" 11	27.50		" 24	27.04
" 12 Sunday			" 31	27.25
" 13	27.50		Aug. 7	28.46
" 14	27.00		" 14	27.62

Monthly Averages

	1915	1916	1917		1915	1916	1917
Jan.	13.60	24.30	29.53	July	19.09	25.66	29.67
Feb.	14.38	26.62	34.57	Aug.	17.27	27.03	....
Mch.	14.80	26.65	36.00	Sept.	17.69	28.28	....
Apr.	16.64	28.02	33.16	Oct.	17.90	28.50	....
May	18.71	29.02	31.69	Nov.	18.88	31.95	....
June	19.75	27.47	32.57	Dec.	20.67	32.89	....

LEAD

Lead is quoted in cents per pound, New York delivery.

Date			Average week ending	
Aug. 8	10.87		July 3	11.57
" 9	10.87		" 10	11.25
" 10	10.87		" 17	10.98
" 11	10.87		" 24	10.29
" 12 Sunday			" 31	10.55
" 13	10.87		Aug. 7	10.87
" 14	10.87		" 14	10.87

Monthly Averages

	1915	1916	1917		1915	1916	1917
Jan.	3.73	5.95	7.64	July	5.59	6.40	10.93
Feb.	3.83	6.23	9.01	Aug.	4.67	6.28	....
Mch.	4.04	7.26	10.07	Sept.	4.62	6.86	....
Apr.	4.21	7.70	9.38	Oct.	4.62	7.02	....
May	4.24	7.38	10.29	Nov.	5.15	7.07	....
June	5.75	6.88	11.74	Dec.	5.34	7.55	....

QUICKSILVER

The primary market for quicksilver is San Francisco, California being the largest producer. The price is fixed in the open market, according to quantity. Prices, in dollars per flask of 75 pounds:

Week ending

Date				
July 17	105.00		July 31	110.00
" 24	110.00		Aug. 7	115.00
			" 14	115.00

Monthly Averages

	1915	1916	1917		1915	1916	1917
Jan.	51.90	222.00	81.00	July	95.00	81.20	102.00
Feb.	60.00	295.00	126.25	Aug.	93.75	74.50	....
Mch.	78.00	219.00	113.75	Sept.	91.00	75.00	....
Apr.	77.50	141.60	114.50	Oct.	92.90	78.20	....
May	75.00	90.00	104.00	Nov.	101.50	79.50	....
June	90.00	74.70	85.50	Dec.	123.00	80.00	....

ORES

Tungsten: Asking prices are as high as $23 per unit for 60% concentrates, with $25 to $26 quoted for some grades. Imports have been larger than expected, or 3823 gross tons for the 12 months ended June 30, 1917. Ferro-tungsten is quoted at $2.25 to $2.50 per lb. of contained tungsten, with both domestic and foreign demand excellent.

Molybdenum and antimony: Molybdenum is quoted unchanged at $2.10 to $2.20 per lb. of MoS_2 in 90% material with offerings readily absorbed. Antimony ore has sold at $2.25 per unit for high-grade material.

Eastern Metal Market

New York, August 8

The entire metal market is inactive and slow, with prices practically nominal in almost every case. Government uncertainty as to purchases and prices continues to halt the trade and to act as a depressing influence. The whole trade is nearly demoralized.

Copper is lower but firm.

Tin is quiet but steady.

Lead is dull and almost nominal.

Zinc is stagnant and unchanged.

Antimony continues inactive.

Aluminum is quiet at unchanged prices.

The steel market continues upset by prospective price-fixing and cost-finding endeavors. Repeated outgivings from Washington of sweeping action intended in the Government's dealings with steel makers and the reiteration of the President's call for a one-price-to-all policy have only added to the uncertainty that is holding back all iron and steel markets. Price changes in the main have been narrow. Buying for the Allies is held up by the steel cost inquiry and more time for the latter will be necessary than was expected. Pig-iron output in July fell off further, due to coke shortage, heat, and humidity. The decrease from June was 1200 tons per day, the same as the June decline from May.

COPPER

The features of the copper market are the strong fundamental conditions and the continued dullness, two elements not usually associated. Were it not for uncertainties at Washington, business would probably be of larger proportions, for certainly the demand is latent somewhere. With the amount of 500,000,000 lb. or more bought by the Allies for delivery to July 1, 1917, practically all of which has been shipped, and with the confident prospect of more copper needed from that quarter, to say nothing of this country's needs, the future of the market is generally regarded as bright; but no one is willing or anxious to buy when the price-situation is where it is. Rumors as to what the Government will pay continually crop up, the latest one placing it at 22 to 23c. per lb. No one knows, and perhaps nothing will be known until the cost investigation is completed, which will be at least two or three weeks yet. The market is a slow and nominal one and has gradually receded again the past week until yesterday both Lake and electrolytic were quoted nominally at 28c., New York. Re-sale lots, the withdrawal of which was the main reason for the market's advance last week, again appeared on August 3 but no sales were reported. Efforts to excite interest in the market about that time were fruitless as reported by some traders. A surprising incident of the week was that the exports of copper for June were only 28,198 tons as compared with 38,512 tons per month for the previous five months. The London quotations were unchanged yesterday at £137 for spot electrolytic and £133 for futures.

TIN

One element of market-uncertainty seems in a fair way to be removed. According to the revenue bill as reported to the Senate last Monday, there is to be no duty on tin and the trade now inclines to regard the imposition of a duty as unlikely. The market the past week has been quiet with not much business reported. Late last week off-grade brands were sold at a substantial discount for prompt delivery, these sales amounting to more than those of Straits tin. On one day business was held up because of the non-receipt of cables. On August 3 there were sales of 100 tons of futures. Early this week the market was quiet on the surface, though one seller reported the disposition of 100 tons of futures, but no other sales were noted, and spot Straits was nominal. The principal activity this week has been in shipments from the East. Arrivals up to August 7 were 510 tons, but the quantity afloat has not been officially reported so far this month and it would not be surprising if this information is to be kept from the public, although no announcement to that effect has been made. The quotation yesterday was 63.62½c., New York, for spot Straits. Yesterday the London quotation for spot Straits was £247, a decline of £1 from that of a week ago.

LEAD

Whether the Government has really placed its order for 8000 tons of lead to cover its August requirements is not known, but many believe that this quantity has been bought at 8c. per lb., as in the July purchase, but that no announcement will be made. The Government's needs dominate the entire future of this market. If they are to continue at 8000 tons per month, they will be sufficient to take up the ordinary slack in the market and absorb about one-quarter of the country's yearly output. A strong market is looked for with production keeping up at full pace. The extent to which transportation and labor troubles may hinder output in the future is a possible important factor. The market the past week has been quiet but firm, and almost nominal at 10.87c., New York, and 10.75c., St. Louis. Labor troubles, which were prominent last week, seem to have been settled in many cases. The leading interest still quotes 11c., New York, but sales have been made below this, and we therefore quote the market at 10.87c., New York. For the present the demand is small.

ZINC

Surprise has been general in the trade the last week over the unexpected action of the Government as represented by the Zinc Committee in asking for bids on 11,500,000 lb. of grade C zinc, instead of fixing a price beforehand and ordering the metal, as was done in the recent purchase of grades A and B zinc. The reason for this move has not been clear. The bids are to open on Friday and considerable speculation exists as to the price to be paid. As high as 9 to 9.50c., St. Louis, is generally expected. Grade C zinc is nearly the same as prime Western except that it is a little better as to some impurities. The report of the Geological Survey for the half yearly consumption of spelter, soon to be issued, is awaited with interest as shedding some light on supply and demand. The market continues in its dull and lifeless condition and quotations are nominally unchanged at 8.50c., St. Louis, or 8.75c., New York, for prime Western for early delivery or as far ahead as September inclusive. There is no demand and sales are of no volume, consumers being unwilling to commit themselves under present uncertain conditions caused by the Government's attitude as to purchasing and price fixing. Those willing to quote beyond September ask from 8.67 to 8.75c., St. Louis, or 8.87 to 9c., New York. Well-informed authorities confidently anticipate heavy buying in the not distant future, accompanied at the same time by unusual railroad congestion due to war-traffic.

ANTIMONY

Chinese and Japanese grades are obtainable at 15 to 15.50c., New York, duty paid, but there is little demand with few sales.

ALUMINUM

No. 1 virgin metal, 98 to 99% pure, is quoted at 50 to 52c. per lb., New York, but demand is very low and the market inactive.

Book Reviews

THE FRONTIERS OF LANGUAGE AND NATIONALITY IN EUROPE. By Leon Dominian. Published for the American Geographical Society of New York, by Henry Holt & Co., 1917. Royal 8vo., pp. XVIII, 375; pl. ix.; fig. 67.

The outbreaks in Mexico some years ago created such impossible conditions for the mining industry that many mining engineers cast about for other fields of usefulness. The analogous Germanic upheaval has produced similar and world-wide effects, but has differed from the Mexican in sometimes supplying opportunities for mining engineers in lines not immediately connected with mining. The book named above is the result of both these causes. Its author, long a member of the American Institute of Mining Engineers and in active practice in Mexico, became connected, after the Mexican disorders, with the American Geographical Society and turned his early training in Constantinople, his extensive knowledge of European and neighboring Asiatic languages and conditions, his world-wide subsequent experience, and his engineer's disposition to find the solution of a stiff problem, into the prosecution of a study which might indicate a way out of the European difficulties. The book, therefore, is shaped by a desire to emphasize such adjustment of political boundaries as will make for stable conditions. The body of the work sets forth the facts of language and nationality; the concluding pages state the author's solution.

After an introduction by Madison Grant of the Geographical Society, the author summarises the ancestral races of Europe and the sources of their languages. He then takes up the boundaries of the French and Germanic languages in Belgium, Luxemburg, Alsace-Lorraine, and Switzerland, which constitutes one phase of the problem. The borderlands of the Italian language afford a second and quite individual one. The Scandinavian and Baltic languages furnish a third. The Slavic question with its Polish, Bohemian, Moravian, Slovakian, Hungarian, and several Balkan ramifications, supplies a fourth. The geographical case of Turkey and its various peoples makes a fifth and concluding one. But brief and incidental mention is made of the peoples of central Russia, White Russians, Great Russians, and Little Russians, although the recent evidence of internal disaffection have brought the last two prominently to the fore, a condition of things not easy for Mr. Dominian to foresee. No mention is made of Spain, Portugal, or the peoples of Great Britain, as the boundaries of the last three were probably not believed to be immediately involved in post-bellum settlements. Still, Portuguese colonies bulk large in that extensive area, euphemistically described as Germany's place in the sun, and the same luminary lights up the Moroccan mountains with their prospective copper and iron mines and raises questions which vitally interest Spain.

In the discussion of the various topics, Mr. Dominian adopts first the genetic standpoint so as to make clear the introduction of the several elements of a local population into their present homes. He uses also the statistical treatment based on census reports, and illustrates the relationships by numerous maps and photographs. Many readers will be surprised to learn the complexity of mixtures which the several European States, as at present constituted, have attained. In treating of Turkey and its position at the crossing of the two great inter-continental overland trade-routes, Europe to Asia, and Europe to Africa, the author is particularly well-informed. Altogether the book is supported by wide reading, and is timely, instructive, and of great interest.

The final solution suggested is a grouping of peoples into autonomous states, according to language and national aspirations. A series of buffer-states between France and Germany is advocated, consisting of Belgium, Luxemburg, Alsace-Lorraine, and Switzerland, but no guarantees are mentioned to replace the discredited 'scrap of paper.' The author supports the transfer of the Trentino to Italy; a Serbian union of the old Serbia and the provinces along the north-eastern Adriatic which are of Serbian kinship; the addition of German-speaking Austrians to Germany; and the independence of Bohemia and of Hungary. The author thus quite unconsciously demonstrates what all observers of intelligence now realize, that the colossal cuckold nation of the present war is Austria-Hungary. Mr. Dominian upholds the union of the Rumanians, an autonomous Poland, a similar Finland, and the restoration to Denmark of those areas where Danish-speaking people predominate. A solution of the tough problem of the Turkish empire is somewhat vaguely suggested by the establishment of Christian rule, but no specific mention is made of any one Christian nation as sponsor, or of several. Asia Minor is described as practically a German colony, but every intelligent man now knows that Germany is frankly pagan. Odin and Thor, or, one might just as well say, Jupiter and Vulcan, or Zeus and Hephaestos, have replaced the conception of the supreme deity, as taught by Christ. Nevertheless, in so far as they go, the suggestions are of interest and value. In the end they must hold their own in the contest with dynastic and national selfishness, in order to insure stable conditions.

J. F. K.

LAWS OF PHYSICAL SCIENCE: A Reference Book. By Edwin F. Northrup. J. B. Lippincott Company, Philadelphia, Pennsylvania, 1917. Pp. 210, index. For sale by MINING AND SCIENTIFIC PRESS. Price, $2.

This is an extremely handy book of reference, prepared by a man competent for the task. In fact, the need of such a collection of all the fundamental laws of physics and chemistry, without having to look them up in a number of different volumes, is one that the author himself experienced in his own work in the Palmer Physical Laboratory at Princeton; it was reasonable, therefore, to assume that other men must experience the same need. It covers the field of mechanics, hydrostatics, hydrodynamics, capillarity, sound, heat, physical chemistry, electricity, magnetism, and light; there is also an index, and a bibliography; moreover, the sources of authority are given with each law and the corresponding working formulae. The book is handsomely printed, and is bound most attractively in flexible leather; the dress is in keeping with the character of the contents.

FATS AND FATTY DEGENERATION. A Physico-Chemical Study of Emulsions and the Normal and Abnormal Distribution of Fat in Protoplasm. By Dr. Martin H. Fischer and Dr. Marian O. Hooker. John Wiley & Sons, Inc., New York, 1917. Pp. 155, subject and author index, ill. For sale by MINING AND SCIENTIFIC PRESS. Price, $2.

The title sounds remote from matters of interest to metallurgists, but science is one, and the principles of chemistry are the same to whatever service we yoke them. In this book, moreover, the physiologic application is separated distinctly from the experimental evidence and scientific conclusions derived from work that may be considered equally apart from direct physiologic associations. In other words the question of the physical character of emulsions was submitted to examination, and the relations of emulsions to the colloidal state of dispersion and of colloidal dispersion-media, were searchingly studied. When it is remembered that Dr. Fischer is one of the eminent colloid chemists of the world, and has been selected by Wolfgang Ostwald as his translator and expositor in English, it will be understood that the data presented in this volume possess the merit of coming from a man qualified to deal understandingly with the phenomena under investigation. We have elsewhere drawn heavily upon the researches of Dr. Fischer in connection with an independent study of emulsions in relation to flotation. The significance of the evidence adduced by Dr. Fischer to show that control of an emulsion is possible only when the whole of the dispersion-medium is employed in the emulsification, or that the two phases must be wholly complementary, therefore leaving no room for a residuum, could not be lost sight of by the practitioner of flotation. It suggested immediately a new view of the important economic question of control, not alone of the emulsion but of its rate of breaking in order to function properly in the pulp at the moment of producing the gas-in-water dispersion that we call foam, so that the mineral might be floated. Whether the metallurgist may care to read the final chapters on Fatty Infiltration, Fatty Secretion, Natural and Artificial Milk, and Mimicry of Mucroid Secretion, or not, he will be intensely interested in the first half of the work that deals with the important question of emulsions. The style in which the book is written is so charmingly lucid and simple and the subject matter so intensely interesting to cultured men that we believe no student will be able to lay it down until he has read it to the last page.

ANNUAL CHEMICAL DIRECTORY OF THE UNITED STATES. Editor, B. F. Lovelace. Williams & Wilkins, Baltimore, Maryland, 1917. For sale by MINING AND SCIENTIFIC PRESS. Price, $5.

The enormous development of the chemical industry in this country has been one notable effect of the War, and one that will permanently influence the subsequent life of the nation. The development within two years has surpassed the understanding of the man in the street. We now really have a chemical industry; therefore the appearance of this chemical directory is most timely, and will prove of great value. It is a good performance so far as it goes; it will doubtless improve with the experience of the publishers. We assume that it is what it pretends to be, a directory, intended for the benefit primarily of those who buy it in order to find makers of chemicals and consumers of materials, irrespective as to whether the listed have paid for the privilege or not. It would not be a directory if it made such a discrimination. We mention this because we find some omissions where we have tested it, especially in regard to certain minor products; and it is the minor things that are the most evasive. Everyone knows where to look for Sherwin-Williams when he is interested in paints, although they are not included in this directory (why?), but it might be harder for him to guess that the Krebs Pigment & Chemical Co., for instance, are buyers of zinc and barite with which to make lithophone—and they are not in the book either. It is, therefore, not complete; it is useful, nevertheless, and we hope it will be still better next year than this.

REPORT OF THE ROYAL ONTARIO NICKEL COMMISSION, with appendix. Printed by order of the Legislative Assembly of Ontario. Toronto, Canada, 1917.

We have already given extended notice of this illuminating volume, both in our editorial columns, and by making abstracts of portions of special interest. It now comes in bound form, ready to take the place it deserves in the library, for it is, in effect, the most exhaustive study of nickel from every aspect that has hitherto been printed in English. It covers the history of nickel mining in Canada, and gives an elaborate account of the geology of the deposits; also the deposits in other parts of the world are fully described, in part as the result of personal inspection by the distinguished geologists of the Commission; and finally the volume deals with the metallurgy of nickel and the alloys that have been studied, so that it constitutes the only existing treatise on the metallurgy of this metal that may pretend to be anywhere near complete. As a reference book it will prove of the greatest value to the metallurgist.

Recent Publications

SANDSTONE QUARRYING IN THE UNITED STATES. By Oliver Bowles. Bulletin 124 of the Bureau of Mines. Pp. 143. Illustrated with half-tone engraving and line cuts. Washington, 1917.

THE MINING INDUSTRY IN THE TERRITORY OF ALASKA IN 1915. By Sumner S. Smith. Bulletin 142, U. S. Bureau of Mines. Pp. 65. One drawing. Index. Washington, 1917.

This volume briefly describes the progress of the mineral industry in Alaska during 1915, reviewing new gold mining districts, gives the mineral production, work at the mine experiment station, new territorial laws, the coalfields, the Government railroad, and gives considerable information regarding the condition in the important older districts. A flow-sheet of the Kennecott copper concentration mill is an interesting feature of the book.

The following publications have been issued recently by the United States Geological Survey:

Professional Paper 97. GEOLOGY AND ORE DEPOSITS OF THE MACKAY REGION, IDAHO. By J. B. Umpleby. Pp. 129, 21 plates, 14 text-figures.

Professional Paper 100-A. THE COAL FIELDS OF THE UNITED STATES. General introduction. By M. R. Campbell. Pp. 36, 1 plate, 3 text-figures.

Professional Paper 108-C. A COMPARISON OF PALEOZOIC SECTIONS IN SOUTHERN NEW MEXICO. By N. H. Darton. Pp. 29, 9 plates.

Professional Paper 108-D. WASATCH FOSSILS IN SO-CALLED FORT UNION BEDS OF THE POWDER RIVER BASIN, WYOMING, AND THEIR BEARING ON THE STRATIGRAPHY OF THE REGION. By C. H. Wegemann. Pp. 6, 2 plates, 1 text-figure.

Professional Paper 108-E. GEOLOGIC HISTORY INDICATED BY THE FOSSILIFEROUS DEPOSITS OF THE WILCOX GROUP (EOCENE) AT MERIDAN, MISSISSIPPI. By E. W. Berry. Pp. 14, 3 plates, 2 text-figures.

Bulletin 660-A. NOTES ON THE GEOLOGY AND IRON ORES OF THE CUYUNA DISTRICT, MINNESOTA. By E. C. Harder and A. W. Johnston. Pp. 28, 1 plate.

Bulletin 666-P. OUR MINERAL SUPPLIES—ALASKA'S MINERAL SUPPLIES. By A. H. Brooks. Pp. 14.

Bulletin 666-T. OUR MINERAL SUPPLIES—CLAY AND CLAY PRODUCTS. By Jefferson Middleton. Pp. 3.

Bulletin 666-W. OUR MINERAL SUPPLIES—BARIUM AND STRONTIUM. By J. M. Hill. Pp. 3.

Part of Mineral Resources of the United States, 1916, namely: SAND-LIME BRICK IN 1916. By Jefferson Middleton. Pp. 2. Part II. Pp. 2. Geologic Folio 205. DETROIT, MICHIGAN, (WAYNE, DETROIT, GROSSE POINTE, ROMULUS, AND WYANDOTTE QUADRANGLES). By W. H. Sherzer. 22 folio-pages of text, 12 maps, 12 plates, 20 text-figures. Price, 50c. retail or 30c. wholesale. Geologic Folio 206. LEAVENWORTH-SMITHVILLE, MISSOURI-KANSAS. By Henry Hinds and F. C. Greene. 13 folio-pages of text, 4 maps, 10 plates, 10 text-figures. Price, 25c. retail or 15c. wholesale.

MAPS. Topographic maps (about 16½ by 20 in.) of the quadrangles named below are now ready for distribution. Price, 10c. each; if included in wholesale orders amounting to $3 net, 6c. MICHIGAN, ST. CHARLES: long. 84°-84°15', lat. 43° 15'-43°30'; scale 1:62,500; contour interval 5 ft. OREGON DIAMOND LAKE: 122°-122°30', lat. 43°-43°30'; scale 1:125,000; contour interval 100 feet.

TOPOGRAPHIC MAP OF THE PLATTSBURG TRAINING CAMP, N. Y., 31 by 63½ in.; long. 73°25'-73°35', lat. 44°34'-44°50'; scale 3 inches to 1 mile; contour interval 5 ft. Price, 20c. retail or 12c. wholesale.

EDITORIAL

T. A. RICKARD, Editor

DEFENSIVE action against Minerals Separation has been taken in Colorado by organizing a Mill Operators Association with a view to ensuring voluntary financial support and legal assistance for the protection of any member of the Association attacked by the patent-exploiting corporation. We presume, however, that the legal contest over patent rights will be concentrated on the big copper companies, which are the most notable objectors to the payment of a flotation royalty.

JOURNALISM and diplomacy alike have been cheapened by the stuff that Mr. Gerard has published in the Hearst and other, better, newspapers. Omitting the Kaiser's telegram, which was intensely interesting and should have been made public by the State Department, not by the ambassador or the syndicate to whom he sold it—omitting that telegram, the rest is no better than kitchen gossip. Most of it is not new and much that is new is trivial. For an ambassador to splash a Hearst paper with large type and larger headings, and thereby spread two-thirds of a column of matter over a whole page, is small business. It is more worthy of a Tammany judge than the envoy of the United States at the Imperial Court of Berlin.

ZINC-PRODUCTION has been seriously restricted in response to the dullness of a waiting market. The Government has been unable to determine its future needs, as indicated from week to week by our New York correspondent, and the smelters themselves have taken the practical step of curtailing the output. A mid-year statement from the United States Geological Survey announces that 14 entire plants were idle on June 30, involving a total of 35,000 retorts, in addition to the retorts engaged in refining prime Western spelter and in re-distilling zinc 'ashes.' The market has been further weakened by a slight increase in the production for the half-year, the amount having been 364,000 short tons, or 13,000 tons in excess of the product of the smelters during the second half of 1916.

IRON ORE movement by the lake fleet from the North-West involved 10,241,633 gross tons in July, which is the largest tonnage on record, surpassing the shipments in August 1916, the former high level, by 391,493 tons. It is estimated that the requirements of the inland furnaces will exceed 58,000,000 tons this year. Accompanying this large output of ore from the mines, a decline in the production of steel is noted, this being ascribed in part to the hot weather. Timidity in the face of uncertain Government influence upon the market is another explanation that commends itself as reasonable when it is recalled that the price of No. 2 furnace pig-iron at Chicago, for example, has risen from $19 per ton on August 1, 1916, to $55 on the same date this year.

PLATINUM is another metal enhanced in value by the War. We hope that the rumored discovery in Alaska will prove true. The price of platinum has risen from $28 per ounce in 1907 to $110 at the present time. Its high melting-point, relative insolubility, malleability, and ductility render it particularly useful for the making of chemical utensils. Much of it has been used in dentistry and jewelry. It is to be hoped that the last will become unpopular; for the diversion of platinum for ornament represents waste. In 1915 the United States produced 2329 ounces of the metal and imported 61,437 ounces, the total consumption being valued at $2,507,183. The world's output in 1914 was 263,450 ounces.

IN the current magazines the heroines are discarding elderly financiers and young slackers in order to plight faith with the heroes in khaki, which goes to show that literature is a criticism of life even in its 5-cent form. We have had calls from young engineers out of work and asking for advice how to obtain an appointment. Any unmarried man between 25 and 35 can get the most honorable kind of employment at this time, and wear a uniform that is the most becoming in the world just now. We read the other day about a young man that went to see Joffre and Viviani when they came to New York. The great enthusiastic crowd did not shout either the General's name or the statesman's, they repeated one name again and again, "France! France!" A great longing came over him to hear that name the rest of his life or to hear millions say 'America!' in just that way.

OUR contemporary 'The Daily Metal Reporter' draws attention to the fact that the producers of zinc ore are importuning the Government to investigate the conditions under which they have received so small a part of the rise in the price of spelter. This rise is now reversed, the price being about one-third what it was two years ago. The reason for it is to be found in the statistics of export, for whereas the United States exported 230,557,691 pounds in ten months preceding June 30, 1915, it exported 423,474,263 pounds in the corresponding period ended June 30, 1917. Imports are decreasing, from 163,526,191 to 157,386,336 in the ten-month periods of the fiscal years 1916 and 1917 respectively. The imports have decreased because the contracts for Australian concentrate have been fulfilled and no more is likely to come because the smelting is being done in Australia. Similar efforts to smelt Australian concentrate in England have come to fruition, with the immediate effect of decreasing the demand for our output of spelter, which therefore is being produced at a tremendous rate in face of a sagging European demand. The laws of economics are incorrigible.

HYDRO-METALLURGY of lead has developed with exceeding slowness. According to Mr. O. C. Ralston, writing on 'Salt in the Metallurgy of Lead' in the transactions of the American Institute of Mining Engineers, the fundamentals of the process of roasting lead sulphide with salt, followed by leaching with brine, were stated 63 years ago by M. M. Becquerel. He foresaw that such a method would be possible by the application of electrolysis to recover the lead from the solution, but at that time no economic electric generator had been developed. Practically the same method has been studied by Mr. Clarence W. Larson, and the details of plant and technique have been developed by him at the Bunker Hill & Sullivan smelter in Idaho. He tells the story on another page in this issue. It constitutes a step forward in the new metallurgy, and opens a splendid opportunity for an improvement widely applicable in the reduction of lead. Mr. Larson shows that the economic treatment of raw ores containing as little as 8% lead is entirely feasible, when a proper 'mix' has been made to ensure the necessary heat-units for successful roasting. The copper and the precious metals also are dissolved and are precipitated before sending the liquors to the electrolytic vats. The process has been so far perfected that it will soon be in operation on a large scale at the Bunker Hill & Sullivan plant at Kellogg.

DISPATCHES from Washington announce that signs of oil have been discovered on the island Angel de la Guarda in the Gulf of California. This news has created excitement in some quarters and threatens to start a rush to the island. It can scarcely prove agreeable to Carranza, since it would precipitate a crisis between the central Government and the semi-independent administration of Governor Cantú. He has maintained his authority in all essential matters throughout the revolution. An oil-boom in Lower California might bring such wealth and power to the recalcitrant Governor as to provoke a political situation of peculiar interest. Indications of oil are known, not only on the island named, but for a considerable distance along the precipitous coast between the island of San José and Point Concepción, and also across the peninsula between the bay of La Paz on the east and Magdalena bay on the Pacific coast. Serious study of these districts has been retarded by the existence of old titles and concessions that made it necessary to deal with owners who invariably imposed prohibitory terms. The new constitution of Mexico, however, has 'nationalized' oil-land, that is, it has placed oil under the same rule as metalliferous minerals. Accordingly rights in petroleum have been severed from the ordinary surface-rights, and, the new law being retroactive, it becomes possible now to locate claims wherever indications of oil are found, and to proceed to acquire a leasehold-title from the Government. The new regulations, originally intended to admit of extracting ready cash from the oil-producers of the East Coast, have incidentally opened the way for the adventurous to explore the Lower Californian area, which so long has interested chance visitors to that remote corner of the world.

USUALLY it is ungracious and unprofitable to criticize the published writings of others, but the bulletins of the American Institute form an important part of the technical writing read by mining engineers and metallurgists. If the writing is bad, it sets a poor example. We have received the advance-proofs of a paper written by four metallurgists any one of whom ought to know better than to send such slovenly writing to the Institute; that four men should sponsor such a product is inexcusable, and that the Secretary should allow it to get into print would be incredible if we were not aware that he is over-burdened with his executive duties. We quote: (1) "It is possible in a 15-min. treatment to send off most of the lead and leave a partially roasted pyrite which can be discharged from the downdraft roaster while still ignited, and quenched with water." The incompletely roasted pyrite does well to emerge as quickly as possible from a roaster that is ignited and that is threatened with drowning. (2) "If zinc is present as zinc sulphide, there is not so much danger of overheating the mass with resulting matting of the iron sulphide, and good extractions of the lead are thus possible." The zinc cannot be present as a sulphide of another metal; the mass is not likely to be heated or over-heated with 'resulting matting,' although matting is an inflammable substance in dry weather; the lead is not 'extracted' until later in these metallurgical operations. The four well-informed but careless gentlemen mean that "if the zinc is not present as a sulphide, there is less danger of raising the heat sufficiently to make an iron matte, and a good recovery of lead is therefore practicable." (3) "The proper roast that will allow this is rather too difficult at present for commercial application, but the idea is a good one with which further work might be done." What is the idea hidden behind this verbal fog? Apparently it is meant that "the kind of roasting ensuring this result is too difficult to be economical, but the idea is worthy of further trial." These three quotations, taken almost at random from one paragraph, may not have been revised to express exactly what the authors meant, for they have not taken sufficient pains to make their meaning clear; we give alternatives merely to drive home the criticism. This is not done in any unfriendly spirit, either to the Institute or the four author-members, but to lay stress on the need for more care in writing. The transactions of the Institute ought to set a good example and gentlemen in responsible positions ought to show their sense of responsibility when they write.

W. H. Storms

With deep regret we record the tragic death of William H. Storms, the editor of our news columns. On Sunday August 19 he was shot fatally by an old miner, George R. Hutchinson, who had been trying to interest him in a prospect in Placer county. Crazed by his confident hope of selling his mine and demented by disappointment when Mr. Storms indicated his inability to examine it, Hutchinson became angry and in the ensuing altercation used the weapon that he carried in his capacity of watchman on the Key Route railway. It was one of those pitiful tragedies for which no reasonable explanation can be made. The news will come as a horrible shock to the many that knew and liked Mr. Storms. He was widely known and respected among the very class of men represented by the slayer. Born at Hackensack, New Jersey, our late associate was 57 years old. When 19 he went to the Black Hills of South Dakota and there acquired his first mining experience. Then he came to California and lived in the Mother Lode region, being connected with such well-known mines as the Wildman and Black Oak. Drifting into journalism he was connected successively with the Pasadena 'Star' and the San Diego 'Union', leaving the city editorship of the latter in 1891. For six years he was assistant to the State Mineralogist. Then he joined the staff of the MINING AND SCIENTIFIC PRESS, becoming editor under Mr. J. F. Halloran in 1902. He became news-editor when Mr. Halloran transferred his control of the paper to Mr. Rickard, at the end of 1905, and remained on the staff until late in 1906, when he accepted the superintendency of the Yellow Aster mine. He engaged in active professional work as a mining engineer until he was appointed State Mineralogist of California by Governor Johnson in 1911. For this post he was particularly well fitted and he would have discharged its duties efficiently if he had not been dismissed in 1913 by the Governor because he protested against the staff of the Mining Bureau being filled with political henchmen, thereby leaving insufficient funds for field-work. Thereupon he resumed his practice as a mining consultant, but in November last year he re-joined our staff as news-editor. A skilful writer, he was the author of two books, one of which, 'Timbering and Mining', is regarded as a most useful contribution to technology. A kindly and humorous man, he was much liked by his associates. Full of all sorts of information concerning mining in the West, he was always interesting to mining men, but it was among prospectors that he was most at home, as his short stories, published by us last year, testify. We join with his many friends in expressing bitter regret at his sudden end and deep sympathy for his wife and children.

'Flotation'

This is the title of a timely book on the metallurgic process that holds the stage at the present time. The authors of 'Flotation' are Messrs. T. A. Rickard and O. C. Ralston. The former, as editor, has done much to make known the fundamental technology of the process, while the latter, as metallurgist in charge of the Salt Lake station of the U. S. Bureau of Mines, has conducted researches as well as collected data on the practice of flotation. This, the latest book on the subject, resembles the one published last year by Mr. Rickard in being in part a collection of articles by many contributors to the MINING AND SCIENTIFIC PRESS. The present volume, however, contains numerous contributions that have not appeared in print before, notably several by Mr. Ralston, of which two are of special value, namely, 'Differential Flotation', and 'Mechanical Development'. As yet the differential treatment for the separation of blende from galena, particularly, has not attracted as much attention here as at Broken Hill, where the Horwood, Lyster, Bradford, and Seale-Shellshear processes have been developed. In the United States the use of froth-flotation in the concentration of copper mineral, particularly chalcocite, has been the outstanding feature, although the success of the process in the treatment of the zinc ore of the Butte & Superior is notable, not metallurgically alone, but because it has furnished occasion for the first big litigation, named after the defendant, Mr. Hyde, and also the third principal suit, against the Butte & Superior Mining Co. The mention of the first litigant recalls the fact that the book has been dedicated, presumably by the senior author, to Mr. James M. Hyde, as "the pioneer of the froth-flotation process in America." Our readers will remember that Mr. Hyde started experimental work in the Black Rock mill, at Butte, in March 1911, and in August he erected a small plant, the operation of which drew upon him the legal attack of Minerals Separation. Disregarding the merits of that quarrel, it is to Mr. Hyde's enduring credit that to his initiative we owed the first demonstration in America of the practicability of froth-flotation; it seems therefore not inappropriate that he should receive this compliment. Returning to Mr. Ralston's contributions to the book, it will be found that the chapter, or article, on the mechanical development of the process reviews the widespread inventive activity marking the last twelve months, for a large number of new machines have been patented and tried. A series of 35 drawings illustrates the diversity of mechanical design. The pneumatic or air-bubbling type, first introduced by Mr. J. M. Callow, is gaining ground as against the blade-agitators of the kind introduced by Mr. E. H. Nutter and others on the Minerals Separation staff. The original machines have been changed and enlarged by some of the big users of flotation, such as the Anaconda and the Inspiration companies, with results not particularly noteworthy. Indeed, we venture to demur against the taking of a machine bearing a man's name and then widening or deepening it, making it of steel or of bronze, or in some other way departing from the original just enough to excuse the dropping of the inventor's name. However, the big companies are not the only ones to do it. The variations on the Callow machine are already numerous, but the fundamental principle of the porous bottom is common to all of them. That suggests the fact that the Towne patents, in which carborundum was the medium, should prove valuable when the courts have become sufficiently enlightened concerning the difference between bubbling and agitation. This will come in time. Another chapter in the book that is likely to be thumb-marked is the one on the 'Disposal of Flotation Products' by Mr. Robert S. Lewis. This appeared in our issue of April 7 and at the time of publication we complimented Mr. Lewis on the care, and the consequent success, of his manner of writing. He has collected a large quantity of useful information and arranged it in logical form, with explanatory notes. The making of concentrate by flotation is only half the battle, as every millman knows. The concentrate must be sold to a smelter or treated on the spot. Mr. Lewis states how it can be done at home, as it were, and how it is done by the big buyers. Whatever the process of treatment, it is preceded by the dewatering of the concentrate, for the thin slime of finely pulverized sulphide is in no condition for transport. Mr. Lewis describes the multitudinous ways in which the water is drained. Since last year the flotation of oxidized ores has become a live subject and the book says all that can be said just now. It is most suggestive. Another phase of the subject that must not be overlooked is the behavior of colloids, on which Mr. E. E. Free has a chapter, dealing with a recondite subject in admirable form. It is much to be regretted that Mr. Courtenay De Kalb's article on the 'Control of Emulsions', appearing in our issue of August 11, was too late to be included in this volume. Much remains to be done in the study of emulsification as a factor in flotation. Here we may draw attention to Mr. Ralston's discussion of the over-oiling effect in his chapter on 'Flotation-Oils'. As to these we have yet much to learn, particularly concerning the difference in effect between soluble and immiscible oleaginous substances. We know that oil and certain kinds of solids, especially metallic sulphides in a fine state of subdivision, stabilize air-films in contact with films of water modified by soluble substances or by adsorbed gel-colloids. This introduces a problem that has not yet been definitely answered, namely, what property in common is possessed by oil and mineral particles in the interfacial relation? The hypothesis explaining froth-flotation is yet inchoate; the practice of the process is ahead of the explanation of it; we are obtaining results long before we know why they are produced; empiricism continues to direct the millman, and the physicist follows him with timid steps. In this new book there will be found eight chapters devoted to theory. Several recognized specialists, notably, Messrs. Wilder D. Bancroft, Joel H. Hildebrand, and Will H. Coghill, have essayed to elucidate the principles of flotation, and Mr. Rickard has contributed 25 pages in which he has endeavored, not without some success, it may be hoped, to interpret the scientific observations of physicists and to render the subject interesting by recording a number of simple experiments, most of which he has tried himself and therefore is able to describe vividly. We welcome the help of such learned doctors as Bancroft and Hildebrand; to them and to their fellow-workers in physical research we look for light upon the many obscure problems of flotation. Here again we take the opportunity to urge a larger measure of co-operation between metallurgists and physicists. The former, of course, ought to know something about physics, but it so happens that the chemical side of metallurgy has absorbed the major share of their attention, so that the sudden growth of the flotation process has taken them at a disadvantage. We need more literature on the physics of the subject, and it must be written as plainly as possible if it is to reach the men that want it, namely, the practitioners. From them, in turn, we ask for more details of manipulation, for those that have studied flotation are keenly aware that manipulation plays an important part in producing successful results within the mill. That is why we welcome such articles as 'Flotation Tribulations' by Mr. Jackson A. Pearce and such clear technical descriptions of mill-work as Mr. Hallet R. Robbins gives in 'Flotation at the Calaveras Copper' mine. None of us knows much about flotation, not even the gentlemen of the Minerals Separation company, and the only way to promote the further advance of this highly fascinating and industrially important branch of metallurgy is to come together in a true spirit of co-operation. That recalls a thought—one not without sardonic humor—suggested by the mention of Minerals Separation. The patentees are scientific men; Mr. H. L. Sulman particularly has proved himself a keen and successful investigator; he and his partner, Mr. H. F. K. Picard, have performed much work of research in the floating of minerals; they and their associates, notably, Messrs. A. H. Higgins, G. A. Chapman, and E. H. Nutter have ascertained much that is worth knowing, but not one of them has contributed voluntarily anything to the knowledge of the subject; on the contrary, they have united not only in keeping secret what they know but also in attempting to place an embargo on the publication of knowledge gathered by others. In consequence, they have seen others discover the main principles underlying the process and they have had to stand aside while others acquired the knowledge that they have withheld, until now the theory of flotation is as well understood, if not better, outside the Minerals Separation office as within its secret portals, and the technology of the process has grown to notable dimensions without their honorable participation in the literature of the subject. Shortly after the first decision in the Miami case, Mr. Nutter said that his company expected to publish a brochure on the technology of the subject as soon as the litigation was ended; we venture to say that any such brochure will have a mortuary interest by the time indicated, and that it is deeply regrettable not only that the Minerals Separation people should have attempted to wrap the process in secrecy but that they should have tried to compel others, such as ourselves, to abstain from giving forth information so urgently needed by the profession and by the industry to which it is devoted. That is why, among other reasons, we are glad to publish another book on flotation.

More Manganese Needed

Whoever can produce manganese today can sell it; that is one metal that does not have to pass through the guarded gateway of a 'trust' to reach a market. The extraordinary demand for manganese in the manufacture of open-hearth steel, and the curtailment of available foreign resources, has led to a crisis. The details of the situation are presented elsewhere in this issue. The imports of ferro-manganese have declined from an average of 7577 tons monthly in 1916 to 2717 tons in June, while our domestic output during the same period has shown an insignificant increase from 18,461 to 21,041 tons per month. The average monthly home-production for the first half of this year has been 19,829 tons and the average imports 4916. At the end of June a total monthly tonnage of slightly less than 24,000 tons seemed to be established, whereas the requirements of the steel industry amounted to 28,000 tons. The shortage is so serious that the Alloy Committee of the Council of National Defense has issued a warning to the steel-makers to adopt every possible means for recovery of manganese by the re-smelting of slags, and the substitution of spiegeleisen as far as practicable also has been suggested. The problem, as stated by our correspondent, is to produce a spiegeleisen of low carbon-content. Such an alloy has not yet been produced in the blast-furnace, and it is doubtful whether it can be done in the electric furnace. Could this be accomplished then a great quantity of manganese ores relatively high in iron could be utilized. In making ferro-manganese the reducing conditions in the furnace are maintained by the use of carbon, a part of the carbon being absorbed and tenaciously held as a manganese-carbon alloy. It is possible, when starting with an ore containing about 48% manganese, to hold the carbon-content close to 4%, and with that amount present over-carburization of the steel is avoided, but spiegeleisen contains about the same quantity of carbon, and the ratio of manganese to carbon therefore is nearly double the quantity that can be tolerated. The proportion of manganese to the melt in the open-hearth furnace cannot be reduced, else there would not be enough to serve the purpose of a 'scavenger' through its reducing effect upon the miscellaneous oxides in the iron that is being converted into steel. If a reducing agent other than carbon could be found for making a low-manganese ferro-alloy in the electric furnace a satisfactory solution of the difficulty might be reached, but the only metals more greedy for oxygen than manganese, that would be completely eliminated as slag or by volatilization, are too scarce and too costly. Therefore the outlook for making a low-manganese alloy from high-iron manganese ore is not encouraging. The present method of utilizing ores of this class is to mix them with those that are quite rich in manganese. In this way the proper proportions can be obtained. This suggests the possibilities of concentration. In many places the common wet methods of concentration have been successfully applied. The log-washer has been used in the Atlantic States, and jigging also has been employed with considerable success. Magnetic separation has long been used in the concentration of franklinite, and it may have been applied elsewhere. It would seem to offer an opportunity for helping to provide an increased output of desirable manganese ores from a number of relatively low-grade deposits. It would be easy to determine experimentally with a large average sample from any particular deposit precisely what amount of merchantable manganese concentrate could be recovered. The minerals highest in iron would be removable by developing a magnetic field of low intensity, and subsequent concentration by fields of greater intensity would give products of decreasing iron-content with a correspondingly higher percentage of manganese. Such plants, of course, would not be practicable without having large quantities of ore developed, but we have reason to believe that suitable deposits exist, both in Eastern and Western States. Fortunately the construction of magnetic separators is relatively simple, so that it would not be necessary to await the delivery of machines from manufacturers. Greater delay would probably be experienced in securing the needed crushing and sizing equipment. With 40% ore selling at 90 cents per unit, or $40.32 per long ton, f.o.b. Chicago, or Pittsburgh, it is evident that an ore with a concentration-ratio as low as 4:1 should prove profitable in almost any part of the country. While the chief demand originates in the East it must not be overlooked that 14 open-hearth convertors are now in operation on the Pacific Coast, and that a great iron and steel plant is being erected in the State of Washington. The continuance of the steel-industry in the West now seems assured. The Noble Electric Steel Company has responded to the call for ferro-manganese by producing nine tons per day, with a single furnace, in its works at Heroult, California, and it has transformer capacity for making 25 tons of alloy daily. It is only the difficulty of securing electrodes from the East that limits the expansion of these operations. The demand for manganese is becoming more insistent on both sides of the continent, and he who contributes in any detail to the solution of the problem is doing a patriotic service to his country. In California psilomelane is found abundantly in the foothill counties from Tuolumne to Nevada, some of the deposits offering much promise.

Crisis in the Manganese Trade

Ferro-manganese receipts from Great Britain are rapidly declining and the situation for the steel industry of the United States is not as hopeful as it was. With losses by submarine sinkings amounting to 2000 to 3000 tons in the past two months, and with a more acute situation in England as regards supplies of ore and of the alloy to meet the increased demands of their own industry, future imports from that country are not likely to renew the rate requisite to meet our need of 28,000 tons per month. The May and June available supplies were less than 23,000 and 24,000 tons respectively. Receipts of ferro-manganese from Great Britain in May and June showed a decided falling off from previous months and rendered the situation as to supplies less reassuring than a short time ago. The following table presents the status of the trade, giving the imports and the domestic production of the alloy as shown by the blast-furnace reports in *The Iron Age*, in gross tons.

	Imports	Domestic output	Available supply
Average per month, 1916	7,577	18,461	26,038
January, 1917	6,211	21,130	27,341
February	6,379	19,942	26,321
March	5,324	18,529	23,853
April	6,846	17,989	24,835
May	2,019	20,722	22,739
June	2,717	21,041	23,758

This reveals that the available supply has been diminishing for some time, particularly in the last two months. The average output of domestic alloy has been 19,892 tons per month for the first half of this year. About two months ago the Alloy Committee of the Council of National Defense, after a thorough canvas of the situation, made the statement that the present consumption of ferro-manganese by the steel industry of the United States was 28,000 tons per month. With our production close to 21,000 tons, and with imports at that time reaching about 6000 tons per month, the available supply seemed approximately adequate, but with imports declining as decidedly as they have done, and with no indication of a change, and taking into account the submarine sinkings already reported, the outlook is not as full of promise as it should be.

Realizing this the Alloy Committee of the Council of National Defense has recently notified the steel-industry of this country that not only must every possible effort at conservation of the alloy be made, but that a more liberal use of spiegeleisen must be introduced where feasible. This unexpected development again brings up the question of means to meet this or a worse contingency. It is generally accepted that metallurgically there is no satisfactory substitute for manganese in the important rôle it plays in steel. It has come to be accepted in practice that only the high-percentage alloy is sufficient. Both ferro-manganese and spiegeleisen are high in carbon. Were lower manganese alloys available having a low carbon-content, they would probably be nearly as satisfactory as the present ones, and would find a wide use.

A solution for the problem, which holds out some hope, is the conversion of our immense stores of valuable manganiferous iron ores into a low-carbon manganese-iron alloy. Blast-furnace reduction does not meet the problem. The electric furnace may perhaps solve it. Important achievements in this field are being accomplished in California. Some metallurgists suggest the value of intensive and prompt research with a view to ascertaining whether such ores cannot be electrically converted into low-carbon alloys. Such a metallurgical process, successfully carried out on a commercial scale, would be of incalculable value to the great steel industry of the United States and perhaps would render us independent of foreign sources of manganese. Its feasibility at least deserves thorough investigation. For high-grade ore, from which to make ferro-manganese, the United States is absolutely dependent on foreign countries. Before the War, Russia, India, and Brazil were the sources of this supply; now Brazil has stood almost alone as our help in greatest need, sending in 1916 over 500,000 gross tons. The dependence of this country upon Brazil or other outside sources of supply is appreciated when it is known that only 80,000 tons of such ore will be available in 1917 from the United States as against a theoretically necessary supply of 700,000 to 800,000 tons per year.

ALTHOUGH the 'permissible explosive' list of the Bureau of Mines is intended for coal mining, the regulations in large part are of general application. An explosive is called 'permissible' when it is similar in all respects to the sample that passed the tests by the Bureau of Mines, and when used in accordance with the conditions prescribed, but even explosives that have passed those tests and are named in the list as permissible, are to be so considered only when used under the following conditions: (1) That the explosive is in all respects similar to the sample submitted by the manufacturer for test; (2) That detonators, preferably electric detonators, are of not less efficiency than those prescribed, namely, those consisting by weight of 90 parts of mercury fulminate and 10 parts of potassium chlorate, or their equivalents; (3) That the explosive, if frozen, shall be thoroughly thawed in a safe and suitable manner before use; (4) That the quantity used for a shot does not exceed 1½ lb. (680 gm.), and that it is properly tamped with clay or other non-combustible stemming. An explosive, however, is not 'permissible' if kept in a moist place until it undergoes a change in character; if used in a frozen or partly frozen condition; if used in excess of 1½ lb. per shot; if fired with a detonator of less efficiency than that prescribed; if fired without stemming; and if fired with combustible stemming. The Bureau of Mines permissible list contains many varieties which would be useful in metal mining, and increase the tonnage broken, particularly where the ground was not so hard as to require the stronger grades of explosive.

The Flotation of Gold and Silver Mineral

By T. A. RICKARD

In the title to this article I have avoided the use of the word 'ore,' because the object of flotation is to float not the 'ore' but the valuable mineral in the ore, leaving the gangue to sink. It is a selective process, based upon the idea that the ore consists of valuable and of valueless components, which must be separated so that the valuable component may be concentrated as cleanly as possible previously to a final treatment in which the metal or metals are extracted and prepared for the market.

Some of the earliest work in flotation, such as that at the Glasdir mine, in 1896-1899, was done on an ore containing gold and silver, but the recovery of the precious metals was incidental to the concentration of the chalcopyrite with which they were intimately associated. Likewise the saving of the silver in the Broken Hill ore was incidental to the recovery of the sulphides of lead and zinc. In such cases—and they are typical—the floatability of gold and silver in the native state, or of their mineral compounds, does not present any special problem because the recovery of the gold and silver follows the concentration of the base-metal sulphides by which they are usually so closely accompanied in ore deposits. However, the floatability of native gold, as of native silver, hardly needs special demonstration here. 'Float gold' has been a bugbear of processes in which water is used, whether in the sluice-box of the gulch or in the stamp-mill on the hillside. Small particles of gold, particularly when flaky, are easily transported on water, as every miner has learned to his sorrow. The platy form of gold, so common in veins, lends itself readily to flotation, if the particles are small, by offering a large surface to the play of surface-tension and to the adhesion of air. The high metallic lustre of gold is a characteristic that experience in flotation would lead us to associate with easy buoyancy in the presence of air and of oil. If the gold is 'rusty,' that is, coated with iron oxide or with manganese di-oxide, or, as more rarely happens, with a film of silica, we should not expect it to float, as we should not expect it to amalgamate or to cyanide freely, until it had undergone such abrasion as would expose a fresh clean surface. Similarly gold in a clayey ore may make trouble for any process in which water is used, but this, like the 'rustiness.' is nothing new and is not peculiar to flotation.

As regards silver, the same general ideas apply. Silver in flaky form is elusive when running water is used, because it is readily floated; likewise when it presents a clean surface it is easily amenable to the guidance of the ascending bubbles. One would expect native silver and those of its compounds that are either highly lustrous, or have a marked cleavage, to float easily. This is a fact. Experiments made at Cobalt* showed a recovery of 92 to 97% for metallic silver, 77 to 89% for argentite, 85 to 87% for pyrargyrite, 78 to 80% for proustite, and 69% for frieslebenite. These experimental results have been confirmed in practice, a recovery of 96% having been made by using oil-flotation to supplement gravity-concentration.† Of manganiferous silver ores, a type familiar to the Mexican miner, it can be stated that when they cannot be cyanided they also cannot be floated. The obstacle probably is the double oxide of silver and manganese. Even preliminary sulphidization appears ineffective because the sodium sulphide will not attack the manganese-silver compound.

As regards the economic gold minerals, namely, the tellurides, they are so lustrous that one would expect them to be eminently floatable. That would apply to the silver tellurides also. Such has been the experience at Cripple Creek, provided, of course, that oxidized ore is carefully excluded from the mill-feed. At the Vindicator mine the practice is to wash the oxidized material out of the ore before it goes to the flotation plant. At the mines on the Mother Lode, in California, where the carbonaceous slate causes trouble by re-precipitation of gold in the cyanide solution, it has been found advantageous to apply flotation after amalgamation.

However, even if sundry gold and silver minerals will float, that does not mean that they can be recovered successfully as a concentrate by the frothing process. Direct floatability would refer to the surface-tension methods, such as those of Wood and Macquisten.

For the success of the older methods, such as that introduced by the Minerals Separation company, employing mechanical agitation, it is necessary that the pulp should contain finely divided mineral able to pass into the oil-water interface and in quantity sufficient to stabilize the air-bubbles by armoring them. Thus, a clean gold-bearing quartz is unsuitable to a machine working on the principle of violent agitation unless, of course, it contain enough gold, so sub-divided by the time it reaches the flotation-cell as to suffice for froth-making. The tailing from a vanner is not as suitable for the agitation-froth process as the pulp before it has undergone concentration on the vanner. Similarly, a gold-quartz ore containing 5% gold-bearing pyrite or other sulphides is better adapted to agitation-frothing

*Canadian Mining Institute. Bull. No. 62. J. M. Callow and E. B. Thornhill.

†At Cobalt the flotation-cell has replaced the slime-table in the McKinley-Darragh mill, while in the Nipissing, Buffalo Mines, and Dominion Reduction mills the cyanidation of tailing has given way to re-grinding and flotation.

than one containing 1% only. This applies to the mechanical stirrer; it does not apply to the pneumatic machine, in which air-bubbles, supplied lavishly, rise quietly through the pulp. In such a machine it is not necessary to have a large proportion of mineral for stabilizing the froth because the plentiful supply of bubbles obviates that requirement. As one bubble breaks, another is ready to take its place, so that the float, or concentrate, does not fall, but is lifted successively until it passes over the lip of the cell.

Another important phase of the subject is the recovery of the base metals associated with the precious metals. In mills using amalgamation and cyanidation, the presence of base-metal sulphides may be so detrimental that an ore containing any considerable percentage of them is likely to be left in the mine. In some cases the presence of base-metal sulphides, insufficient to be a source of revenue, but sufficient to interfere with the milling, has rendered it unprofitable to treat an ore. For such mines the use of flotation comes as a real boon. In the San Juan region of Colorado at this time there is a pronounced growth of productive activity because flotation has facilitated the recovery of the base metals associated with the precious, and so long as the metal markets remain propitious, we may expect a further expansion in this direction. Flotation is superior to any of the older wet processes—amalgamation, chlorination, and cyanidation—in that it will enable the miner to recover not gold and silver only, but copper, lead, and zinc as well.

These general remarks will serve to introduce some analyses of specific conditions—a more satisfactory method of discussing a problem that is economic as well as scientific.

North Star Mill

In the early part of 1916 the management of the North Star Mines, at Grass Valley, undertook a critical analysis* of the milling methods then in use, with a view to consolidating the existing plants and obtaining greater economy of treatment. The ore was being reduced in two 40-stamp mills, each with a cyanide annex, situated at the two main openings of the mine. The combined capacity of the two plants was 110,000 tons per annum, development work during 1915 had been highly successful, and it was anticipated that by centralizing the entire plant at one shaft a considerable saving in operating expense might be made.

The ore was being crushed by 1050-lb. stamps to pass 20-mesh screens, the treatment involving amalgamation in the mortars and on plates, then table concentration, followed by classification into sand and slime for separate cyanidation. The concentrate was re-ground in tube-mills to pass 200-mesh and treated in the slime-plant. The total extraction averaged $10 in gold per ton of ore; of this, $5 was obtained in the stamp-mortars, $3 on the amalgamating plates, and $2 from the concentrate, sand, and slime. The tailing averaged 25 to 35 cents per ton, so that the extraction averaged 97%. To treat 108,000 tons in 1915 the cost of milling was $51,000, for cyanidation $42,000, a total of $93,000, besides the tailing-loss of $30,000—a total deduction of $123,000.

*For this information I am indebted to William Hague, managing director of the North Star Mines.

Whatever process might be adopted, it was deemed advisable to retain amalgamation, since fully 50% of the gold extracted was caught in the mortars and 30% on the plates. It seemed wise to use amalgamation to catch as much of the free gold as possible early in the operations, as a sportsman tries to shoot his bird with the first barrel rather than the second.

The necessary experiments in flotation were made by the firm of Hamilton, Beauchamp & Woodworth, of San Francisco. Samples of plate-tailing were sent to them. The assay-value of these samples ranged from $2.50 to $2.90 in gold per ton. The first tests indicated that the material required re-grinding, to pass 80-mesh, in order to obtain a good recovery in the flotation-cell. When the heading assayed $2.50 the residue from these tests ranged between 10 cents per ton on pulp ground to 200-mesh, and 40 cents on pulp reduced to pass 65-mesh. When the plate-tailing was ground to pass 80-mesh the flotation residue averaged 25 to 30 cents per ton, like the residue after cyanidation.

The treatment of the flotation concentrate was the next step in the investigation. To ship the concentrate to a smelter—the Selby smelter, near San Francisco—would cost $17 per ton, for sacking, haulage, freight, treatment, and losses. This cost, on an ore containing 3½% sulphides, would mean 50 cents per ton of ore. Smelting therefore was not to be recommended. There remained the possibility of cyaniding the concentrate. The ratio of concentration being 30 to 1, the flotation product was worth from $70 to $90 per ton. The residue from the working-tests, when cyaniding the flotation concentrate, averaged $6 per ton. The average consumption of cyanide was 6 lb. per ton. Experiments were made also to learn what effect the flotation-oil had on the precipitation of the gold or on the fouling of the cyanide solution. The results indicated that no difficulty was to be anticipated from the presence of small quantities of oil in the concentrate.

Given the results of these flotation experiments and comparing them with the results obtained in cyanide practice, it appeared that three methods were available for use in the consolidated plant, which was to treat 110,000 tons per annum:

No. 1. To use 60 stamps of 1500 lb. each and crush to 0.04 inch diameter, employing amalgamation, classification, concentration of the sand, the classifier overflow and the concentrator tailing each going by separate conduits to the cyanide plant; the concentrate to be re-ground to 200-mesh and to undergo separate treatment in the cyanide annex, in order to ensure sufficient contact with the solution before being delivered to the slime-plant.

No. 2. To use 40 stamps of 1500 lb. each, crushing to 8-mesh, followed by amalgamation, classification, and re-grinding; to re-grind 70% of the stamp-product in tube or ball-mills to pass 80-mesh, this product to be treated by flotation, re-grinding the flotation concentrate to 200-mesh before cyaniding it.

No. 3. The same as No. 1 except that, instead of cyaniding the slime, to treat everything finer than 150-mesh by flotation and to leach the coarser material with cyanide solution; the flotation concentrate to be cyanided with the re-ground product of the water-concentrators.

Tests made in the mill showed that the 1500-lb. stamp falling 6 inches 105 times per minute would crush a little more than 5 tons per day to 0.04 inch, the screen having an open area of 36%; and also that this weight of stamp, crushing to 8-mesh, would have a duty of 7½ tons, yielding a product 70% of which would be coarser than 80-mesh.

The question of using ball-mills, both as primary and secondary crushers, had to be considered. The fact that 50% of the gold could be saved in the mortars was favorable to the retention of the stamps. Furthermore, in referring to descriptions of South African practice, it was noted that from 5 to 10% less gold was recovered by amalgamation after coarse crushing followed by tube-milling was introduced. If 5% more of the gold in the North Star ore, formerly saved by amalgamation, were thrown into the cyanide annex, the extra loss in cyanidation would off-set the saving of power and supplies to be anticipated from the substitution of ball-mills.

In estimating the operating cost of the 60-stamp mill and cyanide annex, the management was able to supplement its own experience with that of a neighboring plant having a capacity equal to that of the one being planned. In estimating the cost of flotation treatment, the North Star management was less confident, having to depend largely upon the advice of others; but by giving due weight to the evidence available it was possible to come to a fairly trustworthy conclusion, namely:

No. 1 method, applied to 110,000 tons in one year, involved a working cost of $75,000, plus a tailing-loss of $33,000, making a total deduction of $108,000.

No. 2 would require a working cost of $69,000 on the same tonnage and in the same time, plus a tailing-loss of $27,000, plus a loss of $20,000 more (3300 tons of flotation concentrate per annum in which $6 per ton would be left after cyanidation) in the residue of flotation concentrate after cyanidation. The total deduction would be $116,000.

No. 3 required an operating cost of $75,000, plus a total tailing and residue loss of $39,000, making $114,000. In this estimate, as in No. 2, royalty has not been included in the operating cost.

The capital cost to be incurred was estimated at $52,000 for No. 1, $42,000 for No. 2, and $50,000 for No. 3 method.

It was decided that the saving of $15,000 per annum to be made by consolidating the plant, under No. 1 method, was justified, this decision being strengthened by the disadvantage, occurring under No. 2 and 3 methods, of having "to choose between paying a royalty or fighting a patent suit."

THE MELONES MILL, CALAVERAS COUNTY, CALIFORNIA

A Mexican Mill

Next I shall quote figures relating to a silver mine in Mexico. The question had arisen of substituting flotation for cyanidation. Experiments, made by the same firm as had tested the North Star ore, indicated that this Mexican ore was amenable to flotation. The silver sulphides floated readily; the recovery ranged from 70 to 83% of the combined silver and gold, varying according to the proportion of oxidized minerals in the ore—oxidation being a deterrent to flotation, of course. This compares with an extraction of 91% by the existing method, which gives 77% of the metallic content as bullion and 14% as concentrate. But a closer analysis of more detailed data is required to make a trustworthy comparison. The present method of treatment includes crushing in cyanide solution followed by concentration of the sand on Wilfley tables and of the slime on Deister tables, re-grinding the sand and cyaniding an all-slime product. The capacity of the plant is 150,000 tons per annum and the ore assays $10.50 per ton when silver is worth 65 cents per ounce. The ratio of gold to silver in the ore is

10 oz. silver to 0.07 oz. gold. The tailing is worth $1 per ton. The total cost of milling is $1.55 per ton, to which must be added 10 cents per ounce of fine metal for marketing the bullion, this charge including export-tax, expressage, and refining, and 15 cents per ounce of metal for marketing the concentrate, this expense including taxes, freight, smelting, and sacking. The present output yields a bullion 700 fine and approximately 50 tons of $350 concentrate per month.

On the other hand, the flotation plant would cost $50,000 and would produce 100 to 150 tons of $500 to $800 concentrate, to be marketed at a cost of 15c. per fine ounce and leaving a tailing assaying $2.10 per ton. The Mexican export-tax on concentrate is 2% (of gross value) more than on bullion, thus:

	Bullion, %	Ore, %
Federal	5	7
State	2½	2½
	—	—
Total	7½	9½

The final comparisons are as follows:

1. Between table-concentration followed by cyaniding and straight flotation, the saving in cost of treatment by flotation is about equal to the extra recovery by tabling and cyaniding, but the increased expense for marketing the large tonnage of lower-grade concentrate plus lower smelter-returns on concentrate than on bullion represents a loss of 90 cents per ton of ore by straight flotation.

2. Between table-concentration and cyanidation, as against flotation and cyanidation of flotation tailing, the extra recovery is 33 cents (extraction 94%) and the saving in cost is 57 cents per ton, while the additional expense of marketing the concentrate plus lower smelter-returns on concentrate than on bullion is 90 cents, so that the difference is extinguished.

Cyanide is taken at 30 cents per pound, 1.8 to 2 lb. being consumed per ton of ore. Lately the precarious character of the cyanide supply has furnished an argument in favor of flotation. Packing the concentrate on mule-back or freighting it by train, with the uncertainty of getting cars for shipment to a smelter, possibly outside Mexico, are points requiring careful consideration. On the other hand, a load of concentrate is less easily stolen by revolutionists or brigands than a bar of bullion. The smelter deductions are important; usually payment is made on 95% of the silver and $19 (or 91.92%) is paid per ounce of gold in the form of concentrate against 100% of these metals in the form of bullion. This represents a net loss of 6 to 8% on this class of ore. At present therefore it is inadvisable to spend $50,000 in erecting a flotation plant to obtain a result no better than that given by the existing system of treatment, but it may prove advantageous in the future to adopt flotation followed by cyanidation when the flotation product itself can be treated safely and profitably on the spot, yielding bullion. This example will serve at least to emphasize the fact that such problems cannot be settled in the laboratory, and to show that the object of metallurgy is to give the greatest net returns rather than the highest percentage of extraction.

The Melones Mill

Another interesting comparison between cyanidation and flotation is afforded by an investigation made by the Melones Mining Company, which treats a low-grade gold ore typical of this part of the Mother Lode region. The existing plant consists* of 100 stamps weighing 1000 lb. each, and dropping 6 inches at the rate of 107 drops per minute. The stamp-duty, when discharging through a 20-mesh screen, is 5.3 tons. No plate-amalgamation is attempted inside the mortar, but mercury is fed into the battery, and the pulp when discharged passes over the usual amalgamating tables. The ore, a gold-bearing quartz containing pyrite, averages $3.65, from which $1.83, or 50%, is extracted by amalgamation and cyanidation, leaving a 51-cent tailing and 3.4% of concentrate assaying $36 per ton. The further extraction of the gold in the concentrate may be disregarded for the moment. Of the total extraction 43% is obtained as amalgam, 17% as bullion in the cyanide annex, and 40% in concentrate, which also is treated by cyanidation. The pulp from 60 stamps passes from the amalgamating tables to Wilfley concentrators, while that from the new mill of 40 stamps undergoes classification in spitzkasten before being concentrated. It has been observed in the Mother Lode region that the pulp from the stamp-battery does not concentrate so well after hydraulic classification as without such preliminary treatment. Sizing seems to be preferable. On the Wilfley tables the gold-bearing pyrite is recovered as a concentrate; at the same time, by the addition of extra water in the feed-box of the Wilfleys, the sand and slime are separated without the intervention of the usual de-sliming classifiers. The sand, assaying $1.15, goes to the leaching-vats of the cyanide annex; the slime, assaying $1, passes through cone-classifiers, the underflow from which joins the sand in the leaching-vats, while the overflow runs to the slime-plant. This includes Dorr dewaterers and Devereux agitators,† followed by Dorr thickeners and an Oliver filter. A middling, assaying $8 per ton, is made on the Wilfleys; this, after classification, goes to six 'finishing' Wilfley tables, the tailing from which is passed to the sand-plant.

The following facts are pertinent to our enquiry. The Wilfley tables recover 95% of the gold that is so intimately associated with the pyrite as to have escaped amalgamation; such gold-bearing pyrite as escapes into the cyanide annex is either in the form of slime or it is material that has been insufficiently pulverized. The pulp after concentration on the Wilfleys assays $1.12, whereas the mill-tailing, discharged from the cyanide annex, assays 51c. per ton. That represents the residual loss, to which must be added the loss in the treatment of the concentrate. The 'sand' and 'slime' are nearly equal in weight. The extraction of gold from the 'sand' and

*For most of the information in these paragraphs I am indebted to W. G. Devereux, manager for the Melones Mining Company.

†For a description of these machines see M. & S. P., March 3, 1917.

'slime' together is about 55%, that is, 55% of the 31% of gold remaining after amalgamation and concentration. A solution containing ½ lb. KCN per ton of ore is used in cyaniding the slime, and a 4-lb. solution on the sand. The total cost of milling (excluding concentration) is 50 cents per ton of ore.

What can flotation do on this ore? Samples were sent to the Minerals Separation people in San Francisco and they reported thus:

	Weight %	Gold oz.	Recovery %
Heading	100.0	0.06	100.0
Concentrate	1.4	1.83	42.7
Middling	6.2	0.20	20.7
Tailing	92.4	0.02	30.8

that the ore as submitted can be given flotation treatment usefully," but it must be noted that the sample "as submitted" was much too coarse to undergo successful flotation. The extra cost of re-grinding to 200-mesh is a vital factor in the problem. What would be the cost of re-grinding? At a neighboring mine, producing a similar ore, the cost of re-grinding is 25c. per ton. But all the pulp would not have to be re-ground; only the sand, say, 275 tons in all, out of the daily output of 530 tons. At 25c. per ton on 275 tons, the cost per original mill-feed would be 12c. per ton for re-grinding.

A flotation plant to treat 530 tons of such ore would cost $6000 f.o.b. San Francisco, or $10,000 erected; but the necessary re-grinding plant would cost $20.000 more.

THE APP MILL, TUOLUMNE COUNTY, CALIFORNIA

This showed a recovery of 63.4%, including the middling, which, in mill-practice, would be re-treated continuously. However, account is rendered for only 94.2% of the total gold in the heading. The material tested was slime from the Wilfley tables, the screen-analysis showing 98% through 200-mesh. In a later test, made at the same laboratory, a sample of the pulp as it came from the amalgamating tables was subjected to flotation, the result being a failure owing to the coarseness of a large part of the product. After the sample had been re-ground, until only 4% remained on a 200-mesh screen, the flotation machine did as follows:

	Weight %	Gold oz.	Recovery %
Heading	100.0	0.085	100.0
Concentrate	1.7	4.400	88.4
Tailing	98.3	0.010	11.6

The chief engineer (E. H. Nutter) for Minerals Separation reported that this test "indicates very definitely

The items of operating cost are estimated, by Mr. Nutter, as follows, per ton of dry ore:

	Cents
Reagents	6
Labor	5
Power (35 hp. per 200-ton unit)	2
Royalty at 25c. per oz. gold	2
	15

To this must be added the present cost of crushing and amalgamation, which is 20c.; so that the total cost of combined treatment, to the point of making a concentrate, would be

	Cents
Crushing	20
Re-grinding	12
Flotation	15
	47

This compares with the present cost of 50c. Allowing a 90% recovery by flotation, as against the present 86%, on a $3.65 ore, the additional winning would be 14.6c. per ton.

The Melones concentrate, representing 3.4% of the weight of ore, is re-ground, at a cost of 50c. per ton, to pass 200-mesh and is then cyanided, without roasting. The cost of treatment is $5.50 and the extraction is 92%. The cost is 19c. per ton of crude ore and the tailing retains 9½c. in gold.

In making this comparison the treatment of the concentrate is assumed to be the same, whether it be the product from the Wilfley tables or from the flotation-cells. No doubt exists as to the successful treatment of the flotation concentrate, which would not require re-grinding, so that the present cost of re-grinding the concentrate, which is 50c. per ton of concentrate, or 1.5c. per ton of mill-feed, would be saved. Moreover, the concentration would be higher, if one may judge from the results of the test made on the re-ground pulp. In that experiment the concentrate was only 1.7%, but it assayed $88, that is, it was half in quantity and double in richness as compared with the Wilfley product. Apparently the re-grinding had liberated some gold, which had become included in the concentrate; on the other hand, some of the quartz attached to pyrite had been loosened so as to join the rest of the gangue in the tailing. The reduction in the weight of concentrate would decrease the cost of treating concentrate from 19c. per ton to, say, 15c. The higher-grade concentrate would require more careful handling, and the extraction, at the same ratio of 92%, would leave a higher residue, namely 8% of $88, or $7.04, as compared with 8% of $36, or $2.88, making a difference, however, of only 2c. per ton of original ore, owing to the higher rate of concentration by flotation. In the event of adapting flotation to a stamp-mill, such as that of the Melones, it would be advisable to take the pulp from the amlagamating tables to the re-grinding machinery, in preparation for flotation, and not to attempt any table-concentration, because flotation would be better when leaving the pyrite in the pulp than when treating pulp after concentration. It only remains to remark that the Melones treatment might be changed to all-sliming and cyanidation, discarding concentration and separate treatment of the concentrate.†

The cost might be reduced 3c. per ton, but it must be remembered that the flotation figure is only an estimate as against the actual cost by the existing method. The increased extraction might be 15c. per ton. The total gain, of 18c. per ton, would be attractive if confirmed by further experiment, and if the use of flotation did not involve inquisition by, and subservience to, a patent-exploiting company. That undoubtedly, in my opinion, is a deterrent now. If it were a question of erecting a mill on a mine that had no reduction plant, and that produced ore of the kind we have been discussing, it would be rational to adopt flotation, not only for the sake of the small extra extraction but on account of the first cost of plant. The Melones cyanide plant cost $50,000.

†Compare this with the Nickel Plate experience. 'The Nickel Plate Mine and Mill.' By T. A. Rickard. M. & S. P. January 20, 1917.

The Dutch-App Mill

At the Dutch and App, a neighboring group of mines, in Tuolumne county, a conventional Californian mill of 40 stamps of 1050 lb. each, followed by amalgamation and concentration, has been changed from concentration by Wilfley tables and Frue vanners to flotation, so that a closer comparison is possible. I am informed‡ that the cost in the old mill was 74 cents, of which 38c. was for stamping, amalgamation, and concentration, and 36c. was for the transport and treatment of the concentrate. The Californian custom of reporting the cost of milling without including the expense of realizing upon the concentrate is misleading; so also is the practice of ignoring the loss or smelter-deduction from the assay-value of the concentrate. This item should be added to the assay of the mill-tailing in order to ascertain the total loss, and therefrom the total extraction of valuable metal. Thus the total cost in the old mill was 74c.; in the modified mill it was found to range between 88c. and $1 per ton. The modified plant, however, is at best a mill patched-up for the purpose of testing the ore by flotation; therefore, the cost should be much reduced. The tonnage treated was at the rate of 200 tons daily, whereas the proposed new mill will treat 600 tons daily.

The tailing in the old mill, treating daily 200 tons of $3.75 ore, assayed 90 cents (excluding the loss in concentrate-treatment), but when flotation replaced concentration by Wilfley tables and Frue vanners on the same grade of ore, the tailing was reduced to 35c. during the first month of operation, showing a saving of 55c. per ton. The comparative cost of a plant to treat daily 200 tons of ore carrying 5% of concentratable pyrite is estimated under normal conditions as follows:

(1) Twenty stamps of 1250 lb. each, with two re-grinding tube-mills, plates, and vanners—including rock-breaker, $48,000.

(2) Same mill, using flotation in place of vanners, $55,000.

(3) Same mill, concentrating on tables, the sand and slime treated by cyanide, $65,000.

No. 3 does not include the treatment of concentrate, because it is the usual custom to send it to the smelter at Selby. It will be noted that, for a comparison of plant-cost, No. 3 must be compared with No. 2. Flotation has given a slightly higher concentration of the gold-bearing pyrite in the Dutch-App ore; thus:

Old mill, 4½% of $35 to $40 concentrate.

Flotation, 5% of $50 to $65 concentrate.

The cost of transport and treatment is 38c. per ton of ore. The cost of haulage, freight, and smelter deductions is $9.50 per ton of concentrate. The cost of stamp-crushing and ball-mill re-grinding together is 35c. per ton. The cost of flotation by itself is 15c., making 88c. per ton in all for the extraction of the gold.

‡By W. J. Loring, who, I hope, at a later date, will contribute his own testimony on the subject.

When the new mill is treating 500 to 600 tons daily, it is expected to reduce the cost to 68c. and the tailing-loss to 20c. per ton.

By way of further comparison I may quote the cost of milling at an up-to-date plant, using stamps, amalgamation, and table-concentration, namely, at the Plymouth Consolidated in Amador county, also on the Mother Lode. There the cost of milling is 36c. and the concentrate realization 24c. more, making 60c. per ton. The treatment is most satisfactory because, among other reasons, the yield of concentrate is only 1½%. On such an ore flotation would not be as beneficial as on the Dutch-App ore, which contains 5% concentratable gold-bearing pyrite.

The Argo Mill

A tribute to the elasticity of the flotation process is furnished by the use of it in a plant treating custom ores. In the Argo mill, at Idaho Springs, Colorado, it has been found highly advantageous to substitute flotation for cyanidation, after stamp-milling and classification, as shown on the accompanying flow-sheet.* Most of the ore comes through the Argo (formerly Newhouse) adit, which taps the veins of Clear Creek and Gilpin counties. These contain gold and silver associated with pyrite, chalcopyrite, and tetrahedrite, so that it is a question of concentrating the sulphides encasing the precious metals. Much of the ore is of the so-called 'free-milling' kind, that is, it is amenable to stamp-mill amalgamation. The coarse gold is caught on the tables before flotation; the fine gold is caught in the flotation concentrate. Good results are obtained by classifying before flotation, particularly on oxidized ore. A trial on an oxidized gold ore from the Paris mine yielded 86% by flotation and 98% by cyanidation. The mill levies the same charges for treatment as the smelter at Denver and has helped, by its competition, to lower the smelter-rates. Its own concentrate is sold to the smelter. Rens E. Schirmer, the manager, and Jackson Pearce, the metallurgist, informed me that they are able to treat ore assaying $40 to $80 per ton. They quote $10.50 for treatment on a $50, and over, ore, but the bulk of the custom ores that come to this mill range between $9 and $20, gross value, and on such material the milling-charge ranges from $4.75 to $5.50 per ton.

For zinc there is no pay and also no penalty.

For lead they pay the regular smelter rate, namely: "Deduct 1.5% from wet assay. The prices paid per unit for lead ore are based upon a quotation of $4 per hundred pounds, 1c. up or down for each change of 5c. in this quotation, which shall be 90% of the sales prices in New York of the A. S. & R. Co. for common desilverized lead, provided said price does not exceed $4 per hundred pounds. When price does exceed $4 per hundred pounds, the quotation used as a basis of settlement shall be $3.60 per hundred pounds, plus three-fourths the excess of said sales price above $4 per hundred pounds."‡

For copper they pay on the dry assay, that is, 1% deduction from the wet assay, and the Western Union quotation for casting copper, less 6c. per pound.

For silver, they deduct ½ oz. from assays up to 10 oz. per ton, and pay for the remainder at 95% of New York quotation on date of assay. For ore over 10 oz.

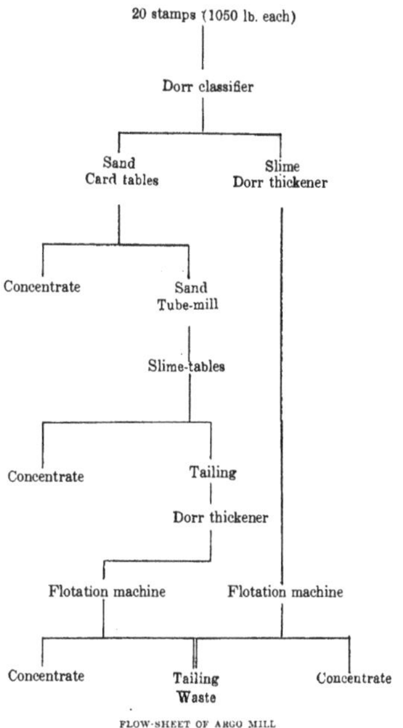

FLOW-SHEET OF ARGO MILL

per ton, they pay 95% of the quotation, without further deduction.

For gold they pay $19 per oz. between 0.05 and 1.5 oz. per ton; on richer ore they pay $19.50 per ounce.

The Argo mill, with its flotation annex, has treated 4 oz. gold ore, 100 oz. silver ore, 40% lead ore, and 5% copper ore at different times. Such high-grade ore is mixed with the lower-grade before being milled. Money has been made even on a $1.25 ore.

It was found that the recovery by flotation was slightly better than by cyanidation; moreover, the cost was lower, owing to the less expense in chemicals. A much simpler flow-sheet became practicable; there is less pumping, less

*See also an excellent article by Jackson A. Pearce. 'Flotation Tribulations.' M. & S. P., Sept. 16, 1916.

‡I quote the exact words in order that the reader may appreciate the hocus-pocus of this smelter method of fixing the price of lead. It is to laugh!

power, and less labor. Hence the adoption of flotation. A further and decisive advantage is the ability to beneficiate the base metals, notably copper, which was a cyanicide. Since flotation was introduced the mill finds a wider scope and is able to command a larger custom. The flotation machine in use is one devised and patented by Mr. Pearce. It is of the mechanical-agitation type, but it consumes less than one-third of the power required by the Minerals Separation machine previously used in this mill. A 6-cell Pearce machine requires only 5.3 horse-power. Two such 6-cell machines are employed, one to treat sand (80% + 200 mesh) and the other to treat slime. The former gives the higher recovery. I watched them at work and can testify that they produced a uniformly good froth and appeared to be operating admirably. Two pounds of an oil-mixture, consisting of two parts of gas-oil to one part of pine-oil (Pensacola Tar & Turp. Co.'s No. 400), is used per ton of ore. The gas-oil comes from Wyoming; it is one from which the lighter gasoline has been removed. The more sulphidic an ore the larger the proportion of oil required. As soon as sulphide particles appear in the tailing, more oil is added. Gas-oil costs 6c. per gal. and pine-oil 32c. laid down. The water is neither acid nor alkaline. Neither acid nor alkali is added. Fresh water is used, there being no return of water from the tailing. By using fresh water the millman avoids fouling of the liquid by an accumulation of colloids. Formerly the overflow from the concentrate-thickener was run back to the flotation-cell but it was found that this closed circuit tended to collect colloids detrimental to a high recovery by flotation.

Experiments have been made with as much as 40 lb. oil per ton of ore; the recovery was slightly higher than when using 2 lb. per ton. What a commentary on the 'critical' point! When using 40 lb. of oil the bubbles are smaller owing to the larger proportion of gas-oil, which was increased to 85% of the oil-mixture. When using 5 to 6 lb. oil and storing the concentrate in a wooden bin, Mr. Pearce noticed that the oil seeped visibly.

Ordinary variations of temperature appear to have no effect. In winter the froth freezes occasionally, so that it sinks, but so long as the temperature is just short of freezing the operation is not affected. In summer the recovery is no higher than in winter. The mill recovers 92% of the gross market-value of the various ores and treats 100 tons daily.

The Portland Mill

As yet scarcely anybody that has tried flotation has discarded it after trial. One example of such a reversal is furnished by the Portland Gold Mining Co. of Cripple Creek, and the reasons for it are interesting.

The ore was dump material assaying $2.25 per ton, carrying half an ounce of silver for every ounce of gold, both minerals being present as tellurides, chiefly calaverite. After treating over 100,000 tons the management decided that flotation was not superior to their older method of treatment by table-concentration and cyanidation, for the following reasons:

(1) Good extractions are obtained by cyanidation when grinding to 20-mesh, whereas flotation calls for grinding to 48-mesh. This extra work costs 10c. per ton, which is a serious item on $2.25 ore.

(2) Cyanide bullion is sold to the Mint, and the small amount of concentrate that is made is so low in silica and so high in iron that it can be marketed at the smelter on easy terms, whereas flotation yields a large amount of silicious concentrate on which the cost of freight and treatment is three times that of marketing the by-product of a cyanidation mill. On this ore the recovery by flotation was found to be inversely proportional to the grade of the concentrate; a high recovery made a large amount of low-grade concentrate; a small amount of high-grade concentrate entailed a poor recovery.

(3) Only $20 per ounce was paid for gold in the flotation concentrate, and nothing for the silver. Cyanide bullion is sold to the Mint at $20.67 per ounce for gold and 95% of New York quotation for silver. Thus 4% more is received for the product from the table-concentrate than for the flotation product. On account of the highly silicious character of the latter it was found that the cost of marketing at the smelter was out of all reason, whereupon it was sent to the custom roasting-cyanide mills at Colorado Springs. This flotation concentrate proved ideal stuff to treat after roasting, but as the silver is not recoverable after the ore has been roasted, the mill could not pay for it.

(4) The royalty payable to the Minerals Separation company.

The principal difficulty was to make a high-grade concentrate and a low tailing concurrently. Amorphous slime rises with the froth, and any effort to prevent it involves a loss of the sulpho-telluride mineral; in short, it is difficult to separate the gangue-slime from the mineral in this particular ore. Any free gold in the ore floats with the telluride and sulphide minerals. Incidentally, it is worthy of note that flotation in cyanide solution was accomplished successfully at the Portland by J. M. Tippett. He had to use caustic soda, in preference to lime, in order to ensure sufficient alkalinity. The ore was ground in the presence of caustic soda and cyanide. Mr. Tippett avoided dewatering, and the consequent loss of cyanide, by establishing a closed circuit. The pulp flowed from the flotation-cells to thickeners and pachucas, and thence to filters. By this method of treatment he was able to obtain a 40c. tailing on a $20 ore. From $17 to $18 was taken off in the form of a 12-ounce flotation concentrate, the tailing from which assayed $2 or $3 per ton and was reduced to 40c. by treatment in a pachuca.

Summary

In most cases the substitution of flotation means the making of a concentrate instead of bullion. The latter is ready for the market and easily handled or transported. The concentrate is bulky and if it has to be shipped elsewhere for treatment the handling of it entails loss, to which must be added freight and smelter deductions. However, these are troubles that can be obviated if the concentrate is treated at the mine. Some difficulty is said to have been caused by the oil retained by the concentrate; it may interfere with cyanidation, as Paul W. Avery testifies,* but we have the statement of E. M. Hamilton† that he has treated several samples of flotation concentrate with complete success. The experience at the Melones mine is to the point. Speaking generally, I infer that if an ore can be cyanided, then its concentrate, obtained by flotation, can also be cyanided successfully. In short, any gold and silver ore that can be amalgamated or cyanided is amenable to flotation, and the resulting concentrate is equally amenable to treatment by cyanidation. In some cases, as with Cripple Creek tellurides and the Cobalt silver minerals, it may be necessary to roast first, but the use of flotation on precious-metal ores need not involve dependence upon a smelter. That is important. On the concentrate made at Cobalt, for example, the total cost of marketing is $38 per ton. A chloridizing roast followed by leaching appears to be the only escape from this exaction. In California, the marketing of concentrate from a Mother Lode mine involves a cost, in freight, treatment, and other deductions at the smelter, of $9 to $15 per ton. This can be avoided, at many mines, by cyanidation on the spot without roasting, at a cost of about $5 per ton. It is remarkable how little enterprise has been shown in this department of ore reduction, despite the obvious saving of money promised by such a departure from precedent.

Thus we see that the substitution of flotation for the older processes of amalgamation and cyanidation is an economic rather than a metallurgic problem. In most of the specific cases discussed in detail it is safe to assume that if the manager were starting today to equip his mine with a mill, he would select the flotation process rather than the older methods, or make flotation a part of his flow-sheet. The scrapping of an existing, and expensive, plant is quite another matter. Usually the simple flotation plant would cost half that of the more complicated cyanide annex. Yet, in conversation with various managers, I have ascertained that the control or assumed control of the basic flotation patents by the Minerals Separation people is a strong deterrent. Most of us do not like to be inquisitioned by the agents of a patent-exploiting company, nor do technical men care to be placed under pledges of professional secrecy to anybody. The royalty on gold is only 25 cents per ounce and on silver 2½%, so that the tax is not onerous, but a tax of any kind is an irritation to most men, particularly when the right of the tax-gatherer is still undecided by the courts of law. If and when this question of Minerals Separation's right to collect a royalty is decided satisfactorily we may expect a wide extension in the application of flotation to ores chiefly valuable for gold and silver.

*'Cyanidation of Flotation Concentrate.' M. & S. P., May 16, 1916.

†M. & S. P., March 11, 1916.

The Thiogen Process

Recovery of sulphur di-oxide from smelter fume has for several years engaged the attention of the Bureau of Mines. These studies have included a critical investigation of the wet thiogen process. The results are given in the Bureau of Mines Bulletin 133, prepared by A. E. Wells. In carrying out the thiogen process the gases are first cooled and cleared of dust and fume. They are passed through an absorption-tower in which the sulphur di-oxide is absorbed in water or mother-liquor. To the solution of sulphur di-oxide is added powdered barium sulphide, when there is precipitated barium sulphite, thio-sulphate, and sulphur; the precipitate is settled and the mother-liquor returned to the absorption-tower. The settled precipitate is then filtered and dried. The elemental sulphur and one-half the sulphur from the thio-sulphite are distilled and the sulphur vapor condensed. The residue, consisting of a mixture of barium sulphite and sulphate is then reduced to sulphide, which is returned to the operation as a precipitate. In the investigations made, all the operations involved in the process were studied, not only on a laboratory scale but on a scale that may be considered semi-commercial, so that the technical possibilities of the process were determined. The results indicate that the process can be carried out successfully. However, in the reduction of the barium to sulphite it was found that there is a tendency for the formation of barium oxide and carbonate, which will react slowly in the precipitation. These compounds tend to increase in amount, as the barium is used in a cyclic operation. The proportion of these insoluble barium compounds can be reduced to a low figure by rapid reduction with pure carbon at a high temperature in a furnace or retort externally heated. In any case, as the soluble barium compounds increase in successive cycles the soluble barium sulphide can be leached out and the insoluble compounds allowed to react with strong sulphur di-oxide solution, resulting in their conversion to sulphite or sulphate, which can subsequently be reduced to the sulphide. The conclusions are that in some places the process can be conducted successfully for the recovery of sulphur and sulphur di-oxide from waste smelter-gas on a commercial scale at a cost of about $1 to $13 per ton of sulphur produced.

I. W. W. Sabotage.—In an interview published in the *Oregon Journal*, J. P. Weyerhaeuser, of Tacoma, manager of extensive timber interests, said: "I recently asked one of the I. W. W. leaders what his people wanted, anyway." "We want your mill," was the reply. Asked what he would do with the mill after he got it inasmuch as none of his organization knew how to run it the response was: "That doesn't worry us; we will have plenty of time to learn to run the mill after we get it." That seems to be the spirit of the crowd. They are out to disorganize and disrupt legitimate business. Do you know that practically all the men in the lumber camps this year are working on 'slow orders'?"

Properties of Vanadium Steel

By G. L. NORRIS

*A prevalent idea credits vanadium with being a powerful scavenger, and ascribes the beneficial effects of its use in steel principally to its action in removing minute residual amounts of oxygen and nitrogen from that metal. It has even been advanced by some that when all the vanadium added has been completely used up in scavenging, and none remains in the steel, all the improvement or beneficial effect it is capable of yielding has been accomplished, despite the fact that there is an increase in the mechanical properties of the steel with increasing amounts of vanadium present, and that vanadium has equally as beneficial effects in steels made under reducing conditions, such as crucible and electric-furnace steels, as in the case of those made under oxidizing conditions, like open-hearth and bessemer-steels. While it is true that vanadium oxidizes readily and will combine with nitrogen, its value as a scavenger is negligible, as there are far cheaper metals that are equally effective, or perhaps more so. The remarkable effects of vanadium on steel are due entirely to its presence in the steel as an alloying element and its influence on the other constituents with which it is in combination. It is found in both the main constituents, ferrite and pearlite, but principally in the latter. Only a few hundredths of one per cent of the vanadium combines with the ferrite. This minute amount, however, appears to increase the strength, toughness, hardness, and resistance to abrasion, of the ferrite. Nearly all the vanadium, however, is found in the pearlite, in chemical combination with the cementite, as a compound carbide of vanadium and iron in the case of ternary steel, and as more complex carbides in the case of quaternary steels.

Vanadium replaces the iron in the cementite or the carbide by increasing amounts until, finally, when the amount of vanadium is about 5%, all the iron is found to be replaced by the vanadium. The vanadium-containing cementite is not as mobile as ordinary cementite and consequently does not segregate into as large masses, but it occurs in relatively minute particles, and, therefore, is more uniformly distributed. It does not, consequently, readily occur as lamellar or thin plates in the pearlite, but in a granular or sorbitic condition. This strong tendency of vanadium to form sorbitic and even troostitic pearlite is doubtless one of the reasons for the mechanical superiority of steels containing vanadium, not only statically but dynamically. Vanadium carbide is not as readily soluble on heating as iron carbide, and consequently vanadium steel requires a higher temperature to dissolve the cementite and to put the steel into the austenitic condition for quenching.

The effect of vanadium on the physical or mechanical properties of steel increases with the percentage of vanadium until about 1% is present, after which there is a decrease, even in the case of quenched steels, and with 3% or more of vanadium the steel is actually softened on quenching except at unusually high temperatures, such as 1300° to 1400°C. Vanadium steel hardenite has a greater thermal stability, that is, power to withstand elevated temperatures without softening or breaking down or without separation of the cementite. This property is responsible for the great improvement in high-speed steels through the addition of vanadium enabling it to stand up longer under the high temperatures developed at the point of the tool in taking heavy cuts at high speed. Percentages of vanadium as high as 3.50 have been successfully used in high-speed steel, and 1.50 to 2.50% are not uncommon, although only a few years ago the percentage ranged from 0.30 to 0.75, and it was thought that the addition of over 1% gave little additional advantage. The improvement in high-speed steel through the use of vanadium has borne an almost direct relation to the percentage of vanadium present, and is considered to be from 60 to 100%. In the case of carbon-vanadium tool-steel, the use of vanadium has proved almost equally beneficial, although at present only about 0.2% of vanadium is used in such steel. It has a wider quenching-range, that is, it can be heated higher without injury; it hardens deeper, retains its cutting edge longer, and is much tougher and stronger. A bar of 1% carbon tool-steel containing 0.25% vanadium, quenched and drawn back at 400°C., will bend 90° without failure, whereas a similar steel without vanadium will bend only about 20° or possibly 30°. Comparative compression tests of such tool-steels, with like tempering or draw-back, gave on 1½-in. cubes, 490,000 lb. for the vanadium steel and 278,000 lb. for steel without vanadium. For battering-tools, such as pneumatic-chisel, setts, calking-tools, rock-drills, etc., vanadium tool-steel possesses marked superiority on account of its combination of hardness, strength, and toughness.

One of the principal applications of vanadium steel has been for steel castings, particularly for locomotive frames. The addition of this small amount of vanadium increases the elastic limit of the annealed castings 25 to 30%, without lowering the ductility. The tensile strength is not increased proportionately in the case of thoroughly annealed castings, but is usually 10 to 15% greater. Vanadium-steel castings require a somewhat higher annealing-temperature than ordinary steel castings. They are also more susceptible to hardening, and therefore should be cooled slowly in the annealing furnace. The annealing temperature should be about 875°C. in this case.

*Abstract: *Iron Age*, July 19, 1917.

THE BUNKER HILL & SULLIVAN SMELTER

Hydro-Metallurgy of Lead-Silver at the Bunker Hill Smelter

By CLARENCE L. LARSON

Although the hydro-metallurgy of lead has been discussed in technical articles and has been the subject of patents, no appreciable results have hitherto been obtained on a commercial scale. During the past three years, however, a serious study has been made of the possibilities of developing some kind of a wet process for lead ores. Although contributions to the progress of the method have been made by others, the Bunker Hill & Sullivan must be credited with the major effort, in which the research work of various earlier experimenters has been utilized. The Bunker Hill & Sullivan investigation has been directed toward the production of bullion from its lead-silver sulphide ore at its plant at Kellogg, Idaho. The process under consideration involves, in the main, four steps: (1) roasting, (2) leaching with saturated brine, (3) direct electro-deposition of the lead from the brine-lixiviant, using soluble iron-anodes, (4) regeneration of the brine-lixiviant.

The rejuvenation of the Holt-Christensen chloridizing-process for copper, silver, and gold ores, which is virtually the Longmaid-Henderson process with new mechanical apparatus, contributed the first item of note to the present study. In the Holt-Christensen leaching experiments at Park City, Utah, lead was detected in the mill-solutions. At the Knight-Christensen mill at Silver City, Utah, Christensen again observed lead in the solutions, and though the lead in the ore treated was usually inconsiderable, a deposition of such as was carried in the solution was highly desirable, since its recovery would be clear profit on the custom contracts. Accordingly Christensen did some experimental work with iron electrodes. This appeared encouraging, but before the results could be verified on a larger scale the mill burned down and that terminated Christensen's active work on lead. In the Park City plant Holt had also carried on some small-scale work, but had reported unfavorably. In August, 1914, I began an examination of the Holt-Dern system of blast-roasting, doing my work in the Bureau of Mines laboratories at the Department of Metallurgical Research, in the University of Utah. In a couple of months my tests verified the Holt-Dern claims as to operating on low-grade copper, silver, and gold ores. In testing the American Flag ore at Park City, a nearly perfect extraction of lead was obtained. I then suggested to D. A. Lyon, the Director of the Station, that this be followed up, and that a study be made on the behavior of zinc ores under similar treatment. He heartily approved, and to him is due credit for steady co-operation in counsel and in experimental work along this line. Considerable investigation followed throughout the year 1914-15, both on lead and zinc, with distinctly encouraging results on the lead ores but with decidedly discouraging results on zinc. During this period a sample of vanner-tailing was secured from the Bunker Hill & Sullivan, and the successful roasting of this material led to my joining the staff of the Bunker Hill & Sullivan Mining & Concentrating Co. in July, 1915.

At this time the Bunker Hill & Sullivan Co. was experimenting on a small scale with a process, an outline of which is as follows: the ore was chloridized in a Dewey hearth-furnace; leaching of the roasted product followed, utilizing saturated brine as a lixiviant; slaked lime was then used to precipitate the lead, after which the lead-precipitate was filtered off and re-dissolved in acetic acid, the resulting solution being electrolyzed between graphite-anodes and lead-cathodes. This work developed many difficulties both in practice and in economics. The roasting was inefficient and expensive; the lime-precipitation was awkward in regulation and operation, and produced a low-grade product; the silver did not precipitate well; the lead-precipitate was difficult to wash free from salt; the electrolysis was expensive and the electrolyte degenerated rapidly. Worst of all, the brine-lixiviant fouled rapidly in cyclic use. These difficulties forced a shut-down of the experimental plant, and much laboratory study followed, resulting in the development of the process that will now be described, and which promises to be satisfactory from an economic standpoint.

Roasting tests on lots of 500 to 1000 lb. were made on various mill-products and on mine-run ore, in a Holt-Dern intermittent type of blast-roaster. This is the type which was used at the plant at Park City in 1914-15 on copper, silver, and gold.* Ore, dry-crushed to the proper mesh, is mixed with fuel if necessary, and with salt if desired, and then with water. This gives the 'roaster-mix'. An essential condition in the roaster-mixes is that the size of mesh to which the ore should be crushed must be such that the particles properly react in roasting. Bearing in mind that brine is to be the leaching agent, it is evident that it is essential to transform as much lead sulphide into chloride and sulphate as possible, with a minimum production of oxide. In my experience two kinds of galena have been observed. That derived from the Eagle & Blue Bell mine in Utah roasts efficiently when crushed to No. 8 mesh, and Bunker Hill ore, if crushed to No. 8 mesh, also will roast in an apparently satisfactory manner except that all of the coarser particles of the galena will be found unchanged in the calcine. It seems that the Eagle & Blue Bell galena decrepitated during the roasting, whereas the Bunker Hill material does not. In the latter case it appears to be necessary to crush through at least No. 20 mesh. The size of the material greatly affects the mechanical conditions of the roasting operation. In order to make the matter more clear a description of the test-roaster is presented. Fig. 1 gives detail drawings of the furnace shaft, and Fig. 2 of the grate used. The roaster is essentially a shaft-column of concrete with an open top for charging and a closed bottom, with suitable doors for discharging the roasted product and for cleaning under the grates. The grate-details are assembled as seen in Fig. 2: the beveled posts are used to support the notched strips, while into the notches are fitted thirty $\frac{1}{2}$ in. by $1\frac{1}{2}$ in. by 30 in. strips. Thus is formed a grating of $\frac{1}{2}$ in. spacings. This is covered with a depth of three or four inches of $\frac{3}{4}$ in. down to $\frac{1}{2}$ in. tailing, in order to prevent the wooden grate from burning. A positive-pressure air-blast is provided under the grates, the 5 in. by $2\frac{1}{2}$ ft. by $2\frac{1}{2}$ ft. space serving as an equalizing chamber. Above the level of the wooden grate with its cover of tailing comes the calcine-discharge door. This is luted air-tight with fire-clay during the period of roasting.

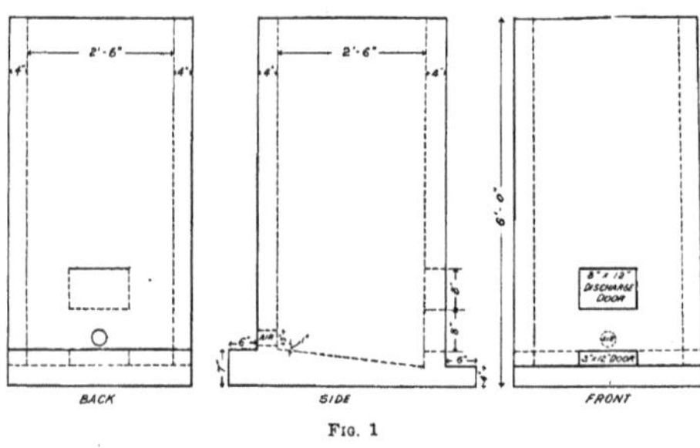

Fig. 1

The method of roasting is as follows: Upon the grate is built a fire of kindling, the air-blast being utilized to burn the wood until only live glowing coals remain. These are spread evenly over the entire area of the grate. The air-blast is now turned off, and the wet roaster-mix is dropped in and spread evenly over the fire to the desired depth, which may vary from 12 to 40 in. The air-blast is again turned on, and the roasting proceeds, 6 to 20 oz. air-pressure being required. When the fire-zone has broken through the entire surface of the charge the calcine is drawn and the roaster is ready for another charge.

Returning to the subject of size of crushing, the finer the mesh the greater the difficulty in getting a charge introduced so as to have uniform air-permeability throughout its entire mass; also, the slower the roasting the lower the capacity per square foot of grate-area. The size of the crushed ore also has an effect on the fuel required to roast the mass, for the finer the particles the more perfect become all the reactions with the metal sulphides, thus increasing the realizable fuel-value of the ore. The fine material also requires a greater height of roast-column and a higher air-pressure than the coarser sizes. With the increased air-pressure the danger of getting blow-holes or chimneys through the charge becomes correspondingly greater. Crushing should be done by rolls rather than by ball-mills because sharp angular particles tend to increase the permeability of the ore-bed in the roaster. The fuel-factor of the roaster-mix is apparently of the utmost importance in the blast-roasting of lead ores. Galena always shows a fuel-value dur-

*Bibliography on Holt-Dern roasting: U. S. Patents on the Holt-Dern Process, No. 1,107,240, 1,113,961, and 2; Bull. A. I. M. E., July 1914, 'Chloridizing Leaching at Park City,' by T. P. Holt; M. & S. P., Aug. 29, 1914, 'Rejuvenating the Chloridizing Roast,' by F. Sommer Schmidt; *Eng. & Min. Jour.*, Feb. 6 and 13, 1915, 'Chloridizing Blast-Roasting and Leaching,' by Glenn A. Keep; U. S. Bureau of Mines, Technical paper 90; *Mex. Min. Jour.*, May 1915, 'The Holt-Dern Process,' by C. L. Larson.

ing the oxidation of the lead sulphide. Pyrite also possesses considerable fuel-value, but other factors of ore-composition may lessen the heating-effect of the galena and pyrite contents. For instance, the gangue may contain inert constituents having high specific heats. Two analyses of ore follow:

	B. H. & S. mine, %	Alhambra mine, %
Lead	8.28	8.04
Lime (CaO)	0.26	0.74
Iron	16.20	21.75
Silica	48.15	32.10
Sulphur	3.26	3.14
Manganese	1.43	1.47
Alumina	2.36	11.00
Zinc	0.98	0.20
Copper	tr.	0.44
Silver, oz.	3.50	8.40

These ores come from adjacent properties. The Bunker Hill & Sullivan ore, with no addition of fuel, tends to roast too hot. The temperature is so high as to sinter the charge pronouncedly, and the result is to cut down the silver recovery from 30 to 50% and the lead recovery from 10 to 20% below the results obtained under suitable conditions. The Alhambra ore requires the admixture of 1½ to 2% of coal-dust in order to blast-roast it successfully. Its high alumina-content is undoubtedly responsible for this fuel-requirement. In order to modify the high temperature produced by an ore that contains too high a fuel-value it becomes necessary to use a diluent. To the Bunker Hill & Sullivan ore has been successfully added some calcine from a previous roast, or a highly silicious mill-tailing containing lead and silver. If the mixing of ores or material carrying mill-products were not possible, inert rock of high specific heat could be used, perhaps coarser in size, so that the calcine might be screened and the coarse particles of diluent returned to the roasting charge. If an ore requires fuel, another ore containing an excess fuel-value, or pyrite, or coal-dust must be added. Rarely will as much as 2% of coal-dust be required. About 2% of pyrite is equivalent to 1% of coal-dust, although pyrite has the more beneficial effect on the reactions desired. The limits of fuel-value in the roaster-mix are (a) maximum, the first indication of sintering, and (b) minimum, a tendency of the roast not to spread evenly to the furnace walls on all sides. Suitable calcine is pulverulent, shows no trace of sinter, and exhibits a distinctive reddish-blue color. Calcine showing deep black particles with a shiny lustre is not good, as it contains lead compounds insoluble in everything except hydrofluoric acid. Sintered material also prevents the recovery of any silver. In the blast-roasting of lead ores, (and this involves low-sulphur content by reason of the type of roasting) it seems advisable to add salt to the charge. The following table indicates the general experience in such tests:

Kind of roast	Neutral brine, extraction		Acidified brine, extraction	
	Lead, %	Silver, %	Lead, %	Silver, %
No salt	70 to 80	0	75 to 85	20 to 30
2 to 7% salt	80 to 90	50 to 65	88 to 95	80 to 90

The ores assayed from 8 to 11% lead and 4 to 10 oz. silver. Muffle-roasting or hearth-roasting of high-sulphur ore has been found to give practically perfect lead-sulphating results on galena without the use of salt. The roasting of carbonate-ore is distinctly different from that of sulphide-ore. I attempted to blast-roast a lead-carbonate ore from the Wilbert mine in Idaho, which has about 6% lead and a trace of sulphur. Chloridizing was first attempted, and after varying every factor in the mix and in the operation, it was concluded that lead oxychlorides had formed during the cooling period. Tests made on cooling roasted calcine in air-tight chambers seems to confirm this theory. Sulphidizing of the lead in this ore was then tried and an efficiency of 80% was attained by an addition of 10% of pyrite.†

The water required in roaster-mix charges varies from 5 to 15%. Fine sizes require more than coarse, and clayey ores more than silicious ones. For the most part water has proved to be a not very important factor in

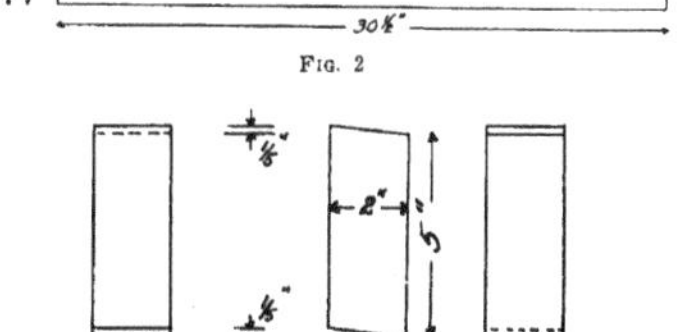

Fig. 2

POSTS FOR WOODEN GRATE

the roasting of lead ores. Care need be taken only to the extent of merely moistening all the slime-particles of the mix. In Hofman's 'General Metallurgy' will be found the detailed reactions occurring in a chloridizing roast. In this discussion I limit myself to the formation of lead chloride, lead sulphate, lead oxide, sodium and manganese sulphates, and insoluble iron oxides. Upon the addition of a leaching solution or of a preliminary water-wash, all chloride of lead is changed to sulphate in the presence of sodium sulphate, so that the solubilities involved in leaching are those of lead sulphate. The Holt-Christensen process utilizes a 22° Baumé brine for a mill-solution in work on copper, silver and gold ores. Solubility-tests on lead sulphate readily show the necessity of using a saturated brine in the leaching of roasted sulphide-ore, or of carbonate-ore if sulphuric acid be used with the brine. A solubility-curve seen in Fig. 3 shows tests made at 15° and with freshly made brine and pure lead sulphate. A maximum concentration is noted of 15 gm. per litre. All work done to date in cyclic leach-

†The Department of Metallurgical Research at Salt Lake City has performed interesting work in the direct acid-brine leaching of carbonate-ores. Some of the early work done at that laboratory is reported in the *Met. & Chem. Eng.*, Oct. 1, 1916, 'Electrolytic Recovery of Lead from Brine Leaches,' by Sims and Ralston.

ing indicates that this will be about the maximum that will be attained. The use of saturated brine in the leaching of calcine involves, in cyclic work, a gradual building up of the sulphates in the solution. This rapidly lowers the possible lead concentration and lessens the speed of leaching. Iron also builds up to about 20 gm. per litre and then maintains a practically constant concentration, but no effect of this iron has been noted. Removal of the accumulating sulphates at suitable intervals, or after each cycle, effects a regeneration of the lixiviant to its original power.

As soon as the fouling of the brine with sulphate was proved, attention was directed toward minimizing the quantity of the sulphates that might go into solution. Tests showed that, from average calcine, about one-half of the total brine-soluble sulphur could be removed by a water-wash without incurring loss of lead or silver.

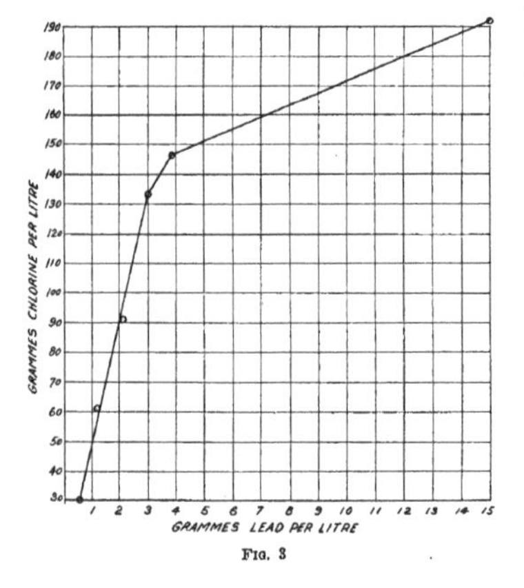

FIG. 3

The economics of this step were such as to warrant the inclusion of a preliminary water-wash as an integral step of the process. Cyclic leaching showed that as soon as the sulphur-content rose above 5 or 6 gm. per litre, the deterioration of the lixiviant became evident and rapidly increased, accompanied by further concentration of the sulphur in the solution. By the time the sulphur-content reached 10 to 12 gm. per litre the dissolving of the lead grew infinitely slow and the resulting solution nearly worthless.

Elimination of sulphur may be accomplished by direct precipitation with calcium chloride, but such a step in most places would not be economically favorable. Slaked lime is used and causes elimination of sulphur according to the following reactions:

$$FeCl_2 + Ca(OH)_2 = Fe(OH)_2 + CaCl_2$$
$$CaCl_2 + Na_2SO_4 = CaSO_4 + 2NaCl$$

The precipitate of calcium sulphate and iron hydroxide must be filtered off. Should any lead remain in the solution subjected to this regeneration it would be precipitated and lost. Consequently this step is utilized only after the lixiviant has been completely stripped of lead. In order to produce a pure marketable bullion it is necessary to remove all the copper and silver from the brine-lixiviant previous to sending it to the electrolytic vats, as any remaining silver or copper would there precipitate upon the depositing lead and upon the iron electrodes. Numerous precipitants have been tried, but technical and economic considerations presage the limitation of the use of finely divided iron, and a short violent agitation. Finely divided lead would also serve, but is rather slow in action. The brine-lixiviant coming from the silver-copper precipitation is electrolyzed‡ directly for its lead-content between iron electrodes. Insoluble anodes showed high current-costs and a deterioration of lixiviant in cyclic use. Because of the low metal concentration of the electrolyte, this being about 15 gm. of lead as a maximum, a very good circulation of electrolyte is required. Series-connections afford current-efficiencies of 60 to 70%, and parallel-connections 80 to 95%. Elec-

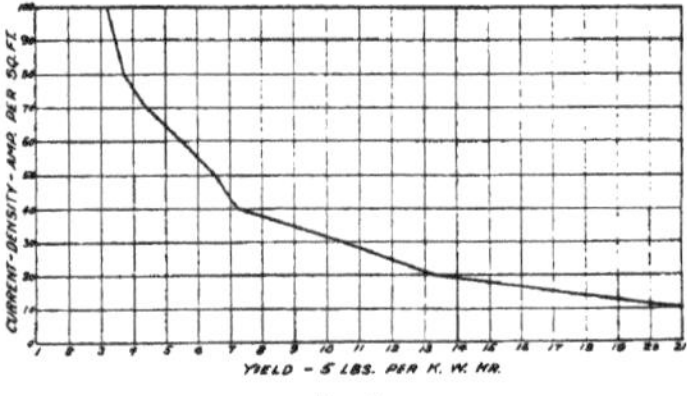

FIG. 4

trolysis is carried on to a lead concentration of about 1 gm. per litre, with but a natural slow decrease in the current-efficiency. If regeneration is to follow, the lead is entirely removed from the solution. The rate of decline of the current-efficiency, however, increased rapidly with solutions containing from 1 gm. of lead per litre down to those entirely free from lead. With air-agitation of the electrolyte, which makes it difficult to attain suitable efficiency and uniformity, it has appeared advisable to remove about half of the lead at current-densities of 50 to 60 amp. per sq. ft. of cathode-surface, and the remainder of the lead at 20 amp. per sq. ft. With a more uniform and efficient agitation of electrolyte, such as may be obtained by a rapid flow, or by rotating electrodes, the above condition does not hold, and one current-density may be taken for the entire period of 'stripping' the solution of lead. Thus the question of current-densities becomes the critical factor in the balance of yield versus plant and interest charges, plus the item of favor to the lower densities that cause less danger of carrying the

‡Sims and Ralston, 'The Electrolytic Recovery of Lead from Brine Leaches,' *Met. & Chem. Eng.*, Oct. 1, 1916.

electrolysis too far. The curve in Fig. 4 shows the relation of the current-densities to the yield in the case of parallel connections and under ideal conditions of circulation of the electrolyte.

Proper circulation of the electrolyte and the use of appropriate current-densities causes a deposit of lead which is fine and semi-crystalline, the material being so soft that it can be pressed between the fingers somewhat like an amalgam. The deposit of lead adheres for the most part quite firmly to the cathodes, but not as a sheet. Failure to provide for a suitable circulation of the electrolyte causes the production of a light fluffy lead which fills up the spaces between the electrodes and floats about promiscuously. This lead is well-nigh worthless, as it cannot be pressed together, and hence cannot be gathered. It is practically an ooze. The sponge-lead is pressed free of solution and converted into briquettes under several tons pressure per square inch, and these briquettes are melted and cast into bullion. If sponge is allowed to air-dry without pressing it rapidly oxidizes, and the melting then must be done with the aid of a reducer.

The iron from the anodes that is dissolved by the electrolysis is the source of the ferrous chloride of which use is made in the regeneration of the lixiviant. As iron begins to settle out at 20 gm. per litre, regeneration must always be used before the iron builds up so far. In the event that iron settles out in the leaching of calcine, regeneration must follow each passage of the electrolytic step by the solution, before being again sent to the leach-vat.

In the process discussed a neutral brine has been used. Roasting transforms some of the lead sulphide into oxide and this is rendered soluble by slightly acidifying the brine-lixiviant. Such acidulation also has a markedly beneficial effect on the recovery of the silver. Acid need be added in small amount only, as, after roasting, but little acid-soluble, or acid-consuming, material can exist. Enough is added to take out the metal and then the leaching is carried to neutrality.

This article necessarily has been limited to a discussion of the general characteristics of a process developed in the laboratory on a small experimental scale. During the past few months about 1½ tons of ore per day has been treated and about 25 tons of lead-bullion has been recovered by it, assaying 999.5 fine. The cost of a commercial plant and the cost of its operation, as well as the question of suitable equipment, cannot be discussed at this time. The preliminary estimates, however, indicate that favorable economic results may be expected.

SILICON increases the tensile strength of pure iron from 38,000 to about 90,000 lb. per sq. in. in annealed samples, when 4% silicon is present. Beyond this the strength rapidly decreases. The elastic limit likewise steadily increases up to 4% silicon. The ultimate elongation and reduction of area are practically unaffected until 2.5% of silicon is reached. From this point the drop is rapid. Beyond 7% silicon forgeability ceases.

Stadia-Reduction Chart

By EARL GLASS

The accompanying chart may be used for computing horizontal distances from stadia-readings and observed angles of inclination, the curves giving the correction to be subtracted. The horizontal distance of the observation will be found at the lower edge of the chart, and the degrees of inclination of the angles of elevation or depression appear at the left. Find the intersection of these two lines, and follow the curve from this point of intersection to the right-hand edge of the chart where the correction will be figure found on the margin. Thus, for a stadia-reading of 100 ft., and an angle of inclination of 8°, the correction is 2 ft., and the horizontal distance is the difference between the reading and the correction, or 98 ft. For a stadia-reading of 250 ft. and an angle of 3°40′, the correction is 1 ft., and the horizontal distance is 249 feet.

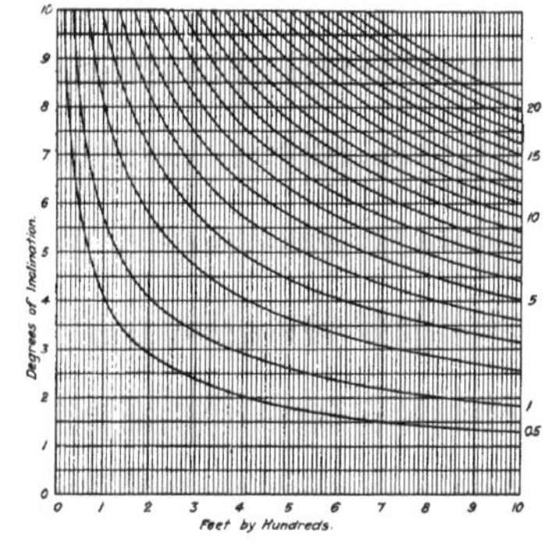

PYRITE on being calcined has a tendency to explode, thus producing a large amount of extremely fine material that prevents free access of the air for further oxidation. This explains the irregularity in de-sulphurizing pyritic ores when burned in the McDougall type of roaster. Recent tests by C. R. Gyzander confirm the suggestion of Lunge, made many years ago, that the explosion is due to occluded carbonic acid gas. Fine crushing before calcination reduces this tendency, and the quantity of dust produced by grinding in the ball-mill is less than that resulting from explosion of the pyrite in the furnace.

ULTRA-VIOLET light is readily absorbed by colloids. A suspension of gel colloid acts as an effective screen for such rays.

Velocity and Discharge in Ditches

By GEORGE HENRY ELLIS

The accompanying charts may be used for computing the velocity and discharge in small unlined earth ditches. The quantities have been calculated by using Kutter's formula, n being taken as 0.0225, which is an average value for earth canals in fair condition. Thus for a canal 6 ft. wide flowing 2 ft. deep, having a slope of 0.0004 or practically 2 ft. to the mile, the velocity is found to be approximately 1½ ft. per second and the discharge 30 cu. ft. per second. For a canal 8 ft. wide, flowing 3 ft. deep, with a slope of 0.0002 or 1 ft. to the mile, the velocity is slightly under 1½ ft. per second and the discharge is 60 cu. ft. per second. Again, given a canal 8 ft. wide, with a slope of 0.0009 or 4½ ft. per mile; required the depth of water to give a discharge of 50 cu. ft. per second. By interpolation, the depth is found to be approximately 1.85 ft. The velocity for this discharge is found by interpolation to be 2.3 ft. per second. The diagram may be used also to solve such problems as, given a canal 6 ft. wide in which it is desired that the velocity be 2 ft. per second, what is the greatest depth that can be permitted without delivering more than 40 cu. ft. per second, and what would be the corresponding slope? The depth is found to be 2 ft. and the slope approximately 0.00066 or 3.3 ft. per mile. Again, given a canal 8 ft. wide flowing 2.5 ft. deep; required the slope necessary for a discharge of 50 cu. ft. per second, and also the corresponding velocity. The slope is found to be slightly under 0.0003 or 1½ ft. per mile and the corresponding velocity, computed by interpolation, is 1.6 ft. per sec. To deliver 60 cu. ft. per sec. it would require a slope of 0.0004 and a velocity of 1.9 ft. per second.

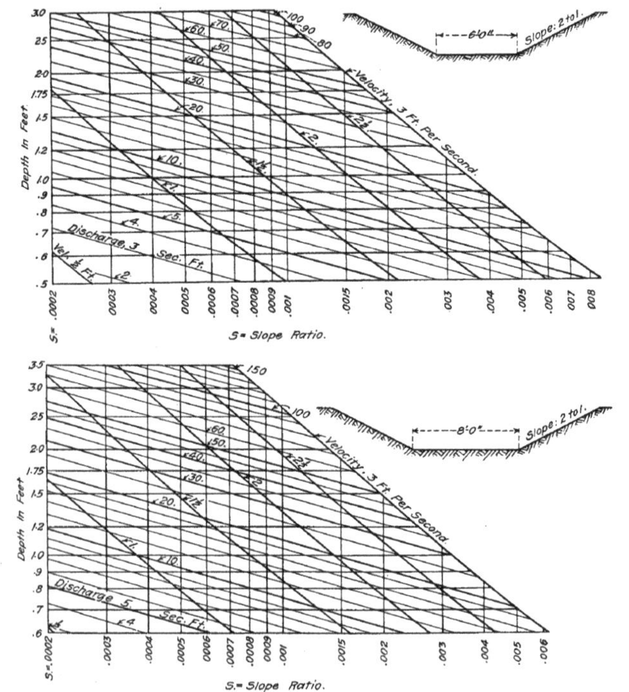

CHARTS FOR COMPUTING DISCHARGE OF EARTH DITCHES.

Detrital Copper Deposits

By W. TOVOTE

Detrital copper deposits, similar in derivation to gold placers are not infrequent in the South-West. From personal observation I know them to exist in the following localities in Arizona:

Castle Dome, near Miami, Gila county.

Ray, Pinal county.

Copper Butte, near Ray.

Copper Basin, near Prescott, Yavapai county.

Copper Creek, near Hillside, Yavapai county.

Mineral Park, near Chloride, Mojave county.

In spite of this wide distribution the general features are the same. They are

1. The source of the copper is not veins, but desseminated pyritic deposits in porphyry or schist.
2. The distribution of the copper is not uniform, but follows certain channels, suggesting mechanical concentration.
3. The copper-conglomerate is accompanied by similar conglomerate showing rusty iron-oxide cement, barren of copper. This iron conglomerate usually covers a far greater area.
4. The copper content is largely in the cement, being concentrated along later fissures and fracture-zones, suggesting chemical concentration and causing the ore to appear sometimes as deposited in a 'gash-vein,' because the ore invariably ceases at a shallow depth.
5. The predominant minerals are chrysocolla and the mixtures of chrysocolla with the oxides of iron and manganese, commonly called copper pitch; azurite and impure cuprite are rare and malachite is exceptional. Both the oxides of iron and manganese occur independently without admixture of copper, but while iron is prominent in the barren breccia, manganese seems to follow closely the copper. Much of the earthy manganese (wad) assays well in copper, without showing the color of copper. This is one reason why those unfamiliar with these deposits under-estimate the grade of the ore.

All the deposits mentioned, with exception of the Copper Creek area, have produced shipping ore, but only the Copper Butte and possibly the Mineral Park are of economic importance. A short sketch of the individual occurrences follows:

CASTLE DOME. Cemented gravel in creek-beds, issuing from the pyritized granite-massive of Porphyry Mtn. Apparently originally arranged in several terraces (lake-beds), but much dissected by recent erosion. Sorted ore as high as 15 to 20% copper has been obtained in small quantities. The deposit is not extensive.

RAY. Small areas of high-grade conglomerate in the canyons crossing the Ray orebody. The conglomerate is angular and deposited in immediate vicinity of its source.

COPPER BUTTE. This is the most notable member of the group. At least 30,000 tons of ore averaging about 4% copper is indicated in one mass right at the surface. Several gash-veins with narrow high-grade streaks are found in the immediate vicinity. The ore is in the pre-dacite conglomerate, named the Whitetail Conglomerate by F. L. Ransome in his Globe and Ray reports. It was described a few years ago in the MINING AND SCIENTIFIC PRESS by F. H. Probert as a 'volcanic neck' and is given as such in Weed's 'Mines Handbook.' How this misconception arose is not clear. This conglomerate extends as a broad belt from the Gila river to Copper Butte, pointing to Ray as its source. The copper-bearing area is surrounded by a broad belt of bright red conglomerate, which will average about 0.5% copper indicating a slight admixture of cuprite. Practically all the pebbles are slightly water-worn fragments of schist with occasional pieces of limestone, quartzite, and diabase. Below the schist-conglomerate lies another conglomerate composed largely of quartzite and again deeper another of limestone fragments, repeating in inverse order the sequence of the Paleozoic sediments exactly as they would be attacked by the agencies of erosion. The conglomerate is overlain by remnants of the dacite flow, and dissected by post-dacite faulting and erosion. Every trace of both dacite and conglomerate has been removed from the area intervening between Copper Butte and Ray, where the erosion has cut down to the diabasic and granitic rocks. The conglomerate belt is apparently a broad river-channel or a lake-bed. Copper-bearing areas recur in it at several places, but none of them is as extended or important as the Copper Butte ground. The ore is principally in the cement, but a few pebbles with copper stain are found also. The Whitetail conglomerate here, as everywhere else, is of variable thickness. It indicates the first lacustral and fluviatile deposition accompanying the subsidence of a sharply dissected land-area during a period of great geologic unrest with frequent shifting of stream-channels and drainage-systems. It is odd to consider that only such a shifting of erosion has saved the Ray orebodies from complete destruction.

The best exposure of the entire deposit, here over 200 ft. thick from limestone-conglomerate at the bottom resting on diabase to schist-conglomerate on the top, and conformably overlain by the dacitic flow, is found on the west slope of Copper Butte peak. The bottom of the dacite flow is marked by the bed of obsidian described by Ransome. The occurrence at Copper Butte has so many features in common with the copper deposits of the Black Warrior and Black Copper mines near Inspiration, that I am convinced that they are similar in origin. Near the Black Warrior the Whitetail conglomerate is lacking, but copper ore is found in the lowest part of the dacite, where chrysocolla, frequently in concentric aggregates, replaces the dacite. The dacite preserves its glassy texture, but is decomposed and crumbling as a rule. While the deposition of the copper is strictly by chemical replacement, chemical action has played an important part in the Copper Butte deposit also. The copper is concentrated most along post-dacite faults of N 30° W strike, traversing the conglomerate. Veins of this strike, with 20% ore, persist below the horizon that generally marks the bottom of the copper-bearing strata in the conglomerate.

COPPER BASIN. Creek-gravel, well cemented and indicating former terraces, now dissected. Derived from the pyritized monzonite, which is the principal country-rock. The conglomerate contains large boulders, usually barren, and cement rich in manganese with iron oxide and copper salts. Azurite is common. The copper has not traveled far and marks practically the lode-croppings. This is probably due to the fact that the ground-water comes to within a few feet of the surface. Good ore has been shipped from here. The conglomerate led to the discovery of the veins and lodes now being worked.

COPPER CREEK. Creek-gravels, mostly with iron cement but containing irregular areas of copper-bearing gravel; all well cemented. Material derived from the pyritized monzonite, in which the Bagdad Copper Co. has proved about 15 to 20 million tons of disseminated ore.

MINERAL PARK. Gravel-beds skirting the base of the masses of pyritized porphyry in Niggerhead and Congress mountains. The surface is much obscured by later barren gravel. The granitic boulders are barren; copper occurs in the cement, apparently in well defined channels. Ore of about 3% copper has been proved to a depth of 25 feet.

In conclusion I would say that none of these copper-conglomerates can be considered a true placer, because in every case the chemical concentration and re-deposition of copper from solution in surface-waters is far more important than the mechanical action of transportation by water. The Copper Butte is nearest to being a true placer; there the apparent source of the ore is over six miles away; the other extreme is Copper Basin, where there is practically no mechanical transportation at all. Copper Butte, for the same reason, is also the only one where copper is found to any extent in the pebbles. Otherwise the genesis is probably simply that copper was leached from the pyritic copper deposits as a sulphate and found a ready precipitant in the cement of near-by gravel deposits. In some cases, where erosion was rapid or oxidation shallow, boulders of ore found their way into the conglomerate together with barren material, so that they give the impression of true placer material, but ore of economic grade will probably never be found in deposits of this type unless due to chemical deposition and concentration. The copper can be derived as well from the gravel itself as from the original source of the gravel.

Montana Mineral Production

A mid-year report by the U. S. Geological Survey on the mines of Montana shows a large output of silver, copper, and zinc, in spite of delays by severe winter weather and by fire in the mines at Butte. As the six-month period closes a labor strike threatens to interfere with the year's record. The copper production of the first four months in 1917 was decidedly higher than during the corresponding period of 1916. The total output for 1916 was estimated at 357,000,000 lb. Next in importance is the silver output, which increases or decreases with the production of copper. In 1916 the production of silver was about 16,000,000 oz. The Anaconda Copper Co.'s smelting plants were averaging 28,200,000 lb. of copper monthly during the first four months of 1917, and the East Butte smelter 1,786,000 lb. Each of these establishments had higher averages in certain months, and may increase their output materially before the end of 1917. The estimated quantity of recoverable zinc produced in the State during 1916 was about 227,000,000 lb. With the increase in the number of shippers to the electrolytic plant this may be exceeded in 1917. At the Butte & Superior the output of the first quarter was not up to the average of 1916, owing to shaft-sinking and changes in the flotation equipment. The monthly average was 13,360,000 lb. of zinc in concentrate, against about 15,000,000 lb. in 1916. For several years the largest zinc producers were the Butte & Superior and the Elm Orlu mines. In addition to these, the Emma, Lexington, and Davis Daly are now large shippers to the Anaconda zinc concentrator. The concentrate is calcined at Anaconda and shipped to Great Falls for treatment at the electrolytic zinc works. The Ophir mill at Butte, operated by the Butte Detroit Copper & Zinc Co., has been converted into a zinc concentrator for custom ore. In Fergus county and at Marysville, in Lewis and Clark county, the output of gold has been increased. The new cyanide mill at Bannock, in Beaverhead county, was recently started, and the first lot of bullion was shipped. At Giltedge another cyanide mill is under construction for the treatment of gold ores, and near Helena one gold mill and a concentrator have been put in commission.

ACID-RESISTING alloys, almost completely insoluble in nitric and other acids, have been made in the Engineering Experiment Station of the University of Illinois by mixing chromium, copper, and nickel with small quantities of tungsten or molybdenum. The results were given in Bulletin 93 of the Experiment Station.

PICRIC ACID may be precipitated from waste liquors by adding sodium bi-sulphite in sufficient quantity to form a high-gravity solution. The picric acid present rises to the top in a flocculent form. The sodium bi-sulphate can be re-concentrated for subsequent use. The method was devised by Henry A. Gardner.

Recent Patents

1,232,783. CONCENTRATOR. James B. Freeman, Los Angeles, Cal. Filed Dec. 23, 1913. Renewed May 2, 1917.

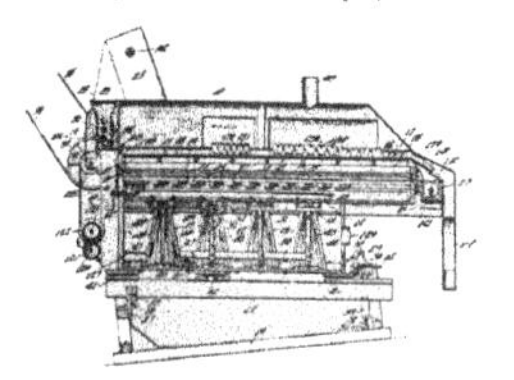

1. A dry concentrator comprising an inclined perforated table provided with diagonal riffles, means for oscillating said table, means for passing air through said table, and means for adjusting said table about an axis perpendicular to the plane of the table.

5. In a dry concentrator, a table having on its upper surface a series of diagonal riffles, a series of low converging walls at the lower end of said table so placed as to form narrow lanes to take all the discharge from said riffles, and thin plates between said walls in said lanes partially covering a series of holes in the table.

1,233,061. FILTERING APPARATUS. William M. Jewell, Chicago, Ill., assignor to Jewell Engineering Company, Chicago, Ill. Filed Nov. 16, 1915.

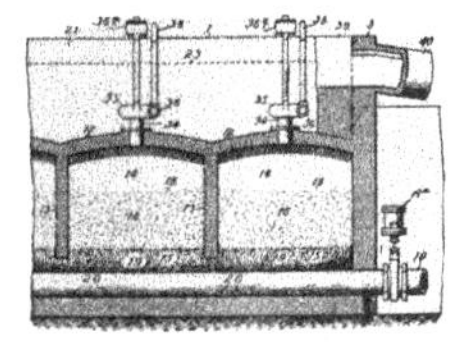

1. A filter apparatus comprising a chamber containing a filter-bed and provided with an inlet for water to be filtered and an outlet for filtered water, a filtered water receiver in communication with said filter and into which said filter discharges by gravity, and suction-producing means for exerting suction-action in said chamber above said filter-bed to draw water from said container upwardly through the filter-bed for washing the latter, the point at which suction is applied to said chamber being above the said water receiver.

5. A filter apparatus comprising a chamber containing a filter-bed and provided with an outlet for the filtered water, depending partitions extending downwardly partially through said bed and forming filter-compartments closed to each other above said filter-bed, and means for producing suction in said compartments above said bed independently of each other, for washing the bed.

1,231,684. REFRACTORY BRICK. Robert H. Youngman, Pittsburg, Pa. Filed May 25, 1915.

As a new article of manufacture, a refractory brick, consisting of calcined bauxite, about seven per cent of magnesite or its equivalent, and about one and one-half per cent of chromium ore.

1,232,805. ART OF FILTRATION. William M. Jewell, Chicago, Ill., assignor to Jewell Engineering Company, Chicago, Ill. Filed Oct. 23, 1915.

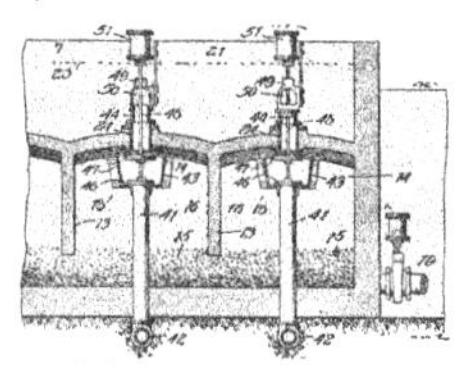

1. In the art of filtration by gravity, the method of washing a filter which consists in producing a flow of water through the filter-mass reversely of the flow therethrough in the filtering operation solely, by suction applied to the filter above the filter-mass.

2. In the art of filtration by gravity, the method of washing a filter which consists in producing a flow of water previously filtered in the filter, through the filter-mass reversely of the flow therethrough in the filtering operation solely, by suction applied to the filter above the filter-mass.

1,231,790. ORE COOLING, MOISTENING, AND FEEDING TABLE. Adelbert Harry Richards, Salt Lake City, Utah, assignor to American Smelting and Refining Company. Filed July 9, 1914.

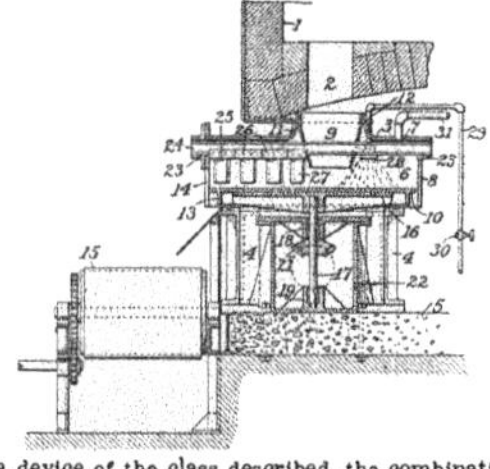

1. In a device of the class described, the combination with a furnace having an irregular discharge of ore therefrom, of a horizontal table, means for rotating said table, a fixed hood inclosing said table to form an inclosed casing, said hood having rabble blades disposed in operative relation to said table to mix the ore carried thereby, a spraying device for directing cooling liquid to the ore on said table, a liquid conduit leading to said spraying device, said hood having an outlet for directing the ore from said furnace onto said table and the side of said casing having an outlet for the mixed and wet ore.

5. In a device of the class described, the combination with a hopper adapted to contain a charge of ore, means for drawing said ore from the hopper in a uniform stream, said means including a rotatably mounted table adapted to receive the ore, a bar disposed above the table, a plurality of rabble blades positioned on said bar, each of said blades having a vane disposed at an angle to the path of movement of the ore on the table and spaced from each other to permit the passage of the ore therebetween, said blades being fastened to one side of the axis of rotation of the table acting on said ore to draw the same uniformly from said hopper toward the edge of the table to spread the same evenly on said table, and a spraying device for subjecting the layer of ore on the table to a cooling treatment on the side of said axis opposite the rabble blades.

REVIEW OF MINING

As seen at the world's great mining centres by our own correspondents.

CRIPPLE CREEK, COLORADO

NEW STRIKE AND A REVIVAL OF INTEREST ON TENDERFOOT HILL.—LESSEES SHIPPING AS USUAL.—ANOTHER STRIKE OF HIGH-GRADE ORE REPORTED IN THE CRESSON CONSOLIDATED.

The most important development of the past week in this district has been the proving of the continuity of value at depth in the Queen Bess mine, a Tenderfoot hill property, by the owners and operators, Hahnewald, Olsen, and associates. Since the property passed into possession of the present owners, who exercised the option of $15,500, given them by the C. L. Tutt estate, Mr. Olsen secured the right to extend laterals from the 500 and 900-ft. levels of the El Paso Gold King mine, south of the Queen Bess, and at least 100 ft. lower down, in Poverty gulch, through the Mollie Kathleen and into Queen Bess lines. The 900-ft. level, extended at a point 200 ft. south of the original ore-shoot, has opened up ore of shipping grade. The depth below surface on the Queen Bess is fully 1000 ft. and is the deepest working on the hill. The discovery has given encouragement to operators and property owners of that part of the district and other properties long idle will now be developed.

The Wide Awake, a Raven hill property, originally owned by the Wide Awake Gold Mining Co. and now owned by the Willshire G. M. Co., a close corporation, is again in the shipping class, Porter Hedge, lessee. Hedge shipped a trial lot last week, 5 tons each of screening and coarse rock. The screenings ran $46 per ton and the coarse rock $27 per ton. A second shipment is being taken out. The property had been shut-down for several years.

The Bertha B. on Raven hill, near the Cresson mine, owned and operated by Carl Johnson of Denver, will commence shipping by the first of the ensuing week. A road is being graded from the ore-house to the loading-station in Eclipse gulch and ore sorters are getting out the first shipment.

The New Boston, a Stratton Estate property on Womack hill, has been brought back to the producing classs. A. E. Gilbert, the lessee, loaded out a shipment last week that returned about $30 per ton.

The El Paso Extension Corporation, whose officers and stockholders are from the Southern States, are operating extensively on the southern slope of Beacon hill. The company has a lease with option to purchase, on the Rocky Mountain and North Slope claims, formerly owned by the Rocky Mountain G. M. & M. Co. and on the Hiawatha, a near-by property, and have started to drive a cross-cut adit to exploit the territory. Al Campbell, former superintendent of the Gold Dollar and Isabella mines, is in charge.

A prospecting permit with lease and option has been secured by Sam W. Vidler, on the properties of the Republic Gold Mining Co., the Janet W., Laura M., and Lester W., containing about 9 acres, on the western slope of Beacon hill, north of the Rocky Mountain and south of the El Paso Gold Mining Co.'s property, and has men prospecting. Frank Hervey Pettingell, president of the Los Angeles Exchange, is president of the Republic G. M. Co. This new work on Beacon hill is being closely watched and, in the opinion of mining men, ore will be found.

Articles of incorporation have been filed for the Ocean Wave Mining Co., which holds a long-time lease on the Ocean Wave on Battle mountain, and has installed machinery and is sinking a new vertical shaft at the line between the Ocean Wave and Little May. Both properties are included in the lease from the Portland company.

Irving T. Snyder, vice-president and general manager for the Vindicator Consolidated Gold Mining Co. and secretary-treasurer of the Golden Cycle Mining & Reduction Co., has tendered his resignations of all offices excepting that of vice-president of the Vindicator company, and has removed with his family to Santa Barbara, California.

The July output of the United Gold Mines Co., made by lessees on the Trail mine on Bull hill, the W. P. H. mine, on Ironclad hill, and the Bonanza on Battle mountain, approximates 750 tons of ore with a gross bullion value of between $57,500 and $60,000.

A report is current in the district that another rich discovery of ore has been made at the 1500-ft. level of the Cresson Consolidated Gold Mining & Milling Co.'s main shaft. The report has not been confirmed officially, but is believed to be reliable.

BRITISH COLUMBIA

LARGE IRON DEPOSITS BOUGHT BY D. C. JACKLING AND ASSOCIATES TO SUPPLY AN IRON AND STEEL PLANT TO BE BUILT AT SEATTLE.

The purchase of iron-ore deposits on Vancouver island, estimated to contain several millions of tons, has been completed by D. C. Jackling and associates. This ore will be used to supply a large steel plant, which is to be erected by them, near Seattle. This confirms the reports regarding the establishment of a steel plant and the opening up of the iron deposits of British Columbia. So far there has been little development on these orebodies owing to the lack of smelting facilities, but large deposits are known to exist and have been the subject of Government reports.

Most of the iron deposits are on Vancouver island, and it was there that the principal option was secured. Other options are held, however, in this Province, and also in the State of Idaho.

A 500-acre site for the steel plant has been obtained near Seattle, where it is proposed to erect blast-furnaces and steel-mills, also a coking plant for the treatment of such by-products as gas, tar, and ammonia.

Construction of the new plant soon will begin and it is expected to turn out pig-iron within a year. As soon as rolling-mills can be equipped ship-plate will be made. Forged billets, car-axles, and tin-plate also are to be included in the output of the plant.

The iron ore involved in this transaction is situated on the Gordon river, on the west coast of Vancouver island, a short distance north-west of Victoria, and almost due north of Cape Flattery. They have been under option since April 1913 to the Alaska-Gastineau Mining Co., which has made the first payment on the $250,000 purchase price.

The iron ore in this group of claims is practically all magnetite, carrying from 60 to 65% iron.

Options on five other large properties in British Columbia, it is said, are also controlled by the Jackling interests, two being on the mainland, not far from Vancouver. In addition to these the same financiers have secured options on three valuable quartz and dolomite claims, one on Vancouver island, one on Quatsino, and one on a small island in Cardero channel.

Representatives of the Alaska-Gastineau Mining Co. announce that at least $100,000 will be spent shortly on bunkers and a wharf at the claims at Gordon river, which are conveniently situated for shipping by water. The development of the claims is to be commenced promptly and it is planned to take out a large supply of ore in readiness for the completion of the smelter.

As a side issue of this transaction there is being perfected a system of producing coke from local coal. It is claimed that this coke can now be produced in British Columbia at a cost of $6 per ton and that it will be superior to the $14 Pennsylvania coke.

No one kind of British Columbia coal will produce a metal. It has been found that by mixing coal from the various Pacific It has been found that by mixing coke from the various Pacific Coast mines, including a little of the Washington coal, a coke equal to if not superior to that produced in Pennsylvania can be made. The larger part of the coal used is from the Nanaimo mines on Vancouver island.

SUTTER CREEK, CALIFORNIA

THE KEYSTONE MINE BEGINNING TO SNOW A MUCH MORE ENCOURAGING BALANCE-SHEET.—THE CENTRAL EUREKA IMPROVING, AND ALL STAMPS DROPPING.—PREPARATORY WORK CONTINUES AT THE OLD EUREKA.

The semi-annual report recently issued by the Keystone Mines Co., of Amador City, shows that the 40-stamp mill has been in continuous operation since the first of the year with the exception of 10½ days lost by order of the California Debris Commission on account of tailing, and about 2 days lost as a result of repairs, clean-up, and electric power being off. During this period, 39,896 tons of rock went through the mill, yielding $14,281 in bullion and 1505.124 tons concentrate, valued at $84,893.12. From this estimated total yield of $99,174.12 must be deducted freight and smelter-charges to the amount of $18,061.64, leaving a net product of $81,112.64, or about $2.03 per ton of ore worked. The company's expenditures for the past six months are segregated as follows:

Operating expenses, mine and mill	$75,200.00
License tax	32.75
Salary of manager, postage, and stationary	1,994.40
Total	$77,227.15

The company's indebtedness of $20,423.43 has not been reduced during this half-year, but conditions in the mine justify the hope that ore of better grade will soon be accessible, and the fact that the old mine has been able to hold its own, despite high costs of material and labor, on ore of such low grade, speaks well for future operations. Work is in progress on the 900, 1000, 1200, 1400, and 1800-ft. levels, the greater part of the ore having come from the 1200 and 1400-ft. levels. The 1800 station has been completed and a cross-cut driven east 550 ft. Upon reaching the vein at a point 500 ft. east of the shaft in this cross-cut, drifts were started north and south, the latter having been extended 100 ft. in ore all the way, conditions being also encouraging in the north drift. Carlton R. Downs, of Sutter Creek, is manager for the Keystone Mines.

For the months of May and June, the Central Eureka Mining Co. reports total receipts of $21,944.57, of which $9068.02 was derived from bullion and $12,571.52 from concentrate, the remainder being insurance re-fund and proceeds of supplies sold. The expenses for the same period amounted to $29,204.02, making a net loss of $7259.45. This is due in part to the extensive development and repair-work carried on in the mine, and to the re-inforcement of the tailing-dam, as required by the Debris Commission. The list of expenditures for the two months includes $4284.78 for development, $4159.87 for shaft and mine-repairs, and $1956.96 for strengthening the tailing dam, while the actual cost of mining and milling the 7760 tons of ore totals less than $15,000. The underground preparations made during May and June for extensive ore extraction enabled the company to start additional stamps during July and the 40 stamps have been in operation for about a month with encouraging results.

Clearing and re-timbering old drifts and repairing the shaft continues at the Old Eureka, and the company has just completed the construction of an up-to-date change-house for their employees.

HELENA, MONTANA

REVIVAL OF MINING ABOUT HELENA.—MANY OLD-TIME PRODUCERS AGAIN IN OPERATION WITH GOOD RESULTS.—GRASS VALLEY DISTRICT AGAIN PROSPEROUS.—THE JARDINE DISTRICT.

It is many years since there has been so much interest in mines and their development in the districts within a radius of 30 miles of Helena. For years the mining industry near Helena has been dormant, but it is stated that within the several months past, more progress has been made in mining in the region about Helena than in a decade before. While the developed mines that are producing and shipping are not many as yet, the number is rapidly increasing.

The Grass Valley district, west of Fort William Henry Harrison, which was practically dead a year or so ago, is now active. The present demand for the metals of the district, principally lead and silver, with some gold, has stimulated development.

The Rock Rose and Looby mines of the Cruse Consolidated Mining Co. are the most prominent of the Grass Valley district. The Looby is on the same vein as the Helena group and parallel to the vein of the Rock Rose. On the latter the vein at 200 ft. depth is 6 to 8 ft. wide, with a two-foot pay-streak on the hanging wall, having a value of $100 per ton, 35% lead, 31½ oz. silver, and 80c. in gold. The average value of the vein is $32 per ton. A station has been cut at the 200-ft. level and the shaft will be sunk to 300 ft., from which point drifts will be run. The inclined shaft on the Looby is down 160 ft. in ore of unknown width, which assays 20 oz. silver and 20% lead per ton. Improvements and new machinery will be placed as development warrants. Headquarters of the company is at Helena, with James J. Cruse, president; E. D. Phelan, vice-president; and R. A. Weisner, secretary.

The Helena mine at Grass Valley, near Fort William Henry Harrison, under the control and management of the Helena Mining Bureau, is making weekly shipments to the East Helena smelter. The ore nets $22 per ton. This property was one of the first in the Grass Valley district and considerable good ore was taken out in the early days. It is located between the Looby and Rock Rose properties.

Gold ore to the value of nearly $4000 per car, is going out from the Scratch Gravel Gold Co.'s property, at the rate of two cars per week. A pump has been installed which will lift 200 gal. per minute.

At the True Friend shaft, a new gasoline hoist has been installed. The shaft is in ore to a depth of 150 feet.

The mining district of Jardine is about seven miles from Gardiner, in a region of crystalline schists, which have been penetrated by rhyolite and trachyte dikes. The principal old mines, that have been idle for a decade, resumed operations during the past winter. The new companies are managed by practical men and engineers. The mines of the district have a record of over $3,000,000. Scheelite, carrying from 50 to 60% tungstic acid, is found in spots throughout the district.

The mines of the Maiden district, in Fergus county, are being worked by lessees. The ore runs about $8 in gold per ton. The McGinnis has produced over $1,000,000; the Spotted Horse, $3,000,000, and the Cumberland, $350,000. William A. Young, lessee of the McGinnis, has transformed the old mill into an up-to-date cyanide-plant, using crusher and rolls. The Spotted Horse is under lease to E. B. Coolidge; and the Cumberland to George Wiezlader. These properties, with the Gilt Edge, are gold producers, with cyanide treatment.

MARYSVILLE, MONTANA

THE DRUMLUMMON MINE AGAIN A PRODUCER.—THE SHANNON DEVELOPING INTO A VALUABLE PROPERTY.—THE OLD CORBIN DISTRICT ONCE MORE IN PROFITABLE OPERATION.—A GREAT PRODUCTION IN THE PAST.

The 80 tons of ore mined daily in the Drumlummon mine, now under control of the St. Louis Mining Co., is milled at the old 25-stamp mill, and then treated by continuous cyanidation. The mill is operated by electric power. William and Charles Mayger are in charge.

The Shannon mine, in the heart of the Marysville district, is surrounded by the great mines of the past. It is developed by a 750-ft. adit, with a hoist at the 500-ft. point, where a shaft has reached a depth of over 300 ft., giving a depth below surface of over 750 ft. The vein is from 4 to 35 ft. wide, with an average of 12 ft. The ore assays about $20 per ton, principally in gold. One hundred tons per day is sent by aerial-tram two and one-half miles to the Piegan-Gloster mill near Marysville, where three 10-ft. Lane mills are operated. Forty men are employed at the mine, which is owned by the Barnes-King Development Co. Thirty samples, taken for 150 ft. along the 500-ft. level, gave average returns of $76.60 per ton. W. R. Price is in charge.

The re-opening of the Blue Bird group, near the Shannon, by the Marysville Gold Mining Co., for which L. S. Ropes is the consulting engineer, has been undertaken, and new development along the foot-wall of the old workings is meeting with encouraging results.

Throughout the Corbin mining district, 35 miles east of Butte, is found the Butte granite, and the fracturing of the rocks is identical with that at Butte. In this district are found the carbonate and sulphide of lead associated with bismuth, chalcopyrite, cuprite, chalcocite, and native copper. The silver ores are the same as those at Butte. Gold is associated with all the ores, ranging from $1 upward per ton. The first discoveries in the district were made in 1864, and since that time the Alta mines on Alta mountain are credited with a gross production of more than $40,000,000 in lead, silver, copper, and gold. The depth of the Alta is 1600 ft. About 100 men are employed at the Alta mines and concentration plant. James Madden is in charge of the development.

On the east side of the Alta mountain is the Boston Corbin property. There are at least 10,000 ft. of workings, including shafts, tunnels, and drifts, to a depth of 1300 ft. in fissure-veins 2 to 10 ft. wide with copper and silver, and 30c. in gold to a unit of copper. The monthly production is 70,000 lb. of copper and 5000 oz. of silver. The concentrating plant has a capacity of 200 tons per day. Eighty men are employed. D. J. Courtney is superintendent.

The Montana Zinc Co., lately leasing the Minnie Shea, which adjoins the Rarus group in the Corbin district, has put in machinery. The 250-ft. adit cut two veins of silver, lead, and zinc ore that assays $78 per ton. The main vein is 6 ft. wide. The vein is in the Butte quartz-monzonite.

The Crystal Copper Co. owns 84 acres in the Cataract mining district, eight miles from Basin. The ore in the Crystal vein is from 30 to 40 ft. wide. Several shafts have been sunk from 60 to 130 ft. The ore runs 3 to 6% copper, 6 to 13 oz. silver, and 80c. in gold, per ton. Walter Harvey Reed is in charge. The Minah mine west of Alta mountain near Wickes has a record of $1,250,000 in silver, lead, and gold. There is much unexplored ground. It is owned by the Pittsburgh Minah Development Company.

The Comet, another old mine with a record of more than $13,000,000 in lead, silver, and gold, is again in operation. The main shaft is down 976 ft., where zinc ores are practically intact as in former days of development.

The Carbon Hill Mining Co. proposes sinking to a depth of 400 ft. and is shipping ore from the 100-ft. level. The property is equipped with two gasoline engines and a hoist. When the limit of gasoline power is reached a 40-hp. steam-plant will be installed.

At Basin is a profitable concentrating plant owned and operated by Max Atwater. He is treating tailing from former concentrating of zinc ores. His payroll is stated to be $15,000 per month. His proposed hydro-electric plant will be able to effect a great saving. Instead of shipping the zinc when the plant is in operation, only spelter will be sent to market.

HOUGHTON, MICHIGAN

HEAVY PRODUCTION OF COPPER FOR AUGUST.—LABOR CONDITIONS SATISFACTORY.—ISLE ROYALE'S GREAT OUTPUT.—CALUMET & HECLA MAINTAINS ITS STATUS.

Michigan's production of copper for August will approximate 26,000,000 lb. if the present rate is maintained. This is a larger amount of refined copper than has been turned out any month this year, excepting January. Production is running close to normal and while there has been an increase in the number of men employed in the Lake Superior district, the total number of working days will not show any increase for the reason that miners and trammers are making so much money that they can afford to take vacations as they please. Likewise some of the men own farms and are working on them. There is no accumulation of copper at the docks, practically all of the August production having been contracted for early last spring. Shipments are being made by both water and rail, the larger amount going by rail, notwithstanding the greater expense. General labor conditions here are excellent, and the men are getting larger wages than ever before in the history of the district and there is work for all. Wise miners, realizing that paid agitators were sent to practically all of the iron and copper mining districts of the United States, figured that these conversationalists will accomplish little in the Michigan copper country, so the best miners applied for work in this district. The I. W. W. gang did not make any headway in its efforts to stir up trouble in the Michigan district and the few agitators who visited this district took good jobs and went to work, or hurriedly left the district.

The Isle Royale is one of the instances of low-grade copper production in this district. When it had become evident that there was only one hope for a mine that produced an ore that averaged but 12 or 13 lb. of copper per ton, and that was by the output of an enormous tonnage of rock, there was little to the future for the Isle Royale except hope. Now the Isle Royale is running on an average of better than 15 lb. of copper per ton. The average for last month was 15 lb. and for August to date 16 lb. While the output of rock is not as large as it will be later it is keeping its own stamp-mill running to capacity and sending 450 or 500 tons daily to the Point Mills plant, and it could send 1000 tons daily to the custom-mill if the miners could be secured. The property is operating profitably. The seventh shaft is down to the ninth level and the most southerly openings are in rock which runs a little better than was anticipated and as good as anything recently opened in any of the Isle Royale shafts.

One shaft of the Champion mine continues to produce rock

from four levels that run better than 50 lb. of copper per ton.

The tonnage from the shafts of the Calumet & Hecla is not yet up to 10,000 tons per day, but it is crowding that figure, and by October more than 9500 tons daily can be counted on. There is practically no change in the average run of the rock. Four months ago there was little enthusiasm for the White Pine property as a permanent producer, but present indications show an improvement, and the application of the improved concentration process has added 30% to the copper recovery from the sand.

The Osceola Consolidated is showing a steady improvement in the value of the rock from the drifts south of No. 6 shaft. The average run of rock from the old Osceola mine has been lower than that produced from the North Kearsarge but the south drift toward the La Salle line, continues to improve and other levels running into the same territory show rock much better than any other found in this mine. Two shafts at South Kearsarge continue to furnish the richest rock that comes from the Kearsarge lode at any point. The total daily output is 3700 tons.

At three points the Quincy is showing an improvement in the grade of rock. The encouraging feature is that the improvement is at the lowest point at which the mine is operating. Quincy is shipping some copper by water and has 1000 tons on the docks at the smelter, all sold for delivery.

PLATTEVILLE, WISCONSIN

New Jersey Zinc Co. Makes a Rich Find.—Criminal Carelessness.—Mineral Point Reviving.—New Mines Opened and Some Old Ones Closing Down.

During July the New Jersey Zinc Co. struck rich veins of zinc ore, 2½ ft. thick, on the Clark-Underwood range, in the Highland district. At Montfort, the Hump Development Co., operating the O. P. David mine, was put out of business by an adverse decision rendered by the Industrial Commission on claims growing out of a dynamite explosion in the change-room at the mine on April 13, 1917, when one man was killed and eight others seriously injured. The David mine was shut-down, and it was said officially no further attempt would be made to mine this property under the present management. The heavy penalty imposed, says the Commission, follows a violation of safety orders, which prohibits the storage of dynamite and other explosives in the change-house or in other buildings frequented by employees.

The Linden district had some set-backs several producers shutting down permanently, among them the Glanville, an old-time producer, and the Wickes, a more recent development. The Weigle mine, of the Saxe-Pollard Co., was shut-down after a determined effort to make it a success and the equipment will be removed to new quarters, where the company is developing new mines. The Stoner Bros. Mining Co. purchased the Wickes equipment and began removing it to its mine.

Mineral Point, the site of a great oxide works, but for several years backward in mine operation, is reviving. The Utt-Thorne Mining Co. is responsible for the awakening in mining. The Melibou mine, Harris mine and equipment, and the Pierson leases are in this company's hands and are producing and shipping. The electric-power company has extended its lines in the northern districts to new mining towns and to several operating companies.

The Mifflin district, including Livingston and Rewey, had a good month through the operations of the New Jersey and Vinegar Hill Zinc companies. The new Yewdall mine shipped a car of ore daily, the new lands adjoining, drilled and now ready for production, will be permitted to rest until the labor conditions are more encouraging. The M. & A. Mining Co. of Platteville, operating the Big Tom mine at Rewey, paid a 25% dividend on July 15, and followed by another of 25% on July 31, making 150% paid since the mine began operating last March. Officials have brought in additional leases where drills are at work and extensions of the main ore-bearing vein are shown on the new tracts leased for mining.

At Dodgeville the North Survey mine with equipment is making regular weekly shipments, something the district has not known in several years, although there are others operating that are well established. The North Survey mine points the way to some excellent deposits in a stretch of country as yet untouched by the prospector.

Platteville, in the geographical centre of the field, fared badly all month, shipments from local producers being light. The Bell Mining Co., after providing equipment and getting down to a shipping basis, pulled up the pumps and quit. Further action will be left to a meeting of stockholders called for August 7. The New Rose Mining Co. paid off its indebtedness during July and laid aside a sufficient amount to declare a 40% dividend. The Block-House Mining Co. (Kistler-Stephens Co. owners) declared a 50% dividend, and since the mine began operations, one year ago, has paid $80,000 in dividends, which is 400% on its capitalization. At the end of the week, July 28, 1100 tons of roasted blende and 1400 tons of crude zinc concentrate was held in bin, 290 tons of 43% zinc ore was made in the last week of July. The Block-House vein has been drilled out for nearly one-half mile ahead of the breast now being broken. The company has drills at work on the McGregor, Heffernan, and Kistler lands, all having remarkable success. The Mann & Harding Mining Co. ships one car of high-grade ore per week. A. W. Plumb, general manager for the Wisconsin Zinc Co., who took charge of the company's holdings July 1, is planning big undertakings.

The Cuba district, noted for the operation of two magnetic separating-plants, was busy throughout the month. Receipts of ore were high and shipments of the finished product were regular. The C. S. H. Mining Co., recently organized by Platteville and Milwaukee men, bought the old Roosevelt plant and had it removed to the G. O. P. mine on the Raisbeck lease, where the contract called for operation August 20. At the National Separators, Cottrell electric precipitators were installed to collect dust and gas-fume. The Stafford Mining Co., working on a new range, put down two shafts and began the shipment of zinc ore. Pumps were installed in shaft No. 1, of the Coulthard mine and two shifts were put to work in shaft No. 2 on the Dall mine.

In the Shullsburg district the Winskill mine, of the Wisconsin Zinc Co. reported indifferently. The Rodhams-McQuitty Mining Co. is on a profitable shipping basis selling high-grade lead and zinc ore. The Oliver-Mulcahy mine, idle in June, shipped two cars of high-grade separator-ore weekly during July. The Rowe Mining Co. supplied a surface-rig to the Milwaukee-Shullsburg mine. The Fields Mining & Milling Co. took the Pacquette farm and mine under option. The owners of the land lacked the capital to develop this find, reported to be rich, and the new owners, successful operators of the Crawhall and Thompson mines, soon will get the new ground into shape for mining.

The Benton district is producing more than one-half of the mine-run output of the entire field. The principal producers are the Wisconsin Zinc Co., the Frontier Mining Co., and the Vinegar Hill Co. Two of the finest mills in the field were recently completed for the Frontier Mining Co. The Vinegar Hill Co. has a new producer in the Meloy mine, shipments the last week of July running better than one car per day. A new mill is completed for the Wisconsin Zinc Co. on the Copeland estate, which will commence operating in August. Another new mill has been completed for the Strawbridge Mining Co. of Milwaukee. The Zinc Concentrating Co., of New York, investigated several properties in the district, seeking a number of low-grade zinc-ore producers to supply a magnetic separating-plant planned for the Cuba station. The New Jersey Zinc Co. has a new producer in the Hoskins mine.

TORONTO, ONTARIO

THE DOME MINES.—CONIAGAS NEGOTIATING FOR NEW PROPERTY. —PRODUCTION AT COBALT INCREASING.—RICH ORE DISCOVERED AT THE HARGRAVES.—LARGE COPPER DEVELOPMENT AT SCHIST LAKE.

The Dome Mines has passed its dividend, the reason assigned being the heavy expenditure required to carry on development work. This caused no surprise in mining circles, as it had been anticipated for some time. The Porcupine Crown, which had been paying quarterly dividends at the rate of 12% per annum, has also suspended dividends for the present, as the shortage of labor renders it impossible to keep development ahead of production. The directors announce that their policy will be to limit production to the amount necessary to meet expenses. Operating profits for the six months ended June 30 were $124,179. Contrary to expectations the production of the McIntyre in July, which amounted to $137,790, was less than in any preceding month of the current year, and the average value of the ore treated, $9.32, was the lowest in seven months. During the first four months of the year the average grade of the ore was $10.43. The main vein, which is being driven at the 1000-ft. level, divides into two branches just over the Jupiter line, one of which is over 12 ft. wide and very high in grade, the remainder averaging $12 per ton. At the Keora the main vein, about 16 ft. wide, has been cut by diamond-drilling 750 ft. below the surface where it averages $6.26 per ton. Vein No. 2, cut at a depth of 615 ft., assays $4.20 throughout a width of 4½ ft., and vein No. 3, which is 2 ft. wide, and was cut at 430 ft., assays $33 per ton. Another high-grade vein has been found at a depth of 380 ft. The Coniagas, at Cobalt, which is operating the Anchorite, is negotiating for the Maidens-Macdonald property adjoining, which was formerly under option to the La Rose. The Kirkland Lake has cross-cut the main orebody on the 700-ft. level, but the width at this depth has not yet been ascertained. The Teck Hughes is sinking to the 600-ft. level on vein No. 1. A small fire in the 'rock-house' on August 6 did damage to the amount of $1000, but caused no serious delay in operating.

At the Croesus, in Munro township, where difficulties have been occasioned by flooding, new pumping equipment has been installed and the flow of water checked sufficiently to permit a resumption of operation. The mill is working steadily on ore from the dump, the gold content of which is about $40 per ton. Production at Cobalt lately has been stimulated by the high price of silver, and shipments have been heavy. During July the Nipissing mined ore of an estimated net value of $272,490, and shipped bullion and residue derived from its own resources and from custom ore of an estimated net value of $295,495. The total bullion recovery for the year to the end of July was $1,765,167. The new flotation plant of the McKinley-Darragh, the machinery for which is being installed, is expected to be in operation within the next few weeks. This will give a total capacity of 500 tons per day, half of which will be taken from the mine and the other half from the tailing dump, to be re-treated by flotation. Improvements have been made in the hoisting facilities to handle the increased tonnage. It has been found that several bodies of low-grade rock, which could not be profitably treated two years ago, are now of economic value, owing to the increased price of silver. At the Hargraves, in driving an old vein, a lens of high-grade ore stated to assay about 5000 oz. per ton, has been discovered. It is more than 3 in. wide, and is accompanied by good ore on both sides. There is a large quantity of low-grade millable ore on the dump which, under present conditions, can be profitably treated. The Adanae has begun stoping, and is sacking ore for shipment.

Development in Rickard township, about 40 miles northeast of Porcupine, where a recent gold discovery attracted a number of prospectors, has been delayed by the refusal of the Provincial Mines Department to grant working permits. It appears that the Abitibi Pulp & Paper Co. is constructing a dam in the neighborhood, which, when completed, may flood a considerable area of the gold-bearing district and mining operations cannot be undertaken until some arrangement for protecting the company's rights can be effected.

The copper deposits of the mineral belt of northern Manitoba are undergoing active development. At the property of the Mandy Mining Co. at Schist lake, from which 3600 tons of sulphide ore was shipped to the Trail smelter last winter, the crude makeshift equipment first in use has been replaced by modern machinery. The shaft is down 90 ft., and when the 100-ft. level is reached driving will be undertaken. It is proposed to take out at least 8000 tons of ore before spring, for shipment next year. At the Flin-Flon the most extensive deposit of sulphide ore so far discovered in Manitoba is being exploited by diamond-drilling. Approximately 6000 ft. of hole has been drilled this year, and over 6,000,000 tons of sulphide ore thereby has been proved.

COBALT, ONTARIO

HEAVY OUTPUT MAKES RECORD PRODUCTION FOR THE YEAR.—INCREASED BONUS BEING PAID.—NEW DEVELOPMENT AND NEW MILLS.—INCREASED PRICE OF SILVER OFFSETS HIGHER COST OF OPERATION.

During the first week of August, Cobalt companies sent out 900,000 lb. of ore. During the month of July a total of 3,621,300 lb. was shipped, which is a new high record for one month during the current year. Bullion shipments for the year so far total 6,100,344.61 oz., valued at $4,614,843.54. Nipissing, with over 3,000,000 oz. was the heaviest shipper; the Mining Corporation, with over 2,500,000 oz., was second. The camp is well supplied with men but their efficiency is low. The increasing cost of production is being offset by the higher quotation for bar silver. The 50c. per day bonus to mine workers when silver averages over 70c. per oz. is being paid, and when the average is over 80c. for any one month an additional 25c. per day bonus will be paid. In July the average was a fraction below 79. The mine workers are now anticipating the extra bonus. It is an interesting fact that an increase of one cent per ounce adds $200,000 to the value of the 20,000,000 oz. annual output from Cobalt district, and an increase of 25c. per day to the 3000 men employed approximates $225,000. Thus a one-cent rise will nearly pay the extra bonus.

The report recently issued by the Kirkland Lake Gold Mines, Ltd., which is controlled by the Beaver Consolidated, of Cobalt, shows approximately 6000 tons of ore in the dumps, valued at $60,000. The ore blocked out underground is valued at $420,000. It is proposed to sink a new shaft 500 ft. west of the present workings, where a 150-ton mill will be erected. The main shaft of the Kirkland Lake Gold Mines is down to the 700-ft. level where the downward continuation of the main orebody was reached recently, and which is similar to the ore at and above the 600-ft. level. The average grade of the ore is around $10 per ton.

The official report of the Kerr Lake output for June was 251,367 oz. silver, making a total of 2,195,485 oz. for the 10 months ended June 30 and indicating approximately 2,700,000 oz. for the 12 months ending August 31.

To the north along the contact, the Chambers Ferland property of the Alladin Cobalt Co. is finding high-grade ore. The Alladin, formerly an English company, will be listed within the next few days on the Standard stock exchange.

The flotation plant at the McKinley-Darragh is nearing completion. A part of the machinery is in place, and within a few weeks the mill will be ready for operation.

THE MINING SUMMARY

The news of the week as told by our special correspondents and compiled from the local press.

ALASKA

(Special Correspondence.)—Twelve men under the direction of W. Krippachne have started development work on the Auk Bay property recently bonded by the Treadwell company. One of the big boilers from the Treadwell Central power-plant is being shipped to Camas, Washington, where it will be used in a pulp-mill.

Lynch Brothers, diamond-drill contractors, have sent a drill and crew to El Capitan, on the west coast of Prince of Wales island, where they will drill the marble deposits at that place. The crew will be in charge of Fred Lynch. The El Capitan marble deposits were formerly worked extensively, but the quarries have been idle for some time. Henrie Brie and associates of Douglas, who are interested in mining claims at Auk Bay, have had men at work constructing a road over which machinery and supplies can be taken to the properties. The exports from Alaska for July, as compiled by the local custom-house, show a marked increase in value, due in part to the salmon shipments and the large shipments of gold. The total value of all exports of Alaskan products, including gold, during the month amounted to $6,357,966. The copper ore shipments amounted to $2,288,708. The value of gold bullion, gold ore, and silver ore shipped was $1,914,011.

Chas. T. Counsell, Federal immigration officer, has arrived at McCarthy, as a representative of the Department of Labor, to investigate the Kennecott strike, report on conditions, and mediate a settlement if possible. The Kennecott imported men are not being interfered with and the plant is running at about a fourth of its capacity. There are 240 men working, most of these are from the Government railroad, canneries, and different points in south-eastern Alaska. The Bonanza mine is idle, but the reserves in the Jumbo have been drawn on to keep the mills running.

Treadwell, August 5.

ARIZONA

YAVAPAI COUNTY

(Special Correspondence.)—The General Development Co. has taken over the Mayer-Belford group of 11 mining claims four miles east of this place, and the Ewing & Hooker mine near Jerome. The finding of high-grade ore on the 600-ft. level of the Arizona Binghampton is evident in the flotation-mill. Large orebodies running from 2 to 3% copper have been found in the Copper Queen mine adjoining the Arizona Binghampton, and a 3-ft. vein of 25% copper ore has been cut in a 40-ft. shaft on the Big Bug Copper Co.'s ground a mile south of the Copper Queen. The Barbara mine is in the hands of Bisbee lessees, John Ross, W. A. Tucker, James Smith, and Lee Hunt, in the same district. Senator Reynolds and Gelora Stoddard have begun the reorganization of the old Stoddard Mining Co. and will do extensive development. High-grade copper are has been discovered in a 50-ft. shaft by O. M. Bywaters, of Roxton, Texas, on a property south-east of the Copper Queen. The Half Moon company will start development.

The Mayer-Belford mine is known locally as the Christmas, having been located by William Belford on a Christmas day. The group is five claims long, covering a strong outcrop of mineralized schist and iron. At several places along the strike, the vein is from 200 to 300 ft. wide and there are zones of it that will run from 2 to 17% copper. A shaft has been sunk 50 ft. It is reported that each owner has received a cash payment and a block of the stock in payment in full for their holdings. The company will give particular attention to the Ewing & Hooker mine, near Jerome. It is reported that $250,000 was paid for this property, a good sized payment having been made, the balance being due at the end of the year. It is also said that work will be commenced on the Christmas mine at Mayer within 90 days. It is reported that the General Development has put up $500,000 so far.

The district around Copper mountain covers an area at least five miles square with many properties yet to be exploited adjacent to the mines already taken over by various companies. Prospecting can be carried on easily, as there is no sedimentary formation covering the mineralized formations, which in many cases can be traced across the entire district.

Mayer, August 15.

CALIFORNIA

The State Water Commission has approved the following applications, granting permits for the appropriation of water for mining and power purposes:

La Grange Mining Co., 10 cu. ft. per second of the waters of Salt creek, tributary to Stuart's Fork in Trinity county for hydraulic mining purposes, the water to be returned to the stream.

G. D. Meikeljohn of Ohama, Nebraska, 0.19 second feet of springs in San Bernardino county for mining purposes, the gold mines proposed to be served being at the south end of Providence mountain. The works consist of a masonry dam 10 ft. high, 30 ft. on bottom and 20 ft. on top, and a pipe-line 4 miles in length, the estimated cost being $5000.

J. Irving Crowell, of Los Angeles, 0.05 cu. ft. per second of Cane Springs for gold mining 15 miles south-west of Rhyolite, Nevada. There is a pipe-line half a mile long. The estimated cost of the diversion is given at $7000.

Southern California Edison Co. of Los Angeles, permit under section 12 of the Water Commission Act, fixing the time for the completion of a diversion that originated prior to the time the act took effect. Amount not to exceed 600 cu. ft. per second of the waters of Kern river, for power purposes. Actual construction is to begin by September 1, 1917, and be completed by July 1, 1922. The estimated cost of completing the project, exclusive of transmission lines, is $4,118,295. The total capacity of the plant is given as 30,000 kw., the total fall to be utilized as 800 ft. and the theoretical horse-power to be developed as 54,545.

Mariposa Mine Association of Alleghany, 0.3 second feet of Steamboat gulch waters in Sierra county, for mining purposes, the water to be returned to the stream after use.

H. S. Williams, of Los Angeles, the waters of a mountain spring in Ventura county for the development of oil-wells.

CALAVERAS COUNTY

(Special Correspondence.)—At the Buffalo gravel mine a Deister concentrator has been placed in the mill to save the fine gold. Preparations are being made to work continuously throughout the coming winter.

Mokelumne Hill, August 14.

(Special Correspondence.)—The Newman mine is mining and crushing pay-ore. New machinery is being installed.

There is a general revival of interest in mining in this district.

West Point, August 14.

Suit for $850,000 was filed August 10 in the United States District Court by E. G. Parlow, C. A. Tarbat, Domingo and R. Roleri, owners of the Finnegan Extension quartz mine, against the Melones Mining Co. It is alleged that the owners of the Melones mine have extended its workings into the Finnegan property and that 170,000 tons of ore, valued at $5 per ton, has been taken out. The property involved in this suit is situated on and near the summit of Carson hill, in the southern part of the county.

Del Norte County

(Special Correspondence.)—Otto Anderson is opening up copper ore on one of his claims in the Shelley Creek district.

The Monumental mine, which has been idle for several years, also on Shelley creek, is to be re-opened by Eastern parties. There is a large tonnage of ore developed running well in gold.

J. C. Kendau and Earnest Rackliff, of Nevada, have been examining the John Griffin cinnabar property on Diamond creek, adjacent to the Ehrman cinnabar mine.

A rich shoot of copper ore has been opened at the Old Crow property, situated in the Monumental district. The vein is 3½ ft. wide. It is owned by F. E. Bauman, of San Francisco, and managed by E. A. McPherson, of Crescent City.

J. N. Britten is cross-cutting a body of copper ore in the Patricks Creek district.

Chrome ore is now being shipped from the Gravlin property at Cold Spring.

Crescent City, August 13.

Mendocino County

(Special Correspondence.)—The Sullivan Machinery Co. is moving its diamond-drill machinery from Yreka to commence new operations after completing the contract with the Gray Eagle company on the Klamath river in the Happy Camp district, where 8000 ft. was drilled for the company, and the result showed that everywhere the ore went down to a great depth, and proved to be rich.

Yreka, August 16.

Nevada County

(Special Correspondence.)—J. H. Batcher, of San Francisco, has purchased the Summit mine, near Nevada City, from the Mountaineer company and will consolidate it with the Mohegan and New England mines. It is planned to rehabilitate the deep New England shaft for working purposes and to build a mill. A large tonnage of milling ore is stated to be blocked out in old workings.

M. J. Brock and associates have begun work on the Oak Tree claim, within the city limits of Grass Valley. The 200-ft. shaft has been repaired to a depth of 40 ft. The claim was worked with good results in pioneer days.

C. K. Brockington, superintendent of the Golden Centre, is rapidly completing arrangements for sinking the 1000-ft. shaft to 2000 ft., where it is expected to recover the rich faulted veins. Approximately $100,000 will be expended.

Several shipments of chrome are going out from the Washington district. The Red Ledge gold mine is producing large quantities of chrome, and the deposit lately opened on Poorman's creek by Peter and Robert Murcateti has a 40-ton shipment prepared.

Nevada City, August 20.

Plumas County

(Special Correspondence.)—Under the management of Albert Burch of the Philadelphia Exploration Co., the Crescent and Green Mountain mines are being cleared of water to the 400-ft. level, and pumps have been provided. Both mines contain gold-bearing quartz and formerly yielded well. They are near Crescent Mills.

Driving of a 550-ft. tunnel is progressing at the Pilot mine, in Genesee valley, to intersect the orebody at a depth of 280 ft. The vein is 6 ft. wide and carries gold, silver, and copper. It is operated by A. P. and J. W. Goodhue.

A deposit of manganese is being prospected in Dixie valley, near Greenville, by the Noble Electric Steel Co. It is near the Indian Valley railroad and some high-grade ore has been exposed.

The Gruss Mining Co., composed of the Gruss interests of San Francisco, has been formed to operate the Genesee, Little Genesee, North East Extension, and Mohawk properties, in the Genesee district.

The Billie Boy Mining Co., of Los Angeles, has taken under bond the Stover and MacKenzie gold-copper properties, two miles east of Lake Almanor. Driving of a cross-cut has begun. Some rich ore has been opened near surface.

Quincy, August 18.

San Bernardino County

(Special Correspondence.)—The discovery of platinum ore is announced as having been made on the 150-ft. level of the West End mine, at Cima. This property has been producing and shipping a large tonnage of lead are, and it is thought probable that considerable platinum was unwittingly shipped before its presence was known. The platinum is said to occur with the lead carbonate. The best ore of this character assays: gold, 4.2 oz.; silver, 5 oz.; platinum-palladium, 7.61 oz.; lead, 47.5%. Copper is thus far absent from the ore, which fact it is thought will make the recovery of the platinum and associated metals easy. The ore occurs in a vein at the contact of limestone and granite.

Cima, August 11.

Shasta County

(Special Correspondence.)—Fifteen electrolytic cells are in operation in the Mammoth Copper Co.'s zinc-plant at Kennett. The first carload shipment of zinc was made this week. A large amount of cadmium also has been produced.

The flotation plant of the Afterthought Copper Co. at Ingot is nearing completion. It is thought the mill will be ready for test about September 1.

Redding, August 16.

Siskiyou County

(Special Correspondence.)—D. M. Watt, of Butte, Montana, and David D. Good, of Ashland, who have been developing the High Grade group of gold mines in the Stirling district on the south side of the Siskiyou mountains since early last spring, have closed down operations and returned to Ashland. This group was purchased by these operators last spring from James Hopwood of Ashland for $10,000 cash after producing rich returns to the original locators. The present owners are expecting to close a deal for the purchase of the property for people of Butte.

W. F. Green of the Boorse, Green & Duck Copper Co. that has holdings in the Ike's gulch is expected in Happy Camp to begin development work on the company's holdings. These properties adjoin the Gray Eagle property on the north-west.

Charles Williams and G. Logan Brown are among the San Francisco mining men who have been examining mining properties in Happy Camp district.

Electric power equipment is being installed on the old Jitson mine near Hornbrook, the group being now known as the Hazel Mining Co.'s holdings. For some months a large force has been employed on the property preparing for development. It is reported that the present owners, with O. Jillson at the head, will spend $100,000 to put this group in the producing list again.

Hornbrook, August 16.

COLORADO

Boulder County

(Special Correspondence.)—James Maxwell, superintendent for the Tungsten Mountain Mines Co., reports a satisfactory season's work. The lessees are making a regular production of a good grade of ore. Development will be continued throughout the winter. The new machinery is giving good satisfaction. The new aerial tramway of the Mojave Boulder Co.'s tungsten mines is completed. At the Huron property, at Eldora, Woodring and Dupont are pushing work. This is one of the highest grade gold properties in Boulder county. Shipments soon will commence. There is now more mining and ore production in Eldora district than for several years past. The improvement is due principally to the smaller mines and the new enterprises that have started this season.

Eldora, August 14.

Summit County

The American Metal Co. and the Molybdenum Products Corporation are opening extensive deposits of molybdenite near Climax, a station on the Colorado & Southern railroad at the summit of Fremont pass, at an altitude of 11,303 ft. There are 300 men employed in mining and construction work. These deposits have been known for years, but it is only recently that the demand and high price for molybdenite have encouraged the investment of the large capital necessary for the development and equipment of the mines. It is proposed to crush the ore and treat the pulp by flotation, reducing the volume from 65 to 1. The first mill will have a capacity of 250 tons daily. The deposits are on Bartlett mountain and in Ten Mile gulch.

IDAHO

Shoshone County

(Special Correspondence.)—The Federal Mining & Smelting Co., operating in the Coeur d'Alene and Wood River regions of Idaho, has declared its regular quarterly dividend of $210,000, payable September 13. This makes the total for the year $930,000, and the grand total to $13,716,695 on the preferred stock. Payments on common stock aggregate $2,708,750, making the total disbursements $16,425,445.

The Tamarack & Custer Consolidated Mining Co. produced 3000 tons of crude lead and concentrating ore in July. The net value was $50 per ton, or $150,000. The net profits of the company for 1916 were $338,746. More than 350 men are employed in the mine and mill.

The Hecla Mining Co. has declared its regular monthly dividend of $150,000. This will raise the total of payments this year to $1,200,000 and the grand total to $8,505,000.

Spokane, August 15.

MICHIGAN

Houghton County

(Special Correspondence.)—Wolverine showed a marked falling off in the percentage of refined copper secured from the ore during July, the pounds per ton being 14.112. In the first two weeks of August there has been a decided recovery, however, indicating an average for August close to the recovery for June, which was 16.516 lb. per ton. The refined output from the Wolverine for July was 363,888 lb. from 564,000 lb. of concentrate secured from 25,771 tons of ore.

The Mohawk produced 41,687 tons of copper ore during July, from which the stamp-mill produced 1,499,100 lb. concentrate and the smelter returned 970,538 lb. of refined copper.

Houghton, August 14.

MONTANA

Fergus County

(Special Correspondence.)—A shortage of labor is interfering with the development and operation of the mines of Fergus county. Many miners have been drafted and others have left to secure the higher wages paid in base-metal mining camps. At Kendall, the Barnes-King is producing about one-half of its usual tonnage. Lessees are working the old Kendall mine with a small force and assessment work is being done on several claims.

The Spotted Horse, Maginness, and Cumberland mines at Maiden are being successfully operated by lessees. A fresh inflow of water last week flooded the Cumberland, but the new pump has handled it and the workings are drained. Lewistown capitalists are planning the development of the War Eagle, on which a body of silver-lead ore carrying gold and zinc has been explored by an adit and cross-cut. Lewistown men are also financing the Nelson group, which has an excellent surface showing.——An adit is being driven on the Sutter group of claims on Armell's creek to strike the copper-bearing orebody in depth. Several carloads of copper carbonate was shipped from these claims last fall. Oswald Lehman has brought suit against the Sutters, claiming prior location of the claims. N. J. Littlejohn is re-opening the Mammoth property near Giltedge. Some of the tailing from the Giltedge mill is being re-treated in a small way by the cyanide process.

The Casofour plant of the U. S. Gypsum Co. is situated 10 miles east of Lewistown. The property is developed by two 500-ft. inclines. The inclines, surface cuts, and churn-drilling have shown that the gypsum bed averages 14 ft. thick, but only 8 ft. will be mined; the rest will be left for a roof, as the overlying rock is treacherous and would require much costly timbering. The property is awaiting the construction of a side-track and the arrival of construction material and machinery.

Progress is being made with the plant of the Three Forks Portland Cement Co., seven miles west of Lewistown. A 14,000-ft. aerial tramway is being installed by Leschen & Sons Co. and is nearing completion. This will bring limestone from the South Moccasin mountains. The Three Forks Co. has spent $8000 on a road to the top of the South Mocassin to haul the tramway and crusher parts to the quarry. The heaviest parts weigh 20 tons. The tramway and cement-plant will be operated by electricity furnished by the Montana Power Co. There will be room for two kilns, but only one will be installed at present. This will be 285 ft. long. The cement will be made by the wet process. The plant will not be in operation before January 1.

A carload of lead-silver-copper ore has been shipped to East Helena from the Baker claims on Wolf creek in the Little Belt.

Lewistown, August 12.

Lewis and Clark County

(Special Correspondence.)—Lewis and Clark county shows more activity and development than at any time in its history. The Drumlummon is again in operation, but by a new company. This change resulted from litigation in which the St. Louis Mining Co. was given, under a judgment, all of the holdings of the Drumlummon company, which included a mill, and the placer and quartz claims, under patent. The 25-stamp mill has been remodeled. The process used is continuous cyanidation. Electric power operates the machinery. The ore thus far milled is from the Drumlummon, taken through the Maskeline tunnel of that property. The stopes contain low-grade material that was left by the old company. The cyanide process has made it possible to work these ores at a profit. William Mayger, organizer of the St. Louis Mining Co., is in charge.

The Barnes-King company is operating the Shannon mine and the Piegan-Gloster group. The Shannon is comparatively a new mine, though the Piegan-Gloster was worked years ago.

In the Shannon a large body of low-grade ore exists in the upper levels. On the 500-ft. level an orebody has been opened that assays high in gold. Low-grade ore is coming from the Piegan-Gloster.

The Marysville Gold Mining Co. is developing a group of claims between Belmont and the Shannon mine, and high-grade ore has been uncovered in many places. The most extensive work is in the Blue Bird-Hickey adit. Duluth, Minnesota, interests are in control. A milling plant is to be built when the development warrants it.

At the Bald Butte mine, near Marysville, prospecting for new orebodies is in progress. The mine has not been a producer for some years. Bald Butte formerly paid $1,500,000 in dividends. It has a complete milling plant.

Marysville, August 14.

NEVADA

Humboldt County

(Special Correspondence.)—Surveys have been completed for an electric railway between the mines and mill of the Rochester Combined company. Construction of the mill building is proceeding rapidly and equipment is being placed in position. The drift from the Kromer shaft has discovered sulphide ore.

Connections have been established between the 800-ft. level and the Friedman adit of the Rochester mine by a raise, and all ore from the 800-ft. level and upper workings is dropped direct to the Friedman adit, and trammed to the mill. On the 800-ft. level the Adams orebody is from 15 to 24 ft. wide, all of good grade. The mill is treating 198 tons of ore per day. The July bullion output totaled $55,355.

R. G. Gillespie, of Pittsburgh, Pennsylvania, has bought seven claims adjoining the Happy Jack group of the Rochester Combined from John Nones and William Merklinger. Rich silver ore is exposed. Gillespie operates the Monarch Consolidated property near Sierra City, California.

Arrangements have been made for an examination of the potash deposits recently discovered near Lovelock, and now under option to Chicago capitalists. If the examination proves satisfactory development will be started and a plant erected.

Rochester, August 19.

Nye County

(Special Correspondence.)—The Tonopah Mining Co. last week produced 1450 tons of ore. The Extension shipped 35 bars worth $63,500, which represents the bi-weekly clean-up. The ore production the past week was 2380 tons. On the 1440-ft. level of the Victor, the Murray vein shows a 10-ft. face.

The clean-up of the Tonopah Belmont mill consisted of 100,918 oz. of bullion valued at $120,000, and 35 tons of concentrate valued at $15,000. The plant at Millers shipped 47,234 oz. of bullion valued at $49,000, and 25 tons of concentrate valued at $10,000. The tonnage for the past week was 2634 tons.

The Rescue-Eula the past week produced 255 tons of ore.

Having finished the station work on the 900-ft. level of the Desert Queen, the Jim Butler is cross-cutting into the Ophir King. This mine sent 800 tons of ore to the Belmont mill at Millers the past week. The company earned $297,863 net during the first six months of the year.

Tonopah, August 13.

NEW MEXICO

Socorro County

(Special Correspondence.)—A strike of high-grade ore has been made on the 1100-ft. level by the Socorro Mining & Milling Co. The drift has advanced 30 ft. in this ore. In the last half of July, 16,000 oz. of bullion was sent to the Mint. The Mogollon Mines Co. has cleaned up 13,000 oz. from the last two weeks' run. A diamond-drill has arrived at the mine, and will be used to prospect the lower levels.

Regular shipments are being made to the custom-mill from the Maud S., Pacific, and Johnson mines. Work has been started on the Wiley group in the White Water district.

Mogollon, August 12.

OREGON

Harney County

(Special Correspondence.)—Lorenz, Lubbinga & Ebling are developing a copper property in the southern end of this county on the Red Horse claim of the Happy Chance group of 13 claims. They have uncovered 3% ore 400 ft. wide. The ore occurs principally in monzonite. There are dikes of amygdaloidal rock. On the Big Jag and Coppertown claims there is 300 ft. of 2½% ore. The ore is copper carbonate, occurring in the amygdaloid. On the Black Dike claim two shifts are cross-cutting a 60-ft. vein. The formation here is andesite. The ore on this claim is copper glance, tetrahedrite, and nickel, similar to the ore of the Sudbury district of Ontario.

Fields, August 12.

Josephine County

The placer mines in this district are shut-down until the rainy season. The Waldo copper mine shut-down August 6. It is reported a sale is pending and changes are to be made.

Waldo, August 8.

(Special Correspondence.)—During July 1600 tons of copper ore from the Waldo district passed over the branch road, 15 miles to Grants Pass, the nearest shipping point. This shipment was from the Queen of Bronze group, and the Pickett creek properties, owned and operated by the New York & Oregon Development Co. The wagon-haul from these properties is from 20 to 30 miles to the railroad. The ore went to the Tacoma smelter.

The chrome tonnage from the same district and over this road was 1200 tons. Shipments of chrome will be increased during August, as the new wagon road to the deposits was not completed until after the first week in July. The chrome shipments also require a long haul to the railroad. All of the shipping chrome deposits adjacent to this new wagon road are being operated by the United States Steel Co., shipments going to Pittsburgh, Pennsylvania.

Tucker and Fife, of Takilma, have taken a lease on the Cow Boy and Lyttle copper claims and are shipping. These properties are controlled by the Queen of Bronze people. The ore runs from 10 to 36% copper, and the shipments amount to two carloads per week.

Collins McDougall, of Grants Pass, has taken a lease on the Meade gold prospect, situated on Jones creek, 5 miles east of Grants Pass. This is a free-milling mine, running $5 per ton in gold. The vein is 12 ft. wide and has been prospected for 1200 ft. Mr. McDougall is negotiating for a 5-stamp mill to be operated by a gasoline engine. He expects to be able to crush 15 tons of ore per day.

The output of the Logan placer mines in Waldo district, which recently closed down for the season, amounts to $40,000 in gold. The yield of platinum was said also to be large but the figures are not available. The mine is operated by George M. Esterly of Seattle. The water supply permits mining for 8 months of the year. The gold is very fine, and accompanied by platinum, also a little osmium and iridium. The deposit is from 10 to 25 ft. in depth. There are three ditches, the water from one being used in the elevator under a head of 325 ft., another is employed in two giants, and the third is used to clear away the tailing from the end of the sluice at the head of the elevator.

Grants Pass, August 9.

(Special Correspondence.)—Chrome mining is increasing near Adams station on the Crescent City-Grants Pass road.

A large group of chrome claims are owned or leased by R. J. Rowen, M. E. Young, and Geo. S. Barton, of Grants Pass, who are pushing development. About 40 adits on the veins have been started, in most of which chrome has been uncovered. The properties are on French hill, seven miles from the wagon road. It is planned to haul the ore by auto-trucks to Crescent City.

Mike G. Womack, of Medford, and M. A. Clark, of Ashland, are developing manganese, two and a half miles west of their recent manganese strike, 10 miles west of the Josephine caves on Buck peak. The ore assays 50% manganese. The nearest shipping point is 22 miles at Wilderville, on the Grants Pass-Crescent City railroad.

C. A. Winetrout has finished the installation of a gas engine to operate the Huntington mill on the Boswell mine three miles from Holland.

Grants Pass, July 25.

SOUTH DAKOTA

Lawrence County

The Homestake Mining Co. has declared its usual monthly dividend of $65,000, payable August 20. Total paid to date $43,786,234.

WASHINGTON

Stevens County

(Special Correspondence.)—Operations in the magnesite properties at Chewelah and Valley, 50 miles north of Spokane, are in progress. Shipments of magnesite from Chewelah station aggregated 9450 tons in July. The net earnings were $37,800 on a basis of $4 net per ton of crude ore. The manufacturing department of the Northwest Magnesite Co. is about ready for operation. The investment represents $260,000, including the buildings at Chewelah that cover five acres, the equipment, and a tramway connecting the manufactory and quarry, which are five miles apart. The plant will be used in the manufacture of magnesite specialities.

Engineering Council

Subsequent to the organization meeting held on June 27, 1917, the Engineering Council has held two other meetings, one on July 13 and one on July 26. It has considered many matters of interest to engineers in general.

There have been appointed standing committees:

1. On Public Affairs: C. W. Baker, G. F. Swain, S. J. Jennings, and E. W. Rice, Jr.

2. On Rules: J. P. Channing, Clemens Herschel, N. A. Carle, and D. S. Jacobus.

3. On Finance: B. B. Thayer, I. E. Moultrop, Calvert Townley, and Alex. C. Humphreys.

Many matters coming before the Council, both from the several founder societies and from the Council's predecessor, the Joint Conference Committee of National Engineering Societies, have been considered and referred to appropriate standing committees for investigation and report.

The Council has also created a War Inventions Committee comprising H. W. Buck, A. M. Greene, Jr., and E. B. Kirby, to co-operate with the Naval Advisory Board and other departments at Washington if desired in the promulgation to engineers of war problems now before the Government and for which there are opportunities for solution by means of inventions. It also created a committee comprising George J. Foran, E. B. Sturgis, A. S. McAllister, and A. D. Flinn, which is to collect and compile such information regarding engineers of the country as will enable the committee to co-operate with the different departments of the Federal Government on request and to assist in supplying the Government's need for engineering services.

Personal

Note: The Editor invites members of the profession to send particulars of their work and appointments. This information is interesting to our readers.

F. L. Sizer is at Sutter Creek, California.

Frank Stock is in San Francisco from Cuba.

C. D. Kaeding was at Salt Lake City this week.

R. A. Kinzie has returned from Bishop, California.

Gilmour E. Brown has returned to Shanghai from Sumatra.

Morton Webber will return to New York early in September.

H. J. Maguire is now superintendent of the leaching-plant at Anaconda.

George A. Tweedy has gone to Honduras to examine the Opetika mine.

Edwin E. Chase and his son will be at Jarbidge, Nevada, until September.

Victor C. Alderson has accepted the presidency of the Colorado School of Mines.

Guy C. Riddell is expected at New York in September on his return from Australia.

Corey C. Brayton is in Idaho, and will go to Utah before returning to San Francisco.

George S. Young has left the Colorado School of Mines, and will reside in San Francisco.

B. H. McLeod, formerly in Mexico, has joined the Royal Flying Corps in the British Army.

H. W. Aldrich is superintendent for the Ladysmith Smelting Corporation, at Ladysmith, B. C.

Drummond MacGavin, now lieutenant in the Field Artillery, has gone to an Eastern military depot.

Fred. B. Ely, who has been in the officers training camp for three months, is now at Superior, Arizona.

Ellard W. Carson, manager of the Oceanic quicksilver mine, is examining properties in north-east Nevada.

E. M. Rabb, superintendent of the Tom Reed mine, has returned to Oatman after a trip to mining districts in Colorado and Arizona.

Henry M. Payne has moved to Pittsburgh on his appointment as assistant to the president of the Bertha and affiliated coal companies.

W. W. Henry, superintendent of the Arps Copper Co., Shasta county, California, has been admitted to the second training camp for officers.

Rowland Feilding, formerly in practice as a mining engineer, has been promoted to lieutenant-colonel in the British Army and awarded the D. S. O.

Walter E. Trent, president, and Dr. L. A. Dessar, secretary-treasurer, of the Louisiana Consolidated Mining Co., are now at the mine in the Tybo district.

F. W. Denton has resigned as manager for the Copper Range Co., becoming vice-president of the company. He is succeded as manager by W. H. Schact.

G. H. Coleman is now superintendent for the Noble Electric Steel Co. at Heroult, California, relieving W. W. Clark, who has held that position for nearly two years.

Nathan H. Jones, engineer for the Spring Valley Water Co., at San Francisco, has received a commission as first lieutenant and is now at Vancouver Barracks, Washington.

Braxton Bigelow, an American mining engineer, serving as captain in the engineer corps of the British Army, was wounded in a raid on the German lines on July 23 and is reported 'missing.'

THE METAL MARKET

METAL PRICES

San Francisco, August 21

Aluminum-dust (100-lb. lots), per lb	$1.00
Aluminum-dust (ton lots), per lb	$0.95
Antimony, cents per pound	15.50—17.00
Electrolytic copper, cents per pound	29.50
Pig lead, cents per pound	12.25—12.50
Platinum, soft and hard metal, respectively, per ounce	$105—111
QuicksilVer, per flask of 75 lb	$115
Spelter, cents per pound	10.50
Tin, cents per pound	60
Zinc-dust, cents per pound	18

ORE PRICES

San Francisco, August 21

Antimony, 45% metal, per unit	$1.20
Chrome, 40% and over, free SiO_2 limit 8%, f.o.b. California per unit	$0.55
Magnesite, crude, per ton	$8.00—10.00
Tungsten, 60% WO_3, per unit	$22.00—25.00
Molybdenite, per unit for MoS_2 contained	$40.00—45.00

Manganese, Carnegie schedule less freight to Chicago. Less than 40% special quotation.

Manganese prices and specifications, as per the quotations of the Carnegie Steel Co. schedule of prices per ton of 2240 lb. for domestic manganese ore deliVered, freight prepaid, at Pittsburgh, Pa., or Chicago, Ill. For ore containing

	Per unit
AboVe 49% metallic manganese	$1.00
46 to 49% metallic manganese	0.98
43 to 46% metallic manganese	0.95
40 to 43% metallic manganese	0.90

Prices are based on ore containing not more than 8% silica nor more than 0.2% phosphorus, and are subject to deductions as follows: (1) for each 1% in excess of 8% silica, a deduction of 15c. per ton, fractions in proportion; (2) for each 0.02% in excess of 0.2% phosphorus, a deduction of 2c. per unit of manganese per ton, fractions in proportion; (3) ore containing less than 40% manganese, or more than 12% silica, or 0.225% phosphorus, subject to acceptance or refusal at buyer's option; settlements based on analysis of sample dried at 212° F., the percentage of moisture in the sample as taken to be deducted from the weight. Prices are subject to change without notice unless specially agreed upon.

EASTERN METAL MARKET

(By wire from New York)

August 21.—Copper is quiet, no movement in the market and quotations are unchanged, being at 28c. all week. Lead market is dull, prices being merely nominal at 10.87 to 10.75c. Zinc is dull at 8.75 to 8.62c. owing to there being no buyers. Platinum remains unchanged at $105 for soft metal and $111 for hard.

COPPER

Prices of electrolytic in New York, in cents per pound.

Date		AVerage week ending	
Aug. 15	28.00	July 10	31.50
" 16	28.00	" 17	30.29
" 17	28.00	" 24	27.04
" 18	28.00	" 31	27.25
" 19 Sunday		Aug. 7	28.46
" 20	28.00	" 14	27.62
" 21	28.00	" 21	28.00

Monthly AVerages

	1915	1916	1917		1915	1916	1917
Jan.	13.60	24.30	29.53	July	19.09	25.66	29.67
Feb.	14.38	26.62	34.57	Aug.	17.27	27.03	....
Mch.	14.80	26.65	36.00	Sept.	17.69	28.28	....
Apr.	16.64	28.02	33.16	Oct.	17.90	28.50	....
May	18.71	29.02	31.69	Nov.	18.88	31.95	....
June	19.75	27.47	32.57	Dec.	20.67	32.89	....

The Granby Consolidated Copper Co. of British Columbia, paid an August dividend of $374,962 and has in cash and copper over $3,000,000. The cost of production in June is stated to haVe been 9.3c. per lb., deliVered at New York.

The 37th quarterly report of the Utah Copper Co., covering the second quarter of 1917, shows that the mills produced 56,403,465 lb. of copper, in addition to which 441,594 lb. was contained in ore shipped direct to the smelter, making a total production of 56,845,059 lb., as compared with 43,060,450 lb. in the previous quarter. The total net profit during the period was $10,563,541.11.

SILVER

Below are given the average New York quotations, in cents per ounce, of fine silVer.

Date		Average week ending	
Aug. 15	86.75	July 10	78.70
" 16	86.75	" 17	80.62
" 17	86.75	" 24	79.12
" 18	86.75	" 31	78.20
" 19 Sunday		Aug. 7	80.50
" 20	87.75	" 14	82.95
" 21	88.25	" 21	87.16

Monthly AVerages

	1915	1916	1917		1915	1916	1917
Jan.	48.85	56.76	75.14	July	47.52	63.06	78.92
Feb.	48.45	56.74	77.54	Aug.	47.11	66.07	....
Mch.	50.61	57.89	74.13	Sept.	48.77	68.51	....
Apr.	50.25	64.37	72.51	Oct.	49.40	67.86	....
May	49.87	74.27	74.61	Nov.	51.88	71.60	....
June	49.03	65.04	76.44	Dec.	55.34	75.70	....

The tendency of the silVer market has been toward greater ease. Recent arrangements of the Indian GoVernment haVe made supplies more accessible in London to meet the demand, which, apart from the requirements of the Indian GoVernment, has been by no means heavy for some time past. The market, however, has been so sensitive that the price has often fluctuated with little apparent cause for moVement. In these circumstances larger supplies will tend to restore normal conditions. On May 17 last attention was called to possible action on the part of the German GoVernment to withdraw silver coins from currency. It is now decreed that two-mark pieces shall cease to be current from January 1, 1918, though accepted by the Government in exchange for paper currency up to July 1, 1918. The following detailed returns of the Indian currency show an increase in the silVer-holding of 212 lacs. The total holding of 2431 lacs on July 22 last, compares with 2651 lacs held exactly a year before.

The stock in Bombay consists of 1700 bars, the same as reported last week. The stock in Shanghai on July 21, 1917, consisted of about 19,700,000 oz. in sycee and $15,100,000, as compared with about 19,900,000 oz. in sycee and $15,100,000 on July 14, 1917.

Quotations for bar silVer per ounce standard, averaged 39.833 cash for July 20 to 26 inclusive.

LEAD

Lead is quoted in cents per pound, New York deliVery.

Date		AVerage week ending	
Aug. 15	10.87	July 10	11.25
" 16	10.87	" 17	10.98
" 17	10.75	" 24	10.29
" 18	10.75	" 31	10.55
" 19 Sunday		Aug. 7	10.87
" 20	10.75	" 14	10.87
" 21	10.75	" 21	10.79

Monthly AVerages

	1915	1916	1917		1913	1916	1917
Jan.	3.73	5.95	7.64	July	5.59	6.40	10.98
Feb.	3.83	6.23	9.01	Aug.	4.62	6.28	....
Mch.	4.04	7.26	10.07	Sept.	4.62	6.86	....
Apr.	4.21	7.70	9.38	Oct.	4.62	7.02	....
May	4.24	7.38	10.29	Nov.	5.15	7.07	....
June	5.75	6.88	11.74	Dec.	5.34	7.55	....

ZINC

Zinc is quoted as spelter, standard Western brands, New York delivery, in cents per pound.

Date		Average week ending	
Aug. 15	8.75	July 10	9.20
" 16	8.75	" 17	9.06
" 17	8.62	" 24	8.60
" 18	8.62	" 31	8.75
" 19 Sunday		Aug. 7	8.75
" 20	8.62	" 14	8.75
" 21	8.62	" 21	8.66

Monthly AVerages

	1915	1916	1917		1913	1916	1917
Jan.	6.30	18.21	9.75	July	20.54	9.90	8.98
Feb.	9.05	19.99	10.45	Aug.	14.17	9.03	....
Mch.	8.40	18.40	10.78	Sept.	14.14	9.18	....
Apr.	9.78	18.62	10.20	Oct.	14.05	9.92	....
May	17.03	16.01	9.41	Nov.	17.20	11.81	....
June	22.20	12.85	9.63	Dec.	16.73	11.26	....

QUICKSILVER

The primary market for quicksilVer is San Francisco, California being the largest producer. The price is fixed in the open market, according to quantity. Prices, in dollars per flask of 75 pounds:

Date		Week ending	
July 24	110.00	Aug. 7	115.00
" 31	110.00	" 14	115.00
		" 21	115.00

Monthly Averages

	1915	1916	1917		1915	1916	1917
Jan.	51.90	222.00	81.00	July	95.00	81.20	102.00
Feb.	60.00	295.00	126.25	Aug.	93.75	74.50	....
Mch.	78.00	219.00	113.75	Sept.	91.00	75.00	....
Apr.	77.50	141.60	114.50	Oct.	92.90	78.20	....
May	75.00	90.00	104.00	Nov.	101.50	79.50	....
June	90.00	74.70	85.50	Dec.	123.00	80.00	....

TIN

Prices in New York, in cents per pound.

Monthly Averages

	1915	1916	1917		1915	1916	1917
Jan.	34.40	41.76	44.10	July	37.38	38.37	62.60
Feb.	37.23	42.60	51.47	Aug.	34.37	38.88	....
Mch.	48.76	50.50	54.27	Sept.	33.12	36.66	....
Apr.	48.25	51.49	55.63	Oct.	33.00	41.10	....
May	39.28	49.10	63.21	Nov.	39.50	44.12	....
June	40.26	42.07	61.93	Dec.	38.71	42.55	....

ORES

Tungsten: The position of tungsten has strengthened considerably in the past week, and the market has been correspondingly firm. No change can be made in quotations, which are $25 per unit for high-grade wolframite, and $22 to $23.50 for material containing slight impurities. The arrivals during the week were small and the aVailable stocks were not sufficient to take care of the inquiries. The domestic and foreign demand for ferro-tungsten is excellent, the prices quoted ranging from $2.25 to $2.50 per lb. of contained tungsten.

Molybdenum and antimony: There is no change in the quotations made last week of $2.10 to $2.20 per lb. of MoS_2 in 90% material, and the supplies are only nominal. Antimony ore has followed the trend of the metal market, and has declined slightly, sales having been made in the past week at $2 per unit for high-grade material.

Eastern Metal Market

New York, August 15.

Uncertainties as to the future, especially as regards the price-fixing policy of the Government, still continue to affect the metal markets, and prices remain practically nominal. Should a definite decision be reached regarding the price that will be paid for all the metal that will be purchased both for this country and the Allies, conditions would undoubtedly improve.

Copper is quiet and prices, which are nominal, remain unchanged.

Tin is dull.

Lead is very dull, but prices are firm.

Zinc prices are unchanged and the market is lifeless.

Antimony is quiet, with no change in prices.

Aluminum still continues inactive.

Increased orders for steel have been placed by the Government, and in spite of the embargoes, the export movement is large. Ordinary domestic business in iron and steel, however, is insignificant. Under pressure from some of its Allies the Government is hurrying its cost-inquiry, but the investigators are still some distance from the end of their work. Without taking a final position on the question of selling to the Allies at the same prices as the Government, steel manufacturers have accepted orders from officials at Washington this week for annealed wire and wire rods for Italy, the prices to be fixed after the Trade Commission's findings are made up. While the general question is in abeyance, it is understood that some steel interests have expressed a willingness to take business from the Government at a price to be fixed later, even though the material is for an Ally.

COPPER

The market has been quiet during the week with no pressure to sell and no inclination on the part of purchasers to buy. The present dullness is attributed solely to the uncertainties of the Government's price-fixing policy. As was the case last week there has been considerable speculation as to the price that will eventually be determined by the Government. While 18c. per lb. has been talked of in the past few days, this is, however, pure conjecture, since no announcement has been made of the price that will be paid, nor is there any indication as to just how soon an official statement will be forthcoming. The market in both Lake and electrolytic is dull, with 28c., New York, the nominal quotation for both. Sales for last quarter have been made by some dealers at 25.50 to 27c., and some spot business at 29c. is also reported. The London quotations yesterday were unchanged at £137 for spot electrolytic, and £133 for futures.

TIN

The throttling effect of price-fixing talk is apparent here also. Trading in tin is rendered more difficult by delays in the receipt of cables from London, due to the additional censorship on this side of the Atlantic. This, in conjunction with that already exercised by the English authorities, frequently causes a delay of several hours. Business for the day is often deferred until the daily cables are received, and sometimes this is late in the afternoon. Last week the market was quiet, coming to a halt on Friday, but Monday witnessed more activity for futures and a fair tonnage changed hands. Nobody is either willing or anxious to buy with the price-situation as it is, consumers, of course, being unwilling to purchase at the present quotations if there is a likelihood of a lower price being established later by the Government. Considerable interest has been aroused by the report from Washington that, in return for the effort made by this Government to obtain the same prices for the Allies as it has paid, the European nations must, in turn, sell to the United States and to its consumers on the same basis. The effect of the adoption of such a policy by England would be to depress the price of tin. Off-grades have been in fair demand. The arrivals of tin up to August 13 were 1810 tons, and there were 4215 tons afloat from the Straits and the United Kingdom. The quotations yesterday, when inquiry was light and the sales correspondingly small, was 62.25c., New York, for spot Straits.

LEAD

The Government requirements of lead for August, September, and October have been allotted to producers according to their capacity for supplying the metal. Unofficially the order is stated to be for 25,000 tons and the price paid, according to the same authority, was 8c. per lb., St. Louis. As is the case with the other metals no business is being transacted except under the spur of necessity. Despite this lack of sales the market remains firm at 10.87c., New York, although the leading interest has not reduced its quotation of 11c. per pound.

ZINC

The Zinc Committee of the Council of National Defense last week opened bids for furnishing 11,500,000 lb. of grade C metal. While no official information has been made public as to the prices paid, it being the intention of the Government to maintain secrecy regarding the purchase of this lot of zinc, it is reported that the great majority of these bids was on the basis of 8.75c. per lb., St. Louis, with a small number at 9c., and that orders for the Government requirements will be placed at these figures. If these prices are correct, it would seem to show that the market is not as strong as was thought last week when bids of 9 to 9.50c., St. Louis, were generally talked of by the trade. A preliminary report on the production of zinc in the United States for the first half of 1917 has been issued by the Geological Survey. According to this report, which is based upon returns from producers of 99% of the total output, the production reached 364,000 net tons, an increase of 13,000 tons over the last half of 1916, and approximately 48,000 tons more than the corresponding period of last year. This output is about 75% of the total amount of zinc produced in the United States during the entire year of 1915. The stocks on hand July 1 are estimated at 33,000 tons, or nearly twice what they were at the beginning of the year. Fully 33,000 retorts, including 14 complete plants, were reported idle on June 30. The market continues sluggish, with no change in quotations from last week. These are nominally 8.50c., St. Louis, or 8.75c., New York, for prime Western for delivery any time before October 1, the prices for last quarter being ½c. higher. Some business has been done for next year at even slightly higher figures.

ANTIMONY

Chinese and Japanese grades are still to be had at 15 to 15.50c. per lb., New York, duty paid. The market is dull and sales are few in number.

ALUMINUM

The quotation for No. 1 virgin metal, 98 to 99% pure, is unchanged at 50 to 52c. per lb., New York. There is little demand, and the market is not active.

Company Reports

OLD DOMINION

The Old Dominion Copper Co. for the year 1916 reports the ore extracted, 152,059 tons, containing an average of 5.88% copper. Mining cost was $6.48. The company produced, including custom ore, 40,776,611 lb. of copper, 223,228 oz. silver, and 4097 oz. gold.

ALASKA TREADWELL GOLD MINING CO.

The annual report of the Alaska Treadwell Gold Mining Co. for the year ended December 31, 1916, is as follows: operating cost for mining was $0.862 per ton milled. Development to the amount of 4222 ft. was done during the year, and 539,940 tons was stoped. A total of 671,378 tons was crushed.

In the 240-stamp mill 217,070 tons was crushed, and in the 300-stamp mill 454,308 tons. The quantity of sulphurets saved was 4397.48 tons in the 240-mill, and 10,720.66 tons in the 300-mill. The cost of milling in the 240-mill was $0.3993 per ton; and in the 300-mill $0.3585 per ton. The total operating cost was $1.3642 per ton, making a total cost of $915,865.54. Minor construction costs amounted to $688.81. The net profit on the year's operations was $658,118.69.

Dividends, No. 115 and 116, paid during the year—5% on the par value of the stock—amounted to $250,000.

ALASKA MEXICAN GOLD MINING CO.

The annual report of the Alaska Mexican Gold Mining Co. for the year ended December 31, 1916, shows the following:

Mining development, 1050 ft., stoping 69,285 tons cost $159,-469.89, at the rate of $0.9088 per ton. The mill crushed 175,-476 tons, from which 4397.54 tons of concentrate was saved, at a cost of $68,946.26, or $0.3929 per ton of ore treated. The expense of treating concentrate was $0.1371 per ton of ore milled, making a total cost for this branch of ore treatment of $24,-058.75. The total cost of mining, milling, and sundry minor expenses was at the rate of $1.5123 per ton, making a total of $665,366.97. The total operating cost, including a small construction charge, was $265,402.97. The free-gold recovered from 175,476 tons of ore crushed was $91,307.88, or at the rate of $0.5204 per ton. Base and copper bars increased this amount by $27,855.01. Concentrate treated, 4397.54 tons, yielded $155,859.64, at the rate of $0.8882 per ton of ore crushed. Total receipts from treatment of ore, $275,022.53. Interest received, $49,183.99. From other sources, $6447.05. Total income from all sources, $330,653.57. Net profit for the year, $64,236.29. Carried forward from 1915, $10,190.09. Total balance on hand, $74,426.38.

CORDOBA COPPER COMPANY

The report for the year ended December 31, 1916, shows a production of 1746 tons of blister copper, on which was realized £205,575 10s.2d. The working-costs amounted to £162,-884 4s.9d., and the income tax absorbed £6064 5s. A dividend of 10% was declared. As a result of the installation of an improved Murex plant for treating the wet middling and slime a higher average extraction of copper was obtained. The combined recovery in the mill by wet concentration in combination with the Murex plant was 75.2% of the assay value of the ore. This is an improvement of 4.7% over the recovery for 1915. The reserve of ore at the end of the year was 142,914 tons, having an average content of 2.39% copper. The mines of the company are situated about 10 miles north-east from the city of Cordoba in Spain. Mining is conducted through four shafts ranging from 820 to 1550 ft. deep. There is in operation a smelter with converters, and the company holds a lease upon a neighboring coal mine. The capitalization is £200,000 divided into 800,000 shares of 5s. each.

UNITED STATES SMELTING, REFINING & MINING COMPANY

The eleventh annual report covering the year 1916 gives as the total net earning of all the companies held by the corporation, exclusive of deduction for depreciation, the sum of $9,737,664.02. After deducting $839,200 for depreciation and exploration the net profit was $8,898,463.92, from which dividends on common stock of $1,492,238.75 were declared. An undistributed surplus of $12,957,454.96 was carried forward. The metals produced during the year, including those derived from custom ores and the output of the Mexican properties were as follows:

	Pounds	Average selling price, cents
Copper	28,888,093	27.297
Lead	103,855,451	6.676
Zinc	64,584,001	12,327
Silver, ounces	11,647,205	65.386
Gold, ounces	129,273	.....

At the Bingham mines improvements on the Niagara tunnel-level, including the installation of a pneumatic-haulage system, has been completed. The haulage-system has proved successful, and a similar installation has been made on the 400-ft. level, from the Galena shaft. The Niagara tunnel improvements have eliminated the necessity of operating the aerial tramway, and it has been dismantled and shipped to the Stowell mine in California. The Bingham mines of the company shipped during the year 109,586 tons of lead ore and 46,017 tons of copper ore. The Utah smelter at Midvale, Utah, has been enlarged by the installation of a seventh furnace, a magnetic separator, and a thaw-house. The magnetic separator is making good recoveries of pyritic material heretofore lost, and the thaw-house is used for thawing frozen ores. A Weatherby electric separator is being set up, and experiments are being conducted on the treatment of the mill-tailing by flotation. The zinc smelters acquired in Kansas during the year were in opertion, but were handicapped by insufficiency of the supply of gas, and a new plant having a capacity of 200 tons per day was purchased at Checotah, Oklahoma. Exploration was continued at the Centennial-Eureka, in Utah, and 51,381 tons of lead ore was shipped. At the same point some work was done on the Bullion-Beck and the Champion properties, and a small production was made, mostly by lessees. The Mammoth copper mine in Shasta county, California, produced 244,445 tons of ore, of which 10,597 was zinc sorted from the copper ore. Other mines in Shasta county operated by the company are the Stowell, the Spread Eagle, Anderson, and Friday-Lowden. The smelter has been operated throughout the year, and has also treated a large tonnage derived from the Balaklala mine. The construction of a plant has been started to recover the metals in the bag-house fume which has been accumulating for some years. The chief feature of this plant consists of a plant for the electrolytic recovery of zinc. The refinery at Chrome, N. J., is being increased by an addition that will give 50% greater capacity. At the Gold Roads mine in Arizona, the ore on the lower levels proved to be too low in grade to admit of being mined at a profit, and the mill has been closed. Further development has revealed some promising orebodies. The Tennessee mine at Chloride, Arizona, is being worked to a depth of 1400 ft., and a winze is following the ore-shoot below that level. The Needles mill treated 47,570 tons of ore during the year. Operations at the Real del Monte y Pachuca, in Mexico, were on a reduced scale, due to lack of supplies. Exploration on the orebodies, however, continued with satisfactory results.

EDITORIAL

T. A. RICKARD, Editor

WE take pleasure in announcing that Mr. George J. Young, lately professor of metallurgy in the Colorado School of Mines and formerly professor of mining in the Nevada School of Mines, has joined our staff as assistant editor.

IT is announced that the Miami Copper Company will not file a petition for the writ of certiorari required in order to appeal from the Appellate Court to the Supreme Court of the United States. Another course of procedure will be adopted.

AMONG the effects of the War is a great demand for bibles. We are glad of that, not only for moral and ethical reasons, but because of the opportunity given to learn good English. If our mining engineers and metallurgists would read the Bible their style in writing might be improved.

SHIPMENTS of scrap-metal have proved an embarrassment in the operation of the railroads in consequence of the large amount of rolling-stock involved. On recommendation from the Council of National Defense many railroads now decline to accept such shipments until evidence of acceptance by responsible buyers has been presented.

JUDGE BOURQUIN of the District Court at Butte has decided in favor of Minerals Separation in the suit against the Butte & Superior Mining Company for infringement of patent. We shall publish the text of the decision as soon as it reaches us. Meanwhile we look forward to attending the trial of this case when it is heard on appeal before the U. S. Court of Appeals of the Ninth Circuit in San Francisco.

SYMPHONY concerts are to be held as usual in San Francisco and the entry for the golf tournament at Monterey is unusually large. A year hence we shall be ashamed to remember either the concerts or the golf tournament. Any man having the health, strength, time, and money to qualify him for a golf championship ought to be engaged in a nobler occupation at this time. To spend money on symphonies when thousands of innocent victims are dying of destitution is a heedless stupidity.

LABOR-UNIONS are criticized for their persistent objection to incorporating themselves and thereby assuming legal responsibility for their acts. This recalls the fact frequently stated by broad-minded lawyers that the chief cause for the failure of justice in the courts is found in the complex trivialities of the rules of procedure, and the lack of a fundamental code. The legal trickster is able to take advantage of petty technicalities to discomfit the poor man that resorts to the law. The first practical step toward social reform is to restore confidence in the tribunals of justice, and that means to take them out of politics.

WE have received a pamphlet issued by the Mellon Institute, in the University of Pittsburgh, giving the text of the two opinions rendered in the U. S. Circuit Court of Appeals in the Miami case. The pamphlet is distinguished by an introduction written by Mr. Raymond F. Bacon, Director of the Institute, in which he dwells upon the importance of the litigation over the flotation patents and lays stress upon an understanding of the legal status of the processes involved. What interests us most, however, is his statement that Judge Buffington's minority opinion commends itself to mining engineers as being a fairer statement of the case than the majority opinion of Judge Woolley. Dr. Bacon says: "Having spent some years in the study of ore-flotation processes and the literature upon the subject, it is my personal opinion that the minority opinion of the Court is more closely in accord with the facts as mining engineers see them, and it should be sustained. I am able to say that all of a large number of neutral mining engineers with whom I have talked have expressed the same view." With this we concur.

THE editor of this paper is a member of a committee, of which Mr. James H. Collins is chairman, to assist in arousing the interest and support of the trade and technical press, and more particularly of their readers, in the effort to economize food under the leadership of Mr. Hoover, who has been appointed by the President to direct the organization created for that purpose. In our issue of August 18 we quoted the food-saving pledge in these columns and we reproduced it on an advertising page, so that any reader willing to sign it could cut it out without mutilating the reading-pages. The pledge will be found in this issue on page 30 of the advertising supplement. Please sign it and ask others to sign a similar pledge. You might also make copies of it and distribute them among your friends. When signed, send them either to us or direct to Mr. Hoover at Washington. The MINING AND SCIENTIFIC PRESS is read, we are glad to think, by the wives of many of the members of our profession. We appeal to them to co-operate in this patriotic work. General Pershing has said that the War will not be won by talking or by subscribing to the Red

Cross; it will be won by our soldiers on the field of battle and by our sailors on the sea; but our gallant men depend upon us to supply them with food and munitions, and our gallant Allies look to us to assist them in making good the shortage of food-stuffs due to the ravages of war. Those that cannot fight—the women and the men either too young or too old, even the children—can all help in so saving food that there will be plenty for those immediately affected by the destructiveness of war. Let us unite in this good work. It is a small service compared to that rendered by the brave men that go to the front, and we should be ashamed not to perform it enthusiastically and efficiently.

THE STATEMENT by Mr. W. H. Whittier, appearing on another page of this issue, emphasizing the peculiar advantages possessed by California for manufacturing electric steel and ferro-alloys, should serve to spur public insistence for a wiser policy in the administration of American, including Western, water-powers. The available energy in California, Oregon, and Washington is equalled only by Norway and Sweden, where Government supervision renders it possible to obtain rates equivalent to $6 per horse-power year. The growth of electric-steel production in this country is enormous, and the output is highly desired because of its superiority. The West Coast should take advantage of its marvelous water-power and develop a great steel industry, but in order to do this rates must come down and more hydro-electric projects must be undertaken. The Pacific Electro-Metals Company is now struggling to obtain a rational rate for its works at Bay Point, near San Francisco, where 10,000 horse-power will be required in the manufacture of ferro-alloys. This is a mere hint of the magnitude of the metallurgical operations that would be started under favorable regulations ensuring cheap electric current.

DANIEL GUGGENHEIM has allowed himself to be interviewed on the price of silver. He predicts a further rise in the quotation to $1 per ounce. Incidentally he mentioned the cessation of production in central Europe as a reason for the scarcity of the metal. Writing in the Gerard, or Tammany-Hearst, style we would say that "as a rule, no large amount of silver is produced in central Europe. Of course, silver is very valuable and any large mines, in consequence, would be controlled by the Yunkers, who are Prussian squires, usually counts of many quarterings. As a matter of fact, most of the silver credited to German production is obtained from Australian ore which is smelted by skilful professors from Freiberg. These metallurgists are hard workers and refuse to drink beer until the silver in the cupel has brightened. This important stage of the process, as a rule, of course, is marked by the fact that the noble technicians acquire a noble thirst, which they alleviate out of mugs made, naturally, out of the silver they produce. This is another reason for the scarcity of silver in the German empire, which was founded by the Elector of Brandenburg many years ago." We commend this valuable information to Mr. Guggenheim's attention. Meanwhile we may be permitted to state that in normal times the production of silver in Germany is about 15,000,000 ounces and that of Austria-Hungary 2,000,000 ounces. The larger part of the German production is won from foreign ores, notably the zinc-lead concentrate shipped from Broken Hill for treatment in Rhenish Prussia, so that the total domestic yield of silver in the two enemy-countries is not more than 10,000,000 ounces out of a world production of 250,000,000 ounces. Apparently, therefore, the interruption of mining in central Europe is not an important factor in accentuating the demand for silver.

IT is interesting to receive confirmation from India of our contention that wolframite is a readily soluble mineral in the zone of oxidation, and that its persistence among the minerals in an outcrop is determined by its associates. If enclosed in quartz, so that the solutions from the decomposing soil cannot attack it, wolframite will remain until erosion exposes it to the atmosphere. Under these conditions it will sometimes endure for ages. Its stability even then is apparently the result of the accident of superior purity, and is due particularly to the absence of iron sulphide. This compound is commonly present and aids in the decomposition of the wolframite. Mr. J. Coggin Brown, in a letter published on another page of this issue, writes that the conditions found in the rainy tropics yield results like those observed in the tungsten deposits of the United States. Also he finds that it is equally true in Burma that wolframite disintegrates in the detritus from the lodes, yielding soluble compounds that leach so rapidly as to preclude the formation of wolfram placers. The same is true of molybdenite, which is generally regarded as resistant to the influences of weathering. On exposure to the air it persists in a manner that gives it another point of resemblance to graphite. It is not directly attacked by alkalies, but in the process of soil disintegration, under the influence of which everything must pass that finally becomes concentrated into a placer, it undergoes complete oxidation. This is not, as some think, because it is carried away by the current when saturation during the seasons of excessive rainfall imparts mobility to the gravel. Whatever may happen to individual flakes, molybdenite as a whole, when contributed by erosion to the alluvial wash of a district, is first oxidized and then dissolved and leached. The process may be studied in the lodes at many points in the high mountains of California and Oregon, and in Ontario, where molybdenite deposits occur under similar climatic conditions. This leads to the further observation that the products of decomposition of vegetal matter play an important part in the solution of minerals in vein-outcrops. The effects in outcrops are frequently analogous to those happening to the same minerals in the soil. The zone of katamorphism possesses a 'grass-root' phase that is perhaps more significant than has been realized. The disintegrating agents are many and infinitely complex compared with those operative at a relatively shallow depth where the quantity of organic substances becomes small.

The I. W. W. and Butte

Some light is thrown on the purpose of this organization by a report made by Mr. Harris Weinstock in 1912 to the Governor of California. The I. W. W. had been making life unbearable to the citizens of San Diego and had exasperated many of them to the point of organizing a counter-irritant, a company of vigilantes, who took the law into their own hands and perpetrated outrages hardly less reprehensible than the ones that they were intended to check. The Governor appointed Mr. Weinstock to investigate. He appears to have discharged his task in a fair-minded way and for that reason his statements are worthy of respect. After examining a number of pamphlets issued under the I. W. W. stamp, he concluded that this anti-social organization teaches and preaches the following:

1. That workmen are to use any and all tactics that will get the results sought with the least possible expenditure of time and energy.
2. The question of right or wrong is not to be considered.
3. The avenging sword is to be unsheathed with all hearts resolved for victory or death.
4. The workman is to help himself when the proper time comes.
5. No agreement with an employer of labor is to be considered by the worker as sacred or inviolable.
6. The worker is to produce inferior goods and kill time in getting tools repaired and in attending to repair-work, all by a silent understanding.
7. The worker is to look forward to the day when he will confiscate the factories and drive out the owners.
8. The worker is to get ready to cause national industrial paralysis with a view to confiscating all industries, meanwhile taking forcible possession of all things that he may need.
9. Strikers are to disobey and to treat with contempt all judicial injunctions.

Mr. Weinstock says that the members of the I. W. W. are not anarchists because they stand for a "co-operative commonwealth." This is funny. To divert such a kindly social term as "co-operative commonwealth" to the characterization of the I. W. W. nightmare is ludicrous. Co-operation involves the surrender of personal privilege, not an insistence upon rampant individualism. Mr. Weinstock says that "if all men and women were . . . to follow their teachings . . . it would simply mean a nation of thieves, liars, and scoundrels." A stronger term might be used for Vincent St. John, responsible for the assassination of Arthur Collins at Telluride in 1903 and now the general secretary of the I. W. W. organization, and for other men identified with the murders and brutalities of the strikes in the Coeur d'Alene and Cripple Creek, such as William D. Haywood, also one of the leaders of these anarchists. It is revolting to the mining engineer to recall for how many years such men at St. John, Haywood, and Moyer have been prominent in making trouble not only in an economic way—for that there is justification—but by inciting men to crime. To return to Mr. Weinstock; he describes how the people of San Diego, inflamed by an irresponsible press, took the law into their own hands, organized a vigilante movement, kidnapped a number of the I. W. W., and subjected them to indignities and cruelties before driving them out of the county. He condemns this lawlessness properly: "It cannot now be said . . . that there was any justification whatever on the part of men professing to be law-abiding citizens themselves to become law-breakers and to violate the most sacred provisions of the constitution." Again he says: "Who are the real anarchists . . . these so-called unfortunate members of 'the scum of the earth' or these presumably respectable members of society?" Similarly, in her recent speech at Butte, Miss Jeanette Rankin referred to the lynching of Frank Little and expressed entire lack of sympathy with both the victim and the murderers. "Lawlessness has no place in organized society," she declared. It is well that a woman should utter this truism. We go further and say that when the law is broken by "respectable members of society" and by men of education they must bear a bigger blame than the ignorant and perverted disturbers of the peace on whom they lay violent hands. So also the briber is worse than the man bribed—we have in mind municipal corruption—because the giver of the bribe usually has a better understanding of the baseness of the act than the receiver. Education has its responsibilities. According to the little Congresswoman it is the 'rustling-card' more than the question of wages that has driven the men at Anaconda to strike. The facts are made clear in the article from our correspondent at Butte appearing on another page. It shows a spirit of fairness throughout and is well worthy a careful reading. Since that article was written the smelter-men at Aanconda have gone on strike and both the Washoe and Great Falls smelting plants have ceased operations. That will compel the closing of many mines and cause 15,000 men to stop work. This action of the Mill and Smeltermen's Union at Anaconda is due to the radical element and represents a repudiation of the agreement made by the union a few days earlier. Here is one of the great obstacles to the recognition of the unions: their bad faith and their lack of any such legal status as would permit action against them. Thus industrial anarchy ensues, and coming as it does at a time of national crisis, it is to be hoped that a Federal investigation will be made without further delay. Miss Rankin said: "War spells sacrifice. No one can escape its far-reaching effects," and so forth. How hollow it must have sounded at Butte. She spoke of the men "whose patriotism causes them to continue to work under conditions that they are daily unnecessarily risking their lives." The less said about patriotism the better. This applies to the other side also, whose patriotism causes them to refuse to sell copper at less than 25 cents per pound when the average price for ten years has been 16.67 cents. One reason for the cat-and-dog fight between capital and labor is indicated by our Butte correspondent, who states that the local press has aggravated the antagonism between the mining companies and their employees by misrepresenting the facts in a malicious manner. That, of course, is one of the underlying causes of social unrest. A sense of justice cannot be developed so long as the press is subservient to either side.

The Food Law as a Price Regulator

Press reports of the recent address by Mr. Frank A. Vanderlip at Chicago did not make clear what he meant by the word 'inflation'. He predicted financial hardship as a result of it, and indicated that the absorption of the railroads by the Government seemed inevitable. As no issues of inconvertible paper money have been made, we here confront a new use of the term; it is not inflation of currency, but an inflated face-value of credits that is already beginning to react on the price-boosters. They are trading in profits, not in the true values of goods delivered. The steel-makers, the copper-producers, the lead trust, and a long list besides, are uncurbed in their efforts to pile up vast credits in the hope that they will be valid without contraction in the days to come when prices of commodities must shrink to a level comparable with average incomes. The railroads, however, are not free to take all that the traffic will bear; they find themselves between the devil of greedy capital, unwilling to divide profits with them, and the deep sea of popular distrust, which will not sanction an increase in rates. This is one phase of the difficulty. Another is the restriction upon food-prices; and yet another the action being taken by the Government to control the prices and deliveries of fuel in order that the people may not freeze next winter and that industry may not be crippled. The rule given by Congress in the 'food bill' is not one for universal application; it singles out specific things, which it calls "necessaries", and in so doing it may work a seeming injustice upon many industries. An all-inclusive measure would have been better; it would be no more difficult to establish bases for equitably adjusting the prices of all commodities than to interpret wisely the extent to which the classification given in the statute might be applied. The words of the law, defining its purpose, are: "to assure an adequate supply and equitable distribution, and to facilitate the movement of foods, feeds, fuel, including fuel-oil and natural gas, and fertilizer and fertilizer ingredients, tools, utensils, implements, machinery, and equipment required for the actual production of foods, feeds, and fuel, hereafter in this act called necessaries". In Section 12 the statute authorizes the President when "necessary to secure an adequate supply of necessaries for the support of the Army or the maintenance of the Navy, or for any public use connected with the common defense," to take over and to operate "any factory, packing-house, oil pipe-line, mine, or other plant or any part thereof, in or through which any necessaries are or may be manufactured, produced, prepared, or mined." As this is a war measure it is unlikely that resistance to its operation could be made by injunction, or that a case arising under it could be decided by the Supreme Court in time to test its constitutionality before the occasion that called it into being fortunately had passed into history. Nevertheless, we feel that it might fairly be questioned whether the text makes "necessaries" include more than "foods, feeds, and fuel." This view has been taken with hilarious gusto by a large number of prominent gentlemen who claim exemption for their purses as well as for their sons. It opens the way to play politics; it invites the unwholesome practice of using 'influence' in the sinister and corrupt sense of the word; it tends to introduce into our administration the evils of bureaucracy. Exemptions by special favor should be made impossible in a republic. In this respect the Food Bill fails most dismally.

It was not our purpose to criticize the food law, because we are duly thankful for any measure with which the Government may set to work effectively in the solution of its mighty problems. Some way will be found to extend it, as circumstances require, for the salvation of the country. The thing that does concern us, however, is the inevitable unfairness that will result, in consequence of which the evils of inflation will be felt in many industries. Regulated businesses will be at the mercy of the unregulated. Their own costs will be inflated with the profit-gas of the free producers through the purchase of supplies. They will not be able to increase wages nor to distribute compensating bonuses. The effect is certain to be disturbing, not alone to finance, but to an already aggravated labor situation. We cannot afford to add to the hardships of those that bear the brunt of industry. We hope the Administration will realize the importance of equalizing the opportunity for every kind of economic effort, and that it will not hesitate to read authority into the statute to bring all industry under control through the regulation of prices. It is a wiser way than to attempt an adjustment through the excess-profits tax. The padding of costs will defy Government scrutiny, and the wrinkles will not be smoothed out; big business will find ways for paying a lesser proportion of its actual profit. The War Industries Board seems to be taking steps to meet the threatened difficulty, and we quote approvingly its recent announcement that a just price means one that "will sustain the industries concerned in a high state of efficiency, provide a living for those who conduct them, enable them to pay good wages, and make possible expansions of their enterprises which will from time to time become necessary as the stupendous undertakings of this great War develop."

DISCUSSION

Our readers are invited to use this department for the discussion of technical and other matters pertaining to mining and metallurgy. The Editor welcomes expressions of views contrary to his own, believing that careful criticism is more valuable than casual compliment.

Sampling Large Low-Grade Orebodies

The Editor:

Sir—I have read with considerable interest your editorial of May 26, and the following contributions by Messrs. T. H. Leggett, Morton Webber, and Albert Burch.

I do not know that Mr. Leggett is right in stating that the mine that cannot be sampled "is a good mine—for the owner to keep," nor is Mr. Burch exactly right while agreeing with Mr. Webber but qualifying what he says that such mines are "freaks" and unimportant. I have had several years experience in both the South-Eastern and South-Western base-metal fields of Missouri. I can safely say that sampling is never undertaken in the underground operations of these mines, which are among the most important lead and zinc producers of the world. That surface drilling is done is true, but the cores are never assayed. Drilling is done to demonstrate the extension of the profitable beds; and the underground development that follows depends solely on ocular observation. This may seem strange, but new-comers who have gone into the field and started underground sampling found they were being misled and drifted into the method found by experience to be the best.

In Mr. Webber's article on the subject entitled 'Latent errors in mine sampling' he refers to this field as belonging to his class five, and I notice the views of H. C. Hoover are also in agreement with Mr. Webber as to the classification of this district.

R. E. Raymond.

New York, August 8.

Steel on the Pacific Coast

The Editor:

Sir—The article on the 'Outlook for Iron and Steel on the Pacific Coast' by Ernest A. Hersam in your issue of July 28 was read with a great deal of interest by the writer, as was also the editorial 'Steel on the Pacific Coast.' This subject is claiming considerable attention in the West and is particularly interesting to me because I just recently completed a year's research work on the iron ores of the North-West at the University of Washington under the direction of the Bureau of Industrial Research of that institution. The work was taken up with the idea in mind of determining, if possible, the feasibility of establishing a steel industry on the Pacific Coast. In order to arrive at any conclusions in this regard it was necessary to investigate the ores themselves, fuel, fluxes, and labor, markets and transportation, and costs of production. The results of this work form the subject matter of 'An Investigation of the Iron Ores of the North-West' which is now in the hands of the printer and is being published by the Bureau of Industrial Research of the University of Washington.

Mr. Hersam's article is well written and shows the result of considerable thought and study. There is one point, however, which is not brought out that I believe should be emphasized. That is the superior quality of iron and steel and ferro-alloys manufactured in the electric furnace. This finer quality of product demands higher prices and of course enables the electric furnace to produce at a higher cost than an open-hearth or blast-furnace. It is said that conditions on the Pacific Coast relative to electrothermic smelting and refining are more nearly parallel to those of Sweden and Norway than are those in any other country. It would seem that this is a phase of the subject of iron and steel that deserves more investigation on the Pacific Coast.

W. H. Whittier.

Ray, Arizona, July 31.

Mill-Tests v. Hand-Sampling

The Editor:

Sir—I have read with great interest the article by Morton Webber published in your issue of July 28, entitled 'Mill-Tests v. Hand-Sampling' in valuing mines, together with the discussion by Albert Burch published in the same issue.

Knowing something personally of Mr. Webber's methods and the thoroughness of his work, and having a high regard for his ability, it occurs to me that he makes two points of fundamental importance with reference to correct mine-valuation that should be elucidated more clearly and in greater detail. I refer especially to his remarks concerning "Pre-determined foot-weight of the stopes as a basis for mill-test", which to me, and I presume to many other engineers, is somewhat novel and lends itself to considerable enlargement. Also I refer to the last paragraph of his article wherein he suggests a "division of the area under examination into mill-test zones." It seems to me this point is provocative of much thought, and I suggest that a more detailed scheme of this point would be of decided and general interest to your readers, and especially to those interested in mine-valuation.

It is a trite saying, of course, that each mine or orebody is a problem unto itself and a proper solution of each individual problem of valuation should be based upon regional and local characteristics, and not upon general or typical characteristics, as is too often the case. All orebodies have not only a developed but a prospective value, and this is a proper element for consideration in valuation. How far and to what extent the prospective element is entitled to consideration depends upon what I term regional and local characteristics, the proper judgment of which introduces the personal element. I should be glad to have the experience and judgment of Mr. Webber and others on this point, and believe it would be of general interest, especially as applying to properties of meager or incomplete development.

I quite agree with Mr. Webber that a rule-of-thumb method of mine-valuation will not apply generally and that terms and conditions of purchase will frequently, and should properly, govern methods of examination. I do not, however, wholly agree with him as to a definite time for mixing a given sample, such as five minutes. It seems to me that the time for mixing should be governed wholly by the conditions involved in each individual case. Is there not danger in many instances of re-concentration by specific gravity by over-mixing? Why would not two or three minutes mixing accomplish all that five minutes will? If five minutes is better than two or three, why not extend the time of mixing to ten minutes or longer? My own experience leads me to believe that as soon as a sample is thoroughly mixed, whether it be in one, two, or more minutes, the operation should cease. Possibly that is a habit acquired in the assay-laboratory, in which conditions are somewhat different than those under discussion.

Referring to the taking of mill-test samples that shall be representative of the average orebody, I find that Mr. Burch, in his discussion, asks the same question that occurs to me, namely, "How can it—the mill-test sample—be made representative?" I will cite a case in point which represented a most difficult problem of valuation either by hand-sampling or by mill-test sampling, the latter by reason of the fact that it seemed impossible from the character of the deposit to take mill-test samples that would be 'representative' of the average ore. The mine in question is in Montana and is a large low-grade deposit, in superficial area approximating 1200 by 1500 ft., in which area the entire mass is more or less mineralized. An assay-plan representing over 700 samples, said to have been taken by a reputable engineer, was submitted and showed an average gold-value of $2.65 per ton. These samples were taken systematically from over the entire surface-area of the deposit and in 5-ft. cuts throughout the length of an adit 900 ft. long driven into the deposit, giving at the face a depth of about 500 ft. The value stated individually ran from less than $1 to over $10 per ton, and while the distribution of the gold was continuous throughout the deposit, it was erratic.

I made a preliminary test by hand-sampling at different points, locating the original point of sampling as closely as possible, both on the surface and underground and taking comparatively small samples. My results showed an average of 92c. per ton. I then made a second preliminary test by hand-sampling, taking much larger samples, mixing thoroughly and quartering carefully, my results showing an average of 79c. per ton. A third preliminary test by hand-sampling was made with an engineer representing a large mining corporation that was considering the purchase of the property. This test was preliminary to a final complete sampling of the property for valuation. He took a definite 100-ft. section of the adit and sampled every 5-ft. section, while I took a definite section of the surface, sampling also in 5-ft. sections. These samples were mixed thoroughly and quartered, the combined results showing an average of between 80c. and 90c. per ton. None of the individual samples showed any close comparison with the individual comparative samples on the assay-plan. Here then was a case of two engineers against one, that one maintaining he was right. To settle the controversy, a mill-test sampling was suggested, but, the problem was how to get mill-test samples that would be 'representative' of an orebody of this character. We decided it could not be done. It is still unsolved. Possibly this belongs to Mr. Webber's Class 5. I should like to have his solution of that problem. It is quite probable that the dividing of an orebody of this character into "mill-test zones" as suggested in Mr. Webber's closing paragraph, which, so far as I know, is original with Mr. Webber, may be a solution of many vexing problems in mine-valuation. May I not, therefore, suggest again that a more detailed scheme of this suggestion by Mr. Webber would be of decided value and general interest?

E. P. SPALDING.

Spokane, August 15.

Solubility of Tungsten Minerals

The Editor:

Sir—In your issue of March 31, Mr. W. H. Emmons classes tungsten as a metal that dissolves very slowly when its ores are exposed at or near the surface and subjected to the action of water and air. In the same number you venture to take issue with Mr. Emmons regarding this conclusion, supporting your views by a number of occurrences that are exceedingly difficult to explain if the natural tungstates of iron and manganese, or both, are regarded as stable ores at all comparable say with cassiterite, for Mr. Emmons places both tin and tungsten in the same division of his classification along with gold (in part), bismuth, chromium, and molybdenum.

In the Tavoy district of Lower Burma, quartz veins containing wolfram and cassiterite are common in and about the granites that have been intruded into a group of ancient argillites and clay-schists, locally known as the Mergui series. The climate is a tropical one, the country is mountainous, and for five months of the year subject to an exceedingly heavy rainfall. Denudation is rapid and there are abundant opportunities for observing the results of general atmospheric causes on tungsten minerals in place. My own results confirm your view that only in a special environment may this metal be classed as one that dissolves slowly.

A large portion of the wolfram produced in Tavoy is obtained by hydraulic methods from detrital deposits, but there are no true wolfram placer deposits or anything resembling them, though the aluvial deposits of the slopes merge into true placers in the valley-bottoms. These often contain cassiterite, sometimes in profitable quantities, but long before the water-sorted gravel is reached the wolfram has ceased to appear, though it comes from the same lodes as the tin ore in which usually it is found in considerably greater proportion. The only wolfram seen in the placers themselves occurs tightly enclosed in unfractured quartz, to which it owes its preservation. Chemical analysis of tin ore obtained by dredging in this district reveals as a rule less than 0.25% WO_3, though an examination of the detrital deposits of the hillsides from which the placers are derived would in most cases show more wolfram than tin. The rapidity with which wolfram disappears in this way is remarkable; this is partly explained by the very perfect cleavage of the mineral, resulting in its disintegration on movement and the production of a comminuted form eminently suitable for chemical decomposition.

Again, observations on the weathered outcrops of the lodes themselves prove that wolfram is readily dissolved. Sometimes nothing is left but the empty cavities once occupied by the crystals; occasionally these are filled with sintery oxides of iron and manganese while nests of yellow tungstite bear witness to the original presence of the mineral.

Wolfram-bearing pegmatites are not unknown here and the kaolinization of their feldspars results in the rapid decay of the wolfram by the formation of soluble alkaline tungstates.

Calcareous rocks, to all intents and purposes, are absent in those portions of the district opened up at present and the formation of the insoluble lime tungstate is rare.

The invariable association of sulphides with the wolfram of Tavoy is not without a bearing on this subject. Iron pyrite always accompanies this mineral, chalcopyrite is common, pyrrhotite and mispickel are not rare. Galena, zinc-blende, molybdenite, and bismutite are found in places. The ready oxidation of the iron and copper sulphides produces solutions that attack the tungstate, while the mechanical effect of their removal from the lodes opens the way for deeper-seated chemical action.

You have remarked that tungsten is a common accompaniment of gold and that it is a persistent associate of high-temperature copper ores. Gold is known to occur in small amounts in the alluvial deposits of Tavoy, but its association with tungsten in the lodes of the district has not yet been proved, probably because the matter has never been investigated. Copper ores are found in small quantities in most wolfram-quartz lodes here. Minerals like topaz, tourmaline, and flourspar are quite subordinate to the sulphides as associated minerals. They are found in small quantities, but appear to have played very little part in the reactions of wolfram genesis.

Thus evidence gathered in the Tavoy district indicates that tungsten must not be classed as a metal that dissolves slowly.

J. Coggin Brown,
Geological Survey of India.

Tavoy, Lower Burma, May 6.

The Extra-Lateral Right

The Editor:

Sir—I have read Mr. Colby's article, 'The Extra-lateral Right,' in your issue of June 2, with a great deal of interest. As a review of the growth and use of the extra-lateral idea from its earliest inception to the present day this paper is a classic and must represent a large amount of painstaking labor and research. As an argument in favor of the retention of the extra-lateral right in our American system of mining law it also presents many good points, and while in his list of conclusions Mr. Colby does not commit himself definitely to the stand that it ought to be retained, still his evident inclination is favorable toward such a course.

The first five of his conclusions, which tend to support his evident belief, are based upon precedent, which, if I may say so, is one of the strongholds of his profession. As an individual, he would be more than human if he could entirely shake off the influence of many years of legal training in which knowledge of and dependence upon precedent in the form of old laws (all human) and court decisions (all human) have formed so large a part. And it is right and proper that, to the end that all men may know with certainty the laws by which they are governed, those laws should rest upon the firm foundation of past authoritative interpretations, but while this principle properly applies to all existing laws, it ought not to govern our action in the matter of making new ones to fit the ever-changing conditions and needs of the people. Therefore, if it can be shown that the extra-lateral right law does not in the best manner possible meet the present and probable future requirements of the majority of the people, it should be discarded and a substitute found just as any progressive manufacturer would 'scrap' an inefficient machine and would install one that would perform his work to better advantage. With a man who approaches the subject in this mental attitude, the first five conclusions, being based on precedent, would have no weight. With the sixth conclusion I heartily agree, namely, that if the extra-lateral right be abolished the principle of discovery must go with it, but is it not true that changed conditions in this country justify the abolition of both, and is it not possible that similar changes in conditions in the older mining countries of the world were largely responsible for the abolition of both as recorded in Mr. Colby's excellent historical sketch? When the principle of discovery and the right of extra-lateral pursuit were incorporated in the miners' rules of this country no way was known for determining the existence of underground orebodies except to search for outcrops; and the discovery of such outcrops was comparatively easy. It was a game at which all men, skilled or unskilled, rich or poor, might play with fairly even chances of success, and to that period and to a similar period in older countries I believe the laws now in existence were well adapted. It was proper that a man who had been successful in discovering the outcrop of a vein should have the right to follow it downward beyond the side-lines of his location, because it enabled a comparatively poor man to take advantage of his discovery without spending unnecessary time and money in the acquisition of protecting side-claims, thus putting rich and poor upon the same footing. Even with the benefit of the extra-lateral right, however, he rarely obtained the full fruits of his discovery because of the pernicious habit that orebodies have of raking one way or the other within a vein in total disregard of end-lines.

But today the conditions surrounding the prospector have changed and the chances of finding valuable orebodies by searching the surface for outcrops are few and far between. The mining geologist has come into the field and tells us that under this broad exposure of gossan there may be a valuable deposit of copper ore or it may be a valueless body of pyrite; that under this flow of lava may be found the extension of the vein-system from the other side of the bill, if it has not been pinched out or been displaced by an unknown fault; or that this great vein which apparently terminated against a slip may be found buried under the débris of yonder valley, if some other calamity has not overtaken it; and based on this indefinite information—indefinite because, unlike the law (?), geology is not an exact science, men spend thousands and hundreds of thousands of dollars attempting to make a discovery. They will not do this, and as a matter of fact do not do it without first acquiring some sort of title to sufficient territory to proteet their discovery no matter which way the vein, when found, may course or dip. And the poor prospector of today, not in spite of the extra-lateral right law but because of it endeavors to protect his meagre discovery by locating enough surrounding territory to prevent the encroachment of neighbors. It is practically impossible today in any new district in the United States to find the outcrop of a vein sufficiently prominent to enable a prospector or even a trained geologist to determine its exact course and dip for any considerable distance. Therefore both classes of mineral explorers, the capitalists who gamble on geological information and the prospectors who hunt for croppings, violate the law by making numerous protecting locations upon which no mineral discoveries are or can be made. These violations of the law are winked at by everyone except the guardians of the national forests, but why not change the law so as to make legal the actions of nine-tenths of the men who make locations under present conditions instead of trying to make nature fit the law based on precedent? To do this would undoubtedly mean applying the solution suggested in Mr. Colby's ninth conclusion.

Regarding the volume and expense of litigation that might arise because of such a radical change in our system as compared to the volume and expense of litigation under the present laws, I do not pretend to the wisdom of Solomon, but believe from the known experience of other countries that the change would be for the better in this respect. In the first place, as fully set forth in a letter from V. G. Hills, published in your issue of June 30, I believe that in estimating the relative importance of extra-lateral litigation Mr. Colby has omitted most of the really heavy items of expense, namely, the sums of money expended by mining companies to avoid such litigation. Secondly, the absence of such litigation in the Spanish-American countries and the British dominions, including even British Columbia, where a new law has been super-imposed upon the old one, which gave extra-lateral rights, is in itself good proof of the benefits that might be hoped for from the abolition of the extra-lateral right and the correlative discovery principle. As regards British Columbia, it is true that some litigation has arisen over the adjustment of the new law to the old, but it must be remembered that mines do not last forever and that in time mines located under the present law will become exhausted and with them the chances for new extra-lateral litigation.

I am in perfect agreement with Mr. Colby's eleventh and final conclusion, namely, that if the law is to be changed, it should be changed only after careful study by a competent commission to the end that it may fit as well as may be with all other laws governnig the disposal of our public domain.

In conclusion, it is reasonably well known, Mr. Editor, that in the past I have derived a considerable revenue in payment for services connected with extra-lateral litigation and in advocating means to avoid such litigation, I am reminded of the story of the railroad conductor who had been caught collecting fares for his own benefit instead of that of his employer. He was called before the superintendent and asked what he had to say for himself. He invited that official to take a ride with him in his motor-car and as they passed through the country he pointed out to him a well-kept farm with good buildings, cattle, and horses. He said, "I own that place and I own this motor-car; I have all I want and if you discharge me and hire a new man he will have to begin at the bottom and acquire them as I did."

ALBERT BURCH.

San Francisco, August 12.

FERRO-MANGANESE is now selling at $400 per ton at Eastern points, this being based on the alloy containing 80% manganese. Quotations for the first quarter of 1918 are $350 per ton. Spiegeleisen is quoted at $80 at the furnace. The indifference regarding the future is interesting in view of the Government's appeal for rigid economy in recovery of manganese from slag.

The Labor Troubles at Butte

By AN OCCASIONAL CORRESPONDENT

The labor situation in the Butte district, while it bears a similarity to the condition prevalent for some months past in most mining and industrial districts of the United States, and while it is the result of all the other labor troubles that have existed in this big mining district and have never been definitely remedied, was undoubtedly brought to a crisis by the Granite Mountain fire.

Under a seemingly placid surface the current of trouble has been flowing for many months. In February last the Electrical Workers Local Union took up with the Montana Power Co. the question of higher wages and

THE FIRE IN THE GRANITE MOUNTAIN SHAFT

better working conditions for its members. Definite answer to these demands was postponed repeatedly, until, on June 5, the Electrical Workers voted to discharge their conference committee and to declare a strike against the Montana Power Co. They demanded $6 per day, and time and a half for Saturday afternoons and Sunday. Before this difficulty was settled the fire occurred in the Granite Mountain shaft of the North Butte Co., as a result of which more than 160 men lost their lives. Immediately following this the miners began to 'lay off' and the seed was sown for the organization of the Butte Mine Workers Union, which took place on June 11. The leaders were Tom Campbell, Joe Shannon, Dan Shovlin, and others prominent in labor circles in the West for many years. The organization appealed to the miners, and a large membership was soon secured.

Almost the first question before the new union was that of affiliation with some national organization, but it has become a principle of this union to remain independent. The electricians working for the mining companies went out in sympathy with the electricians that were striking against the Montana Power Co., and they demanded that the mining companies bring pressure to bear upon the Montana Power Co. to grant the demands of its electricians. The electricians' strike committee recognized the Butte Mine Workers Union, voting to support that union in its demands. Soon afterward the

The paper stuck by the vigilantes on Little. The figures refer to a grave 3 ft. wide, 7 ft. long, and 77 in. deep. The initials refer to labor leaders, the one with the circle meaning the first victim.

Metal Trades Union, blacksmiths, machinists, and boilermakers employed by the mining companies went on strike in sympathy with the electricians, because non-union men were doing electrical work at the mines. However, the Metal Trades Union was unable to recognize the Butte Mine Workers Union, as that body was not affiliated with the American Federation of Labor, of which the Metal Trades Union is a member.

The engineers' local union was approached by a committee from the Electrical Workers Union to call a sympathetic strike, which would close the mines completely. The hoisting-engineers had endeavored about seven years ago to withdraw from the Western Federation of Miners.

During this trouble it was claimed that electricians had taken the places of some of the engineers, and it seems probable that the refusal of the engineers to vote more than moral support was by way of retaliation.

The B. M. W. U. then took a vote on the question of affiliation with the A. F. of L. The proposal was rejected by a majority of over 10 to 1, the principal reason doubtless being the fact that Charles Moyer, former president of the W. F. of M. is now head of the International Mine, Mill & Smeltermen's branch of the A. F. of L. Following this action the Butte local of the I. M. M. & S. union withdrew their charter and joined the B. M. W. U. in a body. This action pulled out all the mill and smelter workers in Butte.

The B. M. W. U. demanded, among other things, a flat wage of $6 per day for all underground employees, regardless of the price of copper, abolishment of the 'rustling-card' system of employment, appointment of a committee consisting half of members from the union and half of members appointed by mining companies to make mine examinations continuously, the trial before another such committee of all men discharged, improvement in general underground working conditions, such as providing manholes in all underground concrete fire-bulkheads.

At all meetings of the B. M. & W. U. one of the conspicuous points noticed is the constant warning by its leaders against the use of violence, and this body has conducted its strike in a manner remarkably free from anything of that sort. The union constantly petitioned congressmen from Montana and the Department of Labor to have working conditions and living conditions in the Butte district investigated by Federal authorities. After considerable delay, W. H. Rodgers of the Department of Labor was sent to Butte. His investigation, extending over a considerable period, and based on information from all possible sources, failed to result in any remedy for the situation, Mr. Rodgers being withdrawn suddenly without making any statement or any attempt at relief.

After about three weeks the Montana Power Co. reached a settlement with its electrical workers, granting their demands. At the same time the operating companies announced their new sliding scale of wages based on the price of copper. This gave the miners $3.50 minimum wage as before, with copper under 15c., $4 with copper at 15 to 17c., and beyond that a raise of 25c. per day for each 2c. increase in the price of copper. The same sliding scale applied to all other crafts, the base rate with copper under 15c. remaining the same as it was for each class of workmen, with the exception of hoisting-engineers, who were given a base rate increase of 50 cents.

During the entire period of trouble in the Butte district the daily newspapers of Butte and Anaconda have continually misrepresented the facts. They have taken advantage of every event in the usual current of city life to arouse antagonism against the union, its sympathizers, and its aims. When events have been reported truthfully, it has been in a sneering and belittling way, which has called down upon the press the condemnation of all fair-minded persons, and has brought public condemnation from the Butte Typographical union. The city and county authorities have denied the unions a place to hold large meetings, even to using the city ball-park, in which nearly all Butte unions are shareholders.

All craft-unions that are members of the A. F. of L. voted an acceptance or rejection of the new sliding-wage scale, as it applied to each particular craft. It was agreed that if a majority of the several unions should accept each its individual scale, the entire proposal would be binding on all crafts affected. In this way it was possible for unions with memberships as small as 11 members to cast as important a vote as a union of 800 members. The majority of the unions voted to accept this scale, and all crafts went back to work. The individual members of the unions that voted to reject the scale were, however, in a majority.

A concession in respect to the 'rustling-card' system was granted with this scale, giving the applicant a 'temporary rustling-card' upon application for a job, the same to be replaced by a permanent rustling-card issued after the necessary information concerning the man had been received by the company. Within a week after the majority of all craft-unions had voted to accept this scale, and had all gone back to work, 90% of the electricians on the hill quitted individually, and, without any action as a union, demanded a flat rate of $6 per day and time and a half for overtime. Probably 75% of them are still out.

On the morning of August 1, Frank H. Little, a member of the I. W. W. executive board, who had been in Butte for a short time, was taken forcibly from his room by six masked and armed men. He was gagged, thrown into an automobile, driven about two miles, and strangled to death by hanging from the Milwaukee railway trestle. While the situation was tense during the following few days, there was absolutely no lawlessness nor attempt at revenge. He was buried on the following Sunday in a Butte cemetery, and escorted to his grave by what was probably the largest procession of men ever seen in Butte. Little's speeches differed radically from the B. M. W. U. leaders' speeches, in that he bitterly denounced the soldiers, attacked the Government for its war policy, and counselled "direct action."

Many of the Montana unions have voted money to help the B. M. W. U. and most prominently the M. M. & S. union of Anaconda. Several weeks ago Campbell, of the B. M. W. U., went to Anaconda to meet the members of the Anaconda union. He and his party were met at the train by the sheriff and deputies and ordered to return to Butte. A few days later he was publicly invited by that union to come to Anaconda and address them, and practically every union man in Anaconda appeared in the escorting party from the train to the Union hall. Last week this union, which had previously voted against accepting the scale proposed by the operating companies, took a strike vote on the question of standing by the agreement of the other unions. The strike vote secured a small majority of the members, but failed to obtain the necessary two-thirds majority required by their constitution.

Miss Jeanette Rankin, congresswoman from Montana, arrived in Butte on August 14, on a visit to her home at Missoula, and stopped in Butte to offer her services in an attempt to bring about an adjustment of the labor troubles of the district. She addressed the B. M. W. U., their friends, and sympathizers at a mass meeting held at Columbia Gardens, about three miles outside the city, on August 18. To quote from her speech: "I am convinced that the demands of labor in this struggle are just, and should be granted . . . If the rustling-card can be defended in theory, its abuses in Butte are reprehensible and cannot be defended . . . Labor has no right to take advantage of unsettled conditions to enforce unjust demands, and capital has no right to take advantage of labor to compel labor to make all the sacrifice simply because we are at war." In speaking of the Government's purchase of 50,000,000 lb. of copper at 16.67c. per lb., when the market-price was 32c., Congresswoman Rankin drew attention to the fact that the operators refused to accept that price for the second 50,000,000 lb. purchased by the Government, and that the price which the Government must pay is being determined by the Committee on War Purchases. Should the committee set the price at 28c. per lb., this one purchase will cost the Government $6,000,000 more than the advertised price. In conclusion, she appealed to the miners to attempt a peaceful settlement of the strike and ever to bear in mind that we are at war.

THE GRANITE MTN. SHAFT ON JUNE 9, 1917, SHOWING THE CROWD OUTSIDE THE FENCE

A Federal investigator in the person of Justice Covington of the Supreme Court of the District of Columbia has been appointed to look into conditions in the Butte district. His appointment, following so closely on the addresses of Congresswoman Rankin of Montana before the House, is taken as a result of that speech. The B. M. W. U. has petitioned the Government to investigate charges of German activity and German financing of the strikers, and the results of Justice Covington's investigations will be awaited with interest.

The main issue at stake seems to be the rustling-card system, instituted after the trouble in 1914; it is in use at all the mines except the Elm Orlu and the Tuolumne, owned by W. A. Clark and the Tuolumne Mining Co., respectively, and at a few small independent properties. While some such system properly used is of undoubted value to an employing concern, still the rustling-card system as applied in the Butte district has been badly abused and employed for political purposes. Since the results desired—efficiency of workmen and employment of a better type—might be obtained in other ways, it is

RESCUE CREW FROM THE TRAMWAY MINE WORKING AT THE GRANITE MTN. SHAFT. THE DOOR BEHIND THEM LEADS TO AIR-TIGHT CHAMBER WHERE THEY STOPPED ON THE CAGE

not clear why this complete concession is not granted. It is certain, however, if every concession asked by the striking workmen be granted, it will be many months before the district can return to the same level as that on which it was working before the Granite Mountain fire, for hundreds of miners have scattered to different parts of the country. If the increase of 50c. per day, which has been forced from the operating companies, had been given on their own initiative, as a share to the workers of the immense profits being made from this unprecedented war-business, it would have gone far in averting the greatest part of the existing trouble.

It is a well-known fact that, in the Butte district, any increase in wages granted to the miners is followed immediately by a much greater proportionate increase in the cost of necessaries. If the operators, instead of cultivating the friendship and support of the merchants and politicians, would seek the goodwill of their own men by controlling to some extent these irritating conditions, it is safe to say that their difficulties with labor would be greatly relieved; and, in this direction, announcement by the operating companies of weekly pay-days should be noted. This will do away with the loan-shark and the assignment-broker. A few such steps as this would make Butte the best city in the world for workmen and would attract the best class of miners instead of 'ten-day' men.

As is clearly shown by the loss caused by workman's compensation through the Granite Mountain fire, the operators themselves would be benefited by a more rigid mine inspection than is afforded by the present system, in which the inspection of all mines in the district is made by one man, who is appointed by the Governor as a reward for political activities. This does not disparage the excellent work of the Safety First departments.

A final word: Looking at the wonderful strides that have been made at Butte in mining methods, machinery, and metallurgy, it is hard to conceive how the district can be so negligent in caring for the welfare of its employees. It is this defect in modern corporate management more than the wage question that causes an irritable spirit and a repetition at frequent intervals of these strikes. The present situation is this: All crafts except electricians are at work, 75% of the electricians are still out, but are not on strike, and from 35 to 40% of the normal number of miners is at work in mid-August.

The figures 3-7-77 on the paper attached to the body of the murdered man were used by the early vigilantes; they signify a grave 3 ft. wide, 7 ft. long, and 77 in. deep. The initials at the bottom are supposed to be those of the labor leaders, the first, with the circle around it, meaning Little, and the others Dunne, Campbell, Shannon, and Sullivan. The shaft to the right of the Granite Mountain, in the third photograph, is the Speculator; this was bulkheaded at 800 ft. at the time of the fire and the draft turned down-cast by the fan, all the rescues being effected through it at 2800 ft. deep. The rescue-team is a reminder of the systematic live-saving work organized by the U. S. Bureau of Mines.

An Industrial Romance

The purchase of the Buffalo Copper & Brass Rolling Mill by the American Brass Co. brings to light an industrial romance which has few parallels even in this land of opportunity. Prior to the War the energies of this concern were chiefly employed in trying to keep down the size of the annual deficit. It was manufacturing a small quantity of roll and sheet copper. Then came the War, which catapulted the company into the front rank of munitions producers, with $60,000,000 of gross business a year and millions of net profits, making it also, next to the American Brass Co., the world's largest consumer of copper and a tremendous user of spelter. William A. Morgan, who left Swift & Co. to take hold of the Buffalo company, secured in Canada and the United States contracts for making brass discs to be used in the manufacture of shrapnel shells. At this juncture a practical brass man, in the person of Russell A. Cowles, brought new blood and new capital into the management. He represents the third generation of a brass-making family. He severed a 22-year connection with the American Brass Co. and its predecessors, invested heavily in the Buffalo company's securities, and later, when a programme of expansion had been decided upon, interested his friend, John N. Willys, sufficiently to invest over $1,000,000 in the Buffalo enterprise. The demand for brass rods for the French and British governments becoming so large and insistent, the company then erected the largest rod mill in the country, all in a period of 90 days. Orders secured from J. P. Morgan & Co. and others showed paper profits more than sufficient to pay for this new mill before a wheel was turned. The rolling-mill capacity was then doubled and Buffalo Copper & Brass became the largest producer of brass discs and cartridge metal, as well as the largest single mill producing brass rods, in the United States. Some idea of the size of this plant may be derived from the fact that the amount of metal cast reached the enormous total of 40,000,000 lb. per month. Furthermore, Mr. Cowles's purchases of copper and spelter during 1916 were probably the largest of any single individual in this country, with the exception of purchases made for the Entente allies. All of this resulted in gross business of $60,000,000 in 1916, for the Buffalo company. From the profits of 1915 and 1916 the company early in 1917 paid a dividend of $1000 per share on the common stock, owned chiefly by Messrs. Morgan, Cowles, and Willys. Following this big disbursement the plant was sold to the American Brass Co. for several million dollars. Mr. Morgan has now become general manager for the Curtiss Aeroplane Co., while Mr. Cowles has assumed the presidency of one of the oldest New England manufacturing concerns, the Ansonia Clock Co., which his father, A. A. Cowles, headed at the time of his death.—*Boston News Bureau.*

SCIENCE is organized common sense.

THE KALATA PLANT OF THE VERK ISETZ CORPORATION, RUSSIA.

Mining and Smelting Copper Ore at Kalata

By F. W. DRAPER

INTRODUCTION. The Kalata mines, belonging to the Verk Isetz Corporation, are situated on the extensive series of copper-bearing schists that lie mainly on the eastern slope of the Ural mountains, running north and south parallel with the axis of the range. The Kyshtim properties, described by E. J. Carlyle in the *Engineering & Mining Journal* of June 22, 1912, mark the southern end of this copper belt. Kalata is about 145 miles north of Kyshtim and between these two points copper has long been produced on the Sissert estates, and since the purchase of these by English capital the energetic campaign of exploration conducted under the direction of their chief engineer, Norman C. Stines, has resulted in the blocking-out by diamond-drilling of large lenses in an entirely new locality. At other points it is also certain that workable deposits exist. About 75 miles north of Kalata, in the southern part of the Bogoslovsk estates, diamond-drilling during the last few months has proved the existence of a large deposit.

The ores along this belt are all heavy sulphides, but their character varies in the different localities. At Kyshtim the copper is quite evenly distributed, the richest part being in the centre of the lenses, gradually diminished toward the sides, while at Kalata the copper is very unevenly distributed. There are low-grade areas also, but they have so far proved too irregular both in horizontal and in vertical extent to make it feasible to break them separately for sale to the sulphuric-acid manufacturers. The Kyshtim ores contain considerable barite and zinc-blende, which are entirely lacking at Kalata, and the gold and silver contents are also considerably higher at Kyshtim than at Kalata. A distinguishing feature of the Kalata orebodies is bands of magnetite in the pyrite, these bands in places being several inches wide. They are particularly noticeable at the southern end.

Both the schist and the ore are much more silicious at Kyshtim than at Kalata. Carlyle quotes the silica content of the Kyshtim ores as 12%, whereas at Kalata it is only about 2%. The fine is slightly more silicious than the coarse, due to admixture of schist from the partings in the lenses. The average analysis of the coarse ore sent to the smelter during the first eight months of 1915 was Cu 2.3%, SiO_2 1.93%, Fe 43.5%, and S 50.0%. We frequently have over 51% S for considerable periods.

The mines were opened very early in the last century and were worked mainly for the pyrite used in the manufacture of acid. Copper was smelted in the neighborhood, but there was no rich gossan on the Kalata lenses and all the available data indicate that to a considerable depth the pyrite was lean, so that probably none of it found its way to the smelting-plant. The name Kalata, derived from the Russian verb 'to strike', is probably suggestive of working conditions in the mines in the early days.

In 1912 the Verk Isetz management, deciding that the grade of the ore then worked was sufficiently high in copper to make its smelting profitable, decided to install a modern smelting-plant, and I was engaged to design and supervise the construction. At that time the mine was practically worked-out down to a depth of 357 ft., but diamond-drilling had proved the continuity of the lenses below this level, and subsequent drilling disclosed large ore-reserves. The principal lens at the present working-levels, at 406, 455, and 560 ft., has an extreme length of 560 ft. and a maximum width of 84 ft., yielding about 3000 tons of ore per foot of vertical depth. There are a few stringers of schist, but otherwise the lens is solid pyrite.

The pitch of the lenses down to 455 ft. was eastward at 70° from the horizontal, but below that level down to the 665-ft. level they stand almost vertical. Besides the principal lens, there are smaller parallel lenses in the hanging wall.

MINING. Prior to 1913, the ore was extracted with close timbering, but since then square sets have been used, except in the small blocks that had been left in the upper levels. The last 7-ft. slice between levels will also be taken out with close timbering. The ore when exposed to the air quickly develops cracks, making a dangerous roof, and timbering must be kept close to the faces. In general, freshly exposed ore must not stand over two weeks without timber-support. Heavy pressure develops so rapidly that filling must not be allowed to fall behind. It is never safe to have over one 7-ft. slice unfilled. For filling we use blast-furnace slag broken from the smelter-dumps, which is cheaper here than breaking rock and hauling it to the shafts. A considerable saving will be made as soon as we complete arrangements to supply granulated slag.

At present there are three shafts. The Marinsky shaft was formerly the main working-shaft, but the upper 140 ft. is in the lode, and the ground around it is now badly caved and the surface is constantly sinking. In July 1912 a fire broke out in the old close-timbered stopes above the 364-ft. level; in consequence the mine had to be closed and flooded, it being impossible to wall-off the burning area. The management then decided to sink a new working-shaft in the hanging wall, and work was started under contract in August, the work being done by the Belgian firm of Prüdan-Nöel, which sank 420 ft. in 139 days. Compared with work in other countries, this was not rapid, but for this part of Russia, where 35 ft. per month was considered good, it constituted a record. The work was done by Russian and Tartar workmen under Belgian foremen. At the extreme northern end of the lenses, the Savinsky shaft connects with the 560-ft. level. This is the main auxiliary shaft, and most of the men and timber enter through it.

All hoisting of ore is through the Kalata shaft, which is now down to a depth of 770 ft. From this shaft the 675-ft. level is just being opened up. The hoisting equipment consists of a direct-connected double-drum hoist, using three-phase alternating current at 2100 volts. The hoist was manufactured by Brown, Boveri & Co. of Baden, Switzerland, and has given satisfactory service. The motor is of the double-commutator type, with Deri connection. The control-standard has a single-hand lever, which controls both the regulation of the motor and the adjustment and operation of the brake. This lever swings in two planes, the longitudinal displacement controlling the motor and the lateral movement adjusting the brake. At the time of writing, the hoist has been in service three and a half years, and the only serious repair required was the re-winding of a burned-out coil on the stator. When purchasing the hoist there was some question as to how long the commutator-rings would last, but no difficulty whatever has been caused by them.

The skips are self-dumping, of the Kimberley type, each capable of hoisting 3½ tons at a speed of three metres per second. The ore falls from the skips on an inclined grizzley with bars spaced 1½ inches, the oversize going to the coarse-ore bins, and the undersize to a revolving trommel with ¾-inch round holes. The oversize from the trommel falls into the coarse-ore bins and the undersize into a separate bin. Only the coarse ore and oversize from the trommel are sent to the furnaces. We can take material smaller than ¾-in. to the furnaces, but the fine wet pyrite blinds the screen if we use smaller holes.

The mines are fairly dry, the water to be pumped amounting to about 250 gallons per minute, but it is strongly acid and contains normally about 0.2 grammes of copper per litre. About 80% of this is recovered by precipitating on scrap-iron. This extraction can be improved, but up to the present it has not been feasible to re-construct the precipitating-plant.

Owing to the acid character of the water, the pumps have to be of special composition. The first equipment in the new shaft on the 455-ft. level consisted of two centrifugals and one three-throw plunger-pump, each capable of handling the entire quantity of mine-water. The plunger-pump was purchased owing to uncertainty as to whether the centrifugals would stand the combination of grit with acid water. Three years' experience has shown that the centrifugal pumps are more economical, as regards both power-consumption and repairs, than the plunger-pumps, and further installations on the lower levels will be entirely centrifugal. The bronze of which the pumps are made is 80% copper, 19% tin, and 1% lead, and it must be free from even traces of zinc.

In starting a new level, a cross-cut is driven from the shaft to the foot-wall, the main haulage-drift is then run on the hanging wall to the north and south with intermediate cross-cuts at intervals of 50 ft. These cross-cuts are connected along the foot-wall, and then raises are made to the level above at the ends of each intermediate cross-cut. These raises usually have two compartments, one for filling and one for a ladder-way or for lowering timbers. Next stoping is started on the slice above the level. The ore is taken out in horizontal 7-ft. slices, and the filling dumped into place from small cars or from wheelbarrows. Owing to the character of the ore, the roof will not stand, and a heavy pressure develops quickly. This often requires the support of the roof either on rock or timber cribs, and somewhat increases the cost of filling.

Drilling is done mainly with Ingersoll-Rand jack-hammers, the only exception being in the raises, where B. C.

21 stopers are used. Practically all stoping is done on contract, based on the footage broken. In 1915 the average wages of a man in the stopes for an 8-hour shift was 1.65 rubles, and the general average of all men above and below ground was 1.34 rubles (a ruble being equal to 50 cents).

The following figures, which are averages for the year 1915, give an idea of the work done:

Consumption of dynamite per cubic foot of ore broken, including that from development, 90-93% nitro-glycerine	0.0245	lb.
Ore produced per drill-shift of 8 hours in stopes	125.5	cu. ft.
Ore produced per stope per day	27.92	" "
Ore produced per man per day in stopes	21.41	" "
Ore produced per man above and below ground per day	8.232	" "
Ore produced per foot of hole drilled in stopes	5.737	" "

These results are below normal because in March we began to use prisoners, who came to us absolutely without experience in underground work, and at the beginning of summer we took Chinese, who, although the contractor agreed to furnish miners, turned out to be woodchoppers and agricultural laborers that had never seen a mine and that did not act as though they ever wanted to. They were from northern Manchuria, which seems to furnish a less desirable type of coolie than other parts of China. This caused considerable decrease of output per man and per stope per day. In January and February, with Russian workmen, the output per stope per day was 42.76 cu. ft.; per man in stopes per day it was 26.57 cu. ft.; and per man above and below ground, 9.536 cu. ft. During the year, the proportion of experienced Russian workmen decreased steadily, the deficit being made up by more prisoners and Chinese, all without previous experience. Under these conditions the output per stope per day had decreased in December to 19.24 cu. ft. per man; in stopes to 16.46 cu. ft.; and per man above and below ground to 7.889 cu. ft. That is, the efficiency of the Chinese and prisoners working with Russian instructors in the stopes after they had had from six to nine months experience was about 70% that of the Russian workmen.

The cost of mining ore for the year 1915 was 3.5988 rubles per long ton, which includes all development, both in the lode itself as well as the shaft-sinking and station-cutting with cross-cuts to the lode. The cost of development amounted to 23.5% of the total. Owing to the War, the cost for the year was about 20% above normal.

THE SMELTER. The site chosen for the plant was close to the main hoisting-shaft, the ore being taken direct from the head-frame bins to the feed-floor by the charge-trains.

Quartz of excellent quality was known to exist at various points within easy reach of the smelter-site, but not in large quantities at any one point. Limestone, also of excellent quality, was to be had in unlimited quantity about 1½ miles away. This scattered distribution of fluxes within a radius of 7 to 8 miles of the smelter led us to utilize horse-transport. The choice was also influenced by the decision of the management not to connect the smelter with the Government railway 5½ miles distant, pending the conclusion of negotiations with the latter for the construction of the necessary branch. This was undoubtedly a mistake even in a country where wages are low. Labor at low wages is not always cheap labor. For light supplies horse-transport in the Urals has been cheap, but for heavy pieces very expensive. It would have been cheaper to construct the branch at our expense. Negotiations, especially with the government-owned railways of Russia, do not move rapidly.

As transportation by horses is much cheaper in winter than in summer, and in summer the supply of available horses is much less, provision had to be made for the storage of fluxes for 6 to 8 months. In order to be able to negotiate with the teamsters at the beginning of winter, it was necessary to have a couple of months reserve still in stock at that time, so that they might see that their immediate services were not indispensible.

After considering various plans, it was decided to install a Bleichert cable-crane having a span of 700 ft., with one movable and one fixed tower, and suitable bins for storing a 24-hour supply of ore and fluxes. Our first idea was to equip it with a grab-bucket, but, after careful investigation, we came to the conclusion that such a bucket would not pick up either coarse ore or limestone. The crane was accordingly equipped with ordinary detachable boats.

The working of the crane, from a mechanical point of view, has been quite satisfactory, but the cost has been higher than we anticipated, partly owing to war conditions and partly to fluctuating demands on the crane. During holiday periods, which are frequent in Russia, the miners will not work, and consequently it was necessary to increase greatly the tonnage handled by the crane. For single days the available storage in the mine sufficed to supply the tonnage needed, but the Russian dearly loves to take several days preparing for a holiday and several days getting over it. However, since the prohibition of the sale of vodka, things have greatly improved. The handling of this extra tonnage necessitated adding men to the regular crew and these men could be obtained only at excess wages, which made it necessary to pay extra wages to the whole crew.

The movable tower has also proved to be of no advantage. The boats can be hoisted only from points directly under the cable, or at most not over seven feet on one side, and with piles of various materials, each of which is being removed at the same time, it is not feasible to keep the edges in a straight line. We have found it cheaper to lay transverse service-tracks, on which the boats are moved by means of transfer cars, than to move one tower at short intervals. These service-tracks also greatly extend the storage-area that can be served by the crane. The cost of transferring material from the stock-piles to the bins was, under pre-war conditions, about 12½ cents per long ton.

The charge-trains, which run either to the shaft-bins or to the cable-crane bins, consist of side-dump cars, each holding two tons of ore, and are drawn by 6-ton single-phase electric locomotives. The charge is dumped directly into the furnaces through doors on the sides, alternating so that ore is charged first on one side, then on the other, the fluxes being charged on the side opposite to the ore. Fuel is charged by shovel through the end doors. The average charge weighs about 20 tons, of which 65-70% is ore. A normal charge consists of 25,200 lb. of pyrite, 7200 lb. quartz, 3060 lb. limestone, and 3600 lb. of secondaries. The quartz carries 91-93% SiO_2, and the limestone 50-51% CaO.

BLAST-FURNACES. There are two water-jackets, each measuring 287 by 56 inches at the tuyeres and 287 by 71 in. at the throat, with an effective height from the tuyeres to the charge-plates of 14 ft. They have two tiers of jackets, the water from the lower set going directly into the upper through flexible-hose connections. This decreases the piping, the quantity of water to be handled by the circulating-pumps, and ensures clean water into the upper jackets, thus removing a fruitful source of burning-out of these jackets on the lower corners and edges. In two years we burned out one upper jacket. Long single jackets are, I think, better than the arrangement I have mentioned, but at the time we planned our furnaces there was difficulty in getting the long jackets made in Russia.

The spout connecting the furnace with the fore-hearth is shown in the accompanying sketch. They are cast from converter-copper in cast-iron forms, and have given us most satisfactory service. The connection of the individual cooling-pipes through stuffing-boxes to the main cast-iron supply and delivery pipes permits the quick cutting out of any one or more burned pipes, these being simply cut off with a hack-saw, and blanks bolted over the stuffing-boxes on each side. We have run spouts for some time with three consecutive pipes burned through, and for some days with six, by keeping a spray of water on the outside. A separate detachable tip, cast from converter-copper, is used, and to facilitate changing this without cutting off the water from the main spout, this tip is provided with separate cooling connections. At first we had much trouble with the burning-out of these tips with our low-grade matte, but we now protect them with a piece of talc cut to shape and laid in a mortar of magnesite mixed with water-glass. These talc blocks usually last 24 hours and the changing of them is a matter of only five minutes. Magnesite brick can be used in place of the talc, but it does not last any longer.

The furnaces can be drained through a small tap-jacket on the side toward the converter-aisle. Thus the contents of the furnace can be poured back into the settler by the crane. In case of spout trouble, the furnace can be operated through this side opening, the slag and matte being poured into the fore-hearth by the crane.

SMELTING PRACTICE. When smelting a green-ore charge, our concentration is 5.25-5.5 : 1, the resulting matte being 12.5 to 13.5% copper. This matte we convert direct in basic-lined 12-ft. converters of the Great Falls type. Under our conditions, the cost of oxidizing a pound of iron is less in the converters than in the blast-furnace, and when we take into consideration the extra cost of re-handling low-grade matte to get it to the furnace, and that it crowds ore off the furnace, with one furnace operation this economy is greater than the extra slag-loss involved in the larger tonnage of converter-slag poured through the settlers.

The average grade of matte which we treated during the first eight months of the year 1915, together with the cost of oxidizing a pood of iron in the blast-furnace and in the converter, is shown in the following table:

		Cost of oxidizing 1 pood (36 lb.) of iron Kopecks (1 kopeck = ½ cent)	
	Grade of matte	(A)	(B)
Month	% copper	In blast-furnace	In converter
January	12.7	10.8	8.23
February	19.5	11.34	11.51
March	19.3	10.29	10.38
April	16.8	9.97	7.80
May	13.7	12.57	11.39
June	11.8	12.71	8.89
July	15.1	19.29	11.82
August	16.3	14.88	7.02

The cost of converting in February and March was abnormally high because we did not have enough matte to keep one converter busy all the time, and labor and fixed charges were correspondingly high. In July the whole plant was stopped for a week for some general repairs, and the cost of these repairs and alterations was charged direct to operations, giving high costs for the month.

The average chemical composition of the slag sent to the dump during the first eight months of 1915 was as follows: SiO_2, 31.04; Fe, 44.14; CaO, 6.31; Cu, 0.33%. The samples are taken every hour and assayed once in 24 hours, a portion proportioned to the tonnage removed to the dump being kept in the laboratory for making up the monthly general sample. Converter-slag is poured through the settlers and is therefore included in the sample of the ore going to the dump. Making correction for this, it gives as the composition of the blast-furnace slag itself approximately the following: SiO_2, 33.0; Fe, 41.5; CaO, 6.7%.

Blast is furnished by a single-stage turbo-blower capable of delivering 35,000 cu. ft. of free air at 60-oz. pressure; this is coupled direct to a three-phase alternating-current motor of 800 hp., taking current at 2100 volts. For measuring the quantity of air delivered to the furnace, a collar is fitted into the supply-main between the filter and the blower, to diminish the area at this point. This causes a difference in pressure in front of and behind the collar, which can be measured directly and from which the volume and weight of air passing can be computed. We have arranged this so that the weight and volume at various temperatures can be read directly on a scale. These readings are taken half-hourly, and from them the amount of air blown in the 24 hours is computed. The consumption of air per long ton smelted varies from 1550 to 2000 kilo., the average for the year 1915 being 1787. This corresponds to about 50,000 cu. ft. of free air at our pressure and temperatures. The average tonnage per day was 530 tons (long), and the fuel consumption 1.66% of the charge, or 2.49% based on the ore. Several times we have been without coke and have tried replacing this with coal, successfully. The furnace runs slower and has a tendency to chill in the focus. This could be avoided if we used coke at intervals to re-establish the necessary temperature in the focus. At the time of writing we have just completed a run of five days under such conditions. The coal available was mainly a coking coal, containing 18% ash with which was mixed a little high-grade coking coal. The slag is a little colder than with coke. The following table gives the results of this run:

Date	Ore smelted, long tons	Coal to charge, %	Copper in matte, %	Silica in slag, %
March 27	466	2.49	13.0	30.1
" 28	468	2.41	12.7	28.3
" 29	503	2.28	17.5	31.1
" 30	505	2.27	15.5	30.8
" 31	513	2.11	19.0	29.1

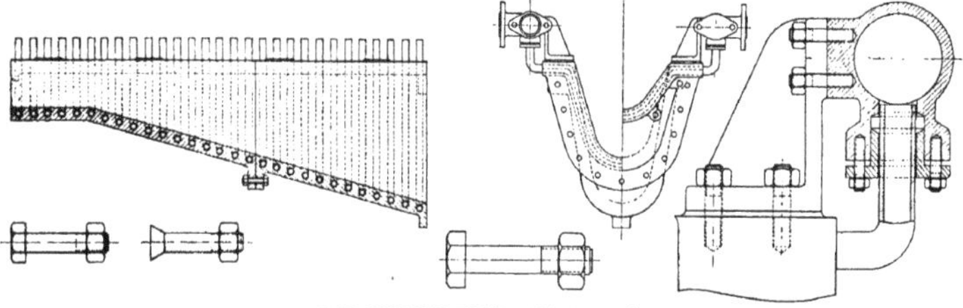

SPOUT CONNECTING FURNACE WITH FORE-HEARTH

The furnace was running normally and was stopped for reasons other than the fuel, although such experience as we have had leads us to think that after a few more days it would have been necessary to heat it by running a day on coke. Our normal average now on coke is about 600 tons, so that the coal causes the furnace to slow up to 80% of its normal rate of running.

Owing to the physical character of our ore, we make but little flue-dust. In 1915, while smelting 169,760 tons of charge, we produced 1050 tons of flue-dust, or 0.6%. The downtakes from each furnace are nine feet in diameter, leading to a dust-chamber with a cross-section of 350 sq. ft., and 150 ft. in length beyond the first down-take. From the dust-chamber the gases escape through a stack 212 ft. high and 10 ft. 2 in. inside diameter at the top. This stack also serves the boiler-plant. With two furnaces in blast, this allows for a speed in the dust-chamber of 250 ft. per minute, and the dust settles quite completely. The gases entering the dust-chamber carry 12.3 grammes of dust per cubic metre assaying 2.6% copper, and the gases entering the stack carry 1.666 gr. per cu. m. assaying 1.2% copper. There is much finely divided sulphur in the gases leaving the dust-chamber.

CONVERTERS. There are two stands of upright converters, 12 ft. diam. inside the shell, lined with magnesite brick. As we handle low-grade matte, we blow from 15 to 20 five-ton ladles of matte before blowing to blister. The corrosion of the lining is mainly at the tuyeres, and repairs are required every 30 to 40 days. The remainder of the lining lasts over a year. With more skilled workmen, the lining can be made to last longer and the campaigns should be not less than two months. In July 1914 all our skimmers were called to the army, and I had to try to continue operations with entirely green men. Two of the available men had been punchers, but knew nothing at all about running the converters or watching the temperature; the other was a blast-furnace foreman, who said that he knew the difference between matte and slag as shown on the hook. So they have had but 1½ years' experience, and it has taken a good deal of sitting-up nights to get even the rudiments of converter-work into their heads. During 1915 the average length of campaign of a converter before patching at the tuyeres was 32.67 days, and the consumption of magnesite brick for repairs was 4568 standard 15-in. brick on a production of 4,400,000 lb. of refined copper. The general work of the converters is given in the following table:

Month	Slag analysis SiO_2, %	Slag analysis Fe, %	Iron oxidized per min., kilo.	Average pressure, lb. per sq. in.	Blowing time, %	Air per ton blister, kilo.	Copper in matte, %
January	20.3	55.2	57.00	11.75	53.60	15,996	19.3
April	21.7	54.1	52.52	11.42	58.13	18,478	16.8
May	23.4	53.2	44.98	11.20	58.91	20,565	13.7
June	22.6	54.4	55.18	10.47	55.86	28,060	11.8
July	22.3	54.2	53.22	11.44	67.15	23,114	15.1

Blast is furnished by a multiple-stage turbo-compressor designed to furnish 5500 cu. ft. of free air per minute at one atmosphere. The compressor is direct-connected to a 540-hp. variable-speed motor taking alternating current at 2100 volts. With the speed variation we can run with any pressure down to seven pounds. At the start we had some difficulty with the compressor pumping, but this was remedied by a valve in the intake. This compressor has proved so satisfactory that we have just installed a second one of similar type.

REFINING. The gold and silver contents are too low to cover the charges imposed by existing copper refineries in Russia, therefore, pending completion of our own plant, the copper has been marketed direct. The refining-furnace is placed in one corner of the converter-aisle in line with the converters, and molten copper is poured directly into it from the ladle. This avoids the expense of casting and makes a saving of time and fuel in the refining. The furnace is a simple wood-fired one of the usual design. As it has to stand for considerable periods between charges, the fuel consumption is greater than it would be if the furnace could be crowded to capacity. Under present conditions the consumption of pine wood is 0.9 cord per ton of refined copper.

POWER-PLANT. The general design of the plant included electrically driven units at all points so that accurate determinations of power consumption could easily be made. For all units consuming considerable quantities of power, separate integrating watt meters are provided, and for the other units the power consumption is approximated by the use of a portable meter, the reading in each case extending over several days.

The boiler-house contains three Babcock & Wilcox boilers, each with a heating surface of 374 sq. m. and equipped with super-heaters delivering steam at the turbines at 330°C. Normally two boilers are in service, the third being in reserve. There is provision for placing another boiler similar in size to those already installed. When the plant was built, economizers were not added, but space was left and they are now being erected.

As planned, the average power requirements at mines and smelter was 1100 kilowatts, and to supply this two Brown-Boveri turbo-generators, each of 1500 kw., were installed. These furnish alternating current at 2100 volts and 50 cycles. Thus one unit is always in reserve. Surface-condensers are used, and the circulating and air pumps are electrically driven, the motor being direct-connected to the one shaft that operates all the pumps.

The power-station is housed in a brick building served by a five-ton hand-operated traveling crane; this building, besides the turbo-generators, contains a Jaeger single-stage turbo-blower for the water-jackets, capable of delivering 35,000 cu. ft. air per min. against a pressure of 60 oz. This blower is direct-connected through a flexible coupling with an 800-hp. motor running at 2950 r.p.m. This unit has now been in continuous operation for three years with no lost time for repairs. It is rather noisy and makes it hard to converse in the room. A second blower of similar type and 38,500 cu. ft. capacity is now being installed. No difficulty with pumping has arisen, even when running with considerably less than half the rated quantity of air going through. Such a blower should be equipped with a variable-speed motor, as otherwise power may be wasted if the pressure required happens to be less than that for which the machine is designed at full speed.

Two multiple-stage Brown-Boveri turbo-blowers serve the converters, each blower delivering 5500 cu. ft. of free air at 15 lb. These are also direct-connected through a flexible coupling with a 450-hp. alternating-current motor operating at 2000 volts. Each blower furnishes air for one 12-ft. Great Falls converter.

The switch-board is placed on a raised platform at the end of the building nearest the generators, and the instruments controlling the main circuits are on a low stand, so situated that the attendant is at all times facing the engine-room with the main panels of the switch-board directly behind him.

In the basement are the circulating-pumps and the fire-pump. This latter is a steam-unit that can deliver into the main water-circuit of the plant in case of accident to the electric equipment. The water-supply for the plant is obtained from ponds left from placer operations about 1½ miles from the power-plant. From this point the water is pumped by motor-driven centrifugal pumps. These pumps are placed in a sunken station so that they are below the water-level in the ponds. There is thus never any difficulty from failing suction, and the pumps are started and stopped from the power-plant, only a watchman being required at the pumps. A man goes once a day to oil and inspect them. The cost of such a station was somewhat higher, but the saving in cost of attendance has more than compensated for this. To decrease the size of the pumping-station and the diameter of the pipe-line, cooling-towers with catch-basins were built at the plant. One of these serves for the jacket-water and the other two for condenser-water.

THE PHOSPHATE LANDS of Idaho lie in the south-eastern part of the State, in Bear Lake, Bannock, Bingham, Bonneville, and Fremont counties. The Government has reserved 966,377 acres in Idaho as workable phosphate land, to which must be added 4080 acres of officially classified phosphate land in the Fort Hall Indian reservation. Idaho appears to contain the largest body of high-grade phosphate rock in the West. In the Georgetown Canyon region are three commercially workable beds, from 3 to 6 ft. thick, that carry 72.9 to 82.3% of tri-calcium phosphate. The main bed averages 5 to 6 ft. thick and contains about 70% of tri-calcium phosphate. Estimates based on detailed studies indicate that the areas examined by the U. S. Geological Survey probably contain 5,200,000,000 long tons of tri-calcium phosphate. These estimates include chiefly the main bed, which contains high-grade rock carrying 65% or more. Large areas of phosphate land have not yet been examined. Rock-phosphate has been quarried and shipped from Montpelier, Bear Lake county.

Metallurgy of Lead in Lower Mississippi Valley

By HERMAN GARLICHS

*The development of the extensive South-East Missouri lead deposits greatly preceded that of the Iowa and Wisconsin deposits. It began about 1720 at Mine La Motte and other localities, and has continued uninterruptedly to the present time. It is estimated that this district produced about 184,000 tons of lead in the year 1916, having a value of approximately $25,000,000. This is a new high record. The South-West Missouri orebodies were hardly known before 1845, and were not extensively developed until 1870. Lead production in South-West Missouri for 1916 was 38,788 tons. Oklahoma produced 14,399 tons and Kansas 2345 tons, making a total for South-West Missouri and the adjoining districts 55,532 tons of lead.

The first metallurgists in Missouri were the Indian and the hunter. The early settler learned to procure his bullets either by melting ore in his camp fire or by throwing pieces of galena on an old stump and depending on the usual 'roasting and reaction' method to obtain the metal. A little more refined method was to arrange two flat stones in the form of a V. Wood and galena were placed in the furnace and ignited. As heat developed, some of the metal was extracted and molded into bullets. The next step in utilizing the lead ore was the log-hearth. All lead in Missouri was smelted in this crude affair before 1820. It was built on sloping ground, and consisted of a hearth of stone, surrounded on the front and two sides by a stone wall. The wall was 7 ft. high in front. The top and rear end were left open and in front an arch or opening was made, forming the eye of the furnace. In front of this a pit was dug in the ground, to receive the molten metal. Large logs were rolled in at the back and made to rest upon ledges formed inside, to raise them from the hearth and to give a draft. These logs filled the entire width of the furnace. Small split logs were then set up around the two sides and the front, and the ore was piled on until the furnace was full. Finally, the mass was covered with logs and fuel until the ore was completely surrounded. A gentle heat was started, which was raised gradually. After 12 hr. the heat was increased and continued for 12 hr. more, 24 hr. being required for each smelting. The furnace treated about 5000 lb. per 24 hr. and the ore yielded about 50% of its metallic lead. A considerable quantity of the ore was not de-sulphurized, and fell between the logs into the ashes, forming a kind of slag, which was called 'lead ashes.'

The ashes were rich in lead and were frequently treated in a furnace of a peculiar construction, called an 'ash furnace.' This was introduced from Virginia and was a crude reverberatory, built of limestone, which lasted 15 to 20 days on continuous work. As the ash or residue was partly oxidized, the charge was immediately reduced at a moderately high temperature. Silica, as chert or sand, was mixed with the charge. In about two hours the furnace was ready to tap. The slag was tapped first, then the lead on the opposite side of the furnace. The total extraction from the two operations was about 75% of the lead in the ore. The next step in the progress of metallurgical operations in Missouri was the reverberatory furnace patterned after those of Carinthia, Silesia, and England. This type of furnace is no longer in operation, although the Desloge Consolidated Lead Co. operated the Flintshire furnace on a small scale until recently. The reverberatory required pure ores and it gave fair results, considering that no attempt was made to recover the fume.

The Backwoods hearth was another type utilized for the reduction of Missouri ores. The early ones were built entirely of stone and the blast was supplied by bellows. An improvement was the water-backed hearth, which permitted continuous work. The Jumbo hearth of Joplin was an enlarged furnace of the same character. Before the recovery of the lead fume, these furnaces extracted from pure ores about 70% of the contained lead, not considering the lead in the gray slag and fume. It was not until E. O. Bartlett, in the 'seventies, installed the bag-house for recovering the fume that the hearth could be considered economic. On account of the purity of the South-West galena, the Lone Elm Smelting Co. (now the Eagle-Picher Lead Co.) volatilized all the lead possible in short shaft-furnaces (slag-eyes), producing a pigment which they called sublimed white lead, being a so-called basic lead-sulphate consisting of $PbSO_4$, with varying amounts of PbO and ZnO, the zinc being derived from the Joplin ores, which carry from 2 to 3% of that metal in the high-grade concentrates.

The cupola process was the next one in line, and differed from the methods just described in that it divided the operations into two distinct steps, each performed in separate furnaces. The concentrate was roasted in the usual reverberatory, 55 to 60 ft. long, at gradually increasing temperatures. By later additions of sand, the roasting concentrate was sintered and partly slagged. The usual amount of sulphur left in the sinter was 5 to 6%. To reduce the sulphur lower meant increased temperature to decompose the lead sulphate, with resulting heavy losses in volatilized lead. The round (Pilz) blast-furnace for reducing the sintered product was later replaced by the rectangular (Raschette) furnace. This allowed a large production of lead, but the total extrac-

*Trans. A. I. M. E., St. Louis meeting, 1917.

ANALYSES OF SOUTH-EAST MISSOURIAN ORES AND RESULTING CONCENTRATE

	Ag, oz.	Cu, oz.	Pb, oz.	SiO_2, oz.	Fe, oz.	Al_2O_3, oz.	CaO, oz.	MgO, oz.	Zn, oz.	S, oz.	Ni and Co, oz.
Ore	0.12	0.06	5.7	5.0	4.1	4.9	25.5	14.2	0.8	2.0	....
High-grade concentrate	0.7	0.13	73.2	1.0	3.5	...	2.6	0.8	0.4	15.0	0.05
Medium concentrate	1.3	0.12	68.6	1.4	4.6	...	3.1	1.4	0.8	15.5	0.06
Low-grade concentrate	1.0	0.30	65.8	0.5	3.1	0.5	4.3	2.8	1.7	13.7	....
Flotation slime	..	0.50	45.0	2.6	4.4	3.1	7.5	4.2	4.0	12.8	....
Flotation slime, high-grade	3.7	0.05	57.8	6.0	2.7	...	2.2	1.4	9.4	45.5	....
Joplin concentrate	..	....	80.2	1.1	1.0	...	0.4	...	1.7	13.3	....

tion was not over 90%, due to losses in sintering, smelting, and re-treatment of matte.

I have passed lightly over these old practices, which are now almost wholly abandoned.

Analyses of the ores mined in South-East Missouri, that are about the average of the district, and also analyses of grades of resulting concentrate, show that the metallurgy is simple, the removal of the sulphur, and the slagging of the dolomite in the blast-furnace being of prime importance. There is a little nickel and cobalt in these ores, especially in those at Mine La Motte and the North American mine near Fredericktown. The nickel and cobalt concentrates with the copper in the matte, but are eventually lost in refining the copper. James W. Neill made a number of successful blast-furnace runs in the 'eighties at Mine La Motte with nickel and cobalt containing copper and lead, producing a bullion, a matte low in nickel and cobalt, and a speiss containing 23 to 24% of nickel and cobalt. He obtained the necessary arsenic from Western argentiferous speisses, recovering the silver and gold therein. Later the North American Lead Co. built a plant to separate nickel, cobalt, and copper by the Hybinette process. This produced matte containing the three metals, which were then separated by a wet method. The plant has been closed for many years, but is now controlled by the Missouri Cobalt Co. and is being remodelled for another campaign. The big improvements which have revolutionized the metallurgy of the Mississippi Valley ore in the last 10 or 12 years are:

1. The introduction of Huntington-Heberlein pot-roasting, preceded by a preliminary roast in a mechanical furnace (Godfrey), subsequent smelting in the blast-furnace, and filtering all fume through cotton or woolen bags.

2. The introduction of the Dwight-Lloyd sintering process which eliminated preliminary roasting for the concentrate, but not for the matte, with subsequent smelting in a blast-furnace with bag-house attachment.

The Godfrey pre-roasting furnaces were later abandoned in the H. & H. process, and the Wedge roaster was adopted at Herculaneum for the pre-roasting of matte. It soon became apparent that elimination of sulphur was still unsatisfactory, and that there was an undesirable, but necessary, dilution of the concentrate from 70 and 65% to 40 and 45% in lead to render the D. & L. machine applicable with a single roast. To avoid these objections, the double roast is being installed by both the Hercnlaneum plant of the St. Joseph Lead Co. in Missouri, and the Federal plant of the Federal Leal Co. in Illinois. A rapid preliminary roast is given on a separate line of D. & L. machines, and this pre-roasted material is then turned over to another set of D. & L. machines (Herculaneum), or to H. & H. pots (Federal) to reduce the sulphur of the finished product to about 2 to 2½% and with its lead content increased to 50 or 55%. It is expected that the double roast will effect the following improvements over present methods:

1. Avoid the large production of matte that locks up lead and requires expensive re-treatment for its recovery.

2. Avoid diluting a 65 to 70% lead-concentrate to 40 to 45%, which means large additions of barren flux.

3. Utilize to the fullest extent the fluxing components of the ore. This means that no limestone should be added to the blast-furnace charge, while some silica and iron will be required. The resulting blast-furnace slag will contain from 6 to 7% MgO with a corresponding increase in capacity for original concentrate.

4. Recover a part of the copper in the concentrate, which metal is at present mostly oxidized and lost in the large amount of slag.

The other important improvement was the mechanically rabbled hearth, known as the 'St. Louis hearth.' This was developed at the Collinsville plant of the St. Louis Smelting & Refining Co., fully described by W. E. Newman. The St. Louis hearth is applicable only to concentrate assaying 68% or more in lead, and therein differs from the H. & H. and D. & L. methods which treat concentrate of varying grades. When concentrate assays over 70%, it is considered that the hearth shows better recovery in lead, and is the most economical method, although the double roast, when perfected, promises to be equally as cheap, with the advantage of being able to treat any grade of concentrate. The peculiar advantages of the mechanical hearth for the higher grade concentrates are:

1. It treats the concentrate undiluted with flux.

2. It is a very efficient de-sulphurizer, expelling 95% of the sulphur.

3. It re-treats all the dust and fume produced, leaving the gray slag as the only product to be sent to the blast-furnace, with a large reduction in slag and matte produced.

4. The hearth makes a total extraction of 82 to 84% of the lead content of 70% concentrate. With concentrate assaying 75 to 80% lead, 90 to 95% represents the total extraction, leaving from 5 to 10% lead content in the gray slag. This practically eliminates any re-treatment of matte.

The mechanical rabbling machine is limited to 68% lead concentrate, and, for this reason, where a lower grade of concentrate is received at Collinsville, the D. & L. machine must be used as an auxiliary. All matte produced by smelting the gray slag of the hearths and the D. & L. sinter at Collinsville can be easily handled raw on the D. & L. machine, so that a Wedge roaster is not needed for pre-roasting.

The St. Joseph Lead Co. is erecting at Herculaneum 12 St. Louis hearths to be used in connection with the double-roast process for the treatment of the excess of high-grade concentrate. In the South-East district, about 75% of the output of the mills is a high-grade product, applicable to hearth-work, the remainder being lower-grade concentrate and flotation slime. In order to sweeten the mixture for the double roast, in other words, to raise its grade, it is necessary to use a proportion of the high-grade concentrate for this purpose. The balance of the concentrate should be worked by the mechanical hearth to obtain the most economical results both as to cost and recovery.

The lead produced from Missouri ore contains few impurities, and when subjected to a single liquation can be used for all purposes except corroding. A typical analysis of Missouri un-desilvered lead is as follows. Ag, 0.0080 (2.4 oz. per ton); As, trace; Sb, 0.0030; Bi, trace; Cu, 0.0800; Fe, 0.0015; Zn, trace; Ni and Co, 0.0080; lead by difference, 99.8995. The Herculaneum and Collinsville plants de-silver part of their lead, recovering about 1 to 2 oz. of silver per ton, and producing an unusually pure lead for the manufacture of white lead and other purposes requiring a very pure lead. The following analysis of South-East Missouri refined, indicates the unusual character of Missouri de-silvered lead: Ag, 0.005; As, trace; Sb, 0.0020; Bi, trace; Cu, 0.0002; Fe, 0.0004; zinc, 0.0004; lead by difference, 99.9965. The difference from Western refined is almost entirely due to the bismuth content (about 0.05 in Western refined) from which the Missouri ores are singularly free. The refining of the Missouri crude lead for corroding purposes follows the usual practice, except that no softening-furnaces are necessary.

At Collinsville the lead from the blast-furnace is poured molten into the drossing kettle. After carefully drossing to eliminate copper as much as possible, it is pumped into a de-silverizing kettle. The hearth-lead is charged as pig directly into the kettle, drossed, and de-silverized. Zinc is added, stirred in, and the zinc-silver-lead-alloy (crust) removed. This first crust is re-worked in succeeding kettle-charges until sufficiently high in silver to be set aside. The resulting retort-bullion (lead riches) assays from 500 to 600 oz. Ag per ton, and represents a concentration of the silver-content of 250 to 300 tons of lead into 1 ton of retort-bullion. The zinc consumption per ton of refined lead produced is about the same in amount as that consumed in refining argentiferous bullion. Special attention must be given to removing copper as much as possible from the lead to be de-silverized. The quantity of zinc-silver-lead alloy made is insufficient to warrant the operations of retort and cupels, and heretofore the alloy has been sent to some outside refinery to be separately distilled in retorts, and the returns are based on the sampling of the various products obtained, such as retort-bullion and dross, spelter, and blue-powder. The assay-results check quite closely. Retorts and cupels are being installed at Collinsville, so as to be able to treat the alloy whenever occasion demands.

A Switch-Tongue

In constructing turn-outs in the line of track underground, either for sidings, or for branches to side workings, most miners desire to avoid the making of frogs. To do this they substitute the tongue-switch. The accompanying sketches illustrate the manner of making and operating the tongue-rail for such a switch. To hold the lug at the end of the tongue in place, a plate having

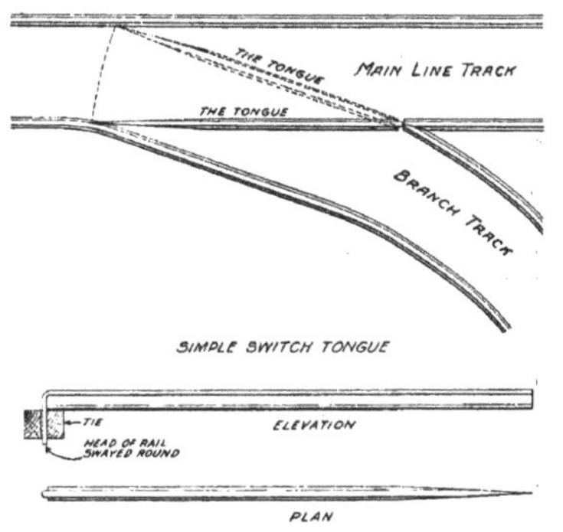

a round hole punched through it should be placed on the upper and lower face of the ties; through this the lug passes vertically. This holds the tongue in position when the pointed end is swung across to join the opposite side of the track. The tongue should be at least 8 ft. long to permit of some flexibility, or the car-wheels are likely to be bound by the rails, causing the car to jump the track. It will be noted that the outer rail of the switch is in a position parallel to the tongue when thrown across the main track, and there is consequently a jog on the outer rail opposite the end of the tongue, and another opposite the swivel-end. Thus the gauge is maintained throughout, and the likelihood of derailing cars is obviated.

COMMERCE between Spain and the United States is growing at a rapid rate. The total value of the imports from Spain in the first six months of 1916 were $18,-553,341 against $25,291,338 for the corresponding period this year, while the exports were respectively $38,960,413 and $55,348,860. Among the receipts from Spain were 11,000 tons of iron ore from the province of Vizcaya in March and April.

Modern Milling Methods Applied to Californian Gold Ores

By ALEX McLAREN

There has been, during past years, a system of gold-milling practised on the Mother Lode region of California that developed in connection with the so-called California mill, a system that through years of continual use became 'bred in the bone' of the mill-men of that region. This type of mill consisted of the stamp-battery as a pulverizing device, preceded by a crusher, and followed by amalgamating-plates and Frue-vanners. When the oxidized ores were being treated, good recovery generally resulted, but the recovery on the sulphides averaged only about 80%. The reason for this low recovery was due to a number of conditions that were mainly mechanical. The excessive quantity of fine material produced in stamping was discharged from the battery with a consistence of six or seven of water to one of solid. This worked out well as far as amalgamation was concerned, but in very few of the mills, if any, was any attempt made to either de-water or to classify the pulp for concentration. It was, instead, run over the vanners with the excess of water that was necessary to use in the battery and with the addition of more water on the vanner. This, of course, is not conducive to good concentration of either classified or unclassified pulp.

The belief was current among many of the old Mother Lode mill-men that they were getting the most from the ore that was possible, and suggestions for de-watering and classifiying were generally greeted with ridicule. A few months ago I was called upon by the management of the Kroromick Mining Co. at Coarse Gold, California, to take charge of its mill with a view to bettering the recovery. It was a modern 20-stamp Californian type of mill; only 10 stamps were in use, followed by amalgamating-plates and two Wilfley tables in series. I operated the plant for eight days without making any changes, meanwhile sampling carefully to determine in just what form the loss occurred. The ore was typical of the Mother Lode, a mixture of 'ribbon-rock,' clay-slate, and quartz, yielding about 60% of its gold-content by amalgamation, the remainder being carried by the lead and iron sulphides. After being churned in a deep mortar and discharged through a 30-mesh screen, more than 40% would pass a 200-mesh screen. This was treated on the Wilfley tables where the excessive amount of water carried the fine particles of sulphide over the tables and into the tailing-launder. Under these conditions the total recovery during the eight days of sampling averaged 75.11%, and the loss was found to be in the material finer than 200-mesh.

The next step was either to attempt to cut down the quantity of fine, or to leave the pulverizing conditions as they were, segregate the slime, and treat it on a slime-table or on some device adapted to such work. For classification the free-settling system was used. This was accomplished in a wooden tank 12 ft. long, 4 ft. deep, 3 ft. wide at the top, and 30 in. wide at the bottom. Six spigots were provided along the side of the bottom of the tank and spaced at regular intervals. A Senn pan-motion vanner was in the mill but had not, for some reason, been used. The coarse particles of pulp were drawn from the first two spigots and treated in a thickened condition on the first Wilfley, the next two spigots on the second Wilfley, and the last two were delivered to the Senn pan-motion concentrator, using it as a slimer, inasmuch as a screen-analysis of the issue from the last

SCREEN ANALYSIS

Screen	Mill head, %	Table head, %	Senn head, %
+ 30 mesh	1.14	3.71	....
+ 40 "	2.60	8.61	....
+ 60 "	10.07	23.77	....
+ 80 "	8.44	12.66	0.39
+ 100 "	13.78	13.24	5.31
+ 150 "	11.55	10.27	12.93
+ 200 "	8.56	6.85	14.00
— 200 "	41.94	19.22	65.63
Screening loss	1.92	1.67	1.84

LAST TWO SPIGOTS TO SENN CONCENTRATOR

Date 1917	Mill head	Plate tail	Saving by amal., %	Table tail	Senn head	Senn tail	Senn recovery, %	Mill recovery, %
March 22	$5.20	$1.60	69.2	$0.60	$2.40	$0.80	66.7	88.47
" 23	6.40	3.00	53.1	0.80	2.60	0.90	65.4	87.50
" 24	8.20	2.60	68.2	0.80	2.80	0.90	89.1	89.09
25	6.40	2.00	68.7	0.40	1.60	1.00	62.5	96.00
26	7.60	4.20	44.8	0.60	3.20	0.80	75.0	89.50
Average	$6.76	$2.68	60.8	$0.64	$2.52	$0.88	71.7	90.11

two spigots showed 65% passing through 200-mesh and 99.61% through 80-mesh. This permitted thickening the pulp, the surplus water overflowing the end of a tank fairly clear. The spigots, being partly closed, permitted a steady flow to the concentrators with a consistence of about 30% solids. Under these conditions, the ensuing five days' sampling showed a recovery of 90.11%, being

TEXAS FLAT MILL, MADERA COUNTY, CALIFORNIA

an increased recovery of 19.9%. It was readily seen that the additional recovery was parly due to a classification, the settling being far from perfect, thus reducing the water-content of the pulp on the tables, segregating the fine and treating it separately on a Senn pan-motion concentrator. With it I found possible a large recovery from the sulphide that was previously lost. It was an exceedingly difficult case, since more than 80% of the concentrate recovered was so fine as to pass through a 200-mesh screen. The average results before making modifications in the mill, taken from the assay-sheets representing a week's run were: heading, $8.05; plate-tailing, $2.73; recovery by amalgamation, 58.8%; table tailing, $1.62; table recovery, 37.8%; average total recovery for the week, 75.11%, the daily recoveries ranging from a minimum of 62.5% to a maximum of 92%. The changes made resulted in a recovery which, as shown by a typical week's run, gave an average of 89.8% extraction, ranging from 86.4 to 94.8%. The accompanying table shows the results of a five days' run, including the Senn pan-motion concentrator. The screen-analysis also reveals the conditions under which the concentrator had to operate.

INTERIOR OF THE TEXAS FLAT MILL

The Texas Flat mine, now owned and operated by the Krornmick company, was discovered in 1853, and was equipped and extensively worked by the Haggin & Hearst syndicate for several years prior to 1890. It has seen long periods of idleness and several of prosperity. The vein is from 3½ to 4 ft. wide in slate, the strike and dip of the fissure conforming generally with those of the inclosing rocks. An adit has been run on the vein 1400 ft. and connects with the incline shaft 260 ft. vertically below the surface. The property was equipped with an electric power-plant, situated on Fresno river, three miles from the mine, which is one mile north of Coarse Gold post-office. An electric hoist has been placed at the adit-level of the shaft, which is 920 ft. deep. A 4-drill compressor and mill with twenty 1500-lb. stamps and concentrators complete the equipment. In 1914 high water in the Fresno river destroyed the dam and did considerable damage to the plant, but power is now supplied by the San Joaquin Light & Power Co. It is said that the ore contains about $4 in free gold, and 0.5% sulphide that assays about $100 per ton. It is claimed that the cost of mining and milling is under $1.50 per ton.

ZIRCON is used chiefly in the manufacture of refractory crucibles and for the refractory lining of furnaces. It is therefore a material for which there may be increasing demand under the pressure that is being put on the steel industry of the United States. Silicate of zirconium, the mineral known as zircon, is found in small crystals mixed with monazite in the sands from which the Brazilian monazite is washed. In the process of recovering monazite from these sands, in Bahia, Espirito Santo, and in a smaller way in Minas Geräes and Rio de Janeiro, zircon and ilmenite are separated. As the ilmenite has no commercial value it is thrown away, but exporters claim that zircon can be marketed at $50 per ton. With the electric separating machines now in use it is practically impossible to remove all the monazite from the zircon, and a residuum of 2 to 6% remains with it. Increasing shipments of zircon may be expected from Brazil provided it is not too heavily burdened with duties and is classified as zircon and not as monazite.

COLOMBIA shipped to the United States during the year 1916 a total of 21,645 oz. of gold, or a value of $432,900. Of this amount 1394 oz. was in the form of cyanide precipitates, the rest being bars and gold-dust.

California Mines in 1917

While the mineral industry of California has shown exceptional activity during the first six months of 1917, gold mining has been suffering a period of depression, according to Charles G. Yale, of the U. S. Geological Survey. He reports that the U. S. Mint at San Francisco, and the local smelters and refiners, received about the same quantity of gold and silver as in the corresponding period of 1916, proving that no advance in output has been made, notwithstanding the opening of new gold mines and the re-opening of old ones. Investigation has shown that thus far in 1917 the expenses of gold production have increased 18% over 1916, due to extra cost of labor and supplies, and increased taxation. The scarcity of skilled and unskilled labor has caused many mine-owners to hang up some of their stamps for a few months this year because they have been unable to hoist the normal quantity of ore. Machine-men have gone to munition-plants and other factories where they obtain better wages than in the mines. Good timber-men are scarce, as are good miners. Adjoining States are offering higher wages than in California, although wages have been raised in this State also. Copper mines give a bonus, based on the value of the metal, and that has also attracted men from the deep gold mines.

The California Debris Commission has been vigorously enforcing its regulations. As a result many deep mines have had to construct dams to impound the tailing so as to prevent it from passing into streams which feed the navigable rivers. All these things have combined to retard the gold-mining industry. During the first half of 1917 both the Kennedy and the Argonaut companies, the largest producers in Amador county, have exceeded a vertical depth of 4000 ft., and at that depth are mining rich ore. This is encouraging to other Mother Lode miners who have not so great a depth. In Calaveras county the indebtedness of the old Royal Consolidated mine at Hodson has been paid off, and re-habilitation of the property has commenced. The Drake group, including the Last Chance south of Angels, is about to be unwatered and worked by Eastern men. There is also activity at Mokelumne Hill, affecting the Safe Deposit, Buffalo, Hector, Lucas, Mokelumne, and Esperanza mines. In El Dorado county the Church-Union and several adjoining claims south of Placerville have been re-opened after a long period of idleness. The Placerville Gold Mining Co. has acquired 23 claims at Placerville, including the Pacific, and some have been unwatered for development. In Mariposa county, the Fremont Grant of the Mariposa Mining & Commercial Co. has been sold to a company which is about to re-open the Princeton, Pine Tree, and Josephine mines. On this tract of 44,000 acres are 71 mines which were once productive but have been idle for many years. The Mount Gaines mine, near Hornitos, formerly a producer, but for some years idle, is being put in shape for re-opening; and the Mary Harrison, near Bagby, is also being re-habilitated. At the Red Bank a flotation-plant has been installed. In Tuolumne county the Eagle Shawmut mine, with its 100-stamp mill and chlorination plant, has been purchased by the Tonopah-Belmont Mining Co., and will again become productive, the main shaft, now down 2200 ft., being deepened to explore the main orebody.

In Inyo county there is much activity around Keeler and Darwin. The Cerro Gordo properties are producing large quantities of zinc. The Darwin Development Co., after proving the efficiency of the flotation-process on the Lane and Lucky Jim mines at Darwin, has purchased the Defiance mine and the water-rights at Coso. In and around Zabriskie there is activity in silver-lead mining. In Mono county the Masonic district has again displayed new life. An ore-shoot of high-grade gold and silver ore has been discovered in the Pittsburg-Liberty mine by lessees. The Serita company has enlarged its milling activities, and has also increased its water-supply for the electric power-plant. In Nevada county the new mill on the Allison-Ranch mine has been completed and is running. This was formerly a heavy producer. The North Star Mines Co. at Grass Valley continues to produce heavily, the richer ore coming from below the 600-ft. level on the incline. This company has recently acquired the Rogers mine near Washington and has started work. More stamps have been added to the Golden Center mill at Grass Valley. The Empire Mines Co. continues to operate the most productive deep gold mine in the State. The Valley View mine in Placer county has been sold to the H. C. Winters Co. of New York. Machinery has been ordered, and an electric-power line is being constructed.

Goldstone is a new district in San Bernardino county, where developments are encouraging. The Goldstone Mining Co. has completed its mill, and the Belmont, Red Bridge, and other gold claims are being developed. In Sierra county work has been resumed on a number of properties in the Downieville region. The Shamrock mine, adjoining the Sierra Buttes, at Sierra City, has been sold to San Francisco men and is being re-opened. The Plumbago and Oriental mines of the Croesus Mining Co. are having extensive development work done; in Siskiyou county, the Dewey mine, after an idleness of eight years, is again in operation, and an electric-power plant has been provided. In the region around Happy Camp numerous gold and copper properties are receiving attention from New York and Philadelphia people. The Bonanza King mine, near Trinity Center, Trinity county, is again producing.

The gravel-mining industry of California. including

dredge and hydraulic operations, is in a more prosperous condition than quartz mining. Some features of the latest style of dredge put into operation this year by the Yuba Consolidated Goldfields in Yuba county, are of interest. The capacity is 350,000 cu. yd. of gravel monthly, with an estimated working cost of 3½c. per cu. yd. The dredge-buckets hold 18 cu. ft. each, and the machine will dig to a depth of 84 ft. below the surface of the pond. This machine will move boulders weighing four tons and will remove roots and trunks of trees embedded in the gravel. The dredge cost nearly $500,000. It has two tailing-stackers, each 220 ft. long, stacking on each side so as to leave a 500-ft. channel. Another dredge will dig a channel 500 ft. wide to comply with the requirements of the California Debris Commission for a 1000-ft. channel. A large dredge, of similar design, to cost $600,000, has been started this year. In Trinity county the Pacific Gold Dredging Co. started operations early this year in the Carrville district. The buckets of the dredge have a capacity of 9 cu. ft., and the machine digs 60 ft. below the water-line. The Trinity Star Dredging Co. is building a wooden dredge on the Trinity river on the Paulsen ranch near Lewiston. The Trinity Gold Dredging Co., owning the old Alta Bert dredge, has contracted for a second dredge with 82 buckets of 18 cu. ft. capacity each. The company owns 3600 acres or dredging ground. The Gardella Dredging Co. is preparing to build a second dredge on Clear creek, near Redding, where it already has one in successful operation. A third dredge is to be constructed by the American Dredge Co. on the Mokelumne river, near Camanche, Calaveras county. In hydraulic mining, activity has been greater than for many years in California. The new big concrete impounding dam across Slate creek in Sierra county will permit hydraulic mining again at Howland Flat, St. Louis, Port Wine, and a number of other old-time placer camps, which have been virtually deserted. Many old abandoned hydraulic mines in the Cherokee and Columbia Hill districts of Nevada county, and other places, are being tested with a view to reopening them and starting the giants. Experiments are being made in drift mines in Butte county, with a view to mining the gravel of the buried rivers by machinery instead of by hand.

The mining of copper, lead, and zinc in California, thus far in 1917, has been more active and prosperous than gold mining. As usual the larger proportion of the copper output comes from Shasta county, although Calaveras and Plumas are large producers. The Afterthought mine at Ingot, Shasta county, after being closed nine years, has resumed activity and is being equipped with a flotation-plant and a roaster. The Minnesota mine, near Keswick, after 20 years of idleness, has been re-opened by the Mountain Copper Co. The same company has also opened the Keystone group of copper claims on Flat creek, in the same vicinity. The Balaklala Copper Co. is active and shipping ore. The electrolytic zinc-plant of the Mammoth Copper Co. has been completed and put in operation. The Arps Copper Co., adjoining the Bully Hill, in Shasta county, has placed orders for hoisting machinery, compressor, and drills, preparatory to deeper development. The Copper King mine, at Clovis, Fresno county, long idle, is under new management and began to ship ore last May to the Mammoth smelter. In Calaveras county a converter has been installed at the smelter of the Penn Mining Co. so that blister-copper will be shipped hereafter instead of matte. Electric power has been substituted for steam throughout the plant. More units are being added to the flotation-plant of the Calaveras Copper Co. in the same county. In Plumas county, the Indian Valley railroad is operating between the Engels Copper mines and Paxton, on the Western Pacific railroad. The capacity of the flotation-plant at the mine has been increased. The International Copper Co. is making large shipments of copper ore from Taylorsville to the Tooele smelter in Utah. The United States Smelting & Refining Co. is increasing its activities in Plumas county. In Siskiyou county the Gray Eagle Copper Co. has repaired its roads in Happy Camp, and has several diamond-drills at work. The Hunters Paradise copper group has been bonded to the Mason Valley Mines Co., with a view to prompt development.

Advance in Electric Steel Production

Great strides have been made within the last three or four years, and particularly last year, in the production of steel in electric furnaces both here and abroad. Great Britain last year more than doubled her output of 1915. Canada stepped into prominence by making more electric steel last year than has ever been made in France, the originator of the electric furnace. Austria-Hungary, under war conditions, also doubled her production last year over the output for 1915. The leading countries producing electric steel last year turned out more than twice as much as was made in the previous twelve months. The extent to which electric steel is being made since the War began is given in gross tons in the order of importance in 1916, as follows:

	1916	1915	1913	1912
Germany	180,335*	129,000	101,755	74,177
United States	169,918	69,412	30,180	18,309
Great Britain	49,256	22,000	......	.....
Austria-Hungary‡	47,247	23,895	26,837	21,556
Canada	43,790	61	......	.....
France	......	......	18,000*	15,922
Total	490,546	244,368	176,772	129,964

*Estimated, metric tons.
†Metric tons.

It is estimated that the electric steel output of the United States is now at the rate of 200,000 tons per year and by January 1, 1918, with the prospective furnaces already contracted, it will reach 300,000 tons.

Recent Patents

1,232,572. MINER'S CARBID-LAMP. Ronchi Lazare, Portage, Pa. Filed Apr. 26, 1917.

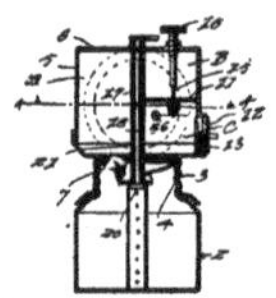

1. In a miner's lamp, a cylinder, heads closing the opposite ends of the cylinder, a partition wall extending vertically through the cylinder, a horizontal partition wall between one side of the first-mentioned partition wall and the cylinder, a carbid receptacle formed by the horizontal partition wall and the vertical partition wall, a water receptacle formed above the horizontal partition wall and carbid receptacle, said vertical partition wall having an opening in its upper edge, whereby communication is established between the water receptacle and the remainder of the cylinder, a valve controlling the flow of water from the water receptacle into the carbid receptacle, a valve extending through the cylinder for controlling the flow of water therefrom, a main carbid receptacle adapted to be secured to the under side of the cylinder, burners mounted on the side of the cylinder, and tubes leading to the burners for conducting the gas generated by the mixture of the carbid with the water to said burners.

1,232,156. PROCESS FOR OBTAINING POTASSIUM CHLORID FROM CERTAIN WATERS CONTAINING BORAX. Noah Wrinkle, Keeler, and Wilfred W. Watterson, Bishop, Cal. Filed Dec. 11, 1916.

2. The herein described method of obtaining potassium chlorid from waters containing the same and sodium borate, sodium chlorid and other salts, embodying first eliminating the other salts and concentrating to obtain a solution high in the three named salts, treating the solution with an acidifying agent to form boric acid, concentrating the solution by heat to deposit the named salts and acid, treating the mixed salts with alcohol to dissolve the boric acid formed and separating the solution thus formed, and washing out the sodium chlorid from the remaining mixture of sodium chlorid and potassium chlorid with clear cold water, leaving the potassium chlorid.

8. The herein described process of treating water containing potassium chlorid, sodium borate, sodium sulfate, sodium carbonate, and sodium chlorid, embodying first treating the liquor with carbon dioxid to transform the carbonate to an insoluble bicarbonate and removing the liquor therefrom, then heating and evaporating the liquor to concentrate it to the saturation point of potassium chlorid at ordinary temperatures, then treating the liquor further with carbon dioxid to remove further sodium carbonate and then cooling to deposit sodium borate and removing the liquor, then further concentrating the liquor by evaporation to the saturation point of potassium chlorid at the temperature employed, then cooling the liquor to deposit the contained salts, removing the liquor to be returned to an original liquor being treated, re-dissolving the mixed salts in clear water, treating the solution with chlorin to form boric acid and sodium chlorid from the sodium borate, concentrating the solution and depositing the salts and boric acid therefrom, removing and treating the mixed deposit with alcohol to dissolve the boric acid and removing the solution, and washing out the salts with clear cold water leaving undissolved the potassium chlorid.

1,232,830. METHOD OF MAKING CATALYTIC MATERIAL. Charles Burrows Morey and Charles Rollin Craine, Buffalo, N. Y., assignors to Larkin Company, Buffalo, N. Y. Filed Feb. 2, 1916.

1. The method of producing a catalyzer consisting of mixing an absorptive non-catalytic material with an insoluble compound of a catalytic agent, removing from the interstices of said absorptive material the moisture and gases contained therein and causing said compound to enter into said interstices to form a secure bond between the same and said compound, drying the mixture and reducing the same.

2. The method of producing a catalyzer consisting of forming a suspension in a liquid of an insoluble compound of a catalytic agent, mixing an absorptive non-catalytic material with said suspension, removing from the interstices of said absorptive material the moisture and gases contained therein and causing said compound to enter into said interstices to form a secure mechanical bond between the same and said compound, drying and reducing the same.

1,232,471. CYANID IN THE FORM OF GRANULES. Fritz Abegg, Perth Amboy, N. J., assignor to The Roessler & Hasslacher Chemical Co., New York. Filed June 23, 1914.

1. As a new article of manufacture, cyanid in the form of a hollow granule.

2. As a new article of manufacture, cyanid in the form of a hollow granule, open at one end.

3. As a new article of manufacture, cyanid in the form of a hollow granule, open at one end, and glazed on both its outside and inside.

4. As a new article of manufacture, sodium cyanid in the form of a hollow granule.

5. As a new article of manufacture, sodium cyanid in the form of a hollow granule, open at one end.

1,232,080. PROCESS OF RECOVERING COPPER. Frederick J. Pope and Albert William Hahn, Douglas, Ariz. Filed July 29, 1914.

1. The improved cyclic method of extracting copper from copper-bearing material whose copper content is soluble in a sulfuric acid solution, comprising the leaching of the material with a solution of sulfuric acid, whereby the copper is converted into copper sulfate in solution, electrolyzing the solution containing the copper sulfate and regenerating sulfuric acid, returning the solution to the ore, and at some point in the circuit of the solution, when it becomes foul withdrawing a portion of the solution from the circuit, predetermined in amount in accordance with the amount of impurities necessary to be removed from the solution as a whole, eliminating impurities from the portion so withdrawn and returning it, purified, to the circuit.

4. The improved cyclic method of extracting a metal from an ore containing it in cases where the metal content is soluble in an acid solution, comprising the leaching of the ore with a solution of such acid whereby the metal is converted into a soluble salt of the metal, electrolyzing the solution containing the salt, returning the solution to the leach, and at a suitable point in the circuit of operations, when the solution becomes foul withdrawing a portion of the solution predetermined in amount in accordance with the amount of impurities necessary to be removed from the solution as a whole, eliminating impurities from the portion so withdrawn, electrolyzing, and returning it purified for further use in the cycle.

1,232,772. METHOD OF AND MEANS FOR RECOVERING METALLIFEROUS MINERALS BY FLOTATION. George Crerar, Spokane, Wash., assignor of sixty one-hundredths to James L. Boyle, Los Angeles, Cal. Filed Mar. 9, 1915. Renewed Nov. 27, 1916.

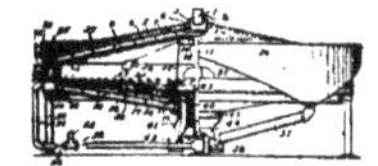

4. The method of recovering metalliferous minerals from ore pulp which consists in converting the pulp into froth; subjecting the froth to the uplifting and drying action of finely divided air currents; causing the froth to flow toward a place of separation meanwhile spreading the froth over a wider area, at the same time deepening the body of froth and subjecting said deepened body to the continuous action of such air currents forcing the surface froth to separation; drawing of frothy material from below the surface of said deepened froth body and the frothy material thus being drawn off.

5. The apparatus for recovering metalliferous minerals from ore pulp treated with a frothing agent which comprises a sloping porous deck, means to supply the treated ore pulp to the upper end of the deck; means to force air up through the deck to froth the pulp thereon; separating means for collecting a body of froth of considerable depth at the lower end of the deck; said means terminating at a level intermediate the levels of the upper and lower ends of the deck; means to force air up through such deep froth body; and means to draw tailings off at the bottom of such deep froth body.

1,232,840. SHOT-FIRING DEVICE. Elmer Porphir, Pittsburg, Kan. Filed May 4, 1916.

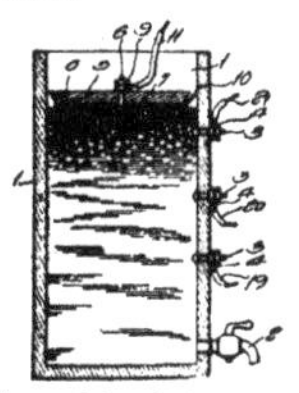

1. A multiple charge firing device comprising a polysided liquid receptacle, a series of vertically spaced contacts passing through one side of said receptacle for connecting current conducting wires, the inner ends of said contacts being in the form of rounded heads, a float movable vertically in said receptacle, a contact plate having its outer end directed upwardly to form a contact for engagement with said series of contacts, and having means for attaching a current conducting wire thereto, said float being of a shape corresponding to the polysided shape of the receptacle to prevent rotation of said float and to thus maintain the contact plate and the series of contacts in alinement, and a valve at the lower end of said receptacle for preventing the escape of liquid and supporting the float above said series of contacts until the charges are to be fired and for then permitting the escape of such liquid.

1,233,118. ALLOY TOOL-STEEL. James Heber Parker, Reading, and Berton H. De Long, Springmont, Pa., assignors to Carpenter Steel Company, Reading, Pa., a corporation of Pennsylvania. Filed June 26, 1916.

1. A tungsten-steel having a base composed of between 30% and 45% of cobalt and between 55% and 70% of iron, said compound constituting the bulk of the alloy and the tungsten between 8% and 20% thereof.

2. An alloy tool-steel composed mainly of iron and cobalt proportioned approximately two of iron to one of cobalt, 8% to 20% of tungsten and .5% to 2.5% of vanadium.

3. An alloy tool-steel composed mainly of iron and cobalt proportioned approximately two of iron to one of cobalt, 8% to 20% of tungsten and .5% to 2.5% of vanadium and 1% to 6% of chromium.

1,232,362. ELECTRIC FURNACE. Claude G. Miner, Berkeley, Cal. Filed Feb. 17, 1917.

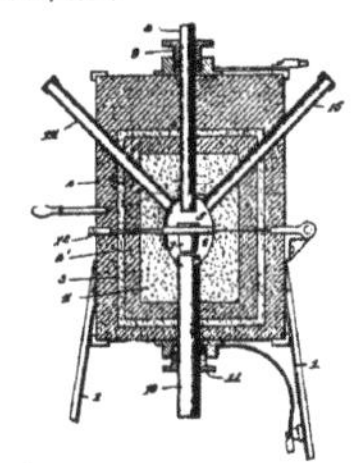

4. An electric furnace, outer and inner spaced walls and heat insulating material in the space between said walls; said furnace formed interiorly of said inner wall with a reaction chamber having the shape of two opposed paraboloids; and a crucible composed of conducting material in the principal foci of paraboloids, said crucible constituting a centre electrode for two arcs in series.

10. In an electric furnace, a body having outer and inner spaced walls and having the space between said walls filled with heat insulating material; said body formed interiorly of said inner wall with a cavity, and a crucible in said cavity, said crucible constituting a common electrode for two arcs in parallel.

1,233,086. ORE-GRINDING MILL. Peter A. MacEachern and Nolen O'Daniel, Douglas, Ariz. Filed Aug. 12, 1916.

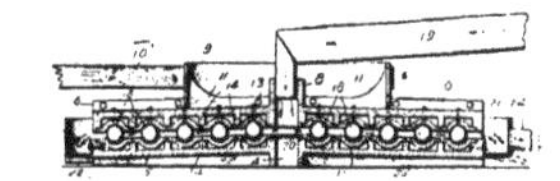

In a mill of the character specified comprising complementary members, each having radial ribs on its outer face, and parallel circular ribs on its inner face, the circular ribs of the two members coinciding, half-round runways placed between the circular ribs of the members and touching the members on medial lines and having outer flanges at their edges to overlap and bear against the ribs, and balls arranged in the runways.

1,228,414. METHOD OF SEPARATING AN AROMATIC SULFONIC ACID AND OBTAINING THE SULFONIC ACID IN SOLID FORM. Louis M. Dennis, Ithaca, N. Y. Filed Nov. 20, 1916. Serial No. 132,339.

3. A method of separating a sulfonic acid of a hydrocarbon of the aromatic series from sulfuric acid and of obtaining the sulfonic acid in solid form, consisting in treating a mixture of the acids with an organic solvent which dissolves the sulfonic acid but not the sulfuric acid and which dissolves an appreciably greater amount of the sulfonic acid at high temperature than it does at low temperature, separating the sulfonic acid solution from the sulfuric acid and cooling said solution.

REVIEW OF MINING

As seen at the world's great mining centres by our own correspondents.

BRITISH COLUMBIA

THE SMELTER AT LADYSMITH BLOWN-IN AFTER AN IDLENESS OF SIX YEARS.—SMELTER RATES COMPARED WITH THOSE AT THE TACOMA SMELTER.

After being closed down for five years the smelter at Ladysmith, on Vancouver island, has been re-opened. The 20,000 tons of ore necessary as a supply before the furnaces could be blown-in has been collected.

This is one of the most important events in British Columbia mining developments since the War started and will go a long way in fostering and maturing the mining activity on the island, as well as in other parts of the province.

One of the great advantages of the smelter will be its arrangements for treating local ores. The plant is of sufficient capacity to handle the entire output of the island for many years to come, and present and prospective owners will be able to proceed with development work in full confidence of having adequate smelting facilities.

It was in 1901 that the smelter, then known as the Tyee smelter, was erected. It was built at Ladysmith because of the advantageous position, cheap transportation rates being available from most parts of the province.

The plant was closed-down in the latter part of 1911 and has remained idle ever since. Last autumn it passed into the hands of the Ladysmith Smelting Corporation, which company has made many contracts for ore that will ensure constant operation.

The president of the company is Colonel Stevenson, who has long been associated with important mining interests in British Columbia and in Alaska. He is one of those controlling the Alaska Corporation, which operates in conjunction with the Ladysmith smelter. H. W. Aldrich is superintendent of the plant.

Naturally there is deep interest in mining circles as well as among business men generally as to the effect the operation of the smelter will have upon the movement of ore into the United States. At present large quantities are shipped to the Tacoma smelter and it appears likely that a considerable proportion of that ore will in future be sent to Ladysmith.

The Ladysmith corporation has announced a treatment charge of $5 per ton. The copper will be paid for on a basis of 100%, but a deduction of 0.4% will be made from the wet assay.

The price paid for copper will be 3c. per lb. less than New York quotations 90 days after the receipt of the ore; 95% of the gold content will be paid for at $20 per oz., but if there is less than 0.1 oz. per ton it will not be paid for; 95% of the silver will be paid for at New York quotations 90 days after sampling, but the ore must assay over 0.5 oz. in order to be paid for.

An advance of 60% of the value of the ore at current prices will be made, but the shipper must pay 7% interest for the advance.

Tacoma, with which Ladysmith is in direct competition, issues two schedules, of which the shipper may make his choice. The charge for treatment is $1.50 per ton, $3.50 less than the Ladysmith rate.

In both Tacoma schedules 95% of the gold and silver content is paid for, but no gold under 0.3 oz. or silver under 1 oz. per ton is paid for. On this point the Ladysmith terms are better.

Under schedule A, of the Tacoma smelter, 100% of the copper is paid for after deducting 1.3 for wet assay. When copper is quoted higher than 14c. per lb. only 75% of the excess is paid for—which means that the smelter claims 25% of all the copper value over 14c., after deducting 1.3 for wet assay and 3c. per lb. in addition.

Under Schedule B, of the same smelter, the same percentages of value are allowed, but preliminary settlement will be made on the basis of copper at 14c. and silver at 55c. Final settlement will be made 120 days after sampling, on the basis of the average New York quotations for a week prior to the expiration of the 120 days.

If the final quotation is in excess of 18c. per lb. for copper the treatment charge is advanced from $1.50 to $2.50 per ton. Any shipments under five tons are charged $10 flat for sampling in addition to the treatment charges and reductions.

A comparison of the Ladysmith and Tacoma rates shows that the former is generally the most favorable.

COBALT, ONTARIO

BULLION OUTPUT.—NIPISSING.—TRETHEWEY SILVER-COBALT COMPANY.

Bullion shipments during the second week of July from the Cobalt camp were heavy; Mining Corporations alone sending out a little over 250,000 oz., and with 113,571 oz. to the credit of Nipissing. Ore shipments were about the average, Nipissing with 4 cars heading the list. For the seven months ended July 31, Nipissing has mined ore of an estimated net value of $1,765,167, made up as follows:

January	$173,988
February	271,527
March	256,953
April	259,082
May	261,663
June	269,469
July	272,490
Total	$1,765,167

It will be noted that the July record is the highest for any one month during the current year, and that since March the output has showed a steady increase.

Unusual light promises to be thrown on the affairs of Temiskaming in that certain sharerholders of that company have suggested that a report, other than that of the management which was comparatively unfavorable, be made by a disinterested party. A special examination by B. Neilly, superintendent of the Penn-Canadian mine, is now under way and will be ready for the shareholders at an early date. It is the consensus of opinion that there will be very little difference shown between the estimate of the management and that of the special report.

A dividend of 5% has been declared by the Trethewey Silver-Cobalt Co., payable August 31 to shareholders of record August 20. The disbursement amounts to $47,272, making the total paid to date by that company $1,159,270.50.

With the price of silver holding well above 80c. per oz., the mine-workers in the silver mines are beginning to count on the extra bonus of 25c., which is to be paid them when the average for any one month is above 80c. per oz. Also, ground which hitherto was considered to be too low grade, or containing only inconsistent amounts, is now likely to come in for serious attention. Already owners of such ground are growing restless, and with the arrival of fall, provided the high quotation for the white metal continues, the number of active properties in the camp will probably increase.

MANHATTAN, NEVADA

White Caps Development.—Union Amalgamated.

The reduction plant at the White Caps mine has reached the final stage of completion and the wheels are practically ready to turn and commence the production of gold bullion.

The weekly report of development: On the 300-ft. level the east cross-cut has been advanced 31 ft., cutting several joints which gave good assay-values. The west cross-cut was advanced 40 ft. in quartzite. On the 400-ft. level the raise on the west orebody was put up a distance of 25 ft. and shows a full face of $30 ore.

A large station is being cut in the east orebody where an electric hoist will be installed to facilitate sinking the winze to the 350-ft. level.

On the 550-ft. level a station is being cut on the north side to accommodate a large station pump and sinking-hoist, also a working-station on the south side. The dimensions of the working-station will be 8 to 15 ft. high, 11 ft. wide, and 50 ft. long. The ground is solid quartzite and will not require timbering.

Work at the Union Amalgamated has been confined principally to cutting a large station on the 600-ft. level to accommodate the new tri-plex sinking pump. The new pump will have a capacity of 125 gallons per minute under a 1400-ft. vertical head and will handle the maximum flow anticipated for several hundred feet additional depth.

A new ore-shoot panning from $75 to $200 has been discovered close to the surface on the top of Mustang hill, at the Mustang mine. The oxidized shoot on the 225-ft. level of the lease workings shows more quartz carrying coarse gold. The sulphide shoot still maintains its high value. An excess of water caused by the late heavy rains has prevented the exploration of the lower workings of the company shaft.

HOUGHTON, MICHIGAN

Increase of Capital Stocks of New Cornelia.—Reports from Mines.

Shareholders of the New Cornelia met yesterday at the offices at Calumet and voted to increase the capital stock from 1,203,800 shares to 1,800,000 shares, the vote being practically unanimous, 1,120,400 shares being present and voting for the proposition. The increase in stock will provide for the purchase of seven patented claims and 52 unpatented claims, comprising 1150 acres, covering the extension of the New Cornelia orebodies and providing additional ore reserves for the New Cornelia, estimated to be 21,000,000 tons of 1.55% ore in extent. Calumet & Arizona owns 76% of the stock of the New Cornelia and this stock all was in favor of the increase in capital. An issue of 200,000 shares will be made and $500,000 in cash will accompany the stock to pay for the new claims. The half million in cash will be made in two payments, $250,000 in six months and the balance in one year, all bearing interest at 4%. The cash with the stock valuation and other costs makes the additional property come to approximately four millions.

Calumet & Hecla. Rock output for July was 270,000 tons and will run close to 300,000 for August. This company has less trouble in keeping workingmen than any other in the district.

Quincy. Rock output for this month will run over 100,000 tons and copper output for third quarter of this year will break previous records, as a larger amount of copper is coming from rich mass opened at 6800 ft., a production 'fatterer' that comes in very handy with the present metal prices.

Keweenaw. Property keeps one stamp-head busy and is slowly adding to its working-force.

Ahmeek. August output of this mine may reach 110,000 tons of rock.

South Lake. Production ought to reach 10,000 tons in August, as 3500 tons already has gone to the mill this month.

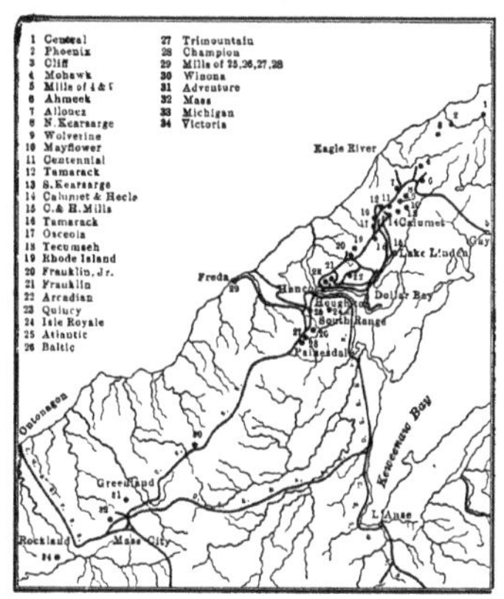

MICHIGAN COPPER REGION

That is better than the total for July and the rock is better in copper content.

Victoria. Four levels are furnishing a good grade of rock for the mill, where improvements are being made that will result in further economies of operation.

Adventure. Parsons Todd and C. D. Hancette of the Adventure management have returned from a visit to the property. Adventure will this month ship 90 cars of copper rock of a good average grade. The shaft now is unwatered to the eighth level and good milling rock has been opened from this shaft on the series of formations, including the Evergreen, the Butler, and the North Butler. The management has decided that it will not be necessary to open the second shaft, as good ground can be economically mined from the present working-shaft.

ELY, NEVADA

Nevada Consolidated.—Miscellaneous Mining.

The Nevada Consolidated is milling about 14,000 tons per day, better than 3000 tons per day coming from the Ruth underground workings, the balance from steam-shovel operatons. Only one-half of the new crushing-plant is in operation. It was found necessary to strengthen and re-inforce some of the structural parts. Several drills are on prospect work and are developing new orebodies, several million tons having been developed during the year.

The Consolidated Copper Mines (old Giroux) has not yet put the second unit of the mill in operation. The underground work, from the New Giroux 1445-ft. shaft on the tenth level is exposing good ore; one body shows 60 ft. in width of 11% copper. Some of the finest specimens of 'ruby' copper, cuprite crystals, and native copper are being found.

The old Ward mine is continuing development and shipments of 30 to 40 tons per day; the company has let a new ore contract, and some expected new trucks will materially increase the output. Osterlund and associates are shipping a car of lead-silver ore from Hamilton; they recently have found 5 ft. of sand carbonate, assaying about $100 per ton. Witcher and Dan McDonald have recently shipped a carload of 75% lead ore from a new discovery known as the Miller property, a few miles north of Sacramento pass, in Snake range, the ore going out to the Deep Creek railroad.

Lacey and Clary have just received returns of a car from their Red Hills mine, in the north-east corner of White Pine county. The ore returned 57% lead and 26 oz. silver per ton, shipment from Currie, on the N. N. R. R.

Several lessees in the Robinson district ship an occasional car of lead-silver ore. H. S. Williams is doing a little work in the tunnel on the Smoky Development Co.'s ground just above Lane City. There is but little movement in tungsten in this county to date. A few men are working on the Minerva and U. S. Tungsten Co.'s properties. The prices do not warrant much development. There are two tonnage propositions which, if sufficiently large milling capacity were obtainable on the ground, would pay at present or lower prices.

There is much unrest with the labor element. A large meeting of the trainmen and associated unions was held last week at Ely.

A new compressor and other machinery are being installed at the Red Top. As soon as the compressor is ready to run, a station will be cut on the 100-ft. level on which the Deming triplex pump will be installed and sinking resumed.

OATMAN, ARIZONA

BULLION PRODUCTION FOR JULY.—MONTHLY PRODUCTION RECORDS OF UNITED EASTERN AND TOM REED MINES.

Gold bullion production during the month of July in the Oatman mining district was the largest ever turned out in the history of the camp. Over $225,000 in gold was shipped to the U. S. Mint in San Francisco from the United Eastern and the Tom Reed, the two producing mines.

The United Eastern treated 7324 tons during the month, of a gross value of $175,898, or $24.01 per ton. This is the highest average value ever attained by the United Eastern. Average extraction for the month was 96.47%, total operating costs $6.68 per ton and net profits $120,802.

For the first seven months of operation the United Eastern has mined and milled 44,889 tons, having a gross value of $969,395 or $21.59 per ton. Total operating costs, including loss in tailing, were $357,455 leaving total net profits of $611,-940. July and August dividends of 5c. per share each, amounting to about $68,000, have been paid. Monthly dividends at this rate will be continued for the rest of the year.

The Tom Reed Gold Mines Co. for July treated 7630 tons of a gross value of $58,000, or about $8 per ton. Total bullion produced was $54,500. Total mining and milling costs were $4.75 per ton, to which should be added about $1 for administration, depreciation, and development. Average extraction for the month was about 97%.

It is expected that both the United Eastern and Tom Reed will be able to maintain this rate of production for the next few months. Within sixty days the United Eastern will have added 50% to the capacity of the mill and expects to treat 10,000 tons monthly.

Active development work on the Big Jim Consolidated and the Red Lion was started within the past few days. On the former, A. G. Keating, the manager, was unable to secure a suitable compressor engine within reasonable time, so he bought a six-cylinder 60-hp. automobile engine recovered from a wreck, installed it in the engine room, connected it with the Imperial type Ingersoll-Rand compressor, and is securing excellent results. Not only is air furnished for the drills, but for a 300-gal. per min. pump as well. At the Red Lion a complete plant is being installed, including an electric generator, which will furnish light for the underground work as it progresses.

PORCUPINE, ONTARIO

HOLLINGER.—MCINTYRE.—DOME.—PORCUPINE CROWN DISCONTINUES DIVIDENDS.—MILL CONSTRUCTION AT SCHUMACHER.

In the gold camp, Hollinger and McIntype continue to occupy their usual strong position. The individual system with regard to a straight wage of $4 per day to all underground employees at this mine appears to be working very satisfactorily and development work is going forward at a rapid rate. As a consequence, Hollinger ore reserves are steadily improving and the physical condition of the mine each day is growing stronger.

McIntyre is also being developed at the maximum. Production also, in sharp contrast to any other gold mine in the country, is at high pitch. It is a remarkable fact that during July the McIntyre treated 15,365 tons of ore, which is a new high record for any one month in the history of the mine, and compares with 14,455 tons in June. During the first seven months of the current year this company has treated 101,449 tons from which $1,011,393 in gold bullion has been recovered.

Dome is treating a large tonnage, but mill-heads are comparatively low, due to the fact that sufficient men are not available to mine and hoist ore from the lower or higher-grade parts of the mine.

Porcupine Crown has decided to discontinue the payment of dividends until a return to more satisfactory labor and supplies. The management, however, intends to conduct an aggressive development campaign, and will operate the mill only at sufficient capacity to pay expenses of operation. The directors have just issued a statement pointing out that during the first half of the current year a total net profit of $120,305.98 was realized, thus leaving a surplus of $305.98 above dividend requirements. To do this, however, it was necessary to allow a falling off in development work, which condition, it was decided, must be avoided, and toward which end further dividend disbursements will only be resumed when sufficient manpower is available to keep up both branches of the work. The total surplus now on hand at the Porcupine Crown amounts to $277,390.91. Mill construction at the Schumacher is being rushed and at this mine operations will shortly be resumed.

SPOKANE, WASHINGTON

PHILIPSBURG BOOM DUE TO DISCOVERY OF HIGH-GRADE MANGANESE ORE.—NORTH-WEST SHIPMENTS TO TRAIL.—MAGNESITE IN STEVENS COUNTY.—REPUBLIC.

A tale of a mining town that has 'come back.' as a result of the War is told by J. E. VanGundy, Montana representative of the United States Steel Corporation, now in Spokane. The town is Philipsburg, which 20 years ago was known as a heavy silver producer. The silver deposits dwindled and the town has virtually been off the mining map for many years. But manganese ore, exposed in large quantities during the silver workings, has brought the camp back to its own. Mr. VanGundy describes conditions: "Philipsburg is experiencing an era of activity surpassing even that of the silver-boom period.

Men without means six months ago have become affluent. Five men are sharing the profits from daily shipments of four carloads worth $2000 net per car. Another man has $75,000 worth of manganese on the dump of an old mine, and will incur no expense except for loading it. Among the sources of production is a shoot opened by a glory hole having an area of 100 by 100 ft., and another 50 ft. sq. at a depth of 400 ft. Operations are proceeding on 10 properties. The deposits of pyroinsite have been pronounced the largest and best in the United States. Government engineers have expressed surprise at the extent and quality of these ores. The ore is classified as a war necessity, and the Government sees that sufficient railroad cars are furnished for its removal. The output of the camp is about 600 tons per day, but this will be increased, it is believed, to 1000 tons per day within the next 30 days.

Eighty per cent of the value of the ore is paid to the producer as soon as his ore is loaded on cars at Philipsburg."

The Florence Silver Mining Co., operating at Ainsworth, B. C., earned $20,000 net in July. Shipments for the month aggregated 320 tons valued at $30,000 with operating expenses of $10,000.

Increased production of all mines in the North-West, due

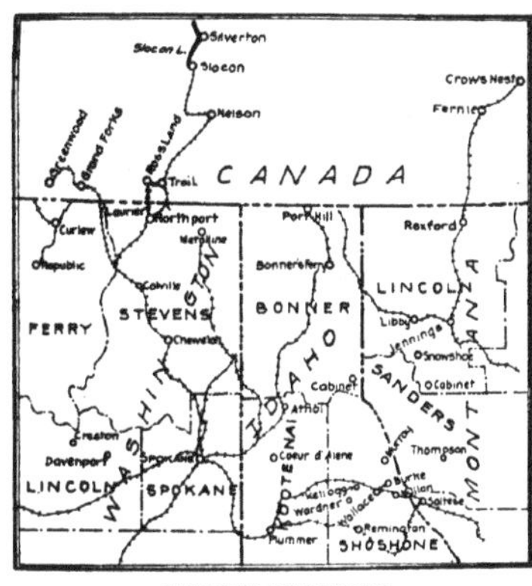

NORTH-WEST MINING REGION

largely to war conditions, with the consequent high price of metals, is indicated by one week's record at the Trail, B. C., smelter. During the seven days ended August 14 the smelter received 7541 tons of ore. Of this total the mines of Washington contributed 1017 tons, the remainder coming from British Columbian properties. The shipments were divided as follows:

Sullivan, Kimberley ... 2791
Iron Mask, Kamloops ... 38
Tip Top, Kashabowa ... 57
Paradise, Athalmer ... 34
Lead & Zinc Co., Metaline Falls ... 72
Le Roi, Rossland ... 1189
Centre Star, Rossland ... 759
Electric Point, Boundary, Wash ... 567
Emerald, Salmo ... 31
Monarch, Field ... 23
Mandy, Le Pas ... 237
Blue Bell, Riondel ... 180
Lanark, Illecillewaet ... 34
Venus, Carcross ... 42
Knob Hill, Republic, Wash ... 145
Emma, Coltern ... 736
St. Eugene, Moyie ... 36
St. Eugene Leasing Co., Moyie ... 35
Standard, Silverton ... 88
Couverapee, Field ... 40
Silver Hoard, Ainsworth ... 16
Slocan Star, Slocan Star ... 77
Sovereign, Sandon ... 34
Quilp, Republic, Wash ... 68
San Poil, Republic, Wash ... 165
Elkhorn, Greenwood ... 14
Retallack, Retallack ... 33

One deposit of magnesite in Stevens county, sufficient to last 37 years, is claimed by Thomas W. Cole, of the American Minerals Production Co., also president of the Spokane Valley & Northern Railway Co., which has let a contract for the building of a railway to the property. The company has a total of 2000 acres. Speaking of the magnesite industry, Cole says: "This is not merely a war affair, for we believe we have in Stevens county the purest magnesite in the world, better than the California product, which is being used now, and better than the Austrian product shipped to America before the War.

"In 1914 the total magnesite used in the United States was about 700,000 tons, of which about half was shipped crude and the other half calcined. With the impetus to the steel trade there probably is 1,000,000 tons per year now used in this country."

Since the resumption in July of shipments from the Knob Hill mine in Republic, Wash., 14 carloads of ore have been sent out, with a net value of $6000. Sinking on the shaft from which a cross-cut will be made at the 200-ft. level to tap the main vein is proceeding at the rate of five feet per day. The vein to be sought from the 200-ft. level of the shaft contains a series of orebodies that have been followed for several hundred feet on the lower adit-level. These shoots have yielded upward of $400,000 gross. Out of this gross production $145,000 has been paid for the property and nearly $100,000 in dividends.

Directors of the Richmond Mining Co. have declared a 2c. dividend, payable September 5. This amounts to $16,800 and is the fourth regular monthly dividend this company has paid. It will make its total disbursements $67,200. The property is in the East Coeur d'Alene. The regular quarterly dividend of $32,000 has been declared by the directors of the Utica Mines, Limited, operating in the Slocan district, B. C. This is at the rate of 2c. per share. Payment will be made September 15. A dividend of $32,000 was disbursed on June 15 last.

PHILIPPINE ISLANDS

BENGUET CON. M. CO. REPORTS OPERATION AND PRODUCTION FOR JUNE AND FIRST HALF OF 1917.

Tonnage to mill, reduced and treated, for June, 1735; for half year, 9381; estimated content, gold, for June, $32,992; for half year, $174,566.80; recorded value recovered, for June, $28,240.13; for half year, $149,876.80; average value ore entering, for June, $19 per ton; for half year, $18.60 per ton; percentage of recovery, for June, 86; for half year 85.9; estimated value bullion, for June, $28,760.66; for half year, $148,224.58. Slag and matte on hand to the value of at least $7500, making actual recovery considerably in excess of recorded.

Two dividends of 5% each have been paid in 1917, and it is the expectation that two more at least will be paid before the close of the year. Physical condition of mine is very good. Stock of principal supplies on hand is sufficient for about a year.

THE MINING SUMMARY

The news of the week as told by our special correspondents and compiled from the local press.

ALASKA

(Special Correspondence.)—The concentrates from the Alaska Jualin Mines Co. and the Chichagoff Mining Co. will soon be treated at the cyanide plant of the Alaska Treadwell Co. The cyanide plant at present is treating the concentrates from the Ready Bullion and the 'clean-up' from the various mills. An appreciable saving to these companies both in freight and treatment costs will result. A large pocket of high-grade ore was recently encountered at the Chichagoff mine. It is stated that for a day and a half the mill was kept busy milling ore which ran from $900 to $1100 per ton and the mill-men were forced to clean the plates every hour during that time. It is also reported that the general average of the ore being developed in the Chichagoff is higher than formerly.

W. S. Peckovich, owner of a group of mining claims at Funter bay, has secured a large amount of mining machinery and will commence the development of his property at once. He has put a crew of 15 men to work and will install the machinery as speedily as possible. H. McGuire of New York has been examining properties on Iron creek, a tributary of the Talkeetna river, near the Government railroad running inland from Anchorage. The showing there is said to be low-grade copper in veins from 5 to 200 ft. wide.

The copper mines in the vicinity of White Horse and Caribou are busy shipping ore. At the present time there is 1000 tons of sacked copper ore from the Vallerie mine on the dock at Skagway awaiting shipment to the Granby smelter. The Pueblo mine, which has been closed since the cave-in last fall, is being re-opened. Sixty men are at work in the mine, the old shaft has been pumped out, and an effort is being made to re-open the mine through the old workings.

Treadwell, August 12.

ARIZONA

Gila County

Inspiration Con. Copper Co. resumes operations at Miami on a limited scale.

Greenlee County

(Special Correspondence.)—A fire was discovered in the Coronado mine of the Arizona Copper Co. All efforts to extinguish it have been unsuccessful. No damage is reported.

Metcalf, August 25.

Maricopa County

(Special Correspondence.)—The mill of the Arizona Molybdenum Co. is operating. It has a capacity of 50 tons and is situated nine miles north-west of Wickenburg. It is producing 1500 lb. of concentrate per day working one 8-hour shift. The capacity is expected to be doubled, and the production of concentrate will be 6000 lb. per day, assaying from 18 to 20% molybdenum trioxide.

Phoenix, August 9.

The Mazatal quicksilver district in the mountains of the same name, about 70 miles north-east of Phoenix, Arizona, is coming into renewed prominence. The Sunflower cinnabar mine has recently shipped 10 flasks of 75 lb. each. The property is being operated by C. N. Sears, who organized the Sunflower Cinnabar Mining Co. in 1913. Sears is installing a 12-nest retort of the Johnson McKay type. The quicksilver resources of the district were reported upon by F. L. Ransome in Bull. 620-F of the U. S. Geological Survey.

Pima County

(Special Correspondence.)—The sinking of the two-compartment 200-ft. shaft of the Hecla-Arizona company in the Twin Buttes district is progressing at the rate of 5 ft. per day and is now down 75 ft. The shaft will be completed to 200 ft. before any driving or cross-cutting is done. Two other shafts will be started within a short time.

Thirty men are employed at the Hecla-Arizona. K. M. Said, the superintendent, announces that plans for a million dollar leaching plant, which it is proposed to erect on the property in the near future, are in Tucson. The proposed plant will serve mines in the Twin Buttes district, the San Xavier, the San Xavier Extension, the Empire Zinc, the Vulcan, the Mineral Hill, the Paymaster, and several others, all of which are now shipping their ore by motor-truck to Tucson, a distance of 22 miles.

The Hecla-Arizona has expended $40,000 in equipment and machinery and there is reason to believe additional money will be forthcoming to carry out the proposed work. The Hecla-Arizona adjoins the San Xavier Extension and is on the same vein in which a strike of good ore was made recently by the San Xavier Extension.

Tucson, August 6.

(Special Correspondence.)—William R. Ramsdell has resigned as president of the Arizona Cornelia Copper Co. The property adjoins the New Cornelia and consists of a number of claims still in the prospect stage.

Tucson, August 14.

(Special Correspondence.)—A big gold strike is reported by the Arizona Consolidated Copper & Mining Co., in the Greaterville district. Maurice J. Fink is in charge. The Arizona Consolidated Copper & Mining Co. is affiliated with the General Mining Co. of Nevada, of which H. G. Humphreys is president. The strike was made on the J. B. Anderson property. C. H. James will examine and report on the property.

The Total Wreck mine, near Pantano, is being operated by E. L. Vail & Co. The Arizona Lead & Silver Co. is operating under the management of Selzig and Shanks. The Sesame Copper Co., operated by Charles Taylor, is cross-cutting on the 100-ft. level.

Tucson, August 16.

Pinal County

(Special Correspondence.)—Two new mills are under construction at the Mammoth mine, for treating tailing that had been formerly worked. The mine is also being operated.

Tucson, August 16.

Yavapai County

(Special Correspondence.)—The 10,000-ft. adit planned by the United Verde Extension, to connect with the 1400-ft. level of the mine, has been changed. The adit as now planned will be 12,500 ft. long and has been contracted for by the Porter Construction Co. of New York and St. Louis. Completion within a year at a cost of over $1,000,000 is proposed. Tunneling has been commenced as well as the sinking of a 500-ft. shaft on the adit line which will give additional working faces. The adit will connect with a five-mile railroad from the United Verde Extension smelter. The construction of the railroad will be commenced at once. At present the United Verde Extension is hoisting and shipping its ore to the smelters at Douglas, Globe, and Humboldt. As soon as the 4-mile railroad from Clarkdale to the smelter site is completed, the erection of the steel work of the 800-ton smelter will commence. Since the labor troubles the shipment from the United Verde Extension mine has averaged 230 tons per day of 25% copper ore.

The flow of water in the Jerome Daisy shaft has proved to be greater than anticipated and cannot be handled by the compressed-air equipment now installed. Arrangements are being made for the installation of electric power to handle the increase. A cross-cut has been started at the 250-ft. level to tap the orebody at a distance of 125 ft. Another block of 50,000 shares has been placed on the market at 60 cents.

The Copper Chief of the Hayden Development Co. is operating with 20 men instead of 40, since the labor trouble in May. Preparations are being made to overhaul the mill and rearrange the milling so as to increase the recovery. Before the strike the mill was treating 130 tons of gold ore per day. The presence of copper lowered the gold recovery.

The Consolidated Arizona Smelting Co. has declared its first 2% dividend. The accumulation of a larger working-capital has delayed the declaration of this dividend. The company has produced approximately 9,730,000 pounds of copper in the first half of 1917. The operating profit was at the rate of 5c. per pound, the gross profit figuring at 65c. per share per annum on the 1,663,000 shares.

Prescott, August 6.

(Special Correspondence).—It is reported that an examination has been completed by an engineer of the Standard Oil, and that he has recommended development in the China Valley district. Several companies are now in the field. It is expected that the Arizona Queen company will start drilling here by September 15. The Arizona Oil Syndicate controls the largest area in the China Valley field, recently having acquired 4000 acres completely surrounding the holdings of the China Valley Oil Co. The China Valley Oil Co. has sunk its first well to a depth of 2000 ft. and is preparing to start a second.

Prescott, August 9.

(Special Correspondence).—The Verde Combination shaft has now reached copper-bearing schist at 545 ft. There is a band of schist 27 ft. wide at 500 ft., which assays 2% copper. The showing in the bottom of the shaft is encouraging.——The cutting of the 600-ft. station at the Calumet & Jerome is finished and cross-cutting toward the orebody has commenced.——The Jerome Copper Co., which holds the Hooker and Ewing group of claims, a mile south of Jerome, is now financed. Negotiations are under way with the Arizona Power Co. for power. A contract has been let for the building of a 6000-ft. road, which when complete will enable the company to bring in its equipment of compressor, drills, steel, and the like. This machinery was bought from a company south of Prescott and is practically new. The development as now outlined will involve driving an 800-ft. adit to intersect the ore at a depth of 450 feet.

Prescott, August 16.

YUMA COUNTY

(Special Correspondence.)—The Black Reef Copper Co. has been financed by Toronto and Niagara Falls capitalists. It controls a property in the Cunningham Pass district of the Harcuvar mountains. The mine is an old partly developed prospect; the main shaft is 325 ft. deep. The shifts have been increased and development is now being pushed.

Yuma, August 14.

CALIFORNIA

Conditions causing the small number of new wells started in the oil-fields in the past two weeks as shown by State Mining Bureau reports, have evidently been altered, as the report this week shows 18 new wells, which is approximately the average shown during this year, and brings the total for the year up to 696. The Midway Sunset fields, as usual, show the major portion of this new development, the number of wells there being 13. The Santa Maria fields show an unusually high number, as 4 wells are being started, 3 of them being by the Union Oil Co. In the fields near Los Angeles only one new well is reported, belonging to the Standard Oil Co., and located in the new Montebello field, which has been developed this year.

There were 21 wells reported as ready for test of water shut-off, 17 deepening or re-drilling, and 4 abandoned, which is about the usual number.

In order to facilitate the securing of labor for fire-fighting on the National Forests, labor agents have been appointed by the Forest Service in Chico, Oroville, Marysville, Sacramento, Fresno, and Bakersfield. These agents are authorized to hire, transport, and provide food for fire-fighters at the request of Forest officers.

The total receipts from all sources on the National Forests in California for the fiscal year 1917 was $436,032.05. In addition to $109,000 turned over to the State, 10%, or more than $43,000, was expended in road and trail construction.

AMADOR COUNTY

(Special Correspondence.)—Operations are in satisfactory progress at the property of the South Keystone Con. Mining Co., between Sutter Creek and Amador City. Development work is under way at the 1000-ft. North Star shaft. A short cross-cut to the east at the 600-ft. level encountered a good-looking vein, 10 ft. wide. Raising and two cross-cuts have proved a width of about 200 ft. A chute has been cut at the 800-ft. level, preparatory to cross-cutting for this vein at that level, and, if conditions warrant, a cross-cut will later be run from the 1000 or lowest level, with the same object. A cross-cut has also been started to the west to cut the veins in the McIntire claim.

The north drift, which was started 300 ft. west of the shaft from the cross-cut, now has a length of nearly 800 ft. and is between 300 and 400 ft. north of the South Keystone's south end-line. Marked improvement is also in evidence in this drift, which in another month's time should reach a point opposite the South Spring Hill mill-site, where there is good reason to expect a rich vein of ore. It is estimated that three months' time will be required to reach the high-grade rock existing at the north end of the South Keystone. At a point in the north drift about 350 ft. north of the South Keystone's south end-line, a cross-cut was started east to encounter the vein that passes through the road about 200 ft. south of the old mill-site, and indications are favorable for the finding of a valuable orebody there.

The property is being opened by Tacoma capitalists, who are sparing no pains to make a success of the enterprise. John A. McIntire, of Sacramento, was formerly the owner of a large portion of this ground and retains an interest in the company.

Sutter Creek, August 19.

SHASTA COUNTY

(Special Correspondence.)—The Mammoth Copper Co. of Kennett has taken a long term bond on the Delta Consolidated Gold Mining Co.'s patented claims on Dog creek five miles west of Delta. The bond is for $150,000, and a substantial payment is to be made in six months. A narrow-gauge railroad connects the mines with the Southern Pacific at Delta. The Mammoth owns the McCourt group of copper mines adjoining the Delta group of gold mines and has been working thirty men on them for a year and shipping the ore to the smelter at Kennett. The smelter needs the gold quartz for flux and can treat the ore for gold more cheaply than is possible by stamp-mill or other process. The Delta mines have produced high assays in gold, but the company has had trouble to secure advantageous treatment. Arkansas capital is largely interested.

At the Old Texas mine in Old Diggings the working-force has been increased from 6 to 35 men.

The Shasta Hills Mining Co. has erected a 5-stamp mill and concentrator on its mines five miles west of French Gulch. The principal mines are the Accident and the Sybil. The management is well pleased with the results of development. J. R. Davis is superintendent and part owner.

At the Little Nellie mine, between Keswick and Iron Mountain, the Pittsburg Mining Co. is working two diamond-drills to test the ledge at a considerable depth below the working tunnel. One hole has been driven 800 ft. to a depth of 1800 ft. below the apex.

William T. Morgan, of Denver, operating the Minnesota mine between Keswick and Iron Mountain, has closed a deal for a bond on the Sharp copper claims near Buckeye. Morgan will start development of the four claims about the first of September.

The Mountain Copper Co. is shipping an average of ten 50-ton carloads of ore per day from the Iron Mountain mine to its smelter and the Standard Oil Co. at Martinez. The output would be greater but for the shortage of miners.

Redding, August 19.

Siskiyou County

(Special Correspondence.)—Ten years ago the Big Bullfrog Gold Bar mine in the Yreka district was closed down after a disappointing run of the new mill and has since been idle; $150,000 had been spent in underground development, but no recovery process at the time would treat the ore. Changed conditions now make it possible to operate. It is fully equipped and in good condition. The re-organization committee, of which F. H. Oliphant, of Los Angeles, is secretary, is giving the former stockholders an opportunity to recover their interest in the form of new stock in a re-organized company on payment of two cents per share exchange bonus, payments to be made through the Continental National Bank of Los Angeles.

D. D. Good and D. M. Watt of Ashland, Oregon, report a new strike at their High Grade mine located on the south slope of the Siskiyou in the Stirling district. They were opening up the second ore-shoot in the new 240-ft. drift and working in 2 ft. of rich ore.

Chrome mining is quite active in the Etna district, and ore is being shipped to Eastern points.

Hornbrook, August 20.

Trinity County

The Federal Chrome Co.'s mines near Peanut are now producing chrome that keeps three auto-trucks busy hauling to Redding. The distance is 73 miles. A round-trip is made in three days. S. B. Hall is superintendent. It costs $20 per ton to haul the chrome to Redding, but it brings $60 per ton in Pittsburgh, to which point all of it is shipped.

IDAHO

The Idaho-Hecla Mining Co. on August 20 declared dividend No. 171, of 15c. per share, amounting to $150,000. The total dividends for 1917 amount to $1,200,000, making a grand total to date of $6,505,000.

MONTANA

Lewis and Clark County

(Special Correspondence.)—The mining outlook is favorable. At Marysville the mill at the Shannon and Piegan-Gloster is in constant operation. Ore in the Shannon mine at the 500-ft. level assays above $75 per ton for a distance of 150 ft. in length. About $40,000 is due on the payments for the Shannon and is expected to be paid within two months. The properties are all owned by the Barnes-King Development Co. C. W. Goodale, president of the company, whose headquarters are in Butte, visited the mines at Marysville last week.

The St. Louis Mining Co. is milling about 80 tons of gold ore every 24 hours. The results are satisfactory to those in charge. The company owns all of the ground held by it originally, and in addition has acquired through the courts all of the holdings of the Drumlummon (Ltd.). Ores treated in the mill by cyanide are taken from old stopes in what was formerly Drumlummon ground.

Immediately surrounding Helena most of the mining work consists of development, although some ore is being shipped. The Scratch Gravel Gold Co. sends to the smelter two cars per week. The ore nets over $3000 per car. The Thomas Cruse Development Co. is having an exhaustive examination of its property with a view to determining the best plan to pursue in future work.

The Helena Mining Bureau has installed a new boiler and is taking ore for shipment from the Helena mine. The ore comes from the 300-ft. level and averages about $35 per ton in silver and lead.

Work on the Rock Rose and Looby ground, which are controlled by the Cruse Con. Co., consists mostly of development. Ore is coming from both the Rock Rose and Looby shaft.

Helena, August 20.

Silverbow County

The strike of the smelter-men at the Washoe smelter at Anaconda has caused the Anaconda Copper Co. to close down its mining and reduction properties. The smelters at Anaconda and Great Falls, together with the mines at Butte, employ 15,000 men.

NEVADA

Nye County

(Special Correspondence.)—The Tonopah Belmont last week produced 2616 tons of ore. The intermediate drift, above the 700-ft. level on the South vein, intersected a fault, but the vein was picked up again, and it shows good assays. The east of the 800-ft. level on the Rhyolite vein has a 3-ft. face of good ore. The west drift of the 1166-ft. level on the Mizpah Fault vein has been discontinued.

The Tonopah Extension the past week sent 2380 tons of ore to the mill. Development in the No. 2 shaft was 57 ft., and 70 ft. was made in the Victor. The east drift of the Victor on the 1440-ft. level has intersected a large body of fair ore.

The Tonopah Mining Co. has discontinued development in the Mizpah Extension, which it had operated for the past three years. The 35 ft. of development at the Sandgrass was confined to the winze, which is going down on the foot-wall of the vein. Development at the Silver Top consisted of 113 ft. The ore production the past week was 2050 tons.

The net earnings of the Jim Butler Mining Co. for July were $53,564, from 3112 tons of ore. The ore production the past week was 700 tons.

The development of the West End is centred on raise 522 of the Ohio, which is being run for air connections. The West End produced 959 tons of ore the past week, while the Halifax sent 51 tons to the mill.

The production of the Rescue-Eula the past week was 128 tons, the North Star 55 tons, and the Cash Boy 105 tons. Lessees at the Montana produced 55 tons, making the week's output at Tonopah 8999 tons, with gross value of $162,000.

Tonopah, August 20.

(Special Correspondence.)—The Louisiana Consolidated Mining Co. is shipping steadily from the Tybo mine, 68 miles north-east of Tonopah, ore that averages $75 per ton, assays showing 20% lead, 50 oz. silver, and $4 gold. From a single opening the company is outputting 20 tons per day, the ore going by auto-truck to Tonopah and thence to the smelters. Five trucks and trailers are already in use and five more are to be added. The ore is coming from the foot-wall vein, which has heretofore been little explored, the previous production of $5,000,000 having come almost entirely from the hanging-wall vein. The foot-wall vein was found by cross-cutting the andesite at the 400-ft. level, 4 ft. of ore at this point averaging $117 per ton. In a raise from the 300 to the 400-ft. level the orebody had an average width of 4 ft. the entire distance. On the 300-ft. level the orebody has widened to 8 feet.

Arrangements have been completed for extending the electric power-line of the Nevada-California Power Co. from Belmont to Tybo, and a mill will be built to concentrate the low-grade ore and make a separation of the lead and zinc. The mill will be erected on the site of the old stamp-mill. The Louisiana company has acquired the property of the Tybo Lead company, which was the west extension of the Tybo. A new shaft will be sunk. Sinking of the main Tybo shaft from the 400 to the 800-ft. level will also begin during the month. The Louisiana company has purchased the Pearl and Zenda mines in Kern county, California, and will start operations there this month. The property is a large low-grade silver-gold deposit, which is worked on the open-cut system. The company estimated that there is 300,000 tons of $5.50 ore in sight and that it can be mined and milled at a total cost of $3 per ton. The present mill will be enlarged to treat 100 tons per day. Later it will be enlarged to treat 250 tons per day.

WHITE PINE COUNTY

An eight-hour day is to be put into effect in the open pits of the Nevada Con. M. Co. A settlement of an impending labor difficulty is avoided by the change.

The Ely School of Mines, a local school for miners, will open September 4.

NEW MEXICO

GRANT COUNTY

(Special Correspondence.)—The Pinos Altos mining district came into existence through the discovery of rich placer gold in the gulches and it was but a short time until a camp sprung up. Placer mining continued profitable for 20 years. Later miners and prospectors began looking for the source of the gold. Lode mining claims were filed on and developed, but the ore proved to be complex-lead, copper, and zinc, with some silver and gold. After several years of operation the old companies suspended work and the camp became quiet. As time went on prices and demand for copper, lead, and zinc began to increase and men of small capital endeavored to mine the Pinos Altos ores; but experience of the past few years has proved that while the district abounds with pay-ore it needs capital which, judiciously expended, will be amply repaid in exploiting these ores. This complex ore is susceptible to separation processes, either by the Huff electro-static, the magnetic separators, or the flotation process, the last being preferable, having given evidence of a greater recovery, as proved by the successful operation of the Cleveland mine and mill of the Empire Zinc Co. The ore is heavy with zinc—a complex ore, treated by the magnetic separators and the flotation process.

Other companies are operating on a small scale, but there is not the activity in the Pinos Altos district that the district justifies. Companies with plenty of capital for initial work and exploiting at depth is what is needed. The small companies now operating are doing a great work in furthering development, but it is a case of opening a vein of ore as soon as possible, suspending development while they get out a few carloads to help pay expenses.

The companies operating in the district besides the Empire Zinc Co. and that are doing good work on small capital are the El Paso Mining & Milling Co., Doyle and Clark, of the Manhattan group, the Calumet & New Mexico Mining Co., and the United States Copper Co. The last began operations on what is known as the Hardscrabble group in January of this year, but suspended operations the middle of May. It is expected they will resume development in August or September. The Calumet & New Mexico Mining Co. has blocked out sufficient ore to warrant a concentrating-plant this coming winter. They ship the high-grade lead and zinc ore to Coffeyville, Kansas. Doyle and Clark, of the Manhattan, are mining and shipping from their Indian Hill shaft and have a crew sinking a shaft at the end of the Manhattan adit in 900 ft., the shaft to be sunk 500 ft. A good tonnage of low-grade ore is developed. The El Paso Mining Co. is mining in the Pacific No. 2 shaft on the Skilacorn vein and occasionally encountering rich gold ore but developing the low-grade, which will be treated in their own concentrator. The Huff electro-static separators and the flotation process are used in this mill.

The Hardscrabble group, covering a dike 1000 ft. wide and over 6000 ft. long, of grano-diorite, and the ore is not complex, no zinc being found. The low-grade ore is ideal for concentrating and the high-grade acceptable to the smelters for its self-fluxing qualities. All ores are hauled to Silver City for shipment by rail, from 8 to 10 miles.

Pinos Altos, August 6.

(Special Correspondence.)—The sinking of the 4 by 7 ft. winze at the end of the 900-ft. adit on the Manhattan property has reached a depth of 350 ft. A station will be cut and driving on the ore will begin this week. Considerable stoping-ground will be opened. The adit cut a good vein of heavy iron sulphide carrying copper, gold, and silver, when in between 450 and 500 ft. The shaft is sunk 8 ft. to the west of the vein which first had a slight dip to the east, later straightened, and now is dipping toward the shaft. The ore is of shipping grade, running between $40 and $50; 150 ft. of stoping-ground has been opened on the 4th level, where mining and shipping will begin this week. The company expects to ship 25 tons or more daily to the Copper Queen smelter at Douglas. The nine-mile haul to the railroad at Silver City costs $2.50 per ton, and the transportation rate to Douglas is $1.75 per ton. Thirty men are employed and the prospects are bright for the fall and winter. Jackhammer drills are used in sinking and driving, and one Ingersoll and one Waugh stoper are being tried out. An Ingersoll-Rand compressor, driven by a 24-hp. Muncie oil-engine furnishes the power. A Fairbanks-Morse hoist, capable of sinking to 500 ft., with a ⅝-in. cable and a 1000-lb. bucket, was installed some time ago and is being operated successfully. The new ore being found in this shaft demonstrates that its value is increasing with depth. The vein averages about 3 ft. in width. Independent of this high-grade ore considerable low-grade milling ore is also being opened up. A custom-mill is much needed in this district. The Indian Hill shaft of the Manhattan group is shipping regularly. R. L. Doyle, of Cripple Creek, is in charge of all work at the Manhattan, and M. S. Clark is in charge of the office.

One shift is driving on ore at the Langston property.—— The El Paso Mining Co.'s mill has not yet resumed operations, but is expected to soon. Shipment of high-grade ore is being made.——The Empire Zinc Co.'s ore is increasing in copper content in the lower workings.——J. W. McAlpine, president of the United States Copper Co., arrived at Pinos Altos this week, and will remain until September 1. It is reported that development of the company's Hardscrabble group will begin about August 15.

Pinos Altos, August 6.

(Special Correspondence.)—Six miles south of Lordsburg a rich silver-lead strike has been made on the Brown-Wiggens-Weldon group, which is being worked by Weldon and Cole. From the bottom of two 100-ft. shafts ore is being taken out that averages $10 in gold, $6 in silver, and 65% lead. This property adjoins the Lister property on the east and is believed to be an extension of the latter. Weldon and Cole are shipping about two tons of high-grade lead ore per day to the smelter at Point Richmond, California.

The El Dorado group of copper properties in the vicinity of Lordsburg was sold by Messrs. Killebrun and Fairly, the owners, to James P. Porters of Lordsburg, who is understood to be associated with prominent mining men. All of the ground between the Bonney mine and the Eight-five mine is under the control of Porters. B. L. Cunningham, assistant geologist of the Southern Pacific railroad, and Mr. Porters have recently made a complete examination of the district.

Lordsburg, August 6.

(Special Correspondence.)—Development work at the mine of the Austin-Amazon Copper Co., in the Burro mountains, totals 2500 ft. of shafts, adits, and cross-cuts. A great part of this work is in payable shipping and milling ore. A cross-cut has been driven through the mineral-zone from the first level which shows an orebody 80 ft. wide, 30 ft. of which averages better than 4% copper. This orebody will be cross-cut at the 200-ft. level. The company expects to make new shipping contracts. Negotiations now pending may result in the erection of reduction works and other improvements by the company. The Janney flotation machines are to be given a practical test here. This machine is especially adapted to the complex ores of the Hanover district. The Republic Mining & Milling Co., of which Mr. Janney is president, will build a 100-ton mill on the company's property at Hanover, which will include the Janney machines. The principal value of the ores in this district is in copper and zinc sulphide. The successful operation of the new mill and the new process will result in a revolution in the milling methods in this district and also in the Pinos Altos district, where the same class of ore predominates.

Hanover, August 14.

SOUTH DAKOTA

LAWRENCE COUNTY

(Special Correspondence.)—The Capital Gold Mining & Milling Co., on Deadwood gulch, one and a half miles west of the Homestake mine and adjoining the Trojan mine in Bald Mountain district, is preparing for active operations on the property, a compressor with 20-hp. McViters gas engine, with 650 ft. of air pipe-line has been installed, and work will be started at once to complete the adit on the south side of the Deadwood creek to get under the shaft in which assays indicate ore worth $24.80 per ton gold. A part of this ore will be shipped to the Trojan mill, until the company has completed the transfer of the machinery of a large concentration plant at Central City, which will be converted into an up-to-date cyanide plant with 350 tons capacity. The company has a vein 110 ft. wide, exposed in the railroad-cut of the C. & N. W. railroad, which shows assays of $3 to $12 per ton. This vein can be traced across the property, which consists of 60 acres. It lies in the centre of the heaviest ore-producing area of the silicious belt. The company has been re-organized under the laws of South Dakota with a capital stock of $1,500,000, the most of which is held by Omaha, Nebraska, business men. R. M. Harrop, a graduate of the South Dakota School of Mines, is secretary and general manager for the company.

Deadwood, August 1.

KOREA

The Seoul Mining Co., operating the Suan Concession in Whang Hai province, reports results for the month of July, 1917: Total recovery, $99,835.

The Holkol mill has been running continuously for several years and was in need of a general overhauling. This caused a shut-down for a considerable part of the month and the results were reduced thereby.

Personal

Note: The Editor invites members of the profession to send particulars of their work and appointments. This information is interesting to our readers.

W. DeL. Benedict is here from New York.

Russell C. McGinnis has obtained a commission in the Army.

Walter H. Wiley was here last week and is now at Tucson, Arizona.

Thomas Kiddie has returned to Los Angeles from British Columbia.

Elbert H. Gary and D. C. Jackling have returned to Seattle from Alaska.

R. A. F. Penrose has been in San Francisco, engaged in patriotic work.

James L. Head is in the Engineer Officers' Training Corps at Fort Leavenworth.

H. Vincent Wallace has returned to Los Angeles from Guanajuato, Mexico.

F. S. Norcross is superintendent of the Copper Mountain mines, near Princeton, B. C.

Herbert C. Hoover has been elected an honorary member of the American Institute of Mining Engineers.

Wilton J. Crook, of Stanford University, is in the Officers Training camp at the Presidio, San Francisco.

B. W. Knowles has been appointed superintendent of the Nickel Plate mine, at Hedley, B. C., in succession to W. Sampson.

F. C. Lincoln, director of the Mackay School of Mines, Reno, Nevada, has been appointed State Assayer and Inspector for the State of Nevada.

John F. Berteling, Michigan representative of the Sullivan Machinery Co., has taken a position as superintendent for the Newport Mining Company.

F. C. French is manager for the Consolidated Mining & Milling Co., of Salt Lake City, and is now in charge of the Golden Fleece mine at Lake City, Colorado.

S. R. Brown has resigned as mill superintendent for the Plymouth Consolidated to accept a position with the Porphyry Dyke Gold Mining Co. at Rimini, Montana.

Francis A. Thompson has resigned as dean of the Washington State School of Mines at Pullman in order to accept a similar position at the University of Idaho, at Moscow.

C. B. Dunster of the Mining Department of Breitung & Co., New York, has succeeded H. L. Kaufman as manager of E. N. Breitung & Co., at Cleveland, and as assistant general manager of the Breitung iron properties.

Erle P. Dudley, construction engineer at the Bunker Hill & Sullivan concentrators, at Kellogg, Idaho, has received his commission of the grade of captain, Engineer Officers Reserve Corps, from the Adjutant General, United States Army.

Judge C. C. Goodwin, Nevada pioneer, well-known editor and author died on August 25 at his home at Salt Lake City. Goodwin was editor of the 'Territorial Enterprise' of Virginia City, Nevada, during the period preceding the early 'eighties. Mark Twain, Bret Harte, Dan De Quille (William Wright), Joseph T. Goodman, and others who later gained fame, were associated with Goodwin. He was the author of 'The Comstock Club', 'The Wedge of Gold', and other Western books. As editor of the Salt Lake 'Tribune' he exercised a wide and useful influence.

THE METAL MARKET

METAL PRICES

San Francisco, August 28

Aluminum-dust (100-lb. lots), per lb	$1.00
Aluminum-dust (ton lots), per lb	$0.95
Antimony, cents per pound	15.50—17.00
Electrolytic copper, cents per pound	28.50
Pig lead, cents per pound	12.25—12.50
Platinum, soft and hard metal, respectively, per ounce	$105—111
Quicksilver, per flask of 75 lb	$115
Spelter, cents per pound	10
Tin, cents per pound	60
Zinc-dust, cents per pound	18

ORE PRICES

San Francisco, August 28

Antimony, 45% metal, per unit	$1.20
Chrome, 40% and over, free SiO_2 limit 8%, f.o.b. California per unit	$0.55—0.70
Magnesite, crude, per ton	$8.00—10.00
Tungsten, 60% WO_3, per unit	$25.00—28.00
Molybdenite, per unit for MoS_2 contained	$40.00—45.00

Manganese, Carnegie schedule less freight to Chicago. Less than 40% special quotation.

Manganese prices and specifications, as per the quotations of the Carnegie Steel Co. schedule of prices per ton of 2240 lb. for domestic manganese ore delivered, freight prepaid, at Pittsburgh, Pa., or Chicago, Ill. For ore containing

	Per unit
Above 49% metallic manganese	$1.00
46 to 49% metallic manganese	0.98
43 to 46% metallic manganese	0.95
40 to 43% metallic manganese	0.90

Prices are based on ore containing not more than 8% silica nor more than 0.2% phosphorus, and are subject to deductions as follows: (1) for each 1% in excess of 8% silica, a deduction of 15c. per ton, fractions in proportion; (2) for each 0.02% in excess of 0.2% phosphorus, a deduction of 2c. per unit of manganese per ton, fractions in proportion; (3) ore containing less than 40% manganese, or more than 12% silica, or 0.225% phosphorus, subject to acceptance or refusal at buyer's option; settlements based on analysis of sample dried at 212° F., the percentage of moisture in the sample as taken to be deducted from the weight. Prices are subject to change without notice unless specially agreed upon.

EASTERN METAL MARKET

(By wire from New York)

August 28.—Copper is easier, being nominal at 26.50 to 26.25c. Lead market is quiet, prices a little lower for the week at 10.67 to 10.50c. Zinc is dull and weaker at 8.50 to 8.25c. Platinum remains unchanged at $105 for soft metal and $111 for hard.

SILVER

Below are given the average New York quotations, in cents per ounce, of fine silver.

Date			Average week ending		
Aug.	22	88.25	July	17	80.62
"	23	88.25	"	24	79.12
"	24	88.50	"	31	78.20
"	25	88.50	Aug.	7	80.50
"	26 Sunday		"	14	82.95
"	27	88.75	"	21	87.16
"	28	88.75	"	28	88.50

Monthly Averages

	1915	1916	1917		1915	1916	1917
Jan.	48.85	56.76	75.14	July	47.52	63.06	78.92
Feb.	48.45	56.74	77.54	Aug.	47.11	66.07	....
Mch.	50.61	57.89	74.13	Sept.	48.77	68.51	....
Apr.	50.25	64.37	72.51	Oct.	49.40	67.86	....
May	49.87	74.27	74.61	Nov.	51.88	71.60	....
June	49.03	65.04	76.44	Dec.	55.34	75.70	....

COPPER

Prices of electrolytic in New York, in cents per pound.

Date			Average week ending		
Aug.	22	26.50	July	17	30.29
"	23	26.50	"	24	27.04
"	24	26.50	"	31	27.25
"	25	26.50	Aug.	7	28.46
"	26 Sunday		"	14	27.62
"	27	26.25	"	21	28.00
"	28	26.25	"	28	26.41

Monthly Averages

	19[illegible]	1916	1917		1915	1916	1917
Jan.	[illegible]	24.30	29.53	July	19.09	25.66	29.67
Feb.	14.38	26.62	34.57	Aug.	17.27	27.03	....
Mch.	14.80	26.65	36.00	Sept.	17.69	28.28	....
Apr.	16.84	28.02	33.16	Oct.	17.90	28.50	....
May	18.71	29.02	31.69	Nov.	18.88	31.95	....
June	19.75	27.47	32.57	Dec.	20.67	32.89	....

Combined production of the four Jackling porphyry copper mines for the first half of 1917 was 223,015,685 lb., larger than ever before in a like period. All of the mines aided in establishing this record with the single exception of Nevada Consolidated, whose half year's total of 36,669,677 lb. was exceeded last year when over 43,000,000 lb. was turned out.

Utah Copper was responsible for the greater part of the gain, although in pounds of new copper produced it was not far ahead of the increase scored by Ray Consolidated. Chino showed a moderate advance over last year.

Production figures in detail for the first six months of the past two years compare (pounds):

	1917	1916
Utah	99,289,787	84,949,462
Nevada Consolidated	36,669,677	43,347,854
Ray Consolidated	47,284,523	35,581,346
Chino	39,771,704	34,366,632
Total	223,015,685	198,245,294

Assuming a 28c. average price received for copper, these companies did a gross copper business of $62,444,491 in the first half of the year, of which perhaps $35,000,000 was saved for net.

None of these companies has been affected by the labor difficulties that have visited practically every other camp in the country.

LEAD

Lead is quoted in cents per pound, New York delivery.

Date			Average week ending		
Aug.	22	10.67	July	17	10.98
"	23	10.67	"	24	10.29
"	24	10.67	"	31	10.55
"	25	10.67	Aug.	7	10.87
"	26 Sunday		"	14	10.87
"	27	10.50	"	21	10.79
"	28	10.50	"	28	10.61

Monthly Averages

	1915	1916	1917		1915	1916	1917
Jan.	3.73	5.95	7.64	July	5.59	6.40	10.93
Feb.	3.83	6.23	9.01	Aug.	4.62	6.28	....
Mch.	4.04	7.26	10.07	Sept.	4.62	6.86	....
Apr.	4.21	7.70	9.38	Oct.	4.62	7.02	....
May	4.24	7.38	10.29	Nov.	5.15	7.07	....
June	5.75	6.88	11.74	Dec.	5.34	7.55	....

A mid-year canvass of lead production by the U. S. Geological Survey, the results of which have been collated by C. E. Siebenthal, shows that the output of domestic de-silvered lead, excluding de-silvered soft lead for the first six months of 1917, was 152,231 short tons, compared with 158,235 in the same period of 1916. Domestic soft lead, including the de-silvered, was 124,292 tons against 117,879 last year, and the lead produced from foreign ores and bullion was 29,539 tons while the amount for the half year in 1916 was only 9453. In view of the scarcity of the metal the increase in lead of foreign origin is encouraging. The greater part came from Mexico. According to the records of the Bureau of Domestic and Foreign Commerce, the total lead imported during the first six months of 1917 was 30,620 tons, of which 22,507 tons came from Mexico and 4569 from Canada. The exports of domestic lead amounted to 29,241 tons and of foreign lead 6066 tons. The lead used in articles exported with the benefit of drawback was 3270 tons. Thus the total exportation of lead amounted to 38,577, and the apparent consumption in this country in the six months was 268,952 tons, being an increase of 38,365 over the same period in 1916. The production of new antimonial lead was 7822 tons, and of secondary antimonial lead 1959 tons, being a marked decrease. The average outside spot price of lead during the six months was 9.9c. per pound, as against 6.9c. in 1916.

ZINC

Zinc is quoted as spelter, standard Western brands, New York delivery, in cents per pound.

Date			Average week ending		
Aug.	22	8.50	July	17	9.06
"	23	8.37	"	24	8.60
"	24	8.37	"	31	8.75
"	25	8.25	Aug.	7	8.75
"	26 Sunday		"	14	8.75
"	27	8.25	"	21	8.66
"	28	8.25	"	28	8.33

Monthly Averages

	1915	1916	1917		1915	1916	1917
Jan.	6.30	18.21	9.75	July	20.54	9.90	8.98
Feb.	9.05	19.99	10.45	Aug.	14.17	9.03	....
Mch.	8.40	18.40	10.78	Sept.	14.14	9.18	....
Apr.	9.78	18.62	10.20	Oct.	14.05	9.92	....
May	17.03	16.01	9.41	Nov.	17.20	11.81	....
June	22.20	12.85	9.63	Dec.	16.75	11.26	....

QUICKSILVER

The primary market for quicksilver is San Francisco, California being the largest producer. The price is fixed in the open market, according to quantity. Prices, in dollars per flask of 75 pounds:

Week ending

Date					
			Aug.	14	115.00
July	31	110.00	"	21	115.00
Aug.	7	115.00	"	28	115.00

Monthly Averages

	1915	1916	1917		1915	1916	1917
Jan.	51.90	222.00	81.00	July	95.00	81.20	102.00
Feb.	60.00	295.00	126.25	Aug.	93.75	74.50	....
Mch.	78.00	219.00	113.75	Sept.	91.00	75.00	....
Apr.	77.50	141.60	114.50	Oct.	92.90	78.20	....
May	75.00	90.00	104.00	Nov.	101.50	79.50	....
June	90.00	74.70	85.50	Dec.	123.00	80.00	....

TIN

Prices in New York, in cents per pound.

Monthly Averages

	1915	1916	1917		1915	1916	1917
Jan.	34.40	41.76	44.10	July	37.38	38.37	62.60
Feb.	37.23	42.60	51.47	Aug.	34.37	38.88	....
Mch.	48.76	50.50	54.27	Sept.	33.12	36.66	....
Apr.	48.25	51.49	55.63	Oct.	33.00	41.10	....
May	39.28	49.10	63.21	Nov.	39.50	44.12	....
June	40.26	42.07	61.93	Dec.	38.71	42.55	....

Eastern Metal Market

New York, August 22.

While prices are lower than a week ago, business is not stimulated thereby, and all of the markets will undoubtedly remain dull and stagnant, pending an announcement by the Government of its decision in the matter of price-fixing. It is almost certain that the situation cannot be allowed to drift much longer. A preliminary report on the cost of producing steel and copper has been completed by the Federal Trade Commission and will be transmitted to the President within a day or two. It is expected that he will devote several days study to the report before taking action, and until then no information regarding the findings of the commission will be made public.

Copper is dull and prices are nominal.

Tin is lifeless and quotations are slightly lower.

Lead is stagnant and prices have declined slightly.

Zinc prices are unchanged, although there are indications of weakening.

Antimony is lifeless with no sales reported.

Aluminum is quiet and prices are nominal and unchanged.

Although the steel trade is awaiting the decision of the Government as to the prices to be paid for ship and munition steel and expects that it will be made shortly, the outlook for the prices of steel for the Allies and the public is not clearer in any respect. The orders of the Government for wire rods and wire for Italy, mentioned last week, have been held up. While this had been accepted by the manufacturers at prices to be determined later, with the proviso that the Government actually should place the order and become responsible for the payment, it was found that no appropriation existed which could be drawn upon to buy steel for a foreign government. As in previous weeks the market has shown no significant changes in prices. Any softening of pig-iron prices is usually found in re-sale transactions. Cases are cropping up in which there was over-buying and not a little Southern iron was sold at concessions for this reason. Embargoed iron for export is also coming on the market and can only be moved at less than recent quotations.

COPPER

Since the President has announced a fixed price for coal the trade realizes that a precedent has been established, in spite of the belief of many that he could not and would not arbitrarily fix prices to be paid for copper and other non-ferrous metals purchased by the Government. Fixed prices for steel, copper, and other materials are expected to be announced shortly, the trade apparently being reconciled to such a drastic policy. For this reason buyers and sellers continue to play a waiting game until the uncertainty as to Government prices is removed. The market has weakened slightly in consequence and the quotation today is about 1c. less than last week. Since hardly any metal is changing hands, and since that which is sold is without exception for actual nearby requirements, this quotation is purely nominal. Both Lake and electrolytic are quoted at 26.50c., New York. A cable from London on Monday showed no change from last week's quotations of £137 for spot electrolytic and £133 for futures.

TIN

Some business in futures has been done, but delays in the receipt of the cables from London are retarding transactions considerably. This was particularly the case last Friday when the interest of consumers was aroused in the shipments to be made from the Straits Settlements late this year, but nothing was done owing to delays in cables. The market has been stagnant since then, although on both Saturday and Monday a slight interest in futures developed, and some business was transacted. Some cheap offerings of tin about to be shipped from England were made last Thursday, but these were not taken. The arrivals of tin to August 20 were 2470 tons, and the quantity afloat on that day from both the Straits Settlements and ports in the United Kingdom was 4215 tons. The quotation yesterday was 61.75c., a decline of ¾c. from the price which has prevailed for the greater part of the week.

LEAD

Domestic demand and business is almost nothing, and as a result the market continues quiet. Not much is known regarding the policy which the Government is likely to follow with respect to lead prices. The price paid for the lots purchased to meet the Government requirements of about 8000 tons per month is announced as 8c. per lb., St. Louis. Producers, it is understood, are not disposed to criticize this price, as the profit made at that figure is a fair one. More lead was offered in the week than could be absorbed, and quotations declined slightly as a result. The price is now 10.75c., New York, the leading interest still continuing to adhere to the 11c. quotation. A few days ago this price, 10.75c., was paid for a small tonnage for export which changed hands, and another good sized lot was sold at 10.50c., New York, in bond.

ZINC

The transactions in zinc are not enough to make a market. Some indications of a slight weakening on the part of small handlers have appeared. As a result of this action, which is attributed either to a desire to stimulate activity, or to dispose of stocks that have been held for some time, prices have declined to 8.62c. in the last few days.

ANTIMONY

There is virtually no market, and prices for Chinese and Japanese grades remain unchanged at 15 to 15.50c. per lb., New York, duty paid.

ALUMINUM

Aluminum remains exceedingly dull. The quotation for No. 1 virgin metal, 98 to 99% pure, is unchanged at 50 to 52c. per lb., New York, this figure being a nominal one.

ORES

Tungsten: Although the market has been firm in the past week the quotations remain unchanged at $25 for high-grade wolframite. Inquiries for forward position continue on a large scale, and several hundred tons has changed hands. Ferro-tungsten is unchanged at from $2.25 to $2.50 per lb. of contained tungsten.

Molybdenum and antimony: Offers of $2.10 to $2.20 per lb. of MoS_2 in 90% material have been made for small quantities now in transit, and the market is practically bare of supplies. Very few offers of antimony ore have been made, as the high freight rates make shipments at the prevailing low prices almost out of the question. The quotation, which is practically nominal, is unchanged at $2 per unit for high-grade material.

MANGANESE shipments from Bahia, in Brazil, have been resumed. The largest deposit is in the municipality of Bom Fim, and is reached via the Central Railway of Brazil. Two shiploads of the ore have recently left the port of Bahia for the United States, one cargo consisting of 4000 tons and another of 4300. Deposits near Nazareth, State of Bahia, have large amounts of manganese developed but lack of transportation facilities makes them unavailable.

INDUSTRIAL PROGRESS

Information furnished by manufacturers

The Process of Spellerizing as Applied to Pipe

The form and chemical composition of matter are well-known to be distinct properties. For example: Fig. 1 represents carbon in the form of coke; Fig. 2 represents the same material in the form of a diamond. The diamond is practically pure carbon, bearing no outward resemblance to carbon in the form of coke, coal, or charcoal. Metals undergo changes almost as remarkable if subjected while hot to certain heat treatment and mechanical manipulations.

FIG. 1. FIG. 2.

Commercially pure iron is produced in several forms; puddled iron, knobbled charcoal-iron, so-called ingot-iron, and soft welding-steel have much the same chemical composition as commercial products, but may differ considerably in physical properties and durability according to the treatment given in the process of manufacture.

Uniformity, both as to chemical composition, density, and character of structure and finish, has been demonstrated to be the most important factor governing corrosion in pipe. The actual chemical composition of the iron or steel has been demonstrated to be of comparatively little importance provided it is not unduly variable in the same piece.

Insofar as pipe is considered, with which this article is particularly concerned, thorough investigation and research have shown that the greatest danger from corrosion occurs to the smaller sizes of pipe owing to the thinner walls. The larger sizes with thick walls are made from heavy plates of such uniform quality that corrosion does not seriously affect them to any appreciable extent. To overcome the tendency to corrosion in the smaller sizes a process has been evolved, known as 'spellerizing.'

Inasmuch as this process is entirely mechanical and does not in any way depend upon skilled labor, beyond keeping up the machinery involved, uniform treatment is assured.

The spellerizing process is applicable to the smaller sizes of pipe—say, 4 in. and under—although it is possible in special cases to spellerize pipe a few inches larger.

As a matter of fact, the process of spellerizing metal may be considered analogous to the kneading of dough from which bread is made. Dough is kneaded to produce a smooth uniform texture; to facilitate the escape of confined gases, which would form air-holes and other irregularities in substance and on the surface, and to make an even grain and fine smooth surface. Much the same results are obtained by spellerizing steel.

In a general way, this illustrates the principle on which the spellerizing process is based, that is, the aim is to make the metal uniform, so that corrosion will be uniform, and not in the form of pitting, for pitting in pipe represents the presence of weak spots.

There are other factors, such as contact with other materials, which are electro-negative to iron, such as carbon or oxide of iron, or electrolysis due to stray currents which will cause local failures no matter how carefully the steel is made.

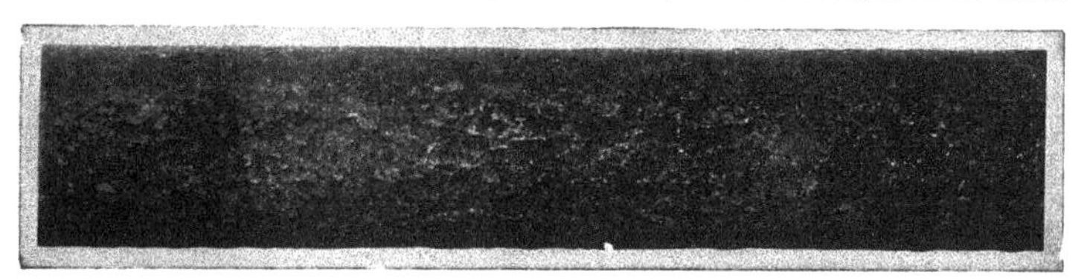

FIG. 4. SAMPLE OF WROUGHT-IRON PIPE IMMERSED IN RUNNING MINE WATER IN WEST PENNSYLVANIA COAL MINE. COMPARATIVE LOSS BY CORROSION. WROUGHT IRON, 100%; SPELLERIZED STEEL PIPE, 73%. (See Fig. 5 below.)

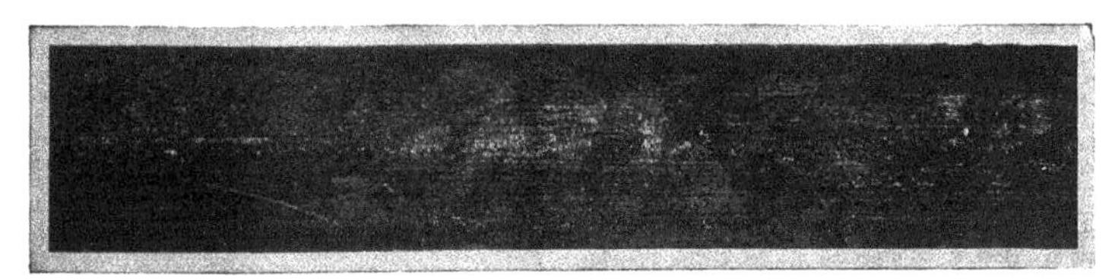

FIG. 5. SAMPLE OF SPELLERIZED MILD STEEL PIPE IMMERSED IN RUNNING MINE-WATER IN WEST PENNSYLVANIA COAL MINE, SIDE BY SIDE WITH FIG. 4. COMPARATIVE LOSS BY CORROSION. WROUGHT IRON, 100%; SPELLERIZED STEEL PIPE, 73%.

Pipe has been made by this process for ten years in increasing amounts. It is significant to note that official records of the American Iron and Steel Institute show that during this period steel-tubes and pipe have increased from 74.3% of the total production in 1906 to 87.9% in 1916.

A few special references to experiences in the use of this process and its influence on steel pipe is of practical interest:

(1) H. J. Macintire, professor of mechanical engineering, in Washington University, in an article in the issue of June 5, 1917, of *Power* (pages 760-761), says, relative to corrosion about the power-plant: "In the case of ordinary steel pipe, mill-scale is always present, and this likewise is electro-negative to the iron. If this scale is evenly distributed, as in spellerized steel, the self-corrosion on its account will be slight; but if it is segregated, then local electrolysis and pitting of the material will result."

(2) Morgan B. Smith, in the October 1913 issue of *Ice*, states with reference to the merits of the spellerizing process: "Steel pipe, which has been treated in such a manner as to eliminate, or at least distribute evenly the mill-scale, may be joined with wrought-iron or cast-iron safely, as a rule. . . . The same stock without the treatment for mill-scale will show a decided tendency to corrode when joined with wrought-iron or cast-iron. The so-called spellerized steel fulfills this condition with respect to the scale."

(3) R. B. Duncan, associated with the United Gas Improvement Co., of Philadelphia, in a paper, 'Installation and Maintenance of Service,' read before the ninth annual meeting of the American Gas Institute, 1914, states: "The steel industry has been developing a new process which, after several years, has given many encouraging results. By this process the steel is treated mechanically and does not in any way depend upon skilled labor, beyond keeping up the machinery involved, hence uniform treatment is assured. This new process is a method of treating metal which consists in subjecting the heated bloom to the action of rolls having regularly shaped projections on their working-surfaces, then subjecting the bloom, while still hot, to the action of smooth-faced rolls and repeating the action whereby the surface of the metal is worked so as to produce a uniform dense texture better adapted to resist corrosion, especially in the form of pitting."

(4) *The Gas Record* of September 23, 1914, page 222, in commenting on Mr. Duncan's paper in regard to the spellerizing process, says: "The consensus of opinion is that modern steel-pipe, particularly if spellerized, is as durable as wrought-iron, and is, besides, cheaper, stronger, and more ductile and more uniform in composition."

(5) Pipe-steel made in 1906 by this roll-knobbling process tested against pipe-steel made in 1897 resulted not only in a somewhat greater loss of weight by corrosion of the latter, but a decidedly deeper pitting of the 1897 steel in 6 months, than occurred in the 1906 steel in 13 months. In comparison with wrought-iron it was found that the two materials lost practically the same weight by corrosion, yet the steel had the advantage of uniform corrosion, since the wrought-iron skelp pitted in 7 months much deeper than the steel did in 13 months. (Prof. H. M. Howe, Am. So. for Testing Materials, 1908.)

(6) New spellerized steel-pipe and new wrought-iron pipe, considered best by the master mechanic of a large western Pennsylvania coal mine, were submerged in running mine water side by side, but insulated from each other. Fig. 4 and 5 show the uniform corrosion of specially worked pipe-steel as compared with irregular corrosion and pitting of wrought-iron pipe.

(7) A. Sang, in a résumé of the question, entitled 'The Corrosion of Iron and Steel' (McGraw-Hill Book Co., New York, 1910), says: "The carefully acquired experience of the largest manufacturers of tubes in the world, which induced them recently to abandon the manufacture of wrought-iron pipe, teaches that the use of steel in place of iron, at least in the United States, for the special purpose of tubing is to be preferred; the tendency of steel to pit is somewhat less than that of iron and it welds at the joint fully as well."

(8) "There is little, if any, difference between the corrosion on the wrought-iron and the corrosion on the steel-pipe. If anything, the wrought-iron is pitted a little deeper, that is, the pitting on the steel pipe is probably more general all over the surface, but the pitting on the wrought-iron pipe is deeper in spots that are affected." (Proc. American Gas Institute, Vol. III, 1908, page 274.)

(9) "While the corrosion was about the same, there was a pitting in the iron that we did not find in the steel, and the steel was corroded more uniformly. From the tests made, I know that the steel pipe is better for such conditions." (Superintendent of one of the largest bituminous coal operations in the Pittsburgh district.)

The evidence might be continued ad infinitum; but it is believed sufficient has been mentioned to show that the tendency of the spellerizing process is to render the surface of pipe uniform and reduce the tendency to corrosion, especially in the form of pitting.

'Smelting-Furnaces and Accessory Equipment'

The above is the title of Bulletin 1417-A of the Allis-Chalmers Manufacturing Co. of Milwaukee. The bulletin contains 48 handsomely illustrated pages, which display and describe blast-furnaces ranging in size from the small water-jacketed cupola, suitable for out of the way places, to the large welded water-jackets, 48 by 192 in., for copper smelters of large capacity. Lead furnaces also are shown, as well as reverberatories, movable forehearths, settling-pots, blowers, and all the usual accessories of the modern smelter. Those interested in this sort of metallurgical equipment may procure a copy of this interesting bulletin by addressing the Allis-Chalmers company in the prominent cities of the United States, and also in foreign countries.

Roller-Chain Drives

The Link-Belt Company has issued two neat little books descriptive of the Link-Belt roller-chain drive, which finds many uses in driving machinery, and particularly in driving motor cars of every description and size. On the Mexican border, after many experiments with numerous varieties of trucks the United States government gave preference to chain-driven trucks. It is also said that in Germany, chain-driven trucks are displacing other types of machines driven by shafts and internal gears. These publications give much useful data on the power required, the load per tooth of the sprocket-wheels, and durability and cost of operation of chain-drives on cars. These booklets will be sent to those interested, upon application to the Link-Belt Co., which has offices in all of the principal cities of the country. The main office is at 39th street and Stewart avenue, Chicago.

A Twenty-Eight Story Hotel

The General Electric Co. is to play an important part in the building of the world's greatest hotel, the Commonwealth, soon to rise 28 stories in the Times Square district of New York, under the supervision of Charles E. Knox. The electrical equipment for power will be sufficient for a city of 20,000 inhabitants.

Not only the electrical apparatus for the lighting, heating, and ventilating of the great building, but the electrical devices for operating and controlling the machinery used in the kitchen, laundry, and refrigeration departments, the pneumatic-tube service, vacuum-cleaner system, and the various pumps necessary in a building of the Commonwealth type will be furnished by the General Electric Co. This company also will furnish the electrical conduits, wires, cables, and incandescent lamps for this greatest of hostelries.

The Hotel Commonwealth will occupy an entire block-front in the heart of New York's shopping and amusement centre. It is to contain 2500 rooms, costing from $1.50 per day (with bath) upward. To be designed, built, decorated, and equipped under a single contract originated by W. J. Hoggson. It will cost a total of $15,000,000 and will be owned and operated on a co-operative basis by over 100,000 shareholders, each of whom may subscribe for from one to ten shares of stock at a par value of $100. No one person will be allowed to subscribe for more than ten shares, and with so great an army of shareholders, it is expected that the Commonwealth will have capacity patronage the moment its doors are opened for business.

Dividends, rebates upon expenditures for food and lodging, discounts through the shopping bureau on purchases at New York stores, and the splendid club privileges of the Commonwealth, are some of the advantages to shareholders.

H. L. Merry, one of the most widely known hotel managers in the country, will manage the Commonwealth. C. M. Ingersoll, the watch manufacturer, is president of the corporation. The enterprise is backed by about 50 firms of national importance. The *New York Globe*, after a thorough investigation of the project, pronounced it the greatest co-operative movement ever undertaken in America and heartily endorsed it.

Outdoor Switch-House Construction

Large power companies are coming more and more to buy portable switching and metering equipments in quantity, in order to be in position to connect-up desirable loads along

Fig. 1.

transmission-lines. They are thus enabled to reach small plants where the load is not large enough to warrant the expense of a sub-station with indoor apparatus. The increasing popularity of this form of equipment has naturally resulted in a steady development in the design and construction of switch-houses, especially with a view to increasing their accessibility.

An open type of construction has recently been brought out by the Westinghouse Electric & Manufacturing Co., that provides the maximum of accessibility. In this switch-house, which is shown in Fig. 1, the oil circuit-breaker is mounted on

Fig. 2.

a specially constructed bracket and the meters are mounted on a slate-slab. The bracket is so designed that it will take different sizes of breakers.

This method of mounting the circuit-breaker has the advantage of being easily accessible for the inspection of wiring, removal of oil-tanks, inspection of contacts, and replacement of fuses protecting the voltage transformers—all very desirable features not possessed by switching-houses employing panel-mounted circuit-breakers.

A ground-mounted switch-house with panel-mounting circuit-breaker is shown in Fig. 2. The circuit-breaker and instruments are all mounted on one slate-slab, and while the construction is not as open as with the bracket-mounted breaker, it has a neat and pleasing appearance.

The tendency of power companies to buy this type of equipment in quantity is well illustrated by the Iowa Light, Heat & Power Co., which has been furnished recently with twelve switch-houses of the ground-mounting type, and the Northwestern Ohio Light Co., of Van Wert, Ohio, with five switch-houses of the pole-mounting type by the Westinghouse Electric & Manufacturing Company.

Duplex Trucks Doing Pioneer Work

"While hundreds of miles of roads have been improved in the United States during the last few years, and as a result, motor-trucks are being used more than ever before for hauling freight from one city to another, there are thousands of miles of highway over which it would be impracticable to haul freight with the average motor-truck," declared H. M. Lee, of the Duplex Truck Co. of Lansing, Michigan. "Seventy per cent of the roads in this country are unimproved and horse and mule teams have always been used for haulage because it was found that motor trucks could be operated only at certain times. As these roads are impassable for the average motor-truck nine months out of the year, the horse and mule teams continue to be used, although many would be replaced with trucks were the roads improved. Because of their tremendous reserve power and their dependability under adverse road conditions, Duplex trucks are being used for hauling freight in many sections of the country where motor-trucks have never been used before. More than half the Duplex trucks in operation today are doing pioneer work, that is, hauling full capacity loads where other types of motor-trucks could not be used efficiently. Turpentine operators in the South-East, oil companies in the South-West, lumber operators in the mountainous districts, pit and quarry owners in the Central States, and many contractors, manufacturers, and jobbers throughout the country, who are required to make deliveries off the pavements, are using them for the first time, because we have conclusively demonstrated that the Duplex will haul full-capacity loads wherever horse and mule teams can be used. The Duplex truck enjoys a unique position because in many instances it does not compete with other motor-trucks. It is performing haulage-work which hitherto has been done exclusively by horse and mule teams." The company is now rushing new factory buildings to completion and the output will be trebled before the first of the year. The manufacture of trucks will be centralized at Lansing instead of operating factories at both Lansing and Charlotte.

Pneumatic Tires for Trucks

A method for reducing costs in mine, mill, and smelter transportation service has been shown since pneumatic tires have developed such great economy on motor-trucks. This was substantiated in a test-run of 4288 miles from Detroit to Mexico and return, under the supervision of Lieut. J. W. O'Mahoney, who was in command of motor-truck company No. 35 on the Mexican border. A Packard 1½-ton truck, carrying a load of over 2½ tons, and equipped with 36 in. by 7 'Nobby' tread pneumatic cord-tires was used for the test, and was run over roads similar to those found in nearly all mining districts. Economy of gasoline and oil, and far greater mileage under extremely adverse conditions were obtained. Lieut. O'Mahoney, who drove the truck, reported in part as follows: "I made up my mind that, in order to give the hardest kind of army transport test, which is the severest that any truck could ever be put to, I would not spare the truck. The amazing part of the trip is that the truck came back to Detroit after covering 4288 miles of the worst roads and 'trail blazing' that I have ever known, in practically as good condition as when it first left Detroit. This fact alone demonstrates that, if a pneumatic tire could be built to stand up under the strain of truck service, truck maintenance could be cut to a minimum. I have seen a great deal of army truck-service along the Mexican border, and with Gen. Pershing's army beyond the border in Mexico where roads are the exception, but I have never heard of such a series of difficulties as these tires overcame. We went through hundreds of miles where the roads were covered with 12 to 14 in. of soft clay, and secured ample traction. We went other hundreds of miles over roads with mixed mud and sharp stone. There were still more hundreds of miles over the sandy Texas desert. When we did not have mud, sharp stones, and hot sand, we were crashing through bridges and fording rock-bottomed streams. Under such conditions, the tires not only stood up well for 'pneumatic,' but did better than 'solids' have done. The average for the entire 4288 miles was considerably more than seven miles per gallon of gasoline. On fairly good country roads, such as are found from San Antonio to Sherman, Texas, and across Kentucky and Indiana, the consumption was about 10.6 miles per gallon. The average mileage per quart of oil for the entire trip from Detroit to Mexico and back was 33.42 miles, which proves that pneumatic tires for trucks do much to reduce the pulling-strain on the motor. I believe these United States 'Nobby' pneumatic tires will give an average of 40% more mileage per gallon of gasoline than 'solids'; that they will reduce the oil consumption from 25 to 30%; and that by taking up road-shock they will lower the depreciation of the truck fully 50%. The 'Nobby' tread-cord truck-tires used in this test were made by the United States Tire Company.

Commercial Paragraphs

W. F. Stevens has accepted a position with DENVER ROCK DRILL Co. Mr. Stevens has just resigned from the H. M. Byllesby Co. at Lovelock.

On August 1 the Kewanee plant of the National Tube Co., at Kewanee, Ill., was sold to the Walworth Manufacturing Co., and on that date the National Tube Co. retired from the fitting business.

The GENERAL ENGINEERING Co. of Salt Lake City, has issued the fourth edition of its Ore Testing Bulletin. This little publication contains a great deal of information of value to those who have ore-testing problems in hand. It will be sent on request.

The WALTER A. ZELNICKER SUPPLY Co. of St. Louis, Missouri, has issued Bulletin No. 221, in which are listed what the company describes as exceptional offerings in rails, locomotives, cars, cranes, pipe, piling, and tanks. The bulletin may be had without cost upon application.

The WESTINGHOUSE ELECTRIC & MANUFACTURING Co., of East Pittsburgh, Pennsylvania, has issued an interesting illustrated publication on the electrification of the Norfolk & Western railroad. It was written by A. H. Babcock, consulting electrical engineer for the Southern Pacific Co. and will be found useful by all who are interested in recent electrical development.

The SPRAY ENGINEERING Co., 93 Federal street, Boston, Massachusetts, has issued Bulletin No. 311, which describes and illustrates the Spraco pneumatic painting equipment, embodying the recent improvements. By means of this device, painting is quickly and inexpensively done without a brush, and with no knowledge of the art of painting. A copy of the bulletin may be had by addressing the company.

MARK R. LAMB is publishing a Revista Semanal, being a folio sheet in Spanish relating to special products of the west coast of South America. It is intended as a market guide for shippers in that part of the world. Quotations are given for the common metals, tungsten, tin, the variety of rubber known as caucho, cotton, hides, and other articles. It is a sheet to supplement the activities of the brokerage house which Mr. Lamb has established at 30 Church street, New York. As the title indicates, it is published weekly. It is attractively printed and illustrated.

THE NATIONAL ASSOCIATION OF PURCHASING AGENTS is composed of the membership of local branches with a present total membership of more than 1000. Active associations exist in New York, Chicago, Philadelphia, Cleveland, Pittsburgh, Baltimore, Boston, Detroit, St. Louis, Columbus, Buffalo, Rochester, South Bend, San Francisco, and Los Angeles. The association is the outgrowth of a movement which originated in New York five years ago. The object is to bring together on common ground purchasing agents generally, to gather and to disseminate information and knowledge of benefit to the profession, to foster and promote friendly relations between members, to secure more uniform methods of routine for purchasing, and the standardization of specifications and classifications, and the dissemination of data relating to the subject of buying. The annual Congress of the Association will be held at Pittsburgh, Pennsylvania, on October 9, 10, and 11. Speakers of national reputation will be in attendance.

EDITORIAL

T. A. RICKARD, Editor

DIVIDENDS of copper companies for the eight months of this year aggregate $113,205,220 as against $81,291,946 in the corresponding period of last year. The Red Cross has received $4,149,683 from these copper companies, but that does not include the gifts of individual shareholders.

WE take particular pleasure in publishing an article from Mr. Walter Douglas. This shows the detrimental effect of increased wages on the efficiency of workmen in the Clifton-Morenci district. Most of this labor is Mexican, and, as most of our readers know, it is not a Mexican habit to be thrifty. Hence the higher wages have had a demoralizing effect in an economic sense.

FROM the latest issue of 'The Missouri Miner' we note that the Missouri School of Mines has answered nobly to the call for service in battle. The number of former students enlisted and commissioned in the Army does honor to this School. The young mining engineers of England, Canada, Australia, and South Africa have done splendid work in the War; those of the United States will maintain the record of our profession.

THE third National Exposition of Chemical Industries opens on September 24 in the Grand Central Palace, New York. It will be the largest exposition of its kind ever held. Three floors of the Grand Central Palace will be occupied by the exhibits of 350 manufacturers of chemical products. Five days will be devoted to papers and symposiums. We hope and believe that this exposition will have a stimulating effect on the chemical industry and serve to mobilize it effectively for the national purpose.

HEARST, it is reported, has presented Marshal Joffre with "a handsomely bound volume containing clippings of news-matter and other references" to himself on the occasion of his recent visit to the United States. We have Mr. Hearst's newspaper's assurance that "the gift is valued intrinsically at several thousand dollars." We trust that the great Frenchman will not be overwhelmed by the costliness of the gift. Mr. Hearst is a gentleman possessed of a great deal of taste, and it is all bad. Fortunately Marshal Joffre does not read **English**.

WE are informed that the American Smelting & Refining Company has brought suit for injunction and specific performance of contract against the Bunker Hill & Sullivan Mining & Concentrating Company. The hearing on an application for injunction is set for October 18 at Portland, Oregon. As our readers are aware, the Bunker Hill & Sullivan company has built, and is now operating, its own smelter near Kellogg, Idaho. The Guggenheim smelter trust claims that in starting to smelt its own output the Bunker Hill & Sullivan is breaking a contract made in 1905. On the other hand, the mining company claims that it is acting in good faith and within the terms of its contract with the smelting company. The suit promises to be both interesting and important. We welcome the prospective disclosure of the sunday-school methods employed in the smelting business.

ALTHOUGH the resilience of magnesite was known to mineralogists, it was overlooked as a factor in construction until the Germans utilized it. In the early days of the War an American company operating magnesite mines in Austria obtained the use of a large number of Russian prisoners, paying them 17 cents per day, this being the cost of subsistence. By aid of these Russian prisoners the American company hastened to dig magnesite as fast as possible, filling boxes and barrels with the mineral, preparatory to shipment overseas; but the German submarine campaign, with the consequent scarcity of ocean freightage, spoiled the scheme; and the magnesite remained at the mine until the Germans seized it, having discovered that it would furnish splendid material for gun-emplacements, on account of its resilience and the quickness with which it sets, namely, in six hours, as compared with the six days required by ordinary cement when in large blocks. The Italians also are using magnesite for their gun-emplacements. Another new use for magnesite is in making flooring or 'deck' for warships, the magnesite being laid upon steel on account of its imperviousness to moisture and heat, besides its resilience, which is a quality of greatest value in many forms of construction.

PUBLISHERS have received the attention of Congress to an embarrassing degree. In the legislative effort to find money for the War it occurred to somebody that the plethoric advertising supplements furnish a suitable target for taxation. As to that we cannot say, being interested. We may inform our readers that publications like ours have been compelled to pay twice the normal price for paper, whereas many big publishers have escaped the imposition owing to long-term contracts with the paper manufacturers. These have exclaimed against the cost of the raw material and have promptly cinched the small man and thereby played into the hands of the big publishers. It must be remembered that the recent attempt to reduce the cost of news-print paper does not affect the publishers of periodicals, such as ours, which use so-called book-paper, a product of better quality and therefore much more expensive. The high cost of paper is a glaring and successful attempt to use the War as an excuse for commercial depredation. As to taxing the profits of publishers, we cannot object to that if every publisher contributes in proportion, but we deprecate the half-accidental manner in which this or that industry is called upon to divide its profits for the national exchequer; obviously everybody, every industry, ought to be taxed equally; every citizen and every business in the country ought to give a fixed percentage for the furtherance of the national purpose, which is to save this country from invasion and to prevent the destruction of the orderly way of living we call civilization.

SULPHUR as a fertilizer is attracting increased interest throughout the country. The remarkable results obtained by Dr. P. J. O'Gara, who recently published an article on this subject in the MINING AND SCIENTIFIC PRESS, has led to a new demand for sulphur. It may be helpful briefly to point out the conditions under which it may benefit the growth of crops. Improperly used it might produce no beneficial effect, and, in the cases of some neutral or slightly acid soils, it would reduce the output. On the other hand, if employed with a proper understanding of the reactions and the bacterial processes involved, it is sure to stimulate the growth of leguminous and cereal crops to a marked degree. The soil must contain an abundance of humus or vegetal matter, must have a slight excess of available lime or else lime must be added with the sulphur, and the percentage of moisture should remain normally about 10%. Below 5 or above 18% of moisture no advantageous results from sulphur can be anticipated. Many of the sulphur deposits of the West, while quite unsuited to treatment for the recovery of refined sulphur, yield material that is admirably adapted for use as a fertilizer, after being ground, on account of the accompanying lime, which serves to protect the soil against acidity. Development of the use of sulphur in this way should create a steady market for the output from some of the low-grade deposits in California, Nevada, Utah, and Texas. Dr. O'Gara showed an increase of over 300% in the quantity of peas harvested from land treated under proper conditions with elemental sulphur.

DESPITE the effort made by the Interstate Commerce Commission to regulate railroad freight-rates and to prevent unfair discrimination in the interest of big combinations of capital, it is noteworthy that injustice is still perpetrated, particularly in the shipment of ores and metals. For example, the rate on blister copper from Garfield (Utah) to Tacoma (Washington), a distance of 1056 miles, is $6 per ton, whereas the rate on pig-lead from Kellogg to Vancouver, a distance of only 510 miles, is $4.50 per ton. The copper is worth $500 per ton, and the lead only $210 per ton, so that the valuation is not allowed to affect the rate logically. This comparison is striking, but a more striking contrast is afforded by the schedule on ore produced within the State of Washington: on ore having a value up to $100 per ton the rate is $9.25 for a haul of 400 miles at the most. Thus a $500 product is carried 1056 miles for $6, whereas a $100 product is carried 400 miles for $9.25 per ton. A still more glaring inequity is disclosed by the fact that $100 ore hauled from Weiser to Tacoma is charged $6.50 on a distance of 566 miles, whereas the $500 copper product from Garfield is carried through Weiser, which is half-way between Garfield and Tacoma, for 50 cents per ton less. Thus a product of one-fifth the value of the copper is charged 50 cents more per ton on a haul of half the length. All this information is abstracted from the published tariffs of the railroad companies, and from the official time-tables, in which the distances between points are given. Any reader can verify our statements for himself. What is the reason for such curious discrepancies? We venture to draw the attention of the Interstate Commerce Commission and of the State Public Service Commissions. Undoubtedly the rates quoted above operate to the advantage of the Guggenheim smelter trust and to the disadvantage of the smaller operators of mines.

Never Again

His Holiness the Pope has been the means of eliciting a frank expression of American policy. To us the latest note of the President is his best; it is short; it is to the point; and it enunciates clearly our object in this war. "We seek no material advantage of any kind." Our object is "to deliver the free peoples of the world from the menace and the actual power of a vast military establishment controlled by an irresponsible government, which, having secretly planned to dominate the world * * * now stands balked, but not defeated, the enemy of four-fifths of the world." Again the President makes the distinction between the people of Germany and the German government. He recognizes the impossibility of dealing with the Potsdam crowd, the gamblers that played with chips of humanity and counted upon bluffing their opponents; and, if they failed, then to let loose the powers of hell carefully prepared for many years. With them we can make no peace because they are outlaws to whom solemn agreements are scraps of paper. The President says, "We cannot take the word of the present rulers of Germany as a guarantee of anything that is to endure, unless explicitly supported by such conclusive evidence of the will and purpose of the German people themselves as the other peoples of the world would be justified in accepting." So far the German people have been more unflinchingly loyal to their rulers than any other people. Signs of an understanding by them of the manner in which they have been exploited by the Potsdam military party are beginning to multiply. When the German people disassociate themselves from the predatory enterprise of their Prussian leaders we shall be able to talk peace; until then any compromise would only afford another opportunity for a more advantageous assault upon civilization. There must be no 'next time.' This horror must not be repeated. 'Never again' is the motto of all the Allies. This is a war to end war. We await the day when the German people will repudiate the trespass, robbery, arson, rape, murder, and slavery that have turned a peaceful neighbor's land —Belgium—into a shambles, a dissolution, and a mark of the beast. We fight that such a crime shall never again be permitted and that American soil shall never again be in any danger of a similar fate.

The Flotation Unpleasantness

Patents have come to be regarded as public nuisances rather than the legitimate reward of inventive genius. That is because few men know how to use their rights under a patent without becoming porcine. During this period of war the litigation engendered by doubtful rights becomes an impediment to the utilization of national resources at a time when unhindered effort is most needed. The fight over the flotation patents is a case in point. It is bad enough that a number of great mining enterprises should be harrassed by the claims of patentees, but a larger number of smaller undertakings are restrained altogether from taking advantage of the foremost ore-dressing discovery of our day—a discovery due to the effort of many men in many places and not to the cleverness of one man or one group of men—because they can neither afford to pay either the royalty or the other exactions demanded nor to pave a way with gold through a succession of courts. The confusion caused by the litigation is due largely to the fact that the Minerals Separation people set out to monopolize flotation before they understood what it was; they made claims that subsequent physico-chemical investigation has proved incompetent as a basis for the most successful application of the process. It is the old story of the so-called 'basic' patent: to be basic it must stand on newly discovered principles, and in the case of flotation the Minerals Separation people have failed to advance a definition that would hold before a jury of physical chemists. They did not know the rationale of the process that they attempted to monopolize and if they ever do come to understand the fundamental principles of the process it will be because many others outside of their organization have contributed to the result.

The foregoing remarks will serve to introduce the announcement, made at Washington, that the American Mining Congress is to urge that one out of three drastic steps be taken to terminate the unpleasantness arising from the Minerals Separation company's campaign to impose a royalty on the production of metals. The first suggestion is, "That the Minerals Separation company shall immediately adopt a liberal policy by the fixing of a fair and justifiable charge for the use of its process and the relinquishment of its apparent purpose of establishing a continuous and burdensome monopoly." This is a pious idea, but how is it to be put into action? The owner of a patent has a legal right to exact even an unreasonable royalty from those whom he licenses to use his patent; moreover, the intent of the law apparently is to give him an unrestricted monopoly in the use of the device or process that he has patented. Who is to curtail the exactions of the Minerals Separation people? The Government, one would suppose, cannot withdraw the privileges it has granted under the law, except by agreement with the patentees; this involves monetary compensation for the rights to be withdrawn. Some such notion underlies the second 'suggestion' made by the Mining Congress; this is, "That the Government, through the Council of National Defense, shall fix a reasonable charge for the use of the process." Here we have the idea of national emergency in time of war and an over-riding of the law in the interest of patriotic service. It can be done, on the principle of *inter arma silent leges,* but any good democrat would hate to see it done. In England they have a law called the Defence of the Realm Act, called 'Dora' for short, a legal and patriotic juggernaut for authorizing the Government to make short cuts toward aiding and expediting the work of warfare; but Great Britain has no written constitution. The enactment for food-control gives the President sundry such powers and if necessary further authority to the same purpose might, and probably will, be granted to him and to the members of his administration. That brings us to the third 'suggestion' of the American Mining Congress, namely, "That Congress shall annul the rights granted by the patent." Nobody will deny the possibility of Congress doing this, but any such legislation would be subject to revision by the Supreme Court. Annulment of patent-rights by Congress would be unconstitutional, presumably, and therefore would be vetoed by the super-legislative authority. We can discover nothing practicable in these suggestions of the Mining Congress, although we share the irritation, common to the entire mining industry, caused by the exactions and embargoes of the patent-exploiting company. The dispatch from Washington says that the 'Journal' of the Mining Congress charges the Minerals Separation company with sundry acts of malfeasance. "It demands that its licensees shall give it the full benefit of any invention or discovery made by them which may be an 'improvement, modification, or addition,' or 'may be useful in carrying out any of the processes thereby protected,' and the right to patent for its own use any invention made by its licensees and further requires its licensees to 'bind their employees to assign or transfer to the licensor any invention made by such employees'." It requires its licensees to pay a royalty "in each and every instance computed on the entire metal value recovered from all ores and material treated by the licensee, provided any portion of such value has been recovered by flotation."

We quote this verbatim because it corroborates much that we have said in these columns during the last two years. The Minerals Separation company insisted that the Britannia Mining & Smelting Company, for example, should pay royalty not only on the concentrate obtained in jigs and tables, before flotation, but also on the ore sorted by hand from a picking-belt in the mill, and shipped direct to the smelter. An impudent effort to bluff the profession was exposed in our issue of February 5, 1916, in which we reproduced the form of contract imposed upon metallurgists in the employ of a licensee, and in the same issue we quoted competent legal opinion concerning the invalidity of this contract. We are glad that the American Mining Congress is taking the matter in hand, for it is one of large moment to the mining industry. We should like to see the Mining Congress take further action in the matter and in order that such action may be effective we suggest that, for instance, an effort be made to expedite the present litigation. When the U. S. Supreme Court decided the Hyde case it was hocus-pocused by the plaintiff's assertions concerning the so-called critical point. This has been stultified by later work on a large scale in various mills. The Miami case is not going before the Supreme Court because the record in that case does not include the latest evidence. The Butte & Superior case will go before the Circuit Court of Appeals and then an effort will be made, whoever wins, to bring the case before the Supreme Court on a writ of certiorari. It is greatly to the interest of all concerned, not the litigants only but the whole metal-mining industry, that the decision of the Supreme Court be handed down with as little delay as possible. If the Supreme Court, after hearing all the latest evidence on the rationale of flotation, decides in favor of Minerals Separation, the only thing to be done will be to float without oil and without the soluble agents covered by the Minerals Separation patents, or to pay the royalty. Meanwhile it is to be remembered that there are other patents besides those of Minerals Separation and the pending litigation does not cover anything like the whole field of possible controversy. If Minerals Separation fails in obtaining a monopoly, it is not likely that others will be encouraged to imitate its exacting methods.

If, however, the Government can be persuaded that the production of the copper necessary for munitions is being hindered by the exactions, or attempted exactions, of the Minerals Separation people, then the President and his advisers may see fit to use the power given them by Congress to curb the patent-exploiters. A condition somewhat similar existed until recently in the manufacture of aeroplanes; the patents owned by rival companies retarded the development of aviation in this country so much as to leave our designers, the originators of the successful heavier-than-air type of avions, far behind the French and Germans. We found ourselves practically without an air-fleet, and almost destitute of trained pilots, at a moment when it became apparent that a strong aerial navy would have turned the tide of war more quickly and certainly than anything else. The prohibitory price of aeroplanes and the inability of either of the two competing manufacturers to utilize all the improvements required for safe and successful aviation prevented such private initiative as would have developed aviation and aviators in this country. The patents stood in the way. Thereupon the Council of National Defense stepped between the rivals and made peace between the Wright-Martin Aircraft Corporation and the Curtiss Aeroplane Company. Co-operation between these two is now assured for the welfare of the country, and for their own good, by means of a cross-licensing agreement giving the advantage of all the patented mechanical ideas to each manufacturer. This establishes a precedent that may warrant action in the flotation unpleasantness. If the Government can regulate the price of copper, why should it fail to regulate the royalty to be paid by the producers of copper for the use of a metallurgic process necessary in the production of copper?

Estimates of Ore-Reserves

Our Canadian friends have engaged in vigorous discussion over a suggestion, made by Mr. R. W. Brigstoke, that the official estimates of ore-reserves as given in the annual reports of mining companies should be verified by an independent engineer. Messrs. John E. Hardman, R. B. Watson, and E. P. Mathewson have joined in condemning this proposal as impracticable. Mr. Fraser D. Reid suggests that mining companies should be compelled to file an affidavit stating exactly how their estimate was obtained. Apparently these Canadian engineers are unaware that several British companies have been in the habit of doing something of this kind. Thus Mr. C. E. Knecht, manager of the Brakpan, and Mr. William McC. Cameron, formerly consulting engineer to the Goerz group, used to furnish an interpretation of the data on which their estimate of ore-reserves was based. Other London companies give such information frankly, although unfortunately the majority not only withhold it but give their estimates in a confusing way. Of course, the shareholders, who are the real owners of a mine, are entitled to have such information and to have it in the most explicit fashion. The whole policy of mine administration, especially in this country, is saturated with the idea that those in control are entitled to inside information, and after them their friends, leaving the minority shareholder in a cold fog of ignorance. We go further and say that not only should vital statements, such as the annual estimate of the mine's resources, be accompanied by all the explanation necessary to make them intelligible and convincing, but we would suggest that the board of directors in their annual report should appraise the mine, stating what in their opinion, based on the advice of the company's engineer, is the present value of the property. They make such a valuation in their own interest, as large shareholders, and we see no reason why they should not give the other shareholders the benefit of their opinion, not as individuals but as trustees for the shareholders.

Increased Wages and Decreased Efficiency in the Clifton-Morenci District

By WALTER DOUGLAS

As an interesting and instructive commentary upon the effect of greatly increased wages on the manual efficiency of a certain class of mine-labor the accompanying diagram speaks for itself.

In the summer of the year 1915, at which time a sliding scale of wages was put into effect by the management of the Detroit Copper Co., the average pay of wage-earners in the mine department was approximately $2.50 per day, which average included a large force of surface-laborers, whose $2 rate served to offset the considerably higher pay earned by the miner and other underground labor. At this time the number of tons mined per man, exclusive of those engaged on development work, was 2.6. For the first six months of the year 1917, with an increase of wages that brought the average daily earning to $4.10, the number of tons of ore mined per man dropped to 2.18; in other words, the cost for labor to mine a ton of ore in the early part of 1915 was 90 cents, while for the first six months of 1917 it increased to $1.90. The miner who earned 36½ cents per hour, based on 15c. copper, during the former period earned 56½c. per hour in the latter period on 28c. copper.

The labor employed in the mine department of the Detroit Copper Co. since the initiation of the company's operations in 1880, has been, and is now, largely Mexican, with a substantial proportion of Spaniards and a few Italians.

In September of 1915 the Western Federation of Miners sent organizers into the district; these, largely owing to the strained relations existing between the United States and Mexico at the time, were successful in inducing all the Spaniards and a number of Mexicans to join their organization. They then demanded recognition of the union, which, being refused, a strike was called. It is unnecessary here to recite the history of this strike, which, through incendiary speeches by the then Governor of the State, assumed the aspect of a revolution, in which all workers who would not join the Federation were deported from the district. No attempt was made by the operators to introduce outside labor, and in February 1916, through the efforts of Federal mediators, Messrs. Myers and Davies, the union surrendered its charter, work being resumed under the same conditions as existed before the strike was called.

At the suggestion of the mediators, the operators consented to the election by the employees of grievance committees to discuss with the management matters relating to wages, conditions of labor, etc. Unfortunately for the future peace of the district, the men elected to these committees were of the most radical type. Prompted by J. L. Donnelly and other officials of the Arizona Federation of Labor, they assumed the right to dictate to the operators not only who should or who should not be discharged, but who should be employed or promoted.

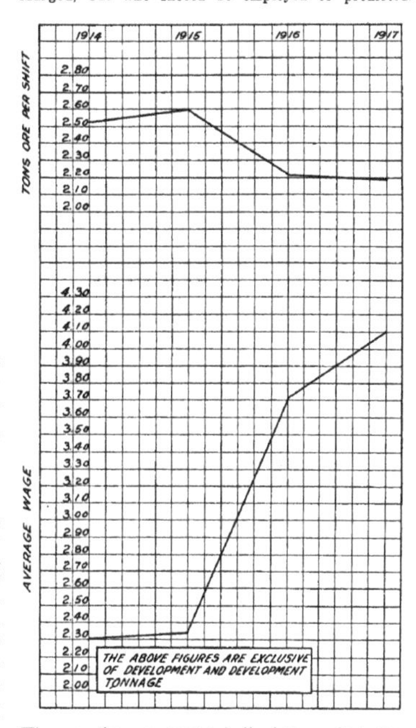

Whenever the management declined to accede to these demands a strike was called either in one department at a time, or all the employees of the companies were called out simultaneously. Nine of these strikes have occurred in the 15 months since operations were resumed. These methods not having been entirely successful, in the opinion of the Federation, in obtaining all their demands, a system of sabotage was introduced underground. This apparently appealed strongly to the class of labor employed, and was effective in reducing the ton-man efficiency. As an example: In May last the management having decided to concentrate its tool-sharpening operations, thereby being enabled to dispense with six or eight Mexican blacksmiths, the grievance committee protested, and not being able to convince the management of the validity of their contentions there was introduced a system whereby thousands of feet of steel was buried in the stope-filling and the company not only put to great expense, but its operations necessarily hindered until an additional supply could be obtained. Discipline underground became impossible and insubordination was rife.

In June last the employees of all the companies operating in the district, through the instigation of the Arizona Federation of Labor, and with its endorsement, demanded the so-called Miami wage-scale. Declining to await the arrival of Mr. Myers, the Federal mediator, they called a strike that has suspended all operations in the district for the past five weeks.

It is an interesting fact that while the operators in the so-called Morenci-Metcalf-Clifton district do not recognize the union—the grievance committees consisting exclusively of employees of the mining companies—yet this district is the most completely unionized in the State of Arizona. The instructions to the grievance committee are issued by the union and the strike was called by it. When it is understood that the tonnage of ore produced per man in the mines in this district is only 2.18 as compared with nearly 13 tons at Miami, the comparison of the wages paid at the two points will appear to be greatly in favor of the Morenci district.

Most readers of this statement will naturally wonder why, if such conditions exist with this class of labor, American miners are not employed, and the wages and efficiency increased. Local conditions have in the past governed this matter, but the main reason at the present time is that there are not enough American miners to man the mines of the State. Then again, the Mexican workmen of the district have been in most cases resident there for many years and are individually a desirable well-meaning class of workmen. They are unfortunately racially susceptible to the socialistic arguments of agitators and easily influenced by oratory of the I. W. W. type, with whose doctrine the Arizona Federation of Labor is permeated. They cannot understand that the union, which persuades them to demand the wages that are paid to the highest class of intelligent miner, does so with the object of eliminating them as an industrial factor in the State's labor forces. The union that sponsored this strike of highly-paid Mexican workmen was the author and active advocate of the notorious 80% law, which prohibited the employment of other than qualified electors or native-born citizens and which was declared unconstitutional by the Supreme Court of the United States.

Cave Deposits of Nitrate

The accumulation of nitrogenous deposits in caves is a well-known fact but the extent to which this sometimes takes place is not fully appreciated. A letter, recently received by the U. S. Geological Survey describes a cave in one of the Southern States that was worked by the Confederacy during the Civil War for potassium nitrate. This cave is said to contain at least 1,000,000 tons of nitrous earth, which, however, contains only 1 to 2% of nitrate. The Survey now states that it seems doubtful whether such material can be profitably used as a source of nitrate salts. The minimum grade of caliche now worked in the Chilean fields contains 12% of sodium nitrate, and though there has been much criticism of the crudeness of the methods employed there, the work is done by cheap Indian labor, and it is doubtful whether leaner material could be worked to advantage here, where the price of labor is so much higher. Several hundred thousand dollars has recently been expended in one of the Western States in testing the proposition to utilize low-grade nitrate. The results have been negative. The nitrate caves in the South were worked during the Civil War by crude methods. Generally the cave-earth was shoveled into iron pots, where it was treated with water and heated over wood-fires to leach out its soluble parts. The liquor was drawn from one pot into another and used for treating fresh material until it became a highly concentrated solution of nitrate salts. It was then drawn off and allowed to cool, whereupon the nitrate crystallized. The remaining liquor was then employed to leach fresh material and the crystals were separated and sacked for use. It is doubtful whether there are any caves of this nature in the West, but one cave at least has been discovered in Nevada from which a considerable quantity of 'bat guano' was marketed.

BRADSTREET'S index-number of commodity prices for July 1 is $16.068, a new high level. It shows an increase of 3.8% over June 1. This upward sweep nearly coincides with that witnessed in England in the same period. Bradstreet's index-number reflects the eleventh consecutive advance noted within as many months, and it also exhibits a rise of 39.4% over July 1, 1916, and of 85.6% over July 1, 1914, just before the outbreak of the War.

FRESNO, Tulare, and Monterey counties have supplied feldspar to California potteries. There are large areas of granitic and metamorphic rocks in the Sierra Nevada mountains which contain pegmatite dikes but the U. S. Geological Survey states that there is little information concerning their commercial availability.

GASOLINE extracted from natural gas in this country increased from 65,364,665 gal. in 1915 to 104,212,809 gal. last year. The quantity of natural gas from which this was derived amounted to 2080 million cubic feet, or approximately 0.5 gal. per 1000 cu. ft., according to the U. S. Geological Survey.

DISCUSSION

Our readers are invited to use this department for the discussion of technical and other matters pertaining to mining and metallurgy. The Editor welcomes expressions of views contrary to his own, believing that careful criticism is more valuable than casual compliment.

Flotation Physics

The Editor:

Sir—Your recent editorial hits the nail on the head when it says that the physical phenomena of flotation must be associated in some way with what we term 'metallic lustre'. Certain ideas related to this had been in my mind for a considerable period, and I propose to let you have them for what they are worth. Metallic lustre means the effect of light on a metal. First we must know what light really is and then see what is its particular behavior with metals. About the middle of the 19th century Clerk Maxwell and others of the more advanced scientists of that day developed a theory of the electro-magnetic character of light. Speaking in terms fitting for use among mining and metallurgical men, this was about as follows: Light corresponds to a movement or rate of displacement of electricity in the ether. This generates a magnetic stress or strain at right angles to it, and the change of magnetic strain in turn generates electric stress, so that there is a progressive transverse wave of these causes and effects transmitted through the ether in somewhat the same way as mechanical stresses and strains combine with dynamic properties to cause a transverse wave to travel along a stretched cord.

Trying to connect this theory with the facts of the transmission of light through material bodies, Maxwell saw that only those substances which—like ether—would not conduct electricity would conduct light. If a substance conducted electricity it would not stand electric strain, because electricity ran from one part of it to another in such a way as to neutralize any strain. Of course, some substances were seen to be opaque, merely because light was absorbed and turned into heat in them, but other substances, such as most metals, reflected a large proportion of the light. This was realized to be due to a reaction or rebound occurring normal to the surface, due to its being the boundary or end of the conducting medium past which no electric strain could pass. The law of equal angles of incidence and reflection was thus obtained and also a law of refraction showing the constant ratio of the sines of the angle of refraction and the angle of incidence.

If that had been all there was to it the scientists of that day might have been happier, but those of today would have had a smaller field for further discovery. The difficulty they met was that transparent substances had a different index of refraction for each different shade of the light of the spectrum; in fact, that was in those days the way they always made a spectrum.

Toward the end of the last century Sir William Thomson, (afterward Lord Kelvin) in the course of his efforts to unravel the mysteries of the atom and the molecule, brought forward a theory that I shall try to explain in simple language. The theory in physics does not necessarily aim to be an ultimate and final explanation, but it is more than a hypothesis, because it explains all well-known phases of the subject to which it relates. It is understood, however, that the discovery of further facts may make necessary a revision of the theory. As it stands the theory serves as an important basis for scientifically correlating all well-known facts in regard to the subject. It is in this sense that the theory of the constitution of the atom, as advanced by Kelvin, is to be understood. It states that the atom is made up of equal amounts of positive and negative electricity. When we understand more exactly what electricity is, we may have to make some modifications of this statement. Negative electricity is, however, presumed to consist of small bodies called electrons, which all carry equal charges or are themselves the charges. Positive electricity is less definitely understood, but is presumed to be a kind of atmosphere constituting the body or nucleus of the atom. The electrons move in or

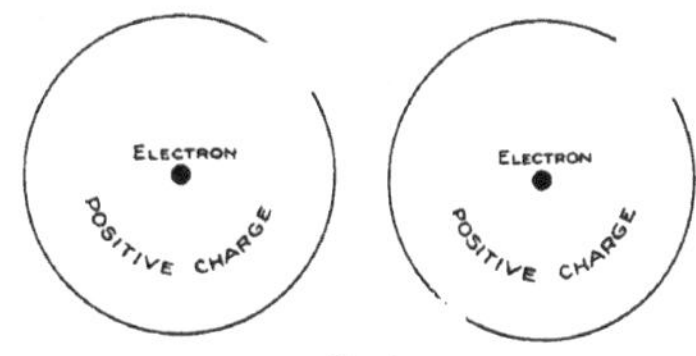

FIG. 1

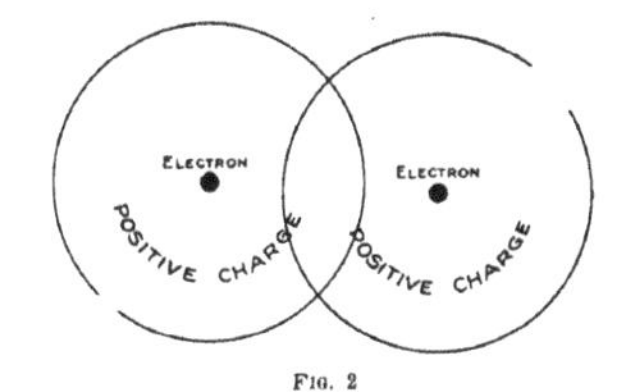

FIG. 2

about the positive charge in orbits, being attracted to it as planets are to a sun, but by reason of the opposite nature of the charge. The atomic weight is presumed to be proportional to the positive charge and is also proportional to the negative charge or the number of electrons taking part in the atomic structure. The details leading to the above conclusions are matters of pure physics, and I shall concern myself here with such deductions from the theory as will help to solve our particular problem.

Suppose each of two atoms (Fig. 1), equal in all respects, to consist of a single electron within a sphere of uniform positive electricity. If they do not touch one another, we know by a general theorem, worked out by Newton for gravitational attraction, that the effect of the repulsion and attraction due to the spheres will be just the same as though their whole effective substance were concentrated in the geometrical centre. Hence there is no resultant attraction or repulsion as between the atoms, as long as they do not touch one another. The theory of gravitational attraction tells us that if the spherical atmospheres lap on one another (Fig 2), part of the repulsion between these positive electric atmospheres is neutralized. The repulsion of the electrons remains the same, and the attraction between the positive and negative charges remains the same. Hence the algebraic sum is no longer zero but in favor of the attraction. The centres of the atoms cannot approach one another indefinitely, however, for the inclusion of one electron in the positive atmosphere of the opposite atom tends to diminish the mutual attraction.

This then constitutes our theory of the force that holds two or more equal atoms together, and accounts incidentally for surface-tension.

An atom must, in general, have more than one eleciron and is believed to have a large number. It would therefore be possible that the inter-lapping of the orbits of the electrons, when revolving in a concentrated charge of positive electricity, would produce a result similar to that I have described.

Following the original presumption of the positive electric atmosphere and supposing a number of electrons in each atom, we may suppose that when two equal atoms are in contact, the electrons pass from one to another of the positive atmosphere. If the two equal atoms are then separated by some outside force, the establishment of equilibrium requires that the same number of electrons remain in each atom.

If, however, the atoms concerned are unequal, we cannot make any such supposition, and it will be possible that when two atoms of different sorts separate from one another, one will carry with it one or more electrons of the other, and thus one atom will have one or more units of negative electric charge and the other atom an equal positive electric charge. These charged atoms are called 'ions'.

Upon the above theory are based all of our modern ideas of the relation between electricity and matter. On the one hand, it has given rise to efforts, such as those of Larmor and J. J. Thomson, to get a more definite knowledge of the constitution of the atom; on the other hand, it has given us a much more exact knowledge of the constitution of solutions, of galvanism, and of chemical reaction in general, and it has been extensively applied by practical chemists in correlating and putting their work upon a more truly scientific basis.

According to this theory of matter, the nature of electric conduction is about as follows: In any metallic solid there are many positive ions. Unlike the ions of a solution, those of a solid metal are fixed in one position. The electric neutrality of the metal as a whole is brought about by there being, within the metal, electrons (negative electric particles) which pass from atom to ion. Under normal conditions there is a continual interchange of this sort, and consequently a continual motion of electrons in all parts of the metal. These motions, however, are such that a balance of electric equilibrium is retained within the metal.

As in the case of cohesion, which you have well described, there must be a surface or boundary condition which retains or reacts upon the condition of stress, or strain, or heat, or charge—or whatever we may call it—within the surface.

This boundary condition (which we might call electronic surface-tension) is well recognized in galvanics. The copper plate of a galvanic cell, for example, must be supposed to have on its surface a latent charge of negative electricity, even before any circuit is made, and before any zinc pole is inserted in the solution or electrolyte. Otherwise we cannot explain how galvanic action can be started. This hypothesis was in fact put forward, quite independently of any modern theory, by the early galvanists.

Now we may better consider the effect of the metallic surface in flotation. Consider the metal immersed in oil. The metallic side of the boundary tends to become charged with negative electricity (electrons). The other side is composed of a dielectric, that is to say, a substance in which there are no ions or loose electrons, and which resists their formation. The molecule of oil is electrically self-contained. The electrons of the surface of the metal are attracted into close contact with the positive portions of the oil atoms or molecules, and without entering them, form a close bond. The electrons of the oil molecules move away as far as they can, by repulsion but cannot leave their own molecules (unless the dielectric rupture stress is reached). This electric displacement is in turn induced in molecules farther from the surface. A bond of adhesion is thus formed at the surface of contact of the metal and oil.

If we now consider the contact of a metal and an electrolyte, we find that the main difference between the metal and the electrolyte is that the ions of the former are fixed, while those of the latter are free to move. Positive ions are attracted to the surface of the metal and stay there, and the corresponding negative ions are repelled from it, initiating a latent charge in the mass of the electrolyte. The neutral dielectric water molecules have no part to play, being neither attracted to nor repelled from the now neutralized metallic surface. The water may therefore be easily separated from the surface. The number of ions necessary to neutralize the latent electrical condition of a face is very small, for a single ion contains quite a large charge.

It might not be altogether out of place to illustrate the difference between the two actions by supposing an electrified conductor placed on one side of a thin piece of glass. On the other side we place an unelectrified conductor of some considerable thickness. The two conductors will then adhere to the glass. This is the case of metal and oil.

To illustrate the opposite effect we allow two differently electrified conductors to come together. Immediately they become equally electrified and begin to repel one another.

The part played by surface-tension should not be overlooked. We have to deal with the combined effect of this and the electrical action we have just described. Oil will wet quartz and water will wet a metallic surface. The electronic action would appear generally to rule. It is all important, however, to recognize the pairing of the unlikes, the solid conductor with the liquid dielectric and the liquid conductor or electrolyte with the solid dielectrics.

It might be useful to know whether the preference of oils, in general, for pure metals does not follow the order of the latent potentials of their surfaces, which is also approximately the order in which their sulphides are precipitated from solution by sulphureted hydrogen. If the adhesive quality of oils has approximately this same order, an apparatus could be devised to find electrolytically the corresponding order of all metallic minerals.

BLAMEY STEVENS.

Mexico City, August 4.

[We are pleased to publish this thoughtful contribution, believing that the statement of different points of view from our readers may lead to interesting and valuable discussion.—EDITOR.]

The Avicaya Mill

The Editor:

Sir—In your issue of June 2, Dr. F. C. Lincoln publishes a brief description of the Pazña tin-mining district of Bolivia, in which he gives a flow-sheet of the Avicaya mill. This flow-sheet is neither complete nor correct, but it would not be fair to blame Dr. Lincoln for this error, since during his visit to the mine all work had been suspended on account of the usual August holidays, and it is difficult to obtain a concise idea of the work that is being done in a mill that is not running. The writer, who has been during part of the time in charge of reconstruction work at the Avicaya mill, thinks it may be of interest to publish the flow-sheet of the mill as it was

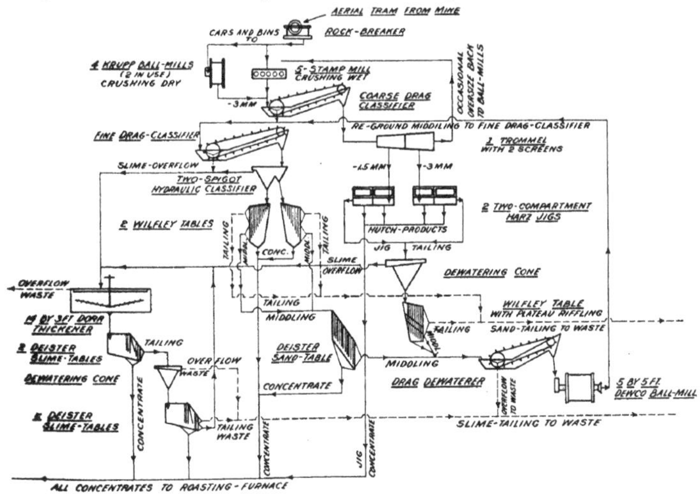

FLOW-SHEET OF THE AVICAYA MILL

operating in the beginning of December of last year when the re-arrangement had been completed.

The tinstone occurs in the ore in small patches and stringers in vugs and on cleavage-planes. It is sometimes of exceptional purity, but more generally intimately combined with iron oxides in the surface ore and with pyrite and arsenopyrite in the ore from lower levels. The gangue itself does not contain appreciable quantities of tin. These conditions call for an initial coarse crushing so as to recover the clean mineral in as large grains as possible and to eliminate the bulk of lean gangue without grinding it unnecessarily fine. The middling from the first concentrating operation contains the mixed grains of tin and iron minerals and must be ground very fine in order to separate these minerals.

The accompanying flow-sheet shows how this has been accomplished in milling practice, the tin-iron middling being kept in circulation over the two Wilfley tables and through the ball-mill until it is either ground fine enough to overflow the classifier and pass to the slime-plant, or has been separated into concentrate and tailing. It should, however, be said that a small part of the mixture is so fine that there is practically no limit to the size of the combined grains, and those grains are found in the middling from the last slime-tables, passing finally into the tailing.

The object of the drag-classifier is to obtain separation of the bulk of the slime in an entirely automatic way that cannot be interfered with by the mill-crew, and if the hydraulic classifiers, which were installed with a view to finish the de-sliming of the drag-belt separated sand, were not correctly operated no very serious losses could result. Instead of classifying the jig-tailing and sending the very diluted overflow of the de-watering cone to the Dorr thickener, it would, of course, have been more practical to install a hydraulic classifier in front of the revolving screen and separate the slime with less water, but there was no height available at this point of the mill.

The mill-ore did not average more than 4 to 5% sulphur, but pyrite and arsenopyrite would accumulate to such an extent in the concentrate that roasting and re-dressing of the product was imperative. The concentrate as produced by the mill would carry 50 to 55% tin, and after roasting it in a small hand-rabbled reverberatory furnace and washing out some iron oxide by hand, it was passed over a small magnetic separator so as to eliminate metallic iron which had entered the tin-product through abrasion of the crushing machinery. The concentrate was finally shipped with about 65% tin, and the tailing resulting from cleaning the roasted concentrate was periodically returned to the Dewco ball-mill.

Since the management has passed into other hands at the beginning of 1917, I am not able to say whether the mill is at present running along these lines.

M. G. F. Söhnlein.

Machacamarca, Bolivia, July 6.

[We are much obliged to Mr. Söhnlein for this interesting information.—Editor.]

Screen Analyses

The Editor:

Sir—I read with some surprise the statement recently made in the U. S. Bureau of Mines Year-Book, as reproduced in abstract in your issue of August 18, in which it is said the finely divided amorphous material in dry pulp must be removed by washing prior to sizing by screens for the purpose of making a screen-analysis that may be depended upon to yield concordant results. In the first place it is a loose use of technical terms to call the fine material adherent upon the larger particles 'amorphous'; it may or may not be amorphous, but it certainly is fine, and that is all that the writer evidently meant. Examination of the more minute particles produced by fine crushing, when made with the aid of a microscope, does not by any means reveal that they exist in an amorphous condition, if by that is meant a non-crystalline condition. In the next place, the writer suggested removing this by washing. The effect of that is often completely to alter the character of the sample. Moreover, I venture to assert that if the Bureau of Mines experimenters will study the pulp after sizing they will discover that the separate larger particles will contain as much adherent fine material as similar particles did in the unwashed pulp. While no means is entirely effective for making clean-sized products in screen analysis, the use of vacuum-cleaning gives altogather the best results, and it is quite the best practice to clean the sized products after screening, then adding the dust collected to the under-size from the finest mesh in the series of screens employed.

John T. Millman.

San Francisco, September 1.

Maintain good order and cleanliness. This is the first step toward effective fire prevention. Wherever excelsior, papers, straw, or other materials are used for packing, keep only a day's supply on hand in a box or bin, lined with tin and provided with a counter-weighted door having a fusible link to insure automatic closing in case of fire. Use standard waste cans as receptacles for such materials as oily or soiled waste, rags, or excelsior. Burn under boilers or elsewhere all such material as is past usefulness every day before closing. Do not permit the accumulation of any waste combustible material near buildings, especially those of wooden construction. Shingles or other refuse from the building, waste paper, old lumber, and empty boxes are sometimes neglectfully allowed to accumulate in such places. Provide metal lockers for the clothing of employees. These minimize the danger of hot pipe in the street coat or a bunch of waste in the overalls pocket. Place dressing rooms in accessible places where fire can be easily controlled.—Nat. Board of Fire Underwriters.

Searles Lake is estimated to contain 20,000,000 tons of potash, an amount sufficient to supply the needs of the country for 15 years.

Geology of the San Sebastian Mine, Salvador

By C. ERB WUENSCH

The San Sebastian mine is situated in one of the numerous volcanic mountains of the coast-range of south-eastern Salvador, Central America, in the Department of La Unión, about 30 miles north of the port of the same name. The mountains, though very rugged, nowhere rise over 2000 ft. above sea-level. Several of the mountain slopes are abnormally steep suggesting faulting. The gulches are sharp, the stream-beds in the vicinity of the mountains contain no gravel, and invariably the solid rock is exposed. The sub-soil is seldom more than 5 or 6 ft. deep. These features are evidences of strong erosive action.

The ore deposit consists of contact-fissure veins carrying gold-bearing pyrite. The veins occur at the contact between a quartz-monzonite porphyry dike and the surrounding eruptives, which are basalt capped by trachyte. In a few cases small orebodies are found in simple contraction-fissure veins in the trachyte. All the rocks are of recent age, probably late Tertiary. During the first volcanic activity the whole region was covered by a flow of basaltic lava, which must have been not less than 3000 ft. thick, which extended irregularly over a great portion of the region. There followed a period of volcanic inactivity during which erosion removed much of the basalt and produced the irregular surface upon which now lies the trachyte. The tuffs, pumice, and glasses belonging to this eruptive period have all been eroded. A second period of vulcanism followed along the same fissures, during which the trachyte was ejected. The trachytitic lava being more acid and hence much more viscous, piled up close to the vent and covered only a small area. The trachyte flow must have been at least 500 ft. thick. At the close of the trachyte flow, after the outer portion had cooled but before the interior had solidified, the quartz-monzonite porphyry was intruded along the original lines of weakness, into the trachyte. The uniform coarse-grained texture of the monzonite shows that it solidified under conditions of slow cooling and pressure. Because of the differences in the coefficients of contraction between the viscous and partly cooled trachyte, the superficially heated fissure walls in the basalt, and the molten monzonite, torsional strains were exerted which produced the irregular fractures, slips, and fissures along which the ore was deposited. After the monzonite intrusion, heated aqueous solutions charged with mineralizers ascended in close proximity to the dike. Reduction of pressure was apparently the important factor in bringing about deposition, for the ore is found along fissures and small fractures. The veins were formed by replacement, sometimes extending to unusual distances from the fractures, and they were not only formed in the trachyte but in the basalt as well.

The structure of the deposit is perhaps the most important feature. By consulting the section on *A-A'*, it will be noticed that the lower limit of the ore-zone is determined by the slope of the basalt-floor upon which the dike rests. Likewise the basalt approaches the vent of the dike from the east. The bottom of the deposit is structurally well defined. Vein I (see section on *B-B'* and the map of the surface) distinctly 'bottoms' on the basalt. This is on the 300-ft. level of the mine at an elevation of 850 ft. The lower portion of the vein contains heavy barren secondary sulphides, and as it de-

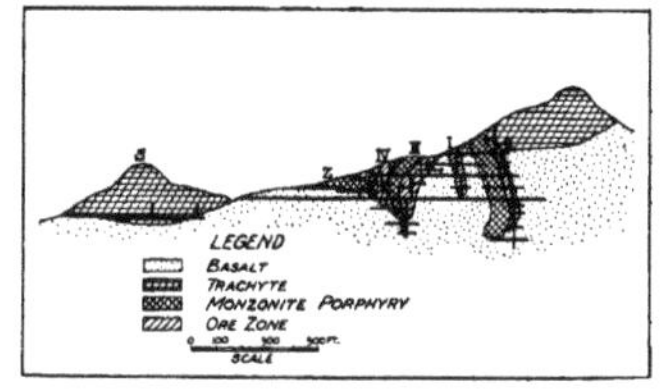

GEOLOGICAL SECTION ON BB: SAN SEBASTIAN MINE

scends in depth mineralization and silicification become poorer and it gradually merges into fresh, unfractured basalt. Likewise vein II terminates against the basalt to the west and continues in depth toward the east to about 50 ft. below the 700-ft. level, below which point no ore has ever been developed. On the 700-ft. level considerable secondary sulphide was found, though ore was found erratically in rich pockets. On the 800-ft. level are some heavy barren secondary sulphides, but a general impoverishment in mineralization and silicification is evident. Vein III was uniformly mineralized in the oxidized zone, but at the 100-ft. level, 50 ft. below the lowest point of the outcrop, when the ore turned into sulphide it became spotty. Rich shoots, however, were found until the 300-ft. level was reached, below which it is universally barren. On the 400-ft. level the vein is heavily mineralized, but with barren secondary sulphide. Veins IV and V behaved similarly. Vein VI is unique in that the ore lies entirely in trachyte and 'bottoms' upon the top of a sill from the dike as shown in the section on *C-C'*. As the dike is approached the sulphide becomes barren, and disappears just before the dike is reached. This ore-zone has thrown light upon the method of fracturing. The fractures are quite irregular. The whole knob of the hill in which these veins are found is composed of slips, cracks, and fractures which strike and dip roughly at 90° to each other as shown in the section on *C-C'*. The majority of the fractures are warped surfaces and not planes. This system of fractures shows a striking resemblance to those which Daubreé produced in his classic experiment with a mass of resin and beeswax. Veins VII and VIII are not of the same type as the others, as has before been mentioned. They are relatively unimportant and have produced only a small amount of ore, averaging about one ounce in gold. The veins are simple contraction-fissures in the trachyte. The ore was apparently derived from the small amounts of gold present in the trachyte and was deposited by vadose waters along the kaolinized walls of the fissures. The streaks are seldom more than small cracks filled with kaolin and granular quartz stained by limonite and hematite. In the vicinity of these veins the trachyte has suffered the type of alteration so common to rhyolites and trachytes, that is, a general silicification of the groundmass. The veins 'pinch-out' within a hundred feet of the surface.

Invariably the ore is confined to fractures or contacts, whereas silicification may have extended for unusual distances into the wall-rock. Frequently there will be a 'horse' of unreplaced basalt adjacent to the dike, or silicification may be poorer at the contact of the dike than a few feet away. In veins I, II, III, and IV brecciation is prominent in the upper portions of the basalt. In veins I and II the brecciated walls have been thoroughly replaced by silica and the open spaces cemented with chalcedony, quartz, and sulphides. In veins III and IV the brecciated boulders of basalt are often largely unreplaced, although in the ore-producing portions of the zone the silicification and cementation of the brecciated boulders is as complete as in veins I and II. On the 600 and 700-ft. levels of vein II the basalt has been thoroughly silicified and evidences of brecciation are rare. Here, however, veinlets of bluish-black quartz are found along small fractures, producing a banded effect. Rich ore has been found in this type of vein. The ore often is found frozen to the contact of the dike. When the pyrite occurs in coarse crystals, or when it is whitish in color the ore is usually lean in gold. When the pyrite is associated with a sooty-black powder, a mixture of chalcocite and manganese di-oxide, the gold content is usually from 2 to 6 oz. per ton. The veins are all very spotty. The upper portions of veins I, II, and III, though spotty, consisted entirely of ore. Some large pockets yielded large amounts of ore containing from 10 to 50 oz. gold per ton, the average being about 3 ounces.

Barite is often associated with high-grade streaks of ore. In a few rare cases, calaverite was found occurring adjacent to a small seam of barite; tellurides, however, are seldom found. Secondary sulphides can usually be recognized by the fact that they replace the wall-rock, whereas the primary sulphides are confined to cracks and openings. When the vein is in the trachyte no brecciation exists. The ore is found only along slips, fractures, and cracks associated with limonite, stained kaolin, or milky-quartz derived by silicification of the trachyte. Propylitization of the basalt is prominent where the basalt is in contact with the dike, the basalt assuming a dull greenish color due to the presence of chlorite, derived by alteration of the ferro-magnesian silicates; pyrite crystals are also developed. From this type, should alteration have advanced further, barren veins of imperfectly silicified material containing streaks of heavy barren secondary sulphide are developed. Some rich ore has been found associated with 'horses' of basalt that have suffered this type of alteration, notably on the 500 and 600-ft. levels of vein II, but where the contact is removed from the known ore-zones, this type of mineralization proves barren. These features of the vein structure point to deposition from hot ascending solutions within the fractured zone, where the pressure was suddenly reduced. Since the primary ore was deposited veins of barren secondary sulphide have been formed, with the development of some ore-shoots, probably by the migration and deposition of some of the gold present in the primary ore.

In many portions of the mine slickensides, striated gouges, and brecciation are present. By careful mapping and extensive development it was satisfactorily demonstrated that, though evidence of faulting was pronounced, it was of minor consequence. In several instances post-mineral faulting could be measured. Nowhere was it more than three feet, and usually less than one foot. Most of the fault-planes were curved surfaces, or consisted of small cracks that extended only a short distance and were then intercepted by another similar fracture. The heavy gouges were usually the result of kaolinization and sericitization of the feldspars by the action of descending sulphuric acid waters rather than of movement and friction. Slight movements are to be expected in a region of such recent volcanic activity. Occasionally slight tremors are still felt. Four years ago an earthquake developed a distinct fracture on the surface of the hillside above the mine. Consequently the striae and slickensides may be due to settling of the rocks to restore equilibrium in sympathy with an eruption from one of the relatively nearby active volcanoes. In vein VI several of the fractures have off-sets of as much as 20 ft. The distinction between faulting and fracturing during cooling, because of torsional stresses, must be recognized. These fractures were formed before the ore was deposited, but even so they make mining difficult.

It was noted that rich ore-shoots were found in the upper portions of the zones of secondary sulphide, and that free gold was often visible in such form as to be suggestive of secondary origin.

Samples of water were collected from various portions of the mine, and each was tested to determine the state of oxidation of the iron, the presence of chlorides, and free sulphuric acid. On a composite sample a complete qualitative analysis was made. The water contained mostly basic iron sulphate, also some ferric and some ferrous sulphate, free sulphuric acid and chlorides. The following elements were also determined: copper, manganese, zinc, nickel, calcium, barium, and magnesium. The surface waters contained a little ferrous sulphate and free sulphuric acid, after it had migrated a short depth, that is, the oxidation of the pyrite had commenced but had not been carried to a point sufficient to form ferric sulphate. A sample near the west end of the 300-ft. level contained both ferric and ferrous sulphates, a little copper and free sulphuric acid; samples from

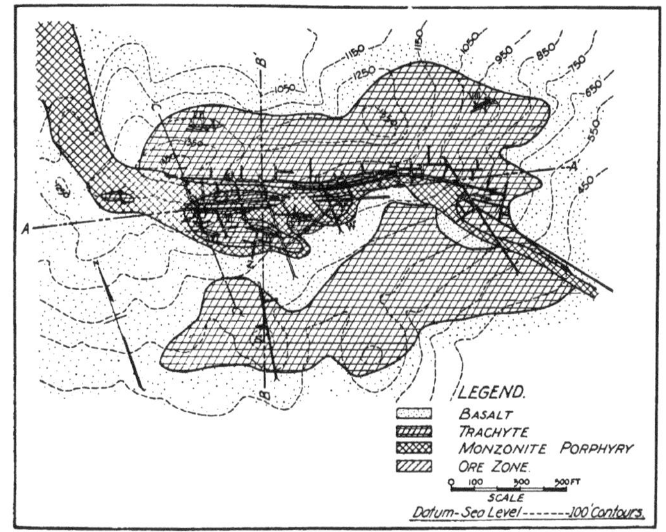

AERIAL GEOLOGY: SAN SEBASTIAN MINE

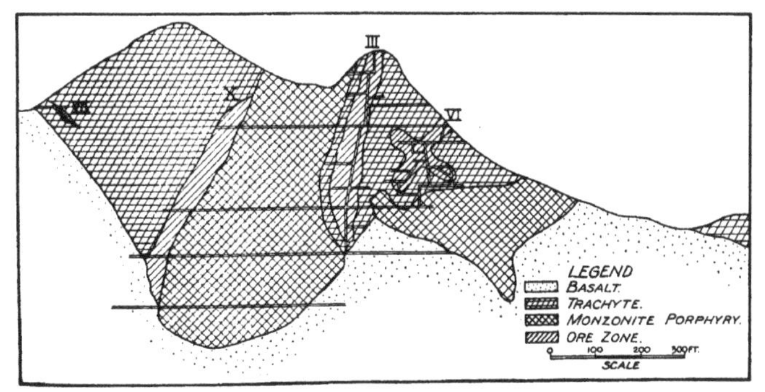

GEOLOGICAL SECTION ON CC': SAN SEBASTIAN MINE

the 400 and 500-ft. levels were extremely acid and contained an abundance of both states of iron and copper; samples from the 600-ft. level showed more ferrous sulphate, less copper and less free sulphuric acid; the same tendency was exhibited by the waters from the 700-ft. level. A spring issuing from the 800-ft. level that was not contaminated by the seepage from above showed only ferrous iron; while copper, ferric iron, and free H_2SO_4 were absent. Unfortunately the tests were not quantitative. The presence of the free H_2SO_4, manganese and chlorides are of paramount importance. These provide solvents for gold. Ferric sulphate is also a solvent, but only in the presence of the manganese and chlorides, while ferrous sulphate and pyrite, as well as other metallic sulphides, are effective precipitants of gold. Judging from the nature of the free gold and its occurrence in several rich pockets on this zone, it is evidently of secondary origin. Specimens polished and examined under a microscope showed small flakes or films on basalt, also associated with the blue-black quartz, or coating the chalcocite and sulphides. In a few instances small vugs containing probably a small quantity of high-grade ore showed comparatively coarse gold intergrown with chalcocite, pyrolusite and pyrite, mixed with minute crystals of quartz, barite, and pulverulent calcite. Gypsum was also frequently present. In the upper levels some copper sulphides and manganese oxides are found. The ores from pockets on the 600 and 700-ft. levels contain more copper than those on the upper levels. Coatings on the abandoned stope-walls of the upper levels show an abundance of chalcanthite, whereas melanterite prevails in the lower levels except in the region of the rich pockets where frequently brochantite occurs together with melanterite.

The ground-water level slopes with the pitch of the basalt-floor. In the western extremity of the deposit sulphides appear at an elevation of 1100 ft., 50 ft. above the 100-ft. level, but in the extreme eastern end the oxidized zone extends within a few feet of the 500-ft. level. At ground-water the ore shows marked enrichment, increasing from two to six times above the gold-content in the oxidized zone. Erosion has removed a considerable portion of the ore-zones, yet no placers have ever been found. The trachyte on this area is exceedingly resistant to weathering and erosion. It undergoes silicification of the groundmass and surficial coloration due to limonite and hematite in the outcrop. Immense talus slopes have developed, and portions have been scattered by the torrential rains. This has developed the abnormally steep slopes so suggestive of faulting. The outcrop was similarly eroded. Before the mine was developed thousands of tons of rich float was gathered and milled. The basalt weathers easily and forms a dark red soil; the monzonite in weathering crumbles to a loose granular soil. The feldspar kaolinizes and often assumes a color similar to the basalt.

The gold in the deposit is intimately associated with the pyrite. In the oxidized zone some of the gold is liberated and can be panned. This is not a safe guide, however, since much of the rich ore will not show free gold. Failure to appreciate this resulted in filling many of the old stopes with ore. Subsequent milling of a number of these fills produced considerable 1 to 2-oz. ore. Owing to the extremely erratic distribution of the gold assaying must be relied upon entirely. Silver is never determined because it is present in only very small amount, the gold-silver ratio being about 10 to 1 in weight. The tailing from the mill contains approximately 0.1 oz. gold regardless of the value of the heading assay. Gold telluride has occasionally been found, but even when the ore is roasted prior to cyaniding, the tailing still contains about 0.1 oz. gold. Excessively fine grinding and prolonged contact with cyanide has helped only slightly. Other minerals of minor importance occurring in the deposit are: tetrahedrite, chalcopyrite, bornite, chalcocite, calaverite, molybdenite, rickardite, marcasite, pyrrhotite, pyrolusite, barite, gypsum, chalcanthite, brochantite, melanterite, and amygdaloidal cavities filled with zeolites such as natrolite, thomsonite, quartz and epidote, with some calcite.

Slides of the basalt from different parts of the mine show olivine, basalt, porphyrite, and augitite. Differences are noted in texture, composition, and color of the basalt at various points, but from their structural relations they are evidently all members of the same flow. The texture of the fresh basalt is compact and fine ophitic texture of the feldspars and the augite is occasionally developed; it is sometimes veined and brecciated, and alteration has produced many secondary minerals. An important alteration of the basalt by contact-metamorphism is found throughout the property. For convenience it has been called locally the 'fine-grained dike.' It varies from a coarse rock resembling the monzonite to a highly silicified rock containing visible striated plagioclase crystals and some scattered flakes of biotite. In a few places it was possible to trace its gradation into fresh basalt. The trachyte consists essentially of orthoclase, biotite, and hornblende in a devitrified groundmass. The biotite and hornblende are not usually visible to the eye; flow-structure is apparent; the texture is uniformly fine-grained and smooth; limonite and hematite stains are prominent, and in the ore-zones the typical silicification of the groundmass is universal. The quartz-monzonite porphyry has a coarsely granitoid texture and hypidiomorphic crystals are well developed, yet there is ample evidence that it was formed close to the surface. In places where it was intruded into fissures through the basalt, glassy boundaries are found. Even in thin sills its texture is coarse-grained. It is locally called the 'granitoid dike.' Just north of the extreme western end of ore-zones I and III at the surface, a small area is found in which the texture is coarsely porphyritic in a stony groundmass not unlike the trachyte, containing sanadine, quartz and biotite in large crystals. The mineralogical composition of the rock is essentially orthoclase, plagioclase, quartz, biotite, hornblende, and magnetite. In the lower portions of the mine, quartz is less abundant and plagioclase predominates, the rock approaching diorite in composition. Such segregations are not uncommon. These features, together with the character of the fracturing, led me to believe that the dike was intruded into the trachyte before it solidified and that its lower portion occupies the throat of the volcano from which the trachyte was ejected. Formerly it was supposed that the dike was intruded along the southern edge of the trachyte plug, but numerous cross-cuts have shown that there is no trachyte plug. All cross-cuts on the lower levels on all sides of the dike have found basalt. Along the contact the dike is frequently kaolinized and sericitized. The trend of the underground circulation is toward the dike. The acidulated waters thus circulating have brought this alteration.

The genetic relationship of the ore deposit to the intrusion of the dike is evident. By referring to the accompanying maps it will be noted that ore-zones I and II; III, IV, and V occur on opposite sides of the different branches of the monzonite dike. It is reasonable to suspect the possibility of ore existing under conditions of similar structure, such as at X, on the north side of the dike; at Z and W, under the flat-lying dike; at Y, on the south side of the dike, and at S, (section on B-B') should the flat-lying dike at Z have extended over this hill. Area X has been explored by four cross-cuts through the dike, two from the 200-ft. level, one from the 100-ft., and the other from a point 200 ft. above the 100-ft. level. One cross-cut from the 200-ft. level entered basalt, but all of the others revealed a large well silicified and somewhat brecciated vein, mineralized only with barren sulphides. It is interesting to note that sulphides are found in the upper cross-cut, at a point 150 ft. higher than the level of sulphides on the south side of the dike in the ore-zones I and III. Since the outcrop of this vein is indistinct and has yielded no payable ore, it is apparent that this is a barren vein and probably was formed by surface waters which derived their silica and iron from the trachyte. The water circulating along this side of the dike is drinkable, in this respect differing from that coming from the known ore-zones. Area Y was explored by a long drift along the south side of the dike on the 300-ft. level, and by two cross-cuts to the south through the dike from the 400 and 500-ft. levels and by two drifts around the western end of the dike. In each case only a propylitized basalt contact-zone was found. Areas W and Z were investigated with like results. A little ore was obtained in these regions but never enough to warrant consideration. In area S a small amount of ore was taken from the summit of the hill, and a long tunnel was then driven along the basalt-trachyte contact, but only heavy barren secondary sulphides were found. The bulk of the ore may have been eroded.

This unique deposit has produced about $15,000,000. There are eight main levels with an elaborate system of intermediate workings. Only one shaft is used in order to hoist ore from the 700 and 800 to the 600-ft. level, at which elevation the mill is situated. There is an 1800-ft. drainage tunnel 30 ft. below the 800-ft. level. Before this was completed the pumping costs were enormous because of the acid water. Because of its erratic character the deposit is peculiarly well adapted to being worked by the cheap, though inefficient, native labor. Seams are followed with careful and frequent sampling. The stringers often widen to an ore-shoot. If the ore-shoot proves large a raise is driven with a machine-drill, but if, as is more often the case, small, the ore is then stoped by hand and removed by 'tenateros,' that is, boys who carry the ore in raw-hide bags, holding about 40 lb. In the broader veins the ore is broken by machine-stopers. In the vicinity of old fills the temperature is excruciatingly hot, hence it is necessary to make numerous underground connections to provide for ventilation. An attempt to estimate the ore-reserves is impossible. From the experience of past years a unique method has been developed. Each month the mine foreman and engineer hold a consultation, and by referring to the stope records of the past month and from their knowledge of conditions, they forecast the amount of ore in each stope, aiming to keep three months ore-reserve available. The object of making the estimate is to obtain a figure for the monthly reports, showing the probabilities. About 2000 ft. of development work is performed each month. On the basis of 3500

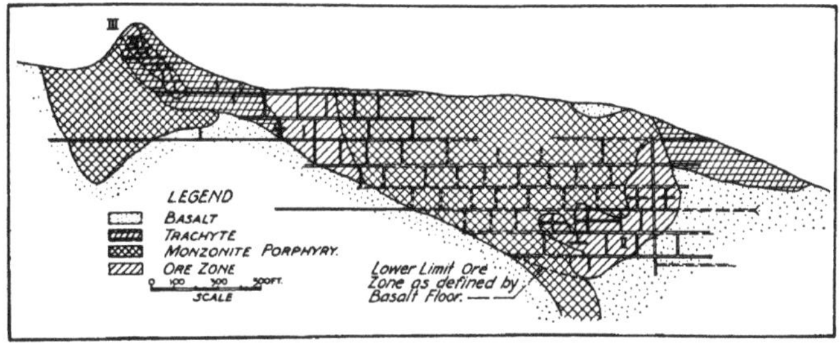

GEOLOGICAL SECTION ON AA': SAN SEBASTIAN MINE

tons stoped and milled, and an equal amount developed, it is easy to see that the deposit is exceedingly erratic. Development is pushed at about 100 working faces. The superintendent, Gustave Scogland, well expressed it: "We get a hat-full of ore here and another there, and by beating the kids on the back we manage to keep the bins full." Although at this writing the end would seem to be in sight, the life of the mine will undoubtedly exceed all estimates. In the process of robbing the known pillars of ore new streaks may be disclosed. Though there are approximately 30 miles of underground workings, no one can say with certainty when the mine will be worked out.

Mineral Output of Idaho

The value of the output in 1916 was $48,767,783. The shipments in the first six months of 1917 from the Coeur d'Alene region, which give an idea of conditions in Idaho, were at the rate of the previous year, which was nearly 44,000 tons of crude ore and concentrate per month. Greatly increased output from the Hercules, Hecla, Morning, and Bunker Hill mines have balanced decreases in the Tamarack & Custer, Stewart, and Last Chance. At the same time the Bunker Hill custom-smelter at Kellogg is nearing completion, and the first ore has been shipped to this new lead plant. The plant at Northport, Washington, continues to treat the lead-product of the Hercules mine at Burke, in addition to custom material. In 1916 the value of the gold produced, coming largely from Boise, Elmore, Idaho, and Lemhi counties, amounted to $1,115,810. There was a decrease in the output of gold and from placers, and an increase from deep mines. Operations at Atlanta, in Elmore county, and at Burgdorf, in Idaho county, are adding to the gold production. The output of silver last year was 12,300,873 oz., valued at $8,093,974. Shoshone county supplied the greater part, and mines at Gilmore, in Lemhi county, and Mackay, in Custer county, constituted most of the remainder. If the present rate of shipment of ore continues, this quantity may be exceeded in 1917, and the value will be greater, as the average price of silver increased from 65.8c. in 1916 to about 75c. in the first part of 1917. The Hercules mine at Burke is by far the largest producer of silver in the State, the gross value of the silver and lead output in 1916 having been $7,278,258. The Hecla, Hercules, Morning, Bunker Hill, Caledonia, and Greenhill-Cleveland produced nearly 75% of the total silver in 1916, and the first three mines are making heavier shipments this year. The copper output in 1916 was 8,478,281 lb., valued at $2,085,657. Over 65% came from Custer county, largely from the Empire property. In Shoshone county, the Caledonia, Richmond, Horst Powell, and National produced copper ore and concentrate. In 1917 the Richmond is becoming a large shipper. In Custer county, activity in the Alder Creek district is notable. The Copper Basin has become a good producer and the Empire is being opened for greater production from the lower tunnel. The output of lead in 1916 was 375,081,781 lb., valued at $25,880,643. This rate has been sustained during the first half of 1917, and, with a better price and a new lead smelter, the shipments may be increased toward the end of the year. The Hercules has materially augmented shipments. The monthly average of crude ore and concentrate in 1916 was over 7200 tons. In Lemhi county, which produced 11,208,985 lb. of lead in 1916, the shipments in the early part of 1917 were less, but at the Idaho Continental, in Boundary county, they increased. The Homestake-Copper Queen property, near Mackay, has become an important lead producer, and the Blue Bird mine, at Clayton, is to have a concentration mill. The shipments of zinc and lead-zinc ore and concentrate in 1916 cotained 86,505,219 lb. of recoverable spelter. An average of about 10,000 tons per month of crude ore and concentrate was shipped. Most of the zinc is from the Consolidated Interstate Callahan mine at Sunset. In the first quarter of 1917, 13,839 tons of zinc ore and concentrate, containing 12,614,475 lb. of zinc, was shipped. This is slightly less than the quarterly average for 1916. The shipments from the Greenhill-Cleveland and Frisco were less, but the Morning, Success, Highland Surprise, Douglas, Rex, and Hercules made the usual contributions. In 1917 the Hecla and Ray Jefferson joined the ranks of the zinc shippers.

THE U. S. Geological Survey is planning to make the desert regions of the western part of the United States more accessible by locating their widely separated watering-places and erecting hundreds of sign-posts to give directions and distances to the watering-places. The project involves also the work of making accurate maps showing the locations of the watering-places, of preparing guides describing them and giving the distances between them, of selecting well-sites, and of developing watering-places (so far as money available will permit) in localities where water is most needed and where the geologic investigations indicate that underground supplies can be obtained. It is expected that this work will help to expedite the discovery and development of the rich mineral deposits in parts of these regions. It will also be valuable in other respects.

FLUORSPAR is a mineral that has not attracted the attention of the Western miner and for obvious reasons, nevertheless, deposits may be discovered. Its importance can be appreciated by the fact that in 1916, 133,651 short tons of 'gravel spar' was produced. The average prices per ton received at the mine were $5.34 for gravel, $7.94 for lump, and $12.38 for ground. The increased demand for fluorspar has come largely from the manufacturers of open-hearth steel, who use the mineral as a flux, but the demand for it in other metallurgic operations and for the manufacture of hydrofluoric acid has been very active. One of the newer uses for fluorspar is as a re-agent in the recovery of potash from feldspar and from portland cement clinker.

Soluble Frothing-Agents

By C. L. PERKINS

There is a great deal of misunderstanding as to the solubility of many of the oils used in flotation. This may be due in part to confusion of the terms 'solubility' and 'miscibility.' If an oil does not apparently mix with water, this is not necessarily an indication of absolute insolubility. It might be well to emphasize the fact that there are widely varying degrees of solubility. Even though a substance dissolves in water only in the proportion of 1 part in 10,000 parts, it is soluble from the point of view of the chemist, and in all the chemist's work with that substance its solubility is taken into account. The flotation engineer and experimenter must learn to look upon solubility in the same way, for it is precisely these slight degrees of solubility which often determine the success or failure of flotation.

An interesting case in point is found in U. S. Patent No. 962,678. Among the soluble frothing-agents mentioned in the claims of this patent are phenol and its homologues. The statement is made that all previous processes of flotation have employed oils which were immiscible in water. Reference is also made to Patent No. 777,274, in which phenol, cresol (a homologue of phenol), and others, are said to be liberated in an insoluble condition in the pulp by the addition of their alkaline compounds to an acid pulp. It is, of course, absurd to consider phenol and cresol as being soluble in one case and insoluble in the other. As a matter of fact, phenol is soluble in the proportion of about 6 grammes in 100 gm. of cold water and much more so in hot water. Cresol is soluble to a much smaller degree, but is completely soluble within the limits practically used in flotation. The figures for the solubility of cresol are variously given as from 0.3 gm. in 100 gm. of water to 3 gm. in 100 gm. Assuming the smaller value, and a 3:1 flotation-pulp, this would mean that cresol might be used in the proportion of 18 lb. per ton of ore and still be entirely in solution.

A recent writer, in commenting upon the decision of the Court in the Miami appeal case, states that the purer grade of pine-oil and creosote are insoluble. Pine-oil is, of course, of varying composition, dependent upon the details of distillation. The refined product, however, still contains soluble substances, being a hydrocarbon complex and hence containing compounds of different properties. The term 'creosote' is loosely used. The chemical composition of creosote obtained from wood is widely different from that of the coal-tar product. The U. S. Pharmacopoeia classes wood-creosote as "slightly soluble." Coal-tar creosote also contains appreciable amounts of soluble material.

According to Mueller,* cresol, or cresylic acid, was formerly the favorite oil for the flotation of copper ores in this country. Its use became less general only as the material became more expensive, and it was found by experiment that cheaper oils and oil mixtures could be substituted for it. The oils replacing it were mainly wood-creosote, coal-tar creosote, and coal-tar itself. The soluble portions of these oils, which are the constituents responsible for their frothing power, are largely cresol and other phenols. In view of the fact that the froths obtained throughout all flotation practice have been largely dependent for their production upon soluble material, and that at least one of the oils extensively used in the beginning was completely soluble in the proportions employed, it is idle to speak of soluble frothing-agents as a distinct class and it is unjustifiable to claim their use as a broad new discovery.

The action of various flotation-oils in producing froth has been discussed by several writers in this journal and elsewhere. Coghill,[1] Hildebrand,[2] and Corliss and Perkins[3] have outlined the considerations governing the formation of a stable froth. Nevertheless, it is apparent, from statements appearing from time to time in the literature of the subject, that the principles involved are still not fully understood by many of those interested in the art of flotation. As the matter is one of considerable importance, especially in connection with the patent situation, it is perhaps worth while to re-state the necessary conditions for the formation of a froth, in order to emphasize the fact that all frothing-agents are to some extent soluble.

The first requisite for a froth is variable surface-tension on the part of the liquid films surrounding the air. In other words, the material of the film must be such that the tension of a freshly-formed surface will differ from that of an older surface. This principle was recognized by Rayleigh,[4] who demonstrated the fact that adsorption requires time for its completion. In working with solutions of saponine he found differences between the tension of freshly formed surfaces and that of slightly older surfaces. These differences were of considerable magnitude and easily susceptible of measurement. The bearing which this fact has upon the stability of a froth is explained as follows:

If a bubble-film starts to thin out, due to the liquid of which it is composed draining away by its own weight, or if it is subjected at any point to a strain or a blow

*Eng. & Min. Jour., Vol. 102, p. 31.
[1]M. & S. P., July 29, 1916.
[2]M. & S. P., Sept. 2, 1916.
[3]M. & S. P., June 9, 1917.
[4]Philosophic Magazine. Vol. XLVIII, p. 321.

which would tend to rupture it, a new surface is automatically produced at that point. The new surface always has a higher surface-tension than the old one, and the result is that the tendency to rupture is resisted. The higher tension of the new surface is immediately equalized by diffusion from within and from adjacent portions of the film. This process is continually going on in a stable film. It is a well known fact that pure liquids will not froth. If low surface-tension were enough in itself to cause a stable froth, many of the pure oils as well as other organic liquids would froth. In order to obtain a froth, it is necessary to add some substance that will produce variable surface-tension. Theoretically, a froth should be caused by materials that raise the surface-tension of water and, as a matter of fact, solutions of many inorganic substances do froth slightly. The reason that they are not practicable for use in flotation is the fact that the differences in surface-tension produced by them are only slight. On the other hand, many of the substances that lower the surface-tension of water do so to a marked extent.

Only materials in solution can impart variable surface-tension to water, because only such materials can diffuse through water with the necessary rapidity. Insoluble oils cannot do so. If they are present in the water in sufficient quantity, they may form films of molecular thickness between the air of the bubble and the surrounding water. Such films are, however, lacking in stability. Colloidal material or other very small solid particles may increase the stability of a froth in several ways. They may serve as points of support for the bubble-films. If the particles are sufficiently closely packed, the friction of one against another may increase the viscosity of the liquid to such an extent that the tendency toward thinning by drainage is greatly decreased. Finally, they may prevent coalescence of the bubbles in the same way that similar materials stabilize an emulsion, that is, by forming an envelope or sac about them, preventing actual contact of one bubble with another.

Depreciation Tables

The accompanying tables, taken from the final report of the Committee on Valuation of the American Society of Civil Engineers give the amount to be allowed for depreciation for different lives and interest rates under the so-called compound-interest or equal-annual-payment method. Thus, for an article valued at $100, having a 20-year life, no salvage value, and interest at 5%, the allowance for depreciation for the first year is $3.02; the allowance for interest is $5, making the total for interest and depreciation, $8.02; the second year the depreciation is $3.18 and the interest on the depreciated value of $96.98 is $4.84; making a total of $8.02 as before. The tables can also be used for articles having a salvage-value at the end of their useful life. Thus assume the original value to be $100, the salvage value $25, and the useful life 10 years. The amount of value which depreciates to zero is $75. By using the ratio in the table and applying it to this amount the depreciation for any period can be determined.

Age, in years.	Interest Rate. 4%		Interest Rate. 5%		Interest Rate. 6%		Interest Rate. 7%	
	Value.	Dep.	Value.	Dep.	Value.	Dep.	Value.	Dep.
0	100.0000		100.0000		100.0000		100.0000	
1	81.5873	18.4627	81.9025	18.0975	82.2604	17.7396	82.6109	17.3891
2	62.3361	19.2012	62.9002	19.0023	63.4563	18.8041	64.0046	18.6063
3	42.3668	19.9693	42.9477	19.9525	43.5241	19.9322	44.0959	19.9087
4	21.5988	20.7680	21.9976	20.9501	22.3959	21.1282	22.7935	21.3024
5	0.0000	21.5988	0.0000	21.9976	0.0000	22.3959	0.0000	22.7935
		100.0000		100.0000		100.0000		100.0000
10-Year Life.								
0	100.0000		100.0000		100.0000		100.0000	
1	91.6709	8.3291	92.0495	7.9505	92.4132	7.5868	92.7623	7.2377
2	83.0086	8.6623	83.7015	8.3480	84.3712	8.0420	85.0179	7.7444
3	73.9998	9.0088	74.9361	8.7654	75.8467	8.5245	76.7314	8.2865
4	64.6308	9.3690	65.7324	9.2037	66.8107	9.0360	67.8648	8.8666
5	54.8869	9.7439	56.0686	9.6638	57.2325	9.5782	58.3776	9.4872
6	44.7533	10.1336	45.9216	10.1470	47.0797	10.1528	48.2263	10.1513
7	34.2144	10.5389	35.2672	10.6544	36.3177	10.7620	37.3644	10.8619
8	23.2538	10.9606	24.0801	11.1871	24.9099	11.4078	25.7421	11.6223
9	11.8549	11.3989	12.3337	11.7464	12.8177	12.0922	13.3062	12.4358
10	0.0000	11.8549	0.0000	12.3337	0.0000	12.8177	0.0000	13.3062
		100.0000		100.0000		100.0000		100.0000
15-Year Life.								
0	100.0000		100.0000		100.0000		100.0000	
1	95.0059	4.9941	95.3658	4.6342	95.7037	4.2963	96.0205	3.9795
2	89.8120	5.1939	90.4998	4.8660	91.1497	4.5540	91.7625	4.2580
3	84.4104	5.4016	85.3906	5.1092	86.3224	4.8273	87.2064	4.5561
4	78.7927	5.6177	80.0260	5.3646	81.2054	5.1170	82.3314	4.8750
5	72.9503	5.8424	74.3930	5.6330	75.7815	5.4239	77.1152	5.2162
6	66.8743	6.0760	68.4784	5.9146	70.0322	5.7493	71.5338	5.5814
7	60.5551	6.3192	62.2681	6.2103	63.9378	6.0944	65.5616	5.9722
8	53.9832	6.5719	55.7472	6.5209	57.4777	6.4601	59.1715	6.3901
9	47.1484	6.8348	48.9004	6.8468	50.6301	6.8476	52.3340	6.8375
10	40.0403	7.1081	41.7112	7.1892	43.3717	7.2584	45.0190	7.3150
11	32.6478	7.3925	34.1625	7.5487	35.6776	7.6941	37.1896	7.8294
12	24.9596	7.6882	26.2364	7.9261	27.5221	8.1555	28.8136	8.3762
13	16.9639	7.9957	17.9140	8.3224	18.8771	8.6450	19.8511	8.9625
14	8.6483	8.3156	9.1755	8.7385	9.7135	9.1636	10.2612	9.5899
15	0.0000	8.6483	0.0000	9.1755	0.0000	9.7135	0.0000	10.2612
		100.0000		100.0000		100.0000		100.0000
20-Year Life.								
0	100.0000		100.0000		100.0000		100.0000	
1	96.6418	3.3582	96.9757	3.0243	97.2815	2.7185	97.5607	2.4393
2	93.1492	3.4925	93.8002	3.1755	94.4000	2.8815	94.9505	2.6100
3	89.5171	3.6322	90.4660	3.3342	91.3455	3.0545	92.1579	2.7928
4	85.7396	3.7775	86.9650	3.5010	88.1078	3.2377	89.1697	2.9882
5	81.8110	3.9286	83.2890	3.6760	84.6758	3.4320	85.9728	3.1974
6	77.7253	4.0857	79.4292	3.8598	81.0379	3.6379	82.5510	3.4218
7	73.4761	4.2492	75.3764	4.0528	77.1818	3.8561	78.8898	3.6607
8	69.0570	4.4191	71.1210	4.2554	73.0943	4.0875	74.9734	3.9159
9	64.4611	4.5959	66.6528	4.4682	68.7614	4.3328	70.7822	4.1912
10	59.6814	4.7797	61.9612	4.6916	64.1686	4.5928	66.2977	4.4845
11	54.7105	4.9709	57.0350	4.9262	59.3003	4.8684	61.4992	4.7985
12	49.5407	5.1698	51.8625	5.1725	54.1398	5.1604	56.3649	5.1343
13	44.1641	5.3766	46.4314	5.4311	48.6697	5.4701	50.8711	5.4938
14	38.5725	5.5916	40.7287	5.7027	42.8715	5.7982	44.9926	5.8785
15	32.7573	5.8152	34.7400	5.9878	36.7253	6.1462	38.7030	6.2896
16	26.7094	6.0479	28.4587	6.2872	30.2104	6.5149	31.9729	6.7301
17	20.4196	6.2898	21.8521	6.6016	23.3045	6.9059	24.7717	7.2012
18	13.8782	6.5414	14.9304	6.9317	15.9843	7.3202	17.0655	7.7062
19	7.0751	6.8031	7.6421	7.2783	8.2250	7.7593	8.8219	8.2436
20	0.0000	7.0741	0.0000	7.6421	0.0000	8.2250	0.0000	8.8219
		100.0000		100.0000		100.0000		100.0000

Seale-Shellshear Flotation Apparatus

In the discussion of flotation as practised in Australia frequent mention is made of the Seale-Shellshear apparatus, but no statement of the details has been published in the technical journals. Accordingly we reproduce from the original patent papers the inventors' description, with drawings from the patent specification. The British patent is No. 10,666, and the date claimed, in accordance with the patent office regulations, is July 30, 1914. The inventors are Harry Vernon Seale and Wilton Shellshear, both of Junction North Mine, Broken Hill. In their preamble they affirm that the invention is specially applicable to "froth-flotation in which ores are subjected to aeration in an aqueous separating medium in the presence of a frothing-agent for the formation of a froth or scum of metalliferous particles . . . as described in prior patents No. 7803 (1905), 2359 (1909), and 21,857 (1910). The specification follows:

"Hitherto the aeration or emulsification of the pulp has been effected by vigorous agitation by means of rotating stirrers, impellers or beaters or by means of centrifugal pumps the driving of which involves the consumption of a considerable amount of power. It has also been proposed in the concentration of ores to pass the ore mixed with water and a frothing-agent through a U-tube so that at the inlet end thereof the pulp sucks in air, which as the pulp descends to the bend of the U-tube becomes dissolved and on rising in the other limb of the U-tube again comes out of solution and attaches itself to the metalliferous particles in the ore. From the U-tube the mixture is discharged into a flotation vessel where the froth floats and is separated. It has also been proposed to mix the powdered ore with water and pass it to a hopper arranged with its lower end beneath the surface of liquid in a spitzkasten, a steam-injector being arranged in the said hopper and having means for the injection of oil and air to the steam-jet whereby the pulp is forced into the spitzkasten and is thoroughly agitated with the air drawn in by the steam, and whereby a froth is formed which is floated and separated.

"The object of the present invention is to provide improved means for the aeration or emulsification of the pulp for the formation of the froth or scum in a thoroughly efficient manner with a great saving in the consumption of power. According to the present invention in the concentration of ores by flotation separation the ore-pulp is fed or delivered into the open top of a pipe or tube projecting beneath the surface of the liquid in the separating-box or vessel with such a flow of material within the said pipe or tube as to ensure the entrapping of the air necessary for the aeration of the pulp by the flow of the material itself.

"In apparatus for carrying out this process for the concentration of ores by flotation separation, there is combined with a separating-box or vessel having a residues-outflow at the bottom, a feed-box open at the top, having an inflow-pipe at the side thereof and a discharge-pipe at the bottom delivering into the said separating-box or vessel below the surface of the liquid,

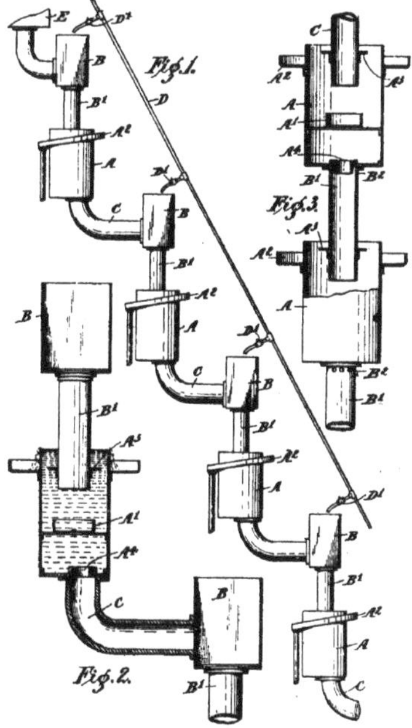

FACSIMILE OF DRAWING IN PATENT

with such a flow of material within the said pipe or tube as to ensure the entrapping of the air necessary for the aeration of the pulp by the flow of the material itself.

"A number of separating-boxes and a number of feed-boxes may be arranged in series one above the other, each separating box having a residue outflow at the bottom leading to a feed-box, and each feed-box having a discharge-pipe at the bottom delivering to the next succeeding separating-box or vessel below the surface of the liquid therein."

Fig. 1 shows the arrangement of the plant with a number of separating-vessels in series; Fig. 2 is a vertical section of a single separating-vessel with a feeding-device; and Fig. 3 represents a modification of the arrangement for effecting the purposes of the invention.

"Each separating-unit preferably comprises a separating-box A and a feed-box B. These feed-boxes B are open at the top and are also of any convenient shape, but we have found that square shape in cross-section well answers the purpose. Each feed-box B is furnished with a depending feed-pipe B^1, which delivers into the separating-box A below the level of the pulp or separating medium therein. The feed-pipe B^1 delivers onto a baffle A^1 arranged within the separating-box A. Each separating-box A is provided with launder A^2 for the float concentrates and a circular baffle A^3 surrounding the feed pipe B^1 for the purpose of maintaining a state of quiescence on the surface of the pulp in the separating-vessel A. The vessel A is also furnished with an outflow A^4 for the unfloated residues which is provided with a removable annular plug or ring whereby the size of the same may be regulated. Leading from the outflow A^4 is a pipe or duct C, which discharges into the next feed-box B of the series. A main water pipe D is provided, having taps or faucets D^1 adapted to discharge into the feed-boxes B whereby the level therein may be adjusted and compensation made for losses occasioned by the withdrawal of concentrates. A common feed-box or vessel E may be provided at the head of the series for the purpose of equalizing the feed so as to ensure a regular flow to the machine.

"In operation the pulp is fed from the equalizing-feed box E into the first open feed-box B, whence it flows by the feed-pipe B^1 into the separating-box or vessel A, being delivered under the surface of the liquid striking the baffle A^1. The float concentrates rise and overflow into the launder A^2 while the unfloated residues flow out through the hollow plug A^4 and the pipe or duct C to the next feed-box B in the series, and so on. The feed-boxes B are open at the top and the flow of material thereinto by the pipe C causes a vigorous agitation. The outflow through the pipes or tubes B^1 from the feed-boxes B has a slight excess capacity over the flow of material therethrough so that a vortex or swirling action is set up which entraps air within the said pipes or tubes B^1 by the flow of material itself thereby ensuring the requisite aeration necessary for the thorough emulsification of the pulp to effect flotation. The head or height of the feed-box B above the separating-box A is such as to regulate the flow of material to ensure the aforesaid entrapping of air. The pulp with its entrapped air discharges immediately on the baffle-plate A^1 by which thorough emulsification is ensured. The outflow through the hollow plugs A^4 is controlled and regulated by adjusting the size thereof so that by having a series of hollow plugs A^4 of definite constant size the whole of the plant is self-regulating providing the equalizing-feed box E at the top is kept at a constant level or other means adopted for maintaining a constant feed. Water or aqueous separating medium is added in the various feed-boxes B by the taps or faucets D^1 to compensate for that removed with the float-concentrates and maintain a constant level within the separating-boxes A. This may be effected by varying the size of the outflow-orifices in the hollow plugs A^4.

"The flotation circuit-liquor which may be hot, is prepared from any water that is readily available, such as underground water, mill circuit-water, or fresh supply-water. A frothing-agent, such for example as eucalyptus oil, is added immediately after the liquor is removed from the residues, when circuit-water is employed, and before it enters the suction of the return-pump which delivers the liquor to the storage-tank. In this way the oil is well agitated with the water by the pump. It is to be understood, however, that the frothing-agent may be added at any stage of the process, for example it may be added to any of the mixing-boxes B. Similarly any acid required in the flotation circuit may be added at any stage of the process as desired. If a further quantity of frothing-agent or acid is required at an intermediate stage of the process, the material from the flotation vessels may be run into a tank where the agents may be added, from which tank it is then passed to a second series of mixing-boxes and flotation-vessels."

BARYTES was produced to the amount of 221,952 tons in 1916, the average price being $4.56 per ton. The leading producing State was Georgia, with an output of 401,295 tons, followed closely by Missouri, with 365,111 tons. The domestic production of barium chemicals in 1916 increased 90% over that in 1915, which was the first year in which a considerable quantity of barium chemicals was made in the United States. The production of ground barytes increased 27% and that of lithopone approximately 10%. White bleached and floated barytes is used in ready-mixed paints and in the rubber and paper industries. Off-color ground barytes is used in preparing colored mixed paints and in several chemical industries. Lithopone is used principally in making rubber goods and as a pigment in 'flat' wall-paints. It is also used in making linoleum, calcimine, and paper. The barium chemicals have many uses, the largest consumption being that of barium binoxide in the preparation of hydrogen peroxide.

SCIENTIFIC RESEARCH on a large scale is being undertaken in Japan, involving the establishment of a great laboratory in Tokio, which has been voted a credit of 2,000,000 yen by the Parliament, in addition to a sum of 2,900,000 yen subscribed by private individuals, and another sum of 1,000,000 yen donated by the Emperor. These funds are to be available in ten annual installments, thus making 590,000 yen or approximately $294,000 per annum.

Precipitating Gold From Coppery Cyanide Solution

By ROBERT LINDSAY

*Cyanide solutions containing copper deposit this metal readily on clean zinc; the copper forms a closely adherent metallic film or plating, which after a time coats the zinc so completely that the precipitation of gold almost ceases, and here the lead-zinc couple is used with advantage for cupriferous solutions. Lead-coated zinc does not precipitate copper so readily from working solutions as clean zinc, and the object aimed at is not to precipitate all the copper, but to keep the amount of copper in solution more or less constant. If the lead coating be too light, too much copper will be thrown down, and more zinc will have to be added to the extractors before the clean-up, resulting in an unnecessarily high consumption of zinc. The practice in dipping is to regulate the quantity of zinc dipped with the strength of the lead-solution so that each 100 lb. of zinc will carry 5 lb. metallic lead. That suits the copper content of the solution here and precipitates about one-half of the copper entering the extractors, while the gold content of the effluent is seldom above 0.03 dwt. even in the cold season. Increasing the cyanide strength as a remedy for bad precipitation is not recommended. It gives temporary relief, no doubt, by keeping copper in solution, but the gradual accumulation of copper in solution detrimentally affects the recovery, and the excess of copper has to be got rid of eventually, either by precipitation or by running some of the solution to waste.

If the soluble copper in the ore should become abnormally high, it will be found advantageous to have uncoated zinc in the first compartment to precipitate a proportionally large amount of copper, leaving the remaining compartments of lead-coated zinc to precipitate the gold and a portion of the remaining copper. The first compartment of coppery zinc is taken out every 4 or 5 days and immersed in dilute acid for a few minutes, and returned to the box along with some new zinc to fill the compartment. There is an increase in zinc consumption, but the gold precipitation is good. The precipitation of copper on zinc is, of course, accompanied by a regeneration of cyanide, thus:

$$Cu_2(CN)_2.4NaCN + Zn = 2Cu + Na_2Zn(CN)_4 + 2NaCN,$$

or

$$Cu_2(CN)_2.6NaCN + Zn = 2Cu + Na_2Zn(CN)_4 + 4NaCN,$$

consequently the cyanide strength leaving the extractor box is always higher than that entering it, as shown in the accompanying table.

	Sand boxes			Slime boxes		
	KCN, %	CaO, %	Copper, lb. per ton	KCN, %	CaO, %	Copper, lb. per ton
Entering	0.032	0.007	0.12	0.022	0.008	0.09
After 1st compt.	0.039	0.008	0.11	0.025	0.011	0.07
After 2nd compt.	0.044	0.009	0.08	0.028	0.010	0.05
After 3rd compt.	0.042	0.010	...	0.029	0.010	...
After 4th compt.	0.040	0.010	...	0.028	0.011	...
After 5th compt.	0.039	0.011	0.07	0.028	0.011	0.04

The tendency of the copper to remain in the stronger solution of the sand boxes will be noted, the precipitated copper there being 41.7% as against 44.4% in the case of the weaker cyanide solution of the slime boxes with a lower copper content on entering. One cubic foot of zinc is allowed per ton of solution per 24 hours. The clean-up takes place twice each month. The top compartment only of each box is taken out, and is treated as usual with sulphuric acid. Bisulphate is not used on account of difficulty of transport and inconvenience in handling, the mine being about ten miles from the nearest railway station at Graskop. When all the zinc has been dissolved, the copper is attacked by nitric acid, or with sulphuric and nitric acids together, which is cheaper than nitric acid alone, thus: With nitric acid only

$$3Cu + 8HNO_3 = 3Cu(NO_3)_2 + 4H_2O + 2NO$$

189 + 504 @ 1 shilling per lb.

= 504s. = £25 4s. for 189 lb. of copper.

With sulphuric and nitric acids

$$3Cu + 3H_2SO_4 + 2HNO_3 = 3CuSO_4 + 4H_2O + 2NO$$

189 + 294 + 126 = 62s. + 126s. = £9 8s. for 189 lb. of copper.

Over 700 lb. of copper is thus dissolved at every clean-up. The acids are added in the proportion of 1 part sulphuric to 0.44 part nitric acid, and the temperature should not be under 60° C. if copper is to be kept out of the bullion. The precipitate, kept in suspension by mechanical stirrers, is tested for the presence of copper after each addition of acid, so as to ensure no more acid being added than is absolutely necessary. When the copper is all in solution the vat is filled with water and a few pounds of size added to assist settlement, which usually takes place in five or six hours. The solution containing the zinc and copper is passed through a sand-filter to catch any fine gold-slime in suspension. The surface of the vat is skimmed every other month and the sand used as flux. The solution now flows through a stoneware pipe to two copper precipitating vats, which are filled with scrap-iron. Any old iron is put into these, fire-bars, fire-doors, worn-out plates, or battery screening. Every four months the vats are completely cleaned and the precipitate screened to keep out coarse iron. It is air-dried, thoroughly mixed, sampled, assayed, and sacked in sugar pockets at a cost of 25s. per ton. The following are the details of a parcel of this copper precipitate sent to England last July, and on which a net profit of £600 was relized:

*Abstract: 'Notes on Treatment of Pilgrim's Rest Ore'; Jour. Chem., Met. & Min. Soc. of South Africa, April 1917.

Dry weight, tons	Assay per ton Au, oz.	Ag, oz.	Cu, %	Total content Au, oz.	Ag, oz.	Cu, tons
8.171	10.99	123.79	40.95	89.802	1011.550	3.346

There is no gold or silver in the copper vat effluent, and only a trace of copper toward the end of the run. It is found that coarse iron gives a much purer product, and by using that exclusively it is hoped to bring the precipitate up to over 60% grade. The copper recovered is equal to 0.13 lb. per ton milled, and agrees fairly closely with the difference between the copper content of original and residue, and is rather less than one-half of the copper content of the ore for the period. The washed gold-slime is filter-pressed and dried. Calcining is not found necessary after using nitric acid. The gold-slime assays about 30% gold and is fluxed with 25% borax, 12½% sand, and 12½% manganese di-oxide. The bullion averages 850 fine. The average total recovery from the plant is 94%, and the cost 5s. 7d.

Portable Electric Hoist

The accompanying photograph shows a portable electric hoist that is in almost constant use by the San Antonio Water Co., of Ontario, California. The hoist is composed of the following: a two-horse farm-wagon, upon which a 3 by 10-in. Oregon pine wagon-frame has

PORTABLE HOIST AND DERRICK

been securely bolted. The housing over the rear wheels contains an induction-motor with accessories and sufficient additional space for a belt and necessary tools. A standard two-ton hoist, belted to the induction-motor, is placed immediately back of the front axle. As shown in the photograph, the hoist is in place, its motor wired to the power-line and its hoist-cable passing over a device, ready to be used in removing the pump when the belt is connected. By bolting an 'A-frame' to the two uprights, near the front axle, and placing a detachable boom, the hoist may be used as a portable derrick with a swinging radius of 10 or 12 ft. This hoist has been used with economy for the following purposes; hoisting tunnel-waste and lowering pipe and concrete for lining tunnels; hoisting large rock from pipe-line trenches and lowering pipe to place; moving earth by the drag-line scraper method; sand-pumping of oil wells; setting pumps, pipe, and machinery at pumping-plants; loading and unloading cars; setting and replacing poles and towers; and performing many minor operations.

Methods of Construction of Concrete and Timber Flume for Cove Power Project

The Cove flume, which is part of the Cove power project of the Utah Power & Light Co., is described in the *Journal* of the Utah Society of Engineers by L. M. Pharis. The project is on Bear river in southern Idaho.

The flume is unusual on account of its size and the details of construction. It has a clear width inside of 20 ft. The length is 5700 ft. and there is a fall of 2 ft. in this distance. The top of the flume is kept level so that the depth varies from 11½ ft. at the upper end to 13½ ft. at the lower end. This allows the water in the flume to assume the static level without spilling. In addition to this there is a spillway 300 ft. long at the lower end to take care of surges when the turbine-gates are closed suddenly, and a wasteway in the canal-section just below the headgate at the upper end to prevent spilling over the flume in case there is a sudden rise in the pond-level.

Instead of constructing the frames or bents of wood, which would have had a comparatively short life, besides presenting some difficult problems in construction, the sills, posts, and tie-beams are constructed of reinforced concrete. The tie-beams are 10 by 14 in.; the posts are 22 in. wide at the top and 27 in. at the bottom. The sills are 2 ft. wide and both posts and sills are 14 in. thick. Each frame was poured as a monolith and the corners are reinforced. The sills are supported on three concrete piers, each 18 by 36 in. cross-section. The height of the piers varies with the topography, from 6 in. to 15 ft. They are carried below the frost-line and have spread footings where the nature of the ground is such as to require them. The frames are spaced 6 ft. centre to centre. Around the inside of each frame is bolted a nailing-plank 5 by 14 in., to which a wooden lining is spiked. This lining varies in thickness from 3 to 5 in., the bottom and lower part of the side-lining being 5 in. and the remainder of the side-lining being 3 and 4 in. The lining is the best grade of Oregon fir and was shipped to the job slightly over-size in width to allow for shrinkage. The planks were then planed just before placing in the flume. This process made the final cut to exact and uniform width and also cut the groove for the spline.

LITHIUM is a silver-white soft metal which forms the oxide or some other salt on exposure to the air, and which decomposes water, forming the hydroxide of lithium; for these reasons metallic lithium does not occur in nature. The U. S. Geological Survey announces that some chemical products derived from lithium minerals have found application for military purposes.

Solubility and Orientation of Molecules in Surface of Liquids

*The structure of the surfaces of liquids, and the relation of solubility to the work done by the attraction of two liquid surfaces as they approach each other, is engaging the attention of physicists more than ever before, due in part to the importance of an understanding of these phenomena in their practical application in flotation. An interesting study of the question was recently contributed to the Journal of the American Chemical Society by William D. Harkins, F. E. Brown, and E. C. H. Davies. Their measurements of the work done in ergs per square centimetre when the surface of two liquids come together to form an interface, show in a striking manner that the film of any liquid in contact with water is composed of molecules oriented so that the active (or polar) group at the end of any hydro-carbon chain is in contact with the water. The double bond of an unsaturated hydro-carbon, and the double bonds in the benzine ring act in this respect like polar groups. It is also shown that the attraction between water and another liquid is one of the important factors in the determination of the solubility of the other liquid in water. Imagine a volume of water divided by a horizontal plane parting and lifting the upper portion. Two surfaces now appear. If the temperature be 20° C. the surface-tension would be 72.8 dynes per centimetre, and the free energy per square centimetre would be 72.8 ergs. The total energy of the two surfaces, each with an area of 1 sq. cm., would be 145.6 ergs. Suppose the two surfaces to be again brought into contact and that this free energy disappears; if, however, one of the surfaces be water and the other benzine, the free surface energy of the two would be 72.6 for the water and 28.98 for the benzine, or a total of 101.78 ergs. The decrease of free energy is seen to be 66.7 ergs when they meet, which is equal to the work done by the two liquids as they approach. Therefore the free energy of the 1 sq. cm. of interface is 35 ergs. The amount of work necessary to separate the liquids after their surfaces have come together, is, however, not the same as the decrease as shown above, since the surface-tension of each liquid is affected by the other. Thus, when mutually saturated at 25° C. the free surface-energies become, water 60.19, plus benzine 27.9, total 88.09; the decrease when they come together is 43.41 ergs. It is seen that the relative attractions exerted by a water-surface upon the surfaces of other liquids may be approximately measured. This applies only to liquids that do not mix; it is not possible to determine the interfacial tensions of extremely miscible liquids, since, in fact, they form no interface. As solubility increases the interfacial tension decreases. For example, the interfacial tension of water with octyl alcohol is only 8.28, with iso-amyl alcohol 4.42, and with iso-butyl it has decreased to 1.76. Therefore it may be affirmed that the interfacial tensions of liquids mutually soluble in all proportions are so small as not materially to affect the calculation of the decrease of free energy.

It is found that if the interfacial tensions between two liquids is small the liquids are miscible. In some cases, however, the surface-tension is not the property that actually determines the solubility of a substance as a molecular disperse system (so-called true solubility); this is shown by the fact that it is possible to make water colloidally soluble in benzine or benzine in water, commonly called emulsification, by adding a substance which causes the interfacial tension to drop to a low value, but in this way such a substance cannot be made to dissolve in water as a molecular disperse phase, that is to say, a true solution of any considerable concentration with a substance of that kind as a solute cannot be formed. The necessary condition for the formation of a molecularly dispersed solution is that as the particles are made smaller the surface-tension must not increase, and this condition is not met by stable emulsions and colloids. As an example of colloid formation an emulsion of benzine in water may be stabilized by sodium oleate. The interfacial tension is either zero, or very low. If the interfacial surface could be removed so rapidly that no time would be given for adsorption the surface-tension would rise to a comparatively large value. Polar liquids are in general mutually soluble; also slightly soluble polar liquids are soluble in other liquids similarly showing slight polarity. The idea that potential differences and the other energy-relations at surfaces, together with certain phenomena of adsorption, colloid formation, and solubility, might be explained on the basis of the polar setting of molecules in surfaces appeared to the authors of the paper some years ago. The adsorption of a substance in a benzine-water interface is greater for any particular salt than it is in the water surface whenever the organic radical is of such a nature that it would be expected to be soluble in the benzine, and this may be a general relationship. This point is now under investigation. It is found that a polar group, such as COOH, SO_3H, OH, CN, NH_2, and others, and their salts, will drag into solution in water a short slightly-polar chain, such as a hydro-carbon, but as the length of the slightly-polar chain increases, the solubility of the com-

*Abstract: Jour. Am. Chem. Soc., Vol. XXXIX, No. 3 and 4.

pound with one end polar and the other slightly-polar rapidly decreases; also, as the length of the slightly-polar, which may be called the insoluble end of the molecule, increases, the adsorption in the surface liquid-vapor from a polar solution increases. The increase in adsorption in such solutions is easily explained by the tendency of water to throw out the insoluble radical, but to hold to the one which is soluble. On this basis there should be an average orientation of such solute-molecules in a water surface with the active or polar end toward the liquid and the slightly-polar end toward the vapor. At an interface between two liquids the polar end would be expected to turn toward the polar liquid, and the slightly-polar end toward the slightly-polar liquid. Also, in the case of a single pure liquid, if it is of the polar type, it would be expected that the polar ends of the molecules would have an orientation toward the inner side of the surface. From this standpoint the ends of the molecules that stand outward in the surface of a pure substance consisting of molecules containing a moderately long paraffine chain attached to a polar group, should not be very different from those in a liquid paraffine, since in both cases the extreme outer surface would be made up of paraffine groups. The positive adsorption in a liquid surface of organic acids, or salts of organic acids, depends on the fact that one end of the molecule is less soluble than the other. The films of solutes positively adsorbed at either liquid-vapor surfaces or liquid-liquid surfaces, frequently become saturated. The rapidity with which this saturation occurs depends upon the nature of the solute, and increases with the length of the insoluble end of the molecule. Thus sodium oleate in water is highly adsorbed, and the surface-tension of the solution decreases with extreme rapidity. After the concentration of the solution reaches the value 0.002 normal the surface-tension no longer decreases, but remains constant up to 0.1 normal or more, proving that the film has become saturated. The stability of emulsoid particles seems to be brought about by orientation of molecules at the interface with the medium of dispersion. The best emulsifying agents, for example, have very long molecules, with a polar or active group at one end of the molecule. For the emulsoid particle to be stable the molecules which make the transition from the interior of the drop to the dispersion-medium, or the group of molecules making up the film, should fit the curvature of the drop. Therefore, the surface-tension of very small drops is a function of the curvature of the surface.

COLLOIDS when precipitated from their solutions or suspensions by an electrolyte yield a precipitate which always contains a definite amount of the ion to the influence of which the precipitation was due, the quantity retained being proportioned to the equivalent of the ion. For example, the precipitate obtained by the action of barium chloride solution on arsenious sulphide sol has the composition $90As_2S_3 + 1Ba$, the barium being present as hydroxide. On washing this precipitate with calcium or strontium salts the barium is replaced by equivalent quantities of these metals. As the coagulum contains the positive ion, which cannot be removed by washing with water, the liquid becomes acid, and the acidity can be determined by titration. The phenomenon is known as adsorption, and is similar in character to the adsorption of gases by charcoal.

Behavior of Aluminum in Cyanide Solutions

Metallic aluminum is used as a precipitant of gold and silver from cyanide solutions, replacing zinc in certain cases. Furthermore, aluminum in a soluble form occurs in certain gold and silver ores. On account of the fact that there is little specific knowledge regarding the behavior of aluminum in cyanide solutions, the reactions occurring between metallic aluminum and the caustic alkalies were investigated by the Bureau of Mines, and a particular study was made of the ore and pulp from the Goldfield Consolidated mine, Goldfield, Nevada, as that ore is one of the best examples of the occurrence of soluble aluminum salts in a precious-metal ore. The aluminum occurs as a sulphate, and is the result of the decomposition of minerals containing aluminum. This work has been completed, and the chemical reactions taking place under the conditions of mill-operation have been established. The more important conclusions reached were that there are two aluminates of calcium formed, namely, a primary aluminate when alumina predominates, and a secondary aluminate when lime is in excess. The primary aluminate is soluble, whereas the secondary aluminate is not; therefore, in mill-operation, if troublesome insoluble aluminum compounds in the precipitate are to be avoided. the amount of lime used in milling must be restricted to no more than that necessary to form the soluble aluminate. It might be explained that lime is universally used as a source of alkali in cyanide work, and also as an aid in settling slime; hence, if aluminum is present, it is important to know the behavior of the calcium aluminates formed in the humid way. Upon the addition of lime during cyanide treatment, first the primary and then the secondary aluminate is formed. The presence of aluminum accounts for the rather high consumption of lime in treating such ores. It was found that, contrary to popular supposition, precipitated aluminum salts did not cause waste of cyanide by carrying down adsorbed cyanide when the sulphates were present. On the other hand, during the pricipitation of cyanide solutions with metallic aluminum, under certain circumstances when an aluminum salt is precipitated, cyanide is lost through adsorption. In connection with this investigation, it was found that too long contact of the solution with the precipitant, whether aluminum or zinc, caused a loss of cyanide, probably through decomposition by the nascent hydrogen evolved.

REVIEW OF MINING

As seen at the world's great mining centres by our own correspondents.

VANCOUVER, BRITISH COLUMBIA

REPORTS FROM MINING DISTRICTS

From the Omineca district comes news of development and progress. Old mines are being improved and new ones are being opened.

In the Rocher Deboule mine, near Hazleton, a shoot of ore has been opened. Development of new ore is satisfactory and considerable tonnage is exposed.

The Delta Copper Co. controls the Delta and Chicago groups, as well as several other locations, in all twenty-five claims. They adjoin the Rocher Deboule on the east. Active work was carried on during last winter on the Highland Boy and the Delta groups under the supervision of G. A. Clouthier, the newly appointed government engineer in charge of No. 1 district. At present the necessary communication with the railway is being completed and work is being concentrated on that while the weather permits. In addition to the known exposures of ore on this company's property a fine showing of copper was found the other day on the Delta group's Lucky Jack claim. Six feet of the vein is reported to run 5½% copper. As soon as communication is established the company intends to actively work the known orebodies, and good results are expected. Harrison Clement is managing this property for the Delta Copper Co. Most of the financing is being done from Edmonton and there is said to be much capital interested.

The Hazleton View and Indian groups, comprising eight crown granted claims, are owned by the New Hazleton Gold-Cobalt Mines. They adjoin the Rocher Deboule company's ground on the west and north. Since last summer a tunnel has been driven on the main gold-cobalt vein on the Victoria claim and a number of high-grade shoots of promising ore have been cut. An interesting feature of this work has been the development of several shoots of molybdenite ore, which have successively shown richer content. They occur quite distinctly from the gold-cobalt ore-shoots, and assays show from 5 to 20% molybdenum. The company is about to install a light aerial tram to take ore from the adit down the mountain to a point from which it can be packed to the railway pending the construction of a wagon-road. Stoping is being commenced on both the gold and molybdenite ores with the intention of making a shipment of each class of ore as quickly as possible. It is understood that the plans of the company for this season include considerable exploration and development work on several parallel veins. They will also do some work on their copper veins.

Adjoining the New Hazleton Gold-Cobalt Co.'s holdings on the west, a group of ten claims is being developed by a Vancouver syndicate.

The Golden Wonder group, owned by W. S. Harris and Denis Comeau, of Hazleton, has been bonded to M. W. Sutherland, who has been conducting development work upon it. The group is situated at the foot of the west slope of Rocher Deboule mountain on the Hazleton wagon-road.

Work is being carried on at the Comeau group, which lies on the west slope of Rocher Deboule mountain, between Hazleton View and the Golden Wonder group. The owners, Denis Comeau and Magnus Johnson, are driving a tunnel on one of the promising outcrops.

The Silver Standard mine is installing a new power-plant, consisting of gasoline engine, dynamo, compressor, boiler-plant, and shop equipment. It will enable operations at the mine to be carried on to better advantage. Erection of a concentrator is now being considered.

Reports from Talkwa indicate considerable mining activity in the surrounding territory. Representatives of a number of well-known mining and smelting corporations are at present in the district examining properties. Among the companies represented are the Granby company, the Consolidated of Canada, the American Smelting & Refining Co., and the Tonopah Mining Co. The Tonopah company has bonded the Wilson Brothers claim near Knockholdt station on the Grand Trunk Pacific railway.

Upon completion of the Babine road the Cronin mine will ship ore and it is expected that the Debenture Mines will also be ready at that time to join the list of shippers.

Prospecting is active on the main branch of the Telkwa river, the Babine mountains, the Sibola country, and to the south of Houston.

There is a good deal of interest in the work of the geological survey of the Hazleton district, which is being carried on by J. J. O'Neill of the Dominion Geological Survey Branch. He is conducting a detailed survey of an area fifteen miles square, embracing all the district immediately surrounding Hazleton and the main working-camps.

A complete index to the forty-two million dollar mining industry of British Columbia is provided in the mineral exhibit at the Vancouver exhibition which opened last Monday. The display occupies the entire upper floor of the forestry building and is under the charge of the Vancouver Chamber of Mines, the honorary secretary of that institution, James Ashworth, having supervised the arranging of the display.

The Dunvegan mines, on the head-waters of Fish river and about ten miles east of Revelstoke, have been purchased by a group of capitalists represented by R. A. Grimes of Calgary. A company is being formed to be known as the Dunvegan Mining Co., Ltd., and it will be capitalized at a quarter of a million dollars.

Development work has been started on the Montgomery group at Downie, which was recently bonded to the Granby Mining Co. of Phoenix.

Ore is now being packed down from the Mastodon mines on Laforme creek, twenty miles north of Revelstoke. A trail from the Columbia river is under construction by the company and as soon as it is finished it will be possible to make shipments.

The Merritt Collieries are installing a compressor and drills at No. 3 mine of the Diamond Vale property. Production is on the increase and work has been carried on steadily since the mine was opened last winter.

The Donahue Mines at Stump lake will resume operations about September 1 next.

On August 29 there will be a special meeting of the stockholders of the Canada Copper Corporation, for the purpose of authorizing the issue of $2,500,000 ten-year 6% first mortgage sinking fund convertible bonds. They will also act on the proposal to increase the authorized capital stock a million shares, of which 833,333 would be held for conversion and the remainder for corporate purposes.

A good strike of ore is reported to have been made at the Ontario mineral claim at Kimberley. A force of eight men is working on the property, which is owned by Sir Daniel Mann, vice-president of the Canadian Northern railway.

It is reported that arrangements have been made for 20,000 ft. of diamond-drilling on the Britannia Mines property on Howe sound.

The Standard Silver-Lead Mining Co. has been engaged in exploration work by diamond-drilling.

There is not likely to be any dividend disbursement by the Rambler-Cariboo Mining Co. in August. Recent shipments to the smelter have not fulfilled expectations.

The Florence Silver Mining Co.'s new 300-ton mill is now handling 220 tons of ore per day and is producing about 32 tons of concentrate.

From five to ten feet of ore has been struck in a drift in the Cork-Province mine. The company owns a concentrator, and when the new ball-mill is in place, shipments will be started.

With a capitalization of $300,000 the Beatrice Silver-Lead Mines, Limited, has just been incorporated to take over the Beatrice Mines from the bondholders, Messrs. K. G. McRae and S. Daney of Ferguson.

The Vancouver Chamber of Mines has secured new and better quarters on the second floor of the Dominion building. They will have about 2500 sq. ft. of floor space, about five times the amount in the present rooms in the Molson's Bank.

JOHANNESBURG, TRANSVAAL

UNDERGROUND WATERS.—EAST RAND PROPRIETARY.—NEW FAR EAST AREAS.—CON. MINES SELECTION.

EAST RAND PROPRIETARY MINES. From what has appeared in previous letters with respect to the tapping of underground springs of water yielding nearly 5,000,000 gal. of water per day at a depth of 4000 ft. in the East Rand Proprietary mines,

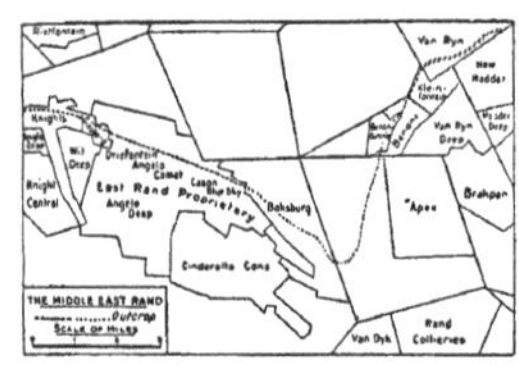

little surprise will be felt that it has been deemed advisable to modify the scheme proposed to prove the southern area, by the driving of two cross-cuts and afterward sinking to the banket. R. C. Warriner, now consulting engineer to the company, has recommended that instead of driving two converging cross-cuts 2000 ft. long from the Angelo Deep shafts and then sinking from the point of convergence for another 2000 ft. to the banket, that a circular shaft be sunk from the surface in the place of the flooded cross-cut. This shaft will be about 6000 ft. deep. It is worthy of notice that the shaft is to be circular in shape, and no doubt the engineer has acted wisely in selecting this shape, seeing that the shaft will be largely used for ventilation and will possibly have to pass through heavily watered ground. It may be pointed out that the Driefontein Deep shaft, situated on the westerly boundary of the East Rand property, and having practically the same strata to pass through as the proposed circular shaft, was stopped by the inflow of water, but not before a heavy loss of life had occurred through the shaft collapsing while in course of being sunk. It will therefore be interesting to watch the progress of this circular shaft, for the Driefontein Deep shaft was rectangular in shape. At the annual meeting of the stockholders of the East Rand Proprietary the chairman pointed out the serious position of the company. The working-profit last year reached £497,165, which was £103,826 less than for the preceding year. The ore milled during the year was 1,939,200 tons, a decrease of 44,400 tons, while the yield was 24s.4d. per ton, a decline of 9s.2d. per ton. The working-cost had risen to 19s.3d. per ton.

NEW FAR EAST AREAS. The Union Government has called for tenders for the exclusive right to work four of the most fancied gold-mining areas on the Far East Rand, and there is every likelihood of keen competition between the various groups to secure the leases. The areas now thrown open for competition consist of the triangular Springs Farm along with a small portion of Geduld, an area amounting to 2050 claims, the eastern portion of Geduld Farm between Cloverfield and Geduld Deep, the two other areas consisting of the remainder of De Reitfontein Farm less the portion already taken up by the Springs Mines.

The Consolidated Mines Selection Co., which controls the Brakpan, Springs, and Daggafontein mines in this neighborhood, is expected to secure at least two out of the four areas, while the Geduld Proprietary Mines is expected to make an equally strong bid for the Geduld area consisting of 2526 claims, and situated alongside their present mynpacht. Had the adjoining Grootvlei mynpacht been considered sufficiently attractive to American capital when invited to participate a year ago, it is thought that the Grootvlei Proprietary Mines might also have competed for this area, but there seems little possibility of any competition now arising from that quarter. Both of the De Reitfontein areas, each consisting of 2236 claims, ought to prove attractive to the controllers of the Springs Mines and should they succeed in obtaining both areas, the Consolidated Mines Selection Co. will be heartily congratulated. The Springs area is estimated to require £800,000 for shaft-sinking and development, and an additional £700,000 for equipment before being brought to the producing stage, whereas the three other areas are each estimated by the Mines Department to require £900,000 for shaft-sinking and preliminary development, and £800,000 for equipment. As regards the probable depth to the banket, the Geduld East area has the advantage, as it is estimated to be 3000 ft. as against 3500 ft. for the Springs area. The depth of the 'reef' on the De Reitfontein east area to the immediate dip of the Springs mynpacht cannot be less than 5000 ft., and on the western area the depth will run from 4000 to at least 5000 feet.

CRIPPLE CREEK, COLORADO

AJAX MINE.—ORE DISCOVERY AT THE C. O. D.—EL PASO.—ROOSEVELT TUNNEL

An important discovery at depth is reported from the Ajax mine on Battle mountain. The property is under lease to the Carolina company. W. S. Black is superintendent. The shaft was recently sunk to 1800 ft. and the main Ajax vein cut at this depth.

A surface discovery of promise has been made by a prospector, Tom Ross of Cripple Creek, on the property of the C. O. D. M. & D. Co. in Poverty gulch. The vein, a new one for the C. O. D., has been opened at the surface and a trial shipment of float and ore from the vein is at a local sampler. The quartz pans freely. The strike was made on the west side-line and close to the El Paso Gold King line. The El Paso Gold King is the oldest producer on the west end.

A new vein has been opened up by the J. H. B. Leasing Co. The discovery was made at a depth of 60 ft. The quartz assays one ounce gold per ton.

Operations have been resumed at the main shaft of the Blue Flag G. M. & M. Co. on Raven hill. An underground air-hoist has been set in place at the bottom station, 1100 ft. from surface, and the shaft is to be deepened 300 feet.

The El Paso Extension company, Al Campbell, superintendent, has started an adit from the south-western slope of Beacon hill through the Rocky Mountain and North Slope claims. The adit will cross-cut at considerable depth the extension of the El Paso, Tillery, and Little May veins of the El Paso vein system.

A 100-hp. electric hoist has been installed at the Wild Horse mine of the United Gold Mines Co., near Midway on Bull hill. The company is cross-cutting and will resume sinking operations. The shaft is 1250 ft. deep.

The Big Toad M. & L. Co., a new corporation, has taken over the Gowdy-Chapman lease on the Dante mine of the Dante G. M. Co., on Bull hill, and the Reid mill, on the property, will be used for the treatment of low-grade mine and dump ores.

The narrow-gauge spur extending to the El Paso Con. Gold M. Co.'s main shaft on the western slope of Beacon hill is being widened to broad-gauge. The property is shipping from 6 to 8 cars of milling ore per week.

The Roosevelt tunnel of the Cripple Creek Deep Drainage & Tunnel Co. was advanced 95 ft. in the first 15 days (single shift) of August.

GOLDFIELD, NEVADA

REPORT OF GOLDFIELD CON. M. CO.

During the month of July 1917 the Goldfield Con. M. Co. produced 18,400 tons from which resulted net realization of $9809.21.

DEVELOPMENT WORK. 1690 ft. of development work was performed at a cost of $7.10 per foot.

OPERATING COSTS. The operating costs of the company were as follows:

	Per ton ore handled	Per ton total ore
Mining:		
Stoping	$2.925	
Development	11.742	
Total mining	3.454	$3.048
Leasing expense	1.884	0.142
Dump moving	0.511	0.022
Transportation	0.087	0.087
Milling	2.534	2.534
Marketing	1.496	1.496
General expense	0.390	0.390
Bullion tax	0.005	0.005
Filter royalty	....	....
Flotation royalty	0.069	0.024
Surface	0.060	0.060
Total operating costs		$7.808
Miscellaneous earnings		0.764
Net operating costs		$7.044
Construction		....
Net costs		$7.044

DEVELOPMENT ON MAIN LEVELS, JULY 31

MOHAWK. On the 450-ft. level, 200 ft. south-east from the shaft, 307-EY sill was cut and produced 18 tons of $71.72 ore.

LAGUNA. On the 600-ft. level, 620 ft. north-west from the shaft, 302-J raise was run and produced 223 tons of $10.19 ore.

CLERMONT. On the 750-ft. level, 430 ft. north from the shaft, 450-B sill was extended and produced 241 tons of $17.05 ore.

On the 1250-ft. level of the Grizzly Bear mine, 250 ft. south-east from the shaft, 802-E sill was extended and produced 384 tons of $12.69 ore.

LESSEES. During the month lessees produced 1386 dry tons of ore having a gross value of $22,436.29, of which the company received net $8505.79.

TONOPAH, NEVADA

MILL AND PRODUCTION REPORTS

The Tonopah Mining Co. during July milled 7452 tons of ore with a value of $109,130, making a net profit of $40,300. During the past week the production was 2600 tons. During the past week the Tonopah Extension Mining Co. produced 2380 tons of ore; 23 bars of bullion, having a value of $51,000, was shipped to the smelter. The bi-weekly clean-up of the West End mill consisted of 31 bars of bullion valued at $52,885. The ore production the past week was 943 tons. The Tonopah Belmont produced 2325 tons of ore in the past week. The Belmont mill shipped 50 bars of bullion and 30 tons of concentrate with a combined value of $109,895. The plant at Millers shipped 25 bars of bullion valued at $49,240.

The production of the Jim Butler the past week was 900 tons, the Rescue-Eula 126 tons, the Montana 78 tons, and the Cash Boy 55 tons, making the week's output at Tonopah 9407 tons with a gross value of $164,622.

BUTTE, MONTANA

SOUTH-WESTERN BUTTE DISTRICT.—HAYDEN-STONE INTERESTS.

In the south-western part of the Butte district the larger mining companies have resumed. This part has been idle for many years, owing to the low price of silver and the large amount of zinc in the ore. With silver at its present price and zinc so readily recovered by milling methods, a great incentive is offered for re-opening the old producers. Among the latest to start active operations are the Clark interests, which are developing the Travonia property. The old Travonia shaft was 300 ft. deep and on the 300-ft. level had about a 1000-ft. cross-cut. As soon as the shaft is unwatered it will be re-timbered and sunk to the 500-ft. level, and then cross-cutting will begin. The veins of the Travonia and adjoining claims owned by Clark are among the richest silver-zinc veins in the district, and were heavy producers in the early days. In addition to the silver-zinc ore, a large body of manganese ore is ready to be mined. This will be exploited by the open-cut or glory-hole method as soon as arrangements can be made for the milling of it. At present there is a large construction crew at work, putting in boilers, a compressor, and a steam-hoist. It has been decided to replace the steam-hoist with an electric hoist as soon as one can be obtained. The Anaconda company is operating the Nettie and the Emma, as well as the Bonanza. The Hayden-Stone interests and J. L. Bruce, of the Butte & Superior, are opening up a group of claims through the Germania shaft. This is one of the early-day producers of high-grade silver-zinc ore. The Davis-Daly company is opening up the Hibernia, and also mining manganese ore on this property. The Butte-Detroit is working the Ophir. The Britannia is being opened by W. L. Creden.

JOPLIN, MISSOURI

PRICES.—MARKET.—PRODUCTION

The Joplin *News-Herald* gives the following summary of the Missouri, Kansas, and Oklahoma zinc and lead region:

AVERAGE PRICES FOR AUGUST

	High	Low	Average
Week ended 4	$75	$65	$70
Week ended 11	75	65	70
Week ended 18	75	65	70
Week ended 25	75	65	70

Average for month, $70.

No change in wage scale.

A noticeable increase in the demand for all grades of zinc ore last week led producers to believe that higher prices might be expected during the week. Although the price for high-grade ore was unchanged at $75 per ton, in carload lots, second grades were stronger, selling for as high as $72.50 per ton in some instances. The range was from $65 to $75, with an average basis price of $70.

Buyers were in the market for practically every pound of ore the producers had in their bins. However, quite a number of operators, with large amounts of ore stored up, held under the belief that the increased demand might lead to better prices. One operator with something like 500 tons of zinc in his bin refused an offer for the lot.

Just what the sudden flurry among the buyers for increased lots of ore was caused by, operators are at a loss to say. One mining man advanced the theory that the strike at the Anaconda Copper Co. in Montana might have caused a shortage in the zinc metal market, and therefore the demand for ore.

Prices for lead were unchanged, choice ores bringing $100 per ton. Calamine was unchanged at from $38 to $40 per ton.

The spelter market is dull and weak following the further decline in price as quoted by the New York metal market. Producers say that a further curtailment of operations will be forced by the decline in the price to $8 per hundred pounds. The mid-year statistics showed that 35,000 retorts, or more than one-seventh of the total, were idle on June 30, although the market in June never went below $9.

Following is a report, in pounds, with values of zinc, lead, and calamine ores produced in cities, towns, and camps in the Missouri-Kansas-Oklahoma district for the week ended August 25:

ZINC AND LEAD PRODUCTION

	Blende	Lead
Miami	7,142,630	408,540
Webb City-Carterville	4,944,330	1,214,540
Joplin	2,596,010	345,190
Duenweg	1,362,950	116,090
Oronogo	1,028,170	247,840
Galena	925,380	326,930
Alba-Neck-Purcell	533,810	......
Lawton	503,630	......
Granby	492,420	......
Wentworth	467,470	......
Baxter	349,230	102,270
Belville	403,400	......
Spring City-Saginaw	175,980	......
Carthage	255,030	......
Cave Springs	149,750	......
St. Louis	96,160	......

CALAMINE, IN POUNDS

Joplin	87,830
Granby	1,356,960
Wentworth	124,920

TOTALS FOR WEEK

	Pounds	Value
Blende	21,425,250	$765,343
Calamine	1,569,710	34,032
Lead	2,761,400	137,800

AVERAGE VALUE PER TON

Blende	$70
Calamine	43
Lead	100

WASHINGTON, D. C.

DIRECTOR MANNING REVIEWS THE OIL SITUATION.—SCARCITY OF WELL-DRILLING SUPPLIES.—INCREASE IN COST OF SUPPLIES.—NECESSITY FOR THE USE OF MUD-LADEN FLUID IN OIL-WELL DRILLING.

In a review of the oil situation in its relation to war needs, Van H. Manning, director of the Bureau of Mines, stated that the situation is critical, for the operators find themselves unable to obtain adequate supplies for the drilling of oil-wells. If they are unable to drill new wells, our present production cannot be maintained, much less be increased, for oil-wells do not maintain a constant production but show a steady diminution from the time they are first completed. Were no wells to be drilled for one year's time, our petroleum production would decrease at least one-fourth.

At the present time drilling is being curtailed because of the difficulty in obtaining supplies. Not only have the costs of oil-well supplies been increased, but it has become impossible to get them in adequate quantities for any price, and the situation is now worse than at any time since the War in Europe started. Perhaps the situation of the producer is best illustrated by the fact that from California to the fields of Pennsylvania the small producing wells are being abandoned in order to use the casing and other equipment of these wells in new wells from which larger productions are expected, thus, wells still capable of small profitable production are being permanently abandoned in the desperate search for well-drilling materials.

For saving, attention is called to a method long advocated by the engineers of this bureau. He refers to the use of mud-laden fluid in the drilling of oil-wells on which method the Bureau of Mines has published Bulletin 134, obtainable free of charge from the Washington office. By this method of well drilling, less casing is necessary and by intelligent application of the methods outlined in Bulletin 134 the same amount of casing can be made to supply a greater number of wells. This is a method that has long been in use in Texas, Louisiana, and California in connection with the rotary and circulator drilling system and in the wells drilled by these methods less casing is necessary than in other fields. During the last few years, the use of mud-laden fluid has been demonstrated to be entirely feasible in wells drilled by the standard-tool system in the oilfields of the Mid-Continent district and farther east, and that it possesses desirable features not to be overlooked by the operator. Where casing is difficult to obtain the saving that can be effected is of first importance. One of the largest operating companies in the Mid-Continent district is using the method to great advantage.

The Bureau of Mines maintains a staff of engineers and practical drillers who have made a special study of this process, and will gladly see that an expert is sent to consult with the operators in any district who show themselves sufficiently interested.

THE MINING SUMMARY

The news of the week as told by our special correspondents and compiled from the local press.

ALASKA

(Special Correspondence.)—Rich telluride ore has been struck in the Marble mine. Molybdenum discoveries are reported. The Martin mine is working a full crew. The Gold Cord Co. has developed 12 ft. of $36 ore. A 2400-ft. aerial tram, connecting with the Independence mill, will be installed by the same company. The Independence M. Co. is restricting its work to development. The Rae-Wallace Co. has purchased the Ike Rosenthal group and will install a 50-ton mill and a 1600-ft. aerial tramway. A motor truck will be used to transport supplies.

Willow Creek, August 11.

By the end of 1917 the main-line grade of the Alaska railroad will reach mile 210. The Engineering Commission will build a pier at Anchorage. The Tanana Valley Railroad has been taken over by the Commission.

The discovery of platinum in the gravel of the Kahiltna river by Dr. Herdschel C. Parker of New York has resulted in the assignment of four Government experts to study and report upon the situation. Alaskan gold miners report platinum in the Ketchikan, Koyukuk, Fairhaven, and Christochina districts.

ARIZONA

Owing to the unsettled conditions existing in Arizona mining districts local meetings of the A. I. M. E. will be indefinitely postponed.

Maricopa County

(Special Correspondence.)—The Mendota Mining Co. has purchased the Yorba Mining Co.'s property in the Harquahala district. Work has been commenced on the road between the property and Aguila, the railroad point. Development preparations are being made.

Wickenburg, August 18.

Mohave County

(Special Correspondence.)—It is announced that N. P. Moerdyke and E. M. Rabb of the Tom Reed Gold Mines Co. have resigned. Assistant superintendent W. B. Phelps will take Mr. Rabb's place temporarily at least. Mr. Moerdyke was elected president of the company a year ago last April, for the purpose of adopting a progressive management and to carry on a programme of development. Mr. Rabb was employed as superintendent. During the past year important projects have been carried to completion, including the development of the Aztec orebody for a distance of over 1000 ft. One hundred thousand tons of ore has been developed in this working. It averages about $8 per ton and with rising costs of mining and milling, economy will be necessary for profitable operation.

It is expected that the new management will discontinue development and institute a policy of retrenchment. The new mill is handling ore at much less cost than the old, the principal problem to be solved being that of mining.

F. M. Graff, president of the Graff Coal Co. of Pittsburgh, a large stockholder in the Oatman Combination property adjoining the Aztec claims of the Tom Reed, thinks so well of the property after a recent examination of it that he will finance further development on an extensive scale.

The Gold Ore Mining Co. has leased a portion of the Gold Road mill and will treat about 50 to 75 tons of ore per day. The Gold Ore is about one mile north of the mill and a tramway will be installed to convey the ore. About 2000 tons of ore was shipped from the Gold Ore to the mill when the latter was in operation a year ago last spring.

Oatman, August 28.

(Special Correspondence.)—The Payroll is sending 35 tons of ore daily to the mill at Needles. The New Tennessee is opening up a good body of ore in the north drift, 240-ft. level. New Mohawk shipped a car of gold-silver ore to Selby's on August 18. Schenectady is hoisting high-grade ore from the north and south drifts, 175-ft. level. A carload is soon to go to the smelter. Copper Age struck 14 in. of shipping ore in the new adit at 700 ft. The same vein is being opened in the south drift upper tunnel. The mill has been started up. The furnace department is well under way to completion. The Keystone mill is expected to resume this month. Several thousand tons dump ore is ready to be milled. The mine will then be unwatered and operated. The Rattlesnake mine has a carload of silver-lead ore in its bins ready for the smelter. The Diana has shipping ore on the 200-ft. level. The Bullion Hill shaft is nearing the 300-ft. level. A station is to be cut and the shaft continued to the 400-ft. level where cross-cutting will be started. The Union Metals Co. has arranged with the owners of the Golconda mine to run a 600-ft. cross-cut from the latter's 1100-ft. level to intersect the vein at a depth of 800 feet.

Chloride, August 24.

(Special Correspondence.)—The Red Lion Mining Co. is about to start development work. Arrangements are being made to sink a 500-ft. shaft and to develop the property. The Rico property is being systematically developed. The Gold Ore Co. will construct a 10-stamp mill and a cyanide plant for the treatment of 50 to 75 tons of ore per day. A big strike has been made 700 ft. in the main adit of the Copper Age (Arizona Ore Reduction Co.). The ore is high grade and carries gold, silver, lead, copper, and zinc. The mill will be ready for operation in a week or 10 days. Ore is picked and shipped at present. A strike of silver ore is reported by prospectors in the Wheeler wash district of the Hualpai mountains. A silver-gold strike has been reported recently in the old Weaver mine. The mine is situated 28 miles north of Chloride and is under lease to Parry and Bonsall.

Kingman, August 18.

The Gold Ore Co., at Goldroad, is preparing to sink its shaft to a depth of 700 ft. A two-mile tram has been purchased and will be installed.

Pinal County

(Special Correspondence.)—A contract for the treatment of the Mammoth mine tailing has been let to S. G. Musser and associates of Los Angeles. The erection of a $20,000 flotation mill for preliminary testing has been ordered. The Ray silver-lead property is shipping 20 tons of ore daily. The Curry Mining Co. has been formed and has received authorization to dispose of 200,000 shares. The company has taken over the William Curry group of seven claims in the Silver King district. William Curry will be in charge.

Florence, August 17.

It is reported officially that the adit of the Fortuna mine at Superior has cut two feet of rich ore. The first class assays 492 oz. silver per ton and 14% copper; the second class assays 196 oz. silver and 11% copper.

SANTA CRUZ COUNTY

(Special Correspondence.)—Development is being pushed at the Idaho mine in the Oro Blanco region. Copper ore carrying gold and silver is being taken out. There is 3000 tons of concentrating ore on the dump. The erection of a small mill is being considered.

Nogales, August 18.

YAVAPAI COUNTY

(Special Correspondence.)—The result of an examination of the Jerome Copper Co.'s holdings at Jerome and Moyer has been the purchase of the controlling interest in the Jerome Copper Co. by the General Development Co. The Moyer-Bedford mine, locally known as the Christmas mine, and the Moyer holdings of the Jerome Copper Co. are included. W. Tovote, an engineer formerly with the Copper Queen at Bisbee, has taken charge of the Loma Prieta Co. in Copper basin. The Verde Hub has engaged J. E. Wallace, an experienced diamond-drill man of Michigan, to prospect the property.

Prescott, August 18.

YUMA COUNTY

(Special Correspondence.)—The Swansea mine, under lease, is shipping 30 cars of ore per week. The new three-compartment shaft is down 125 ft., and it is expected that ore will be reached at 300 ft. The completed shaft will be 500 ft. deep, and will be used to develop orebodies already discovered. It is reported that the Clark copper interests will construct a new smelter at Swansea during the coming winter.

Bouse, August 18.

CALIFORNIA

During the week ended August 25, fourteen new wells were started, bringing the total since the beginning of the year up to 710. Thirty-five wells were reported as ready for test of water shut-off, 12 deepening or re-drilling, and three abandoned.

In the Whittier field two new wells are reported, one on the Murphy-Coyote lease producing 10,000 bbl. daily and the other on the Baldwin-Montebello lease producing 800 bbl. daily.

AMADOR COUNTY

(Special Correspondence.)—The Argonaut Mining Co. at Jackson has just completed installing a large pumping system. The water is pumped from its lowest level at 4300 ft. to a reservoir at the 2000-ft. level of the main shaft, and from this reservoir to the surface. Both large pumps are electrically driven.

The Keystone mine is said to have made a $1500 profit in July. The improvement is due to systematic development on the lower levels, particularly the 1200 and 1400. Good ore is also being mined on the 1800-ft., or lowest, level.

A rich strike in the Hardenburg mine on the Mokelumne river, south of Jackson, is reported during the week. The find was made on the 1500-ft. level in the north drift, the assays from a well-defined vein ranging from $16 to $7600 per ton in gold. The ore is being saved by hand panning and sluicing instead of milling. This property is operated by the W. J. Loring company.

The South Eureka company has directed its attention to the Oneida mining property, which it purchased a few years ago to work in connection with its Sutter Creek holdings. Prospects have proved discouraging at the South Eureka mine. Ore of good grade has been in course of extraction from the Oneida property for several months and the mill on that ground is being worked to capacity.

Sutter Creek, August 24.

BUTTE COUNTY

Prospecting for chrome, manganese, and quicksilver is active in the vicinity of Paradise, and chrome ore is being shipped by the Union Chrome Company.

ELDORADO COUNTY

(Special Correspondence.)—The main deep channel containing gold-bearing gravel on the Rising Hope mine, three miles east of Placerville, was recently tapped through the bedrock by a 14-ft. raise from the face of a 2000-ft. adit. The gravel contains gold and is cemented, like concrete. The gold is coarse in size. From a gold-pan of gravel, gold $2 in value was obtained.

A peculiar feature at the point tapped, 200 ft. or more below the surface, was the absence of water and a subterranean cavern between the lava-capped roof of the ancient river and the bed of gravel. The cavern was full of foul air.

The thickness of the gravel in the channel is 12 ft. Crosscuts will be run across the channel to the rim on each side, and drifts will be extended up and down stream to open a large tonnage before installing a stamp-mill. A revolving barrel has been in operation at the mine for some time reducing the loose bench gravel. The manager, George W. Engelhardt, states that as soon as sufficient tonnage of the cemented gravel has been developed a 10-stamp mill will be installed.

Placerville, August 29.

INYO COUNTY

L. D. Gordon, president and general manager for the Cerro Gordo Mines Co., reports that the same rich ore-shoot that has been worked down to the 200-ft. level had been caught on the 700-ft. level. The next step will be to develop this orebody on the 500-ft. level and later on the 1100-ft. level. The company is now making a net profit of $50,000 per month and has more than $200,000 in its treasury.

NEVADA COUNTY

(Special Correspondence.)—The Empire Mines Co. is installing a 2500-cu. ft. Ingersoll-Rand compressor of the Rogler type. It will be driven by a 500-hp. motor. At present 100 drills are in operation, and it is probable that 20 to 30 more will be added. Development below the 4600-ft. level continues to uncover ore of excellent grade; a 60-stamp mill is operating at capacity; 400 men are employed.

The shaft of the Allison Ranch has been dewatered to a point below the 700-ft. level and preparations are being made for mining in the deeper workings. The 20-stamp mill is supplied with good-grade ore from the third and fourth levels; 65 men are employed.

Preliminary surface work has been started at the Ford and Franklin mines, owned by Salt Lake capitalists, and adjacent to the Allison Ranch mine. The group is believed to contain extensions of the Allison Ranch vein-system.

The Washington Asbestos Co., Oakland capital, has purchased an extensive area of asbestos-bearing ground in the Washington district, a short distance from the Fairview mine. The deposit is about 150 ft. wide and has been traced for 5000 ft. The old 20-stamp Fairview mill has been acquired and a tram built from mine to mill. Arrangements have been made for the addition of special machinery to the mill for sorting. It is estimated that the product averages 10% of fine-quality asbestos. John D. Hoff, of Oakland, is president, and R. E. Conrad secretary-treasurer.

Grass Valley, August 25.

Junk from old mining plants continues to be gathered up wherever available. Old cable is being sold to logging camps at prices ranging from $25 to $50 per ton.

The 20-stamp mill, compressor-house, and change-house of the Blue Tent mine, owned by the California Powder Co., was destroyed by fire. The loss is estimated to be $15,000.

ORANGE COUNTY

Eighty-three wells are being driven in the Fullerton oilfield. There are 679 producing wells with a daily output of 43,485 barrels.

SHASTA COUNTY

(Special Correspondence.)—The Balaklala Copper Co. (National Copper Co.) has just paid its second dividend. The amount is 40c. per share. The mine is shipping 200 tons of ore per day to the Mammoth smelter at Kennett.

The Accident Gold Mining Co. of French Gulch has been disincorporated by order of the Superior Court. The Shasta Hills Mining Co. owns all the claims formerly held by the Accident.

Kennett, August 27.

The strike at the Balaklala has spread to the Mammoth mine at Kennett. Seventy-five Italian miners, being denied a $5 daily wage, struck. At the Iron Mountain mine 50 miners and other employees, bringing the number up to 200, are on strike. At the Stowell and Shasta King mines 50 men are out. Poor sanitation and a demand for an increase of one dollar in the daily wage are the principal points at issue. No violence has been reported. It is asserted that 1200 miners are out.

SISKIYOU COUNTY

(Special Correspondence.)—The Homestake mine, owned by R. S. Taylor, and located on Taylor creek near Snowden in the Salmon river mining district, was recently equipped with a quartz mill and began operations last Monday. The ore is taken from the stope to the mill without rehandling. A 5½-ft. Huntington mill is contemplated. George A. Foster of the Oro Grande Mining Co. is operating the McKeen, the Trail Creek mine, and the Facy mine near Callahan. A crew of 20 men is employed and 30 tons of ore per day is mined from the McKeen mine. J. E. McFadden of the Siskiyou Dredging Co. of Oak Bar is at Callahan looking after his mining interests.

William Koerner of the Grey Eagle copper mines at Happy Camp reports that the air-compressors at the mine would be ready for operation the first of September, and that they are short of miners for underground work.

Hornbrook, August 25.

TEHAMA COUNTY

(Special Correspondence.)—The Tedroc Chrome Co. is preparing to ship chrome ore to Red Bluff from its mines on Tedoc mountain, 52 miles west. The company built 12 miles of road from the mines to the county highway at a cost of $27,000. Seven 7-ton auto-trucks have arrived and will be put at work, day and night, about September 5. Four thousand tons of chrome must be shipped on a Government order before snow falls.

Red Bluff, August 27.

TRINITY COUNTY

(Special Correspondence.)—B. C. Clark of Sutter Creek has been made superintendent of the Globe mine at Dedrick, superseding Henry Sheafe, who was summoned to duty at Fort Leavenworth, Kansas. Sheafe is a captain in the engineering corps. Thirty men are employed at the Globe on development work.

Dedrick, August 27.

TUOLUMNE COUNTY

(Special Correspondence.)—A small crew of men is busy at the Keltz, making surface improvements. Mining operations will be superintended by G. E. Cheda. The Keltz is an old mine and produced quite liberally years ago. The Western Exploration & Mines Co. has acquired a group of mines near Italian Camp and will soon begin active mining operations. The Black Oak, for several weeks past operated with a greatly reduced force, has added just six men. Good ore was recently uncovered in the bottom level. The Rosendale mine, on Table mountain, near Jamestown, is looking well and the operators are pushing development. Development of the Driesam mine, at Arastraville, has been resumed with a small force of men.

Sonora, August 24.

COLORADO

BOULDER COUNTY

(Special Correspondence.)—The price of tungsten is satisfactory. Producing mines are increasing their output and the mills are operating at full capacity.

Eldora, August 31.

(Special Correspondence.)—Andrew Anderson has opened a body of high-grade silver ore in the old Caribou camp.——The Boulder Natural Gas Co., that has been incorporated for the purpose of drilling another well alongside the old Arapahoe well, has ordered materials and expects to commence operations next week.——Morris & Burr have just shipped 1400 lb. tungsten ore from their lease on the Wheelman property at Ferberite.——The aerial tram from the Good Friday mine to the canyon road at Boulder Falls will soon be in operation. The mine now produces steadily.——The tungsten mill of the Long Chance Mining Co., on Beaver creek, is operating day and night.——The Mountain Mines Co. is extending the Windy adit.——The Elsie mine is making preparations to deepen its present shaft.——The Fourth of July mine near Arapahoe peak is operating under the management of W. S. Harpel.

Eldora, August 18.

The Boulder Tungsten Production Co. is preparing to produce tungstic acid and ferro-tungsten. The output of high-grade ore and concentrate is continuing while the refining plant is approaching completion.

LAKE COUNTY

The Ibex and Garbutt mine on Breece hill resumed operation with a decreased working-force. A $4 per day wage scale has proved insufficient to attract men. Work has been resumed on the Bull's Eye property on Iron hill by C. Jarbeau of the Julia M. Company.

High-grade ore is being shipped from the Penrose shaft.

The Porter and Fairview properties are producing a combined output of 3000 tons per month.

Small orebodies have been discovered at the Penn properties on Breece hill. Lessees are operating on the New Monarch mine in Evans gulch.

In Leadville a strike of 2000 miners, pumpmen, and others has been discontinued and 80% of the workers have resumed their tasks, at a compromise wage scale of $4 per 8-hr. shift.——The Yak Tunnel Co. expects to be shipping at the rate of 300 or more tons per month. The Down Town M. Co. is shipping 400 tons daily.

Preparations are being made on an extensive scale to mine the low-grade molybdenum ores near Climax by the American Metal Co. and the Molybdenum Products Corp. The ore contains less than 1% of molybdenum.

PARK COUNTY

(Special Correspondence.)—The London Mining Co. has entered into a contract with the Pawnee Mining Co., for a mill-site, dumping-ground, and the use of the first 600 ft. of the latter's adit, in South Mosquito gulch. The work of enlarging the Pawnee adit from 4 by 6 ft. to 7 by 7½ ft. was begun in the middle of June. The surface improvements will cost $25,000. Haulage in the adit will be done with storage-battery locomotives. Electric power will be furnished by a steam-driven generator. Later a mill will be built and a hydro-electric power-plant installed.——The Louisiana-Colorado Mining Co. has secured a lease and option on the Dolly Varden mine on Mount Bross and will begin the construction of a 2-mile aerial tram. There is an estimated tonnage of ore on the dump of 75,000 tons, for the treatment of which a concentration and flotation mill will be erected. Etienne A. Ritter is manager.

Alma; August 20.

SAN JUAN COUNTY

During the month of August, 25 mines shipped 130 cars to outside smelters and 10 mines shipped 63 cars to the Silver Lake custom mill. The total production, including 30 cars shipped to the North Star mill, is 223 cars.

MICHIGAN

(Special Correspondence.)—No difficulty is anticipated in the shaft-sinking for the Old Colony-Mayflower to a depth of 1000 ft., where the broken zone is expected. Sinking is progressing rapidly.——The Michigan copper district has furnished 1200 men for the army and navy service, including the number taken in the first selective draft. This is a larger proportion than any county in the State, although the county, which produces little outside of copper, a war munition, could claim exemption for 75% of the men who have volunteered.——The length of life of Ahmeek is conservatively figured at 30 years, at a rate of production of rock at 1,800,000 tons per year.——The production of rock at the Franklin never runs below 1000 tons per day and often reaches 1400. It is hoped that ground will be broken by September 1 for the Seneca shaft. Surface buildings are going up rapidly. Several innovations in shaft-sinking, all calculated for speed and efficiency, will be inaugurated at the Seneca.——August production of Adventure will be 4000 tons.——Seventy-five cars of rock have gone to the mill from the South Lake shaft and 30 more will be sent down before September 1.——Installation of electric tramming at the Mohawk will materially benefit, since labor shortage has been a serious problem for some time. The quality of rock continues high and will compensate for diminution in tonnage.——Rock at the Phoenix is running 13 pounds of refined copper per ton, compared to 12.6 in June and 9.3 for the first six months of the present year. This betterment is due to mill improvements. The average tailing loss was six pounds for six months of this year. It is now close to four pounds. The mineral returns per ton are 18 lb.——Ahmeek's newer lower openings are in a grade of rock well up to the average and production will be increased in the expectation of securing new men. Dividends will be increased unless the price of copper goes down.—Three months ago the White Pine of the Calumet & Hecla was looked upon dubiously, as the shafts had exhausted the rich rock. Having returned the development investment, doubt as to its profitable operation on low-priced copper market was expressed. The recovery of large tailing losses by the flotation process and the discovery of richer ore in development make the outlook more favorable.

The Isle Royale Copper Co.'s mine has increased its rock shipments to the Point Mills stamp-mill, where the surplus from its own mill is handled. Over 800 tons per day is treated at the Point Mills, while the Isle Royale mill is milling its regular tonnage. Osceola is falling off and is 300 tons below its daily average for the first two weeks of August. Allouez, Franklin, Copper Range, Quincy, and Calumet & Hecla are making their average, but others are showing a slight falling off.

Houghton, August 27.

NEVADA

CLARK COUNTY

The Yellow Pine M. Co. declared a dividend of $50,000 for August.

ESMERALDA COUNTY

(Special Correspondence.)—A working station has been cut on the 1750-ft. level of the Atlanta mine, from which point a winze will be sunk to a depth of 2000 feet and development begun on the 1900 and 2000-ft. levels. The richest ore found in the Atlanta was opened on the 1750-ft. level, and A. I. D'Arcy, the manager, believed that the vein will extend into the shale and underlying alaskite. Shipments of high-grade copper-gold ore are being made to custom smelters.

The Jumbo Extension Co. has started prospecting new ground in the south-western section of the Poloverde claim and south-eastern end of the Velvet. Exploration of the latter is proceeding from the 1017-ft. level, and the Poloverde will be explored from the 830, 932, and 1017-ft. levels. The company is also driving from the Velvet shaft to intersect a large orebody recently opened at a depth of 650 ft. by the Goldfield Consolidated.

Shipping ore is being mined in the raise from the 300-ft. level of the Great Bend, the vein extending to this point from the 160-ft. level. High-grade material has been reported, and several shipments have been sent out recently that averaged $65 per ton in gold.

The main drift on the 500-ft. level of the Red Hill-Florence is advancing on a 4-ft. vein of profitable ore, and raise No. 2 is also exposing good ore. The vein is a part of the Florence orebody.

The Goldfield Merger Co. is concentrating work on the 1350-ft. level, where, 400 ft. from the Merger shaft, a deposit of copper-silver-gold ore has been uncovered, which averages $20 per ton. Small shipments are being made. It has been officially stated that the Jumbo Extension vein does not pass into Merger ground.

Goldfield, August 25.

HUMBOLDT COUNTY

Rochester Mines Co. reports 25 ft. of ore on the 900-ft. level. Average daily tonnage is 180.

The old Michigan and Nevada mine in the Humboldt range will be re-opened.

The Yerington M. & C. Co., operating the Adelaide mine, is making shipments of 5% ore at the rate of 200 tons per month.

The American M. & E. Co. has purchased the 'Rye Patch Mines.'

LINCOLN COUNTY

The Prince Consolidated is shipping 300 tons per day. A sulphatizing and flotation process is being successfully used. Diamond-drilling is being used for exploration.

MINERAL COUNTY

The Nevada Rand Co. has decided to construct a 10-stamp mill.

NYE COUNTY

(Special Correspondence.)—One of the heaviest bullion shipments made by the Tonopah Belmont went out last week. From the Belmont mill 100,918 oz. of silver-gold bullion and 35 tons of concentrate were shipped, and the Millers plant sent out 47,234 oz. and 25 tons of concentrate. Satisfactory reports have been received from the Surf Inlet and Eagle-Shawmut gold properties.

Shipments have been resumed from the large orebody on the 1200-ft. level of the Halifax, the product going to the West End mill. Cross-cutting from the 12th and 17th levels is proceeding, but without particular change.

Eight sulphur claims near Cuprite have been acquired by George Wishart, E. Campbell Douglass, and capitalists of New York and London.

Tonopah, August 24.

The White Caps mill was started on August 22. In the mine development is proceeding on the 300, 400, and 500-ft. levels. Additional work is reported on the White Caps No. 2 shaft.

A gold strike is reported on Tulicha mountain in the south-eastern end of the county.

The joint shaft being sunk by the Extension and Zanzibar companies is making satisfactory progress. Foundations for compressor and motor have been constructed. The machinery is on the ground and bunk-houses and dwellings are being constructed.

The Manhattan Con. M. Dev. Co. is sinking from the third to the fourth level. A block of ore 130 ft. slope distance between levels, 50 ft. in length and 28 ft. in average width, has been exposed. The average value is estimated to be $30 per ton.

WASHOE COUNTY

The Le Duc Mining Co. filed articles of incorporation for the conduct of a general mining business. The directors are C. H. White, J. P. Jones, and C. J. Hogan.

NEW MEXICO

SOCORRO COUNTY

(Special Correspondence.)—James P. Porters and W. T. McCaskey, of Lordsburg, interested in the Tyndale Copper Co., have taken over the New Golden Bell mine and mill at Rosedale, about 30 miles west of San Marcial on the Santa Fe railroad. The ore is gold, averaging about $20 per ton. About 15,000 tons is on the dump. The 50-ton mill is in good repair. The main shaft of the property is 430 ft. deep, with nearly 5000 ft. of drifts and cross-cuts. The vein varies from 5 to 14 ft. in width. H. H. Shelley, of Lordsburg, will take charge and commence work immediately.——Another sale in the Rosedale district was that of the Rosedale mine to the Rosedale Mines Corporation of Chicago. Harry Carey, of Denver and Chicago, is interested in the company. The property consists of two patented claims with a shaft 732 ft. deep and 5400 ft. of other workings. It is claimed that the mine has 33,000 tons of ore blocked. A new 100-ton mill will be erected immediately.

Socorro, August 12.

(Special Correspondence.)—The recent strike on the 1100-ft. level of the Champion mine has increased in width and value. The ore is of a heavy sulphide and is high-grade. This is apparently the top of a new orebody and is another strong indication that the ores in this district go deep. Mr. Kelly and associates are opening up placer ground on Mineral creek. Considerable work has been done and rockers are being used.

The Oaks Co., operating the Deep Down mine, has completed the surface installation and is ready to begin sinking. Driving is in progress to connect with the Maud S. mine, also being worked by this company.

During the week examinations have been made on the Bearup properties in the west side of the district. The owners have shipped some of the highest grade ore yet taken out of the district.

Mogollon, August 16.

(Special Correspondence.)—The Empire Zinc Co. and the Ozark M. Co. are operating in the Kelly mining district in the Magdalena range, twenty miles west of Socorro. One hundred lessees and 500 men are at work. Monthly shipments totaling 2500 tons of copper ore, 1000 tons zinc, and 2000 tons of lead ore are made. Two concentrators, one at Kelly and the other at Magdalena, are dressing the ores. The Empire Zinc Co. is developing the Lynchburg and the old Kelly mine and is constructing dwellings for workers and officers.

Magdalena, August 23.

(Special Correspondence.)—The Oaks Co. opened new ore in the Deep Down and Maud S. mines during the week. Ore shipments have been doubled. Production for August for the district will amount to nearly 12,000 tons. Mines and mills are working to capacity. The improved conditions of the road between Mogollon and Silver City has resulted in a large increase in the number of trucks and a large part of the freight is now handled entirely by auto-truck.

Mogollon, August 28.

PLANT OF THE EMPIRE ZINC CO., KELLY, NEW MEXICO

OREGON

CLATSOP COUNTY

Serious forest fires are raging in 15 counties. Greatest damage is reported from the Minain, Deschutes, and Cascade reserves. The majority of the fires were found to be of incendiary origin.

JOSEPHINE COUNTY

(Special Correspondence.)—Robert Grimmett, who located chrome deposits near Holland, is now shipping the ore to Grants Pass.

Chrome deposits near Wolf creek are being developed by Dr. Reddy of Grants Pass.

Th Waldo copper mine, at Takilma, is again in operation with a crew of 40 men under same management.

Grants Pass, August 13.

(Special Correspondence.)—Three motor-trucks have arrived at Grants Pass for the California Chrome Co.'s mine near Selma. Two were of 4-ton each, the third of 1½-ton capacity. Four more 4-ton trucks will follow. The distance from the mine to the railroad is 21 miles, with grades ranging as high as 20%. H. L. Egan accompanied the outfit.

It is stated that 10,000 tons of ore is in sight, half of which the contractors expect to move before the rainy season arrives. Building of the truck-road from the mine, the loading-point on the Grants Pass-Coast road, 15 miles south-west of Grants Pass, cost $32,000. The loading and unloading is accomplished by gravity. The chrome tonnage over this railroad for August will be 100% over July shipments.

R. L. Thompson, representing Seattle people, has taken an option on the Osgood placer mine, a mile south of Waldo. This property is owned by F. H. Osgood of Seattle. A shipment of 23 tons of the bedrock has been made to Seattle. It is said to be rich in gold and platinum.

A small crew is employed at the Neil-Success group of gold quartz mines on Fidlers gulch 7 miles west of Kerby preparing for the winter's run. The mine is operated by water-power. The property is under the direction of Fred Furth of Seattle. The group consists of the Neil and Mood properties, which have been producers, the ore being ground in arrastras. The underground works are extensive. A 50-ton rotary mill was recently installed on the properties. Water furnishes power from 4 to 6 months annually.

Grants Pass, August 14.

(Special Correspondence.)—More ore-bunkers are being built at Waters Creek, the terminus of the California & Oregon Coast railroad, to accommodate the increasing ore production from the Takilma district.

Grants Pass, August 20.

(Special Correspondence.)—W. C. Williams, representing the Chemical & Alloy Ores Co. of Cleveland, Ohio, is buying chrome ore, tungsten, and other minerals in this region.

E. A. McPherson of Grants Pass has uncovered a rich shoot of copper ore at the Old Crow mine in the Monumental district, owned by F. E. Bausman of San Francisco. The vein is 3½ ft. wide on a 100-ft. drift in porphyry, assays $7 in gold, and $4 in silver. Negotiation is being made for its sale to Eastern men.

It is rumored that the old Monumental, which has been idle for a number of years, will soon be operated again by Eastern capital. There is a large tonnage of ore running $10 to $40 in gold.

It is reported that the Monkey Creek mine is soon to be an antimony producer again.

The Diamond Creek cinnabar mine, owned by W. J. Ehrman, is being equipped with a furnace.

John Griffin, of Kerby, accompanied by J. C. Kendau and Ernest Rackliff of Reno, Nevada, recently inspected the Griffin cinnabar mines on Diamond creek, with a view to purchasing.

R. R. Horner, of the United States Bureau of Mines, has been in southern Oregon mines the past week. He is giving special attention to the black sand deposits.

Grants Pass, August 20.

MALHEUR COUNTY

A cinnabar discovery in the Owyhee desert has been made by T. H. Murphy and W. S. Brezt.

TEXAS

Buildings at a cost of $175,000 are being erected for the Texas School of Mines at El Paso.

WYOMING

The United States government on October 10 next will offer for lease to the highest bidder 15,000 acres of land for oil and gas exploration at the superintendent's office of the Shoshone agency at Fort Washakie. All tracts are in 160-acre units. Leases will be made for 20 years; cash bonus per acre, 20% at time of sale and balance on notification of approval; cash rental $1 per acre in advance; one-eighth royalty; drill and complete rig on each tract within one year; limit of 2500 acres to one buyer.

Personal

Note: The Editor invites members of the profession to send particulars of their work and appointments. This information is interesting to our readers.

F. L. SIZER is in Arizona.

R. A. F. PENROSE is at Los Angeles.

CHARLES HARDY has been here from New York.

E. M. HAMILTON is on his way to Pachuca, Mexico.

JOHN J. CRAIG is in the training camp at Fort Riley, Kansas.

W. E. SIMPSON has returned to Cobalt from western Quebec.

H. KENYON BURCH has returned to Los Angeles from Arizona.

H. L. SMALL is in the training camp at Plattsburg, New York.

R. M. NICHOLS is in France 'with Uncle Sam,' we are informed.

E. H. GARY and D. C. JACKLING have returned to San Francisco.

JOHN W. MERCER was in San Francisco last week, on his way to Arizona.

WADE L. LEWIS is with the 18th Regiment of Railway Engineers in France.

GEORGE S. BLAIR is at the Officers Training Camp near Chattanooga, Tennessee.

RALPH W. STONE, of the U. S. Geological Survey, was in San Francisco this week.

LOUIS DOUGLAS, son of JAMES S. DOUGLAS, has a commission in the Field Artillery.

NELSON DICKERMAN has returned to San Francisco from a visit to Placer county.

ALBERT L. WATERS has returned to Los Angeles from Detroit and Grand Rapids, Michigan.

T. S. DUNN is now captain in the 304th Regiment of Engineers at Camp Meade, Maryland.

F. R. WEEKES has been engaged in examination work in the Trout Lake district, British Columbia.

HARRY J. SHEAFE has been given a commission as captain of engineers and is now at Fort Leavenworth.

WILLIAM G. REYNOLDS has gone from Anchorage, Alaska, to the Officers Training Camp at Fort Sheridan, Illinois.

JAMES E. MOORE, lately at Goldfield, is now at the Reserve Officers Training Camp at the Presidio, San Francisco.

JAMES S. DOUGLAS sailed from New York on September 1, on his way to France, where he will work for the Red Cross.

BAYLIES C. CLARK, of Sutter Creek, has been appointed superintendent for the Globe Mines, in Trinity county, California.

HORATIO C. RAY has received a commission as first lieutenant in the Engineer Officers Reserve Corps and has been ordered to Washington.

H. H. NICHOLSON has resigned as superintendent for the Plinco Copper M. & M. Co. and is examining mines at Twin Lakes, Colorado.

HAROLD COGSWELL, formerly in Amador county, California, is now first lieutenant with the 316th Regiment of Engineers at American Lake, Washington.

F. N. NELSON, superintendent for the Colorado Fuel & Iron Co. in the Hanover-Fierro district, New Mexico, has completed his work and will leave for an extended vacation in Michigan and Minnesota.

The SAN FRANCISCO SECTION OF THE A. I. M. E. will meet on Tuesday, September 11, at the Engineers Club at the usual hour. A discussion on the labor problem in its relation to the mining engineer will be started by T. A. Rickard.

THE METAL MARKET

METAL PRICES

San Francisco, September 4

Aluminum-dust (100-lb. lots), per lb	$1.00
Aluminum-dust (ton lots), per lb	$0.95
Antimony, cents per pound	17.00
Electrolytic copper, cents per pound	28.50
Pig lead, cents per pound	10.25—11.25
Platinum, soft and hard metal, respectively, per ounce	$105—111
Quicksilver, per flask of 75 lb	$115
Spelter, cents per pound	10.5
Tin, cents per pound	60
Zinc-dust, cents per pound	20

ORE PRICES

San Francisco, September 4

Antimony, 45% metal, per unit	$1.20
Chrome, 40% and over, free SiO_2 limit 8%, f.o.b. California per unit	$0.55—0.70
Magnesite, crude, per ton	$8.00—10.00
Tungsten, 60% WO_3, per unit	$25.00—28.00
Molybdenite, per unit for MoS_2 contained	$40.00—45.00

Manganese, Carnegie schedule less freight to Chicago. Less than 40% special quotation.

Manganese prices and specifications, as per the quotations of the Carnegie Steel Co. schedule of prices per ton of 2240 lb. for domestic manganese ore delivered, freight prepaid, at Pittsburgh, Pa., or Chicago, Ill. For ore containing

	Per unit
Above 49% metallic manganese	$1.00
46 to 49% metallic manganese	0.98
43 to 46% metallic manganese	0.95
40 to 43% metallic manganese	0.90

Prices are based on ore containing not more than 8% silica nor more than 0.2% phosphorus, and are subject to deductions as follows: (1) for each 1% in excess of 8% silica, a deduction of 15c. per ton, fractions in proportion; (2) for each 0.02% in excess of 0.2% phosphorus, a deduction of 2c. per unit of manganese per ton, fractions in proportion; (3) ore containing less than 40% manganese, or more than 12% silica, or 0.225% phosphorus, subject to acceptance or refusal at buyer's option; settlements based on analysis of sample dried at 212° F., the percentage of moisture in the sample as taken to be deducted from the weight. Prices are subject to change without notice unless specially agreed upon.

EASTERN METAL MARKET

(By wire from New York)

September 4.—Copper is dull, being nominal at 26.00 to 25.50c. Lead is quiet, prices being a little lower at 10.50 to 10.30c. Zinc is stagnant and prices nominal at 8.25c. all week. Platinum remains unchanged at $105 for soft metal and $111 for hard.

SILVER

Below are given the average New York quotations, in cents per ounce, of fine silver.

Date			Average week ending		
Aug.	29	89.75	July	24	79.12
"	30	90.75	"	31	78.20
"	31	90.75	Aug.	7	80.50
Sept.	1	90.75	"	14	82.95
"	2 Sunday		"	21	87.16
"	3 Holiday		"	28	88.50
"	4	93.67	Sept.	4	91.12

Monthly Averages

	1915	1916	1917		1915	1916	1917
Jan.	48.85	56.76	75.14	July	47.52	63.06	78.92
Feb.	48.45	56.74	77.54	Aug.	47.11	66.07	85.40
Mch.	50.61	57.89	74.13	Sept.	48.77	68.51	....
Apr.	50.25	64.37	73.51	Oct.	49.40	67.86	....
May	49.87	74.27	74.61	Nov.	51.88	71.60	....
June	49.03	65.04	76.44	Dec.	55.34	75.70	....

COPPER

Prices of electrolytic in New York, in cents per pound.

Date			Average week ending		
Aug.	29	26.00	July	24	27.04
"	30	25.75	"	31	27.25
"	31	25.50	Aug.	7	28.46
Sept.	1	25.50	"	14	27.62
"	2 Sunday		"	21	28.00
"	3 Holiday		"	28	26.41
"	4	25.50	Sept.	4	25.65

Monthly Averages

	1915	1916	1917		1915	1916	1917
Jan.	13.60	24.30	29.53	July	19.09	25.66	29.67
Feb.	14.38	26.62	34.57	Aug.	17.27	27.03	27.42
Mch.	14.80	26.65	36.00	Sept.	17.69	28.28	....
Apr.	16.64	28.02	33.16	Oct.	17.90	28.50	....
May	18.71	29.02	31.69	Nov.	18.88	31.95	....
June	19.75	27.47	32.57	Dec.	20.07	32.89	....

LEAD

Lead is quoted in cents per pound, New York delivery.

Date			Average week ending		
Aug.	29	10.50	July	24	10.29
"	30	10.40	"	31	10.55
"	31	10.40	Aug.	7	10.87
Sept.	1	10.30	"	14	10.87
"	2 Sunday		"	21	10.79
"	3 Holiday		"	28	10.61
"	4	10.30	Sept.	4	10.38

Monthly Averages

	1915	1916	1917		1915	1916	1917
Jan.	3.73	5.95	7.64	July	5.59	6.40	10.93
Feb.	3.83	6.23	9.01	Aug.	4.62	6.28	10.75
Mch.	4.04	7.26	10.07	Sept.	4.62	6.86	....
Apr.	4.21	7.70	9.38	Oct.	4.62	7.02	....
May	4.24	7.38	10.29	Nov.	5.15	7.07	....
June	5.75	6.88	11.74	Dec.	5.34	7.55	....

ZINC

Zinc is quoted as spelter, standard Western brands, New York delivery, in cents per pound.

Date			Average week ending		
Aug.	29	8.25	July	24	8.60
"	30	8.25	"	31	8.75
"	31	8.25	Aug.	7	8.75
Sept.	1	8.25	"	14	8.75
"	2 Sunday		"	21	8.66
"	3 Holiday		"	28	8.33
"	4	8.25	Sept.	4	8.25

Monthly Averages

	1915	1916	1917		1915	1916	1917
Jan.	6.30	18.21	9.75	July	20.54	9.90	8.98
Feb.	9.05	19.99	10.45	Aug.	14.17	9.03	8.58
Mch.	8.40	18.40	10.78	Sept.	14.14	9.18	....
Apr.	9.78	18.62	10.20	Oct.	14.05	9.92	....
May	17.03	16.01	9.41	Nov.	17.20	11.81	....
June	22.20	12.85	9.63	Dec.	16.75	11.26	....

QUICKSILVER

The primary market for quicksilver is San Francisco, California being the largest producer. The price is fixed in the open market, according to quantity. Prices, in dollars per flask of 75 pounds:

Week ending

Date					
Aug.	7	115.00	Aug.	21	115.00
"	14	115.00	"	28	115.00
			Sept.	4	115.00

Monthly Averages

	1915	1916	1917		1915	1916	1917
Jan.	51.90	222.00	81.00	July	95.00	81.20	102.00
Feb.	60.00	295.00	126.25	Aug.	93.75	74.50	115.00
Mch.	78.00	219.00	113.75	Sept.	91.00	75.00	....
Apr.	77.50	141.60	114.50	Oct.	92.90	78.20	....
May	75.00	90.00	104.00	Nov.	101.50	79.50	....
June	90.00	74.70	85.50	Dec.	123.00	80.00	....

TIN

Prices in New York, in cents per pound.

Monthly Averages

	1915	1916	1917		1915	1916	1917
Jan.	34.40	41.76	44.10	July	37.38	38.37	62.60
Feb.	37.23	42.60	51.47	Aug.	34.37	38.88	62.53
Mch.	48.76	50.50	54.27	Sept.	33.12	36.66	....
Apr.	48.25	51.49	55.63	Oct.	33.00	41.10	....
May	39.28	49.10	63.21	Nov.	39.50	44.12	....
June	40.26	42.07	61.93	Dec.	38.71	42.55	....

ORES

Tungsten: Firmness characterizes the market with demand reported as good. High-grade concentrate, 60%, is held at $23 to $26 per unit with $25 reported as paid. Ferro-tungsten is held at $2.30 to $2.60 per lb. of contained tungsten with both domestic and foreign demand excellent.

Molybdenum and antimony: There is no change in quotations. Molybdenum ore is scarce with $2.10 to $2.20 per lb. of MoS_2 asked for 90% concentrate. Antimony ore is also hard to get.

PRICE FIXING AND TAXATION

The National City Bank of New York's September circular says in part: Price fixing and the new measures for taxation are still factors of uncertainty in business calculations. The authorities will closely supervise the movement of the wheat crop out of first hands, through the process of manufacturing into flour, and through distribution to consumption, determining the profit of every handler.

The price of coal has now been fixed by an order of the President, varying with the different fields. The price of steel is under inquiry and may be authoritatively fixed for government use. There is every reason to believe that the authorities will be guided in this policy by what they conceive to be the best interest of the public, and that they will seek to make prices that are fair to producers, but the whole policy is an experiment, and unless it is carried to extremes and the results are very pronounced, it will never be known whether the experiment is successful or not. It is not by any means certain that prices made artificially low are the most desirable. Prices have an important function in bringing supply and demand into equilibrium, adjusting them to each other.

The Senate still has the revenue bill under consideration, debating proposals to increase the levies upon incomes and profits. The advocates of extreme levies upon profits insist upon viewing book profits as cash in hand, but in the present condition of industry they are in actual use, and necessary in most cases to the effectual conduct of the industries. Bethlehem steel is now going to the public market with its securities, to raise money to carry on and enlarge its business. Enlargement of industrial capacity concerns the public as well as the proprietors, since war supplies are produced. The enlargements are made upon a level of costs probably 50% above normal, and nobody knows what these investments will be worth after the War. The profits which Senators are naming are in many instances paper profits, which stockholders have not seen and never will see, because the shrinkage in the value of inventories and plant investments will cause them to disappear.

A favorite argument for high taxes has been that the British government has exacted 80% of the profits in excess of pre-war level without unfavorable effects. Since this 80% rate has been in effect only a month it is too early to report on the effects.

Eastern Metal Market

New York, August 29.

There is but one word with which to describe the condition into which all metal markets have fallen—stagnation. This was induced and perpetuated by the Government's policy of regulating prices and it will continue until definite adjustments occur as a result of known bases on which to proceed. Announcement of prices, at least of copper, are expected within a week or so. Quotations generally are nominal.

Copper is dead, lower and nominal.

Tin is quiet and largely nominal.

Lead is dull but steady.

Zinc is stagnant and lower.

Antimony is stationary and inactive.

Aluminum is weaker.

Government announcement of steel prices is known to be close at hand, but prices have receded slightly on their own account. At Pittsburgh active business has been done almost uniformly at the expense of prices. Plates have fallen to 8c. per pound, a decline of $20 per net ton; bessemer and basic steel-making pig irons have declined $2 and $4 per ton respectively and in finished steel recent high prices have been decidedly reduced. The trade is at sea as to the extent to which the readjustment will go. At a recent important meeting with tin-plate makers in Washington the food administrator showed that there was an indicated shortage of about 2,000,000 boxes, with a greater one in prospect for 1918.

COPPER

Inactivity could scarcely be more pronounced than the present state to which the market has fallen. The trade itself scarcely knows just where the market stands and inquiry among dealers is almost useless so far as ascertaining real basic conditions. No parallel to present dullness can be recalled. The entire situation is the result of uncertainty regarding the action of the most important factor in the situation—the Government, which will buy not only for itself but for its Allies as well. Various rumors have been let out to influence the market but without effect. One important rumor is to the effect that the Government's price will be close to 25c. and not less than 20c. per lb. It is mere guess, for no one knows. Events, however, are moving more rapidly and it is evident that a decision or an announcement is near. The appointment of a dictator for copper will probably be followed by definite news as to price. It is believed that cost data are sufficiently known to warrant a decision. Serious strikes out West cause a disturbing influence and the court decision against the Miami Copper Company in its flotation suit was a bear influence yesterday in copper stocks. Quotations are purely nominal and nearly guesswork. Yesterday both Lake and electrolytic were quoted at 26c., New York, for August and September delivery, with 24.50 to 25.50c. the quotation for last quarter. The London market continues unchanged at £137 for spot electrolytic and £133 for futures.

TIN

Dullness, even more pronounced than in past weeks, pervades this market, and this is also true abroad. There is no desire to buy except where needs are vital. With Government uncertainties predominant, consumers are very cautious. Supplies of spot Straits tin seem light with the demand also light. An August 22, small sales, amounting to about 25 tons, were reported, some being offered as low as 61c., New York. More tin was offered than there were buyers. On August 23, there were sales of 50 tons of Straits tin at 61.75c., New York, but on the next day there was no business and no inquiry. On Monday of this week, August 27, the market was very dull with more sellers than buyers. Spot Banca tin was sold at 58.75c., New York, and there were also on that day fair sales of December shipment from the Straits at 56.62½c. Yesterday the market was lifeless. There was no inquiry except from dealers or consumers who desired information as to the state of the market. Yesterday Straits tin was quoted at 61.50c., New York, as compared with 61.37½c. on the 27th and 61.75c. an the 22nd and 23rd.

LEAD

With the price of 8c. per lb. paid by the Government for its recent purchases still in the minds of consumers, it is not surprising that buying is at a low ebb. The market does not improve. Demand is less than at any previous time and prices have receded. Yesterday the New York quotation was 10.50c., with the St. Louis price at 10.40c. A continuation of the present inactive condition is looked for until definite action is taken by the Government. This is expected soon in all metals. Buyers are extremely cautious and the demand for manufactured goods is reported to have fallen off. The rather radical prices put on coal also tend to influence consumers of metals to expect lower prices. Yesterday the quotation in the outside market was 10.50c., New York, at which price small sales of carload lots were made. The price of the American Smelting & Refining Co. still stands at 11c., New York. Sales of re-sale carload lots of lead for September-October delivery are reported to have been made yesterday at 10.37½c., New York. Some interest has been taken in the first mid-year report of the U. S. Geological Survey, just issued, on the output of refined lead. This is put at 2,306,062 tons, or 20,495 tons in excess of one-half of the 1916 production. The increase is due almost entirely to a gain in the output of lead from foreign ores and bullion of 20,086 tons. Exports decreased 19,231 tons for the half year but the apparent domestic consumption is around 38,265 tons.

ZINC

The market is 'very sick', as one dealer puts it; 'it has typhoid and other complications.' It is without life and stagnant, no one desiring to buy and no sales of any consequence being reported. One broker reports the sale of a carload at 7.75c., St. Louis, but in general transactions are so few that quotations are varied and there is no firm price fixing based on sales. They are entirely nominal and depend on the viewpoint of the holder or prospective purchaser. Some are holding firm to a quotation of 8.25 to 8.50c., St. Louis, or 8.50 to 8.75c., New York, while others are as low as 8.25c., St. Louis, or 8.25 to 8.50c., New York, for early delivery. Probably a fair average nominal quotation is 8.12c., St. Louis, or 8.37c., New York. Later or fourth quarter deliveries are from ⅛ to ¼c. per lb. higher. Any pressure exerted, as evidenced in the last week or so, has resulted in recessions and the market is in no condition to withstand it. Unless foreign buying in particular materializes, no great improvement is looked for and it is a matter of comment that Great Britain and France have been able to go so long without further purchases. Large buying is only slowly accumulating and will be in full swing ere many weeks. The removal of some of the Government and other uncertainties will clarify the situation.

ALUMINUM

No. 1 virgin metal, 98 to 99% pure, continues inactive with the quotation now generally acknowledged as 47 to 49c. per lb., New York. The market is weak on re-sale offerings and no demand.

ANTIMONY

The market is without interest and lifeless. Chinese and Japanese grades are held at 15 to 15.50c., New York, duty paid.

Company Reports

BUTTE & SUPERIOR MINING COMPANY

The net mining profit for the year amounted to $8,792,131.23, to which is added miscellaneous income of $81,314.62. That is equivalent to $30.57 per share on the 290,197 shares outstanding at the close of the year. During the year two quarterly dividends of 75c. each per share and two of $1.25 each were declared. In addition extra dividends amounting to $4 per share were paid. The contract of the company with the American Zinc, Lead & Smelting Co. for the treatment of its concentrate at Caney and Dearing, Kansas, which was for two years, expired on March 27, 1917, and from that date shipments have been made to the American Metal Co. The former contract was based on the smelting only of the concentrate, the metal being sold by the Butte & Superior company, but the new contract is for the sale of the concentrate based upon current published market quotations. The price for spelter during the year 1916 was 12.63c. Development disclosed importat extensions of the orebodies on practically all levels, from the 700 to the 1800-ft. Developments on the 1000, 1100, and 1300-ft. levels show a continuance of the orebodies easterly beyond the end-line of the Black Rock claim for a maximum distance of 500 ft., thus leading to the expectations that profitable orebodies may be found to the east of the ore-shoots on the 700, 800, and 900-ft. levels. On the 1500-ft. level a continuation of the orebodies disclosed above has been found. On the 1600-ft. orebodies of great width and of good grade have been exposed. These are practically continuous for a distance of 850 ft., varying from 16 to 65 ft. wide. The total development during the year was 10,845 ft. of driving, 4832 ft. of cross-cutting, 3015 ft. of raising, and 2649 ft. of shaft-sinking. The total length of underground workings amounts to 102,445 ft. The ore reserves are estimated at 1,044,850 tons containing an average of 17.7% zinc, and 5.65 oz. silver per ton. The ore mined and hoisted during 1916 was 626,803 tons, at an average cost of $4.50 per ton, exclusive of taxes. The average content of the ore produced during the year was 15.55% zinc and 6.57 oz. silver per ton. The mill-recovery averaged 93.1% of the zinc content, being an increase of nearly one per cent over the results obtained in 1915. The principal additions to the mill consisted of further facilities for settling slime and concentrate, so as to recover a larger quantity of the mill-water; also the filtering plant was remodeled to admit of more economical handling of the increased tonnage. The metal content of the output is summarized as follows:

	Gold, oz.	Silver, oz.	Copper, lb.	Lead, lb.	Zinc, lb.
Zinc concentrate	5,539,600	3,699,950	1,861,713	9,955,691	181,624,842
Zinc concentrate, average assay	0.0323	21.544	0.542%	2.898%	52.876%
Lead concentrate, 5470 tons	226.734	150,060.70	38,556	4,099,977	1,523,363
Lead concentrate, average assay	0.0415	27.433	0.352%	37.477%	13.925%

ESPERANZA LIMITED

The political conditions are reported to have been more satisfactory during the year 1916, although short stoppages of work were caused by strikes of the Mexican miners. The mill started in April and was operated continuously thereafter. The ore treated amounted to 113,921 dry metric tons, and the reserves on January 1, 1917, were 111,723 dry metric tons having an estimated content of 61,086 oz. gold, and 243,477 oz. silver. Development work amounting to 8900 ft. was done on the San Rafael vein and its branches on the upper levels. The ore reserves should yield a profit of about $650,000 U. S. currency. The profit and loss account for 1916 shows a net return of £8412, to which is added a balance from the last account of £4818. Out of this an income tax was paid of £9638, and the balance of £3592 was carried forward,

RAY CONSOLIDATED COPPER CO.

The report of operations for the second quarter of 1917 shows a total income of $3,679,898, disbursement to stockholders $1,577,179, and a net surplus of $2,102,719. The total production was 22,255,598 lb., or an average of 7,418,533 lb. per month. During the quarter 892,200 tons of ore, averaging 1.66% copper, was milled. The recovery was 75.48%. On the sulphide mineral it was 85.23%. The underground development was 19,211 ft. The milling cost was 74.69c. and the mining 98.257c. per ton. The average cost of the copper was 11.272c. per pound. The quarter's profit is equivalent to $2.33 per share on a selling-price of 27.562c. per pound of copper. The quarterly dividend amounted to $1 per share; of this 50c. went to earnings and 50c. to capital distribution. Total dividends and capital distribution to the end of the quarter amount to $12,085,458.

CONSOLIDATED ARIZONA SMELTING CO

The annual report of the Consolidated Arizona Smelting Co. for the year ended December 31, 1916, shows the following:

At the Blue Bell mine, during the year, 4819 ft. of development work was done at a cost of $13.11 per foot, or $0.84 per ton of ore produced. This work developed 319,590 tons of ore at a cost of 20c. per ton of such new ore developed. The output of the Blue Bell mine was 75,070 tons of ore, containing an average of $1.70 per ton in gold and silver, and 8.278% copper, which was a distinct improvement over the value of the ore in 1915. The cost of Blue Bell ore delivered at Humboldt was $3.565 per ton.

At the De Soto mine a long cross-cut was extended west under the Whale outcrop, and a drift was run 252 ft. along the mineralized zone. The total development and exploration work amounted to 1814 ft., which cost $17,301.40, or $9.53 per foot. This was at the rate of 50c. per ton for ore produced during the year, and proved up 72,882 tons of new ore, against which the charge for development would be $0.237 per ton. The De Soto production amounted to 34,382 tons of ore of an average value per ton of $1.58 in gold and silver, and 3.375% copper. The De Soto ore is highly silicious and none of it is suitable for direct smelting. The sulphide ore concentrates readily, but the silicious ore from the upper levels, not being amenable to concentration, is used as converter lining and flux at the smelter. It is expected to produce 50,000 tons during 1917. The cost of delivery of ore at Humboldt was $3.358 per ton, being about the same as that of the Blue Bell ore. The estimate of De Soto ore reserves at the beginning of 1917 was 103,500 tons available, containing $1.50 per ton in gold and silver, and 3.2% copper.

The concentration mill treated 80,804 tons of ore during the year. The cost of milling, including flotation royalty, was $1.51 per ton. A recovery of 80.3% of the gold, 79.9% of the silver, and 92.5% of the copper was made.

The ores now being mined come from the deeper levels of the mine and are generally more basic in character and contain a higher percentage of combined sulphides, but with less oxidation, hence a better recovery is obtainable, but a slightly lower grade of concentrate results, with a lower ratio of concentration.

The smelter at Humboldt during the year smelted 103,748 tons of ore and concentrate, the production being 5972 oz. of gold, 150,012 oz. of silver, and 11,989,139 lb. of copper, which was about double the output of 1915. Of the above, 6,159,639 lb. of copper was produced from custom ores. The cost of receiving, crushing, tramming, storing, and sampling ores and concentrate amounted to $0.539 per ton of material smelted. The cost of roasting and smelting to matte, including copper supervision and general charges, was $3.944 per ton treated.

Value of copper, gold, and silver produced	$3,686,535.24
Total cost of production	2,813,770.53
Balance	$ 872,764.71
Interest charges and mortgage bonds	13,750.00
Net operating profit	$ 859,014.71

Mining Decisions

OIL WITHDRAWALS—PRIOR DISCOVERY

When during the two weeks preceding the Presidential withdrawal of oil lands of May 6, 1914, oil locators had their property examined, made a lease, and the lessee made an agreement and ordered material for drilling wells on the premises, and thereafter pushed the wells to completion diligently at an expenditure of over $22,000, the locators through their lessees were at the date of withdrawal "bona fide occupants" in the "diligent prosecution of work leading to the discovery of oil or gas," and were not affected by the withdrawal order. Affirmative testimony by the claimants that oil had been found in test wells in August, 1913, in such quantities as to justify further work toward discovery was given more weight than merely negative testimony of the government's witnesses to the effect that more than a year later no evidence of oil could be found in such test holes.

United States *v.* Ohio Oil Co. (Wyoming), 240 Federal, 996. January 31, 1917.

OIL WITHDRAWALS—GROUP DEVELOPMENT THEORY

Locators of a group of oil placers made a contract for the exploration of one of their claims with a view to developing oil thereon, and if oil should be developed, to exploring the remaining claims in the group. Before oil was discovered on the first claim the withdrawal order of September, 1909, was made. The locators claimed that the work being done on the first claim was for the benefit of the remaining locations, and that they should be protected as persons who were at the date of the order "in diligent prosecution of work leading to discovery of oil or gas." Held, that the group development theory applies only to annual assessment work done after a discovery is made and the location perfected, and that the claims on which work was not actually being done at the date of the order were thereby withdrawn.

United States *v.* Stockton Midway Oil Co. (California), 240 Federal, 1006. January 5, 1917.

OIL LEASE—DIVISION OF ROYALTIES

When several owners of adjoining tracts unite in a single lease thereof to a third party for oil and gas purposes, royalties must be divided in the proportion that the area of the tract owned by each co-lessor bears to the total tract leased, regardless of the ownership of the tract or tracts upon which the wells are actually drilled.

Lynch *v.* Davis (West Virginia), 92 South-eastern, 427. May 22, 1917.

Recent Publications

MAGNESITE. By Hoyt S. Gale, U. S. Geological Survey Bull. 666-BB.

PORTLAND CEMENT. By E. F. Burchard, U. S. Geological Survey, Bull. 666-S, 5 pages.

GRAPHITE IN 1916. By H. G. Ferguson, U. S. Geological Survey. Mineral Resources of 1916.

CARBON MONOXIDE POISONING IN THE STEEL INDUSTRY. By J. A. Watkins; Tech. paper 156, U. S. Bureau of Mines.

ADVANCED FIRST-AID INSTRUCTIONS FOR MINERS; U. S. Bureau of Mines. In this little pamphlet first-aid instructions are standardized for the first time.

LIMITS OF COMPLETE INFLAMMABILITY OF MIXTURES OF MINE GASES AND OF INDUSTRIAL GASES WITH AIR. By G. A. Burrell and A. W. Ganger; Tech. paper 150, U. S. Bureau of Mines.

THE WET THIOGEN PROCESS FOR RECOVERING SULPHUR FROM SULPHUR-DIOXIDE IN SMELTER GASES. By A. E. Wells. Bulletin 133, U. S. Bureau of Mines. A critical study of the process on a laboratory scale demonstrates that the technical operation of the process can be carried out successfully.

THE UTILIZATION OF PYRITE OCCURRING IN ILLINOIS BITUMINOUS COAL. By E. A. Holbrook; Circular No. 5, Eng. Exp. Station, University of Illinois.

THE THRALL OIL FIELD. By J. A. Udden and H. P. Bybee; ozokerite from the Thrall oil field, by E. P. Schoch; bulletin of the University of Texas, No. 66.

ANNUAL YEAR-BOOK OF THE MICHIGAN COLLEGE OF MINES, Houghton, Michigan. A catalogue of information with courses.

PHOSPHATE ROCK IN 1916. Mineral Resources of the U. S., 1916. Part 11, pp. 29-41; published August 13, 1917; prepared by Ralph W. Stone.

SLATE IN 1916. Mineral Resources of the U. S., 1916. Part II, pp. 61-72; published August 11, 1917; prepared by G. F. Loughlin.

THE COAL FIELDS AND COAL INDUSTRY OF EASTERN CANADA. By Francis W. Gray. Bulletin No. 14, of the Department of Mines of Canada, Mines Branch. Pp. 67. Ill., pocket-map, and index.

This publication gives a general description of the coal-fields, the mines, and plants. Also a geological and historical, as well as economic description of the coal regions of eastern Canada.

THE IRON-ORE DEPOSITS OF VANCOUVER AND TEXAS ISLANDS. By W. M. Brewer. Bureau of Mines report, Bulletin No. 3 of the British Columbia Department of Mines. Pp. 41. Maps and sketches. Victoria, B. C., 1917.

This report deals with the history, bibliography, geography, topography, and geology of the iron-ore deposits on the islands off the mainland of British Columbia. It describes the numerous deposits, the character of the ores, and makes an estimate of the amount of ore available—about 13,000,000 tons.

'THE CANADIAN MINING MANUAL', 1916-1917. Edited by Reginald E. Hore. Mines Publishing Co., Toronto. 445 pp. Ill. Price, $3.

This is the year-book published by the 'Canadian Mining Journal', of which the author is editor. The book is what it pretends to be: a trustworthy compilation of information concerning mining enterprises in Canada. Besides giving necessary details concerning mining companies operating in the Dominion, it furnishes reliable information concerning the various mineral products of the country, the distribution of the metals, with maps and other illustrations, including a number of portraits of the leading mine managers. It is an attractive and useful volume.

EDITORIAL

T. A. RICKARD, Editor

HOW the members of our profession are responding to the call to arms is indicated by the list of the members of the Institute now in military service as given in the current bulletin of the society. The list could be much enlarged by obtaining the names of members that have been fighting with our Allies during the earlier stage of the War.

JUST before the War the German government endeavored to persuade the British government to intervene jointly in Mexico, but the suggestion was declined, asserts Mr. James Keeley, editor of the Chicago 'Herald.' During the Spanish-American war Germany organized an effort to aid Spain against the United States and it was England that killed the scheme. We hope to see Mexico restored to order by aid of the United States without the help, but with the goodwill, of England and France.

DISCREPANT prices for chrome ore are caused, in large part, by the different uses to which this mineral product is put. Three classes of ore are recognizable: (1) that assaying 40 to 45% is used by the manufacturers of steel and of refractory materials, (2) that assaying not less than 48% is used in the manufacture of chemicals, and (3) that assaying 51% or more is used for special chemicals, such as the bichromates of potash and soda. There is plenty of the low-grade chrome ore, which therefore is cheap; there is a scarcity of ore assaying 51%, and therefore it is costly. Another factor in making prices is the reliability of the seller; many buyers have discovered to their loss that supposed producers of chrome failed to 'deliver the goods' after receiving a payment on account.

SPEEDY re-opening of mines in Chihuahua is expected by General Francisco Marguia, now that protection against Villa's depredations is assured. But the General says nothing about Carranza's official depredations on the mines, in the form of taxation and requisitions of a most burdensome kind. Perhaps the Carranza government will acquire better manners as the need of a loan becomes felt increasingly. Persistent rumors come from New York that a Mexican loan of $150,000,000—dollars, not pesos—is requested. But Carranza has suspended the constitution that he claimed to be defending, and Mexico is today under the uncertain control of a pompous dictator. He has decreed that no transfers or deeds to property can be made to aliens unless they renounce their citizenship rights and thereby lose the protection of their own flag—whatever it may be worth. Unfortunately no healthy opposition to Carranza exists in Mexico today, only bands of wandering bandits; but cartridges and munitions of war generally are becoming scarce, so that the confusion of misrule is becoming abated. Carranza's violent friendship for German propagandists and exploiters is waning and he may find it useful to make friends with Uncle Sam in order to borrow a little money, on conditions.

CALIFORNIA's mineral resources are playing an important part in the organized industry of the War. The State Mineralogist, Mr. Fletcher Hamilton, and the staff of the Mining Bureau are aiding the consumers of useful minerals by supplying information concerning the localities of supply within the State. It ought to be known to the public that one of the best sources of information in all matters relating to the occurrence and distribution of minerals in California is the office of the Bureau in the Ferry building, at the gateway of San Francisco. Another invaluable agency for the dissemination of such information is the office of the U. S. Geological Survey, in the Custom House building, where Mr. Charles G. Yale is ever ready to assist the investigator and operator of mines.

PRODUCTION of copper decreased 50 million pounds in July, owing chiefly to the cessation of production at the Inspiration, Miami, Arizona Copper, Old Dominion, Shannon, East Butte, and Greene-Cananea mines. It is noteworthy that the New Cordelia began to produce in June, and in July yielded 2,100,000 pounds. The Mason Valley and the Ohio copper companies have resumed production this year and are getting into their stride—not a 7-league gait, but sufficiently rapid to make a considerable contribution of copper. The wonderful little United Verde Extension mine yielded 33,548,000 pounds in the first seven months of this year. Apparently the consumption of blister copper has been greater than the capacity of the refineries, so that the recent decrease in blister copper is not an unmitigated evil.

ON another page we publish a statement made by Mr. Henry D. Williams, chief counsel for Minerals Separation, in the course of which he tells the public what his clients expect to get out of their claims against sundry alleged infringers of their patents. This statement is interesting and that is why we print it. Such as it is, it shows what Minerals Separation expects to get out of the American metal-mining industry. We hope the expectation will not be fulfilled, for it would represent a depressing burden on useful enterprise. Mr. Williams has very little to say about the appeal in the Butte & Superior case. This will come before the court of the

Ninth Circuit, in San Francisco, the court that gave a decision against Minerals Separation in the Hyde case. No reason exists for believing that Judge Bourquin's decision in the Butte & Superior case will not be reversed in San Francisco. The litigation is not so near an end as Mr. Williams so confidently anticipates.

MAGNOLIUM is the name given by British and American experts to a metal that is an alloy of magnesium, aluminum, and other elements. It is a war discovery. When one of the zepellins brought to earth in England was examined it was remarked that it had a framework remarkably flimsy for a machine that had carried over four tons in men and material. Upon further examination it was ascertained that the framework was made of a metal of unusual strength and lightness. It was four times lighter and yet four times stronger than the best steel. The analysis was sent to the United States and the metal is now being made for the manufacture of American aeroplanes. A few days ago we held a piece of it—about the size of a pill-box, being a section from a rod of the metal—and it was so light as to induce the illusion that it must be hollow. The magnesium that it contained was derived from Californian magnesite.

UNDER 'Discussion' we publish three interesting contributions. Mr. C. M. Eye writes from the Philippines to express his opinion concerning the Colorado School of Mines fiasco. As an alumnus of the School he has no liking for an unnecessary ventilation of the scandal, but he is impelled to express agreement with the conclusions previously published in these columns. Before now he will have learned that Mr. Parmelee is no longer President of the School and he will learn shortly of the manner in which the Golden trustee has been permitted to make confusion worse confounded. The one certain cure for these deplorable conditions is to place the governance of the School more in the hands of its alumni and less in the hands of small politicians, newspaper reporters, and other people wholly unfit to act as the directors of an educational institution. The next letter, from Mr. E. E. Fleming, who writes from Alaska, adds to the value of Mr. Harold Power's recent article by furnishing interesting comment on deep-placer mining and by giving details of practice in the Nome district. The third contribution comes from Mr. John B. Hastings, an old friend of ours, a veteran engineer, to whom the word 'sampling' is as the smoke of battle to an old warrior. His remarks on the subject will be much appreciated, more particularly the description of the method of sampling introduced by him in the Centre Star and War Eagle mines at Rossland.

INCIDENCE of taxation on excess profits is causing amateur economists in London, as in New York, to realize sundry fundamental truths in mining finance, for instance, that a mine is a wasting asset and that dividends represent, in large part, only a return of capital. Discussion has arisen regarding the taxing of the profit earned by the two South American subsidiary enterprises of the Oroville Dredging Company, to which we referred in our issue of August 18. Both of these subsidiary companies, the Pato and the Nechi, began to produce gold after the War began, so that the profit made now is compared with the absence of profit in 1913, when these undertakings were in their development stage. If all the profit now declared is to be regarded as 'excess,' an injustice will be done. Therefore the directors are endeavoring to make an equitable arrangement with the Government. Again, the tin-dredging companies on the Malay peninsula are benefiting from the rise in the price of the metal that they produce. It has been suggested that the gain due to the increase in the price of tin should go to the financing of the War, after making deductions for the extra freight-rate, insurance, and other increased expenses caused by the War. In Nigeria, where likewise dredging supplements sluicing operations, an arbitrary figure of 13% was fixed as a reasonable profit. What should be the reasonable profit from dredging operations? Such metal-bearing ground is more clearly a wasting asset than even a vein deposit, because the latter affords promise or prospects in depth, whereas the resources of an alluvial deposit are usually known fully at the start and any change from the original estimate of returns is commonly on the wrong side. We would regard 20% as a proper return on a dredging enterprise, that is, we would regard it as a poor 'investment' if it failed to yield that much.

Silver

Silver is nearing the dollar mark. The latest quotation—96½ cents per ounce—is the highest since 1893, when the Indian mints were closed to free coinage, but it is by no means the highest since the Civil War, as has been asserted by careless commentators. During that war the price was fairly steady at about $1.34 and it was not until 1885 that it fell below the dollar, rising subsequently as high as $1.21 in 1890, when the Sherman Act authorized the purchase of 4,500,000 ounces per month so long as the price did not exceed $1.2929 per ounce. The repeal of that Act and the closing of the Indian mints started a debacle that culminated on July 29, 1915, when the price reached 46.87 cents. During the 20 years preceding 1916 the price averaged 58 cents and in 1916 it was 65.78. Between the low figure of two years ago and the present quotation there is an advance of 50 cents or slightly more than 100%. Even the recent high quotation does not reflect the whole advance, for a premium of as much as 4½% has been paid, at Toronto, on purchase of the metal. Here, in San Francisco, a premium of 4 cents per ounce has been paid on purchases for export to China. This condition of the market is due to causes that are recognized. Not only has the production decreased, but a growing demand has been created by the War. Twenty years ago the world's production of silver was 186,000,000 ounces, of which the

United States contributed 58,500,000 ounces and Mexico 41,400,000. In 1916 the world's total was 172,400,000 ounces, of which the United States contributed 72,800,000 and Mexico 35,000,000. Thus at the end of the 20-year period the production was actually less than at the beginning, this being due chiefly to the disorderly condition of Mexico, where production declined from the maximum of 87,000,000 ounces in 1911 to 35,000,000 ounces last year. In 1911 the world's production was at its highest, namely, 254,214,000 ounces. During the 20-year period the United States has shown a steady increase in production, the ratio of her yield to the world total increasing from 37 to 42%. Since 1905 Canada has become an important source of supply. Whereas in 1896 Canada produced only 3,200,000 ounces, in 1916 she produced 25,500,000 ounces. The outstanding fact, however, is the collapse of silver mining in Mexico, which now produces only 40% of her previous maximum. Turning from supply to demand, it is noteworthy that the United States exported $53,172,000 worth of silver in the fiscal year 1916, of which $41,032,000 went to England, mostly for re-shipment to China and India. In 11 months of the current fiscal year the United States has exported $75,000,000 worth. The need of silver coinage among the European belligerents and the expansion of trade in the Far East were the principal factors in improving the market for silver; more recently the declaration of war by China has intensified the normal drift of the metal to that country. In face of these growing demands there is a decrease of production; therefore the high price is fully explained. What it means to the fortunate producers of silver is indicated by a few individual statistics of production:

	Production last year oz.	Increased profit at 96 cents per ounce
Anaconda	10,624,747	$3,187,424
Butte & Superior	4,126,938	1,238,081
Bunker Hill & Sullivan	1,910,281	573,084
Kerr Lake	2,036,962	611,098
Nipissing	4,044,668	1,213,400
Tonopah	1,387,557	416,267

Of these, one is a copper mine, one a zinc-lead mine, another a silver-lead producer, and the other three yield plain silver ores. Most of them produce some gold also. The copper ores of Montana average $1.05 per ton in gold and silver, those of Nevada 44 cents, and those of Arizona 42 cents per ton. The lead ores of Utah yield an average of $9.23 per ton in the precious metals, and those of Idaho $3.03. Whereas the production of silver used to come mainly from silver-lead ores, an increasing proportion is now being won as a by-product in the course of copper mining, and in periods marked by a highly favorable copper market this silver is a clear gain. In normal times two-thirds of the silver production of the world is consumed in the arts. Now an abnormal proportion is being used for silver coinage, for the convenience of paying the soldiers of the big armies and for replacing the gold that is being hoarded by governments of the belligerent countries. India continues to be a sink for silver; that country absorbs fully half of the world's output. The cinema trade is consuming an increasing proportion of silver. Altogether the outlook is sufficient to prompt renewed search for deposits of the metal. Idle mines and old dumps should invite examination in the light of these improved conditions, for, although labor and cyanide are high, the market for the associated base metals is also favorable.

Geologic Eccentricities

In this issue we publish an interesting article by Mr. Horace B. Patton, Professor of Geology in the Colorado School of Mines, describing the peculiar occurrences of extremely rich ore uncovered in the Cresson mine at Cripple Creek. This is accompanied by such an outline of the local geologic conditions as will afford evidence upon which other experienced observers can form their own hypotheses to explain the eccentricity of ore deposition. The present writer happens to have visited the Cresson mine and is in a position to appreciate Professor Patton's description of the two bonanzas discovered previously to May this year. The dominant structural feature on the 1200-ft. level is a line of fracture that follows a dike of basalt, two to four feet wide, which has been traced from the surface downward. This dike in its passage through the andesite-breccia is accompanied by a distinct selvage, but the extent of the dislocation caused by the dike is not measurable. The distribution of the orebodies discovered in the mine appears also to be related to basalt dikes. On the 1200-ft. level a mass of broken rounded fragments of basalt adjoins the main line of fracture and there is the suggestion of a volcanic vent. Vugs are numerous. One of them is a pear-shaped chamber 50 feet high and 35 feet in greatest width. This cavity, according to Mr. Richard Roelofs, the able manager of the mine, was filled to a depth of ten feet with fine particles of ore, and the walls were encrusted with two feet of rich ore consisting of quartz and calaverite, much of it in beautiful crystals. The telluride extended into the wall-rock, but the vug itself constituted the core of the orebody. From this curious cavity $1,250,000 worth of ore was removed, most of it being screened and sacked without further sorting or concentration. The orebody was found to extend irregularly alongside the dike up to the 300-ft. level, but it did not reach the surface and therefore had no outcrop. On the 600-ft. level the ore was on the east side of the dike, changing to the west side at about 1100 feet from surface. The quartz in the vug was agatized and sintery. This peculiar orebody has several points of similarity with that uncovered in the Anna Lee claim of the Portland mine. It is in the same dike, which in the Portland ground is of the same thickness, namely two to four feet. The Anna Lee 'chimney,' according to Mr. R. A. F. Penrose, was 12 to 15 feet in diameter and reached from the surface to 875 feet in depth. It was nearly vertical. The rock was intensely kaolinized and brecciated; Mr. Penrose considered it to be "the neck of a hot spring or fumarole." Two other extraordinary enrichments of possibly analogous origin in the Cripple Creek district may be mentioned: the Klondike Stope at 700 feet in the Portland and the Jewel Chamber at 400 feet in the Independence mine. Ore deposits of this type were described by S. F. Emmons in his report on the Silver Cliff district. There the Bull Domingo mine had an irregular chimney identified with a volcanic neck of brecciated rock, the ore consisting of concentric shells of silver-bearing galena, blende, and various spars, encrusting fragments of different kinds of rock torn by violent thermal eruption and filling a vent, 40 by 100 feet in horizontal section, traversing syenite. Another equally remarkable deposit was found in the adjacent Bassick mine. This was a vertical chimney of agglomerate, mainly andesite, within a similar agglomerate. In cross-section the shape was elliptical, the two dimensions being 30 by 100 feet. Gold and silver were contained in concentric shells of sulphides—of lead, zinc, antimony, copper, and iron—covering fragments of rock of varying size, up to three feet in diameter. Emmons came to the conclusion that the deposit marked a dying phase of volcanic activity, that the agglomerate occupied "the neck or crater of an old volcano," and that the precipitation of valuable minerals was due to fumaroles, or gaseous emanations. Such eccentric orebodies suggest that geologic action occasionally departs from the uniformitarian processes on which our theories are largely based and that the scientific observer must leave place for catastrophic ideas in his interpretation of nature. Of course, the activities by which such results are achieved are on a scale relatively small as compared with the major movements of geologic development, but they are an essential part of the great work that has made the earth's crust as we now find it. Turning to the economic phase of the subject, it is well to note that the remarkable orebodies in the Cresson mine gave no signs of their position to the prospector, for no outcrop existed. The ore was found by exploration underground, itself started on such vague indications as are plentiful in a mineralized area like that of Cripple Creek. This suggests that diamond-drilling on a systematic scale, or some similar method of research, should be tried in such districts, for it requires but little imagination to see that the finding of ore at best is haphazard. The experienced miner will say, quite properly, that to seek for a small mass of such ore as constitutes one of the Cresson bonanzas is like looking for a needle in a haystack, but he will recognize the fact that such drilling, to be a useful guide for working underground, need not find the orebody itself, only the fringes or signs of it. If the Cripple Creek massif—the mass of the ancient volcano—were made of glass we should be astonished to see how much rich ore had been missed despite the numerous holes made by man. Our system of exploration even when most systematic is fortuitous. It remains to record the fact that in 1916 the Cresson mine produced 74,435 tons, averaging $32.26 per ton and yielding $2,401,434 in gold and silver. The net income of the company operating the mine was $1,634,171 or 68.05% of the gross output.

'Near-Ore'—A New Term

In the many discussions arising out of the attempt to define 'ore,' it has been apparent that the unwillingness to abstain from applying the term to unprofitable vein-filling or lode-matter was due in part to a dislike of the word 'waste.' Every mine contains rock that is not an asset and yet is not negligible when an engineer is appraising the mine. The Mexican has three terms, *metal* (or *mena*), *mineral,* and *tepetate;* the first of these means ore, that is, rock that can be exploited to economic advantage; the third is plain waste or country-rock; the second lies between, it may be raised to a level of economic value if a better metallurgic method is introduced, or a new system of mining applied, or a larger scale of operations adopted; it may be lowered to the level of economic worthlessness by adverse conditions, such as an increase in the cost of labor, higher smelter-rates, or a lower extraction in the mill. For this intermediate product we need some new technical term. We venture to suggest 'near-ore.' It may seem humorous, but it is better to smile when using it at first than to apply 'ore' to stuff that yields no profit and be compelled at last to grin on the wrong side of your face. There is 'near-beer,' some of which is prohibition tipple and some so near to the real thing that arrests have been known to follow the sale of it. 'Near-ore' is good plain English and tells its meaning unblushingly. 'He laughs best who laughs last' is a proverb that like most proverbs is equally true if reversed. 'He laughs best who laughs first' applies to the engineer who smiles at this new term and then makes a seriously truthful distinction between what contributes to the profit of a mine and what does not. The shareholders of a mine in the Transvaal are told, for example, that among the resources of their property is 2,000,000 tons of 4-dwt. banket labeled 'unpayable ore.' Of course, 2,000,000 tons of anything is impressive, but, as a matter of cold fact, that reserve of 2,000,000 tons is worth less than 30 cents, because the cost of winning the 4 dwt. of gold is at least $5 per ton. In course of time—with an advance in metallurgy, or by selecting a more intelligent board of directors, for example—this banket may leave a margin of profit over the expense of exploitation. So the stockholders of a mine in California are told that it contains a large tonnage of 'low-grade' ore, meaning quartz containing so little gold as to yield no profit now, nor for several years perhaps—possibly not until the Greek Kalends. Why not call it 'near-ore'? Then the shareholder, and anybody else interested in knowing the facts, will understand that it is a potential asset. At present he thinks of his millions of tons of 'unpayable ore' as something that gives value to his shares, and in the end the laugh is against him—and a sardonic laugh it is, nothing like the kindly smile that first greeted 'near-ore.'

DISCUSSION

Our readers are invited to use this department for the discussion of technical and other matters pertaining to mining and metallurgy. The Editor welcomes expressions of views contrary to his own, believing that careful criticism is more valuable than casual compliment.

Colorado School of Mines

The Editor:

Sir—I have read your editorial on the Colorado School of Mines, in the issue of June 16, commenting on the recent trouble there, and as an alumnus of that institution, I wish to express my appreciation of your frank but true criticism.

There never has been, to my knowledge, the strict enforcement of discipline at the School that makes for the highest success and the best training for the students on the threshold of a professional career. During my days at the School, in the early 'nineties, enforcement of discipline was sporadic; we were allowed to do as we pleased for a period until there was some sort of outbreak, often innocent enough in itself, and then the wrath of the faculty would be visited upon us; there would follow strikes, lock-outs, etc., the board of trustees would take a hand, the papers would be full of it, encouraging by undue publicity the rebellious spirit of the student body; but finally every one would get together, a truce would be patched up, and things would run along loosely until another occasion arose, usually a year or two later. I imagine from what I have heard and read that much of the same thing has continued through the various administrations since. No doubt much of the trouble is due to the system under which the School is conducted, involving as it does, governance by a politically appointed board, making for good administration if the board happens to be composed of capable men, and if not, the reverse. The board has always been weak in the two particulars you point out, several members being local men with no qualifications for the job, besides being beholden to the residents of the town, whereas other members that do happen to be capable are too occupied to give the work sufficient attention. As you point out, the appointment of capable alumni is the best hope, and in my opinion the majority of the members should be chosen from among the alumni.

The next requisite is a president and faculty not afraid to outline a rigorous system and to enforce it without fear or favor. In the present crisis the president, at least, shows commendable firmness, and it is to be hoped, for the good of the School, that he may continue to do so, but the facts as recited indicate that discipline was very lax indeed, for most of the students were drunk and disorderly and mishandled a member of the faculty. I am glad to say that this sort of thing did not occur in my days at the School. We had more respect for the members of the faculty and drunkenness was uncommon, though not unknown. Now, I am satisfied that most of the trouble that has occurred at various times at the School is due to the feeling among the students that they may *en masse* do almost anything and get away with it, for they feel that the faculty and the board of trustees back of them are so anxious to have a good attendance and make a good showing that they will not dare to take any action that would cut down the attendance materially. Therefore the students rely on simply standing together and forcing the faculty and trustees to yield, in other words, to use the weapon that many of them have had used against them in later life by the Western Federation and others, such as, a strike, with its attendant picketing, intimidation, etc. Now, this is not right, and the only way to correct it is to insist on the right of the president and faculty to make the rules, and to abide by them when made, whether it results in the class-rooms being full or empty. The time for the trustees and faculty to settle any differences of opinion regarding the advisability of regulations is before they are announced. If the board of trustees does not approve of what the faculty wants, then it is up to them to get a faculty that is in accord with them, and if the faculty is dissatisfied with the stand of the board, it is up to them, individually and collectively, to resign. But having once agreed on a programme for the governance of the School, it should be distinctly understood that the rules must be obeyed, and that any action by the faculty in carrying this out will have the unqualified support of the board. I admit that it is difficult to carry this out with a State-supported and politically run institution; but until it can be done, there will always be dissensions.

I hope with you and, I believe, with the most of the alumni, that president Parmelee will stand firm in the present crisis, and will not allow any student to return to the School without proper apology, and with the distinct understanding that he and the faculty, not the student body, are running the School. Otherwise it were better for the School to close up until such time as it can be properly conducted, with a student body having a decent respect for authority, and it were better for the students to go into the Army and receive much needed training in obedience, for until they learn this, they will never be of much use in the great industrial army that digs and delves in the hills and smelts in the valleys.

Of course, no one can question the right of the student to leave an institution at any time that he thinks he is not getting a square deal, though in doing so without just cause, he is doing an injury to a community that has furnished him, up to that point, with the start of an engineering education at considerable expense to itself, but I should think that no other school would be keen on receiving an insurrecto from Golden. I am quite sure that in the largest school of all—that of experience—we who are pretty well through the course would hesitate to take on a man whom we know had helped to disrupt affairs at his alma mater, especially at a time when the needs of the country are calling for more technically trained men. My advice, therefore, to the men who have left the School, is to ask for re-instatement, express regret for the action taken, cut out the booze, and work for the great cause, as well as for the alma mater and their own future professional careers.

C. M. EYE.

Bagui, Philippine Islands, July 21.

Block-Stoping and Timbering in Deep Placer-Mining

The Editor:

Sir—The article appearing in your issue of August 11, by Harold T. Power, describing timbering in deep-placer mining, was read by me with keen interest, because in 1904 when in charge of the Wild Goose Mining & Trading Co.'s operations at Nome, Alaska, we were confronted with the same problem, but solved it a little differently.

The pay-streak, known as the Mattie channel, was discovered on the south slope of King mountain, on the Anvil-Dexter divide. The first discovery was made in 1900 on the Anvil side of the divide, at 400 ft. elevation above Anvil creek, where it was worked by open-pit methods, rockers being used to recover the gold. Water for the rockers was hauled from Anvil creek and the ground was so rich that as high as $1200 was realized from a rocker in one day. As the channel crossed the divide at right angles and the surface rose rapidly from the point at which it was discovered, where Anvil creek had cut it out, only a small area could be worked by open-pit methods; deep mining was unavoidable. Water was delivered afterward to this pay-channel by the Wild Goose pumping-plant on Snake river; 300 miners inches of water being pumped through pipe and flume, a distance of 10 miles to an elevation of 750 feet.

As the ground was all thawed, from the surface down, timber had to be continually used in the workings. The price of timber at that time was $60 per M plus the local freight from Nome to the mine. This made the item of timber a costly one, and the supply was hard to get.

The bedrock was a micaceous schist, varied by slate and lime. About 1½ to 2 ft. of bedrock and 2 ft. of gravel immediately above the bedrock, was run through the sluice-boxes.

The method of working was as follows: The channel, which ran nearly straight, making it easy to follow, was divided into sections 600 ft. long and a shaft was sunk mid-way between the end-lines of each section, care being taken to sink the shafts on the lower side of the channel and not nearer than 25 ft. from the lower rim, so that caving would not affect the alignment of the shafts when the channel was worked out and always permitted our shaft-houses, bunkers, and strings of sluice-boxes to be on solid ground. The shafts were 6 by 8 ft.; they had a single compartment, with a man-way for ladders and pipes. They were from 69 to 140 ft. deep; 140 ft. being the depth to bedrock on the crest of the divide.

A drift was run from the bottom of each shaft at right angles to, and toward, the middle of the channel. From the shaft this drift was wide enough for a double track, for a distance (usually about 40 ft.) sufficient to give a side-track for holding empty cars. When the middle of the channel was reached the drift was split and continued lengthwise with the channel a distance of 300 ft. each way, thus covering the section of 600 ft. into which the channel had been divided.

At each end of the main gangway the drift was T'ed by a cross-cut to each rim. The channel was from 80 to 110 ft. wide. At first we tried breasting as described by Mr. Power, using 10 by 10 in. and 12 by 12 in. posts, placing them on 3 ft. centres, but we soon abandoned this method as we could not recover any timber when once placed and it proved dangerous. One breast or stope, after it had been opened clear across the channel, caved in a rush to the surface, and we not only lost the breast and all the timber but nearly lost the men who were working in the face.

We then devised the method of block-stoping. This proved successful, permitting us to save all our timber, using it over and over again, only now and then losing a stick. To show to what extent we succeeded, we worked out the entire channel in one 600-ft. section, losing only 5000 ft. B.M.; for we waited until fall, when the frost set the gravel walls of the shaft. We even pulled the shaft-timbers. This method may have been used elsewhere before, but it was original with us at the time and was adopted afterward throughout the Nome district in all the deep mining in thawed ground. Frozen ground did not present any difficulties, as no timber was required.

Our greatest difficulty was caused from swelling bedrock in the main gangways. To overcome this trouble we used the methods described by Mr. Power.

In block-stoping we commenced at both ends of each T and took out a block of ground (each marked No. 1 on the diagram), never more than 10 ft. square, using 8 by 8 in. posts, set on 3 ft. centres, with 3 by 12 in. caps 1½ ft. long, never permitting the caps to cross two posts. When the gravel was so loose that it ran, we filled the spaces with false lagging. When this 10-ft. square block was out and the bedrock cleaned up, we pulled every post (using double or triple block and tackle) except the row against the two solid walls of unworked ground on two sides of the block. Behind this row we lagged solid with 2 by 6 in. lagging, against the unworked walls. Usually the roof caved as fast as we pulled the posts and the block was immediately filled. Then we took out all blocks marked No. 2 and worked back toward block No. 1 (which we had just left) until we struck the wall of lagging, cleaned up the bedrock, and pulled as before, repeating until the main gangway was reached.

In the meantime new cross-cuts had been cut from the main gangway to the rims, leaving a pillar 20 ft. wide of unworked ground to make a new T. The blocks were worked out as before. We kept these cross-cuts just ahead of the stopers. This method permitted us to work in four places in the mine, besides the cross-cuts, and was sufficient to keep the hoist on a single-compartment shaft busy.

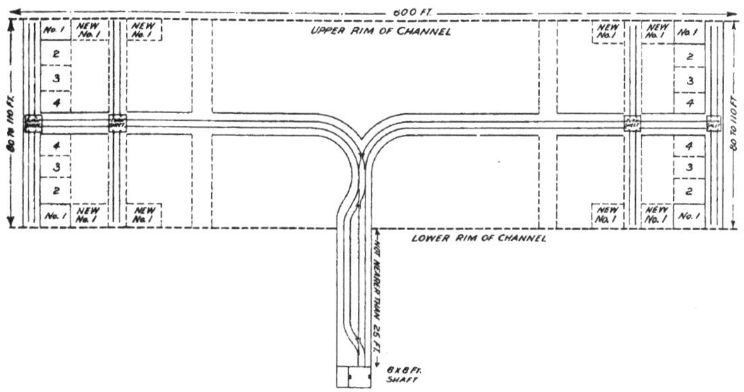

DIAGRAM OF MINE-WORKINGS, SHOWING SYSTEM OF EXTRACTION

We gradually worked back toward the shaft outward, and, as before stated, recovered practically all our timber and never lost a man, the secret being that we always worked under solid ground, and moved ground so quickly that it had no chance to get heavy and take weight.

E. E. FLEMING.

Chichagoff, Alaska, August 7.

Sampling of Large Low-Grade Orebodies

The Editor:

Sir—I think that most engineers have mutual confidence in each other's estimates of tonnage and value when succinctly stated, and particularly if after reading the report they approve of the way in which the mine was opened to allow sufficient exposures for sampling.

The development necessary for estimation of ore can perhaps be judged by the tonnage that may be allowed to each sample. Even in the larger mines it is hardly justifiable to take samples more than 10 ft. apart or more than 10 ft. of the vein-width in one sample. On this basis in the simplest of cases, suppose a vein 20 ft. wide, opened by an adit with an incline on the vein 250 ft. deep, accessible on one side, enabling samples to be taken seven feet wide, and three drifts, one at 50 ft., one at 125 ft. down, and one at the bottom of the shaft, each 200 ft. long, and each with three cross-cuts, one at the shaft, another 100 ft. from it, and the last at the face, enabling 6-ft. samples to be taken across the vein in the drifts and two samples of 10 ft. each in the cross-cuts. It is unnecessary to say how the samples should be taken, or how many cuts to each one. According to the weight of ore, this gives roughly for each sample 750 tons for quartz and 1100 tons for heavy sulphides, so that on a 20-ft. vein with fair development any estimates based on 1100 tons per sample for quartz and 1600 for heavy sulphides should be satisfactory. With a 5-ft. vein the samples should not be over six feet apart, and only 200 tons for light and 300 for heavy ores could be allowed, while on a 60-ft. vein opened 750 ft. deep it could easily run up to 2500 and 5000 tons for light and dense ores respectively. If it were necessary to gauge the value of such a property by a mill-test, I would slab off the end of the shaft, and the roofs of the drifts and cross-cuts, sending the material in an unmixed stream through properly arranged bins to, say, five stamps, with coarse screens with only the width of the screen-frame for discharge, discarding amalgamation in the battery; and I would take screen-samples for assay every half-hour, by-passing the pulp with a pump, if necessary, to an adjoining battery for the metallurgical test. The Butte & Superior was a mine opened in the manner described above, only much more so, and to the 1600-ft. level. Just after it was acquired by Hayden, Stone & Co. I was allowed to study it underground with the thorough assay-maps on hand, for a prospective investor. My conclusions on tonnage and assay-value agreed with the reported estimates, which seem to have been well realized.

Mill-runs and the past product of mines, as we all know, can only be applied intelligently after thorough study of the property. In 1901 I spent three days on a large irregular deposit that had paid $3,000,000 net and estimated there was $4,000,000 more net left above the bottom level, but as the price was $7,000,000 and the missing $3,000,000 plus the profit had to come out of new ground, I thought it not advisable to proceed with an expensive examination. No statement had been made by the owners as to reserves, and they accepted my conclusions without chagrin. It is a good mine yet, and results have been slightly more in favor of the owners than was predicted.

On the other hand, a few years ago I was asked to report on the ore-reserves and future development of a gold mine, on the wisdom with which $250,000 had been spent on the new surface plant, and on the best way to spend $250,000 more to hurry it to completion. After reading a report by the engineer employed to advise concerning the original investment, I said that even though the conclusions stated happened to be true they were entirely unwarranted by the evidence set forth. From the previous output of the mine—80,000 tons yielding $600,000—which had about ended at the time of his visit, and some erratic samples, he placed the ore-reserves at 1,000,000 tons of $7 ore, with probabilities of several times that amount. He preferred the mill-results to sampling, it being one of those mines that could not be sampled, the metal-content being too uneven. After a flying visit to the property, a competent sampler was chosen by the investors at my request. He found that everything of value had been stoped and that all the remainder, including levels under the old stopes, averaged $1.25. I was asked to re-sample, but I told them that the 1000 samples taken were uniform—only a few going over $4, the highest $7—the descriptions of the samples were clear, and I felt sure the $1.25 was correct.

With tedious care correct samples can sometimes be taken with a small pick. Nearly 25 years ago I sampled the Chainman deposit at Ely; this was a large body of hard quartz and iron, including magnetite, several hundred feet long, 20 ft. wide. For the 200 to 300 ft. it was developed, oxidation had disintegrated the vein into rounded boulders, from the size of a marble to six feet in diameter. They hung together without timbering, but cutting through them made large samples. Years after, in Alaska, Charlie Lane said, 'What did your Chainman samples go?" "$4.50," I said. "You were right, ours went $6," he replied.

Chalcopyrite can be spotted. Preparing to sample the Mamie mine, on Prince of Wales Island, Alaska, the superintendent pointed to the irregular bunches of yellow chalcopyrite in the black magnetite and the impossibility of correct sampling. Two-foot holes were put every five feet along the sides of the levels, and blasted on canvas. Along the croppings 3-ft. holes were drilled, loading them just enough to crack the material so it could be barred. Then miners broke the pieces with sledge-hammers, on platforms, and I coned and quartered it, till the samples were small enough to be carried away. After the ore was stoped it was pleasant to see in the U. S. Geological Survey's report that it had averaged about 2%, and was within 0.1% of my sampling.

Regarding the 'high' sample, unless its inclusion is well warranted I take care of it individually. I think that on the whole estimates of ore-reserves are slightly optimistic, probably because every effort has been made, and wisely, to uncover the best places. Mines are not bought solely on their reserves; the future possibilities must be considered. A mine should be young and strong enough to compensate for any disappointments occurring in blocks of ground already opened. In the War Eagle and Centre Star mines at Rossland, in the first few years, the samples averaged 30% of quite high assays; the remainder were low, say, under $10 gold or the cost of mining and treatment. I gave a table indicating this fact in the third annual report, and the local papers were quite shocked. These samples were taken after each round of blasting. It was all machine-drilling, before the face of a drift or a slope had been picked down, a necessity in such intensely hard rock. In the assay-office each sample was divided by weight into its component parts, so that in a composite sample it could have read one-eighth quartz, one-eighth spar, one-eighth chalcopyrite, one-eighth arsenopyrite, one-eighth pyrrhotite, one-eighth country-rock (included in vein), one-eighth silicified foot-wall, one-eighth dike. Each was assayed separately in order to study the metallurgical problem and for general knowledge. When beginning operations I took the samples and made the division myself in presence of the samplers, afterward the older men taught the newcomers. Soon A. A. Cole, now president of the Canadian Institute of Mining Engineers, had charge of the laboratory, and consequently of the divisioning. In spite of the great difference in value of the samples, as they were really only from faces three feet apart, as soon as the ore was mixed by stoping, it was uniform in the 100-ton lots sampled, about $20 gold plus copper and silver. Sometimes I would feel quite desperate at the uneven assays, fearing what the average outcome would be, but each month the smelter-returns and mine-sampling agreed.

The mention of mill-tests always makes me wonder how precautions are taken against salting and recalls stories of moth-eaten $5 gold-pieces that have been found in mortars.

JOHN B. HASTINGS.

Los Angeles, August 31.

THE U. S. GEOLOGICAL SURVEY estimates that nominally six pounds of magnesite was formerly used for every ton of steel made by the basic open-hearth process, but not more than half a pound per ton is now being used, and at some steel plants cheaper and less satisfactory refractories have been substituted for magnesite.

The Cresson Bonanzas, at Cripple Creek

By HORACE B. PATTON

INTRODUCTION. Nothing seems to appeal so quickly to the popular imagination as the finding of rich gold ore. This, of course, is due to the limitless possibilities of sudden wealth. From the time when Stratton first made his memorable strike in the Cripple Creek district and the Independence mine was made to pay for itself 'from the grass-roots' this district has been conspicuously in the public eye. Since the first discovery many other 'strikes' have been heralded in the public press. Most of these have come to nothing, but the successes have been sufficiently frequent to keep the public expectant. As the years have gone by, however, and mining has been carried to greater depth, startling discoveries have been rare.

Such was the opinion in December 1914, when the Denver papers announced the discovery of a veritable 'bonanza' in the Cresson mine. What gave this announcement more than ordinary significance was not alone the fact that it proved to be well-founded but that the strike was made at a depth of over 1200 ft. below the surface. This gave an impetus to deep exploration.

During the three years since this rich deposit was struck there have been many announcements of additional rich 'strikes' in the Cresson mine, most of which have been born of newspaper imagination. Still it remains true that there have been two important discoveries since the first find. One was made in the summer of 1916 and the other about the middle of July this year. The last one has attracted less attention than most of the others, yet it bids fair to equal, or even surpass, the first. The importance of it lies in the fact that a vein has been opened up not only of great richness but of a character entirely different from that discovered in 1914. It is this latest discovery and what it may develop that justifies at this time the publication of this article.

THE CRESSON MINE, CRIPPLE CREEK

THE DISTRICT. Cripple Creek lies in Teller county, Colorado, west of Pike's Peak and 25 miles from Colorado Springs as the crow flies, but 50 miles by rail. It lies at an elevation that ranges from 9400 to 10,800 ft. above sea-level. The area within which important mines are situated is quite limited, being roughly elliptical in outline and measuring about three by five miles.

Geologically the Cripple Creek mining district is a fine example of a deeply eroded volcanic dome. The floor through which the igneous masses broke consisted of pre-Cambrian granite, which had itself been injected into gneisses and schists and which caught up and enclosed fragments of these schistose rocks. Some of the included fragments measure a mile or more in their greatest dimension. Volcanic activity began with a series of tremendous explosions that rent the rock and developed a crater of about the dimensions of the present active mining centre, that is, 15 square miles. This crater was filled in part by falling fragments of the pre-Cambrian rocks but mainly by the shattered igneous material ejected by successive volcanic outbursts.

Periods of violent explosive activity seem to have alternated with periods of quiescence during which massive lava was forced into and through the volcanic ejectamenta that filled the central crater and that at times poured over the surface in the form of lava streams. In this way the brecciated area became injected with numerous and irregular dikes and small bosses. A continuation of this process built up an extensive volcanic cone, which grew to an unknown but considerable height above the present surface. Erosion has removed a large portion of the volcanic pile that rises above the general level of the pre-Cambrian floor and has given the surface its present mountainous character.

The igneous rocks of the district were originally described as definite types of andesite, phonolite, rhyolite, and basalt, but later researches by officers of the U. S. Geological Survey have shown that most of these rocks are to be considered as intermediate types between phonolite and andesite, of which latite-phonolite predominates.

THE ORES. Gold is the principal metal mined in the Cripple Creek district, and it occurs almost exclusively combined as a telluride. Where oxidation has removed the tellurium it has left the gold in a spongy condition, known locally as 'rusty gold'. The gold tellurides have been described under the names sylvanite and calaverite, the former being the name in common use. These two tellurides both contain gold and silver, but in different proportions. Sylvanite is supposed to have the composition $(Au, Ag)Te_2$, in which the two precious metals are present in equal parts. This gives the following percentages: Te 62.1, Ag 13.4, Au 24.5, or a proportion of gold almost twice that of the silver. As a matter of fact, however, the Cripple Creek ores give a much higher percentage of gold and a correspondingly lower percentage of silver than that required by the formula for sylvanite. Analyses of pure telluride show approximately 40 to 41% gold and less than 3% silver. Therefore calaverite, the telluride of gold, predominates.

Pyrite is abundant, and other sulphides are associated with the tellurides. The customary gangue-minerals are quartz, chalcedony, dolomite, and fluorite.

THE VEINS. Lindgren and Ransome in their treatise* on this district state that the ore deposits are of two types, (1) lodes or veins, (2) irregular replacements; but that the two cannot be kept sharply distinct, as replacement may modify the development of distinct veins, and minute veins may traverse replacement bodies. Ore-shoots are likely to be associated with fissures. This is especially true of the smaller veins, which are likely to show richer mineralization at the junction of two veins. Fissuring is not so characteristic of the larger ore-shoots.

The telluride deposits are supposed to have been formed by heated waters making their way to the surface from a considerable depth. In this connection Lindgren and Ransome say: "The view that most of the water and most of the substances contained in the veins were given off by intrusive bodies slowly cooling at considerable depth and were forced up through the upper part of the volcanic mountain as soon as the formation of the fissures allowed them to rise is considered more plausible. The waters ascended in the deeper part of the volcano with comparatively greater velocity on the fewer fissures here available. Nearing the surface they spread through a larger space in a more complicated fissure-system. The speed became checked and the conditions for precipitation improved. Deposition and the chemical action of the country-rock changed the composition of the solutions, and a mingling with fresh ascending waters, possibly also with atmospheric waters, induced further precipitation. In this matter are explained the smaller amount of ore deposited in depth and the richness and abundance of ore nearer to the old surface. The proportion of the volcano removed by erosion may have contained still richer deposits."

THE CRESSON MINE is situated at an elevation of approximately 10,000 ft. in Arequa gulch at a point about half a mile north-east of the small town of Elkton. It is owned by a company incorporated in 1895. The mine was worked by the company or by lessees without any noteworthy results until 1905. Since that date it has been under the direction of Richard Roelofs, to whose efficient management is largely due the profitable development of later years.

In co-operation with H. J. Wolf,† I have published a fuller account of the great Cresson strike of 1914 than can find room in the present article. The following quotation will suffice to give an idea of the geology of the mine.

"The Cresson shaft is close to the contact of the breccia of the central part of the district and an intruded mass of latite-phonolite. The rock of the mine is breccia, with the exception of one or two dikes. According to the statement of Mr. Roelofs there occurs within the breccia, which ordinarily is a firmly cemented rock, a local area of irregularly elliptical horizontal section in which the breccia has been greatly shattered so as to produce numerous cracks and minute cavities throughout the mass. This area of shattering extends in a north-easterly direction for a length of about 1000 ft. and a breadth of 500 ft. It extends down to and below the 12th level. This condition of this fractured mass is such as to suggest some local explosion that produced an unusual shattering of the rocks. Inasmuch as the breccia of the Cripple Creek district is normally highly indurated and as the spaces that must at first have existed between the fragments have been filled by the cementing materials, it does not permit a ready circulation of mineralizing waters except where the breccia has been subsequently shattered, as in the case of this elliptical zone

*'Geology and Gold Deposits of the Cripple Creek District, Colorado', by Waldemar Lindgren and F. L. Ransome, U. S. G. S., Prof. Paper No. 54.

†H. B. Patton and H. J. Wolf, 'Preliminary Report on the Cresson Gold Strike at Cripple Creek, Colorado'. Colorado School of Mines Quarterly. Vol. IX, No. 4, 1915.

of fracturing in the Cresson mine. Undoubtedly this condition is responsible for the occurrence of the ore-shoots.

"The comparatively loose or porous structure of the shattered rock seems to have furnished a ready line of circulation. In fact, two such lines of circulation developed, one along the south-east and the other along the north-west side of the fracture-zone, leading to the formation of two extensive ore-shoots. The accompanying sketch, Fig. 1, shows the position of these two ore-shoots, as disclosed on the 12th level. The larger ore-shoot forms a roughly elliptical body 500 ft. long on the south-east side, with a greatest width of 150 ft. This shoot becomes narrower above the 12th level and has been followed as far as the 3rd level. Its dip is that of the wall of the fractured area of which it forms an integral part.

"On the other sides of the fractured area is the other shoot also elliptical in outline. This likewise runs up to about the 3rd level. It measures 250 ft. by 35 ft. in greatest dimensions.

"These two ore-shoots contain telluride of gold fairly well distributed throughout the mass. The gold content, as will appear later, is not uniform, but it does not follow visible veins. Usually the telluride is invisible. In places it may line the sides of some of the numerous small fractures. It is probably more likely to occur on the face of the fractures, even though it may not be present in sufficient quantity to be visible, but according to Mr. Roelofs it is to be found also inside the breccia fragments. The ore occurrence in the Cresson differs somewhat from that in other properties of equal importance in that the ore-shoots were of relatively trivial importance for the first 350 ft. from the surface. The ore improved in quantity and quality below this level and from the 6th to the 12th level the two main ore-shoots have been continuous. The ore has improved in quality with depth, and near the 12th level there has been a decided increase in the width of the larger ore-body."

THE FIRST BONANZA. In the latter part of November 1914 a raise from one of the cross-cuts driven in the larger of the two ore-shoots on the 12th level suddenly broke into a large cavity or chamber at a point five or six feet above the floor of the level. The chamber was opened near the lowest part. The floor and lower part of the walls consisted of a soft, porous, almost chalk-like, whitish material. Apparently the entire chamber was at one time lined one or two feet thick with this material, which was so soft that it could readily be shoveled. From the upper part of the chamber it had become loosened and had fallen, accumulating to a thickness of several feet on the floor. All of this white lining assayed high in gold, and some of it astonishingly high. As sacked and shipped it yielded from $10,000 to $16,000 per ton.

Behind the porous lining the walls of the chamber consisted of the ordinary breccia of the large ore-shoot. The fragments composing the breccia had been extensively altered by thermal waters, and solution-cavities had been formed between them. Crystals of calaverite thickly studded the surfaces of these solution-cavities and penetrated the substance of the rock-fragments. Close to the wall of the large chamber the impregnated breccia showed varying but extremely high gold contents approaching those of the chamber-lining. On working outward from the chamber it was found that the rock became less impregnated with calaverite until at, say, five or six feet, the assay-value was not greatly above the average of the main ore-shoot.

The chamber as first opened may be said to have had the shape of a pear, the narrow portion being above. Its dimensions were approximately 23 ft. long, 13½ ft. wide,

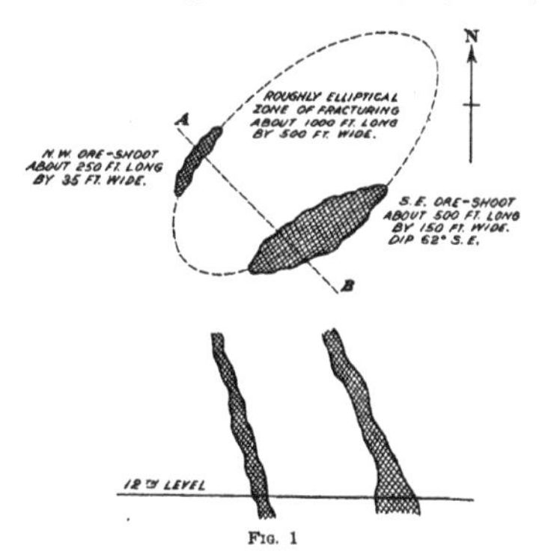

FIG. 1

and 40 ft. high. When the richer portions of the breccia had been worked out the greatly enlarged chamber had a much more irregular shape, and extended downward 10 or 12 ft. below the floor of the 12th level.

The soft lining, when closely examined under a magnifying glass, was seen to consist in part of minutely crystalline matter and in part of an earthy substance. The crystalline material was celestite, the sulphate of strontium. The earthy portion, on account of admixture with the celestite, was not definitely determined. It looked like kaolinite. The gold content of this lining consists of calaverite, which occurs in small crystals thickly scattered throughout the white gangue. The great richness of this material would hardly be surmised from its appearance. The actual amount of gold obtained from this interesting deposit cannot be given accurately because much of the high-grade ore was mixed with low-grade ore before shipping, and for the further reason that the gold content decreased gradually from the walls of the chamber outward, so that no sharp line could be drawn between the chamber-deposit and the ore of the larger shoot in which this deposit was formed. It may be stated safely, however, that at least

$1,200,000 net was taken from the lining and walls of this cavity.

Chemical analyses of the gold telluride occurring in the chamber-walls correspond with the mineral calaverite, the proportion between the gold and the silver ranging from 8 to 18 for gold to 1 for silver. The gangue-minerals with the calaverite in the wall-rock were chalcedony, quartz, dolomite, and celestite. Recognizable quartz was rather rare, as the silica occurred almost entirely in the form of a spongy chalcedony. Fluorite, so common in association with quartz in the veins of Cripple Creek, was missing.

As to the origin of this rich pocket of ore, it may be presumed that the cavity represents an enlarged portion of a passage along which heated mineralized waters were working their way to the surface from deep-lying igneous masses. That the conditions under which the ore was deposited in this case were unusual may be inferred from the fact that large cavities are not common in the Cripple Creek district, and, when found, are almost invariably barren. Such a barren chamber has, in fact, been encountered in this same mine.

ORE AT LOWER LEVELS. It was naturally to be expected that a rich chamber of the character just described must have been fed by mineralizing waters coming from a considerable depth; therefore a connecting vein ought to be found beneath the chamber proper. The ore was traced from the chamber downward for a distance of a little less than 30 ft. below the 12th level, but efforts to trace it were unavailing. Furthermore, the two main ore-shoots that had been followed downward also suddenly ceased a few feet beneath the 12th level, with the exception of a comparatively small finger-like extension that has been proved to extend as far as the 14th level and deeper. A large amount of exploration was done on the 13th level in a vain effort to find a continuation of the rich ore-chamber or of the two regular ore-shoots. Mr. Roelofs has stated that for a year and a half the outlook was extremely discouraging. Failing to find any considerable orebody on the 13th level, development was continued on the 14th for a long time without better results. Eventually persistent search was rewarded.

THE SECOND BONANZA. Failing to find ore on the trend of the regular ore-shoots the manager prospected east of where the main shoot should have been found. This was done largely by means of diamond-drill holes bored, some at an angle of 45°, others on the horizontal, and by driving cross-cuts. In this way, in the summer of 1916, an orebody was struck 85 ft. east of the natural course of the main ore-shoot and in a part of the brecciated area that elsewhere had not been found to carry workable ore.

This new orebody measured 150 by 50 ft. The longer axis was roughly parallel with that of the main ore-shoot. I am unable to state to what extent, if any, this ore differed from that of the main ore-shoot, but I understand that, with the exception of the core, it was similar to that of the two ore-shoots outside the above-described chamber. This core was in the shape of a narrow strip 30 ft. long, with its major axis nearly parallel with that of the larger body in which it lay, and it appeared to be a specially enriched part of that larger body. No information is at hand as to the output from the larger body nor as to the richness of the high-grade core except that $250,000 was taken from it.

THE THIRD BONANZA. Further development has extended the workings to the 16th level. On this level until recently no ore had been struck. On the 15th level some ore had been obtained from the finger-like extension of the main shoot and from a downward extension of the orebody just described. The third find of high-grade ore is the most promising of them all.

Early in July of this year, in driving a cross-cut on the 14th level in ground east of the main drift along the line of the larger ore-shoot, a vein of quartz was struck; this, on being followed, showed ore and, farther on, rich ore. In this vein the gold occurs as sylvanite in comparatively large crystals. The gangue consists of quartz, massive and in crystals, together with purplish fluorite and much pyrite. It contains none of the celestite or dolomite or chalcedonic silica that characterizes the gangue in the great chamber on the 12th level. The vein has a tendency to split into several irregular and parallel smaller veins or stringers that afford a strong contrast with the ordinary breccia of the wall-rock; it has been followed in continuous high-grade ore for 102 ft., ending in a breast of low-grade ore. At the time of my visit practically no stoping had been done, but a drift had been driven on a vein averaging 4½ ft. wide. The total material broken in this drift, consisting of country-rock with included veins and stringers, averaged $500 per ton. In places the vein is astonishingly rich. At the time of my inspection stoping had just begun and the vein exposed in the stope was of the kind that every miner fondly hopes some time to drive his pick into. In spots, several inches wide, the vein had the appearance of almost pure calaverite. At one particular point it would have been possible to cut out a rectangular block measuring four and a half inches thick and a foot square that would contain 50% calaverite, the rest of the vein being white quartz.

The significance of this strike lies not merely in the extraordinary richness of the ore, but in the fact that this is the first distinct and strong vein struck in this mine, and that its character is entirely different from that of the first strike of high-grade ore in the big chamber; it is also entirely different from the ore of the two main stopes that have furnished most of the ore taken from the Cresson; lastly, it was struck on the 14th level and there is every reason to expect that the vein extends downward as well as upward to a considerable, if an unknown, distance. As it was found in a part of the ground in which ore has not heretofore been suspected, the discovery encourages deeper prospecting; all the more so as this vein appears to have been fed from an entirely different source from that of the other orebodies.

It is too early to venture an estimate of the money to be taken from this newly discovered vein, but $500,000 would be a conservative guess.

Figures as to the average production from the Cresson mine for any one period, of, say, a month or a year, are misleading on account of the irregular production of high-grade ore; still it may be of interest to know that the mine has produced during the last 12 months an average of 8000 tons net per month with an average assay-value of $30 per ton. Since the discovery of the great chamber the mine has produced $3,400,000 in smelter returns.

Financial Assistance to Families of Soldiers and Sailors

The faulty working of the pension system has emphasized the necessity for carefully considering the methods of furnishing financial assistance to the families of soldiers and sailors. Obviously the task is so great that the Government alone can assume it. A bill representing the careful study of the whole problem of possible dependency as a result of the War has been simultaneously introduced in both houses of Congress.

The bill is an amendment to the act establishing the Bureau of War Risk Insurance. It adds to the duties of this Bureau the administration of family allowances, allotment to dependents, compensation for death or disability, and life insurance for men in the military or naval service. Secretary McAdoo states that the cost for the first year for the complete activities of the Bureau will be $176,150,000 and for the second year $380,500,000. The main purpose of the bill is to grant reasonable government indemnity against the losses and risks incurred in the discharge of a patriotic duty and in the performance of an extraordinarily hazardous service to which the Government has called and forced the citizen. It aims to accomplish these ends by granting a reasonable measure of indemnity against risk of loss (1) of support of the bread-winner, (2) of life and limb, (3) of present insurability at ordinary rates. The risk of dependency, in the case of an enlisted man's family, is indemnified by allotment of part of the pay of the enlisted man, supplemented by a family allowance granted and paid by the Government.

Secretary McAdoo illustrates the working of the system in the following example: "A private gets $33 per month for service abroad. If he has a wife and two children he must allot to them at least $15 out of his pay. The Government supplements this by giving the family an allowance of $32.50. This family's minimum monthly income, therefore, would be $47.50. The father can allot as much more as he pleases. If there is another child, the Government will allow $5 additional. If the man should have a father or mother actually dependent upon him, he can secure an allotment of $10 per month from the Government for the parent by alloting $5 more of his pay. Thus, the private with a wife, three children, and a mother actually dependent upon him, by giving $20 out of his $33 per month, would get from the Government $47.50 per month, giving the family an income of $67.50 and still leaving the man $13 per month for spending money. If there are more children, or if there is also a dependent father, the Government would grant as much as $50, over and above the man's own allotment."

Provision is made whereby men that do not allot their half-pay fully may deposit the remainder at 4% compound interest. The risk of disability to officers, men, and nurses is indemnified by compensation analogous to the Workmen's Compensation Act. Total disability is compensated on the basis of a proportion of the pay, from the minimum of from $40 to $75 per month up to the maximum for the higher officers of $200 per month. Compensation for disability is granted in proportion to the completeness of the disability. Medical, surgical, and hospital treatment is included. Provision although not at present detailed, is made for the rehabilitation and re-education of the injured. Risk of non-insurability at ordinary rates is provided for by Government insurance in amounts from $1000 to $10,000 for total disability or death. The excess cost due to the increased mortality and disability risk should be borne by the Government. The bill places the responsibility where it obviously belongs—upon the National Government. Local relief obligations will be assumed by the Red Cross. Clearly the purposes of this Act are to prevent families from suffering want while the bread-winners are fighting for their country.

THE possible recovery and utilization of pyrite occurring with bituminous coal is suggested in circular 5 of the Engineering Experiment Station of the University of Illinois by Professor E. A. Holbrook of the department of Mining Engineering. A series of experiments conducted on a commercial scale has shown that, where pyrite occurs in sufficient quantities to justify its recovery, a 50-ton plant may be designed to serve a single mine or group of mines which will yield a profit of about $75 per day, or $1.50 per ton of raw pyrite, with an initial capital cost of about $18,000. It is pointed out that the market for domestic pyrite is active at present and that prices as high as $8 per ton are offered. The supply of Spanish pyrite, which normally is adequate to meet the demand in the United States, has been cut off while the extent of the uses for the mineral has increased. The circular referred to presents a discussion of the composition of pyrite, a detailed review of present market conditions, a description of the machinery required in a preparation plant, and a statement of the methods of operations to be employed. The circular is illustrated with charts and photography, and also with drawings showing suggested plant arrangements. Attention is called to the fact that the processes outlined have been developed upon a working-scale. Copies of the publication may be had by addressing the Engineering Experiment Station, Urbana, Illinois.

Purity of Selected Copper Made in Converters

By HENRY F. COLLINS

*Reference has been made in a former paper (Trans. A. I. M. E., 1915) to the bottom-process in converters as practised in my work in Spain. For this I can claim no priority, since, although evolved independently, the same method had been previously employed by other metallurgists in Australia. Soon after the outbreak of the War in 1914, the Huelva Copper Co. was led to employ this method in order to save part of the gold content and to facilitate the utilization of the product in Spain for making brass for munitions, as well as bronze for coinage and other alloys, without further previous refining. It is noteworthy that practically the whole of the output produced by this method has been saleable upon the basis of 'best selected' or at most with a deduction of £1 or £2 from the full 'best selected' quotation, and as regards freedom from impurities the copper is considered as quite equal to the 'best selected' quality for making alloys. Particular testimony to this fact has been afforded by its utilization for coinage purposes at the Madrid mint. For rolling, the copper is of course unsuitable, on account of gas-pores and because the slight excess of oxygen renders it brittle, so that for this purpose it must be re-melted and 'brought to pitch.' In order that the copper as cast from the converters may be pure enough for brass making, it must be very slightly overblown. The operation is, however, one of some delicacy, and great care must be taken not to overshoot the exact point, since too much oxygen would be injurious to the quality of the resulting brass. The reaction between residual combined sulphur and dissolved oxygen is still proceeding when the operation is finished, but, on account of chilling, the charge cannot be left in the converter long enough to admit of settling down to tranquil fusion. When employing small converters holding charges of from 15 to 30 cwt. of copper, no difficulty was, however, experienced in pouring direct from the converter, without serious chilling, into small ingot-molds holding about 44 lb. apiece instead of the much larger pigs of 140 to 220 lb. into which converter copper is ordinarily cast. The ingots are dumped into water-tanks in the ordinary way for 'best selected' copper, and show the same characteristic rose-red color. Their upper surface is, however, marred by blisters from the expulsion of SO_2 upon solidifying, and, as already stated, the excess of oxygen, which it is necessary to introduce in order that the impurities may be sufficiently expelled, makes the ingots too brittle for rolling.

In order to show the exceptional freedom from deleterious impurity of the copper produced from highly arsenical zincy and leady ores by the selecting process, two analyses are appended, each of which is the average sample of a lot of from 20 to 50 tons and represents the average of a large number of charges mixed together. Two analyses are also given of the 'bottom,' in order to illustrate the extent to which the impurities are separated.

*Inst. Min. & Met., 1917.

	Selected copper		Bottom copper	
	%	%	%	%
Copper	99.4400	99.4100	96.7800	97.9700
Silver	0.0860	0.1060	0.1522	0.1724
Gold	0.0015	0.0010	0.0146	0.0206
Cobalt	0.0440	0.0256	0.1680	0.2410
Nickel	0.1322	0.0966		
Iron	0.0150	0.0353	0.5500	0.5600
Arsenic	0.0029	0.0031	0.0058	0.0058
Antimony	0.0016	0.0012	0.0045	0.0030
Bismuth	0.0017	0.0007	0.0016	0.0030
Sulphur	0.0860	0.0364	1.4600	0.7770
Lead	0.0888	0.0253	0.1170	0.0370
Zinc	0.0096	0.0240	0.0890	0.0290
Insoluble slag	trace	trace	0.5600	0.0510
Oxygen and loss	0.0907	0.2348	0.0973	0.1302
	100.0000	100.0000	100.0000	100.0000

At the works of the Huelva Copper Co., additional refining is omitted, but reverberatory refining of converter-selected copper is, however, carried out at a well-known Australian works, when copper is being handled which is too low in gold and silver content to justify electrolytic refining of the whole of the copper. The bulk of the precious metals present (particularly the gold) being first concentrated into 'bottoms' for electrolytic refining, the 'selected' copper produced by the remainder of the converter-charge is 'poled' and brought to pitch in a reverberatory refining-furnace, from which it is cast into ingots of the usual 'best selected' type. These are known as 'fire-refined' copper, and have been recognized recently by the London Metal Exchange under that title as of equal quality to 'best selected.' In order to show its purity, the following analyses are given, each of which represents the average sample of a large lot as shipped.

	Fire-refined copper		
	%	%	%
Copper, Silver	99.5800	99.6000	99.6200
Gold	......	......	......
Nickel and cobalt	0.3390	0.3300	0.2800
Iron	0.0045	0.0040	0.0030
Arsenic	0.0038	0.0040	0.0030
Antimony	trace	trace	trace
Bismuth	trace	trace	trace
Sulphur	....	trace	trace
Selenium and tellurium	trace	trace	trace
Lead	0.0006	0.0050	0.0050
Zinc	......	......	......
Tin	trace	......	......
Insoluble	......	......	......
Oxygen by difference	0.0721	0.0570	0.0890
	100.0000	100.0000	100.0006

The purity in copper, silver, and nickel combined is, in the three cases, respectively 99.92, 99.93, and 99.90%, the impurities ranging from 0.07 to 0.10%.

Cyanidation of Flotation Concentrate

By JAMES G. PARMELEE

INTRODUCTION. Any analysis of the problem of cyaniding flotation concentrate that will help to bring about an economy in the recovery of silver must interest all metallurgists. The present price of silver is high owing to the demand created by war conditions. An increase in the amount now produced will not materially affect the price, but if the cost of production can be lowered the net returns will be higher. A prime factor is a lower cost of freight to the smelters. The distance of a concentrating-plant from the smelter, then, is the main proposition. If situated a few miles away it is possible to have the flotation-concentrate treated and obtain reasonable returns. On the other hand, concentrators a hundred miles or more distant must meet such a high freight charge that the net profit becomes smaller. Therefore, to overcome this, the flotation concentrate must be cyanided on the property. As a result of the success attained by flotation in the past four years the cyanidation of the concentrate is at present a subject of intense interest to the metallurgical world.

The cyaniding of flotation concentrate is difficult, owing to the fact that two-thirds of the oil added to the ore usually appears in the concentrate. Finely divided metallic gold in the form of flotation concentrate has been cyanided in several cases[1] without much difficulty while the cyaniding of flotation concentrate containing silver as argentite (Ag_2S) or tetrahedrite ($4Cu_2S.Sb_2S_3$ approximately) has been unsuccessful. Certain minerals are difficult to cyanide in any case, and the oil as retained by the sulphides tends to increase the difficulty. Some metallurgists claim that oil has no effect on the cyanidation of flotation concentrate. J. E. Clennell[2] says: "Starting with a 0.204% KCN solution, and using various oils, my lowest percentage of KCN remaining after agitation was 0.108%. This plainly shows less than a 0.3% consumption of KCN, and also that the oil has no deleterious effect." Hugh Rose[3] says: "The raw flotation-concentrate cyanides with no more difficulty than a gravity-concentrate. The cyanide consumption maintains more or less the same ratio as in the crude ore, namely, 4 gm. of NaCN per gramme of silver. In this test an extraction [meaning 'recovery'] of 98% of the gold and silver was obtainable." The ore that Rose used for this test was notably free from base metals and the silver was in the form of argentite (Ag_2S).

In comparison with this, E. A. Herson says:[4] "Were it not for the trouble of the oil, the treatment of ore by a combined process of flotation and cyanidation would promise great economy. Cyanide treatment in many cases would be a direct process for recovering the valuable metal of the concentrate without the costs involved in shipment and in smelting. A low extraction is, however, the result."

The hindrance to high extraction is apparently due to the physical action of the oil, which forms a film-sack or coating around each metallic particle. This coating is thin, but nevertheless may prevent the access of the cyanide solution. It is evident that an oil must be chosen, not alone for its flotative action, but also for the influence it may have upon the extraction by cyanide.

As the cyanide solutions must be used cyclically, serious difficulties are apt to occur. This is ably expressed by P. A. Avery,[5] who says: "Especially is this true in the precipitating end of the plant. It seems that a combination of oil, alkali, cyanide solution, and zinc-dust bring about the precipitation of a gelatinous compound that clogs the zinc-press cloths, causing pressure to rise beyond dangerous limits. This case is cited to show what might be expected, if certain flotation-oils are not removed before cyaniding. Not only does oil affect the cyanidation of the flotation concentrate but it also affects the zinc-box precipitation. The same reaction would take place in a zinc-box, for zinc is zinc whether it be thread or dust. A nasty oily precipitate might finally accumulate causing a decrease of the inflow and a drop in the precipitative efficiency of the zinc."

In the present investigation a series of experiments was made to determine first, the effect of the presence of oil in cyanidation, upon an ore otherwise amenable to cyanide treatment; second, the effect of the presence of oil on the cyanidation of the sulphides, that is, the flotation concentrate; and third, the possibilities of removing the oil.

EFFECT OF OIL. For investigation of the first point a silicious ore, assaying 20 oz. silver in the form of silver chloride was obtained and reduced to 200-mesh. To the pulp was added water and oil, thoroughly mixing in a Janney testing-machine. Potassium cyanide and calcium oxide were then added to the solution, which was agitated for 16 hours in inverted glass chimneys fitted for air-agitation of the pachuca type. A blank test without oil was run at the same time to obtain a comparative extraction. The oils used in this test and in those that follow were, crude pine-oil, pine-tar creosote, wood-creosote, coal-tar creosote, and special resin-oil. Three separate

[1] M. & S. P., November 29, 1915.
[2] M. & S. P., May 13, 1916.
[3] Trans. A. I. M. E., Vol. CXVI.
[4] Met. & Chem. Eng., Vol. XIV, p. 675.
[5] M. & S. P., Vol. CXII, p. 661.

groups of tests were conducted in which varying percentages of oil were used. The results of the effect of oil upon an ore amenable to cyanidation show that oil has no noticeable effect upon an ore when no sulphides are present.

In the following table constant factors were: weight of pulp, 100 gm.; 500 cc. water; 0.5% KCN; 1% CaO; 16 hours agitation; through 200 mesh:

TABLE I

Trial No.	Oil— Kind	%	Silver— Head., oz.	Tail., oz.	Extraction, %
1.	None	..	20.0	3.3	83.6
2.	Crude pine-oil.....	0.27	"	3.4	83.0
3.	Pine-tar creosote ..	"	"	3.5	82.4
4.	Wood-creosote	"	"	3.8	80.5
5.	Coal-tar creosote ..	"	"	3.9	80.1
6.	Crude pine-oil	0.67	"	2.9	85.2
7.	Pine-tar creosote...	"	"	3.3	83.6
8.	Wood-creosote	"	"	3.2	84.4
9.	Coal-tar creosote ..	"	"	3.1	84.6
10.	Special resin-oil....	"	"	3.1	84.6
11.	Crude pine-oil.....	2.70	"	5.6	71.4
12.	Pine-tar creosote...	"	"	3.4	83.0
13.	Wood-creosote	"	"	4.5	77.5
14.	Coal-tar creosote...	"	"	3.1	84.6
15.	Special resin-oil...	"	"	3.5	82.4

EFFECT OF OIL IN PRESENCE OF SULPHIDES. To note the effect of oil upon sulphides it was desirable first, to cyanide raw concentrate with and without oil, and then to cyanide the flotation concentrate, after which a comparison could be made. The pulp used was from a lead-zinc tetrahedrite ore containing silver. This tetrahedrite has in the past yielded poor recovery in cyaniding. Therefore low results would naturally be expected. Ordinary table and flotation-concentrate were obtained as usual. Oil was added, in all cases save one, on the table-concentrate, after which the pulps with oil and those without were cyanided for 18 hours in the lamp-chimney cells.

The results of these tests show that the presence of oil makes an average difference of 20% in extraction between raw concentrate with and without oil, and 15% difference between the raw concentrate without oil and the flotation concentrate. It is noticeable that the heavy oils give a lower extraction than the light oils. This would be expected from their adhesive qualities and therefore such oils should be avoided if flotation is to be used prior to cyanidation.

In the following table the constant factors were: weight of pulp 50 gm.; water 250 cc.; KCN 0.5%; CaO 1%; agitation 18 hours; and oil 1%.

TABLE II

Trial No.	Oil— Kind	Mesh	Silver— Head., oz.	Tail., oz.	Extraction, %
	Table concentrate				
1.	None	100	40.3	18.0	55.0
2.	Crude pine-oil.....	"	"	21.3	46.6
3.	Pine-tar creosote ..	"	"	21.9	45.2
4.	Wood-creosote	"	"	26.8	33.6
5.	Coal-tar creosote ..	"	"	30.3	24.4
	Flotation concentrate				
6.	Crude pine-oil	150	68.8	36.4	46.2
7.	" "	"	"	38.3	44.2
8.	" "	"	"	38.3	44.2
9.	" "	"	"	42.0	40.1
10.	" "	"	"	40.3	42.6

REMOVAL OF THE OIL. When oil is present in small quantities in some materials it can be saponified and washed away from the substance in the form of soap. Furthermore, when the amount of oil is appreciable it can be removed by simple oil-solvents. Roasting expels or destroys the oil, but when silver is present the roasting tends to form an insoluble silver compound. A chloridizing roast is not open to this objection. Clennell and Butters[6] refer to experiments in which roasting was tried. The ore that Clennell was investigating contained gold and silver, but the silver was not considered because of the smallness of the quantity. Clennell says: "As results obtained by direct cyanidation of flotation concentrate did not prove encouraging we turned our attention to roasting the concentrate previous to cyanidation. Three tests were carried out with the percentages noted below:

	Extraction Gold %	Silver %
(1) Oxidized roast, water-wash and cyanided...	97.0	5.0
(2) *Oxidized roast, acid-wash and cyanided....	98.8	11.1
(3) Cyanide flotation roast, acid-wash and re-cyanide	64.9	10.9

*Concentrate.

The most favorable results were found when roasting at a low temperature; under these conditions a maximum amount of copper was extracted by washing with water, and the highest extraction of gold with a minimum consumption of cyanide.

From the last statement it is seen that gold can be recovered easily by cyanidation after the pulp has been roasted at a low temperature. As said before, the silver in an ore can also be recovered by cyanidation after the flotation concentrate has been given a chloridizing roast at a low temperature.

SOLVENTS FOR OIL. The following oil-solvents have been tried. Hot water, alcohol, ether, alkali, soap, benzine, gasoline, carbon bi-sulphide, carbon tetra-chloride, and acetone were used to determine the solvent action on various oils. Into sets of glass beakers small pieces of clean 'steel' galena were placed; oils of different quality were then added to each set of beakers, which were allowed to remain stationary for 15 minutes, that is, about the same length of time that the ordinary particle of sulphide in flotation remains in contact with the oil. Solvents were then added in a similar manner and the results obtained are given in Table III.

CYANIDING FLOTATION CONCENTRATE WITH OIL REMOVED. Alcohol and gasoline gave the best results in dissolving the oil and were therefore used with the flotation concentrate in the next series of tests. The concentrate was visibly cleaner after washing than that which had

[6] M. & S. P., Nov. 29, 1915.

TABLE III

Solvent	Solvent effect with different oils			
	Crude pine-oil	Pine-tar creosote	Wood-creosote	Coal-tar creosote
Hot water	None	None	None	None
Alcohol	Complete	Complete	Complete	Complete
Ether	Partly soluble	Partly soluble	Partly soluble	Partly soluble
NaOH (15%)	Incomplete	None	None	Incomplete
Gasoline	Complete	Complete	Complete	Complete
Acetone	Partly soluble	None	Partly soluble	Partly soluble
Benzine	" "	Partly soluble	" "	" "
Carbon bi-sulphide	" "	None	" "	" "
Hot soap solution	None	None	None	None

not been washed. This material was then weighed and cyanided as before. The results show a similarity in the recovery, namely, 60%, with both alcohol and gasoline. The increase of 5% over the result by cyaniding table concentrate in this case is a result of finer grinding, the pulp being ground to pass 150-mesh.

Constant factors used throughout are: weight of concentrate used 50 gm.; water 250 cc.; crude pine-oil 1%; KCN 0.5%; CaO 1%; agitation 18 hours; fineness 150-mesh:

TABLE IV

No.	Oil-solvent	Silver Head., oz.	Silver Tail., oz.	Extraction %
1	Alcohol	68.8	24.2	64.5
2	"	68.8	27.8	59.5
3	Gasoline	61.9	24.9	59.4
4	"	61.9	24.6	60.0

CONCLUSIONS. (1) Oil has no effect upon the cyanidation of ore in which the metals are not in the form of a sulphide. This shows that the trouble, in cyaniding flotation concentrate, is due to the oil-film on the sulphide particles. Furthermore, it is found that the oil has no effect upon the cyanide solution.

(2) The presence of oil in flotation concentrate hinders cyanidation because the sulphide particles are enveloped in a jacket of oil that prevents a good extraction by the cyanide.

(3) The film or covering of oil may be removed by some solvent such as alcohol, or gasoline.

(4) Heavy oils, such as pine-tar creosote, should be avoided, owing to their insolubility.

(5) Roasting at a low temperature will suffice for removing oil from a flotation concentrate containing gold, but, when treating flotation concentrate containing silver, a chloridizing roast at a low temperature must be used.

LATERIZATION is defined by J. M. Campbell as the process by which certain hydroxides, principally those of ferric iron, aluminum, and titanium, are deposited within the mass of a porous rock near the surface. Unless a rock contains uncombined alumina in the form of hydroxide it cannot be regarded as lateritic. Lateritic constituents are deposited in porous rocks between maximum and minimum vadose-water level only in places near the surface where oxygen can gain free access. Lateritic constituents are believed to be deposited in limited quantity in suitable situations in temperate climates. Rock, at the time laterization commences in it, usually contains no iron.

Determination of Tungsten

*To one gramme of the finely powdered wolfram in an 8-oz. beaker add 10 cc. strong hydrochloric acid, and agitate carefully to prevent caking. Add more acid (about 90 cc. in all), cover with a watch-glass, and boil briskly until the volume of acid is reduced to about 5 cc. Allow to cool, add 5 cc. strong nitric acid, and digest at nearly a boiling temperature for from five to ten minutes. Dilute with water to about 100 cc., allow to settle, and filter through a 9-cm. filter, wash with water, working so as to obtain as little tungstic acid as possible on the filter. To the beaker add about 10 cc. distilled water and 10 cc. dilute ammonia, adding the latter in small quantities down the sides of the beaker. With a rubber-tipped rod remove any deposit from the sides of the beaker into the alkaline liquor, raise to the boil, and stir well. If the suspended matter does not readily subside, boil again until the desired result is attained. Allow to settle, and while still hot filter through the original paper into an 8-oz. flask. Wash the beaker two or three times with small quantities of distilled water, decanting each time from any heavy mineral that may be present. Complete the washing of the filter-paper, and reserve the cover-glass, filter-paper, and beaker. Evaporate the solution of ammonium tungstate in the flask to a small volume, transfer carefully to a weighed platinum dish, and continue the evaporation to dryness on a water-bath. Ignite the dish and contents gently at first, and more strongly afterward. Weigh as tungstic acid. To the beaker containing the heavy residues add 5 cc. hydrochloric acid, cover with the original watch-glass, and boil till nearly dry; add a few cc. of nitric acid, and heat again for ten minutes. Dilute with water, allow to stand until the deposit has subsided, filter through a small paper, and wash the beaker and paper several times. Wash the filter-paper, also the original filter-paper, with a little hot dilute caustic-soda solution, followed by small washes of distilled water, and collect the alkaline liquors and washings in a beaker. If the watch-glass is stained with tungstic acid, dissolve with a few drops of the soda solution and add to the contents of the beaker. Raise the contents of the beaker to boiling, add about 2 gm. ammonium nitrate, and stir well while boiling. Filter through one of the papers previously used, and wash a little. Nearly neutralize the filtrate with dilute nitric acid, and a few cc. of mercurous nitrate solution, stir well, filter on a small ashless paper, wash, ignite, and weigh as tungstic acid. Add this weight to the weight of the tungstic acid in the platinum dish. Fne grinding is essential. One gramme is a convenient quantity, but with experience $2\frac{1}{2}$ grammes may be taken without requiring any modification of the quantities given in the assay. The complete decomposition with the boiling hydrochloric acid takes about an hour.

*A method described by H. W. Hutchin before the Cornish Institute of Engineers, and published in 'The Mining Magazine,' August 1917.

Simple Methods of Finding Density and Weight of Solids in Mill-Pulp

By R. B. KILIANI

One of the simplest methods of finding the specific gravity or density of pulp is to use a 2000 cc. glass graduate with a sheet-iron top for cutting the sample as shown in Fig. 1. By knowing the tare or weight of the graduate, it is a simple matter to determine the weight of pulp in the graduate by weighing the whole thing on a scale reading in grammes. The specific gravity of the pulp can then easily be found by dividing that weight in grammes by the volume of the pulp in cubic centimetres. Knowing the specific gravity of the dry ore, the percentage of solid in the pulp can be found by the formula:

$$p = \frac{G}{G-1} \times \frac{a-1}{a}$$

Where p = per cent of solid; G = specific gravity of the dry ore; a = specific gravity of the wet pulp.

Another less accurate method for finding the specific gravity of a pulp and the per cent of solid present in it, capable of being used where there is not room below the discharge-spout or bell of the mill to insert the graduate, is to use a bucket about 10 or 12 in. diameter by 6 or 8 in. deep, with a narrow spout on one side for cutting the stream of pulp. A sketch of this bucket is given in Fig. 2.

A rod is indicated in the centre of the bucket; it serves as a gauge of the height to which the bucket should be filled. The volume of the bucket to the top of this rod can be found by finding the grammes of water required to fill it to that point. The weight of the wet pulp can then be found by weighing the bucket and the pulp, and the specific gravity of the pulp may be calculated as before.

Having ascertained the specific gravity of the pulp and the per cent of solid, the tons of solid in the amount of pulp handled in each 24 hours can be found by one of two methods. The first is to run the entire discharge of the mill into a barrel or tank for a definite number of seconds, taking the weight of the tank and pulp, and deducting the weight of the barrel, thus giving the pounds of pulp per second. From this the tons of pulp per 24 hr. can easily be figured, and, again, from this and the per cent of solid, as previously determined, the tons of solid handled per 24 hr. can be estimated.

Another method, which avoids weighing the pulp, is to use a tank below the launder leading from the discharge-end of the mill, with a gate to divert the pulp-stream into the tank at will. The gate should be so made that practically no time is lost in diverting or stopping the entire stream of pulp. This box or tank can be made to hold any desired quantity, and may be allowed to fill for any pre-determined number of seconds. By knowing the internal dimensions of the tank, the cubic feet of pulp can be computed from the depth of the pulp-level below the top of the tank, this distance being measured with an ordinary rule. If the box is not level, it is best to fill it a few inches deep with water, so as to have a level bottom. This will also cover a gate or valve in the bottom for emptying the tank, and will avoid errors in computation due to such a condition. The actual depth of the pulp poured into the tank in taking the sample is then measured easily. The tons of solid handled per 24 hr. may be found by multiplying the volume of pulp in cubic feet by the number of seconds in 24 hr., by 62.5 (the weight in pounds of 1 cu. ft. of water) times the specific gravity of the wet pulp, times the per cent of solid, divided by the number

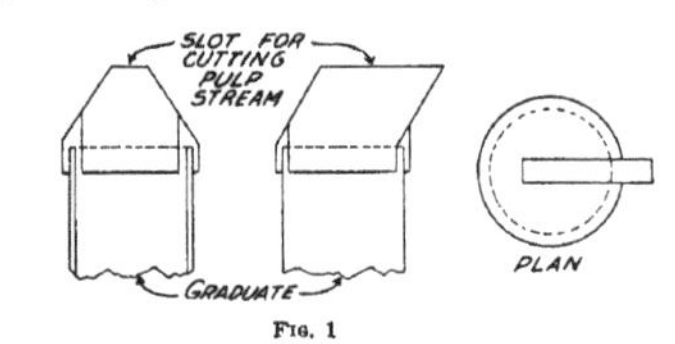

FIG. 1

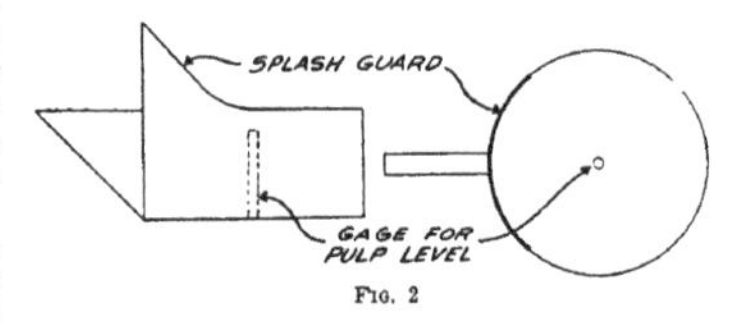

FIG. 2

of seconds during which the sample was flowing, times 2000 (the number of pounds in 1 ton). This can be expressed as follows:

$$\text{Tons per 24 hr.} = \frac{\text{Vol. in cu. ft.} \times 86{,}400 \times 62.5 \times \text{sp. gr. of pulp} \times \text{per cent solid}}{\text{Time of sample in seconds} \times 2000}$$

The use of this formula can be simplified if a standard number of seconds is always used for the time of taking the sample, say, for example, 5, 10, 15, or 20 seconds. The tons of solid per 24 hr. then will be equal to a constant, times the depth of pulp in the tank in inches, times the specific gravity of the wet pulp, times the per cent of solid. Tons per 24 hr. = constant × depth of pulp in inches × sp. gr. of pulp × per cent of solid.

The constant in this formula will be equal to the length of the tank in inches, times the width of the tank in inches, times the number of seconds in 24 hr., times 62.5, divided by 1728 (the number of cubic inches in 1 cu. ft.), times the constant period in seconds during which the sample was taken, times 2000; or, expressed in an equation:

$$C = \frac{L \times W \times 86{,}400 \times 62.5}{1728 \times T \times 2000}$$

Where C = the constant; L = length of tank in inches; T = time of taking sample in seconds; and W = the width of the tank in inches.

Graphite

By L. W. BROOKS

The use of graphite in the industries of the world has been constantly progressing, not only in the amount used, but in the number and variety of its applications. Beginning in the middle of the sixteenth century through its utilization in the manufacture of lead pencils, followed perhaps a century later in the manufacture of melting pots, its sphere of economic adaptation has so broadened that today it occupies a distinct and important place in the world's industrial scheme.

The words graphite, plumbago, and blacklead, are practically synonymous, in that they refer to the same chemical substance. 'Blacklead' was probably the original term, and a comparative one, because it indicated the color of the streak made by this substance as compared with that made by metallic lead. Evidently it was so employed at an early period for writing or lining on paper. From the word blacklead the word plumbago naturally developed through the Latin. Graphite is derived from the Greek; and all these names have reference to its most remote use, namely, that of making visible markings on paper. Although these terms are synonymous, there have arisen special applications for them, thus, we import plumbago from the island of Ceylon, and, before the War, blacklead came from Germany, Austria, and Italy; at the same time, we export graphite from this country to the other countries of the world. There are lead pencils, plumbago crucibles, and graphite lubricants; blacklead stove-polish, plumbago foundry-facings, and graphite paint. This confusion of names may seem to be somewhat misleading, yet there is considerable method in the nomenclature.

Graphite may be regarded as occupying the middle place in the carbon trinity, charcoal and the diamond being the other forms, but it has its own individual characteristics, which make the number of its useful applications much greater than that of either of the others. Graphite occurs naturally in two forms, the amorphous and crystalline, each form being subject to variations due to the incidents of its occurrence. Amorphous graphite does not occur pure, but is always associated with other earthy materials, the character of which have great bearing on its use. Crystalline graphite shows great variations also, because of the distortion of its crystals during the process of formation. Crystalline graphite occurs both massive and with its particles disseminated through a containing rock; thus, Ceylon graphite occurs in large masses of crystals, while that coming from American deposits usually occurs as small laminated flakes disseminated through containing rocks. These differences of formation are made use of in the proper selection of graphite for its various uses.

At the present time the annual production of graphite for the eentire world probably exceeds 100,000 tons, and of this amount probably 50% is of the crystalline variety and the remainder of the amorphous kind. Of the crystalline graphite, more than 80% comes from the island of Ceylon, and is amorphous, and also probably the same relative proportion comes from Korea. The remainder of the crystalline graphite is supplied almost wholly from the mines of Madagascar, New York, and Alabama. The Mexican graphite is of the amorphous kind as well as that from Rhode Island, and Canada. The localities mentioned furnish practically all of the graphite used, but it must not be understood that graphite is of rare occurrence. It is widely distributed, and the Mineral Resources of the United States, published annually by the Geological Survey, makes mention of production in the States of Alabama, Michigan, Pennsylvania, California, Colorado, Montana, South Dakota, and Alaska. It has been stated that there is probably as much carbon, occurring as graphite, north of the Mohawk river, New York, as there is south of it in the form of coal, but while it is so generally distributed it is rarely found so associated as to be available as a source of supply.

The first importation of Ceylon graphite into this country was made in 1829, by Joseph Dixon, of Salem, Massachusetts. Two years previous he had experimented with graphite from the State of New Hampshire, as a substitute for German blacklead, or pot-lead, as it was sometimes called, in the manufacture of crucibles. The success of his experiment was so pronounced that he secured a small shipment of Ceylon plumbago, samples of which he had previously seen brought back by sailors to the New England ports, for at that time Salem was a thriving centre of the East India trade. This was the first application of crystalline graphite in the manufacture of crucibles, which branch of the industry now absorbs somewhat more than one-half of the world's output of graphite. The range in the prices of graphite is greater than the range of the quality. The inferior grades of Korean blacklead sell as low as $20 per ton in original packages in this market, while during the past year, crucible grades of Ceylon plumbago have sold higher than $700 per ton; between these limits, certain grades of the Mexican amorphous graphite have brought higher prices than some of the inferior grades from Ceylon; the percentage of carbon in low grades of German and Austrian blacklead is as low as 40%, while fine grades of Ceylon sometimes run as high as 98% of carbon. The Ceylon product is imported directly by steamers in packages having net weights of about 600 lb. This is graded into lump, chip, and dust, the first grade consisting of the larger and softer pieces, while the last is much finer, and, of course, carries the bulk of the earthy impurities, while chip is a grade lying between the two. These grades are again sub-divided according to their brightness or 'boldness', as it is called, and their softness. About 50% of the Ceylon product is imported into this country, while England takes a smaller amount, the remainder being distributed between France, Japan, and Russia. The total output of graphite is consumed in something like the proportions indicated in the following table:

	%
Crucibles	55
Stove-polish	15
Foundry-facings	10
Paint	5
Lubricants	5
All others	10

The last item includes graphite for powder-glazing, electro-typing, steam-packings, and other minor uses. Of course, an exact apportionment would be quite impossible, and the figures are given merely as an indication of individual opinion. The uses to which graphite is put depend on the physical characteristics which it possesses, none of its uses, except that of a foundry facing, involving a chemical reaction. These physical properties are its infusibility at temperatures below that of the electric arc, its great capacity for absorbing and transferring heat, its comparatively high electrical conductivity, and its softness which may be defined as unctuousness. The latter expression may sound peculiar, but it describes the quality of yielding on contact with other surfaces, and that is the reason for its use in lead pencils, lubrication, powder-glazing, and stove polish, for example. It adheres readily to any surface with which it comes in contact and is highly polished by the slightest friction.

The use of graphite in the manufacture of lead pencils, which was shown to be the oldest, is now its most common one. The use of the lead pencil itself is universal. The first lead pencils were made from blacklead mined near Barrowdale, Cumberland county, England, and the first mention of them occurs in Conrad Gessner's work on 'Fossils,' dated in the year 1565. The method of manufacture was to cut strips of blacklead from larger pieces of mineral and insert them in grooves cut in small bars of wood. The product of this mine became extremely valuable, and it is stated that at one time its product sold for as much as 30 shillings per pound. Laws were promulgated, making robbery of blacklead pits a felony; and military escorts were furnished for the carts on their way from the mines to the works. This method of manufacture continued until Conte, of Paris, in 1795, devised the method of manufacture which is now universally followed. In this process the graphite and the clay are ground to the finest possible state of division, mixed, filtered, and formed into cakes by means of hydraulic filters; it is then again mixed by repeatedly forcing it through plates perforated by many minute holes. It is then placed in hydraulic presses and forced through dies into the shape and size required. As it issues from the press, it resembles nothing more strongly than a long, round, black cord. It is laid out straight on boards, and, when dry, is cut into proper lengths. It is then packed into plumbago crucibles and fired in kilns. Red cedar is the wood most generally employed in pencil-making, although poplar is sometimes used for the cheaper grades, and for slate-spencils. The cedar logs are sawed into small slabs of the proper length for a pencil, and of a sufficient width for four, five, or six pencils. These are grooved lengthwise, the grooves being exactly the diameter of the lead which they are to receive. The leads are laid in the grooves and another similar block is glued firmly to it. The resulting slab is then run through shaping machines, which cut each individual pencil from the larger block. The pencils are then ready for varnishing, polishing, stamping, and whatever else is to be done to them. The grade of hardness depends upon the relative percentage of clay contained in the graphite mixture, the larger the amount of clay the harder the grade. Colored and slate-pencils are made in much the same manner, other pigments being substituted for the blacklead. For the manufacture of lead pencils the amorphous type of graphite is required. It is supplied principally from the mines of Austria and Mexico. The Mexican product has come into the market for this purpose only within the last few years, but it has rapidly taken the place of graphite from other sources. The pencil-making industry, although not consuming a great quantity of blacklead, is of considerable size, probably 15,000 people being employed in their manufacture in this country, Germany, and Austria.

As previously stated, the making of graphite crucibles absorbs more graphite than all the other uses combined. In this country, in England, and in France, the crystalline variety, mostly from the island of Ceylon, is employed almost exclusively in the manufacture of crucibles. In central Europe, Germany and Austria, blacklead is largely used for this purpose, sometimes without any admixture, but also in connection with the crystalline form in varying proportions. The use of plumbago in the manufacture of steel-melting crucibles in England is rapidly increasing, notwithstanding the existence of an easily available supply of good refractory material suitable for the manufacture of crucibles from clay.

Clay crucibles for steel-melting ordinarily last one heat, while plumbago crucibles will serve an average of five to eight heats. The difference is still greater when melting copper and composition metals, plumbago crucibles often lasting for thirty heats, treating 150 lb. at a time.

In outline the manufacture of crucibles is as follows: the components of the mixture, in a plastic condition, are put into a revolving mold, the interior of the mold having the contour of the outside of the crucible; a rib having its forming edge shaped like the inside of the crucible and attached to the end of a lever, is drawn down inside of the revolving mold forcing the plastic mass into the desired shape. The crucibles are then dried thoroughly, this operation requiring from one to six weeks, depending on their size and the thickness of the walls. They are next set in kilns and fired, the temperature of the kilns being sufficient completely to eliminate the combined water from the clay. The crucibles are now ready for use, although the precaution should be taken, immediately before use, to anneal them sufficiently to drive off any hygroscopic moisture that may have been acquired during shipment or storage. The function of the plumbago in a crucible is somewhat obscure, but the principal one is that of its capacity to absorb and transmit heat. It is also infusible at temperatures short of that of the electric arc. The plumbago crucible is remarkable for its ability to withstand great shock, due to sudden changes of temperature. A crucible at white heat may be thrown into cold water without any apparent damage. Of course, this is an extremely severe test, and if repeated would eventually injure the crucible. When well made, crucibles give way through actually wearing out, that is, the repeated formation of slag on the outside of the crucible during successive heats, gradually reduces the thickness of the walls until they become too thin to carry the weight of the metal put into them. The forms of crucibles vary with special applications, for example, for liquid brazing, as applied to the manufacture of frames for bicycles, automobiles, and the like, a crucible, rectangular in plan, with the bottom sloping from the ends to a depth of eight to ten inches at the middle, is set in a brick furnace. Brazing metal is melted in the crucible and covered with a fluxing agent. The frame to be brazed is dipped into the molten metal for a minute or more, until the metal has had time to run into the farthest crevices of the joint. Previous to immersion, the parts of the frame near the joints are coated with a mixture of graphite with some quick-drying varnish like shellac. This coating prevents the adhesion of the metal to the surfaces so protected. Square and rectangular crucibles are also used for carbonization of electric-light filaments and for annealing purposes. Trays for 'blueing' and tempering screws and small machine parts, steel ladle-stoppers and nozzles, and retorts for the distillation of lead and zinc, are also made from the same materials as the graphite crucibles.

The practice of facing molds, in which castings are to be made, with some carbonaceous material, is general. The material used is usually anthracite, charcoal, or graphite. The reason for its use is to prevent the adhesion of the iron to the sand of the mold. The principle of its use for this purpose is as follows: the air contained in the mold, and which is carried in by the stream of molten metal, furnishes oxygen for the combustion of the carbonaceous material of which the facing is composed, so that a condition obtains analogous to that of the spheroidal state of a drop of water on a hot surface; thus the iron is effectually prevented from coming into actual contact with the sand, so that when the casting is removed it will be found to be covered with a thin crust which will easily peel off, leaving the iron smooth and clean. In order to secure perfect results, two conditions must prevail, namely, first, the facing must adhere perfectly to the mold-surfaces. The hot metal coming in contact with the sand, dries out the mold, and if the facing has not been properly compounded, it will be washed away in front of the advancing metal, hence it is necessary to have a certain percentage of clayey material mixed with the facing to prevent it 'running' before the metal; second, it must be slowly combustible. If the facing burns quickly, trouble is likely to ensue from: (a) too great a volume of gas to be readily vented, causing 'blows' and 'cold shuts'; the latter term being applied to those cases where the iron has not filled the mold, and (b) where combustion is too rapid, it is not likely to endure during the entire time that the metal is in the fluid condition, so that while at first the spheroidal condition exists, it ceases before the metal is solidified, thus giving opportunity for adhesion.

Castings which have been made in connection with the use of a facing well suited to the particular case are superior because of the finer surface-texture, and of the ease of cleaning and the lessened tendency to dull the cutting-edges of machine-tools. Plumbago is the one material that combines in a greater degree than any of the others, the requisites necessary to a good facing. It contains no volatile matter whatever, and it burns evenly and slowly, so that a smaller quantity may be used. It has in addition another quality; it can be polished, giving the smoothest possible surface to the mold.

It is probable that the most interesting use of graphite, viewed from a mechanical standpoint, is that of a lubricant. This is a comparatively new application, very little of it having been employed for this purpose earlier than 25 years ago. Much has been claimed for it as a lubricant, without pointing out the manner in which it operates. The crystalline graphite is the only form suitable for lubricating purposes. The amorphous kind is always associated with impurities; accordingly, when finely ground, the particles are liable to be compacted into a mass. The crystalline graphite on the other hand will not pack in this way. While any pure, soft, crystalline graphite is valuable for lubricating purposes, the laminated form is the one specially adapted for the purpose. The true function of graphite as a lubricant is to change the character of solid friction, the co-efficient of friction of iron on iron being changed to that of iron on graphite, which may be taken safely at 60% less. The laminated form is most suitable for lubricating purposes, because it adheres so closely to the surfaces with which it comes into contact. The laminated or flake form, as it is prepared for market, is irregularly circular in shape, having an average diameter of 1/40-inch. The thickness of these flakes varies from 1/500 in. to a thickness too small to be accurately determined by a micrometer caliper; it is certainly less than 1/4000 in. From this it may be appreciated how closely these thin flakes become attached to friction-surfaces. Graphite must not be considered as a substitute for oils or greases in lubrication. It would be foolish to claim that any solid friction could be as low as that of the fluid friction of mineral-lubricating oils or greases, but there are cases where the latter fail and others where the admixture of graphite is of the greatest benefit. As with all good things, discretion must be used in applying it. The greatest error consists in using it in too great quantities. The broad flat flakes will stand a large amount of wear before being destroyed. Furthermore graphite may not be used in connection with oil, where the operation of the oiling device is based on the capillary principle, because the graphite will clog the tubes and entirely interfere with the lubrication. One valuable feature of graphite in lubrication is that a bearing so lubricated is not liable to 'seize' while running hot. The oil may burn out, but the graphite comes into play and effectually prevents adhesion of the friction-surfaces. Graphite also has a number of special uses, such as in air cylinders, type-setting machines, lace machinery, where oil stains are to be avoided, on piano and organ-actions, and other places where the use of oil must be avoided. Perhaps no more remarkable illustration of the lubricating qualities of graphite can be cited than its effect on tile-roofs, near the sorting-compounds in Ceylon. These tiles are set at an angle less than that of repose; the winds carry the dust from the plumbago compounds and drive it into the crevices of the tile-roofs, and so change the angle of friction that without warning the roof comes sliding down.

An exceedingly important use to which graphite is applied is the process of electro-tpying, in which it serves a double purpose. The process of electro-typing as applied to the art of printing consists in making a wax-impression of a type-form or a half-tone cut or engraving, and then depositing copper thereon. Graphite is applied to the face of the wax-case to prevent the adhesion of the type-form. For this purpose it must be very fine, and of such a character that it will not pack in masses that would cause minute blotches on the mold. Its second function is to supply a conductive coating to the surface of the mold. This is done by highly polishing the surface with graphite by means of fine hand brushes. Graphite for this purpose should be of the greatest purity and fineness. When a case so prepared is immersed in the electro-plating vat, a deposit of copper begins to form at the connecting points and extends over the whole surface that has been polished with the graphite. This may take from 20 to 40 minutes. In some cases, where the finest results are not wanted, and where time is to be saved, a chemical deposit of copper is first precipitated on the form. This is done by pouring a solution of sulphate of copper over pure iron filings upon the surface of the mold. Copper is deposited on the surface, but only where there is a graphite coating. Such a coating reduces the time required to deposit the coating in the electro-plating vat, but the quality of the electrotype so made is not so good. Many attempts have been made to find a substitute for graphite in this process, but so far without success.

Considerable quantities of crystalline graphite are used for glazing gunpowder, in order to prevent the absorption of moisture, which would result in caking. This is done in a large tumbling-barrel, the powder and graphite being tumbled together for a number of hours until the correct polish has been obtained.

Stove-polish is also one of the common articles in which graphite is used in large quantities. It consists of two components, the graphite which imparts the polish, and the clay binding-material which holds it in place. Amorphous blacklead can be used without any admixture, but the addition of crystalline graphite gives a more brilliant but grayer lustre, and one that is more enduring.

Within the past decade the use of carbon as a pigment in the paint industry has steadily increased, and the form of carbon usually employed is graphite. The great durability of paint made from lamp-black and linseed oil is well known, but it possesses the great disadvantage of drying with extreme slowness. Paint made with graphite as a pigment seems to have all the durability of lamp-black with the advantage of drying in a reasonable time without the addition of excessive 'dryer.' As a pigment graphite is absolutely inert, producing no injurious effect on the oil, and it is not subject to any change as the result of exterior conditions. One peculiarity of graphite paint is the extreme ease with which it may be worked out under the brush. This is due to the same quality that makes it a good lubricant. The workman finds himself spreading it out properly in spite of himself, it being actually easier to spread it than to apply heavy coats. For the same reason, it is possible to use a larger volume of graphite than of heavier pigments, without materially reducing the spreading capacity of the paint. Amorphous graphite is not as well suited for paint as the crystalline form, because of its earthy impurities. Of the crystalline varieties, the laminated is superior, for the reason that the surface of the oil exposed to destructive influences is least. The action of the brush in spreading the paint causes the flakes to lie in their natural position, that is, flat and overlapping each other, so that only the edge of the oil-film is exposed.

Crystalline graphite is used for polishing the bottoms of racing-craft, and more than once has been brought into requisition in preparing large steamers for their speed-trials. Graphite is also used to furnish body to hold the dye, in the manufacture of felt hats, the amorphous kind being used for this purpose.

A close examination of an incandescent electric lamp will reveal two small black lumps of graphite at the point where the filament joins the platinum wires. A certain kind of amorphous graphite, ground into a fine powder and made into a paste, is applied to this connection and baked hard.

Rods, plates, and pieces of different forms, and having a great range of electrical resistance, are made from mixtures of graphite with some non-conducting binder.

Thus two pieces of the same size may have conductivities in the ratio of 1:100,000. This material is used as a substitute for wire where high resistance is desired in a limited space.

Finally graphite is used as a substitute for red-lead on threaded joints, gaskets, and similar connections. Its advantages for this purpose are that the mixture never sets hard as does red-lead, and the joints may be readily opened, even years after they have been put together; also, due to the lubricating quality of the graphite, the pipes may be screwed up at least a half-turn more on an average.

Mention should be made of the method for the artificial production of graphite devised by A. G. Acheson, of Niagara Falls. The process consists in the production of carborundum in the electric furnace, and then, by prolonged heating, decomposing it and setting carbon free as graphite. Thus made it has the gray color and lustre of the natural crystalline form, but it lacks the same smoothness and softness when pulverized, probably due to the fact that the transformation has not been complete. This process is used successfully for the graphitization of electrodes and brushes, first formed in the ordinary way from coke, with such other materials added as are necessary for the production of carborundum. The durability and efficiency of electrodes are greatly increased by this treatment.

A Useful Chart

THE accompanying chart may be used for computing the velocity-discharge in a V-ditch, based on Kutter's formula. where n is taken as 0.0225, which is a safe average for well-made earth-ditches in fair maintenance. The chart was constructed by George Henry Ellis. From the chart, for example, a V-ditch flowing 1 ft. deep and having a slope of 0.001, or 5 ft. per mile, the velocity is 1 ft. per second and the discharge 1.5 cu. ft. per second.

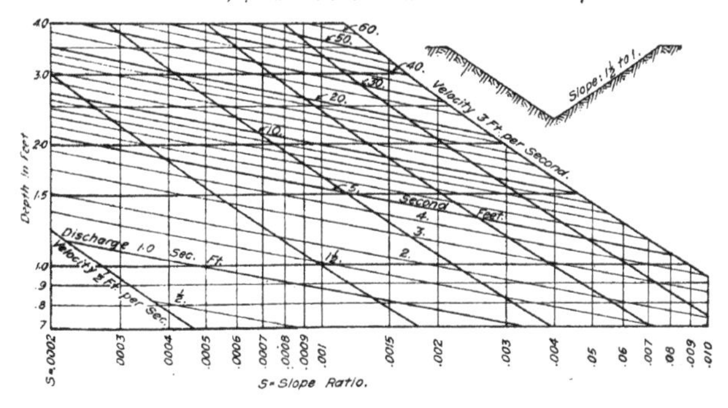

VELOCITY RATE AND QUANTITY DISCHARGE FOR V-DITCHES

First-Aid Instructions

*Be calm. Take command and give orders.

Find location of the injury.

Know what you want to do and do it.

Keep onlookers away from the patient.

Look for red spurting blood and check it by tourniquet or by pressure of finger over blood vessel.

Look for shock; if present, lower head of patient, apply blankets and wrapped hot-water bottles; and give aromatic spirits of ammonia in water, if patient is conscious.

Look for fractures; never remove a patient, unless absolutely necessary, until splints have been applied.

Place bandage compress over compound fracture before applying splints.

Cover all wounds with bandage compress and bandage.

The fingers or instruments should not touch a wound.

Remove a foreign object from a wound, if you do not have to put your fingers into the wound or touch the edges of the wound.

Exclude air as quickly as possible from burned surfaces by using picric acid gauze or other material.

Leave reductions of dislocations or fractures for the surgeon, except dislocation of jaw or finger.

Only part of your work is completed when the patient is ready for the stretcher.

Unnecessary or rough handling of a patient may undo all your work.

Slowly place patient on stretcher, avoiding jerky movements, and carry him to safety.

Don't touch a wound with your fingers or any instrument.

Don't put an unclean dressing or cloth over a wound.

Don't allow bleeding to go unchecked.

Don't move a patient unnecessarily.

*From Advanced First-Aid Instructions for Miners, U. S. Bureau of Mines.

Heavy Blow to Mining Companies

This is the heading of an article in the 'Boston News Bureau', in the course of which that paper gives a statement made by Henry D. Williams, counsel for the Minerals Separation company. We reproduce it because it is interesting, but with the warning that it represents the hopes of the patent-exploiting corporation rather than a conclusive statement of the case. Mr. Williams says:

The effort of the Miami Copper Co. to reach the Supreme Court having been abandoned, two proceedings follow as mere matters of form and inevitably. One is a permanent injunction against the Miami Copper Co. restraining further infringement of the three patents in suit, the first patent for flotation with the use of a small quantity of oil, the second patent for flotation with the use of a solution frothing agent dissolved in the ore pulp, and the third patent for flotation with the use of phenol or cresol in the cold and without acid.

In my opinion this will absolutely prevent the use of the flotation process in any form, with Callow cells now installed, or with any other contrivances. The comfort which Mr. Callow has derived from the opinion of the Philadelphia Court has been solely based upon what that court said as to the oil patent. That court said as to the solution patent that the claims "are not confined to a particular device or a particular degree of agitation" and that the means for bringing about the agitation are "described in terms that are wide and inclusive," and then, quoting the specification itself:

"The air or other gas is to be 'liberated in, generated in, or effectively introduced into the mixture,' in order that the ore particles may come into contact with the gas and as a result may float to the surface in the form of a froth or scum which can be separated afterwards by any well known means. The object of introducing the air or other gas into the mixture is such agitation of the pulp as will produce the desired froth, but the claims are not confined to a particular device or a particular degree of agitation."

I may further add that the difficulties which the Philadelphia Court experienced as to the limitation of the first or oil patent to some particular degree of agitation were all swept aside in the Butte case and shown to be directly contradicted by the very terms of the patent itself.

As to the accounting, the plaintiff is entitled to all of the profits attributable to the infringing acts. The Miami Copper Co. added flotation at the tail end of its plant to recover metal from material formerly thrown away. All of that recovery is therefore due to flotation, and Minerals Separation, Ltd., is entitled to all of the profits of that recovery.

The Miami company has filed in court sworn reports covering the period from October 5, 1916, to the first of August and showing that the total value of the concentrates recovered by flotation during that period was $2,765,672. They commenced to use flotation in December, 1913, and thereafter put in their pneumatic or Callow plant, and commenced to operate this in August 1914. We roughly estimate that their total recoveries in flotation amount to $5,000,000, that the costs of flotation were very small indeed, and the difference between the actual cost and the value constitutes the profits due to infringement to which Minerals Separation are entitled.

The accounting will proceed before William H. Mahaffy, the clerk of the court, who has been appointed master by consent of parties to conduct these proceedings. He will summon the defendant before him and proceed to a complete examination of their accounts for the purpose of determining the profits due to infringement. He will have full access to all of their accounts, and can call before him any of their officers and employees to obtain the necessary information. The accounting will be comparatively simple.

The master's report will be filed, and after approval by Judge Bradford an appeal can be taken to the Court of Appeals, which will, however, only consider questions involved in the accounting. This decision will be final and unappealable, although the losing party will have the right to ask the Supreme Court to call the case up for review.

In the meantime, however, the Miami company will be under injunction, and the most that it can do is to pile up in dumps, subject to deterioration, the valuable tailings from its water concentration plant now treated by flotation and go back to the conditions which prevailed in the plant before flotation was adopted. The patents in suit expire respectively November 6, 1923, June 28, 1927, and June 9, 1931. Until these patents expire the Miami company will be absolutely bound by the injunction and can use flotation only with the consent of the owners of the patents.

In the Butte & Superior suit the sworn reports filed in court by the defendant covering the period from November 1, 1913, to date, show profits of about $20,000,000 due to flotation. In this instance, the costs of operating the flotation plant appear in the statements and show that they are considerably less than 6% of the value of the concentrates.

The Utah, the Chino, and the Ray proved their flotation operations with complete tabulated statements in their efforts to help the Butte & Superior Co., so they have supplied evidence of their infringement and its extent. The monetary values have not yet appeared, but will undoubtedly be very large. The fact that the Nevada Consolidated has also infringed was also proved.

Under the patent laws the patentee is entitled to all of the profits due to the infringement. If, for any reason, the profits cannot be determined, then he is entitled to damages, and the best measure of damages is the royalty usually charged to licensees. Obviously a court will not compel a patentee to take from an infringer merely the compensation willingly given to him by licensees except as a last resort in the event that the profits due to infringement cannot be determined.

Mining in Utah

By L. O. HOWARD

Last month I referred to the measures being taken to relieve the threatened coal shortage. Following the appeal to the Federal authorities and the meeting of railway officials with local boards, the principal coal carrier agreed to furnish more cars to Utah mines, asking only for time in which to move the accumulation of through-freight already consigned to it, after which coal was to receive preferential treatment. The recent damage to track by wash-outs has been repaired. The railroad has started to gather all sorts and conditions of cars for the coal traffic. The effort is rather belated. Meanwhile coal mines have been forced to operate far below normal, organizations have become scattered, labor has been employed elsewhere, and the return to full capacity is bound to be slow.

The railroad now charges $1.60 per ton on lump coal from Carbon county to Salt Lake City, and $1.35 on slack, and has petitioned the State Utilities Commission for permission to increase the rate to $1.75 and $1.50 respectively. Local commercial organizations are opposing the increase as does the city of Salt Lake, asking in turn for a reduced rate, $1.30 and $1.20. The city further charges that present rates are excessive, and that discrimination in the distribution of cars has been one of the principal causes of the threatened shortage. Concerning this discrimination it is alleged that coal mines served by the Denver & Rio Grande railroad in Colorado have been given preference in the distribution of cars, and that the rate from the mines to points on the Los Angeles & Salt Lake route beyond Salt Lake is only $1.25 per ton.

The railroad in its petition admits the discrimination in rates, but bases its rights on an old tariff that has never been repealed.

The coal resources of Carbon county have been expanded several-fold in the last few years. New mines are opened at frequent intervals. Railroad facilities have not maintained equal progress.

In August the coal companies demanded transportation for 14,500 tons of coal per day. The average now being hauled is about 7000 tons. The discrimination in distribution of cars has also been made the basis of a suit before the Interstate Commerce Commission on the part of the independent companies in Utah. A hearing has been set for October 5. The purpose of the suit is to compel railroads serving Pacific Coast points, especially California, and the Middle West, to reduce rates so that Utah coal may enter these markets. California has a rate of $5.75 on Utah coal, and a reduction to $4 is asked so that Utah coal may take the same rate as that from Wyoming. Damages are demanded on the ground that some Utah companies have received more than their share of cars at the expense of the independent companies.

The completion of the Utah Railroad from Carbon county to Provo is looked upon as the solution of the transportation difficulty. This road is under construction by a subsidiary of the United States Smelting, Refining & Mining Co. The track and yards are nearly completed. War conditions have hampered the delivery of cars and locomotives. Those ordered last November have not been delivered. Upon investigation it is found that the Federal government has ordered the builders of locomotives to fill an order of 300 for France, then 1000 each for France and Russia, before making any domestic deliveries. The question was then taken up with the War Department and an order secured that, following the completion of the 300 engines for France, 9 be sent to the Utah Railroad. Of the 2000 cars ordered, not over 1500 are expected before November 1. The Utah Railroad will not be in a position to help the situation effectively until winter. The D. & R. G. is also prevented from getting much-needed equipment.

The fixing of prices on bituminous coal by the President has created many new problems. The outstanding feature is the favoring of consumers of domestic grades at the expense of factories, smelters, and other industrial plants. The average price received at the mines will be about the same. The increase in the price of slack at the mine from $1.50 to $2.35 offsets to a certain extent the decrease ordered on the larger sizes. Run-of-mine has sold for $2.35. The Government's price is $2.60. For points outside the State the prices have been 50 cents to a dollar higher. All are now on the same level. The net result is to benefit outside consumers as against those inside the State, especially industrial consumers, and to reduce the price to the householder a few cents per ton.

The Workmen's Compensation Act has been in force two months, but some questions as to its interpretation are still unsettled. One of the first questions to arise has been the right of operators to ensure in companies charging a lower rate of insurance than that asked by the State Industrial Commission for State insurance. The Scranton Leasing Co. of Tintic took out a policy on July 1 ensuring 30 workmen in a company that charged $5 per $100 of monthly payroll. The Commission was notified and on July 26 the policy was filed. It was rejected by the Commission, which stated that the policy could not be received "since the rate is $5.59 per $100 of employee's payroll and the policy provides a rate of $5 per $100 of employee's monthly payroll." The Commission asserts that under the law it has the right to take this stand; further, the policy was a participating one that the Commission cannot accept. The leasing company claims that the Commission's power is limited to fixing the rate for insurance by the State, and has applied to the Supreme Court of Utah for a writ of mandamus compelling the Commission to re-instate the policy. This move may be regarded as but another attack on the excessive rate for metal mines established by the Commission.

Another question concerns the right of mining companies to deduct a certain sum per month from each employee's wages for the maintenance of a co-operative hospital. One provision of the Act prevents any deduction from employee's wages for any purpose. The issue is clean-cut between the legal prohibition of any co-operative arrangement, however much desired by both parties, and the unconstitutionality of a provision that prevents the execution of a previously arranged contract between employer and employee. The miners are said to favor the old arrangement, since better attention and conveniences are available at the co-operative hospitals. The individual commissioners are also said to favor the old system, believing that the employee receives services that he could not afford to pay for otherwise. A friendly suit may be instituted to test this particular provision of the Act.

The lead producers met late in August to discuss the request of the Government for one-fifth of the production in August, September, and October, at 8 cents per pound. A committee was appointed to investigate the probable cost of this lead in view of the high wages for labor and increasing cost of supplies. The effect of the new taxation will also be considered. Some of the producers think that if a possible loss is to be taken on this business the stockholders of the companies concerned should be first consulted.

The smelters have been so congested with lead ore that during the second week in August the American Smelting & Refining Co. requested shippers to suspend shipments for a week and then to go slowly until the congestion could be relieved.

Once more has the Utah Copper set a new record for loading at Bingham. On one day late in August there was loaded 47,600 tons of ore, exceeding the previous mark by 1600 tons. During August the average was 37,500 tons per day. Reports for the first quarter were recently issued by the Ohio Copper Co. They show net earnings of $180,122 for the period. A 3000-ton flotation plant is under construction, 1500 tons of which may be ready in January. A saving of 80% is anticipated, against the present recovery of 47%.

In July the Park City mines shipped a record tonnage of 8449 tons. August tonnage has exceeded this, although final figures are not available. The California-Comstock mine, since resumption of operations on July 20, has shipped regularly and new discoveries of ore near old workings are of frequent occurrence.

The Big Four Exploration Co., one of the enterprises founded on the large tailing-dumps below Park City, has not yet been able to show a profit. The mill was described in the MINING AND SCIENTIFIC PRESS of September 25, 1915. The plant was not a success and has been repeatedly remodeled. At a recent meeting of stockholders, directors, officers, and creditors, the affairs of the company were found to be in a most unsatisfactory state. No trial-balance has been cast since last December, but liabilities are estimated at over a quarter of a million dollars. Although, after the mill resumed operations in April, the manager signed a statement that profit was being earned at the rate of $10,000 per month, it has been shown that the net result bf operations from April until the mill closed early in August was a loss of $10,000. At no time has a profit been earned, yet a dividend was paid. The stock was boosted to $2 and is now quoted at 5 cents, with a 10-cent assessment in the offing. Assets of the company consist of a mill, reported to have cost $300,000, and $40,000 to $50,000 worth of concentrate and ore, also a few hundred thousand tons of tailing, value unknown. Naturally there is much dissatisfaction at the turn in the affairs of the company, and so far responsibility for the unfortunate state of affairs has not been fixed. A committee was appointed to investigate all phases of the fiasco, as well as to test the tailing for a suitable process, to report at a later date. This is the end of an ambitious project to treat the hundreds of thousand tons of tailing scattered over a space four miles long, 500 ft. wide, and 2 ft. deep, involving the use of drag-line excavators, steam-shovels, and industrial railways.

Shipments from Tintic in July were 38,600 tons, all smelting ore. This district has shipped nearly 300,000 tons in seven months, 32 shippers being represented. This is a remarkable showing for a district that has been producing ore for nearly 50 years, considering further that all of this ore has gone to the smelters. In August nearly 40,000 tons was shipped.

In Little Cottonwood the orebody of the Columbus-Rexall continues to improve. Ore was first found along the bedding of the limestone, and was two feet thick. In 50 ft. it thickened to 10 ft. Since the fissure is about 50 ft. ahead it is expected that it will continue to increase in thickness up to that point. Fifteen teams are hauling ore to the smelter at Murray. Eight cars had been shipped up to the end of August. The Michigan-Utah settled for 59 cars of ore in August, although there was difficulty in getting cars. The Alta-Germania reports the beginning of shipments in a small way. The Peruvian Consolidated reports the opening of four feet of $75 ore. All the companies are handicapped by the shortage of teams and the inadequacy of railroad equipment. Some investigation of the molybdenite deposits in Little Cottonwood is being made. These deposits have been known since early days, but lack of a steady market and the metallurgical problem have made their exploitation unattractive.

In Big Cottonwood the Cardiff is sinking a shaft from its 600-ft. level; this has been in ore for 500 ft. The company is shipping 125 tons per day, but is having many difficulties in transportation.

In Dugway district the Metallic Hill, Golconda, Bertha, Laurel, and Cannon groups are being developed, and the last two have started shipping. In Deep Creek the Western Utah is shipping 275 tons per day from its Gold Hill properties. The company shipped 4760 tons in August. The output of the Pole Star, Woodman, Spotted Fawn, and Silver King brought the total for the district to over 5000 tons. At Ferber the Big Chief group has been developed to the shipping stage. The shaft is now down 45 ft. in $75 silver-lead ore.

REVIEW OF MINING

As seen at the world's great mining centres by our own correspondents.

LEADVILLE, COLORADO

INFLUENCE OF HIGH SILVER PRICES.—RESUMPTION OF WORK IN MINES.

With the high silver-market prevailing and indications pointing to a continued demand for the white metal, mining in the Leadville district is resuming the activity that marked operations prior to the strike. With the exception of two companies, the Yak M. M. & T. Co. and the Empire Zinc Co., all of the large mines have resumed and have produced a combined output of 51,000 tons in August. The Yak company is operating on a small scale as compared to that before the labor trouble, but it is stated by company officials that plans are now being perfected for new development. The extension of the main adit north into the Vega ground, previously developed through small winzes, is being considered. Such an undertaking would open an area that is known to contain ore.

The Empire Zinc Co., operating the Robert Emmet and McCormick properties in Stray Horse gulch prior to the strike, has abandoned operations for the present. These mines were recently purchased from the Small Hopes-Boreel Mining Co. for $750,000 and are considered to be the richest lead-zinc and silver producers in the district. The Mikado shaft of the Iron Silver Mining Co. is now producing lead-zinc sulphide and zinc carbonate at the rate of 150 tons per day. This output will be increased as the orebodies in the lower portions of the R. A. M. property are being developed. Electric-driven centrifugal pumps will be installed during the month. A new ore-house and boiler-room is now under construction at the Mikado and will be completed in October.

The Greenback mine in Graham Park is again operating at full capacity. At the Wolftone shaft, the Western M. Co. is shipping 200 tons per day of zinc carbonate and lead-zinc sulphide. In the lower workings several large bodies of sulphides were discovered which have been under water for many years. These sulphides are rich in zinc and lead and some carry silver. They are being blocked out by drifts and the output will be increased during the next month. It is reported that a few new orebodies of carbonate ore have been opened by recent prospecting.

Cramer & Co., leasing a large tract on Carbonate hill owned by the Star Con. M. Co., are undertaking a new enterprise of great interest to local mining men. A new 200-ton mill, for the separation of lead from the low-grade lead-iron ores occurring in large bodies in the Star properties, is being erected and the first unit will be in operation by October 1.

Mining at the Star properties has been resumed. The Cramer lease is shipping 200 tons per day of high-grade manganese ore. This tonnage is being hoisted through the Star, No. 5, and Ladder shafts. At the Waterloo No. 1 and No. 2, zinc carbonate is being produced by the Cramer interests and at the Yankee Doodle and Catalpa a large tonnage of iron is being shipped. M. L. Buchanan & Co., leasing from the Star Consolidated, are extracting 100 tons of manganese ore per day from the Porter shaft. Recently large bodies of silver-bearing iron ore have been developed and the lessees are preparing to double their output by adding 100 tons of the silver ore to the present production. The operators of the Seneca have discovered several large deposits of manganese ore on the upper levels. High-grade lead-silver ore has also been found in this mine. Shipping will be started during the present month at the rate of 50 tons per day. The output will be increased.

The Louisville mine on Iron hill, owned by E. A. Hanifen and W. O. Reynolds of Denver, and for many years considered the richest silver producer in the Leadville district, is now operating on a larger scale than ever before and is producing 200 tons per day. Stopes of silver-bearing ore, carrying as high as 6000 oz. per ton, have been opened in past years. These stopes are now being re-worked and quantities of low-grade ore are being recovered. New orebodies have also been discovered recently. Preparations are being made to increase the output to 300 tons per day. A shortage of railroad cars has interfered with the attainment of larger shipments.

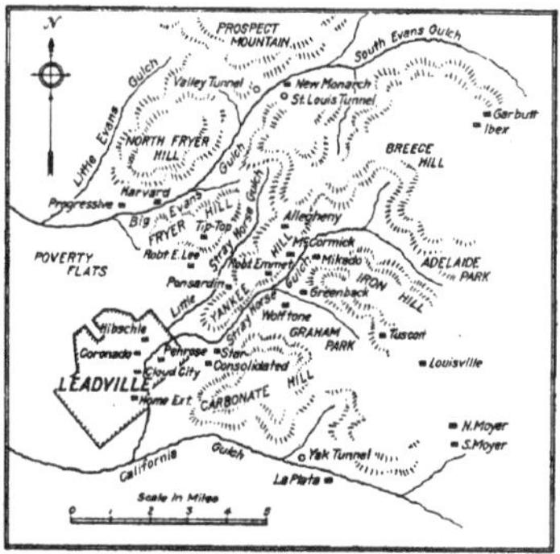

LEADVILLE AND VICINITY

M. A. Nicholson and associates, operating the Tarsus mine on Yankee hill, have discovered several orebodies of high-grade lead-silver ore and iron oxide assaying high in silver. The discovery was the result of a well planned prospecting campaign. The orebody was opened on the 335-ft. level. A 4-ft. vein of lead-silver ore is being mined on this level. Shipping will start with an output of 50 tons per day. The Ponsardin mine, in which a rich orebody of lead-zinc sulphide-ore was discovered two years ago, is again in ore. A blanket-vein of silver ore, 30 ft. thick and assaying from 45 to 75 oz. per ton, has been recently discovered. The vein dips slightly to the north-east into a large undeveloped area. The mine is under lease to W. E. Bowden & Co. An output of 100 tons of lead and zinc ore per day has been maintained during the past two years.

Lessees at the First National mine in Iowa gulch have recently made a rich strike on the bottom level of the shaft at a depth of 330 ft. A 6-ft. vein of ore, assaying 60% lead and 40 oz. silver, has been cut, and shipping has commenced. The ore-shoot is reported to be well defined.

TONOPAH, NEVADA

WAGE AGREEMENTS.—DEVELOPMENT.—PRODUCTION.

Labor day was observed here, and all the mines and mills were closed down, but no public demonstration was held. In November 1916 the miners and mill-workers asked for an increase in wages, justifying the request by the price of silver and the increase in the cost of living. The mine-operators' agreement to a 50-cent increase per day while silver remained at 70c. or above was accepted. The increase commenced December 1916. The continued rise in the price of silver and the cost of living has prompted the miners and mill-workers again to suggest another increase in pay. They ask for a minimum wage of $4.50 per day for muckers and $5 for miners, the present wage, and an increase of 25c. per day for each 5c. increase in the price of silver above 70c. Under their proposal the wage at the present quotation of silver would be $5.50 for muckers and $6 for miners. The mine-operators will give an answer to the proposal on September 5, and a satisfactory arrangement will be reached. For the week ended September 1, the development and production follow.

The Tonopah Extension Co. milled 9476 tons of ore during July, making a net profit of $36,225. During the past week 49 ft. of development was made in the No. 2 shaft and 120 ft. in the Victor. The south cross-cut on the 1680-ft. level has been started. It is expected to intersect the Merger, O. K., and Murray veins. The weekly production was 2380 tons.

The Tonopah Belmont during the past week produced 2431 tons of ore. A raise from the south cross-cut on the 900-ft. level intersected the South vein and exposed a width of 21½ ft. showing fair assays.

The Tonopah Mining Co. shipped 25 bars of bullion valued at $47,500 from the plant at Millers. Increased development has resulted in 127 ft. at the Silver Top, 65 ft. at the Sandgrass, and 19 ft. at the Mizpah. A station is being cut at the 1500-ft. level of the Sandgrass and a north cross-cut will be started. The production during the past week was 2350 tons.

In the Jim Butler mine, cross-cut 754 is being driven to intersect the 648 vein. Raise 2 on the 850-ft. level intersected a vein of low-grade quartz. The weekly production was 800 tons.

The West End Con. M. Co. continues to develop the Ohio by raises and cross-cuts. Raises 513, 517, and 523 are all in ore, while 555, intermediate, is being driven to intersect the vein. The usual weekly progress was made in 717 D.3 to the east, 717 D.2 to the west, and 680 D.2. The production during the past week was 1109 tons.

The Monarch Pittsburg has commenced stoping on the 900-ft. level, as the raise, that was recently connected, gives ventilation. The Rescue-Eula produced 193 tons during the past week, the Montana 48 tons, the Cash Boy 53 tons, and miscellaneous 10 tons, making the week's production at Tonopah 9374 tons with a gross value of $164,045.

TORONTO, ONTARIO

RELEASE OF MUNITIONS WORKERS.—MINING DEVELOPMENT.

The release of 75,000 munitions workers, caused by the curtailment of the work at a number of munitions plants, will materially relieve the labor situation in mining districts.

The Hollinger Consolidated has discontinued its weekly statements and little information of its operations is obtainable. The central shaft is in operation. At the Schumacher M. & M. Co. operations have been resumed. The addition to the mill, bringing the capacity up to 300 tons per day, will not be completed before October. The McIntyre, at present the only dividend-payer in the Porcupine district, has declared another 5% dividend. At the Newray a strike has been made in the cross-cut on the 400-ft. level. The vein is 5 ft. wide and assays $20 per ton. The Dome Lake is stoping ore from No. 3 vein on the 400-ft. level. Diamond-drilling indicates ore in this vein at a depth of 600 ft. Operations at the McRae mine have been suspended after sinking a 100-ft. shaft. The Whelpdale, comprising 160 acres near the Hollinger, is being developed. A number of strong veins have been found and a shaft has been sunk 20 ft. Assays show $7 gold per ton. The Beaver Consolidated has paid the third installment of the purchase money for the Kirkland Lake, leaving $75,000 to be paid. Work on the Black claim, at Kirkland Lake, has been temporarily suspended owing to the closing of the plant from which compressed air was supplied. At the Wright-Hargraves the shaft is almost 300 ft. in depth. The vein, cut at a depth of 200 ft., was persistent and carried a high gold content. The United Kirkland, a newly organized company, has obtained a group of claims known as the Dodge-Ellis in the Kirkland Lake district. Surface work on the Canadian Kirkland has discovered 17 veins, some of which give high assays. The Trethewey has ordered a Callow flotation plant of a capacity of 100 tons per day. It has 60,000 tons of tailing on hand. The McKinley-Darragh has cut a new high-grade vein, averaging four inches in width, on the 350-ft. level and has intersected the extension of the Cobalt Lake fault-vein on the 400-ft. level. The Provincial has been re-opened under the management of John Redington. A shaft is being sunk at the south-west end of the area. The La Rose Consolidated is sinking an open-pit between two old veins, from which the high-grade ore has been extracted, and is mining low-grade ore, of which a large tonnage is available. Owing to the high price of silver it can be profitably mined. At the Adanac a series of rich veins has been developed, one of which, from three to ten inches wide, has been proved for 125 ft. and carries high-grade ore. The Seneca-Superior, which ceased operations in May 1916, has made its final annual report, showing a realizable surplus of $7709. During its activity it produced $2,191,280 and paid dividends amounting to $1,579,817, equal to 326% on the capital issued.

CRIPPLE CREEK, COLORADO

AUGUST OUTPUT.—DIVIDENDS.

The production of the Cripple Creek district for the month of August totalled 102,269 tons, with an average value of $11 per ton and a gross bullion value of $1,137,323.96, an increase of 15,122 tons and $16,328.80 as compared with July. The local plants treated 53,880 tons of low-grade mine and dump-ore with a gold content of $2.11 per ton, from which was recovered $112,783. The ore disposition and values are:

	Tons treated	Av. value per ton	Gross bullion value
Golden Cycle, Colo. Springs...	33,000	$20.00	$660,000.00
Portland, Colorado Springs...	10,688	20.32	217,180.16
Smelters, Denver & Pueblo....	2,600	55.00	143,000.00
Portland, Victor mill, Cripple Creek district	19,500	2.24	43,680.00
Portland, Independence mill, Cripple Creek district.....	34,380	2.01	69,103.80
Rex Mill, Ironclad lease, Cripple Creek district........	1,700	2.00	3,400.00
Rubie mill, Kavanaugh & Co., Cripple Creek district.....	400	2.40	960.00
	102,269	$11.12	$1,137,323.96

Dividends paid during August were limited to the monthly distribution of the Golden Cycle M. & R. Co., 3c. per share, total $45,000, and the Cresson Con. G. M. & M. Co., 10c. per share, total $122,000. Both will pay their regular monthly dividends at the same rate on September 10.

The August output from mines of the United Gold Mines Co. was the heaviest on record, 3200 tons, a gross value of

$65,800. The mines are operated under a liberal leasing system. The Trail mine on Bull hill produced 81 cars, 2800 tons of an average value of $20 per ton. The W. P. H. mine, Ironclad hill, was credited with 9 cars; the Bonanza, Battle mountain, 3 cars; and one car was shipped from the Deadwood mine on Bull hill. The Granite Gold Mining Co. mined and shipped from its Dillon mine on Battle mountain, during the month of August, 2360 tons of ore of an average value of one ounce gold per ton. This company will pay its second bimonthly dividend of one cent per share, $16,500, on September 15. Lessees of the El Paso Consolidated Gold Mining Co., 10

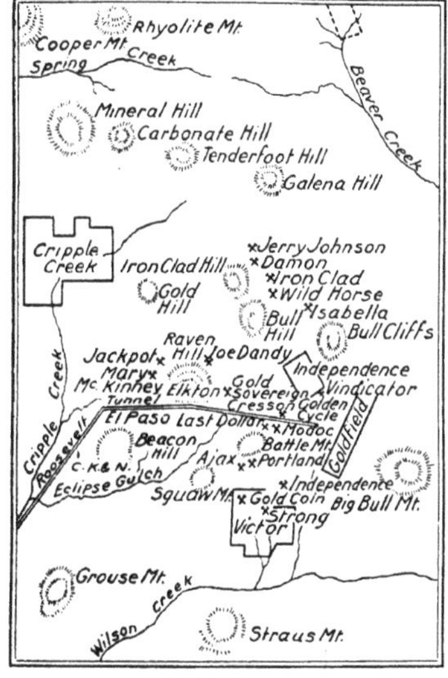

THE CRIPPLE CREEK DISTRICT

in number, shipped from its Beacon Hill mine 530 tons of $20 ore to the Golden Cycle mill in August. The Jerry Johnson mine, of the Jerry Johnson Mines Co., operated by two sets of lessees, shipped 350 tons of ore in August. The Caley lease ore averaged $30 per ton. The Cripple Creek Deep Leasing Co.'s ore, mined below the Caley block, averaged $20 per ton. Shipments by the Komat Leasing Co., from the Victor mine on Bull cliffs, are leaving bi-weekly and 10 cars to date ranged from $8 to $20 per ton. The mine is managed by Math Korf of Cripple Creek and is owned by the Smith-Moffat Mines Co. of Denver. The Caledonia, sold by the late W. S. Stratton to a London syndicate and now owned by M. Doran of Denver, is again shipping. H. M. Hardinge of Cripple Creek, lessee, is mining and shipping a good grade of mill-ore. Machinery, hoist, and a boiler have been hauled to the Roanoke mine on Mineral hill. A leasing company headed by William McPhee will operate the mine. The Millasier Mines Corp., owning and operating the Clyde on Battle mountain, has sunk the main shaft to the 1000-ft. point and at 125 ft. additional depth will cut a station and commence development. The company, a close corporation, is controlled by Michigan parties, represented by Frederick G. Lasier, of Detroit.

The following directors were re-elected at the annual stockholders meeting of the Cresson Con. G. M. & M. Co. held at Colorado Springs on September 4: A. E. Carlton, Leslie G. Carlton, Cripple Creek; Charles M. Macneill, Spencer Penrose, E. P. Shove, Colorado Springs; Irving T. Snyder, T. B. Burbridge, Adolph F. Zang, and C. K. Boettcher, Denver. A. E. Carlton will remain at the head of the company as president.

HOUGHTON, MICHIGAN

GOVERNMENT COPPER PRICE.—LABOR SCARCITY.—AUGUST OUTPUT.

The cost of copper production at the mines of the Lake Superior region will be higher in 1917 than it was in 1916 by at least 10% on the average and higher by 20% in specific instances. The prospective Government price is disturbing producers. A price of 25c. per pound will reasonably permit continuance of operations. A 20c. price will close down a number of smaller mines. Costs last year were higher than they have been, but this year everything that enters into the cost of production has advanced. Labor is not efficient.

The Winona mine has completed arrangements for its coal supply, securing Illinois coal at a fair price. Winona has difficulty in keeping sufficient men on the pay-rolls. Operations will be suspended if the price of copper is not fixed above 20 cents.

Effort will be made to have the Seneca shaft and the collar under cover so that sinking may continue uninterruptedly during the winter of 1917-18. The shaft collar will be of concrete and steel will replace timber in many places in the shaft. The surface-plant buildings all have modern equipment.

Cost at the Hancock Consolidated is 20c. per pound. A substantial reduction can be expected only when the working-force is doubled and the rock output increased.

A decrease in Wolverine's next dividend is anticipated. One of the problems that has come up in the past two years is the presence of a large amount of epidote, the specific gravity of which causes it to lower the grade of the concentrate. No. 4 shaft now has reached a depth of 4500 ft. Its north openings are in a good grade of rock.

Four levels, the 23, 24, 25, and 27, are in rock that is up to the average of the Victoria, the 11th level being in poor-grade rock. The company has 300 men employed and could use 500. The production is under 100 tons of refined copper per month.

Southerly levels of the old Osceola Con. mine continue in a grade of rock better than the average, but these levels are the only openings in this part where the showing is particularly good. The tonnage from the two shafts operating here is under 14 pounds per ton. An 18c. copper prise, at present costs, would necessitate closing down the mine.

Butte miners are returning to the Lake Superior region.

Four electric trams are in operation at the Allouez and production is maintained notwithstanding the shortage of tram-labor.

Tonnage at the Ahmeek is 3700 per day as compared with 4000 tons last week.

Estimated rock outputs for August are: Calumet & Hecla, 244,400; Centennial, 11,600; South Lake, 5185; Isle Royale, 70,000; Osceola, 100,000; La Salle, 15,615; Hancock, 32,800; Mass, 16,000; Superior, 10,000; Ahmeek, 103,400; Wolverine, $28,980; Mohawk, 48,090; Franklin, 27,045; Allouez, 48,465 tons.

The most noticeable increase in August is at the Stanton mines and at Allouez. Mohawk sent more rock to the mills in August than in July and it was better grade. Rock from the southerly openings is showing an improvement in quality and now averages better than 20 pounds without much selection, while the run of the mine is approximately 19 pounds. Wolverine increased rock output by 3000 tons in August and the grade of rock was half a pound better.

Calumet & Hecla is constantly increasing its output of copper from the sand of Lake Superior. One-eighth of its total output comes from the old Torch Lake sand and is costing less than six cents per pound.

Silver is coming to be more and more important to Quincy as a source of income, not so much because of the rise in the price of silver, but because of the increasing amounts that are found in the lower openings. In the early days of the Quincy 'picking' silver was one of the important branches of work in the stamp-mill. Not until recently did the silver commence to reappear again to an appreciable extent. Calumet & Hecla's lower conglomerate workings likewise are showing more silver right along.

No. 4 shaft, North Kearsarge (Osceola Consolidated), has resumed operations after a short shut-down. The shaft continues in a grade of rock which is low in copper, not much better than that which is secured from the old Osceola lode. There is reason for the belief that there will be an improvement when the shaft intersects the Wolverine and Allouez ore-shoot.

At the South Lake, a higher copper-content is showing in the second shaft. The daily output is 200 tons and the August output 85,000 lb. refined copper. Coal-dock improvements at the Calumet & Hecla are nearing completion. The steel work on the addition to the leaching plant at Lake Linden is nearing completion and this part of the plant will be in operation early in the spring.

MONTERREY, MEXICO

TAXES.—RESUMPTION OF MINING.—TRAFFIC CONDITIONS.

An effort on the part of the government authorities to enforce the collection of the exorbitant taxes that have been placed upon mining properties promises to have a disastrous effect upon the industry, according to advices that have been received here from many of the larger producing districts. It is stated that in a great number of instances the owners of mines have temporarily abandoned their properties rather than comply with the Government's demand for taxes. It is claimed that these taxes are almost equivalent to the confiscation of the properties. In cases of Americans and other foreigners who own mines, they have made protest to their respective governments against the alleged oppressive taxes. In several of the districts the Government has already taken over some of the mines upon which taxes have not been paid, and permission has been given to Mexican miners to work the different properties for their own profit. It is said that this practice has become quite general in the State of Sonora.

The Alvarado Mining Co., which is an American-owned concern, is preparing to resume extensive development operations of its mines in the Parrall district, State of Chihuahua. Its holdings include the famous Palmilla mine which was discovered by the peon, Pedro Alvarado, who amassed a big fortune from the property before he sold it to the Alvarado Mining Co. The operations of the company have been conducted intermittently under some trying conditions during the long revolutionary period. Assurances have been given by General Francisco Murgia, commander of the military division of the State of Chihuahua, that ample protection will be given to the employees and officers of not only the Alvarado Mining Co. but of all other concerns that resume mining operations in that State. It is reported that the large smelter of the American Smelting & Refining Co. at Chihuahua will be placed in operation again in the next few weeks. Orders have been placed for coke and when these shipments arrive, the smelter will be blown-in. Some of the mines in the Santa Eulalia district near Chihuahua have recently resumed operations and it is anticipated that there will be no lack of ore to keep the smelter supplied. The Peñoles Copper Co. and other German-owned mining corporations in Mexico are unusually active at this time in producing oil and in enlarging their mineral holdings. According to advices received here the ore output of the mines in the Mapimi and adjacent districts where the German investments are very large, is greater than it has been for many years. Several promising mines in the Guanajuato district have been recently acquired by German interests and similar investments have been made by men of wealth of that nationality, who already have large business interests in Mexico.

The traffic condition of the railroads of Mexico has improved considerably during the last few weeks and the smelters which are in operation are not meeting with the difficulty which they formerly had in obtaining ore-shipments promptly. The equipment is being gradually increased by the re-building of old freight cars and the reconstruction of the motive power.

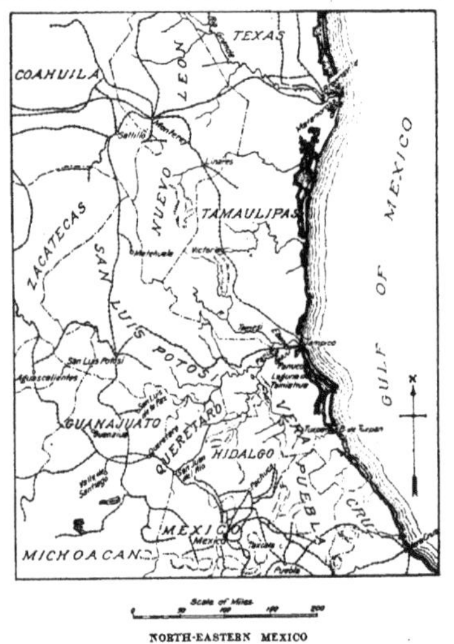

NORTH-EASTERN MEXICO

Conditions in the State of Durango, which for a long time were too dangerous on account of the operations of bands of brigands and revolutionists to permit the carrying on of mining development, are said to have improved very much within the last few weeks. The division of the National Railways of Mexico which runs between Torreon and Durango will soon be open for regular traffic, it is announced, after a suspension for more than three years due to the unsettled condition of the country through which it passes. With the restoration of traffic on this division, it is expected that the smelter of the American Smelting & Refining Co. at Velardeña will resume operations. In the San Dimas and other districts which are far removed from railroad transportation, being in the heart of the mountains of Durango, some of the mines have been operated without interruption during the revolutionary period. These properties have enormous quantities of ore and bullion ready for shipment.

THE MINING SUMMARY

The news of the week as told by our special correspondents and compiled from the local press.

ALASKA

(Special Correspondence.)—B. L. Thane, manager for the Alaska Gastineau Mining Co., states that the rumor of the contemplated shut-down of the mine is without foundation and that as long as the men can be obtained to handle the reserve-ore the property will be operated. The problem confronting the company is a shortage of labor. The campany needs 200 more men and can use 250 more than are now employed.

The Gypsum mine on Chichagoff island is again in trouble. Excessive quantities of water, which for a time were under control, are now flowing into this mine and the management is confronted with a difficult problem.

The Mohawk Mining Co., of the Willow Creek district, has let a contract for the driving of a second adit, 200 ft. long. The new adit will be started 1000 ft. below the upper adit.

H. Embhard, of Anchorage, has acquired the Pearl group of claims on Archangel creek, in the Willow Creek district, and will let a contract for 150 ft. of development. The vein of free-milling ore is three feet wide and assays $56 per ton. A vein of high-grade molybdenum ore from 8 to 36 in. wide has been found by the Hercules Mining Co. near the headwaters of Archangel creek in the Willow Creek district. The company is preparing to operate on a moderate scale.

Sulphur from the volcanic deposit near Bales Landing, Stepovak bay, 50 miles east of Sand Point, will be mined by the Standard Metals Co. of Los Angeles, and a crew of men will sack the sulphur preliminary to its haulage by two auto-trucks to a loading point.

Treadwell, August 26.

ARIZONA

Charles F. Willis, of the State Bureau of Mines, is arranging extensive courses in mining and first-aid. M. A. Allen, of the Bureau, will assist in the search for rare minerals.

Cochise County

(Special Correspondence.)—The Mascot mine, recently acquired by the American S. & R. Co., is shipping 100 tons of ore per day. Additional men are being employed and new machinery installed.

Willcox, August 25.

Greenlee County

(Special Correspondence.)—The Rival Mining Co., operating in the Twin Peaks district, is rapidly sinking its new shaft. The shaft is 60 ft. deep.

The Carlisle Mining Co., at Steeplerock, is operating a new mill. The mill is treating 100 tons of ore per day and concentrating 4 into 1. Concentrate shipments are being made regularly.

Clifton, August 25.

Maricopa County

(Special Correspondence.)—A deposit of manganese ore has recently been discovered by J. D. Peggs of Tucson, 18 miles south-west of Sentinel in the Eagle Tank mountains. The ore analyzes 44% manganese, 0.018% phosphorus, and 4% silica. The orebody is said to be 10 ft. wide and 1500 ft. long.

Phoenix, August 26.

Mohave County

(Special Correspondence.)—The Mary Bell strike, two feet of ore assaying $125 per ton in gold, silver, and lead in the Chloride district, continues to improve. The Rattlesnake mine has a carload of ore valued at $60 per ton ready for shipment to the Selby smelter. There are places in the vein where 18 in. will average $100 per ton in gold, silver, and lead. The shaft is down 160 ft. Both the north and south drifts are developing good ore. The New Mohawk has started shipping. The Payroll continues to ship 35 tons of ore per day to the Needles mill. The Schenectady continues to open up rich silver ore. It has been announced that a 600-ft. cross-cut will be started from the 1100-ft. level of the Golconda mine to explore the 800-ft. level of the Fredonia vein of the Union Metals Co. In the shaft, 60 ft. in depth, are two veins of 14 inches, which assay $50 per ton. The Union Metals Co. also owns the Golconda Annex, which has opened up some very rich surface ore.

Kingman, August 25.

The Golconda, in Union basin, is being run in a most efficient manner by John D. Wanvig, a graduate of the Michigan College of Mines. At present the deepest workings are at 1400 ft., where an ore-shoot 1000 ft. long in a 5-ft. vein is being worked. The pay-streak is from two feet to four inches. It contains brown zinc-blende, with silver and gold, the ore averaging 12% zinc, with $8 in precious metals. This mine is doing well. The mill is being remodeled.

The Chamber of Commerce of Chloride offers to send to interested parties reports on the standing of mining enterprises in the Chloride district whenever possible.

Pinal County

(Special Correspondence.)—The Magma Chief Copper Co. is re-opening an adit for the purpose of developing a deposit of manganese ore. The manganese ore has been contracted for by a New York company.

Florence, August 25.

Yavapai County

(Special Correspondence.)—There is much activity in the Hillside district. At the Golden Burro mine the mill is operating steadily. Eight teams are employed hauling copper ore from the King mine to Hillside station.——A. E. Anderson, owner of the Silver Butte group in the Thumb Butte district, has arranged the transfer of his property to a group of Los Angeles financiers and a company will be organized. The property has over 1000 ft. of development. The Silver Con. Belt M. Co. is employing 12 men in shaft-sinking. It is expected that ore shipment will commence in September.

Prescott, August 25.

CALIFORNIA

Colusa County

The 100-ton amalgamating plant of the Cerise G. M. Co., Charles Austin, superintendent, was destroyed by fire September 3. The plant was valued at $20,000 and contained a rock-breaker, 6-ft. Hardinge ball-mill, Senn amalgamating tables, and a 50-hp. gas-engine.

Plumas County

(Special Correspondence.)—The Great Western Power Co. has completed its electric transmission line from Las Plumas to Veramont, a station on the Indian Valley rairoad, and will soon start the delivery of power to the Green Mountain, Crescent, and other mines near Crescent Mills. The line will also supply the Engels copper group in the winter. An extension of the line to the mines near Genessee is expected.

Th farmers of the northern end of Indian valley have appointed a committee to confer with E. E. Paxton, manager for the Engels Copper Co. The farmers assert the tailing escaping from the mill-pond is ruining irrigation ditches and valuable land and demand remedial measures.

The Droge gold property, near Greenville, has been taken under option and lease by Eastern people and will be operated under the management of John W. Dailey. The Droge is equipped with an electric plant.

J. T. Buel, representing Marin county investors, has acquired the Quigley gravel mine, near La Porte. Prospecting has been in progress for several months and a wide body of blue gravel has been discovered.

Quincy, September 3.

SACRAMENTO COUNTY

The Pilliken mine, near Folsom, is mining and shipping chrome ore. Two trucks are used in transporting the ore to Folsom.

SAN BERNARDINO COUNTY

The Mohave Annex Tungsten M. Co. has completed and is operating a 20-ton mill on wolframite ore. Ore, 10 to 15% tungstic acid, has been developed.

SHASTA COUNTY

(Special Correspondence.)—The Clipper group of mines in Big Backbone district, west of Kennett, was bought at a sheriff's sale by C. C. Hartman.

The Forest Queen mine at Gibson, near Lamoine, is shipping a carload of chrome every three days. The Highland Lake mine at the same place is also shipping chrome in small amounts.

The El Oro company of Oroville is prospecting the Menzel ranch across the river from Redding.

The Afterthought Copper Co.'s flotation plant at Ingot was started September 1. The company has spent $300,000 upon its plant. Twenty miners are employed at the mine and shipments of ore are being made to the flotation-plant over the half-mile railroad. George L. Porter is superintendent.

Redding, September 4.

The managers of three copper companies and the strikers' committee of eight reached no agreement at a meeting, September 5. The miners insist on an increase of one dollar per day and the offer to arbitrate wage-differences was rejected by the miners. Twelve hundred men are on strike.

SIERRA COUNTY

(Special Correspondence.)—The adit at the Wehe gravel mine, on St. Charles hill, has extended 750 ft. and is in soft ground. It is believed that the face of the adit is under the gravel deposit and a raise will soon be started. W. W. Casserly is superintendent.

The Kieffer brothers have purchased a 40-hp. engine, hoist, and duplex pump for the Gibraltar mine. The shaft is to be sunk to 310 ft. The gravel-channel was intersected at a depth of 160 ft., but the pitch of the bedrock necessitated additional depth. The gravel mined has been of excellent grade.

The air-raise at the City of Six quartz mine has been completed and the lower adit is being extended. The mine is owned by Henderson & Hodgkinson.

The main working adit of the Monarch, near Sierra City, is in 1650 ft. and is expected to intersect the main orebody within 750 ft. Upon its completion the adit will be used as the main outlet. The vein averages $10 per ton in free-milling ore.

Work is proceeding at the Tightner, Mariposa, El Dorado, North Fork, Wisconsin, and other mines in the Alleghany-Forest field. Attention is being devoted to the development of milling ore.

Downieville, September 1.

SISKIYOU COUNTY

(Special Correspondence.)—The Salmon River Mining Co. has completed arrangements for extensive work at its hydraulic mine near the Forks of the Salmon. O. H. Poor has been appointed manager. D. P. Doak and associates of San Francisco are interested. The small stamp-mill, placed in operation at the Homestake mine a few weeks ago, is operating steadily and making a good gold-extraction. It has been decided to install a Huntington mill to increase the tonnage. Ore is delivered to the plant from the lower adit workings by a 1200-ft. tramway. Enough ore has been developed to keep the mill running two years. R. S. Taylor of Yreka is manager and owner. The eight-stamp mill at the Know-Nothing gold mine, near Happy Camp, has been completely overhauled. A large compressor has been placed in position and mining will in future be done with machine-drills. The water pipe-line has been improved and the mill will be operated by water-power. Sufficient ore is stated to be exposed to insure operation of the mill for three years. The property is controlled by W. B. Beall.

Arthur Coggins and associates are developing extensive chrome deposits near Dunsmuir, and a promising molybdenite deposit near Lamoine. A concentrating plant will be erected to treat the low-grade chrome. The high-grade is being shipped to the Carnegie Steel Co. and other Eastern manufacturers. Molybdenite is shipped to Denver, Colorado.

Yreka, September 8.

COLORADO

CLEAR CREEK COUNTY

The Primos Exploration Co. has begun the operation of its Urad molybdenum mine near Empire. The work is under the management of Charles A. Chase. E. A. Strong will be superintendent. A mill will be constructed and will be in operation by the first of 1918. The first unit will have a capacity of 50 tons. The mine was developed two years ago.

TELLER COUNTY

(Special Correspondence.)—The strike made on the 2000-ft. level of the Portland Gold Mining Co.'s No. 2 shaft has been confirmed by the officials at Colorado Springs. In keeping with its conservative policy, however, no mention has been made of the rich core in the vein, a streak from eight inches to a foot wide, that assays several hundred ounces per ton. It was admitted in the official statement that 100-oz. assays were frequent.

The Leora V, on Newton hill, has been purchased by Harry Newton, president of the recently incorporated Victory Gold Mining Company.

The Elkhorn, on Carbonate hill, owned by the Star of the West company, has been leased to the Geraldine company. The mine has produced a small tonnage of high-grade ore, but has long been idle. The Zoe mine on the eastern slope of Beacon hill, controlled by James F. Burns, has been brought back to production by lessees.

A surface discovery of promise has been made on the Flying Cloud on the eastern slope of Bull hill. The property is owned by the Gold Pinnacle company, and the strike was made by a lessee, Tom Roberts of Goldfield.

W. A. Dewitt, of Lamar, Colorado, president of the Geraldine M. M. & D. Co. and leasing on the Elkhorn, Carbonate hill, will install a steam-hoist and a compressor. The shaft, 450 ft. deep, is to be developed.

Cripple Creek, September 1.

IDAHO

The Con. Interstate-Callahan Mining Co., operating in the Coeur d'Alene, shipped ore having a net value of $737,115 during the second quarter of the current year, according to a statement from John A. Percival, president. To this is added $6385 in miscellaneous receipts, making the total net value of $733,500. A deduction of $324,513 for operating costs leaves a profit of $408,086; less $69,530 for improvements, leaves a surplus of $339,456 for the quarter. The ore mined was 43,298 tons, and the ore milled 35,474 tons. Its average content was 22.4% zinc, 2.3 oz. silver, and 6.5% lead.

The Big Creek Leasing Co., operating the Yankee mine in the Evolution district, has declared a dividend of 50c. per share, making a disbursement of $2500, the capitalization being 5000 shares, par value $10. Shipments were first made early in 1916 and have continued in increasing volume. The ore is said to have high silver-content.

A dividend of 10c. per share has been declared by the Black-hawk Leasing Co., operating on Government gulch near the property of the Coeur d'Alene Dev. Co., a subsidiary of the Stewart Mining Company.

The gross receipts of the Richmond M. M. & R. Co. in the Coeur d'Alene for August will be $38,750. The rate of production was a car per day. The average value of the ore is $1250 per car, according to the average of settlements on shipments in July. This is after the deduction of freight and treatment charges. The operating cost will be $7000 for August, which leaves a net operating-profit of $31,750.

The enhanced price of silver will increase the annual output of the Hecla to $1,500,250. The ore averages 160 lb. of lead and 4.55 oz. of silver per ton. The output is 328,500 tons per annum. The company pays a monthly dividend of 15c. per share. The last monthly dividend, August 2, brings the total dividends to $5,505,000. Shipments of lead-silver concentrate are being made to the Bunker Hill smelter at Bradley.

Serious forest fires are reported in five of the national forests of Idaho and Wyoming. Reports indicate the probable origin as incendiary.

The Yukon Gold Co.'s dredge on Prichard creek is nearing completion.

The mill of the Bluebird mine, United Mines Co., in Little Smoky district, will be ready to operate by September 15. It will have a capacity of 100 tons per day.

MONTANA

The Intermountain Copper Mining Co., operating near Iron Mountain, Montana, will receive $38,000 net for its August output according to Edward Evans, president.

Lewis and Clark County

(Special Correspondence.)—The prediction is made that the Barnes-King Dev. Co., operating mines in Fergus county and at Marysville, will produce gross returns of $100,000 for the month of August. The principal production is expected to be made by the Shannon mine at Marysville. The orebody is exposed at a depth of 500 feet.

The St. Louis M. & M. Co., operating in Marysville, is milling 100 tons per day from the Drumlummon mine. The ore is reduced by a 25-stamp mill and a cyanide plant.

At the Bald Butte mine a new compressor has been installed and work is to be carried on.

David Blake is shipping ore from the Larson claim. The ore is silver-lead and is consigned to the East Helena smelter.

In the Grass Valley district six claims are in operation and shipments are going to the smelter at East Helena. The ore is silver-lead.

An important discovery of silver-lead ore has lately been made in the Wickes district. The find was made in the Maryland mine of the Comet-Rumley group. A cross-cut at a depth of 75 ft. was extended a distance of 85 ft. before encountering the hanging wall. Six feet of solid ore on the foot-wall was exposed. The Baltimore mine in the same neighborhood is shipping ore to the smelter.

In the Mt. Washington mine, which is developed by an adit 3000 ft. long, 15 ft. of galena ore is exposed at a depth of 800 ft. The output from the adit is one car per day. The Alta Mines Co. is developing a large tonnage. Old adits have been re-opened and a cross-cut has been extended from the bottom of an 800-ft. shaft beneath the old workings.

Helena, September 1.

NEVADA

Elko County

(Special Correspondence.)—Preparations are being made for further tests of the shale deposits near Elko to determine their gasoline-content, and for the perfection of a process invented by A. C. Crane. It is stated that the Crane process facilitates the recovery of a maximum amount of gasoline and other oils from shale.

The Michigan group of lead-zinc claims on Battle creek, on the east slope of the Ruby mountains, has been taken under lease and option from Frank Shoot by Backman, Blewett, and Woodward. The group is equipped with a 50-ton mill and arrangements are being made to operate it and to develop the deposit. The mill product assays 20 to 33% zinc, 11.5% lead, and 6 oz. silver. Bids for the construction of a road into Jarbidge from Twin Falls, Idaho, are being advertised for in Twin Falls newspapers, and it is stated an endeavor will be made to have the road completed before winter.

Elko, September 12.

Esmeralda County

The Southwestern Mines Co., S. H. Brady, manager, expects to have its new 200-ton mill in operation by October 1. The shaft of the company's mine, at Hornsilver, has reached a depth of 500 feet.

Humboldt County

The north drifts on the 800 and 900-ft. levels of the Rochester Mines Co. are advancing in ore. The bullion production for the first 20 days in August was valued at $36,500. It is expected that repairs to the tramway will be completed by September 8.

At the Rochester Combined Co. development is proceeding. A compressor and Dorr cyanide equipment have been received and installation is under way.

Seven Troughs is producing ore from the 700 and 750-ft. levels. Preparations are being made to haul the ore to the Darby mill for treatment.

Lincoln County

(Special Correspondence.)—The Greenwood lease on the Amelgamated Pioche has opened a large body of silver, lead, and gold ore on the 400-ft. level of the Amalgamated shaft. Ten tons of shipping ore is being mined per shift. A compressor has been installed. It is reported that the Amalgamated management is preparing to drive a cross-cut on the 1200-ft. level.

The Virginia-Louise Mining Co. proposes a bond issue of $60,000 to finance further development and equipment of its holdings. It is stated that 200,000 tons has been blocked out and that indications are favorable for the development of a larger tonnage of silver, lead, and gold ore.

H. M. Lansdowne will arrange for the resumption of work at the Pioche Mines group, two miles north. The mine is stated to contain large bodies of good ore. The scarcity of labor is interfering with the re-opening.

Activity is evident on the east side of the Ely range. The Ely Valley and Tulloch are producing, and lessees on the Pioche Metals group are developing good ore. The high price of silver has stimulated work and promises to cause the re-opening of several mines that have long been idle.

It is expected that the sampler at Jean will be moved to Salt Lake in the near future and its capacity increased to 600 tons per day. Negotiations are in progress and J. B. Jensen, the manager, is conferring with Utah producers. If the plant is moved, favorable freight rates to local producers are anticipated.

Pioche, September 2.

The Prince Con. M. & S. Co. will pay another dividend of 5c. per share, in the aggregate $50,000. The total dividends disbursed amount to $550,000. Diamond-drilling has been in progress for three months.

MINERAL COUNTY

(Special Correspondence.)—The Mason Valley Co. will place the second furnace of its Thompson smelter in commission as soon as sufficient coke is on hand to insure steady operations, which will be within 30 to 60 days. A regular supply of custom ore has been guaranteed by operators at Luning, and in nearby districts. This includes copper, silver, and gold producers. The first furnace is operating on ore from the Mason Valley, Nevada Douglas, and Bluestone mines. High-grade copper ore has been intersected 150 ft. from the 1000-ft. level of the main shaft on the Calavada group, formerly known as the Giroux. Cuprite and tetrahedrite occur. The ore is the richest yet discovered in this field and the deepest point to which Luning development has been carried. Several stringers of rich ore have formed an orebody near No. 8 station in the Hahn adit of the Luning-Idaho, and work is proceeding to determine its extent. Arrangements have been made for shipments from the McDavitt shaft and Conte lease as soon as the second furnace of the Thompson smelter has been blown-in. Driving is progressing to develop rich silver-copper ore, outcropping on the Sophie, Monarch, and other claims. The sinking of a main working shaft will be started in the near future.

No. 1 shaft of the Copper Mountain mine, near Rand, has intersected the ore-zone, and drifts from the 150-ft. cross-cut are developing shipping-ore. The shaft on the Miller-La Patt lease has been depened 150 ft. and ore is being shipped. Shafts No. 2 and 3 are being sunk to reach the orebody. The property is operated by the Jumbo Extension Co. of Goldfield. C. Koegel, superintendent of the Nevada Rand mine, expects to intersect the main orebody on the 450-ft. level. Ore has been exposed on the 150, 180, and 250-ft. levels. The company has decided to install a 25 to 50-ton mill. Conditions at Reservation Hill are improving. The Mountain View Co. is operating its small mill on a good grade of gold ore.

Luning, September 8.

NYE COUNTY

(Special Correspondence.)—The preliminary work of the White Caps mill is satisfactory to the management. The feature which caused doubt was whether the temperature of the roaster could be regulated and held at the point where the refractory compounds, found in the White Caps ores, could be volatilized without fusing. The results have proved that the temperature of the roast can be maintained within 20 degrees of the desired point. It is officially announced that the first test-run on a −10-mesh product, after roasting and passing through the leaching plant, yielded an extraction of 87%. With this knowledge, after changes have been made in the crushing plant and by the addition of a trommel, already installed, to size the product, it is expected that an extraction of 90% will be attained in the mill. It already has been stated that the average will be 120 tons per 24 hours. It takes two hours for the ore to pass through the roaster and 72 hours is required in the leaching department of the mill. It is expected that the roasting time will be reduced. The first mill-runs have been made on ores lower than the average. Ten men are employed in the mill. In the mine the development for the week totals 146 ft. A large double-drum hoist has been delivered and will be installed.

Since the resumption of mining east of the working-shaft of the Union Amalgamated Mining Co., a total of 50 ft. of new development has been accomplished. The maximum water-flow encountered has been 100 gal. per minute and is being handled by the triplex station pumps.

The working-shaft of the Manhattan Consolidated M. & D. Co. has now reached a depth of 45 ft. below the third level. The formation encountered in the shaft is a hard, dark-blue, calcareous shale, the soft shale, struck at a point 75 ft. above the third level, having passed out of the shaft 20 ft. below the level. Until the shaft has been sunk 100 ft. below the third level, no further work will be undertaken in the mine, as the development of the east orebody on the third level has been sufficient.

The joint shaft of the White Caps Extension and Zanzibar is being sunk in record time. It is a three-compartment shaft and has now reached a depth of 122 ft. The average speed in the sinking is maintained at six feet per day. Jack-hammer drills are used in the shaft. The size of the shaft is 6 by 16 ft. The Chicago pneumatic air-compressor is in place near the shaft. The compressor has a capacity of 314 cu. ft., sufficient for drilling and pump purposes.

The surface machinery recently purchased by the Red Top Mining Co. has been received and much of it has been already installed. The Cameron sinking-pump is ready to use. The two-stage Ingersoll compressor is in place and ready for operation.

Manhattan, August 20.

STOREY COUNTY

Milling-ore is being developed by the Jacket Surface Tunnel of the Jacket-Crown Point-Belcher M. Co. A plant of 125 tons capacity per day is being erected. The equipment consists of a rock-breaker, Hardinge steel-ball mill, Dorr thickener, pachuca agitation tanks, and Oliver filter. Air-lifts are used to transfer the pulp. The Jacket Con Co. is controlled by the Sturgis interests.

OREGON

CURRY COUNTY

(Special Correspondence.)—The Chetco Mining Co. of Harbor is spending $35,000 in erecting new machinery and building roads. The mine is situated on the summit of Mt. Emery, 12 miles east of Harbor, at an elevation of 3200 ft. The vein of free-milling ore is 60 ft. wide.

Harbor, September 4.

JOSEPHINE COUNTY

(Special Correspondence.)—The decline in the price of copper has closed many of the copper properties in the Waldo district. The miners and equipment from these properties are being diverted to the production of chrome and other minerals. Every effort is being made to develop the production of manganese in Josephine and Jackson counties.

The Pittsburg-Oregon M. & M. Co. was incorporated last week. The office of the company is to be at Grants Pass and a general mining and ore-marketing business is planned.

With the burning of the C. O. Power Co.'s sub-station in the Greenback district north of Grants Pass, the Greenback mine will suspend operation for ten days.

Grants Pass, September 4.

TEXAS

BREWSTER COUNTY

(Special Correspondence.)—Scherer and Whall have a claim-area of 800 acres near Shumla from which they have shipped 100 tons of manganese ore. They propose a weekly shipment of the same tonnage. The ore contains from 21 to 40% manganese and from 7 to 25% iron. It is shipped to East Chicago, Illinois.

Langtry, August 16.

(Special Correspondence.)—The large deposit of fuller's earth situated near here has been leased by J. C. Melcher to F. G. Lewis and C. C. Knight of Wichita Falls for a period of 25 years. It is stated that the lessees will install machinery and equipment. The mineral substances in this deposit is said to be of high grade.

O'Quin, September 1.

(Special Correspondence.)—The Texas Nitrate & Fertilizer Co. has been incorporated for the purpose of developing a large deposit of 'nitratine' or soda niter in Presidio county in one of the most arid parts of the upper border region. It is stated that the discovery was made by Charles Mitchell of Houston.

The deposit covers an area of two miles long, one-half mile wide, and a depth of 5 to 12 ft. Samples taken from various parts of the deposit give an analysis showing nitratine amounting to 11 to 45%. It is the theory of geologists who have visited the deposit that it was formed by an ancient guano bed which was covered by volcanic eruptions. The company has obtained a lease upon 2700 acres of land and is preparing to begin development operations. Associated with Mr. Mitchell are D. A. Gregg of Houston and others.

Alpine, September 1.

UTAH

The Associated Companies, a company exclusively furnishing workmen's compensation insurance for coal mines, has entered the Utah field. E. C. Lee, with headquarters at Pittsburgh, and E. H. McCleary, inspector in charge of the western field, have spent 10 days inspecting mines in Utah. Negotiations are practically completed with the State Industrial Commission governing the methods of work of the companies. The base rate of the Commission is $7.81 per $100 of pay-roll.

UTAH

A mine complying with the merit safety standard of the Associated Companies can obtain 60% of the base rate or $4.69.

The Eldorado G. M. & M. Co. has been operating since May in opening an old property 6 miles north of Ogden.

The Golden Reef Con. M. Co., in Beaver county, is developing a promising property. D. P. Rohlfing of Salt Lake City is consulting engineer. A new vertical shaft has been sunk to a depth of 450 ft. On the 400-ft. level the vein is 100 ft. wide.

The Cardiff M. Co., in Big Cottonwood canyon, is cutting a station at the 225-ft. level. A daily shipment of 125 tons of lead and silver ore is being made.

Stockholders, officers, directors, and creditors are endeavoring to disentangle the affairs of the Big Four Exploration Co. It is asserted that frequent changes were made in the mill and the recovery was unsatisfactory. The liabilities are $258,000 and the chief assets, a new plant, stated to have cost $300,000, $40,000 to $50,000 worth of ore and concentrate, and the Ontario tailing dump.

Beaver County

The Golden Reef, operating four miles north of Frisco, has opened up a body of galena ore, which assays 30 oz. silver per ton and 32% lead.

The Carbonate Center, operating near Frisco, has encountered good grade galena ore. Electrical-driven equipment, supplied with current from the Beaver River Power Co., has recently been installed. A concentrating plant has been designed and a part of the construction material is on the ground.

A compressor has been purchased by the Indian Queen and cross-cutting in the adit is planned. Lessees are mining lead and copper ore from the upper workings.

The Imperial Mines Operating Co. has been organized to lease the Imperial group of claims, situated three miles west of Newhouse. A compressor and other necessary equipment has been ordered. The wagon road at Newhouse has been repaired. The organizers of the company are H. H. Adams, J. J. Brouher, E. E. Meyers, E. H. Street, and S. H. Fotheringham.

The Utah Leasing Co., working the tailing-dump at Newhouse, is treating 700 tons daily. The concentrate is shipped to International.

Grand County

A. M. Rogers is authority for the statement that from 600 to 800 tons of 2% uranium ore is awaiting shipment from Dry valley. Two 2-ton trucks will transport the ore to the railroad.

Juab County

August ore-shipments from the Tintic district reached 928 cars. The important shippers were the Dragon Con., Chief Con., Iron-Blossom, Mammoth, Centennial-Eureka, Grand Central, Eagle & Blue Bell, Gemini, Gold Chain, and Colorado Consolidated.

The north drift of the Connolly mine at Eureka has encountered a good body of ore assaying from $55 to $65 per ton in lead, gold, and silver. The drift was run from the 400-ft. level in the new shaft, and is one of the important developments in the Eureka district. The new Connolly shaft, projected to go 1000 ft., was sunk to the 400-ft. level before cross-cutting. The drift was extended north 100 ft., and intersected an ore-bearing zone. Several shipments, amounting to nearly 100 tons, have been made.

The Huebner property, recently sold to the Eureka-Utah Mining Co., of Salt Lake, is shipping regularly. Since acquiring this property the company has been working in ore of good grade that has now developed into carbonate and galena ores carrying gold and silver. Shipments run $65 per ton in car lots. The property is owned by the Eureka-Croesus Mining Co. of New York.

Summit County

The August shipments of Park City mines totaled 10,919 tons. Production of over one thousand tons per month was made by the Silver King Coalition, Judge M. & S., Daly West, Silver King Con., and the Ontario. The Silver King Con. declared a quarterly dividend of 15c. per share, aggregating $105,000 and bringing the dividend total to $1,422,765.

Tooele County

A discovery of high-grade bismuth ore in the Pole Star mine has been reported.

WASHINGTON

Stevens County

Jesse M. Hall, secretary, reports a net smelter return, from three cars of ore from the Gladstone Mountain mine, of $1778.69. The cars contained 102 tons.

The earnings of the Electric Point mine in July were estimated at $60,000. This is sufficient to pay a dividend of 7c. per share. The surplus is $300,000. The ore reserve is estimated at $1,500,000.

BRITISH COLUMBIA

(Special Correspondence.)—Four new shippers are shown on the statement of ore shipments to Trail last week. There are 34 shipping mines and the ore movement totaled 8327 tons, as compared with 7541 in the preceding week. The year's total tonnage to the end of the week is 224,279 tons. The mines which have shipped for the first time this year are the Gray Copper and Lone Bachelor at Sandon in the Slocan district, the Prince Henry near Greenwood and the Kamloops Agencies at Kamloops. Their shipments were respectively 37 tons, 7 tons, 18 tons, and 7 tons. Of the 34 shippers, 15 are in the Slocan and Ainsworth district, 2 in Rossland, 3 in the East Kootenay, 3 in the Boundary country, 1 in Nelson, and 7 in the United States. The heaviest shipper was the Sullivan mine at Kimberley.

Operation of the new mill at the Surf Inlet gold mine commences September 1 and it is intended to operate at full capacity, 300 tons per day. The Surf Inlet mine has been bonded to the Tonopah-Belmont Co., a new company having been organized with a capital of $2,500,000.

Out of a total of $1,500,000, the Tonopah-Belmont interests expended $1,300,000. They have taken an option on the Pugsley mine, owned by the Princess Royal Gold Mines, Limited.

Vancouver, September 1.

The Slocan Star Mines, Ltd., operating near Sandon, B. C., shipped 775 tons of zinc-bearing ore and 185 tons of lead concentrate in the past three months. The zinc ore yields $25 per ton after the deduction of freight and treatment charges, and the lead $110 to $125 per ton. The 775 tons of zinc concentrate had a net value of $19,375 and the lead a value of $20,350. The total value of the shipments is $39,725. In addition the company has concentrates in storage valued at $10,000.

Shipments of 26 cars of ore with a net value of $68,351 are reported by the Highland Valley M. & D. Co., operating near Ashcroft, B. C., since beginning the production of ore late last year. Crude-ore shipments, made in 1916, total four cars containing 114½ tons, according to Fred Keffer, president of the company. The shipments of concentrate began in January, since which time 22 cars having a value of $66,198 have been made.

GUATEMALA

The Torlon Mining & Smelting Co. has been organized at Huehuetenanga, Guatemala, for the exploration and development of mines in the vicinity of that city. The general manager is Santiago Molina. Members of the board of directors are Santiago Molina, Teodoro M. Recinos, Federico Morales, Joseph Allan, and Juan F. Prem.

MEXICO

The mine-owners and operators of the State of Durango, Mexico, are requested to organize a Chamber of Mines, which shall serve as a registry of all operating mines and their assets. The chamber is expected to make regulations for the welfare of the members, and every owner of a mining claim or of a possessory right in a claim is entitled to representation and a vote at the meetings. On the face of the announcement to the mine-owners there is nothing to indicate an official link between the proposed chamber and the State Government except that the object is to benefit the individual mine operators and principally the commonwealth that requires the utmost activity upon all known deposits of minerals. The call for the organization is signed by attorney Alberto Terrones Benitez, of the city of Durango.

Personal

Note: The Editor invites members of the profession to send particulars of their work and appointments. This information is interesting to our readers.

DONALD F. CAMPBELL, of Seattle, is here.

C. H. MUNRO and JOHN F. NEWSOM are in New York.

W. J. ELMENDORF is in the Copper River district, Alaska.

ROBERT E. CRANSTON is in Alaska on professional business.

G. W. CUTTING has a commission in the Field Artillery, U. S. Army.

F. G. CLAPP has gone on a trip to Texas and other parts of the West.

F. B. FORBES has come home from Nicaragua to serve in the Army.

E. W. WESTERVELT is superintendent for the Hobson Silver-Lead Co., at Ymir, B. C.

GEORGE J. CARR has resigned as superintendent for the Yuba Consolidated Gold Fields.

ANDRES SALAS, a Peruvian mining engineer, is visiting flotation-mills in the West.

OLOF WENSTROM, of Boston, passed through San Francisco on his return from Arizona.

HOWARD D. SMITH has been examining the Jerome Verde property at Jerome, Arizona.

L. WEBSTER WICKES has received a commission as Captain in the Engineers Reserve Corps.

DONALD F. FOSTER holds a commission in the British army and is now stationed in Egypt.

DEAN B. THOMPSON, now with the Mason Valley Mines Co., was in San Francisco this week.

F. C. FRENCH has received a commission as Captain of Engineers in the Officers' Reserve Corps.

IRVING A. PALMER has been appointed professor of metallurgy in the Colorado School of Mines.

J. PARKE CHANNING is in the Michigan copper region, visiting the Naumkeag exploration operations.

SCOVILL E. HOLLISTER has been appointed resident manager for the Leviathan Mines Co., at Yucca, Arizona.

FRANK R. WICKS has succeeded W. H. JANNEY as mill-manager for the Chino Copper Co., at Hurley, New Mexico.

MARK L. SUMMERS, formerly dredge-foreman, has been appointed superintendent for the Yuba Consolidated Gold Fields at Hammonton, California.

A. G. LANGLEY, formerly with the Copper Queen Consolidated, is now resident engineer in the Revelstoke district for the Provincial Government of British Columbia.

EUGENE H. BARTON was born at Bartonville, Vermont, February 22, 1853, and died at his residence in Los Angeles on August 12, 1917. After graduating from the University of Louisiana, he went to Germany, where he took a two years' post-graduate course at Heidelberg. Upon his return to the United States, he joined a civil-engineering corps engaged in railroad construction in Wyoming and Colorado. Later he went to Bodie, California, where he built and operated the railroad running from Bodie to Mono Mills. Later he was in charge of the operation of some of the largest mines at Bodie. In 1887 he became chief engineer for the La Grange dam of the Turlock irrigating system. Upon the completion of the dam, he resided at Sonora, in Tuolumne county, California, where he opened a civil-engineering office. About 1902 he went to the Yellow Aster mine at Randsburg, as engineer and later became superintendent of the property. Early in 1907 he became mine-superintendent for the Ray Consolidated Copper Co. at Ray, Arizona, but left there when mining operations were greatly curtailed owing to the panic of 1907, since which date his time has been devoted largely to general engineering practice.

THE METAL MARKET

METAL PRICES

San Francisco, September 11

Aluminum-dust (100-lb. lots), per lb	$1.00
Aluminum-dust (ton lots), per lb	$0.95
Antimony, cents per pound	16.50
Antimony (Wholesale), cents per lb	15.25
Electrolytic copper, cents per pound	28.50
Pig lead, cents per pound	11.25
Platinum, soft and hard metal, respectively, per ounce	$105—111
Quicksilver, per flask of 75 lb	$115
Spelter, cents per pound	10
Tin, cents per pound	60
Zinc-dust, cents per pound	20

ORE PRICES

San Francisco, September 11

Antimony, 45% metal, per unit	$1.20
Chrome, 40% and over, free SiO_2 limit 8%, f.o.b. California per unit, according to grade	$0.60—0.70
Magnesite, crude, per ton	$8.00—10.00
Tungsten, 60% WO_3, per unit	$23.00—25.00
Scheelite, highest grade, per unit	$26.00
Molybdenite, per unit for MoS_2 contained	$40.00—45.00

Manganese, Carnegie schedule less freight to Chicago. Less than 40% special quotation.

Manganese prices and specifications, as per the quotations of the Carnegie Steel Co. schedule of prices per ton of 2240 lb. for domestic manganese ore delivered, freight prepaid, at Pittsburgh, Pa., or Chicago, Ill. For ore containing

	Per unit
Above 49% metallic manganese	$1.00
46 to 49% metallic manganese	0.98
43 to 46% metallic manganese	0.95
40 to 43% metallic manganese	0.90

Prices are based on ore containing not more than 8% silica nor more than 0.2% phosphorus, and are subject to deductions as follows: (1) for each 1% in excess of 8% silica, a deduction of 15c. per ton, fractions in proportion; (2) for each 0.02% in excess of 0.2% phosphorus, a deduction of 2c. per unit of manganese per ton, fractions in proportion; (3) ore containing less than 40% manganese, or more than 12% silica, or 0.225% phosphorus, subject to acceptance or refusal at buyer's option; settlements based on analysis of sample dried at 212° F., the percentage of moisture in the sample as taken to be deducted from the weight. Prices are subject to change without notice unless specially agreed upon.

EASTERN METAL MARKET

(By wire from New York)

September 11.—Copper is stronger and higher at 25.25 to 26.25c. Lead is dull and prices are a little lower at 10 to 10.25c. Zinc is quiet and nominal at 8.25c. Platinum remains unchanged at $105 for soft metal and $111 for hard.

SILVER

Below are given the average New York quotations, in cents per ounce, of fine silver.

Date			Average week ending		
Sept.	5	95.62	July	31	78.20
"	6	95.62	Aug.	7	80.50
"	7	96.62	"	14	82.95
"	8	96.62	"	21	87.16
"	9 Sunday		"	28	88.50
"	10	97.62	Sept.	4	91.12
"	11	98.62	"	11	96.80

Monthly Averages

	1915	1916	1917		1915	1916	1917
Jan.	48.85	56.76	75.14	July	47.52	63.06	78.92
Feb.	48.45	56.74	77.54	Aug.	47.11	66.07	85.40
Mch.	50.61	57.89	74.13	Sept.	48.77	68.51	....
Apr.	50.25	64.37	72.51	Oct.	49.40	67.86	....
May	49.87	74.27	74.61	Nov.	51.88	71.60	....
June	49.03	65.04	76.44	Dec.	55.34	75.70	....

COPPER

Prices of electrolytic in New York, in cents per pound.

Date			Average week ending		
Sept.	5	25.25	July	31	27.25
"	6	25.50	Aug.	7	28.46
"	7	25.75	"	14	27.62
"	8	26.00	"	21	28.00
"	9 Sunday		"	28	26.41
"	10	26.25	Sept.	4	25.65
"	11	26.25	"	11	25.83

Monthly Averages

	1915	1916	1917		1915	1916	1917
Jan.	13.60	24.30	29.53	July	19.09	25.66	29.67
Feb.	14.38	26.62	34.57	Aug.	17.27	27.03	27.42
Mch.	14.80	26.65	36.00	Sept.	17.69	28.28	....
Apr.	16.64	28.02	33.16	Oct.	17.90	28.50	....
May	18.71	29.02	31.69	Nov.	18.88	31.95	....
June	19.75	27.47	32.57	Dec.	20.67	32.89	....

Advices from Boston state that a purchase of 60,000,000 lb. of copper was made at 25c. per pound. This sale was made direct to the United States government, which acted as intermediary for account of the Allies. The hope is entertained that by virtue of the Administration's 'one-price-for-all' policy the Government will permit a price of 25c. for our own war requirements. The copper people expect to hear from Washington definitely this week. It is understood there are certain interests in Washington, heretofore prominent in Wall Street, who are laboring to secure a quotation of not over 22c. per pound. Notwithstanding heavy wage increases, high cost of coal and other materials, the copper-producing industry is confronted with the most formidable strike situation which it has ever been called upon to face. All of which lends force to the argument that anything less than 25c. for copper will be entirely out of line with $2.20 wheat.

All activities of the Anaconda Copper Mining Co. in Montana have ceased, this condition applying both to mines and smelters. Copper production will be affected to the extent of 30,000,000 lb. monthly, and the output of high-grade spelter will be materially encroached upon. Earnings, of course, will be reduced to a minimum. Anaconda operates three copper refineries. Two of these in Montana were shut-down during the early days of the strike and all smelter product diverted to the Raritan refinery in New Jersey, which has been able to maintain operations, although at less than normal. The Washoe smelter has since closed, leaving as the active factors in this company's operations the Tooele smelter in Utah and the Raritan works. The International company's smelter in the Globe, Arizona, district has been closed for two months, thereby removing from monthly production approximately 36,000,000 lb. of copper.

LEAD

Lead is quoted in cents per pound, New York delivery.

Date			Average week ending		
Sept.	5	10.25	July	31	10.55
"	6	10.25	Aug.	7	10.87
"	7	10.25	"	14	10.87
"	8	10.25	"	21	10.79
"	9 Sunday		"	28	10.61
"	10	10.25	Sept.	4	10.38
"	11	9.87	"	11	10.20

Monthly Averages

	1915	1916	1917		1915	1916	1917
Jan.	3.73	5.95	7.64	July	5.59	6.40	10.93
Feb.	3.83	6.23	9.01	Aug.	4.62	6.28	10.75
Mch.	4.04	7.26	10.07	Sept.	4.62	6.86	....
Apr.	4.21	7.70	9.38	Oct.	4.62	7.02	....
May	4.24	7.38	10.29	Nov.	5.15	7.07	....
June	5.75	6.88	11.74	Dec.	5.34	7.55	....

ZINC

Zinc is quoted as spelter, standard Western brands, New York delivery, in cents per pound.

Date			Average week ending		
Sept.	5	8.25	July	31	8.75
"	6	8.25	Aug.	7	8.75
"	7	8.25	"	14	8.75
"	8	8.12	"	21	8.66
"	9 Sunday		"	28	8.33
"	10	8.12	Sept.	4	8.25
"	11	8.25	"	11	8.21

Monthly Averages

	1915	1916	1917		1915	1916	1917
Jan.	6.30	18.21	9.75	July	20.54	9.90	8.98
Feb.	9.05	19.99	10.45	Aug.	14.17	9.03	8.58
Mch.	8.40	18.40	10.78	Sept.	14.14	9.18	....
Apr.	9.78	18.62	10.20	Oct.	14.05	9.92	....
May	17.03	16.01	9.41	Nov.	17.20	11.81	....
June	22.20	12.85	9.63	Dec.	16.75	11.26	....

QUICKSILVER

The primary market for quicksilver is San Francisco, California being the largest producer. The price is fixed in the open market, according to quantity. Prices, in dollars per flask of 75 pounds:

Week ending

Date					
Aug.	14	115.00	Aug.	28	115.00
"	21	115.00	Sept.	4	115.00
			"	11	115.00

Monthly Averages

	1915	1916	1917		1915	1916	1917
Jan.	51.90	222.00	81.00	July	95.00	81.20	102.00
Feb.	60.00	295.00	126.25	Aug.	93.75	74.50	115.00
Mch.	78.00	219.00	113.75	Sept.	91.00	75.00	....
Apr.	77.50	141.60	114.50	Oct.	92.90	78.20	....
May	75.00	90.00	104.00	Nov.	101.50	79.50	....
June	90.00	74.70	85.50	Dec.	123.00	80.00	....

TIN

Prices in New York, in cents per pound.

Monthly Averages

	1915	1916	1917		1915	1916	1917
Jan.	34.40	41.76	44.10	July	37.38	38.37	62.60
Feb.	37.23	42.60	51.47	Aug.	34.37	38.88	62.53
Mch.	48.76	50.50	54.27	Sept.	33.12	36.66	....
Apr.	48.25	51.49	55.63	Oct.	33.00	41.10	....
May	39.28	49.10	63.21	Nov.	39.50	44.12	....
June	40.26	42.07	61.93	Dec.	38.71	42.55	....

MOLYBDENUM

The New York market for molybdenum is unchanged and the price quoted is $2.10 to $2.20 per pound for 90% material.

ANTIMONY

The New York market continues dull and lots have been offered by jobbers at 15⅝c. per pound, the nominal quotation for spot duty-paid antimony. Antimony ore remains unchanged with very little ore reaching this market. Needle antimony spot is quoted at 10c. per pound.

MANGANESE

Manganese ore coneinues fairly strong at $1 quoted on a basis of 48% ore or over.

FINISHED STEEL PRICES

It is the common belief in the steel trade that the average price for finished steel during the June quarter was 34¼c. per pound, or $65 per ton net. Recent prices have been about 5¾c. Should the average price for steel be settled on at $65 per ton there is room for a 50% decline from present prices.

Eastern Metal Market

New York, September 5.

The metals are all lower, with quotations in nearly each case nominal. Business is slacker than in many years and the market as a whole is nearly demoralized. Buyers are not anxious to buy and sellers are filled with uncertainty as to the future of prices because of Government price-fixing and are unwilling to commit themselves into the future. The result is a stand-off market.

Copper is lifeless and nominally a little weaker.

Tin is dull and featureless but steady.

Lead is stagnant and softer.

Antimony is quiet and lower.

Aluminum is dull and unchanged.

In the iron and steel market repeated postponement of action in arriving at some conclusion regarding price-fixing reveals the magnitude of the task. Additional data have been asked for from producers in the past week. The pig-iron semi-finished steel and rolled products markets continue to drift. Statistics for the August pig-iron output show how small has been the success of furnace-men to increase production. The output was 3,247,947 tons, or 104,772 tons per day against 3,342,438 tons in July or 107,820 tons per day—this in spite of the fact that five new modern furnaces were started in August.

COPPER

Stagnation bordering on paralysis characterizes the copper market. Even the market in copper stocks has the disease and is experiencing a decided slump. The air is saturated with rumors regarding the price to be paid by the Government and its Allies, but the question is still undecided, at least so far as the public is concerned. Anywhere from 20 to 25c. per lb. is mentioned, but it is not even known yet what the final price is to be on the 60,000,000 lb. bought by the Government some months ago. Added to these disconcerting factors are labor troubles limiting production and law-suits cutting in on profits. It is no wonder that buying is paralyzed and that fear seizes the market. There is no buying reported, except spasmodic lots here and there which some consumer needs badly. Small quantities of prime Lake are reported sold at about 26c., New York, with September electrolytic changing hands on August 27 at 25.75c., New York, and October at 25.50c., New York, but these sales are unconfirmed. Prices in general are distinctly nominal. Yesterday both Lake and electrolytic were quoted nominally at 25.25c., New York, with variations from this, on either side, of ½c. For the last quarter about 24 to 25c. seems to be the nominal range. There will be sold on the New York Metal Exchange at auction tomorrow 448,000 lb. of standard electrolytic for prompt delivery, 448,000 lb. for September, 448,000 lb. for October, and 448,000 lb. for November delivery. London prices remain unchanged at £137 for sp t electrolytic and £133 for futures.

TIN

Activity is only spasmodic and the market is featureless. General basic conditions in other metals sympathetically affect the tin market. Inquiry and buying was fairly active on August 29 and 250 tons of Straits tin was reported sold at 61.50c., New York. Other sales on that day embraced metal, ex-steamer 'Foyle,' soon to arrive, at 60.87½c., New York; Straits tin, due in September, at 60.50c. and December shipment from the Straits at 57.12½c. A fair inquiry in futures developed on August 31, but no sales of consequence were noted. August 30 was practically a holiday, due to the city's testimonial to the New York National Guard and the market was neglected, though metal was offered at 61.75c., New York, with few buyers. After another holiday on Monday there was a little inquiry yesterday which developed into sales of 100 to 125 tons, mostly spot with some sales of tin on the seas. The quotation yesterday was 61c., New York, that for August 30 and 31 having been 61.75c. The London market yesterday was £243 10s. per ton for spot Straits, a decline of 10s. from last week's price. Since Monday the London market has declined 30s. for all grades. Arrivals this month, including September 4, were 325 tons, with 4015 tons reported afloat.

LEAD

The lead market is the only one really distinguished by a feature. This was the announcement by the American Smelting & Refining Co. of a reduction in its quotation which was lowered $10 per ton or 50c. per lb. to 10.50c., New York.

The unusual circumstance attending this change was the statement that the company is willing to book orders without any string attached. The trade looks upon this as considerable of a change and as pointing to a liberal supply of lead. Whether the leading interest can make sales at this price is to be watched with interest. General conditions are not altered but prices are. In the outside market also quotations have declined with some selling reported at 10.37½ to 10.25c., New York. Yesterday the outside market broke to 10.25c., New York, at which price at least one 100-ton lot probably changed hands. Buyers and sellers seem as far apart as ever, taking the market as a whole, and buying in no case is on a large scale. Buyers continue to hold off and no activity is expected until confidence in general business conditions is renewed. Recently October metal has been offered at 10.25c., St. Louis, and November at 10c. London quotes £30 10s. per ton for spot lead and £29 10s. for futures.

SPELTER

In the absence of any sustained demand the market continues stagnant and weak. Prices, which are almost entirely nominal, are unchanged at 8c., St. Louis, or 8.25c., New York, for prime Western. For the last quarter, where quotations are obtainable they range from ⅛ to ¼c. higher. The market is very difficult to judge. Some sellers quote as low as 7.75c., St. Louis, for September delivery, with others quoting from 8 to 8.25c. and above. There are some who are in a weakened position and are selling or quoting under 8c., St. Louis, but the amount of business done at these or any other figure is very small. The market is demoralized. Sellers as a rule will not part with metal at near or under the cost of production, nor will buyers enter the market when price-fixing uncertainties are the dominant factor. It is certain, however, that consumption is still going on here and abroad and that a demand is accumulating which will some time express itself in renewed buying on a large scale. This is the only ray of hope.

ANTIMONY

Chinese and Japanese grades are lower in the absence of any pronounced demand. Duty paid, New York, the quotation now stands at 14.50c. per pound.

ALUMINUM

The market is weak and unchanged at 47 to 49c. per lb., New York, for No. 1 virgin metal, 98 to 99% pure.

ORES

TUNGSTEN. The market is firm and fairly active with demand from domestic and foreign sources strong. Prices are unchanged at $23 to $26 per unit for 60% concentrate. Fair sales are reported. Ferro-tungsten is quoted at $3.30 to $2.60 per lb. of contained tungsten with demand continuing good.

MOLYBDENUM AND ANTIMONY. Conditions and quotations are unchanged from those reported last week.

DOLLAR silver is no longer a dream. Shades of 1893! The whirligig of time brings strange revenges. We shake hands with old friends in Colorado and Nevada.

GUESSES concerning the price to be fixed by the Government for copper range all the way from 20 to 25 cents per pound. We shall be curious to learn what the Government experts have ascertained to be the average cost of production.

HOW the War has stimulated our foreign trade in base metals is indicated by the fact that exports of zinc for the fiscal year ended on June 30 amounted to 504,294,665 pounds, as against only 4,560,000 pounds during the fiscal period that ended just before the War began. Copper exports for the twelve months just closed were 1,021,501,398 pounds, which compares with 711,342,146 pounds in 1916. By the way, is it not time to talk in units of tons, not pounds?

IT is our pleasure to reproduce an article contributed to the Institute by W. J. Sharwood. Keen observation and wide experience in cyanidation give special value to his writings. Zinc-dust precipitation is his subject this time. The de-oxidation of pregnant solutions, the avoidance of a low temperature, the control of alkalinity and free cyanide, the use of the zinc in as fine a state of subdivision as possible, and the thorough contact between solution and precipitant are some of the points on which he lays stress. While recognizing the accuracy of knowledge possessed by this metallurgist, may not a hard-worked editor express appreciation of the accuracy of Mr. Sharwood's writing. It is a rare pleasure to be able to publish an article that needs no revision, only half a dozen hyphens and the deletion of a couple of verys. A scientific observer that writes so accurately is one whose research commands confidence.

OBJECTION has been taken to a paragraph appearing in our issue of September 8 concerning the incidence of the flotation royalty on the ore treated in the mill of the Britannia Mining & Smelting Company. The clause in the contract with Minerals Separation specified the royalty as payable on "crude ore milled." It was claimed by Mr. E. H. Nutter, representing Minerals Separation, that this included all ore entering the mill, whereas Mr. J. W. D. Moodie, manager for the Britannia company, held that it applied only to the feed going to the flotation machines. Mr. Nutter waived any claim on the ore removed on the picking-belt for direct shipment to the smelter, but insisted that it applied to all the ore fed to the rolls after the shipping ore and the waste had been sorted out. We were incorrectly informed when stating that this sorted ore was made liable for royalty. In consequence of the difference of opinion between Mr. Nutter and Mr. Moodie, the clause was subsequently amended so as to make the royalty payable no longer per unit of copper, but per ton of ore, tailing, or slime treated by use of any of the patents owned by Minerals Separation.

PLEDGES to save food and thereby to support Mr. Hoover's patriotic effort have been sent to us at a gratifying rate. We give the pledge this week on page 38 of the advertising supplement. If you are willing to economize in order to aid the national cause, sign the pledge and mail it to us; we will forward it to Mr. Hoover at Washington. To those that have signed it we venture to say that you are on your honor to fulfil your promise. It is not much we stay-at-homes are doing. Surely we can forego a little butter, a little white bread, and a little beef for the sake of the cause to which our young men have dedicated their lives. The present writer is not of those that think "Food will win the War," as a foolish heading in our local evening paper stated recently. Not by talk, nor by subscribing to the Red Cross, nor even by Hooverizing, shall we win the decision by which the world is to be made free from Prussian outrage, from "brutality organized to a system," from "cruelty reduced to a science." The victory over "the beast with the brains of an engineer" will be won by brave and intelligent men completely equipped with the scientific implements of war; the men that go forward under the folds of the flag to save American soil from the devastation that has blackened France and the piracy that has violated Belgium—they are the ones that count. Yet we can do something, and the best that we can do is the least that we should do. We can supply the money and the food, by paying taxes cheerfully, by buying the national bonds, and by economizing in food. This last everybody can do in his and her way, saving for the sake both of our soldiers and of the hungry peoples in Europe victimized by invasion. Therefore eat the perishable food-stuffs produced locally and release as much as you can of the food-stuffs that can be transported to distant points—to the trenches and to the broken hearths across the water. Sign the Hoover pledge and live up to it conscientiously.

THREE mining engineers have reached the grade of Lieutenant-Colonel during the War, and we are proud of them: Rowland C. Feilding, Peter N. Nissen, and Ralph Stokes. Colonel Feilding was captain in a yeomanry regiment when hostilities commenced, so that he was already started on the road to promotion. He has won the D. S. O., or Distinguished Service Order, which is only given for acts of especial courage, intelligence, and military value. Colonel Nissen likewise is a D. S. O. He volunteered at the close of 1914 and soon made a name for himself by inventing the Nissen hut, a half-cylinder of corrugated iron. The stamp and the hut have established his honorable fame. Colonel Stokes has won the Military Cross on the battlefield; he enlisted in the Royal Engineers in the very first days of the War, crossing from New York to London in early August of 1914. He went to the front a corporal, but soon won a commission. As Controller of Mines he directs the big sapping and undermining operations, such as those that enabled the Messines ridge to be captured. While referring to such good men and true it is a pleasure to mention Mr. Kenneth A. Mickle, an Australian metallurgist whose name we have quoted more than once in connection with flotation research. He is Captain in the Artillery and also a D. S. O. These men have greatly honored our profession.

WE note with interest that the 'Boston News Bureau' has been obtaining information concerning the flotation litigation from Mr. Henry D. Williams, chief counsel for Minerals Separation. In our last issue we quoted an interview with that distinguished attorney. In a later issue of the 'News Bureau' we find Mr. Williams laying stress on the iniquity of the supposed infringers and stating that "there is no escape from their liability for the past even should they now discontinue the highly profitable use of flotation." They won't. Flotation includes processes not covered even by the numerous patents of Minerals Separation. Mr. Williams quotes Judge Bourquin's explanation of the Supreme Court's decision as if the Judge at Butte were the court of final resort. The Montana decision will be reviewed by the Appellate Court in San Francisco. This is a most intelligent court and already versed in flotation technology. In another issue the 'News Bureau' quotes the terms of the Anaconda-Inspiration contract with Minerals Separation and estimates that both pay the minimum royalty of 4 cents per ton, which, in the case of the Inspiration, would amount to about $212,000 per annum. The Jackling 'porphyries' are mentioned as "vigorous contestants" of the Minerals Separation claims and are credited with producing 22,000,000 tons of ore per annum. Supposing—anybody may suppose—that an arbitrary royalty of 8 cents is charged against the copper companies and 12 cents against the Butte & Superior, then the five Jackling companies would have to pay $1,782,753 per annum—so says the 'News Bureau,' disclaiming any "official basis" for assuming any of the royalties instanced, and suggesting that this $1,782,753 is "by no means a burdensome total." It may or may not be "burdensome," but $1,782,753 will pay for a good deal of litigation and might be incurred cheerfully to avoid the embargoes and impositions that unfortunately are tied to a Minerals Separation license. Moreover there is a glimpse of ebony in this wood-pile, for the 'News Bureau' says: "Of course, the above figures are outside of any adjustment yet to be made on account of damage claims." As to the kind of damages that Minerals Separation expects, it is noteworthy that this same source of gossip, the 'News Bureau,' refers, in a third issue, to the American Zinc, Lead & Smelting Company, which is said to be liable for damages. The directors of this company are undecided, says the 'Bureau,' "whether to attempt a cash settlement, which could be had, we are told, for not over $500,000, or to persist along with the Jackling porphyries in the litigation." It looks as if the 'Bureau' had been informed concerning that succulent $500,000 by Mr. Williams, for in the next paragraph counsel for Minerals Separation is quoted as saying that suit has not as yet been brought against the American Zinc company and if the company has "not given material assistance to the companies which have contested the validity of the Minerals Separation patents," then this company will be allowed graciously to buy a license and settle for its past infringement. There is nothing mean about these M. S. people, they will allow you to pay them liberally and in return they will permit you to discover for yourself how to treat your ore.

Chuquicamata

It is a pity that such a pleasant mouthful of a name should have been set aside by Mr. Daniel Guggenheim when he formed the Chile Copper Company to exploit this remarkable deposit of ore. The Chuquicamata has become one of the romances of mining. In 1911 Mr. Claude Vantin brought the project to Mr. A. C. Burrage, who in turn interested the Guggenheims. In 1912 these enterprising gentlemen organized a development company to undertake systematic drilling and other exploratory work, which proved so successful that on April 16, 1913, a $95,000,000 corporation was formed to absorb the $20,000,000 development company. Immediately thereafter the capital of the corporation was increased by the issue of $15,000,000 worth of 7% convertible bonds, to which the public was invited to subscribe, while the Guggenheims and their friends retained all the common shares; in short, having demonstrated the value of the property, the promoters allowed the public to complete the development and equipment on an investment basis, while they, the promoters, retained 86% of the capitalization. At that time it was stated that 100,000,000 tons of 2½% copper ore had been developed and twice as much was 'possible.' The deposit had been found to have a length of 12,000 feet, a maximum width of 700 feet, and an extreme depth of 1000 feet. By that time the metallurgical treatment of the ore had been worked out by Mr. E. A. Cappelen Smith. It is noteworthy, however, that it was Mr. Vautin's recognition of the true mineral character of the ore that indicated the economic value of the deposit. Samples of the ore had been listed in many museums as atacamite, the oxychloride of copper, whereas it really was brochantite, a basic sulphate of copper named after the French mineralogist Brochant de Villiers. These two minerals are much alike, both being emerald-green with a vitreous lustre; and an occasional admixture with chalcanthite, or 'blue vitriol,' tended to confirm the confusion. The recognition of the predominant copper mineral as brochantite was of supreme importance because this sulphate is readily soluble in cold dilute sulphuric acid even after only a coarse crushing. That is the basis of the existing metallurgy at Chuquicamata. A combination of leaching and electrolysis was immediately outlined. Two difficulties presented themselves; the chlorine in the salt of the surficial portion of the deposit promised to yield chlorine, which would foul the electrolyte; the other difficulty was to find a suitable anode. A magnetite anode of German manufacture was selected; this broke too easily, but the War prevented further purchases, so that a substitute had to be found. Duriron is now being used satisfactorily, except that a small proportion of iron goes into solution, accumulating sufficiently to spoil the electrolyte, a part of which is rejected in order to maintain favorable conditions. In the starting-plant copper anodes are used with cathodes of sheet-lead, on which 7 to 10 pounds of copper is precipitated in 24 hours. The copper thus deposited is stripped and used as a cathode in the main precipitation-plant in conjunction with the duriron anode. This goes into solution slowly, lasting about seven months; it is a mixture of iron and silicon, and is cast in the company's foundry by aid of an electric furnace. The same material is used in the pumps that move the copper solutions. Acid is made on the spot, but this will be obviated shortly because sufficient acid is made incidentally when dissolving the brochantite. The original estimate was 10 pounds of sulphuric acid per ton of ore treated. Owing to the relative purity of the metallic product from electrolytic precipitation, the smelting plant is small, consisting only of two reverberatory furnaces for melting the cathode copper and removing the impurities by oxidation, the skimmings being treated in a blast-furnace. To the metallurgist the most striking feature of the local practice is the crushing, which is not carried beyond ⅜ inch, this sufficing for effective leaching. Thus the main expense and difficulty incidental to most milling operations—namely, fine grinding—is escaped. The original scheme was to break the ore at the mine so that all would pass a 14-inch grizzly. This was drawn by a caterpillar feeder to a 4-inch grizzly, the oversize from which went to McCully gyratory crushers and the undersize to secondary crushers, thence to trommels and Garfield rolls, supplemented by Symons disc-crushers, delivering to belt-conveyors. The feeder and conveyors proved a nuisance on account of rapid wear. Now two large jaw-crushers are followed by 8-inch grizzlies, thence to the secondary crushers. It is proposed to substitute three-stage disc-crushers for the rolls and the small gyratory crushers. The leaching plant contains six square concrete vats each of 10,000 tons capacity; they are charged by a system of 48-inch belt-conveyors and discharged by electrically operated clam-shell buckets that load the residue into a hopper-bin for loading in a train of cars. The leaching covers five periods, each of 24 hours, for charging, upward saturation, downward leaching, washing, and discharging. The solution contains 2½ to 5% sulphuric acid, according to requirements. Many troubles, mechanical and metallurgical, such as are inevitable to so big an undertaking, have been overcome successfully. The staff gives the major credit to Mr. Cappelen Smith. The original appraisement of the venture is to be credited to Mr. Pope Yeatman and the exploratory development to Mr. Fred Hellmann. Mr. H. C. Bellinger is now in charge. The enterprise has enhanced the reputations of all these men, besides giving the Guggenheims the opportunity for one of the most remarkable financial coups on record. According to a recent estimate the ore proved amounts to 350,000,000 tons averaging 2.23% copper, of which 2% is considered recoverable. The total cost was estimated, on a production of 10,000 tons per day, at $2.10 per ton, equivalent to 5¼ cents per pound of copper. The ore now being treated averages 1.7% and the total cost is 6.76 cents per pound of copper, this higher figure being due largely to conditions created by the War.

Our Oil-Supply

The declaration of war against Germany evoked many expressions of patriotic endeavor, some of them of a fussy and feminine kind, more particularly the organization of an untold number of committees upon which all sorts of people have found place. Washington is cluttered with committees. Every State has lots of them, many being mere safety-valves for a vicarious patriotism that accomplishes nothing except to let off steam and thereby obscure the realities of the national crisis. In California we have had our share of this infliction. It becomes all the more agreeable therefore to make note of one committee that has promptly and efficiently performed the particular duty for which it was assigned;

we refer to the Committee on Petroleum appointed by the Governor on May 9 and requested "to obtain the facts relating to the production, distribution, and utilization of petroleum" with a view to making such recommendations as might aid the proper disposition of an important natural resource. This committee consisted of Mr. Max Thelen, who is President of the State Railroad Commission, Mr. Eliot Blackwelder, who is Professor of Geology in the University of Illinois, and Mr. D. M. Folsom, Professor of Mining in Stanford University. So expeditiously did these gentlemen discharge their task that their report was transmitted to the Governor on July 7, barely two months after the date of their appointment. We have read this report of 190 pages, with maps and statistical tables; it is a surprisingly complete and convincing statement. There can be no question concerning the timeliness and trustworthiness of the information. The 'Conclusions and Recommendations' are comprised within 13 pages, which are so much to the point that we regret being unable to quote more than a part of them elsewhere in this issue. One-third of the oil produced in the United States comes from California; but the production of this State is decreasing, while the consumption is increasing, so that the entire available storage of fuel-oil will be exhausted by June 1919. The question is how to stimulate production and to conserve the produce. A dearth of equipment hinders drilling; it is suggested that the Federal government take steps to expedite the making and the delivery of such equipment. Much of the unexhausted oil-land is unavailable for exploitation owing to litigation with the Government, notably the most desirable area, known as Naval Reserve No. 2; but there are considerable tracts of oil-bearing ground that could be opened to exploitation if the Attorney General were to sanction some arrangement whereby the rights of the claimants were protected. One obvious measure of conservation is to place an embargo on all exportation of Californian oil outside the United States, although this is a step so drastic and likely to injure established industry so greatly as not to be taken without careful enquiry. The substitution of coal for oil wherever practicable should be supplemented by more liberality in the distribution of hydro-electric energy for fixed plants. This is a matter within the province of the State Railroad Commission, through its rate-fixing power. As Mr. Thelen is chairman of that commission, a remedy is not far to seek. The report acknowledges the cordial assistance given to the Committee by Mr. E. J. Justice, special assistant to the Attorney General of the United States and in charge of the litigation arising over the withdrawn oil-lands. Mr. Justice has died since this acknowledgment was made; he is succeeded by Mr. Henry F. May, whom we happen to know well. Mr. May is a lawyer of ability and distinction, a graduate of Harvard and a highly respected citizen of Denver; he is a man of rare intellectual honesty and peculiarly fitted to see that right is done in the many legal complexities that have developed out of President Taft's order. We hope he will be able to expedite a decision in these cases, so that the oil industry may be relieved of an incubus.

Another Mexican Crisis

Events of the highest importance are transpiring in our relations with Mexico. So far-reaching are the arrangements now being made that it may be affirmed that a new era has begun. The first public announcement of the changed attitude of Carranza's government was given in a dispatch from Washington on September 12. It stated that the last steps in the formal recognition of Carranza as president of the Mexican Republic had been taken by the United States. Although the authority of Carranza as *de facto* head of Mexico was recognized many months ago, followed by accrediting an ambassador to represent us before the provisional government at Mexico City, the status of our international relations was not permanent. It was still uncertain whether the Mexican Government would assume a friendly attitude toward us in our conduct of the War. The German influence south of the Rio Grande was a serious menace, and the circular letter from Carranza to the other Latin-American countries proposing an embargo upon commerce under the plea of perfect neutrality was logically construed as indicating a hostile feeling toward us in our struggle with the Central Powers. It was inevitable, however, that Mexico could not long maintain economic independence of the United States unless Germany should promptly achieve a decisive victory. At that time the ruthless submarine campaign had just begun, and it is easy to understand that the representations of German agents may have made the possibility of success by their country against the Allies seem plausible. To take the most generous view possible of the evident animosity of Mexico at that time, it is conceivable that Carranza, seeking to play safely so as to be *persona grata* to the winning side, should choose an enigmatic position. As the War went on, and the submarines failed to cut off American supplies from France and England, the needs of Mexico required a change of policy. This became manifest chiefly through the persistent efforts of Carranza to obtain a substantial loan from this country. The dispatch from Washington on September 12, previously referred to, definitely declared that no loan would be made by the United States, but that the question of "a loan still is under consideration between the Mexican Government and a group of American bankers." The fact that important negotiations are in progress is revealed between the lines of this semi-official announcement. From sources close to the governments of both countries it may be accepted as practically certain that not only has the United States obtained guarantees that admit of an endorsement of such a loan, but that a tripartite agreement has been reached in regard to it between this country, England, and France. It is understood, furthermore, that the diplomatic engagements entered into have settled many difficult questions concerning claims for indemnity that these powers have preferred against Mexico. It will thus serve to smooth a wide expanse of troubled water, and to remove possibilities of international friction that have been freely discussed, arising from an assumed responsibility on the part of this country on account of its insistence that European powers should not intervene in Mexico for the protection of their nationals. Apparently the administration at Washington has discovered in the demands of Mexico for a loan some brilliant opportunities for promoting cordial relations between a great group of interested nations. It was definitely stated by parties in touch with the negotiations now pending that the loan would be made before the Mexican Independence Day on September 16. That would have enabled Carranza to utilize his success in a spectacular manner at a psychologic moment for creating a new spirit of popular friendship for America. Evidently there has been a postponement. But a loan appears assured. When it is arranged his own personal pride in this achievement will naturally be shared by the Mexicans themselves, so that it will go far toward producing an era of good feeling advantageous to them and also to our mining and commercial activities in that country. The amount of the loan is said to be $250,000,000 United States currency. Of that amount $150,000,000 is to be available for the Government in covering its deficit, which has been accumulating since last spring at the rate of nearly ₱5,000,000 per month. It will also enable the Mexican government to pay interest on its bonds, now in default, and to pay its soldiers, without which Carranza could not count upon a loyal army to maintain his authority. The remaining $100,000,000 is to be used for rehabilitating the National Railways. The deplorable conditions of these roads is shown by the fact that practically no freight is moving under Government administration on the National lines. Indeed, there is scarcely any rolling-stock remaining that is in condition for service. The American Smelting & Refining Co. is handling its freight over the Mexican railways by means of its own equipment, having in operation 20 locomotives and 400 freight-cars. The subsidiary companies of the American Metal Co., known in Mexico as the 'M. & M.' (La Compañia de Minerales y Metales, S. A.) is doing the same thing, as is the well-known house of Danti Cusi in Mexico City, and other lesser operators. The arrangement under which this traffic is maintained constitutes an example of hard bargaining that is extraordinary. The individual operators provide their own staff and crews, working in co-operation with the staff and crews of the National Railways, and they pay full freight-rates to the Government despite the fact that the goods are hauled in cars drawn by locomotives that all belong to the shipper. A natural result of this anomalous condition is to force shippers from small mines into the position of beneficiaries of the great smelting companies, through whose favor alone can they hope to market their ores. Since the smelters are primarily interested in moving the output of their own mines, the independent producer is willing to pay handsomely for the privilege of shipping his ore. It means extra profits for the smelter, but the difficulties of the small operator are chargeable primarily to the abnormal conditions that follow as the aftermath of a devastating revolution.

Returning to the loan, it is rumored on good authority that the vast sum of $100,000,000 for improvement of the National Railways is not to be turned over to the Government for disbursement, but that the money and the roads themselves are to be placed for a considerable period in the hands of a committee representing the holders of the National Railways of Mexico bonds, exclusive of the Mexico North-Western Prior Lien 6% bonds. It would seem logical that such a string should be tied to the money that will be advanced by bankers already directly and indirectly interested in the earlier bond-issues, although this restriction is denied by a prominent official of the Carranza government. The National Railroad Prior Lien 4½% bonds, due in 1926, are outstanding to the amount of $23,000,000. The interest due in January 1914 was paid by three-year 6% notes due January 1, 1917. Interest was defaulted in July 1914. The National Railroad First Consolidated 4% issue, due in 1951, amounts to $24,740,000 outstanding. The interest due in April 1914 was paid with 6% notes but the interest was defaulted in October 1914. The National Railways of Mexico Prior Lien 4½% bonds, due in 1957, are a first mortgage on 131 miles of road, a second mortgage on 3650 miles, and are a further lien on 2400 additional miles of track, covered by underlying bonds. Interest was defaulted on these bonds in July 1914. The National Railways General Mortgage 4% bonds, due in 1977, are a junior issue guaranteed as to principal and interest by the Mexican government. Interest on these was also defaulted in October 1914. The Mexican Government bonds represent an indebtedness of $140,830,211, the interest on which is also in default. The only other direct obligations of the Mexican government are $30,000,000 of 6% notes issued in 1913, and two internal loans. It will be seen that a correct application of the funds now to be placed in the hands of the Mexican government would re-establish Mexico's credit so far as relates to the payment of interest on her obligations, and would suffice for a material rehabilitation of her railway system. It would not, however, sustain the Government through the period that must elapse before the revenue from her reviving industry could balance her current expenses. Therefore it is thought that she will soon apply for an additional loan. It is certain that the Government is anticipating these future needs, and the difficulties that may be experienced in obtaining further accomodation on the heels of her first success in Wall Street, by undertaking to float a 'patriotic loan' in imitation of our Liberty Loan. It is to be nominally a voluntary offering by the people from the poorest laborers to the richest alien corporations. In effect, however, it will be obligatory, at least so far as the owners of the larger enterprises are concerned. Arbitrary schemes to recuperate the national treasury are by no means ended, and it is not to be expected that new investment of foreign capital will be made in Mexico until the country is in a sound financial condition. In its present extremity it must have recourse to excessive taxation, and that involves the necessity of building upon the ruins of the old rather than upon new industrial foundations. Those having property will be compelled to operate in order to save it; Carranza has frankly adopted the principle of use as the basis of valid ownership, and under the leasehold system this principle can be more easily invoked than under the American plan of issuing title to mining property.

The Mexican government has undertaken to rid itself of the burden of its fiat money in a manner that is effective even though it lack something in the quality of strict justice. The uncounted millions of currency reeled off from the presses during the heat of the revolutionary struggle were treated as a mere expedient entailing no obligation after having served its immediate purpose, but the later issues of Constitutionalist currency, engraved and printed by the American Bank Note Co., are recognized as representing at least some claim for redemption. Because these cannot be easily counterfeited they are called 'infalsificables', and it is chiefly because of this that the Government does not undertake to wholly repudiate them. Accordingly they are being retired, but in a unique manner. Specie payment is now obligatory in the Republic, and even the 'infalsificables' are not legal tender, but the Government has created a market for them by requiring that all duties, taxes, and imposts of every nature that are payable in coin, must be accompanied by an equal amount, in face-value, of this currency. It means nominally the payment of double taxes, but in reality the 'infalsificables' sell at the rate of about $8 U. S. currency for every ₱1000. Over ₱500,000,000 of this currency was originally issued, and the amount still outstanding is unknown. Thus the burden of revolutionary finance is being quickly and comfortably thrown aside, leaving the future untrammeled. With silver selling at 98⅝c. per ounce the Mexican dollar is quoted at 80c., and two Mexican silver half-dollars are worth 65 cents. Meanwhile the gold peso has an exchange value of 54 cents. The coining of silver in Mexico is proceeding rapidly and the Government is purchasing the output of many of the larger silver mines, especially from those in the Pachuca district. Until recently American coins, both gold and silver, were received in Mexico at a discount of 7½%. This made them desirable from a speculative standpoint, since they could be melted and the metal sold at a profit. The order by President Wilson on September 7, prohibiting the exportation of United States coin, will put an end to this practice.

It is generally conceded that Carranza is now seriously endeavoring to re-establish law and order, and to develop industry. The personal independence of the governors of the various States that threatened the future of the country until recently has been curtailed, in many cases by the voluntary co-operation of the local autocrats. This is another favorable indication of a return to stable government. The only States nominally Constitutionalist that persist in arbitrary local self-government are Yucatan and Oaxaca. Zapata still holds the greater part of Morelos and Guerrero, including the ports at the mouth of the river Brazos. A portion of Michoacan is also dominated by the Zapatistas. Villa remains in the field near Parral, and frequently interrupts service on the Central Railroad between Chihuahua and Torreon. The German propaganda is apparently waning in Mexico, although the sentiment of the people still inclines toward the Teuton. It is thought that public sympathies will be given a twist in favor of America and the Allies when the announcement of the loan is officially made. The activites of the American Metal Co., through its Mexican subsidiaries, have been credited with significance in connection with the pro-German movement, although the enormous investments made by this company will ultimately yield profitable results, no matter what turn affairs might take. It has purchased mines in all parts of the Republic, and is doing a huge business, including the operation of the coal mines of Aguijita in Coahuila, the old Madero smelter at Torreon, the Saltillo smelter, and the rcently constructed smelter at Monterrey. Up to the present time very little ore is coming to the United States from any sources in Mexico. The principal shipper is the Potosi Mining Co., operating in the Santa Eulalia district near Chihuahua. It is sending an average of 250 tons of lead ore daily to the El Paso smelter. This concern is a subsidiary of the company that owns the Britannia mine in British Columbia. The local manager is Mr. E. S. Plumb, and the financial agents are Moore & Schley of New York. It has operated without intermission throughout the revolution, a distinction that few mines in Mexico can claim. The American Smelting & Refining Co. is operating its smelters at Monterrey, Matehuala, and Aguas Calientes, and is working mines at Reforma, Monclova, the Bonanza in the Mazapil district, Matehuala, San Luis Potosi, Tepazala, Aguas Calientes, and Angangueo. It is complying with the 8-hour law, but not with the 7-hour law for night work. The workmen are generally dissatisfied, having had their expectations of an I. W. W. Utopia aroused by the new constitution. The strike at Monterrey has been settled, but the men are arrogant and difficult to control. Government officials are said to be reasonable, and to display a spirit of helpfulness in maintaining order and in quelling the rebellious tendencies of the laborers. In Sonora Governor Calles has openly declared that he will stand between the mining companies and the Constitution, but the labor problem is fraught with grave difficulties, and the problem before the Government in restoring favorable conditions for operating industries is most serious. There are many storms to be weathered before the new industrial era in Mexico can be fully established, but the outlook is improving.

FIG. 1. SURFICIAL RESULTS OF THE SLICING SYSTEM OF MINING

Miami: The Mining of the Ore—II

By T. A. RICKARD

PROSPECTING. When once the existence of the orebodies had been demonstrated by shaft-sinking it became economical and expeditious to ascertain the size and copper content of them by drilling. For this purpose the churn-drill proved most serviceable. The holes were drilled without any casing until the caving of the sides of a bore prevented progress, usually soon after ground-water was reached. The casing was then inserted and drilling resumed with the next smaller set of tools. At the start the hole was given a diameter of 10 inches and it was usually finished with a 6-inch tool, although in some instances it became necessary to employ a 4½-inch tool to reach the required depth.

METHODS OF MINING. Three systems of mining are in use at Miami; all have points in common, but, for the sake of differentiation, they may be designated

(1) Top-slicing,

(2) Shrinkage-stoping, with caving of pillars,

(3) Under-cutting, with caving of the orebody as a whole.

The first two methods are exemplified by the Miami mine, and the third has been adopted in the Inspiration mine. The size and the position of the orebodies relative to the surface are as follows:

	Miami		Inspiration
	Captain orebody	Main orebody	Joe Bush orebody
Depth of top of ore, ft......	130	380	115
Depth of bottom of ore, ft..	530	1,000	385
Maximum area of ore, sq. ft.	213,700	770,000	544,500

From the Tinkerville ridge, 200 yards west of the Miami No. 4 shaft, one can see how the surface has settled above the slicing-stopes in the Miami mine. These stopes cover an area extending 325 ft. south, 425 ft. north, 625 ft. east, and 25 ft. west of the old Red Rock (or No. 2) shaft, which penetrated the first big orebody. The ore has been excavated or 'sliced' for a maximum thickness of 200 ft., between the 220 and 420-ft. levels. The consequence is a steady subsidence of the overlying surface, the sinking amounting to 70 ft. The area of subsidence—its edges particularly—is marked by large cracks, as is shown in the accompanying photograph (Fig. 1). The big crack to the right does not extend to the main shaft, as might appear; it turns sharply to the left. The hill on which the little houses stand is solid ground. Just beyond or westward of the Red Rock shaft—now identified by the ruined water-tank—a hole made by 'piping' breaks through the cap-rock from a shrinkage-stope below. Such piping, or running of ground from the surface into the mine, never happens when employing top-slicing. The 'pipe' mentioned is connected with the old shrinkage-stopes in the south-west orebody that were completed some time ago and will not be considered in this article.

From the dump of the old Captain (or No. 1) shaft one can see the surficial results of shrinkage-stoping plus caving at the west end of the Miami mine in the Captain orebody, and of the under-cutting plus caving system at the east end of the Inspiration mine in the Joe Bush orebody. The two properties are contiguous, as shown by the map in my previous article. From the boundary looking eastward, so as to face the Miami end, the observer notes (as seen in the second photograph, Fig. 2) that the cracking and slipping of the ground on the Captain claim is at right angles to the strike of the orebody; the removal of ore underground has produced a step-like subsidence without disintegrating the surface, which is settling gently. This is due to good control of the ore while it is being withdrawn through the raises and the establishment of a horizontal movement, as will be explained later. Looking the other way, toward the Inspiration, one notes how the step-like slips cease abruptly at the boundary, from which westward there extends a V-shaped pit, about 75 ft. deep, having its longer axis along the strike of the orebody, as is shown in another photograph, Fig. 3. This is the hole made by the caving of the surface above the Inspiration stopes; and it serves to distinguish, superficially, the method of excavation in use in that mine. The method is adapted to the conditions, as we shall see; and that is the best that can be said of any method. The 'cap', or overburden of leached and impoverished rock, is disappearing into the open pit and mingling with the sulphide ore of the deeper ground, diminishing its richness but permitting it to be mined extremely cheaply. A steady widening of the pit follows the drawing of the overburden.

Looking farther westward, another pit is detected; this marks the top of the caving ground on the Copper Wonder claim, a part of the Inspiration property. The hills are becoming scarred, but, at their worst, they do not suggest the big holes being made underground. On the south slope of the Inspiration's big pit, I was shown the frame of the testing-plant in which the original wet-milling experiments were made and the first 15-ton flotation plant tried. But that is another story.

SLICING. This method of mining is applied to the Miami main orebody, which is mostly in schist. The contact of this schist with the overlying conglomerate follows the line of a fault, the vertical throw of which has not been determined, but it is at least 2000 ft. A dike of granite runs through the orebody but it is mineralized just like the schist, and as richly. This tongue of granite connects with the massive intrusion constituting the matrix of the Captain orebody.

Stoping is in progress from 380 ft. below the surface to the 570-ft. level. On that level the orebody, which is in schist, is 700 ft. wide from east to west and 1100 ft. long from north to south, and averages 2.4% copper. The first thing to do was to reduce the orebody to a horizontal plane and then form a mat to serve as the cover to the subsequent slicing. Therefore the work began with placing square sets across the full width of the ore at a shallow level, followed by stopes that were protected by stulls at the back. Thus, in starting the slicing system, the first step is to remove the peaks and irregularities by ordinary square-set stoping, reducing the top of the orebody to a floor large enough to serve as a unit. See Fig. 4. The square sets are blasted as soon as the ore has been removed, so that the timber may form part of the 'mat', which is the cover of shattered timber that serves as a roof above the horizontal stope, or 'slice', in which men are at work.* See Fig. 5. The next step is to stope a depth of 10 ft. under the upper floor, which is supported by round posts and caps, the mat acting as an overhead lagging and also as a bridge or roof between the broken cap and the next slice, a stope 10 ft. high over an area 250 ft. square. The broken 'cap' follows each successive slice as the excavation is progressively extended downward. Access for men, timber, and supplies is obtained through a central two-compartment raise. From this 'supply-raise' four drifts at right angles are run to the extreme limits of the 250-ft. block. These drifts are as high as the slice is deep—namely, 10 ft. Then stoping is started at the extreme corners of the block, the excavation retreating as rapidly as possible toward the central raise, so that the drifts may be kept open for the passage of men and material until the moment when all the ore has been removed. See Fig. 6. The ground settles as the ore is under-cut. No attempt is made to keep open a stope any larger than 50 ft. square; as soon as such a stope has been made it is the practice to blast the timbers in order to cause a subsidence of the mat and broken overburden, and promote a complete settlement of the ground to the slicing-floor before another stope is opened adjacent to the preceding excavation. If a stope is extended beyond the 50-ft. limit it will be necessary, in order to prevent dangerous caving of the back, to build numerous bulkheads. To avoid both the danger and the expense, it is found advisable to cave the stope, start another 25 ft. away, and then advance toward the former. See Fig. 6 again.

When about to blast the timbers, the foreman marks (with chalk) those that are giving the largest measure of support and that therefore when shattered will liberate the most weight. Into each post marked by the foreman a 1½-inch hole is bored by a power-auger; this is loaded with a 1¼-inch stick of dynamite, primed with an electric-blasting cap, connecting either with a blasting-battery or the lighting-circuit. It is customary to 'shoot' 100 posts in an area of 2500 sq. ft., or about two posts out of three, on the average. This proportion, however, varies widely according to local conditions. The blasted timber falls with the unblasted into the mat and serves in turn to form a roof for the next slice underfoot.

In the early stages of slicing no effort is made to recover timber because it is necessary to accumulate enough to form a thick mat, but once the mat has reached a sufficient thickness, about four feet, varying however with the physical character of the cap, it becomes advisable to recover a portion of the timber. The small pieces of wood, broken by the movement of the ground, pass with the ore to the crusher at the surface, where they are thrown to one side for the use of the employees, who are glad to have them for domestic fuel.

When a raise is choked, so that the ore hangs, it is customary to blast by means of a dynamite cartridge or two—sometimes five or six—tied with string to a stick 1 by 2 inches in section and 10 ft. long, as shown in Fig. 7. Strong cord, cut to length, is provided for this purpose. If this were not done, the men would use fuse as a binder. The cartridge, of course, has fuse attached to

*Mr. Maclennan showed me a piece of 2-inch plank that had been in the 'mat' for 10 months; it had become compressed to less than half its original thickness.

FIG. 2. SUBSIDENCE OF GROUND ABOVE SHRINKAGE-STOPES

FIG. 3. OPEN PIT ALONG STRIKE OF OREBODY FORMED ABOVE CAVING UNDERGROUND

it. The fuse is 'spit' before the stick is shoved into the raise.

The maximum length of wheelbarrow work is 25 ft., the chutes being 50 ft. apart. Ore that is not more than 10 ft. from a chute is shoveled into it, the distance of shoveling varying according to the strength, reach, and willingness of the man doing the work.

SHRINKAGE. This is the system applied to the Captain orebody, which extends vertically 400 ft. and has a horizontal section of 250,000 sq. ft., namely 500 ft. across and 500 ft. along the ore-belt. The Captain shaft was sunk in a small body of schist entirely surrounded by granite; at 170 ft. this shaft passed into the granite, being continued to a depth of 205 ft., where it was stopped. Subsequently the shaft was re-opened and sunk to 500 ft. in order to serve the orebody found by churn-drilling from the surface. The Captain claim covered 4,000,000 tons of ore. No wonder that the engineer at this shaft comes to his work in a motor-car. I saw it standing, exposed to the Arizona sunshine, beside the shaft-house.

The shrinkage system consists essentially of a series of narrow vertical stopes and partitions; part of the ore being removed, and the remainder settling, while mining operations retreat horizontally, instead of vertically downward as in the 'slicing' system. Stoping is started at a depth determined mainly by the shape of the bottom of the orebody, but also by a consideration of the tonnage to be extracted from that orebody. In ground of the character of the Captain orebody there is no known limit to the depth of subsidence. From this base-level the narrow stopes (10 to 20 ft. wide by 500 ft. long, spaced at 25-ft. centres, thus leaving partitions between them of from 15 to 5 ft. thick) are extended upward until the ore overhead begins to cave. Work in the Captain orebody began on the 270-ft. level, the top of the ore being at 70 ft. below grass-roots, so that 200 ft. of backs had to be removed. About 20% of ore is stoped in order to induce the remainder to break away; in other words, five tons is mined by nature for every ton that is mined by man.

Stoping on the 270-ft. level began at right angles to a series of parallel drifts 25 ft. apart, centre to centre, at the western or Inspiration end of the orebody. This stoping was extended eastward in the form of a succession of stopes 25 ft. apart—narrow rooms, about 12 ft. wide, 500 ft. long, and about 100 ft. high. See Fig. 8. In order to allow for the extra space taken by broken ore as compared with ore in place, it is necessary to remove 35% of the broken ore in each vertical slice. As soon as this stoping had reached a safe distance, the withdrawal of broken ore began at the innermost or extreme west end, with the object of establishing a plane of contact between broken ore and 'cap' at an angle of 60° west. When this had been done, the withdrawal of ore was continued, retreating eastward in the wake of the former stoping. These operations started the caving in the upper part of the orebody—the part untouched by the miner—such caving being facilitated by the fact that the first stope, in the extreme west end, was carried up high enough to ensure the settling of the ground from the grass-roots. When this was being done some fence-posts and barbed wire were drawn through the chutes at the 270-ft. level; their presence in the broken ore looked ominous of an undesirable dilution with waste, but the effect was intentional; it indicated that one essential factor had been established; it proved that the ground was caving all the way to surface, creating a void at the western end so as to cause a horizontal movement of the broken ore toward it, thus accomplishing the two purposes essential to the success of the method, namely:

(1) The overturning and breaking of the partitions left between the vertical slices of stoping, thereby ensuring the withdrawal of this remaining unstoped ore.

(2) Checking the formation of 'pipes,' or runs, through the broken ore to the surface, with consequent dropping of cap-waste into the withdrawing-chutes.

Thus dilution of the ore and interference with the withdrawal of the adjacent ore were prevented. The effect of a 'pipe' may be inferred from the fact that although only 6 ft. in diameter it may reach all the way to the surface, 250 ft. above.

The stopes were carried upward for about 100 ft. only, or scarcely half-way through the back. By the time they had reached that far upward an inspection of the ground overhead, by means of sub-levels entering from the side, indicated that the overlying ore would break by subsidence without further mining. The condition of the unworked ground overhead is watched carefully by means of the access given through these sub-levels, which are 100 ft. apart horizontally and about 30 ft. apart vertically. In order to trace the movement of the broken ore during the withdrawing operations, several thousand blocks of wood were marked and introduced into the stopes or they were placed in sub-levels where the stopes had not reached sufficiently high. The blocks were placed at points 50, 100, and 150 ft. above the 'draw-off,' or withdrawal, level. A differently shaped block was used at each of the three levels mentioned and on each block its latitude and departure were inscribed, using a cutting-tool called a 'scribe.' The blocks introduced at 50 ft. upward were 6-inch cubes; those at 100 ft. were 6 by 6 by 8 in.; and those 150 ft. up 8-in. cubes. The blocks came out almost intact; they were not so much bruised as encrusted with finely crushed ore. By knowing the exact position of the chute through which these blocks were subsequently withdrawn the movement of the ore accompanying any block could be traced. By this means it was proved that a horizontal movement of about 20 ft. westward had been established.

The raises are placed at intervals of 6¼ ft. on alternate sides along the withdrawal-drifts, which themselves are 25 ft. apart. Thus the raises tap the broken ore at 12½-ft. intervals in every direction, so that each raise serves for a block 12½ ft. square. The chute itself is 32 inches wide.

When the 'cap' starts to run through a chute and the records show that the expected tonnage of ore has been

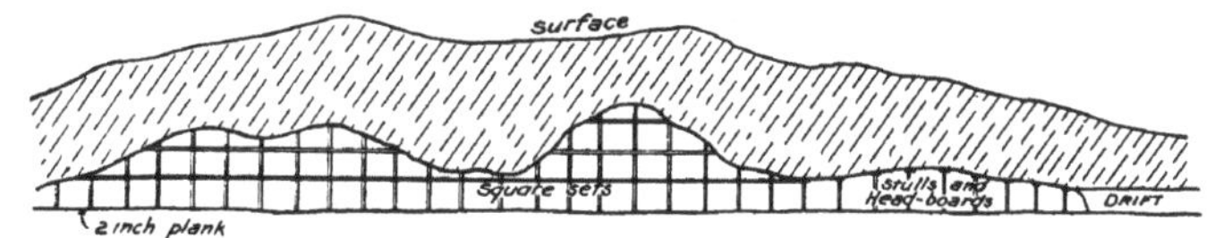

FIG. 4. PREPARATORY WORK

withdrawn, the opening is blocked and the chute is removed to the opposite side of the drift, in a space left between the sets. Usually it is possible to draw an additional 150 tons of ore before the waste comes down, so that the ultimate distribution of chutes is only 6¼ ft. apart east and west, but 12½ ft. apart north and south. This multiplicity of chutes may seem excessive to those inexperienced in this kind of mining, but, of course, the object is to withdraw the ore as cleanly and as completely as possible, excluding any waste. In practice it was found that in first starting to withdraw ore the 'cap' might begin to run through a chute before 25% of the estimated tonnage had been drawn, but that was stopped as soon as the horizontal movement of the broken ore was established, as previously explained. A careful record is kept in order to measure the ore withdrawn at any chute and this showed that the cap had come down prematurely and that further ore remained to be drawn through this particular chute. By sealing this chute and by drawing from the surrounding chutes it became practicable to break the walls of the pipe of waste that had been formed through the broken ore and resume the withdrawal of ore after a small tonnage of waste—15 to 20 tons—had been removed. As J. G. Flynn, the efficiency engineer, remarked: "Once in a while we have a chute running red before time." The cap consists of red schist, which affords a contrast with the gray granitic matrix of the orebody. Fig. 9 shows how a pipe of waste spills into chutes, *B* and *C*, on either side, blocking the withdrawal of broken ore. It should be clear that in this system of mining, when the withdrawing-period has been reached, there are no raises remaining above the withdrawal-level. Such chutes as *A*, *B*, and *C* in Fig. 9 are simply openings in the back of the level through which the broken ore is drawn. This broken ore lies directly upon the drift-timbering. The horizontal movement, of which mention has been made, prevents the establishment of such a pipe as is shown in Fig. 9 by breaking the continuity of it.

The records shown to me by F. W. Maclennan (McGill '98), the mine-superintendent and assistant-manager of the Miami mine [to whom I am indebted for the details of my description of these mining methods], show that

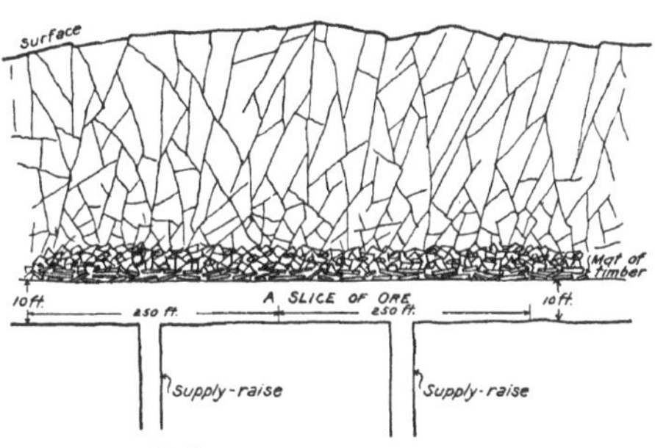

FIG. 5. THE MAT OF TIMBER ABOVE THE SLICE

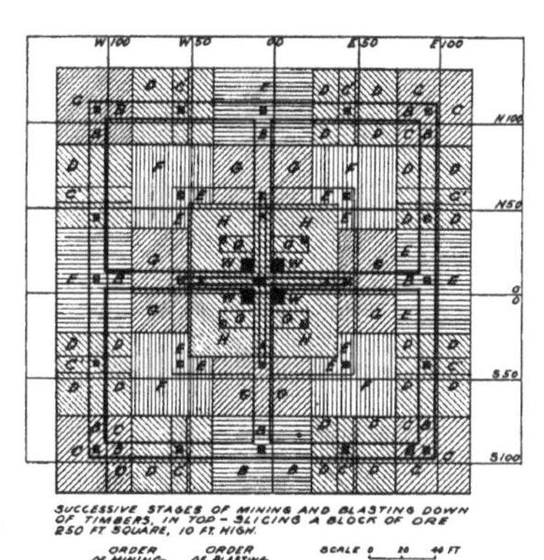

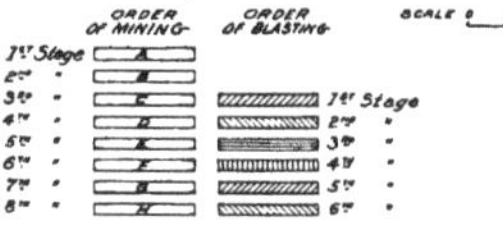

FIG. 6

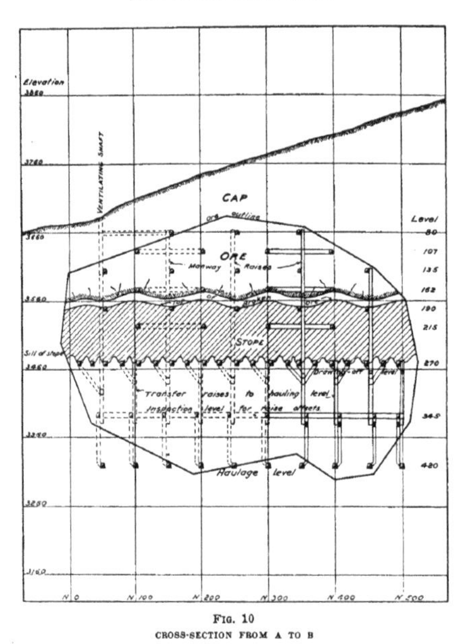

FIG. 10

CROSS-SECTION FROM A TO B

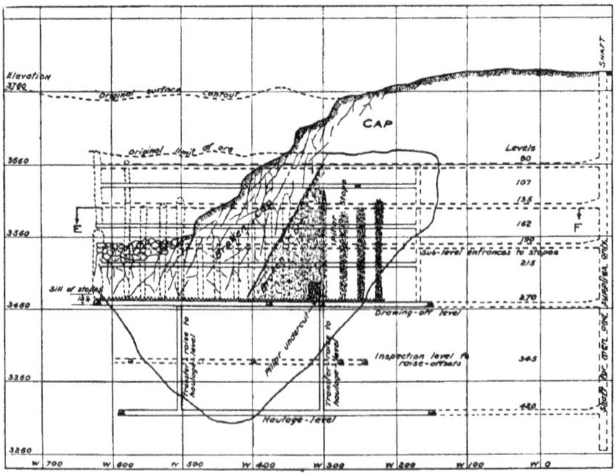

FIG. 11

LONGITUDINAL SECTION OF SHRINKAGE SYSTEM

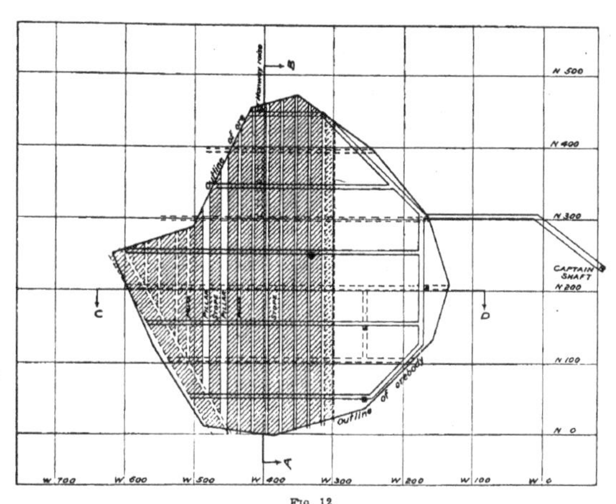

FIG. 12
HORIZONTAL SECTION THROUGH E F

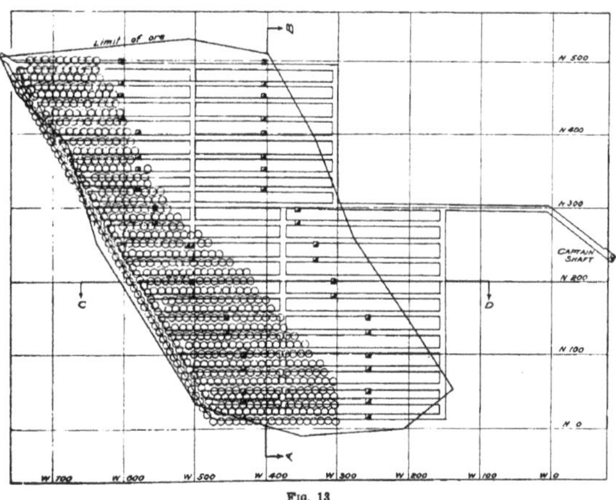

FIG. 13
PLAN OF DRAWING-OFF LEVEL, SHOWING WHERE CHUTES HOLE INTO THE STOPES

more ore has been withdrawn from the ground completely stoped than was expected. This gratifying result is due probably to an under-estimate of the original height of the orebody and to an admixture of waste. The assay of the ore drawn compared with the sampling of the orebody shows a decline in grade from 2.05 to 1.90% copper. The smallness of the decrease indicates that the dilution of ore by waste has been relatively slight, and that the operation has been highly effective.

The accompanying diagrammatic plan and sections will serve further to explain this system of mining. See Fig. 10, 11, 12, and 13. The longitudinal section (Fig. 11) was prepared for me by Mr. Maclennan to show the progress of operations to date. It will be noted that the pillar immediately west of W 300 has just been undercut and the pillars east of this point are left intact in order not to throw weight on the withdrawal-drifts until such time as it is necessary to break these pillars for the purpose of extracting what ore they contain. In this way the length of drift in which expensive repairs are necessary is only the base of the triangle between W 400, W 300, and the top of the orebody.

The terms 'chute' and 'raise' must not be used indiscriminatingly. Mr. Maclennan tells me that he and his staff make "a hard and fast distinction" between them. He writes: "The term 'raise' is applied to an upward opening either through solid ground or through broken ground, but in either case it is an opening with well-defined walls which it is our intention to maintain. The term 'chute' we apply to the arrangement at the opening into a drift or other working through which ore may be passed and controlled, whether this opening is at the bottom of one of our transfer-raises from a haulage-level such as is in use in our top-slicing or whether it is the opening directly to the bottom of a mass of broken ore, such as the chutes on our drawing-off level in the Captain orebody. To make this more clear a parallel may be drawn from a water-system. If we imagine a tank set 20 ft. above the ground and from it a pipe at the lower end of which there is a tap, then this pipe may be likened to a 'raise' and the tap to a 'chute' in the top-splicing system, whereas to be comparable with the chutes in the shrinkage system of mining in our Captain orebody, the tap would have to be placed in the bottom of the tank itself." This distinction, of course, is one that should be maintained. By restricting a technical term to a definite use we give it a precise meaning. Many descriptions of work underground are rendered confusing by the reader's inability to ascertain whether a 'chute' is meant to indicate the chimney in the ground or only the gate to that chimney. Let us call the chimney a 'raise' and the gate a 'chute.'

TIMBER. While underground I noticed an aromatic fragrance due to the use of cedar. This cedar comes from Coos Bay, in California, and is called the Port Orford cedar. It is used in timbering permanent ways, because experience has shown that the life of Oregon pine (the Douglas fir) is only four or five years, whereas evidence goes to show that the cedar will last at least three times as long. The adoption of it is an economy because it saves the re-timbering of the main level during the time it is required for haulage. Steel and concrete are used wherever, in the principal openings, the ground is heavy or where several openings converge. At such points it is customary to construct concrete piers with steel caps of 8-inch H-section. Steel was tried in one of the withdrawal-drifts, but it was not successful, because not even steel will hold the weight of settling ground and any repair is more difficult where steel is fixed than where sets of timber have been placed. At one time steel caps were bolted to the posts, but the movement of the ground sheared them off; since then spreaders—plank two inches thick—have been used under the steel cap. Various timber-protectives have been tried, such as (1) creosote, both the full-cell (using 12 lb. creosote per cubic foot of wood) and the open-cell (using only 6 lb.) method; (2) burnetized timber, subjected to heat, vacuum, and then impregnated with zinc chloride at the rate of 6 lb. per cubic foot. This second method is best adapted to timber not exposed to water, for zinc chloride is soluble. 'Date-spikes,' nails made of galvanized iron, with the date impressed upon their heads, are driven into the timber in order to furnish correct information concerning their length of service in the mine.

The timbering of No. 4 shaft has been creosoted. Once done, this is satisfactory, but the job itself is unpleasant, because the creosote makes the wood slippery and the fume is hard on the tear-glands, like the lacrimosal shells used today on the battlefield. In 1910 when the Giroux shaft was being sunk in the Ely district the men struck on account of the unpleasant effects of the creosote, so that this treatment of the timbers was abandoned after ten sets had been put in place.

VENTILATION has not been ignored. At the 570-ft. level a blower drives air into the top-slicing stopes, through one set of raises, while another set of raises is connected with an exhaust-fan placed on the surface at the top of a special air-raise, thus ensuring a constant circulation of fresh air. The top-slice is hard to ventilate because there is no opening through the back of it; it is large laterally; the air is warmed by the heat, partly chemical and partly mechanical, generated in the mat of timber overhead. In 1914 a fire started in No. 4 shaft, at about 150 ft. below the collar, owing to short-circuiting of electric cables within a 3-inch iron pipe. The heat of the electric arc fused the pipe and set fire to the shaft-timbers. Now the electric conductors are carried into the mine through a cased churn-drill hole of 8-inch diameter. In these mines at Miami, as in all mines, one is reminded of the fact that the consumption of air by the drills is an insignificant part of the output of a compressor, as likewise is the cost of power used in underground electric haulage when compared with the total cost of haulage, it being slightly over 1%. Comparisons based on the consumption of either air or electricity for such purposes in order to prove the efficiency of a particular type of machine are too often fallacious. The difference of consumption between one machine and another is negligible compared with the use in other directions or the wastage.

LABOR. The daily output of the Miami mine is 6300 tons; this is done by 760 men, so that the output per

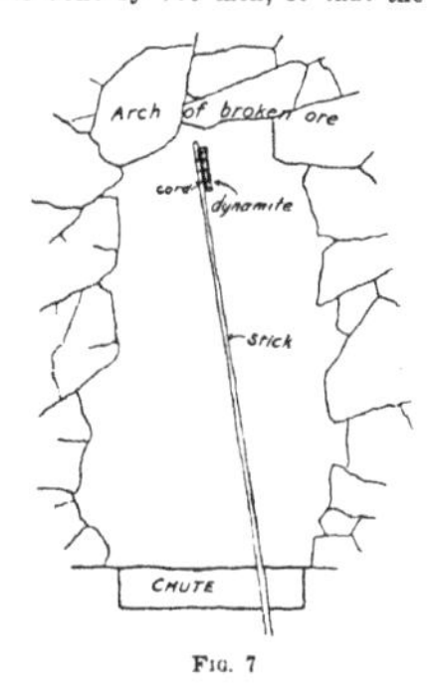

FIG. 7

man is 8.3 tons. In the shrinkage-stopes the output per trammer is 40 tons and per drill-shift 1905 tons. Once caving begins in the stopes, drilling is needed only to 'bulldoze,' or break the excessively large pieces. In the

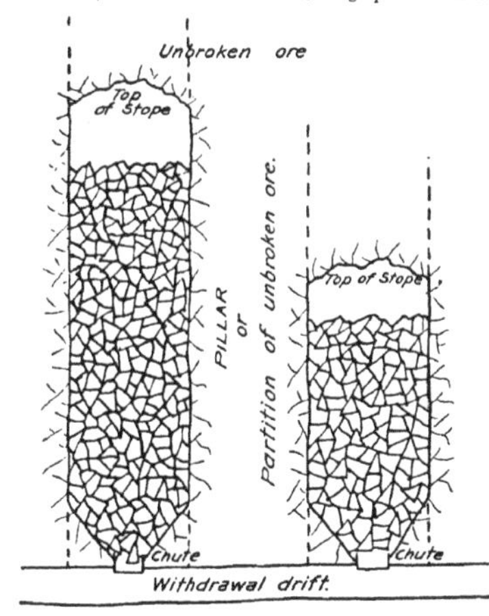

FIG. 8

top-slicing stopes the output per miner is 47.2 tons, and per shoveler 16.4 tons. The consumption of powder is 0.31 lb. per ton of ore and of lumber 7 ft. board-measure. At the surface 13 men are employed in the timber-framing shop, 8 in the blacksmith-shop, and 11 in the machine-shop, making 32 in all. The cost of dynamite, fuse, and caps is 0.7c. for shrinkage and 6.22c. per ton of ore for top-slicing. The cost of timber, lagging, boards, planks, and poles is 4.28c. for shrinkage and 22.54c. per ton for top-slicing. Shrinkage requires 229 drill-shifts per month and top-slicing 1453 drill-shifts in the same period. A drill-shift is worth $2.496 in air, drill-repairs, sharpening, pipe-line, machine-oil, and purchases of drill-parts, but not including the wages of the machine-man. The total underground labor cost is $130,000 per month, of which $102,000 goes to stoping;

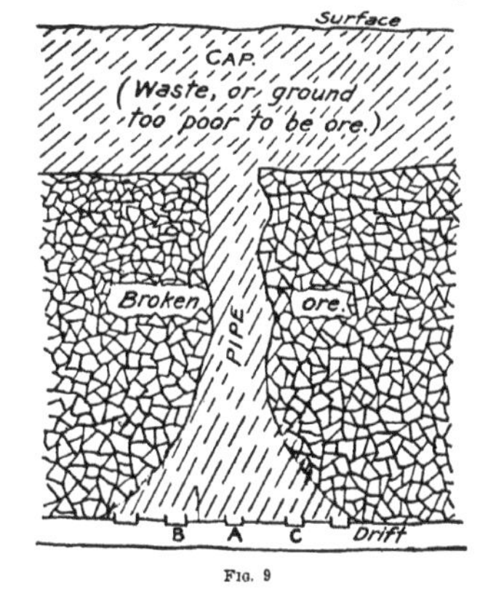

FIG. 9

800 men are employed underground and of these 630 are in the stopes. Based on ore delivered to the haulage-level, the output of ore by slicing is 8 tons per man and of shrinkage 21 tons.

The working force for the whole mine consists of 4 foremen on salary, 30 bosses on day's pay, 98 machine-men, 330 shovelers and chute-men, 216 timber-men and helpers, 40 miscellaneous, 65 in connection with electric haulage on the main level, and 30 hoisting, this last including engineers, firemen, oilers, cagers, skip-tenders, and top-landers, the grand total being 809.

(To be Continued)

THE United States is the largest producer and consumer of talc in the world, according to the U. S. Geological Survey. The softness, absorptive capacity, difficult fusibility and solubility, and electric resistance of talc make it one of the useful minerals in the arts and industries. Its principal military use is to prevent sore feet among marching soldiers. Although the War has stimulated production it has not greatly increased the price.

Method of Filing Drawings

By ALBERT G. WOLF

The several methods of filing drawings in mine-offices may be classified as follows:

1. Horizontally and flat { (a) In drawers. (b) On trays.

2. In vertical filing-cases.

3. Rolled { (a) In drawers. (b) In open racks. (c) In pigeon-hole compartments. (d) In tin cylinders.

4. Bound, on arch bill-files or by wooden strips on a rack, like newspapers in a library.

5. On spring rollers.

The first of these methods, that of filing flat in drawers, is in common use; it is considered the best by some engineers and the worst by others. It has the advantages of keeping the drawings flat and clean, and the sheet desired can be easily found. Some of the disadvantages, such as the difficulty of getting a drawing out from under a pile of others, or back again, without damage, or having to remove all drawings above it, can be minimized by making the drawers fairly shallow, and by having two or three sizes of drawers for different sizes of drawings. Weights can be used, if found necessary, to keep the corners of maps from being caught when moving the drawers in and out. The one disadvantage that cannot be overcome is the large floor-space required. If the set of drawers, however, be only table high, the top will serve for desk-room, and a well-finished set of drawers is not an unsightly piece of furniture in the office.

The tray system has the advantage that as the trays are closer together, many more of them than drawers can be put in the same space, hence, fewer drawings will be placed on each tray, and a closer subdivision of titles is possible. The set of trays can be enclosed in a cabinet with hinged doors, making the whole as dust-proof as any book-case. The use of trays, of course, presupposes that the drawings are absolutely flat. Drawings that have been rolled can be placed in large envelopes or containers, like those supplied with vertical filing-cases. Fig. 1 shows in elevation the method of constructing a set of trays. The horizontal dimensions may be anything desired up to four feet without the trays sagging materially. Two or three sets of trays of different sizes for various sized maps may be used, as with drawers.

Vertical filing-cabinets have the great advantage of large capacity for the small floor-space required. They also keep the maps clean, and the large number of compartments makes closer classification possible. The disadvantages are such as to make the use of the vertical cabinet undesirable when floor-space for drawers or trays is available. Lifting the large pockets, containing many drawings, in and out of the case is an inconvenience in itself. The small maps slip down, losing themselves between the larger ones, making it difficult to find them. The pockets, if heavily laden, wear out rapidly. If the drawings are all bound together at the top, instead of using the pockets, much trouble is involved in removing one of these drawings from the set. These vertical cases are furnished in only relatively large sizes, so that for small drawings they are a decided disadvantage.

Rolled maps in drawers take up a great deal of room, unless several maps are rolled together, in which case it is difficult to find the sheet wanted. All rolled maps, of course, have the disadvantage that they are harder to handle than maps kept flat. Very large maps, however, can only be filed without folding when rolled. Open racks are of two forms: one, consisting of vertical strips on the walls, with pegs projecting upward at an angle a little less than a right angle (Fig. 2). This method is hardly suited to filing a large number of maps because of the wall-space required, difficulty of finding a map, and unsightliness. A use for this rack is found in closets and vaults, and especially to support such drawings as are filed in tin tubes. The second form of open rack consists of a rectangular skeleton frame, as shown in Fig. 3. The numerous cross-pieces furnish support for smaller drawings. The over-all dimensions may be anything desired. The pigeon-hole compartment method is a variation of the skeleton rack, with the disadvantage that short drawings get pushed back into the compartments and are difficult to get out. Both forms have the objections inherent to methods of filing rolled maps; they do not keep the maps clean, and are not adapted to filing a large number of maps, or drawings, without crowding or rolling many maps in the same roll. The skeleton form has the advantage that maps of any width (length of roll) can be filed without having an unduly large frame.

Filing rolled maps in tin cylinders is adapted to only a small number of maps, but is an excellent method of protecting (from dust, moisture, and other injury) tracings of large maps that are brought up-to-date periodically for the purpose of making prints. As only a few maps are filed in this manner, the cylinders are conveniently placed on the peg-rack shown in Fig. 2.

Method 4, that of binding the drawings together and suspending in some manner, is one more adapted to grouping prints that are constantly referred to in the office, rather than as a method of filing original drawings or tracings, as its use necessitates the removal of many drawings before the one desired can be extracted, and the drawings are not protected from dust, etc. Two methods of binding are suggested here: the first, using the arch bill-file with a backing in proportion to the size of maps filed, and for large maps, two sets of arches on the backing instead of one; the second, by using two strips of wood, say, ½ by 2 in., held tightly together by bolts and thumb-nuts (Fig. 4B). These groups of maps are then suspended by means of the binding strips on racks (Fig. 4A).

The last method, that of using spring-rollers, can hardly be considered alone as a method of filing, but is an excellent way to proteet a few large general maps, such as a contour or geological map of a State, claim-map of a district, etc., to which constant reference is made. The maps are then as convenient for reference as the ordinary flat wall-maps, and less wall-space is consumed.

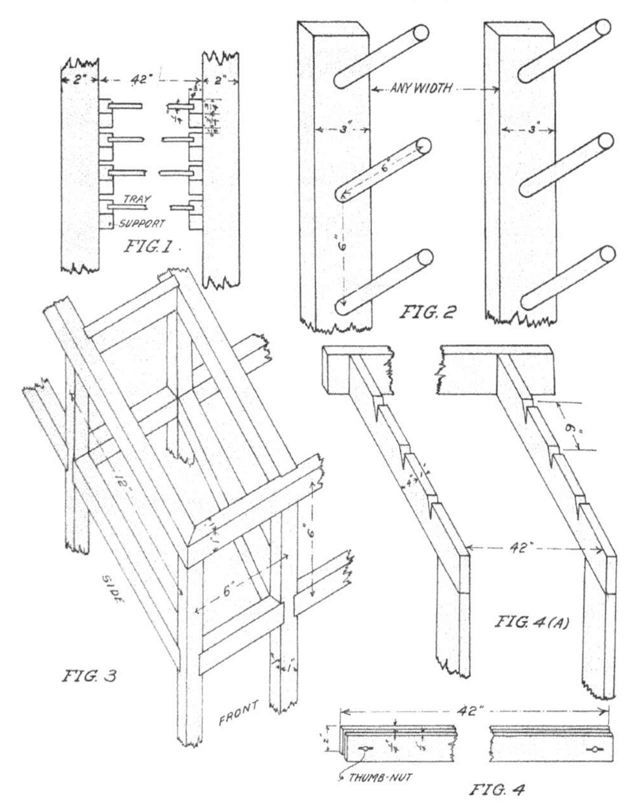

At most plants, no one system of filing will suffice unless the number of drawings is very small. Wherever any one of the systems of flat filing is used, it must be supplemented by filing in rolls the larger tracings that must not be folded.

Zinc-Dust as a Precipitant in Cyanidation

By W. J. SHARWOOD

*In the cyanide process, gold and silver are dissolved from crushed ore as double alkali-metal cyanides, from which they may be precipitated by such positive metals as sodium (amalgam), aluminum, or zinc, or by electrolysis. Two extreme conditions may be noted. Some works, especially slime-plants practising decantation, use a relatively large volume of solution—possibly 4 or 5 tons per ton of ore—nearly all of which may require precipitation, so that the solutions handled are of much lower value per ton than the ore. On the other hand, in some leaching-plants it is possible to extract with very little solution, and to percolate some of this more than once through the charge before precipitation, so that the solution to be precipitated may be much less than half the weight of ore, and proportionally richer. Solution intended for further use need not have all its precious metal removed, but any that has to be thrown away should be impoverished as far as is economically possible.

In spite of certain advantages possessed by other precipitants, zinc in some form has been almost universally used. In some of the first attempts to utilize cyanides as gold solvents, a 'piece of plate of zinc' was suggested as a precipitant, but extension of surface was early recognized as a desideratum.

Macarthur and the Forrests adopted a 'metallurgical filter' of zinc shavings, turned from disks or rolled sheets. They had previously experimented on other forms of zinc, and Macarthur records having tried zinc-dust, which had been long known as a general reducing agent. It had also been known and used as a precipitant of precious metal from plating and photographic solutions, and comminuted zinc had been patented for recovering copper and other metals from ore-leaches. Other inventors proposed the making of zinc powder by the attrition of balls of zinc, and by similar means, during the passage of a stream of gold-bearing solution.

Sulman claimed the use of zinc-dust, or fume, in a special apparatus, effecting a more or less regular feed of dust and solution by means of intermittent siphons; the mixture or 'emulsion' rising with diminishing velocity through an inverted cone in which it deposited most of its burden, and being clarified by passing a baffle-box and finally a cloth filter. This system was used at Deloro, Ontario.

The first cyanide plants in the United States to use zinc-dust were the Mercur (Utah), treating coarsely crushed oxidized or roasted ore, the Drumlummon (Montana), and Delamar (Idaho) tailing-plants, followed by the Homestake (South Dakota). All these treated large tonnages and used filter-presses to catch the precipitate; the first three, and the Homestake when first installed, used square presses of the Johnson type, but differing in size and design. Nearly all the larger cyanide installations on the American continent now precipitate with zinc-dust, and the Merrill triangular press has almost entirely superseded other forms for this purpose. Zinc-shaving is still used almost exclusively in South Africa, and in the smaller plants elsewhere. Aluminum-dust is used to a limited extent, chiefly on Canadian silver ores.

ZINC-DUST. The early method of applying zinc-dust was to fill a vat with pregnant solution, agitate, add the dust, and pump through the filter-press. The zinc was first introduced by stirring the body of solution mechanically and sprinkling the dust on the surface. A variation was the use of compressed air at the vat-bottom, through a small central cross or coil of perforated pipe. In a few seconds this set the solution in violent motion, and the zinc-dust was then scattered on it with a shovel. This was at best a disagreeable job, and the introduction of oxygen at the precipitating stage was opposed to chemical theory. Elimination of air, however, and cautious sifting of the zinc-dust over the surface before pumping, showed imperfect precipitation—for instance, some tests gave only 75% of the gold precipitated as against 96% with thorough agitation by air.

To obviate the use of dry dust, various attempts were made to add it in the form of an 'emulsion' or suspension in water or in cyanide solution.

Bosqui's zinc-dust feed-system has two relatively small tanks, alternately filled and emptied by means of a tilting launder that actuates a counter and throws a measured charge of zinc into the tank to be filled; the zinc is continuously stirred by jets of solution on a revolving agitator.

Mr. Mills has a set of vacuum filter-frames submerged in a vat through which the mixture of zinc and solution is circulated by a centrifugal pump, while settling is overcome by revolving rakes. At one plant such precipitate as settled on the tank-bottom was allowed to accumulate there, while only the suspended portion went to the filter-press. Two grades of precipitate were thus produced, the settled material being decidedly lower in value and harder in texture.

Elsewhere, when the solution was pumped nearly to the tank-bottom, a man entered the vat in rubber boots and swept settled material toward the pump-intake, a small amount of fresh solution being used to assist the sweeping out. In spite of this sweeping there was a tendency for hard lime-zinc scale to accumulate on the bot-

*A paper to be presented at the St. Louis meeting of the A. I. M. E. in October.

tom, and to some extent on the tank-staves, so that after six months there might be several thousand dollars' worth of solution thus tied up in a pair of large vats, only recoverable by periodical 'scaling' in which hammers and chisels were used. Pumping a large tank to the press might occupy two to six hours, and one might expect re-solution to take place during this period when all the zinc was added at the start. Repeated tests showed that, although the precipitation by mere agitation was far from perfect, very little re-solution took place during pumping, but the part played by the zinc accumulated in the press was evidently an important one.

The amounts directly precipitated ranged from 20 to 90%, while after passing the press 92 to 98% of both precious metals had been removed, but only 1 to 40% of the copper. Recently the positive removal of oxygen from the entire body of pregnant solution before adding zinc has been carried out on a working-scale. Long-continued tests show that, by subjecting the solution to a vacuum during pumping, a considerable economy is effected, both in the amount of zinc-dust consumed, and in the acid required for the subsequent refining. Patent has been applied for in connection with this modification of the process.

Turbid and cold solutions, extremely low in alkalinity and free cyanide, present the most unfavorable conditions for precipitation. The quantity of zinc-dust required per ton is not proportional to the precious-metal content, depending largely on the amount used in side-reactions.

CONTINUOUS PRECIPITATION. In adding zinc-dust to a moving stream, as it is termed in the Merrill patent, a uniform feed is essential to secure maximum efficiency, the volume added per ton being often so small that any diminution becomes temporarily fatal to precipitation. Feeders of various forms have been designed and used. One of the first and most satisfactory consists of a slow-moving horizontal belt on the surface of which is spread a charge of zinc-dust in a layer of uniform width and thickness. The zinc usually falls into a small mixing cone, through which an auxiliary stream of solution passes, carrying the zinc by a small pipe to the pump-intake. A slow drip of lead acetate or nitrate, or of strong cyanide solution, may be added here to facilitate precipitation. Many ingenious elaborations have been devised to secure uniformity in the fall of zinc and the flow of auxiliary solution and chemicals. Formerly the zinc and auxiliary solution in the mixing cone were continuously agitated by a small jet of compressed air. According to data published by Clark† a saving of about one third of the zinc was effected by eliminating this air-agitation of the zinc-feed, and preventing air from being drawn into the mixing cone.

Another good feeder has a cylindrical roller or pulley slowly revolving at the lower end of a hopper, conveying a narrow ribbon of dust, the thickness of which is controlled by an adjustable slot; or an auger-like horizontal screw may remove the zinc from a similar hopper. All such hopper-fed devices require jarring mechanism to prevent the zinc-dust from bridging. A miniature tube-mill has been introduced as a mixer to smooth out irregularities in the zinc-feed.

FILTER-PRESSING. Solution intended to be precipitated and thrown away may be run by gravity to a press at a low point; if for further use, it is generally pumped to a press at a considerable elevation, from which it falls to a storage-tank. If the distance between intake and press is considered insufficient to allow of complete reaction, it is increased by leading the pipe in a zigzag line. A zinc-press with gravity feed, the clear effluent elevated by a pump to storage, has the advantage over a pump-fed press in that zinc and precipitate are kept out of the cylinders and valve-chambers; the claim is, however, made that the pump-feed gives better precipitation with a given proportion of zinc.

A close filtering medium is necessary to retain zinc-dust and the extremely fine precipitate obtained from low-grade gold solutions. In early practice the medium was sometimes paper between two cloths, or a single thickness of chain-cloth—a rather expensive fabric. Chain-cloth or cheaper heavy canvas may be covered with a light cheap twill, which is removed and burned at each clean-up to recover adhering precipitate, while the heavy backing is occasionally washed or treated with hydrochloric acid to remove limey accumulations. Another plan is to use two thicknesses of medium-weight cotton-twill, the outer being taken off at each clean-up and either washed or burned. The other then becomes the outside cloth and a new or washed cloth is put under it for the next run. One square foot of net filtering surface for 1.5 tons (say, 50 cu. ft.) solution per day, or 6 tons per hour for 100 sq. ft., may be taken as a conservative ratio: with clear solutions a press may be run at double this rate for long periods. Colloidal suspended matter soon increases the pressure and reduces the pumping rate.

Two-inch distance-frames are suitable for a press used on gold solutions, 3-in. or 4-in. for silver. When a press is opened, most of the cake readily falls into the wheeled tray placed beneath and the remainder is removed by scrapers. It is normally soft but sometimes caked hard as a result of oxidation or the presence of calcium carbonate; this condition can often be controlled by excluding air from the solution.

It was early found advisable, and is still the custom, when starting up a zinc-press after a clean-up, to add to the first charge 50% or more zinc in excess of the weight normally required; this is gradually diminished in successive charges, until the regular amount has been reached. If the zinc-feed is cut too low at any time, no effect may be noted until several charges have been thus treated, then the 'barren' assays rise suddenly and it becomes necessary to add a considerable excess for several charges, until normal working has been restored.

†'The Mining Magazine' (April 1910), Vol. IV, p. 289.

Apparently a certain excess of zinc must be maintained in the press, or re-solution takes places there to some extent. This is evident when a press stands idle for several hours; the effluent samples, caught during the first few minutes of pumping, will be abnormally rich—occasionally richer than the pregnant solution entering the press.

FINENESS OF ZINC-DUST. The virtue of zinc-dust as a precipitant is explicable by its fine state of division or, what amounts to the same thing, its extended surface. Waldstein's patent claimed that the zinc oxide, invariably present, formed a beneficial galvanic couple with the metal, while Sulman's process involved the preliminary removal of oxide by a solvent. In early practice it was noted that dust containing 1 or 2% of lead was more effective than purer samples, and this was confirmed by laboratory tests on synthetic alloys. Cadmium seems to have but little effect. That fine division is the main factor is indicated by the fact that the finest unoxidized metallic zinc, made by grinding sifted filings, and levigating in absolute alcohol, can be made even more effective as a precipitant.

To get some idea of working conditions in using zinc-dust on the large scale, the particles may be assumed to be equal spheres of a diameter which we may take as 0.0001 inch. One pound of this assumed zinc-dust will then contain 7544 million particles and will expose 1650 sq. ft. of surface, which is much greater than the surface of an equal weight of shaving. In precipitating gold, the practical minimum of zinc-dust is probably between 0.1 and 0.2 lb. per fluid ton of 32 cu. ft., or 50 to 100 parts per million of solution. One tenth of a pound, uniformly distributed through a ton of solution, would give some 13,650 particles per cubic inch, spaced at an average distance of about 0.042 inch. If we could substitute particles of half the diameter, the same weights of zinc-dust per ton would give eight times as many particles per cubic inch, at half the distance apart. These considerations make evident the desirability of obtaining zinc-dust in a fine state of division, and the bad effect of stray coarse short or agglomerated masses. In strong solutions, rich in silver, it is probable that the economic limit of fineness would soon be reached; with gold solutions, is seems unlikely.

EFFICIENCY OF PRECIPITATION. Zinc-dust is sometimes valued by the percentage of zinc actually in the metallic state, estimated by the reducing effect on ferric sulphate or chromic acid; or the difference between total zinc and zinc as oxide may be taken as metallic. As a guide to its precipitating value this is insufficient, and the fineness, or the speed of reaction, must be considered. As a rule, a good zinc-dust will nearly all pass a sieve of 200 meshes to the inch, and in some zinc-dusts now obtainable very little over 1% is retained by a 300-mesh sieve. In either case, the coaser portion is merely accidental material—crystalline aggregates, a few shots, or foreign matter.

A precipitating efficiency of 100% assumes pure zinc, 1 atom of which should precipitate 2 atoms of silver or its equivalent in gold or copper from the double cyanide. At this rate 1 unit weight of zinc should precipitate 6.03 units of gold, 3.30 of silver, or 1.93 of copper, but in actual practice the results are much lower. Applied to commercial dust, the laboratory test usually shows an efficiency value of 30 to 60%. Some extremely fine and pure 'artificial' zinc-dusts, prepared by re-distillation of spelter, have shown laboratory efficiencies of 75% or more, and pure electrolytic zinc-powder is claimed to be equally good. Otherwise the efficiency is generally higher when lead is present to the extent of a percentage or two. Distinctly coarse or granular zinc preparations generally show low results.

Judged by the actual precious metal precipitated, the working efficiency on a large scale may be a mere fraction of 1% in the case of low-grade gold solutions. A considerable amount of zinc is always wasted in side-reactions, such as the evolution of hydrogen, reducing dissolved oxygen, or precipitating copper and lead. With rich silver solutions the efficiency may approach 50%, especially if no attempt is made to recover the last traces of silver.

A practical example of mixed precipitation may be taken from the records of the Drumlummon tailing-plant, covering three seasons, or 24 months of treatment of sandy tailing by leaching, the silver being largely in excess of the gold.

Sand treated	290,000 tons		
Solution precipitated	390,000 tons		
	Lb. Av.		
Precipitate obtained	104,000		
Zinc-dust used	113,600		
Metallic zinc in dust	102,240		
Zinc remaining in precipitate	40,300	= 39.4%	
Zinc dissolved	61,940	= 60.6%	
Gold in precipitate	2,380	= 0.39%	efficiency
Silver in precipitate	16,000	= 4.75%	efficiency
Copper in precipitate	7,800	= 3.93%	efficiency
Total efficiency (Au, Ag, Cu)		9.07%	efficiency

Percentage efficiency is calculated by dividing the weight in pounds by the electro-chemical equivalent (Zn 32.7, Au 197.2, Ag 107.88, Cu 63.5, and Pb 103.6, under these conditions), dividing the quotient by the number of equivalents of zinc used, and multiplying by 100. The metallic zinc in dust has been taken throughout at the approximate figure of 90%.

Another instance may be taken from the published results‡ of a months run (May, 1911) of Sand Plant No. 1 of the Homestake Mining Co., in which the silver is about 1% of the gold value. Some lead was added to the solution as nitrate and recovered in the precipitate.

The gold showed 1.18% efficiency, the silver, 0.73%, and the copper 0.19%, in the weak solution, making a total efficiency of 2.1%. The same total efficiency (Au, Ag, Cu) in the low solution was only 0.71%.

That such low efficiencies are tolerated in gold extraction is explained by the fact that, when 'low solution' is going to be thrown away, it is obviously worth while to extract the last two cents worth of gold recoverable, if this can be done at the expense of one cent

‡Trans. Inst. Min. & Met., Vol XXII, pp. 142, 149.

for zinc, making a fair allowance for refining cost. With zinc at 14.5c. per pound, 1 oz. Troy costs one cent. If this ounce is used to precipitate 0.001 oz. of gold (2c.) from a ton of waste solution the practice is defensible, although the actual chemical efficiency attained is less than 0.002%.

DUST v. SHAVING. The practical efficiencies obtained with zinc-dust have been, generally speaking, about the same as with zinc-shaving, and the accumulation of zinc in the solutions is about the same; greater variations occur between two plants using the different precipitants on similar ores.

The dust process involves a more expensive installation than the zinc-shavings, but has the advantage of greater compactness and cleanliness, and involves less labor in maintenance and cleaning up as well as less risk of theft. The periodical clean-up is absolute, while a hold-over of several thousand dollars' worth of precious metal commonly occurs with zinc-shaving, and makes it impossible to compare the actual with what is often called the 'theoretical' recovery. After a destructive fire, precipitate in a filter-press has been found intact, while zinc-boxes have entailed great difficulty in the attempts to recover their contents. At a gold-plant a press occupying a floor-space of 5 by 14 ft. can easily carry a month's accumulation of $40,000. The comparative cost of the two systems at any time depends, of course, upon the wage-scale and the relative prices of zinc-dust and spelter.

COMPOSITION AND TREATMENT OF PRECIPITATE. Except in the absence of coarse fibres in the former, there is no essential difference in the composition of zinc-dust precipitate and that obtained with shaving. Both contain metallic gold and silver, usually co-precipitated as an alloy, sometimes amalgamated with mercury or alloyed with copper or with zinc itself—some zinc being apparently re-precipitated electrolytically with the precious metal. Any lead in the solution is also thrown down, while any lead and cadmium in the zinc-dust remain with the undissolved zinc, which may form 20 to 70% of the dry weight of precipitate. Calcium carbonate, moisture, and fine-ore particles make up the total, with sometimes calcium sulphate or zinc ferrocyanide and basic cyanide. Lead, mercury, and silver may be present either as metals or sulphides, the latter condition resulting from sulphur compounds in the solution. Blowing air through the press to dry the precipitate causes rapid oxidation of zinc, and consequent heating. Dilute sulphuric acid, followed by thorough washing, removes most of the zinc and lime, and more or less copper, while lead, cadmium, mercury, and silica remain with the precious metals. The mercury may then be removed by drying and heating in a retort with a little lime. Imperfect washing leaves zinc sulphate, and this, like calcium sulphate or sulphides, yields a matte in the subsequent refining. Hitherto no way has been found to utilize the zinc-sulphate solution, impure and generally saturated with calcium sulphate, which is obtained in the acid treatment of precipitate.

Acid treatment, or lead refining and cupellation, or a combination of the two, is necessary before melting low-grade precipitate. Richer gold precipitate, and that from most silver ores, may be fused with a suitable flux, directly or after roasting, yielding fairly fine bullion.

German Cyanide

James W. Gerard, recently American Ambassador at Berlin, states: "Another commodity upon which a great industry in the United States and Mexico depends is cyanide. The discovery of the cyanide process of treating gold and silver ores permitted the exploitation of many mines which could not be worked under the older methods. At the beginning of the War there was a small manufactory of cyanide owned by Germans at Perth Amboy and Niagara Falls, but most of the cyanide used was imported from Germany. The American-German company and the companies manufacturing in Germany and in England all operated under the same patents, the English and German companies having working agreements as to the distribution of business throughout the world. The German Vice-Chancellor and head of the Department of the Interior, Delbruck, put an export prohibition on cyanide early in the War and most pig-headedly and obstinately claimed that cyanide was manufactured nowhere but in Germany. Therefore, he said, if he allowed cyanide to leave Germany for the United States or Mexico, the English would capture it and would use it to work South African mines, thus adding to the stock of gold and power in war of the British Empire. It was a long time before the German manufacturers and I could convince this gentleman that cyanide sufficient to supply all the British mines was manufactured near Glasgow, Scotland. He then reluctantly gave a permit for the export of 1000 tons of cyanide, and its arrival in the United States permitted many mines there and in Mexico to continue operations and saved many persons from being thrown out of employment. When Delbruck finally gave a permit for the export of 4000 tons more of cyanide the psychological moment had passed and we could not obtain through our State Department a pass from the British."

PRACTICALLY every industry requires the use of sulphur in some form, and the United States, to be industrially independent, must therefore have an adequate supply. The record of the last three years, according to a recent statement issued by the U. S. Geological Survey, shows that this country has enough sulphur to support its present industries even under the requirements imposed by the drain of a large foreign demand. The two main sources of sulphur are the native mineral and the sulphides. Each year at least 300,000 long tons of native sulphur and 1,250,000 long tons of sulphides are used, principally for making sulphuric acid.

Zinc-Burning

By W. R. INGALLS

*The manufacture of zinc oxide directly from the ore is one of the most important contributions that America has made to the metallurgy of zinc. Hitherto this has been done chiefly for the production of zinc and zinc-lead pigment under the name of the Wetherill process. The principles involved are capable of broader application and are discussed under the term 'zinc-burning'. In zinc-burning, either in blast or reverberatory smelting, the gangue of the ore is scorified and drawn off as slag while other metallic minerals are reduced to matte or metal. F. L. Bartlett first employed the blast-furnace and used an ore-column 18 inches high. In certain blast-furnace practice in the iron industry zinc-burning has been effected as an incident in the smelting of iron and spiegeleisen. The reverberatory offers advantages over the blast-furnace because there is no shaft to be clogged by accretions. Early attempts abroad to use the reverberatory were not commercially successful. Frederick Laist, at Anaconda, Montana, first used the reverberatory successfully for zinc-burning on a large scale. This was done early in the present year, for the extraction of zinc in the residue of the leaching from electrolytic zinc extraction.

The Wetherill grate-furnace, unlike a blast-furnace, produces a clinker and at the same time volatilizes the zinc and lead in the ore. Its operation is difficult and requires much skill where a pure pigment is desired, but for concentrating zinc no great skill is necessary. The process is so simple that attention is being increasingly directed to it. Its application is indicated for (1) the concentration of low-grade zinc ore, especially calamine, thereby saving freight and treatment charges on worthless gangue, and for (2) the recovery of zinc in the leached residue from electrolytic zinc works. It is possible that it may be found to be most economical in the electrolytic process to roast the ore, burn it on Wetherill grates, and leach the fume.

Whether the Wetherill grate or the reverberatory furnace is used will depend on the nature of the gangue, the precious-metal content of the ore, and the commercial and metallurgical conditions that prevail. Lead is even more easily burned out of an ore, either in Wetherill or reverberatory furnaces, than zinc. The fume from a lead-zinc ore will contain both lead and zinc. Sulphur in such an ore will cause the presence of sulphur in the fume. Silver in some ores will largely go into the fume and in others it will remain in the residue. The conditions determining the behavior of silver are not well understood. No hard and fast rules can be laid down as to choice between the Wetherill grate and the reverberatory. As yet the data on the subject are insufficient.

Reverberatory smelting for zinc-burning follows the same lines as for the treatment of copper ore. Sufficient carbon must be mixed with the charge to reduce the zinc oxide. Great excess over the theoretical amount is unnecessary, nor is it necessary that the reverberatory fuel be greatly in excess of what is required in copper smelting. The critical factor is the slagging of the zinc, but, by the maintenance of a neutral or reducing atmosphere the zinc content of the slag may be reduced to low figures and the extraction of zinc as fume may compare favorably with that obtained by Wetherill grates. It is important, once the oxide of zinc is reduced, to give it the best chance possible to be volatilized, and to allow oxidation above the bath, not in it. Good work in the reverberatory may make a slag with only 3% zinc and this compares well with the zinc content of the cinder made in the best zinc-burning.

In the use of labor and fuel the large reverberatory of the Anaconda type is superior to the Wetherill. With a smaller reverberatory the difference will be less. For the small plant treating ores containing no precious metals and of an essentially acid or basic type, the Wetherill grate is most advantageous. With ores of a self-fluxing nature or with appreciable precious-metal content the reverberatory has the advantage, since smelting can be accomplished coincidently with the expulsion of the zinc.

*Abstract of paper to be read at the St. Louis meeting of the A. I. M. E., October 1917.

MAGNESITE is reduced to magnesia either in 'dead-burned' or sintered form, or in what is known as 'caustic calcined' form. Dead-burned or sintered magnesite has been so strongly heated that essentially all its carbon di-oxide and moisture have been driven off and most of the shrinkage taken up. In this condition it is chemically very inert—that is, it is not subject to attack or disintegration even under extreme heat. The caustic form is not so thoroughly calcined; it still retains 1 or 2% of carbon di-oxide and is thus a product more like ordinary caustic lime in its properties, although not quite so active chemically. Caustic magnesite 'slacks' when exposed to the air, re-combining with moisture and carbon di-oxide. Combined with calcium chloride it forms a distinctive cement known as sorel or oxychloride cement, which is much favored by builders for floors and other places where tile or special finish is desired. This is probably the most important use to which pure caustic calcined magnesite is put, although it is used also for making liquors in which wood pulp is digested to make paper, as well as for other purposes.—U. S. Geological Survey.

COMPARING the mineral output of Japan in 1916 with that of 1907, the 'Japan Times' states that within 10 years a remarkable increase has resulted. Copper increased by 160%; lead, 269%; and iron, 194%. This has been accompanied by higher prices, so that output-values show increases as follows: Copper, 238%; lead, 561%; and iron, 340%. The aggregate output-value of zinc, antimony, and phosphorus mines shows an increase of 172%. Compared with the output of 1912 the increase is 124%.

Report of Committee on Petroleum

Some of the Conclusions and Suggestions

UTILIZATION. Far beyond the borders of California, her petroleum and its products play a vital part in commerce and industry.

California fuel oil operates the Panama Canal; the greater part of the steam railroads of Washington, Oregon, California, Nevada, Utah, Arizona, and New Mexico; the steamship lines along the Pacific coast from Mexico to Alaska and across the ocean to Hawaii; the artificial gas plants of California, Oregon, Washington, Nevada, Arizona, and Hawaii; the mines and smelters of California, Nevada, and Arizona; the cement plants and sugar refineries of California; and a substantial portion of the manufacturing, industrial and agricultural enterprises of the Pacific Coast States of the Union.

California fuel oil supplies the Pacific coast fuel requirements of the United States Navy and Army and of the various state and municipal governments.

Radiating from California, north, west, and south, California fuel oil, to a considerable extent, operates the railroads and industries of western Canada; the sugar refineries and railroads of Hawaii; and the railroads, steamship companies, mines and smelters of the west coast of Central and South America.

The products of California petroleum, such as kerosene, gasoline, distillates, lubricants, and road oils meet the requirements of Arizona, California, Nevada, Oregon, Washington, Alaska, and Hawaii. California kerosene is shipped in enormous quantities to China, Japan, India, and Australia and to the western coast of Central and South America. California distillates and lubricants are sold in nearly every important State of the United States, as well as in England, Canada, and Australia.

Important as is the part which California petroleum and its products have heretofore played in the industrial life of the nation during times of peace, much more important is the part which the State should play and will play, if possible, in meeting the emergency in the supply of fuel oil, gasoline, and lubricants, created by the War.

PRODUCTION. California produces between one-fourth and one-fifth of the world's supply of petroleum. One-third of the entire supply of the United States is produced in California.

The year of California's greatest production was 1914, in which year 103,620,000 bbl. were produced. Production fell in 1915 to 89,570,000 bbl., and increased slightly in 1916 to 91,820,000 barrels.

Notwithstanding the fact that a net average of 52 new wells was added to the producing wells of California during each month from January to May, 1917, inclusive, and the fact that drilling during these months was more active than during any other period in the last three or four years, the daily production was slightly less in May than in January.

Unless the measures hereinafter recommended for increasing production are adopted, it is improbable that the production in 1917 will exceed the 1916 production by more than 2,000,000 bbl., if at all.

From the data available to us, we are convinced that the public cannot hope for further substantial increases in California's annual petroleum supply by the discovery of important new fields or of large extensions of existing fields.

CONSUMPTION. The consumption of California petroleum in 1916 was 104,930,000 bbl., being more than 13,-100,000 bbl. greater than the production. The excess was taken from storage.

Consumption in 1916 outran production an average of 1,100,000 bbl. per month and 35,650 bbl. per day.

During the first five months of 1917, consumption outran production 5,415,000 bbl., being 1,083,000 bbl. per month and 35,860 bbl. per day.

Due to normal increase in consumption as well as the additional extraordinary requirements of the War which are already being felt, it is doubtful whether consumption in 1917 can be reduced below the consumption of 1916, notwithstanding the substitution of other fuel or power and the further possibilities of conservation pointed out in this report.

Consumers of California petroleum will shortly face a condition of decreasing production and increasing demand. This condition points inevitably to the necessity of developing other sources of fuel or power.

STORAGE. Crude oil stocks in California have fallen from 57,147,000 bbl. on December 31, 1915, to 44,036,000 bbl. on December 31, 1916, a reduction during the year of 13,100,000 bbl., or 23%.

Standard Oil Company reports that during the first five months of 1917 the field and pipe line crude oil stocks of all companies were further depleted as follows:

1917	Storage depletion in barrels
January	976,036
February	1,031,960
March	854,333
April	1,197,475
May	1,355,318
Total	5,415,122

The total remaining storage on June 1, 1917, is reported to have been slightly in excess of 38,000,000 barrels.

A portion of the crude oil in storage cannot be utilized

because it is located below the outlets of tanks and reservoirs or is being used for the operation of oil pipe lines or for other reasons. Of the total stocks on June 1, 1917, not in excess of 32,000,000 bbl. were available for use. Of this amount, 12,000,000 bbl. were refining oil and would yield approximately 7,000,000 bbl. of residuum. On June 1, 1917, there were available from crude oil stocks approximately 27,000,000 bbl. for fuel.

If the present excess of consumption over production, amounting to an average of 1,083,000 bbl. per month, continues, the entire available storage of California fuel oil will be exhausted by June 1, 1919.

If a margin of safety of 10,000,000 bbl. of fuel oil is maintained, and if the present relationship between production and consumption continues, the margin of safety for fuel oil will be reached by September 20, 1918.

If consumption is materially increased, as seems likely, both because of normally increased requirements, as well as the extraordinary requirements of the War, or if production decreases, as seems likely unless the relief herein recommended is given, both the margin of safety and the complete depletion of all California stocks will be reached considerably prior to the dates indicated.

The principal railroads of California, with the exception of the Southern Pacific Company, have made the necessary arrangements to meet their requirements for at least one year.

At the present rate of production by Kern Trading & Oil Company, the Southern Pacific Company's fuel oil bureau, bearing in mind also the purchase of fuel oil by the Southern Pacific Company, including 1,000,000 bbl. bought from Union Oil Company and not as yet drawn on, and bearing in mind also the Southern Pacific Company's consumption of fuel oil, the Kern Trading & Oil Company's storage of fuel oil will be exhausted by December, 1917, unless the recently augmented drilling operations of Kern Trading & Oil Company increases the Southern Pacific Company's production and unless the Southern Pacific Company effects a substantial saving of fuel oil by converting to coal those portions of its system which are located in proximity to the coal fields of Washington, the Rocky Mountain States, and New Mexico. If the receipts of fuel oil by the Southern Pacific Company decrease or its consumption increases, the depletion of its stocks will occur before December 1, 1917.

If the Kern Trading & Oil Company's storage should be exhausted, it would be necessary for Southern Pacific Company to enter the market to purchase oil from general stocks in storage, which amounted on June 1, 1917, to slightly over 38,000,000 bbl. These stocks are owned principally by Standard Oil Company and Union Oil Company. If these companies should be unwilling to sell to the railroads fuel oil from their storage, we assume that the Federal government would have the right to commandeer the stocks and to compel their delivery for the operation of the railroads as long as the stocks hold out. Such action, if on a large scale, would necessarily deprive other industries of petroleum and its products.

CONSERVATION. Field losses of petroleum in the California fields have been almost entirely eliminated and the amount of fuel oil used in field drilling and pumping has been largely reduced by the substitution of natural gas and electric energy. The operators are now generally taking steps to reduce the remaining use of approximately 8500 bbl. of fuel oil daily by the installation of jacks and electric motors. The use of fuel oil in the fields for pumping and drilling cannot be entirely eliminated.

The principal petroleum refineries of California are working on improved processes of refining. The amount of crude oil which is being refined is increasing and a proportionally larger amount of gasoline and lubricants is also being secured. The result has been a large surplus of kerosene which it has been necessary to export to the Orient, Australia, and Central and South America.

Mexican petroleum was substituted early in 1917 for California petroleum amounting to approximately 2,750,000 bbl. annually, heretofore sold by Union Oil Company in Chile. A considerable portion of the remaining 3,000,000 bbl. of California fuel oil sold in 1916 on the west coast of Central and South America, including the Panama Canal, can likewise be saved by the substitution of Mexican petroleum from the fields of Tampico and Tuxpam. By reason principally of transportation difficulties, Mexican petroleum will not be available, during the War, as a further substitute for more than 3% of California petroleum. After the termination of the War and the resumption of normal transportation conditions, we may assume that Mexican petroleum will play an important part in the commerce and industry of a considerable portion of the Pacific coasts of North, Central, and South America.

Coal cannot be substituted for California fuel oil to any substantial extent during the War because of present difficulties in the production and transportation of coal. Approximately 1,000,000 bbl. of California fuel oil will be saved in the ensuing year in the North-West by the substitution of coal for California fuel oil by the Oregon Short Line and other industries. The Los Angeles & Salt Lake Railroad Company and the Western Pacific Railroad Company are converting a portion of their systems in Utah and Nevada from California fuel oil to coal produced in the Rocky Mountain States. The Southern Pacific Company and the Atchison, Topeka & Santa Fe Railway Company can also gradually convert from fuel oil to coal those portions of their systems which are in proximity to the coal fields of the North-West, the Rocky Mountain States, and New Mexico. Apart from what has already been accomplished and the further possibilities herein indicated, there is little possibility of further conversion from California fuel oil to coal during the War, unless the conditions surrounding the production and transportation of coal materially change.

Powdered coal has been successfully used in cement plants, stationary boilers or power plants and metallurgical furnaces. One cement plant in the North-West has recently converted its plant from California fuel oil to powdered coal, resulting in a saving of 84,000 bbl. of California fuel oil annually. Apart from a possible slight additional saving in the North-West, it is not reasonable to expect that powdered coal will be further substituted for California fuel oil during the War.

Hydro-electric energy has already been substituted to a considerable extent for California fuel oil in industrial and agricultural uses, but the difficulty in securing copper and other material and the disturbance of existing conditions are such that large additional savings of California fuel oil by the substitution of hydro-electric energy cannot be anticipated during the War. A small saving of California fuel oil can be effected, during the War, by the further substitution of electric motors for fuel oil in the California oil fields and by such interconnection between the systems of various electric companies as will eliminate or reduce the necessity of maintaining steam-electric plants. After the termination of the War and the restoration of normal industrial conditions, we may expect that hydro-electric energy will play an increasingly important part as a substitute for fuel oil in all the Pacific Coast States in which such energy is available.

Natural gas has already been substituted to almost the entire extent of its supply, for fuel oil in the California petroleum fields, and for fuel oil and artificial gas for higher industrial and domestic uses. The maximum production of natural gas in the California petroleum fields has been reached and will shortly decline so that it is not to be anticipated that natural gas will, to any substantial extent, further replace other forms of fuel.

We conclude that during the War some further saving of California fuel oil is possible by elimination of losses and substitution of other forms of fuel or power, but that no large saving can be effected without very serious impairment to the efficiency of the transportation systems and industries of the Pacific coast.

The great importance of gasoline and lubricants must be recognized and steps should be taken as soon as reasonably possible to the end that no more unrefined petroleum is burned by any railroad company or other industry. No part of California petroleum should be thus burned except the residuum left after refining.

If a sufficient amount of fuel oil cannot be secured from Mexico after the War, the railroads and other industries of the Pacific coast must gradually make arrangements to use other forms of energy, such as hydro-electric energy or powdered coal.

THE REMEDY. The remedy which imperatively presents itself in view of the emergency created by the War is the prompt and substantial increase in the production of California petroleum.

While we do not desire to minimize the results which can be accomplished and should be accomplished by the further diminution of field losses, the higher use of petroleum and its products and the substitution of other forms of fuel or power, during the War as well as thereafter, the cardinal fact remains that the only means which will be effective in a large way to meet the present emergency is a prompt substantial increase in production.

We estimate that this increased production should amount to more than 35,000 barrels per day and that such increased production cannot reasonably be expected before June 1, 1918.

Each difficulty standing in the way of prompt and substantial increased production must be quickly solved, if the increase is to forestall a serious industrial crisis.

INCREASED PRODUCTION—MATERIAL. The necessary increased production cannot be secured unless large additional amount of oil-well casing, drill-stem pipe, and other oil-well material are promptly brought into California.

The oil-well supply-houses report that they can fill no orders for complete drilling-outfits in addition to those already taken.

The larger oil companies have on hand or have heretofore placed orders for enough material to complete their 1917 drilling operations as heretofore planned, but have no material for additional drilling beyond such plans. Many small operators report that they would drill if they could secure the necessary material, but that it has been impossible for them to secure such material, even at the high prices now prevailing.

Receiver Payne reports that he is willing to drill, if authorized by the Federal court, but that he does not know where he could secure the necessary material unless the Federal government should take the necessary steps to assist the California producers.

In our opinion, the only way to meet the situation, in view of the existing conditions, is to have the Federal government direct the manufacturers of oil-well supplies to devote sufficient capacity of their plants to supply the requirements of oil producers in California and other sections of the United States and to direct the railroads to transport such supplies promptly.

INCREASED PRODUCTION—LABOR. Over 80% of the laborers who are employed in the oil-fields and refineries are skilled men, most of whom it would be difficult to replace. If any considerable number of these men are taken from their present employment it will be impossible to increase the production of petroleum in California and difficult to maintain the present production.

About 3% of these men have already left their employment and have volunteered for service in the various branches of the Army and Navy. About 32% of these employees have registered for the draft.

We suggest the advisability of drawing this situation to the attention of the Federal government.

INCREASED PRODUCTION—THE LAND. In order to secure an increase of 35,000 barrels per day in the production of California petroleum, drilling must be done on the land which will yield the largest production in the shortest time by the expenditure of the smallest amount of drilling material and labor.

Upflow Clarification of Solution

By A. W. ALLEN

When in South Africa in 1913 I devised a conical clarifier shown in Fig. 1 to be used in connection with an experimental counter-current continuous-agitation cyanide plant. The apparatus worked so satisfactorily that the idea was subsequently elaborated as a zinc-box clarifier, as shown in Fig. 2. In both cases the filtering medium was composed of small pebbles, small stone, gravel, and sand, in the order named. An apparatus on a working scale with filter-cloth in place of the sand-filter (Fig. 3), supported between coarse wire-screen, was then considered, but the idea was not followed up. I took out a provisional patent for a clarifier of this type at Pretoria in March 1913, but abandoned any further interest in the matter on hearing that the system was in operation in the United States. No details were forthcoming, however, but I gathered that something similar had been used for clarifying boiler-feed water.

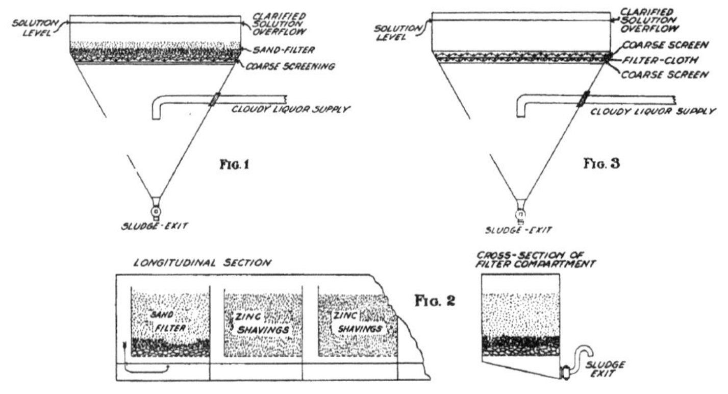

FIG. 1 FIG. 2 FIG. 3

It would be of interest to know what work has been done along these lines by metallurgists, particularly cyanide operators, and whether as a result some modification of the idea is applicable as a cheap substitute for the ordinary sand-filter.

In the type of clarifier illustrated herewith the filtering medium is cleansed by shutting off the flow and discharging through the sludge-exit. This causes a reversed flow that frees the filter of much of the accumulated solid.

The disadvantages of the conical clarifier lie in the fact that an accumulation of solid on the sides of the cone is inevitable. The disadvantage of upflow clarification through sand is the small capacity of the filter-bed area.

THE most important magnesite development of the year 1916, according to the U. S. Geological Survey, is the opening in eastern Washington of large deposits of a coarsely crystalline magnesite that is like marble or dolomite in texture but is essentially magnesite in composition. This material is now being shipped at the rate of several hundred tons per day, and calcining furnaces are in course of erection to prepare magnesia for use in making refractory material and, it is said, also for use in cement mixtures. Coming at a time when the sources of supplies abroad are cut off, the discovery of these deposits appears to be most fortunate. Apparently authentic reports indicate that the deposits are large and that they will afford a supply of uniform character by relatively cheap methods of mining. It is perhaps too soon to say just how well the material is suited for refractory or other uses, but the present indications are that it may be adapted to some of these uses.

THE National Board of Fire Underwriters advise that open lights or flame of any character should never be permitted for use in the presence of light combustible or volatile inflammable materials, or where inflammable dust is liable to be present; incandescent electric lights in such localities should be of the keyless socket pattern and enclosed in wire guards, with operating switch situated in an apartment separated from the inflammables. If possible, keep oils outside of the main buildings in a separate oil-house. Never keep the main supply of light inflammable oils, such as gasoline or kerosene, inside of the main buildings except in nominal quantities. Illuminate the oil-house safely and so brightly that there will never be a temptation for an employee to light a match in order to see while drawing oil. Catch oil drip in metal pans; never use sawdust or other combustible material to absorb it. Excellent oil-cabinets are made which drip back into the main tank.

PLATTEVILLE, WISCONSIN

ORE-PRICES.—PRODUCTION.—MINING.

Mining development and equipment are being seriously retarded at all points because of the lack of labor. For five consecutive weeks the price of zinc ore remained on a base of $70 per ton for standard 60% blende, and ranged down to $65 for medium and second grades. The limitations on the inferior grades were partly removed so that ore of 52% zinc content was given consideration on the basis of the published quotations. The first part of the month low-grade producers, except where protected by contract, found difficulty in the marketing of their product. Later a better disposition on the part of refiners secured an outlet for low-grade zinc-ore producers but the greater quantity of low-grade product was delivered to the Mineral Point Zinc Co. The high-grade product was in good demand all the month. The buying of the Grasselli Chemical Co. was a feature of the business during August. A few producers mining lean deposits suspended operations in the Montfort and Linden districts.

Lead-ore producers who had been receiving from $110 to $130 per ton found it difficult to reconcile the gradual decline in the price of lead ore with the prices ruling in the pig-lead markets and preferred to hold. The first two weeks of August witnessed a few sales but the bulk of the ore, estimated at 1500 tons, was held in reserve. The price of lead-ore fell until it reached $100 per ton. A fair portion of the reserve was worked off before the offerings dropped under $100 per ton. The ore was held back as the price approached $90, and shipments ceased. The Federal Lead Co., represented by N. H. Snow, secured the bulk of the ore offered for sale. A few cars produced by the New Jersey Zinc Co. was shipped to the company's smelter at Palmerton, N. J. The estimates made earlier in the month on reserve lead ore were considered too high, but lead ore came in quantity from districts that were overlooked and careful estimation at the close of the month showed that at least 1000 tons was in reserve. The bulk of the lead ore is obtained as a by-product in the wet concentration of zinc ore, and a greater output of zinc ore is invariably associated with an increased output of lead ore.

The recovery of pyrite was confined to that obtained at reduction plants where electrostatic separation is used. The price for the average Wisconsin pyrite may be put at $5 per ton. Shippers have found cars available for loading rather restricted during the month. Where loading facilities were immediately available deliveries were steady but smaller than heretofore.

No encouragement was given producers of carbonate-zinc ore in the northern part of the field. The price offered Wisconsin miners would be on the basis of $25 per ton, 40% zinc. Most of the Wisconsin carbonate-zinc ore, termed locally 'dry-bone,' contains 30% zinc and brings the miner about $15 per ton. Occasionally higher-grade ore is found and the scale is adjusted up or down on the basis of $1 per unit. Thousands of tons of this class of ore are available for mining and are used by the principal buyer, the New Jersey Zinc Co., for the manufacture of zinc oxide. The New Jersey Zinc Co. has recently completed arrangements that will give the company tonnage from the mines where this ore is obtainable.

The mine production, shipped during August, follows:

District	Zinc, lb.	Lead, lb.	Pyrite, lb.
Benton	25,268,000	768,000	....
Mifflin	8,862,000	152,000	....
Linden	3,008,000	84,000	600,000
Galena	2,888,000	90,000	....
Platteville	1,776,000	......	....
Hazel Green	1,696,000	60,000	....
Highland	1,470,000	126,000	....
Shullsburg	1,254,000	394,000	56,000
Dodgeville	984,000	70,000	....
Potosi	726,000	......	....
Mineral Point	434,000	......	....
Cuba City	....	......	3,642,000
Total	48,366,000	1,744,000	4,298,000

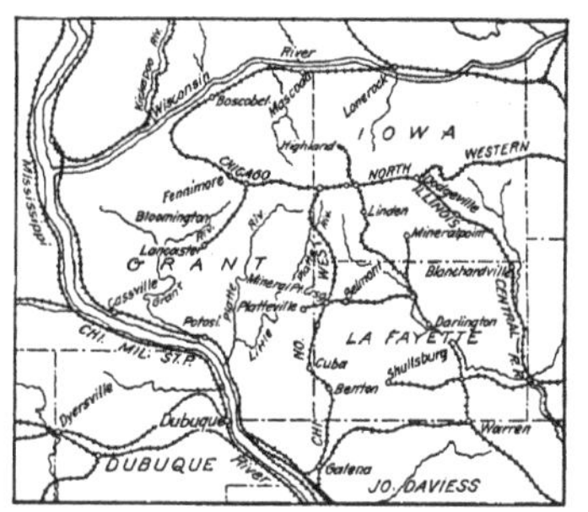

WISCONSIN ZINC REGION

Shipments of high-grade dressed zinc ore from ore-dressing plants were made to smelters as follows:

	Pounds
Mineral Point Zinc Co.	6,064,000
Benton Roasters and Wisconsin Zinc Roasters	5,196,000
National Separators	4,584,000
Linden Refining Co.	496,000
Total	16,340,000

The sales and distribution of zinc ore during the month of August are as follows: To Mineral Point Zinc Co., 6515 tons; Grasselli Chemical Co., 4796 tons; Wisconsin Zinc Co., 4495 tons; Vinegar Hill Zinc Co. to National Separators, 4657 tons; American Zinc Co., Hillsboro, Ill., 3256 tons; Linden Zinc Co., 1537 tons; Matthiessen & Hegeler Zinc Co., La Salle, Ill., 755 tons; Illinois Zinc Co., Peru, Ill., 702 tons; American Metals Co., 667 tons; Benton Roasters, 374 tons; Lanyon Zinc Co., 205 tons; Edgar Zinc Co., 151 tons; American Zinc & Lead Smelting Co., 117 tons; total 28,227 tons.

In the Highland district a rich strike by the Mineral Point Zinc Co. was surveyed on the range for the Saxe-Lampe Mining Co. and prospect work with a drill showed that the range carried into the Saxe-Lampe ground, giving new life to the latter. Underground development was hastened to connect both finds. For the Mineral Point Zinc Co. much new ground was opened up on the Red Jacket mine at Centerville and an increased output is now assured.

No important developments were undertaken in the Montfort district. At Linden, the Saxe-Pollard interests are engaged in developing two new mines, one on the Treloar and one on the Kickapoo lease. Two shafts, one-half mile apart, are being driven to ore-level. After cross-cuts have been driven the management will consider the advisability of supplying a complete surface-plant midway between the two mines to afford milling-facilities. New ground was developed on the Optimo No. 3 mine. Another producer in the district, the Lucky Five Mining Co., is holding several hundred tons of zinc ore, both hand-cobbed and concentrate. The North Survey Mining Co. is developing new deposits. The Mineral Point Zinc Co. operated continuously, producing concentrate, and shipping large quantities of high-grade zinc-oxide. In addition one car of 66% sulphuric acid per day was shipped. Utt-Thorne Mining Co. brought several leases up to production.

In the Mifflin district the New Jersey Zinc Co. and the Vinegar Hill Co. are making a steady production. The New Phoenix mine promises to be a regular shipper. Other producers are the Grunow, Peacock, Royal, Lucky Six, Biddick, Big Tom, and Squirrel mines. Prospecting in new ground is active.

The Platteville district made little showing in August. The Block House Mining Co. made the principal production. The New Rose Mining Co. was organized under L. N. Piquett to operate the Old Mexico mine. A power and milling plant, $15,000 in value, is reaching completion on this mine.

Two magnetic zinc-ore dressing plants are in operation in the Cuba district. The C. S. H. Mining Co. is completing a concentrator. In the Shullsburg district large low-grade deposits have developed in the Winskill mine. The McDermott Mining Co. is producing lead ore.

Shipments of crude ore from the Benton district were at the rate of 65 to 70 cars per month during the first half of August, but were somewhat lower in the second half. Shipments of high-grade ore were small in the first and large in the second half-month. Price declines hastened the shipments. The Domestic Mining Co. opened a new producer and will install a new mill. The Frontier Mining Co. completed a power, mining, and milling-plant on the Hird mine.

The new Wilkinson Mining Co., on the George Wilkinson farm, is driving for a deposit that was discovered by drilling. The Grand View Mining Co., composed of Indianapolis mining men, established a surface-rig here one year ago and was successful in discovering a deposit of carbonate-zinc ore. The poor market offered compelled the company to suspend operations and the plant has been closed. Several cars of high-grade carbonate-zinc ore concentrate were sold to the American Zinc & Lead Smelting Co. The largest production for the month came from the Wisconsin Zinc Co., the Frontier Mining Co., and the Vinegar Hill Zinc Co. The production of the Meloy mine, a recent Vinegar Hill development, was delivered to the Grasselli Chemical Co. The New Jersey, the Fox, the Penna. Benton, and the New Hoskins mines failed to maintain the production-record usually attained by them.

The Hazel Green district showed further decline. The Cleveland Mining Co. made an indifferent showing. The Vinegar Hill Zinc Co. is completing a new 200-ton milling plant at the Jefferson mine. In the Potosi district the Wilson Mining Co. maintained a normal shipment of one car of high-grade wet concentrate per week. The Tiffany Zinc Co. withheld ore from shipment until the latter part of the month when it sent several cars to the Mineral Point Zinc Co. It carried a moderate reserve at the close of the month. The Mudd Range Mining Co. re-built the Dinsdale mill and completed its underground development.

BUTTE, MONTANA

New Company Organization.—Labor Situation.

The Big Foot Mining Co. has just been organized in Butte. The company is incorporated for 600,000 shares with a par value of $5. The object of the company is to develop seven claims in the Big Foot district in Jefferson county, 12 miles from Boulder. The property has not been operated for 20 years, and at that time was operated for the silver in the ore. The ore contained silver, lead, and zinc, and therefore at that time had to overcome a large penalty, which made operation unprofitable. There are several shallow shafts, ranging in depth from 50 to 200 ft., in the strike of the main vein. All of these shafts will have to be de-watered, and where this has been done, fair assays have been obtained.

Labor Situation.—On September 6 a strike vote was taken by two metal trade unions, the machinists and the blacksmiths. Both failed to get the needed majority to declare a strike. The machinists, who, according to international rules, must have a majority of three-fourths to strike, voted 133 for and 85 against. The blacksmiths, who needed a two-thirds majority to strike, voted 54 for and 54 against a strike. The result of the voting seems to indicate that the independent companies will continue operating. All of the electricians who walked out from the mines some time ago have returned to work. Independent companies in Butte are working now, witn the exception of those who ship their ore to the Anaconda company's smelters at Anaconda and Great Falls. An early settlement of the labor situation is expected at the smelters. The concentrator owned and operated by M. W. Atwater, at Basin, was forced to shut-down, due to the labor trouble in Anaconda and Great Falls. Work will be resumed as soon as the smelter commences operations.

COBALT, ONTARIO

August Shipments.—Mining Corporations.—Nipissing Mines.

The total ore shipments from the Cobalt district for the month of August amounted to 49 cars. There is a slight shortage of labor, but in general operations are satisfactory. During the first six months of the current year the price of silver averaged 75.44c. per oz. With the price of silver over 95c. per oz. and with an average of 85c. during August, it is evident that the average for 1917 may be above 80c. per oz. Such being the case, the value of this year's production, probably 20,000,000 oz., will reach $16,000,000.

Mining Corporation and Nipissing Mines are striving for first place among the shipping mines of the district. During the first half of 1917, 2,000,000 oz. of silver bullion was shipped by the Mining Corporation. The Nipissing production is about the same. The two companies are sending out almost one-half the total production of the district. The affairs of the Temiskaming are now much in the public eye, due to the controversy between Max Morganstern, a heavy shareholder of the company, and the present mine-management. A special report has just been made by B. E. Neilly, manager of the Penn Canadian mine, in which it is shown that there is from 400,000 to 450,000 oz. silver in the available ore. The report compares with that recently issued by the management, and which was apparently the starting-point in the difficulty. Morganstern has threatened to oust the present board of directors, claiming to have control of sufficient stock to do so. It appears to be probable that another report will be made by a representative of Morganstern. The Neilly report just issued, dealing, as it does, with only 'positive ore,' leaves room for conjecture as to whether additional estimates can be made on 'probable' or 'indicated' ore.

PORCUPINE, ONTARIO

HOLLINGER.—SCHUMACHER.—MCINTYRE.

Mining operations in the district are more satisfactory than for several months past. The Hollinger production is understood to be greatly improved and the output for the current year, despite the adverse conditions in the labor market in the early summer, will be almost on a par with that of the preceding year. The milling facilities at the gold mines of northern Ontario, taking into account the grade of ore at each of the producing mines, are sufficient for the production of $18,000,000 per annum. For the first half of 1917 the gold output amounted to $4,586,941. Working-forces are steadily improving in numbers, and there is an improvement in efficiency. There is now no talk of striking, and with the close of the harvesting season in the West labor will be obtained more easily.

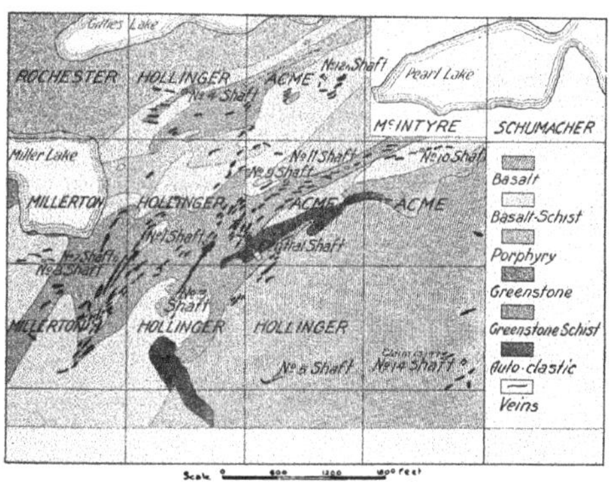

PORCUPINE DISTRICT

The directors of the Schumacher mine have announced their intention of sinking the main shaft to a depth of 1000 ft. The position of the mine would appear to warrant deep mining as it lies adjacent to the Hollinger, as also does the McIntyre. The tongue of porphyry, along the contact of which the Hollinger is finding highly productive orebodies, extends in an easterly direction, partly on the McIntyre and partly on the Schumacher. The McIntyre long drift on the 1000-ft. level is on the north contact of the porphyry belt. The Schumacher proposes to go to a similar depth and prospect the south contact. Not long ago a diamond-drill, operating in the McIntyre, tapped the downward continuation of the north contact at a depth of over a quarter of a mile and the orebody was of uniform width and contained high-grade ore. It is the opinion of mining men of the district that operations will be successful along this contact to a considerable depth.

The mill of the McIntyre-Porcupine is now running and production is understood to be between $6000 and $7000 per day. Despite the adverse conditions during the past twelve months, the management of this mine has not only been able to bring the mill up to capacity, but has also been able to add materially to the ore-reserves and to break a very large tonnage of milling ore in excess of that possible to handle with the present milling facilities. It is now believed probable that a decision will shortly be made for the adding of further equipment which will bring the milling-capacity up to 1000 tons per day. The average grade of the McIntyre ore is $12.50 per ton, and with an output of 600 tons per day the gross bullion-input will be $7500 per day.

At the Porcupine Crown, since the suspension of dividend disbursements, the policy of the management has been to carry on milling operations at a rate sufficient only to pay the expenses of development work. Development is being carried on, and with a return to normal condition in labor supply and the cost of material, a period of production is at hand.

The installation of a mining-plant at the Anchorite is progressing, and this property will soon be classed among those active in the development stage in this district.

With ore-reserves at the West Dome Con. officially estimated at over $2,000,000, and with reports of arrangements about to be made for the installation of a new mill, this property would appear to be about to occupy a good position among the smaller mines of Porcupine.

LEADVILLE, COLORADO

MANGANESE ORE.—RE-SMELTING SLAG.

At the Bohn shaft on East Second street near the Penrose of the Down Town Mines Co., a discovery of manganese ore has been made by a company of lessees headed by Peter Horrigan. The orebody, which was encountered on the 440-ft. level, has been developed by driving and raising. The present dimensions are 12 ft. high, 44 ft. long, and 15 ft. wide. The top and end of the shoot have not yet been reached, and it is estimated that the deposit will prove to be from 200 to 300 ft. long and more than 100 ft. high. Compared with other deposits of manganese ore that have been opened in the district, the Bohn strike exceeds them in the size of the orebody and the purity of the ore. Assays show the ore to average 40% of manganese. None of the assays are below 35%. The material contains 8% silica and 14% iron.

Driving and raising will be continued until the walls have been reached. The fact that the orebody lies along the eastern side of the Cloud City fault is considered to indicate favorable conditions for the occurrence of a large deposit.

Operations at the Bohn were started in January, when Horrigan and associates secured a lease from the Down Town Mines Co. The property had been idle for 25 years, and offered a difficult problem to the lessees as the shaft was caved from a point 85 ft. below the collar to the bottom-depth of 500 ft. The lower level was reached late in July, requiring seven months to reclaim the shaft. Development was started at 440 ft. A drift was carried to a point 300 ft. north. A 30-ft. raise intersected the bottom of the orebody at this point.

Preparations are being made on the surface to handle an output of 100 tons per day.

An important project for the recovery of metal from old dumps has been launched and is being worked out at the old Harrison reduction works. A company known as the Harrison Recovery Co. has been organized by Fred J. Johnson, James Fyfe, and Fred F. Furnas. It is the purpose to repair the blast-furnaces at the Harrison works and re-smelt the slag-dumps which assay from 6 to 300 oz. silver per ton. Both lead and copper are present in small amounts. One of the furnaces has been repaired and is ready for operation. An incline-track, equipped with a 2-ton skip, has been constructed from the old dump to the top of the furnace. After completing preparations, 200 tons per day will be smelted by the furnace. Other furnaces will be put into operation as rapidly as they can be overhauled.

Data compiled from tests on the slag-dump show that from each 500 tons of slag 50 tons of matte, containing 3000 oz. of silver, will be recovered. A market for the product has already been secured at the plant of the Ohio & Colorado S. & R. Co. at Salida.

The Denver & Rio Grande Railway Co. is rapidly completing the standard gauging of the line to the mines in Big and South Evans gulches and Breece hill. The work has reached to within two miles of the Ibex and is expected to be completed during the next two weeks. A large tonnage of ore will be handled over this line when finished. Lessees on the Silent Friend mine in South Evans gulch have opened a large body of lead-zinc sulphide ore and have started shipping at the rate of 50 tons per day. A small shoot of zinc-carbonate ore has also been uncovered on the upper levels. The Matchless mine on Fryer hill is to be operated again. Edward Huter, formerly superintendent at the Square Deal mine at Frisco, representing Denver capitalists, has secured a lease on the mine and has started the work of re-timbering the shaft. It is planned to sink through the parting-quartzite into the ore-zone existing between the porhpyry and the white limestone.

HOUGHTON, MICHIGAN

Calumet & Hecla and Minerals Separation.—Expenditures for Electrical Installations.

The decision of the U. S. Supreme Court on the validity of the claims of the Minerals Separation company will without doubt induce other companies to use flotation in lessening tailing-loss. While conditions at other properties do not offer the same opportunities of success as at the White Pine and the Calumet & Hecla, they are at least encouraging. At the Calumet & Hecla the flotation process was introduced months ago under an arrangement with Minerals Separation which permitted operation on a royalty basis contingent upon the outcome of the law-suits. At the termination of the controversy Calumet & Hecla concluded their contract and as a result flotation will be applied on a larger scale in the mills at Lake Linden. A recovery of 70% has been made in the treatment of slime that resulted from the amygdaloid tailing. The leaching process will be continued but used on material wherein the copper particles are coarser than 200 mesh. At the White Pine, one of the subsidiary properties of the Calumet & Hecla, flotation is successfully applied.

At the Wolverine the vein continues in the foot-wall in No. 4 shaft, due to the dip of the formation at the bottom, the 43rd level. This is the mining-limit in the shaft. The copper content at the bottom is better than it has been for four levels and the laterals are expected to open a grade of rock up to the old-time Wolverine average. Last year the 39th and 40th levels extended to the boundaries, but the three levels below have yet to reach their full length. The 40th level has not been exhausted. Wolverine continues to obtain 40% of its copper from the vein in the foot-wall in the old stopes.

Rock from No. 1 and 2 shafts of the Ahmeek is better than 14 lb. per ton, and the average from the North Ahmeek is better than 20 pounds.

In addition to maintaining a fair shipment of high-grade rock, the South Lake is finding mass-copper. While it is not large, it is larger than the 'barrel copper' and goes directly to the smelter.

The enlistment of miners is more than compensated for by the influx from Butte. These are mostly Cornish miners, many of them trained in the tin mines. The working-force employed here is over 22,000.

An estimate places the amount of money expended and represented in contracts by Lake Superior mines for electrical equipment, being installed or about to be installed, at $1,000,000. Much of this is for underground-tramming. Calumet & Hecla is now installing a 10,000-kw. turbine at the mill at Lake Linden and is scrapping two steam-engines to make room for it.

The Tamarack, now owned by the Calumet & Hecla, has copper in its tailing. In 1913 the company started a reclamation plant to handle the conglomerate sand. The building will now be finished and the machinery installed. It will be patterned after the Calumet & Hecla's re-grinding plants.

Laterals at 220 ft. in the White Pine Extension open up 1800 ft. of mining-ground, all in copper-bearing rock. At the 440 good rock extends over 400 ft. in both directions from the shaft.

CRIPPLE CREEK, COLORADO

Production.—Development.—Milling.

The August production from the properties of the Gold Dollar Con. Mining Co., on the eastern slope of Beacon hill, totalled 12 cars, or about 325 tons of ore, ranging in value from $15 to $18 per ton. Al Osberg is mining ore from a 14-ft. vein at present being developed on the 600-ft. level of the Mable M. shaft.

The Camp Bird M., L. & P. Co., John Nichols superintendent, operating the mines of the Rose Nicol Gold Mining Co., on Battle mountain and Bull hill under a six-year lease, has made an important discovery. The lateral extending into the Rose Nicol from the 1000-ft. level No. 2 shaft of the Portland Gold Mining Co. has entered an ore-shoot within the Rose Nicol claim. The ore is being stored underground until an ore-house can be constructed. The Rose Nicol shaft is down 720 ft., and is being sunk as rapidly as possible to connect with the Portland lateral.

Hahnewald, Olsen, and associates, operating the Gold Sovereign M. & T. Co.'s mine on Bull hill, are doing lateral work on the new 1600-ft. level. The lessees are mining and shipping two cars of ore each week from the shoot under development on th 1100-ft. level. The ore is of average milling-grade.

The Queen Bess mine, on Tenderfoot hill, owned and operated by August Hahnewald, Olsen, and associates, produced 14 cars or about 500 tons of ore in August of a value in gold between 1½ and 2 oz. per ton.

The local milling-plants of the Portland are now treating approximately 2000 tons of low-grade mine and dump-ore per day. The Victor mill is treating 600 tons, and the remodelled Stratton's Independence mill 1400 tons per day. The average grade of this ore will not exceed $2.25 per ton.

The Roosevelt tunnel of the Cripple Creek Deep-Drainage & Tunnel Co. was advanced 155 ft. in August. The flow is constant at 5450 gal. per minute.

LORDSBURG, NEW MEXICO

Concentration Plants.

Concentration plants in which flotation will be used are to be immediately erected by the 85 Mining Co. and the Lawrence Mining Co. The capacity of the mill at the 85 mine, according to plan, will be 300 tons per day. The capacity of the mill for the Lawrence has not been determined. Final tests will decide the mill capacity of the latter mine. Both mills will be available for custom ores. The plant at the 85 mine will be fitted with separate ore-bins, and will also have a custom-sampler.

ALASKA

(Special Correspondence.)—Alaska uses 600,000 bbl. of fuel-oil per annum and, in the absence of coal and oil development, must import from California. Falcon Joslin, recently returned from Washington, stated that there is little hope for the modification of the leasing regulations governing the coal and oil-lands in Alaska. Oil-lands were withdrawn from entry seven years ago and development was stopped. Efforts to secure the cancellation of the withdrawal-order were unsuccessful. The coal-leasing act has proved inoperative. The threatened oil-shortage in California and the high shipping rates combined have made the fuel situation acute.

Livingston Wernecke, geologist for the Alaska Treadwell, has been examining property on Chichagoff island. W. J. Rogers, manager of the Mt. Andrews copper mine at Hedley, has been examining mines. The Mt. Andrews is shipping 1000 to 1200 tons per month of high-grade copper ore to the Tacoma smelter.

Treadwell, September 2.

(Special Correspondence.)—A discovery of importance near the head of Talkeetna river has been made by W. M. Foster and B. Whittridge. Samples from the find assay from $2 to $40 per ton. F. Spengler and B. T. Spaulding, engineers, are on their way to the discovery.

The Cache creek dredge has had a successful season and will operate until the middle of October. H. Parker is prospecting for platinum in the Mulchatna. A reported discovery of gold ore at Snug Harbor was discredited.

Willow Creek mines have made a light production for the season. There is much new development.

The Hercules Mining Co. is developing a molybdenum deposit on Reid creek. Sulphide and oxide-ore of molybdenum occur in quartz and diorite.

Anchorage, August 1.

ARIZONA

COCHISE COUNTY

(Special Correspondence.)—The Denn mine has shipped a car of ore. It is expected that shipments will continue regularly. One hundred and twenty-five men are employed.

Tombstone, September 8.

GILA COUNTY

(Special Correspondence.)—The Copper Hill Co. has just received an oil-burning engine and other machinery for the extensive development work planned. Since April this company has shipped seven cars of high-grade copper and molybdenum ore.

Globe, September 6.

GREENLEE COUNTY

(Special Correspondence.)—The strike in the Clifton district continues and has apparently reached a deadlock. There is a general exodus of both American and Mexican miners. In order to break the deadlock, Mayor Coty of Clifton, at the suggestion of the citizens, called a mass meeting of citizens and miners. After a general discussion of conditions, a proposition was agreed upon for submission to the company-managers. The labor organizations ratified this plan and agreed to abide by the decision of the board. A deputation has now presented the proposition to the mine-managers. Their decision has not yet been made.

Clifton, September 4.

MARICOPA COUNTY

(Special Correspondence.)—The Arizona Gypsum Co. has been organized by Phoenix people. The company plans the construction of a plant at Phoenix for the manufacture of gypsum products. The propertty owned by the company comprises 4160 acres near Wickenburg.

Phoenix, September 4.

MOHAVE COUNTY

(Special Correspondence.)—The Elkhart Extension Mining Co., financed by A. A. Barton and H. M. Russell, of Los Angeles, has taken over the Argyle property and preparations are being made for development. An important strike of silver ore has been made at the New Mohawk in the east drift from the bottom of No. 2 shaft. Five feet of silver ore is exposed. The New Tennessee vein maintains its size and value, there being three feet of ore in the face of the north drift of the 240-ft. level.

Kingman, September 4.

PIMA COUNTY

(Special Correspondence.)—According to Roos and Tovote, engineers, the Twin Buttes Arizona Mining Co. has opened up a promising prospect. This property consists of three claims, 26 miles south of Tucson. The ore won in development assays in copper 11.5%, $1.25 in gold, and 8½ oz. per ton in silver. Two cars of ore have been shipped from the Pandora mine in the Mineral Hill district. The Pandora is under lease to D. C. Sullivan. Shaft No. 2 of the San Xavier mine in the Twin Buttes has been shut-down and work commenced at shaft No. 6 on the west side of the property. The Hecla Arizona Mining Co. of the Twin Buttes has shut-down temporarily.

Tucson, September 3.

YAVAPAI COUNTY

(Special Correspondence.)—The Verde Combination broke shaft-sinking records for the Verde district in August by sinking 163 ft. In the shaft, now at 700 ft., a station will be cut and cross-cutting started to the schist-porphyry contact, 200 ft. north. It is expected that mining operations will commence at the Jerome Copper Co. in two weeks. Manager George Kingdon announced that the United Verde Extension shipped 4,500,000 lb. of copper in August. The work on the United Verde Extension haulage-adit is proceeding steadily but the work on the shaft is temporarily at a standstill.

Prescott, September 4.

YUMA COUNTY

(Special Correspondence.)—The Arizona Manganese Mining Co. is making arrangements to ship ore. A contract has been let to a Los Angeles firm to construct a river-boat and a barge of 30 tons capacity. They will be used to transport ore from the mine to Parker. The mine is 22 miles north of Parker and near the river.

Yuma, September 4.

CALIFORNIA

DEL NORTE COUNTY

(Special Correspondence.)—The cinnabar mine on Diamond creek is in operation and quicksilver shipments will soon be made. The Diamond Creek copper mine has been recently examined for Eastern parties.

Crescent City, September 4.

KERN COUNTY

(Special Correspondence.)—The Yellow Aster Co. is installing a new crushing and screening plant, and completing improvements to the mill. Equipment is arriving and mine operations are being conducted along broader lines. Several small gold mines are producing steadily, and tungsten mining is active at several places. The Black Hawk Tungsten Co. is

sinking a shaft on the Leonard claim, near the Churchill deposit. Small veins have been encountered but sinking will be continued to depth.

Randsburg, September 15.

SHASTA COUNTY

(Special Correspondence.)—The flotation-plant of the Afterthought Copper Co. has been in operation for several days and is treating a heavy tonnage from the upper levels of the mine. Ore is delivered by the Ingot & Afterthought railroad. The old dump has been purchased by G. C. Taylor, who is sending sorted ore to the Mammoth smelter at Kennett.

The Leonard cross-cut from the main adit of the Shasta-Belmont has cut a vein of copper ore for a width of 13 ft. A second cross-cut will be driven 70 ft. ahead of the Leonard lateral to intersect the vein, and the adit continued to intersect the main deposit.

Indications are good for the resumption of work at most of the mines in the copper region within a few days. The Mountain Copper Co. has over 100 men on the payroll and the Mammoth company expects to operate the Mammoth mine in a few days.

Redding, September 16.

At a mass meeting of miners the striking miners repudiated the agreement made by their committee earlier in the day. The miners are holding out for an increase of $1 per day in wages. It is expected that a settlement will be made in the immediate future.

SIERRA COUNTY

(Special Correspondence.)—At the Plumbago mine 76 men are on the payroll and the 15-stamp mill is crushing ore of good grade. Sufficient quartz is exposed to ensure a long run and it is stated that operations will be conducted during the winter.

At the Twenty-One a large bunk-house and other buildings are under construction. Mine developments are proceeding with a crew of 35 men. The mine lies near the Tightner group. Small pockets of rich ore are being found occasionally in the Tightner mine and ore conditions are good. Under A. Hall, superintendent, 60 men are employed and the 10-stamp mill is in operation.

Rich ore has been encountered in the shoot under development in the North Fork mine, at Forest, and arrangements have been made to develop to additonal depth. It is reported that a mill will be installed in the spring. George F. Stone is manager.

There is prospective activity in the Howland Flat, Scales, St. Louis, and contiguous districts, and several hydraulic-mining companies are preparing for operations. The impounding-dam on Slate creek will enable many mines to operate for the first time in a generation. Los Angeles capital has been active in this field during the past year.

Alleghany, September 17.

SISKIYOU COUNTY

(Special Correspondence.)—Arthur Coggins, of Sacramento, is mining chrome on Little Castle creek near Dunsmuir. He is planning a concentrating-plant to reduce the low-grade ore. The plant will be erected at once. Last year the output was shipped to Niagara Falls. The present output is being shipped to the Carnegie Steel Co. and the Electro-Metallurgical company.

C. T. Loftus, of Dunsmuir, and Arthur Coggins, of Sacramento, are shipping from their Boulder Creek molybdenite property, near Lamoine, to the Henry E. Wood Ore Testing Co. at Denver.

Hornbrook, September 6.

Denial is made of the shut-down of the High Grade mine. It is stated that development is in progress and that a 250-ft. adit has been completed during the summer. The owners are D. M. Wall, of Butte, Montana, and D. D. Good, of Ashland, Oregon.

COLORADO

BOULDER COUNTY

(Special Correspondence.)—The Nil Desperandum, situated in the Sunshine district, will be re-equipped with electric pumps, shaft-houses, and hoist. The propertty has been a producer in the past. Operations will be resumed. The Royal Gem Mining Co. of Caribou is preparing to re-open its property. New Orleans capitalists are back of the enterprise. C. E. Wenzel will be superintendent. Woodring & Dupont of the Huron mine are preparing for the winter's work by erecting necessary buildings.

Eldora, September 5.

(Special Correspondence.)—The Longfellow and adjacent mines have been consolidated and a general plan for deeper development is being inaugurated. The Bush, Alice, and other mines are developing and mining ore. G. F. Stringer, president of the Mohave Boulder Tungsten M. Co., is supervising the re-construction of the milling-plant.

Eldora, September 10.

MONTANA

Work was resumed on September 17 by the Anaconda Copper M. Co. in both mines and mills. At the smelting plants more men applied for work than could be accommodated.

August operations of the Barnes-King indicate 2239 tons, bullion production $22,800, from the North Moccasin mine and 4586 tons, estimated bullion production $103,000, from the Gloster and Shannon mines. The balance due, $44,940, upon the Shannon mine is to be paid at once. The total payment including the balance just mentioned will be $228,000.

The Anaconda G. M. Co. announces that operations will be resumed on September 17. The union accepts the new scale offered by the company in July.

NEVADA

ESMERALDA COUNTY

(Special Correspondence.)—The Goldfield Consolidated Co. is still experiencing trouble in marketing its flotation-concentrate, due to the unwillingness of the custom-smelters to receive a refractory gold product. As a consequence the production and earnings of the company have declined. It is reported here that the directors are considering the erection of a refining-plant for the treatment of the concentrate. Copper-gold ore is again being produced from the 1250-ft. level of the Grizzly Bear mine, where a large body of medium-grade ore has been developed. Much of the richer ore is being drawn from the 600 workings of the Laguna and the 450 level of the Mohawk.

The Jumbo Extension is preparing for the sinking of a deep winze from the 1017-ft. level to prospect the shale and alaskite below.

The Atlanta company has installed on the 1750-ft. level a large compressor, pumps, electric hoist, and other equipment, and will immediately sink a winze to the 2000-ft. level for the purpose of prospecting. Regular shipments of selected ore are being made to custom-smelters, and the company is said to be earning fair profits. The new raise from the 160-ft. level of the Great Bend, about 35 ft. east of the main stope, has advanced 60 ft. in high-grade ore, the working-face disclosing a width of 4½ ft., assaying $150 per ton. Cross-cutting and driving is in progress on the 300-ft. level, and raises from the level have disclosed good ore. Small shipments are being made, and the management expects to increase the output.

Goldfield, September 18.

HUMBOLDT COUNTY

The Nevada Packard has granted its employees a 10% bonus of their present wages. The bonus applies to the technical staff as well as the laborers and will be in force as long as silver is quoted at 93c. or more per ounce. The present minimum wage is $4.95.

NYE COUNTY

(Special Correspondence.)—The development of the White Caps Mining Co. totaled 161 ft. for the week. The West claim is being prospected. The mill is operating steadily and the anticipated tonnage is being handled.

In the Manhattan Consolidated the shaft has been deepened 108 ft. below the third level. After constructing a sump, the development of the 400-ft. level will be started. No water has been encountered in the shaft.

The Union Amalgamated has completed the development of its orebodies upon the 600-ft. level. The total development for the week was 62 ft. Option to purchase the controlling interest in the company has been exercised and the first payment made upon the stock. It is understood that Whitman Symmes will take charge as consulting engineer.

The working shaft of the Extension and Zanzibar has penetrated the White Caps limestone. After additional depth it is presumed that lateral development will be extended to the White Caps fault. The hoist and compressor-houses have been completed.

The Amalgamated Extension has made arrangements to resume operations. The shaft will be continued well below the water-level before additional lateral work is attempted. The equipment of the company consists of a hoisting and compressor plant.

Manhattan, September 13.

(Special Correspondence.)—The Tonopah Mining Co. during the week produced 1900 tons of ore. Operations at the Sandgrass shaft have been suspended to re-timber the shaft.

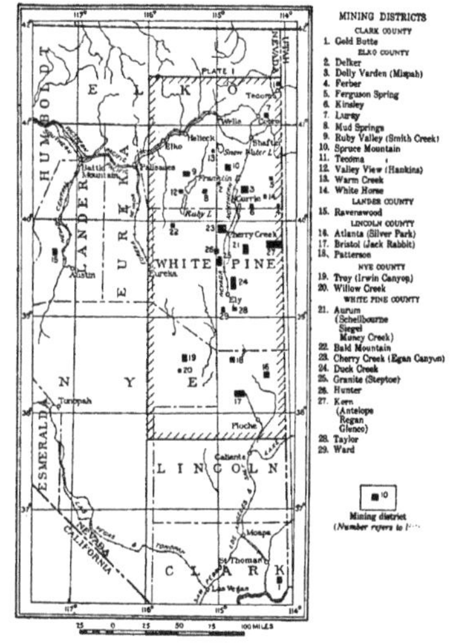

EASTERN NEVADA MINING DISTRICTS

The exploration department was discontinued September 1. During the past five years the Tonopah Mining Co. has acquired an interest in the Tonopah Placers Co., at Breckenridge, Colorado, the Canadian Mines Co., in Manitoba, Canada, the Eden Mining Co., in Nicaragua, the Tonopah Nicaragua Co., the Mandy Mining Co., in Canada, and the Tunkey Transportation & Power Co., in addition to the Desert Power & Milling Co., at Millers, Nevada, and the Tonopah & Goldfield railroad.

The Tonopah Extension's auxiliary power-station, consisting of a steam turbine and a 400-kw. generator, is ready for operation and is to be used in the Victor territory. Development, consisting of 193 ft., was made in the Victor and No. 2 shafts. The tonnage for the week was 2380.

The Tonopah-Belmont produced 2184 tons for the week.

The Jim Butler Tonopah Mining Co. produced 750 tons. Driving on the West End vein from the 816 winze shows a face of low-grade ore.

The bi-weekly shipment of bullion from the West End mill consisted of 31 bars, valued at $55,930. The tonnage was 909, from the Ohio and West End shafts, and 48 tons from the Halifax. The 522 raise in the Ohio made connection with the level. In development work, drifts 525, 526, and 520, and raises 523 and 524 showed good widths of milling ore. Conditions in the 717 drift on the east and west side are unchanged.

The Rescue-Eula produced 125 tons, and the Cash Boy 50 tons, making the week's production at Tonopah 8356 tons with a gross value of $146,230.

Tonopah, September 11.

By a vote of 402 to 317 the mine and mill employees of Tonopah accepted the proposition of the mine operators to grant an advance of 50c. per shift for all who work 20 shifts in one month. A part of the workers voted not to work for a less advance than 75c. per shift regardless of the time employed.

NEW MEXICO

SOCORRO COUNTY

(Special Correspondence.)—A new strike was made on the 300-ft. level of the Johnson mine during the past week. This property mills a good grade of ore but the recent find has opened three feet of ore exceeding the average in value.

The Oaks Co. is constructing new ore-bins at the Maud S. mine to provide for an increased production. The bins are arranged for serving either burro or auto-truck.

The recent discovery of high-grade ore in the lower level of the Socorro mine continues to show a good width and value. It is on the 1100-ft. level and, as it is the deepest ore so far opened in the district, it is causing satisfaction.

The Pacific mine production for the past month was increased over the normal.

Mogollon, September 11.

OREGON

JACKSON COUNTY

(Special Correspondence.)—The Blue Ledge copper mine is again an object of inquiry. Its sale is forecasted to the English syndicate that owns the Lady Smith smelter at Vancouver, B. C. The copper ore from this mine with that from the Alaskan mines makes an economic flux. Sixteen motor-trucks are employed in hauling ore from the mine to the railroad at Jacksonville.

Andrew Jeldness of Medford, who has been developing the Bloomfield copper mine in the Blue Ledge district, has uncovered a body of rich ore and will start shipping. C. J. Fry of Medford is developing several promising copper veins in the Blue Ledge district.

Operators are active in the Gold Hill district investigating manganese deposits. Recent tests from the Gold Hill iron mine have proved that this ore will produce a low-carbon manganese-iron alloy. Several mines in the Meadows district, north of Gold Hill, will be developed to prove the extent of their manganese deposits.

Medford, September 12.

(Special Correspondence.)—The Blizzard gold mine, two miles north of Gold Hill, was sold yesterday to J. W. Wakefield of Medford, Oregon, and his associates, who are Eastern people. The new owners have been investigating the district and selected this mine. They will re-open the old workings and expect to erect a 5-stamp mill. The vein, in a porphyry and slate-contact, contains free-milling ore, three feet wide. Ore averaging $40 per ton was reduced in local mills when the mine was in active operation.

Another strike of specimen ore has been made on the old Elashia Ray mine, three miles north of Gold Hill. It is operated by J. W. Davies and associates of Sacramento.

Gold Hill, September 12.

JOSEPHINE COUNTY

(Special Correspondence.)—Five auto-trucks are being used to haul ore from the California Chrome Co.'s mine to Waters creek. The haul is 21 miles and 50 tons is shipped per day. The destination of the ore is Niagara Falls, N. Y. At the mine 21 men are employed.

Chrome-deposits are being developed by R. C. Pehely and R. W. Kitterman on Gray Back mountain.

Grants Pass, September 4.

(Special Correspondence.)—C. Long and F. Nelson are developing a copper deposit in the Preston Peaks district.

Chrome ore is to be concentrated at the Dorothea Chrome mine in Coyote creek. Crushers, stamp-batteries, classifiers, and standard tables comprise the mill equipment. The ore contains 30% chrome, associated with serpentine, and it is expected that a 65% concentrate will be made.

Grants Pass, September 4.

(Special Correspondence.)—The Logan placer-mine situated on the Grants Pass-Crescent City highway, two miles northwest of Waldo, has been sold to George M. Esterly and associates of Seattle, Washington. The purchase price is $140,000. The mine is one of the oldest in the State. Its early history is contemporaneous with the gold-rush to Jacksonville. An option on the property was taken by the new owners ten months ago, and it was operated by them during the past season, the output being $50,000 in gold and platinum. The property will be operated on a much larger scale. The water-supply permits mining for eight months of the year. The placer-gold is fine, and is accompanied by platinum, together with osmium and iridium.

Grants Pass, September 4.

(Special Correspondence.)—The Waldo copper mine, two miles from Waldo, has been sold to the American Exploration Co. of Grants Pass. The purchase price was $135,000. The property has been in operation 12 years, but only during the last two years has it been worked with modern machinery. A mill of 50-ton capacity is used to concentrate the ore. Recent production totals $300,000 worth of copper. The ore is a massive chalcopyrite associated with pyrrhotite and pyrite.

Grants Pass, September 7.

WASHINGTON

The Northwest Magnesite Co., operating at Chewelah, shipped 11,800 tons of crude magnesite in August. The calcined product aggregated more than 500 tons. The September production is expected to be 15,000 tons of crude mineral and 1500 to 2000 tons of the calcined product.

Arthur L. Lyddane, of Washington, D. C., and C. Edwin Oyster, of Berkeley, California, representatives of the Federal Trade Commission, are in the Coeur d'Alene investigating the cost of producing lead. They are visiting the principal lead-producing districts of the United States. Mine-owners have contended that 8 cents per pound is not enough for this metal while cost of production is so high. This is the price the Government is now paying.

The third dividend of the Loon Lake Copper Co., operating 45 miles north of Spokane, amounting to $14,513, was paid September 11.

Grant county has a new industry being developed at Quincy by the American Mill & Produce Co. A short time ago this concern bought the lease on a half-section of State land containing infusorial earth, and shipments of the earth are being made to different parts of the United States.

FERRY COUNTY

For the week ended September 15 the mines of Republic camp have shipped 1250 tons of ore to the smelters, as follows: Quilp lessees, 600 tons; Lone Pine mine, by the Northport Smelting & Refining Co., 500 tons; Knob Hill mine, 150 tons.

The Northport S. & R. Co. has added to its working force at

the Lone Pine mine for the purpose of increasing ore shipments to 150 tons per day.

A discovery of silver-lead ore near Swan lake, 20 miles west of Republic, has been reported and a number of locations have been made.

At the Drummer mine, two miles south of Curlew, the National Silver-Lead Co. has installed a compressor plant for the purpose of extending the main adit, now in 700 ft. and carrying out plans for other development through the fall and winter.

In the Keller district, at the Golden Crown mine a large quantity of ore is on the dump ready for shipment and the ore-body is being developed. The ore is rich and shipments will soon be made after the return of J. C. Davis, manager, from Pittsfield, Illinois. The mine is owned by an Illinois company.

Republic, September 14.

CANADA

BRITISH COLUMBIA

(Special Correspondence.)—The heavy demand for copper for war purposes has greatly stimulated the copper mining and smelting industry of British Columbia. Recent returns give the production by the smelters for 1916 at 65,379,364 lb. of fine copper, valued at the average New York price for the metal at $17,764,494. This is an increase in quantity of 8,460,959 lb. and in value of $7,948,994 over the output of 1915. The production for the current year will show a still greater increase, owing to the large expansions which are being undertaken by the producing companies. The Granby Consolidated, which during the year ended June 30, spent approximately $1,000,000 for improvements, has plans in hand which will necessitate even a larger outlay. It has acquired a high-grade coal property, one seam of which shows 20,000,000 tons of coking coal, and is constructing a by-product coke plant which will supply its coke requirements at a low cost. The Canada Copper Co. is opening up a large mine near Princeton, B. C., in which the ore is estimated at 10,000,000 tons. The power to operate this mine will be supplied by the West Kootenay Power & Light Co., a subsidiary of the Consolidated Mining & Smelting Co. of Canada, and the ore will be, for some time at least, smelted and refined at the plants of the company at Trail.

Toronto, September 11.

(Special Correspondence.)—The Granby Consolidated has declared the usual quarterly dividend of 2½%.

It is reported that Eastern interests have acquired 1,400,000 of the 1,600,000 outstanding shares of the Utica mines. The mine is in the western part of the Ainsworth district, West Kootenay.

At the smelting plant of the Canada Copper Corp., at Greenwood, 60 men are employed. A car of blister copper was recently shipped to Trail.

Vancouver, September 7.

The Ladysmith smelter after being closed down for five years has been re-opened. Announcement has been made that the treatment-charge will be $5 per ton. All of the copper will be paid for with the exception of 0.4%. The price paid for copper will be three cents per pound less than New York quotations, 90 days after the receipt of the ore; 95% of the gold content to be paid at $20 per oz. (under 0.1 oz. not to be paid for); 95% of the silver content to be paid for at New York quotations; 90 days after the receipt of the ore (under 0.5 oz. not to be paid for).

MEXICO

The Mexican government threatens to confiscate the properties of the American S. & R. Co. in the State of Chihuahua unless work is started at once. The properties have been idle for two years. It is said that a syndicate of wealthy Mexicans has offered to take over the smelting company.

PERSONAL

Note: The Editor invites members of the profession to send particulars of their work and appointments. This information is interesting to our readers.

H. P. GORDON is in Idaho.

MORTON WEBBER is at Salt Lake City.

H. F. A. RIEBLING is at Gold Basin near Chloride, Arizona.

J. B. TYRRELL is examining mining property in Newfoundland.

WARD ROYCE is Captain of Engineers at Fort Leavenworth, Kansas.

FRED. SEARLS, JR., is in service with the Engineer Corps in France.

HARRY H. HANNAH has moved from Los Angeles to Lida, Nevada.

COURTENAY DE KALB is at Bisbee, after a visit to the El Paso smelter.

L. WEBSTER WICKES has moved from Los Angeles to Landore, Idaho.

M. C. DRAKE is at the Officers Training Camp at Fort Sheridan, Illinois.

ROY KING has returned to California, after spending four years in Colorado.

ELLSWORTH H. SHRIVER is Lieutenant in the 23rd Engineers at Camp Nevada, Maryland.

LAWRENCE COMAN is in the New National Army as a private at Ft. Leavenworth, Kansas.

H. H. BURHANS is First Lieutenant with the 316th Engineers at Camp Lewis, Washington.

STUART RAWLINGS is now assistant manager for the Cerro de Pasco Mining Co., in Peru.

FALCON JOSLIN has returned to Fairbanks, Alaska, after a visit to New York and Washington.

MORTIMER NORTH is Lieutenant of Engineers and is raising a company of miners to go to France.

T. C. ROBERTS, engineer to the United Verde mine, at Jerome, is in New York on his way to South America.

JOHN BALLOT, chairman of Minerals Separation Ltd., has returned to New York from Houghton, Michigan.

NORMAN D. LINDSLEY is First Lieutenant in the Engineer Officers Reserve Corps at Ft. Leavenworth, Kansas.

THOMAS M. SMITHER is First Lieutenant in the Engineer Officers Reserve Corps at Fort Leavenworth, Kansas.

FAYETTE A. JONES, formerly president of the New Mexico School of Mines, has returned from Alaska to Albuquerque.

HORACE V. WINCHELL passed through San Francisco on his return from Japan, where he examined the Ashio copper mine.

JOHN G. KIRCHEN, general manager for the Tonopah Extension Mining Co., has moved his headquarters to Reno, Nevada.

E. E. FREE has been given a commission as Captain in the Ordnance Department and has been detailed to inspect munitions.

G. R. DE BEQUE is Lieutenant in the Engineer Corps of the American Expeditionary Force and is now on his way to France.

HUGH MCNAIR, son of F. W. MCNAIR, president of the Michigan College of Mines, has been decorated in France for special bravery in the Ambulance Corps.

Harry S. Richards, First Lieutenant of Infantry, son of James Richards, superintendent of the Isle Royale mine, Houghton, is on his way to France for special duty.

J. E. SPURR has resigned as advising engineer to the Tonopah Mining Co., on the decision of the company to drop its exploration department, which has accomplished its object.

Mineral Output of Nevada

The mine production of gold, silver, copper, lead, and recoverable zinc in Nevada in 1916 is given by the U. S. Geological Survey in the accompanying table. Nevada has made steady progress in mineral production since the discovery of Tonopah. Nine counties in the State have an annual production of more than $1,000,000 in value. The four most important are White Pine, Nye, Clark, and Esmeralda. Almost two-thirds of the ore-tonnage is derived from copper and the remainder from gold, silver, lead, and zinc mining.

MINERAL PRODUCTION OF NEVADA IN 1916
(Advance figures by V. C. Heikes, U. S. Geological Survey)

County	Number of producers	Ore.	Gold. a	Silver a	Copper	Lead	Recoverable zinc content.	Total value.
		Short tons.	Fine ounces.	Fine ounces	Pounds	Pounds	Pounds	
Churchill	20	110,943	13,734.15	1,386,524	9,609	8,902		$1,199,221
Clark	85	103,374	14,912.66	404,788	573,088	8 405,796	29,400,537	5,235,275
Douglas	5	84	4.31	38	5,981			1,585
Elko	76	40,729	13,649.20	268,156	1,771,689	1,098,911	963,438	1,099,363
Esmeralda	59	389,351	129,871.78	170,677	1,391,352	408,534	17,680	3,169,833
Eureka	36	22,082	5,551.12	143,686	79,066	2,804,981	34,935	426,972
Humboldt	95	109,683	19,264.42	869,424	135,180	75,734		1,008,792
Lander	41	16,099	9,765.02	210,539	576,493	199,039		495,947
Lincoln	23	162,153	1,659.50	559,045	806,938	10,356,926	2,002,105	1,583,574
Lyon	47	31,422	3,945.76	36,080	3,491,390			964,188
Mineral	96	266,183	34,300.58	431,130	2,986,361	270,734		1,746,064
Nye	100	564,450	121,213.92	8,868,267	6,023	135,355		8,351,855
Ormsby	3	107	16.45	484	134			691
Storey	24	63,187	23,371.75	285,827	903	11,057		672,196
Washoe	28	5,061	1,698.59	55,750	174,621	220,243		129,951
White Pine	41	4,056,257	35,945.00	:147,110	93,107,945	1,641,066	24,494	23,260,917
Total, 1916 ...	779	5,941,165	428,904.21	13,837,525	105,116,813	25,637,278	32,443,189	b49,946,424
Total, 1915 ...	799	5,118,620	551,683.01	14,459,840	68,636,370	16,637,277	24,376,450	34,551,436

a Includes placer production.
b Average value of metals: Gold, $20.6718 per ounce; silver, $0.658 per ounce; copper, $0.246 per pound; lead, $0.069 per pound; zinc, $0.134 per pound.

Recent Decisions

MINERS' LIENS—ALASKAN LAW CONSTRUED

Under the Act of the Alaskan legislature of April 30, 1913, a mine owner who leases his claim and fails to post non-responsibility notices or record the lease is liable to have his mine taken under foreclosure of the liens of miners and material-men employed by the lessee.

Spalding *v.* Martin (Alaska), 241 Federal, 372. April 9, 1917.

OIL LEASE—FAILURE TO DRILL OFFSET WELLS

Where the lessor in an oil and gas lease sues the assignee of the lessee for damages resulting from his failure to drill wells and save the lessor's ground from being drained of oil through wells drilled in adjacent territory, he must prove the assignment and transfer of the lease, operations under the lease by the lessee, negligence in failing to offset wells driven on adjoining territory, and the actual damage in loss in royalties to the lessor resulting therefrom.

Steel *v.* American Oil Development Co. (West Virginia), 92 South-eastern, 410. April 17, 1917.

OIL LEASE—EVIDENCE OF EXTENSION

When a mineral lease by its terms expires within a definite period unless oil or gas is discovered, in which event it is to be extended so long as either can be produced in paying quantities, the fact that after the expiration of the period named the lessor executed a deed purporting to be subject to the lease, and the lease itself was assigned, constitutes no evidence, in an action for rent, of an extension of the life of the lease. Nor is the fact that such has not been released of record evidence of its still being in force; nor does its execution create a presumption of possession by the lessee.

Ashgrove Lime & Portland Cement Co. *v.* Chanute Brick & Tile Co. (Kansas), 164 Pacific, 1087. May 12, 1917.

ACT OF 1866—SURFACE PATENTED LIMITS APEX

A patentee under the Act of 1866 is entitled only to so much of the vein or lode as apexes within the surface for which he has elected to apply for patent, even though if he had kept his location upon which the patent was based good through performance of annual labor he might have claimed under the location a greater length of vein with sufficient surface ground to work it. Although the Act of 1866 is indefinite as to the grant of surface ground and apex, the election of a locator under that Act of the amount of surface he will take in his patent delimits also the amount of apex he gets under that patent, and the patent is void as to any part of the apex attempted to be included which lies outside the limits of the selected surface.

Whildin *v.* Maryland Gold Quartz Mining Co. (California), 164 Pacific, 908. May 17, 1917.

METAL PRICES

San Francisco, September 18

·lb. lots), per lb	$1.00
lots), per lb	$0.95
pound	16.50
), cents per lb	14.75
cents per pound	28.50
ound	9.50
ard metal, respectively, per ounce	$105—111
of 75 lb	$115
und	10
	60
ound	20

ORE PRICES

San Francisco, September 18

l. per unit	$1.10
er. free SiO_2 limit 8%, f.o.b. California	
ng to grade	$0.60— 0.75
r ton	$8.00—10.00
per unit	$23.00—28.00
for MoS_2 contained	$40.00—45.00

schedule less freight to Chicago. Less than 40%

und specifications, as per the quotations of the Car-
ule of prices per ton of 2240 lb. for domestic man-
freight prepaid, at Pittsburgh, Pa., or Chicago, Ill.

	Per unit
manganese	$1.00
manganese	0.98
manganese	0.95
manganese	0.90

n ore containing not more than 8% silica nor more
s, and are subject to deductions as follows: (1) for
8% silica, a deduction of 15c. per ton, fractions in
each 0.02% in excess of 0.2% phosphorus, a de-
nit of manganese per ton, fractions in proportion;
s than 40% manganese, or more than 12% silica, or
subject to acceptance or refusal at buyer's option;
nalysis of sample dried at 212° F., the percentage of
le as taken to be deducted from the weight. Prices
without notice unless specially agreed upon.

EASTERN METAL MARKET

(By wire from New York)

per is quiet and nominal, remaining at 26.25c. all
and weaker, ranging from 9.87 to 8.87c. Zinc is
25 to 8.37c. Platinum remains unchanged at $105
11 for hard.

SILVER

average New York quotations, in cents per ounce.

		Average week ending		
	98.62	Aug.	7	80.50
	98.62	"	14	82.95
	100.50	"	21	87.16
	100.50	"	28	88.50
		Sept.	4	91.12
	102.50	"	11	96.80
	103.50	"	18	100.70

Monthly Averages

916	1917		1915	1916	1917
.76	75.14	July	47.52	63.06	78.92
.74	77.54	Aug.	47.11	66.07	85.40
.89	74.13	Sept.	48.77	68.51	....
.37	72.51	Oct.	49.40	67.86	....
.27	74.61	Nov.	51.88	71.60	....
.04	76.44	Dec.	55.34	75.70	....

COPPER

in New York, in cents per pound.

		Average week ending		
	26.25	Aug.	7	28.46
	26.25	"	14	27.62
	26.25	"	21	28.00
	26.25	"	28	26.41
		Sept.	4	25.65
	26.25	"	11	25.83
	26.25	"	18	26.25

Monthly Averages

16	1917		1915	1916	1917
30	29.53	July	19.09	25.66	29.67
62	34.57	Aug.	17.27	27.03	27.42
65	36.00	Sept.	17.69	28.28	....
02	33.16	Oct.	17.90	28.50	....
02	31.69	Nov.	18.88	31.95	....
17	32.57	Dec.	20.67	32.89	....

ds have been declared: Nevada Consolidated, $1
ated, $1; Chino Copper Co., $2.50; Butte & Su-
Copper Co., $3.50 per share.

Anaconda's production last month of 11,175,000 lb. was the smallest in four years. The properties had been operating at 33 1/3% capacity through July and greater part of August. Washoe smelter closed August 25, following a walk-out of the men. Great Falls refinery followed, and the mines shut-down, so that activity of the company in Montana has practically ceased.

Exports of copper for the fiscal year ended in June amounted to 1,021,-501,398 lb., as against 711,342,146 in 1916. Exports of zinc were 504,204,665 lb., as against only 4,560,000 lb. in the year ended just before the War began.

It is announced that the sale of copper to the Allies in August amounted to 77 million pounds at 25c. per pound.

LEAD

Lead is quoted in cents per pound, New York delivery.

Date			Average week ending		
Sept.	12	9.87	Aug.	7	10.87
"	13	9.75	"	14	10.87
"	14	9.50	"	21	10.79
"	15	9.00	"	28	10.61
"	16 Sunday		Sept.	4	10.38
"	17	9.00	"	11	10.20
"	18	8.87	"	18	9.33

Monthly Averages

	1915	1916	1917		1915	1916	1917
Jan.	3.73	5.95	7.64	July	5.59	6.40	10.93
Feb.	3.83	6.23	9.01	Aug.	4.62	6.28	10.75
Mch.	4.04	7.26	10.07	Sept.	4.62	6.86	....
Apr.	4.21	7.70	9.38	Oct.	4.62	7.02	....
May	4.24	7.38	10.29	Nov.	5.15	7.07	....
June	5.75	6.88	11.74	Dec.	5.34	7.55	....

ZINC

Zinc is quoted as spelter, standard Western brands, New York delivery, in cents per pound.

Date			Average week ending		
Sept.	12	8.25	Aug.	7	8.75
"	13	8.25	"	14	8.75
"	14	8.25	"	21	8.66
"	15	8.25	"	28	8.33
"	16 Sunday		Sept.	4	8.25
"	17	8.37	"	11	8.21
"	18	8.37	"	18	8.29

Monthly Averages

	1915	1916	1917		1915	1916	1917
Jan.	6.30	18.21	9.75	July	20.54	9.90	8.98
Feb.	9.05	19.99	10.45	Aug.	14.17	9.03	8.58
Mch.	8.40	18.40	10.78	Sept.	14.14	9.18	....
Apr.	9.78	18.62	10.20	Oct.	14.05	9.92	....
May	17.03	16.01	9.41	Nov.	17.20	11.81	....
June	22.20	12.85	9.63	Dec.	16.75	11.26	....

QUICKSILVER

The primary market for quicksilver is San Francisco, California being the largest producer. The price is fixed in the open market, according to quantity. Prices, in dollars per flask of 75 pounds:

Date			Week ending		
			Sept.	4	115.00
Aug.	21	115.00	"	11	115.00
"	28	115.00	"	18	115.00

Monthly Averages

	1915	1916	1917		1915	1916	1917
Jan.	51.90	222.00	81.00	July	95.00	81.20	102.00
Feb.	60.00	295.00	126.25	Aug.	93.75	74.50	115.00
Mch.	78.00	219.00	113.75	Sept.	91.00	75.00	....
Apr.	77.50	141.60	114.50	Oct.	92.90	78.20	....
May	75.00	90.00	104.00	Nov.	101.50	79.50	....
June	90.00	74.70	85.50	Dec.	123.00	80.00	....

TIN

Prices in New York, in cents per pound.

Monthly Averages

	1915	1916	1917		1915	1916	1917
Jan.	34.40	41.76	44.10	July	37.38	38.37	62.60
Feb.	37.23	42.60	51.47	Aug.	34.37	38.88	62.53
Mch.	48.76	50.50	54.27	Sept.	33.12	36.66	....
Apr.	48.25	51.49	55.63	Oct.	33.00	41.10	....
May	39.28	49.10	63.21	Nov.	39.50	44.12	....
June	40.26	42.07	61.93	Dec.	38.71	42.55	....

ANTIMONY

The ore seems to be a little lower at from $1.70 to $2 per unit for the best material.

MOLYBDENUM

Concentrate continues unchanged at $2.10 to $2.20 per lb. of contained MoS_2. Not much material is being offered here.

MANGANESE

As high as $1.25 to $1.30 per unit, seaboard, has recently been paid for high-grade (40-50% Mn) ore from India. This is a record price to date.

Eastern Metal Market

New York, September 12.

A better feeling generally pervades the metal markets. There has been more inquiry in some cases than in weeks, particularly for copper and zinc. The prospect of an early settlement of the copper controversy is having a stimulating influence on the market. In the absence of large buying many quotations are still nominal.

Copper is firmer but nominal.
Tin is steadier and higher.
Lead is in poor demand and lower.
Zinc is more active but unchanged.
Antimony continues lifeless.
Aluminum is quiet and no stronger.

Government purchases make up the greater part of the current business the last week in the steel market. Other consumers seem to believe that the market will continue to decline and are in no hurry to buy. Any readjustment of iron and steel prices is very slow and the situation in which the Government finds itself is more complicated if anything. It is significant that representatives of some of the Entente Allies are again negotiating actively on their own behalf and independently, at least temporarily, of the War Industries Board. Sales of re-sale pig-iron are more frequent, due possibly to unconfirmed rumors that Government prices on pig-iron are to be lower.

COPPER

It is at last possible to report a better tone to the copper market after three or more months of dullness and stagnation. The principal cause for this is the official announcement of the purchase by the authorities at Washington of 76,654,000 lb. of copper for the Allies at a price understood to be 25c. per lb. This means an expenditure of nearly $2,000,000. There remains yet to be decided the price and the requirements of our own Government. Negotiations looking to a decision of this point are now active at Washington and a decision is expected today. It is hoped by some that the same price will be decided upon, but it will not be surprising if 20 or 22.50c. per lb. is fixed. It is certain, however, that a decision of this long drawn out matter will be of incalculable benefit to copper and the non-ferrous metals in general. There has already appeared in the trade a more active inquiry in the last few days, a direct result of the allied purchase. Domestic consumers are on the look-out and ready to cover their accumulated needs just as soon as a price level seems assured. Actual sales have so far been few, or almost negligible, despite rumors of large domestic buying. Quotations have certainly stiffened but they are still quite nominal. Yesterday both Lake and electrolytic were quoted at 26.25c., New York, for early delivery with last quarter firm at 25.50 to 26c. Without doubt better days are ahead for the copper market and an active market is felt by many to be not far distant. The London market continues unchanged at £137 for spot electrolytic and £133 for futures.

TIN

The market last week, after our letter was written, continued nominal and stagnant. There were more sellers than buyers. On Friday a little better tone developed with inquiries appearing for lots of 20 to 25 tons, but no sales were reported. This week thus far the market has shown more life. On Monday, while quiet, there developed a better tone, aided to some extent by fair sales of small volume. Yesterday, Tuesday, demand was still better, moderate sales being recorded aggregating 125 to 150 tons. The prospects are that this better feeling will continue and the trade will not be surprised if the tin market picks up after a long dullness. Arrivals to September 11 inclusive have been 700 tons with 5400 tons reported afloat. The London market has declined 10s. per ton in the past week and now stands at £243 for spot Straits.

LEAD

The American Smelting & Refining Co. has again contributed the feature to a more or less lifeless market. It repeated on Wednesday last what it did the week before—lowered its quotations $10 per ton or ½c. per lb. to 10c., New York. Two such recessions in two weeks are not of a nature to stabilize a market. There appears to be a decided lack of confidence which pervades the entire situation. There is an indication that consumption of manufactured goods is falling off. Sellers lack initiative and fear to press the market, realizing that such pressure will only accelerate the decline. The entire tone is lower and new inquiry is absent. This was the status at the close of last week. Since then the markets, particularly the outside ones, have declined and free offers have been made at 9.50c., St. Louis. This is reported to have resulted in a fair business. There are hints of fresh inquiries from Canada. With 8c. per lb. still in the minds of purchasers as the Government price, continued hesitation is to be expected, supported as it is by the conditions here related.

ZINC

Sentiment is decidedly better and it is believed that the market has recently touched bottom in the stagnant weeks of the immediate past. A turn for the better is believed to be at hand or at least to have started. This is based on the manifest increase in inquiry in the last week. Dealers report that in the week ended Saturday there was more inquiry and more business done than in many weeks and that so far this week, two days, inquiry largely exceeded anything in several months. It must be acknowledged, however, that the sales referred to have not been of large enough volume to greatly affect quotations which yesterday were 8c., St. Louis, or 8.25c., New York, for early delivery. For future delivery, or last quarter, about ⅛ to ¼c. above these prices is asked, but commitments beyond October are not eagerly entered into. The apparent cleaning up of at least part of the copper situation, referred to in this letter, is also a stimulating market influence in zinc. The fact also should not be lost sight of that production has been cut down from 30 to 40% in the entire country owing to some producers being obliged to operate at a loss under present price and labor conditions. 'Under the rule' on the New York Metal Exchange there was sold last week 200 tons of prime Western at 7.50c. in bond and 300 tons at 7.80c. in warehouse, New York, for immediate delivery.

ANTIMONY

There seems to be but little demand for antimony and Chinese and Japanese grades are quoted at 14.50 to 14.75c. per lb., New York, duty paid.

ALUMINUM

The market is dull with No. 1 virgin metal, 98 to 99% pure, quoted at 46 to 48c. per lb., New York.

TUNGSTEN

Prices continue to range from $23 to $25 per unit for tungsten ores with $26 asked for some high-grade ores. A fair amount of business is reported. Ferro-tungsten is a little stiffer with a minimum of $2.50 and up to $2.80 asked. Demand continues good from all sources.

A World's Series of base-ball games sounds pathetically ridiculous in these times. It will sound like an insult a few months hence.

SOLDIERS are proverbially sparing of words. Great soldiers are laconic. When General Pershing placed his wreath at the foot of Lafayette's tomb, he said simply: *Lafayette, nous voilà!* Thus the historic debt is to be paid. "Here we are, Lafayette!"

FIXING of prices for steel and iron, involving a reduction ranging from 43 to 70% for various products, is an epochal event; it is fraught with tremendous consequences both now and for the future of industry in this country. Incidentally it should prove a great boon to the manufacturers of machinery.

RUBLES are down to 12 cents as compared with a normal parity of 50 cents. The Mexican dollar is worth more dead than alive. The mark is not quoted. Exports of gold are under embargo. Silver is nearing $1.10 per ounce. The new Liberty bonds will be offered next month. Are you saving your money to invest in the second issue of the Liberty Loan?

MOLYBDENUM has proved a great help to the 75-mm. gun of the French army. This wonderful field-gun is lined with steel containing 9 to 10% molybdenum. It is reported that 2500 shots can be fired before re-lining becomes necessary. Molybdenum steel is also used most successfully in the trench-helmet, which is light and shrapnel-resisting, although useless against the rifle-bullet.

THE statistics of gold production in China for 1915 have been given officially. They are belated, but none the less welcome. The production was 200,000 ounces, of which 120,000 is credited to Manchuria, 60,000 ounces to Outer Mongolia, and the remainder to the provinces of Chihli, Hunan, Shantung, Kansu, Turkestan, Szechuan, Tunnan, Kiangsi, and Honan. It is interesting to compare this total of 200,000 ounces with the estimates published by various statisticians: the 'Mineral Industry' says 177,744 ounces; the 'Mining Magazine' says £755,000, or 182,500 ounces; while the 'Engineering & Mining Journal' credits China "and others" with $3,675,000, say, 183,750 ounces. These guesses were not at all bad.

IMMIGRATION records for the fiscal year ending June 30 show that 295,403 aliens entered this country and 66,277 departed, the net gain being 229,126 as against a net gain of 915,142 in 1914. It is interesting to note that the foreign countries making an increased contribution to our population last year as compared with 1914 were France, Spain, Portugal, and Mexico. In 1914 there came 251,612 immigrants from southern Italy; last year only 35,154. Similarly 138,051 Hebrews and 122,657 Poles came in 1914, as against 17,342 and 3109 respectively last year. The German immigration declined from 79,871 to 9682 during these two years.

TUNGSTEN is fairly steady in price. It is believed in well-informed quarters that a quotation of about $20 will be maintained even when the War ends, owing to the demand for railroad work in Europe, the tungsten-steel being used for punching rails before attaching the fish-plates and, of course, in all sorts of high-speed tools. The dumps of 1 to 2% tungstic acid that were exploited when the boom began are now depleted, as are also other surficial deposits of tungsten ores. The cost of production has increased so that no quick return to a pre-war price is at all likely. At present the production in this country is about 5000 tons and the importation 4000 tons, the imports being nearly equal to the exports of ferro-tungsten and tungsten powder, so that the production just takes care of the domestic consumption.

COMMISSIONS from Russia are plentiful. Such work is agreeable at this time. We hope that their efforts will be successful in stimulating American interest in Russian trade and industry, but it will all go for naught until order is restored and property rendered secure. The I. W. W. element is too prominent just now to warrant anybody engaging in new business in Russia. Alexander F. Kerensky is a sincere patriot, we do not doubt, but he is too much of a visionary and a talker to re-integrate the uneducated millions left suddenly without a recognized leader. A democracy is not made in a day. China and Mexico afford unfortunate examples of the disorder following the destruction of despotism and bureaucracy. If Mr. Kerensky is supported by the best of the military element he may succeed, but before order is established some of the forms of democracy must be dropped temporarily.

WE are asked to state that a fund is being raised to assist British and Canadians in California enlisting for service at the front. The allowance made by the British government is insufficient and it is desired to help the wives and mothers of soldiers by making a grant of $10 to $30 per month. Twelve hundred such men have volunteered for service from California and are now with the British army in France. It is hoped that some of our British and Canadian mining friends, unable to engage in active service, will contribute to this useful fund. Subscriptions should be sent to Mr. John Bishop, at 433 California street, San Francisco.

A LITTLE bird in Montana brought us the following story. It was a bird of the most veracious—not voracious—type. A new kind of concentrating table had been tested in the mill, and this table had yielded a tailing so poor in copper that everybody was delighted, especially the inventor, Mr. Emil Deister. Just then Mr. John Ballot, chairman of Minerals Separation, enters on the left, as they say in the play-book. He is surprised to learn that water-concentration can do as well as froth-flotation, on this particular ore at least. Whereupon he says to Mr. Deister: "Why are you so foolish as to sell your machine outright? Why don't you put it out on a royalty?" "No, thank you," says Mr. Deister, "I am quite content to make a decent profit and retain the goodwill of my customers, rather than harvest a crop of lawsuits." And when the little bird said that he cocked his eye in a knowing way.

ANNOUNCEMENT is made that the exploration department of the Tonopah Mining Company is to be discontinued and the energies of the management are to be concentrated on the development of the mining properties acquired through the exploration department during recent years. It is stated that Mr. J. E. Spurr, the distinguished economic geologist, has resigned as advisory engineer, having accomplished the purpose for which his services were retained. He has done well for the Tonopah company. In five years the total expenses of his department were $261,108, and during that time "the net value of the results" was $7,858,601. Apparently this means that successful business to that amount was transacted by acquiring properties or purchasing stock in operating companies.

IN the article describing mining methods in the Inspiration mine it is stated that technical men, that is, graduates from mining schools, are employed to supervise the tapping of chutes. This is a sign of the times. As hand-labor is decreased, giving way to automatic devices, the proportion of unskilled labor diminishes and the value of scientific supervision appreciates. These methods of mining, in which nature is called upon to assist so largely as to reduce human effort to the mere removal of the ore shattered by gravity, place a premium upon intelligence and thus afford increasing opportunities for the services of technical men. Brain becomes more essential than muscle.

ON another page we comment on Judge Bourquin's decision in the Butte & Superior suit. We are glad to see that our contemporary in New York finds reason for concluding that the Court has not "viewed the case in the light of common sense." Hitherto our contemporary has shown no proper appreciation of the exacting nature of the demands made by Minerals Separation; now it uses "hoggishness" to describe the demands of the patent-exploiting company. It is well that the New York paper should have awakened, even tardily, to a realization of the real position of affairs, so menacing to the welfare of the mining industry. Here we may record the fact that the District Court at Butte, on September 17, ordered the Butte & Superior company to give bond for $10,000 to cover the cost of appeal. A stay of injunction was granted for 30 days and thereafter until final determination on appeal, provided the defendant filed a bond for $2,500,000 and either paid into court all the profit made each month on all material milled and treated by the process or filed a satisfactory bond each month to secure such amount. The case should be heard on appeal in San Francisco before the close of this year.

COPPER has been priced by the Government at 23½ cents f.o.b. New York, this price being subject to revision four months hence. The War Industries Board, in whose hands the matter lies, has stipulated that the producers shall not reduce the wages now being paid, they shall sell their copper to the Allies and to the public at the price now fixed by the Government, and they shall follow the directions of the Board for the distribution of the metal, with a view to preventing it from falling into the hands of mere speculators, and finally that the producers shall pledge themselves to maintain production "at the maximum so long as the War lasts." The Board acknowledges the "admirable spirit" shown by the producers in giving information and in meeting the wishes of the Government. At the same time the point is made that the high price of 36 cents attained by copper would have been exceeded by this time if the Government had not taken the matter in hand. We believe this to be true and we believe that the regulation of the price at a reasonable figure is to the best interests of the copper-mining industry. A warning is sounded that any mining company failing "to conform to the agreement and price" may be disciplined by having its mine and plant taken over by representatives of the Government and having them operated on the National account. No figures of cost have been disclosed. It is probable that the information recently gathered on the subject will be kept by the Government officials to guide them in the fixing of prices from time to time. The publication of any attempted average would only excite undesirable controversy. The announcement of the price at 23½ cents has given satisfaction on Wall Street and it will, we believe, meet with general approval, as being just and expedient.

IN our issue of July 7 we published some notes concerning submarines and their capabilities. We are glad to know that the American Committee of Engineers in London re-published them in pamphlet form for the use of those interested in devising ways of circumventing the assassins of the sea. In this issue we reproduce a further bulletin prepared by the Naval Consulting Board. We commend the information to our readers. If any of them have good ideas on the subject they should address themselves to the Secretary of the Board, Mr. Thomas Robins, at 13 Park Row, New York. Since we last wrote on the subject nothing else that is important has been made known, at least nothing authentic. The most promising was the unofficial statement that listening devices have been developed to the point where the presence of a submarine at a distance can be detected sufficiently accurately to assist the chase by destroyers and other useful craft. On September 18 a wild story came from Washington announcing that Mr. Thomas A. Edison had invented a means of deflecting the course of a torpedo so as to render it innocuous, but this was denied promptly both at Mr. Edison's laboratory and by the Naval Department; yet, although the yarn was punctured promptly, it was used by one of our local organs of misinformation as the subject of an editorial expatiating upon the wonderful discovery. Country papers will reproduce the misleading comment of this careless scribe and another myth will be added to the many created by the reckless purveyors of news.

Sampling Large Low-Grade Orebodies

For the sake of reference we stick to the above heading, under which an interesting technical discussion was started and has been continued with a good deal of spirit. In this issue we publish another contribution from Mr. Morton Webber, who, in reply to friendly comment, lays emphasis on the employment of the mill, not to sample ore in the mine, but to ascertain the factor of error in ordinary moil-sampling. The object of sampling is to ascertain the yield to be obtained when the mine is exploited on a large scale. This ordinary sampling often fails to do, as every engineer, from the youngest to the oldest, is likely to learn to his cost at some time or other, be he ever so careful and skillful. The reason for the failure is due to an unknown and variable factor of error. Likewise the mill-test by itself has failed to be an unerring guide. At best the ordinary mill-test served as a sequel to the hand-sampling and only confirmed the estimate applicable to a particular and small part of the mine. Now comes Mr. Webber with the excellent suggestion to use the mill-test as a means of determining the sampling-error, which when known can be used at once to correct the results obtained by moiling. We commend his idea and hope that members of the profession will give it serious consideration—also criticism. Commonly the mill-test is used either as a sampling method or as a metallurgical investigation. For the moment we disregard both these functions and concentrate our attention on a new use of the mill. When it is used for the purpose of checking the sampling of various blocks of ore it does not matter whether the recovery, or the extraction as the case may be, is complete or not, all that is needed is to ascertain accurately the valuable content of the tonnage treated. To obtain this it is not necessary to ensure high metallurgic efficiency but to ascertain closely the loss in the tailing, so that it may be added to the recovery, for the purpose of knowing precisely the assay-value of the mill-feed. This permits a choice of milling method and points to one in which the sampling of the tailing is likely to be least inaccurate. Therefore for a gold mine it is better to use a small cyanide mill than a stamp-battery. Here it may be suggested that actual milling is not necessary, merely a rock-breaker and rolls for the purpose of large-scale sampling, in other words, a small sampling-mill, not a metallurgical plant. The motto of the mine-valuer should be 'It is better to be sure than sorry.' We know by experience that too much care can hardly be devoted to the crucial operation of sampling. Mining has suffered in reputation, as a profitable business, more by errors of appraisal than by the interplay of depravity and gullibility contributory to scandalous performance in promotion. The disappointments at Juneau have been due, not to chicanery, but to an ineffectual attempt to ascertain the real gold contents of large masses of low-grade ore. Let us learn by that experience. As we have said more than once, the worst of all waste is the waste of experience. To prevent it we ought to pay close heed to honest blunders. Frank discussion helps to elicit the essential truth, on which all successful technical operations are based. We invite members of the profession to contribute further to this intensely practical discussion of a highly important subject. We are glad to print a letter from Mr. E. E. Chase, in which he draws attention to one fruitful source of error, namely, the insufficient reduction of samples before they are subjected to assay.

The Butte Decision

In this issue we publish the complete text of the decision handed down by the District Court of Montana in the case of Minerals Separation against the Butte & Superior Mining Company. We delayed publication of this decision owing to errors appearing in the text as first printed; and it will be noted that even now the citations are incomplete. There seems little excuse for such imperfections, which, however, are more important to the lawyer than to the layman. To the latter the decision will seem to be poorly expressed, the language of it turgid, the exposition of the physics of the process unconvincing, and the reasoning hardly worthy of the reputation of the Court, for Judge Bourquin is reputed in Montana to be not only an upright but an unusually clever man, despite the lack of a conventional education. The Court states that the ambiguity and obscurity characterizing the flotation process are due "to the inability then and now to know and explain all its laws and principles." This statement has not become less true since Judge Bourquin has added to the legal literature of the subject. The word 'all,' of course, is redundant. Scientific men will be unable to explain 'all' the laws and principles of any metallurgic process until the Greek kalends. If the patent explains clearly the modus operandi, as is insisted by the Court, why was the process not applied successfully in this country until seven years after the grant of patent? The "skilled operator" had to perform a great deal of experimental work, usually without the aid of the Minerals Separation people, before he could put the process to practical use in the mill; and where the Minerals Separation people did assist, the desired result has not always been hastened; sometimes it has been retarded. Note how they stated, as late as 1912, that certain ores of copper, particularly those containing chalcocite, were not amenable to flotation.

Judge Bourquin's decision turns, as was expected, on the definition of the so-called 'critical proportion,' the phrase expressing a supposed limitation to the quantity of oil efficacious in froth-flotation. By accepting this supposed limitation the Supreme Court was, we believe, misled into an erroneous inference. Judge Bourquin places his own interpretation on the dictum of the highest court; he argues that "small deviations from the predetermined amount," namely, the critical proportion of oil, are injurious to the operation of the process, and in saying so he sets aside the evidence of work done successfully with proportions of oil that make the use of 'critical' ridiculous. Yet he acknowledges that "the process can be fairly successfully operated with 1% or more of oil," insisting, however, that "the excess of oil is useless, wasteful, and harmful," and merely a means of escape from infringement. Is it wrong to try to escape from the liability for infringement and to use more than the proportion of oil said to be essential to the working of a patented process, when by using more oil one can prove that the patent has been granted on a claim that is intrue in fact? The Supreme Court sustained the patent because it is based on the use of a proportion of oil that is "critical and minute," and yet if it is proved by large mills treating thousands of tons daily that the proportion need be neither 'critical' nor 'minute,' it does not matter, says Judge Bourquin, even though the Supreme Court rejected the claims in the patent that failed to specify the proportion of oil more closely than by the phrase "a small quantity." Here we come to the most unsatisfactory part of the decision—one exhibiting a curious lack of logic. Judge Bourquin quotes the Supreme Court's decision, limiting the patent to the use of "a fraction of 1%" of oil, and invalidating the claims in which this exact proportion is not specified; yet immediately after quoting this dictum he says: "It seems clear neither patent nor decision undertakes to say the process depends upon less than 1% of oil or is inoperative with 1% or more of oil." Elsewhere he remarks: "There is suspicion that with experiments, as with figures, can be done anything for or against, without impropriety in the operator." This jibe at the technical witnesses is out of place. Obviously there is impropriety, of the grossest kind and of a kind to be severely reprobated by a judge, in using experiments or figures with intent to deceive. But experiments and figures are not the worst implements of deception; to words is given an even greater scope for making confusion, as the Court is well aware. Referring again to the express limitation of the process to "a fraction of 1%" of oil by the Supreme Court, Judge Bourquin says: "With the later knowledge of this suit, it is doubted that such would be the decision now." We shall see in due time whether the District Court of Montana is correct in this interpretation. As regards the use of a particular kind of violent agitation, to which the Supreme Court confined patent 835,120, Judge Bourquin has a variant notion. He refuses to distinguish between "applied agitation," that is, "by beating air into the mass" with blade-impellers, and "self-agitation," such as is "set up by the air particles themselves in merely rising through the mass" of pulp. It is not necessary to comment on this phase of the process to anybody that has compared the working of a Minerals Separation machine with that of a Callow cell.

The decision is sweepingly in favor of Minerals Separation and in accord with the same judge's decision in the Hyde case—a decision reversed by the Appellate Court in San Francisco. Judge Bourquin asserts that "the great mass of new evidence herein is but cumulative of the Hyde suit." We have read the record in both suits and have found much that seemed different to an important degree. This latest decision of the courts is calculated further to undermine public confidence in the judicial interpretation of the patent law. Not only has it become manifest that it is against the community interest to grant the monopoly of a process to patentees themselves ignorant of the underlying principles of their supposed invention, and therefore undeserving of a blanket right to collect royalties from those that introduce vital improvements, but it is clear that contests over the validity of patents and the infringement of them should be tried by judges having special qualifications, such as a scientific training, including a thorough familiarity with the fundamental principles of chemistry, physics, and mechanics. If all such cases could be settled on questions of law, the need for scientific acumen would not arise, but it is obvious that in these flotation suits the judges are asked to swim in waters wherein heretofore some of them have not waded, in waters by which they have been scarcely wetted.

Sampling Large Low-Grade Orebodies

The Editor:

Sir—If the sampling discussion has not closed I would like to suggest from my own experience that the spotty gold mine that "cannot be sampled," or on which two engineers cannot agree as to value, owes its supposed character to the different methods of assayers in reducing the sample to a pulp. Let us assume that the hard-working conscientious engineer has followed the plan I always adopt, namely, maximum length of channel-cut, 5 feet; weight of sample, 50 lb. per 5 ft.; if wet or sticky, then thoroughly dried on canvas in the sun; then all crushed to pass a ½-inch screen-opening. I carry a specially made sieve for this purpose, separating the fine so as not to exhaust energy on stuff already fine enough. Next, the ore is thoroughly mixed and cut down to a 2-lb. sample, most of which will be so fine as to pass a 20-mesh screen. Now—and here is the whole burden of my song—what does your assayer do with this two pounds over which you have spent an hour of human energy? Have you ever looked at his reject? I have, and my blood has curdled on occasions to find the reject looked no different from the original two pounds. I have even suspected at times that he simply spooned out half of an assay-ton. But the right kind of an assayer puts that whole two pounds through a 50 or an 80-mesh screen and I have proved absolutely that in no other way can a spotty gold ore be properly mixed or the two halves of a sample be made to check. There should be standardization among assayers or the whole careful sampling work of an engineer may be spoiled at the finish.

I am one that believes that almost any mine can be hand-sampled and results obtained on which an engineer will then know what to do or what to advise, provided the samples taken are large enough and near enough together and the assayer looks after his end of the business. But I did think I was up against it when I attempted to hand-sample the Ebner mine, in Alaska, a 40-ft. lode in which the highest gold-contents were distributed along parallel diagonal streaks. I tackled it, however, obtaining a $2.18 average. A large mill-test from the same locality gave $2.22. Was it careful work or merely a coincidence? I cannot say.

For the mines that really cannot be sampled, a good stock of common sense may be substituted. For instance, I was called upon to pass an opinion, after two weeks examination, on one of the big lead mines of southeastern Missouri. Flat beds 40 ft. thick. A year's time might have furnished some valuable data by sampling, but it would have also required the 'human fly' to do the work. I got around it by selecting a typical section, cleaning the face and taking samples of the different layers and at the same time fixing mentally in my mind the amount of 'shine' each layer afforded. Then, after the samples were assayed, I visited the place again and further rubbed into my mental vision the relation of 'shine' to assay-value, and found that I could carry this fairly well through the entire miles of workings. The Lake mine in Michigan was a harder one. All mass copper and no mill-runs available. The question to be solved was this: Was the general run of the mine a 2% or a 5% ore as claimed? No one can sample 'mass' copper, and I was baulked for a moment, but solved it pretty well by figuring that a 2% native-copper content meant a chunk nearly the size of a man's head for every ton broken. Then, by hanging around the stopes and drift-faces for a week, I found I could estimate the tonnage broken and the approximate visible copper, all of which was in large pieces.

EDWIN E. CHASE.

Denver, September 5.

Mill-Tests v. Hand-Sampling

The Editor:

Sir—I have read the appreciative criticism by E. P. Spalding of my article of July 28 on this subject. He requests that I furnish a detailed description of my suggested method of dividing a large orebody into mill-test zones with the object of ascertaining the zone-sampling error, the obtained error to be used as a corrective factor for hand-sampling. In answer to Mr. Spalding's principal question, "How can the mill-test be made representative," I desire to say that in the examination of large ore deposits I do not consider a mill-test to be representative individually, but only as a link in a chain of mill-tests supplemented by corrected hand-sampling.

The origin of the discussion to which I was asked to contribute was the discrepancy on the ore of the Alaska Gastineau between the results of mill-tests and the outcome of subsequent full-sized operations. I have no knowledge of this particular case beyond what I have read; therefore I would not be justified in making suggestions as to how this property might have been sampled. I will, however, describe a method designed by me for the examination of a large low-grade gold deposit in Central America. This property was optioned by a client who owned an operating gold property equipped with a mill. Part of this mill was devoted to sampling the property under option. The deposit was divided into zones with the object of making a mill-test of the ore of each zone. The mill-test was not intended to sample the ore of the zone as a whole; its use was solely to determine the sampling-error of the ore at the point subjected to mill-test. It was then intended to estimate the tonnage, grade, and continuity of the ore by hand-sampling to be corrected by the sampling-error factor for each zone as disclosed by a zone mill-test. The property proved undesirable when subjected to this method of examination and the option lapsed. This case may not fully illustrate the result of a 'combination method' with the subsequent large-scale stoping of full-sized operations. I believe, however, that had we been guided by either hand-sampling or mill-tests alone, very much more money would have been spent on the venture than was actually involved. In fact, the loss was very small considering the size of the project. The mill-test disclosed a sampling-error averaging about 40c. on an expected mill-head of $3.50. This was roughly 40% of the expected operating profit.

In dealing with the principles involved in this example, I contend that a mill-test from a particular part of a large ore deposit will represent only the particular spot from which the ore has been removed. In the case of a low-grade simple-treatment gold ore, it is logical to assume that the sampling-errors obtained at numerous points of the area under examination will shade into each other more or less gradually. A reliable estimate can be made of the local sampling-error by obtaining the average of successive samplings of the stope-faces as the ore is removed for a mill-test and comparing this result with the recovery plus the tailing-loss. If, therefore, such mill-tests were made from various openings at predetermined intervals over the supposed ore-reserves, it is logical to assume that the factor applied to the hand-sampling of the intervening spaces would constitute a reliable corrective and would reflect what should be expected when the intervening spaces (the ore-reserves) were mined. To facilitate explanation, I submit two diagrams in further response to Mr. Spalding and also to some other engineers who have requested a detailed scheme. Fig. 1 represents an area *A, B, C, D,* supposed to be an orebody of unknown grade under process of appraisal by the ordinary mill-test method, whereby a relatively large portion of ore obtained from a relatively small number of places is mined and treated. I submit that if the diamond-shaped areas in black are sufficient to determine the sampling-errors at these particular points of the mine, then continuing to mine and mill the shaded portions will not furnish further data on what should be expected in the area *E, F, H, G.*

Fig. 2 represents what I describe as a 'combination method,' whereby I use the mill-test solely as a *sampling-error indicator,* and I fall back on hand-sampling thus corrected to estimate tonnage, continuity, and grade. The area, *A, B,* C, *D,* is divided into eight mill-test zones as shown, and the black diamond-shaped portions represent mill-tests of moderate size, whereby only sufficient ore is treated to disclose the sampling-error for each particular place. For example, in zone *A, E, K, O* an imaginary sampling-error of 10% is assumed to have been disclosed by the mill-test. The crosses indicate hand-sampling corrected by the zone-sampling error. I have gone into further details in zones *I,* P, *C, L,* and a similar refinement would be applied to the other zones. In the zone *I, J, M, L* there will be observed a dotted imaginary square, *R, S, U, T* and *H,* W, Z, *X* in zone *J,* P, *C, M.* The object of this is as follows: As the hand-sampling leaves the place where the mill-test was made in the zone *I, J, M, L* and approaches the mill-test for zone *J, P,* C, *M,* it is logical to assume that hand-sampling will reflect the influence of the other zone. Therefore, the sampling within the areas *R, S, U, T* will be corrected by a sampling-error of 10%, as shown in the diagram, but on passing the line *S U* it should then be corrected by a factor of 14%, this being the average of the two zones. As sampling passes the imaginary dotted line *H X,* a sampling-error of 18% should then be introduced. Further refinement in shading from one zone to the other could be interposed. I question, however, its utility under large-scale conditions.

To anticipate the criticism that an increase in the number of mill-tests entails increased trouble in cleaning-up and that there might be difficulty in forming accurate estimates because of the gold retained by the plates, it should be clearly understood that this method of sampling a large low-grade ore deposit is not intended as a metallurgical experiment; but purely as a 'combination method' of mine-sampling. In other words, the extraction obtained by the mill-test is unimportant. If amalgamation is used and the plates properly scraped after the clean-up, they should carry a uniform or suspended quantity of precious metal. This complication should therefore not be insurmountable. What is fundamental is the comparison between the successive sampling of the stope-faces with recovery plus tailing-loss. If 60% is recovered by the mill and 20% is known by careful assay to be retained in the tailing-pile, the unaccounted difference of 20% represents the sampling-error; which is the information wanted. When sufficient ore has been treated to obtain this factor, little will be gained by continuing to mine and mill more ore from this particular spot. I consider, however, that a small cyanide mill is better adapted to this method of combination sampling, in which the object is to determine the head, recovery, and tailing-loss without reference to metallurgical efficiency. As it is vital to sample the tailing accurately, as part of the calculation to obtain the sampling-error, the simplicity in sampling sand in leaching-vats will illustrate why I favor a small cyanide plant as a sampling-error indicator.

In order to make my suggested 'combination method' clear to some of the younger members of the profession, I take the liberty of summarizing as follows: In the majority of cases, in dealing with the sampling of large

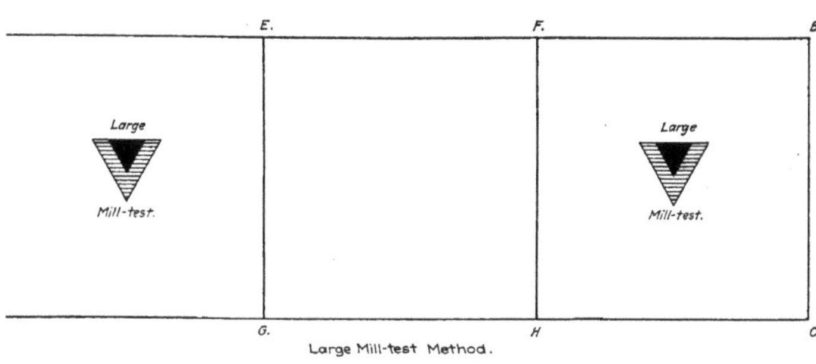

FIG. 1

grade ore deposits, the estimation of ore-reserves t be based on hand-sampling. This can be cor-ed by a series of mill-tests employed for the sole pur- as a sampling-error indicator. Once this error is ined for a given place, the mill-test should terminate. esh mill-test should be made elsewhere to obtain the pling-error for another zone. In other words, the pling-error, which is employed as a factor to correct l-sampling for additional areas, should be based on tests conducted within these areas. The size of each vidual zone will depend on the nature of the ore. his way it may be possible to divide any ore deposit zones to be governed by the engineering expediency e particular case. The principles of this 'combina- method' will apply to best advantage in mines of uniform mineral characteristics. The greater the variation, the shorter must be the distance between one mill-test point and its neighbor, and the reverse. By 'mineral characteristics' I do not mean necessarily the assay of the ore in gold, but the proportion to gold content of friable base sulphides which so often constitute a carrying vehicle for the precious metal.

With regard to Mr. Spalding's further observations: the 'foot-weight' method should not be confused with the method of mine-sampling described above. In the Central American case the method is designed for the sampling of a large low-grade ore deposit, where it is of basal importance to attain the greatest accuracy. In the instance quoted an error of 40c. equalled 40% of the expected operating profit. The mill-test, as explained, in the estimation of a large low-grade deposit, is intended solely as a 'sampling-error indicator.' In my sampling

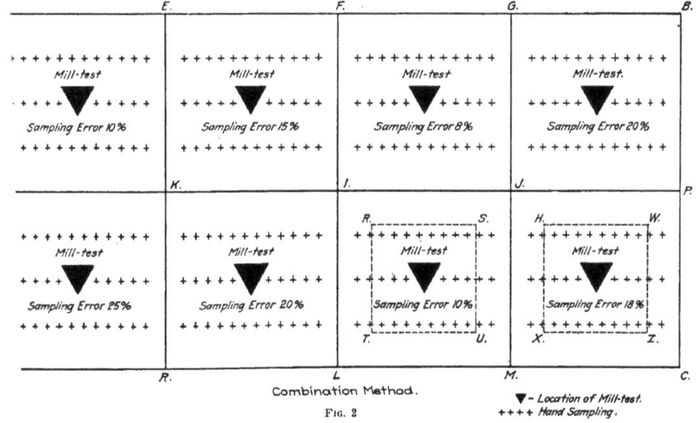

FIG. 2

of the Red mine, to which Mr. Spalding refers, the use of a 'pre-determined foot-weight' of the stopes was designed for an entirely different purpose. In this case the mill-test was used not only as a device for ascertaining the sampling-error, but as a mine-sampler, and also as a metallurgical experiment. In order that the mill-test may perform these three functions, the test must be based on a representative, or 'composite,' shipment of the mine as a whole. Therefore, each stope should furnish its proportional weight, depending on its width and length. How this was accomplished in the case of the Red mine has been described in detail. It will be evident that such a method could not be applied to a large low-grade ore deposit owing to the doubt that would exist as to the intervening spaces. To apply the foot-weight system to such a deposit at a sufficiently short interval to avoid dangerous gaps would entail prohibitive expense.

MORTON WEBBER.

New York, September 11.

Water and Mines in Paradise

The Editor:

Sir—On page 251 of your issue of August 18, there is a reference to the work of the Paradise Irrigation District and a statement that a clash has arisen between the farming and mining interests. It is also said that mining land absolutely useless for farming is included in the district and the owners have the privilege of paying tax without being allowed to have the water. The following is a correct statement of the matter.

It is true that the Irrigation District water will not be sold for mining purposes. The water is needed for bona-fide orchard development, and the people of the district are taxing themselves to obtain water for that purpose only. The mine-owners, who have made informal application to purchase water, have never had the use of this water in the past and are not in any way being deprived of any flow which they have heretofore used. The district, by constructing a large reservoir, will impound storm-water without interference with the legally established and existing rights to the perennial flow. It seems difficult to understand wherein the mining interests are being injured thereby.

Concerning the tax on mining land, there are two answers: First, the owners of all land have an opportunity to appear before the Board of Equalization each year and have the tax-rate adjusted in case their land is actually not susceptible to benefit from the irrigation system. None of the owners of the land referred to has so appeared at any time in the past, but all have paid the tax levied without protest. Certainly a technical journal is not the place to make such a protest when no protest has been made before the board charged by law with the duty of hearing such protest. Second, the land in question includes 20 acres which *is* susceptible to profitable orchard development, as stated before an officer of the district by the owner thereof. The tax levied against this land is necessarily averaged over all land in the tract, although the valuation given on strictly mineral land is but $1 per acre and the annual tax 2 to 4 cents thereon.

To make the matter short, it is not true that the district is levying an unwarranted tax upon mining land, it is not true that the owners pay tax without being allowed to have the water, and it is not true that the mines are being deprived of water to which they have a right for mining use.

IVAN E. GOODNER.
Chief Engineer,
Paradise Irrigation District.

Paradise, Cal., August 27.

[The statement to which exception is taken was not made editorially but by our local correspondent at Paradise.—EDITOR.]

GEORGE A. BURRELL and Alfred W. Ganger of the U. S. Bureau of Mines state that the limits of complete inflammability of mixtures of combustible gases and air are:

	%
Gasoline vapor	1.5 to 6
Ethane	2.5 to 5
Methane	5.5 to 14.5
Natural gas	5 to 12
Acetylene	3 to 73
Artificial illuminating gas	7 to 21
Hydrogen	10 to 66
Carbon monoxide	15 to 73
Blast-furnace gas	36 to 65

In general the limits are slightly lower when ignition occurs from the bottom upward than when it occurs from the top downward.

THE National Board of Fire Underwriters publish the following rules for fire protection: (1) Learn the factory safety rules and observe them. (2) If you discover a fire, give the alarm promptly. Do you know how to do this? Ask to be shown. (3) Don't smoke where it is not permitted. (4) Never drop a lighted match, cigar, or cigarette; be sure that it has no spark left before throwing it away. (5) Report suspicious strangers seen about the plant. (6) If you notice any unusual smoke, the overheating of any machine, or any other accident, notify the foreman at once. (7) Carry your precautions into your own home; keep your house and yard free from rubbish, and help others to do the same. Where would your job be if this plant should burn?

THE relative importance of the principal metal-mining States can be estimated, in a general way only, by the weight in pounds of high explosives used in 1916:

Michigan	29,691,122	Nevada	7,378,443
Missouri	24,875,757	Washington	6,546,844
California	13,942,000	Alaska	5,761,280
Arizona	13,002,555	Utah	5,133,090
Minnesota	10,220,182	New Mexico	3,061,951
Montana	9,162,385	Oregon	2,578,689
Colorado	8,826,030	South Dakota	2,272,840

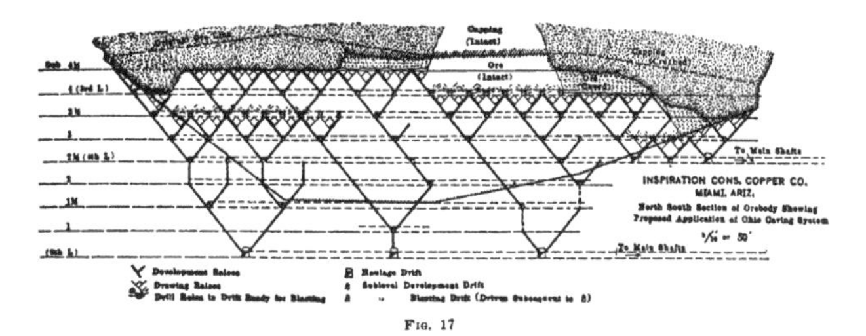

FIG. 17

Miami: The Mining of the Ore—III

By T. A. RICKARD

Before selecting a method of mining for extraction of the 100,000,000 tons of low-grade ore estimated to be available in the Inspiration mine, it was decided to make a series of experiments.† For this purpose a box with glass sides was constructed, and in it was placed crushed ore capped by red waste, so as to be able to observe the effect of drawing ore through miniature chutes. From these experiments it was deduced that: (1) Chute-centres for drawing caved ore should be placed as near together as the ground will allow; (2) ore should be caved in columns of maximum height; (3) when cap-rock appears at the chutes, no more should be drawn unless the bottom slice is being mined. The method of caving finally selected was a modification of that introduced at the Ohio Copper Company's mine in Utah by Felix McDonald, who also supervised the introduction of the method in the Inspiration mine. As Mr. Lehman says, "the method consists, essentially, of under-cutting the ore (taking out a horizontal slice), allowing the ore above to cave and crush, and drawing off the crushed ore through small inclined raises, driven under the caved ore, into main inclined raises that lead down to the haulage-drift chutes." The various stages of the operation are illustrated in the accompanying diagrams, Fig. 14, 15, 16, and 17, taken from Mr. Lehman's excellent description.

After the shafts have been sunk, the haulage-levels are extended under the ore at intervals of 100 ft. Then inclined raises, at an angle of about 50°, are started at intervals of 25 ft., and chutes are built as soon as these are up to 10 to 15 ft. The raises are continued to the first sub-level, where sub-drifts, parallel to the haulage-level, are driven, intersecting the series of raises. These are then extended to the next sub-level, where again another set of drifts is started to connect with the series of raises. See Fig. 14. These operations are continued until the first under-cut has to be made. This will be as far below the top of the ore as it is intended to start the caving. When all is ready, the sub-level drifts on the first under-cutting level are 50 ft. apart, as also on the sub-level immediately below, whereas on the lower sub-levels the drifts are 100 ft. apart. Each drift is connected with raises at intervals of 25 ft. Cross-drifts, at intervals of 150 ft., serve to complete the system of drifts on each sub-level.

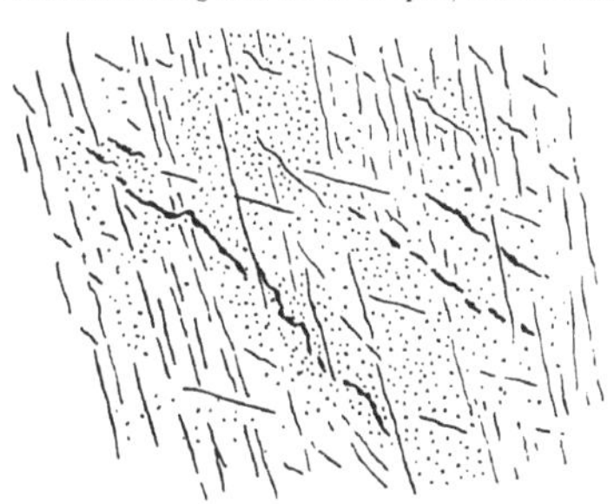

A CHALCOCITE ENRICHMENT IN SCHIST

The next step, before stoping begins, is to drive more sub-drifts on the under-cutting level, so as to have the drifts 25 ft. apart centre to centre. See Fig. 15. These last drifts meet branch-raises at 12½ or 25 ft. intervals as may seem necessary; they are called 'finger' raises and serve to tap the ground completely.

The under-cutting of the ore is started from a cross-

†'Ore-Drawing Tests, Etc'. By George R. Lehman. Trans. A. I. M. E., September, 1916.

drift on the boundary of the section to be mined. In retreating, as shown in Fig. 16, deep holes (8 to 10 ft.) are drilled into the pillars between. The broken ore is drawn through the finger-raises, in which ordinary board chutes are built 4 or 5 ft. below the sub-level, so that the blasting of the under-cutting holes will not destroy them. The ore may begin to cave as soon as it is under-cut, but in hard ground the caving may be postponed until the under-cutting has receded.

Then commences the withdrawal of the ore. Lumps too large for the chutes are blasted. The chute-tappers work in pairs. One climbs through the grizzley sub-raise, opens the chute-gate, and draws the ore, while the other pries the ore through the grizzley on the sub-raise below. From here the ore falls into the main raises and thence into the haulage-chutes, where it is loaded into 5-ton cars. Mr. Lehman, from whom I have borrowed the foregoing details, gives the cost of mining as 60 cents per ton, including a charge of 20c. for amortization of development expenditure.

The mining practice of the Inspiration is essentially a method of caving; it presents a near approach to underground excavation without the aid of either timber or powder. Only two feet, board measure, of timber, and only a quarter of a pound of explosive, is used per ton of ore. On the Joe Bush claim the ore reached to within 50 ft. of the surface, so that steam-shoveling was considered as a method of excavating, but the objection to it was the narrowness of the orebody and the fact that it was bounded on the north side by a steep hill. Thus, in order to uncover the ore it would have been necessary to remove too much of the bill that flanked the orebody. At the start an arbitrary depth of 35 ft.—the equivalent of a sub-level—was taken as the limit of caving; this interval was increased to 70 ft.; now, where opportunity offers, it is 200 ft., with confidence that it will prove practicable. Indeed, experience to date is such as to warrant the expectation of caving to a depth of 500 ft., in an orebody that had such a vertical thickness. In one section of the stopes 5% more copper than was estimated and 29% more tonnage were obtained, owing to dilution of the ore by admixture of the locally enriched cap, although 93% of the copper and 104% of the tonnage represent the current rate of recovery. The cutting-out stope previous to caving is 8 to 10 ft. high. It represents a horizontal slice 10 ft. thick across the full width of the orebody. In the northern end of the orebody—in the Socker claim—an area 175 ft. square was under-cut in the schist before subsidence began. The schist stands better than the granite.‡ Where a stope is wholly in granite a stope 50 ft. square will slough by the time the excavation is completed. In schist a roof three times as large may hold. The granite crushes more easily than the schist, largely on account of less secondary silicification, for where silicification is thorough the ground is brittle. Granitic country, being granular, contains more moisture than schist and this is a factor in aiding movement. The blasting of large pieces—'bulldozing'— is a relatively unimportant feature of mining in the Inspiration. Obviously the greater the depth of subsidence the more the opportunity for a natural breaking of the ore.

Each chute is sampled by itself before the ore from the eight chutes of a system mingles at the main raise. The orebody has been sampled by churn-drilling. All main raises and drifts are groove-sampled every 5 ft., so that a fairly accurate knowledge of the copper tenor of the material below the first undercutting-level is obtained before such ground is mined. In general, churn-drill samples are the only guide as to the copper content of the material lying above that level.

Machine-men do their own shoveling on the undercutting-levels, where the drift connects with a raise every 12½ ft. When sub-driving between two raises 25 ft. apart, as on the grizzley-level, the miner shovels only enough to make room for his drilling-machine, and a shoveler finishes the job. Indeed, the success of the Inspiration practice depends largely upon the branching system of raises, which obviates the necessity for shoveling. The chute-tappers average 132 tons per man, the shovelers, or 'muckers' as they are needlessly called, 138 tons per man. Except in the haulage-drifts, shoveling is intermittent, and the men designated for that work also clear tracks and carry supplies up the raises. In this kind of mining common labor becomes an insignificant part of the whole operation. Nature is induced to do more than her share. For instance, in the Inspiration mine, producing 17,000 to 21,000 tons of ore daily, only 129 machine-men shifts, and 29 machinemen-helper shifts, per day are required.

'Piping', or the descent of waste in a channel formed through the broken ore, is governed largely by the manner in which the chutes are drawn. If one chute is drawn too long it is bound to lead to the tapping of waste prematurely, that is, before the available supply of broken ore has been withdrawn. When two chutes are drawn, each will tap an inverted cone, of 70°, in the broken ore; this is followed by cap-rock right up to the surface and this waste will fill the lower part of each cone, replacing the ore that has been withdrawn. The cap will descend until it reaches an angle of rest. The distance between the chutes must be small enough to afford as short a base as possible for an upright cone or A-shaped mass of broken ore, which tends to remain between the V-shaped openings made above the chutes, and thus fails to be withdrawn. Each chute of the finger-raise system is drawn individually and so furnishes evidence of the progress of ore-withdrawal and of the lowering of the cap on the ore. The making of minor pipes is inversely proportioned to the attention given to the proper tapping of chutes. A small pipe if ignored will soon become magnified. The man that taps the chute is prone to draw from the one that runs most freely because he is held responsible for a given tonnage. He is not easily supervised because he climbs into the

‡For much of this information, I am indebted to C. E. Arnold (California '06), chief mine engineer to the Inspiration Consolidated Copper Company.

raise and can withdraw a lot of waste before he can be scolded. If caught in the act he has plenty of excuses.

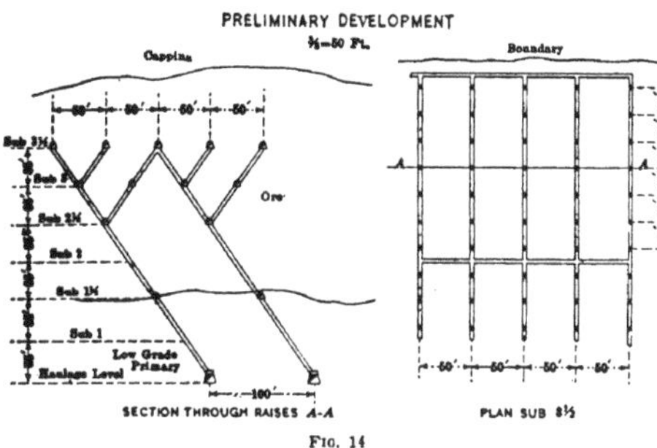

Fig. 14

Apparently the best way to prevent this is to increase the number of bosses—what are called 'boss chute-tappers', and to employ technical men for the purpose. As a rule these chute-tapper bosses appear to be of the same race ('nationality' is the term in general use) as the crew under them. The raise-engineers, who are technical men, are also employed in this work, being assigned charge of given sections of the mine, but their primary duty is to see that the finger-raises are properly placed, and in so doing they are exceedingly helpful to the shift-bosses. However, they are supposed to instruct the boss-tappers when to stop drawing certain chutes as soon as

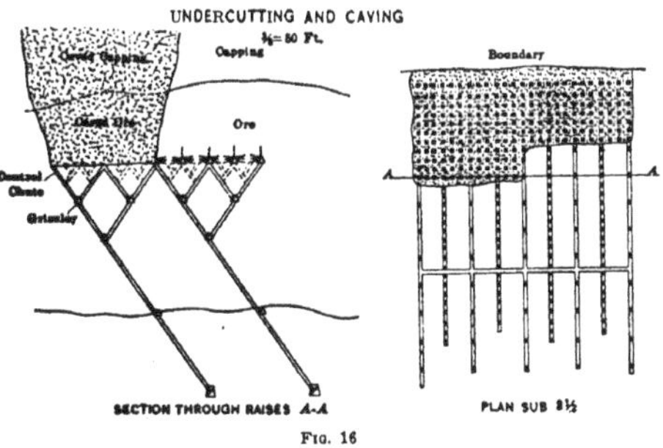

Fig. 16

they (the engineers) see that waste is coming, but they have no power to discipline these bosses. The regular shift-boss is the first man able to dismiss a chute-tapper boss for negligence. Raise-engineers frequently graduate to shift-bosses. The idea is being considered of organizing a 'finishing' crew for each section, to take charge of the last stage of the ore-withdrawal; then there would be less excuse for drawing waste. Fine waste will run under a big chunk of ore that has become 'keyed' or wedged. It becomes necessary to blast the big piece and start it moving. The idea is to promote uniform subsidence immediately the chute is tapped and to prevent big pieces from obstructing the regular descent of the broken ore. On the other hand, it is no part of the method to induce a uniform vertical subsidence over a wide area or to cause the surface to descend uniformly (although this was the idea when the mine was first opened), because the subsidence in steps tends to

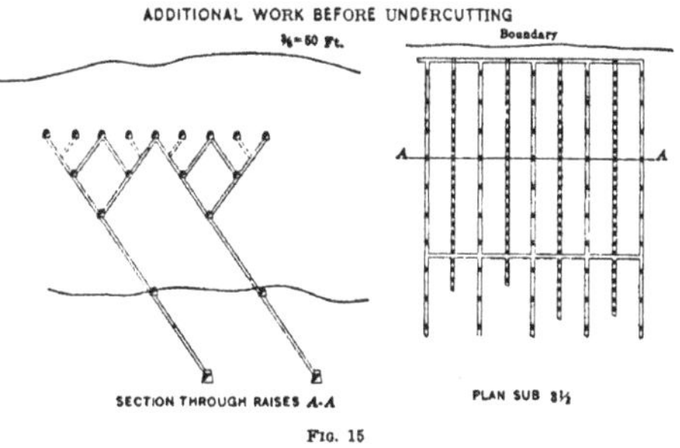

Fig. 15

break any 'piping', as is clearly shown in the Captain ground, where the horizontal movement is a most useful factor in preventing such piping. Moreover, if an attempt were made to lower the surface over a large area, the widespread pressure would necessitate the maintenance of an excessive number of openings and compel a wide distribution of workmen, entailing extra supervision. Theoretically, to ensure the highest extraction of ore, it might seem best to draw from a complete series of chutes close together under a large area of ground, drawing a given tonnage from each chute, and repeat the operation so long as the entire area was subsiding bodily, but such a method would involve an abnormal cost for repairs, maintenance, and supervision.

The lowest cost of mining consistent with safety is the

aim of operations in both mines. The top-slicing of the main orebody in the Miami is logical on account of the relatively high grade of the ore, which it pays to extract cleanly. The shrinkage system in the Captain claim and the caving in the Joe Bush provoke comparison because they are applied to different parts of the same orebody. It remains to be seen which will prove the better. As yet neither has been carried to a definite conclusion. Churn-drill sampling indicated that the Inspiration orebody as a whole contained 1.72% copper. The mill-feed last year averaged 1.55%. The Miami's Captain ore was estimated at 2.05% and the mill shows 1.9%. The Miami 'slicing' ore, estimated at 2.40%, loses practically nothing in grade by mining by top-slicing and brings the combined total output of the mine up to an average grade of 2.15% copper in the mill. This has been the average grade for the past three years. On the other hand, the Inspiration yields 18,000 tons with 1130 men and the Miami 6000 tons with 810 men. At what point is it economical to sacrifice grade to tonnage? That question cannot be answered without a minute inquiry not only into local conditions, but into the financial policy of each company, for the rate at which a mine is exhausted should be regulated by the value of money, that is, whether it is more profitable to take it out rapidly and re-invest the dividends or to take it out slowly as the dividends are required. The rate of exploitation should also be regulated in accordance with the anticipated market for copper.

MOST of the refractory magnesite that has been in general use has peculiar and distinctive properties that are not found in the magnesite deposits of the common type. The value of this refractory material depends not only on its resistance to the corrosive action of heat and metallic slags, but also on the permanence of the forms in which it is put into the furnace. This permanence is due to a natural bonding which tends to make the loose crushed material cling together under furnace heat and thus makes brick forms molded from it more durable. Bricks and granular furnace bottoms made of magnesite that lacks this bond break, and the magnesite floats off on the fluid molten metal and is lost in the slag. Thus, though magnesite that contains a small percentage of iron may be somewhat less resistant to extreme heat than a purer form, the slight fusibility given to the material by the iron tends to hold it in place. For this reason, in part, a type of magnesite so far found only in Austria and Hungary has been the principal source of the refractory magnesia used in this country. The purer magnesite from Greece, California, and elsewhere is used in making plaster or cement or material for other relatively minor uses.—U. S. Geological Survey.

AMERICAN railroads have achieved the feat of adding to their freight service, in the short space of two years, an amount equal to the total traffic of Great Britain, France, Russia, Germany, and Austria combined.

Platinum

James M. Hill, of the U. S. Geological Survey, states that people are apt to think carelessly of platinum as a good setting for precious stones; the metal is, in fact, indispensable to many of our industries. Dishes and utensils of platinum are absolutely necessary in chemical laboratories, and on their laboratories all great industries depend for guidance. Alloys have been developed for some parts of the ignition systems of internal-combustion engines, but no substitute for platinum has been found for certain delicate parts of these systems. Platinum or allied rare metals have wide application in many instruments of precision used in the physical testing of all kinds of materials. Probably the most important use of platinum at present is in the contact process of making concentrated sulphuric acid, an essential commodity to many industries aside from its use in the manufacture of munitions.

The United States alone apparently uses about 165,000 fine ounces of platinum a year. The known supply of platinum is small; possibly 5,000,000 oz. has been produced in the world to date. Estimates based on the official figures of production from Russia since 1843 (which are taken as 25% low), and on the assumption that Russia has supplied 95% of the world's output, indicate that the total quantity of crude placer platinum produced in the world since 1843 has been less than 4,632,000 troy ounces, or about 159 short tons.

Probably 500,000 oz. of this total output of platinum is in use as a catalyzing agent. The quantity manufactured in the form of chemical and physical equipment cannot be readily ascertained, but may be roughly estimated at 1,000,000 oz. Electrical devices may have required 500,000 oz., but perhaps 250,000 oz. would cover that use. The jewelry industry has probably consumed 1,000,000 oz., and the dental industry also about an equal quantity.

Realizing the urgent necessity of increasing the domestic production of the metals of the platinum group, the U. S. Geological Survey has enlarged its field of investigation of the domestic platiniferous deposits. At the request of the Secretary of Commerce the Geological Survey will also take a census of all unmanufactured and scrap platinum in the United States that can be considered as an immediately available supply. It is confidently hoped that this census will show that the United States has a sufficient reserve of platinum to meet all the immediate needs of industry. The use of platinum in heavy articles of jewelry will be curtailed, if the resolutions adopted by the Jewelers' Vigilance Committee are accepted by the manufacturing jewelers, as is to be expected in the present circumstances. This action will release a considerable supply of metal for the more urgent needs of chemists, as advocated recently at the St. Louis meeting of the American Chemical Society.

THE LIBERTAD MINE

SANDSTONE AND LLAMAS

PLAZA IN COROCORO

VIEW TOWARD GUALLATIRI, CHICO

Corocoro Copper Mines

By FRANCIS CHURCH LINCOLN

Corocoro is the important copper district of Bolivia, and, next to Lake Superior, is the largest producer of native copper in the world. Although these deposits have been known and worked since the time of the Incas, the early operations were on a small scale. It was not until 1873, when the Compañia Corocoro de Bolivia, a Chilean corporation, consolidated a number of the smaller properties, and began large production. In 1909, a second and larger combination was made when the Corocoro United Mines Co. took over most of the small mines. According to Lester W. Strauss,* the district produced 100,000 tons of copper from 1873 to 1912, by which time it was shipping at the rate of 5000 tons of copper per year. In 1913 the Arica-La Paz railroad was completed and in the following year a branch line was extended to Corocoro, with the result that oxide and sulphide ores previously untouched were shipped in large quantities. The production for 1915 exceeded 10,000 tons of copper.

The city of Corocoro is the capital of the province of Pacajes in the department of La Paz and has a population of 15,000, consisting mostly of Indians. It lies on the Bolivian plateau at an elevation of 13,210 ft. above sea-level, 70 miles east of the Chilean border, and about mid-way between the main and the western ranges of the Andes. The Bolivian government has constructed a four-mile branch railway connecting Corocoro with the station of Tarejra on the Arica-La Paz railroad, the latter being a metre-gauge line from the part of Arica in Chile to La Paz. Prior to the completion of rail con-

*'The Mining Magazine', Vol. VII, p. 207.

nection in 1914, copper concentrate from Corocoro was packed or carted to Nazacara on the Desaguadero river, loaded upon steamers for transport to Guagui on lake Titicaca, trans-shipped to steamers for Puno, Peru, and there transferred to trains of the Southern Railway of Peru, which completed the journey to the Peruvian port of Mollendo.

Corocoro is situated in a group of low hills composed of shales, sandstones, and conglomerates ranging in age from Triassic to late Tertiary. These sediments have a general ferruginous color, and are also gypsiferous. A number of the beds of late Tertiary sandstone, lying at short intervals apart, are white in color, instead of the prevailing red, and these contain copper. White copper-bearing beds in reddish sediments are by no means confined to the Corocoro deposit, but occur elsewhere in Bolivia and in Peru. Indeed, these deposits form a belt having a length of 240 miles. Thus, to the south-east of Corocoro, deposits are said to occur at Chacarillas and at Turco, while to the north-west I have examined beds near Guaqui, in Bolivia, near the Peruvian boundary in the vicinity of Desaguadero, and at Tacasa, Peru.

Structurally, the Corocoro district consists of a faulted anticline, striking northerly, and with the copper-bearing beds dipping away from the fault in both directions. The dip varies with locality and with depth, being as low as 35° at the Carmen mine and as high as 80° near the fault at Corocoro. The ore-bearing sandstones, which dip to the west, are locally termed 'vetas' (veins), whereas those dipping eastward are known as 'ramos', that is, branches. The strike of the fault is nearly parallel with that of the 'vetas' and the dip is slightly steeper, but the sedimentary series containing the branches has been rotated in such a manner that on going northward along the fault successively lower members of the 'ramos' series outcrop to the east. For example, near the town of Corocoro, the 'vetas' come to the surface and the 'ramos' are only found in depth, but a mile and a half north of the plaza the 'ramos' begin to outcrop. Apparently the veins and the branches are the severed portions of the same group of copper-bearing sandstones, but they present mineralogical differences that will be noted later.

The Corocoro district has a length of about 11 miles, extending northerly from the Guallatiri claim on the south, through the town of Corocoro and the Libertad and Carmen mines, to the Grau property on the north. The maximum width of the district is about half a mile. The beds of sandstone have been explored and found to contain ore to a depth of over 1600 ft. These copper-bearing sandstones are closely spaced, being separated by beds of shale. They vary in thickness from one inch to 25 ft., averaging 4 ft. The fault-plane itself formerly contained a large orebody, which may possibly have been drag from the 'vetas' and 'ramos', but this is now worked out. The intensity of the mineralization varies in different ore-bearing beds and in different parts of the same bed, and in no instance does the entire bed consist of ore. The copper-bearing beds are generally white in color, and when they contain red patches they are always free from copper. The white sandstone is much more friable than the red, as a result of having been leached of its original cementing material and of having had this replaced in part only by ore. The mineralization is confined to the filling of the interstices between the grains of sand.

The principal ore-mineral is copper. In the case of the branches this has been oxidized near the surface to malachite, azurite, and chrysocolla, such oxidation extending on an average to a depth of 100 ft. While the 'vetas' also contain native copper in depth, chalcocite occurs near the surface on a zone averaging 300 ft. in vertical extent, while above this the oxidized ore—consisting mainly of the basic copper sulphate, brochantite—has an average depth of 50 ft. Therefore, the 'vetas' present a reversal of the usual occurrence in which native copper turns to sulphide in depth, because there the sulphide turns to native copper, making a most interesting problem for the economic geologist. In a polished section containing both native copper and chalcocite that I examined, these minerals occurred as a cementing material between the grains of sand, and were intergrown in such a manner as to indicate contemporaneous deposition. Other ore-minerals of minor importance are cuprite, domeykite, native silver, and galena. The principal gangue-mineral, besides the quartz of the sandstone, is gypsum; and small amounts of hematite, barite, and celestite also are found. A mineralogical analysis of the ore from the Carmen mine gave the following result: brochantite 12%, chalcocite 3.5, quartz 73, gypsum 7.5, and hematite 4%.

Except for the altitude, the conditions affecting mining at Corocoro are favorable. The climate is cool throughout the year, and the district is healthful. The summer rains begin near the end of December and continue until April. They are by no means continuous, although they sometimes last for several days. The remainder of the year is dry. Indian labor is employed, the wages paid being, in bolivianos, miners 1.80 to 2.20; helpers, 0.80 to 1.20; surface men, 1.40 to 1.80; women, 0.70 to 1.20; machinists, 3.50 to 4; blacksmiths, 3 to 4.50; carpenters, 3 to 4.50; mechanics, 5 to 10. The value of a boliviano is normally 39 cents, but it has recently been as low as 33c., so that the cost of labor is low. Contracts for drifts are let at from 10 to 30 bolivianos per running metre, and shafts are sunk at the rate of 15 to 30 bolivianos per metre. Little timbering is done, as lumber is costly, but dry-wall supports, laid with gypsiferous shale, cost only 6 to 7 bolivianos per metre. Power was formely produced by burning 'taquia,' the Indian name for llama dung, under steam-boilers. This costs about 3c. per horse-power hour, but Diesel engines burning Peruvian oil are now being introduced, reducing the cost to about one cent per horse-power hour.

The Corocoro United Mines Co. and the Compañia Minera de Corocoro control the district, although there are a few small independent mines, of which the Libertad is a good example. The Remedios mine of the Compañia Minera has a vertical depth of 500 metres, while the Vizcachani shaft of the Corocoro United is 375 metres deep. The head-frame and rock-house at the latter shaft are shown in an accompanying photograph. The output of this mine is 300 tons per day. The hoisting-engine is a 200-hp. Robey. The efficiency at this altitude is only 45%, and in spite of this the uneconomical 'taquia' is still employed in conjunction with Australian coal. The ore is mined by overhand stoping, black powder and Du Pont dynamite being used as explosives. The sulphide ores are sacked and shipped. Owing to lack of cars, immense quantities of the sacked ore accumulates at times.

The native copper ore is concentrated to an 85% copper concentrate locally known as 'barrilla'. There are five concentrating mills in the district: the Libertad, the San Francisco, the Guaychuni, the Guallatiri Grande, and the Guallatiri Chico. All of the mills suffer from

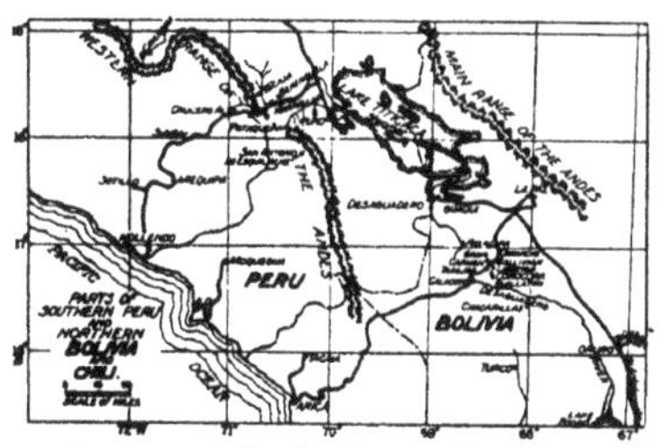

scarcity of water. The San Francisco is a 200-ton mill treating the ores of the Compañia Minera, while the Guaychuni, Guallatiri Chico, and Guallatiri Grande are 200-ton, 180-ton, and 60-ton plants respectively, engaged in treating ores of the Corocoro United, and all four mills use the same water in turn. Notwithstanding this difficulty the Corocoro United is considering the erection of another mill farther down-stream to treat the low-grade sulphide ore by flotation. The Libertad mill and the view from the Guaychuni mill looking toward the two Guallatiri plants are shown in the photographs. Crushing is done by Chilean mills, and concentration is effected by jigs and tables, the coarse middlings being re-ground in Chilean mills. The cost of producing and marketing the Corocoro 'barrilla' ranges from 6½ to 7½c. per pound of copper.

THE use of high explosives for mining purposes as compared with other purposes is presented in the table for two groups of mining States.

	Alaska, Idaho, Oregon, and Washington, lb., 1916	Arizona, California, and Nevada, lb., 1916
Mining other than coal	8,563,685	25,527,705
Coal mining	464,210	2,000
Railway and construction work	3,573,626	3,386,302
All other purposes	7,418,382	5,407,046

Chromite

The importance of chromite as a war supply in the manufacture of armor-plate, armor-piercing projectiles, stellite for high-speed tools, and automobile and other special steels can scarcely be over-estimated, according to the U. S. Geological Survey. The chief sources of supply during the last few years have been Rhodesia, New Caledonia, Turkey, and Greece, and the imports in 1916 were 114,655 long tons.

The greatly increased trade, especially in steel, and the consequently larger demand for chromite have stimulated the search for it in the United States, as shown by the increase in production, the amount sold in 1915 being only 255 tons against more than 47,000 tons in 1916. In Maryland and Wyoming there has been only a small production, but in the Pacific Coast States, especially California, the advance in the output has been remarkable. In Oregon the production was more than 3000 tons; in California it was nearly 44,000 tons.

It is evident that for some time to come California will furnish the chief domestic supply. With a lively demand and good prices deposits farther from lines of transportation will be worked.

An interesting feature of future chrome production lies in the fact that T. W. Gruetter has recently established at Kerby, Oregon, a custom plant for concentrating black sand to win its gold and platinum. The black sand of the Klamath mountains usually contains a considerable amount of chromite, and it is believed that by adding magnetic separators to Gruetter's plant to remove the other minerals from the tailing, sufficient chromite may be obtained from the black sand in areas of chromiferous serpentine to make the operation financially successful. The process will evidently yield a high grade of chrome ore, which may be suitable for special uses.

There is considerable chromite in Cuba, but scarcely anything is known of its occurrence in Mexico or Central and South America.

Prospecting for chromite may disclose other supplies, and the most profitable deposits will be those in areas of serpentine that are adjacent to cheap rail or water transportation or connected with it by good roads. Cheap concentration may in places improve the grade of the ore available for profitable mining.

IN a technical paper of the U. S. Bureau of Mines, Dr. J. A. Watkins states that one of the earliest symptoms complained of by those daily exposed to small quantities of carbon monoxide is persistent and distressing headache, accompanied at times, when the headache is severe, by nausea and vomiting. A peculiar feature of this symptom is that headache often does not begin until some time after exposure, that is, after the employee has left work, but in most instances it affects the individual the whole time he is at work, becoming more severe as the day wears on. Attacks of giddiness are common.

The Butte & Superior Case

Opinion of Judge Bourquin, U. S. District Court of Montana, in Minerals Separation v. Butte & Superior Mining Company

This is trial on the merits of the suit reported in 237 Fed., 401. It involves the patent and claims of the Hyde suit, wherein the Supreme Court held the patent valid, but some claims invalid. The issues are as in the Hyde suit, viz.: Novelty, invention, infringement, and in addition defenses of unreasonable delay and defects in disclaimer of the invalid claims, and estoppel by reason of statements by plaintiffs' counsel to the Supreme Court in arguing the Hyde suit. The evidence herein is that submitted during 25 days and also the record in the Hyde suit. So far as heretofore known the nature and history of the discovery and invention (a process of ore-concentration by air-flotation) are fairly set out in reports of the Hyde suit, of the Miami suit, and of foreign suits cited in a foot-note. This suit is an important contribution and yet it discloses that though the use of the process is very wide, extensive, and growing, its simplicity, economy, and success still surprise and gratify the metallurgical world and its laws or principles of operation still interest and puzzle the scientists. "In the beginning it was very little knowledge and mostly guesswork, and since then there has been every year a little more knowledge and still a great deal of guesswork," testifies one of defendant's experts, Prof. Bancroft of Cornell, a physical chemist of note, acquainted with the process since 1906 and lecturer upon it since 1912. Though speaking for himself alone, the learned doctor's estimate might well be applied to all, practical layman and expert scientist alike.

At the same time, though heretofore somewhat ambiguous and obscure, present knowledge warrants the conclusion that the gist of this remarkable and valuable process and the actual discovery and invention are that whereas theretofore in ore-concentration air had been used in desultory and fugitive bubbles as a makeshift incident of and supplement to oil and skin flotation, air can be made to do all the work by creating in water-ore pulp modified by a suitable oily contaminant an infinitude of bubbles. It is the first of its kind, and the patent sufficiently discloses it and methods to those skilled in the art.

Ambiguity and obscurity were as much due to the extreme simplicity of the process as to the inability then and now to know and explain all its laws or principles. The tendency was to attach prime importance to reduction in amount of oil used when in fact this is but a necessary incident (for which there are substitutes if not equivalents) to the creation of the infinitude of bubbles that do the work. Despite this tendency and to overlook the simple and obvious, the patent fairly clearly sets out the various ways and means to create this infinitude of bubbles and that they do the work.

The tests to determine which kind and amount of "oily substance yields the proportion of froth or scum desired," that flotation is "mainly from the inclusion of air bubbles," the froth, the agitation, all are so many guides in the patent, pointing the skilled operator to and including the infinitude of bubbles and the degree of agitation and amount of soap or oil to produce such bubbles, as surely as the word "crystallization" points to appropriate temperature in Commercial Co. *v.* Co., 135 U. S., 189, and the words "uttered sound" by the "human voice," to articulate speech in the telephone cases, 126 U. S., 531.

Of the new evidence herein are learned dissertations upon the philosophy of the process—upon the philosophy of bubbles, the heart of it, by Professors Bancroft of Cornell and Taggart and Beach of Yale, and Doctors Sadtler of Philadelphia and Grosvenor of New York. From these it is gathered that the mere introduction of particles of air into a liquid does not create bubbles, but that they are created by subsequent agitation, either applied or self-agitation. Air particles introduced into pure water are incapable of creating bubbles. The reasons are the surface-tension of the water and the lack of viscosity to create a sufficient film about the air particles, compel the escape of the air particles into the atmosphere and no bubble is formed. Some soaps and oils possess the quality to lessen this surface-tension of water and to give or increase this necessary viscosity. Their addition in appropriate quantity to water enable air particles introduced therein to create bubbles. Rather, the meeting and co-action of water, oil, and air create a film composed of all three and which surrounds the air particles. This film is more viscous than the mass of the water, and rising to the surface, the tension of which (and of the films) has been reduced by the oil, maintains itself as an air bubble. This quality of oil is of first importance in the process. Another of lesser importance and which all oils possess is the "preferential affinity for metalliferous matter over gangue." Of lesser importance, because it is now known (and patented) that given another contaminant than oil but which possesses the like bubble-making quality though not the said "preferential affinity," the process is equally successfully worked. Air also possesses this "preferential affinity," and in view of the foregoing it well may be that the capture as well as the flotation of the metallic particles is more due to the great volume of air than to the infinitesimal oil. That in this process air without oil cannot capture and retain the metallic particles, seems due to its inability to create bubbles without oil. And why this capture in any case, is still of the unsolved phenomena of the process. On the other hand, water has a preferential affinity for gangue over metalliferous matter. That is, it wets the former more readily than it does the latter. And this contributes to the process in that oil and air displace water from the surface of metalliferous matter more easily and quickly than from gangue, and so more readily capture and float the former than the latter. At the same time, despite these preferential affinities, in successful operation of the process the bubbles generally float more gangue than metal, more in quantity but not in proportion, and why is also unsolved.

There are "critical proportions" of any oil used in this process, perhaps not a sharp divide, but rather a broad one. For the amount of oil to produce sufficient and efficient bubbles must depend on many other factors, viz: the working-cell space, amount of water, degree of agitation, kind and amount of ore, and perhaps on occasion amount of metallic content, kind of oil, etc. For example, if a ton of ore be agitated in a lake of water, doubtless a lake of oil will be necessary to create sufficient bubbles to capture the metal in the ore. But with bona fide operations in a good workmanlike manner—with the proportion of space, water, agitation, etc., such operations and manner dictate, the range in amount of oil will be narrow and well within 1% on the ore. These "critical proportions" are like those known to and solved by every child with its pipe and bowl of suds. Too little soap, the bubbles are few, small, fragile, and break quickly. Too much soap, they flow from the pipe in a torrent, are heavy, and refuse to float. The right amount of soap, the "critical proportions," his bubbles are large, detach readily, and float high, far, and for long. So is it with the bubbles in this process. With excess oil but not enough to defeat bubbles altogether, though of fair aspect to the eye the bubbles will not do the work. In the excess oil in the films the metallic particles do not cling but swim or slide to the bubble's lower surface, "neck off"—detach, and sink. The untechnical workman recognizes there are "critical proportions" of oil, and small deviation from the predetermined amount in the feed, whether more or less, manifests itself to him in the appearance of the froth and poorer results; and he knows and remedies the error in oil.

Metallic content of ore seems of little importance—sometimes seems to require oil inversely. For example, a local operator with the process upon ore from the same vein as defendant's, uses 0.7 lb. of oil per ton of ore of 11.23% zinc content, making 50.59% concentrates with 94% recoveries, and in the same plant uses 2.83 lb. of the like oil per ton of tails of 0.97% copper content, making 9.085% concentrates and 0.266% tails. It is apparent it is the air and not at all the oil that floats the mineral, noting that in the first of this example 211 lb. of zinc are floated by air bubbles in the creation of which only 0.7 lb. of oil is used. How the air particles are introduced into the pulp is immaterial. For introduced, they are still particles and not bubbles. Agitation subsequent to introduction is vital and alone can convert air particles into water-oil-air bubbles. It is this subsequent agitation that within the claims of the patent agitates "the mixture until the oil-coated mineral matter forms into a froth" or "to form a froth." And it is all one, whether this be applied agitation or self-agitation—the agitation set up by the air particles themselves in merely rising through the mass and thereby coming in contact with both water and oil, all co-acting to form bubbles which capture the metal. The mineral particles, either oiled before or by contact with bubbles, attach to and enter the viscous film of the bubbles. The particles also increase the viscosity of the bubble films, armor them and increase their stability, perhaps as stays that decreasing the area of unsupported surfaces, increase the latters' ability to resist rupture.

The great mass of new evidence herein is but cumulative of the Hyde suit. The only new publication is the 'California Journal of Technology' detailing a suggestive but rather misleading and abandoned experiment, sufficiently referred to and disposed of in the Miami suit.

There is much evidence that progress in the process and methods of operating it, now discloses that with some ores and some oils or mixtures of oils, the process can be fairly successfully operated with 1% and more of oil. This is really admitted by plaintiff and is taken as proven. But it is also proven practically without conflict that in all the operations with this process not to exceed 0.2% of oil is used, save by defendant and others in like situation and only since the decision in the Hyde suit and solely to avoid infringement; that some oils are effective and more are ineffective to operate the process; and that the excess oil used is useless, wasted, and harmful.

But the defendant contends that this evidence demonstrates the process lacks novelty and invention, and that because of it the record is substantially different from the Hyde suit, the decision there should not control here, and the patent is and ought to be held invalid. This is without support in the patent and Hyde decision.

In describing the invention the patent refers to Cattermole and says that the patentees "have found that if the proportion of oily substance be considerably reduced —say to a fraction of 1% on the ore," after vigorous agitation the metallic particles rise to the surface in a froth; that the proportion thereof varies considerably with different ores and different oils, and so it is necessary to test "to determine which oily substance yields the proportion of froth or scum desired." An example of a particular ore and oil is of oil "say from 0.02% to 0.5% on the weight of ore," wherein on cessation of agitation "a large proportion of the mineral present rises to the surface in the form of a froth or scum which has derived its power of flotation mainly from the inclusion of air bubbles introduced into the mass by the agitation, such bubbles or air-films adhering only to the mineral particles which are coated" with the oil which has "a preferential affinity for metalliferous matter over gangue." It adds that the minimum of that oil "may be under 0.1% of the ore, but this proportion has been found suitable and economical."

The claims are (1) for "oily liquid * * * to a fraction of 1% on the ore," (2) for oleic acid "to 0.02—0.5% on the ore," and (3) for "a small quantity of oil." These last were held invalid. In upholding the patent the Supreme Court says that "as described and practised" the process consists "in the use of an amount of oil which is 'critical' and minute as compared with the amount used in prior processes, 'amounting to a fraction of 1% on the ore,' and in so impregnating with air the mass * * * by agitation * * * as to cause to rise to the surface * * * a froth * * * which is composed of air bubbles with only a trace of oil in them, which carry in mechanical suspension a very high percentage of the metal"; that "it differs so essentially from all prior processes in its character, in its simplicity of operation, and in the resulting concentrate, that we are persuaded that it constitutes a new and patentable discovery"; that the facts are not overstated by Liebmann that "the present invention differs essentially from all previous results. It is true that oil is one of the substances used but it is used in quantities much smaller than was ever heard of, and it produces a result never obtained before. The minerals are obtained in a froth of a peculiar character, consisting of air bubbles which in their covering film have the minerals imbedded in such manner that they form a complete surface all over the bubbles. A remarkable fact with regard to this froth is that, although the very light and easily destructible air bubbles are covered with a heavy mineral, yet the froth is stable and utterly different from any froth known before, being so permanent in character that I have personally seen it stand for 24 hours without any change having taken place. The simplicity of the operation, as compared with the prior attempts, is startling. All that has to be done is to add a minute quantity of oil to the pulp to which acid may or may not be added, agitate for from 2½ to 10 minutes and then after a few seconds collect from the surface the froth which will contain a large percentage of the minerals present in the ore"; that the Court is convinced "that the small amount of oil used makes it clear that the lifting force which separates the metallic particles of the pulp from the other substances of it is not to be found principally in the buoyancy of the oil used, as was the case in prior processes, but that this force is to be found, chiefly, in the buoyancy of the air bubbles introduced into the mixture by an agitation greater than and different from that which had been resorted to before and that this advance on the prior art and the resulting froth concentrate so different from the product of other processes make of it a patentable discovery as new and original as it has proved useful and economical"; that the Court agrees with the House of Lord's decision that the process is not one before described but a new method in which flotation is by the "buoyancy of air bubbles"; that tests to determine the necessary "amount of oil and the extent of agitation," and "the range of treatment within the terms of the claims," satisfy the law; but that while the patentees "discovered the final step, precedent investigations were so informing that this final step was not a long one and the patent must be confined to the results obtained by the use of oil within the proportions often described in the testimony and in the claims of the patent as 'critical proportions,' amounting to a fraction of 1% on the ore, and therefore * * * the patent is valid as to claims No. 1, 2, 3, 5, 6, 7, and 12 * * * but * * * invalid as to claims 9, 10, and 11." Those held invalid are those heretofore referred to as (3). It seems clear neither patent nor decision undertakes to say the process depends upon less than 1% of oil or is inoperative with 1% or more of oil.

It is true that in the beginning and during the Hyde suit the patentees inclined to so believe, or at least believed better results would be obtained with a fraction of 1% of oil. Perhaps limited investigation and experience with few ores and oils justified the belief. Indeed, all experience to date, plaintiffs', defendant's, strangers', demonstrates that with any ore and any efficient oil, less than 1% of oil gives better results, all circumstances considered. The "critical proportions" referred to seem absent, in terms, from the patent, and ought not to be adversely inferred in disregard of construction in favor of the patentee where the patent is ambiguous. The patent describes oil "considerably reduced," and refers to a "fraction of 1%" by way of example. And though some claims limit oil to such fraction, and a limited range within it, others are for "a small quantity" and for that reason held invalid by the Supreme Court. With the later knowledge of this suit, it is doubted that such would be the decision now. It is to be observed that this limitation of the patent indicates the Supreme Court believed the process might be operative with 1% and more of oil, and contemplated that this would not defeat the patent but might affect infringement. If the patent is limited to the use of a fraction of 1% of oil, that the process can be operated with 1% and more is not material to validity though it may be to infringement. For if a patentee limits his claim voluntarily or because he does not know the extent of his invention, he abandons the excess and the patent is valid to the extent of the claim. If it be conceded that new evidence might warrant and demand that a Trial Court hold invalid a patent by the Supreme Court held valid, such evidence must be unequivocal, clear, and convincing, in quality and quantity that inspires confidence and produces conviction that the patent is invalid and that the Supreme Court would so determine beyond a reasonable doubt. Not only does it fail here, but it strengthens the conviction that the patent is valid.

The disclaimer to conform to the Supreme Court's decision that claims 9, 10, and 11 are invalid, was filed 107 days after said decision and after mandate but before expiration of time for rehearing. It was timely filed. In substance it fairly conforms to the language of the decision, disclaiming "from claims 9, 10, and 11 * * * any process of concentrating powdered ores excepting where the results obtained are the results obtained by the use of oil in a quantity amounting to a fraction of 1% on the ore." The parties differ in its interpretation even as they do in respect to the decision. Written words, not oral claims, control. The patent claims included what the patentees were entitled to and more. The decision pointed out the excess. The patentees disclaim the excess. They can safely rely upon the decision, and a disclaimer conforming to the language of the decision is sufficient. The patent valid, defendant admits infringement from before suit commenced to January 7, 1917, and denies infringement thereafter. During the period of admitted infringement it applied the process to some 1,598,000 dry tons of crushed ore, the tailings from water concentration, of mineral zinc content by yearly averages as follows: 1913, 15.14%; 1914, 14.14%; 1915, 13.66%; 1916, 12.89%. Oil used by yearly averages is as follows: 1913, 5.58 lb.; 1914, 2.22 lb.; 1915, 1.49 lb.; 1916, 1.43 lb. The concentrate grade by yearly averages is as follows: 1913, 47.60%; 1914, 53.03%; 1915, 54.62%; 1916, 53.83%; and recoveries (apparent) likewise, are for 1913, 80.03%; 1914, 86.08%; 1915, 90.18%; 1916, 92.63%. Progress is indicated by leaner ores, less oil, higher grade of concentrates, greater recoveries, all coincident with advancing time.

It is noted that the process is responsible for advance in methods, devices, and machines for its operation. To briefly describe defendant's during infringement admitted, the water-concentrate tailings, oil added, flowed to the head of a pyramid machine of seven cells in series, each cell containing a revolving perpendicular spindle and horizontal blades, and having two opposed spitzkasten. The agitation was very violent. The tails from each cell flowed by gravity to the cell immediately below, those from the last cell flowing to Callow air cells which produced froth-middlings returning to the head of the pyramid, and tails, to waste. The first three cells of the pyramid produced froth rougher concentrates which flowed to cleaners, and the other cells produced froth middlings which returned to the head of the pyramid. The rougher concentrates passed through five cleaner-cells for washing. This produced concentrates which flowed to and through five re-cleaner cells, and tails which returned to the head of the pyramid. The re-cleaners produced final concentrates, and tails which returned to the head of the cleaner. Commencing January 7, 1917, the defendant's methods are as before, save pneumatic as well as mechanical agitation is employed in the lower four cells of the pyramid, some spitzkasten are blocked, an additional cleaner operation is used and from which for some unexplained reason 8.65% zinc tails go to waste, and oil in amount 1% and more is used. Defendant not very insistently claims results for this latter period are more profitable than for the former, but plaintiffs' analysis (neither denied nor criticized and beyond both) of defendant's reports and tabulations makes manifest the fact is otherwise to the extent of about $1.75 per ton of ore—an enormous loss on 45,000 tons monthly. There is considerable like testimony in reference to operations by other infringers. However, coming as it does after very large operations, investigations, and experiments of several years, after the Supreme Court's decision in the Hyde suit, it is incredible that use of 20 pounds of oil per ton results in more profit than one and a half to four pounds per ton. If it does, these great concerns would not have waited to discover the fact and employ it, until after the said decision. Defendant practically admits that it now uses the present amount of oil merely to avoid the patent. It is done, says its counsel in argument, "because the Supreme Court has said it is not at liberty to use less than one-half per cent," and "out of abundance of caution" it uses "more than 1%." The evidence likewise persuades. If the excess oil were effective and useful and not inert, useless, and harmful, it would be without the claims of the patent, would be of that the patentees abandoned to the public, and would involve no infringement.

Plaintiffs somewhat laboriously argue that though the Supreme Court held that the claim for use of "a small quantity of oil" to produce froth is too broad and so invalid, yet since the court identifies the invention by the "results obtained," though confining it to the "results obtained" by the use of oil "amounting to a fraction of 1% on the ore," the import of the decision is that if the results obtained by operation of the process with oil in amount 1% and more on the ore are like and the same results obtained with a fraction of 1% of oil, it is within the patent and is infringement. This is more ingenious than sound, and would deprive the decision of effect. The Court does not confine the patent to the *like* or *same* results obtained, but to *the* results obtained by the use of a fraction of 1% of oil on the ore. It is believed, however, that the Court employs the word "use" in its or the ordinary sense of beneficial service. Patent law is not concerned with the useless, and a valuable result sought is not "obtained" by but *despite* the use of an excess of an essential ingredient, which excess render no or ill service. From the evidence it appears the larger part of the oil used by defendant and all in excess of a fraction of 1% on the ore, if not inert is ineffective, wasted, and injurious to the process and results. Before January 7, 1917, defendant used only pine-oil and about 1.43 lb. per ton of ore, with excellent results. Since said date it uses a mixture of 20 to 24 pounds of oil per ton of ore, made up of 18% of pine-oil, 12% of kerosene oil, and 70% of fuel-oil, with poorer results. The kerosene and fuel-oil are petroleums. As before stated many oils are ineffective to operate the process, and that is because they have not the quality that contributes to bubble-making. What this quality consists of, wherein it lies, does not appear. With these ineffective oils, agitation will not produce froth and so there is no flotation of the metallic particles. One of defendant's witnesses testifies that in the laboratory and plant of the Utah Copper, one thousand oils have been tried, of which but two mixtures give satisfaction. Petroleums seem generally ineffective by the evidence of both parties, though some of defendant's witnesses testify to sometime successful experiments with them. Incidentally, there is suspicion that with experiments, as with figures, can be done anything for or against, without impropriety in the operator. Some petroleums are used in limited quantities, but always in combination with a recognized bubble-making oil, and only, it is said, for a somewhat bubble-stabilizing effect. Defendant's present mixture of oil contains more pine oil on the ore than it used alone before January 7, 1917. The other factors the same, it is obvious the excess petroleums in the mixture are responsible for the poorer results.

Defendant uses the patent process, uses plaintiffs' invention of ore-concentration by air-bubble flotation, uses the same elements in the same combination in the same way with the same function to the same but poorer results; and exceeding the patent claims in reference to one ingredient (oil), uselessly, wastefully, and injuriously and merely with intent to avoid the letter of the patent, does not avoid infringement. The addition of the excess oil no more adds to or changes the process, no more avoids infringement, than would the addition of milk or other useless substance not a part of the process. The excess oil exercises either no function or less efficiently exercises the sàme function in the same way as the limited oil, and to the same but poorer results. To secure to patentees their invention, the law looks quite through mere devices and forms, to the substance of things. And if in substance the invention is taken, if the thing that does the work is taken, all devices to evade the letter of the patent avail nothing to escape the consequences of infringement. Neither principle nor authority to the contrary, is cited or known to the Court.

In the matter of estoppel affecting infringement, it suffices to say neither in pleading nor proof do the elements of estoppel appear. Although the evidence is of great volume and the arguments of relative length, all have been carefully considered, but require no additional reference herein.

The patent is valid in respect to all claims involved. Defendant throughout has infringed and now infringes all said claims save 5, 6, and 7, and decree accordingly.

BOURQUIN, J.

August 25, 1917.

VICTOR C. HEIKES, of the U. S. Geological Survey, states that there were 527 producing mines in Arizona during 1916. The ore produced was 17,033,810 tons and the metals derived therefrom: gold, 192,801 oz.; silver, 7,212,039 oz.; copper, 721,833,169 lb.; lead, 27,062,087 lb.; zinc, 18,220,863 lb. The total value of the metals was $190,806,170.

Prices of Graphite

*Ceylon graphite of the better grades is more largely used by crucible makers than domestic flake-graphite and, even in normal times, commands a higher price. During the last three years the tremendous increase in crucible manufacture has caused a demand for the Ceylon product greatly in excess of the available supply and the price has constantly increased. The sharp increase of price for all grades of Ceylon graphite in 1916 over that for previous years is in part due to the fact that a much larger proportion of the highest-grade product was imported. In March 1917 the prices of Ceylon graphite, ex-dock, New York, were approximately as follows: No. 1 lump, 28 to 30c. per pound; No. 1 chip, 19 to 21c.; No. 1 dust, 11 to 12c. The prices paid for the highest grade of domestic graphite at the mines in 1916 ranged from 10 to 16c. per pound. The New York prices varied greatly according to the demand. In the summer of 1916 the price was as high as 18c. per pound. At the end of the year increased imports from Ceylon reduced the price, and in March 1917 domestic graphite of good grade could be bought for 12 or 13c. per pound. The prices paid at the mines for the highest grade product, in cents per pound, have been as follows: 1911 and 1912, 6 to 7; 1913, 6 to 8; 1914, 6½ to 8; 1915, 7 to 10; 1916, 10 to 16.

The sharp increase in the value of the Ceylon graphite imported in 1916 is due largely to the fact that, owing to the high freight-rates and scarcity of ships, the higher-grade material formed a larger proportion of the total.

Domestic amorphous graphite brought widely varying prices according to the grade of the product, but in general not greatly in advance of prices before the War. Amorphous graphite from Chosen averaged about $45 per ton on the New York market, a price due largely to the prevailing high freight rates.

Assessment Work on Mining Claims

Congress has passed a resolution exempting officers and enlisted men who have been or may be mustered into the military or naval service from the $100 worth of assessment work required by law upon unpatented mining claims that may be in their possession at the time of enlistment. It is necessary for the enlisted owner to file in the office, where the location is recorded, a notice of his muster into the service of the United States and of his desire to hold his mining claim under this resolution. This must be done before the expiration of the assessment year during which the claim-owner is mustered. The exemption is for the period of enlistment.

IN 1915 there was a production of 23,224 short tons of antimonial lead of a value of $3,665,736, and 5364 tons of antimony, valued at $2,373,760.

*Abstract from Mineral Resources, U. S. Geological Survey.

The Submarine and Kindred Problems

*The thousands of suggestions and plans presented to the Naval Consulting Board for assisting the Government in the present emergency indicate the patriotic fervor of the mass of our citizens. The Board makes a careful examination of every proposal presented. To facilitate this work, by suggesting the elimination of impractical ideas, the Board calls to the attention of those who desire to assist it some of the popular misconceptions as to certain fundamental principles which are most frequently misunderstood by the layman.

ELECTRO-MAGNETS AND MAGNETISM. The electro-magnet, the magnetic-needle, permanent magnets, and magnetism have been carefully studied for many years; and the laws governing their application may be found in any book on the subject. Although these laws are generally known, and applied in a practical manner, in a multitude of devices in common use, even the man of wide experience will be astonished at the limited range of practical effect of electro-magnets of large size. For instance, the magnets used in our manufacturing plants for lifting heavy masses of iron or steel are designed to exercise maximum magnetic effect, and for operation require a very considerable amount of electrical energy; yet a magnet which can lift twenty tons, when placed in contact with an iron plate of that weight, will not lift a two-inch cube of iron or steel if separated from it a distance of two feet. Therefore proposed devices which depend on the attractive power of magnets for their operation in deflecting or arresting torpedoes, mines, or submarines, must be governed by the simple laws of magnetism. A torpedo weighing approximately 2500 lb., and traveling at a speed of 25 to 45 miles an hour, will not be deflected to any practical degree by any known application of magnetism; and it is not believed that an enemy torpedo, mine or submarine will ever be found in a position to be interfered with effectively by any electromagnetic means, however powerful.

There is a general misconception regarding the 'electrification' of water and the atmosphere. There is no known method of 'charging the sea with electricity,' or 'shooting a bomb of electricity,' or of 'charging the atmosphere with electrocuting current.' Suggestions along these lines should show that the writer has made research in the laws governing the application of electrical energy and should contain sufficient proof of their feasibility to insure serious consideration. On the other hand, applications of the transmission of electrical energy by means of alternating or pulsating currents—as used in wireless systems, for example—belong to a different class of electrical development. Inventive genius is rapidly improving apparatus of this type for the sending and receiving of signals and messages, and the possibility of valuable results in this field is unlimited.

PROTECTION AGAINST SUBMARINE ATTACK. This subject, which is occupying the public mind as is no other, divides itself into a number of problems, the most important being the following:

(a) Means of discovering the approach of a hostile submarine and locating it so as to permit of prompt action for combating its attack.

(b) Protection of cargo-carrying ships by nets, guards, and screens.

(c) Protection through decreasing the visibility of vessels.

(d) Methods of destroying or blinding a hostile submarine.

Submarines, to operate most effectively, must approach within close range of the vessel which is intended to be torpedoed. The installation of offensive weapons on the merchant marine has increased the necessity for the utmost care being exercised by the submarine commander in remaining unseen by the officers on the vessel to be attacked.

Reports from abroad indicate that in many cases submarines must have remained along certain lanes of travel for periods extending into weeks of waiting with the expectation of torpedoing certain vessels. Under certain favorable conditions, where the waters are less than 200 ft. in depth, a submarine might lie at rest on the bottom, and if equipped with sensitive listening devices attempt to detect the approach of a vessel. As soon as this evidence was secured the submarine might come to the surface for a quick observation by means of the periscope and in this manner obtain the proper aim which would be required to register an effective hit. In case the water is more than 200 ft. in depth a submarine must be kept in motion to obtain steerage way in order to hold its proper depth of submergence. This speed may not exceed 4 or 5 miles per hour, but to remain submerged, and at the same time unobserved, the water must be at least 60 ft. deep.

The latest type of submarine which is being used abroad has a surface speed of at least 17 knots per hour and a submerged speed of probably less than 10 knots. The superior gun-fire from the merchantman which has been properly equipped would make it necessary for the submarine commander to obtain his observations, such as would permit accurate aiming of the torpedo, during the very brief interval of time required to come to the surface for observation through the periscope and to again submerge. If running near the surface, the periscope might be raised, a quick observation taken,

*Bulletin No. 1, issued by the Naval Consulting Board of the United States.

and lowered again within 30 seconds. If, however, the submarine is on the surface and hatches uncovered, from one to four minutes will be required to completely submerge, depending upon circumstances.

A submarine of recent type probably has a total radius of action of as much as 8000 miles when traveling at a moderate cruising speed of from 10 to 11 knots, and may remain away from its home-base for as much as one month, without requiring either fuel or other supplies during this period. This type of submarine may have as many as three periscopes, two conning towers and two rapid-fire guns attached to the upper portion of its hull. The vessel is steered by very efficient gyroscopic compasses, which are unaffected by extraneous magnetic or electrical influences.

A general understanding of the capabilities of the modern submarine for offensive operations will make it easier to appreciate the importance of the three problems which follow:

MEANS FOR DISCOVERY. The aeroplane. When the condition of sea and air are favorable, a submarine is readily discernible from an aeroplane flying at a sufficient height even though the submarine be submerged to a considerable depth.

While aeroplanes have thus been used successfully in the English Channel, they are unable to fly far out to sea where the submarines are now most active. Mother ships for carrying and launching aeroplanes might be used in this connection, but there are only a small number of such ships in operation and the construction of others under present conditions is necessarily a slow process.

Various sound-recording devices, intended to locate surface-vessels, submarines, and even moving torpedoes, are now being carefully tested. Water is an excellent conductor of sound, and the development and improvement of such apparatus offers a promising field for inventive endeavor to those who possess adequate scientific training and laboratory facilities.

Many devices are suggested which depend upon optical means of detection, such as special forms of telescopes and field-glasses to be mounted on ships, or on scouting vessels. Many special forms of searchlights and projectors have been suggested. The fact that a moving torpedo leaves in its wake a stream of air-bubbles caused by the exhaust-air from its propelling engines, offers, under favorable conditions, one means for discovering the approach of a torpedo. This evidence is, however, difficult to detect in a rough sea or at night, and, furthermore, the bubbles do not reach the surface of the water until after the torpedo has traveled onward a distance of from 50 to 200 ft. toward its target. The dragging of trawls, or nets, by special guard-boats, not only with the view of locating submerged submarines but also to sweep up floating and stationary mines, is frequently suggested. Under certain conditions this operation is practicable and effective.

It will be seen that each of the above methods, however useful, has its limitations, and scientists and inventors should apply themselves not only to the task of improving these, but also of finding supplementary methods and devices.

PROTECTION OF CARGO-CARRYING SHIPS BY NETS OR SCREENS. Many designs of such devices are suggested, and most of them are intended to be attached to the hull of the vessel to be protected. Many other suggestions along these lines, and differing only in some of their minor characteristics from the foregoing, have been received by the Board. Up to the present time not one of these proposals involving screens of any kind has received the approval of the Navy Department or of the Merchant Marine. The principal objections offered to these devices are that they are heavy, difficult to hold in position, unmanageable in a heavy sea, and that they interfere with the speed and with the ability of the vessel to manoeuver. The undeniable evidence which has been accumulated during the past few months of submarine activity has demonstrated that the immunity of a vessel to submarine attack is dependent very largely on its speed and also its manoeuvering ability. The percentage of vessels having speeds of 15 knots or more which have suffered from submarine attack is very small, while the losses of slow vessels, whose speed is less than that of a submerged submarine, is practically one hundred per cent of those attacked. Many of the suggested devices would prevent the launching of life-boats or rafts from the vessel to be protected. It is barely possible, however, that there may be developed some form of this general plan which will be found practicable. In no other field have so many suggestions or so many duplicate inventions been presented to the Board.

PROTECTION THROUGH INVISIBILITY. The point of look-out on a submarine being close to the water, the position of a vessel at a distance can only be determined by observing its smoke, which floats high in the air. Improved smokeless combustion is therefore desirable. Relative invisibility may also be afforded by methods of painting. Suggestions as to any other methods of reducing the range of visibility will be of interest.

DESTRUCTION AND BLINDING OF THE SUBMARINES. A rapid-fire gun is effective when the submarine is seen within accurate range of the gun; but the target is so small that it is difficult to hit.

The powerful effect of any submarine explosion on all neighboring bodies provides a simple means of destroying or crippling an undersea boat. Once it has been even approximately located, the setting-off of a heavy charge of high explosive, well submerged in the vicinity of the submarine, will bring about this result.

In certain areas, a quantity of heavy, black petroleum or similar substance which will float on the surface of the water has proved an effective means of clouding the optical glass in the periscope's exposed end.

Under favorable conditions of wind and position, many vessels have saved themselves from torpedo attack by the production of a smoke-screen. This may be formed either by incomplete combustion of the oil used for fuel by most naval vessels, or it may be created by burning chemicals, such as phosphorus and coal tar, or mixtures in which both of these and other materials are used.

After hiding itself from the submarine in a cloud of dense smoke, the vessel, if possessed of sufficient speed, may be able by a quick manoeuvre to change her position and escape before the submarine is able to discharge a torpedo.

MINES. Ever since the first use of gunpowder in the prosecution of war, mines and torpedoes have received great attention both from the warrior and the inventor. Mines are either fixed or floating. The fixed or stationary submarine mine is fired by contact, electricity, timing device or fuse. Such mines, which are extensively used by all navies, are rugged in design and may contain large charges of explosives. They are placed in position by especially equipped mine-laying vessels. Such a mine is provided with an anchoring device.

Floating mines differ from fixed mines in that they are unanchored, and, unless guard boats are at hand to warn friendly vessels of their proximity, may be as dangerous to friend as to foe. Such mines must be, according to laws of war, designed to become inoperative within a few hours after being set adrift.

TORPEDOES. The modern submarine torpedo is about 20 inches in diameter and 20 ft. in length: is self-propelled: is not steered by magnetic means: and keeps a fairly accurate course for several thousand yards at an average speed of more than 30 miles an hour. Its weight is approximately a ton and a quarter; and, when travelling at normal speed, possesses great momentum—in fact, in one case, when the high explosive charge in the 'warhead' failed properly to detonate, the body of the torpedo penetrated the steel hull of the ship attacked. Torpedoes are also provided with means to more or less effectively cut through screens, nets, or guards placed in their path.

A torpedo is propected from a submarine or other vessel by means of a special form of tube or gun. A small charge of gunpowder or compressed air is employed to start the torpedo, after which—if of the usual self-propelling type—it is driven through the water by its own compressed-air motor, the air being supplied from a strongly built reservoir within the body of the torpedo itself. The torpedo is kept upon its course by a gyroscope steering mechanism, which is immune to outside magnetic disturbances.

The detonation of the torpedo is accomplished through a mechanism placed within its warhead; and if the torpedo is either abruptly diverted from its course or is checked in its forward motion, the firing device, which is operated by arrested momentum rather than by any form of a projecting firing-pin, instantly ignites the heavy charge of explosive contained within the warhead. The explosion, if it takes place within twenty feet of the vessel, will usually rupture the ship's plating, because of the terrific blow transmitted through the water from the point of the explosion to the ship's side. The depth at which a torpedo travels may be regulated and is usually between 12 and 15 ft. below the surface.

CONFINING THE SUBMARINES. The question as to why submarines are not destroyed before they reach the open sea is a most natural one, and the best answer which it is possible to give, according to the officers of our Navy and those of the foreign commissions who have visited this country, is as follows:

The submarine bases are very strongly protected by land batteries, aeroplane observers and large areas of thickly mined waters extending to such distances that the largest naval gun cannot get within range of the bases. In spite of these protections, there is now going on a continuous attempt on the part of the Allied navies to entrap or otherwise defeat the submarines as they emerge from the protected areas. Nets are laid and as promptly removed by the enemy, whose trawlers are in turn attacked by our destroyers. The design of these nets and the detailed arrangement of their fastenings and attachments offer a broad field for invention, but it should be remembered that they must be capable of being used in waters in which there is a tidal current running from two to five miles per hour. Many suggestions for 'bottling up' these bases have been offered. but, as will be realized, it is not desirable to publish information which would indicate even in the smallest degree this country's plans.

SHIPS AND SHIPBUILDING. Many suggestions are made for ships of unusual form to provide for safety in case of a torpedo or mine exploding near or against the hull. Most of these plans are an elaboration of the usual watertight bulkhead construction now required as structural design for all modern ships.

The multiplicity of watertight compartments in any hull design tends to add to the vessel's safety.

The modern tank steamer used to carry fluid cargoes, such as petroleum products or molasses, is a good example of this design, which has been in general use for many years.

The explosion of a nearby submarine mine or torpedo frequently tears great rents in the ship's plating, in some cases opening a jagged hole ten feet or more across, but the destructive effect on the hull of a ship caused by the explosion of a mine or torpedo may be greatly diminished by special hull construction.

From recent experiments and experience, it has been demonstrated that the average merchant steamer may be seriously damaged by the explosion of a torpedo thirty-five feet away from its hull, but the destructive effect at any given distance from the point of explosion depends to a large extent upon the design and condition of steel framing and plating. A vessel with a strongly built hull can withstand an explosion that would destroy a weaker vessel.

COMMUNICATIONS should be addressed to

THOMAS ROBINS, Secretary, Naval Consulting Board,
13 Park Row, New York, N. Y.

The Flotation Process

By E. P. MATHEWSON

*This process has revolutionized the metallurgy of copper. Recently constructed plants, costing millions of dollars, have been discarded or scrapped and the flotation process introduced. Notable examples of this are the Washoe reduction works at Anaconda and the plants of the Utah Copper Co. at Garfield, Utah.

It took considerable time to convince the copper metallurgists of the country that the flotation process was a success, but as soon as a satisfactory demonstration was made, which occurred after numerous failures, these metallurgists took up the new process with avidity, scrapped their old plants, and re-built to adopt this modern system of concentration.

The MINING AND SCIENTIFIC PRESS of San Francisco, under the able direction of T. A. Rickard, has published an immense amount of information about flotation and has given out the following figures, which are astounding when one considers that the application of the process on nothing more than an experimental scale is only a few years old:

Treatment	Tons per annum
Flotation	30,000,000
Copper smelting	26,000,000
Gravity concentration	25,000,000
Gold and silver milling	13,000,000
Lead smelting	5,500,000
Copper leaching	2,000,000
Zinc smelting	1,000,000

Of the figures given above, that for copper leaching is perhaps a little low, but it is so much smaller than the figure for flotation that it does not signify. Prior to the adoption of the flotation process in copper concentration the losses were seldom less than 20%, whereas now they are seldom over 8%. The figures published by the Anaconda Copper Mining Co. indicate that the concentration loss in the most modern gravity-concentration plant in existence was 17%; and now with flotation the loss is given at a trifle over 4%. Other large establishments can show equally amazing results. The savings are now so great and the economies so extensive, due to the introduction of the process, that the so-called hold-up by owners of patents on the process cannot possibly cripple the users of the process, even if exemplary damages be allowed by the courts.

The slime problem, the bugbear of copper metallurgists for a generation, has been solved by flotation. Many copper metallurgists will recall the numerous experiments conducted with a view to recovering the values from slime and the enormous sums of money expended in putting up plants that recovered only 60% of the values from these slimes and which were considered marvels in their day. These plants, like the concentrators, have gone into the discard and now from the most pernicious slimes, savings of 90% of the values by means of the flotation process are quite common.

Strange to say, the process is a rule-of-thumb development. The practical application of the process was first made on an extensive scale with zinc. Copper metallurgists never imagined in those days that the process would ever be applied in their specialty, but today the tonnage treated by the process is largely made up of copper ores, zinc long ago having taken a secondary position.

Each particular ore seems to require a special treatment. This is because no satisfactory theory has as yet been evolved regarding the process. Many students are working on the problem with some promise of success, but at this writing we know less about flotation itself than we do about the true nature of electricity, while the results of the flotation process are as well known as the effects of electricity. Naturally a process creating such enormous profits and handling such large tonnages has produced a great amount of litigation. We have had cases involving millions of dollars tried in various courts and the end is not yet. However, as indicated above, there is money enough in the process to pay for all the litigation and still leave a good return in the investments.

*A paper presented at the Boston meeting of the American Chemical Society.

Average Price of Silver in New York, 1865-1916

Year	Price	Year	Price
1865	$1.337	1892	0.87
1866	1.339	1893	0.78
1867	1.33	1894	0.63
1868	1.326	1895	0.65
1869	1.325	1896	0.68
1870	1.328	1897	0.60
1871	1.325	1898	0.59
1872	1.322	1899	0.60
1873	1.297	1900	0.62
1874	1.278	1901	0.60
1875	1.24	1902	0.53
1876	1.16	1903	0.54
1877	1.20	1904	0.58
1878	1.15	1905	0.61
1879	1.12		
1880	1.15		Cents
1881	1.13	1906	66.79
1882	1.14	1907	65.32
1883	1.11	1908	52.86
1884	1.11	1909	51.50
1885	1.07	1910	53.48
1886	0.99	1911	53.30
1887	0.98	1912	60.83
1888	0.94	1913	57.79
1889	0.94	1914	54.81
1890	1.05	1915	49.68
1891	$0.99	1916	65.78

Quality of Tungsten Ores

Frank L. Hess, in a recent bulletin of the U. S. Geological Survey, states that inquiries were addressed to firms known to be reducing tungsten ores, asking what, for their purposes, was the relative desirability of the tungsten-ore minerals, the impurities most hurtful, and the limiting percentages of impurities that would be accepted. Replies were received from eight firms. Of these firms, one reduces its ores by fusing with sodium carbonate, leaching with water, separating tungsten tri-oxide by hydrochloric acid, and reducing the tri-oxide to a metallic powder; two reduce the ores by other wet chemical processes; two use both the sodium carbonate fusion process and direct reduction in an electric furnace; two use an electric furnace only; one uses processes in which the ores are first treated with wet chemicals and reduction is then completed in an electric furnace. Another firm, the Crucible Steel Co., has in use a number of processes, part of which are covered by the Johnson patents.

By the sodium-carbonate fusion process only powdered metallic tungsten is obtained. One of the other wet chemical processes produces powdered tungsten and another makes powdered ferro-tungsten. The electric furnaces produce only ferro-tungsten.

Most of the processes used for reducing tungsten from its ores also partly or wholly reduce nearly all the metallic and some other impurities in the ores, and these impurities are carried with the tungsten into the steel to which it is added. For such use iron makes no difference, but a number of other elements are not wanted, either because, like copper and phosphorus, they are detrimental to the steel or because, like manganese, if they are wished in the steel they can be added more advantageously in some other way. Objectionable impurities found in tungsten ores are antimony, arsenic, bismuth, copper, lead, manganese, nickel, tin, zinc, phosphorus, and sulphur. Few of these occur in large quantity in ores found in this country. Copper is perhaps the commonest hurtful impurity, and therefore most is said about it, but ores from some foreign countries contain nearly all the impurities mentioned. During the early part of 1916 tungsten ores were so eagerly sought that nearly all offered were bought with little objection to impurities, but under more normal conditions consumers are much more particular.

The wet chemical processes give more opportunity to get rid of impurities than the electrolytic process, so that companies using wet chemical processes are, as a rule, though not uniformly, least particular about the ores they buy. Two of the firms that use wet chemical processes buy tungsten ores almost without regard to the impurities present, but one objects to more than 2% copper, and both buy ores containing as little as 20% WO_3.

Only the one firm mentioned is known that does not object to copper in any grade of ore. Another will take cupriferous ore if 'the content of WO_3 is sufficiently high.' The others either will not take copper-bearing ore when other ores are to be obtained or set limits of 0.2 to 2% copper and not less than 50% WO_3, except that one firm will take ores that carry 5% more copper, for such a percentage will pay for separation. Two companies take ores without regard to impurities other than copper, provided the content of WO_3 is sufficiently high. Most of the companies object to tin, sulphur, phosphorus, antimony, arsenic, bismuth, lead, and zinc, two of them object to manganese, and one to nickel. The last company referred to set extreme limits of 0.25% for phosphorus, 0.25% for nickel, 6% for manganese, and a trace of arsenic.

As to the different tungsten ore minerals—ferberite, wolframite, hübnerite, and scheelite—two companies using wet chemical processes reported that they made no discrimination; a company using both processes reported that it made no discrimination if the ores carried more than 60% WO_3; one company uses ferberite and scheelite and will not use wolframite or hübnerite; another prefers scheelite but will take any tungsten ore mineral; a user who does not make steel and whose product does not enter into steel also prefers scheelite. Three others gave their estimates of the comparative values as follows (the estimates being stated in the same order): If ferberite can be bought at $7 per unit then wolframite is worth $7, $6.30, $6.25; hübnerite, $6.50, $5.60, $6.25; scheelite, $6, $6.60, $6.50.

So far as can now be learned, the foreign buyers are quite as various in their demands as the domestic users and are in general more strict in the limits set, and they also demand a purity of 65 to 70% WO_3, which means loss in concentration, for ores cannot ordinarily be concentrated to so high a percentage without great waste in slimes.

Brokers are naturally ruled by the consumers to whom they sell and make the same restrictions as to quality of the ores bought.

In a recent paper W. Geo. Waring discusses the zinc ores of the Joplin district and gives some interesting notes of the occurrence of rare elements in zinc ores. He states that in 1915, a very observant watchman at the zinc smelter of the Bartlesville Zinc Co., in Oklahoma, discovered globules like mercury that had oozed out in warm weather upon the drusy surfaces of the cakes of lead residuum that had been taken out of retorts used for the redistillation of crude spelter at a temperature of about 350° C. below the highest temperature attained in the ore distillation. He collected a handful of the liquid and submitted it to F. G. McCutcheon, the chief chemist, who found that it gave the reactions of both gallium and indium, and it was later proved to be an alloy of those elements. In 1916, G. A. Buchanan, chemist for the New Jersey Zinc Co., succeeded in proving quantitatively the presence of the extremely rare metal garmanium in Joplin ore, as well as in some products of zinc metallurgy. Still more recently Waring found thalium in the zinc ores of the Webb City-Carterville district. Practical methods for the extraction of the rarer elements from zinc ores have not been developed. Waring suggests that no particular difficulty lies in the way of separating the cadmium, thallium, and indium, along with copper, etc., in metallic form from the leach liquors of the electrolytic process by cementation upon granulated spelter, while gallium and germanium would concentrate in the electrolyte.

Prospecting for Platinum

James M. Hill, of the U. S. Geological Survey, gives the following useful suggestions for prospecting. Platinum is not known in all countries, but that all the deposits are known is highly improbable. In fact, platiniferous deposits in southern Spain have been discovered recently bý a geologist working on the assumption that the place to look for platinum is in areas of basic igneous rocks similar to the typical Russian rocks. Systematic search, based on similar assumption, in all countries may lead to the discovery of new deposits of platiniferous gravels of economic importance, but a note of warning should be sounded. In Russia such systematic work has been in progress for many years and although new deposits have been found they are not so numerous nor so rich as the hope of the explorers anticipated.

Platinum, both as the native metal and in the form of sperrylite, (platinum arsenide) has been found in basic igneous rocks in several places in the world but not in commercial quantities. The search for the source of platinum in rocks is therefore not likely to be of particular value in obtaining an immediate supply of the metal from lode deposits; it may, however, develop near-by gravel deposits. As already stated, it is confidently believed that any platiniferous ores of commercial grade will yield to the regular analytical treatment in the hands of competent assayers.

The placer deposits containing platinum are all, so far as known, in the vicinity of areas of basic igneous rocks, and it would seem that in any search for new deposits of platiniferous gravels the first step is the search for outcrops of peridotite, pyroxenite, dunite, and serpentine. When areas of these rocks have been found, the gravels of the streams rising in them should be washed to ascertain if platinum is present. The heavy concentrates found in gravels carrying platinum are usually rich in chromite and olivine. The character of the rock particles often gives a clue to the source from which the gravels were derived.

Crude platinum, as it occurs in placer concentrates, is ordinarily a silvery white metal which could be confused only with silver and possibly pieces of iron or steel. It can be distinguished from both of these metals, as they are soluble in dilute nitric acids; crude platinum can be dissolved only in concentrated aqua regia. In some placer deposits the grains of platinum are coated with a dark film and somewhat resemble the grains of the dark minerals chromite, magnetite, or ilmenite, from which they are separated by careful panning, as the specific gravity of platinum is greater than that of any of those minerals.

Platinum will not amalgamate with quicksilver alone, but will amalgamate if sodium is added. In ordinary quicksilver amalgamation the flakes of platinum float on the surface and can be removed. If sodium amalgam is used, the platinum may be separated from gold by agitating the amalgam with water uutil all the sodium is used up to form sodium hydroxide; then the platinum will come out on the surface of the amalgam, provided, of course, the amalgam is sufficiently liquid. Platinum has a hardness of 4 to 5, and can be scratched with a knife. It is so malleable that it can be pounded without heating into very thin sheets. It is practically infusible; the grains cannot be melted together, as can particles of gold.

A relatively simple chemical test can be made to determine the platinum. The metallic particles are dissolved by boiling in concentrated aqua regia and the resulting solution is allowed to remain on the stove till dry. The residue is dissolved again in hydrochloric acid and evaporated by boiling till the solution is thick but not quite dry. This mass is dissolved in distilled water and a few drops of sulphuric acid and of potassium iodide solution are added, which, in the presence of platinum, causes the solution to turn a characteristic wine-red, if much of the metal is present, or to a reddish pink in the presence of small quantities of platinum.

The test outlined above is fairly delicate, but it cannot be used to detect traces of platinum in the presence of large quantities of iron or other elements.

A second test may be applied to the aqua regia solution after the resolution in hydrochloric acid outlined above. In this test potassium chloride is added to the solution, which precipitates yellow crystals of potassium platinic chloride, if platinum is present.

These tests are comparatively simple and positive when made on single grains, but they cannot be relied upon when various other elements are present in the material tested. It is, therefore, recommended that their use be restricted to grains of a single mineral picked from the concentrate obtained by panning a sample of either rock or gravel.

THE COMPOSITION of zinc ore concentrate is given by W. Geo. Waring in a recent paper before the A. I. M. E. From numerous analyses the range in average composition of first-grade zinc-blende concentrate is: zinc, 42.10 to 63.09%; iron, 0.9 to 5.97%; lead, 0.06 to 2.86%; lime, 0.15 to 1.64%. For second-grade concentrate the range is: zinc, 48.62 to 58.37%; iron, 2.40 to 6.63%; lead. 0.04 to 3.83%. Galena jig-concentrate ranges in lead content from 73.7 to 81.9%, and table-concentrate from 53.7 to 72.5%.

CRIPPLE CREEK, COLORADO

CRESSON MINE—ROOSEVELT TUNNEL

A. E. Carlton, president of the Cresson Con. G. M. & M. Co., in an interview announced the discovery of a rich gold-bearing ore-vein, believed to be the extension of the 14th level shoot, on the 15th level of the main shaft. Mr. Carlton said: "It looks as if we have encountered the rich vein that we uncovered some weeks ago on the 14th level, now in the 15th. The ore is running around $200 per ton and there seems to be an extensive body of it, although we are not yet certain of its extent. However, the recent developments in the mine make it look good."

Light production is resulting from the work in progress at the main shaft of the Elkton Con. M. & M. Co. on Raven hill. Ten groups of lessees are active, and the output is limited to three cars per week of milling-grade ore. The lessees are doing much prospecting, but no new orebodies have been discovered. Dump lessees are shipping from 2 to 4 cars per week of low-grade ore.

The Komat Mining Co., Math Korf, president and general manager, has re-timbered the shaft of the Victor mine, on Bull Cliffs, to the 900-ft. level, and has started to pump the seepage water from the bottom level, at 1000 ft., preparatory to exploitation at that depth. Since shipments commenced, the leasing company has delivered 15 cars of ore, six during the present month. Settlements have ranged from $14 to $18 per ton. The property is owned by the Smith-Moffat Mines Co., of Denver.

The Big Toad Gold M. & M. Co., E. A. Chapman, manager, operating the Dante mine on the south-western slope of Bull hill, under lease from the Dante Gold Mining Co., is engaged in overhauling the Reid mill, recently acquired by purchase, and will handle low-grade mine and dump-ore.

The Raven & Beacon Hill G. M. Co. has extended the lateral from the Roosevelt tunnel of the Cripple Creek Deep Drainage & Tunnel Co. into its ground, 17 acres of the Arequa townsite on Raven and Beacon hills, a distance of about 225 ft. The vein under development is a strong one, but the assays have been low. The work is being watched with interest.

The Roosevelt tunnel was advanced 78 ft. in the first 15 days of September and the heading is now in a dense breccia. The ground is breaking well. The heading is now in the Big Banta claim of the United Gold Mines Company.

The Acacia Gold Mining Co. is engaged in cutting a station at the new 1450-ft. level of the South Burns shaft. Shipments from the upper levels continue and the company and lessees are shipping about 100 tons of ore per week, ranging from $18 to $30 per ton in value.

The following officers and directors were elected at the annual stockholders meeting of the Blue Flag Gold M. & M. Co., held recently at Denver: J. F. Erisman, president and general manager; Joseph J. Gunnell, vice-president; J. S. Pigall, secretary-treasurer; W. B. Warren, assistant secretary and treasurer; directors, Joseph J. Gunnell, J. S. Pigall, J. M. Heaton, D. A. Schaffnit, J. M. Burkhart, J. F. Erisman, and George W. Duke. The company has resumed operations on Raven hill and is sinking the main shaft from the 1100-ft. to the 1400-ft. level. The Blue Flag G. M. & M. Co. is also operating the Laurium group in Summitt county, and the Yellow Jacket placer in Idaho.

VANCOUVER, BRITISH COLUMBIA

ELK VALLEY COAL AREA.

While it is an undoubted fact that the coal mines of Vancouver island are capable of much greater production and that there will probably be no great difficulty in supplying future demands for some time to come, considerable importance is attached to new and undeveloped coal areas that have been the subject of recent investigations. British Columbia's coal consumption is increasing, the supply of fuel oil is threatened and if cut off would have to be replaced by an additional million tons of coal per year. New industrial plants are springing up, commerce is increasing, and after the War there seems reason to anticipate a greater development than ever.

In view of these facts it appears not at all unlikely that the undeveloped coal resources of the Province will be utilized. A report just issued by the Geological Survey of the Dominion Department of Mines states that practically the whole of the Upper Elk valley is occupied by a coal basin and that the thickness of workable coal is large. The coal of the Upper Elk valley is similar to that of the Crowsnest coalfield, which is mined at Michel and Fernie. It is bituminous and is a good coking coal. The coal occurs interbedded with sandstone and shale, and the series as a whole is referred to the Kootenay formation. The thickness of the formation in Green Hills is approximately 3500 ft. This is much greater than that reported for the same formation in other areas.

In Elk valley the first coal is in the Green Hills, which reach an altitude of 3000 ft. above the valley. Several coal-seams of workable thickness outcrop. A section measured shows 89.5 ft. of coal in twelve seams. The formations strike with the trend of the ridge and can be seen outcropping for several miles to the north. The next coal-outcrops are about seven miles farther north at what is locally known as the Canadian Pacific Railway headquarters' camp. Here are two coal-seams, 10 and 15 ft. thick, separated by 125 ft. of sandstone and shale, on the west slope of the Green Hills. Prospect adits have been driven on them.

Aldridge creek cuts across the strata and a number of coal-seams are exposed. Prospect adits have been driven in six seams. The combined thickness of the six is not less than 50 ft. One seam is 15 ft. thick. Two and a half miles north of Aldridge creek at Weary Creek camp an adit has been driven across the same measures and is reported to have cut seven workable coal-seams of an average width of nine feet. One seam at the surface measures 15 ft. in width. Opposite Weary Creek camp on the west side of the Elk river, the coal measures are well exposed on Bleasdell creek. About 20 seams of coal are in evidence, varying in width from a few inches to 30 ft. Several are of workable thickness.

North of Weary creek the valley narrows and for the last 12 miles before the summit is reached the limestone mountains rise steeply on both sides. A thick coal-seam is reported to occur along the river five and a half miles north of Weary creek, but was not seen by the geological survey party.

On ascending Fording river the first coal is encountered at

Ewin creek, a tributary from the east, 12½ miles north of the junction of Fording and Elk rivers. On the south side of Ewin creek the Imperial Coal & Coke Co. has driven adits on six seams, three of which are more than 10 ft. thick. Prospect-pits on the hills indicate several more seams. These lie on the east arm of the syncline between the north arm of the Wisukitshak range and the main range of the Rocky Mountains. They were not followed southward, but to the north they follow the ridge east of the Fording for 13 miles, then cross the Fording and connect with the coal on Aldridge creek, as previously described. Prospect-drifts have been driven on three seams where they cross the Fording river, but were so badly caved that the thicknesses could not be measured.

The information that has been compiled thus far by the geological survey was obtained during a hurried trip through the valley. There were no prospectors in the field from whom guiding information could be obtained and most of the time was spent along the main trails and at the easily reached prospects.

The mining of the coal can be accomplished without serious difficulties. The methods employed on the pitching seams of the Crowsnest district can be applied here. The valleys of Elk and Fording rivers offer easy routes for railways and connections can be made with the Canadian Pacific and Great Northern railways at Michel or with the main line of the Canadian Pacific railway to the north. At present a good wagon-road leads from Michel up Elk river to Weary Creek camp, a distance of approximately 45 miles, and the other parts of the district can be easily reached by pack trail.

MARIPOSA, CALIFORNIA

Hornitos District.—Princeton.—Merced River District.

A good deal of fresh activity is reported. In the Hornitos district the Mt. Gaines is walling off the water on the 900-ft. level and extracting ore on the 500. The water now being pumped is delivered to the Ruth Pierce, where a full crew is employed and 30 tons of ore is being stamp-milled daily. The Castanet mine is sinking the main shaft and will build a small mill to treat its small output of high-grade ore. On the Mariposa grant a good deal of work is being done by lessees. Of these, one party is operating the Long Mary mine and mill, but operations are restricted by the amount of electric power obtainable from the Bagby plant. The lessees on the Queen's Specimen, near Bagby, have finished a 300-ft. tunnel and cut ore 9 ft. wide. It is said to be rich. The Trabuco boys are milling ore from the upper levels of the old Pine Tree and it averages over $20 per ton for the 10 tons per day. The dumps on the Pine Tree and Josephine are being hauled to Bagby and milled there in the old Mariposa company's mill. It is reported that 240 tons averaged $5.20 per ton. About 300 tons remains to be treated. The Princeton mill is being prepared to treat the output of the Mt. Bullion district. The new company has made the first payment of $75,000 on the purchase of the Mariposa grant, but a lack of working capital appears to restrict new operations. In the Merced River district, the White Oak is being operated by A. E. Rau-Roesler of New York. A new electrical equipment is being erected; a hoist and new compressors have been bought. The intention is to sink 250 ft. and drive 1000 ft. on the vein, which is 4 ft. wide and averages $8 per ton in gold. The Original mine, on the Merced river, is working full time in both mine and mill, paying dividends of $2000 per month during the current year. The Mountain King, after much trouble in getting a crew, is operating full time, milling ore from the 1400-ft. level that averages $7 per ton for 60 tons per day. In the Coulterville district, the Champion is getting ready to add an electrical equipment. The San Joaquin Power Co. is running a line to this mine and new equipment is to be purchased. Meanwhile extremely rich ore is being mined on the 300-ft. level. The White Gulch Mining Co., operating the Virginia mine, is employing 22 men and extending the drift on the 700-ft. level. Ore for the 10-stamp mill, treating 30 tons daily, is being taken from the 300 and 500-ft. levels. A new hoist has been erected and the enterprise is being operated with notable efficiency. The Gray Eagle has installed pumps and is getting ready to operate. The Mary Harrison is idle.

AUSTIN, TEXAS

Deep Potash Exploration.

George Otis Smith, director of the U. S. Geological Survey, in responding to an inquiry as to what research work that department was doing in Texas and the South-West, makes particular reference to the potash explorations that are being conducted. He says:

"The search for potash in the Panhandle or north-west Texas will be continued. The well now drilling at Cliffside, seven miles north-west of Amarillo, for the purpose of proving the presence or absence of potash salts in the Permian formations of that area has now reached a depth of over 1700 ft., without, however, encountering potash in important amounts. It is hoped that the hole may be carried to a depth of 2300 ft. or more. Meanwhile the studies of the rocks in the Red Beds basin of north-eastern New Mexico, northern Texas, and western Oklahoma with the object of determining the centres of greatest saline deposition, so far as that is possible from surface criteria, will be continued by N. H. Darton. The study of the occurrence of salt and potash deposits in the Permian Red Beds of Europe encourages the hope that somewhere in the great Red Beds area of eastern New Mexico, western Texas, western Oklahoma, western Kansas, and eastern Colorado and Wyoming deposits of potash salts, probably lenticular in shape and restricted in geographic extent, but perhaps profitably workable or even as valuable as those in Germany, may somewhere be found. An earnest realization of this hope is to be seen in the tremendous thicknesses of rock-salt reported in some of the wells drilled for water and oil. It is a misfortune tnat chemical tests for potash were not made of the saline deposits and of the brines encountered in most of these wells. All the strata, including the gypsum, anhydrites, and clays, as well as the salt-beds cut by the drill at Cliffside, have been chemically tested on the spot by R. K. Bailey, chemist, who also has from time to time visited other wells operating in eastern New Mexico and western Oklahoma for the purpose of testing the drill-samples taken from the wells and held for his inspection through the courtesy of the operating companies. It may be that a number of wells will be bored by the Government or by private interests seeking oil or water before potash salts in profitably workable beds are finally discovered in this great basin. However, the conditions of deposition, including the character of the sediments, the precipitation of vast amounts of salt, gypsum, and anhydrite, and even the conditions of shallow water and aridity were so similar contemporaneously in the Stassfurt district of Germany and in the great Red Beds region of the West that the presence of potash deposits in the latter may be tentatively predicted on the law of probability."

Manganite is a black brittle massive or crystalline mineral with a reddish-brown or nearly black streak. It has a specific gravity of 4.2 to 4.4, a hardness of 4, and a theoretical content of 62.4% metallic manganese and 10% water. It occurs generally with psilomelane in alternating layers in botryoidal masses. In such masses it consists of numerous needles that radiate from the walls toward the middle of the vein. Psilomelane is a black or steel-blue mineral of specific gravity 3.7 to 4.7. Its chemical composition is not definite.—U. S. Geological Survey.

ARIZONA

COCHISE COUNTY

(Special Correspondence.)—Machinery is being overhauled and buildings repaired at the Central Butte mine, three miles from the Commonwealth. The work is in preparation for de-watering the 180-ft. shaft for further sinking. W. J. Anson of Los Angeles examined a tungsten property near Dragoon and purchased an interest in it. Machinery is now being shipped from Los Angeles and operations will start at once. The Arizona Binghampton output for September is expected to be in excess of the previous month. A new unit was recently added to the mill, increasing the capacity to 250 tons per day. A carload of high-grade ore per day is being shipped. Diamond-drills will be used for prospecting the lower levels. The Nelson mine is now producing and shipping 125 tons of ore per day to the plant of the Bradshaw Development Co. As a result of four years of development much ore is in reserve. The ore is complex and is treated by flotation.

Bisbee, September 15.

The Shattuck-Arizona, Bisbee, is developing deposits of lead-zinc sulphides. A concentration-plant will be constructed.

GILA COUNTY

(Special Correspondence.)—Operations in the Globe-Miami district are rapidly approaching normal. The Old Dominion has 700 men working. Miami has just about as many, and half of its mill is in operation. Four sections of the Inspiration mill are in operation. The International smelter expects to resume. The Ray Hercules has been able to push its construction work with such rapidity that the mill will be in operation by the first of the year. The initial capacity of the mill will be 1500 to 1800 tons per day. There is a reserve of 10,000,000 tons of ore. The Sulphide Copper Co. has been organized to develop 37 claims 12 miles from Globe in Russell gulch.

Globe, September 17.

GREENLEE COUNTY

(Special Correspondence.)—It is rumored that the Phelps-Dodge interests are contemplating the purchase of the Arizona Copper Co., now controlled by Scotch capitalists.

Clifton, September 10.

MOHAVE COUNTY

(Special Correspondence.)—The Rico Mining Co., Robert W. Wilde, president, leased and bonded the C. O. D. silver mine. The old mill is to be re-constructed and the mine re-opened. W. J. Hennessy is superintendent. In the New Mohawk, a 4-ft. vein of silver ore was cut in No. 2 shaft, 150-ft. level. A larger hoist will be installed and sinking resumed. The New Tennessee has resumed sinking and the shaft will be continued to the 500-ft. level. Four feet of milling-ore has been opened in the north drift, 240-ft. level. The Copper Age mill is in operation. The mine continues to develop ore-reserves on the upper adit-level. Final payment was made on the Tuckahoe. The shaft is down to 250 ft. On the 200-ft. level of the Diana an orebody four to six feet wide and averaging $20 per ton in gold and silver was opened. The Bullion Hill shaft is down 280 ft. Cross-cutting will begin at the 300 level. Several feet of ore was discovered in the south drift of the Schenectady upon the 175-ft. level. Ore is being broken in both drifts for shipment. The Rattlesnake is still mining ore of shipping-grade. The Chloride Queen has installed new hoisting machinery.

Chloride, September 6.

(Special Correspondence.)—The C. O. D. mine, consisting of 15 claims, has been purchased by the Rico Con. Mining Co., which is headed by Robert W. Wilde, of the Desert Power & Water Co. The C. O. D. mine is 12 miles north of Kingman. The property has been worked by a number of lessees in the past. Work has been started on re-modeling the mill and preparations are being made for the unwatering of No. 2 shaft, 350 ft. in depth. The New Mohawk has reached 60 ft. in shaft No. 3, and rich silver ore has been encountered. An assay of ore, not including a high-grade streak, gave a return of $53. The sorting-tables have been completed at shaft No. 2, and all ore is being sorted. The final payment has been made on the Tuckahoe. The new shaft has now reached a depth of 200 ft. Lessees, operating the Cornish property which adjoins the New Mohawk, have sunk the shaft to 80 ft. Samples assay $7 in gold and several ounces in silver per ton.

Kingman, September 14.

ARIZONA

NAVAHOE COUNTY

(Special Correspondence.)—Coal declarations have been filed at the U. S. Land Office at Phoenix. Nineteen have been filed by men from Holbrook.

Holbrook, September 15.

PIMA COUNTY

(Special Correspondence.)—The Mile Wide Copper Co. has entered into a contract with the Sasco smelter and will ship four cars of ore per week. The ore will average 8% copper. The development has reached 325 ft. in depth and cross-cuts have been extended at each 100-ft. level. The orebody is from 22 to 28 ft. wide. The first payment on another group of four claims has been made and a lease on the Copper Plate claim taken by the same company. The Cornelia Copper Co. has contracted with the Southern Sierras Power Co. for 2000 kw. to be delivered at Ajo. A line will be constructed from Yuma to Ajo, the contract calling for power delivery within 18 months. The Dos Cabezos Gold Ridge Mining Co. is developing ore at the Gold Ridge mine. The company contemplates erecting a mill. The Mascot mine of the same company shipped 75 cars of ore in August. More machinery is being installed at the Mascot and a larger force of men is being put on.

Tucson, September 17.

A 50-ton custom smelter will be built at Ajo by H. R. Haines, I. J. Tolinski, and J. W. Mayes.

The New Cornelia Copper Co., Ajo, will install several additional precipitation tanks at its leaching-plant.

The Hecla-Arizona Copper Co., Twin Buttes, is preparing plans for a large leaching-plant.

PINAL COUNTY

(Special Correspondence.)—The Silver King has now been equipped for pumping, and dewatering has commenced. The management expects to be able to start shipping ore by October 15. Developments on the Fortuna Consolidated continue to improve, and it is expected that shipments of copper ore will soon be made. Work has been resumed on the Camelback Gold & Silver Mining Co.'s property. E. F. Gordon of the Primos Chemical Co. is making an examination of the Mohawk mine in the Mammoth district.

Florence, September 15.

The Ray Hercules M. Co. contemplates the installation of electrical equipment, a crusher, and a 1200-ton concentrating plant.

YAVAPAI COUNTY

It is stated that the General Development Co., Jerome, has appropriated $500,000 for the purchase and installation of machinery and the development of its property.

The Copper Queen Mining Co., Mayer, will construct a flotation-plant.

(Special Correspondence.)—Operations by the Arkansas & Arizona company have ceased. The controlling interest is held by the Goodrich-Lockhart Syndicate. Pumping and re-opening the property was started a year ago and $250,000 spent in development and improvement. A complete report of the affairs of the United Verde Copper Co. has been forwarded to the Corporation Commission by Eugene M. Barron, who was appointed manager recently. The directors felt that if the commission approves its present plan of raising more money they will make the mine a copper producer. The Shannon Copper Co., which took over the Yaeger Canyon mine, 8 miles south of Jerome, has opened up ore in cleaning out the old workings. A $30,000 electric plant is being installed. The power-line of the Arizona Power Co. crosses the Yaeger property.

Jerome, September 10.

The Pocahontas Copper Co., W. H. Skinner, president, is contemplating the erection of a flotation mill at its mine, two miles south-east of Mayer.

YUMA COUNTY

(Special Correspondence.)—The shaft of the Black Giant Mining Co. was sunk 160 ft. in August. The equipment used was a 25-hp. Fairbanks-Morse hoist, Ingersoll compressor and jack-hammers. The property was taken over in April by D. W. Hale and associates of Mobile, Alabama. W. Tovote, of Tucson, is making an examination of the property.

Yuma, September 15.

(Special Correspondence.)—Lessees on the Bullard property in the Cunningham Pass district shipped their first car of ore last week. They are said to have a moderate tonnage of high-grade ore. An adit on a vein from which 80 cars of ore had been shipped was extended to a length of 400 feet.

The Wenden King is operating on a double-shift basis and is sinking a shaft. The Black Reef is sinking and has almost reached the 500-ft. level. At the Black Giant, air-drills are being used and the Gosson vein will be prospected at a depth of 500 ft. Boston people have taken over a mining property near the Socorro mine. They have organized a company which is called the Harqua Hala Ridge M. & M. Company.

Wenden, September 16.

CALIFORNIA

AMADOR COUNTY

(Special Correspondence.)—The Bunker Hill Mining Co. at Amador City has a flume in course of construction, extending on trestles from the mill or slime plant to the tailing dam on Rancheria creek. The flume will carry the waste from the mill to and along the line of the dam, and the material itself will thus be used to strengthen the concrete restraining-barrier.

The California Clay Corporation, which has acquired a large tract of land near Ione and has plans made for the development of the clay deposits there, has been hampered by financial conditions, but active operations are expected at an early date.

Pending the completion of plans for handling its large deposits of low-grade ore, the South Eureka Co. laid off its employees, with the exception of those necessary to keep the mine open and in repair. The Oneida mine is not affected.

Sutter Creek, September 15.

A copper mine west of the Allen copper mine is being re-opened by F. W. Ruhser, J. Strohm, and A. Huberty of Jackson.

Wm. Hooton and J. R. Tiernan have leased a tailing-dump near Forest Home and will begin the erection of a treatment plant.

NEVADA COUNTY

The Golden Center mill is receiving ore from the 500 and 700-ft. levels and will resume operations within a short time.

PLUMAS COUNTY

(Special Correspondence.)—Under B. R. Binns, as manager, Nevada people have begun work on a group of copper claims adjoining the Engels mine. A camp has been established and work will soon start on a 3000-ft. adit. The Engels Copper Co. is steadily increasing the capacity of its flotation mill and developing areas of outside ground.

The Juneday Mining Co. has arranged for the resumption of work with F. D. Searight as manager. Cleaning out of the old tunnel has begun and by the end of fall development will start. The Juneday is a gold-quartz mine near Crescent Mills.

The Noble Electric Steel Co. has completed a 1½-mile road from the Burch manganese mine in Round valley to the recently acquired group of manganese claims above Indian Falls, and is about to start regular shipments. From its Braito mine this company is shipping two cars of manganese ore per week, and is prospecting numerous properties in the vicinity of Indian Falls and Greenville. From the Crystal Lake manganese group, on the summit of Mt. Hough, Allen & Robinson are shipping several tons per week to the Heroult smelter, and opening one of the best deposits yet found. The property is five miles east of Indian Falls and is owned by Myton & Kloppenberg of Quincy.

Indian Falls, September 26.

The Engels Copper Co. proposes to file an application before the Commissioner of Corporations for permission to issue 167,686 additional shares of capital stock. The company will use the money derived from the sale of the stock to defray the expenses of the first unit of the new mill and the tramway and for the construction of another 750-ton unit which will be built early in 1918. A two-mile adit, a three-compartment shaft, and the construction of a hoist are also planned.

SAN BERNARDINO COUNTY

The Valley Wells smelter will soon be in operation; the delay in the completion of the plant being due to the congested condition of the railroads, and the failure to deliver some of the machinery on time. The plant will treat custom-ores as well as ore from the Copper World mine which is being operated by the same company. There are many thousands of tons blocked out in the mine and the old dumps are being worked over. Enough ore is in sight to run to capacity for three years. This company has taken over the Francis mine for its supply of sulphides.

A shipment of platinum ore from the West End mine was made to the Pacific Platinum Works in Los Angeles, the assay of this shipment was: gold 1.56 oz., silver 5.10 oz., lead 54.3%, platinum 3.57 oz. per ton. Arrangements are now being made for the marketing of the ore. In the development now being done for determining the extent of the platinum ores, a body of high-grade lead ore was opened. A 10-ton dry-concentrating plant is to be erected for the treatment of the lower grades of ore. The Standard mine is making regular shipments of high-grade copper ores. This property, idle 10 years, shows promise of again becoming one of the large producers of the district.

Calico, a once noted silver-mining district, is showing signs of activity under the stimulus of the prevailing high price of silver.

SHASTA COUNTY

The El Oro Dredging Co. is prospecting the Menzel ranch. The acreage lies north of Redding, directly across the Sacramento river.

The Gardella Dredging Co. is constructing No. 2 dredge for operation on ground adjoining the area now being worked by No. 1.

The Old Diggings mining district is active and gold-bearing ore is being shipped to the Mammoth smelter.

The Shasta Hills Co. is developing the Sybil mine. A 5-stamp mill and concentrator has been erected.

The miners strike is practically over. The Mammoth, Iron Mountain, Balaklala, and Hornet mines are in operation with reduced forces.

TUOLUMNE COUNTY

(Special Correspondence.)—The Wonder mine has been purchased by E. G. Blake, and work started in the 50-ft. shaft. The vein has an average width of 3 ft. and assays are from $4 to $108 in gold per ton.

Chrome ore is being shipped from the western part of the county, most of it being loaded on cars at Jamestown and Chinese Camp. Buyers are in the field almost every week. A few stamps of the Shawmut mill are once more in operation, after being hung up for several months. For some time work has been confined to the development of the property and to making improvements.

The McAlpine mine, near the southern boundary line of the county, is being worked and it is said that the outlook is much more promising than at any time in the past. The shaft is to be deepened, and the operators are now seeking additional men. The return from the gravel milled and washed at the Buckeye mine, on Table mountain, since the beginning of milling and washing operations three months ago, is reported to have been fair. A flow of water was tapped in the mine recently, causing the men to rush to surface to save their lives. An air-compressor and other machinery have been installed at the Keltz, and under the direction of G. E. Cheda, superintendent, development work will be pushed.

A two-stamp mill has been installed on the A. C. Smith claim near Chinese Camp. The claim was recently leased to Pearce, Rowe, and others. The lessees report the finding of rich ore. The shaft at the Yosemite mine, adjoining the Rawhide, has attained a depth of 500 ft. After cutting a station a cross-cut will be driven.

Sonora, September 15.

COLORADO

BOULDER COUNTY

(Special Correspondence.)—The lessees on the Scroglund mine are steadily producing tungsten ore. The Boulder County Tungsten Production Co. has let a 700-ft. contract to O'Day Bros. The company has completed its refining plant and is now directing its attention to the exploitation of its mineral area. The Huron mine, president R. T. Shaw of Holton, Kansas, and G. M. Woodring of Eldora, manager, is producing rich gold ore and is planning a milling plant at an early date. The Consolidated Leasing Co., of which C. E. Kohler is president and manager, has perfected plans for the erection of a mill. The company has large orebodies on Spencer mountain. The Ludlow mill at Lake Station is being completed, and is expected to be ready for operation in the near future.

The Primos Chemical Co. is planning the erection of a mill for the treatment of molybdenum ores in the Daily district.

Eldora, September 15.

TELLER COUNTY

The plant of the Lost Dollar mine is being re-built and mining operations are expected to begin late in November.

IDAHO

Investigations of the treatment of ores which was begun this summer by the U. S. Bureau of Mines in co-operation with the School of Mines of the University of Idaho is expected to stimulate metallurgical progress in the North-West, according to C. A. Wright, head of the Bureau of Mines at Moscow, Idaho, and Francis A. Thomson, dean of the School of Mines. The co-operation has been made possible by generous contributions from some of the mining companies in the Coeur d'Alene district.

The work at Moscow is to be carried out in the School of Mines' laboratories under the direction of C. A. Wright, with Thomas Varley, superintendent of the North-West Bureau of Mines station, as supervising metallurgist. The staff at Moscow, in addition to Mr. Wright, consists of an analyst and two research men.

Special emphasis is to be placed upon differential flotation as applied to ores containing lead, zinc, and iron minerals. It is also planned to study processes and methods of treatment that might be applied to the low-grade ores which cannot be profitably concentrated by the methods now in use.

The Consolidated Interstate-Callahan Mining Co. produced 15,005 tons of ore in August. An average of 14,400 tons per month was made for the quarter ended in June. Shipments during August aggregated 5226 tons of zinc and 951 tons of lead product, a total of 6177 tons. The product contained 47% zinc, 46% lead, and 18 oz. of silver per ton. Before concentration the ore averaged 21.5% zinc, 7.2% lead, and 3 oz. of silver per ton. The installation of an additional mill-unit will be completed soon. The unit is equipped with flotation-cells and is expected to handle greater quantity of ore and increase extraction.

The Richmond M. M. & R. Co., operating in Shoshone county, has declared its regular monthly dividend of $16,800, at the rate of 2 cents per share. Payment will be made October 5. The total of disbursements is $84,000.

The Douglas Mining Co. of Wallace has declared a dividend of one cent per share. The Douglas mine is under lease by the Anaconda C. M. Co. which pays a royalty of $3 per ton of ore shipped.

The Con. Interstate-Callahan M. Co. has suspended its dividend for tne current quarter. The war tax is the chief reason for withholding the dividend. The earnings for the third quarter were $461,000, an equivalent of one dollar per share.

The Armstead mines on Lake Pend Oreille will be developed by the Silver Hill Mining Co. The installation of a compressor and a concentrator is contemplated.

LEMHI COUNTY

C. W. Sherwood of Salmon will install a cyanide plant on the Gold Dyke. Plans have also been made to install an electric power-plant on the Salmon river to supply current for the electrolytic treatment of the zinc ores of the Gold Dyke.

MONTANA

DEERLODGE COUNTY

At the Washoe smelter, four sections of the concentrator are in operation. One blast-furnace, 5 reverberatories, and No. 2 roaster building are also in operation.

LINCOLN COUNTY

The Snowstorm Consolidated, at Troy, has been milling ore at the rate of 7000 to 8000 tons per month during June, July, and August, according to Leo Greenough, president and manager. Production has been curtailed owing to a shortage of underground labor but conditions in this respect are improving.

SILVER BOW COUNTY

Resumption of work at the Anaconda is steadily taking place. Over 4000 miners are at work.

NEVADA

CLARK COUNTY

The Los Angeles Platinum Co. is constructing a plant at Los Angeles for the treatment of ore from the Boss mine.

HUMBOLDT COUNTY

The Seven Troughs Mining Co., at Fairview, is extracting high-grade ore and shipping to the Darby mill at Mazuma. It is said that the erection of a mill is being considered.

MINERAL COUNTY

The Pilot Copper Co. has resumed work on a 200-ton leaching plant.

The Luning-Idaho company will start shipments from the McDavitt shaft as soon as the second unit is put into operation at the Thompson smelter.

Rich copper ore has been struck in the Calavada on the 1000-ft. level.

It is stated that the Nevada Rand Co. has definitely decided upon the construction of a 10-stamp mill.

NYE COUNTY

(Special Correspondence.)—The Tonopah Belmont during the past week shipped 49 bars of bullion valued at $97,823 and concentrate valued at $12,400 from the Belmont mill, and 22 bars of bullion worth $44,200 and 31 tons of concentrate worth $12,400 from the plant at Millers, making a total of $166,823 for the two mills. Raise 44 on the 900-ft. level continues on the South vein with a 4-ft. width. The intermediate east drift on the 1100-ft. level reached the foot-wall of the Belmont vein. South cross-cut 20 on the 1166-ft. level intersected a branch of the Mizpah fault vein, and it shows good assays. The tonnage of the past week was 2692 tons.

The Tonopah Extension, during the last part of August, shipped 25 bars of bullion valued at $54,000. The station on the north side of the Victor shaft on the 1680-ft. level is complete, and a cross-cut into the unexplored ground to the north will soon be started. In the No. 2 shaft, 24 ft. of development was made, and in the Victor 120 ft. On the 1440-ft. level of the Victor, the Murray vein has a width of six feet and the O. K. three feet. The tonnage the past week was 2380 tons.

The Sandgrass territory of the Tonopah Mining Co. is still idle due to the re-timbering of the shaft. In the Silver Top. 54 ft. of progress was made and in the Mizpah, 29 ft. The Burro and the South Dipping veins both show a 4-ft. face of milling ore. The tonnage the past week was 2200 tons.

The West End mill shipped 28 bars of bullion valued at $54,457 for the first half of September. In the Ohio territory of the West End, cross-cut 527 has been started. Cross-cut 680, No. 2 of the West End, has continued to advance under the vein; 717 D. No. 2 is advancing in low-grade vein material. At the Halifax 1704 and 1256 cross-cuts have advanced at their usual rate.

The south cross-cut from the 900-ft. level of the Desert Queen shaft of the Jim Butler Tonopah Mining Co. cut a small stringer of mineralized quartz. Raise 368 has a 4-ft. face of fair ore on the Wandering Boy vein. The tonnage the past week was 900 tons.

The Monarch Pittsburg Mining Co., since the visit of president H. C. Brougher, has started a more active campaign of development. Raising on the vein from the 850-ft. level will be started to develop the orebody.

The Rescue-Eula produced 128 tons of ore the past week, the Cash Boy 40 tons, and the lessees at the Montana 327 tons, making the week's production at Tonopah 9521 tons with a gross value of $166,617.

Tonopah, September 18.

(Special Correspondence.)—Tonopah operators have agreed to pay a minimum wage of $5 per day to miners, and have waived the original stipulation of 20 days of continuous employment before a worker could command the increase of 50c. daily. The offer of the operators was accepted by the men by a vote of 402 to 317. In some quarters dissatisfaction is expressed but the more conservative miners and millmen are in favor of the agreement.

It is reported that the MacNamara mine will resume work in a few days and that the mill will be operated on ore from the upper workings, where a large tonnage is blocked out. This product was formerly unprofitable, but with $1 silver can be mined to advantage. It is also rumored that the Montana company is contemplating resumption of operations at its mine and mill, and that other companies are planning early activity.

The Round Mountain Mining Co. is repairing the Fairview mill and expects to start work at the Fairview mine. The mine adjoins the Round Mountain group and was acquired by the company several years ago. Activity prevails at the Round Mountain quartz and placer mines. L. D. Gordon is president and manager.

Several properties in outside districts have started ore-shipments to Tonopah mills. They are the Louisiana Consolidated, at Tybo, the Confroth, at Bellehellen, and other old producers. Production has been greatly stimulated by the advance of silver and from present indications all the Tonopah mills will be taxed to capacity to take care of shipments.

Difficulty in securing motor trucks is interfering with shipments, and mule-teams are also difficult to secure. Manager Joseph Confroth of the Confroth company reports that he has been unable to secure enough trucks, as all the manufacturing companies are engaged in supplying Government demands.

Prospecting of silver deposits is active at Gold Mountain. H. D. King of Tonopah has uncovered small shoots of rich ore on the Jew and Jewess claims and is preparing for deeper work.

Tonopah, September 24.

(Special Correspondence.)—The Tonopah Extension Co. has satisfactorily tested its new auxiliary electric-power plant. It develops 500 kw. and will be used to operate the Extension mill, the pumps in the Victor shaft, and other equipment whenever the usual source of electric power fails. Cross-cutting to the south from the 1680-ft. level of the Victor shaft has begun, the object being to pick up the Murray, North Merger, and other veins. Another station is being completed on the north side of the shaft preparatory to cross-cutting from this point.

Claude Marvin, Walter Vickers, Charles Kielhofer, and other Tonopah men are developing a group of claims in the Gold Reef district, south of Tonopah. A shaft is down 75 ft. and is in ore showing free gold. At the surface the vein is 14 ft. wide.

The rise of silver has stimulated the re-opening of properties at Barcelona. The Consolidated Spanish Belt Co., F. Pike, superintendent, has driven an adit 350 ft. and has discovered ore. A compressor, machine drills, and other equipment have been installed. The shaft of the Stimler mine, at Cottonwood, is down 445 ft. and the ore is being tested. A six-drill compressor and 40-hp. semi-Diesel hoist are in operation and, on the 165-ft. level, 12 men are stoping ore. Cross-cutting from the bottom of the shaft is about to begin. Harry Stimler is superintendent.

Tonopah, September 16.

The Tonopah Mining Co. has discontinued its exploration department. J. E. Spurr stated recently that, at a total cost of $261,108, the exploration department had yielded results to the extent of $7,858,600.

ORMSBY COUNTY

(Special Correspondence.)—The Univeda Gold Dredging Co. has taken a five-year lease on the Brunswick and Merrimac property of 400 acres and is preparing to install a dredge to recover the gold and silver from mill-tailing in the Carson river near Brunswick, six miles from Carson City.

The Bessemer Con. Mining Co. has completed preparations for mining iron deposits in Brunswick canyon, four miles south of Brunswick. The upper adit is in 200 ft. and has cut several iron-bearing veins, of which the main orebody shows a width of 17 ft. of ore. Adit No. 2, 160 ft. below No. 1, is in 500 ft. and a raise is being extended to develop the main vein. A five-ton motor truck will be employed to haul ore to the railroad station for shipment to San Francisco.

Orin H. DuBois is developing silver ore in the Washington-Nevada, 10 miles north of Carson City.

Carson City, September 27.

WHITE PINE COUNTY

(Special Correspondence.)—The Nevada Consolidated is mining 13,000 tons of ore per day. Being unable to renew satisfactory contracts for fuel oil, a plant is being erected at the smelter for the storage and crushing of coal to dust, which is to be used for smelting and the power-plant. It is being constructed of concrete, as structural steel cannot be obtained. It will cost $400,000, and will be completed at the end of the year, when the oil contracts expire. Several steam-shovels at work on overburden have been laid off.

The State and Federal authorities requested that all their labor must be on an 8-hour basis, and this went into effect on September 1. Heretofore most of the open-pit mining was done on 9, 10, and 12-hour shifts. During the last two months the company has paid all employees 50c. per day more than the scale controlled by the price of copper; at the pit the men are paid for nine hours, although working only eight hours. The men are charged $25 per month for board, no change having been made for several years either in food-quality or price. The men are also furnished rooms, with heat and lights free; married men are given houses at very low rentals.

Robert Linton has been made manager and consulting engineer of the Consolidated Copper Mines Co., at a recent directors meeting. A smelter-site has been selected by Linton near the old Giroux shaft and construction will be started next year. Equipment for the stripping of the Morris orebody is coming, and work will be started within 60 days. The development in the Alpha continues in high-grade ore. Shipments from the 10th level of the Giroux shaft assays from 10 to 12%.

The Delkir mine, in Elko county, 29 miles west, is shipping copper ore from Currie station. Additional ore will be mined but, after the developed ore has been exhausted, further mining is doubtful as the vein is badly faulted, and the continuation has not been discovered.

The Ward mine, S. B. Elbert of Denver in charge, is making preparations for regular shipments. It is 18 miles south of Ely. Two White auto trucks and trailers and three 'cub' tractors and accompanying trailers, manufactured by the J. I. Case Plow Co., have been received and will be used for haulage.

The lessees on the Hunter mine, 18 miles south-west of Cherry creek, have been unable to make a satisfactory engagement with the Eastern owners and have stopped work.

Hamilton, 45 miles west of Ely, is shipping lead-silver ores by way of Eureka. Lessees from the Robinson district are shipping a car of high-grade lead ore from a new lead mine on top of Snake range, north of Sacramento pass. They have a large orebody, of a 20% concentrating ore. The Red Hills mine is mining and preparing to ship another car of lead-silver ore.

Three cars of ore are being shipped from the Lucky Deposit mine at Aurum. The copper sulphide ore assays 7 to 8% copper and $6 in gold and silver per ton. It is hauled by auto truck 56 miles.

Ely, September 18.

UTAH

JUAB COUNTY

The Tintic Drain Tunnel Co., Provo City, has been incorporated and proposes the construction of a 5½-mile adit from Goshen valley to a point near Silver City. Mine drainage and the development of water for irrigation are the objectives sought by the company.

SALT LAKE COUNTY

Bingham Mines Co. has declared a dividend of 50c. per share, amounting to $71,500.

The Utah Apex Mining Co. has declared a dividend of 25c. per share. The aggregate sum is $264,000 and the total dividends to date amount to $851,125.

SUMMIT COUNTY

A dividend of 25c. per share has been declared by the Judge M. & S. Co. The total dividends paid amount to $1,900,000.

The Daly West has declared a dividend of 10c. per share on 150,000 shares.

The Silver King Coalition will pay a dividend of 15c. per share on 1,250,000 shares.

Preparations are being made to work on the Utah-Park through the Ontario drainage-adit.

The August production of Park City totalled 10,217 tons.

WASHINGTON

Official announcement was made by Bart L. Thane that a steel plant together with subsidiary enterprises would be constructed on Puget Sound. At present the preliminary engineering work is being done but actual construction will depend on more favorable economic conditions.

The Electric Point declared a dividend of three cents per share, payable September 30. The mine is earning at a greater rate and has a surplus. This is the fourth dividend and the total disbursements are $71,370.

CANADA

BRITISH COLUMBIA

(Special Correspondence.)—The Canada Copper Corporation has increased its force at the Copper Mountain mine to 250 men. Bunk-houses, dwellings, water and sewer systems are being constructed. Funds to the amount of $2,500,000 have been raised by the sale of bonds which were underwritten by Hayden, Stone & Co. of New York.

Charles Camsell, of the Dominion Geological Survey, is at Copper Mountain engaged in geological investigation.

A 50-ton concentrator will be built on the Bowena mine, Bowen island. The King Solomon mine near Greenwood is shipping 15% copper ore. Deposits of good-grade ore have been discovered on the 750-ft. level of the Erin, near Kamloops. A mill and compressor-plant will be installed at the Victor mine near Fort Steele. The vein on the 8th level of the Slocan Star has widened to six feet.

The mill of the Surf Inlet gold mine is in satisfactory operation. Enlargement of its present capacity of 250 tons per day is looked for. A 96% recovery is being made and concentrate assaying $100 per ton is produced. The concentrate is shipped to the Tacoma smelter.

Vancouver, September 14.

The Fisher Maiden mine, 1½ miles from the Slocan Star, in the Slocan district, has been bought by Barney Crilly, an operator of British Columbia, on a bond and lease for three years, at a reported price of $100,000. Jules L. Prickett is president of the corporation, known as the Fisher Maiden-Troy Mines, Limited.

The Hedley Gold Mining Co. has declared its regular quarterly dividend of 3% and an extra dividend of 2%. This is at the rate of 50c. per share on 120,000 shares issued.

The Ladysmith smelter will be enlarged at once at a cost of $2,000,000. The plant now handles 400 tons of copper ore per day and the extensions will increase the capacity to 1200 tons.

The Con. M. & S. Co. received at its plant at Trail, during the first seven days of September, 10,864 tons of ore from 31 mines.

KOREA

The Seoul Mining Co., operating the Suan Concession in Whang Hai province, Chosen, reports a recovery of $139,305 from their operations during the month of August.

MEXICO

The subject of making spelter in Mexico has been frequently studied by many companies and individuals in the past, and all have decided that owing to the high cost of fuel, such an industry would not be payable. It is interesting to note, therefore, that an experimental plant has been erected and that a working plant is in course of erection. The International Ore Co., at Saltillo, in Coahuila, whose manager is F. E. Salas, is constructing two furnaces of 300 retorts each, having a capacity estimated at 40 to 45 tons per day, and a production of 15 tons of spelter per day. The furnaces will be fired with crude oil, and the gases from them will be used under the boilers of the power-plant. The total cost of the above plant is estimated to be $90,000. That spelter will be produced by the above plant there is no doubt, but whether the business will be profitable or not with the present prices, remains to be seen.

PERSONAL

Note: The Editor invites members of the profession to send particulars of their work and appointments. This information is interesting to our readers.

F. G. COTTRELL is here.

J. F. THORN has returned to San Francisco.

R. A. F. PENROSE has returned to Philadelphia.

J. H. MEANS is in the Manigotagon district of Manitoba.

F. L. SIZER is examining mines near Deming, New Mexico.

RUSH M. HESS has gone from Arizona on a visit to New York.

ASKIN M. NICHOLAS has arrived at New York from Australia.

D'ARCY WEATHERBE is in Japan, on his return from Siberia.

EDWARD L. DUFOURCQ is manager for the Teziutlan Copper Co. in Mexico.

J. R. FINLAY has returned to New York from the Lake Superior region.

NORMAN L. CALDER is with the Royal Engineers at Sandwich, in England.

W. A. PRICHARD, manager of the Pato and Nechi mines, in Colombia, is here.

H. W. BAKER, of Boston, is examining mines in the Kirkland Lake district of Ontario.

J. E. THOMPSON has returned to New York after a visit to the Californian oilfields.

RUSSELL WOAKES, son of ERNEST R. WOAKES, has won the Military Cross in Mesopotamia.

HEATH STEELE has been appointed chief engineer to the American Metal Co., at New York.

FRANK L. HESS, of the U. S. Geological Survey, was in San Francisco on his way to Arizona.

H. B. PULSIFER has become Professor of Metallurgy in the Montana School of Mines at Butte.

ERLE P. DUDLEY has received a commission as Captain in the Engineer Officers Reserve Corps.

ARTHUR B. FOOTE, superintendent of the North Star Mines, was in San Francisco during the week.

W. H. WHITTIER is Lieutenant with the Engineer Officers' Reserve Corps at Fort Leavenworth, Kansas.

A. E. RAU-ROESLER has returned from Arizona and is operating the White Oak mine in Mariposa county.

C. D. KAEDING, manager for the Dome Mines company, at Porcupine, has returned to New York from San Francisco.

J. C. HOUSTON had to resign as superintendent for the Dome Mines owing to ill health and is now examining mines in British Columbia.

E. COPPÉE THURSTON has resigned his appointment with the Bunker Hill smelter to enlist in the service of the American Red Cross in France.

ARTHUR L. WALKER, Professor of Metallurgy in Columbia University, has been retained as consulting metallurgist by the War Department.

HENRY M. PARKS has resigned as Dean of the School of Mines at Corwallis, Oregon, to become Director of the Oregon Bureau of Mines and Geology.

CLARENCE WOODS has resigned as superintendent for the Louisiana Con. M. Co., at Tybo, Nevada, to accept a position with the A. S. & R. Co., at Blue Hill, Maine.

F. E. DOWNS, formerly mill-superintendent for the Rex Con. M. Co., at Wallace, Idaho, is now at Monterrey, Mexico, where he holds the same position with the Compañia de Minerales y Metales, S. A.

Production of Ontario Mines

Returns received by the Ontario Bureau of Mines from the smelters, refineries, and mines of the Province for the six months ended June 30, 1917, are summarized herewith.

Product	Quantity 1916	Quantity 1917	Value 1916	Value 1917
Gold, oz.	235,060	228,673	$4,822,740	$4,586,941
Silver, oz.	10,267,743	10,073,787	6,188,269	7,584,439
Cobalt (metallic), lb.	121,817	162,250	103,677	237,004
Nickel (metallic), lb.	13,933	45,864	5,899	19,073
Nickel oxide, lb.	} 410,408	5,495	} 204,638	1,648
Cobalt oxide, lb.		153,498		175,308
Other cobalt and nickel compounds, lb.		122,076		15,879
Molybdenite, lb.	12,631	36,777	13,075	47,942
Lead, lb.	......	912,934	......	114,955
Copper ore, tons	922	1,543	14,368	45,688
Nickel in matte, tons	20,651	20,230	10,325,766	10,115,000
Copper in matte, tons	11,426	10,381	4,207,620	4,152,400
Iron ore (exported), tons	......	24,332	......	85,135
Pig-iron, tons	......	40,968	......	715,912
			$25,886,052	$27,897,322

It will be noted that the figures are for pig-iron produced from Ontario ore only. Nickel and copper in matte have been valued at 25 and 20c. per pound, respectively, and copper was valued at 18¾c. per pound in 1916.

GOLD. It was anticipated that the production for the half-year would show a decline as compared with the same period in 1916, owing to labor troubles and shortage in the Porcupine district. Nearly all the mines, including the Hollinger and Dome, have been developing their orebodies and increasing their milling capacity in preparation for the time after the War when labor will be more plentiful and operating costs decreased. In the meantime production and dividends have been curtailed. New producers are Gold Reef and Tommy Burns at Porcupine, Teck-Hughes at Kirkland Lake, and Miller-Independence at Boston Creek. Mines, producing 5000 oz. or more gold, in order of importance, were Hollinger, McIntyre, Dome, Porcupine Crown, Tough-Oakes, Schumacher, and Porcupine V. N. T.

High prices for silver, which averaged 75.44c. for the half-year as compared with 62.53c. for the same period of 1916, have stimulated production from Cobalt. The advance in value has offset the increased cost of mining. If the Miller-Lake O'Brien continues shipping at the same rate throughout the year, Gowganda will show a record production for 1917. The increase is attributed to the high-grade vein discovered in the summer of 1916. The Hargrave mine is now shipping regularly. The National, formerly the King Edward mine, is a new shipper. The Mining Corporation of Canada (Cobalt Lake and Townsite City mines) shipped over 2,000,000 oz. in the half-year. Shippers of 500,000 oz. or more were Nipissing, Kerr Lake, O'Brien, Beaver, and Coniagas. Silver recovered from gold ores totalled 38,492 oz. and from copper ores 646 ounces.

NICKEL-COPPER. The production of nickel-copper matte at Copper Cliff and Coniston shows a small decrease as compared with the same period in 1916, owing to shortage of labor. Assays of samples of nickel-copper matte for their precious-metal contents were made for the Royal Ontario Nickel Commission by Ledoux & Co. of New York. Platinum and palladium were found in quantities varying from 0.32 to 1.97 oz. per ton of matte. These metals are quoted at $100 per ounce. The British American Nickel Corporation has announced that its new electrolytic refinery will be built at the Murray mine, and will have an initial capacity of 5000 tons of nickel per annum. The Port Colborne refinery of the International Nickel Co. will produce 7500 tons of nickel, and provision is made for quadrupling the capacity.

COPPER. Shipments for the half-year came from three sources, the Tip Top mine near Kashabowie, the Hudson Copper Co. at Havilah, and the Kenyon Copper Co. at Massey. The last-mentioned operates the Massey mine, where a 100-ton Callow flotation-mill is producing 20% concentrate. Shipments from the Bruce mines are included under nickel-copper. The Port Arthur Copper Co. at Mine Centre is erecting a concentrator and will be shipping soon.

The A. I. M. E. Meeting at St. Louis

How to increase the supply of manganese ore for the American steel industry in the war emergency is one of the principal subjects to be discussed at the 115th meeting of the American Institute of Mining Engineers during the week of October 8 to 13. Prominent engineers from many parts of the country will attend the sessions which are to be held in mining districts of Illinois, Missouri, Kansas, and Oklahoma.

It is impossible for the United States to supply the steel industry with enough manganese if the present practice of using manganese is maintained. This is the word that D. F. Hewett of the U. S. Geological Survey has given the Institute. As a result, the Institute officials have informally issued a call for some 6000 members to present suggestions for methods of using these ores. This is only one of the subjects in connection with the development of the country's mineral resources for greater war service that is to be presented at the mining engineer's meeting. An effort will be made along several lines to co-operate with the Government and the mining interests with the expectation of increasing production, conserving the present supply, and improving the methods of using the country's minerals.

There were only two shippers of manganese ore in the United States in 1913 and last year this number had grown to 53. This year the rate of production is about three times that of last year and certain States have increased their output to five times that of 1916. Even with this increase it is stated that the war demand already calls for a greater increase in production. Manganese is chiefly found in Montana, Virginia, Arkansas, and California.

The mining engineers will open their convention at St. Louis on October 8 with a patriotic meeting and will journey to neighboring points in Illinois, Kansas, and Oklahoma by special train. Technical sessions will be held relating to the work of the individual districts at every important centre that is visited. An elaborate social programme has been arranged by the engineers of the St. Louis district.

Potash in 1916

A comprehensive review of the potash industry of the U. S. is given by Hoyt S. Gale of the U. S. Geological Survey. The domestic production of potash for 1916 totaled 9720 tons of a value of $4,242,730 at the point of shipment. The significance of this production is commented upon by Gale, who states that much of the present production of potash is maintained by the war prices and cannot be continued after these prices fall. Some of the projects may be permanent, but unfortunately their output is yet small. The only exception, perhaps, is Searles Lake, which now bids fair to produce enough potash to take the edge off the market by supplying the most urgent requirements for home consumption, such as those of the chemical industries, but the operation of the Searles Lake deposits is hardly yet under way. So far as is now evident, there is probably no domestic source adequate to supply more than a small percentage of the domestic consumption, but the slight promise should not by any means exclude all hope of larger success.

The activity of private initiative in the development of possible sources of potash is increasing. Interest in the possible recovery of potash as a by-product is very keen, and some of these sources, such as the dust of cement and blast-furnaces, seem to offer much promise of development on an extensive scale, but are not yet far enough advanced to warrant confident prediction that they will ultimately supply the country's whole needs. Perhaps the real hope of making the country independent of foreign supplies lies in the possibility of discovering, in association with some of the very extensive and little-known domestic deposits of salt, a supply of potash that may compete with the foreign supply on equal terms. There are valuable deposits of potash in the Permian in central Germany, in the Oligocene (Tertiary) in Alsace, the lower Miocene (Tertiary) in Galicia, and in the supposed Eocene or Oligocene (Tertiary) in north-eastern Spain, all accompanied by deposits of common salt, dolomite, and gypsum, and it would be strange if deposits of this sort were confined entirely to Europe.

Governmental agencies have served and are still serving a useful purpose in aiding development.

The imports of potash salts into the United States are given in considerable detail and from Gale's tables the quantities given in the succeeding summary have been taken.

IMPORTS OF POTASH SALTS

	1912		1916	
	Quantity (tons)	Value	Quantity (tons)	Value
Chloride	241,665	$7,229,121	1,299	$348,961
Sulphate	49,119	1,783,846	1,693	81,684
Carbonate, crude	3,813	234,868	341	113,413
Nitrate	3,256	202,899	5,769	1,519,375
Caustic	4,845	370,506	242	16,694
Cyanide	363	109,434	1	803
Other salts	8,429	761,611	1,243	271,518
Kainite	573,413	2,386,362	40	1,173
Manure salts	192,368	1,797,057	1,249	21,273

The prices per metric ton of German potash salts delivered at Eastern ports before the beginning of the War were:

	1912-13	1914
Muriate of potash (80% KCl; 50% K_2O)	$38.05	$39.07
Sulphate of potash (90% K_2SO_4; 48% K_2O)	46.30	47.57
Manure salts (20% K_2O)	13.30	13.58
Kainite (12.4% K_2O)	8.25	8.36

Since the beginning of 1915 the prices of these salts have risen enormously. During 1916 potash-bearing fertilizer, on the basis of water-soluble potash, has commanded a price ranging from $3 to $3.50, and even reaching $5 to $6 per unit.

Recent Publications

NOTES ON THE GEOLOGY AND IRON ORES OF THE CUYUNA DISTRICT, MINNESOTA. By E. C. Harder and A. W. Johnston. Bulletin 660-A, U. S. Geological Survey. The bulletin is a review of the more important features relating to the occurrence of iron ores in the Cuyuna district.

TUNGSTEN MINERALS AND DEPOSITS. By Frank L. Hess. Bulletin 652, U. S. Geological Survey. The bulletin makes known the general facts about tungsten, the minerals in which it is found, and the nature of the deposits. Methods for the approximate determination of tungsten are given. Excellent illustrations are a conspicuous feature.

THE DE SOTO-RED RIVER OIL AND GAS FIELD, LOUISIANA. By G. C. Matson and O. B. Hopkins. Bull. 661-C, U. S. Geological Survey.

ENRICHMENT OF ORE DEPOSITS. By William Harvey Emmons. Bull. 625, U. S. Geological Survey. The paper is an amplification of an earlier Survey bulletin on the enrichment of sulphide ores (Bull. 529). It is a discussion of representative deposits, especially of the paragenesis of their ores and of the principles that underlie the process of enrichment. To Western metal-miners the material of this bulletin should be of special value.

STRUCTURE OF THE NORTHERN PART OF THE BRISTOW QUADRANGLE, CREEK COUNTY, OKLAHOMA—PETROLEUM AND NATURAL GAS. By A. E. Fath. Bull. 661-B, U. S. Geological Survey.

CHEMICAL RELATIONS OF THE OIL-FIELD WATERS IN SAN JOAQUIN VALLEY, CALIFORNIA. By G. Sherburne Rogers. Bull. 653, U. S. Geological Survey.

ECONOMIC GEOLOGY OF GILPIN COUNTY AND ADJACENT PARTS OF CLEAR CREEK AND BOULDER COUNTIES, COLORADO. By Edson S. Bastin and James M. Hill. Prof. Paper 94, U. S. Geological Survey.

GEOLOGY AND ORE DEPOSITS OF THE MACKAY REGION, IDAHO. By Joseph B. Umpleby. Prof. Paper 97, U. S. Geological Survey.

REPORT ON MINING OPERATIONS IN THE PROVINCE OF QUEBEC DURING THE YEAR 1916. By the Mines Branch of the Dept. of Colonization, Mines and Fisheries.

THE RUSTLER SPRINGS SULPHUR DEPOSITS. By E. L. Porch, Jr. Bull. 1722, University of Texas.

REPORT OF THE FEDERAL TRADE COMMISSION ON ANTHRACITE AND BITUMINOUS COAL. Government Document, Washington, D. C.

THE USE OF THE PANORAMIC CAMERA IN TOPOGRAPHIC SURVEYING. By James W. Bagley. Bull. 657, U. S. Geological Survey. The methods described have been used in Alaska. Camera and plane-table offer advantages in cost and speed of execution of topographic surveys. The methods outlined should be of interest to mining engineers.

MANGANESE DEPOSITS OF THE CADDO GAP AND DE QUEEN QUADRANGLES, ARKANSAS. By Hugh D. Miser. Bulletin 660-C., U. S. Geological Survey.

POTASH IN 1916. By Hoyt S. Gale. Mineral Resources of the United States, U. S. Geological Survey.

COKING OF ILLINOIS COALS. By F. K. Ovitz. Bulletin 138, U. S. Bureau of Mines.

ABSTRACTS OF CURRENT DECISIONS ON MINES AND MINING. By J. W. Thompson. Bulletin 152, U. S. Bureau of Mines.

EXTRACTION OF GASOLINE FROM NATURAL GAS BY ABSORPTION METHODS. By George A. Burrell, P. M. Biddison, and G. G. Oberfell. Bull. 120, U. S. Bureau of Mines.

METAL PRICES

San Francisco, September 25

Aluminum-dust (100-lb. lots), per lb	$1.00
Aluminum-dust (ton lots), per lb	$0.95
Antimony, cents per pound	16.50
Antimony (wholesale), cents per lb	14.75
Electrolytic copper, cents per pound	28.00
Pig lead, cents per pound	8.50— 9.50
Platinum, soft and hard metal, respectively, per ounce	$105—111
Quicksilver, per flask of 75 lb	$105
Spelter, cents per pound	10
Tin, cents per pound	61
Zinc-dust, cents per pound	20

ORE PRICES

San Francisco, September 25

Antimony, 45% metal, per unit	$1.10
Chrome, 40% and over, free SiO_2 limit 8%, f.o.b. California per unit, according to grade	$0.60— 0.75
40-45%	$0.75
45-50%	$0.80
50% and over	$0.90
55% and over	$1.00

Chrome prices show a tendency to remain stationary. With the advent of winter, when the haulage conditions are less favorable than at present, prices will be more erratic and are likely to go higher.

Magnesite, crude, per ton ..... $8.00—10.00

Manganese: The Eastern manganese market continues fairly strong with $1 quoted on the basis of 48% material.

Tungsten, 60% WO_3, per unit 28.00

Molybdenite, per unit for MoS_2 contained ..... $40.00—45.00

EASTERN METAL MARKET

(By wire from New York)

September 25.—Copper is inactive and nominal, remaining at 26.25c. until Friday; then Government price at 23.50c. the remainder of the week. Lead is steady and unchanged, remaining at 8c. all week. Zinc is quiet and firm at 8.37 to 8.50c. Platinum remains unchanged at $105 for soft metal and $111 for hard.

SILVER

Below are given the average New York quotations, in cents per ounce, of fine silver.

Date		Average week ending	
Sept. 19	105.50	Aug. 14	82.95
" 20	106.50	" 21	87.16
" 21	108.50	" 28	88.50
" 22	108.50	Sept. 4	91.12
" 23 Sunday		" 11	96.80
" 24	108.50	" 18	100.70
" 25	108.50	" 25	107.66

Monthly Averages

	1915	1916	1917		1915	1916	1917
Jan.	48.85	56.76	75.14	July	47.52	63.06	78.92
Feb.	48.45	56.74	77.54	Aug.	47.11	66.07	85.40
Mch.	50.61	57.89	74.13	Sept.	48.77	68.51	....
Apr.	50.25	64.37	72.51	Oct.	49.40	67.86	....
May	49.87	74.27	74.61	Nov.	51.88	71.60	....
June	49.03	65.04	76.44	Dec.	55.34	75.70	....

LEAD

Lead is quoted in cents per pound, New York delivery.

Date		Average week ending	
Sept. 19	8.00	Aug. 14	10.87
" 20	8.00	" 21	10.79
" 21	8.00	" 28	10.61
" 22	8.00	Sept. 4	10.38
" 23 Sunday		" 11	10.20
" 24	8.00	" 18	9.33
" 26	8.00	" 25	8.00

Monthly Averages

	1915	1916	1917		1915	1916	1917
Jan.	3.73	5.95	7.64	July	5.59	6.40	10.93
Feb.	3.83	6.23	9.01	Aug.	4.62	6.28	10.75
Mch.	4.04	7.26	10.07	Sept.	4.62	6.86	....
Apr.	4.21	7.70	9.38	Oct.	4.62	7.02	....
May	4.24	7.38	10.29	Nov.	5.15	7.07	....
June	5.75	6.88	11.74	Dec.	5.34	7.55	....

COPPER

Prices of electrolytic in New York, in cents per pound.

Date		Average week ending	
Sept. 19	26.25	Aug. 14	27.62
" 20	26.25	" 21	28.00
" 21	23.50	" 28	26.41
" 22	23.50	Sept. 4	25.65
" 23 Sunday		" 11	25.83
" 24	23.50	" 18	26.25
" 25	23.50	" 25	24.41

Monthly Averages

	1915	1916	1917		1915	1916	1917
Jan.	13.60	24.30	20.53	July	19.09	25.66	29.67
Feb.	14.38	26.62	34.57	Aug.	17.27	27.03	27.42
Mch.	14.80	26.65	36.00	Sept.	17.69	28.28	....
Apr.	16.64	28.02	33.16	Oct.	17.90	28.50	....
May	18.71	29.02	31.69	Nov.	18.88	31.95	....
June	19.75	27.47	32.57	Dec.	20.67	32.89	....

The Chile Copper Co. operating the Chuquicamata mines, reports a production of 22,573,314 lb. copper in the second quarter of this year and a net income during this period of $3,030,143. The tonnage treated was 741,371 and the average grade 1.68%. The net recovery of copper was 82.9%. The cost at the plant was 8.53c. per pound, but the total cost of copper delivered at New York was 12.22c. per pound.

ZINC

Zinc is quoted as spelter, standard Western brands, New York delivery, in cents per pound.

Date		Average week ending	
Sept. 19	8.37	Aug. 14	8.75
" 20	8.37	" 21	8.66
" 21	8.50	" 28	8.33
" 22	8.50	Sept. 4	8.25
" 23 Sunday		" 11	8.21
" 24	8.50	" 18	8.29
" 25	8.50	" 25	8.46

Monthly Averages

	1915	1916	1917		1915	1916	1917
Jan.	6.30	18.21	9.75	July	20.54	9.90	8.98
Feb.	9.05	19.99	10.45	Aug.	14.17	9.03	8.58
Mch.	8.40	18.40	10.78	Sept.	14.14	9.18	....
Apr.	9.78	18.62	10.20	Oct.	14.05	9.92	....
May	17.03	16.01	9.41	Nov.	17.20	11.81	....
June	22.20	12.85	9.63	Dec.	16.75	11.26	....

QUICKSILVER

The primary market for quicksilver is San Francisco, California being the largest producer. The price is fixed in the open market, according to quantity. Prices, in dollars per flask of 75 pounds:

Week ending

Date			
Aug. 28	115.00	Sept. 11	115.00
Sept. 4	115.00	" 18	115.00
		" 25	105.00

Monthly Averages

	1915	1916	1917		1915	1916	1917
Jan.	51.90	222.00	81.00	July	95.00	81.20	102.00
Feb.	60.00	295.00	126.25	Aug.	93.75	74.50	115.00
Mch.	78.00	219.00	113.75	Sept.	91.00	75.00	....
Apr.	77.50	141.60	114.50	Oct.	92.90	78.20	....
May	75.00	90.00	104.00	Nov.	101.50	79.50	....
June	90.00	74.70	85.50	Dec.	123.00	80.00	....

TIN

Prices in New York, in cents per pound.

Monthly Averages

	1915	1916	1917		1915	1916	1917
Jan.	34.40	41.76	44.10	July	37.38	38.37	62.60
Feb.	37.23	42.60	51.47	Aug.	34.37	38.88	62.53
Mch.	48.76	50.50	54.27	Sept.	33.12	36.66	....
Apr.	48.25	51.49	55.63	Oct.	33.00	41.10	....
May	39.28	49.10	63.21	Nov.	39.50	44.12	....
June	40.26	42.07	61.93	Dec.	38.71	42.55	....

ORES

Tungsten: Considerable business is reported at prices ranging at the usual quotations and the market is strong. Ferro-tungsten is quiet at unchanged prices. Exports of the alloy and the metal for the fiscal year ended June 30, 1917, were 1,784,306 lb., a record never exceeded before by this country or probably any other.

Molybdenum and antimony: It is reported that small transactions have been put through at $2.20 per lb. of MoS_2 for 90% material. Low-grade ore, containing 50% MoS_2 has gone at $1.35. Antimony ore is unchanged and dull.

CYANIDE

Sodium cyanide, 96-98% (51-52% cyanogen) is quoted at 37c. per pound at New York.

PRICES OF IRON AND STEEL

On September 24 the War Industries Board made known the new schedule as arranged with the producers. These prices will hold good until January 1, 1918, when they will be revised. The new prices follow:

Steel bars at Pittsburgh and Chicago, $2.90 per hundredweight.

Coke (Connellsville), price agreed upon, $6 net ton, a reduction of 62.5%.

Steel plates, basis Chicago and Pittsburgh, price agreed upon, $3.25 per hundredweight, a reduction of 70.5%.

Pig-iron, price agreed, $33 gross ton; recent price, $58 gross ton; a reduction of 43.1%.

Steel bars, Pittsburgh and Chicago basis, price agreed upon, $2.90 per hundredweight; recent price, $5.50 per hundredweight; a reduction of 47.3%.

Steel shapes, basis Chicago and Pittsburgh, price agreed upon, $3 per hundredweight, a reduction of 50%.

Eastern Metal Market

New York, September 19.

The better feeling, noted last week, has given place to one of disappointment, principally due to indecision in the copper price-fixing matter which was fully expected last week. All the metals are dull and transactions are few and small in number.

Copper is dead and nominal.

Tin is fairly firm and in occasional demand.

Lead is lifeless and much lower.

Zinc is steady and firmer.

Antimony is quiet but higher.

Aluminum has declined.

The steel market is marking time waiting for price fixing and in the meanwhile prices decline. Re-sale pig-iron is weakening this market and generally the market is $2 below last week's level. The fuel situation is causing considerable anxiety in the steel industry and has become serious since the fixing of the $2 base price for soft coal. Some Eastern makers of plates and shapes are facing a shut-down because of a shortage of gas-coal and the Government may have to supply coal to some iron and steel producers from whom it has bought material. There is dissatisfaction with present war buying methods and manufacturers who recently ceased to act on advisory committees because of legal questions are not likely to resume the relation.

COPPER

After a brief experience of a better feeling and what promised to be a more active market last week, due to the large purchase for the Allies at 25c. per lb., the copper market has again become stagnant and lifeless. The fact that the decision, which was fully expected last week, as to the price for Government copper, has been delayed as it has, has caused keen disappointment and brought out strongly a realization that the conflict of opinion at Washington is hard to smooth out. Various rumors continue to fill the market, the latest today being that 24c. will probably be the level for Government purchases. Any level would be better than no decision which thus far has demoralized trade and interfered with manufacturing and business in general. Quotations based on actual sales for any position are hard to find, buyers being disinclined to commit themselves. For early delivery we continue to quote electrolytic copper nominally at 26.25c. per lb., New York. The average quotation for last quarter is nominally 26.50 to 27c., New York. The London market is unchanged at £137 for spot electrolytic and £133 for futures.

TIN

The tin market has passed through an advance since last week but yesterday was at a lower level than the high point of the week. As high as 62.50c. was paid last Friday, September 14, but yesterday the market was 61.75c., New York. Since our last letter there has been but one day on which any transactions of volume were recorded. This was on Wednesday, September 12, when quite a business was done very quietly, made up largely of tin afloat to the United States, of prompt shipment from London, and of metal coming overland to New York from Vancouver. These sales resulted in a better tone. For the rest of the week, however, the market was dull and inactive. Some excitement was caused by an inquiry from a big steel company for 25 tons of spot tin and for 25 tons for delivery each month for the rest of the year. Spot Banca tin sold on Thursday, September 13, at 60c. in store, New York, but spot Straits was scarce with occasional 5-ton lots changing hands. Due to the sudden and unexpected reduction in the price of lead on Thursday, further buying of tin was checked and the market eased off. Yesterday and Monday, this week, the situation has been stale and lifeless with no sales reported. An easier tone has developed, due to freer arrivals and to sellers being more anxious to sell. Arrivals up to and including yesterday were 1635 tons as compared with only 700 tons a week ago. The quantity afloat yesterday was 4600 tons. The London market has advanced £3½ in the week to £246 10s. for spot Straits.

LEAD

No such record has ever been made in lead price reductions than has taken place in the last week. On top of the two reductions of ½c. each by the leading interest, previous to last week, the same producer has made two cuts of 1c. per lb. each in the past week. The first of these was on Thursday, September 13, when its price was put at 9c. per lb., New York. At that time it was regarded as the greatest alteration in price ever made by the big company. On top of this, late yesterday came another reduction of 1c. per lb. to 8c. This is the level at which sales to the Government have been made in recent months for its monthly needs. Outside producers have met the cut in each case so that yesterday the general quotation was 8c., New York, or 7.87½c., St. Louis. When the first cut of 1c. was made a plausible reason advanced in the trade was the fact that refiners had received ores, concentrates, and bullion faster than sales of metal. Under such conditions there has been a decided abstention from the market, some sellers have been anxious to sell when the market was around 9.50 to 9.80c., New York. The fixing of 8c. is regarded as a really bright spot for it is believed this will be the bottom since it is the Government price.

ZINC

A distinctly better feeling pervades the market and the entire situation is stronger than for some time. The improved sentiment, noted last week, has not lost any force and the belief is firmer among many dealers that the bottom was reached the past month or two and that, unless some unforeseen event changes present conditions, an extensive buying movement is in the making which will be in force before many weeks. It is pointed out that prices are low, actually below cost, that consumption has gone on, perhaps at a diminished rate, but in large volume, and that production has been curtailed 30 to 40%. A continuation of the war for one year at least, as now seems probable, will result in increased needs from the Allies and from our own Government. While inquiry is better, it is modest but is improved over that for August. The market is stronger and quotably higher with September delivery at 8.12c., St. Louis, or 8.37c., New York, and October at 8.25c., St. Louis, or 8.50c., New York. Better than this is reported to have been done. Exports of zinc for the fiscal year ended June 30, 1917, were 66,108,586 lb., exclusive of ore and dross. This compares with 45,867,156 lb. for the fiscal year of 1916 and with 21,243,935 lb. for the same period in 1915. Bids have been asked by the Government on 1000 tons of grade B zinc for extended deliveries, bids to be opened September 24. The lowest quotation on this has been 11.50c. per lb., some demanding as high as 12c.

ANTIMONY

The market fell to 14c. or under since our last letter at which levels fairly large sales are reported to have been made. An advance has taken place since and 15 to 15.50c. per lb. is now quoted for Chinese and Japanese grades, duty paid, New York, with the market dull and lifeless.

CONGRESS has passed an act authorizing prospecting for potash salts on the public lands of the United States, except "adjacent to Searles Lake." Prospecting permits for two years and covering not more than 2560 acres each are to be issued and the patenting of not more than one-fourth of the land embraced in the prospecting permit is to be allowed. The Act stipulates that the deposits upon unsurveyed land adjacent to Searles Lake "may be operated by the United States or may be leased by the Secretary of the Interior under the terms and provisions of this Act." We give the full text on another page. This legislation has been urged as an emergency measure and as a war measure by the Secretary of the Interior and by the Council of National Defense.

THE president of the Butte & Superior Mining Co., Mr. Bruce MacKelvie, has stated that the Butte decision in the flotation suit was not unexpected, but he looks forward to the review of the case in the Circuit Court of Appeals. He considered that Judge Bourquin's requirement of a bond of $2,500,000 was excessive and likely to compel a cessation of production at a time when metal is needed for the national service. We are glad to say that the judicial order has been modified since he spoke, so that the Butte & Superior company is now only required to deposit its monthly profits with the Court. He made another significant remark: "It is to be hoped that while the interests representing the Minerals Separation Company in this country are operating under the form of an American concern, the consistent attack on this company in the press as well as in court, particularly since the outbreak of the War in 1914, is not another way of reducing the available munitions supply for this country." This hints at German malice. We have heard the same suggestion from other quarters, but, prejudiced as we are against Minerals Separation, we do not hesitate to say that to the best of our knowledge such an idea is mistaken. It is true the firm of Beer, Sondheimer & Company acted as agent for the Minerals Separation and it is also true that this firm was placed on the British black list, but we have no reason to believe that the German affiliations of Beer-Sondheimer played any part in the selecting of that firm as the American agency for Minerals Separation, Ltd., represented now by the M. S. North American Corporation. We have ascertained that the Beer-Sondheimer company is an American corporation controlled by Messrs. Elkan and Frohnknecht, natives of Germany but legally American citizens. Mr. John Ballot, the chairman of Minerals Separation, is an Afrikander, a Dutchman born in Cape Colony; his close associate in the business, Mr. J. H. Curle, is a Scot of pre-eminent loyalty, who has devoted all his time since the War began to patriotic service; the next director, Dr. S. Gregory has been described to us as of Austrian extraction, but we are informed now that he is a Scot by birth, although of Armenian origin; Mr. W. W. Webster is English, and so is Mr. J. H. Krohn, his German origin having been eradicated except as to name. That disposes of the directors of Minerals Separation. The metallurgists, the firm of Sulman & Picard, are British, the junior partner being a graduate from the Royal School of Mines. In this country Mr. E. H. Nutter is a native Californian and a graduate of Stanford university; Messrs. Higgins, Chapman, and Wilkinson are English, Mr. E. B. Malmros is a Swede. Finally, to cinch the matter, the chief counsel, a most important person in such a business as that of Minerals Separation, is Mr. Henry D. Williams, a Yankee of the eighth generation. There may be some enemy-alien spots on Minerals Separation, but the core is British.

GENERAL satisfaction has been expressed with the price at which copper was fixed by the War Industries Board. It is not so low as to discourage production, even by the poorer mines, and it is not so high as to check consumption for useful purposes. The fixing of a price removes the premium on 'spot' copper and brings the market to a uniform level. The statistics of cost collected by Government officials have not been made public; so guesses continue to be made. One authority places the cost for a number of the bigger mines between 10 and 13 cents; another objects to the assumption that a price of 23½ cents allows a general profit of 10 cents per pound. Wages are now relatively higher, for the sliding scale is not to operate down-hill. Even if the Government had not stipulated for a maintenance of the existing wage-level, it is unlikely that the rate of pay could have been decreased without wide-spread labor troubles. Sliding scales are unfair arrangements, because both parties do not equally play the game. Miners in Arizona are now getting $5.15, whereas the sliding scale on the basis of 23½-cent copper would call for $4.75 per shift; in Montana the miners are being paid $5.25, whereas under the scale they could be receiving only $5 per shift. However, none of us will grumble at that, if only they will do their work and join with the managers in a continuously active production of one of the sinews of war.

ON the 27th ult. Dr. R. L. Wilbur, the president of Stanford University, who has been one of Mr. Hoover's most valued coadjutors on the food administration, delivered an address before the Commonwealth Club. It is unfortunate that such utterances from such men are rarely given verbatim in the local press, whereas columns of piffle are printed with painful regularity. Dr. Wilbur said many things likely to provoke serious thought. For example, he stated that the enemy, Germany, controlled 235 millions of people and if that control over these people and their resources continued after the War, our country would be faced with a long and bitter struggle. He asked his hearers to imagine what would happen if the western battle-front 'cracked,' if the English and French armies were defeated; that line of defence meant as much to us as if it were at San José. Our naval and military efforts would not count immediately, although splendid efforts were being made, but there was one thing that we could do now most effectively to assist in the winning of the war, and that was the conservation of food. Our immediate task was to save and set aside enough food to supply the soldiers and the dependents of the soldiers of our Allies, to ensure that they were sustained, otherwise they would fail in winning the victory without which our national effort would be inconclusive. The placing of an embargo on exports to the neutral countries, rendered doubly effective by our control of the coal necessary for ships departing from our ports, was one of the most important steps taken by the Government. We were rationing the neutrals, he pointed out, and under such restrictions as to prevent them from sending supplies across their borders to the enemy. That alone was worth a million men, for that many would have to be withdrawn from the battlefield in order to produce the supplies that Germany needed for the successful continuance of warfare. Dr. Wilbur spoke enthusiastically concerning the fine spirit and true patriotism shown by the leaders of agriculture and other basic industries in this country and at the same time warned us against the widespread German propaganda, signs of which were to be met at every turn, trying to hinder and to cripple the work of the National Government, in the control of food as in other directions. A carefully considered programme for the homes and public eating-places has been arranged by the Food Administration. If our people live up to this programme we can see the solution of our food problem and that of our Allies. Appealing to his hearers for support in the conservation of food, he said: "Three times each day everyone of you must decide whether you will play Germany's game or America's."

The School of Experience

In this issue we publish another of those biographical records from which so much of human interest can be learned. Our subject, Mr. Denis Mathew Riordan, is a veteran whose kindliness of heart and wide sympathy with all that concerns the profession have kept him young amid the vicissitudes of a varied engineering career. That career did not start in the conventional way; he did not graduate from a school of mines or a university; he did not have an indulgent uncle to give him a job as soon as he wanted one; on the contrary, he had a struggle to reach even the lowest rung of the ladder that leads to promotion and success. A carpenter and the son of a carpenter, he began life with but little prospect of achieving his present place as an honored member of a highly technical profession. Perhaps he was fortunate in the time of his debut, for the Civil War was followed by a remarkable era of mineral exploration and industrial development. He followed Greeley's advice and came West, to grow up with the country that offered many chances to the intelligent and energetic; but first he did his duty as a good citizen by serving as a soldier. In these days when 'slackers' are not far to seek, it should be more inspiring than amusing for our young men to read of the boy that tried five times to enlist before he succeeded in getting into uniform. Those that know Mr. Riordan need not be told that the soldier-boy of 50 years ago is a keen patriot today and as eager to serve the national cause now as he was then. We met Mr. Riordan in Europe when he went thither on the 'Tennessee' to distribute the funds for American refugees and we recall that he avowed himself 'neutral,' but he used the word in an ironical way that left no doubt about his feelings on the subject—feelings strengthened by later events. He served a various apprenticeship: soldier, brake-man, shipping-clerk, carpenter, millman, miner, all of these occupations he tried before, at the age of 22, he obtained a position of responsibility as superintendent of a lumber-yard; but the call of the mine was insistent; he returned to Virginia City; and when 30 years old he won his first real promotion, becoming superintendent of a stamp-mill at Bodie. He tells the interviewer how he taught himself in his spare time, overcoming not easily, but slowly and laboriously, the lack of special training. We were not at Bodie in 1878, but from another we learn that Mr. Riordan was "a live wire" in those exciting days; he was ambitious and speculative; he was fond of hunting, proving himself a good scout but a poor shot; our informant remarks that "he had to shoot oftener and tramp farther than most of us, but he always got birds." Apparently he was persistent in small as well as big things. He was keenly interested in politics and local affairs, as he is now in national and world affairs; his mental horizon has been enlarged, but it is full of movement as heretofore—never provincial, he asserts the best right of civilozed man: to make all knowledge his patrimony. When working underground on day's pay he began to pick up scraps of mineralogy and geology. As a young man he had the good sense and the good fortune to make friends among the leaders of the profession and to learn from them by daily association more than books could tell. The reading of reports and technical writings was supplemented by an intelligent inquisitiveness. He was never afraid to acknowledge ignorance and always quick to ask questions. Here is a hint to our young men. Nothing is more foolish than the false assumption of knowledge, the intellectual 'bluffing' that checks the acquirement of accurate information. To realize ignorance and to be eager to correct it is the mark of the real student. Later, in his first experience with the General Electric people, Mr. Riordan was true to form; he refused to undertake work for which he considered himself unqualified; he was glad to avail himself of the assistance of specialists. His saying is true that a mature man, if intelligent and honest, knows his capabilities and his limitations, and is wise if he recognize them. He knew his own and that made him a safe adviser. Our readers will note with keen interest the names of the mines he selected as likely to become profitable enterprises. Any man might be proud of such a record. It was the result of native sense, keen observation, and wide information. Again there is a lesson for our young friends: the proper study of mankind is man, to know men is better than to devour books, the understanding of one's fellows is capital of an indestructible kind. The interview also suggests that men differ greatly in their capacity to be helped by their surroundings; some sink to the bottom, others swim to the top. "Men live their future now. They determine by today's behavior and aspiration the strength or weakness that will tomorrow honor or shame them." Such is the moral of this story; but no maxims can give to a man the genius of friendship. This our friend possesses, and it is the very crown of life. He has the social gifts of quick understanding and willing co-operation; more particularly the happy knack of making congenial persons known to each other, for work as well as pleasure. His friends know him to be a charming letter-writer and a winning teller of stories either by mail or across the table. The equanimity shown by him in times of trouble is largely the product of an essential sanity and of a philosophic temperament disciplined by a keen sense of humor. That salt is the savor for making palatable even the unpleasant happenings of checkered days. And one thing more this mining engineer possesses, and shares with his fellows freely: an innate kindliness that has found fruition in acts of service to the less fortunate whose trails he has crossed. Many that read this will jump to affirm what we have just said, for in his trail across the waste places of the earth where mines are found there grow for him many flowers of grateful remembrance.

Labor v. Mine Production

No student of affairs is able to explain the present labor situation by attributing it all to industrial discontent. Civilization involves discontent, because it also involves an advance in the standard of living, and thus the unfortunate thread becomes interwoven with the warp of progress. These things are inseparable from our economic life and will remain so until a wiser generation has found the way to cure a thousand defects in our systems of social organization. The demon of discontent we have ever with us, but the widespread labor troubles that now impair our efficiency in grappling effectively with the dominant issues of war are wholly exceptional. They are so manifestly a reaction from abnormal conditions that one is forced to consider them all, directly or indirectly, a part of hostile propaganda. It is significant of hidden motives that the same kind of interference with the operations of government appeared instantly in Brazil when her relations with Germany became strained. The phenomenon has been repeated in the Argentine republic at the moment when the senate and deputies of that country have accepted the Teutonic challenge, despite the restraint of President Irigoyen's cautious diplomacy. The recognition of international influences in these disorders necessarily requires that other considerations than those of mere economic adjustment must weigh strongly with our Government in restoring a safe working basis to meet the existing crisis. We are at war, and the production of raw materials and their manufacture is as vital a factor in the military programme as the conscription and training of soldiers. The methods of war must be as relentlessly applied to industry as to the discipline of the Army that is to do the actual fighting for the preservation of our country.

The sudden demands for an enormously increased output from mine and factory have intensified the peril of the situation. A real shortage of labor is apparent, and at a time when industry is crippled by this shortage the labor-unions continue to ask and to obtain a wider recognition of the eight-hour day. We have always been in sympathy with the movement for a shorter working day, and would be glad to see it generally accepted in principle, but this is an inopportune moment to attempt the carrying of that reform into effect. If we are to cope successfully with the strains to which the Nation is being subjected in the present struggle, it is certain that every true patriot must lengthen rather than curtail the amount of his daily service. He must be ashamed to take 16 hours rest out of 24 while his country calls for aid. If our statesmen, our legislators, our committees for defence were to trim their daily labor to eight hours by the clock, we would speedily be in the way of submitting to the same brutal invasion of our rights that has crushed and ruined Belgium. That means that every man would go to work at the point of the bayonet, that his choice of labor and hours would be derided, that he would receive only the bare necessaries of life from his military taskmasters. If any of us are so unpatriotic as not to respond willingly to the call of duty, if we must render our quota to the common good under compulsion, then it had better be under the orders of a peace-loving democracy that offers the assurance of future blessings and of a continuous social uplift. The labor-unions should take a long look ahead; in the free institutions of a popular government guaranteeing unrestricted and equal franchise, they should recognize the safe foundation for a social structure on which the principles of larger equity shall be more perfectly established. This end is surely not to be anticipated through a military autocracy committed to the policies of ruthless national and racial aggrandizement. The time has come when the menace of the future must force those entrusted with the government of this free republic to remind us that in time of war our enjoyment of the liberties of the civil law is by sufferance only.

We are not yet short of metals, but production must be maintained at an undiminished level or perilous difficulties will confront us. It is probably true that our manufacturing capacity is insufficient to handle all the iron and copper and zinc that is being mined, but that capacity is being increased, and these materials are absolutely essential to the conduct of the War. We shall need more and more of the manufactures into which they enter as the War progresses. Engineering enterprises more gigantic than have ever been seen are planned, and the longer the War lasts the more extraordinary will it become as a battle between giant machines, mechanical monsters that will make the creations of H. G. Wells' imagination seem tame and commonplace. We are becoming so accustomed to the extraordinary that we are no longer astonished, and we fail to realize how tremendous are the mechanical appliances that this war is calling into existence, but the fact is that the demand for all the metals will display from this time forward a rapidly rising curve. In the face of this it is disquieting to note a dispatch, couched in terms of jubilation, that the mines at Butte are working at 50% of their normal capacity. A pitiful fifty per cent! That is not the result of patriotism; it is the fruit of a misnamed organization, the Industrial Workers of the World. Again we read another dispatch from Clifton, Arizona, announcing that the shortage of labor is so severe that the great mines there cannot attempt to operate after September unless immediate relief is obtained. At Bisbee the Copper Queen and the Calumet & Arizona are struggling to maintain normal production with forces reduced by 40%. The managers have not so stated, but it must be apparent to every miner that the effort to maintain the regular output with 60% of the customary number of employees, when no changes in mining methods have been made, can mean only one thing, the curtailment of development work. That may be tolerated for a brief period in the face of critical necessities, but it may not long persist without provoking financial trouble and crippling the output of the property. Mines, such as those at Clifton and Bisbee, are not developed by the drill, and the ore is not available for stoping without time-consuming underground preparation. The same cry of a labor shortage comes from Jerome, and at Ajo barely half the needed number of laborers is available, so that this plant, perhaps the most perfectly automatic of any copper property in America, is seriously hampered in its operation.

It is surely not patriotism that has drawn the men from these great mines; it is the I. W. W. and the pernicious influence that has spread from that centre of infection. It has bred an indifference to responsibility that duplicates in a remarkable degree the spirit that led to disintegration in Mexico when the weak Madero proved unable to administer a democracy that was ready to lapse into the extravagance of anarchy. To put it plainly, the working people do not take their contribution to the common welfare seriously. They are insubordinate to those whose duty it is to supervise them and to see that an economic result is forthcoming; when chided for inefficiency or for lagging, they reply that the union, not the foreman, sets the standards for them. Duty has dropped out of the scales, and the managers no longer manage; from necessity they are declining into mere persuaders, coaxers, cajolers of labor. The big business of the world is not done in that manner and never will be. The old Americanism, the sort that developed workingmen into great builders and organizers and producers, was founded on the principle of duty and of obedience to those whose function it was to manage the work in which they took their necessary part. The old Americanism did not shirk; it did not drop the hammer on the stroke of the clock; it regarded work as an expression of dignity, not the sign of servitude. It made freemen who loved their country because it represented liberty and opportunity and knew no such thing as caste. When the unions turn human labor into a mechanical operation to be run so many hours per day at a pressure governed by a wage-meter, and liken service for a stipend to slavery, the spirit of the old American has gone and the changed conditions call for new methods. We must save free institutions for the laborers whether they appreciate it now or not. After the country is safe we may proceed with the process of civic evolution. With one great steel plant of the United States Steel Corporation paralyzed by a strike, with the building of ships hampered by union opposition, with the output of copper threatened at the great sources of supply, we are driving toward disaster. The Government is dealing with the problem of regulating prices, which means reduction of prices and a better economic equilibrium, but price-regulation downward only hastens the producer toward the closure of his mines if he is harrassed at the same time by shortage of labor and impairment of service. It is not surprising that the Clifton mines have announced their determination to close. The same thing will become necessary at many places unless more effective measures are taken promptly by the Government.

Physics of Flotation

The Editor:

Sir—In your recent valuable contributions to the discussion of the physics of flotation, I notice that you continue to lay great stress on the importance of lowering the surface-tension of water, in order to produce stable bubble-films.

An interested person, after reading through your articles is very liable to conceive the idea that the normal S. T. of water is too great for the tensile strength of the film, which therefore ruptures and allows the bubble to burst. Is not this an extremely misleading conception? My own opinion is that the actual intensity of the S. T. has very little, if anything, to do with the stability of liquid films. This opinion was considerably strengthened after noticing that a fine froth of beautiful silvery bubbles was produced by violently agitating some 'foul' quicksilver. Granting that the contaminant, whatever it was, lowered the S. T. somewhat, still it must have been several times as great as that of pure water.

Looking at the question from a purely theoretical point of view, is it not permissible to regard the S. T. of a liquid merely as the reaction from the cohesion of its molecules? Starting with this hypothesis, it is only necessary to apply the fundamental physical principle that "action and reaction are equal and opposite," to arrive the deduction that the intensity of the S. T. can have nothing to do with the rupturing of a liquid film.

Several enlightening articles on the theory of flotation have appeared in the technical press during the last few months, but any theory which depends largely on the presence of oil seems to me to be out of order, since, in practice, flotation plants have been operated successfully without oil, and it is quite a simple matter to prove in the laboratory that bubble attachment and levitation can take place when oil or oily substances are absent.

In attempting to apply the straight surface-tension theory to actual flotation-plant practice, a difficulty is immediately encountered. This theory, as presented in several excellent articles, always takes into account three surface-tensions, whereas in a flotation-plant, at the commencement of operations, there are only two in existence, namely, gas-liquid and solid-liquid. Before considering the other, namely, solid-gas S. T., is it not desirable first to describe how it is brought about? Yet I have failed to find an explanation in the writings of any investigator. If the casual observer thinks it is only necessary to bump bubbles against solids in order to produce a solid-gas interface, let him try bumping a bubble, held in the ordinary bubble-holder, against a clean glass bead, and see what success he has in getting the air into contact with the glass. On the other hand, if he thinks that the bubble and the solid must be forced together in order to produce the solid-gas interface, let him move a bubble, ever so carefully, into the vicinity (say, within 1/100 of an inch) of a grain of shot coated with paraffin-wax, and try to pass it by without a solid-gas interface being formed. I think the result of these two simple experiments will prove that there is something which needs explaining before it is logical to start to explain flotation phenomena on the basis of three surface-tensions.

The explanation is probably so simple that none of the investigators have thought it worth while to put it into print, but in this connection, it is as well to remember that flotation is still very much a millman's problem, even if it has ceased to be a physicist's.

Holkol, Korea, August 18. H. HARDY SMITH.

Pneumatic Flotation

The Editor:

Sir—In your article in your issue of May 17, entitled 'History of Flotation', and which has since been repeated in your and Dr. Ralston's book 'Flotation', you have omitted, in detailing its history in the United States, to make mention of the pneumatic flotation plant erected by me for the National Copper Co. at Mullan, Idaho. The construction work on this plant was started on August 4, 1913, and went into operation about April 10, 1914. Its capacity was 500 tons per day.

The flow-sheet employed was published by me in my 'Notes on Flotation', American Institute of Mining Engineers, February 1916. An early description by Ernest Gayford also appeared in the 'Engineering and Mining Journal' about June 27, 1914. A large number of people visited it in operation, including many prominent metallurgists, mining engineers, and managers. The plant contained ten Callow pneumatic cells.

I would like again to point out that this was a most significant installation, since it was the first plant employing pneumatic flotation in any form, either in this or any other country, and the phenomenal developments that have since that time taken place, in this form of flotation, date from the original disclosures made in this plant. The flotation process employed at this plant was the forerunner of the present flotation practice of the Inspiration, the Miami, Magma, Arizona Copper Co., Calaveras, and the many other users, both large and small, of pneumatic flotation which have since followed this lead.

I shall be glad, therefore, if, in justice to me, you would give the necessary prominence to this fact, as I do not think the history of flotation in America is truthfully complete without some mention of it.

Salt Lake City, September 25. J. M. CALLOW.

[We take pleasure in publishing this criticism and correction. We accept Mr. Callow's statement as a part of the history of the process.—EDITOR.]

Magnetite and Copper

The Editor:

Sir—If your far-distant readers are to take part in your discussion columns, their contributions must necessarily be belated, but, unless chiefly controversial in character, they should be of interest, and seldom quite out of date. That is my excuse for this letter.

In your issue of March 3, W. H. Storms, commenting on the occurrence of magnetite with copper and iron sulphides, mentioned by you in the issue of February 24, says: "I am under the impression that this association is more common than is generally supposed." I think he is right. In Australia the association is frequent, and occurs in widely separated localities, in several well-known mines in Southern Australia, in New South Wales, and in Queensland. In Queensland, in the Dawes range at the back of the port of Gladstone, there extends for over 50 miles a contact-metamorphic series of garnet-rock, crystalline limestones, magnetites, and hematites, containing iron and copper sulphides. Weathering, under a tropical sun, has made spectacular outcrops where the deep gullies, intersecting the series, cut across the copper-bearing portions of the garnet-rock and magnetite, and these outcrops have led, several times during the last forty years, to short-lived copper-mining 'booms'.

The garnets and magnetites carry, also, small quantities of gold. Having occasion, years ago, to examine several of the so-called copper mines of that district, and, the associated minerals being new to me, I searched all the literature available for enlightenment, and found, to my surprise, that the association is world-wide.

Apparently, where the magnetite may be considered subsidiary to the pyrite, there is a concentration of copper, and the mixture of magnetite and sulphides proves higher in copper than the average of the clean pyritic mass. Some of our biggest mines show this feature, and I think the presence of magnetite in large pyritic ore-bodies may be considered a favorable feature for copper. But in the massive magnetite bodies the copper occurs as veins running diagonally across the lenses of magnetite. Kemp, in his 'Ore Deposits', refers to the Cornwall mine, in Pennsylvania, as showing that characteristic, and in Europe there are similar instances. The deposits in the Dawes range show the same feature, but the veins are nowhere large enough to make a profitable copper mine; and, I gathered that, except where the magnetite is mined as an iron ore as well, the copper veins, even in Europe, have not been worked commercially.

It would appear, therefore, that copper and iron sulphide, forming distinct veins in massive magnetite, are not likely to become important mines.

In your issue of April 28 you refer editorially to the 'daylight-saving' movement. That passed through this country, and a 'Daylight Saving Act' was passed and put into force, last year, in the Commonwealth. The experience was not generally approved, and the Act is to be repealed. This is contrary to European experience, where its adoption last year has been followed again this year. General opinion in Australia was in favor of the Act being passed, from which much benefit was anticipated. After trial, general opinion is against it. The explanation appears to be that the advantage, or otherwise, is all a question of latitude. Australia, as a whole, is too near the equator. The disapproval was loudest in Queensland, and diminished as one went south. Probably the United States lies so far south of the latitude of European countries that the advantage found by the latter, in the adoption of a special 'summer-time', would hardly exist in the former. If one lives in a land where the sun rises at 3 a.m. in mid-summer, the question assumes a different aspect from what it has to those who live where sunrise is never earlier than 5 a.m.

EDGAR HALL.

Silver Spur, Queensland, August 31.

POLITICAL ECONOMY. 'The Financial Times', London, has this to say: Mr. H. C. Hoover, who when we first knew him was one of the shyest and most taciturn directors who ever faced a shareholders' meeting, is now getting quite a forceful speaker. In one of his first addresses as United States Food Controller he told the following effective story: "Germany's claim that she imports nothing, buys only of herself and so is growing rich from the war is a dreadful fallacy. Germany is like the young man who wisely thought he'd grow his own garden stuff. This young man had been digging for about an hour when his spade turned up a quarter. Ten minutes later he found another quarter. Then he found a dime. Then he found a quarter again. 'By gosh,' he said, 'I've struck a silver mine,' and, straightening up, he felt something cold slide down his leg. Another quarter lay at his feet. He grasped the truth: There was a hole in his pocket."

ELECTROLYTIC ZINC is gaining favor for galvanizing on account of the thin tough coating that this very pure spelter gives. It is also preferred in high-grade manganese bronze, aluminum alloys, sheet-brass, and die-cast alloys. The process results in the accumulation of residues rich in cadmium, but the demand is small, and metallurgical experimenters are urged to develop new uses for this metal.

D. M. Riordan and the School of Experience

AN INTERVIEW. BY T. A. RICKARD

Mr. Riordan, your name is Irish?

It is. My father was Irish, and his forbears as far back as our records go. My mother was English. I am an American, having been born in New York state. Your question reminds me of the time my youngest grandchild came to her mother, crying, when she was about three years old, and amid sobs managed to say: "Ain't, ain't, m-my grandfawther of Irish deecent?" Her mother responded soothingly that I was. Alicia's rejoinder was: "W-w-well, that's what I said, but Muriel says he is just plain Irish."

In what town were you born?

In Troy, New York, on June 26, 1848.

What was your father's occupation?

My father was a carpenter, of the kind the old country produced when they served seven years as apprentices before they were even allowed to become journeymen. He could do almost everything with tools in wood, except carving.

What was your early education?

I did not get much early education, but what I did get was in common schools, intermittently. I left school for good at the age of ten, and have never seen the inside of a school-house since, except as a visitor. I have worked to help out the family finances and to make my own living, continuously from the age of ten, in Chicago at first.

You speak of Chicago. Your parents evidently moved from New York while you were a child.

They did. My father decided to move to Chicago before any railroad had been completed, and therefore went around the lakes, by boat. Our family landed at Chicago in October, 1852. We lived there until I went into the Army.

You first left home then to enter the Army?

Yes, in 1864. But I started to enlist in 1861; I enlisted five times, but was taken out of the Army each time by my father, until the fifth enlistment. On the fifth I succeeded in getting into the field, in my 16th year.

Did you see much of the Civil War?

Not a great deal. I served with the Army of the Cumberland in the second separate division, mostly on detached scouting service in Missouri and Kentucky, and after being discharged I re-enlisted, which made the sixth enlistment, serving in Tennessee and Georgia, and was finally discharged at Memphis in 1865, after Lee surrendered.

Were you in any battle?

The only engagement of importance in which I participated was the battle of Nashville.

Were you wounded?

I was knocked down by a bullet, which struck the brass plate of my cartridge-box belt, but was not injured.

After the War, what did you do?

I went into various occupations as opportunity offered, the first job being that of a freight brake-man on the Rock Island railroad.

You were getting near the occupation of engineering?

Well, you might look at it that way; I found that there was a decided reluctance to hire a man who had been in the Army. I never knew why, but assumed from the nature of the inquiry, as well as the manner of it, that most employers had the idea that a man who had been a soldier would not work at any regular occupation unless he had to. Therefore, I refrained from mentioning the fact that I had been in the Army until after I got a job and had shown my willingness, at least. I was particularly anxious to have this job of brake-man, although the work was hard, the hours long and irregular, and necessitating exposure to all weathers, because my 'bunkie' in the Army had moved to Davenport, Iowa, which was the end of my run, and I thus gained frequent opportunities for a pleasant visit with him.

How long did you serve as brake-man?

Probably four or five months. My father objected to the occupation, both on account of there being no likelihood of advancement, and on account of the danger.

What did you do then?

Upon leaving the railroad I succeeded in getting employment with Palmer & Leiter, which firm subsequently became Field, Palmer & Leiter, and then Marshall Field & Co. I worked as shipping-clerk with them through these changes, until the fall of '68, when I made up my mind to go West, being then 20 years old. At this my father again demurred, and offered to stake me to a store in Chicago if I would remain and cared to follow merchandizing as an occupation. But my experience in the Army and my natural bent for out-of-doors made me determined to strike out for the West. I liked to tackle Nature, rather than human nature; and the building of a railroad, the subduing of a forest, or the opening up of a mine, were even then dreams in my mind. So I went west in January, 1869, my first stop being at Bear River in Wyoming, whence I worked on the Union Pacific as a carpenter—being handy with tools, as the result of my father's training—until I reached Echo City in Utah. There six of us decided to leave the railroad and go to Salt Lake City, inasmuch as the approaching completion of the trans-continental railroad was in sight. We went to Salt Lake, and after spending a couple of weeks there, all being carpenters and having our tools with us, we flipped up a half a dollar to decide whether we would go east or west from there. I did the flipping, and West won. So we bought a six-horse team and a wagon and started for White Pine, a booming mining camp in eastern Nevada. We passed around the southern end of the great Salt Lake and through Tooele across the Steptoe valley, which was then a quagmire, in which we spent three bitter days, thence through Egan canyon, until we made our way into Hamilton, White Pine district, Nevada. This was in 1869.

And there your mining began?

Only in a temporary way. I started to work as a carpenter at $9 per day. Miners were being paid from $5 to $6. Everything was high; to illustrate: flour was worth $36 per hundred, lumber $300 per thousand feet, and all other things in proportion. It was a typical raw frontier mining camp in mid-winter. Most of the miners lived in tents or dug-outs, and a reckless good-natured lot they were.

What were the principal mines in the locality?

Up to the time I left there, there was nothing developed that could be called a 'mine'. There were innumerable holes all over Treasure Flat, anywhere from 10 to 50 ft. apart, but each claiming "1500 ft. on this lead, lode, or vein," and it was not at all unusual for a shot put into one hole to throw rocks that fell fairly into the next 'mine'. The ore was principally silver, and some of it ran as high as $20,000 per ton.

It was not long before you went into mining yourself?

I did some prospecting on Treasure hill, but nothing that could be called 'mining'. I worked principally at carpentry. I left White Pine early in May, 1869, to go to Virginia City, where I arrived with $1.60. I walked from Hamilton to Elko, something like 180 miles, and also from Reno, with my blankets on my back, to Virginia City, a distance of 21 miles, and on my arrival at Virginia City I took a glass of beer. This cost me 10 cents, and was all I ate that day. I then divided the remainder of my fortune into a three-day grub-stake, eating one meal each day while looking for work. I hunted faithfully for a job all the way from Virginia City down Six-Mile canyon to the Devil's Gate, and in all the quartz-mills down Seven-Mile canyon, but at that particular time there had been a fire in the Yellow Jacket group and the town was filled with idle men, many of whom were disgusted White Piners; others had come from the railroad-camps, the trans-continental railroad having been recently completed.

So you had a poor show?

A mighty poor show; but when my last half-dollar was under my belt, I decided, after consultation with Phil Smith, the foreman of the Kentuck mine, to go to Washoe Valley, which I did, afoot, and got a job in Dall's mill at Franktown. I worked in that mill at different jobs, from feeding battery to amalgamator, but did not learn assaying at that time. In the fall of 1869 I went back to Virginia City, to work underground in the Chollar Potosi, of which I. L. Requa, father of M. L. Requa, was then superintendent. I worked in that mine until I was again persuaded to change my occupation, going into a store-office, and later, through the friendship of Capt. T. M. Hart, who was time-keeper at the Chollar Potosi, I was brought into a conference with H. M. Yerington, and within two hours installed in charge of the business of the Carson lumber-yard, in which duty W. O. Mills, a nephew of D. O. Mills, was my predecessor.

This was at Carson City?

Yes. My duties covered not only the Carson lumber-yard, but grew into the handling of the mills, large and small, that were cutting lumber at Lake Tahoe. When I first went to Carson, the Virginia & Truckee railroad had recently been completed from Carson to Virginia City. The rails and locomotives were hauled by horses from Reno to Carson. I can recall an argument between H. M. Yerington and William Sharon (who was then the agent of the Bank of California at Virginia City), when Yerington was importuning Sharon to support his recommedation that the road buy another engine. They had two engines, but Mr. Yerington insisted that they should have a reserve engine, and should have their trains so arranged as to start one from each end of the road practically simultaneously. Mr. Sharon demurred strongly, but Mr. Yerington's recommendations finally prevailed with Mr. Mills. At the time of this argument the railroad was being run without even a telegraph-line, but before I left the road it was running 40 trains per day and was easily paying 100% per annum.

You are now referring to the Comstock boom, I presume.

Yes, to one of them. From 1870 to 1872 there was a boom on the Comstock. Those who were there at the time will recall that Crown Point stock went from about $2.50 to about $2000 per share; and the shares of other companies, notably the Con. Virginia, Kentuck, Belcher, made almost equally phenomenal gains.

Did you speculate yourself?

I had no money for speculating. But only a few days ago I met Mr. A. M. Ardery, who was my chum, is now vice-president of the Virginia & Truckee and Carson & Colorado railroads, and is still my friend; he reminded me of the time we decided that we would put $20 apiece from one month's salary into Crown Point stock. This would have bought us 16 shares at the price then prevailing. We flipped up a $20 piece to decide the matter, the flipping decided against us, and we did not buy.

How long did you remain at Carson?

A couple of years. Then I went back to Virginia City and worked in the Hale & Norcross and other mines as a miner.

What wages did you get?

Four dollars a day.

What did board and lodging cost?

Board cost $8 per week, and lodging anywhere from $10 to $20 per month. Many of us 'batched' and did our own cooking. I was one of three, all old soldiers and members of the same G. A. R. post, that batched in a cabin on the divide between Virginia City and Gold Hill. I had not been long in the mine, however, before Mr. Yerington met me one day and asked if I wasn't tired of such dangerous and heavy work, and if I didn't want to come back to my old position at Carson. This resulted in my doing so, and subsequently I was appointed station-agent at Mill Station in Washoe Valley, where I had a large flume under my direction, in addition to my routine work as agent. From there Mr. Yerington brought me back to his office, and although I had no title ent, I served as assistant-superintendent of the Tioga and the Bulwer, and looked after all of the underground work of the Syndicate, and did all of the office-work for the three companies. My superiors were S. B. Ferguson and Warren Rose. My old foreman, John F. Parr, who lives in Berkeley now, and has since been in Alaska, Siberia, South Africa, and other mining regions, and is now mining in Tuolumne county in this State, was one of the most resourceful men I ever knew. We were 120 miles from a railroad. We had odds and ends of six different plants collected from among the idle mills of the then deserted camp of Aurora, in Nevada, and put together in an old mill-building. Mishaps of all kinds were continually occurring, but no excuse would be accepted for allowing the mill to shut down, and we prac-

OLD-FASHIONED PRAIRIE-SCHOONER TYPE OF TRANSPORTATION. HAULING SUPPLIES FROM REDDING TO THE LA GRANGE MINE

I acted as private secretary for him until he asked me one day how I would like to go to Bodie.

When was that?

That was in '78. I don't know whether Mr. Yerington knew I had previously had quartz-mill experience, but I was put in charge of the Syndicate mill at Bodie. There was no job in the mill that I couldn't do with my own hands, but I recognized my deficiencies in that I had no knowledge of assaying or metallurgy. I decided to correct this, and made an arrangement with a Freiberg graduate, who had an assay-office in Bodie, to teach me how to assay for gold, silver, and copper. In pursuance of my ambition to learn, I performed my routine duties until half-past nine or ten in the evening, then went up to Bodie, about a mile and a quarter, and worked until eleven or twelve, as the case might be, then back to the mill, and to bed, and was usually on deck again at half-past five. In addition to my duties as mill-superintend-tically had to run it on 'rawhide and wire', and did. Shoes and dies were obtained from Pittsburgh at a cost of about 15 cents per pound. Wood cost $12 per cord. Other supplies in proportion. We had an old 40-ft. two-flue boiler, and an old marine engine, 18 by 40, slide-valve; yet with all these handicaps we managed to keep our milling cost down pretty low—to within $6 per ton, and sometimes as low as $5.

What was your total cost per ton?

We ran mostly on custom ores.

Then you made a profit?

Oh, yes. We received $12 per ton for doing the work. During this time we had one rather unusual run on ore from the Bodie mine, and from this run we produced about a million dollars in six weeks. We seldom could run more than two of the 5-stamp batteries at a time, and frequently only one, because the ore was so rich that it kept the tanks, pans, and settlers full of amalgam. I have frequently known the amalgam to cling to the pan, shoes, and mullers so as to slow down the engine until it would stop on the centre. Any mill-man who lives near enough to his mill to have it become a part of himself will realize that if a stamp goes wrong, or a belt breaks during the night, it will bring him up out of a sound sleep into the middle of the floor standing, almost before he has time to recognize what the sound is. Therefore, any old-fashioned millman will recognize the fact that an unusual sound caused by one of the stamps going wrong would cause me to hurry on my clothes and go out to the mill to find that sometimes the stamp was not striking ore, neither was it striking iron, but that the bottom of the mortar had become filled up to and above the level of the dies until the material overflowed onto the dies themselves, and that the stamp was falling on malleable gold, deadening the sound. Under such circumstances I have removed the screen and filled a Wells Fargo express-box full of gold from one mortar. Pretty good ore!

This ore came from the Bodie Consolidated?

Yes. We got 900 pounds of gold amalgam one day.

How much of this was gold?

About one-third was metallic gold. During the run on this Bodie ore, I now remember sending a message to Mr. Yerington, who was treasurer of the company, announcing the recovery for that day. He made the remark, "Riordan is probably excited. He speaks of 900 pounds of gold amalgam, but; of course, he means ounces. But that's good enough." I did not mean ounces; I meant pounds.

What was the yield of the ore per ton?

It averaged about $1000 per ton; but some of it ran as high as $50,000 per ton.

Was it a pocket?

It proved to be a pocket; and it was opened up quite accidentally, against orders. The Bodie company had sunk a shaft to the depth of 250 ft. on the south end-line of the Standard mine, with a horse-whim. At that depth they started a cross-cut and intersected some 17 veins. The principal vein, which was the Standard main vein, was very low-grade at that point; but one of the night-foremen, finding some pieces of quartz showing free gold in a vein about 18 inches wide, afterward called the Burgess vein, decided, without orders, to start a drift south on that vein. In 27 or 28 ft. this drift widened out to about 9 ft. and was filled with gold-bearing quartz, the ore running thousands of dollars per ton.

What was the later history of the Bodie mine?

That I cannot tell without making fresh inquiries, because I left Bodie in the fall of 1880 and went to Arizona. I only know that before this phenomenal run the stock was selling at about 50 cents per share, and that it increased in value on the market to $50 per share, and before the 'pocket' was exhausted it began to pay dividends at the rate of $5 per share per month. I went down to Arizona as the result of a proffer of a position as superintendent of a mine in what is now, I think, Cochise county.

Where was the mine?

It was in the southern end of the Santa Rita mountains. Prior to my leaving Bodie, a number of samples of the ore were sent to me there for assay, together with maps. The samples went within $1 or $2 of the amounts shown on the assay-certificates that accompanied them, and some of them ran over $150 per ton. When I went down there to look the property over, I found that all of the maps and descriptions were accurate. But I also discovered that the ore never came from that mine. That fact had not been stated. So I refused to accept the superintendency of the property.

How old were you then?

About 32.

What did you do next?

The next thing I did was to spend several months with Fred F. Hunt, who is now an analytical chemist in New York City, and who was then covering the South-West in a search for possible copper opportunities for the Orford Nickel-Copper Company, with Thomas A. McElmell, an ex-naval officer, who was a member of the American Institute of Mining Engineers. We three were cabin partners and to a certain extent prospecting partners, especially McElmell and myself, although we had no formal agreement. I decided to take up a property adjoining the Copper Queen at Bisbee, and two business men from Iowa and myself formed a partnership and put some $40,000 into this property in the search for copper ore.

You had saved money from your salary?

I had saved money from my salary and I had made several lucky turns in mining stocks, at both Virginia City and Bodie.

So you were a capitalist?

In a small way. I was always willing and ready to take chances and have been flush and broke perhaps twenty times, and expect to be broke again a few times before I pass in my checks. This copper mining venture at Bisbee not proving profitable, I tackled another prospect in the Silver Bell district, north-west of Tucson, a venture in which one of my previous partners and Senator Norwood, of Georgia, now deceased, were principals. The enterprise languished and finally expired for want of capital.

What has since then happened to the prospects that you tackled both at Bisbee and Silver Bell?

The prospect at Bisbee I do not believe ever developed into a mine, but the prospect at Silver Bell district was afterward developed by E. B. Gage and Frank Stanton, associated with Frank M. Murphy, into the Imperial Copper Co., which, I understand, has been acquired by the Guggenheims. From Silver Bell I went to Prescott, and there became associated with the then Governor of Arizona, F. A. Tritle, whom I had known previously at

Virginia City. Governor Tritle, F. F. Thomas, and myself at one time had the United Verde property under option, but we had to lay down our hands for the want of $14,000. This was before it had been examined by W. A. Clark. While associated with Tritle I made examinations of properties in the South-West and down into northern Sonora, as far as Altar. Subsequently, and perhaps growing out of our intimacy, I was asked by the Governor to take the agency for the Navajo Indians in north-eastern Arizona, because rumors had reached the trading-posts of discoveries of valuable minerals—gold and silver, lead and copper—on the reservation. The Governor wanted somebody in whom he reposed confidence to go up there and take charge of the Indians, as agent for the Government; and while in that capacity to ascertain whether the rumors were true.

Before we proceed further, I shall ask whether you had been trying to supplement your practical experience with the reading of technical books?

When I was a 'cub' miner, in Virginia City, a working miner who knew five or six different kinds of rock was regarded as being something of a geologist. For instance, if he knew granite, limestone, quartz, porphyry, slate, and sandstone, he was regarded as being a fellow who was observant and a good judge of rock. I was always desirous to know the reason of things. In working in the lower levels of the Comstock in bad air I have had occasion to notice that the heat of the water coming from the face of the drift was sufficient to cook eggs. I wanted to know what made that water hot. On inquiry among men who had had a technical education, I learned that it was caused by the chemical reactions in the rocks themselves when exposed to oxidation.

What books did you have?

Usually none, because of my wandering life. I did manage to get hold of Dana's 'Mineralogy', and occasionally I would have my attention called to some book on geology by such men as Fred Hunt, who knew; and I made it a habit to make a humble exposure of my ignorance, so that I never failed to elicit information from any one who was near me. Another thing that helped me was that, for a number of years, the members of the U. S. Geological Survey, many of whom I knew personally, including Major Powell, Arthur Davis, Prof. Thompson, and Prof. Hiller, camped in that part of the South-West where I happened to be, and I got the benefit of their criticisms and their observations, in many cases extending back 25 years, while sitting around camp-fires, and the information they imparted would be etched on my memory. In '82 I became a member of the American Institute of Mining Engineers, and learned more through my acquaintance with such men as Dr. R. W. Raymond, Anton Eilers, George W. Maynard, E. G. Spilsbury, John Stanton, and Charles Macdonald. Some of the younger men with whom I made friends were Arthur S. Dwight, Karl Eilers, Dr. Spencer, and Dr. Hess of the Geological Survey, and such engineers as John B. Keating, Herbert R. Hanley, and others with whom I was in close association both in discussions over practical operations, as well as technical methods.

BULLY HILL, LOOKING SOUTH-WESTWARD OVER THE RISING STAR CLAIMS

Then, Mr. Riordan, you got your technical equipment from nature and from men, rather than from books?

Exactly so. Or, to put it in another way, I got it from hard knocks and absorption. My books were few.

You have given me the impression that you read a great deal now.

Yes, I do now, but then I had neither the time nor the facilities. Some sixteen years ago I was asked by the General Electric Company to analyze for them 700 or 800 mining reports. I have never quite known how they came to hear of me. But I took up the task, and out of the 700 or 800 badly mixed manuscripts I picked seven mines that I thought were worthy of investigation and recommended that they get some high-class engineer to make the investigations; but, instead of doing this, they gave me authority to choose the engineers and have the investigations made. For a number of years I had a free hand in the making of examinations in various parts of the continent, and naturally in company with these engineers, analyzing the facts they ascertained and giving me the reasons why, I was bound to get some of the knowledge they had acquired by study as well as by observation, but which I could not have got in any other way. Then when I found I lacked any information in any particular I would not hesitate to ask where I could get the original authority or the original source of information on this or that point. If I heard an ore spoken of as 'bornite', when I did not know what that mineral was, I made it my business to go and hunt it up. And thereafter that was fixed in my memory. Habitually for 15 years I read reports and articles on mining subjects from two to three hours each day, besides the work I did during office-hours.

Thank you. That answers my question. Now we will go back to your acceptance of the agency on the Navajo reservation.

I probably spent more time on horseback scouring all over the reservation than any agent who preceded me, and possibly more than any who has been there since.

How long were you there?

I was there about two years and a half, from early in 1882 until the middle of '84.

What did you find?

I did not find anything that could be called a hopeful prospect, but I did find an ingenious system among the Indians of getting samples of rich ore from the San Juan region or elsewhere in Colorado, taking a sample as large as one's head, breaking it into smaller pieces, and giving the pieces to a number of different Indians to show to prospectors. And this in turn developed a carefully devised scheme for awakening the interest of any 'tenderfoot' or even 'old-timer' whose ambition might be aroused by showing him a sample of this rich ore and describing in a rough way the course of a vein along a hillside and the kind of rocks on both sides of the vein. For giving this information the Indian usually had himself and party well supplied with flour, bacon, coffee, or anything he wanted that could be obtained at the trading-post. Then he would return to his own 'stamping-ground' and a few days later another Indian would come down with another sample and describe a mine in the same locality. Probably a month later the first fellow would return. Every time the Indian messenger showed up he was listened to eagerly and went back loaded with the goods that he wanted. Then the white man would name a time when the Indian would be prepared to take a party up there and show them the locality, and the day would be set and this man would come down from the heart of the reservation to Gallup, or some other point, meet the white man there, and two or three hours afterward another Indian was likely to come tearing in with his horse all covered with foam, with the word that the hearts of the Indians in that vicinity had grown 'bad' and that it was not safe for the white man to go into that region to investigate. They would work that scheme with ingenious variations, until I exploded the bubble. Governor Tritle's idea was, if any valuable mineral was found, to negotiate with the Indians for a right-of-way for a road from the most available point on the railroad or the nearest white man's wagon-road, and thus avoid any conflict, setting aside the district for mining purposes with an amicable arrangement that was fair to the Indians. But, of course, the failure to discover valuable minerals in quantity spoiled everything.

Then?

Edward E. Ayer, of Chicago, who was probably the largest tie and telegraph-pole dealer in the world, and who had a saw-mill at Flagstaff, telegraphed me while I was at Washington with a party of Indians visiting the Secretary of the Interior, to call and see him on my return journey. His first question was: "Do you know anything about a saw-mill?" I said, "Not a damned thing." He said, "You are just the man I want." That resulted in my running the lumber business until Ayer invited me to come to Chicago at $10,000 a year as manager of a part of his business and a half-a-dozen superintendents under me. This was in 1887. I told him I would rather make a success of that old mill out at Flagstaff and get $1000 a year than to be in Chicago at $10,000. I said I would make a success of the business if it was humanly possible, or leave my bones in the pile of saw-dust. So he sold me the mill, and the only security I could give him was my bones and clothes. Inside of two months I lost $18,000 through the failures of customers; inside of two months more I started in helping to build what was then known as the Arizona Mineral Belt Railroad; I trusted the company to the extent of $40,000, and then the silver panic of '93 came, and the Atlantic & Pacific Railroad went into the hands of the receiver, owing me the $40,000. The lumber business proved a trial all right; but it never failed. It had taken me over 11 years to work out problems that I thought I could solve in three. Even at that time labor controversies were not unknown; and while we had no serious troubles I decided to put into effect a profit-sharing plan that I had thought over for years, the idea having been suggested by Dr. Edward Everett Hale. It would be too long a story to tell you now, but some day we may have a chance to talk it over. It is sufficient to say that my effort ended in a heart-breaking conspiracy among the beneficiaries, and rather than endure the continuing disappointment, I disposed of my remaining interests, at a sacrifice, and left that region, completely cured and freed from illusions on profit-sharing in this country under present social conditions. In leaving Flagstaff I had been encouraged to go to the South, where E. P. Ripley, president of the Santa Fe, and his associates, had an uncompleted railroad from Macon to Savannah.

What did you do there?

I went over the portion of the road that had been completed from Macon to Dublin, which was then earning a profit over operating expenses, but not on the bonded issue. It was a good road, if backed up.

What was your work?

It was in the nature of preliminary investigation, for the purpose of advising the owners and making up my mind whether I would take the presidency of the road. However, I had not gone far before I discovered that while Mr. Ripley, Mr. Soper, and their associates had apparently unqualified confidence in my ability to perform the physical task of completing the road, their chief desire to have me accept the presidency was based upon their belief that I could finance the completion; inasmuch as I had previously succeeded in getting some heavy undertakings of a similar nature financed—that is, heavy for me.

What did you decide to do?

I decided not to accept the presidency, although I told them plainly that if the road were financed I would undertake to complete it and accept the responsibility of its direction. I left. Then I started on a horseback reconnaissance of the Southern States, from Washington to Tampa on the east, and as far west as Texas, and in the doing of this I spent nearly two whole years.

Making a mineral exploration?

Not necessarily, although mineral investigations were included in some cases. For instance, I examined tracts of land embracing some 50,000 acres on which there was poplar, pine, and the other merchantable woods of the Piedmont region, but on which it was also claimed there was gold, silver, copper, and coal. Among other tracts I examined was one of 25,000 acres, at the instance of Abram S. Hewett. This property had a wonderful growth of poplar on it, but it also had two horizons of coal that were probably fifty times as valuable as the timber, while Hewett's proposition was to buy the tract based upon the stumpage value of the poplar alone.

Did Hewett buy it?

He did. I may say that I was introduced to him by Dr. R. W. Raymond, who is one of the staunchest and most loyal of friends. Two years of this work brings me to 1898, the year of the Spanish War. I went to Tampa during the preliminary rumblings and was there the night the Maine was blown up, on my way to Cuba to see things for myself; but I was stopped. My family was in Europe at this time, and I had been keeping in mind the possibility of a home in the South, but decided to give it up and return to the West. Upon my return to Arizona, I went to Tucson, and while there was invited by Epes Randolph and Eugene S. Ives of Tucson to become an investor in the King of Arizona mine, north-east of Yuma. At that time the four principal owners and stockholders were Randolph, Ives, Blaisdell of Yuma, and S. Morgan Smith, a manufacturer from Pennsylvania. The mine had been discovered by a man named Eichelberger and taken over under option from him by Randolph and Ives. Smith was an ex-minister from the South, and was the principal person who put real money into the venture, besides myself. I was elected president of the company, and had for manager a nephew of Marshall, the discoverer of gold in California, Thos. Marshall Irwin, a bright young fellow. He died while in our employ, in Arizona.

What happened to the King of Arizona?

We were not a happy business family. I was not long in discovering that each of the three Arizona directors entertained a very poor opinion of Smith, and of each other, except as they could combine forces to do up Smith without coming within the grasp of the statute law. And each expressed an opinion of the others, as I do when I do not want to be offensive and describe a man as a stew-

THE BULLY HILL SMELTER, SHASTA COUNTY, CALIFORNIA

ard. For a good mouth-filling definition of a steward I would refer you to Kent's answer in the second scene of the second act of 'King Lear', to the question of the steward: "What dost thou know me for?" I wrote to each one of my associates repeating what he had said about the others and asking him if my recollection of his characterization was correct. In each case my correspondent replied that my memory was not at fault, whereupon I called a meeting of the board at which all were assembled and placed the correspondence before them.

Were any of them armed?

It is fair to assume that at least two of those present were heeled. But nothing happened other than that they became less acrimonious, and after some interchanges requested me to see if I could not do something to bring about the sale of the property. Then I told them they would either have to nominate other representatives on the board or secure another President and General Manager.

What happened?

They decided to give me an option on the property at a sum which was a good round figure for such a property at that time, but its subsequent development and production has apparently justified the price they then asked. However, I did not succeed in disposing of the property for them, and while I was in New York on other matters, in 1900, I was in the old Engineers Club on Fifth Avenue when I was asked by a representative of the General Electric Company to meet some of the directors with a view to examining into the feasibility of their searching for copper-mining opportunities. After a preliminary conference I was invited to examine a mine for which they had negotiated and upon which they had made a preliminary payment. This mine was in New Hampshire, not far from Berlin. I suggested that they send a high-class engineer to make the examination, and this resulted in A. R. Ledoux undertaking the task. When Dr. Ledoux was chosen I assumed that this would result in my not going, as I was not under definite engagement with these people; but they asked me to accompany Dr. Ledoux, which I did. While we were at the mine together, our investigations were independent of each other. After his return Dr. Ledoux made an elaborate report on the property. I made none; at least none in writing, and openly declared that I would not attempt to make a mining report either as confirmation or as correction of the report of a man of Dr. Ledoux's standing. I did not regard myself as competent. My habitual term, in speaking of myself, was that I was a sort of a saw-mill metallurgist, and my experience has taught me that I could not truthfully claim to be anything more. When a man has reached the age of 48 or 50, if he is honest with himself, and intelligent, he usually has made a pretty accurate mental analysis of himself. Others may flatter him and regard him as a very much higher-grade man than he does himself. Still others may have a depreciated opinion of him; but he ought to know what he is, himself. And while I am bound to know that a great many men of the highest standing in the profession regard me as a competent man, nobody knows as well as I do myself how keenly I recognize my deficiencies. My training was of the most haphazard character, as you must have recognized already, and entirely the result of some previous experience growing out of my being plunged into the vicissitudes of the frontier under conditions which compelled me to undertake the running of a lumber-yard this year, building a piece of railroad the next, running a quartz-mill the following year, and perhaps appraising timber-land the next.

What happened to the mine in New Hampshire?

My principals took my advice and forfeited the payment they had made, but I never made any report on the property. After Dr. Ledoux's report had been received, one of the directors came to me and asked me whether I intended to make a report myself, and I said, "No, I do not regard myself as competent to submit a report upon a mine which is intended either to confirm or to correct the report of an engineer of Dr. Ledoux's professional standing." Then the director asked me, "Will you tell me this? Will you tell me what you think of that mine? And will you tell me what you would advise us to do?" I said, "I will do that. I think the mine is not worth a damn. I advise you to forfeit the payment you have already made and make the best settlement you can with the entrepreneurs." The promoters of the mine were Dr. Edward Peters and a man named Elliott.

Did that lead to your being retained by the General Electric Company?

It did. There were turned over to me 700 or 800 reports of mines of all kinds and descriptions. I was asked to go through them carefully, and to select some one or more that I regarded as being promising. This resulted in my choosing six or seven—among which were the Shannon, in the Clifton district; another in the Cananea district; the entire Shasta county, California group; the Granby at Phoenix, B. C.; the Britannia on Howe Sound; the Wall property, in Bingham canyon, now the Utah Copper; and two or three others—the Tezuitlan and the Inguaran, in Mexico.

Was it a question whether the General Electric people should purchase shares in companies operating these mines, or whether it should purchase the mines and operate them themselves?

The idea was that the General Company would operate the mines themselves.

How many mines did the General Electric Company proceed to operate and how did you select those they did operate?

When I gave the directors who had the matter in charge, notably, C. A. Coffin, Robert Treat Paine, and Gordon Abbott, the list of mines and districts in which I thought the possibilities were sufficiently good to warrant an investigation, they asked me if I would undertake to secure engineers to make the examinations. I told them I would, so we selected W. Lawrence Austin, Thos. Marshall Irwin, T. S. Mathis, who had previously made some examinations under the supervision of Ellsworth Daggett and Hinsdill Parsons; Matthew J. Walsh, E. Gybbon Spilsbury, Martin Schwerin, John H. Mackenzie, and others. Usually, especially in the case of mines situated in our South-West, I knew somebody in the district who was familiar with local conditions and upon whom I could call for a preliminary opinion.

Would you give the names of those who helped you in this way?

They were J. C. Smith, Capt. Berry of Globe, M. P. Freeman of Tucson; Charles Akers of Phoenix; T. G. Norris, Hugo Richards, and others at Prescott; and one or two each in Bisbee and Tombstone whose names will not come to me on the instant; and several old Comstockers whom I knew in Nevada.

THE LA GRANGE HYDRAULIC MINE

What mines did your company acquire?

Upon the technical recommendations of Mr. Austin—and my endorsement, of course—our people decided to take up a number of mines and prospects in and about Hanover, New Mexico—the Ivanhoe group—and something like a hundred claims stretching along the railroad in what are known as the Upper and Lower Basins, in the vicinity of Fierro and Hanover, were put together under the name of the Hermosa Copper Company. My principal aim in making these recommendations was to establish a basis for the acquisition of the Santa Rita property, which was then owned by J. Parker Whitney, of Boston. In the early days of the Hermosa I had an exhaustive examination of the Santa Rita property made by John M. Sully. This examination cost us considerably over $10,000, and included over 4000 samples from various parts of the property, and I unqualifiedly recommended that our people take the property over, but they failed to do this. The property is now owned by the Chino Copper Company, which has something like 80,000,000 tons of developed ore, according to the latest estimates; and the man who made the examination for us is the manager, Mr. Sully.

What happened to the Hermosa?

We started on the recommendations of Frank H. Probert, who confirmed all of Mr. Austin's previous recommendations, and decided to take the property over; but it did not respond to our work as was hoped. It is now being operated in a rather small way by lessees, and recently some of our people have been recommending that they rehabilitate the property and resume operations.

What other properties did you acquire?

The only other property we acquired was the Bully Hill in Shasta county, California.

This has done well, has it not?

It did well, very well, until the Government began to make trouble about the smelter-fume. The Bully Hill proved to be a good property, but my original recommendation was not the acquisition of the Bully Hill alone, but the combining of it with the Mammoth, and my still earlier recommendation was to take up the whole Shasta group of copper mines.

Why was your recommendation ignored?

That I am unable to say. But there was always a strong feeling among some of our important directors against the company going into mining at all, and a whispered negative would outweigh the strongest positive recommendation.

When did you acquire the Bully Hill?

In 1905.

Were your headquarters in California?

I kept my headquarters in New York, but spent from two-thirds to three-fourths of my time in the field.

Aside from your mining activities for the General Electric Company did you engage in any on your own account?

I did. As a matter of fact, I was always engaged more or less in mining activities in what might be regarded as a small way. But sometimes these ran into considerable amounts of money for one of my means. I was always 'grub-staking' prospectors, all over the mining regions with which I was familiar. I directed some drift-gravel mining in Calaveras county, with Matthew J. Walsh as superintendent, after the investigation staff of the General Electric Co. was cut down to the minimum; and I also became identified with hydraulic mining in Trinity county, California, at the La Grange mine. I was treasurer and managing director of this property for a number of years, becoming its president after the death of Robert Mather, who went into the venture when he was president of the Rock Island. Pierre Bouery, a French engineer, was the manager of the property, and a mighty good manager he was. This venture proved profitable to all interested; subsequently the property was sold to English capitalists for whom Baron de Ropp was the managing director in this country. We took out over a million and a half dollars during our operating period of some seven years.

I have never been entirely out of mining operations for nearly 50 years; some of them treated me very nicely, some of them not so agreeably, but on the whole I have no cause for complaint. I put just about as much energy, just as much intelligence, and usually more money into the losing venture than into those that proved winners; and as I look back now, taking the whole series of ventures, by and large, I cannot recall one that I greatly regret having made.

Does your connection with the General Electric Co. continue?

It does. Some five years ago I resigned my position as president of the various subsidiary companies that had been formed and recommended that John B. Keating, who was the manager at Bully Hill, be elected in my stead. Mr. Keating is a trained metallurgist, a fine mining engineer, and an exceptionally good business man. It has always been my habit, and one out of which I derived a great deal of pleasure, to boost the younger men. I never consciously wanted to be the head of anything. I would rather be the trusted right-hand man to some man whose character and ability commanded my respect, than be the boss. And usually I preferred to have a younger man in the position of the executive head, rather than occupy that position myself. But from the time I began work with a pick and shovel and wheelbarrow, even where there were no more than two of us together, and even if the other man was older than myself, the boss always seemed to look to me for a lead. But when Mr. Keating took charge of the Bully Hill and other properties, as president, he and his assistant, Herbert R. Hanley, started to solve the Bully Hill fume problem by electrolysis instead of smelting. That work has been going on for over three years now, and is just about completed.

You are going to erect an electrolytic zinc refinery?

Our people are not, that is, not immediately, but the Mammoth company has passed favorably upon our process, after an elaborate examination by the heads of their metallurgical department.

Can you tell me what process you are going to adopt?

In most essentials, I judge, although I have not seen the Anaconda process in operation, it is similar to the method devised by Laist and Frick, although in some important respects prior to that. Mr. Hanley, who was the assistant-manager at Bully Hill, was the man who elaborated the process and who found out the way to do things which were sometimes baffling, and, under Mr. Keating's supervision when he happened to be there, they worked out the methods to the point where the process seemed sufficiently developed to invite the Mammoth people to examine it with a view to the erection of a plant.

By the way, Mr. Riordan, I met you in London when you came over to aid the refugees at the beginning of the War.

I remember. I was deputed by the Secretary of the Treasury to accompany the officials of the Treasury department to Europe, the entire party on board the 'Tennessee,' on which we sailed, being composed of representatives of the State, War, Navy, and Treasury departments.

You were bringing money to aid American refugees in Europe?

Yes. We took over $1,500,000 of Government money, and something like $3,000,000 of money that was hastily got together by the leading banks of New York, Boston, and Philadelphia, for a similar purpose.

I remember your telling me how nearly the 'Tennessee' became the victim of a floating mine.

Yes. It was a close call, although we did not know it at the time. The Dutch government provided one of the boats which were patroling the coast to accompany us far enough out to sea to do the honors and to note the direction in which we sailed. The last time I went back to England from the Hook of Holland, it was reported that the boat that escorted us, after turning around to leave us, was sunk by a mine and went to the bottom. It is fair to assume that we had a narrow escape.

You have now settled in California?

I believe I have. I opened an office here last year, but I have been such an Arab on the face of the earth that for years I have not dared to count on a permanent residence. I have had to travel, until within the last year, probably 25,000 miles per annum for 20 years; and have wandered from the mines of Cariboo to Bolivia, and habitually crossed the Continent from three to five times a year. I probably traveled 30,000 miles in Arizona alone, on foot, on horseback, and by buckboard—most of it before the railroads entered the Territory.

Have you any opinion regarding the future price of copper?

My opinion is that the price of copper will hardly drop below 20 cents during the five years following the end of the War.

Potash Production

More potash has been produced during the first six months of 1917 than was made during the entire year of 1916, according to the U. S. Geological Survey. In terms of available potash (K_2O), the production from natural salt and brine amounted to 7749 tons; from alunite and dust from cement-mills and blast-furnaces 1867 tons; from kelp 2143 tons; from industrial wastes 2153 tons; and from wood-ashes 111 tons. The aggregate is 14,023 tons, and the value at the point of shipment $5,864,039.

The Nebraska alkali lakes still lead, having yielded about one-third the entire production. There are now at least four important operators in this field.

The production from Searles Lake, California, would undoubtedly be materially assisted by passage of the legislation now before the House of Representatives dealing with the leasing of potash-bearing lands. Continued uncertainty as to the status of titles to this property has hampered the development of this important deposit.

No production is reported from feldspar or other silicate rocks, but considerable quantities of potash salts and potash-bearing fertilizers were obtained from the dust in cement-mills and blast-furnaces.

The production from kelp was about 15% of the total, as it was in 1916.

The prices quoted range from $3.50 to $6 per unit, a unit meaning 1% potash in a ton of the material as marketed.

The figures given indicate that the production for 1917 will exceed 25,000 tons of potash or two and one-half times that made in 1916. This is about 10% of the average normal yearly consumption of the country before the War.

MONAZITE is essentially a phosphate of cerium, lanthanum, and didymium, including small and varying quantities of thorium, silicon, and other elements. The mineral is yellow, varying in shade and tint from a light greenish yellow to dark honey-brown; it is somewhat translucent to opaque, and it is characterized by a resinous or greasy lustre. It shows no distinct cleavage, and is brittle, breaking with an uneven fracture. In the original rock it may show sharp edges and plane crystal faces, but as found in sand it is in rounded grains, intimately mixed with other heavy and resistant minerals, the most abundant of which are zircon, magnetite, ilmenite, and garnet. The monazite sands as found usually have to be concentrated, as the sands of this country usually carry less than 50% monazite. A compilation of analyses of the mineral monazite—that is, the pure selected mineral—has been made to serve as a basis for comparison of the value of monazite from different localities. The percentage of thoria present in the sand from which this monazite was selected can be obtained by multiplying the percentage of thoria in the pure mineral by the percentage of monazite in the sand.—U. S. Geological Survey.

Drop in Pressure of Compressed-Air Hose

By WALTER S. WEEKS

Last year W. W. Sprague and E. H. Wisser, under my direction in the Mining College of the University of California, made a large number of tests to determine the drop in pressure that compressed air undergoes in passing through a 50-ft. length of commercial air-drill hose. These tests were made for the purpose of furnishing to engineers data that might aid them in selecting a hose.

Tests were made on $\frac{3}{4}$-in. and 1-in. hoses, which were lent by different manufacturers. The drop in pressure, with the same amount of air passing through, varied somewhat with different types of hose. The charts given on the opposite page may be considered representative of an average hose. The pressure marked on each curve is gauge-pressure at the entrance of the hose. The drop shown on the charts includes the drop through the hose-fittings, which in each case consisted of two couplings and two spuds. An approximate test was made on the drop through the fittings alone, and the drop was so slight that it would have been a waste of time to pursue it further.

FIG. 3. APPARATUS FOR TESTING HOSE

In addition to the data given by the charts a few more points may be of interest. The buyer should make sure that the hole in the hose is of the specified size. The hose with the most 'give' has the least drop in pressure, because when it is under pressure the diameter of the duct air with less drop in pressure than more rigid hoses. which have a layer of rubber between the wraps of duck, apparently have more expansion, and consequently conduct air with less drop in pressure than more rigid hoses. Wire-wound hoses are very rigid. They have the additional disadvantage of becoming closed when struck by a falling piece of rock. The inside of a wire-wound hose is often corrugated by the tight winding. These corrugations increase friction and constrict the opening.

If a 50-ft. hose is coiled in four or five coils the drop in pressure will be increased by about 17%.

Manufacturers rate their drills in free-air consumption at sea-level. The free air on the charts is at sea-level, so the rating given by the drill-maker is used to find the drop in pressure. The effect of the altitude at which the drill is used may be neglected.

The method of conducting the tests was briefly this: Referring to Fig. 3; air was supplied to a receiver (1) through a reducing-valve (2), which maintained the pressure at any desired figure. One end of the hose was on the receiver and the other end on a tank (3), which was fitted with a low-pressure nozzle (4), a water manometer (5), and a thermometer (6). From each end of the hose a small copper tube (7) led to a mercury manometer (8). The air passing through the hose was controlled by a valve (9) leading into the tank.

From a chart the water-gauge was determined; this signified that a given amount of free air was passing through the nozzle. The valve in the tank was set so that this pressure was maintained in the tank. The difference of level in the mercury manometer gave the drop in pressure with a given amount of air flowing through, with a given pressure in the receiver.

POTASH occurs principally in the rocks and soils of the earth, where it is very widely distributed. Chemical analyses of rocks and soils show that potash (K_2O) forms from 2.65 to 3% of the entire earth's crust. Most of it occurs in combinations that are insoluble or only slightly soluble in water, and so is not considered 'available' for use either in agriculture or in many large industries that utilize salts of potassium. From 10 to 15% of certain minerals, some of them, such as feldspar, very common, consists of insoluble potash. Potash occurs as chloride or possibly sulphate in ordinary sea-water; in smaller proportions in the waters of most streams and lakes, and in larger proportions in many natural brines. —U. S. Geological Survey.

Chemistry

Next to the work of creation itself, possibly the most marvelous achievements which have been wrought in material things on this planet have come as a result of the work of the chemist. Indeed, chemistry would seem to come nearer to truly creative work in material things than anything else of which man knows. Leading in all human progress is the chemist. Whether it be dealing with the soil, increasing its fertility and enlarging the output of foodstuffs; whether it be in extracting fertilizers from the air with which to save civilization in the years to come from starvation by lack of food; whether it be in the creation of explosives or the making of dyes, the production of medicines or the thousand and one other things which enter into every phase of human activity, chemistry is the dominating power. In olden times—and indeed by many people of today—chemistry is sometimes associated with apothecary shops. For many years people thought that the chemist was a druggist or apothecary, and there are still some people in the land who associate these interests and do not realize the broader work of the chemist in every line of human endeavor. Without the chemist there could be no agricultural advancement; without the chemist there could be no manufacturing growth; without the chemist there could be created none of the drugs and medicines to alleviate human suffering. It is, however, largely due to the utilization of chemistry in the creation of industries, agricultural and manufacturing, to which the world is now giving its greatest attention.

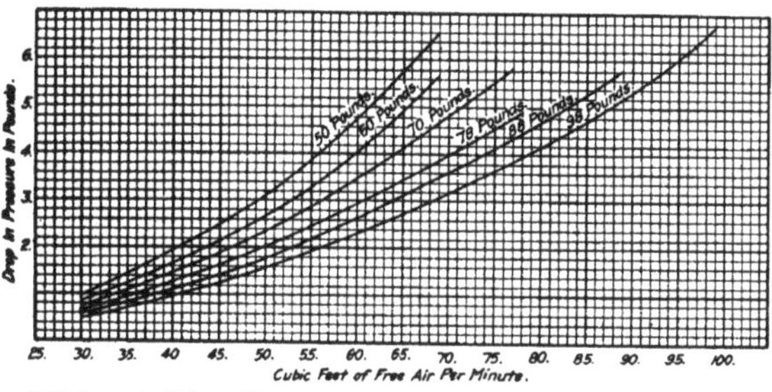

Fig.1. Curves for 50 Feet of ¾ Inch Woven Hose with Sullivan Fittings Showing Drop In Pressure Plotted against Air Delivered at Different Pressures.

We have entered a world war which is in reality a war of chemistry and mechanics, largely of chemistry. When Germany invaded Belgium it awakened the world to a realization of the fact that this country, as others, had for years been dependent upon German chemistry for myriads of things which should have been produced at home. We were instantaneously cut off from a supply of potash, though we should have searched the world for potash or produced it from by-products, as we are now beginning to do. We were cut off from dyes, though there are more resources in this country for dye-making than Germany ever dreamed of having. We had permitted ourselves to be handicapped, and had we at that time been forced into war our aid would have of necessity been very small, because chemistry had not made it possible for us to do the things which are now essential to maintain war. Within the last three years, however, the chemists of this country have almost been born anew or, rather, a new birth has come to chemistry. Thousands of chemists have concentrated their attention with untiring zeal upon the opening up of new avenues for producing the things which heretofore were brought from other lands. They are seeking to make this country independent of Germany, to make it self-reliant and self-contained.—From the 'Manufacturers Record.'

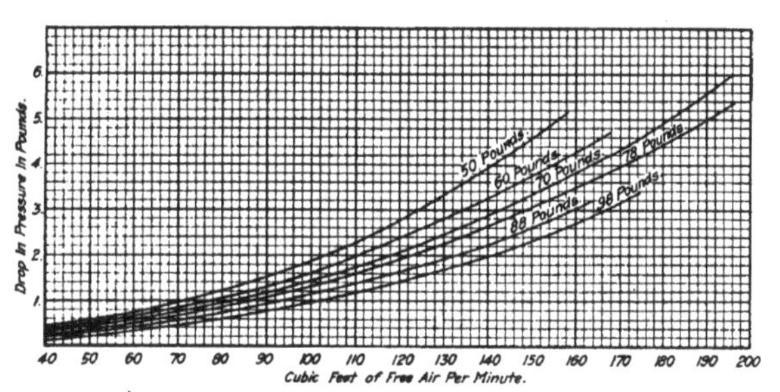

Fig.2. Curves for 50 Feet of 1-Inch Wrapped Hose with Sullivan Fittings Showing Drop in Pressure Plotted against Air Delivered at Different Pressures.

Mining Practice in the Joplin District

By H. I. YOUNG

PROSPECTING. *Where ore is found at a depth of 100 ft., or more, the prospecting is done by churn-drilling. Machines driven either by steam or gasoline are used. The drilling is started with a 6-in. bit and the size of the bit is reduced, if necessary, as the hole is advanced. It is customary to drill holes from 100 to 400 ft. deep, but this is governed entirely by the position and formation of the ground. In drilling sheet-ground, close spacing of the drill-holes is unnecessary, but in narrow orebodies the holes are placed from 15 to 50 ft. apart.

After the orebody is reached, cuttings are taken every two or three feet, samples being sometimes obtained by pouring the water off the cuttings and taking a portion of the coarse material for assay. It has been found, however, that the sludge removed with the water may contain mineral. To overcome this, cuttings from each two or three feet are put in a container and after the water has evaporated, a sample for assay is taken by means of an automatic device.

Records of cuttings are kept in glass jars or glued on large cards. Drilling at present costs from $1.25 to $1.50 per foot. In normal times, the drilling cost ran from 75c. to $1 per foot.

DEVELOPMENT. After proving the property by drilling, vertical shafts are put down to the orebody. The dimensions of the one-compartment shaft vary from 4 by 5 ft. to 5 by 7 ft. in the clear, the two-compartment shaft is from 5 by 10 ft. to 7 by 12 ft. in the clear. Recently a few three-compartment shafts have been sunk. The shafts are usually put down near the edge of the orebody, so that very little driving is required to open up the ground. The shafts are usually cribbed with 2 by 4 or 2 by 6-in. sawed timbers, which extend several feet below the top of the solid rock. The cribbing is lined vertically with 7-in. boards, which help to keep the shaft free from small rocks and water and permit faster hoisting. The cost of sinking a 5 by 7-ft. shaft, where from 100 to 500 gal. of water per minute is handled, to a depth of from 200 to 250 ft., is from $25 to $30 per foot. Usually two or more shafts are sunk on each project. The shafts are connected by an air-drift 5 by 7 feet.

In the Granby, Aurora, and Galena districts where ore is found at from 40 to 100 ft., the practice is to sink small shafts, very little drilling being required to prove the shallow orebodies. To increase the tonnage where working-faces are limited, a 7 by 8-ft. drift is extended ahead of the working-faces for 300 to 600 ft. and the ground opened up so as to permit the use of several machines. Bore-holes are necessary for ventilation where this method is used.

MINING METHODS have been developed to suit local conditions. The orebodies, which are 8 to 40 ft. thick, lie in a horizontal plane and are usually found on one level. In parts of the district, however, two distinct ore-horizons exist, and as many as four separate ore-bearing strata, these being separated by relatively thick layers of hard barren rock, the several ore-seams being mined independently. Where the orebody is thicker than 16 ft., underhand stoping is practised, especially in the hard-ore sheet-ground. Pillars are left at frequent intervals, the distance between pillars and their thickness depending on the nature of the ground and the thickness of the orebody. The ore is usually found in a hard flint formation, known as the Grand Falls chert, which lies between beds of limestone. The limestone formation is known as the Mississippian, and is overlain by the Pennsylvanian series of shales and sandstones. A comparatively small amount of timbering is required in the mining operations, as there is a cap of flint from 1 to 5 ft. thick. This makes a good roof. In some cases, the flint decomposes into what is known as 'cotton rock.' It is soft and white, and pillars of it must be spaced closely, and the roof arched. In sheet-ground from 10 to 20% of the ore-formation is left as pillars. These are sometimes trimmed, many of them being removed after the orebody has been worked out.

DRILLING AND BREAKING. Where the orebody is over 16 ft. thick, it is customary to employ the 'heading and stoping' method of breaking the ground. The heading is supported on 6 or 7-ft. posts. The face is irregular, which is necessary for good breaking. From five to seven holes are drilled at each set-up, and the heading is advanced from 15 to 20 ft. ahead of the stope. If the stope is more than 10 ft. high, a 'splitter' is drilled horizontally so as to relieve the stope-hole. In high stopes more than one splitter is used. Stope-holes are pointed below the horizontal in order to maintain a level bottom.

Where the orebodies are less than 16 ft. thick, the 'heading and overhead stoping' method is employed. A 10-ft. post is used to support a bottom heading, and from five to eight holes are drilled per round. The holes vary in depth from 10 to 12 ft., from 75 to 150 tons of ore being broken per round. After the bottom heading has been advanced from 10 to 25 ft., holes are drilled in the overhanging ore, which is blasted down cheaply. A comparison between these two methods is shown in Fig. 1.

Two types of air-drills are used, the 'piston' and the 'hammer.' In sheet-ground, where the dust is an important consideration, the hammer type of water-drill

*Abstract of a paper to be read at the St. Louis meeting of the A. I. M. E., Oct. 1917.

has been used with success, showing a drilling efficiency 25 to 40% more than the piston-machine. Drills are operated with air at a pressure of from 85 to 100 lb. High-carbon hollow drill-steel is used with hammer-drills. Holes are started with a 2¾-in. bit and finished with 1½-in. It is necessary to change the size of gauge in each two feet drilled, and it is sometimes necessary, in exceedingly hard ground, to use several steels to drill the two feet. Usually the drill-steel is re-sharpened in an underground blacksmith-shop. Mechanical sharpeners are preferred.

Table 1 shows the cost of drilling in a mine where the faces are from 10 to 16 ft. high, for four months of 1917:

TABLE 1

Tonnage broken 136,272

	Amount	Cost per ton of ore
Machine-men	$9,630.40	$0.0706
Machine-helpers	7,940.45	0.0583
Drill-repairs	3,505.14	0.0257
Drill-steel	2,627.29	0.0193
Sharpening drill-steel	3,194.86	0.0234
Oil for machines	207.78	0.0015
Compressed air	6,168.63	0.0453
Air-hose and fittings	703.43	0.0052
Total	$33,977.98	$0.2493

BLASTING. After drilling, the holes are chambered and prepared for powder. Both ammonia and gelatin dynamite (30 to 40%) are used, in sticks of from 1 to 1¼ in. diameter. No. 8 caps have been found to give the best results. Where hoisting is done on double-shift the powder-crew works on the second shift, and the blasting is done after the shift, so several hours is given for clearing the smoke. It is necessary to employ skilled labor, both on account of the high cost of explosives and for the safety of the men. The holes are loaded with charges varying from 25 to 150 pounds.

The cost of breaking is from 15 to 30c. per ton, according to the height of the working-faces. The items shown in Table 2 are for a 12-ft. working-face.

TABLE 2

Tonnage 136,272

	Amount	Cost per ton of ore
Labor	$4,006.20	$0.0294
Powder	17,914.00	0.1314
Fuse	782.33	0.0057
Caps	217.62	0.0016
Total	$22,920.15	$0.1681

These costs are on powder purchased by contract, the present market price being 70% higher.

SHOVELING. After the ore is broken it is necessary to drill large pieces with a jack-hammer, and blast them. The usual practice is to shovel into tubs, which are commonly known as 'cans,' holding from 800 to 1200 lb. of ore. These are 30 by 30 in., 30 by 32 in., or 32 by 32 in., and are trammed to the shaft on small trucks. The shovelers use short-handle No. 2 scoops, an average of 20 tons per shoveler per shift being obtained. All shoveling is done on contract, and shovelers earn from $3 to $7 per 8-hr. shift. Both the Thew steam-shovel and the Myers-Whaley shoveling-machine have been used underground to good advantage, but without a decrease in cost. However, if labor continues to become scarcer, it may be necessary to adopt this method to ensure steady production.

A number of mines have substituted cars for 'cans.' The cars hold from 1500 to 2000 lb. of ore; they are easier to shovel into, as they are less high than the cans. They are also more stable, which permits faster tramming. The cost of shoveling ranges from 20 to 25c. per ton.

TRAMMING. Several different methods are employed: by hand, by mule, by motor, and by rope. The selection depends on the length of the haul, the tonnage, and the grade. The tracks are from 15 to 24-in. gauge, the rail varying in weight from 12 to 30 lb. per yard. The trucks are made of wood or steel mounted on plain wheel and axle or on roller-bearings. The roller-bearing truck has been found economical. Where the haul is 200 ft. or less, the shovelers run the cans to the shaft, but for any distance between 200 and 400 ft. additional labor is required. Where a large tonnage is required, as in the sheet-ground mines, and the distance from the shaft to the working-face is from 400 to 1500 ft., mule-haulage is advisable. A mule will haul 100 tons 1500 ft. in an 8-hr. shift, at a cost of from 5 to 6c. per ton. In the large sheet-ground mines, where 1000 tons is hoisted in an 8-hr. shift, and ore must be hauled 2000 ft. or more, motor-haulage is necessary. The 6-ton gasoline locomotive is much used and does not spoil the ventilation. Rope-haulage is employed where the grade is excessive.

The costs in Table 3 show a comparison of gasoline-locomotive and mule-haulage for four months of 1917. The distance on locomotive haulage is 1750 ft. and on mule-haulage 700 feet.

TABLE 3

GASOLINE-LOCOMOTIVE

Tons hauled 136,272

	Amount	Cost per ton
Engineer	$1,883.62	$0.0138
Brakeman	1,659.12	0.0122
Oil for locomotive	239.27	0.0018
Gasoline for locomotive	1,458.10	0.0107
Repairs	2,063.39	0.0151
Car-couplers	2,241.65	0.0164
Total	$9,545.15	$0.0700

MULE

Tons hauled 81,145

	Amount	Cost per ton
Mule-driver	$1,868.70	$0.0230
Maintenance mules	406.68	0.0051
Car-couplers	964.56	0.0119
Car-greaser	470.37	0.0058
Total	$3,710.31	$0.0458

HOISTING. Three methods are used; cans, cages, and skips. The can is best adapted to the depth of hoisting and to the tonnage of the average Joplin mill. Both steam and electric hoists are in use. The hoisting cycle is as follows: A trammer runs a loaded can to the side of the platform, the hoisting-cable is unhooked from the platform and attached to the load by the tub-hooker; this operation is done quickly, and without any signals to the hoist-man, who now picks up the loaded can, which swings over the platform, where the tub-hooker steadies it in the centre of the shaft. It is raised to the top of the derrick, where the hoist is placed. Here the hoist-man hooks a tail-rope, which is fastened in front of the shieve-timbers, to a ring in the bottom of the can. The hoisting-cable is slacked, and the weight is taken by the tail-rope, causing the can to dump into the mill-hopper. The hoisting-cable is taken up and the tail-rope unfastened so as to allow the can to descend in the shaft. With practice a hoist-man and tub-hooker will become so expert at their respective operations that as many as 150 cans per hour can be hoisted from a depth of 250 ft. The record for the district is 1071 cans in an 8-hr. shift. Where two compartments are used for hoisting, the hoists are placed side by side at the top of the derrick, and the shaft is divided for its entire depth to prevent the cans from bumping.

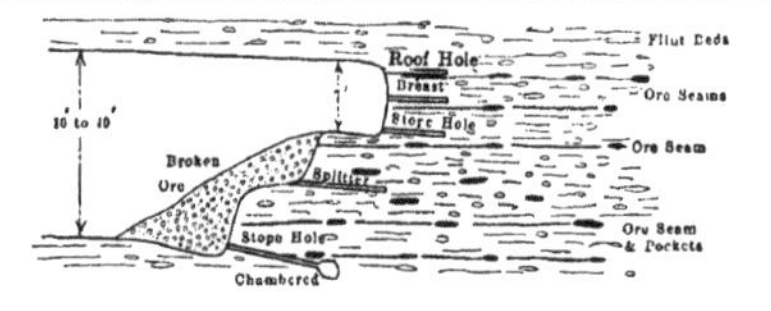

HEADING & STOPE METHOD OF UNDERHAND STOPING

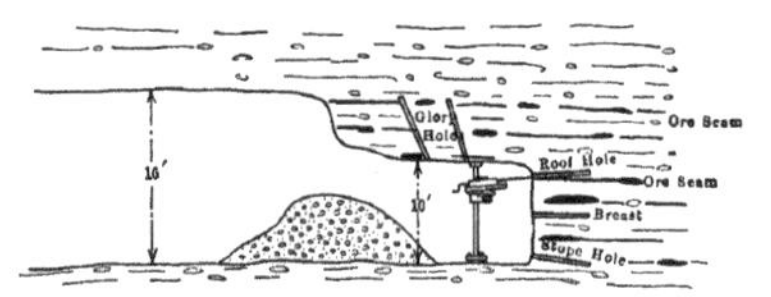

OVERHEAD STOPING IN SHEET GROUND

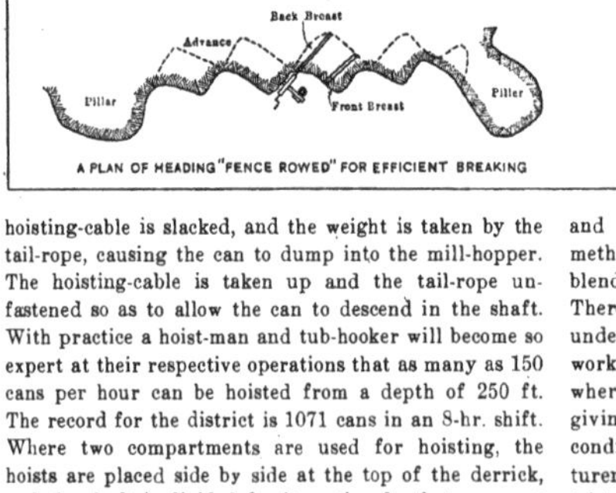

A PLAN OF HEADING "FENCE ROWED" FOR EFFICIENT BREAKING

In order to do the shoveling on the day-shift, and to secure a large tonnage, many of the operators hoist from several shafts, and tram the ore on the surface of the central mill-hopper by means of inclined surface-trams; some few aerial trams are used, but the level topography and short hauls have made the inclined tram most popular.

The cost of hoisting is between 6 and 7c. per ton; surface tramming costs 1c. per ton. Table 4 shows a distribution of hoisting-cost for four months in 1917:

TABLE 4

Tons hoisted 136,272

	Amount	Cost per ton of ore
Hoisting-engineer	$2,316.07	$0.0170
Tub-hooker	2,115.57	0.0155
Cable	720.50	0.0052
Hoister-repairs	429.92	0.0033
Hoist-derrick repairs	110.85	0.0008
Shaft-lacing repairs	99.16	0.0007
Steam for hoister	2,744.92	0.0202
Slickers for hooker	140.30	0.0010
Oil and waste	73.16	0.0005
Total	$8,750.45	$0.0642

PUMPING. The serious problem is not the amount of water to be raised, but the acid character of the water. It is seldom that over 1000 gal. per min. has to be pumped at any one mine. In sheet-ground the mines are connected so that many do no pumping at all, a central drainage company being formed. Many types of centrifugal pumps are used; the Pomona and Texas are favored for surface plants, the Texas being used for unwatering mines, while the large-geared duplex or triplex pumps are usually found in the underground pumping-stations.

COSTS. There is no uniform system of accounting. At present operating costs, which include mining, milling, pumping, miscellaneous, and administrative, are from $1.20 to $1.75 per ton of ore mined. On account of the high prices of supplies and the scarcity and inefficiency of labor, costs are gradually increasing.

MAGNETIC SEPARATION. C. E. Siebenthal of the U. S. Geological Survey states that roasting and magnetic separation continues to be the general method of eliminating the iron from the marcasite-blende ores of the upper Mississippi Valley district. There were nine roasting-plants in operation and one under construction in 1915. Some of these did custom work in addition to handling the output of the mines where located, and some did wholly custom work. By giving a very light roast, the pyrite is separated in a condition suitable for sale to sulphuric-acid manufacturers. An electro-static separator was built in this district in 1908 and operated until destroyed by fire in 1911. Though successful it was not re-built, possibly because it was more expensive than the roast-magnetic plants which also do good work. Moreover the atmospheric conditions in the district are not the best for such a process.

Potash Exploration

Full Text of the Act Authorizing Exploration for Potassium on Public Lands

Be it enacted by the Senate and House of Representatives of the United States of America in Congress assembled, That the Secretary of the Interior is hereby authorized and directed, under such rules and regulations as he may prescribe, to issue to any applicant who is a citizen of the United States, an association of such citizens, or a corporation organized under the laws of any State or Territory thereof, a prospecting permit which shall give the exclusive right, for a period not exceeding two years, to prospect for chlorides, sulphates, carbonates, borates, silicates, or nitrates of potassium on public lands of the United States, except lands in and adjacent to Searles Lake, which would be described if surveyed as townships twenty-four, twenty-five, twenty-six, and twenty-seven south of ranges forty-two, forty-three, and forty-four east, Mount Diablo meridian: Provided, That the area to be included in such permit shall not exceed two thousand five hundred and sixty acres of land in reasonably compact form.

Sec. 2. That upon showing to the satisfaction of the Secretary of the Interior that valuable deposits of one or more of the substances enumerated in section one hereof have been discovered by the permittee within the area covered by his permit, the permittee shall be entitled to a patent for not to exceed one-fourth of the land embraced in the prospecting permit, to be taken in compact form and described by legal subdivisions of the public-land surveys, or if the land be not surveyed, then in tracts which shall not exceed two miles in length, by survey executed at the cost of the permittee, in accordance with rules and regulations prescribed by the Secretary of the Interior. All other lands described and embraced in such a prospecting permit from and after the exercise of the right to patent accorded to the discoverer, and not covered by leases, may be leased by the Secretary of the Interior, through advertisement, competitive bidding, or such other methods as he may by general regulations adopt, and in such areas as he shall fix, not exceeding two thousand five hundred and sixty acres, all leases to be conditioned upon the payment by the lessee of such royalty as may be specified in the lease and which shall be fixed by the Secretary of the Interior in advance of offering the same, and which shall not be less than two per centum on the gross value of the output at the point of shipment, which royalty, on demand of the Secretary of the Interior, shall be paid in the product of such lease, and the payment in advance of a rental, which shall be not less than 25 cents per acre for the first year thereafter; not less than 50 cents per acre for the second, third, fourth, and fifth years, respectively; and not less than $1 per acre for each and every year thereafter during the continuance of the lease, except that such rental for any year shall be credited against the royalties as they accrue for that year. Leases shall be for indeterminate periods, upon condition that at the end of each twenty-year period succeeding the date of any lease such readjustment of terms and conditions may be made as the Secretary of the Interior may determine, unless otherwise provided by law at the time of the expiration of such periods, and a patentee under this section may also be a lessee: Provided, That the potash deposits in the public lands in and adjacent to Searles Lake in what would be if surveyed townships twenty-four, twenty-five, twenty-six, and twenty-seven south of ranges forty-two, forty-three, and forty-four east, Mount Diablo meridian, California, may be operated by the United States or may be leased by the Secretary of the Interior under the terms and provisions of this Act: Provided further, That the Secretary of the Interior may issue leases under the provisions of this Act for deposits of potash in public lands in Sweetwater County, Wyoming, also containing deposits of coal, on condition that the coal be reserved to the United States.

Sec. 3. That in addition to areas of such mineral land to be included in prospecting permits or leases the Secretary of the Interior, in his discretion, may grant to a permittee or lessee under this Act the exclusive right to use, during the life of the permit or lease, a tract of unoccupied non-mineral public land not exceeding forty acres in area for camp sites, refining works, and other purposes connected with and necessary to the proper development and use of the deposits covered by the permit or lease.

Sec. 4. That the Secretary of the Interior shall reserve the authority and shall insert in any preliminary permit issued under section one hereof appropriate provisions for its cancellation by him upon failure by the permittee or licensee to exercise due diligence in the prosecution of the prospecting work in accordance with the terms and conditions stated in the permit.

Sec. 5. That no persons shall take or hold any interest or interests as a member of an association or associations or as a stockholder of a corporation or corporations holding a lease under the provisions hereof, which, together with the area embraced in any direct holding of a lease under this Act, or which, together with any other interest or interests as a member of an association or associations or as a stockholder of a corporation or corporations holding a lease under the provisions hereof, or otherwise, exceeds in the aggregate in any area fifty miles square an amount equivalent to the maximum number of acres allowed to any one lessee under this Act; that no person, association, or corporation holding a lease under the provisions of this Act shall hold more than a tenth interest, direct or indirect, in any other agency, corporate or otherwise, engaged in the sale or resale of the products obtained from such lease, and any violation of the provisions of this (sec) tion shall be ground for the forfeiture of the lease or interest so held; and the interests held in violation of this provision shall be forfeited to the United States by appropriate proceedings instituted by the Attorney General for that purpose in the United States district court for the district in which the property or some part thereof is located, except that any such ownership or interest hereby forbidden which may be acquired by descent, will, judgment, or decree may be held for two years, and not longer after its acquisition.

Sec. 6. That any permit, lease, occupation, or use permitted under this Act shall reserve to the Secretary of the Interior the right to permit for joint or several use such easements or rights of way upon, through, or in the lands leased, occupied, or used as may be necessary or appropriate to the working of the same, or of other lands containing the deposits described in this Act, and the treatment and shipment of the products thereof by or under authority of the Government, its lessees, or permittees, and for other public purposes: Provided, That said Secretary, in his discretion, in making any lease under this Act may reserve to the United States the right to dispose of the surface of the lands embraced within such lease under existing law or laws hereafter enacted, in so far as said surface is not necessary for use of the lessee in extracting and removing the deposits therein: Provided further, That if such reservation is made it shall be so determined before the offering of such lease; that the said Secretary, during the life of the lease, is authorized to issue such permits for easements herein provided to be reserved.

Sec. 7. That each lease shall contain provisions deemed necessary for the protection of the interests of the United States, and for the prevention of monopoly, and for the safeguarding of the public welfare.

Sec. 8. That any lease issued under the provisions of this Act may be forfeited and canceled by an appropriate proceeding in the United States district court for the district in which the property or some part hereof is located whenever the lessee fails to comply with any of the provisions of this Act, of the lease, or of the general regulations promulgated under this Act and in force at the date of the lease, and the lease may provide for resort to appropriate methods for the settlement of disputes or for remedies for breach of specified conditions thereof.

Sec. 9. That the provisions of this Act shall also apply to all deposits of potassium salts in the lands of the United States which may have been or may be disposed of under laws reserving to the United States the potassium deposits with the right to prospect for, drill, mine, and remove the same, subject to such conditions as to the use and occupancy of the surface as are or may hereafter be provided by law.

Sec. 10. That all moneys received from royalties and rentals under the provisions of this Act, excepting those from Alaska, shall be paid into, reserved, and appropriated as a part of the reclamation fund created by the Act of Congress aproved June seventeenth, nineteen hundred and two, known as the reclamation Act, but after use thereof in the construction of reclamation works and upon return to the reclamation fund of any such moneys in the manner provided by the reclamation Act and Acts amendatory thereof and supplemental thereto, fifty per centum of the amounts derived from such royalties and rentals, so utilized in and returned to the reclamation fund shall be paid by the Secretary of the Treasury after the expiration of each fiscal year to the State within the boundaries of which the leased lands or deposits are or were located, said moneys to be used by such State or subdivisions thereof for the construction and maintenance of public roads or for the support of public schools.

Sec. 11. That the Secretary of the Interior is authorized to prescribe necessary and proper rules and regulations and to do any and all things necessary to carry out and accomplish the purposes of this Act.

Sec. 12. That the deposits herein referred to, in lands valuable for such minerals, shall be subject to disposition only in the form and manner provided in this Act, except as to valid claims existent at date of the passage of this Act and thereafter maintained in compliance with the laws under which initiated, which claims may be perfected under such laws: Provided, That nothing in this Act shall be construed or held to affect the rights of the States or other local authority to exercise any rights which they may have to levy and collect taxes upon improvements, output of mines, or other rights, property, or assets of any lessee.

Sec. 13. That the Secretary of the Interior is hereby authorized and directed to incorporate in every lease issued under the provisions of this Act a provision reserving to the President the right to regulate the price of all mineral extracted and sold from the leased premises, which stipulation shall specifically provide that the price or prices fixed shall be such as to yield a fair and reasonable return to the lessee upon his investment and to secure to the consumer any of such products at the lowest price reasonable and consistent with the foregoing: Provided, That such lease issued under this Act shall also stipulate that the President shall have authority to so regulate the disposal of the potassium products produced under such lease as to secure its distribution and use wholly within the limits of the United States or its possessions.

Passed the Senate August 10, 1917.

Attest: James M. Baker,
Secretary.

The Supply of Gasoline

By MILTON A. ALLEN

A shortage of gasoline has been threatened recently and it has been announced that if the consumption was not reduced voluntarily during the Wár, legislative regulation would be necessary.

Prior to the War the United States produced 65% of the world's petroleum. Since then the supplies of Russia, Rumania, and Austria have been cut off, thereby placing the burden of production upon the United States. Besides this there is the ever-increasing domestic demand.

In 1914 the marketed production of American petroleum was 250,000,000 barrels and this increased to 281,000,000 bbl. in 1915. Our estimated resources are 5,500,000,000 bbl., or 20 years' supply at the present rate of consumption. The future source of gasoline to supply the increasing consumption and to take the place of our depleting resources is a serious problem. Most of the crude petroleum yields only a small proportion of gasoline and burning naphtha or oils that lend themselves readily to being manufactured into gasoline by cracking. The Mexican, Californian, Texan, and some Mid-Continental oils are of this class. Gasoline must therefore be supplied by the high-grade paraffine-base oils. The best of these yield only 15 to 25% of refined gasoline and naphtha.

The Appalachian and parts of the Mid-Continental fields yield the bulk of these paraffine-oils, but this represents less than 40% of the total crude petroleum output in the United States. From this it is seen that our gasoline supply is limited and restricted. During the year ended June 1915 the United States exported 244,000,000 gal. of gasoline and naphtha that could have been diverted to home consumption. It will be interesting to consider the possible source of gasoline other than crude petroleum, casing-head gasoline, and cracked gasoline.

The sources from which oil can be obtained in quantity for making motor-spirit are natural bitumen, highly volatile coal (lignite and cannel), and shale.

Prior to the discovery of oil in this country in the 'fifties, the cannel-coal deposits of Kentucky were mined and the coal distilled as it is in Scotland today. The oil yielded excellent lighting-oil or kerosene, lubricating oils, and tar. Similarly the albertite deposits of Canada were worked and distilled. Both these industries were forced out of the market when natural petroleum was produced in quantity.

For many years the Scottish oil-shales have been distilled at low temperatures (1000° F.). The richer shales, from which as much as 40 gal. per ton was obtained, are now exhausted; the average yield is now between 20 and 25 gal. only. In the process of distillation used, a large yield (45 lb.) of ammonium sulphate is obtained. The small yield of gas is scrubbed and used in the distillation process and for power. The crude oil is refined, yielding naphthas, burning-oils, gas-oils, lubricating-oils, and paraffine-wax. In spite of the increased price of chemicals, the improvements in retorts and in recovery, together with the installation of electric power have enabled the industry to operate at a profit. It is a fact, however, that in depth the Scottish shale yields less oil.

Large quantities of natural bitumen, similar to albertite, are found in Utah, Oklahoma, and Colorado. It is estimated that there are 32,000,000 tons of uintaite in the known veins in Utah alone. Analyses of albertite and uintaite are given herewith:

	Albertite	Uintaite
	%	%
Volatile	60.0	56.5
Fixed carbon and ash	39.5	43.0
Sulphur	0.5	1.5

Experiments made in Scotland under working conditions showed a yield of 134 gal. of oil per ton of albertite. The yield of uintaite should approach that of albertite.

The market demands for cannel-coal are limited, hence the present output is small; it is used to enrich gas in gas-works and as a grate-coal in some localities. For this reason large areas of cannel-coal in Ohio, Kentucky, West Virginia, Indiana, and Tennessee have not been developed. These coals vary in volatile content from 35% to 54%, and in many instances they approach 60%. It is estimated that there are millions of tons of workable cannel-coal in Kentucky alone.

An analysis of a typical Kentucky cannel-coal shows

	%
Moisture	2.4
Volatile	50.3
Fixed carbon	43.3
Ash	3.4
Sulphur	0.6

This coal holds considerably less ash than the European cannels, hence the greater possibilities of disposing of the coke by-product if the coal is distilled like the Scotch shale. The coal residue could also be burned in producers and electric power generated on the side.

Experiments in distillation at low temperature of high volatile cannel-coal show that it yields from 35 to 60 gal. of dry tar-oil, the amount depending on the percentage of volatile in the coal and the temperature at which the destructive distillation is conducted. These cannel-coal tar-oils are high in paraffine, and the heavier fraction could readily be cracked to yield a high percentage of motor-spirit.

For many years in England cannel-coal was used in gas-works as an enricher; but if large quantities were employed the tar by-product was difficult to sell to the tar-distillers because of its high percentage of paraffine. This bears out recent experiments, except that the percentage of paraffine hydro-carbons is considerably increased when the coal is treated at a low temperature as in the Scotch shale distillation. The advantages of cannel-coal distillation for tar-oils over the Scotch shale would be that the residue or coke could be used either for power purposes on a large scale, sold directly, or ground and made into briquettes. Ammonium sulphate would be a by-product as in the Scotch shale industry and the yield would probably be large. Experiments on albertite yielded 65 lb. ammonium sulphate per ton.

Apart from the tar-oils the gas when scrubbed yields benzene and toluene and some light oils. The benzene and toluene could be mixed with some of the light oils similar to the use to which casing-head gasoline is put. Benzol has been used in France as a motor-spirit for many years.

In many of our central, northern, and western States, there are hundreds of square miles of lignite. Canadian lignite deposits, it seems, are inexhaustible. Lignite is useless there as a fuel because of its low heating-value. If the lignites are subjected to low-temperature distillation they will yield from 30 to 40 gal. of dry tar-oils, depending on the volatile content of the lignite and the temperature of the destructive distillation. The residual coke can be used similarly to that from cannel-coal.

In Nova Scotia, New Brunswick, and in our western States are large areas in which oil-shale is known to exist. These shales resemble the oil-shales of Scotland. Experiments made on shales from Nova Scotia and New Brunswick yielded from 40 to 65 gal. oil per ton, the ammonium-sulphate by-product varying from 67 to 110 pounds. The following is an analysis of these shales:

	%
Moisture	1.00
Volatile	45.32
Fixed carbon	1.29
Ash	50.69
Sulphur	1.70

In the Carboniferous system of Kentucky above the cannel-coal deposits are bands of shale similar in composition; also there are separate beds of workable shale. These separate shales show the following analysis:

	%
Moisture	0.9
Volatile	43.0
Fixed carbon	13.1
Ash	42.5
Sulphur	0.5

A yield of 75 to 85 gal. of oil has been obtained from French shale deposits having the following analysis:

	%
Volatile	44.4
Fixed carbon	25.0
Ash	30.6

No sulphur exists in the oil. This resembles a high-ash cannel-coal.

From the results given, it will be seen that the shales in almost all cases give a high yield of oil. Prior to the War the oil from the Scotch shale was not cracked for motor-spirit, but there is every reason to believe that this now is being done.

In crude experiments in the cracking of lignite tar-oils, after the extraction of the light naphthas, a yield of 40% of motor-spirit has been obtained. It can be expected that the tar-oils of cannel-coal and the shale-oils, which carry a larger percentage of paraffine-oil, will yield a larger percentage of motor-fuel on cracking, and will be more amenable to such treatment. The processes now being used in the cracking of petroleum in this country are the Burton, Hall, and Rittman. The Hall process is being applied successfully in England and Russia.

The War has created a demand for benzol and toluol that did not exist before. Therefore, they were not recovered in the by-product plants. It is doubtful if these products will fall below the price of gasoline after the War because of their value as motor-spirit. Gasoline will probably not go lower.

Considering the low costs of mining and treatment of the Scotch shales by the uneconomical method practised, it is difficult to see why the shales, natural bitumens, cannel-coals, and lignites would not yield large quantities of gasoline profitably at the present price, and at the same time develop resources that are idle for want of profitable use, by the adoption of a modification of the most recent continuous retorts of the Woodall-Duckham or Glover-West type to suit low-temperature work, together with a modern cracking plant.

A NEW METAL as light as aluminum and as strong as steel is now made in the electric furnace. It possesses unusual anti-corrosive properties also. The U. S. Government is testing a pump made of this new metal, for pumping sea-water. It is said to be a chemical union of aluminum with other elements and not an alloy. The name of the new product is 'acieral', probably from the French word *acier*, steel, with the *al* to signify aluminum, or 'steel-aluminum.' It is understood to contain about 92 to 97% aluminum, with 3% nickel and other metals. The chemical reaction is obtained by the use of a secret flux. Its color is white and it has a compact texture similar to the best steel. It is as sonorous as bell metal and has a good conductivity for use in electrical work. Castings made from it show high tensile strength, and when rolled into rods or sheets its strength nearly doubles. It can be cast in sand or die cast, with or without pressure, and it flows well when being cast. It may be forged hot or cold, and can be annealed, drawn, or rolled into shapes. It is resistant to all acids, hot or cold, except hydrochloric, and it resists oxidation by the air when damp, at all temperatures, even when salt is present. Its greatest use is for aeroplanes and automobiles. With aluminum selling at 60c. per lb., the new metal will sell at $1. It was discovered by M. de Montby.

SUTTER CREEK, CALIFORNIA

OLD EUREKA MINE.—WILDMAN-MAHONEY.—SOUTH EUREKA.

At the Old Eureka mine, preparations for active development are proceeding satisfactorily, and the portions of the shaft below the 1600-ft. station that had but one compartment, has been enlarged and re-timbered to a point 1900 ft. from the surface, leaving only 260 ft. to reach the bottom. By means of this work the entire depth of the shaft will have three

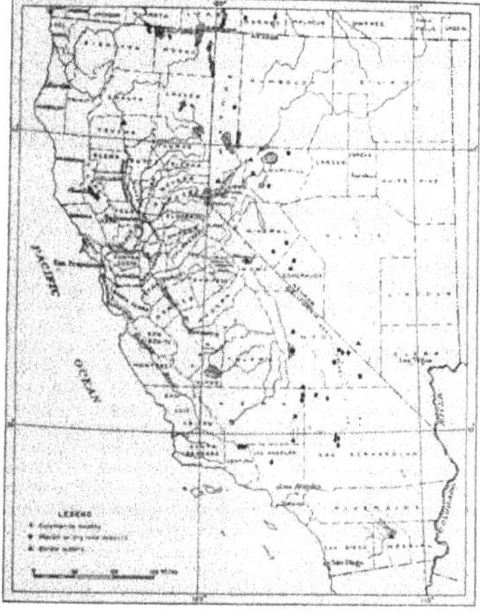

MAP OF CALIFORNIA

compartments and the large skips can be used from the surface all the way. Three important levels have been re-opened with encouraging results, namely, the 800, 1200, and 1700. On the 1200-ft. level, about 265 ft. north of the shaft on orebody 30 ft. wide, assaying about $25 per ton, shows in the face of the drift and gives indication of being the same vein as that on the 800-ft. level. Ore of good grade has been found on the 1700-ft. level. Some of the drifts and cross-cuts are caved, but in most parts of the mine the old timbers appear to be in an excellent state of preservation. Where deemed expedient, parallel drifts and cross-cuts will be run near the caved ground in preference to re-opening it.

Michigan capitalists, several of whom are interested in the Old Eureka mine, have made payment of part of the purchase price of the Wildman-Mahoney-Lincoln group and the Lincoln Con. Mining Co. will pay its stockholders a 10% dividend. This company has made application to the Corporation Commissioner for authority to distribute $60,655.50 to stockholders, their financial statement showing $57,000 cash on hand and $68,000 in marketable bonds. The group of mines adjoins the Eureka on the north and could be economically operated under one management. The Wildman-Mahoney and Lincoln mines are credited with a production of over $7,000,000 and have been opened by means of four shafts: the Lincoln, 2000 ft. deep; the Mahoney, 1200 ft.; the Wildman, 1400 ft., and the Emerson, 619 ft. On the Wildman 1400-ft. level a vein 160 ft. wide, assaying $3.50 per ton, is said to have been discovered shortly before work ceased in the Old Eureka. The expensive upkeep of the Wildman shaft was a large factor in preventing the successful mining of this orebody. The Emerson vertical shaft was sunk in diabase 1000 ft. east of the Wildman shaft with the object of cutting the vein at 2300 feet.

The South Eureka shaft is likely to be deepened. The working-force has been gradually reduced from 350 to less than 20 men. Those now on the pay-roll are engaged in keeping the shaft in repair, operating pumps, and getting the mine in readiness for sinking. While large deposits of low-grade ore are in reserve, it has become necessary to go deeper for sufficient good ore to mix with the low-grade. For some time, negotiations have been pending for the sale of the property and it is not yet known whether future development work will be done by the present company or by one of the companies that have recently investigated the mine.

CRIPPLE CREEK, COLORADO

CRESSON CON. AND GOLDEN CYCLE DECLARE DIVIDENDS.

The directors of the Cresson Consolidated have declared the regular monthly dividend of 10 cents, payable on October 10. A. E. Carlton, the president of the company, announced a net profit for August of $220,000, and ore-reserves estimated at $4,500,000, with ore in transit and cash in bank to the amount of $1,280,000.

The monthly dividend of the Golden Cycle company, of 3c. per share, amounting to $45,000, has been declared.——The Montana vein is reported cut at the bottom or 1650-ft. level of the Dillon shaft of the Granite mine, and ore of marketable value has been exposed. The company will deter development until such time as the lateral cuts the Bobtail vein, which extends into the Dillon from the Portland estate and is producing heavily on the 1550 and 1450-ft. levels of the Dillon claim.

The Hahnewald brothers, lessees of the El Paso Gold King, in Poverty gulch, were shipping again this week. New orebodies have been opened up in the deeper levels and sinking is now proposed. The El Paso Gold King shaft is 1000 ft. deep and it is proposed to sink a 250-ft. lift for a starter. Edwin Gaylord, lessee of the Forest Queen mine on Ironclad hill, is sinking below the 600-ft. level. He continues to ship from the caved ore between the 400 and 600-ft. levels; this ore averages $15 to $20 per ton.——The new Hurst shaft on the Ocean Wave, on Battle mountain, is being sunk with two shifts and the work will continue to 500 ft. The property is under lease to the Ocean Wave company.——The Unity Leasing Co., operating the Nightingale of the Stratton estate, is sinking a new vertical shaft on Raven hill. The Nightingale extends to Bull hill. The south end of the claim adjoins the Maggie mine of the Cresson and the Gold Soverign, both producing mines.

A depth of 500 ft. has been attained in the shaft on the Longfellow No. 2 of the Stratton estate and a station is now being cut at this depth. The property is under lease to the Excelsior company, which has opened up ore on the 500 and 600-ft. levels of the Golden Cycle mine. A cross-cut will be driven to this orebody from the 500-ft. level of the Longfellow. ——The Dig Gold Mining Co., M. B. Burke president and general manager, has installed an electric hoist and compressor on the Alpha and Omega group on the southern slope of Gold hill and is now engaged in sinking the shaft from the 150-ft. point.

SPOKANE, WASHINGTON

ENTERPRISE LEASED.—CALEDONIA AND STANDARD DECLARE DIVIDENDS.

With silver at over a dollar and prospects of further advance the silver-producing mines are in clover. The Inter-state-Callanan and the Bunker Hill mines of the Coeur d'Alene may be taken as illustrations. The former earns from 3 to 20c. per ton with each cent of advance. This means that it receives $1.59 to $10.60 more per ton at present prices than two years ago. The Bunker Hill & Sullivan produces an average of 22 oz. per ton or $11.66 more than two years ago, and the Hecla 26 oz. per ton of ore shipped, or $13.78 per ton. Hecla would receive $371,000 more for its annual output than two years ago. These profits are not net, for the cost of production has been increased.

The profits of British Columbia mines are greater. The crude and concentrated ores of mines in the lead belt contain from 30 to 90 oz. silver per ton. The price received is $15.90 to $47.70 more than the average price of two years ago. The mines of the dry belt in British Columbia average 200 oz. silver per ton. This makes the gross value more than $205 per ton for silver alone, an increase of $106 over the price of two years ago and an increase of $74 for the current year.

With a score of teams and trucks, each hauling 10 tons of ore at a time, the American Minerals Production Co. is rushing shipments of magnesite from its quarries at Valley, Stevens county, Washington, without waiting for the railroad to be built into the district. The railroad will be standard gauge and have a length of 16 miles. Steel has been delivered for the first 4½ miles. A contract for the work is being executed by the General Construction Co., of Spokane. It will be fulfilled early in January. The line will be operated between Valley and the quarries. A connection will be made at Valley with the Great Northern. Grading has commenced.

Net earnings of the Slocan Star Mines, in the Slocan district, B. C., for August were $4500, according to a preliminary estimate. This is about the same as reported for July. Shipments in September aggregated 110 tons of zinc and 40 tons of lead up to September 14. The carload shipped this month contained 60% lead and 100 oz. silver per ton. The Enterprise mine, in the same district, which Finch & Campbell of Spokane sold for $375,000, has been taken over by a new company on a lease and option for $125,000. The new company is the Seattle-B. C. Lead Silver Co., organized by H. F. Millard, of Valdez, Alaska. Finch & Campbell worked the property for many years, but a high percentage of zinc made the ore less profitable than was expected. The Spokane men sold the property to the British Columbia Mines Co., of London, which worked it profitably by hand-sorting. This company is now selling to Millard's corporation. It is said that the mine has produced about $600,000 in the past. Mr. Millard estimates the ore in reserve at 40,000 tons; this will sort 10,000 tons and produce 3,891,000 lb. lead, 4,337,000 lb. zinc, and 182,400 oz. silver.

Another car of high-grade antimony ore, from the Pine Creek district, has been shipped to the smelter by the Coeur d'Alene Antimony Mining Co. Pine Creek is one of the few districts in the United States where antimony ore is produced. Its production is small, but better prices for the metal have caused two or three properties to be prospected.

An expenditure of $300,000 for the purchase of the Independence mine in the Wood River district, Idaho, and $200,000 more for its development hinges on the close of negotiations by the Federal Mining & Smelting Co. with the owners. An option signed at Denver recently gives the Federal a right to take possession on November 15. In the meantime all the ore that can be broken will be shipped by the owners. The agreement must be ratified by officials of the Federal company and confirmed by stockholders of the Independence company. The owners of the Independence are Mrs. H. J. Allen of Hailey, and Dr. Harper of Chicago, who own three-fifths, and Charles F. C. Rutan of New York and others, who own two-fifths.

The Caledonia Mining Co., in the Coeur d'Alene, has declared its regular monthly dividend of $78,150, disbursement to be made on October 5. This payment will be at the regular rate of 3c. per share. It will bring the total of payments to $2,678,941, of which $781,500 will have been paid this year. The directors of the Standard Silver-Lead Mining Co. have declared the regular quarterly dividend of $100,000, payable on October 15. This disbursement will be the third of $100,000 made this year at the rate of 5% quarterly on the capitalization of $2,000,000. The forthcoming disbursement will bring the total to $2,700,000, or $700,000 more than the par value of the issue.

HOUGHTON, MICHIGAN

HANCOCK CONSOLIDATED.—CALUMET & HECLA.

Operations at the Wyandot continue to be encouraging. The lode is 30 ft. wide, strong, and persistent. Another mill-test will be made when the bins are filled.——Production from the Hancock Consolidated for September will be 250,000 lb. Electric tramming on the lower levels of No. 7 Quincy opens up some of the richest rock this mine can handle. It may be necessary to put an auxiliary shaft down to mine some 600 acres of particularly good lode. In No. 2 shaft the openings are getting into more settled ground, having passed through the disturbed conditions of the Pewabic.——The Wolverine property is recovering 67% of refined copper from its mineral and averaging under 17 lb. per ton. Hardly any waste-rock is hoisted to surface. Last year the total linear feet of sinking, driving, and raising was 1924. While total costs last year were 12c. per pound, that figure has been increased 40% since.——Development on the Butler lode at the Lake mine are so encouraging that a separate shaft will be sunk. In the main shaft the openings all are in good ore.

A careful survey of the Michigan copper district indicates that no mines will be shut-down because of the Government price for copper. The majority of the mines here will be operated to their full capacity. There is a shortage of labor and not much hope of any substantial increase excepting such as comes in the fall and winter after the completion of farm work. Costs will be higher this year than ever before in the history of Lake Superior mining, in spite of the fact that there is no unnecessary construction. The Calumet & Hecla, for itself and subsidiaries, is making extensive improvements, particularly at Lake Linden. The Mutual plant at Tamarack city is one notable instance of improvement for subsidiaries. So, also, is the new coal-dock with its modern equipment. Quincy and Isle Royale are completing a number of new homes. The Copper Range and the Mass companies did some sensible and practical work along this line early in the year.

Lake Superior copper output is likely to be 30,000,000 lb. less than last year.

TORONTO, ONTARIO

McINTYRE REPORT.—SCHUMACHER RESUMES WORK.—INCREASED ACTIVITY AT HOLLINGER AND KIRKLAND LAKE MINES.—HEAVY SILVER SHIPMENTS FROM COBALT.

Remarkable progress is shown in the McIntyre, at Porcupine, by the report covering a period of 15 months, ended June 30. Despite labor shortage and increased mining cost, ore to the value of $1,954,793 was treated at a profit of $725,-790. The mining cost per ton was $2.99 and the milling cost 88c. The most important feature is the increased value of the ore-reserve. The estimated value of this on June 30 was $4,943,000, as compared with $2,247,000 on March 31, 1916. During August the ore milled made a new high record, being 15,410 tons averaging $9.52 per ton, with a production of $141,394. The Schumacher resumed operations last month. Approximately 3400 tons of ore was treated during the month ended September 15 with a recovery of $23,000. Most of the new milling equipment is now in place and the complete plant is expected to be in operation by the end of October. It will have a capacity of 200 tons per day. The Dome is operating at great disadvantage on account of labor shortage, the mine force having been reduced to 200 men. The underground equipment has been improved; now it has a capacity of nearly 4000 tons per day, double that of the mill. The Hollinger Consolidated shows increased activity, and the mill is stated to be working close to capacity. Reports conflict as to the results of the development work in progress, some big figures being current as to the additions to the ore-reserve. At the McEnaney, formerly known as the Hollinger Reserve, a good grade of ore has been cut in the winze sunk from the 200 to the 500-ft. level. The ore above the 300-ft. level shows a gold-content of $13 per ton. A mill will shortly be built with a capacity of 75 tons. The Coniagas, at Cobalt, has exercised its option on the Maidens-MacDonald mine, which will be worked in connection with the Anchorite. A mill is being built on the Premier Langmuir barite-silver property, about 15 miles south-east of Porcupine, which is expected to be in operation by the end of the year. It will produce about 30 tons of barite per day. A shaft is down 100 ft., and some driving has been done. Considerable native silver, which will be recovered as a by-product, occurs along the walls of the vein.

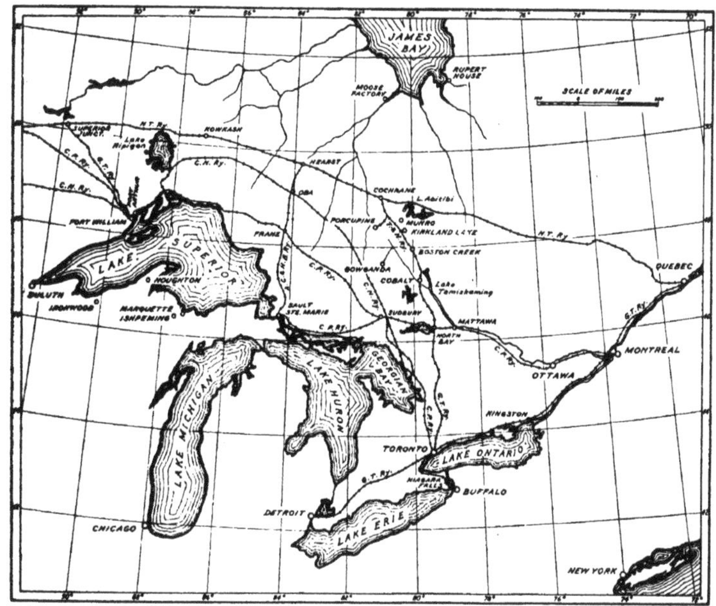

MAP OF ONTARIO AND THE GREAT LAKES

The Tough Oakes, at Kirkland Lake, has 126 men on its payroll, and during August the mill treated 3400 tons of ore. At the Lake Shore the foundations for a mill are being set; as the machinery is on the ground, it should be completed in about a month. Development is in progress at a depth of 400 ft. A company under the title of the Ontario Kirkland is being organized with $1,500,000 capital to operate the Hurd mine, on which five promising veins have been opened. The Orr claims are being developed by the Kirkland Porphyry, a newly formed company. A shaft is being sunk and tree gold is showing.——The Lucky Cross at Swastika is being wound up. The property has been taken over by the principal bondholders, who will form a new company.

Heavy shipments of ore are being made from Cobalt, where the improved silver market has led to the opening up of properties that formerly were unprofitable. Among these are the Silver Queen, containing a large body of low-grade ore, the Provincial, and the National. On the last of these a flotation-plant is treating 100 tons daily. Underground work will be resumed. The Kerr Lake, during the year ended August 31, produced 2,586,532 oz. silver, being the highest yearly production in its history. The net profit is estimated at $1,166,358. At the Peterson Lake the plant lately in operation on the Susquehanna lease has been removed to the Gould lease, where the 200-ft. shaft has been unwatered preparatory to resuming work. The shareholders of the Ophir have authorized the increase of the capital from $1,000,000 to $1,500,000, which will enable development to be continued for at least a year.

There is renewed activity in prospecting in the Portage Bay and Maple Mountains silver district, where some good surface showings are reported. It is hoped that the high price of silver will induce speculators to furnish the capital necessary for the investigation of many promising claims in these areas.

MANHATTAN, NEVADA

EXTENSIVE DEVELOPMENT AT WHITE CAPS AND UNION AMALGAMATED

The most important work in the White Caps has been the driving of the south cross-cut from the 500-ft. level to catch the downward extension of the east orebody. During the past week the work has been in limestone. Assays from the face have run as high as $12.20 in gold. A slight turn has been made in the direction of the cross-cut, to carry it deeper into the limestone. There has been some surprise that the east orebody has not been already exploited from the bottom level, but a glance at the map of the underground workings will show that the edge of the orebody, if continuing from the fourth level downward at the same pitch and dip, would be several feet from the face of the south cross-cut, and the heart of the ore would be at least 40 ft. away. The hanging-wall cross-cut on the third level has been extended 24 ft., and now is out a distance of 58 ft.; the 310 raise has been put up 15 ft.; the cross-cut has been extended 18 ft. from the fourth level. The south cross-cut on the fifth level is in 257 ft. The footage for the week was 137. The daily tonnage for the mill is still about 100, but the grade is being raised. A large trommel has been erected below the rolls and a closed circuit formed, so that all the products reaching the roaster pass through a 6-mesh screen. The new Case oil-burners that are being tested in the roaster have given satisfaction. The entire roaster will be equipped with these burners.

In the Manhattan Consolidated mine rapid work has been made at the fourth station and the station-timbers are all in place. A cross-cut south has been started from the third level to develop the downward extension of the orebodies, proved to the south and west. The most important orebodies are known locally as the Mushett vein and the Kendall vein. From the fourth level a drift east has been started to expose the orebody, already well developed at the 200 and 300-ft. levels. An underground survey shows that the downward extension of the east orebody should be reached 250 ft. from the shaft.

The operations in the Union Amalgamated for the week are as follows: from the 600-ft. level the north cross-cut on fissure advanced 12 ft.; the east drift from No. 7 north cross-cut east 25 ft. Owing to the distance of the work from the shaft and also to atmospheric conditions, recent work has been considerably retarded by gas. A blower is now being installed, and as soon as this is in operation development work will show a marked increase. The face of the north cross-cut on the fissure is now in low-grade quartz and should shortly cut the north orebody. The east drift has exposed a mill-grade of ore, the shoot being 80 ft. long and No. 11 north cross-cut has been started to prove the extent of the ore toward the footwall of the vein. No. 9 south cross-cut east has cut the hanging wall of the Swanson orebody and some rich ore has been developed, though the orebody is broken up at the point where struck. Owing to the fact that the recent work has exposed so many faces of mill-ore, it has been necessary to relieve the congestion of the mine-bins and during the week the mill-bins have been filled with this ore.

The joint operating shaft of the Extension and Zanzibar companies has now reached a depth of 130 ft. During the week 21 ft. has been added to its depth, entirely in limestone.

MAYER, ARIZONA

BIG LEDGE.—COPPER QUEEN.—ARIZONA BINGHAMPTON.

The Big Ledge Copper Co. announces that construction work is to be resumed by October 1 at the smelter. The second furnace will have a capacity of 400 tons, and the total furnace capacity for the two furnaces will be 600 tons per day. A third furnace, two converters, and one reverberatory are also to be installed at a later date. Three shifts are to be put on two of the company's mines, the Butternut and the Henrietta. On the 200-ft. level of the Butternut, the orebody is said to be 50 ft. wide. The company plans the erection of a flotation-mill at the mine. A wire-rope tram has been purchased for the Butternut mine. The orebodies in the Henrietta are not as large as in the Butternut but are higher grade. The company has increased its capital stock to $15,000,000 and on September 27 will complete the purchase or consolidation with the Great Western Smelter Co. The policy of the company will be to encourage the small mine-owners by providing a market for their ores.

The Copper Queen Co. is developing more ore in the lower workings of the mine. Claude Ferguson, mine superintendent, has outlined the system of development. The two deepest shafts have been unwatered and extensive old workings are being thoroughly sampled. The shaft at adit No. 2 is to be sunk 200 ft. deeper. It is estimated that 25,000 tons of milling-ore has been blocked out.

The Arizona Binghampton Co. has purchased a diamond-drill equipment and will prospect the lower workings with cross-holes before running mine cross-cuts. The second unit at the flotation-mill has brought the capacity to 250 tons. The high-grade ore that was being sent to the Humboldt smelter is now being run through the mill with the lower grade and satisfactory results have been obtained.

The district is active, especially in the vicinity of Copper mountain. There is a demand for silver properties.

MEXICO

AMERICAN SMELTING & REFINING CO.

According to C. L. Baker, general manager, there is no immediate prospect of the smelter of the A. S. & R. Co. at Chihuahua being re-started. Mr. Baker said: "A little construction work is being done at Chihuahua, but it is not with a view to the re-opening of the smelter, and the idea circulated that Mexican officials have offered the protection of General Francisco Murgia's forces is all poppycock. The only reason the smelter is not being run is lack of coke. No Mexican officials have approached me with the suggestion that the Mexicans take over the smelters and reduction works is not to be considered. I have no statement to make concerning the smelter situation. The work of the American Smelting & Refining Co. will continue as before. We have three smelters operating at Monterrey, Matehuala, and Aguas Calientes."

ALASKA

(Special Correspondence.)—The Rainy Hollow lode properties and the Porcupine gold placer properties that are being operated by Col. W. L. Stevenson, under option, are doing well. Ore taken out of Rainy Hollow is said to assay $260 per ton, mostly in silver, although the copper value is about $60. The hydraulic plant on the Porcupine river is said to be producing about $1000 per day. The Rainy Hollow properties were optioned from a dozen claimants on condition of development, and this has been in progress all summer. The Porcupine placer was secured on option from the Porcupine Mining Co., a corporation organized by Col. Horvey Conrad several years ago. Martin Nolan, representing Col. Stevenson, is in charge. ——The Kennecott mine is now employing 450 men.——The Mother Lode mine has produced about $5,000,000 worth of ore this year. At present the company is erecting a steam-electric plant at the mouth of McCarthy creek, and is building an auto-truck road up the creek 14 miles to the lower terminal of the tramway. They are also building a 250-ton mill. The present production of the Mother Lode is about 250 tons per day, and the working force is about 200 men.——Pat Bonner has a bond on the Erickson property, which shipped two carloads of high-grade copper ore to the Tacoma smelter this spring. Mr. Bonner expects to build an auto-truck road from McCarthy to the mine. The mine is 36 miles south-east of McCarthy creek on the Nizina river. There are several hydraulic gold mines in the Nizina district that are working crews of from 10 to 150 men. The principal operators are Frank Kernan, George Max Esterly, and Howard Birch.

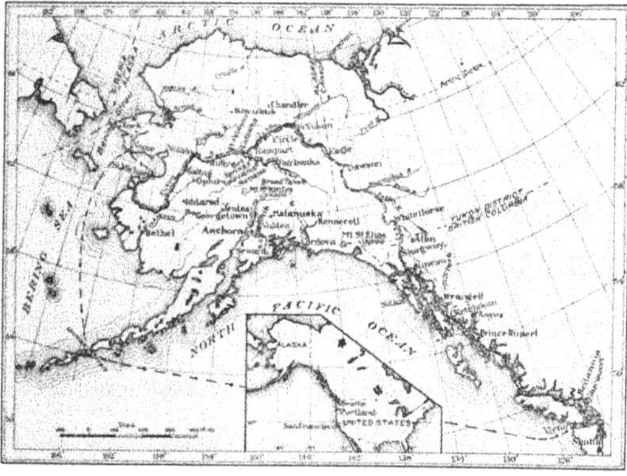

MAP OF ALASKA

The War Eagle mine on the Kuskulana river, owned by the Chitina Kuskulana Copper Co., is being developed by driving two adits, one on the west end of the War Eagle and one on the east end of the Calcite claim. The working force is about 45 men. The plant consists of a 120-hp. boiler, a 100-hp. engine, a 65-kw. generator, and an air-compressor. The ore is chalcopyrite. The company intends to build a spur railroad from the main line, a distance of about eight miles.——The Alaska Copper Co., operating on Nugget creek near the Kuskulana river, is one of the high-grade properties of Alaska. The company has at present a working force of 100 men. Howard W. DuBois is consulting engineer. An auto-truck road is being built from Strelna to the mine, a distance of 18 miles, and the company expects to ship ore soon. The machinery consists of two combustion tandem engines, and a compressor with a capacity of about 500 cu. ft. free air per minute.——The North Midas copper and gold mine is employing 20 men and is operating in the east adit on a vein of high-grade ore. The machinery consists of a semi-Diesel combustion tandem engine, and compressor. The company expects to be shipping ore this year.——The Great Northern Development Co., at Copper Mountain, on the Kuskulana river, has completed 8000 ft. of development work, opening up a large low-grade property. E. F. Gray is general manager. He expects to install a 5000-ton flotation plant, work to commence in 1918. There are 165 claims in this property and a majority of the patents have been issued.——The Big Horn mine, on Big Horn mountain, owned by L. C. Dillman, is being developed by an adit driven on the ore. T. W. Lynch is in charge of the property.——The Climax mine, owned by London & Cape Co., is being developed by two adits. This property consists of 14 claims and was the second property to be patented in the Copper River district. The ore outcrop is 40 ft. wide and runs from 3 to 8% copper. The property joins the east end of the War Eagle.——Hubbard & Elliott Co., on Elliott creek, has a property being developed by a long cross-cut.

Treadwell, September 16.

(Special Correspondence.)—The Alaska Agricultural Industrial Fair held here September 3, 4, and 5 was a decided success, and hereafter will be held annually. The Willow Creek mining district was well represented in the mining section. One of the most interesting exhibits was a quantity of bornite brought from the Kashwitna district by four prospectors, William Trout, Dolf Smith, Anton Stander, and Fred Kelly. Besides bornite the ore contains quartz showing gold in leaf and wire form. The vein is said to be in granite. A number of prospectors have left for the scene of the discovery. ——An expedition carrying 10 men has left here for Snug Harbor to investigate an alleged mineral discovery.——There is a strong possibility that owing to increased cost of materials and abnormal conditions generated by the War the Government will suspend operations on the main line of railroad, but the branch line to the Chicaloon coalfields is nearing completion and will be finished shortly.

Anchorage, September 11.

ARIZONA

(Special Correspondence.)—Of the eleven large smelters in the State, seven are now operating at full capacity, one is reaching full capacity rapidly, while three are idle.

E. L. Jones, Jr., of the U. S. Geological Survey, has been investigating manganese deposits of the State during the past month.

More than 1800 men have come to Bisbee since the beginning of August.

The Oatman mines produced $228,000 during August, the best return for a single month to date.

Oatman, September 29.

COCHISE COUNTY

(Special Correspondence.)—It is announced by Col. T. N. Seger, president of the Golden Reef Mines Co., that fresh exploration is to be undertaken. Samples taken over a distance of 300 ft. on 1000 ft. of outcrop give an average assay of $9.92 per ton.——Owing to the advance in the price of silver, shipments of ore from the Tombstone district are increasing. Considerable manganese ore is also being mined and shipped.——A rich strike of silver ore is reported in virgin ground at the Commonwealth mine, which has been idle for some months. ——The Central Butte Mining Co., of which A. H. Struthers, of Douglas, is manager, is making an excellent showing and is expected to begin production soon. At the Middlemarch mine, seven miles from Pearce, 40 men are now employed; ore is being hauled to Pearce by motor-truck.

Tombstone, September 20.

GILA COUNTY

(Special Correspondence.)—The International smelter is now working at full capacity and handling all the concentrate from Miami and Inspiration. Miners are returning to work daily, and in 30 days the mines will be working full shift. ——W. W. Reese, principal owner of the Columbia mine, at Grass Valley, California, accompanied by engineers, has been looking over the Greater Miami property with view to purchase.

Globe, September 22.

MARCIOPA COUNTY

(Special Correspondence.)—The Blue Ridge prospect, at Wickenburg, owned by John Grace, has been sold to Fraser & Ridgley. The Blue Ridge adjoins the Monte Cristo mine.

Phoenix, September 20.

MOHAVE COUNTY

(Special Correspondence.)—The new Steffy mill has commenced to treat 3000 tons of Altata ore, supposed to average $11 in gold, silver, copper, and lead.——An important strike of ore has been made in the Emerson tunnel of the Chloride district. This tunnel is now in 406 ft., and 150 ft. below the surface it opened up good ore for its entire length.——Operations have been resumed at the Merrimac mine. A compressor and engine have been ordered, and a contract calling for 300 ft. of sinking and cross-cutting will be let in a few days. The firm of Sill & Sill, engineers and ore-testers, have taken samples for treatment.——Fred W. Sherman, general manager of the Keystone mine, expects to have the re-constructed mill operating within a few days. New machinery has been added to the plant. It is intended to treat the dump ore for several months and in the meantime the mine will be unwatered. Prior to the closing of this mine from 70 to 80 men were employed.——Two Oatman companies, the Gold Road Bonanza and the Oatman United, are calling meetings of their shareholders to change the articles of incorporation so that the stock can be made assessable. It is hoped that the extra money raised in this way will enable the companies to complete their development.——Work is expected to be started soon on a tunnel to open up the Diamond Joe mine.——The shaft of the Leviathan mine is now down to the 200-ft. level and is in ore 5 ft. wide averaging 5% molybdenum. It is the intention to sink this shaft to 250 ft. and cross-cut a parallel vein. The Arizona Moss Back Co. has been organized to develop and work the old Moss Back mine, which was last worked in 1901. The present 330-ft. shaft will be sunk to 530 ft. The property became idle on account of litigation, but in the spring the mine was unwatered and examined, resulting in the formation of the new company. There is ample water one mile north of the property for mining and milling.

Kingman, September 18.

A large body of molybdenite ore, assaying from 5 to 12% MoS_2, has been struck in the Jackman mine, Hualpai mountains. The property is under option to the Willys Automobile Company.

PIMA COUNTY

(Special Correspondence.)—The control of the Daily Arizona Consolidated Copper Co. has been acquired from W. H. and John Daily by W. H. Martin and associates, of Indiana. The company is to be re-organized and $100,000 spent in development. Machinery is on the ground. The ore mined during development, which is expected to be 50 to 100 tons per day, will be shipped. Diamond-drilling is being continued on the Little Ajo. About a mile west, on the property of the Arizona Mines Co., which adjoins the New Cornelia, diamond-drilling is also being carried on. On the latter property are good showings of ore, some of which has been shipped. The New Cornelia company has increased its copper output this month.

Tucson, September 21.

PINAL COUNTY

(Special Correspondence.)—It is reported that the shares of the Silver King Extension, from which several hundred thousand dollars was taken out in the early days, are in demand because this property adjoins the Fortuna mine, in which the recent strike has been made. Development work has commenced in the Camelback mine. A contract has been let for the first 100 ft. of adit to intersect a lode in which a shaft has been sunk to a depth of 40 ft. The company owns 12 claims. Ore in good quantities has been shipped, running 270 oz. in silver and $26 in gold. At a recent meeting of the Arizona Ray Copper Co. the name was changed to the New Arizona Ray Copper Co., and the capital increased from one to five millions. It is proposed to sell 100,000 shares at 75 cents for further development, which will include drilling and sinking.

Florence, September 21.

YAVAPAI COUNTY

(Special Correspondence.)—F. P. Taylor is in Prescott for the purpose of opening the Comanche mine, in the Weaver district near the Hassayampa river. This property was forfeited by Mr. Taylor while on military duty at Fort Whipple. Development work will be commenced early in the fall. Assays made for Mr. Taylor 35 years ago ran from 60 to 300 oz. in silver.——It is reported by E. H. Oldham, of the Yeage Copper Co., that an excellent showing of manganese ore has been found on the company's property in Yeage canyon. Sinking was started until 35 ft. was reached, when water stopped the work. Samples show 40% manganese.

Prescott, September 21.

(Special Correspondence.)—A remarkable body of copper ore is being opened into by the Big Ledge Copper Co., at the Butternut mine, two miles north-west of here. At a depth of 300 ft. the vein is 48 ft. wide; the average value being 4% copper. On the 425-ft. level a cross-cut is being run, and so far 15 ft. of the vein has been traversed, the copper-content increasing to 10%. The finding of this ore in such a large body settles the question of tonnage for the company's smelter here; it is now to be rushed to completion. Dwight L. Woodbridge, the consulting engineer of the company, is due from Duluth. Plans are made for a large flotation-mill at the Butternut mine to treat the ore that has already been developed in the upper workings.

The Blue Bell mine, owned by the Consolidated Arizona Copper Co., is stoping 40 ft. of ore on the 1000-ft. level, and the Arizona Binghampton and Copper Queen companies are working on orebodies that will average 15 ft. in width above the 400. These mines are not contiguous, but they have much the same character of ore, principally chalcopyrite with some chalcocite on the lower levels.

Driving will begin in a few days on the new 1200-ft. level in the Blue Bell mine. The shaft has been sunk 200 ft. deeper from the 1000-ft. level. It is expected that the orebodies that are now furnishing 300 tons daily for the Humboldt smelter will be cut at this new level. Chalcocite has been coming into the ore at the 1000-ft. level recently, giving reason to believe that a higher grade of ore will be found on the 1200 level.

Mayer, September 28.

CALIFORNIA

DEL NORTE COUNTY

(Special Correspondence.)—Manganese claims have been located in the Shelley Creek district by G. W. Gravlin of Grants Pass, Oregon. Some high-grade copper ore is being taken out at the Old Crow copper mine. E. A. McPherson is manager.

Crescent City, September 24.

EL DORADO COUNTY

Owing to the increase in the price of silver, the once famous camp of Calico is beginning to show renewed activity. A new cyanide plant is treating 400 tons of tailing per month. It is estimated that there is 200,000,000 tons of tailing on the old dumps. A number of mining men have visited the locality and it is probable that other plants will be erected.

SHASTA COUNTY

(Special Correspondence.)—The Noble Electric Steel Co. has arranged to operate a second furnace on manganese ore, which will bring the output of ferro-manganese to 400 tons per month. The company is turning out an 80% product and for the present will confine its attention to the manufacture of this material. Shipments of manganese are being received from several points. Recent developments in properties in the vicinity of Greenville and Indian Falls, Plumas county, have proved encouraging.

The Pittsburg-Mt. Shasta Co. is operating the 15-stamp mill at the Little Nellie mine on gold ore, and shipping copper-bearing ore to the Kennett smelter. The latter ore is drawn from the lower levels, and diamond-drills indicate wide bodies of copper ore to an average depth of 800 ft. The property lies near the Mountain Copper mine.

The Craig Mining Co. has arranged to prospect placer deposits near Igo with diamond-drills. William McKee is in charge. The gravel channels have been worked in the past with indifferent results, although rich ground has been found. With a modern dredge it is believed much of the ground can be mined profitably.

The East Side belt continues to show growing activity. The Afterthought Copper Co. is treating 300 tons of ore daily at its flotation plant and expanding its work in the mine.——The Bully Hill group is undergoing aggressive development and shipping to the Kennett smelter.——The Arps group continues to develop well.——Ore conditions in the Shasta-Belmont have improved to a point where early shipments are expected. Several small properties are being worked vigorously and two important deals are reported pending.

Kennett, September 30.

SIERRA COUNTY

(Special Correspondence.)—Under the management of D. E. Hayden, the Wisconsin Mining Co. is installing an electric hoisting and pumping plant. Underground work has been discontinued pending completion of surface improvements. The company has developed some excellent ground, and the outlook is considered satisfactory.

At the adjoining North Fork mine the main incline and drift have been re-timbered and development has been resumed. North and south cross-cuts will be driven to determine the length of the ore-shoot recently uncovered, from which rich ore is being mined. Prospecting for a second shoot has begun. George F. Stone is superintendent. The mine is near Forest.

With two shifts the main adit of the City of Six group is advancing rapidly toward the lode. It is at present in 350 ft., with the vein showing harder rock and better defined walls. A cross-cut has been started to seek the hanging wall, and other cross-cuts will be driven. This mine is near Downieville.

The Twenty-One Co. has erected a bunkhouse capable of housing 50 men, a boarding-house, and other structures at its property near Alleghany, and has begun building a blacksmith-shop and several cottages. Considerable new work has started. Thirty men are on the payroll.

Downieville, September 30.

SISKIYOU COUNTY

(Special Correspondence.)—The Dewey gold mine, nine miles west of Gazelle, resumed operations this summer, after being closed for ten years. It is equipped with a 10-stamp mill, run by electricity. The vein is from 10 to 20 ft. wide and is opened by a 900-ft. drift. The new improvements include nine miles of power-line. G. C. Smith, of Los Angeles, is superintendent.——The Jillison gold mine near Hornbrook has closed-down until next spring.——The Siskiyou Dredging Co. is operating successfully on Greenhorn creek.——The Woods & Brown company is developing the Sunset group of copper claims near Slater Butte on Cade mountain.

Hornbrook, September 27.

COLORADO

The Platon Rey Leasing Co., which leased the Margin & Silver King claims, shipped a car of high-grade silver ore this week. This is the first shipment, but the dead-work being completed, regular shipments may be expected henceforth.

A large body of manganese ore, assaying 35 to 40% Mn, has been discovered in the old Bohn mine at Leadville. The orebody was struck on the 400-ft. level, and a level has been driven 40 ft. in solid ore and a raise put up 12 ft. without reaching the limits of the deposit.

BOULDER COUNTY

(Special Correspondence.)—The plant of the Mojave-Boulder Tungsten Co. is completed and in operation. This company acquired the Bracken group, and has been developing it actively. The April Fool and Good Friday give promise of becoming good properties. H. Brower, the superintendent, estimates that with the force now employed, he can produce 20 tons daily for many years.

The Slides, an old mine of the early days near Boulder, has a dump that is being worked over by the Gold Hill Concentrating Co., namely, George Teal, Charles Gustafeson, Robert Kimoch, and Clifton Barr. The ore is being treated by the flotation process with satisfactory results.

A strike has been made in the Boulder Falls section on the Catastrophe mine, belonging to Henry Lawrence of Boulder. The vein ranges from 10 to 2 ft. in width and averages over 40%. A ton of tungsten is produced daily. This is considered to be one of the best strikes of the season. Maurice Beckman, a mining man, has moved from Clear Creek county to Boulder and is now operating a lease on the Yellow Pine mine. The Evans group near Summerville, consisting of eight patented claims, are being cleaned out and re-timbered. On Gold Hill, George Kirkbridge has a group of 10 patented claims. He has leased the Cash, St. Joe, and three others to Denver parties. Active operations are in progress. The Myrtle mine belonging to D. S. Clark on Gold Hill has been leased to Jim Gouyde and Oscar Wallon. Regular shipments of gold ore are being made. W. F. Murphy, from Iowa City, Iowa, is building a mill on the Little Johnny mine at Salina. He has a new process. The old Victoria mine at Summerville, with a big silver-producing record, has been modernized in equipment and is maintaining a steady output. Frank M. Pickard is superintendent. The Hudson mill in the same district has added a Dorr agitator and another flotation unit. The Huron people at Eldora are getting out some good gold ore. The Consolidated Leasing Co. is getting material hauled for the new mill. C. E. Dewitt, superintendent of the Wolf-Tongue Tungsten mill, reports everything running to fullest capacity.

Eldora, September 22.

NEVADA

EUREKA COUNTY

(Special Correspondence.)—At the Marne mine, Harris and Finn have completed sinking their shaft to the 150-ft. level and are making their first shipment from that place. The Eureka Croesus shipped nine cars this week from the new strike on the 400-ft. level which is reported to run about $60 per ton.——Work has been commenced by J. H. Rogers as manager for a new company on Mineral hill immediately south of the Richmond property to try to discover the fault that is supposed to have enriched the orebodies on Ruby hill now owned by the Richmond Eureka Mining Co., a subsidiary of the United States Smelting Co.——C. F. Wittenburgh of Tonopah came in a few days ago and made arrangements to commence work on the Cyanide mine on Adams hill. Two outfits in the Philipsburg district, about 18 miles north-east of Eureka, have about three carloads of ore out but are held up owing to lack of teams. Several shippers have been notified by the smelters to curtail shipments on account of congestion of ores at the smelters, scarcity of labor, and coal for smelting. Four trucks and a trailer are hauling ore to this place for shipment from Hamilton in the White Pine district, about 30 miles east, making about two trips in three days, averaging five tons per load.

Eureka, September 24.

HUMBOLDT COUNTY

(Special Correspondence.)—An exploratory drift from the 900-ft. level of the Rochester mine last week cut an orebody 40 ft. wide. The vein has since been tapped by ten more cross-cuts and shows high-grade ore. The discovery, which is beyond the old foot-wall, opens interesting possibilities in an entirely new section of the property. On the 800-ft. level the Adams vein has widened so that it has been found necessary to maintain a central pillar of ore to support the roof. The main stope is 20 ft. wide and is in solid ore. The winze from the 900-ft. level is progressing in ore assaying $17 silver and $4 gold, with better ore as the foot-wall is approached. The mill is treating 211 tons daily, averaging about $13 per ton. A shipment of bullion valued at $31,300 was made on September 21.

With the exception of motors for the tubes and ball-mills, and trolley-locomotives, all equipment for the 300-ton mill of the Rochester Combined is either on the ground or has been shipped. Sinking of the Happy Jack shaft has advanced to the 200-ft. level, where a station has been cut, and north and south drifts started. Excellent ore is developing in the north drift and west cross-cuts on the 100-ft. level. New work has been begun from the Kromer and Shepherd adits, with encouraging results.

The orebody lately intersected in the La Toska mine, at

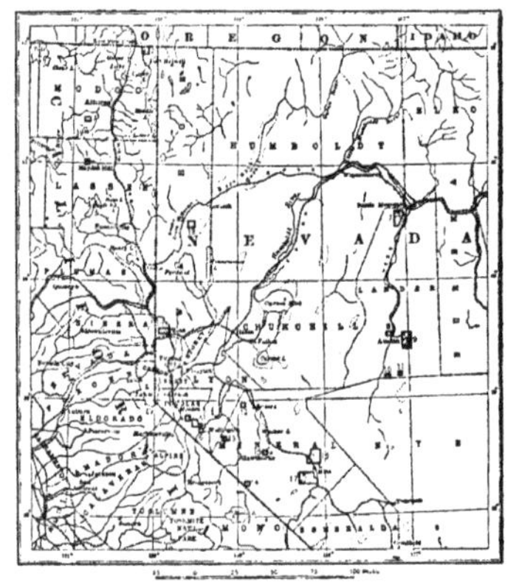

MAP OF PART OF NEVADA

Wright's canyon in the Humboldt range, is yielding ore running high in silver and a little gold. It was intersected about 3000 ft. from the adit-portal, after a search of six years. The lode is six feet wide and lies between walls of limestone and rhyolite.

Humboldt now stands second among the counties of Nevada in point of productive mines, and is steadily increasing its yield of gold, silver, copper, and lead. Prospecting is active, and the general outlook is encouraging.

Rochester, October 1.

NYE COUNTY

(Special Correspondence.)—In the Silver Top shaft of the Tonopah Mining Co. 136 ft. of development was performed and 15 ft. in the Mizpah shaft. Re-timbering of the Sandgrass is completed and stoping has been resumed. The raise driven on the Mizpah vein, showing a 4-ft. face of mill-ore, has developed considerable tonnage. The production for the week was 2400 tons. The Tonopah Extension shipped 20 bars of bullion, valued at $46,000, which represents the bi-weekly clean-up. Development in the No. 2 shaft amounted to 124 ft. The Murray vein on the 1260-ft. level is 8 ft. wide. The Victor was developed 119 ft. On the 1440-ft. level 613 raise is being driven in 6 ft. of mill-ore, while 1503 raise on the 1540-ft. level is in 3 ft. of ore. The south cross-cut on the 1680-ft. level was driven 61 ft. during the week. The production was 2380 tons.

The Tonopah Belmont has announced the purchase of the Alta, St. Louis, and Palmyra mines, near Telluride, Colorado. The large tonnage of complex ore will be treated by the Jones-Belmont flotation process. Raise 44 from the 900-ft. level has found some ore, while raise 90 on the 1100-ft. level has developed an orebody on the Rescue vein. On the 1166-ft. level the Mizpah fault-vein has been intersected 4½ ft. wide. The plant at Millers shipped 21 bars of bullion valued at $43,908, while the Belmont mill shipped 58 bars of bullion valued at $122,650, and 31 tons of concentrate valued at $12,000. In August 10,790 tons of ore was milled for a net profit of $121,695. All the stored silver of the Belmont company has been sold. The production last week was 2445 tons.

The West End Con. Mining Co. produced 905 tons from the West End and Ohio shafts. Sampling in the West End has developed a large tonnage of low-grade ore that can be exploited if the price of silver is sustained. At the Ohio, 512 cross-cut has reached the foot-wall of the vein. Owing to the high price of silver the McNamara Mining Co. has been able to resume operations after a shut-down of three years. The underground workings were found to be in excellent condition, so stoping was commenced immediately. A crew of about 20 men has been added. A few minor repairs have been made in the mill, which will commence operation on September 28. Development is contemplated in addition to the regular stoping.

The Jim Butler Tonopah Mining Co. produced 3403 tons of ore in August, realizing a profit of $46,740. The production last week was 750 tons. The Buckeye Belmont has commenced development on the Buckeye-Eagle vein. The Rescue-Eula produced 171 tons of ore, the Montana 71 tons, the North Star 54 tons, the Midway 94 tons, the Cash Boy 40 tons, and miscellaneous 58 tons, making the week's production at Tonopah 9388 tons with a gross value of $164,290.

Tonopah, September 26.

NEW MEXICO

(Special Correspondence.)—The Gold Eagle group has resumed operations and will start milling shortly.——The recent strike on the 300-ft. level of the Johnson mine is developing well. The Socorro Mining & Milling Co. operates this property and is arranging to increase shipments to the mill.——During the week the Mogollon Mines Co. has opened up a quantity of good ore in the upper workings that had been overlooked by former operators.——The Oaks Co. is shipping from each of its properties, namely, the Maud S., Deep Down, Eberle, and Clifton.

Mogollon, September 25.

Obituary

James F. Burns died at his home in Colorado Springs on September 22 at the age of 64. He was one of the locators of the Portland mine at Cripple Creek. He was associated with W. S. Stratton and James Doyle in the development of the mine, making a large fortune. From 1895 to 1905 he was president of the Portland Gold Mining Company.

E. M. de La Vergne, a pioneer of Cripple Creek, formerly general manager of the Elkton Consolidated, superintendent of the United Gold Mines, and locator of the El Dorado claim, the first recorded location in the district, died at his home in Colorado Springs, on September 18, at the age of 71 years. De La Vergne was superintendent for five years of the Old Man mine at Camp Fleming, and the Black Hawk mine at Silver City, N. M. Returning to Colorado he managed the Orient mines at Lawson, and the first assay-certificate signed by Professor Lamb of Colorado College in December 1890 is believed to have been the first authoritative record of gold in the Cripple Creek district.

PERSONAL

Note: The Editor invites members of the profession to send particulars of their work and appointments. This information is interesting to our readers.

J. B. Tyrrell has been to Newfoundland.

Benjamin Rezas, of Mexico City, is here.

A. A. Arluck has been drafted into the National Army.

Blamey Stevens has left Mexico and is now at Brooklyn.

H. Vincent Wallace has returned to Los Angeles from New York.

Drummond MacGavin has arrived in England on his way to France.

T. A. Rickard has gone to St. Louis, to attend the meeting of the Institute.

Frank W. Royer, of Los Angeles, was in San Francisco during the week.

F. R. Wolfle has gone from Spokane to the Florence mine at Ainsworth, B. C.

Roy Starbird, recently at the Perseverance mine, has returned from Alaska.

Charles W. Merrill has gone to Washington to join the Food Administration.

F. S. Kirkland is consulting engineer at the Yankee John mine in Shasta county.

N. A. Stockett has enlisted in the 164th Depot Brigade at Camp Funston, Kansas.

M. M. Johnson is consulting engineer at the Silver King mine in Shasta county.

Thomas Neilson, metallurgist from Siberia, will sail for Manila on professional business.

Irving A. Palmer has been appointed professor of metallurgy in the Colorado School of Mines.

F. M. Leland, of New York, formerly manager of Balaklala mine at Kennett, is at Redding, California.

Courtenay De Kalb has returned to San Francisco from a journey of observation in southern Arizona.

Fletcher McN. Hamilton, State Mineralogist of California, is attending the Institute meeting at St. Louis.

L. E. Parker, formerly of Montana, is general manager of the Silver King mine, near Redding, California.

Homer Hamlin has opened an office as consulting engineer and geologist at 1103 Central Bdg., Los Angeles.

G. P. Watson, metallurgist at the Braden Copper Co., arrived in San Francisco last week on his way from Chile.

Frederick G. Clapp, chief of the Petroleum Division of the Associated Geological Engineers, is in Kansas and Oklahoma.

Charles F. Williams, formerly with the Tonopah-Belmont company, has been commissioned Lieutenant in the Engineer Corps and is at Fort Leavenworth, Kansas.

A. C. Beatty has offered his London house as a hospital for American officers under the supervision of the Columbia Hospital Unit.

P. K. Lucke, formerly consulting mining engineer for the Cia. Minera de Penoles and Cia. de Minerales y Metales, has resigned, and will resume independent practice at 621 Bedell Bdg., San Antonio, Texas.

E. G. Snedaker and Andrew Newberry have been commissioned Lieutenants in the Engineer section of the Officers Reserve Corps and are now at Vancouver barracks, Washington.

The St. Louis meeting of the American Institute of Mining Engineers opens on October 8. An excellent programme has been arranged, including visits to the zinc and lead districts of Missouri and the oil-fields of Oklahoma. A special session of the War Minerals Committee will be held.

THE METAL MARKET

METAL PRICES

San Francisco, October 2

Aluminum-dust (100-lb. lots), per pound	$1.00
Aluminum-dust (ton lots), per pound	$0.95
Antimony, cents per pound	17.00
Antimony (wholesale), cents per pound	14.75
Electrolytic copper, cents per pound	23.50
Pig-lead, cents per pound	8.25—9.25
Platinum, soft and hard metal, respectively, per ounce	$105—111
Quicksilver, per flask of 75 lb	$105
Spelter, cents per pound	10.50
Tin, cents per pound	61
Zinc-dust, cents per pound	20

ORE PRICES

San Francisco, October 2

Antimony, 45% metal, per unit	$1.10
Chrome, 40% and over, free SiO_2 limit 8%, f.o.b. California per unit, according to grade	$0.60— 0.75

Chrome prices show a tendency to remain stationary. With the advent of winter, when the haulage conditions are less favorable than at present, prices will be more erratic and are likely to go higher.

Magnesite, crude, per ton	$8.00—10.00

Manganese: The Eastern manganese market continues fairly strong with $1 quoted on the basis of 48% material.

Tungsten, 60% WO_3, per unit	28.00
Molybdenite, per unit MoS_2	$40.00—45.00

EASTERN METAL MARKET

(By wire from New York)

October 2.—Copper is inactive and nominal, remaining at Government price at 23.50c. all week. Lead is dull and steady, remaining at 8c. all week. Zinc is quiet and easier at 8.50 to 8.37c. Platinum remains unchanged at $105 for soft metal and $111 for hard.

SILVER

Below are given the average New York quotations, in cents per ounce, of fine silver.

Date		Average week ending	
Sept. 26	106.50	Aug. 21	87.16
" 27	101.50	" 28	88.50
" 28	97.62	Sept. 4	91.12
" 29	96.62	" 11	96.80
" 30 Sunday		" 18	100.70
Oct. 1	95.12	" 25	107.66
" 2	93.62	Oct. 2	98.49

Monthly Averages

	1915	1916	1917		1915	1916	1917
Jan.	48.85	56.76	75.14	July	47.52	63.06	78.92
Feb.	48.45	56.74	77.54	Aug.	47.11	66.07	85.40
Mch.	50.61	57.89	74.13	Sept.	48.77	68.51	100.73
Apr.	50.25	64.37	72.51	Oct.	49.40	67.86	....
May	49.87	74.27	74.61	Nov.	51.88	71.60	....
June	49.03	65.04	76.44	Dec.	55.34	75.70	....

COPPER

Prices of electrolytic in New York, in cents per pound.

Date		Average week ending	
Sept. 26	23.50	Aug. 21	28.00
" 27	23.50	" 28	26.41
" 28	23.50	Sept. 4	25.65
" 29	23.50	" 11	25.83
" 30 Sunday		" 18	26.25
Oct. 1	23.50	" 25	24.41
" 2	23.50	Oct. 2	23.50

Monthly Averages

	1915	1916	1917		1915	1916	1917
Jan.	13.60	24.30	29.53	July	19.09	25.66	29.67
Feb.	14.38	26.62	34.57	Aug.	17.27	27.03	27.42
Mch.	14.80	26.65	36.00	Sept.	17.69	28.28	25.11
Apr.	16.64	28.02	33.16	Oct.	17.90	28.50	....
May	18.71	29.02	31.69	Nov.	18.88	31.95	....
June	19.75	27.47	32.57	Dec.	20.67	32.89	....

Copper producers express general approval of the plan to pool output of the country. It is stated on highest authority that prompt arrangements to this effect will be put in operation by the Government, regardless of possible appointment of a special board of administration. Smaller producers will share an apportionment of trade with all consumers, and on the other hand larger producers will share pro rata in whatever allotment is diverted to domestic consumption.

The owners of leading American copper shares will receive in September in dividends an estimated total of $27,232,000. This compares with disbursements in August of $11,467,000 and of $26,389,000 in September a year ago. In the first nine months of 1917 these copper companies will have paid to their stockholders the huge total of $140,437,443, compared with $107,681,058 in the same period of last year.

A canvass of the larger copper producers in New York discloses great uncertainty in the copper market, notwithstanding price-fixing by the Government for the next four months. Both producers and consumers are 'up in the air,' due to lack of details in connection with the carrying out of the proposed plans for handling the copper market, and this condition will continue until Washington furnishes more detailed advices as to what can be done and what should not be attempted under the new order of things.

LEAD

Lead is quoted in cents per pound, New York delivery.

Date		Average week ending	
Sept. 26	8.00	Aug. 21	10.79
" 27	8.00	" 28	10.61
" 28	8.00	Sept. 4	10.28
" 29	8.00	" 11	10.20
" 30 Sunday		" 18	9.33
Oct. 1	8.00	" 25	8.00
" 2	8.00	Oct. 2	8.00

Monthly Averages

	1915	1916	1917		1915	1916	1917
Jan.	3.73	5.95	7.64	July	5.59	6.40	10.98
Feb.	3.83	6.23	9.01	Aug.	4.62	6.28	10.75
Mch.	4.04	7.26	10.07	Sept.	4.62	6.86	9.07
Apr.	4.21	7.70	9.38	Oct.	4.62	7.02	....
May	4.24	7.38	10.29	Nov.	5.15	7.07	....
June	5.75	6.88	11.74	Dec.	5.34	7.55	....

ZINC

Zinc is quoted as spelter, standard Western brands, New York delivery, in cents per pound.

Date		Average week ending	
Sept. 26	8.50	Aug. 21	8.66
" 27	8.50	" 28	8.35
" 28	8.50	Sept. 4	8.25
" 29	8.37	" 11	8.21
" 30 Sunday		" 18	8.29
Oct. 1	8.37	" 25	8.46
" 2	8.37	Oct. 2	8.43

Monthly Averages

	1915	1916	1917		1915	1916	1917
Jan.	6.30	18.21	9.75	July	20.54	9.90	8.98
Feb.	9.05	19.99	10.45	Aug.	14.17	9.03	8.58
Mch.	8.40	18.40	10.78	Sept.	14.14	9.18	8.33
Apr.	9.78	18.62	10.20	Oct.	14.05	9.92	....
May	17.03	16.01	9.41	Nov.	17.20	11.81	....
June	22.20	12.85	9.63	Dec.	16.75	11.26	....

QUICKSILVER

The primary market for quicksilver is San Francisco, California being the largest producer. The price is fixed in the open market, according to quantity. Prices, in dollars per flask of 75 pounds:

Week ending

Date			
Sept. 4	115.00	Sept. 18	115.00
" 11	115.00	" 25	105.00
		Oct. 2	105.00

Monthly Averages

	1915	1916	1917		1915	1916	1917
Jan.	51.90	222.00	81.00	July	95.00	81.20	102.90
Feb.	60.00	295.00	126.25	Aug.	93.75	74.50	115.00
Mch.	78.00	219.00	113.75	Sept.	91.00	75.00	....
Apr.	77.50	141.60	114.50	Oct.	92.90	78.20	....
May	75.00	90.00	104.00	Nov.	101.50	79.50	....
June	90.00	74.70	85.50	Dec.	123.00	80.00	....

TIN

Prices in New York, in cents per pound.

Monthly Averages

	1915	1916	1917		1915	1916	1917
Jan.	34.40	41.76	44.10	July	37.38	38.37	62.60
Feb.	37.23	42.60	51.47	Aug.	34.37	38.88	62.53
Mch.	48.76	50.50	54.27	Sept.	33.12	36.66	....
Apr.	48.25	51.49	55.63	Oct.	33.00	41.10	....
May	39.28	49.10	63.21	Nov.	39.50	44.12	....
June	40.26	42.07	61.93	Dec.	38.71	42.55	....

Charles Hardy writes on September 25 from New York:

Tungsten has been exceedingly active during the past week and considerable business has been done. The tonnage turned over is heavy, yet some of the buyers would have bought more if the price-fixing for steel had not made them hesitate before committing themselves. As it appears now, the prices fixed for steel leave high-speed steel and tool-steel out of consideration and further business is now looked forward to in tungsten. Off-grade ore containing copper though high in tungstic acid which could not be sold a few days ago at $21 was sold this morning at $23, which would put the price for highest-grade wolframite at $25 or better. The demand for tungsten has received further impetus by heavy inquiries on the part of the motor industry where aeroplanes and tractors are called for by the Government. Scheelite remains unchanged at $26 basis 60% WO_3.

The molybdenite market continues firm at $2.20 per pound on the basis of 90% MoS_2. Lower-grade material has been sold in proportion.

Small business has been done in prompt antimony around 15¼c. per pound for November delivery. Prices remain for prompt around 15½c. The market continues practically lifeless.

The manganese market remains the same with $1 still quoted on the basis of 48% material.

Consolidated Interstate Callahan Mining Co. has passed its dividend. In explanation, President Percival says: "The directors decided that the uncertain outlook, due to the War, and the increased cost of production, together with the necessity of providing for heavy war taxes, estimated under the provisions of the statute as now drawn at approximately $300,000, which must be paid out of 1917 earnings, make it desirable for the company to create a strong reserve in order to be properly protected against any emergency."

Eastern Metal Market

New York, September 26.

The entire metal market has a better aspect due to the fact that some important uncertainties that have been hanging over the market for so long and causing extreme unsettlement have at last been determined; at least a definite price for copper has been named by the Government, affording a basis for future negotiations, although the market for copper at present is at sea. Lead has also fallen to the Government price-level.

Copper is inactive and nominal.

Tin is firm and strong.

Lead is steadier and firm.

Zinc is quiet and fairly firm.

Antimony continues inactive.

Aluminum remains unchanged.

The long expected Government announcement of steel prices came on Monday and with unexpected suddenness. The steel world, excepting those on the inside of the negotiations, has not yet recovered, nor does it really know where it stands. Pig-iron has been fixed at $33 per gross ton, but no grade of iron is mentioned, and they vary in prices. Coke has been fixed at $6 per ton, but there are two kinds of coke. Steel bars are placed at $2.90 per 100 lb., or $58 per ton; structural shapes at $3 per 100 lb., or $60 per ton; and plates at $3.25 per 100 lb., or $65 per ton. Whether private business has to be done at these figures is still uncertain, and at the present writing the trade is 'up in the air' and at a standstill.

COPPER

The sensation of the week has been the announcement by the Government of its price for copper, 23.50c. per lb. This appeared last Friday, September 21, and was reported as applying to both Government, foreign, and domestic business for a period of four months. Generally speaking the price is distinctly satisfactory, and is higher than many expected. It is deemed eminently fair and conducive to the maintenance of production at a high rate. There is no question that a good, and in some cases a high, profit can be made at the new price. This is the general opinion in the trade. So much for the price. The market, however, at present does not know just where it stands. There is no market quotation unless that of the Government be adopted; there is no business for producers, and dealers refuse to quote; there is considerable inquiry, but producers are generally saying that they have no copper to sell at the Government price at present, while the smaller dealer is not informed whether he may sell at a higher price than 23.50c. per lb. without getting into trouble. Many small wholesale dealers have high-priced copper that they sell in 5 and 10-ton lots. Until some definite information as to the exact status of both large producers and small dealers is forthcoming business is practically at a standstill. It is not known to what extent domestic business will be ruled by the Government price, as it is a fact that legally the Government cannot fix any prices except for its own needs. In the absence of any transactions for any purchases, our quotation is necessarily nominal at the Government price of 23.50c. For delivery in the first quarter 26c. is quoted by some, but consumers are uncertain what the condition as to price-fixing will be then, and consequently are not anxious to enter the market. The general feeling, however, while chaotic at this writing, is that the big problem will work out for the benefit of all concerned. It is recognized that the Government's main object is to cut out speculation.

ZINC

The market has not changed greatly from the distinctly better tone reported last week. Demand is not large and has apparently eased off. In the last two weeks a fair business has been done at the prices already quoted. It has been well distributed, but not large in volume. For the most part it has been for fairly early delivery, business for the later positions being of small consequence. Producers and dealers are not inclined to commit themselves far ahead. Today the market is steady, but is a little easier, if anything, with quotations at 8.25c., St. Louis, or 8.50c., New York, for September and October delivery, with last quarter held about ⅛ to ¼c. higher. Reports are to the effect that demand from the galvanizing and brass interests is not large. It is expected, however, that the latter phase of the trade will demand large amounts of zinc as the War progresses.

TIN

The tin market is higher than a week ago when our last report was sent. This is due to several causes. The principal one is a report that an embargo in shipments from the East was to be put into effect in October. This has not been confirmed, but nevertheless it has a stiffening influence. Another factor is an advance in Atlantic war-risks due to a fear of submarine activity on this side. The fact that the London market has advanced decidedly has been a factor also. Straits tin advanced there to £3 5s. per ton for spot delivery with spot standard up to £3 and future standard up £2 15s. Yesterday spot Straits on this side was nominal at 62.50c., New York, but the market is nervous and inquiry is lacking. Tin arrivals have been 2285 tons so far this month, with 4500 tons afloat. No large volume of business is reported.

LEAD

Unquestionably the lead market is gradually assuming a more stable aspect. The fact that the independent and Trust price are at the level at which the Government is obtaining its supplies is a stabilizing influence. There is more inquiry, and the market is steady. It is believed that the market quotation of 8c., New York, or 7.92c., St. Louis, can be and is being slightly shaded by some small interests, but, on the whole, what business is being done is mostly going at 8c., and it is likely that the bottom has been reached and that an upward swing is in prospect or certainly a firm and stable market at near the present level.

ANTIMONY

The market continues in its dull and lifeless condition with quotations for Chinese and Japanese grades unchanged to 15 to 15.50c., New York, duty paid.

ALUMINUM

No. 1 virgin metal, 98 to 99% pure, can still be obtained at last week's quotations of 41.50 to 42.50c. per lb., New York. The market is no stronger and demand is still of small volume.

ORES

Tungsten: The market has been active with considerable business reported at the prevailing prices of $23 to $26 per unit. Ferro-tungsten is unchanged at $2.30 to $2.60 per lb. of contained tungsten.

Molybdenum and antimony: Firmness characterizes the molybdenum market at $2.20 per lb. of MoS_2 in 90% material with lower grades in proportion. There is nothing to report in the antimony-ore market. Quotations are unchanged.

Recent Decisions

OIL LEASE—SOLD SUBJECT TO DELIVERY CONTRACT

Where one purchases an oil lease with knowledge that his vendor has contracted that the oil produced on the property shall be sold and delivered to a pipe-line company at a price named in the sale-contract, he is bound by such contract.

Simms v. Southern Pipe Line Co. (Texas), 195 South-western, 283. May 17, 1917.

POLLUTION OF STREAMS FROM MINES

Where a State statute requires the draining of working places in mines into adjacent streams or water courses, the State cannot prosecute a mining company for pollution of a stream resulting from such drainage even though an indictment would lie at common law.

Commonwealth v. Kingston Coal Co. (Kentucky), 194 South-western, 1038. May 25, 1917.

MINING CORPORATION—WHEN NOT DOMICILED

A New York corporation engaged in mining in Alaska is not "doing business" in Washington because it is maintaining an office there for buying supplies and forwarding them to Alaska, such as will enable a Washington court to obtain jurisdiction of it on a cause of action arising in Alaska, by service of the complaint and summons on such purchasing agent.

Macario v. Alaska Gastineau Mining Co. (Washington), 165 Pacific, 72. May 19, 1917.

OIL LEASE—INJUNCTIVE RELIEF DENIED

An oil lessee who has drilled one "dry hole" but has not thereafter proceeded within the time set, with diligence and good faith, to drill other wells to develop the property will be denied injunctive relief to prevent the lessors from enforcing a forfeiture of his lease.

Advance Oil Company v. Hunt (Indiana), 116 North-eastern, 340. May 29, 1917.

COAL LEASE—CASUALTY CLAUSE CONSTRUED

A coal-mine lease provision that minimum royalties need not be paid during strikes and other unavoidable casualties beyond the lessee's control while the mine is closed due to such casualties, includes shut-downs caused by failure to get railroad-cars after the exercise of due diligence in the attempt to obtain them, but is inapplicable to shut-downs caused by machinery troubles or poor market conditions.

Bennett v. Howard (Kentucky), 195 South-western, 116. May 25, 1917.

MINING OPTION—ACCEPTANCE

Where an optionee under a mining bond goes on the property solely to examine and prospect with a view to determining whether or not he will buy, such actual entry and possession do not constitute an exercise of the option or render the optionee liable for payments under the option, even though in accordance with a parol understanding the optionee remains after the time for acceptance has elapsed in order to complete the annual assessment work for the owner.

Johnson v. Clark (California), 163 Pacific, 1004. March 13, 1917.

MINER LIENS—FAILURE TO POST NOTICE

Where a co-owner of a group of mining claims permits his co-tenants to go into possession of the same and purchase materials and supplies for and employ labor in the development of the claims, without posting any notice of non-responsibility for them, he cannot estop lienors from establishing their liens merely by a subsequent suit and judgment quieting his title to the entire group against his former co-owners.

Bishop v. Henry (Oregon), 165 Pacific, 237. May 29, 1917.

Book Reviews

DESCRIPTIVE MINERALOGY. By William Shirley Bayley. 12 mo. Pp. 542. Ill. D. Appleton & Co., New York and London. For sale by the MINING AND SCIENTIFIC PRESS.

This volume is not intended as a book of reference; but rather to give a comprehensive view of modern mineralogy. The minerals selected for description, therefore, are not the most common ones, nor those that occur in greatest quantity, but those that are of scientific interest, of economic importance, and illustrations of some principle employed in mineral classification. The book is divided into three parts and appendices, and is sub-divided into 23 chapters. The first part, consisting of two chapters and 25 pages, deals with composition and classification of minerals, their formation and alteration. In the second part, in which are 19 chapters and 25 pages, we find descriptions of the minerals selected. They have been grouped together from a chemical rather than a physical standpoint. Thus, for example, we find sulphides, tellurides, selenides, arsenides, and antimonides in one chapter; chlorides, bromides, iodides, and fluorides in another; phosphates, arsenates, and vanadates in another, and so on. The third part is devoted to blow-pipe analyses and characteristic reactions of the more important elements and acid-radicals. This occupies two chapters and 16 pages. The four appendices, covering 32 pages, give a guide to the descriptions of minerals, a number of exceedingly useful well-arranged tables, and a bibliography that is far from complete. The book is well illustrated with 40 photographic reproductions and over 200 drawings. In all, nearly 700 minerals are described, and well described. The descriptions deal with their chemical qualities; physical qualities; crystalline form; action before blow-pipe on charcoal and in closed tubes, with and without reagents; effect of acids and other reagents; synthesis; occurrence and origin; place where found; and uses. The giving of the synthesis of minerals is unusual and goes to show the advancement of science in this direction. The chapter devoted to blow-pipe analyses, though not attempting to be exhaustive, is distinctly good, while in that given to characteristic reactions of the more important elements and acid-radicals there are some excellent tests not only for the common but for many of the rare elements. The tests, too, are well chosen, simplicity being the key-note when possible. The tables in the appendices, as we have said, are excellent, but with regard to them the author hoists a danger signal that should be heeded. He says: "The most serious objection to the use of determinative tables lies in the danger that the student will feel, when the name of the mineral is obtained, that the object of his search is at an end, whereas their true aim should be to lead him to such a thorough study of the mineral that there will remain no doubt in his mind as to its real nature." He advises further that the tables are intended only to lead the student to the text in the main body of the book where the true distinction between the different species is to be found. The tables divide the minerals into two classes, namely, those with metallic lustre, which are always opaque on their thinnest edges, and those without metallic lustre, which are transparent in splinters and on their thinnest edges. There are 14 double-column pages of these tables. Nine pages of tables arrange the minerals according to their principal constituent. Thus, under cobalt we find: cobaltite, erythrite, glaucodite, linnaeite, smaltite. Finally, four and a half pages are given to a list of minerals arranged according to their crystalline form. The type is good, the cuts are clear and sharp, and the book is a convenient size. Anyone interested in mineralogy should have it on his shelf. F. H. M.

COST of the War to date is estimated at about 100 billion dollars, of which 25 billions, or one-quarter, has been incurred alike by England and Germany, leaving 50 billions as the money spent by the other belligerents. France and Russia are credited with 20 billions apiece.

BEGINNING on October 8, the 115th meeting of the American Institute of Mining Engineers will be held at St. Louis. An attractive programme has been arranged and much has been done to ensure the success of this gathering of the profession at the home-town of the president of the Institute, Mr. Philip N. Moore.

F. A. VANDERLIP is a type of the enlightened patriot. As president of the National City Bank of New York he has exhibited the highest intelligence and public spirit in many directions, particularly since the beginning of the Great War. Now he resigns all his regular duties in order to assist Secretary McAdoo in financing the Liberty Loan. His services, which are gratuitous, should prove of great value to the National purpose.

QUESTION has arisen whether the Government's price for copper applies to sales made for the period beyond the four months specified in the agreement with the producers—for it was an agreement, not a peremptory dictation. It is said that 26 cents is asked for copper sold for the first quarter of 1918 instead of the uniform price now prevailing of 23½ cents. We expect that the War Industries Board will answer this question equitably by applying the present price to forward sales also, otherwise the intent of the agreement would be stultified. In these matters all concerned must forego any quibbling and do their best to live up to the spirit of the agreement, which is such regulation of the market as is helpful to the national purpose.

OUR people are asked again to provide the sinews of war. The second Liberty Loan must be as successful as the first, for we have the money and we have the obligation. Since the issue of the first loan our soldiers have begun to go forward to the battlefield and the youth of the country has been called to prepare for active service. Those that cannot serve in that way can aid by contributing the money needed to bring the War to a victorious conclusion. This 4% Government loan is the most gilt-edged of all securities; it is particularly suitable as an investment for professional men. While the loan is sure to be a success, it is the duty of every citizen to help to crown the issue with an over-subscription to prove to all the world, particularly the hostile part of it, that our people are supporting the Government without reservation.

FLOTATION is being improved in a multitude of details, and some of the most interesting opportunities for betterment seem to lie in the peculiar properties of many chemicals both for modifying the physical state of the flotative medium and for cleansing and altering the surfaces of the minerals to be floated. This subject is treated at some length by Messrs. O. C. Ralston and L. D. Yundt elsewhere in this issue. They raise a question concerning the supreme importance of a change of surface-tension in the water that will renew interest in the investigation of this phenomenon. Above all they show how wide is the field for new discovery in the application of chemistry to the process. They are justified in their assertion that the art of flotation is in its infancy.

EFFORTS to control the price of silver are being made in defiance of the inexorable laws of supply and demand. The Indian government has prohibited both the import and export of silver, in order to protect its currency, endangered by the temptation to melt rupees into bullion. This will not prevent the conversion of coins into feminine jewelry nor the demand of prosperous workmen for silver in preference to paper. In India, as in France and England, the increase in wages calls for more silver coin, the prejudice against paper money being persistent. The recent eccentricities of the silver market, a big rise being followed by a sudden precipitous drop, are not surprising, considering the changes of value so suddenly created. Stores of silver have been brought into the market for realization at the high price prevailing. The 350,000,000 ounces of silver represented by minted dollars stored in the United States treasury against outstanding silver certificates represents a potential regulator of the market, for it may become necessary in the interest of the United States and of our Allies to prevent an excessive appreciation of silver. Such control can be exercised easily.

USERS OF CYANIDE will welcome the article on the synthesis of sodium cyanide by Dr. G. H. Clevenger, which we print in this issue. It gives the results of an interesting series of experiments in the production of this chemical by direct fixation of the nitrogen of the air in the presence of a catalytic agent, following the remarkable method recently developed by Prof. John E. Bucher of Brown University. The process is revolutionary in its simplicity and economy, and possesses the merit of being adaptable to operations on a relatively small scale. Mr. Clevenger has shown the difficulties that will be met in practice and the way to overcome them. He has also tested the synthetic cyanide which he produced in the extraction of gold and silver from ores, with parallel tests employing the cyanide of commerce. The results are entirely satisfactory. The article is accompanied by a tentative flow-sheet which will be useful to those that may wish to follow the subject further.

A REFUGEE from Mexico gives a picture in this issue of the tribulations of the American citizen, deserted by his country, in a land rent by revolution. We need to be reminded many times of the low regard that our official policy engendered for American rights abroad. What then happened in Mexico had much to do with inspiring contempt for us in Germany, and it emboldened the Kaiser to assail us in ways that ended by drawing us into the War. We are now setting forth to redeem ourselves, and one splendid fruit of this great struggle must be to exalt the dignity of American citizenship, and to make it a guarantee of safety and just treatment among all nations. Conditions are said to be improving in Mexico, and perhaps our 'refugee' could now return without fear of the abuses formerly suffered, but it is too early to count on a change of heart in that country. We must first give proof of our prowess, and of our determination to make the world everywhere safe for Americans. After it has been made safe for Americans it will be quite a safe place for democracy.

SEVERAL members of the Mining & Metallurgical Society of America have written to us asking for aid in selecting the mining engineer to whom is to be awarded the Society's medal, which this year is to be given for "distinguished service in the administration of mines." This self-imposed task of awarding a medal every year is a gracious act, but it must be embarrassing at times. We could mention a hundred men worthy of a medal under the specification quoted, but we should dislike very much to be compelled to select one super-eminent in this branch of professional work. If a medal must be given we would like to see it awarded to some modest fellow on whom the rewards of wealth and prominence have not been bestowed already. But then his "service" would not be of a "distinguished" kind, it may be urged. So the medal will probably go to one already possessed of such prizes as befall the successful. Nothing succeeds like success. It would be a nice thing if it were understood that the medal could not go to any member of the Society, which then would become a splendid jury for the awarding of these marks of distinction and thus have one more reasonable excuse for existing.

A New Copper District

On another page we publish an account of a visit to the newly discovered copper deposits in Manitoba. This is written by Mr. Walter Karri-Davies, well known in South Africa as one of the two organizers of the Imperial Light Horse, which gave such splendid service in the Boer war. The name of Major Karri-Davies is part of the history of South Africa and we need say no more. The article is interesting not only as coming from a distinguished man but as giving information concerning a new copper-mining district in the North. Sundry descriptions of the locality have appeared previously in print. Mr. J. A. Campbell, the Commissioner of Northern Manitoba, has published an account of a trip he made thither last July. In that account he does not fail to descant upon the scenic beauty of the region. He refers to the Mandy mine, now controlled by the Tonopah Mining Company, but he makes no statement regarding its value. Concerning the Flin Flon, which was the object of Major Karri-Davies' trip, Mr. Campbell says that 6000 feet of drilling proved 6,000,000 tons of ore and that this "will develop into the greatest orebody of its kind in America." We fear that his opinion will not carry much weight after such a statement, for the largest body of ore of that kind is so vastly bigger than 6,000,000 tons as to make the comparison ludicrous. However, this healthy prospect justifies his next suggestion that the Flin Flon and Mandy ventures "will result in opening up a number of other claims in the district where mineral deposits have been shown to exist, and which, owing to their remoteness and the amount of money involved in handling them under existing conditions, make their development by present owners now out of the question," in short, that the prospects justify a vigorous exploration of the district by those possessed of sufficient capital. Another report worthy of note is by Messrs. R. C. Wallace and J. S. Delury of the University of Manitoba. These geologists describe the formation of masses of copper-bearing sulphide ore, containing gold and silver, in sericitic schist near the contact with intrusive granite, presenting conditions, as our readers know, often associated with pyritic bodies of lenticular shape, when of magmatic origin. They mention seeing at Flin Flon "an almost solid mass of chalcopyrite and zinc-blende, 25 feet wide in the centre, the sulphides grading into pyrite on both sides." Besides the copper lodes there are gold-bearing quartz veins. The finding of these on the shore of Herb lake incited the first prospecting in the region.

These also show lenticular masses of some extent in which the gold is associated with arsenical pyrite. The presence of tourmaline marks them as contact deposits, and we cannot say that the description is prepossessing. They remind us too much of the Lake of the Woods. However, if they contain enough gold to yield a profit, we need not dispute over the character of the geologic environment. The best evidence of a gold mine is gold not mixed with too much quartz or other valueless rock. Undoubtedly the discovery of large bodies of copper-sulphide ore in this part of Canada is a notable event. It is likely to lead not only to the development of one or two big mines, but it will furnish a new point of departure for intelligent exploration in a vast region full of possibilities. Those that have been in this glaciated country, one of inundated *roches moutonées*, as the geologist sees it, are well aware of the difficulties presented to the search for veins and lodes underneath the mantle of moss and drift that covers the surface. Moreover, so large a part of the region is covered by shallow lakes that the search is further hindered by another obstacle even less easy to circumvent, although in the Lake of the Woods district and on Silver Islet, in Lake Superior, the miner sank shafts even under the water. We believe that the Canadian government will be wise in assisting bona-fide exploration in northern Manitoba, both by giving the scientific aid of its Geological Survey, as already done, and by encouraging the building of branch lines of railway. In some cases transport by water can be facilitated by blasting the rocky bottom of rapids and by building short tramways at portages between the lakes. The presence of the precious metals in the copper lodes is of particular significance and encourages the expectation that a local mining industry of some importance may be established. American capital is participating in the development of the region and we venture to add that the Canadian people at this time are particularly friendly to enterprise originating from this side of a frontier on which no fort has been built or is ever likely to be built. Cordial co-operation is assured.

An Austrian Melchizedek

Count Czernin has spoken, so that we now know somewhat more of the mental aberrations of the Central Powers. It is not our purpose to treat his utterances with levity, for he speaks with the authority of his Emperor, and the Austrian throne is in the same position as "me-too" Platt when Roscoe Conkling committed a certain irreparable and famous political blunder. What Count Czernin says regarding the peace of the future reflects the mature ideas of Kaiser Wilhelm as to what he would have the outer world believe. Dr. Michaelis recently treated us to a masterly expression of Prussian contempt for the military power of the United States; he brusquely read us out of the list of worthy foes, which may serve perhaps to render us more fully conscious of some of our defects which may have looked larger than they are when viewed through the Teutonic microscope. Taking us in the large we will disappoint this microscopic estimate of the German chancellor. The after-dinner speech of the Austro-Hungarian Minister of Foreign Affairs, however, is not all bluff; it represents a point of view that will affect many people whose sentiment occasionally proves toxic to their practical patriotism. When Count Czernin tempers his announced policies in pretended accord with the spirit of Pope Benedict's recent appeal, he is playing on a very ancient and historic chord that has not yet wholly lost its resonance. Although the Holy Roman Empire as an actual force in world-politics faded away long before the Napoleonic era, the formal abdication of this imperial dignity by the house of Austria took place barely a century ago. The Emperor at that time possessed a sufficient sense of humor and self-respect to insist no longer on a form that had no substance. Nevertheless the Empire of Austria has continued to occupy a peculiar position in the affection and confidence of the Papal councils. When the Austrian minister ventures to announce what purports to be an interpretation of the Pope's letter we may assume that the construction he puts upon that document is not entirely without sanction.

Coming at the same moment with the revelations of the duplicity of Bolo Pasha and Count von Bernstorff, the attitude of this mouthpiece of the Huns as an advocate of a peace to be established on broad humanitarian principles is interesting. Likewise Bolo Pasha and Count von Bernstorff were preaching the doctrine of love and amity among men, and at the same time buying newspapers to promulgate falsehood in order to befuddle and mislead the world into acceptance of a 'German peace.' Bolo Pasha was the arch peace propagandist, encouraging well-meaning sentimentalists to impede, in the name of humanity, every sane effort to prepare for the protection of our country, and now we find that he and the former ambassador from Germany were distributing a vast 'slush fund,' a nasty name that suits the nasty purpose to which it was applied. Our good people who used to preach the doctrine of submission and national impotence so earnestly should now be filled with shame and remorse. Their folly has helped to plunge us into a war from which we would have been saved by preparedness. Count Czernin, however, is clearly of the opinion that the pacificists have not fully recovered from their delusion, and he poses as the modern Melchizedek. In his speech at the dinner given by the Hungarian minister, Dr. Alexander Wekerle, he insists over and over upon the willingness of the Central Powers to come to an agreement for peace upon the basis of disarmament and arbitration. This is the fundamental idea in his announcement. As a corollary to it is, specifically, naval disarmament, although he would not extend this to the narrow or territorial waters, whereby the high-seas would become severely modified by application of the principle of the *mare clausum*. Conceding these two basic doctrines, he argues that "every ground for territorial guarantees disappears," which is a unique way of proposing to defend the principle of nationalism. He might have stated this thesis more boldly had he been convinced of the essential benefit of nationalism as a factor in human development; but he was thinking of the anger of an outraged world that might disrupt the two empires that have filled the earth with destruction and sorrow. Finally he proposes the relinquishment of economic war . . . by the opponents of Germany and Austria! This is to trifle with the one great principle that could insure the peace of the world. He would apply it to his enemies, not to the Central Powers which have not forgotten in the midst of war to make preparation with increased factories and an augmented merchant marine to go forth under the shield of Teutonic protectionism to the commercial conquest of the Earth after Mars has sheathed his sword. Disarmament on land, disarmament on the high-seas but not in the *mare clausum*, absence of territorial guarantees, and abandonment of the economic war by the enemies of the Teuton only, is the "pacific and moderate programme" set before the world by the spokesman for the Kaiser under cover of Austria's historic pretensions to second, as a temporal power, the spiritual manifestoes of Rome.

Mr. Hoover and Sulphuric Acid

Great Britain, by order of the Minister of Munitions, has fixed the price of super-phosphate, based upon the percentage of tri-basic or mono-calcium phoshate present. The maximum prices are $22.50 per long ton for super-phosphate containing from 15 to 16% of the soluble salt. This shows that the British farmer is worse off than the American, since the current Atlantic seaboard quotation for acid phosphate is from $16 to $18 per ton. Meanwhile crude unwashed Spanish pyrite, averaging 48 to 52% sulphur is worth 16 cents per unit of 20 pounds at Philadelphia and Baltimore, a price that has remained stationary since March last, although the quotation on acid phosphate has risen 34% within that period. The difficulty of obtaining pyrite is, therefore, not as real as it has been represented, despite the sudden flurry at the beginning of the ruthless submarine campaign. Another interesting circumstance is that sulphur has likewise sold quite steadily at $35 per ton in the open market for many months, and the demand, though large and insistent, has not justified the erection of expensive plants to produce it in large quantity from the abundant low-grade deposits in various parts of the country. It is difficult to penetrate the mysteries of commodity markets, but it is quite certain that no real shortage of raw material for making super-phosphate exists. Enormous deposits, actually stupendous deposits, of high-grade phosphate rock lie untouched and undeveloped in Utah and Idaho; the great smelters within range of these supplies are just beginning to convert their waste fume into sulphuric acid, and are doing it cautiously with full realization that the demand for fertilizer is not sufficiently acute to warrant the building of elaborate plants. Florida phosphate rock averaged about $3.45 per long ton before the War, and is now quoted at from $3.25 to $6 according to grade. The average has not increased in proportion to the advance in price for the finished product. California has an enormous developed tonnage of pyritic ores that will assay more than 48% sulphur and 1% copper with small amounts of gold and silver, and these could be loaded on vessels at San Francisco bay at a cost that would compare not unfavorably with Rio Tinto pyrite put on ship at Huelva. The Calumet & Arizona Mining Company has such great quantities of pyrite developed in its Junction and Briggs shafts at Bisbee that it could produce 1000 tons per diem of material that would average higher than 45% sulphur and 2% copper, with a fair assay value in gold and silver, if the Council of Defense deem it important enough to arrange an emergency railroad-rate for the sake of hurrying this acid-making material to the producers in the East. We have heard of the struggle of the acid manufacturers to obtain sulphur-burners to replace their pyrite roasters, in order to maintain the supply of sulphuric acid for munitions and fertilizers, and we also hear of restrictions upon the exportation of sulphur; and 60° sulphuric acid, furthermore, is still selling at about $25 per ton, although it formerly sold at $6 and less when Spanish pyrite was laid down at the works for 9.75 cents per unit of sulphur. The elevation in the price of Spanish pyrite has been 64% against a corresponding rise of 316% for the acid. Here is a trade mystery that needs explaining. The world measures its advancement in the arts by its consumption of sulphuric acid, as emphasized by Disraeli and other statesmen and thinkers of long ago. It is as true a measure as can be found, and the selling price of this essential may also serve as a measure of commercial values in other industries. In this instance it may prove to be a convenient measure of inflation, and may merit the serious attention of the price-regulators. The cost of fertilizers is vitally related to the cost of food, to the cost of labor, and in logical sequence to the cost of producing metals, and Mr. Hoover may become exercised over the fact that the farmers, shrinking from the enormous prices, are adopting cheaper makeshifts in fertilization. These may be useful in their way, but they would be more valuable as a sure reliance if combined with plenty of the old dependable acid phosphate. The farmers are even skimping the land, putting it on short rations, so to speak, as well as the tiller with his meatless and wheatless days. 'Hooverizing' is excellent as a means of economy and for the cure of over-worked livers; a less bilious people will do better work in overcoming the unspeakable Teuton, but we must not apply Hooverizing to the soil. Evidently there is no need, if the powers possessed of the proper authority look carefully into the question of fair prices for sulphuric acid and super-phosphate, now that the locked treasure of potash in Searles Lake has been set free by recent legislation, so that this necessary may be abundantly supplied.

Mill-Tests v. Hand-Sampling

The Editor:

Sir—I have read with interest the article by Mr. Morton Webber on this subject appearing in your issue of July 28. I was particularly struck with the care in the example cited called the Red mine in obtaining a representative sample for a mill-test that would reflect the proper proportion that each stope of the mine would contribute, if operated under large-scale conditions. As Mr. Webber points out, it is impossible to break down the backs of a stope in a mine under examination so that each stope will contribute its proper portion to the mill-test. He states one stope will break down 50% over its proportion, while another would break 100% in excess of its proportion. He states it is here most mill-tests or car-shipments to a smelter (for the same purpose) fail. In this I thoroughly agree. The method devised by Mr. Webber depending on a "predetermined foot-weight" is, I think, the most accurate and practically convenient method that could be devised. The intention to break more rather than less than the required foot-weight is also an important point, the idea being, as explained, to prevent going back and tampering with the face to obtain the deficient portion.

My object in contributing to the subject is because I am a believer in the mill-test if properly conducted, but my experience is that extremely few mill-tests are properly conducted; for the reason that the above fundamental requisites are not usually considered. There are many text-books on the moil-sampling of mines, but I know of no standard work on the mill-test as a mine-sampler. Many of us, at one time or another, have shot-down various stopes in a mine and flattered ourselves we were 'sampling' the property, when as a matter of fact we did not, because we failed to realize the fundamental necessity of the 'foot-weight' as described in Mr. Webber's article.

Referring to the last paragraph of the article where reference is made to the mill-test as an indicator of the sampling-error for hand-sampling large low-grade deposits, it is not clear how such a method should be devised—as the detail is not outlined.* If, however, it was possible to divide the mine into zones using the mill-test as the sampling-error indicator for various zones, as described by Mr. Webber, an obvious advance would be gained, and if practicable, would no doubt have prevented the failure of the Alaska Gold property, which, I understand, was the inception of your editorial and the interesting and constructive outcome that followed in your paper.

CHARLES BENNETT.

London, September 3.

*This was explained in detail by Mr. Webber in our issue of September 29, 1917.—EDITOR.

Secondary Zinc Deposits

The Editor:

Sir—In his interesting and clear-cut article, appearing in your issue of May 19, Dr. Nason advances a theory of enrichment of secondary zinc deposits by the subtraction of gangue-material. This clarifies some puzzling problems encountered in the study of oxidized zinc ores. I know that Dr. Nason is particularly qualified to write on this subject and that it was chiefly through him that the great zinc mines of the Empire Zinc Co. at Hanover, New Mexico, were discovered. Nevertheless, I do not believe that his theory of residual zinc deposits is applicable to the zinc deposits of Hanover, as he states in the article.

The zinc ores of the Hanover district are found in the sedimentary metamorphic zone, surrounding a quartz-monzonite-granite batholith. The sediments are heavily garnetized, and all the primary ores of the district are zinkiferous garnets, the alteration varying from slight to complete metamorphism. It is evident that the zinc accompanied the other volatile garnetizing agents and is a constituent part of the rock-mass.

There appear to be two horizons of garnetization, (1) the upper portion of the crystalline lime and the blue lime immediately above it, and (2) the lower portion of the crystalline lime and the basal sediments underlying that. The crystalline lime is approximately 150 ft. thick and is about 200 ft. above the basal granite. Late evidence, namely the finding of zinc-sulphide ore at a depth of 200 ft. in the Republic mine, where the sediments are known to be thin, lends weight to the theory that garnetization and mineralization will extend down to the basal granite.

The oxidized orebodies are found immediately below the upper garnet. This formation is usually the crystalline lime. When precipitated under these conditions the ore is always a carbonate, as are the ores of the great Antonio mine, discovered by Dr. Nason. But in one important instance, the Pewabic mine of the Hanover Copper Co., a stratum of silicious shale lies between the zinkiferous garnet and the crystalline lime; and here the oxidized ore is the hydrous zinc silicate.

It is my contention that the oxidized ores of this district are a direct result of the decomposition, by meteoric waters, of the upper zinkiferous garnets; that the zinc, instead of being residual, has been the first metal to migrate, being re-precipitated by the first precipitant encountered; and that the character of the ore, whether carbonate or silicate, is controlled by the character of the precipitant.

It appears to me that this theory is borne out by the conditions in the field. The first indication of an oxidized orebody, not actually exposed by the erosion of the decomposed garnet, is a low-grade residual iron gossan, which shades off quickly into a soft decomposed hornblendic material, undecomposed portions of which show zinc and often lead sulphides. The decomposed hornblende usually contains small amounts of oxidized zinc and lead. Underlying this decomposed garnet in the formerly barren crystalline lime are the secondary zinc ores. They are peculiarly free from all impurities. Inasmuch as the primary ores usually contain lead, iron, etc., and the oxidized ores are free from them, it is evident that they are the result of the selective leaching by descending solutions.

In one remarkable instance, the Bond mine of the Hanover Copper Co., the garnet in decomposing gave up its silica as well as its zinc. Crystalline lime underlies the garnet at this point. The resultant orebodies are enclosed in a true 'honeycomb' quartz, each cell filled with zinc. This ore effervesces weakly upon the application of hydrochloric acid, showing carbonate, but along the walls of the cell the ore has the radial crystalline structure of calamine. I have never found this ore to assay above 33% zinc, but when exposed by erosion the zinc leaches out, leaving the quartz behind as light as pumice. Below this pumice-like quartz, in the crystalline lime, the purest and highest grade of smithsonite ever found in the district has been mined.

Dr. Nason refers to the carbonate orebodies that abut against a low-grade zinkiferous garnet in the Antonio mine. I have studied this particular condition and I am convinced that it is due to faulting, there being absolutely no shading of one formation into the other. The change is abrupt and clearly defined. Along the course of the fault a little oxidation occurs in the garnet, but these ores are always low-grade and impure, while the abutting carbonate orebodies are of an exceptionally fine grade and quality.

In these carbonate orebodies of the Antonio mine and in the other oxidized ores of the district, residual pebbles and boulders of the replaced formation are common. These are always absolutely barren and unleached. The attacking solutions have come from without and not from within. A boulder presenting the appearance of solid ore is often found to have only an outer shell of ore, a second shell of gypsum crystals and ore, and a core of crystalline lime. Gypsum crystals are always most plentiful (except where found in vugs) where an orebody has floored on a solid bed of the lime, which lies in a manner to check or completely impede the descending solutions. The presence of gypsum crystals in the oxidized orebodies seems to indicate that the zinc came down in the form of a sulphate and that the result of its reaction with the lime was zinc carbonate and calcium sulphate.

In the Maggie mine, five miles north of Hanover, in Little Shingle canyon, a very low-grade zinc carbonate changes within a few feet to sulphide. The surface ore assays from 3 to 8%, and the sulphide from 10 to 25% zinc. The gangue here instead of being zinkiferous garnet is an almost unaltered soft lime-shale, and the conditions are ideal, apparently, for the operation of the theory of segregation of the zinc through the leaching of the gangue-matter; yet the converse is true.

During the past two and one-half years I have studied every known zinc deposit in this district, and am convinced that the theory of replacement of the limestone by descending zinc-bearing solutions, whereby a barren formation has been transformed by constant enrichment to commercial ore, is the only theory that accounts for the results.

CLYDE M. BECKER.

Hanover, New Mexico, September 25.

Mining Laws

The Editor:

Sir—In making recommendations for needed changes in the mining laws at present in force in these islands, I have been impressed by the lack of a proper classification of claims. I am therefore proposing the following classification, which I believe covers the field.

(1) LODE CLAIM: This shall be understood to apply to all lands containing veins or lodes of quartz or other rock in place, bearing gold, silver, mercury, tin, lead, zinc, copper, tungsten, or other valuable metals or the minerals thereof, in workable or appreciable quantities.

(2) PLACER CLAIM: This shall be understood to apply to all lands containing deposits of an alluvial nature, bearing gold, silver, platinum, or other precious metals in the native state, or containing minerals of tin, tungsten, copper, lead, zinc, or of other metals, susceptible of recovery by washing or concentration from such alluvium.

(3) FUEL CLAIM: This shall be understood to apply to all lands containing or suspected of containing deposits of coal, oil, peat, bitumen, asphalt, natural gas, or other hydro-carbon substance capable of being used commercially for fuel or in manufacture.

(4) QUARRY CLAIM: This shall be understood to apply to all lands containing stone or earth or earthy minerals of commercial value, such as building stone, limestone, iron ore, manganese ore, salt, sulphur, infusorial earth, guano or other phosphates, soda, potash, nitre, and all other non-metallic mineral substances of commercial value, including cement rock, sand, and gravel.

C. M. EYE.

Baguio, Philippine Islands, August 24.

VILLISTAS ON THE MARCH

THE RAVAGES OF WAR

Recollections of Mexico

By A REFUGEE

Now that we are safe on this side of the Rio Grande, where neither Carrancistas, Zapatistas, Villistas, nor any other kind of 'istas can touch us, even though we tell the whole truth about them and their unhappy regimes, we can indulge in a few reminiscences. Since the revolution began in 1910, the life of the American residents of Mexico is best described in the words of that little verse that goes something like this:

"In again, out again,
On again, off again,
Home again, Flanagan!"

For in the last five years as many orders have come from Washington for its citizens to "get out" of Mexico, as though "getting out" of your home was the merest trifle in the world. The only way to appreciate just how we felt when these famous warnings came is to imagine what it would mean to you to be ordered periodically to leave your home and business, knowing that when you returned you would find, at least, your interests neglected, if not ruined. Caretakers are only human, and get tired of a continued performance; in time they begin to feel that they own what they were left to care for. And yet that is what the citizens of the United States living in Mexico have been forced to do repeatedly: to leave their all. If we had known even how long we would have to remain away, we might have arranged our affairs more advantageously. Often we left, thinking that it would be necessary to remain away but a few days, until this or that trouble blew over. But the uncertainty of safety would oblige us to stay away for months at a time. We could make no plans, as our future was so uncertain, always expecting that the American government would take some steps to safeguard our lives and properties, and that we would be able to return to our homes. Surely those who blaze the way for American trade and commerce deserve protection. One of the hard parts, during these enforced absences, was to maintain a second home in the United States while our houses in Mexico stood vacant. After becoming discouraged with watchful waiting, we would return to take our chances once more in Mexico until the 'next time;' which was never far. Remedies for our situation have never been wanting. A favorite one is "Why don't you sell out all your interests and leave Mexico for good?" But who will buy? The wealthy Mexicans are all living in exile, and the Carranza government has decreed that no more foreigners may acquire property in Mexico. Even if we could realize on our Mexican holdings, there would be difficulty in finding any one rash enough to venture now into that disordered republic, in spite of its vast resources. And it is one easy for anyone accustomed to doing business in Mexico, where different modes, customs, and laws obtain, to adapt himself to new ways in this country or to stand up against the competition of established firms. We have even been advised to live with relatives, until such time as Mexico is normal. Though they too urge us to flee when danger threatens, we would be precious welcome, as anyone who has ever tried this expedient can testify!

Then again we are asked "What do you gain if you do save your property by endangering your life?" Most of us feel that there is small choice between being shot if we stay or of starving if we go.

Advice is never lacking, but no relative and no government offers to pay our expenses or set us up in business. We are simply advised to go, at any cost, no matter what the nature of our business, the limit of our income, or the size of our family. After the first few warnings from Washington came to naught, we decided not to heed them in the future, so when the next order to leave came, we refused to heed our consul, though he notified us daily of some special refugee train from farther south that was passing through our town. Finally the consul was recalled; believing then that intervention was at hand, and that the last train was leaving, we made up our minds to go too just one hour before train-time. How we rushed around, throwing things into trunks and suit-cases, food into baskets, filling bottles with water, for well we knew from past experience that there would be nothing on the train. There was no time to call a *coche*, so we rode down to the station atop our trunks in an express-wagon. So great was our haste, we left the house with our shoes unbuttoned, but one of us had the presence of mind to grab a button-hook, as we rushed out! We reached the station just as the conductor called out *vamonos*, which, being interpreted, means 'All aboard.' But often when instructions came to leave the country, there was nothing to go in but box-cars, and many of us have made the trip thus, perched on a suit-case but glad of the chance to be able to ride at all. They could hardly be called *de luxe*—the trains in which we have traveled in our various flights from Mexico.

Even on those rare occasions when we were fortunate enough to get a Pullman or day-coach, there were never any lights, or water, or anything but the car. Imagine a pitch-dark car at night, without a porter or conductor even. As to bed-clothes and food, we carried our own, and always had to provide for several weeks, as we never knew how long we might be on the road. Mere special and passenger trains were side-tracked at any old place in order to accomodate the many military trains on the road. But we did not waste time pitying ourselves; we considered ourselves lucky to get away with a whole skin. There were times when warnings came to quit the country, that the train service was so poor that the refugees would have to wait at the station several days for a train, going home to eat and sleep, but coming back to their post until a train happened to come along. There was no regular schedule, and the station-agent either would not or could not tell when a train was due. These exoduses had their amusing as well as pathetic sides. One time a Scotchman was commissioned by the foreign consuls to go to the outlying districts in our State to notify all foreigners living there to come into town in order to leave the country on a special train that was to go on a certain day. After advising those living near-by, the Scotchman decided to save himself a long horseback ride by telephoning the news to a distant American mining camp, from the little village he was in. There was only one long-distance telephone and that was in charge of the Mexican military authorities, and they were not supposed to know until after the exodus. Nevertheless the Scotchman asked permission to use the telephone and his request was granted, but on condition that his business be transmitted in Spanish. Now the Scotchman's knowledge of Spanish was very limited, consisting of several words only; besides, the nature of his message was rather delicate to have the military authorities know. Possibly that is why they insisted upon the Spanish, having surmised the situation. However, the poor Scotchman used all the Spanish at his command, but the American mine-owner at the other end of the line failed to get his meaning. Finally, to the relief of the distracted message-bearer, the soldier on guard relinquished his vigilance long enough for the Scotchman to get in two good old English words, "Beat it." Needless to add, it did not take the Americans at the mining camp long to get his meaning and act on his suggestion.

Once, having been asked by a friend in the United States to explain the exact procedure of getting the Americans out of Mexico, every time relations between the two countries were strained, the inquirer replied, in all seriousness, "Oh, I thought that the President wrote each one of you a note."

A mildly pathetic incident happened just before our last flight from Mexico. A foreign-refugee train from farther south was detained for days by military operations, in our home-town in the north. As the refugees had several little babies and children and had been on the road so long, in crowded day-coaches, the American residents of the northern city invited the refugees to their homes that they might rest, refresh themselves with baths, etc. Of course, the refugees dared not leave their train without some guarantee from the military authorities that their train would not be ordered to move during their absence. But this the authorities refused to do, so the tired travelers spent many more weary hours in their crowded cars, and instructions to pull out came several days after.

In our various flights to the border we have had many a thrill. The last time we fled from the wrath of Villa, suddenly our train came to a stop, away out in the desert, where there was no sign of human habitation. As we were close to one of the bandit's favorite hiding-places, we were thoroughly alarmed to see a large body of horsemen approaching the train. The soldiers that the military authorities had sent to guard our train grabbed their guns as they jumped to the ground and the male passengers, also heavily armed, took positions at the doors and windows of the car, ready for any emergency. The captain of our escort went bravely to challenge the men. Imagine our relief to learn that they were Government scouts and not bandits coming to hold up the train, as we had feared. Our train carried no lights, not even a headlight, but as darkness came on we felt more and more apprehension until the last Mexican town was reached and only the international bridge lay between us and safety. Across the river on all sides we heard the slogan, 'Remember Columbus; remember Santa Ysabel.' But we were trying to forget.

Some of the refugees have established themselves in business in this country, but many are idle, depleting their small fortunes, the fruit of years of labor in Mexico. And the exiled Mexicans, whose only crime was the possession of wealth, dare not return to their native land. Their confiscated houses and lands have not been returned to them, as Carranza promised. Their beautiful homes have continued to be used as barracks, stables, and even abattoirs, exactly as in the days when Villa ruled. Though we hoped for better things when Carranza came into power, the situation remains the same. When the City of Mexico was under Carranza's rule and we in the North were in Villa's dominion, a friend wrote "While

COMMISSARY OF THE MEXICAN ARMY

MACHINE-GUNS IN CAR AND SOLDIERS TO GUARD REFUGEE TRAIN

you are praying for Carranza to come into power there, we are just as earnestly wishing for Villa here, for while Villa may not be any better than Carranza, he cannot possibly be any worse." All industries are paralyzed, chiefly because Carranza has made such a fiasco of finance. Had he allowed the old Federal bills to circulate instead of throwing them out, thousands of people would have been saved from ruin. Instead, he reeled off his own fiat money, as a printing-press makes handbills. And although it had no backing, Carranza declared each paper dollar worth ten cents gold. He, of course, had dreams of confiscating mines, oil-wells, etc., to back his currency. To appreciate the havoc wrought by the repudiation of the Federal money, imagine if you can, a president of the United States declaring null and void the legal tender of the country. Outside of a few native products, such as corn and beans, most articles of food and clothing come from the United States, so the merchant must pay gold for his goods. How then can he continue to do business in Mexico while he is obliged to sell for worthless paper money, as he has had to do the past few years? Just now silver is again in circulation. Of course, if he could charge paper or silver money in proper proportion to his outlay, it would be another story. But the Government has made decrees forbidding exorbitant prices. They fix the value regardless of the merchant's purchasing price and the gold duty they have been known to charge on imported goods, even during times when it was a crime for a private individual to refuse the paper money. The reason given for forcing the merchants to sell at impossible prices is to help the poor. How they have been helped, the spectacle of the country in ruins, financially, industriously, and literally is the most significant answer. The poor were a thousand times better off when there was honest labor for everybody, and when their wages were paid in real money.

The poor, as well as the rich, the native as well as the foreigner, all have suffered at the hands of the brigand governments of the past five years. The only ones who have benefited are the military grafters themselves. We are tired of war, we do not want either internal strife or outside intervention. All we hope is that any faction strong enough to preserve peace and order may attain ascendency, so that we may return to our abandoned homes and neglected businesses, exiles and refugees no longer.

> "Rough-piled, far-flung, unending, range on range
> And still beyond all wrapped in purple mist,
> Are mountains dimly beckoning."

IN AN ADDRESS by W. S. Kies at the Chemical Exposition in New York he stated that in 1914 this country exported, in round numbers, chemicals, drugs, and dyes to the amount of $27,000,000. In the fiscal year 1917 these exports amounted to $185,000,000; but this does not tell the full story because the list of exports does not include the great group called 'explosives,' which is so closely allied to the chemical industry as to be in fact a part of it. In explosives the value of the exports grew from $6,000,000 in 1914 to $820,000,000 in 1917. Under this class were listed cartridges, dynamite, and gunpowder. Under the heading 'other explosives,' the exports grew from $1,000,000 in 1914 to $420,000,000 in 1917, showing that in the industries closely allied with chemistry the growth has been quite as striking as in chemicals proper. In the articles which may be considered as strictly chemicals, we exported in 1914 $500,000 worth of acids; in 1916 $24,000,000, and in 1917 approximately $55,000,000 worth. Our dye-stuffs exports have grown from $333,000 in value in 1914 to $12,000,000 in 1917, and these latter figures do not include foreign dyes re-exported. A few other illustrations may be of interest. Our exports of soda salts and preparations grew from practically nothing in 1914 to $18,000,000 in 1917. The importance of the chemical industry is evideuced by comparison of the amount of capital and the value of the products turned out with that of other leading industries. According to the census of 1915, the value of capital of the group known as 'chemicals and allied industries' was $723,000,000 in 1914. The capital of the automobile industry in 1914 was $408,000,000, and the capital of the flouring-mill industry was $380,000,000. The amount of capital invested in the group distinctly classed as chemicals was, in 1914, $224,000,000, which represents an increase of nearly 800% since 1880, and the value of the products was $158,000,000, or 70c. per dollar of capital invested.

A New Canadian Mining District

By WALTER KARRI-DAVIES

In Australia a new chum takes to the bush and a sundowner goes to town to knock down his check. In South Africa a roineck treks across the veldt and a backvelter inspans for the nearest dorp to nacht mahl. In Canada a greenhorn hits the trail and an old-timer goes to town to have a blow-out.

In taking these jaunts, thoughtful kindness on the part of all concerned is important. The old-timer is just as apt to require friendly guidance when he steps into his boiled shirt at Ritz-Carlton as the greenhorn when he steps into his canoe in north-western Canada. It would be surprising if it were otherwise. One finds it easier to forget than to remember that this is true.

It is a little difficult to say where one first hits the trail. A Pullman sleeper and a dining-car can no more be called the 'trail' than the new stern-wheel river steamer 'Nipewin' with its comfortable berths, excellent food, and good attendance. These luxuries are better described by saying that "we struck it rich." In our case we might fairly say we hit the trail when we started from Sturgeon landing. It was there we took to our feet and launched our canoes.

We started from Winnipeg, Canada, by train for The Pas at 9 a.m. on the 6th of August and arrived at our destination at 8:30 next morning. The Pas is a frontier town, but it is clean and comfortable. Take a map of North America, find Lake Winnipeg, picture yourself standing on the north-west corner of the lake facing west. Walk 90 miles due west and you will come to the place where Sir John Franklin's sailors wintered. You may enter the little church and see the seats that his sailors carved, all of which are in good preservation. This is The Pas, in Manitoba, on the bank of the Saskatchewan river.

At 1 p.m. on August 9 we started down the river on the 'Nipewin' of the Ross Navigation Co., Captain H. H. Ross in command. The Captain belongs to the cultured pioneer type. In normal times he spent part of his life shooting pheasants in England and the rest of the year shooting rapids on the Saskatchewan.

It was from my comfortable shake-down on the deck of the 'Nipewin' at about midnight that I first saw the northern lights. They are too beautiful to describe; all I can do is to write my impression as I wrote it for my children. It seemed to me as if the great torch-lights of eternity were being flashed by The Almighty to light the fairies on their joyous way from the stars and planets to Heaven, where they arrived bright and happy, and the angels tucked them in and kissed them good-night; then the lights went out and all was peace.

Next day we arrived at Cumberland House, which is the oldest Hudson Bay trading-station in Manitoba, having been established in 1774 by Samuel Hearne. Crossing Cumberland lake and Namew lake we arrived at Sturgeon landing at the mouth of the Sturgeon river at 5:30 p.m. on the 11th. This trip can be done in 16 hours, but we were delayed owing to the necessity of tuning up the engines on a maiden voyage.

The name Sturgeon landing expresses all there is to say about this spot. From here last winter the Mandy mine transported about 3600 tons of high-grade copper ore over the ice from the mine on Schist lake, about 65 miles away, the ore being loaded on barges and towed to The Pas, whence it was sent by rail to the smelter at Trail, B. C., 1200 miles away. I understand that the ore assays about 22% copper, 7% zinc, $2 gold, and about 3 oz. silver per ton. A heap of about 1500 tons of ore was still awaiting shipment. Messrs. H. C. Carlisle, D. M. Haynes, and C. B. Morgan, together with 100 men and 92 teams of horses were responsible for this creditable piece of work of hauling the ore in the depth of winter over the ice.

At 5:45 a.m. on August 12 we sailed forth from Sturgeon landing, our party consisting of Messrs. Jack F. Hammill, Henry C. Perkins, John H. Black, William Wallace Mein, the brothers Dan and John Mosher, myself, and four Indians, together with three canoes, tents, mosquito-nets, eiderdowns, food, fishing-tackle, guns, knapsacks, etc. The natives poled the canoes up the rapids with two of the party to help. The rest of us tramped for nine miles, the walking being over moss, which is called 'muskeg' in Canada. A two-horse wagon carried most of our equipment and one canoe, and we got on the wagon to cross Sturgeon river. Mr. Black took the opportunity of examining the bed of the river to see if it was suitable for a railway-bridge. He decided it was, without delaying us. After walking another two miles, we came to Goose river, where we loaded everything into the canoes and sent the wagon back.

On the way up the river we landed and lunched, and then crossed Goose lake, landing on Gooseberry island on the north side at 3:30, when we camped for the night, having a swim in the lake before turning in. The island gets it name from the wild gooseberries growing on it. Starting at 5:45 on the 13th, we paddled most of the day, making several portages. Then we crossed Athapapuskow lake, a most beautiful sheet of water with many lovely islands, which look like botanical gardens. The formation in places was limestone, and in other laces schist and quartzite. At 5:30 p.m. we arrived at ie Mandy mine on Schist lake, a distance of about 65 iles from The Pas. We camped for the night on the ainland near the island on which the Mandy is situ- ted. At the mine, which is owned by the Tonopah ining Co., we found Messrs. Graham and Kennedy in shore of the island. Reynolds went into the bush a short way and pulled up some muskeg, under which he found copper pyrite. Jackson had said to his mate Reynolds when they first went out prospecting, "Now, if we find anything worth while, I want it to be called the Mandy, as it is a name my wife fancies"; so when Reynolds saw

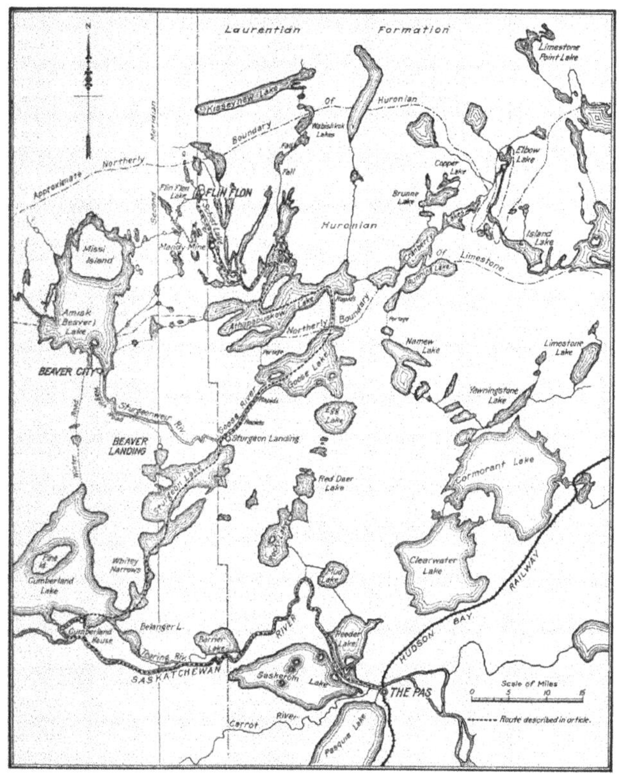

PART OF MANITOBA, SHOWING NEW MINING DISTRICT

irge, and they extended the courtesies that real min- men always offer each other the world over.

The history of the Mandy mine is interesting. A dis- ery of copper had been made on the mainland, and the point of the island, claims being staked by the sher brothers. Two other prospectors, F. C. Jackson Sidney Reynolds, later paddled their canoe to the the copper ore he called out to Jackson, who was still in the canoe, "Well, here's your Mandy for you." The claim was pegged and registered in Jackson's name. He sold it to the Tonopah company for a 15% share in profits on condition that they develop it. Now he de- nies that Reynolds had any interest, and the case has been heard in court, but no decision has yet been given.

I do not know what the legal position is, but what I do know is this, that when a man goes out prospecting, and discovers something of value, he should benefit by his discovery and it should not be in the power of any man to take this benefit away from him entirely.

The Mandy is a glory-hole out of which 3600 tons of ore has been taken. I understand about 20,000 tons of similar ore has been proved in the lode by diamond-drilling, besides 100,000 tons of disseminated ore that will require concentrating. In any case, there are some millions of dollars to be taken out of this property and I sincerely hope that those benefiting will recognize the moral, if not the legal responsibility to Reynolds, as I am satisfied that his story is substantially true.

A trip in the canoe and a portage on the 14th brought us to Flin Flon lake, on which is situated the large copper, gold, silver, and zinc deposits—now under option to David Fasken of Toronto. Several million tons of ore have been proved already by diamond-drilling. Here we found George Scarfe of Nevada City in charge. The name of Flin Flon was given to this property by one of the discoverers, who had read a book of this name, the hero having found a mine, a hole in the ground, out of which he eventually got untold wealth. The name will probably be changed, as it sounds a little too much like a gay lady robed in a modification of Highland costume, and it is thought that the name, like the costume, is not quite suitable to cover a proposition of such large dimensions. The North-Western Mining & Smelting Co. might be a more suitable name. It is too early to say what place this mine will take in the world; more boring is necessary. All that can be said at present is that it has held up to all that could be expected, and if it answers to its present promise it will be one of the big mines of Canada.

The people of the Province of Manitoba and the Dominion government will no doubt take steps to furnish the necessary transportation facilities. Power also will have to be developed. The new wheat railroad to Port Nelson will benefit if it has back freight from a thriving mining district.

The orebody is about 200 ft. wide; it has been proved by diamond-drill for a distance of 2000 ft. and to 600 ft. deep. The ore is wider southward where the last drill-hole went through it. Drill-holes show about 2% copper and 5% zinc, with $2.50 in gold and 1¾ oz. silver per ton. It is recognized by those directing the enterprise that a larger tonnage of such an ore will have to be proved before it warrants the big expenditure necessary to develop a productive mine.

Having spent a few days at the mine and enjoying the hospitality of Mr. and Mrs. Scarfe and Mr. and Mrs. Jack Callenham, we retraced our steps, carrying the impression that a new and important addition has been made to the mineral resources of Canada.

CAUSTIC-SODA burns should be treated with acetic acid diluted to 2%. Stronger acid must not be used. This may be followed by treatment with carron oil.

Production of Pyrite

Never before was so much pyrite produced or imported by the United States as in 1916. The increase in production is to be attributed mainly to an increased yield from old mines rather than to the opening up of new deposits. The domestic production was 423,556 long tons, valued at $1,965,702, which is about 30,000 tons more than was produced in 1915 and was valued at about $290,000 more than the ore produced in 1915. The total consumption of domestic and imported pyrite amounted to about 1,670,000 long tons. In addition to the pyrite ores returns from acid manufactures show that 577,045 long tons of domestic copper-bearing sulphide ores, 196,404 tons of foreign copper-bearing sulphide ores, 531,625 tons of domestic zinc-sulphide ores, and 92,002 tons of foreign zinc-sulphide ores were treated in the United States in 1916 for their sulphur as well as for their metallic content. Of the pyrite 218,000 long tons, valued at $1,245,000, came from the Appalachian region, including the States of New York, Virginia, South Carolina, and Georgia; 36,000 tons, valued at $93,000, came from the region east of the Mississippi, including Pennsylvania, Ohio, Indiana, Illinois, and Tennessee; 23,000 tons, valued at $60,000, came from Missouri, Wisconsin, and South Dakota; and 145,000 tons, valued at $565,000, came from California. The quantity of crude pyrite was about four times the amount of concentrate sold, the record showing that 336,000 tons of lump ore, valued at $1,588,000, and 87,000 tons of concentrate, valued at $377,000, were produced. The importation of pyritic ores was the greatest in the history of the industry, being 1,244,662 tons, valued at $6,728,318.

BRICK from ordinary coal ash is one of the newer industries. A company has erected a plant in the East for making fire-brick and building brick from this material. Several thousand of such brick have been made for experimental and test purposes, and they are reported as having proved better than the usual fire-brick in the market. This is said to be due to the fact that coal ash possesses greater heat-resisting qualities than fire-clay. This has been recognized for a long time. It is stated, however, that heretofore brick makers have found it impossible to secure a suitable binder. The new process was invented by Earl V. Wagner, formerly a chemist for two large railroad companies. The new brick is lighter than commercial fire-brick. Last August the oil-fuel combustion chamber of the National Paving Co.'s plant at Milburn, N. J., was lined with the new ash-brick, and they gave three times the service of the brick formerly used, when subjected to a temperature of 3000° Fahrenheit.

SULPHUR in Sicily is sold subject to Government regulation of price. Many grades are produced, from the refined article to material as low as 65%. The refined sulphur is now selling for 46.03 lire per quintal of 220.46 lb., which is equivalent to $88 per short ton.

The Synthetic Making of Sodium Cyanide

By G. H. CLEVENGER

*Early in March, after the appearance of the article by John S. Bucher,† covering his work upon the fixation of nitrogen, I undertook a series of experiments to ascertain whether a synthetic process of cyanide formation employing carbon, a sodium salt, and the nitrogen of the air, together with a suitable catalytic agent, would be of such a nature as to give promise of developing into a commercial method for the production of sodium cyanide. It will be noted that the experiments follow closely those outlined by Bucher, with the exception of my proposal to leach the cyanized briquettes directly with mill-solutions rather than to attempt to produce a refined product.

Briquettes were prepared from the various mixtures according to the method suggested by Bucher. The iron and carbon were ground together for a period of about four hours in a large Abbé jar-mill. The sodium carbonate was then added, and the grinding continued until a thorough mixture resulted. The finely ground and intimately mixed constituents were moistened with hot water, and brought to a stiff paste in a water-jacketed kettle to avoid setting. A small household meat-chopper was remodeled in such a manner that the material fed to it, instead of being ground and torn to pieces, was forced straight out through holes 5/32 in. diameter. This gave briquettes which, when dry, were approximately ½ in. to 1 in. long, depending upon how they broke during handling. These were placed in shallow layers in iron pans, and dried on a hot-plate. The furnace used for heating the iron pipe containing the briquettes consisted of rings of arch fire-brick held together by hoops of heavy iron wire. Two gasoline burners supplied the necessary heat. The temperature of the briquettes was determined in the first experiments by means of a Le Chatelier thermocouple and a Siemens Halske millivoltmeter, but later it was found more satisfactory to use a Féry radiation-pyrometer, for the reason that the wires of the couple were rapidly destroyed by vapors which penetrated the iron protecting-tube, and, moreover, it was found to be rather difficult to introduce the couple into the charge in a satisfactory manner. In the first experiment an attempt was made to use a 4-in. gas-pipe, but in the later experiments it proved more convenient, on account of the small quantity of briquettes available, and the slow rate of heat-penetration with the larger pipe, to use 6-ft. lengths of 1¼-in. pipe.

Producer-gas was used throughout the tests as a source of nitrogen. This was made by blowing a carefully regulated blast of air through incandescent wood-charcoal contained in a 6-ft. upright section of 4-in. gas-pipe. The pipe was arranged with a grate at the bottom for supporting the charcoal, and a large removable plug at the top for charging and cleaning. This small producer worked satisfactorily throughout all the tests, except in the first experiment, when some difficulty was experienced on account of an attempt to use coke instead of charcoal. It was found difficult to keep the coke burning in this small apparatus, and ashes and clinker interfered also, whereas with the charcoal the light ash formed could be blown off readily at intervals by opening the large plug at the top of the pipe. The body of the pipe was jacketed by slipping over it a section of 10-in. light sheet-iron pipe, the annular space between the two pipes being filled with sand. The general arrangement of the producer and the cyaniding furnace is shown in the accompanying drawing. In making a run the briquettes were charged into a length of the 1¼-in. gas-pipe, which had previously been fitted at the bottom with a perforated supporting-disc. After the furnace had reached the proper temperature this was screwed into position so as to make a connection with the producer. After allowing sufficient time for the pipe and the contained briquettes to reach the proper temperature, the gas from the producer was turned on and the cyanizing continued for the period stated in each experiment. Upon the conclusion of each test the pipe was unscrewed, lifted from the furnace, and closed at both ends by means of caps which prevented the entrance of air. The contents were not discharged from the pipe until cold.

EXPERIMENT 1. Briquettes consisting of two parts coke, two parts iron-filings, and one part sodium carbonate were charged into a 5-ft. section of 4-in. pipe. This was placed in the furnace and heated to the temperature recorded in the accompanying log. When this experiment was made the difficulties of operating the small gas-producer had not been overcome, so that a 2-ft. column of coarsely crushed coke was placed at the bottom of the pipe, and air was passed through the charge instead of producer-gas. The briquettes were separated from the coke by means of an iron screen.

Time	Temperature, degrees C.	Remarks
1:00		Started burners.
2:00	600	Bottom of pipe red. Added a 2-ft. column of coke.
2:15	650	Added a 1½-ft. column of briquettes.
2:40	500	Increased heat.
2:50	600	Upper 2 ft. of pipe dull red. CO flame at mouth of pipe.
3:03	740	Slow stream of air passed through. CO flame increased.

*Published by permission of the Director of the Bureau of Mines.

†Jour. Ind. & Eng. Chem., March 1917.

Time	Temperature, degrees C.	Remarks
3:20	860	CO flame at mouth of pipe showing more sodium at temperature increases.
3:30	900	
3:40	950	
3:50		Air line out.
4:00	1920	
4:15	1030	

Upon the conclusion of the experiment it was found that the bottom of the pipe had melted, due to the intense heat produced by the combined effect of the burners and the combustion of the coke inside the pipe. The briquettes contained in the tube were sampled, and titrations for cyanide were made upon samples as below:

Sample	NaCN in equivalent of KCN,‡ per cent
1 (top)	0.15
3 (middle)	0.45
5 (bottom)	0.41

‡Throughout this article cyanide percentages are expressed in terms of potassium cyanide. The actual percentage of sodium cyanide present in any particular case would be lower, and may be calculated by dividing the figure given by 132.8%.

EXPERIMENT 2. In this an attempt was made to heat a section of the 1½-in. pipe electrically, but despite the fact that it was insulated on the outside by pulverized fire-brick, it was impossible to raise the temperature, with the current available, above a maximum of about 650°. However, producer-gas was passed through the pipe containing a charge of briquettes of similar composition to those used in experiment 1, but careful analysis of samples taken from various parts of the charge failed to show any cyanide. Failure to form cyanide was doubtless due to the low temperature.

EXPERIMENT 3. 950 grammes of briquettes of the same composition as used in experiment 1 was charged into a 1½-in. pipe. When the charge had reached about 900°C. a slow stream of producer-gas was passed through, and was maintained for 20 minutes after the charge had reached a temperature of 1060°. The total time of passing gas through the charge was approximately one hour. As soon as the gas was turned off the thermo-couple was removed. The pipe was taken from the furnace and allowed to cool. While it was cooling, the air had access to the pipe, as the opening of the thermo-couple was not closed. A small amount of water inadvertently found its way into the pipe, causing some decomposition. Titrations for cyanide upon three samples taken from various parts of the mass were made; the results are given below:

Sample	NaCN in equivalent of KCN, per cent
1 (top)	0.5
3 (middle)	9.5
5 (bottom)	0.075

EXPERIMENT 4. 950 grammes of the same briquettes as those used in experiment 1 was charged into a 1½-in. pipe. The temperatures throughout the test were as follows:

Time	Temperature, degrees C.	Remarks
1:15		Started burners.
2:00	900	Briquettes added, 950 gm.
2:30	920	
2:50	1020	Producer-gas turned on.
3:10	1070	
3:30	1090	
4:00	1100	
4:25	1100	Producer-gas turned off.

Upon conclusion of the test the pipe was removed and immediately capped and allowed to cool in the air. Samples were taken at various points in the charge and were found to contain the percentages of cyanide noted below:

Sample	NaCN in equivalent of KCN, per cent
1 (top)	10.2
2 (next to top)	18.5
3 (middle)	5.3
4 (next to bottom)	trace
5 (bottom)	trace

EXPERIMENT 5. 1772 grammes of briquettes was charged into a 1½-in. pipe which was placed in position after heating the furnace to 1000°C. During this test the gas issuing from the tube containing the briquettes was passed through a 0.3% solution of sodium hydroxide in order to determine whether there was any volatilization of cyanide.

Time	Temperature, degrees C.	Remarks
10:15		Pipe placed in position and burners started.
10:45	1020	
11:00	950	Producer-gas turned on.
11:20	1030	
11:30	1040	
11:45	1050	
12:00	1090	Producer-gas turned off.

Upon completion of the test the pipe was removed from the furnace, capped at both ends, and allowed to cool to room temperature. Titrations for cyanide were run upon samples from various parts of the charge (samples 1-5) and also upon a general sample (6) divided from a mixture of the whole charge.

Sample	NaCN in equivalent of KCN, per cent
1 (top)	0.075
2 (next to top)	4.1
3 (middle)	16.9
4 (next to bottom)	5.1
4 (bottom)	0.11
6 (general)	4.2

The total weight of the briquettes after treatment was 1500 gm. No cyanide could be detected in the sodium hydroxide solution through which the exit-gas bubbled. This would indicate that the loss of cyanide by volatilization under the conditions of this experiment was negligible.

EXPERIMENT 6. Briquettes composed of two parts coke, two parts roll-scale, and one part sodium carbonate were used in this test. The original weight of the briquettes as charged into the 1½-in. iron pipe was 15 gm. In drying the briquettes containing roll-scale it was noted that even when the pan in which the briquettes were placed to dry did not rest directly upon the hot-plate, but upon a thick sheet of asbestos, and when every precaution had been taken to dry at a low heat, the briquettes would suddenly become red hot and glow for some time as the drying progressed. The black color of the original briquette was then transformed to brown. Although considerable care was exercised in drying the briquettes containing iron oxide, this difficulty was not entirely overcome. The exit-gas was passed through a 0.3% solution of caustic soda, as in the previous tests.

Time	Temperature, degrees C.	Remarks
1:30		Pipe placed in position and burners turned on.
2:00	960	
2:06	1020	Producer-gas turned on.
2:20	1040	
2:40	1070	
3:06	1080	Producer-gas turned off.

Upon shutting off the producer-gas, the pipe was immediately removed from the furnace, capped, and allowed to cool to the room temperature. Cyanide was found in the various samples of briquettes as indicated:

Sample	NaCN in equivalent of KCN, per cent
1 (top)	7.8
2 (next to top)	19.7
3 (middle)	6.4
4 (next to bottom)	0.03
5 (bottom)	0.035
6 (general)	5.4

In this test the sodium-hydroxide solution through which the exit-gas bubbled showed a trace of cyanide, which indicated that there had been slight volatilization of cyanide. The weight of the briquettes upon completion of the experiment was 1000 grammes.

EXPERIMENT 7. Briquettes were made up from two parts of sage-brush charcoal, two parts iron filings, and one part sodium carbonate; 975 gm. of these which were somewhat bulkier than those previously used being charged into a section of 1½-in. pipe. After the furnace had reached a temperature of 1000°C., the pipe was introduced, and producer-gas passed through. The temperature at various stages of the test was as below:

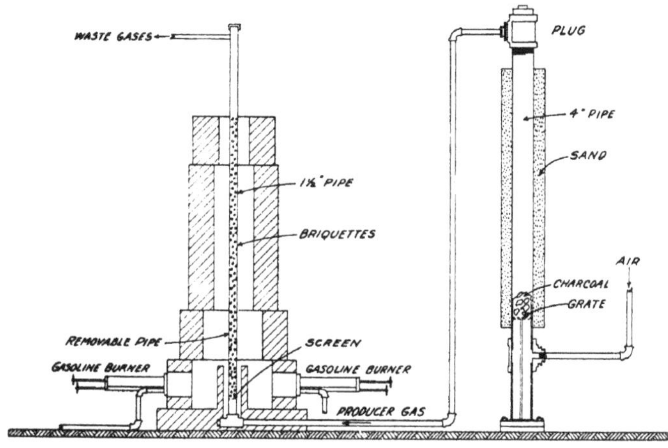

APPARATUS FOR MAKING SODIUM CYANIDE

Time	Temperature, degrees C.	Remarks
3:10		Pipe placed in position and burners turned on.
3:35	1020	
4:00	1070	
4:20	1070	
4:35	1050	Producer-gas turned off.

Upon conclusion of the test the pipe was removed, capped, and allowed to cool to the room temperature. Various sections of the charge showed cyanide as indicated below:

Sample	NaCN in equivalent of KCN, per cent
1 (top)	0.05
2 (next to top)	1.45
3 (middle)	6.60
4 (next to bottom)	3.70
5 (bottom)	0.12
6 (general)	4.3

No cyanide was found in the caustic-soda solution

through which the exit-gas bubbled. The briquettes in the bottom portion of the charge did not disintegrate and fall into powder, as was the case in the other experiments. The weight of the briquettes at the end of the experiment was 768 grammes.

EXPERIMENT 8. Briquettes were made from two parts coke, two parts Swedish iron ore (magnetite), and one part sodium carbonate. The weight of the briquettes which was charged into the 1½-in. pipe was 1208 gm. In this test the thermo-couple was damaged, so the temperature readings were low, and as a consequence the bottom portion of the pipe became overheated and melted through at one point. Producer-gas was passed through the tube for a period of one hour.

Sample	NaCN in equivalent of KCN, per cent
1 (top)	12.2
2 (next to top)	8.2
3 (middle)	0.3
4 (next to bottom)	trace
5 (bottom)	0.4
6 (general)	4.6

A trace of cyanide was found in the caustic-soda solution through which the exit-gas bubbled, indicating slight volatilization of the cyanide. The weight of the briquettes after treatment was 923 grammes.

EXPERIMENT 9. Briquettes were prepared consisting of two parts of oak-charcoal, two parts roll-scale, and one part sodium carbonate, 1585 gm. of these briquettes being charged into the 1½-in. pipe. Producer-gas was passed through the charge for a period of one hour. Temperatures as given below were determined by the Féry pyrometer:

Time	Temperature, degrees C.	Remarks
1:20		Pipe placed in position and burners turned on.
2:20	1040	Producer-gas turned on.
2:40	1050	
3:20	1070	Producer-gas turned off.

On completion of the test the pipe was removed from the furnace, immediately capped, and allowed to cool. Over one-third of the charge at the bottom of the tube was found in a pulverulent condition. Cyanide determinations were made upon various samples, as recorded below:

Sample	NaCN in equivalent of KCN, per cent
1 (top)	trace
2 (next to top)	13.0
3 (middle)	21.4
4 (next to bottom)	0.12
5 (bottom)	0.68
6 (general)	9.50

The weight of briquettes after treatment was 1089 gm. No cyanide was found in the caustic-soda solution through which the waste-gas bubbled.

EXPERIMENT 10. Briquettes similar to those used in the earlier tests, namely, coke, iron filings, and sodium carbonate, amounting to 1700 gm., were charged into the 1½-in. pipe. Arrangement was made in this case for passing the producer-gas through an emulsion of lime, with the idea of removing such CO_2 as still remained in the gas. The temperatures of the containing pipe as indicated by the Féry pyrometer were given below:

Time	Temperature, degrees C.	Remarks
4:30		Pipe placed in position and burners turned on.
5:10	970	Producer-gas turned on.
5:40	1030	
6:10	1070	Producer-gas turned off.

The total time of passing producer-gas was one hour. Cyanide determinations as tabulated below were made upon samples taken from various parts of the charge:

Sample	NaCN in equivalent of KCN, per cent
1 (top)	0.10
2 (next to top)	13.60
3 (middle)	28.30
4 (next to bottom)	1.70
5 (bottom)	trace
6 (general)	10.40

The total weight of briquettes after treatment was 1007 gm. A small portion of the charge at the bottom of the tube was unavoidably lost in removing the pipe from the furnace. No cyanide was found in the caustic-soda solution through which the exit-gas bubbled.

The highest maximum cyanide-yield was realized in this test, and a much higher total yield was obtained than in the preceding tests where the conditions were the same except for the removal of the carbon-dioxide from the producer-gas. This test would seem to indicate that complete removal of the carbon-dioxide is advantageous. However, it is possible that the small amount of water-vapor absorbed by the producer-gas in passing through the lime-emulsion is also advantageous. The opinions expressed in 'The Cyanide Industry' by Robine, Lenglen, and Le Clerc, appear to differ upon this point, some investigators claiming that a small amount of water-vapor is advantageous, while others claim that it is disastrous. Further experiments will be necessary to settle this point.

TREATMENT OF GOLD AND SILVER ORES. The question at once arose as to whether the cyanide leached directly from the briquettes could be utilized for the economic

Sample	Time agitated, hr.	Cyanide consumed, lb.		Percentage extracted			
				Gold		Silver	
		Comm'l.	Synthetic	Comm'l.	Synthetic	Comm'l.	Synthetic
B composite	24	0.48	0.66	98.8	97.6	74.6	8.3
B "	48	0.53	1.26	98.8	98.2	80.1	12.5
E "	24	0.40	0.78	95.9	94.7	73.7	55.3
E "	48	2.36	1.02	95.2	96.5	72.5	57.6
Average		0.95	0.93	97.2	96.7	75.2	33.4

Sample B assayed, gold, 1.66 oz.; silver, 1.72 oz. per ton. Sample E assayed, gold, 1.70 oz.; silver, 4.48 oz. per ton.

TEST WITH VIRGINIA CITY ORE

Time agitated, hr.	Cyanide Before test	Cyanide After test	Cyanide Consumed per ton	Alkalinity Points before	Alkalinity Points after	Assay of tailing Gold, oz.	Assay of tailing Silver, oz.	Percentage extracted Gold	Percentage extracted Silver
Synthetic cyanide									
24	0.101	0.08	1.26	47	67	0.02	1.98	96.0	74.5
48	0.101	0.079	1.32	47	30	0.01	1.35	98.0	82.5
Commercial cyanide									
24	0.1005	0.0785	1.32	51	43	0.01	1.35	98.0	82.5
48	0.1005	0.074	1.59	51	17	0.01	0.67	98.0	91.3

extraction of gold and silver in the well-known cyanide-process. In order to determine this point, the tests described below were made upon ores of different types. A small lot of briquettes was leached with tap-water and the strong-solution thus produced was diluted with water to produce a working-solution containing 0.123% of cyanide in terms of potassium cyanide. Agitation-tests employing this solution were made upon two samples of a Nevada gold-silver ore. The tabulation above shows the results obtained, together with the extraction previously made from the same samples when employing a solution of commercial cyanide such as is ordinarily used in the treatment of gold and silver ores.

The silver extraction with the briquette-solution is lower, particularly with sample B. The average extraction of gold is slightly lower, but the difference could easily have arisen in the tests, as each series was made independently by different men. In this case no attempt was made to purify the briquette-solution before applying it to the ore. Qualitative tests had indicated the presence of sulpho-cyanate, ferro-cyanide, and alkaline sulphides in the briquettes, but there was considerable variation in the proportion of these impurities present in the different samples. The lower silver extraction obtained in this series of tests with the briquette-solution may have been due to the presence of a small amount of alkaline sulphide, since no effort was made to remove it from the solution prior to use. In order to more thoroughly test the briquette-solution for ore-treatment a portion of the briquettes from Test 9 was leached with tap-water and the resulting solution diluted with water to a working-strength of 0.101% cyanide in terms of potassium cyanide. This solution was found to contain a small amount of ferro-cyanide and only traces of sulpho-cyanate and alkaline sulphides. A few drops of lead acetate was added to assure the absence of alkaline sulphides, and lime was used to give a protective alkalinity of 47 points. A solution was also prepared from commercial sodium cyanide, such as is commonly used in the treatment of gold and silver ores, containing 0.1005% cyanide in terms of potassium cyanide. This solution was similarly treated with a few drops of lead-acetate solution, and with lime to give a protective alkalinity of 51 points. The aim was to have both solutions in as nearly the same condition as possible. Parallel tests were made, as noted below, on three different types of ore, the only difference, within commercial limits, being the source of the cyanide used. A silver ore from the Presidio district in Texas, assaying 23.18 oz. silver per ton and but a trace of gold, was ground to pass a 100-mesh screen, and samples were treated with both solutions by agitation for periods of 24 and 48 hr. Lime of a very low grade, containing only 38.6% CaO, was added at the rate of 15 lb. per ton.

A silver-gold ore from Virginia City, Nevada, assaying silver 7.72 oz. and gold 0.50 oz. per ton, was then ground to pass a 100-mesh screen and was given the same treatment as in the case of the Presidio samples, the same grade of lime being used as in the former test.

Next a gold-silver ore from the Jarbidge district of Nevada, assaying silver 2.08 oz. and gold 1.65 oz. per ton, was ground to pass a 65-mesh screen and was treated similarly to the Presidio samples, the same lime, in the proportion as before, being added.

With the exception of the silver in the Ophir ore the results upon the two solutions check within the ordinary limits of experimental error. However, in this case the difference might have arisen through irregularity in the sample or through an error in assaying. On the whole it would appear from these small-scale tests that the synthetic cyanide is about as effective for the treatment of gold and silver ores as the commercial cyanide ordinarily used. Nevertheless, cyclic tests should be made, covering a considerable period of time, in order to determine whether harmful impurities would accumulate in the solution. A tentative flow-sheet is offered that shows how the proposed synthetic method of cyanide-formation might be made to fit into the operation of a cyanide-

TEST WITH PRESIDIO ORE

Time agitated, hr.	Cyanide Before test	Cyanide After test	Cyanide Consumed per ton	Alkalinity Points before	Alkalinity Points after	Assay of tailing Gold, oz.	Assay of tailing Silver, oz.	Percentage extracted Gold	Percentage extracted Silver
Synthetic cyanide									
24	0.101	0.0655	2.13	47	9	...	3.73	...	83.9
48	0.101	0.064	2.22	47	16	...	3.29	...	85.8
Commercial cyanide									
24	0.1005	0.051	2.97	51	2	...	3.50	...	84.9
48	0.1005	0.065	2.13	51	11	...	3.50	...	84.7

Grammes lime added, 0.75, or 15 lb. per ton ore.

Time agitated, hr.	Cyanide Before test	After test	Consumed per ton	Alkalinity Points before	Points after	Assay of tailing Gold, oz.	Silver, oz.	Percentage extracted Gold	Silver
Synthetic cyanide									
24	0.101	0.0915	0.57	47	45	0.03	0.61	98.2	70.6
48	0.101	0.0940	0.42	47	27	0.02	0.60	98.8	71.2
Commercial cyanide									
24	0.1005	0.092	0.51	51	33	0.02	0.54	98.8	74.0
48	0.1005	0.081	1.17	51	19	0.02	0.66	98.8	68.3

plant. Inasmuch as the cyanide solutions employed in cyanidation are very dilute, and as the increase in the strength of cyanide necessary at the end of each cycle to restore the original working-strength of the solutions is small, the conditions for applying the idea are considered to be favorable. Although the final cyanide-product is of a low grade, the fact that the bulk of impurities (carbon and iron) are comparatively insoluble renders it possible to introduce, through the suggested method, high-grade cyanide into the mill-solution in a very simple manner. It would appear that the formation of sulpho-cyanates and soluble sulphides could be avoided by using raw material free from sulphur. Traces of soluble sulphides, when they do occur, can readily be removed by the use of a lead salt. Excessive formation of ferro-cyanide apparently could be avoided by rapid leaching of the freshly cyanized product. The presence of ferro-cyanide would invariably be expected when using iron as a catalytic agent, but it is not anticipated that the concentration of the mill-solutions in the presence of the zinc would increase to the point where it would cause trouble.

As indicated in the flow-sheet it is proposed to use a portion of the barren solution for dissolving the cyanide from the briquettes, thus giving a strong solution with which to bring up the strength of the whole mill-solution. The barren solution is used rather than a solution containing the precious metals in order to avoid premature precipitation of the gold and silver by the finely divided iron and carbon in the briquettes. The spent cake would be returned to the mixer, where, after adding carbon, and as much of the catalytic agent as might be necessary, together with a sodium salt, it would be converted into fresh briquettes which, after being dried and having had producer-gas passed through them, would again be leached. The reagents required would therefore be carbon, sodium carbonate, nitrogen from the air, and replacement of so much of the catalytic agent as might be necessary to replace the losses. The proposal to leach the cyanized briquettes with mill-solution avoids the difficulty of recovering alkaline cyanides in the solid purified form, which, according to Williams,* has proved a stumbling block with many processes of cyanide manufacture, particularly those in which atmospheric nitrogen has been used.

The most vital point in the application of the process is the furnace in which the combination of the elements to form cyanide is effected. It is assumed that in heating the charge it would be necessary to keep it out of contact with the products of combustion. It therefore becomes necessary, unless electric heating be used, to put the charge in a metallic container of refractory material, heated from the outside. The difficulties arising from the use of refractory material for this purpose would be those of cracking, of slagging with substances in the charge, and of slow transfer of heat. On the

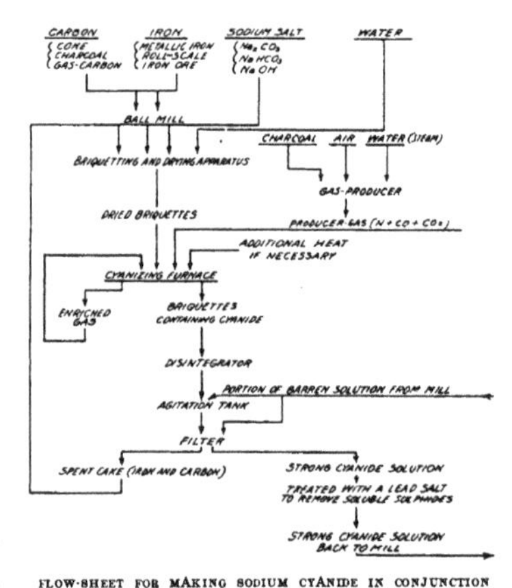

FLOW-SHEET FOR MAKING SODIUM CYANIDE IN CONJUNCTION WITH MILLING

other hand, metallic containers are open to the objection that they might be subject to rapid destruction through oxidation and burning from the outside at the temperature necessary to bring about the reaction. The rate of heat-penetration into a charge of this kind is slow; therefore, if the time-factor is to be kept down to a reasonable limit, and the overheating of the containing tubes is to be avoided, it would be necessary to use a container of comparatively small section. So far as we know, this should be made preferably of iron. Increased capacity of the individual unit could be obtained by making the tube elliptical in form, and a sufficient number of these units could be mounted in a suitable furnace to give the desired capacity. They should be interchangeable and removable, so that when they failed they could readily be

*'The Chemistry of Cyanogen Compounds and their Manufacture and Estimation.' Blakiston's Son & Co. (1915), p. 302.

taken out and replaced. In the experiments cited the iron pipes rapidly oxidized during cooling, forming a thick layer of scale, a pipe rarely lasting longer than three tests. It is claimed by the inventor, however, that if the containers are not allowed to cool, and particularly if the furnace gases are maintained at a reducing composition, and furthermore if the cyanized briquettes are discharged as fast as formed, into special receptacles for cooling, the life of the container is greatly prolonged. When electrical heating is adopted it would appear possible to surround the outside of the containers and heating-elements with a reducing-gas or nitrogen. Assuming that a product containing 10% of cyanide is formed (which I have by no means definitely proved to be possible in continuous commercial operation), and that the furnace is properly designed, this method of cyanide formation would appear to be feasible, even though the loss of containers be high, particularly under conditions of high cost of cyanide, such as have obtained for some months.

Attention is called to the fact that a small fund only and a very brief time were available for making these tests. Therefore, in the absence of complete data, my discussion of the subject and of the conclusions reached must be looked upon as being of a purely tentative character. I am informed, however, that the Nitrogen Products Co., the owners of the Bucher patents, is rapidly placing the process upon a commercial basis for the production of high-grade sodium cyanide.

Largest Ingot Mold Ever Made

The largest ingot mold ever made has just been poured by the Bethlehem Steel Co. at South Bethlehem, Pennsylvania. Its dimensions are 15 ft. 7 in. high, with a mean diameter of 91½ in., its shape being octagonal and corrugated. At its thinnest section or point of corrugation the thickness of the metal is 15 in., and the heaviest section is 20¾ in. A mold of this size is made so that steel can be poured into it to form ingots large enough to be reconverted into 16-in. and 18-in. guns. A steel ingot poured in this mold will weigh 300,000 lb., and from it tubes and jackets for the large guns will be forged. Before the mold can be used large steel bands, 12 in. wide, must be shrunk around each end of the casting after it has cooled. Because of the necessity of pouring this large mold quickly bessemer pig-iron was melted in three large open-hearth steel furnaces and the melted iron suspended in ladles over the mold at one time. A continuous runner from the ladles was made so that the iron was thoroughly mixed before entering the mold. It took 340,000 lb. of pig-iron to pour the casting, not counting the 10,000 lb. needed later to fill up the sink-heads so as to take care of the natural shrinkage. The mold had to be left covered in the sand for a considerable time to completely cool. Two large 100-ton cranes were used to lift it from its pit and prepare it for transferrence to the open-hearth department where it will be used regularly for casting ingots.

Non-Ferrous Metal Exports

Exports of non-ferrous metals for June were $59,918,666, against $64,148,723 in June 1916, a decrease of $4,230,057, or 7%. Exports for 12 months totaled $820,818,850, $414,033,182 in 1916, and $164,747,501 in 1915. The average rate of export on the basis of 12 months ended June 30 has been more than four times the corresponding figures for 1915, which was a representative period just before the war movement started on a large scale.

The following gives totals for the six metals:

	1917	1916
Copper and manufactures	$30,860,643	$24,379,526
Brass " "	21,565,239	31,929,451
Zinc " "	3,602,070	5,161,552
Lead " "	1,859,031	1,301,128
Alum. " "	1,278,043	488,007
Nickel " "	753,640	889,059
June total	$59,918,666	$64,148,723

Exports of the copper group in 1917 increased in value 223% compared with corresponding figures for 1915. Exports of zinc and zinc manufactures were three times greater than those of 1915 and aluminum and manufactures in 1917 were nearly six times those of 1915. Comparative statistics of the exports of the six principal non-ferrous metals for the 12 months show:

	1917	1916	1915
Copper and manufactures	$322,284,174	$173,946,226	$99,558,030
Brass " "	383,291,964	164,876,044	20,544,559
Zinc " "	66,108,586	45,867,156	21,243,935
Lead " "	16,563,290	13,823,004	9,004,479
Alum. " "	20,299,982	5,664,349	3,245,799
Nickel " "	12,270,854	9,876,403	11,110,699
Total, 12 months........	$820,818,850	$414,033,182	164,747,501

PAINT can be removed from iron and steel surfaces by the following process as employed by the U. S. Coast Artillery Corps which employs it to clean the exterior surfaces of big guns and carriages. One pound of concentrated powdered lye is dissolved in six pints of hot water, and lime is added to make the solution thick enough to spread. It is then applied, freshly mixed, with a brush, and allowed to remain until almost dry. When it is stripped off it will take the paint with it. In case the paint is thick and old it may be necessary to wash the surface and apply a second coat of the emulsion. After the old paint has been entirely removed and before a new coat of paint is applied, the metal should be washed with a solution consisting of one-half pound washing soda, or sal soda, dissolved in eight quarts of water, after which, of course, the parts should be thoroughly dried.

IMPORTATIONS of sulphur in 1916 were practically the same as for the last five years, amounting to 22,235 long tons, valued at $404,784. The exports, however, increased nearly 350% over the exports in 1915, amounting to 128,755 long tons, valued at $2,505,857.

A Simple Head-Frame

At small mines it is usually necessary to equip and operate at the least possible expense, as there generally is little money available to be wasted in unnecessary equipment or in doing work that has not some direct and immediate value. Among the structures that small mines most commonly require is a head-frame designed to meet all requirements, and one that may be made of a minimum amount of timber, easy to construct, and therefore inexpensive. As the shafts of most new mines and prospects are sunk on the vein, and as the vein is seldom vertical, inclined shafts are more common than those that are vertical.

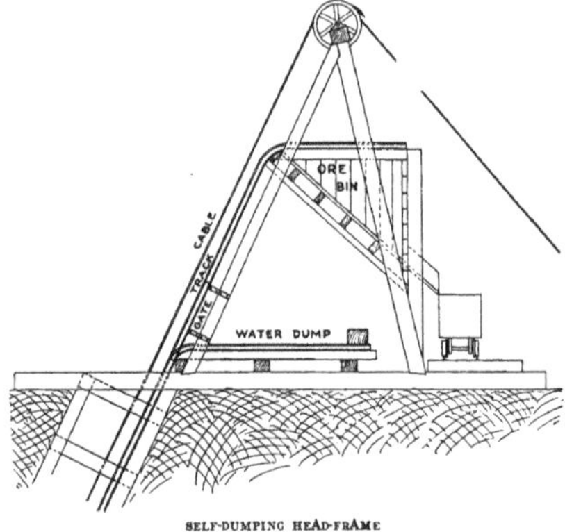

SELF-DUMPING HEAD-FRAME

The accompanying illustration is that of a head-frame designed for a small mine, having an inclined shaft. Usually miners select timbers for head-frame construction that are much heavier than necessary. It is a very common thing to see a head-frame not more than 30 or 35 ft. high, built of 12 by 12-in. timbers, when the greatest stress it will ever be called upon to meet will be, perhaps, weight of a skip or bucket, loaded with rock or ore, with perhaps two or three men, the whole weighing less than 3000 lb. A properly constructed head-frame of 8 by 8-in. timbers will be strong enough for this load with a large factor for safety, if properly designed and built. The structure here illustrated is composed of as few pieces of timber as possible to meet the necessities of the work, which requires provision for an ore-bin in the frame and a dump for both ore or rock, and one for water. The water-dump is arranged just above the collar of the shaft, as shown, where a segment of the track on either side is provided with hinges and forms a gate which may be swung back, when it is desired to dump water or tools from the skip.

When the skip comes up loaded with ore the top-ı closes the gates and the skip descends into the shaf the pleasure of the engineer.

The skip is of the ordinary pattern, without spe arrangement of extra wheels, or other dumping de\ When the skip arrives at either the lower or the uF dump, the forward wheels follow the curve in the t and run out horizontally. The engineer continue hoist, taking care not to overwind, and as the lower] of the skip is lifted by the draw-bar, it is tilted forw and the load dumps automatically. As soon as the gineer releases the brake the rear wheels of the ı again drop onto the track and the skip runs down shaft. The whole thing works automatically, one man at the engine may do all the top-w when necessary. Not only is this device i pensive and simple in operation, but it has t demonstrated to be perfectly safe at many m in California.

The accompanying drawing is incomplete a is intended merely as a suggestion, for the rea that each superintendent has his own ideas ab the construction of ore-bins and other structu about his mine. However, the ore-bin may secured to the main timbers of the head-frame it may be provided with individual posts to s port it.

This frame is pyramidal in shape, the m members tapering, and meeting at the apex, w the sheave-bearings are placed, the sheave i running between the right and left-hand side the structure. If for any reason a rectang frame is preferred it must be made some\ higher than here shown and then the sheave in boxes secured to two upright timbers, that mortised into the cap and into a tie placed a three feet below the cap. This will give the neces clearance for dumping the skip in the upper bin.

THE EXPECTED has happened in the spelter indu Confronted with operating at a loss, a number of plants have closed and still others have inaugu drastic curtailment. At the mid-year 14 zinc plant become idle, their capacity being 35,000 retorts, w probable yield of about 120,000 tons of spelter pe num. Both the Caney and Deering smelters of American Zinc, Lead & Smelting Co. have closed. have a combined capacity of about 38,000 tons of s per year. It is understood that one of the Gr smelters, also controlled by American Zinc, will shortly. Other smelters have curtailed heavily an plant, controlled by one of the largest factors, has down. On June 30 there had accumulated 35,000 according to the Government compilation. Prod believe that, despite the growing curtailment since time, unsold stocks have been piling up. Domest mand has fallen flat; Government requirements l means call for today's production and the export ment has subsided materially.

Chemicals Used in Flotation

By O. C. RALSTON and L. D. YUNDT

INTRODUCTION. *The development of the flotation process in a few years to a stage where over 100,000 tons of ore is treated every day in the United States, and half as much more in the rest of the world, threatened, for a time, to upset the market for various products such as pine-oil, coal-creosote, and coal-tar. Incidentally, flotation has also opened a market for other chemicals that formerly could not be used in the Western inter-mountain districts.

ACIDS. Almost the only acid used for flotation work is sulphuric, which costs less than most of the other acids. No one knows why the addition of sulphuric acid to certain pulps improves the flotation. We do know that it often produces a froth containing less gangue, and also that higher recoveries are made. Further, it can reduce the amount of oil necessary to produce the same frothing effect. However, the addition of acid to an ore-pulp sometimes proves fatal to good flotation, especially with copper ores in which a part of the copper has been oxidized by weathering, as at Inspiration, Arizona. The trouble may be due to the formation of copper sulphate. The oxidized copper minerals dissolve easily in dilute sulphuric acid solutions, whereas the natural copper sulphides are not attacked.

When the addition of sulphuric acid causes an improvement in the flotation of a particular ore, its exact function is difficult to explain, and theories are conflicting. The Potter-Delprat process depended on acid to develop bubbles of gas through reaction with carbonate minerals (such as calcite) in order to form a froth, and bubbles of air were not beaten into the pulp. Consequently it was natural that some of the early theorists claimed that the addition of acid caused an increased number of gas bubbles to form in the pulp and so facilitated the flotation of all the desired mineral. The fallacy of this argument has long been recognized.

Fanciful electro-static theories have been advanced to explain flotation and the statement has been made by one theorist that the addition of an electrolyte to the water of the pulp allowed the electro-static charges on the gangue-particles to be conducted away while the charges on the particles of the desired minerals were insulated by a surrounding film of oil. This theory was soon exploded, as it rested on a misapprehension as to just what caused charges to appear on suspended particles. The surfaces of small particles suspended in a liquid usually tend to adsorb more ions of one electric sign than of the other sign and hence carry apparent electro-static charges. The thing to be noted, however, is that these charges on the particles are not conducted away in the presence of electrolytes, but their existence is dependent upon the presence of ions in the solution. The surface of oil droplets and of air bubbles likewise adsorb ions from solutions, and a certain parallelism between the charges on these various particles of matter and the conditions of good flotation has been observed. Just what is the inner connection has never been explained. It is an interesting fact that one mill in Colorado was reported to receive pure water from melting snow and that flotation of its ore in this water was unsatisfactory until the water was contaminated with an electrolyte, sulphuric acid proving acceptable among other things.

Bancroft[1] calls attention to the fact that the acid may not react because of the replaceable hydrogen atom but by cutting down the concentration and consequently the adsorption of the hydroxyl ions. The connection that he wishes to draw is that "in acid and neutral solutions air seems to be adsorbed by organic liquids much more readily than by water," and he thus derives a reason for the attachment of the air bubbles to the oiled sulphides. The preferential adhesion of the oil to the valuable sulphide minerals in the presence of gangue-minerals is well known.

The surface-tension theorizers have claimed that the addition of the proper electrolytes, like sulphuric acid, so alters the surface-tension of the water that frothing can take place. This may be true without conflicting with the colloidal theory, but we would call attention to the small amounts of acid that produce a marked effect on flotation and the small resulting changes in surface-tension of the water. Small amounts of electrolytes are sometimes known to produce profound changes in colloids and we feel more inclined to favor the colloidal theory.

Nevertheless it is not right to claim that the replaceable hydrogen atom has nothing to do with the improvement in flotation results when acid is added. Allen and Ralston[2] have called attention to the fact that if any sample of ore is ground dry it is usually necessary to use acid in the flotation, whereas it is often possible to grind this same ore in a wet condition in a pebble-mill and obtain the same flotation results without the use of acid. The natural inference is that the dry grinding has caused oxidized films to form on some of the particles of natural sulphides and that the sulphuric acid was needed

*A paper read before the American Chemical Society, at New York, September 1917.

[1] Wilder D. Bancroft: 'Ore Flotation', 'Met. & Chem. Eng.' Vol. XIV, p. 631 (1916).

[2] M. & S. P., January 1 and January 8, 1916.

to dissolve these films, leaving a clean unaltered surface for the action of the oil and the bubbles. This, of course, is a special case and does not make any exception to the colloidal theory.

The amount of acid used is usually the minimum that will produce a given effect. It varies from two or three pounds of sulphuric acid per ton of ore (mixed with three to five tons of water), to as high as one or two hundred pounds of acid. When a large excess of acid is necessary we feel that it must be dissolving off oxidized films, while small amounts of acid are probably effective in causing selective-adsorption effects. Acid seems to be necessary in the flotation of sphalerite and iron pyrite. Occasionally a pyritic copper ore needs acid. The Anaconda slime constitutes an example. The present tendency among flotation men is to find substitutes for acid.

COPPER SULPHATE. So much has been said about the "flotation mystery" involved in the improved results obtained at the Mascot, Tennessee, plant of the American Zinc, Lead & Smelting Co., that a word should be said here. There is no doubt that the use of less than one pound of copper sulphate per ton of ore, suspended in over four tons of water, has caused much higher recoveries of the sphalerite in their ore and allowed cleaner work. The same effect has been noticed in many other mills treating zinc-sulphide ores, although an improvement in the flotation is not necessarily the invariable result of adding copper sulphate to zinc-sulphide ores.

At Mascot, the large-scale plant was unable to reproduce the results obtained in the laboratory. At last it was realized that the bronze castings used in the laboratory machine might account for the difference, and the hanging of a plate of copper in the mill-pulp while it was passing through the large flotation machines corrected the difficulty, and led to the discovery of the effect of salts of copper on the flotation of sphalerite.

Copper sulphate is deleterious in the flotation of silver ores[3] and of copper ores, its effects being overcome by precipitation with such a reagent as hydrogen sulphide, sodium sulphide, or sodium carbonate. Neither silver minerals nor copper minerals precipitate copper ions from solutions, but it is known that natural zinc sulphide reacts with copper-sulphate solutions, precipitating copper sulphide; in fact, this reaction is used in the Huff electro-static separation process for separating zinc sulphide from minerals of the same specific gravity and likewise poor conductors, such as barite. By soaking the mixture of minerals in a dilute copper-sulphate solution a thin imperceptible film of copper sulphide is formed over the zinc-sulphide particles, and when dry these act like good conductors and can be caused to jump off the electro-static machines, leaving the barite behind.

The suggestion is inevitable that in the case of improved flotation of sphalerite the copper going into solution is precipitated as copper sulphide on the surface of the zinc-sulphide particles, making them capable of better flotation. Why the modified particles float better is the real mystery. Whether the increased electric conductivity of the surfaces of the particles has anything to do with it, or whether copper sulphide has some inherent property of being more easily 'oiled,' is hard to say.

Some other effects not previously recorded have been reported to me by F. G. Moses, one of the engineers of the General Engineering Co. He had observed the appearance of the tailing from the flotation machines treating a sphalerite ore, when it was passed over a slime-table (as a pilot to indicate the work of the flotation machine by the size of the concentrate streak produced). Before the addition of copper sulphate the slime in the tailing seemed impalpable, but afterward it had the appearance of curdled milk (flocculated). However, in a second mill the slime in the tailing still seemed to be deflocculated. In both instances the finely ground zinc sulphide in the concentrate seemed to be flocculated and in the second case the grade of the concentrate was raised considerably by passing it through a drag-classifier from which some of the deflocculated gangue-slime could overflow. This observation may prove important. I have often noticed the flocculated condition of the concentrate in flotation-froths but had attributed it to the oil that supposedly collected the material into little flocks. Evidently the copper sulphate could assist in this flocculation, for it is well known that heavy metal ions are strongly adsorbed. In colloid chemistry it is usual to find small amounts of such chemicals producing striking effects and it is probable that the extremely small amounts of copper sulphate involved in the case under discussion are sufficient only for some colloidal reaction, be that what it may.

The condition of the concentrate and gangue particles, with regard to their degree of flocculation, has not been sufficiently investigated; hence we are confronted with the necessity for determining just what becomes of the copper introduced into such ore-pulps. Is it precipitated as copper sulphide or is it adsorbed?

LIME AND ALKALIES. Of late there has been a tendency to use various alkalies as addition-agents in flotation, with the result that in many places they are competitors of acid. Of course, the best thing to do, if possible, is to find an oil mixture that will not call for any addition-agent, but as long as the use of these further addition-agents increases the extraction of the desired mineral or improves the grade of the concentrate enough to more than pay for the amount of addition-agents used, they will not be displaced. The use of alkaline addition-agents has advantages in lessening the corrosion of the machinery, and often diminishes the amount of oil necessary. With alkaline pulp the oils can be added to the pulp ahead of the final grinding machinery so that the tubes or ball-mills can be caused to disseminate the flotation-oil before the pulp reaches the flotation machinery.

Few people have reported any observations on the phenomena produced by the use of these addition-agents.

[3]M. & S. P., December 18, 1915. 'Effects of Soluble Compounds of Ore on Flotation.'

yond the improvement of the metallurgical results. iere is no doubt that soluble heavy-metal salts are ecipitated by the use of these various alkaline sub- ınces and it is also well known that these soluble :tal compounds are often deleterious to flotation, so at their removal by precipitation is imperative. It is o known that in a number of instances it has been ssible to float a 'tarnished' ore by the addition of rious alkaline sodium salts where the same flotation- s did very little good when used on the ore alone. If dium sulphide were the alkali used this would not be rprising, as it could be easily seen that the sodium lphide would react with copper carbonates and other idized minerals, forming sulphides and thus causing eir flotation, but when sodium hydroxide brings about e same results it makes it seem possible that the thin ms of oxidized material are dissolved off the faces of e natural sulphide particles. This would be possible th zinc, copper, or lead ores, since the hydrates of ese metals are somewhat soluble in sodium alkaline lutions. Instances of improvement of the grade of ork with ores containing each of these metals are ıown. I have not heard of many instances of the im- ovement of the flotation of pyrite by the use of such dition-agents. British Patent 5856 of 1914, granted Minerals Separation and DeBavey's Processes Austra- ı Proprietary, Ltd., states definitely that in the flota- ın of such tarnished ores the water in the circuit ould be distinctly alkaline toward phenol-phthalein. dium carbonate is the alkali preferred.

The use of caustic soda has resulted in some degree of tisfaction, especially with the disseminated copper s. At the September 1916 meeting of the American stitute of Mining Engineers, during the discussion of e of the papers, it developed that the Old Dominion ll at Globe, Arizona, was using caustic soda to the ount of one pound per ton of ore. According to the tallurgist, W. B. Cramer, this yielded a much less ous froth, which lay more flatly in the flotation chine, was more easily broken down, and ran more dily in the launders, so that not so much water was uired. This, in turn, permitted better practice in the rworked Dorr thickeners. Further, the recovery was ewhat increased, the grade of the concentrate was ed, and hence less 'insoluble' was left in the concen- e. At still another mill there also resulted a reduc- in the flotation-oil necessary.

is noticed that such alkalies produce a weaker and e mobile froth. The effect of the addition of alkali ball of stiff clay lying on a board is well known. op of alkali will cause the clay to flatten out, whereas same amount of water does not. The alkali deflocсu- ı the clay so that it 'liquefies.' So also the alkali s to deflocculate the gangue-slime, and therefore any able mineral particles entrained in the flocks will be ated and become available for flotation. In conse- ce less flotation-oil and agitation will be necessary e attempt to break up the flocks and bring the oil ntact with the sulphide minerals. Not so much air being necessary to bring the mineral particles to the surface, there will not be such a large proportion of mechanically entrained gangue in the concentrate.

Of the various alkalies used, caustic soda, sodium silicate, sodium carbonate, sodium sulphide, and lime are the principal. Sodium silicate is a more powerful de- flocculent than sodium hydrate, possibly owing to a 'protective' action of the colloidal silicic acid formed in dilute solutions. It was used for a while with success in the flotation plant of the Utah Leasing Co., but was later displaced by sodium carbonate. Our own experi- ence with many different types of minerals, at the Salt Lake station of the Bureau of Mines, has been that sodium carbonate is the most desirable alkali as it does not tend to deflocculate the gangue slimes so perma- nently, allowing easier thickening and filtration of the flotation concentrate and tailing.

A surprising claim is made in U. S. Patent 1,203,341 of October 31, 1916, granted to Allen C. Howard of the Minerals Separation Co. He claims that if alkaline substances are made up into a strong solution and the proper amount dripped into the first cell of a series of flotation machines a better result is obtained than where the circuit-water is already alkaline to the same extent. Attempts to check this claim by experimentation in our laboratory have failed to justify it. The patent says that in the mill of the Caucasus Copper Co., in southern Russia, this effect is quite marked. We fail to under- stand just how this can be so, and strongly suspect that the patent is actually attempting to claim the use of these alkalies in the manner in which they are most naturally used, in the form of a strong solution, the proper amount of which is dripped into the pulp at the most convenient point in the flow-sheet. To avoid in- fringing the patent the alkali would have to be pre- viously mixed with all the mill-water before the ore was ground in it, which might prove more expensive and less easily performed.

Lime is not such a desirable addition-agent as the alkaline sodium compounds; and while it may cause de- sirable effects when added in small amounts, an excess is often harmful. Lime and most other calcium compounds are usually flocculators of gangue-slime, rather than de- flocculators as the alkaline sodium compounds are. This may explain the difference between the two, although it has been hinted often that we know too little about the degree of flocculation of the ore-pulp during flotation. It is probable that in the cases noticed where lime im- proves the flotation it functions in precipitating other less desirable soluble impurities, such as the iron sul- phates. At the Miami mill it was found that when lime was added to the pulp before entering the ball-mills an increased recovery was obtained, whereas the addition of lime at the head of the flotation-cells resulted disas- trously. Supposedly the lime had to be allowed time enough to react with all the soluble or semi-soluble im- purities of the pulp before going to the flotation ma- chines, so that there would be no raw caustic lime to cause trouble.

Sodium sulphide and calcium sulphide or poly-sulphides are used in the flotation of oxidized ores of lead and of copper. Hydrogen sulphide is also in use for the same purpose in the flotation of copper-sulphide ores. The oxidized minerals are 'sulphidized' by these chemicals and can be floated in the same manner that the natural sulphides can be floated. However, there is no objection to using sodium sulphide or the other sulphides to precipitate soluble heavy-metal salts from solution. Hence these soluble sulphides can serve a double function. Three mills are now in operation using sodium sulphide in the sulphidizing of lead-carbonate ores and several copper mills use it for oxidized copper compounds. The method is fairly satisfactory because only small amounts of the sodium sulphide are used, the lead-carbonate particles being merely coated superficially with a film of lead sulphide that can be oiled and floated. As little as three pounds of sodium sulphide per ton of ore containing 15% lead is successful in causing the flotation of over 90% of the lead minerals. Manganese di-oxide or basic iron sulphates or lead-peroxide minerals tend to use up the sodium sulphide before the lead carbonate is satisfactorily sulphidized. Since such minerals are fairly common in the oxidized ores of lead the method can be applied to only those ores which are free from such constituents. Hydrogen sulphide is in use in the mill of the Magma Copper Co., at Superior, Arizona, sulphidizing copper carbonate.

Of all the alkaline compounds mentioned above, lime is the cheapest, but not the most desirable. Sodium carbonate is not only highly desirable but also can be obtained more cheaply than most of the other sodium compounds. In fact, most of the flotation work which calls for the use of sodium carbonate is situated in the inter-mountain region not far distant from the alkaline lakes and natural trona deposits of the arid and semi-arid regions. An outlet for this natural soda has long been desired. At the present time most of the soda in use by flotation-mills is shipped from the East and is much purer and more expensive than the work demands.

Sodium sulphide also costs too much when delivered into the inter-mountain country from the Middle West or the East. It can be made from nitre-cake or from sodium sulphate by reduction. Sodium sulphate is available in a fair degree of purity from deposits on the bottom of Great Salt lake or from Searles lake in California.

ORGANIC ACIDS like tartaric acid, citric acid, etc., and their salts, have been mentioned in British Patent 17,327 of 1914, granted to George A. Chapman, one of the Minerals Separation engineers. Argol, the crude tartrate of potassium, was added to the pulp in both the Anaconda and the Burro Mountain mills, which are treating copper ores. The argol was introduced into the pulp as it entered the tube-mills, in amounts ranging from one-tenth to one-fourth of a pound per ton of ore. Both of the ores mentioned are noted for the high percentage of colloids present, due to clay-like minerals. The flotation results were said to be greatly improved. The effect of the tartaric acid seems to be a slow flocculation or coagulation of the gangue-slime. These organic acids and their salts are known to have low dissociation constants and the theory advanced in Chapman's patent is that these reagents do not flocculate the gangue-slime so quickly, but that the sulphide minerals escape entrainment in the flocks so that they are collected by the oil and air into the froth before flocculation of the gangue. To quote the words of the patent, "the effect of electrolytes upon colloids (as is known) depends *inter alia* upon the valency and upon the ionizing power or dissociation constant of the electrolyte. Tartaric acid is divalent and as compared to ordinary mineral acids it has a low dissociation constant. Citric acid has similar properties but is trivalent. Many other electrolytes such as organic acids and their salts also have similar properties as electrolytes and the choice of a suitable electrolyte for the purposes of this invention can be readily determined by a preliminary test."

This is a most interesting theory, but I have noticed that citric acid, which is mentioned above, usually does not flocculate gangue-slime but rather exerts the reverse effect. It is probable that the effects of organic acids in flotation will be found to be as erratic as their adsorption effects, commonly known to colloid chemists.

SILICO-FLUORIDES. British Patent 4938 of 1914, granted to H. L. Sulman and Minerals Separation, calls for the use of silico-fluorides as a substitute for acid. It is claimed that these compounds give the desired results. The complex fluorine salts have rather unusual ionization and other chemical constants, and hence it might easily happen that their effects on flotation would be unusual and hard to predict. The diversity of the various compounds mentioned as having been patented by the Minerals Separation Co. shows how far their engineers and chemists have gone in developing the process.

A test of Broken Hill tailing was made in London tap-water without the addition of potassium silico-fluoride and another test was made with this addition-agent to the extent of 3.4 lb. per ton of ore. The oil used in each test amounted to 2 lb. of eucalyptus-oil per ton of ore. The data on the two tests were as follows:

	Heading %	Concentrate %	Tailing %	Recovery %
ORDINARY FLOTATION				
Zinc	19.0	29.0	14.4	56.0
Lead	7.0	9.3	2.5	70.2
WITH SILICON-FLUORIDES				
Zinc	19.0	36.1	4.8	86.2
Lead	7.0	14.7	1.6	88.5

The patent also makes a good suggestion that the silico-fluorides can be prepared by adding hydrofluoric acid to the ore pulp. This produces soluble silico-fluorides of the bases in the pulp. The main value of a substitute for sulphuric acid would be for ores containing a great deal of calcium carbonate or other acid-consuming materials in the gangue.

CALCIUM SULPHATE is a compound which has been added to ores containing colloidal gangue, although its success has been somewhat erratic. It was once used in one of the Broken Hill mills and its effect was that of an electrolyte. It is sparingly soluble, so that there can never be a high concentration of its ions in solution and it is hence more or less equivalent to the tartrates in providing ions rather slowly, so that, supposedly, the flocculation of the gangue-slime will not entrain particles of the desired mineral.

Sodium manganate and sodium chromate are two oxidizers recommended for differential flotation of two flotative minerals. They seem to deaden one or the other of the flotative minerals so that it will not float. The exact mechanism of this deadening action is not known, for while chromates will deaden galena particles in the presence of sphalerite, manganese compounds have the reverse effect. They are efficient in this type of work, but there is a wide-spread impression among metallurgists that they are too expensive. To be sure, the market prices on the chemicals usually sold are rather high, but extreme purity of these compounds is not essential. There is no apparent reason why they should not be made by fusing local manganese or chromium ores with sodium carbonate. Manganese ores are usually available in zinc and lead districts, whereas the chromium ores are harder to obtain. Mill-managers who have considered only potassium permanganate of the usual high commercial purity might well give their attention to the more cheaply prepared sodium manganate.

SULPHUR DI-OXIDE, SODIUM SULPHITE, SODIUM THIOSULPHATE. These reagents are inhibitors of flotation, if used in sufficient amount. If sulphur di-oxide is applied to an ore-pulp the first effect is one similar to that obtained when sulphuric acid is added to the pulp, an improvement in the flotation. However, if the sulphurous acid is allowed to act on the ore long enough to form sulphites the result will be non-selective flotation and the gangue will be floated with the desired mineral particles. The great use of such compounds has been found in differential flotation, the basic idea being to add just enough of one of these inhibitors to deaden the surfaces of one kind of flotative minerals while another group is only slightly affected and can still be floated. Mixtures of sphalerite, galena, pyrite, and chalcopyrite or any two of them usually need separation and in case they need to be finely ground to be liberated from each other mechanically differential flotation methods are in demand.

CHLORIDE OF LIME is a reagent said to be good for differential flotation of sphalerite in the presence of pyrite.

SODIUM CYANIDE. In most cyanide-mills it has been found that the cyanide left in the tailing is deleterious to flotation. E. J. Atckison has informed us that in some tests made by him with the ore of the Amparo Mining Co., of Etzetlan, Mexico, this effect was noticed unless the amount of cyanide present was less than 0.007% on the weight of the ore or 0.0018% in the mill-water, under which conditions improved flotation results were obtained over what could be done in water containing no cyanide.

At the mill of the Ohio Copper Co., in Utah, mine-water containing a rather high percentage of the sulphates of iron and similar impurities is used in the flotation and the addition of a small amount of cyanide has been found to greatly improve the flotation.

It is probable that in both these instances complex double cyanides are formed with the solutes in the mill-water and it may even be that some of the most undesirable impurities are precipitated by the cyanide.

CONCLUSION. The use of certain chemicals in the flotation concentration of ores has been described and theories of the action of these chemicals have been explained. The use of chemical addition-agents in ore-pulps during flotation is only in its infancy and as the process is better understood operators will make greater use of chemical addition-agents which will allow them to obtain the highest economic results. The possibilities of such applications are almost unlimited and it is probably along lines of this kind that some of the great advances in ore flotation will be made.

ACCORDING to Hoyt S. Gale, of the U. S. Geological Survey, two establishments, intended primarily for separating and refining potash salts from the water of Great Salt Lake, have been built, and one of them was in more or less regular operation in 1916. They are the Utah Chemical Co., whose plant is at 'Potash Siding' on the branch of the Salt Lake Route from Salt Lake City to Saltair, and the Salt Lake Chemical Co., whose plant is at Grants, a station on the Western Pacific Railroad about 30 miles west of Salt Lake City. The Utah Chemical Co. was organized to recover potash as a by-product from the waste liquors of the salt works near which its plant stands, and the Salt Lake Chemical Co. was designed to handle the lake water directly without consideration of the possible utilization of the salt it contains. The water of Great Salt Lake is a strong solution of sodium chloride and other salts, and it contains potash in about the same proportion to its other constituents as does sea-water. It carries about 20% of dissolved salts, however, whereas sea-water carries only 3.5%. The proportion of potash in the natural water is very small—less than 0.5% of the water and its contained salts—but when this water is concentrated by evaporation, the first salts to crystallize out are those which are nearest saturation in the solution—that is, common salt and sodium sulphate—some of the minor constituents, like potash, tending to remain in solution, and to collect and concentrate in the mother liquor. This general order of crystallization has determined the methods now used to separate the potash from both Salt Lake water and sea-water. The dried salt resulting from the evaporation of the lake water contains 3.16% potassium chloride, 75.91% sodium chloride, 10.92% magnesium chloride, and 9.52% sodium sulphate. Calcium sulphate and carbonate are present in small amounts.

TONOPAH, NEVADA

TONOPAH MINING CO.—TONOPAH EXTENSION.—WEST END CON.

The Tonopah Mining Co. during the first part of September shipped 30 bars of bullion, valued at $61,500. In the Silver Top 199 ft. of development was done, in the Mizpah 10 ft., and in the Sandgrass 35 ft. Stoping continues on the Burro No. 4 vein on the 540-ft. level in 3½ ft. of ore. A north cross-cut has been started on the 1500-ft. level of the new winze at the Sandgrass. Last week the production was 2550 tons.——The Tonopah Extension milled 9721 tons of ore during August, resulting in a net profit of $32,444. At the Victor shaft 144 ft. of progress was made. On the 1440-ft. level the 605 east drift advanced on a 3-ft. face of ore, while 613 raise advanced on a 6-ft. face. On the 1680-ft. level the south cross-cut advanced 65 ft. Cross-cutting has been suspended on account of the large flow of water encountered. At the No. 2 shaft 115 ft. of progress was made. West drift 503 on the 1350-ft. level continues to advance on ore 3 ft. wide. The production for the past week was 2380 tons.——The production of the Tonopah Belmont was 2454 tons. The north-east drift 8005 on the 800-ft. level on the south vein has a 4-ft. face of ore. On the 1100-ft. level, raise 89 on the Rescue vein shows a 3-ft. face of ore, while the east drift on the Belmont vein continues in 4 ft. of ore.——The West End Con. Mining Co. is cross-cutting and rising from 415 east drift on the south vein in the hanging wall of the West End vein, which carries good ore. Raise 680, No. 4, holed through into stope 647, No. 4, facilitating mining operations. At the Ohio shaft drift 526 shows a good face of ore. Cross-cut 512 reached the foot-wall of the vein, and shows excellent ore. Cross-cut 555, No. 1, has also reached the foot-wall of the Ohio vein. Last week's production was 1095 tons.——The Halifax-Tonopah continues to drive on the 1256 and the 1704 cross-cuts, while 1018 cross-cut on the 1000-ft. level has been started to the east.

The McNamara Mining Co. has its mill in operation, and stoping has been resumed on the 200, 300, 350, 375, 500, and 700-ft. levels. Development on the 800-ft. level will be started shortly.——The Jim Butler-Tonopah Mining Co. continues to make its regular dividend of 10% semi-annually. It has at present $84,000 in excess of that required. On the 800-ft. level from the 816 winze, east drift No. 3 and west drift No. 1 continue on the McNamara vein on low-grade quartz. The production for the past week was 850 tons.——The raise from the 1050-ft. level of the North Star property holed into the 950 main level. A drift has been started on one of the veins cut by the raise.——The Midway Mining Co. has commenced stoping on the 800-ft. level, where a 2-ft. vein was recently opened. The output for last week was 63 tons.——The Rescue-Eula produced 127 tons of ore, making the week's production at Tonopah 9519 tons with a gross value of $166,582.——The new 200-ton mill of the Silver Mines Co. at Hornsilver will be started soon, according to S. H. Brady, the manager. The incline shaft has reached the 500-ft. level, while development on the levels above the 400-ft. level has provided sufficient ore to operate the mill at capacity.

The Kanrohat in Jefferson canyon, Round Mountain district, Nye county, has been taken over by S. H. Brady & Co. through a bond. The property is developed by ten adits and four shafts which have proved considerable ore. The ratio of silver to gold is about the same as at Tonopah, 100 to 1. The property is equipped with a 150-ton mill which is in excellent condition.——On the Life Preserver, at Tolicha, the owners, Zeb Kendall and George Wingfield, are sinking four inclined shafts on the veins. Two parallel veins of quartz, varying from 10 to 30 ft. in width, in a rhyolite fault-zone, constitute the ore.——The Lone Mountain Lead & Zinc Mining Co. near Tonopah, is making regular shipments of ore, containing 25% lead, 10 oz. silver, and 3% copper per ton, to the Salt Lake smelter.

Tonopah, October 3.

JUNEAU, ALASKA

TREADWELL.—LABOR SUPPLY.—GASTINEAU.

Of the properties operated by the Alaska-Treadwell group, the Ready Bullion, at the southerly end of the group, is the only one that is making any production. This mine contains the best ore found in any of the Treadwell properties, which is indeed fortunate, as it is the only one not flooded through the caving at the surface. There are now about 100 men employed underground and if the management could secure them it would have at least 250 men at work. The consolidated company is conducting exploration work on the Red Diamond property, which lies on the west side of Douglas island, and it is likely that if the Red Diamond should prove valuable the ore will be trammed across the island underground and hoisted and milled at the Ready Bullion.——The three other mines of the Treadwell group, namely, the Treadwell, Seven Hundred, and the Mexican mine, are entirely filled with water and completely closed. The only work being done is the dismantling and cleaning-up of the mills. The Alaska-Treadwell consolidation is exploring several properties in the North-West. In addition to the Red Diamond, just mentioned, the company has a crew at Auk Bay, about 10 miles up the channel from Juneau, exploring the Brie property. This is a low-grade gold deposit similar to those of the Juneau mainland. Another property on the Stikine river above Wrangell is being tested.

The entire district is handicapped by the lack of good mining labor. This is due principally to the high wages paid at the copper mines above Cordova, at Ellamar, and Latouche, and also to the higher wages paid by the Government to the men on the new railroad from Anchorage, as well as to the higher wages and better conditions of living in the base-metal mining camps of the States. All the gold mines of the world are working under a great disadvantage at the present time, it being often more profitable for the operators to close-down than to purchase supplies and labor at the extraordinary high prices now prevailing. This unusual prosperity in base-metal mining has led many men to engage in prospecting during the past summer and it is likely that the scarcity of men will be somewhat relieved when winter comes.

The scale in the Juneau district has been and still is $3.50 per day for machine-men and $3 per day for muckers. In order to meet the advanced scale of wages in the copper districts the mine-operators have introduced the plan of giving contracts on all mining with the exception of stoping, at such prices and with such conditions that the contractors are able to pay their men 50 cents more than the regular wages paid by the companies. This is a good policy for all concerned, as it makes it possible for the contractor, thus offering a bonus of 50 cents per day to his men, to get more work out of a man per dollar than it would be possible in any other way. It likewise serves to keep the men from wandering to other districts, and it proves of advantage to the workmen, because the cost of living has not been advanced as much at Juneau as it has in the districts where an increased wage-scale has been publicly announced and put into effect. That the local business-men have not sent their prices skyward as has been done elsewhere is shown by the fact that the company boarding-houses of the Juneau district have continued to charge $30 per month for board and room since the War began. In general, it may be said that the collapse at Treadwell has cut off fully 75% of the business at Treadwell, 50% at Douglas, and 20% at Juneau. As an illustration of the loss suffered by the town of Douglas through the disaster, it may be mentioned that a house built last year in that town, and containing $1200 worth of lumber, was recently sold for $400, which included the lot. The loss at Juneau is principally on the sale of the better grades of merchandise and fancy goods that are not carried in stock in the stores of Treadwell and Douglas and in the saloon business, which drew considerable trade when miners from Douglas island, tired of the monotonous struggle, wanted a little vacation. As Alaska goes prohibition next January, the loss of the saloon business is only advanced a few months. The disaster has weeded out incompetent business interests that were struggling along on a precarious footing, and it has stimulated the fishing industry by giving a local market for the fisherman, who formerly had to take his product to the canneries near Ketchikan. The business-men of Douglas, particularly, in order to make up for the loss of trade, are catching small ice-floes, which float down Gastineau channel, crushing them, and boxing the fish packed in this ice for direct shipment to Seattle.

ALASKA JUNEAU MINE

The Alaska Juneau Gold Mining Co., whose property is situated on the mainland near the town of Juneau, is now employing about 150 men, and could use 300 men to good advantage. The Alaska Gastineau Gold Mining Co., which is the Jackling-Hayden-Stone promotion, also on the mainland about two miles south of Juneau, has about 300 men employed underground and should have at least 600 men in order to operate to full capacity. The Alaska Gold Belt Mining Co., which was a promotion of Makeever Bros. of New York City, drove a 1600-ft. adit and did considerable diamond-drilling on their property in 1915. A snowslide in the winter of that year removed the surface plant, and as the owners would not take stock in part payment for the property and the operators did not make any discoveries of importance, the option was allowed to lapse and the property reverted to the original owners, Captain Harry Lott and R. P. Nelson. It is likely that development work on this property would have been continued further if money had been easy to secure, and the War had not advanced the prices of supplies and equipment. The Ebner mine, also in the Juneau belt, is being operated by the U. S. Smelting, Refining & Mining Co., employing only 10 men. This company is secretive about this work. At the Eagle River mine, 35 miles north of Juneau, there is no work in progress. The Chichagoff on the ocean side of Chichagoff island, south-west of Douglas island, is the only mine in the district that has a full crew. This property produces a comparatively high-grade ore, and is fitted with a 30-stamp mill. W. R. Rust, formerly of the Tacoma smelter, Mr. Wallace, and Mrs. de Graff are the principal owners.

The Jualin Alaska Gold Mining Co. is employing only 50 men. The Kensington group, 65 miles north of Juneau, operated by the Kensington Mining Co., which is affiliated with the Alaska Gastineau Mining Co., would be operating under normal conditions and would employ about 300 men. This means that the district immediately tributary to Juneau, which should employ about 2100 miners, is now giving employment to about 600 miners, or less than one-third of its employing capacity.——Herschel C. Parker, of New York, who has been in charge of the examination of placer ground on the Kahiltna river, reports that platinum is generally distributed in the wash of streams in the mountains about Mt. McKinley.

ALASKA JUNEAU G. M. CO. MILL

VANCOUVER, BRITISH COLUMBIA

CANADA COPPER.—LADYSMITH SMELTER.

Placer ground, covering some 800 acres situated between the falls on Perry creek and Saw Mill creek, is now owned by the Wild Horse Creek Gold Placer Mining Co., of which Mayer Clapp of Cranbrook is president. Preparation is being made to develop this ground next spring.——Slocan and Ainsworth mines shipped over 27,000 tons of ore to Trail this year.——Henry Krumb and W. H. Aldridge of New York and John M. Turnbull of Vancouver arrived recently at White Horse, Yukon. They have examined the Engineer mine at Atlin for Eastern capitalists. The Canada Copper Co. has contracted for a supply of electric power from the West Kootenay Power & Light Co. It will be used at the mines and concentrating plant that the company intends to establish near Copper mountain in the Similkameen district.——The Tonopah-Belmont Development Co. has instructed F. W. Heller, manager of the Surf Inlet mine, to examine the Dorothy Morton property with a view to re-opening it.——Stone is being quarried on Haddington island for construction of the new Canadian Northern station at Vancouver.——Magnetite assaying 67% iron has been discovered on King island.

Reports of progress at Alice Arm, the new mining camp near Prince Rupert in northern British Columbia, continue favorable. Considerable work has been done and some important new discoveries have been made. A strike of silver has been made at the North Star, which joins the Dolly Varden lode. It is about 20 ft. wide and the ore averages 70 oz. silver per ton.——A good showing is being made on the Moose property, owned by Don Cameron and Chris Nelson. An average of assays taken across a 16-ft. orebody runs 200 oz. in silver. In some places the lode is from 50 to 60 ft. wide. This property adjoins the Silver Hoard. George Coalback, A. McPhailand, and Pat Marley have discovered an orebody 100 ft. wide that has been traced for 3000 ft.——The Granby Consolidated has bonded several properties in Alice Arm district and has two diamond-drills at work. Several men are employed at the foot of the glacier on a copper prospect owned by Gus Pearson and A. Davidson.——The light railway to the Dolly Varden mine is progressing rapidly. Five miles of steel has been laid and 15 miles of grading completed.——From 1000 to 1200 tons of ore per month is being shipped from the Marble Bay mine to Tacoma. The mine is operated by the Tacoma Steel Co.——W. Trelour, G. Brister, and D. Weir are operating the Copper Queen on lease and are shipping steadily. The last shipment of 60 tons, from this mine, returned $62 per ton. Good ore was struck on the 500-ft. level.——The Loyal group near Blubber bay, owned by E. W. Watson and H. A. York, has shipped ore running 7½% copper and 4 oz. silver to the Ladysmith smelter.——The Pacific Steel Co. recently shipped 400 tons of 67% iron ore from West Redonda island to Irondale for experimental purposes.

Reports come from Dawson City that much staking has been done on the newly discovered placer creeks of Seymour and Kitchener creeks and their tributaries. One miner claims to have got 2c. per pan for four miles on Seymour creek. Bottom has not been reached.

Converters are to be added and other improvements made to the Ladysmith smelter and the company will make blister instead of shipping matte to Tacoma.——With the blowing-in of the Ladysmith smelter there is a good prospect of the re-opening of a number of properties on Mount Sicker on Vancouver island, which have been closed for years.——The Donohue Mines Co. is preparing to resume work at the Stump Lake mine.——The Kamloops Silver Mines Co., which has been incorporated by Smith Curtis, will operate the old Homestake silver mine.——The last shipment from the Blue Grouse mine, at Cowichan lake, to the Ladysmith smelter, averaged 8% copper.——A fine showing of silver has been uncovered accidentally in excavating for a boiler foundation at the Silver Hoard. A carload of this ore, running 130 oz. silver, is ready for shipment.——A short tramway is being constructed on the New Hazleton gold and cobalt property from the mouth of the adit to the camp. Molybdenite is being taken from the mine and a shipment is to be made in a short time.——The Standard mine at Silverton is milling ore; the bins at the mine are full and the mill is running at full capacity.

The Granby Con. M. S. & P. Co. has purchased 350 acres of coal lands near Cassidy Landing, Nanaimo mining division, and will start active development work immediately. Coal mined from this property will be sent to Anyox.

MONTERREY, MEXICO

NEW BULLION AND MINERAL EXPORT LAW

American and other foreigners owning gold and silver mines in Mexico are greatly exercised over the recent decree issued by President Carranza, which requires that exporters of mineral ores or concentrate containing more than six grammes of gold per ton shall re-import in gold bars for coinage, or in foreign gold, an equal value to the gold content of the ore or concentrate shipped out; and that exporters of silver in bars or ore or concentrate containing more than 50 gm. of silver per ton must re-import in gold bars, or other foreign gold, 25% of the value of the silver bullion or ore shipped out.

It is claimed by these foreign mining interests that the putting into effect of this decree will practically mean the confiscation of their properties. It is common rumor here that German influences are behind this drastic move on the part of the Mexican government, and that its real purpose is that American-owned mines in this country may be forced into the hands of the coterie of Germans that have been active during the revolutionary period, and even more so of late, in acquiring rich mines in the different districts.——The recent decree places the foreign owners of gold and silver properties between two fires. If they do not work their mines they are subject to confiscation by the Mexican government, and if they do work them under the existing requirements it will mean bankruptcy to many.

The following is an official translation of the provisions of the decree without attempting to give the full text: First. The general prohibition to export gold or silver Mexican coins. Second. From September 28 the exportation of gold in silver bars is absolutely prohibited. Third. Exporters of mineral ores or concentrates of all kinds containing over six grammes of gold per ton will have to re-import in gold bars for coinage, or in national or foreign gold coins, an equal value to the gold content, as per assay of the minerals or concentrates, exported. Fourth. Exporters of silver in bars or mineral ores or concentrates, containing over 50 gm. of silver per ton must re-import in gold bars for coinage or in national or foreign coins 25% of the value of the silver content of the bars, mineral, or concentrates. Fifth. To ascertain the amount of the gold to be re-imported, the silver will be valued at the rating made by the treasury department for the collection of duties on the date of the exportation of the ore. Sixth. Exporters of silver in bars or minerals, and gold or silver concentrates, must give bond to the custom house at the point of export, or to the treasury department, for the amount of gold they must re-import. Seventh. Gold re-imported must be delivered to the mint for coinage within 30 days from date of introduction into Mexico. The mint will only charge coinage expenses. Eighth. Re-importations of gold must be made within 10 days of the exportation which they cover, it being understood that failure to make the re-importations will be punished by the forfeiture of the bond.

Protests against the decree are being prepared by American mine-owners, and will be forwarded to the United States government at Washington.

ARIZONA

COCHISE COUNTY

(Special Correspondence.)—Plans have been completed for the erection of a 450-ton mill, including flotation, at the Shattuck mine. This plant will treat silver-lead ore. It is expected that the mill and necessary development will cost $500,000.

Bisbee, September 28.

GILA COUNTY

(Special Correspondence.)—The Silver Peak claim of the Sulphide Copper Co. has been leased to some of the stockholders. The Northwest Inspiration Copper Co., adjacent to the Inspiration, has been carrying out preliminary work for the past nine month. A road has been completed and such surface equipment as hoisting-plant, compressor, power-plant, and head-frame have been put in place. It is expected that sinking will be commenced shortly.

Globe, September 27.

GREENLEE COUNTY

(Special Correspondence.)—The companies and strikers in the Clifton-Morenci-Metcalf district are still at a deadlock. On October 1 the companies were supposed to resume operations and it was thought that the differences between them and the union would be settled. The companies wished the men to sign up as individuals and not as an organization. The men refused to do this, so all efforts on the part of the companies to adjust matters have been dropped.

The companies are now bulkheading all portals and mine entrances and there is every indication of a long shut-down. The three companies in the district, Arizona Copper, Phelps-Dodge, and Shannon, hope to complete the work of closing the properties by October 10.

The fire which was discovered August 21 in the Coronado mine, one of the largest mines owned by the Arizona Copper Co. at Metcalf, is still burning. The fire is between the 655 and 600-ft. levels. The work of bulkheading between the 500 and 700-ft. levels is nearly completed. When this is finished the burning area will then be flooded in an effort to extinguish the fire.

Metcalf, October 5.

MOHAVE COUNTY

(Special Correspondence.)—With the advance in the price of silver practically every mine and prospect in the chloride district is being worked. The Oatman Gold Mining Co. has been organized and has taken over the Oatman United Mines Co., share for share. The new company's stock is assessable, and an assessment of 1c. per share ha been levied to obtain $17,000 for development work. J. J. Connolly reports that some fine ore has been developed on the old Byron Collins mine at Cobalt. The main orebody will assay $100 in gold. Ten sacks of ore taken from the winze is expected to yield $2000. Fifty tons of ore, averaging 65 oz. silver and 16% lead, has been shipped from the Granite Point silver-lead mine.——J. J. Robertson and associates have taken a lease on the Bella Union group of six claims. Machinery has been ordered; development will proceed pending its arrival.——The Mineral Development Co., operating the Golden Hammer and other claims in the Mineral Park, is removing the entire equipment purchased recently from the Arizona Gold Star at Oatman. As soon as this machinery, consisting of 40-hp. gas engine, 9 by 8 compressor, and machine drills, is in place active development will commence.——The Hackberry mine is employing 30 men in development work. About 35,000 tons of ore is blocked out. Two carloads per week is being shipped, running 50 oz. in silver, 10 to 15% lead, and $6 in gold.——It is reported that recent development has opened the main orebody of the Roadside mine, 25 ft. of $12 ore is said to be proved, streaks of the ore will run $100 per ton.

Kingman, September 27.

(Special Correspondence.)—W. B. Twitchell, of Phoenix, Arizona, is preparing to start operations on the Jack Pot mine. A good deal of work has been done on this property, but it has become necessary to add hoisting machinery. In the old shaft there is a vein of gold and silver ore, averaging $100 per ton. A new shaft will be sunk.——A carload of machinery, including a 50-hp. gasoline engined and an air-compressor, has arrived and is being erected on the Merrimac.——The lessees of the Mary Bell have shipped a car of ore. This property has one of the strongest croppings in the district. The Rico Consolidated Mines Co. is erecting a new hoist on the Rice shaft of its property. This company has secured the C. O. D. mine.——The Emerald Isle plant is turning out 1200 lb. of electrolytic copper every day. Twenty cells are now in operation and 30 more are to be added. The Copper Age mine is driving its adit from the east and the west ends with all possible speed. When finished the adit will be 1300 ft. long.

Chloride, September 28.

CALIFORNIA

PLUMAS COUNTY

(Special Correspondence.)—A copper orebody 10 to 12 ft. wide has been exposed in the Madero mine, 25 miles from Portola, and operated by the U. S. Smelting, Refining & Mining Co. Arrangements have been made to increase developments, and to maintain work throughout the winter. Shipments will be made in the spring to Utah smelters. E. H. Greenwood is manager.

The Walker Copper Co. has 160 men at work; it is pushing a 16-mile electric-power line to obtain current from the Great Western system. Thirty-two men went on strike October 1, demanding a flat wage increase of 50c. per day. Miners are paid $4, tool sharpeners $4.50 to $5, carpenters $4.50, and surface labor $3.50; board is $1 per day. Mine developments are reported satisfactory. The flotation plant is treating 75 tons per day. Concentrate is shipped to the Tooele smelter, Utah. W. V. Hart is manager.

More miners have been employed at the Seneca Consolidated and Shaffer gold properties, and underground work at both mines is being pressed. Because of the difficulty in securing equipment, the management has decided to postpone construction of the cyanide plant until next spring. J. J. Reiley is manager.——The Plumas Grass Valley Co. has found rich gravel on the south fork of the Feather river. J. F. Buel is superintendent. On the Dinsmore group, R. A. Edgar and associates are prospecting promising gravel. Work has been suspended for the winter at the Monitor mine.

The Noble Electric Steel Co. is surveying for an aerial tram from its Braite manganese mine to Indian Falls, an approximate distance of 1½ miles. At present ore is hauled by teams to Crescent mills. Several manganese deposits in this field are attracting attention.

Portola, October 10.

COLORADO

BOULDER COUNTY

(Special Correspondence.)—The recent silver-strike at Caribou is turning out well. The size of the vein has increased from 2 ft. at the surface to 5 ft. at a depth of 30 ft. The high-grade streak is 12 in. wide and carries 41% lead and 200 oz. silver. The rest of the vein averages $40 in silver and lead. The Comstock mine, leased to Hawkins & Co. by Gray & Perry, made a shipment of high-grade ore last week. The No Name, Silver Point, Poorman, Seven-Thirty, and Sherman are shipping regularly. The Up-to-Date and the Caribou have been shipping steadily for four years. A car of silver ore from the Up-to-Date at Caribou gave 140 oz. silver and 1.75 oz. gold.——A mill is to be erected at the Huron mine, Eldorado, this fall. ——H FrankHadley, superintendent of the Gold Hill, Boulder county, has resigned and will take charge of the Iron Mask mine at Breckenridge, Colorado.

Eldora, September 29.

TELLER COUNTY

The production of gold ores by the mines of Cripple Creek during the month of September totalled 97,511 tons, with a gross bullion-value of $1,028,554. The average value of the ores treated was reduced to $10.54 per ton, due to the heavy treatment of low-grade ores at the local mills. The two Portland plants, the Independence and the Victor, milled 56,425 tons, with a general average of but $2.04 per ton. This was dump-ore and ore-house reject. The treatment at the several plants was as follows:

	Tons	Value	value
Golden Cycle M. & R. Co., Colorado Springs	28,000	$20.00	$560,000 00
Portland G. M. Co.	9,786	21.62	211,573.32
Portland G. M. Co., Independence Mill, Cripple Creek district	37,100	2.05	76,075.00
Portland G. M. Co., Victor Mill, Cripple Creek district	18,325	2.02	37,016.50
Rex M. & M. Co., Kavanaugh lease, Cripple Creek district	1,400	2.00	2,800.00
Rubie mill, Bull Hill, Cripple Creek district	350	2.40	840.00
	97,511	$10.54	$1,028,554.82

Dividends paid in September were confined to the monthly distributions of the Cresson Consolidated Gold Mining & Milling Co., 10c. per share, $122,000; and the Golden Cycle Mining & Reduction Co., 3c. per share, $45,000, a total sum in dividends of $167,000.

The September production from the Dillon mine of the Granite Gold Mining Co. totalled 72 cars, or between 2400 and 2500 tons. The ore mined on company account averaged slightly below $20 per ton. Lessees on the Dead Pnne, Upper Granite, and Gold Coin mines, were shipping, but returns are not yet at hand from the mills for all September shipments. ——The Forest Queen mine on Ironclad Hill, active under lease to Edwin Gaylord, of Cripple Creek, produced 12 cars, or close to 400 tons of ore, during September, having an average grade of about $15. This mine is owned by parties in California and Colorado.——The properties of the Gold Pinnacle Mining Co., on the eastern slope of Bull Hill, near Cameron, have been leased to Mrs. J. E. Dunn. A company, not yet recorded in Teller county, has been organized by Mrs. Dunn, and it is reported that the new concern has been strongly financed.——Work commenced on Monday of this week on the Elkhorn, a property long idle, but at one time a producer. The Elkhorn has been leased to the Geraldine Mining & Milling Co., and it is said machinery will be installed. The mine is on the south-eastern slope of Carbonate hill, at no great distance from the city of Cripple Creek. Water-level has been reached at a depth of 1973 ft. in the main shaft on the Ajax mine on the south-eastern slope of Battle mountain at an elevation of 8135 ft. above sea-level. The Ajax is under lease to the Carolina company. This company commenced operations 10 months ago and has sunk the shaft 500 ft. since taking charge, besides cutting stations at the several levels.

Cross-cutting has commenced at the 1100-ft. level of the new vertical Frankenberg shaft on the Ocean View claim of the Modoc Mining & Milling Co., by the Modoc Consolidated Mines Co., which is now operating the property. Shipments will shortly be made from the mine.

The properties of the former Bull Hill & Straub Mountain Gold Mining Co., comprising 16,613 acres, contained in the Accident, Bull Domingo, Anna May Wells, Hannah Britt, Mae Clasby, Baby Dora, and Midnight lodes, were sold at public auction on October 1 to the United Gold Mines Co. The purchasing company owned 70% of the Bull Hill & Straub Mountain Company.

Deep development has commenced from the bottom of the winze sunk to a depth of 1500 ft. from the bottom or 1250-ft. level of the Lee shaft, by the Isabella Mines Co. on Bull Hill. The East Victor will be shortly cross-cut and another cross-cut is headed for the Buena Vista vein, 220 ft. west from the winze and about 400 ft. from the Lee shaft.

MONTANA

LEWIS AND CLARK COUNTY

Silver-lead ore is going to smelter from three mines in Grass Valley district and three more are expected to be shipping soon. The Helena mine is in the lead with 15 carloads to the smelter at East Helena. In Scratch Gravel Hills several properties are being worked profitable. The best producer is the Scratch Gravel Gold; $60-ore is being taken from the 300-ft. level west. The mine is equipped with twin 80-hp. boilers.——The Thomas Cruse Developing Co. is shipping ore from the 500-ft. level.—— The Blue Bird is mining and shipping copper-silver ore.—— The Katie mine is producing a high-grade lead-copper product. ——The leading producer in the Marysville section is the Shannon, under operation by the Barnes-King company. The Shannon, together with the Piegan-Gloster in Marysville camp, and the North Moccasin mines at Kendall, produced $125,000 in bullion during August.——The St. Louis Mining Co. is running its stamp-mill on gold ore drawn from the old Drumlummon stopes.——The Bell Boy mine is again in operation.

A good strike of gold ore is reported from Bald Butte mine. A new compressor has been added to the mining equipment. Rich ore has been found in a lead parallel to the old Bald Butte.

The Eclipse-Argo Co. is sending four car-loads of 30% copper ore from its Hell Gate mine monthly.——The Economy Mining Co. is unwatering a 300-ft. shaft.——The Mt. Washington mine, at Wickes, is shipping a car of lead ore per day. This output is to be increased to two cars per day. The mine is just opened to a depth of 800 ft. by tunnel. The orebody at the face is 14 ft. wide.

NEVADA

HUMBOLDT COUNTY

(Special Correspondence.)—F. H. Vahrenkamp and associates of Salt Lake City and New York have arranged for the erection of a flotation plant at the Rye Patch mine and the addition of a flotation unit to the old mill of the Inlay group. The mill at the latter property will also be remodeled. The mines produced heavily 40 years ago and are said to contain large quantities of silver ore.

Re-construction of the recently-burned mill at Sulphur is proceeding, and Arthur Crowley, the superintendent, expects to start sulphur production within 30 days.

Discoveries of rich silver ore in the foot-hills of the Humboldt range, 12 miles south-east of Lovelock, are attracting interest. On the Badger group, M. D. Faretta and partners have been working five months and have opened a 5-ft. vein to a depth of 30 ft. It has been traced on surface for 200 ft., and ore eposed at 25 points. The ore averages $40, and 50 tons is ready for shipment.

The Rochester Mines Co. has connected the 800 and 900-ft. levels and the 800 to the 650-ft. levels by raises. All ore is now handled through the Friedman adit, from which the buckets of the aerial tramway are directly loaded to the ore-bins. The winze from the 900-ft. level is down 50 ft, and as soon as the 1100-ft. level is reached development of the Adams lode will be commenced. The mill is crushing 217 tons per day. The management estimates the September earnings at $70,000.

Lovelock, October 10.

NYE COUNTY

(Special Correspondence.)—Litigation has developed over the first claims located at Tolicha. David Ward of Goldfield has secured a court order restraining Edward Yeiser and J. J. Jordan from disposing of the Life Preserver claims to Zeb Kendall, on the grounds that Ward is entitled to a third interest by reason of a grubstake agreement. Despite many sensational stories, the deposits of the Tolicha field are not high-grade, but contain gold-bearing quartz of good milling character, so far as development has progressed. Prospecting is going on at fully twenty claims. Water is obtained from Monte Cristo springs, three miles distant, and from springs in the mountain. It is thought sufficient water can be developed in the gulches 2000 ft. from camp. The district is 48 miles by wagon-road from Goldfield, and an automobile stage operates between these points.

Whitman Symmes, of Virginia City, and Reno associates, have acquired control of the Union Amalgamated group at Manhattan and are arranging for development as soon as the final payment is made. The Wittenberg mill will be moved close to the mine and a three-compartment shaft sunk. Stamps will be replaced by ball-mills. On the 800-ft. level the orebody has been penetrated 22 ft. by the cross-cut with no indication of the hanging wall. Ore averages $14 per ton. The main vein has been intersected in the cross-cut from the 550-ft. level of the White Caps mine, Manhattan. The ore is assaying from $24 to $28 per ton. The roasting-plant is now working satisfactorily and the entire mill is stated to be effecting a high metal extraction.

The Tonopah Belmont Co. has acquired a three-fourths interest in the Alta, St. Louis, and Palmyra gold properties, near Telluride, Colorado, and a mill and Jones-Belmont flotation process, is being erected. Besides gold the ores contain silver, copper, and lead. The subsidiary company will be known as the Belmont-Wagner Co. From the Belmont mine 2450 tons is going weekly to the mill.

The MacNamara company has resumed work on the levels from the 100 to the 600 points, and will soon start extensive operations on the 800. With silver above 90c. most of the ore exposed can be mined and milled profitably. The mill has been overhauled and is ready for operation. J. L. Joseph, president of the company, is personally directing affairs.

Tonopah, October 9.

NEW MEXICO

SOCORRO COUNTY

An important discovery has been made during the week in the east 300-ft. level of the Mogollon Mines Co. While the management has not given out the details, it is understood that there is now 8 ft. of ore at the breast of the drift that will average between $30 and $40 per ton. As this is approaching the orebody opened in the north end of the Deadwood, it is likely that it will be extensive. The Oaks Co. has opened a new orebody in the Maud S. mine. This is being developed from two sides and daily samples are running as high as $30 per ton. This section has not been develiped before.

The Socorro Mining & Milling Co. is ordering supplies for a narrow-gauge road to run between the Johnson mine and Fanny mill. This will be used to transport ore to the mill from all the holdings in the west end of the district.

All of the principal mines in the district are now on an eight-hour basis. While men are rather scarce, working conditions remain normal.

Mogollon, October 2.

OREGON

JACKSON COUNTY

(Special Correspondence.)—Marshall mine, eight miles east of Wolf Creek station on the Southern Pacific, was operated many years for gold. Chrome ore was recently discovered in the mine, which is now being worked for that mineral. The ore is crushed and concentrated on standard tables. The ore is said to contain 30% and the concentrate 65% of chromic oxide.

Medford, October 6.

JOSEPHINE COUNTY

(Special Correspondence.)—John Hampshire, of Grants Pass, and associates are considering the purchase of the Waldo copper mine, one mile east of Takilma. The property consists of 480 acres of patented ground and 12 unpatented claims, and 325 acres adjoining the Queen of Bronze and the Lyttle mine, both of which are owned by the new purchasers. Col. Draper operated this property between 1900 and 1906 and the ores were treated at the Takilma smelter. Litigation, however, on account of the sulphurous fumes caused its closing down. The total output has been about $300,000. The price is said to be $135,000.

Grants Pass, October 6.

SOUTH DAKOTA

LAWRENCE COUNTY

(Special Correspondence.)—The concentrating plant of the Custer Peak M. & M. Co., near Roubaix, has been completed and is now treating ore from the 300-ft. level. The new equipment consists of ten stamps, Harz jig, ball-mill, and four Wilfley tables. The ore carries native copper, gold, and silver. ——The machinery for the concentrator of the Spokane Lead & Silver Co., of Custer, has arrived. Several shipments of high-grade material have been made to the smelter.——The Refinite company, at Ardmore, has nearly completed its new plant for the treatment of kaolin. The mineral is found in large deposits in that district and is readily recovered. The capacity of the plant is estimated at 10,000 lb. of dried product per day.——The south-western portion of South Dakota is being prospected for oil, and already several companies have been formed to develop the Ardmore field. The Hat Creek Co. has leased a drill and will commence work near the centre of the dome. The Shiloh company holds leases on land adjoining the Hat Creek and a Des Moines company will operate a few miles north of Ardmore.

The Black Hills Tungsten Co., of Hill City, is re-modeling its plant. The mine is being put in shape for a regular production. The concentrator will handle custom ore. Emil Anderson, formerly in charge of the Homestake tungsten plant, is superintendent.

The Trojan company has leased the property of the Republic in Black Tail gulch, and is making regular shipments to the mill at Trojan.

The Crawford Mica M. & M. Co. has re-opened the Old Mike mine near Custer and is shipping mica at the rate of one car per week. A new kerosene engine and a compressor have been erected.

The Homestake company is changing its plant to use electric power for everything except the hoists. The boilers at the several shafts will be removed and steam will be furnished to the hoists from the central plant. The Nordberg hoist at the B and M shaft has been placed in commission and the six-ton skips are removing the ore at the rate of 1400 tons per shift.

Lead, October 4.

UTAH

JUAB COUNTY

(Special Correspondence.)—The car shortage is again making itself felt, especially in the Tintic district; which, by the way, is one of the foremost mining districts of Utah. Forty men were laid off at the Mammoth mine on this account.

Mr. Rawson, the manager for the Tintic Delaware Mining Co., says that ore-hauling will start within the next few days, as the new road to the Tintic Delaware is now practically completed. The Tintic Delaware is at West Tintic; it was held for a number of years by the Ekker Brothers and only recently attracted the attention of Mr. Rawson and his associates of Salt Lake. Recent assays show that the ore is worth $65 per ton, the predominant metal being lead.

The mines of the Tintic district should be reaping the benefits of the excellent metal market. The smelters and the railroads should be in position to handle the product of the mines, but instead we find that conditions, which have been very bad for several months, are getting worse. Most of the big mines of this district are being held back to a point where they can make little or no money, while some are unable to ship any ore at all. Here, in Utah, the smelters have not kept pace with the rapid strides of the mining industry and, to make matters worse, the railroads are making but a pitiable effort to handle the freight business.

The Utah Power Co. has made arrangements for the construction of a seven-mile line connecting the Tintic Standard property with the new drain tunnel, which is to be driven by Jesse Knight and associates.——The Tintic Drain Tunnel Co. has filed articles of incorporation with the county clerk of Utah county. The company is organized to conduct a general mining business, including the construction and operation of tunnels. The company's principal place of business is Provo. The capital stock is $150,000, in shares of par value of 5c. each, based on the valuation of 300 mining claims along the route of the Tintic drain tunnel in the south-east end of the Tintic mining district. The following are the directors and officers: Jesse Knight, president; W. Lester Mangum, secretary and treasurer; John S. Smith, E. F. Birch, Henry Barney, and C. W. Reese.

At the Iron King property, in the eastern end of the district, the work of cutting out a large station on the adit-level, where the engine is to be placed, is practically completed. This engine, operated by air, will be used in sinking the first few hundred feet of the Iron King's new working shaft, but before the sinking is started a portion of the raise, which furnishes an opening through to the surface from the adit, is to be enlarged and re-timbered. The officials of the Iron King are fast getting everything in readiness for the extensive development which is to be taken up there. The management intends to thoroughly prospect the ground at a depth of several hundred feet.

The mill, which C. E. Loose and associates have erected at the Big Indian property in the La Salle range of mountains in Grand county, will be in commission within the next few days. Preston G. Peterson of Provo is on the ground looking after the milling operations for the Big Indian company.——The Central Eureka was this week listed on the Salt Lake Exchange, receiving an opening bid of 2c. per share. The company is capitalized for $15,000 with 1,500,000 shares.

Mammoth, October 1.

PERSONAL

Note; The Editor invites members of the profession to send particulars of their work and appointments. This information is interesting to our readers.

William Kent is here.

Charles S. Herzig is here.

Howard D. Smith is at New York.

M. J. Falkenburg is at Anchorage, Alaska.

Horace V. Winchell is in San Francisco.

F. G. Clapp is a Corporal in the Veteran Corps of Artillery of the State of New York.

Kwong Wu, engineer for the Chinese government, sailed on October 11 on his way home.

George L. Porter, president of the Afterthought Copper Co., at Ingot, California, has resigned.

Rowland King is in San Francisco as company clerk of the second contingent from Spokane.

R. S. Burdette has been commissioned Captain in the Ordnance Section of the United States Reserve.

R. E. Tremoureux is First Lieutenant in the Engineer Section of the O. R. C. at Camp Lewis, Washington.

Paul M. Paine has left the Honolulu Oil Co. at Taft, California, and is now with the Gypsy Oil Co. at Tulsa, Oklahoma.

George E. Farish of New York, together with R. R. Leslie and U. A. Fritschi, sailed on October 2 on the 'Peru' for Salvador.

Pierre R. Hines has received a commission as First Lieutenant in the Engineer Officers Reserve Corps and is stationed at Fort Leavenworth, Kansas.

Hugo Gillespie has left the Clayburn Company to become superintendent of the Drumlummon copper mine, south of Prince Rupert, British Columbia.

E. K. Soper, formerly in charge of the department of mining at the University of Idaho, has been appointed Dean of the Oregon State School of Mines at Corvallis.

W. H. Aldridge accompanied by Henry Krumb and John M. Turnbull of Vancouver has arrived at White Horse, Yukon. They have been examining mines in the Atlin mining division, British Columbia.

Capt. W. H. Landers and Capt. O. B. Perry have been assigned to the Chief Engineers Office in the War Department, at Washington. They are preparing specifications for the mining machinery to be used by special regiments that will be sent to France.

A dinner will be given in San Francisco on October 25 in honor of Ira N. Hollis, under the auspices of the joint engineering societies, consisting of the American Society of Mechanical Engineers, of which body Mr. Hollis is president, the American Institute of Mining Engineers, the American Chemical Society, the American Society of Civil Engineers, and the American Institute of Electrical Engineers. Mr. Hollis is now president of the Worcester Polytechnic Institute, and was formerly in charge of the engineering department of Harvard University. He was long connected with the Navy, and was active in the construction of the White Squadron. As an authority upon marine architecture his address in this city will possess peculiar interest and significance.

Royal Canfield Peabody died on September 17. He was connected with the Combustion Engineering Corporation.

THE METAL MARKET

METAL PRICES

San Francisco, October 9

Aluminum-dust (100-lb. lots), per pound	$1.00
Aluminum-dust (ton lots), per pound	$0.95
Antimony, cents per pound	17.00
Antimony (wholesale), cents per pound	14.75
Electrolytic copper, cents per pound	23.50
Pig-lead, cents per pound	8.25—9.25
Platinum, soft and hard metal, respectively, per ounce	$105—111
Quicksilver, per flask of 75 lb.	$105
Spelter, cents per pound	10.50
Tin, cents per pound	61
Zinc-dust, cents per pound	20

ORE PRICES

San Francisco, October 9

Antimony, 45% metal, per unit	$1.10
Chrome, 40% and over, free SiO_2 limit 8%, f.o.b. California per unit, according to grade	$0.60— 0.75

Chrome prices show a tendency to remain stationary. With the advent of winter, when the haulage conditions are less favorable than at present, prices will be more erratic and are likely to go higher.

Magnesite, crude, per ton	$8.00—10.00

Manganese: The Eastern manganese market continues fairly strong with $1 per unit Mn quoted on the basis of 48% material.

Tungsten, 60% WO_3, per unit	26.00
Molybdenite, per unit MoS_2	$40.00—45.00

EASTERN METAL MARKET

(By wire from New York)

October 9.—Copper market is chaotic and prices nominal at 23.50 all week. Lead is dull and easy at 8 to 7.95c. Zinc is quiet and unchanged at 8.50c. all week. Platinum remains unchanged at $105 for soft metal and $111 for hard.

SILVER

Below are given the average New York quotations, in cents per ounce, of fine silver

Date			Average week ending		
Oct.	3	93.62	Aug.	28	88.50
"	4	92.12	Sept.	4	91.12
"	5	91.12	"	11	96.80
"	6	90.62	"	18	100.70
"	7 Sunday		"	25	107.66
"	8	91.12	Oct.	2	98.49
			"	9	91.72

Monthly Averages

	1915	1916	1917		1915	1916	1917
Jan.	48.85	56.76	75.14	July	47.52	63.06	78.92
Feb.	48.46	56.74	77.54	Aug.	47.11	66.07	85.40
Mch.	50.61	57.89	74.13	Sept.	48.77	68.51	100.73
Apr.	50.25	64.37	72.51	Oct.	49.40	67.86	....
May	49.87	74.27	74.61	Nov.	51.88	71.60	....
June	49.03	65.04	76.44	Dec.	55.34	75.70	....

COPPER

Prices of electrolytic in New York, in cents per pound.

Date			Average week ending		
Oct.	3	23.50	Aug.	28	26.41
"	4	23.50	Sept.	4	25.65
"	5	23.50	"	11	25.83
"	6	23.50	"	18	26.25
"	7 Sunday		"	25	24.41
"	8	23.50	Oct.	2	23.50
"	9	23.50	"	9	23.50

Monthly Averages

	1915	1916	1917		1915	1916	1917
Jan.	13.60	24.30	29.53	July	19.09	25.66	29.67
Feb.	14.38	26.62	34.57	Aug.	17.27	27.03	27.42
Mch.	14.80	26.65	36.00	Sept.	17.69	28.28	25.11
Apr.	16.64	28.02	33.16	Oct.	17.90	28.50	....
May	18.71	29.02	31.69	Nov.	18.88	31.95	....
June	19.75	27.47	32.57	Dec.	20.67	32.89	....

The output of the Miami Copper Co. for September was 1,900,000 lb. as compared with 250,000 lb. in August and nothing in July. The June output was 5,340,000 pounds.

Anaconda's production of only 2,800,000 lb. for September was practically negligible as it compares with 29,400,000 lb. for September a year ago. With the mines operating today at only 55% it is doubtful if October will result in more than 20,000,000 lb. Last year the company, including the output of the Great Falls smelter, produced 331,900,000 lb. With the production for the second half of the current year curtailed, as it will require the balance of the year before the properties are in full working order, the total output of refined copper will be not more than 262,000,000 lb. This means a falling off of about 70,000,000 lb. from last year's returns.

It would seem safe to say that the price received for the copper this year will be as high as last, which was 25c. With a 15c. cost the profits for the year, not including the new source of revenue from zinc, and before deducting excess profits taxes, should amount to approximately $11 per share. This will compare with about $22 for 1916.

LEAD

Lead is quoted in cents per pound, New York delivery.

Date			Average week ending		
Oct.	3	8.00	Aug.	28	10.61
"	4	7.95	Sept.	4	10.38
"	5	7.95	"	11	10.20
"	6	7.95	"	18	9.33
"	7 Sunday		"	25	8.00
"	8	7.95	Oct.	2	8.00
"	9	7.95	"	9	7.96

Monthly Averages

	1915	1916	1917		1915	1916	1917
Jan.	3.73	5.95	7.64	July	5.59	6.40	10.93
Feb.	3.83	6.23	9.01	Aug.	4.62	6.28	10.75
Mch.	4.04	7.26	10.07	Sept.	4.62	6.86	9.07
Apr.	4.21	7.70	9.38	Oct.	4.62	7.02	....
May	4.24	7.38	10.29	Nov.	5.15	7.07	....
June	5.75	6.88	11.74	Dec.	5.34	7.55	....

ZINC

Zinc is quoted as spelter, standard Western brands, New York delivery, in cents per pound.

Date			Average week ending		
Oct.	3	8.50	Aug.	28	8.33
"	4	8.50	Sept.	4	8.25
"	5	8.50	"	11	8.21
"	6	8.50	"	18	8.29
"	7 Sunday		"	25	8.46
"	8	8.50	Oct.	2	8.43
"	9	8.50	"	9	8.50

Monthly Averages

	1915	1916	1917		1915	1916	1917
Jan.	6.30	18.21	9.75	July	20.54	9.90	8.98
Feb.	9.05	19.99	10.45	Aug.	14.17	9.03	8.58
Mch.	8.40	18.40	10.78	Sept.	14.14	9.18	8.33
Apr.	9.78	18.62	10.20	Oct.	14.05	9.92	....
May	17.03	16.01	9.41	Nov.	17.20	11.81	....
June	22.20	12.85	9.63	Dec.	16.75	11.26	....

QUICKSILVER

The primary market for quicksilver is San Francisco, California being the largest producer. The price is fixed in the open market, according to quantity. Prices, in dollars per flask of 75 pounds:

Week ending

Date					
Sept.	11	115.00	Sept.	25	105.00
"	18	115.00	Oct.	2	105.00
			"	9	105.00

Monthly Averages

	1915	1916	1917		1915	1916	1917
Jan.	51.90	222.00	81.00	July	95.00	[illegible]	[illegible]
Feb.	60.00	295.00	126.25	Aug.	93.75	[illegible]	115.00
Mch.	78.00	219.00	113.75	Sept.	91.00	75.00	112.00
Apr.	77.50	141.60	114.50	Oct.	92.90	78.20	....
May	75.00	90.00	104.00	Nov.	101.50	79.50	....
June	90.00	74.70	85.50	Dec.	123.00	80.00	....

TIN

Prices in New York, in cents per pound.

Monthly Averages

	1915	1916	1917		1915	1916	1917
Jan.	34.40	41.76	44.10	July	37.38	38.37	62.60
Feb.	37.23	42.60	51.47	Aug.	34.37	38.88	62.53
Mch.	48.76	50.50	54.27	Sept.	33.12	36.66	61.54
Apr.	48.25	51.49	55.63	Oct.	33.00	41.10	....
May	39.28	49.10	63.21	Nov.	39.50	44.12	....
June	40.26	42.07	61.93	Dec.	38.71	42.55	....

ALUMINUM

The market is quite nominal at 41.50 to 42.50c. per lb., New York, for No. 1 Virgin metal, 98 to 99% pure.

ORES

Tungsten: Prices continue unchanged at $23 to $26 per unit for 60% concentrate, depending on the quality of the ore, scheelite being the best and held at $26. Inquiries are appearing from European buyers but there is a question whether exports will be permitted. Ferro-tungsten is quoted at $2.30 to $2.60 per lb. of contained tungsten, but probably considerable can be obtained at the lower figure.

Molybdenum and antimony: Some business is reported in molybdenum ore at $2.20 per lb. of contained MoS_2 in 90% concentrate and this continues the quotation. In antimony ore the market is dull with nothing to report.

The total value of the mineral production of the country in 1916 was more than $3,470,000,000, increasing $1,076,200,000, or 45%, over the $2,393,800,000 recorded for 1915, and exceeding the former record year (1913) by more than 1,000,000,000, according to preliminary figures compiled by the United States Geological Survey. Practically all the minerals shared in this increase, gold being the only one of the more important products that showed a decrease in value, though silver and anthracite showed decreases in quantity but increases in Value.

Eastern Metal Market

New York, October 3.

All the metals are unusually dull and inactive. The markets await the solution of perplexing problems arising from Government control of industrial activity. Producers in many cases do not know what course to pursue, and buyers are uncertain as to the future of prices and demand.

Copper is lifeless and nominal.

Tin is featureless and lower.

Lead is steadier but inactive.

Zinc is again idle and a little weaker.

Antimony is dull and unchanged.

Aluminum is quiet.

Price-fixing and its effect are uppermost in the minds of the steel trade. The market has come to a standstill, buying and selling being practically out of the question. There have been daily sessions this week of steel committees of manufacturers to work out prices on the full line of products in proper relation to those already fixed. Next week the President will announce prices in various grades of pig-iron, on semi-finished steel and a long list of rolled products. In all the confusion it is clear that the new prices will not figure much in general business for weeks. The pig-iron output in September again fell off, the total having been 3,133,954 tons for 30 days, or 104,465 tons per day against 3,247,947 tons in August, or 104,772 tons per day.

COPPER

The copper market is still flat on its back. The resuscitation which was expected to follow the fixing of a Government price has failed to materialize. Indeed the situation seems to be worse than before the price of 23.50c. per lb. was announced. Until then some buying was possible; now practically nothing can be sold. Producers in general, especially the large ones, are unwilling to sell to ordinary domestic consumers at any price until the needs of the Government and its Allies are definitely known. Meanwhile contracts are being filled at prices agreed upon when made, which are probably in most cases above the present new price. There has been no known move to change contracts. Whether all grades of copper, such as Lake, electrolytic, arsenical, and casting-copper are to be maintained at the new price is still a perlexing question. It is admitted that consumers needing small wholesale lots are obtaining them from jobbers and other sources at from 26 to 28c. per lb. We continue to quote the Government price of 23.50c. per lb., but it is entirely nominal, as no transactions at this figure are reported. It was rumored yesterday that an important meeting of producers and consumers was in progress which would probably result in important news affecting the solution of many of these vexing problems. Copper exports in August are reported to be 42,285 tons, making a total of 342,904 tons to September 1, or only 40,000 less than the large total for 1913, and exceeding those for 1915 and 1916. The London market has declined to £125 for spot electrolytic and to £121 for futures.

TIN

The tin market is at a standstill. It is both featureless and monotonous. The entire week, since our last letter, has been quiet. Yesterday and Monday were some of the dullest days experienced in a long time. Cables have been delayed repeatedly, in most cases coming at least a day late. The market has declined from 61.75c., New York, on September 26, to 61.25c. on September 28, and to 60.50c. on both October 1 and 2. There is an entire absence of demand. Yesterday some spot, believed to be a small quantity, was pressed for sale at 60.50c, New York, but takers were not eager. It is regarded as having been a 'distress' lot. Last Friday, September 28, was virtually a holiday (Jewish), and there was no desire to do business. Some little was done, however, including principally 25 tons of December shipment from the Straits at 57.25c, and 25 tons of January shipment at 57.12½c. September 27 was a quiet day, the sales totalling 100 tons, of which 25 tons was spot-metal at 61.62½c., and 50 tons December-January shipment from the Straits at 57.35c. On Friday September 28 there were a few sales of spot tin at 61.25c., New York. At the close of October 2 there were 60 tons of tin arrivals and 4500 tons reported afloat. The London market is lower at £245 10s. for spot Straits. Total tin deliveries for September are reported to be 5402 tons, of which 3302 tons came from Pacific ports to the East. Stocks of tin on landing on September 30 were 3297 tons.

LEAD

It can safely be said that the lead market is steadier, and that it has largely recovered from the shock due to the sudden decline of 3c. per lb. in less than as many weeks. The decline is regarded as checked, but there is little activity. Buyers are still awaiting developments. Sellers certainly want orders, but seem unwilling to make concessions much below 8c. per lb., New York, which is still the quotation of the American Smelting & Refining Co. A moderate business has been done at 8c., and some small concessions have been made on a few large lots, but extended buying is absent. It is reported that some export sales have been made and that they are large, but no particulars are available. An explanation of the recent fall of lead is that pig-lead had to come to the present 8c. level in order to save the white-lead industry, because substitutes are making encroachments in the manufacture of paint. Lead is quoted in London at £30 10s. per ton for spot, and £29 10s. for futures.

ZINC

The more confident and cheerful tone that has prevailed for the last two or three weeks has disappeared gradually in the last few days. The market is easier and lower, with demand very quiet, and sales of small consequence. Quotations have eased off a little until yesterday prime Western for October delivery was held at 8.12c., St. Louis, or 8.37c., New York, with about ⅛ to ¼c. higher asked for November and December delivery. Large producers for the most part continue to ask 8.25c., St. Louis, and 8.42 to 8.50c., New York, for October. Sales for October delivery have been made at 8.37 to 8.42c., New York. The only interesting transaction since our last letter has been the awarding by the Government of 1000 tons (2,000,000 lb.) of grade B spelter (99.50% zinc) to the lowest bidder, which was the American Metal Co. Its bid was 10.67c. per lb. as against a range of 11 to 12.12½c. from some other companies. It is to be recalled in this connection that early in September the Government fixed 12c. as its price for this grade of zinc, on 2,750,000 lb. it was to buy, but later withdrew the order for the material, deciding to go into the open market. The present sale is part of this order. The transaction has not had a stimulating effect on the market, but rather the contrary. The Government is also asking bids on 320 tons of prime Western, and awards are expected any day.

ANTIMONY

There is no change since a week ago. Chinese and Japanese grades are still obtainable at 15 to 15.25c. per lb., duty paid, but demand is at a low ebb.

GREAT BRITAIN has increased its output of steel from 7,000,000 tons per annum before the War to 10,000,000, and facilities are being provided that will admit of a production of 12,000,000 tons next year.

ASSESSMENT-WORK on mining claims, with the exception of oil, will not be required for the years 1917-18. The bill exempting claim-holders in general, in addition to those enlisted in the Army and Navy, from this obligation imposed by the mining law substitutes a lieu service for assessment work. We will publish the full text of the Act in a later issue.

UNDOUBTEDLY the demand for various metals and minerals in all the belligerent countries is stimulating governmental interest in the mining industry, and to that extent it will prove useful even when the War is ended. In every country a systematic examination of its mineral resources is being made under official instigation and from it there will come enlarged knowledge, as well as a more sympathetic interest in mining.

MEXICAN labor is becoming scarce in Arizona and New Mexico. Part of the difficulty experienced at the mines in those States from shortage of workmen is due to the repatriation of large numbers of refugees who had sought freedom from conscription at home during the revolutionary turmoil. Fleeing from one war they rushed into another, and the call for registration by our Government produced an exodus. German sympathizers encouraged the Mexicans to believe that men of all ages would be drafted, thus intensifying the labor stringency in the South-West.

VILLA is still active, despite the denials of the Mexican government. Only a few days ago Col. Miranda, with four survivors of a sanguinary encounter with Ildefonso Sánchez of the Villa party, escaped across the border near Candelaria and surrendered to a troop of the Sixth Cavalry. Such episodes indicate a vigorous opposition that should not be suffered to exist if Carranza would have us believe in the stability of his administration. A government that seeks a loan of hundreds of millions should remove the doubt as to its ability to apply this in the economic redemption of the country by an effective campaign against these revolutionaries. Carranza has long displayed a tender regard for the personal freedom of Villa that never has been satisfactorily explained.

SPAIN has undertaken to establish Government laboratories in all the provinces of the country for making commercial analyses as well as for assaying and metallurgical testing of mineral products. The services of these laboratories will be available to the public at cost. A notably increased activity in the development of Spanish mineral deposits has resulted from the demands occasioned by the War. Despite the milleniums during which the Iberian peninsula has contributed to the world's supply of copper, lead, and iron, there remain undeveloped resources that promise to become important.

CORRECTION of proof often leads to the duplication of error. In this way the meaning of an author may be obscured. Such a fate befell the article on 'Drop of Pressure of Compressed-Air Hose,' by Mr. Walter S. Weeks, published in our issue of October 6. The last few lines at the bottom of the first column on page 504 should have stated that "The hose with the most 'give' has the least drop in pressure, because when it is under pressure the diameter of the hose is slightly increased. The more expensive hoses, which have a layer of rubber between the wraps of duck, apparently have more expansion, and consequently conduct air with less drop in pressure than more rigid hoses."

GYPSUM is selling for use as land plaster at an average rate of $7 per ton, unsacked, in carload lots, at most main-line points in the United States. The raw gypsum is ground so that about 60% will pass a 100-mesh screen. Gypsum occurs widely scattered throughout the country in deposits that often are not capable of being developed on a large scale, but which could be utilized for the local agricultural demand. The erection of suitable mills is not very costly, and the same plants would be available also for grinding limestone as a dressing to cure acidity and to loosen up soils heavy with clay. Many chemical plants in the country are utilizing their grinding equipment for producing ground limestone, which now sells at about $4 per ton at the larger distributing centres. It is generally pulverized so that 60% will pass a 200-mesh screen. An increase in the use of pulverized limestone is a marked tendency of recent agricultural practice.

CHEMICAL industry received prominent advertising at the recent exhibition in New York. A brief comment upon the presentation made is furnished us by Mr. M. W. von Bernewitz, which is printed in this issue. From this it will appear what great forward strides are being taken in applied chemistry by this country. We might take a divergent view from that expressed by our correspondent concerning the propriety of utilizing such an exposition to reveal the resources of the southern States. It is surely as important to make known available raw materials as to show what advances have been made in utilizing them. The object of the Chemical Exposition was to stimulate national development. The Southerners may have occupied a large amount of space, and if it appeared disproportionate in size it would indicate that other districts had neglected a valuable opportunity.

SURREPTITIOUS exportation of tungsten continues. Important arrests have been made in New York of parties engaged in this illicit trade, and some of the metal was recovered. The work of the Federal authorities in breaking up the ring responsible for this violation of the President's proclamation is said to be equivalent to the sinking of ten submarines. We can easily believe it. Shipments of tungsten, molybdenum, and nickel to Scandinavian ports for transmission to Germany have been continuous for the past three years. Every conceivable device has been employed for evading the British restriction upon contraband. Travel to Scandinavia increased enormously, and semi-manufactured cotton and rare metals were regularly taken in the baggage; women went oftener to Norway and Sweden than was seemingly necessary or safe, and a spirit of chivalry protects a lady's baggage from impertinent scrutiny; barrels of pork were not all meat, and much of the tobacco was incombustible. Americans of high standing were among those accused of falsifying invoices, and some of them were active buyers of tungsten for certain well-known German firms that apparently had millions available for accumulating these supplies in their warehouses. Members of these firms were prominent 'pacifists,' and openly announced that they were hoarding the metals to prevent their use for purposes of war, although these same concerns were on the British lists as violators of the embargo. Evidently the war, in their eyes, was a crime only when directed against Germany. Furthermore, these firms were active in trying to secure the passage of a bill to place a high protective tariff upon tungsten at a time when it would have been useful to them in helping to prevent the importation of this article from British ports, after it was found necessary for England to relieve the market here in the interest of speeding the manufacture of munitions and machinery.

Grinding Ore for Flotation

In the working out of the problems presented in the fine grinding of ores for flotation there exists, perhaps, a larger opportunity to reduce the costs of milling than is to be found in any other branch of the art. Fine grinding is beset with difficulties, and therefore the experience at every important plant, when studied and discussed by a competent technologist, is notable. For this reason the article on the milling of the ore at Miami, by Mr. T. A. Rickard, which appears in this issue, will make an exceptionally wide appeal to mining engineers. The subject is treated in great detail, and the able metallurgists responsible for the economic results obtained have been frank in their confession of the difficulties met, and have supplied elaborate data showing how they were overcome. At Miami, as at many other great mills that have been erected since mammoth treatment-plants have become the vogue, the original estimates of capacity for the crushing department were not realized in practice. The reason lies not in lack of engineering skill on the part of the designers, but in the special difficulties presented by the ore drawn from any particular deposit. A test-mill suffices to indicate the general plan of treatment that will lead to success, but the scheme of a pilot-plant is rarely capable of being followed exactly in the works finally erected for reduction on a larger scale. Moreover, the products obtained when milling thousands of tons daily seldom conform in character and proportion to those derived from a plant treating only a few hundred tons. At Miami the mill had been running for a considerable period before the most desirable method of crushing was ascertained, and extensive changes in plan and equipment were then made. Even now further changes are in prospect, for, as Mr. Rickard pertinently remarks, "a progressive management knows no such thing as completion."

The comparison at Miami lay between rolls, Chilean mills, and Hardinge mills. The Hardinge in this case yielded the more advantageous product. Rolls are perhaps the most efficient granulators made, but their special province is restricted to crushing dry unless the ore is exceptionally free from clayey, or, more broadly speaking, colloidal substance. On the other hand, while rolls tend to choke, as they did at Miami, from too large a percentage of primary colloids, they do not produce as much secondary colloidal material as results from the wet-grinding of ore in other types of comminutors. This is partly for the reason that the time of contact is brief, partly because the method of applying the crushing force involves no shearing, and finally because trituration is entirely avoided. It is interesting to note that the ball-mill, as exemplified by the Hardinge machine in the trials made at Miami, yielded appreciably less pulp of minus 200-mesh size than the Chilean mill. This indicates less abrasion. Although pre-eminent as a comminutor, the ball-mill nevertheless appears to be essentially a granulator. It is a moot point, since practitioners differ in their opinions, but it seems that the effect in a ball-mill is due mainly to what has been called 'radial crushing,' in which respect it resembles the action of rolls, differing, of course, in many respects in the mode of application of that principle. Crushing by impact no doubt takes place to some extent in ball-mills, especially in those adapted for relatively coarse crushing, but the load of balls at any moment is an exceptionally rigid mass, and is caused to move only by disturbance of the equilibrium when the mill is revolving. The result, then, is a tendency on the part of each ball to rotate around its neighbor. The ore-particles gripped between the balls are reduced as if caught between the cylinders of opposing rolls. Within the mass of balls there can be no impact of one against another, and the moment that shearing might come into play it would be a case of ball cutting ball rather than of the abrasion of ore-particles. Whatever be the explanation of the crushing-effect in the ball-mill, it is evident that less slime is produced than when the work is done in a Chilean mill. It is not a question of relative merit as crushing devices, but of the kind of product desired by the metallurgist for securing the best possible recovery of metal, and the difference may reside to a large extent in the peculiarities of the ore undergoing treatment. If clay were an abundant constituent, and it were important to eliminate this as rapidly as possible from the crusher, it is conceivable that the Chilean mill would show superior adaptability.

The object at Miami, as at all plants where ore is ground for flotation, is to produce as large an amount of exceedingly fine material as possible. Mr. Rickard points out that the decrease in the proportion of sulphide remaining in the tailing at Miami accurately follows the curve of decreasing size of the particles "until below 200-mesh the recovery of sulphide is seen to be reasonably complete, and the cause of the loss in the tailing becomes due predominently to the inability to save the oxidized mineral." Mr. Gottsberger states that his aim is "to produce the maximum amount of material between 60 and 200-mesh." The product finer than 60-mesh shows 60% under 200-mesh, and although 30.9% of the total pulp, or 45.5% of the minus 60-mesh, is classed as 'slime,' it is certain that an extremely large proportion of this is actually granular, and hence would be specially well adapted to yield a high recovery of the sulphide by flotation. The significance of this is that sulphides reduced so fine as to approximate the condition of suspension-colloids, as distinct from gel-colloids, are peculiarly amenable to successful treatment by flotation, since they are readily taken up and held in the interface between the oil and water. The presence of gel-colloid, apart from any influence it may exert in emulsification, is a deterrent, shrouding the mineral particles, and preventing them from reaching the interface where they would be held and lifted to the surface by the bubbles. The action of the ball-mill, as shown in the operation of the Miami plant, is interesting in this connection. It has yielded a higher degree of comminution with the production of a less amount of slime.

An American Peace

The new Liberty Loan is making progress, but it is being subscribed at a rate that reveals a thoughtful sacrifice by many patriotic investors. This is really better than tempestuous enthusiasm, for emotion is fickle, while serious thought develops a purpose that will reach through to the end. The people who are using paper and pencil to figure out how much they can spare are getting into their systems the inflexible determination of the old patriots who knew no compromise with the principles of liberty set forth in the Declaration of Independence. These are the kind of men who will translate their convictions into action, who will literally see it through, and not finish by writing an interesting letter, as did Mr. Britling. Secretary McAdoo gave the people an energizing doctrine when he said, "I do not believe in compromise with enemies." Neither will the men and women who are giving their sons for the contest and then following that supreme sacrifice with a further yielding of ease and comfort by figuring how much they may abstract from their incomes in order that the Government may sustain the young men who have gone to deliver the iron message of America to the German people. No; there will be no compromise! No politician would dare to misinterpret the wills of these men and women who are making sacrifices for America and American rights, by trifling with the Teutonic scourge; the War must go on until that is forever obliterated. This means more than a 'German peace,' more than acceptance of treaty pledges from a German nation. America will see it through, and will pay the bills with all the resources at its command. It is one of the most useful things for the revitalizing of that old-fashioned American spirit, which nearly had been emasculated by the pacifists, that the Liberty Loans are issued as popular loans, not subject to the manipulations of Wall Street, but available at par, without financial jugglery, to every citizen. It introduces a personal relationship to the Government that carries with it the sense of personal responsibility and the feeling of personal control that are inseparable from the act of paying the cost of national administration consciously and directly. That is the peculiar merit also of the direct tax, with which our people will soon become familiar as never before; and after the national experience with popular loans, and personal contributions, and bitter warfare for our principles, we may turn out a better democratized people, and prove in the future less open to the chicanery of politicians who thrive on general ease-loving habits and political indifference.

This loan is working for our political uplift, and it is being subscribed with an intelligent sense of its high significance. Nevertheless, the forces of newspaperdom are operating still against the spirit necessary for strengthening our military programme and for making it effective. Perhaps they are doing it unconsciously; they forget a higher duty in performing their function as purveyors of news; the evil is apparent on both sides of the Atlantic. The news agencies of England are equally reprehensible. The dispatches from London, based upon messages that tell of German uprisings and mutinies and the threatened degradation of high officials, have not modified the poison that was sent from Berlin to blind and befuddle the world and to beguile us into inaction. Can anyone be so simple as to suppose, if he but pause and think it over, that any information derogatory to Germany, indicative of internal treachery and the peril of imperial collapse, could be published in that country and allowed to find its way through the tightly contracted censorship of an autocratic military power into the countries arrayed against it, unless every word had been carefully prepared with a view to its psychologic effect? If this were not so, then we would be obliged to admit that the Kaiser was puerilely advertising the tottering of his authority. It is too soon for that, though it must come in due season. The entry of America into the War has listed the downfall of Prussianism among pre-destined events. We may consider it stupid to send out such dispatches, whether true or false, for they must in turn exert an influence upon the German mind. The Teutonic leaders have displayed a vast amount of childish simplicity in their misjudgment of the intelligence of other people, and also sometimes of their own. The stupidity of these dispatches may be partly intentional, but it is not wholly so. It would be well if our news-gatherers should so handle the misinformation transmitted from Berlin through Amsterdam as to make clear the intentional as well as the unintentional stupidity in them.

The purpose of the game is to arouse anticipations of an early peace, so as to slow down our preparations. After months of exasperating performances of this character, and after repeated warnings by prominent men returning from Germany that such events would be heralded in order to make our military activity seem unnecessary, President Wilson has at last urged the people of this country to disregard these rumors. He should go further; he should request the newspapers to pass dispatches from such untrustworthy sources through the hands of officials who, by proper editing, would render them innocuous. No newspaper is so un-American as to disregard the expressed wish of the President in any matter pertaining to this War. We are ready to buy bonds, we are determined to sustain the Government, we are sending our sons to the front, but on account of the misinformation disseminated through the press we are talking a great deal too much about an early peace. Our present supremely urgent business is war, and we had better talk about war. An arctic explorer who should start from his base on the northward trudge over the forbidding snow-fields, and should persistently talk about the final homecoming, would certainly fail of the inspiration that would carry him through to the Pole. One way to help win the War is to insist upon grim determined fighting. Very likely the Germans are tired of the War; beyond doubt Kaiser Wilhelm would have been a less blustering war-lord in 1914 had his foresight been as good as the hindsight now possible over a fruitless triennium of slaughter; but Germany has not yet been sufficiently punished, and our soldiers are needed at the front to make the correction permanent. The dispatches from Berlin must be read in the light of the Teutonic estimate of Americans as silly fools, and until the Kaiser sends a plea for mercy direct to the Allied Governments we must recognize that he still hopes for some form of a 'German peace.' We ought to print on every bond "No compromise!" and then go on until we can dictate an American peace.

Copper Producers Combining

Copper production has fallen off. No other result could follow from the double embarrassment of obstrep erous labor and an indecisive Government. The decreas of 45,000,000 pounds of copper per month during Augus and September is not surprising under these circum stances. The slump will become more precipitous unles wise steps be taken promptly. The action of the coppe producers, under the marshalship of Mr. Joseph Clen denin of the American Smelting & Refining Company in arranging what the market denominates a 'pool,' is a necessary step in the direction of self-preservation. Th copper producers have borne the brunt of increasin difficulties, and the outlook promises more stress o weather. As we pointed out recently, in discussing th labor situation, ceaseless agitation, interrupting opera tions and lessening the will and efficiency of the worke coupled with an actual shortage of men, is a matter o national concern in view of the fact that the demands o this Government and of our European Allies hencefo ward will exert a tremendous pressure upon the capacit of the copper mines. Certainly co-ordination of effort requisite if the problem is to be met with hope of solvin it. The committee on distribution, which has just bee formed in representation of the producers, has frankl admitted that what it proposes to do, and must do to sa the situation, is technically in violation of the Sherm Act. In these times of mammoth combination of indu try, under Government sanction and by its direct inte vention, for the sake of directing production and stabili ing prices, it is almost a shock to be reminded that Sherman Act stands lonely and forgotten on the statu books. We imagine that the Government will find so convenient means by which it may adapt this committ as a useful wheel into its War-mechanism, and we ho in the interest of plain people who must buy a f things made of copper and brass for general use, t the proposed committee of consumers, being organized co-operate with the producers' committee, may rece official recognition. Surely in some manner it should come known whether outsiders may or may not h copper, and whether they may have it at 23.5 cents at some other price. The Government needs to t decisive measures or empower those to do so who are b informed in regard to the conditions of production a the needs of the domestic market.

Mill-Tests v. Hand Sampling

he Editor:

Sir—Contributions to the MINING AND SCIENTIFIC RESS on the subject of mine-sampling appear continually and certainly few, if any, subjects in the technology of mining are more worthy of attention. The llowing are some of the impressions of a man who has een more engaged in the management and development f mines than in inspecting and reporting on them, but ho has also done and seen a considerable amount of ch work.

Much has been said as to methods to be adopted in mpling, that is useful and necessary. These are the tcome of applying experience and common sense. It ems to me, however, that there is a tendency to overate the applicability of particular methods. I submit at knowledge of methods is not sufficient in a man appraising the value of a mine. Probably no competent ngineer would say that it is. I submit, however, that e mere knowledge of methods falls much further short f fulfilling the requirements than appears to be genrally accepted. I affirm that the primary requirement appraising the value of a mine is an understanding f the deposit itself. This diagnosis having been made, correctly as available data and the ability of the man ermit, the method to be adopted should then be a matr for consideration in the light of experience, of comon sense and of the facilities and money available for e work.

There has been much discussion as to the respective erits of bulk-sampling, that is, sampling by means of tually treating certain tonnages of the ore, as against mpling by means of considerable numbers of small mples taken according to certain methods. Personally have no doubt as to the usefulness of treating large nnages of the ore, except in certain circumstances. ose circumstances apply in the case of ore formations a very homogeneous character and on which other ines have already, by actual operation, shown the aracteristics of the ore under treatment, and have also oved the efficacy of certain methods of sampling to rive at the metallic content of the ore. Obviously the praising of such mines is a simple matter and can be ected by the application of rule of thumb methods of npling, and at no great cost compared with the size the deposit sampled, supposing, of course, that they developed to a reasonable extent. Such cases are not se that call for discussion; nor are they the ones which present difficulties or call for initiative or a very deep understanding of ore deposits. It is the case of deposits outside of such circumstances of which I now write. To deal, then, with the appraising of deposits where the problems presented are fresh ones, it will be agreed that, from the point of view of appraising the character of the ore metallurgically, that is, what the recovery and the cost of treatment will be, the treatment of considerable tonnages of ore, taken from the mine so as to be as fairly representative of the whole deposit as may be feasible, is invaluable. It may be costly to effect, and in many cases impossible. If it can be done it is undeniably most useful, as the recovery and the cost of treatment are fundamental factors in the value of the ore, which value is the profit obtainable from it.

Even, however, from the point of view of ascertaining the metal content of the ore, the treatment of certain tonnages judiciously chosen is most valuable. For instance, information given by the treatment of the whole of the ore coming from particular headings such as drifts and winzes, accompanied, of course, by continuous sampling at the treatment-plant, is certainly much more complete and reliable than that obtained by sampling the sides by any method based on a number of small samples. In the one case the whole of that portion of the deposit represented by the particular headings actually has its full metal-content accounted for. In the other case the same portion is merely sampled by methodical scratching around it. Whether all the circumstances would justify such treatment of the ore, or of ore broken from selected spots, if it be unavailable from the headings, is another matter.

Every mine or deposit should be judged on its own merits as to the manner in which it should be sampled and to what extent sampling should be carried out. In diagnosing cases for this purpose many factors come into play, the degree of regularity in the distribution of the metal and the facilities for carrying out bulk-sample tests, the history of the mine, and the like. The mine record is most informing if known in reasonable entirety, and especially if the reasons underlying the scheme of past operations are known. In some mines I have preferred sampling the trucks or kibbles as they have come from development work to a sampling of the faces. In these cases the treatment of all the ore from the development was not feasible. In other mines I have found that sampling of the development-faces by the usual method of cutting strips across gave satisfactory results. The former method is feasible only while development is progressing and is not to be effected by an engineer whose time is limited to a visit. Dump-sampling may or may not partly take the place of such sampling, according to circumstances. Where it is decided that face-samples are to be utilized for valuation, I submit again that there is no rule to guide the method of taking them. The degree of homogeneity in value and in hardness, the variability in width, the degree of regularity of outline, and so on, must be carefully considered in deciding on a plan. The stereotyped taking of sections across the run, or what may be considered the run of the ore at certain spaces apart, by no means meets all conditions. To give examples of different circumstances which would indicate special methods of sampling would be too lengthy. My contention is that it is the man, and his general grasp of essentials, including the significance of results from sampling when obtained, that is the main thing, and not knowledge of methods, as such.

I will, however, say a word regarding the face-sampling of ore of a kind I have often seen quoted as presenting a difficult sampling problem. I allude to cases where the ore is variable in hardness and where either the harder or the softer part, more frequently the latter, is much richer than the other. Most engineers have dealt with such ores. The difficulty is that in cutting out a section the jarring causes a lot of the softer material to fall from the ore alongside of and behind the section. The problem is to get the hard and the soft portion of the ore represented in proper proportions in the sample. I would suggest in such a case that before the sample is taken the face should be subjected to a jarring similar to that which it receives during sampling, that a wide strip be cut, and that a rectangular receptacle, say a box, of the same or of a less width than the strip, be used. If such a box be used, one side being kept parallel to the length of the strip, and care being taken that no part of the box is held outside of the limits of the strip, then little or none of the ore shaken from outside and alongside of the strip by the jarring should go into the box. These precautions practically dispose of error due to softer ore from the part outside and alongside of the strip taken. With regard to the amount of softer ore which would fall from behind the strip, this should be counterbalanced by the amount of the same which had previously been jarred from the strip. As a matter of fact this effect is achieved to some extent in the work of making the excavation, the face of which is being sampled. The explosions and picking will have jarred out much of the softer ore, leaving on the outer surface a depth composed of a larger proportion of the harder ore than was the case originally. This last fact and the additional use of preliminary jarring has, I think, been often overlooked in considering the correctness of sampling in ore of the above description.

I have seen advocated a method of sampling faces by means of boring holes in them. In ore with the metals fairly well distributed, this should give reasonably accurate results. The few inches at the collar will be unduly represented and in a fairly long hole the amount of drillings would be 50% greater per inch a few inches from the collar than at the bottom. The conditions thus differ considerably from the rules laid down for sampling faces in sections, if all the drillings from a hole go into one sample. This can be met by the division of the drillings into separate samples as the boring proceeds, say into sections of one foot each.

This method has been recommended, or stated as having been used, for cases where the hardness varied and the metal-content of the ore varied with the hardness. Where the softer part of the ore is of a crumbly nature, this method would result in considerably more of the softer part going into the sample than should be allowed. The jarring of the drilling would shake out a quantity of the softer ore from parts around the hole. In any case the hole would generally cut somewhat larger in the soft parts than in the hard parts. The effect of jarring out crumbly ore from around the hole would apply particularly if percussion drills were used. If done by hammer drills or by hand-drilling the effect would be less, but might still be considerable. In this case one would expect more accurate results from rotary drills, if it were feasible to use them. Even then any crumbly parts might be too largely represented in the drillings. Cores would probably represent such parts insufficiently.

I have never myself utilized this method of sampling and cannot speak from experience as to its reliability. I have utilized the boring of holes for exploratory purposes, to see what ore existed beyond the exposed faces, but that is a different thing. This was seeking general information, not results upon which to calculate values. I am satisfied that the difficulties of securing a representative sample of the immediate spot being taken can practically always be overcome by intelligent consideration of the problem, though perhaps with trouble and expense. In irregular deposits any scheme of sampling must be recognized as giving at the most only reasonably accurate indications of the average value of the exposed faces, and that the unexposed parts may be richer or poorer than the sampling indicates. It comes back to the fact that sampling is only one of the instruments used by a man with a sound grasp of his work, certainly an essential instrument, but in itself insufficient to indicate the probable worth of a mine. The intelligence and experience to observe and interpret other data and to recognize the most probable meaning of sampling results is necessary. Perhaps some will think this so obvious as to be unnecessary to say. I believe that it is not obvious to all, and is so important that it is worth the saying even at the risk of superfluity.

H. R. SLEEMAN.

Perth, Western Australia, July 17.

FACING-SAND suitable for foundry use, especially the finer grades that may replace the imported French sand, is the subject of a paper by C. P. Karr of the U. S. Bureau of Standards.

VANADIUM combines with both oxygen and nitrogen in molten steel at high temperature, eliminating them.

Miami, Arizona: The Milling of the Ore—IV

By T. A. RICKARD

CRUSHING-PLANT. A part of the Miami mill went into operation on March 15, 1911. It was designed by H. Kenyon Burch (State College of Washington, '01). From the No. 4 shaft the ore is delivered to the coarse-crushing plant, which, as planned originally, consisted of two No. 7½ (Kennedy) gyratory crushers, breaking the mine-run to 2-inch cube, without return, the crushed ore passing to two 4 by 10-ft. trommels equipped with manganese-steel screen-sections having ¾-in. round holes. The oversize of these trommels was delivered to two 54 by 24-in. Traylor rolls, intended to reduce the ore to ⅛-in. cube.

While this arrangement seemed ideal on account of its simplicity, it failed to produce the degree of fineness desired. An improvement was brought about by making several changes. A small grizzley was placed before the gyratory crusher so as to remove the fine stuff, already sufficiently reduced. The chief fault, however, of the coarse-crushing plant was the inability of the rolls to furnish a uniform product. When the faces of the rolls became worn they were so corrugated as to yield a large proportion of oversize. With a 2-inch feed such corrugation seemed inevitable. The only way to obtain a suitable product was to re-grind. Therefore the product of the two 54 by 24-in. rolls is now (see Fig. 1) elevated to two 4 by 7½-ft. trommels with ¾-inch holes, the oversize from which goes to two 42 by 16-in. spring-rolls, the product from which is returned to the trommels, making a closed circuit, so that only the undersize from the trommels goes into the mill. Today the product from the coarse-crushing plant contains only 1% that remains on a ¾-inch screen and 10.9% that remains on a ⅛-inch screen. The crushed ore is delivered through chutes leading from the trommels to a traveling belt, which already has carried over 7,000,000 tons. It was made by the New York Belting & Packing Co. It has a width of 30 inches and moves at the rate of 411 ft. per minute. For 256 ft. it has an incline of 16°30′ and then delivers the ore to a horizontal belt 329 ft. long, passing over the mill-bins. The distribution of the ore is effected by a tripper. A weighing-machine (the Merrick weightometer) keeps a record of the supply.

ORIGINAL MILL. The mill is divided into six sections, each of which is provided with a circular steel bin of 1000 tons capacity. In May 1911, three months after the start, the first three sections were in operation, treating 39.657 tons during that month. From each bin the ore was fed by a shaking feeder to 42 by 16-inch rigid rolls designed by Mr. Burch, who had provided a device whereby one roll-face moved across the other so as to prevent corrugation. It failed mechanically because the pressure upon the two faces in contact was so great. At this stage the reduction was supposed to be to ⅛-in. cubes. This result was nearly obtained, only about 2% failing to pass through a ¼-in. round hole. The rigid rolls, however, proved too light for the work required, and on account of the high cost of maintenance were replaced by an extra heavy type of spring rolls specially designed by the Traylor Engineering Co. The original flow-sheet provided for a wet feed to these rolls and for

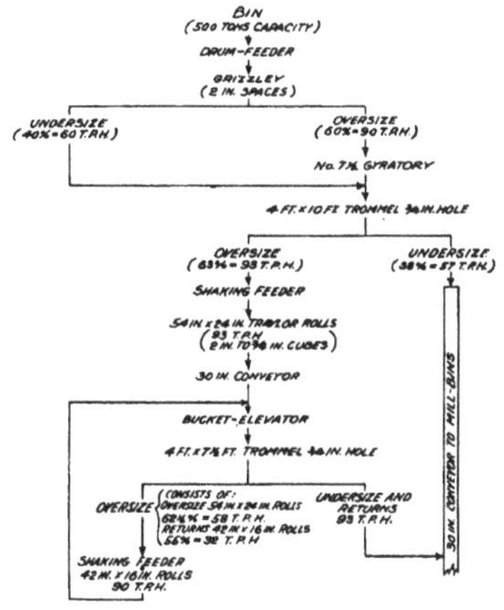

FIG. 1. FLOW-SHEET OF ORE-CRUSHING DEPARTMENT

the removal of some of the fine by means of a shaking launder, fitted with a perforated steel plate where water was added, but it was found in practice that the rolls did better work on a dry feed, so that later the water was added below the rolls.

The product from the rolls was divided into two parts, each of which passed over a duplex Callow screen, having openings of 0.026 inch, equal to about 26-mesh. See flow-sheet, Fig. 2. The undersize from each screen went to a drag-belt dewaterer, while the oversize passed to a 6-ft. (Saturn) Chilean mill provided with a screen having 0.024-inch openings, equivalent to 28-mesh. The product of the Chilean mill joined the undersize from the Callow screen at the dewaterer, the overflow from which went to the slime-plant while the sandy product from the dewaterer went to a 10-spigot launder-classifier equipped with the Richards hindered-settling device. Each of the 10 spigots served one No. 2 Deister sand-table. The middling from the first six tables on each side was fed to a 6-ft. Hardinge pebble-mill, the product going to two No. 2 Deister tables, which re-treated this re-ground middling from the first Deister tables. The middling from the last four Deister tables on each side was re-concentrated, without re-grinding, on a No. 3 Deister slime-table. That finished the sandy product; we turn now to the treatment of the slimy product.

The overflow from the dewaterer and the overflow from the launder-classifier joined in their passage to one 8-ft. cone-tank. The underflow (or spigot product) was divided into six parts and fed to as many No. 3 Deister slime-tables. The overflow was distributed to eight 8-ft. cone-tanks, the thickened product from these being treated on eight No. 3 Deister slime-tables. The overflow was used as returned water. The middling from all the slime-tables was re-treated without re-grinding on two No. 3 slime-tables.

Such was the scheme of water-concentration. The results were as follows: From a feed averaging 2.46% copper there was obtained a concentrate of 39.69% and a tailing of 0.688%. Thus the recovery was 71.8%—a result fully up to the standard reached at that time—six years ago—on a disseminated-chalcocite ore.

CHANGES. The screen of the Chilean mill was changed to 0.029 and later to 0.037 (say, 18 mesh), partly to gain capacity and partly to promote durability. As B. Britton Gottsberger, the manager, said in his report of January 1, 1913: "It was found necessary to sacrifice metallurgical efficiency for mechanical reasons." Coincidently No. 4 section was equipped with three sets of 42 by 16-inch rigid rolls, instead of Chilean mills; but the experiment proved a failure, because the capacity of the rolls was insufficient when crushing wet. After the failure of the rolls in No. 4 section, an 8 ft. by 22-in. Hardinge mill was tried in competition with the Chilean mills. This was in July 1911. The immediate result was a telegram stopping the shipment of another type (Evans-Waddell) of Chilean mill already ordered and the purchase of three Hardinge mills for section 6, besides 10 more Hardinge mills, one to be placed in section

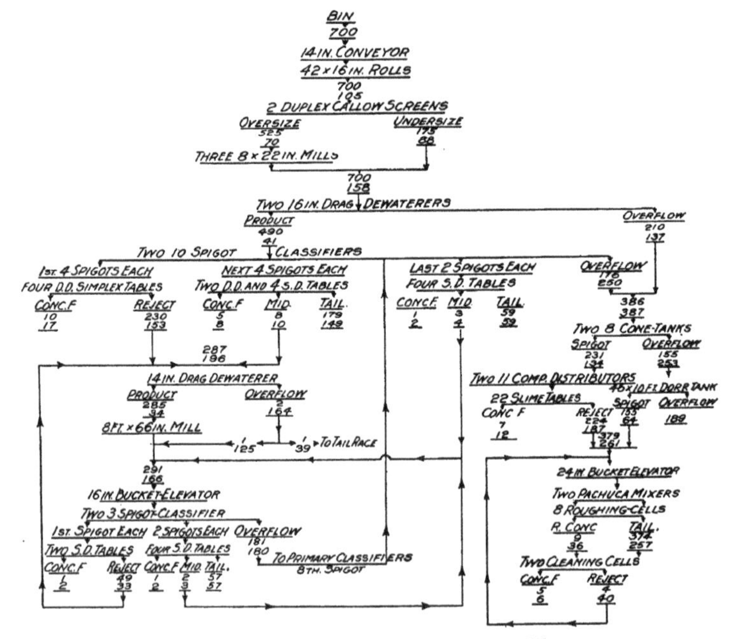

FIG. 3. FLOW-SHEET OF SECTIONS 1, 2, AND 4. SEPTEMBER, 1915

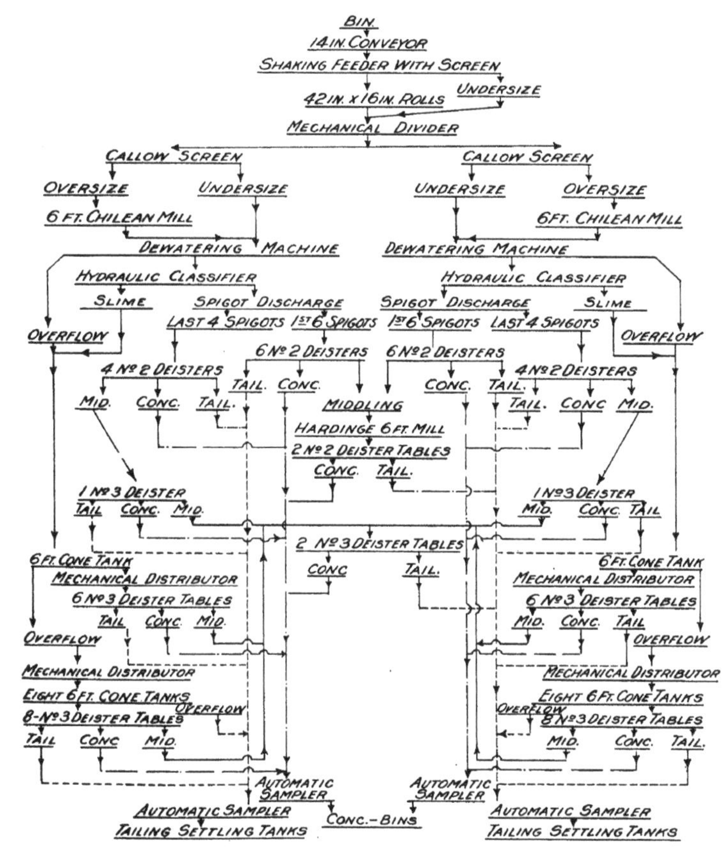

FIG. 2. FLOW-SHEET OF ONE CONCENTRATOR UNIT. MARCH, 1911

4 and three each in sections 1, 2, and 3. In the report already quoted, Mr. Gottsberger gave a tabulation of screen-analyses showing the work done by the Hardinge and Chilean mills, both taking the Callow-screen oversize from the rolls; I quote it herewith:

	Hardinge mill		Chilean mill	
Mesh	Feed	Product	Feed	Product
+ 4	12.9	0	13.9	0
+ 10	47.3	0	47.5	0
+ 20	26.8	0.2	22.9	2.3
+ 30	5.0	3.2	5.2	11.8
+ 40	0.8	4.9	0.9	6.7
+ 60	0.8	13.8	1.0	11.4
+ 80	0.4	10.4	0.5	6.7
+ 100	0.3	8.6	0.4	5.4
+ 150	0.3	8.0	0.5	6.3
+ 200	0.5	10.0	0.7	7.2
— 200 sand	0.8	10.0	1.2	10.6
— 200 slime	4.1	30.9	5.3	31.6
— 200 total	4.9	40.9	6.5	42.2

Mr. Gottsberger said: "Inspection of these shows at a glance that the Hardinge mill as a grinding-machine for the purpose we aim at is the more efficient. Our aim is to produce the maximum amount of material between 60 and 200-mesh, avoiding sizes both above and below these limits. By this standard it is found that the Hardinge mill produces 22.1% plus 60-mesh in comparison with 32.2% for the Chilean. Of material between 60 and 200-mesh, the Hardinge mill produces the larger amount, showing 37% in comparison with 25.6% for the Chilean, while on material under 200-mesh, the Hardinge mill here shows the lower amount—40.9% as against 42.2. It is thus evident that as a grinding-machine the Hardinge mill is far in advance of the Chilean." Further, he pointed to the fact that where the mill-sections equipped with Hardinge mills needed only 172 gallons of water per section per minute, the sections containing Chilean mills required 290 gal. per min. Another feature was the effect of the fine slime upon the cone-tank overflow. The lower cone-tank overflow carried 19.9 tons of solid per section per 24 hours where Hardinge mills were at work, while the Chilean sections averaged 75.1 tons. As regards capacity, Mr. Gottsberger stated: "In the case of the Chilean mill, as the size of the product depends upon its passage through a screen, the capacity can be increased by forcing the machine, without any change in the size of the material delivered. This gain in capacity, however, is limited, and is accomplished only at the cost of an abnormal power-consumption. In the case of the Hardinge mill, for a given quality of feed, only a certain amount can pass through the machine and this will be governed by the quality of the product desired. The capacity of the Hardinge mill will vary possibly to a greater extent than the Chilean with different percentages of fine material in the feed. We have found in practice that with two Saturn mills, a section will handle about 480 tons per day with a power-consumption of about 150 hp. By overloading the machines, between 500 and 600 tons per day for two mills can be obtained, but at this rate the power-consumption will rise to 175 hp. or over. In comparison with this we find that with three Hardinge mills and our present intermediate-roll product, a section-tonnage in excess of the 500 tons per day can be obtained with a consumption of about 140 horse-power."

OXIDIZED ORE. Thus was brought about a series of mechanical changes. The capacity of the mill was increased from 500 tons per unit to 700 tons. The copper in this ore was in two diverse conditions, sulphide and oxidized. This fact influenced not only the water-concentration, but the froth-flotation that came at a later stage. At the time of which I am now speaking, at the end of 1912, the copper was divided between the sulphide (chalcocite, with a little chalcopyrite) and the oxidized minerals (chrysocolla, malachite, and azurite) in the proportion of 83:17. Of the total copper recovered by milling, 93% was in the form of sulphide and 7% oxidized, whereas the loss in tailing was distributed between 61.7% sulphide and 38.3% oxidized. In short, the recovery of sulphide copper was 76.8%, whereas that of the oxidized was only 28.8%. Thus the success of the milling varied inversely to the proportion of oxidized mineral in the feed. Of the sulphide, 66% remained on the sand-floor, as against 36% of the oxidized copper. On the slime-floor the ratio was 34% of sulphide to 64% of oxidized. It was evident therefore that in the course of the successive crushing operations the oxidized minerals tended to collect in the slimed products, where the recovery of them was difficult, and, in fact, of relatively small consequence. On the other hand, of the oxidized copper going to the sand-floor, 58% was recovered, as against 12% on the slime-floor. This relation was confirmed by the assays of the tailing, showing that of the total loss of copper in oxidized form, 21% had to be debited to the sand-floor and 79% to the slime-floor.

A test made between a Wilfley and a Deister multiple-deck table showed a recovery of 46.4% on pulp assaying 1.785% copper by the Wilfley, as against 46.1% on a feed of 1.578% by the Deister. The tailing from the Wilfley contained 0.954%, whereas the Deister tailing contained only 0.875% copper. An experiment was made in lengthening the cylinder of a 6-ft. Hardinge mill from 22 to 38 inches. This proved advantageous. One long mill re-ground as much as two short ones. It was found advisable to re-grind the sand in a Hardinge mill, following this by classification before re-treatment on Deister tables. A test-run showed that the tailing could be reduced to 0.47%. Thereupon, a new flow-sheet was planned; but shortly afterward, in June 1913, the possibility of using flotation as an auxiliary process had to be considered, in consequence of the experiments made at the Inspiration mine. These experiments were confirmed at the Miami mine in January 1914, when an agitation-machine was used with encouraging results.

RESULTS. Before proceeding to describe the introduction of flotation, I shall quote figures indicating the results obtained during the past five years:

HORIZONTAL BELT AND TRIPPER DELIVERING CRUSHED ORE INTO MILL-BINS

ROLLS AND HARDINGE MILLS, HEAD PULLEY OF DRAG DEWATERER AND HEAD END OF CLASSIFIERS

	Tonnage	Feed, %	Concentrate, %	Tailing, %	Recovery, %
Jan. 1913..	92,615	2.33	37.13	0.699	71.57
Jan. 1914..	102,497	2.31	39.97	0.747	70.05
Jan. 1915..	60,115[1]	2.22[2]	40.03	0.655	71.67
Sept. 1915..	121,713	2.21[3]	44.16	0.462	79.94
Jan. 1916..	128,866	2.19[4]	41.64	0.628	72.47
Mch. 1916..	141,272	2.09[5]	42.40	0.533	75.48
Jan. 1917..	171,882	2.15[6]	43.14	0.592	73.48

[1]Curtailment due to the War. [2]Oxidized, 0.363%. [3]Oxidized, 0.251%. [4]Oxidized, 0.33%. [5]Oxidized, 0.33%. [6]Oxidized, 0.247%.

During this period the cost of milling was 65.86, 57.22, 49.96, 57.94, and 58.91 cents for the years 1912, 1913, 1914, 1915, and 1916 respectively. Flotation was introduced in August 1914. The first result was to increase the capacity of the mill rather than the recovery, although that also was improved. The effort to get more ore through the mill was intensified by the rise in the price of copper, so that the tonnage treated in January 1917 was nearly 70% more than was milled in January 1914, whereas the recovery during those three years improved only 3.43%. During this interval the cost per ton of ore increased considerably, from 49.96 to 58.91 cents, because the higher profit justified a more hurried exploitation, but the cost is now being reduced, having been 51.57c. for the first three months of the current year.

FLOTATION. In June 1914 the attention of the Miami management was called to the Callow machine and in August of that year a flotation unit was added to section No. 5 of the mill. This new equipment consisted of four Callow roughing-cells and one cleaning-cell. Meanwhile preparations were made for a re-modeling of the mill in accordance with a new flow-sheet, shown in Fig. 3.

The capacity of the steel bins was enlarged by enclosing the spaces between the original circular bins, thereby increasing the total storage to 10,000 tons, as against an original capacity of 6000 tons. Three Hardinge mills (8 ft. diam. by 22 in. length of cylinder) were added to the grinding-machinery, replacing two Chilean mills. The launder-classifier was retained, but the first six spigots were made to feed three double-deck Deister sand-tables, the other four spigots feeding four Deister No. 2 single-deck tables, as previously. The reject from the first two double-deck tables on each side of the section, together with the middling from the first six tables, was re-ground in an 8 ft. by 66-in. Hardinge pebble-mill. The product from this was elevated to a 3-spigot launder-classifier feeding three sand-tables.

The overflow from the dewaterer and from the classifiers passed as before through an 8-ft. cone-tank (one on each side of the section), the spigot-product of which was distributed to 11 No. 3 Deister slime-tables, while the overflow from the cone-tank went to a 46 by 10-ft. Dorr thickener. The thickened product and the reject from the slime-tables passed together to a bucket-elevator and thence to the Callow flotation machines—eight roughing-cells and two cleaning-cells for each section.

That was the flow-sheet as designed three years ago; it is the flow-sheet of the entire mill today, except that the flotation concentrate, with the concentrates from the sand-tables and slime-tables—that is, all the concentrates—used to go to a series of circular tanks provided with filter-bottoms whereby the concentrates were subjected to a vacuum, reducing the moisture to 14%. In January last, one Dorr thickener (60 by 12 ft.) and one 12-ft. Oliver filter were added; so that now the flotation concentrate and the slime-concentrate are drained in one thickener and one filter, whereas the sand-concentrate still goes to the old filtering-tanks. During April 1917 the distribution of concentrates was as follows: Sand-tables, 3809 tons; slime-tables, 1093 tons; flotation plant, 1674 tons per day. The concentrates from these three sources are of almost equal grade, ranging from 41.94 to 42.79% copper.

STEEL BALLS were tried, in the place of pebbles, during April 1915; the use of these in all the Hardinge conical tube-mills was adopted throughout the plant in the August following. A comparison made at that time showed:

	Balls	Pebbles
Tons treated per day per section	814	650
Percentage of material larger than 48-mesh in the product	14.8	20.7
Tons of + 48 material in the product	120	135
Tons of — 48 material in the product	694	515
Tons of — 48 material produced per hp-day	4.63	3.43

The same power was consumed in each case, namely, 150 hp. Two mills ran with balls against three with pebbles. The chief gain, obviously, was the increased grinding capacity. Two-inch 'manganoid' balls, made by the Jeffrey Manufacturing Co., are now used in the Hardinge mills; the size of the balls should vary with the maximum feed—the finer the feed the smaller the balls.

FURTHER CHANGES will be made in the mill, for a progressive management knows no such thing as completion. Experiments having demonstrated that iron balls are more effective than pebbles for grinding in Hardinge mills, the ore will be fed from the bins into an 8 ft. by 30-in. Hardinge mill supplied with 4½-inch balls. The product will pass to a Dorr duplex 4½-ft. classifier, the overflow from which will be of finished size (-48 mesh) and will be sent direct to the flotation-cells. The sand from the classifier will be re-ground in two 8 ft. by 30-in. Hardinge mills supplied with 2-in. balls. The latter will complete the grinding to 48-mesh in closed circuit with two 6-ft. Dorr duplex classifiers. The overflow from the classifiers passes to flotation machines of the pneumatic type, the total area of filter-bottom per mill-section of 1000 tons capacity being 431.2 sq. ft. in the roughing-cells and 107.8 sq. ft. in the cleaning-cells. The air-pipes will enter from below and the flotation unit of each mill-section will have 5000 cu. ft. of air per minute, under 5-lb. pressure, or three times as much air (per 1000 tons of dry ore) as is used today. Three Oliver filters will be provided for the entire mill (of six sections), two filters constituting an operating unit and one being held as a spare.

SINGLE-DECK DEISTER TABLES, ELEVATOR-HOUSING, AND SECONDARY CLASSIFIERS

FLOTATION-CELLS; 'ROUGHERS' ABOVE AND 'CLEANERS' BELOW

The object of the crushing and grinding is to liberate the valuable mineral from its encasing gangue so as to obtain the full effect of differential treatment, whether by gravity or flotation. The rolls tend to crack without grinding; on the other hand, the Chilean and Hardinge mills, in which the ore is exposed to a variable measure of abrasion, are better fitted to induce the parting of mineral from gangue. The defects of the Chilean mill are its small capacity, high power-consumption, and excessive use of water; moreover, it produces an excess of slime, although the sand-tailing is cleaner. In 1915 the average sand-tailing for the whole mill was 0.52% copper, whereas that of the Chilean mills was only 0.44%; but on referring to the total tailing-loss it is recorded that for the whole mill it was 0.48%, while for the Chilean mills it was 0.52%; this being due, as suggested above, to the excessive slime produced by these machines. Fine grinding is desired by the millman because it is his aim to liberate the valuable mineral. In water-concentration the production of slime was a deterrent to fine grinding because it entailed the loss of metallic mineral; thus the millman was between the devil of insufficient reduction and the deep sea of excessive slime. When, however, he was given the use of flotation the bugbear of slime was removed, because the froth floats the finest mineral most completely.

The ore from the Captain orebody of the Miami mine, and from the extension of that deposit into the Inspiration ground, contains a large amount of secondary quartz, which is readily disintegrated into angular sand; indeed, chunks that seem at first so solid as to be unbreakable without the use of a hammer, if held in the clenched fist and manipulated often will granulate and yield much sandy material when under no greater pressure than that developed by the hand. To this peculiar physical condition of the ore is due in no small part the success of the caving-systems of mining here used, and more conspicuously the cheapness of crushing and grinding the ore in the process of milling. It also helps to explain the striking success of the Hardinge mill in granulating so large a tonnage, and it renders more comprehensible the lower capacity and higher proportions of slime in the pulp coming from the Chilean mills, since the function of abrasion or shearing performed by a Chilean mill is excessive as compared with that in a ball-mill, as elucidated by De Kalb[1] long ago.

OXIDIZED ORE. The copper in the tailing after flotation is distributed as shown by the following screen-analysis:

Size Mesh	Weight, %	Total copper, %	Oxidized copper, %	Sulphide copper, %
28	1.4	0.61	0.18	0.43
35	4.3	0.58	0.14	0.44
48	8.1	0.52	0.14	0.38
65	9.7	0.45	0.15	0.30
100	12.9	0.38	0.14	0.24
150	12.1	0.40	0.15	0.25
200	5.9	0.40	0.23	0.17
—200	45.6	0.83	0.68	0.15

[1]'Ore Dressing.' Robert H. Richards. Vol. III, page 1326.

This shows that the coarse sizes contain relatively more sulphide, owing to incomplete liberation of the chalcocite. Going down the scale of size, one finds a decreasing proportion of sulphide and a relative increase of oxidized mineral, until below 200-mesh the recovery of sulphide is seen to be reasonably complete and the cause of loss in the tailing becomes due predominantly to the inability to save the oxidized mineral, whether on tables or in the flotation-cell. This point was emphasized when discussing the earlier flow-sheet. The oxidized ore contains a variable proportion of silicate with the carbonate. The silicate accompanies the gangue during concentration and it is also difficult to leach.

A testing-plant of 100 tons capacity has been operated for the past year as a pilot for the flotation operations and of late for experiments looking for a suitable method for the treatment of mixed sulphide-oxide ore.

FROTH-FLOTATION is always interesting; at least, I can say that my feeling of wonder when I see the froth at work is still strong. It is a fascinating process. In the Miami mill only four-tenths of a pound of oil is used per ton of ore, or, more accurately, 2.7% of the weight of chalcocite in the feed to the flotation-plant. This oil is a mixture containing 50% coal-tar, 45% coal-tar creosote from the Barrett Co., and 5% steam pine-oil. It is added to the tailing from the Deister slime-tables on its way to join the stream of the spigot-product from the Dorr thickener.

On the roughing-cells, or first flotation treatment, the froth looks like wool; it has a dark tint, which varies, of course, with the supply of air and mineral. The surface is pitted as the bubbles break, to be replaced by others rising from below. The froth in these pneumatic machines is flimsy and evanescent but voluminous and continuous, so that the mineral is borne steadily to the surface and over the lip of the 'cell' or compartment. Any excess of oil shows itself in smooth streaks that form lines of depression along the surface of the froth, killing it. This is the 'over-oiling' effect. If the supply of air or oil is insufficient the froth does not overflow, the mineral particles drop, to meet other bubbles, rising again, falling, finally agglomerating and dropping to the bottom.

In the roughing-cell the froth is 6 inches deep; in the cleaning-cell, 16 inches. Two-inch bubbles suffice for the 'rougher,' but a few bigger bubbles, up to 6-in. diam., are liked as an evidence of slight over-oiling. In the 'cleaner' the size of the bubbles is not so important: an average of 2½-in. is liked. The air-inlet valves of the rougher are kept wide open; those of the cleaner only one-third open. However, the supply of air to the cleaner is relatively the greater because the flow of pulp is less. The roughing-compartments determine the loss in the tailing; the cleaning-compartments regulate the grade of the concentrate. The grade of this product is determined by the rate at which the froth overflows, and this in turn is controlled by the supply of air and oil. The higher the grade of concentrate, the richer the tailing; therefore the metallurgist aims to maintain the economic balance, that is, the most profitable result.

An air-washing device is to be attached to the inlet of the blower with a view to removing the dust in the air, thus preventing or retarding the choking of the canvas that serves as a porous medium on the bottom of the cell. I saw several pachucas, or Brown agitators, which played so fateful a part in the litigation with the Minerals Separation company.[2] They are no longer in use.

THE FROTH-CONCENTRATE is washed with the slime-table concentrate down a 6-inch pipe and a launder into a Dorr tank, 12 by 60 ft., in which it is thickened from 7:1 to 1:1, or half-liquid and half-solid by weight. Two Dorr tanks are now in use, but four will be needed when the new flow-sheet is put into effect. From the spigot of the thickener the concentrate runs in thick pulsations along a launder inclined at two inches per foot to the Oliver filter. An addition of two pounds of slaked lime per ton of concentrate is made in order to enable the concentrate to hold its moisture, preventing the water from rising to the surface of the concentrate when it is loaded subsequently in cars on its way to the smelter. Two 12-ft. Oliver filters are in place (one in use) and a third is being erected. They work under a vacuum of 23 lb. per square inch. The tank under the filter serves for storage. The surface of the filter is kept clean by scrubbing it every three weeks with acidulated (HCl) water. This increases the efficiency of the machine as a dewaterer. It works admirably. Invented originally[3] as a continuous cylindrical filter for use in cyanide mills, it has proved of great service in flotation plants. From the filter the concentrate is conveyed on a traveling belt to a railroad-car.

The sand-table concentrate is easier to drain. It goes to a separate annex containing 12 circular filter-bottom tanks, in which it is dewatered to 4½% moisture. It is discharged through the bottom of these tanks upon a traveling belt, which conveys it to the railroad-cars.

It will have been noted that ordinary gravity concentration is applied to the coarser part of the pulp before sending the slime to the flotation annex. A screen-analysis of the flotation-feed shows that

1% remains on a 48-mesh screen
12% remains on a 100-mesh screen
62% passes through a 200-mesh screen

DISTRIBUTION OF SOLID AND WATER FOR ONE SECTION OF MILL

		Tons solid 24 hr	Ratio Solid to water	Tons water 24 hr	Gal water per min
Roll Product		700	1 : 0.9	630	105
Callow screen	Oversize	525	1 : 0.8	420	70
	Undersize	175	1 : 3.0	528	88
16-in. Drag	Product	490	1 : 0.5	246	41
	Overflow	210	1 : 3.9	822	137
Primary classifier table products	1st 4 Spigots Conc	10	1 : 10.2	102	17
	1st 4 " Reject	230	1 : 4.0	918	153
	Next 4 " Conc.	5	1 : 8.6	48	8
	" " Mid.	8	1 : 7.5	60	10
	" " Tail.	179	1 : 5.0	894	149
	Last 2 " Conc	1	1 : 12.0	12	2
	" " Mid.	3	1 : 8.0	24	4
	" " Tail.	59	1 : 6.0	354	59
Secondary classifier table products	1st Spigot Conc.	1	1 : 12.0	12	2
	" " Reject	49	1 : 4.0	198	33
	Last 2 " Conc.	1	1 : 12.0	12	2
	" " Mid.	2	1 : 8.0	18	3
	" " Tail.	57	1 : 6.0	342	57
14-in. Drag	Feed	287	1 : 4.1	1176	196
	Product	285	1 : 0.7	204	34
	Overflow	2	-	1098	164
Sand-elevator	Feed	291	1 : 3.4	996	166
Classifier-overflow	Primary	176	1 : 8.5	1500	250
	Secondary	181	1 : 6.0	1080	180
Upper cone-tanks	Feed	386	1 : 6.0	2322	387
	Spigot	231	1 : 3.6	804	134
	Overflow	155	1 : 9.8	1518	253
Dorr tank	Spigot	155	1 : 2.5	384	64
	Overflow	–	–	1134	189
Slime-floor table products	Conc	7	1 : 10.3	72	12
	Reject	224	1 : 5.0	1122	187
Flotation products	Net Feed	379	1 : 4.0	1506	251
	Tailing	374	1 : 4.1	1542	257
	Cleaner conc	5	1 : 9.6	48	8
	Cleaner feed	9	1 : 24.0	216	36
	Cleaner reject	4	1 : 60.0	240	40

FIG. 4

SUMMARY

		Tons solid 24 hr	Ratio Solid to water	Tons water 24 hr	Gal water per min
Material to reclaiming water plant	Sand-tailing	295	1 : 5.4	1590	265
	Flotation-tailing	374	1 : 4.1	1542	257
	14 in drag overflow	1	–	234	39
	Dorr tank overflow	–	–	1134	189
	Total	670	1 : 6.7	4500	750
Material to concentrate plant	Sand-concentrate	18	1 : 10.3	186	31
	Slime-table conc	7	1 : 10.3	72	12
	Flotation conc	5	1 : 9.6	48	8
	Total	30	1 : 10.2	306	51
Total	All products	700	1 . 6.9	4806	801

ADDED WATER IN GALLONS PER MINUTE

	Per section	Entire mill
To roll product	105	630
To Callow screen oversize	53	318
To 16-in drag dewaterer	20	120
To 14-in. drag dewaterer	2	12
To sand-floor dressing-water	212	1272
To slime-floor dressing-water	65	390
To hydraulic classifiers	330	1980
To flotation wash-water	14	84
Total	801	4806

FOR ENTIRE MILL

	Gallons per min	Ratio solid to water	Tons water per day	Gallons per ton ore	% total water
New water added	1087	1 : 1.55	6522	373	22.6
Water reclaimed from tailing	3416	1 : 4.89	20496	1175	71.1
" " " conc.	303	1 : 0.44	1818	104	6.3
Total	4806	1 : 6.87	28836	1652	100.0

FIG. 5

Pulp that will just pass 48-mesh would be unsuitable. Most of the + 100 is gangue. Ideally the finer the pulp the better for flotation. The recovery is almost restricted to the - 200 mineral. The coarser mineral is saved on the tables. Thus the mill-man produces a slimy concentrate and 'passes the buck' to the smelter-man. The loss is in ore that has been insufficiently ground, that is, insufficient to release the chalcocite from the gangue. The loss in - 200 material is constant, being due to the imperfection of the apparatus and not to the

[2]See editorial 'The Miami Appeal.' M. & S. P., June 23, 1917.

[3]By E. L. Oliver (Cal. '00), in the North Star mill, at Grass Valley, in 1908.

condition of the mineral. The mill-feed is comparatively uniform in its content of copper during any month, thus in three months it was:

Feed, %	Tailing, %
2.29 to 2.03	0.52 to 0.43
2.28 " 2.09	0.55 " 0.54
2.39 " 2.13	0.56 " 0.44

When the oxidized copper is crushed, it is ground finer than the gangue, passing readily into the slime and thence into the flotation-cells, where it is lost.

Flotation, like ordinary gravity concentration, makes a high recovery on the sulphide copper, but is powerless to save the oxidized mineral. Thus on a 2% ore containing 0.5% oxidized and 1.95% sulphide, the tailing will be 0.5%, of which 0.3 will be oxidized, so that the saving on the sulphide is 88% and on the oxidized only 40%.

Another factor is the grade of the ore. If 6000 tons of 3% ore were milled, it would be impossible to make a low tailing because, with a crushing-plant designed to treat 2% ore, the ore would not be crushed sufficiently fine to liberate the particles of chalcocite. Likewise the tables and flotation-cells would become over-crowded. If, on the other hand, 6000 tons of 1% ore were treated, there would be ample scope for all the apparatus and the tailing-loss would decrease, but the output of copper would also diminish to such an extent as to curtail the profit sadly.

The results for the first three months of 1915 and 1917 compare as follows:

		Feed				
	Tons per day	Total copper, %	Oxide copper, %	Concentrate %	Tailing, %	Recovery, %
1915	2,979	2.222	0.428	39.67	0.695	69.97
1917	5,641	2.116	0.256	43.35	0.638	70.87

Out of the entire mill-feed

30% is treated by flotation alone
40% is treated by gravity alone
30% is treated by gravity and flotation

To R. B. Yerxa (Mass. Tech. '03) I am indebted for much information concerning this department, of which he has charge. The oil is now added at the launder at the foot of the elevator that lifts the pulp to the flotation-cells. It is inadvisable to add the oil to the ore when in the Hardinge mill, because that machine is so far away from the flotation-cells that the oil would lose much of its effectiveness, by dilution with fresh water and by aeration, in its passage through the other apparatus.

Water is now added as the pulverized ore emerges from under the rolls and passes into the Callow screens. These are of 30-mesh; they pass about 28% of the feed into the undersize. To be efficient they need to be sprayed, but a spray cannot be used without entailing such an increase in the proportion of water as would be too much for the Hardinge mill.

The water used in the mill is reclaimed from the tailing-ponds and thickeners. Thus 75% of the mill-water is recovered. It shows a marked tendency to froth. No alkali or acid is added. The slight natural alkalinity of the water corresponds to $\frac{4}{10}$ lb. acid per ton. Note may be made here of the quantity of wood, in the form of splinters, that enters the mill with the ore in consequence of the slicing system of mining. This is a nuisance, which is mitigated by arresting the wood on screens distributed all over the mill, particularly where there is any break in the launders.

The ratio of concentration is 28:1. The grade of concentrate (42% copper) is suited to the operations of the smelter, for it leaves enough silica to slag the iron. It is the present aim of the manager of the Miami mine to make a 45% concentrate, with 20% insoluble, and a 0.25% tailing on a 2.15% ore containing 0.2% oxide. The good work already done augurs well for further accomplishment.

Official List of Technical Men

The Bureau of Mines has completed a census of mining engineers, metallurgists, and chemists, showing that 7500 men engaged in mining and 15,000 men in the chemical industries have been classified according to the character of work in which each claims proficiency. It is not the purpose of this census to enable the Bureau to act as a clearing house for technical men in obtaining commercial positions. The reasons for taking this census are obvious. The War is one in which chemists and engineers play a greater rôle than ever before. The products of the mines, furnaces, factories, and chemical plants are being so rapidly consumed that the highest possible skill is required to keep pace with the destruction. The problem is to get the best qualified specialists for each place. Men with a knowledge of sanitation are essential to the health of the soldiers at the training camps; those with a knowledge of pyrotechnics in the manufacture and use of signal devices; telephone and telegraph operators are essential for systems of rapid communication, without which valuable time may be lost; mining engineers, under military control, will be of the greatest assistance in planning and directing sapping operations, in digging trenches, in the erection of concrete and steel supports for trenches, dug-outs, and tunnels behind the lines, and in planning systems of ventilation and drainage for such excavations; in re-habilitating or re-developing wrecked coal and iron mines taken from the enemy; and in increasing the output of minerals for military uses. The experience of some mining engineers in the use of oxygen mine-rescue apparatus might be invaluable in the exploration of underground saps or dug-outs. As it is important to keep the lists up to date, the registered engineers and chemists are requested to advise the Bureau of Mines of any change of address. Communications should be sent to the Director of the Bureau at Washington. Engineers and chemists who have not yet registered should do so at once.

Flotation of Lead and Zinc in the Joplin District

By C. A. WRIGHT

From flotation tests made on the lead and zinc ores of the Joplin district it was found that it is relatively easy to float the sulphides. The ore consists mainly of zinc sulphide (sphalerite), with smaller quantities of galena. The gangue consists mainly of flint. In Joplin there are certain conditions, however, such as the quantity of floatable material available without resorting to further crushing, and the possibilities of finer grinding, that must be met, and the problems are to be worked out to insure success on a commercial scale. Due to these conditions which confront the millman in the treatment of these ores by flotation, the operators in the district have been somewhat reluctant in letting themselves be carried away by the possibilities of a greater saving by this means of concentration. Although flotation may not at present hold as important a place in the treatment of the Joplin ores as at many of the larger copper and zinc mills of the West, it is believed that before long many of the Joplin mills will have small flotation units for saving a large proportion of the metal in the fine now going to waste. Already five or six mills are using flotation successfully, while several others are still in the experimental stage.

In the Joplin district the mills are small, their capacity being 100 to 500 tons of ore treated in 10 hours. They are equipped with crusher, rolls, jigs, and tables. At most of them the material that reaches the jigs has passed a one-half inch trommel. The middling from the jigs, locally called 'chats,' are re-crushed for further treatment on sand-jigs or tables, or, at some mills, are returned to the roughing-jigs. Middling from the tables is seldom re-crushed. Therefore the first point to be considered is how much material now being produced is fine enough to be amenable to flotation; in other words, what percentage of the total tonnage is crushed to a suitable size and what proportion of this size can be saved for re-treatment. To show more clearly the quantity of material available and of suitable size for flotation, a simple mill-test is submitted herewith. The ore treated was from a 'hard-ground' mine, better known as 'sheet-ground' ore, consisting chiefly of a flint gangue with a small percentage of zinc, and practically no lead.

Ore treated, tons	1,994
Concentrate produced, tons	35.5
Tailing, tons	1,958.50
Solids in overflow from settling tanks, tons	34.73
Tailing without overflow, tons	1,923.77
Concentrate (metallic zinc), per cent	59.4
Tailing without overflow (metallic zinc), per cent	0.41
Overflow (metallic zinc), per cent	5.01
Zinc (metallic) in concentrate, tons	21.087
Zinc (metallic) in tailing without overflow, tons	7.887
Zinc (metallic) in overflow, tons	1.740
Total zinc in ore treated, tons	30.714
Zinc-recovery (21.087 ÷ 30.714) × 100, per cent	68.6

The results of the mill-test shown in table 1 indicate a zinc-recovery of 68.6%, which is somewhat higher than the average zinc-recovery throughout the district. From the screen-analysis of the mill-tailing, not including the overflow from the settling-tanks, it is seen that 2.31% passes a 65-mesh screen, a product suitable in size for flotation. The tonnage treated during this test was 1994, giving 44.44 tons of material passing a 65-mesh screen. The calculated assay of this size of material is 5.24% zinc, and the total amount of zinc in the 44.44 tons figures 2.328 tons. Assuming that the overflow is of suitable size, 34.73 tons of solids is present assaying 5.01% zinc, which gives 1.740 tons of zinc. The sum of

TABLE 1. RESULTS OF A MILL-TEST IN TREATING SHEET-GROUND ORES

Mesh	Screen-analysis of mill tailing, not including overflow: Opening mm.	Opening in.	Weight of screen product %	Cumulative weight %	Assay, zinc	Per cent of total zinc	Cumulative per cent of total zinc
..	13.33	0.625	6.89	....	0.30	5.6	...
3	6.680	0.263	32.08	38.97	0.22	19.3	24.9
6	3.327	0.131	26.75	65.72	0.20	14.6	39.5
10	1.651	0.065	18.40	84.12	0.19	9.5	49.0
20	0.833	0.0328	9.07	93.19	0.30	7.4	56.4
35	0.417	0.0164	2.70	95.89	0.56	4.1	60.5
65	0.208	0.0082	1.80	97.69	1.27	6.3	66.8
100	0.147	0.0058	0.44	98.13	2.93	3.5	70.3
150	0.104	0.0041	0.39	98.52	3.35	3.6	73.9
200	0.074	0.0029	0.21	98.73	5.70	3.3	77.2
-200	0.074	0.0029	1.27	100.00	6.55	22.8	100.0

Calculated assay .... 0.37 Assay of original ore 0.41

these two products gives a total of 4.068 tons of zinc in the material suitable for flotation, or a tonnage of 79.17, with a calculated assay of 5.14% zinc. Provided that this material would be available, and that 80% of the zinc could be recovered by flotation, there would be a saving of 3.256 tons of the metal. The total zinc contained in the 1994 tons of ore treated was 30.714 tons. The quantity saved in the concentrate was 21.087 tons. Including now the additional amount of zinc, 3.256 tons, that it would be possible to recover from the 79.17 tons of floatable material in the 21.087 tons of zinc, there results 24.343 tons of zinc saved out of a possible 30.714 tons, or a zinc-recovery of 79.2%. In other words, the percentage of zinc-recovery would be increased from 68.6 to 79.2, or an additional saving of 10.6% of the total zinc in the ore.

The above figures bring out the possibilities of a marked saving by flotation on a relatively large tonnage. Unfortunately, the average daily tonnage treated by the mills of this district is considerably less than the tonnage treated during the time the mill-test was being made. It is reasonable to believe, however, that the mills treating 500 to 1000 tons of ore per day could well afford to install a small flotation unit at a profit, the size of the plant depending upon the quantity of fine material produced. In the case of soft-ground deposits there is generally a much larger percentage of fine produced in milling the ore, so that there would be a sufficient tonnage of fine coming from a mill treating 200 to 300 tons per day for a 20 to 30-ton flotation unit. The next step in determining the possibilities of flotation for the ores of the Joplin district is the profitable limit of grinding to a size suitable for flotation. This is one of the main difficulties with respect to flotation, as the gangue consists mostly of flint. Therefore, in case there is not sufficient material available for flotation, the problem resolves itself into one of fine grinding rather than to the floating of the mineral. Many kinds of fine-grinding mills have been tried in this district with somewhat discouraging results, due to the high abrasive action of the flint. In fact, fine grinding of Joplin ores, either by ball-mills or other means, is still a problem to be solved. There are certain products, however, from the various stages in the concentration of the ores, that would be less resistant and would respond more readily to finer grinding. Some of these products are: (1) 'chats' or included mineral particles from the jigs; (2) tailing from the tables treating fine; (3) all table-middling that needs to be re-crushed before further treatment; and (4) possibly the oversize often discarded from the sand screens. The important factors to be considered for the treatment of these products by flotation are the cost of fine grinding, the quantity of material to be re-ground, and the zinc content. If the material contained 1% metallic zinc, equivalent to 2% of the 50% zinc concentrate, and the concentrate was worth $50 per ton, the material would have a value of $1 per ton, or 80c. per ton on a basis of 80% zinc-recovery. For material containing 2% zinc these values would be $2 per ton and $1.60 for an 80% recovery. In other words, it is possible to figure from the zinc-content and the possible zinc-recovery what value can be placed on the material to be crushed and treated. The cost of fine grinding would depend upon the type of grinding-mill used and on the character and coarseness of the material. The greater the tonnage treated the less will be the cost of grinding and subsequent treatment. The possibilities are, therefore, greatly increased when one considers the increased tonnage available by finer grinding. In addition to the increased tonnage of floatable material secured by finer grinding, slime or fine sand at present being treated on slime-tables could be well taken care of by flotation, and a higher zinc-recovery effected than is usually possible on slime-tables.

In view of the demand for high-grade zinc concentrate by smelters, zinc ores containing a comparatively large proportion of lead and iron sulphides would require either differential flotation or re-treatment of the concentrate on tables. A preliminary treatment of the floatable material on tables to eliminate the lead and iron has been suggested, but this would require extra tables and its efficacy is doubtful. Re-treatment of the concentrate on tables seems to be the most feasible method in separating the lead and iron sulphides from the zinc-blende, and could be accomplished at a small cost, using a small diaphragm-pump operated in closed circuit to return the tailing from the tables to the flotation system. This method of re-treatment gives satisfactory results in some Western mills. The froth from the concentrate is largely 'killed' by spraying it with water, or by some other means, before it reaches the table. The ease with which the froth is killed depends largely upon the oil-mixture or frothing-agent used in producing the froth. The composition of the gangue in the ores of the Joplin district differs somewhat and the mineral content varies. Consequently, though many available oils give acceptable results, no one oil or fixed quantity of oil under the same conditions will give the best possible results for all ores. The best oils or mixtures of oils must therefore be determined by experiment on each ore.

The Bureau of Mines, in co-operation with the Missouri State Geological Survey, started flotation experiments in the Joplin district in the latter part of the year 1914, and continued until the spring of 1915. In the first experiments different oils were tried in varying quantities to decide which oil was best suited to the material treated. When this point had been determined the same tests were repeated with the addition of sul phuric acid in varying quantities. The effect of tem perature was then considered. It was found from th preliminary results that the sulphides of this district ar apparently not difficult to separate from the gangue, an that a fairly good grade of concentrate could be secure by the use of roughing and cleaning-cells. Oils with coal-tar base gave a high recovery but a low-grade co centrate, a considerable amount of gangue being e trained in the froth. Other oils gave a higher-grade co centrate but not as good an extraction under like co

TABLE 2
Acid roughing-tests

Test No.	Acid, lb. per ton	Oil, lb. per ton		Weight of froth, gm.	Zinc %	Zinc recovery %
1	7.36	1.37	S. S. No. 4 pine-oil	105.5	35.9	88.8
2	14.72	1.37	S. S. No. 4 pine-oil	96.5	39.3	88.8
3	36.8	1.37	S. S. No. 4 pine-oil	95.5	38.9	87.0
4	73.6	1.37	S. S. No. 4 pine-oil	100.0	38.8	91.0

TABLE 3
Roughing-tests

Test No.	Kind of oil and lb. per ton	Weight of froth, gm.	Zinc %	Zinc recovery, %	Remarks
5	0.5 S. S. No. 4 pine-oil	104.5	36.5	89.5	
6	1.0 S. S. No. 4 pine-oil	86.5	43.2	87.5	
7	2.0 S. S. No. 4 pine-oil	103.5	36.2	87.8	
8	4.0 S. S. No. 4 pine-oil	94.0	39.7	87.5	
9	1.0 S. S. No. 4 pine-oil	120.5	31.6	89.2	Cold (20°C.)
10	2.5 crude cedar-oil	93.5	40.1	87.8	Rapid separation
11	1.8 S. S. No. 8 pine-oil	73.5	49.6	85.5	Very rapid
12	1.7 wood-creosote	93.0	40.8	89.0	Strong froth
13	2.0 S. S. gas-creosote	78.0	46.8	85.8	Strong froth, rapid
14	1.7 eucalyptus-oil	110.0	34.2	88.2	Good frother

ditions. The best results were obtained with wood-creosote and pine-oil. As the coal-tars and some petroleum products are good collectors and the pine-oils and some other oils from wood distillation are good frothing-agents, it would seem that a mixture of these oils, such as 80% coal-tar and 20% pine-oil or wood-creosote, or combinations of these, would be best suited to the treatment of the Joplin ores by flotation.

To verify the data obtained from this preliminary work, slime produced from one of the sheet-ground mines near Webb City, Missouri, was shipped to Salt Lake City, Utah, where the Bureau of Mines in co-operation with the University of Utah is conducting flotation experiments. The results given in tables 2, 3, 4, and 5, from experiments performed by O. C. Ralston and G. L. Allen, showed that it is fairly easy to float the sphalerite from the gangue by using warm solutions and about 1 lb. per ton of any suitable oil, either from wood or coal distillation, and that acidity, although it does not seem to be necessary, allows the froth and tailing to separate more quickly. Cold solution gave as high an extraction as warm solution, but the grade of the product was not as good.

RESULTS IN FLOTATION OF JOPLIN SHEET-GROUND SLIME

Conditions of tests: Assay of slime tested, 8.54% Zn, 0.51% Pb, 0.7% Fe. Charge in each test 500 gm., with 1500 to 2500 cc. H_2O. Temperature of pulp, 60°C., except in test No. 9, which was run at 20°C. Total time consumed in agitation for each test was about 18 minutes.

CLEANING-TESTS

Samples of each froth from roughing-tests were taken and the remainder saved for cleaning-tests. Combined roughing-froths made enough material for two cleaning-tests, the results of which were as follows:

TABLE 4
Cleaning-test No. 1

Froth used, gm.	Assay of head: Zinc %	Assay of head: Lead %	Product, gm.	Assay of product: Zinc %	Assay of product: Lead %	Extraction, %
500	39.8	1.3	366.5	51.7	2.35	95.3

TABLE 5
Cleaning-test No. 2

Froth used gm.	Assay of heads: Zinc %	Assay of heads: Lead %	Weight of product: Froth gm.	Weight of product: Middling gm.	Weight of product: Tailing gm.	Assay of products: Froth Zn %	Froth Pb %	Middling Zn %	Middling Pb %	Tailing Zn %	Tailing Pb %
479	39.8	1.3	346	47	77	51.5	2.04	13.5	2.66	1.0	0.4
Total recovery, zinc and lead						93.2		3.3		4.0	

The oils used were: Sunny South No. 4, pine-oil of General Naval Stores Co., New York; crude cedar-oil, from Idaho cedar swamps, Government work; Sunny South No. 8, special pine-tar oil, General Naval Stores Co., New York; beechwood creosote, specific gravity 1.078, Merck & Co.; Salt Lake gas-creosote, fraction 500-612°F., Salt Lake Gas Works; eucalyptus oil, Atkins, Kroll & Co., San Francisco.

The above tests, No. 1 to 4, given in table 2, were to determine the effect of various quantities of acid on the percentage of recovery, the grade of the concentrates, and also to obtain a sharp line of separation between the yellow froth and white tailing, the quantity of oil used being constant. The results indicate that the quantity of acid used has little effect upon either the grade of the concentrate or the percentage of recovery, but it was found that the line of separation between the concentrate and tailing, as seen through the glass slide of a laboratory machine, is much more quickly and better defined by the use of a large quantity of acid. The next four tests, No. 5 to 8, table 3, were made with varying amounts of oil, some acid having been added to allow the experimenters to see the froth and tailing separate more quickly, and to clean the surfaces of the sulphide minerals. No distinct effect was observed except that one pound of oil per ton of ore treated seemed to give the highest grade of concentrate. Test No. 9 was run to parallel test No. 6, except that a cold pulp at 20°C. was used in place of the warm pulp at 60°C. in the other tests. The lower grade of froth proves the greater ability of froth and tailing to separate from each other at elevated temperatures. The other tests, No. 10 to 14, were made with various kinds of oils. After sampling the froth-concentrates they were mixed, and two tests (tables 4 and 5) were made using the flotation cell as a 'cleaning' unit. The previous tests had been performed with the cell as a 'roughing' unit with a high recovery of zinc as the end in view. When cleaning the roughing-froth a concentrate of high zinc-tenor should be the aim, but these tests were carried to nearly complete removal of the zinc, so that only a small amount of middling was left for returning to the roughing-cell and the concentrates obtained were perhaps not quite as high in zinc as could be secured by returning a large quantity of middling. A scheme making use of roughing and cleaning-units of flotation cells seems essential in order to get a high recovery and to produce concentrate of acceptable grade.

In the experiments made in Joplin, the ratio of solids to water in the pulp treated ranged from 1 to 3 up to 1 to 7. In practice the most favorable ratios for different ores would, of course, have to be determined by experiment. In general, a mixture of fine and slime requires a denser pulp, whereas for a mixture consisting wholly of slime (finer than 200 mesh) a thinner pulp is desirable. Thickening of the slime from the mills of the Joplin district would be necessary and could be accomplished by the use of Dorr thickeners or some other device that would give the flotation machines a uniform feed of constant density. The addition of acid may not be absolutely essential for all the ores of the district, although the tests made locally showed that a small quantity of acid is desirable, especially in cleaning the roughing-concentrate.

During the past year the following companies have installed successful flotation units, varying in capacity from 25 to 100 tons per day: American Zinc, Lead & Smelting Co. at both Carterville and Granby, Missouri; Picher Lead Co., at Picher, Oklahoma; Mining & Flotation Co., at Joplin, Missouri; and the Oronogo Circle Mining Co., at Oronogo, Missouri. Other operating companies are still in the experimental stage, but will probably soon be listed among those using the flotation process.

Utilizing Pyrite in Coal

Illinois coals contain from 1 to 6% of sulphur. The outlook for recovering a large part of this is the subject of a most suggestive bulletin by Joseph E. Pogue, published by the Illinois State Geological Survey. The sulphur in coal-seams is present chiefly in the form of the brass-yellow pyrite or marcasite, called commercially 'coal-brasses' and termed by the miners 'sulphur.' This material occurs as small seams, lenses, and nodules scattered through the coal, and is generally looked upon as an objectionable impurity, because it must be removed before the coal is marketed. As a matter of fact, however, this so-called impurity is a raw material of value for the manufacture of sulphuric acid, and its recovery, in those mines in which it is now thrown aside, affords a good opportunity for turning a troublesome waste-product into profit, and establishing a new source of revenue. There is a possibility that some mines, abandoned because of the unusually high sulphur content of the coal, may be re-opened and successfully operated as pyrite producers, with coal as the by-product. A few Illinois coal mines for a number of years have been recovering the pyrite and selling it to chemical works, but the output of the State has never exceeded 27,000 long tons per year. This figure falls far short of the possibilities of the State. With a côal production of 60,-000,000 tons per year and with pyrite worth about $8 per ton, the Illinois coalfields may greatly extend the pyrite yield. There are four sources of domestic material suitable for the manufacture of sulphuric acid. The first of these is included in the many small deposits of pyrite and related sulphides scattered through the Eastern States. These were formerly unable to compete with the Spanish ore, which was brought to this country cheaply as ballast by Mediterranean freighters. The second is formed by the fume of many smelters, chiefly in the Western States, formerly allowed to go to waste, but now beginning to be turned into acid. The third is the large native-sulphur deposits at Sulphur, Louisiana, and at Bryan Heights, Texas, which yield a very pure product, too costly under normal conditions to compete with pyrite, but which may be utilized to meet the pyrite shortage during the present emergency. The fourth source is the pyrite accompanying coal-beds, now largely wasted.

SHIPMENTS of high-grade manganese during the past six months aggregate 28,345 tons, or nearly 10% more than the tonnage for the whole 12 months of 1916. Deliveries are reported from ten States. In Montana alone the shipments since last January exceed the shipments for the whole United States in 1915. The U. S. Geological Survey estimates that 80,000 tons of the high-grade ore will be produced in this country during the current year. This, however, is less than 20% of the demand of the steel industry, based on present consumption. The output of lower-grade manganese ore is likewise record breaking.

Mechanical Ventilation for Metal Mines

By GEORGE RICE

*Mechanically produced ventilation of mines has been regarded, until the last few years, as necessary only in coal properties. The haphazard ventilation of metal mines is in striking contrast to the high order of engineering in other departments of metal-mining operation. Ventilation of coal mines, on account of the inflammable gases given off from the strata, or because of the loss of oxygen from the air through absorption by the coal, made the positive 'coursing' or circuiting of the air-current a necessity. This was recognzied a century ago after serious explosions of fire-damp had occurred, although coal-mining operations were on a very small scale at that time. Notwithstanding the slow development of positive ventilation in metal-mining operations, mechanically produced ventilation of such mines received an early start. Agricola in his book describing the mining practice of the 16th century in the Harz mountains repeatedly refers to ventilation. The following are some quotations taken from the Hoover translation of Agricola's book:

"I will now speak of ventilating machines. If a shaft is very deep and no tunnel reaches to it, or no drift from another shaft connects with it, or when a tunnel is of great length and no shaft reaches to it, then the air does not replenish itself. In such a case it weighs heavily on the miners, causing them to breathe with difficulty, and sometimes they are suffocated, and burning lamps are also extinguished. There is, therefore, a necessity for machines * * * which enable the miners to breathe easily and carry on their work.

"These devices are of three genera: The first receives and diverts into the shaft the blowing of the wind, * * * The second genus of blowing machine is made with fans, and is likewise varied and of many forms, for the fans are either fitted to a windlass barrel or to an axle. * * * Blowing machines of the third genus, which are no less varied and of no fewer forms than those of the second genus, are made with bellows, for by its blasts the shafts and tunnels are not only furnished with air through conduits or pipes, but they can also be cleared by suction of their heavy and pestilential vapors."

Agricola's drawings show many varieties of ventilating devices; some are dependent, as indicated in the above quotation, on the effect of the wind, while others are the predecessors of the centrifugal fans. Several of the drawings show large feathers set radially in the axle; others show wooden blades or paddles arranged as in the later Guibal fan. Apparently, however, the proper arrangement for admitting the air to the fan at the axle had not then been discovered. In these old drawings, the air appears to enter a slot on the circumference of the casing and is discharged through a wooden conduit at another place on the circumference. Presumably the action of the fan was improved through the leakage of the wooden sides, so that accidentally it may not have been as inefficient as it appears. The drawings seem to indicate that the use of brattices in shafts and ventilating-doors was understood, although there is no indication that a circuit of air was maintained, away from the shaft or beyond where a shaft intersected drift-workings, except by wooden pipes.

Little advance seems to have been made in metal-mining ventilation for the 300 years thereafter, until compressed-air drills came into use; then the exhaust was regarded as furnishing sufficient air in the working-place. When the drill is not running, if the atmosphere is oppressive in the heading or stope, or is foul after blasting, the compressed air is allowed to blow direct from a pipe-outlet. Undoubtedly in small metal mines, or those in which there is no considerable amount of noxious gas given off, or where workings are a short distance from the drift-mouth or shaft, this may be sufficient. In most metal mines, owing to the ore formation generally having considerable length along the strike, and depth vertically or on the dip, but generally no great width, the difference in temperature between the lower workings and the outside atmosphere generates air-currents when secondary connections have been made to the surface.

Inflammable gases are rarely found in metal mines, and generally there is a little absorption of oxygen by the ore or enclosing walls, although in mines in which there is much timbering it has been found that there is considerable absorption of oxygen by the timbers due to bacteriological action. Nevertheless, thousands of samples of air that have been gathered by the Bureau of Mines in metal mines indicate that, except after blasting, it is rare when the air contains a considerable amount of impurity. Occasionally, as in the Cripple Creek district, metal mines contain large amounts of pure nitrogen gas residual in the strata. Such mines can only be worked by the employment of a plenum-pressure system of ventilation. In general, however, the chief air-impurities in metal mines arise from blasting.

In the case of long drainage-tunnels, the rule is to ventilate with fans and air-pipes, and it is obvious that it is an economic advantage to quickly ventilate the heading after blasting so that the miners may get back to their work as soon as possible.

In great orebodies, such as are found at Butte, in the

*Paper presented at the 6th Annual Safety Congress of the National Safety Council.

Lake Superior iron and copper mines, in the copper mines of the South-West, and especially in the deep and extensive gold mines of South Africa, there has been a gradual introduction of mechanical ventilation apart from any dependence on compressed air at the faces.

Attempting to ventilate by means of compressed air when there is a considerable volume of noxious or poisonous gases present, as from blasting, is extremely inefficient both in sweeping away the smoke and gases which it tends merely to stir up, and from the important standpoint of mechanical efficiency in using air which is usually compressed uneconomically to a high pressure, and liberating it without doing mechanical work against atmospheric pressure. The only remedy is to arrange as far as possible for circuiting of the air, supplemented by 'booster' fans.

It is hardly worth while to argue the need of having two means of exit from a mine, as this is the requirement in most State metal-mining laws. Probably over 95% of the metal mines of the country which have reached a permanent basis of ore-production, have two exits. Where there are two such exits one naturally becomes the up-cast and the other the down-cast. Unfortunately, however, these conditions sometimes change with the season and with such changes the natural ventilation underground is insufficient. Hence there is necessity for the use of fans. In very deep mines, such as in the copper mines of the Lake Superior district, normally the up-cast and down-cast conditions do not reverse, and the natural main up-casting current is very strong. It then becomes a problem whether or not in the interest of safety to prevent reversal of the current in the case of a mine-fire in the down-cast, it is not best to install a large fan, which may never be needed except for fire-protection. However, in the great majority of mines a strong non-reversing current due to the difference in temperature in the up-cast and down-cast shafts cannot be obtained. This being the case it is highly important, both for fire-protection and for a positive circuit of air through the mines that fans shall be used.

The great variety of conditions found in metal mining in the United States, varying from typical veins to the great lenticular and irregular iron-ore deposits of the Lake Superior district, makes it impossible to lay down as exact a procedure as is usually regarded necessary in planning coal-mining development. Nevertheless it will be generally conceded that a great deal more might be done in the future than in the past if preliminary planning of a ventilating system is borne in mind, particularly in developing low-grade deposits.

While metal-mining laws and modern practice require two exits and recognize the necessity of air down-casting in one shaft or winze, and returning through another shaft or winze, adit, or tunnel, the necessity for provisions for circulating the air through the mine has not been well recognized, with the result that in some of the great mines of the United States the ventilation is of the most haphazard character. Often I have had occasion to observe that the main fan placed underground is ineffective because part of the air-discharge has been circuited round and round. Therefore, even when the percentage of impurity in the air might be low it did not take care of one of the most important features of deep mine ventilation, namely, to keep the temperature below that of the mine-walls. It is not always a question of depth or natural heat of the strata, but where large masses of timber are employed, especially in the caving-system employing a timber-mat, the temperature is frequently far greater immediately under the mat than it is in the deeper levels.

It is believed that if the same degree of care were taken in laying out in advance the ventilating system that is given to laying out the levels, drifts, cross-cuts, and raises, for extracting the orebody, a great deal of the difficulty of a physiological nature now experienced by the miners would be prevented. These troubles are not simply those of temporary discomfort of the workers, but their health is likely to be affected, making them liable to contract diseases, lessening their working efficiency, and thus decreasing the earning power of the mine. Ventilation therefore comes to be a matter of economic importance. It is difficult to speak of specific mines or districts without making unfortunate comparisons, but that the problem is acute will be readily admitted by most mining authorities.

Another phase of the physiological problem concerning workers in hot mines, is not alone in the high temperature of the air but in the fact that the air is not in motion. The experiments of Dr. Haldane, of Oxford, a celebrated physiologist, brought out strikingly that in still hot air the body is enveloped with a film of air at or near the blood-heat and the skin is not cooled by evaporation of the sweat, but if the air is in motion, unless it is saturated, which it seldom is, evaporation follows, and thus the skin is cooled. I recall a case where, in a certain stope, there was no movement of the air, and the miners complained of the heat and from time to time went into a nearby cross-cut where there was a strong current of air and were comfortable, although in the cross-cut the temperature was actually several degrees hotter.

The great variety of mining conditions found in the United States prevents formulating an exact specification for a ventilating scheme which would fit all conditions, but in order to bring out discussion the following general principles are mentioned which may be of guidance in planning a ventilating system for new mines or in adapting the arrangement of old mines as far as possible:

(1) There shall be a permanent up-cast and permanent down-cast shaft.

(2) The down-cast shaft shall be, as far as practicable, the shaft in which the men are hoisted and lowered. It may or may not be the main ore-hoisting shaft, the object being that, in case of fire, the main hoisting arrangements for the men will be available, and, if practicable, this shaft and its landing shall be concreted or otherwise fire-proofed, and back of each landing there

shall be a fire-proof door to protect the men from the workings beyond.

(3) The main shaft shall extend to the lowest working so that the fresh air may go to the bottom of the mine and then ascend on either side through the workings to the up-cast shaft; sufficient splits of air to be taken off on the different levels to provide for proper ventilation; the return air to be taken around the down-cast shaft by cross-cuts or concreted bridges.

(4) The main ventilating fan should be placed on the surface; the fan to be of fire-proof construction and with fire-proof housing; this to be a reversible fan. Generally the most convenient arrangement would be to place it near the collar of that shaft which is not the main hoisting shaft. Of course a tunnel or slope mouth would be equally as good. This is to avoid an air-lock; therefore, the fan would normally be an exhaust-fan, but as previously stated it should be reversible to provide for emergencies.

(5) The fan should be of sufficiently large size to take care of the largest volume of air which is likely to be necessary for at least 5 or 6 years of development, and it should be able to deliver and to draw air at a pressure in excess of that which could be produced by maximum natural causes, that is, the maximum motive head of the air column, due to the maximum difference which may obtain, between the temperature of the mine-air and of the outside air, in the coldest or the warmest days of the year. In general, the main fans at metal mines are entirely inadequate, and fans such as are employed in large coal mines should be installed in metal mines where the workings are expected to be of large extent and to require many men underground. By this is meant fans which will discharge from 100,000 to 300,000 cu. ft. of air per minute against a pressure measured by 2 to 6 in. of water-gauge. On account of the fire-risk it is not good practice to place the main fan underground, and moreover there is usually a liability of the air short-circuiting, that is, going around and around and not being discharged to the surface.

(6) 'Booster' fans are necessary in exploratory drifts, and in running winzes or raises and in high stopes. New devices in the way of canvas pipes and hanging arrangements have been developed recently that make admirable auxiliaries for reaching the face of headings and winzes, but wherever possible, it should be planned, as in coal mining, that a continuous circuit of air be obtained after it has been established that bodies of ore exist that sooner or later will have to be blocked in order to give adequate ventilation throughout.

A number of the greatest fire disasters in metal mines would have occasioned a much lower loss of life had there been continuous circuits of air, which would have enabled the miners to reach the in-taking air. The great merit in fire-proof down-cast shafts and the landings adjacent to these is to enable the escape of the men who could keep in fresh air by traveling against the in-taking current. On the other hand, if an unfire-proofed down-cast takes fire, as at the Cherry coal mine or the North Butte copper mines, it nearly always means disaster, and the only remedy then is a prompt reversal of the fan, which if accomplished in time might save the men. If this cannot be done, or if all passages for escape are already filled with smoke, the only thing left for the men is to brattice or bulkhead themselves off in some heading free from smoke, marking with chalk where they have gone, for the subsequent direction of the rescue-parties. While the cost of providing adequate ventilation is considerable, whether it be in metal mines or in coal mines, it must be regarded as a vital necessity for deep mining. Anyone who has visited deep or hot mines, such as those of the Comstock Lode or at Butte, must realize the need of providing the best ventilating system possible in order that deep mining may be carried on at all. Where nitrogen gas is found, or where spontaneous fires occur in rich sulphide ores, as in some Montana and Arizona mines, or in mines where sand or slime filling cannot be accomplished, it is necessary to erect fire-walls and put in a plenum-pressure system of ventilation so that the gases may be forced back into the crevices in the strata or behind the fire-walls.

Chilean Nitrate

Government purchases of Chilean nitrate for this country will be co-ordinated through the purchasing committee of the War Industries Board, under the immediate supervision of Mr. Baruch, so that there will be no competitive bidding for this material. This should simplify the problem and make it feasible to secure the best possible terms. The price of Chilean nitrate on board ship in Chile has greatly increased within recent months without justification. Shipping rates also have advanced. The increase in price is due in part to unfounded statements regarding the demand for Chilean nitrate for munitions and for fertilizer, especially in connection with the $10,000,000 recently appropriated by Congress to purchase nitrate of soda. Actually the demands of this country for nitrate of soda will be smaller than heretofore. The Navy already has placed contracts for this material to satisfy its needs for the next twelve months. The War Department announces that practically all the sodium nitrate which it has been planned to procure has been contracted for. Under a recent authorization of Congress there is available an appropriation of $10,000,000 to be used, at the discretion of the President, to secure nitrate of soda and to supply it to farmers at cost for cash. The quantity which may be purchased for fertilizer use under the special appropriation of Congress will not be an addition to the quantity normally used. Any quantity purchased by the Government and sold to farmers will simply replace part of the quantities heretofore supplied to them through private agencies.

BRASS on being melted suffers by oxidation of the zinc and copper. E. D. Frohman states that this may be prevented by covering the metal with charcoal.

The Chemical Exposition

By M. W. von BERNEWITZ

Fully 100,000 people, of whom 25% were men of special scientific training, received inspiration at the third National Exposition of Chemical Industries held at the Grand Central Palace, New York, during the week of September 24. At no time was the place crowded, but those in attendance appeared bent upon studying what was presented. There were meetings of scientific societies, moving pictures illustrating various industries, and over 400 booths containing an aggregation of apparatus used in chemical manufacturing; therefore visitors could not complain of a lack of variety. There is no doubt that the show was a success, which gives encouragement for promoting another in 1918. Broadly speaking, the exhibits might be classed under eight groups; for example, dyes and colors, instruments, acid-proof apparatus and connections, glass, porcelain, and silica-ware, refractories, metals, ore-treatment apparatus, insulating materials, and by-products. The first of these was shown in about 25 booths, the products being in bottles and in various cloth, silk, and leather materials. Dye manufacturers were probably the most prominent of any exhibitors. Instrument-makers were in force, showing the latest ideas in balances, control, electrical, liquid, temperature, and vacuum-measuring devices, some of which were to be seen in operation. Measuring temperature by radiant heat was of interest. A dozen firms showed chemicals, including explosives. A demand is increasing for acid-proof apparatus, such as evaporators, heaters, pumps, and pipe connections. These were shown as special iron compounds, enameled iron, lead, and earthenware, some of the last being cleverly made. There was also acid-proof cloth. Considerable laboratory glassware was on view, and porcelain for all purposes. Silica-ware, both transparent and opaque, was keenly examined. This, as most chemists know, is made from pure quartz. It will stand great variations in temperature without cracking, is proof against all acids save hydrofluoric and phosphoric, and is superseding platinum for many purposes. Refractories included ores and brick. Several well-known smelting concerns had excellent exhibits of electro-spelter, copper, and lead. Much interest was taken in the spelter, the latest metallurgical triumph on a large scale. Millmen had a chance to see in operation ball-mills, concentrators, flotation machines (the odor of the oil sufficing to direct one to the place), and filters. Cyanide (30%), made from cyanamid, was on view. It is blackish and reminds one of the salt produced 20 years ago. The 52% salt was also shown. Smoke-farmers could see how their damage-suits are going to lapse by reason of the well-known precipitator. Several types of smoke-helmets were demonstrated in a practical way. Insulating materials included rubber, vulcanite, bakelite, and brick. Tungsten producers had a chance to see 99% acid (WO_3), a canary-yellow powder worth $1.50 per pound. By-products were shown by numbers of manufacturers. Smelters displayed gold, silver, platinum, bismuth, tellurium, and the like, all recovered during the refining of base metals. Perhaps the most interesting by-product exhibit was one arranged like a family tree, showing actual specimens of the derivatives from coal, including coke, tars, oils, colors, disinfectants, drugs, explosives, and many other things.

It would be unkind to indulge in much criticism of such an exposition, but as an unbiased spectator during four days some defects were apparent. The ramifications of chemistry are far-reaching, but a line should be drawn somewhere; and such exhibits as 50-ton evaporators, large mixing, drying, and centrifugal machines, elevators, trucks, corn products, engineering firms, Japanese chemicals, milling and treatment-machinery for mines, and excessive space occupied by the Southern States and railroads showing their mineral resources, it would seem might better be shown elsewhere, and the space so occupied filled with matter pertaining to the real chemical industries. A number of the large chemical firms was conspicuous by being absent. These should be in evidence at future expositions.

Sulphuric-Acid Production

Under the conditions imposed by the War an enormously increased demand has developed. The production in 1916, expressed in terms of acid of 50°B., was 5,642,112 tons, valued at $62,707,369, to which must be added 443,332 tons of acid of strengths higher than 66°B., not convertible into acid of 50°B., valued at $10,806,757. This includes by-product acid, that is, acid produced at copper and zinc smelters. The production of acid from this source in 1916, expressed in terms of 60° acid, was 1,069,589 tons, valued at $12,158,266, and 92,802 tons of acid stronger than 66°B., valued at $1,941,661. Sulphuric acid was produced at 211 plants. In the production of all grades of acid the following amounts and kinds of sulphur ore were used:

ORE USED IN MAKING SULPHURIC ACID, LONG TONS

	Domestic	Foreign	Total
Sulphur	261,574	10,903	272,477
Pyrite and pyrrhotite	324,602	1,154,550	1,479,152
Gold and silver-bearing pyrite	1,673	....	1,673
Copper-bearing sulphides	568,116	196,404	764,520
Zinc-bearing sulphides	531,625	92,002	623,627

Included in the pyrite and pyrrhotite are all the sulphide ores used which were not treated further for their copper, lead, zinc, gold, or silver content. Much of this material contains small amounts of these metals.

RUBBER has been decreasing in price for many months. 'Up-river fine' has gone down from 81c. per pound in 1916 to an average of 65¾c. for the month of September 1917. Of the other chief materials used in rubber-tires, cotton is now worth 27c., zinc oxide 10½c., and lithophone 6½c. per pound.

CRIPPLE CREEK, COLORADO

CRESSON CONSOLIDATED.

The following circular letter to stockholders from A. E. Carlton, president of the Cresson Consolidated Gold Mining & Milling Co., accompanying the October dividend checks, is self explanatory: "Under date of September 20, 1917, Mr. Noble, consulting engineer of your company, reports net profits on fully developed ore-reserves, as of September 1, 1917, of $3,652,-562. The ore shipped in August yielded a net return of $220,-221.27 above cost of operation. Your company has cash in bank today of $1,255,003.84, and, in addition, about 30 cars of ore in transit and at the mill, of an estimated value of $40,000. New ore of a net value of $17,883 was added to reserves during the month. Mr. Noble reports regarding the new workings as follows: On stope 1417A a winze 25 ft. deep has shown continuous high grade to the bottom, and a drift has now been started westerly. In drift 1517, directly below 1417, 45 ft. of ore of good grade has been opened. A raise from this is showing improving values in going up. The ore cut in 1602 has been developed 50 ft. in length. Many good streaks carrying sylvanite ore are observed coming into the main vein, and it is probable that a greater width than now disclosed will be opened when air-connections permitting more active work are made. In drift 1609, 50 ft. of ore is shown and in 1608 a length of 43 ft. The ground between these two will probably be found to be mineralized when stoping is started. Mr. Noble has not as yet figured in the reserves any of the orebodies now under development on the sixteenth or lowest level. Distribution No. 60 of 10c. per share has been declared from funds received by your company from the sale of ore-reserves acquired prior to March 1, 1913, check for which is enclosed."

Richard Roelofs, superintendent of the Cresson company, has submitted his resignation and is reported to have been paid a handsome cash bonus. His contract with the company had still two years to run. He will be identified with the McNeill Carlton interests. He will be succeeded as superintendent by A. L. Bloomfield, for many years associated with the Golden Cycle Mining & Reduction Co. The change will take effect on November 1.——The Vindicator Consolidated Gold Mining Co. has declared its regular quarterly dividend of 3c. per share, amounting to $45,000, payable October 25, to all stockholders of record on October 15.——The Portland Gold Mining Co. will pay its regular quarterly dividend of 3c. per share, $90,000, on October 20 to stockholders of record on October 10. With this distribution Portland stockholders will have received $10,957,080 in dividends. The company is now making ready to accept and treat custom ores. Scales are now being set in at the new sampling-plant at the Independence mine, and low-grade ores will hereafter be treated at this mill within 30 days. The rate for treatment has not been announced.——The September production from the South Burns mine of the Acacia Gold Mining Co. was limited to about 200 tons of ore. Two cars mined on company account averaged $25 per ton. Lessee's shipments of four cars averaged $18 per ton.——The Cripple Creek Deep Leasing Co. and F. T. Caley, operating leases on the Jerry Johnson mine on Ironclad hill, owned by the Jerry Johnson Gold Mining Co., mined and shipped seven cars of ore running from $18 to $25 during the month of September. The Roosevelt tunnel, of the Cripple Creek Deep Drainage & Tunnel Co., was advanced 166 ft. during September. The heading is now in the Rose lode mining claim on the north-western slope of Battle mountain belonging to the Rose Nicol Gold Mining Co. The flow from the tunnel has fallen to 5123 gal. per minute. Practically all of the corporations operating on company account in this district have adopted the bi-monthly pay-day. Employees will now be paid on the 10th and 25th of each succeeding month.

HOUGHTON, MICHIGAN

THE 1917 PRODUCTION.—LABOR CONDITIONS.—ISLE ROYALE.

If the copper production from the mines of the Lake Superior district reaches a total of 235,000,000 lb. for 1917 there will have to be a substantial improvement in the output of all mines from now until January 1. The production for September is estimated at 18,000,000 lb. The production for October may show an increase of 1,000,000 lb. over that amount, but the fact remains that the whole district is likely to show a decrease of close to 30,000,000 lb. over 1916. Last year this district produced 263,000,000 lb.——The daily output of ore from the Calumet & Hecla increased in October from 8000 to 9000 tons and before the middle of the month the mine should be producing 9600 tons. This mine is more fortunate than the smaller ones and seems likely to have no labor difficulty ahead. During the spring and summer trammers were not only difficult to secure but hard to keep, and all the mines were constantly changing this form of labor. Now the situation is different; the men are coming in seeking work every day; moreover, they are sticking to their jobs and show no desire to quit. Miners are plentiful.——Production of ore from the Mass Con. has increased 10% in the past month and now runs 600 tons per day.——The Ahmeek is handling an average of 85 cars or ore, a reduction of 15% since mid-summer.——The Old Osceola, of the Osceola Con., is shipping 650 tons, a 15% increase, while the North and South Kearsarge are handling 2600 to 2800 tons per day.——The Wolverine has reduced ore shipments from 1200 to 950 tons, owing to accident. An ov r-hoist put the skip, loaded with ore, through the top of No. 4 shaft-house and hauled it over to the compressor.——The St. Paul railway, in conjunction with Copper Range, may build a spur to Michigan.——No. 5 shaft at the Quincy is showing a sensational amount of mass and barrel copper that is going directly to the smelter. This runs 75% copper.——Production at the Wolverine is normal again, repairs having been completed at No. 4 shaft. Ore shipments are running 40 cars per day.——There has been a slight falling off in the Franklin ore shipments lately; production is now running about 850 tons.——Calumet & Hecla authorities deny the rumor that the company is increasing its holdings of Isle Royale stock. This sets at rest the story, but does not in the least divert attention from the fact that no Calumet & Hecla subsidiary is attracting more favorable attention than this property. Isle Royale is a pronounced success. The management is working the mine in the only way possible for such a low-grade ore, namely by handling large quantities. Isle Royale reached its maximum production in 1916 when the output was 12,412,111 lb., compared to 9,342,10C in 1915, 6,601,235 in 1914, and 4,158,548 in 1913. When it is realized that the output of copper from this mine is four times what it was four years ago, some idea of the way this property has been developed may be judged. Isle Royale owns $160,134 worth of stock in the Lake Milling, Smelting & Refining Co. as well as $32,000 worth of stock in the Lake Superior Smelting Co. at Dollar Bay.——The Cherokee shaft has been sunk 420 ft., but cannot go much deeper until a new rope is secured. Good looking copper ore, some mass, some amygdaloid, and some epidote, have been found.——The second unit in the leaching-plant at the Calumet & Hecla mills at Lake Linden is being built rapidly.——A flotation unit at White Pine, to handle 600 tons per day, will be in operation at this subsidiary of the Calumet & Hecla as soon as possible.

PLATTEVILLE, WISCONSIN

ZINC AND LEAD SITUATION.—GRASELLI CHEMICAL CO.

A leading smelter representative, when asked for information regarding the spelter situation, blamed spelter manufacturers for the perplexing and unsettled conditions in the spelter market. He pointed to the agreement of copper smelters to establish a price to the Government on copper and declared that had zinc smelters used the same tactics, spelter would have occupied a more favorable position. The willingness of the American Metal Co. to accept 7.75c. per lb. on 5500 tons of spelter, declared this authority, upset the whole brew, as it was shown only too plainly the company was willing to accept a loss on the entire lot in order to work off this quantity of metal. Prices on zinc ore were maintained during the first part of September on the identical basis reported for the month of August, namely, $70 per ton base for 60% and top-grades of refinery ore, with the range down to $65 per ton for seconds and lower-grades. A break in prices was seen in the third week, the published figures remaining the same, but local correspondents in the different camps handed in reports that showed declines were imminent, the first decline carrying the base down to $68 per ton. The last week found top and standard ore at $67.50 per ton, with the range down to $63 and the latitude of assays restricted to apply on grades down to 54% zinc. The disposition of producers and refiners to hold for better prices, based on some vague belief that such a contingency was possible, gave way to a calm resignation that prompt marketing was the best policy and shipments of finished ore were much heavier during the latter half of the month. On the lower grades of zinc-concentrate independents found a ready market with the Mineral Point Zinc Co., which did a heavy business at its reduction plants at Mineral Point all through the month. Price offerings were out of tune entirely with the base that carried on the higher-grade ores, but one operator admitted that even at the prices submitted outputting still remained profitable. Powder, steel, and all mining supplies have advanced from 40 to 100%, with the increase in the cost of labor about double that obtaining two years ago. Shovelers are receiving from 8 to 12c. per can of 1000 lb. and some earn as high as $6 on a single 9-hour shift.

The buying of ores by the Grasselli Chemical Co. of Cleveland was a feature of the month's business, although the total tonnage secured was under the amount shown for the month of August. This company secures most of its mine-output under contract arrangement, but inquiry revealed the fact that sellers were dissatisfied and it is highly problematic what future arrangements well be made, as several important contracts are about to expire. High prices for zinc ore seem destined to become past history; the Cleveland corporation has given ample evidence of recent date that more attention will be given to the finished product obtained at refining plants. Should this programme be realized, increased facilities for zinc-ore dressing will be established promptly. Operators running on lean deposits are face to face with conditions that require extraordinary executive ability and it has been intimated in reliable quarters that several producers will be under the painful necessity of shutting down entirely if present costs of operating continue and prices for zinc ore descend to lower levels.

An item fraught with much interest to mining men is found in the annual tour of mine inspection paid to this field during September by leading officials and mining engineers of the New Jersey Zinc Co. The majority found in favor of continuing the company's policy in this field, and a liberal buying-up programme has been recommended. Mineral lands known to possess merit for mining purposes and mines now in operation and under the control of small independent corporations will come in for early inspection and negotiations designed to transfer ownership and control to the big Eastern zinc syndicate. The liberal attitude of the New Jersey Zinc Co. in dealing with land owners and mining men in the past insures further exploitation in this field that lends encouragement at this time.

Of the several separate establishments in the field only two were content during the month to market iron pyrite in any appreciable quantity; the Linden Zinc Co., with one plant at Linden and another at Cuba and the National Separators at Cuba. Prices showed no change and the total deliveries for the month were less than half that usually reported. At the National Separators a considerable quantity of fine obtained in electro-static separation was carried over. At the reduction works of the Mineral Point Zinc Co. such residue as was obtained was promptly converted to the uses of the acid-making department.

Deliveries of zinc ore, lead ore, and pyrite were made during the month of September by districts as here shown:

District	Zinc lb.	Lead lb.	Pyrite lb.
Benton	19,716,000	192,000	....
Mifflin	6,252,000	86,000	....
Galena	3,492,000	......	....
Linden	2,786,000	62,000	....
Shullsburg	1,896,000	......	....
Platteville	1,754,000	......	....
Hazel Green	1,650,000	46,000	....
Highland	900,000	146,000	....
Potosi	616,000	50,000	....
Mineral Point	294,000	......	....
Dodgeville	264,000	......	....
Cuba City	....	......	3,520,000
Total	39,620,000	582,000	3,520,000

Shipments of high-grade blende from refining plants in the field to smelters were made during September as follows:

	Lb.
Mineral Point Zinc Works	5,548,000
Wisconsin Zinc Roasters	3,438,000
National Zinc Separators	3,192,000
Linden Zinc Refiners	1,008,000
Benton Roasters Co.	518,000
Total	13,704,000

The gross recovery of crude concentrate from mines for the month aggregated 20,104 tons; net deliveries out of the field 11,707 tons.

Sales and distribution of zinc ore for September showed the following accounting: Mineral Point Zinc Co. 197 cars, 7335 tons; Grasselli Chemical Co. 85 cars, 3134 tons; National Separators 80 cars, 3381 tons; Wisconsin Zinc Roasters 80 cars, 3231 tons; American Zinc Co. 56 cars, 2287 tons; Linden Zinc Co. 40 cars, 1452 tons; Matthiessen & Hegeler Zinc Co. 24 cars.

1148 tons; American Metal Co. 16 cars, 676 tons; Illinois Zinc Co. 12 cars, 511 tons; Lanyon Zinc Co. 10 cars, 364 tons; Benton Roasters 9 cars, 369 tons; and Edgar Zinc Co. 8 cars, 294 tons.

Interesting developments occurred in all of the camps of the field during the month. In the Highland district the New Jersey Zinc Co. was engaged with a large force of men and mules over narrow-gauge tracks in turning gutters and dumps filled with waste of mining operations conducted two and even three generations ago. The sludge recovered proved profitable, assaying as high as 15% zinc. The Mineral Point & Northern Ry. Co. is meeting the increased traffic over its 34 miles of roadway by the laying of 90-lb. rail. In the Linden district the Optime company is showing heavy returns from mine No. 3. Spring-Hill Mining Co. is driving an adit to connect with an orebody struck by borings on the Ross lease. The Milwaukee-Linden Development Co. completed a new shaft on the Thompson prospect but was driven out by heavy overflow, and heavier pump equipment is being supplied. Another shaft is being sunk on the Kickapoo lease. Producers of low-grade zinc ore received during the month an average of $17 per ton for 25% zinc concentrate; $24 for 30%; $35 for 40%; $68 for 60%. The Linden Zinc Refining Co. ran steadily all the month, making recoveries in excess of 60% zinc. Drilling outfits were engaged on the Rolling lease for the Optimo Mining Co. Labor was well paid, breast-men receiving $3.50 per shift; muckers $3; shovelers 8c. to 12c. per can (1000 lb.); hoist-men $3.25; foremen $125 to $175 per month; machine men $4 per shift; blacksmiths $3.50; dynamiting crews $3 each; teamsters $6 per day of 8 hours. There is a labor shortage.

The Mifflin failed to show as well as for August by an even 1000 tons of concentrate. The principal producers were the Yewdall and Coker mines. A new producer called New Phoenix made regular shipments; the Big Tom mine increased its deliveries and paid a monthly dividend of 25%. The Grunow, Peacock, Lucky Six, Biddick, and B. M. & M. mines operated steadily all month and made shipments at intervals. The Billings mine, a new development opened on the Bourett farm, made its first shipment.——The Block-House Mining Co., paid a 100% dividend during the month, making 570% since January 1, 1917. Additional leaseholds secured by the Block-House Mining Co. adjoining the present producer have been exploited by drills with gratifying results.

MANHATTAN, NEVADA

WHITE CAPS.—MANHATTAN.—UNION AMALGAMATED

Since last week's report, the east orebody has been exposed on the fifth level of the White Caps. The hanging-wall side of the ore was first struck by the south cross-cut, although the cross-cut was entirely in ore. After driving for 25 ft. along the vein, the course was changed to cut the ore at right angles toward the foot-wall. This cross-cut has now been extended 15 ft. and there is no sign of the foot-wall. The first sign of calcite which has been prominent in the heart of the vein in the upper levels was noted on October 4. As mentioned in last week's report, the main portion of the orebody is still farther ahead of the cross-cut from the fifth level, as is distinctly shown on the maps of the underground workings. However, judging from the size of the vein already exposed, with neither wall yet uncovered, it is practically a certainty that the bottom workings are going to prove the east orebody considerably larger in size than on the fourth level. Each foot of work along the vein on the bottom level is adding many tons to the known ore-reserves in the mine. Between the third and fourth levels the ore-reserves in the east orebody alone adds over a million dollars to the value of the White Caps, and the present development should more than equal this amount. The weekly amount gained from the various levels in the property, brought up to date, is as follows: From the third level, the hanging-wall cross-cut gained 30 ft., now a total of 74 ft. A new drift east has been started during the week and 12 ft. has been made. From the fourth level, raise 405 is breaking the records for rich ore, seven feet next to the foot-wall assaying $35. This raise has been extended 26 ft. and is now up a distance of 135 ft. from the level. From the fifth level, the south cross-cut has been advanced 37 ft., and is now 292 ft. from the station. The total for the above work shows 105 ft. made since last report. The same average daily tonnage of ore has been maintained for the mill. In the White Caps mill a set of rolls has been added to equipment above the roaster for secondary grinding, to bring the entire product to under 6-mesh size before passing into the roast. A clean-up of the zinc precipitate in the zinc press has been started so that before long the first gold bars should reach the mint from the milling operations.

CALUMET-NEW MEXICO SHAFT AT PINOS ALTOS

Driving east from the fourth level in the Manhattan is progressing with two shifts. The face of the drift is now 50 ft. from the shaft. The east orebody should be struck at 250 ft. From the third level, the south cross-cut has been extended from the point where development work ceased some months ago, and now the face of the cross-cut is 150 ft. from the point in the east drift where the cross-cut was started. A raise will shortly be extended from this cross-cut to the second level and thus assure ventilation. The raise will also serve to develop the high-grade orebody already proved from the second level to the surface.

The footage made in the Union Amalgamated during the past week is reported as follows: From the 600-ft. level, the north cross-cut on the fissure made 19 ft., No. 11 north cross-cut east made 30 ft., the south cross-cut on fissure extended for 27 ft.; the total footage made in above workings being 76 ft. The face of the north cross-cut is in quartzite. The north orebody should be found in the continuation of this work in 10 to 15 ft. Gold is now showing in the seams found in the face. No. 11 north cross-cut is still in ore of low grade and considerably broken. This cross-cut has been extended through the Earl ore-shoot and has proved the orebody to be at least 30 ft. from hanging to foot-wall. About 10 ft. of the ore showing in this work is high-grade. The south cross-cut on the fissure has demonstrated a large tonnage of ore. The continuation of this cross-cut will cut the hanging-wall contact in about 20 ft. It is expected that another valuable orebody will be exposed, as the present work should cut the extension of the orebody developed along the contact from the 350-ft. level of the Earl.

VANCOUVER, BRITISH COLUMBIA

GRANBY CON. DIVIDEND.—PRINCETON MINING AND DEVELOPMENT

In the fiscal year ended June 30, Granby Consolidated Mining, Smelting & Power Co. earned $5,452,796 from all sources, or $36 per share. This compares with $25 earned in the preceding fiscal period.——With the starting of the new mill at the Surf Inlet the gold production of British Columbia will be increased by about $1,000,000 per year. This is equal to an increase of 25% on the present output of lode gold. The mine has 11,000 ft. of development work done, the main, or 550-ft., level is 3000 ft. long, and develops two veins. From this level a shaft has been sunk to a depth of 285 ft., from the bottom of which levels have been driven 600 ft. Above the main level, intermediate levels have been driven at 430 and 320 ft. respectively. The west vein has been developed by five and the east vein by four levels. The west vein averages 18 ft. and the east vein 5 ft. on these levels. The orebodies occur as lenses along the course of the shear-zone, one lens having a length of 700 ft.——The main working-level is equipped for electric haulage, the cars being dumped into the crusher-bins. The ore averages $11 per ton in gold and 385,320 tons has been developed. The mill is of a type entirely new to this country; ball-mills and tube-mills replacing the customary stamps; the tailing is treated by flotation. The concentrate will be shipped either to Tacoma or the Selby smelter. While the mill has a present capacity of 300 tons per day it is probable that after the War it will be increased to 1000 tons, and the concentrate will be treated at the mine as is being done at Hedley.

The Princeton Mining & Development Co. has been incorporated to develop the Copper Farm group, recently acquired by F. F. Foster and associates from Col. Stevenson. The capital of the company is $150,000 and the head office is at Princeton. F. F. Foster, president; Perley Russell, secretary-treasurer; and Guy Murphy form the directorate. The property is about five miles below Princeton and is traversed by the Great Northern railway. It is fairly well developed; assays of the ore run from $33 to $174.——Nearly 500 tons of concentrate and crude ore has been shipped from the Florence silver mines at Ainsworth during the month of September.

COBALT, ONTARIO

BEAVER CONSOLIDATED.—MCINTYRE.—PORCUPINE CROWN.

The outlook at the silver mines at Cobalt continues exceedingly bright, ore shipments for last week totaled 18 cars containing 1,250,000 lb. as compared with 1,500,00 lb. during the preceding week. It seems likely that September shipments will be the highest during the year.

Development at the Beaver Consolidated is resulting in the recovery of high-grade ore. One ore-shoot at the 600-ft. level, 130 ft. long, though narrow, contains 3000 oz. per ton. This company is also operating the Kirkland Lake Gold and has paid $200,000 toward the purchase of the mine and spent $100,000 in development. Only $70,000 remains owing.

It is estimated that ore to the value of $500,000 is blocked out. The capital of the Buffalo has been reduced from $1,000,000 to $750,000. The shareholders are to receive 25c. and a share of the new issue for each share of the old, the par value of which will be 75c. instead of $1. This disbursement of $350,000 makes a total of $3,037,000 paid to the shareholders in this company. The oil-flotation plant at the Buffalo is treating the huge pile of old tailing at a fair margin of profit.

Developments at the 400-ft. level of the McKinley-Darragh has revealed considerable ore. Further cross-cutting is under way, and one of the main veins showing at the 300-ft. level is expected to be cut at any time. The deal for the old Alexandra property is still pending, and will probably not be decided before the first week in October. The Chambers Ferland continues to yield fair quantities of ore from the zone just above the contact.

The Genessee is being explored by cross-cut just above the contact at a depth of 550 ft. in the hope of picking up extensions of the Chambers Ferland veins.——The O'Brien mine is yielding $100,000 per month, and the ore-reserve is several years ahead of production.

The main drive on the 1000-ft. level in the McIntyre is now about 200 ft. on Jupiter ground. The Jupiter shaft, 1800 ft. farther east, has reached a depth of 840 ft. The management plans to connect this long drift with the Jupiter shaft which is being sunk toward that level. A force of 230 men is employed and there appears to be no reason why the present interim dividends of 5% quarterly should not be maintained. The annual meeting is being held at Toronto.

The Hollinger is now employing nearly 1000 men. Extensive development is being done, and it is expected that the ore-reserves at the end of 1917 will be $40,000,000. The centralization of work has improved conditions considerably and will cheapen the cost of production.

SPOKANE, WASHINGTON

INTERSTATE-CALLAHAN.—ELECTRIC POINT.

War taxes, which must be paid out of this year's earnings, will cut heavily into mine dividends of the Inland Empire. The first example is given by the Consolidated Interstate-Callahan Mining Co., operating in the Coeur d'Alene, whose directors suspended the dividend for the current quarter. The president of the company, J. A. Percival, says:

"The directors decided that the uncertain outlook, due to the War, and the increased cost of production, together with the necessity of providing for heavy war taxes, make desirable the creation of a strong reserve, the reserve to become a protection against any emergency. The war tax is approximately $300,000, estimated under the provisions of the statute as now drawn. It must be paid out of the 1917 earnings." Other strong corporations of the Coeur d'Alene are preparing to do their bit for Uncle Sam, according to reports. The work of arriving at figures is being facilitated by a corps of government statisticians. On a basis of 20%, mines of the Coeur d'Alene would contribute $2,400,000, if the earnings this year were the same as those of 1916.

Washington mines contributed ore worth $65,000 to the smelter of the Consolidated company at Trail, B. C., the last week of September. The output aggregated 1086 tons, divided as follows:

	Tons
Quilp, Republic	200
Knob Hill, Republic	96
Valley Mining Co., Valley	32
Electric Point, Boundary	637
Metaline, Metaline Falls	93
Whitford, Chewelah	28

Nearly 500 tons of concentrate and crude ore was shipped in September from the Florence Silver Mining Co. at Ainsworth, B. C. A dividend of $118,960, making the total to date $166,530, has been declared by the Electric Point Mining Co., of Northport, Wash., and $30,000 will remain in the treasury after the disbursement is made. The mine from which these profits were taken was discovered only two and a half years ago. The latest disbursement is at the rate of 15c. per share on the issue of 793,500 shares. Three cents of it is the regular quarterly dividend and 12c. an extra.

ALASKA

(Special Correspondence.)—K. I. Fulton, of the Talkeetna Mining Co., operating in the Willow Creek district, is in town with the first clean-up of the company amounting to $2000. The clean-up is the result of a 140-hour run in a Denver quartz mill. Gold from Cook Inlet district sent to the U. S. assay offices at Seattle for the fiscal year ended June 30, amounted to $406,000.——The Willow Creek Development Co., a local company formed to exploit mines in the Willow Creek district, has elected the following officers and directors: J. E. O'Reilly, Charles Herron, Leopold David, George M. Campbell, and A. E. Lathrop. Capital stock is $100,000. The stock is being placed on the market at 50c. per share.——L. H. French has bonded a number of claims in the vicinity of the Mabel mine and will begin active development in the spring.——A rain and wind storm on September 11 and 12 caused great damage to Seward by destroying the lighting-plant and also a number of buildings and bridges. The Crow Creek Mining Co., at Glacier Creek, estimates its damage at $100,000. Two miles of track on the government railroad was washed out near Mile 34, and some machinery was destroyed. Pearson camp lost its entire clean-up.

Anchorage, September 15.

(Special Correspondence.)—J. C. Murray, in charge of the Cache Creek, where a dredge was started in the fall of 1916, is in from the camp, and reports a good season, although forced to suspend work earlier than anticipated, on account of excessive rains which raised the water in the streams and inundated much adjacent ground. He reports that the many independent operators have had a satisfactory season, but lays stress on the necessity for roads to lower the present excessive cost of transportation. However, Ross Kinney is making a preliminary survey for the Alaska Road Commission with a view to building a wagon road from Talkeetna to Cache Creek. A trail will be built probably this winter.——Frederick Spengler, an engineer who recently returned from a reconnaissance in the Indian Creek district, 50 miles up the Talkeetna river, reports that section as presenting an attractive field for the exploitation of smelting ores which are in proximity to hydro-electric power and fluxing materials, and which have the further advantage that a railroad could be easily and cheaply constructed to them.——Herschel C. Parker, of New York, who has been in charge of the examination of placer ground on the Kahiltna river, reports that platinum is generally distributed in the wash of streams in the mountains about Mt. McKinley, and indications are that the combined gold and platinum is sufficient to justify the exploitation of these gravels on a big scale. K. B. Mertie, government geologist, has been instructed to investigate, and is now on the ground.

Anchorage, September 24.

ARIZONA

COCHISE COUNTY

(Special Correspondence.)—The milling machinery of the Commonwealth Mining & Milling Co., part of which has been sold to the United Verde Copper Co., at Jerome, is now being removed. The Commonwealth will erect more suitable machinery immediately. Owing to the high price of silver the old tailing is being treated by A. T. Smith and associates. Cross-cutting has been started on the 300-ft. level to prove in depth the recent surface strike. Samples from this strike, taken over a width of 10 ft., assayed $25 per ton.

Bisbee, October 6.

MARICOPA COUNTY

(Special Correspondence.)—A number of men from Mesa and Tempe have acquired a lease on the Buckhorn and Boulder mine. The shaft has been re-timbered. A company is to be formed to finance the equipping and developing of the property.

Phoenix, October 5.

MOHAVE COUNTY

(Special Correspondence.)—Fire of undetermined origin destroyed the mill of the Union Basin Mining Co. at Golconda, 15 miles north of Kingman. The loss is estimated at $100,000; zinc ore was treated at the mill. The company employed 300 men. ——At a meeting of the Amalgamated Gold Mining Co. at Oatman it was decided to sink 200 ft. deeper, to the 500-ft. level, and carry on development.——Sinking to the 500-ft. level and driving on the Leland property will commence shortly. Work is progressing on the Gold Bond mill, which is being repaired by the Gold Ore Mining Co. The Gold Ore shaft is being re-timbered and preparations are being made to mine and mill 100 tons per day. Machinery, consisting of 100-hp. hoist, compressor, and 500-ton milling plant, is expected at Kingman for the Frisco Gold Mine Co. The providing of these improvements resulted from the discovery of a large quantity of $10 gold ore in the Gold Dome claims.

Kingman, October 6.

PIMA COUNTY

(Special Correspondence.)—A petition to permit the optional sale of the G. B. McAneny placer mine to A. J. Taylor of Denver for $200,000 has been made by his trustees.——Suit against T. Childs, alleging damages of $75,000 for refusal to deliver the Iron Reef, Apex, and Rambler claims of the Cardigan property at Ajo, has been filed in the Superior Court by H. E. Fredrickson.——As the result of a rich strike of lead and copper ore the Black Prince group, 87 miles south-west of Tucson, in the Quijotoa mining district, has been leased.——The San Xavier Extension has commenced shipping galena and chalcopyrite from the 200-ft. level.

Tucson, October 6,

PINAL COUNTY

(Special Correspondence.)—The Ray Consolidated Co. has subscribed $1,000,000 to the second Liberty Loan.——It is announced by H. S. Bryan that a contract has been let for the erection of a 50-ton mill for the United Vanadium Development Co. near Kelvin Junction. Part of the machinery is now being shipped from Denver. It is estimated that there is sufficient ore on the dumps to operate the mill for two years. ——A rich ore-shoot in the west drift on the 105-ft. level of the Gila Development Co. has been found. The ore is free-milling quartz, a sample taken over three feet assayed $133 per ton. The Cave Spring Co. has been doing extensive development work during the past nine months, and plans are now under way for the erection of a mill and cyanide plant. Pending the erection of a mill all high-grade ore will be shipped.

Florence, October 5.

YAVAPAI COUNTY

(Special Correspondence.)—The Venture Hill has resumed sinking. This property has been closed-down since the strike last May. When work ceased good showings of chalcopyrite and bornite had been uncovered in the shaft. Machine drilling has commenced in the adit of the Jerome Copper Co.'s property, which was formerly the Ewing and Hooker group. It is proposed to reorganize the Jerome Victor Copper Co., under the name of the West United Verde Copper Co., with a capital of $5,000,000. John H. Banks of New York has been engaged as consulting engineer.

Prescott, October 6.

(Special Correspondence.)—The management of the Big Ledge Copper Co. announces that construction work will be commenced on October 1 on the smelter, and three shifts are to be put on at the Butter Nut and Henrietta mines. A 50-ft. vein, all milling ore, has been opened on the 200-ft. level of the Butter Nut. The company also expects to handle custom ore.——It is expected that the copper output of the Arizona Binghampton property will be between 300,000 and 400,000 lb. for the month. The first unit of the mill has been operating a year, the second unit was started a month ago. The company plans the sinking of a larger shaft in order to increase the production. A new shoot has been discovered on the 200-ft. level.——The Copper Queen mine, which adjoins the Arizona Binghampton on the east, is planning extensive development in view of the proposed flotation plant which is to be erected at the mine. This company already has some 35,000 tons of ore blocked in the upper levels. Some of the old workings have been unwatered and sampling is being done. The stopes in the mine are 20 ft. wide. Work is being resumed by the Half Moon company immediately south of the Copper Queen. A new hoisting plant has been erected. The Pocahontas has resumed operations; a flotation plant is to be erected.——The Barbara mine, leased by Bisbee men, is driving an adit to tap an orebody recently opened by them in a shaft 300 ft. above. It is understood that the report is favorable on the Harvard and Yale, Iowa, and Celebration mines, which were examined during the first part of the month by a Chicago engineer.——The Ventura Hill Co. has arranged to raise $25,000 for development. Sinking in the 200-ft. shaft will be continued.——The Jerome-Anaconda Copper Co., with a capital of $2,000,000, has been formed to develop 32 claims, belonging to T. Bresnahan, in the Yeager Canyon district.——The Green Monster Co. has just erected a new hoist at the Dorothy May shaft.——The Jerome Verde will erect a 750-cu. ft. compressor. Last month this company drove 550 ft. of development work. Driving will be started on the 600-ft. level of the Verde Combination. Cunningham Smith and Black of El Paso are drilling for oil at China valley.——L. P. Morgan of New York has just completed an examination of the Lucky Five in Copper Basin and has reported favorably.

Prescott, September 28.

YUMA COUNTY

(Special Correspondence.)—C. Flynn and associates have just purchased from H. B. Hanna of Bouse what appears to be a large deposit of manganese, situated in the north of this county.

Yuma, October 5.

(Special Correspondence.)—The water in the shaft of the Silver King mine has been lowered 220 ft., only 30 ft. remains to be unwatered. Timbering and repairing are following the unwatering closely. Systematic development is to be carried on prior to starting the mill. For this work a 50-hp. Fairbanks-Morse engine, a 12 by 10 Ingersoll-Rand compressor, a hoist, Cameron pump, and 25-hp. West Coast engine have been erected. An electric-light plant for lighting the surface buildings has been completed. Three motor-trucks are in service hauling supplies. The road from the mine to the railroad siding, three miles away, has been completed. From two drifts in the upper workings some good ore has been obtained, assaying as high as $100.——The Queen Creek Copper Co. has uncovered a lode in the bottom of the 300-ft. shaft, which assays $100 in gold and 15% copper. This company is controlled by Scotch capital.——Fire starting from an explosion of gasoline consumed the hoist and compressor buildings of the Mammoth Development mine, at Mammoth, and burned its way down the shaft. The men escaped through another shaft. This mine had just reached a dividend-paying basis. It is planned to continue operating part of the mill on 10,000 tons of dump-ore. The Old Collins shaft will be opened and used until re-equipment is finished.

Florence, September 29.

CALIFORNIA

CALAVERAS COUNTY

(Special Correspondence.)—At the Safe Deposit mine in Old Woman's gulch preparation is being made to turn on the water for a winter's season. About 700 in. of water will be played against the bank and the gravel hoisted from the pit by hydraulic elevators. The debris dams were completed some months ago and were passed by the Debris Commission engineers.——Reports from the Mokelumne group of mines, operated by Max Muller of Los Angeles, are satisfactory. A Joshua Hendy ball-mill was erected last fall. The mill is running two shifts and the company is preparing to sink 500 ft. The force employed at present is 35 men.——The Buffalo gravel mine in Chili gulch has unwatered the shaft and re-timbered it. The mill will be operating at capacity shortly. ——Shortage of Miners is hindering work in this section, but the outlook is good. A rich body of ore has been struck at Gold Cliff. It is reported, also, that the Lightner mine will be re-opened shortly.

Mokelumne Hill, October 1.

NEVADA COUNTY

(Special Correspondence.)—The orebody lately uncovered on the 500 and 700-ft. levels of the Golden Centre is developing well and the management believes it persists to the surface. Unwatering of the lower levels is proceeding steadily and, as soon as conditions permit, the vein will be sought at a lower depth. The shaft will be sunk to 2000 ft. and the Peabody group developed.

The Grass Valley Consolidated Co. has put in a centrifugal pump on the 500-ft. level of the Allison Ranch mine and is rapidly unwatering the levels below this point. The pump has a capacity of 500 gallons per minute. Ore for the mill is being drawn from the 500 and 700-ft. levels. The 20-stamp mill is running one shift.——The smallpox quarantine on the mine and plant of the Columbia Consolidated, near Washington, has been raised. In the Ocean Star section of the group, excellent ore is being mined and sent to the mill. The shaft is down 300 ft. and good ore is exposed in the lower workings. ——The Washington Asbestos Co. has overhauled the old Fairview mill and will soon have it operating on asbestos. The flume-line has been repaired and everything around the mine placed in shape for a steady output. The asbestos is stated to be of fine quality; it will be shipped to San Francisco. Oakland and San Francisco people are interested chiefly.

The Black Bear Mining Co. is pressing work steadily at its gold property in the Rough & Ready district and is developing some promising ore. William Bucholz has been re-elected president and manager, and M. Shaver secretary.

Grass Valley, October 15.

SHASTA COUNTY

(Special Correspondence.)—The Afterthought Copper Co. at Ingot has shut down its oil-flotation plant while alterations and enlargements are being made. The operation for over a month demonstrated the success of the process on the zinc-copper ore. The shut-down will be for only two or three weeks. One hundred men are empolyed in the mine.

The Yankee John Development Co., operating the Yankee John mine, four miles west of Redding, contemplates building a cyanide plant. It has condemned its cannon-ball mill. An oil-flotation plant is contemplated. The company is making extensive improvements, having just put up a new head-frame and built a new shaft-house, boarding-house, and assay office. T. E. Graff of Redding is superintendent.

Redding, October 11.

COLORADO

CLEAR CREEK COUNTY

(Special Correspondence.)—The Primos Chemical Co. has resumed work on its molybdenum properties at Camp Boericke, near Empire. A contract has been let to S. Knowles to drive the main adit at the base of the mountain, 2000 ft. The adit is now in 500 ft. Grading is under way for a 200-ton mill. The milling machinery will consist of crushers, ball-mills, with tube-mills for re-grinding, followed by a series of flotation units. A. R. Nuimer is in charge of the work. An aerial tramway from the upper workings will be erected to serve the mill until the adit is completed, when electric hauling will be used.——The Central Colorado Power Co. is putting in a power-line to the mines to supply light and power for all purposes.——The Argo Leasing Co. is erecting a 50-ton concentrating mill to treat the silver-lead-zinc ore recently opened up.——The Sceptre Tunnel Co. has completed the re-timbering of its adit and the raise to the Sunburst adit is nearly finished. Several of the old stopes contain good ore. An aerial tramway from the mine to the railroad at the base of the mountain, a distance of 6000 ft, is to be erected. An electrically-driven air-compressor and drills will be added. J. H. Lewis is manager.

Georgetown, September 29.

IDAHO

BONNER COUNTY

The No. 3 adit of the Armstead Mines, Inc., on Lake Pend Oreille, was advanced 366 ft. in September. This is believed to have made a record for the district. It increases the length to 1748 ft. and adds 607 ft. since the middle of August. The adit was driven 26 days in September, or at the rate of a little more than 14 ft. per day of 24 hours, operations having been interfered with on two of the remaining days by the employment of the crew in fighting a forest fire. Two drills were run on 19 days and one drill on 7 days. Two machine-men and a helper, four or five muckers, and a motorman were employed on each shift. One man prepared the powder for all shifts and performed other work, and the steel was sharpened by one crew, using a Leyner machine. The muckers extended ventilating pipe, the air and water lines, track, and ditch.

MONTANA

SILVER BOW COUNTY

(Special Correspondence.)—The North Butte Mining Co. has declared a dividend of 25c. per share on 430,000 shares of the outstanding capital stock of the company; the Barnes-King Development Co., a dividend of $40,000, being 10c. per share, and the Butte-Bullwhacker Mining Co. a dividend of $10,000, being 1c. per share.——At a meeting of the stockholders of the Montana-Canadian Oil Co., held at Butte this week, the stock, which is owned mainly in Montana, was turned over to C. F. Kelly and associates of the Anaconda Mining Co. The action, which is described as a sale, amounts more nearly to a merger of the Montana-Canadian with more powerful financial interests; the Montana-Canadian stockholders receiving their pay in stock in the new company, which will be organized as a foreign corporation. In taking over the Montana-Canadian, Mr. Kelly and his associates are said to have in mind plans involving an expenditure of between $10,000,000 and $15,000,000 for development of the gasfield at Sweet Grass, Montana, and construction of pipe-lines to Great Falls, Helena, Butte, and Anaconda. The gas is to be used for lighting and heating purposes, and also at the smelters at Great Falls and Anaconda. The new corporation will be organized by C. F. Kelly and will be capitalized at $15,000,000. The oil and gasfield, which will come under the control of the Anaconda Copper Mining Co., comprises 120 square miles near the town of Sweet Grass. There are two flowing wells with a capacity of 8,000,000 cu. ft. per day; one well being 1850 ft. deep, the other 2500 ft. In this field 1100 to 1300 ft. of shale is found on top of the gas-sand. It is said that this field is an exension of the Bowisland range, in which is found wells that have been flowing for 28 years.

Butte, October 6.

NEVADA

LANDER COUNTY

(Special Correspondence.)—The Gold Top Mines Co. is erecting a small mill at its group of gold-mining claims near Galena, and expects to have it in operation by November 15. Sufficient ore is exposed to insure a long period of production.——The Nicholas Mining Co. is preparing plans for a cyanide mill to treat its silver-lead ores.——The Antimony Silver Mining Co., controlled by Salt Lake people, is operating a small mill on gold ore.

A 3-ft. vein, assaying $100 in silver and lead, with a little gold, has been opened by John Clifford, John T. Martin, and O'Toole Bros. near Washington canyon, south of Austin, in the San Juan district. An examination of the deposit is being made on behalf of Eastern capital by J. L. Bryson.

Austin, October 14.

MINERAL COUNTY

(Special Correspondence.)—Several properties in the Luning field will soon be shipping to the enlarged sampling-plant, just completed at Hazen by the Western Ore Purchasing Co. The plant has a capacity of 900 tons daily, and, wherever possible, has been equipped with automatic machinery.——The R. B. T. Co. has arranged for a shipment of two carloads of $40 gold-silver ore to the Belmont mill, Tonopah, and expects to ship steadily henceforth. From 5000 to 7000 tons of $20 ore has been opened on the Cevita and Abe Lincoln claims in addition to a large quantity of shipping product averaging $40 per ton. Connections have been established between the Meyer adit and Edith shaft, and driving is proceeding on the main vein, which has been cut at various points along its strike. At the adjoining Luning-Idaho group teams are being assembled for transportation of copper ore to Luning, and from there it will be sent to custom smelters by rail. The shipping product ranges from 8 to 25% copper and carries some silver and gold. The Stockham adit has advanced 100 ft. in low-grade oxidized ore and penetrated the sulphide zone at a depth of 100 feet.

Some excellent silver-lead ore has been uncovered in the June Bug. The property is controlled by W. E. Casson and George Hedges of Carson City, and San Francisco associates. The Packsaddle group of gold-silver claims has been examined by San Francisco people. Good ore is developing in the Mountain View and the small mill is running steadily.

On the 180-ft. level of the Nevada Rand, six to eight feet of $80 silver-gold ore has been opened by a cross-cut in the foot-wall. Driving for the main orebody is proceeding from the 350-ft. level.

Luning, October 14.

NYE COUNTY

(Special Correspondence.)—The Liberty Mining Co. is erecting a 10-stamp mill at its gold property near Willow Creek. Milling-ore has been exposed at several points, and small quantities of high-grade quartz. Driving of an 80-ft. adit has begun to tap the main ore-shoot.

Tonopah, October 13.

(Special Correspondence.)—The Tonopah Mining Co. the past week produced 2200 tons of ore. At the Silver Top, Sandgrass, Mizpah, and Red Plume shafts 225 ft. of development was made. At the Silver Top shaft stoping on the smaller veins furnished considerable tonnage.——Cross-cutting south on the 1680-ft. level at the Victor shaft of the Tonopah Extension Mining Co. has not been resumed on account of the flow of water encountered last week. The vein that was intersected appears to have a flatter dip and a different strike from the Merger vein, but may prove to be that vein. The ore assays about $20. In the Victor territory, 167 ft. of development was made and 68 ft. in No. 2 shaft. Last week's production was 2380 tons. The bullion shipment for the second half of September consisted of 25 bars valued at $53,500.

The Tonopah Belmont will dismantle the reduction plant at Millers. A part of the equipment will be sent to the Eagle-Shawmut mine in California, and the remainder to the Belmont-Wagner Mines Co. at Telluride, Colorado. A sampler and conveying system are being added to facilitate handling of custom ore by the Belmont mill. Cross-cutting from the 888 east drift has been started to pick up the faulted segment of the Shoestring vein. The intermediate east and west drifts from stope No. 3 on the Belmont vein continues on a full face of fair-grade ore. The tonnage the past week was 2320 tons. ——The West End mill shipped 31 bars of bullion, valued at $56,333, which represents the clean-up for the last half of September. In the Ohio territory, cross-cut 512 continues in high-grade ore. The hanging wall of the vein has not been reached. Drift 526 continues in excellent ore, while drift 525 is in low-grade material. Raise 529 is in the foot-wall of the vein. The shipment the past week was 967 tons.

As the Jim Butler Tonopah Mining Co. ships its ore to the Millers plant of the Tonopah Belmont, the Belmont mill will treat the Jim Butler ore when the plant at Millers is dismantled. The shipment the past week was 700 tons.

The McNamara Mining Co. has confined its stoping operations to the 700-ft. level, stope 1, where a 5-ft. face of ore is exposed, and stopes 1 to 4 on the 300-ft. level.

Stoping on the raise from the 950 to the 1050-ft. level in the North Star has progressed in good ore. About 50 tons of ore was sent to the West End mill.——Last week the Rescue-Eula produced 133 tons of ore, the Montana 239 tons, and miscellaneous 5 tons, making the week's production at Tonopah 8944 tons with a gross value of $156,520.

At the Kanrohat property in the Round Mountain district,, recently purchased by S. H. Brady & Co., the stamp-batteries are being replaced by ball-mills and combined cyaniding and flotation will be used. Jay A. Carpenter will be in charge of the mill operations.

In the issue of September 29 it was erroneously stated that the adit at the Con. Spanish Belt Co. has discovered ore. Only a few small stringers of low-grade quartz have been cut by this adit.

Tonopah, October 15.

WASHOE COUNTY

(Special Correspondence.)—A recent discovery was made at the bottom level of the Wall Street mine at Luning when a short cross-cut was driven into the foot-wall and a 4-ft. body of high-grade ore was opened which averages 30% copper. The company has a force of 75 men, and is doing extensive development as well as producing ore.——A winze has been sunk 20 ft. on the high-grade orebody opened up on the 1000-ft. level of the Calavada Copper Co. with no change in the character of the ore. A cross-cut is being run from the 1100-ft. level, to get under this ore, and a connection will be made by putting up a raise.

The main working adit of the St. Patrick mine, owned by the Kirchen Mines Co., is in 750 ft. and has found high-grade shipping ore, although from surveys made of the ore-shoots in the upper workings it was not expected to reach this ore for another 25 ft. Near the face of the adit a raise has been put through and connections made with the upper workings; this will do away with hoisting to the surface from the upper levels. Development has been started on a 6-in. stringer that was found in a body of iron gossan 250 ft. from the mouth of the adit. In 10 ft. this stringer has widened to 3 ft. of high-grade ore which assays from 10 to 12% in copper.

Reno, October 11.

NEW MEXICO

SOCORRO COUNTY

(Special Correspondence.)—The Socorro M. & M. Co. clean-up for the last half of September amounted to twenty-one 100-lb. bars of bullion. Twenty men are now at work on the road to Whitewater power-plant, which is being put in condition for heavy freight.

The governor, state engineer, and others have made an inspection trip over the new wagon-road between Mogollon and Socorro and it has been decided to put a large force of men to work and complete this road.

The Oaks Co. is now mining and milling $17 ore on the Deep Down mine, one of the central group. This adds another shipper to the district and makes the fourth property of the Oaks central group that is maintaining regular production. Auto-truck transportation to mill is under consideration, as there are not enough burros in camp to handle the production.

Mogollon, October 9.

OKLAHOMA

OTTAWA COUNTY

A subterranean cavern lined with lead and zinc ore has been struck by the Admiralty Zinc Co., in its mine at Douthat. The cavern is 75 by 20 by 12 ft. and presents one of the most spectacular sights seen in many a day.

The Montreal Mining Co., at Douthat, is shipping 350 tons of ore per week. This is an old mine that has been re-opened by Canadian capitalists who have equipped it with a modern plant. Comfort for the men, in the way of shower-baths and comfortable bunk-houses, is a pleasing feature.

OREGON

DOUGLAS COUNTY

(Special Correspondence.)—The Elkhart quicksilver mines, nine miles south-east of Yoncalla on the Southern Pacific railroad, on which $75,000 was expended in development and plant fifteen years ago, was taken over early in 1917 by E. P. Perrine and H. L. Marsters of Roseburg. The plant has been reconstructed and the mine is being re-opened. The ore is said to run 10 to 12 lb. of quicksilver per ton.

Roseburg, October 6.

JOSEPHINE COUNTY

(Special Correspondence.)—J. M. Finch and T. P. Johnson are mining chrome ore near Kerby and will begin shipping soon. Their property is west of Kerby, across the Illinois river, and, there being no bridge at that point, they are endeavoring to get as much ore as possible before the rains come.

Kerby, September 26.

UTAH

Earnings of Utah Apex Mining Co. for the three months ended with August were as follows:

	Receipts	Expenses	Net
une	$88,345	$60,652	$27,692
uly	144,928	83,033	61,894
ugust	215,932	110,013	105,919
Total	449,206	253,699	195,506

The consistent increase in tonnage mined, which is now veraging approximately 550 tons of crude and milling ore per ay, has not been secured through the sacrificing of development work, the ore-reserve today is in excess of the amount in ight at the beginning of the year.

Cash on hand on August 31 amounted to $449,000, not including $100,000 American Foreign Securities 5% bonds and 30,000 United States Liberty Loan 3½% bonds.

The May Day mine at Tintic is showing interesting development on the 700-ft. level, according to J. C. Dick, the manager. n a level being run in the east end of the mine, the vein has widened to about three feet, showing a streak of ore averaging 5 oz. silver and 20% lead per ton.

The mines of Alta were never more active than at the present time. The output of six mines alone is close to 250 tons er day. The Sells, South Hecla, Michigan-Utah, Columbus lexall, Emma, and Peruvian are all contributing largely to he camp's output. The Emma Con. is producing 65 to 70 tons er day, the ore running $35 to $80 per ton.

TOOELE COUNTY

Approximately 5000 tons of ore was shipped from Gold Hill tation in the Deep Creek district during the month of September. Of these all but five were sent out by the Western 'tah Copper Co; the Pole Star shipped three cars, the Garrison Ionster one, and one came from the Copperopolis. Most of he ore shipped by the Western Utah Copper averaged between 5 and 5% copper, with 5 to 6 oz. of silver per ton. The Pole tar company shipped two cars of copper-silver ore and one of smuth and copper. The Copperopolis shipment was second-'ade ore which carried about 9% copper, 10 oz. silver, and 0.50 gold per ton. The shipment from the Garrison Monster as low-grade lead ore.

Ores carrying rare metals have been reported by the Semi-ole Copper and Wilson Con. companies of the Deep Creek strict. The shipment included 20 tons of molybdenite ore d 10 tons of tungsten ore from the Seminole property and tons of bismuth ore from the Wilson Consolidated.

WASHINGTON

FERRY COUNTY

(Special Correspondence.)—The ore shipments from the public to smelters during the week ended September 28, re as follows: Knob Hill mine 150 tons; Lone Pine, by rthport Smelting & Refining Co., 650 tons; Quilp mine, by sees, 550 tons; Last Chance mine, by Lone Pine Consolied Mining Co., 90 tons.——In the Last Chance mine a drift in progress from the bottom of the new 500-ft. shaft to nect with the old stopes. The old workings are full of ter. The drift is within a few feet of old workings.——In Lone Pine 50 men are employed.

Vork has been delayed in the Knob Hill mine by a shortage listillate for the hoist and compressor. Eighteen men are ployed on two shifts. The cross-cut from the new 200-ft. ft is within 40 ft. of the main Knob Hill vein. When this ut a raise will be carried up to the old stope and the cross- will be continued to the rich and smaller vein.——The kings in the Princess Maud mine below the adit-level are g drained.

Republic, October 6.

PERSONAL

Note: The Editor invites members of the profession to send particulars of their work and appointments. This information is interesting to our readers.

J. S. Brown is in San Francisco.

James M. Hyde has returned to San Francisco.

L. D. Ricketts is at Santa Barbara, California.

C. F. Tolman, Jr., has been at Grass Valley, California.

Thomas Neilson is on his way to the Philippine Islands.

W. deL. Benedict has returned to New York from San Francisco.

Phil Lauzon has joined the 164th Depot Brigade, and is at Camp Funston, Kansas.

J. S. Harris has joined the 164th Depot Brigade, and is at Camp Funston, Kansas.

Harvey D. Miller has joined the 164th Depot Brigade, and is at Camp Funston, Kansas.

Richard Roelofs has resigned as manager for the Cresson Consolidated Gold M. & M. Company.

K. Ikeda, chief metallurgist for Fujita & Co., passed through San Francisco on his arrival from Japan.

George Bronson Rea, of the 'Far Eastern Review', is in San Francisco and is likely to open an office here.

E. L. Hermann has become assistant superintendent of the Hancock Consolidated M. Co., at Hancock, Michigan.

Myron L. Fuller has completed a summer's fieldwork in West Virginia for the Associated Geological Engineers.

W. G. Haldane has resigned from the Colorado School of Mines to accept a position as superintendent of a potash plant in Nebraska.

P. K. Lucke has resigned as consulting engineer for the Cia. de Minerales y Metales in Mexico, and has opened an office at San Antonio, Texas.

Charles S. Herzig was married on October 13 in San Francisco to Miss Florence F. Upmeyer of Minneapolis. Mr. Herzig and his bride have gone to Salt Lake City.

Erle P. Dudley, Captain at the Engineers Training Camp, Vancouver, Washington, has been relieved from duty and been assigned to the 20th Regiment, U. S. Engineers, at Washington, D. C.

Courtland E. Palmer, for a number of years one of the most active operating and consulting engineers in this country, died at his summer home in Arlington, Vermont, September 21. Mr. Palmer was born in New York in 1857. He was educated at Loyola College, Baltimore, at the Johns Hopkins University, and subsequently graduated from Columbia School of Mines, in 1878. Starting in the engineers corps of the Pennsylvania company of Pittsburgh in 1879, he later became assistant engineer of the New York Board of Health. In 1884 he removed to Colorado where he filled many important positions. He operated well-known properties at Aspen, Rico, Cripple Creek, and other districts in Colorado. He was general manager and vice-president of the Mollie Gibson Con. M. & M. Co. at Aspen, Colorado, during the time this property made its famous production. In 1903 he was appointed general superintendent of mines of the Guggenheim Exploration Co., and of the American Smelters Securities Co., and later became consulting engineer of the American Smelters Securities Co., which post he filled for several years. In 1910 Mr. Palmer retired from general professional work and devoted his energy to the mining interests in which he was personally concerned. He left a wide circle of friends who recognized his accuracy in details, his sterling integrity, and soundness of judgment.

Recent Decisions

Claim Lines Defined—Extralateral Right

A mining contract required the conveyance by the Quilp company of "all orebodies of every description lying east of the west line of the Quilp claim extended down vertically and lying south of the north end-line of the Quilp mining claim extended down vertically and which may apex within the lines of the Surprise lode mining claim." See Diagram. The Surprise vein extended diagonally south-easterly on the dip beneath the Quilp claim.

Held, the north end-line of the Quilp claim, as referred to in the contract is the line *CD* (the Surprise claim being senior

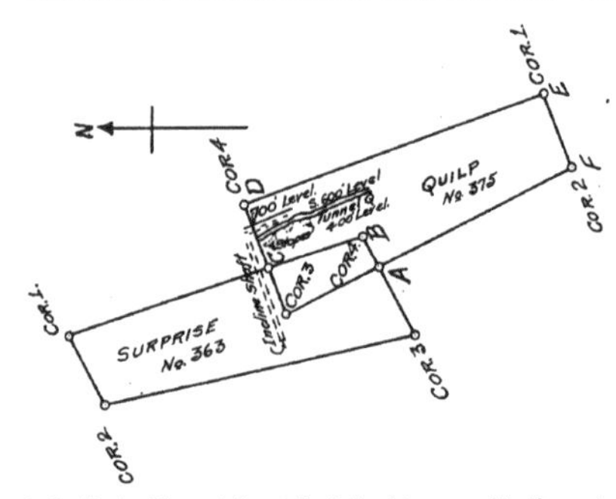

in location). The west line of the Quilp claim as used in the contract refers to the line *FABC*. No rule exists in mining law against broken side-lines. The end-lines alone must be parallel. It follows that the Quilp company must convey all orebodies apexing in the Surprise claim and extending easterly of the west side-line of the Quilp and south of the north end-line as defined above.

Quilp Gold Mining Co. *v.* Republic Mines Corporation (Washington), 165 Pacific, 57. May 19, 1917.

Review of the Alice Case

Sale to the Anaconda Copper Mining Co. of Montana of the Alice Gold & Silver Mining Co. of the same State was upheld on October 1, 1917, bly the U. S. Circuit Court of Appeals. The decision sustained by the U. S. District Court of Montana, upon an appeal made by the minority of the stockholders of the Alice Gold & Silver Mining Co. The property of the Alice Gold & Silver Mining Co. consists of the Alice, Magna Charta, Curry, Valdemere, Rooney, Hawkeye, Reef Fraction, Magnolia, Plover, Sankie West, Rising Star, Cotton Wood, Boston, Wood Yard-Gusset, Walkerville, Midnight, Roy Walker, Blue Wing, and Neptune lode claims, three-fourths interest in the Sankie East and Paymaster claims, the Alice mill-site, and a great many lots. This property is in the northern part of the Butte mining district. After the Anaconda company originally had made the purchase of the Alice Gold & Silver Mining Co., suit was brought by Peter Gedes, Joseph R. Walker, and other minority stockholders to have the same annulled and a new sale ordered. After a hearing Federal Judge George M. Bourquin gave his decision, in which he said: "It is apparent that both parties contemplate that in the event of no greater bid than the upset price, it shall be not optional but obligatory upon the Anaconda to pay Alice stockholders entitled thereto, in money. The Court does not attempt to define who is entitled to the benefit of these proceedings. Of course, when a fund comes into the hands of a Court from corporate liquidation all stockholders share therein and are properly noticed to appear, but here is no such case. Here, if no sale is made, all stockholders are not entitled to be paid in money, but only those who did not consent to or acquiesce in the sale, or who are not otherwise estopped. It is their duty to come in, to, in effect, take a place as litigants, and surely not the Court's to seek them out and solicit them to do so. The suit is pending nearly four years, ample time for any of them to appear. At the same time, in view of the nature of the suit, the special relief sought, and the perhaps unforseen determination, it is believed a limited time should still be allowed for any stockholder entitled to appear and prove his right to share in the relief granted."

When the first sale was made the stock named in the price was the old stock of the Anaconda. The present stock represents twice the par value of the old stock. Under the terms of Judge Bourquin's order directing a re-sale of the Alice Gold & Silver Mining Co., there being no bidder at a special sale before John Lindsay, November 10, 1915, the property reverted to the original purchaser, the Anaconda company. The consideration given by the Anaconda for the property was 30,000 shares of Anaconda stock, the value of which was placed at $1,500,000 by the Court, and the assumption of the debts of the Alice company, amounting to about $40,000. The lowest price which could have been accepted by the master under the Court's decree was $1,904,391.07, had there been any bidders. Previous to the sale the Anaconda company placed in the hands of the master a certified check on the Daly bank, of Butte, Montana, for $190,439.11, the 10% of the lowest price allowed, and which was required as a condition precedent to any bid being made or received. No other deposit was made. When 'no sale' was announced, the check was returned to John Gillie, general manager for the Anaconda company.

Extra-Lateral Right—Burden of Proof

The presumption is against the validity of the extra-lateral claims of plaintiff to orebodies underlying the surface boundaries of another claim until he proves by a preponderance of evidence the continuity of his vein and the identity of said orebodies as part of that vein. Not only are the defendants *prima facie* entitled to all ore beneath the surface of their claims, but, as a working hypothesis, it is fair to assume, in the absence of a contrary showing, that the vein or veins there found will continue to extend upward at the same angle as exhibited below. In any event, ore presumptively belonging to the defendants, because beneath the surface of their claims, cannot be taken from them on the mere opinion of witnesses for plaintiff that the ore has its apex in the plaintiff's claim. The presumption in favor of the overlying surface owners "is not overturned by speculative conjecture or intelligent guess." Plaintiff's failure to furnish substantial evidence of the identity of the vein by working down from their apex to the disputed territory held fatal.

Barker *v.* Condon (Montana), 165 Pacific, 909, May 28, 1917.

Coal Lands—Surface Support

The owner of coal lands conveyed the surface, reserving the right to remove the coal without liability for damage to the surface. Subsequently he conveyed the coal in place without expressly granting the right to remove it without liability for surface-support. Held, that the grantee of the coal was liable for the support of the surface.

Penman *v.* Jones (Pennsylvania), 100 Atlantic, 1043.

METAL PRICES

San Francisco, October 16.

Aluminum-dust (100-lb. lots), per pound	$1.00
Aluminum-dust (ton lots), per pound	$0.95
Antimony, cents per pound	16.00
Antimony (wholesale), cents per pound	14.75
Electrolytic copper, cents per pound	23.50
Pig-lead, cents per pound	7.25— 8.25
Platinum, soft and hard metal, respectively, per ounce	$105—111
Quicksilver, per flask of 75 lb.	100
Spelter, cents per pound	10
Tin, cents per pound	60
Zinc-dust, cents per pound	20

ORE PRICES

San Francisco, October 16.

Antimony, 45% metal, per unit	$1.10
Chrome, 34 to 40%, free SiO_2, limit 8%, f.o.b. California, per unit, according to grade	$0.50— 0.60
Chrome, 40%, and over	$0.60— 0.75
Magnesite, crude, per ton	$8.00—10.00

Manganese: The Eastern manganese market continues fairly strong with $1 per unit Mn quoted on the basis of 48% material.

Tungsten, 60% WO_3, per unit	26.00
Molybdenite, per unit MoS_2	$40.00—45.00

EASTERN METAL MARKET

(By wire from New York)

October 16.—Copper is unchanged and nominal at 23.50c. all week. Lead is easy and lower at 7.80 to 7c. Zinc is dull and easier at 8.50c. all week. Platinum remains unchanged at $105 for soft metal and $111 for hard.

SILVER

Below are given the average New York quotations, in cents per ounce, of fine silver.

Date		Average week ending	
Oct. 9	89.62	Aug. 28	88.50
" 10	88.87	Sept. 4	91.12
" 11	88.25	" 11	96.80
" 12 Holiday		" 18	100.70
" 13	86.34	" 25	107.66
" 14 Sunday		Oct. 2	98.49
" 15	86.25	" 9	91.72
" 16	85.34	" 16	87.01

Monthly Averages

	1915	1916	1917		1915	1916	1917
Jan.	48.55	56.76	75.14	July	47.52	63.06	78.92
Feb.	48.45	56.74	77.54	Aug.	47.11	66.07	85.40
Mch.	50.61	57.89	74.13	Sept.	48.77	68.51	100.73
Apr.	50.25	64.37	72.51	Oct.	49.40	67.86	....
May	49.87	74.27	74.61	Nov.	51.88	71.60	....
June	49.03	65.04	76.44	Dec.	55.34	75.70	....

COPPER

Prices of electrolytic in New York, in cents per pound.

Date		Average week ending	
Oct. 10	23.50	Sept. 4	25.65
" 11	23.50	" 11	25.83
" 12 Holiday		" 18	26.25
" 13	23.50	" 25	24.41
" 14 Sunday		Oct. 2	23.50
" 15	23.50	" 9	23.50
" 16	23.50	" 16	23.50

Monthly Averages

	1915	1916	1917		1915	1916	1917
Jan.	13.60	24.30	29.53	July	19.09	25.66	29.67
Feb.	14.38	26.62	34.57	Aug.	17.27	27.03	27.42
Mch.	14.80	26.65	36.00	Sept.	17.69	28.28	25.11
Apr.	16.64	28.02	33.16	Oct.	17.90	28.50	....
May	18.71	29.02	31.69	Nov.	18.88	31.95	....
June	19.75	27.47	32.57	Dec.	20.67	32.89	....

LEAD

Lead is quoted in cents per pound, New York delivery.

Date		Average week ending	
Oct. 10	7.80	Sept. 4	10.38
" 11	7.70	" 11	10.20
" 12 Holiday		" 18	9.33
" 13	7.60	" 25	8.00
" 14 Sunday		Oct. 2	8.00
" 15	7.50	" 9	7.96
" 16	7.00	" 16	7.52

Monthly Averages

	1915	1916	1917		1915	1916	1917
Jan.	3.73	5.95	7.64	July	5.59	6.40	10.93
Feb.	3.83	6.23	9.01	Aug.	4.62	6.28	10.75
Mch.	4.04	7.26	10.07	Sept.	4.62	6.86	9.07
Apr.	4.21	7.70	9.38	Oct.	4.62	7.02	....
May	4.24	7.38	10.29	Nov.	5.15	7.07	....
June	5.75	6.88	11.74	Dec.	5.34	7.55	....

Lead started the year in New York at a price a little over 7.5c. per lb., but immediately began to rise, reaching 10.5c. late in February and early in March. A decline carried the price down to 9.25c. early in April, but this was followed by a rise reaching 12.25c. in the middle of June. The first half of the year closed with lead at 11.25c. The average price of lead for the period was 9.9 cents.

At the average New York price the value of the lead produced from domestic ores the first half of 1917 was $54,751,000, and the value of that produced from foreign ores was $5,849,000, a total of $660,600,000, as compared with a total value for the year of $78,816,000 in 1916 and of $51,705,000 in 1915.

One result of the increasing imports of foreign lead is that a larger quantity of foreign lead is 'in stocks in bonded warehouse,' of which there was 19,517 tons on June 30, as compared with 12,369 tons at the close of 1916. Other stocks of lead are comprised in the ore that has necessarily been gathered preparatory to smelting at the new Kellogg smelter of the Bunker Hill & Sullivan Mining Co. and at the Deming plant of the Empire Smelting & Refining Co. The increasingly high price of silver and the high price of lead have greatly stimulated prospecting in the Western States and have led to the opening of new deposits and to the re-opening of silver-lead mines that were abandoned on account of the low price of silver. Though not adding to the sum total of domestic lead supply, the decreased use of lead pigments owing to their high cost leaves that much more of the domestic supply available for use as metal. The annual report of the largest producer of white lead says that the quantity sold in 1916 was the smallest for many years.

ZINC

Zinc is quoted as spelter, standard Western brands, New York delivery, in cents per pound.

Date		Average week ending	
Oct. 10	8.50	Sept. 4	8.25
" 11	8.50	" 11	8.21
" 12 Holiday		" 18	8.29
" 13	8.50	" 25	8.46
" 14 Sunday		Oct. 2	8.43
" 15	8.50	" 9	8.50
" 16	8.50	" 16	8.50

Monthly Averages

	1915	1916	1917		1915	1916	1917
Jan.	6.30	18.21	9.75	July	20.54	9.90	8.98
Feb.	9.05	19.99	10.45	Aug.	14.17	9.03	8.58
Mch.	8.40	18.40	10.78	Sept.	14.14	9.18	8.33
Apr.	9.78	18.62	10.20	Oct.	14.05	9.92	....
May	17.03	16.01	9.41	Nov.	17.20	11.81	....
June	22.20	12.85	9.63	Dec.	16.75	11.26	....

The price of spelter at St. Louis started at 9.75c. per lb. and after receding to 9c. at the middle of January it rose gradually to 10.75c. early in March. The price dropped to 8.9c. late in April, rose gradually to 9.4c. by the end of May, and closed the period at 9c. even. The average price of a pound of spelter at St. Louis for the first six months of 1917 was 9.8c. The foregoing prices are those of the ordinary commercial grades of spelter. High-grade spelter suitable for cartridge spinning has been in so great demand that it has brought a good premium. At the average price for immediate delivery at St. Louis the value of the spelter produced from domestic ores during the six months, if reckoned as prime Western grade, was $61,062,000 and that of the spelter produced from foreign ores was $9,721,000, a total of $70,783,000. Most spelter is sold for future delivery, and in the preceding years of the war the average price has been considerably lower than the average spot quotations. For the first six months of 1917, however, spelter for future delivery has brought within a half cent of spot spelter and in May and June commanded a small premium over spot spelter. Taking into consideration the high grade of the spelter sold, the average selling price of spelter must have been above the average quotation for spot spelter, and the total value indicated above must be somewhat under the real figures.

QUICKSILVER

The primary market for quicksilver is San Francisco, California being the largest producer. The price is fixed in the open market, according to quantity. Prices, in dollars per flask of 75 pounds:

Week ending

Date			
Sept. 18	115.00	Oct. 2	105.00
Sept. 25	105.00	" 9	105.00
		" 16	100.00

Monthly Averages

	1915	1916	1917		1915	1916	1917
Jan.	51.90	222.00	81.00	July	95.00	81.20	102.00
Feb.	60.00	295.00	126.25	Aug.	93.75	74.50	115.00
Mch.	78.00	219.00	113.75	Sept.	91.00	75.00	112.00
Apr.	77.50	141.60	114.50	Oct.	92.90	78.20	....
May	75.00	90.00	104.00	Nov.	101.50	79.50	....
June	90.00	74.70	85.50	Dec.	123.00	80.00	....

TIN

Prices in New York, in cents per pound.

Monthly Averages

	1915	1916	1917		1915	1916	1917
Jan.	34.40	41.76	44.10	July	37.38	38.37	62.60
Feb.	37.23	42.60	51.47	Aug.	34.37	38.88	62.53
Mch.	48.76	50.50	54.27	Sept.	33.12	36.66	61.54
Apr.	48.25	51.49	55.63	Oct.	33.00	41.10	....
May	39.28	49.10	63.21	Nov.	39.50	44.12	....
June	40.26	42.07	61.93	Dec.	38.71	42.55	....

Eastern Metal Market

New York, October 10.

The metal situation as a whole has not shown any life during the past week and all the markets are dull and weaker with the exception of tin. Uncertainties continue the dominant factor. What appears to be a settled matter today is reversed tomorrow and there is no stabilizing influence nor any basis for settled business.

Copper is chaotic and nominal.

Tin is more active and higher.

Lead is dull and weaker.

Zinc is again stagnant and nominally lower.

Antimony is quiet and unchanged.

Aluminum is inactive and lower.

New buying of steel is negligible still. The entire trade is awaiting the announcement of agreed prices on at least a hundred different steel products, and this may come this week. In the meantime prices continue to fall, 1000 tons of billets having been sold at $55 for October delivery, a decline of $5 per ton.

COPPER

The copper situation has been decidedly chaotic and still is somewhat so. The principal cause of this has been the virtual commandeering by the Government of all supplies of copper. This has been brought about by the fact that producers have not only been told not to sell at even the Government price of 23.50c. per lb., but also not to deliver any metal already contracted for. The object of this has been to conserve supplies until it shall have been determined how much copper will be needed by the Government and its Allies. It is now stated that delivery on contract may proceed since it is found that supplies over and above these needs appear reasonably ample. The report states that the requirements of the Government and its Allies will call for 120,000,000 lb. per month for the last quarter, or 360,000,000 lb. to January 1, 1918. While production in August and September was low, or an average of only 140,000,000 lb. for each month, against an average of 185,000,000 lb. per month for April and May, the estimated output for the last quarter is 200,000,000 lb. per month, leaving at least 80,000,000 lb. per month for domestic consumers, which is regarded as sufficient. A 'pooling committee' of producers, which has been holding daily meetings in New York recently, has arranged to meet a similar committee of consumers, and it is expected that the statistical data that will result from this meeting, and the better understanding of the entire situation as to supplies, and the requirements of the trade, will act as a stabilizing influence and clear the atmosphere. Business has been at a standstill. Small lots to needy consumers have gone at 26 to 28c. per lb. and higher, and in no case has lower than 26c. per lb. been quoted for first quarter metal. The London market is unchanged at £125 for spot electrolytic, and £121 for futures.

TIN

The tin market is the only one which has exhibited any life in the last week so far as sales are concerned. It has been fairly active until yesterday and Monday when actual buying eased off though inquiry was fairly good on Monday. Last week, on October 3, an inquiry for 125 tons appeared from one consumer but no sale is known to have resulted. Another consumer reported that he had bought 200 tons for prompt and nearby shipment, although apparently there were no sellers at that time. The situation was a mixed one, with leading sellers all reporting the situation dull. On October 4, however, sales of 300 tons by one or two sellers were reported, all for future delivery. Again, on October 5, there was a fair business in futures, with transactions totalling about 200 tons, also futures. Today the market is quiet and dull. The feature of the week since our last letter has been a distinct tightening in the present situation, especially as regards shipments from England and the Straits settlements. This may mean a hardship later for buyers of spot-metal, because the names of persons to whom the tin is to go is now required. Yesterday the quotation was 61c., New York, the Monday price having been 61.12½ cents.

LEAD

There has been only one bright spot in a very dull lead-market, which has been stagnant and uninteresting since our last report; this has been a demand of considerable proportions from England and Canada. Outside of this the demand has been as light as it is possible for it to be, with buyers holding off. Sellers are desirous of orders and have made concessions. Sales for November delivery have been reported as low as 7.50c., New York, with bids for December delivery at 7.37½c. The New York quotation yesterday had fallen to 7.80c. or 7.65c., St. Louis, but little business even at this figure was reported. It is evident that some consumers are expecting a lower price than 8c. to be fixed by the Government and a reduction of the American Smelting & Refining Co.'s quotation of 8c. would not be a surprise. It is an unconfirmed report that producers have agreed to furnish the Government October metal at less than the contract price of 8c. per lb. if the published quotations should show an average of less than 8c. for that month. This has had an unsettling influence.

ZINC

The market is exceedingly depressed. Filled with hope that the bottom had been reached two or three weeks ago, the market rose to 8.50c., New York, and a fair buying move was started, and it was not then expected that zinc would again fall to the 8c. level or lower; but today zinc is decidedly weaker. The quotation, regarded as nominal by some large producers, is now 8c., St. Louis, or 8.12c., New York, for prompt and October delivery, but demand is absolutely lacking. For delivery in November and December about 8.12c., St. Louis, or 8.37c., New York, is asked, but sales for this position are inconsequential. It is stated, but we have been unable to confirm it, that metal as low as 7.75c. has been offered by one producer. Today this offer is said to have been withdrawn.

ANTIMONY

There is no market. Demand is at a standstill. Chinese and Japanese grades are obtainable at 15c. per lb., New York, duty paid.

ALUMINUM

No. 1 virgin metal, 98 to 99% pure, is quoted at 40 to 42c. per lb., New York, but the market is inactive.

ORES

Tungsten: Quotations for the various grades of tungsten ore are unchanged, and quite a little business is reported as having been done at $23 to $26 per unit for 60% concentrate. European inquiry has appeared. For ferro-tungsten $2.35 per lb. of contained tungsten was recently bid on an inquiry, but was not accepted. Quotations range from $2.30 to $2.50 per pound.

Molybdenum and antimony: Molybdenum is held at $2.20 per lb. of MoS_2 in 90% concentrate, with $2.15 asked for the 85% grade, and a small volume of business is reported. There is nothing to report in antimony but quotations are unchanged.

AMONG the appeals to the mining public is a post-card issued by the Idaho Mining Association, calling upon those engaged in mining to show their patriotism by investing (in the purchase of Liberty Bonds) the amount saved through the enactment of the law exempting mining claims from assessment work during the current year and in 1918. That is a good argument. Gentlemen of the pick, it is up to us! Let us help to forge the sword that shall thrust the blond beast back into his lair.

UTAH COPPER is feeling the effects of war conditions and some not unnatural anxiety has been shown among its thousands of shareholders. The excess-profits tax is estimated to call for a contribution of $2,250,000, but the company is reported to have $15,000,000 worth of copper on hand in process of treatment, besides $19,000,000 in cash, so that even a large contribution to the National exchequer will not break its back. The big mining enterprises of the country are proving real pillars of strength not only in furnishing the metals requisite for munitions but as sources of revenue. We believe that the big-scale men in control of them will make no effort to evade their responsibilities, but, on the contrary, that they will respond to their duty with a proud heart.

SUBSCRIPTIONS to the Liberty Loan are being made at an accelerating rate as the time shortens. Our readers will have received various appeals to do their share. Many are the slogans sounded to stimulate patriotic participation. One of the best is that suggested by the Governor of California: "Go across with guns or come across with funds." Those of us that cannot perform the supreme duty of manhood can at least do the smaller service, loaning our money for the national purpose. We do not give the money, we simply exchange it for gilt-edged bonds at 4%, the safest and best investment, in a financial sense, that is available to the thrifty and loyal citizen. The President has set aside next Sunday as a day of prayer for the success of American arms in the War, in accordance with a resolution of Congress. Let us pray for the success of our great campaign, which is no chivalrous adventure but a vital effort to save our country from brutal aggression and organized piracy. Let us pray, but let us also give the outward and visible sign of fervor by furnishing the money urgently needed to complete the great task undertaken by our forces on land and sea.

MEXICO, according to an agreement made between Señor Ignacio Bonillas, the Mexican ambassador at Washington, and the United States Treasury department, has promised to remove the restrictions on the export of metals whereby a counter-importation of an equal amount of gold for all gold bullion exported was required and 25% on all the silver imported. In return, any money due on the balance of trade will be paid in gold. Under this agreement Mexico will receive $8,000,000 in gold during October, $5,000,000 in November, and $2,000,000 in each month thereafter, with a distinct understanding that Mexico must not ship the gold to other countries. Apart from facilitating the business of mining companies, the arrangement appears to be in the line of consummation of the plans to aid Mexican finance and thereby hasten the restoration of order.

PREFERENTIAL flotation is the subject of an illuminating article by Mr. W. Shellshear, printed in this issue. It has the particular merit of dealing with actual practice, and it represents the fruits of long experience by a devotee of the art. Mr. Shellshear is interested in the theoretical aspects of the problem, and he offers pertinent suggestions concerning the recondite causes for some of the effects observed in practical work, but he contents himself mainly with presenting the facts that have been learned empirically. His discussion of the details of the application of sulphur di-oxide as an agent for preferential flotation is the most exhaustive of any that has hitherto been published. It is interesting to be reminded that, when floating galena with the aid of sulphur di-oxide, no oil or organic substance is required, and he calls attention to the importance of sodium carbonate as a frothing-agent. Of special significance in connection with the possibilities of discarding oil and soluble organic material is his statement that the function of the frothing-agents is to intensify the physical conditions that promote flotation rather than to produce them. This serves to accentuate the need of a more thorough investigation of the physical principles upon which the process depends. The nearer we approach to an understanding of these fundamental conditions the closer will we come to emancipation of the flotation method from the impositions that now hamper the mineral industry.

THE supply of sulphuric acid has been taken under consideration by the Food Commission, and Mr. Charles W. Merrill has been placed in charge of the investigation. The United States Geological Survey has recently completed an examination of the sulphur resources of Louisiana and west Texas, and we understand that a pyrite survey will follow. No lack of raw material exists, and a competent organization, such as will now be effected, should render it available for the acid-makers. In case of extreme need the fume now going to waste at many smelters in the West could be utilized, but that entails further difficulties in obtaining the necessary tank-cars, and would throw new burdens upon the over-taxed railway systems of the country. It would be more simple and expeditious to draw upon the developed resources of pyritic ores in the mining districts of the Western States than to erect new acid-works and prepare for transporting the acid. Large tonnages of these ores can be had on short notice. Meanwhile, the abundant outcrops in the famous pyrite belt of Virginia should be promptly developed. Orebodies have been opened at intervals from the James river northward to the Potomac, and the indications point to enormous undeveloped resources. The proximity of this zone to the centres of sulphuric-acid manufacture gives it great importance at this time. The pyrite from Virginia contains about 3% of zinc and some copper. On this account it is of peculiar value for mixing with zinc-blende concentrate, whereby it can be cheaply roasted, primarily for acid-making, leaving a cinder from which the zinc is leached for use in the manufacture of lithophone. Many of the pyritic copper ores now penalized by smelters on account of the large zinc-content might also be utilized for acid-making, afterward extracting the metals from the cinder by leaching methods. These possibilities will not be overlooked by the committee over which Mr. Merrill will preside.

Russia

Mr. Horace V. Winchell contributes a timely article to this issue on 'Russia in War-Time.' We are all vitally interested in the fate of that huge country, for our own prospects of an early solution of the difficulties confronting us depend upon an unshaken Russian defensive. Mr. Winchell brings first-hand news of supreme importance to every American, for he can be relied upon as a competent and trustworthy observer. The picture which he presents is that of a country at the mercy of a rabble, for the most part good-natured and not given to malicious excesses. The central authority is unable to bridle the wild multitudes bent on enjoyment of liberty in the spirit of a debauch. The tendency to graft permeates all classes, and offers the largest resistance to effective co-ordination of the nation's resources and strength. Nevertheless, one splendid fact stands forth pre-eminent above the seeming dissolution of civic organization. There is a national purpose in the hearts of the people, and that is to withstand the onslaughts of Germany. Were this not so, the efforts of Kerensky would have been in vain, and the troops of the Kaiser would now be swarming into the rich fields of middle and southern Russia. There is a genuinely effective Russian resistance, despite the ceaseless dispatches telling of anarchy. That circumstance alone is proof that the Russian leaders have found a rallying point for the people, and are in control not only in name but in fact. If Germany could penetrate the eastern front, and scatter the Russian forces guarding even 50 miles of the long frontier, the resources of the richest territory in Europe would be at her command, and the War might continue for a decade. Indeed, with Russia conquered, the world might be forced to compromise and to accept an inconclusive peace. It is evident that Kerensky is maintaining a firm défence, and that also means that he has succeeded in organizing the resources of the country so as to bring up the needed munitions and supplies for the army. In the face of this, the attack upon Riga loses much of its importance. Success in the north, at the threshold of winter, when the Frost Giants come to the aid of Russia, is an equivocal advantage, and time is thus afforded for putting the country in order and preparing to hold back the Teuton from Moscow and the South. If the health of Kerensky does not weaken under the terrible strain there is hope that he may accomplish his purpose, with such aid as America can lend in money and material.

A Pro-American Policy

Adverse criticism of the Secretary of War and of other officials at Washington for allowing the obstructionist tactics of un-American advisers to frustrate the intent of Congress as expressed in the bill authorizing the construction of a great government nitrate plant, has borne good fruit. We hasten to congratulate Secretary Baker upon having overcome the obstacles that were placed in his path to obscure the facts and to induce a hesitant policy in order to fritter away time and opportunity as well as the funds appropriated. W[illegible] confess to sympathy for every high official of the Ad[illegible] ministration at this time of battling interests. The Wa[illegible] rages around them in subtle and confusing ways, th[illegible] force of which an outsider can but dimly surmise; ever[illegible] conceivable device is employed to divert them fro[illegible] courses that would most benefit the Nation; and if th[illegible] doubts cast before a departmental secretary achieve [illegible] more than to work delay, a gain has been scored by th[illegible] agents of the Enemy. At last, however, decisive actio[illegible] has swept away the pernicious net of plausible scientif[illegible] objections that had been woven about the nitrate projec[illegible]

The 'Nitrate Division' of the Ordnance office has stated that the original restrictions requiring that the site selected should be "somewhere in south-west Virginia" has been removed, and that the plant will be erected at Muscle Shoals in Alabama if a site can be secured at a reasonable price. We may count upon the people of Alabama attending to that side of the matter, even if they have to purchase riparian rights from private owners and present the land to the Government. The boards of trade of the chief Alabaman cities have gone on record regarding that question. The interesting point that has appeared in connection with the recent announcement of the Ordnance office is that the entire appropriation of $20,000,000 will be concentrated on the erection of an efficient plant as soon as the selection of a site at Muscle Shoals shall have been made. The original purpose of the Act will thus be conserved. The constituents who urged their congressmen to pass that bill were men who realized the dependence of this country upon Chile for the bulk of our fixed nitrogen, and they insisted, as a preparation against the emergency of war, that a substantial effort be made to eliminate this weakness from our military and economic situation, and they had no thought of delays in carrying the project into effect, nor of expending a large part of the sum appropriated in further technical experiments. The bill was meant to strengthen our defences immediately at a particularly vulnerable point. If processes were to be tested a special bill, carrying its separate appropriation, should have been introduced for the purpose. With our supply of nitrate for munitions and agriculture dependent on an easily assailable 7000-mile ocean highway, it is clear that our diplomacy must be tempered by recognition of that fact. The action of the War Department in this case is now in the direction that will help to stiffen the pro-American policy of the Government.

The St. Louis Meeting

The recent convention of the American Institute of Mining Engineers at St. Louis was a success so great as to surprise even its sponsors and so complete as to stultify the small group of small men in south-eastern Missouri that withheld the support to have been expected from all loyal members of the organization. We leave the recalcitrants to their chagrin and hasten to congratulate the President, Mr. Philip N. Moore, also Mr. H. A. Buehler, the chairman of the local executive committee, Mr. Walter E. McCourt, the secretary of that committee, Mr. H. A. Wheeler, chairman of the excursions committee, and Mr. Victor Rakowsky, chairman of the Joplin-Miami committee. To these gentlemen and to their numerous aides the success of the meeting was due, backed by an attendance larger than any recorded previously outside the headquarters at New York. Space will not permit us to give a full account of the proceedings, so we shall dwell upon the outstanding feature of the convention, namely, the patriotic meeting held on the afternoon of Monday, October 15. This was called to hear a report of the War Minerals Committee, but it developed into a forceful expression of loyal co-operation in the National effort by the Institute as a whole. Letters of regret were read from Sir Robert Hadfield, who referred to the passage of the American troops through London as "the greatest day in Anglo-Saxon history;" from Dr. L. D. Ricketts, mentioning his slowly recovering health as the only reason for being absent; from Mr. H. C. Hoover, who used his opportunity to make a plea for "saving, serving, and sacrifice." At the suggestion of the chairman, Mr. Moore, the audience honored the Food Administrator by standing during the reading of this letter. Thus early in the proceedings the audience showed itself keenly enthusiastic, and its enthusiasm was sustained throughout the session. Mr. Moore explained the objects of the meeting. Each speaker was introduced by him in felicitous terms, epigrammatic as those of a chancellor bestowing an honorary degree. Mr. H. M. Ami, the French-Canadian geologist, was present as a representative of the British legation at Washington. He spoke eloquently concerning the solidarity of the three great Allies, likening the French to the polished mica, the British to the gritty quartz, and the American to the interlocking feldspar that together composed the strong granite of our battle-front. He was followed by Colonel E. De Billy, the head of the French commission, an accomplished soldier, commanding a flow of scholarly English, which he used most effectively. He pointed at the fact that not armies but nations were in the field, along a line of trenches stretching for 750 kilometres and so continuous as to render outflanking impossible. Whereas previous battles, such as those in Manchuria, during the Russo-Japanese war, ceased when supplies of ammunition were exhausted, in the present war the organization of industry has assured a steady supply that guarantees uninterrupted fighting. Service at the front has undergone specialization "as in a well-conducted factory." The democratic life in the trenches has developed a fellowship that will have enduring results in helping to settle political difficulties. Fully 80,000 enlisted men have received commissions in the French army. Colonel De Billy then referred to the rehabilitation of the coal mines in northern France and acknowledged the offer of assistance already made by American mining engineers. When he ended his speech, the chairman, Mr. Moore, called for three cheers for France. They were given with gusto. Next came Mr. Fedor F. Foss, a mining engineer representing the Russian Ministry of Commerce, now on a special mission to the United States. He dwelt upon the extraordinary difficulties of his country. When hostilities began the mining and metallurgical societies of Russia combined and offered their united services to the Ministry of War, in vain. The various departments of War, Navy, and Commerce were of the opinion that any concerted effort to develop the national industries would be too slow to count in the War; the results were deemed too remote;

so nothing was done, until the Galician retreat in 1915. By that time the authorities had shown themselves bankrupt of ideas and the People's War Industrial Committee had been organized all over Russia largely under the leadership of Alexander Goutchkoff, who became Minister of War after the revolution, only to be driven out of office by the extremists. Now, however, the whole country is covered by a net of enquiry created by the War Industrial Committee in its systematic effort to mobilize the resources of Russia. This speech was received with obvious interest and sympathy. The chairman then explained the various efforts made by our own people to mobilize our metal and mineral resources; he spoke of the many good citizens "milling around" Washington, of governmental machinery that was designed for tasks other than those of war; he asked his hearers to make allowances, to be tolerant of early errors; he asserted that the work was being done better every day, that it should be a matter of pride to the mining profession that the one committee in which the Institute was officially represented had done so well, namely, the War Minerals Committee. He called upon Mr. F. W. De Wolf, the State Geologist of Illinois and Assistant Director of the Bureau of Mines, who gave a summary of the various committees and councils organized for war preparations, explaining the correlation of Federal and State activities, and the gradual adjustment of relations between volunteer agencies and established bureaus, all tending unmistakably toward patriotic unity of purpose. A clearance of supernumerary groups was already apparent, he said, and the definite crystallization of policy out of a previously fluid condition. He spoke with deep seriousness and was heard with sustained interest. Mr. Moore, when introducing Mr. W. J. Westervelt, the chairman of the War Minerals Committee, laid emphasis on the fact that in the files of the members of the Institute there is recorded an accumulation of knowledge concerning mineral resources such as is obtainable nowhere else; he urged those present to volunteer whatever information they could, especially concerning the distribution and supply of the minerals most needed at this juncture. Mr. Westervelt explained how the American Association of State Geologists started the campaign for co-operation by organizing a committee on mineral imports and how subsequently the directors of both the Geological Survey and the Bureau of Mines were approached successfully, so that finally a larger committee was formed by the appointment of representatives from the State Geologists, the Survey, the Bureau of Mines, the Institute, and the Mining and Metallurgical Society, thus constituting what is now known as the War Minerals Committee. Mr. Westervelt described the scope of the work, he appealed for information, particularly concerning sources of pyrite and manganese. It was necessary to stimulate domestic production, in order to replace foreign supplies now shut off by shortage of shipping and other causes. He suggested a modification of the technique of sulphuric acid manufacture so as to beneficiate deposits of pyrrhotite for the purpose. Then came Mr. David White, of the Geological Survey, an organization, as he said, possessing the largest body of information on American mineral resources. As part of the assistance given to the War Department he instanced the mapping of the Mexican frontier and the coastal regions. This mapping is proceeding at the rate of 100,000 square miles per month. A school for training topographic engineers has been organized and a detachment of them has been sent already to France. The engraving department of the Survey has reproduced countless charts for the Navy and the War Colleges, as well as others to be studied by the men in training for aviation. Engineers formerly engaged in foreign exploration have given the Government their unpublished maps for the use of the Allies. Data concerning water-supply near camps and exploration for the alloy-metals represent other branches of activity. Every effort is being made to obtain a domestic output of all classes of necessary minerals in order to release vessels for the carriage of food and munitions. Mr. A. G. White, of the Bureau of Mines, secretary of the War Minerals Committee, outlined the scope of the work outside the principal metals of industry; he referred to the fact that in 1916 we received 800,000 tons of manganese from Brazil, as against a domestic consumption of 900,000 tons, whereas in 1917 our own production of high-grade ore was only 80,000 tons. We received 1,225,000 tons of pyrite from Spain in 1916, as against only about 400,000 tons to be imported in 1917. The next speaker, Mr. W. O. Hotchkiss, the State Geologist of Wisconsin and secretary of the Association of State Geologists, was introduced as the one chiefly to be credited for starting the good work on War minerals. He said, "War is not an objective thing; it is my war and your war; if you realize that, you will furnish all the information that you have for the benefit of all of us in winning this war; we are tackling the biggest job we have ever undertaken; our enemy is utilizing all his resources; we are less organized than he—by patient plodding research he has built up a complete structure for waging a campaign meant to place us at the feet of an arrogant autocracy." This evoked loud applause. Mr. Hotchkiss proceeded to state that the Government will take a quarter of our lead production; the Navy department has advertised for paint in excess of 10% of our entire production of mercury. Each man has his responsibility to do his share. Let every engineer do so even before he is called upon. Mr. Moore, as chairman, closed the meeting in a few appropriate remarks and asked for the names of men ready to go to Washington on special service.

This is a sketchy outline of the proceedings at a meeting that will long be remembered by those privileged to be present. Throughout it was marked by sustained interest and keen enthusiasm. At later sessions of the War Minerals Committee a large mass of useful information was given and recorded for the benefit of those in authority at Washington. We believe that much of it will prove of first-rate value. Beyond the mere collection of useful data there was evoked a feeling of patriotism and of national service that has done more than anything else in its history to justify the right of the Institute to speak for the profession.

Economic Disturbance of Silver

The disappearance from view of the tremendous totals of commercial transactions through the balancing of one credit against another, until only a small fraction remains to be settled with actual money, leaves the public quite unprepared to think in the proportions of the trade-volume when expressed in terms of currency. Suddenly the country sees Government securities mounting into the billions, and these represent a measure of commodities absorbed for the purposes of War. The great buyer in this case is not like ordinary buyers; Uncle Sam, engaged as he now is in the property and life insurance business for his people, is a consumer only. In business the buyer stands in the position at one and the same time of consumer and producer; he assembles supplies and he again issues them, having served either to benefit them in passing through his hands, or, presumably, to benefit the public as a distributor. In either case there is an income and an outgo of commodities, the difference between these being a charge for services rendered. When Uncle Sam buys copper and steel and nitrate and toluol, or beans and corn and potatoes, the goods simply disappear. When goods go into the Quartermaster's department it is the same as when they enter the wage-earner's kitchen; they have reached the hands of the ultimate consumer. Therefore, just as the wage-earner pays in coin and bills for the necessaries of life, so Uncle Sam must issue exchangeable currency for the things consumed in this vitally important insurance business in which he is now engaged. Having no material output to offer, the item becomes a debit acknowledgment from Uncle Sam to the producer that covers, not a balance, but the entire bulk of the goods taken. Hence, of necessity, the vast issues of Liberty Bonds stand as a promise to pay, not protected by a metallic reserve, but based on service rendered in protecting the national existence and on the hope of future peace and normal prosperity they are to promote. Thus we suddenly see a large part of the totals of trade massed before us in an aggregate of securities that soar into unbelievable and incomprehensible billions. For one thing, it is helping us to appreciate the magnitude of trade and the immensity of the interests that we have to protect. On the other hand, timid men are beginning to worry about the volume of metallic currency available, and a clever manipulation of the silver market, based upon a real augmentation of the demand both here and abroad, which forced the price more than 100% above the level of a year ago, awoke the slumbering ogre of bi-metallism to again disturb the foundations of exchange. It is fortunate for the financial sanity of the world that the quotations fell as soon as they did. Primarily our sympathies are with the metal-producer. It is to his interests that we are devoted; but it is not to the advantage of any silver miner to have his output so misused as to unsettle national credits by adding the strain of an uncertain medium of redemption, in consequence of which he would have to buy his supplies with a depreciated currency. As a fact, we suppose that the cost of producing silver has grown so that a doubling of its former value in terms of gold would represent an increase no more unwarrantable than the enhanced market value of the other metals. At 83 cents we believe the silver-producer is proportionately underpaid, and we would not be surprised to see another upward movement of the market.

To the bi-metallists, who so promptly appeared with their old fallacy, which the people of this country once rebuked with such unmistakable emphasis, it should be pointed out that if all the silver of Potosi and Cerro de Pasco once more could be collected into a national treasury, it would but feebly suffice to guarantee the national issues that would be needed to care for a credit-balance involving almost the total volume of the original credit. We must, then, if we can, think in terms of commodities, and recognize the dollar, whatever its substance, as but a token after all, assumed as a convenient evidence of an obligation to deliver merchandise and service. The commodities of the nation, and the labor of the people in producing them, are at the command of the Government. It must have them, and will have them; but the hardship upon the people will not be lessened by complicating the problem of ultimate liquidation through disturbing the monetary standard by adding the variables inherent in the value of silver to the lesser variables that exist in the purchasing power of gold.

Recently Prof. Irving Fisher has been quarreling with the gold standard, and fundamentally he is correct in saying that the true measure of exchangeable value is a unit derived from a commodity-index number. That is what money is always meant to accomplish; it represents a token of materials and services delivered in trade. Men have fixed upon gold because it is beautiful, heavy, extraordinarily indestructible, and because it has never been found in quantities that would admit of its general use as an article of utility on an important scale. The family plate is made of silver and not of gold—for a reason that does not depend upon the whims of governments and financiers, but upon the fact that there is enough silver available to go round, and to increase the brilliancy and joys of weddings for a long time to come. It has been said that rising prices meant a superabundance of gold, but this fails to explain the abnormal conditions that had developed even before the War, when foods rose in value for a decade while the scale of metal-prices continued almost unchanged. It is certain that a deluge of gold would produce the effect of depreciation of its purchasing power, but we are in no danger of discovering another Rand and Klondike over-night. Silver, however, would soon swamp the treasury at a statutory price of one dollar per ounce. If gold is going down, there is no need of increasing the destructive momentum by adding the weight of silver. Actually the limit of productivity for the mines of the Transvaal is now in sight; not many years will elapse before the annual output of gold will seriously contract; the gross production today is lessening, and the relative yearly addition to the world's supply is far below the increase in the volume of business transacted. We are not at all inclined to think that depreciation is the result of the existing accumulation of gold reserves; it is more likely to be found in the changed proportions between producers of raw materials and those engaged in manufacturing and other occupations that add nothing to the fundamental commodity-stocks of commerce.

The recent advance in the price of silver, which began in April, was a logical response to a growing use in coinage. The European nations were employing larger quantities, and the Oriental demand was more insistent than it had been in the last few years before the outbreak of hostilities. Upon our entry into the War it became necessary for the United States to coin more silver to be used abroad in connection with our military operations. It was more convenient than to use gold, and it kept our stock of that metal at home. The upward tendency of the silver market was deliberate but steady until the latter part of August, when the quotation stood at 90.75 cents. From that date the market was distinctly speculative, and trading became wild and unreasoning. A price of 108.50 cents was unjustified, and the sharp recoil has recorded the uncertainty of dealers. Regulation of the price would undoubtedly benefit the producers, but this could not be done except through an international agreement without affecting the gold standard and thereby deranging prices to an even greater extent than at present.

Exemption From Assessment Work

It is a pleasure to be able to print in full herewith the text of the Act exempting locators and owners of mining claims from the obligation to perform the customary annual labor for holding their claims. The exemption is clean-cut and absolute, without qualification of any sort, and includes not only the present year, but also the year 1918. As approved the bill was designated 'Public Resolution No. 12, 65th Congress (S. J. Res. 78), and it became law by the signature of the President on October 5. As originally introduced simultaneously in both houses of Congress the measure required certain other performances for the benefit of the Army and Navy, or for the National treasury, in lieu of the regular assessment work on the mining claims. The bill was accepted in committee during the last days of the session, and many congressmen even were unaware that it had finally passed both the Senate and the House. The difficulty experienced in trying to obtain a reliable and authentic statement may be judged from the fact that we received a positive announcement as late as October 15 from the private secretary of a distinquished senator that the bill had failed. The text is quoted in full from an official copy printed by the Government:

"Joint Resolution To suspend the requirements of annual assessment work on mining claims during the years nineteen hundred and seventeen and nineteen hundred and eighteen.

"*Resolved by the Senate and House of Representatives of the United States of America in Congress assembled,* That in order that labor may be most effectively used in raising and producing those things needed in the prosecution of the present war with Germany, that the provision of section twenty-three hundred and twenty-four of the Revised Statutes of the United States which requires on each mining claim located, and until a patent has been issued therefor, not less than $100 worth of labor to be performed or improvements to be made during each year, be, and the same is hereby, suspended during the years nineteen hundred and seventeen and nineteen hundred and eighteen: Pr*ovided,* That every claimant of any such mining claim in order to obtain the benefits of this resolution shall file or cause to be filed in the office where the location notice or certificate is recorded on or before December thirty-first, of each of the years nineteen hundred and seventeen and nineteen hundred and eighteen, a notice of his desire to hold said mining claim under this resolution: *Provided further,* That this resolution shall not apply to oil placer locations or claims.

"This resolution shall not be deemed to amend or repeal the public resolution entitled "Joint resolution to relieve the owners of mining claims who have been mustered into the military or naval service of the United States as officers or enlisted men from performing assessment work during the term of such service," approved July seventeenth, nineteen hundred and seventeen.

"Approved, October 5, 1917."

It should be said that any claim-owners who already have performed the statutory annual labor for the current year, either purposely or incidentally in connection with development work that would apply directly or through benefit conferred on adjacent claims, would do well to file affidavits to that effect, stating the amount and cost of the work done, as usual, since it advances the position of the claim toward patentability, and there is a bare possibility that such work later might be made to inure to the further benefit of the claimants through legislation covering a period beyond 1918. It is recommended that such affidavits be filed in addition to the notice of desire to hold the claim under the Exemption Act, and that this is to be done whether the State mining regulations make affidavits of annual labor obligatory or merely permissive.

The last paragraph of the new law expressly states that it neither amends nor repeals the previous resolution relieving enlisted men from the necessity of doing annual labor. This reservation is because the exemption granted by the two bills does not cover identical periods of time; for the general public the relief applies to 1917 and 1918 only, while for those who have joined the colors it extends throughout their term of service.

Russia in War Time

By HORACE V. WINCHELL

The total area of the Russian Empire is 8,600,000 square miles. Its population is estimated at 181,000,000. The total area of Europe is 3,800,000 square miles and the population about 450,000,000. The climate of Russia is continental and the variation between the summer and winter temperatures is great. The mean January temperature at Moscow is 54° lower than that of July. Central Germany has a temperature-range of 34°. The snow lies for many months over a large portion of the vast Muscovite territory. The annual rainfall in northern and central Russia is moderate, being from 20 to 24 in., but quite sufficient for crops, being distributed throughout the year. As one goes south the rainfall decreases rapidly, averaging from 8 to 12 in., and in the extreme south there is very little precipitation in the fall and winter, so that the period of vegetation is too short to permit the growth of trees. Thus we find the forests of northern Russia contrasted with the steppes of southern Russia, and this contrast makes itself felt in every department of economic and national life. The forest district begins at the Arctic circle, where the trees are coniferous, consisting largely of spruce, fir, pine, and birch, and extends southward to the latitude of Moscow and the upper waters of the Volga. The mixed forest region covers central and western Russia as far south as the Black Sea. In the Caucasus I saw the finest forests of white oak, white ash, Persian walnut. and beach, together with beautiful fir and hemlock.

The population of the coniferous belt varies from 1½ to 36 per square mile. In Finland, which is somewhat warmer, the population is 21 per square mile, but agriculture and stock raising are comparatively unimportant. The population of the mixed forest region is from 50 to 150 per square mile, exclusive of Moscow. The population of the Baltic region is from 60 to 80 per square mile. while Poland, which is extremely fertile, has a dense agricultural and industrial population amounting to 288 per square mile. The great agricultural region of Russia is found in the steppes. This is largely known as the 'Black Earth' region. It has a moderate rainfall and produces all kinds of grain, and large crops of sugar beets. The population is from 120 to 230 per square mile.

PEOPLE. Just as varied as the physical features of the country are its inhabitants. The different peoples have through long years of conquest become united to form one nation, but they are still far from being welded into one race. There are wide discrepancies in language, physical type, customs, dress, and religion. The two main races in Russia proper are the Indo-Europeans in the south-west and the Finns in the north-east, the Finns being an offshoot from the Mongolian race. Russians or Slavs of Aryan or Indo-European descent amount to about 75% of the population of European Russia. They are members of the Greek Catholic church, of which several sects exist, and they are divided into three stocks. The Great Russians, numbering about 52,000,000 in European Russia, live in territory from which they have driven the Finns, with which people they are somewhat mingled, and thus occupy northern and central Russia, including the north-eastern part of the 'Black Earth' region, and also territory in eastern and south-eastern Russia from which they have driven the Tartars. Physically they are blond, blue-eyed, and vigorous, with broad shoulders and bull necks, often somewhat clumsy, and with a strong tendency to corpulency. Their character has been influenced not only by a long history of subjugation to feudal despotism, but also by the gloomy forests, the unresponsive soil and the rigorous climate, and especially by the enforced inactivity of the long winters. In disposition they are melancholy and reserved, clinging obstinately to their traditions, and full of self-sacrificing devotion to Czar, Church, and feudal superior. They are easily disciplined, and so make excellent soldiers, but they have little power of independent thinking or of initiation. The normal Great Russian is thus the mainstay of political and economic inertia and reaction. Even the educated Russian gives apparently little response to the actual demands of living. He is more or less the victim of fancy and temperament, which sometimes lead to a despondent slackness, sometimes to emotional outbursts. Here we have the explanation of the want of organization, the disorder, and the waste of time, which strikes the Western visitor to Russia. This pessimistic outlook finds expression in the word which is forever on Russian lips—'Nitchevo.' 'it doesn't matter.'

The White Russians number about 6,000,000, and probably derive their name from the white color of their clothing. They occupy the provinces of Minsk, Mohilev, Vilna, Vitelsk, and Grodno. They are the poorest and least advanced of the three Russian stocks.

The Little Russians, numbering 20,000,000, are settled in the 'Black Earth' district or Little Russia proper, and in the Ukraine, which includes the provinces of Kiev. Poltara, Kharkov, and Tchernigov. They have also spread into Galicia and north-eastern Hungary, and have colonized in other directions. They are slender and dark. with the emotional southern temperament, and speak a dialect different from the other Russians.

The Cossacks are not a distinct stock, but are descended from the refugees and outlaws that occupied frontier districts between the settled and the nomadic tribes. They were afterward organized as a frontier militia and as a light cavalry. The Cossacks are found in the valley of the Don, in the Urals, and in Siberia.

About 8,000,000 Poles are found in western Russia. Other races included within this vast nation are the Letts, 1,400,000; the Lithuanians, 1,200,000; Germans, 2,000,000; Swedes, 370,000; Rumanians, constituting the bulk of the population in Bessarabia, 1,000,000; Bulgarians and Greeks, who are fairly numerous in southern Russia; Jews, 5,100,000, until recently not permitted to live in either Great Russia or East Russia, speaking a German dialect mixed with Hebrew, but also familiar with the Russian language; Mongolians, of whom there are said to be 9,000,000 in Russia proper, and many more in Siberia; 170,000 Kalmucks of Mongol blood professing the religion of Lama; the Ural Mountain people, 5,400,000, consisting of eastern Finns, which includes the Ugrians, the Permiaks, the Syryenians, the Samoyedes, the Volga Finns, the Votyaks, and the western Finns, amounting in all to nearly 4,000,000, including the true Finns, the Esthonians, the Lapps, the Tchudes, and the Livonians. There are also Mohammedan Turks, Tartars of various branches, Bashkirs, Kirghizes, and many others, including Georgians, Lesghians, Daghestanians, and more besides. It will thus be seen that, if in America we have a melting-pot of nations, in Russia there is a racial volcano from which eruptions may be expected for a long time to come.

ECONOMIC CONDITIONS. Russia, by virtue of its geographical situation, its natural features, and its history, is a land of raw products, although in the past 50 years the manufacturing industry has developed rapidly. Fully 80% of its population lives by agriculture, yet only 26.2% of the total area is under tillage, and 15.9% is devoted to gardens, meadows, and pastures. Forests cover 28.8%, and 19.1% is wholly barren. The manufacturing industry is as yet of minor importance. There is a great lack of native capital and of competent workmen. The products are limited to articles of common use for distribution in Russia and in central Asia. In Poland the predominant industry is the manufacture of cotton and other textiles. One important industrial district in central Russia extends on the south from Moscow to the coalfields of Kaluga and Tula, on the north to Tver and Yaroslavl, and on the east to Ivanovo and Vladismir. The chief manufactures here are textile and metal wares, together with wooden articles. The iron industry of southern Russia, where coal and iron are found together in the Donetz basin, has already materially decreased the imports from foreign countries, and a few years before the War it developed a small export trade. This district now has a production twice as great as the old and famous iron district of the Urals, where charcoal and lignite were used for fuel.

There are manufacturing establishments producing cotton, metal wares, machinery, and chemicals in the cities on the Baltic, such as Petrograd and Riga. In Finland the great abundance of timber and water-power support wood-pulp, paper, and other industries, and in the steppe-region are flour mills and beet-sugar factories.

The railroad system of European Russia is nearly as large as that of Germany. Considering mileage alone, 25,447 miles were open for traffic on January 1, 1912, but the ratio of mileage to area was one mile of railroad for every 59 square miles of territory, which is less than for any other country except Norway. The United Kingdom has one mile of railroad for each 5.3 square miles; Germany one mile for 5.8, and the United States one mile for every 15.7 square miles. The Russians use a broader gauge than the other railroads in Europe and this difference is a serious obstacle to through traffic.

The people of Russia, including the Jews and nomads, are divided into four classes: nobles, officials, clergy, and peasants, the latter including the laborers. The really sharp distinction, however, is that between the great mass of the people on one side and the hereditary and official nobility and the burgess class on the other. Alongside of the admirable achievements in all spheres of intellectual activity, we find also a great deal of merely outward imitation of Occidental forms with a tendency to rest content with a veneer of Western culture and a stock of Western catch-words. Side by side with the unquenchable desire for scientific knowledge, which shuns no sacrifice, and is constantly drawing new elements from the lower classes, there is only too often a total inability to put into practice and to make efficient use of what has been learned. Fancy and emotion are much more widely developed in the soul of the Russian than true energy and joy in creation. The upper classes are noted for their luxury and extravagance, and for their reckless gambling, their better side showing itself in their unlimited hospitality. The lower classes live in unspeakable poverty and destitution. Beggars are numerous and troublesome, especially in the vicinity of churches. At the present time conditions of living are most difficult. With the declining value of the Russian ruble, which is normally worth 51c., and which is now quoted at about 14c., with its actual purchasing-power still less, together with the extremely high prices asked for whatever small stocks of supplies still remain in the hands of the merchants, it is a wonder how the ordinary individual can support himself and his family. A pair of boots costs from 100 to 200 rubles, and can be obtained with difficulty even at such prices. Food of all kinds is scarce in the larger centres of population, and can be obtained only in limited quantity by means of bread-cards, sugar-cards, and the like. A lemon or a small apple costs from one to two rubles. A Russian pound of strawberries, which is nine-tenths of a pound avoirdupois, costs from two to four rubles. These strawberries, by the way, are the finest I have ever eaten. I saw canteloupe selling at 40 rubles; cucumbers, of which the Russians are very fond, at one to two rubles each. No white bread was to be had in even the best hotels in Petrograd and Moscow for a number of weeks before my departure. The Russian black bread seemed to be composed of tar and cobblestones. I broke two teeth upon this locally esteemed Russian delicacy.

The procuring of a railroad ticket is a matter of extreme difficulty, and it was not an unusual experience to wait two or three weeks, sending a man each day to stand in line at a total expense of perhaps 100 rubles, to find your space in the train occupied by soldiers who had taken possession and were going home for a vacation, or returning to the front, or deserters who had paid no fare whatever. It was even the experience of myself and wife to be obliged to stand up in the corridor outside of a first-class compartment all night long, while nine soldiers occupied, free of charge, the space for which we had waited weeks and had paid full rates.

The entire Russian people, men and women, are mobilized for war and are under military orders. Every person over the age of 18 must have a passport, and the men all wear uniforms and are drafted for military duty. Thus Russia has under arms the greatest number of men ever assembled in one country. From 15,000,000 to 18,000,000 men have been taken out of industrial life, and, regardless of their usefulness to the nation in the fields or factories, are now being supported in idleness. The quantity of man-power now rusting and doing nothing but eating and talking is appalling. During the last three months of my sojourn in Russia I saw hundreds of thousands of soldiers, but not once during that time did I see a soldier drilling, although I passed hundreds of barracks and drilling grounds, nor was my heart gladdened once by the sight of a squad of these men working upon the roads or engaged in useful occupation. It is true that many of the idle soldiers find ways of earning money. They act as porters at railway stations, asking ten times the normal fee, and perhaps absconding with the baggage entrusted to their care. They are the only purveyors of food and clothing when there is absolutely none to be had from the merchants. More than once we were able to buy from some soldier butter, bread, sugar, and other commodities when none could be had at the stores. I was informed of an instance in which a railway bridge was washed away, the rails and ties remaining,

PETROGRAD AND BRIDGE ACROSS THE NEVA

and a board-walk laid to permit the foot-passengers to cross. One lady engaged two soldiers to carry her baggage across. They agreed upon a price of 10 rubles. Upon reaching the middle of the stream the soldiers informed the lady that if she did not immediately pay them 100 rubles they would throw the baggage in the river. She finally compromised by paying them 65 rubles, all the money on her person, and got her baggage across. I had a similar experience with some horse-packers in the Caucasus. I engaged them for a week at a certain price. When we were a few days out they demanded more, threatening otherwise to leave me in the mountains without horses. I promised to accede to this, but when we reached the end of our journey I paid them the price originally agreed upon and no more.

It is not, however, the common people alone who are thus biassed as to their understanding of the sacredness of a contract. The Government owns the telegraph service and many of the railways. The managers of these public utilities are Government officials; yet it was the universal practice for the telegraph companies to exact three times the regular rate for a telegram, saying that there was great congestion and that a telegram must be rated as 'urgent' in order to receive attention. After accepting triple payment these Government officials would perhaps put a postage stamp upon the telegram, or would pay no further attention to it whatever. Occasionally telegrams from Petrograd to the Caucasus would come through in 10 days. Occasionally they arrived after three weeks and frequently not at all. Likewise with the railway officials. I paid several hundred rubles to have my baggage transported by railway from Petrograd to the Caucasus. On arriving at the Don river, where a bridge was out, I found thousands of soldiers fighting for a chance to cross upon the small steamers, and there was no possibility of getting my baggage across unless I took it myself. I was told that the bridge might not be repaired for a month, but that the baggage would eventually be forwarded. Having heard so many tales of the loss of baggage I took my trunks from the possession of the railway company on the west side of the river Don, and at an expense of 300 rubles and after much labor, I got it across the river and presented it to the officials of the same railway on the eastern side of the river. These officials informed me that having once taken my baggage from the custody of the railway, it was again necessary to pay several hundred rubles to get it carried to its destination.

I could relate many instances of Governmental graft that came within my own experience or that of my friends. Every merchant who has succeeded during the last three years in getting merchandise ordered for the Government or for private consumption from Archangel or Vladivostok to Petrograd will tell of innumerable instances in which he has been obliged to pay extortionate graft to the railway officials before he could get a car in which to forward his freight. At the time of leaving Russia factories were closing, and all kinds of business was rapidly coming to a standstill. The workingman has no conception of a limit to what he may demand. Pay has been increased and hours shortened again and again. Efficiency, which was never great, has fallen off to such an extent that no one can operate at a profit, notwithstanding the high prices commanded by his products. Take for example the cost of mining coal in the Donetz basin; before the War this may have been at most three rubles, or $1.50 per ton. At the present time the cost is not less than $8 gold, and is probably nearer $12. Manufacturers in this district are losing money. They are required by the Government to continue operating, and as a result they are borrowing from the Government to meet their payrolls.

Politically the situation in Russia is decidedly mixed. There are many different parties and no one can say which will secure the adherence of the mass of the people. These parties include the following:

First: Anarchists.

Second: Quite similar to the anarchists, but denying this appellation, since the object of the anarchists was to establish a Government without laws and primarily to overthrow the Czar's régime, which has been accomplished, are the Maximalists or Bolsheviki. These Bolsheviki are extreme socialists who demand confiscation of all kinds of property without compensation and the carrying on of all business by the Government.

Third: Minimalists or Mensheviki. These are socialists whose programme includes the taking over of all forms of business by the Government, and involves Government ownership of all lands now in private estates, but they do not demand that this be done without compensation to the present owner.

Fourth: The Social Revolutionists, whose programme is still somewhat more moderate.

Fifth: The Labor Group allied with the previous two, but concerned more with labor questions than with the question of land ownership.

Sixth: The National Freedom party, the 'Narodniki', or Liberal party, of which Miliukoff, Lvoff, and others are members.

Seventh: The Octobrists, or Monarchists, who believe that Russia is not yet ready for a democracy and who wish an autocracy to be re-established.

Eighth: The extreme right or supporters of the old Czar, Grand Dukes, and the rest.

Today the Socialists are in power, but not the most radical section. The Bolshevikis are the most dangerous and make the most noise. They are the disturbing element; they consent one week to the authority of the Provisional Government and the next week, because this government does not act in all respects in accordance with their demands, they revolt against it. The Liberal party has been forced into the background. Its members are some of the finest men in Russia, but they are persons of markedly Russian type, lacking in both determination and executive ability. These men had for years spasmodically striven to educate the people and to bring about political reform. When the revolution was accomplished they were appointed Ministers under a Provisional Government, and had it not been for the strong socialistic tendencies and propaganda of the soldiers and workingmen, there would have been established in Russia under their guidance one of the most advanced democracies of the world. Miliukoff is a splendid character. He was lecturer on international law at the University of Chicago. He has traveled. He is a magnetic speaker, and is one of the finest men in Russia today. Fearless and honest, he, it was, who at the risk of his life, denounced in the Russian Duma the former Prime Minister Stuermer who was accused of being a German spy and a traitor to Russia, and who is said to have died in Petrograd on the eve of his trial for treason. The former Minister of War, Sukhomlinoff, has been condemned to life imprisonment for betraying his country; and Protopopoff, former Minister of Railways, is still awaiting trial. These men were accused of being in the pay of Germany, and are said to have been directly responsible for the loss of hundreds of thousands of Russian lives.

Now, as to the future: Americans are interested to know what can be expected of Russia, first in the present War; second, as a land of opportunity after the trouble. It is my belief, although I realize that no one can speak with any assurance, that there is at present no general sentiment in Russia for a separate peace with Germany. The majority of the people do not favor it. I believe further that, although disorganized, Russia will continue to hold a line which will require the presence of large German and Austrian forces in opposition. I do not look for any great and general offensive by Russia, certainly not this year. No country in the throes of a revolution can at the same time vigorously conduct a foreign war. The wonder is that she has done so well. I do not believe Germany will get to Petrograd, nor will she capture large stores of food in Little Russia; but even the taking of Petrograd would not mean Russia's downfall or surrender. Russia is too great. There is too much territory and, as someone has said, "Russia is like a feather-bed—the further you push in the more you smother." Moreover, the Russians themselves have not that sentimental and affectionate feeling for Petrograd that they have for Moscow. Russia is no place today in which to set on foot new industrial enterprises. Labor is too scarce, too inefficient, and too expensive. Titles are insecure until we know whether or not the socialistic programme is to be carried out to its full extent. Taking a broad view, I should say that Russia is a land of opportunity. The favorable and unfavorable factors may be summarized as follows:

Among the favorable factors I would specify, (1) governmental and political, (a) the country is now on a democratic basis. It is the general belief that it will remain so and will become an advanced republic; the Octobrists and Czarists are in the minority, and only the unexpected failure of the efforts now being made to preserve the freedom of the people will make it possible for the monarchists to organize a successful counter revolution; (b) there are able men in Russia, men of character and of patriotism, and some of these are conducting its affairs. Kerensky is one of the strong figures in pubilc life today, and in determination and fixed endeavor he is a super-Russian. So long as he remains the guiding spirit the Russian people will be in excellent hands for the solving of their problems and the working out of their destiny; (c) the United States is looked upon as a sincere ally. The Russian people are proud to feel that they are politically on a par with us, and they wish to show us that they are worthy of our trust; (d) the War is sure to result favorably for us and for them. Whatever aid we, as a nation, may render to Russia will be appreciated, and our investments will be repaid tenfold in the satisfaction derived from the performance of a duty to a struggling people and in the industrial opportunities which will surely develop; (e) the old laws which were frequently impracticable and discouraging to foreign investors are now being revised and simplified. (2) Practical consideration of the favorable side of the case needs to take into account, (a) that the great territory of Russia and Siberia is the largest undeveloped domain. Other things being equal and favorable, it presents the largest field for exploration and the investment of capital; (b) it is in the north temperate zone, and covers a wide range of climate and latitude. Its products are as varied as those of the United States, both in mineral and agricultural wealth. Its population, which is already nearly twice as great as ours, is rapidly increasing; (c) whatever the form of its future government, the country is now on the eve of great development, and will furnish a market for all forms of products, especially those which can be manufactured and produced within its borders; (d) the people are friendly by nature, and especially friendly to Americans. They are also moderate, and will, I think, with a little education, be fair in their dealings with us; (e) prohibition prevails throughout the land, and there is no disposition to make a change in that respect; (f) although there is need of money for large undertakings, there seems, nevertheless, to be an abundance for the purchase of every-day necessaries and luxuries; (g) Russia no longer trades with Austria nor with Germany, which latter country formerly supplied the greater part of the Russian imports; (h) she no longer wishes to trade in a large way with England.

The unfavorable factors are the following:

(1) Governmental and political; (a) The present Government, which is frankly provisional, will exist only so long as it holds the confidence of the soldiers and the people. The organized force behind it has limited power for compelling obedience to its mandates. Under the guidance of patriotic men who seek only the common good, it finds itself obliged to decide between the selfish and conflicting demands of the various classes of society, many of whom, especially the peasants and laboring men, have not the slightest conception of the meaning of the words 'freedom' and 'democracy'. The ship of state is in perilous waters, and requires a skilful skipper. Among the temporary ministers are representatives of all classes with the possible exception of the peasants, and an occa-

sional clash of interests is unavoidable. The socialists demand the distribution of lands formerly held by the Cabinet and now in large estates (two different varieties of large holdings); the laboring men clamor for concessions and privileges which they believe are enjoyed by laborers in other countries; the tribes are demanding freedom and even rising in revolt to obtain it, although lacking entirely in the ability to govern themselves; and the soldiers have an idea that 'freedom' means that they can have everything without price, and, often in large numbers, are riding on street-cars and railroad-trains without paying fare; and while there is rejoicing over what has been accomplished, there is also unrest and apprehension for the future. (b) The concessions which have already been made to the Poles and Finns, without requiring any guarantee of future support and allegiance, are felt by some to constitute a menace. The same may be said of the priviliges thus early granted to the workingmen. (c) The old Russian laws, most of which remain in force, are cumbersome and inapplicable to business. Until they are revised, new enterprises will of necessity be undertaken with great caution. (d) The Jews, who are returning to Russia in great numbers, and who have long exerted an influence beneath the surface, are now becoming more outspoken and often arrogant. Their presence will be valuable in a way, for they are already acquiring property which formerly they were forbidden to hold, and will not favor the confiscation of real estate, but if they become too bold and prominent disturbances will arise from existing racial antipathies.

(2) Practical Difficulties. (a) The most obvious difficulty in the way of business at present is the lack of transportation. The railway systems are totally inadequate. Rolling stock is out of order, and little effort seems to be made to repair it. Repair shops are needed, but even if provided there would be no mechanics to fill them. There is a shortage of every conceivable commodity all over the Russian dominions. It is impossible to get freight forward. Most of the available cars and engines are needed for transporting troops and war-supplies. Hundreds of thousands of tons of freight have been lying for months at Vladivostok without means of moving it to the western centres of population. Nothing but war-supplies may be brought over the railways from Archangel and Mourman. There are but few wagon-roads, and those very bad. The railroad from Sweden is almost useless for handling freight since the neutrality laws forbid the bringing of supplies that way. What is most needed just now is 10,000 American railroad men and mechanics to build shops and repair the rolling-stock already on hand and to operate the Siberian road properly, and thousands more of new cars and hundreds of locomotives from the United States; but to such a programme there is an objection which will be mentioned later. Transportation now is frightfully slow, and this applies as well to passenger traffic. The American railroad commission is doing much to improve the situation, but it is still bad. (b) In the next place there is a lack of men for almost every kind of work, and when they can be hired they are arbitrary, expensive, indifferent, and unreliable. (c) There is an inordinate number of holidays, about 30 under the old régime, each lasting from a day to a week, and during these festal occasions laborers of all kinds refuse to work, stores are closed, the post-office goes out of business, hotels and restaurants refuse to serve meals, and industry is out of joint. (d) The Russian workman cannot be said to be particularly industrious or efficient. He has little initiative and less executive ability. For ages he has been dependent upon his master, has earned little more than his keep, and has not learned to think for himself. His ideas now are greatly confused, and he spends a considerable portion of his time discussing with his mates the wonderful thing that has happened and what to do. The results of his deliberations are sometimes grotesque indeed. Whatever the outcome, the present effect is low efficiency and poor discipline. (e) In these days when the people see property of various kinds being confiscated by the new Government, they are apt to believe that there is no limit to possible confiscations. There is a desire on the part of the proletariat to take from those that have and give to those that have not; and there is fear on the part of those who have that they will not long be permitted to own and enjoy that which they possess.

In spite of the inability of the Russians to furnish all the supplies needed for the army and at the same time provide ample supplies of food, fuel, clothes, and other necessaries for the non-belligerents, there is an insuperable reluctance to permit Americans or anyone else to perform this service for them. In their pride and in their desire to develop their own resources they object to the introduction of foreign labor or manufactured products. Thus they are in dire need of railways, but in their charters for new roads they always require the use of rails and other structural materials of Russian manufacture, even though it is quite apparent that these cannot possibly be obtained. They need skilled men in their shops and factories and for operating their railroads, but will not permit them to be imported even for the time necessary to put things in order. Under these circumstances it will be next to impossible to render them the greatest service within our power.

In summing up the Russian situation I will quote a well-known Moscow merchant now in this country: "Day after tomorrow Russia will be all right, but there may be a long day and a couple of dark nights in the interval."

To the business man I would say: "Keep your eyes on Russia, but your money out of it for the next few years." To the younger men, students of business and of engineering: "Study the language of Russia, become familiar with its history, its geography, its people. The time is near when such knowledge will be valuable and will command high wages and will offer rich rewards in the opportunity to take an active part in the development of a great nation."

Skip-Changing Devices at the Butte Mines

By THEODORE PILGER

The continued development of greater orebodies in the Butte district and the increasing average depth from which the ore is hoisted have made necessary the adoption of more advanced methods of handling both neath a single-deck cage on which the station tenders ride. The fact that all workmen at Butte are lowered on 'company time,' that the shafts are already pushed to the limit of capacity for hoisting ore, and that all the

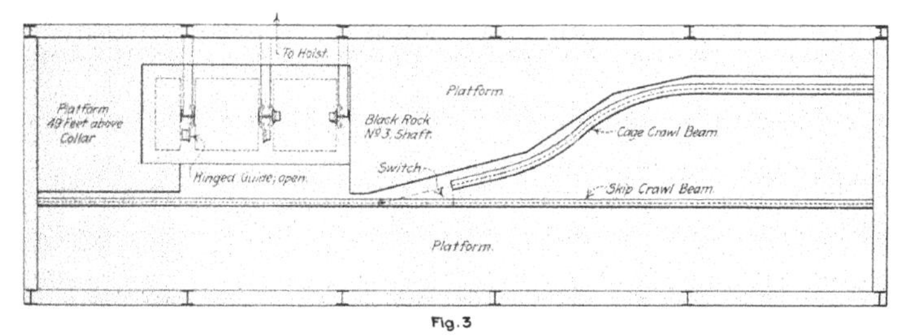

Fig. 3

men and ore, to secure greater capacity and to save time. At first a single-deck cage was used, for all purposes. Then a cage was added below, then another, and another. As the mines continued to enlarge, some more shaft repairing that can be accomplished has to be done between shifts, makes important every minute that can be saved in the handling of men. This condition led to the development of changing devices which would admit

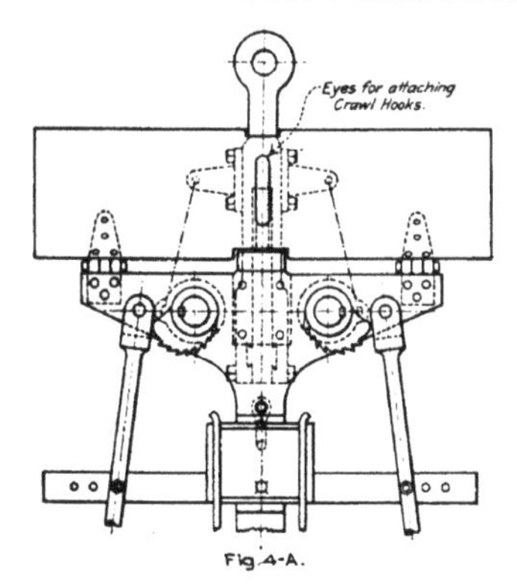

Fig. 4-A.

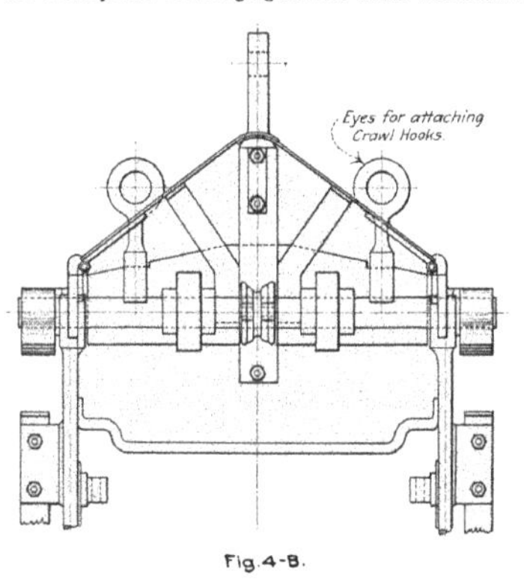

Fig. 4-B.

rapid method of hoisting than could be done by hoisting mine-cars and tramming them to the ore-bins had to be devised. The mechanism used for this purpose in the vertical shafts at Butte is the skip with trunnion-wheels and half shoe, hung by a cross-head and bail from be- of a rapid change from skips to multi-deck man-cages, and the reverse. There are several systems of these skip-changing devices now in use in the Butte mines, namely, the underground-chamber method, such as is employed at the Anaconda and St. Lawrence mines;

the head-frame chamber system, as seen at the Leonard; and the overhead-crawl system in use at the West Stewart and the new shaft of the Black Rock.

In the old incline-shafts the changing device was as follows: The skip to be taken off was pulled up high in the head-frame, a section of the guides below the skip was then swung to one side, and the skip, which ran on wheels, on being lowered, would be run upon a horizontal carriage that had in the meantime been pushed on tracks over the collar of the shaft. The cables were then detached, and the carriage carrying the skips was run back upon the tracks from the hanging-wall side of the shaft. Other carriages for transporting man-cages would be run over the collar of the shaft, the cables attached, and the cage hoisted into the head-frame. The guides then swung back into place, were locked with pins, the empty carriage was withdrawn from above the shaft, and the operation was complete.

This plan could not be used in vertical shafts because the angle at which the cages must be removed was too great. The advantage of the method was that the skips and cages resting on these carriages could easily be run directly to the shops for repairs, which can more easily be made when the cages are in a horizontal position than when suspended or lying at an angle. As the inclined shafts were gradually discarded on account of their general inconvenience, great cost of maintenance, and the great proportion of power wasted in hoisting, the underground-chamber method was developed in connection with the vertical shafts, and still is in general use in the older mines of the district. The essential parts of this system are an underground-chamber containing two sets of inclined guides in each compartment, radiating from a point above the collar of the shaft (Fig. 1) at which point the swinging guides in the hoisting compartment are pivoted. This chamber is just below the collar and is usually on the side of the shaft through which hoisting is done. It is timbered all round, and is deep enough below the bottom end of the swinging guides to accommodate the longest set of cages that would be employed at the mine.

The swinging guides, about five feet longer than the longest cages, are locked in place at the bottom when they are in regular hoisting position. They are built of two angle-irons placed back to back several inches apart, riveted on a plate with maple guides for fillers (Fig. 2). The bottom end is pulled out to match the inclined guides which extend downward into the chamber and carry the skips and the cages. The swinging guides are pulled over the chamber by means of a compressed-air cylinder and are stopped at the proper place to match the inclined guides by lugs provided for that purpose. As these guides are usually pivoted just below the skip-dump in the head-frame, and from 20 to 30 ft. above the collar of the shaft, it is necessary to have doors in the platform at the collar, so placed that they can be opened to allow the guides to swing at the bottom. This door is also opened and closed by means of a compressed-air cylinder.

The procedure in making a change from skips to man-cages in the underground-chamber system is as follows: The station tenders, on coming to surface, spot the skips so that they are hung between the swinging guides. They then raise the door at the collar of the shaft, go below, and, having unlocked the swinging guides, pull them over at the bottom until they match the inclined guides which extend downward into the empty chamber. They then signal the engineer to lower until the skips are in their resting positions. Having withdrawn the king-bolt, the cables are pulled over and attached to the man-cages and the swinging guides are placed to match the inclined guides on which the man-cages are resting. They now signal the engineer to hoist until the man-cages are in proper position on the swinging guides, then push the swinging guides by means of the air-cylinder into position in the hoisting compartment, lock them, and the cages are ready for use. When the new head-frame of the Leonard mine was erected, the device was changed in so far that the chamber was built into the head-frame above ground and below the ore-bins. This was possible here because the skip-dump was more than 75 ft. above the collar of the shaft and because the ore-bins are set back about 20 ft. from the shaft. This arrangement is exactly the same as the underground-chamber method in every other particular.

The latest method which has been installed at the No. 3 shaft of the Black Rock mine, and which is an improvement on the method in use at the West Stewart mine, where the idea originated, consists of swinging the skip from the hoisting compartment upon an electric-power crawl which runs on the bottom flanges of an I-beam, placed parallel to the long direction of the shaft and about six feet from it (Fig. 3). This beam is suspended under structural steel frame-work as shown in the accompanying illustration. It is high enough above the platform level to give clearance for four decks of cages, and is below the skip-dump. It branches at one end, having a switch that can be operated from the platform delivering to two similar I-beams from which the spare cages and skips are suspended. A separate crawl is provided for each skip and for each cage, and these crawls are run into place as required. The crawls, as shown in the illustration, are provided with two pieces of steel-cable about seven feet long, on the ends of which are fastened the hooks used to support the cages and skips when taken off the hoisting cable. The skips and cages each have two eyes (Fig. 4A and 4B) fastened to the cross-head and extending through the bonnet into which the crawl-hooks are fastened. The guides are hinged for a distance of about five feet longer than the length of the four-deck cages. These guides are built up of two angle-irons with the corresponding flanges turned together and the space between filled with the usual maple guides. They are all hinged on a rod extending the full length of the guide and are locked every 12 ft. simultaneously (Fig. 5).

The method of changing skips with this device is the following. The station tenders, on coming to surface signal the engineer to lower the skip below the hinged guides. These are then opened and the skips raised until the eyes on the cross-head of the skip can be reached with the hooks attached to the crawl. They are lowered until the weight of the skip, now hanging from the crawl instead of from the hoisting cable, pulls the skip out of the shaft compartment. The king-bolt is withdrawn from the skip, which is then run on the crawl to its receiving side track. The cage is now run from its receiving side track over a switch to the main beam, the hoisting cable is attached, and the engineer is signalled to hoist until the weight is taken off the crawl,

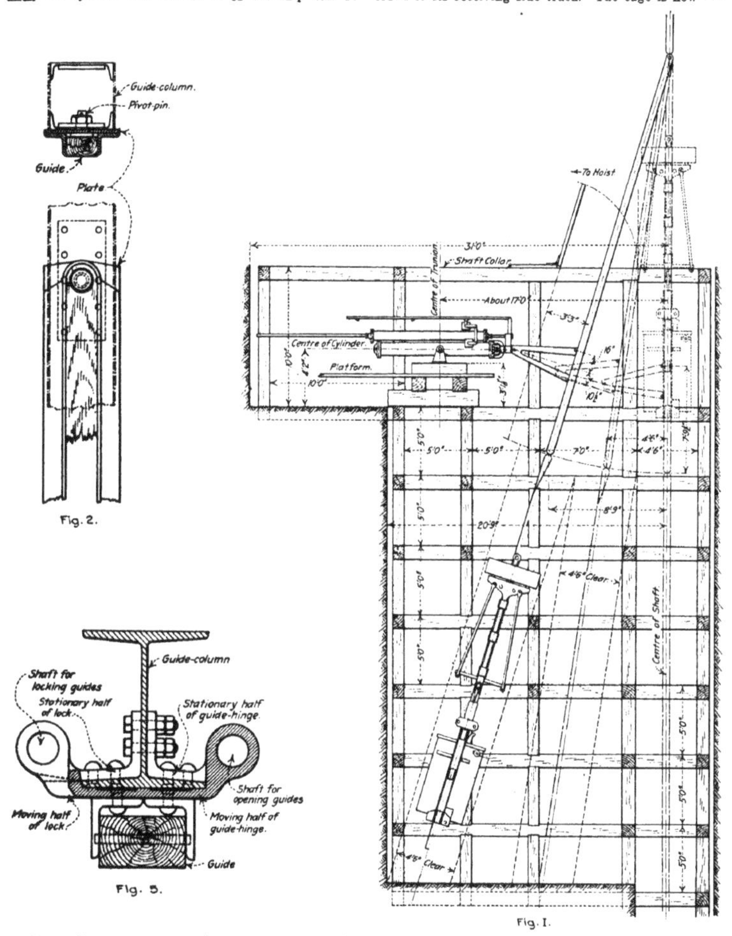

when the cage swings of its own weight into proper position in the shaft. It is then lowered to the guides, the hinged guides are closed into position and locked, and the change is complete.

The advantages of the underground-chamber system over a surface-change system are that it can be used readily on low head-frames, that it requires no surface-space, and the station tenders, who are perhaps wet through from riding the shaft, do not need to make the change in the wind and cold. Its disadvantages lie in the fact that it is almost impossible to make any satisfactory examination of the condition of the skips and cages, and that it is hard to discover defects, or to make repairs in a dark dirty hole underground. Furthermore, this chamber, being at the collar of the shaft, is generally in loose shifting ground, perhaps in an old dump, and the cost of holding the chamber open and keeping the shaft in good shape is very high. Also, in both the underground-chamber system and the head-frame chamber system, the skips and cages stand at such an angle that repairing is difficult. Another source of great trouble has been the compressed-air cylinder used in pulling back the swinging guides to match the inclined guides in the chamber. The connection to the bail which pulls these guides has a swivel-joint and the cylinder itself is on a rocking base, but nevertheless the piston is continually buckling and binding. The air-cylinders often freeze and cause a great deal of delay as well as danger from fire in being thawed. The door in the platform binds, particularly in winter, and often has to be opened with bars. Finally, the heavy skips and cages in passing over the swinging guide, which is fastened only at top and bottom, hammer the ends of the guide and cause it to buckle.

In the new surface-change system the greatest advantage lies in the facility with which examinations and repairs can be made on the cages and skips above ground and while they are held in a vertical position. If it is necessary to take any to the shops they can be lowered from the crawl directly upon the trucks. If it be desired for any reason to support them from the bottom they can be blocked up. The switching device will allow for the storage of extra skips and cages without the delay of hoisting which was incident to the underground-chamber method. The principal advantage lies in the fact that the taking off and changing of cages does not occupy any of the regular hoisting time because the new cages can be raised to the crawl and stored for emergency without the use of the hoisting engine.

The underground-chamber method will cost 50% more to install than the surface-rig, and furthermore, the expense attached to the up-keep of the surface-change system will be very little, while the underground-chamber requires constant repairs.

POTASSIUM NITRATE is reported to have been found recently as a constituent in dark gray arenaceous shale in Presidio county, Texas.

Photo-Micrographs in Color

Lantern slides representing photo-micrographs of sections should closely resemble the appearance of the section itself. This can be attained by making the print in stained gelatine instead of by the usual photographic process. The procedure for making such a print is as follows: Lantern plates (Seed or Standard plates are satisfactory) are sensitized by bathing for five minutes in a 2½% solution of ammonium bichromate containing 5 cc. of strong ammonia to the litre, the temperature of the bath being not above 65°F. The plates are then rinsed for two or three seconds in clean water, drained, and dried as uniformly as possible, the plates being kept in the dark during drying. The sensitized plates are then exposed through the glass under the negative to the light of an arc-lamp, the average exposure being about three minutes at 18 in. distance. Printing cannot be done by daylight, or sharp images will not be obtained. The exposed plates are then developed by rocking in trays of water at about 120°F. until all soluble gelatine is removed. Under-exposure is indicated by the high-light detail washing away, and over-exposure by the film being insoluble to too great a depth. The plates are then rinsed in cold water, fixed in hypo, and washed free of the hypo. They are then ready for staining. The staining is done with a 1% solution of dye containing 1% of acetic acid, the dye being selected to stimulate most closely the original stain of the section, the time of dyeing being chosen so that the necessary depth is obtained. When sections stained with two different colors are being photographed negatives are made through suitable color-filters and are then dyed in the two stains and placed face to face so that a two-color slide is obtained. For example, suppose that a section is stained red and green. Two negatives are made on pan-chromatic plates, one with a red filter, which will cause the green to appear as clear spaces in the negative and will not record the red, and the other with a green filter, which will record the red and not the green. The slides made as described from these in bichromated gelatine are stained—that from the red negative with the original green stain, and that from the green negative with the original red stain.

The filters required can be chosen from the set prepared for photo-micrography under the name of Wratten M filters. The choice of the filter is decided by visual trial under the microscope, the filters chosen being those which most nearly absorb one color and transmit the other. Thus, photographing a section stained with Delafield's hæmatoxylin and precipitated eosin, the *A* filter (red) shows no trace of the eosin and gives a good strong negative of the hæmatoxylin. The *B* and *C* filters are used together for the other negative, giving a blue-green color and recording the eosin and hæmatoxylin both fully, and from these two negatives positives are made and stained with a blue and a red dye. Prints thus made beautifully reproduce the original colors.

Recovery of Converter-Fume at Tooele, Utah

By L. S. AUSTIN

The installation and operation of a Cottrell plant at Garfield, Utah, was described by W. H. Howard in 1911.* This consists of seven units containing in all 2520 five-inch pipes, 10 ft. long, and was to precipitate the dust and fume from an average of four Pierce-Smith barrel-type converters, each yielding 25,000 cu. ft. per min. of converter-gas, calculated at standard temperature and pressure. Tests made at a smaller experimental plant upon gases from the blast-furnace, roasters, and reverberatory furnaces, indicated a recovery of not less than 95% of the dust and fume. Mr. Howard states, however, that the results obtained were upon gases particularly favorable to the process, and that it does not follow that all smelter-smoke can be treated so successfully. At the International smelter at Tooele, Utah, a small Cottrell treater for two or three years has been in successful operation for collecting the dust from the Dwight-Lloyd sintering machines, and last year a similar installation was built to recover converter-fume. O. M. Kuchs describes‡ a bag-house intended for collecting the lead and zinc fume formed in the converting of leady copper-matte coming from the silver-lead blast-furnace of these works. This was made possible by carefully watching the temperature of the escaping gases, which carried enough volatilized lead and zinc to neutralize the sulphurous anhydride (SO_3) fume produced by converters; but, for those converters when used on straight copper-matte this was not practicable, and to take care of some dust and the abundant fume from this source the more enduring Cottrell apparatus was installed.

COTTRELL TREATER AT INTERNATIONAL SMELTER, TOOELE, UTAH

The converter-house at the Tooele plant has a single row of five converters extending in a row from north to south, the three converters at the north end, treating copper-matte, delivering the fume to a flue going southward to the new Cottrell treater, whereas the two southern converters, which treat leady matte, blow to a paral-

*Trans. A. I. M. E., Vol. XLIX, 540.

‡Trans. A. I. M. E., Vol. XLIX, 579.

lel flue which leads north to the bag-house above mentioned. There is provided a branch pipe from the treater-flue, so that, in case of need, part of its gases can be diverted to the bag-house. The course of the converter gases is by the long bottom flue of 81 sq. ft. cross-sectional area to the distributing flue, thence by two branches to either of the two Cottrell units, each consisting of 110 pipes, upward through these pipes, thence by a down-cast flue of 45 sq. ft. area to a No. 12 Socorro suction-fan, and to the 6-ft. diameter stack. If desired the flow of gases may be by-passed directly to the stack, at the same time shutting off the treaters. Either unit of the Cottrell treater may be cut out from the flow for cleaning or repairing. Just before going to the treater at the distributing flue atomizing water sprays are inserted, by which water is sprayed into the moving gases for the purpose of wetting and cooling them. Cooled they certainly are, but the dust refuses to take up the moisture to any extent, and, when deposited on the treater-pipes, it appears as a bulky, fluffy, dry product, difficult to wet. Through each pipe, from an insulated frame above, an axial wire passes to another frame beneath the pipes. Thus the high-voltage current, of 50,000 to 70,000 v., has to jump across the 6-in. interval between the wire and the inside of the 12-in. pipe. In so doing it drives the dust particles to this inner surface. Here it attaches itself, forming a gradually thickening coating. Some dust also clings to the central wire. During each shift one of the units is cut out by switching off the electric current; the dampers leading to the unit are closed, and, by jarring the pipes, the dust is caused to fall into the hopper-shaped bottom of the unit, where it may be drawn off. The cleaning being completed, the current is turned on and the dampers re-opened to resume operation. The product as obtained is light, fluffy, dry, bulky, and, when withdrawn, is difficult to wet. Due to these qualities it is an unsatisfactory material to treat. It may be either briquetted, or added as one of the ingredients of the charge sent to the Dwight-Lloyd sintering machines.

It will thus be seen that the purpose of the installation is to pass the gases upward through a nest of tubes, each tube taking its share of the total flow. It is during the brief period of upward movement that the work of removing the fume must be done, and it is the phenomena of dust-separation that is now to be considered. At night, looking downward through a tube, one notices a glow, called the aurora, created by the high-tension current. A similar glow may be noticed at the insulators at the towers of a high-tension transmission line, at the poles of a static electric machine, and in the atmospheric phenomenon of the aurora borealis. W. W. Strong, at a recent joint meeting of the American Institute of Mining Engineers and the American Electrical Society, discussed the theoretical aspects of electrical precipitation of fine material, as a result of these observations, and his diagram, which is a cross-section of a tube, merits careful examination. It shows the central wire and the pipe upon which the deposit is received. Upon passing a positive electric current the gases, oxygen and hydrogen respectively, are ionized, and the negative ions attach themselves to the dust-particles and are driven to the pipe by the electric field. The particles with the positive ions go to the wire from a smaller region of plus space. To clear the deposit on the wire a striking-rod

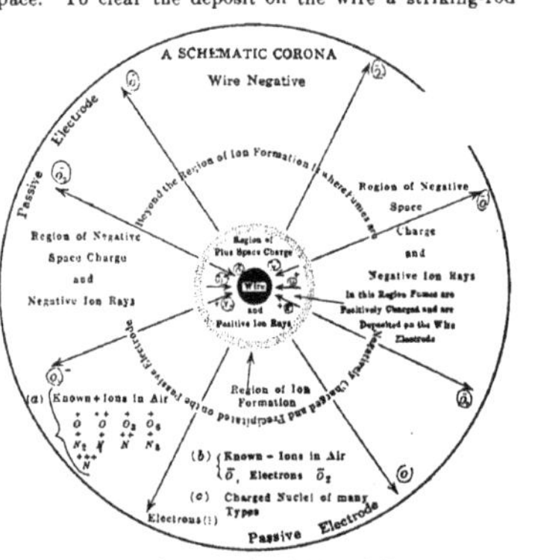

SCHEMATIC CORONA IN COTTRELL TUBE

is run just below the lower wire-guide bars and just above the weight which stretches the wire. This rod strikes the wire, jarring off its attached load.

The treater is lined with ⅜ to ½-in. sheets of 'transite,' an asbestos-board made of asbestos and cement, pressed together under heavy pressure. The dampers are 'transite' sheets ⅜ in. thick.

SULPHONIC ACIDS are prepared by D. Tyrer in conformity with a patented process in which he first makes the mono-sulphonic acids of benzine and other hydrocarbons, boiling below 200°, by passing the vapor of the hydro-carbon through H_2SO_4 heated to 100°. By this method the H_2O formed is steadily removed from the reaction mass. In the preparation of disulphonic acids, the current of vapor is stopped when a given amount of the mono-sulphonic acid has been formed. More H_2SO_4 is added and the temperature is raised to the favorable point. The unsulphonated portion of a hydro-carbon, treated by this method, has been freed from its impurities, notably from sulphur compounds, and can be condensed in a purified form.

TRIPOLI or diatomaceous earth was produced to the extent of 2721 short tons in 1916. The chief centres of production were California, Connecticut, Maryland, Massachusetts, Nevada, New Hampshire, New York, and Washington. The market value of tripoli is about $9.50 per ton. Large deposits are being developed in Maryland and Virginia.

Preferential Flotation

By W. SHELLSHEAR

By preferential flotation is meant the separation by flotation of one or more sulphides from each other. It is to be regretted that there is still much confusion of terms in describing flotation work. The American nomenclature which applies to selective flotation, indicating the separation of sulphide minerals from non-sulphide gangue, as distinct from preferential flotation or the separation of one or more sulphides from each other, however, well meets the case, and its adoption should be universal. These terms will need to be modified to some extent to cover the flotation of oxides and carbonates, but these are special cases. It is intended, rather, to deal here with facts gained from experience in operation and conclusions derived therefrom than to attempt to formulate a theory to account for the action of the different processes described.

Although the possibilities of preferential flotation were hinted at by Cattermole and others, it is only in recent years that processes have been devised that have made it a commercially profitable operation. The first of this type, which has proved successful in large-scale operations, was the Horwood process, in which certain sulphides have their surfaces rendered immune to flotation by roasting, leaving the surfaces of other sulphides unaffected. The success of this process largely depends, therefore, on careful attention to the detail of roasting, such as the range of temperature, and the rate of feed. Large-scale operations, however, have demonstrated that, with modern mechanical furnaces, these factors can be kept well under control, and it is difficult to understand why this process is not in use for the treatment of many mixed sulphide ores, especially those containing lead, zinc, and iron.

It had long been recognized in the Broken Hill mills that when ore finely ground to a pulp was subjected to aeration, either by exposure to an air-surface, as in tabling operations, or else by a fall into a spitzkasten, cone, or similar apparatus, the galena and silver sulphide floated more or less imperfectly on this surface in preference to the zinc sulphide. This flotation was undoubtedly due to lubricating oil in the mill-circuit. Samples of the frothy scum from mill dewatering cones assayed in some cases as high as 60% lead, and 50 oz. silver per ton, but in all cases the scum was thin and represented only a small percentage of the sulphides present. Milling experience had thus demonstrated that there were some inherent properties possessed by the sulphide minerals, which were more marked in the cases of the lead and silver sulphides. Advantage was not taken of this fact, however, till Lyster proved that by controlling certain factors in the flotation treatment, use could be made of the relative differences between the natural properties of the various sulphides to effect their separation from each other. This was the first important step in the evolution of preferential processes. These distinctive properties of the individual sulphides vary somewhat with the different classes of ores. Thus, for example, they are found in a more marked degree in calcitic Broken Hill ores than in the rhodonite or garnet sandstone variety. For this reason, even in apparently similar types of ore in the same locality, working conditions, as far as preferential flotation is concerned, need to be modified and adjusted to suit each particular case.

The controlling factors for preferential flotation of this type are the nature and amount of the agitation, variation in temperature, and proportioning of the medium used, the control of the air-supply during agitation, and the proportion of mineral salts in solution. It is believed by some that the preferential action in processes of this type is due to poor flotation. For instance, one writer, referring to Lyster's process, hints that it is probably due to imperfect conditions, the high grade of the lead being caused by only the finest particles of the lead floating, and that therefore the residues probably are high. This is a mistaken idea as there are large-scale plants working on the Lyster principle, obtaining lead concentrate assaying as high as 60% in that metal while the residues averaged less than 2% lead. The secret of their success is the scientific adjustment of the controlling factors previously mentioned. Although these distinctive physical properties of the sulphides, by reason of which preferential flotation of this type was rendered possible, were present in a rather marked degree in slime freshly crushed from crude ore, it was found that if the slime were allowed to settle and accumulate in dumps these properties were seriously altered. This was probably due to oxidation and to concentration of salts in solution on the surfaces of the sulphides. Especially was this the case if the sulphide particles had dried from exposure, because it was then extremely difficult to obtain preferential flotation at all along these lines. As a rule if a dump-slime is treated on the Lyster principle a poor grade of 'float' is obtained, consisting of mixed lead and zinc sulphides.

It was the continued research in the treatment of dump-products that resulted in a great forward leap in preferential work, the discovery of certain reagents or mediums, which, although used only in small quantities, could temporarily wet, or render incapable of flotation, certain sulphides, leaving the floatable properties of other sulphides unimpaired. It was found possible, by using these reagents in conjunction with various methods of flotation, to preferentially separate one or more sulphides from each other in an efficient manner, whether these products were freshly crushed or had been stacked in dumps. The fact that these reagents cause only temporary wetting, this effect being more complete when the reagent is still active as a medium in the solution, enables a sharp line of distinction to be drawn between these processes and those of the Horwood type, in which the faces of certain sulphides, such as zinc and iron, are permanently altered by the roasting action, rendering them for the most part permanently incapable of flotation.

Another type of process in which the Horwood effect is produced by chemical means is the ferric-chloride method. The cost of chemical reagents generally makes a process of this type uneconomical.

The action of deadening reagents is extremely difficult to understand, as every medium at present known appears to act in a different way. Under certain conditions in a mixed lead-zinc sulphide-slime sulphuric acid will cause a preferential flotation of lead sulphide. This method was worked for some months as a commercial process treating the Central mine slime-dumps. Sulphuric acid in this case was added to the slime-pulp in the cold, no oil being used, although a certain amount of the soluble constituents from the oil used for the zinc flotation was returned with the circuit-liquor. In most cases the addition of fresh oil disturbed the preferential action. After floating the lead sulphide, using machines of the sub-aeration type, the zinc sulphide was floated by the addition of heat, more acid, and eucalyptus oil. This process is only workable in the case of certain Broken Hill slime. In other cases, under practically similar conditions, a mixed sulphide-float is obtained. Even in a favorable slime the preferential action is upset by an excess of sulphuric acid.

In some cases, when experimenting with this method, one can obtain a float, the upper portion of which consists of lead sulphide and the lower portion of zinc sulphide, one sulphide gradually merging into the other. This process was discovered by Hebbard and Harvey, to whom also is due the successful development on a large scale of the under-driven suction-aeration machine for flotation.

The action of reducing gases, for example, sulphur di-oxide (Bradford's patent), especially if generated in the solution, is entirely different. Sulphur di-oxide when used in treating a mixed sulphide ore deadens each sulphide in succession according to their relative flotation properties. Thus, when treating an ore containing iron, lead, and zinc sulphides by this process, conditions may be so regulated that each sulphide may be floated in succession, the conditions being practically controlled by the amount of sulphur di-oxide gas in the pulp. An excess of sulphur di-oxide will, in most cases, render all the sulphides present temporarily immune to flotation, and in this case the surface will merely be covered by a mass of thin white bubbles. Although a fine adjustment of conditions is required to float the sulphides in succession as described, in actual practice it is not difficult to float the lead and the iron sulphides as one product, leaving the zinc sulphide in the residue. It is interesting to note that once a sulphide is wetted it is practically impossible to re-float it as long as the sulphur di-oxide gas is in the solution. To float the wetted sulphide, heat and sulphuric acid usually are required. Although this process would appear to be delicate, the working conditions are not difficult to control in a well-designed plant, as proved by the fact that several large-scale plants are working successfully at Broken Hill using a modification of this method.

The action of sulphur di-oxide appears to be unaffected by the nature and amount of the salts in solution, within reasonable limits, and the amount of gas required is usually so small that its chemical action, if any, on the surfaces of the sulphides must be very slight. In some cases, to apply this process, the gas is passed directly into the solution, while in others it is generated therein by the addition of sulphuric acid and certain salts, such as sodium hypo-sulphite.

As another instance of the action of these agents I may take the case of potassium permanganate and potassium bichromate. Both are powerful oxidizing agents, and yet each acts in a different way. Potassium permanganate when added in small quantity to a lead-zinc sulphide slime will wet the zinc sulphide, and, providing conditions are correct, the lead sulphide may be floated as a high-grade product. The addition of sulphuric acid at normal temperature to the lead residue will cause a rapid and practically complete separation of the zinc sulphide from the gangue. This proves that the surfaces of the zinc sulphide are only temporarily affected. An excess of permanganate in the primary treatment will result in both sulphides being deadened and their subsequent flotation is usually difficult, even with heat and sulphuric acid. On a slime of a similar type potassium bichromate will cause a preferential flotation of the zinc sulphide, leaving the lead sulphide in the residue.

With this reagent the conditions required are usually more intense. Thus, in most cases a 0.2 to 0.5% solution of the reagent is required at a temperature of 120° to 140° F., whereas the potassium permanganate works at normal temperatures. In some cases a long digestion of the pulp with potassium bichromate is reqired to effect the deadening of the lead sulphide. Although this paper deals more particularly with the preferential flotation of lead-zinc sulphides, there are other sulphides which can be preferentially floated with these reagents. Potassium bichromate, for instance, will cause a preferential flotation of copper pyrite from iron pyrite.

In connection with the processes previously mentioned, the term wetting one sulphide in preference to another has been used. This term is employed by Bradford in describing his preferential processes, and as far as one can see, no other term adequately meets the case. When agitating a lead-zinc ore with one of these reagents in a glass vessel maintaining correct conditions, it will be found that one sulphide will be in a flocculent condition, and the other in separate grains similar to the constituents of the gangue. If the glass vessel with its contained pulp be now connected to a vacuum apparatus, and the air above the surface of the solution be evacuated, gas bubbles will be plainly seen attached to the lead sulphide, but none to the zinc. It therefore seems reasonable to conclude that the function of these agents is to rob one or more of the sulphides of gas, leaving the others with sufficient gas on their surfaces to admit of their subsequent flotation.

The sulphur di-oxide process points to a gradual displacement of gas from the various sulphides in rotation, according to their flotation properties, especially when it is considered that this gas is a powerful reducing agent. Possibly other agents of this class act in the same way.

To obtain a preferential flotation of zinc sulphide, leaving the lead sulphide in the residue, it is obvious that, as the flotation properties of galena are stronger than those of zinc-blende, the preferential agents must act in a different way. It will be found that in almost all cases of preferential lead flotation heat is not required, and only comparatively small quantities of reagents are necessary. This is not the case with the preferential flotation of zinc sulphide, since in all known cases strong solutions and high temperatures are necessary, indicating that a chemical action takes place on the surface of the lead sulphide. Thus the bichromate action is probably due to the formation of lead chromate on the surface of the sulphide, and experiment has proved that in Bradford's salt process a small percentage of lead is in solution as lead chloride. The fact, however, that the lead sulphide can be floated afterward in this process by the use of potassium permanganate and other salts indicates how comparatively slight must be the chemical action. It therefore seems probable that the flotation of any sulphide will be hindered in some degree by the presence in the solution of a salt which will chemically react with it. It was first pointed out in the case of Minerals Separation *v.* Elmore that, if virgin sulphides from certain localities were washed in ether and the ether afterward removed and evaporated, a residue was left consisting of a hydrocarbon oily in nature mixed with free sulphur which could be plainly seen in crystalline form. This has been verified by similar work on Broken Hill sulphides, but it has also been proved that some sulphide ores contain more natural oily residue than others. Whether there is any relation between the percentage of oily residue, the amount of adsorbed gases, and the floatative properties, remains to be proved. Experiments up to the present have indicated that dumps containing a relatively high percentage of this oily residue float more readily than those which do not. From the very small quantity of oil in a frothing agent necessary to produce efficient flotation, it seems possible for the adsorption of gases on the sulphides to be accounted for if merely an infinitesimal trace of oily matter be present in this composition.

Although the principles underlying preferential processes are only imperfectly understood, by following routine conditions of working obtained by experiment in the laboratory and on the large-scale processes of this type, they may be made commercially profitable. A scientific control of such factors as rate of feed and its regularity, quantities of reagents, and the like, is, however, vital.

Regularity of feed is essential. To call attention to the importance of this factor would seem unnecessary were it not for the fact that in many cases it is not given sufficient prominence. It is important in all metallurgical operations, but even more so in the case of preferential processes, because, without regularity of conditions, it is not practicable in many cases to work these methods at all. One must aim at getting feed and conditions of mechanical running as perfect as possible, the main axiom of all flotation-work being that without smooth running the accurate adjustment of metallurgical conditions is virtually impossible. In working a plant for the treatment of a mixed lead-zinc ore preferentially, irregular conditions will spoil both the lead and the zinc products, and neither will be of marketable value.

Control of aeration is also vital. Bradford was the first to discover that aeration could be controlled by a valve on the suction-pipe of a centrifugal pump, the air thus sucked in being atomized by the action of the impeller. Owen's patent shows a further development of this principle, the aeration in this case being controlled by a pipe connected to the bottom of an ordinary agitating vessel, the required air being sucked in and atomized by the agitator. It was found that by working on these lines the ordinary spitz-box used in the Hoover machine was unnecessary, the aeration and flotation being conducted continuously from the bottom and top of the same machine. In this type of flotation-machine there are three distinct zones of action: (*a*) the aeration-zone, where air is sucked in by the impeller; (*b*) the atomizing-zone, where the air is completely broken up by the aid of baffles which also check the rotary motion of the liquid; and (*c*) the flotation-zone, a still zone on the top of the vessel where the flotation takes place. Similar control of aeration may be effected on either of these machines by replacing atmospheric air with air under pressure.

In considering the action of frothing-agents it must be said that it has now been definitely proved that, to float certain sulphides, oil, in the sense of an organic agent of any description, is unnecessary. Thus, when preferentially floating galena by the sulphur di-oxide process no oil or organic agent as a general rule is required. There is at the present time a commercial plant giving a high recovery of lead with this process without the use of any medium of this description. In some cases oils or organic frothing-agents have been replaced by soluble inorganic frothing-agents, such as sodium carbonate. The function of the frothing-agents in most flotation processes is to intensify the physical conditions which promote flotation rather than to produce them. When experimenting with a mixed sulphide ore with a preferential agent, if a strong stream of atomized air be passed through the agitated pulp, the color of the sulphide which will float will be seen in the froth as it breaks at the surface of the solution before any oil or frothing-agent has been added at all. In the cases of most preferential processes the addition of certain unsuitable oils or frothing-agents will completely nullify the preferential action. For instance, the mixing of the pulp with an exceedingly small amount of oils of the wrong type before the addition of the preferential reagent will generally result in no separation of the sulphide minerals whatever. It is extremely difficult to separate two or more sulphides which have been floated with oil as a mixed concentrate in a previous operation, although this can be done with a process where intense conditions are used, as in Bradford's salt-process.

Although, as previously stated, the action of frothing-agents is to intensify the conditions of flotation, one of the many problems of preferential flotation is to use oils which will assist the flotation without disturbing the preferential action. As an example of a preferential process in which oil is not used, the sodium carbonate process may be taken. Freeman discovered that by treating mixed lead-zinc sulphide slime with sodium-carbonate solution varying in strength from 1 to 10%, according to the ore treated, employing an iron vessel with ordinary conditions of agitation and aeration, the lead sulphide floated preferentially. The sodium carbonate in this process serves a double purpose, being a preferential reagent and at the same time a soluble frothing-agent. The addition of any oil is generally deleterious to the successful working of the process, but in any case it is unnecessary, as the sodium carbonate usually gives a firm coherent froth. Although this process works quite successfully at normal temperatures, the grade of the lead concentrate is not affected by heating the solution up to 120°F. In a copper machine, on the other hand, sodium carbonate has a quite different action, for, if the temperature of the solution be raised to 160°F., a preferential flotation of zinc sulphide is obtained, leaving the lead sulphide in the residue. Even at normal temperatures the preferential flotation of the lead is imperfect if any copper is present as a part of the treatment-machine.

This is due to the fact that flotation conditions as a rule are materially altered if there is present in the solution even a trace of metallic salt such as a salt of copper, or mercury. If quantities of these reagents up to 2 or 3 lb. per ton be added to a pulp containing zinc sulphide or a sulphide of a similar nature, and the pulp be subjected to flotation-treatment, after flotation has taken place there is usually no trace of copper present in the solution. From this fact one is led to the conclusion that the action of these metallic salts is similar to the action of copper sulphate on scrap iron, or, in other words, that the sulphides become coated with metallic copper or mercury, as the case may be. Bradford was the first to discover the properties of these metallic salts, and in 1913 he patented the use of reagents such as copper sulphate to exalt the flotation effect. Another interesting example of the action of copper salts is afforded by Bradford's salt-process. As is well known, this process consists in agitating a mixed lead-zinc ore with sulphuric acid in a salt solution varying from 7 to 10%, and heating the solution to about 140°F. The zinc sulphide floats preferentially, leaving the lead sulphide in the residue. After the zinc sulphide has been floated off, the lead sulphide may be floated by the addition of copper sulphate to the solution used for the flotation of the zinc, this being practically a continuous process. There seems to be large scope for this process, as it makes an extremely clean zinc concentrate, and yields a high recovery of zinc.

Different processes require different machines to obtain the best results, and it is only by experiment on any particular ore that the best type of machine in any case can be determined. For instance, Bradford's salt-process will work successfully in the ordinary Hoover machine, whereas processes of the Lyster type usually work to advantage in the combination of pump and spitz-box machine illustrated in Lyster's patent. Permanganate of potash and agents of this type work to the best advantage in a sub-aeration machine, but they will give good results in the ordinary Minerals Separation apparatus. Every type of machine, however, usually requires different quantities of reagents and different conditions. Although I have not seen the Callow machine tried in this particular class of work, it seems to be suited for a number of these processes, and there is no reason why it should not work successfully. The action of metallic salts clearly demonstrates that the metal used in these machines, especially in laboratory machines since commercial machines are usually made of iron or wood, is important, and for consistent results the laboratory machine must approximate the large-scale apparatus in its details.

As in ordinary flotation so in preferential treatment in order to obtain good results two distinct phases are necessary. First, the pulp is brought into a physical condition for efficient flotation, and, second, aeration takes place, enabling the sulphide-mineral that is to be floated to be separated as a froth. In the first place conditions are to be so adjusted that the mineral to be floated shall assume a clotted form, or, in other words, bound together in minute masses in the pulp. This clotted appearance is due to formation of primary gas-nuclei, and only under certain conditions will these nuclei form. If this primary phase is incomplete, aeration will cause a thin froth which is troublesome to handle and the recovery will be low.

In preferential flotation when deadening one sulphide the other, if conditions are correct, will be in a clotted condition. If aeration be now applied an efficient flotation will take place. If, however, physical conditions are such that the action of the preferential reagent is imperfect, no amount of aeration will give good results, regardless of the manner in which it is applied.

The action of oil or frothing-agents appears to b

mainly to increase the size and maintain the stability of the primary nuclei already referred to.

Every ore appears to have distinctive physical properties as far as its floatability is concerned, and consequently the method of treatment of two different ores is rarely the same. This is particularly the case with lead-zinc sulphide ores. The varying composition of zinc-blendes from various localities to a large extent accounts for this. The properties of blende, as far as flotation is concerned, appear to be influenced by the amount of iron in its composition. Thus the zinc-iron blende, marmatite, having an approximate composition of 5ZnS. FeS, has quite different properties from a true 'resin blende' of the composition ZnS, and even resin blendes differ to some extent in their properties. It is obvious, therefore, why one preferential process will work successfully on one class of sulphide ore and fail on an ore of a similar type. This has been previously illustrated in the case of the sulphuric-acid process, and other examples might be cited. Silver sulphide follows lead sulphide so closely in milling operations that for a time it was believed that the two sulphides were in chemical combination. Flotation has absolutely disproved this, since different preferential processes will in many cases give the same grade of lead concentrate from the same ore, but the concentrate will in some cases be high in silver and in others low. Indications show that silver sulphide is one of the most easily floated of all sulphides. The sulphur di-oxide process seems to be peculiarly adapted to the flotation of silver, and in many cases on obtaining a lead concentrate preferentially by this process the recovery of the silver is higher than that of the lead.

Preferential flotation is yet in its infancy, but most of the processes mentioned have worked or are in operation on a large scale and are not merely laboratory conceptions.

To test an ore for this class of flotation requires a great deal of experience and observation, as there are so many variable factors involved, and the disregard of minor details will usually affect the success of the treatment seriously. Preferential flotation has come to stay, and there are many fine-grained complex ores which can be made commercially profitable by these methods. Mixtures of zinc and iron sulphides, fine-grained lead and zinc sulphides, and other ores of this type, are practically dependent on these processes for commercial separation. The scope for preferential flotation is, therefore, a large one, and these methods will soon be quite universally adopted.

DOUBLE ACID-PHOSPHATE is a new product evolved by the Bureau of Soils of the Agricultural Department. The object is to provide a richer fertilizer, thus reducing the consumption of sulphuric acid direct. Phosphoric acid is obtained by electric smelting of phosphate rock,

Manganese in Furnace Slags

*In the production of certain alloy steels the effect of superheating the working slags results in the separating of the most readily reduced slag constituent. Thus in making manganese-steel in the open-hearth furnace the primary slag covering the molten pig-iron charge was a rich ferrous slag of the composition approximately $FeO.Fe_2O_3 = 80\%$, and $SiO_2 = 20\%$, fusible at about 1300C. Into this a pyrolusite ore containing

	%
MnO_2	85
$CaO + MgO$	5
SiO_2	10

was gradually added. The MnO_2 becomes MnO on superheating, the FeO being changed to Fe_2O_3 by the oxygen liberated from the MnO_2. It was then found that the FeO and Fe_2O_3 in the slag, if increased by further additions of 'hematite ore' (manganese) and lime, caused a rapid reduction of the manganese from the slag, manganese entering the liquid metal. When the slag contained over 50% manganese that element entered the metal, whereas if the slag was adjusted so as to contain above 50% iron, then iron was reduced from the slag in preference to manganese. The slags higher in MnO were more fluid than the rich FeO slags, and were apparently fusible at a lower temperature.

The refining of alloys may be illustrated in the case of the elimination of manganese in the puddling or open-hearth steel processes. Thus the reaction between the slag constituent $FeO.Fe_2O_3$ and the alloy constituent Mn_3C in the pig-iron is of the type:

$$FeO.Fe_2O_3 + Mn_3C = 3MnO + CO + 3Fe.$$

Manganous oxide displaces the iron oxides in the slag, the manganese in the pig-iron being replaced by pure iron from the reduction of the iron oxides. Superheating the slag renders the $FeO.Fe_2O_3$ more active, while that of the molten pig-iron increases the instability of the Mn_3C; CO gas is liberated; and the slag is enriched by the entrance of MnO at the cost of the iron oxides $FeO.Fe_2O_3$. In such actions the active oxides $FeO.Fe_2O_3$ are the reducing agents. They are least stable and hence in the condition to react with the other unstable compounds present when by superheat their stability is decreased. Possibly at the moment of exchange it is the ferrous oxide which reacts with the Mn_3C in a reversible relation, thus:

$$3FeO + Mn_3C \rightleftarrows 3MnO + Fe_3C.$$

Superheating the FeO in an oxidizing atmosphere leads to the formation of Fe_2O_3. On the other hand, liquid Fe_2O_3 in contact with molten iron-alloys is readily converted to FeO providing a reducing or neutral atmosphere is maintained in the working chamber.

TUNGSTEN is being found in quantities ...

Japanese Copper Production

By KENZO IKEDA

The Japanese copper output last year was about 90,000 avoirdupois tons, mainly produced in the form of electrolytic refined copper. The important smelting works and their contribution of metals are as follows:

Smelter and company	Copper production, tons	Annual capacity of blast-furnace, tons
Hitachi, Kuhara	17,000	400,000
Ashio, Furukawa	16,000	120,000
Besshi, Sumitomo	11,000	250,000
Kosaka, Fujita	10,000	400,000
Inujima, Fujita	4,000	100,000
Abeshiro	4,000	......
Ikuno	3,500	
Osaruzawa	3,000	
Ogoya	2,500	
Others	10,000	
	90,000	

The amount of copper mentioned above is produced from copper deposits which may be classified into the following four groups:

1. Fissure-filling deposits, or true veins, the majority being in Tertiary rocks and the mineralization caused by circulating solutions, represented by the Ashio mines. The largest known vein is 3000 ft. long, averaging two feet wide, and is 3000 ft. deep, occurring mainly in a liparite intrusive through Paleozoic slate, sandstone, and quartzite. The ores are composed of rich copper pyrite in a silicious gangue.

2. Pyritic bedded deposits in crystalline schists or Paleozoic rocks, such as the Besshi mines, the rocks being chloritic, graphitic, and Piedmontite schists. The deposits consist of pyrite and some magnetite, the copper content being 3.5%. The largest lode explored up to the present time is 500 ft. long, 4 ft. wide, and 3000 ft. deep, the dip being nearly 49° The form is nearly that of a sheet.

3. Complex sulphide deposits, or simple metasomatic replacements in Tertiary sedimentaries, but with no limestone beds, and in volcanic rocks, this class being represented by the Kosaka mines. The country rocks are liparite dikes and prophylitic andesite intrusives through liparite tuffs. The orebody is 1600 ft. long, 400 ft. in maximum width, and 400 ft. deep, massive in form, with ores of three kinds, namely, pyritic or yellow ores, complex sulphides or black ores, and those with a quartz gangue, called silicious ores, the average mixture having 2% copper, 20% iron, 20% sulphur, not taking into account the sulphur contained in baryte, 22% SiO_2, 20% baryte, 6% zinc, and 1.3% lead, carrying some silver and gold. The mines are being worked by open-cut chiefly.

4. Metamorphosed metasomatic deposits, or so-called contact-deposits, in Archean, Paleozoic (mainly), Mesozoic beds, or older sedimentaries, and rather acidic plutonic rocks, as seen at the Kamaishi mines. The form and size are very irregular as compared with the other groups, and they are confined mainly to the margins between the Paleozoic clay-slates and limestones, and the intrusive granites and diorites. The copper deposit is associated with large adjacent accumulations of magnetite ore, so that the mine produces material that is treated for its iron as well as for copper.

These four classes of copper deposits are in different zones, each having characteristic features, so that usually it is possible to tell to what class a certain orebody in a district belongs, and, moreover, a fair conclusion may be hazarded from previous study and experience as to what extent it may be possible to develop the mines. More than 90% of the total copper output, considered as to its origin, is distributed as follows:

	%
From fissure-filling deposits	50
" bedded deposits	30
" metasomatic deposits	14
" contact-deposits	6

The remaining 10% of the copper production comes from miscellaneous ore deposits.

The methods of treatment followed are in outline as follows: Ores from fissure-filling and contact-deposits are subjected to concentration and are then smelted in blast-furnaces by semi-pyritic smelting. The coke on the charge is from some 8 to 16%.

Ores from bedded, as well as from metasomatic deposits are usually smelted direct in blast-furnaces by pyritic smelting. The coke consumed amounts to about 2 to 5%. The blast-pressure employed is 1.5 lb. per sq. in. as a maximum, but generally is kept close to 1 lb. The ash-content in the coke is from 20 to 25%. The ores treated in these smelters are obtained chiefly from mines belonging to the same company.

I may cite another example of a pyritic smelter, namely, the Inujima, which was recently designed by myself. The smelter consists of one blast-furnace 180 in. long for matting the ore, another furnace 110 in. long for matte concentration, and 14 Mabuki hearths for making blister copper. The smelter is for treating custom ores only, the charges being composed of a mixture of nearly 40 different kinds and sizes of ore. The blast-pressure is about 0.8 lb. per sq. in., and the amount of blast supplied per minute is about 500 cu. ft. of free air for each foot in length of the hearth. No coke is used in the normal charge, but coal fed at the tuyeres as auxiliary fuel amounts to about 3%. The average copper-tenor is nearly 2.6%, the amount of fine in the charge being 35%. This is subjected to a preliminary stamp-briquetting, or to pot-roasting, according to the nature and degree of fineness. The Mabuki hearth may be called a Japanese copper converter. It is rather primitive, and is very simple in construction, but it requires some well-trained workmen. In addition to the furnaces in regular daily operation in this plant there is a spare 160-in. blast-furnace and six extra Mabuki hearths.

LEADVILLE, COLORADO

DEVELOPMENT AT BIG EVANS.—DOWN TOWN MINES.—HILL TOP.

The discovery of an extensive shoot of low-grade sulphide ore on Prospect mountain has disclosed many important geological facts and has opened the way for intelligent development. In a cross-cut that is being driven from the 300-ft. level of the Silver Spoon shaft, in Big Evans gulch, north, under Prospect mountain, Warren F. Page, manager for the Luema Mining Co., has cut an extensive ore-shoot. The cross-cut has penetrated it for 120 ft. and is still in ore. The shoot occurs in the big Monarch fault-fissure, which has been identified with all the important orebodies in the Big Evans district, and is believed to extend to the west from the fault on the white-lime contact. This has not been determined to date, as no development, other than advancing the breast, has been done. The ore is stated to be a low-grade sulphide of a character entirely new to the Big Evans area. It carries gold, silver, and lead. Small bunches of extremely high-grade lead-silver ore have been found in the deposit, indicating that larger shoots of the rich ore will be found as development proceeds. The greatest importance attached to the discovery at this time is the fact that it proves the continuation of the regular Leadville formations under Prospect mountain. Following the theory of the pioneers, mining men have regarded Prospect mountain as a barren territory and have believed that the ore-bearing formations of the district extended no farther north than Big Evans gulch. This theory has now been exploded. The presence of ore on the white-lime contact proves that conditions in this area will be similar to those on Iron and Breece hills, and opens to development a territory greater in area than both of them. Ore has now been found in the upper and lower contacts of Prospect mountain. Last year, Mr. Page completed the Valley tunnel 1600 ft. into the mountain, and cut an immense body of iron oxide near the breast. A winze put down a depth of 100 ft. at this point remained in ore throughout. The conditions found made it impracticable to operate through the tunnel, and early last spring the driving of the cross-cut from the Silver Spoon shaft was started to get under this deposit. Although driving proceeded along the big Monarch fault, the cutting of ore before nearing the shoot opened above was not expected. The sulphide deposit was encountered a distance of 600 ft. north from the shaft.——Mr. Page states that the breast will be pushed ahead until reaching a point under the winzes, regardless of what conditions are found. Then cross-cuts and raises will be driven into the orebody to determine its extent and value.

A new leasing company with M. M. McGinnis, a well-known Leadville mining man, as manager, has secured a sub-lease from the Down Town Mines Co., on the P. O. S. property on Carbonate hill. Work is now under way repairing the surface buildings and installing machinery. The P. O. S. is one of the properties in the Down Town area, just south of the Penrose. It is believed to be rich in silver-bearing iron and manganese ore. The property, at one time a heavy producer of iron and silver, has been idle for twenty years.——The A. V. shaft, near the old Harrison reduction plant in California gulch, is now being examined and plans completed for erecting hoisting and pumping machinery. The A. V. is in the iron and manganese zone that extends west from the Carbonate property through the P. O. S., Bon Air, Bohn, and Home Extension, and at one time was an extensive producer. The present lessees hope to discover large bodies of high-grade manganese ore.

C. H. Weaver, of Toledo, Ohio, who for some time was associated with the late W. B. Anderson in operating the Anderson adit near Birdseye, has organized a company of Eastern capitalists for the purpose of developing several properties in the Leadville district. The company has purchased the Anderson adit and has changed its name to the Ohio. Operations have been resumed and a contract to drive the main adit 1000 ft. has been let. The adit is now 1700 ft. long, following a south-easterly course through Prospect mountain toward Leadville. This distance is almost entirely in the Webber grit with a few porphyry dikes. The addition of the 100 ft. is expected to bring the breast through the grit and blue lime and into the first contact. The recent discovery of ore through development carried on by Warren F. Page at the Silver Spoon shaft adds materially to the prospects of the Ohio project. Mr. Page's cross-cut is following a course a little west of north and at a greater depth than that attained by the Ohio adit, but if continued far enough the two projects will cross each other. Ore has been found by Mr. Page both in the blue-lime and the white-lime contacts, proving that these formations continue into Prospect mountain. A slight rise in the formations beyond the Page ground will bring the ore-zones directly in the path of the Ohio project. Both of these enterprises are of great importance to the Leadville district and are being closely followed by local mining men.

Mr. Weaver and associates have also secured a lease and bond on the old Hill Top property in Park county, just over the range from Iowa gulch. Plans have been completed for extensively prospecting the property by means of a diamond-drill in the hope of discovering a lower contact or any parallel ore-shoots which may exist beyond the big vein that for years made the Hill Top a famous producer of lead and silver ore. Bodies of marketable low-grade ore are also expected to be found in the old workings. Supplies for the winter are now being brought in.

COLORADO SPRINGS, COLORADO

CREEDE EXPLORATION CO.—WAGON WHEEL GAP FLUORSPAR MINE.

Things are livelier at Creede than they have been in many years. The increased prices of silver and lead have been partly responsible for this, but the advent of the Creede Exploration Co., a subsidiary of the American Smelting & Refining Co., which has a lease on practically all the ground on the Amethyst vein below the level of the Wooster adit as well as some of the Commodore mine above that level, has had more to do with the increased activity than anything else. This company is now unwatering the Amethyst shaft below the adit-level and expects to start unwatering the Commodore shaft as soon as the necessary machinery [illegible] be done, the extraction of the ore below the adit will be begun. The company has built a power-house near the Humphreys mill which will supply power for the Commodore workings. The power for the Amethyst shaft comes from the city plant at Creede.——The Mineral County M. & M. Co. is mining a little ore from the Happy Thought ground, which is being treated in the Humphreys mill.——The Del Monte & Chance Mining Co. is operating the Last Chance-Del Monte-New York property from which about 300 tons per month is shipped.——The lessees on the Bachelor are shipping about 500 tons per month.——Collins & Co. are said to have opened up a large body of ore of good grade on the Quintette at Sunnyside. They have already started shipping.——The smelters have refused to receive more than a limited quantity of low-grade ore, which has greatly restricted the output of the district. ——The fluorspar mine at Wagon Wheel Gap is producing 1500 or 2000 tons per month. About 40 men are employed at present and more will be put on as soon as additional bunk-houses can be built. A track has been built from the lower terminal of the tramway to the railroad, a distance of nearly two miles, and the spar is hauled in cars of 3000 lb. capacity by horses and dumped into the railroad cars; a great improvement over wagon haulage. An air-compressor is being erected at the mine. The underground development has been very satisfactory. The vein is over 20 ft. wide in some places. This mine was opened primarily to supply flux for the Colorado Fuel & Iron Co.'s plant at Pueblo, but the product is now being shipped to steel works all over the country and some, under a guarantee of 98% purity, to chemical works.

MAYER, ARIZONA

Arizona Binghampton.—Black Canyon.—Big Ledge Copper.

The output of the Arizona Binghampton mine, east of Mayer, is steadily increasing. The first six days of October showed a daily output from the flotation mill of 12,500 lb. of copper in a concentrate averaging between 25 and 27% copper. The mill-head averages 4% copper. On this basis, the output of the company for this month will be better than 375,000 lb. of copper. The management announces that at the 500-ft. level an extension of the main orebody from the 400-ft. level has been cut, showing 25 ft. of 9% and 15 ft. of 3% copper. The new strike was made in a raise from the 600-ft. level. It is expected that this same orebody will be broken into at any time on the 600. This strike of ore is considered to be one of the most important ever made in the Mayer district, as it proves another portion of the great Jerome belt.——On the adjoining property, owned by the Copper Queen Co. under the management of Claude Ferguson, a good sized vein of low-grade copper ore has been found 200 ft. below adit No. 2 in a cross-cut run from this winze. Extensive development work is in progress in this property.——The Half Moon Co. has added a gasoline-hoist on the shaft on the next property south of the Copper Queen and Arizona Binghampton. This shaft is to be sunk 500 ft. A number of strong copper veins outcrop here. ——Ore is expected at any time in the adit being driven by the Red Metal Mining Co. The property is being developed by Bisbee capital. The adit is being driven to strike a known vein that has been opened by a shaft.——Thorough testing of the ore of the Pocahontas mine is being made with a view to building a flotation mill. The mine is two miles from Mayer. The company has recently been financed by Eastern capital through the president, W. H. Skinner.

The Black Canyon Mining Co. is opening a large body of silver-lead ore in a property four miles south of Turkey station. A drift has been run on a vein that shows 500 ft. of continuous ore, 12 ft. wide at the face. Kansas City capital is interested.——A deal has been closed recently by Bisbee investors for a copper property 12 miles south of Mayer. The ground was owned by J. L. Dillon and associates. The property is to be extensively developed by an adit.

Anderson and Birch, owners of a silver-lead mine one mile west of Humboldt, shipped a carload of high-grade silver ore this week. A flotation mill with a capacity of 75 tons per day is treating $15 ore. The Silver Belt company on the north is developing an extension of the same orebody.

Work is to be resumed within a few days on all the properties of the Big Ledge Copper Co. The officials of the company will arrive from Duluth. The size of the orebody recently found in the Butternut at the 425-ft. level, is increasing. The grade continues at 8% copper. The company's smelter is to be completed by the first of the year. By then the company expects to have its two principal mines, the Butternut and Henrietta, opened sufficiently to run its smelter at full capacity.

MAMMOTH, UTAH

Big Indian.—United Tintic.—Empire.—Marshall Lake

Mammoth mine has paid the following dividends this year: May 12, 5c., or $20,000; May 29, 25c., of $100,000; June 29, 25c., or $100,000; August 29, 25c. or $100,000. The dividend paid on October 12, of 10c., will bring the total this year up to 90c. per share, or $360,000, making the grand total to date $2,780,000.

The copper-reduction mill at Big Indian is now ready to commence operations. The ore-bins are full. During the past week practically every piece of machinery in the mill has been tested and found to be in satisfactory condition. The copper-solution tanks are being steamed and are already watertight. Preston G. Peterson, secretary and manager for the Big Indian company, who is at the camp, stated that he believes the mill may be producing copper by October 15, that all the machinery will be properly adjusted, and the preliminary troubles overcome. Sulphur for the manufacture of acid is being freighted to Big Indian, and ample supply is already on hand. At first 100 tons of ore per day will be milled, but when everything is running smoothly the plant will be operated at capacity, namely, 300 tons. Fifty feet of work will give the Tintic Standard company a connection between the new shaft and the old workings. This connection will mean much to the Standard, giving better ventilation of the workings and more economical ore extraction. This week the Tintic Standard will send out about six cars of ore.——George H. Dern, manager for the company at Silver City, met with an accident while driving his auto near Hailey, Idaho, last week, and broke three ribs.

A meeting has been called at Salt Lake City for the stockholders of the United Tintic Mines Co. to consider the consolidation with the South Standard Mining Co. It has been suggested that the properties of the two companies be consolidated either by increasing the capital stock of the South Standard to 1,300,000 shares at 10c. par value, or by forming a new corporation with capital stock of 1,300,000 shares at 10c. par value. In case the first method of re-organization is favored one share of South Standard will be issued for two shares of the United Tintic Mines Co., and the balance of the stock placed in the treasury.——The Iron King Consolidated Mining Co., C. E. Loose, president, Preston G. Peterson, secretary, has filed amended articles of incorporation with the clerk of Utah county. The amendment increases the capital stock from 1,000,000 to 2,000,000 shares of the par value of $1 each. Of this, 1,000,000 shares have been subscribed to the original incorporators and other stockholders, and 1,000,000 shares will be placed in the treasury to be disposed of by the directors for working capital.

Good progress is being made in the development of the Empire mines, at Mammoth. Two cross-cuts are being driven by the Walter Fitch Jr. contracting company, one toward the eastern portion of the property at a depth of 1800 ft., and the other toward the west at 1500 ft.——Driving is still in progress on the 2000-ft. level of the Colorado Consolidated mine, and it is understood that the management is preparing to sink on the fissure which has been followed for some distance on this level. Aside from this work there is the usual amount of leasing being done in the Colorado mine.

A gold brick weighing 122 lb. and worth $24,424 was shipped recently by L. T. Holte, from his mine in the Marshall Lake district, 58 miles north of McCall. This brick represents the clean-up for one month. Mr. Holte bought this mine in November 1915; since then it has produced $300,000 in gold, and during the present year at the rate of $1000 per day. For the month of August Mr. Holte cleaned up $60,000. The mill has a daily capacity of 40 tons. From 40 to 60 men are employed.

WILLOW CREEK, ALASKA

RAE-WALLACE.—GOLD CORD.—FERN AND GOODELL.

Most of the promising prospects and mines are on the tops of rugged sharp mountains, and transportation of the ore to the reduction works is by aerial tramways. Weathering goes on rapidly and slides change the face of the hills and uncover new outcrops daily; talus slides make it necessary in many places to run in open-cuts to expose the formation, but the district as a whole is one where ore is present everywhere. The wagon-road from the railroad-town of Wassile, over which all supplies must come to this district, is being patched up again this year and, while such work is not of a permanent character, it improves conditions considerably. It is to be hoped that the Alaska Road Commission will give this district more attention, so that it will not cost the operators all they make to get in supplies and machinery. Willow Creek has produced more than $3,000,000 and has paid from the surface. Freight costs $80 to $100 per ton from Seattle, and over the so-called wagon road $2.75 per 100 lb., making the cost of getting materials almost as much as if the district were 100, instead of only 16 miles from a railroad. High-grade ore, which local miners call a telluride, has been found in a number of places on Fishhook creek. A grab-sample of the ore found on the Rae-Wallace property gave $1800 per ton, and arrangements have been made to ship out 150 sacks. This property was purchased early in the season from William Martin for $10,000, and from the present outlook the purchase price will soon be won from the ground.——Four miles below the camps now in operation the country grades out to large flats and rolling hills, heavily timbered, and it is only a question of time until a large central plant will be erected to work all the properties through a long tunnel, doing away with the present excessive mining costs. Ample power can be secured from the Little Susitna river and, with mine-timber near, it will be an ideal location. At the present time all ore assaying less than $20 per ton is discarded. As no suitable timber is found near the mines, it must be brought up early in the year over This vein has been traced over the mountain for nearly a mile, and miners on the other side, on the Archangle creek, claim that they have the northern extension. A 2350-ft. aerial tramway has been erected; the ore is run to the Independence mill for treatment. A mill will be built next year. The owners are Charles Horning, Charles Bartholf, the Isaac brothers, and Byron Bartholf.——The property adjoining the Gold Cord on the south is owned by Kemp and Bartholf, and five strong veins, averaging $72 per ton, have been developed by open cuts.——The Brooklyn claims, owned by Milo Kelly and associates in Fishhook creek, has shown up well; the main vein is a continuation of the Independence.——William Martin of the Alaska Free Gold Mining Co. has had a successful season, despite the labor troubles brought on by I. W. W. agitators, which forced a close-down early in the year. A vein of good ore has been discovered behind the present mill.

The Independence Mining Co. has done only development work this year, and, as it should be one of the biggest producers in the section, the output for the district will suffer. High-grade ore has been found in many places.

The Rae-Wallace company, which has a bond on the Skyline group, is developing the mine and erecting a 50-ton Gibson mill. A 1600-ft. aerial tramway will be necessary to bring the ore down from the mine. A mill, bunk-house, and assay-office will be built during the winter. The main vein is near the top of the mountains directly above the mill-site. The ore is free-milling and carries $35 per ton. The original discovery of high-grade was a 4-in. vein, but, in stripping this, four more parallel veins were uncovered, running from $1800 to $5000. From the general pitch and strike of this, it is thought by many to be a continuation of the Mabel vein. This venture is backed by Wallace, Idaho, capitalists.

INDEPENDENCE MINE, WILLOW CREEK

At the Fern and Goodell, at the head of Archangle creek, a 100-ft. adit exposes a strong vein. The ore averages $15. At the surface this ore was a free-milling white quartz but changed with depth to a blue sulphide ore.——Adjoining the Fern and Goodell property on the south is the Talkeetna mine. Though little development work has been done, a mill and aerial tram have been

and the striking of high-grade this year makes this one of the best little producers in the country. A Denver mill is in operation on this property; aerial tramways bring ore from the mine.——Molybdenite has been found in many places, but no development work has been done.

COBALT, ONTARIO

NIPISSING.—MCKINLEY-DARRAGH.—KIRKLAND LAKE.

During the first week of October, 11 Cobalt companies together shipped 29 cars, containing 2,184,919 lb. of ore. This is by far the highest record this year, and, in fact, is the first time in recent years that the aggregate has amounted to more than 2,000,000 lb. during a single week. The total shipment during September was 63 cars, containing 4,600,082 lb., thus leading by a considerable margin all preceding months of the current year.——The monthly report of development at the Nipissing mine by Hugh Park, manager, states that during September the company mined ore of an estimated value of $349,258. The total ore mined from the Nipissing during the nine months ended September 30 amounts to $2,406,541. No new veins were found during the month, but all stopes continued to produce satisfactorily, and nearly all of the ore-shoots are extending their previously known limits.——McKinley-Darragh is in a strong financial position according to a statement sent out with the last dividend checks. Cash in bank amounts to $180,687.96 and ore in transit and at smelter amounts to $98,700. This with ore at the mine ready for shipment brings the total to $392,887. Dividends paid by this company equal 229% of its issued capital. The new oil-flotation plant will be completed by the end of this month.

The Mining Corporation of Canada is producing at the rate of 5,000,000 oz. per year and by the end of the current year will have 28,000,000 oz. to its credit. The company has acquired control of the old Waldman property and preparations for development are being made.

In the gold camps the spirit of optimism is running high, and the declaration of peace in Europe will spell for Porcupine and for Kirkland Lake the beginning of an era of prosperity that will eclipse all past records in the annals of precious-metal mining in northern Ontario. The amount of ore-reserves at the various mines is estimated at $80,000,000.

In the Kirkland Lake area, the Tough-Oakes and the Teck-Hughes are operating their mills at full capacity. The mill at the Lake Shore mine will be completed by January, adding to this camp. The Kirkland Lake Gold Mines, which is owned by the Beaver Consolidated of Cobalt, will also soon have a mill, excavation for the foundations having already commenced. Owing to defective machinery, underground operations at the Wright-Hargreaves mine have been suspended; 10 men are employed on surface making foundations for a new and more powerful plant.

MADRID, SPAIN

FUEL COMMISSION APPOINTED.—GUADALCANAL COAL.

The reform in administration of the mining department of the Government has been productive of excellent results. Explorations for copper and lead are active in many provinces where there was practical stagnation a year or more ago. A commission has been appointed to make further studies of metalliferous prospects as well as to investigate the resources of potash salts withdrawn by the Government from location in the provinces of Barcelona and Lérida. The Minister of Fomento has also appointed a technical commission to make a detailed study of the oil prospects in the vicinity of Cádiz, where a number of wells have already been drilled. In addition, this commission will study the coal resources of Asturias and the underground waters of Almeria. The strike of the coal-miners in Asturias, which was declared on August 13, tied up the output of this vital necessary for a time, but an adjustment has been made by which it is hoped that further difficulty will be avoided. The demands of the workmen were not excessive, and were in no sense intended to cripple the industry.

Under the leadership of Viscount de Eza an association of the coal producers of Asturias is being effected, which is intended not only to further the development of coal mining in this large and important basin, but to regulate the industry in such manner as to meet the exigencies of trade after the War. It is understood that guarantees of some sort will be sought from the Government, before any considerable developments are undertaken, it being pointed out that while the difficulty of obtaining coal from abroad has resulted in the extraordinary development of the native resources, so that the indications are that the Spanish coal mines will meet the domestic demand in 1918, they will be laboring under a difficulty in competition with foreign sources of supply after the War, unless some means of protection is assured. Viscount de Eza affirms that it would be unfair to burden manufacturing industries in Spain with a higher cost by imposing a duty upon imported coal for the sake of protecting a new and struggling industry. It is thought that recourse will be had to a direct bounty upon the domestic product sufficient to protect the coal miners against cut-throat competition from abroad. Under these circumstances it is believed that a large development of the coking industry will be undertaken, and the Government has already proposed to erect an experimental plant for coal-washing in order to assist in the development of these deposits. In conjunction with the coal-washing, the Government engineers propose the distillation of the slate washed from the coal in order that the valuable combustible materials may be conserved.

A report was recently issued through the Geologic Institute written by Señor Mallada regarding the coal deposits at Guadalcanal in Sevilla. It is a very brief description of the geology of the district, and it also touches upon the relations of this deposit to those of Córdoba and Badajóz. It is believed that the coal deposit, which lies parallel to that of Bélmez, that is north-west and south-east, passing approximately about 3½ miles north-east of Guadalcanal, is connected with the coalfields of Fuente del Arco. The geologists of the Institute are working upon the problem in order to construct a geological cross-section of these coalfields for which data at the present time are not available. Analyses of the coal from the mine of San Epifania, near the Arroyo of Cortijo Viejo de Castelló are as follows:

	Samples		
	1	2	3
	%	%	%
Volatile hydro-carbon	36.60	36.80	38.00
Fixed carbon	49.07	48.54	50.00
Ash	14.33	14.66	12.00
Coke (by test)	63.40	63.20	63.00

A large part of this coal lies within 15 metres of the surface, and much of it is not deeper than three to five metres. The field lies nearly horizontal, and it is thought it will be possible to operate by first stripping the over-burden. The Government has undertaken to fix the rates of transportation on coal from the coalfields of the country.

In the Spanish zone in Morocco several concerns are operating mines of hematite, lead, and zinc. The Cia. Españolo de Minas del Rif exported from the Port of Melilla 6000 tons of hematite, and the Cia. del Norte Africana has shipped 87,957 tons of lead and zinc ore. The zinc occurs in the form of calamine, and the lead as galena. Development of mines in Morocco is hindered by inability to obtain bottoms in which to ship the output.

ALASKA

(Special Correspondence.)—W. Martin, of the Alaska Free Gold, is in town with the bi-weekly clean-up of the mill. He reports the mill, which is a Lane Chilean, as producing $100 per day.——Theodore Pilger, from Butte, has been making examinations in the Willow Creek district.——Joseph Morris, representing Spokane people, who has been prospecting on a group of copper claims on Iron creek, 40 miles from Talkeetna, is in town on his way south. He brought a large number of samples that have been assayed in the local office with gratifying results. He says the district is one of great possibilities, numerous well-mineralized outcrops exist that have never had a pick or a shot put into them.

Anchorage, October 4.

(Special Correspondence.)—Lynch brothers, diamond-drill contractors, recently completed a three-year contract at the Perseverance mine and have sent the machine and crew to Kennecott, where two machines are now working. During the term of the contract at Perseverance over 75,000 ft. of diamond-drill holes have been run. The Lynch brothers now have machines working at Jualin, Ellamar, Latouche, El Capitan, and two at Kennecott.

The trestle over the cave at the Treadwell has been started; most of the piles are in place.

Treadwell, October 7.

Orders have been received from Washington to close the Federal Mine Inspection Office at Juneau at once, for an indefinite period. John A. Huff, who has been in charge of the office for the past three months, has been instructed to report to Washington at the earliest possible moment. Sumner S. Smith, the Federal Mine inspector, is at present superintending the opening of the Government coal mine on Eska creek for the Alaska Engineering Commission. The records of the office will be sent to him there.

ARIZONA

GRAHAM COUNTY

(Special Correspondence.)—Since last December considerable development work has been done on the Silver Cable claim, part of the holdings of the Grand Reed Mining Co., which has been leased to T. Parks of Tucson. This property has a small well-defined vein of molybdate of lead. A road to this property has been completed and grading for a 100-ton mill commenced.

Safford, October 11.

GREENLEE COUNTY

(Special Correspondence.)—The smelter of the Arizona Copper Co. is the only active plant in the Clifton-Morenci district. It is operating on custom ores. The mines are being closed for an indefinite period, irrespective of the decision of the Federal inquiry, which is expected to take place in a week or so.

Clifton, October 12.

The Mexican and Spanish miners who had voted to return to work at the mines at Clifton have voted to rescind their remain idle. Sheriff Slaughter, who is holding 70 Mexicans of the 400 who marched on Clifton from Morenci, contradicting the statement recently given out by Gutierrez, the Mexican consul, in which the latter said the Mexicans were going peaceably to Clifton to go to work. The sheriff said the advance party of Mexicans told him and his deputies they were going to run the 'scabs' and Americans out of Clifton.

MARICOPA COUNTY

(Special Correspondence.)—A syndicate, formed last May to re-open the Buckhorn and Boulder mines in the Goldfield district, is now forming a company to further develop the property. In cleaning out and re-timbering the old shaft and workings high-grade carbonate ore was found, and also a lode of chalcopyrite beneath the oxidized zone. Cross-cutting has proved the ore to be over 15 ft. wide. While sinking the shaft at the Abe Lincoln property, a body of chalcopyrite ore has been opened at the 96-ft. level which appears to be the main orebody. It is planned to sink to a depth of 500 ft. A compressor and other machinery have just arrived at the mine and orders have been placed for an electric pump.

Phoenix, October 11.

"While the labor commission is in Arizona it will endeavor first to settle all existing labor troubles in the mining districts by mediation," stated William B. Wilson, secretary of labor and head of the special commission appointed by President Wilson to investigate and act in labor troubles of all kinds throughout the West, which began its labors at Phoenix, October 5. Charles H. Moyer, of Denver, president of the Mine, Mill, and Smelter Workers union, arrived on the same train that brought the commission. Members of the commission are William B. Wilson; J. L. Spangler, of Pennsylvania; Verner C. Reed, of Colorado; John H. Walker, of Illinois; and E. P. Marsh, of Washington. Felix Frankfurter is secretary, Max Lowenthal, assistant secretary, and R. B. Horner, clerk.

MOHAVE COUNTY

(Special Correspondence.)—The new hoisting plant at the Chloride Queen has been completed and sinking of the 100-ft. shaft another 200 ft. will commence immediately. Development work at the United Northern will be resumed as soon as the new machinery, now on the property, is in place. Bids are being received for sinking the shaft of the Oatman-Golconda from the 25 to the 425-ft. level. This property consists of six claims lying between the Pioneer and Leland, both of which have been producers in the past.

Oatman, October 15.

PIMA COUNTY

(Special Correspondence.)—The controlling interest in the Mile Wide Copper Co. has purchased the Wakefield-Belmont group of 19 claims in the San Xavier district and development work is to commence immediately. Increased production in the San Xavier district has prompted the continuation of the Twin Buttes railroad to the Bush, Vulcan, and Mineral Hill properties. These properties are now hauling to Tucson.

Tucson, October 13.

Superior district announces that actual development will start in a couple of weeks. A road has been completed and buildings are being erected. It is announced by the Queen Creek Copper Co. that, in the inclined shaft which is being sunk to the 500-ft. level, good gold-copper ore has been opened from 300 to 325 ft., the present bottom of the shaft. The Magna Chief Copper Co. is making preparations for diamond-drilling to commence November 1. A body of silver-lead ore, exceeding any yet developed, has been opened up on the Cincinnati claim of the Ray silver-lead mine. Machinery which arrived recently is being erected. There are about 85 men working at the mine and 285 burros packing the ore.

Florence, October 9.

YAVAPAI COUNTY

(Special Correspondence.)—When the new addition now being made at the United Verde crushing plant and smelter is completed, which will be early next summer, it is expected that the United Verde will double its present output of 6,000,000 lb. of copper per month.

Jerome, October 16.

YUMA COUNTY

(Special Correspondence.)—Mr. Ormsby, superintendent of the Wenden copper mine, has recently reported a strike of high-grade ore in a cross-cut from a shaft. The Black Giant has added another shift of men. The Rainier Development Co. has re-organized as the Ranier Mines Co., having taken over claims of the Harqua Hala Ridge Mining Co. Lessees on the Bullard property have a force of fifteen men working. They are using an Ingersoll compressor and two jack-hammers. The Wenden King has closed down temporarily until repairs can be made on its 50-hp. Fairbanks-Morse gasoline hoist.

Wenden, October 18.

CALIFORNIA

COLUSA COUNTY

The erection of the new machinery to replace that destroyed by fire at the Cerese mine is nearly completed.

DEL NORTE COUNTY

The Cold Spring chrome property, owned by G. W. Gravlin of Grants Pass, has been in operation for several months. Although extensive prospecting has been done the mine has produced only a few hundred tons of ore.——Chrome mining is still increasing in the Low Divide and French Hill districts, many tons being shipped daily. William Hawkins has located a number of chrome deposits in the Low Divide district and is shipping ore to San Francisco.

ELDORADO COUNTY

John C. Evanis is mining one ton of 40% chrome ore daily from a deposit about three miles north-west of Clarksville. He is doing the mining alone and gets $16 per ton for ore at the mine, from where auto-trucks haul it to Folsom.

INYO COUNTY

Important changes are in progress in the mills at Tungsten. The Standard company, with L. E. Porter, superintendent and general manager, will have a new Marcy mill in operation within the next two or three weeks. All but one part of the machinery is now on the ground. In the meantime the old plant is running steadily while construction of the new mill is going on, there being room to place the new machinery without interfering with the old for more than three or four days in the final stages. The new mill will bring the capacity up to 200 tons per day.

Base Metals Mining Co., at Big Pine, is shipping 300 tons of silver-lead ore per month to the U. S. Smelter Co. at Salt Lake.

NEVADA COUNTY

The Columbia Consolidated has completed the construction of a 20-stamp mill at the Ocean Star mine, near Washington.

SHASTA COUNTY

(Special Correspondence.)—A carload of ore per day has been shipped for the last month from the Star shaft of the Bully Hill mine to the smelter at Kennett. The proceeds help to continue the development work, on which $75,000 has been spent already by the Arnstein interests. The company is contemplating putting in a flotation plant at Copper City.

Kennett, October 15.

(Special Correspondence.)—Six hundred thousand brick used in constructing furnaces and roasters for the Balaklala Copper Co. at Coram have been sold to a Redding builder. The Balaklala has a long-term contract for treatment of its ore at the Mammoth smelter, at Kennett, three miles away. ——J. A. Wilson, with the consent of the Superior Court, has bonded the land of the Drew minors near Gas Point for dredging purposes. Prospecting with a Keystone drill will begin shortly.——The Mammoth, Mountain, and Balaklala are running, though with crews smaller than before the strike. The Mammoth has resumed work at its outside mines—the Stowell, Spread Eagle, Friday-Lowden, and Sutro. The Mountain Copper Co. has resumed work at Hornet, Iron Mountain, and United States mines, the last being near the Balaklala.

Redding, October 17.

A second manganese furnace has been completed at Heroult, increasing the capacity from 200 to 400 tons of ferro-manganese per month. Ore is being shipped from a number of points, and several new deposits are being worked.

SIERRA COUNTY

(Special Correspondence.)—W. A. Loftus, W. A. Sage, and other Los Angeles capitalists have purchased the huge concrete restraining dam recently erected on Slate creek, and acquired 90,000,000 ft. of virgin gravel channels in the vicinity of St. Louis, Scales, Howland Flat, Poverty Point, Port Wine, Whiskey Diggings, and other old hydraulic-mining camps. A large adit will be driven to Poverty hill through which tailing from the several properties will be run to the restraining dam. The dam is 50 ft. high and is so situated that its height can be easily increased to 150 ft. The company is incorporated under the title of the Loftus Blue Lead Gold Mining Co. Steel flumes will be used in place of the usual ground-sluices. ——A new deposit of gravel has been uncovered in the Herkimer mine at Howland Flat, showing considerable coarse gold and occasional nuggets. Supplies have been shipped in and operations will be pushed steadily throughout the winter season. G. De Brettville is manager.——High-grade ore containing arsenical pyrite has been intersected 220 ft. from the adit-portal of the High Commission group. Water is hampering mining; work probably will be suspended until the spring. The claims are owned by John Reid and Sheriff Johnson.—— Mining in the north end of Sierra county is active, but shortage of skilled miners is keeping down production. Around Sierra City a number of mines are producing steadily, notably the Sierra Butte and Monarch Consolidated. At the latter driving of the new main working adit is proceeding steadily. Prospecting is active above Downieville, and in the Alleghany-Forest district the situation is satisfactory.

Downieville, October 21.

TRINITY COUNTY

(Special Correspondence.)—The development of the Silver King, four miles west of Redding, is attracting the attention of mining men. L. E. Parker, of Butte, Montana, after a thorough examination, organized the Silver King Mining Co., of Salt Lake, at San Francisco, with Robert Sloan as president. Over 1200 ft. of development has been done. The silicious copper ore is mined from an altered andesite dike 70 ft. wide and more than a mile long. The ore is found on both the hanging and foot-wall and runs 33 oz. silver to a unit of copper. Two cars are being shipped every ten days to the smelter at Kennett. L. E. Parker is general manager and M. M. Johnson, consulting engineer.——The Barron Gold mine, nine miles east of Ashland, Oregon, has been purchased by H. J. Sallee of Redding from the Byron L. White estate. The vein is 16 ft. wide, and contains gold, silver, and antimony in paying quantities. Development consists of a 200-ft. shaft, 750 ft. of levels, and a 5-ft. winze. Sixty men are to be employed with John Kemple as superintendent. The ore will be shipped to Kennett smelters.

The 10-stamp mill, at the Strode mine, near Carrville, is being run continuously. The owners have spent $40,000 on development and plant in the last 14 months. H. L. Stewart, one of the owners, is superintendent.

Redding, October 15.

TUOLUMNE COUNTY

(Special Correspondence.)—The Italian Camp group of mines, situated above Phoenix lake, has been acquired by an Eastern company. G. L. Caulfield is superintendent; a few men are getting things in shape preparatory to more extensive work. It is stated that the company will re-open the old Belleview, an adjoining property that has been idle 25 years.——Twenty men are employed at Black Oak, near Soulsbyville. The mill has been operated intermittently on ore from the 1400 and 1900-ft. levels, with satisfactory results.——A new body of rich ore has been uncovered on the 810-ft. level of the Confidence mine. A 100-ton roller mill is being erected.——A shaft is being sunk at the Erin Go Braugh, situated north of the Golden Rule, near Stent. The mine is held under bond and lease by W. E. Booker and Ed McGinn. The shaft, which is 100 ft. deep, exposes some good ore. A pumping plant has just been put in.——The erection of a mill on the Ruby mine, in the Groveland district, is contemplated by A. Steinmetz, owner of the property. A considerable quantity of $30 ore has been taken out in development, and the vein is expected to average $8 per ton. The main adit is in 190 feet.

Sonora, October 17.

COLORADO

BOULDER COUNTY

(Special Correspondence.)—John L. Bell has made a big strike on his properties near Sunset. Mr. Bell is one of the oldest miners and prospectors in Boulder county, with several big producing mines to his credit.——J. D. Hume, of Post City, Texas, is at Boulder attending a meeting of the stockholders of the Tungsten Mountain Mining Co., of which he is a heavy stockholder.——The old camp of Ward is having a real boom. The Big 6, White Raven, Modoc, Utica, Grand View, and scores of others are busy. The price of silver has given a new impetus to Ward, gold and silver properties vying with each other for supremacy.——The old Independence mine above Eldora will be re-opened; work has already been started, with Henry Martin in charge. The property belongs to the Rockwell Bros., of Denver.——Several auto loads of men interested in the Huron mine were down this week looking at the property. Active operations will be continued all winter, and steady shipments made. Guy M. Woodring has been in Denver completing arrangements for extensive improvements.——C. E. Kahler, of Columbus, Ohio, president of the Consolidated Leasing Co., returned home for a few days. The machinery for the mill has arrived, and it will be completed as soon as possible. W. F. Harper is manager.——Gunter, Dirkis, and Ludlow will work their property near Hesse this winter.—A. R. Scofield of the Dixie has opened up a fine body of high-grade ore.——John A. Wilson has purchased his partner's interest in several claims at Woodlawn.

Eldora, October 9.

(Special Correspondence.)—The Moore Mining, Milling & Developing Co. is doing a big amount of development work on the Livingston dike at Sugar Loaf. The company is composed of Boulder men and has been operating in this district for the past three years. The Nil-Desperandum group of mines in the Sunshine district has purchased the Inter-Ocean property. It has also enlarged and increased its milling plant and is working on a large scale. It is heavily financed. This property was one of the old producers in the early days and is among the famous old properties of this telluride camp. S. M. Brandt is manager of the property.——The Tungsten Mountain Mines Co. was greatly pleased with the report of Mr. Clark, the manager, given at the annual meeting of the stockholders. Increased development work will be conducted.——Around Lakewood centres the big producing mines of the tungsten belt: The Clyde mine, operated by the Wolf-Tongue company, and the Gale. The Vasco is doing a big business. The Primos Co. is operating crews on Conger, Lonetree, Corkscrew, Beddig, 18, and 4. The milling capacity has been doubled during the past year. Other improvements are also being made. Some of the most productive private leases are the Rake-Off, Lily, Old Donnelly, Spider-Leg, Bob Cat, and No. 19 on the Coon patent.——The fifth annual meeting of the Colorado Metal Mining Association will be held at Denver next January. The value of the metal output in Colorado for 1916 was $49,200,675.——The big silver excitement at Caribou still continues, and another strike is reported. Miners and prospectors are rushing in and all available ground is being taken up. Several sales have been reported.

Eldora, October 15.

LAKE COUNTY

(Special Correspondence.)—The Derry Ranch Gold Dredging Co. paid its eighth dividend on October 1, and the total dividends up to that date amounted to 100%. It has produced $278,301 up to that time, and the ground averages 28c. per yard. This is the second year of operation. The property has a life of five more years, in which it is expected that the present yield will be maintained.

Leadville, October 20.

IDAHO

SHOSHONE COUNTY

The second stack at the big smelter of the Bunker Hill & Sullivan M. & C. Co., at Kellogg, has been blown-in; the company has been operating steadily for several months with one furnace; a third one, completing the plant as now planned, will be blown-in about November 1, giving the smelter a capacity of 600 tons of ore or concentrate per day.——The Hecla Mining Co., operating at Burke, has declared its regular October dividend of $150,000. This disbursement will be on a basis of 15c. per share on the 1,000,000 shares issued. It will increase the total payments this year to $1,500,000 and the grand total to $6,895,000. The company reported a net profit of $1,681,059 last year. It ranked third in the district, being exceeded only by the Hercules and the Interstate-Callahan in the order named.——The first unit of a mono-rail road, which will transport lead-silver concentrate from the mill of the Highland-Surprise mine to the end of the spur which will be built this winter up Pine creek, in the Coeur d'Alene, has just been completed. The concentrate is now being hauled by wagon, at a cost of approximately $4 per ton for the nine miles from the mine to the O.-W. R. & N. The mono-rail will cut the cost to 25c. per ton. The road consists of a light steel rail spiked to the top of an 8 by 10-in. timber, supported by wooden uprights and braces. Below the top timber are wooden guide rails. The locomotive has for its motive power a converted Ford engine attached to the driving wheel by chain and sprocket. On either side, low enough to throw the centre of gravity of the car below the metal rail, are bins, each of which will contain a ton and a half of concentrate.——A dividend of 3½c. per share has been declared by the Tamarack & Custer, one of the big mines of the Coeur d'Alene, making the disbursement $53,287. Stock of record October 15 will participate. This is the third dividend the company has paid. In May and again in April of last year it paid 2c. per share, makink disbursements totaling $71,050.

MONTANA

LEWIS AND CLARK COUNTY

(Special Correspondence.)—The old Honey Comb, Mt. Pleasant, Blue Bird, and Hickey claims are being developed.——The Barnes-King Development Co., owner of the Shannon, paid a dividend of 10c. per share on 400,000 shares. This is the first dividend paid under the present management. The company is working the Shannon and the Piegan-Gloster mines at Marysville and the North Moccasin group, together with the original Barnes-King.——The Economy Mining Co., which is operating 15 miles east of Helena, has unwatered the 300-ft. shaft and is mining a high-grade gold-copper ore. Five mines are shipping ore from Grass Valley district and a like number from Scratch Gravel hills.

A. B. Woolvin has been at Helena consulting with his engineer, L. S. Ropes, who has charge of a large group of gold-mining claims in the Marysville district. The claims are known as the Marysville Gold Syndicate group, but were formerly called the Marysville Gold Mining Co.; they were secured at the instance of M. L. Hewitt, who interested Maurice Eisenberg, of New York. After spending considerable money Mr. Eisenberg transferred the entire holdings to A. B. Woolvin and associates of Duluth, Minnesota. Considerable development has been done and the showing is fair.——The Amalgamated Mines Co., a new corporation, has acquired the Free Coinage group in Lump gulch and has located 25 additional claims. The aim is to unwater the Free Coinage which is about 350 ft. deep. No work has been done on the property for 25 years. The company also owns ground in Grass Valley, California, and in Nevada.——The Economy Mining Co., operating in Mitchel gulch, 12 miles east of Helena, has unwatered its 300-ft. shaft by electric power, lately put in, and has resumed mining.——The Mt. Washington mine near Wickes continues to send out one 50-ton car of ore per day. The ore is taken from the vein at the face of the adit, which shows 14 ft. of galena ore.——Grass Valley and Scratch Gravel hills are sending a considerable tonnage to the smelter every week.

Helena, October 15.

NEVADA

HUMBOLDT COUNTY

The completion of the 350-ton mill of the Rochester Combined Mines Co. has been delayed by the difficulty in obtaining some essential machinery; it will probably not be started before January or February. The company is centring its activities on the Happy Jack shaft, Shepard adit, and Kromer winze, work having been discontinued on the Link adit, and many of the outlying claims.——The Packard mill is crushing 100 tons per day. The new Walter open-cut, as well as the Margrave cut, is producing a quantity of good ore, mined entirely by 'glory-holing.' Mill-ore is being produced from the 160 and 120-ft. levels, which are being rapidly developed. Forty-nine men are employed.——The sulphide orebody, cut by the Rochester Mines Co. several months ago, continue to produce excellent ore, with an increased ratio of gold to silver. The mill at Lower Rochester is treating 200 tons per day, and the monthly clean-up averages $65,000. The Packard-Rochester district has produced over $2,225,000 in the past four years. Aside from the $68,000 paid in dividends by the Packard, and payments made to original claim-holders in the district, stockholders have received no returns.

An announcement of much importance to the district was recently made by the Nenzel Crown Point Co. The United States Smelting, Refining & Mining Co. has contracted to take 100 tons of ore per day from the Nenzel Crown Point, Lincoln Hill, and Buck-and-Charley, the greater proportion to be furnished by the Crown Point. The silicious ore is desirable as a flux and it is understood that a very low treatment-charge will be made. The ore will be shipped to the Kennett smelter. This will mean the rehabilitation of the Nevada Short Line railroad running from Oreana, on the Southern Pacific, to Rochester. The Nenzel Crown Point has blocked out a large quantity of low-grade silver ore. The Buck-and-Charley and Lincoln Hill are small properties at Lower Rochester.——Some good ore was recently found on the Hoffman and Beecham claims adjoining the old Relief mine, five miles south-east of Packard. The Relief was one of a series of phenomenally rich but shallow silver-gold deposits opened along the Humboldt Range in the 'sixties and 'seventies of the last century.

Lower Rochester, October 11.

NYE COUNTY

At the White Caps the past week has been devoted almost entirely to breaking ore for the mill, the average production of 100 tons per day being maintained with ease. Practically the only development work has been on the fifth level, the east orebody having been exposed during the week for an additional 32 ft. The total length of the ore so far is 65 ft., although each day's work is adding an average of six feet to this length. The level on the ore has been kept along the foot-wall of the vein, so that the dimensions between the walls cannot be determined. The orebody yields $30 per ton; next to the foot-wall assays run as high as $65. Car samples taken on October 10 for the entire day's hoisting gave an average of $38 per ton. The vein is as strong and well defined in the present face on the fifth level as anywhere in the mine.——The week's work at the Amalgamated is as follows: From the 600-ft. level the north cross-cut on the fissure has been advanced 16 ft., No. 11 north cross-cut east 14 ft., and the south cross-cut on fissure 34 ft. The total for the week is 64 ft. The recent development work in the mine has demonstrated conclusively that the two orebodies which had been mined in the Earl shaft from the 350-ft. level to the surface have come together at the present 600-ft. level. The average assay-value of the orebody, eliminating all samples which assayed higher than $17, is $14 per ton. There is still one more orebody known as the Swanson ore-shoot which remains to be proved from the present bottom level, and should this be of the same size and richness as proved in the Earl upper workings, it will yield over 8000 tons of $20 ore. The working shaft on the Earl is to be straightened, enlarged to double-compartment with manways, and timbered with eight by eight. From the present bottom of the shaft, 350 ft., it is to be sunk, and a raise started from the 600-ft. level of the Amalgamated workings, to make connection with the shaft bottom. This work is to be carried forward at once and the requisite underground surveys are being made. Owing to the fact that a complete hoisting-plant is in place at the Earl shaft in addition to the plant worked by the company several hundred feet farther west on Litigation hill, there will be no delay in the carrying on of this work, and development will be continued as well from the 600-ft. level in exploring new ground, to pick up the downward extension of some important orebodies which have yielded well. It is the intention of the management to remove the mill from its present position and re-build it on Litigation hill, convenient to the Earl working shaft. When this is done several important changes in construction will be made to reduce the operating cost and increase the tonnage. In all probability the stamps will be replaced by ball-mills, as the Litigation Hill ore is soft and friable and ideal for such treatment.

The week's work at the Manhattan Consolidated has been devoted to driving the east drift from the fourth level and the south cross-cut from the third level. The east drift is 90 ft. from the shaft. The formation in the face of the drift has changed into much tighter ground, being now a black hard lime-shale. The point where the east orebody should be found is still about 160 ft. from the face.

The Manhattan Red Top will shortly have the heavy flow of water, that has caused so much delay in shaft-sinking, within bounds. The management announces that a large Cameron sinking-pump has been shipped from San Francisco and should reach the mine before the end of the week. The quartz lease recently granted the Red Top by the Union Amalgamated Mining Co. is an important concession to the latter, as it gives access to some promising ground which on the surface shows several rich stringers of ore that can be easily reached from the Red Top's working shaft.

The Manhattan Morning Glory is prospecting in an inclined shaft, about 150 ft. east of the old Pine Nut shaft. This work is 90 ft. from the surface; some rich quartz stringers are being followed.

NEW MEXICO

Fire completely destroyed the plant of the Empire Smelting & Refining Co., situated a mile and a half east of Deming, on October 4, entailing a loss of $20,000, about half of which is covered by insurance. The fire started when a pipe carrying gasoline broke while the engineer, Horace Deull, was heating a burner under the oil-burning engine. Flames quickly spread to all parts of the structure and could not be stopped by the fire-department from Deming, which had to pump water from a reservoir near the smelter. Mr. Deull was slightly burned. L. C. Baker is the manager and W. H. Seamon the superintendent of the company, which is financed by Eastern men. The plant smelted lead-silver-gold ores and was doing a thriving business on ores from the districts surrounding Deming. The plant will be re-built at once.

OKLAHOMA

OTTAWA COUNTY

Protection for a mining plant in the shape of a barbed-wire fence 8 ft. high, charged with electricity, is something new for this district, but is what is planned by the W. E. Merrill Mining Co. property north of Quapaw. The plant is nearing completion and is one of the most up-to-date in the district. Besides the thoroughly modern mill, the company has erected a 10-room hotel for the accommodation of its workers, and a club-house for its officials, with a concrete floored garage that will hold ten automobiles.

The equipment at the Merrill plant comprizes three 7-cell rougher-jigs, with cells 36 by 42; one cleaner-jig, with seven cells of the same size, and a 5-cell sand jig. Power is furnished by Fairbanks-Morse oil engines, one of 150-hp. being used to drive the compressor, two of 100-hp. to drive the mill and hoist, and one 75-hp. for the sludge department. There are 18 sludge tables, a Dorr thickener and classifier, and a set of 42-in. high-speed rolls for re-grinding. There is also a 90-ft. classifying screen, which is an unusual feature for a sludge-room in this district. The mill is virtually completed and it is expected that the plant will be ready for operation by the first of November.

OREGON

JACKSON COUNTY

(Special Correspondence.)—There is evidence of steady activity and increasing interest in the mining and prospecting in Jackson county. The government price on copper has served to quell the excitement about that metal, but the established price is quite sufficient to keep the present plants active. Clark & Webb, of Medford, have leased the Gold Ridge gold and silver mine on Kanes creek, three miles south of Gold Hill, owned by T. C. Norris, of Medford. They have a crew operating a 3-stamp mill and developing a 20-ft. vein that averages $100 per ton in gold and silver.——M. G. Womack, of Medford, and M. A. Clark, of Ashland, representing Trinity county, California, people, have taken a lease on the old Red Oak gold mine, three miles south-west of Gold Hill, on Galls creek. This property produced $40,000 from the 60-ft. level 25 years ago, but since then it has been idle owing to litigation. Recently it fell into the hands of John Ralls and Claude Lawrence, local miners, who will re-open the 300-ft. level. The vein is three feet wide. W. M. Cowley, president of the Cowley Investment Co., and Howard H. Startzman, both of Seattle, have been inspecting the Copper King copper mine on Grave creek, 20 miles south-west of Gold Hill. They announce that their re-organization plans have been adopted. The mine is owned by the United Copper Co. of Seattle; it is fully equipped and there is a large orebody said to run 36% copper. A new wagon-road from the mine to Rogue River, eight miles west of Gold Hill, is being made.——Earl M. Young, of Rogue River, has sold his manganese property near Wimer, 12 miles west of Gold Hill, to Seattle steel people, who will develop the mine. The ore occurs in small veins running through serpentine.

Gold Hill, Oregon, October 9.

UTAH

UTAH COUNTY

The directors of the Iron Blossom Mining Co. met at Provo October 10 and declared the usual dividend of 5c. per share or $50,000, payable October 25 to stock of record October 15. This brings the company's total up to $3,110,000 and makes a total for the year of $250,000, a highly gratifying result. The mine was reported in excellent condition.——The directors of the Dragon Consolidated posted the usual dividend of 1c. per share or $18,750 with similar dates, making this company's total $75,000, all of which has been paid this year. This property also was reported in good shape.——Earnings of both mines were said to have been affected by the car-shortage, although development work had resulted in piling up a considerable quantity of ore which would be shipped just as soon as cars were available.

The Rico Wellington has been unable to ship on account of scarcity of labor and car-shortage, but this has not affected the condition of the mine, which was said to be in good condition. ——The directors of the Miller Hill Mining Co. have levied an assessment of 2c. per share on the company's capital stock, delinquent November 12, sale day November 30. This gives the company a total of $14,000 for further development work on the property in American Fork canyon.

The Mineral Flat Mining Co. has also levied an assessment of 1c. per share, delinquent November 12, sale day November 30. This will raise the sum of $10,000, which will be expended in development.

An assessment of ½c. per share has been levied on the capital stock of the Tintic Central Mining Co. The assessment becomes delinquent November 15, sale day December 8. The assessment is to cover costs of running the drift through the Iron Blossom into the Tintic Central on the 1700-ft. level of the Iron Blossom and on the 1500-ft. level of the Tintic Central. The drift has been extended into Tintic Central ground 140 ft., and is 500 ft. lower than any previous work on the property.

BOLIVIA

The deputies in session at La Paz have for some time been discussing a new project of law, which would place a tax upon copper, whether produced for treatment as raw ore or in the form of concentrate. Congress is also considering new treaties with Germany.

BRAZIL

The State Government of Minas Geraes is expected within a short time to impose a duty of 8% on all manganese exported. On account of this, energetic efforts have been made to export as large quantity as possible before such a tax might apply. In the month of August manganese ore transported by the Central railroad exceeded 50,000 tons, which is a quantity without precedent in the Republic.——Edwin Morgan, U. S. Ambassador to Brazil, has announced that the United States will send to Brazil at once 120 engineers, specialists in different technical branches, as an industrial commission to study the resources of the country, but with special reference to the possibilities of developing the timber resources of Brazil. The commission will also study the question of transportation, which, at the present time, is a great impediment to developing the mineral resources of several States.

For some time past capitalists from Rio de Janeiro have been producing coal in the form of briquettes made at the mines of Barra Punta, situated in the State of Paraná. These are being used with success by railroad companies and manufacturers.

CANADA

BRITISH COLUMBIA

The Trail smelter has issued a notice stating that in future no ores containing over 4% of zinc will be accepted. Practically all the silver and lead ores in British Columbia carry a larger proportion than this; the smelter used to allow shipment of ore carrying 8% without imposing a penalty. The effect of the new order was marked on the Vancouver Stock Exchange by a slump in practically every British Columbia mining property listed. In anticipation of the continued need of the Imperial Munitions Board, 15,000 tons of lead ore in addition to about 2000 tons of lead matte had accumulated at Trail. The board has now informed the management of the smelter that, owing to a reduction in shrapnel making, it is unable to take more than 1000 tons of lead per month until January 1. After that date the requirements of the board are uncertain. In order to meet the situation a reduction in the shipments became necessary. Shipment of lead ore from the Sullivan mine has been stopped. By adopting the policy of accepting only that ore containing 4% zinc or less the smelter will be able to put more ore into the furnace and in that way clean up the large accumulation on hand.——Negotiations are pending for the amalgamation of the interests of the Little Bertha and Pathfinder properties on the North Fork.——The Aberdeen mine on Ten-Mile creek has been re-opened. T. J. Corwin, the manager, has just returned from the coast, where he made financial arrangements for developing the mine.

MEXICO

(Special Correspondence.)—Notices have been posted throughout the camp at Cananea that all difficulties between the Cananea Copper Co. and the Mexican government have been adjusted and that the mine and smelter will resume operation shortly.

Cananea, October 15.

PERU

William C. Gastes has obtained a concession from the Government for the exploration and ultimate denouncement of gold placers near the headwaters of the Rio Inanbari and its principal tributaries. The concession grants Mr. Gastes three kilometres on each side of the main axis of flow of the river Inanbari and two and a half on each side of the axis of flow of the main tributaries. The islands and high banks of the rivers are included in the concession. It is understood that the concession restricts the rights of the concessionaire to operations conducted upon an important scale, which involves considerable engineering work.

PERSONAL

Note: The Editor invites members of the profession to send particulars of their work and appointments. This information is interesting to our readers.

F. W. Bradley is at Kellogg, Idaho.

R. M. Atwater was at Joplin last week.

John A. Dresser, of Montreal, is at the St. Francis hotel.

Victor Rakowsky, of Joplin, was in New York this week.

C. A. Romadka passed through San Francisco on his way to Eureka, Nevada.

George W. Bryant, formerly at Guanajuato, Mexico, is now at Joplin, Missouri.

Clyde T. Griswold is spending a vacation at home after a summer in Wyoming.

Henry Hay, formerly in South Africa and Chile, is now at Santa Ana, California.

Ernest Gayford, of the General Engineering Co., Salt Lake City, has been to Alabama.

Lloyd C. White, mill-superintendent for the Mountain Copper Co., is in San Francisco this week.

Harvey S. Mudd is at Washington, serving as assistant secretary to the War Minerals Committee.

S. B. Weed has become engineer to the Keweenaw company, succeeding G. M. North, now in the Navy.

Hoyt S. Gale is inspecting nitrate deposits in southern California for the U. S. Geological Survey.

H. W. Edmondson has been appointed Captain in the Engineer Reserves and is at Osage City, Kansas.

J. H. Devereux has become a member of the firm of Wilkens & Devereux, with offices at 120 Broadway, New York.

James J. Norton has been appointed ore-dressing engineer with the U. S. Bureau of Mines at the University of Minnesota.

T. A. Rickard has returned from St. Louis, where he attended the meeting of the American Institute of Mining Engineers.

W. A. Argall has joined the 148th Regular, Battery C., U. S. Field Artillery, and is at Camp Green, Charlotte, North Carolina.

R. B. Earling and Ferdinand Meinecke, Jr., have joined the 166th Depot Brigade, and are at Camp Lewis, American Lake, Washington.

Warren D. Smith passed through San Francisco on his way from the Philippine Islands to the University of Oregon, at Eugene, Oregon.

James F. Kemp has joined Dorsey Hager and M. Bates, making the firm of Hager, Bates & Kemp, petroleum geologists, at Tulsa, Oklahoma.

John C. Greenway, general manager for the Calumet & Arizona Mining Co., has joined the Army and expects to leave for France shortly.

Lawrence Addicks and A. F. Kuehn sailed from Vancouver on October 25 on their way to the Bawdwin mines of the Burma Corporation.

H. C. Miller is superintendent of the Round Valley Tungsten Co., at Bishop, California, succeeding Cooper Shapley, who is now general manager.

A. W. Fahrenwald has resigned the position of professor of metallurgy with the New Mexico School of Mines to accept an appointment with the A. S. & R. Co. at El Paso, Texas.

Pearson John Claudet died recently in London. He was the metallurgical partner in the well-known firm of F. Claudet, Ltd., assayer to the Bank of England. He survived his elder brother Arthur by only a few years, and although he had been in ill health for some time, his death was sudden and unexpected.

METAL PRICES

San Francisco, October 23

Aluminum-dust (100-lb. lots), per pound	$1.00
Aluminum-dust (ton lots), per pound	$0.95
Antimony, cents per pound	16.50
Antimony (wholesale), cents per pound	14.75
Electrolytic copper, cents per pound	23.50
Pig-lead, cents per pound	6.75— 7.75
Platinum, soft and hard metal, respectively, per ounce	$105—111
Quicksilver, per flask of 75 lb	100
Spelter, cents per pound	10
Tin, cents per pound	60
Zinc-dust, cents per pound	20

ORE PRICES

San Francisco, October 23

Antimony, 45% metal, per unit	$1.10
Chrome, 34 to 40%, free SiO_2, limit 8%, f.o.b. California, per unit, according to grade	$0.50— 0.60
Chrome, 40% and over	$0.60— 0.75
Magnesite, crude, per ton	$8.00—10.00

Manganese: The Eastern manganese market continues fairly strong with $1 per unit Mn quoted on the basis of 48% material.

Tungsten, 60% WO_3, per unit	26.00
Molybdenite, per unit MoS_2	$40.00—45.00

EASTERN METAL MARKET

(By wire from New York)

October 23.—Copper is unsettled and nominal at the Government price of 23.50c. all week. Lead is inactive and lower at 7 to 6.50c. Zinc is Mislese and steady at 8.25c. all week. Platinum remains unchanged at $105 for soft metal and $111 for hard.

SILVER

Below are given the average New York quotations, in cents per ounce, of fine silver.

Date			Average week ending		
Oct.	17	85.87	Sept.	11	96.80
"	18	84.33	"	18	100.70
"	19	83.87	"	25	107.66
"	20	83.50	Oct.	2	98.49
"	21 Sunday		"	9	91.72
"	22	83.00	"	16	87.01
"	23	82.50	"	23	83.85

Monthly Averages

	1915	1916	1917		1915	1916	1917
Jan.	48.85	56.76	75.14	July	47.52	63.06	78.92
Feb.	48.45	56.74	77.54	Aug.	47.11	66.07	85.40
Mch.	50.61	57.89	74.13	Sept.	48.77	68.51	100.73
Apr.	50.25	64.37	72.51	Oct.	49.40	67.86	....
May	49.87	74.27	74.61	Nov.	51.88	71.60	....
June	49.03	65.04	76.44	Dec.	55.34	75.70	....

COPPER

Prices of electrolytic in New York, in cents per pound.

Date			Average week ending		
Oct.	17	23.50	Sept.	11	25.83
"	18	23.50	"	18	26.25
"	19	23.50	"	25	24.41
"	20	23.50	Oct.	2	23.50
"	21 Sunday		"	9	23.50
"	22	23.50	"	16	23.50
"	23	23.50	"	23	23.50

Monthly Averages

	1915	1916	1917		1915	1916	1917
Jan.	13.60	24.30	29.53	July	19.09	25.66	29.67
Feb.	14.38	26.62	34.57	Aug.	17.27	27.03	27.42
Mch.	14.80	26.65	36.00	Sept.	17.69	28.28	25.11
Apr.	16.64	28.02	33.16	Oct.	17.90	28.50	....
May	18.71	29.02	31.69	Nov.	18.88	31.95	....
June	19.75	27.47	32.57	Dec.	20.67	32.89	....

Secondary metal is a term used by the U. S. Geological Survey for metal btained from scrap, sweeping, skimming, dross, and refuse, as opposed to rimary metal that is obtained from ore. The total quantity of secondary opper recovered in 1916, on the assumption that the brass re-melted had n average copper content of 70%, was 350,000 tons, of which 52,212 tons 32,425 tons more than in 1915) was recoVered by plants refining pri-ary metals and the remainder by plants treating only secondary ma-rials. The copper produced by smelters of the latter class includes 74,100 ns of pig-copper, 14,000 tons of copper in alloys other than brass, and 10,000 tons of copper in re-melted brass. These figures indicate an in-ease for 1916 of about 20,000 tons of pig-copper, 113,750 tons of copper brass, and a decrease of about 2000 tons in alloys other than brass. least 175,000 tons was recovered from clean scrap made in the course manufacture of copper and brass ware, so that less than 175,000 tons as obtained from ashes, cinders, and scrap, or from material that had tually been used and discarded. From the reports received it is quite ident that there was an increase in the quantity of old scrap brass and pper smelted and that there was a much larger proportionate increase in quantity of clean new scrap re-melted. The increase in the quantity scrap brass used amounted to 162,000 tons, and of this increase at least 0,000 tons was clean new punchings, filings, and clippings. The de-ase shown in the quantity of secondary copper in alloys other than brass undoubtedly due to including alloys containing tin (which properly should be termed bronze) under the classification brass. The value of the copper, both as metal and in alloys, is computed at the average yearly price of 24.6c., the average sales price of all marketable grades of new metal.

LEAD

Lead is quoted in cents per pound, New York delivery.

Date			Average week ending		
Oct.	17	7.00	Sept.	11	10.20
"	18	6.87	"	18	9.33
"	19	6.87	"	25	8.00
"	20	6.50	Oct.	2	8.00
"	21 Sunday		"	9	7.96
"	22	6.50	"	16	7.52
"	23	6.50	"	23	6.70

Monthly Averages

	1915	1916	1917		1915	1916	1917
Jan.	3.73	5.95	7.64	July	5.59	6.40	10.93
Feb.	3.83	6.23	9.01	Aug.	4.62	6.28	10.75
Mch.	4.04	7.26	10.07	Sept.	4.62	6.86	9.07
Apr.	4.21	7.70	9.38	Oct.	4.62	7.02	....
May	4.24	7.38	10.29	Nov.	5.15	7.07	....
June	5.75	6.88	11.74	Dec.	5.34	7.55	....

ZINC

Zinc is quoted as spelter, standard Western brands, New York delivery, in cents per pound.

Date			Average week ending		
Oct.	17	8.25	Sept.	11	8.21
"	18	8.25	"	18	8.29
"	19	8.25	"	25	8.46
"	20	8.25	Oct.	2	8.43
"	21 Sunday		"	9	8.50
"	22	8.25	"	16	8.50
"	23	8.25	"	23	8.25

Monthly AVerages

	1915	1916	1917		1915	1916	1917
Jan.	6.30	18.21	9.75	July	20.54	9.90	8.98
Feb.	9.05	19.99	10.45	Aug.	14.17	9.03	8.58
Mch.	8.40	18.40	10.78	Sept.	14.14	9.18	8.33
Apr.	9.78	18.62	10.20	Oct.	14.05	9.92	....
May	17.03	16.01	9.41	Nov.	17.20	11.81	....
June	22.20	12.85	9.63	Dec.	16.75	11.26	....

QUICKSILVER

The primary market for quicksilver is San Francisco, California being the largest producer. The price is fixed in the open market, according to quantity. Prices, in dollars per flask of 75 pounds:

Date		Week ending			
Sept.	25	105.00	Oct.	9	105.00
Oct.	2	105.00	"	16	100.00
			"	23	100.00

Monthly AVerages

	1915	1916	1917		1915	1916	1917
Jan.	51.90	222.00	81.00	July	95.00	81.20	102.00
Feb.	60.00	295.00	126.25	Aug.	93.75	74.50	115.00
Mch.	78.00	219.00	113.75	Sept.	91.00	75.00	112.00
Apr.	77.50	141.60	114.50	Oct.	92.90	78.20	....
May	75.00	90.00	104.00	Nov.	101.50	79.50	....
June	90.00	74.70	85.50	Dec.	123.00	80.00	....

TIN

Prices in New York, in cents per pound.

Monthly Averages

	1915	1916	1917		1915	1916	1917
Jan.	34.40	41.76	44.10	July	37.38	38.37	62.60
Feb.	37.23	42.60	51.47	Aug.	34.37	38.88	62.53
Mch.	48.76	50.50	54.27	Sept.	33.12	36.66	61.54
Apr.	48.25	51.49	55.63	Oct.	33.00	41.10	....
May	39.28	49.10	63.21	Nov.	39.50	44.12	....
June	40.26	42.07	61.93	Dec.	38.71	42.55	....

MOLYBDENUM AND ANTIMONY

Reports are that $2.25 has been paid for high-grade molybdenite, while for 90% concentrate $2.20 is the prevailing quotation. The antimony ore market is unchanged.

The production of secondary antimony, of which all but 80 tons was recoVered in alloys, increased from 3102 short tons in 1915 to 4480 tons in 1916. The Value assigned is arbitrary and is based on the aVerage yearly price for ordinary brands of antimony published by the 'American Metal Market.' The regular smelters reported the recovery of 528 tons of antimony contained in antimonial lead scrap, an increase of 52 tons. The principal materials refined or re-melted which contained antimony as an alloy were hard lead dross, babbitt, solder, pewter, and type metal. The imports in 1916 of antimony as metal in ore or as oxide or salts amounted to 11,928 tons, and the recoVery from secondary sources was equal to 38% of the imports. The production of antimony from antimony and antimonial lead ores of domestic origin in 1916 was about 4500 tons, or about the same quantity as that recoVered from secondary sources.

CHROME

Quotations in this market are hard to obtain but it is stated on good authority that chrome-ore is hard to buy and that not less than $1 per unit, f.o.b. New York, for 45 to 50% ore would have to be paid to obtain California grades.

Eastern Metal Market

New York, October 17.

All the metals are inactive. There has been no buying except in tin and this has not been large. Price-fixing has been reported as about to be extended to lead and zinc.

Copper is at a standstill and nominal.

Tin is steady with some business done in futures.

Lead is unsettled and weaker with prices considerably reduced and they may go further.

Zinc is stagnant and fairly firm at mostly nominal prices.

Antimony is dull and unchanged.

Aluminum is inactive and lower.

Outside of Government orders business has been light in the steel market. Sales of steel have been made at both the fixed prices and higher, the latter, in the case of 1000 tons of billets at $50, representing largely the closing of options, made before the price announcements. Signs point to the consummation of sufficiently numerous sales before long to establish a public market at the agreed figures.

COPPER

The general uncertainty which has prevailed for some time is still with us. What appeared to be a definite or early clearing of the atmosphere last week has not materialized. Various reports are circulating but definite information is not available. Joint separate meetings of committees representing the producers and consumers are being held daily with the object of unraveling the tangle regarding distribution. It is probably known what the Government and Allies' requirements are for the last quarter, but how much is to be left for general consumption is not made public if indeed it is known. It is probably true that the producers are now permitted to deliver metal on old contracts, but no copper can be sold at the new Government price of 23.50c., which we continue to quote as nominal. The intricate problem will probably be solved soon, but until then business is practically at a standstill in order to conserve the Government's interests and that of the Allies. Small lots to needy users continue to go at about 28c. per lb. The London market is quiet and unchanged at £125 for spot electrolytic and £121 for futures.

TIN

The week has been generally quiet, with no features of special interest. On the holiday, Friday, October 12, there was no market, and previous to that, on October 10 and 11, practically no business was done, except the sale of 25 tons of February shipment from the Straits at 57c. On Saturday, October 13, good inquiry for November, December, and January shipment from the East appeared and two lots of 25 tons each for January and February shipment from the Straits went at 57.37½c. On Monday, October 15, further good inquiry appeared. Two lots of 50 tons each of January and February shipment from the East went at 57.62½c. In all, about 300 tons of futures were sold as well as some nearby metal. Yesterday the market was quiet, with buyers showing but little interest, though there were fair sales quietly made of January-February shipment. In the entire period since our last letter there has been practically no business in spot-tin. This has come to be regarded as in a specialty-class, for shipping permits are hard to obtain unless the buyer's name is stipulated. Since last week the price has not fluctuated outside of 60.62½ to 61c., New York, the latter having been the quotation yesterday. Arrivals to October 15, inclusive, have been 1540 tons with 4520 reported afloat. The London market is strong, having advanced £2 since a week ago to £247 10s. per ton yesterday.

LEAD

There have been two important developments since the one mentioned in our last letter to the effect that the Government's October requirements would be furnished below the agreed price of 8c. per lb. if the quotation for the month of the 'Engineering & Mining Journal' showed an average below 8c. The new developments include the report of price-fixing by the Government of pig-lead at 6c. or under, which is unconfirmed, and the unexpected and sudden announcement of a reduction of 1c. per lb. by the American Smelting & Refining Co., bringing its quotation to 7c., New York. While a reduction was looked for, its proportions exceeded expectations. Four months ago lead was selling at 12c. and buying was good. This is the fifth decrease in price since August 29, bringing the price to the lowest level it has reached in a year. It is not surprising that the market is now unsettled, with buyers uninterested and expecting a further drop, influenced by Government price-fixing. The reason for the decline has been few offerings of round lots for any delivery this year at concessions below the 8c. price, but no buying resulted. We quote the market at 7c., New York, but offerings under this are already rumored.

ZINC

Into a market that has been on the verge of nervous prostration, and that has been lifeless and stagnant for the greater part of several months, is now injected price-fixing by the Government. It is reported, but unconfirmed so far, that "price-fixing of lead and zinc by the War Industries Board is imminent as a result of investigations into both the lead and zinc industries by the Federal Trade Commission. Complete figures on the cost of production of the two commodities have been sent to President Wilson and the Navy Department. After consideration has been given to the data an announcement from the White House probably will be made." This has caused a dampening rather than an inspiring effect. The market continues dull with little interest or demand. It is, however, fairly firm with prime Western for early delivery quoted at 8c., St. Louis, or 8.25c., New York, and ⅛ to ¼c. per lb. now asked for latter positions. Sales, however, are not of consequence. The apparent firmness is due largely to the fact that quotations are very near to the cost of production. Another factor by no means making for an upward trend in the market is the mid-year report of the U. S. Geological Survey on spelter. This shows that up to June 30 inclusive, smelting capacity was still being enlarged despite the unsatisfactory market conditions. The total number of retorts on June 30 was 233,050, an increase in the half year of 13,632, with about 15,000 additional building or planned.

ANTIMONY

Chinese and Japanese grades are quoted at 15c. per lb., New York, duty paid, but demand is poor and the market is dull.

ALUMINUM

No. 1 virgin metal, 98 to 99% pure, is lower, being quoted at 38 to 40c. per lb., New York, but there is no demand and prices are nominal.

ORES

Tungsten: There is no change in this market. Scheelite continues to command a higher price than other ores, $26 per unit on 60% concentrate being the quotation which has been realized in recent moderate sales. From $23 to $25 is quoted on other ores. Inquiry is reported good. Ferro-tungsten is unchanged at $2.30 to $2.50 per lb. of contained tungsten.

TYPOGRAPHICAL errors on our advertising pages are not debitable to the editorial department, but we venture to draw attention to such an error on the first page of our issue of September 29, where the horse-power required by the Marcy mill is given as 255, whereas the copy said 225.

COPPER dividends declared during the first nine months of this year make a total of $139,000,000, as compared with $108,000,000 in the same period of last year. The biggest contributor is the Utah Copper, which paid $17,869,390 in the first three quarters of 1917. The Anaconda paid $15,153,125, and the Kennecott $13,054,-616. No other company paid over ten millions, but the Phelps-Dodge group distributed $9,900,000, the Inspiration $7,387,294, and the Calumet & Hecla group $7,500,000.

EXEMPTION from assessment work on mining claims under the recent Act of Congress applies equally to lode and to placer claims. The notice of desire to take advantage of the Act for exemption does not need to be in the form of an affidavit, but may consist merely of a letter to the official recorder authorized by the State law to record notices of location of mining claims in any district or county, giving in the letter such a description as will unquestionably identify the claim. If the other facts can be accompanied by a citation of the book and page of record, all question as to the identity of the property will be obviated.

ON October 25 Dr. Ira N. Hollis, chairman of the National Engineering Council, president of the American Society of Mechanical Engineers, and president of the Worcester Polytechnic Institute, was entertained at dinner in San Francisco, and on that occasion delivered an address on 'The moral influence of engineering and efficiency.' We hope to publish the text of this address in our next issue; meanwhile we venture to remark, as on the occasion of Dr. Wilbur's recent address before the Commonwealth Club, that the serious public utterances of our leaders of thought receive no attention from the daily press, whereas the vaporings of cheap notorities are given ample space. More than that, the engineer, who is the king-pin in our organization to wage successful war, is not recognized as being at least as important as the talkative lawyer or the nondescript politician, two types that embarrass our national efficiency at this crisis.

SPELTER exports from Atlantic ports during the first seven months of this year amounted to 255,005,-289 pounds, as against 177,993,047 pounds in the same period last year and 142,051,969 pounds the year before. At the end of this seven-month period, however, exportation decreased considerably, that for July being the least for the year, at 19,769,286 pounds, as against 29,097,748 pounds in June and 61,912,499 pounds in May. England continues to be the chief importer, taking 98,201,360 pounds in seven months. Canada took 17,242,373 pounds. The contracts made with the allied governments are helping our spelter industry, but while the market is regulated by the abnormal demands of war, it is bound to be subject to vicissitudes, and we expect therefore that the Government will fix a price with such limitations of trading as will prevent speculation.

ANNOUNCEMENT is made that the American Zinc, Lead & Smelting Company has adjusted the penalties for infringement and damages claimed by Minerals Separation, agreeing to pay the latter $250,000 during the next 12 months, this payment being divided into $50,000 cash and two sums of $100,000 in six and twelve months respectively. The smelting company becomes a licensee of the patent-exploiting company. This agreement was made against the advice of Mr. D. C. Jackling, who has resigned from the directorate of the American Zinc, Lead & Smelting Company. The sum to be paid by the latter is just half of the figure mentioned by the Minerals Separation people a month ago. We sympathize with Mr. Jackling in his protest, which we believe to have been made in the interest not only of the company of which he was a director, but also of the mining industry as a whole. As regards the Butte & Superior Mining Company, it is satisfactory to note that the order signed on September 17, whereby the company was required to file a bond for $2,500,000 and to deposit monthly all the proceeds of its operations, less actual expenditure, has been modified, so that now the company is enjoined from disposing of its assets until the final termination of the suit and payment of judgment. The company is given permission to apply to the local court for leave to incur extraordinary expenses for betterments. Thus the filing of a bond for $2,500,000 is obviated, the Minerals Separation people stating that they consented to the change because they wished to avoid any procedure "that would restrict the output of metals in war-time." The order of the Court provides further that the Butte & Superior company shall file monthly reports of its operations and shall monthly deposit in Court the actual receipts, less actual expenses, "or a bond in the amount thereof that they will be forthcoming for application upon any profits and damages finally awarded to the plaintiffs." Preparations for the appeal are being made, and the defendant has good reason to expect that Judge Bourquin's decision will be reversed.

OUR compliments to Os Estados Unidos do Brazil! "The land of bloodless revolutions" has gone to war! She may not send a great army to Europe, though she could train and equip more men than Italy has lost in her terrible reverse in the Julian Alps. The problem would be to transport them and to maintain them when there without straining an already overburdened trans-Atlantic food-service. She will, of course, dispatch a body of troops so that the Germans may see another banner adding to the blend of national colors like a menacing aurora over her frontiers, and they will be good fighters, too—those Brazilian standard-bearers. The man that is slow to take up arms proverbially proves a worthy foe, and Brazil has a record of some relentless and effective fighting against Rosas in the Banda Oriental and against Lopez on the banks of the Paraguay. What Brazil's entry into the War chiefly means as an aid to those now struggling to save the Anglo-Latin civilization is a loosening of international relationships, so that the important contributions of Brazil to the stock of needed supplies may be forwarded without the limitations imposed by an observance of neutrality. Moreover, Brazil might render a notable service by mustering a great industrial army to plant more cotton and sugar in Ceara and Maranham, and to raise more grain in Parana, and to improve transportation facilities for the metals needed from Minas Geräes.

Minerals Separation

It is apparent from recent issues of the 'Canadian Mining Journal,' the 'Northern Miner,' the Toronto 'World,' and other papers across the border, that the company claiming a monopoly of rights to the flotation process has been under the fire of journalistic shrapnel. The 'Canadian Mining Journal' has maintained a dignified and correct attitude throughout this campaign, with the result that it has been attacked by its less responsible contemporaries. A charge has been made that Minerals Separation, more particularly its American subsidiary, is under German control. We stated the facts succinctly in these columns in our issue of October 6. Nothing has happened since to call for a revision of the statement then made. The connection, as agents at New York, of Beer, Sondheimer & Co., a German firm placed temporarily on the black list of the British government, is the chief evidence on which Minerals Separation is charged with being a German corporation anxious to hinder the production of war-metals in the United States and Canada. Since December 1916 the American business of Minerals Separation has been in the hands of the Minerals Separation North American Corporation, the directors of which are Messrs. John Ballot, S. Gregory, and Frank Altschul, the last holding a commission in the American army. We infer nothing from the naturalized citizenship of this or that director, nor do we think that the efforts to separate the identity of the British parent company from the American subsidiary is of any importance, but we do feel absolutely confident that Mr. J. H. Curle, well known to readers of our paper, and a man of unimpeachable loyalty, as also of unquestionable honor, could be no party to any attempt to interfere with metal production in the interest of the Enemy. Amid the various charges of the Canadian press and the explanations published by Messrs. Ballot and Williams, we place our reliance squarely on the good faith of Mr. Curle, who is a director of Minerals Separation and a close friend of Mr. Ballot. Mr. Curle's association with the enterprise gives us confidence in Mr. Ballot personally and in Minerals Separation as a corporation. We believe that the papers at Cobalt and Toronto are barking up the wrong tree. We agree with them that Minerals Separation with its attempted patent-monopoly, is a pestilential nuisance to the mining industry, but we see no good in beclouding the issue with false charges. The 'Canadian Mining Journal' brushes aside the mis-statements of the daily press and reviews the position in an eminently sane fashion. We appreciate our contemporary's reference to ourselves. Our attitude is now prejudiced for good reasons, but we did not come to the conclusion that Minerals Separation's claims to a monopoly were unjustified by the facts of discovery without first making an honest and thorough inquiry. The patent-mongering company has tried to bluff the metallurgical profession in America by imposing an invalid agreement on technicians employed by its licensees, it has tried to exact extortionate royalties, and it has endeavored persistently to stifle the publication of information concerning the technology of the process. The decisions of the American courts have been, in the main, favorable to Minerals Separation, but the end is not yet. Meanwhile the patent laws of Canada afford better protection to the public than our own and we expect that any attempt to impose an unreasonable royalty can be defeated. It is to be hoped that the Mines Department of the Dominion will give serious attention to the question, not only in the furthering of metal production, but in the vi[illegible] st of the mining industry. Owing to the closer ıl exercised by the Canadian government over ; royalties and over mining regulations, especially ı time of war, it ought to be practicable to bring inerals Separation people, whether in New York ıdon, into a sensible frame of mind and to make arrangement with them, whereby the mining in- ' of Canada may be relieved of an incubus. We hat the Canadian Mining Institute and the 'Cana- lining Journal' will unite in pressing the subject the attention of the Canadian government with a :o granting relief to the mining industry of the ıion and preventing a further development of nnical, extortionate, and stifling control upon the : of base metals.

:e writing the above we have received a telegram he 'Northern Miner' stating that Mr. Frank Coch- who was Minister of Lands, Forests, and Mines in io before entering the cabinet of the Dominion ıment, stated at Cobalt on October 24 before a the War' convention: "It will be my business to r Robert Borden [the Premier] on the question, view to bringing them [the M. S. people] to their and, for the benefit of the mining community, I ry to get him to cancel their contracts." He in- l the meeting that, if, after a thorough investi- , it were found that the control was not vested :man alien enemies, it would be the duty of the ıment to see that the royalty imposed was so able that it would in no way embarrass the in- or in any way retard the development and pro- ı of Canadian metals. In this connection he sug- that the Government would appreciate the advice rators in Canada as to what, in their estimation, be considered a fair and reasonable royalty. This summary as telegraphed by our courteous con- ary at Cobalt. We are glad to see that attention ; given by the Canadian government to the ques- d we hope that the mining profession in Canada ;tir itself to co-operate with the proper depart- authorities. As to a fair royalty, five cents per ırdless of quantity would yield a handsome in- r the patentees without being burdensome to the 's of mines. The Anaconda and Inspiration es, with others in that group, are paying 4 cents on an aggregate output in excess of 30,000 tons Ve do not believe in a royalty based on tonnage in principle it is unreasonable, and in practice hard on small mining enterprises. Some official of the royalty would be a great relief to the ı mining industry, and we wish it were prac- ı this side of the border, but we suggest that dian government might do even better, by buy- isputed patent-rights, and then either make a f them to the mining industry of the Dominion arge a small royalty, sufficient to represent 4% rchase. Such a step would serve at once as a mulant to mining development.

Making a Career

Mining makes demand upon a great variety of talent, and the successful application of that talent is conditioned by the vicissitudes inherent in the adventurous search for mineral wealth; so that the career of the mining engineer is not without a romantic element. He may climb from a small opportunity and attain to a big achievement; he may begin in a frontier camp and end in the seats of the mighty; above all, he may start under a big handicap and yet become a winner in the long race of life. This is true of other professions, particularly in this land of opportunity, but in mining it is linked to the great adventure—the opening of an Aladdin's cave under the crust of the earth. Several of the most notable men in our profession have achieved distinction despite the absence of those early educational advantages that they themselves have not failed to give to their own sons. We confess to sincere admiration for the force of character that can surmount an obstacle so fundamental. In this issue we publish an interview with a distinguished mining engineer in whose life natural abilities backed by tenacity of purpose have triumphed over the lack of a conventional education. Mr. W. J. Loring is a native Californian whose childhood was spent amid the muffled thunder of the stamp-mills. From the first he knew that men make holes into the ground to find the golden ore, which they then smash into powder in order to separate the gold from the worthless quartz or slate with which the precious metal is enveloped. Early he realized that the purpose of mining was to make money, and early he was impressed with the essential virility of the men that directed the technical operations. His reference to James Parks will be appreciated along the Mother Lode, for that worthy mine-manager was a notable figure in the foothill country and represented a type of man never common, combining good judgment, ready initiative, and sterling integrity. At an age when most boys have just begun their schooling Mr. Loring learned to hold a drill and to do the small work about a mine, following this with similar labor in a mill. One that has broken ore with a hammer and fed it with a shovel into a stamp-mortar will appreciate the automatic crushers and feeders of a later period. Personal experience in the benefits of mechanical development must stimulate an intelligent interest in labor-saving devices. Thus, in the years that followed, Mr. Loring helped to test and use various appliances that release human muscle for more effective service in the mine and mill. As a young man he worked with a progressive group of men. His experience in reducing the output from a pocket of specimen ore, in the Tryon mine, was useful in preventing him thereafter from becoming excited at the sight of free gold. By that time, when he was barely 20, he had attained a position of considerable responsibility and had the satisfaction of taking his part in the erection of the Utica mill, which was destined to extract several millions in gold. His chief, Charles D. Lane, was one of the last of the old type of Californian miners; from him he got the zest for mining adventure, the prospector's love of underground exploration. As Charlie Lane used to say: "I am digging all the time, and praying like hell." He knew that he needed the aid of luck—every miner does—and if he went too often to spiritualism instead of science, like his associate Alvinza Hayward, yet he had that saving common-sense that is closely akin to science and through it he was saved from many of the stupidities into which spirit-rapping might have led him. Fortunately, Mr. Loring acquired the practical sense of these old Californians without being attracted by their mystical vagaries. In due time a call came for service in a larger and more distant field. He went to Australia, on the invitation of the Hoover brothers. It is a far cry from the young engineer in that mine-manager's office in Western Australia to the Food Administration's building at Washington, but the qualities manifested by Mr. H. C. Hoover 15 years ago are those that he is now using to such great purpose. Our readers will like Mr. Loring's reference to his former associate, through whom he obtained the great opportunity of his career. It enabled him to prove his mettle. Then, as now, he realized that the object of mining is profit, not low cost per ton or high extraction, except in so far as these are contributory to the main result. His early training made him a keen millman, but he realized that the major economies were to be made in the mine itself. He studied the cost-sheet, and kept his eye on the payroll, for 60% of the total cost of mining is labor. The success scored by him at the Sons of Gwalia marked him as a capable administrator. Here we may refer to the joint management of the Australian mines under the direction of Bewick, Moreing & Co. in 1903. We refer to this simply to lay stress on the inadvisability, to put it mildly, of such division of control. The experiment has been tried so often that it is no longer an experiment, but an assured failure. The reader will note Mr. Loring's part in starting the smelting of the ancient Chinese slag-dumps at Bawdwin, now identified with the important enterprise of the Burma Mines company. The story of his reconnaissance at Porcupine, in its early stages of development, is likewise interesting. We ask the younger men to note again the injury done by malaria. In Mr. Loring's case it was slight, but the mention of it gives us the opportunity once more to advise young men not to sell their birthright—good health—for a mess of pottage—an increase of pay. We would like to dwell on Mr. Loring's remarks concerning the British administration of mining companies, but space will not permit. The publicity given to operations and the protection given to shareholders are admirable, but the red tape, figure-head direction, and flux of responsibility are points of obvious weakness in the British system. After engaging in professional work around the world, Mr. Loring has returned not only to his native land but to the very mining region in which he made his start. By his successful re-opening of several mines on the Mother Lode he has proved that a proverb can be read backward as well as forward, for he is held in honor eve his own habitat. He has retained the qualities ingrained, the love of mining itself; "he likes to d as Charlie Lane said; he is keen on milling, now su mented by flotation; he is kind to workmen and is by them, so that he can get the best out of them; climbing the ladder of professional success he has not the sense of comradeship and the sympathetic touch makes the ideal employer and manager. We note own regret that he missed the educational opportun that he has taken care to give to his promising son, b this day his judgment of a mining venture is better his technical description, which exemplifies the d nance of the practical over the theoretical in his m development. The same utilitarian sense may be acqu by men of full technical training that, in the spir ambitious apprenticeship, have not hesitated to learn business details of mining after graduation from col The point is that Mr. Loring does not belong to that of so-called self-made man whose boast is that he has pensed with technical training. The question a whether such men as Philip Argall, J. H. Macke D. M. Riordan, and W. J. Loring, successful engin whom we have interviewed and who have told us they overcame the lack of a conventional educa would have done better if they had been better equi at the start. The answer cannot be made confide The handicap spurs the runner, the absence of eq ment and patronage necessitates special effort, u which the best qualities of the man may have the chance to grow. If similar force of character could been developed along an easier trail, it is likely tha educational help would have been turned to good count. To a mining engineer a college training is a cut; by aid of it he can learn more rapidly and t more logically than the man not so assisted. He often because his essential manhood has not been tr with the severity that marks the life of his less fa comrade, who has to depend entirely upon native i ligence, pertinacity, courage, and opportunity. It is to judge a man by what he has achieved. We respe man that makes the most of his opportunities. It is than delightful, it is inspiring, particularly to the y to see men overcome the limitations of their environ and prove that strong character is not to be held by the fortuities of place and surrounding. Possibl consciousness of limitations is less oppressive in a ing camp, for instance, than in a great city. The who deals daily with a great variety of person equally educated has a better chance of becoming scious of mental power, and is encouraged there forget his own handicap; he is encouraged to ove that handicap and to place himself on a level with that have undergone a special education, to find, persevere, that their accidental advantage over h less decisive than the natural capacity with whi was born. Thus mental power and essential v enable men to conquer circumstance and achieve purpose, which is to be happy and effective.

Mill-Tests v. Hand-Sampling

The Editor:

Sir—I have read with great interest the 'combination method' of sampling large low-grade orebodies devised by Morton Webber which appeared in your issue of September 29. I think all engineers who have studied the latent difficulty in estimating what the true mill-heads of a large deposit will be when the ore is mined, will feel that Mr. Webber's system marks an important advance in sampling and in the general technology of the subject. I am one of the losers in the Alaska Gastineau venture, but I decidedly agree with what you say editorially, that the loss of over $15,000,000 of public money was "not due to chicanery, but an ineffectual attempt to ascertain the real gold content of large masses of low-grade ore. Let us learn by experience." After considering the matter carefully, it is difficult for me to conceive how Mr. Webber's method, if applied to this case, would have failed at a comparatively early date to disclose what would have happened when the supposed ore reserves were stoped on a commercial scale. Mr. Webber's system seems to have as its object the utilizing of the chief benefits of moil-sampling and mill-testing, and the elimination of their relative defects.

The defect of moil-sampling, as is well known, is that it has a latent sampling-error which, prior to Mr. Webber's method, could not be estimated until the ore was mined, or in other words until 'after the horse was gone.' The defect in the mill-test is that it only accurately samples the particular place from which the ore is removed. To use a chain of mill-tests to determine the sampling error, as in Mr. Webber's method, and to correct the hand-sampling of the ore-reserves by using the mill-test as a "sampling-error indicator" for various zones, is in my judgment a radical improvement. If this system was properly applied to any large low-grade ore deposit, it would be difficult for me to understand how responsible people could be misled. The main thing would be to properly apply the system to a particular case. This would require judgment and experience, but that it could be applied I am confident, the principal item being expense. The mining profession had to have, as I suppose, the 'rap over the knuckles' of the Alaskan experience, and I do not think any ill feeling has resulted. I think sufficiently well, however, of the 'combination method' that there should be no excuse for a second failure. The question, however, that requires further clarification in my judgment, is the size of the zone that should be corrected by the sampling-error mill-test conducted within it. Mr. Webber states, "the size of each individual zone will depend on the nature of the ore. In this way it may be possible to divide any ore deposit into zones, to be governed by the engineering expediency of the particular case. The principles of this 'combination method' will apply to the best advantage on mines of fairly uniform mineral characteristics. The greater the variation the shorter must be the distance between one mill-test point and its neighbor, and the reverse." While the excerpt here quoted deals with the principles that should govern the size of the mill-test zone, the means of determining the size is open for discussion. Obviously the personal factor must be introduced. On thinking over the most practical method of deciding upon the size of the zone, I suggest if it is intended to put in a 'chain' of mill-tests, I should recommend that the variation of the sampling-error between one mill-test and its neighbor should be the governing factor on deciding on the area of the zone. Whether the sampling-error is large or small should not be a factor in this decision. A mine indicating a fairly uniform sampling-error of 40% could stand obviously much larger zones to each mill-test than a mine where the sampling-error fluctuated from zero to, say, 20%. So, I think, we may take it that the degree of sampling-error is not the deciding factor in the area of a zone, but the fluctuation in the sampling-error between one mill-test point and its neighbor.

To apply this method to meet the underlying principles I will suppose a main development-drift being driven along a large low-grade orebody. At a certain point a sampling-error mill-test 'A' is instituted from the roof of the drift. The sampling-error here is found to be 20%. Another point, say, 400 ft. from the first test, is subjected to a mill-test, which we will call 'B,' and an error of 25% is obtained. I feel confident that a factor for the average of ∠ and B would properly govern the hand-sampling of the intervening space. If a third test 'C' was instituted 400 ft. farther along the drift and the error proved to be 40%, showing a violent fluctuation from B, a mill-test should then be made between B and C, dividing the intervening distance into two blocks of 200 ft. This is probably what Mr. Webber had in mind when he stated, "It should be possible to divide any ore deposit into mill-test zones depending on the engineering expediency of the individual case." I feel, however, that it is most important that the application of his system should be discussed. In former cases, where the mill-test has been employed as a means of sampling, and not exclusively as a sampling-error indicator, as in the 'Webber method,' a large amount of ore was mined and treated from a relatively small number of places. This meant special expense and exploration for the purpose of sampling. Also the expense of a mill was entailed before it was known if the mine would be a success or otherwise. The numerous small mill-tests used as sampling-error indicators in the 'combination method' could be made from the development openings of the deposit, and much expense would be avoided.

The great benefit, in my judgment, of Mr. Webber's method is that the actual mill-heads of large-scale operations can be determined as the property under exploration is being opened up. All former systems failed to do this. A large investment of capital was involved before this was ascertained. Therefore a method that can disclose the true mill-head for each zone as it is developed is unqualifiedly a radical step in the sampling of large low-grade ore deposits.

New York, October 1. R. E. RAYMOND.

The Editor:

Sir—While reading the valuable contributions from Morton Webber, and others, on mine-sampling methods, the thought occurs to me, as it has done heretofore, that there is not enough common ground for the mine-sampler and the one who excavates the ore. The miner is depended on to make good the computations, estimates, surveys, sampling, and other preliminaries to his final labor. Sampling, whether by moil or mill-test, is mining in miniature. In taking his samples, making his measurements and estimates, the engineer should visualize what will happen when the stoping is done, whether or not the miner will take out the ore that was sampled in a way represented by the sampling-process. The same stope under different administrations will produce entirely different grades of ore, depending, first, upon the industry and skill with which the workers perform their task, and, secondly, upon the character of the ore. If the ore is short, and labor is scarce and incompetent, the ore, excepting what is carelessly lost by reckless blasting and imperfect methods, will go into the chutes accompanied by a lot of waste, both from waste inclusions in the vein and from the wall-rock; on the other hand, if the ore breaks big and clean, the careful miner is glad to be able to sort out the refuse, as the waste helps to support his stope, and to prepare it to properly care for the next blast and to prevent scattering and loss. The ore from such an administration will not only be of higher grade, but more actual metal will be won. Again, ore slabbed off from the sides of drifts by machine-drills will not give the same mill-feed as material from the same orebody when caved in large masses, mixed unavoidably with waste from included bodies and the wall-rock. My point is that ample regard must be had, by the sampling engineer, for the conditions under which the ore actually will be produced. Small-scale mining, or sampling, must not assume conditions of perfection not attainable in every-day work.

Kellogg, Idaho, October 4. STANLY A. EASTON.

Water and Mines in Paradise

The Editor:

Sir—In your issue of September 29, 1917, Ivan E. Goodner, the chief of the Paradise irrigation district, takes exception to some correspondence sent in by me to the MINING AND SCIENTIFIC PRESS for a previous issue. In the interest of your periodical, whose correspondents are generally reliable, I must take exception to some of Mr. Goodner's remarks. They are not only inaccurate but untrue. After an opening remark or two the statement is made that "The following is a correct statement of the matter." My understand of the word 'correct' is possibly different from that of the young engineer. The first statement is that "None of the owners of the land referred to has so appeared at any time in the past, but all have paid the tax levied without protest." How about the Mineral Slide Mining Co., in Sec. 10, T. 23 N., R. 3 E., M.D.M.? Mr. Goodner not only protested, but forced withdrawal of their property from the irrigation district. Statement No. 2 is: "The land in question includes 20 acres which is susceptible to profitable orchard development, etc." The land in question includes 54 acres, not 20, and the tax levied was $5 per acre instead of $1. Next year's tax will be $1. The young engineer should correctly inform himself of these things before rushing into print. As for "profitable orchard development," I am not a farmer, but if a lava scarf at an angle of from 35 to 65° will support a "profitable" orchard, I have that thing to learn.

The last paragraph of the statement is as follows: "To make the matter short, it is not true that the district is levying an unwarranted tax upon mining land; it is not true that the owners pay tax without being allowed to have the water, and it is not true that the mines are being deprived of water to which they have a right for mining use.'" In answer I will cite one case, that of the Lucky John mine in the N. ½ of Sec. 11, T. 23 N., R. 3 E., M.D.M.; 54 acres of this property was included in the irrigation district on which the tax has been paid. As no water may be used for mining purposes, it is obvious that the Lucky John mine will get no water. The miners are being deprived of water, as witness the damage suit now pending in the court of Butte county between the Mineral Slide Mining Co. and the Pacific Gas & Electric Co., that sold only their "right, title, and interest" (not grant, bargain, and sale) to the Paradise Irrigation Co. The two ditches that supply Paradise with water were built originally by mining men, Delaplain, Nickerson, Moody, and others, for mining purposes only, and later sold to the Oro Light & Power Co., the original owners retaining certain rights.

J. D. HUBBARD.

Paradise, California, October 2.

Methods of Mining

The Editor:

Sir—The reading of your series of articles on the Miami methods of mining was a pleasure and brought to my mind the thought that the evolutionary steps that have made possible some of the mining methods used in the winning of large copper deposits might be of interest to mining engineers. Unfortunately, it is not possible at this time to trace the historical evolution, and indeed it would be an extremely difficult matter, since the historical record is fragmentary and spread through the mining lore of a number of countries. What I have to contribute is more the piecing together of older methods in an endeavor to show how the present practice could have been developed logically from a consideration of the essential features of well-known methods.

The room-and-pillar method is a system evolved from the primitive plan of mining coal in irregular chambers. As applied to thick coal seams it enabled a goodly proportion of coal to be won, but when applied to narrow seams the tonnage output was greatly restricted. Under difficult roofs the method was objectionable. These two considerations undoubtedly caused the development of the long-wall system. The application of this method taught mining engineers that it was not necessary to support the roof where it was extremely difficult to do so and that the better way was to allow it to subside. The use of the method gave skill and familiarity with working under heavy roofs. The winning of all the coal on the 'first working' was a particularly desirable feature. With the application of the room-and-pillar method to deeper and deeper coal beds the necessity for winning a greater proportion of the coal, in order to distribute the investment in deep shafts and more expensive hoists and other equipment over a larger tonnage, became apparent. The greater earth-pressures caused engineers to use wider pillars and narrower rooms; this naturally reduced the proportion of coal won. Under he stimulation of these necessities the bord-and-pillar method was evolved. In this method the coal bed is ivided into large square pillars by a rectangular system f galleries. Each pillar was mined by a modified longall method. Instead of attempting to carry a longall face, by mining progressively from the 'gob' to e 'solid,' slices were removed and the roof caved before removing the next slice. The initial cut was made ose to the gob, but with a protecting pillar between. ie coal was then mined up to the gob, props being used give temporary support. When the limited area was ned the props were blasted out or removed and the of caved. Fig. 6 on page 421 of your article illustes a sequence similar to that used in mining a pillar coal by the plan just described. Top-slicing is thus ced back to the methods of mining coal. The evolu- of the timber mat came in response to the need of a hod for mining extensive thick orebodies. The bord-pillar method could be applied to the first slice, but limited in the thickness of the slice by the height of the available props. Working under loose ground was effected by fore-poling. This was applied to the mining of the second slice, and with succeeding slices the miner undoubtedly discovered that the caved timbers of the slice above gave considerable support, becoming a distinct advantage. It was a simple step to the recognition of the desirability of forming a thick mat of timbers as a protection against sudden runs and the control of the top pressure. When these essential ideas had been developed, systematic application followed as a logical evolution and in the application of top-slicing on the Mesabi and at Miami the method reaches its full development.

The beginning of shrinkage-stoping can be traced back to overhand stoping. Either the substitution of waste-filling and the attack of the back from the top of the waste-pile, or the piling up of the broken ore upon the bottom of the stope and the use of the pile as a platform, must have suggested to the miner the filling of the stope with broken ore as a substitute for supporting-timbers and working-platforms.

The application of shrinkage-stoping was in response to a demand for a method that could be applied to wide orebodies and that would avoid the expensive square-set system or waste-filling. A primitive method of mining magnesite came under my observation some years ago and is of interest in this connection. The magnesite occurred in a network of small veins in a vertical zone of serpentine 40 ft. wide. The stope was opened out on an adit-level for the full width of the zone. The miners broke down all of the material and selected the white magnesite, leaving the serpentine as a filling and also as a platform from which to reach the back. Enough serpentine was trammed out to give a working-space between the back and the top of the pile. The method was undoubtedly a wasteful one, you will say, but it shows how the miner makes use of the obvious and most direct means even at the expense of valuable mineral lost. The combination of top-slicing and shrinkage-stoping in the systems used in the South African diamond mines illustrates the readiness of the engineer to invent new ways out of old ones.

Block-caving can be traced to the method of under-cutting an earthen bank and to under-cutting a coal seam in long-wall mining. In a stone-quarry the under-cutting of the toe of the bench and the wedging of the overhanging rock-mass may be considered as a further development. The excavation of a high bench, 150 to 200 ft. high, by first driving galleries into the bottom and then cross-galleries in order to support the mass on pillars, is similar to the methods used in underground block-caving. The reduction of the size of the pillars and the final removal of them by blasting parallels the case in most of its details. While both of the above methods of quarrying are obsolete, in the application of underground block-caving in the Pewabic mine the method just described was followed, with the additional feature that the ends of the block were cut off by open stopes. At Cananea the block was surrounded by shrinkage-stopes; and in this example we have the first combination of the two methods. The application of shrinkage-stopes separated by narrow pillars to the bottom of the block in place of a system of pillars and galleries was the next development and is described in your article. In the Pewabic mine the caved block was removed by fore-poling drifts driven into the loose pile. Necessarily a large amount of shoveling ensued and the application of a system of chutes and the substitution of chute-loading for shoveling, all of which had been worked out in shrinkage-stoping, was to be expected. At the Ohio mine the system of collecting-chutes was worked out and a similar system was adopted at the Miami. Thus the method at Miami represents a combination of advantageous features developed in other mines. This is good engineering.

GEORGE J. YOUNG.

San Francisco, October 21.

Mill-Tests v. Hand-Sampling

The Editor:

Sir—Bearing in mind that the object of sampling generally is to ascertain as closely as possible the metallic contents of some specific orebody, and bearing in mind that the expense of doing this must be as small as possible, we find ourselves between the two extremes, that of taking out the whole block and putting it through mill or smelter, or that of taking a single small band-sample to represent the whole mass of ore. Obviously neither will be done, some middle course will be followed by which we shall secure the maximum information at the minimum expense.

How much information we gather, to what expense we go, whether we trust entirely to moil-samples, to mill-samples, or to a combination of both of these, will depend upon the many factors entering into the problem of examination, its object, and the results as exposed from time to time.

We can say little of the merits of Mr. Webber's method or that of Mr. Jones or Mr. Smith unless we have a concrete case on which to apply it, some case where it is particularly applicable or particularly inappropriate, but this we can say: any suggestions that make for increasing accuracy in mine-sampling are appreciated. We store them away in the backs of our minds and some day they become useful in a particular case.

Mr. Webber's method, though he speaks of a mill-sample, is, as he himself admits, not a mill-sample at all. What is it then that he does? He samples a face with a moil as closely as the nature of the ore warrants, he puts in a light round of holes, breaks out a foot in depth, samples again, breaks out another foot, and repeats the sampling and breaking once or twice more. All of the material shot-down he crushes fine enough to pass a mill-screen and samples the crushed material thoroughly. He is then able to compare the average result of a large number of moiled samples from one point in the mine with a single sample taken a foot deep from directly under his moiled trenches.

If the large sample shows a different percentage of metal from the average of the moiled samples, he corrects all of the moiled samples in that part of the mine by a corresponding plus or minus factor. This will be a safe rule to follow, only when the engineer's good judgment tells him to what area he will be safe in applying such correction. It will not be a panacea for all large low-grade deposits. Every deposit is a problem in itself, presenting some new factors, requiring ingenuity, skill, and experience to solve satisfactorily.

Whether the idea is new or not I do not know, but it is new to me, and yet I have done pretty nearly the same thing on more than one occasion. I recall a rather low-grade, wide, oxidized silver-bearing vein in Mexico. I had moiled a series of trenches for 100 ft. in the back of a drift, found them to check with a previous sampling, but had some fear lest the moiling was shattering an excessive proportion of brittle oxidized silver-bearing minerals. As there were several thousand feet of drifts and cross-cuts to be sampled, I selected two such places as mentioned above and shot-down from the roof of each of the two 100-ft. sections between 125 and 150 tons. These two large samples were crushed and quartered as in any ordinary sampling-mill except that the first breaking was down to about ¼-in. mesh. The results happened to check quite closely with the average of the moil-samples, so closely that I thought they required no correction. If they had been different, I might have used such a factor to correct my moil-samples as Mr. Webber used. In this instance it could have been done to advantage had there been an appreciable difference.

My personal experience with large low-grade deposits has not been so extensive as to warrant my laying down any rules. It so happens that in my own work I have been fortunate in having such large and low-grade deposits as I have examined and sampled by moil check out fairly well in the mill. My own troubles have been with the small higher-grade and erratic veins. Here I have found serious discrepancies, but they were not such as to have been reduced by Mr. Webber's method or any other method short of taking out all the ore.

Permit me, through you, to express my thanks to Mr. Webber for the suggestion he has made, a suggestion that may be very useful on some future occasion in my sampling career.

F. F. SHARPLESS.

New York, October 8.

METAL industry in California is increasing greatly under the stimulus of the War. The chrome output in 1916 jumped from a value for the previous year of $38,044 to $717,244, and the contrasting figures for magnesite were $283,461 and $1,311,893 respectively. Quicksilver also grew from $1,157,449 to $2,003,425, and manganese from $49,098 to $274,601.

THE SONS OF GWALIA MINE, IN WESTERN AUSTRALIA

W. J. Loring. A Californian Engineer

AN INTERVIEW. By T. A. RICKARD

Mr. Loring, you are a native of California, are you not?

Yes; I was born near Half Moon bay, in San Mateo county, California, on March 6, 1869.

Was your father a mining man?

No; my father was a lawyer, and came from Illinois. On his side my people were French and English, and on my mother's side they were Dutch and Scotch.

What schooling did you have?

A common-school education in Amador county.

Then your father moved to Amador county?

Yes; the family moved to Amador county in 1879.

So you were brought up among the sound of stamp-mills?

Yes; there were a number of stamp-mills in operation at Amador City, where we lived, at that time.

Have you any early recollections?

I recall James F. Parks, Sr., who was underground foreman of the Keystone mine; and I also remember distinctly when he left the Keystone and took charge of the Kennedy mine. As a boy, his long legs impressed me, and I often wondered whether I would ever occupy as exalted a position as Mr. Parks. My family remained in Amador City for about three years. At that time the Original Amador operated a 40-stamp mill, and I remember when it was closed down. The Bunker Hill was also operating a 40-stamp mill on Rancheria creek. As a boy I was impressed by the activity around these mines, and particularly the noise of the hoisting-plants. When I was about 12 years of age a man named Collins, who was then employed as a miner in the Keystone mine, owned a claim called the East Keystone, which happened to be close to my home; his assessment work consisted of driving a cross-cut into the side of a hill. It was here that I began my mining career, turning a drill and doing such work as a boy of twelve could perform when not at school.

You found that more interesting than books?

Mainly for the reason that I was paid at the rate of 50 cents per day to begin with, then 75 cents per day, and my pay-day was every second day.

When did you leave school and begin your career?

From Amador City my family moved to Plymouth, in the same county, and when not at school I was employed intermittently breaking ore with a hammer and shoveling it into the feeders in the old Empire mill. There were no rock-breakers except one at the South shaft, but this was usually out of order.

When did you begin regular work?

At the age of 14, at Plymouth. My first responsible job was taking care of 16 Tulloch feeders in the old Empire mill, belonging to a property which is now part of the Plymouth Consolidated. From the feeders I was gradually promoted to be assistant-amalgamator and concentrator-man, until the mine took fire on January 26, 1888, when the property was closed down by the owners. These were Alvinza Hayward, Walter Hobart, Sr., and their Eastern associates.

What did you do next?

John E. Reaves, who was foreman of the Empire mill, took a fancy to me as a boy, and when operations ceased at Plymouth, he was transferred to the Utica mill, at Angels Camp, then under option to Alvinza Hayward and the elder Walter Hobart. Within a month after his arrival at Angels Camp he secured for me a job in a 10-stamp mill owned by George Tryon, Walter Tryon, Thomas Hardy, and James McCreight, at Albany Flat in Calaveras county. I was then 19 years of age. The Tryon mine, as it was then known, was producing very rich ore—so rich that it was possible to pound out $200 to $500 per day in a hand-mortar. The ore that was stamped yielded from $30 to $60 per ton.

So you must have got credit for being a good millman?

Evidently I was considered a good millman because when I first went to work I was put on the day-shift and remained for one week, when I changed to the night-shift; but Walter Tryon, the superintendent, thought so well of the improvements that I had made that he only allowed me to work by night for three shifts, putting me back on the day-shift, where I remained until the property was purchased by the Utica Mining Co. in the summer of the same year, namely, 1888. The mill was operated by a 52-ft. overshot-wheel, and, as is usually the case, water became short in the summer and the mill closed down until the winter rains. I was transferred to the Utica mill, then containing 20 stamps and under the superintendency of Charles D. Lane. Soon afterward the company decided to add 40 more stamps. During the re-construction of the original 20-stamp mill and the addition of the new 40 stamps, I was employed in placing machinery and doing whatever was necessary to assist in the erection of the plant. I remember distinctly having personally set the first tappets and also pounded on the first shoes that were used in the new mill on the Utica. I did that with my own hands.

How long did you remain at the Utica?

Until August 4, 1901. During the 13 years I rose to the position of assistant-superintendent of the Utica under various superintendents, including Charles D. Lane, his son Thomas T. Lane, E. L. Montgomery (formerly superintendent of the Plymouth Consolidated), Theo. Allen, and L. W. Shinn. I occupied the position of head amalgamator until April 30, 1894, when John E. Reaves was taken sick and died. I was then promoted to the position of mill-superintendent, in charge of 160 stamps, during which time the bonanza orebody in the Utica mine was discovered and worked, producing $203,000 in one month.

What made you leave the Utica?

W. C. Ralston, with whom I had become acquainted, offered me the position of superintendent of the Melones mine at Robinson's Ferry, as it was then called. He offered me a considerable increase in salary, which I accepted, taking up my new duties in August, 1901. It had been decided, prior to my acceptance of the superintendency, to erect 60 of the 120 stamps then lying on the property, and to thoroughly equip it and put it into operation. During the following six months 60 stamps were erected, a dam was built in the river, together with a flume and ditches for the transmission of water to drive the 60 stamps and auxiliary plant. In January 1902 I received a letter from T. J. Hoover, who was then at Bodie, Mono county, asking me if I would consider an appointment as superintendent of a mine in Western Australia. This offer came from his brother, H. C. Hoover, who was then a partner in the English mining firm of Bewick, Moreing & Co. The terms of the offer were $6000 per annum, and the contract called for two years and traveling expenses. This offer I declined. Later I was offered $9000 per annum, with traveling and other expenses, which I accepted. I left San Francisco for Western Australia on February 9, 1902, going direct to Sydney, overland by train to Adelaide, and thence by steamer to Perth. From Perth I went by train to Kalgoorlie, where I met Mr. Hoover and W. A. Prichard, who had previously managed the Keystone mine at Amador City, and whom I had known at that time. Mr. Prichard recommended me to Mr. Hoover. This was my first acquaintance with Herbert Hoover.

What impression did he make?

He impressed me first with his youthful appearance, for he was then only 27 and looked younger; but after several days' contact with him, he impressed me as being a very capable young engineer. He had a unique way of getting at the root of things without preliminaries, avoiding unnecessary detail. Then, as now, I do not believe that he cared for details. He appeared to have the faculty of getting at the essentials by eliminating non-essentials. I was entertained by Mr. Hoover and enjoyed his company for several days, prior to leaving for the Sons of Gwalia mine, which I had been engaged to superintend. Before leaving Kalgoorlie, Mr. Hoover told me that in his opinion 20 of the 50 stamps on the Sons of Gwalia should be closed down for a time to allow the mine to be thoroughly developed. I asked his permission to examine the mine before coming to a decision as to what should be done, to which he agreed. After looking over the situation I came to the conclusion that the mill could be run at full capacity and great improvements made throughout the whole scheme of operations. I had the Californian dislike of hanging up stamps, and I also believed that the mine was capable of producing sufficient ore to maintain the mill at full capacity.

What was the result of your recommendations?

I took charge of the mine on April 11, 1902, and on that day the mine payroll had 814 men on it. The mill was treating 8000 tons monthly, and the working-cost was 35s.6d. per ton; the output barely covered the cost of operating and development, being in the neighborhood of 50s. per ton. The mine had an ore-reserve of 60,000 tons when I took charge, and at the end of the year the reserve had been increased to 120,000 tons; a bank overdraft of £10,000 and an outstanding debt of £10,000, had been liquidated; in addition £46,750 had been remitted to London. The number of men on the payroll had been reduced to 420 and the working-cost reduced to 21s., the over-all cost becoming 35s. per ton.

How long did you remain at the Sons of Gwalia?

For about two years. On January 15, 1903, W. R. Feldtmann resigned as general manager for Bewick, Moreing & Co. in Western Australia, and Mr. Prichard and myself were appointed joint general managers in his place. The firm was then operating 16 mines in Western Australia, half of them in and about Kalgoorlie and the rest scattered. Among the latter was the Sons of Gwalia, and it was these scattered mines that I took charge of; while Mr. Prichard attended to the Kalgoorlie group.

So you had much travel over the desert?

Yes; I traveled about 30,000 miles a year among the

A GROUP OF WESTERN AUSTRALIAN ENGINEERS

Top row, left to right: G. W. Borrowe, W. E. Simpson, F. G. T. Nicholas, V. S. Allen, J. V. Jukes, W. R. Bawden, C. W. Skrine, E. A. Griffith, D. D. Sinclair, E. Graham Price, E. Günther.

Middle row: A. Sheard, H. A. Shipman, W. W. Johns, P. Ledoux, W. W. Barton, W. H. Turner, H. Lawrence Read, T. Maughan, Henry Cribb.

Bottom row: Norbert Keenan, W. J. Loring, F. A. Moss, G. M. Roberts, F. A. Govett, G. A. Touche, R. Hamilton, W. A. Prichard, J. W. Sutherland.

THE IVANHOE MINE, AT KALGOORLIE, WESTERN AUSTRALIA

various mines that were in my division. Some of the traveling was done under great discomfort, owing to the heat and dust. We did not use the motor-car until 1906. It required 64 horses to transport me on one of my round trips.

How long did the joint general-managership last?

It lasted 16 months. Mr. Prichard then resigned and I was left in full charge in Western Australia until 1906, when I was appointed general manager of the whole of Bewick, Moreing & Co.'s interests in Australasia. This included mines in Victoria, Queensland, New South Wales, Western Australia, and New Zealand. My headquarters were then at Melbourne.

When did you become a partner?

I took over Mr. Hoover's interest in July 1908, when Mr. Hoover left the firm; and soon afterward I took up my residence in London, so as to be with my senior partner, C. Algernon Moreing.

When did you go to Burma?

I arrived in Burma in that year, 1908. The purpose of my visit to Burma was to look over the properties of the Burma Mines, Ltd., which owned the Bawdwin silver, lead, and zinc mines. The Burma Mines company was organized to treat 110,000 tons of slag lying in the jungle, but the intention was not to re-open the mines, at first. It was necessary to build a railway 52 miles from the Government railway in order to transport the slag to the smelter at Mandalay. This smelter was operated for two years, during which time I made yearly visits to the property, each time becoming more impressed with the possibility of opening up an enormously rich lead and silver mine below the old Chinese workings. The history of this mine had been traced back to 1320, and during my several visits I found remnants of high-grade silver-lead ore mixed up with the slag; and judging by the enormous excavation in the mountain it appeared to me that large bodies of ore must have been extracted at an unknown period. After smelting operations had commenced, we started development work on a small scale, and almost immediately high-grade ore was encountered in the form of pillars and remnants in the old workings. This was in 1910. After operating the smelter at Mandalay at excessive cost, due to over 200 miles of rail haulage, it was decided to move the smelter to Namtu, 12 miles by rail from the mine and 2500 ft. lower than the old excavations. I recall my connection with this enterprise now with keen pleasure, seeing that it has become one of the great mines of the world; and, curiously enough, the later developments and expansion have been done largely under the direction of my former chief, H. C. Hoover.

Where did you go next?

On my return to London from Burma, late in 1908, I paid a visit to the Gold Coast of West Africa, visiting most of the principal gold mines in that country.

And you escaped malaria?

No; I did not. I suffered from malaria for two years after my return from the Gold Coast, and do not believe that I have ever fully recovered from the effects of it.

I hope that the next mine that you inspected was in a healthier climate?

It was. It is said that I have the distinction of being the first engineer to visit the Porcupine goldfield in the early days of its boom—in January 1910. I went to this district, which is in Ontario, to examine a mining claim that had been taken under option in my senior partner's name by Henry Van Cutsem, who accompanied me on my journey to Porcupine. I went from Toronto to North Bay, and from there to mile-post 222, now called Kelsey, arriving there at 11 o'clock at night with the thermometer below zero. We were dumped into the snow with a train-load of prospectors, food, and mining implements of all sorts. My party consisted of the brothers Henry and Noah Timmins, their two partners, and A. T. Budd and Henry Van Cutsem. The recent death of Van Cutsem I regret greatly. He was a delightful companion on that trip and I shall never forget his many kindnesses to me. We arrived at the Hollinger mine after 15 hours of sledding over a rough snow-road at nine o'clock at night. Our road passed across Porcupine lake, which was frozen two to three feet thick. We arrived there at about 8 p.m.; the night was cold and clear, the temperature ranging around 40° below zero. When we were half-way across the lake the horses refused to go forward, and upon looking ahead much to our amazement we saw wolves about a quarter of a mile away. We had only one gun in the party, this being a 32-calibre revolver. Messrs. Timmins, who knew all about wolves, would not let us shoot at them. After a consultation, it was decided to divide the party, one part to make a detour around the wolves, hoping by some means to frighten them away and thus leave a free passage for the team and sleigh. Upon making the detour we closed in upon the wolves and the nearer we approached the less did they appear to move. Upon closer investigation we found that what we had taken for wolves was a broken-down sleigh loaded with goods, which had been scattered about the snow, and which appeared to move about in the star-lit and snow-clad whiteness of this northern night. The following day an examination, such as was possible, was made of the Hollinger and surrounding claims, then covered with snow and ice. The outcrop could be traced; in many places it stood as high as seven or eight feet above the surface. I saw the first shot fired in the outcrop, which afterward became the main shaft of the Hollinger mine.

What made you see wolves?

The optical illusion. The eye sees motion when everything is perfectly still.

Have you found any difference between English and American company methods?

American companies would do well to adopt some of the methods employed by English companies. One distinctive feature is the publicity given to results obtained and to general conditions at the mine. This enables a comparison to be made, thus avoiding the possibility of the management living in a fool's paradise. Publicity also keeps the management up to date, by inviting comparisons.

THE FIRST SHAFT SUNK ON THE IVANHOE LODE, AT KALGOORLIE

I am also a believer in the British custom of annual meetings of shareholders; it gives every shareholder an opportunity to hear from the directors a statement as to the many points of interest. Under the American system a manager or president considers himself too busy to give much consideration to the distribution of information to shareholders, whereas under the English system this is compulsory. The system is often called, by us Americans, 'red tape.' While it is burdensome at times at any rate it furnishes a record and there is less likelihood of serious mistakes being made.

W. J. LORING AT THE DOME MINE, PORCUPINE, IN JANUARY 1910

Have you ever attended a meeting of shareholders that made any particular impression upon you?

I have attended many meetings, all of which to my mind were most useful, and some of which greatly amused me.

What particular meeting amused you?

I have one in mind that I shall never forget. This was the annual meeting of a company operating in Rhodesia. The property consisted of a concession covering many thousands of acres, upon which were several mines. The parent company sold mines to subsidiary companies, ran mines itself, farmed and raised cattle, was interested in all the business that a company of this nature could develop. In the early days the company had a bright future; as time went on reverses came, the shares fell, and shareholders became discouraged. This resulted in much mud-slinging at the annual meetings. I was present at one of these meetings, which was called for 1 p.m., this hour being fixed for the purpose of eliminating certain shareholders who thought more of their stomachs than they did of their business. However, the meeting was well attended and was presided over by a fine-looking old Colonel. His address to the shareholders dealt with the minutest detail of operations, and consumed over two hours. During this time such shareholders as had other engagements were compelled to leave, while a few went to sleep, but the Chairman rambled on, finally coming to the end of his speech and proposing the adoption of his report and statement of accounts. He asked the shareholders if they had anything to say before the resolution was put to the meeting; on this invitation six men in the front rows were on their feet at once. Each of the six in turn critized the directors in the plainest language for mismanagement. While each speaker was trying to see which could get the floor first and say the worst about the directors, an old gentleman—sitting next to me in the back row of the hall—who had armed himself with reports and statements of accounts for several years past, and who had become nervous during the long and tiresome speech of the Chairman, and had tried to gain the floor as each previous speaker resumed his seat, finally made up his mind to remain standing, and wait until all the others had finished vilifying the directors. The Chairman looked over the heads of the meeting and said: "Gentlemen I have heard the remarks of Mr. Blank and I can only say what he has said passed like water off a duck's back; if there are no further comments I will put the resolution to the meeting." Whereupon the old gentleman, who had been standing all this time, became quite excited, waved his hands (grasping papers) in the air like a wind-mill and said in a loud tone of voice: "Hold on—hold on— what do you suppose I have been standing here for, for the past hour, I want you to know, Sir, what I think of you and your co-directors,"—and he certainly did tell them what his opinion was, which would not look well in print.

Where it is possible for shareholders to meet directors of companies and critize or give expression to their satisfaction for good work, there is certainly a chance for shareholders and directors to become better acquainted than under the American system, thus, I think, benefiting the company generally.

What about your experience in the Hollinger deal?

I have already stated how I went to Canada early in 1910 for the purpose of inspecting the Porcupine goldfield. I could see, after spending a few weeks at Porcupine and Toronto, that deals in mining claims were made in the most reckless fashion. Nothing less than spot cash would do. A group of mining claims was offered to me, at my room in the King Edward hotel, Toronto, for $90,000 cash; as no reports had been made, and no work had been done upon the property I naturally turned it down. That same property was sold within two hours for $140,000 cash, the deal being completed the same day. So long as a man had four stakes in the snow he could sell a claim for a tidy sum of money. The Hollinger, at the time of my visit, did not embrace the property that is now known as the Hollinger mine, but only a part of it, for the part next to Pearl lake was offered to M. J. O'Brien, who afterward gave up his option, so that the Hollinger people were able to amalgamate the O'Brien property with the Hollinger. However, feeling sure that Porcupine would turn out some good mines, I decided to return to London and advise my partner, C. Algernon Moreing, to form a company, with sufficient cash capital so that Mr. Moreing could go to Canada with at least £100,000 at his disposal to be used in any way that he saw fit. This was done, the Northern Ontario Exploration Company, with a capital of £500,000, being formed. Then £100,000 was called up and Mr. Moreing proceeded to Canada when the weather moderated sufficiently to allow an examination of the surface. I left for Australia and upon reaching Port Said I received a cable from our London office stating that Henry Van Cutsem, who had remained in Canada after my departure, had secured an option on the Dome mine, on behalf of my firm, under the following terms: We were to furnish sufficient cash to develop and equip the mine, the character of the development and equipment being specified. Out of the first profits obtained we were to reimburse ourselves for all of the moneys so expended, plus interest, after which the profits were to be divided, 60% to the owners and 40% to my firm. I thought so well of the Dome property that I cabled my firm strongly recommending that the terms be accepted, provided the property included the big outcrop from which the mine took its name, with enough ground to protect its dip and lateral extent; and I suggested that a reliable surveyor be engaged to check the lines. I proceeded on my way to Australia, expecting the deal to go through, but, much to my disappointment, when I reached Colombo a cable was awaiting me stating that owing to our heavy interest in the Maikop oilfield it had been decided not to exercise the option on the Dome property. Mr. Moreing eventually proceeded to Canada, to exercise his own judgment as to the purchase of other prospects. He was met at New York by Henry Timmins, of the Hollinger mine. By this time the Hollinger had been incorporated into a large company and Timmins's partners had also acquired a large number of other claims in the Porcupine district. It was decided between Messrs. Moreing and Timmins that a number of the Timmins holdings, outside the Hollinger, should be sold to a new company, which was to be formed by our firm, headed by Mr. Moreing. The terms and conditions of the sale need not be recited, but I may say that the Timmins people received a fairly large sum of money, together with a considerable share in the new company, which was eventually called the Ontario Porcupine Goldfields Development Co. This company was formed in London and the Northern Ontario Exploration Company took an interest in it. Mr. Moreing visited Porcupine and inspected the Hollinger mine, which had then reached a depth of 200 ft., where rich ore was being exposed. On the day of Mr. Moreing's departure from Canada, Mr. Timmins agreed to sell 50,000 Hollinger shares to the Northern Ontario Exploration Co. on the understanding that these shares should be taken to London and a market made for them there. My firm was appointed transfer-agent for the Hollinger company. The price per share, I believe, was $4. Mr. Moreing arrived in London in due time and the developments in the Hollinger continued most satisfactorily. As is well known, a market cannot be made in shares by holding them in a safe; the price of the Hollinger shares actually moved up, and the Northern Ontario Exploration Co. assisted the market by disposing of its shares and generally dealing in the shares of the company, finally disposing of its holdings at a considerable profit. The sale of the shares necessitated the signing of the share-certificates

W. J. LORING ON HIS WAY TO MINES ON THE GOLD COAST OF AFRICA

TERMINAL OF RAILROAD TO THE BAWDWIN MINES, IN BURMA. ON THE PLATFORM ARE W. J. LAKELAND, MRS. LOVELL, M. W. MORRIS, MRS. LORING, AND GERARD LOVELL

by my firm, acting as transfer-agent. These certificates finally found their way to Canada and we—as a firm—were severely criticized for selling our shares to the Canadians at a considerable profit to ourselves. As a matter of fact, my firm never owned a solitary Hollinger share. This I say with considerable bitterness, as I believe we, personally, should have been large shareholders in it. We could not have made a market in the shares of the Hollinger company without selling, neither could the transfer be made without our signature, acting as transfer-agents in London for the Hollinger company. We stood the brunt of all the mud that was slung, without having made one solitary cent out of anything Canadian.

Have you—after your long absence from California—anything to say regarding your operations and conditions generally on the Mother Lode?

I have a high opinion of the Mother Lode as a mining region; I believe that many improvements can be made, especially on the metallurgical side. It is not necessary for me to express any opinion regarding continuity in depth, as there are mines on the Mother Lode that have already reached nearly 5000 ft. on the dip, with a splendid showing in the bottom. I believe that the application of flotation to the treatment of certain types of Mother Lode ores will simplify the treatment, as well as improve the extraction of the gold contents of the ore. My belief is based upon results obtained both in an experimental way and on a practical scale. I believe amalgamation to be unnecessary, in many cases, as the results that I have obtained have proved that on a $5.24 ore, 33% of which is free gold, a 98.3% recovery can be obtained by straight flotation. The tailing-loss is 10c. per ton. I do not believe that flotation can be applied to all ores on the Mother Lode, but I believe that important developments in the treatment will follow shortly after we have completed our mill, which is to be erected on the Dutch-App mines, in Tuolumne county.

You must be pleased with the development of the Plymouth Consolidated?

I am proud of the fact that I was personally responsible for the re-opening of the Plymouth Consolidated mine, which has up-to-date produced a profit, over all expenditures, of $630,937 in a period of 36 months. As you are aware, I worked at the Plymouth as a boy and remained at the mine until it closed down, on account of fire, in January 1888. It was then operated by Alvinza Hayward, Walter Hobart, Sr., and their Eastern associates.

Mr. Loring, you are married?

I am; I was married to Miss Marie Ellen Everhart, at Angels Camp, California, on December 12, 1888. My wife has followed me in all of my wanderings.

You have a son?

Yes, my son Edward Amos Loring is a partner in our firm and at the present time is in London attending to the engineering department of our business.

You are still an active member of Bewick, Moreing & Co., although you live in San Francisco?

I am, I am pleased to say.

Why have you changed your residence from London to San Francisco?

The change is only a temporary one, I hope. Before the beginning of the War I came over from London to see the new mill at the Plymouth Consolidated begin operations; the intention being to proceed from California to Japan, Korea, China, and back to London through Siberia. I visited the Far East, but, owing to the War, decided not to return to London through Siberia, but to return to the United States; and as my partners were willing to take care of the London business, during my absence, it was decided that I should remain in the United States for the time being, to build up a branch of our business in this country. I have been enabled to start several promising enterprises, financed by my personal Boston friends, and these will keep me here for some time.

When were you in Korea?

The latter part of 1914. I visited Korea for the express purpose of looking over the properties controlled by my friend, Mr. H. Collbran. He has developed a unique mining proposition in Korea, on the Suan Concession, and is at the present time operating what is known as the Suan and Tul Mi Chung mines. Since my visit he has opened another mine which gives promise of being a very good producer. Fortunately, his rights antedate the Japanese occupation and he is able to operate under their friendly protection.

What advice would you give to the younger men, or to their fathers, in regard to the preparation for a mining career?

My advice would be the same as I gave to my son when he was preparing to enter the university at Sydney, and that was to select certain lines of study that in his opinion would be most useful to him and to avoid a smattering of many subjects; also, as far as possible, to mix with his university education the practical side of mining during vacations. The result has been that while my son is barely 27 years of age he has considerably more experience than many mining engineers 10 years his senior. I most certainly advise young folks to gain as much education as possible, and not leave their future to the mercy of experience as a teacher. Experience is as necessary as schooling, but it is slow and uncertain. Some men have made a mark in the world without a college education, but it is becoming more difficult as time goes on to accomplish great things without a special training. After a man has reached middle age and has had varied experiences he may be considered educated. Much depends upon the man and much depends upon the circumstances surrounding the man as to his success or failure in life. An educated young man has a great start over an uneducated young man.

Principles Governing Zinc-Ore Deposits

By FRANK L. NASON

It is not my aim to discuss synthetic considerations only, but to discuss from an analytic standpoint also the effect observed on fissures, contacts between sedimentaries, contacts between sedimentaries and eruptives, including dikes, sills, and batholiths, and contacts between eruptives, in relation to the origin and genesis of zinc ores. For many years the term 'true fissure vein' has been a catch-phrase of promoters, and it has proved alluring to men with money to put into mining. The proof lay in the results of development; if an economic orebody were found, it was a true fissure, otherwise not. Instances of fissure veins bearing zinc that have proved highly remunerative are the Butte & Superior in Montana, the Interstate-Callahan in Idaho, the Broken Hill in Australia, and the Bawdwin in Burma. Fissure veins which are known to exist, but which generally are not economic, are found widely in Tennessee, Virginia, and in St. Lawrence county, New York. Many other examples could be cited. In Butte, Montana, the Butte & Superior and the Elm Orlu are probably part of the great Rainbow lode on the western end of which are the Alice and the Moulton. This lode, like the copper lodes farther south, is a great fracture in the monzonite or granite batholith. The vein in the main evidently is not a clean sharp break, but seems rather to be a fault or fissure-zone of brecciation. The zone was the channel for an intense circulation of water. The term 'intense' is deliberately chosen, since the extensive leaching of the nore soluble salts from the original rock seems to warant it. Furthermore, the silica of the original web ppears to have gone largely into solution, and to have een re-deposited in the more or less colloidal gangue of he lead and zinc sulphides. The assumption that the ilica gangue was derived largely from the silica of the rushed rock is somewhat arbitrary but not wholly so. udging from the fact that the productive area of the ein widens and pinches on the strike while the lode atter does not, as a rule, follow the varying dimensions the vein; and that where the vein and the colloidal ngue is thin or disappears, rotted rock with grains of ica takes its place, the conclusion is inevitable that tense circulation in the lode was confined to channels the crushed zone.

A study of other fissures in the Butte district affords ditional confirmation. Many fissures, even in the productive area, wholly lack the silica crusts, the vein-matter isisting entirely of rotted kaolinized feldspars, the ginal quartz grains remaining unchanged. Other ns in the district, fissures in every sense of the word e for the presence of minerals in economic quantities, have colloidal silica developed in varying amounts. Another striking example of a crushed zone in which varying intensity of mineral circulation is to be noted is in the Animas Forks district in Colorado. The vein boldly outcrops for two or three miles. At one end is the Sunnyside, at the other the Gold Prince mine. Between these two the vein is 50 ft. thick or more. There are found small pockets or lenses which hold zinc sulphide in colloidal or even crystallized quartz. The mass of the vein, however, consists of partly digested fragments of the country rock, feldspar, quartz, and epidote, seemingly carried to a stage of incomplete solution resulting in a pasty mass, fragments of the original rocks, fairly sharp in outline having pasty rims and cores wholly unaltered. The Interstate-Callahan is undoubtedly a fissure or fault. For the most part the exposed portion of the vein which was being worked on my visit showed one wall to be of quartz monzonite and the other of Pritchard slate. Here the vein-matter seemed to be derived wholly from the slate. Both in the Butte and in the Coeur d'Alene districts, minor veins or fractures were noted, some mineralized, others not; with clean sharp walls and no filling of rotted rock.

In eastern Tennessee I have traced undoubted fissures, though not continuously, for ten miles or more. The fissures on the Goin and Caldwell farms, to be specific, were clean and sharp, and were from a fraction of an inch to 18 in. thick. The vein-matter below the influence of weathering was frozen to the wall-rock on the contact and the country rock was wholly unaltered and unmineralized. The depth on the fissures showing these conditions was proved by diamond-drills to a depth of 200 to 1200 ft. below sea-level. Another characteristic of these narrow but persistent fissures was the continuous mineralization. Where the fissures were exposed by washing off the surface soil, however thin the fissure, it was found to be filled from wall to wall with either zinc and lead or with zinc alone. That these were true fissures is shown by two facts: first, the fissures were followed downward through dolomite into slate; and second, the fissures cut across the line of strike of the country-rock. In four widely separated localities, Lead Mine Bend, The Welch, Lynch, and Russell farms, the fissures displayed additional characteristics. Between parallel fissures the ground was considerably brecciated, and in the New Prospect mine the breccia was extensively mineralized with both lead and zinc. In the other three localities, while strong mineralization appeared in the surface-breccia, no development work has been done to prove depth. In the townships of Macomb and Rossie,

St. Lawrence county, New York, a remarkable line or series of fissures is developed. This was extensively worked for lead seventy years or more ago. There can be no doubt as to the nature of these deposits. The fissures are vertical; they cut across the line of strike at an angle of 60°; they traverse the white limestone and the gneisses which are interstratified with the limestone. They are sharp-walled, that is, no mineralization extends into the walls, nor are the wall-rocks changed, even when they consist of limestone, nor are the limestone or gneiss-fragments altered in the slightest degree. There are no metamorphic minerals either in the veins or in walls; these are as conspicuous by their absence as in the fissures in Tennessee. In both, however, fluorite is present in limited quantities. The fissures in Rossie and Macomb differ in their main essentials from the zinc deposits which so far as now known extend from Sylvia lake to the Northern Ore mines at Edwards. The Rossie and Macomb are cross-country fissures, those of the Edwards belt are seemingly strike-faults or fissures, parallel to the dip as well as to the strike. In Macomb and Rossie the galena and blende are in a calcite gangue and are mainly free from pyrite, the blende being remarkably pure; in the Edwards belt serpentine is the principal gangue mineral; galena is subordinate to the blende and both to iron pyrite, while the blende is so heavily charged with iron as to be highly magnetic. In both Rossie and Macomb the gneiss is frequently seen to be in close contact with the limestone, yet no extension of zinc or lead along the contact-plane, even near the vein, is to be observed.

The large zinc deposits in the brecciated Knox dolomite at Mascot, and the Austinville zinc and lead mine in Virginia are types of zinc deposits where fissuring or strike-faulting is deduced from the surroundings rather than directly observed. The Mascot mine is on a line of brecciation at least 70 miles long. At Mascot the breccia is plainly a crushed bed or series of beds of dolomite. The assumption of faulting is based wholly on over-thrust faults a few miles distant, where the dolomites are seen resting on the Marysville limestone series. Although there are numerous places along the line of the Mascot breccia-zone where blende is observed, Mascot is the only spot where the veins have been proved to be economic. In other places, despite the extensive brecciation, absolutely no trace of lead and zinc is to be found. In Mascot a remarkably pure blende is the only mineral present in a gangue of bitter spar (dolomite). The bitter spar is the secondary filling of the interstices in the brecciated Knox dolomite, the fragments of which can be readily recognized. No blende penetrates these fragments, which, as a rule, are wholly unmodified. At Austinville, Virginia, the mine is indirectly recognized as a fissure. The host is here in places an impure dolomite banded in a striking way with thin ribbons of gray and blue. On the south of the mines the same ribbon-rock appears in contact with the pure dolomite. As the ribbon-rock at the river dips under the pure dolomite, both dipping at an angle of about 45°, the existence of a fault is the obvious conclusion to explain the appearance of the ribbon-rock south of the mine. Aside from the probable main north-east fault with a throw of something like 3000 ft., there are numerous other faults which break the strata into great blocks. Along the fault-planes and many of the bedding-planes the rock has been brecciated and the interstices largely filled with lead and zinc sulphides. In parts of the mine iron pyrite occurs in great masses to the entire exclusion of lead and zinc. While not a perfect simile, the blocks of faulted and broken limestone seem to be wrapped about with a blanket of lead and zinc sulphides. The mineral solutions seem, in places, to have etched the bedding planes and the fault walls, but nowhere do they seem to have found their way into the mass of the dolomite blocks.

In the ribbon-rocks along New river, for more than a mile, lead and zinc sulphides occur in the moss of the rock in a manner to suggest crenulation of the strata, and the small amygdaloidal cavities thus formed are filled with ore. In the Austinville deposit no secondary metamorphic minerals are observed save occasional traces of fluorite.

The La Bufa deposit at Charcas, Mexico, structurally resembles the Austinville property in that, while the mine is evidently opened on a fault-fissure, the fault is inferred from contorted strata rather than from direct observation. In some ways it differs as radically from the Austinville mine as in other ways it closely resembles it. For example, the local metamorphism is intense, as manifested in the abundance of garnet, epidote, wollastonite, and hematite. In places the entire mass of rock is silicified, apparently with incipient wollastonite and garnet. There are many other examples which have come under my professional observation, but the above are types which are widely scattered from Salvador through Mexico and the United States, into British Columbia, and from the Rockies to the Appalachians. Out of the hundreds examined only twenty-two produce zinc ore on a very large scale, or are worked mainly for the zinc content. Of these, one each is in Salvador and Guatemala, six in Mexico, eleven in the Rocky Mountain region of the United States, and three in the Appalachian region, exclusive of the zinc mines in New Jersey.

From the above brief descriptions two significant facts stand out: First, that small, thin, sharp fissures, extensive both longitudinally and in depth, are characterized by unmodified walls, the vein content, consisting of nearly pure mineral, that is 90% or more; second, that in crushed zones in which the sulphides of lead and zinc occur in shoots of various dimensions from very low-grade ore, the mineralized gangue is mainly silicious, the shoots being pinched in places, and that where pinching occurs the zone continues with full width but then contains kaolinized feldspar in which is found the unchanged grains of the original quartz; and finally that wide intervals occur between the shoots, where the zone is filled with rotted rock having no trace of mineralization. The two facts here cited appear to be of great value in passing judgment on a prospect. A thin vein, even with considerable swells on its exposed strike, will in all probability retain the same characteristics to whatever depth it is followed. In all probability there will be no crushed zone developed with depth, and unless the vein as exposed on the surface is rich enough to raise the stoping width to a profitable milling grade, the prospect will prove economically worthless. There are fundamental reasons for such a conclusion. In a sharp break the walls present a smooth plane of attack for the mineral solutions; the corrosive action must necessarily be slow and long-continued before the fracture can be sufficiently enlarged to permit of an economic volume of ore; the flow will be checked by the gradual filling of mineral or it will cease altogether. Again, in crushed zones the interspace between wall and wall will be filled with fragments and powder of the country rock. In passing judgment on a prospect of this nature three important possibilities will have to be considered: first, a general feeble flow of circulating waters, due to lack of high initial pressure, or, owing to the extremely fine comminution which may prevent circulation, however strong the initial pressure; second, capillary flow with a gradual enlargement of the channel, thus forming 'shoots'; third, coarse crushing in places affording open hannels which may rapidly enlarge through solution of he finely comminuted rock, thus forming large and ich shoots, or the result may be interstitial mineralization between the large crushed fragments. To discuss in etail **the foregoing phases would** require **too much** pace; **besides, experienced examining engineers prob**ably in one form or another use the same criteria. For the inexperienced man it will be sufficient to state, as in the case of thin sharp-walled fissures, conditions exposed for a considerable distance along the strike will not be apt to change essentially in depth. Further, the conditions under which ore-shoots are formed should warn the inexperienced against the lure of 'extensions,' even with a rich mine on the adjoining claim. There are such extensions, of course, as seem to be the two contiguous claims in Butte, the Butte & Superior and the Elm Orlu. The Tonopah Extension is another instance. Examined in detail, however, it is more than probable that in both of the above instances the adjoining mines embrace the same shoot.

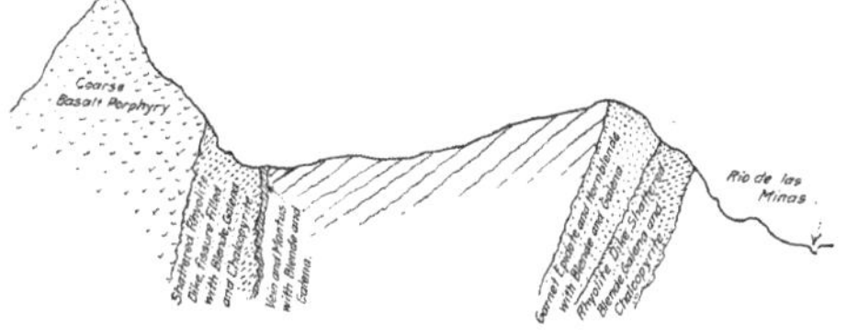

NORTH-SOUTH SECTION, ALOTEPEQUE MINE, GUATEMALA

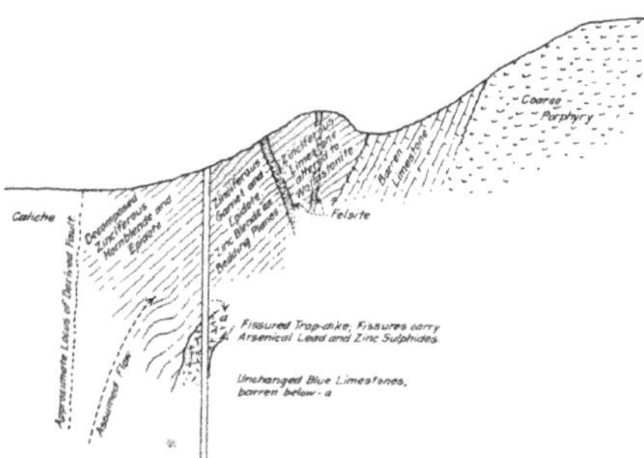

SECTION THROUGH BUFA MINE, CHARCAS, MEXICO

Contact deposits between sedimentaries is a subject that may be discussed briefly. 'Mantos,' or blankets of zinc ore, are common in the northern zinc-producing states in Mexico. In the San Matias district near Saltillo, one manto about one metre thick had been followed down on a slope for about three hundred metres. On the south side of the mountain was a long wide zone of highly metamorphosed limestone with massive garnet abundantly developed. It is more than likely, though not proved, that the manto here was derived from an undeveloped, even unexplored, fissure less than one-half mile to the south. It can be readily understood that seepage, or even a considerable flow of mineral solutions, would follow between bedding planes, especially of lime-

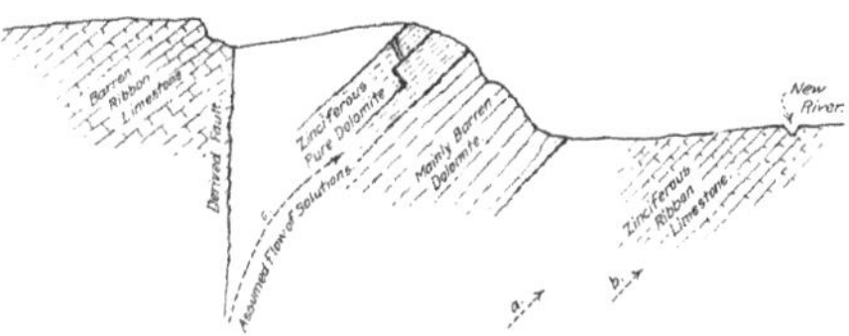

SECTION THROUGH AUSTINVILLE ZINC MINE, VIRGINIA

stone, where the source would be in the crushed highly metamorphic fissure zone. As has already been mentioned, fault-spaces between large blocks of limestone in Austinville, Virginia, are heavily mineralized. In one tunnel of the mine successive limestone strata, wholly unchanged and unbroken, have zinc and lead sheets in bedding-planes, from one-half inch to three inches or more thick. It is probable that the bedding-plane mineral will enrich the barren limestone to milling grade. Between 500 and 700 ft. farther in the tunnel, the heavily mineralized fault-zones are developed. The derivation of the mantos from the fault-zones here seems to be self-evident.

Contacts between sedimentaries and eruptives and contacts between eruptives, to the tyro in mining, and even to the recently graduated mining engineer, are quite as alluring as 'true fissures'. Instances where mining engineers and mining geologists of experience and reputation have fallen victims to the 'contact' theory have often come under my observation. In one instance, over $20,000 was spent in diamond-drilling to exploit the contact under the umbel of an assumed laccolith, in spite of the fact that miles of exposed sills showed unmineralized contacts between them, and their limestone hosts were at no great distance from the position of the drills. It is but fair to say that there is a decided difference of opinion on this subject among writers well qualified to pass judgment. One noted English geologist states unequivocally that "Whether or not an increase of temperature has been connected with the deposition of ore-veins, it is abundantly evident that the occurrence of igneous rocks has no connection with it." On the other hand, an equally well-known American authority on ore deposits, under the caption 'Deposits Related to Igneous Activity', states that* "Several important groups of ore deposits must therefore be considered apart from the ordinary processes of metamorphism. This especially applies to those deposits of the rarer metals, which stand in the closest genetic connection with the eruption of igneous rocks."† My experience compels agreement with the first citation. At the Interstate-Callahan mine, a stringer of zinc-lead was followed for about one mile in the adit. At this distance a trap dike was cut and immediately on the other side of it the great vein was picked up. In Butte, Montana, aplite dikes frequently cut the copper veins. The copper content ceases at the contact but is picked up beyond. The Silver Pick vein in Mt. Wilson, Colorado, is a fissure in a fine-grained diabase. Though primarily the mine yields gold, were the vein larger it would have been an important zinc producer. Mining operations confirmed the surface operations. The diabase dike was intruded into a coarse granodiorite, probably monzonite. In the above two instances of contacts between older and younger igneous rocks, they evidently affected the mineralization not at all. In western Kentucky dikes accompany the faults. In these faults fluorite is extensively deposited, and this mineral, in turn, is impregnated with sulphides of zinc and lead. The influence of the eruptives on the formation of fluorite and blende would seem to be negatived by extensive faults in eastern Tennessee and in south-west Virginia, where zinc and lead occur accompanied by fluorite in small amounts, with one locality in which the fluorite constitutes about 50% of the gangue. In St. Lawrence county, New York, extensive fissures with galena and blende have some fluorite, but on intrusive igneous rocks, unless the black gneisses are conceded to be intrusive. Even so, however, these apparently could have had no influence since the lead-zinc fissures cut through limestone and gneiss in an unbroken line. In Guatemala a great zinc-lead series of deposits is frequently cut by dikes and by what are apparently sills, but only when the dikes and sills are fractured and these fractures filled with galena, blende, and, to a lesser degree, with chalcopyrite, are the intruded limestones mineralized with galena and blende. They then develop great masses of garnet, hornblende, and epidote. Practically the same phenomenon accompanies the zinc deposits in Charcas, San Luis Potosi, Mexico, and Hanover, New Mexico. Numerous similar instances could be cited.

Near Pembroke, Maine, small lenses and veins of blende occur in a fissured trap; in two places in New Hampshire lead and zinc are associated with igneous dikes, but, on the other hand, near Lincoln, in the same State, is a great crushed fault-zone in gneissoid rocks where not only does argentiferous galena occur with zinc-blende, but many of the rock-fragments have their feldspars converted to a greasy mineral, probably sericite, or else have them changed to epidote. No intrusive of any kind was observed. In the Triassic formation, between Northampton and Amherst, and also near Mount Tom, veins are found in the sandstone and are filled with barite accompanying a considerable amount of galena, blende, and chalcopyrite. No eruptives occur near the veins. Almost numberless citations of zinc and lead localities can be made but the summary would leave the question where it now is, balanced. The number of zinc and lead localities without accompanying eruptives would not be outweighed by those in which igneous rocks occur. The assumption of either fissures or eruptives as the primal cause of such deposits would be purely arbitrary. There are miles of barren fissures just as there are miles of barren contacts between eruptives and eruptives, and contacts between eruptives and sedimentaries. The factor common to both is an open channel or rocks prepared in such manner that channels can be opened in them. This leads to the brief consideration of another common factor, namely, rock disturbances, or rather preparation of the rock. It is as impossible to conceive of an igneous intrusion without disturbance as it is to conceive of a mineral deposit without such a forerunner. It is suggested that a fissure filled with an eruptive is as much a vein and even a fissure vein, as though it were a fissure filled with a gold, silver, lead, or zinc-bearing gangue.

*Jukes, 'The Students' Manual of Geology', p. 372.
†Lindgren, 'Mineral Deposits', p. 73.

Even the existence of high temperatures, evidently a powerful factor in many mineral deposits as in many instances of rock metamorphism, may be no more inherent in eruptive rocks than in veins or in the metamorphism of rocks, but the temperature of all may have been derived from the same source, that is, the conversion of the enormous energy required to move and fracture huge rock-masses into nearly or quite corresponding heat equivalents.

In conclusion, I wish to disclaim the idea that this paper introduces any new factor into the consideration of zinc and lead deposits. What I do feel is that, in the honest zeal of investigation, some evident factors in the problem have been selected, to which undue value has been generally attached. Science is no doubt advanced by this method, for counter-checks are certain to follow, but these rarefied discussions are certain to be misinterpreted and misapplied in practical mining. The safe course for the practical miner is to bear in mind that no secondary mineral deposit can be formed without rock disturbance; that in spite of this, deposition of mineral rarely or never is co-extensive with the rock preparation. The fundamental principles governing zinc-ore deposits are fissuring or faulting, with consequent open channels for circulation of the mineral-bearing solutions. All other accompanying phenomena may be safely disregarded by the practical miner.

Fused Bauxite for Furnaces

In a recent issue of 'La Cerambique,' M. N. Lecesne describes the following method of preparing bauxite for furnace lining and crucibles: A mixture is prepared of one part of anthracite and three parts of bauxite of an average composition of about 60% alumina, 10% iron sesqui-oxide, and 10% silica, moist as it comes from the quarry. The mass is charged into a furnace lined with refractory material, preferably fused bauxite, and previously heated with anthracite, air being blown through the charge at an initial pressure of about 40 in. water gauge, rising afterward, according to the depth of the charge, though 160-in. pressure is the usual maximum. The temperature rises quickly and the sudden vaporizing of the water in the bauxite causes it to break up and to granulate, while the aluminum carbide produced burns and raises the charge to a point at which the silica is volatilized, and the iron, reduced by the action, is expelled by the air-blast as a shower of sparks. The air-blast is continued in order to burn off the surplus anthracite and cool the fused mass, which can then be discharged from the furnace and ground, mixed with crude or calcined bauxite as a binder, molded, dried, and fired in the same way as calcined bauxite. If the fusion is performed quickly, say, in about three hours, the resulting mass will be highly porous, besides containing a sufficiency of unconsumed anthracite to facilitate grinding and briquetting. The fused bauxite will give a refractory material stated to be capable of withstanding a temperature of over 2000°C. if mixed with some of the same material freed from iron by magnetic separation, and with water and quicklime, the latter forming a binder of silicate of lime. Mixed with bauxite the product is hard but porous, and forms a suitable lining for reverberatory or other furnaces heated with liquid or gaseous fuel.

Uses of Stellite

The malleable alloys, as prepared by Elwood Haynes, are composed almost entirely of cobalt and chromium, though the proportion of the constituents may vary from 10 to 50% chromium, with a corresponding variation in the cobalt. These alloys are all hard, and while they may be scratched by the file, none of them is practically workable by this means. They cannot be machined, but some of the softest may be drilled by means of a hard-carbon steel drill or a drill composed of hard stellite. They resist nitric acid almost perfectly, even when boiling, particularly if the chromium content is over 15%. They forge with difficulty at temperatures ranging from 750° to 1200°C. They have been forged into tableware, such as spoons, forks, knives, and ladles, surgical instruments, pocket knives, dental instruments, evaporating dishes, crucible supports, lamp-stands, jewelry, including finger rings, cuff buttons, and scarf-pins. These malleable alloys are all slowly attacked by either hydrochloric, sulphuric, or hydrofluoric acid, but are nearly immune to all chemical combinations, as well as the fruit acids. As the evaporating dishes made of this metal take a bright polish, and can be made with a comparatively light section, they will prove suitable for evaporating many chemical salts to dryness, and are particularly suitable for boiling the caustic alkalies. Substances may be evaporated to complete dryness in these vessels without any danger of breaking the vessel, since the tensile strength of the alloy exceeds 100,000 lb. per sq. in., and it also shows considerable elongation before rupture. When a vessel made of this material is struck by a hammer, it emits a clear musical tone, and continues to vibrate for a considerable length of time. The vessels retain their lustre in the chemical laboratory under practically all conditions, since they are not affected in the slightest degree by sulphuretted hydrogen, ammonium chloride, or other vapors. Furthermore, they are practically immune to acid vapors. They give excellent results in the form of lamp-stands, supporting-rings, triangles, and other laboratory apparatus. When these are heated to full redness, they become covered with a deep blue-black film, which does not change either in weight or appearance by repeated heating, and as no scale ever forms, these articles retain their weight, stability, and smooth surface indefinitely under all sorts of use. They can be subjected to temperatures up to 1200°C., and still retain a considerable amount of strength. In fact, the stellite alloys possess the highest 'red hardness' of any of the alloys yet discovered.

War-Steel Production

*Statistics have shown that the steel-ingot production in the United States in July was at the rate of 44,000,000 tons per year. An excellent authority compiled a statement last August revealing that the existing ingot-capacity was 49,500,000 tons, and that an additional 3,800,000 tons capacity was under contract, to be completed within six months. Finished rolled-steel production amounts to 76% of the ingot-production, hence the output for finished rolled steel was at the rate of 33,000,000 tons in July, at which season the production is generally about 5% below the year's average. Allowing for a considerable shortage of coke, even after the by-product ovens under construction are completed, we should be able to count upon 48,000,000 tons of ingots and 36,000,000 tons of finished rolled steel. The 11,500,000 tons of war-steel already included would be 32% of the output. The percentage is likely to be under rather than over 32%. If, under the stress of labor shortage, transportation difficulties, and other circumstances growing out of the War, the output should be restricted more than is here allowed for, there is good reason to believe that the commercial requirements in steel will be even more restricted, so that the essence of the argument is not affected, namely, that there is no serious danger of a decided shortage of commercial steel.

The distribution by the Government of orders on behalf of its Allies succeeds to such buying as these Allies have hitherto been doing on their own account. Last year's exports of steel under orders from the Allies amounted to about 5,200,000 tons. The steel consumed in making shells, machinery, and all other steel manufactures, has been estimated at about 1,800,000 tons, making a total of 7,000,000 tons of rolled steel for 1916 involved directly and indirectly in the export trade. The production was 30,500,000 tons, and thus 22,500,000 tons was left for the domestic trade. That is approximately the same amount as was consumed in 1912 and 1913, the best years before the War. Of the 7,000,000 tons involved in exports one may guess that something like 5,000,000 tons was included in exports to the countries now our Allies. The present war-steel programme absorbs the tonnage, whatever it was. There is every reason to expect that, as long as steel is scarce in the United States, the Government will not permit exportation to neutrals, hence, for the purpose of this argument, such exports may be neglected. The result is that, if we accept the figures of 11,500,000 tons of steel for the war programme in 1918, and the total production at 36,000,000 tons, there is left 24,500,000 tons for domestic commercial consumption, against 22,500,000 tons in 1916. It is difficult to see how this consumption could possibly increase, there being so many directions in which it is visibly decreasing. The greatest pinch will come to ship-plates. The production of mills that can roll such plates was only about 1,800,000 tons in 1916. The largest figure yet set upon American shipbuilding in 1918 is that used by the British controller of shipping, who recently authorized the Associated Press to say that he would like to see 6,000,000 tons of shipping built in the United States in 1918. He set the amount of steel involved at 3,500,000 tons, which is a rather high proportion of the gross tons of steel to the gross register of the vessels. Deduction should be made for wooden ships in the programme of expansion for the merchant marine. There would be required, then, say 2,500,000 tons of ship-plates for our shipyards. This we can furnish, and some plates in addition for foreign yards, possibly 3,600,000 tons in all, or double our production of this class of plates last year, but certainly there will be no plates, except those not available for shipbuilding, for ordinary commercial purposes.

*Abstract: 'The American Metal Market'.

Secondary Enrichment

The theory of ore enrichment may be briefly stated as follows, according to W. H. Emmons, who has prepared Bulletin 625 on this subject recently issued by the U. S. Geological Survey: Mineral deposits that are exposed to weathering, that is, to the action of the atmosphere, of rain, frost, and other agencies, break down and form soluble salts and new minerals that remain unchanged under surface conditions. In the course of weathering, however, the ore of each metal behaves in its peculiar way. Deposits of iron, aluminum, manganese, and some deposits of gold and other metals may be enriched near the surface by the removal of valueless material. On the other hand many of the valuable metals, including copper, silver, and less commonly gold, zinc, and lead, go into solution and are carried downward by rain water and re-deposited, and thus parts of the orebodies may be enriched. Most of the rich or 'bonanza' orebodies of copper, many of silver, and some of gold and other metals, have been formed in this way. Bulletin 625 describes in detail the geologic conditions favorable and unfavorable to the formation of enriched mineral deposits and gives a summary of the criteria by which such deposits may be recognized. The natural chemical processes by which enrichment is produced are fully described and the behavior of each metal is considered separately and is illustrated by descriptions of practically all the known valuable enriched orebodies.

CORUNDUM mining has been revived at Corundum Hill, North Carolina, by the Hampden Corundum Wheel Co. of Springfield, Massachusetts. Emery is also being produced in Westchester county, New York, the quantity shipped amounting to 15,282 tons. The excessive use of abrasives on account of the increase in steel manufactures during the War has again brought these natural products into demand. For years the artificial abrasive, carborundum, has kept natural corundum out of the market, being more cheaply produced. Electric power at Niagara, however, is now more usefully employed in other manufactures not otherwise obtainable.

Chrome Deposits of Alaska

By W. P. LASS

The deposits of chrome ore in Alaska have, until recently, been rather overlooked by the industrial companies of our country and it was not until the rise in the price of chromite that local operators have undertaken to market the ore. The more important known deposits are on the extreme far point of Kenai peninsula, while the zones of mineralization may be traced eight to ten miles inland. Although no systematic prospecting has been done to any great distance from the edge of salt-water, there are nevertheless several areas where important deposits of chromite have been explored. Notable among these are the deposits at Red mountain, described in the U. S. Geological Survey Bulletin 587.

I was called upon to examine these deposits in the fall of 1916, and although shown several exposures containing from a few hundred to a thousand tons of high-grade ore, they were not deemed worthy of development by local capitalists who, in turn, had to depend upon selling their ore in a very unstable market. Should the present demand for chrome ores continue, however, throughout the coming year a more thorough prospecting and development of this belt will take place.

CHROME MINING ON THE KENAI PENINSULA, ALASKA

Although the conditions for marketing have not justified large expenditure in this region, there are deposits, nevertheless, so close to salt-water as to make the shipment of this ore well worth while. The accompanying photograph shows workmen removing the top, quarry fashion, from a high-grade deposit of chromite. Approximately 1000 tons of ore, containing from 46 to 49% chromium oxide, was removed from this deposit during the past summer. No development work was done to determine the extent of the deposit, and there remains approximately 1000 tons of ore in sight above the line of high tide. The ore between high and low tide is 18 to 20 ft. wide and contains from 48 to 52% chromium oxide. No development plans for the working of this deposit below the line of high tide have been made. Some shipments of this ore were made by barges to the docks at Port Graham and Seldovia and thence by freighters to Seattle, but the greater part of the shipments were lightered direct from the beach to steamers anchored about 300 yd. off shore. The present freight rate on 100-ton shipments to Seattle is $3.50 per ton. From Seattle to Atlantic Coast points the railroad rate is $12 per ton. As this material is essentially a war-product, no difficulty has arisen in obtaining cars, and satisfactory deliveries to Eastern points have accordingly been made.

Calcination of Magnesite

Operations have been begun at the reduction plant of the Piedra Magnesite Co. at Piedra, Fresno county, 30 miles east of Fresno, California. The equipment has a total capacity of about 60 tons of calcined ore or twice that amount of crude ore per day.

The plant and its equipment represents an investment of $55,000. Thomas J. Curran, manager for the American Magnesite Co. of Porterville, is general manager for the Piedra Magnesite Co., W. B. Phillips of Porterville is superintendent, and J. W. Thomas treasurer.

The rotary kiln at the plant is 83 ft. long and 8 ft. diameter and weighs 80 tons. The cooling-tower is 60 ft. high. The red-hot ore is carried to the cooling-tower from the kiln by a 125-ft. chain-bucket elevator. The ore is fed into a gyratory crusher from which it is lifted to a large revolving screen where the fine ore is screened and sent through a chute to the ore-bin, which has a capacity of 75 tons, and from which the ore is fed through a dust-chamber into the kiln. The ore that fails to pass through the screen is returned to the crusher. All ore passing into the kiln is reduced to pea-size and is calcined quickly. The machinery about the plant is operated by electricity.

The Piedra Magnesite Co. owns eleven different mining claims, but at present is working only seven or eight of them. The reduction plant, which is on the banks of Kings river, is about half a mile from the quarry. The ore is hauled to the plant on motor trucks. About 125 men are employed at the plant and quarries. All the ore of the company has been contracted for by the American Refractories Co. of Pittsburgh, Pennsylvania.

Gold Output of Colorado

These statistics of the U. S. Geological Survey are interesting; they prove that the Leadville district, in Lake county, is still a dominant factor in the mining industry of Colorado, the silver-lead production of the early days being supplemented now by an important output of gold and of zinc. Cripple Creek places Teller county far in the lead of the gold-producing districts. Next comes San Miguel, which includes the Telluride district. Creede is identified with Mineral county and Aspen with Pitkin county; both show a great decline from the early prosperity. The old mining centres of Clear Creek, Gilpin, and Ouray are still pegging away bravely. It will be noted that the grand total is $5,773,978 higher than in 1915.

PRODUCTION OF GOLD, SILVER, COPPER, LEAD, AND ZINC AT MINES IN COLORADO IN 1916

[Compiled by Charles W. Henderson.]

County.	Producing mines.	Ore sold or treated (short tons).	Gold.	Silver (fine ounces).	Copper (pounds).	Lead (pounds).	Zinc (pounds).	Total value.
Baca	1	5		50	2,772			$715
Boulder	81	33,011	$119,299	292,824	64,707	864,333		387,534
Chaffee	35	69,358	185,050	100,749	1,001,455	3,016,899	4,744,985	1,341,695
Clear Creek	100	94,220	428,931	462,141	621,732	4,295,725	2,572,575	1,527,096
Custer	14	2,245	6,309	36,971	44,004	123,536	10,970	51,455
Dolores	13	6,398	7,426	77,280	419,500	588,333	182,306	226,497
Eagle	32	105,149	96,036	222,126	112,610	1,517,362	28,438,052	4,185,294
Fremont	11	1,734	786	4,529	101,041	31,710		30,810
Gilpin	87	38,913	453,259	126,553	557,317	521,334		709,603
Grand	2	2		134	760			275
Gunnison	18	10,419	31,553	29,023	84,679	313,217	1,964,873	356,386
Hinsdale	18	377	1,346	10,030	16,248	75,638	12,575	18,847
Jackson	1	61	95	199	6,752			1,887
Lake	96	477,240	1,720,440	2,931,281	2,621,675	21,719,392	76,785,567	16,082,059
La Plata	36	1,602	31,653	29,187	12,024	6,551		54,268
Mineral	9	38,103	31,124	373,956	13,138	2,295,087	240,575	471,017
Moffat	5	25	063	41	9,033			3,212
Montezuma	3	86	1,402	193	3,118	116		2,304
Montrose	2	197	10	1,132	100,008			25,357
Ouray	52	111,192	491,175	803,461	444,081	2,339,029	69,015	1,299,737
Park	15	3,005	234,299	13,231	22,698	330,609	47,560	277,749
Pitkin	12	144,330		577,863	28,931	17,519,275	162,574	1,617,966
Routt	4	517	179	237	32,142			8,242
Saguache	20	3,338	8,024	48,959	92,581	255,449		80,640
San Juan	46	146,128	438,628	502,342	1,615,167	7,285,304	4,014,403	2,207,116
San Miguel	20	428,651	2,072,393	812,041	581,427	6,126,551	1,098,485	3,319,676
Summit	55	65,117	673,891	120,207	14,581	1,688,637	13,940,948	2,741,177
Teller	64	945,820	12,119,550	79,804				12,172,061
Total, 1916	852	2,697,243	19,153,821	7,656,544	8,624,081	70,914,087	134,285,463	49,200,675
Total, 1915	821	2,737,020	22,414,944	7,027,972	7,112,537	68,810,597	104,594,994	43,426,697
Increase or decrease	+31	−39,777	−3,261,123	+628,572	+1,511,544	+2,103,490	+29,690,469	+5,773,978

The average prices for metals for the calendar years 1915 and 1916 were:

	1915.	1916.
Silver, per fine ounce, at New York	$0.507	$0.658
Copper, per pound, electrolytic, at New York	.175	.246
Lead, per pound, at New York	.047	.069
Zinc (spelter), per pound	.124	.134

The average price used for spelter is not the market price at St. Louis, but the average selling price of zinc as shown by canvass covering all sales.

Magnalium

Reference was made recently in our editorial columns to this remarkable alloy, used in the making of aeroplanes. We find that Frank L. Hess, in his report on 'Magnesium in 1915,' published by the U. S. Geological Survey, quotes a description of E. F. Law, an English authority:

Commercial magnalium contains only a small percentage of magnesium, and, although a large number of alloys are manufactured and sold under the name of magnalium, few of these, if any, appear to contain more than 2%. On the other hand, they all contain a variety of other metals, more especially copper, tin, nickel, and lead.

The importance of magnesium as a de-oxidizer must not be overlooked, for it is even more readily oxidized than aluminum; and its beneficial influence on aluminum is in no small degree due to its power of freeing that metal from dissolved oxide.

The three magnalium alloys most commonly used in this country [England] are described by the makers as *X*, Y, Z. Of these, *X* is intended solely for castings where strength is of primary importance; *Y* is used for ordinary castings; and *Z* is intended for rolling and drawing.

As regards alloy *X* it has been stated by Barnett that it contains 1.76% copper, 1.16% nickel, 1.60% magnesium, and small quantities of antimony and iron.

Alloy *Y* is somewhat similar in composition, except that it contains no nickel, but small quantities of tin and lead.

Alloy *Z* contains 3.15% tin, 0.12% copper, 0.72% lead, and 1.58% magnesium.

The tensile strength of ordinary castings with alloy *Y* varies from 8½ to 10 tons per square inch, and that of rolled samples of alloy *Z* varies from 14 to 21 tons per square inch.

The alloys work well, and excellent screw-threads can be cut. The speed of working is about the same as that of brass, and the tools should be lubricated with turpentine, vaseline, or petroleum. Alloy *Z* is exceedingly ductile, and can be spun and drawn into the finest wires. For these operations vaseline or a mixture of 1 part stearine and 4 parts turpentine has been found suitable. In drawing tubes or wire the alloy must be annealed by heating and cooling suddenly. Slow cooling produces hardening. Magnalium is very little affected by dilute acids and can be employed with perfect safety for the manufacture of cooking utensils and all culinary appliances.

OIL may be more economically burned, according to W. A. Janssen, by pre-heating the air, then mixing it in definite proportions with the oil, creating a combustible gas which is forced into the furnace under positive pressure. This results in perfect combustion since exact regulation of the air is possible.

MAMMOTH, UTAH

RARE-METAL ORES.—TINTIC STANDARD.—DRAGON.—IRON BLOSSOM.

According to Magnus Benson, an old Nevada-Utah prospector, some rare ores have been found in western Boxelder county. The ores are those of molybdenum and tungsten. Mr. Benson and his associates have taken up the Blue-Bell group of seven claims located 16 miles north-west of Terrace station. This is in the extreme north-western corner of the State, near the Nevada-Idaho line. The formation there is mainly a shattered granite with the ores occurring in prominent fissures. One of the principle fissures in the Blue-Bell group shows on the surface for over 4000 ft. This fissure is from three to four feet wide where opened on the surface and is quartz-filled. The rich ore occurs in pockets and is easily sorted. A shaft will be sunk on the fissure; there are also good tunnel-sites. The owners expect to ship a carload of this rare ore from Terrace at an early date. Associated with Mr. Benson are George W. Parsons and J. A. Lynch. At Bovine, six miles distant, tungsten ore is being mined for shipment this fall. Ogden people are making preparations to start development soon. While the embargo on the ores at the mines to the A. S. & R. Co. has been extended from October 1 to October 15, it was stated yesterday by a mining official that the embargo had been lifted for a period of five days, beginning this week. This will give the mines that have been holding back for three weeks an opportunity to rush in considerable of their accumulated stock. There has been no embargo at the plant of the U. S. Smelting company for more than three weeks.——Directors of the Uncle Sam Con. M. Co. met on Monday and levied assessment No. 2, consisting of 1c. per share. The delinquent date is November 9 and sale date December 10. Fred C. Dern is secretary.——Directors of the Tintic Standard Mining Co. met on October 9 and declared the second dividend of 2c. per share, which means a payment of nearly $24,000. The distribution will be made on October 25 to stock on record October 20. The initial dividend of $24,000 was paid on June 7. This makes a total to date of $48,000 on the 1,200,000 shares issued. E. J. Raddatz, the general manager, said that the company made a payment of $12,000 in cash on some of the newly-acquired property. During the past two weeks the mine has shipped 11 carloads of ore, or about 550 tons. The connection between the new shaft and old workings will be made shortly, only 20 ft. of ground intervening, and a raise of 22 ft. This will make connections at a point from the old 1500-ft. level by incline to the 1650-ft. level, which is 1250 ft. vertically. The drift from the new shaft is on the 1260-ft. level. This has passed out of the shipping ore now and is making for the point of connection.——The director of the Dragon Con. at the regular monthly meeting at Provo on October 14 declared the usual quarterly dividend of 1c. per share, payable on October 25 to stock on record October 15. This is the fourth dividend and brings the total distributed to $75,000. The first dividend was paid on January of this year. The Tintic record of shipments of ore shows that the Dragon leads all other mines of the district by a big margin. A comparatively small amount of higher-grade ore was marketed at the same time, so the car-lots run high. The September total for Dragon amounted to 184 carloads: August, 254, July 145. The total for the first six months of this year was 876 carloads.——An assessment of half a cent was levied by the Tintic directors at Provo on October 14. This became delinquent November 15. Sale day is December 8. It is stated that the object of the levy is to liquidate the company's share of the drift from the 1700-ft. level of the Iron Blossom. This drift is now about 140 ft. inside of Central territory. Boston officials give out the information that the Utah Metal & Tunnel Co. has made a contract with the International Smelting Co. whereby the latter will purchase all the company's product for a period of five years. The price will be the average price for the week preceding the receipt of the ore. The company's engineers estimate that there is sufficient milling ore developed for more than three years' supply, and the low-grade ore is increasing so rapidly that the mill capacity must be still further enlarged or a new mill built at the other end of the tunnel in the middle canyon. The new additions to the mill at Bingham were completed October 1, and will go into operation immediately; this increases the capacity of the mill by 50%, making it 400 to 450 tons per day.——Directors of the Iron Blossom Mining Co. at Provo have declared the regular quarterly dividend of 5c. per share, payable on October 25 to stock on record October 15. This will bring the 1917 dividend up to $250,000, or 25c. per share. The stock sold down to 66c. after this dividend announcement. This distribution will bring the total Iron Blossom dividends up to $3,100,000 or at the rate of $3.10 per share. Last year the company distributed 35c. per share, or $350,000; 33c., or $330,000, in 1914; and 40c., or $400,000, in 1913.——Iron Blossom shipped 75 carloads of ore in September, 85 in August, 82 in July, and 761 during the first half of this year. While it is admitted by officials that the recent extensive prospecting on deep levels in various parts of the mine have been disappointing, it is felt that at any time new orebodies may be discovered, as Iron Blossom still has a large unprospected territory. During the first nine months of 1917, the Utah mines have paid dividends aggregating $22,252,680, compared with $24,529,430 last year. The total dividends paid by these 21 mines amount to the gigantic aggregate sum of $120,314,060.——Theodore P. Holt, superintendent for the Tintic Milling Co., at Silver City, is expected back from Canada during the coming week. Mr. Holt went to Cobalt some months ago for the purpose of building some of the new Holt-Dern roasters. The Holt-Dern roaster, first used at a mill at Park City and then at the Tintic mill, is an invention of Mr. Holt and George H. Dern of Salt Lake City.——Shipments of first-class ore from the mines of Tintic this week totaled 172 cars, estimated at 8000 tons, and valued at $200,000. Dragon Con. continues to head the list, with Iron Blossom and Chief Con. running neck and neck for second place. Ore was shipped from 19 mines.——Cecil Fitch, superintendent of the Chief Con. mine, states that actual mining operations will soon be in progress through the shaft of the Plutus company. The Plutus is controlled by the Chief. The officials of the Chief have decided that the old Plutus shaft could be used to advantage in the development of the large tract of mineral ground which the latter company owns, and consequently the work of putting this shaft in shape was started. The shaft, down to a depth of 400 ft., had to be repaired from top to bottom and new hoisting machinery put in.

All of this work has been completed and shafts and drifts on the 400-ft. level are being unwatered. There is no big flow of water at that depth, but during the many years that the property has been idle the water has accumulated.——A new hoist, secured from the May Day company, is being moved to the Eureka Bullion company in the east end of this district. The latter company's shaft has reached a depth of 545 ft. and better hoisting equipment is necessary to continue the sinking operations. The hoist which has been secured will be good for 1000 ft. or more, according to John Bestlemyer, the superintendent. Manganese is now coming in at the bottom of the Eureka Bullion shaft, and the indications are said to be more favorable than at any time since work started.——The Mammoth mine dividends for 1917 amount to $360,000, all of which has been paid out since May 1. The last distribution of $40,000, or 10c. per share, will be paid on October 12.

Thomas L. McCarthy, M. E. Price, and F. E. Butler, of Salt Lake City, have taken a lease on the manganese claims of the Chief Con. Mining Co. at Homansville, and the Ellener, Cromar, Nelson, and Morgan, in the Erickson district, and work has been started. The ore which is now being mined carries about 35% of manganese. The drift at the Zuma is still following the stringer of lead-silver ore, cut some weeks ago. Since the work was stopped in the main drift and the cross-cut, driven along the fissure, the ore has been followed a distance of over 100 ft., the value remaining fairly constant. ——The Chief Con. has purchased $100,000 worth of Liberty Bonds; this company took $50,000 worth of the first issue.—— Sinking has been started at the North Beck mine next to the Eureka mines. An electric hoist and air-compressor have been added to the equipment.——The Utah Zinc Mining & Milling Co., B. F. Fleiner, superintendent, has uncovered some high-grade lead-silver ore at several points on its property; about 150 tons of lead ore, carrying 35 to 40%, is ready for shipment. ——After the sinking of a winze to 100 ft. below the adit, driving is now in progress at the O. K. Silver property, near Indian Springs. The mine is controlled by the Kearns and Keith interests of Salt Lake; some two years ago it produced some high-grade silver-lead ore.——Two carloads of bullion will be shipped from the Tintic Milling Co.'s plant at Silver City before November 1, according to George H. Dern, the manager; the plant is now shipping at the rate of more than a carload of bullion every month. The ore being treated is mainly from the Dragon Consolidated, Iron Blossom, and Colorado; a small quantity comes from other neighboring Tintic mines. The mill is treating 200 tons per day. The mill-product is shipped to the refineries at Newark, New Jersey; it contains 90% copper and from 600 to 2000 oz. in silver.

SUTTER CREEK, CALIFORNIA

OLD EUREKA.—HARDENBURG.

The Alpha Lode mining claim, sometimes called the West Eureka property, has been acquired recently by the Old Eureka Mining Co., which purchased it from William J. McGee. This strip of ground adjoins the Old Eureka on the west and was first worked in the early 'seventies by D. D. Reaves and associates. McGee's title to it was acquired through foreclosure-proceedings. The ground is considered to be valuable, particularly to the Old Eureka Co. It has been opened by shallow shafts and an adit, and sufficient work has been done to demonstrate its worth. It will doubtless be worked in conjunction with the Old Eureka mine, as some of the Old Eureka orebodies crop there. In addition to the property purchased from Hetty Green, and last worked about thirty years ago by Alvinza Hayward and others, the company has purchased two large ranches to the east for a mill-site and tailing-dump. The principal stockholders are said to be the new owners of the Wildman, Mahoney, and Lincoln properties, which group adjoins the old mine on the north. The bottom of Old Eureka shaft is being enlarged to provide for three compartments.—— Failure to find enough ore to justify further expenditure is the reason the W. J. Loring company has discontinued operations under its option at the Hardenburg mine, south of Jackson. The property reverts to the former owners, who are the same men interested in the South Eureka property at Sutter Creek. The rich find reported in the Hardenburg mine recently proved to be only a small pocket. Large low-grade orebodies are in sight, but the cost of handling them under present conditions is prohibitive.——Conditions at the Central Eureka mine are improving. The value of the mill-heads shows an increase and there is prospect of the present 3c. assessment putting the property in a workable condition.

HOUGHTON, MICHIGAN

A RETROSPECT.—OSCEOLA CON.—MOHAWK.—CALUMET & HECLA. COPPER RANGE CO.—SOUTH LAKE.

The production of copper for October from the Lake Superior district will exceed that of September by at least 2,000,000 lb. if the present rate of production is maintained. The decided falling off in production in September occasioned considerable surprise and a good deal of comment from shareholders. In comparison with the copper production in other mining districts in the United States the showing was exceptionally good, and, in this connection, it must be remembered that the slump in output is not entirely due to shortage of labor but to a more important factor. A few months after the War started the price of copper began to advance. Just at the outbreak of the War it was low, sales being made at 10c., and there was a considerable accumulation of metal on hand. The wiser buyers for the Allies and the wiser domestic consumers immediately began to buy heavily at the low price. Then the buyers for the Allies all jumped into the market, and the price began to advance sensationally. There was a brief respite when foreign war requirements were all placed in one hand for purchasing, but the price continued at an abnormal figure. Confronted with the certainty of 'fat' profits, mine managers in the Lake Superior district were faced with two problems. One was constantly increasing costs; the other was whether it was desirable to mine the richest ore in their property, making all the possible profit, and take every advantage of the high price, or to mine just average ore, maintain advance openings to the limit of expectations, and even to send to the mill considerable ore that would show a profit with copper at 25c. but could not be mined with the metal at 11c. There are plenty of arguments on both sides of the question. A few of the mines in this district began mining copper that ran 50 lb. per ton and made wonderful profit. In fact they figured that never again would the opportunity present itself to get such a result from a copper mine. And the surprising thing about it was the way results kept up. Month after month the looked-for slump did not come; now that it has, there is surprise and dismay. Yet the result is but the natural outcome of the policy which has plenty of justification in profits accrued and in dividends paid. But most of the 50-lb. copper has been cleaned up, and even if tonnage is maintained, a lower percentage of copper is all that can be expected unless some richer stopes are opened in newly-developed territory. The reduction in output of two or three Lake Superior properties, due to the fact that the 'fat' places are cleaned up, is going to count as a considerable factor in the reduction of copper output from this district in 1917.

Directors probably will not accept the resignation of John C. Greenway, manager for Calumet & Arizona, who is now a Major of Engineers in the United States Army, but will grant him leave of absence during the War and select one of the directors, either Mr. Campbell, the Secretary, or E. C. Congn to act as managing director during his absence.——The duction of the Osceola Consolidated dividend to $2 per quar- r is based on the preparation for the super-tax, for reduced ofits owing to the lower price of copper, and a desire to con- rve a treasury-surplus until the metal situation becomes ttled. There is a great shortage of trammers at No. 4 North arsarge.——Notwithstanding the high cost of construction rk, all of the mines here are repairing the residences of ployees and erecting new ones; Champion is to erect 25 w homes for miners; Trimountain 15. At the Quincy stamp- ll new dwellings are being built and the old ones improved. There is a slight improvement in the small amount of open- g work being done in Mohawk No. 1 shaft. Last year there s no betterment and the mineralization was considerably low the average of the mine. For the purpose of exploring formation to the east and west of the Mohawk lode, cross- ts were driven on the 21st level at a point 670 ft. south of No. haft. The cross-cut west, or in the hanging, has now reached listance of 740 ft. and the cross-cut east 654 ft. Several amyg- loids were intersected, none of which shows sufficient en- uragement for further work. The Wolverine sandstone was ind in the east cross-cut 350 ft. from the foot-wall of the hawk lode.——The 17th and 18th levels south of No. 3 shaft the Ahmeek have been extended to the Mohawk boundary: are in fair ore. Last year the five levels just above were all tended to this same boundary line. The shaft is down close the 22nd level. The Fulton fissure crossed this shaft at the h level.——One third of the production of copper which the lumet & Hecla secures from the three shafts it operates on Osceola lode, at the old mine, comes from the foot-wall pes. This lode, particularly in the Calumet & Hecla prop- y, has peculiar difficulties. It is utterly impossible to mine ross from foot to hanging wall because the lode varies in dth from 10 to 160 ft. The copper content is so lean, running tle better than 13 lb. per ton, that there is but one practical thod of handling it, namely by following the copper course carefully as possible. For this reason operations in shafts 14, and 15 require miners of unusual ability and shift- sses who use discretion in avoiding poor ore.

Annual report of mines inspector of Houghton county for r ended October 1 showed total number of underground n to be 16,423. This compares with 16,250 in 1916, 16,005 1915, and 12,954 in 1914. It will be seen, then, that, with the hue and cry of shortage of men and the fact that every e needs more men underground, the working-figure is ter than the average. The percentage of casualties was 29. Of the 48 men killed in the mines, Champion lost n and Calumet six.

t the present rate, production at Mass Con. ought to run 9,000 tons of ore in October. The Evergreen lode is fur- ing some ore, but Butler continues to be the largest pro- r. Recent openings are encouraging.——The No. 4 North rsarge of Osceola Con. is getting into the best grade of ore d anywhere in the mine. The ore tonnage at old Osceola ch is maintained, but the total for October will not be h over 100,000 tons.——North Ahmeek now furnishes 40% he Ahmeek total which is likely to be 110,000 tons this th.——The ore coming from South Lake shows an average 7 lb. per ton, all of which goes to Winona mill.——The er Range railway plans to take over haulage for Wolver- nd Mohawk in two months. Ore production at Wolverine wn to 34 cars per day.——Mohawk may reach an output ,000 tons in October. The copper content averages 19 lb. ton.——Long drifts at the La Salle toward old Osceola good copper. The output is close to 15,000 tons per h.——The Centennial September output was 10,000 tons. difficulty of securing men for this mine and long tram makes the cost high.——Repairs to the old shaft-hoist e Franklin are completed, and hoisting is now at the ar speed. The output for September was 23,000 tons compared with 27,000 in August. The old shaft has a full crew, but more men could be worked at No. 2.——Allouez is one of the few Lake Superior copper mines that will show a substantial increase in output for 1917.——No new ground is being opened at Superior and the tonnage is declining steadily. The Houghton Copper, next adjoining, confines operations to exploring from Superior's openings.

The three mines of the Copper Range Co. continue to produce copper at pretty close to normal rates, the daily tonnage running 3600. In September the three mines sent to the mills over 100,000 tons of ore, averaging 30 lb. per ton. This tonnage is slightly under the average of the properties covering a period of ten years. During 1916 the three mines combined took out 1,187,119 tons of ore. The Champion mine continues to be the richest and largest producer, turning out more ore and more copper than the Trimountain and Baltic combined. In fact the Champion produced last month more than 60,000 tons of ore and the Baltic 26,000 tons, while the Trimountain turned out only 16,000 tons.——South Lake produced 3500 tons of ore last month, which was rich in copper. The ore was crushed at the Franklin mill at Point Mills. Beginning with October 1, however, the South Lake ore will be crushed at the Winona mill. This arrangement saves railway haulage. The September output is 1500 tons under that of August.——Surprising though it may seem, Adventure is getting out more copper ore than Lake, the tonnage in September being 6000, compared to 5500 for Lake.——Arrangements have been completed whereby R. R. Seeber, who has been superintendent of Winona for a number of years, will hereafter operate the property on tribute, guaranteeing profits to the company.

Foundations for the machinery for a power-house at the Seneca have been laid. It is hoped that the shaft-house will be enclosed before the winter, so that sinking can start before the new year.

TONOPAH, NEVADA

TONOPAH EXTENSION.—TONOPAH BELMONT.—WEST END CON.—JIM BUTLER.

During the month of September the Tonopah Mining Co. milled 9800 tons of ore averaging $12.70, from which a net profit of $35,060 was obtained. During the previous week 2600 tons of ore averaging $10.13 was milled.——In the Silver Top 130 ft. of development was done, 85 ft. in the Sandgrass, and 7 ft. in the Mizpah. Last week the production was 2350 tons. ——The Tonopah Extension Mining Co. has opened up the vein on the 1680-ft. level of the Victor; it averages 5 ft. of good ore. In the Victor 115 ft. of development has been done and 76 ft. at the No. 2 shaft. At the latter, on the 1260-ft. level, stoping continues on the Murray vein on a 6-ft. face and on a 5-ft. face on the O. K. vein. The Victor raise 631 shows a 5-ft. face of ore. On the 1540-ft. level, 1515 east drift cut a 2-ft. vein, and the 1503 raise continues on a 3-ft. face of ore. The production for the past week was 2380 tons.——The Tonopah Belmont Development Co. milled 10,160 tons of ore during September, making a net profit of $102,742. On the 1000-ft. level, raise 72 on the Rescue vein shows an increase in the width of the vein and also an increase in value. On the 1166-ft. level, west drift 21 on the Mizpah Fault vein shows a 3-ft. face of fair ore. Branch B of the Mizpah Fault vein, cut last week, shows a 2-ft. face of fair ore. Last week's production was 2429 tons.——The fine showing of high-grade ore in the Ohio vein of the West End Consolidated Mining Co. continues. The ore exposed in the 512 and the 555 cross-cut No. 1 proves the vein for nearly 700 ft. on its dip to the east and 500 ft. along its strike, north and south. As the vein has been exposed within 50 ft. of the MacNamara company's end line, there is a possibility of the Ohio vein being found there. Raises 528 and 529 have both reached the hanging wall of the vein, exposing

7 ft. of fair-grade ore. Drift 531 to the south continues to open up the high-grade orebody. Drift 530 is being run north to prove that area. Drift 717 No. 2 at the West End shaft has been resumed for connection on the west side, and 717 cross-cut No. 3 has been discontinued. The mill clean-up for the first half of October consisted of 31 bars of bullion, valued at $49,458. The output the last week was 1063 tons. At the Halifax Tonopah Mining Co. 1256 cross-cut intersected a vein of quartz carrying gold.——The Jim Butler Tonopah Mining Co. during the month of September milled 3467 tons of ore, resulting in a net profit of $55,000. On the 200-ft. level raise 367 on the Wandering Boy vein shows a 2-ft. face of good ore. Between the 300 and 400-ft. level the north cross-cut intersected the Wandering Boy vein exposing 1½ ft. of low-grade quartz. Raise 6 on the 700-ft. level cut the hanging-wall branch of the West End vein. It shows 18 in. of fair-grade ore. The production for the past week was 500 tons.——At the MacNamara Mining Co. stoping on the 200-ft. level shows some excellent ore, while the other stopes show fair-grade ore. The production the past week was 315 tons.——Last week the Rescue-Eula produced 132 tons, the Montana 116 tons, and the North Star 56 tons, making the week's production at Tonopah 9341 tons with a gross value of $163,467.

COBALT, ONTARIO

BEAVER CONSOLIDATED.—HOLLINGER CONSOLIDATED.—MCINTYRE-PORCUPINE.

An aggregate of 16 cars of ore in a single week was the new high record set by the Mining Corporation during the second week of the current month. The approximate weight of the ore was a little more than 500 tons. Not only at the Mining Corporation, but at every other producing mine in the camp the fixed policy appears to be to maintain production at the highest possible point. Results of operations below the diabase sill at the 1600-ft. level at the Beaver Consolidated are encouraging. This is the deepest working at Cobalt and will have a material bearing on the future of a number of neighboring properties. In order to get below the sill, a shaft had to be sunk through approximately 1000 ft. of barren diabase. Early in the year a vein containing considerable high-grade ore was found. Stoping operations, however, have been disappointing, the grade of the ore not holding as good as that first found. This week a winze has been started on the high-grade ore-shoot, and the development of the next few months will be watched with critical interest.——The Cobalt Silver Bird has been re-opened recently. The property has been classed as a wild-cat and has never received a thorough test. However, it is now in the hands of the O'Brien mine and will be thoroughly explored.——The production at the Kerr Lake is being maintained at around 7000 oz. of silver per day. Working over unusually wide stopes the tonnage also is well maintained.——From the Nipissing mine the production is understood to approximate $10,000 per day. Practically all ore-shoots are extending beyond their previously known limits and the ore-reserve at the end of the year is expected to compare favorably with that of the preceding year.

The number of mine-workers in the Porcupine field is increasing. The increase appears to be the result of the high rate of wages being paid and the release of workers from harvesting and threshing in the West. The Hollinger Consolidated and the McIntyre-Porcupine are both in a favorable position, and the rate of production is increasing. Development at the 200-ft. level on the Miller-Middleton side of the Hollinger is understood to be greatly exceeding expectations, and, in fact, has constituted one of the big surprises of the year. Though the wide body of ore on surface was low grade, it has been found at the 200-ft. level to be above the average in the mine.——The development campaign at the Porcupine Crown is being continued with favorable results, energy being directed largely to this one branch of the work. Were the working forces to increase materially, the Porcupine Crown would at once re-commence milling operations at full capacity. In the meantime, however, production is being maintained at a rate sufficient to cover expenses, and development work is going ahead at a record rate.

The Porcupine V. N. T. orebody at the 600-ft. level is approximately 20 ft. wide and has an average gold content of over $11. Driving operations are under way and the ore-reserves of the mine are increasing.

TAMPICO, MEXICO

DEVELOPMENT OF THE OILFIELD.

Oil is the magnet that is attracting large sums of American and British money to Mexico. Although foreign investments have already wrought a wonderful transformation in the physical aspect of the Gulf coast to the south and west of Tampico, this improvement has only begun, judging from present indications. The influence of the remarkable industrial development that is in progress in Tampico and the coast oilfields is having its beneficial effect upon every class of citizens. An enormous amount of industrial improvement is under way and in contemplation at Tampico and adjacent oil-producing territory. According to reliable estimates, approximately $60,000,000 gold will be expended in the near future in the construction of new refineries, oil-pipe lines, wharves, storage-tanks, and pumping-plants. It is stated that plans have been prepared for the immediate construction of twelve new refineries here. The plant of the Aguila Oil Co., which is owned by Lord Cowdray of London, England, and associates, will, it is said, be among the largest in the world. Practically all the large oil-producing and refining companies in the United States and many in Europe have extensive holdings in the Tampico district and are preparing to build refineries either at Tampico or in its immediate vicinity. Although the oil of the different fields in this territory is of a heavy grade, modern refining methods are capable of obtaining from it good yields of gasoline, leaving the unrefined portion available for fuel. According to recent estimates, the wells now completed, nearly all of which are capped, are capable of a total production of about 10,000,000 barrels per day. This available output is more than forty times the quantity that is now being used. Construction has already been started on several new pipe-lines, pumping-plants, and storage-tanks, and material for new refineries is being received by every boat from the United States.——The well of the Mexican Petroleum Co. at Cerro Azul is keeping up its regular flow of 15,000 bbl. through the pipe-line to Tampico. Although this well has been flowing for several years and has given a total output of more than 57,000,000 bbl., it showed by a recent test when the throttle was taken off that it still has a daily capacity of 257,000 bbl.——The Mexican Petroleum Co., E. H. Doheny of Los Angeles, president, has several producing wells in the Tampico district that are almost as large as the one at Cerro Azul. The British government is not taking any chances on the oil-wells of Lord Cowdray and associates being destroyed, in order that the supply of fuel-oil from that source for the British navy may be cut off. The exportations for the British navy by the Cowdray syndicate are all delivered by means of ocean-loading pipe-lines placed some distance out from the mouth of the Tuxpam river, near the port of Tuxpam, about 120 miles south of Tampico. The source of this oil-supply is from fields adjacent to Tuxpam. In order to be protected in case of any possible emergency, the British government entered into a contract with the Mexican Petroleum Co. whereby the latter is required to keep 40,000,000 bbl. of oil in storage at Tampico. At a cost of more than $2,000,000 the Mexican Petroleum Co. constructed the necessary steel tanks to take care of this enormous quantity of oil. Some idea of the oil-resources of this company may be had by the statement that an emergency supply of 40,000,000 bbl. for the British government was placed in storage in remarkably short time and is now available for any demand that may be made upon it. But for the difficulty in obtaining material for the construction of storage-tanks, the different companies would by this time have added enormously to their oil-stocks. The iron and steel plant at Monterrey, which was erected many years ago by a syndicate of Italians at a cost of $10,000,000, is preparing to construct a department for the manufacture of steel plats and other materials that enter into the erection of oil-storage tanks. The Mexican government is receiving a large revenue now from the oil industry. The military protection which the independent revolutionist General Paliaz is affording to several of the larger oilfields, to the south of Tampico, and particularly those from which the Cowdray syndicate draws its supply, is still being given, much to the satisfaction of the foreign oil interests, notwithstanding that they have to contribute heavily toward the cost.

GLOBE, ARIZONA

SETTLEMENT OF GLOBE-MIAMI STRIKE.

William B. Wilson, Secretary of Labor, head of the Federal commission, announced on October 22 the plan whereby normal production will be resumed by the mines in the Globe-Miami district. Under the orders of the commission the strike must be called off immediately and all men who desire to return to work must report to the companies within five days. The following are the terms of the agreement:

"First—Each company will recognize a workmen's committee representing the employees, said committee to be selected from and by men actually in the employ of the company. This committee shall consist of four men representing the different departments in the same manner as the existing workmen's committee. Election shall be made by secret ballot. The method of conducting this election shall be as a majority of the workmen shall agree upon and it shall be held as hereinafter provided for. The first election of workmen's committees under this arrangement shall be held on the last Saturday of December and the term of office of the committeemen elected at that time will be from January 1, 1918, to June 30, 1918. Thereafter the election shall be held on the last Saturday of each June.

"Second—The workmen's committee shall have no jurisdiction over individual grievances until the workman has done all in his power to bring about an adjustment of the same with his foreman.

"Third—When an individual grievance is brought to the attention of the workmen's committee it shall first endeavor to bring about an adjustment between the workman and the head of the department in which he works. Failing in that it may take the grievance up with the next official in authority and in like manner to other superior officers until an adjustment is reached or the matter has been brought to the attention of the general manager. No man shall be discharged or discriminated against in his work because he does or does not belong to a union. Disobedience of the established rules of the company or of orders for the carrying out of such rules shall be cause for discharge or suspension.

"Fourth—Grievances of a general character taken up by the workmen's committee shall first be presented to the official of the company having jurisdiction of the same and if an adjustment is not reached, may be carried to the next superior officer and in like manner on through to the general manager, unless an adjustment is sooner reached.

"Fifth—If the workmen's committee and the management are unable to mutually adjust the difference in dispute they may submit the same to an arbitrator, whose decision shall be final and binding upon both parties. There shall also be two alternate arbitrators, who shall act whenever from any cause the arbitrator is unable to act. The arbitrator agreed upon is Joseph S. Meyers, and the alternates are Hywell Davies and Judge George W. Musser. When any dispute affecting non-union men is presented to the arbitrator for decision, they may select such person as they desire to present their case to the arbitrator.

"Sixth—When any dispute affecting a union man is submitted to the arbitrator for decision, the union to which that man belongs may, with the approval of its international officials, select one of its members to present the case to the arbitrator. When any dispute affecting non-union men is presented to the arbitrator for decision, they may select such person as they desire to present their case to the arbitrator.

"Seventh—All men now on strike who report for duty within five days after the acceptance of this arrangement shall be re-employed without discrimination as soon as places can be found for them, except those who since the beginning of the strike have been guilty of utterances disloyal to the United States, or who are members of any organization that refuses to recognize the obligation of contracts, and the fact of such disloyal utterances or membership shall be determined by the arbitrator here'n provided for. In re-employing workmen preference shall be given to married men and those with dependents. The placing of striking workmen reporting for duty shall be handled as a district problem.

"Eighth—The workmen's committee may make such investigations of the hospital department on behalf of the employees as it may from time to time deem necessary and make such recommendations to the trustees relative to improvements as it may find desirable.

"Ninth—It is understood that this machinery will take the place of strikes or lock-outs during the period of the War, and no other method for regulating relations between employers and employees shall be substituted except by mutual agreement."

TORONTO, ONTARIO

BONUS FOR COBALT MINERS.—KIRKLAND LAKE.

For some time past the iron and steel industries of Canada have been considerably handicapped by the shortage of raw material. Much construction work has been indefinitely delayed by the difficulty in procuring steel, for which the munitions plants have the first call, and the newly-developed shipbuilding industry is seriously imperilled. The situation has been brought to a crisis by the embargo recently placed by the United States government on many lines of iron and steel products, for which Canadian manufacturers look to the United States as their principal source of supply. On September 28 representatives of the leading firms using iron and steel held a conference, at Ottawa, with Sir George Foster, Minister of Trade and Commerce, at which, as a preliminary measure, a committee was appointed to ascertain from manufacturers their immediate requirements, with a view to making representations at Washington to secure a relaxation of the embargo, at least so far as regards the material needed for the munitions and shipbuilding industries. It is proposed that the limited supplies obtainable should be conserved and co-ordinated under government supervision, for which purpose an Iron and Steel Controller may be appointed with duties analogous to those of the Fuel Controller.——A representative gathering of Cobalt mine managers recently passed resolutions requesting the Government to investigate the validity of the patents of the Minerals Separation Co., with a view to their cancellation.——The general basis for a bonus directly relative to the price of silver—instead of a wage increase—to be paid to the mine-workers of Cobalt is 25c. per day on 60c. silver; 50c. on 70c. silver; and 75c. on 80c. silver.

Some of the mines pay a $1 bonus on 90c. silver and $1.25 when silver is at $1, and it is understood that others will pay $1 when silver stands at that figure. The number of men now employed at the camp is approximately 3000, and the sum total that will be thus paid out is estimated at $78,000 per month.

The Kirkland Lake district is coming into increased prominence and attracting much attention from investors. At the Teck-Hughes the mill is running at full capacity, treating 80 tons of ore per day. The usual early difficulties in starting have been surmounted and the mill is operating satisfactorily. The winze from the 400-ft. level is down 150 ft. and when the 600-ft. level is reached, a point under the main shaft will be struck by a cross-cut and the workings connected by a raise. The company has laid out its own townsite and will erect ten houses immediately for its employees. The central shaft at the Kirkland Lake is down 20 ft. The main working has reached a depth of 700 ft. An adit run at the 400-ft. level toward the centre of the claim has proved the orebody for a distance of 600 ft., and it is expected that this orebody will be encountered by the central shaft. The Canadian Kirkland No. 2 vein is, at a depth of 25 ft., the whole width of the shaft and has a gold content of $8 per ton. No. 1 vein has been uncovered for 400 ft. and is well mineralized.——A shaft is being put down on the Lucky Baldwin from the 100 to the 300-ft. level; the ore taken out is good-grade, carrying free gold.——Difficulties from encountering a water seam at the Croesus mine, Munro township, proved more formidable than at first anticipated, but have been overcome, and development and milling operations are now normal. The mill-head is the highest of any in northern Ontario, being over $40 per ton. ——At the Buff Munro, one mile from the Croesus, two shafts are down 50 ft. each; a mining plant is being brought in.

OATMAN, ARIZONA

GOLD ORE.—MOSS BACK.—OATMAN GOLCONDA.

Mining operations in the Oatman district have been revived and speeded up to a marked degree during the past week. Following the hottest summer ever known in this section of the country, the beginning of the pleasant fall season witnesses a dozen important new development-projects. Summer weather in the Oatman district is usually not too hot for conducting mining activities, but the past summer has been exceptional.——Chief among the new undertakings announced recently is the work now being commenced on the Gold Ore. Control of this partly proved property has been assumed by H. E. Teter & Co., who developed the Big Jim and sold it to the United Eastern. An entirely new plant has been purchased and will be erected at the mine; the shaft will be lowered 200 ft. to the 730-ft. level; the orebody will be opened up at the 350, the 530, and the 730-ft. levels. A new road will be built to the Gold Road mill one mile distant, and at least 100 tons of ore will be mined and transported per day. The Gold Road mill has been leased for one year and will be operated on ore from the Gold Ore. The capacity of the mill is 300 tons per day and it is possible that some of the ore now being developed on the Gold Road property may be milled there. A. C. Werden, who located the Gold Ore, will be retained as manager, with A. G. Keating as consulting engineer. It is estimated that 30,000 tons of ore is already developed in the mine. Over 2000 tons was shipped to the Gold Road mill, while that plant was in operation, returning an average of $14.50 per ton.——Development work has been resumed on the United Northern property, between the Gold Road Bonanza and the United Eastern. A shaft 550 ft. deep was sunk on this property last year and cross-cuts will be driven at the bottom. Development work will be resumed shortly on the Gold Road Bonanza, where driving along the vein at the 550-ft. level will be pushed. The drift has been advanced for a distance of 400 ft., following a stringer averaging two feet wide.——Activity has been resumed in the vicinity of the historic Moss mine, where gold was first discovered in this district in 1864, according to tradition, by John Moss, a scout and guide employed in the U. S. Army. The old Moss property itself will be further developed by the company, of which A. C. Werden, of the Gold Ore, is president. The immense Moss vein, 200 ft. wide, can be traced over a long distance on the surface; diamond-drilling has revealed it at a depth of about 350 ft., and its further exploration should prove to be one of the most interesting projects in the district.——At the Moss Back the new owners are building a road to the property and preparing to sink the shaft 200 ft. deeper, to the 500-ft. level. The Girard property is being explored by sinking a shaft on the vein to a depth of 300 ft.——In the southern part of the district, a shaft 425 ft. deep is being sunk on the Oatman Golconda, between the Pioneer and the Leland. This mine is controlled by F. M. and H. E. Woods, of the Nellie and the Lexington-Arizona. Development work on both the Leland and the Pioneer will be started in the near future.

McCARTHY, ALASKA

MOTHER LODE COPPER MINE.—FRESH PROSPECTS.

The Mother Lode Copper Mines Co. has completed its new power-plant at McCarthy, built an auto-truck road from McCarthy to its lower camp, 15 miles away, and erected a concentrator unit at the lower camp which started running on October 10. It is expected that within a week regular shipments will begin; these, according to previous estimates, will amount to 30 tons of concentrate per day, carrying 50% copper and 3 oz. silver. To secure this product 200 tons of ore will be mined and delivered by the 6000-ft. aerial tramway into a gyratory crusher, reducing it to ½-in. mesh, thence through a trommel screen, where it is sized, thence to a Woodbury jig, that makes a high-grade product only. The tailing from the jig is saved for re-treatment in a permanent mill that is to be built next summer. The power-plant just completed at McCarthy generates 240 volts, stepped-up by the two transformers to 16,500 volts, then stepped-down again to 440 at the lower camp and to 220 volts at the mine. For this purpose five 500-kva. and two 100-kva. transformers are used, two 500-kva. transformers being used at the power-plant, two at the lower camp, and two 100-kva. at the mine, providing for changing from two-phase to three-phase; the fifth transformer is kept as a spare one. The telephone line is now in operation to the mine. The wagon-road, also just completed, is substantial and good, and it is graded either on the gravel bars of McCarthy creek or in rock side-cuts. In crossing McCarthy creek there are three double-truss bridges and 12 single-truss deck-bridges, besides 16 log bridges and 34 culverts. A two-story building 60 by 40 ft. has been erected at the lower camp. The office building is of two stories, 30 by 36; the ground floor contains a general office, the president's office, and commissary; there are four bed-rooms on the upper floor for the use of the officers. At the mine 212 ft. of development work has been driven during the past summer, making in all upward of 8000 feet.

A new and important copper strike is reported to have been made during the past summer by M. B. Vaugh and M. V. Fox near the head of the McLaren, in the Susitna water-shed, about 50 miles from Yosts' on the Government Fairbanks-Chitina trail. The location will be about 100 miles from the Government railroad, when it is completed through to Fairbanks. The vein is said by the locators to be in amygdaloid greenstone, is from 2 to 10 in. wide, maximum width shows on the surface for about 40 ft., while the vein can be traced for 300 ft. The samples of the ore shown by the locators, which they claim to be a fair average of the vein, is a good-grade chalcopyrite, about 20% copper content.

ALASKA

(Special Correspondence.)—W. W. Logan, who for the past season has been superintendent of the Gold Bullion cyanide plant, came in from the mine and sailed for San Francisco on the 'Alameda'. Under Mr. Logan's management the Gold Bullion, which is the farthest north of any cyanide plant in America, handled the greatest tonnage and had the most successful season since it started. Henry Landers is in town from the Gold Bullion with a clean-up from that property. The Gold Bullion was the first producer in the Willow Creek district. It is under lease to a company of Canadian capitalists, who, it is reported, will make a substantial payment on the property this fall. The company is arranging to work a crew of men all winter in developing the mine.——The Willow Creek Development Co. has made a shipment of ore to Falkenburg & Laucks, of Seattle, who will make a thorough test and design a mill for treating it.——The Government railroad has already borne fruit in rendering the Willow Creek district accessible. It can be reached now from Anchorage in a day, while a year ago the journey required four days. This has resulted in increased activity in the district and bright prospects for the coming year.——The Cache Creek placer district will be rendered accessible when the railroad reaches Talkeetna, and the Alaska Road Commission completes the road from that place to McDougall, which is now being laid out by Ross Kinney.——Hershel Parker, who has reported the discovery of platinum in commercial quantity in the gravels of the Kahittna river, left for New York on the 'Alameda' tonight.——M. A. Ellis of the Thunder Creek Mining Co. is in town attending to the transportation of supplies for his company's operations next year. He reports a very successful season and is preparing for extensive operations next year.

Anchorage, October 8.

Louis C. Dillman, a Seattle capitalist, announced on October 20 his sale of the West Over group of copper mines in southwestern Alaska to the Alaska-Chitina Exploration Co. of Seattle for $416,000 in bonds and cash. F. A. Seiberling, president of the Goodyear Rubber Co. of Akron, Ohio, is reported to be interested in the purchasing company.

ARIZONA

COCHISE COUNTY

(Special Correspondence.)—Sinking has commenced in the Cochise shaft of the Copper Queen company to develop more water to operate the proposed new concentration plant planned to treat 15,000,000 tons of low-grade ore, the tests for which are now nearly complete. A large tonnage of chalcocite ore has been developed as a result of several years' drilling operations in the granite porphyry mass of Sacramento hill.

Bisbee, October 12.

(Special Correspondence.)—The Arizona Commercial Mining Co. has completed a lot of repair work and is now producing at about 75% normal. Additions are being made to the pumping and hoisting equipment.

Globe, October 12.

GRAHAM COUNTY

The Copper Reef mine, in the Stanley district, has recently been doing new development work that is proving productive. The mine has been opened by a long adit, and by winzes. A new orebody was disclosed by a cross-cut 200 ft. long, driven under the direction of the superintendent, Leonard W. French. Shipments of ore averaging about 12% copper are now being made through San Carlos.

MOHAVE COUNTY

(Special Correspondence.)—Lessees have taken 3000 oz. silver ore from the White Hills property, 30 miles north of Chloride. This property has been idle for some time. The same lessees took out a small fortune from this property some years ago. Molybdenum ore, estimated at 4% sulphide, has been uncovered at a depth of 285 ft. in the shaft of the New Tennessee mine. Traces of this ore came in at the 270-ft. level.——H. B. Tetor & Co., developers of the Big Jim mine, have assumed control of the Gold Ore property. The present 520-ft. shaft will be sunk another 200 ft. and levels driven at 200-ft. intervals. The ore will be treated at the Gold Road mill, a mile away, until the Gold Ore mill is ready to handle it. It is estimated that the Gold Ore has some 30,000 tons of milling ore developed.——Preparations are being completed for resuming work on the United Northern property. A new engine and compressor have been purchased, and, as soon as erected, the 500-ft. shaft will be sunk another 100 ft. and drifts started.

Kingman, October 20.

(Special Correspondence.)—The Keystone mill, comprising a Blake crusher, a ball-mill, an Aikin classifier, 10 roughing and 3 cleaning flotation-cells, a Dorr thickener, settling-tanks, and drying-bins, is working perfectly, and turning out a concentrate averaging 100 oz. in silver, $7.50 gold, and 6% copper. At present the mill is running one shift per day on dump-ore.——The New Mohawk has erected a 35-hp. hoist on shaft No. 2, where a vein of high-grade silver ore has been cut.——The Schuylkill mine has started to ship ore. The shaft is down 800 ft. and levels are being driven at 400 and 800 ft., respectively. The Schenectady mine is driving on the vein at the 200-ft. level.

Oatman, October 19.

A. M. MacDuffee, mineral commissioner of this county, has been gathering specimens for exhibit for the fairs at Prescott and Phoenix. He has visited nearly every mine in the district.

The Jamison mine, controlled by Lauzier & Walcott of Butte, Montana, Charles McKinnis of Wallace, Idaho, and F. Kays of Kingman, Arizona, is being developed rapidly; the main adit, which is driven on the vein, is about 800 ft. long and has been in ore from the start.——The Golden Hammer has a good showing of high-grade ore on the dumps. New hoisting machinery is being erected. The orebody has been proved for 600 ft. without a break, and is 5 ft. wide. The ore runs about $40 per ton. This property is under bond and lease by W. B. Twitchell of Phoenix and F. C. Smith of Chloride.

In the George Washington mine, owned by the Washington-Arizona Copper Co., of which Elmer D. Reece is president and manager, the development consists of an adit 960 ft. long exposing the vein the entire distance, with several stopes, a winze 100 ft. deep, two 50-ft. raises, and several cross-cuts. Two other adits farther up the hill show milling ore; these adits with the cross-cuts will measure about 2000 ft. The mine is well timbered, electric lighted, and equipped with modern machinery. The mill has just been put in operation.

The Town mine is shipping a couple of cars of ore to the Needles smelter, which is reported to average $50 per ton.

PIMA COUNTY

(Special Correspondence.)—Engineers have examined the Vulcan property in the San Xavier-Twin Buttes district for possible Eastern purchasers.——The northern part of the property of the Swastika Copper & Silver Mining Co., in the same district, has been sold for $125,000. Negotiations are pending for the sale of the four remaining claims.

Tucson, October 21.

PINAL COUNTY

(Special Correspondence.)—The Climax Hill shaft of the Troy Arizona Copper Co. is down 377 ft. and indications point to an early penetration of the sulphide zone. F. C. Armstrong, president of the Arizona-Hercules Copper Co., has taken an option on the Pinal Consolidated mine, a silver-lead property, formerly known as the Silver Bell.

Florence, October 22.

YAVAPAI COUNTY

(Special Correspondence.)—It is reported that 20% ore is being mined on the Jerome Verde, 10 ft. north of the United Verde extension line. Ore from this area is now being hoisted through the Edith shaft of the United Verde Extension.

Prescott, October 21.

Some of the richest gold ore ever seen in Yavapai county has been found on the Rosa Mary group of five claims owned by John Revello. How much of this ore will be found on the property is a matter of conjecture, but Mr. Revello recently extracted 39 lb. of the richest ore and sent it to the Selby smelter, and he received the firm's check for $1704.20. The fortunate owner also put three tons of what he calls second-grade ore through an arrastra and recovered $620. There is also a quantity of ore which has assayed from $12 to $15 per ton. The property is situated in the Crown King district, near the famous Monte Carlo property.

YUMA COUNTY

(Special Correspondence.)—The Mohican Copper Co. has completed 512 ft. of sinking and 1200 ft. of driving. The people controlling the Mohican property have taken over the Groundstone group of gold claims three miles from Alamo springs.

Yuma, October 12.

CALIFORNIA

ELDORADO COUNTY

(Special Correspondence.)—Preparations are under way to erect a stamp-mill on the Montezuma lode mine, on the Mother Lode, in the Nashville district, 11 miles west of south from Placerville. The Montezuma, which is developed by a shaft 1000 ft. deep and drifts north and south on the vein at each 100-ft. level below the 500-ft., is owned by J. Cam. Heald, Mayor of Nashville. The Havilla lode mine, situated just south of the Montezuma, is being re-opened and the stamp-mill is being re-constructed. San Francisco capitalists are financing the reviving of the Havilla.——The Washington & California Gold Mines Co., of Seattle, is starting operations on its gold quartz properties, which are on the main Mother Lode of California, a short distance from the Montezuma mine. A 5-ft. vein of high-grade quartz, showing gold, was recently uncovered in sinking a shaft on one of the claims owned by the company.——Farther north on the Mother Lode, near the Martinez group of gold quartz claims, Antony Crafton is developing a group of quartz claims, which he has under lease and bond for Seattle capitalists from Grant Busick, an old-time miner. This property contains a fair tonnage of low-grade gold ore.

Robert G. Hart, mineralogist, who is making a systematic investigation of the Placerville Quadrangle for ores containing rare metals, left this morning for a three-day trip to the north part of Eldorado county to examine and sample a deposit of high-grade molybdenite.

Placerville, October 25.

FRESNO COUNTY

(Special Correspondence.)—Operations at the reduction plant of the Piedra Magnesite Co., at Piedra, 30 miles east of Fresno, began September 21. The construction of the plant, which is one of the most up-to-date of its kind on the Pacific coast, was begun early in July. The plant has a capacity of about 60 tons of calcined magnesite per day, which is equal to a little more than twice that amount of raw material. The plant represents an investment of approximately $55,000. Thomas J. Curran, manager for the American Magnesite Co. at Porterville, Tulare county, is general manager for the company. J. W. Thomas of Porterville is treasurer and W. B. Phillips, also of Porterville, is superintendent. The calcining kiln is of the rotary type, 83 by 8 ft., and weighs about 80 tons. The cooling-tower is 60 ft. high and the mineral is carried to the tower from the kiln by 125-ft. chain-bucket conveyors. The mineral is first fed into a heavy gyratory crusher from which it passes through a screen into the ore-bin and then into the kiln. The screen allows only mineral that is finer than a large pea to pass into the kiln, thus ensuring rapid and thorough calcination. The large pieces are returned to the crusher. All machinery about the plant is run by electricity. The Piedra Magnesite Co. owns 11 mining claims and employs more than 100 men. The mine is about half a mile from the plant which is on the bank of the Kings river. The mineral is hauled from the mine to the plant on auto trucks. The output has been contracted for by the American Refractories Co. of Pittsburgh, Pennsylvania.

Fresno, October 24.

PLUMAS COUNTY

The Walker mine and mill are being worked to capacity once again, after a strike started by I. W. W. agitators. There are 200 men on the payroll at the mine and 30 men making a connection with the electric line of the Great Western Power Co., 15 miles away.

SAN BERNARDINO COUNTY

The California Supreme Court has upheld the Superior Court of San Bernardino in the case of R. Waymiro *v.* the California Trona Co. This confirms the title of some 20,000 acres, known as Searles Lake, in the California Trona Co. The decision has no bearing on the Government case against the Trona company, which is still pending. The Government has charged that the company is an alien corporation, because most of the stock is owned by the Consolidated Gold Fields of South Africa, represented by Lord Brabourne and Baron de Ropp, and therefore should not be granted a patent to the lands. Searles Lake, a lake during the rainy season only, contains both potash and borates, and, on account of the great need for the former for agriculture and munitions, the case is exciting considerable interest.

TRINITY COUNTY

In a recent clean-up from his mine on Coffee creek, Patrick Holland found three nuggets, valued at $1155, $600, and $155 respectively. These are the largest nuggets that have been found in the county for a number of years. The rest of the clean-up amounted to $600 and contained several nuggets over $50 in value.

PIEDRA MAGNESITE PLANT, FRESNO COUNTY, CALIFORNIA

SHASTA COUNTY

The Shasta Hills Mining Co. is erecting a five-stamp mill at the Sybil mine, three miles west of French Gulch.

COLORADO

BOULDER COUNTY

(Special Correspondence.)—A number of old mines are being re-opened in the Caribou district, and some new discoveries have been made.——F. G. Albrecht, one of the pioneers of this camp, has opened up a fine body of high-grade silver ore.——Guy M. Woodring, vice-president of the Huron mine at Eldora, has purchased an air-compressor for the property; it will be put in this week.——A new mill is being erected by the Consolidated Mining Co.——The Good Morning adit at the Wall Street has tapped a large vein of low-grade silver and gold ore. F. H. Hammond is the foreman. An electric-drill outfit has been put in.——W. W. Degge, who built a refinery at Boulder, has returned from the oilfields of Wyoming, and will again start business at the oilfields of the Boulder company.——The Great Britain & Emerson property at Wall Street has been taken over by a new company, which will push operations on a large scale.——The Doss adit at Wall Street is getting out some rich gold ore; steady shipments are being made.——A large force of men is employed at the Fairfax mine, at Salina.——The Melvina mine, west of Salina, an old-time producer, will be opened up by A. Murphy, who has just arrived from Arizona.

Eldora, October 22.

SUMMIT COUNTY

The Tonopah Placers Co. shipped a gold brick valued at $8100 on October 10 to the United States mint at Denver. It was the result of a previous 10 days run of one of the company's three gold dredges.——The Wellington Mines Co. produced in September about four carloads of heavy lead concentrate in addition to its regular output of zinc concentrate. The company issued dividend checks to the amount of $100,000 on October 1.——The Carbonate, on Mount Baldy, shipped three carloads of lead carbonate ore during the first ten days of October. The ore, besides running high in lead, carries considerable gold and silver. The property is operated through an adit several hundred feet above the timber-line.——The Iron Mask mine, operated by the Mid-West company, shipped a carload of carbonate ore last week to a smelter. The Iron Mask is operated through a main adit 1500 ft. long. At 1300 ft. from the portal, in a winze, sunk below the adit-level, a 3-ft. body of lead ore has been opened.

Paul Burdette, lessee of the Ella on Mineral Hill, shipped a carload of lead-silver ore last week. The Ella is controlled by the Wellington-Crescent Co. The Minnie mine, also controlled by this company, is being worked under lease by E. E. Miller and associates.——A discovery of wolfenite has been made by George Dunnigan on Peak No. 9 of the Ten Mile range, opposite the Breckenridge townsite.

IDAHO

SHOSHONE COUNTY

The Hecla mine, at Burke, Idaho, which is paying monthly dividends of $150,000, will soon be improved by the addition of storage bins, a sorting-plant, and a change-room for 500 employees. The bins will accommodate the daily production from the mine; they are designed to make possible a continuous flow of ore through the sorting-plant and mill when delays occur elsewhere. The Hecla is steadily extending its levels and increasing its production and reserve. The ore-reserve is estimated to be four to five years in advance at the current rate of output. Of the 1000 tons hoisted daily from the lead-silver bodies about 75% is ore.

An October dividend of $78,150 has been declared by the directors of the Caledonia mine in the Coeur d'Alene, bringing the disbursements to stockholders for the current year to $859,650 and the total since the mine entered the dividend-paying class of $2,753,150. The surplus on August 31 was $719,183. The net earnings during September were $117,000, and it is estimated that the surplus at the present time is in the neighborhood of $800,000.

MISSOURI

JASPER COUNTY

The structural portion of the new 200-ton mill of the N. J. & D. Mining Co., situated on a lease of the Walker land, just north of Duenweg, is rapidly nearing completion. A boiler has been set and machinery is being erected in the new concentrator. Construction work was begun early in September and the members of the company plan to have the mill completed within a month. The shaft is being shaped up for operation by the time the plant is ready to run. The company has sunk a new shaft and opened up some good lead and zinc ore. Below the 160-ft. level 20 ft. of ore was proved by drill-holes. C. E. Laplant is superintendent and part owner of the mine.

MONTANA

LEWIS AND CLARK COUNTY

(Special Correspondence.)—The Amalgamated Silver Mines, a Montana corporation, has secured the Free Coinage claims in Lump gulch, together with the Alma and Muskeegan claims, and has, in addition, located 25 claims adjoining. Three shafts, 180, 200, and 350 ft. respectively, are available. The present work consists in opening and deepening the 180-ft. shaft on the Free Coinage. A hoisting and pumping plant has been installed and the shaft re-timbered to water-level, previous to sinking another 100 ft. Silver ore to the value of $12,000 was mined from this shaft in the past. The ore averaged 115 oz. per ton and carried over 20% zinc. The company is composed of New York and Montana people and is capitalized at $1,000,000 in dollar shares. The veins are in granite, 10 miles from the smelting works and 1½ miles from the railway.

In the old Park district, Broadwater county, considerable work is being done. New buildings are going up. A new road is being made to connect with the main wagon-road from Hassel to Townsend. The ores contain silver, lead, and gold. The ground is very wet and requires machinery to control the water.——The Helena mine, operated by the Helena Mining Bureau, has shipped its twentieth carload and has a large tonnage blocked on the 300-ft. level, the present depth of the mine. ——The Economy Mining Co. has unwatered its 300-ft. shaft and is taking out high-grade gold ore. The plant is operated by electric power.——Negotiations are pending for the treatment of a large tonnage of manganese ore from Senator Clark's mines at Butte. The New York-Montana Engineering & Testing plant at Helena wants the ore for concentration, but nothing will be decided until Senator Clark returns to Butte.

Helena, October 26.

NEVADA

HUMBOLDT COUNTY

(Special Correspondence.)—The gross earning of the Rochester Mines Co. for the quarter ended September 30 totaled $158,883, and net profit was $64,690; 15,908 tons of ore was milled, averaging nearly $10 per ton. The mill is now treating 217 tons of ore per day with most of it coming from the Adams vein on the 800 and 900-ft. levels. On the 800 the vein has been opened for 500 ft. and ranges from 12 to 32 ft. wide. On the 900 it has been developed for 200 ft. and is from 13 to 40 ft. wide. The ore on both levels ranges from $8 to $35 per ton. Mill-ore is under development on the 250 and 650-ft. levels, and from the 900 a winze to open the Adams vein on the 1100 is progressing. A bullion shipment worth $34,000 was made October 5.——The Nevada Packard Mines Co. is preparing to sink a shaft to the sulphide zone from the main adit which opened ore at a depth of 200 ft. Work has been resumed in the Kromer-Hampton section and ore averaging $10 per ton is being mined. A considerable tonnage of $8 to $10 ore has been opened in the Griffin and Margrave cuts.——Construction of the 300-ton mill of the Rochester Combined Co. has been nearly finished, but delayed shipments of electrical equipment will prevent ore-crushing until December. Excellent ore continues to be mined from the 100 and 200-ft. levels of the Happy Jack shaft, and from the Shepherd and Kromer adits. Prospecting of outlying claims is proceeding.——The Nenzel Crown Point and Buck & Charley have arranged with the Nevada Short Line railway for steady shipments to custom smelters. Some ore will be sent to the Hazen sampling-plant, and negotiations are proceeding for shipments to the Mammoth smelter at Kennett, California. The Nevada Short Line has purchased a new locomotive and is preparing to handle shipments for several companies.——The Bloody Canyon antimony mine and adjacent holdings near Unionville have been acquired by the Pioneer Extraction Co. of San Francisco. Ore will be treated in a mill, being constructed at Mill City. Besides a large tonnage of milling-ore, considerable shipping material has been uncovered. Homer Wilson is president and general manager. ——The Darby mill, at Mazuma in the Seven Troughs district, has been purchased by George W. Forbes and Brooke Hartley and will be operated on custom ore, besides treating the product of the Hero and Wildcat mines at Farrell, which are leased by the mill-owners. Arrangements have been made to crush ore from the Fairview property, Vernon, and from the large dumps at the Mazuma Hills group.——The silver district recently discovered eight miles south-east of Lovelock is attracting favorable attention. A shipment of high-grade ore will be made from the Badger group, operated by M. Farretta. A dozen promising claims are being prospected and rich ore exposed at several points.

Lovelock, October 22.

The American Exploration Co., through F. H. Vahrenkamp, its president, announces that it has just closed a contract with the Magnolia Metals Co. of New York, aggregating $6,000,000 for antimony to fill government orders. These call for shipments of at least 500 tons of antimony per month and as much more, up to 7500 tons, as it is possible for the company to produce. As a start toward filling the contract, preparation is being made to re-open the Imlay properties at Imlay, and a 200-ton mill is nearing completion.

MINERAL COUNTY

Definite plans for the development of its property comprising the old Moho group of claims in this county were announced recently by the Western Silver Mines Co., a concern recently incorporated in this State. The property is six miles from the line of the Southern Pacific railroad, and not far from the town of Mina. It comprises ten lode-mining claims and a mill-site and, according to those interested, shows four separate veins, one of which, the Moho, has been opened for a distance of approximately 4500 ft., while another, the Shoemaker vein, has been opened for about 1500 ft. These veins are said to show an average width of 4½ ft. that runs $9.43 per ton.

NYE COUNTY

(Special Correspondence.)—The Nevada California Power Co. has completed its large auxiliary electric-power plant and given it a thorough test. The generators are operated by steam and the plant will enable the company to supply local mines with abundant power should storms demolish the transmission lines between here and Bishop, as happened last winter. Completion of the plant also insures steady work at the Great Western and West Tonopah mines. Both properties are being explored by a long cross-cut, which is being driven by the Western Tonopah company, composed largely of Boston and Butte capitalists.——The employees of the White Caps and White Caps Extension companies at Manhattan are on strike, following refusal of the companies to grant the Tonopah wage-scale, a minimum of $5 per day. The strike affects mines and mill. The Morning Glory, Consolidated, and Union Amalgamated granted the wage asked and are operating with a full crew. Some operators maintain they will close for the winter rather than grant the demands of employees.——The Bullfrog district again claims attention. The Sunset company is vigorously working the old Tramps Consolidated group and recently found ore assaying $20 per ton at a depth of 450 ft. The vein is a part of the Denver and is five feet wide. With the mill-ore occurs a streak assaying $270 per ton.——The Mayflower company has been re-organized as the Consolidated Mayflower Mines Co. Ore assaying $6 to $50 per ton is going to the 15-stamp mill, operated by a 35-hp. Fairbanks-Morse engine. An eight-drill Ingersoll-Rand compressor is driven by a 90-hp. Western gas-engine. Leasing operations have begun on the Montgomery-Shoshone and other properties. In the Pioneer section considerable activity is reported.

Tonopah, October 14.

NEW MEXICO

SOCORRO COUNTY

(Special Correspondence.)—The Socorro M. & M. Co. is increasing mill tonnage; the past week being one of the largest so far made. October output to date is several hundred tons ahead of normal. Ties for the new narrow-gauge railroad to the Johnson mine are being cut and delivery will begin next week.——The Oaks Co. has increased its development and ore breaking on the Central group. A considerable tonnage is awaiting shipment to mill.——Business men of Clifton, Arizona, are working to secure the connection of Clifton and the Mogollon district by a State highway. This will shorten freight and passenger service and is receiving encouragement from Mogollon people.——Miners are scarce generally through-

FRONT VIEW OF MERCURY FURNACE AT CHISHOLM MINE, OREGON

BACK OF MERCURY FURNACE AT CHISHOLM MINE

out the district and the increasing work offers employment for good men.

The local bullion production for the first 15 days of October amounted to about 33,000 oz., containing one-third gold, two-thirds silver.——Development on the east 900-ft. level at the Mogollon mine is steadily improving. This drift is entering the same orebody that was developed on the 700-ft. level.——The new strike in the Maud S. mine on the Oaks Co. property is being opened from two sides and the value and width of the ore continue to increase.——Fifteen men are working on the Whitewater road for the Socorro M. & M. Company.

Mogollon, October 23.

OREGON

JACKSON COUNTY

(Special Correspondence.)—Herbert Brewitt of Tacoma, Washington, representing investors of that city, has taken a lease on the Truit manganese deposits in the Eagle Point district, 20 miles north-east of Gold Hill. The lessees have done considerable development work on the property and ordered machinery for a concentration plant, and 40,000 ft. of lumber from a mill-building. C. F. Daugherty, examiner and buyer for the Noble Electric Steel Co. of San Francisco has been examining manganese properties in this district during the past week.——Clark & Webb of Medford have leased the Gold Ridge gold and silver mine, three miles south of Gold Hill, on Kanes creek. The lessees are operating a three-stamp mill and making further developments on a 20-ft. vein that averages $100 per ton in gold and silver. The vein is in a gabbro-porphyry contact and is free-milling.——The owners of the Copper King copper mines on Graves creek have re-organized and operations have been resumed at the mine. New equipment will be added and a wagon route from the mine to Rogue River, nine miles west of Gold Hill, is under construction. A large body of ore has been uncovered recently.

Earl M. Young of Rogue River has sold his manganese property, 12 miles west of Gold Hill, on Evans creek, to Seattle steel people, who will develop the mine. The ore occurs in small seams, running 35 to 50% manganese. Operations have been resumed on the Nellie Wright gold mine, two miles east of Gold Hill, after a close-down of two months pending reorganization. The 25-ton Beers mill will be replaced by a larger and more modern one. R. M. Wilson is the lessee, representing San Francisco investors. Development work is progressing favorably on the Cheney, Simmons, Ray, and Haff gold mines, two miles north of Gold Hill. This group is being operated by J. W. Davies for Sacramento people.

Gold Hill, October 22.

(Special Correspondence.) — The first modern quicksilver furnace has been completed and is in operation in this county. It is situated 12 miles north of Gold Hill, at an elevation of 2500 ft., on the slope of the Umpaqua mountains in a heavily timbered district. The mines are owned and operated by W. P. Chisholm of Gold Hill, and consist of 40 acres, which have been mined for cinnabar since 1878, but the ore has been reduced in only a crude way.

The vein, which strikes N. 53° W., has been exposed by adits for 2000 ft.; the greatest depth attained is 75 ft. It occurs along a granite-sandstone contact, where the granite is in part pegmatitic. The mineralized zone is from 100 to 200 ft. wide; it is not a well-defined vein, but is mineralized along an irregular contact. The ore, or mass, contains cinnabar, native mercury, pyrite, gold, zinc, silver, and a heavy black mineral resembling meta-cinnabarite. Samples taken from all the adits assay from $5 to $6 per ton in gold, 5 oz. in silver, 2.5% zinc, and 1% mercury. The cinnabar appears all through the ore and also in the hanging and foot-walls in the form of seams and kidneys. The seams are from 6 to 20 in. thick and average from 17 to 70% mercury.

The initial run of ten tons of ore in the new furnace produced three flasks of quicksilver; 1200 lb. of this ore was taken from a rich seam that runs as high as 70% quicksilver. With recent developments in the mine there is 5000 tons of ore in the block. The furnace, a 12-pipe Johnson and McKay, is placed within 300 ft. of the main adit.

Gold Hill, October 25.

JOSEPHINE COUNTY

(Special Correspondence.)—A deposit of chrome ore has been discovered on the old McGrew road, 15 miles south-west of Waldo, by J. H. Gregg, and is being hauled to Waters Creek for shipment.——The chrome claims located on Canyon creek, near Kerby, a few weeks ago by J. M. Finch and T. P. Johnson have proved to contain a large deposit of high-grade ore. The ore will be hauled to Waters creek.——A carload of chrome ore has been shipped from the Falls Creek district by W. E. Gilmore of Kerby. The ore was packed several miles on horses.

Kerby, October 26.

TEXAS

The Toyah Valley Sulphur Co., Seaton Keith, president, J. A. Daniel, secretary, and Ira Fisdale, treasurer, all of Houston, has been organized to develop a 1200-acre sulphur property.

UTAH

SUMMIT COUNTY

A deposit of sulphur has been discovered in a mine, nine miles south-west of Coaldale, that was originally opened by Boston capitalists for potash and alum. Some of the ore runs as high as 95% sulphur.

Copper Share Dividends

In the first nine months of the year there was paid out to copper shareholders approximately $139,000,000, against $108,-000,000 in the same period of last year..

The principal copper-share dividend payers and the amounts distributed throughout the first ten months of the past two years appear in the following tabulation:

	1917		1916	
	Amount per share	Total 9 months	Amount per share	Total 9 months
Allouez	$9.00	$900,000	$4.50	$450,000
Shattuck-Arizona	3.75	1,312,500	3.50	1,225,000
Osceola	18.00	1,730,700	11.00	1,037,650
Ahmeek	12.00	2,400,000	8.50	1,700,000
Granby	7.50	1,124,886	5.00	749,924
Isle Royale	5.00	750,000	2.00	300,000
North Butte	2.25	967,500	1.75	752,500
Anaconda	6.50	15,153,125	5.00	11,656,250
East Butte	1.00	411,000	....	....
Mohawk	20.50	2,050,000	17.00	1,700,000
Inspiration	6.25	7,387,294	3.25	3,841,392
United Verde Ex	2.10	2,205,000	0.50	525,000
Miami	7.25	5,416,644	4.25	3,175,362
United Verde	13.50	3,850,000	9.00	2,700,000
Shannon	1.00	300,000	....	....
Mass Consolidated	3.00	291,951	1.00	97,317
Calumet & Arizona	9.00	5,556,708	6.00	3,704,472
Greene Consolidated	2.00	2,000,000	2.50	2,500,000
Greene-Cananea	6.00	2,930,574	6.00	2,930,574
Champion	38.40	3,840,000	43.40	4,340,000
Cerro de Pasco	4.50	2,909,997	3.00	1,999,998
Copper Range	7.50	2,888,901	6.50	2,503,714
Quincy	15.00	1,650,000	11.00	1,210,000
Calumet & Hecla	75.00	7,500,000	50.00	5,000,000
Utah Consolidated	3.00	900,000	2.25	675,000
Kennecott	4.70	13,054,616	4.50	12,499,101
Centennial	1.00	90,000	1.00	90,000
Wolverine	7.50	450,000	6.00	360,000
Utah Copper	11.00	17,809,390	8.50	13,808,165
Chino	7.90	6,524,850	5.75	5,002,385
Ray Consolidated	3.20	5,046,972	1.75	2,759,977
Nevada Consolidated	3.15	6,298,289	2.25	4,468,780
Phelps-Dodge	22.00	9,900,000	20.00	9,000,000
Magma	1.60	384,000	1.50	360,000
First National	0.40	240,000	0.25	150,000
Consolidated Arizona	0.05	83,150	....	....
Old Dominion of Maine	7.00	2,053,471	8.50	2,493,497
Old Dominion of New Jersey	...	....	8.50	1,377,000
United Globe Mine	...	....	51.00	1,173,000
Total		$138,511,518		$108,336,058

PERSONAL

Note: The Editor invites members of the profession to send particulars of their work and appointments. This information is interesting to our readers.

H. H. WEBB is at the Fairmont hotel.

DAVID W. BRUNTON is on a visit to New York.

C. C. BROADWATER is on his way to New York.

GARDNER WILLIAMS has returned to Washington.

GEORGE E. FARISH sailed on October 2 for Salvador.

L. D. RICKETTS is making a tour of inspection in Arizona.

C. H. DOOLITTLE is here, having motored from Salt Lake City.

A. G. HARBAUGH is here from Belleville, in the Candelaria district.

JOHN A. DRESSER, of Montreal, has gone to El Paso, thence to Tennessee.

NORMAN PICOT sailed on the 'Sierra' from San Francisco for Melbourne.

E. H. HOAG has returned from San Francisco to Cosala, Sinaloa, Mexico.

CURTIS WEBB has returned from Morococha, Peru, to serve in the Aviation Corps.

CHARLES E. PRIOR has obtained a commission as First Lieutenant in the Army.

W. C. RUSSELL, general manager for the Caribou Mining & Milling Co., is at Denver.

C. E. ARNOLD, engineer to the Inspiration Consolidated Copper Co., is here from Miami.

HAROLD C. CLOUDMAN has been commissioned a Captain in the Engineer Corps, U. S. Army.

THOMAS CARLYON has resigned his position with the Keweenaw company and retired to Houghton.

G. L. SHELDON recently made a professional trip through Eureka, Lander, and Nye counties, Nevada.

R. D. PARK, of the United Verde mine, has received a commission in the Engineer Corps of the U. S. Army.

WALTER R. VIDLER has returned to Colorado from professional work near El Paso, Texas, and Nogales, Arizona.

H. VINCENT WALLACE is at Tucson; he is now consulting engineer to the Guanajuato Development Company.

JAMES F. BROWN has been commissioned in the Ordnance Department of the Army, and is stationed in Illinois.

L. O. HOWARD has been appointed Dean of the School of Mines of the State College of Washington, at Pullman.

I. J. RUSSELL has resigned as superintendent for the United Gold Mines Co., and has gone to the oilfields of Oklahoma.

W. N. THAYER has returned to Cincinnati from the Wyoming oilfields and has left for the Vermilion district in Minnesota.

D. L. MCCARTHY, superintendent for the Granite Gold Mining Co., is seriously ill at the Glockner Institute, Colorado Springs. JOHN BAILEY is acting in his stead.

J. H. WINCHELL, JR., has resigned as engineer for the Deming Mines Co. to become junior land-classifier for the U. S. Geological Survey, and is stationed at Sterling, Colorado.

A. C. BARKE has become engineer to the U. S. Mining, Smelting & Refining Co.'s exploration department, in succession to ALBERT ROBERTS, who has joined the Minerals Separation staff in San Francisco.

HENRY A. WENTWORTH has been appointed a vice-president of the American Zinc, Lead & Smelting Co. He will take charge of the Exploration Department recently formed for the investigation of new properties and processes.

ALBERT BURCH has been elected president of the California Metal Producers Association for the current year. G. W. METCALFE is first vice-president and ROBERT I. KERR secretary-treasurer, with EDWIN HIGGINS as safety and efficiency engineer.

THE METAL MARKET

METAL PRICES

San Francisco, October 30

Aluminum-dust (100-lb. lots), per pound	$1.00
Aluminum-dust (ton lots), per pound	$0.95
Antimony, cents per pound	16.00
Antimony (wholesale), cents per pound	14.75
Electrolytic copper, cents per pound	23.50
Pig-lead, cents per pound	5.75— 6.75
Platinum, soft and hard metal, respectively, per ounce	$105—111
Quicksilver, per flask of 75 lb.	100
Spelter, cents per pound	9.50
Tin, cents per pound	60
Zinc-dust, cents per pound	20

ORE PRICES

San Francisco, October 30

Antimony, 45% metal, per unit	$1.10
Chrome, 34 to 40%, free SiO_2, limit 8%, f.o.b. California, per unit, according to grade	$0.50— 0.60
Chrome, 40% and over	$0.60— 0.75
Magnesite, crude, per ton	$8.00—10.00

Manganese: The Eastern manganese market continues fairly strong with $1 per unit Mn quoted on the basis of 48% material.

Tungsten, 60% WO_3, per unit	26.00
Molybdenite, per unit MoS_2	$40.00—45.00

EASTERN METAL MARKET

(By Wire from New York)

October 30.—Copper is unchanged and nominal at 23.50c. all week. Lead is more active and firm at 6.25 to 5.60c. Zinc is dull and weaker at 8.25 to 8c. Platinum remains unchanged at $105 for soft metal and $111 for hard.

SILVER

Below are given the average New York quotations, in cents per ounce, of fine silver.

Date			Average week ending		
Oct.	24	82.50	Sept.	18	100.70
"	25	82.50	"	25	107.66
"	26	83.25	Oct.	2	98.49
"	27	83.25	"	9	91.72
"	28 Sunday		"	16	87.01
"	29	84.75	"	23	83.85
"	30	90.62	"	30	84.71

Monthly Averages

	1915	1916	1917		1915	1916	1917
'an.	48.85	56.76	75.14	July	47.52	63.06	78.92
'eb.	48.45	56.74	77.54	Aug.	47.11	66.07	85.40
(ch.	50.61	57.89	74.13	Sept.	48.77	68.51	100.73
.pr.	50.25	64.37	72.81	Oct.	49.40	67.86	....
(ay	49.87	74.27	74.61	Nov.	51.88	71.60	....
une	49.03	65.04	76.44	Dec.	55.34	75.70	....

Samuel Montagu & Co., of London, report that the silver market has een inert, and in the absence of any important buying orders the moveent of prices has been again retrograde. The prohibition of exports to rtain neutral countries, as from October 8, is hardly likely to have much fect upon the market except so far as any additional restrictions tend hamper business. The return of the price to a lower level is in favor the Indian government, which is undoubtedly in a position to use rge quantities of silver when it can be secured advantageously. The dian currency returns record an increase of one crore (10,000,000 rupees) the holdings of silver.

The High Commissioner for Canada has stated that "Since the discovery silver in Cobalt in 1903 there has been produced approximately 266,-0,000 oz. of silver, valued at $143,000,000. It is estimated that this ar's production will have a valuation of $15,000,000. If this be callated at the average price of the first six months of this year, the ight should equal about 19,600,000 oz., implying a reduction of about 400,000 oz., or about 25%, on the Government estimate of 26,000,000 for 1916.

COPPER

Prices of electrolytic in New York, in cents per pound.

Date			Average week ending		
	24	23.50	Sept.	18	26.25
	25	23.50	"	25	24.41
	26	23.50	Oct.	2	23.50
	27	23.50	"	9	23.50
	28 Sunday		"	16	23.50
	29	23.50	"	23	23.50
	30	23.50	"	30	23.50

Monthly Averages

	1915	1916	1917		1915	1916	1917
	13.60	24.30	29.53	July	19.09	25.66	29.67
	14.38	26.62	34.57	Aug.	17.27	27.03	27.42
i.	14.80	26.65	36.00	Sept.	17.69	28.28	25.11
	16.64	28.02	33.16	Oct.	17.90	28.50	....
	18.71	29.02	31.69	Nov.	18.88	31.95	....
e	19.75	27.47	32.57	Dec.	20.67	32.89	....

hile Copper Co. states that reduced production in September of 3,294,-000 lb., as against 8,036,000 lb. in August, was due to inability to secure deliveries of Californian and Mexican fuel-oil, and to the shortage of Chilean coal, which conditions persist.

LEAD

Lead is quoted in cents per pound, New York delivery.

Date			Average week ending		
Oct.	24	6.25	Sept.	18	9.33
"	25	5.50	"	25	8.00
"	26	5.50	Oct.	2	8.00
"	27	5.50	"	9	7.96
"	28 Sunday		"	16	7.52
"	29	5.60	"	23	6.70
"	30	5.60	"	30	5.66

Monthly Averages

	1915	1916	1917		1915	1916	1917
Jan.	3.73	5.95	7.64	July	5.59	6.40	10.93
Feb.	3.83	6.23	9.01	Aug.	4.62	6.28	10.75
Mch.	4.04	7.26	10.07	Sept.	4.62	6.86	9.07
Apr.	4.21	7.70	9.38	Oct.	4.62	7.02	....
May	4.24	7.38	10.29	Nov.	5.15	7.07	....
June	5.75	6.88	11.74	Dec.	5.34	7.55	....

ZINC

Zinc is quoted as spelter, standard Western brands, New York delivery, in cents per pound.

Date			Average week ending		
Oct.	24	8.25	Sept.	18	8.29
"	25	8.12	"	25	8.46
"	26	8.12	Oct.	2	8.43
"	27	8.12	"	9	8.50
"	28 Sunday		"	16	8.50
"	29	8.12	"	23	8.25
"	30	8.00	"	30	8.12

Monthly Averages

	1915	1916	1917		1915	1916	1917
Jan.	6.30	18.21	9.75	July	20.54	9.90	8.98
Feb.	9.05	19.99	10.45	Aug.	14.17	9.03	8.58
Mch.	8.40	18.40	10.78	Sept.	14.14	9.18	8.33
Apr.	9.78	18.62	10.20	Oct.	14.05	9.92	....
May	17.03	16.01	9.41	Nov.	17.20	11.81	....
June	22.20	12.85	9.63	Dec.	16.75	11.26	....

QUICKSILVER

The primary market for quicksilver is San Francisco, California being the largest producer. The price is fixed in the open market, according to quantity. Prices, in dollars per flask of 75 pounds:

Week ending

Date					
Oct.	2	105.00	Oct.	16	100.00
"	9	105.00	"	23	100.00
			"	30	100.00

Monthly Averages

	15	1916	1917		1915	1916	1917
Jan.	.90	222.00	81.00	July	95.00	81.20	102.00
Feb.	19.00	295.00	126.25	Aug.	93.75	74.50	115.00
Mch.	[illegible].00	219.00	113.75	Sept.	91.00	75.00	112.00
Apr.	.50	141.60	114.50	Oct.	92.90	78.20	....
May	8.00	90.00	104.00	Nov.	101.50	79.50	....
June	[illegible].00	74.70	85.50	Dec.	123.00	80.00	....

TIN

Prices in New York, in cents per pound.

Monthly Averages

	1915	1916	1917		1915	1916	1917
Jan.	34.40	41.76	44.10	July	37.38	38.37	62.60
Feb.	37.23	42.60	51.47	Aug.	34.37	38.88	62.53
Mch.	48.76	50.50	54.27	Sept.	33.12	36.66	61.54
Apr.	48.25	51.49	55.63	Oct.	33.00	41.10	....
May	39.28	49.10	63.21	Nov.	39.50	44.12	....
June	40.26	42.07	61.93	Dec.	38.71	42.55	....

Charles Hardy says: Manganese ore remains in good demand. Such ore as reaches our market from abroad finds ready buyers. The basis for 48% material is still $1, and chemical ore of good quality is sold in large quantities at prices of about $100 per ton.

High-grade chrome still remains firm at 90c. per unit f.o.b. Western Coast shipping point and the market is considered firm.

The magnesite market continues practically the same with $40 per net ton f.o.b. California quoted for high-grade calcined magnesite. Small business, however, has been reported.

Californian virgin quicksilver is still quoted at $100 per flask, but very little spot material is available. Mexican and recovered is sold at the regular discounts from the Californian virgin price.

The United States Geological Survey is gratified to report a new source of the much-needed war mineral, chromite. J. T. Pardee, of the Survey, found it in serpentine while hunting for platinum and manganese on Cypress island near Anacortes, Washington. Some of the chromite has recently been shipped to an alloy company at Tacoma, and it is probable that on account of its nearness to tide-water the Cypress Island locality may become an important producer of chrome ore.

Eastern Metal Market

New York, October 24.

The markets continue inactive. Uncertainty pervades nearly every department and buying is only from hand to mouth.

Copper is still unsettled and nominal at the Government price.

Tin is firm and steady but there is little demand.

Lead has fallen again and there is almost no buying.

Zinc continues stagnant with no change.

Antimony is in poor demand and a little lower.

Aluminum is quiet and unchanged.

More active buying is apparent in the steel market at the new prices, especially in pig-iron. Further announcements of new prices are not expected, but it is probable that the trade will adjust itself to the new levels on other commodities not already stipulated. It is felt generally that a broader and more active market, as well as a healthier one, is in prospect.

COPPER

The entire situation continues unsettled. Reliable or official information concerning the market is extremely difficult to obtain. The producers' and consumers' committees are making no announcements, but it is believed they are rapidly working out a satisfactory solution so far as large producers and consumers are concerned. The main object is to conserve supplies so that war-needs may be adequately met, but how much is to be left after these are satisfied is not known to the public. Some predict a shortage; others an over-production. It is believed that large consumers, especially if they have Government work to carry out, are getting supplies on old contracts. The position of the small dealer and of the consumer that usually buy small quantities is difficult. The statement is made that sales by such dealers at a price higher than the Government figure are subject to confiscation. If this has any basis in fact, it will put small dealers and consumers in a serious position so far as continuing their business is concerned. It is true, however, that small lots of copper have sold as high as 30c. per lb., New York. It is also a fact that copper in large quantities cannot be bought at the Government price nor at any other price. The entire market is stifled and inactive. We continue to quote the official rate of 23.50c., New York, as nominal for all grades of copper. The London market is unchanged.

TIN

The tin market has been firm and steady since our last letter, though little business has been reported. Prices have fluctuated closely on either side of 61.50c., New York, on each day. Yesterday the quotation for spot Straits was 61.62½c., New York. The last three days of last week were very dull, for the good demand that appeared earlier has flattened out. On Monday and yesterday the market was nearly nominal, although yesterday a moderate business was done quietly in futures. Late in the day a fair demand for futures developed which may result in business later. Cables continue delayed, interfering with trading. The only interesting feature of the market is the fact that indications point to an easier situation in the shipping-permit question. Nothing definite is known about this, but it is believed that both the American and British governments have come to an understanding regarding the matter and that, as a consequence, consumers that have bought ahead are to get their metal more easily than in the past. It will be the small consumers, accustomed to buying in a small way, that will be hard hit. It is recognized in upper circles that the American tin-plate maker must have tin with which to make cans, and this has helped to bring about an understanding. Arrivals to October 23, inclusive, have been 1910 tons, with 4300 tons afloat. The London market is 5s. under that reported last week, or £247 5s. per ton for spot Straits.

LEAD

An already weak and declining market has been further unsettled by the announcement last Thursday, October 18, of a reduction by the American Smelting & Refining Co. of ½c. per lb. in its price, bringing its quotation to 6.50c., New York. This was not a great surprise in view of similar reductions recently and in view of the fact that independents were making efforts to sell under the trust price. This has continued at the new level, and yesterday the outside market was quoted at 6.25c., New York, or 6.12½c., St. Louis, for early delivery. The market may be quoted today at 6.40c., New York, for prompt; 6.25c., New York, for October and November; and 6c., New York, for December delivery. Buyers are ordering only what they need and what they can get hold of quickly. It is believed the bottom has been nearly reached. An interesting instance is that, in the last week, fair amounts of lead in transit have been sold, but they did not command a premium as is usually the case.

ZINC

The situation is discouraging and in some respects mysterious. Demand now is almost negligible, but a year ago, with the United States not yet in the War, the Allies were buying large amounts of zinc for their needs of the following year and American brass-makers were heavy purchasers for the filling of shell-contracts for the Allies. Now, with the United States in the War, and with its demands for ammunition for its own use expected to be large, orders for zinc are few and far between. Added to this is an inconsequential demand from galvanizers. The entire market is stagnant, with sales of small consequence so far as concerns the establishing of quotations. The nominal price for prime Western is 8c., St. Louis, or 8.25c., New York, for early delivery, with offerings reported at 7.87½c., St. Louis. It is probable that desirable inquiries would bring out a concession from 8c., St. Louis. For later positions from ⅛ to ¼c. per lb. higher is asked.

ANTIMONY

In a lifeless market the quotation for Chinese and Japanese grades has declined to 14.75c., New York, duty paid, for early delivery, but sales are of small proportions.

ALUMINUM

The quotation for No. 1 virgin metal, 98 to 99% pure, is unchanged at 38 to 40c. per lb., New York, with demand of small proportions.

ORES

Tungsten: The market is only fairly active, with quotations unchanged at $23 to $26 per unit in 60% concentrate. About 100 tons of scheelite is said to have been sold at $26 per unit, the prevailing quotation. Ferro-tungsten is quoted at $2.35 to $2.50 per lb. of contained tungsten.

Molybdenum and antimony: The market is unchanged. Fair sales of molybdenite are reported at $2.20 per lb. of MoS_2 in 90% concentrate. There is nothing to report in antimony ore.

THE COMPOSITE-METAL PRICE is based on current selling-prices of each of the common metals (spelter, lead, electrolytic copper, and tin) multiplied by the ratio of average consumption, added together and reduced to one pound. The proportions of the metals in the order named are 2.5, 4, 3, and 0.5. The composite price October 24 was 14.2387c. per lb., as compared with 20.0562c. in February.

AMONG the peaceful advances of man we may record the completion of the Australian transcontinental railway, the last link between Kalgoorlie, in Western Australia, and Port Augusta, in South Australia, having been closed. Now there is continuous rail connection between all the States of the Commonwealth from Perth to Brisbane. The last portion of the line—1060 miles—was started in 1912; the completion of it was delayed by natural difficulties and by an under-estimate of cost. A sparsely inhabited and unproductive region had to be traversed, so that materials of construction and food had to be hauled as much as 400 miles. Labor strikes proved another hindrance to speedy work. The journey from Brisbane to Perth, 3467 miles, or 215 miles more than the distance between New York and San Francisco, will take five days.

THE origin of saltpetre has been a subject of discussion by scientists for many decades. In this issue we publish a suggestive article by Mr. Hoyt S. Gale, of the U. S. Geological Survey, in which he attributes the formation of natural nitrate salts to the action of nitrogen-fixing bacteria. There is much warrant for this explanation of the phenomenon. A French biologist many years ago wrote an elaborate treatise on the Chilean deposits, in which he advanced the theory now advocated by Mr. Gale. It is certainly plausible in view of the extraordinary nitrogen-fixing function of the azotobacteria that are found in all soils containing humus. The presence of abundant organic matter, the absence of acidity, and the need of magnesia and of certain chlorides are essential features, however, and such conditions do not always exist on the exposed cliffs where nitrates are sometimes discovered. The bacterial theory is sound, and fully accounts for most of the small deposits known in this country, but evidently other principles are operative, and in places may be important.

WE draw attention to our New York correspondent's remarks, under 'Eastern Metal Market,' concerning the selling of copper. His statement is confirmed by information coming from other sources. The market for copper has passed into the hands of a so-called Producers Committee consisting of representatives of the United Metals Selling Company, American Smelting & Refining Company, the Phelps-Dodge Corporation, and the Calumet & Hecla group. This semi-official body calls for monthly reports from producers and consumers; it is obtaining the most complete information concerning the industry. No orders for shipment of refined copper can receive attention unless countersigned by this committee. Selling agencies have been asked to show orders on their books, the amount of copper sold and the names of their customers, together with the scheduled date of delivery. Mr. Edwards Mosehauer, of the United Metals Selling Company, is in charge of the Producers Committee's office at 61 Broadway, New York. It is presumed that all this is being done by authority from Washington, but we would like to be assured that such is the fact. The small producer and the small consumer may be forgiven for feeling apprehensive of this organization, which concentrates so much power and information into the hands of a few big corporations.

EFFORTS now being made by representatives of the Mexican government to raise money at Washington, in order to buy corn for the famishing peons are likely to have good results in more than one direction. It is reported that the Administration will not listen to the request unless Mexico shall align herself with the Allies, but this is a crude way of stating the case; what is meant is that we have no food to spare except for our Allies, and that therefore Mexico will knock in vain at our granary until she comes as a partner in our great enterprise. General Obregon asserts that four or five million Mexicans are on the edge of starvation, and he may not have explained that lack of food will drive the miserable victims of misrule into a new revolt against Carranza's government. It is suggested that Secretary Lansing will tell the Mexican envoy that nothing can be done until Mexico breaks with Germany, and puts an end to the propaganda of the Enemy not only in Mexico City but along the border. Of course, it is not in the fitness of things that we should go out of our way to feed a country that chooses to support the German at this time, but we see something bigger to be done, namely, to win good-will and thereby promote friendly relations with the people of Mexico. If the United States were to supply corn, or other food, for the starving peons, and if the distribution of such food were done by the hands of Americans, under the direction of the Food Administration, we would have a chance to prove to the Mexican people that the gringo is a good fellow, after all, and that he was willing to help them at a time when Carranza and his marauding generals were helpless to prevent famine. This would lead to the establishment of such a friendly understanding as would go far to settle the Mexican problem.

ELSEWHERE we refer to the importance of the coal-fields of northern France now held by the invader. Part of this industrial region has been regained from the Enemy, for it has been reported that 10% of the producing capacity of the Lens district is in the hands of the French. Of course, the invaders have pillaged and destroyed the area under their control and the equipment of the mines has been wrecked with Hunnish thoroughness. The system of mining practised in this coalfield is the long-wall, not the pillar method. It is likely that the Germans will have blasted the linings of the shafts so as to admit water and quicksand, the sinking having been done in part by the freezing process. Some of our readers are aware that a plan has been set afoot to assist our French allies by organizing a group of American mining engineers to aid in the re-opening of these coalfields. The work has been taken in hand by a committee appointed by the Mining and Metallurgical Society; it is headed by Mr. C. R. Corning and includes Messrs. A. R. Ledoux, R. V. Norris, E. T. Conner, and S. C. Thomson. Conferences have been held with the French Ambassador at Washington and with the French Commission now in this country. We wish our friends every success in this useful work; the abilities of our profession could not be applied to better advantage.

NEWS comes from Johannesburg that the Anglo-American Corporation has been organized at Johannesburg with a capital of $5,000,000, for the purpose of tendering for the mining areas to be placed on offer by the government of the Transvaal. The directorate includes Mr. W. L. Honnold, a distinguished American mining engineer, formerly identified with the Consolidated Mines Selection and Brakpan Mining companies; also Mr. Charles H. Sabin, president of the Guaranty Trust of New York, and Mr. H. C. Hull, formerly Mininster of Finance in the South African government. The chairman is Mr. Ernest Oppenheimer, of Kimberley. Whether this new corporation is to issue a loan in America in behalf of the Union government or to engage in mining, we do not know. Any active competition in tendering for the exploitation of Government areas on the Rand has been killed, in large measure, by an agreement between the leading 'houses' controlling the major operations of South African mining, so that these financial powers will not compete with each other. The question arises, will General Botha and his administration regard this combination in restraint of trade with a favorable eye? Unfortunately the interest shown by the Lewisohns, through Mr. W. W. Mein, early in 1916 has come to nothing and the American participation in Rand mining has, as a Cornishman would say, "just naturally petered out."

WE are informed that the directors of the American Institute of Mining Engineers have appointed Mr. E. K. Judd, assistant professor of mining in Columbia University, to be editor of the bulletins and transactions of the society. This is a step much to be commended. The present secretary of the Institute, as our readers know, is an able executive, but his multifarious duties have rendered it impracticable for him to do the work of an editor, so that the publications of the Institute have exhibited a lack of revision, forcing a contrast with the work of the old régime, under which the editing was done so thoroughly by Dr. Raymond, leaving much of the executive work to his assistants, notably to Mr. Theodore Dwight and Dr. Joseph Struthers. We have deplored the un-edited texts that have been published by the Institute, because it seemed to us that a bad example was being set to the profession; therefore we record Mr. Judd's appointment with keen pleasure. His early editorial experience was shared with the present writer, in another place; so that we are in a position to know something of his aptitude for such work. He is editor of the 'Columbia School of Mines Quarterly'; more recently he has acted as editor for the Mining and Metallurgical Society, but that was not a test of his ability, because the junior organization does not publish much and because its chief contributors have been gentlemen not without experience in technical writing. We wish Mr. Judd every success at his new post.

Miners Advance!

A telegram from Major O. B. Perry informs us that the Engineer Corps of the United States Army has been authorized to raise, by voluntary enlistment, a special mining regiment to consist of six companies of 250 men each and to be known as the 27th Engineers National Army. This regiment is now being recruited, the first company having been formed and placed in training at Camp Meade, Maryland. The regiment is to be composed entirely of picked men representing various mining districts throughout the country and comprising all the trades and occupations incident to mining. Each company will contain representatives of every branch of mining work so that it can operate as a unit. The men most wanted are experienced miners, both hand and machine drillers, shovelers, trammers, timber-men, track-layers, pump-men, blacksmiths, tool-sharpeners, electricians, machinists, carpenters, surveyors, time-keepers, cooks, shift-bosses, mine-foremen, and top-men. That is the order in which they were named to us. The work to be done is purely military in character, and is what is known as 'first line' work; it consists in the preparation of underground shelters for the fighting troops and the placing of explosive mines. Such work requires a high degree of skill in rapid tunneling and involves the hanlling of all sorts of material, from clay and chalk to the hardest rock. In addition to the regular engineering equipment, each company will be provided with special tools, such as tunneling and boring machines, drills, compressors, hoists, and lighting sets. Although the chief function of the 27th regiment will be underground mining, the men will be trained to fight as well as to dig; an engineer officer of the Regular Army will be placed in command, the other officers being selected mainly from the mining engineers that have already volunteered and undergone training. Major Perry and Captain W. H. Landers, two mining engineers well known in the West, have charge of the enlistment. Applications should be addressed to the Commanding Officer, 27th Regiment, Office of the Chief Engineers, Washington.

This announcement should meet with a ready response from the younger members of the profession and from the men that are working with them in the mines. It is a gratifying compliment to the mining fraternity and offers an exceptional opportunity to perform patriotic service under favorable conditions. This will not be the first time that a regiment of miners has played a noteworthy part in history. During the Indian mutiny, in 1857, when the residency at Lucknow was defended by the troops under Sir Henry Lawrence, the regiment that held the fort was the 32nd Duke of Cornwall Light Infantry, known as the Cornish regiment, because it was recruited in the old mining county of Cornwall. They made a victorious stand, largely because they were able to countermine the sapping of the Sepoys. As the chronicler says, "they took kindly to mining." We shall be surprised if some of the American descendants of those Cornishmen do not find a way into the 27th Engineers of the National Army. Brothers of the pick, this is your chance. Assert your manhood. You have been the pioneers of industry in the waste places of the earth. Now it is your privilege to save civilization from the onslaught of organized hell! Enroll yourselves not only to keep the world safe for democracy but to keep your hearths and homes safe from piracy on land and sea. Miners, follow on!

The Spoils of War

Recent discussion of the terms of peace has overlooked an item of some importance. Among the objects of the German invasion of France and among the reasons for the tenacious holding of the invaded territory we cite mineral resources of the greatest interest to the mining engineer. Most of our readers are aware of this fact without knowing the details. Among the public statements of German purpose none is more representative than the manifesto of the six industrial associations dated March 20, 1915, and presented to the German chancellor in May of that year. The six societies signatory to this document represent the non-socialistic classes in the German empire; they include the manufacturers of Saxony and the peasant proprietors of Würtemberg; they represent the economic interests of Germany. In this manifesto emphasis is laid on the necessity for holding Lorraine, "including the fortifications of Longwy and Verdun." Iron ore is not specified, it being taken for granted that the implication would be understood, for the next paragraph reads: "The possession of large supplies of coal and, in particular, of coal rich in bitumen, which is found in great quantities in the basin of northern France, is decisive, for the result of the War, at least to as great an extent as the iron ore." A glance at a map showing the distribution of coal in France makes clear the importance of the northern region. According to the committee appointed by the International Geological Congress of 1910 to report on the coal resources of the world, Germany has 409,975 million tons of bituminous coal in reserve, whereas France has 12,680 million tons; of which 9000 million tons are in the northern coalfields of the Valenciennes and Pas de Calais basins. In 1912 the mines of the Valenciennes coalfield yielded 26,139,948 tons and employed 123,698 workers. The names of the chief mining centres have been made known by the operations of war; from west to east they are Marles, Lens, Douai, Valenciennes. Thence the coal-seams extend across the northern border into Belgium, where the names eastward again are resonant of war, Charleroi, Chatelet, Namur, Liège, and Visé. Near the last of these mining centres the coalfield leaves Belgium and crosses the border into Westphalia. In Belgium there are two other coalfields, those of Dinant and Campine, but much the most important is the one that follows the valley of the Meuse from Namur to Visé. It is estimated that a reserve of 3000 million tons of coal remains in this Belgian coalfield. Here we have one reason for the German invasion and occupation; but the intention to seize the iron resources of France was a more important object. The deposits of the famous 'minette' ore extend into Luxemburg, Lorraine, Belgium, and France. Germany is in temporary possession of the whole of this mineral wealth. The ore is low-grade, namely, 36%, but the deposit is of such big dimensions that it can be mined cheaply. The ore-zone is 8 miles wide and 40 miles long, and as much as 60 feet thick. In France the principal orebodies are in the Nancy and Longwy-Briey basins. What the iron deposits of Lorraine and those of the adjoining part of France covering the Briey basin mean to either Germany or France is shown by the statistics of production in 1913. From Alsace-Lorraine, Germany obtained in that year 21,000,000 tons of iron ore, out of her total production of 28,600,000. In the same year the Briey basin, now occupied by the German invaders, produced 12,700,000 tons out of the total French production of 21,700,000. Thus three-quarters of the iron ore mined in Germany before the War came from the French territory annexed in 1871, and if the lost provinces were regained Germany's output of iron ore would decrease to 7,600,000 tons, while France's would increase to 42,700,000. On the other hand should Germany remain in possession not only of Alsace and Lorraine, but also of the

Briey basin, her output would increase to 41,300,000, while France's would dwindle to 9,000,000 tons. If therefore we consider the coal and the iron resources existing in the French area occupied by the German troops and add to that the iron resources of Lorraine, we find that on the possession of these disputed territories depends the future industrial development of both Germany and France, but more particularly of our gallant Ally.

Foods, Metals, and Labor

A summary of the economic situation in the United States appearing in the October circular emanating from the National City Bank of New York predicts prosperity throughout the coming year on account of the enormously increased purchasing power possessed by the agricultural population. The farm-values of all foods this season, according to the forecast of the Department of Agriculture, reach the extraordinary total, in round figures, of 21 billions. This is 47% in excess of 1916. On the other hand, products from the farm for this year and 1916 combined are considerably below the output for the preceding biennial period. The shortage is clearly due to the same inability to obtain labor that has curtailed the production of metals. Therefore the outlook for next season is viewed with alarm by many economists. It must be remembered that a continuous increase in the manufacturing population has been observed for decades, and that this movement has been enormously accelerated since the outbreak of hostilities in Europe. For a time the mines did not suffer, and before the War it was noted with interest that farm-products had advanced greatly in price, while the metal-market remained nearly stationary. Now the mines are beginning to feel the shortage of labor. Part of this has been the direct result of enemy propaganda, but part of it also has come from natural causes. The pressure to manufacture the supplies needed on so huge a scale for maintaining our Allies beyond the sea and for equipping our own Army and Navy, has called into existence a manufacturing capacity in excess of anything hitherto witnessed in this country, and altogether out of proportion to the quantity of raw materials issuing from our mines and farms. This is an economic problem of magnitude that must be considered and dealt with wisely or we shall be placed under a strain that may in the end weaken our resistance. From Arizona comes news that the special conciliators representing the President have secured the adherence of the mine-operators to an agreement recognizing the principle of the grievance committees, and establishing a central employment-bureau in order to facilitate the manning of the mines. This is hoped to aid in holding laborers by making it cheaper and easier to find work. It will be interesting to see whether this attempt to overcome the scarcity of workers at the individual properties will have an influence on the district as a whole by attracting a larger number of men. The outcome may throw light on the possibility of stemming the tide of labor now pouring into manufacturing operations to the detriment of our fundamental producing-power. Without considering problems of industrial management involved in the settlement of the strike at the Arizona mines, the precedent established for meeting the demands of labor unions throughout the West may incidentally involve a wider control of the economic situation.

The Komspelter Region

A recent visit to the zinc-mining districts on the western flank of the Ozark uplift suggests the need for the coinage of a comprehensive name to cover this most interesting and important region. For a time it was called the Joplin district, because the zinc-lead deposits were traced south-westward from that famous mining centre in Missouri. The exploratory work and mining development of the last two years, however, have proved that ore deposits, more important than those of Joplin, exist in the north-eastern corner of Oklahoma, at Commerce and at Picher, near Miami. The mines are not at Miami, but a few miles north of that budding town, the importance of which has been magnified by a systematic campaign of publicity, including the employment of a professional booster, a gentleman whose rhetorical exuberance may be likened to that of a gusher in the neighboring oilfield of Cushing. We admire the enterprise of the Miami towns-people and regret to seem cold toward their energetic propaganda; but it will be evident to anybody not intoxicated by the local boom that Miami is an unsuitable name. It is already the name of an older and more famous copper district in Arizona. To give the same name to two important mining centres is only to invite confusion detrimental to both. Moreover, the ore-belt extends not only far beyond the Miami district, but outside the State of Oklahoma, otherwise the 'Oklahoma zinc region' would be a suitable name. The programme of the recent Institute meeting used the compound 'Joplin-Miami' to cover the mining industry of this part of the Middle West, but even that will not do, because Joplin is relatively unimportant, save for the fact that it boasts the best hotel within reach of the mine-operators. On the other hand, the principal ore-belt is now being traced northward into Kansas, so that it is impracticable to localize the mining activities by using the name of any two or more towns or of any one or two States. We need a comprehensive name, adequate to cover not only the past productiveness of Joplin and the present output of the Commerce-Picher-Miami groups of mines, but also the future development of the promising area that extends to Waco, in Kansas, and even farther north. Therefore we suggest 'Komspelter.' This links the initials of the three States of Kansas, Oklahoma, and Missouri to the trade name of the metal zinc. We hope that this suggestion will be considered favorably by those whom it may most concern.

Mill-Tests v. Hand-Sampling

tor:

ly attention has been called to an article, under e caption, by Morton Webber, in your issue of er 29.

sts have long been used in one way or another k on hand-sampling, usually under the assump- t the mill-test, however crude and unscientific, rect results and hand-sampling did not. It is usual mill-test gives two factors, the value of and the recovery. Mr. Webber, however, limits test to the point of establishing the value of the e and he attempts to accomplish this object in a d round-about manner. "The object"—he in- —"is to determine the head, recovery, and tail- without reference to metallurgical efficiency. As l to sample the tailing accurately, as part of the on to obtain the sampling-error. the simplicity ing sand in leaching-vats will illustrate why I small cyanide plant as a sampling-error indi-

Mr. Webber really requires is a sampling-plant give directly the value of the crude ore and k *mill*-sampling against *lot*-sampling, a not un- method of procedure. However, instead of mechanical sampling of ore *lots* he introduces ith its many sources of error, not the least of the actual tailing-loss, and assumes that "If recovered by the mill and 20% is known by say to be retained in the tailing-pile, the un- l difference of 20% represents the sampling-

turn down hand-sampling on any such state- method or assume the 20% difference above as a error is quite unsound.

re many sources of metal-loss outside the tail- n fact, I cannot recall a case where the metal ing plus the metal recovered equaled the metal de ore milled. Nor could I conceive of any est on small lots of ore, based on cleaning up ted plates, in the manner indicated.

mples can be cut down, pulverized, and as- minimum error, and if cut in the mine with usually give commercial results. Lot-samples ried out similarly, the lots passed through a nd the correct value of the ore determined xing in metallurgical errors and the sampling ng of various mill-products with their attend- ancies.

The examples given by Mr. Webber of a low-grade gold deposit in Central America is illuminating; the assumed inerrant mill-test we are told "Disclosed a sampling-error averaging about 40c. on an expected mill-head of \$3.50." No approved system of sampling should be questioned by such a statement or condemned on such a mill-test.

In common with, I believe, most engineers, I have had all my sampling done on a basis of pounds per foot of sample cut and pointed out the importance of this procedure in a paper on mine-sampling, 14 years ago.

PHILIP ARGALL.

Denver, October 19.

King of Arizona Company

The Editor:

Sir—My attention has just been called to an interview with Mr. D. M. Riordan published in your issue of October 6. The statements therein contained with reference to certain of the directors of the King of Arizona Co. are so far at variance with facts that I am inclined to think you have misquoted Mr. Riordan. If you have not misquoted him, his memory has failed him in the past 18 years and in reminiscence he paints pictures of things that did not exist.

There was never at any time a disagreement among Messrs. Ives, Blaisdell, Riordan, and myself at any meeting of the board. The minutes evidence the fact that the only disagreement that ever occurred at any meeting of the board was between Messrs. Riordan and Smith. That was at a meeting held at Yuma, Arizona, on June 27, 1899, and grew out of a report and recommendation submitted by Blaisdell and adopted as the policy of the company by the votes of Riordan, Blaisdell, Ives, and Randolph. Smith declined to vote and made the following statement, which was spread upon the minutes of the meeting:

"Mr. Riordan is responsible for the opinion of Mr. Blaisdell as reported. Mr. Blaisdell's views as reported were not Mr. Blaisdell's views until today. Lest the stockholders reading these minutes should regard this policy as emanating from Mr. Blaisdell and thereby acquire an improper opinion of Mr. Blaisdell's ability as manager I object to the form of the motion."

This was not a serious matter, but I recall now, as my memory is refreshed by reading the minutes, that Mr. Riordan's feelings seemed unduly wounded by Mr. Smith's criticisms. Riordan was the president of the company, Blaisdell the manager. The policy suggested by Blaisdell was sound and even if Riordan had influenced him to make the report in the form made, it was not only his privilege to do so, but his duty. However, from that time forward, Mr. Riordan seemed to lose interest in the affairs of the company and only attended one more meeting of the board, which was held on July 21, 1899. At this meeting the only business transacted was a report from me to the effect that I had borrowed for the company $20,000, giving the terms, etc., all of which is spread upon the minutes. The following resolution concludes the meeting of the board:

"On motion of Mr. Riordan, seconded by Mr. Ives, the action of Mr. Randolph was ratified and approved. On motion of Mr. Ives the meeting then adjourned."

The next meeting of the board was on August 13, 1899. This meeting Mr. Riordan did not attend but sent in his written resignation as president of the company without assigning a reason. Thus ended Mr. Riordan's connection with the company.

Now, if two of the directors of this company were "heeled" as Mr. Riordan puts it, at any meeting of the board, it must have been at the June 27th meeting, and those two directors must have been Riordan and Smith, but if they were "heeled" there was no evidence of the fact at the time.

Riordan's statement in your interview to the effect that there was a combination among the directors "to do up Smith" is utterly false. On March 7, 1899, the King of Arizona Co. entered into a contract with Smith and Blaisdell, as a firm, to build a cyanide plant and do certain other work in and about the mine, for which they were to receive, when completed, a stipulated number of shares of the stock of the company. In view of the fact that these two men were to become large stockholders of the company upon the fulfilment of their contract, they were, simultaneously with the signing of the contract, given each a share of the company's stock and elected members of the board of directors that they might have the fullest knowledge of all transactions of the company. Blaisdell was also made manager of the mine that he might conduct its affairs upon the ground at the same time that he was engaged in erecting the cyanide plant in fulfilment of the contract of Smith and himself. It was of the utmost importance that the work undertaken by Smith and Blaisdell be completed on contract time, for the reason that the company required the proceeds from the plant in order to meet certain deferred payments on the mine. In the progress of this work it became apparent that Smith and Blaisdell would not complete their contract on time or anywhere near it, and it was the opinion of the directors that this work was being purposely delayed by Smith. When still later it became evident that Smith and Blaisdell would not complete their work on contract time, a forfeiture was declared and the company took over the unfinished plant and completed it. At the time it did so, it offered to issue to Smith and Blaisdell shares of the company's stock in the ratio that the work already performed by them bore to the entire contract. Blaisdell for himself accepted this offer and stock was forthwith issued to him. Smith declined to accept and brought suit against the company. This suit was decided in favor of the company in the trial court at Yuma and was later affirmed by the Supreme Court at Phoenix. Smith appealed to the Supreme Court of the United States and while the case was there pending the matter was compromised and stock issued to Smith in accordance with the compromise agreement. Smith later sued to rescind this latter contract, employing a new firm of lawyers. The firm first employed by Smith sued him attaching the stock, which was later issued to them by order of the court and they drew dividends upon it from the beginning of the dividend period to the time that the mine ceased to operate. So much for S. Morgan Smith's connection with the enterprise.

Your interview makes Mr. Riordan state that Mr. Smith was the principal person who put "real money" into the venture besides himself (Riordan). It is not known to the undersigned just how much money Smith spent upon his contract. Mr. Riordan's investment in the company was $5000 "real money." The undersigned invested $10,000 "real money" on the day the mine was optioned, and thereafter procured for the company or invested in its shares from time to time very much larger sums of money. There were nine stockholders each of whom put more "real money" into the property than Mr. Riordan, and some of them very many times more "real money."

The company records here in Tucson and the court records in Yuma and in Phoenix will verify all of the statements herein made, and they are open to inspection by anybody at any time.

EPES RANDOLPH.

Tucson, Arizona, October 25.

Estimating Ore

The Editor:

Sir—Given an orebody 800 ft. wide, 1600 ft. long, that rises above the surface from 100 ft. on one end to 800 ft. at the other end, all being quartz and exposed; with assays from many parts from zero to $19 per ton, and with several general samples weighing from 200 to 350 lb., showing from $5 to $8; mill-tests on 1½ tons, 3 tons, 10 tons, 4 tons, and 40 tons, averaging better than $5 gold per ton saved on the plates; I wish to know whether more than two well-known experienced engineers can agree upon the proper methods to sample this with a view to purchase. The above is an actual orebody and information is wanted.

G. L. SHELDON.

Ely, Nevada, September 30.

[The general samples and mill-tests do not mean much until we know who did the sampling. The assays "from zero to $19" mean even less until we know how many inclined toward zero and how many approached $19 per ton. The accuracy of any sampling will depend upon the distribution of the gold; whether it be evenly distributed or whether it be confined to veinlets or patches in the mass of quartz. If the supposed ore rose 800 ft. above the surface it should be possible to sample it by driving one or two cross-cut tunnels, and then sample the sides of the tunnels in 5-ft. sections.—EDITOR.]

A Criticism

The Editor:

Sir—I wish to express my appreciation of and say that I am greatly pleased with the new 'headpieces' with which you are introducing the various departments of the magazine. They are cleverly chosen and well drawn, with one exception, and without that you would never have heard that I liked them.

Being not only a miner but a motorist, I wish to enter a vigorous protest, in the name of 'Safety First' against allowing your artist to employ a chauffeur (in the Review of Mining) who will stop on over a 15% grade with a 'flat' less than a car-length ahead. A slight slippage of the brakes, considering the position he is standing in, might dislocate Mr. B. A. Stockholder to such an extent that all of the arguments of Mr. A. Geologist would be of no avail. If that point of view is absolutely necessary, make the 'prospect' walk.

LESTER S. GRANT.

Stent, California, October 19.

[The point is well taken.—EDITOR.]

Flotation Physics

The Editor:

Sir—I have noted with interest in your issue of September 8, a discussion by Mr. Blamey Stevens of 'Flotation Physics.' In the course of the experimental work I have done in flotation, a line of reasoning very similar to that presented by Mr. Stevens has often suggested itself to me, and I should like to say a word or two in that connection.

In the first place, if Mr. Stevens' hypothesis be correct, we should expect that the best conductors would float best. It has been my experience that native copper or silver are almost impossible to float unless filmed with sulphide or something of the sort, although copper and silver metals are both excellent conductors.

Secondly, Mr. Stevens argues that the latent negative charges on the surfaces of conductors will attract and repel oil-films and ions outside the surface. As he previously states in his article, a distributed charge of electricity will act upon bodies outside as if it were concentrated at its geometrical centre. It would appear to me that the latent surface charge would act in this manner, and would be equal to and opposite in sign to the residual charge in the interior of the conductor. Acting from the same centre, their combined effect outside would be zero. This I have always understood as the reason for saying that the charge is latent.

For considerations such as the above, I have always felt that a theory of this sort did not offer as logical a picture of the mechanism of flotation as might be desired. In fact, I might go further and say that my present opinion is that it is not advisable to explain flotation in terms of electro-statics. My reason is that some of our leading physical chemists say that ordinary laws of electro-statics do not apply at molecular distances.* Consequently, a scientifically acceptable electro-static theory must be so complicated as to be practically useless to the average engineer or millman. We are also told, however, that chemical reactions, adhesion, adsorption, surface-tension, and many other phenomena, are fundamentally due to electro-static or electro-magnetic attractions and repulsions between the various atoms and molecules involved, possibly somewhat as Mr. Stevens outlined in the first part of his discussion. This being the case, we should be able to express the attractions and repulsions occurring in flotation in terms of chemical affinities. In doing so, we not only make our theories similar to those used by many scientists in explaining condensation and evaporation, sublimation, crystal structure, and many other apparently physical phenomena, but we are using terms and laws far better understood among metallurgists than many now found in flotation discussions. This idea was first suggested to me last winter by Mr. Schwartz, in much the same language he uses in his letter published in your issue of July 7, 1917. I have collected data since then which makes it appear that a satisfactory theory of flotation can be so presented in terms of chemical affinities. Lack of time and space prevent a full discussion of this at the present time, but I hope to find an opportunity soon to present my ideas in detail.

JAMES A. BLOCK.

Hayden, Arizona, October 25.

[This interesting letter should provoke discussion. We invite flotation experts to take part.—EDITOR.]

CALCAREOUS MARL used for soil-sweetening was reported to the U. S. Geological Survey, in 1916, from Arkansas, California, New York, Pennsylvania, South Carolina, and Virginia, according to returns compiled by G. F. Loughlin. The output was 35,588 short tons, valued at $107,768, or $3.03 per ton. Individual reports gave prices ranging from $1 to $3 per ton in the Southern States and from $4 to $6 per ton in the Northern States and California.

PLATINUM production in Colombia is expected to increase materially during the coming year. Prospecting has shown that the metal occurs in the stream gravels and in the high gravels for long distances on the Atrato and San Juan rivers in the Department of the Chocó. The most important operations are those of the South American Gold & Platinum Co. along the Condoto river, and the numerous native washings in the vicinity of Istmina.

*G. N. Lewis, Jour. Am. Chem. Soc., Vol. 38, p. 773 (1916), and Irving Langmuir, Vol. 38, p. 2237 (1916).

Origin of Nitrates in Cliffs and Ledges

By HOYT S. GALE

*The prominence given to the subject of nitrates and the necessity of establishing a domestic nitrate supply both for munitions and for agricultural use, has naturally directed attention to the possibility of finding workable deposits of these salts within our own country. Nitrates exist in small amounts in many places, and nothing is more natural than that the prospector should attempt to follow these indications to their source, in the hope of discovering some large body of material containing a sufficient percentage to justify its exploitation. It appears that many well-informed people are being misled by small amounts of these salts, found in many places, so that it would be well, while this campaign of exploration is going on, to give thought to the peculiarities of the occurrences of nitrate in nature, and the way in which it is known to be formed. Consideration of this point may eliminate much unnecessary work and expenditure in prospecting. Turning to an encyclopedia we find: "Saltpeter (or nitrates) occur but seldom in strata, being for the most part a product continually formed by the action of nitrifying bacteria upon decomposing protein in the presence of oxygen." Protein is the name given in organic chemistry to a class of compounds found in animal and vegetable tissues made up of the elements carbon, hydrogen, and oxygen, together with nitrogen in chemical combination, the class being distinguished from other organic compounds by the presence of the combined nitrogen. Nitrogen in this form is generally called organic or albumenoid nitrogen. Protein is regarded as one of the most essential constituents of food for man and animals. It is represented by the albumen of the egg, the gluten of the wheat, and certain parts of meat and blood, as well as various grains and seeds. Although protein is food for animals, and is generated in plants, it is not directly available as food for plants that require nitrogen in the form of a nitrate, that is, a salt of nitric acid. Nitrates are essential to the growth of plants, and are present in all fertile soils, in which the nitrates are formed mainly by the action of nitrifying bacteria.

These nitrifying bacteria are present almost everywhere that conditions are favorable. They are found in practically all soils, and nearly everywhere on the surface of the earth. They have been found on the surfaces of rock masses high in the mountains. They become active as do other more familiar ferments, as for instance the yeast in dough, under proper conditions of warmth and moisture and with access to the oxygen of the air, and they become less active or quite inactive in the absence of any of these favoring conditions. The farmer treats his soil to exactly the conditions leading to the best results, and one of the principal results is that the nitrifying bacteria become active and supply nitrate for the growing plants. There must be organic matter or humus present to support this action, as well as other conditions.

*By permission of the Director of the U. S. Geological Survey.

Large quantities of commercial nitrate salts are produced today by taking advantage of bacterial activity. During 1916 more than 2500 tons of saltpeter came from India to the United States, to be used mostly in munition manufacture. Nitrate is produced regularly in India, and in some other countries, by the cultivation under exceptionally favorable conditions, of organic wastes including manure, but chiefly the filth that accumulates about the dwellings of a none too sanitary population. This refuse is collected in small yards, is turned over from time to time so that it shall have proper aeration to promote bacterial activity, is protected from the leaching of rains during the wet season, is rendered slightly alkaline as one of the conditions favoring the nitrifying action, and is then periodically taken up and leached to extract the nitrate salts that have accumulated in it. The same earth or soil is worked over and over, but organic refuse is continually added, and care is taken to keep the compost slightly alkaline.

The same thing that is being done artificially in India is going on in many natural accumulations of compost, either manure or the general refuse of animal life where animals are herded together. Cave-earths were worked as a source of saltpeter during the Civil War and at other times in the history of the United States. There is no doubt that the nitrates in the caves originated in an exactly similar way. Caves or cavernous ledges are the natural refuge of all sorts of animal life, including insects, birds, reptiles, and many of the larger animals. In these recesses they leave, not only extrementa, but bits of their food, hair, bones, flesh, and even grains, mixed in a soil that is often light and porous, and may be filled with twigs and dried leaves. This is probably stirred by the coming and going of the cave denisons. Parts of these recesses are often damp with ground-moisture, or with wind-blown storm-water or mist, and when damp and at the same time warm they are in an ideal condition to promote the activity of the nitrate-forming bacteria. Thus it is easy to account for the accumulation of saltpeter or nitrate salts in caves, and in the cavernous recesses of rock-ledges.

Nitrate salts are so extremely soluble in water that they readily dissolve in ground or surface-moisture. They are continually forming in soils and as continually being washed away, so that there seldom exists any considerable accumulation of nitrates in soil. In caves, too, the nitrates are largely washed away in the same manner. In places more protected from such action the accumulations found have been formed. Even these are doubtless variable, the product found at any one time being the difference between what has been formed or brought in and what has been carried off. The ready solubility of the nitrate salts accounts for one other factor in the accumulation of such deposits, namely the migration of the salts from the soil or place of formation and its recrystallization on the walls of the rock, in crevices, and even in the cavities of a porous rock. Almost everyone is familiar with the tendency of a soluble salt to take up moisture in a moist climate, and to creep up the sides of the container, often spreading considerable distances. Nitrate formed within a cavern will thus creep wherever the air or the ground-moisture can take it, and will crystallize where the evaporation of the dissolving moisture may leave it. Such evaporation can not go on in the depths of the earth away from access to the air, so that the accumulation of such crystalline crusts can not be looked for at a distance from the surface nor where the air can circulate.

The foregoing is an adequate explanation for most of the occurrences of nitrate salts that have been reported from many parts of the United States. Very often the nitrate is found in direct association with cave-earth rich in decomposing organic material, and this earth will frequently give high results on testing for nitrate. Visible organic matter is not always plainly associated with the deposits of nitrate salts, but the connection may almost invaribly be traced. With this explanation it is seen that a source for the nitrate is found which is active only at the surface, and does not account for the formation of deposits in depth away from access to the air. This accords with the facts observed in all cases where attempts to exploit the deposits have given opportunity to compare the surface with the underground evidence. There are many variations which will lead the prospector to think that his case is different from the type described. For instance, how is one to account for the fact that a test for nitrate may often be obtained on a barren hillslope, in a soil just below the surface layer, especially in certain peculiar clay soils? The explanation is, however, the same as in the more definite cases cited above. Even these apparently barren soils contain much organic matter, especially at the surface. Bits of animal refuse and vegetation are sufficient to show fixed nitrogen almost anywhere that a sample may be taken. In a semi-arid climate, the soluble salts formed in the soil are not so completely washed out as in moister climates, and not infrequently they collect in the soil in considerable quantity. If the soil be a 'gumbo' clay it is quite easy to imagine that a surface may seal over very effectively almost as soon as it becomes wet, and although the soluble salts in the surface may be washed away, a part will be carried down. As the surface dries, it cracks, and letting in the air, the drying of the underlying layer is promoted as a sort of secondary process. Thus, a lower layer rich in the soluble salts ordinarily characteristic of such soils is formed. This layer has, at a number of places, been found to contain nitrate in small amounts associated with sodium sulphate, chloride,

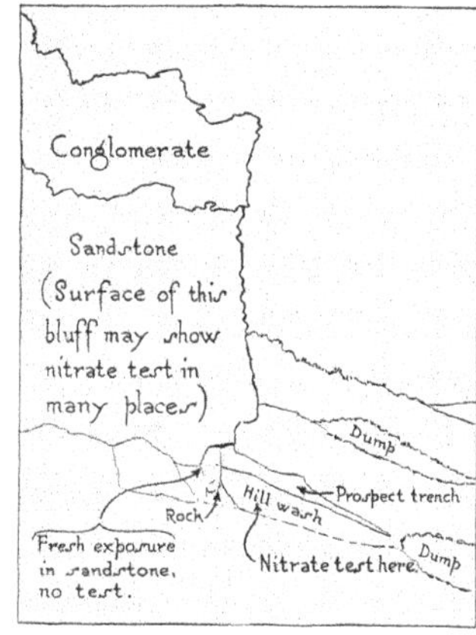

PROSPECTING FOR NITRE, VALE, OREGON

and possibly carbonate. The nitrate found in such places may be explained as the accumulation of ordinary soil-nitrate in the sub-soil, where it happens to be under such conditions that it is not more completely washed away.

The accompanying illustrations are taken from prospects on nitrate deposits of the type mentioned and will serve to elucidate the features described.

The question is often asked, how were the great deposits of nitrate in northern Chile formed, and is it not possible that something of the same sort might be found in this country? The origin of the Chilean nitrate deposits has been widely discussed and it can not be said that a positive answer has yet been given. However, the deposits in Chile are not unlike the hillside occurrences of nitrate-bearing soils referred to above, and it is not unlikely that they may have originated in the same way. The caliche, or nitrate-bearing layer in the Chilean deposits, lies just below one or two definite layers of what is essentially a surface-leached soil-zone. The soil in this case happens to be largely a mixture of broken rock, as it is the wash that lies between the bedrock of the hills and the almost flat fill in the floor of the inter-montane valleys. Rains in this region are so exceedingly rare that the soluble salts are not leached as they would be in almost any other country. Fogs are prevalent, which serve to dissolve the more soluble salts at the surface, and to carry them down. Ground-moisture is found at the shallow depth which is doubtless being constantly drawn toward the surface by capillarity and evaporated by the heat and dryness of the atmosphere during the day. Between these two influences the soluble salts become concentrated in a zone which is usually not far below the surface. The deposits are also localized regionally, as well as horizontally, and this may perhaps be accounted for in the same way, chiefly as a result of the movement of ground-moisture. In spite of the many fanciful explanations that have been given to account for the origin of the Chilean nitrate, it seems not unlikely that scientists will in the end come to an agreement that these salts were produced in the same way that they are being formed all over the world, and that the extraordinary accumulations in Chile are principally the result of the unusual climate at the place where they are found. As such a climate does not exist anywhere in the United States it seems unlikely that considerable deposits of this type will be found here. It is possible that some of the lower-grade accumulations of soil-nitrates might be turned to account if they could be found in sufficient quantity to make their exploitation worth while.

JAPAN exported goods in 1916 of a value of 1,127,468,000 yen, being an increase of 69.8% over the previous year. Of this amount 30% came to the United States. The imports had a value of 340,228,000 yen, or 8.6% more than in 1915, and 27% of this trade-movement represented shipments from this country. The present output of pig-iron in Japan is 1,000,000 tons per year.

Lost Mines

Every mining camp has its story of romantic discovery. Those of us who have shared the bivouac of the prospector or the warmth of the stove in a frontier hotel have heard of men that won wealth at a stroke. The poor prospector that went out as Bill or Dick is promoted to Mister or Colonel. As one old fellow expressed it: "A man can hustle, and starve may be, while trying to find a mine, but let him make a strike and they'll want to give him an oyster supper when he ain't hungry." There is scarcely a mining community in the West but has its mythical "lost mine." There was the Blue Bucket in Oregon, the Lost Cabin in Idaho, the Nigger Ben in Wyoming; there were others even more famous, such as the Breyfogle and Gunsight, both lost mines in the Death Valley region; but the most famous of all was the Pegleg, which was said to have been found by a prospector called Smith. He had lost a leg and stumped around actively enough by aid of a wooden substitute. His "stamping ground," as they say, was among the sand dunes and rock ridges of the Colorado desert, in Imperial county, California. According to the story, Smith was journeying from Yuma to Los Angeles when he was overtaken by a wind that drifted the sand across the trail. When the storm had passed he climbed to the top of one of a group of three small knolls in order to ascertain his way. When he reached the top of this knoll he was astonished to find the surface covered with pieces of dark-blue quartz so rich in gold as to be nearly solid metal. So astounded was he by his discovery that he forgot to take his bearings and hastened down the hill to make his way as quickly as he could toward Warner's pass. As soon as he could get a burro and some supplies he returned to his treasure-trove, but it was nowhere to be found. For days he searched the desert unsuccessfully. The days became weeks, the weeks lengthened into months, the years slipped by, and yet he failed to find that golden-crested hillock. He made many journeys, varying his direction, changing his area of search. Finally, he lived in a dug-out and mated himself to an Indian squaw at the foot of the hill now called Smith mountain. He never found his mine and died with his quest unfulfilled. In his last days he became garrulous and told the secret to other prospectors, who, in turn, started on a still hunt for the golden treasure. They too failed. The newspapers of the West have retailed the story in various guise. Scores of men have searched for the Pegleg mine and many have died in the effort; for example, four men in one party that went out from San Diego in 1895. Perhaps you would like to know where it is, as described by Smith's squaw long after Pegleg had found a resting-place in the one prospect-hole to which every man comes at last: 30 miles southwest of Smith mountain not far from the Butterfield trail, which is the old overland route between Yuma and Los Angeles, at the summit of one out of three little knolls. But two of those knolls may have been sand-dunes that the wind shifts as it lists.

THE INSPIRATION MILL, WITH THE INTERNATIONAL SMELTER IN THE BACKGROUND

Miami, Arizona: The Milling of the Ore—V

By T. A. RICKARD

Flotation constitutes an important part of the operations in the Miami mill, but in the Inspiration mill it is the whole thing. The Inspiration is the largest all-flotation plant in existence. It treats 19,000 to 20,000 tons per day; this is equivalent to about 13¼ tons of ore yielding ½ ton of concentrate per minute.

Entering the building, I was impressed by the bigness of it and then by the relative simplicity of the mechanical arrangement. Next I noticed the smell due to the aromatic constituents of the coal-tar used in the flotation department, and then the heat. It was a warm day outside, but upon entering the mill the temperature seemed appreciably higher. The idea came to me that this was due to the heat generated by the grinding of the ore, but a thermometer placed in the classifier-overflow, that is, the finished product of a ball-mill in combination with a Dorr classifier, showed a temperature of 70°F. as compared with 66°, which was the temperature of the water mixed with the ore in the ratio of 3:1. Making allowance for the relative specific heats of the ore and the water, it is evident that the rise in temperature caused by the grinding is small, namely, only about 3 degrees. A further heating of the pulp is due to the introduction of warm compressed air into the flotation machines, but this is balanced largely by the cooling effect of evaporation while the air is passing through the machines, so that this item also is negligible. Thus I had to conclude that the heat noticeable in the mill was due to the moisture contributed by the air passing through the flotation-cells and by the evaporation of the pulp in the launders and other parts of the mill. In short, I had stepped from an Arizonan atmosphere into one resembling that of Florida.

Each of the 20 sections into which the mill is divided treats nearly 1000 tons of ore per day, but it occupies a width of less than 17 ft., and a length of 273 ft. On the day of my visit the mill produced 708 tons of concentrate containing 29.52% copper and discharged a tailing assaying only 0.13% sulphide copper. The average sulphide-copper content of the tailing for the whole month was still lower, namely, 0.104%.

A complete flow-sheet of the mill is given herewith (Fig. 1) and on a smaller diagram the only change, made recently, is shown (Fig. 2). This is the addition of a small flotation-plant to re-treat the middling from the tables.

Each section includes two Marcy mills (8 by 6 ft.),

their axes parallel with the flow of the pulp. This pair of grinding-machines is separated by two duplex Dorr classifiers (6 by 27 ft.), the underflow from each classifier being returned to the nearest ball-mill for re-grinding, forming a closed circuit. The overflow from the classifiers runs over a screen, so as to remove chips of wood in the ore, and then direct to the flotation-plant.

This consists of several kinds of machines, one section being provided with a Hebbard machine, four with Callow cells, and the other 15 sections with a somewhat different type of porous-bottom machines designed on the spot. These last, called the Inspiration machines,[1] are the biggest of their kind, for each one of them is able to treat the pulp of an entire section, namely, 1000 tons of dry ore and 3000 tons of water. First the pulp, with its emulsified oil, goes to the roughing-machine, which in its later form is made of steel, 51 ft. long and 11 ft. wide over all, being subdivided into two double series of cells each 4 ft. 3 in. by 3 ft. At the end of each series of eight compartments the pulp can overflow, so that the level of the pulp in the preceding compartments is maintained at a constant level. The air that makes the froth is admitted above a solid cast-iron bottom and below a false bottom that serves as a porous medium. This consists of canvas protected on its upper side by an iron grate. From this first flotation machine the concentrate goes to a 'cleaner' of similar design but smaller. It has six double compartments 3 ft. square. The consumption of air is 4000 cu. ft. per minute per section, at a pressure of 4.5 lb. per square inch. The power consumed in supplying this air is 87.5 kw. per section or 2.1 kw-hr. per ton treated, to which about 10 kw., or 0.24 kw-hr. per ton, must be added for the pumping of cleaner-tailing.

The tailing from the cleaner is pumped back to the rougher. Meanwhile the tailing from the rougher goes to a drag-classifier, which separates it into sand and slime, the first being re-classified by Deister hydraulic machines, the sandy product from which is concentrated on 11 Deister double-deck tables. The slime from the various classifiers joins the tailing from the Deister tables as a final reject, which is treated by Dorr thickeners in order to reclaim as much of the water as possible.

Water is added to the coarsely crushed ore, or mill-feed, when it enters the ball-mills. The Dorr classifiers produce overflow-pulp of a 3:1 consistence. This is regulated by a float that responds to the density of the pulp. The float is attached to a thin wire connecting with a lever that swings between two contact-wires connected with an electric mechanism, which, in turn, operates a butterfly valve, thus admitting of stopping the flow of water in accordance with the rise or fall of the float. The various concentrates, from the Deister tables and from the flotation machines, are sent to Dorr thickening-tanks (ranging from 60 to 80 ft. diam.), which provide 40 sq. ft. of settling-area per ton of concentrate. On top of these tanks is the famous notice, "Keep your dog away." It appears that vagrant dogs jump on the cake of slimy concentrate formed over the surface of the thickener, thinking it solid, with a result disastrous to the dog and annoying to the millman.

The spigot-discharge (1.65 solid to 1 water) from the thickening-tanks goes to Oliver filters, of 12-ft. diameter and 12-ft. face. Six of these are in position, and are required to treat the entire output of the mill, namely, 700 tons of concentrate per day; it is intended to add two more. They reduce the moisture to 17%, but the percentage varies with the proportion of insoluble material in the concentrate.

At the time of my visit the flotation tailing from one section was going direct to the Deister tables, without classification; but it appears that classification is slightly the more favored. In the sections using the Callow cells, the overflow from the drag-classifiers is subjected again to flotation. The middling from the tables was re-floated experimentally with such success that the mill-practice is about to be modified accordingly. Dr. Rudolf Gahl, who has charge of the technical operations, is unwilling to believe that the last word has been said, for he knows that flotation as an art is still in its beginning. One of the subjects being studied, as at the Miami, is the recovery of the oxidized mineral. Whereas the recovery of the copper sulphide is 90%, that of the oxidized mineral is less than 25%. Leaching experiments are in progress, the oxidized copper being converted into sulphate, which is precipitated by crushed limestone. The results are said to be encouraging.

The total cost of milling, including crushing and grinding, is less than 40 cents per ton under normal conditions. For the first half of the year it averaged 46 cents on account of high wages (sliding scale based on the price of copper) and high cost of supplies. The cost of operating the flotation department is just under 6 cents per ton, not including royalty. An average of 1¼ lb. of oil is used per ton of ore. This means about 12½ tons of oil per day in the mill. A more significant figure is the quantity consumed in proportion to the flotation-concentrate produced: this is 49 lb. per ton. The oil is fed into the scoop-boxes of the ball-mills. Formerly a mixture of 95% crude coal-tar and 5% pine-oil was used, but it has been found that since the proportion of reclaimed water to fresh water used for milling purposes has increased, it is possible to reduce the quantity of pine-oil to the vanishing point. Nothing but coal-tar was used for considerable periods. Dr. Gahl used to think[2] that a slight admixture of a wood-distillation product was needed to produce a multiplicity of small bubbles for the saving of the minutest mineral particles, but he now finds it is not necessary, if the consumption of fresh water is kept low. When work was first started at the

[1]Excellent descriptions of the mill and its operations will be found in the September 1916 bulletin of the A. I. M. E. 'History of the Flotation Process at Inspiration,' by Rudolf Gahl. 'Mine and Mill Plant of the Inspiration Consolidated Copper Co.,' by H. Kenyon Burch.

[2]Op. cit. supra.

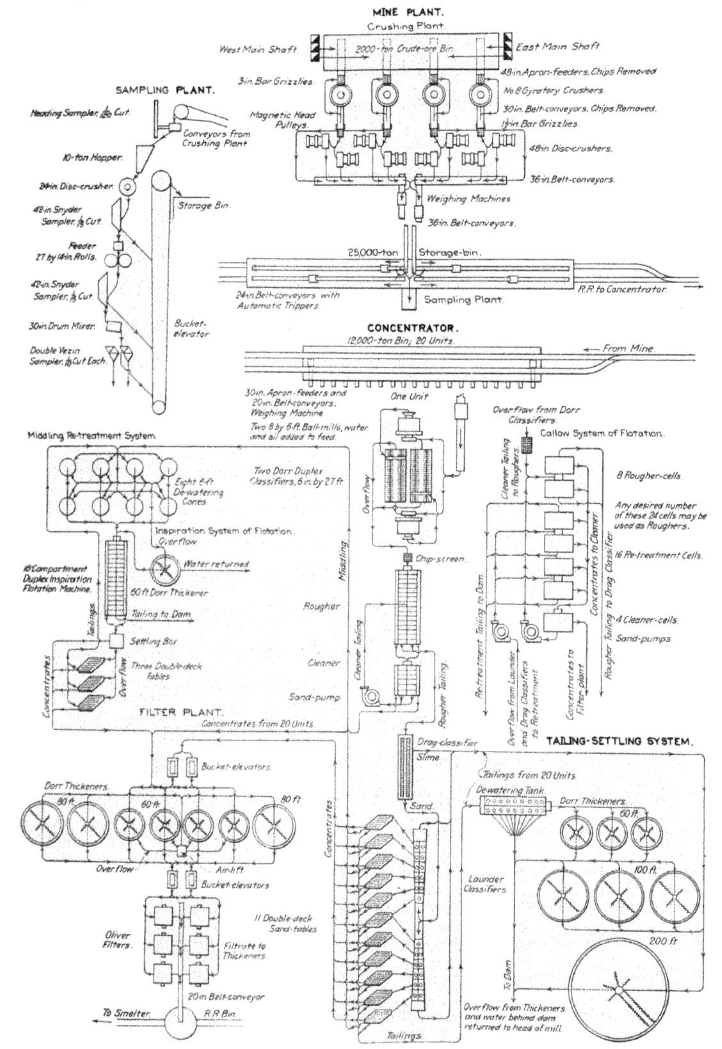

FIG. 1. FLOW-SHEET OF THE INSPIRATION MILL

Inspiration it was the practice to use one pound of flotation-oil consisting essentially of cresol at 30 cents per gallon (now 60c. per gal.); at the present time 1¼ lb. coal-tar at less than 6c. per gal. does the work. These prices do not include freight. This marks a distinct advance in the economics of flotation, even if the physical explanation of the change is yet less assured than the value of the money saved.

The ore that comes to the mill is hauled 1.6 miles, from the mine, for 1½c. per ton. The mill-bins hold 12,000 tons and are of the suspended type. They are made of sheets of steel in a catenary curve, which represents the most economical shape for a bin of this type. The mill is served by three cranes, one of 60 hp. and two of 40 hp. A broad-gauge incline runs along the side of the building, so that parts of machinery needing repair are carried on a flat car to the machine-shop, where they are lifted by another 40-hp. crane.

The history of the design of this big mill, and of the metallurgical processes conducted within it, is well worthy of American enterprise. In order to go back to the first suggestion of flotation, I must record the fact that on December 6, 1911, Edward H. Nutter, the representative of the Minerals Separation company in San Francisco, received a sample of chalcocite ore, sent to him by I. L. Greninger, from the Joe Bush dump. A test on that sample yielded a 9% concentrate and a 0.4% tailing on a 1.4% ore. Other samples were tried, with equally poor results. Nothing more was done until November 1912, when another series of tests was made, using the slide machine. On a 2% ore, a 15% concentrate was obtained, with an 85% recovery, and a 39% concentrate with a 55% recovery, according to Mr. Nutter. The various tailings ranged between 0.11 and 0.2% copper. These tests were made by F. A. Beauchamp, also in the Minerals Separation laboratory at San Francisco. A telegram stating that a 15% concentrate could be obtained on an 87% recovery was mutilated on its way to New York and was caused to state that a 50%, not 15%, concentrate could be made. This telegram caused considerable elation to the parties concerned, notably Dr. S. Gregory of Minerals Separation and W. D. Thornton of the Inspiration company. The immediate result was the erection of a small testing-plant by the Minerals Separation company at the mine in December 1912. In describing the big hole[3] made by the underground mining operations on the Joe Bush mine, I mentioned the remains of this 50-ton plant in which the first local experiments on Inspiration ore were made by representatives of the patent-owning company, notably I. L. Greninger and G. A. Chapman.[4]

The test-mill consisted of two sets of rolls, one Chilean mill, one Hardinge pebble-mill (6 ft. diam. by 12-in. cylinder), a Richards hindered-settling classifier, two Deister tables, and one M. S. flotation machine of the so-called standard type. This mill ran for eight hours each day and treated about 10 tons in that time. It was in use from January to June 1913. Good results were obtained, but at first it was intended to float only the tailing from the gravity-concentrator. The testing of the various kinds of ore existing in the mine was one of the principal purposes of this test-mill. To this end 10-ton lots from various places were supplied and investigated in order to ascertain their suitability for flotation. It was ascertained that chalcocite in schist was easily floated, whereas any kaolinized granite, such as was found along fault-fractures, produced a colloidal slime that diverted the oil from its proper function. To overcome this obstacle to a good recovery, it was suggested by Mr. Chapman that the oil be added to the pebble-mill on the theory that it might come in contact with the mineral particles at the instant when they exposed fresh fractures.[5]

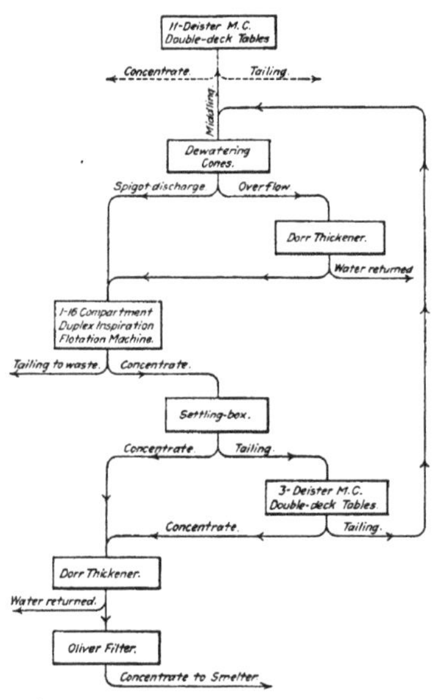

FIG. 2. FLOTATION PLANT FOR RE-TREATMENT OF TABLE-MIDDLINGS

In order to check various doubtful points on a bigger scale, it was decided to build a larger testing plant. This was designed by H. Kenyon Burch and started in Janu-

[3]See previous article. Page 418. M. & S. P., September 22, 1917.

[4]Lewis R. Wallace, subsequently superintendent of the International smelter at Miami and still later general manager for the Andes Copper Mining Co., represented the Inspiration company in these tests.

[5]'Flotation of Chalcocite Ore.' M. & S. P., March 18, 1916.

LOOKING DOWN AT SECTION-UNITS OF THE INSPIRATION MILL

THE SAME. MARCY MILLS, FLOTATION MACHINES, AND DORR DRAG-CLASSIFIERS

ary 1914. It was given a capacity of 600 tons, because C. E. Mills, the general manager, figured that a plant of this size would at least pay its own way and be large enough to take care of the ore furnished by the development work in the mine. Running the ore from the development work through the test-mill would give an excellent chance for checking the mine-sampling and for determining in a final manner the suitability of the ore for flotation. At this time Dr. Gahl began to take part in the technical operations.

Grading for the existing mill, also designed by Mr. Burch, was begun in August 1913. The erection of steel was begun in March 1914 and the mill was completed in February 1916.[6] It likewise was intended mainly for gravity-concentration, and a contract for structural steel had been made when the general manager, Mr. Mills (to whose good sense and mature experience so much of the success of this great copper-mining enterprise is due), decided that an all-flotation plant was amply justified. Each section, with its two Marcy mills, was expected to grind 600 tons per day, but this anticipation has been much exceeded, as already recorded, for the capacity of some sections is now above 1000 tons per day. It was found also that the capacity of the flotation units could be increased commensurately. Thus the mill, despite a capacity 66% greater, occupies a floor-space of 50% less than was originally designed. It will be noted that much care was taken in experimenting before a large plant was built, but, once the experiments appeared conclusive, the erection of the big mill was undertaken with the least possible delay. This record of achievement does credit to all concerned.

The concentrator report for May 12, the day of my last visit, showed:

Tonnage (total time)	19,348.9
Average number of sections running	19.469
Average tonnage rate per section	993.8
Assay of feed	1.30
Screen-anlaysis of feed (on 48-mesh)	3.8
Oxide copper in feed	0.28
Flotation concentrate	38.43
Table concentrate	14.30
General concentrate (from filters)	29.70
Insoluble in general concentrate (from filters)	32.4
Moisture in general concentrate (from filters)	17.5
General tailing	0.35
Oxide copper in general tailing	0.26
Recovery of copper	73.96
Recovery of copper sulphide	91.47

BORAX should be free from arsenic and from sodium carbonate. Much of the refined borax of commerce contains, by weight, more than four parts of arsenic per million. Such borax should not be used as a food preservative. Likewise sodium carbonate is an injurious adulterant except when the borax is to be used as a water softener. In any case, borax adulterated with sodium carbonate should be so declared, and should sell at a correspondingly reduced price.

[6]This refers to 18 sections of the mill, two more were added later and brought into service during April 1917.

British Standard Alloys

Interesting details of the composition and permissible variation in content of the alloys cast at the Royal Mint, London, are contributed by W. J. Hocking in a paper read at the recent annual general meeting of the Institute of Metals. The metals are generally procured in the form of fine ingots, and are alloyed in the proportions shown in the following statement:

Standard gold, 91⅔% gold, 8⅓% copper.
Imperial silver, 92½% silver, 7½% copper.
Coinage bronze, 95% copper, 4% tin, 1% zinc.
Cupro-nickel, 75% copper, 25% nickel.

In addition to the fine ingots and alloys, the average charge, to the extent of about one-third of the total, consists of scrap-metal from the various processes of manufacture, returned for melting. The bars cast are about 2 ft. long, but differ in width and thickness according to the denomination of coin desired. The width varies from 4 inches in the case of bars for bronze coins, to 1¼ in the case of those for three-pence pieces; and the thickness from ¾ in. for cupro-nickel bars to ⅜ in. for bars for coinage bronze. In another department the bars are rolled to the thickness of the coin required. The average weight of the different classes of coinage-metals cast annually during the last five years was about 2000 tons, or a mean rate of a little over 7 tons (7000 kg.) for the working day. In melting gold and silver for coinage, great care must be taken to secure the correct proportions of metals in the alloys as finally turned out, in accordance with the legal requirements. The limits of variation from exact fineness are narrow, and are specified in the Coinage Act of 1891 as two parts per thousand for gold and four parts per thousand for silver. The variations permitted in practice are much less than these, and the necessity that the bars cast should be uniform in composition tends to restrict the size of the charges. The volume of the charge is mainly determined by the convenience of stirring it when molten and before pouring. Gold and silver are isolated in their respective pots until they are reported by the assayer to be either suitable or unsuitable for coinage. The usual charge of standard gold is 2800 oz., or 87 kg., and of silver 6000 oz., or 187 kilogrammes.

HOOKWORM infection is reported in the case of 444 miners out of 1440 examined in the Mother Lode district of California, according to a circular just issued by the U. S. Bureau of Mines. The investigation was made by Dr. James G. Cumming, director of the Bureau of Communicable diseases of California, in co-operation with Joseph H. White, sanitary engineer for the U. S. Bureau of Mines. Of those infected, 91 have taken treatment and have been freed from the infection.

TIN-PLATE will be produced on a larger scale next year than ever before. The demand is leading to the erection of many new tinning plants.

Factors in the Production of Electrolytic Zinc

By R. G. HALL

*An article on the subject of electrolytic zinc no longer needs to be preceded by an apology. The production of zinc by electrolysis is past the laboratory stage and has become an economic factor of importance in the United States.

The various methods of procedure to be followed for the greatest economy of manufacture are fairly well known, consequently a detailed description of any method is no longer necessary or desirable. Assuming that, for the present at least, the electrolytic production of zinc will be confined to localities having developed water-power, a glance at the powers developed in the United States with their relation to known zinc ore-bodies of economic size, will not be without interest.

In the North-West are water-powers, developed and undeveloped, in Washington and Montana. These must bear a close relation to the rapidly growing zinc-production of the Coeur d'Alene and the Butte districts. The power of this territory is relatively low-priced, as compared with the cost of developing power from coal, and the near neighborhood of large bodies of somewhat complex zinc ores makes these water-powers of more than usual importance.

One of the large complex ore deposits of the West is in Shasta county, California, in the territory of the Northern California Power Co., and developments to utilize it in zinc reduction are under way. In Utah are developed water-powers in connection with such ore deposits as the Tintic and the Park City districts. Evidently the relation between the power and the value of the ore deposits is sufficiently close to justify further development, and such are already undertaken. Farther :o the east is power development at points in Colorado n the neighborhood of the large and valuable complex ores of the Lake, Summit, and Eagle County districts, as vell as those of the San Juan. In Arizona and New Mexico there is not, so far as I know, sufficient development of low-cost power to make attractive any particular ffort to develop electrolytic zinc. Passing east from the Rocky Mountain region there is no further development f power on a large scale short of the Mississippi Valley. Here is the plant at Keokuk, Iowa, producing ower, which, while perhaps not as low-priced as some of he North-Western powers or the older Niagara Falls ower, when considered in connection with its economic osition it may be regarded as a point of considerable dvantage. Farther east is the power development at iagara Falls, but from all accounts this is already iken up to the fullest extent and no power is available, at least none in large quantities for electro-metallurgical work.

The electrolytic refining of copper requires power not alone for the electrolytic tanks, but it also requires large quantities of coal for heating the solutions with steam or otherwise, as well as coal of special quality for cathode furnaces. In the electrolysis of zinc solutions we do not have this requirement because, in general, the solutions do not require heating, but on the contrary require occasional cooling, and the temperature for melting the zinc is so low as not to necessitate a large amount of coal for the cathode furnaces.

The consumption of acids and chemicals in electrolytic-zinc production is relatively small. No matter what kind of ore is used as a supply of zinc, unless it should be an oxidized ore with considerable quantities of soluble carbonates other than zinc, the amount of sulphuric acid lost in a cycle is a small factor, and in a modern electrolytic plant the use of other chemicals than those for laboratory purposes may be considered as nil.

The relation of freight charges to grade of ore is an important one and requires to be considered more in connection with the general ore-supply than by itself. It is to be noted, however, that the basing point for lead and zinc, the principal metals with which we are concerned in this inquiry, is the Mississippi Valley, consequently the value of the refined product must be considered only in connection with some delivery point having a freight rate basis which can be compared to the rate from East St. Louis.

At the present time, of course, with much of our metals going for export, this factor is considerably complicated, but for the purpose of this paper we shall consider the cost of the metals f.o.b. Mississippi Valley points, or equivalent to East St. Louis basis.

The requirement for the electrolysis of zinc is zinc sulphate and zinc sulphate only. Most other elements found in the solution are harmful to the electrolysis of zinc sulphate. I have never found any that I could confidently say were beneficial. This requirement seems relatively simple, but when it is understood that this process is to be applied to the complex of ore minerals coming from available zinc mines, it will be understood that the production of pure zinc sulphate of standard strength is not the easiest problem in the world, yet it is only this problem that has stood in the way of the manufacture of zinc electrolytically for so long a time. The factors necessary to the production of a solid coherent plate of electrolytic zinc were well known, but it

*Abstract: Bull. A. I. M. E., St. Louis meeting.

is only recently that we have been able to produce this pure zinc sulphate on a commercial basis and have been able to obtain the electric current at such a cost as to make the production of electrolytic zinc a possibility.

Inasmuch as sulphide of zinc as it occurs in nature is not soluble in dilute sulphuric acid for the direct production of zinc sulphate, it is necessary to roast sulphide ores in order to eliminate the sulphur and convert the zinc to an oxide or sulphate. The roasting of zinc ore for sulphuric-acid leaching presents a different problem from the roasting of the same ore for retort-smelting. In the case of retort-smelting it is necessary to eliminate the sulphur to the greatest possible extent. Sulphur remaining behind either as sulphate or sulphide, beyond a certain amount, is harmful. For the solution of zinc in sulphuric acid, however, only the sulphur remaining behind as sulphide, and not always all of that, is harmful or causes a loss of metal. Zinc that has been roasted to sulphate is, of course, not harmful and is even desirable, as the sulphuric acid combined therewith helps to make up for the mechanical and other losses in the subsequent process. Another difference in the two methods of roasting arises from the fact that in the roasting of ferruginous zinc ore for retort-smelting little attention is paid to that complex compound of zinc and iron which is generally formed in a roaster from such ores and is usually referred to as a zinc ferrate. In roasting Colorado ores for the production of spelter, I have found as much as one-third of the total zinc in the ore to be insoluble in a relatively strong solution of sulphuric or hydrochloric acid. A material such as this would be fatal if formed in quantity when the ore is to be subjected to leaching with dilute sulphuric acid containing only from 6 to 8% H_2SO_4. Fortunately, however, the conditions which produce the ferrate are not such as need necessarily prevail in the roasting of ores for leaching with sulphuric acid. The production of this material seems to be a function of temperature, fineness of ore, and of time. In roasting for the production of spelter the temperature is usually raised at the end of the roast for the purpose of decomposing the sulphates and the elimination of the last possible amount of sulphur, but in a roast for leaching with sulphuric acid, the presence of sulphates is an advantage rather than an objection, and consequently the high temperature is not necessary. When roasting a complex concentrate such as that from Colorado and the South-West, containing only about 35 to 40% zinc, most of the balance of the metal-content being iron, there is great danger that an undue proportion of the zinc will be insoluble. On roasting zinc ores for the complete elimination of sulphur, the loss of fine material carried out mechanically, and of zinc, lead, and silver volatilized, is appreciable. I had the doubtful pleasure of smelting several thousand tons of some of the earliest concentrate produced by flotation in Montana. The concentrate was roasted in the standard type of Hegeler kiln, and by the time it had dropped seven times from hearth to hearth, and leaked out of the doors and other parts of the kiln, by no means all the original concentrate found its way into the retort. Subsequent practice has improved this, but the roasting of a zinc-flotation concentrate in a multiple-hearth kiln is not conducive to sweetness of disposition in the metallurgist.

When leaching the calcine with sulphuric acid, other metals than zinc may go into solution. Contrary to the general belief, the solution of iron is not a difficult one to deal with, as this element is rather easily removed. In case the concentrate contains copper, a considerable amount of that metal will be dissolved with the zinc. This removes the copper from the place where it ought to stay, namely, with the lead and silver, and takes it where it should not be, that is, with the zinc. Consequently it must be removed by a separate treatment. Other soluble metals and metalloids are treated in the familiar standard manner which needs no description. The handling of the residue from the sulphuric-acid leach is, however, of considerably more interest. The economically valuable metals in it are copper, lead, and silver, with some gold, and the method of smelting must be adapted to the metal of principal value, whether copper or lead. If the lead is high, as is frequently the case, the material will go to a lead-smelting plant where the copper is of secondary consideration. In many cases, notably in the South-West, the copper is the metal of principal value, and under these conditions it may pay to send the material to a copper smelter and sacrifice the lead. The losses in leaching are slight in the case of lead, silver, and gold, but may be considerable in the case of copper, depending upon the subsequent treatment of the zinc sulphate.

Considering the ore as produced from the mine and carried straight through to the production of the finished metals along the lines above indicated we have: (1) When the ore is concentrated for the purpose of roasting and leaching with sulphuric acid, it would be desirable that a concentrate should be made as high in zinc and as low in iron as possible. The loss in leaching will be directly proportional to the volume of tailing remaining and also somewhat proportional to the amount of iron in the calcine. (2) For an ore high in zinc and low in iron, there will not be much difficulty in roasting to a point that will make possible an economic extraction of the zinc. (3) A calcine containing 68 to 70% of zinc oxide, equivalent to about 55% metallic zinc, would leave a residue of about 35% when this residue contains 10% zinc; whereas, a calcine containing 40% zinc would leave a residue of nearly 60% when the residue contains 10% zinc. In the first case, there would be a loss of only about 6% of the original zinc, while in the second case there is a loss of approximately 15% of the original zinc, and with such a low-grade concentrate the loss is likely to be even higher. It is evident, therefore, that, as the grade of concentrate so roasted and leached is lowered, the losses in zinc may rise rapidly, not only relatively but absolutely, and this loss will become greater as the percentage of iron increases in

[th]e residue. (4) Again it will be seen that only in [th]ose cases where the source of current is relatively near, [o]r where the concentrate can be made high-grade, will [it] be permitted to follow this method of handling. Com[p]arison of the total freight paid on the contents of a [5]0% zinc concentrate from Montana, landing 85% of [th]e product in New York, is of interest as showing the [a]pproximate distribution of the freights paid and the [r]elative amount of freight in each case.

[2]000 lb. ore to Mississippi river	$7.00
5% moisture	0.35
[8]50 lb. metal to New York	1.44
	$8.79
[8]50 lb. metal, Montana to New York	4.31
Difference in freight paid on 1 ton 50% ore	$4.48

From the difference of approximately $4.48 per ton [o]f ore shown above, in favor of electrolysis at or near [th]e mine in Montana, must be deducted such local [fr]eights as may be paid in case the power is not obtain[a]ble at the mine itself. This may decrease the difference [a]s much as $1 per ton. All of these conditions have to [a] large extent been fulfilled in the Montana and Idaho [fi]eld, in that there can be made a high-grade concentrate [w]ith a high efficiency of concentration, and that there is [p]ower close to the ore-supply, so that in no case does [tr]ansportation become a serious matter. The balancing [o]f these factors is of the greatest importance for success [in] the production of electrolytic zinc.

The ores of Colorado are of a complex character which [d]o not admit of a high ratio of concentration or of a [go]od recovery. The cost of milling is relatively high [an]d the recovery of all the metals contained is low as [co]mpared with the modern practice in simple ores. I [ap]proached this problem, therefore, with the idea of [el]iminating to the greatest possible extent this prelim[in]ary concentration, as many of the ores in Colorado are [al]ready what one could consider, in other industries than [zin]c, relatively high-grade. Where the ores as mined [co]ntain too much waste, or are too low in valuable [me]tals, these can readily be concentrated to a product [co]ntaining all the metals in a form which, though value[les]s for the retort-process, are easily treated by the [me]thod proposed. That is, the ores may contain from [..] to 20% zinc with varying quantities of lead, silver, [an]d gold. It was necessary, therefore, to find a method [of] handling these. The problem was to find which [wo]uld produce a high-grade simple concentrate, enabling [suc]h concentrate to be transported to a central refining [pla]nt without incurring high freight charges, and then put this concentrate into such shape that it would be [rela]tively easily separable into its constituent metals. [Con]centration smelting, with the volatilization of the [zin]c and the lead, was consequently decided upon. Most [of] the ore-supplies of central Colorado today are of [suf]ficiently high grade to permit of direct smelting in [the] blast-furnace, whereby the principal part of the [pre]cious metal is collected in a copper matte and the zinc and lead are driven off as a fume and collected in the bag-house. This is the process as carried out at Florence, Colorado, today. The physical character of the ores must be such as to permit charging into the blast-furnace, or in the case of concentrate, it must be put into this physical condition. Again, the ore-charges must be balanced in such way as to give a slag that will carry the minimum of zinc to waste. The Colorado ores are of such varied character as to permit of the fulfilling of both of these conditions. The ores are charged without preliminary roasting; and with low copper content on the charge a first matte is generally made carrying sufficient metal to bear shipment directly to the refinery. Part of the silver in the charge is carried into the matte and the balance goes over with the volatilized metals into the bag-house. The volatilization of the zinc and the cleanness of the slag still leave much to be desired and interesting work in this direction remains to be done. Concentrate produced in this way will contain only those metals and metalloids that are volatile under the conditions of the blast-furnace, consequently it is a relatively simple mixture of oxide of zinc and sulphate of lead, with some sulphate of zinc and a small amount of silver. In this connection there is necessary a word of acknowledgment of the pioneer work performed on these lines by F. L. Bartlett in his smelter at Cañon City, Colorado, for a number of years.[1] An intimate acquaintance with Bartlett's work which I had the opportunity of obtaining some years ago really forms the groundwork for the practice in Florence, Colorado, at the present time. In such concentrate as this the zinc is readily soluble in dilute sulphuric acid, and the lead and silver are quite insoluble; consequently the purification becomes simple. The physical character of the material is such as to make it easily handled for shipment and easily handled in the leaching-vats. The metal content is high, and shipment is in the direction of consumption of the metals. This point must be borne in mind in connection with refining at Keokuk.

The residue of lead and silver is relatively small when figured back to the tonnage of the original ore handled, and being a finely divided heavy material, it is washed with considerable ease reasonably free from zinc sulphate. The treatment of this for its lead presents no great difficulty, as the only metals of economic value are lead and silver.

From the above it appears that, considering the character of the concentrate possible with these ores, and the situation of the power available for the electrolytic process, only some such method could be employed. It is true that the loss of zinc in the blast-furnaces is high, extremely high when compared with metallurgical work on other metals, but the endeavor is as far as possible to make the milling and smelting operations one, so that one loss covers everything, and I am inclined to think that, under the conditions, to have a big hole in one pocket is perhaps no worse than to have little holes in

[1]Trans. A. I. M. E. (1893), 22, 661.

three or four. The freight on the zinc concentrate from the smelting point to the refinery becomes relatively small when figured back to the original ore, and in any event the concentrate is moving toward the market. Should such concentrate be moved in any other direction, the freight charge would become a direct charge on the cost of production which would not be redeemed in any way.

The cost of electric current in the production of electrolytic zinc is fortunately directly proportional to the metal recovered, rather than to the tonnage of ore treated. For this reason it becomes profitable to use it on a 20% ore or concentrate from Colorado, where it would not be profitable to use it under present conditions on a 60% concentrate from the Joplin district. Furthermore, when the actual cost of current per ton of ore is figured on a 20% ore it becomes, relative to the other charges incurred, only a small factor. In such case it would probably not exceed 25% of the whole cost of producing the metal and shipping to point of consumption, when based on the present power cost in most of the Eastern territory. The actual power consumed in the production of metallic zinc will be for most cases about 1½ kw-hr. per pound of zinc in the electrolytic tanks. Additions to this, of course, must be figured in the auxiliary power for the plant, transformer losses, conductor and other losses, all of which will be familiar to an electrochemist, so that altogether, probably, when working on central-station current 75% of the current purchased will go into the electrolytic zinc on the basis above mentioned. Costs on this basis need not be calculated, as each district will be a law unto itself, and power-contracts will be based on load-factor as long as central-station power is used. Central-station power is usually received as high-tension alternating current, and for the electrolytic tanks has to be stepped down and rectified. For the auxiliary power it has to be stepped down and may or may not be rectified, as the purchaser pleases. Load-factors in electrolytic work are usually high, approaching 100%, consequently the cost of horsepower per year can be figured almost directly into a kilowatt-hour cost; that is 0.1c. per kilowatt-hour is approximately equal to $6.50 per horse-power year at 100% load-factor.

WOLFRAMITE is reported from the State of Rio Grande do Sul, Brazil. The mineralized area has been explored to a depth of from 8 to 9 metres and is found in over 8000 hectares of land. The property is 10 miles from the city of Lavras, and on account of the lack of railway facilities the ore has to be hauled by cart from that point to Lavras, whence it is shipped to Porto Alegre. Shipments from this property up to the outbreak of the War had been made regularly to Germany, the Krupp works having contracted for the entire output. In the years immediately preceding the War the exports from this mine are said to have averaged more than 1000 tons of 60% ore per year. The property could produce from two to three tons daily.

Concentration Practice in South-East Missouri

By A. P. WATT

*The problem of concentrating the disseminated lead ore of south-east Missouri is extremely simple. The economic mineral is galena and the gangue dolomite. The ore assays from 4 to 6% lead, 80% of which is recovered in a 70% concentrate. The ore is crushed through 10-mm. and sized on 2-mm. screens; the oversize is jigged, while the undersize is tabled, and the slime treated by flotation. The district is the largest lead producer in the world, the output of metallic lead for the year 1915 being 183,906 tons. Over 20,000 tons of ore per day is being treated. The management of the different properties is progressive and always open to suggestion. The fact, however, must not be lost sight of, that the special problem in the district is peculiar, as the average ore contains but 80 to 120 lb. of a metal normally selling at about 4c. per lb. The low price of the metal limits the treatment-system to one of low cost. An elaborate treatment is neither logical nor advisable. An engineer first visiting the district may consider that the method of treatment could be improved and, in some cases, this may be the case. On the other hand, the method of treatment followed is largely one developed by evolution and, consequently, is probably not far from correct. However, the ultimate method of treating the ore is a problem that has not yet been solved. The methods of concentration used at the different properties are practically identical, differing not in principle, but in detail. This is because of the similarity of the ore in the different properties.

The usual method of treating the ore is as follows: It is given an initial crushing in gyratory breakers, later finished in rolls to pass the first screen, which is 9 or 10-mm. round-hole punched plate. Water is first added to the undersize of the first limiting screen, the pulp passing to the second limiting screen. This size varies to some extent, the usual size, however, being from 1½ to 2 mm. Punched plate with round holes is used entirely on the second limiting screen, woven wire not being used for screening. The oversize of the 2-mm. screen passes to Hancock jigs which yield concentrate, middling, and tailing. The concentrate is either a finished product or else it passes to a re-treating plant. The tailing, in general, is an end product. The middling is generally crushed in high-speed rolls, but much work is being done in an endeavor to replace them with regrinding mills; Allis-Chalmers, Hardinge, Marathon, and Marcy mills being now tested for the purpose of regrinding middling. It is desirable to crush the jig-middling through a 1½ or 2-mm. trommel. Formerly the re-ground middling was joined with the original feed, but at present the usual practice is to keep them separate. The discharge from the middling rolls, or regrinding mills, is sent to the middling elevator, which discharges to the middling-screen. This is generally a trommel covered with 1½ or 2-mm. punched plate, the oversize of which may be sent to a middling-jig or to the jig treating the original feed. There are two undersize products to treat on the tables, the original undersize and the middling undersize. They may be combined, but are generally treated separately; the latter course is the logical one. The original undersize is comparatively small in amount, but contains much galena and little water. The contrary is true of the middling undersize; here the tonnage is large but of low grade, and this product contains all of the circulating water in the middling system.

The method of treatment given each of the two products smaller than 2 mm. is practically identical. The product may be classified, each spigot going to tables, or it may be de-slimed in any of a large number of de-sliming machines. In this latter case, which is the more common, the de-slimed spigot is treated on Butchart tables. In any event, whether the table-feed is classified or simply de-slimed, the sand tables make a final tailing, a concentrate that may be final or re-treated, and a middling. This middling is circulated, re-tabled or reground, and generally joins the main stream of middling at the middling elevator. The table middling contains considerable pyrite which builds up in the system, and this 'iron', as it is locally called, is now being ground through 80 mesh in ball-mills and tabled, the tailing going to flotation.

The overflow of the de-sliming apparatus or the classifiers may pass directly to the flotation plant, or else be first sent to the slime-table department for treatment before going to flotation. If the latter course be taken, the overflow of the de-sliming cones is settled, the spigot is tabled, yielding a concentrate and a tailing. The concentrate is a finished product and the tailing is combined with the overflow of the settling tanks in the slime-table department and sent to flotation.

At the flotation plant the slime is settled in Dorr tanks, the overflow being sent back to the mill-water system. The spigot product, thickened to 20% solids, goes to the flotation machines, of which both agitation and pneumatic types are used. Creosote is the usual frothing-agent employed. The flotation concentrate is not re-treated, while the tailing generally receives further treatment. Re-treatment of the tailing generally consists in sending the primary tailing to pneumatic machines,

*Abstract of a paper read at the St. Louis meeting of the I. M. E., October 1917.

which make a final tailing, and a concentrate that may be final or may be cleaned later. The general flotation-concentrate is settled in Dorr tanks, the spigot product is treated by Oliver filters, the cake from which contains 50% lead and 14% moisture. It is loaded into box-cars and shipped to the smelter, but separate from the jig and table-concentrate. The general mill-tailing is stacked by conveyor-belts, as the topography of the ground does not permit of disposal of the tailing by gravity alone. The tailing from the flotation plant is either pumped to the 'chat' pile to be used as flushing water, or else sent to a slime-settling pond. For operating and metallurgical reasons, the mill-concentrate is kept separate and distinct from the flotation-concentrate.

No crushing is done underground in the district, all being done at the mills, with the exception of a few cases where coarse-crushing is done at certain shafts, the crushed ore then being transported to the mill. The usual practice in the district is to centralize the crushing plant at the mill and, as would be expected, the use of the gyratory type of crushers is almost universal, although one of the older plants still retains the jaw type.

A typical installation would be as follows: The run-of-mine ore is fed from the crude-ore bin to the primary gyratory crushers. They are usually either No. 6 or No. 7½ of the standard makes. The ore is generally fed direct without preliminary screening of the fine. The feed may be regulated by arc-gates operated by air or by the usual type of air-operated gates. The use of steel-pan conveyors in the capacity of crusher feeders is becoming common practice, as with a slow-moving pan conveyor a very large gate-opening can be permitted, thus preventing 'choke-ups' due to large rocks. The pan conveyors prove satisfactory for this purpose, being operated either by an adjustable ratchet and pawl or by a variable-speed motor. The discharge of the crushers passes to trommels, the oversize in general passing to the fine-crushing gyratories. At one plant, however, the discharge from the first gyratory passes, after screening, to 48-in. Symons horizontal disc-crushers. At another plant, the discharge from the first gyratory crushers, after screening, passes to Symons vertical disc-crushers, which are capable of making a reduction from 3 in. to approximately ½ in. This arrangement, of course, simplifies the crushing and screening plant.

There is nothing of particular interest about roll-practice in the district. The usual type is the high-speed roll, although two mills are still operating the Cornish type. The high-speed rolls are of the standard makes, the largest rolls in the district being 54 in. diameter. Rolls that crush sizes larger than 10 mm. are operated dry, while rolls crushing material finer than the first limiting screen are operated wet. The usual practice for dry-crushing is as follows: The ore is crushed to approximately 1½ or 2 inches in the crushing plant, then conveyed to and stored in the mill-bins. From here, after screening over 10 mm., the oversize is fed to the rolls, the roll-product passing to the elevator raising the original roll-feed. In other words, the elevator, screen, and rolls are in closed circuit, the rolls crushing from 1½ or 2 in. to 10 mm. Graded crushing is not used in the lead belt, although graded roll-crushing is practised at one plant in the Fredericktown district, all roll-practice at that property being wet, due to milling conditions. Graded roll-crushing gives much better operating conditions than were formerly obtained by the use of a closed circuit, principally because the roll-spacing could be better adjusted. Both manganese and rolled-steel roll shells are used.

All crushing and screening is done dry in sizes larger than 10 mm. There are many screening devices used but the trommel is the usual type, and for heavy work it is supported on rollers at the feed end, the discharge end being supplied with a gudgeon. For 10-mm. screening, the usual type of shaft and spider trommel is used. The old type of conical trommel, as well as the hexagonal trommel, is used at one plant, but both of these types, as well as the cylindrical trommels, are being replaced by the flat screens of the Ferraris type. These are to be found at two mills where they have replaced or are replacing the cylindrical trommels. The flat screen has many advantages over a trommel, since the screening efficiency is higher and the wear on the punched plate is much less than that occurring on a trommel, and, for the space occupied, the flat screen has a much larger capacity than a trommel. The flat screen will also handle a greater volume of water than a trommel of the same screening capacity. The impact screen, as manufactured by the Colorado Iron Works Co., is being tested at one mill in the district. Flat stationary screens follow the coarse crushers in one plant, and another is using a rotary grizzly for removing the undersize of 4 in. before feeding the original ore to the primary crusher. As the finest screening in the district is done on the second limiting screen, corresponding to a 2-mm. trommel, it will be seen that the screening problem is not serious. About 80% of the screening is done with the standard trommel, and the remainder mainly on the Ferraris type of flat screens.

JIGGING THE ORE. The ore when crushed through 10 mm. readily yields commercial tailing, as at this size it contains much dolomite that is free from lead. This clean dolomite, however, gradually progresses to the condition of middling where the galena is finely disseminated. When crushed through 10 mm., the first free lead of any great importance occurs on 4.699-mm. screen. The jig-tailing will contain from 0.7 to 0.8% lead.

The Hancock jig is one of the noteworthy appliances in the district, since jigging in the lead belt is now limited entirely to the use of the Hancock jig. This belongs to the type of jig having a movable screen. The screen tray rests in the tank which is made of 4-in. lumber. The tank is 25 ft. long, 4 ft. 2 in. wide, and 10 ft. high. It is divided by five headers into six compartments, the headers being placed at any desired position. The products from the first two compartments are concentrate, the third spigot may yield a concentrate or a product for circulation, depending on operating conditions. The fourth and fifth hutches yield middling for crushing, while the tailing is discharged over the end of the tray into the last compartment. A chain drag of the Esperanza type is usually built into the last compartment, which drags the tailing up over the end of the jigs. The tray, as built, is 20 ft. 6 in. long and 3 ft. 2 in. wide, outside dimensions, built of lumber. One company has replaced the wooden tray by one of steel. A steel tray has many advantages over wood, being lighter, and, due to the increase in effective screen-width, the capacity of the jig is increased about 15%. The feed to the Hancock is generally crushed through 10 mm. on 1½ or 2-mm. screens. Very efficient work is done on this range of sizes.

In the past this district has made use of classifiers for preparing table-feed, the usual grade treated being the undersize of a 1½-mm. screen. Nearly every type of standard classifier has been used, as well as the usual flock of home-made ones. The introduction of the Butchart table has done much to simplify the question of classification, and in practically every plant the classifiers have been either entirely displaced or largely replaced by apparatus for de-sliming only, so that the classification problem is not serious. In the past, use has been made of classifiers for preparing Harz-jig feed, and two plants for a while endeavored to use classifiers for preparing the feed to the Hancock jigs. This arrangement was not noted for its satisfactory results, and was replaced by screens. The present practice may be discussed under three separate headings, namely, classification, de-sliming, and settling. For all practical purposes, we can say that classification, in the usually accepted sense of the term, is not used in the district, preparation of table-feed being done by de-sliming apparatus. Much use is made of de-sliming appliances for preparing the table-feed. The problem is not very serious as, due to the large margin in difference between the specific gravity of the gangue and the galena, extremely efficient work is not of great importance. The de-sliming apparatus should eliminate from the spigot-product the galena too fine to be saved on the tables, and the overflow should contain a minimum of sand and lead that can be efficiently treated on the tables. This size is usually set at 150 mesh. The 2-mm. undersize is the largest now being de-slimed. Work has been carried on with undersize from the 3-mm. trommel with satisfactory results, so far as classification is concerned. De-sliming is accomplished by means of Dorr classifiers, drag classifiers, Akin classifiers, and de-sliming cones. The Dorr and drag classifiers, in some cases, send the sand-product direct to tables. In other cases, the sand is treated first in classifiers, or de-sliming cones, before tabling. At one plant the undersize from the 2-mm. screen is sent to a de-sliming cone, the spigot of which discharges to a belt-drag, which in turn yields a sand-product for tabling. Other plants de-slime the screen undersize in de-sliming cones, the spigot-product going direct to tables. The Delano cone is used at several plants, and is giving excellent results.

The problem of table treatment is simple and, due to the great difference in specific gravity between the galena (7.4) and dolomite (2.85), close classification is not essential. In the past, the usual method employed was to classify the feed and treat it on Wilfley tables, the usual size treated being the undersize from a 1½-mm. trommel, but at present the table-feed is de-slimed and generally treated on Butchart tables. The usual size of feed to the tables is the undersize from 1½ or 2-mm. screens. The table-concentrate will average about 77% lead, while the tailing will contain about 0.3%. Probably the largest size of original feed to be tabled is the undersize from a 3-mm. screen. This yielded good results by tabling, but the wear of the riffles and of the linoleum was so great as apparently to offset the advantages gained by tabling this size. A Butchart table was used for the test, and with 90 tons of feed per table, the tailing assayed between 0.4 and 0.5% lead. The object of endeavoring to table this coarse feed was to remove the fine sizes from the jigs and, furthermore, to remove from the jigs as much of the galena as possible, thus reducing the loss due to abrasion. The excessive wear of the table-tops necessitated replacing the 3-mm. screen with a 2-mm. plate. The number of tables to each Hancock jig varies to a great extent. At one plant there are supplied three tables for each jig, at another plant there are ten tables for each jig. This ratio varies according to screen sizes and methods of middling and grinding. The average ratio is one jig to seven tables.

Re-grinding of middling is receiving a great deal of attention in the district. Two types of middling are being re-crushed, the jig-middling and the table-middling. The problem of crushing the jig-middling will be first considered. In the past, the jig-middling has generally been crushed in rolls, but now one company is using Chilean mills for crushing the fifth hutch-product of the Hancocks, and one company has just discontinued the use of Huntington mills. Rolls are used in all plants for crushing middling, and are exclusively employed for this purpose in five plants. Fine-grinding mills are being introduced at all the plants in order to determine the best mill adapted to this purpose. The Huntington and Chilean mills are not desirable. The result obtained by crushing the middlings in rolls is not satisfactory, and every attempt is being made to replace the rolls and other crushing machines with one of the newer types of fine-grinding mills.

The first company to introduce a ball-mill into the district for the grinding of jig-middling was the Desloge Consolidated Lead Co., which began operating a 6-ft. by 22-in. Hardinge mill in March 1916. Since then many different types of mills have been introduced, and fortunately the choice of the different companies was not the same, so at present the following types of re-grinding mills are operating: Allis-Chalmers, Hardinge, Marathon, and Marcy mills. An Allis-Chalmers ball-granulator is operating in competition with a 6-ft. by 22-in. Hardinge at one plant, and at another plant an 8-ft. by 30-in. Hardinge is running in competition with a 4 by 9-ft. Marathon. The Marcy mill as yet is not directly competing with any other mill, as is true of the 4 by 10-ft. Marathon.

The feed to the fine-grinding mills is generally from the fourth and fifth hutches of the Hancock jigs, but one plant is crushing the jig-middling and the oversize of the 10-mm. screen in the ball-mill. The usual practice, however, is to take as feed to the re-grinding mill, the jig-middling which passes 10 mm. and stays on 2 mm. The fifth hutch of the jig will contain from 1.5 to 2.5% lead, and the fourth hutch may assay from 8 to 20%, and it is generally desirable to crush the middling to pass a 1½ or 2-mm. trommel. The product should be similar, if possible, to that produced by roll-crushing, as the minimum of slime is desired. The feed to the re-grinding mills will assay from 3 to 8% lead, and after crushing through a 1½ or 2-mm. screen, an economic tailing and a clean concentrate can be produced and a middling for re-grinding in a separate circuit. By economic tailing is meant a product that cannot be further treated and yield a product. Test-work has shown that where a jig-middling is crushed through a 0.208-mm. aperture (65 mesh) practically 100% of the lead is free. It is thus evident that extremely fine grinding is not necessary nor desirable. The ideal product would be one that would just pass the limiting screen, 1½ mm. or other size, and have a minimum of material finer than 65 mesh. If a tailing can be made at 1½ mm., it is evident that it is not desirable to produce any larger tonnage below this critical size than is necessary. At the other extreme, no product is desired finer than the lower critical point at which all the lead is freed.

The term 'slime' has considerable latitude in the district and refers to the product passing a limiting screen of some definite size. Formerly 200 mesh was adopted as the common limit, but the introduction of large-tonnage tables has caused this definition to be slightly changed. These tables treat a 'de-slimed' feed, all material finer than 150 mesh preferably having been eliminated. It is thus desirable to overflow material finer than 150 mesh from the de-sliming apparatus, and this overflow is termed slime, since it is all minus 150 mesh. Some companies, however, still retain 200 mesh as the limiting screen, and references in this article will consider 200 mesh (0.047 mm.) as the limiting size. At present slime is treated both by gravity methods and by flotation. It may be said that reciprocating tables are the only means now used for slime-treatment. These tables give excellent results for the treatment of material within their range of efficiency, as a table should make a satisfactory saving of galena in sizes coarser than 300 mesh and also do excellent work on the undersize of 300 mesh. A test on a slime-table treating the overflow of classifiers showed the following results: In the feed to the tables, 94% of the lead was through 200 mesh. The resulting concentrate assayed 76% lead, 89% of the concentrate being through 200 mesh. Of the tailing loss, 98.5% was through 200 mesh and the screen-analysis showed this loss to be with minus 300 mesh. The recovery was 65%. When flotation was introduced, the tendency was to eliminate the slime-tables and to treat the entire slime-tonnage directly by flotation, but the pendulum is now swinging back, which is the logical course. The usual practice now is to table as much of the slime as possible. The slime-tables never make a finished tailing, the material in every case being treated by flotation, but tabling has served to lower greatly the grade of the flotation-feed. Table-treatment serves to yield a concentrate very much cleaner than is possible with flotation, while the smelting charge on the table-concentrate is lower and the tonnage shipped is less than if the slime had been treated directly by flotation. Some of the plants send the overflow of the de-sliming apparatus direct to the flotation-plant, but such a method means a rich feed. Other plants settle the classifier overflow and give it a table-treatment, and by this means at least 65% of the lead is saved in the form of a table-concentrate, assaying fully 75% lead. One plant is even re-treating by further tabling the middling from the slime-tables. Another, by careful slime-table treatment, is sending a feed to the flotation-plant containing but 2.25% lead, which feed, before treatment by slime-tables, contained 6% lead.

The Federal Land Co. did the pioneer work in this district on flotation, and it solved the problem of floating the fine galena after it had been considered impracticable. This company had the first operating flotation-plant, and knowledge there gained was freely given to other operators in the field, so that now flotation-plants are found at every mill in the district. Roughly, the flow-sheet used is as follows: The slime is settled in Dorr tanks, the spigot-product, containing 1 part of solid to 4 of water, is treated in flotation-machines, with crude wood-creosote as the frothing-agent. No acid is used. The concentrate is a finished product, while the tailing is treated on tailing-machines. The introduction of flotation has not greatly altered the metallurgy of the ores. Flotation is not replacing gravity concentration except for the very fine sizes, and economically never can. Fortunately the physical nature of the ore is such that an economic tailing can be made on the jigs and tables, and this fact precludes the possibility of flotation ever encroaching upon the field of gravity concentration. The logical field of flotation here is the treatment of galena finer than 300 mesh, as efficient work can be done with tables on coarser sizes than this. It will thus be seen that the field for flotation is sharply defined, being limited to the treatment of slime-products only, no attempt being made to displace gravity concentration by flotation on the sizes coarser than 200 mesh.

Scheelite is preferred by smelters of tungsten in this country, and therefore it commands a superior price to wolframite and hübnerite. The reverse is the case in England, where the iron tungstate holds a more favorable position in the market.

LEADVILLE, COLORADO

IRON PYRITE PROSPECT.—MANGANESE MINES.

A. S. Harvey of Los Angeles has secured an option to purchase for $50,000 eighteen placer-claims comprising 2100 acres in the Arkansas valley 10 miles south-west from Leadville. He plans to construct a dredging plant on the ground that will cost $50,000. The ground has been held by Lake county for several years for delinquent taxes which total approximately $50,000. Construction on the dredging plant will be undertaken early next spring and will be completed before the end of the year. Examination during the summer resulted in the finding of good ground. Mr. Harvey, who is a former Leadville resident, will have D. A. Cannon and others associated with him in the enterprise. The dredge will be the second to enter Lake county; the boat now operated by the Derry Ranch Gold Dredging Co., a mile north-west from the new site, is the only one in the district.——Because of the huge sum owing to Lake county in back taxes on mining property, the Board of County Commissioners has instituted a new plan that is expected to result in the payment of a large proportion of the delinquent money. The board has decided that offers to purchase property for back taxes, or a part of them, will be considered and accepted under the same ruling now governing real estate. Heretofore no reduction on taxes due on mining claims would be made by the commissioners, and the result is that the county is out several hundred thousand dollars. Under the new plan, it is believed that numerous offers will be made to secure title to old claims for a fair part of the taxes now due on them, thereby enabling the county to realize a portion of the money already owing and at the same time place these redeemed claims in the active assessable property of the county. The new ruling is also expected to cause an increase in development and be of general benefit to the entire district.——Leadville mines containing iron pyrite with a high-sulphur content are soon to have a good market for their ore, according to Gustavus Sessinghaus, a Denver mining engineer. For several years, Mr. Sessinghaus has been engaged in the purchase of ores used in the manufacture of chemicals; and he is an authority on the subject. In the Leadville district there are several large mines in which extensive deposits of pyrite have been opened. These bodies are practically free from any metals that would make them valuable and therefore have remained undeveloped for years. The Ibex on Breece hill, the Adelaide in Adelaide park, the Louisville, Moyers, Tucson, and R. A. M., on Iron hill, the Greenback in Graham park, and several properties developed through the Yak tunnel have large deposits of pyrite uncovered in the lower workings within easy access to development. Mr. Sessinghaus states that a considerable tonnage could be extracted from these properties without delay. A slight reduction in the freight-rate on pyrite between Leadville and Eastern points where the ores are used will help local production, and an effort is now being made to have pyrite classed with low-grade lead and zinc ores. It is stated that this change in rating would be sufficient to make mining of pyrite from Leadville mines profitable.

The production of manganese in the Leadville district is rapidly becoming an important factor. Recently several large bodies of ore averaging from 30 to 40% manganese have been discovered, the two largest being at the Grey Eagle and the Boben properties on Carbonate hill. In the former, two large shoots of the ore have been opened, showing the material to contain 30 to 38% manganese, 1 to 6% silica, and 21% iron. At the Penrose, Northern, Seneca, and Nisi Prius, new bodies of manganese have been discovered, and there are a number of big deposits opened at the Star properties. A slightly better market will cause a heavy tonnage of this ore to be shipped from the district.——The latest strike of importance is that at the Hayden shaft on the May Queen and Hibernian claims on Yankee hill. The May Queen Leasing Co., headed by David Harris and P. C. Madscen, has opened a body of zinc carbonate that ranks with the big discoveries made at the Wolftone several years ago. A shoot 110 ft. long, 29 ft. high, and 20 ft. wide has been opened from the 150-ft. level of the shaft. The ore averages 26% zinc. Development in the shoot still continues and the limits of the ore have not been reached.

M. J. Nicholson, superintendent of the Western Mining Co., operating through the Wolftone shaft on Carbonate hill, states that for several days the property has not been shipping its normal output because of the car-shortage. Large storage-bins have been erected on the surface to hold the ore when it is impossible to secure the necessary cars. In this way the property has maintained normal development. Mr. Nicholson admits that the problem is becoming serious, and that it may be necessary in the near future to reduce operations at the Wolftone, as it has been at other properties.

ELY, NEVADA

A SURVEY OF THE DISTRICT.

The Nevada Consolidated is running at full capacity; it, as well as the railroad, is having trouble to secure sufficient coal for current demands.——The Copper Mines Co. has almost 700 men on its payroll; it is doing a large amount of surface work and erecting buildings and new machinery.——Development work is proceeding at the Alpha from the Giroux shaft. ——The Ward mine, 18 miles south of Ely, is having difficulty in transporting its ore. A large amount of ore is ready for shipment but the company is unable to get it to the railroad, 60 tons per day being all that it is possible to move. A narrow-gauge railroad seems to be the only solution to the difficulty. ——Williams & Thayer shipped a carload of carbonate ore from the Robinson district; it is expected that it will run 40% zinc.——The Hamilton district is making a better output than for many years past; a large amount of ore is sacked and piled up at the various mines, but it is impossible to secure teams or trucks. Truckage to Kimberley, 35 miles away, costs $16 and to Eureka, about the same distance but with less grade, $14.

The old camp of Eureka is showing some life. Julius Huebner, in charge of the California, is adding a compressor and a boiler. Some Salt Lake people leased an undeveloped piece of ground last spring and are making shipments of ore running $150 in lead-silver. They have an excellent showing. ——It is believed that if the railroad facilities were better, the camp might return to its old-time prosperity, but it is impossible to handle any tonnage. Only a few properties are working at or near Austin. Eastern people have an option on the old Austin Consolidated. The mill-records of this mine show a production of almost $20,000,000, and the deepest workings are only 600 ft.——Salt Lake people have taken an option on the old Ophir silver mine in the Toyaba range. There is a vein of good size, and the old dumps indicate that there have been extensive operations in the past.——The old Jefferson mine, six miles up Jefferson canyon, south of Round Mountain, has been purchased for $165,000, it is reported, of which $75,000 has been paid. A new mill has been completed and recently a trial-run was made. Fifty men are employed. ——John Seaver is working his silver property, and occasionally makes shipments of good-grade cobbed silver ore.——Berg Bros. of Round Mountain have madé a discovery recently. A 12-ft. shaft 6 by 8 ft. has been sunk in solid ore that assayed $240 per ton.

PORCUPINE, ONTARIO

HOLLINGER. — MCINTYRE-PORCUPINE. — DOME MINES. — MINING CORPORATION.

The Hollinger Consolidated equipment has a capacity for handling between 2700 and 2800 tons per day as compared to 1700 and 1800 during 1916. During 1916 this company paid $3,126,000 in dividends; the ore-reserve is several million dollars in excess of what it was a year ago, so it is evident that with anything like an adequate supply of labor this greatest of Canadian gold mines will set a still higher standard of achievement.——The tonnage being treated at the McIntyre-Porcupine is greater than ever before, an average of over 500 tons per day is being maintained. The production of bullion is being maintained at the rate of $140,000 to $150,000 per month and the dividend requirements of 5% quarterly, or about $186,000, is being more than earned. The development at the 1000-ft. level is being pushed forward unremittingly and, since the estimate of $5,000,000 in ore-reserve on June 31, a considerable quantity of ore has accumulated.—The decision of the Dome Mines to curtail production during the winter months has caused neither surprise nor alarm. It was made on account of the shortage of labor. In a general way, the mine is in a better condition than ever before. The facilities for handling ore are understood to be far in excess of present mill-capacity. The ore-reserve of nearly $15,000,000 is sufficient evidence that shortage of ore had no bearing on the decision to curtail milling operations.——At the 600-ft. level of the Porcupine V. N. T., excellent results pertain. The 100-ton mill is running at almost capacity, and the improvement in the supply of labor is being felt in a marked degree at this mine.

Mining companies of Cobalt are producing silver bullion at the rate of approximately $60,000 per day, as figured on an average price for silver during September. The weight of the output is about two and one-half tons per day or at the rate of 900 tons of silver annually.——Mining Corporation continues to lead all other mines in point of production, while Nipissing easily holds the lead in point of ore in reserve. Nipissing is in the happy position of having considerable ground as yet unexplored. Mining Corporation is carrying out a policy of acquiring additional property, having taken over the Waldman about two weeks ago and only this week it has purchased the old Alexandra property, situated near the Bailey Cobalt. ——In the lower levels of the Chambers Ferland mine, the results continue to be satisfactory. Some fairly high-grade ore is being found just above the contact, and considerable low-grade ore is being broken.——The sinking of a winze at the 1600-ft. level of the Beaver Consolidated, the deepest working in the Cobalt district, is going ahead with highly encouraging results and, owing to its bearing on other properties, is commanding a great deal of attention.——The profit from the Cobalt mines for 1917 will probably compare favorably with the banner years of the camp. The output for the year will aggregate at least $15,000,000. The annual payroll, due to the high wages and large bonuses being paid the mine-workers, will be not far short of $4,500,000, while the cost of material will be about $3,000,000, thus making in all about $7,500,000 for operating expense and leaving a net profit of about an equal amount. In view of the high price of supplies and cost of labor the showing made is extremely satisfactory.

TONOPAH, NEVADA

TONOPAH EXTENSION.—TONOPAH BELMONT.—WEST END CON.—DESERT QUEEN.

Last week the Tonopah Mining Co. produced 2850 tons of ore. In the Silver Top 123 ft. of development was done, in the Mizpah 21 ft., and in the Sandgrass 111 ft., making a total of 255 ft. On the 240-ft. level of the Silver Top stoping continues on a 3-ft. face on the Valley View vein, while on the 440-ft. level the vein is only 2 ft. wide. On the 540 and 640-ft. levels the Burro vein shows a 3-ft. face, as does the South Dipping vein. On the 1440-ft. level of the Sandgrass the raise in the Upper Sandgrass vein shows about 4 ft. of ore.——The Tonopah Extension Mining Co. shipped 20 bars of bullion valued at $40,000, representing the clean-up for the first half of October. In the No. 2 shaft 95 ft. of development has been done, and 125 ft. in the Victor. In the No. 2 shaft, cross-cut No. 510 on the 1350-ft. level cut the Merger vein, exposing 4 ft. of ore. At the Victor on the 1440-ft. level, the Murray vein is being stoped on an 8-ft. face. Raise No. 1503 on the 1540-ft. level continues in mill-ore. On the 1680-ft. level development in the south cross-cut shows the Merger vein, recently intersected, to have a width of 6 ft. Cross-cutting has been delayed in order to cut a pump station to handle the excess water. The production for the past week was 2380 tons.

The Tonopah Belmont Development Co. shipped 37 bars of bullion from the Belmont mill and 23 bars of bullion from the plant at Millers. The shipment represented the bi-monthly clean-up and was valued at $104,684. On the 700-ft. level, raise No. 5 from No. 760 west cross-cut reached the faulted segment of the South vein, exposing 3 ft. of fair ore. Raise No. 6 on the Mizpah Fault vein has a full face of fair ore. The east drift on the 1166-ft. level on the Mizpah Fault vein is being driven on a 2-ft. face of good ore. The last week's production was 2491 tons.——The West End Consolidated Mining Co. continues to push development on the Ohio vein. Drift No. 531 continues in high-grade ore and drift No. 555 cross-cut No. 1 also maintains its excellent showing. Stoping on the Ohio vein furnishes a large tonnage of good ore. At the West End shaft No. 717 drift No. 2 continues to make good progress and raise No. 677 has reached the foot-wall of the vein. The output the last week was 1118 tons.——In the Desert Queen section of the Jim Butler Tonopah Mining Co. unfavorable mining conditions have caused a discontinuance in the development of the Ophir King on the 900-ft. level. Development has been resumed in an upper level. At the Desert Queen shaft on the No. 625 intermediate, a cross-cut has been started toward the vein on the South fault. The production for the past week was 1050 tons.——The MacNamara Mining Co. has commenced development on the 725-ft. level. Drift No. 18 shows a 2-ft. face of fair ore and the north drift has a full face of fair ore. Stope No. 2 above the 200-ft. level shows 5 ft. of ore. Stoping continues in the 300-ft. level in stopes No. 1 and 4. The production the past week was 406 tons.—— The Cash Boy Consolidated Mining Co. intersected the footwall branch of the vein, which is about 6 ft., on the 1700-ft. level. Driving will be started on the hanging-wall branch of the same vein. The output for the past week was 55 tons.—— Last week the Rescue-Eula produced 128 tons, the Montana 53 tons, and miscellaneous 47 tons, making the week's production at Tonopah 10,578 tons with a gross value of $185,115. ——At the Silvermines Corporation at Hornsilver a station is being cut on the 500-ft. level of the incline shaft to facilitate development of the large orebody exposed at that point. A new ore-shoot has been found to the west of the 400-ft. level station.

MANHATTAN, NEVADA

WHITE CAPS.—UNION AMALGAMATED.—MANHATTAN CONSOLIDATED.

The east orebody of the White Caps has been opened up from both the fourth and fifth levels. On the fourth level the orebody has been proved to have greater dimensions between walls than was known previously, and on the fifth level it is found to be longer than on the one above. As shown on the upper levels the orebody is continuous up to the east fault where it is cut off. This fault will not be reached on the fifth level until 120 ft. has been driven along the course of the ore. So far there has been no development work to demonstrate the width of the east orebody on the fifth level, the drift being run along the foot-wall. During the past week 35 ft. has been added to the east drift on the ore. Already the east orebody has been exposed for a length of 115 ft. and when continued to the fault mentioned it will have a demonstrated length of at least 235 ft. The car samples for the past three days have run $28, $53, and $56 per ton. A sample taken close to the foot-wall showed $80 in gold. On the fourth level the east sill is being put in ready to start stoping operations, by the shrinkage system. The cut being made across the ore for the sill-floor has now been extended 32 ft. and neither wall shows as yet. The foot-wall is known to be back at least six feet beyond the sill-floor and the hanging wall will be at least as far, so it is safe to state that at this point the east orebody is 45 ft. between the walls. The ore exposed in the present work has a higher average value than that shown in the drift originally run along the east orebody. On the third level, the east drift from the foot-wall cross-cut has been extended 11 ft. during the week, making a total of 27 ft. It has now been discontinued, as the stoping operations from the fourth level will make use of the waste rock from this drift for back-filling after connection is made. From the fifth level the west cross-cut has been started to extend to the shaft orebody. The White Caps hoist-house is being connected with the various levels by flash and electric-bell system, and a mine telephone is being added. The bunk-house for the White Caps employees will shortly be completed.

During the week the Manhattan Consolidated has had 45 ft. added to the length of the east drift from the fourth level, the face now being 200 ft. from the shaft. The east orebody should be cut within another 50 ft. From the third level the south cross-cut has been extended 25 ft. for the week, only one shift being employed. A decided change is showing in the face of the east drift from the 400-ft. level. Although too soon to be sure, probably the mud fault has been encountered.

The development for the week to October 24 at the Amalgamated is as follows: from the 600-ft. level the raise on the fissure vein has been advanced 20 ft., making a total of 41 ft.; the shaft cross-cut has been advanced 32 ft. The raise on the fissure has cut across the Mushett ore-shoot for 25 ft. with no hanging wall in sight as yet. Owing to the present programme of the company to stop all development and concentrate the full force on the enlargement of the Earl shaft and connect same with the present bottom level of the mine, the fissure raise will not be advanced any further at present. The shaft cross-cut is now directly under the Earl shaft, but will be extended an additional eight feet before the raise is started, to allow room for an ore-chute. When this is in place a raise the full width of the shaft will be put up to connect with the Earl shaft at the 350-ft. level. Work in straightening and enlarging the Earl shaft is proceeding favorably. The surface equipment at the Earl has been placed in first-class condition and the shaft has already been enlarged to a full two-compartment shaft for a distance of four sets below the surface. Three shifts will be employed in this work and it will be pushed forward with all expedition now that the preliminaries are finished. The Amalgamated mill is employed on ore extracted during the recent development work in the Earl orebodies. The mill is to be dismantled when this has been completed and it will be re-erected on Litigation hill close to the collar of the company's permanent working-shaft.

CHLORIDE, ARIZONA

A GENERAL SUMMARY OF THE DISTRICT.

The greater number of the operating mines in Chloride are situated in the main range, east of town. The ores contain silver, lead, zinc, and copper, together with a little gold. To the west of town, the gold content runs higher, and lead and zinc are less prominent.——The Tennessee mine, which is practically in the town of Chloride, is being worked to a depth of 1600 ft. It has been a producer of lead-zinc ore for many years and at the present time is shipping 150 tons per day to the concentrator at Needles.——The Schuylkill mine, operated by Frank Garbutt of Los Angeles, is being actively developed and has a large amount of high-grade lead ore blocked out.——The Elkhart mine and the Elkhart Extension have been shut-down.——The Chloride Queen group is in operation and showing good ore.——The Hercules property is in active operation with a good crew.——The Schenectady, operated by the Telluride & Chloride Mining & Leasing Co., has developed a large body of rich silver-lead ore.——The Bullion Hill mines are in course of development; an electric plant has been erected.——The New Tennessee is working a full force.——One of the most notable properties of the district is the Copper Age, operated by the Arizona Ore Reduction Co. This is undergoing extensive development and a modern mill is being erected.——The Hidden Treasure, belonging to the Chloride Mining Co., is shut-down, pending re-organization.——The Emerson property, operated by E. M. Bind, is working a small crew of men; it is expected that ore-shipments to the Steffy custom mill will be made shortly.——A 300-ton flotation plant has been erected at the George Washington group. This mine was purchased recently from the Moyle brothers.——A 50-ton flotation plant was constructed last year at the Keystone Con. M. & M. Co., and has recently been re-modeled by Charles Sherman, the superintendent, with satisfactory results. Mr. Sherman is testing this mill upon low-grade ores from the old dump, and is concentrating about 50 tons into one, making a high-grade flotation-concentrate that carries about 5% of copper, together with silver and gold. The operation is interesting from the fact that the raw ore contains only 0.2 to 0.3% of copper which is nearly all saved by flotation.

To the west of the town of Chloride the Puzzle group is in active operation under John J. Robertson, for the New Mohawk Mining Co. The Mohawk Mining Co., an Eastern corporation which lately took over the old Puzzle, Moose, and Columbia groups, has done considerable development at each mine.——The Diana, under management of George F. Beveridge, has developed some high-grade gold and silver ore.——The Jamison, owned by Coeur d'Alene parties, is making good on the east and west drifts of the 300-ft. level; the vein is 6 ft. wide and has a value of $15 per ton.——The New Jersey Mining Co., E. Mingle, manager, is working on a big body of gold ore.——The orebody of the Emerald Isle Copper Co., near the entrance to Mineral Park canyon, is mainly a silicate and carbonate of copper; it is opened to a depth of 25 ft. by pits and shallow shafts, and assays 2.75% of copper. The ore is treated by a leaching process and the copper is recovered electrolytically.——The Union Metals Mining Co., to the south of the Golconda, is developing a lead-silver-gold ore having a value of $50 per ton. A cross-cut to a second level has cut a vein of silver-lead ore. Occasional shipments are being made to Selby. J. P. Ryan, of Chloride, is the superintendent.—— The face of the 600-ft. adit on the Mary Bell mine on Cerbat mountain, owned by J. P. Ryan and others, shows 20 in. of dense sulphide ore. It is proposed to open this mine to an increased depth of 500 ft. by driving from the same lode in the Payroll mine.

MAMMOTH, UTAH

TINTIC SHIPMENTS.—IRON BLOSSOM.—CHIEF CONSOLIDATED.

For the week ended October 19, the shipments of first-class ore from the mines of the Tintic district totalled 203 carloads. These are estimated at 9500 tons, and valued at $250,000. This is compared with 172 carloads the previous week. The temporary lifting of the embargo on ores had the effect of increasing the weekly tonnage by 31 carloads. The camp is now in shape to add largely to these figures if cars can be secured and the smelters will accept all the ore the mines are capable of producing. The principal shippers this week were: Dragon Con. 45 cars, Mammoth 30, Iron Blossom 25, Chief Con. 23, Centennial-Eureka 15, Gemini 14, and Colorado Con. 14 cars. The combined shipments of the other mines was 37 cars, making a total of 203.

Preparations for extensive mining operations are progressing favorably at the property of the Tintic Drain Tunnel Co. on the slope to the south-west of Elberta and it will not be more than a few weeks before the new compressor is running. The machine is being placed on its foundation, and material for the line which will give the Tintic company electric power has been received. The company has six new buildings and a large amount of timber and other material on the ground. Miners are already busy enlarging the face of the adit and placing it in shape for fast work when air is available for the drills. This new corporation was formed by Jesse Knight and associates for the purpose of driving an adit into the south end of the Tintic district to drain the Knight properties and other mines at a depth of more than 2000 ft. This undertaking will require years of labor and the expenditure of a large sum of money, but the matter has been investigated by Mr. Knight and there is no doubt but that it will be carried to completion. The local mining paper states that "The local officials of the Knight mines are unable to account for the recent slump in the price of Iron Blossom stock unless it can be attributed to the lighter shipments, due to the smelter embargo and car shortage, and a corresponding decrease in the mine's earnings." An official report from the office of the Iron Blossom company at Provo states that the treasury contained more than $160,000 when the directors were in session last week and the recent $50,000 dividend was declared. This is not a bad showing when it is understood that every mine in this district has been forced to slow down during the past few months, none being allowed to move its regular amount of ore. Stories that the Iron Blossom would hardly be able to continue the payment of dividends are ridiculous. It is true that the heavy shipments of the past few years have materially decreased the mine's ore-reserve but there is quite a long life ahead of the property even if no new ore is found, and in Tintic the only mines that fail to find new ore are those that limit development work. There has been a slight improvement in the car situation during the past few days, so far as the Knight mines are concerned, and the Dragon is once more sending out a large tonnage, this fluxing ore being in demand at the various Utah smelters. ——At a recent meeting of the directors of the Chief Con. M. Co., a 10c. dividend, or approximately $88,000, was ordered to be paid. This is the regular 5c. dividend and an extra dividend. Books will be closed to the transfer of stock on October 20, and the dividend checks will be in the hands of the shareholders on November 5. This dividend will bring the total up to $747,784.——In 1913 the Mammoth Mining Co. paid under protest a tax of $43,214.58, levied by Juab county on a dump, the material in which was taken out of the mine between 1876 and 1890. The appeal is now before the Supreme Court.

A contract has been let for sinking the No. 1 shaft at the Iron Blossom from the 1000 to the 2100-ft. level. It has been some time since the Iron Blossom company did any development work on the 1900-ft. level which is the lowest point in the No. 1 workings. The vein had been cut there, however, and owing to the fact that the limestone on the 1900-ft. is silicious the management has decided to sink a couple of hundred feet, in the hope that a less silicious limestone will be found. Two men, Dan McKenzie and William Sharp, lost their lives in the No. 1 shaft of the Iron Blossom mine on October 25; their bodies being blown to pieces by the explosion of a round of holes that they had just loaded; they were lighting at the time of the accident. Both men were experienced miners and it is hard to understand why they remained in the shaft until the shots went off. The Iron Blossom shipped 21 cars of ore this week; the shares sold at 65c.——At a stockholders meeting of the Golden Key, held at Provo on October 29, the company's property was sold to Jesse Knight for $1500. It consists of two claims in the Tintic mining district on the course of the Knight Tintic Drain Tunnel, the owners of which will take it over. The Golden Key has ceased to exist; the stockholders will be paid about one cent per share on the capital stock.——A winze is being sunk from one of the drifts in the Zuma mine on the 500-ft. level where the ore had been followed for nearly 100 ft. While this winze has been sunk but a short distance there has been improvement in the showing.——Local people have organized the Eureka Metallurgical Co. and it is understood that they are planning to erect and operate a custom milling plant similar to the Argo, Canyon City, Colorado. R. V. Smith, one of the well-known metallurgists of the State, is working out the process to be used; flotation will be one of the chief factors. Mr. Smith has secured patents on several new features which will be used in the plant. The company at present has a small experimental plant near the Bullion Beck mine and the tests made at this plant have been satisfactory. The company is capitalized for $50,000 divided into shares of the par value of 10c.; the officers and directors are: M. A. McCrystal, president; R. V. Smith, vice-president; Edward Pike, secretary; Thomas McCormick, Jr., treasurer, and A. W. Larson. All of the incorporators are residents of Eureka.

The Chief Con. mine is marketing little or no ore as a result of the embargo and the shortage of railroad cars, but there will be no suspension of development work. For several weeks the Chief has been able to ship a limited amount of lead ore, notwithstanding the fact that the smelters had placed an embargo on the dry ore from this mine, but early this week the United States Smelting company served notice that its plant was overstocked with lead ore and that no more would be accepted. This leaves the mine in a rather bad predicament with a market only for a limited amount of dry ore at the International smelter, at Tooele. In addition to the development work in the main portion of the Chief, this company is spending a lot of money on new work at its Homansville tract at the Plutus. The old Plutus shaft has been re-timbered and equipped with the necessary machinery and the workings on the 400-ft. level are receiving attention. The drifts were in bad shape and it has been necessary to remove a lot of caved ground and put in new timbers. The present plans call for some new work on the 400-ft. level of the Plutus before sinking the shaft below this point. According to N. W. Roberts, superintendent of the Iron King, everything will be in shape for sinking the new shaft within a few days. Men are engaged in enlarging the raise which connects the adit with the surface; this will be the upper part of the company's new working shaft. Sinking will be started from the adit-level below the raise; temporary machinery will be placed in the adit.——The Ohio Copper flotation mill at Lark is handling between 2000 and 2500 tons of ore per day. The concrete foundations for the enlarged plant are all in and the machinery is on the way. The first units of the flotation plant are giving satisfaction. It is expected eventually to have all the ore finally treated by the flotation process, which is said to be giving about 85% recovery as compared to 40 to 50% in the old mill.——The directors of the Mammoth moth on November 2 declared another dividend of 10c. per share, which calls for the sum of $40,000 to be distributed on November 8. The books will close on November 3, and re-open on November 8. This will bring the 1917 distribution up to $1 per share, and the total distribution to $2,820,000. The Mammoth shipped 36 carloads of ore in October, 33 in September, 73 in August, 80 in July, and 399 in the first six months of this year.

HOUGHTON, MICHIGAN

New Cornelia.—Adventure.—Quincy.—Isle Royale.

After an investment of practically $15,000,000 New Cornelia, controlled and managed by the Calumet & Arizona, has become a profitable producer. While it has yet to pay back the money invested, it gives every promise of success, and illustrates the way Calumet & Arizona interests do big things. The production of copper from this property for October will approximate 3,000,000 lb. The output for the past four months follows: June, 1,318,896; July, 2,267,328; August, 2,180,090; September, 1,608,850; making a total of 7,375,164. The falling off in September was due to a combination of circumstances, all of which were overcome before the first of October. Two kinds of ore come from the New Cornelia, one requires to be leached and the other to be smelted. The copper recovered by leaching during the last four months amounted to 6,739,162 lb., while 646,002 lb. was recovered by smelting. The Calumet & Arizona owns 76% of the stock in the New Cornelia. Six weeks ago the Ajo Consolidated was acquired by a stock payment. The outstanding first-mortgage bonds amounting to nearly $4,000,000 are owned, practically all, by the shareholders of the New Cornelia. The management believes that 40,000,000 tons is a conservative estimate for the low-grade ore in the New Cornelia. An estimate is made that 20,000,000 tons were acquired by the absorption of the Ajo Consolidated. Practically all of the New Cornelia is low-grade steam-shovel ore. There have been irregular shipments of 3% high-grade that goes directly to the smelter. The company has created a world-wide market for old tin cans and is advertising for them all over the country, paying $15 per ton for the old tin plate. These cans, together with old iron scrap, are used as a precipitant of copper from part of the solution. Tin cans used in his apparently ludicrous manner secured 160 tons of copper ast month.

The application of the Government's plan for price fixing for munitions of war has operated in the Lake Superior copper istrict with unfortunate results in but one case. The susension of operations at the Adventure property was caused irectly by the fact that with copper at 23½c., it is impossible, 'ith its limited operations, for the mine to work at a profit. he suspension of work at Adventure is unfortunate as the rade of the ore was improving in nearly all directions.——The chest ore-shoot ever opened at the Quincy is on the west ein at the 65th level. Masses and barrel work, some running pieces weighing ten tons and necessitating special apparatus handle them underground and at the shaft, are being found. his comes from a vein 60 ft. wide; it is opened on the strike r nearly 100 ft. and continues in richness, practically all the rock go'ng directly to the smelter.——Nothing more heard of the $54 price that was advertised as the figure the alumet & Hecla intended to pay for the outstanding Isle oyale stock. The impression, created at the time the rumor was given publicity, that it was put out for stock-market purposes, continues to prevail here. In the meantime the general condition of the property indicates that its intrisic worth is greater now than at any time in its history, both as to possible earnings and future possibilities as shown by its newer openings.——Despite passing the dividend Mass Consolidated will increase its production in October and November. Shipments have been increased 150 tons per day during the past few days and now average 700 tons.

TORONTO, ONTARIO

Mineral Separation.—Cobalt Shipments.—New Discoveries.

Complaints are being heard that the Canadian government is slow to investigate the validity of the patents of the Minerals Separation North American Corp., covering the flotation process. This is only to be expected, as the Government is being re-organized by the admission of several Liberals to office, thus forming a Union in place of a Conservative administration and involving a general re-distribution of portfolios. Hon. Martin Burrell, formerly Minister of Agriculture, becomes Secretary of State and Minister of Mines, in place of Hon. Arthur Meighen. As a general election will be held on or about December 19, the Government will be fully occupied with important matters, so it will not be surprising if action in connection with the claims of the Minerals Separation Corporation is delayed considerably. While not abandoning the accusation that the corporation is under German control or influence, the Cobalt mine-owners are basing their case mainly on the ground that the process as covered by the patents is not suited to Cobalt ores and was only rendered available through modifications made by Cobalt metallurgists.

Cobalt ore-shipments for September exceeded all previous records with a total of 4,737,811 lb.; this included 59,100 lb. from the Miller-Lake O'Brien. The heaviest shipper during the month was the Nipissing with 1,830,000 lb. During September the Nipissing mined ore of an estimated value of $349,258 and shipped bullion from its own and custom ores of an estimated net value of $346,948.——The Mining Corporation of Canada, which has been re-organized, so as to make it distinctively Canadian, the largest portion of the stock having formerly been held in England, has recently made remarkable progress. On September 9 it had a balance at the credit of profit and loss of $3,519,768 as compared with $2,447,582 on December 1, 1916, and the cash in the bank amounted to $1,338,117, an increase of $491,750. Two dividends, aggregating $933,778, have been paid during the current year.——The Kerr Lake is being re-organized by the formation of a new company under the laws of Canada, with the same capitalization as the old company. This is done in order to escape double taxation, as at present the company is liable for taxes in the United States as well as in Canada.——At the McKinley-Darragh development work is being done on the 400-ft. level of the Cobalt Lake fault vein. A width of six feet of low-grade ore has been shown along a 40-ft. drift. The La Rose has sunk 160 ft. on the Violet property, with the expectation of cutting a rich vein coming in from the O'Brien, adjoining. At the Beaver the high-grade vein found on the 1600-ft. level is showing improvement. It has been stoped for 100 ft. and a winze is being put down.

The number of laborers available for the Porcupine mines is steadily increasing as the result of the high wages paid and the return of workers from the harvest fields. The Hollinger Consolidated is pushing development on the Millerton property with highly encouraging results, the large orebody, which was low-grade on the surface, becoming high-grade on the 200-ft. level. The Acme is being worked on the 800-ft. level with satisfactory returns. At the Dome the development drift run on the 700-ft. level to cut a new orebody approximately 119 ft. wide, indicated by diamond-drilling, has passed above the deposit. Another drift will be run on the 800-ft. level to strike it.

An important discovery has been made in the Kirkland Lake district by the cutting of a vein 15 ft. wide in the 340-ft. level at the Kirkland Elliott. This mine is situated in the extreme western section of the camp and the find proves the extension of the mineralized area further than was previously known. At the United Kirkland the shaft is down 25 ft. with the vein showing a width of four feet. Operations at the Wright-Hargraves have been retarded by the breaking of the compressor. The new mill of the Lake Shore is expected to be in operation by January. It will have a capacity of 80 tons. The main shaft is down 420 feet.

A gold discovery has been made in Eby township, about three miles west of Swastika. The vein found is reported to be 30 ft. wide, showing considerable visible gold. There are extensive outcrops of quartz in the neighborhood and many prospectors have gone to the new find. Reports of gold discoveries in other localities have obtained currency, but require verification.

ASPEN, COLORADO

DEVELOPMENT OF PARK TUNNEL.

The Park Tunnel, Mining & Milling Co. has been formed to develop an immense area in the Tourtelotte Park mining district. The company was organized on July 26, under the laws of Colorado with a capital stock of 150,000 shares at a par value of $1. The officers of the company are Robert Shaw, president; W. R. Foutz, vice-president; Frank M. Yates, secretary and treasurer; these together with William H. Cornwall and Edward Turner form the board of directors. These men have spent more than a year acquiring through purchase and by lease, 20 full claims in the Park district; and are now in possession of a tract that gives them entire control of this rich district. A number of the properties in the combination, including the Camp Bird, Iowa Chief, Edison, Silver Star, Best Friend, Justice, and Jenny Lind, were at one time among the best mines in the Aspen district and are credited with a combined output of $1,000,000. Prior to the panic of 1893, which closed down nearly all the mines in the Aspen district, there were more than 1000 miners working at the properties in Tourtelotte Park; and scores of four and six-horse teams were engaged hauling the rich silver ore from the mines to the railroad at Aspen. Since then the high cost of transportation for supplies and movement of ore, high smelter charges, and greatly increased cost of production have combined to restrict development in the Park to a few small concerns operating under leases. Now, after 25 years has elapsed since the abandonment of the mines in the Park, several factors important to mining development have placed these old properties on a splendid footing. Great advancement has been made in the scientific treatment of ores, improved machinery has increased efficiency in production, electrical appliances have cheapened the cost of mining, and freight rates and smelter charges have been greatly modified. With silver again commanding a good price, the possibilities of development in this district are exceedingly promising.

The company proposes to develop this large area through an adit with the portal near the head of Keno gulch. Driving 3000 ft. in a south-easterly direction will place the breast directly under the Last Dollar workings and will give the adit a depth of 700 to 800 ft. Cross-cuts are to be driven both north and south from this point to thoroughly explore the ground. From the portal, a tram is to be constructed to the railroad siding near the Newman adit for handling supplies and ore. The Park district was examined in 1897 by J. E. Spurr of the U. S. Geological Survey and his report was published by the Government in 1898. In it he stated that the contact fault in the Last Dollar, Minnie Moore, and Justice is remarkable for the immense amount of low-grade ore formed along it. The men responsible for the re-opening of this large area are confident that their project will prove to be one of the biggest and most important enterprises that has been launched in the Aspen district for many years.

CRIPPLE CREEK, COLORADO

OCTOBER PRODUCTION AND DIVIDENDS.

The production of gold ores by the mines of the Cripple Creek district for the month of October, as compiled from the reports of the mill managers and smelter representatives, totaled 98,297 tons, with an average value of $10.44 per ton and gross bullion value of $1,026,933.70.

The labor shortage both in the district and at Colorado Springs, where the Portland and Golden Cycle mills are situated, reduced the output and treatment. The treatment at the several plants was as follows:

Plant and location	Tons treated	Average value	Gross bullion value
Golden Cycle M. & R. Co., Colorado Springs	26,000	$20.00	$520,000.00
Portland G. M. Co., Colorado Springs	10,582	22.20	234,920.40
Smelters, Denver and Pueblo	2,755	55.00	152,625.00
Portland G. M. Co., Independence mill, Cripple Creek	38,750	2.04	79,050.00
Portland G. M. Co., Victor mill, Cripple Creek district	18,610	2.03	37,778.30
Rex M. & M. Co.'s mill, Ironclad lease, Cripple Creek	1,600	1.60	2,560.00
Total	98,297	$10.44	$1,026,933.70

The October dividends were as follows: Cresson Consolidated Gold M. & M. Co., $122,000; Golden Cycle M. & R. Co., $45,000; Portland Gold M. Co., $90,000; Vindicator Con., $45,000; a total sum of $302,000 in dividends.

On November 10 the Cresson Con. and Golden Cycle companies will pay their regular monthly dividends of 10 and 3c. per share, respectively, amounting to $122,000 and $45,000.—— The Roosevelt tunnel of the Cripple Creek Deep Drainage & Tunnel Co. was advanced 161 ft. during October. The heading is in a dense black syenite and is still in the property of the Rose Nicol Gold M. Co., on the north-west slope of Battle mountain. The flow has fallen to about 5000 gallons.

The Cripple Creek Deep Leasing Co. and F. R. Caley, lessees of the Jerry Johnson mine of the Jerry Johnson Mining Co., produced 8 cars of ore during the month of October. The Deep Leasing company billed out 5 cars and the Caley lease 3 cars. The average value of the first was $20 per ton with Caley shipping ore ranging from $30 to $50 per ton.—— Twelve sets of lessees on the property of the Elkton Con. M. & M. Co. shipped 18 cars or about 500 tons of 1-oz. ore to the mill of the Golden Cycle M. & R. Co., during October.—— Hahnewald and associates have resumed shipments from the El Paso Gold King mine; the first producer in the west end of the district. New orebodies have been opened up by the lessees at the 1000 and 800-ft. levels.——Carnduff and Duncan, lessees on the Dead Pine mine of the Granite Gold Mining Co., mined and shipped 16 cars of ore of $16 to $18 grade during October. The lessees are operating from the 1000-ft. level of the Gold Coin shaft.——The Mt. Rosa Mines Co., recently incorporated, has taken over the holdings of the Rexall Gold Mining Co. which has been adjudged bankrupt. The properties embrace the Mt. Rosa placer on which is situated the town of Victor, portions of the Gold Coin and Spicer veins, the Rosa Lee, Mt. Rosa, and Adams veins. The new company will shortly block out its holdings for leasing purposes.

ALASKA

(Special Correspondence.)—William Martin of the Alaska Free Gold Mining Co. is in town with the semi-monthly clean-up. The mill is running on high-grade ore that has been discovered recently. Mr. Martin expects to keep operating until November 1 when cold weather will necessitate closing down the mill for the winter. The development of the mine will continue.——The necessity for a central plant that will supply power the year round is being felt in the Willow Creek district. Conditions are ideal for the development of hydro-electric power, and during the past season conditions have been investigated by an engineer representing Colorado investors, who, it is said, are considering the proposition.——George Anderson, who during the past season has been exploiting the Little Gem mine in the Willow Creek district, reports the adit in 65 ft. with six inches of specimen ore. He is displaying a gold button worth $17 which is the result of a test run of 100 lb. of ore.——The Cache Creek Dredging Co. has 125 tons of supplies at McDougall, which have been brought there by river steamer, and are awaiting transportation to Cache Creek.——Thomas Riggs, Jr., member of the Alaskan Engineering Commission, and who is in charge of the Fairbanks end of the Government railroad, is at Anchorage conferring with the chairman of the commission.

Anchorage, October 15.

ARIZONA

COCHISE COUNTY

(Special Correspondence.)—J. H. Hunt & Co. have acquired the old tailing-dump of the Grand Central mine, one of the big producers of the early mining days, one mile east of Fairbanks. This firm will re-treat the tailing as soon as the necessary machinery has been erected. In the meantime a spur track is to be built connecting with the Benson-Nogales branch of the Southern Pacific. The concentrate will be shipped to El Paso or Deming, New Mexico.

Bisbee, October 25.

GILA COUNTY

(Special Correspondence.)—P. G. Beckett, for several years manager of the Old Dominion mine, has resigned his position and will take another position with the Phelps-Dodge people. W. C. McBride, who has been assistant general manager for the Detroit Copper Co., will succeed Mr. Beckett.——The Old Dominion Copper Co. has offered a $50 Liberty Bond as a bonus for all employees who remain with them continuously for one year. The bond must be paid for in installments by the employee, and at the end of the year he receives the bond plus the money he has paid and interest.——The New Dominion Mining Co. is shipping several hundred tons of ore per month; considerable new machinery has been added. The present equipment is sufficient to sink 1500 ft. and arrangements are now complete for deepening the 800-ft. shaft. A winze is being sunk in the main sulphide zone, which has been opened on the 800-ft. level. All ore is being shipped to the Old Dominion company.

Globe, October 25.

MOHAVE COUNTY

(Special Correspondence.)—The Dalseil Mining & Milling Co. of New York has purchased the Twins mine in the district north of Kingman. Two officials are on the ground, making arrangements for active work. A 25-hp. hoist has been ordered and other orders for machinery will be placed as development warrants. It is planned to unwater the old 330-ft. shaft, re-timber it, and commence sinking. It is claimed the old workings contain 6000 tons of known ore.——Active mining operations have commenced at the Merrimac. The power-plant and compressor are working well. Sinking at the new shaft has commenced.——As a result of the find of molybdenum ore in the Tennessee and the Silver Union group, active prospecting is being done on the extension of these orebodies in adjacent properties. Some samples have been taken which run as high as 12.77%.——J. D. Wanvig, Jr., general manager for the Union Basin Mining Co. at Golconda, announces that a new mill, to replace the one destroyed by fire, will be constructed as soon as the design and details have been worked out.

Kingman, October 26.

PIMA COUNTY

(Special Correspondence.)—Work has begun on the Yellow group of claims close to the Copper Prince property in the Twin-Buttes district. The lease on this property is held by Ross and Tavote of Tucson.——Cross-cutting on the 300-ft. level has proved the orebody at that depth on the Mile Wide Copper Co. in the Tucson mountains. This orebody has been cross-cut at the 100 and 200-ft. levels, and has opened ore so favorable that the power-plant at the Copper King shaft is to be doubled by the erection of a 100-hp. Diesel engine and a compressor. Another 250-hp. engine has been ordered for the Margareta claim, to sink a shaft and open up the original workings.——The diamond-drill hole at the Little Ajo is down 1500 ft. and is still in rhyolite containing native copper. It is expected that the monzonite will be penetrated soon. Another hole will be started as soon as the first hole is completed.——Diamond-drilling at the Arizona Cornelia Mines Co., formerly known as the Pittsburgh, has been discontinued. Two leases are in operation at the Ajo Cornelia, both of which are shipping ore to the Copper Queen.

Tucson, October 27.

SANTA CRUZ COUNTY

(Special Correspondence.)—The Duquesne Mining & Reduction Co. is shipping about 200 tons of ore per day. This company has sold its steam-power plant to an El Paso wrecking company. Electric power is now being used at the mill.

Nogales, October 26.

YAVAPAI COUNTY

(Special Correspondence.)—Machine-drilling has started at the Jerome Copper. The adit, now being driven, is in 105 ft., the original work being done by hand. It is estimated that the contact is in 800 ft., the present adit giving 600 ft. of backs when complete.——G. Kingdon, manager for the United Verde Extension Mining Co., reports that owing to the large amount of development work that is being done the shipments of ore have not been increased above 325 tons per day. The main haulage-level is in about 2000 ft. The air-shaft connecting this adit is down 200 ft.——The concreting of the Edith shaft is progressing rapidly.——Sinking is now going forward at the shaft of the Dundee Arizona Copper Co.——The south cross-cut on the 700-ft. level of the Verde Combination has penetrated the Yavapai schist, well mineralized with chalcopyrite.——The Mildred Mining Co., in the Weaver district, is awaiting labor troubles to be settled before resuming work. ——The Arizona Del Rio company will have a drill in operation shortly.——A deal has been closed whereby Nevada people will exploit a tract of 640 acres situated two miles to the north of the China Oil company.——Thomas C. Byrd, of Coalinga, and M. L. Bryant, of Taft, have secured a section of land. It is estimated that 120,000 acres in this field has been taken up.

Prescott, October 27.

CALIFORNIA

ELDORADO COUNTY

(Special Correspondence.)—A systematic investigation of the Placerville fields is being made by Robert G. Hart, to search for rare metals. Near El Dorado a large deposit of molybdenite is reported; the investigator is sampling the deposit. Other sections of the county will also be given attention.——Arrangements are being made to erect a stamp-mill on the Montezuma mine at Nashville, 11 miles south-west of Placerville. The shaft is 1000 ft. deep with levels at every 100 ft. below 500 ft. In former years considerable rich ore was mined near surface. W. J. Loring and George Wingfield are said to be interested in the property. J. C. Heald is principal owner. ——The Washington & California Mines Co. is developing a promising group a short distance from the Montezuma. A 5-ft. vein of free-gold ore has been found in the new shaft, and preparations are being made for more comprehensive work.——San Francisco people have taken a bond on the Havilla and are reconstructing the mill.——Seattle capital is financing the rejuvenation of the Martinez property. Large reserves of low-grade ore are exposed.

Placerville, November 5.

CALAVERAS COUNTY

(Special Correspondence.)—The Buffalo Gravel mine, three miles south of Mokelumne Hill, in Chili gulch, is using electricity to run its entire plant. A Fairbanks-Morse 15-hp. hoist lifts the gravel from a 200-ft. vertical shaft into the mill building. The mill is a 3-stamp triple-discharge Merrill, with ½-in. steel-mesh wire screens; from there the ore goes over riffles to a Neil jug, which separates the fine material from the coarse; the fine going to a No. 2 Deister concentrating table, which separates the precious metals and black sand from the gravel. The black sand assays $1200 per ton, being rich in platinum, osmium, and others of the rare metals. The gravel averages $2.20 per ton, and its average thickness is six feet; 4200 ft. of drifts and cross-cuts has been driven. At the bottom of the shaft is a Dow electric pump, driven by a 10-hp. motor. Ed. Westbrook, Jr., late of Alaska, is superintendent, and has designed a novel scheme of getting a large body of gravel with the minimum of powder. The bedrock is a soft clay-slate and he proposes to undercut it with a fine jet of water flowing through a nozzle under a 455-ft. head and then break down the gravel from above.

The Fischer quartz mine, five miles east of Mokelumne Hill, has been acquired by San Francisco capitalists, who have assembled material on the ground to commence operations. There is a 110-ft. shaft on the property, and a 60-ft. drift on a pay-shoot, averaging 4 ft. wide and assaying $9.20 in gold. The building of the head-frame for the shaft is to start on November 1. At present there are four men employed, but three shifts will be put on as soon as the work warrants it.

Mokelumne Hill, October 27.

NEVADA COUNTY

(Special Correspondence.)—The old Washington district, above Nevada City, is producing considerable gold and chrome, and has started to yield asbestos. The Washington Asbestos Co., controlled by Oakland capital, has placed 10 stamps of the old 20-stamp mill at the Fairview mine in operation and is shipping high-quality asbestos to Oakland. The mill has been equipped with special separating machinery, making two products. Shortage of water is hampering work, but as soon as the fall rains set in the mill will be operated steadily. John D. Hoff is president and general manager.——At the Columbia Consolidated group 20 stamps are dropping on ore from the Columbia and Ocean Star mines. The product averages over $10 per ton in gold. At the Ocean Star shaft much new work has been begun. A hoist and other equipment is being erected at the Snow Point, four miles above Washington, and the sinking of a shaft will start shortly. Chrome and gold are being extracted from the Red Ledge. Several chrome deposits are receiving attention and small shipments are being made.——More machine-drills have been placed in commission in the Empire mine and development of new territory is proceeding below the 4400-ft. level. The 60-stamp mill is running steadily, crushing good ore from the Empire and Pennsylvania properties. The monthly gold output is reported to approximate $100,000.——Suit for $20,000 against the South Yuba Mining & Development Co. has been filed in the Superior Court by Luella B. Kyle, on grounds that the breaking of defendant's Diamond Creek flume seriously damaged the Eagle and other claims owned by plaintiff. The properties are near Washington.——Mining is quiet around Nevada City, due partly to high cost of materials and labor. The North Star Mines Co. continues to operate the Champion group, and several small properties are working along narrow lines.

Nevada City, November 5.

SHASTA COUNTY

(Special Correspondence.)—The flotation plant of the Afterthought mine at Ingot is showing good results. In a recent trial 300 tons of ore produced 144 tons of concentrate that assayed 44% zinc and 8% copper. The result exceeds expectations.——A new five-stamp mill on the Sybil mine four miles west of French Gulch is under construction by the Shasta Hills Mining Co. All the machinery is on the ground.——The Noble Electric Steel Co. at Heroult is operating two furnaces on ferro-manganese ore. The output of ferro-manganese is from 10 to 14 tons per day. The alloy is worth $325 per ton in the markets. G. F. Coleman is the new superintendent. A force of 125 men is employed.——Shipments of ore from the Iron Mountain and Hornet mines, owned by the Mountain Copper Co., are being made at the rate of 1000 tons per day. The company has a full crew; the strike having been overcome.

Redding, November 1.

SIERRA COUNTY

A verdict for $100,000 has been rendered by a jury in United States Judge Frank H. Rudkin's court against the Twenty-one Mining Co. in favor of the Sixteen-to-One Mining Co., which sued the former for $280,000 for gold alleged to have been taken from its vein. The companies operate adjacent properties in this county. In a counter suit, the Twenty-one company asked for $155,000, declaring its part of the vein had been entered. The trial occupied three weeks. Bert Schlesinger, attorney for the Twenty-one company, obtained a thirty days' stay of judgment, announcing that the verdict would be appealed.

TRINITY COUNTY

(Special Correspondence.)—The new dredge under construction at Trinity Center for the Estabrook Gold Dredging Co. will be by far the largest one in the county. The hull will be 165 by 70 ft. and the buckets will have a capacity of 20 cu. ft. The company's saw-mill is ripping out lumber for the hull. J. G. Souther is the new superintendent. He superseded G. B. Hodge, who was called by the draft.

Carrville, October 25.

COLORADO

TELLER COUNTY

(Special Correspondence.)—The following dividends have een paid by Cripple Creek mining companies: Cresson Consolidated Mining & Milling Co., 10c. per share, $122,000, monthly lividend, paid October 10. Golden Cycle Mining & Reduction o., 3c. per share, $45,000, monthly dividend, paid October 0. Portland Gold Mining Co., 3c. per share, $90,000, quarerly dividend, paid October 25. Vindicator Consolidated Gold Iining Co., 3c. per share, $45,000, quarterly dividend, paid ctober 20.

The Millasier Mining Co., owning and operating the Clyde, n Battle mountain, has reached a depth of 1125 ft. in the main haft and will sink to the 1300-ft. level.——The Modoc Consolidated Gold Mines Co., operating the Modoc mine adjoining he Clyde property on Battle mountain, commenced shipping his week. Lessees are also making a slight production from he old incline shaft on the Ocean View, from the Bull Hill nd of the claim.——A ground tramway has been constructed

NEW FLOTATION PLANT AT AFTERTHOUGHT MINE, SHASTA COUNTY, CALIFORNIA

the Eclipse mine, of the Queen Gold Mining Co., and ore again moving from the property. The company is controlled J. T. Milliken of St. Louis, and is a close corporation.——ere is a shortage of men in the district.

The net earnings of the Vindicator Con. G. M. Co. for the ird quarter of the year totaled $49,000, according to the atement of G. S. Wood, the president, in a circular letter dressed to stockholders and mailed with the checks for the tober dividend. The reduction in earnings, as compared th the first two quarters of the year, was due to the inased cost of supplies and of labor in the district. Leasing erations, too, from which the company has derived a subntial income, have been largely curtailed for the same son.——A drift has been started from the Roosevelt Tunnel el, to prospect the Funeral or Anna Lee dike, and is being ried north-west through the Big Banta claim of the United ld Mines Co., toward the main shaft of the Cresson Condated G. M. & M. Co. This dike has produced heavily in Portland estate to the south-east. The heading of the adit still in the Rose claim of the Rose Nicol G. M. Co. on the th-east slope of Battle mountain, and is in hard syenite.—— Modoc Consolidated Mines Co. has been loading ore of to $30 grade during the present week, from the old inclined ft workings. The ore was taken out in course of development at the 11th and 15th levels. The new vertical Frankenberg shaft will soon be in commission, when the production from the property will be materially increased; leasing operations will be carried on through the old inclined shaft.——Lessees of the Elkton Consolidated Mining & Milling Co., John Barnard and associates, have opened up rich ore on the Henley vein, at the third and fourth levels of the main Elkton shaft. A streak about four inches wide assays as high as 25 oz. gold, but the lessees are working about 3-oz. ore.——The output from the Elkton mine for October will total nearly 15 cars, produced by lessees who are also shipping from the Tornado mine on Raven hill.——The Camp Bird Mining, Leasing & Power Co., operating the Rose Nicol mine under a six years' lease, has ceased operations at the 1000-ft. level of the Portland No. 2 shaft, and is confining work to sinking the main Rose Nicol shaft.——The Unity Mining & Leasing Co. is sinking a new shaft on the south end or Raven Hill half of the Nightingale claim; it has attained a depth of 450 ft. On November 1 the manager, H. M. Hardy, will increase the working force to two shifts and will sink to a depth of 800 ft. The Nightingale adjoins the Maggie mine of the Cresson Consolidated G. M. & M. Co., and the Unity shaft is but a short distance from the Cresson main shaft.——The Excelsior M. D. & E. Co. has sunk the Longfellow shaft to the 575-ft. level. This company, previously operating through the Golden Cycle shaft, has ore opened up on two levels that are reached from the 500 and 600-ft. levels of the Longfellow shaft. The property is owned by the Stratton estate.——The annual stockholders' meeting of the Catherine Mining Co. for the election of a board of five directors will be held at Victor on November 13. The company has a lease and bond on the properties of the Last Dollar Gold Mining Co. on Bull hill.

Cripple Creek, October 25.

IDAHO

SHOSHONE COUNTY

The interest of the Bunker Hill & Sullivan in the Star-Morning case was frankly disclosed by Stanly A. Easton, the manager, during cross-examination on November 2. Mr. Easton on direct examination told of his investigation of the properties involved, and expressed the opinion that the apex is along the lode crossing the Morning and Evening Star claims. On cross-examination he was asked if the Bunker Hill & Sullivan was not interested in the outcome of the case at bar. Frankly stating that it was, Mr. Easton explained that the Federal Mining & Smelting Co. was a subsidiary of the American Smelting & Refining Co., which sought to control the lead ores of the district; that if the Federal company was permitted to mine the ores from the Star property, that tonnage would be diverted to the A. S. & R. Co., which was even now attempting through action at bar to control the Bunker Hill & Sullivan tonnage, and to prevent treatment of that tonnage in the company's own smelter.

MONTANA

LEWIS AND CLARK COUNTY

(Special Correspondence.) — Marysville and Bald Butte, which are practically one district, are both looking well. New ground is being developed at the old Bald Butte mine, west of the main range from Marysville. The old mine was worked to a depth of about 600 ft. when the ore appeared to run out and work ceased. Since then good ore has been found in a parallel vein, 700 ft. south. To open this vein at depth a cross-cut is being run from the old shaft at 400 ft. The cross-cut has been driven 500 ft. through extremely hard rock by hand-drilling; this work is to be concluded with air-drills. A compressor driven by electric power has been erected. A shaft is being sunk on another vein, two feet wide, south-west from the main shaft.——The St. Louis is milling ore from old stopes in the Drumlummon mine.——The Barnes-King is working the Shannon orebodies at the head of Rawhide gulch.——A new 300-ton electrically-driven ball-mill has been erected at the Porphyry Dike, replacing a 20-stamp mill.——The Helena, Cruse Con., and Rock Rose are shipping galena ore from Grass Valley.

Helena, October 28.

SILVER BOW COUNTY

(Special Correspondence.)—It was announced yesterday in Boston that the Goldsmith mining claim, one of the early silver producers in the Butte district, was purchased by the Crystal Copper Co.; the deal having been consummated on October 23. The Goldsmith claim is in the north-western part of the Butte district; it consists of 17 acres, and is credited with a production of $1,500,000 in silver from ore extracted above the 367-ft. level. The silver occurs frequently in native form and some of the finest specimens in the Butte district have come from this property.——The Crystal Copper Co. backed by Boston people, is operating in the Cataract district, 11 miles from Basin. Regular shipments of copper are being made to the Washoe smelter at Anaconda. Walter Harvey Weed is the managing director of the campany and associated with him as resident director is Paul Gow, a prominent mining man of this district. The consideration involved was not made public, but it is rumored that it is in the neighborhood of $250,000. The Goldsmith property was formerly owned by C. W. Ellingwood, a prominent pioneer resident of Butte.——According to a report issuing from the local office of the Tuolumne Copper Co., a big strike was made at the Main Range property. A drift that is being driven easterly on the 700-ft. level of the Spread Delight fissure has cut a zone of high-grade copper ore; a breast 5 ft. wide runs 12.5% copper and 26.75 oz. silver. This ore was found at a distance of approximately 260 ft. east of the main cross-cut on that level. As this drift progresses it gains depth by reason of penetrating a high hill, and a maximum depth of about 1200 ft. will be attained. The Tuolumne has rights on the strike of this fissure for a distance of 1100 ft. beyond the point of the present breast. The driving of a raise from the 700-ft. level of the Main Range mine to the 500-ft. level is proving the continuation of the orebody and a considerable tonnage of ore is being blocked out. A feature of the ore development at this point is the uncovering of native-silver ore which is being found in fairly large quantities.

The erection of an air-compressing and hoisting plant, capable of operating to a depth of more than 2000 ft. and of hoisting 1000 tons per day, is under way at the Sinbad shaft, the principal working shaft of the Main Range property. This is being enlarged to three compartments and will be sunk from the 700-ft. level to the 1100-ft. level.——Sinking is under way at the Colusa-Leonard shaft of the Tuolumne company, which is 600 ft. west of the Sinbad shaft. These two mines are being connected with cross-cuts for ventilation, safety, and economy.

Butte, October 26.

NEVADA

ESMERALDA COUNTY

(Special Correspondence.)—Ore averaging $16 per ton in gold and copper has been opened in the alaskite belt by the Jumbo Extension Co. This is the first time ore of a profitable grade has been found below the shale in Goldfield. The vein is a fissure and has been followed through shale from the overlying latite. It was intersected from a 75-ft. winze sunk from the 1017-ft. level and has an incline of about 30°. A drift will be extended from the 100-ft. level of the winze to explore the vein at that point. J. K. Turner is superintendent.——At a point 200 ft. south-east of the 500-ft. level of the Silver Pick mine, ore assaying $20 per ton in gold has been found. A strong vein is indicated and the management is centring work at this point.——Ore conditions are improving in the Spearhead and Kewanas mines; a small shipment will be made from the former shortly.——The winze from the 1750-ft. level of the Atlanta has passed the 1900-ft. point and a cross-cut has been started to seek the vein in the alaskite. As soon as the 2000-ft. level is reached another cross-cut will be started. The manager states the best ore yet found in the Atlanta is exposed on the 1750-ft. level, with indications favorable for persistence of the vein into the underlying alaskite.——The Goldfield Consolidated is operating its cyanide plant on oxidized ore from the upper workings and on tailing; it is treating 500 tons per day. The flotation unit is treating copper-gold ore from the deep workings of the Grizzly Bear and Laguna mines, but difficulty in disposing of the flotation concentrate to custom smelters continues to hamper production of this class of material.——It is reported that good ore is being developed in the Surcease mine near Oroville, California, which the Consolidated holds under bond and option.——The new mill of the Silvermines Co. at Hornsilver is ready for service and will start with a capacity of 150 tons per day. At a depth of 510 ft. the shaft entered sulphide ore, and a station is being cut on the 500-ft. level. Water is delivered to the mill through a pipe-line 8½ miles long. S. H. Brady is manager.

Considerable activity is reported from Talicha, the new gold camp near the Nye county line.——Work is being pushed at the Life Preserver, Periscope, and other properties, and high-grade milling-ore is exposed. Recent reports of bonanza discoveries are denied by operators, who state the ore is of good milling character but not particularly rich. At the Life Preserver group, under bond to George Wingfield and associates, four shafts are being sunk on the main vein.

Goldfield, November 3.

MINERAL COUNTY

(Special Correspondence.)—An interesting development has taken place at the St. Patrick mine of the Kirchen Mines Co. at Luning which shows the intense action of the copper-bearing solutions that formed the orebodies in that mineral belt. Before starting a stope on high-grade ore, recently cut in the main working level, a cross-cut was run into the foot-wall and this shows the monzonite wall-rock to be heavily impregnated with copper, assaying 2 to 4%.——The first carload of ore to be shipped under the new contract by the Pilot Copper Co. left Luning on October 29. The Pilot company has secured one of the most favorable contracts that the smelters have made in recent years; enabling it to ship 100 tons per day of a much lower grade of ore than had previously been shipped and yet make a good profit.——A rumor is current that a large adjoining acreage, with a fine ore showing, will be added to the Wedge Copper Co. The plan also includes the building of a large reduction plant at the mine.

Luning, November 1.

WASHOE COUNTY

(Special Correspondence.)—Though the ore that previously has been mined from the No. 2 vein at the 300-ft. level of the Nixon Nevada has been unusually rich, assaying 20 to 40% copper, with a substantial gold and silver content, there has been a decided improvement during the week and much that is now coming to the surface will assay 75% copper. This brings the first-class ore now in the bin to an average of 50% copper. Two shifts are at work extending the north drift on the No. 1 vein, from which the bulk of the production of the

Nixon Nevada mine up to this time has come. The first ore-shoot found in this drift was 200 ft. long and averaged 28.47% copper and $19.26 in gold and silver per ton. Two shifts are extending the cross-cut which will tap the No. 4 vein, the biggest of the seven, at a depth of 800 ft. on its dip.

Reno, November 1.

NEW MEXICO

SOCORRO COUNTY

(Special Correspondence.)—Friday October 26 a fire started about 8 p.m. in a newly-built oil-burning furnace in the blacksmith shop of the Socorro M. & M. Co.'s plant. This quickly spread to the head-frame and crusher-plant and through the conveyor runway to the upper end of the mill, the upper section of which was destroyed as well as store-houses, machine-shops, assay laboratory, custom ore-bins, and tramway terminal. The fire was controlled, and that section of the mill below and including the Pachuca tanks was saved. This included about $30,000 in gold and silver. The main shaft burned down but a few sets as the draft was up-cast. All miners escaped without accident through the Little Charlie adit. The loss amounted to between $200,000 and $300,000; it was covered by insurance. This month was to have been a record breaker in the tonnage output and the recently favorable development of this property will likely hasten the re-building of the plant. William Childs, the president, and Mr. Cox are expected this week.

Mogollon, October 30.

OREGON

JACKSON COUNTY

(Special Correspondence.)—The 1000-bbl. cement factory of the Portland Beaver Cement Co., on the outskirts of Gold Hill, began operations last week with a crew of 85 men. This factory was recently completed at a cost of $700,000 furnished by local and Portland investors.——The State lime board has taken over the J. H. Beeman limestone deposits lying on the opposite side of Rogue river from the cement factory, and will equip the quarry for furnishing the Oregon farmers with limestone fertilizer at cost in excess of $20,000 at once. This property is 1½ miles from Gold Hill, and the output will be delivered by aerial tramway to the Southern Pacific railway. ——M. G. Womack, L. R. Bigham, and George Thrasher, of Medford, are developing a promising quartz gold deposit three miles south of Jacksonville in the Poormans creek, which is a rich placer district. The vein is in a greestone-porphyry contact, and 10 ft. wide.

The Manganese Metal Co., of Tacoma, has purchased the J. H. Tyrell ranch in the Lake Creek district for $18,000 cash. This ranch contains the most extensive and promising manganese deposits in this new mining district, 20 miles east of Gold Hill. The new owners have taken options on two adjoining ranches. Three carloads of mining machinery are on the way from Tacoma to equip the mines. Charles W. Scott, representing the company, is at Lake Creek, and is hiring men and teams preparatory to moving the machinery from Eagle Point. ——W. F. Sears of Gold Hill, who recently purchased the Larsen ranch on Kanes creek three miles south of Gold Hill, is preparing to re-open an old gold-quartz vein on the property, and equip it with a small stamp-mill and pumping plant.

Gold Hill, October 29.

JOSEPHINE COUNTY

(Special Correspondence.)—The Kerby Queen copper mine in Hanfort gulch five miles north-east of Waldo is to be re-opened at once. The property is owned by D. W. Collard of Kerby and is developed by a 900-ft. adit on the lower level and a 300-ft. on the upper level. There are many thousand tons of 4% copper ore on the dumps. The high-grade ore being shipped averages $6 per ton gold.——The Collard, Moore & Collard chrome mine, near Kerby, will begin to concentrate ore next week. The concentrator was erected last spring. Over 2500 tons of chrome ore has been shipped from this mine and there are many thousand tons of high-grade ore in sight.

A two-stamp mill has been erected at the Abbott and Williams gold mine on the Illinois river near Selma. The ore runs $17 gold.——I. L. Thompson of Seattle has made the second payment on the Osgood placer mine, one mile south of Waldo. Preparations are being made for extensive development work. The bedrock in this mine carries platinum.——The Del Norte Claimholders Association has commenced to grade eight miles of wagon road between Waldo and its copper claims in the Preston Peak district. Supplies have arrived and work is being directed by J. T. Gilmore and A. C. Hoffman.

Grants Pass, November 1.

UTAH

TOOELE COUNTY

Another carload of copper ore has been shipped from the Frankie mine by the Woodman Mining Co. The main adit in this mine, which is about 2½ miles south from Gold Hill, on the Western Utah Copper, is in 200 ft. This gives a vertical depth of about 120 ft. The portal is in granite, but the adit has been in low-grade ore for 72 ft. The ore occurs on the contact between the granite and limestone. A fair sample of the entire body shows that it carries 2% copper, $1.20 gold, and 2 oz. silver. On one wall is a streak of three or four feet of richer ore. This samples up to 15 and 20% copper, $1.50 to $2 in gold, and 2 oz. silver. At this point a winze has been started.

PIUTE COUNTY

Howard S. Chappell, president of the Mineral Products company, has returned to Salt Lake from the East and announces that the Marysvale potash plant, which burned to the ground on October 25, will be immediately re-built. Already he has placed orders with Eastern firms for approximately $150,000 worth of machinery and is now negotiating for an ample supply of lumber which will be placed on the ground at once and work will be started without unnecessary delay.

Mr. Chappell states that after the mysterious fires of last spring it was extremely difficult to secure insurance on the plant, but that they finally succeeded and the recent loss is fully covered.

SALT LAKE COUNTY

Officials of the Boston Development Co., operating the well-known Maxfield mine in Big Cottonwood canyon, are highly optimistic concerning the future of that property. Their enthusiasm is based on recent developments which have disclosed, at a distance of 1100 ft. in the adit, well-defined streaks of high-grade ore, some of which assays 62% lead, 5 oz. silver, $22 in copper, and $6 in gold, while the average of the ore found has an estimated value, it is said, of $165 per ton in these metals. The fissure on which the miners are now working is said to be highly mineralized and it is thought by those conversant with geological conditions in that section to be leading to the bedding from which thousands of dollars worth of high-grade mineral was taken not many years ago.

A. Boulais, the superintendent, was in from the mine a day or two ago and brought with him a magnificent sample of the ore from which much is expected in the near future. A force of nine men is busily engaged pushing development and if expectations are realized shipments from the Maxfield will be in order at no distant date.

WASHINGTON

FERRY COUNTY

Renewal of activities in Republic camp has developed a shortage of help. About 100 men are now employed in the mines and the services of many more could be used. About 50 are employed on the Lone Pine, owned by the Northport Smelting & Refining Co., 18 on the Last Chance of the Lone Pine Consolidated Mining Co., 18 on the Knob Hill, 5 on the Quilp dump, and several on the American Flag at Sheridan.

SPOKANE COUNTY

A shipment of 5720 lb. of tungsten ore, said to be worth $6000 and the largest shipment ever made from Spokane, was consigned to New York a few days ago. It was composed of several small lots, assembled from different sources.

STEVENS COUNTY

Computations of the quantity of magnesite in the deposits of this county have been made recently by the U. S. Geological Survey. Single deposits are estimated to contain over 1,000,000 tons within 100 ft. from the surface, while the deposits of the county are estimated at over 7,000,000 tons. Four companies are operating in the county, namely the Northwest Magnesite Co., the American Mineral Product Co., the U. S. Magnesite Co., and the Valley Magnesite Co. The combined output of these companies amounts to about 700 tons per day. Only a small part of this is calcined material.

CANADA

BRITISH COLUMBIA

Urgent appeals are being received by the Hon. W. Sloan, Minister of Mines for British Columbia, asking him to intervene in the silver-lead situation created by the decision of the Trail smelter not to handle ores containing more than 4% zinc. Unless immediate relief is obtained, many properties will be obliged to close down. In a telegram to the Provincial Premier, who is now at Ottawa, the Minister says that practically all the silver-lead mines of the Province, with the exception of two or three, will be closed down. He proceeds: "It will probably relieve the situation if the order in council, prohibiting exports of nickel, zinc, and lead, if lifted, so far as the United States is concerned. This matter is most serious and disastrous to our mining industry. Would suggest that you take it up at Ottawa, with a view to getting a definite understanding as to munitions orders and also as to lifting the embargo on exports to the States. The United States should now be in a position to handle our surplus lead and zinc. I cannot too strongly emphasize the alarm of our mine operators; and I am receiving urgent wires from all over the interior, demanding immediate action."

A 'get-together' week will be held by the Vancouver Chamber of Mines from October 29 to November 2. There will be demonstrations and lectures, and the whole affair is being arranged so that it will be of general interest to the public.

Alexander Hardie who died at Sanova recently, is one of the last of the early pioneers of British Columbia. Born in Scotland in 1831 he left for the Californian goldfields in 1854. A few years later he came to the placer-fields of the Fraser river and finally settled in Camerontown in the Cariboo. There he located the Caledonian claim, which has yielded nearly a quarter of a million dollars.

Heavy purchases of mining properties in the Camp McKinney and Fairview districts of British Columbia have been completed by the Consolidated Mining & Smelting Co. of Canada, and preparations are being made for production. About 2000 acres in each camp is involved but the price is withheld. Camp McKinney and Fairview, situated in the neighborhood of Keremeos, B. C., and Molson, Wash., are valuable to the great operations of the Consolidated company because of the silicious-fluxing ores they contain. These will become a large asset in the operation of the smelter of the company at Trail. Included in the acquisitions is the old Cariboo mine in Camp McKinney, famous for its yield of gold. Both camps, about 20 miles apart, are traversed by veins of quartz carrying gold. ——Another strike has taken place at Coal Creek, and nearly 900 men are idle. The dispute seems to centre around No. 1 East mine where the men demand an investigation of the yardage on No. 1 level and that an award be made by the Director of Coal Operations.

PERSONAL

Note: The Editor invites members of the profession to send particulars of their work and appointments. This information is interesting to our readers.

W. B. Dennis is in New York.

F. W. Bradley is at Portland.

Hoyt S. Gale is at Los Angeles.

J. C. Hopper is at Silverton, Colorado.

Horace V. Winchell is now at Butte.

G. M. Esterly, of Waldo, Oregon, is in Alaska.

T. J. Jones has returned to London from Russia.

Malcolm Maclaren has returned to London from Peru.

Corey C. Brayton has returned to San Francisco from Salt Lake City.

H. B. Bateman, now Captain in the British Field Artillery, is in East Africa.

E. S. Bastin is lecturer on geology at the Massachusetts Institute of Technology.

Frank Lawrance has left Shasta county and is now residing in San Francisco.

G. M. Clark has been elected president of the South African Institution of Engineers.

E. E. Hardach, who returned from Johannesburg last June, is residing in San Francisco.

W. B. Phelps is superintendent of mines for the Tom Reed company at Oatman, Arizona.

John Russell is now assistant superintendent for the A. S. & R. Co. at Monclova, Mexico.

W. J. Elmendorf has returned from a three-months trip in the Copper River district, Alaska.

Rodolphe L. Agassiz, president of the Calumet & Hecla, has returned to Boston from Michigan.

R. B. Watson, general manager for the Nipissing Mining Co. of Cobalt, is at Phoenix, Arizona.

Byron M. Johnson has been appointed engineer to the A. S. & R. Co.'s mines at Matehuala, Mexico.

T. P. Holt has returned to Tintic, Utah, from Cobalt, Canada, where he is erecting some roasters.

W. G. Anderson has resigned as manager for the Ore Chimney Mining Co. at Northbrook, Ontario.

C. E. Stuart, formerly in Mexico, is Captain in the Engineer Officers Reserve Corps at Fort Leavenworth, Kansas.

Donald G. Campbell, of Campbell, Wells & Elmendorf, is examining mines on Vancouver island, British Columbia.

B. C. Austin has resigned as manager for the Mountain King Mining Co. and will open an office as consulting engineer in San Francisco.

Bruce Marquand, formerly mill-foreman for the Broadwater Mills Co. at Park City, Utah, has taken a similar position with the U. S. S. R. & M. Co., at Silverton, Colorado.

Horace V. Winchell, W. H. Wiley, C. K. Leith, and Oscar H. Hershey have been testifying, as experts, in the Bunker Hill & Sullivan v. Star-Morning case, at Wallace, Idaho.

H. DeWitt Smith has resigned as assistant manager at Kennecott for the Kennecott Copper Corporation to accept a position as superintendent of the mining department of the United Verde Copper Co., at Jerome, Arizona.

Junius F. Cook, until recently consulting mechanical engineer to S. Neumann & Co., the Consolidated Mines Selection Co., the Brakpan and Springs mines, passed through San Francisco last week on his return from the Transvaal to New York, where he expects to reside.

The Sulphur Committee of the War Industries Board was in Texas this week. The committee consists of J. Parke Channing, J. W. Malcolmson, A. B. W. Hodges, P. S. Smith of the U. S. G. S., and W. O. Hotchkiss of the University of Wisconsin.

METAL PRICES

San Francisco, November 6

Aluminum-dust (100-lb. lots), per pound	$1.00
Aluminum-dust (ton lots), per pound	$0.95
Antimony, cents per pound	16.00
Antimony (wholesale), cents per pound	14.75
Electrolytic copper, cents per pound	23.50
Pig-lead, cents per pound	6.25—7.25
Platinum, soft and hard metal, respectively, per ounce	$105—111
Quicksilver, per flask of 75 lb	100
Spelter, cents per pound	9.50
Tin, cents per pound	62
Zinc-dust, cents per pound	20

ORE PRICES

San Francisco, November 6

Antimony, 45% metal, per unit	$1.10
Chrome, 34 to 40%, free SiO_2, limit 8%, f.o.b. California, per unit, according to grade	$0.50— 0.60
Chrome, 40% and over	$0.60— 0.75
Magnesite, crude, per ton	$8.00—10.00

Manganese: The Eastern manganese market continues fairly strong with $1 per unit Mn quoted on the basis of 48% material.

Tungsten, 60% WO_3, per unit	26.00
Molybdenite, per unit MoS_2	$40.00—45.00

EASTERN METAL MARKET

(By wire from New York)

November 6.—Copper is unchanged and nominal at 23.50c. all week. Lead is strong and higher at 5.75 to 6.25c. Zinc prices are quoted at 8 to 6.87c. Platinum remains unchanged at $105 for soft metal and $111 for hard.

SILVER

Below are given the average New York quotations, in cents per ounce, of fine silver.

Date			Average week ending		
Oct.	31	90.12	Sept.	25	107.66
Nov.	1	89.75	Oct.	2	98.49
"	2	88.87	"	9	91.72
"	3	88.75	"	16	87.01
"	4 Sunday		"	23	83.85
"	5	88.75	"	30	84.71
"	6 Holiday		Nov.	6	89.25

Monthly Averages

	1915	1916	1917		1915	1916	1917
Jan.	48.85	56.76	75.14	July	47.52	63.06	78.92
Feb.	48.45	56.74	77.54	Aug.	47.11	66.07	85.40
Mch.	50.61	57.89	74.13	Sept.	48.77	68.51	100.73
Apr.	50.25	64.37	72.51	Oct.	49.40	67.86	87.38
May	49.87	74.27	74.61	Nov.	51.88	71.60	....
June	49.03	65.04	76.44	Dec.	55.34	75.70	....

LEAD

Lead is quoted in cents per pound, New York delivery.

Date			Average week ending		
Oct.	31	5.75	Sept.	25	8.00
Nov.	1	5.85	Oct.	2	8.00
"	2	6.00	"	9	7.96
"	3	6.10	"	16	7.52
"	4 Sunday		"	23	6.70
"	5	6.20	"	30	5.66
"	6	6.25	Nov.	6	6.02

Monthly Averages

	1915	1916	1917		1915	1916	1917
Jan.	3.73	5.95	7.64	July	5.59	6.40	10.93
Feb.	3.83	6.23	9.01	Aug.	4.62	6.28	10.75
Mch.	4.04	7.26	10.07	Sept.	4.62	6.86	9.07
Apr.	4.21	7.70	9.38	Oct.	4.62	7.02	6.97
May	4.24	7.38	10.29	Nov.	5.15	7.07	....
June	5.75	6.88	11.74	Dec.	5.34	7.55	....

TIN

Prices in New York, in cents per pound.

Monthly Averages

	1915	1916	1917		1915	1916	1917
Jan.	34.40	41.76	44.10	July	37.38	38.37	62.60
Feb.	37.23	42.60	51.47	Aug.	34.37	38.88	62.53
Mch.	48.76	50.50	54.27	Sept.	33.12	36.66	61.54
Apr.	48.25	51.49	55.63	Oct.	33.00	41.10	....
May	39.28	49.10	63.21	Nov.	39.50	44.12	....
June	40.26	42.07	61.93	Dec.	38.71	42.55	....

The U. S. Geological Survey reports that the secondary tin recovered in the United States in 1916 was equal to about 24% of the tin imported, as metal or as oxide, into the United States during the year, 69,055 short tons. Secondary tin recoveries increased from 13,650 tons, valued at $10,554,180, in 1915, to 17,400 tons, valued at $15,131,040, in 1916.

COPPER

Prices of electrolytic in New York, in cents per pound.

Date			Average week ending		
Oct.	31	23.50	Sept.	25	24.41
Nov.	1	23.50	Oct.	2	23.50
"	2	23.50	"	9	23.50
"	3	23.50	"	16	23.50
"	4 Sunday		"	23	23.50
"	5	23.50	"	30	23.50
"	6	23.50	Nov.	6	23.50

Monthly Averages

	1915	1916	1917		1915	1916	1917
Jan.	13.60	24.30	29.53	July	19.09	25.66	29.67
Feb.	14.38	26.62	34.57	Aug.	17.27	27.03	27.42
Mch.	14.80	26.65	36.00	Sept.	17.69	28.28	25.11
Apr.	16.64	28.02	33.16	Oct.	17.90	28.50	23.50
May	18.71	29.02	31.69	Nov.	18.88	31.95	....
June	19.75	27.47	32.57	Dec.	20.67	32.89	....

Anaconda's production for October will amount to about 22,000,000 lb. The zinc plant at Great Falls is now operating practically at capacity and the output will be increased from 20 to 30% by the first of the year through completion of four additional furnaces for roasting of zinc concentrate and high-grade ore. The Washoe smelter is now operating at three-fourths normal. By request of the Government arsenic production is being pushed and from cleaning of smelter flues at Great Falls 1000 tons will be obtained by December 1.

ZINC

Zinc is quoted as spelter, standard Western brands, New York delivery, in cents per pound.

Date			Average week ending		
Oct.	31	8.00	Sept.	25	8.46
Nov.	1	8.00	Oct.	2	8.43
"	2	6.87	"	9	8.50
"	3	6.87	"	16	8.50
"	4 Sunday		"	23	8.25
"	5	6.87	"	30	8.12
"	6	6.87	Nov.	6	7.42

Monthly Averages

	1915	1916	1917		1915	1916	1917
Jan.	6.30	18.21	9.75	July	20.54	9.90	8.98
Feb.	9.05	19.99	10.45	Aug.	14.17	9.03	8.58
Mch.	8.40	18.40	10.78	Sept.	14.14	9.18	8.33
Apr.	9.78	18.62	10.20	Oct.	14.05	9.92	8.32
May	17.03	16.01	9.41	Nov.	17.20	11.81	....
June	22.20	12.85	9.63	Dec.	16.75	11.28	....

The receipts of spelter at St. Louis since January 1, 1917, as compiled by Eugene Smith, secretary of the Merchant's Exchange, were 10,879,510 slabs in 1917, as against 5,390,130 in 1916; and shipments were 5,981,400 slabs in 1917 and 4,001,810 in 1916.

QUICKSILVER

The primary market for quicksilver is San Francisco, California being the largest producer. The price is fixed in the open market, according to quantity. Prices, in dollars per flask of 75 pounds:

Week ending

Date					
Oct.	9	105.00	Oct.	23	100.00
"	16	100.00	"	30	100.00
			Nov.	6	100.00

Monthly Averages

	1915	1916	1917		1915	1916	1917
Jan.	51.90	222.00	81.00	July	95.00	81.20	102.00
Feb.	60.00	295.00	126.25	Aug.	93.75	74.50	115.00
Mch.	78.00	219.00	113.75	Sept.	91.00	75.00	112.00
Apr.	77.50	141.60	114.50	Oct.	92.90	78.20	102.00
May	75.00	90.00	104.00	Nov.	101.50	79.50	...
June	90.00	74.70	85.50	Dec.	123.00	80.00	....

Makers of ferro-chrome used in manufacture of high-grade steels and other products have received instructions from War Industries Board that pending a better situation in the chrome-ore supply, henceforth their product must be devoted to Government purposes and be subject to Government control. Automobile manufacturers and munition makers are the largest users of alloy and other high-grade steel, and it may be judged from the ruling issued that the use by automobile companies except on Government work will be made secondary to that by the munition makers. Automobile manufacturers are of the opinion that there will be sufficient supplies of both high-grade steels and ordinary steel to satisfy Government wants and also allow substantial allotments to the making of automobiles. A conference of automobile men and of representatives of the Government board was held on November 2.

The Carnegie Steel Co. has raised the price which it will pay for domestic manganese ore to $1.20 per unit delivered at its furnaces for ore running 50% and higher in manganese.

Eastern Metal Market

New York, October 31.

The markets as a whole are in better condition than for some time, since two of the metals are stronger instead of all being dull and weak.

Copper is unchanged and nominal at the Government price.

Tin is more active and considerably higher.

Lead is much stronger and more active, with a recovery in prices after another recession.

Zinc continues inactive and is lower.

Antimony is nominally lower.

Aluminum is inactive and has slightly declined.

In the steel market, prices have been dropping toward levels which are regarded as marking the new maxima. No further price-fixing is now announced, but adjustments on commodities not already fixed will be agreed upon by those interested. The whole movement indicates a broad spirit of confidence in the steel-makers' representatives. Buying of pig-iron is not very active on account of the unwillingness of furnaces to accept additional orders.

COPPER

So far as official information is concerned the copper market remains unchanged. It is apparently completely in the hands of the Copper Producers' Committee which is stringently regulating deliveries of metal to consumers with a view to conserving supplies for the Government and its Allies. It is impossible to confirm reported sales by large producers at the Government price of 23.50c. per lb., New York, which they continue to quote. It is stated that for first-quarter delivery sales have been made at the 23.50c. price, but subject to any change the Government may then make. The actual business of the committee, according to an authority of weight, is done through the United Metals Selling Co. and the American Smelting & Refining Co., which have become practically the business agents, the former representing the United States government and the latter the foreign countries. On orders received these agents requisition the several producers to fill orders according to information as to their ability to supply the material desired. The two agents render bills to the governments and then remit to the various producers. The position of the small dealers and consumers is still unsolved so far as official information is concerned. Proceeding on the assumption that sales of small lots can be made safely at above the official price, the same as in the case of steel jobbers, copper continues to change hands, but the realized price is reported now to be more nearly approaching the Government level. The London market is unchanged at £125 for spot electrolytic and £121 for futures.

TIN

The market for spot tin has become more active, due to an acute shortage and to some doubt as to the ability of small buyers to obtain further supplies. The market has advanced 4.37½c. per lb. in the past week, until yesterday spot and nearby tin was quoted at 66c., New York. After a quiet day on Wednesday, October 24, when about 100 tons changed hands, the new activity started on Thursday with the appearance of good inquiry for early delivery but with few sellers. Outside the occasional spot inquiry, however, the market was for the most part quite uninteresting. Buyers of spot, however, became nervous, and this helped other positions so that on Friday, October 26, sales of at least 500 tons of all brands for all positions were made. The fact that the license question is not as easy as was expected has had its influence on the sharp rise in spot. Cablegrams also continue to be delayed daily, and often do not arrive in time to conduct business. On Monday a buyer of 50 tons could only obtain small lots, even bidding up to 65c., New York. The spot situation became still more acute until yesterday, when spot sold in small lots at 66c., New York, with plenty of buyers but a lack of sellers. Arrivals up to and including October 29 were 2060 tons, with the quantity afloat 4300 tons. The London market for spot Straits has advanced over £9 per ton in the week, the quotation yesterday standing at £256 10s.

LEAD

The continued decline in the lead market reached its culmination last week when, on October 24, the American Smelting & Refining Co. reduced its price again 1c. per lb., bringing its quotation to 5.50c., New York. This was lower than had been expected and the trade was thrown temporarily into consternation. Previous to this, each cut by the leading interest had been met by a cut on the part of the independents. The new price was evidently regarded as rock-bottom and its establishment evidently had the effect intended. Since then inquiry has been large, and consumers, whose stocks were probably very low, have come into the market. By the end of last week, Saturday, prompt and November shipment was selling from 5.50 to 5.60c. in large quantities. So far this week the market has continued to gain in strength and sales are reported as having been large. By yesterday independent and other sellers had withdrawn from the market, with no one left to supply the demand except the leading interest, and with as high as 5.87 to 6c. asked. Yesterday sales were heavy, as high as 5.75c., New York, having been paid, with some reporting 5.87c., New York, as having been obtained. An advance very soon by the American Smelting & Refining Co. is not unlikely.

ZINC

The market is devoid of interest and is sluggish to a degree. It has gradually fallen during the past week, the decline being measured by about ¼c. per lb. It is due, without doubt, to the weakness in the lead market. Demand is of small proportions. In the absence of any volume of business the quotation, as judged by or based on the few sales reported, is about 7.75c., St. Louis, or 8c., New York, for prime Western. At this price one or two lots of 100 tons each have been sold within the last two days. It is also reported that a few small lots have gone at 7.62c., St. Louis, or 7.87c., New York, but not all sellers are willing to meet this price. It is still a matter of comment that galvanizers and brass-makers still remain largely out of the market, nor is there any apparent desire on the part of consumers to stock up for the winter in spite of the fact that railroad congestion may interfere with later deliveries.

ANTIMONY

There is no demand, but quotations for Chinese and Japanese grades are nominally lower than last week at 14.25 to 14.50c. per lb., New York, duty paid.

ALUMINUM

No. 1 virgin metal, 98 to 99% pure, is a little lower, but demand is of small proportions. It is quoted at 37 to 39c., per lb., New York.

ORES

Tungsten: Quotations are unchanged at $26 per unit in 60% concentrate for scheelite, with $23 to $25 per unit for other ores. One dealer reports an active market with about 60 tons of scheelite sold in the week. The minimum price for ferro-tungsten is $2.35 per lb. of contained tungsten.

Molybdenum: For high-grade molybdenite $2.20 per lb. of MoS_2 in 90% concentrate has been paid.

LEACHING of copper ores continues to engage the attention of progressive metallurgists. Elsewhere in this issue is printed a suggestive article on this subject by Mr. A. E. Drucker. Descriptions of the leaching practice at Bisbee and Ajo will appear in proximate issues of this paper.

WE regret to record the death, at the early age of 50, of Dr. E. F. Roeber, editor of 'Metallurgical and Chemical Engineering,' New York. Born in Germany, he came to this country when 27 and soon made himself a leader in electro-chemical research. An industrious, capable, and kindly man, he won an honorable position in technical journalism and also the personal regard of a wide circle of friends.

ONE of our local organs of misinformation undertook to define for its readers the various Russian factions. The Maximalists, we were told, constituted a party that insisted on "the maximum programme of extreme Socialism." That seemed the limit of radicalism, but apparently it was not, for we were informed immediately that the Bolsheviki were "Maximalists raised to the n-th degree." The plain truth is that 'Bolsheviki' is the Russian equivalent for 'Maximalists,' which is a curtailment of Maximumalist, the meaning of which is obvious.

STATISTICS issued by the U. S. Geological Survey indicate a decided decline in the domestic production of spelter, the output having decreased from 180,569 tons in the first quarter to 132,700 tons in the third quarter of the current year. Stocks are increasing. The number of idle retorts on September 30 represented 28% of the total, and included 18 entire plants. It is evident that an unprofitable production has been maintained by several smelters during recent months in order to retain their labor force in the expectation of obtaining Government orders, which have not come on the scale expected, so that a curtailment of operations has been forced.

WE publish two short articles on manganese. One we owe to the Geological Survey and the other to the Bureau of Mines. They supplement each other. Both are timely. Philipsburg, formerly associated with the highly profitable mining of silver, as represented by the famous Granite Mountain mine, is proving an important source of manganese. So also is the Lake Superior region, heretofore associated with copper and iron. If steel-makers can modify current practice so as to utilize ferro-alloys containing moderate proportions of manganese the pressure to secure high-grade manganiferous ore will be relieved. Converter-slags relatively rich in manganese are now being utilized to some extent by being charged into the blast-furnace, and the Mayari ores of Cuba, which have a considerable manganese content, are also employed with success. The question would seem to involve chiefly an improvement in technical methods to meet a condition that did not exist while sources of high-grade manganese ore were open.

UTILIZATION of liquid oxygen to form an explosive in connection with carbonaceous material has been revived by Mr. George S. Rice of the Bureau of Mines. It is proposed to erect an experimental plant under Government auspices, in co-operation with the producers of the liquefied gas, who have shown great interest in the project. In a recent discussion before a meeting of the Mining and Metallurgical Society of America, Mr. Rice stated that shortage of materials commonly used in making explosives had led the Germans to adopt liquid air for saturating cartridges of carbonaceous matter to be employed for blasting coal in the mines of Upper Silesia. It was claimed that the efficiency of this makeshift explosive was equivalent to dynamite at 12 cents per pound. Mr. Rice is using a mixture of various carbonaceous substances placed in a cheese-cloth container along with a detonator; the cartridge is then dipped into liquid oxygen, and although the best grade of oxygen at present obtainable is composed of only 55% of that element, a strength equal to about three-fourths that of 40% dynamite is realized. The function of the oxygen, of course, is to completely and instantaneously oxidize the carbon, the suddenness of the effect being such as to bring it within the realm of detonation. A large amount of information regarding explosives made with liquid oxygen is available, as this subject was extensively investigated about 18 years ago by a distinguished chemist, Sir James Dewar, in England. An experimental plant for its manufacture was also under consideration, if not actually erected, at Boston by a subsidiary of the British corporation. The investigations then made indicated favorable possibilities for the oxygen explosive, though some difficulties were experienced. Financial troubles prevented Dewar from continuing his efforts to commercialize the invention.

THE University of California is served by a publicity agent that is remarkable for an exuberance detrimental to accuracy. Twice a statement has been given to the local press asserting that the Research Council of the State Council of Defense had discovered a way of making cyanide so cheaply as to be of immediate assistance to the citrus industry of California and also to the gold mines. We asked Professor John C. Merriam, the chairman of the Council, and he informed us that the statement was not warranted, but that a great deal of useful work was being done, some of it of a promising kind. Such undisciplined publicity is injurious to the University and stultifying to those concerned. On the other hand, it is true that the Bucher catalytic process for the production of cyanide has opened the way for successfully manufacturing it with relatively inexpensive equipment. We published, in our issue of October 13, the details of an ingenious adaptation of Bucher's idea by Mr. G. H. Clevenger, in which he used a few pieces of gas-pipe, an assay-furnace type of gasoline-burner, and a few other odds and ends, with which he made sodium cyanide in sufficient quantity to perform laboratory leaching-tests on gold and silver ores. These experiments were suggestive of a method by which cyanide could be produced in small plants under the direction of clever chemists and mechanics.

Labor and I. W. W.

Under 'Discussion' we publish a genuine, and therefore interesting, plea in behalf of labor. It is true that most laborers think only of getting more money for their work and that most employers think only of getting more work for their money; that is why there is so much friction. Neither attitude is honest, in so far as each tries to get more than he gives. We agree with the writer of the letter, Mr. J. F. Harrington, that goodwill and fair dealing furnish a basis on which employer and employee can co-operate to better advantage; but he has ignored the chief advantage of the contract system as adopted in mines. It is not to substitute the contractor as a driver instead of the foreman as a supervisor; the idea is that if the men are paid for extra work, they will do it; the contractor is one of themselves, merely the partner in whose name the contract is made. In the mines with which the present writer was professionally connected the system of contracting was a means of developing team-work among a small party, usually four, six, or eight, joined in an effort to drive a level or sink a shaft at an agreed price and with such speed as would yield them considerably more than day's pay. Of course, the best way to get a square deal is to give one, and the quickest way to get euchred is to try to trick the other fellow. As to the I. W. W., we cannot discuss them patiently during this time of War, nor with any pleasure at any time. The I. W. W. is the problem of the hobo; it is anti-social, subversive of all government, including "government of the people, by the people, for the people," an ideal that we shall not willingly let die. We have no fear lest the brave men that come home from the trenches will subvert existing institutions in favor of the sentimental savagery of the I. W. W. "Liberty and justice" are implicit in the American idea, certainly, but we do not connote them with Haywood and St. John or any of the other irresponsible conspirators that have prostituted unionism in their own interest and that of a small group of anarchists. We believe in collective bargaining, in unionism; we believe that the laborers should combine lawfully to protect their interest, just as the capitalists combine in corporations and syndicates to conduct their business advantageously; but the chief cause of labor unrest is the fact that so many of the most intelligent men do not join the unions, whereas too often they are dominated by ignorant hot-heads or by rank outsiders, who employ the unions to blackmail the companies. If that element, usually non-resident, the Buttinskis of industrial unrest, as bad as the Bolshevikis, could be excluded and the unions composed of all the employees organized in self-protection, and not for an economic hold-up, there would be fewer strikes, better feeling, and a larger prospect of developing true Americanism, which means an equality of opportunity under the law.

The Russian Crisis

Conditions in Russia are engaging the attention of most of us. We are glad therefore to publish a letter from a Russian engineer, Mr. L. A. Mekler, criticizing the recent article by Mr. Horace V. Winchell, whose contribution, in our issue of October 27, was particularly timely and interesting. It must be confessed that the logic of events appears to be against Mr. Mekler and on the side of Mr. Winchell; the optimism of the one has been stultified by recent happenings, whereas the pessimism of the other has been sadly justified. For the moment Nicholas Lenine and his international group of I. W. Ws are in control, proving the culmination of German penetration and propaganda. The Russian people are cutting a sorry figure and it is unlikely that they will come to their senses until they realize that they must either hang together or be hung separately. The unfortunate feature of the Russian revolution is that it went much further than was intended by those originally responsible for it. The moderate Liberals, men like Lvoff, Milyukoff, and Rodzianko, aimed at constitutional reform and had no expectation that all constituted authority would be swept suddenly into the discard of political chaos; they hoped to open the doors for democratic government, but when they pushed what seemed to be such a door of progress they fell headlong into the dark void of undisciplined revolt; instead of preparing the way for a democratic change of government they found themselves in the midst of a mob of talkative visionaries and traitorous anarchists. The Russian public has been fed on false analogies with the French revolution. That historic uprising happened among the best educated people in Europe; the Russian revolt destroyed the government of a people that, taken as a whole, is the least educated in Europe. About 80% consists of ignorant peasants. France had a population of only 30 millions in 1789, and the revolution was accomplished in Paris, which then had only 600,000 inhabitants. At that time, the swiftest communication was effected on horseback; last March in Russia the post and telegraph brought 180 millions within the range of the great event. Again, whereas Paris was the heart of France, Petrograd is only the stomach of Russia; it is a cosmopolitan city, poisoned by German intrigue, and out of sympathy with the rest of the country. As soon as the revolution occurred, the Russian soldiers at the front were told that the land was about to be distributed to the people and that they had better hurry home to get their share. They deserted by the thousand, they seized trains and scurried homeward. The story of a division of the land was one of the many lies spread by German agents, such as Lenine and Trotzky. No such allotment of land characterized the French revolution, nor did the French armies win battles so long as the death penalty for desertion was held in abeyance; it was not until 1793, when discipline was restored by making death the punishment for desertion, that the military power of France asserted itself victoriously. In short, the Russian crisis is a sad exhibition of plain ignorance supplemented by German intrigue, made effective by an international group of anarchists masquerading as socialists. They are not that, they are anti-social. Most of the leaders of the Maximalists have assumed Russian names, but they are not natives of Russia any more than our I. W. W. element is American; it is not even a distinctly foreign element; it is a non-national group of irresponsibly destructive vagrants to whom representative government and liberal ideas are just as alien as a military despotism or a predatory bureaucracy. We may pause to note the curious fact that the Prussian who hates a socialist at home, and supresses him as much as he can, is quite willing to support the same kind of irreconcilable in Russia; so also while claiming to be model Christians at home they aid the Turk in killing Christians in Armenia. 'To hell with principles' is the expression of Prussian policy. Thus today the world sees the strange spectacle of the Prussian, who has developed social organization to a repulsive tyranny, co-operating with Maximalists, to whom the disorganization of a political catch-as-catch-can is a desirable consummation. Those that aim to spiritualize the human relationship and to develop the ideals of representative government can join heartily in fighting the combined cohorts of the I. W. W. and of

Food for the War

California has taken the lead in signing pledges to economize food. This is as it should be, for many of the officials of the Food Administration come from this State, and the chief of it is a man of whom California may well be proud. We have printed pledges on advertising pages of several former issues and we do so again on page 8 of the advertising annex of this issue. It has been a pleasure to forward a large number of signed pledges to Mr. Hoover, who, we are authorized to say, appreciates keenly this evidence of support from the members of his own profession. Interest in food-thrift is growing, but it is not nearly as keen as it ought to be. In the course of a recent transcontinental journey to St. Louis the present writer noticed that the ideas of the Food Administration had not as yet made much impression on the dining-car service of the Pullman company. It ought to be possible for the Controller to compel the Pullman company to set a better example. Only in one dining-car, between Kansas City and St. Louis, did we note a sincere effort to follow the suggestions emanating from Washington. In some instances hooverizing was being used by the Pullman people as an excuse for raising prices, serving half-portions at full rates, for example. The idea of using perishable and local products, so as to consume the minimum of those that can be shipped to a distance or of those that have undergone long transportation, was not in evidence. For example, trout, easily procurable in Colorado and Utah, was served for 75 cents, whereas white-fish from the Great Lakes or salmon from the Pacific Coast cost only 50 cents. The prices should be reversed. A prodigal display of white bread is still common; more bread than is wanted is served, to be fingered or broken, and then thrown into the garbage-can. The people of the United States waste more than enough food to feed the whole population of Belgium. To waste food today is a crime—a mean and disgraceful act, arguing no sympathy with the millions now famishing in the countries devastated by war. We discussed the subject on the train with one or two young men and found a not infrequent irresponsiveness, a careless disregard of the fateful crisis. One young man, for example, asserted his right to eat as much meat as he liked three times a day, because, said he, there was plenty of meat, and other food, in the country. Such cases of callous—and callow—detachment from current history may be rare—we would like to think so—but there are many people that refuse to see the necessity for the United States feeding the peoples across the Atlantic. We have tried to explain to our young friends—and some of the older ones—that the United States is fighting not only to keep the world safe for democracy and to assist the victims of autocratic piracy, but to preserve inviolate our own hearths and homes, to prevent our land from coming under the heel of a ruthless conqueror, to maintain the right to live and let live as against the 'verboten' and the 'goose-step' of

Prussian militarism, with its associated 'frightfulness,' and above everything to prevent the destruction of that spiritualizing of the human relationship that is the essence of our kind of civilization. Some of our young men are willing to die for this cause—some have died already—thousands are preparing to face the supreme sacrifices of virile manhood; many of our older men, and women also, are giving all their time and much of their money to the service of the flag and for the furtherance of the national purpose; therefore the least that the rest of us—the run of mine—should do is to save food, for the feeding of our own soldiers, for the sustenance of the gallant men that have been fighting during three long years for what is now our cause, for the nourishment of the peoples that cannot cultivate the soil productively either on account of the numbers of their men in the armies or because the battalions of death have despoiled, devastated, and destroyed their fields and gardens. Every man, woman, and child can help in this work—which is vital. It is no mere bit of sentimentalism; it is no fad of Mr. Hoover; it is no red-tape of the Government; it is an integral part of our warfare against organized hell. The man or woman lacking the qualities immediately useful in war—youth, strength, brains, or money—can help to save food and so assist those more capable than themselves. It is the least that we can do, and the least that we can do should be our very best to help and promote the national purpose—the purpose of the world's welfare, as we understand it. One practical suggestion we give: let every one of us cease to eat 'hot cakes' made of wheat flour. It is one of the most extravagant ways of consuming wheat. To most people the 'hot cakes' at the close of breakfast are an extra, merely the gastronomic tamping of a full charge. Most of us will be better without them and if all agree to abstain from eating them we shall effect a great and useful economy—and not only in wheat flour, but in butter also. If you must fill the chinks, use cakes made of corn or rice, or bran muffins instead. This will seem trivial to some of our readers. It is not. The sinfulness of squandering wheat flour can be made clear by a few statistics. Our Allies will produce 393,770,000 bushels of wheat this year; their normal consumption is 974,485,000 bushels; the shortage is about 580,000,000 bushels, besides a shortage of 673,000,000 bushels of other cereals. The United States will have a crop of 668,000,000 bushels of wheat, as against a normal consumption of 590,-304,000, leaving a surplus of only about 80,000,000 bushels, while we shall have a surplus of about 1,000,000,000 of all other cereals. Obviously we must save wheat, substituting other cereals in our dietary. It is a fact determined by physiologists and confirmed by experience in Belgium that wheat bread is indeed 'the staff of life,' the fundamental food of man. If any reader desire to become better informed let him read Mr. Vernon Kellogg's article on 'Patriotism and Food' in the November issue of the 'Atlantic Monthly.' As he says, this association of patriotism and food is not sordid: "It can be as fine as the spirit of democracy and as ennobling as the struggle for democracy. If we cannot organize our effort in this world-crisis by the individual initiative, spirit, and consent of the people, then democracy is a faith on which we cannot stand." It is as certain as the coming of tomorrow's dawn that if we do not make good in this war we shall be shamed to all eternity.

The Phosphate Outlook

No apology is due for devoting space to fertilizers, for, apart from the fact that a large proportion of the raw material is obtained by mining or quarrying, and belongs to the mineral industry, we are at war, and it is essential that we and our Allies should be fed. We have devoted considerable space to occasional reviews of the nitrate, sulphur, and potash problems. A glance at the phosphate outlook, then, will complete the review of the important commercial fertilizers. Unlike nitrates, sulphur, and potash, the supply of phosphates is ample, not only for present needs but for many decades, and probably for centuries. The United States is the world's greatest producer of phosphate rock. Florida leads with an annual contribution of two and a quarter million tons or 75% of the total domestic output. Deposits exist in Tennessee, South Carolina, Kentucky, and Arkansas, while an immense belt, discovered in 1906 and since mapped by the U. S. Geological Survey, extends from Salt Lake City to Helena. A survey of these deposits of phosphate rock has disclosed a tonnage greater than any yet discovered in the world. Unfortunately, however, there is a pucker in the persimmon. Phosphate rock is insoluble, and the vegetal kingdom requires its nutriment in liquid form. Previously phosphate rock has been treated with sulphuric acid to convert the insoluble basic phosphate of lime into a soluble acid phosphate, readily assimilable by plant life. Owing to a shortage of bottoms in which to convey pyrite from Spain, temporary embarrassment in the production of sulphuric acid resulted, and at the same time the call from munition-plants reduced the amount of acid available for making fertilizer. Attention has been called to pyrite and pyrrhotite deposits in the eastern States, which were unable to compete with Spain in pre-war times, and we understand that more energetic development of these has been undertaken. Also the western States are beginning to develop larger resources of these minerals. The sulphur deposits of Louisiana and Texas are producing all that can be transported. Meanwhile it has been found that when phosphate rock is fused in an electric furnace with silica and a small quantity of charcoal, coke, or coal, the phosphoric acid is eliminated by volatilization in a very pure state, and can be collected in fume-chambers, leaving a silicate of lime that can be cast into bricks for paving-blocks. The process would seem to be costly, even if the blocks could compete successfully with concrete as a paving material, but with cheap hydro-electric power it may be feasible.

Russia in War Time

The Editor:

Sir—In the last issue of your journal, there appeared an article by Mr. H. V. Winchell, on 'Russia in War Time,' in which the author after extensive statistical data and comments on the present political, economical, and other conditions, gives his views on the possibilities for American enterprise and capital in that country.

Before commenting on the article, I would like to say a few things with regard to this kind of article in general, and articles about Russia, in particular. Russia has never had a square deal in the United States, because most of the information concerning Russia is given out through the Sunday supplements of the so-called 'popular' press, in cheap novels of no less popular character, and by travelers who, passing through the country in an express train, without knowledge of the language and people, in most cases, would, on their return home, write a book, with such titles as 'Russia from Within,' or 'The Spirit of Russia.' Not only titled or semi-titled travelers succumbed to this folly, but men of sound judgment and great powers of observation, being struck by the peculiarities of the people and the country, could not withstand the temptation to describe and 'explain' what impressed them most.

Destructive criticism is easier than constructive comment, and it is only human nature to make sweeping generalizations. In the result, the picture is anything but what it ought to be, and should have been, if a more thorough investigation had been made, and the first impression not given too much weight.

Knut Hamsun, the great Scandinavian writer, a great judge of human nature, was one victim of this psychological condition, and Mr. Winchell appears to be another, Mr. Winchell having the disadvantage of describing Russia in a period hardly understood even by natives of the country. Many a Russian would not have done better under the conditions confronting Mr. Winchell. He could not help but see some of the things he saw. But facts and figures will say whatever one requires of them. It is the inner meaning and not the outer expression of things that is the more important. Being a native Russian and having the good fortune of having both a Russian and an American college education, I am enabled to appreciate both points of view, and would, therefore, like to explain some of the facts stated in Mr. Winchell's article.

In coming to Russia, an American usually lands in Vladivostok, the only Russian port open to both the military and non-military freight and passenger traffic. No wonder he meets congestion, disorder, and general turmoil. He proceeds on the trans-Siberian railway, the only artery feeding the front and the rear, and usually congested, even in peace time. If there was a congestion on some of the railroads in this country where, according to Mr. Winchell's own statistics, there are three times as much railroad tracks per mile, and where there are many routes from one point to another, is it anything but natural that the service on this Russian trans-continental railroad is not what one would expect under normal conditions?

At the end of the journey, the traveler, already brought to the highest pitch of nervousness by his uncomfortable experience on the journey, finds himself in Petrograd, settled in the most expensive and most crowded hotel in the city. After a day or so of discomfort, he is ripe for complaints, and sees only the bad side of things. Overwhelmed by the large numbers of rubles asked for commodities, he forgets that a ruble is only 14 cents, and that many commodities being imported are bought by traders on the dollar, not the ruble basis, and have to be sold accordingly.

It is an economic fact that in an isolated community, commodities eventually equalize in their relative price and therefore even home-made products will climb up to their proportional cost. The big Russian cities, now that the system of communication is out of order, are isolated communities. Recognizing this fact and the depreciation of the ruble, and comparing the prices obtaining in the large Russian cities with the prices of, say Paris, which has much better transportation facilities, one can see that "the devil is not as black as he is painted."

In cities of smaller population, and at inland localities of Russia, conditions are different. Food is plentiful, the exorbitant costs of Petrograd are almost unknown, and even luxuries can be obtained at moderate prices. My brother, now attending the University of Tomsk, the capital of Siberia with about 100,000 population, lives comfortably on 150 rubles (about $22) per month. In many Siberian and some Russian cities, even at present, meat can be obtained at 6 cents per pound.

From all this it follows that the conditions of living and prices in Russia generally (not the large congested cities that constitute about 3% of Russia) are as good as, if not better than, in other warring countries.

The industrial situation is bad, as might be expected when all the man-power and skilled labor are withdrawn. The new government realizes this and is already withdrawing all skilled men from the army and placing them where they will be most useful. To increase the number of skilled and trained men, the Government is issuing a call for all Russian engineering and medical students abroad to come home to complete their education, full credit being given to them for the work done abroad. This is a radical departure from the practice of the past, when even a graduate of a foreign university usually had to begin his work over again in Russia almost from the freshman year.

Labor troubles occur frequently in Russia, but so they do in other countries where the position of labor is much better, and where people have had a long experience in organization and self-government. The present misunderstandings between labor and capital are temporary. Things are already straightening up and much quicker than expected by the most optimistic of those who knew better than to expect an immediate settlement of problems confronting the world for the last 50 or 60 years.

Many local labor organizations, especially those removed from Petrograd, the chief source of trouble, are withdrawing from the influence of the central body and doing their best to stabilize conditions. Recently when an order for a railroad strike as a protest against the Moscow conference was called by the Petrograd Council, practically all local unions disregarded the order, and only the Moscow-Petrograd non-military traffic was interrupted. As to the closing of industries or inability to run plants without Government subsidies, it is mostly due to purely local factors of transportation, supply, management, and the pre-revolution relations between Capital and Labor, the latter being sometimes the most important.

It takes a man of great adaptability to concede suddenly the fact that labor is something more than chattel, and that pre-war conditions of work and wages are no longer possible. On the other hand, the laborer when made to understand his power and given an opportunity to exercise it, will be the more aggressive, the worse his conditions were before the revolution. Where there were fair labor conditions and where the management did not act hastily or resorted to drastic measures the worst of misunderstandings were settled to the satisfaction of both parties.

I know of a case where the workmen appropriated a munitions plant from the owner, and appointed him general manager, in recognition of his good management of the men, as well as his executive ability. Appreciating the phychology of the moment, the owner did not shut-down the plant, and now after things have been thought and talked over he is still its owner and operates it to his own, the employees, and the Government's satisfaction.

The political condition is improving, in spite of all the reports in the press, and the so-called eye-witnesses. These reports certainly have a foundation to them; but while one sees and hears the soap-box orator, and the demonstrations, which are growing smaller in number and importance, he hardly can notice, without especially looking for it, the silent constructive work.

Just now Russia is like a bottle of good wine that was kept tightly corked for centuries. The first to appear is the foam and the longer the time of 'seasoning', the more foam must escape. But there is good wine and plenty of it in Russia. The silent creative work of the best people of the country is felt already. The Government grows stronger every day. The extremist is losing his influence over the Council of Workmen and Soldiers and the Council has not as much power as it had a few months ago. Just today I read in the papers that the Government proclaimed martial law in Finland, and refused any financial help to Ukraina, the provinces that, supported by the Russian extremists, insisted on becoming independent immediately. Could a weak government do this?

The fear that private property will be abolished in Russia may be dismissed. The Russian peasant demands land, that is, he wants to become a proprietor himself. It is hardly probable that he will vote against his own desire. There may, and probably will be, a distribution of crown-lands and other large land-holdings, and the adoption of the leasing system in mineral development, but that is as far as it may go.

Graft among officials, so common during the old regime, is dying out wherever new men are substituted for the old 'tchihovnik' and Mr. Winchell must have had queer luck to witness so many cases personally.

Summing up my statements, I would like to say that in studying Russia and Russian conditions, especially at the present time, one must not be impressed too much by what he sees or hears. Russia is a land of great possibilities and their realization is much nearer than one may expect. If the United States wants to help Russia and to profit by this help, it must start now, when there is need for-help and when the field is clear. Germany will not give in easily. She is ready to step in any moment to re-establish her economic ante-bellum status in Russia. Do not give up Russia as hopeless, for the present. The successful person is he who does the thing while other people are deciding whether or not to do it, and in nine cases out of ten, 'the early bird gets the worm.'

L. A. MEKLER.

Berkeley, California, October 30.

Freight-Rates on Drill-Steel

The Editor:

Sir—A vivid object has come under my observation in relation to another thrust at the high cost of mining. While the West Coast enjoyed the competition of ocean transportation from the Atlantic ports, the mines were able to secure a freight-rate on mine drill-steel of 45c. per 100 lb. During this time the railroads were making a rate of 80c. on the minimum car of 40,000 lb., and Pittsburgh, Pa., had a rate of 60c. on an 80,000-lb. minimum. When the railroads asked the Interstate Commerce Commission for an increase of 15% in transcontinental rates, during the latter part of 1916, the rates on mine drill-steel, classified as bar-steel, were increased to 90c. on the 40,000-lb. minimum, and 85c. on the

50,000-lb. minimum, with a 75c. rate on the 80,000-lb. carload. Less than carload rates were $1.50 per 100 lb. During all of this time the classification was ambiguous, and with the water competition there never was any attempt made toward a close inspection or classification of mine drill-steel other than as bar-steel. The last transcontinental west-bound freight tariff, No. 1-P, issued February 20, 1917, effective April 16, 1917, placed the 20 and 25-ton minimum on bar-steel under the classification of structural iron and steel, which left only the 80,000-lb. minimum at the 75c. rate on bar-steel, under the iron and steel classification. Under the structural iron and steel articles, item 1514, containing the two lower minimums (40,000 and 50,000 lb.), specifies .... 'bars, including corrugated and twisted bars.'' Item No. 1522, under the heading of iron and steel articles which carry the 80,000-lb. minimum, specifies bar-iron or steel, including corrugated or twisted bars. The only difference between the first two minima and the third is that the first two are under the sub-heading of structural iron and steel, and the third under iron and steel articles. While it is easy to recognize the distinction, the fact remains that the railroads have allowed mine drill-steel to come through under the structural iron and steel classification, which contains the two lower minima, and have been billing it at the corresponding rates when shipped as bar-steel, without any attempt at close inspection, thus establishing a long precedent in favor of the 20 and 25-ton carload rate. From my observation, it all depends on the classification placed by the inspectors, and as cruciform drill-steel is a corrugated steel and has been so construed, the point upon which the railroads are now deciding is a direct discrimination against this class of steel to that which it formerly enjoyed. Carloads of steel have been received where an endeavor has been made to assess the 75c. rate on the 80,000-lb. minimum, and the railroads have corrected their assessments to apply to their 85c. or 90c. rate on the lower minimum, according to the quantity which the car contained over the 20 or 25-ton carload. Just to show how ambiguous the classification is: 999 agents throughout the West Coast, if required to give the rate on bar-steel (as there is no other classification for mine drill-steel other than bar-steel), would quote one of the three minima to which I have above referred. This is also a direct discrimination against the mills' entire output of which, or at least 90% of it, is mine drill-steel. Unless they ship in 80,000-lb. cars they would have to compete with mills that ship all grades of steel, including a portion of mine drill-steel. This would make a rate of $1.50 against 75c. It seems to me that mine drill-steel should be descriptive so that it would be impossible for any one firm to make up a car including mine drill-steel, thus securing a lower rate in discrimination against another firm shipping all mine drill-steel and having to pay the higher rate. The Interstate Commerce Commission has decided that so far as present rates go, the 20 and 25-ton cars to the Pacific Coast will be eliminated from the tariff, which automatically increases the rate from 75c. to $1.50 per 100 lb. in amounts less than 80,000 lb. Some action should be taken on the part of the mining companies to secure a hearing before the Interstate Commerce Commission to secure for themselves an equitable rate on mine drill-steel, and not permit the railroads at this time to take advantage of such ambiguities for setting up an arbitrary tariff. As things now are the lack of proper classification causes an assessment of $600 on a 50,000-lb. carload of steel instead of $425 as formerly. The Atchison, Topeka & Santa Fe have the matter under advisement with the Transcontinental Freight Bureau at Chicago, and the outcome of the decision is awaited with great interest.

Trusting that the facts I have presented will arouse action by mining people that will secure proper representation to prevent an arbitrary increase in freight rates.

H. D. STALEY.

San Francisco, October 22.

Tungsten

The Editor:

Sir—It has occurred to me that many shippers of tungsten ores do not realize the position of the tungsten market regarding quality, and it might be useful to give an idea of what the market requires. Different buyers make different specifications. The main division as to quality depends upon the method of treating the ore, either by a chemical process, by an electrical process, or by a semi-wet process. Prior to the War, most tungsten ores were treated by the chemical process, but at the moment of writing, the preponderance of users in the United States reduce the ores by electrolytic methods.

In the electrolytic process, copper, tin, and phosphorus are the most objectionable impurities, and it is almost impossible to sell to anyone employing the electrical process any material containing in excess of 0.05% tin, 0.05% copper, and 0.04% phosphorus; thus ore that is free from these impurities is put in the highest class, and naturally commands the top-market price. Ore that does not meet these specifications is therefore called 'off-grade,' and while in a rising market it is sometimes possible to sell ore with impurities slightly above the percentages given, it is exceedingly difficult to sell it at good prices in a sluggish or steady market.

In the chemical process, tin or copper or even phosphorus can be eliminated, and therefore, the chemical-process buyers are not so particular as to the impurities, but inasmuch as the number of these producers is limited, all the ore coming on the market and containing tin, copper, and other impurities is offered to these few producers, who naturally find an abundance, and are not prepared to pay top-market prices. As a matter of fact, they have been able to make their price, and while for high-class ore the market is often a seller's market, for off-grade ore it is almost invariably a buyer's market. The buyer's idea of price governs the market, otherwise off-grade ores cannot be got rid of and will ultimately reach the buyer still cheaper than if his price is accepted as and when the lots come on the market.

Inasmuch as supply and demand are at the present moment balancing themselves, it is easily understood that buyers even for the chemical process prefer to take ores from which they have not the trouble of extracting the impurities. Ores containing deleterious matter have to be sold at heavy discounts from the regularly established price. I have handled a considerable quantity of tungsten ores amounting to thousands of tons already this year and have not yet had a single lot of tungsten ore unsold, though some of them assay as high as 11% tin. Of course, where tin is contained in the ore in such a large percentage, it would be advisable to have it extracted either at the port of shipment or in New York, if arrangement to that end could be made. Such extractions would prevent a sale prior to the result of the extraction being known.

CHARLES HARDY.

New York, November 1.

A Plea for Labor

The Editor:

Sir—I had a mind not to subscribe for your paper at all, yet after thinking about it over-night, I believe I shall do so. On page 561, issue of October 13, in an article from Treadwell on the labor-supply, I read: "This is a good policy......to get more work out of a man." Can the gentleman who penned those lines really have any just cause for complaint if the men adopt a method that makes it possible to get more money out of the company "than can be gotten in any other way?" Is it not because of the extra 50 cents that more work is done and not because of "the policy" of contracting? Could not the foreman take the place of the contractor, and in that case would not the increase amount to just 50c. anyway? To be sure! But the idea is that the contractor will do the driving, relieving the company of this disagreeable phase of getting the work done. Not long ago, while talking to a very fair conscientious man about the kicking against working conditions he had heard among the miners, I remarked that I believed in giving them everything that was fair, and perhaps a little bit more, to be sure; then, if they still kicked, discharge them instantly, and with a clear conscience. I said, "Make a man's work as easy to do as possible, for the easier he can do his work, the more work he will do." The gentleman said "The more work he *must* do!" Like the writer of this article, he believes that force is necessary; he does not give the miner credit for ambition, a sense of justice, or the desire to give the company a square deal. The late Mr. Taylor, consulting engineer to the Midvale Steel Co., said: "By getting the goodwill of the worker, he will do more work for you voluntarily than all the bosses on earth can force him to do." Very well! Mr. Emerson lays much stress on his thirteenth principle, 'The principles of personal efficiency'; it is the square deal, he says. "The best way to get the square deal for yourself is always to give it to the other fellow." Language was made to conceal thought, as some one has said. It requires a very brainy man, a skilful talker, to make me believe what he does not believe himself. To try to convince a man you have perfect confidence in him, when in the back of your head is that deep-seated idea that force is necessary to compel what may be called a fair day's work for a fair day's wage, is almost impossible. Perhaps it is so, but if it is, why blame the I. W. W., if he is convinced of the truth of it too? Enclosed find clipping from the Los Angeles 'Times,' 'Edgerton Discusses Federal Ownership,' especially paragraph headed 'Transition Period.' The proposition is this: either men can be governed by reason, by justice, and by fairness, or they cannot. It is either the American idea of government or the German. The battle royal between these two conflicting ideas is on, and must be fought to a finish ere the sword is sheathed in Europe. Should the War last two or three years longer, I expect to see American army and navy officers in command of the Allied military and naval forces; an American field-marshal and an American admiral outranking all other admirals. General Goethals did a very bad thing for the private control of big interprise when he decided to turn down the contract method on the Canal, and gave the socialists cause to rejoice. Another thing, the officers of all the armies at war are climbing out of the trenches into uniforms with shoulder-straps, consequently the armies are more democratic than ever. When reforms are demanded, backed up by a victorious army, democratic to the core, what will happen? Why, Sir, there would be nothing to do but get out of the way of that reform and that army pushing it along. Have you ever noticed how much good can be done by a kind word? How it can put courage and hope into an almost broken heart? Did you ever try having confidence in people and telling them of it? Have you ever, as Daniels has done, when reviewing the findings of a court-martial, suggested to the Judge Advocate, "Put yourself in his place." Do you remember that the first workingman was a beaten brother, a slave, conquered in battle, probably with a war club or a stone hatchet, tamed by hunger and blows to such an extent that he obediently worked for his conqueror to preserve his life? What wonderful ideas have caused the abolition of such conditions, and how far have we gotten from the first master and servant. This great advancement is not due alone to the struggles of the slave to be free; a great deal of credit is due to the minds that were imbued with lofty ideas of justice. That's what Americanism means. Liberty and justice to the oppressed!

J. F. HARRINGTON.

Chloride, Arizona, October 28.

[Mr. Harrington refers to a news-letter from Treadwell, Alaska. Our correspondents are allowed to express their own views concerning local events; we do not always agree with what they say; in short, the editor is not responsible for every opinion expressed in this paper. We discuss this letter on another page.—EDITOR.]

Hydro-Metallurgy of Copper Sulphides

By A. E. DRUCKER

INTRODUCTION. The hydro-metallurgical treatment of table and flotation concentrate on the spot in a good many cases has proved to be more profitable than transporting to a smelter. The smelter generally gains at the shipper's expense mainly through the extraction of the metal and the moisture deduction. A large number of mining companies that in the beginning shipped gold and silver-bearing concentrate to a smelter, paying in some cases exorbitant freight rates and smelting charges, finally adopted a hydro-metallurgical process at the mines with the result that large additional profits were realized. The saving on drying, bagging, carting, and

ACID-PROOF DORR THICKENER, MERRIMAC CHEMICAL COMPANY

transporting ranged from $5 to $20 per ton plus a 3 to 7% increase in the recovery of the gold. A saving also was made in certain cases on the cost of treatment. The following list includes some important companies that have made this change:

	Tons treated	Metal contained
Oriental Consolidated, Korea	2400	gold
Alaska Treadwell, Alaska	3000	"
North Star, California	...	"
Esperanza, Mexico	...	"
Waihi, New Zealand	500	silver-gold
Frontino & Bolivia, Colombia	150	gold
Goldfield Consolidated, Nevada (preliminary roast)	2400	gold-silver
Ivanhoe, Australia (preliminary roast)	1950	gold
Golden Horseshoe, Australia (preliminary roast)	2000	"
Oroya-Brownhill, Australia (preliminary roast)	1000	"
Lake View Consols, Australia (preliminary roast)	1800	"

The cost of treatment per ton at the above plants varies from $2 to $6, while the recovery of precious metals is from 90 to 97%. The successful treatment on the spot has been developed during the past ten years. There is equally as great a field with a similar, acid-leaching, treatment in the case of copper sulphides, and I believe that hydro-metallurgy will become eventually as important to the copper industry as cyanidation has been to gold and silver miners. Present-day smelting methods of extracting copper from table and flotation concentrate will be gradually replaced, except in a few special cases, by combined roasting and leaching, producing electrolytic copper direct in marketable form.

In the consideration of the acid-leaching process the first factor is the character of the concentrate. The table and flotation concentrates obtained from the porphyry copper ores at Bingham, Globe, Ely, and Miami should be amenable to acid-leaching after a preliminary roast. This concentrate contains 10 to 28% of copper, 16 to 23% iron, 22 to 32% sulphur, and 18 to 28% insoluble, and with the exception of the Bingham concentrate there is no appreciable amount of gold and silver.

TREATMENT. The following is a general outline of the treatment which I suggest for the porphyry sulphides or similar material amenable to acid-leaching. No attempt has been made to go into the details of roasting, leaching, and electro-precipitation. As a general rule porphyry concentrate contains practically no zinc, lead, silver, or gold, and therefore is particularly suited to roasting and acid-leaching. Copper is one of the most readily soluble of all metals, and most easily precipitated electrolytically. Small amounts of zinc present would not affect the operation materially. The zinc problem has been satisfactorily solved at Anaconda. Where gold is present to the extent of $2 to $3 per ton of concentrate, as in certain cases at Bingham, Utah, acid-leaching would not recover this, and it is questionable if smelting could be replaced advantageously.

ROASTING. A multiple-hearth muffle roasting-furnace has been successfully employed at low temperatures for sulphate-roasting, and in some cases as high as 80% of the copper was rendered soluble in water as sulphate. The total soluble copper has been increased to over 95% by the use of a weak acid-solution. If the sulphides are roasted so as to make about 50% of the copper soluble in water, the remaining copper may be extracted with the regenerated acid resulting from the electro-deposition of this metal. For every pound of copper deposited 1.54 lb. of acid is regenerated. With such regenerative methods no additional acid need be provided and no fouling of solutions is likely to occur. The Chile Copper Co. offers one example. The sulphurous gases from a roasting-furnace have been applied direct to the wetted crude ore or as a spray to ore pulp, producing the necessary acid solvent. This is a means of dispeusing with the usual expensive acid-making plant. The sulphatizing of the wetted roasted sulphides with sulphur gases, followed by leaching with the regular solutions from the plant, is one cheap and effective way of getting the copper into solution. The elimination of sulphur from the sulphides is accompanied by the evolution of considerable heat generated during its oxidation. Concentrate containing as low as 15% sulphur has been roasted to a certain point in multiple-hearth furnaces without the addition of fuel preliminary to smelting. On concentrate high in sulphur, that is, from 20 to 35%, the roasted material may be reduced as low as 7 to 9% sulphur without the use of fuel. Reducing this 7% down to 3 or 2% requires extra fuel.

For a sulphatizing roast, the temperature should not exceed 600 to 650°C. Copper sulphides are not particularly sensitive to high temperatures, and if sintering or fusion occurs, it is impossible to get a satisfactory extraction. The product, when well roasted for acid-treatment, will contain from 3 to 4% of sulphur as sulphide, containing practically all the copper in the form of sulphate or oxide, and all the iron in the insoluble ferric state. The Wedge multiple-hearth type of furnace is being successfully used for the chloridizing roast of silver and copper ores. The furnace may be fired with oil, gas, coal, or wood, if a more thorough roast is desired for hydro-metallurgical work. In the case of fuel oil, it can be injected into the lower finishing hearths as desired. If solid fuel be used, the enclosed fire-box type may be employed. The fuel consumption with wood or coal usually varies between 9 and 16% of the weight of the concentrate, when roasting to less than 1% sulphur remaining in the form of sulphide. In roasting sulphides down as low as 7 to 10%, fuel is not ordinarily necessary. At Kalgoorlie, Western Australia, the following results were obtained on concentrate:

Company	Sulphide sulphur % Raw	Sulphide sulphur % Roasted	Wood used weight %	Furnace used
Lake View	25-30	0.35	9.8	Edwards
Oroya-Brownhill	35	0.20	16.0	Merton
Ivanhoe	27	0.11	14.0	Edwards

COUNTER-CURRENT ACID-LEACHING PLANT, DAVISON CHEMICAL COMPANY

MECHANISM FOR CONTROLLING LEACHING-VATS, DAVISON CHEMICAL COMPANY

I believe that with a modern Wedge or McDougall furnace concentrate or copper sulphide could be de-sulphurized to 1% sulphide sulphur with a consumption of 6 to 8% by weight of wood or coal. At the Goldfield Consolidated, Nevada, nine gallons of crude oil is required per ton of concentrate, or 3.3% of the weight. The raw concentrate contains about 19% S, and the roasted product 0.15% S. The cost of roasting sulphides varies from 25c. to $1 per ton, depending upon the ton-

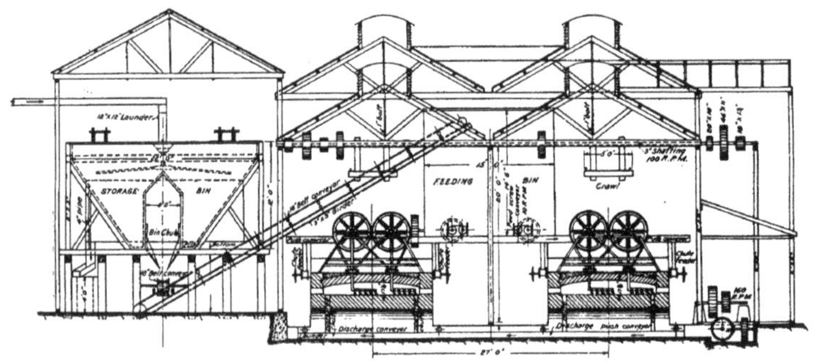

ROASTING-PLANT, GOLDEN HORSESHOE ESTATES CO., SHOWING EDWARDS FURNACES

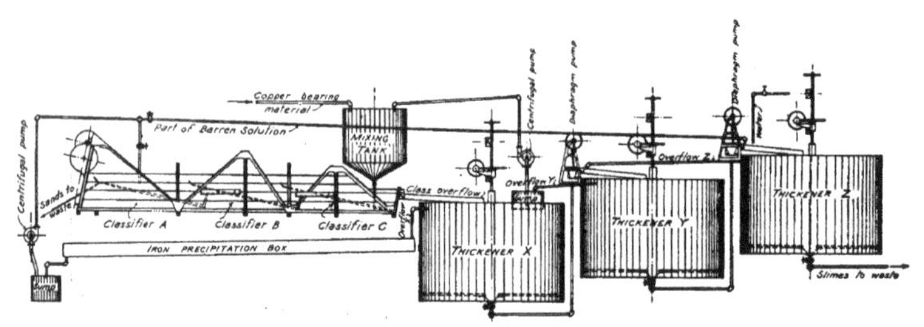

PLANT FOR LEACHING COPPER CALCINE FROM PYRITE-BURNERS

nage, the degree of roasting, the type of furnace, the fuel used, and other local conditions. At Butte and Garfield costs have been as low as 25 to 35c. per ton for incomplete roasting in McDougall furnaces, reducing the sulphur down to 7% prior to smelting. I see no reason why the cost of this preliminary operation to acid-leaching with a well-designed roasting-plant to treat 100 to 200 tons of sulphides per 24 hours would not average about 50c. The cost of a modern roasting-plant of the Edwards, Ridge, or the Wedge and McDougall multiple-hearth type, will vary between $600 and $800 per ton of daily capacity for de-sulphurizing.

Leaching and Precipitation. After roasting the sulphides, the copper should be in the form of sulphate and oxide, although some sulphide will remain, as a complete roast in this particular case is not desired. The roasted material is then treated by the ordinary sulphuric-acid leaching process. The oxide of copper is readily soluble in weak-acid solution, and to a certain extent by the ferric sulphate formed by air-agitation of the pulp and by the electro-deposition of copper from the solutions. Copper sulphate is practically all soluble in water, especially when some acid is present. After a proper roast it is an easy matter to get the copper into solution very completely, but, as all metallurgists know, the most difficult part of the process is to precipitate the metal efficiently and economically in a marketable form. Most of the improvements in the future will be in the electrolytic deposition of the copper from these complex solutions. Considerable advancement has been made with this at Anaconda and at Chuquicamata, Chile. If there is lime in the original ore, it will not be present in the concentrate in injurious amounts. The same is largely true with other impurities which usually cause a loss of acid and also contaminate and foul the electrolyte, causing trouble in the electrolysis and producing impure copper. I realize that it is not all smooth sailing with the acid-leaching process and especially with electric precipitation, but I do say that the problems are no more complex and difficult than with the cyanide-leaching process, which during the past thirty years has been perfected to a high point. There are no such complex solutions to deal with as compared with those resulting from cyanide leaching of mixed sulphides.

At the present time the electrolytic process for the recovery of copper from acid solutions offers economic possibilities, especially when one considers the present cost of re-melting blister copper into anodes and refining these. The regeneration of the acid solvent to a large extent is a most important consideration. Sulphate solutions, resulting from acid leaching in practice, will require a voltage of about 1.5 to 3, with a current density of 5 to 12 amp. per sq. ft. In practice it has been found that one to two pounds of copper will be deposited per kilowatt-hour, regenerating the acid.

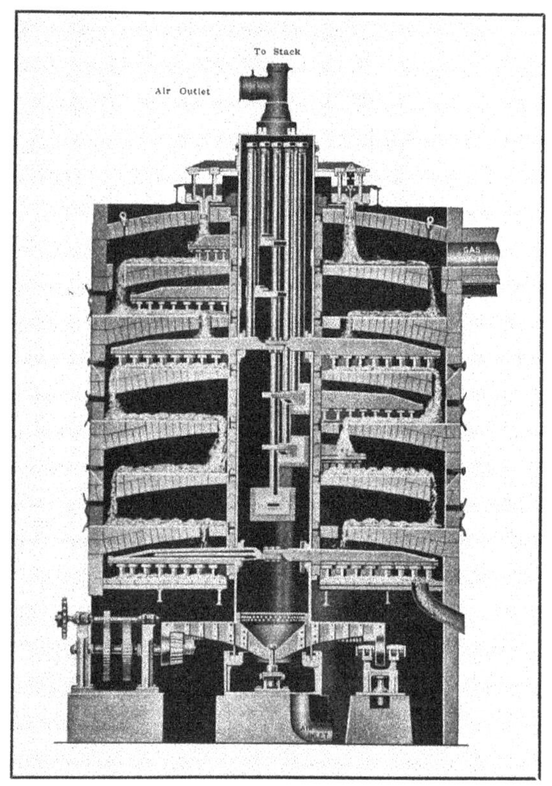

WEDGE ROASTING-FURNACE

With concentrate containing considerable slime and material finer than 200 mesh, the ordinary sand-percolation treatment in vats would not give a satisfactory extraction. It has been found on the Rand, and in other places, that a sand-charge containing more than 2 or 3% of slime will not give the best result. The presence of slime is detrimental when it goes above that amount, and the extraction will gradually fall as the amount of slime increases. The most efficient and profitable method for leaching roasted concentrate finer than 150 mesh would be air-agitation of the pulp of a consistence of 1 to 2 or 1 to 3, in Dorr agitators followed by dilution with the 5 to 10% acid-solution, and with counter-current decantation in a series of Dorr thickeners. A water-wash would be used in the last thickener. I think that there would not be any need of a vacuum or pressure-filter at the end of this operation for removing the last traces of dissolved copper. This part of the operation would be identical with the latest practice in cyanide counter-current decantation, which has proved so successful on gold and silver ore. For treating the heavy material, coarser than 150 mesh, we have the Dorr agitator, followed by the multiple-deck washing classifier, which is suitable for counter-current leaching and washing. A four or five-deck machine could be used for this purpose, or several standard Dorr classifiers operating in series. All machinery used in leaching must be acid-proof.

Costs of construction of a plant vary largely in different parts of the United States, depending on the cost of material, transportation, labor, and other local conditions. A plant consisting of Wedge furnaces, acid-recovery equipment, Dorr classifiers, agitators, and thickeners, and with clarifying-filters and the necessary electro-precipitating plant, would cost from $1500 to $1800 per ton of concentrate treated per 24 hours. The cost for treatment per ton of concentrate would vary between $3 and $5. Copper extraction would vary from 90 to 95% or more, depending on the degree of roasting, time of treatment, and the percentage of copper in the concentrate. There is reason to believe that, within the next few years, hydro-metallurgical methods, with electrolytic deposition, will play an important part in extracting marketable copper direct from concentrate on the spot. Smelting methods gradually will be replaced to a large extent by the more economical wet methods. Copper, zinc, gold, and silver-bearing concentrate are nearly all amenable to the wet methods after roasting. In many cases the marketing concentrate costs in the aggregate from $10 to $40 per ton. This includes freight, sampling, smelter-charges and deductions, electrolytic refining, and selling. Under such conditions the leaching of copper offers great possibilities at a cost that should not exceed $3 to $6 per ton.

DOMESTIC prime white floated barytes now brings $30 per ton, and off-color grades sell at $22 to $24. The foreign competition is practically eliminated on account of lack of shipping facilities.

Mining Dividends and Income Tax

An income-tax ruling just promulgated by the Treasury Department will interest stockholders in a number of mining companies that have adopted the plan of declaring dividends out of reserve set aside to meet depreciation and depletion of property. The ruling follows:

"Referring to the practice of certain corporations of declaring dividends out of reserve set aside to meet depreciation and depletion of property, and of advising stockholders that such dividends represent a distribution of capital assets, your attention is directed to the ruling made herein as follows:

"All such dividends received by stockholders declared out of such reserves, accumulated subsequent to March 1, 1913, constitute income to the stockholder under the act of September 8, 1916, and must be accounted for in returns of net income.

"A stockholder's investment is in the stock of a corporation. If he disposes of his stock for more than its fair market value on March 1, 1913, or its cost if acquired since that date, the profit realized must be returned as income; if he disposes of it at a loss, the loss sustained is deductible from gross income within the limits of the taxing act. In computing the profit or loss sustained, there must be taken into account dividends paid from reserves accumulated prior to March 1, 1913, which were not returned as income for the year in which received, under the provisions of the act of September 8, 1916."

The gist of this ruling is to subject to the income tax those dividends that have been called a distribution of capital assets. Hitherto it had been supposed in some quarters that dividends representing distribution of capital were not subject to the tax. Such dividends have been declared by Utah, Chino, Nevada Consolidated, Ray Consolidated, and St. Joseph Lead companies.

AN OCCURRENCE of nickel ore near Perto da Villa de Livramento in Minas Gerães, Brazil, is the subject of a paper by Horace E. Williams, published by the Serviço da Agricultura e Mineralogico do Brazil. The property is 261 kilometres from Rio de Janeiro, and 55 kilometres north-east of Pico de Itatiaya, in the municipality of Turvo. The region is composed mostly of ancient gneissoid rocks, while the chief mountain-mass in the vicinity consists of nepheline syenite. This is intruded by a basic rock in which the chief constituent is olivine, and it is in this rock that the nickel is found. The basic intrusive has been extensively altered to serpentine, and the outcrop contains the nickel as garnierite, much of which will assay 3½%, and some of it as high as 8 to 15%. Mr. Williams attaches minor importance to the deposits so far explored as a basis for a considerable nickel industry, but recent explorations have indicated that this serpentinized belt may yield important amounts of chrome ore, with nickel ores as a valuable accessory.

COBALT oxide is in good demand, and is selling at $1.50 to $1.60 per pound.

Proportions for Cement Mortars and Concretes

By J. A. KITTS

*THE VOID THEORY. As an initial step in the determination of the laws of concrete mixtures, it is more important that one combination of materials be studied thoroughly in a great number of progressive proportions with enough test-specimens of each mixture properly to reduce the errors of the average results, than it is to study many combinations of a large number of materials with only a few mixtures and a few test-specimens of each combination.

TABLES I AND II. PROPERTIES AND SCREEN-ANALYSIS OF SAND IN WITHEY'S TESTS

Sand No.	Weight per Cubic Foot, lb.	Specific Gravity.	Voids, per cent.	Silt, per cent.	Absorption, per cent.
Standard	104.1	2.65	37.0	...	....
Sd. 1	106.2	2.66	36.5	3.0	0.19
Sd. 2	106.9	2.74	38.2	1.3	0.40
Sd. 3	101.6	2.66	39.2	0.8	0.27
Sd. 4	98.9	2.68	30.5	1.2	0.04
Sd. 5	91.2	2.67	45.2	0.5	0.17
Sd. 7	106.2	2.78	38.6	1.6	0.41
Sd. 8	105.3	2.70	36.4	1.5	0.18
Sd. 9	105.5	2.75	38.0	0.7	0.51
Sd. 10	120.3	2.77	27.9	7.7	0.20
Sd. 11	108.7	2.72	35.0	0.4	0.11

Sand No.	Percentage by Weight Passing Sieve Nos.							Uniformity Coefficient.
	10	20	30	40	50	74	100	
Standard	100.0	100.0	0.0	....	....	...	...	...
Sd. 1	86.4	65.5	38.0	25.0	9.1	5.4	2.6	3.0
Sd. 2	81.2	70.9	61.4	52.4	34.5	17.9	9.7	2.6
Sd. 3	91.6	73.7	39.7	26.4	13.7	4.2	1.2	2.4
Sd. 4	100.0	99.5	98.9	88.8	62.5	22.1	8.0	1.8
Sd. 5	100.0	99.9	99.5	98.1	67.6	18.2	5.7	1.8
Sd. 7	67.7	44.5	26.0	17.8	11.2	3.8	2.2	4.7
Sd. 8	88.0	72.5	55.3	36.6	16.1	3.6	1.9	2.6
Sd. 9	65.8	33.7	13.9	10.3	4.9	2.4	1.6	3.7
Sd. 10	69.7	48.5	34.0	26.2	17.4	8.7	6.5	6.9
Sd. 11	79.0	47.7	28.8	12.0	4.3	0.7	0.5	3.4

With a certain combination of cement, sand, and rock, we find that an arbitrary 1:2:4 mixture shows greater strength in proportion to cost than a 1:3:6 mixture, and that a 1:2½:6 mixture is more efficient than either. It is probable that no one of these arbitrary mixtures is the most efficient and it is apparent that any number of arbitrary mixtures may be made and tested with the common result of producing more confusion than knowledge of the laws of mixtures. The arbitrary method of proportioning has been barren of positive results. Some other methods must be used if we are to obtain accurate knowledge of the laws governing mixtures.

*Paper presented at the annual meeting of the American Society for Testing Materials.

The void theory of proportioning has been known for a number of years, but has not been given proper consideration, and the fundamental principle has not been appreciated. This lack of consideration has been because the theory, as generally known, defined the proportions for maximum strength rather than for maximum efficiency or maximum strength in proportion to cost. A broad statement of the void theory is as follows:

The proportion for maximum efficiency of a mortar or a concrete is controlled largely by the proportions of voids in the respective aggregates.

TABLE III. VOLUMETRIC PROPORTIONS FOR SIMPLE WEIGHT PROPORTIONS

Sand No.	Weight Proportions.			
	1:2	1:3	1:4	1:5
Sd. 10	1:1.77	1:2.66	1:3.54	1:4.42
Sd. 2	1:1.98	1:2.97	1:3.96	1:4.94
Sd. 11	1:1.99	1:2.98	1:3.97	1:4.97
Sd. 7	1:2	1:3	1:4	1:5
Sd. 9	1:2	1:3	1:4	1:5
Sd. 8	1:2.05	1:3.07	1:4.10	1:5.12
Sd. 1	1:2.08	1:3.12	1:4.17	1:5.20
Standard	1:2.11	1:3.17	1:4.23	1:5.27
Sd. 3	1:2.13	1:3.20	1:4.27	1:5.33
Sd. 4	1:2.23	1:3.34	1:4.46	1:5.57
Sd. 5	1:2.41	1:3.62	1:4.82	1:6.03

This aspect of the void theory is the basis of the following considerations.

MORTARS. A mortar is a mixture of sand, cement, and water in various proportions. The question of scientific interest is: What determines the proportions for maximum efficiency?

Sands vary in physical, chemical, and mechanical structure, causing a variation in the specific gravity and percentage of voids, and the last two properties cause a variation in the weight per cubic foot or aggregate specific gravity. An important consideration in the study of a large number of mortars from a large number of aggregates is that of comparison. A study of the characteristics of various sands will show that mortars are not comparable either in arbitrary weight-proportions or in arbitrary volumetric proportions. What then determines the conditions for comparison of the mortar from one sand with that from another? An analysis of the results indicated in M. O. Withey's test of mortars throws considerable light on the two preceding questions.[1] In these tests mortars were made in 1:2, 1:3, 1:4, and 1:5 weight proportions and the following tests made: Unit tensile strength; unit compressive strength; leakage of water through specimens 2 in. thick, with pressures of 10 and 40 lb. per sq. in.; density; yield; and compressive strength in proportion to cost. Tables I and II show the physical and mechanical characteristics of eleven of the sands used in these tests.

TABLE IV. RATIO OF VOLUME OF CEMENT PASTE TO VOLUME OF VOIDS IN SAND FOR SIMPLE WEIGHT-PROPORTIONS

Sand No.	Weight Proportions.			
	1:2	1:3	1:4	1:5
Sd. 10	2.03	1.35	1.01	0.81
Sd. 11	1.44	0.96	0.72	0.58
Sd. 2	1.43	0.96	0.72	0.57
Sd. 9	1.39	0.93	0.70	0.56
Sd. 7	1.37	0.91	0.68	0.55
Sd. 8	1.34	0.90	0.67	0.54
Sd. 1	1.32	0.87	0.66	0.53
Standard	1.28	0.85	0.64	0.51
Sd. 3	1.22	0.82	0.61	0.49
Sd. 4	1.13	0.75	0.56	0.45
Sd. 5	0.91	0.61	0.46	0.37

Tables III and IV show the variations of the volumetric and void conditions common to simple weight-proportioning.

Table III is computed by the following equation:

$$\frac{\text{Aggregate volume sand}}{\text{Aggregate volume cement}} = \frac{\text{Weight proportion sand}}{\text{Weight proportion cement}} \times \frac{\text{Aggregate specific gravity cement}}{\text{Aggregate specific gravity sand}} \quad (1)$$

$$\text{Aggregate specific gravity}^{2} = (1 - \text{proportion of voids}) \times \text{specific gravity} \quad (2)$$

$$= \frac{\text{Weight, in lb. per cu. ft.}}{62.5}$$

$$= 110/62.5 = 1.76 \text{ for cement}^{3}$$

Table IV is computed by the following equation:

$$\frac{\text{Volume of cement-paste}}{\text{Volume of voids in sand}} = \frac{\text{Aggregate specific gravity of sand}}{\text{Aggregate specific gravity of cement} \times \text{weight proportion of sand} \times \text{voids in sand}} \quad (3)$$

$$\frac{\text{Volume of cement-paste}}{\text{Volume of voids in sand}} = \frac{1}{\text{Volume proportion of sand} \times \text{proportion of voids in sand}} \quad (4)$$

Table III shows that the volumetric proportions correspond to the 1:2 weight proportions vary from 1:1.77 to 1:2.41, the 1:3 weight from 1:2.65 to 1:3.62 volume, 1:4 weight from 1:3.54 to 1:4.82 volume, and the 1:5 from 1.4.42 to 1:6.03. Table IV shows the proportions of voids in the sands filled with cement-paste varying from 0.91 to 2.03 for the 1:2 weight proportion, from 0.61 to 1.35 for the 1:3, from 0.46 to 1.01 for the 1:4, and from 0.37 to 0.81 for the 1:5 proportion. Nothing could better illustrate the fallacy of the practice of comparing work-sands with standard sand in 1:3 weight proportions.

Neither is there a basis of comparison in a fixed volumetric proportion. In 1:3 volumetric proportions the proportions of the voids in the sands filled with cement-paste would vary from 0.73 to 1.20 for the eleven sands as shown in Table V. It must be admitted, from both scientific and from practical considerations of a mortar, that an important function of the cement is to fill the voids in the sand; and it cannot be expected that a mortar in which the volume of cement-paste is equal to only 73% of the volume of voids in the sand, is comparable with one where the cement-paste is equal to 120% of the voids.

TABLE V. RATIO OF VOLUME OF CEMENT PASTE TO VOLUME OF VOIDS IN SAND, FOR VOLUMETRIC PROPORTIONING

Sand No.	Volumetric Proportions.			
	1:2	1:3	1:4	1:5
Sd. 5	1.10	0.73	0.55	0.44
Sd. 4	1.25	0.84	0.63	0.50
Sd. 3	1.31	0.87	0.65	0.51
Standard	1.35	0.90	0.68	0.54
Sd. 7	1.35	0.91	0.68	0.55
Sd. 1	1.36	0.91	0.68	0.55
Sd. 8	1.37	0.92	0.69	0.56
Sd. 9	1.39	0.93	0.69	0.56
Sd. 2	1.42	0.95	0.71	0.57
Sd. 11	1.43	0.95	0.71	0.57
Sd. 10	1.79	1.20	0.90	0.72

Assume, for example, that six mortars of a certain sand and cement are made in which the ratio of the volume of cement-paste to the volume of voids in the sand has, respectively, the values 0.50, 0.75, 1.00, 1.50, 2.00, and 3.00. It is reasonable to anticipate that the strength of the mortars will increase at a rapid rate until the voids in the sand are filled with cement-paste, after which the strength will increase at a lesser rate until the sand-particles are so widely separated by the cement-paste that the strength of the mortar will approximate closely the strength of the neat cement. In other words, as the cement-content is increased, the rate of increase of strength is greater before the voids are filled than it is after the voids are filled. Also in some finite proportions the strength of the mortar approximates the strength of the neat cement. Fig. 1, which shows the compressive and tensile strengths and the strength in proportion to cost for sand No. 10, plotted against the void-filled ratio, illustrates these assumptions very well. This sand, unfortunately, is the only one of this series of tests which had one mixture (the 1:4 by weight) where the volume of cement-paste was approximately equal to the volume of voids in

[1]M. O. Withey, 'Tests of Mortars Made from Wisconsin Aggregates,' Proceedings, American Society for Testing Materials.

[2]In an article on this subject published in WESTERN ENGINEERING for August 1915, I called this "apparent specific gravity," for want of a better term.

[3]110 lb. of cement makes 1 cu. ft. of normal-consistence paste

the sand, and which had more than one mixture where the volume of cement-paste was greater than the volume of voids in the sand.

Fig. 2 shows the compressive strengths plotted

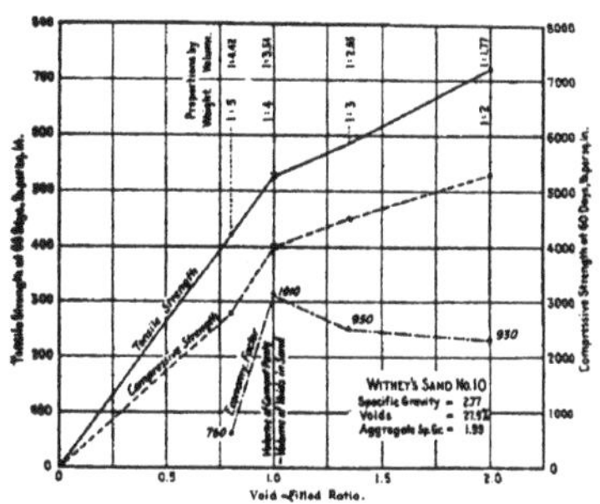

FIG. 1. VARIATIONS IN STRENGTH OF MORTAR FROM SAND NO. 10 IN PROPORTION TO VOID-FILLED RATIOS

against the void-filled ratio. The curve of averages supports the assumption as to the rate of increase of strength. The curve is fairly uniform, and would appear to indicate that the void-filled ratio has a similar effect on all the sands. This, then, appears to establish the principle that the properties, strength, and other characteristics of mortars, are properly comparable on the basis of the void filled ratio.

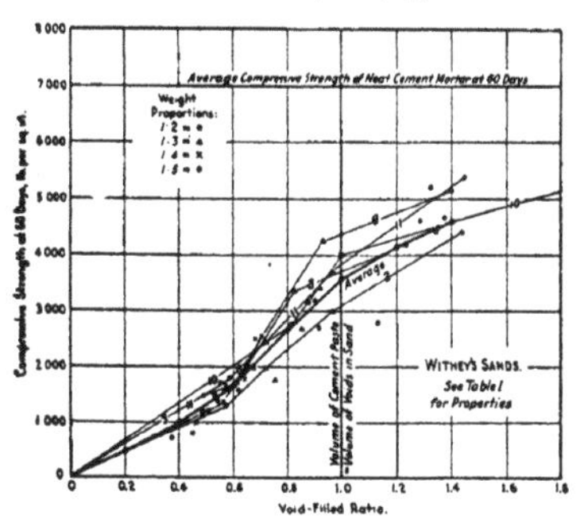

FIG. 2. VARIATION IN STRENGTH OF MORTARS IN PROPORTION TO VOID-FILLED RATIOS

The final and most important consideration of a mortar is the strength in proportion to the cost. This may be expressed as an 'economy factor,' equal to the compressive strength in pounds per square inch divided by the cost of the mortar in dollars per cubic yard. This factor may be expressed by the following equation:

$$\text{Economy factor} = \frac{\text{Compressive strength (lb. per sq. in.)}}{\text{Cost of mortar (dollars per cu. yd.)}} = \frac{\text{Compressive strength} \times \text{yield}}{C_s + \frac{P_c \times C_c}{P_s}} \quad (5)$$

in which P_c and P_s are the volumetric proportions of cement and sand, C_c and C_s are the costs, in dollars per cubic yard, of cement and sand, and the yield is based on the volume of the sand as unity.

This factor is plotted in Fig. 10 of Withey's paper,

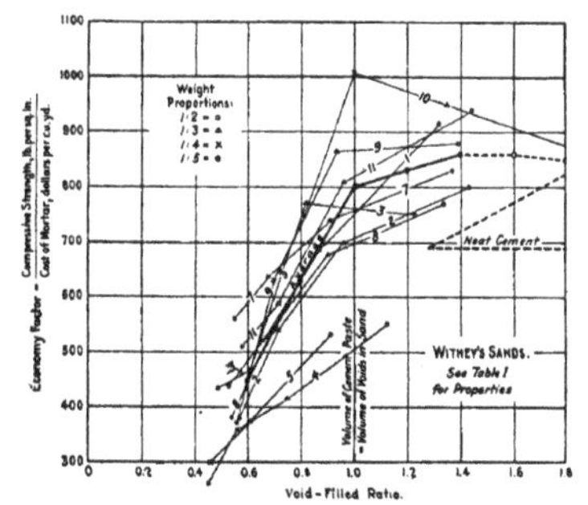

FIG. 3. VARIATION IN ECONOMY FACTOR IN PROPORTION TO VOID-FILLED RATIOS

mentioned above. The cost represents costs of materials only, cement being estimated at $1.50 per barrel and sand at $1.25 per cubic yard. In Fig. 3 these factors for all eleven sands have been plotted against the void-filled ratio. The average compressive strength of the cement used in these tests was less than 7000 lb. per sq. in. at 60 days. The economy factor for the neat cement mortar would then be approximately 7000 ÷ 10.14 = 690.[4] The highest economy factor shown on the curve of averages is 860. This and the curve of averages indicate that the economy factor decreases when the void-filled ratio is somewhat in excess of 1.5. The results of these tests, therefore, indicate that the most economical mixtures lie between proportions giving a void-filled ratio from 1 to 1.5.

As an illustration, Fig. 3 shows that the most economical proportions for sand No. 10 is that in which the volume of cement-paste is equal to the volume of voids. From Fig. 1, it is seen that this proportion is 1

[4]It should be noted here that this value for the 'economy factor' of neat-cement mortar, as well as the other factors in Fig. 3, is based upon the assumption that 1 bag of cement gives 1 cu. ft. of cement-paste. If, as has been done in this paper, it is assumed that it requires 110 lb. of cement to make 1 cu. ft. of neat-cement paste, the cost of a cubic yard of cement-mortar would be $11.85 instead of $10.14, giving an 'economy factor' of 590.

cement to 3.58 sand by volume, giving a tensile strength at 60 days of 525 lb. per sq. in., a compressive strength of 4000 lb. per sq. in., and an economy factor slightly over 1000.

The equations for economical mixtures, as indicated by this series of tests, may therefore be written:

$$\frac{\text{Volume of sand}}{\text{Volume of cement}} = \frac{1}{(1 \text{ to } 1.5) \times \text{proportion of voids in sand}} \quad (6)$$

or

$$\frac{\text{Weight of sand}}{\text{Weight of cement}} = \frac{\text{Aggregate specific gravity of sand}}{(1 \text{ to } 1.5) \times \text{aggregate specific gravity of cement} \times \text{voids in sand}} \quad (7)$$

If the properties of mortars from all sands vary with the variation of the void-filled ratio, the leakage, density, and yield should show similar effects for all sands. Fig. 4 shows the leakage of the various mortars plotted against the void-filled ratio; and this is another proof that the void-filled ratio is the proper basis of comparison of the properties of mortars.

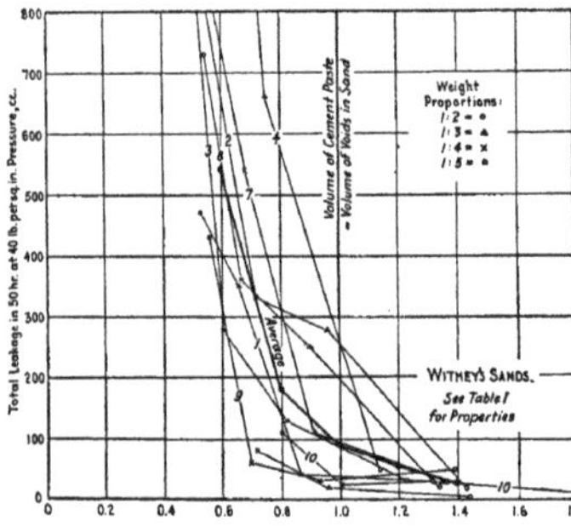

FIG. 4. VARIATION IN LEAKAGE OF MORTARS IN PROPORTION TO VOID-FILLED RATIOS

CONCLUSIONS REGARDING MORTARS. 1. Sand-cement mortars are not comparable in simple weight proportions because of the wide variation in the corresponding volumetric-proportions and the variations of the void-filled ratios.

2. Sand-cement mortars are not comparable in simple volumetric proportions because of the wide variations of the void-filled ratios.

3. The void-filled ratio has a general effect upon the strength, permeability, and economy of a mortar and undoubtedly affects the density and yield.

4. An important function of the cement-paste is to fill the voids in the sand.

5. Sand-cement mortars are properly comparable on the basis of the void-filled ratios.

6. The economical proportions for sand-cement mortars depend upon the void contents of the sands and may be expressed by equations 6 and 7.

7. The economy factor expresses the relative efficiency of mortars and may be determined by equation 5.

8. I do not find any general relation of silt-content, uniformity coefficient, and absorption to the efficiency of sands.

CONCRETE. Where concrete is made from either bank-run gravel, or a prepared concrete mix, and cement, it would be reasonable to expect the same general principles to apply as those previously stated for mortars. Where mortar is added to a coarse aggregate, such as rock, it is reasonable to assume that an important function of the mortar is to fill the voids in the rock.

A scientific test to determine the laws of concrete mixtures would be first, to determine the mortar of maximum density and the yield of mortar; second, to make progressive proportions of the mortar with the rock, using values of the ratio of the volume of mortar to the volume of voids in rock equal to 0.50, 0.75, 0.90, 1.00, 1.10, 1.25, 1.50, 2.00, and 3.00 with a view to determining that proportion giving maximum efficiency; and finally, to make additional progressive proportions, maintaining the ratio of fine to coarse aggregate of the 'efficient' proportion just determined, but increasing the amount of cement to 1.1, 1.2, 1.3, 1.4, 1.5, 2.0, and 3.0 times that used in the mortar of maximum density. The most economical proportions would be determined by the following equation:

$$\text{Economy factor} = \frac{\text{Compressive strength in lb. per sq. in.}}{\text{Cost of concrete in dollars per cu. yd.}} = \frac{\text{Compressive strength} \times \text{yield}}{C_r + \dfrac{P_s \times C_s}{P_r} + \dfrac{P_c \times C_c}{P_r}} \quad (8)$$

P_c, P_s, P_r being the volumetric proportions and C_c, C_s, and C the costs in dollars per cubic yard of cement, sand, and rock, respectively. The yield is based on the volume of the coarse aggregate (rock) as unity.

There are no test-data extant regarding the economical proportions of concrete. Some tests have been made by the U. S. Bureau of Standards using the void theory of proportioning in a limited way. The results of these tests will be analyzed in the following discussion and may throw some light on the subject.

The economy of any mixture depends largely upon the local cost of materials. If the cost of cement is twelve to fifteen times that of the aggregates, the economy of lean mixtures will be greater than if the cost of cement is comparatively lower. In Table VI, the economy factor for several 1:2:4 mixtures is compared with that of several void-theory mixtures, the factors having been computed by me from the results shown in Tables 8 and 12 of the Bureau of Standards paper.[5] The void-theory mixtures may be expressed by equations 9 and 10, the void-filled ratio (1 to 1.5) being 1.1 in every case both for the mortar and for the rock.

[5]Technologic Paper No. 58, U. S. Bureau of Standards.

e cost of the aggregates was assumed at $1.20 per)ic yard and the cost of cement at $12 per cubic rd. The yield of the various mixtures is necessarily umed as unity, as it is not possible to approximate yield closely by calculation.

It would appear from a study of this table, that the d-theory mixtures compare favorably with 1:2:4 xtures, from the standpoint of efficiency. The two nbinations of granite show the 1:2:4 mixtures as most efficient. However, there are too few combina- ns to be conclusive.

In Table VII the economy factor is shown for various nbinations of aggregates in three mixtures, 1:2:4, 3:6, and a void-theory mixture. In 33 comparisons, ratios for the most economical mixtures are as fol- rs:

oretical	1:2:4	1:3:6
18	15	
30		3
	30	3

The ratio of cement to total aggregate is 1:8.12 for theoretical mixture, compared to 1:6 and 1:9 for 1:2:4 and 1:3:6 mixtures. The fact that the 1:9)itrary mixture is much less efficient than an average 8.12 theoretical mixture and that the 1:8.12 theoret- mixture is somewhat more efficient than the 1:6 trary mixture would appear to indicate that there me basis for the void theory.

he 1:2:4 mixture generally appears to be an effi- t one and the explanation may be the efficiency of mortar. As the voids in sands are seldom below or above 50%, the ratio of cement-paste to voids the 1:2 mortar varies from 1 to 1.5, which corre- ds to the limits previously determined for efficient mortars. Rock seldom shows less than 35% or more than 50% of voids, so that in a 1:2:4 mixture the volume of mortar will vary from 1 to 1.5 times the volume of voids in the rock. A 1:3:6 mixture is seldom an efficient one because a 1:3 mortar is seldom efficient, few sands showing less than 33% voids.

TABLE VI. ECONOMY FACTORS COMPARING 1:2:4 MIXTURES WITH VOID-THEORY MIXTURES

Kind of Aggregate.	Volumetric Proportions.	Age at Test, weeks. 4	13	26	52
	1:2.4:4.47	570	720	900	930
	1:3:6	620	690	750	...
	1:2:4	700	750	960	...
	1:2.4:4.45	470	570	680	980
	1:3:6	410	530	600	800
	1:2:4	450	500	660	880
...ne......	1:2.4:5.46	800	1000	1050	900
	1:3:6	790	810	1010	1030
	1:2:4	800	820	...	1040
	1:2.8:6.8	660	780	840	1000
	1:3:6	570	750	820	880
	1:2:4	530	...	830	1340
	1:2.4:5.81	670	900	930	710
	1:3:6	490	730	940	970
	1:2:4	540	820	810	970
....	1:2.4:4.92	470	730	750	920
	1:3:6	390	440	530	580
	1:2:4	540	830	730	980
	1:2.4:7.0	640	820	790	...
	1:3:6	430	540	610	...
	1:2:4	700	...	930	...
....	1:2.4:6.2	960	930	1110	1230
	1:3:6	440	590	680	880
	1:2:4	880	1030	1080	1190
	1:2.2:6.4	750	930	980	...
	1:3:6	410	550	660	...
	1:2:4	700	...	970	...
....	Void Theory[a]	661	817	893	984
	1:3:6	503	624	719	850
	1:2:4	656	805	875	1050

[a] Table 13, *Technologic Paper No. 58.*

TABLE VII. COMPARISON OF THE ECONOMY FACTORS OF CONCRETES PROPORTIONED BY DIFFERENT METHODS

Kind of Aggregate.		Age at Test, weeks. 4	13	26	52	78	104
		1:2:4 MIXTURES—2 COMBINATIONS.[1], [4]					
Granite.......	Minimum.....	490	630	700	990	....	....
	Average......	510	730	710	1010	....	....
	Maximum.....	540	830	720	1040	....	....
		VOID-THEORY MIXTURES—2 COMBINATIONS.[2], [4]					
	Minimum.....	410	510	590	880	....	....
	Average......	440	620	670	900	....	....
	Maximum.....	470	790	750	920	....	....
		1:2:4 MIXTURES—20 COMBINATIONS.[1]					
Limestone.....	Minimum.....	280	380	450	780	1140	1280
	Average......	570	680	750	990	1260	1410
	Maximum.....	830	820	940	1360	1520	1500
		VOID-THEORY MIXTURES—20 COMBINATIONS.[2]					
	Minimum.....	300	390	410	470	1210	....
	Average......	600	760	800	890	1330	1710
	Maximum.....	890	1020	1050	1440	1590	....
		1:2:4 MIXTURES—10 COMBINATIONS.[1]					
Gravel........	Minimum.....	200	300	390	440	....	....
	Average......	550	600	740	810	....	....
	Maximum.....	870	1040	1090	1190	....	....
		VOID-THEORY MIXTURES—8 COMBINATIONS.[2], [3]					
	Minimum.....	290	420	460	530	....	....
	Average......	560	680	740	820	....	....
	Maximum.....	1080	930	1110	1230	....	....

[1] From Table 8 of *Technologic Paper No. 58.*
[2] From Table 12 of *Technologic Paper No. 58.*
[3] The two last combinations of Table 12 have been omitted, since they were not common to both mixtures.
[4] Two combinations in common to both mixtures used.

CONCLUSIONS REGARDING CONCRETES. 1. An important function of a mortar in concrete is to fill the voids in the coarse aggregate.

2. The efficiency of a concrete mixture depends largely upon the efficiency of the mortar.

3. For economical proportions the volume of cement should be equal to or greater than the volume of voids in the sand but should not exceed $1\frac{1}{2}$ times the voids in the sand, and the volume of mortar should be equal to or greater than the volume of voids in the coarse aggregate but should not exceed $1\frac{1}{2}$ times the voids in the coarse aggregate.

4. The equations for economical mixtures may be stated as follows:

$$\frac{\text{Volume of sand}}{\text{Unit volume of cement}} = \frac{1}{(1 \text{ to } 1.5) \times \text{proportion of voids in sand}} \qquad (9)$$

$$\frac{\text{Volume of rock}}{\text{Unit volume of cement}} = \frac{\dfrac{\text{Volume of sand} \times \text{yield of mortar}}{\text{Unit volume of cement}}}{(1 \text{ to } 1.5) \times \text{proportion of voids in rock}} \qquad (10)$$

5. The economy factor expresses the relative efficiencies of concrete mixtures and may be determined by equation 8.

GERMANS are said to be exploiting the Servian copper mines, of which one, the Bor, is of importance. They have also developed a source of copper in the Kielce district of Poland, where the existence of copper deposits had been known to the Russians but ignored. Meanwhile the copper mines of the Caucasus are being exploited on the Russian side.

Leaching and Purification of Zinc Sulphate

By K. B. THOMAS

In the roasting of zinc concentrate in a Matthiessen-Hegeler kiln, a considerable quantity of dust is deposited in the flues that lead from the kiln to the dust-chamber, and in the dust-chamber itself; at times this amounts to an equivalent of 2 or 3% of the green ore charged in the furnace. The amount of flue-dust deposited in the flues of any type of furnace, with any kind of ore-charge, is mainly dependent on three factors, namely, (1) moisture-content of ore when charged, (2) fineness of the ore, (3) draft condition on the top hearth of the furnace. In the roasting of concentrate we have a very fine ore to handle, approximately 75% through a 200-mesh screen, and 90% through a 100-mesh screen; also we are using an ore which on account of its fineness needs a greater draft in the furnace than when roasting coarser ores. These factors account for the high percentage of dust formed. The large amount of dust produced, and the high zinc-content, necessitated the recovery of the zinc, and as practically all of the zinc was in the form of a sulphate, smelting was eliminated, and leaching was adopted, and found to be profitable as well as practical in every respect.

The analysis of a composite of different ores in use on furnaces is as follows:

	%		%
Zinc	45.5	Alumina	2.0
Iron	7.0	Manganese	1.0
Copper	2.5	Lime	1.0
Lead	1.0	Sulphur	27.0
Silica	11.5		

The analysis of a similar composite-sample of flue-dust showed:

	%		%
Zinc	31.0	Silica	4.5
Iron	4.0	Manganese	0.6
Copper	1.8	Lime	0.3
Lead	1.3	Sulphur	17.5

The three following methods were used in the leaching and purification of the zinc-sulphate content of the flue-dust. The first two methods were tried on a small experimental scale only, and the third on a commercial scale.

FIRST METHOD. Take 95% by weight of flue-dust and 5% of roasted ore from the furnace; add an equal amount of water by weight, and boil for two hours. The following reactions take place:

$$Fe_2(SO_4)_3 + 3ZnO = 3ZnSO_4 + Fe_2O_3$$
$$CuSO_4 + ZnO + H_2O = ZnSO_4 + Cu(OH)_2$$

The ZnO must be slightly in excess to give the above reactions. As the lead, silica, and traces of silver are already in an insoluble form, upon filtration of the above liquor there results a solution of zinc sulphate, with traces of sodium salts derived from the original flue-dust, and also a small amount of manganese sulphate and aluminum sulphate. During the concentration of this liquor, preliminary to crystallization, the traces of sodium salts are thrown down in the form of scale on the heating coils.

SECOND METHOD. Roast the flue-dust at a dull-red heat until the following reactions are complete:

$$2FeSO_4 + \text{heat} = Fe_2O_3 + 2SO_2 + O$$
$$CuSO_4 + \text{heat} = CuO + SO_2 + O$$

The roasted dust is cooled and then treated with an equal weight of water, heated for one hour, and filtered. This gives approximately the same quality of liquor as the first method.

THIRD METHOD. The flue-dust is placed in a lead-lined dissolving tank with twice its weight of water, kept hot with a lead-pipe steam-coil, and agitated with a mechanical stirrer for 8 to 10 hours. Then add sufficient soda ash to give the mixture a basic reaction, thereby precipitating the iron and manganese and neutralizing any free acid that is present. Next treat the mixture with chlorite of lime to oxidize the balance of the iron and manganese, and add more soda ash to precipitate these. Sodium sulphide is then added to precipitate any traces of copper not removed by the soda ash. Zinc-dust may be used to remove the last traces of copper, but it has a tendency to reduce some of the iron and drive it back into solution. The mixture is then filter-pressed, using a press built of wood as far as practicable, with lead-pipe connections.

Concentrate the liquor in a lead-lined tank with a lead-pipe steam-coil until it registers 52° or 53° Baumé. Then crystallize in lead-lined tanks.

Dry the crystals for 20 minutes in a centrifugal machine with a brass or copper basket. This method gives a crystal pure enough for the lithopone manufacturers.

The residue from the filter-press contains all the silver, gold, copper, and lead, and is in a form acceptable by the refineries for the recovery of these metals. The manufacturing cost of producing zinc sulphate from flue-dust, making no charge for the flue-dust, but including maintenance of plant and all other items, is from $20 to $25 per ton.

KALGOORLIE, in Western Australia, began its career as a gold-mining district when Pat Hannan found gold-bearing cement in the desert country 25 miles east of Coolgardie on June 14, 1893. The deposit was the sort of concentration made by wind and rain in an arid region. The gold was extracted by dry-blowing. In the course of this superficial mining a quartz vein was struck in the bedrock, and it served as the basis for a wild-cat flotation called the Great Boulder Proprietary. While working on the low-grade quartz vein the shallow workings broke into a rich lode, the top of which, being soft, had been eroded and covered by detritus. This proved the vein that made the Great Boulder one of the great gold mines of the world. To the end of 1915 it had produced $49,092,333, and distributed $25,000,000 in dividends, but it began life as a puling wild-cat.

The Manganese Deposits of Philipsburg, Montana

The shortage of manganese brought about by the effect of the War on oceanic transportation and the derangement of the foreign manganese industry have stimulated production in many parts of the United States. Philipsburg, unknown a few years ago as a commercial source of manganese, has taken first rank as a domestic source of ore suitable for making ferro-manganese. For the few months prior to an examination by Joseph B. Umpleby of the U. S. Geological Survey in August of this year, the daily production was about 500 tons and this output will probably be maintained for several months. We are indebted to Mr. Umpleby for the information that follows.

All of the manganese mines are situated in the foothill area east of the town of Philipsburg, and at no point more than 2½ miles from it. The deposits are scattered irregularly over an area about 1½ miles north and south by about 1 mile east and west. Wagon-roads from 1 to 2 miles long lead from the mines to the Northern Pacific railroad. In August 1917, shipments were being made from 14 deposits, and several others were being tested.

The manganese deposits are enclosed in magnesian limestone and fall into four general classes, which grade into one another: (1) Concentrations of manganese oxides on the hanging-wall side of silver-bearing quartz veins; (2) replacements of certain beds of limestone where crossed by fissure-zones; (3) irregular replacements adjacent to the grano-diorite contact; and (4) replacements of certain beds up to 1000 ft. from the igneous contact, where fissures have not been recognized.

Deposits of the several groups seem to run about the same in manganese, but the representatives of the third group contain a greater amount of waste, and the only member of the fourth class is high in silica. In preparing the ores for shipment, waste is picked out by hand and the entire product passed over screens to reduce the silica content. The screening from one of the larger mines runs 32 to 40% manganese and 20 to 28% silica, whereas the lump ore averages 40.5% manganese and 15% silica. Most of the ore shipped from the district runs about 45% manganese, less than 2½% iron, and about 15% silica, with low phosphorus. Material of this grade is worth about $30 per ton f.o.b. Philipsburg.

An effort will be made to concentrate the screening and the ore from several silicious deposits known to be of considerable size. To this end the Philipsburg Mining Co. is now building a concentrator, and several other operators are investigating the problem. Experiments show that it is possible to make a highly desirable concentrate, but that loss in the tailing is heavy.

In a locality like Philipsburg, where extraction follows development closely, it is exceedingly difficult to form an idea of the tonnage of ore available. Some of the larger orebodies will probably yield from 10,000 to 40,000 tons each, and it is possible that a few of them may greatly exceed this estimate. It also is likely that a number of the small prospects will add considerably to the total production of the district. In view of the present meagre development, however, it seems most reasonable to estimate that the Philipsburg deposits contain 135,000 tons of ore suitable for making ferro-manganese, and that the amount may be 350,000 to 400,000 tons. Under prevailing market conditions this ore should be recoverable. The manganese carbonate, rhodochrosite, appears on the dumps of some of the old silver mines, but as the workings are now inaccessible and no data are available concerning the quantity of such material, it is not included in the estimate. An incomplete analysis of a typical piece of the rhodochrosite follows: Mn 41.39, Fe 0.22, SiO_2 0.24%. Such material is highly desirable and it is possible that considerable deposits of it may be found below the oxidized ore.

It is expected that a more complete report will be issued by the U. S. Geological Survey on this subject within the next few months. Several geologists of the Survey familiar with metalliferous deposits have been making systematic examinations of manganese and manganiferous ore-bearing localities during this season with a view to appraising the available resources in manganese and to aid where possible the mining industry in extending exploration for these ores and thus further stimulate production.

Effect of Sulphides on Cement

Cement is materially affected by the presence of sulphides, whether in the original material or in the water employed for mixing. J. C. Witt has conducted elaborate experiments on this subject under the auspices of the Bureau of Science at Manila, Philippine Islands. His conclusions, printed in the Philippine Journal of Science (Vol. 9, sec. A, No. 6) are that, (1) The setting time of cement is greatly modified by the presence of sodium sulphide. With low concentrations the set is retarded, but, after reaching a maximum, further additions accelerate the set. In general, the cements highest in iron content are most sensitive to this influence; (2) there is a decided decrease in tensile strength. The percentage-loss varies with the concentration of the sulphide and with the iron content of the cement. The briquettes appear normal in every other respect, except in color. There is no cracking nor distortion of any sort; (3) in most cases sulphide may be present in concentrations up to 1 gramme per litre without causing the tensile strength to fall below the U. S. Government specifications; (4) certain results indicate that a colloid is formed by the action of sodium sulphide on the iron in the cement; (5) based on the results of both chemical and physical observations, the following explanations of the decrease in tensile strength are offered: (*a*) The precipitated colloid forms films of inert material through the cement and interferes with the cohesion; (*b*) When the colloid is precipitated a portion of the dissolved calcium hydroxide is removed from solution. Since the latter substance is a very important factor in the strength of cement, the strength will be lowered when some of it is removed; (6) It is probable that other factors influence the effect of sulphide on cement. Among these are the fineness of the cement, the temperature at which it is mixed, the proportion of water used, and the amount of dissolved calcium hydroxide.

Manganese Investigations

On account of the increasing difficulty of obtaining imported manganese and the vital importance of the metal in the production of the steel, the U. S. Bureau of Mines is making an investigation of the possibilities of relief through a more extensive use of domestic ores. Since the steel-industry is dependent on manganese and because the problems of manganese-ore concentration are similar to those in the case of iron ore, these investigations have been assigned to the Lake Superior station, recently established at Minneapolis.

The Lake Superior region is an important producer of manganiferous iron ores; it contains large reserves of these ores, and can mine and ship much larger amounts if metallurgical practice can be so modified as to create a better demand for them. For many years steel-makers have relied on manganese alloys imported from England, and on manganese ores from India, Russia, and Brazil. These ores are high in manganese and are suitable for making 80% ferro-alloy, which has been demanded in constantly increasing quantities in proportion to the growth of the open-hearth process. With the low ocean freight-rates that prevailed during the past 10 or 20 years, it was well nigh impossible for producers of domestic manganese ore to compete with imported ore. Consequently the manganese industry of the United States languished. There was little incentive for exploration and development and mining was carried on without system. With the outbreak of the War in Europe in 1914 the manganese industry faced new conditions. The entrance of Turkey into the conflict shut off shipments of Russian ore from the Caucasus. A growing scarcity of shipping made it increasingly difficult to transport the material from the usual sources of supply and the great rise in ocean freight-rates increased the cost at the very time when the producers in the United States faced an unprecedented demand. Until last year this demand had little effect on production because of the difficulty of interesting capital in an industry that might die with the ending of the War. A clearer and broader conception of the causes of high manganese-prices and a belief in the likelihood of these causes existing for some time after the War has partly removed that fear. A still greater difficulty in obtaining domestic supplies is the fact that manganese-ore of high grade, suitable for making 80% 'ferro,' is scarce in the United States, while manganiferous iron ores are much more abundant. These have not been in much demand by the steel-industry in the past and it is essential to investigate by research on a large scale the possibilities and limitations involved in their use. It is this work that the Bureau of Mines desires to undertake, and the sooner the facts can be determined the more valuable they will be. Suggestions and constructive criticism will be greatly appreciated.

The Bureau plans to assemble and tabulate as much information as possible concerning the character and available tonnage of known domestic ore, the current practice in mining and concentration, and the present methods of manufacturing manganese-alloys and of using them in steel. When compiled, this material should afford a starting point for original researches dealing with related problems.

The possibility of concentrating various types of manganese ores will be studied; fundamental thermo-chemical data relating to manganese and its compounds will be determined, and the equilibria of the fundamental reactions between iron and manganese compounds will be investigated. Also, attention will be given to the development of substitutes for manganese in steel-making, to the possibility of using alloys that are not employed in what is now considered standard practice, such as alloys containing between 20 to 80% manganese, and those containing manganese with other constituents, as well as the possibility of increasing the manganese-content of basic pig-iron in order to lessen the amount of the alloy added at the end of a heat. Part of this would be left as residual manganese in the bath-metal. Another problem will be the production of manganese-alloys in the electric furnace where power is cheap. The Bureau has been conducting a series of experiments at Pittsburgh on the production of alloys in the electric furnace from slags high in manganese. In the laboratories of the Lake Superior station, which co-operates with the University of Minnesota, concentration tests of ores will be made. Similar investigations in other alloys will be undertaken at laboratories throughout the country.

THE TECHNICAL commission of Colombia has recently concluded its studies for the completion of the Cauca Valley railroad from kilometre 63 south of the city of Cali to Popayán, a distance of 86 kilometres. The maximum grade will be 2.3%, and the minimum radius of curvature will be 80 metres. The estimated cost is $3,412,000. The construction of this road will open to development a territory rich in gold placers and in mines of silver. Popayán was a famous mining centre in the early days, and great fortunes were made. The remoteness of the region has retarded development of the mines by modern methods. The climate is salubrious, in this respect differing radically from miasmatic valleys in the lower Cauca. The railroad will open easy communication from the port of Buenaventura on the Pacific coast.

ALUMINUM sulphate is now quoted at 3¾ to 4¼c. per pound.

Effect of Mouthpieces on Flow of Water

An elaborate series of tests was recently made by Fred B. Seely on the effect of mouthpieces on the flow of water through a submerged short pipe. These experiments were made with the advantage of the excellent hydraulic equipment of the Engineering Experiment Station of the University of Illinois, and the results have been issued as Bulletin No. 96 from the Station. The data are given in detail, and are reduced to mathematical expression, accompanied by tables and curves. The conclusions reached are the following:

The preceding discussion has shown that the losses accompanying the flow of water depend largely upon the state of its motion, and this in turn is influenced by many factors, the effects of which often can be but roughly estimated. While the results of the experiments tend to define the range of such effects for certain conditions of flow, additional experiments would be necessary to establish all the inferences that have been suggested. The conclusions here given, however, seem justified:

(*a*) As applying to conditions likely to be met in engineering practice, the value for the head lost at the entrance to an inward-projecting pipe, that is, without entrance mouthpiece and not flush with the wall of the reservoir, is 0.62 of the velocity-head in the pipe ($0.62 \frac{v^2}{2g}$) instead of $0.93 \frac{v^2}{2g}$, as usually assumed. To put it in another form, the coefficient of discharge for a submerged short pipe with an inward-projecting entrance is 0.785 instead of 0.72 as given in nearly all books on hydraulics. Further, the lost head at the entrance to a pipe having a flush or square entrance is 0.56 of the velocity-head in the pipe ($0.56 \frac{v^2}{2g}$) instead of $0.49 \frac{v^2}{2g}$ as usually assumed. In other words, the coefficient of discharge for a submerged short pipe with a flush entrance is 0.80 instead of 0.82 as given by nearly all authorities.

(*b*) The loss of head resulting from the flow of water through a submerged short pipe when a conical mouthpiece is attached to the entrance end, may be as low as 0.165 of the velocity-head in the pipe ($0.165 \frac{v^2}{2g}$) if the mouthpiece has a total angle of convergence between 30° and 60° and an area of ratio of end-sections between 1 to 2 and 1 to 4, or somewhat greater. In other words, the coefficient of discharge for a submerged short pipe with an entrance mouthpiece, as specified above, is 0.915.

(*c*) The loss of head which occurs when water flows through a submerged short pipe having an entrance mouthpiece varies but little with the angle of the mouthpiece if the total angle of convergence is between 20° and 90°, and if the area-ratio is between 1 to 2 and 1 to 4, or somewhat more. The loss of head for any mouthpiece within this range would be approximately 0.20 of the velocity-head in the pipe ($0.20 \frac{v^2}{2g}$). There is, therefore, little advantage to be gained by making an entrance mouthpiece longer than that corresponding to an area-ratio of 1 to 2. Thus, an entrance mouthpiece, with a total angle of convergence of 90° and the length of which is only 0.2 of the diameter of the pipe, gives approximately $0.20 \frac{v^2}{2g}$ for the loss of head.

(*d*) The amount of velocity-head recovered by a conical mouthpiece when attached to the discharge end of a submerged short pipe depends largely upon the angle of divergence of the mouthpiece, but comparatively little upon the length of the mouthpiece. This is true for lengths greater than that corresponding to an area-ratio of 1 to 2 and for total angles of divergence of 10° or more. The amount of velocity-head recovered decreases rather rapidly as the angle of divergence increases from a total angle of 10° to 40°. At or near 40° the amount of velocity-head recovered rather abruptly falls to approximately zero.

(*e*) A conical discharge mouthpiece, having a total angle of divergence of 10°, and an area-ratio of 1 to 2, when attached to a submerged short pipe, will recover 0.435 of the velocity-head in the pipe, which is 58% of the theoretical amount of recovery possible.

(*f*) The amount of velocity-head recovered by a diverging or discharge-mouthpiece, when attached to a submerged short pipe, is considerably more when a converging or entrance-mouthpiece is also attached than it is when the entrance end of the short pipe is simply inward-projecting, that is, with no mouthpiece attached. This excess in the velocity-head recovered diminishes rather rapidly as the angle of the discharge-mouthpiece increases, and it becomes zero for a discharge-mouthpiece having a total angle of divergence of approximately 40°. This increase in the velocity-head recovered is probably due to the effect of smooth flow in the pipe as the water approaches the discharge-mouthpiece. The smooth flow allows the mouthpiece to recover more of the velocity-head in the pipe than when a more turbulent flow exists; this increase amounts to as much as 33% in the case of the discharge-mouthpiece having a total angle of divergence of 10° and an area-ratio of 1 to 2.

While these conclusions are drawn from experiments on the flow of water through a particular short pipe having various entrance and discharge conditions, it is felt that the results of the experiments are applicable in a general way to a large variety of cases in engineering practice where the contraction and expansion of a stream of water occurs.

The deductions made are capable of use in connection with the loss of head which occurs when a stream contracts or expands under differing conditions of flow and they show the marked effect that turbulence of flow may have upon the amount of head lost, and they also have a direct bearing upon problems in hydraulic practice that involve the contraction and expansion of a stream in flowing through passages. Comparatively little experimental work has been done hitherto to determine the value of conical mouthpieces of various angles and lengths in reducing the lost head at the entrance to and discharge from a submerged pipe, particularly for mouthpieces of the sizes and proportions comparable with those met in engineering practice. The need for such experiments was apparent. The minimizing of the lost head due to the contraction and expansion of a stream may be of considerable importance in many hydraulic problems; for example, the intake of a pipe, particularly when the pipe is of short length and of large diameter, the suction and discharge-pipes of a low-head pump, the reduction or expansion from one pipe to another of different diameter or of different shape, the passages through a large valve, the passages through locomotive water-columns, the draft-tube to a turbine, the connection from a centrifugal pump to a main, the sluice-ways through dams, the slat-screens at head-gates, culverts, and short tunnels, jet-pumps, the Boyden diffuser as formerly used for the outward-flow turbine, the Venturi meter, the suction and discharge pipes of dredges, and the guide vanes and runner of a turbine. Losses due to this cause are difficult to estimate and easy to overlook. Even where such losses are in themselves of little consequence as compared with other quantities involved, they may have an important influence upon subsequent losses on account of the turbulent motion started by the contraction or expansion. The efficiency of a drainage-pump or other low-head pump, for example, may be increased by an entrance-mouthpiece on the suction-pipe because it allows the pump to receive the water in a smoother condition of flow. It is well known that a turbine must receive the water from the guide-vanes without shock if, in the subsequent flow through the runner, the energy of the water is to be absorbed efficiently by the turbine. The loss of head through a Venturi meter may be considerably increased if the meter is placed too short a distance downstream from a valve, elbow, or other obstruction or cause of disturbance in the pipe. The friction-factor for a pipe following an obstruction or bend may be changed by the disturbance thus caused; the lost head at the entrance to a pipe, particularly when projecting inward, may be more than that ordinarily assumed for a tube three diameters long. There is but little definite knowledge on the subject of the effect of abnormal conditions, and it offers a large scope for investigation. The fact that a comparatively small change in the form of the blades of a turbine-runner may result in a large effect on the efficiency of the turbine should prove suggestive when estimating the probable effect of turbulent flow in less severe or critical cases. It is also worth mentioning in this connection that the recent advances in turbine design have been due largely to the attention given to the approach-channels to the guide-vanes and to the design of the draft-tube.

The flow of water usual in engineering practice is rather turbulent. The general equation of energy, or Bernoulli's theorem, so generally used in hydraulics, applies only when the particles of water move with uniform velocity in parallel stream-lines. Although this condition of flow seldom occurs, satisfactory analyses may often be made by using an average velocity and introducing empirical constants. A very slight change in the conditions under which flow takes place may cause a large difference in the behavior of the water. There is always danger in extending the use of experimental data or empirical constants to apply to conditions of flow different from those under which the data were obtained.

A Regiment of Miners

The following letter is being sent to mine-managers:

War Department,
Office of the Chief of Engineers,
Washington.

Subject: Recruits for Mining Regiment.

1. The Engineer Corps of the United States Army has recently been authorized to raise a regiment of Miners for service in France. We inclose herewith a brief description of the service, the character of men wanted and blank applications for enlistment.

2. Your name has been given us among others as one who is an employer of miners and who would be willing to help us in securing the men wanted. There are, no doubt, in your district, some men who sooner or later are going to enlist for the war and to whom this Mining Service may appeal. We ask that you aid us by sending these men to us now.

3. We appreciate that we are asking you to make a sacrifice in sending men to us at this time, but the total number of men wanted is not large, some 1500 in all, and if each Mine Manager or Superintendent will send us a few men, the desired quota will soon be raised. Information recently received from General Pershing indicates that these men are urgently needed.

4. We request that you distribute the inclosed information and enlistment blanks to such men as you are able to reach who are eligible and likely to be interested. In addition, we would like to have the names and present addresses of skilled men who have already been drafted from your district.

5. We wish to thank you in advance for your co-operation.

By order of the Chief of Engineers:
O. B. Perry,
Major, Engineer Officers' Reserve Corps.

Address reply to
The Commanding Officer, 27th Engineers,
Office of the Chief of Engineers,
War Department, Washington, D. C.

De-oxidizing of steel by the use of carbohydrates is being tried as a means of economizing manganese. Desulphurization is also said to result, and only traces of carbon are added to the steel. The substances used are sugar, starch, and cellulose. The purer the carbohydrate the better the result obtained. It is said that wood-shavings serve the same purpose.

War-Tax on Canadian Mines

Taxation of mining companies in Canada under business-profits war-tax or income-tax is of such conservative character as to have received almost unanimous endorsement by mining-company executives. Not only do Canada's war-taxation measures leave profits untouched up to 7%, but so generous is the allowance for exhaustion of mines that not more than 20 companies, probably fewer, will pay any Federal war-tax in 1917. This bears out the contention of the Minister of Finance that Canadian mining corporations have more freedom to divide earnings than similar corporations in most other countries. Out of the great annual dividend totals from the Cobalt and Porcupine districts the Ontario government gets just $300,000, and the whole mining revenues of the British Columbia government in a year are not above $150,000. The Federal finance department reckons on a revenue from mining sources under war-taxation of only a few millions. It is not possible to set forth the profits of various Canadian mining companies and calculate on the amount of their 1917 levies. The Finance department has no standard practice. While the United States government stipulates that not more than 5% of the gross output in any one year shall be allowed for exhaustion of property, the Canadian authorities have left themselves free to place that percentage anywhere between 2 and 15. In the opinion of the Finance Minister the United States standard is too low for many metalliferous mines that may have an average life of only 8 to 10 years. Some Canadian coal-mining companies are not being allowed even a 5% rate for depreciation of property. With some silver and copper mines enjoying depreciation allowances of 10 to 15% of gross output, it is plain that the total of Canadian mining companies liable to payment of war-tax will not come above 20. In 1915, exclusive of nickel companies, only 24 mining companies in the Dominion paid dividends.

The following statement of Sir Thomas White, minister of finance, may smooth out some misunderstandings as to method of calculating mining companies' capital for taxation purposes. Seven per cent on 'capital' is free; what that 'capital' is has been expressed by the Minister thus:

"For the sake of taxation, you have your capital, your reserve or rest account, and your accumulated profits, substantially representing net capital of company invested in the business. But in case of mining companies you will find many anomalies such as a company incorporated originally with $250,000 capital. Property today may be worth $5,000,000. A holding company has been created holding stock in original company and dividends of 15% or 20% are being paid on $5,000,000. Taxation applies to underlying company, but regard is had to amount of fully paid-up capital, values of reserves, rest, and accumulated property—the three put together representing value of the mine. Again, where a company bought a property at $100,000, and spent $150,000 in plant, a discovery of rich ore ran valuation of the mine to a million dollars. In opinion of the department the capital is $250,000. Reserve, rest, and accumulated profits make up the balance. Capital of that mine for purposes of taxation is taken at $1,000,000."

In 1915 no war-tax was collected, but as the initial Canadian measure of war-taxation, called the Special War Revenue Act, was retroactive, companies earning above 7% on capital were obliged to pay 25% of surplus profits on two years' operations. These deductions were usually charged against first accounting period in 1916, but exact amounts paid are by no means ascertainable in case of all mining companies' balance-sheets, war-tax being included with other extraordinary items. The business war-tax of 1916, and its amendments, greatly increased the load of taxation by taking 50% of profits between 15% and 20%, and 75% of profits above 20%. Latter tax applies only to accounting periods falling within 1917. It must be emphasized that the business-profits war-tax does not supersede original special war-revenue tax taking 25% of profits in excess of 7%. Both measures work in harmony.

ALUMINUM cables reinforced with steel have come into use lately for long-span high-voltage electric transmission-lines. The co-efficient of expansion of aluminum is about one-third greater than that of copper, so that, in a climate where the extremes of temperatures are considerable, it is necessary to provide for the expansion and contraction of an aluminum cable by giving it a greater sag between supports than would be necessary with a copper cable. This was feasible in the days of short spans and low voltages, but now that the reverse condition commonly exists it has been found necessary to reinforce aluminum cables with a steel-strand; the steel to provide the strength and the aluminum the electrical conductivity. Thus in a seven-strand cable the centre strand is of steel, heavily galvanized to protect it from atmospheric corrosion, and the six outer strands are of aluminum. Such cables can be made with any number of strands, the ratio of steel to aluminum being arranged to fit the particular need for which the cable is required. Steel reinforced cables are being used in a number of transmission-lines on this continent.

PLACER MINING in the Kantishna district, Alaska, is done by the open-cut method. The upper part of the gravel is sluiced off within a foot or two of bedrock, then shoveling the gravel into the sluice-boxes by hand. Most of the miners complete the ground-sluicing early in the spring when the water is abundant, but a few have built automatic dams, and are thus enabled, by alternately storing and releasing the water, to carry on the operation during the period of low-water.

WOLFRAM exports from Tavoy, Burma, according to reports from Rangoon, amounted to 202 long tons in June, compared with 104 tons for the same month in 1916.

Safety-Orders for Explosives

We reproduce recent orders issued by the California Accident Commission relative to the storage and use of explosives, this being known as order No. 1105. These rules for the use of blasting materials have a highly economic importance aside from their desirability in the prevention of accident. Much loss of time and efficiency in mining results from disregard of the simple precautions set forth in this circular. It is always agreeable to the public when a 'safety first' requirement is found also to have the desirable effect of increasing revenue.

(1) Magazines shall not contain tools other than a wooden mallet and wedge or a phosphor-bronze chisel.[1]

(2) When supplies of explosives or fuse are removed from a magazine, those that have been longest in the magazine shall be taken first. Packages of explosives shall be removed to a safe distance from the magazine before being opened, and such package shall be opened only with wooden or bronze tools. Empty or broken boxes, paper, and rubbish shall not be allowed to accumulate within 100 ft. of a magazine. Such boxes, paper, and rubbish shall be removed to a safe distance (at least 500 ft.) and burned or otherwise disposed of.

(3) Packages containing explosives shall not be opened in a magazine.

(4) Magazines shall at all times be kept clean and dry. Before repairing or altering a magazine, explosives shall be carefully removed and the magazine cleaned thoroughly. Nails or screws must not protrude from any part of the interior of a magazine.

(5) No detonators shall be taken into a magazine containing other explosives.

(6) No detonators shall be transported in the same receptacle with other explosives.

(7) Detonators shall not be removed from original container except as they are used for capping fuses.

(8) Explosives shall not be thawed in a magazine where other explosives are kept.

(9) Explosives shall not be thawed in any device other than a room, box, or other receptacle heated by exhaust steam, hot water, manure, or electricity. If steam or water be the agent employed, the stove, boiler, or other primary source of heat shall not be nearer to the thawing-room than 10 ft. Explosives shall not be thawed by direct contact with steam. If electric current be the thawing agent, the current shall not be brought within 5 ft. of the explosive to be thawed, and in no case shall explosives while being thawed be exposed to a temperature higher than 80° Fahrenheit.

(10) Explosives shall not be placed or left within 5 ft. of live electric wires.

(11) Explosives shall not be kept exposed to the sun any longer than is necessary. Explosives deteriorate under the direct heat of the sun.

(12) No person shall remove explosives from a magazine without the written or verbal orders of the superintendent or foreman of the job.

(13) Fire-extinguishers or hydrants and hose shall be provided and kept ready near all powder-magazines for use in case of fire outside the magazine.

(14) Smoking in a powder-magazine or while handling powder, or entering a magazine with an open light is prohibited.

Order 1106: Fuse.

(1) No fuse shall be used for blasting that burns faster than 1 ft. in 30 sec. or slower than 1 ft. in 55 sec. Only fuse of sufficient length shall be used so as to allow the men to retire to a point of safety before firing a hole or round of holes.

Under no circumstances shall fuse less than 2½ ft. long be used.

(2) Caps and electric fuses are extremely sensitive, hence it is forbidden to carry them in one's pocket or store them in buildings except a magazine as specified in these Orders.

(3) The use of oil or grease to waterproof joints between cap and fuse is forbidden. This practice causes misfires. The use of a compound sold by powder manufacturers for waterproofing, such as roofing paint, celakap, etc., is recommended.

(4) In capping fuse, at least 1 in. shall be cut from the end of each coil of fuse to be used in blasting. (This will prevent damp fuse-ends from getting into the cap.)[2]

(5) Only a crimper[3] shall be used for attaching fuse to blasting cap. Crimping with the teeth or a knife is strictly forbidden. Crimpers shall be provided and kept in good repair ready for use.

(6) It is forbidden to use fuse that has been hammered or injured by falling rocks or from any other cause. (Such injury may increase the rate of burning, or may render the fuse entirely useless.)

(7) In cold weather, fuse shall be warmed slightly before uncoiling, to avoid cracking the fuse.

(8) The hanging of fuse on nails or other projections, which cause a sharp bend to be formed in the fuse, is prohibited.

[1]At the San Francisco meeting, the omission of the clause following the word 'tools' was suggested.

HARDWOODS, such as oak, beech, birch, hickory, ash, and locust have a comparatively high fuel-value, and one cord of the dry wood is equivalent to a ton of bituminous coal. Pine, hemlock, douglas fir, sycamore, and soft maple require two cords to equal a ton of coal. The fuel-value is related to the weight, 4000 lb. of wood yielding as much heat as one ton of coal. Resinous woods, however, are better than those which do not contain that substance.

THE gold reserve of the United States is approximately $3,000,000,000. Before the War it was about $1,900,000,000. The significance of the increase in our gold reserve lies in the fact that there are, in round figures, only eight and a half billions of coined gold and gold bullion in the world. The world's credit-structure rests upon that foundation.—'Amer. Met. Market.'

LEADVILLE, COLORADO

NEW STRIKE AT FANNY RAWLINGS.—MANGANESE ORE.

At the Fanny Rawlings property on Breece hill a rich shoot of silver ore has just been cut by development carried on by Ray McConihay, who recently succeeded Robert W. Coates as manager. The shoot occurs in a big fissure-vein which traverses the entire property, running almost due north and south. The ore has been opened from three levels—the second, intermediate, and third. The second and third levels are being developed by winzes on the vein, while in the intermediate work is being done by raising. A force of men has been put to work on the first level also, and indications are that the shoot will be found there. The ore is a good silver-bearing sulphide. A high-grade streak in the vein assays 0.51 oz. gold, 330 oz. silver, and 6% lead. The main shoot assays 80 oz. silver. Mr. McConihay has shipped a four-ton lot of the high-grade ore, which was hauled from the property to the smelter, but he has been unable to ship any of the lower-grade material because the Denver & Rio Grande has not yet completed the standard-gauge siding to the property which extends from the main-line on Breece hill. Work on the siding is now in progress and will be finished this week. The surface bins, with a capacity of 100 tons, have been filled with ore and the underground chutes are now being used to store the material. Mr. McConihay estimates that he will be able to maintain a steady output of 50 tons per day, and in addition will ship regular lots of sacked high-grade ore. The discovery was made while developing two large bodies of zinc sulphide lying on either side of the fissure. The vein is a continuation of one of the big fissures that has produced rich gold ore.

The strike at the Fanny is regarded as an event of great importance by mining men operating in the Breece Hill district. It is the first evidence of the existence of high-grade silver ore in this part of the district, and opens new possibilities for development in many of the old properties including the Ibex. Ore assaying as high as 100 oz. silver has been found in the Ibex, but not in large bodies.——M. J. Nicholson, superintendent for the Western Mining Co., has secured a lease on the Big Four property, adjoining the Fanny Rawlings, and has started a small force of men to work cleaning out old drifts and putting the mine in shape for operation. New surface machinery will be erected this week. Prospecting for the extension of the new Fanny vein will be first undertaken.——A company of Leadville men, composed of T. V. Gallagher, James T. McMahon, H. M. Shephard, F. H. Sinclair, and T. J. Morrissey, leasing on the Nisi Prius Extension in Iowa gulch, has resumed operations and is shipping 50 tons of manganese daily. The ore is being marketed through Gustavus Sessinghaus who has been active in the district recently investigating deposits of manganese and pyrite. The orebody at the Nisi Prius is now 100 ft. long, 65 ft. wide, and 35 ft. high. The limits of the shoot have not been reached at any point. The ore contains 33% manganese.——Lessees at the Bohen shaft on Carbonate hill, where an immense body of manganese has been opened, have made another discovery that greatly increases the dimensions of the deposit. In a new drift, driven from the shaft 25 ft. below the level on which the ore is now being developed, the same shoot has been cut. The ore at this point contains less manganese but considerably more silver.

This week has witnessed the arrival of greatly improved conditions regarding the car shortage. At the Breece Hill properties in particular, have the operators been able to ship more regularly and move a heavier tonnage than for some time. Box-cars, coal cars, gondolas, and every other class of rolling stock that can be secured is being put into service here.

Production from the Penn properties on Breece hill has been stopped until the Denver & Rio Grande completes remodeling the siding from narrow to standard gauge. This work has been under way for several weeks and will be finished by the middle of the month. The Penn combination produces a large output of iron and gold ore.

TORONTO, ONTARIO

INTERNATIONAL NICKEL.—HOLLINGER.—MCINTYRE.—KIRKLAND LAKE.

The Canadian Copper Co., the subsidiary of the International Nickel Co., has paid into the treasury of the Province of Ontario, a check for $1,366,892, for taxes for the last two com-

THE M'KINLEY-DARRAGH MINE

pleted years of the company's operations under the retroactive taxation bill adopted at the last session of the Legislature. The tax, while levied on the Canadian Copper Co., is computed from the earnings of the International Nickel Co. and based on the value of the nickel matte exported from Sudbury for refining, which is determined by deducting from the finished production the cost of refining and marketing. The present tax is 5% of the profits up to $5,000,000, with an addition of 1% for each additional $5,000,000 of profits, should they ever reach such a figure.

The new milling equipment of the Hollinger Consolidated has been completed after considerable delay caused by adverse conditions, especially the lack of efficient labor. It is now being tried out, and is expected to increase the present milling capacity of 1800 tons by approximately 1000 tons. There is little prospect, however, of its being operated to anything approaching capacity for some time, as further interferences with the inadequate labor-supply are inevitable, owing to the conscription measure now in process of enforcement. The expenditure in connection with the new equipment, including the extra power and the equipping of the new central shaft, is estimated at about $1,000,000.——The financial statement of the

Dome Mines for the six months ended September 30, shows a gross income of $701,810; operating and developing costs $534,-575; and net profits $167,234, which, added to the surplus of March 31, makes a total of $864,285. Deductions of $141,164 for depreciation, $27,415 for war tax, and $100,000 for dividends, leave a surplus of $595,706. While these figures do not indicate that present operations are yielding a profit, it is realized that the conditions responsible for the decreased output are merely temporary, as development has largely increased the ore-reserve and under normal conditions production will be greatly increased. The main working shaft will be sunk shortly to the 1500-ft. level.——The Schumacher is treating 180 tons of ore per day, the mill-heads being nearly $7 per ton. Operating costs being maintained at about $4 per ton, substantial profits are being realized.——The McIntyre mill is running to capacity and treating about 500 tons per day of ore running $7 per ton or more. The Jupiter shaft is being sunk to the 1000-ft. level. The big vein is believed to cross the whole of the Jupiter property, and to enter the Plenaurum adjoining, on which the McIntyre holds an option.——The Coniagas, of Cobalt, which recently acquired the Anchorite, has let a contract for the sinking of a 3-compartment shaft to the depth of 500 ft.——The Newray has been taken over on option by the Crown Reserve and Dominion Reduction companies. This property, formerly known as the Rea, comprises 320 acres and was at one time regarded as a very promising mine, as it has paid dividends to the amount of $120,000 in 1914. The main vein faulted and has not since been re-found, but other veins of potential value have been discovered.

An important deelopment in the Kirkland Lake camp is the cutting of a high-grade orebody, 12 ft. wide, on the 325-ft. level of the Elliott-Kirkland, which indicates a considerable extension westward of the auriferous zone. It is of special importance to the Kirkland Lake mine as evidencing mineralization of the entire width of that property. The Burnside, adjoining the Tough-Oakes, has been taken over by the Aladdin-Cobalt. A number of veins on the property have been opened up. A depth of 600 ft. has been reached in the winze put down from the 400-ft. level of the Teck-Hughes. A station is being cut and a cross-cut will be run to connect with the workings from the main shaft.

An approximate estimate of the production of silver from the Cobalt mines for the 10 months ended October gives a total of 17,000,000 oz. of a value exceeding that of the total output of 1916.——The new mill of the McKinley-Darragh is practically completed. It will have a capacity of 500 tons per day, 250 of which will be tailing, there being quite a large accumulation that can be profitably treated by flotation. The ore-reserves have lately been increased by the cutting of several new veins on the 450-ft. level.——The Buffalo flotation-mill is treating old tailing at the rate of about 400 tons per day.——The annual report of the Temiskaming & Hudson Bay for the year ended August 31, shows an income of $190,992, operating expenses $96,043, and net profit $94,949. The ore-reserves are estimated at 107,614 oz.——The Coniagas has passed its dividend due November 1, the directors considering a conservative policy advisable in view of heavy expenditures in the acquisition and development of Porcupine gold claims.

TONOPAH, NEVADA

TONOPAH MINING CO.—TONOPAH EXTENSION.—JIM BUTLER.—MACNAMARA.

The Tonopah Mining Co. shipped 23 bars of bullion, valued at $39,300; during the previous week 2563 tons of ore averaging $12.07 was milled. In the Silver Top 79 ft. of development was done, 103 ft. in the Sandgrass, and 8 ft. in the Mizpah. On the 500-ft. level of the Mizpah the stope on the vertical vein continues in a 2-ft. face of ore. A considerable tonnage from the old dumps is being sent to the mill. Last week the production was 2850 tons.——On the 800-ft. level at the Tonopah Belmont, raise No. 16 shows 5 ft. of good ore, while raise No. 17 shows 4½ ft. of medium-grade ore where it cut the faulted segment of the South vein. Between the 800 and 900-ft. levels, the east and west drifts on the Belmont vein, there is a 4-ft. width of medium-grade ore. On the 900-ft. level a cross-cut is being driven to pick up the faulted segment of the Rescue vein. Last week's production was 2415 tons.——The Tonopah Extension Mining Co. during September milled 8692 tons of ore, making a net profit of $17,866. At the Victor 106 ft. of development was done and 216 ft. in the No. 2 shaft; on the 1440-ft. level, 604 east drift shows a 3-ft. width of ore, while raise 631 has a 5-ft. face of ore; on the 1540-ft. level winze 1501 is going down on ore. The production the last week was 2380 tons.——The West End mill shipped 33 bars of bullion, valued at $58,193, which represents the final clean-up for October. At the Ohio shaft the ore in drift 530 has increased in value, while drift 531 continues in high-grade ore, proving a large body of ore; on the 555 intermediate, drifts have been started to prove the vein. At the West End shaft raise 677 continues on the vein; drift 717 No. 2 continues to make good progress; raise 623 No. 7 has been started to prove the hanging wall. The output the last week was 913 tons. At the Halifax Tonopah on the 1200-ft. level, cross-cut 1256 has made good progress since it uncovered a vein of low-grade quartz.——At the Jim Butler Tonopah Mining Co. on the 625 intermediate of the Desert Queen shaft, a vein has been cut showing a full face of ore. On the 200-ft. level of the Wandering Boy shaft, the faulted segment of the Wandering Boy vein, recently cut, shows a 4-ft. face of fair ore. The production for the past week was 800 tons.——The MacNamara Mining Co. made its first shipment of bullion since resuming operations. The shipment was valued at about $10,000, resulting from the treatment of 903 tons of ore. A raise from the 725-ft. level to prove the extension of the Ohio vein, has cut a silicified zone. Development on the 600-ft. level will soon be started. The output the past week was 476 tons.——Last week the Rescue-Eula produced 195 tons, the Montana 105 tons, and the Midway 89 tons, making a total production at Tonopah of 10,223 tons with a gross value of $178,902.

COBALT, ONTARIO

CROWN RESERVE.—BUFFALO MINES.—DOME REPORT.—HOLLINGER CON.

During October Cobalt made another record shipment, aggregating 91 cars or 6,360,658 lb., against 4,357,610 lb. for September. The movement among Cobalt companies to secure new silver prospects has become general. The Gowganda field is attracting considerable attention, and already a number of deals has been consummated. Among those interesting themselves in Gowganda are the Mining Corporation and the La Rose.——The Crown Reserve has secured an option on the Newray property at Porcupine. It is proposed to increase the capital of Newray to $3,000,000. Some 1,800,000 shares will be under option to Crown Reserve at a price said to be 45c. per share. The Crown Reserve agrees to spend at least $6000 per month in development.——The Mining Corporation of Canada and the Nipissing are now occupying the centre of the stage in Cobalt. The Nipissing is producing upward of $10,000 per day and the Mining Corporation is exceeding that amount. The aggregate from these two mines for the current year will approximate nearly 10,000,000 oz. of silver.——The Buffalo Mines is in a strong physical condition. A new vein carrying considerable rich ore has been opened recently. The huge pile of old mill tailing is being treated at the rate of about 400 tons per day in the company's oil-flotation plant and a very satisfactory recovery is made. Taking them as a whole, the mines of Cobalt are experiencing a period of exceptional prosperity.

The new milling equipment at the Hollinger Consolidated is completed. This addition will add a capacity for 2800 tons per day to the Hollinger mill and make possible a production of $9,000,000 annually.——The McIntyre Porcupine is maintaining production at full capacity. Over 500 tons per day is being milled; the mill head ranges from $7 to $10 per ton. The physical condition of the McIntyre is satisfactory; the net earnings are large.——The Schumacher is making a far larger profit than at any previous period in its history.——It is officially announced that the Dome Lake ore-reserve approximates over 9000 tons and contains $82,008. The average grade is $9.03 per ton.——A contract has been let by the Coniagas company of Cobalt for the sinking of a three-compartment shaft to a depth of 500 ft. on the Anchorite property. The Anchorite is developing well, and is now generally looked upon as an established mine. It is owned by the Coniagas, which has a dividend record of more than $8,500,000 to its credit.

HOUGHTON, MICHIGAN

VICTORIA.—MASS CONSOLIDATED.—OSCEOLA CONSOLIDATED.

Omitting the consideration of the higher wages, the highest in the history of the Lake Superior, the important factors that enter into the cost of copper production have advanced at such a startling rate that mining costs during 1917 will be higher here than at any time within 30 years. Coal is costing the average independent producer 90% more than it did last year, when it was higher than it had been for 25 years. Powder is 100% higher and no contracts can be obtained, as the Government has the first call on all output. Steel rails, which are 110% over the prices of a year ago, are hard to obtain. It is difficult to get carbide even at $115 per ton and the miners may have to go back to the old-fashioned candle.

The management of the Victoria has decided to continue operations during the winter. The directors visited the property last week. The cost of production during recent months has been above 23½c. However, it is believed that by suspending some construction work now under way, dropping underground development, and confining mining to ore of known merit, the Victoria can continue through the winter at a profit. ——Mass Consolidated is shipping 1000 tons of ore per day. The working force is better than it has been for some time, both as to numbers and general efficiency. The physical conditions underground are good.——South Lake is operating only one shift.——In considering the deep cut in the last dividend of Osceola Con. the shareholders must not lose sight of the fact that the future of this corporation depends on the North Kearsarge and on the Old Osceola. The South Kearsarge is gradually becoming more and more a negative proposition. Pillar mining has been going on at both No. 1 and No. 2 shafts, and from these sources a total of 650,000 tons of ore can be secured. Last year over one-third of the South Kearsarge ore came from the foot-wall, cleaning out old stopes and surface stock piles. These sources were not to be drawn upon this year, except the foot-wall workings. Altogether the output of South Kearsarge for 1917 is going to be considerably less than it was last year and the expectations for the year ahead are not good. While the South Kearsarge shafts are gradually getting near their limits it must be remembered that this branch of the mine has for years been the richest of all. Last year the ore showed better than 17 lb. per ton and the year before it exceeded 16 lb. At present the North Kearsarge ore is yielding slightly better than 12 lb. per ton and the Old Osceola rock 12¾ lb. per ton. It is difficult to operate these two branches at a profit under existing conditions. The Old Osceola, while it continues to open a good grade of ore in the cross-cuts toward La Salle, is confronted with extremely high mining costs. The long distance that this ore has to be trammed makes it exceptionally difficult.

SPOKANE, WASHINGTON

STATE METAL-MINING ASSOCIATION.—HECLA DIVIDEND.—VALLEY MAGNESITE.

The Washington State Metal Mining Association has been formally oranized at Spokane, with George Turner, former U. S. Senator, president; Thaddeus S. Lane, first vice-president; Conrad Wolfe, Frank T. McCollough, Jerome L. Drumheller, George S. Bailey, and R. E. M. Strickland, executive committee. The secretary-treasurer will be elected by the executive board. Purposes of the association the the promotion of the mining interests, the consideration of taxes and industrial insurance, and a closer relation of all persons identified with mining, including the prospector, the miner, the mine-owner, and the investor. The association will encourage the use of safety appliances designed to reduce the number of injuries and the losses in industrial insurance, encourage a diligent search for rare metals and minerals, and aid in building the mining in-

COEUR D'ALENE REGION

terests by giving reliable information. The bureau through which this information is to be disseminated will be operated in conjunction with the bureau of any other State. County divisions will be formed with a vice-president in each county that has mining interests. The State organization will meet annually and the executive committee monthly. The new body will co-operate with the Northwest Mining Association, which is broader in its territorial scope and which has social, educational, and legislative objects.——Enough mineral to maintain a production of 300 tons of crude material per day for 15 years has been proved at the quarries of the Valley Magnesite Co., 65 miles north of Spokane, according to Irving Whitehouse, a director of the company. A third kiln has been started, increasing shipments to a carload of calcined material per day. Equipment has arrived for eight additional kilns.——Following several discussions of the war-tax on mines the Northwest Mining Association adopted resolutions expressing belief that the "law in its effect on the mining industry is unequal and unjust and in some cases oppressive and confiscatory. Neither this association nor the mine-owners object to any tax, however onerous, in aid of the present righteous and most necessary war, but it is only common justice that inartificial standards be corrected so that the tax upon mining be equalized with that of other industries and that the mines, as between themselves, be put upon an equal and common footing in the burdens imposed upon them." The next convention of the association will be held at Spokane in February.

The Hecla Mining Co. has declared a November dividend of $50,000. The regular rate of disbursement has been $150,000 per month, or at the rate of 15c. per share. The Hecla is the first of the big mines of the Coeur d'Alene region to cut the dividend rate because of uncertainty of the amount of excess war profits the company would be called upon to pay. The latest disbursement will increase the total of distributions to $6,855,000, of which $1,400,000 will have been paid this year.—— What is said to be a record in driving has been made at the Armstead mines on Lake Pend Oreille, in the Idaho panhandle. In 29 days of October the main adit on the property was advanced 488 ft., maintaining an average of 17 ft. per day. Ordinarily a rate of 12 to 15 ft. is considered fast progress in north-western mines. Everything essential to scientific mining is being used in driving at the Armstead. The track is equipped with 20-lb. rails, which have been laid as carefully as if for a steam-railroad. Storage-battery locomotives are used in hauling and the ore-cars are equipped with roller bearings. The machine-drills are of large size and power. A bonus is paid for each foot over 300 driven per month.

PLATTEVILLE, WISCONSIN

ZINC, LEAD, AND PYRITE SALES.—PLATTEVILLE, CUBA, AND BENTON DISTRICTS.

Prices for zinc ore were well maintained during the first half of the month, namely, $67 per ton base for premium and top grades with the range down to $62 for the inferior grades down to 54% zinc. The first appreciable break came about October 20, when the top quotation receded to $62 per ton base with the range down to $57. Exchange quotations held to these figures, but local correspondents in camps where refineries are situated and deliver top-grade ore showed a wide discrepancy in figures, proving that the published figures did not entirely reveal inside buying conditions, Linden at one time quoting $56 per ton for 60% blende. The closing days of the month found the base on premium grades back to $65 per ton, with the range for the lower grades down to $59 per ton.

The lead producers adroitly managed to evade current quotations for ore all the month and the shipment made of six cars in all in no wise reflected offerings published in trade and metal-market journals. The determination to keep out of the market did not at first seem well taken, as the production was by far the best shown for any month of this year, and a reserve had accumulated in the field estimated at 1500 tons. It appears that miners have confidence in later demands and are willing to carry lead ore over until next spring if necessary, as most of the ore secured in this field is obtained through wet concentration of zinc-lead ore. There are few exclusive lead-ore producers in the Wisconsin field.

Shipments of iron pyrite were small but steady during the month. The demand improved, but there was no appreciable change in price. Spot prices for sulphuric acid ruled as follows: 60°B., $20 to $25 at maker's work; 66°B., $30 to $35.

Shipments of carbonate zinc ore were confined to deliveries made by the Mineral Point Zinc Co., on which quotations did not apply. The standing offer of $25 per ton base of a 40% zinc content found little favor with miners and less of this class of ore is being mined than at any time in the history of the field. A scarcity of miners in the northern districts is responsible at present for many good producers being idle. This ore is used exclusively in the manufacture of zinc oxide for paint, heavy shipments were made daily from Mineral Point to army cantonments.

The deliveries of ore, reported by districts, from October 1 to 27 were:

Districts	Zinc Lb.	Lead Lb.	Pyrite Lb.
Benton	20,506,000	262,000	....
Miffin	6,838,000	......	....
Galena	4,272,000	170,000	....
Linden	2,020,000	88,000	508,000
Hazel Green	1,626,000	......	....
Shullsburg	1,355,000	......	....
Platteville	1,006,000	......	....
Highland	840,000	......	....
Potosi	348,000	......	....
Dodgeville	346,000	......	....
Cuba City	240,000	......	4,910,000
Mineral Point	224,000	......	....
Total	39,621,000	520,000	5,418,000

The gross recovery of crude concentrate from all mines totalled 19,964 short tons; the total net deliveries to smelter were 13,001 tons. While there was about the same recovery of crude ore for September the net deliveries exceeded August by 1300 tons.

The shipments of high-grade refinery ore made during October were: Mineral Point Zinc Co., 5,218,000 lb.; National Separators (Cuba), 4,506,000; Wisconsin Zinc Roasters, 3,986,000; Linden Refinery Co. (Linden), 330,000; Benton Roasters, 426,000; a total of 14,466,000 pounds.

Distribution and sales of zinc ore were: Mineral Point Zinc Co., 5950 tons; Grasselli Chemical Co., 3081; National Separators, 3078; Wisconsin Zinc Co., 3265; American Zinc Co., 1996; American Metal Co., 1440; Linden Zinc Refiners, 882; Benton Roaster Co., 707; Matthiesen & Hegeler Zinc Co., 634; Illinois Zinc Co., 513; Lanyon Zinc Co., 338; Edgar Zinc Co., 338 tons.

In spite of bad weather, which retarded mining operations considerably, a good showing was made in all branches of the industry. Lack of sufficient labor to man mills was responsible for the shutting down of two mills in the Highland district. Forces were kept at work underground and shipments were maintained without interruption. The Saxe-Lampe mine resumed regular deliveries after a period of idleness due to the opening of recently-discovered deposits. The Linden camp was more active than usual. A new shaft was bottomed in ore on the Treloar mine for the Milwaukee-Linden Development Co. The Spring-Hill Mining Co. broke into new deposits on the side of the old Mason mine. The Optimo Mines No. 3 and 4 produced steadily and a new shaft was put down on No. 4; a mill and power-plant is to be built on the site of No. 4. Roos Bros. made improvements to power and milling plant. Stoner Bros. Co. re-milled old dumps found on the Anaconda mine, securing a high-grade concentrate.

The zinc oxide and acid departments of the Mineral Point Zinc Co. were active all the month. The two reduction plants operated constantly and the shipment of high-grade blende to smelters was heavy. Considerable far-western low-grade calcined blende and carbonate-zinc ore from Mexico were received at the works. The North Survey Mining Co. of Dodgeville was the only active ore producer in this district.

The Miffin district made a good showing both in output and deliveries. The two Coker mines and the Yewdall claimed the lion's share of attention. The Grunow Mining Co. made regular deliveries. The M. & A. Co., operating the Big Tom, shipped two and three cars of zinc ore per week and paid a 25% dividend. Lucky-Six, Biddick, and Peacock were regular shippers.——The Block-House Mining Co. shipped frequently and the close of the month found 20 cars of refinery ore held in bin, for which bids were solicited in open market. The company paid a 100% dividend on October 15. This concern has paid $134,000 in dividends since starting, only a year ago, on a capitalization of $20,000.

ALASKA

(Special Correspondence.)—William Martin, of the Alaska Free Gold, left for Seattle on October 16. He states that plans are under way to finance the construction of a hydro-electric power-plant to deliver power all the year and supply all producing mines, which, under present conditions, are only able to operate about 120 days in the year.——Steel on the main line of the Government railroad will be laid to Montana creek within the week, when the track-laying gang will be laid off for the winter.

Anchorage, October 22.

(Special Correspondence.)—A suit to recover $16,500 has been filed against the Eagle River Mining Co. The company is controlled by Georgia capital and has been operating the mine at Eagle River for a number of years. Last summer the work was closed down and the property has been idle since. B. L. Thane, C. C. Whipple, and the Alaska Gastineau Mining Co., with various Juneau merchants, appear among the plaintiffs.——A rich discovery of gold is reported from Good News Bay in the Kuskokwim district. A stampede is on from Bethel.——The Government will let a contract for mining 200 tons of coal from the Nenana fields for test purposes.

Treadwell, October 26.

The Westover group of copper mines in south-western Alaska is said to have been sold to the Chitina Exploration Co. of Seattle, for $516,000.

The Gold Bullion mine, in the Willow Creek district, has closed down after a successful season, and for the first time in several years will not employ any men during the winter months. The mine has been under lease by a syndicate of Canadian capitalists and has been managed by A. T. Budd. Mr. Budd and his crew of workers came to Anchorage on October 26 and left for Toronto two days later. The Canadian syndicate, Mr. Budd said, had surrendered its lease, although the mine has paid a handsome dividend as the result of this season's operations. The majority of stock in the company owning the mine is held by Frank Bartholf, of Denver, and the property will be operated next season by the company owning the ground.

Walter McRay, a deep-sea diver of Tacoma, Washington, proposed to the copper mining people at Cordova that he recover the ore that had been dropped overboard in loading ships for one-half its value. They told him he was welcome to try, and hoped he would be lucky enough to make fair wages. Hundreds of tons of copper annually has been lost overboard during the loading of ships at that place, and no previous effort has been made to recover it. An official report states that McRay has so far earned a trifle over $5000 per month as his share.

ARIZONA

COCHISE COUNTY

(Special Correspondence.)—The Phelps-Dodge corporation is employing all available forces in the Tombstone district in mining manganese ores.——L. D. Ricketts, consulting engineer for the Calumet & Arizona, will take charge during Major Greenway's absence.

Bisbee, November 1.

Complete agreement has been reached between the President's Mediation Commission and the mine operators of the Warren district. The plan, which will go into effect immediately, provides that no man shall be refused employment because he does or does not belong to a union, and makes provision for adjustment of disputes through grievance committees composed of men actually employed by the companies. Disputes that cannot be settled by the committees to be adjusted by Federal mediators whose decision shall be final.

GILA COUNTY

(Special Correspondence.)—The Old Gibson mine at Bellevue has been sold to a syndicate of Eastern capitalists who have organized the Gibson Consolidated Copper Co. The company proposes to erect a 300-ton mill. The Gibson mine has a record of $2,100,000 in smelter returns.——The Porphyry Copper Co. has resumed operation.

Globe, November 2.

GREENLEE COUNTY

(Special Correspondence.)—The Federal Board of Investigation, which has been in the Clifton-Morenci-Metcalf district, has brought about a settlement and has adjusted the differences between the copper companies and the miners. The companies announced on November 3 that the strike was settled and that operations would be resumed. The miners are applying for work as individuals. The conditions upon which the mines are opening are practically the same as before the strike.——Owing to the general overhauling of the mill at Morenci, the largest mines of the Arizona Copper Co. at Metcalf will not be operated for some time. Only high-grade ore will be mined for the present.——A few of the Americans who left to seek employment during the strike have returned to the district and many more are expected. As but few Mexicans left, labor shortage is not expected to be serious.

Metcalf, November 5.

MOHAVE COUNTY

(Special Correspondence.)—It is reported that leading stockholders of the Washington-Arizona Mining Co. will begin work on the Lady Anne group of claims, just south of the Washington-Arizona.——The Jamison mine is opening up some good gold-silver-copper ore on the 300-ft. level. This ore will run $15 in gold and 6% copper.——The Golconda Annex of the Union Metals Co. will drive under the Black Butte to develop the surface showings.——The Distaff mine is being unwatered.——A rich strike of ore has been made in a winze sunk on the Prince George mine by the Arizona Butte company. The ore is 4 ft. wide and averages 128 oz. of silver. This ore is well below the old workings; an adit is being driven to cut the orebody at a deeper level.——The output of the Tom Reed mine has been increasing, the bullion output for the first half of October being $34,000.

Kingman, November 2.

PIMA COUNTY

(Special Correspondence.)—E. Cripen of Tucson has under bond and lease the Shacklette-Upshaw group of four claims in the San Xavier district.——G. W. Warren and associates of New York have purchased under bond and lease a group of eleven claims in the Twin Buttes district from Mike Serasio. Drill-holes have shown ore carrying 11.37% copper and 7 oz. silver. Contracts for 2000 ft. of bore-holes are to be let immediately.——The Sahuarita smelter is being rehabilitated by lessees from the Pioneer Smelting & Mining Co., and it will be ready by the end of the year.——G. Ulter and Bisbee associates have taken over the old Gunsight mine. The old shaft will be re-timbered. It is claimed there is an orebody some 200 ft. wide carrying 4% lead..——Grading for the spur-track connecting the smelter and leaching-plant is being done.——An old vein of $20 free-milling gold ore has been uncovered in the old Mammoth mine in the Casa Grande district.

Tucson, November 1.

YAVAPAI COUNTY

(Special Correspondence.)—Operations at the United Verde Extension are proceeding rapidly. One blast-furnace of the new smelter will be blown-in next month. The branch of the railroad that is to convey the ore from the main adit to the smelter is rapidly approaching completion.——A six-drill compressor and 60-hp. hoist are to be erected at the Jerome-Superior.——A syndicate of Jerome and Prescott investors has secured an option on a large part of the Verde Apex stock and undertaken the financing of a company. This property is about one mile to the south of the United Verde.——Recent developments in the shaft sunk on the line of the Socrates claim of the Venture Hill company are encouraging.——The New York Verde has taken over the claims of the Victor Copper Co. in the Verde district and the Maynard district of Mohave county.

Prescott, November 2.

(Special Correspondence.)—The Arizona Binghampton Co., locally known as the A B C company, has gone through 31 ft. of high-grade copper ore in the new cross-cut on the 600-ft. level, most of the ore being rich enough to ship to the smelter. The cross-cut on the 600-ft. level has cut a strong vein of ore; it is a continuation of the same vein that has been found on all of the upper levels, and settles the question of the permanency of the orebodies.

The Patton copper mine, situated four miles east of Mayer, has been taken over by the newly-organized Rio Tinto Copper Mining Co. W. D. Mahoney and P. J. McIntyre, of Jerome, are the principals of the new company. The property is well developed and considerable high-grade ore has been exposed. The company will be sufficiently financed to carry out extensive development work.——The Copper Mountain Mines Co. is working a force of a dozen men in the old Stoddard mine at the north end of Copper mountain. A gas hoist has been erected and one of the old shafts is being re-timbered. This company is a re-organization of the old Stoddard Copper Co. which operated extensively a quarter of a century ago. The new company is capitalized for $5,000,000. The old mine is one of the best developed on Copper mountain; many shoots of high-grade and milling ore are exposed; these were developed when only rich ore could be handled economically. The close affiliation of this company with the Arizona-Binghampton, a mile to the north, will mean that the experience of developing the lower workings of the Binghampton will be used in the development of the old Stoddard mine. Gelora Stoddard is president and general manager of the new company, Senator Reynolds of New York vice-president, and M. A. Pickett, of Phoenix, is secretary and treasurer.——There are indications of ore in the adit that is being run on South Copper mountain by the Monte de Cobre Copper Co. This is the mine that was taken over last summer by W. A. Tucker, John Ross, and James Smith, of Bisbee. The adit is being run to tap a strong vein of copper ore on which a shaft has been sunk for 60 ft. The formation has changed and some chalcopyrite has appeared in the last drillings.——The announcement made by the Black Canyon Mining Co. that a flotation-mill is to be erected at the mine, five miles below Turkey, means a great deal to this silver-lead camp. The mine was originally located by J. D. Thompson, of Prescott, eight years ago. Thompson partly developed this Black Canyon mine and then sold it to the present Kansas City owners, who have run a drift about 800 ft. on the main orebody opening a strong vein from 8 to 16 ft. wide. There is about 300 ft. of good ore above the drift. W. A. Moses is president, A. L. Harroun vice-president, Edna Harroun secretary, and Claude Baker general manager.

Mayer, November 10.

CALIFORNIA

NEVADA COUNTY

(Special Correspondence.)—The North Star Mines Co. has re-constructed a 40-stamp mill at the Champion mine and enlarged the cyanide plant. The deep Providence shaft has been splendidly equipped and converted into the main working outlet. Work will be pushed more vigorously in the Nevada City mine, and in other claims that have been developed to comparatively shallow depths.——The shaft of the Allison Ranch mine, four miles below Grass Valley, has been unwatered to a point below the 800-ft. level and the management is preparing to repair additional levels and start deeper development work. The mill is now running with two shifts, most of the ore coming from the 400 and 700-ft. levels. Several promising blocks of ground are being worked by lessees.——With possible exception of the Red Ledge mine, all properties in the Washington district have suspended chrome shipments for the winter because of unsatisfactory road conditions. In the Red Ledge a large body of chrome has been found in an open-cut and has been quarried to fair depth. It is pronounced the best chrome deposit yet discovered in Nevada county. Upward of 6000 tons of high-grade chrome has been shipped from properties above Nevada City in the past six months.——Copper mining is active around Smartsville with several claims producing. From the Dibble group, owned by William Dibble of Grass Valley, regular shipments are going to the Kennett smelter. San Francisco people are inspecting the field. Most of the ore carries gold as well as copper.

Nevada City, November 12.

SHASTA COUNTY

Confirmation is lacking of the extensively reported discovery of a 7-ft. seam of coal on Beegum creek in the south-west part of this county.

SIERRA COUNTY

(Special Correspondence.)—Preparations are being completed for an active season at the Brandy City hydraulic mine, near Brandy City. Flumes and ditches have been improved, the restraining dam strengthened, and much new ground prepared for hydraulicking. Four to six giants will be operated as soon as sufficient water is available.——All surface work has been completed at the North Fork and Wisconsin mines near Forest, and underground mining has been resumed at both properties. The Wisconsin has been equipped with one of the best electrical plants in the county. Although under separate management, both mines are worked from the same main adit. Driving on the new vein is proceeding in the North Fork, the work being conducted from the shaft, which has been thoroughly repaired.——An electric hoist and compressor have been shipped from Nevada City for the South Fork mine, near Forest. A promising quartz deposit has been opened near the main adit of this property and is showing ore of excellent grade. Preparations have been made for sinking as soon as hoist and compressor are ready for service, meanwhile driving for the Bald Mountain gravel channel will continue.

Downieville, November 10.

R. G. Matzene and Mrs. Matzene, of Syracuse, and Richard

Phelan, of Oakland, have bonded the Hilda mine at Sierra City. The Hilda was equipped some years ago as a hydraulic mine, but the California Debris Commission stopped operations because there were no restraining dams. In future the mine will be worked as a drift-mine.

SISKIYOU COUNTY

(Special Correspondence.)—While some prospecting and staking of claims throughout this county is going on, but little actual work will be attempted on new properties until next season. The suspension of assessment work leaves but few mines working at present.——W. R. Beall, of the Know Nothing mine near the Forks of Salmon, expects to resume operation with new machinery in two weeks. He has been handicapped in not being able to get the new equipment delivered.——The Mercury Company of America, under the management of Eugene C. Belknap, will soon begin operation with new equipment on its Cowgill cinnabar mines, at Gottville, near the head of Beaver creek.——Henry Musgrave and Albert Hadley are taking out considerable high-grade chrome ore at Lime gulch in the Greenhorn district. F. Le May is also operating on chrome deposits near their mine.——T. K. Anderson of Gottville, operating the Pilot Knob gold mine on Empire creek, is erecting a new gasoline engine to run a 5-stamp mill.——W. B. MacAdams of Yreka will operate hoists and pumps on his placer mine near Gottville by water-power this winter. J. R. Clute and W. J. Beagle are operating a 2-stamp mill on a rich gold deposit near Gottville.——All the placer mines through this district are in shape for activities as soon as the rains start.

Hornbrook, November 9.

COLORADO

BOULDER COUNTY

(Special Correspondence.)—J. C. Clark has secured control of the Grand Group Silver mines at Caribou; he is erecting machinery on the Grand County shaft, which is about 300 ft. deep. The Grand Group consists of seven claims and is one of the oldest properties in the district. It has produced heavily and is reported to have shipped some of the richest ore ever produced at Caribou. It adjoins the famous old Caribou on the north. Extensive development work will be done in re-opening the old ore-shoots and exploring for new deposits.——William Wilson, who has a lease on the Wolcott group, just above the Lucky Two, has one of the largest bodies of tungsten ore in Boulder county. The high-grade ore runs from 28 to 35% and the low-grade from 2 to 4%. Heavy shipments are being made and the property is being worked to its full capacity. Among other lessees on the Wolcott holdings are the Sellers Bros., who are operating an extension of the Eureka vein; Yates and Starkey in the Blue-Jay adit; and Harris, Tripp, and Premier, the Tip-Top, which is an extension of the Eureka. Regular shipments from these properties are being made.——A big body of free-milling gold ore has been discovered at the Huron property.

The Dana Mining Co. has been organized with headquarters at Boulder. It is giving most of its attention to the development of the old Hoosier mine at Summerville. James E. Simpson, of Boston, is president; Captain Coan, of Boulder, secretary-treasurer; and A. S. Sloan, manager. The Hoosier was one of the principal producers of the early days.——W. R. Baker, the veteran miner at Caribou, was in Boulder a few days ago. He is greatly pleased with the revival in silver mining and the good showing which has been made in the mines of the Caribou district. James Cowie, president of the Cowie Coal Co., reports that its new mine near Gorham should be ready to begin production within the next two weeks.——W. F. Harpel, manager for the Consolidated Leasing Co., at Eldora, is pushing work on the company's new mill.

Eldora, November 5.

LA PLATA COUNTY

On November 1 the plant of the American Smelting & Refining Co., at Durango, closed. F. C. Gilbert, the manager, has made the following announcement.

"In retiring from the business we beg to express our appreciation for the co-operation which the shippers of the district have extended to our venture.

"Great progress has been made in the art of milling the San Juan ores and we believe that the problems which seemed so difficult two years ago are now practically solved. Since we started in the business the custom-mill idea has spread and we are convinced that the interests of the camp will be well taken care of by the small mills located at favorable points. The increased tonnage of both concentrates and crude ore has necessitated an increase in the roaster-capacity at the Durango plant. We predict for 1918 the biggest tonnage in fifteen years."

SUMMIT COUNTY

The property of the Michigan Mining & Milling Co. on East Sheep mountain near Kokoma has been shipping steadily

PART OF COLORADO

to the smelter at Leadville. The ore runs from 20 to 30% lead and carries from 12 to 15 oz. of silver. The mine has been opened by two adits, the lower one being now used as the main working-adit. The mine is equipped with an air-compressor and an underground hoist draws ore from the long incline driven on the orebody. All machinery is driven by electric power furnished by the Colorado Power Co. Edward Moir is president, S. H. Dunlop manager, and John M. Moir and G. E. McKim are directors.

TELLER COUNTY

(Special Correspondence.)—Dividends in the sum of $183,500 were paid on November 10 to stockholders of the following Cripple Creek mining companies: Cresson Consolidated $122,000, Golden Cycle Mining & Reduction Co. $45,000, both monthly dividends, and the Granite Gold Mining Co. $16,500, a bi-monthly distribution.——Recent development at the bottom or 1600-ft. level of the shaft of the Cresson company has exposed a huge body of ore of an average value of one ounce gold per ton. Ore at this point in the levels above has been too poor to pay for shipment.——The October production from properties of the United Gold Mines Co., the Trail, W. P. H., and Bonanza mines on Bull and Ironclad hills and Battle mountain, totalled 84 cars with the content closely approximating 3000 tons, averaging $21 per ton. The Trail mine was the heaviest producer with 67 cars shipped by three sets of lessees.——The properties of the Requa Gold & Silver Mining & Milling Co., a corporation controlled by the United Gold Mines Co., the charter of which has expired by limitation, will be offered for sale at the Court House in Cripple Creek on November 30 by G. R. Lewis, trustee appointed by the district court of Teller county. The properties consist of the Trail mine on Bull hill and the Big Banta, Old Ironsides, and Lost Fraction on Battle mountain containing 25,338 acres, U. S. Mineral Survey No. 7812. An important discovery has been made in the bottom level of the Modoc inclined shaft at a depth of 1150 ft. An ore streak, on the west vein about 4 in. wide of tetrabedrite or gray copper, assayed 88 oz. gold and 100 oz. silver per ton.

Austin T. Holman, formerly superintendent of the Golden Cycle mine, and more recently identified with the J. T. Milliken interests in Alaska, has been appointed superintendent of the Granite G. M. Co., to succeed the late D. L. McCarthy.——E. M. Rapp, a graduate of the Colorado School of Mines, has been appointed superintendent of the Ajax mine for the Carolina Mines Co., lessee.——The hoisting-plant at the Ajax was wrecked last week, when the skip passed beyond control of the engineer and crashed to the bottom of the shaft. The damage to plant and shaft has been estimated at $10,000.——J. T. Alexander of Elkton has been appointed superintendent for the Queen Gold Mining Co., operating the Eclipse mine on the south-western slope of Battle mountain. The Queen is controlled by J. T. Milliken. John Lynn, late superintendent of the Queen, has left Colorado for Los Angeles and will remain with the Milliken interests in California.——The properties of the Sedan Gold Mining Co., on Galena hill, the Anna Belle, Fourth of July, and McGee lodes, containing 13.738 acres, will be sold under a judgment issued from the district court of El Paso county, in the sum of $19,861.65 in favor of A. B. Crane, on December 8.——The district mines are working short handed and with excessive cost of mining supplies, few leases are active, and production is maintained by operation of the larger corporations and their lessees operating under the split-check system.

Cripple Creek, November 8.

IDAHO

SHOSHONE COUNTY

A mining case that will occasion wide interest throughout the West and which involves hundreds of thousands of dollars began November 7 in Federal Court, Judge Wolverton presiding, the litigants being the American Smelting & Refining Co., a New Jersey corporation, and the Bunker Hill & Sullivan Mining & Concentrating Co., an Oregon corporation. An injunction is sought by the American company to restrain the Bunker Hill company from making other disposition of its ore and concentrate than the delivery at the smelter of the plaintiff corporation. The Oregon corporation owns the Bunker Hill mine, of the Coeur d'Alene district of Idaho.

A 25-year contract, entered into between the two companies, and which has half its term yet to run, provided for the delivery of all ore and concentrate from the Bunker Hill properties to the American smelter. About a year ago the Bunker Hill company began the construction of a large smelter of its own near the mine. The American corporation alleges that to divert any portion of the contracted supply would work great injury to its business, inasmuch as the Bunker Hill ore contains certain elements that have been counted upon for the smelting of gold and silver ores received from other mines. The plaintiff sets forth that it would be injured, not only by the loss of business, but by the retarding of its other smelting operations, owing to a shortage of the Bunker Hill ore, which contains a low percentage of zinc and is needed to reduce the proportion of zinc in other ores received for smelting.——The profits of the Bunker Hill company during the first half of the contract period have been $200,000 per month, it is alleged, while the mining company has paid to the American Smelting & Refining Co. approximately $15,000,000.

The Federal Mining & Smelting Co., operating in the Coeur d'Alene and Wood River regions of Idaho, reports earnings of $112,422 in September, after all deductions. This is an increase of $16,455 over the net earnings for September 1916, which were $95,967. The earnings in July were $259,728, June $276,533, May $217,016, April $204,581, March $151,664, and January $77,791. The net profits for the year to September 30 last were $1,368,930, as compared with $720,596 in the corresponding period in 1916. Three dividends of 1¾% each, amounting to $209,757 each, have been paid on preferred stock during the year. An addition of 350,000 tons, by estimate, has been made to the ore-reserves by deepening a shaft and workings in the Coeur d'Alene region during the summer.

MONTANA

LEWIS AND CLARK COUNTY

(Special Correspondence.)—The Marysville Gold Syndicate, which is developing numerous mining claims east of the Shannon and which owns the Bald Mountain and Belmont claims, has entered into arrangements with the St. Louis Mining & Milling Co. at Marysville to mill 30 tons of ore per day. The ore will be taken from the Honeycomb and Mt. Pleasant veins. This arrangement is regarded as only temporary, as the syndicate purposes building a plant of its own in the spring. A. B. Woolvin, who was connected with several large mining deals in Montana, is the main promoter of the enterprise. The capital is being furnished by Duluth, Minnesota, people. ——The Barnes-King Mining & Development Co., whose principal mines are in Marysville camp, is producing at the rate of about $100,000 per month.——The Helena mine, in Grass Valley, has sent out ore to the East Helena smelter, for which it has received $18,000. The ore was mined from the 300-ft. level.——The Rock Rose and Cruse Con. mines have lately shipped two carloads of lead-silver ore to the smelter.

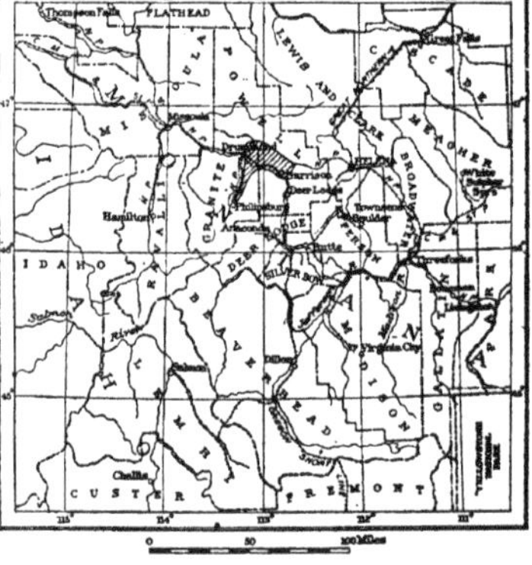

MONTANA

Two carloads are being shipped from the Coffee-Byrnes claims, Grass Valley. Part of the ore will run from 400 to

500 oz. of silver per ton. The other half is lower grade, running about $50 per ton.

Helena, November 5.

NEVADA

CLARK COUNTY

A check for $7070, exclusive of the freight, which amounted to nearly $1100, has been received for one car of ore which the Boss company shipped to Irvington, New Jersey, several weeks ago. In future the ore from this mine will be sent to the Pacific Platinum Co. at Los Angeles, which started operations late in October.

ELKO COUNTY

The Holden Mining & Milling Co., whose stock is owned by the Stewart Mining Co. of Idaho, is operating the Commonwealth, Nevada, McVystal, Surprise Fraction, Diana or 16 to 1, and Extra or May Queen mines in Tuscarora. At present 16 men are working and the force is being increased. A new mill is to be erected this winter; the treatment will be ball-mill crushing followed by electro-cyaniding. The directors of this company reside in New York. J. D. Hubbard is general superintendent at Tuscarora.

MINERAL COUNTY

(Special Correspondence.)—For the year ended October 30, 488 carloads of ore, aggregating 25,672 tons, and of an estimated value of over $1,000,000, was produced in the Luning district. One carload averaged 42% copper and 12 oz. silver, or about $275 per ton at the then prevailing prices. Little, if any went under 5% copper. The principal shippers were the Kirchen Mines Corporation, Wedge Copper, Calavada Copper, Pilot Copper, Congress Copper, Luning-Idaho, Wall Street, Fermina Sarrias estate, R. B. Todd Mining Co., and the Iriquois Copper Co. About 20 companies are operating in the district and upward of 100 lessees have been at work during the past year. The deepest working, namely 1100 ft., is at the Calavada on the 1000-ft. level and it is the best showing in the Luning copper belt. The orebody here is 10 ft. wide and will average 15% copper. A strike has been made on the 260-ft. level at the Giroux, the vein is over 10 ft. wide and will average 10% copper.——In addition to the high-grade shipping ore, all the mines have orebodies of good mill ore that should pay well with a local reduction plant. V. C. Alderson, president of the Colorado School of Mines, reports that the Pilot company alone has 298,100 tons of ore blocked out on four sides that will average 2.51%, and he believes that many times that amount will be developed. It is probable that a big plant will be built in the near future at Luning.

Luning, November 10

NYE COUNTY

(Special Correspondence.)—The Cash Boy company has arranged to ship regularly to the Mammoth smelter, at Kennett, California; the ore will average $25 per ton in silver. The cross-cut from the 1700-ft. level has cut the foot-wall section of the vein, disclosing a width of six feet. East and west drifts open the vein on its strike. Good ore continues to be found in the main upper workings.——The Belmont company is adding conveyors and sampling equipment in its Tonopah plant and will be ready to treat custom ore before the end of December. The Millers plant will be dismantled. On the 700-ft. level the faulted segment of the South vein has been uncovered. Approximately 2491 tons is shipped weekly to the mill. The general manager, Frederick Bradshaw, has gone to British Columbia to inspect the Surf Inlet mine.——The West End company is concentrating work on its California claim where the Ohio vein has developed into the principal ore-producer of the property. On the 500-ft. level the cross-cut has shown the vein to be fully 15 ft. wide, and the ore of excellent grade. A drift is advancing in a full face of ore and a cross-cut from the 555 intermediate has entered the orebody. The management is preparing to develop the vein at further depth.——The Nevada Ophir Mining Co. has resumed work at its silver mine near Millet, with R. J. Williams as superintendent. The property has been idle for 40 years; it is stated to contain large quantities of medium-grade ore. It is planned to work the property through a series of adits.——An ore-shoot showing considerable free gold has been tapped on the 400-ft. level of the Manhattan Consolidated. It was uncovered in a cross-cut from the 175-ft. east drift, 200 ft. from the shaft. Developments are proceeding from several points on the third and fourth levels.

Tonopah, November 10.

STOREY COUNTY

Encouraging reports continue to come from the north end of the Comstock, especially at the Union Con. mine, where further important developments have been made the past few days. On the 2300-ft. level, from the north drift an east cross-cut, which is being extended in the direction of the hanging wall of the vein, has broken into good ore, and the last car-samples show an average of $25 per ton. There is also an improvement working south from the east cross-cut on the 2400-ft. level, where the last 16 cars of ore taken out in the advance gave an average of $35.50 per ton. Three feet in the face has given samples of $110 per ton, and three feet next to it $10.

Developments are also being made in the Ophir mine, where work is being carried on to the south from No. 1 cross-cut on the 2100-ft. level. This drift was started to follow a 3-in. quartz stringer, which has now widened out to 14 in., and assays $18 per ton. In the Mexican mine on the 2900-ft. level, the cross-cut is showing more quartz with some clay seams, the formation heretofore being in hard ground. The Sierra Nevada has started to clean out and repair the east cross-cut from the main north drift on the 2500-ft. level, which leads to an ore-discovery made in the early days and which is to be further investigated, as reports show little or no work done there. Work continues in the 2600-ft. raise.——The Andes is working on the 350-ft. level at the stope and south, extracting some low-grade ore.——Alpha and Exchequer are prospecting on the 200-ft. level from the Imperial shaft.——The Jacket Crown Point-Belcher Co. on Gold hill is placing the mill in shape; part of the plant is in operation.

NEW MEXICO

SOCORRO COUNTY

(Special Correspondence.)—The output for October on Pacific mine amounted to 390 tons.——The Mogollon Mines Co. is now handling custom ores and greatly relieving the congested milling situation of the district. During the week the Oaks Co. made regular shipments from the Maud S., Deep Down, and Eberle mines of the Central group. Another burro train was added to the freight service.——The Socorro M. & M. Co., which recently lost the upper portion of the plant by fire, is pushing plans for immediate re-building. The shaft collar is being re-timbered and building material is expected. The filters, refinery, agitators, and pumps were uninjured and the regular clean-up has been made. October was to have been the largest production the company has had and the recent splendid ore opened in the lower levels makes the delay doubly unfortunate.——All mines are crowding in their winter freight and both teams and trucks are working to capacity.

Mogollon, November 6.

SOUTH DAKOTA

LAWRENCE COUNTY

(Special Correspondence.)—With the resumption of work at the Silver Queen property, the Galena district is once more taking on the aspect of old times when daily shipments of high-grade lead-silver ores were common. Within the past month ten cars of ore were shipped from the Silver Queen to the smelter at Denver. This material was all taken from the dumps and netted the owners satisfactory returns. Work has been started in the mine and several high-grade bodies of ore are being developed.——The New Puritan Co. is adding machinery to both mine and mill at Strawberry Gulch and the 100-ton cyanide plant will be in operation soon after the first of the year. For the greater part of the present year the mine has been undergoing development.——Excavation for the 10-stamp mill and concentrator to be erected in Spruce Gulch below Deadwood, by the Deadwood Zinc & Lead Co. is nearly completed. The ore, which is exposed in several places on the property, is a heavy sulphide containing lead, zinc, gold, and silver.——At a meeting held at Deadwood on October 12 there were present R. R. Horner, a mining engineer in the U. S. Bureau of Mines, and Thomas Varley, superintendent of the Experimental Station at Seattle. These men have been sent to the Black Hills by the Federal branch to investigate the tin deposits of that district. The purpose of the meeting was for the Government officials to meet the mining men. While it has been known for some time that there were large deposits of low-grade tin ores in South Dakota, the development and production of the metal has been limited. Within the present year 20 tons of tin concentrate averaging about 50% metal was produced in the Hill City district and sent to the Eastern market. Most of the ore was taken from the Cowboy mine.——O. M. Lane of Keystone continues the production of spodumene at the rate of one car per week. The Rheinboldt Met. Co. is taking out the same amount of amblygonite weekly. At the Summit mine west of Hill City, the Dakota Continental Copper Co. has commenced diamond-drilling to further develop the ground. The shaft is down over 200 ft. and several hundred feet of cross-cut work has been completed since the first of the year.

Lead, October 26.

CANADA

BRITISH COLUMBIA

The production of the Phoenix mines of the Granby Consolidated Mining, Smelting & Power Co., in the Boundary district, averaged 13.52 lb. copper, 0.177 oz. silver, and 0.027 oz. gold per ton, according to the annual report. The development-cost averaged 16c. per ton and 12.2c. per foot. Diamond-drilling amounted to 6502 ft., making the total diamond-drilling on this property 119,122 ft. The cost per ton of ore on cars was $1.17. The wage bonus amounted to 15.7c. per ton; the increase in cost of supplies over pre-war prices was 11.2c. per ton; workmen's compensation assessment, 1.5c. per ton; making an increased cost of 28.4c. per ton over pre-war prices, but this item was reduced by a 0.7c. decrease in railway freight resulting from the opening of the Kettle Valley railway. The Phoenix mines have produced 12,825,426 tons of ore, leaving reserve estimated at 3,274,996 tons, and carrying 17 lb. copper and 75c. in gold and silver per ton. Production for the last fiscal year was $672,292 tons of ore. The company's staff and mine force at Phoenix have contributed to the War 45 men and $42,198.35 in money, of which 75% went to the patriotic fund and most of the remainder to the Red Cross fund. The contributions during the past year amounted to $19,440.07. Of the company's employees enlisting, five are reported to have been killed, seven wounded, and two made prisoners of war. The Granby company has taken working options on a number of properties in the Kamloops and Skeena districts and already has a force of men doing development work on them.

KOREA

The Seoul Mining Co., operating the Suan Concession in Whang Hai Province, reports the recovery of $140,725 for October.

PERSONAL

Note: The Editor invites members of the profession to send particulars of their work and appointments. This information is interesting to our readers.

A. W. Allen is at Denver.

Arthur Winslow is here from Boston.

Theodore Piloer has returned to Butte.

J. W. Mercer is at Mokelumne Hill, California.

William C. Russell, of Denver, has gone to New York.

James W. Neill is visiting San Francisco, from Pasadena.

H. C. Wilmot is on a professional trip through Arizona and New Mexico.

George H. Sexton, of New York, was in San Francisco this week.

C. W. Laws has returned from Butte to Esqueda, in Sonora, Mexico.

Edwin W. Mills has gone to Peking, on his return from Siberia.

William Motherwell has returned from Nelson, B. C., to San Francisco.

J. B. Harper, manager of the Jerome Verde mine, at Jerome, Arizona, is in New York.

Philip L. Foster has been given a commission as Captain in the Signal Corps, U. S. Army.

Frederick H. Jackson sails on November 20 on his return to Sydney, Australia, from Mexico.

Frank Langford, of Eureka, California, is with the Andes Exploration Co., at Santiago, Chile.

Henry R. Putnam passed through San Francisco this week on his return from Shanghai to New York.

C. A. Rodegerdts has resigned as construction engineer with the Atolia Mining Co. and is now in San Francisco.

J. D. Hubbard, of San Francisco, is general superintendent for the Holden Mining & Milling Co. at Tuscarora, Nevada.

W. A. Meloche, mining engineer to the Jualin Alaska Mines Co., has resigned, to serve in the 115th Engineers, U. S. Army.

W. O. Borchert, of the New Jersey Zinc Co., passed through San Francisco on his way from Los Angeles to Salt Lake City.

F. Jenkin, for many years mine-superintendent and assistant-manager for the El Oro company, at El Oro, Mexico, has resigned.

Gerard W. Williams has been awarded the Military Cross for gallant service at the Somme front. He is Captain in the Royal Engineers.

Bancroft Gore, formerly Professor of Metallurgy in the Montana School of Mines, has left Butte to go to the Naltagua smelter at El Monte, Chile.

F. L. Ransome will take charge of the work on quicksilver for the U. S. Geological Survey in order to release H. D. McCaskey for administrative work.

F. H. Hayes, formerly chief geologist with the Detroit Copper Co., has been appointed Major in the 79th Infantry Brigade, and is now at Camp Kearny, California.

C. B. Clyne has resigned as constructing engineer to the Demming Mining Co., in Idaho, to accept a position with the Mine & Smelter Supply Co., at Denver.

Horace B. Patton, for 24 years Professor of Geology and Mineralogy in the Colorado School of Mines, has opened an office as consulting geologist at Golden, Colorado.

M. Hirokawa, R. Hamasaki, G. Ansai, and S. Miyake, of the Metal Mining Department of the Mitsubishi Company, in Japan, are making a tour through the mining districts of the West.

E. J. Carlyle and M. R. Hull, respectively metallurgical and mechanical engineer for the Sissert Mining Co., of Petrograd and London, have come to San Francisco from the Ural Mountain region, on their way to Salt Lake City.

METAL PRICES

San Francisco, November 13

Aluminum-dust (100-lb. lots), per pound	$1.00
Aluminum-dust (ton lots), per pound	$0.95
Antimony, cents per pound	16.00
Antimony (wholesale), cents per pound	14.75
Electrolytic copper, cents per pound	23.50
Pig-lead, cents per pound	6.50— 7.50
Platinum, soft and hard metal, respectively, per ounce	$105—111
Quicksilver, per flask of 75 lb	100
Spelter, cents per pound	10.00
Tin, cents per pound	65
Zinc-dust, cents per pound	20

ORE PRICES

San Francisco, November 13

Antimony, 45% metal, per unit	$1.10
Chrome, 34 to 40%, free SiO_2, limit 8%, f.o.b. California, per unit, according to grade	$0.60— 0.70
Chrome, 40% and over	$0.70— 0.80
Magnesite, crude, per ton	$8.00—10.00

Manganese: The Eastern manganese market continues fairly strong with $1 per unit Mn quoted on the basis of 48% material.

Tungsten, 60% WO_3, per unit	26.00
Molybdenite, per unit MoS_2	$40.00—45.00

EASTERN METAL MARKET

(By wire from New York)

November 13.—Copper is unchanged and nominal at 23.50c. all week. Lead is quiet and firm at 6.25 to 6.37c. Zinc is dull and steady at 7.87 and 7.75c. Platinum remains unchanged at $105 for soft metal and $111 for hard.

SILVER

Below are given the average New York quotations, in cents per ounce, of fine silver.

Date		Average week ending	
Nov. 7	86.62	Oct. 2	98.49
" 8	86.12	" 9	91.72
" 9	86.12	" 16	87.01
" 10	86.12	" 23	83.85
" 11 Sunday		" 30	84.71
" 12	86.12	Nov. 6	89.25
" 13	86.00	" 13	86.18

Monthly Averages

	1915	1916	1917		1915	1916	1917
Jan.	48.85	56.76	75.14	July	47.52	63.06	78.92
Feb.	48.45	56.74	77.54	Aug.	47.11	66.07	85.40
Mch.	50.61	57.89	74.13	Sept.	48.77	68.51	100.73
Apr.	50.25	64.37	72.51	Oct.	49.40	67.86	87.38
May	49.87	74.27	74.61	Nov.	51.88	71.60	....
June	49.03	65.04	76.44	Dec.	55.34	75.70	....

Samuel Montagu & Co., London, in their letter of October 11, state that the Bank of England gold reserve against its note-issue shows a decrease of £169,810, as compared with last week's return. In order to prevent the hoarding of silver currency caused by the increased value of silver, the Italian official gazette publishes a decree stating that the one or two lire (nominally 25 lire = $5) silver pieces may be changed for paper money as from November 1 next to December 31. The silver currency will then cease to have a monetary value, and persons found in possession of silver pieces over 10 lire in value will be fined from 60 to 1000 lire. The sterling amount, however, involved is not large. The U. S. Mint report gave the total stock of silver Italian coin in 1913 as $22,400,000, or 64c. per capita. It is reported that a bill has been introduced in Denmark for the issue of small money made of iron. We hear that hundreds of tons of Mexican silver dollars have been imported into the United States in the few months prior to September for melting and re-sale as bullion.

COPPER

Prices of electrolytic in New York, in cents per pound.

Date		Average week ending	
Nov. 7	23.50	Oct. 2	23.50
" 8	23.50	" 9	23.50
" 9	23.50	" 16	23.50
" 10	23.50	" 23	23.50
" 11 Sunday		" 30	23.50
" 12	23.50	Nov. 6	23.50
" 13	23.50	" 13	23.50

Monthly Averages

	1915	1916	1917		1915	1916	1917
Jan.	13.60	24.30	29.53	July	19.09	25.66	29.67
Feb.	14.38	26.62	34.57	Aug.	17.27	27.03	27.42
Mch.	14.80	26.65	36.00	Sept.	17.69	28.28	25.11
Apr.	16.64	28.02	33.16	Oct.	17.90	28.50	23.50
May	18.71	29.02	31.69	Nov.	18.88	31.95	....
June	19.75	27.47	32.57	Dec.	20.67	32.89	....

Possibility of an upward revision in the price of copper, in so far as 'high' producers are concerned, is discerned in the report of the mediators in the Clifton-Morenci copper strike. The following excerpt from the findings of the mediators would seem to indicate that where it can be shown that under the advanced wage scale a fair profit is not left to the producer, a higher than the 23.50c. rate may be granted: "If the administrator should recommend a wage increase and such a wage scale allows a fair profit to the companies under the existing price of copper, the President's commission shall at once promulgate such new wage scale and the company shall pay all such increases in wages as of the first day of the return of men to work. If, however, such a wage scale recognized by the administrator does not leave a fair profit under existing prices of copper, the President's commission shall recommend to the President an increased selling price which will yield a fair profit, and the wage increase recommended by the administrator shall not be made effective until such selling price has been obtained."

ZINC

Zinc is quoted as spelter, standard Western brands, New York delivery, in cents per pound.

Date		Average week ending	
Nov. 7	7.87	Oct. 2	8.43
" 8	7.87	" 9	8.50
" 9	7.75	" 16	8.50
" 10	7.75	" 23	8.25
" 11 Sunday		" 30	8.12
" 12	7.75	Nov. 6	7.42
" 13	7.87	" 13	7.81

Monthly Averages

	1915	1916	1917		1915	1916	1917
Jan.	6.30	18.21	9.75	July	20.54	9.90	8.98
Feb.	9.05	19.99	10.45	Aug.	14.17	9.03	8.58
Mch.	8.40	18.40	10.78	Sept.	14.14	9.18	8.33
Apr.	9.78	18.62	10.20	Oct.	14.05	9.92	8.32
May	17.03	16.01	9.41	Nov.	17.20	11.81	....
June	22.20	12.85	9.63	Dec.	16.75	11.26	....

As a result of a canvass by U. S. Geological Survey to ascertain production of spelter in the United States during third quarter of 1917, C. E. Siebenthal estimates, on basis of returns of over 95% of the production, that output of primary spelter from domestic ores was 132,700 short tons and from foreign ores 23,900 tons, a total of 156,600 tons, as compared to an average of 180,569 tons per quarter during first half of 1917, 175,502 tons per quarter during last half of 1916, and 158,226 tons per quarter during the first half of 1916. Stocks of spelter at smelters September 30 amounted to 47,186 tons, compared with 33,147 tons June 30, and 17,598 tons January 1. The number of idle retorts September 30 was 64,626, or 28% of total number of retorts, and included 18 entire plants, several of which are being dismantled. One electrolytic plant also will possibly be dismantled.

LEAD

Lead is quoted in cents per pound, New York delivery.

Date		Average week ending	
Nov. 7	6.25	Oct. 2	8.00
" 8	6.37	" 9	7.96
" 9	6.37	" 16	7.52
" 10	6.37	" 23	6.70
" 11 Sunday		" 30	5.66
" 12	6.37	Nov. 6	6.02
" 13	6.37	" 13	6.35

Monthly Averages

	1915	1916	1917		1915	1916	1917
Jan.	3.73	5.95	7.64	July	5.59	6.40	10.93
Feb.	3.83	6.23	9.01	Aug.	4.62	6.28	10.75
Mch.	4.04	7.26	10.07	Sept.	4.62	6.86	9.07
Apr.	4.21	7.70	9.38	Oct.	4.62	7.02	6.97
May	4.24	7.38	10.29	Nov.	5.15	7.07	....
June	5.75	6.88	11.74	Dec.	5.34	7.55	....

The total value of lead concentrates of all classes sold or treated by producers in the Central States in 1916 amounted to $26,070,323, against $15,424,244 in 1915. The value of all zinc concentrates sold increased from $30,157,854 in 1915 to $39,311,185 in 1916.

QUICKSILVER

The primary market for quicksilver is San Francisco, California being the largest producer. The price is fixed in the open market, according to quantity. Prices, in dollars per flask of 75 pounds:

Week ending

Date			
		Oct. 30	100.00
Oct. 16	100.00	Nov. 6	100.00
" 23	100.00	" 13	100.00

Monthly Averages

	1915	1916	1917		1915	1916	1917
Jan.	51.90	222.00	81.00	July	95.00	81.20	102.00
Feb.	60.00	295.00	126.25	Aug.	93.75	74.50	115.00
Mch.	78.00	219.00	113.75	Sept.	91.00	75.00	112.00
Apr.	77.50	141.60	114.50	Oct.	92.90	78.20	102.00
May	75.00	90.00	104.00	Nov.	101.50	79.50	....
June	90.00	74.70	85.50	Dec.	123.00	80.00	....

TIN

Prices in New York, in cents per pound.

Monthly Averages

	1915	1916	1917		1915	1916	1917
Jan.	34.40	41.76	44.10	July	37.38	38.37	62.60
Feb.	37.23	42.60	51.47	Aug.	34.37	38.88	62.53
Mch.	48.76	50.50	54.27	Sept.	33.12	36.66	61.54
Apr.	48.25	51.49	55.63	Oct.	33.00	41.10	....
May	39.28	49.10	63.21	Nov.	39.50	44.12	....
June	40.26	42.07	61.93	Dec.	38.71	42.55	....

Eastern Metal Market

New York, November 7.

The markets are all dull, with sales confined to small lots. Two of the markets continue to gather strength, but the others are inactive with a tendency to become weaker.

Copper continues unchanged at the Government price and is nominal.

Tin is scarce and continues to advance.

Sellers of lead are disinclined to part with their metal, and the market is higher and stronger.

Zinc continues in poor demand and is lower.

Antimony has nominally declined.

Aluminum is weaker and inactive.

Further maximum prices on certain steel-products were announced by the Government on November 5, but they are in line with those previously fixed. The Government announces that on products not yet covered by official schedules manufacturers have agreed to adjust prices promptly. Further price-fixing is not looked for unless a general revision should be made later. Ill-advised agitation of contract-abrogation from a Government source has restricted commercial buying of steel. The feeling is that the present price-basis will not be changed on January 1, and it may be that there will be no change during the first half of 1918. Government shell-steel now under inquiry amounts to 1,500,000 tons, of which 300,000 tons was alloted last week. Pig-iron production increased in October to 3,303,038 tons, or 106,550 tons per day, as against 104,465 per day in September.

COPPER

There is little to be said about the copper situation beyond what was contained in this letter in the past week or two. The general wholesale market is bare of sales at any price, and the distribution of supplies is entirely in the hands of the Copper Producers' Committee. Information of an official nature is impossible to obtain. Consumers are reported as being taken care of if they have Government work, and it is even stated that electrolytic copper in quantities has been delivered at the Government price to consumers doing urgent Government business but having no contracts for copper. The relation of consumption to production is a matter of speculation, some contending that the supply is not enough. Government and Allied needs are receiving first consideration. The jobbers' position continues a hard one. Small lots to needy small consumers are being daily sold at prices above the Government price of 23.50c. per lb., which we continue to quote as nominal. No interference with this practice has so far been reported. Stocks in such hands, however, are rapidly dwindling, and their replenishment at any price, or at a price to insure a profit, is problematical. A solution of this problem is hoped for in the near future. The London market is unchanged at £125 for spot electrolytic and £121 per ton for futures.

TIN

The strong market reported last week has continued to grow stronger, due to the scarcity of spot-tin, which continues the controlling factor. Late last week a buyer desiring 15 tons could not obtain it. A prime cause is the permit situation, which continues stringent, confining purchases only to buyers whose names are given. The position of the small consumer continues to tighten, and the future of his supplies is problematical. The quotation for spot Straits on October 31 was 66c., New York, with the same price for November 1 and 2. On Monday, November 5, however, a small lot sold at 68c., New York. Outside of the spot situation the market has been quiet. On Monday moderate sales of futures were reported, but on October 31, and November 1 and 2 there were sales of fairly large quantities of all grades of tin, mostly for future shipment, but these were not made public. Arrivals up to and including November 2 were 160 tons, with 4300 tons reported afloat. The London market is strong and higher. It has advanced £4 per ton in the last week, standing on November 3 at £260 10s. for spot Straits. The average price for tin for the month of October was 62.24c. per lb., New York.

LEAD

In the large buying movement that followed the radical lowering of prices reported last week, probably 20,000 tons of lead changed hands. This included large quantities for France and Canada. The Government has not closed for its November needs, it is believed, but report has it that it is inquiring for a quantity for two or three months' supply. The market continues strong and higher. Prices have advanced almost daily until on Monday the quotation in the outside market was 6.25c., New York, or 6.12½c. per lb., St. Louis, for early delivery. The American Smelting & Refining Co. also advanced its price on Monday, November 5, ½c. per lb. to 6c., New York. Demand has been active in the last few days, but sales have been small, due to an absence of most sellers from the market. The leading producer has not been a free seller recently, and has even restricted the purchases of its regular customers. There is little metal available from either the independents or the leading maker. A higher market is looked for because buyers are anxious to buy, and sellers are refraining from selling.

ZINC

Stagnation continues to dominate the zinc market. There is no life and no demand. The low sellers are the only ones parting with their metal and some of these have done so for November delivery at 7.50c., St. Louis. Others have sold at 7.75c., St. Louis, while some refuse to consider anything less than 8c., St. Louis. Brass makers and galvanizers continue out of the market, and the few orders taken do not represent any broad trading. The quotation for prime Western for November delivery is about 7.62½c., St. Louis, or 7.87½c., New York. All sales made have been of small lots only. For December delivery about ⅛c. per lb. higher is asked, with 7.75 to 7.87c. per lb., St. Louis, asked for first quarter.

ANTIMONY

The antimony market continues inactive and dull. In the absence of demand we quote the market easier and nominal at 14 to 14.25c. per lb., New York, duty paid, for Chinese and Japanese grades for early delivery.

ALUMINUM

No. 1 virgin metal, 98 to 99% pure continues to sag and is now 36 to 38c. per lb., New York, for nearby delivery with demand of small proportions. Export inquiries from Italy have appeared.

ORES

Tungsten: There has been no change in the quotations for the various grades of ore, scheelite commanding $26 per unit in 60% concentrate with $23 to $25 per unit for the other grades. The market is fairly active. Ferro-tungsten continues to be quoted at $2.35 per lb. of contained tungsten as the minimum price.

Molybdenum and antimony: The market for both these ores in this district is quiet. The quotation for high-grade molybdenite is unchanged at $2.20 per lb. of MoS_2 in 90% concentrate. There is nothing to report in the antimony ore market.

MANY inquiries have been made in recent months concerning the details of the process for making inc oxide. The question is fully answered by Mr. leorge E. Stone in an article which we re-print this 'eek from the transactions of the Institute. Mr. Stone, s chief engineer for the New Jersey Zinc Company, is n authority on this as on other branches of the metal- irgy of zinc.

TACOMA is now able to boast the highest chimney in the world. The stack of the copper smelter is 571 et high. It is constructed of paving brick, which is itable for the purpose because this material is able to ithstand a high compressive strain, moisture, and acid. he height is advantageous because it maintains the fluent gases at a high temperature and so permits them travel far before precipitation. The dilution of fume proportioned to the square of the height of a chimney.

HONOR to Mr. Gompers! The president of the American Federation of Labor has taken a stand the iportance of which can be gauged by imagining him have done the opposite. At this juncture the patri- ism of the leader of organized labor means everything the nation. We note that C. H. Moyer, president of e International Union of Mine, Mill, and Smelter orkers has taken the cue. No strikes are to be toler- ed during the continuance of the War. The decision of immense importance. It remains for all employers recognize this patriotic action and to meet it in the ne spirit. Such solidarity will assure our soldiers d sailors unfailing support in the completion of our at task.

EVERAL industrial companies at Chicago have united to issue a circular stating their intention not issue Christmas cards this year, but, instead, to set le for the Red Cross or other relief organizations the ney usually spent in this way. We commend this de- on heartily. Both in business and in private life we ht to discontinue the sending of Christmas cards. s is no time for wasting time, money, or paper-pulp such trifles. The same is true of the larger propor- of little presents, outside the family, that go into discard within a few weeks after their arrival. ead, let us restore the good old fashion of writing letters, with some feeling for the amenities of the epistolary art. We must be soberly thoughtful this Christmas, realizing the national crisis and the world crisis, more particularly the suffering and destitution of millions whom it is our duty and privilege to help.

THE premier goldfield of the world, the Witwatersrand, is feeling the pinch just like any small mining district anywhere else. We are informed that some of the poorer mines are likely to be shut-down in order to conserve supplies for those that are more productive. The scarcity of shipping has interfered with the importation of materials, so that a real difficulty exists in continuing the large-scale operations for which the Rand is famous. Old shoes and dies, together with other discards, are being utilized. Every bit of scrap-iron is being re-melted. At the Robinson this is being done by the Heroult method of electric smelting. Old boilers and compressors have become valuable, and have been rescued from the junk-pile. The electrification of the mine and mill plants caused a lot of machinery to be scrapped; this is now being used to good purpose. Thrift has become compulsory. The British government early in the War took steps to ensure a supply of cyanide, steel, and other necessaries for the mines at Johannesburg, because the production of gold was a source of strength to the Empire. Now that the credit of the United States has been thrown on the side of the Allies, the gold from the mines is not so important and it might seem advisable to save shipping for the transport of men and food across the Atlantic. The cessation of gold mining in South Africa, however, is unthinkable because it would bring widespread destitution to the communities now dependent upon this industry not only in the Transvaal and Rhodesia directly but in other parts of South Africa indirectly.

CANADIAN papers continue to attack Minerals Separation vigorously. On November 7 Mr. George Chapman met the Mine Managers Association of Cobalt at a special session and delivered an address in which he replied to the charge that his principals were pro-German. Apparently he was not successful, for the president of the Association, Mr. B. Neilly, stated that although his remarks were interesting they were not convincing, and that the managers "as a body were equally sincere in their belief that Minerals Separation was German-controled." This we learn from the 'Daly Nugget,' which, however, has not lost its head in the controversy, recognizing that no proof has been forthcoming that Minerals Separation is an enemy-alien. On the other hand, our contemporary raises proper objection to the attempt at a monopoly shown in the contract enforced upon its licensees by Minerals Separation, in which the latter calls upon its victims "to promptly communicate to the licensors [M. S.] every invention or discovery made or used by them which may be an improvement, modification, or addition to any of the inventions specified in the letters patent within this license or may be useful in carrying out any of the processes thereby protected or any addition thereto or modification thereof whether patentable or not which the said licensees may use or be or become possessed of." Truly, Minerals Separation is not Prussian, but some of its methods appear to have been learned at Potsdam.

OUR readers are aware that the price of paper has risen about 100% during the last two years. We refer to 'book-paper,' such as that on which this is printed. The publication of technical journals has been burdened by the increase of expenditure due to this cause and we have taken a keen interest therefore in the inquiry made by the Federal Trade Commission created by resolution of the United States Senate on September 7, 1916. This commission has issued a pamphlet detailing "the findings of fact, together with conclusions and recommendations with reference to the book-paper industry." The report can be summarized in a few words. It has been ascertained that "the advances in the prices of book-paper in 1916 were excessive and not justified either by the increase in cost or by the changes in conditions of supply and demand. The advance was brought about in part by the activities of the members and secretary of the bureau of statistics." This bureau is an association of manufacturers that combined to advance prices "without fear of competition." While ostensibly organized to compile statistical information, the bureau "has been devoted to encouraging members to increase their prices." In short, the book-paper manufacturers have used the exigencies of the War to establish a cinch and to conspire to make an excessive profit by combining in restraint of legitimate trade. The statement made by some of them that there was a shortage of 'bleach,' or chlorine, was not a verbal inexactitude; it was a plain lie, like some other statements made by them to cover their tracks. If more chlorine be needed, the way to get it is quite easy. Electro-chemists habitually complain of the difficulty experienced in disposing of the chlorine evolved as a by-product in many chemical processes. For example, chlorine is an obstacle to the development of the electrolytic manufacture of caustic soda on a large scale. In the States of California and Maine, to examine the American conditions no more closely, large quantities of hydro-electric power are available at the switchboard between peak-loads, and salt water is handy; yet the paper-makers put up the price of paper with a weak apology that chlorine cannot be obtained, and at the same time the Government is commandeering the output of the alkali-makers, causing a caustic-soda famine, while power is going to waste that could be used in the production of both commodities. The Commission recommends "as a war emergency measure that all print-paper mills and distributing agencies operate on Government account during the continuation of the War, and that the total product be pooled in the hands of a governmental agency so that it may be equitably distributed at a fair price based upon cost of production and a reasonable profit per ton." We would like to see action taken under the Sherman Act for the prosecution of this successful effort to break the law against unlawful conspiracy in restraint of trade. We mention this matter not only because it affects us but because it is an example of a large number of instances in which the War is being used to victimize the public. If a man does not fight for his country and if he does not subscribe to the national fund, at least let us see to it that he is not given an opportunity to make business depredations on the community under cover of the extraordinary conditions created by the War.

IN a recent issue we referred to the Anglo-American corporation that has been organized at Johannesburg, London, and New York to finance business in South Africa. Primarily the purpose is to provide capital for the exploitation of new leases adjacent to the properties controlled by the Consolidated Mines Selection Company, with which Messrs. Walter McDermott and W. L. Honnold, among others, are identified. The names of the New York directors suggest that the Morgan and Rockefeller groups are participating. This is true only indirectly, the American interest in the enterprise having been subscribed in the name of Mr. William B. Thompson, a gentleman that needs no endorsement from any other financial powers. We are informed that this group has made a successful tender for the south De Rietfontein lease, the north lease going to the Barnato Bros. The Transvaal government declined tenders made for two other leases, one of which, the west De Rietfontein, will be re-advertised shortly. The proportion of profit payable to the Government as royalty on the south De Rietfontein lease is 25%, whereas that payable on the Barnato tender will be nearly double, making the latter an unattractive scheme. The lease secured by the Anglo-American group, through the Mines Selection, will be incorporated later in the Springs Mines, which, with the Brakpan, constitutes part of the valuable property already controlled by this group through its Johannesburg and London affiliations. We understand that the Anglo-American Corporation may take part in financing a loan for the South African government. Whatever be the outcome of this new departure, it is most interesting as an evidence of American participation in the mining industry of the Rand.

Thanksgiving

The President has called upon our people to observe next Thursday as our annual day of thanksgiving, and the Governor of the State has issued a similar proclamation. Indeed, there is much for which we may be thankful; to be alive to see the greatest events in human history; to be in a land safe from invasion at a time when ten countries are feeling the tread of an enemy; to have national wealth and prosperity when many nations are humiliated by poverty and failure; to have plenty of food when millions are going hungry; to be in the enjoyment of peaceful homes when thousands of hearths are smoking ruins; to be free when others have been enslaved by a savage conqueror; to be free when millions are prisoners in the hands of the enemy; to be confident of our freedom when other peoples are in fear of subjugation; to be able to fight for freedom for all the world as once before we fought for our own; to have the sinews of war, the material, and the machinery for making the weapons, munitions, and ships needed for successful war; for the gallant men that are able and willing to battle for our cause on land, on sea, and in the air; for the patriotic spirit that has united all parties, professions, and classes in a common endeavor; for all of these we are thankful, but most of all for the awakening of this nation to a purpose greater than material success, to the discovery of a soul, undismayed and unconquerable, that shall guide us through this dark hour to final victory and enduring peace.

Recruiting Labor

On another page we publish a bitter protest, on the part of an engineer directing the work of a gold mine in Mariposa county, against the attempt to induce miners to leave their employment in California for the sake of the higher wages offered at a copper mine in Arizona. Our protestant even imputes a low code of professional ethics to the young engineer that has been recruiting in his neighborhood and also to the manager of the Inspiration mine. We happen to know all the persons concerned in the dispute. Mr. C. E. Mills, the manager of the Inspiration, has a reputation that will defend him better than anything we need to say, and his energetic representative, Mr. C. E. Arnold, will be pardoned if he fail to recognize the description of himself appearing in Mr. S. E. Rau-Roesler's letter. The question cannot be discussed on that plane; let us consider the fundamental point, as to whether it is fair to induce men to leave one company in order to work for another. Usually it is not done in the home district because such action causes ill feeling, and the managers of neighboring mines try to live on good terms with each other. It is extremely annoying to have your neighbor steal your cook, for example, and it is not done in the 'best society,' whatever that may mean. The amenities of neighborliness are transgressed by such a depredation. However, Mrs. Jones of Jonestown would not hesitate to beguile the chef in the employ of Mrs. Smith of Smithtown, supposing that the ladies had not a bowing acquaintance. So Mr. Mills' representative thought it proper to invade Mariposa county and secure workmen for Miami, not thinking of the annoyance he might cause the manager whose preserve he would violate. It looks like a competition, not pleasant to the victim, but justified by the exigencies of business. Personally, we believe that miners on the Mother Lode might well hesitate to exchange the foothill country of California for the desolation, however picturesque, of Miami. No married man should be in a hurry to transplant his family from a garden to a desert. Moreover, we would like to ask Mr. Arnold how the cost of living compares? But a broader question arises, should the gold mines be depleted of their labor in order to stimulate the mining of copper. The idea obtains that this nation, being now engaged in war, needs copper more than gold. For the moment the stock of gold in the Federal Reserve banks is plethoric, but it is not a superfluity, for it will be needed to sustain the enormous credits that we are supplying our Allies for the furtherance of the War. Moreover, is it wise to injure the old settlements around the gold mines in order to invigorate the new camps about the copper mines? Certainly the State of California will resent any steps taken to injure its established industry for the sake of strengthening that of another State, even such a splendid member of the Union as Arizona. These are knotty problems. On the whole, we believe that there should be as little disturbance as possible of one industry for the sake of another, and that while recruiting for labor on the part of agents from Arizona is in no wise nefarious, it is objectionable.

The Lady or the Tiger?

Villa's attack upon Ojinagá the other day, revives interest in Mexico. Under the strain of vital and catastrophic events in Europe, we have been inclined recently to ignore the country south of the Rio Grande. By preference we avoid slang, but it is inevitable, in this connection, that we should recall Villa's apt remark when the European storm broke in 1914, "the Kaiser's war will make mine look like thirty cents." It was true; the diversion of attention from "our pet bandit" made it more difficult for him to finance his operations, until the stock fell so low that he could not have sold the entire issue in the revolutionary market for the proverbial 'thirty cents.' The European war helped Carranza to establish himself because it also assisted President Wilson to pursue his chosen Mexican policy less hampered by adverse opinion. Lately the revolutionists have shown renewed strength, which is prejudicial to the interests of Mexico and in consequence to our own country. Political stability and advancing prosperity in Mexico have always been the sincere wish of the people of the United States, and at this time of international struggle it is more earnestly desired than ever. It may be amusing to pet bandits when no serious national enterprise is afoot, but today we care nothing for the momentary enthusiasms of the time when we had leisure to view with interest the spectacular episodes of romantic Mexico. It was natural then to applaud first one and then another of the players as each scored in the game. Today we wish that Villa and Zapata and Carranza would form a truce and settle down to orderly living. Carranza might properly enough offer a sweeping amnesty, without any strings tied to it, as an inducement to 'bury the hatchet.' We are constrained to say that after the high-sounding verbiage of the decrees that were promulgated with this ostensible purpose has been stripped down to the inner fact, it is certain that no political opponent of the 'supreme government' in Mexico would find sufficient guarantee of safety to induce him to risk his life in the hands of his erstwhile enemy. That has been a fault of every faction that has risen to power; Madero committed it, and so did Huerta, and Carranza has done no better. The real balm that healed the wound at the end of our Civil War was the generous amnesty granted, so complete in its scope, and so unequivocal in its terms, and so dependable because it was an Act of Congress and not a decree by a political chief, that men could accept it and then forget the matter.

Venustiano Carranza just now has such an opportunity as seldom comes to any man in this world. In calling attention to it we may remind him that Porfirio Diaz also had the opportunity of a glorious immortality in the archives of the world, and he sacrificed it. Today he is remembered as a man of achievements; but his character is under discussion, and his personal weaknesses are brought forward for examination. He failed of being a Washington, against whom it is held a sacrilege to recall his human shortcomings. After being the master-builder of a devastated country Diaz should have crowned his work by stepping down, as did Washington, ready to help and counsel if need be, but eager to become a passenger in the ship of state, leaving another master on the bridge. On September 16, 1910, a host of friends of the valiant old Mexican Cid watched and waited for the great stroke that should give him a place among the immortals, but the advice of Ramón Corral and other sycophants robbed him of the proper fruit of his long and conscientious services to the country. Carranza is also at the parting of the ways. He may choose a seat of honor in the temple of history or one of ignominy and regret in the outer darkness that follows the sunset of his opportunity. It lies before Carranza to re-build the shattered fortunes of Mexico faster than Diaz ever dreamed of doing; if he should so elect, he could make himself first in the hearts of his countrymen, and be held in his presidential seat by the pressure of universal acclaim. In a single year he can do more for Mexico than Diaz did in thirty. The opportunity is to follow the lead of the great nations that have made common cause against the Teutonic scourge. By joining the allied nations he becomes one of us, a friend, a coadjutor, working in harmony with us, sharing what his people can offer, and welcome to all the financial and other help that America and England and France can render. Financial and industrial relations in Mexico would then be equalized with the rest of the world, and her share of prosperity would grow great in proportion to the magnificence of her natural endowments. Mexico has all to gain by accepting the favor of the world. It would be as an industrial contributor, adding to the supplies of metals and foods and cordage that her participation would be most active, and she would grow enormously rich. This is the opportunity, the way to stifle revolution, to silence political foes, to bring back the exiles as friends ready to help in the upbuilding of the 'patria.'

There is something more to be said, however. In the first place, amnesty that is frank and conclusive must be granted; something far more frank than has yet been spread upon the statute-books of Mexico. Next, President Carranza must remember that the revolution was not meaningless; that it stood for two quite definite ideas, namely, agrarian reform and "Mexico for the Mexicans." Up to date no true agrarian reform has been achieved by Carranza, and it must be achieved if he is to become in any measurable degree a father to his people. It is a knotty and difficult problem; any man who knows the complicated antecedents of the situation as it existed in 1910 will admit that; but it is not insoluble; nor will the solution of it necessitate doing serious damage to the interests of the landed proprietors of record. Finally, the issue between the Mexican and the 'gringo' must be met, and that is a much simpler matter than agrarian reform. We believe heartily and completely in the doctrine "America for the Americans;" our Congress and our State legislatures have long expressed the national feeling, sometimes a bit peevishly by restrictions upon land tenure, but these usually have been erased from the statute-books by adverse court decisions. The thing, however, that our lawmakers have not done was to grant monopolistic concessions and to peddle them among foreigners. We venture to suggest to President Carranza that while his onslaught upon the concessions of the old regime may have lacked something of strict legality, it was to be expected as a logical result of the revolution, with possibility of the ultimate compensation that ever crowds on the heels of confiscation; but his own hand is becoming free with these iniquitous dispensations, on which no secure economic life can be established. By nature a concession is obnoxious to the people; it is provocative of unrest and revolution. It is, at best, a crude instrument for promoting investment and for inviting foreign capital. The United States has been a favorite field for foreign enterprise, and such participation has accelerated the growth of the nation without threatening either the industrial or political rights of the people. The foreigner has come and helped us to develop America for the Americans, just as he will help develop Mexico for the Mexicans if the rulers pursue sound policies in legislation and in administration of the laws. Carranza is in the arena with the world looking on; he must open one door or the other; one leads to the tiger's cell; and the other to .... ?

Recruiting Labor From the Gold Mines

The Editor:

Sir—Your articles on methods employed at the Inspiration mine have proved of much interest to me, and I am sure that all the members of the profession appreciate them. I also beg to draw attention to methods employed by the Inspiration, and especially by members of the engineering corps of that company, which have occasioned some remark and would seem to denote a considerable lowering of the ethical bars presumably observed by members of the profession elsewhere.

It is a matter of record that the pay of miners on the Mother Lode is at present considerably below that paid on the copper scale at Miami. It is also a matter of record that there is at present great unrest and considerable dissatisfaction among miners all over the country. There is an unquestionable shortage among operating mines of men suitable to do the work. The Mother Lode has been fortunate in most instances because there is a distinct class of men who have always followed gold mining, who know the Mother Lode conditions thoroughly, and who have been in consequence fixtures on the Lode. Generally they have been dependable, though they have grown exacting in some ways. It is not going too far to say that these men were satisfied with their wages, their working conditions, and their surroundings.

The Inspiration has barred the I. W. W. agitators, after a rather disastrous experience. The Inspiration, however, does not hesitate to send their own peculiar brand of I. W. W. agitators to the Mother Lode, to coax men from our already-undermanned properties, with promises of higher wages, to make them dissatisfied with conditions on the Lode, and generally to spread the gospel of discontent.

Unfortunately, and this is the reason why I draw your attention to the matter, the man in charge of the agitation is C. E. Arnold, consulting engineer. We, on the Lode, going our own way in happy content, had no idea that the profession was honored by members of this un-American crew. It would as soon occur to us to turn agitator and body-snatcher as it would occur to us to go to Miami to disturb adjusted working conditions among the men there. However, I am credibly informed that inducements with regard to wages were made to certain of the men employed at my mine, that their fare was posted with the railroad, and that letters were sent them by Mr. Arnold urging them to go to Miami. Remember that these men were and are employed, have had no occasion to be discontented, and are given the best of food and living conditions in a climate almost too ideal for mining.

Either Mr. Arnold is not aware of the fact that there are certain things an engineer cannot prostitute himself to undertake, or his pretentions of belonging to the profession are without fact, or the price offered by the Inspiration for miners was too great an inducement for his small ethics to withstand. At any rate, even the best interpretation of his actions does not reflect credit on himself, on the Inspiration in particular, and on the profession in general.

This protest, I believe, is directly in line with your exposition of Inspiration methods, and also I think conforms to the better standards among the men of the profession that your paper stands for.

S. E. Rau-Roesler.

Bagby, California, November 8.

[We discuss this subject on the editorial page.]

King of Arizona Company

In our issue of November 10 we published a letter from Colonel Epes Randolph taking exception to a reference made to him in our interview with Mr. D. M. Riordan, appearing in the issue of October 6. When publishing Colonel Randolph's letter we omitted the concluding paragraph, which read as follows:

"It is my wish that you publish this letter in the next issue of the Mining and Scientific Press, giving it the same prominence that you gave to the libelous Riordan interview. Also I request that you wire me upon receipt of this letter, at my expense, whether or not you will comply with this request."

We replied in the affirmative and published his letter, minus this paragraph. Colonel Randolph insists that this paragraph should not have been omitted; we demur, but we publish it now rather than prolong the controversy. The paragraph was omitted because it formed no part of his statement; it was clearly in the way of an instruction. Of course, the term 'libelous' is one that we cannot accept. If Colonel Randolph's feelings were hurt by anything said in the interview, we regret it; so does Mr. Riordan, who states that he had no more feeling of animosity in his mind when he was answering my questions than if he were to say that the Oaks would 'do up' the Seals, or that Yale would 'do up' Harvard. He says that he felt at the time, 18 years ago, that undue rigor was being manifested in expressions to him in regard to Smith and they conveyed to his mind a sense of severity that he did not feel was quite deserved. That was all.

Further he says: "The quotations from the minutes of the meetings of the board of directors of the King of Arizona are no doubt correct. I have no means at hand of verifying them. But in our meetings, as in all the meetings of boards of directors in which I have taken part, there is a great deal of discussion that is never spread on the minutes. Things are said back and forth that are not made a part of the permanent record. At the time I was talking to you I was trying to crowd into the insistent activities of a busy day such answers to your questions as came into my mind on the spur of the moment, and with no thought of unkindness toward any human being. I think if Messrs. Smith, Blaisdell, and Ives could be got together in a room, some, or all of them, would verify the substantial correctness of the picture I was trying to present of *my* activities, which, as you know, I was reluctant to have published at all, because I did not believe there was anything in my recollections that could be of particular interest. I think Colonel Randolph is taking this matter entirely too seriously, and I am sorry that he has done so.

"With reference to my remark about 'real money,' to which Colonel Randolph objects, it seems singular to me that the company records would not 'show how much money Mr. Smith spent on his contract,' but at the time I invested my money I had it not only from Mr. Ives, but, I believe, from Messrs. Blaisdell and Eichelberger that Colonel Randolph had put only 'a little money' into the venture, and had given his notes for the balance. This I am stating off-hand, from recollection, and am willing to stand corrected if I am in error as to what was then stated to me. But I am now prepared to accept Colonel Randolph's statement that he invested $10,000 in real money on the day the mine was optioned; and if $10,000 is, in his estimation 'only a little,' I congratulate him; $10,000 looked to me like a very respectable sum of money then, and I have a good deal of respect for that amount today; even when many people talk as glibly about billions of dollars as if it were a wheelbarrow load of potatoes. But I certainly did have it verbally from Messrs. Randolph, Ives, and Blaisdell that Mr. Smith had put a considerable amount of money into the venture, or had pledged himself to do so, and at the time they spoke of this to me they were looking to Mr. Smith as the principal person who was putting in 'real money,' at that time, and was expected to continue doing so.

"If I have injured Colonel Randolph, or anybody else, in mind, body, or estate, I would (if need be) walk from here to Tucson and back again to make due, full, and satisfactory reparation. In trying to give you a composite picture fished out of the dusty pigeon-holes of a more or less battered old memory, I chatted away conversationally as I would if we were riding together on a buckboard across the desert, or climbing on horseback some mountain-trail. It was the merest recital, entirely devoid of malice or animosity. Colonel Randolph, at the time I first met him, had a high reputation as an engineer, and all I have known of him since confirms this estimate of his professional equipment and his achievements. His plan to utilize the waters of the Colorado in the Grand Cañon was a splendid one, and should be, and I believe will be, some day put into practical service. His subsequent plan, carried through to a successful conclusion, to control the waters of the Colorado, after the Government engineers had failed to do so, should, and, I believe, will stand as an enduring monument to his resourcefulness and daring. So that all my predilections were in favor of avoiding the saying of objectionable things about him, or anybody. I think if Colonel Randolph were present in the room when you were questioning, and heard my answers just as I gave them, he would have been inclined to treat them in the good-natured way I did, and laugh at them instead of taking umbrage."

This closes the controversy.—EDITOR.

Feeding the Mexicans

The Editor:

Sir—Your suggestion regarding the distribution of food to the starving peons in Mexico is, I believe, entirely correct. The same conditions existed in Nicaragua after the long campaign resulting in the overthrow of President Zelaya. The neglect of agricultural work during the revolution and a series of crop failures had resulted in a severe scarcity of food. This was being relieved by importation of corn, beans, flour, etc., from the United States, Salvador, and other countries. The short-lived Mena revolution, by cutting off these supplies, left the people in a state of destitution. When the revolt was ended by the intervention of the United States, the two thousand American marines and blue-jackets with their officers took over the civil administration of the municipalities, established law and order, and organized the distribution of food to the native population. In spite of their natural antipathy to the 'gringo', heightened by his interference in their internal affairs, the natives were compelled to admire and respect the honest and efficient administration of our soldiers.

Such a distribution of food in Mexico would do much to establish in the mind of the peon the idea that we wish him well, and that, rather than wishing to engage in a war of conquest, we are sincerely desirous of making the world safe for democracy.

J. A. PARKER.

Mina, Nevada, November 12.

NICKEL is in great demand for Germany, especially for making tubes for torpedoes. It is evident that she had not taken precautions to obtain a sufficient store of this metal, as she did in regard to copper, zinc, tungsten, and some others. It is now reported that the Christiania nickel works are to resume operations, under a contract to supply Germany with 100 tons of nickel per month. Exportations of nickel from Norway to Germany had been suspended for a considerable time.

Heap-Leaching of Copper-Sulphide Ore

By COURTENAY DE KALB

Large-scale leaching-tests have been conducted for a period of nearly eight months at the Copper Queen mines at Bisbee, Arizona, following work for three years in precipitating copper with scrap-iron, in preparation for handling the disseminated sulphide ore recently developed in Sacramento hill. The orebody, as tested by drilling, and further confirmed by underground exploration, has shown a large tonnage, some of which is of a grade suitable for smelting without previous concentration. The larger volume, however, with a tenor averaging about 1½% copper, will be treated by flotation. In addition, large tonnages that assay no higher than 0.9 to 1% copper will be available. This lower-grade ore will be leached, and it is proposed to treat it in heaps, taking advantage of the reactions that occur through oxidation when the mass is treated with the requisite amount of water. The principle, of course, is the same as that long utilized at the Rio Tinto mines in Spain, but the details whereby the system is made adaptable to American conditions have been evolved by Joseph Irving, who is in charge of the operations at Bisbee. The novel features introduced into the process are found chiefly in the precipitation, which will be considered first. As now conducted, taking into account the existing costs of supplies and labor, the entire process has demonstrated a recovery of 80% of the copper in the ore, at a cost of 20c. per ton for transporting the ore from the mine to the leach-yard, building of the heap, and disposing of the tailing; 20c. for leaching; and 60c. for precipitation, including labor, power, maintenance, and the cost of iron consumed, charging current rates for the scrap-metal. The ore treated in the experimental heaps has averaged 1.25% copper, yielding 20 lb. cement-copper per ton of ore. The cement is obtained as an unusually high-grade product, assaying 60% copper. The cost of the whole operation, therefore, is only 5c. per pound of copper recovered. To obtain the total cost of the copper there must be added the cost of mining, shipping of the cement-copper to the refinery, melting, poling, and casting.

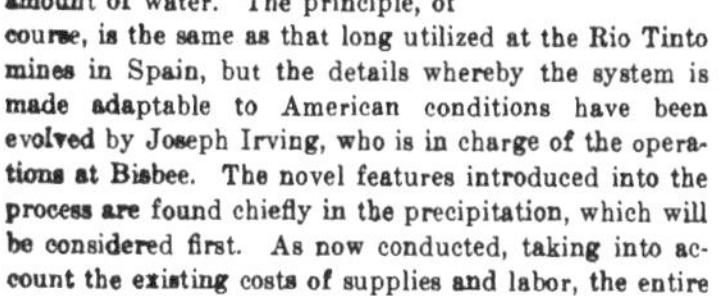

COPPER PRECIPITATING-PLANT, BISBEE

PRECIPITATING-VATS AND LAUNDERS

It will be seen that the precipitation is the most expensive step in the process. When applied on a larger scale this cost will be reduced, of course, and improvements are even now being made that will effect further economy. At the present time an average of about 400,000 gal. of solution is passed through the precipitation-vats daily. Only 70,000 gal. of this comes from the leach-yard; the remaining 330,000 gal. consists of water from the Czar shaft, which contains about 63 grains of copper per gallon. The experiments in precipitation were conducted on the mine-water for a period of nearly three years before 'heap-leaching' was tried. The possibility of economically reducing the ferric to ferrous salts by the use of Mr. Irving's pyrite filter was thus proved, and this paved the way for successfully treating the low-grade sulphides developed by drilling the disseminated granite-porphyry ore on Sacramento hill. As a preliminary to the precipitation of the copper from the mine-water it is essential to eliminate the solids carried in suspension. The larger portion of the solid matter is removed by diverting the water through shunt-tanks where the checking of the current causes it to settle. Most of what remains is then caught in the pyrite filter.

The arrangement of the shunt-tank is illustrated in the accompanying sketch (Fig. 1), where *F* is the flume carrying the mine-water, *S* the diversion-launder, with a gate *G*, which may be changed so as to close either the flume *F*, or the diversion-launder *S*. *T* is the tank, 32 ft. long, 8 ft. wide, and 6 ft. deep. The tank is provided with a discharge-door, *D*, near the bottom, for flushing out the accumulated sediment. About 400 cu. ft. of sludge is settled from the mine-water in three months, and it assays from 0.3 to 1% copper.

FILTRATION AND REDUCTION. After the coarser material has been settled the waters are conducted to the filtration-vats at the head of the precipitating-plant. As originally designed it was intended to use six vats, in each of which would be placed approximately 40 to 45 tons of pyritic ore to serve as a filter. The vats are round and are made of 3-in. redwood. As now constructed, the solutions are caused to flow downward through vertical box-pipes at the side, and then to pass upward through the filter-bed, discharging at the opposite edge of the vat into a 2 ft. by 10-in. launder. The filter is composed of pyritic ore, from ½ to 1 in. diameter, with some fine material, assaying about 2% copper. The bed is 2½ to 3 ft. deep, and is supported upon a wooden grating resting on 4 by 6-in. beams. The solution flows through each filtration-vat at the rate of 100 gal. per minute, and the filter retains 75% of the solid matter still in suspension. In addition, the pyrite converts the ferric salts in the solution into ferrous salts, thereby saving iron in the subsequent precipitation. It does more than this; the rapid corrosion of iron in the precipitation-tanks when ferric salts are abundant would add to the amount of corroded iron-scale, which becomes detached and accumulates in the cement-copper. It also provides acid which maintains a 'sweet' condition of liquor in the precipitating-tanks, and prevents settling of basic iron salts, thereby improving the recovery. An indication of the influence of the pyrite filter in accomplishing the reduction of the ferric salts is shown by the fact that after introducing the filter the grade of the precipitate recovered was raised from less than 40% to an average of 60% copper. Economies follow from such a change in assay-value, not only in the handling, drying, sacking, and shipment, but in the subsequent melting at the refinery.

Each filtration-vat is provided with a large door on the level of the filter-bed, through which the accumulated sediment may be flushed at intervals; it is also used for discharging the spent filter. The mud is washed out once a month. If allowed to collect for too long a time the particles of pyrite become coated with slime and the chemical action of the filter ceases. It is also desirable at intervals to turn the material that constitutes the filter-bed in order to expose fresh surfaces to the action of the solutions. Mr. Irving is now preparing to modify the arrangement of the ore-filter by raising the tank, using a 3-ft. filter-bed of screened pyritic ore and filtering downward. This will permit a percolation of 500 gal. per minute. The filtered solution will flow by a siphon from the lower part of the vat to the top of a precipitating-vat set two feet lower. The sediment caught in the filter will be flushed from time to time without removing the ore-bed. The action of the pyritic filter is represented by the following equation:

$$2FeS_2 + 11Fe_2(SO_4)_3 + 12H_2O = 24FeSO_4 + 12H_2SO_4 + S$$

Although this explains what takes place, and is in accord with the deposition of elemental sulphur in nature, as often found in pyrite casts resulting from such a reaction, the reducing effect of the filter is undoubtedly more complicated than the above equation would indicate. The promptness of the reaction is the interesting feature of Mr. Irving's application of the pyrite filter. Nishihara* has shown that many common minerals yield a relatively prompt reduction of tenth-normal ferric sulphate. For example, he obtained a sharper and more immediate effect with kaolinite than with pyrite, although after more prolonged contact the larger reducing effect was obtained with an equal weight of pyrite. Pyrrhotite was far more active, the reduction, expressed in terms of cubic centimetres of $KMnO_4$ used in titration, being four times greater than for pyrite. The various common ferro-magnesian silicates found in the gangue of many ores also exert a not inconsiderable reducing power. Chalcopyrite likewise displays a reducing action, and, as might be expected, marcasite is more pronounced in its effect than pyrite. Moreover, the reaction in all cases becomes more energetic when copper sulphate is present in the solution, and this is the case in the solutions derived from copper mines and from the heap-leaching of copper ores. It should be considered, also, that the pyritic ores from different mines, and perhaps from different parts of the same mine, would be expected to show important variations in reducing power, since refined analysis proves that pyrite rarely, if ever, corresponds to the formula FeS_2; it is a somewhat indeterminate solid solution of FeS in FeS_2. The effect of this on the ability of pyrite to reduce ferric sulphate in waters coming in contact with it, is to be taken into account, and upon this probably depends the continued action of the filter, which must represent a

*'Economic Geology,' Vol. IX, pp. 743-757.

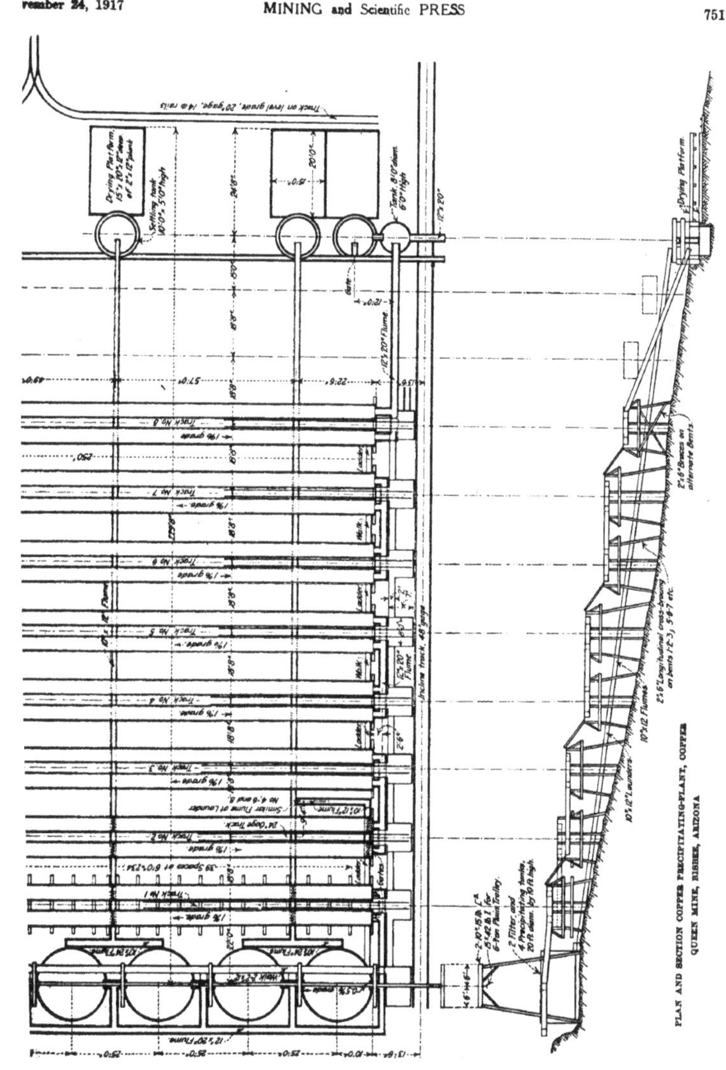

PLAN AND SECTION COPPER PRECIPITATING-PLANT, COPPER QUEEN MINE, BISBEE, ARIZONA

penetration and breaking down of the pyrite particles rather than a mere superficial reaction.

Another feature in the economic use of pyrite as a reducing filter in the preparation of copper-bearing solutions for precipitation is the fact that the filter itself gradually becomes depleted of its copper. Ferric sulphate reacts with either cupric or cuprous sulphides, reducing the iron salt to the 'ous' state and yielding, respectively, copper sulphate, and copper sulphate plus cupric sulphide. The mine-water at the Copper Queen gives an extraction of 50% of the copper in the filter in five months, lowered from the original assay of 2.02% to 0.6% in the discarded filter-bed, and at the end of a year the filter has become too inactive to yield economic results. The mine-water varies materially in its load of metallic salts at different times of the year, depending on the rainfall. A typical analysis shows about 21 grains of copper per gallon, equivalent to 0.04%, about the same quantity of ferrous iron, and approximately 16 grains of ferric iron, or 0.03%. The amount of copper in the dry season often increases to double the above quantity. A fair example of the composition of the solutions entering the reduction-vats and of the outflow to the precipitation-plant is given below:

REDUCING EFFECT OF PYRITE FILTER ON SOLUTIONS

	Copper	Ferrous iron	Ferric iron	Free H_2SO_4
	Grains per gallon			
Entering vat	20.98	13.32	15.66	6.00
Leaving vat	39.64	25.10	1.74	20.00

The solutions represented by the foregoing analyses consist of mine-water only, and the reduction shown is the average at the present time.

PRECIPITATION. The clarified and reduced liquors after filtration are conducted to the precipitation-vats, of which four are now in service. These are of the same size as the two vats that contain the pyrite filters, namely, 20 ft. diam. by 10 ft. deep. They are similarly fitted with false bottoms to support the scrap-iron employed for precipitation. By the present arrangement only the mine-water passes through the filters, since the leach-liquor from the ore-heap has been subjected to the same action in the process of leaching through the pyritic ore. The liquor from one filter-vat passes through only one precipitation-vat, and the overflow thence goes to a cascade of precipitation-launders below. The mine-waters going to the second filter-vat overflow to another precipitation-vat, and the outgoing liquor unites with the leach-liquors from the ore-heap, and thence goes through two more precipitation-vats placed in series, the final overflow being then delivered to the head of the system of precipitation-launders. The course of the solutions may be more readily traced by means of the accompanying flow-sheet.

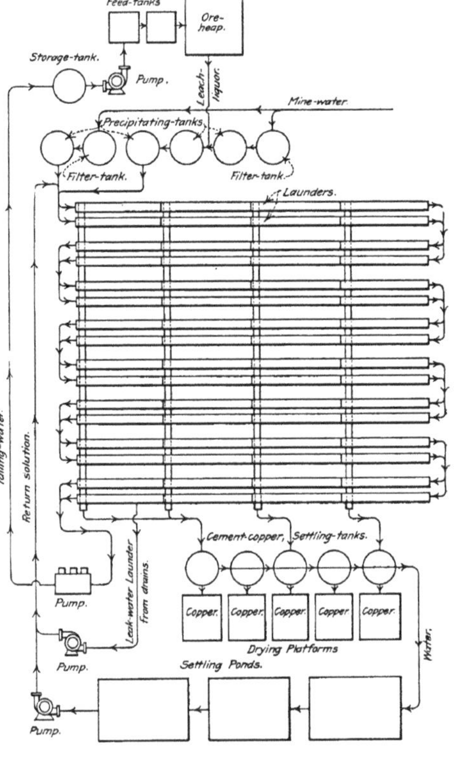

SOLUTION FLOW-SHEET

Better precipitation is obtained in the vats than in the launders, and this fact will lead to some important modifications of the system when carried into effect on a larger scale after the Sacramento Hill mine begins to deliver a heavy tonnage of ore to the leach-yard. It is found that a single vat of the size now used will extract on an average 40% of the total copper-content from the solution when flowing at the rate of 100 gal. per minute d carrying 10 lb. copper per 1000 gal. The iron in e large round vats consists of cumbrous scrap from the ine and smelter, such as old cars, pumps, compressor rts, worn-out converter-shells, old flues, and the like. e solution is conducted through a down-take to the

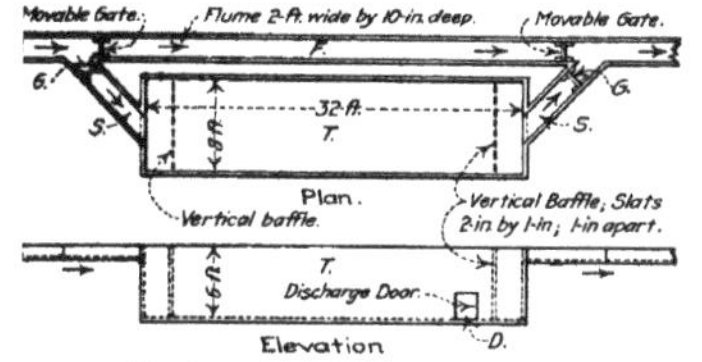

FIG. 1. SETTLING-BOX FOR MINE-WATER

ttom of the vat, and flows upward through the false ttom and the scrap-iron. The cement-copper is washed wn through the false bottom with tailing-water from e precipitation-plant, and is discharged into launders iderneath that carry it to settling-tanks below. The

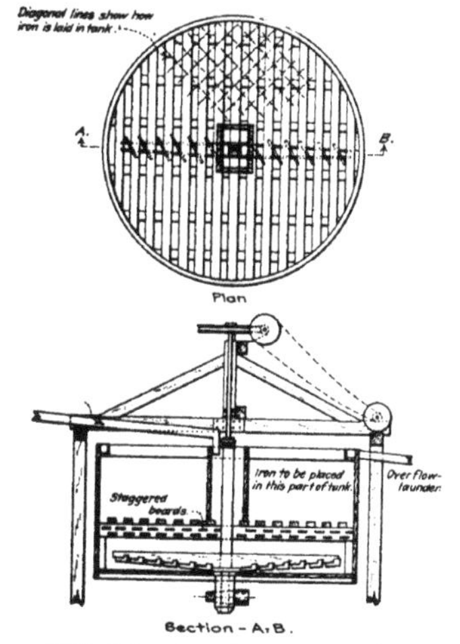

FIG. 3. DORR MECHANISM FOR PRECIPITATION-VAT

ng-water, as will be more fully explained later, is iped to the ore-heap above, and the main through h these spent liquors is pumped is so placed that ich-pipes may be utilized for drawing off what water quired for washing the cement-copper from either recipitating-vats or the launders.

nsulting the plan it will be seen that the precipi-tation-launders, consisting of eight pairs, are set parallel to the line of upper round vats. The launders in each pair are on the same level, and cascade into the next pair below. Except when cleaning up, the stream of solution is divided between the two launders in each

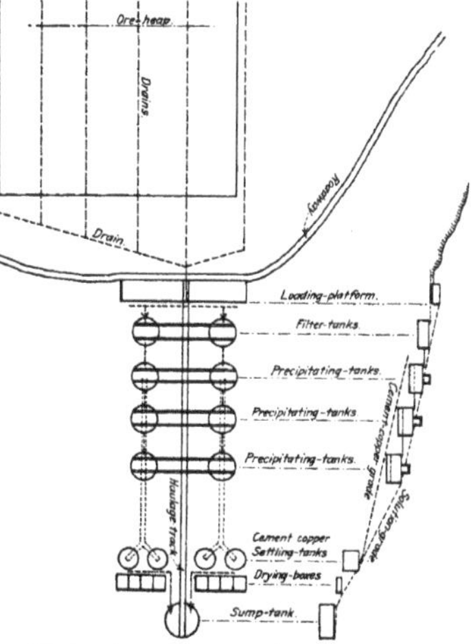

FIG. 2. PLAN FOR BURRO MOUNTAIN LEACHING-PLANT

pair. The process of cleaning up is continuous, so that one launder in the system is always cut out, and during this process its neighbor carries the full stream. The

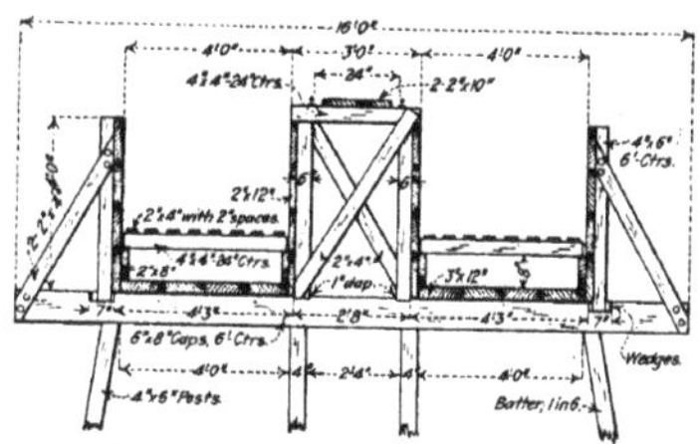

FIG. 4. SECTION ON PRECIPITATION-LAUNDERS

launders are each 250 ft. long and 4 by 4 ft. in cross-section, divided into four compartments. There is a narrow downtake at the head of each compartment to deliver the solution at the bottom. Smaller sizes of scrap-iron are used in the launders, resting on a false bottom made of slats, 4 in. wide and set 2 in. apart. This false bottom gives 6-in. clearance to admit of thoroughly hosing out the cement-copper. The copper is discharged into launders underneath that take it to the settling-tanks and ponds. The tailing-water used in the clean-up is taken from 2-in. bronze pipes connected to the large main leading from the pumps to the leach-yard. All the launders are built of 2-in. Oregon pine and the bents and bracing that support them are of the same material. Considering the weight carried the construction seems light, 4 by 6-in. posts being used in the bents, and 2 by 4-in. material for bracing. The leakage is small, but some solution unavoidably escapes, and this is preservative against decay, although the cellulose suffers chemical changes that leave the wood soft, and with greatly lessened transverse strength. No trouble, however, has been experienced by collapse, and the structure has stood with very little repair and no renewals for three years. This is interesting, because in addition to the dead load there are vibrations from moderately heavy traffic in running trucks loaded with iron for filling the launders. A tramway is provided between each pair of launders; it is so arranged that the truck can be run off at any level from a platform-car operated by cables on an inclined track extending from the scrap-pile at the railroad siding below to the upper round vats, following the grade of the system of precipitation-launders.

On the ground underneath the precipitation-plant ditches serve to drain escaping solutions to launders that lead to a lead-lined Buffalo centrifugal pump with a 2-in. discharge, which returns the leak-water to the precipitating-system. The cement-copper is delivered by launders placed at right angles to the precipitation-launders, into a series of five circular redwood settling-tanks, discharging one into another, the inflowing stream being delivered at the centre, and the outflow over the periphery falling into surrounding circular launders. The settling is fairly complete, and the thickened product is flushed from the round tanks into large square shallow drying-vats where it evaporates in the Arizonan sun. The dried cement has a lustrous red color, and assays 60% metal. The overflow from the tanks is further settled by circulating through a series of three ponds, the collected sludge assaying from 10% copper in the first pond to 5% or less in the last. All this cement copper is shipped to the smelter at Douglas. The clarified liquor from the final settling-pond is called 'tailing-water.' It contains an insignificant amount of copper, which, however, would involve an appreciable loss if allowed to go to waste. The tailing-water has the following composition:

Copper Gr. per gal.	Ferrous Iron Gr. per gal.	Ferric Iron Gr. per gal.	Free H_2SO_4 Gr. per gal.
2.32	150.80	26.68	3.48

The tailing-water is pumped to the feed-tanks above the leach-yard, and is used for wetting the ore-heap, thus returning, enriched in copper, to the precipitation-system. It is handled by a geared 3-piston bronze Aldrich pump, driven by a 20-hp. Allis-Chalmers motor taking current at 220 volts. The quantity of solution is 300 gal. per minute, delivered against a head of 100 ft. The main that carries the returned tailing-water is an 8-in. Crane wood-lined pipe. The leak-liquor, which is returned to the head of the precipitation-system by lead-lined centrifugal pumps, is carried in a 3-in. cement-lined pipe. The cement lining is about ½ in. thick, and the coating is done at the Copper Queen shop at Bisbee. The entire operation of the plant, consisting of filtration, precipitation, cleaning, handling scrap-iron, drying and sacking cement-copper, pumping, and general maintenance, requires a force of eight men and one foreman. The iron consumed amounts to 1½ lb. per pound of copper precipitated. The theoretic quantity required is 0.88 lb. iron for each pound of copper, but there is a mechanical loss, and a certain amount of ferric sulphate still remains in the solution after passing through the pyrite filter. Every unit of ferric iron calls for half as much metallic iron in its conversion to ferrous sulphate in the precipitation-boxes. The free sulphuric acid also consumes iron. Therefore the recovery of one pound of copper with an excess of only 0.62 lb. of iron over the theoretical requirements, in which is included the mechanical loss as well as the loss from free sulphuric acid in the solutions, shows how important a rôle is performed by the pyrite filter when it is considered that the original mine-water carries nearly pound for pound of copper and ferric iron. Without the pyrite filter this ferric iron alone would dissolve a half pound of metallic iron for each pound present as a ferric salt in the solution.

It has been pointed out that the precipitation in large round vats has been shown to be more efficient than in launders. In the new plant that will be built to handle the solutions from heap-leaching on a large scale when treating the Sacramento Hill low-grade ore, such vats will entirely supersede the launder-system. The same system may be installed at the Burro Mountain mine, the tentative plan being shown in outline in the accompanying sketch (Fig. 2). A further improvement is proposed, in accordance with a design by Joseph Irving, Jr., which consists in adapting the Dorr mechanism to a precipitating-vat. As will be seen in Fig. 3, the stirrer is placed below the false bottom, and the solution is fed through a vertical wooden conduit surrounding the spindle of the machine, which may be a solid or a perforated cylinder. The stirrer serves to impart a movement to the solution, causing the copper, as fast as it is reduced by the scrap-iron resting on the false bottom, to become detached. As the cement-copper settles it is swept to a pocket in the bottom of the tank, from which it can be drawn off at intervals.

HEAP-LEACHING. The ordinary reactions in the oxidation of sulphides of iron and copper have long been utilized as a ready means for leaching low-grade ores in which the copper is present either wholly as a sulphide or as a mixture of sulphide and oxide. Although the last word has not been written on the subject, the principles are quite well understood; nevertheless, the application of heap-leaching has led to frequent disappointment, partly because of the difficulties arising from ferric sulphate in the solutions, and from lack of familiarity with the details of management in leaching ore-heaps. Therefore the work done at Bisbee is of great interest, since it has established an economic line of treatment that will be employed both at the Burro Mountain mine in New Mexico and at Bisbee on a larger scale. Indeed, the possibility of adapting the method as a means for profitably extracting the copper from the 1,500,000 tons of the lower-grade ore developed in Sacramento hill has been an important factor in the plans for beneficiating that unique deposit.

It may be desirable to re-state the reactions responsible for the solution of copper in moistened ore-heaps, and to point out that no inconsiderable merit in the system lies in the possibility of recovering much of the copper that may happen to be present as an oxide. The very low-grade copper-sulphide bodies usually are found in fairly well-defined zones, either on the borders of higher concentrations of such ores, or on zones through disseminated orebodies where oxidation has been more rapid in the reactions that determine enrichment. These two cases are found in Sacramento hill, and in each of them a portion of the copper has been converted into the oxide. Carbonates of copper, however, are practically absent. The reactions on which the solution of the oxide depends may be represented by the action of cupric sulphate upon cupric oxide, yielding basic copper sulphate, thus:

TOP OF ORE-HEAP, SHOWING TERRACES AND BASINS

BUILDING ORE-HEAP FOR LEACHING

$$CuSO_4.5H_2O + 3CuO = CuSO_4.3Cu(OH)_2 + 2H_2O$$

Acid generated in the oxidation of the sulphides will also directly attack both cuprous and cupric oxides, as follows:

(a) $Cu_2O + 2H_2SO_4 + O = 2CuSO_4 + 2H_2O$

(b) $CuO + H_2SO_4 = CuSO_4 + H_2O$

On the other hand, there is danger of reducing cuprous oxide to metallic copper in the presence of ferrous sulphate and free acid:

$$Cu_2O + 2FeSO_4 + H_2SO_4 = Fe_2(SO_4)_3 + H_2O + 2Cu$$

This tendency, however, should be offset by the solubility of freshly precipitated copper in dilute acid and in ferric-sulphate solutions at the temperature present and with the time available in a slowly oxidizing ore-heap, according to the reactions:

(a) $Cu + H_2SO_4 + O = CuSO_4 + H_2O$

(b) $Cu + Fe_2(SO_4)_3 = CuSO_4 + 2FeSO_4$

The reaction yielding the cuprous sulphate found in Nature as the mineral dolerophanite may also take place between cupric sulphate and cupric oxide:

$$CuSO_4 + CuO = (Cu_2O)SO_4$$

It should be pointed out that the danger of producing metallic copper is always considerable, and deserves attention on the part of the superintendent with a view to obviating as far as possible the conditions favorable for such reactions. Metallic copper may result also from the action of ferric sulphate upon cuprous sulphide, as seen below:

$$Cu_2S + 4Fe_2(SO_4)_3 + 4H_2O = Cu + CuSO_4 + 8FeSO_4 + 4H_2SO_4$$

Mr. Irving admits the possible formation of some

metallic copper in the leaching of the ore-heap, but it has not been definitely ascertained what proportion of the copper in the leach-tailing, which contains about 20% of the original copper, is in the form of metallics. It would be an interesting inquiry, since it is difficult to believe that any considerable amount of sulphide could escape the action of the solutions after a year of treatment under conditions carefully maintained to secure the highest possible degree of oxidation.

The following equations are offered by Mr. Irving as representative of the reactions that take place in the ore-heap. The first stages are those that produce the active reagents, namely:

$$(1)\quad FeS_2 + H_2O + 7O = FeSO_4 + H_2SO_4$$

$$(2)\quad 2FeSO_4 + H_2SO_4 + O = Fe_2(SO_4)_3 + H_2O$$

The function of the ferrous sulphate and sulphuric acid has already been discussed. The ferric sulphate is the compound that attacks the copper sulphides, and when a proper balance exists the ferric sulphate should be wholly reduced to the 'ous' state; furthermore, the effect of the pyrite in reducing the ferric salt is operative as explained in connection with the pyrite filter. Accordingly a well-managed heap should yield leach-liquor comparatively free from ferric sulphate, thus rendering precipitation on scrap-iron economical. To effect this desirable result the heap would have to be rich in pyrite. The further reactions on which the leaching of the copper sulphide depends are the following:

$$(3)\quad Cu_2S + Fe_2(SO_4)_3 = CuS + CuSO_4 + 2FeSO_4$$

$$(4)\quad CuS + Fe_2(SO_4)_3 = CuSO_4 + 2FeSO_4 + S$$

The latter reaction is probably more generally accomplished in the presence of abundant oxygen without the liberation of elemental sulphur, as expressed in the equation:

$$(5)\quad CuS + Fe_2(SO_4)_3 + H_2O + 3O = CuSO_4 + 2FeSO_4 + H_2SO_4$$

In making ready to build the heap a site was chosen in an arroyo or gulch on the side of Sacramento hill, out of the way of flood-waters. No special preparation of the ground was attempted; the space was merely well-covered with wash from the hill, and the slope was about 8°. The heap was built of run-of-mine ore from the Sacramento Hill deposit, and contained 9487 dry tons, the largest lumps being 10 in. diam. The upper surface of the heap has an area of 12,000 sq. ft., the vertical depth at the lower edge is 25 ft., and at the upper edge it is 5 ft. A vertical depth greater than 25 ft. is inadmissible, owing to the danger of channeling. A gradual and uniform subsidence of the leach-water is of prime importance. The top of the heap has two levels or terraces, in order to facilitate inspection so as to obtain information regarding the percolation. The difference of elevation between the two terraces is 4 ft. Each terrace is subdivided into basins 15 ft. square, separated by ridges of ore about 12 in. high. The basins, furthermore, are so arranged that the fresh leach-water will overflow from one to another.

On first flooding the heap it absorbed 360,000 gal. of water, or 38 gal. per ton. The solution exudes from all sides of the heap until the surface becomes sealed by a crust of basic salts. After a little time the heap swells, and in places may burst out and then must be repaired by building bulkheads of the larger blocks of ore. The floors of the basins also become clogged by a deposit of basic iron salts, which checks the filtration. When this occurs they are dug up, and material from below is brought to the surface and again leveled preparatory to a resumption of leaching. One laborer on each shift is required. He serves as heap-tender, looks after the measuring or feed-tank, which supplies the leach-water, and takes care of a centrifugal pump that throws the water from a storage-tank to the feed-tank. As previously explained, the tailing-water from the precipitation-plant is elevated by an Aldrich pump, through an 8-in. wood-lined main, to be used for leaching the ore-heap. It is delivered direct to the measuring-tank, which has accumulators and an automatic tripper for discharging a volume of 450 gal. every five minutes. The average delivery of leach-water to the heap is 70,000 gal. per day of 24 hours. The amount varies somewhat, depending upon observed conditions of percolation and outflow, and is regulated by the superintendent and by the heap-tender. The rate of delivery of leach-liquor from the foot of the heap may not correspond exactly to the rate at which the fresh water is added, but the weekly averages will nearly balance each other. As an example of these discrepancies, on three successive days the following quantities of feed and outflow were observed:

PERCOLATION

	Feed-water to ore-heap Gal.	Leach-liquor outflow Gal.
(a)	29,580	38,800
(b)	27,000	31,000
(c)	30,000	23,000
Total	86,580	92,000

Accurate samples of feed-water and leach-liquor are taken daily and assayed. The subjoined table gives representative analyses of the feed-water and the corresponding leach-liquors, in a series of four days.

SOLUTIONS TO AND FROM ORE-HEAP

Pounds per 1000 gal.

	Copper		Ferrous iron		Ferric iron	
	Feed-water	Leach-liquor	Feed-water	Leach-liquor	Feed-water	Leach-liquor
1	0.20	5.18	0.33	0.66	0.33	0.50
2	0.44	8.82	0.33	0.33	0.50	1.00
3	0.44	7.01	0.33	10.44	0.50	1.82
4	0.20	7.00	15.33	9.83	1.00	2.00

The last line in the table represents the present reactions with special reference to the formation of ferrous and ferric iron, and indicates a desirable result.

The speed of percolation is about $\frac{3}{4}$ ft. per hour, so that the solutions take approximately 20 hours to pass through the ore-heap. The rate of extraction is calculated from the daily volume of solution and its assay. The average of the samples of ore taken daily while building the ore-heap gave the following result:

	%
Copper	1.3
Silica	60.7
Iron	10.5
Lime (CaO)	1.2
Alumina	12.1
Sulphur	9.9
Silver	0.12 oz.
Gold	none

Thus the total copper in the heap amounted to 246,662 lb., and in a period of five months 25% of the total copper has been extracted. Taking the recovery from the solutions in precipitation at 96%, this means an actual recovery of 59,198 lb. of copper, or nearly 400 lb. per diem. The cost of transporting the ore to the leach-yard and building the heap was 10c. per ton, and the average cost for leaching has been $10 per day. It is possible that it may require two years to complete the leaching, giving an extraction of about 80% of the original copper, or approximately 189,435 lb. recovered as cement-copper.

Applying the reactions which occur in the direct leaching of ores in heaps, it has been proposed by Mr. Irving to crush copper ores for flotation in a solution of ferrous and ferric sulphate. The result anticipated is the solution of the oxidized copper compounds, thereby increasing the total recovery, and incidentally cleaning the sulphide particles from films of oxide so as to facilitate concentration in the flotation machines. The solutions would then be passed through the filtration plant to recover the dissolved copper. The outcome of these experiments will be awaited with keen interest.

Michigan Copper-Tailing

Utilization of waste-tailing sand from the old mines in northern Michigan is attracting attention. Some of this tailing, such as that at the old Cliff mine, in Keweenaw county, now controlled by the Calumet & Hecla and its associated companies, are said to be richer in copper than that now yielding millions of pounds of copper yearly by the process applied for the treatment of the Calumet & Hecla tailing. The old sand-piles were deposited from 20 to 50 years ago, when milling was by no means a highly developed art, so that a great deal of copper that would now be recoverable from the ore was discarded with the waste. The Calumet & Hecla has proved that its sand can be re-worked, after being dredged from Torch lake, re-ground, and run through the leaching plant, at a cost of not more than 6c. per lb. for the copper recovered. The Tamarack, Quincy, Franklin, and other sand-piles contain great amounts of recoverable copper. Some of these have already been tested.—'Daily Metal Reporter.'

SPANISH CRUDE, unwashed, fine pyrite is still quoted at 16c. per unit of contained sulphur, and Spanish washed lump pyrite ore commands 16½ cents.

Oregon Mines in 1917

Although receipts from Oregon at the mint and smelters in San Francisco show a decrease in the output of gold of $75,000 for the first five months of 1917, as compared with the corresponding period of 1916, it is not an indication, according to Charles G. Yale, of the U. S. Geological Survey, that the mines of the State are less prosperous. Confirming this opinion, the copper output of Oregon increased 794%, so that the total yield of metal in 1916 increased $939,969 or 47% over 1915. A steady increase in the copper yield is apparent this year, and the placer mines are also increasing their gold yield. The deep mines, however, are not producing as much gold as in 1916. As the copper output augments, so does that of silver. The Homestead Iron Dyke mine in Baker county, re-opened after years of idleness, is now by far the largest copper and silver producer in the State. In June last a new orebody was discovered on the 850-ft. level. A large concentrator has been constructed, and the flotation process is used. Other important deep-mine producers in Baker county are the Baker Mines Co., the Cornucopia Mines Co., and the Commercial Mining Co. The largest placer in the State is the Powder River Dredge Co., operating two dredges in Cracker Creek district. The Laclede mine, near Baker, has been re-opened this year; and the old Ben Harrison, at Sumpter, once a large producer, has been sold to Idaho men, who intend to make extensive development. In south-western Oregon the old Blanco, a beach-sand deposit in Curry county, between Port Orford and Langlais, is being equipped with new machinery to treat 500 tons of gold-platinum sand daily. The Nellie Wright mine, near Gold Hill, Jackson county, was recently sold to Salt Lake people, who are installing compressors and drills. The mine has a 25-ton Beers mill. The Cheney, Simmons, Ray, and Haff group of quartz claims, in the same county, has been sold to Californian capitalists, who are starting a new 1200-ft. adit. In Josephine county the Grayback copper group has been leased to owners of the California-Oregon Coast Ry., who also own the Queen of Bronze and the smelter at Takilma. Development on a large scale has been commenced. The Queen of Bronze is working 60 men and producing a good grade of copper ore. In Canyon district, Grant county, the Empire Gold Dredge Co. began operations in 1916, and has continued during 1917 with successful results. This is the third dredge operating in Oregon. There is considerable activity in the hydraulic mines, particularly in Baker, Grant, Jackson, and Josephine counties.

PAINT-MATERIALS are coming into greater demand, owing to an energetic movement under the slogan 'Use more paint,' undertaken by the National Paint, Oil & Varnish Association. Although having a selfish motive the propaganda is based on sound principles of conservation, and a nation-wide educational campaign will expound the function and use of paint as a preservative.

Leadville Manganese Resources

Leadville, early famous for gold, and later for silver, lead, and zinc, has recently come into prominence as a source of manganese ore suitable for making spiegeleisen. For more than forty years 'black iron' has been mined for fluxing purposes from several of the Leadville mines, but only since the scarcity of manganese incident to the War became acute has any of it been diverted to the steel industry. At present about 500 tons per day is being shipped as manganiferous iron ore from the district, one of the large steel plants deriving most of its supply from this source. The ore, which averages from 20 to 25% manganese, 24 to 30% iron, and 10 to 15% silica, is reduced to the 33% spiegeleisen that is now being used with excellent results in both the bessemer and open-hearth practices.

"The manganese deposits at Leadville," according to J. B. Umpleby, of the U. S. Geological Survey, who recently examined them, "are distributed along the eastern edge of Poverty Flat, the middle and lower slope of Carbonate hill, and in Iowa gulch. One deposit of undetermined size has recently been opened on the west slope of Iron hill near the old McKean incline. At about this same topographic position is the McCormick shaft from which 2000 tons of low-grade ore (14% Mn) was produced.

"Few of the deposits crop out at the surface. Prospecting for them is determined by the records of earlier mining and by the occurrence of manganese in waste material on the dumps. In general the ore occurs in the blue limestone below the old lead stopes, but there are numerous exceptions to this, some of the old mines that produced very little lead or zinc being among the most important sources of manganese ore.

"In August 1917 ten mines were shipping, and since that time at least two others have entered the list of producers. Cramer & Co., operating the Star group, is the largest shipper in the district, with a daily output of about 250 tons. The other mines, most of them operated under lease, are shipping from 20 to 50 tons per day.

"Hand-sorting is a considerable item in the cost of mining as silicious nodules are common and seams of clay traverse most of the orebodies. In places also ground is being re-worked for the third or fourth time, and old timbers, waste dumped into abandoned stopes, and large areas of caved ground, must be contended with. On the other hand at every mine, except one, it is possible to dump the ore directly into the railroad cars. The large number of shafts available also serves to lower costs by reducing underground haulage to a minimum.

"One of the large deposits is in the Fair View or All Right No. 1 mine, where a mass of fairly homogeneous ore has been shown to be 120 ft. thick. Its areal extent has not been proved, although workings penetrate an area 200 by 75 ft., and similar ore is found 275 ft. west in the old Fair View shaft, 400 ft. north-west in the Jason lease, and 300 ft. north in the All Right shaft. The ore as shipped runs 20% manganese, 30% iron, and 15% silica. The distribution of workings about this deposit and the fact that it is situated in an area not previously mined extensively, affords a better basis for estimating tonnage than in most of the Leadville manganese mines. It is by no means certain, however, that it contains more ore than two or three other deposits in the district.

"Any attempt to estimate the tonnage of manganese ore in Leadville is extremely hazardous. Old stopes, not recorded on mine-maps, are continually being broken into. Furthermore, the small cost of installation, together with the numerous faces of favorable appearance, discourages extensive development in advance of extraction. It is the opinion of the operators that the present rate of production might be doubled and maintained for a period of at least two years. This would call for about 750,000 tons of ore of shipping grade, but after an examination of all the producing mines, one is led to think that more than twice this amount may be produced if market conditions become slightly more favorable. It is reported that the present price for ore of between $4 and $5 per ton f.o.b. Leadville, leaves little margin of profit and does not encourage extensive development. Without doubt the rate of production can be increased rapidly for every reasonable advance in the base price."

LEGISLATION affecting mines in the Philippine Islands has engaged serious attention during the past year. The tax placed on the gross output of mines has not affected the mining industry as much as was anticipated. No company was forced to stop operation because of its application. The Bureau of Science has furnished information during the year on the wharfage tax. There can be no doubt that this tax retards mineral exportation, and it remains for the Legislature to decide whether the mineral resources of the Islands should be developed or should be conserved. The division of mines of the Bureau of Science is in a better position to know the needs of the mining industry and the defects and needed changes in the mining laws than any other department of the Government, and it is studying the question. Some of the old Spanish laws in regard to locating and working claims could be adopted to advantage. Under the present system a great deal of ground is being held for speculation to the detriment of development.

CEMENT is now being made from a by-product obtained in the manufacture of beet sugar, according to 'La Revue' of Paris. The first step in sugar-making from beets is to boil them, and a thick scum rises to the top. It has been customary to discard this, but analyses show that it contains large quantities of carbonate of lime. The amount of calcium carbonate recoverable from 100,000 tons of beets is about 6000 tons, and by adding to this the requisite amount of clay it has been found that, on being burned, it yields an exceptionally high-grade portland cement.

Oxide of Zinc

By GEORGE E. STONE

*The method for making oxide of zinc direct from the ore was invented and developed at the works of the New Jersey Zinc Co. at Newark in the middle of the last century. The process was invented by Burrows, who had not the ability, either financial or technical, to work out the economic details. This was done by John P. Wetherill, whose name is commonly attached to the process. The grate-bars used are also frequently called Wetherill grates, which is a misnomer because they were in common use for boiler-firing at the time and he never claimed originality for them. The development of the process was due to the efforts of the New Jersey Zinc Co. to find a profitable means of working the ores from Franklin and Sterling Hill, New Jersey. These are a mixture of franklinite, willemite, and zincite, containing about 20% zinc. The first attempts were to make spelter. These were not successful, owing to the low grade of the ore and the fusibility of the residue. The next attempt was to make oxide in large muffles and reverberatory furnaces. This succeeded, although the cost of operation was high, the recovery low, and the quality of the product uncertain. In 1855, the new process was patented and has been in successful operation ever since. Essentially, it consists in spreading a mixture of coal and ore on a body of burning coal on a perforated grate and blowing an excess of air through the grate. The zinc is reduced in contact with the coal, volatilized and burned by the excess of air in the upper part of the furnace and in the flues. It is then carried to the bag-rooms by the excess air and products of combustion, which are forced through the flues by fans. In its main features, the process is the same today as at the time of its invention, but the details have been so modified that it would hardly be recognized by its originators.

The process is applicable to all the oxidized ores of zinc and to roasted sulphides, provided the gangue is not so fusible as to leave a residue that is impervious to the blast. In many cases the ores contain impurities that make it impossible to produce an oxide of a color good enough to be available as a pigment. These impurities are volatile metals that form colored oxides or sulphides. Cadmium is one of the worst, because it is very volatile and its dark brown oxide and bright yellow sulphide both have strong tinctorial powers, so that small fractions of a per cent seriously injure the color of the oxide. Lead, which is one of the commonest impurities in zinc ores, also injures the color, though to a much less degree, as it usually forms basic sulphates which are nearly white. On other accounts lead is often objectionable, particularly when the oxide is to be used for the manufacture of rubber goods. Sulphur may occur in oxide as sulphides, sulphates, or as sulphurous anhydrides, the first and last of which are objectionable, the first on account of probable injury to the color, while the last is believed by many paint manufacturers to have a bad effect on the grinding properties. Opinion on the latter point is by no means unanimous. Sulphates, if soluble, have very objectionable qualities for outside paints as they are leached from the coat of paint, leaving discolored spots. Chlorides, in appreciable quantity, are rare, but if present would be open to the same objection.

*Abstract: Trans. A. I. M. E., St. Louis meeting, October 1917.

The fuel must be one that does not produce a black smoke, as this would ruin the color of the oxide. In the West, where anthracite is expensive, semi-bituminous coals, low in volatile constituents, are sometimes substituted, but they are never as good. Coke alone does not work well, as it causes a very intense local heat that makes the residue impervious to the blast. It is occasionally used as a part of the charge-fuel. With silicious ores, limestone is sometimes added, but in general it does more harm that good, owing to its tendency to make the residue too fusible. The ores and coal are usually received at the works in fine enough condition for use without further crushing. Nearly all the ores consist of concentrates or roasted products inevitably finer than necessary. As small sizes of coal are suitable, and are cheaper, they are invariably used. If fluxes are used they are crushed to about the size of the ore and coal. There should not be a great difference in the size of the different materials, or it will be impossible to mix them properly.

The Palmerton plants of the New Jersey Zinc Co., in Pennsylvania, are the largest and best equipped oxide plants in the world. The general arrangement of the two plants is similar, but the east plant has fewer and larger bag-rooms and differs somewhat in details. At both plants the raw materials are delivered on the trestle at the south of the plant and are either placed in bins or stocked in piles. The ore and coal are taken from the bins to the mix-house by lorries. The mixed charges are delivered to the bins above the furnaces and distributed by traveling cranes. Each block of furnaces is a separate unit, having its own blower, exhaust-fan, and bag-room. The oxide is trucked from the bag-rooms to the packing-room and the final packages skidded or trucked to the storehouses.

The ore and coal are delivered on a double track, steel and concrete trestle, 22 ft. high, with Brown tangential bins below the track next the furnace-room. Materials for current use are dumped directly into the bins and the surplus is stocked back of the trestle by a bridge-crane of 233 ft. 9 in. span equipped with a 10-ton Brown grab-bucket. This equipment also returns materials from the stock-piles to the bins when necessary. The charges are drawn from the bins to a weighing lorry, which has two hoppers, each containing a furnace-charge. These are carried to Ransome concrete-mixers placed at one end of the furnace-room. The mixed charges are elevated by a pair of counter-balanced skips to bins above the furnaces. From the bins the mixed charge is taken to the furnaces by cranes each having three hoppers holding one-third of a furnace-charge each, and each hopper is placed on a scale so that the charge is equally divided between the three. At the east plant all materials are delivered on a single-track, steel and concrete, trestle 47 ft. high, down the centre of the storage-yard. The materials are taken from this by a bridge-crane of 48 ft. 8½ in. span with a 2-yd. bucket and placed in elevated bins next the furnace-room. The ore and coal are stocked and reclaimed by the traveling bridge and by locomotive cranes on the high trestle. The total storage capacity is 200,000 tons of ore and 175,000 tons of coal. The fuel and the various ores are taken by bridge-cranes to small bins placed over the mixers. The materials are dropped from these bins into weighing-hoppers and from these into the Ransome mixers. The mixed charges are elevated to bins over the furnaces and then distributed by cranes much as at the east plant, except that there are no scales on the cranes or lorries. This arrangement of doing all of the weighing at the mixer on a fixed hopper is more satisfactory than weighing-lorries, because the latter are difficult to keep in adjustment.

The furnace-rooms at the two plants are similar, the main differences being that at the west plant the furnaces are carried on a concrete foundation and at the east plant on the floor. In both cases the furnaces are elevated sufficiently to allow the residue to be dropped into hoppers below the floor and then into cars. The framing of the buildings is quite different, the east plant being arranged to give more room for the charge-cranes. The west plant furnace-room contains 34 blocks of furnaces and the east plant, 26. All the blocks are alike, consisting of four furnaces 19 ft. 6½ in. by 5 ft. 11½ in. Each furnace has independent blast and flue-connections and three charge-openings in the roof surmounted by hoppers. Each furnace works three charges in 24 hr. The different furnaces of the block are charged alternately at 2-hr. intervals in regular order. Charging in this way, there is always one furnace starting, one nearly finishing, and two working strongly, and the volume and temperatures of the gases to be handled are nearly uniform at all times.

When the furnace is worked off, the dampers in the flues are closed, cutting it out from the bag-room. The blast is shut off and the working-doors taken down. Three men work on the furnace at once, each having a separate door. The loose material on top of the charge is raked off and dropped on the floor close to the furnace. The clinker is then broken by heavy slice-bars and raked into hoppers under the floor. The furnace-men now thoroughly clean the side and back walls from any clinker that has fused to them. When all of the old charge has been removed and the furnace cleaned, the grate is covered as rapidly as possible with a thin layer of coal, the doors closed and a light blast turned on. The coal is lighted by the radiation from the hot arch and it is important that it should light quickly and evenly. When the bed of coal is burning brightly, the charge from the hopper is dropped upon it and leveled. The doors are now closed, the blast turned on, and the furnace left to itself with occasional inspection to stop any blow-holes that may form. With the large charges that are now worked considerable time elapses before zinc vapor appears; when it does, the dampers in the flues are reversed, sending the gases and oxide to the bag-room.

The blowers for the oxide furnaces are direct-connected to motors, the wheels of the blower being placed on extensions of the motor-shaft. They are placed in small houses outside the furnace-room, two in each house. Each furnace has two outlet pipes, usually 20 to 22 in. diameter which lead up to the drum-pipe overhead. The drum has an outlet in the middle of its length leading to the exhaust-fans. Formerly the gases passed through a brick settling-tower between the furnaces and the exhaust-fan, but this is now usually omitted. The exhaust-fans are extra heavy plate-fans with special Hyatt roller-bearings and water-cooled shafts, belted to motors. Aside from the shafts, bearings, and extra thickness of plate, they are of the ordinary types. For convenience in attendance, a number are placed in one building. The west plant has four fan-rooms, three containing 8 fans each and one containing 10 fans. At the east plant there are two with 10 fans each and one with 6. From the exhaust-fans to the bag-rooms, in each case, run round pipes 3 ft. diameter, made of No. 16 sheet-iron and provided with numerous clean-out openings. In all cases, the pipes are arranged to give as nearly as possible the same length from furnaces to bag-rooms. This is done partly to equalize the friction in the different lines, but mainly to secure proper cooling of the gases before reaching the bag-room.

The method of collecting oxide in bags was first patented by S. T. Jones in 1852 and has been patented by others many times since. The bag-rooms at Palmerton vary considerably in size, the first two built holding bags enough for four blocks, while the latest contains the bags for 20. The larger rooms are more economical in labor and do not cause inconvenience provided the ventilation is equally efficient. The bags used are of heavy closely woven cotton sheeting, 22 in. diameter and about 40 ft. long. In the majority of cases the gases enter a header at the top of the room and from there pass to a series of parallel pipes from the bottom of which the bags are connected to hoppers, four bags to each, with a single collecting-bag tied to an outlet at the bottom of each hopper. At intervals the gases are shut off from the two lines of pipe connecting with a single line of hoppers, and the bags shaken. This pair of pipes is again connected to the main header and the next two lines shut off and shaken. The frequency with which the bags are shaken depends upon the relative area of grate and bags, and the amount and quality of the oxide being made. The proper ratio of grate-surface to bag-surface has been a subject of much discussion. It varies in practice between 1:80 and 1:130. The amount actually necessary depends mainly on the efficiency of the ventilation of the buildings and the frequency of shaking. A well-ventilated room will consume no more muslin with a ratio of 1:80 than a poorly ventilated one with a ratio of 1:130. Recent experiments indicate the probability that a well-ventilated bag-room with a mechanical shaker operating at short intervals will allow a great reduction in the necessary bag-area. The collecting-bags are removed from the hoppers every 24 hr., or oftener if necessary, and taken to the packing-room. This is done by a trucking gang that removes the filled bags and replaces them with fresh ones, trucking them by hand to one side of the bag-room, where they are coupled together and the train drawn to the packing-room by a motor.

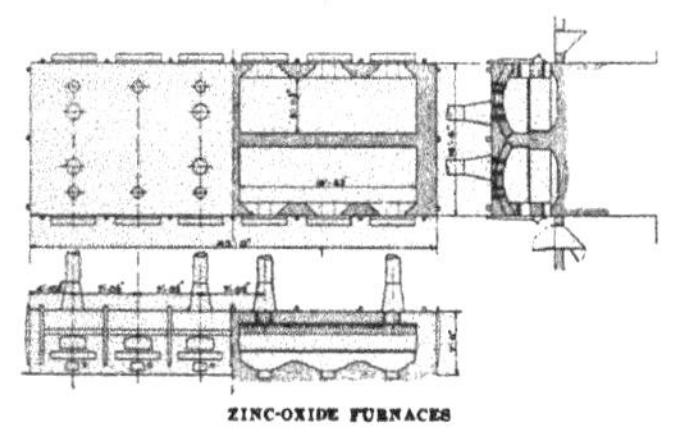

ZINC-OXIDE FURNACES

The packing-room is a four-story building of mill construction. The trucks are hauled into the ground floor and then taken one by one on a platform-elevator to the upper floor. There are a number of openings in this floor leading to bolters on the floor below. Around each opening is a wide shelf on which the bags are placed while their contents are emptied into the bolters. The third floor contains nothing but the bolters, and a room for sewing and making bags. The bolters are of the ordinary types used in flour-mills. The packing-machines are arranged in four lines on the second floor. They are also of a type largely used for flour. There is usually one packer below each bolter and fed directly by it. In some cases, owing to the lack of room, one bolter feeds two packers. The oxide is packed partly in paper bags holding 50 lb. each, partly in barrels with 30-in. staves and 19½-in. heads, holding 300 lb., while material for export is packed in special barrels and weights to suit the different markets. The barrels are kept in a storage-room where the hoops are driven by a machine and then nailed by a second one, and from here they are passed to the packers. All barrels are weighed empty and filled, the weights adjusted, and the barrels headed. They are passed to the ends of the building and rolled on skids to the store-houses. The second floor of the packing-room is high enough above the ground-level to allow sufficient slope to the skids to carry the barrels to all but the most distant store-houses. The paper bags are loaded on trucks and taken to the store-houses on the same skidways.

Power for both plants is supplied from a central station between the oxide and the spiegel furnaces. The boiler-room is equipped with Edgemoor boilers of 536 hp. each, which can be fired with coal or with the waste-gases from the spiegel furnaces. The engine-room parallels the boiler-room and is about 20 ft. north of it, and contains three horizontal cross-compound blowing-engines for the spiegel-furnaces; four direct-current generators, two of 500 and two of 600 kw., each direct-connected to a horizontal cross-compound Corliss engine; three Allis-Chalmers turbo-generators, two of 2000 and one of 4000 kw., 6600 volts, and three phases; two motor-generator sets used as a reserve for both alternating and direct-current lines, as either part can be used as a motor to generate current of the other kind, if needed. In addition, the engine-room contains the exciters, condensers, and pumps.

The ores used at Palmerton consist almost entirely of concentrate from the company's mines at Franklin Furnace and at Sterling Hill, New Jersey. Three of these products are smelted for oxide, namely, franklinite, 'half and half,' and dust. The composition of these is about as follows:

	Franklinite %	Half and half %	Dust %
Zn	18.2	18.4	17.6
Fe	35.0	15.2	20.1
Mn	12.8	12.2	8.9
SiO_2	3.6	12.6	7.2

The franklinite is worked by itself and the residue smelted in blast-furnaces for the production of spiegel. The 'half and half' and dust are mixed, and the residue wasted, as it contains too little Fe and Mn and too much SiO_2 to be a profitable smelting material. Franklinite is the easiest to work and gives the best product. The 'half and half' and fine are mixed and worked together. The charges per furnace average:

	Ore tons	Coal tons
Franklinite	2.62	1.62
Half and half and fine	1.95	1.36

The recovery for franklinite averages about 86% and for the 'half and half' and fine about 78%. The recovery and quality of product are controlled by the selection of the ores, the ratio between ore and coal and of both to the grate-surface, and, what is of great importance, the balance of the blast and exhaust and the proper proportioning of both to the charge being worked. Where leaded ores are treated the charges and recoveries vary more so that average figures are of little value. In general, with such ores, the charges are little lighter and the recoveries lower than for the New Jersey ores. The recovery of lead is usually better than of zinc, roughly as 9 to 8. The output of oxide is generally estimated as equalling the combined Zn + Pb in the ore, the gain in O and SO_3 nearly balancing the losses in Zn and Pb.

The product is divided into five grades, two for the use of paint manufacturers, two for rubber, and an off-grade that is re-worked either for oxide or spelter. For the paint trade, the grading is entirely by color, samples being daily rubbed down with oil and compared with standards. For the rubber trade, slight variations in color do not make so much difference, and the grading in this respect is not so close, but freedom from lead and absence of small hard particles is insisted upon. The variation in composition of the different grades is very slight. The material as shipped uniformly contains 99% or more of ZnO, the principal impurities being: SO_3, 0.25 to 0.33%; moisture, 0.10 to 0.50%; and PbO, 0.05 to 0.30%. The principal use of zinc oxide is as a pigment and for this purpose its purity of color and freedom from discoloration by gases and atmospheric conditions, fineness, and uniformity, make it particularly useful. As a pigment it has the disadvantage of slow drying and a tendency to become unduly hard in time. The general concensus of opinion among paint manufacturers is that a mixture of pigments is better than any single one. In these mixtures, zinc oxide is an almost invariable constituent because it prevents 'chalking' and gives a surface that retains its color. Next in importance is its use in rubber goods, in which it is employed as a pigment to produce white goods. It is excellent for this purpose, when practically free from lead. If this were the only effect it might be replaced by cheaper materials, but it also greatly increases the tensile strength of the resulting rubber. A mixture of pure rubber and sulphur when vulcanized has a tensile strength of about 2000 lb. per square foot and an elongation of 960% of its original length. The effect of adding zinc oxide is to increase the tensile strength but diminish the elongation as shown below:

ZnO %	Tensile strength Pounds per sq. in.	Elongation %
25	2400	720
35	2400	700
45	2700	680
55	2500	620
65	2000	540
75	1300	400

It will be seen that up to a content of about 60% the ZnO increases the tensile strength of rubber. No other material tested has this effect.

As the supply of lead-free ores is limited, and as the demand for ZnO is increasing, much 'leaded' oxide is being made. The process is the same, but the furnace-charges are usually lighter and the recovery lower. In most cases the furnaces used are double-ended with one or two working-doors at each of the opposite ends. This makes the furnace easier to clean, as there is no back-wall, which is the most difficult part from which to cut accretions. On the other hand, this type of furnace cools more when charging and does not, as a rule, light up as quickly. Which type is better is still an open question. In making 'leaded' oxide, it is customary to have a brick combustion-chamber over, or close to, the furnace. This insures sufficient time of contact at a proper temperature to cause the lead to be converted into basic sulphate. This is essential, as all of the oxides of lead are strongly colored and injure the color of the product, while the basic sulphate is very nearly white and does not have this effect. In many cases the bag-rooms of the works making 'leaded' oxide have no over-head distributing pipes, the hoppers being connected and used for this purpose. Leaded oxide is divided into several grades, depending on the amount of lead it contains. This is always stated as the amount of neutral sulphate ($PbSO_4$) equivalent to the total lead present, although it is always present mainly as a basic sulphate. It forms an excellent material for the manufacture of mixed paints. So far this grade has not proved satisfactory to the rubber manufacturers. It is impossible to give the production of oxide of zinc, as the statistics are not published by the Government. Such figures as have been published are not comparable, because the grouping of the different products is not always the same. The increase has been very rapid, especially within the last three or four years.

Paints and Painting in the Tropics

Arrangements have been made at the Miraflores purification plant in the Panama Canal zone for a central laboratory for the inspection and testing of paints used on canal structures and equipment. The purpose is to make a study of formulas and methods of application and determine the best processes of protecting woods and metals in varying circumstances in the tropics. Expenditures by the Canal administration for paint materials amounted to $127,173.44 in 1914; $255,366.56 in 1915; and $229,463.58 in 1916. There are five paint-grinding machines at the Balboa shops supplying paint for the general needs of the canal, and one at Pedro Miguel lock, which does work for all locks. The problem of painting has never been solved, although practically every kind of paint that is made has been tried on the Isthmus, and tests have been made in coating with concrete and with metal sprays. There are now between 15 and 20 steel plates coated with various mixtures exposed at the lower end of the Miraflores locks for tests under weather and water, in intermittent and constant submersion. The determination of the most effective protection is to be made the subject of special study.

FERRO-MANGANESE is sold in England for domestic use at £25 per long ton under government control of prices, but for export a basic price of £80 per ton is taken. Consequently British ferro-manganese sells for $325 to $375 per ton in American sea-board markets. Meanwhile 'prompt domestic,' of American manufacture, brings $275, and fourth quarter prices are approximately $300 per ton.

Flotation Tests With Hardwood Oils

By R. E. GILMOUR and C. S. PARSONS

*The most promising wood oils were tested at Cobalt under commercial conditions. Three hardwood oils were given a thorough trial at the Buffalo mines in comparison with a standard pure pine-oil (G.N.S. No. 5). They are as follows:

F.P.L. No. 26, heavy hardwood creosote-oil.
F.P.L. No. 27, acid (creosote) oil.
F.P.L. No. 29, ketone residue.

These oils are quite uniform in the crude state as produced and can be considered standard flotation-oils. The hardwood oils used in these large-scale tests were supplied from the plant of the Standard Chemical, Iron & Lumber Co. of Canada, Ltd., at Longford, Ontario.

Samples of these three oils when examined at the Forest Products Laboratories showed 8.55%, 5.25%, and 2.55% respectively, of water. The aqueous fractions of the first two were distinctly acid and strong enough to corrode ordinary iron metal-containers. The water content of the ketone residue was natural. Although the hardwood oils can be used with the small percentage of water present mostly as an emulsion, the removal of the aqueous fraction seems strongly advisable, not only on account of the above objections but by reason of the fact that the acetic acid and its homologues forming the acid content are essentially of no flotation value. The higher aqueous content of heavy hardwood creosote oil No. 26, compared with acid oil No. 27, is due to its lighter specific gravity. The mixing of these two oils in the proportions in which they are produced commercially would be advisable, since the value in flotation of each oil is equally satisfactory. In this way a better gravity separation of the water content could be effected than would be possible with oil No. 26 alone. The yields of these two oils, No. 26 and 27, are respectively 0.486 and 0.284 imperial gallons per cord. Figuring on 500 cords of hardwood distilled per day, this means that 385 gal. of this Canadian wood-oil product is available. For about 2000 tons of silver ore treated daily at present, only slightly more than 25% of this oil supply can be consumed in the Cobalt area. Should the demand for these hardwood creosote-oils become greater than the supply, it is interesting to note that oil No. 25B, according to laboratory results, is equally satisfactory. As yet hardwood creosote-oil mixture No. 31, comprised of oils No. 25, 26, and 27 in the proportions produced, has not been tried in actual milling at Cobalt. Although small-scale laboratory experiments on the creosote-oil mixture in the crude state show good comparative results, the indications are that after the removal of the fraction that distills below 150° C., and that has a gravity below 0.89, the mixed residue corresponding to oils No. 25B, 26, and 27, would give excellent results with Cobalt ores. Treating oil No. 31 in this way by a preliminary distillation would also remove all the aqueous content and would give a desirable water-free flotation-oil. The commercial supply of this crude hardwood-creosote mixture after such treatment would be over 500 gal. per day in Canada.

*Abstract: Bull. Can. Min. Inst., Nov. 1917.

The proposed specifications for hardwood creosote-oils are as follows: Oils No. 26 and 27 with 17.5% and 10% respectively, distilling below 150° C., including 8.55% and 5.25% aqueous content, have been found to give satisfactory flotation results when used in the crude state. Calculating to the water-free oil basis this means that these two crude oils contained respectively 9.8% and 4.9% of moisture-free oil distilling below 150° C., according to the standard method of fractional distillation for the examination of crude oils proposed. Taking these two crude oils as a standard, and keeping in mind the hardwood creosote-oil mixture No. 31, with the aqueous and lighter oil-fractions removed, as suggested above, it would seem that the general specifications as given below would be fair restrictions to propose for the buying of hardwood creosote-oils, to be used as selective frothing-agents for Cobalt ores. (1) The aqueous content should be less than 5% of the crude oil as received. (2) Not more than 10% of the water-free oil shall distill below 150° C. (corrected temperature), and with specific gravity below 0.89 at 15° C., according to the standard method of fractional distillation outlined. (3) At least 65% of the water-free oil should distill between 200° C. and 265° C., where the gravity shall be between 0.975 and 1.065. (4) Not more than 15% of the water-free crude oil shall remain as a hard pitch or coke-residue. These specifications would not apply to ketone residue, which is somewhat different in nature from the creosote-oils. Where a high-grade concentrate is not desired, and the aim would be to lift considerably more mineral-bearing gangue in order to reduce the tailing-assay, considerably more pitch residue than indicated in the fourth specification may be used. This, however, may be accomplished by adding hardwood tar as such, containing over 70% pitch residue, to the total oil-mixture.

LARGE-SCALE TESTS. The tests carried on during 42 continuous shifts were made in large Callow cells without interruption of the general procedure. During the first 26 shifts the standard mixture of pine-oil and coal-tar products was used. In the remaining 16 shifts the hardwood-oils were substituted for pine-oil in the following order: Heavy hardwood creosote-oil No. 26 in 6 shifts, acid oil No. 27 in 5 shifts, and ketone residue No. 29 in the last 5 shifts.

TESTS WITH PINE-OIL MIXTURE

	%
Oil-mixture—Pine-oil (G.N.S. No. 5)	20
Coal-tar creosote	60
Coal-tar	20

Oil-mixture per ton of dry ore, 1.06 lb.

Shift	Ore treated tons	Silver, oz. per ton Heading	Tailing	Middling	Con-centrate	Recovery %
7-3	248	6.2	0.9	8.4	381.0	85.5
3-7		6.8	1.0	20.0	417.5	85.3
7-3	327	9.0	1.0	9.8	392.0	88.9
3-7		8.6	1.0	10.0	480.0	88.4
7-3	316	8.6	1.0	10.4	421.0	88.4
3-7		8.8	0.9	12.8	528.0	89.8
7-3	283	8.2	0.9	8.0	355.0	89.0
3-7		7.8	0.9	10.0	365.0	88.5
7-3	265	7.8	1.0	8.0	481.5	87.2
3-7		9.0	0.9	14.9	473.1	90.0
7-3	297	8.2	0.9	9.0	465.0	89.0
3-7		7.8	0.9	11.6	447.0	88.5
3-7	247	9.0	1.0	16.0	462.5	88.9

TESTS WITH HARDWOOD-OIL MIXTURE

	%
Oil-mixture—Hardwood acid (creosote) oil (F.P.L. No. 27)	40
Coal-tar creosote	50
Coal-tar	10

Oil-mixture per ton of dry ore, 1.06 lb.

Shift	Ore treated tons	Silver, oz. per ton Heading	Tailing	Middling	Con-centrate	Recovery %
3-7	187	12.0	1.1	92.0	900.0	90.8
7-3	283	9.6	1.2	15.2	320.0	87.9
3-7		7.0	1.0	22.8	732.5	85.8
3-7	125	8.8	0.9	10.6	435.0	89.9
7-3	80	8.8	0.8	10.6	344.0	91.1

The tables speak for themselves. These tests gave more than 20% higher recovery than were obtained in small-scale laboratory tests. The operation of the flotation machines needed no special attention during the tests on the hardwood creosote-oils No. 26 and 27. When using ketone residue, however, care had to be taken to keep out traces of foreign oil such as lubricating-oil which had a deleterious action on the somewhat sensitive though heavy froth.

CONCLUSIONS. 1. Crude hardwood-oils, the supply of which is abundant, can be substituted for pine-oil as a selective frothing-agent in the commercial flotation of Cobalt silver ores.

2. Heavy hardwood creosote-oil, F.P.L., No. 26 and acid (creosote) oil F.P.L. No. 27, give equally satisfactory results and a mixture of these two oils in proportions as produced is recommended as a standard hardwood flotation-oil.

3. Ketone-residue in the crude state can also be considered a satisfactory flotation-oil for practical use on Cobalt silver ores.

Hardwood creosote-oils have been tested on a laboratory scale with copper ore from Bruce mines, Ontario. From ore assaying 7.9% copper a mixture of 40% heavy hardwood creosote-oil, F.P.L. No. 26, 54% tar-creosote and 6% coal-tar gave 93.1% recovery, leaving tailing with 0.8% copper. Other hardwood-oils should give equally satisfactory results. Ketone residue has proved satisfactory for the flotation of low-grade copper-nickel ore in the Sudbury district and a commercial demand for the oil has already been effected as a result of this investigation. Hardwood creosote-oils, especially light hardwood creosote-oil, F.P.L. No. 25, has promising qualities for the treatment of the lead-zinc and the copper ores of British Columbia. Further investigations on the flotation problems of other Canadian ores, especially those from British Columbia, are now being conducted in the ore-dressing division of the Mines Branch, which has secured a complete collection of the Canadian crude wood-oils and their fractions.

Life of Cast-Iron Pipe

Cast-iron, especially in the form of pipe, seems to last longer than any other form of steel or iron. Some remarkable records have been reported. There are records of pipe in use in Europe for over 200 years. In this country, and particularly on railroads, such long-time service is not recorded because of the limited time that these pipes have been in use. In the early years of American railroads, water-service was given with pipes of less than 3 in. diameter and the use of cast-iron pipe in consequence was limited to cases where 3-in. or larger pipe was used. Instances of installations of cast-iron pipe were recently made the subject of study on the Illinois Central railroad. The life of several of these lines has been traced to the present time. The most interesting is that of a pipe-line laid at Centralia, Illinois, in 1855. This was of 4-in. cast-iron pipe approximately 10,000 ft. long. After 12 years of service it became inadequate for the increased consumption, and upon examination it was found to be heavily encrusted. Efforts were made to improve its capacity by cleaning it and relaying a part of it with clean-out boxes at intervals of 100 ft. This provided only a temporary relief and the following year about 5000 ft. of the line was taken up and re-laid with 8-in. pipe, which is still in service after being in the ground 49 years. The 4-in. pipe relieved at Centralia was re-laid in 1868 at Ramsey, Illinois, where it remained in service until 1903, when it again became too small for the demand and was replaced by 6-in. pipe. Since that time the old 4-in. pipe has been used at various places, principally as drains. Another installation of which authentic records are available is one of 7500 ft. of 4-in. cast-iron pipe at Little Wabash river, laid in 1857. This remained in service until July 1893, when it was replaced. The old pipe was stored at Centralia until 1895, when portions of it were re-laid at that place and at Effingham and at Big Muddy. The installations at the last-named places have since been removed but the pipe at Centralia is still in service. No sign of paint can be found on the pipe at the present time and apparently it never had been painted.

TONOPAH, NEVADA

TONOPAH BELMONT.—WEST END CON.—MACNAMARA.

The Tonopah Mining Co. shipped 34 bars of bullion, valued at $62,400; this represents the final clean-up for October. During the previous week 2700 tons of ore, averaging $13 per ton, was milled. In the Silver Top 57 ft. of development has been done, and 112 ft. in the Sandgrass. At the Red Plume shaft stoping continues on the Red Plume vein on 3 ft. of ore. Last week the production was 2500 tons.——The Tonopah Belmont Development Co. shipped 40 bars of bullion and 27 tons of concentrate from the Belmont mill, and 29 bars of bullion and 20 tons of concentrate from the plant at Millers. The value of the shipment was $121,930. On the 800-ft. level raise No. 16 shows 3½ ft. of good ore, and raise No. 17 from the north-east drift No. 8006 shows 6 ft. of excellent ore. On the 900-ft. level north-east cross-cut No. 9019 is being driven to intersect the South vein. On the 1000-ft. level a raise is being driven to cut the extension of the North vein. Last week's production was 2438 tons.——The Tonopah Extension Mining Co. shipped 23 bars of bullion, valued at $46,500. In the Victor 101 ft. of development has been done, and 121 ft. in the No. 2 shaft. At the No. 2 shaft on the 1350-ft. level the development in the intermediate east drift No. 504 exposed a 2-ft. face of ore, in the west drift No. 510, 3 ft. of ore, and 3 ft. in the east drift No. 510. The stopes on this level were the Murray vein 5 ft., O. K. vein 3 ft., and Merger vein 3 ft. On the 1680-ft. level of the Victor, drifts have been started to the east and the west to prospect the Merger vein which was discovered recently. The flow of water at this level has slackened. The production for the past week was 2380 tons. ——At the Ohio shaft of the West End Consolidated Mining Co. cross-cut No. 506 is in foot-wall quartz. Drift No. 526 is being driven on a high-grade hanging-wall stringer. A sill stope is being started in drift No. 530. Drift No. 531 continues to show a full face of high-grade ore. No. 454 drift No. 2 shows a full face of fair ore. On the 555 intermediate level, drifts No. 1 and 3 show a full face of good ore, while raise No. 3 is being driven for the purpose of a connection. At the West End 623 raise No. 7 has exposed considerable ore.. The output last week was 1154 tons. Development on the 1700-ft. level of the Halifax Tonopah will be resumed soon. The tonnage the past week was 54 tons.

The MacNamara Mining Co. is treating 70 tons of ore per day. On the 300-ft. level stope No. 1 is producing fair ore from both the foot-wall and the hanging wall. Stope No. 2 shows a 6-ft. face of good ore. On the 625-ft. level good-grade ore is being broken on a 6-ft. face in stope No. 1. Development on the 725-ft. level consisted in advancing drift No. 18 on a 6-ft. face of quartz. The output last week was 487 tons.

The Jim Butler Tonopah Mining Co. produced 850 tons of ore last week. The recent shipment of bullion and concentrate from the Millers plant of the Tonopah Belmont company represents the production of the Jim Butler for the last half of October. The value was $29,837.——At the Cash Boy the hanging-wall drift on the 1700-ft. level was extended 25 ft. in ore, assaying $22 to $60 per ton.——The Great Western Consolidated Mining Co. will resume operation as soon as the Nevada-California Power Co. can furnish uninterrupted power service. The auxiliary plant of the power company is nearly completed.——The week's production at Tonopah was 9863 tons with a gross value of $172,602.——The addition of a compressor and air-drills at the Kanrohat property, in the Round Mountain district, has facilitated development.

MANHATTAN, NEVADA

WHITE CAPS.—MANHATTAN CON.—UNION AMALGAMATED.

The mine work covering the week ended November 8 at the White Caps is as follows: from the 300-ft. level, raise 303, being extended in the foot-wall of the shaft orebody for mining purposes, has broken into the orebody, and shows ore aver-

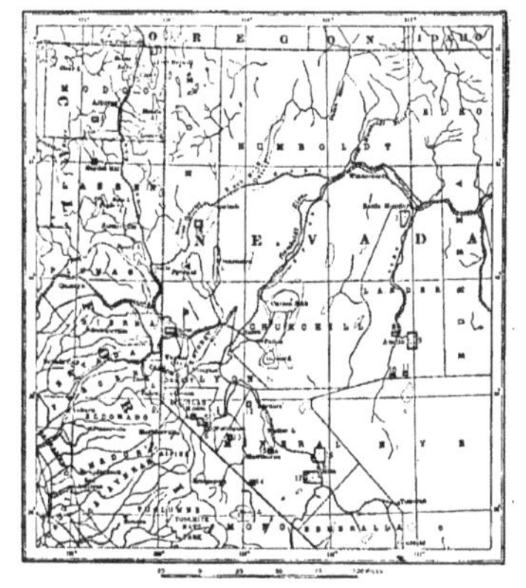

NEVADA

aging two ounces in gold. Four feet of progress is reported; the raise will be discontinued and a stope started from it. On the 300-ft. a sump is being cut east from the shaft beyond the east fault for the purpose of obtaining a supply of water for domestic purposes, as there is a spring of pure water at this point. When this work is completed the sump will be concreted, a pipe-line extended to the shaft, and the water pumped to the surface and placed in a cistern, to supply the White Caps and White Caps Extension. From the fifth level, south cross-cut No. 502, from 501, has been discontinued. A raise has been started which will be extended to connect with the fourth level. This raise will be entirely in ore. The west cross-cut, No. 503, being extended to prospect for the shaft orebody, has made 28 ft. through an area of shale and quartzite. No. 504 south cross-cut has just been started to cross-cut the east orebody, and 14 ft. has been made, all in ore. The south cross-cut, No. 501, running on the east orebody is now 399 ft. from the station, the last 167 ft. being in ore. A full face of ore shows, with a foot of same averaging two ounces in gold. The distance made driving east is 36 ft. The new work started is the 500-ft. sump, which is being excavated preparatory to the resumption of sinking the shaft one more lift. Thirty cubic yards has been taken out from this sump. The summary for the week shows a total of 48 cu. yd. of excavation and 82 ft. gained. In the White Caps mill over 125 tons is crushed per day. The Butters filter, which is one of the improvements to be added, is now being made in the Tonopah Extension workshop, the complete change in the mill is expected to be finished in two weeks. The clean-up of precipitates from the mill has been shipped to the Tonopah Extension for treatment.

The east orebody of the Manhattan Consolidated has been exposed from the fourth level during the past week. The east drift had been extended 200 ft. from the shaft. After 10 ft. had been made with the cross-cut the orebody was cut and since then 20 ft. has been made across the orebody with no signs of hanging wall in sight as yet. The late developments have proved the orebody to be continuous from the surface to the 400-ft. level with incline distance of over 450 ft. The management plans to immediately continue the shaft-sinking for an additional 200-ft. depth, as soon as the east orebody has been developed from the fourth level.——The Union Amalgamated has confined itself to shaft-sinking during the week. Two shifts in the raise being extended from the 600-ft. level to connect with the Earl shaft have been employed and the distance made is now 37 ft. above the level. This raise should hole-through into the shaft bottom in about twelve days. After connection is made the completion of the enlargement work on the shaft will be greatly facilitated as the waste can then be dropped to the 600-ft. level and raised to the surface from the Bath shaft. Three shifts have been employed in the Earl shaft, and at present the shaft has been enlarged and timbered for a distance of 84 ft. The last 36 ft. has been in very treacherous ground.——In the Manhattan Red Top the shaft has been sunk 60 ft. No material change in the formation has been noted. The flow of water has increased to some extent but is being handled without much trouble. The total depth of the shaft is 167 ft. vertical.——The operations by the White Caps Extension and Zanzibar companies in the sinking of the shaft for the benefit of both companies have been maintained steadily. The three-compartment vertical shaft has reached a depth of about 370 ft. It is intended to sink the shaft to about 390 ft. and from that point the first station will be cut and cross-cuts extended to explore the formations for orebodies.

PLATTEVILLE, WISCONSIN

VINEGAR HILL.—GRASSELLI CHEMICAL CO.

In the Cuba district, noted for the operations of the National Zinc Ore refinery, operations were uninterrupted and shipments of high-grade ore were heavy from both this and the Linden Zinc refinery. The raw-ore deliveries to the National Separators all came from the Vinegar Hill mines, showing that this was the heaviest producing group in the field. A portion of its output was sent to the Grasselli Chemical Co., the National Separators being unable to treat the entire output. The C. S. H. Mining Co. has completed a new mill and shipped concentrate to the Benton Roasters. The Connecting Link Mining Co. bottomed two shafts in ore on the old Dall range. The Big Eight Mining Co. is assembling material for a new mill on the Raisbeck mine. Drills were kept busy proving spar ranges on the Gritty-Six lease for William Raisbeck, former owner of the Black Prince mine. The Connecting Link was incorporated with a capital of $60,000.

The Benton contingent of zinc miners made heavy deliveries to the Grasselli Chemical Co. The leading producers were the Wisconsin Zinc Co., Frontier Mining Co., Vinegar Hill Zinc Co., New Jersey Zinc Co., and the Fields-Thompson mine. The Vinegar Hill Zinc Co., after much underground development at the Martin Kittoe, and Blackstone mines, is in much better shape and increased deliveries were made the second half of the month. The new Meloy mine is showing an improved grade of ore, at times exceeding 50% zinc. The Wisconsin Zinc Co. met with exceptional success in work at the Thompson, Champion, and Longhorn mines. The New Jersey Zinc Co. had three good producers in the Fox, Penna-Benton, and Hoskins mines. The Fields Mining & Milling Co. more than doubled the zinc output from the John Thompson mine. Operating conditions will be greatly benefited by the completion of a new branch railway connecting with the main line of the Chicago & Northwestern railway.

CASPER, WYOMING

BIG MUDDY.—WEST SALT CREEK.

Developments in the Wyoming oilfields within the past few months have been almost sensational enough and important enough to form the basis for a boom of old-time proportions. Old-timers, followers of the boom-towns—Leadville, Cripple Creek, the Klondyke, Nome, Goldfield, Oatman, and Tulsa—unite in saying that Casper is the best of all. The growth of Casper and the development of surrounding country are based on a firm foundation. Oil is coming to the Casper refineries from a dozen fields, meaning that railroads, refineries, and all allied businesses are employing full crews and enjoying prosperity. Casper is being built rapidly and substantially. It claims a population of close to 14,000 and the fact that it has a capacity for housing only one-half that number has caused congestion. Basement rooms, attics, woodsheds, and private garages have been pressed into service for sleeping quarters, and hotels with cots in halls and extra beds in rooms turn away guests every night.

Big Muddy gave Casper its great impetus about a year ago. This field is on the railroad 18 miles east of Casper. Deep drilling has been fruitful of remarkable results. Four distinct oil-bearing sands have been found within reach of the drill. The Shannon sand, at 1000 ft., is spotty. At many points it has been found water-bearing, but the Producers company recently finished the best well in the field in this sand. The well makes upward of 1000 bbl. per day. The Kinney sand, at 1900 ft. has been an abundant producer where tapped, but it does not appear to be of great extent. The same may be said of the 'stray' sand, found in the east part of the field around 1600 feet. But the best sand of all is the Wall Creek, found at 3000 ft. Twelve wells have so far been driven to this sand and all but one are flowing. The oldest, which was started just a year ago, is making better than 200 bbl. per day. Deep wells on Big Muddy are expensive to drill. Cable drills do not operate satisfactorily in the soft and crumbly shales, but the big rotary rigs have been found efficient. So far the fastest time made in completing a deep well is 47 days. On the other hand some holes started months ago still are unfinished. Big Muddy has been connected with Casper by a pipe-line operated by the Illinois Pipe Line Co., and the daily production of the 80 wells in the field is bought by that company on a base price of $1.15 per barrel. Standard Oil interests are dominant at Big Muddy. The Producers Oil Co., a subsidiary of the Texas company, has made good progress in developing its holdings in this field and now has a number of producing wells.——West Salt Creek has been the sensation of the year in the Casper district, for Big Muddy was assured, though not proved, in 1916. This district embraces a territory west of the Government's Salt Creek Naval Reserve, and lying entirely outside of the escarpment. In this field three sands are within reach of the drill. The first, Wall

Creek, a wonderfully productive sand in the main Salt Creek, here is found to be water-bearing, but beneath it are two other sands, the first of which yields oil, and in the second, which soon will be tapped by the Bessemer company, oil has been found along fault-planes in the shale. Since the Hjorth well brought in oil in the second sand early this year, this district has been the scene of much activity. Probably 75 rigs have been erected and fully as many more ordered. The long haul from Casper has delayed development. West Salt Creek is 50 miles from the railroad over one of the worst wagon-roads that trucks and teams ever tried to negotiate. Some idea of the road may be had when it is stated that the regular rate for freighting from Casper to the field is $30 per ton.

All of the Salt Creek production goes to the Midwest refinery at Casper, through two pipe-lines. The Midwest Oil Co. owns or controls nearly all of the production of Salt Creek and buys the balance on a basis of $1.15 per barrel. The field produces about 15,000 bbl. of crude oil per day. A casing-head gasoline plant is just being completed that will materially add to this company's revenues. To convey this volatile product to Casper a line of 4-in. pipe, with welded joints, has been laid.

The Lander field is being rapidly developed, and, with Grass Creek, Elk Basin, and Greybull, is contesting for third place in point of production. Partly developed fields, where either oil or good indications have recently been found, are Powder River Station, Tisdale, Notches, Poison Spider, Lost Soldier, Buffalo Basin, Thornton, and Old Woman Creek. Development has been started on Midway, Buck Creek, Big Sand Draw, Government Draw, Castle Creek, East Salt Creek, Rumford, Emigrant Gap, North Casper Creek, Bates Hole, Big Hollow, and a score of others.

HOUGHTON, MICHIGAN

AHMEEK.—ALLOUEZ.—OSCEOLA CON.—CENTENNIAL.

The October output of copper from the mines of the Michigan district was the largest since March, and November promises to show a further increase. Three or four of the mines are again working normal forces and others are increasing their crews steadily. By December, Michigan mines may again produce at the rate of 20,000,000 lb., despite the loss of hundreds of the best miners who have gone into the service of the Government. Ahmeek will probably show a production of over 30,000,000 lb. for the present year, compared with a little better than 24,000,000 lb. last year. This is one of the few mines that will show a larger output in 1917 than 1916. Ahmeek has the advantage of youth, rich material, good management, and fine treatment of employees. Young miners are glad to get work at the North Ahmeek particularly. The tonnage of ore is increasing daily and the production is averaging better than 22 lb. per ton. This mine did not follow the policy of sending down all of its richest stuff when prices were high, a few months after the War started, and the result is that today it has as rich an ore-reserve as any mine in the district.——Allouez, another Calumet & Hecla subsidiary, will come close to a production of 10,000,000 lb. this year, if the increase for November and December comes up to expectations. This mine reasonably can be expected to produce at this rate for at least ten years to come. The grade is 18 lb. per ton, and, while this may be maintained for several years, it cannot be expected to continue indefinitely.

The future of the Osceola Consolidated depends on further economic changes that will make possible the profitable operation of a grade of ore lower than anything handled in the past. There can be little expectation of any increase in the proportion of copper secured in the North Kearsarge shafts with greater depth, unless the history of every ore-formation in this district changes, as the copper content of the Lake Superior ores do not increase with depth.——Production at the Centennial is not heavy, but the yield for the present year will be close to 2,250,000 lb., an increase of at least 750,000 lb. over 1916. This is due to an improved quality of ore and a slight increase in tonnage. The Centennial now is into all of the Wolverine richness that its cross-cuts have been seeking.——Because of the delay in securing a right of way for its railway the Seneca has been handicapped considerably in preparing for shaft-sinking. At the present time, however, the Seneca has a fine surface prospect and shaft-sinking will commence in a short time.

MAMMOTH, UTAH

CALIFORNIA-COMSTOCK.—HECLA DIVIDEND.—ORE SHIPMENTS.—CHIEF CONSOLIDATED.

The entire Tintic mining district is hard hit with the embargoes and car shortage, the total shipments of first-class ores being only 127 carloads with an estimated weight of 6000

UTAH

tons and valued at $165,000.——J. C. Jensen, general manager of the Park City mine, reports that during the month of October two important strikes were made at the California-Comstock. One of these strikes was made on the 300-ft. level of the Comstock. In a drift started to the south-east from the bottom of the winze a full face of a splendid lead-silver-zinc ore was uncovered, giving promise of a large tonnage. The other strike was made in the bottom of what is known as the 250 lead-zinc stope, where two feet of high-grade ore was exposed.——An assay of the ore gave 55.8% lead, 4.14% copper, 14.2 oz. silver, and 9.1% zinc. The equipment and buildings at the property have been overhauled and repaired during the last few months so that everything is in shape for continued operations regardless of weather conditions.——With the raising of the embargo on lead-zinc ores the Rico-Argentine's production will be greatly increased. In October it shipped 875 tons to the smelters. The marmatite ore is netting about $300 for each small carload; it carries 5% copper, 6 oz. silver, and a little gold. During October 49 carloads of ore containing 35 tons each were shipped from the Michigan-Utah mine at Alta.——A rich strike has been made at the Lehi-Tintic, 2100 ft. from the mouth of the adit. The face is fully 600 ft. deep and carries 6 oz. of silver, 28% lead, and $2 in gold. After driving the adit for 2000 ft. the Empire vein was cut, and it was near the hanging wall that the strike was made.——The Hecla Mining Co. declared a dividend of 5c. per share, payable November 25 to stockholders of record November 5. This is a reduction of 10c. per share from the previous disbursement.

The second dividend of the Mammoth Mining Co. within one month was declared on November 4 amounting to 10c. per share and totalling $40,000.——The shaft being sunk on the Eureka-Bullion property, under the direction of John Bestlemyer, is down 570 ft. and is progressing satisfactorily. Sinking has been hampered by the loose ground during the last 100 ft.——Electric power is now being supplied to the North Beck Mining Co. by the Utah Power Co., and the work of sinking the three-compartment shaft is progressing rapidly. The new company is operating under the direction of E. J. Raddatz. The machinery includes a 40-hp. compressor and a 25-hp. hoist, both electrically driven.——The ore shipments during the week ended November 10 totaled 120 cars as against 127 cars for the previous week. The mines are still greatly hampered for lack of transportation facilities and the output therefore is being materially curtailed. Following are the shippers:

	Cars
Dragon Consolidated	35
Iron Blossom	17
Centennial-Eureka	17
Mammoth	11
Chief Consolidated	10
Tintic Standard	5
Grand Central	5
Gold Chain	5
Eagle & Blue Bell	4
Empire Mines	3
Victoria	3
Gemini	3
Chief manganese lease	2
Ridge Valley	2
Bullion Beck	2
Sharp	2
Eureka Hill	1
Colorado Consolidated	1
Minnie Moore lease	1
Total	129

The Chief Con. earned 76c. per share for its stock in 1916 and closed the year with a $477,295 surplus, or 54c. for each of the 884,000 outstanding shares. Its income for the year was $1,391,869 and its operating expenses, including construction, $719,766, leaving a net profit of $672,103. The company paid $176,481 in dividends, or 20c. per share, purchased the properties of four mining companies and other groups of claims adjoining its own ground, comprising altogether about 600 acres, for $228,360, and added $267,261 to its previous treasury surplus. Not more than 10 to 15% of the ground owned by the Chief company has been developed, and yet from its own workings more than $3,000,000 worth of ore has been sold to the end of last year, approximately half of which was net profit. Since its organization the company has paid $562,000 in dividends, invested more than $400,000 in additions to its mining property, and put $130,000 into equipment, besides accumulating $477,295 cash surplus.

MONTERREY, MEXICO

NEW MINING LAWS.—REVOLUTIONARY OPERATIONS DELAY MINING.

According to advices received by mine-owners in Mexico, the Federal Government, through the Department of Fomento, is drafting a new mining law. It is stated that the measure will differ in many material respects from the existing act and that it will not contain some of the drastic decrees that have been promulgated by President Carranza. The tax on mining properties will be reduced and the oppressive restrictions that have operated to the detriment of many foreign mine-owners will be removed. The drafting of the new bill will be finished soon and submitted to Congress for enactment into law.——The revival of revolutionary and brigandage operations in different parts of the country is serving to cause further postponement of the plans that were on foot for the re-opening of various mines. It is stated that conditions in the State of Chihuahua have grown steadily worse during the last two or three weeks, due to the re-entrance of Francisco Villa into the military field. A number of bands of armed men claiming to be followers of Villa are operating in the vicinity of Parral, Santa Barbara, and other mining districts not only of the State of Chihuahua, but also in those of Durango and Zacatecas. Several independent bandit leaders have sprung up in different parts of the country and are making it again unsafe for mining men to attempt to resume work upon their respective properties. Railroad traffic is also much disturbed by reason of this renewal of activity on the part of armed opponents of the existing Government.

American and other foreigners who own mines in the State of Chihuahua were recently served with official notice by General Francisco Murgia, the military commander of that State, that if they do not resume the operations without further delay, their properties will be confiscated by the Government and turned over to Mexican laborers, who have been employed in them.

COBALT, ONTARIO

CHAMBERS FERLAND.—MCKINLEY-DARRAGH.—HOLLINGER CON.

It is probable that the value of the silver production at Cobalt for the current year will approximate $16,000,000 and, with the exception of 1912, when the value of the production was $17,408,935, will be the highest in the history of the camp.

At the lower workings of the Chambers Ferland mine development and exploration will be commenced at once. This company has an extensive area in which little exploration work has been done. It has $200,000 on hand and in British government war bonds and about $100,000 in bullion.——The McKinley-Darragh is understood to be yielding approximately 75,000 oz. of silver per month. The favorable development at the 400-ft. level has added materially to the life of the mine.——Encouraging results are being obtained in the 310-ft. level at the Adanac where driving toward the north is going on. The vein maintains a width of about five inches and contains good-grade ore, the best of which is being bagged.——The Nipissing is maintaining production at the rate of about 10,000 oz. per day.

The improvement in the labor market at Porcupine continues, although nothing like pre-war efficiency is being obtained. The production of approximately $750,000 per month from the camp under existing conditions plainly illustrates what may be expected when things become normal again.——The Hollinger Consolidated and the McIntyre-Porcupine alone are understood to be turning out upward of $500,000 per month, and the Schumacher is producing at the rate of about $35,000 per month.——The tendency throughout Porcupine camp is to devote energy toward development. Ore-reserves are being built up considerably in excess of production which would appear to point toward increasing prosperity.

ALASKA

(Special Correspondence.)—D. C. Stapleton, superintendent of the Gypsum mine, has recently purchased a boiler and pumps from the Treadwell company. The machinery will be erected at Gypsum, and an effort made to pump the mine free from water. Several weeks ago the large leak, which has been a source of trouble for years, broke out again and the property was flooded.——Development work on the Red Diamond group of claims, under option to the Treadwell company, has been discontinued.——Train service on the newly-constructed trestle went into effect today. R. G. Wayland, superintendent for the Alaska Treadwell company, is examining mining property in the vicinity of Wrangel.

Treadwell, October 22.

(Special Correspondence.)—Advices from the Land & Industrial Department of the Alaskan Engineering Commission are to the effect that machinery is on the way from San Francisco to operate leasing units No. 10 and 11 in the Matanuska coalfield, consisting of 1440 acres, near Chickaloon, leased to Lars Nelland, of Oakland, California. Two cars of coal from Chickaloon reached Anchorage last week. This coal was from Unit 12, consisting of 480 acres on the Chickaloon river, which was withdrawn by an executive order of June 17, 1917, for the Alaskan Engineering Commission, to furnish coal for the construction of the Government railroad. The coal tested by the navy in 1914 was from this unit and the completion of the railroad to, and transportation of coal from, this point is accomplishing one of the foremost purposes for which the railroad is being constructed. During the construction of the railroad to Chickaloon, the Commission used coal from the lower Matanuska field on Moose creek, some 40,000 tons having been shipped from there to Anchorage. The mines in the Willow Creek district have closed for the winter.

The Alaskan Engineering Commission's dredge 'Sperm' has been put into winter quarters after having been assembled and successfully tested in the mouth of Ship creek, where it will operate in the spring. The hull, a converted whaler, is 140 ft. long, 41 ft. beam, 7½ ft. depth of hold, and draws a trifle over 4 ft. of water. The ladder well is 20 ft. wide and 40 ft. long. The ladder is 100 ft. long, built of wood, and will dig at a depth of 70 ft. This great depth was made necessary because of the extremely high tide on Knik Arm, the maximum being over 34 ft. Power is furnished by coal under two Scotch marine boilers of 250 hp. each. The main engine is a compound-condensing, rated at 750 hp., built by the Ideal Engine Works of Springfield, Illinois. The centrifugal pump has 18-in. suction and discharge, and is 96 in. diam. There is a 5-drum hoisting-engine; feed, condenser, and underwriters' pumps, electric light plant, and steam capstans. The spuds are 100 ft. long, of Washington fir. The digging-ladder is equipped with a cutter specially designed to work in the tenacious glacial silt, which has built up the delta at the mouth of Ship creek. All the machinery, excepting this cutter, and the main engine, was built by the Willamette Iron Works of Portland, Oregon, and was formerly in use on a dredge owned by the Portland, Spokane & Southern railroad, and wrecked in a storm off Cape Flattery in the winter of 1916. The machinery was salvaged and the present dredge is the design of William Gerig, and has been erected under his supervision. It has a rated capacity of 6000 cu. yd. per day, and will be used to dig a channel from the present ship anchorage and a harbor which will occupy the silted flat and tortuous channel at the mouth of Ship creek.

Anchorage, November 2.

Willow Creek mines have made only a light production for the season. There is much new development.

The Hercules Mining Co. is developing a molybdenum deposit on Reid creek. Both the sulphide and oxide of molybdenum occur in quartz and diorite.

G. C. Martin, of the United States Geological Survey, who recently reached Fairbanks from the lignite coalfields, 57 miles south of Nenana, said he had completed investigations preparatory to offering the lands for lease. Dr. Martin stated that this work is supplementary to the survey of the main part of the coalfields, which was finished last year. At that time the coal on the west bank of the Nenana river was not examined, but this has now been done. The leasing offer of the mining units as provided by law will soon be made, for after the survey of the general land office the law requires that the land be divided into blocks best suited for economical mining. The coal lands in the Bering River and Matanuska districts have already been offered for lease in this manner.

THE 'SPERM'

ARIZONA

COCHISE COUNTY

(Special Correspondence.)—Machinery for sinking the 70-ft. shaft of the Arizona Bisbee Copper Co. is being hauled to the mine. Walter H. Weed has reported on the property. Until recently this company has been a close corporation, but now shares are being placed on the market to raise money to carry on development. Ample water has been piped to the property.——The Copper Chief Mining Co. is preparing to commence operations. A large compound compressor and hoist have been erected. Prior to closing down last January this company shipped two carloads of ore per day.——Some promising ore has been opened up recently on the lower levels of the Denn Arizona Copper Company.

Bisbee, November 10.

GILA COUNTY

(Special Correspondence.)—The management of the Globe Dominion Copper Co. has decided to continue sinking its two-compartment shaft, which is now down 800 ft., to the 950-ft. level before cross-cutting. This property adjoins the Old Dominion on the east and south, and consists of some 25 claims. Considerable machinery and surface equipment is already in place.

Globe, November 9.

MOHAVE COUNTY

(Special Correspondence.)—It is reported that the New Mohawk has again opened up a rich streak of silver ore.——The Tennessee has opened up a large body of ore on the 1170-ft. level.——The Henry Gold Mining Co. has been re-organized and has made its stock assessable in order to raise money to commence development work.——The United Eastern has added a heating system in the mill, so the production will not be curtailed during the winter. A splendid body of gold-silver-lead ore is being opened on the 800-ft. level of the Schuylkill. A shipment of 500 tons will be made shortly for experimental treatment at Needles.——It is reported that the Hill Top Metal Mines Co. has made a big strike in its upper workings.

Kingman, November 10.

PIMA COUNTY

(Special Correspondence.)—The Whitcomb property in the San Xavier district, which consists of the Wedge group and the Prosperity, has been purchased under option by the Magnate Copper Co.——An extensive deposit of tungsten ore has been discovered in some old gold workings near Ajo.——The Daily Arizona Consolidated Copper Co. has been re-organized and sufficient money raised to develop the orebodies on the Copper Princess claim. Recent developments have blocked out about 75,000 tons of ore and shipping is in progress.

Tucson, November 10.

PINAL COUNTY

(Special Correspondence.)—A rich strike of copper carbonate ore has been made in the 300-ft. level of the Queen Creek Copper Company.

Florence, November 9.

SANTA CRUZ COUNTY

One of the most important decisions that has been rendered in this State, for mining men, was handed by Judge O'Connor, of Nogales, recently, when he decided that the $8000 worth of tungsten ore, which had been re-plevined by Barry & Barry of Nogales and Boyle & Pickett of Douglas, attorneys for the N. A. Hart interests, was the property of Mr. Hart and his associates, of whom J. S. Douglas of Douglas is one. The ore in question was re-plevined from Wells Fargo company, who accepted it from ore dealers of Nogales for shipment. Systematic stealing of high-grade tungsten ore from the La Cruz mine, at Sonora, has been going on for the past two years, and all attempts to detect the thief or to recover any of the ore have been of no avail until the settlement of the case before Judge O'Connor.

YAVAPAI COUNTY

(Special Correspondence.)—A two-foot vein of chalcopyrite assaying 9% copper, 1.5 oz. silver, and $1.20 gold has been cut on the Cherry claim of the Green Monster Mining Co.——Work has commenced on the main drift of the Crown King. The drift is to be extended another 100 ft.——The Arizona Copper Queen Co. has commenced shipping to the Humboldt smelter.——The Yellow Jacket and Empire groups of gold claims in the Groom Creek section have been taken over by St. Louis capitalists and development work has been started.

Prescott, November 10.

CALIFORNIA

ELDORADO COUNTY

A 60-ton mill is being built at Stockton to the design of Burr Evans for the Cincinnati mine. The mill has several novel features, the crushing being done by eight 15-in. chilled-iron balls rotating on a grooved tread in a five-foot pan. It will be worked by an oil engine.

NEVADA COUNTY

The North Star Mines Co. is preparing to put a full force of men at work underground in the Champion mines on Deer creek. The work of re-building the 40-stamp mill has been completed, and it will be kept running to capacity. Not only have the 40 stamps and equipment been entirely overhauled and re-built, but a concrete concentrating floor has been finished and the cyanide plant has been enlarged.

Ostrom & Co. has run an adit 600 ft. long into the hillside to tap the gravel-channel and has struck it several feet below bedrock. The gravel shows free gold, but the adit is being driven ahead to a point where it is expected that better ore will be found. As soon as the winter rains set in the washing of gravel will be started, and it is expected that it will be continued all winter.

PLUMAS COUNTY

(Special Correspondence.)—The Walker Copper Co. has arranged to erect a Marcy ball-mill and other equipment which will increase the capacity of its flotation-plant from 75 to 200 tons per day. All equipment is now electrically operated. The sinking of a 340-ft. incline shaft has been started from the third level, the shaft being sunk at an angle to escape a water-pocket disclosed by the core-drills. The shaft is designed to open a new section of the mine and will carry development well within the sulphide zone. V. A. Hart is manager.——Norris & Noyes of San Francisco are developing an extensive chrome deposit near Swayne and shipping a carload per day to Niagara Falls. The ore is of excellent grade and the deposit is one of the largest yet discovered in California. Motor trucks are employed to haul ore from the mine to Swayne.——The development of manganese deposits in the Crystal Lake district and near Indian Falls is proceeding steadily, and several companies are making occasional shipments to the electric smelter at Heroult. Most of the ore under development is of excellent grade.——Development has begun at the Juneday group of gold claims, near Crescent Mills, under management of F. D. Searight. The old adit has been thoroughly repaired and opening of new ground started. Good ore is reported to be showing in the drifts.——The Uncle Sammie Mining Co., composed of Oakland people, has started preliminary work on a group of copper claims near the Engels mine. Good ore has been found near surface.

Portola, November 19.

SIERRA COUNTY

The Dolan Bros. and A. G. Bell have closed their mine on Rock creek. A considerable amount of development has been done during the year, and there is a quantity of rock broken and ready for crushing in the spring.

SISKIYOU COUNTY

(Special Correspondence.)—L. M. Prindle and H. G. Ferguson, of the U. S. Geological Survey, and C. A. Logan, of the State Mining Bureau, spent several months the past fall examining the platinum deposits in Siskiyou county. They have awakened the placer miners to the appreciation of the value of the metal and extra efforts will be exerted toward its recovery in the future.

F. B. Hall of Reno, Nevada, has purchased the Simas quartz mine in the Greenhorn district and will begin operation soon.——Wike & Mathis of Sawyers Bar have purchased the George D. Carter placer mine, five miles below Sawyer Bar. The mine is fully equipped for operation.——A. E. Dench of Seattle has taken a lease on the Osgood placer mine, near Yreka, and is equipping it with a large force of men.——Theodore Knackstedt of Etna will operate the Bloomer placer mine near the Forks of Salmon during the coming season.——John Boyle is operating the Home Stake mine in the Etna district; he reports a satisfactory clean-up from the last run.——Cleveland Barry of Etna has taken charge of the Bonally mine.

Hornbrook, November 16.

(Special Correspondence.)—The Grey Eagle Copper Co. has 50 men at work and has provided quarters for 30 more. Additional air-drills have been purchased; all underground work is being done by contract. William Korner is manager.——Representatives of Eastern capital are examining large low-grade gold deposits in this county, including a property recently opened by W. P. Fisher of Humbug.——Alex MacCartney and associates of Seattle are developing the Taylor Lake group of gold claims above Etna Mills. So satisfactory has recent work been that the operators are making preparations for development along broader lines, and have shipped in a large quantity of supplies.——The old Pilot Knob group, near Gottville, is reported to be developing well. Work has been in progress several years, and a large tonnage of medium-grade gold ore has been blocked out. A five-stamp mill is running steadily and a gasoline engine has been erected to furnish power. T. K. Anderson is managing owner.

Etna Mills, November 19.

As the great dam of the California-Oregon Power Co. across the Klamath river at Copco, 45 miles from Klamath Falls, is completed, the tunnel which carried the waters of the stream during the construction period was closed on November 13, and the immense reservoir formed in the channel of the river by the dam is filling with water. It will require nearly four weeks to fill the reservoir, which stretches up-stream for five miles. The bed of the river was excavated to a depth of 125 ft. to get a solid foundation for the great weight of masonry and reinforced concrete in the large dam, which rises to a height of 125 ft. above the bed of the Klamath river. The dam is 95 ft. wide at the base and 500 ft. long, and the cost of the work will exceed $1,500,000. For three years over 300 men worked on the project. The company plans to generate 26,000 hp. under present conditions, and an equal amount can be developed by the addition of other units to the power plant.

YUBA COUNTY

The new $500,000 dredge, No. 17, was launched at Hammonton on November 4. Fully 500 people from Marysville and this vicinity were present to witness the launching of the big steel dredge, which is one of the largest in the world. The dredge, which will be put to work immediately, was built by the Yuba Manufacturing Co. of Marysville, under the supervision of Paul E. Morse. It was built for the Yuba Consolidated Gold Field Company.

COLORADO

BOULDER COUNTY

(Special Correspondence.)—W. S. Harpel, superintendent for the Consolidated Leasing Co. west of the Huron, has a force of men at work on the new mill, which is being pushed as rapidly as possible.——J. A. Wilson has opened up a big body of high-grade ore in the Lost Lake mine near Eldora.——A. B. Clarkson of St. Louis and H. C. Newton and Otto Victor of Eldora, have exposed a vein on the Lone Pine lode in the Caribou district. The property is in the vicinity of the Boulder County mine.——The Up-to-Date mines at Caribou are producing a high-grade ore from the shoot recently opened in the adit.

Eldora, November 10.

TELLER COUNTY

The Ella W., on Tenderfoot hill, is the scene of a new discovery made by Johnstone and Tomkins, who two years ago opened up the first ore on the claim and sold their lease for a good round sum. The prospectors have made their latest discovery on the south end of the Ella W. and from shallow trench workings have taken out between 15 and 20 tons of ore assaying two ounces in gold per ton. The property is owned by the Gumaer estate.——The north 400 ft. of the Wilson mine of the Free Coinage Gold Mining Co., on Bull hill, has been leased by the Free Coinage Consolidated Mines Co., holding control of the first-named corporation, to Leon F. Le Brun, of Cripple Creek, a miner who some years ago operated on the same ground. The lease runs for a two-year term and the lessee, who is already mining milling-grade ore, will pay a graded scale of royalties on ores marketed. A meeting of the shareholders of the Free Coinage Gold Mining Co. will be held at the company's office at South Altman, on December 15 for the purpose of electing three directors.——The Excelsior Mining & Milling Co., F. O. Wiborg of New York president, holding a long-time lease on the Longfellow mine of the Stratton estate, is constructing a new, up-to-date, steam-heated ore-house, and erecting a seven-drill Imperial Rand compressor. E. J. O'Flaherty, the superintendent, has just finished sinking the shaft to the 600-ft. point and will drive for connection with the Golden Cycle mine workings, extended into the Longfellow, where a good grade of ore is exposed in both the 500 and 600-ft. levels.——The mill of the Rex Mining & Milling Co., on Ironclad hill, will be closed down for the winter months as soon as the clean-up now in progress by Thomas Kavanaugh is completed. The Rex mill, operating under lease, has during the past year treated approximately 8500 tons of ore. This ore is quarried out from the hillside.——The high cost of mining supplies is reducing materially the number of miners employed on leased properties.

Cripple Creek, November 16.

Richard Roelof, general manager for the Cresson Consolidated Gold Mining & Milling Co., reports an ore-reserve of $3,652,562 on August 1, as against $4,130,318 on the same date in 1916. During the year 101,384 tons has been shipped, yielding $2,359,663.88 net.

IDAHO

SHOSHONE COUNTY

Several stringers of high-grade copper ore were encountered recently in the lower adit of the Big Creek Mining Co. The adit lacks 50 ft. of a point under the upper workings where it is expected the lode will be cut. The lower adit has reached a point 2300 ft. from its portal and several hundred feet below a point in the upper workings from which shipments of high-grade ore have been made. The work is being directed by G. Scott Anderson.

MISSOURI

JASPER COUNTY

The new 150-ton mill on the property of the N. J. & B. Co., north of Duenweg, was started on November 12. James H. Johnston, of Carthage, is manager and one of the principal stockholders. The property has been exclusively drilled and some promising ore has been found.

MONTANA

SILVER BOW COUNTY

(Special Correspondence.)—The week ended November 3 showed a remarkably steady increase in the production and in the number of men employed in the Butte mines. The Anaconda company is hoisting an average of 12,000 tons of ore per day and has more than 8500 men on its payroll in this district. These figures show that the company is now employing more than 80% of its normal crew. Men are returning to the mines so rapidly that plans are under way to open two large properties now undergoing repairs. A great many men are returning from small properties that have closed down for the winter. Owing to the fact that this district has given up over 2000 men for federal service, and that 1650 more men have left for unknown reasons, it is considered that the mining companies are doing wonderfully well.——The increase shown by the other operators, namely, Butte & Superior, Clark interests, East Butte, and Davis Daly, is in the same proportion as the Anaconda company. The North Butte Mining Co. is confining most of its efforts to the repair work at the Granite Mountain shaft.

Butte, November 1.

A dividend of 10c. per share was paid by the Barnes-King Development Co. on November 15, amounting to $40,000. This is the third dividend declared by the company, the others being in March 1916, $30,000, and in June 1916, $30,000, making a total of $100,000.

NEVADA

MINERAL COUNTY

(Special Correspondence.)—The Nevada Rand mine has started shipments to the Hazen sample-works, the initial consignment averaging $77 per ton in silver and gold. The shipping ore is taken from the new vein on the 180-ft. level, where four to seven feet of ore is being broken in the stope. Driving from the 450-ft. level is being pressed to cut the new vein and open the main orebody, which has been tapped on the 150, 180, and 250-ft. levels. S. E. Montgomery recently stated the reserve contained 35,000 tons of ore.——The Copper Mountain Co., controlled by the Jumbo Extension Co. of Goldfield, has purchased the Miller-La Patt lease and resumed shipments. Ore of shipping grade is reported in bottoms of shafts No. 1 and 2, and No. 3 is being sunk rapidly to the ore-zone. The management states that sufficient ore has been developed to warrant erection of a small concentrator. It is probable the company will be re-organized on an assessable basis.——The Mogul Mining Co., operating at Rawhide, has arranged to sink its 3000-ft. shaft an additional 700 ft. It is stated that a six-foot vein has been opened on the 300-ft. level, assaying $9.80 per ton, the discovery being made on the Regent claim.——Heavy shipments of copper ore are going to custom smelters from properties near Mina, and several new mines are nearing the productive stage. The Ray Consolidated recently entered the shipping list.

Mina, November 20.

(Special Correspondence.)—The Pilot Copper Co. at Luning is taking out 18 tons of 5% copper ore per day from the Anderson group.——With the blowing-in of the second unit of the Thomson smelter, the Kirchen Mines Corporation expects a demand for the big body of iron ore, carrying a small proportion of copper, which is near the surface of the St. Patrick mine. Heavy shipments of this ore were made last spring but were discontinued when the smelter ran short of coke. Regular shipments of high-grade ore are made from the St. Patrick.——The last three samples assayed from the bottom of the Siri shaft of the Wedge Copper Co. returned $17.37, $23.50, and $14.81, respectively, per ton. Good progress is being made by the contractor on the No. 2 adit which opens the vein 200 ft. below the No. 1.——The second carload of silver ore went out from the R. B. Todd mine last week and a carload of copper ore will go from the Luning-Idaho mine next week.——The Congress Copper Co. is confining most of its efforts to development work.

Luning, November 13.

NEW MEXICO

QUAY COUNTY

(Special Correspondence.)—A deposit of potash has been discovered upon the property of Red Peaks Copper Co. near Tucumcari. Arrangements are being made by the company to start development and to erect the necessary plant for extracting the mineral.

Tucumcari, November 12.

SOCORRO COUNTY

(Special Correspondence.)—A large gang of workmen are re-timbering the upper section of Little Fanny shaft. In the meantime miners are using the Little Charlie entrance. The Socorro company has taken over the saw-mill of the Mogollon Lumber Co., has put on its own men, and is getting out timber for re-building the head-frame and upper portion of the mill recently destroyed by fire. A 2100-lb. clean-up was made during the week. The Mogollon Mines Co. has been running its mill to full capacity and steadily increasing the ore-reserve. About 3000 lb. of gold and silver was sent to mint during October.

The Oaks Co. is maintaining regular daily shipments from

NEW MEXICO

its Central group. The Maud S., Deep Down, Clifton, and Eberle mines are all being developed and shipments are being made regularly from each.

Mogollon, November 13.

OKLAHOMA

OTTAWA COUNTY

Good results have been obtained by drilling on the Thomas land, four miles south-west of Baxter Springs on the Oklahoma side. The fourth hole drilled on the lease showed a 112-ft. orebody. In this hole ore was found at the 48-ft. level and continued to the 160-ft. Assays of the borings between the 56 and 72-ft. level ran from 42 to 56% zinc and 7 to 17% lead.

Two separate mills are being built by the Hawkins Lead & Zinc Co. on the site of the old Lancaster mine, at Sunnyside, one mile east of Lincolnville. One is a regular mill of 200 tons capacity, while the other is a modern tailing-mill, which will be used to work over the big tailing-pile left by the original Lancaster plant, which was burned down some years ago. The plants are both strictly modern and will be equipped with oil engines. In the tailing-mill there will be high-speed Cornish rolls for re-grinding. The company intends to work the level that was opened ten years ago by the original company, and also to sink to a deeper level.

If plans do not go awry construction work on a new 800-ton mill of the Brown & Head Zinc Co., with headquarters at Joplin, will begin within two weeks on a lease of the Cooper land situated on the Kansas side north-west of Picher. It was planned to begin building the mill in October, but delay in obtaining building material and machinery made it necessary to postpone the work. Shaft-sinking is being carried on; both shafts are down near the ore-level. An almost uniform orebody was shown from 175 to 255-ft. levels; 25 drill-holes were sunk on the lease before shaft-sinking was begun. The company's lease is situated just south-east of the new 800-ton mill of the United States Smelting company. J. E. Head, of Tulsa, is chief owner of the property. Don D. Molloy, of Joplin, and W. W. Marsh, of Webb City, are in charge of field operations.

A company of Texas and Oklahoma capitalists has purchased the concentrating plants of the St. Louis Smelting & Refining Co. and the lease of the 160-acre tract on which they were situated and additional leases on 1240 acres, immediately to the east, in the heart of the Missouri-Kansas-Oklahoma zinc-lead district. The two concentrating plants are the largest and best equipped in the district. Extensive drilling has been done on a large part of the acreage purchased, which, it is said, has proved an immense orebody running from Century in a north-easterly direction toward Baxter. Among the capitalists interested in the purchase are C. C. Slaughter, head of the Slaughter cattle interests in Texas, Walter Morris and L. P. Gamble of Dallas, and O. D. Halsell of Oklahoma.——R. P. Sharp of Miami and H. M. Ferguson and associates of Mangum, Oklahoma, have purchased the Sambo mine at Lincolnville from the National Bank of Commerce of St. Louis for $150,000. The property consists of a 40-acre well-developed lease and a 300-ton concentrating plant. Recent operations have proved an orebody of 75 by 100 ft. The new owners will increase the mill to 500 tons capacity. Another mill is to be built on the lease of the Buffalo Mining Co. The No. 1 mill is situated at Quapaw and has been in operation for only a short time. The No. 2 mill will be built on the Red Skin lease in Kansas to the west of the property of the Blue Mound Mining Company.

TEXAS

HUDSPETH COUNTY

(Special Correspondence.)—The Sierra Blanca Mining Co., which has a capital stock of $1,000,000, has begun the development of its mine seven miles west of Sierra Blanca. The ore from the present workings runs more than 14 oz. silver and 52% lead per ton. The stock of this company is owned principally by William W. Crosby of Eagle Pass, and J. W. Atkins of Shreveport, Louisiana. Several other promising prospects in the same section of the Quitman mountains where this mine is situated are being developed. Regular ore shipments are being made from some of the properties and plans are on foot for the building of one or more ore-reduction mills. It is stated that besides silver and lead, some of the ores contain gold, zinc, and copper.

Sierra Blanca, November 14.

UTAH

The report of the Chief Consolidated, covering the period from July 1 to September 30, shows 13,075 tons dry ore mined, yielding, after smelting, transportation, and sampling charges were deducted, a total of $414,850. The net profits, after paying all charges, was $69,041 for the three months. The report shows $478,566 cash on hand. The total development work was 5098 ft. This includes 579 ft. at Pinyon Peak and 584 ft. in the Plutus. The metal content of the ore mined was as follows: Gold, 2458 oz.; silver 382,302 oz.; lead, 2,083,587 lb.; and zinc, 194,508 lb. The assays of the ore were 0.188 oz. gold per ton, 29.28 oz. silver, 10.86% lead in lead ores, and 36.29% zinc in zinc ores. The average gross value was $45.98 per ton. The smelting, freight, and sampling-cost was $14.25 per ton, leaving an average net of $31.73 per ton. The dry tons mined are classified as 9592 lead, 3215 other ore, and 268 tons zinc ore, a total of 13,075 tons.

SALT LAKE COUNTY

The Utah Metals Co. has declared a dividend of 30c. per share, payable December 10 to stockholders of record November 30. This means a payment of $207,476 on the company's outstanding stock of 691,588 shares and swells its dividend total to $892,598. It is the second dividend declared this year and is a reduction of 20c. per share in the amount distributed last February, when the company paid 50c. As is generally known the Utah Metals Co. operates at Bingham, where it controls nearly 4000 acres of valuable ore-bearing ground. In the year ended December 31, 1916, the company produced 6,301,670 lb. of lead, 1,761,520 lb. of copper, 388,757 oz. of silver, and 17,934 oz. of gold, from which it earned a net profit of $528,737. Its possessions include the ground formerly owned by the Bingham-New Haven Copper & Gold Mining Co., equipped with a concentrating plant, and it has several large orebodies which are said to warrant a heavy production for years to come. The company paid its first dividend of 50c. in August 1916.

WASHINGTON

FERRY COUNTY

The shipments of the Knob Hill company totaled 500 tons in October, according to John Byrne, president, who has concluded an inspection of the property. The net smelter returns, mainly for gold, averaged more than $400 per car of 40 tons. The greater part of this ore came from above the main adit, but the shaft-workings, where development has been started, contributed in a small way. Ore was reached on the 200-ft. level of the shaft-workings soon after a drift was started from the cross-cut. It was lost for awhile, but found again at 135 ft. The body has a width of 5 to 24 in. and contains $5 to $58 per ton. A raise will be made to connect with a 130-ft. winze, sunk from the adit, as soon as the winze is freed of water. The ascent will be about 100 feet.

KING COUNTY

Formal announcement was made on November 16 that the plant of a newly-organized steel manufacturing company, financed for the greater part by California capitalists, will be built on Puget Sound and that work will soon begin on the first unit, requiring an investment of $9,000,000. The full development of the project will require $25,000,000. The announcement was made by Bartlett L. Thane, general manager for the Alaska Gastineau Mining Co., who is now at Seattle. Associated in the project are W. H. Crocker, Herbert Fleischhacker, S. F. B. Morse, and other San Franciscans. Eighty per cent of the money for the beginning of the work has been subscribed in California, most of it in San Francisco. It is expected that the plant will be in operation within a year and a half.

STEVENS COUNTY

The Northwest Magnesite Co. produced 19,000 tons of magnesite in October, of which 13,000 tons was calcined; the remainder was shipped in the crude form and placed in reserve at the plant near Chewelah. The shipments of calcined magnesite averaged 180 tons per day in October and is proceeding at the same rate this month. The building of a calcining plant and of a tramway are nearing completion. Two of the metal kilns, each about 130 ft. long, have been in operation many weeks and the third will be in commission by December 1. The tramway, which is to be five and one-half miles long, will be completed about the same time. The cost of this and other equipment and of improving the property otherwise was estimated at $250,000 when work was started several months ago. The freight charges for the movement of magnesite are about $100,000 per month. R. S. Talbot, president of the company, has returned from the East.

The Thorp Iron Co. has bought material for the construction of two brick kilns to be used in calcining magnesite on its property, 14 miles west of Valley. The company has nine claims in the magnesite belt. Part of the group gives promise of a yield of iron, for which it was acquired. The property is a mile or two from a railroad in process of construction from Valley.

MEXICO

According to G. M. Seguin, Mexican consul general at El Paso, the Alvarado Mining & Milling Co. has resumed operation of its Parral properties. Four hundred men are working and this force will soon be doubled. The San Francisco del Oro corporation, a British syndicate, with offices in New York, has also partly resumed operation of its property. Twelve carloads of mining machinery were shipped through the Juarez port to this corporation on November 6 and will be transported to its properties in the Parral district. This company expects to have fully 500 men employed in the Parral district within the next 10 days. Preparations are being made by the American Mines & Smelter Co. to resume the operation of mines in the Magistral, Durango district. The American Smelting & Refining Co. expects soon to finish the construction of its smelter at Chihuahua. Mr. Seguin said that General Francisco Murguia had promised to give protection to all mining corporations that resume operation. He has sent detachments of troops to the Alvarado and San Francisco del Oro properties in the Parral district.

According to the Boletin Financiero y Minero de Mexico, of October 10, the International Ore Co. of Saltillo, State of Coahuila, is constructing two furnaces, each one containing 300 retorts, with a capacity of approximately 40 to 45 tons per day and a daily production of zinc amounting to 15 tons. The furnaces will burn crude petroleum and the waste gases will be used under the boilers. The cost of this project is estimated at $90,000 American currency.

Water and Mines in Paradise

In our issue of November 3, we published a letter from Mr. J. D. Hubbard, on 'Water and Mines in Paradise.' An error of punctuation appeared in the sentence: "How about the Mineral Slide Mining Co. in Sec. 10, T. 23 N., R. 3 E., M.D.M. Mr. Goodner not only protested, but forced withdrawal of their property from the irrigation district." This statement should read: "How about the Mineral Slide Mining Co. in Sec. 10, T. 23 N., R. 3 E., M.D.M., Mr. Goodner? They not only protested, but forced withdrawal of their property from the irrigation district."

At the annual meeting of the WESTERN METALLURGICAL AND CHEMICAL SOCIETY, held at Chicago, October 1 to 3, Baron C. B. Smith, London, was elected president; R. H. Burgess, Chicago, vice-president; John M. Meyerburg, Oakland, treasurer; and Roy Franklin Heath, Billings, Montana, secretary. A re-organization of the society was announced by the chairman. An annual and a quarterly bulletin will be published by the society. The clause restricting the membership to 200 was eliminated from the by-laws, while the initiation fee was reduced from $50 to $15. The yearly membership dues were reduced from $25 to $10. A resolution was adopted unimously, urging President Wilson to use every effort and power within his reach to defeat "the Prussian chief and his powder gang." A committee was appointed by Mr. Smith to undertake an investigation regarding professional ethics. Reports covering the various departments were read and discussed.

PERSONAL

ote; The Editor invites members of the profession to send particulars of their work and appointments. This information is interesting to our readers.

GEORGE S. BUTTERWORTH has returned to New York from Venezuela.

G. BROKENSHIRE has enlisted in the Engineering Corps of the U. S. Army.

E. A. JULIAN, of the Goldfield Consolidated, has returned from New York.

BERNARD MACDONALD has returned from Antofagasta, Chile, to South Pasadena, California.

FREDERICK BRADSHAW, manager for the Tonopah-Belmont Development Co., is in Colorado.

G. W. LAURIE, of the firm of Drucker & Laurie, New York, has joined the 23rd U. S. Engineers.

BENJAMIN REZAS has gone to the Philippine Islands to join the staff of the Colorado Mining Company.

G. A. JOSLIN has returned to Salt Lake City, having spent the summer in the Bay Horse district, Idaho.

FRITZ MELLA has resigned from the employ of the A. S. & R. Co. and is now on his way to Santiago, Chile.

J. L. PHILLIPS, formerly manager of the Inguaran mine, in Mexico, was here from Los Angeles during the week.

L. R. DAVIS, from Wyoming, and R. V. AGETON, from Butte, are commissioned in the engineer section of the Air Service.

W. C. WEBSTER, secretary and general manager for the Nichols Copper Co., New York, has resigned, and expects to take a rest.

BEN. B. THAYER, having recovered from a recent illness, has gone to Butte on his periodical trip of inspection from the Anaconda Copper Company.

Obituary

LEO VON ROSENBERG, Baron Leo Franz Seraficus Georg Orsini Von Rosenberg Lipinsky, was born on June 29, 1858, at Klaugenfurt, Austria. He died at New York on October 6, 1917. He received his education in the best schools in Austria and at Freiberg. Coming to the United States with his father in the early 'eighties he shortened his name to Von Rosenberg, by which he was thereafter known. His first work in this country was as a draftsman. Most of the illustrations in the important book on 'Tunneling,' by H. S. Drinker, were prepared by him. In 1887 he published a book on the Vosburg tunnel, profusely illustrated and giving a complete description of the construction of that important work. Later he published a work on the Musconetcong tunnel. His great work in publishing was 'The Washington Bridge'; probably no engineering work has been more thoroughly and accurately described and illustrated. At the time he was engaged upon this publication he also was occupied in preparing maps and reports for various mining companies. Many of these reports were printed in pamphlet form for distribution among the stockholders and others; and aside from intrinsic value were remarkable for artistic make-up. He practised as a consulting mining engineer and obtained an honorable reputation for such work. Among the mines upon which he either made reports or issued publications may be mentioned the Playa de l'Oro, Cerro de Pasco, Umbria, Russell, Osol Condo, Victor, Gold Coin, Vindicator, Fortuna, and Golden Cycle. An engaging and kindly man, he made many friends in the West and his death will cause wide-spread regret.

METAL PRICES

San Francisco, November 20

Aluminum-dust (100-lb. lots), per pound	$1.00
Aluminum-dust (ton lots), per pound	$0.95
Antimony, cents per pound	16.00
Antimony (wholesale), cents per pound	14.75
Electrolytic copper, cents per pound	23.50
Pig-lead, cents per pound	6.50— 7.50
Platinum, soft and hard metal, respectively, per ounce	$105—111
Quicksilver, per flask of 75 lb.	100
Spelter, cents per pound	9.50
Tin, cents per pound	65
Zinc-dust, cents per pound	20

ORE PRICES

San Francisco, November 20

Antimony, 45% metal, per unit	$1.10
Chrome, 34 to 40%, free SiO_2, limit 8%, f.o.b. California, per unit, according to grade	$0.60— 0.70
Chrome, 40% and over	$0.70— 0.80
Magnesite, crude, per ton	$8.00—10.00

The demand for magnesite is confined almost entirely to the calcined product.

Manganese: The Eastern manganese market continues fairly strong with $1 per unit Mn quoted on the basis of 48% material.

Tungsten, 60% WO_3, per unit	30.00
Molybdenite, per unit MoS_2	$40.00—45.00

EASTERN METAL MARKET

(By wire from New York)

November 20.—Copper is unchanged and nominal at 23.50c. all week. Lead is quiet and firm at 6.50c. all week. Zinc is dull and steady at 7.87 to 8c. Platinum remains unchanged at $105 for soft metal and $111 for hard.

SILVER

Below are given the average New York quotations, in cents per ounce, of fine silver.

Date		Average week ending	
Nov. 14	86.00	Oct. 9	91.72
" 15	85.75	" 16	87.01
" 16	86.75	" 23	83.85
" 17	86.50	" 30	84.71
" 18 Sunday		Nov. 6	89.25
" 19	86.50	" 13	86.18
" 20	85.25	" 20	85.62

Monthly Averages

	1915	1916	1917		1915	1916	1917
Jan.	48.85	56.76	75.14	July	47.52	63.06	78.92
Feb.	48.45	56.74	77.54	Aug.	47.11	66.07	85.40
Mch.	50.61	57.89	74.13	Sept.	48.77	68.51	100.73
Apr.	50.25	64.37	72.51	Oct.	49.40	67.86	87.38
May	49.87	74.27	74.61	Nov.	51.88	71.60	....
June	49.03	65.04	76.44	Dec.	55.34	75.70	....

The silver market has undergone very little change. The fluctuation was limited to about one cent per ounce, and on November 13 the official quotation was 96⅛c. per ounce.

LEAD

Lead is quoted in cents per pound, New York delivery.

Date		Average week ending	
Nov. 14	6.50	Oct. 9	7.96
" 15	6.50	" 16	7.52
" 16	6.50	" 23	6.70
" 17	6.50	" 30	5.06
" 18 Sunday		Nov. 6	6.02
" 19	6.50	" 13	6.35
" 20	6.50	" 20	6.50

Monthly Averages

	1915	1916	1917		1915	1916	1917
Jan.	3.73	5.95	7.64	July	5.59	6.40	10.93
Feb.	3.83	6.23	9.01	Aug.	4.62	6.28	10.75
Mch.	4.04	7.26	10.07	Sept.	4.62	6.86	9.07
Apr.	4.21	7.70	9.38	Oct.	4.62	7.02	6.97
May	4.24	7.38	10.29	Nov.	5.15	7.07	....
June	5.75	6.88	11.74	Dec.	5.34	7.55	....

COPPER

Prices of electrolytic in New York, in cents per pound.

Date		Average week ending	
Nov. 14	23.50	Oct. 9	23.50
" 15	23.50	" 16	23.50
" 16	23.50	" 23	23.50
" 17	23.50	" 30	23.50
" 18 Sunday		Nov. 6	23.50
" 19	23.50	" 13	23.50
" 20	23.50	" 20	23.50

Monthly Averages

	1915	1916	1917		1915	1916	1917
Jan.	13.60	24.30	29.53	July	19.09	25.66	29.67
Feb.	14.38	26.62	34.57	Aug.	17.27	27.03	27.42
Mch.	14.80	26.65	36.00	Sept.	17.69	28.28	25.11
Apr.	16.64	28.02	33.16	Oct.	17.90	28.50	23.50
May	18.71	29.02	31.69	Nov.	18.88	31.95	....
June	19.75	27.47	32.57	Dec.	20.67	32.89	....

Not until the current year have exports of copper in a single month run above 100,000,000 lb. In January clearances were 114,960,000 lb. August figures, just at hand, show that in that period there was shipped abroad from this country 101,129,000 lb. of copper, making the fifth month of 1917 in which the 100 million mark has been exceeded. Up to the first of September there had been shipped 778,389,561 lb. of copper, against 521,976,630 lb. in the same months last year and 433,205,804 lb. two years ago. In addition to these large copper tonnages the shipments of brass—two parts copper to one part spelter—have broken all previous records.

It is understood that the Anaconda Copper Mining Co. has been increasing its holdings in shares of the Inspiration Consolidated Copper Co. around the recent low prices. Anaconda has been carrying for some time 200,000 shares of Inspiration upon which it has been drawing dividends at rate of $8 per share. During the past two years a strong surplus has been built up by Anaconda, and Wall Street believes a part of this has been invested in additional amounts of Inspiration stock. Both companies have managements closely allied, with John D. Ryan the dominant factor.

ZINC

Zinc is quoted as spelter, standard Western brands, New York delivery, in cents per pound.

Date		Average week ending	
Nov. 14	7.87	Oct. 9	8.50
" 15	8.00	" 16	8.50
" 16	8.00	" 23	8.25
" 17	8.00	" 30	8.12
" 18 Sunday		Nov. 6	7.42
" 19	8.00	" 13	7.81
" 20	8.00	" 20	7.98

Monthly Averages

	1915	1916	1917		1915	1916	1917
Jan.	6.30	18.21	9.75	July	20.54	9.90	8.98
Feb.	9.05	19.99	10.45	Aug.	14.17	9.03	8.58
Mch.	8.40	18.40	10.78	Sept.	14.14	9.18	8.33
Apr.	9.78	18.62	10.20	Oct.	14.05	9.92	8.32
May	17.03	16.01	9.41	Nov.	17.20	11.81	....
June	22.20	12.85	9.63	Dec.	16.75	11.26	....

Australia has ceased the shipment of zinc-bearing material to the United States, the movement having been diverted in great part to England where smelting facilities have been greatly enlarged during the past two years. Originally Australian product was sent to Belgian and French smelters controlled by German metal interests, but the outbreak of war altered this programme, and in the early days of the European struggle large quantities of zinc were brought to this country for treatment.

QUICKSILVER

The primary market for quicksilver is San Francisco, California being the largest producer. The price is fixed in the open market, according to quantity. Prices, in dollars per flask of 75 pounds:

Week ending

Date			
Oct. 23	100.00	Nov. 6	100.00
" 30	100.00	" 13	100.00
		" 20	100.00

Monthly Averages

	1915	1916	1917		1915	1916	1917
Jan.	51.90	222.00	81.00	July	95.00	81.20	102.00
Feb.	60.00	295.00	126.25	Aug.	93.75	74.50	115.00
Mch.	78.00	219.00	113.75	Sept.	91.00	75.00	112.00
Apr.	77.50	141.60	114.50	Oct.	92.90	78.20	102.00
May	75.00	90.00	104.00	Nov.	101.50	79.50	....
June	90.00	74.70	85.50	Dec.	123.00	80.00	....

The market for Californian quicksilver remains at $100 per flask of 75 lb. net for October shipment from the West. Mexican mercury is quoted at $95 to $96 per flask, while there is no quotation for recovered mercury, but probably could be sold around $93 to $94.

TIN

Prices in New York, in cents per pound.

Monthly Averages

	1915	1916	1917		1915	1916	1917
Jan.	34.40	41.76	44.10	July	37.38	38.37	62.60
Feb.	37.23	42.60	51.47	Aug.	34.37	38.88	62.53
Mch.	48.76	50.50	54.27	Sept.	33.12	36.66	61.54
Apr.	48.25	51.49	55.63	Oct.	33.00	41.10	....
May	39.28	49.10	63.21	Nov.	39.50	44.12	....
June	40.26	42.07	61.93	Dec.	38.71	42.55	....

Owning the only tin smelter in this country, the American Smelting & Refining Co. should, early in 1918, be in position to furnish from its New Jersey plant 800 tons of electrolytic tin monthly to consumers in the United States. Sufficient construction headway has now been made to offer assurance that by January the Guggenheim's tin smelter should be up to its capacity. In October the A. S. & R. Co. turned out about 600 tons of electrolytic tin, representing the only domestic production in the United States. This month's contribution should be about 650 tons, December's output 750 tons, while January is expected to show an 800-ton yield. The raw material for the smelter comes from Bolivia in the form of about 60% tin, which has to be reduced to marketable forms and shapes. Despite its richness in practically all minerals the United States has not yet produced a tin mine of value. In 1916 there was found but 278,000 lb. of tin in this country, against 204,000 lb. in 1915.

MAGNESITE

Little business has been reported on magnesite, although the market remains practically the same.

MANGANESE

The November schedule for furnace ore remains at $1.20 per unit for high-grade material, which is an advance of about 10c. over last month's schedule.

Eastern Metal Market

New York, November 14.

The markets are generally inactive, but fairly steady and firm. There is little incentive to buy under present conditions.

Copper is nominally unchanged at the Government price, with no line on actual conditions.

Tin has been soaring, reaching the highest price ever recorded, due to great scarcity.

Lead is quiet but firm and steady.

Zinc displays a better tone and is firmer.

Antimony is in poor demand and is nominally lower.

Aluminum is unchanged and dull.

The steel trade is adjusting itself to the new regime with less friction than was expected. New prices have again appeared, including bar-iron, boiler-tubes, nuts and bolts, boat-spikes, wire-rope, pipe, skelp, etc. Little finished steel is being sold for general commercial needs, most mills being booked for three to six months in advance. The Government's requirements for ship-plates will probably be 100% of ship-plate production. The pig-iron market is quite active, with most furnaces sold up through the first half of 1918.

COPPER

The entire situation is 'up in the air' and shrouded in mystery so far as definite and authentic information is concerned. Whether sales for early delivery in large volume have been made to consumers having Government contracts at the Government price of 23.50c. per lb., or at any price, is not known, though statements are appearing in print to that effect. In the absence of facts we continue to quote the fixed price as nominal. There has been no illumination on the perplexing situation of the small dealer and buyer, but it is understood some sales in this field are being made at higher than the Government price. The market is absolutely uninteresting. Statistics as a rule are dry, but it is significant to note, as affecting the general situation, that copper exports for the first nine months of this year have been larger than for any whole year of 12 months on record. To October 1, the total is reported as 370,843 long tons as compared with 327,310 tons in the whole of 1916, and 360,229 tons in 1914.

TIN

The scarcity of tin, spot Straits, Banca, and Chinese, is so great that prices have reached levels never before recorded. Yesterday spot Straits sold at 73.50c. per lb., New York, in a transaction involving 5 tons, and later another 5-ton lot was offered at 73c., New York. The market on Monday, November 5, was 68c., New York, as reported in our letter last week, but it has advanced continually since. On November 7 it was nominal at 71c., but sales were made at 70c. on both November 9 and 12, with the quotation standing at 73c., New York, yesterday. Another interesting feature was the Government's action yesterday. Failing to obtain any tin on an inquiry for 15 to 25 tons of spot Straits for the Washington Navy Yard, the required material was requisitioned from an incoming steamer which had 250 tons on board, already sold to consumers. What price will be paid has not been divulged. The stiff permit-requirements in England are the prime cause of the present scarcity because shipping-permits are issued only to consumers whose names are divulged, so that buyers of less than 25-ton lots are extremely anxious. On the whole the market is quiet and inactive, buyers abstaining as far as possible until the situation clears. On November 7 a sale of 25 tons of Eastern shipment in October was sold at 64.25c., and on November 8 shipment from the East at sellers' option in February, March, and April went at 61.25c. Arrivals up to November 12 inclusive have been 830 tons, with 4100 tons reported afloat. The London market has had a large advance of £13 10s. in the week, spot Straits being quoted yesterday at £274 per ton.

The Tin Committee has been in Washington in the last week and while the result of the deliberations has not yet had any effect on the market, it is understood that an agreement has been reached, which is expected to relieve the situation.

LEAD

The American Smelting & Refining Co. advanced its price again on Wednesday, November 7, from 6 to 6.25c. per lb., New York, while the outside market had already gone to 6.50c., New York, or 6.37½c., St. Louis. These prices are the present levels with transactions of small volume. This is natural as a result of the large purchasing of the last two or three weeks, noted a week ago. There is but one weak point in the situation which otherwise is strong and firm. It is the fact that there were a few sellers who abstained from the market in its recent large activity and who are consequently probably able to offer some metal at concessions. It is reported that lead has been offered in the last day or two at 6.35c., New York.

ZINC

The prospect of large Government purchases of shells in the near future has imparted a better tone to the market, and it is firmer. One report is to the effect that the Government has purchased or is about to buy 33,000,000 shells of the 3-in. size. This will require 33,000,000 lb. of zinc for making the brass shell cases. It is true that meetings are being held jointly by the War Industries Board and the zinc producers, and definite announcement of plans is expected soon. It is generally thought that the question of price-fixing is being seriously considered with the probability of some definite levels to be agreed upon. While small lots of prime Western have been sold recently at anywhere from 7.50 to 7.87c. per lb., St. Louis, the lower prices have disappeared in the last day or two, and the market is quotable at 7.75c., St. Louis, or 8c., New York, with the market quiet but firm.

ANTIMONY

The market is lower because of poor demand and large imports. Chinese and Japanese grades are quoted at 13.75 to 14c. per lb., New York, duty paid.

ALUMINUM

There is no change, and the market is inactive and quiet at 36 to 38c. per lb., New York, for No. 1 virgin metal, 98 to 99% pure.

ORES

Tungsten: One dealer reports that South American supplies are scarcer, due to restrictions resulting from the Trading with the Enemy Act. The usual quotation for the various grades of ore are firm, with a tendency to advance. Ferro-tungsten is obtainable at $2.35 per lb. of contained tungsten.

Molybdenum: Molybdenite continues to be quoted at $2.20 per lb. of MoS_2 in 90% material and some business has been done at this figure.

Antimony: The market is quiet and nominal at $1.60 to $1.75 per unit.

Chrome: The market for high-grade chrome ore remains practically the same with 90c. per unit quoted f.o.b. Californian shipping points. Chrome over 60% is scarce and practically all the shippers are mining a low-grade product.

FOR the convenience of prospectors and other locators of mining claims we print the legal form to be used by those desiring to claim the benefit of the Act of Congress suspending the requirement of annual assessment work for the year 1917. It will be found on page 17 of our advertising department.

IT is announced that the 200 employees of the Mason Valley Mines Company, at Thompson, Nevada, have agreed to subscribe one day's pay per month to a fund that is to be used "to promote the welfare of American and Allied soldiers at home and abroad." This is a good move. In Canada every man on the payroll of the mining companies gives one day's wage each month to the Canadian Patriotic Fund, which is used in caring for returning soldiers and for the dependents of those at the front.

UNDER 'Discussion' we publish another interesting letter on mine-sampling by Mr. L. A. Parsons. The importance of elaborating a reliable method of sampling large bodies of low-grade ore, particularly in gold mines, is emphasized by the collapse of the share-value of the Alaska Gold Mines, which enterprise has suffered from a frankly acknowledged error in ore-valuation due to an ncorrect basis of sampling. The shares that were quoted ıt 40¾ in April 1915 are now standing at 2¼. Even the onds have depreciated 42%. Thus an original capital nvestment of $10,500,000 is imperiled. Obviously it is vorth while to ascertain the most reliable way of sam-ling such masses of low-grade ore as those near Juneau.

WE publish a detailed description of the International smelter, at Miami, in this issue. This metallurgical lant is particularly interesting because it is the one melter devoted entirely to the treatment of concentrate, hiefly flotation concentrate, a mill-product presenting eculiar features. The tonnage treated daily is classified ı follows: the Miami supplies 211 tons, of which 93 ons is flotation concentrate and 118 tons is table con-ntrate; the Inspiration supplies 668 tons, of which 500 ns is flotation concentrate and 168 tons is table con-ntrate; the Old Dominion supplies 282 tons, of which ! tons is flotation concentrate and 230 tons is concen-ate from tables and jigs. Thus the flotation concen-ates constitute 55.55% of the total ore-supply. Undoubtedly the coarser products from jigs and tables help the smelter superintendent in overcoming some of the difficulties inherent in handling the slime-product of the flotation-cells.

ON page 36 of our advertising department we print a form of application for enlistment in the 27th Engineers, which is the regiment being formed for special mining service in France, as announced in our issue of November 10. We have received requests for these forms of application and we feel sure that this way of circulating them will prove useful. Several companies of the regiment are already in training at Camp Meade, Maryland. The officers will be mining engineers that have undergone preparation in the officers' training camps. The colonel will be a Regular Army engineer officer. The enlisted men will be the men from the mines, more particularly drillers, mechanics, and handy men of all kinds. Again we commend this service to our friends of the pick and gad.

THE weekly review of the metal market as viewed by our correspondent at New York shows that speculation in copper was killed, as was intended, when the Government fixed the price of this metal at 23½ cents per pound. In a recent review it was stated that "the relation of consumption to production is a matter of speculation." Some light is thrown on the subject by the latest quarterly reports of the so-called 'porphyry' copper companies. These show that whereas the production of the Utah, Nevada, Chino, and Ray companies was maintained steadily, the sale of copper was withheld after September 21, the date on which the Government fixed the price at 23½. The stocks of copper at the end of the third quarter of the year are valued at 13¼ cents as compared with the estimate of 28 cents on which profits were based for the second quarter. Why the arbitrary fixation at 13¼ cents? Is that a possible estimate of cost? The average cost incurred by the four companies during the third quarter of 1917 is given as 11.68 cents per ton; this compares with an average of 7.57 in 1915; but costs are figured in a less camoufleurant way in these serious days! Undoubtedly the placing of the Government's hand on the copper market had the effect of causing many of the large producers to hold back their metal with the idea of decreasing the available supply. It is suggested, by interested parties, that the expansive output required for our warfare will not be obtained unless the producing companies are encouraged, by a rise in the official price, to exploit lower-grade ores. We want to see the copper-mining industry maintained on a profitable and prosperous basis, of course, but we are frank to say that we view with suspicion the placing of the Government's buying in the hands of such organizations as the United Metals Selling Company and the American Smelting & Refining Company, both of which are so heavily interested in the production and sale of the metal. The public is entitled to have the statistical facts as ascertained by official inquiry. It would do much to clarify the position if the Government would make a frank statement concerning the production and consumption of copper in the country as ascertained in the course of the investigation that preceded the fixation of price last September.

ACCORDING to statements made by congressman Julius Kahn, a new draft-bill will be introduced on the re-assembly of Congress in December. He suggests that the age-limit may be extended to 35 or 40 years, and also that young men from 18 to 21 may be called for training. This would appear wise. It is certain that most men reach the fulness of strength and endurance between the ages of 30 and 40. Moreover they develop a maturity of judgment that would make them more efficient and resourceful as soldiers. The younger men would benefit by military training and constitute a reserve-line that would increase the military strength of the nation. At the same time, we believe that consideration should be given to the growing need of a large body of technicians, and that any future bill should provide exemption for all students enrolled in technical colleges and universities until they may have completed their course of instruction. Unless attention be directed to this matter we shall presently experience a shortage of capable engineers, chemists, and physicians that will seriously cripple the country not only industrially but in its preparations for effective conduct of the War.

SILVER has been steady in price lately, after a rise and a fall that must have excited many of our readers. Letters from them indicate a not unnatural anxiety over the market position. One correspondent was perturbed by our suggestion that the Government might use its stock of silver to regulate the price. Another correspondent raised vigorous objection to the Government's purchases of Mexican pesos for re-coinage. The recent narrow range of quotations gives color to the statement, made at Washington, that the governments of the United States and Great Britain have entered into an agreement whereby the entire American output, including that of Canada, is to be purchased on joint account for the purpose of establishing a 'corner' and regulating the price at a figure somewhere between 80 and 85 cents per ounce. A later dispatch from London says that a portion of the silver to be acquired, as above stated, is to be placed at the disposal of the Indian government, in order to pay for Indian products required in the United States. There is nothing in all this to intimidate the miner; on the contrary, the fixing of the price of silver at any such figure as 80 to 85 cents is greatly to his benefit, and much better than a hysterical market, fluctuating wildly, and affording no basis for steady operations in mine and mill.

DO we need gold—more gold? That is the question submitted by Mr. Lester S. Grant in our discussion department this week. It is a question that deserves an answer. If our gold reserve is adequate, if indeed, as some are claiming, it is excessive, and is responsible for a lessened purchasing-power of the dollar and therefore accentuating the pinch of high prices, then a war-measure discriminating against freight and labor for gold mines might be a justified application of the utilitarian doctrine of doing the greatest good to the greatest number. On the other hand, if our mountainous obligations, incurred by virtue of the stress of war, should at any moment of military anxiety become the weapon of unprincipled men to scare the timid multitude into seeking any available amount of cash for their Government bonds, a calamity might be precipitated unless there were a gold reserve of sufficient dimensions to admit of emergency legislation to utilize it in refunding under some plan of redeemable certificates. Our feeling is one of satisfaction when contemplating that great and growing reserve of bullion in our national Treasury, but we are ready to stand corrected by specialists in finance.

REFERENCES have been made, in the course of the Minerals Separation controversy, to the Canadian patent law. We have obtained a copy of the Canadian Patent Act of 1906 and find that, under Section 38, it requires "the construction or manufacture of the invention patented" in Canada within two years after the grant of patent; and there is the further condition that the invention cannot be imported "after the expiration of twelve months from the grant of a patent or an authorized extension of such period," except under special license. We note that, under Section 44, any person may apply to the Commissioner of Patents for a license to make, use, or sell the patented invention, in the event of the patentee refusing "to grant licenses to others on reasonable terms." It is also provided that the existence of one or more licenses shall not be a bar to an order by the Commissioner for, or to the granting of, a license on any application." If the patentee "refuses or neglects to comply with such order within three calendar months" then "the patent and all rights and privileges thereby granted shall cease and determine, and the patent shall be null and void." Apparently therefore the Canadian law is such that Minerals Separation will find it advisable to behave circumspectly. It is a pity that our own patent law does not provide for similar disciplinary action.

American Smelting v. Bunker Hill

In our issue of September 8 we recorded the fact that ie American Smelting & Refining Company had brought iit against the Bunker Hill & Sullivan Mining & Con- ntrating Company for infraction of a contract that ecified the disposal of the Bunker Hill ore and mill- roducts. This suit was started in August and was eant to prevent the Bunker Hill company from smelt- g its own ore in its own smelter, which has been built ear the mine at Kellogg, Idaho, and had begun opera- ons in July. A petition was presented by the American melting company to the U. S. District Court at Portland, regon, asking for a temporary injunction against the ntinued operation of its smelter by the Bunker Hill mpany pending the trial of the suit. This petition me up for hearing before Judge Wolverton at Port- nd on November 7. A decision may be expected ortly.

The bill of complaint alleges that on March 20, 1905, contract was signed between the Bunker Hill company id the Tacoma Smelting company; and that by this ntract the Bunker Hill agreed for 25 years to ship all ore that contained between 30 and 75% lead to the nelter at Tacoma; all the products of the mine and mill at assayed outside these specified limits being open to irchase by the Tacoma smelter on payment of the best ing rates and terms. The contract provided also that e products specified should be subject to competitive tes after the first five years. On the day following at on which this contract was ratified, the American nelters Securities Company, a Guggenheim organiza- n, acquired all the stock of the Tacoma Smelting com- ny, which in June 1917, but not until then, assigned e contract itself to the American Smelting & Refining mpany. Our readers will remember that in March 05 the American Smelters Securities Company, ough Mr. Bernard M. Baruch, bought not only the coma smelter but also the Selby plant, near San Fran- co, and thereby the Guggenheims prepared the way their control of the lead-smelting industry. Two irs earlier the American Smelting & Refining Com- ıy had bought the Everett smelter, in Washington, l in 1905 this property also was transferred to the elters Securities Company. In 1912 the Tacoma plant converted into a copper smelter and ceased to treat l ores. As we understand the point at issue between litigants, the Tacoma smelter was entitled to, and nd to take, all the Bunker Hill ore, concentrate, and ie assaying between 30 and 75% lead, but the smelter the option of purchasing the outside products only best going market rates and terms." The idea that Tacoma smelter was entitled to all the products of Bunker Hill mine is asserted by the defendants to lirectly contrary to the express letter of the agree- t. It is also claimed by the defendant that the option hese outside products lapsed with the rest of the con- t when the latter was assigned to the Smelters Se- ties company, in which the American Smelting company owned only 17 million dollars worth of stock, out of the total of 77 millions. It is obvious, however, that the ore-contract was intended for the benefit of the smelter trust, which acquired the control of the Tacoma smelter on the day after the contract was signed. The Guggenheims bought the control of the smelter from the principals of the Bunker Hill company. In effect, the plaintiff company asserts that it was the real purchaser of the defendant's controlling interest in the Tacoma smelter; that the purchase was made with the idea of securing all the ore produced by the Bunker Hill for a period of 25 years, so that the Bunker Hill people would not go into the smelting business again for that length of time, and that the main reason for the new Bunker Hill smelting enterprise is the cupidity aroused by the abnormal price of lead due to the War. To this the defendant company replies that the plaintiff was not the real purchaser of the Tacoma smelter; that all the Bunker Hill product was not covered by the 25-year contract; that the Bunker Hill people incurred no obligation to abstain from smelting operations; that cupidity played no part in the decision to build the smelter at Kellogg, this step having been taken before the War raised the price of lead; and that one of the reasons for building the smelter was the fact that the Guggenheim combination had so suppressed competition in ore-buying that the Bunker Hill company could find no market for such of its products as were not covered by the 25-year contract, namely, ore and mill-products assaying outside the limits of 30 and 75% lead. It is also charged that the Guggenheims during the first five years of the Tacoma contract killed the benefit of competitive terms on the medium-grade products specified in the contract. The other legal defences of the Bunker Hill company may be summarized as follows: The 25-year smelting contract was made between the Bunker Hill company and the Tacoma company; for seven years the latter received part of the Bunker Hill ore at its Tacoma plant and delivered the remainder to the plants of the American Smelting company. In 1912 the Tacoma company ceased to operate its lead plant and since that time all the Bunker Hill ore has been diverted to the plants of the American Smelting company. There has been no contract between this company and the Bunker Hill, but in June 1917, after the Bunker Hill smelter had been completed, the American Smelting company procured an assignment of the Bunker Hill's Tacoma contract and is now seeking to enforce that contract in its own name. The Tacoma company, with which the contract was made, has no plant in which it can treat the Bunker Hill ore and therefore is not damaged if such ore is smelted at Kellogg. The Bunker Hill company contends that the contract could not be assigned without its consent, because the contract imposed financial obligations on the Tacoma company and the Bunker Hill company has a right to say that it will not look to any other company for a performance of those obligations. The U. S. Supreme Court has held that a smelting contract such as this cannot be transferred by one company to another.

In any event, it is a fact that the contract was amended in 1907, two years after it was signed, so that the Tacoma company agreed during the life of the contract to pay for the lead in the Bunker Hill ore at a price made by the American Smelting company in New York City on the day of shipment. In other words, the Tacoma company and the Bunker Hill company agreed to let a third party fix the price for the lead each day, and settlements were made accordingly. Now the arbiter, or price-fixer, claims to have taken an assignment of the contract and seeks the aid of the Court to compel the Bunker Hill company to deliver lead ore to him—the plaintiff—the American Smelting & Refining Company—during the next 12 years at the price to be fixed by him or it. That is asserted by the defendant to be indefensible. Undoubtedly the sore spot in the controversy is the killing of competition—of the competition for which provision was made in the contract of 1905—and the establishment of a 'cinch,' or monopolistic control, by the smelter trust not only over the Bunker Hill output but over that of the whole Coeur d'Alene region.

On another page we publish the affidavit made by Mr. Edgar L. Newhouse in support of the complaint lodged by the Guggenheim company. We have omitted some of the elementary technology, such as is needed apparently to instruct the Court but would be tiresome to our readers. We have also ventured to correct the affiant's relative pronouns, substituting 'that' for 'which' in such sentences as seemed to require the change. In our next issue we shall publish the larger part of Mr. F. W. Bradley's affidavit, and subsequently we hope to give our readers the counter-affidavits of the two leaders in the controversy, believing that these statements made by distinguished members of the profession will prove of particular interest at this time.

Mineral on Agricultural Land

Many prospectors will sympathize with the position taken by Mr. Leonard G. Blakemore in his discussion of the interference of agricultural titles with the development of mineral resources, printed elsewhere in this issue. He maintains that the discoverer of a mineral deposit should be entitled to locate and to hold it for active development against any owner of surface rights, subject, of course, to adequate recompense for appraisable damage done to the interests of the aforesaid owner; he would also have the United States copy the Australian system of granting prospectors' licenses, carrying with them the right of entry upon the estates of agricultural owners, and of valid location when mineral is found. With all this we confess to cordial sympathy in principle, and we are pleased to endorse Mr. Blakemore's suggestion that engineers respond with discussion of the problem. It is a broad and important question, and should command serious attention. The West still holds a large area of public land, and it might be well to assert the ancient principle of the inalienability of minerals from the sovereign state so far as relates to lands that may be patented in future. Retroactive laws are contrary to the spirit and practice of our democratic institutions, and are, moreover, contrary to our fundamental law. Therefore some other remedy must be sought for the unlocking of the buried treasure on lands patented in good faith in accord with previous Acts of Congress. Ingenious methods of taxation sometimes prove a key that will unlock such difficulties. To undertake the unsettling of rights secured by title, however, would be to trifle with the stability of those sacred principles of possession that our Constitution is supposed to guarantee. If a patent once granted under the great seal of the United States can be thus lightly scorned, there would be nothing that a majority in power might not take away; it would be to invite anarchy under the guise of democratic form and procedure. We have already said that we believe in the essential justice of Mr. Blakemore's argument, but the correct basis of political evolution is growth by variation, not change by radical excision. It depends on the antecedents of present law as to what may be permissible. The custom of the world was once that the metals belonged to the king; that was a universal reservation, but that principle was not affirmed in the United States with reference to the public lands. The Government, in fact, undertook to discriminate between agricultural and mineral land, and to classify them accordingly. It was an almost impossible task, and has worked much hardship. What Mr. Blakemore had in mind, when he affirmed that the United States owns the minerals in lands to which agricultural patents have been issued, is probably the fact that in many of the Eastern States, where the laws of property were derived from Colonial charters, the ancient sovereign prerogative is asserted; the present statutes, in other words, are drawn in the light of their origin. In the State of New York a discoverer of mineral is protected in a half-interest to the deposit found and also in its control for exploitation; he may file his claim with the Secretary of State, and forfeiture is possible only through failure to develop and work the mine continuously. This we believe applies to gold, silver, copper, zinc, and lead. In North Carolina the fee-holder has the prior right to the mineral under the land, and may convey it separately from the surface right; but it does not automatically follow the title on conveyance. Unless specifically assigned the mineral-right reverts to the State and may be claimed by a subsequent discoverer. Likewise in Spain, which may be cited appropriately as the oldest and greatest mining region of Europe, the sovereign right exists unimpaired, and a discoverer may obtain a leasehold under which to develop and operate. The same is true in Mexico and in all the Spanish countries, where sometimes mining claims are located in populous cities. It is an old principle and a sound one; nevertheless retroactive laws may not be considered in free America, the country that has ever insisted on individualistic right. Without cause founded on delinquency we may not take from a man that which the law has given him.

Mill-Tests v. Hand-Sampling

The Editor:

Sir—I have followed with interest in your columns the discussion on sampling large low-grade orebodies, a discussion which has lately been somewhat narrowed to a comparison of the relative merits and pecularities of mill-tests and hand-sampling. This narrowing is a good sign; it points, as does the text of the succession of articles and letters, to something approaching agreement among your various correspondents as to the basic principles of good practice. The discussion has thus performed a most useful service in clearing some of the fog; but by showing how much your various skilful contributors agree in fundamentals it has served to accentuate the differences in details. It is about some of these details that I am writing to you today. A close agreement on these will probably never be reached, as the details of good practice are as varied as the orebodies that call them forth; but a further discussion of them may lead to a better understanding of the possible consequences resulting from their variation.

In your issue of September 29, Morton Webber describes a method of using the mill-test for determining the factor of error in hand-sampling. I am glad to see from a previous article of Mr. Webber's that he places his main reliance on hand-sampling, and uses the mill-test only as an aid in interpreting his results. Thus far we are in agreement. His 'Class 5,' a type of orebody for which, in a previous article, he makes an exception to this general rule, is a type extremely difficult to sample correctly, and, in the unusual cases he has cited, great accuracy is impossible by any economic method. Nevertheless, I believe this class does not constitute a true exception, and that the independent engineer will be better able to estimate the possibilities of such an orebody, as of a more uniform one, if he bases his opinion on hand-work alone rather than on mill-tests alone. This, however, is a digression. As a general principle, the checking of hand-samples by mill-tests, if carefully done according to some such method as that outlined by Mr. Webber, gives information that cannot fail to be valuable. I would hesitate, however, to support the argument that an exact sampling-factor can be thus determined; rather the information obtained will be valuable in a general and not too precise way, as an aid in estimating the reliance to be placed on the hand-sampling, and the variations and trend of the sampling-factor. I am ignoring the function of determining the most advantageous metallurgical process; for that purpose a mill-test, or at least a test on that scale, is practically always necessary. I am confining myself to the best means of determining the tenor of the orebody, pure and simple.

For the combination method as outlined by Mr. Webber, it is not necessary that the mill-test be representative of the entire zone from which it is taken. The best that can be expected is that it will be representative—or rather, show the entire mineral content—of the tonnage actually milled. This fact is clearly stated by Mr. Webber. The tenor obtained for the small tonnage milled is, to all intents and purposes, exact. In order to obtain an accurate sampling factor it would be necessary for the channel-samples, cut from "successive ...... stope faces as the ore is removed for a mill-test," to be as representative of the small tonnage of ore milled as the mill-test itself. But this cannot be so. Please note I say 'representative,' not 'accurate.' Singly or in small groups, hand-samples are not supposed to be representative. It is only in large numbers that their individual variations merge in an average that is reliable. This basic fact, a clear understanding of which underlies all successful hand-sampling, is frequently overlooked. The average of all the samples taken on an orebody may thus be close enough to its actual tenor for the business in hand, and yet small groups of these samples, segregated according to zones, may not be truly representative of the parts of the orebody from which they are taken. This fact becomes more noticeable as the orebody becomes more erratic, a truth recognized by Mr. Webber when he states that "the principles of this 'combination method' will apply to best advantage in mines of fairly uniform mineral characteristics."

To refer to Mr. Webber's Fig. 2, if it is assumed that the area *A, B, C, D* indicates the whole of the orebody, then if the orebody is moderately erratic—the sampling of which difficult type we are discussing—it could hardly be expected that the samples within the area *A, E, K, O* would be representative of the tonnage indicated by this area. Still less would the samples from the little triangle represent the, compared to the whole deposit, minute tonnage milled. The sampling-factor obtained would then be partly due to the inherent sampling-factor, and partly to the fact that the comparatively few hand-samples would fluctuate widely according to purely accidental variations. The factor thus obtained might or might not hold good for the rest of the zone. The whole thing comes down to the fact that hand-sampling depends for its accuracy upon the averaging of a *large number* of (sometimes) widely divergent results. It is the safety created by this large number of samples more than any other factor, more than the tonnage represented by each sample, more than the tonnage per foot of development, on which the hand-sampler depends, and which makes his results reliable. To grade a deposit carefully on samples in sufficient numbers to be reliable, and then to alter this grade according to results obtained from a small proportion of these samples, is to introduce the weak link of the chain, to throw overboard much of the safety so arduously acquired.

One more point. To my mind, a large part of the sampling-factor so consistently obtained at various mines is due to an erroneous estimate of the waste that will be unavoidably broken with the ore. Waste from hanging, waste from foot, waste from included 'horses'—it is so easy to under-estimate these, to assume that the 'hanging' and 'foot' will stand well, that the horses can be left as pillars, or at least broken separately from the ore. Sometimes I think it is almost impossible to be sufficiently liberal in this respect. And on this element Mr. Webber's method would throw no light. In the absence of tests practically on an operating scale, the engineer must still estimate the amount of dilution according to his unaided judgment.

All of which leads me to state that I believe Mr. Webber's method will be valuable in a deposit where it is economically feasible, provided it is used by a highly experienced man purely as an aid to the broad interpretation of his sampling; but that on account of the relatively few hand-samples compared with each mill-test, and the inability of such small tests to give results which include dilution with waste, its use to determine mathematically a sampling-factor might be extremely dangerous. For many of the most knotty problems, the very problems where the sampler is most sadly in need of help, its use would be impossible owing to the utter lack of agreement between the sampling-factors obtained. I remember one deposit for which a half dozen sampling-factors, each obtained by comparing between 1500 and 2000 hand-samples with 30,000 tons milled, were hopelessly contradictory. On more uniform deposits an apparent agreement might mask an error greater than that in the hand-samples alone. Although I should be glad to have such information for any deposit I had examined, I am not sure in what manner I should use it; such a weapon is two-edged. I shiver to think of the results that might follow its hasty application to any but the most uniform deposit, and it is methods of sampling the erratic perverse deposit with which this discussion is largely dealing.

L. A. PARSONS.

New York, November 1.

[We are glad to publish this further thoughtful contribution from Mr. Parsons to the discussion, and we hope that other engineers, experienced in the sampling of large low-grade gold mines, will contribute. It seems to us that Mr. Webber's suggestion, apart from its immediately practical value, should continue to serve as a subject for useful discussion.—EDITOR.]

Prospecting Conditions in California

The Editor:

Sir—The prospector is the pathfinder and pioneer in many countries. Some cynical person will exclaim, "Yes, but he is so for his own benefit in the discovery of minerals, mainly gold." That leads to another question, what made this great State of California, if not gold and the prospector of '49 who discovered it? Therefore I contend he is the maker of trails, the pioneer of the vast areas now populated with teeming thousands, and reaping the benefits of his hardihood and privation. Gold still abounds in California, but under what conditions does the prospector now start on his trip? He sallies forth, as in the past, and devotes days and months to the search for something worth while, and if he does find a vein, or a gold channel, he is fortunate if he is not chased off by a bull-dog, or by some individual holding the ground under a patent as agricultural land, even though a jack-rabbit might be lucky to get a living on it. Under these agricultural patents thousands of acres are held in California that are purely mineral lands and fit for nothing else. Under the laws of the United States all mineral belongs to the Government. If the prospector decides to talk the matter over with the owner he is held up for a prohibitive sum to acquire something the patentee should have no right to sell if the mineral laws were so adjusted as to give the prospector the primal right of discovery. Upon the discovery of mineral land held under an agricultural patent the surface rights and such ground as needed for mining purposes should be purchasable at a fair price, or secured through a court of arbitration, appointed for that purpose. There are hundreds of prospectors in the State who know of promising mineral deposits, but are barred from locating them because the owner of the agricultural patents will not allow prospecting done on their lands. Thus Uncle Sam's mineral deposits lie idle and undeveloped, when they could be made a source of revenue to the country! In Australia they had the same evil but eradicated it through laws that made minerals the property of the State and by issuing leases only for 21 years, which are renewable at the option of the lessee, but here an owner of patented land can lock it up for a lifetime simply by paying small taxes upon it. In Australia, if one person desires to farm the surface he pays a farming license; if another desires to mine in the same ground he pays a miner's license, so the Government may receive two revenues from the same piece of land. The lessee, moreover, must keep a man nine months in the year upon the property and mine and develop it. A miner who locates a claim has to work eight hours a day upon his claim. If he runs out of provisions or money he can acquire a leave of absence from the district commissioner for one, two, or three months. He then posts his notice to that effect beside his discovery monument to prevent 'jumping' the claim. The land can be exploited and need not lie idle as in California today. A miner's license in British Columbia costs $5 per year and enables the prospector to locate not more than one claim on any creek or hill and not more than two in the same district. He may, by purchase, however, hold any number of claims; but he cannot prospect or mine without such a license. There should be a thousand men in the hills of California searching for mineral where there is one today. The production of gold is decreasing in California. The placers are gone, but the 'leads, veins, and lodes' remain mostly untouched, because they are on patented agricultural land. This is true of most of the great Mother Lode. When a person applies for an agricultural patent he has to swear that no mineral exists, to the best of his knowledge, upon said land. If the patentee discovers that mineral exists upon this land he immediately tries to dispose of what he swore did not exist. The poor prospector that stumbles upon mineral and tries to acquire a right to it after doing a lot of work upon it is stopped when it is discovered that he is on patented land. The owner then wants $20,000 down, $10,000 more in 30 days, and so forth. That is the average result, and the prospector heaves his 'chew' into a distant spot and departs in disgust. With a miner's license in his pocket, if such existed here, he would be able to locate the mineral deposit and exploit it, paying a reasonable sum per acre for any surface rights required for mining purposes. These things should be made the subject of agitation wherever and whenever the A. I. M. E. meets: they should be brought to the attention of congressmen by engineers and miners; they should be brought to the attention of senators in Washington, until a reform is made, and land that is now bringing in practically nothing to Uncle Sam would be made to yield revenue in the near future, for the benefit of the whole State, directly or indirectly, though giving employment to miners, in the distribution of dividends, and in the payment of taxes.

LEONARD G. BLAKEMORE.

Mokelumne Hill, California, October 31.

Do We Need Gold?

The Editor:

Sir—Why is Gold Mining? Is not the question a a pregnant topic?

I have read your editorial on 'Foods, Metals, and Labor,' in the issue of the 10th inst., and also the circular of the National City Bank, to which you refer. From perusal of their monthly circulars, and other sources, it is my understanding that our present reserve of gold is ample for the needs of the War. If such is the case, should we, in view of the national emergency, be allowed to continue mining gold? It is my belief that this question is in the minds of many, particularly those following the branch of the mining industry in question. Would not the airing of current views on the subject, through the columns of your valued periodical, be of sufficient worth to warrant the space?

The reasons against the present mining of gold are obvious. The industry employs labor, consumes metals, explosives, fuels or power, and foods in the production of a metal not directly employed in actual warfare. All of such consumption would, of necessity, find its way into other channels. It is equally obvious that to suddenly stop gold mining, even temporarily, would work great hardship, not to say ruin, on thousands of individuals, especially in camps like Cripple Creek, where the mining of gold is the sole reason for the existence of the community. But War does that very thing. If the national need of gold is less than the national need of what the industry consumes, then the industry should cease.

As one who, ever since becoming self-supporting, has followed gold mining as a means of livelihood, I would appreciate greatly having the views of those who, like yourself, have a much wider view than I am permitted. And I would especially call your attention to the fact that if such 'regulation' is even remotely in the minds of our government officials, then the quicker that it is brought to the attention of the industry, the better may we prepare ourselves for the blow.

LESTER S. GRANT.

Stent, California, November 16.

Missed Holes in a Wet Shaft

The Editor:

Sir—Can you tell me the cause of missed holes in a wet shaft? The primers are made with the regulation fuse and standard blasting caps. The cap and fuse are then painted with P & B paint after being crimped with what the foreman says is the best crimper made. Still we have missed holes even when two primers are used in a hole. I noticed the primers hanging up in the shaft-house with the caps down, and I thought that perhaps the paint itself got into the cap before the fuse was sent down the shaft, or perhaps that the paint penetrates the fuse. The fuses are made whenever the hoist-man finds the time. We are having missed holes so regularly that it is getting on my nerves.

J. F. HARRINGTON.

Chloride, Arizona, November 22.

[The causes of misfires are so many that it would be mere chance to hit upon the difficulty experienced in any particular case. A critical investigation of every detail will usually reveal the trouble. An article on this subject will appear in our next issue, when a reply will be made to this query.—EDITOR.]

THE BUREAU OF MINES has been working on a method for utilizing the waste sulphur gas at smelters. The SO_2 is the most abundant constituent, and this is an excellent solvent for oxides of copper and zinc. Among the problems involved in the use of this gas as a lixiviant is that of obtaining it from the smelter in a sufficiently concentrated form, and also the precipitation of the metals from the sulphurous-acid solution.

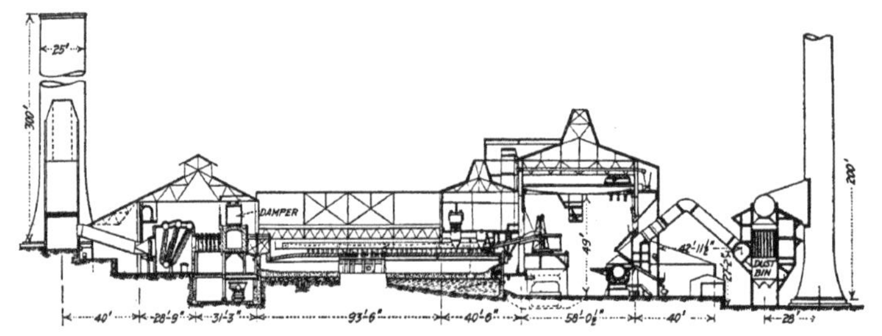

FIG. 4. CROSS-SECTION OF THE INTERNATIONAL SMELTING CO.'S PLANT

Miami: The Smelting of the Ore—VI

By T. A. RICKARD

The various concentrates made in the mills of the Inspiration and Miami companies are reduced to blister copper at the neighboring smelter, a plant belonging to the International Smelting Co. The distance from the Inspiration mill is 8000 ft. by rail and from the Miami mill 17,500 ft. This smelter is the first to be designed for the particular purpose of treating flotation concentrate. Of the total charge of ore (1200 tons) treated daily 75% is such material. The Inspiration supplies 668 tons daily; the Miami, 211 tons. Of this, 68% is flotation product and 32% table concentrate. Besides the concentrates from the two mills at Miami, the smelter also receives daily about 50 tons of flotation concentrate and 230 tons of table concentrate from the Old Dominion mill at Globe.

At the time of my visit the International smelter was treating 1500 tons per day, but a new reverberatory furnace was under construction, to increase the capacity by 600 tons more. Of the 1500 tons of charge, 20% is 'secondaries' and flux; 80% is 'ore,' that is, concentrate. The 'secondaries' include converter-slag and matte-skulls, some of which go back to the converters. A small tonnage of oxidized ore from the Live Oak workings of the Inspiration goes direct to the converters, where it serves as a silicious flux. The total of oxidized copper ore averages 116 tons per day. The limestone and the oxidized copper ore go to the receiving-bins and thence to the coarse-crushing plant, from which they are conveyed to the storage-bins.

The flotation concentrate arrives in 60-ton hopper-bottomed railroad-cars; it carries 17% water, but a little lime added at the mill is most useful in preventing the moisture from working to the top. For the haulage of this concentrate a special car has been designed. It is tight and easy to unload. A longitudinal opening or slot, 22 inches wide, extends along the bottom, on each side of which is a shelf, so that wooden slats (6 in. wide and 26 in. long) can be stretched across the opening, to close it. These slats overlap, as is shown in Fig. 1, which is borrowed, together with Fig. 2 and 3, from a paper con-

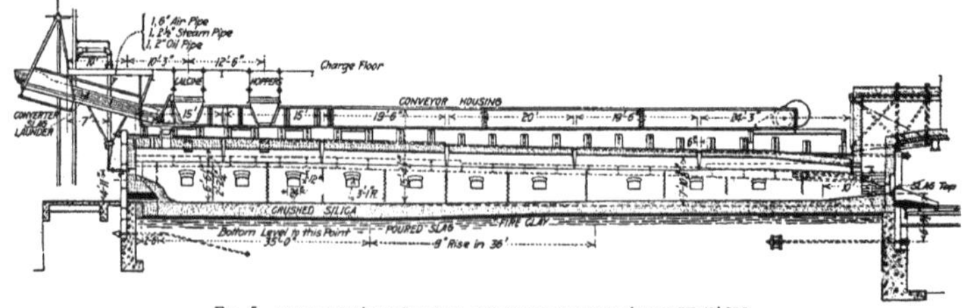

FIG. 5. LONGITUDINAL SECTION OF ONE OF THE REVERBERATORY FURNACES

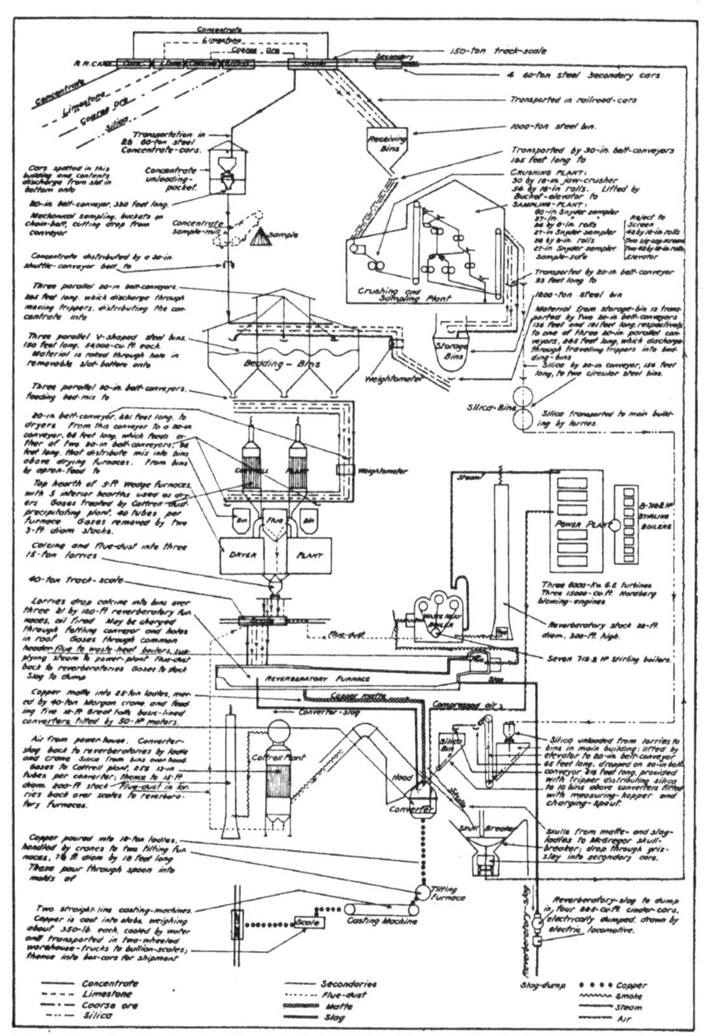

FIG. A. FLOW-SHEET EXPLAINING THE SMELTING OF FLOTATION CONCENTRATE AT THE INTERNATIONAL SMELTER, MIAMI, ARIZONA

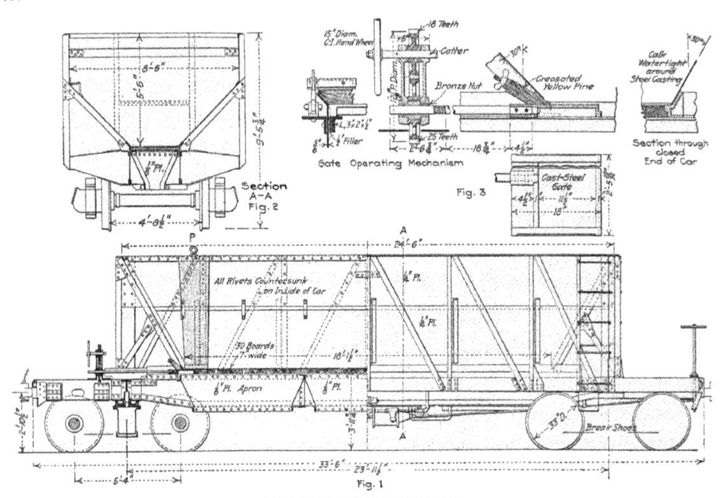

SIDE ELEVATION OF CONCENTRATE-CAR

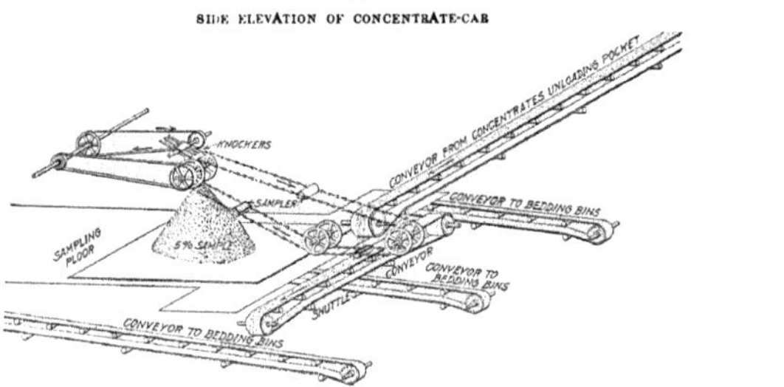

FIG. 6. ARRANGEMENT FOR SAMPLING CONCENTRATE

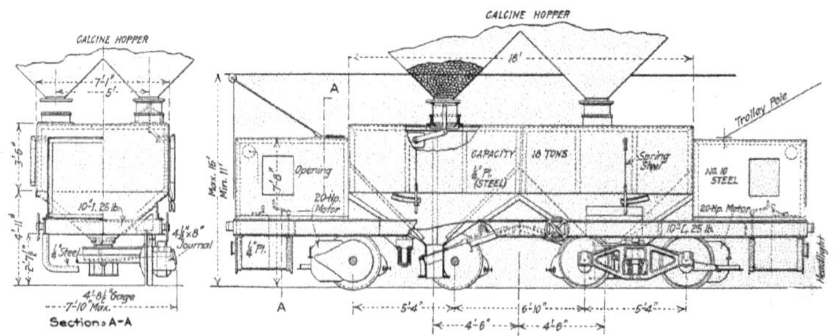

FIG. 7. END AND SIDE ELEVATIONS OF CAR FOR DRIED CONCENTRATE

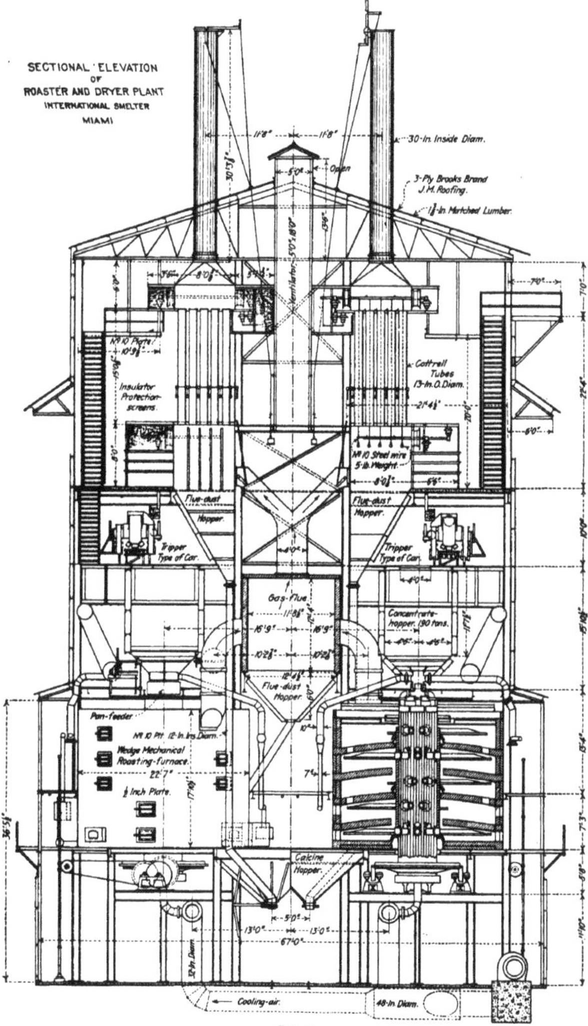
SECTIONAL ELEVATION
OF
ROASTER AND DRYER PLANT
INTERNATIONAL SMELTER
MIAMI
30-In. Inside Diam.
3-Ply Brooks Brand J.M. Roofing.
Cottrell Tubes 13-In. O. Diam.
Insulator Protection-screens.
Flue-dust Hopper.
Tripper Type of Car.
Gas-flue
Concentrate-hopper. 190 tons.
Pan-feeder
Wedge Mechanical Roasting-furnace.
½ Inch Plate.
Calcine Hopper.
Cooling-air.
48-In. Diam.

FIG. B

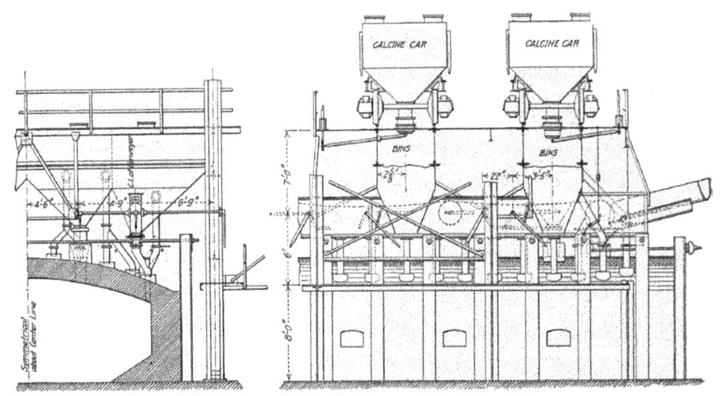

Fig. 8. SECTION AND SIDE ELEVATION SHOWING MECHANISM FOR THE REVERBERATORY FURNACES

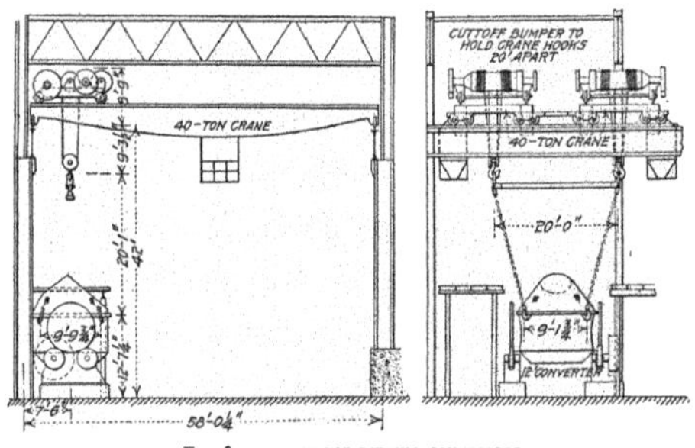

Fig. 9. METHOD FOR LIFTING CONVERTERS

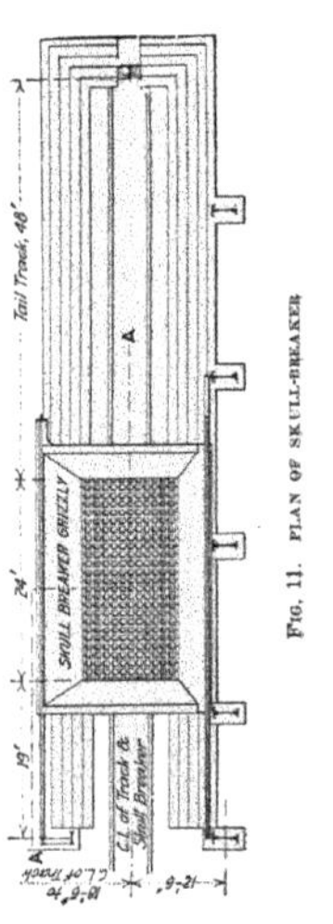

Fig. 11. PLAN OF SKULL-BREAKER

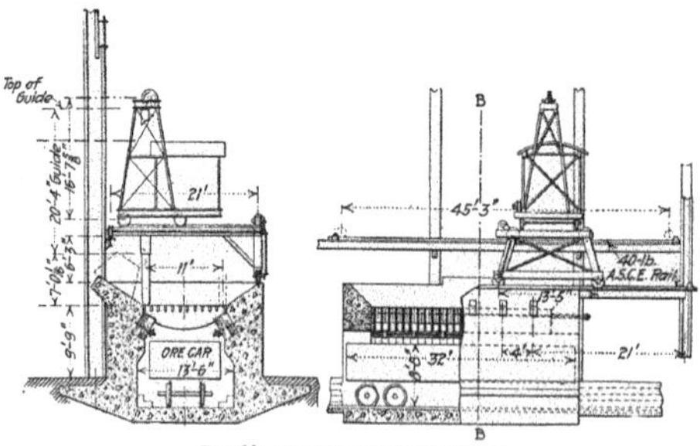

Fig. 10. SECTIONS OF SKULL-BREAKER

tributed by A. G. McGregor to the Institute.* The slats on the narrow bottom of the car are pushed tightly together by means of a steel plate that is made to slide forward by a screw operated by a hand-wheel at one end of the car. On this steel plate rests a slightly tapering steel plug, at the top of which is a lug p by which it can be lifted. Upon arrival at the smelter the plug is lifted by a chain-hoist on a trolley overhead, the plate is moved back by the screw and wheel, an opening 18 in. wide is made, and the concentrate begins to fall out of the car. The concentrate has to be poked with long chisel-pointed bars, worked by men standing on top of the car. When a little of the bottom has been cleared, the slats are shifted sideways successively, allowing the concentrate to fall upon the conveyor running underneath the track.

The concentrate is sampled for moisture by running the galvanized-iron spout of a 'coffee-pot' up the face of concentrate at seven intervals along three vertical faces in each car. This is done while the car is being unloaded. The sampling for assay is done by 1½-inch augers, making 24 bores. This applies to flotation concentrate and to slime-table products. The sand-table concentrate is sampled for moisture as above described, but the sampling for assay is performed automatically by a bucket-chain, which takes two tons out of each 50-ton carload. Mr. McGregor describes the sampling of the concentrates and remarks that the method in use works satisfactorily on table concentrate but not on the flotation product, because the stickiness of this material causes it to form large chunks, which fall on the conveying-belt and give it an uneven load. Since he wrote this the effort to sample the flotation concentrate automatically has been abandoned, but his description of the practice, as applied to the table concentrate, is quoted herewith:

"It will be noted that the conveyor from the unloading-pocket discharges onto a shuttle-conveyor (a tripper on the former conveyor could have been substituted for the shuttle-conveyor, but it would not have handled the sticky concentrate as well). Buckets attached at their ends to chain-belts are arranged to cut through streams of concentrate as it is discharged onto the shuttle-conveyor. The chains are driven and supported by sprocket-wheels. The arrangement is such that at the highest point in their travel the buckets are in an inverted position. At this point they pass under a revolving shaft which has a number of knockers, made of 6-in. belting, attached to it. These knockers slap the bottom of the buckets, jarring loose any concentrate tending to stick to them." See Fig. 6 and 13.

The traveling belt at the bottom of the unloading-pocket is 22 inches wide and delivers the concentrate to any one of three belts passing over the bedding-bins, which are 150 ft. long and have a capacity of 12,000 tons. Each of the three belts has a tripper, operated by motor and movable as desired over the bins. 'Secondaries' and limestone, both in a crushed condition, enter on another belt, which delivers its burden upon the belts running over the bins just before the concentrate begins to arrive. The slimy concentrate splashes as it is discharged upon the belts and off them, making an even feed impracticable. The bins have V-shaped bottoms closed by 28-in. slats covering a narrow longitudinal opening. Again the descent of the material has to be assisted by shoveling or hoeing; it discharges upon a 20-inch belt delivering to a 22-inch belt that runs to the top of the building in which the drying is done. The concentrate falls into steel bins discharging through apron-feeders. Here likewise the concentrate has to be pushed from the top by the use of long bars. It falls upon the top hearth of the furnace. Of these Wedge furnaces there are six, and four more were under construction at the time of my visit in May.

Each furnace has five hearths, besides the one on top. The interior diameter is 20 ft. 4 in.; the shell is 22 ft. 7 in., and the total height is 17 ft. 10½ in., so that each drying-chamber is 39 inches high. Heat is supplied by three fish-tail burners in each furnace. The lining is 12 in. thick. Each rabble makes one revolution in 55 seconds and the total period of drying is 1½ hours. The ore is turned by four arms, each carrying five 6 by 14-in. rabbles. The furnace treats 50 tons per hearth in 24 hours. It is air-cooled. After passing through the rabble-arms the air passes up the central shaft to the top, whence it is conducted to the fire-boxes connecting with the two lower hearths. Thus heated air is provided for the combustion of the oil by which the firing is done. The treatment is drying, not desulphurization; the elimination of sulphur is confined to the small particles that are suspended in the air and come in contact with the flame. The decrease of sulphur is not known but it is very slight, from an original percentage of 19.52 in the feed to the dryer. During 1916 the average sulphur was 18.72%. Owing to a change in the charge consequent upon the receipt of Old Dominion concentrate there is a tendency to increase the sulphur content and decrease the coper content, the result being to lower the grade of the matte produced in the reverberatory. An experiment was made with the use of a steel I-beam as a rabble-arm in these Wedge furnaces, instead of the ordinary air-cooled cast-iron arm, but it was not successful, owing to the bending of the arm when hot. This heating was due to the settling of flue-dust, which became incandescent and prevented the rabble from being cooled by the air. Then an A-shaped cover of concrete was built over the rabble, so as to expedite the shedding of the flue-dust. This was an improvement over unprotected arms. The use of I-beam arms, however, is being discarded.

During the first six months of 1917 the product from the dryers averaged as follows:

	%		%
Silica	17.73	Iron	17.15
Sulphur	19.73	Alumina	6.42
Copper	26.47	Lime	3.83

The Cottrell plant for arresting the dust consists of 12 units of 20 tubes each; it is placed immediately above

*'Features of the New Copper Smelting Plants in Arizona.' Trans. A. I. M. E., Sept. 1916. Fig. 4 to 15 also are borrowed from this article.

FIG. C. POURING SLAG FROM A CONVERTER

FIG. D. A GROUP OF CONVERTERS, TWO ARE POURING COPPER

FIG. E. CASTING MACHINE DISCHARGING SLABS OF COPPER

FIG. F. TRIMMING-FLOOR, SHOWING ACCUMULATED BLISTER-COPPER

the drying-furnaces, as shown in Fig. B. The lap-welded steel tubes are of 13 in. outside diameter and 15 ft. long. They are cleaned by knocking every two hours. The dust drops when thick enough, but if allowed to accumulate it may fill the tube to such a thickness as to cause disruptive discharge of electricity across the tube. The dust is extremely fine and is detached by a rap from a series of hammers suspended on a bar, so that the movement of a lever causes them to strike the pipes. A potential charge of 75,000 volts is maintained. The lavish sparking of the rectifiers suggests the intense potential of the electric energy in use. These rectifiers receive an alternating current from the transformers and in turn convert it into a pulsating direct current. The electric current is distributed within the pipes either by No. 1 jack-chain or No. 16 nichrome wire. For those unfamiliar with this useful apparatus, I may explain that the principle of the Cottrell tube is the electrification of suspended particles by the discharge of high potential electricity into the gases and dust escaping from smelting furnaces. The individual particles acquire an electric charge by which they are caused to be attracted by the opposite unlike polarity of the tube. The first step in the operation is discharge of free electricity from the wire or chain into the gases; these charges of electricity are immediately condensed upon the particles of dust, which then become repelled by the wire, which is of the same sign, and attracted by the walls of the metal tube, which are of the opposite sign.

The dried concentrate drops from the Wedge furnaces into hoppers that load 20-ton cars, which are moved by an electric trolley to the reverberatory building. There the dry concentrate is discharged into hoppers, each holding one car-load. See Fig. 7.

The transfer of the powdery concentrate from the drying-plant to the reverberatory building calls for precautions against dusting. This is accomplished by means of a car of special design. My description is borrowed from Mr. McGregor. In the top of the car are four sliding sleeves that are spaced so as to correspond with the discharge-openings of the hoppers at the drying-plant. When the car is in position for loading, the sleeves are forced against the hoppers by means of levers on the side of the car. The lever is made of flat spring-steel so that the sleeve can be pressed into firm contact with the flange at the bottom of the hopper and held in place by the lever becoming fixed by a notch in a quadrant on the side of the car. The spring handle ensures sufficient pressure between the sleeve in the top of the car and the flange on the bottom of the hopper so as to make a tight connection, regardless of the small variations in distance between them or of any lost motion in the play of the lever on the sleeve. A vent of closely woven wire is provided in the car. Fig. 8 shows the corresponding sleeves attached to the top of the hoppers at the reverberatory furnaces. When the car arrives from the drying-plant a lever operated from the charge-floor forces the sleeve against the flange on the discharge-spout of the car. The lever is engaged by a catch while the car is being emptied. A close contact between the sleeve and the spout is ensured again by a helical spring, so that lost motion, wear of the levers, and variations of distance do not cause leakage. See Fig. 5.

Each of the hoppers feeds a drag-chain conveyor that supplies a reverberatory furnace, of which there are three.† The drag-chain runs within a rectangular sheet-iron conduit (3 ft. high by 1½ ft. wide); this scrapes the dried material forward and feeds it through openings distributed along the side of the furnace at intervals of 30 inches, except at the extreme front, or skimming end. The feed is admitted where desired by withdrawing a slide in any one of a series of spouts leading through a 6-in. pipe into the furnace. Formerly the feed entered along the centre of the back of the furnace, but this involved excessive loss by dusting. The firing itself is done by five 1¼-inch round burners using Californian crude-oil, at the rate of 0.75 bbl. per ton of dry charge. A little silica, for fettling, is introduced at the front of the furnace. At the time of my visit Old Dominion flue-dust was being fed; this is dry and self-fluxing; it is so fine as to be unsuitable for transport on open belts. Each reverberatory receives about 16 tons of the dried product per charge. Besides this, the furnaces handle all flue-dust produced by the plant as well as all the hot converter-slag. Thus composed the charge of 20 tons contains

	%		%
Silica	18.0	Iron	21.7
Sulphur	16.2	Alumina	6.1
Copper	23.9	Lime	2.3

The increase of iron received from the converter gives the slag-forming ingredients, without which the charge would be too infusible to permit the separation of matte. De-sulphurization in the reverberatory is at the rate of 28%.

The slag from the reverberatory contains

	%		%
Silica	41.3	Iron	32.9
Sulphur	0.4	Alumina	14.0
Copper	0.75	Lime	5.4

The slag is skimmed into cars of 225 cu. ft. capacity that take it to the dump. Four cars of slag are usually skimmed at a time. The cars are hauled to and from the dump by a 24-ton electric locomotive and are dumped electrically by means of a motor mounted on each car.

The matte is tapped as it accumulates; it runs along a miniature ditch made of banked silicious ore (without the aid of any metal sheeting) into a steel ladle of 25 tons capacity. This is lifted by a 40-ton Morgan electric crane and poured into a converter lined with magnesite brick. The magnesite brick has deteriorated sadly since the beginning of the War, the consequence being that the brick crumbles easily, owing to lack of uniformity in the bonding substance. The matte averages

	%
Sulphur	24.2
Iron	23.2
Copper	49.7

†Another has been built since my visit.

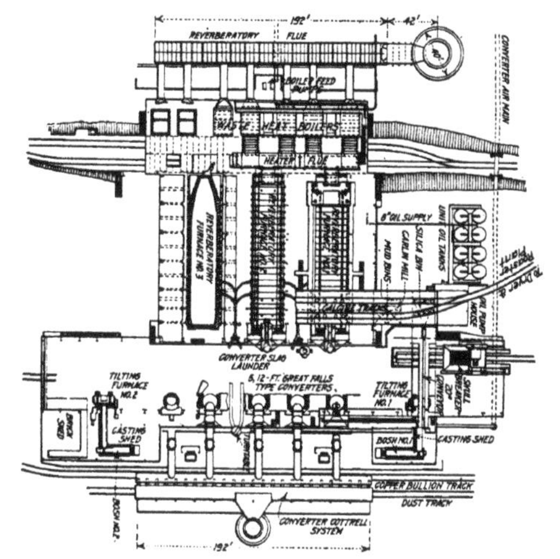

FIG. 12. PLAN OF THE INTERNATIONAL SMELTER

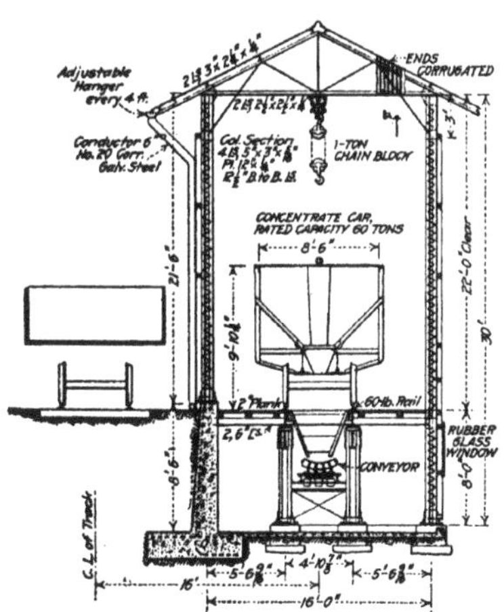

FIG. 13. SECTION THROUGH UNLOADING POCKET

Silica, in the form of silicious copper ore, is added to the converter as a flux to remove the iron in the matte. This flux is fed into a bin through a measuring-hopper. The charging is regulated from the ground-floor by means of a system of levers. Matte-skulls, scrap-copper, and the general clean-up material are fed into the converters. Of these there are six, with five stands.

The average length of blow is 3 hours 9 minutes. The air used is 112,000 cubic feet per ton of iron and sulphur in the matte. Copper is poured at the end of each blow; the slag is removed as it accumulates, say, two or three times during the blow, depending upon the proportion of old slag and other easily fusible material that may have been added.

The converter-slag contains

	%		%
Silica	20.2	Iron	44.7
Sulphur	2.0	Alumina	4.1
Copper	4.0		

This goes back to the reverberatory.

The copper from the converter is poured into a ladle and transferred to a tilting brick-lined furnace heated by an oil-burner, using 1.75 gal. per ton of copper. The pouring of the copper from this furnace, or 'kettle,' is effected by an electric motor (as is the tilting of the converters), the periphery of the furnace resting on wheels actuated by the motor. The copper is poured into a clay-lined spoon and thence into the molds of the straight-line casting-machine, which consists of a series of molds on an endless chain. Four of them are filled and emptied each minute. The molds are made (at the plant) of blister copper with a cast-iron splash-plate, 12 by 12 inches, cast into the bottom. The splash-plate prevents the hot copper from melting the cold copper and so burning a hole through the bottom of the mold. The copper carried by the casting-machine is cooled in its passage by a spray of water and is detached by being pried with a bar in the hands of an attendant. Each slab of copper weighs 250 to 260 lb. Two at a time are taken on a two-wheeled truck to the trimming-floor, where the 'fin' or fringe, is removed previous to weighing. See Fig. E, F, and H.

Thus 333 tons of blister copper, assaying 99.5% is made daily. In the first 15 days of May 5000 tons of copper was produced. In April, owing to the embargo on shipments to Galveston, a stock of copper valued at $1,250,000 was accumulated. The copper goes by train to El Paso, thence to Galveston, there to be loaded on vessels of the Morgan line, which take it to New York, on the way to the refinery at Perth Amboy.

The breaking of 'skulls,' or hard crusts of matte and slag, is done by dumping them upon a heavy cast-steel grill, the base of which supports the frame of a pile-driver. This consists of a hammer that descends between guides; it is mounted on a trolley and actuated by a motor-driven hoist, the whole mechanism being placed on a traveling bridge so that the hammer can be dropped on any part of the grill or grating, as shown in Fig. 10.

The smelter employs 350 men on average, the number of names on the payroll ranging between 348 and 515. The lowest pay is $2, to which is added the bonus based on the price of copper. In May the bonus was $1.65, and in April $2, so that the men received $3.65 and $4 in these respective months.

For most of my information and for other personal courtesies I am under pleasant obligations to Lawrence Ogilvie Howard (McGill '00), the superintendent of the plant. I give his name in full because there are three other professional men with the same initials, besides the Secretary of the American Association for the Advancement of Science. Mr. Howard, however, has achieved his own identity as a metallurgist and has added to the reputation of his alma mater, McGill University, already associated with the successful careers of E. P. Mathewson, W. A. Carlyle, W. H. Howard, and the Hamilton brothers.

Standing still for a moment at one end of the converter-aisle I watched the busy scene. On the left, through the open side of the building and on a higher level, I can see a reverberatory furnace, the serried loop-holes illuminated by the heat inside. A stream of red-hot matte is issuing from the furnace and flowing along the quartz-lined conduit on the upper floor. From the little spout at the lower end of this conduit, the matte is pouring into a ladle. When full this

FIG. G. GENERAL VIEW OF INTERNATIONAL SMELTER

FIG. H. POURING COPPER INTO THE CASTING MACHINE

so much storage over the drying-furnaces; it would have been better to work out the details so that the charge would have been landed onto the furnace-hearth before stopping it in its passage from the bedding-bins. Another thing that we learned from our experience at the International plant is that the dry unroasted concentrate has a much steeper angle of repose than ordinary 'calcine.' As a consequence the hopper-bottoms in our calcine-cars and the bottoms of the calcine-hoppers under the furnace are too flat for the best results.''*

When dried this product becomes dusty, of course. Here the Cottrell tube proves a great boon; without it the loss of valuable metal would be distressing. Mr. Howard was kind enough to give me a demonstration of the saving made by the Cottrell plant. He ordered the electric current to be turned off; whereupon a thick brown smoke issued from the stack above the dryers.

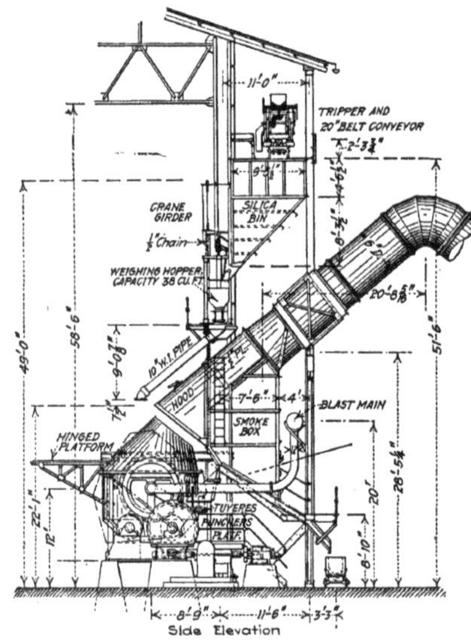

FIG. 14. CONVERTER PLANT

This smoke contains 35% copper, so one can infer the loss that would ensue under such conditions.

Once the concentrate has been melted, in the reverberatory furnace, the special risk of loss ceases. The losses are in the reverberatory slag and in the fume going up the converter-stack. A Cottrell plant has been added to control the latter. Despite the difficulties caused by the flotation concentrate, the smelting operation is highly efficient. I am not at liberty to state the exact extraction of copper, but I can say that the Miami company receives 95% of the copper in its concentrates. Transportation costs $4.50 per ton, this being the original freight to Cananea. The charge for treatment at the smelter is $3 per ton. A penalty of 10c. per unit is made for concentrate assaying less than 35% copper, and 6c. per unit is allowed for each unit of iron, besides 8c. for each unit of insoluble. The total cost of freight, smelting, converting, and refining on the copper in the Miami concentrate as shipped from the mill is 2¼ cents per pound.

*From a letter to the author.

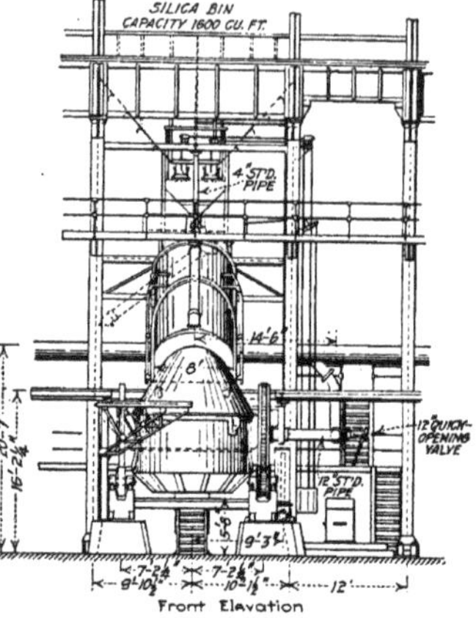

FIG. 15. CONVERTER PLANT

SPELTER has been selling in a depressed market for a long time, and the price has been falling. It is now said that the Government is about to place large orders for 75-mm. shells. A semi-official intimation of the quantity fixes it at 33,000,000. Only the highest-grade spelter will be used in the brass cases, and it is estimated that these will require 80,000,000 lb. of zinc. This will increase the demand for electrolytic zinc, and will be of special advantage to those that have been fortunate enough to establish plants of this character. At the same time comes announcement of an extension of the manufacture of zinc sheets. The difficulty in securing a sufficient quantity of iron sheets for roofing has encouraged the demand for zinc for this purpose, and has induced many manufacturers to experiment in the making of suitable substitutes from soft spelter. Recently the American Zinc Products Co. at Youngstown, Ohio, has started a plant for this purpose.

American Smelting v. Bunker Hill

ABSTRACT OF AFFIDAVIT BY E. L. NEWHOUSE

E. L. NEWHOUSE, being first duly sworn, on oath, says:

I am 52 years of age, and have all my life been engaged in the mining and smelting business, principally the latter. I was educated technically at the School of Mines, Columbia University, graduating in the year 1886. My professional career has been substantially as follows:

In the year 1886, after graduating, I served as assayer and chemist at the Argentine plant of the Consolidated Kansas City Smelting & Refining Co. Later, for the same company, I was mining engineer, with headquarters at El Paso, Texas. Later I was in charge of the mines and ore-buying business of the same company in the Sierra Mojada mining districts in Mexico. In 1888 I built a smelter and operated the same for the Villa Dama Co. at Nuevo León, Mexico. In 1889 I served as superintendent of the Philadelphia smelter at Pueblo, Colorado. Later I served as buyer and manager of said plant. I came to New York in 1903 as director of the American Smelting & Refining Co. In 1907 I was elected vice-president of the American Smelting & Refining Co. and was placed in charge of their smelting operations. In 1915 I was elected senior vice-president of the plaintiff company.

Since my graduation in 1886 I have been continuously engaged in the mining and smelting business as above set forth, and since 1888 exclusively in the smelting business. For ten years last past I have been a vice-president of the American Smelting & Refining Co. in charge of smelting operations, and have particularly had under my jurisdiction supervision over all contracts made for the acquisition of ores used in our smelting business.

Metals are seldom found in nature (except placer gold) in a pure state. They are nearly always in combination with various other metals. The separation of these metals necessitates the processes known as smelting and refining.

The different combinations of the metal in ores from different mines necessitates different combinations in smelting. Even in the same mine the ore changes. The proportions of silver, of lead, of zinc, of iron, of silica, may vary widely, and any of these changes in the ore are important factors in smelting.

The freight-rate to the smelter and to the final market is an important factor; hence the mine seeks to enrich its product by eliminating the waste and saving in freight. This led to the development of the process of concentration under which, by grinding the ore and agitating it in water and sometimes other chemicals such as oil and acid, the waste matter is eliminated and a product containing a larger percentage of the valuable metals is obtained. These products are known as 'concentrates' and 'slimes.'

This concentrating process commercially is an important factor both to the mine and the smelter. There is always a loss of metals in concentrating, hence the mine must study the amount of the metal-losses in concentrating as compared with the additional cost of shipping a larger amount of product with an increased freight and smelting charge. The smelter, on the other hand, as a rule, prefers a larger tonnage, but on the whole finds a product, such as defendant's, most advantageous when containing around 50% of lead.

When I was a young man there were few large smelters in the country. It was customary for every important mining district to have a smelter. Such a smelter was usually small in capacity and did what is now considered very expensive and inefficient metallurgical work. Their losses of metals and expense of operations were simply huge. The first thick cream of the richer deposits was being skimmed off throughout the country as district after district was discovered. As soon as the ore in a district justified, a smelter would be established, the expectation being that with the exhaustion of the ores of the mine the smelter would be abandoned and its total cost would be charged up to profit and loss. While the general principles of smelting have not been materially changed, wonderful advances have been made in methods of operation, principally along the line of larger plants, more perfect apparatus, and especially a better combination of ores permitting of better metallurgical results, which means saving more of the metal contents of the ore. Where twenty years ago there were few smelting plants that had cost in excess of three to four hundred thousand dollars, there are now few lead-smelting plants in which there has not been invested at least a million dollars and in many of them a million and a half. The importance of this change in the smelting business is not generally understood outside of the trade, and it has rested upon the following economic reasons:

Smelting is essentially a chemical process carried out on a commercial scale. There must be a combination of the charge of lime, iron, and silica. This is necessary in order to produce a chemical compound by the application of heat, in which compound the true separation of the precious minerals may take place. Gold and silver ores cannot be smelted profitably alone. The amount of gold and silver, especially of gold, is so small that it would be lost in the process. There must, therefore, be a 'collector,' that is, some metal of relatively large volume in the charge which has a chemical affinity for gold and silver. Lead is such a collector. In the smelting furnace the lead gathers to itself the gold and silver, and through its greater specific weight sinks to the bottom of the furnace. There it is drawn off. This product is known as lead bullion, being a combination of lead, gold, and silver, with some impurities, such, for example, as antimony and arsenic.

Any one mine, or the mines of any one district, rarely, if ever, have the necessary proportions of lime, iron, and silica to enable it to be smelted alone. It follows, therefore, that the smelter must secure the deficient substance from some other course. The main basis of the business of every smelter will be certain ores—the ore of the district or the ore which naturally goes to that smelter because of the lower freight-rate, and the largest tonnage it has under contract. Assume, for example, that this ore is deficient in iron, that is, it requires the addition of iron in some form to make the proper proportion of lime, iron, and silica in the slag. Now, if this iron can be obtained in a combination with precious metals, it is much more economical and commercially better for the smelter than if iron alone (such as scrap-iron) is added to the charge. Hence a smelter which is short of iron will look around for an ore that carries an excess of iron in combination with precious metals. It will treat such an ore for a less amount than it would an ore containing no iron, because everything it saves out of treating is that much saved as against using barren iron ore or scrap-iron. Therefore, every smelter has its own problems, which vary from time to time with the characteristics of the larger part of its ore. If the ores naturally tributary to it have not enough lead to form a collector, but furnish a large quantity of gold and silver ores, then that smelter is short of lead and can pay more for a lead ore to afford the necessary collector. If short of a silicious ore, it will pay more (that is, it will smelt for a lesser charge) an ore containing an excess of silica, and so on for lime and iron. For this reason, it is difficult to compare the different rates of different smelters for the same ore, since the necessity of each smelter for some substances (lead, iron, silica, or lime) may compel it to be seeking that particular ore as a necessary component of its smelting charge and not alone as a source of profit. If a smelter does not make the right metallurgical combination of lime, iron, and silica, or has insufficient lead, its metal-losses in the process will be very great.

In order for a smelter to plan its combinations, it is necessary for it to have its ore-supply known and contracted for. Hence, with the increased size of smelters' the custom grew up of making long-time contracts. Having a long-time contract, the smelter was better able to supply its capacity, could better arrange for its metallurgical needs, and for other supplies such as coke, fuel, etc. When I assumed charge of the operations of the plaintiff company, I worked persistently along this line of securing long-time contracts. I found that most mines also found it to their advantage for their own reasons of gauging their production, etc., to make a long-time contract, particularly with a responsible smelter which would take its entire production. Prior to the organization of the plaintiff company, there were no smelters in the country that were capable of handling an output from one mine in the Coeur d'Alene as high as five to seven thousand tons a month, which is the average output of the defendant, and as a result the Coeur d'Alene mines had several contracts with a number of smelters. These contracts being of short duration, the mine never knew when it would be unable to dispose of its ore and hence might have to curtail its output. Having a long-time contract for an unlimited amount, a mine is always inclined to increase production during times of high prices and to curtail production during periods of low prices, much to their financial advantage.

The principle of long-time contracts has worked out to the advantage of both mine and smelters, and it has revolutionized the smelting business, since now, instead of there being a great number of small smelters doing poor metallurgical work in the vicinity of each mining district, the smelting business has been consolidated at advantageous points selected largely with relation to freight-rates and ore-supply. These are large and expensive plants and the plaintiff has believed itself justified in constructing such plants because its business has been based upon long-time contracts and it thus felt sure of an ore-supply.

The reciprocal advantages to a mine and smelter of the certainty of selling its output on the part of a mine and having a definite supply of ore on the part of the smelter had been felt in the trade and in the Coeur d'Alene district had led to a tendency for either the smelter to become interested in the mine or the mine in the smelter. The Bunker Hill company in order to assure itself a definite output had bought an interest in the Tacoma smelter about the year 1899. This smelter was relatively a small smelter and capable of taking care of or handling about one-third of the production of the Bunker Hill company, but it served its purpose of enabling the company to be sure of a sale of at least that amount of its production. The balance of its production prior to the contract at issue had been sold by the Bunker Hill company to various smelters. Thus, in 1905, when the contract at issue was made, the Bunker Hill company had five smelting contracts with five different smelting companies outstanding: the Tacoma Smelting Co.; the Selby Smelting & Lead Co., at San Francisco; the Ohio & Colorado Smelting Co., at Salida, Colorado; the Pennsylvania Smelting Co., at Carnegie, Pennsylvania, and this plaintiff.

The contracts with the Tacoma Smelting Co. and this plaintiff were at the same price. The contract with the Selby Smelting Co. was $2 per ton less advantageous to the mine.

The negotiations for the Bunker Hill contract are truthfully set forth in the complaint herein, which I have read and which I believe to be true in its entirety.

I inspected the Tacoma smelter personally; am familiar with smelters, their value and their efficiency. A smelter equally efficient, equally well located. and equally pable of treating the same tonnage as the Tacoma elter could have been erected at the time for about the n of $400,000 (I do not, of course, include those im-ovements in the Tacoma smelter for which we particu-ly paid). As an outside figure, we calculated that 00,000 was the full reduplication value of the Tacoma int, and in paying $4,000,000 for the stock of the coma company, the plaintiff figured that we were pay-; $3,500,000 more than we could reduplicate that elter for. The Bunker Hill people, of course, realized s and the negotiation for the Tacoma smelter was con-cted on the basis of buying a going concern that had established business, which was the main asset trans-rred and acquired. The transaction was analogous in inciple to buying out a well-established newspaper or a ll-established and profitable store. In either case the ice paid is generally far more than the reduplication lue of the actual tangible assets. It is the clientele, the od-will, the established position of the business which res it, as did the Tacoma smelter in the preceding year 1904, when it had a similar contract with the Bunker ll and the three Alaska companies, as a basis of its siness, had made certain profits which when capitalized stified the price; but, of course, the price would not be stified unless we were reasonably sure of the continua-n of the business, and to insure this continuation we pulated as a part of the Tacoma transaction that con-cts Exhibits B, C, D, and E in the complaint should executed and become a part of the assets of the Tacoma elter. Having the business thus assured, we felt justi-d in paying the price and this situation was thoroughly derstood by the defendant at the time. Hence the pulation that the contract, Exhibit B, should be made, thorized by the stockholders and deposited in escrow to delivered if the deal went through and to be returned the defendant company if the deal did not go ough. Bernard M. Baruch was acting in that transac-n as agent of the American Smelting & Refining Com-ny.

f these four contracts, thus executed, the Bunker Hill tract (Exhibit B) was by far the more important for following reasons:

1). Because the annual tonnage produced by the nker Hill was roughly nearly three times the tonnage m the three Alaska companies.

2). The Bunker Hill contract was for 25 years, while three Alaska contracts were only for 5 years and reafter could only be held if this company met the of treating the Alaska ores by the cyanide process. was well understood by all concerned that the prob-ity was that we could not meet cyanide competition o ores of the character of those Alaska ores, and this, act, proved to be the case. At the end of five years cost to the Alaska companies of treating their product he cyanide process was so much less than we could d to treat it that we were forced to give up handling r business. The total tonnage shipped in that five s by the three Alaska companies was about 138,000 , while the total tonnage shipped by the Bunker Hill company for the same period was about 400,000 tons. We all of us understood at the time that the probability was that the three Alaska contracts would only last for the five-year period and this plaintiff so figured it in estimating the value of the respective contracts as a part of the assets of the Tacoma smelter.

The three Alaska companies and the Bunker Hill company are all controlled by the same group of individual stockholders. For many years they have been under substantially the same management. For many years F. W. Bradley, president of the defendant Bunker Hill company, has also been president of the three Alaska companies. For the purposes of the trade the four companies and the Tacoma smelter were all considered as representing one interest, although there were, of course, divergences of actual stock ownership in each of the companies.

The essence and the spirit of the agreement between this plaintiff and the defendant in purchasing the Tacoma smelter was that for a 25-year period the Bunker Hill company should continue to treat the output of its mine substantially as it had been treating it and would ship that entire output when so treated to this plaintiff which had the privilege of routing it to such smelter as it saw fit. From the beginning of that contract in 1905 down to the breach complained of in the complaint, the defendant substantially did this.

When the amendment to the contract (Exhibit B-4) expired in the year 1915 a phenomenal condition existed. The present European war had broken out. During the first few months of the war the lead industry was partially paralyzed, but it soon became apparent that the war necessities (coupled with the condition of anarchy then existing in Mexico and labor difficulties in Spain which shut off the supplies of lead from those important lead-producing countries) would lead to higher prices for lead, zinc, and copper.

As, under the contract (Exhibit B) this plaintiff shares equally with the defendant in any increase in the price above $4.10 per hundred, it became apparent that the amount received by it on account of smelting defendant's ore would be correspondingly increased, and to deprive the plaintiff of this benefit stipulated for in the contract, the defendant demanded a radical change in the contract, depriving the plaintiff of such benefit, and failing to secure such change, set out to build a smelter at its mine in Kellogg, Idaho. With the completion of that smelter, defendant began to divert ores, concentrates, and slimes that should have been shipped to us under the contract and treat them in that smelter. Instead of as heretofore continuing to make its mine-product into normal grade and shipping to us, and instead of treating the product from its mine as it had treated it during the preceding 25 years of the history of the mine, the defendant began to treat the ore in a different method and manner and to ship the same to its own smelter.

Defendant has intimated directly and indirectly that it intended hereafter to treat all the product from its mine in its own smelter. It has threatened and is carrying out the threat of diverting its mine-product from this plaintiff to its own smelter.

This conduct on the part of the defendant inflicts tremendous and irreparable losses upon the plaintiff in the following respects:

(*a*) The amount of profit received by plaintiff in the purchase and smelting of ore under said contract with the defendant depends in part upon the current prices of lead and silver which fluctuate from day to day, and the amount of damages that plaintiff would sustain by the failure of defendant to perform its said contract would, if the same be calculable at all, vary practically with each shipment, because of the fluctuations in the market prices, and hence would necessitate a vast multitude of suits, possibly a new suit for each day's shipment since the profit on each day's shipment, if ascertainable at all, would vary with the price of lead and silver on each day.

(*b*) A smelter can only approximately estimate the damage that is sustained by reason of the failure of a particular mine to ship ore, because of the fact that the ores of each mine are commingled with ores from other mines in order to get the necessary flux, the cost of smelting being distributed to the entire charge, composed of ore from several mines, making it always difficult and at times impossible to ascertain what would have been the cost of smelting if any particular ore had been omitted and some other ore from another mine or other flux substituted therefor.

(*c*) The ore of the defendant, because of its high lead content, is desirable for the smelting of what are known as 'dry' or 'silicious' ores. Such leady ore is distinctly limited in quantity. Few new lead mines have been discovered in the United States in recent years and old lead mines have their output already contracted for. If plaintiff does not receive from the defendant the mine-product that it is entitled to receive under its said contract, and that the defendant has been shipping to plaintiff pursuant thereto, plaintiff will be unable to smelt a large part of the dry and silicious ores that it has under contract to purchase and smelt, and the plaintiff will be unable to buy additional silicious and dry ores. Its smelting capacity will be reduced and it will be impossible for it to carry on its business to the extent that it has heretofore arranged, and for which its several plants are adapted to carry on said smelting business, all of which have been arranged and based in reliance on the output to be furnished by the defendant company, and they will be unable to use the large amounts of limestone, fuel, and supplies purchased to smelt said tonnages. If plaintiff be not furnished with defendant's ores, plaintiff will be required to discontinue the smelting of much dry and silicious ore that it now has under contract and be liable for damages therefor. All of such expenses, losses, and damages, it will be impossible to estimate.

(*d*) Though the life of every mine is uncertain, affiant is reasonably well-informed as to the facts, and upon said facts is of opinion that the mineral-contents of the defendant's said mine are such that at the rate of output prevailing since the making of said contract defendant's mine will continue to furnish the same output for the remainder of the term of said contract, to wit, until February 1930. Plaintiff relying upon said contract and upon its promised performance, has spent large sums of money in enlarging its smelters and equipping them with the most modern devices, so as to be able at all times to handle the product of the defendant's mine pursuant to said contract. If plaintiff does not get defendant's said product, not only will there be an impairment of plaintiff's ability to purchase and smelt other ores which it could and would have purchased were it receiving defendant's product, but the large plants and equipment constructed by plaintiff for the purpose of handling the product of defendant's mine and in which plaintiff has invested several millions of dollars, will become greatly impaired in value, and the amount of the damage resulting therefrom it will be impossible to determine.

(*e*) Plaintiff, relying upon said contract and the continued performance thereof by defendant, has also greatly increased its refining capacity in order to refine the lead bullion derived from the defendant's mine, and plaintiff's enlarged refining capacity will become idle and greatly impaired in value, and the amount of the damage resulting therefrom it will be impossible to determine.

(*f*) Plaintiff has built up a large and valuable business in the handling and selling of lead relying upon its said contract with the defendant. This business plaintiff will not be able to supply adequately if it is deprived of the product of defendant's mine. The damage resulting to plaintiff's said business from the breach of said contract in being deprived of the supply of lead promised thereby will be very large in amount and will be impossible to determine.

(*g*) If plaintiff be required to bring suit at law for the diversion or failure to ship each shipment or lot of ore that may be marketed or otherwise treated or disposed of by defendant elsewhere than with plaintiff, or even if such actions be begun periodically, a great and intolerable multiplicity of actions will result. The defendant in the normal conduct of its business, has been accustomed to ship to this plaintiff each day from 150 to 250 tons of ore, slime, and concentrate in from five to ten railroad-cars. The price of lead and silver often fluctuates from day to day, silver especially having small fractional changes per ounce, which, however, involve considerable sums of money when applied to the quantity shipped by the defendant. The amount of lead and silver in each car may also vary, being affected by many factors, such as the changing contents of lead and silver in the original ore produced from the mine in various parts thereof, the necessarily varying efficiency of the concentrating process, and other factors. Thus the amount of plaintiff's damage might change and vary with each car on each day since plaintiff's profit depends in part not only upon the changing price, but also on the amount of lead and silver it may recover in excess of 90% of lead and 95% of silver. If plaintiff be required to wait until the expiration of said contract in 1930, in addition to the losses and damages and dissipations of assets of the defendant hereinbefore mentioned, plaintiff will be unable by reason of lapse of time, the death and disappearance of witnesses, and the loss and destruction of evidence, to prove its several causes of action with anything approaching certainty or likely to accomplish the ends of justice.

(*h*) Said contract like most contracts for similar ores calls for payment for 90% of the lead and 95% of the silver. The remaining 10% of lead and 5% of silver is not paid for, in accordance with a custom based upon the unavoidable loss of these metals in the processes of smelting and refining. A smelter sometimes loses more than 10% of the lead and more than 5% of the silver. On the other hand, if it can make its ore-contracts and combinations successfully, and have the benefit of a high degree of skill and attention on the part of its metallurgists, it may save more than 90% of the lead and more than 95% of the silver and hence increase its profits. It is impossible to prove accurately what percentages of lead and silver in any particular ore is saved, since only accurate knowledge of the percentage saved from the ore of the various mines commingled to constitute the charge in the furnace can be had. If the smelter can rely upon ores being furnished under contracts, it can make its shipments to each smelter conform more to the requirements of each class of ore, and hence it is more likely to save more than 90% of the lead and 95% of the silver of the total ores treated, but it cannot say with accuracy how much of this saving (which means increased profit) is properly assignable to each particular ore treated. Thus plaintiff has made a profit out of this contract by reason of its ability to save more than 90% of the lead and 95% of the silver, but it could not prove in a court in an action for damages exactly what per cent of lead or silver it has been or might hereafter have been able to save, and hence could not show with accuracy its profit in this connection. The quantities of lead and silver saved also fluctuate widely from time to time and are dependant upon many uncontrollable contingencies, such as judgment and skill of the metallurgist, unexpected changes in quality of ore, coke, limestone, etc., unavoidable delays in the shipment of supplies, changes in weather conditions, particularly at East Helena, and many similar factors, which cause the profit made by plaintiff in handling the defendant's ores, to vary widely from month to month and from time to time, entirely dependent of the rise and fall of silver and lead values.

(*i*) There is a definite limit to the amount of lead ores, concentrates, and slimes to be had upon the market, while from time to time and at various times there is a large demand for such products for the purpose of using the same in smelting other ores, and also directly for the value of the mineral lead, and if plaintiff should be deprived of the said products promised by defendant in its said contract, it will be unable to procure the same elsewhere, and the amount of damages and losses arising from such inability it will be, if not impossible, to estimate, but such damages and losses will be great and in fact irreparable and necessitate the closing of several furnaces.

(*j*) Unless a temporary injunction is granted preventing the defendant from diverting its product elsewhere and compelling it to ship the same to this plaintiff, the business of the plaintiff will be irreparably injured for each and all of the reasons hereinbefore stated, which apply to a final injunction. I wish particularly to emphasize the disturbance of the business of the plaintiff which will occur if these ores are lost. It means not alone the loss of this particular ore, but it means the loss in smelting of a large quantity of other ores which the plaintiff can only handle if it has the output of the defendant with which to smelt them. This will be particularly true at our East Helena and Colorado plants. At East Helena we need the defendant's output, which is free from zinc, to mix with the zinky outputs from other mines. If we do not have it, our operations at East Helena will be seriously curtailed. In Colorado we will have to abandon the smelting of considerable tonnages of dry ores, especially those from the Aspen and Creede district. Many of these tonnages we have under contract.

If a final injunction be granted in this case, a great and irreparable injury would be inflicted on this plaintiff if during the interval between the present and the times of a final injunction the plaintiff is not able to smelt these additional ores which require the Bunker Hill ore for smelting purposes. It would have to let that business go. The mines producing this ore would make other arrangements, and if under contract would undoubtedly sue us for damages, and when the final injunction is granted and we could again resume that business, we would find that it was lost and gone to our competitors. I am informed by counsel that it will probably be three to six months before this case can be tried upon its merits. Now, to lose the defendant's ores for from three to six months will entirely upset all our arrangements for that length of time. These arrangements, as I have shown, have been based upon the supposition that we were going to receive the defendant's output.

We entered into the Bunker Hill contract in good faith. We paid a large consideration, as I have heretofore shown. We have arranged our business on the supposition that the contract would be carried out in good faith. We have increased our plants. We have entered into contracts for dry ores. We have entered into contracts for supplies. We have incurred large liabilities and spent large sums of money on the assumption that the defendant would carry out its contract for the life thereof. We cannot now change this situation. If defendant does not ship to us, our business will suffer. We will be made liable for damages on account of failure to live up to our contracts, and we will lose a large amount of business. How much we cannot tell.

The defendant is not in this position. It has not built up a business. It has just started. It knew when it started to build its smelter that we would bring this law suit. We notified the defendant to that effect, both orally and in writing.

Wasting Ore-Values

Litigation recently begun between American Smelting & Refining Co. and Bunker Hill & Sullivan Mining & Concentrating Co. in the Coeur d'Alene district of Idaho brings to light losses of millions in metal-values through false treatment of Bunker Hill & Sullivan ores over a period of 34 years. It is alleged that when Bunker Hill ores were milled and concentrated by water a colossal blunder was made. These ores contain nearly 8 oz. of silver per ton with 14% lead and 25% iron carbonate. Figuring lead at 13% and silver contents at 7 oz., each ton of ore contains 260 lb. of lead and 7 oz. of silver. Fully 35% of lead and silver and nearly all the iron carbonate were lost in water concentration, whereas were these ores treated by direct smelting as they came from the mine, only 7% of the lead and 4% of the silver would have been lost in slag. On an output of 1000 tons of ore per day, 242,000 lb. lead and 6720 oz. silver should have been recovered by direct smelting. Taking five cents as the average price of lead and 65c. the average for silver during past 34 years that Bunker Hill has been operating, and figuring output of mine at 6,000,000 tons, property should have produced 720,000 tons lead and 40,000,000 oz. silver since it began operations, or a gross output of $98,000,000. Estimating cost of production at $6 per ton of ore, cost would be $36,000,000. This leaves $62,000,000 for dividends, whereas actual dividends disbursed have been but 25% of this sum.

But the actual results were far different. Instead of the 720,000 tons of lead, which should have been recovered by direct smelting, Bunker Hill got paid for only 450,000 tons. About 230,000 tons was lost through water concentration and the smelting company deducted 50,000 tons for losses in slag. It is estimated that through failure of the Bunker Hill management to build up an up-to-date smelting and refining plant at the start of mining operations 34 years ago, imperfect recoveries through water concentration have cost the stockholders between $30,000,000 and $40,000,000. Experts have figured that the greater part of 280,000 tons of lead and 15,000,000 oz. of silver have been lost in the Coeur d'Alene river and a company has now been formed to recover this metal.—'Boston News Bureau.'

The above statement is interesting, and true in part. The Bunker Hill ore never averaged as high as 8 oz. per ton and 14% lead; the average has been about 4½ oz. silver and 11% lead. The total of dividends has not been $15,500,000, but $20,370,000, besides $6,500,000 of operating profit that is represented by cash, securities, plant, and property. In addition the mine has an eight years supply of ore developed and ready for exploitation. It is true, however, that the Bunker Hill company would have been better off if several years ago it had built a smelter to replace the water-concentration plants. On the other hand, it is not correct to suggest that the Bunker Hill could have had its own smelter from the start; not only was this rendered impracticable by the lack of railroad facilities, but it would have been inadvisable until the progress of the metallurgical art rendered it profitable to erect such an up-to-date plant near the mine as could compete with other established lead smelters in the West.

Deterioration of Coal in Storage

An interesting bulletin, No. 136, has been issued by the Bureau of Mines, on 'The Deterioration of the Heating Value of Coal During Storage,' by Horace C. Porter and F. K. Ovitz. It is elaborately illustrated and accompanied with a large amount of tabulated data, from which we believe further conclusions might be drawn with the aid of graphic analysis. The experiments in question have been under way for several years. In fact, they are an outgrowth of the interesting work begun in 1909, under the supervision of the late J. A. Holmes, at that time chief technologist of the U. S. Geological Survey. The object has been to determine the deterioration in coal as a result of storage, a subject of great interest to all power-users who are not so fortunate as to have hydro-electric power at command nor to be able to burn fuel oil. There is given what appears to be a rather remarkable statement regarding which additional explanation would be interesting, that is, that the experiments were made upon small lots "in order to make the tests of maximum severity." Open-air storage was made for the most part in 4-ton to 12-ton lots; though in one case 100 tons of run-of-mine coal were thus stored. Tests were chiefly made with smaller lots in barrels or boxes submerged in fresh water or in salt water. A general summary is presented which states that the tests show that the amount of deterioration in the heating value of coal during storage has commonly been over-estimated. "Except for the sub-bituminous Wyoming coal no loss was observed in outdoor weathering greater than 1.2% in the first year or 2.1% in two years. The Wyoming coal suffered somewhat more loss, 2.3% in the first year, and as much as 5% in three years." This is so greatly at variance with the general experience of engineers that it would seem to be desirable that further experiments be made on a larger scale. It has long been recognized that the deterioration is not excessive when the mass is sufficiently small to prevent increase of temperature from the oxidation of the coal and of the contained sulphur compounds. In consequence it has been the custom of engineers to resort to different methods of ventilation of coal piles in order to check such deterioration. The deductions would seem not to be convincing in the light of the data presented, unless supplemented by confirmation from tests involving much greater quantities.

THE rise of silver has been so unfavorable to American employees in China whose salaries are paid on a gold basis that a movement has been started for an adjustment that will offer relief. For some time salaries have been worth only half their former buying value.

PETROGRAD, RUSSIA

GENERAL CONDITIONS.—KYSHTIM, SISSERT, TANALYK, RIDDER, SPASSKY.

Since the revolution industrial conditions in Russia have become almost impossible. The transportation department, particularly the railroads, are so badly disorganized that they cannot even carry the grain from the districts in which it is grown to the centres of population in which it is consumed. This is due not so much to lack of cars and motive power as to the low conductivity of the workingmen. After the revolution everything was put on a 8-hour basis and pay on a 12-hour basis. Two months later wages were increased an average of probably 100%, making the cost of an 8-hour day twice as great as a 12-hour day before the revolution. At first the workmen took the stand that the productivity would not be decreased, but practice has shown clearly that the output per man has fallen from 20 to 50%. Necessarily, the unit-cost of all products has gone up. Before the revolution nearly everything was produced on the piece-work system; now, however, the workmen will not allow this kind of work any longer and they insist wholly on day's pay. From statistics now being published it seems that there are roughly 800,000 idle freight-cars in Russia; most of these are idle wholly because they need repair. It cannot be said that there is a shortage of workmen. Owing to the inability of the railroads to transport the necessary food, fuel, and other raw materials, many of the works in Petrograd and other large centres are either completely closing down or cutting down their production, with a consequent dismissal of staff and men. As a result of chaotic conditions, the cost price of all metals and all manufactured material has increased from five to ten times over that which existed before the War. Up to now the authorities have taken no active steps to curtail this disorganization and I personally think that they will not be able to do anything in that way until they have established a firm government. On the other hand, I fear that we can expect no real help from Russia, either at the front or the rear, until the people themselves realize that they must do a full 8-hours work for their former 12-hours pay. Owing to these conditions the production of iron and copper has materially decreased and I doubt if the total copper production for 1917 will be much more than 60% of that for 1913. Kyshtim is seriously affected and will face a crisis within the next two months. The management has refused to meet the workmen and the workmen claim that they will make no concessions. The Bogoslovsk works have got into financial difficulties. In round figures, they have already this year lost 7,000,000 rubles. On the first of January they had 11,000,000 rubles in bank and on the first of July they had a debt of 4,000,000 rubles; in other words, they had spent, in addition to their ordinary receipts from sales, about 15,000,000 rubles. They have notified the Government that they cannot go on and actual steps have already been started to close their works. I may add that these are the largest works in the Urals, and their idleness would mean a great curtailment of production.——I understand that Bogoslovsk in the last two years has developed a large body of low-grade copper-bearing pyrite, and active steps are being taken to equip a pyritic smelting plant, to which end N. I. Truschkoff is being sent to America.——The Sissert company has completed the prospecting of the Degtiarsky

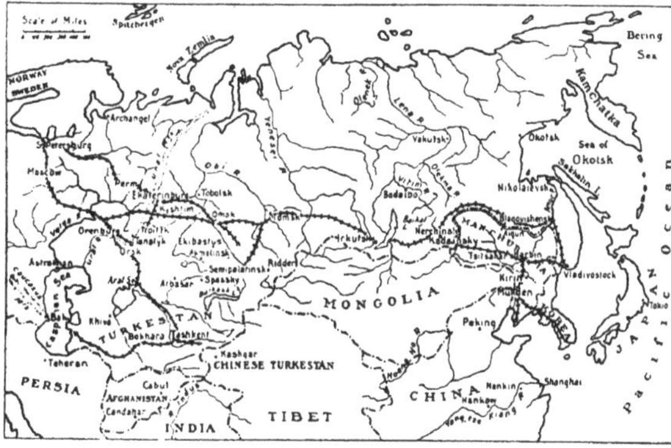

THE RUSSIAN DOMINIONS

mine and is now making the detailed plans for a suitable smelter. However, here again, owing to the conditions, all thought of the erection must be postponed.——The Verk-Isetz property, which adjoins the Sissert on the north, seems to be about the only one of the Ural companies that is in a strong position financially. This is mainly due to the good work done during the last few years by F. W. Draper. It was under his direction that this company's smelter was designed and built, and later the mine itself was developed, equipped, and brought up to its present production.——The Tanalyk seems to just about keep even. As soon as one of its shallow deposits gives out a new one is found, but, on the whole, this enterprise does not seem to have a brilliant future.

In the Altai the Ridder has been seriously handicapped by the flooding of the main workings. I understand that the railroad connecting this mine with the Irtysh river is now completed and that the production of concentrate is being delivered to the barges on the river.——The zinc-plant at Ekibastus has been in operation for nearly a year, while the lead-plant was just about to be blown in when acute labor troubles developed.——The Altai concessions of the Russian Mining Corporation have been taken over by a Russian company called The Altai Mines. Although there is a large body of rich ore in the Zyrianovsk mine, no actual steps toward putting up a considerable reduction plant are being taken, owing to the political conditions. Drilling is being done actively and with satisfactory results.——During the last year an entirely new mine, the Byelousovsk, has been developed by drilling, and something over 1,000,000 tons of copper-zinc ore is now considered as proved. In a prospectus recently issued locally by this company, it was shown that 1,400,000 tons of ore had been proved with a net recoverable value of 28,000,000 rubles.——The Atbasar apparently is doing nothing at all; first it had difficulty with labor and later, owing to certain Government regulations regarding the mobilization of the Kirghese, these, to escape military duty, deserted.——I understand that Spassky is again producing, but on what scale, I do not know. There are unconfirmed reports that considerable development of porphyry deposits of copper-bearing ore have been found in these steppes. I believe that this region offers a promising field for other valuable copper deposits.

The platinum production of the Urals has dropped materially, in spite of the high price that the Government is now paying. As is generally known, all the platinum must be delivered to the Government, which is now paying 136,000 rubles per pood; this works out at 314 rubles per ounce for 83% platinum.——The two Bucyrus dredges ordered by the Nicola Pavdo Co. last year are still at Vladivostok and I doubt if they will be in operation before the end of 1919. I am not aware of any other new construction in the platinum region.

The production of gold is declining. As you know, the greatest single producer in Russia is the Lenskoi; as far as can be ascertained, its production for the financial year ended September 31 will be slightly in excess of that for the previous annual period. The Government price on gold is now 11.50 rubles per zolotnik as against 5.51 in normal times. However, as it is not compulsory to deliver gold to the Government, there is a large trade in this product in the open market and the present price at Petrograd for fine gold is about 22 rubles per zolotnik. The dredge ordered by the Lenskoi company is about ready for delivery, but owing to the chaotic conditions in Russia and to the experience of the Pavdo people, it has been decided to store this dredge somewhere in the United States until it can be more definitely foreseen just what time would be necessary to transport and construct it on the Lena.

During this year we have had in Russia, as you probably know, Charles Janin, T. B. Joralemon, H. V. Winchell, J. C. Ray, and others, looking up business for new American clients, that is, they are new to Russia. I do not know what success they have had, but it only seems necessary to say that they have all left Russia.

TORONTO, CANADA

VIPOND-NORTH THOMPSON.—TECK-HUGHES.—NIPISSING.

The labor situation at the Porcupine camp shows a marked improvement, and the forces at the leading mines are being steadily increased. The enforcement of conscription has had little effect, as a large proportion of the laborers are either alien enemies or other unnaturalized foreigners, and are not liable to be drafted. Production has apparently taken an up-ward turn, and with increased milling facilities is expected to show a large increase in the immediate future.——The shareholders of the Newray have ratified the arrangement giving the Crown Reserve a controlling interest in the company and increasing the capital stock from $1,500,000 to $3,000,000. Maurice Summerhayes, who has been for some years manager of the Porcupine Crown, has been appointed manager.——The management of the Davidson has arranged for the extension of the electric-power line through its property, thus securing power for the new mill, which should be in operation about the end of the year. The Davidson has lately bought up a number of claims in north-east Tisdale, and now controls an area of about 500 acres.——At the Vipond-North Thompson the shaft will be put down from its present depth of 600 ft. to the 800-ft. level, and arrangements have been made for the exploration of the Thompson-Krist property, adjoining, by running a cross-cut at the 400-ft. level of the Vipond.——Two promising veins have been cut in diamond-drilling on the Bilsky claims, under option to the Apex. The drill-hole is down over 1000 ft.——The southern end of the West Dome Consolidated is to be developed from the Dome Lake workings. Some of the Dome Lake veins continue into the West Dome.

The report of the Teck-Hughes, of Kirkland Lake, for the year ended August 31, shows an adverse balance of $105,456, and a net impairment at the end of the fiscal year of $261,926. The extension of the plant to a capacity of 110 tons per day will cost $30,000, and substantial sums are also needed to carry on operations. A plan to secure the necessary funds is being considered.——Operations at the United Kirkland are encouraging. The shaft is down to the 65-ft. level; it was sunk on a vein which dipped out of the shaft, but two other good veins were cut. It is proposed to tap them by cross-cutting from the 100-ft. level.——Charles O'Connell, for some years manager of the Tough-Oakes, has resigned his position to manage the Boston Hollinger property at Boston Creek; he is succeeded at the Tough-Oakes by D. A. Angus, lately of the Right of Way at Cobalt.——A new gold discovery in the Lightning River district, south of Lake Abitibi, has attracted many prospectors and, according to the last account, about 75 claims had been staked. Several tons of quartz taken from the discovery-vein are stated to assay high. Rock exposures extend over a large area, the basalt formation predominating, with diabase, porphyries, and granites. The field is accessible from Ramore, a railroad station on the Temiskaming & Northern Ontario railway, some 30 miles distant, but the winter will probably check the rush.

The Canadian government has taken action in the matter of the patents on the flotation process held by the Minerals Separation North American Corporation, and has ordered an investigation by the Patents Branch of the Department of Agriculture into the complaint that the corporation was imposing excessive charges for royalties upon the Cobalt mine-owners. The question of whether the corporation is under German influence or control does not come within the scope of the inquiry.——During October the Nipissing mined ore of an estimated value of $306,167, and shipped products from Nipissing and customs ore of an estimated net value of $335,386. Extensions of two branch veins were found at 73 shaft, and the results of the stoping of veins 96 and 102 exceeded expectations. The Chambers-Ferland has found a rich ore-shoot 6 to 8 in. wide on No. 15 vein at the 400-ft. level. The ore is reported to assay 5000 oz. per ton.——The La Rose Consolidated proposes to follow the example of other companies liable to taxation both in Canada and the United States, by dissolving the American holding company and so escaping the United States tax. It also will reduce its capitalization from $6,000,000 to $1,500,000.

Plans are being made for the development of electricity at Big Bend falls to supply the Gowganda silver-mining camp and the new district of Matachewan. The falls are capable of supplying between 5000 and 6000 hp., which would be ample for the needs of the district to be served. The project for the development of power at Hanging Stone falls at the south end of Gowganda lake was abandoned owing to financial difficulties. Gowganda has lately attracted increased attention as a coming mining centre owing to the success of the Miller Lake-O'Brien mine, which is now producing silver at the rate of about 8400 oz. per month.

LEADVILLE, COLORADO

THE IRON SILVER GROUP.—FRYER HILL DEVELOPMENT.

The Iron Silver Mining Co. has leased the South Moyer mine to William Carson, who for several years has been employed by the company as superintendent. The South Moyer is one of the original properties in the Iron Silver combination and has been a rich producer of lead, silver, and zinc ores. Up to the first of the present year, it has been the centre of the company's activities. With the purchase of the immense tract of ground including the R. A. M., Pyrenese, and Mikado, which the Iron Silver company is now operating through the Mikado shaft, development in the Moyer has been gradually abandoned. The big ore-shoots have been practically exhausted and it was not apparent that development would discover new ore. It is stated that many small bodies of ore are still to be found in the property, making it an alluring venture to a leasing company such as Mr. Carson has organized.——At the Mikado, the Iron Silver company is now producing a steady output of 200 tons per day; half of this is being extracted by the Commerce Leasing company which controls the upper ore-zones where large bodies of zinc carbonate have been opened. The Iron Silver has uncovered large deposits of high-grade zinc-lead sulphide in the lower workings and is steadily blocking out ore. ——Following more than a year of systematic development which included the draining of a large area through the Harvard and Jamie Lee shafts and the sinking of the latter 250 ft. through the parting quartzite into porphyry and white lime contact, the driving of numerous drill-holes, and the advancement of long drifts from the shaft bottom, the Leadville Unit has abandoned work in the lower levels of its property on Fryer hill. No ore deposits were found in the second contact where they were expected. The company is now developing the upper portions of its property where many bodies of zinc and lead carbonate have been uncovered. The company also has taken charge of the Denver City property on Yankee hill and is developing it. The Unit is now shipping 50 tons of ore per day. This tonnage will be increased as soon as the railroad siding to the Jamie Lee is completed.——Two of the famous Matchless shafts on Fryer hill are now active. Edward Huter, manager of Matchless No. 1 lease, which is held by a company of Denver men, has completed re-timbering the shaft to its bottom, a depth of 246 ft., and has re-opened some of the old workings, where two small bodies of zinc carbonate have been found. A small tonnage of ore is now being shipped. The shaft is to be sunk another 100 ft. which will put it into the lower ore-zone.——On Matchless No. 2, William J. Burke and associates have prepared to re-timber the shaft and sink it into the sulphide contact. Machinery is being assembled at the property and work will begin shortly.——F. E. Kendrick, J. C. Strong, William V. Dolan, and John Brady have organized a leasing company and secured control of the Dunkin property on Fryer hill. A new shaft is to be sunk to reach bodies of lead and zinc carbonate that are believed to exist within a depth of 200 ft. Hoisting machinery and surface buildings are being erected. This is the first new shaft to be put down in the district this year.——A. A. Garrett and associates have secured a lease on the Pittsburgh property between the Ponsardin and May Queen on Yankee hill and are erecting machinery at the Blonger shaft. Large bodies of zinc and silver ore have been discovered in the May Queen and Ponsardin making the Pittsburgh one of the most promising ventures undertaken in the district this year.

MAMMOTH, UTAH

EAGLE & BLUE BELL.—KNIGHT'S MINES.—NORTH TINTIC.

C. E. Loose returned to Provo on November 14 after a visit to his mining properties in the Tintic district. He reports the Gold Chain and the Grand Central in good physical condition, although both have suffered from the embargoes of the smelting companies and the shortage of railroad cars. As the mines have been unable to ship anywhere near their regular amount of ore, the working forces have been cut down and only half of the regular number of miners are on the payrolls. Considerable ore is piled up at both mines.—— Probably no mine in the Tintic district has suffered more as a result of the smelting companies' embargoes than the Eagle & Blue Bell mine which has been unable to ship during 60 days out of the past 75 days. Had it not been for this serious trouble the Eagle would no doubt have closed the year with the best dividend record of any of the Tintic mines, instead of coming second to Mammoth. A considerable amount of new ore has been developed during the past few months. Whenever a drift strikes an orebody the work must necessarily be stopped, as there is no storage room at the mine. The underground workings of the Eagle are in excellent shape for a heavy production when conditions change and the smelters are once more ready to take the product. An immense amount of high-grade ore is awaiting development.——At the No. 1 shaft of the Iron Blossom sinking is progressing satisfactorily. The management intends to sink the shaft to the 2100-ft. level, from which point extensive prospecting will be done. The Colorado Con. company is now driving a drift on the 1100-ft. level to develop a new section of its ground. A drift was sent over into that part of the Colorado from the Crown Point, and the drift which has just been started will cut the same ground 100 ft. deeper.——At the Dragon the No. 1 shaft is being deepened; it is down about 1500 ft., which is equal to 2000 ft. in the Iron Blossom, adjoining the Dragon on the north. A pump has been added to take care of the small flow of water which is lifted through the 1000-ft. level and allowed to flow into the loose formation that exists there. The Iron Blossom continues to send a good tonnage of low-grade ore to the Tintic Milling Co. and the mine is also shipping some higher-grade ore.——In order that work may be carried on below water-level in the Chief Con. mine, pending the question of the new pumping plant, a 500-gal. tank is now being used to bail the water from the sump below the 1800-ft. level. About 80,000 gal. of water per week is being lifted from the mine. One of the largest and richest orebodies which the Chief Con. people has developed in recent years is that which it is now following below the 1800-ft. level.——There is considerable activity among the properties at North Tintic, the busiest being the Lehi Tintic and the Eureka King. The shaft of the latter is down about 250 ft., and when it reaches the 300-ft. level some cross-cutting will be done.

ASPEN, COLORADO

DEVELOPMENT AT THE HOPE.

George W. Tower, Jr., and Frederick W. Foote have completed an examination of the property of the Hope Mining, Milling & Leasing Co., and have submitted their report to the officers of the company with recommendations for future development. They advise the driving of an adit easterly from Castle creek into Richmond hill for exploring the Little Annie and Wilmington mines, together with the territory between these properties. This project would reach a depth of 1500 to 2000 ft. and should expose several orebodies in new territory. They state that the workings from this adit, if carefully directed, will cut three known ore-zones; the Little Annie fault, the Silver fault, and the Contact fault. The opening of these zones would assure a sufficient tonnage of ore to justify the erection of a modern concentrating plant. The Hope adit is now 8000 ft. long, six years having been devoted to its advancement; and it is the big enterprise of the Aspen district. The objective of the bore is the territory underlying the Little Annie mine on Richmond hill where good ore was extracted during the early years of the Aspen district and where big ore-shoots are now known to exist. The main shoot in the Annie is opened at various points for a total horizontal distance of 600 ft. and to a depth of 300 ft. In some places it has shown a width of 60 ft. This shoot lies along the Little Annie fault. The Silver and Contact faults, the chief mineralized zones of the district, have not been cut in the Little Annie; and the reaching of these zones through the Hope is the one incentive to the continuation of the enterprise. The Hope company controls a 100-ton mill at the mouth of the Little Annie adit above the Hope adit. The mill is in good condition and contains a crusher, rolls, and 10-stamp mill in the crushing department; screens, Harz jigs, Wilfley tables, cones, and an 80-hp. steam plant for driving the machinery.——The silver-lead ore of the Little Annie contains much barite, and the process of concentration now employed is not adaptable to its treatment. The company also owns a mill-site at the portal of the Hope adit, and, after the ore-zones have been reached, a flotation plant will be erected. Flotation has been tried on Aspen ore similar to that in the Little Annie with good results. The officers of the Hope company are: Charles O'Kane, president; H. W. Clark, secretary; and J. B. Stitzer, treasurer.

TONOPAH, NEVADA

TONOPAH EXTENSION.—TONOPAH BELMONT.—MACNAMARA.—JIM BUTLER.

The Tonopah Mining Co. during the previous week milled 2700 tons of ore averaging $10.50. In the Silver Top 67 ft. of development was done and 127 ft. in the Sandgrass. On the 1100-ft. level of the Sandgrass raise No. 1163 continues in ore. Last week the production was 2600 tons.——The Tonopah Extension Mining Co. shipped 21 bars of bullion, valued at $45,858, which represents the first November clean-up. At the Vctor 85 ft. of development was done and 164 ft. at the No. 2 shaft. At the latter shaft raise No. 453 on the 1260-ft. level advanced 31 ft. in ore. On the 1440-ft. level of the Victor east drift No. 604 advanced on 4 ft. of ore, while raise No. 612 continues on low-grade quartz. On the 1680-ft. level the 1600 south cross-cut is being driven toward the C. K. and the Murray veins. East drift 1600 on the Merger vein continues on a 5-ft. face. The production for the past week was 2380 tons.——On the 700-ft. level of the Tonopah Belmont, development has been resumed in east drift No. 721 and west drift No. 722 on the Shoestring vein. Raise No. 16 on the 800-ft. level continues in good ore as well as raise No. 17 on the same level. On the 900-ft. level drift 9021 on the South vein shows 2 ft. of medium-grade ore. Raise No. 72 on the 1000-ft. level, which was started from a stope on the Rescue vein, shows a 4-ft. face of fair ore. The last week's production was 2440 tons.——The West End mill shipped 29 bars of bullion, valued at $51,100. The Ohio cross-cut No. 506 has reached the vein, exposing excellent ore, and drifts have been started. Drift No. 531 has been discontinued for the present and a winze is being sunk on the vein. The stope on the 530 drift continues in good ore. Drift No. 1 on the 555-ft. level continues in good ore. Stopes 504, 508, 510, 511, 522, and 523 are all showing good faces of ore. The West End shaft has been re-timbered without interference with the regular work. Drift 717 No. 2 shows a face of low-grade quartz. The output the past week was 1003 tons. The Halifax Tonopah produced 55 tons.——The MacNamara Mining Co. shipped 8 bars of bullion, valued at $14,700. On the 500-ft. level a raise cut the extension of the Ohio vein exposing 12 ft. of ore. Driving has been started toward the west. On the 300-ft. level the raise out of the winze continues in good ore. On the 725-ft. level stope No. 1 shows 8 ft. of good ore, while stope No. 2 shows a 5½-ft. face. The output the past week was 500 tons.——The Jim Butler Tonopah Mining Co. is cutting a station for diamond-drill work on the 800-ft. level of the Wandering Boy. On the 200-ft. level of the Wandering Boy shaft drift No. 370 has been extended on a 3-ft. face of low-grade quartz. At the Desert Queen shaft raise No. 651 shows a 4-ft. face of good ore. The production for the past week was 850 tons.——The Cash Boy produced 100 tons of ore during the past week, which was sent to the Keswick smelter. With the extension of the drifts on the hanging wall and foot-wall on the 1700-ft. level, a considerable tonnage has been developed.——The east drift on the 1050-ft. level of the Rescue-Eula shows a 2-ft. face of ore. Winze 952 is being sunk on 2½ ft. of good ore.——On the 800-ft. level of the Midway Mining Co., a vein showing 3 ft. of good ore was exposed. A vein showing 15 ft. of low-grade ore was also found on the same level.——The Montana produced 144 tons and miscellaneous 21 tons, making the week's production at Tonopah 10,093 tons with a gross value of $176,628. The treatment of the tailing 'sweeps' has been suspended for the season. These 'sweeps' average from $25 to $30 per ton, and approximately 5000 tons was treated during the summer. A greater part of the tailing was treated at the Belmont plant.

HOUGHTON, MICHIGAN

CALUMET & HECLA.—COPPER RANGE CON.

The Calumet & Hecla is shipping an average of 9500 tons of ore per day, and 21 stamp-heads are in use. The Ahmeek shipments are now at the rate of 4600 tons per day. The Mass Consolidated shipments are over 900 tons per day and will be 1000 tons within ten days, as the labor force is being increased. The Allouez shows a slight increase this month and is shipping 2100 tons per day.——The Wolverine also is picking up slightly, producing 37 cars daily, compared to 32 last month.——The South Lake shipped 6000 tons of ore last month.——The Lake shipments were 7300 tons.

The three Copper Range Con. mines, Trimountain, Baltic, and Champion, last month produced 120,000 tons of ore. Of this total Champion contributed 65,000 tons, Trimountain 25,000, and Baltic 30,000 tons. At present Champion is producing 2200 tons daily, Baltic 1300, and Trimountain 900 tons. At Champion this ore-tonnage is considerably lower than the average for last year when it ran about 78,000 tons per month. Underground situation at all properties continues fair; the Champion as usual is producing more than the combined tonnage of the other two and showing a much higher grade of ore. It is the opinion of the best mining men in the Lake Superior district that there has been a smaller increase in costs at the Copper Range mines than at any in the district, except, possibly, the Ahmeek.

ALASKA

(Special Correspondence.)—On November 18 the 'Mariposa,' belonging to the Alaska Steamship Co., was wrecked on her last voyage of the season to Anchorage from Shakan, Alaska. She struck the rocks of Strait island, 50 miles due west of Wrangell, at night and sank a few hours later. No lives were lost. On October 8, 1915, the 'Mariposa' went ashore near Bella Bella, in south-eastern Alaska, but was re-floated successfully. At one time she plied between San Francisco and Australia for the Oceanic Steamship Company.

THE 'MARIPOSA' ASHORE NEAR BELLA BELLA, ALASKA

Anchorage, November 22.

Considering the disappointing exhibit made by the Alaska Gold Mining Co. for the three months ended September 30, it is not surprising that the stock should have found a new low level. The average value of the ore during the three months was only $1.01 per ton, and the final net profit was $8735, which is wholly insufficient to provide a bond interest of $37,500. Alaska Gold represents a capital investment of $10,500,000, $3,000,000 of which is in bonds and $7,500,000 in $10 shares. ——The Alaska lode-gold mines have shown a decrease in both production and ore value during 1916. The production amounted to $5,912,736, as against $6,069,023 in 1915, while the average value fell from $2.75 in 1915 to $1.70. The placer mines, on the other hand, showed an increase from $10,480,000 in 1915 to $11,140,000 in 1916. This increase is due chiefly to the newly developed Marshall and Tolovana districts.

ARIZONA

COCHISE COUNTY

(Special Correspondence.)—A. Young, secretary of the Green-Cananea, has gone to Mexico City where he will remain for a month. It is believed that he has gone there to take part in the negotiations for resumption of operations in Mexico by American mining interests.

Douglas, November 17.

A. G. McGregor of Warren has purchased a large business block in that city and will establish offices and laboratories there. Mr. McGregor has designed many of the large smelting and leaching plants in the State. In future he will make Warren his headquarters.

GILA COUNTY

(Special Correspondence.)—The stockholders of the Miami Copper Co. have received notice that dividend No. 19 of $1.50 per share and $1 per share extra, declared April 2, 1917, and paid May 15, 1917, and dividend No. 20 of $1.50 per share, and $1 per share extra, declared June 29, 1917, and paid August 15, 1917, were paid out of the surplus and profits which accrued and were earned prior to March 1, 1913.

Miami, November 24.

MARICOPA COUNTY

(Special Correspondence.)—The new mineral building at the State fair grounds at Phoenix was sufficiently completed to hold the mineral exhibit, which was by far the best display of Arizona minerals yet exhibited.

Tucson, November 17.

PINAL COUNTY

The report of the Ray Consolidated Copper Co. for the quarter ended September 30 shows a production of 22,255,598 lb. of copper, against 21,656,342 lb. for the previous quarter. During the quarter 901,300 tons of ore, averaging 1.63% of copper, was milled, and 16,159 ft. of development work done. The mining cost was $1.169 and the milling cost 88.46c. per ton. The average cost of production was 11.27c. per pound. The regular quarterly distribution of $1 per share was paid on September 30 and a dividend, of 20c. per share, was paid on July 25. The combined amount totalled $1,892,614.

YUMA COUNTY

(Special Correspondence.)—Walcott and Donnely are taking out high-grade ore from the Bullard lease. They have shipped a number of cars. Jones and Hodges, lessees of the

Vindicator property, have shipped 30 tons of ore which gave a return of 20% copper and one ounce gold.——Fleming and Gilland, on the Davis property, are running an adit on a vein to connect with a winze; they are taking out a heavy glance which will run more than 40% copper. The adit is now in over 100 ft. This vein parallels the Black Giant lode and runs through the Ranier property.——The Black Giant is working two shifts in a drift on the Gossan vein. Shaft-sinking has been resumed at the Black Reef. It is intended to sink to the 800-ft. level. The old Socorro company is removing its steam plant to California.

Wenden, November 21.

CALIFORNIA

CALAVERAS COUNTY

(Special Correspondence.)—Work at the Calaveras Consolidated mine, near Melones, is proceeding most satisfactorily. The property is owned by the Calaveras Consolidated Syndicate, for which W. J. Loring is manager. The controlling owners are said to be Messrs. Clark and Coolidge of Boston. The property covers 8400 ft. along the Lode, which has been trenched and sampled systematically, disclosing several profitable orebodies, the downward continuation of which has been proved by an adit 400 ft. below the surface and 2160 ft. long. A raise, in ore all the way, connects this adit-level with the surface. Furthermore, a shaft has been sunk at a point 950 ft. from the entrance of the adit, and is now 288 ft. below that level. Drifts are being extended in ore. It is estimated that 241,000 tons of ore is blocked out, and the average assay-value is $3.50 per ton. A 3-stamp mill for testing this reserve began work in April, and has been crushing six tons per stamp. A new 10-stamp mill began operations in the middle of August, and has continued the test with satisfactory results. In October the two mills crushed 2942 tons for a yield of $6242 in gold. An additional 10 stamps is now being built. The new plant will include two Hardinge mills and a Minerals Separation flotation unit, followed by Oliver filters. The combined mills will be capable of treating 200 tons per day, and it is expected that the tailing-loss can be kept down to 25c. per ton. The whole plant is driven by electricity supplied by the Sierra & San Francisco Power Co. Included in the property is the famous Morgan mine, once owned by James G. Fair and now owned by his daughter, Mrs. Vanderbilt. The Calaveras property is immediately west of the Melones, which operates 100 stamps, and has been constantly productive for many years, down to a total depth of 3000 ft. It is considered likely that the Calaveras Consolidated will develop into one of the most important mines in the foothill region of California.

Melones, November 22.

SANTA CLARA COUNTY

Maurice Dieches has been appointed by the Supreme Court temporary receiver for the Quicksilver Mining Co. The company was organized with a capital stock of $10,000,000, which was issued and is outstanding, to operate the New Almaden and other mines. The listed assets total $98,000 and the liabilities $94,000.

SHASTA COUNTY

(Special Correspondence.)—Bituminous coal has been found at Beegum Creek, near the south-western boundary of this and Tehama county. The discovery was made by D. Noble and Frank Wild of Knob, who have taken up 320 acres. Two seams have been uncovered, the upper one is four feet thick and the lower one eight feet. The 10-ft. parting between the seams carries several thin layers of coal. The coal has been used for welding iron in a blacksmith's shop at Redding.

Redding, November 21.

COLORADO

PUEBLO COUNTY

(Special Correspondence.)—Eighty acres of land has been secured at Pueblo, by the Norcross Chemical Co. as a site for its plant where acids and industrial chemicals will be manufactured. The company proposes to erect at once a 50-ton sulphuric-acid plant and a unit known as the Zurcher unit. The office of the company is in the Ideal building, Denver.

Pueblo, November 23.

NEVADA

WASHOE COUNTY

(Special Correspondence.)—John M. Baker, of Alderson, Baker & Baker, of Boston, has examined the properties of the Nixon Nevada and Washoe mining companies for Eastern clients. Mr. Alderson made an examination of the Nixon Nevada about a year ago and in his report expressed the opinion that the numerous veins, from which rich ore has already been shipped, came from a common source and will therefore come together at depth.

Reno, November 21.

WHITE PINE COUNTY

The thirty-third quarterly report of the Nevada Consolidated Copper Co. shows a copper production of 20,217,673 lb. for the three months ended September 30, as against 20,817,356 lb. for the preceding quarter. During the quarter 1,014,031 tons of ore, averaging 1.46% copper, was milled, 78% of which was supplied from pits and 22% from the underground workings of the Ruth mine. The cost of production was 11.78c., against 11.02c. There was a deficit for the quarter of $311,051 after payment on July 25 of a Red Cross dividend of 15c. per share amounting to $299,918, and after payment on September 29 of a dividend of 50c. per share and a capital distribution of 50c. per share amounting to $1,999,457, or a total distribution of $2,299,375. There was set aside $172,144 for plant and equipment depreciation and $61,000 for ore extinguishment, leaving a net debit to earned surplus for the quarter of $544,196.

TEXAS

LLANO COUNTY

(Special Correspondence.)—Development has just been started on a deposit of graphite near Llano. The property is owned by the Dixie Graphite Co.; Warren May is manager. The equipment consists of a 100-hp. oil engine, a 50-ton crusher, flotation plant, and dryer. From the time the ore enters the crusher until it comes out a finished product, ready for shipment, it is not handled; every step being taken automatically by machinery. The process is so complete that only four men are required to run the plant.

Llano, November 20.

UTAH

SALT LAKE COUNTY

(Special Correspondence.)—For the three months ended September 30 the Utah Copper Co. produced 54,999,468 lb. of copper, against 56,403,465 lb. for the previous quarter; 3,439,400 tons of ore averaging 1.306% of copper was treated, and the average cost of production was 10.86c. per pound. The regular dividend of $3.50 per share was paid on September 30 and a Red Cross dividend of 50c. per share.

Salt Lake City, November 22.

MEXICO

(Special Correspondence.)—American officials and employees of the American Smelting & Refining Co. in the State of Chihuahua have been ordered to come to the border and eight of the principal officials already have arrived at El Paso. This was announced on November 21 by C. L. Baker, general manager of the company's Mexican interests with headquarters at El Paso. This was done as a precautionary measure and work has been stopped at the Chihuahua smelter pending developments in Mexico.

El Paso, Texas, November 23.

A. I. M. E., San Francisco

The local section of the American Institute of Mining Engineers will meet on Tuesday, December 4, at the Engineers Club. Mr. Henry F. May, Special Assistant to the Attorney General of the United States, will explain the litigation over the withdrawn oil-lands. Mr. Fletcher McN. Hamilton, State Mineralogist, and Mr. J. H. G. Wolf will take part in the discussion, which promises to be unusually interesting and informing. A large attendance is expected. A simple steak-dinner will be served at 6:30 for $1 and the meeting will begin at 7:30 promptly.

War Activities

ALBERT BURCH has been called to Washington by the Bureau of Mines for consultation in the formulation of rules and regulations governing the act relative to the use, distribution, and manufacture of explosives.

THE CALIFORNIA METAL PRODUCERS ASSOCIATION has petitioned the appointment of M. L. Requa, a mining engineer at present employed as the principal assistant to Herbert C. Hoover, as an additional member of the advisory committee appointed by the President for the purpose of devising rules for the guidance of the Internal Revenue Department in the administration of the provisions of the new war-tax law.

Because of the great amount of war work being done by the Bureau of Mines, the Secretary of the Interior has asked six of the prominent chemists of the country to act as an advisory board to the Bureau. The members of this board are: WILLIAM H. NICHOLS, General Chemical Co., New York, chairman of the board; H. P. TALBOT, head of the chemical department of Massachusetts Institute of Technology, Boston; WILLIAM HOSKINS, consulting chemist, Chicago; H. P. VENABLE, head of the chemistry department, North Carolina University; E. C. FRANKLIN, professor of chemistry, Leland Stanford University; and CHARLES L. PARSONS, chief chemist of the Bureau of Mines. The board will discuss and advise upon gas warfare research, the minerals especially needed for munitions, and the recently enacted law for the regulation of explosives.

Obituary

ARTHUR LAKES died on November 21 at Nelson, B. C., at the age of 73. Born in Cornwall, England, he was widely known in the West as a mining geologist and the author of the 'Geology of Western Ore Deposits.' From the start of the Colorado School of Mines until 1895 he was Professor of Geology in that institution, and in the course of his life he contributed largely to the descriptive geology of Colorado. He is survived by a son, Arthur Lakes, Jr., who has followed in his footsteps as a geologist, but is now in training for a commission at the Presidio, San Francisco.

GEORGE T. HOLLOWAY died on October 24 in London. A graduate of the Royal School of Mines, he established himself in practice as assayer and metallurgist, in 1886. Recently he served as the British government's representative on the Ontario Nickel Commission and was chosen chairman by his associates. Crippled early in life, he made a courageous and successful struggle, becoming a useful citizen, a resourceful metallurgist, and a professional man of the highest standing. He was a familiar figure at the meetings of the Institution of Mining & Metallurgy, of which he was a vice-president and to the discussions of which he was a notable contributor. At the Mining & Metallurgical Club, the Chemical Society, and the Institute of Metals he will be sadly missed. In Canada as well as England he had a wide circle of friends by whom his memory will be honored.

PERSONAL

Note: The Editor invites members of the profession to send particulars of their work and appointments. This information is interesting to our readers.

HENNEN JENNINGS is at the Fairmont hotel.

F. LESLIE RANSOME, U. S. Geological Survey, is here.

OSCAR H. HERSHEY is examining mines near Mina, Nevada.

R. E. CRANSTON has returned to San Francisco from New York.

HOWARD D. SMITH, on his return from Arizona, has gone to Ely, Nevada.

J. T. BROWN, until recently assayer at Cripple Creek, is in San Francisco.

ALFRED SCHWARZ has moved his offices to 1401 Broadway, Joplin, Missouri.

W. R. LINDSAY is superintendent for the Tungsten Mines Co. at Bishop, California.

R. A. KINZIE sailed for Mazatlan on November 27, and will be absent one month.

ATKIN & MCRAE announce the removal of their laboratories to 1008 South Hill street, Los Angeles.

STANLY A. EASTON, superintendent of the Bunker Hill & Sullivan mine, was in San Francisco this week.

EDWIN O. HOLTER and H. P. HENDERSON, of the Jerome Verde company, have returned to New York from Jerome.

A. L. NOLAND has accepted the position of mine superintendent for the United Gold Mining Co. at Granite, Oregon.

E. WALKER, formerly chief chemist for the Butte & Superior, is now metallurgist for the Como Con. Mining Co., at Dayton, Nevada.

WALTER BRIGGS, formerly with the Tennessee Copper Co. and other copper enterprises, is residing temporarily at Berkeley, California.

JOHN W. BULL has resigned, after serving 15 years as superintendent of the Afterthought mine at Ingot, to be succeeded by JOHN LIND.

S. F. SHAW is in charge of operations of the Montañas property in Nuevo León for the American Smelting & Refining Co., which is now under lease.

FORBES RICKARD, JR. has been commissioned second lieutenant in the reserve corps of the Regular Army and expects to go to France on December 15.

CHARLES H. WHITE, recently professor of mining in Harvard University, has opened an office as consulting geologist in the Hobart building, San Francisco.

O. E. JAGER, formerly superintendent of the Cerro de Pasco smelter, in Peru, is in England, where he is on the staff of one of the Government-controlled lead smelters.

GEORGE H. GARREY passed through San Francisco last week after having completed a geological examination for the Tonopah Belmont Development Co. in Colorado.

R. E. ADAMS, superintendent of the Velardeña unit of the American Smelters Securities Co., has returned to Mexico after an absence of several months in the United States.

H. DEWITT SMITH has resigned as assistant-manager at the Kennecott mine, Alaska, to become superintendent of the mine department of the United Verde, at Jerome, Arizona.

ALLEN H. ROGERS, SYDNEY H. BALL, and LUCIUS W. MAYER have formed a partnership as mining engineers, with offices at 42 Broadway, New York, and 201 Devonshire street, Boston.

JAMES G. PARMELEE, research metallurgist for the Granby Con. M. & S. Co., of Anyox, B. C., has accepted a fellowship in metallurgy at the University of Idaho, at Moscow, in connection with the U. S. Bureau of Mines.

METAL PRICES

San Francisco, November 27

Aluminum-dust (100-lb. lots), per pound	$1.00
Aluminum-dust (ton lots), per pound	$0.95
Antimony, cents per pound	16.00
Antimony (wholesale), cents per pound	15.25
Electrolytic copper, cents per pound	23.50
Pig-lead, cents per pound	6.50— 7.50
Platinum, soft and hard metal, respectively, per ounce	$105—111
Quicksilver, per flask of 75 lb	110
Spelter, cents per pound	9.50
Tin, cents per pound	70
Zinc-dust, cents per pound	20

ORE PRICES

San Francisco, November 27

Antimony, 45% metal, per unit	$1.10
Chrome, 34 to 40%, free SiO_2, limit 8%, f.o.b. California, per unit, according to grade	$0.60— 0.70
Chrome, 40% and over	$0.70— 0.85
Magnesite, crude, per ton	$8.00—10.00

In anticipation that the Government is likely to fix a price for calcined magnesite, buyers are holding off. There is practically no demand for the crude mineral.

Manganese: The Eastern manganese market continues fairly strong with $1 per unit Mn quoted on the basis of 48% material.

Tungsten, 60% WO_3, per unit	30.00
Molybdenite, per unit MoS_2	$40.00—45.00

EASTERN METAL MARKET

(By wire from New York)

November 27.—Copper is unchanged and nominal at 23.50c. all week. Lead is quiet and firm at 6.50c. all week. Zinc is inactive and easy at 8c. all week. Platinum remains unchanged at $105 for soft metal and $111 for hard.

SILVER

Below are given the average New York quotations, in cents per ounce, of fine silver.

Date		Average week ending	
Nov. 21	85.25	Oct. 16	87.01
" 22	84.62	" 23	83.85
" 23	84.62	" 30	84.71
" 24	84.62	Nov. 6	89.25
" 25 Sunday		" 13	86.18
" 26	84.25	" 20	85.62
" 27	84.25	" 27	84.00

Monthly Averages

	1915	1916	1917		1915	1916	1917
Jan.	48.85	56.76	75.14	July	47.52	63.06	78.92
Feb.	48.45	56.74	77.54	Aug.	47.11	66.07	85.40
Mch.	50.61	57.89	74.13	Sept.	48.77	68.51	100.73
Apr.	50.25	64.37	72.51	Oct.	49.40	67.86	87.38
May	49.87	74.27	74.61	Nov.	51.88	71.60	....
June	49.03	65.04	76.44	Dec.	55.34	75.70	....

Samuel Montagu & Co. say: There have been several causes for nerveless character of the silver market; among these are the facts that the accumulation of orders has been satisfied by an accession of supplies, the difficulty of shipping to neutral countries, the strengthened position of the Indian currency silver reserves, and the freer sales from Mexico.

ZINC

Zinc is quoted as spelter, standard Western brands, New York delivery, in cents per pound.

Date		Average week ending	
Nov. 21	8.00	Oct. 16	8.50
" 22	8.00	" 23	8.25
" 23	8.00	" 30	8.12
" 24	8.00	Nov. 6	7.42
" 25 Sunday		" 13	7.81
" 26	8.00	" 20	7.98
" 27	8.00	" 27	8.00

Monthly Averages

	1915	1916	1917		1915	1916	1917
Jan.	6.30	18.21	9.75	July	20.54	9.90	8.98
Feb.	9.05	19.99	10.45	Aug.	14.17	9.03	8.58
Mch.	8.40	18.40	10.78	Sept.	14.14	9.18	8.33
Apr.	9.78	18.62	10.20	Oct.	14.05	9.92	8.32
May	17.03	16.01	9.41	Nov.	17.20	11.81	....
June	22.20	12.85	9.63	Dec.	16.75	11.26	....

The largest zinc-smelting plant in the country has curtailed operations to the extent of two-thirds of its capacity. This smelter has 13,440 retorts, of which only about 4500 are in use this week. The U. S. Geological Survey reports that 64,026 retorts were idle on September 30, and that since then 20,000 have gone out, making a total of 85,000 retorts idle and but 145,000 in operation at the present time. Surplus stocks of spelter amount to over 100,000,000 lb., and it is because of this situation that it is quite improbable that the Government will fix a price for spelter as it has done in copper.

COPPER

Prices of electrolytic in New York, in cents per pound.

Date		Average week ending	
Nov. 21	23.50	Oct. 16	23.50
" 22	23.50	" 23	23.50
" 23	23.50	" 30	23.50
" 24	23.50	Nov. 6	23.50
" 25 Sunday		" 13	23.50
" 26	23.50	" 20	23.50
" 27	23.50	" 27	23.50

Monthly Averages

	1915	1916	1917		1915	1916	1917
Jan.	13.60	24.30	29.53	July	19.09	25.66	29.67
Feb.	14.38	26.62	34.57	Aug.	17.27	27.03	27.42
Mch.	14.80	26.65	36.00	Sept.	17.69	28.28	25.11
Apr.	16.64	28.02	33.16	Oct.	17.90	28.50	23.50
May	18.71	29.02	31.69	Nov.	18.88	31.95	....
June	19.75	27.47	32.57	Dec.	20.67	32.89	....

LEAD

Lead is quoted in cents per pound, New York delivery.

Date		Average week ending	
Nov. 21	6.50	Oct. 16	7.52
" 22	6.50	" 23	6.70
" 23	6.50	" 30	5.66
" 24	6.50	Nov. 6	6.02
" 25 Sunday		" 13	6.35
" 26	6.50	" 20	6.50
" 27	6.50	" 27	6.50

Monthly Averages

	1915	1916	1917		1915	1916	1917
Jan.	3.73	5.95	7.64	July	5.59	6.40	10.93
Feb.	3.83	6.23	9.01	Aug.	4.62	6.28	10.75
Mch.	4.04	7.26	10.07	Sept.	4.62	6.86	9.07
Apr.	4.21	7.70	9.38	Oct.	4.62	7.02	6.97
May	4.24	7.38	10.29	Nov.	5.15	7.07	....
June	5.75	6.88	11.74	Dec.	5.34	7.55	....

QUICKSILVER

The primary market for quicksilver is San Francisco, California being the largest producer. The price is fixed in the open market, according to quantity. Prices, in dollars per flask of 75 pounds:

Week ending

Date			
Oct. 30	100.00	Nov. 13	100.00
Nov. 6	100.00	" 20	100.00
		" 27	110.00

Monthly Averages

	1915	1916	1917		1915	1916	1917
Jan.	51.90	222.00	81.00	July	95.00	81.20	102.00
Feb.	60.00	295.00	126.25	Aug.	93.75	74.50	115.00
Mch.	78.00	219.00	113.75	Sept.	91.00	75.00	112.00
Apr.	77.50	141.60	114.50	Oct.	92.90	78.20	102.00
May	75.00	90.00	104.00	Nov.	101.50	79.50	....
June	90.00	74.70	85.50	Dec.	123.00	80.00	....

TIN

Prices in New York, in cents per pound.

Monthly Averages

	1915	1916	1917		1915	1916	1917
Jan.	34.40	41.76	44.10	July	37.38	38.37	62.60
Feb.	37.23	42.60	51.47	Aug.	34.37	38.88	62.53
Mch.	48.76	50.50	54.27	Sept.	33.12	36.66	61.54
Apr.	48.25	51.49	55.63	Oct.	33.00	41.10	....
May	39.28	49.10	63.21	Nov.	39.50	44.12	....
June	40.26	42.07	61.93	Dec.	38.71	42.55	....

There is practically no change in the tin situation. The Tin Committee has not given out a report, but there is reason to believe that some satisfactory arrangement has been made. The nominal price is 75c. per pound.

Charles Hardy reports no change in the tungsten situation. A small amount of ore has arrived from South America. European buyers are still in the South American market and continue to pay much higher prices than are being paid in New York. Highest-grade wolframite has brought $25, but sellers are holding for a higher figure. Scheelite is still strong and has been handled at $26. It is, of course, understood that the price varies with the tungstic acid content and with the proportion of tin and copper in wolframite and of tin, copper, sulphur, and phosphorus in scheelite. Ores containing high proportions of impurities have been sold as low as $16 to $17 per unit.

Gold production of Australia for first six months of 1917, compared with 1916 and 1915, was as follows:

	1917	1916	1915
Total fine ounces	747,331	850,728	1,005,173
Total value	$15,447,332	$17,584,548	$20,777,133

A turn in the tide and an increase in gold production can hardly be expected until labor conditions improve.

Eastern Metal Market

New York, November 21.

All the markets are quiet and most of them devoid of feature or interest. Outside of copper, which is not ruled now by supply and demand, buying is of small proportions marketwise.

Copper continues to be quoted nominally at the Government fixed price for all brands.

Tin has reached record heights on small sales and is very scarce for spot delivery.

Lead is quiet but firm.

Zinc is dull with a tendency to weakness.

Antimony is nominally unchanged.

Aluminum remains stationary and unchanged.

Fuel scarcity and transportation difficulties hinder full output in the steel industry, which is adapting itself to war conditions with some difficulty. Although it is two months since price-fixing began, market results are scarcely appreciable except in pig-iron, where sales at the fixed prices have been large. In rolled steel there has been some shifting of consumption as well as some reduction. The Government's order for 1,500,000 tons of shell-steel is being allotted, shipments of which are to be finished by June 15, 1918.

COPPER

There is no longer any such thing as a copper market. A representative of a large dealer said to me yesterday that the public must get rid of the idea that there is such a thing; that one might as well try to call a ball square. The copper 'situation' is absolutely a controlled one, and very tightly controlled at that. Distribution is evidently going along smoothly so far as relates to those having old contracts and delivery is being made in such cases at unchanged prices, but consumers having Government contracts are the ones having the first call. It is acknowledged that firms having Government business that involves the use of copper are receiving the metal at the Government price of 23.50c. per lb., but that the amount is of little consequence marketwise. It is probable that both Lake and electrolytic are being sold for first-quarter delivery at 23.50c. also. How much copper there is to supply the needs of everyone is not known publicly. The jobbing situation is not yet cleared up. Sales of copper bought before the price was fixed are being made at 26 to 28c.—this is acknowledged. It is also admitted that sales of Lake Copper have been made to such jobbers at 23.50c. for delivery in a reasonable time, but at what price they may dispose of this is still unsettled. A satisfactory adjustment of this problem is expected in view of the fact that steel jobbers have been informed at what prices they may conduct their business at levels above the large market. We continue to quote, as nominal, the Government price of 23.50c. per lb. for both Lake and electrolytic.

TIN

Outside of spot tin the market has been inactive and bare of nterest. Spot metal, however, has continued to 'sky-rocket,' ind yesterday it put in the shadow its former high record. 3mall lots were sold yesterday at the highest ever recorded, '7c. per lb., New York, after selling at 74c. on Monday and 73c. n Wednesday, Thursday, and Friday of last week. The carcity continues pronounced, with no relief in sight. As reated before, it is due entirely to the permit situation which Jlows no metal to be shipped from England unless the conumers' name is specified, thus stripping the market of spot naterial. The Tin Committee has again made a trip to Washngton, but no announcement of a decision has been made, hough it has been expected for some time. The entire market has been held up by an apparent lack of action by this committee, though some are inclined to think that a snag has been struck in the negotiations. No relief is in sight, and even though a solution were reached at once, it would take three weeks or a month for supplies to reach the market. Consumers or holders of tin are not allowed to sell, so no relief is possible from this source. Sales that have been made at the prices mentioned have been in small lots only, or just enough to establish a market. Arrivals this month, including yesterday, have been 1185 tons, with 4100 tons reported afloat. The London market continues to advance. Yesterday spot Straits sold there at £279 10s. per ton, an advance of £5 10s. over sales a week ago.

LEAD

There is very little to be said about the lead market. The situation is strong, with producers well sold up for November and December, and consumers fairly well cared for. Demand has gradually declined until now it is light. No demand is expected after the heavy sales of the past few weeks, and conditions are entirely normal. There is no pressure to sell, and the market is simply drifting. The outside market is firm at 6.50c., per lb., New York, or 6.31¼c., St. Louis, at which prices sales of small amounts are recorded. The American Smelting & Refining Co. continues to quote 6.25c., New York.

ZINC

The market has again fallen into a lifeless state after a period of a few days of some animation and buying. It was short-lived, and the market is again weak and inactive. A fair business was done last week, some say at prices as high as 8.12½c., New York, but now sellers are willing to part with their prime Western at 7.75c., St. Louis, or 8c., New York, and are doing so, although the actual amount sold is not large. There is little demand, and it is unquestionable that most large consumers are in need of spelter. The market is quoted at 7.75 to 7.87½c., St. Louis, or 8 to 8.12½c., New York, for prime Western for early delivery. Little is heard about price-fixing, although the deliberations of the new Zinc Committee may result in something along these lines. This committee was appointed last week, and is made up of representatives of leading producers. It is conferring with Government representatives on the spelter situation, but the results are not yet available for the public. It is hoped that there will be a clearing of the atmosphere.

ANTIMONY

Demand continues of small proportions, and the market is nominally unchanged at 13.75 to 14c. per lb., New York, duty paid, for early delivery.

ALUMINUM

Inactivity continues to rule this market. No. 1 virgin metal, 98 to 99% pure, continues unchanged at 36 to 38c. per lb., New York, for early delivery.

ORES

TUNGSTEN. There is a fair demand for the various ores at the usual quotations, though South American ores are harder to obtain. Scheelite still commands $26, or even higher prices, with other grades at $23 to $25 per unit in 60% concentrate. Ferro-tungsten is lower, small lots having been sold at $2.25 per lb. of contained tungsten.

MOLYBDENUM AND ANTIMONY. Molybdenite is quoted at $2.15 to $2.25 per lb. of MoS_2 in 90% concentrate on transactions of small proportions. Antimony ore is nominal at $1.50 to $1.75 per unit.

Dividends From Mines, United States and Canada

Company	Metal	Shares issued	Par value	Paid in current year	Total to date	Latest dividends Date	
Ahmeek Mining Co., Michigan	copper	200,000	$25.00	$2,400,000	$8,450,000	July 10, 1917	$4
Alaska Treadwell, Alaska	gold	200,000	25.00	none	15,785,000	May 28, 1916	0
Allouez, Michigan	copper	100,000	25.00	900,000	1,700,000	July 3, 1917	Amount
Anaconda, Montana	copper	2,331,250	50.00	4,662,500	122,554,375	Feb. 26, 1917	2.00
Atlantic, Michigan	copper	100,000	25.00	....	900,000	Feb. 21, 1905	0.50
Barnes-King, Michigan	gold	40,000	5.00	30,000	60,000	June 1, 1916	0.07½
Brunswick Con., California	gold	395,287	1.00	23,717	203,315	Sept. 15, 1917	0
Bunker Hill & Sullivan, Idaho	lead	327,000	10.00	490,500	18,980,250	Mar. 3, 1917	0
Caledonia, Idaho	l.s.c.	2,605,000	1.00	234,450	2,678,941	Oct. 5, 1917	0
Calumet & Hecla, Michigan	copper	100,000	25.00	7,500,000	139,250,000	Mar. 22, 1917	25
Cardiff, Utah	s.l.	500,000	5.00	375,000	500,000	Dec. 1916	0
Centennial C. M. Co., Michigan	.....	90,000	25.00	90,000	180,000	Mar. 20, 1917	1
Central Eureka, California	gold	396,900	1.00	....	921,323	Feb. 20, 1915	0
Cerro Gordo, California	l.s.z.	1,000,000	1.00	175,000	175,000	Oct. 15, 1917	0
Champion, Michigan	copper	100,000	25.00	6,014,541	17,514,541	Nov. 14, 1916	6.
Chief Con., Utah	.....	884,020	1.00	44,201	576,719	Feb. 5, 1917	0.[illegible]
Chino Copper, New Mexico	copper	869,980	5.00	7,177,335	13,875,330	Dec. 31, 1916	3.00
Continental Zinc Co., Missouri	zinc	22,000	5.00	22,000	385,000	July 2, 1917	0.50
Con. Arizona, Maine	.....	1,663,000	5.00	83,150	....	Aug. 15, 1917	0.05
Copper Range, Michigan	copper	289,363	100.00	2,170,148	21,280,545	1916	.
Cresson Con., Colorado	gold	1,220,000	1.00	1,098,000	5,331,362	Oct. 10, 1917	0.
Dome Mines Co., Ltd., Ontario	silver	400,000	10.00	300,000	1,500,000	June 1, 1917	0
Dragon Consolidated, Utah	.....	1,875,000	1.00	56,250	56,250	July 25, 1917	0
Eagle & Blue Bell, Utah	.....	893,146	1.00	....	535,888	Dec. 28, 1916	0
East Butte C. Co., Montana	copper	411,000	10.00	411,000	411,000	Jan. 29, 1917	1
Electric Point M. Co., Washington	.....	793,750	1.00	166,688	166,688	Sept. 30, 1917	0.[illegible]
Federal Mining Co., Idaho	.....	60,000	100.00	....	2,705,335	Jan. 15, 1909	1
Fremont Con., California	gold	200,000	2.50	12,000	280,000	Mar. 20, 1917	0
General Development Co., New York	.....	120,000	25.00	720,000	4,403,917	Sept. 1, 1917	2
Good Springs Anchor, Nevada	z.s.	550,000	1.00	16,500	158,285	Mar. 15, 1917	0
Golden Cycle M. & R., Colorado	gold	1,500,000	1.00	450,000	8,088,500	Oct. 10, 1917	0
Goldfield Consolidated, Colorado	gold	3,559,148	10.00	none	28,999,831	Oct. 30, 1915	0.
Granite G. M. Co., Colorado	gold	1,650,000	1.00	33,000	319,500	Sept. 1, 1917	0
Hecla Mining Co., Idaho	s.l.	1,000,000	0.25	1,350,000	6,655,000	Sept. 1917	0.
Hollinger Consolidated, Ontario	.....	4,920,000	5.00	738,000	8,194,000	Apr. 23, 1917	0
Hope Mines, Utah	.....	482,260	1.00	none	4,697	Dec. 4, 1915	0
Horn Silver, Utah	l.s.z.	400,000	1.00	40,000	....	....	0
Inspiration Consolidated, Arizona	copper	1,181,967	20.00	7,387,294	13,571,410	July 30, 1917	2
Iron Blossom, Utah	l.s.g.	1,000,000	0.10	200,000	3,050,000	July 25, 1917	0
Iron Cap, Arizona	copper	145,030	10.00	20,459	....	Jan. 1, 1917	0
Iron Silver, Colorado	.....	500,000	20.00	....	5,150,000	Dec. 31, 1916	0.
Isabella Mines, Colorado	gold	2,528,331	1.00	....	767,783	Dec. 30, 1916	0
Isle Royale Copper, Michigan	copper	150,000	25.00	750,000	1,350,000	July 31, 1917	1
Jim Butler, Nevada	.....	1,718,021	1.00	171,802	687,208	Feb. 1, 1917	0.50
Judge M. & S. Co., Utah	.....	480,000	1,00	240,000	1,830,000	July 1, 1917	0.[illegible]
Jumbo Extension, Nevada	.....	1,550,000	1.00	77,500	....	June 30, 1916	0.05
Kennecott Copper Corp., Alaska	copper	2,786,856	no par value	6,827,926	23,148,209*	June 30, 1917	1.50
Kerr Lake M. Co., Ontario	silver	600,000	5.00	450,000	7,170,000	Sept. 15, 1917	0.25
Knob Hill, Washington	gold	1,000,000	1.	5,000	75,000	Mar. 1, 1917	0.00½
Liberty Bell, Colorado	gold	133,551	5.	106,841	2,100,028	June 30, 1917	0.20
Lower Mammoth, Utah	.....	1,000,000	1.	....	67,000	Dec. 15, 1915	0.01
Magma Copper Co., Arizona	copper	240,000	5.	384,000	1,104,000	Sept. 29, 1917	0.50
Mammoth Mining Co., Utah	g.s.c.	400,000	0.	....	2,420,000	Dec. 20, 1916	0.10
Mary McKinney, Colorado	gold	1,309,252	1.	....	1,182,398	Oct. 26, 1916	0.01
Mary Murphy, Colorado	g.s.l.z.	370,000	5.	25,067	93,106	May 1, 1916	0.07
Mass Con. M. Co., Michigan	copper	100,000	25.	300,000	500,000	Aug. 15, 1917	1.00
May Day, Utah	.....	800,000	0.	56,000	300,000	Dec. 23, 1916	0.02
McIntyre-Porcupine, Toronto	.....	3,610,283	1.	541,542	....	Sept. 29, 1917	0.05
Miami Copper Co., Utah	copper	747,114	5.	5,229,798	14,925,582	Aug. 15, 1917	2.50
Mogollon Mines Co., New Mexico	g.s.	355,682	5.	none	173,352	Oct. 1915	0.05
Mohawk, Michigan	copper	100,000	25.	1,700,000	5,575,000	Aug. 1916	10.00
Moscow M. & M. Co., Utah	s.l.c.z.	900,000	1.	40,500	107,980	Aug. 1917	0.02
Nat. Zinc & Lead, Missouri	z.l.	500,000	1.	20,000	210,000	May 15, 1917	0.02
Nevada Consolidated, Nevada	copper	1,999,457	5.	4,298,832	31,772,690	July 25, 1917	0.15
Nevada Douglas, Nevada	.....	979,438	5.	....	125,000	Jan. 1, 1913	0.12½
Nevada Hills, Nevada	gold	1,065,687	5.	159,853	586,128	Aug. 28, 1917	0.05
Nevada Wonder, Nevada	.....	1,408,409	1.	....	915,205	Nov. 21, 1916	0.10
New Idria, California	quicksilver	100,000	5.	100,000	2,330,000	Mar. 31, 1917	1.00
Nipissing Mines Co., Ontario	silver	1,200,000	5.00	600,000	600,000	Oct. 20, 1917	0.50
North Butte, Montana	c.g.s.	430,000	15.[illegible]	1,075,000	....	Jan. 29, 1917	0.75
North Star, California	gold	250,000	10.00	....	5,087,040	Dec. 1916	0.50
Optimo Mining Co., Wisconsin	zinc	500	100.00	....	40,000	Sept. 1915	0.05
Osceola Con. M. Co.	.....	96,150	25.00	1,730,000	16,217,975	July 31, 1917	6
Phelps-Dodge Corp., New York	.....	450,000	100 00	9,900,000	....	Sept. 28, 1917	8.00
Pittsburgh-Idaho, Idaho	lead	821,306	1	65,524	364,389	Apr. 14, 1917	0.06
Plymouth Con., California	gold	240,000	4	115,200	350,640	Aug. 14, 1917	0.24
Portland Gold M. Co., Colorado	gold	3,000,000	1.	90,000	10,687,080	Jan. 20, 1917	0.03
Prince Consolidated, Nevada	s.l.	1,000,000	2	50,000	425,000	Dec. 23, 1916	0.05
Quincy Mining Co., Michigan	.....	110,000	25	1,100,000	25,187,500	Sept. 24, 1917	5.00
Ray Consolidated, Arizona	copper	1,577,179	10	1,577,179	8,931,100	Dec. 30, 1916	1.00
Rico Wellington, Colorado	.....	1,000,000	0.	39,600	39,600	July 25, 1917	0.04
Round Mountain, Nevada	gold	1,320,630	1.	....	363,965	Aug. 1913	0.04
Shannon Copper Co., Arizona	copper	300,000	10	150,000	1,050,000	Feb. 1917	0.50
Shattuck Arizona, Arizona	copper	350,000	10	875,000	5,512,500	Apr. 1917	1.25
Silver King Coalition, Utah	.....	1,250,000	5	750,000	14,709,885	Apr. 1917	0.15
Silver King Con., Utah	.....	700,000	1	313,287	1,422,705	Oct. 1917	0.15
St. Joseph Lead, Missouri	lead	1,409,466	10.	704,733	12,382,097	June 15, 1917	0.25
St. Mary's Mineral, Michigan	.....	160,000	25.	1,920,000	....	Aug. 30, 1917	2.00
Stratton Cripple Creek, Colorado	.....	20,000,000	0.	....	445,000	Feb. 1915	0.14
Superior Copper Co., Michigan	copper	100,000	25.	100,000	200,000	Apr. 30, 1917	1.00
Temiscaming Mining Co., Toronto	silver	2,500,000	1.	300,000	2,000,000	Oct. 24, 1917	0.
Tonopah Belmont, Nevada	gold	1,500,000	1.	187,504	8,580,540	Jan. 1, 1917	0 ½
Tonopah Extension, Nevada	g.s.	1,282,801	1.	320,692	1,912,412	Apr. 2, 1917	0
Trethewey Silver Cobalt, Toronto	silver	1,000,000	1.	50,000	1,161,998	Aug. 31, 1917	0.
Uncle Sam Con., Utah	silver	750,000	1.	....	470,000	Sept. 20, 1911	0.
Union Basin M. Co., Arizona	zinc	83,535	0.	4,177	292,372	May 4, 1917	0.
United Verde, Arizona	copper	300,000	10.	300,000	1,350,000	Mar. 1, 1917	1.
United Verde Extension, Arizona	copper	1,050,000	0.	....	3,255,000	Aug. 1, 1917	0.
Utah Apex Mining Co., Utah	s.l.	528,200	5.	132,050	604,225	Jan. 2, 1917	0.
Utah Consolidated, Utah	copper	300,000	5.	900,000	11,175,000	Sept. 25, 1917	1.
Utah Copper Co., Utah	copper	1,624,490	10.	19,493,880	52,215,777	Dec. 31, 1917	3.
Utah Metal & Tunnel, Utah	.....	691,588	1.	345,794	688,266	Feb. 15, 1917	0.
Utah-Missouri, Utah	zinc	10,000	1.	52,000	62,000	Sept. 18, 1917	2.
U. S. Smelting, Refg. & M. Co	.....	351,115	50.	1,728,900	27,480,292	Apr. 14, 1917	1.
Vindicator Con., Colorado	gold	1,500,000	1.	45,000	3,622,500	July 25, 1917	0.
Wasp No. 2, South Dakota	gold	500,000	1.	....	649,465	May 15, 1916	0. ½
Wellington Mines, Colorado	l.z.	1,000,000	1.	200,000	1,450,000	Apr. 7, 1917	0.
West End Con., Nevada	.....	2,000,000	5.	100,000	700,000	Oct. 24, 1916	0.
White Knob Copper, California	gold	400,000	10.	60,000	245,000	Aug. 25, 1917	0.
Wilbert M. Co., Idaho	.....	1,000,000	1.	10,000	70,000	May 15, 1917	0.
Wisconsin Zinc, Wisconsin	zinc	925,000	1.	20,000	80,000	Mar. 1, 1917	0.
Wolverine & Arizona, Michigan	copper	118,674	10.[illegible]	none	53,403	Dec. 15, 1915	0.
Yak M. M. & T., Colorado	.....	1,000,000	1.	70,000	2,837,685	Mar. 31, 1917	0.
Yellow Aster M. Co., California	gold	1,032,100	1.00	35,000	1,245,789	July 20, 1917	0.03
Yellow Pine M. Co., Nevada	z.l.s.	1,000,000	1.00	90,000	1,753,008	Mar. 31, 1917	0.[illegible]

(Abbreviations: g = gold. s = silver. c = copper. l = lead. z = zinc.)

Note: Companies not included in the above list are requested to submit details. Changes in capitalization and new dividends will be entered on receipt of the information. *In current year additional dividend in stock, $2,090,239.

HOTELS are announcing New Year's Eve entertainments, "bigger and grander than ever," as a Sacramento hostelry says in an advertisement. This is no time for the usual year-end orgy and we hope that good citizens and thoughtful Americans will take pains to express disapproval.

READERS of this paper will note with pleasure that several firms paid for advertising space in which publicity was given to the raising of a miners regiment, namely the 27th Engineers. The following firms are to be credited with this display of patriotic spirit: Ingersoll-Rand Company, Bemis Brothers Bag Company, Holt Manufacturing Company, New York Engineering Company, Mine & Smelter Supply Company, and Broderick & Bascom Rope Company.

SOME time ago we challenged a statement by a well-known geologist in which wolframite was mentioned as a particularly insoluble mineral and hence found abundantly in the outcrop of tungsten veins. Our criticism was confirmed from India, and this week we print in our discussion department a note from Mr. Newton B. Knox giving further examples of the ready solubility of wolframite, especially in the presence of kaolinizing feldspars, in the tin-tungsten mines of Galicia, Spain. It is the free alkali that dissolves the tungsten, and the compounds are highly soluble in water. Alteration to tungstite is rare, and copiapite resulting from oxidation of the pyrite present is often mistaken for it.

WE have referred previously to the efforts made by German firms prominent in the rare-metal market to keep from American manufacturers the necessary materials for producing special steels, such as those employed for high-speed lathe-tools. The attempt to corner the market, feebly excused by the transparent lie of moral scruples against war, developed two surprises: the first was an unsuspected abundance of tungsten, revealed by more careful search under the stimulus of soaring prices; the other was the action of England, which in response to the intolerable situation produced by the high cost of these special steels, released large stocks of tungsten to this country. The German pacifists thereupon sought control through obtaining options on every newly found deposit, or, when this was not possible, by working delay with tempting offers for the output on contract, followed by time-consuming counter-propositions and subtle negotiations over the terms. Meanwhile tungsten was being shipped surreptitiously to the Scandinavian countries for transmission to Germany. Three of the accomplices in this game have been convicted recently in New York, and have been sentenced to pay a heavy fine and serve a year in a Federal prison. These are Fritz Oerundal, Waldemar J. Adams, and R. J. Collin. They were caught red-handed, and the shipment of a considerable quantity of powdered metallic tungsten was narrowly averted.

LETTERS come to us frequently from prospectors, and other worthy developers of mines, asking us to assist them in selling their property or in promoting a mining company. It should be known to our readers that this is quite outside our province. Similarly, specimens of ore are sent to us with the request for an opinion. At one time we maintained a department for such work, which was done for us by a professor in the University of California, but we found it necessary to discontinue the arrangement. Assayers and mining engineers are available for this work. We urge our prospector friends to refer such matters to the nearest assayer or engineer, whose name and address can be obtained by consulting the Professional Directory appearing in each issue of this paper.

DECEMBER 2, which was planned by the Russian I. W. Ws and the Prussian militarists to be the day of a general armistice for the furthering of German domination, was the anniversary of Napoleon's coronation and the battle of Austerlitz. Those events of a previous century should serve to remind the Allies of the fact that Napoleon's domination of Europe was due largely to the obtuseness of the Prussians, Russians, and Austrians in failing to co-operate, allowing Napoleon to smash them in turn, whereas if they had fought unitedly they could have broken Napoleon's power ten years sooner, and thereby spared Europe untold miseries. It was England's bulldog tenacity, and navy, that turned the tide of war then, as now. Reading the records of that era one realizes also that, after all, Napoleon and his marshals waged war like gentlemen and sportsmen as compared with the princes and generals that are leading the Germans to defeat at this time.

MAGNESIUM is used in the new alloy 'magnalium,' not 'magnolium,' to which we referred in a recent issue. In making magnesite cement for gun emplacements, the liquid binder is magnesium chloride. Magnesium is now made, we are informed, by direct reduction from the hydroxide—an important advance in the metallurgy of this metal, the use of which has been intensified by the War. It is used in shrapnel shells to create a vivid smoke by day and a brilliant light by night; it is also used in the star-shells and bombs employed to illuminate the battlefield. Metallurgically magnesium is of great service as a reducing agent, preventing oxidation of aluminum and the other metals with which it is mixed.

PLATINUM is regularly recovered as a by-product in gold-dredging along the Pacific Coast. It has been found also in the beach-sands of California and Oregon. In 1911 the platinum coming from these sources amounted to 511 fine ounces, of which about 90% was derived from the gravel deposits. In 1912 the production was nearly the same. Since then the price of platinum has enormously increased, and it is surprising that so little attention has been given to improvements in methods for collecting the platiniferous sand. We suspect that the unwarranted hopes raised a few years ago by some experimenters, who obtained the public ear, may have acted as a deterrent. Many concentrating plants were erected on the beaches of Coos and Curry counties, Oregon, and near Crescent City, California. The results were not brilliant, nor yet altogether unpromising. The subject is brought forward again in a most practical manner by Mr. James W. Neill in an article that we publish in this issue. He explains the causes for the small recovery of platinum hitherto obtained in the treatment of the sands from gold-dredging, and records a series of tests that should encourage others to introduce similar refinements in the operation. He points out that it might be possible, by following such methods, to produce $150,000 worth of platinum yearly from this State alone.

RAILROAD congestion may be overcome in part by local efforts in the construction of highways suitable for motor-trucks, according to Judge R. S. Lovett, chairman of the War Industries Board. This suggestion is of special significance, coming as it does from the former president of the Southern Pacific Company. He ought to know the limitations of the railroads better than most men, and he understands how much more rapidly motor-trucks may be built than rolling-stock for the railroads. He may have had in mind, also, the remarkable experience of the Long Island railroad. It controls a large and populous area supplying enormous quantities of garden produce to New York. The acquisition of this line by the Pensylvania system was regarded as a brilliant stroke; but good highways have almost stripped it of local freight. The business of the island is chiefly conducted by motor-trucks, and the railroad is reduced to dependence upon passenger traffic. Evidently Judge Lovett, through his previous connection with the Southern Pacific, has seen the advantage of the long haul so clearly that he does not fear the invasion of other and more flexible means of transportation in densely populated districts where the short haul preponderates. The highest operating costs per ton mile of freight carried in this country are in New England, and the reason is obvious. The plan to relieve the railroads by construction of more highways as an emergency war-measure is commendable, and its results will prove beneficial for all time to the public, to the makers of motor-trucks, and to the railroads themselves by increasing their operating profits.

Another Plea for Labor

We publish another letter written from the point of view of the laboring man. Such letters are welcome, and they are a distinct compliment to this journal in that they indicate the belief of the miner that we endeavor to be fair in discussing the vital problem of industry. We should like to see such letters as those of Mr. J. F. Harrington and Mr. 'Miner' answered by one or more mine-managers. A friendly discussion before a neutral chairman, as it were, would be useful in promoting a better understanding between those chiefly concerned. The difference between our own opinions, as previously expressed, and those of Mr. 'Miner' are due to the different directions from which we come to the subject of debate. We are well aware that the contract system is often abused. In our issue of July 7 we said that "the contract system is fair to all concerned and it is the only system that upholds the self-respect of the working-man; we believe, however, that it is not always fairly applied and that it then becomes vicious. Here we come to the root of the whole matter; fair dealing as between men." The system sustains the miner's self-respect, because when working on contract he can feel that he is working for himself, on his own account, being paid in proportion to his effort. The trick of shifting a contract crew as soon as they are making big money is a dishonesty of the cheap and nasty kind; the superintendent that does such a thing is little better than a thief; and he is inefficient, because such tactics fail in their purpose, which is to lower the cost of operations. When the superintendent plays such tricks the best men leave the mine. Where properly used the contract system is a means not of lowering the pay but of raising it, by offering a premium on results—not necessarily such excess of exertion as to kill a man's vitality. In a similar way the 'piece-work' system is abused. Employees are paid, say, so much per hundred units when a thousand will represent a reasonable day's wages; if they do 1250 they receive 25% more pay; they are encouraged to raise their

›utput to 1500 units, whereupon 1250 is made the basis ›f the wage-scale, and they are driven to an over-exertion :bat is killing to those of average strength. It is to this ystem of driving that Mr. 'Miner' refers. We agree :hat it is bad economically, as well as inhuman. So also ı manufacturer will appoint an agent on a small salary ›lus a commission basis; as soon as the agent has estab- ished a good local business the manufacturer dismisses :he agent and substitutes a direct representative, who :akes all the benefit of the pioneer work done by the ›revious agent, but at a lower cost to the manufacturer. Such practices are dishonest in spirit, however legal, and ɛend to undermine that good faith which is essential to ɔonest business and honorable living. In the end such nethods are demoralizing to all concerned, and more ›articularly to the democratic ideal on which this Re- ›ublic is based. Let us return to our immediate subject. Our own experience has been that the contract system, when fairly administered, attracted the best miners, ›ut to establish it successfully it was necessary to set uch prices as would permit the men to earn considerably nore than the standard day's pay. It does not matter how nany dollars a miner gains, but how many dollars it costs o break ground per foot of cross-cut or drift, or per on of ore stoped. Only an inexperienced and stupid uperintendent will cut the price as soon as the men make ı little extra money, that is, lower the rate per foot when ıe lets the cross-cut or drift at the beginning of the fol- owing month; for I do not suppose him to be so dis- ıonest as to break his contract with the men, no matter ıow much money they may be making, by discontinuing rork during the month for which the contract is let. The trouble is that many superintendents do not know ow to judge ground, so they let contracts at figures that re considerably too low or too high; in fact, they make guess, and if they guess to the disadvantage of the ompany they try to remedy the blunder by stopping ork on the misjudged contract. A competent superin- ındent can judge ground better than the men—that is ıe reason why he is 'boss'—and he should use his ıperior judgment in an effort to be fair. If he does, ıe men will have confidence in him and when in places ıe ground changes suddenly it will be recognized on ɪth sides as a factor of error that is corrected in the ng run so as not to work persistently to the advantage disadvantage of either the men or the company. As shirking, to which Mr. 'Miner' refers, shirking is ver a virtue, even for your brother's sake; we demur that idea. A self-respecting man will use his strength d intelligence to good purpose; he will delight in being ective. To slack in order to make it easier for a weaker lazier comrade is an idea that does not appeal to us all. The first thing a truly honorable man desires to be honest with everybody; to deliver the goods; to 'e a fair day's work for a fair day's pay. The work- n is worthy of his hire; the employer is entitled to a en amount of labor in return; it is an unwritten con- ct between two honest men. Slacking is not self-preservation; it is self-degradation; when it becomes a principle of conduct it degrades the cause of labor, just as the breaking of the unwritten code of good faith in the fulfilment of a contract degrades the cause of capital- ism; both alike stultify honest industry.

The Flotation of Silver Minerals

Flotation is predominantly associated, in the United States, with the metallurgy of copper, and after that with the concentration of zinc and lead. The recovery of silver has not been featured, so that it is well to fur- nish information concerning the work done at Cobalt. We are enabled to publish a timely and useful article by Mr. W. E. Simpson, previously honorably identified with cyanidation in Mexico and other countries. In his article he records the fact that the Minerals Separation people discouraged the application of flotation to the silver ores of Cobalt. This agrees with our own informa- tion. It was from Mr. J. M. Callow that the successful initiative came. In March 1915 he made his first test, and in the September following he sent one of his technicians with the necessary equipment to establish an experi- mental plant for the Buffalo Mines Company. He re- crushed the tailing from the old mill, reducing the pulp to a product that left only 5% on a 200-mesh screen. In June 1916 another experimental plant was erected under Mr. Callow's directions at the McKinley-Darragh-Sav- age, where the overflow-slime was subjected to flotation, with satisfactory results. These experiments led to the Buffalo Mines Company building a 600-ton plant, which is yielding a 1½-ounce tailing on a heading assaying 8 ounces per ton, and giving a concentrate containing 350 ounces of silver per ton. The general practice of the district has been described by Messrs. Callow and E. B. Thornhill in Bulletin No. 62 of the Canadian Mining In- stitute. It will be noted that at Cobalt the flotation process does not replace, but supplements, ordinary water-concentration. The new process serves chiefly to save the friable silver minerals and fine metallic silver that escape the older processes. Whereas fine grinding causes loss from the jigs and tables, it favors the success of treatment by flotation. This is also the story of cop- per metallurgy. By the way, we venture to draw atten- tion to the fact that it is unscientific to distinguish be- tween flotation and table-concentration by speaking of the latter as a 'gravity' process. Flotation is not inde- pendent of gravity; on the contrary, the rising of the bubbles, bearing their freight of pulverized mineral, is an operation in which gravity is the chief agent; only it happens to be upward, instead of downward, because the mineral-laden air-bubble weighs less than the water it displaces. Fortunately, the major part of the silver in the ores of Cobalt is present in native form, which is quickly responsive to flotation, if untarnished. We sug- gest that this proviso explains the discrepancy between experimental work and large-scale milling, particularly on old tailing-dumps. Mr. Simpson rightly emphasizes the relatively low cost of a flotation equipment and the relatively small space that is occupies, while he recognizes the disadvantage, as compared with cyanidation, of an unfinished product, namely, a concentrate that has to undergo further treatment before merchantable bullion is forthcoming. On the other hand the concentrate produced by flotation is purer than that obtained on table-machines, that is, it contains less arsenic, nickel, and cobalt—this, however, appearing to interfere with the extraction of by-products from which the smelters win an incidental, but most satisfactory, profit. In fact, the marketing of the flotation product has been the chief obstacle to the general adoption of the process at Cobalt. The smelters have levied arbitrary charges, which, with the railroad freight, have made a distressingly big hole in the gross value. Evidently the future extension of flotation depends upon the development of a successful local treatment for the concentrate. Mr. Simpson gives an outline of the methods that have been tried, both with and without previous roasting. We look forward to publishing further details as experimentation advances to a successful issue. This necessary work, we regret to add, is hindered by the threats of the Minerals Separation company and by the litigation incidental to an attempted collection of royalties. On this phase of the subject we have dwelt frequently in these pages. Suffice it to say that we hope the mine-operators of Ontario will succeed in obtaining the protection of the Canadian government so that the process-mongers may be brought to a reasonable state of mind and be prevented from exercising a tyrannical embargo.

Potash as a By-Product of Cement

When the orange-growers of southern California brought action against the cement-makers of Riverside county in order to compel them to abate the nuisance of a fine fume, which volatilizes in the air and settles upon the foliage, causing injury to the citrus trees, they little thought they were doing a service to the owners of the plants and, incidentally, to the country at large. With the option of either abating the nuisance or of closing their plant, the Riverside Portland Cement Company installed the Cottrell apparatus, which effectively prevented the escape of fume into the atmosphere. In the fume thus arrested from 4 to 12% of potash was found. It is not so long ago that Dr. Wilhelm Friedrich Ostwald announced to the German nation that they held the trump card against any post-war trade-war, for by cutting off the only known large deposit of potash from the enemy—that of Strassfurt—Germany could dictate what they should and what they should not eat; potash being, it is needless to add, one of the prime essentials of plant life and, therefore, a necessary constituent of productive soil. Leaving aside the question of post-war trade-war, potash has become an urgent necessity, and some of the best brains in this country have been at work to find a means of supplying the want; and here, through the veriest accident, one might say, a considerable source of supply has been discovered.

To make things more clear it will be well to consider roughly for a moment the manufacture of portland cement. The necessary constituents are lime, silica, alumina, and oxide of iron. In this country the lime is provided generally in the form of limestone and the other constituents as clay or shale. Most clays contain a small proportion of potash, especially when they are derived, as they often are, from the decomposition of granite containing potash feldspar; but other substances, such as feldspar itself, equally as well as clay or shale, would provide silica and alumina for the cement and, at the same time, would give a much larger proportion of potash to be saved as a by-product. The limestone, and the clay, shale, or feldspar, are ground finely, and thus intimately mixed are burned to a clinker in a large rotary kiln, and it is in this burning that the greater part of the potash becomes volatilized and is eliminated from the cement-clinker as fume. Experiments at Riverside have demonstrated that the addition of a small amount of fluorspar to the raw 'mix' greatly assists the volatilization of the potash, and by this means, and by careful attention to other details, the extraction of potash has been so improved that 70% of that present in the 'mix' is leached from the fume by water and converted to sulphate, in which state it is marketed. Care has been taken, also, to so select the raw material that it shall contain the greatest quantity of potash compatible with the manufacture of a high-grade cement. The efforts of the management have been crowned with such success that today the Riverside plant presents the curious anomaly of being not a cement plant that produces potash as a by-product but a potash plant producing cement as a by-product. In other words, potash is the main product sought at the Riverside plant, and, in order to make as large an output of that substance as possible, more cement is being made there than is needed to supply the demand. The present output of the plant is about 5000 barrels of cement per day, and six pounds of potassium sulphate is recovered for every barrel of cement made. The surplus cement will be stored in the form of clinker until the end of the War, when potash prices will tumble and cement will again be the principal product. It is pleasing to note, too, that although the evolution of the process has been made possible by the abnormal price of potash, it is claimed that potash can be made at a profit when that material reverts to pre-war prices. The operations at Riverside have not escaped the attention of cement-manufacturers in other parts of the country. Many cement-makers have added dust-saving plants, and the only reasons to prevent their becoming standard is the high cost of structural material and the lack of assurance that the present price of potash will be maintained. Based on the production of plants already in existence, the estimate has been made that if dust-saving devices were adopted by all the cement-plants in the country it would result in the production of about 135,000 tons of potassium sulphate per annum.

A Plea for Labor

The Editor:

Sir—Your answer to J. F. Harrington in the November 17 issue aroused my interest, and I feel it my moral duty to say a few words in comment. True, I am not trained to produce articles grammatically correct, and in literary style, as I have grown gray in the occupation with the shovel, pick, hammer, drill, machines, and all that; and those playthings are not conducive to such qualities. Yet I feel confident that the meaning of these lines will be understood by the esteemed editors, regardless of the grammatical errors and faulty English of this missive.

You say, in your editorial, that contracting is a means of developing team-work, etc., at an agree price, and with such speed as to yield the worker more than day's wages. Mark well the word 'speed'! Anyone who has had some experience will know that the extra speed almost invariably is coaxed, cajoled, coerced, or driven out of the men's backs. The extra speed, I say, is the result of the extra energy exerted by the men; in fact, the efficiency engineer's problem seems to begin and nd right thre. This is not what he has been taught in college, as I understand it, but that is the way it works in practice—so with the contracts. The amount more than day's wages is nearly always at the price of evor-exertion, and when the workman's' energy is fagged out, he soon falls short of the mark, or, slowing up in one way or other by failing health, he, like the mule, is turned out to grass, with the difference that the mule is turned into its owner's pasture, and the man is turned into the public feeding-ground. They have to foot the bill when picking along the roadside for him is no longer possible. True, if he is knocked out suddenly, he is taken to the county hospital, usually fairly well taken care of, but that is not the case, however, when the workman suffers from general deterioration of health. But—back to the contract again—I have been told time and again that contracts are for the purpose of raising the pay for good men, and to get work done more speedily. Granting the latter to be true, while the former is only nominally so, or I may say, only temporarily so, it is more often told for the purpose of deceiving, which is equivalent to a lie. Knowing to my regret that by using all the ingenuity I possessed, and with intimate co-operation and team-work, and exertion of energy enabling me and my partners to get more than day's wages for a short time, if the time or distance or both to be driven were specified by written contract, it would last that long; otherwise if no written contract existed, we would be told that the place where we worked would be shut-down, and we would either be laid off indefinitely or sent to some remote corner of the mine to ork for day's pay, or else a 'hard' contract, obviously for the purpose of keeping us away from the former place where we worked on contract, that place meanwhile being let on contract at a much lower figure, usually at such a price as would give day's pay by doing the same amount of work as we did when we worked there. If the new team failed to come up to the scratch ,they would get fired and another lot would be given a contract, only to meet the same fate. This would go from bad to worse until the ground changed to relieve the situation, or else the bosses would raise the price a trifle, so that the contractors could by utmost exertion make the standard day's pay. This is the way the 'contract system' works. Not raising the pay of good men, but lowering the pay by raising the amount of work for the day's pay at the rate prevalent in the district. I can name the mine and the efficers, etc., where a number of definite cases occurred similar to this, but that is not necessary to illustrate the point in this argument; besides it is in the past. I am convinced that this is the way it works in general in all the big mines where contracting is practised, and I will pit my experience, to wit, in Alaska, British Columbia, Washington, Idaho, Montana, South Dakota, Colorado, Arizona, Old Mexico, and some of the continent of Europe against any argument, no matter how nicely put up or well delivered, to the contrary, a sto the result, namely, doing more work for the same pay, for that is what it finally amounts to. You may say, with the majority of others, that it is the men's own fault for setting the pace. True, but if you are candid you will also admit that any normal person, not spoiled by adverse experience, delights in doing and getting results, and this trait should not be manipulated to the possessor's injury. But alas, he soons learns that it is to his own and his fellow's detriment; to the benefit of whom? Is it not a natural consequence then that the worker will try to do less than his best; in fact, self-preservation compels him; so shirking becomes a virtue, in consideration for his fellows, who may be less fortunate as to physical endurance and aptitude.

No, my dear Mr. Editor, most contracts and bonuses can consistently be compared to a cartoon appearing in one of the comic papers some time ago, where a negro hired a team of mules to haul a load to town. Wanting to economize on mule-feed, he conceived the ideà not to feed the mules, but to tie a bundle of hay on a stick, and nail that stick to the wagon-tongue, so the bundle would be in front, and barely out of reach of the hungry animals. Eventually they would, by hard pulling. flatten the collars enough so they could reach and get a bite of the hay once in a while. I venture the opinion, Mr. Editor, that those animals did team-work toward getting that hay, and further, that the negro, swelling with pride of superiority, would prod the animals to see if possible to get more speed out of them. The animals in their state of temper would exercise the heaven-born right, and use their natural weapon of defense.

As to the I. W. W., I have no argument with you; suffice it to say that an article written by Professor Parker of Washington University, in the 'Spokesman Review,' a few days ago, hit the nail fairly on the head. He says the I. W. W. is a symptom rather than a disease.

Wallace, Idaho, November 24. A MINER.

[The writer of this letter states that it was not intended for publication, but that, if we choose, we may publish it. We do so because it is the sincere expression of opinion from a thoughtful man. In discussions of this kind we care little about grammatical errors or lack of style; these are small matters as compared with sincerity. The one man to whom these columns are closed is the one that talks through his hat or adopts a pose in order to get into print. Mr. 'Miner' is heartily welcome to the space taken by his letter; he is as sincere as Mr. 'Editor' and he knows as much about his subject. The article to which reference is made above is 'The I. W. W.', by Carleton H. Parker in the November issue of The Atlantic Monthly.' Evidently it was reproduced by the 'Spokesman Review.' We agree with our correspondent that it hits the nail on the head.—EDITOR.]

Solubility of Tungsten

The Editor:

Sir—In your issue of September 1 there appeared an interesting letter by Mr. J. Coggin Brown on the 'Solubility of Tungsten.' Mr. Brown refers especially to the Tavoy district. The occurrence of wolfram here, at the Phoenicia mines, tends to support your and Mr. Brown's views regarding the rapidity of the solubility of wolfram over cassiterite. In the Phoenicia mines, which are in Galicia, north-western Spain, the ore occurs primarily in three parallel quartz veins cutting across chlorite-schist which has been intruded by a series of pegmatite dikes varying in width from one to twenty-five metres. The minerals of the ore vary from two parts of cassiterite and one of wolfram to equal amounts of each mineral. Wherever the veins pass through the dikes more cassiterite than wolfram is encountered; indeed, in those dikes where the feldspars are highly kaolinized only crystals of cassiterite occur. The crystals vary from those of a beautiful ruby-red, about the size of a mustard-seed, to large gray pyramids and prisms about the size of walnuts and looking not unlike the old-fashioned rock-candy. Some wolfram is found in the granite, but as a rule the veins crossing the schist yield a larger proportion of wolfram than those passing through the granite. It would appear that in the kaolinization of the feldspars the wolfram has been dissolved. On this mine many surface workings have the appearance of extreme old age and are locally attributed to the Phoenicians. On the dumps of these ancient workings—dumps of surprisingly limited extent—more wolfram than tin is found; due possibly to the fact that wolfram was previously discarded as waste. It is, however, interesting to note that wolfram on the surface is most frequently found in the form of large single detached crystals quite free from quartz. These crystals may be the reject of ancient cobbings, or, what is more likely, they have decayed out of their quartz matrix and represent the beginning of chemical decomposition.

I have never found single detached crystals of cassiterite. Let me say, in passing, that the wolfram here occurs as a very pure mineral, our concentrate averaging 73% of WO_3. At the Mawchi mines, in the South Shan States, Burma, a tin-wolfram property, my samples of the pits in the detrital deposits below the veins showed a greater proportion of tin over wolfram than samples taken from the veins themselves. Unfortunately my notes relating to these samples are not here and I am unable to give exact figures.

NEWTON B. KNOX.

Noya, Coruna, Spain, October 14.

Magnetite and Copper

The Editor:

Sir—Referring to the note by Mr. Edgar Hall on the occurrence of magnetite in copper ores, in your issue of October 6, I would say that in looking through the Kalata copper mine in the Ural mountains in 1911, I found some black and yellow banded ore. This, on closer examination, proved to be magnetite with iron and copper sulphides arranged in layers, all these minerals apparently being original. The chief sulphide was pyrite with perhaps 2% of copper in the form of chalcopyrite.

San Francisco, November 30. H. W. TURNER.

THE reaction producing basic copper sulphate was stated in the article on 'Heap-Leaching of Copper-Sulphide Ores,' MINING AND SCIENTIFIC PRESS for November 24, 1917, p. 755, as $CuSO_4 + CuO = (Cu_2O)SO_4$. That was incorrect as regards the form of presentation of the molecule. The compound produced is the basic cupric sulphate, $CuO \cdot CuSO_4$. [DE KALB.]

SPELTER purchases by the United States government since April have amounted to 58,128,500 lb. For grade A the price, f.o.b. New York, has ranged from 11¼ to 13½c. per pound, for grade B from 11 to 13c., and for grade C from 8.75 to 9c. About 40% of the spelter used by the Government is of grade B, under requisitions from the Army and Navy departments. It has been proposed by the Zinc Committee to restrict the production of the metal as a means for steadying the market.

THE DOMINION REDUCTION WORKS, AT COBALT, USING THE CALLOW FLOTATION SYSTEM

Flotation at Cobalt, Ontario

By W. E. SIMPSON

SYNOPSIS. Flotation has proved a useful auxiliary in the treatment of silver ores, and is now in operation at practically every producing mine in the Cobalt district. Mines originally equipped with plants for gravity concentration have bettered their recovery as much as 5 to 15% through the addition of flotation-units, while the added cost is from 5 to 15 cents per ton. With the all-sliming process (grinding in cyanide solution), extensive experiments indicate that when recovering the refractory minerals by flotation, the usual revenue is maintained or improved, and that the consumption of cyanide is reduced one-third. The chief difficulty with flotation lies in the disposal of the concentrate, marketing to distant smelters being expensive and local treatment not yet having proved satisfactory. Threatened litigation by Minerals Separation, Ltd., is also seriously embarrassing metallurgical progress in the Cobalt district.

HISTORY. The earliest application of the flotation process to the treatment of silver ores at Cobalt dates from 1910 when some small-scale tests were made in the laboratory at the Coniagas mine, to note the effect, if any, of violently shaking representative samples of mill-pulp to which had been added a few drops of oil. No commercial importance was attached to the results obtained in these simple experiments and the matter remained in abeyance so far as that mine was concerned until quite recently.

The next attempt was made in 1914, when a former employee of the Minerals Separation company constructed an experimental unit at the Temiskaming mine and demonstrated the feasibility of profitably treating the fine tailing, then being run from the concentrating mill to the waste-pile. The litigation ensuing over the use of flotation acted as a deterrent to the continuance of experiments.

While these tests were being conducted, a sample of slime, representative of what may be now called flotation-feed, was sent by the Cobalt Reduction Co. to its consulting engineer in London to determine whether the flotation process could be introduced successfully in the Cobalt district. The tests were made in the laboratory of Sulman & Picard, metallurgists for Minerals Separation, Ltd., and the results were as follows: The sample assayed 5.5 oz. silver per ton, and the products obtained were a concentrate assaying 43.5 oz. containing 57.11% of the silver in 7.3% of the weight of the original slime, a final tailing assaying 1.99 oz. per ton with 24.88% of the gross content left in 69.15% by weight of the slime, and a middling containing the rest. Alf. Tellman, who signed the report, concluded by saying: "I believe that if the slime can be successfully treated by cyanide you will be able to make more profit than with flotation."

The serious adaptation of the flotation process to the Cobalt ores really started when T. R. Jones, manager of the Buffalo Mines, after conducting an extensive experimental campaign, installed the first treatment plant to operate on a commercial scale in October 1915. This unit employed the Callow type of machine and had a capacity of from 50 to 75 tons per day. So satisfactory were the results that additional plant was immediately erected, bringing the total capacity to 600 tons per day in September 1916.

The first flush of success led to the statement by enthusiastic operators that flotation would entirely replace the gravity method of concentration, completely displace the cyanidation of Cobalt ores, and revolutionize the established practice of metallurgy. Further experience, however, has called for a modification of these views. A satisfactory recovery of the silver-bearing minerals by flotation is only obtainable from material in a fine state of subdivision and the tendency now is to apply the process to the treatment of slime and to such portions of tailing as may be sufficiently rich to warrant the additional expense of fine grinding.

THE ORE. The Cobalt orebodies are aggregations of veins formed, apparently, by the deposition in regular succession of various minerals in zones of fracture in the country-rock, which may be either slate or conglomerate. No actual walls define the limits of these orebodies, but the gross width of mineral matter extractable at a profit may reach 20 or 30 ft., although the veins individually are seldom more than a few inches wide. A geological diagnosis indicates the deposition of the minerals in the following sequence: Smaltite, niccolite, calcite, argentite, native silver, and, lastly, bismuth; copper also occurs in traces. The ore, therefore, is decidedly complex.

About 90% of the silver of the district occurs in the form of the native metal, a fact that did much to simplify metallurgical operations in the early days, and permitted many mining prospects to develop into wealthy producers without any capital outlay beyond that required for the purchase of a few tools. This metallic silver frequently is found in slabs six to eight inches thick and weighing thousands of ounces, but much of it occurs in a state of subdivision so minute as to necessitate pulverization of the ore to pass a 200-mesh screen in order to obtain a satisfactory extraction. According to laboratory experiments by J. M. Callow and E. B. Thornhill* this fine native silver floats more readily than any other silver-bearing mineral. Actual full-scale work in the mills, however, fails to support this statement.

EARLY MILLING. The first mills erected at Cobalt were equipped with crushers, stamps or rolls, jigs, and gravity tables, two products being obtained, a concentrate assaying several thousand ounces per ton, that was sold to the smelters, and a tailing that was stored and is now the principal material available for treatment by the flotation process. Later plants were equipped for both gravity concentration and treatment by cyanide. At this date (October 1917), owing to the success achieved by the Buffalo Mines, every mill is either equipped with a flotation annex or has one under consideration.

APPLICABILITY OF FLOTATION. In reaching a decision regarding the adoption of a process and the installation of a plant, two factors exert an influence in favor of

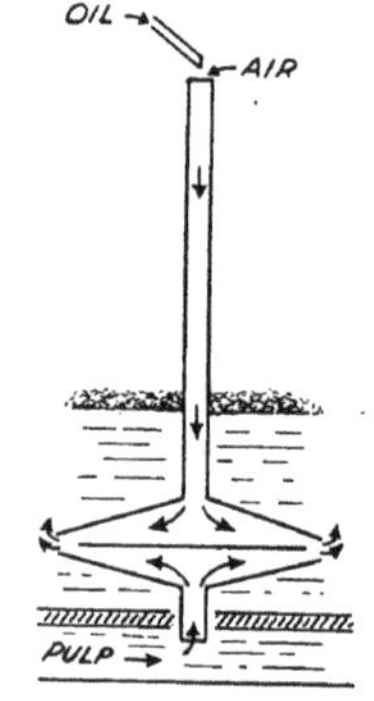

FIG. 1. THE OBOCH IMPELLER

flotation at Cobalt more than in mining districts generally; these are (1) the winter is long and heating is costly, and (2) many of the mines appear to have already passed the zenith of their prosperity, their ore-reserves being now narrowly limited in extent. The most desirable process for the Cobalt mines, therefore, is one that can be housed in the smallest building and installed with the lowest capital outlay. Experience has proved that a flotation plant can be erected in a space less than one-fourth that required for either a group of gravity-tables or a cyanide-plant of equal capacity. In a modern mill for treating 100 tons per day it is estimated that a saving in capital expenditure of about $20,000 can be effected by the substitution of the new for either of the older methods. The working profit also favors the newer method. The chief handicap to flotation is its inability to produce a finished article, that is, one easily marketable, such as high-grade bullion. Flotation in reality is a method for concentrating valuable mineral into small bulk, the ratio of the weight of material treated to that of the product being, in the Cobalt district, approximately between 50 and 100 to 1. This concentrate either must be treated locally or sold to distant smelters, whereas if cyanide bullion is produced no further treatment is necessary.

*Bull. Can. Min. Inst., June 1917.

FLOTATION-OILS. Owing to the difficulty in floating the arsenides and other refractory minerals with which the silver is associated, the oil-mixture to be used must fulfil the conditions of being a collector, a frother, and a powerful adhesive. The Cobalt blend of oil consists of 20% pine oil, 70% creosote, and 10% tar. The last does not have the selective action on the valuable minerals that could be desired, its function being to adhere tenaciously to the mineral and improve the recovery, although this is at the expense of the grade of the concentrate produced. The consumption of oil throughout the district varies from ½ lb. to 2 lb. per ton of material treated.

FLOTATION APPLIANCES. The Callow system of flotation was the first to operate commercially in the district, and, as the results obtained were satisfactory, the pneumatic cell rapidly became popular. The Callow appliance has been so often described that a repetition here is unnecessary. Recently a machine, the invention of a local engineer, F. O. Groch, has made its appearance, and indications are that it possesses merits that should entitle it to consideration. Externally it bears a striking resemblance to the Minerals Separation machine, but the method of operating is quite different. It may be described as a number of impellers operating in a V-box 14 ft. long by 5 ft. wide, divided transversely and longitudinally into compartments, the partitions being slotted or submerged so as to compel the pulp to take a zigzag course, both vertically and horizontally, in its passage through the machine. The impellers, six in number, combine the functions of centrifugally agitating the pulp, diffusing the oil, and sucking air into the mass during agitation, all accomplished in one operation. The impeller consists of a hollow vertical shaft to which is attached at the lower end a duplex centrifugal 'runner,' similar in design to a pair of centrifugal pumps separated from each other by a disc. During the operation this is submerged in the pulp. The discharge from the 'runner' is at the periphery and the pulp-intake to the lower part is from below through the hollow hub, while the upper part has its inlet for oil and air through the hollow shaft. In operation the pulp is admitted into the first compartment at the bottom of the V-box on the under side of a false bottom through which the hollow hub of the centrifugal 'runner' protrudes. There it is transferred from the lower to the upper side of the false bottom by being sucked up through the hollow hub, to be subsequently discharged in a horizontal plane at the periphery. This violent ejection creates a vacuum that reacts in such manner as to suck into the vortex everything in its neighborhood, including the contents of the upper part of the 'runner.' As this upper part has its inlet from the top of the hollow shaft the result is that a lavish supply of air is drawn

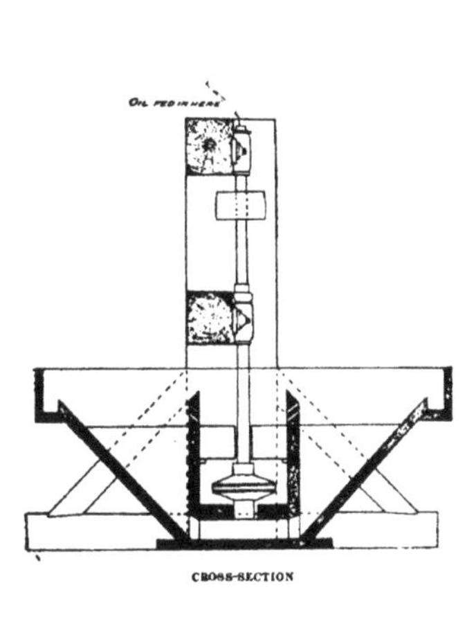

FIG. 2. THE GROCH FLOTATION MACHINE

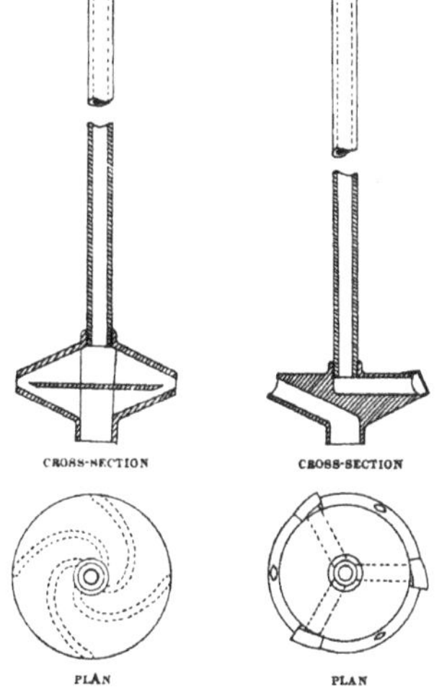

FIG. 3. THE GROCH IMPELLER FIG. 4. THE RUTH IMPELLER

into the impeller and discharged into the pulp in the form of an infinite number of minute bubbles. Theoretically this is precisely what is required for ideal flotation. A working model of the Groch machine was recently shown at the National Exhibition of Chemical Industries, New York, and the discovery was then made that, as so often happens, another operator, Joseph P. Ruth, of Denver, had independently hit upon practically the same idea and was gaining considerable success in the West with his improvement. Illustrations of the impellers of both Mr. Ruth and Mr. Groch are here shown for purposes of comparison. See Fig. 1, 2, and 3.

In the Groch machine, the oil is admitted, drop by drop, through the hollow shaft of each impeller, where it falls on the rapidly revolving disc, and is, as it were, atomized and shot into the pulp along with the minute air-bubbles. By feeding different oils into the pulp through the impellers in series so that the minerals in an ore are brought into contact with oils of increasing viscosity it is claimed that preferential selection is obtainable. The idea is worthy of further investigation, and it is possible that a decided improvement can be obtained over existing methods.

THE CHEMICAL COMPOSITION of the concentrate recovered by flotation differs greatly from that obtained by the gravity method of treatment on tables, for the reason that the physical properties, specific gravity, and flotability of minerals bear no relation to each other. Arsenides, with which the silver is invariably associated, are difficult to float, while chalcopyrite, a frequent constitutent of the ore, floats readily. Comparative analyses follow:

	Gravity concentrate	Flotation concentrate
Silver	1000–2000 oz. per ton	300–500 oz. per ton
	%	%
Copper	0.3	1.0
Arsenic	25–40.0	0.5
Nickel	3–6.0	0.25
Cobalt	5–7.5	0.25
Iron	10.0	9.0
Sulphur	5.5	3.0
Alumina	7.0	18.0
Lime	5.5	6.5
Silica	15.0	55.0

This difference in composition was not anticipated at the time when the introduction of the flotation process was first taken under consideration, so that the situation developed in an unexpected manner. Several smelting companies that had been particularly keen to purchase the gravity concentrate absolutely refused to tender for the flotation product. Its extreme fineness and its deficiency in nickel, cobalt, and arsenic, from which at present valuable by-products are being obtained, is doubtless the cause of its unpopularity.

MARKETING THE CONCENTRATE. The greatest drawback to the flotation process, so far as Cobalt is concerned, has been the high cost of shipping and treating the product. The American Smelting & Refining Co. is practically the only purchaser able to accept any large quantity of the material. This buyer has fixed an arbitrary charge for treatment, and makes payment on 95% of the assay-value, which results in a total cost to the mining companies of approximately $40 to $50 per ton of concentrate exported. Assuming a concentrate assaying 500 oz., with silver at 70c. per oz., based on shipment from Cobalt to Denver, the details of cost are as follows:

		Per ton of concentrate
Steam-drying to approximately 10% moisture, coal at $7.50 per ton		$2.50
Sacking and loading		1.35
Sacks and wire-ties (allowing five trips per sack; original cost of sack, 15c.)		1.05
Haulage from plant to cars at Cobalt, about		0.75
Freight, Cobalt to Denver		14.20
Treatment-charge at smelter		10.00
Extraction-loss, 5% on 500 oz. silver at 70%		17.50
Incidentals:		
U. S. revenue, per car	$1.20	
Assaying	6.00	
Customs' charge	2.50	
Representative at sampling	7.00	
Total	$16.70	0.42
Total		$47.77

LOCAL TREATMENT OF CONCENTRATE. Two methods of treatment locally are being tried, the one necessitating roasting of the concentrate, the other operating with this material in its raw state. The roasting process consists in placing the concentrate, with 15% moisture, in a specially constructed furnace, patented by the Holt-Dern organization, where it is slowly oxidized without allowing any flame to reach the surface. The charge is a mixture of gravity and flotation-concentrate in the ratio of 1:4, giving a self-roasting mixture, the fuel being supplied from the sulphur, antimony, and arsenic present. The furnace may be described as a 7 by 9 ft. rectangular brick-lined chamber 4½ ft. deep, into which during operation the charge is fed wet at regular intervals, a corresponding amount of the roasted product being simultaneously withdrawn underneath through the patented grate with which the furnace is equipped. Thus a blanket of wet concentrate remains continuously on top, minimizing the loss by dusting and volatilization. The contained moisture is really a saturated solution of common salt from which the chloridizing effect is derived.

The roasted material is leached with brine and the silver precipitated on scrap-copper. The copper is later recovered by precipitation on scrap-iron. The resulting silver, in the form of a cement, is treated with acid and melted.

The cost of this treatment is not yet ascertainable as many difficulties are being experienced that will require ingenuity to overcome. The brine eventually becomes foul and the customary high-grade bullion, which Cobalt operators consider as their standard for export, is not obtainable except at a prohibitive cost for labor and acid.

Leaching by brine may be replaced later by agitation

in cyanide solution, thereby eliminating some of the difficulties, but at an increased cost. For this work a thorough roast would be essential and this is scarcely obtainable in any furnace where the charge remains motionless during the whole operation. Results so far indicate that the charge has a tendency to sinter. There remains the alternative of dealing with the concentrate in its unroasted state. Here, as in so many other matters, local metallurgists have shown remarkable enterprise, and hopes are entertained that a process, already installed at the Cobalt Reduction Works for the treatment of gravity concentrate, may prove applicable to the flotation product. This process consists of intensive tube-milling, first in water with 25 lb. of calcium hypochlorite per ton of concentrate treated. The operation exerts an oxidizing and chloridizing effect, which is completed in about 24 hours. Then follows a period of water-washing and agitation in a 0.25% cyanide solution for 48 hours, during which time about 95% of the silver-content is extracted. Filtering then follows, and precipitation from the solution is effected by the sodium-sulphide method with the production of bullion 996 fine.

This method applied to flotation-concentrate has so far failed badly. The oil-mixture used for the recovery of the concentrate is of a decidedly tarry and adhesive character; consequently the difficulty of attacking the oil-enveloped particles with the cyanide is even greater than that experienced generally in dealing with flotation concentrate. A vigorous campaign of experimental research is being started and it is confidently expected that the existing difficulties will be overcome.

Recovery and Cost of Operation. The litigation, with which the Cobalt district is being threatened, prohibits the publication of authoritative details for fear that they may be used subsequently in law-court proceedings. It may be stated, however, that, as a general rule, the recovery varies from 75 to 90% of the silver in the material treated by flotation and the cost of the actual operations from the time of receiving the slimed feed to that of discharging the finished concentrate is roughly 20c. per ton, divided as follows:

Item	Cents
Oil	4
Labor	9
Power	6
Repairs	1
	20

The estimated cost of treating the accumulated sand-tailing of the Buffalo Mines by flotation, and of converting the concentrate recovered into actual cash, is given in detail by Robert E. Dye in a paper read before the Canadian Mining Institute in March 1917, together with those obtained in previous operations by sliming in cyanide. The particulars are as follows:

	Flotation cents	Cyanidation cents
ading tailing by steam-shovel	4	4
be-milling and classifying	40	50
Heating and lighting	3	10
Overhead expense	14	14
Thickening, agitating, etc.	..	16
Chemicals	..	48
Precipitation and miscellanies	..	17
Refining	..	3
Flotation	8	..
Drying and loading concentrate	5	..
	74	$1.62
Smelter-charge, including freight and losses	83	..
Total, per ton	$1.57	$1.62

A 90% recovery is obtained by flotation, and a similar extraction by cyanidation to which intensive tube-milling is a contributing factor. The working cost for both systems is practically identical, but there are possibilities of reduction in the cost of treating flotation concentrate that do not exist in the competitive method. No charge is shown in the foregoing figures for amortization and depreciation, and as the quantity of material available for treatment at the Buffalo Mines is not over 350,000 tons, the deciding factor of capital expenditure is overwhelmingly in favor of flotation. A promising avenue for research in Cobalt is being found in the endeavor to obtain a combination treatment of sliming in cyanide followed by flotation. It has been discovered that, if the customary strong cyanide solution be used, subsequent treatment by flotation is abortive, and that the flotation recovery is increased as the cyanide solution used in the initial treatment is weakened. The probable reason for this is that strong cyanide, in attacking the silver-bearing minerals, pits their surface, thereby lowering their flotability. Experiment shows that a reduction in the strength of the usual working solution from 0.25% NaCN to 0.10% lowers the extraction by that chemical from the regular 90% to 75% of the silver in the ore, but it leaves the residue in such a state that, with suitable water-washing, a further 75% of the content is readily recoverable by flotation. This modification is showing a saving of over one-third of the cyanide ordinarily consumed.

Minerals Separation. In harmony with its attitude elsewhere, the Minerals Separation company, through its subsidiary organization, the North American Corporation, has threatened proceedings against all users of flotation in the Cobalt district, so as to collect, if possible, the royalty of 2½% of the gross value of the whole concentrate recovered, according to the usual demands of this patent-exploiting company. The success of flotation at Cobalt is due entirely to local enterprise, therefore this demand is resented bitterly. The indications are that a legal fight is to follow, and a campaign has been started to enlist Government action "with a view to having the patents annulled." Amid the legal turmoil, metallurgical progress is being seriously handicapped, the free exchange of ideas has been completely stopped, and an embargo is being placed on valuable information.

It is sincerely hoped that an equitable settlement may be obtained at the earliest possible moment.

CONCLUSION. Flotation is undoubtedly destined to play a part in all milling operations in the Cobalt district, although its scope will not be as extensive as was at first anticipated. As a competitor to the sand-table it has not met with the success gained in other localities for the reason that the valuable minerals are difficult to float and are easily recoverable by gravity-methods. For successful flotation, the arsenic, nickel, and cobalt minerals, with which much of the silver is associated, must be reduced to a fine state of subdivision in order to conform to some flotation law in which the ratio of surface-area to mass is an important factor. Fine crushing is expensive; consequently the best field for flotation lies in the treatment of primary slime. On this material, it has already completely superseded the slime-table, the additional revenue being greatly in excess of any additional cost incurred. One concentrating plant treating 150 tons of ore per day has been able to add to its recovery from 200 to 300 oz. of silver in concentrate daily, the additional revenue being directly attributable to the introduction of flotation for the treatment of slime. The approved type of gravity mill for Cobalt, therefore, should contain jigs for the extraction of the coarse metallic silver, gravity tables for the treatment of sand, and flotation for the slime.

In the cyanidation of Cobalt ore, certain inroads by flotation have been made, but this results not so much from the actual benefits derived from the use of flotation as from the disadvantages both temporary and permanent associated with the older system. During the War, a shortage of cyanide has been frequently threatened and the finding of some substitute has become imperative. No actual shortage in Canada has, so far, been experienced but the possibility of having to suspend operations for this reason has acted as an incentive to the development of flotation.

Cyanidation lacks elasticity, especially in respect of the cost-sheet. A fixed consumption of cyanide per ton of ore treated has to be incurred regardless of the value of the material being cyanided. This means practically a fixed limit of cost with no margin for fluctuations in revenue. The ores of Cobalt are notoriously erratic in value, consequently, on many mines, often owing to a fall in grade, cyanidation is actually being conducted at a loss. With flotation, the chief item of cost is directly associated with the treatment of the valuable concentrate produced, consequently the total cost of the whole operation fluctuates directly as the value of the ore being treated. This is a most favorable factor. A combined cyaniding and flotation plant should therefore give an improved recovery, effect a saving in cyanide, and automatically harmonize the cost of treatment with the value of the ore being milled.

Flotation has already been allotted a useful place on practically every producing mine in the district and its sphere of operation is bound to increase greatly as soon as a more advantageous method of disposing of the product has been devised. The probabilities in this matter seem to favor treatment locally.

The capital outlay for a flotation annex, with storage-tanks, filter, buildings, etc., but no crushing appliances, is roughly $50 to $60 per ton of daily capacity. This, in itself, gives the process an advantage over all competitors.

The flotation of silver ores is now a well-established success and the results being obtained from researches in the Cobalt district should prove of much interest to operators in other countries particularly Old Mexico, where the argentiferous minerals are less refractory and can be much more readily treated than those of New Ontario.

Road-Building Over Wet Soft Ground

Wagon-roads built to reach mines in the mountains are frequently impassable in places on account of the existence of wet areas that constitute weak links in the chain of connection between the mine and the shipping-point at the railroad. To overcome these bad spots, a method that has been thoroughly tested by the U. S. Government may be applied; this is known as the burned clay method. The roadbed is plowed as deeply as possible. Furrows are then dug across from ditch to ditch, extending through and beyond the width to be burned. For a 12-ft. roadway the transverse furrow should be 16 ft. long, extending 2 ft. on each side. Across the ridges between the furrows, which should be about 4 ft. apart, a course of cord-wood is laid longitudinally to form a series of flues. These should be laid so that the pieces will touch, forming a floor. Another layer of wood is thrown irregularly across this floor in crib-formation, with spaces left between, in which clay is piled. The lumps of clay placed on the crib-floor should be coarse enough to permit a draught for easy combustion. A third course of wood is laid, in the same manner as the first, on top of this, and each opening and crack is filled with brush, chips, bark, small sticks, or other combustible material. A top layer of clay is placed over all this, being the material dug from the side ditches. The top layer should be from 6 to 12 in. thick. The best results are obtained by firing all the flues of a section simultaneously, and maintaining combustion as evenly as possible. A supply of light dry kindling should be at hand to prevent the fire from dying down at any one place. The firing should be started on the windward side. If the combustion is too rapid, it may be checked by banking the mouth of the flue with clay. When sufficiently cool the roadbed should be brought to a high crown before rolling, in order to allow for compacting of the material. The finished crown should have a slope toward the sides of about ½ in. to the foot. The cost is estimated at approximately $85 for each 300 ft. A road treated in this way will not become plastic when wet; it will remain hard, and will carry heavy loads. Clay clinker burned in this way is used on many railroads for ballast, particularly in Illinois, Indiana, and in some of the Southern States.

THE YOSEMITE DREDGE, AT SNELLING, CALIFORNIA

Recovery of Platinum in Gold-Dredging

By JAMES W. NEILL

During the 22½ months preceding July 1, 1917, the Yosemite Dredging & Mining Co.'s 3¾-ft. dredge, operating on the Merced river, near Snelling, California, shipped to market a concentrated 'black sand' from the clean-ups for the following returns:

Months	Yards	Black sand, lb.	Platinum, dwt.	Value	Per yard	Receipts, total	Per yard
9	617,500	921	365.25	$1734.95	0.281	$1913.10	0.310
4.5	308,750	622	172.25	732.05	0.237	795.30	0.260
9	617,500	1011	285.75	971.55	0.157	1078.05	0.177
22.5	1,534,750	2554	823.25	$3438.25	0.222	$3786.45	0.248

In addition to the platinum, there was recovered 18.75 dwt. osmiridium valued at $24.75 total, and 369 dwt. gold worth $369. This indicates a value per pound of sand shipped of $1.48. Prices of platinum received were $3.40, $4.25, and $4.75 per dwt., the highest figure covering the latest shipment. The refiners, Shreve & Co. of San Francisco, state that this product carries a higher proportion of platinum to osmiridium than any received from mines in the State, other shipments carrying as high as 50% of osmiridium.

If the 65 million yards of gravel dug annually by dredges in the State of California all yielded platinum in this same proportion, namely 0.25c. per yard, it would amount to $150,000 per annum. I am informed that there is some platinum in all the gravel mined, the amount varying with locality, but also more probably, with the methods used to save the metal. On our Yosemite dredge the following method is used to produce the sand described in the table above. The dredge uses Hungarian riffles of 1 by ⅜-in. angle-iron, set one inch apart. At clean-up time these are removed, the gravel loosened, and shoveled to the top end of the table, or sluice, then streamed down with water from a hose, stirring meanwhile with rake, or hand, to get rid of the pebbles and lighter sand. After this operation there remains on each table a small pile of sand and amalgam, which is scooped into an enameled bucket and dumped into the long-tom box, usually at the rate of one small bucket to each table.

This long-tom is of wood, 5 ft. long by 2 ft. wide and deep, set at an angle of 1½ in. per foot. At its lower end a slot delivers the pulp into a bucket set in the upper end of a sluice-box, to catch the amalgam and quick.

THE LONG-TOM AND SLUICES FOR GOLD-SAVING

silver. The overflow from this bucket drops upon the sluice, whence it passes over a shallow quicksilver trap, and then over five feet of riffles set in the sluice; thence the material flows to waste overboard. The sluice below the quicksilver traps is 10 in. wide by 6 in. deep, therefore the capacity of the riffles is small.

All the material from the clean-up is handled in this way, and everything from the long-tom passes over these riffles. When all has been worked up, the riffles in this sluice are lifted, placed in the long-tom box, and washed down again, to remove as much quicksilver as possible, but this time the launder is left without riffles, and all the sand is washed into a galvanized-iron tub, passing on the way through a 20-mesh screen held under the spout by one of the operators.

The coarse material saved in this way is picked over by hand, usually yielding a few small nuggets, half quartz and half gold, the rest being chips of iron, lead, shot, etc., all of which goes back onto the tables. This sand, so produced, is stored, dried, sacked, and finally shipped, yielding as shown in the table given above.

It might naturally be supposed that with the rather crude methods described there will be some loss of platinum and no doubt there is, but I have panned the tailing flowing from this sluice many times, with either entirely negative or very low results; hence I am led to believe that the losses are negligible. Nevertheless, another 5-ft. length has been added to this launder, thus doubling the riffle-area.

Whereas the specific gravities of gold and platinum are practically the same, as also are the particles in size and shape, the conditions governing the recovery of the metals on the dredge are entirely different. The particles of gold are covered by the quicksilver and soon become agglomerated into lumps that are covered by the little pools of 'quick' gathered in the riffles, pockets, and stops, or adhering to the metallic surface in less exposed places. The platinum, not having any affinity for quicksilver under ordinary conditions, is dependent solely on its gravity to keep it in the riffles (like, of course, the coated or rusty gold); hence there must be a considerable loss of very fine or very light grains of platinum. Those familiar with dredging know well that the spaces between the riffles, supposed to form pockets in which there may be some 'riffling' action or some space for metals to settle, quickly become filled with pebbles, the interstices between these being filled with sand and the whole then thoroughly cemented with mud, until it is not possible to make a dent in the surface with a finger or thumb. It seems a miracle that any metal is retained under such conditions! Ordinarily such a surface is all that is offered to the platinum for a resting place.

The Yosemite dredge is equipped with ten Neill jigs, one set at the lower end of each of the first four tables on each side, the other two being set near the distributor, in the sixth table, and handling sand from both the fifth and sixth tables in order to get the necessary fall. These jigs are of the smallest size, having a screen-area of only two feet along the travel of the gravel; they are fitted with an 8-mesh screen and use one to two inches of cast-iron shot-bedding, and deliver a hutch concentrate all 20-mesh or finer. This jig-product is passed over stationary silvered copper plates of varying length, placed in launders below the regular tables, from five to only two feet, according to the space and fall available. All deliver the pulp into one straight sluice which empties over the stern. This latter has a fall of only half an inch per foot, which is too flat, but that is all the fall available. The upper portion of this sluice has a flat bottom, without riffles, but the last five or six feet has riffles, consisting of inch-boards through which one and one-half inch holes have been bored spaced at 5-ft. intervals.

This apparatus yields $200 to $500 per month in amalgam from the plates, and a considerable amount of quicksilver from the riffles below the plates. These riffles were only lifted and the material washed down in the long-tom twice per month at clean-up times. This arrangement of the plates, and riffles to handle the jig-concentrate, is not that originally designed for the dredge, as was illustrated in my article* in a previous issue of this paper (November 28, 1914). That called for elevating all the concentrate to the upper floor, settling in spitzkasten, flow over 3 by 8-ft. plates one on each side of boat, long launders below the plates in which by riffles and stops to detain the platinum and mercury; or, if found necessary, there was sufficient fall for the placing of a Wilfley table. This equipment was not re-installed when the boat was re-fitted, the present arrangement allowing us to dispose of the two sand-pumps and the costs of a man per shift to watch the apparatus; in other words, the later equipment is the application of commercial metallurgy as against theoretical metallurgy.

In order to determine the amount of sand-concentrate produced by these jigs, and its possible value, I took an ordinary galvanized-iron tub holding 17 gal. of water, arranged a movable spout for the end of the plate-tailing launder, and then proceeded to take samples of the tailing, timing the flow until the tub was full, then allowing a short time for the sand to settle, pouring off the water and slime, and weighing the sand. This sampling was continued during several days at intervals of 15 minutes, first on one side of the dredge, then on the other. In this way one complete cut across the pond was sampled, the product weighed, the amount of water measured, and a sample obtained for assay. The tabulated result indicated a flow of 100 gal. water per minute from all 10 jigs, and 42,000 lb. of wet sand in a 22-hour day, or 35,037 lb. dry weight. The sample was sent to the General Engineering Co. of Salt Lake City, and by them run over a laboratory Wilfley table returning to me the concentrate. This test resulted as follows:

	%
Concentrate	1.81
Middling	7.25
Tailing	90.94
	100.00

*See also 'Gold-Saving on Dredges,' by Howard D. Smith. M. & S. P., August 5, 1916.

he concentrate was sent to Shreve & Co., who refined r me, and reported half an ounce of platinum metals ton.

his, therefore, would indicate

$35031 \times 1.81 = 634$ lb. concentrate per day

$634 \times 30 = 19020$ lb. per month

$\frac{2000}{19020} = 9.5$ tons at 0.5 oz. platinum or 4.75 oz. at, say, $90 per ounce.

hus an apparent loss of $427 per month is shown, nst a total caught and saved in the riffles of about per month.

aturally, it would not be fair to expect exactly such lts each and every month, as the yield of platinum ls would be nearly proportionate to that of gold. month during which this test was made being about rmal month in dredge-yield, one might reasonably for an average as indicated.

ollowing this test, I had several tests made on the rial remaining on the riffles in the launders below plates, and in every case these have proved of con- able value, in one case 600 lb. of the sand yielded lb. of concentrate carrying 6c. per pound in plat- a and gold, in the ratio of 4.5 : 1.

e next question was how to handle the material, would a Wilfley table work on the dredge during operating time? The consensus of expert opinion that it would not, but, in spite of this I purchased a size Wilfley and set it up on the upper deck of the ge. This table has a standard timber frame but eck is half the normal size, the total weight being t 2100 lb. The upper deck in our dredge is of light en construction, sloped rather sharply to the sides, placed upon the floor two 4 by 8 in. timbers, 12 ft. set them level fore and aft and cross-ship, and d the table-frame to this, placing no underpinning pports below. The motor was set on one of these s and was belted to a 20-in. pulley, which gave us equired speed, 325 revolutions per minute. The supply comes from the priming pump, delivered 50-gal. barrel, and thence by ¾-in. pipe to the feed ash-water troughs on the table.

sand from the jigs is elevated by a Krogh sand- into a V-bottom tank 4 ft. square with peripheral ow, the flow of the sand to the table being regulated ans of molasses cocks, and the overflow water going wash-water barrel.

trary to expectation, the table, under normal con- s of digging, runs perfectly. Raising or lowering gging-ladder changes the table-level and interferes he discharge of the concentrate, but this is readily ied by a wedge under the tail end of the table- and as the ladder is not often moved this creates iculty.

frequent and often rather violent surging of our oat, which digs into pretty stiff ground and bed- times, does not interfere with the operation at ves of water and some light sand pass lengthwise table, but the streak of concentrate once formed clings to the surface and holds its line in spite of these surges. One may cut a greater or lesser streak of concentrate as desired, and this will preserve itself without change of table-inclination or of cutting-launder for an hour, or hours at a time. Naturally, in this case one does not look for the precision in concentration desired in copper or lead-mining work; nor is it necessary. The streak of concentrate shows at its upper edge a fine white line of minute globules of floured quicksilver, and with this, or detained by some black sand just along side of it, the particles of gold and platinum, which are practically all contained in the upper few inches of the concentrate streak. The free quicksilver in large globules travels along the ends of the table-riffles and finally lands in the middling-compartment, some of it, probably 'lazy' from contained gold, goes into the concentrate-compartment.

Preliminary work with the table demonstrated the necessity for an efficient quicksilver-trap on the feed end of the table, so as to get rid of this element in the concentrate, and relieve the operator of the need of close watching to prevent its loss. Also we found it necessary to install a screen to take out some oversize pebbles, and such shot as gets through broken screens.

Zinc From Low-Grade Complex Ores

During the past two years many processes for the treatment of the oxidized ores of zinc have been investigated at the Salt Lake City station of the Bureau of Mines. In the treatment of most ores the proposed hydro-metallurgical processes have been found to present serious difficulties. On the other hand, zinc-sulphide ores are at present being treated by such processes with considerable success. Owing to the difficulties in attempting to apply hydro-metallurgical processes to oxidized ores of zinc, an igneous concentration process was tested, with encouraging results. The process consists in blowing a blast of air through a mixture of oxidized zinc ore, coke, and limestone. As a result the zinc oxide was drawn off as a fume, which could have been collected in a bag-house, or by electro-static precipitation. The ore from which the zinc has been volatilized forms a slag in the furnace, and is tapped off in the usual manner. The commercial application of such a process is important, as at present there is no process being used that permits the successful concentration of low-grade oxidized ores of zinc. Such ores have been accumulating in all of the important mining districts of the United States, especially in Utah, where there are large deposits.

METALLIC MAGNESIUM is now worth about $4 per pound. This price is the great deterrent to its use in the alloy with aluminum, known as magnalium, which was previously described in the MINING AND SCIENTIFIC PRESS. Electro-chemists are urged by the Council of National Defense to endeavor to solve the problem of producing it cheaply.

Blasting Troubles

Blasting-practice offers many chances for exasperating errors that result in loss of time, labor, and supplies, even if no more serious consequences follow. Every little while an epidemic of some special difficulty will break out, and the foreman may have a long hunt to ascertain the cause. At one mine in Mexico persistent trouble was experienced with 'stinkers' and misfires. The investigation was finally extended to the magazine, which was a low frame structure built into the hillside and covered with dirt. It was summer-time, and the atmosphere was hot and moist; the magazine was poorly ventilated, and the manager had put in store a large supply of 40% dynamite, sufficient for the operation of the mine for a year. The boxes of powder had never been turned while in the magazine, and the high temperature had promoted drainage of the nitroglycerine toward the lower side of the cartridge. Nitroglycerine had leaked out; the boxes were saturated with it; and the homogeneity of the explosive had been destroyed.

Deterioration of explosive through moisture when old damp tunnels are used for storage is well-known. The objection is not to the tunnel but to the moisture. Even at moderate temperatures the moisture facilitates the seepage of the sodium nitrate from the 'dope.' This salt is mixed with the cellulose, or other absorbent, in order to furnish oxygen for its combustion, thus removing what otherwise would be a cause of retardation of the velocity of the detonating wave. It is a very deliquescent salt, much more so than potassium nitrate, which is too expensive to be employed for this purpose. In consequence of the readiness of the nitrate to absorb moisture from the air, it exudes from the cartridge and forms saline crusts on the wrapper and on the box. As soon as it has begun to exude the homogeneity of the powder is gone; the nitroglycerine will then more quickly drain from the dope; the deterioration of the powder will be impaired, and 'stinkers,' misfires, and low-energy explosions result. It is cheaper to provide a dry storage-place than to use a ready-made magazine that is damp. If a man does not choose to determine the moisture in some of the simple ways that every chemist knows, he is, at least, likely to possess a cultivated sense of discrimination between a damp and a dry cigar; let him leave a cigar in the place proposed to be used as a magazine, and in 24 hours he will know what to do, for a place that proves to be a good humidor for cigars is too damp for the economical storage of powder.

Another fertile cause of trouble arises from the improper handling of fuse. At a mine in New Mexico an experienced foreman, sinking a shaft on contract, was in despair over what he called "a perfect spree of misfires." He should have made an advance of three feet per day, but he was gaining only six inches instead. He called on a well-known mining engineer for advice. The fuse was inspected, a few lengths burned, a few caps exploded, and the dynamite carefullly investigated. The engineer then asked to have the day's primers made up; he watched the miner cut a few lengths of fuse for which purpose the ordinary combined crimper and fuse-cutter was used. The engineer asked to see the tool, and found that the blades were dull and the joints loose. The result was that the instrument pinched the fuse while cutting it, thus crowding the powder-stream and leaving a thick ridge of the asphaltic paste where the powder should spit into the fulminate. Side-spitting was the result, so that only an occasional fuse would set off the cap. That day he cut the fuse square across with a sharp knife, and there was not a single mis-fire in the round of shots.

The Rusting of Iron

When we consider the enormous economic waste due to the corrosion of iron, a waste which the Bureau of Mines estimates to be in the neighborhood of 1,000,000 tons of that metal per year, to say nothing of the amount of zinc, other protective metals, and paint annually applied in an effort to diminish this waste, there is abundant reason why time and money should be spent in a thorough study of the problem. The Bureau rejects the electrolytic theory that has been so universally accepted, on the ground that pure iron, that is to say, the purest iron obtainable, rusts and pits as badly and often worse, than the impure product. The work of the Bureau during the last year has been devoted to the study of surface influences, particularly to the influence of rust once formed on the progress of further rusting. The investigation has shown that rust is a factor in the progress of rusting; that there is a reverse polarity according as the rust is dry or wet; that the iron underlying freshly-formed rust is electro-positive to the surrounding metal, and consequently goes into solution; and that dried rust assumes a negative polarity which tends to reverse if it becomes thoroughly wetted. What is considered to be a most important finding is that wet rust does not act as an electrode but as a semi-permeable diaphragm that makes the underlying iron electro-positive, and this promotes its solution. These findings would seem to support rather than to reject the electrolytic theory. That rust begets rust was accepted, if we remember rightly, by John Percy, but he accounts for it as a chemical rather than a physical process. He considered the rust to be a continual conveyer of oxygen from the atmosphere on the one side to the iron on the other; thus, on the outside, the reaction

$$2FeO + O = Fe_2O_3$$

was taking place, while on the inside the reaction would be expressed by

$$Fe_2O_3 + Fe = 3FeO$$

That both chemical and electrical reactions take place in the rusting of iron is beyond doubt, but which is cause and which is effect is not yet proved.

American Smelting v. Bunker Hill

AFFIDAVIT BY F. W. BRADLEY

F. W. Bradley, being first duly sworn, on oath says: I was educated as a mining engineer and have been continuously engaged in the business of mining for the last 34 years. My first connection with defendant was in 1890 as assistant-manager, when construction was beginning on the first of its present concentrating plants. I served in that capacity at Wardner, Idaho, until 1892, when I was made manager. I then served defendant as manager at Kellogg, Idaho, until 1897 when I was made president, and have served ever since in that capacity from offices at San Francisco.

In 1898 I became a director in the Tacoma Smelting Company; and in 1900 was made president of that company, in which capacity I served until June, 1905, when I resigned on learning that the Federal Mining & Smelting Co. (under the same control as the Tacoma Smelting Co.) had begun an extralateral-right suit against the defendant. In 1900 I was appointed consulting engineer for the Alaska Treadwell, Alaska Mexican, and Alaska United gold-mining companies. I served these companies in that capacity until 1911, when I was made their president, in which capacity I am still serving these Alaska companies. Before the year 1890 I was and ever since have been actively engaged in mining at many other places and with many other companies.

The building up of defendant's property has been a long slow process. Because of disappointments in ore-values at the beginning, the work could only be carried along as the mine earned sufficient profit to pay for it; besides the policy of keeping the development work well in advance of ore-shipments finally exposed sufficient ore to excite the cupidity of neighbors, which resulted in a vast amount of extralateral-right litigation. Ore had to be found and put in sight from time to time, so that the proper and necessary equipment of the property could be planned and provided for.

The ore in defendant's mine has been made payable under past conditions by the mineral galena carrying from one-third to one-half of an ounce silver to each per cent of lead. When pure, galena consists of 86.6% lead and 13.4% silver. In addition to gold and silver, some galena in the Bunker Hill & Sullivan mine carries such other impurities as salts of zinc, copper, antimony, bismuth, etc. The country-rock of the Wardner district is composed of layers of quartzite, shale, etc. The galena occurs irregularly scattered in the shape of stringers and bunches in veins that traverse the country-rock. These veins in turn have been broken up, dislocated, thrown, and twisted by many faults, the Wardner district being one of the most excessively faulted areas of similar extent in the world. The galena is often disseminated throughout the vein-matter, which in many cases consists of masses of iron carbonate, and it is also at times found disseminated in the country-rock itself. In either of these cases of dissemination, it is both wasteful and unprofitable to concentrate such ores under methods of concentration heretofore in use because of the severe tailing-losses.

The disappointment at the beginning, referred to above, was that it required a greater number of tons of ore from the mine to produce one ton of a product rich enough to ship at a profit than originally figured on. That is, as ore was found, it did not average as much per ton in lead and silver as originally expected. There never has been from the beginning any concern as to the sale and marketing of ore. In my time with the Tacoma smelter, it was always possible to make contracts for the sale of lead and silver; and we have found it possible to make sales of the lead and silver products of our new smelter at Kellogg. The primary and great concern was to find ore in payable quantities and then to protect the ore so found from predatory litigation. After finding and securing payable orebodies, the next problem has always been the concentration of the ore into sufficiently clean shape from the barren vein-gangue and country-rock to pay for freighting the cleaned or concentrated product to distant custom lead-smelters, which always stood ready, until June, 1915, to buy the ore when and as prepared for shipment.

In those years of smelter competition, representatives of the different smelting companies made it a practice of calling at the Bunker Hill & Sullivan office, at Kellogg, Idaho, for the purpose of soliciting shipment of ore to their many plants at varying freight and treatment rates, depending upon the quantity of silicious ores being shipped to them, refining-rates, freight-rates, etc. This followed the original practice of auctioning the ore for sale to the highest bidder, and resulted in the making of many concurrent and over-lapping contracts for any quantity of ore that we had to spare and for periods of time varying from six months to one or five years; and in practically all cases provided that the Bunker Hill company would have the benefit of any reduction in freight-rates, and also be paid for any excess in iron. Such smelting companies were the Globe Smelting & Refining Co., operating at Denver, Colorado; the Pennsylvania Smelting Co., operating at Carnegie, Pennsylvania; the Ohio & Colorado Smelting & Refining Co., operating at Salida, Colorado; the Selby Smelting & Lead Co., operating at Selby, California; the Puget Sound Reduction Co., operating at Everett, Washington; the Philadelphia Smelting & Refining Co., operating at

Pueblo, Colorado; the Colorado Smelting & Refining Co., operating in Colorado; the Tacoma Smelting Co., operating at Tacoma, Washington; the United Smelting & Refining Co., operating at East Helena, Montana; the Pueblo Smelting & Refining Co., operating at Pueblo, Colorado; the Omaha & Grant Smelting & Refining Co., operating at Omaha, Nebraska, and Denver, Colorado; the Chicago & Aurora Smelting & Refining Co., operating at Chicago, Illinois; the Consolidated Kansas City Refining Co., operating at El Paso, Texas; and the American Smelting & Refining Co., now known as the Trust, in control of all but two of the foregoing mentioned smelting plants.

Each of these smelting companies had what I called its 'personal equation.' That is, some of them were more honest in the weighing, sampling, assaying, settling, etc., than others were. They had different ways of determining moisture, lead contents, etc., which ways were all calculated to be in favor of the smelter and against the shipper.

The first annual output from defendant's concentrating plant, the building of which was begun in 1890, was made with a loss in concentration of over 30% of the lead and over 33% of the silver. From that time on, it has been a constant effort and study not only to lessen the tailing-loss in concentration, but also to increase the lead content of the concentrates, so as to reduce the amount of worthless material upon which freight and smelting charges have had to be paid. In this effort, three concentrators have been built in addition to the first one of 1890. Of these four concentrators, the first one has been partly abandoned with its remaining portion partly modernized, the second one has been completely abandoned, the third one is being experimented with in the effort to work out a wet process for making a high-grade lead concentrate, and the last, or fourth one, is still being changed and modernized from time to time, and meanwhile has reduced the losses in concentration to eleven and a fraction per cent of the lead contents and to twenty-two and a fraction per cent of the silver contents. The wet process of concentration, now being commercially developed, promises to still further reduce these losses, to make a concentrate of approximately pure lead, and to recover in marketable shape some of the other valuable contents of defendant's ores.

Most of the old-time custom smelters were situated at such a great distance from Kellogg that a considerable cost was involved for freight. So, in 1898, the defendant company purchased an interest in the Tacoma smelter for the purpose of assisting in building up and putting that smelter on its feet; because freight-rates to that point were more favorable than the freight-rates at that time to other smelting-points, when taking into consideration the fact that Tacoma enjoyed a favorable water-freight rate on lead bullion to the Selby refinery at San Francisco, and then on refined lead from Selby by the Isthmus of Panama to New York. Tacoma thus not only had a great natural advantage in freight because both of its nearness to Kellogg and of its situation for ocean competition; but the 'personal equation' of the Tacoma smelter had always been good. That is, it had gained a reputation among ore-shippers of returning fair weights, making correct assays, and in general making an honest settlement for all ore contracted for.

But by 1898 lack of working capital had rather put the Tacoma smelter out of the way of receiving a sufficient volume of ore to enable it to operate at a profit. Although this lack of working capital had resulted in the Tacoma plant running short of sufficient custom ores, still it had a splendid tide-water frontage which had been filled in with slag and a good upland acreage, and above all, a good name. At the same time that defendant purchased an interest in the Tacoma smelter, the Alaska Treadwell, Alaska Mexican, and Alaska United companies purchased a similar interest, ample working capital was arranged for, and contracts were made that afforded the Tasoma smelter 2000 tons per month of a very desirable iron-gold concentrate from the Alaska companies and an equal tonnage of lead products from the defendant, and at the same rate of profit per ton. This was all figured out at the time as a matter of equity, because the Alaska companies were paying for a third interest, the defendant was paying for a third interest, and individuals had already paid or were paying for the remaining interest, all of which payments were for the purpose of bettering and adding to the plant. This deal was tentatively started in 1898 in New York City by W. R. Rust, the then principal owner of the Tacoma smelter, securing the general approval of H. C. Perkins, for the Alaska companies, and the general approval of D. O. Mills for the defendant, these two men agreed to loan all working capital that might be needed in case the respective managements of the mining companies approved of the general arrangement. The details were then worked out on behalf of the Alaska companies by Thomas Mein, their then consulting engineer; on behalf of the defendant by myself; and on behalf of the individuals by W. R. Rust, many of whom had no interest whatever in the mining companies.

The Alaska companies and the defendant in purchasing into the Tacoma smelter in this way, provided that they would not only secure honest treatment for their respective ores and a fair share of the profits in smelting, but also funds for enlarging the smelting plant to enable it to do a large general custom business; as it was expected that the partnership of the mining companies with the Tacoma smelter would create such a prestige for the smelter as to enable it to secure many other ore-contracts. This prestige was especially successful in the case of the Alaska companies, as the Tacoma smelter virtually had from this start a monopoly on much of the copper ore shipped from Alaska. In fact, the tonnage of these shipments became so great that the resources of the Tacoma smelter were taxed in increasing its capacity fast enough to care for it, the expenditures on plant by January, 1905, having totaled some $800,000. This constantly growing and profitable copper tonnage eventually resulted in the building by the Tacoma smelter of the first and only copper refinery on the American shore of the Pacific Ocean.

With the prestige of this dividend-paying mining-smelter partnership, the Tacoma smelter not only secured an inside track on all Alaska smelting-ore shipments, but also South American, Mexican, and Oriental smelting ores gravitated to it. In addition, the Tacoma smelter was securing the shipment of silicious and other smelting ores from Nevada, Washington, Oregon, Idaho, Montana, and British Columbia. All this outside business had become so profitable that by January 1, 1905, the smelter was estimated as earning a profit of $550,000 per year with nearly $700,000 profitable annual earning capacity in sight.

Adjustments had been made from time to time in the Bunker Hill contract with the Tacoma smelter so as to continue the equity between the Bunker Hill and Alaska companies because of their respective interests in the smelter. One of these adjustments was made after one of the individual stockholders in the smelter threatened to bring suit on the ground that the smelter was not making as much profit per ton from defendant's ore as it was making per ton from the Alaska companies' ore.

The negotiations that culminated in the sale of the Tacoma smelter were begun in January, 1905, with W. R. Rust in New York City, by the officers of the plaintiff. Rust secured the signature of D. O. Mills (one of the large stockholders of the Tacoma company) to a document providing for the sale of his shares in the Tacoma Smelting Co. W. R. Rust then proceeded to San Francisco in the effort to secure the signature of myself, and of H. H. Taylor, the then president of the Alaska companies, to the same document. We would not sign the document presented, but afterward did sign an amended document, because D. O. Mills had committed himself to the sale and also because we had learned that the controllers of the plaintiff were also buying into the Selby refinery, and could thus embarrass the Tacoma smelter in the refining of its lead bullion. The sale of the stock held by the Alaska companies and by the defendant in the Tacoma Smelting Co. was neither sought nor recommended by Mr. Taylor or myself, but for the reasons stated we asquiesced in making the 25-year contracts for the mining companies, but upon the basis that the mining companies after a fixed five-year period would be given the benefit of competition. The purchasers refused at first to consent to any clause providing for competition, contending that they were taking the risks that the mines might play out; that silicious ores might become unavailable and that lead ores might become more plentiful. The last preceding annual report of defendant showed less than five years' ore in reserve, and plaintiff's agents made no examination whatever of defendant's mine.

Early in March, 1905, the negotiations had reached a point where the purchasers of the Tacoma smelter proposed that a smelting contract with defendant be drawn up on the following basis:

1st. Verbiage of then existing contract between plaintiff and defendant should be followed as near as possible.

2nd. The then existing contract between the Tacoma smelter and defendant to continue for 25 years.

3rd. Defendant to terminate its contract with the Salida and Carnegie plants.

4th. Defendant's contract with Selby not to be renewed.

5th. All ore above that containing the 27 to 37 (at option of defendant) tons metallic lead per day to be shipped to the smelter at Tacoma, to be settled on the terms as then existing between defendant and plaintiff for a period of 25 years, except that if after April 11, 1914, the plaintiff should pay for a majority of its ores in the Coeur d'Alene a better rate than its then existing contract with defendant, then after that date the same better rates should apply to tonnage in excess of any shipped to Tacoma and should change accordingly from time to time during the remainder of the contract, but in no case to be worse for the defendant than its then existing contract with the plaintiff, which did not cover any ore of 40% and under and 70% and over.

The negotiations continued until March 29, 1905, when it was agreed that competition was to become effective after the fixed 5-year period expiring June 1, 1910, for all ores, concentrates, and slimes between 30% and 75%, including the tonnage to the Tacoma smelter as well, with the understanding that defendant was always to receive thereafter at least as good rates as those provided for in the fixed 5-year period and in addition thereto was to receive from time to time any and all benefits established by competition.

It was finally agreed that the essence of the contracts with the mining companies was to be a continuation of the then existing terms for five years, with the provision that after the expiration of the fixed 5-year period, the Alaska companies and the defendant were each and all to obtain the benefit of competition, and paragraph 12 was inserted in defendant's contract (Exhibit B to the complaint) and paragraph 7 was inserted in the Alaska companies' contracts (Exhibts C, D, E to the complaint). It was recognized that the Tacoma smelter was the only practicable outlet for the concentrates of the Alaska companies, and that if they secured any competition at all they would have to create it; but nothing whatever was said at the time about the cyanide process, and it was not in my mind and not talked of by any of the negotiators that it would be feasible or possible for the Alaska companies to create their own competition. This was only accomplished after some three years of experimental work begun in 1908. This work resulted in the following notice:

"San Francisco, Nov. 4, 1910.

Tacoma Smelting Company,
Tacoma, Washington.

Gentlemen:

We inform you that we are now able to treat the products of our mine in such manner that the profits will be greater than by selling you the concentrates, in that we can treat the concentrates at Douglas Island, Alaska, for a cost not exceeding $3 per ton, and will save $2.89 per ton on freight, insurance, and other charges, and the extraction will be 98% and considerable value in cyanide residues for their iron and sulphur contents, and we would have the immedate use of money for the concentrates, which would be equivalent to interest on their value for from 30 to 60 days.

We hereby notify you that if, after a period of three months from the date of this notice, you refuse to accept the concentrates agreed to be delivered under our agreement dated March 20, 1905, and to make returns and render accounts therefor at such prices as will give and yield us as great net returns for such concentrates as can be obtained by us by our present method of treating them at Douglas Island, without cost of freight, insurance, and other charges, and loss of interest and loss of iron and sulphur contents, we will, upon the expiration of said notice, forthwith terminate such agreement and cease to ship and deliver to you any concentrates under it.

We inform you that we are keeping an account of tonnage for final adjustment with you in event that you refuse to take the concentrates as stated in our agreement and in this notice.

Very respectfully,

ALASKA TREADWELL GOLD MINING CO.
(Signed) H. H. TAYLOR,
President.

With the Everett smelter secured by the Federal Mining & Smelting Co., and with the control of this company secured by the same control that was securing the Tacoma and Selby smelters, I knew that a great monopoly was being built up in the lead-smelting business and that the contract for defendant's ore then being negotiated must provide for defendant being able to market its entire output; but I was willing to have the exclusion of the products of 30% and under and 75% and over because a low-grade lead product was always especially sought after by near-by smelters and the special reservation of ores carrying 27 to 37 tons of metallic lead per day to the Tacoma smelter was to be sure of maintaining a lead-smelting point at Tacoma. Also, a high-grade lead product could stand shipment to a distance at which competitive smelters might be reached.

The idea of providing for the same yearly average analysis and lead assay as the ore shipped in any one year of the preceding 12 years was to be sure of being able to deliver to the smelter any possible mixtures of ore. That is, if an ore carrying 30% of lead and under could not be sold locally and if an ore carrying 75% in lead and over could not be sold at a distance, then these two products could be mixed to fit the average of any one of 12 preceding years. This was further provided for by the clauses reserving the options as to making shipments to any smelting plant; and as to lead contents of ores, slimes, and concentrates wherever the same might be shipped; and also by the clause reserving the right of putting different classes of product together to form a shipment.

At the time of the Baruch-Tacoma deal, defendant had five outstanding contracts as follows:

	TACOMA SMELTING CO.	OHIO & COLORADO SMELTING CO.	PENNSYLVANIA SMELTING CO.	SELBY SMELTING & LEAD CO.	AMERICAN SMELTING & REFINING CO.
Quantity and character of tonnage	Ore and concentrates containing 27 to 37 tons daily at option of defendant.	Concentrates shipments unlimited, defendant to cease at any time.	Highest grade of ore and concentrates not less than 600 tons per month and not more than 1000 tons per month.	Ore and concentrates maximum 1000 tons per month and minimum to contain 200 tons lead per month.	Ores, slimes, and concentrates to be of a lead-assay value of between 40 and 70%, unless the smelting company should give its written consent for shipment to it of ores, slimes, and concentrates varying from that lead-assay standard; 40 tons daily minimum; maximum unlimited.
Life	Two years from June 1, 1904, plus 5 years additional contingent.	Two years from Feb. 2, 1904.	May 9, 1904, terminable 30 days' notice from either party.	Jan. 1, 1903, for 5 years.	Five years from April 11, 1904.
Payment	90% of lead contents at 90% of N. Y. price up to $4.10 plus ½ of the excess above.	100% of full fire-assay value. Per unit 25-40%, 40c. 40-45%, 44c. 45-50%, 52c. 50-55%, 56c. 55-60%, 57c. 60 and over 60c. at lead basis of $4.25. Variation 1c. per unit from prices given for every 5c. variation from $4.25.	90% of contents at full N. Y. price.	90% of lead contents at full N. Y. price.	90% of the lead contents at 90% of N. Y. price when 4.10 or less plus ½ excess when over 4.10.
Freight and treatment charges	When gross values not $50, $16 plus or minus ½c. per lb. of lead variation from 53% lead.	Gross value $50 and less, $10 p.t. $50-60, $12 p.t. $60-100, $14 p.t.	$21.50 released to $60 gross valuation based on $12 freight-rate.	$18 based on $8 freight-rate.	$8 for treatment plus railroad tariff-rate to Denver or Pueblo whatever the destination.

Large shipments of crude ore were being made to the smelters at Tacoma, Everett, and East Helena; and the average freight and treatment rate under all above contracts was better for defendant than the average rate for ores, slimes, and concentrates of between 30% and 75% lead as fixed in the Tacoma smelter 25-year contract.

The Tacoma smelter in July, 1905, arranged to take 3000 tons per month of our crude ore at Tacoma, 1200 tons per month at Everett, and 1300 tons per month at East Helena.

The Tacoma company a short time later desired to change the clause that permitted us to ship crude ore to the Tacoma smelter, as well as that permitting us to ship slime and any other product, amounting to 37 tons of metallic lead daily, and they offered concessions in this regard provided the Bunker Hill would change contract Exhibit B so as to make it cover the defendant's entire product. I declined to consent to changing the contract in this respect.

About this time a controversy arose over our option under the original 25-year contract to make deliveries at any smelting plant of any character of product. We insisted that we had the option to determine the lead contents of ores, slimes, and concentrates wherever the same might be shipped. The contract so provided. We considered this option vital because of what I have already described as the 'personal equations' of the different smelting plants. We had it in our power to ship different grades and character of products to the different plants according to their different degrees of honesty in determining weights and lead contents and in making payments.

Late in the year 1905 a controversy arose over our shipments of high and low-grade concentrates containing less than 30% in lead and over 75% in lead to the Ohio & Colorado Smelting Co. and to the Pennsylvania Smelting Co. This controversy extended until March or April, 1906, when the right of the Bunker Hill company to ship to outside smelters material containing 30% and less and 75% and more in lead was clearly established and conceded as shown in the exhibits attached to defendant's answer. From the year 1906 until the year 1915 the Tacoma company met competitive bids made by other smelters for our high and low-grade products and the parties recognized the construction which the parties in the year 1906 placed upon the clause of the contract pertaining to high and low-grade shipments to outside smelters.

In the year 1908 it was arranged that an examination of defendant's mine and concentrating plants should be made by Pope Yeatman, one of plaintiff's consulting engineers, who reported among other things, as follows:

• • •

"CONCENTRATION. The crushing stations and concentrators are situated in the valley just beyond the mouth f the Kellogg tunnel. There are now three mills. The ld one of a nominal capacity of 900 tons per day; the ew or south mill of about 600 tons per day averages or the entire year including delays, and a tailing plant West mill) estimated to be able to treat 900 to 1000 ons of tailing.

"At the present time there is being operated the new lant, a part of the old plant, the other portion being verhauled, after which the part now in operation will also overhauled and the tailing-plant which is being perated in connection with the old plant, taking the ner material for final treatment. When repairs and terations are completed in the old plant, each one of e plants will be self-containing and absolutely independent and able without the others to handle its own particular feed.

"An excellent description of the new plant by the superintendent, Gelasio Caetani, can be found in the 'Mining and Scientific Press' of January 15, 1910."

• • •

"The new mill is thoroughly up to date and well-designed and the old mill will also be brought into excellent condition. In the last few years considerable improvement has been made in concentration over the old operations and while costs have been increased, partly due to the 8-hour law going into effect and also to the greater number of machines used to better extraction, the increased saving has more than justified the greatest cost." • • •

"The products produced are first, hand-sorted ore, done principally in the stopes, but also on belts at the crusher-stations; second, coarse concentrate; third, middling high in carbonate of iron, but for which there has been a market at the smelters; fourth, slime concentrate. The utilization of the middle products has made higher extraction possible than would have been the case had it been necessary to separate these." • • •

Toward the end of February, 1910, the question was raised of arranging the new prices after the expiration of the 5-year period on June 1. W. R. Rust wanted the new arrangement to cover the high-grade product at the rates then in force, but I claimed that defendant was entitled to better rates. At a meeting in San Francisco with Judge Lindley, Edward Brush, E. L. Newhouse, and Karl Eilers, I asked for the rates that plaintiff was then granting the Hercules and made the contention that plaintiff was not buying the majority of Coeur d'Alene ores, as Tacoma and Selby were outside smelters according to the wording of the contract, and that therefore defendant should get the benefit separately of every item in the outside contracts that would better the corresponding items in the 25-year contract. I finally covered all of my contentions to the specific claim of a betterment of $1.70 per ton. This contention resulted in the fourth supplemental agreement of November, 1910, which was the result of a compromise in splitting my claim to 85 cents per ton, without prejudice; but this was to cover only our products of 30% and under and 75% and over which were left free for outside smelters. This agreement was made retroactive to begin June 1, 1910, and to expire five years thereafter, on June 1, 1915.

In June 1912 I notified the Tacoma company that under a misapprehension of instructions the shipments theretofore made to the Tacoma smelter were being sent to East Helena at a saving to the smelting company of about $2.25 per ton. I claimed either the benefit of this freight reduction or else the right of shipping to the Tacoma plant material to contain a maximum of 37 tons of metallic lead per day. At the same time I also notified the Tacoma company that the competitive freight and treatment rate on the high-grade product was $15 per ton and not $18.50, which the Tacoma smelter was then charging us. With these notifications, I thereupon placed the specific carrying out of our contentions with our attorney, M. A. Folsom, and our manager, S. A. Easton, who thereupon secured the competitive rate for the high-grade output for an unlimited tonnage to last until June 1, 1915, in consideration of which we waived for the time being, without prejudice, either our right to ship up to 37 tons metallic lead daily to Tacoma or the $2.25 difference in freight-rates between shipping to Tacoma and to East Helena.

Toward the end of 1912, it became apparent that plaintiff was limiting all its ore-contracts so as to have them expire on or before June 1, 1915. The uncertainty of competition after June 1, 1915, caused me in November 1914 to authorize M. A. Folsom to endeavor to reach some understanding with Tacoma in advance; and in December 1914 I notified the plaintiff that Mr. Folsom had full and sole authority to act for defendant in endeavoring to reach some understanding for the disposal of our ores after June 1, 1915. These negotiations extended for several months without result. It was then that defendant decided to build its own smelter and construction contracts for that purpose were entered into in September 1915.

Plaintiff's letter signed by Edward Brush, which is set forth in the complaint, warning us not to erect a smelter, was dated New York, November 22, 1915; but we had then already let our contracts, after being unable to secure any other outlet for our product. In spite of the great revival in industry, war prices for lead, and demand for ores, we could secure no bids whatever for any of our product. Therefore it was absolutely necessary for us to continue with our plans to construct a smelter in order to provide some market for our high and low-grade products; for the ores of a grade too low to ship to a distance; for the disseminated ores that it would not pay to concentrate before smelting; for the disposal of sufficient quantity of material to yield 37 tons of metallic lead per day, which had been refused at Tacoma, and to provide competition for the remainder of our ores should the plaintiff herein continue to deprive us of the competitive rates provided in clause 12 of our contract.

Defendant accordingly erected a smelting and refining plant near Kellogg, Idaho, where it is now working satisfactorily on lead ores some of which carry 8% zinc. Kellogg is the centre of one of the important lead-mining districts of the United States and is the logical point for a lead-smelting plant. Defendant's smelter will not only be a help to old Coeur d'Alene mines, but will stimulate the prospecting for and discovery of new mines which will probably be developed as fast as old ones are worked out. This prediction is based on the fact that since the making of the Tacoma contract the following comparatively new Coeur d'Alene mines have become productive: Consolidated-Interstate Callahan, Green Hill-Cleveland, Marsh, Ontario, Stewart, Tamarack & Custer, Caledonia, Sierra Nevada, and Alhambra.

Also, defendant's mines for the 12 years, 1905-1916 inclusive, have averaged about three times as much tonnage of shipping product as they averaged in the 12 years prior to 1915. All this increased lead production from the Coeur d'Alene district has resulted in the following lead production for the State of Idaho, as compared with the total production of the United States:

	Idaho tons	United States. tons
1894	33,308	213,650
1895	31,638	235,822
1896	46,662	257,487
1897	58,627	282,169
1898	59,142	302,148
1899	52,154	298,047
1900	85,444	367,773
1901	79,654	371,032
1902	84,742	367,892
1903	99,590	368,939
1904	108,854	393,452
1905	99,027	388,307
1906	117,117	404,746
1907	123,292	413,389
1908	98,464	396,564
1909	103,747	446,909
1910	109,951	470,272
1911	117,355	486,979
1912	127,780	480,894
1913	137,802	462,460
1914	177,827	542,122
1915	160,680	550,055
1916	170,059	571,134

The foregoing tabulation discloses that the Idaho production for the last 11 years has increased at about the same rate as has the production for the whole United States. On the other hand, the production of Missouri has, for the last 11 years, increased twice as rapidly as has the production for Idaho. One reason for this is because of late years the controllers of plaintiff have secured control of lead-mining property in Missouri and largely increased the output of such property. In this connection, plaintiff's annual report for the year ending December 31, 1916, says:

"Due to the falling off in the lead ore coming to the company from the Coeur d'Alene mines, attention has been given to increasing the supply of lead from other sources. About three-quarters of a million dollars have been expended in enlarging the milling and smelting works, and extending the development of mines of the company in Missouri. While the production, from this source, at the beginning of 1916, was approximately 42,000 tons of pig-lead per annum, this has now been increased to 66,000 tons. Constant attention is being given to the increasing of the supply of lead ore at the disposal of the company."

Defendant's last annual report shows ore reserves blocked out totalling 3,453,146 tons, with bottom level of defendant's mine in the best orebody that it has ever enjoyed. As this is eight years' life at the present rate of output, with geological conditions in favor of many years of additional life, defendant is fully justified in increasing the efficiency of its plant in order to fortify itself against periods of low prices for lead. This increase of efficiency is economically necessary.

CRIPPLE CREEK, COLORADO

Cresson Con.—Golden Cycle.—Excelsior.—Pride of Cripple Creek.

Meetings will be held during the ensuing week by the directors of the Cresson Consolidated Gold Mining & Milling Co., and the Golden Cycle Mining & Reduction Co., when in addition to the declaration of the regular monthly dividend, payable December 10, it is confidently expected the directors of these companies will pass upon the payment of an extra, or Christmas, dividend. The Cresson company has paid 11 consecutive monthly dividends of $122,000 each this year, and the Golden Cycle company 11 dividends of $45,000 each. A. E. Carlton of this city is president of both corporations.——Active operations will be resumed by the Catherine Mining Co., on the Last Dollar mine on Bull hill, under bond and lease to the Catherine company. At the annual stockholders' meeting at Victor, the following officers and directors were elected: C. R. Street, president; H. C. Eddy, vice-president; John A. Gallagher, secretary and treasurer; C. Harding and W. Bigby directors. All reside at Chicago. Charles Walden of Victor was re-appointed general manager.——The Excelsior Mining, Milling & Electric Co. has purchased the six-drill electrically-driven compressor formerly in use at the Sitting Bull shaft of the Hondo G. M. & M. Co., and the machinery has been removed to the Longfellow mine on the eastern slope of Bull hill. The Hondo company has ceased operations.

The Pride of Cripple Creek, on the eastern slope of Ironclad hill, has been leased by the Pride of Cripple Creek Gold Mining Co. to E. H. Beebe, L. S. Cox, and Harry Nelson, well-known mining men of this district. A 30-hp. electric hoist has been erected and operations have commenced. The Pride is in close proximity to the Jerry Johnson, W. P. H., and Forest Queen mines, all producing at this time.——A block of the Cresson property, situated on the north end of the Sadie Bell and Draper lode mining claims, above the 1100-ft. level of the Cresson shaft, has been leased to Evan Williams, Edward Olsen, and J. Anderson of this district, who are operating the block from the 1450-ft. level of the Jackson shaft of the Gold Sovereign Mining & Tunnel Co. on Bull hill.——The Hahnewald Leasing Co., operating the Gold Sovereign mine, is pushing development on the 1450-ft. level of the Jackson shaft, and the drift on the main Sovereign vein is showing well.——The Big Toad Mining & Milling Co., operating the Dante mine, adjoining the Gold Sovereign M. & T. Co., on Bull hill, has opened up a strong body of milling-grade ore.——The Rex mill on Ironclad hill has been closed down for the winter months.——A telegram received from Clark G. Mitchell, director and manager, Isabella Mines Co., announced his voluntary enlistment with the U. S. Aviation Signal Corps at Washington, D. C.

HOUGHTON, MICHIGAN

The Price of Copper.—Seneca.—La Salle.—Baltic.

The production of copper from Lake Superior mines for 1917 is estimated at 225,000,000 lb., provided nothing unforeseen intervenes.——There is much speculation in the district as to what action the Government will take relative to the price of copper at the conclusion of the present agreement which ends January 1. That there can be a further reduction in the price seems out of the question, as it would mean suspension of more mines or a cut in wages. The mines now are paying good bonuses to their employees, and while no official announcement has been made, there is every reason to believe they will continue to do so. Few of the mines are working full forces, and all of them could use more men to advantage. Should the Government decide upon a lower price for copper there could be but one logical outcome, namely, Government control of the mines. This would necessitate, not the operation of the mines by the Government, but a supervision over general conditions with a stipulated percentage of profit allowable to the shareholders. Such a profit would undoubtedly be based on amount of copper produced, as all energy would be directed to a large output. Lake Superior mining men believe that stimulation to production might be secured by increasing the price to 25c. per pound, but they hardly expect any such action.

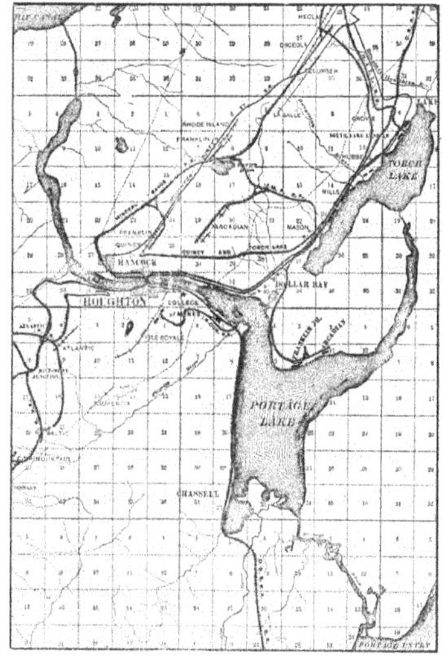

MICHIGAN COPPER BELT

While the official announcement says that the Keweenaw has shut-down only for the winter, and while it may re-open in the spring, the fact is that the Keweenaw is close to the end of the rope. Its copper content has been about 12 lb. per ton and the tonnage limited at that. It has been a long hard pull and the shareholders have stuck with tenacity to their hopes for better things. The present company was organized by the late Charles A. Wright and, as originally planned, it contemplated operations on a large scale.

A 10,000-kw. turbine has been shipped and will be set up at Lake Linden power-plant of the Calumet & Hecla.

The costs at Franklin are now 17c., despite high wages and costs for material. The sinking at No. 2 shaft will be under way this week. The production from No. 1 shaft is better than 1100 tons per day.——Production of White Pine for 1917 will be over 4,000,000 lb., about equal to that of last year.——The accumulated stock-pile at the Wyandot will go to the Winona stamp-mill next week.——The Franklin has made various improvements at the mill, which facilitates operations and ensures economies. The latest is the longer trestle, permitting 16 cars.——Production at the Hancock for November will be 340,000 lb. The mine could use an increased force of one-third underground. There is no indication of any immediate betterment in the labor situation.——The Seneca has practically completed housing in its surface plant, which is one of exceptional capacity. Shaft-sinking will start on January 1; the shaft-house is completed; railroad track being laid; and hoisting-engine foundation was laid early in the month; boilers are now being erected.——The future possibilities of development of rich Osceola lode extension from No. 6 shaft Osceola into La Salle territory are bright. Ore of more than ordinary quality is now showing, but mining is difficult, because of the great depth and long distance from the shaft.——The No. 7 shaft at Isle Royale is being sunk below the ninth level. Three levels south are in over 600 ft. in average ground, not quite up to the quality of No. 5 shaft.——The Baltic mine is coming into good ore again in the No. 4 shaft at the 32nd level. At No. 5 shaft, the best ore is coming from west vein. The best producing shaft both in quality and quantity is No. 2, half of the output coming from this shaft. Practically all of the tailing from the three Copper Range mines, Baltic, Trimountain, and Champion, are being re-ground.

MAMMOTH, UTAH

A. S. & R, Co. Buys More Ore.—Scotia.—Plutus.—Scranton.

The Emma Consolidated property is in a splendid condition. The winze, which is now 45 ft. below the third level, has cut good-grade ore running 100 oz. silver and 5% lead. Recent analyses show the ore to contain tungsten and molybdenum, one assay running 26% tungstic acid.

Certificate of amendment to articles of incorporation of the Monarch Mines Co. has been filed with the county clerk. The capital stock is increased from $40,000 to $62,500, in shares of the par value of 5c. Charles Ohram is the president and J. W. Storrs secretary of the company.

The Six Mile Mining & Milling Co., operating in the West Tintic district, has made an important strike in a new shaft. The vein is over 5 ft. wide and carries from 25 to 160 oz. silver and 60% lead. The shaft was sunk as a test, and as soon as ore was found a cross-cut was started from the 200-ft. station in the adit which is expected to cut the orebody at a vertical depth of 350 feet.

Rapid speed is being made in sinking the main working-shaft on the property of the North Beck Mining Co., in the North Tintic district. Sinking started on October 17, and now the shaft is nearing the 250-ft. level.

The Tintic Standard is mining ore from the new workings north of the new shaft. The orebody is large and assays 70 oz. silver, $6 gold, 10 to 12% lead, and 2 to 3% copper. This new work proves the orebody to have a length of over 400 feet.

While the American Smelting & Refining Co. has notified some of the local mines, among them the Eagle & Blue Bell, that the embargo has been lifted, notice has been given that only a limited tonnage of ore will be accepted until conditions change. The Eagle will probably be allowed to ship three carloads of ore per day, which is less than half what the mine should be sending out. Other Tintic properties which have been shipping to this smelter will receive similar treatment and officials of the smelting company believe that by holding the mines down to a smaller tonnage it will not be necessary to shut them off periodically as they have done in the past. The general outlook for the mines of this district is more promising than it has been for some time, although a weak lead market is a serious drawback, most of the mines being heavy lead producers.——Walter James, owner of the Scotia mining property in the West Tintic district, has sold his interests to Dunn & Fabian, of Salt Lake City. At the time the Chief Con. was operating the Scotia it took out $100,000 worth of lead-silver ore, paying the Scotia company in royalties $20,000. The ore was followed to a depth of 400 ft. by a winze sunk some distance below the 150-ft. level, where the ore was not lost. The new owner will continue the winze and will prospect about 300 ft. of undeveloped ground between the main workings and an old glory hole from which M. S. Sammon quite recently took a couple of shipments of good ore. James McNelis is to be the superintendent.

According to Preston Peterson, superintendent, the Big Indian mill has been fully demonstrated, and as soon as minor repairs are made, the mill will be started. The payroll of the company will be increased to at least 200 men.——W. D. Loose, local manager for the various Loose properties, states that a drift is being sent out to cut the vein in the Central Hill ground at a depth of 285 ft. Nearly 100 ft. above this point the company did some driving with encouraging results. The Central Hill adjoins the Mammoth, Grand Central, and Plutus.——Work is being carried on in five different points at the Plutus, according to officials of the Chief Con. Co., which now controls it. After driving into the Plutus from one of the lower levels of the Victoria a raise is being put up to thoroughly prospect a body of quartz. Another section of the Plutus is being developed from the old Tetro shaft, which recently was cleaned out and re-timbered. This shaft has a depth of 400 ft. and the showing is so promising that extensive driving will be done before sinking the shaft to a greater depth.——Mining operations at the Scranton, in the North Tintic district, have been slowing down somewhat during the past month as a result of the decline in the price of lead, most of the ore taken from the mine being low grade. For the past two years lessees have been busy in the Scranton and have shipped a large amount of ore; the output running near 100 tons per week. Announcement of another assessment of 1c. per share on the stock of the Eureka Lily Mining Co. has been made. This assessment will be delinquent on December 19 and the sale day is January 15.——The Copper Leaf officials have decided to handle the shaft work on company account when the present contract expires. The shaft is now down to a depth of 400 ft. The ground is hard and sinking slow.——On November 24, when the Chief Con. was crowding things to get out as much ore as possible under its old contract, which has since expired, it came close to establishing a new record for fast hoisting, bringing out of the mine 415 mine cars of ore in an 8-hour shift.——Shipments of first-class ore from the Tintic district show an increase of 11 cars, the total being 230 cars as against 219 the week previous. This week's shipments are valued at $275,000. The car shortage has ended, at least for the present. The principal shippers this week were: Chief Con., 50 cars; Dragon Con., 42; Iron Blossom, 26, Mammoth, 16; Gemini, 16; Centennial Eureka, 14; Grand Central, 11; Gold Chain, 10; Eagle & Blue Bell, 8; Tintic Standard, 7; Bullion Beck, 5; Colorado, 5; other mines, 20 cars.——The Eureka Standard Mining Co. has given to the Eureka Standard lidated Mining Co., for 350,000 shares of the latter, quit-deeds to the Mielich or Montana group of claims.——Bullion Mining Co. has withdrawn the 1c. assessment, ly levied.——Work at the mines of the Western Utah sion Copper Co. is being pushed under more favorable ions and the showings in all the workings are better a week ago. The ore in sight has been increased and high-grade ore is being mined from a new opening on the it vein.——Sinking is to be resumed immediately by the itar Copper Co. in the Victor shaft, and the work will be d until a depth of 500 ft. is reached. The shaft is 150 ip and from that point driving to the west and north is way.——High-grade ore carrying copper, silver, and s still being followed from the bottom of the 75-ft. winze Copperopolis mine of the Bamberger-Dunyon. A drift ng sent to the south in the fissure containing the rich at continues to show strength.——A new discovery of as made this week by the Woodman Mining Co. when was started on the west side of the ridge from where ain operations of the company have been carried on. re carries 8 to 9% copper.——A contract to sink the ight No. 4 shaft 100 ft. has been let by the Deep Creek r Mining Co. to Oscar Carlgren of Gold Hill, who has a whim to the property and is arranging to work two The shaft is about 20 ft. deep and follows a small vein ng a good-grade copper ore.

MONTERREY, MEXICO

DESTRUCTION OF GUANAJUATO POWER-PLANT.

rican mining men have recently fled from Guanajuato ape the possibility of falling into the hands of a large of Mexican bandits who are operating in that section. mericans state that this great mining district is now in e chaotic condition than it has been at any time since tion and brigandage began in Mexico. The most serious of the present situation in the States of Guanajuato, acan, Jalisco, and San Luis Potosi grows out of the attack that forces of brigands made upon the great electric plant of the Guanajuato Power & Light Co., d at El Duro, State of Michoacan. They destroyed the es and other valuable machinery and put the entire plant commission. Not contented with this destructive work, ndits cut and carried off large sections of the power-ission cable. The power company owns three generat-ants; all connected into one system with a number of tions, by means of which power is conveyed and dis-d over a wide area of central Mexico. The mines and ls of the rich Guanajuato district, the irrigation plants plateau region of Jalisco, the flour-mills and smelters of Calientes and portions of San Luis Potosi, as well as lustrial establishments and towns of these several were dependent wholly upon the Guanajuato Power & o. for their service. As a result of the partial destruc-the plant and the severing of the transmission lines by , the mines and other industries have been compelled -down, and it will probably be several months, even he most favorable conditions, before the damage can be l.——The bandits in the Guanajuato region are com-by Inez Chaves in the guise of revolutionists. They mmitted robberies within the city of Guanajuato and lly hold that place within their grasp.——El Duro, he central hydro-electric plant of the power company ted, is 120 miles from the city of Guanajuato. The is composed of Americans, most of its stock being y Colorado Springs men. It was through this develop-cheap electric power that a great revival of mining as in the Guanajuato district was brought about some years ago. This cheap power has also enabled the ion of large areas of land by means of irrigation and has been in this and other respects of enormous benefit to the people of that part of Mexico.

COBALT, ONTARIO

McKINLEY-DARRAGH.—HOLLINGER.—McINTYRE.—DOME.

The McKinley-Darragh is meeting with excellent results at the 400-ft. level. Considerable high-grade ore is being found and there is a large tonnage of low-grade in sight. The company has declared its regular dividend of 3%, payable January 1, to shareholders of record December 8. This makes a total

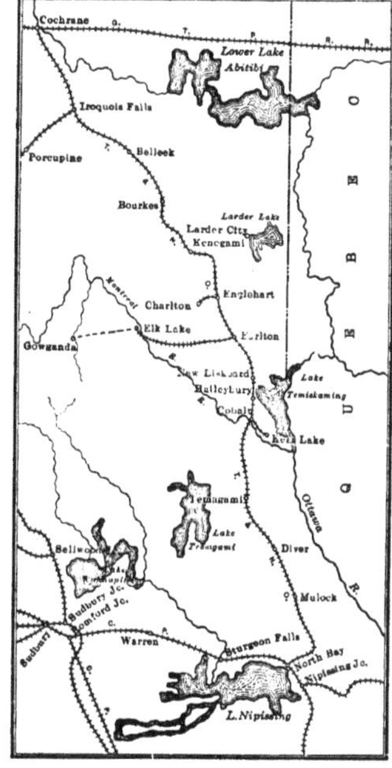

COBALT AND PORCUPINE DISTRICTS

for the year of $269,723, or an aggregate of $5,146,197, which is equal to 229% on the company's capital.——The Coniagas is yielding 108,479 oz. of silver per month. The Ankerite and the Maidens-McDonald property, which are controlled by the Coniagas, are developing satisfactorily.——The Miller Lake-O'Brien mine of Gowganda, which, like the O'Brien mine at Cobalt, is under the control of M. J. O'Brien, Ltd., a twenty million dollar corporation, is rapidly taking a place of honor among the best mines of northern Ontario. The production from the Miller Lake is averaging about 84,000 oz. per month.

The sensational developments at the Hollinger Consolidated during recent weeks, have revealed an enormous orebody at the 400-ft. level. At the 200-ft. level this body was 60 ft. wide and high-grade. At the 400-ft. level the width has increased to

70 ft., and the average gold content is understood to be $27 per ton. When it is remembered that previously the average grade throughout the mine was less than $9 per ton, the importance of a 70-ft. body of $27 ore can be realized.——The McIntyre-Porcupine continues to hold its strong position. A net profit of $70,000 per month is being maintained and with a steadily increasing ore-reserve and better facilities for the handling of its ore, the McIntyre company should be able to continue the present rate of dividend disbursements. This mine, in point of production, is second only to the Hollinger. ——The Dome Mines is not being operated at full capacity. The shortage of labor and the low grade of the ore, preventing its being handled during times of high costs, have forced curtailment. However, the ore-reserve appears to give abundant assurance of the intrinsic value of the mine.——The Porcupine V. N. T. is in full operation and results are better than ever in its history. The 100-ton mill is running at normal capacity and the grade of ore is nearly $10 per ton. Recent developments at the 600-ft. level of this mine have been good.

DEADWOOD, SOUTH DAKOTA

IMPROVEMENTS AT TROJAN CYANIDE MILL.

Two important purposes are being served by the improvements now being made at the mill of the Trojan Mining Co., at Trojan, namely, increased recovery and greater capacity. When these changes are completed the flow-sheet will be about as follows: the crusher product will go to two 6-ft. Chilean mills, one of which will discharge to the closed circuit of a ball-mill and Dorr classifier; the sand to leaching-vats, and the slime to a series of thickeners and agitators, employing counter-current decantation, with final wash and de-watering of the slime on Portland filters. Fine grinding, or 'all-sliming,' will not be attempted, for experiments have demonstrated that fine sand yields as economical an extraction as slime—that is, the extra expense of reducing the ore to slime is not offset by the increased recovery effected. This conclusion is based upon the value of the mill-feed, which is $4 per ton, and probably would not apply to ores worth twice as much. It has been determined that a 60-mesh sand yields a satisfactory recovery of both gold and silver, and with the purpose of grinding as much as possible of the ore to that fineness and at the same time producing a minimum of slime, the new flow-sheet has been designed.

The cyanide solution is mixed with the ore in the proportion of 3.9:1 at the Chilean mills, of which there are two, of the Trent-Monadnock type. From one of the mills the pulp is taken direct by elevator to the classifying department, while the other discharges to a Dorr classifier, operating in closed circuit with a Denver Engineering ball-mill. The discharge from the ball-mill enters the same classifiers; sand from this is returned to the ball-mill, and the slime enters the main classification department, being carried there by a steel-cased elevator and a series of launders. A feature of the ball-mill is a combined scoop and centre feed which permits the feeding of either ore or sand as desired. The scoop has a radius of 38 in. The 20-in. trunnion-discharge can be closed by plates to give any height of discharge. The lining was designed in the Trojan office and eliminates a large number of the bolts ordinarily used. Between the motor and the mill pinion is a shock-absorbing gear, designed by the Denver Engineering Works.

As has been noted, the pulp coming to the main classifiers is comprised of material that will pass a 60-mesh screen. An attempt is made to separate this into pulp larger and smaller than 150 mesh; the larger going to the sand-treatment vats and the smaller to the slime department. Naturally it is impossible to make a clean separation—the sand always containing more or less slime, and the slime some fine sand—but the theoretical point is adhered to as closely as possible. In the classification department are a 16-ft. Weigand and two Dorr duplex classifiers. The Dorr machines are of the ordinary type. All of the details of the slime-treatment have not been worked out. The department includes 5 vats, 40 by 14 ft., equipped with Dorr thickeners. Two of these will probably be used as primary units, to de-water the classifier product, while the others, together with agitating-tanks, will form the counter-current decantation system. They are equipped with Colorado Iron Works diaphragm-pumps for handling the thickened slime. From the counter-current decantation system the slime will go to three 12-ft. Portland filters, now being erected. These machines will de-water and wash the slime. Among other important additions to the plant is the doubling of the precipitating facilities. Three additional steel zinc-boxes are being added. Another tank with filter-leaves is to be put in for clarifying thickener-overflows before precipitation. The mill is expected to treat 400 to 500 tons daily. It does a custom business, for which it is well equipped, as receiving-bins, sampling department, and laboratory are all well adapted to the careful handling of custom business. Recently the Two Johns property was purchased and is being developed and equipped. At Blacktail the company has secured the Republic mine under a long-term lease; this may develop into a large producer.

TONOPAH, NEVADA

TONOPAH MINING CO.—TONOPAH BELMONT.—WEST END CON.—

The Tonopah Mining Co. during October milled 11,400 tons of ore, averaging $9.43 per ton, resulting in a net profit of $17,075. During the week ended November 24, 2550 tons of ore, averaging $8.50 per ton, was milled. In the Silver Top 84 ft. of development has been done, 26 ft. in the Red Plume, and 129 ft. in the Sandgrass. On the 340-ft. level of the Silver Top, stoping continues on the Valley View vein on a 3-ft. face, while on the 640-ft. level the stope on the Burro vein shows a 3-ft. face. On the 1300-ft. level of the Sandgrass the west drift on the Merger vein continues in quartz. A drift to the east has been started. Last week the production was 2750 tons.——The Tonopah Extension Mining Co. during October milled 9393 tons of ore, resulting in a net profit of $13,288. In the No. 2 shaft 111 ft. of development was done and 103 ft. in the Victor. At the No. 2 on the 850-ft. level, raise No. 814 advanced 15 ft. on a 2-ft. stringer of ore. On the 1260-ft. level raise No. 453 advanced 20 ft. on a 3-ft. face of ore. On the 1680-ft. level of the Victor 1600 south cross-cut advanced 7 ft., while 1600 east drift advanced 7 ft. on a 3-ft. face of ore, and west drift 1600 continued on a 4-ft. face of ore. The production for the past week was 2380 tons.——The Tonopah Belmont Development Co. milled during October 10,087 tons of ore, resulting in a net profit of $51,876. On the 800-ft. level raise No. 16 shows a 4-ft. face of ore, while raise No. 17 continues on a 6-ft. face of excellent ore. North-east cross-cut No. 8013, which started from the Occidental vein, cut the South vein exposing 2 ft. of medium-grade ore. At this level the South vein was intersected by cross-cut No. 8014 at a point farther east, showing 3 ft. of low-grade ore. On the 900-ft. level the north-east cross-cut, after picking up the faulted segment of the Occidental vein, was continued to find the faulted segment of the South vein. On the 1000-ft. level raise No. 72 on the Rescue vein broke into the level above. The raise proved the vein to be 4 ft. wide, consisting of medium-grade ore. Last week's production was 2372 tons.——The West End Consolidated Mining Co. continues development at the Ohio shaft. Raise No. 3 on the 555-ft. level made connection with the level above. Drift No. 1 continues in fair ore. On the 500-ft. level the excellent showing, where the 506 cross-cut intersected the vein, continues in the drifts started last week. A the West End shaft drift 717 No. 2 continues on a face of low-grade quartz. The output for the past week was 1211 tons.

ALASKA

he annual meeting of the Alaska Mining and Engineering iety was held on October 17, 1917, at Thane, Alaska, the ety being the guests of the Alaska Gastineau Mining Co. party was met at Thane at 4 p.m. and conducted through shops, power-plant, and mills of the Gastineau company. er the sight-seeing trip, dinner was served in the mess-hall. nager G. T. Jackson, president of the society, welcomed members and guests. Papers, descriptive of the practice the Gastineau mill, were read by E. V. Daveler, R. Hatch, E. Garlock, Fred Hodges, and Emil Gastongway. Officers e elected for the ensuing year, as follows: John Richards, sident; Earl Daveler, vice-president; C. K. White, secre- r; W. S. Pullen and C. E. Davidson were elected members he executive committee.

ARIZONA

COCHISE COUNTY

Special Correspondence.)—The Shattuck-Arizona is grad- the mill-site on Denn ground and putting in concrete ndations. The mill will have a capacity of 400 tons per . The Shattuck production of copper for October was ,000 lb. and there is now every indication that the year's duction will reach 13,000,000 lb.——The Denn Arizona is pping a car of ore from development work every other day. s is chiefly from the 1300 and 1400-ft. levels.——Henry enzuela reports the discovery of hübnerite ore on his ms 13 miles south-east of Benson, near the American gsten Co. property.

isbee, November 23.

GILA COUNTY

Special Correspondence.)—The production of the Superior- ton has reached 80 tons of ore per day. Recent develop- ts on the 1000-ft. level continue favorable. Driving has extended east for 80 ft. in copper glance ore that assays 11% copper. It is proposed to develop this ore on the and 1400-ft. levels.

obe, November 22.

MOHAVE COUNTY

pecial Correspondence.)—The first shipment of concen- from the flotation-plant of the Washington Arizona has made. The adit is in 1000 ft. and the shaft is being from the 100 to the 250-ft. level.——Sinking has com- ed in the Gold Ore shaft; the equipment is electrically n.——Some rich ore is being taken from the 150-ft. level e New Mohawk.——The development at the Hackberry has opened up ore to the value of $3,000,000. F. W. Sher- has been engaged to design and erect a 300-ton mill. For mber of months the Hackberry has been shipping high- ore.——The United Eastern main shaft is almost com- it being down 1060 ft. The lowest level will be 1090 ft. The company has declared a 5c. dividend per share for nber. The ten months of operation have yielded gross ns of $1,512,679; total net profits were $971,359, of which 00 has been distributed in dividends. The Keystone mill erating on dump ore. The Tom Reed production for er was 7500 tons of ore averaging $8.04 per ton, the break-down of the mill curtailing production for about 10 days.——The Middle Golconda is now shipping silver ore.

Kingman, November 24.

PIMA COUNTY

(Special Correspondence.)—Belmont group of claims, in the San Xavier district, has been taken over by the Reineger-Freeman Mining Co. The erection of a mill at an early date is planned. The New Cornelia's production of copper for October was 3,628,800 pounds.

Tucson, November 17.

PINAL COUNTY

(Special Correspondence.)—The cross-cut on the 300-ft. level of the Queen Creek Copper Co. has exposed a 20-ft. vein of copper ore that assays 14% copper. The company has closed a contract to drive a 300-ft. adit in a claim which has an excellent showing of silver ore, 2000 ft. from the shaft.——The new vertical shaft of the Arizona Coronado Copper Co. has cut the orebody at 128 ft. in depth. The orebody is 13 ft. wide and consists mainly of copper glance. Driving east and west for a distance of 208 ft. has opened up ore all the way.

Florence, November 23.

YAVAPAI COUNTY

(Special Correspondence.)—The existing contract for hoisting all Jerome Verde ore in the Edith shaft of the United Verde Extension ceases on December 23. A new contract is being arranged by which the ore from the main top orebody will be trammed through the United Verde Extension to the Columbia shaft of the Jerome Verde.——The Black Canyon Mines Co. in the Turkey Creek district will erect a 200-ton flotation plant.——The Big Ledge Copper Co., which recently took over the Great Western smelter, has arranged to add two new converters and a reverberatory furnace, and to handle custom ores. With the new additions it is expected that the capacity of the smelter will be 800 to 1000 tons of ore per day. ——It is reported that a recent strike in the Nevada mine of the Groom Creek district has opened up 14 inches of 800-oz. silver ore.——The Garford Syndicate, which in the early part of the year acquired the Copper Hill group in Copper Basin, has taken over the Robinson mines under a bond agreement. Recent development has opened up a considerable tonnage of zinc, lead, and copper ore.——The Arizona-Binghamton Co. has cross-cut 31 ft. of high-grade copper ore on the 600-ft. level. This ore for the most part is rich enough to ship.——The Patton Copper mine, four miles east of Mayer, has been taken over by the newly organized Rio Tonto Copper Mining Co.——The Copper Mountain Mines Co. is re-timbering one of the old shafts of the old Stoddard mine, and a gas hoist has been erected.——A two-foot vein of diorite schist, well mineralized with chalcopyrite, has been followed a distance of 30 ft. in the east drift from the Gorge adit of the Green Monster. This ore is said to run 8 to 10% copper.

The following appears in the issue of November 17: 'Operations at the United Verde Extension are proceeding rapidly. One blast-furnace of the new smelter will be blow-in next month.' An error was made in transcribing the notes; the sentence should have read: 'One blast-furnace will be blown-in next March.'

CALIFORNIA

AMADOR COUNTY

(Special Correspondence.)—After several months' work in enlarging the Old Eureka shaft below the 1600-ft. station to conform to the size of the shaft above that point, the bottom, 2065 ft., has been reached. A small opening served the purpose of the former management in prospecting to this depth and it has been a tedious task to remove the refuse from this narrow shaft, widen the opening for more than 400 ft., and timber it throughout. The three compartments are now in use from collar to bottom and sufficient work has been done around the 2000-ft. station to justify active development work there. Some prospecting will be done on the ore now exposed before continuing the shaft to greater depth. Large supplies of timber and lumber are being delivered.——During the last week, some ore of good grade has been developed in the stopes above the 3350-ft. level of Central Eureka mine. Assays of this range from $4 to $50 per ton; much of it shows free gold. Thirty of the forty stamps in the mill are in operation on ore from the 2540, 2690, 3350, and 3425-ft. levels. Shortage of labor accounts for the mill running under capacity.——Prospecting continues at the South Keystone Consolidated mines near Amador City. This property, which adjoins the South Spring Hill property, is being worked through the North Star shaft.

Sutter Creek, November 22.

ELDORADO COUNTY

(Special Correspondence.)—The Breala copper mine, situated one mile west of Breala, will begin shipping copper ore shortly. The manager, Victor E. Bonnefoy, states that at the 70-ft. level, in an old drift, he has opened a vein of sulphide copper ore that is fully 18 ft. wide; an average sample of the ore across the vein gave assay returns of 4.5% copper, $2.53 gold, and $1.11 silver.

The old 65-ft. shaft on the copper mine at Missouri Flat, in section 15, 4½ miles west of Placerville, has been unwatered. The manager states that the workmen have found in the old workings a vein about 1½ ft. wide which contains chalcopyrite and bornite ore that runs 55% copper. After the orebody is further developed the owners intend to erect a mill.

Placerville, November 16.

(Special Correspondence.)—Frank Dittmar and W. Pittenger have examined the Livingston property at Fairplay, in the interest of Montana capital. It is owned by Mrs. A. L. Livingston of Oakland, and was formerly worked profitably, being closed on the death of the present owner's husband.——Suit involving title to the Salmon Falls property has been filed in the Superior Court. Seven plaintiffs and seven defendants are named.

Placerville, November 27.

NEVADA COUNTY

(Special Correspondence.)—Under management of H. Gerdetz, San Francisco people have placed the Blue Lead mine, at You Bet, in shape for extensive operations. A restraining dam and sluices have been erected in Missouri canyon. An abundant water supply is available and hydraulicking will start in a few days. Some of the gravel is excessively hard and will be blasted before monitors are brought to play on it.——The mill of the Golden Center Co. is again running on good ore from the 700-ft. level, and deeper work is being resumed as rapidly as unwatering of the shaft progresses. It is believed that the pumping plant is now sufficiently large to effectively handle all water, and the management expects to continue deep mining throughout the rainy season.——A deposit of graphite has been uncovered on the Black Quartz mine, near Washington, owned by C. S. Waite and H. N. Rawson. The deposit is four to six feet wide, and assays about 26% carbon.

Grass Valley, November 26.

SHASTA COUNTY

(Special Correspondence.)—The Greenhorn mine near Tower House has just made a shipment of 50 tons of copper ore to the smelter at Kennett, as a test run.

Redding, November 30.

The Afterthought mine and flotation plant were closed down November 30 on telegraphic orders from the company's headquarters at St. Louis. One hundred and fifty men are laid off. Only a small crew will be kept at experimental work through the winter. The shut-down is due in part of the winter weather and in part to the returns from the oil-flotation plant. J. T. Robertson, general manager for the company, speaking of the oil-flotation plant, said: "The plant has not proved as satisfactory as hoped for, but it is confidently expected that the metallurgical problem will be solved by the experiments to be carried on this winter." The oil-flotation process has proved

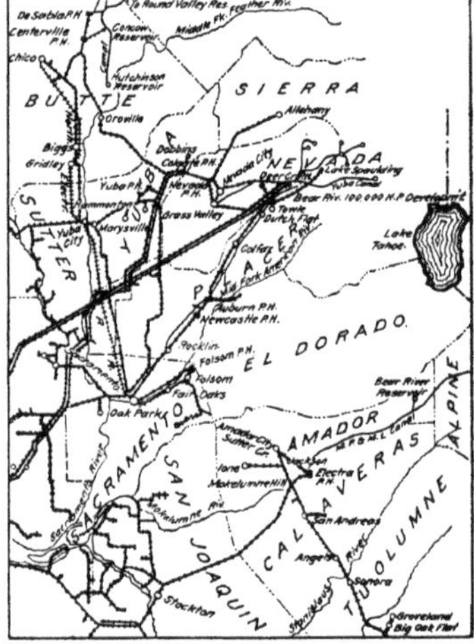

PART OF CALIFORNIA

successful in other copper mines, but at the Afterthough the problem is a little different from that encountered elsewhere. The copper ore is heavily charged with zinc, making it impossible to smelt by the ordinary blast-furnace process. In the oil-flotation plant zinc and copper concentrates were recovered separately, and carload shipments were made to reduction works at Tacoma for copper and in Oklahoma for zinc. The Afterthought company has spent over $250,000 on improvements this season. Ingot, which had been dead ever since the mine and smelter shut-down on January 1, 1908, again became a lively mining camp. Mr. Robertson predicts that the mine and plant will start up in the spring. In the meantime this thrifty town will be depopulated, for there is nothing but the mine and flotation plant to keep it going.

TRINITY COUNTY

(Special Correspondence.)—The Jennings Gold Mining Co. has erected a 250-hp. motor on its claim five miles up Trinity river from Lewiston. The motor will pump water into a reservoir on the hillside, where sufficient pressure will be given to operate a hydraulic giant near the river. Twelve men are employed under Fred. Dorman.——At the Balew mine near Lewiston two small giants will be operated all winter—one on gravel, the other on the dump. V. P. Demars is superintendent.

The Pacific Gold Dredging Co. has dismantled its dredge, moved it from Morrison gulch, and set it up at the mouth of Coffee creek.

Lewiston, December 1.

COLORADO

BOULDER COUNTY

(Special Correspondence.)—The old Smuggler mine, north of Jamestown, has been taken over by Denver people and capitalized at $1,000,000. The Smuggler comprises six lode claims, on which considerable development has been done. Its past record gives it a high rank among the best properties of the district. W. E. Scott, of Gilpin county, has been given the management.——The Mitchell Bros. and T. W. Grouch of Lakewood, who have leased the Corkscrew No. 2 on one of the Primos company's properties, have been shipping tungsten ore, and will continue operations during the winter.——The Golden Age mine above Springdale has been taken over by Barnhill & Son and Salt Lake associates. Thomas Collins is superintendent. The mine will be re-opened and the old dump will be treated by a new process.——The old Longfellow mine, north of Jamestown, which was a heavy producer in the early days and which has been re-opened recently, has been making steady shipments of high-grade silver ore. It will be tapped at depth by an adit, which will be started this week by the owners, William Brown and associates.——The Smuggler Oil & Mining Co., in addition to its interest in the Smuggler mining property at Jamestown, will operate wells in the Brannon field north of Douglas, Wyoming. W. G. Bigelow of Denver is president, and John Rockwell secretary and treasurer.——The activity in mining at Caribou still continues. The recent new strike has ore that runs $1700 in silver and lead. Many properties are producing heavily, and steady shipments from this old-time camp are going on.

Eldora, November 20.

SAN JUAN COUNTY

(Special Correspondence.)—The Sunnyside Mining & Milling Co., a subsidiary of the United States Smelting, Refining & Mining Co., has its new 500-ton mill at Eureka under roof. A new aerial tramway from the mine to the mill is also under construction. About 150 tons per day is being treated in the old Sunnyside mill, where an excellent zinc-lead separation is being made by preferential flotation. Sodium manganate, manufactured at the plant from rhodonite, is used as a flotation reagent. The cascade flotation machines, which replaced the Hyde machines, have been discarded in favor of Minerals Separation apparatus. The Huff electrostatic plant has been dismantled.

The Genessee-Vanderbilt, Yankee Girl, and other properties near Red Mountain, which were recently under lease to James M. Hyde, have been taken over by the Red Mountain Mines Co., a subsidiary of the Mary Murphy Mines Co. Work is being done through the Genessee adit and the old steam compressor there is being used until a new electric-driven compressor can be erected. The equipment at the Joker adit is being overhauled preparatory to starting work there. The operations are being directed by George C. Collins of Denver, with Warren C. Prosser, a local engineer, as superintendent.——A fine body of shipping ore has recently been developed in the Lackawana. An aerial tramway is being built from the mine to the Silverton Northern track, a short distance above town.——The Hamlet is shut-down while some changes are made in the mill. It is expected that operations will be resumed before the first of the year.——The Peerless San Juan mill has been undergoing extensive repairs and is expected to start soon.——The Iowa-Tiger, Pride of the West, Mayflower, Highland Mary, and Gold King are all shipping regularly. The Trilby has a small shift engaged in sinking a shaft.——It has been an unusually open autumn. There is, as yet, hardly any snow on the ground.

Silverton, November 23.

A syndicate composed of Isaac Untermyer, Willard P. Ward, and W. L. Fleming has arranged to underwrite 200,000 shares of the stock of the Vernon Mining Co. for the purpose of increasing the capacity of the mill to 120 tons per day which is estimated to give the company net earnings of about $1000 per day at the present prices of metals. The company is capitalized for 1,000,000 shares with a par value of $1 per share. The mine has been partly opened to a depth of 300 ft., which is still above the water-level; it is said to show 30,000 tons of ore with an average value of $16 per ton in gold, silver, and copper. The property covers 450 acres and contains three veins, one of which is the Mogul vein on which is the well-known Sunnyside mine, recently purchased by the United States Smelting, Refining & Mining Co., at a price reported to be $7,000,000.

IDAHO

SHOSHONE COUNTY

A dividend of 50c. per share has been declared by the Big Creek Leasing Co. at Kellogg, and was paid December 3. The disbursement is $2500. This dividend is the fourth, the first having been made about six months ago. It will raise the total of distribution to $10,000. The profits are derived from ore produced on a section of the Yankee Boy mine on Big creek, east of Kellogg.——At a meeting of the Columbia section of the A. I. M. E. held at Kellogg, S. S. Fowler of Blondel, B. C. was chosen section chairman for the coming year; J. C. Haas of Spokane was elected vice-chairman, and L. K. Armstrong of Spokane was re-elected secretary. The next session of the section will be held at Spokane during the mining convention in February.——The regular monthly dividend of 3c. per share was paid on December 5 by the Caledonia Mining Co., operating in the Coeur d'Alene district. The disbursement was $78,150, bringing payments for the current year to $859,650, and the grand total to $2,835,241.——A dividend of 50c. per share has been declared by the Consolidated Interstate-Callahan Mining Co. of Wallace, to be paid on January 2. The amount is $232,495, which will bring the total payments of the company to $6,500,860. The last dividend was for $1 per share, disbursed on June 30. The dividend for the quarter ended September 30 was passed so that a surplus might be built for war tax requirements. It is assumed that the reduction to 50c. per share was made with the same end in view.

The meeting of the Idaho Mining Association assembled at Wallace, Idaho, November 24, for the purpose of considering business, political, and industrial matters, affecting the mining industry of the State of Idaho, and incidentally of the Nation, connected therewith, refers to the recent War Income Tax Law enacted by Congress and declares that the mining industry should bear its just and equitable proportion of all expenses of government, federal, state, and municipal, under which it is protected; and that it should and does stand ready to meet all taxes imposed upon it by constituted authority, and particularly those taxes imposed by the War Revenue Law passed by the Federal government, and it willingly offers to support all War burdens imposed by the Government to as great an extent as those that may be imposed upon any other industry, even to the limit of the confiscation of all its property, if that be necessary, to bring about a successful termination of the War now being so justly and patriotically prosecuted by our country for humanity.

But this association declares that all should bear the War Tax, levied upon all alike, under a fixed principle, without discrimination, to the end that all shall be equably taxed.

MONTANA

LEWIS AND CLARK COUNTY

(Special Correspondence.)—Low-grade manganese ore is being concentrated at the plant of the Montana Testing & Engineering Co. at Helena. The ore comes from Philipsburg, west of the range, and is below 35%. Arrangements have been perfected to receive the same character of ore from W. A. Clark's mines in Butte. Several carloads of concentrate have already been sent to iron works in Pennsylvania.——The ball-mill at the Porphyry Dike has been completed, and the pipeline and pumping stations for running the tailing from the mill to the Boulder water-shed are in place. The mine is closed and may not be operated before next spring owing to a shortage of water.——The Montana Peerless Co. has taken over the old Peerless Jennie property and is preparing to develop it. The mine has produced no ore for years, but shipments from near the surface in the early '70s ran up to $1000 per ton. The ore was shipped to Swansea in Wales for smelting.——The Helena Mining Bureau is sending out two carloads per week from the Helena mine.——The Shannon mine, of the Barnes-King Co., is producing well. The total production for all its mines amount to $100,000 per month.——The Marysville Gold Syndicate has arranged to mill 30 tons per day from the Blue Bird-Hickey adit.——There is a probability that the Thomas Cruse Development Co. and the Scratch Gravel Gold Co. may consolidate. They have been operating adjoining properties in Scratch Gravel hills.

Helena, November 27.

NEVADA

ESMERALDA COUNTY

(Special Correspondence.)—The Goldfield Consolidated Co. has taken a bond and option on the Curtz Consolidated group of mines near Markleeville, California, for a reputed price of $500,000. The orebodies are large and contain gold, silver, and copper. The Curtz Consolidated Co. has been working for several years and a large ore-reserve is stated to be exposed. The Consolidated is also developing the Surcease gold property near Oroville, California, and is examining gold properties in California, Nevada, South America, and other fields. Small profits are attending operation of the Goldfield Consolidated mines, due partly to difficulty in disposing of flotation concentrate. Large bodies of low-grade copper-gold ore are being developed in the lower levels of the Laguna, Grizzly Bear, Clermont, and other properties. The Aurora Consolidated is reported to be earning a small monthly profit. ——Goldfield companies are steadily extending the use of the leasing system. Numerous blocks of ground in the Consolidated, Florence, Sandstorm-Kendall, and other mines are being worked by lessees with encouraging results, but as work advances it is evident that the rich deposits near surface were largely exhausted by the companies. The hope of the future is based on possibilities attending deeper work in the Atlanta and Jumbo Extension, where promising discoveries have been made within the shale. At the Atlanta the winze is nearing the 1900-ft. point, where a cross-cut will be extended to seek the Atlanta vein, which yielded high-grade ore on the 1750-ft. level. The new vein recently intersected at a depth of 1830 ft. carries little profitable ore and is apparently an off-shoot from the Atlanta vein. The winze is expected to enter the alaskite-shale contact at an approximate depth of 2000 feet.

The Mason Valley Co. hopes to place the second furnace of its Thompson smelter in operation before the end of December. The first furnace is operated largely on ore from the Mason Valley, Bluestone, and Nevada-Douglas mines, and a heavy tonnage from the Nevada-Douglas and several properties in the Luning district is assured for the second unit.——The Nevada-Douglas reports net monthly earnings in excess of $20,000.——More than 20 companies are shipping from the Luning field and others will start production as soon as the Thompson smelter is ready for their ore.——The June Bug Co., of San Francisco, is developing the Scorpion group of silver-lead claims near Reservation and expects to make a shipment soon. The Reservation Hill Co. has a shipment of high-grade silver-lead ore ready.——The Sebastopol, Pack Saddle, Mountain View, and other companies in the Reservation field are active.

Goldfield, November 25.

LINCOLN COUNTY

(Special Correspondence.)—The Springfield Mining Co. has re-timbered the shaft of the Ely Valley to the 400-ft. level. On the 130-ft. level good ore is showing; a shipment of gold-silver ore will be made soon. Below the 450-ft. point a vein containing copper has been prospected to a limited extent. M. Reel is manager.——The Virginia-Louise Co. is preparing for operations along broader lines. A large hoisting engine has been acquired and a head-frame, bunkhouse, and other structures are being built. Development has been in progress upward of a year and profitable ore has been exposed.——The Prince Consolidated is shipping 40 tons of ore per day.——The Davidson Mining Co. controlled by the Prince Consolidated, is increasing operations on the Division group. Cross-cuts are being extended from the 416-ft. incline and 150-ft. vertical shafts; the orebody has been cut at several points, showing a width of 70 ft. Ore is of the same grade as exposed in the Prince Consolidated and Virginia-Louise.

Pioche, November 25.

MINERAL COUNTY

(Special Correspondence.)—At the Congress mine a big vein is being developed; it is 26 ft. wide and averages 3% copper, while the soft monzonite of the hanging wall will go from 1.5 to 2% of that metal for a width of 60 ft. The whole 86 ft. should pay well if treated on the ground.——The Luning Idaho did not make the expected shipment this week as the Thompson smelter failed to blow-in its second unit, on account of a shortage of coke. Shipping-ore is coming from the Conte winze; the Stockham adit is now in 150 ft.——The R. B. Todd Co. is sinking on the Abe Lincoln claim, from which a shipment was recently made; the shaft is in ore of the average grade.——Two more leases have been let on the Anderson group of the Pilot company. The company is sending out a large tonnage under the new contract of the ore buyer, which permits the marketing of a lower grade of ore at a profit.——A carload of ore that will average 8% copper went out from the Siri shaft of the Wedge mine this week. A contract was let for an additional 100 ft. in the No. 2 adit. The proposed merger of the Wedge and some good surrounding mines has not been consummated as yet, but it is believed that the deal will go through.——The Kirchen Mines Co. continues to ship steadily from its St. Patrick mine, the ore of late running 9% copper. What is known as the 'picture stope' on the lower level is attracting much attention on account of the beautiful coloring of the high-grade ore.——E. C. Watson has returned from an extended trip and has taken a lease on the Anderson group of the Pilot company.

Luning, November 21.

(Special Correspondence.)—The east drift on the 290-ft. level of the Wall Street mine at Luning is now out 30 ft. and for that distance two feet of the ore averages 24% copper. The other workings are also looking unusually well, and Mr. Gerhardi, the manager, is making arrangements to handle a much increased production.——E. H. Wedekind, at one time manager of the Balaklala mine and smelter in Shasta county,

California, has been engaged as consulting engineer for the Wedge Copper Co. From a preliminary examination he expresses himself as well pleased with the Wedge and says that work already done has opened up two good ore-shoots from which ore of excellent grade is being produced.——The Kirchen Mines Co. has opened a three-foot silicious vein in the monzonite, separate from the big vein on the lime-monzonite contact, which gave 6% copper as the net return of two carloads shipped.——The Pilot Copper Co. has found a rich deposit away from the big vein on the contact. This is in the lime hanging wall and the return from one carload was 8% copper, 4 oz. silver, and 0.04 oz. gold.——The Luning Idaho is saving ore from a vein that has been proved for a length of 4500 ft. within the company's property lines.——J. C. Brumblay, Reno representative of the United States Smelting company, has been looking over the Luning district for the past week.

Luning, November 29.

NEW MEXICO

SOCORRO COUNTY

(Special Correspondence.)—The Oaks Co. has taken over the active operation of the Pacific mine. The company is largely interested in the ownership of the mine, and plans extensive development. Until the aerial tramway is again in commission burros or auto trucks will be used in moving ore to mill. The Central group is producing daily and the recent developments on Maud S. and other mines of the Oaks Co. are increasing the ore-reserves.——Thirty 100-lb. bars are now being shipped per month by the Mogollon Mines Co. besides several tons of high-grade concentrate. Both mill and mine are working to capacity.——The firm of Cleaveland & Weatherhead, owners of the Deadwood Sunburst mines and mill, has merged its partnership into the Deadwood Mining Company. The ownership remains the same and plans are being made to open up the property in the near future. The mill is modern and arrangements will be made to enlarge the mine equipment.

Mogollon, November 27.

OREGON

JACKSON COUNTY

(Special Correspondence.)—Mine operators in this district anticipated a relief from the shortage of help as soon as the miners working in the lumber camps returned for the winter, but the new supply is insufficient to meet the loss of the miners of the draft age who are enlisting in the service of their own choice before the second draft is called. The Blue Ledge mine lost seven men in one day last week.——H. H. Leonard, lessee of the Bowden mine, three miles south of Gold Hill, is now operating in the 120-ft. level on a 3-ft. vein of rich ore. He also has taken a lease on the Yellow Jacket, an adjoining property, owned by Thomas Hagen, who is now his superintendent.——W. A. Douglas, a local mine operator, has taken a lease on the G. Danielson quartz gold mine on Galls creek, three miles west of Gold Hill. This property is on old producer, but has been idle for a number of years. The lessee will re-open the old drift and extend it on the vein into adjoining property owned by himself and Thomas Dungey.——R. M. Wilson is equipping and will operate the Nellie Wright mine and mill, three miles south of Gold Hill, which has been closed down the past 90 days.——The Manganese Metal Co. of Tacoma has received three cars of machinery for its manganese property in the Lake Creek district, 20 miles east of Gold Hill. Shipments will be made from Eagle Point.

Gold Hill, November 24.

VIRGINIA

LOUISA COUNTY

The Virginia Lead & Zinc Corporation has completed its standard-gage railroad from Sulphur Mine in Louisa county to Valzinco in Spotsylvania county, a distance of 10 miles. It is constructed in accordance with specifications of the Chesapeake & Ohio railway, with which it connects at Mineral, Virginia, as a first-class branch-line, adapted to the heaviest rolling-stock. The corporation is operating the Valcooper zinc mine and mill on the line of the new road, three miles north of Sulphur Mine. Its principal mines, however, are at Valzinco, where a large tonnage of high-grade zinc-lead ore has been developed on a wide vein which has been opened up to a depth of 350 ft. for a length of over 1200 ft. on the strike. Other veins, equally promising, have been prospected. The completion of the railroad will be signalized by extensive development. The company bids fair to become an important factor in the Eastern zinc-situation. The president is Berkeley Williams of Richmond, and the manager is J. H. Batcheller. The mine offices are at Mineral and at Jones Store, Virginia.

WASHINGTON

STEVENS COUNTY

Renewed interest in Deer Trail camp has been aroused by the discovery of valuable ore, missed by pioneer operators a quarter of a century ago, in the old Queen mine, under operation by the Silver Basin Mining Co., according to Charles H. Goodsell, deputy U. S. mineral surveyor. "The body has a width of four feet and a content of 200 oz. of silver per ton," said Mr. Goodsell. "It was found by breaking into a false wall, alongside of which a drift had been run 200 ft. on the 125-ft. level many years ago. Residents are enthusiastic over the discovery. High-grade copper ore is being moved every day from the Turk mine and the camp seems to be coming back, assisted by higher levels for the metals. Other old properties await the touch of the developer."——Mr. Jarvis, the receiver, has sold the Copper King property at Chewelah, to H. S. Wales of the Northwest Magnesite Co. for $75,000. It is said the debts of the Copper King aggregate about $45,000, which will leave a considerable amount for the stockholders. Mr. Wales has been connected with development of the magnesite industry in this county from the first operations. The Copper King lies beyond the United Copper property and is well developed. It is understood Mr. Wales will make arrangements for the development of the property.

CANADA

BRITISH COLUMBIA

The strike of employees at the smelter of the Consolidated Mining & Smelting Co., Trail, B. C., has resulted in a suspension of operations at several large and small properties in British Columbia and a reduction in the number of employees at others. Ore is being accumulated at some mines where the storage capacity is sufficient.——Employees of the Trail plant struck for an eight-hour day. About 400 were working on a nine-hour basis, according to estimates. The number of smelter employees affected is 1500. The company has given notice that it will receive no more ore at present. The suspension interrupts the smelting of 8000 to 10,000 tons of ore per week. The first week of this month, in which the receipts were 8996 tons, may be taken as an illustration. The names of the mines and their contribution in that period were as follows: Blue Bell 183, Centre Star 856, Couverapee 69, Delphia 11, Emerald 40, Emma 615, Electric Point 350, Gem 6, Galena Farm 50, Iron Mask 32, Jessie Blue Bell 3, Josie 353, Knob Hill 151, Lucky Jim 134, Lucky Thought 54, Le Roi 2490, Mandy 29, Metaline Falls 84, Payne 21, Quilp 536, United Copper 144, Rambler 72, P. Schuter, 7, Slocan Star 40, Surprise 88, Standard 79, Sullivan 2224, Triune 28, Tip Top 137, P. Higgins 7, Van Roi 103 tons. When relief may be expected is uncertain, both employer and employee having taken firm stands in the controversy. More than half the ore shipped to Trail is from mines owned by the Consolidated. The smelting company has lead for three months at the recent rate of demand, according to an estimate made early in the fall. Its coke is being diverted to the Granby Consolidated Mining, Smelting & Power Co. The number of miners affected is large. It approaches that at the smelter.

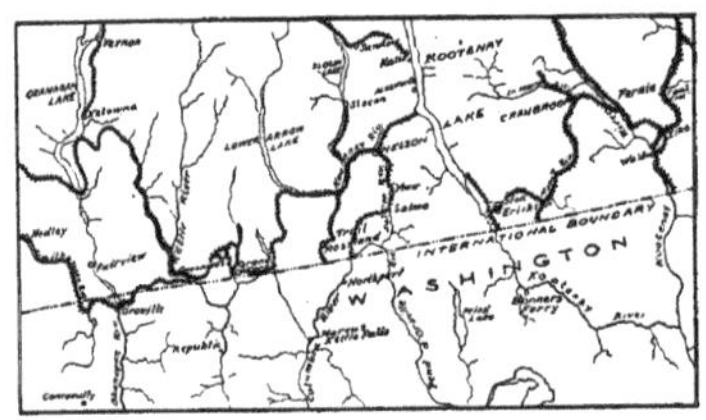

MAP SHOWING POSITION OF TRAIL

The Granby Consolidated Mining, Smelting & Power Co. produced 3,259,974 lb. of copper in October. The Anyox plant produced 2,391,800 lb. of this quantity and Grand Forks 868,174 lb. A comparison of production shows a decrease of 800,000 lb. in October from November 1916, and a decrease of more than 500,000 lb. from August. The decrease at Anyox has been in recent months. The reduction results from changes and improvements being made in the Anyox plant. Two converters are idle, pending the completion of changes, and a steam-driven power plant is being built. The surplus matte is being shipped to Grand Forks for refining. This increases the credit of the latter plant while reducing that of Anyox. The steam plant is expected to make possible the operation of the smelter at the average capacity, or near it, in the winter months, when there has been a reduction because of limited water-power.

The Grand Forks plant was idle in May and June 1916, because of a shortage of coke. Preparations for the erection of buildings for the men to be employed in the development of its recently acquired coal properties are being made by the Granby at Cassidy's landing, near Nanaimo, where its work and wharves will be centred.

Montana Section A. I. M. E.

This section held its regular fall meeting on November 16, at the Silver Bow Club, Butte, when 68 members and guests attended the dinner and technical session. The programme included 'Notes on the St. Louis Meeting,' by C. W. Goodale, with discussion by J. H. Warner, Alexander Leggatt, N. B. Braly, and E. B. Young. 'Work of the War Minerals Committee,' by E. B. Young. 'The Manganese Situation,' by R. H. Sales. 'Fire-Proofing of Mine-Shafts,' by E. N. Norris, with discussion by Messrs. Berrien, Nighman, Braly, Daly, Dunshee, and Carrigan. 'Fire-Extinction by Hydraulic Filling,' by C. L. Berrien and C. E. Nighman. The meeting closed with interesting talks by B. B. Thayer and Spencer Miller, both members of the Naval Consulting Board, and Robert Sticht of Tasmania.

W. B. JEFFREY died at Los Angeles on October 29. At one time he was manager of the Pinos Altos and of the Avino mines, in Mexico. He was a capable and trustworthy engineer, of high character and engaging personality. One of his sons, R. H. Jeffrey, is now manager for the Mazapil Copper Company, at Saltillo, Mexico.

THE CANADIAN MINING INSTITUTE has moved to 503 Drummond Bdg., 511 St. Catherine street, West, Montreal.

PERSONAL

Note: The Editor invites members of the profession to send particulars of their work and appointments. This information is interesting to our readers.

CHARLES BUTTERS is at Washington.

E. M. HAMILTON has returned from Pachuca, Mexico.

V. G. HILLS is examining mines at Chloride, Arizona.

J. W. MERCER visited Denver on his return to New York.

W. A. HUNTER was in San Francisco from Eclipse, California.

F. L. SIZER has moved his office to 1006 Hobart Bdg., San Francisco.

R. B. MCGINNIS has moved his office to Mills Bdg., San Francisco.

W. A. HUNTER, of Eclipse, Plumas county, was in town during the week.

J. H. MACKENZIE returned to San Francisco from Juneau, Alaska, last week.

D. C. JACKLING has returned to San Francisco from a periodical tour of inspection.

GERALD SHERMAN, manager of the Copper Queen, was in San Francisco during the week.

J. I. THOMAS, recently in the Afterthought mill, near Redding, has gone to Ely, Nevada.

O. H. REINHOLT, of San Diego, has been examining a manganese prospect in Lower California.

H. HUNTINGTON MILLER has been examining properties in the vicinity of Tucson, and is now in New York.

RUSH T. SILL, of Los Angeles, has just returned from a business trip to Mexico City and Pachuca, Mexico.

N. I. TRUSCHKOFF, mining engineer to the Bogoslovsk Mining & Trading Co., Petrograd, is in San Francisco.

G. J. STEELE, sergeant in the Military Police, has returned from France as instructor to the American army.

H. W. MACFARREN has been commissioned Captain in the 116th Engineers and is now on his way to France.

JAMES L. KEELYN, of Pasadena, has received a commission as Captain at the Training Camp of the Presidio, San Francisco.

J. W. WHITEHURST has received a commission in the U. S. R., having been in training at Fort Sheridan for the last three months.

F. W. DENTON, director and consulting engineer of the Copper Range interests, has gone to Boston, where he will reside during the winter.

C. H. COOPER, assistant superintendent of the Calumet & Hecla smelters at Hubbell, has been appointed a Captain in the Coast Artillery, U. S. R.

JOHN W. RICHARDS, formerly at Spokane, has been commissioned First Lieutenant in the National Army and is now at American Lake, Washington.

A. W. STEVENS is on his way to Nicaragua, having been appointed mine-superintendent of the Leonesa mine, for the Mina Leonesa, Ltd., of London.

W. G. FARNLACHER and FRANK J. HOENIGMANN have been commissioned First Lieutenants in the Infantry and are to go on active duty on December 15.

LOUIS GAY, for ten years foreman of the Mammoth mine near Kennett, has resigned and will live retired at Redding. RICHARD J. MINEAR succeeds him.

STILLMAN BATCHELLOR, recently engineer for the Globe Mining Co., California, is now associated with CHESTER A. FULTON, mining engineer, at Havana, Cuba.

COREY C. BRAYTON and E. R. RICHARDS have temporarily closed their San Francisco offices in order to give more time to their operations at Midvale, Utah.

ARTHUR P. WATT has resigned from the St. Louis Smelting & Refining Co., and has been engaged as metallurgist by the Missouri Metals Corporation, at Mine La Motte, Missouri.

METAL PRICES

San Francisco, December 4

Aluminum-dust (100-lb. lots), per pound	$1.00
Aluminum-dust (ton lots), per pound	$0.95
Antimony, cents per pound	16.00
Antimony (wholesale), cents per pound	15.25
Electrolytic copper, cents per pound	23.50
Pig-lead, cents per pound	6.50— 7.50
Platinum, soft and hard metal, respectively, per ounce	$105— 113
Quicksilver, per flask of 75 lb.	115
Spelter, cents per pound	9.50
Zinc-dust, cents per pound	20

ORE PRICES

San Francisco, December 4

Antimony, 45% metal, per unit	$1.00
Chrome, 34 to 40%, free SiO_2, limit 8%, f.o.b. California, per unit, according to grade	$0.60— 0.70
Chrome, 40% and over	$0.70— 0.85
Magnesite, crude, per ton	$8.00—10.00

The magnesite market remains practically the same; considerable calcined and raw magnesite has been offered, but no business to amount to anything has been reported.

Manganese: The Eastern manganese market continues fairly strong with $1 per unit Mn quoted on the basis of 48% material.

Tungsten, 60% WO_3, per unit	26.00
Molybdenite, per unit MoS_2	$40.00—45.00

EASTERN METAL MARKET

(By wire from New York)

December 4.—Copper is unchanged and nominal at 23.50c. all week. Lead is quiet and steady at 6.50c. all week. Zinc is dull and easy at 8c. all week. Platinum sells at $105 for soft metal and $113 for hard.

SILVER

Below are given the average New York quotations, in cents per ounce, of fine silver.

Date		Average week ending	
Nov. 28	84.25	Oct. 23	83.85
" 29 Holiday		" 30	84.71
" 30	84.25	Nov. 6	89.25
Dec. 1	84.25	" 13	86.18
" 2 Sunday		" 20	85.62
" 3	85.37	" 27	84.60
" 4	85.37	Dec. 4	84.70

Monthly Averages

	1915	1916	1917		1915	1916	1917
Jan.	48.85	56.76	75.14	July	47.52	63.06	78.92
Feb.	48.45	56.74	77.54	Aug.	47.11	66.07	85.40
Mch.	50.61	57.89	74.13	Sept.	48.77	68.51	100.73
Apr.	50.25	64.37	72.51	Oct.	49.40	67.86	87.38
May	49.87	74.27	74.61	Nov.	51.88	71.60	85.97
June	49.03	65.04	76.44	Dec.	55.34	75.70	....

ZINC

Zinc is quoted as spelter, standard Western brands, New York delivery, in cents per pound.

Date		Average week ending	
Nov. 28	8.00	Oct. 23	8.25
" 29 Holiday		" 30	8.12
" 30	8.00	Nov. 6	7.42
Dec. 1	8.00	" 13	7.81
" 2 Sunday		" 20	7.98
" 3	8.00	" 27	8.00
" 4	8.00	Dec. 4	8.00

Monthly Averages

	1915	1916	1917		1915	1916	1917
Jan.	6.30	18.21	9.75	July	20.54	9.90	8.98
Feb.	9.05	19.99	10.45	Aug.	14.17	9.03	8.58
Mch.	8.40	18.40	10.78	Sept.	14.14	9.18	8.33
Apr.	9.78	18.62	10.20	Oct.	14.05	9.92	8.32
May	17.03	16.01	9.41	Nov.	17.20	11.81	7.76
June	22.20	12.85	9.63	Dec.	16.75	11.26	....

COPPER

Prices of electrolytic in New York, in cents per pound.

Date		Average week ending	
Nov. 28	23.50	Oct. 23	23.50
" 29 Holiday		" 30	23.50
" 30	23.50	Nov. 6	23.50
Dec. 1	23.50	" 13	23.50
" 2 Sunday		" 20	23.50
" 3	23.50	" 27	23.50
" 4	23.50	Dec. 4	23.50

Monthly Averages

	1915	1916	1917		1915	1916	1917
Jan.	13.60	24.30	29.53	July	19.09	25.66	29.67
Feb.	14.38	26.62	34.57	Aug.	17.27	27.03	27.42
Mch.	14.80	26.65	36.00	Sept.	17.69	28.28	25.11
Apr.	16.64	28.02	33.16	Oct.	17.90	28.50	23.50
May	18.71	29.02	31.69	Nov.	18.88	31.05	23.50
June	19.75	27.47	32.57	Dec.	20.67	32.89	....

The Government fixed price remains at 23½c. Electrolytic copper is quoted 25⅝c. per pound, figuring the 5% as arranged for the jobber's benefit, which, of course, is the only way to look at the question of price fixing, as dealers and jobbers would not have the benefit of the 23½c. price before January.

Copper scrap is in excellent demand and bringing higher prices. There is a firm tone for brass scrap. The demand for heavy-lead scrap is excellent, the price being on a close parity with the virgin metal. The small quantity of tin scrap available is bringing fancy prices. Zinc scrap is not in demand.

LEAD

Lead is quoted in cents per pound, New York delivery.

Date		Average week ending	
Nov. 28	6.50	Oct. 23	6.70
" 29 Holiday		" 30	5.66
" 30	6.50	Nov. 6	6.02
Dec. 1	6.50	" 13	6.35
" 2 Sunday		" 20	6.50
" 3	6.50	" 27	6.50
" 4	6.50	Dec. 4	6.50

Monthly Averages

	1915	1916	1917		1915	1916	1917
Jan.	3.73	5.95	7.64	July	5.59	6.40	10.93
Feb.	3.	6.23	9.01	Aug.	4.62	6.28	10.75
Mch.	4.	7.26	10.07	Sept.	4.62	6.86	9.07
Apr.	4.83	7.70	9.38	Oct.	4.62	7.02	6.97
May	4.04	7.38	10.29	Nov.	5.15	7.07	6.38
June	5.75	6.88	11.74	Dec.	5.34	7.55	....

QUICKSILVER

The primary market for quicksilver is San Francisco, California being the largest producer. The price is fixed in the open market, according to quantity. Prices, in dollars per flask of 75 pounds:

Week ending

Date			
Nov. 6	100.00	Nov. 20	100.00
" 13	100.00	" 27	110.00
		Dec. 4	115.00

Monthly Averages

	1915	1916	1917		1915	1916	1917
Jan.	51.90	222.00	81.00	July	95.00	81.20	102.00
Feb.	60.00	295.00	126.25	Aug.	93.75	74.50	115.00
Mch.	78.00	219.00	113.75	Sept.	91.00	75.00	112.00
Apr.	77.50	141.60	114.50	Oct.	92.90	78.20	102.00
May	75.00	90.00	104.00	Nov.	101.50	79.50	102.50
June	90.00	74.70	85.50	Dec.	123.00	80.00	....

TIN

Prices in New York, in cents per pound.

Monthly Averages

	1915	1916	1917		1915	1916	1917
Jan.	34.40	41.76	44.10	July	37.38	38.37	62.60
Feb.	37.23	42.60	51.47	Aug.	34.37	38.88	62.53
Mch.	48.76	50.50	54.27	Sept.	33.12	36.66	61.54
Apr.	48.25	51.49	55.63	Oct.	33.00	41.10	62.24
May	39.28	49.10	63.21	Nov.	39.50	44.12	74.18
June	40.26	42.07	61.93	Dec.	38.71	42.55	....

TUNGSTEN

The fact that Europe is buying considerable ore in South America has had some influence on the market and prices for wolframite and other grades have advanced to $25 to $26, though scheelite is unchanged at $26 per unit in 60% concentrate. Ferro-tungsten has been sold in small lots at $2.25 per lb. of contained tungsten and this is the minimum quotation.

MOLYBDENUM AND ANTIMONY

Molybdenite is unchanged and firm at $2.15 to $2.25 per lb. of MoS_2 in 90% material. There is nothing to report as to antimony ore.

Charles Hardy says: Fair quantities of concentrate have changed hands since my last report. European buyers are still in the South American market, especially in the Argentine and continue to pay high prices. Highest grade wolframite ranges from $24 to $26, the latter price having been obtained for the very highest grade, containing no impurities such as tin and copper. Considerable quantities of wolframite containing impurities have been sold at prices ranging from $23 up, according to grade. Scheelite remains firm at $26. Most of the sellers, however, are holding for higher prices.

The molybdenite market remains steady at prices ranging from $2.15 to $2.25 per pound MoS_2 contained, basis 90%.

The antimony market has apparently raised one-half cent per pound to 14¼c. duty paid, New York; shipment continues at 13½c. c.i.f. New York in bond. The above prices are all nominal.

The present schedule for manganese remains at $1.20 per unit for high-grade ore.

The tin position has felt no relief, the price being on Straits tin at 80c. per pound. It was expected that some good might come from the new Government regulation as covered by the American Iron and Steel Institute, Bulletin No. 1, which was issued November 22, but the market has advanced rather than eased; 99% tin is offered at 66½ cents.

The Trust price of lead remains at 6¼c. per pound, although sale of the first quarter lead has been made at 6½c., New York.

Spelter prices are prompt New York. Prime Western ranges from 7⅞ to 8c. It is apparent that the Government is not going to fix a price for the present at any rate on prime Western spelter, the law of supply and demand being left to work out the situation.

Very little high-grade chrome ore has been offered and it is reported that the quantities of 42 to 45% is also very scarce. The market remains at 90c. per unit, quoted f.o.b. California shipping point.

Eastern Metal Market

New York, November 28.

The markets are all inactive with two of the leading metals under Government control, or practically so, tin now following copper in this respect.

Copper is unchanged, with quotations nominal at the Government price.

Tin is scarce and nominally higher.

Lead is inactive but the tone is firm.

Zinc is dull but no weaker.

Antimony is a little stronger and higher.

Aluminum is dead and unchanged.

In the steel market there is an increasing demand from both Government and domestic consumers, with the latter buying for needs other than war necessities. There is no improvement in blast-furnace or steel-works operations, those at Youngstown being at not more than 50 to 75% of capacity, while at Pittsburgh the decline is less. It is expected that the situation in Russia will result in the release of locomotives and possibly cars to relieve the congestion in this country. The shortage in pig-iron is not relieved, and the fuel situation is unimproved, with the possibility that the Government may have to call on coke-producers to furnish coke to steel and iron-producers to keep them running so as to furnish steel ships, shells, and other war-equipment.

COPPER

The situation has been somewhat cleared by an announcement last week determining the jobbers' position. It is now officially declared that small dealers may dispose of their holdings at a 5% profit for cash, with interest at 6% chargeable for time. This refers officially to the copper bought at the 23.50c. Government price, but, as no such metal is now obtainable from producers, it is inferred that metal now in jobbers' hands can be parted with on at least a 5% profit, or for any price that can be obtained. While the latter practice in many cases has been going on for some time, and while some small dealers have continuously refrained from selling, all are now offering their metal, and sales of small lots are going at 27 to 28c. per lb. After January 1, when metal bought at the Government price is delivered, it will sell at 24.675c. per lb., and no more in jobbing lots. While the present arrangement is a hardship to some, it is felt that ultimately it will result in the elimination of the speculator—a decided boon to the entire trade. In general the trade is strictly a regulated one. Your correspondent had an interview yesterday with a member of the Copper Producers' Committee. The whole situation was made clear. The Government's needs come first absolutely. Its orders for copper will mount into large figures, though specifications on these orders as yet are not coming in very fast. Consumers having Government work are awarded metal first, even though they have no copper on contract, in which case it is being delivered at the Government price of 23.50c., which we continue to quote as nominal. Such metal is, however, only for the consumer's own consumption. Copper is being delivered on old contracts at the original price for first-quarter delivery, even for use in commercial lines, but with the understanding that if the Government require the copper the contract will be cancelled or postponed. General satisfaction seems to prevail, and no one legitimately needing any copper seems to be without it. It is hoped, but not certain, that there will be enough copper to meet all needs, but some believe the supply will not be adequate.

TIN

There have been two important features since my last letter, but no market. In fact the week has been an eventful one. First in importance is the official announcement that the American Iron and Steel Institute will act for the War Industries Board so as to completely control the tin market. In its hands will be the distribution of all tin. To it all imports will be consigned. No one will be permitted to do any business until they have signed blanks strictly defining their rights. A pamphlet has been issued giving details. The arrangement is similar to the existing one by which the British Consul at New York regulates the trade. The new control, however, will be more stringent, as ultimately defined by the United States government. It is not yet effective, however, as some details are still to be worked out, but it is expected that it will be soon. The other feature of interest is the report, probably true but not confirmed by your correspondent at first hand, that the Government has commandeered all tin in bonded warehouses and probably in all others. When its own needs are satisfied the remaining metal will be released to its owners. From a market point of view there has been no market nor is one possible under present conditions. The spot-situation is no better. Attempts to buy spot Straits yesterday and Monday, and on last Thursday and Friday, at 80c. bid, and last Wednesday at 78.50c. bid, were without results. The market is, therefore, quoted nominally at 80c., New York. Some consumers are short of tin, and if they have Government work the Government can get it for them; but there is no incentive to buy tin. If one does buy it he may not get it, and if he does the Government may appropriate it with a possible loss to the buyer. The entire situation is at present unsettled. Arrivals to November 27, inclusive, have been 1310 tons, with 4100 tons reported afloat. The London market continues to advance, spot Straits yesterday being sold at £284 per ton, an advance in the week of £4½ per ton.

LEAD

The market is devoid of interest or feature. Price-levels remain unaltered from those of last week, with the outside market at 6.50c., New York, and 6.37½c., St. Louis, and with the American Smelting & Refining Co. quoting 6.25c., New York. The tone of the market is firm and steady at these levels. Demand is of small proportions, and sales have been light. It is purely a drifting market. New developments must come to arouse interest. Consumers are well covered, and producers and dealers well filled with orders. Some imported lead from Mexico has been done at 6c., New York, in bond.

ZINC

The market is still lifeless and without interest. It continues to drag along in an aimless way. There seems no factor operating to cause a demand, although galvanizers are said to have come into the market recently to a greater extent than for some time, but this does not mean much. The quotations are unchanged at 7.75c., St. Louis, or 8c., New York, for prime Western for early delivery, or even for delivery through January. These quotations are being shaded slightly by certain sellers whom some dealers characterize as 'weak sisters.' Nothing is heard of further Government shell-orders so far as orders for zinc are concerned, and very little is being said about the results of the new Zinc Committee's negotiations with the Government.

ANTIMONY

There has been a little better demand which has rendered the market slightly stronger, and the quotation now stands at 4 to 14.50c. per lb., New York, duty paid, for early delivery, for the brands available, mostly Chinese and Japanese.

ALUMINUM

The market is without interest, with No. 1 virgin metal, 98 to 99% pure, unchanged at 36 to 38c. per lb., New York.

THE title 'Water and Mines in Paradise' suggests a poetic conception of the miner's future state, but, unfortunately, it is the heading to an unpleasant controversy suggesting the wrangles of lost souls in Hades.

IT is not often that discussion comes to hand under the stimulus of an article published two years previously. This week such a compliment is paid by Mr. William P. Daniels, in his communication on the extra-lateral right, indicating the vitality of the subject and the aptness of the examples constituting the original contribution by Mr. L. F. S. Holland.

LONDON appears to take a hopeful view of conditions in Mexico. At the recent annual meeting of the Santa Gertrudis company the chairman, Mr. F. W. Baker, said that he believed the years of depression were near an end and that "conditions in Mexico from a political and financial point of view are steadily improving." We hope this anticipation may be fully justified.

JERUSALEM has been captured by the British. Shades of Richard Coeur de Lion! The latest crusade has won its goal and the Moslems are ousted. Now we may look forward to the consummation of the Zionist plan, whereby the Jew may come to his own and Lower Broadway may become less congested. The Saracen is beaten but the Hun is still triumphant; but it is not too much to say that the ideals of Christianity are flouted less at Jerusalem than at Berlin.

EXPORTATION of gold in bars, or of gold and silver coin, from Mexico was prohibited by a decree of President Carranza dated September 27, 1917, accompanied by another forbidding the payment of taxes in American banknotes. The Mexican government was wholly within its rights in taking these steps, but it is perfectly clear that it was done in a spirit of reprisal because of the order by the United States against the exportation of gold, which in our case was a necessary war-measure. That Mexico has further designs looking toward restrictions upon the disposal of the output of mines in that country is shown by the criticism of the operations of American companies at Santa Eulalia, Chihuahua, in a report by an officer in the Department of Mines. He takes the ground that the recent prosperity of that district is wholly fictitious, and that it signifies the depletion of rich Mexican mines for the benefit of "adventurous foreigners." We admit that an American undertaking to do business in Mexico under present conditions must necessarily possess a venturesome disposition, and we also note that the Mexican official making these unpleasant suggestions bears the name of Honigmann.

IN referring to the article on the new copper district in northern Manitoba, by Major Karri-Davies, in our issue of October 13, we did not mention the publications of the Canadian Geological Survey describing this region. Those interested in the subject will find that Mr. E. L. Bruce, more particularly, but Mr. A. MacLean also, have published several summary reports, together with a memoir and a special report on this part of the Canadian mineral domain. These papers can be obtained by writing to the Department of Mines, at Ottawa.

WE note, with keen pleasure, that Mr. Robert M. Raymond, now Professor of Mining in the Columbia School of Mines, has been nominated a vice-president of the American Institute of Mining Engineers. Many others will share this pleasure. The nomination of Mr. Sidney J. Jennings as president for the ensuing year is in accord with general expectation and is most gratifying. He deserves the honor and should prove an admirable chief. The latest bulletin of the Institute records the proceedings of the Patriotic Meeting at St. Louis, on October 8. We gave a summary of the speeches in our issue of October 27, but we commend the full report to our readers.

WE publish abstracts of the reply-affidavits made by Messrs. Edgar L. Newhouse and F. W. Bradley in the suit brought by the American Smelting & Refining Company against the Bunker Hill & Sullivan Mining & Concentrating Company. They refer to other affidavits that we cannot find space to print. For example, Mr. Stanly A. Easton, the manager for the Bunker Hill company, testifies that on June 24, 1915, in Mr. Newhouse's room in the Samuels hotel, at Wallace, there was a conversation "in relation to the ore situation" and on that occasion Mr. Newhouse stated that "it would be fruitless for the Bunker Hill company to expect to receive from other smelting companies as good or better rates and terms for its high-grade and low-grade ores than would be given by the Tacoma Smelting Company, as there was a better understanding among the various smelting companies who purchased ores in the Coeur d'Alene district than there had been for many years and that a very friendly and co-operative spirit existed among them, and when the representatives of any certain smelter was engaged in negotiating for the purchase of ores it was the present policy of the other companies 'to look out of the window'." It is fair to add that Mr. Newhouse demurs to the accuracy of this statement. This is but the attack and parade of the fencing-bout that precedes the real battle, but it indicates the line of defence.

ORDER of deposition of minerals in a vein is a matter of great practical importance to the engineer when occasion arises to pass upon the desirability of developing a prospect. Every examining engineer is required to do some of the work of an economic geologist, because the functions of the two have not yet become fully differentiated. To state it otherwise, the specialist in economic geology is seldom commissioned to examine a prospect; most of that work devolves upon mining engineers, and it is usually the engineer who requests the assistance of the geologist as the development proceeds. Therefore, a scholarly paper such as that which we publish in this issue by Dr. F. N. Guild, on the relations of the silver minerals, will make a wide appeal. It will help in the determination of the primary or secondary character of the ore in cases where oxidation is not in evidence. It may not be important to ascertain this relationship in every case, but where it is necessary to the appraisal of a prospect, it is vital. The article is also noteworthy in its revelations of the limitations and proper use of the reflecting microscope.

DECLARATION of war against Austria will affect a large number of mine-laborers in the South-West and elsewhere. Some of them have been among the more refractory units during recent labor disturbances. Their status now as enemy-aliens will compel them to act discreetly, for they will no longer be subject to the general laws—State and Federal—but liable to internment or other disposition by executive order of the President. This is one of the regrettable consequences of the War, in so far as it checks the assimilation of sundry European elements in our population, creating foreign islands in the body politic of the nation. Another regrettable consequence is the loss, for military service, of a number of Slovenes and Jugo-Slavs that are technically Austrians but yet willing to fight for the cause. However, the time has come when he who is not with us is against us, and although we may lose the services of some of these loyal soldiers we shall decrease the danger of the spying and sabotage done by enemy-Austrians that have been playing the German game.

NECESSITY mothers opportunity. This has once more been proved in Spain. Inadvertently we did her an economic service when we relieved her from her colonial incubus. Ceasing to dissipate her energies in exploiting peoples across the seas, she then turned upon her own domestic resources of brain and brawn and raw material. Having been shown about the same time the road to sanity in national finance by that splendid Spaniard of the Cervantian type, Don José Echegaray, she has advanced in a more progressive spirit ever since. The War likewise has brought its hardships and its benefits. More than before has it been necessary to develop her native industrial opportunities. Iron smelting has grown rapidly. Furthermore, the search for coal, which had been known to exist in the Peninsula without exciting large interest in the past, has led to remarkable discoveries and to constantly enlarging operations. Spain is proving to possess assets in fuel that ensure a prosperous future. Every week new extensions of the coalfields are reported, and the Minister of Fomento has entrusted this new and growing industry to a special committee known as the Consorcio Carbonero, at the head of which is Don José M. de Madariaga, who is also president of the Council of Mines.

TECHNICAL hand-books in growing number are being issued from the Government bureaus. The U. S. Geological Survey has just produced one that gives a simple and direct account of all methods having the approval of present-day specialists in the drilling of wells. We refer to Water Supply Paper No. 257. The author is Mr. Isaiah Bowman, which means, as in the case of all such governmental treatises, that Mr. Bowman has brought together the conclusions of many men, and that he has enjoyed facilities for gathering information, because he represented the Survey, that would not have been accorded to many private individuals. We mention this excellent manual in this way because the Bureaus are invading the province of the author and publisher of technical books to a greater degree than ever. The projected activities of the Bureau of Mines, for example, will involve the continued publication of treatises on metallurgy, chemistry, fuels, and other subjects, as a result of original research and elaborate investigations into current industrial practice. Evidently, in future, the independent contributor to technical literature will perforce find a field apart, and this may become one of criticism, of careful weighing and checking of results in the discussion of the more minute details of economic operations. The function of departmental technical bureaus will necessarily become limited to the broader principles, in short, to the presentation of standard methods, of which Mr. Bowman's booklet on well-drilling is an excellent example.

The Threat to Gold Mining

It has been intimated, from Washington, that the Priority Board, of which Mr. R. S. Lovett is director, will discriminate against the mining of gold by declaring railroad freightage for this branch of industry to be non-essential to the conduct of the War. If such a ruling should go into effect, we may expect next that an embargo will be placed on the delivery of explosives to gold mines—a step that would terminate gold-mining activity in hard rock, leaving only dredging as a source of the monetary metal. This is a serious matter—tremendously serious—not only to an important phase of Western industry but to the National welfare. We submit that present conditions call insistently for a generous treatment of gold mining. It is true, the value of the gold held in this country is estimated at $3,272,500,000 as compared with $1,871,600,000 when hostilities commenced in Europe, but the obligations incurred by the United States since our own declaration of war have been so great as to decrease radically the proportion of gold to indebtedness, despite the influx of gold from outside. Whereas the National debt was a little more than a billion dollars when we started to make war, it is now being increased at the rate of over 20 billion dollars per annum. Are we to diminish the foundation of our financial security at this time of world-crisis? When England went to war she took steps at once to protect her gold-mining industry by making the arrangements necessary to ensure a supply of cyanide, steel, and other materials required by the mines in South Africa, Australia, India, and Canada. It may be said, truly, that England knew that she would need every available ounce of gold to make payment for the war-materials and food-stuffs to be imported from abroad, whereas the United States can obtain such necessaries from within her own borders, so that gold is not required in anything like equal degree for the consummation of the big purchases made at Washington. Our own Government can buy all that it needs from our own people on credit; it issues bonds and notes that circulate freely within the national family; a domestic circulation is established and the day of reckoning is postponed. But a day of reckoning will come; the United States is not, any more than any other member of the civilized world, an island in the midst of an untraveled sea; the inflation of credit cannot proceed uninterruptedly; we must be prepared for the inevitable deflation; and when that day comes the nation must be able to stand not on the shifting sand of fiat paper, but on the firm foundation of a gold reserve adequate to maintain stability. We suspect that the bureaus and committees at Washington have yet to compare notes, in order to crystallize policy into far-sighted action. It is imperative that no such serious step as an embargo on gold mining should be taken until the facts have been well considered. We venture to suggest that this careful consideration is lacking. Just as the Controller of Fuel and the Controller of Food were in disagreement recently concerning the priority to be given to the transport of fuel and food, respectively, so also different departments of the Government appear to be at variance in regard to the need of gold. When the shipment of steel plates to Japan, which has been building ships for use by the Allies, was stopped by order from Washington, there began a withdrawal of gold from the United States to Japan. In return for steel we had been receiving Japanese products, such as silk and soya-bean oil, in exchange, without trenching upon the reserve of gold. As soon as this international exchange was stopped, it became necessary to export bullion. Again, a number of employees of the Baldwin Locomotive Works, in the service of the Government, were about to go to Russia in performance of duties immediately contributory to the success of the War; as Russian exchange had collapsed, it was impracticable to do business with paper, so these engineers and mechanics were furnished with gold amounting, in the aggregate, to $10,000, which was distributed among them on the eve of their departure from Seattle. The port authorities stopped the removal of this gold to Russia. Further, we are informed that the Californian management of gold mines in Salvador was accorded facilities for the shipment of supplies on the express understanding that the bullion would come to this country. A consignment of candles, for use underground, was held up under the embargo placed upon the export of tallow until the destination of the bullion output of the mine was made clear. These three examples appear to indicate clearly that one or more departments of the Government are awake to the real economic position. We submit that the freightage incidental to gold mining is so small in proportion to the value of the product that the proposed embargo is illogical. For example, a gold mine, such as the Plymouth, in Amador county, producing concentrate that is shipped to a smelter, requires one car inward and three outward per month on an annual production of $1,000,000 worth of gold. The group of mines at Grass Valley contributes $3,000,000 per annum in gold while using only two cars per month on inward freight, all the gold being shipped as bullion in parcels so small as to weigh only five tons per annum. Coal, or other fuel requiring transport by rail, is not consumed by our gold mines, their machinery being actuated by hydro-electric power. The smallness of the freightage incidental to gold mining is an outstanding feature of the industry; it appears to be one of which the members of the Board of Priority are not aware. We commend it to their serious attention. Further, we deprecate the recruiting of men in the gold-mining districts for employment at the copper mines. An idea prevails in some quarters that copper and lead are more necessary for warfare than gold. We demur to that; and we believe that thoughtful financiers will do so likewise. For example, we wrote to Mr. Otto H. Kahn, of Kuhn, Loeb & Co., a man conspicuous today for his intense loyalty as also for his keen intelligence, and he replied that in his opinion "the production of gold is a necessary war operation." Are we going to discard one of the advantages we hold over the Enemy? Of the world's gold production 60% is produced under the British flag, 20% is produced in the United States, and 5% in Russia; thus 85% of the total is in the hands of the Allies. Of the remaining 15%, only 2% is controlled by the Enemy. Here we have an economic factor that, if properly used, should go far to strengthen our arms. Is it not foolish to forego this advantage?

Harnessing a Volcano

To utilize the movement of the tides has long been the fancy of the mechanical genius; to convert the heat of the volcano into economic units has been the dream of the physicist. It is interesting to record a successful effort to put the more visionary of these ideas into action. A consular report from Florence describes experiments made near Volterra, in Tuscany, where volcanic steam issues from fissures in the ground along the Cornia river. Early in the 19th century a Frenchman, François de Larderel, analyzed the steam and discovered that it contained enough boracic acid to warrant economic extraction. He built a plant close to the largest fissure and laid the foundation for an industry that is now identified with the town of Larderello. The vent is covered and the steam is condensed into water from which the boracic acid is precipitated in crystalline form. The evaporation of the liquor and the subsequent drying of the crystals are aided by the heat of steam issuing from the fissures. Likewise the offices of the company and the houses of its employees are heated by the same natural agency. Several years ago it was decided to use the surplus steam from these volcanic fissures to drive a small horizontal engine. This was done by utilizing the heat of the steam to vaporize water. When the low-pressure turbine was invented, another experiment was tried. Borings showed that no diminution of steam-pressure from the vents was caused if the borings were of sufficient distance—not less than 50 feet—apart. Bores were made to a depth at which a pressure of two to three atmospheres and a temperature of 310° to 380°F. could be obtained. Through pipes in the bores the steam is carried from underground into tubular boilers, the heat of which is used to turn pure water, derived from other sources, into steam, which, in turn, actuates the engine driving a dynamo. The electricity thus generated is transmitted and sold to neighboring towns. At present three sets of turbine-alternators of 4000 horse-power each produce a three-phase 50-cycle 4000-volt current, which is transformed to 16,000 volts for transmission to the near-by towns or to 40,000 volts transmitted to more distant points. Although utilization of natural energy is on a scale relatively small, it is immensely suggestive. Man has bridled the wind to his travel, he has tamed the waterfall to do his service, and made the lightning carry his word. As yet the ocean has defied his discipline and the volcano has mocked his control. In the tidal motions and in the volcanic fires he sees two great forces that remain outside his authority. The lunar attraction is not wasted, so it may seem to him, for twice it performs its cleansing function on earth's shores, but the subterranean heat is to him only the touch of devastation. Shall we be able to put it to beneficent purpose? Before all the coal is consumed shall we not learn how to generate light from the dying moon, through the service of the tides, and how to win heat from the volcanic glow that comes from the earth's warm heart?

Flotation

Our contemporary at New York publishes a belated criticism of the recent flotation decisions. This is written by Mr. R. C. Canby, who was an expert for the Miami Copper Company in its suit with Minerals Separation. Two facts emerge from his lengthy contribution: One is that the author was making experiments with the porous carborundum wheel, covered by the Towne patent, for the Teziutlan Copper Company, at a time precedent to his tests with the Minerals Separation machine at the Miami mine. His early work was based on the Elmore vacuum process, not patent 835,120. Another fact is the discarding of the Pachuca tank, more properly called the Brown agitator, in connection with the Callow machine, at Miami, just before the hearing of the Miami suit at Wilmington. We cannot understand why Mr. Canby and the other experts acting for the Miami Copper Company should have made the blunder of using the Pachuca, or why, having once used it, they did not make it clear to the Court, then instead of now, that the Callow machine, or pneumatic process, was effective without the aid of any sort of violent agitation. If they had, the decision of the Court would have been different, there is good reason to believe. Again, why did they not demonstrate that the Elmore vacuum patent was the 'prior art' from which they derived the pneumatic method. It seems to us that the Miami case was badly managed. Also the Hyde case; for the demonstration that the 'critical' proportion of oil was a fallacy—that good results could be obtained in the mill when using more than 1% of oil—was not made until after the Supreme Court had given its decision; indeed, it was demonstrated within a few weeks thereafter—too late. The Anaconda company did the mining industry a poor service by hastening to make its contract with the Minerals Separation people, but, on a purely business basis, this contract aided the Inspiration company, which had previously and too eagerly adopted the Minerals Separation machine, because the Anaconda contract superseded the Inspiration contract and gave the two companies a low royalty on the basis of their combined tonnage. Today the Minerals Separation machines in the Inspiration mill have been replaced by machines of the Callow type, except for one Hebbard (a modified M. S. machine) out of the 20 flotation-cells in use. This in itself is pretty good evidence, as Mr. Canby suggests, of the American contribution to the flotation process. Is it not a fact also that patent 835,120 was seven years old before the Minerals Separation people applied it to the concentration of copper ore in the United States?

Extra-Lateral Right

The Editor:

Sir—My attention has recently been called to a discussion of extra-lateral rights in your issues of December 18, 1915, and February 12, 1916, the reply to the inquiry of E. B. Durham in the latter article being of peculiar interest because the question arose with us some years ago, and our attorney, the late George C. Redd, advised us that the triangle *M L G* in Fig. 6 [herewith reproduced] did not belong to any surface claim; that it could not be located on the surface; but that anyone reaching the triangle by any working could locate it in such working. Mr. Redd stated that he had searched for authorities and decisions but had been unable to find any; that he had consulted a number of attorneys, among them Judge Morrison, and that all agreed with him. If there have been decisions definitely settling the matter, as stated by the attorney who replies to the inquiries for you, it would be interesting to me and probably to others, if you would cite them. It would seem to me that the opinion of Mr. Redd ought to be correct, and that it would be, by long odds, the most equitable settlement of the matter.

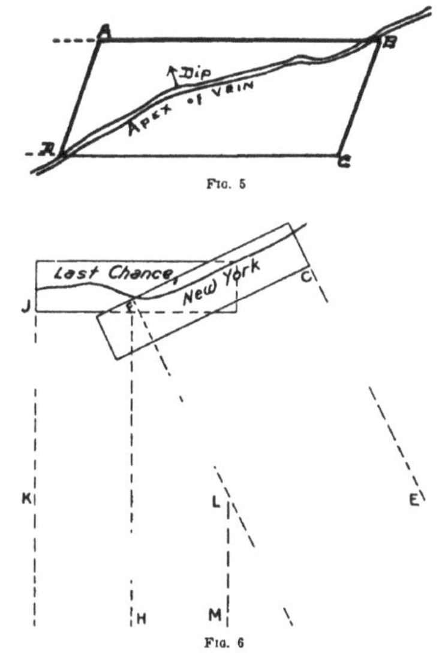

FIG. 5

FIG. 6

WM. P. DANIELS.

Denver, Colorado, October 21.

[The case to which our correspondent refers, taken from the article which he cites, is the following:

"A. Where the vein cuts both side-lines, are there any limits to the application of the rule that the original side-lines become the end-lines that limit the extra-lateral rights? Suppose in Fig. 5, the vein crossed line *A B*, near *B* and the line *D C* near *D*, and dipped north, that is, parallel to the line *B C*; then if the rule held, the extra-lateral rights would be nearly on the strike and also the width of the claim might exceed the maximum of 600 ft. provided by the U. S. law.

"B. Where there is a conflict between two claims, the rule given by Lindley in the next to the last paragraph seems reasonable. If Fig. 6 is drawn with the boundary lines extended, as per sketch herewith, in accordance with the rule given, there is a triangle *M L G* on the vein, with its apex *L* far underground, which has no owner. How can ownership be acquired to this piece? My idea is that additional claims would have to be located along the outcrop of the vein adjoining the Last Chance with their end-lines parallel to the line *L M*; these might have no surface rights and their extra-lateral rights would not commence until the vein had passed beyond the rights of the New York claim."—EDITOR.]

Water and Mines in Paradise

The Editor:

Sir—Under the above heading an article signed by J. D. Hubbard appears in your issue of November 3, in which he takes exception to certain statements made by Ivan E. Goodner, Chief Engineer of the Paradise Irrigation District, in an article appearing in your paper some time ago, and saying that the remarks made by Mr. Goodner are not only inaccurate, but untrue.

Mr. Hubbard writes: "The first statement is that 'None of the owners of the land referred to has so appeared at any time in the past, but all have paid the tax levied without protest.' How about the Mineral Slide Mining Co. in Sec. 10, T. 23 (22), R. 3, M.D.M? Mr. Goodner not only protested, but forced withdrawal of their property from the irrigation district."

I wish to say that as the Mineral Slide Mining Co.'s property has never been in the district, there has never been, nor could there be, any withdrawal from the district, forced or otherwise.

Mr. Hubbard further writes: "The land in question includes 54 acres, not 20, and the tax levied was $5 per acre instead of $1."

The tax levied was 20 cents per acre, as Mr. Hubbard well knows. He may perhaps remember that I was present when he paid it. The total tax on the 54 acres was $10.80.

Next year, Mr. Hubbard says, the tax will be $1 per acre. This statement is not true either; this tax is 4 cents per acre, a total of $2.16 for the entire tract. The valuation of the tract in question is 1/60 of that placed on the best land without improvements. This shows that the assessor has not refused to consider the quality of the land. So far no water has been furnished to anybody. When the system is completed, Mr. Hubbard, or whoever the owner of this 54 acres may be, may feel certain that he will receive the full amount of water to which his $2.16 tax entitles him.

G. C. BILLE, Secretary,
Paradise Irrigation District.

[This closes the controversy.—EDITOR.]

A Method of Mining

The Editor:

Sir—Would you please state the best and quickest way to mine the ore and to remove the waste in a block

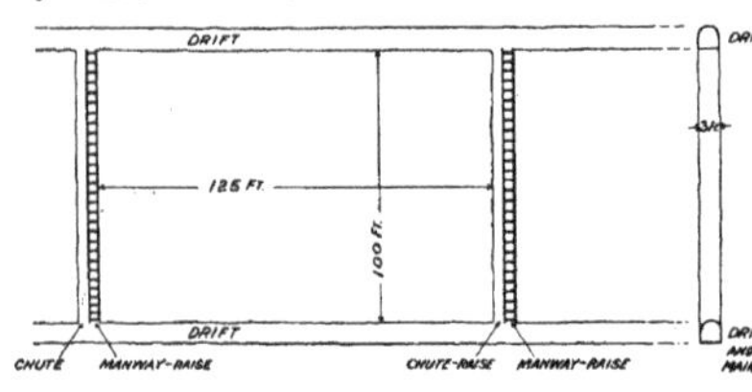

of ground such as is indicated by the accompanying sketch? The vein is three feet wide and we are drilling by hand, using five-eighths of an inch steel. We pay one yen, or 50 cents, per shift of eight hours. The hardness of the vein material is equal to that of average quartz, and the country-rock, which is liparite, is only moderately hard. The vein dips at 75°. The ground stands well, but it would exert considerable side pressure over any large stope, that is, one over 100 ft. long. There is little or no rock for filling, unless specially broken in cross-cuts. The ground to be stoped is from 1000 to 1500 ft. below the surface. The main drift is 5 by 7 ft. inside timbers. There is considerable water, but the vein-filling does not fall out. The waste is often consolidated by the water if left too long broken in the stopes. If you will kindly indicate the approved American way of mining such a block of ground under such conditions I shall be much obliged.

TAKEO IGAWA.

Ashio, Japan, September 2.

The Editor:

Sir—I have noted the inquiry of Mr. Takeo Igawa headed 'A Method of Mining,' and have the following to suggest:

In formulating a system for stoping the particular ground referred to, one should keep in mind two additional essential points, which are not referred to in the inquiry, namely, the value of the ore and the supply of timber.

Assuming that the material carries sufficient valuable metal to be classed as 'ore,' and that timber is easily obtainable at a reasonable cost, I would suggest that, having regard to the fact that the walls are stated to exert considerable side-pressure over a large opening, the ground should be taken out on vein square sets; that alternate sets be filled with waste (the waste to be taken from half-raise cross-cuts in the hanging wall), these cross-cuts to be made at points where waste is required, and that whatever waste is broken when the vein is shot down shall be sorted and used as filling. Thus the ore can be mined clean, and the walls supported, which I believe would suit the requirements of such a block of ground as is described by Mr. Igawa.

If the walls will stand it, it might be found cheaper to simply open the level with ordinary tunnel-sets, placing chutes at intervals of 25 or 30 ft. These sets may be cribbed upward, as the vein is removed, and the stopes filled with waste from hanging-wall cross-cuts. By this method the stope would be filled completely with the exception of the ore-passes and manways, which can be distributed in accordance with the desires of the engineer, having for their purpose the convenience of passing the ore to the level below without having to use anything but a shovel in the stope.

If this method be adopted, the walls will not be supported between the top of the filling, and the vein can be left intact.

W. J. L.

San Francisco, November 13.

The Editor:

Sir—Answering the inquiry of Mr. Takeo Igawa as to the best methods to handle his mine, I would say that two methods are available.

I would advise opening the mine with levels not over 100 ft. apart, starting from the top of the orebody, or at the 1000-ft. level. If the ore-shoots are long, I would advise driving raises at intervals of 75 to 100 ft. along the shoot. These raises should be well timbered to take the strain of the walls; in this way, the stopes can be worked just the same as we worked the small stopes in the Cripple Creek district, which was by stulling and an occasional wall-plate to hold any piece of ground that was blocky.

The other method I have in mind would be to leave pillars of ore at regular intervals. Work the mine on the open-stope method and leave pillars of ore about every 75 ft., the pillars to be from 25 to 50 ft. long; then when all the ore has been removed outside the pillars on a level, start at the back of the drift and draw the pillars. The wall-rock being liparite, which is really rhyolite, should stand well; where blocky ground is encountered, wall-plates can be used back of the stulls. This last method is only advisable where timber is expensive.

As Mr. Igawa is driving his drifts 5 by 7 ft. inside timbers, these will have to be at least 7 by 8 ft. outside timbers; this will give him 32 cubic feet of waste for every foot of drift. This waste he should hoist to the upper levels and dump into the stopes as filling. If later he finds it best to fill the stopes further, this can be done by a series of holes drilled into the hanging wall, shooting it down directly into the stopes.

As his vein is only 3 ft. wide, I would advise that he do not bother with filling the stopes at all. The ground will not break away enough to endanger the mine at any time. The walls are of such a nature that they are not at all dangerous, at least, this is what we found to be the case at Cripple Creek. After they have been left open for a long time, the rhyolite will swell and crush the stulls and commence to fall out in blocks, but the waste-rock that he has dumped into them from his drifts will act as a cushion on top of the drift-timbers.

He cannot use the shrinkage system at all as the water will tend to pack his ore in the stopes and make t very difficult to remove.

To sum up, I would open the mine by the open-stope nethod with levels not to exceed 100 ft. apart, stulling he stopes and putting in wall-plates wherever the round showed a tendency to be blocky and not bother o fill the stopes. With his low cost of labor, he might ind it an advantage to run his drifts at intervals of 0 ft., which would give him such low stopes that he ould easily get out all the ore before any of the wall egan to come in. In this way, he would be well out f the level when the hanging wall did cave, and as ; cannot come over three feet, this is not enough to ndanger the mine in the least.

A. W. S.

Piedmont, California, October 28.

[This appears to be a simple case. In California the round would be excavated for a stoping-width equivlent to the thickness of the ore—3 ft.—by means of the ammer-drill, shooting down on platforms supported by ulls. The ore would be dropped through chutes, either ibbed or lined with 2-inch boards, delivering to the rs on the level below. The stulls used during drilling will protect the men at work and hold the ground long enough to permit of removing the broken ore. After that the stoped area can be allowed to cave. If the rock swells it will be unnecessary to fill.—EDITOR.]

Control of Emulsions in Flotation

The Editor:

Sir—The paper on 'Control of Emulsions in Flotation' by Courtenay De Kalb, in the MINING AND SCIENTIFIC PRESS of August 18, has been read with great interest. It seems to me that the author is setting out with the theory that the mineral is to be oiled; that the oil is first to be emulsified and then delivered to the mineral in the proper state of division. My conception of the happening in the process is somewhat different. The formation of hydrophilic colloids in the flotation process I believe is an absolute necessity; without the formation of these colloids there is no flotation. The presence of oil, however, seems to be unnecessary. I am daily conducting tests and am obtaining results without any oils. In order to analyze the condition which may exist in the flotation process we must, of course, examine the art and come to the conclusion whether there is one flotation process or whether there are two flotation processes. Without referring to legal matters it seems that the Minerals Separation Co. at least takes the stand in accordance with their patents, that there are two flotation processes; one which uses as its medium an immiscible substance and another which uses as its medium an absolutely soluble substance. The patent 962,678 when read carefully leaves no question as to the assertion that the soluble frothing-agents act while in solution, and a disclaimer contained in the patent emphasizes this. In accordance with this theory then we would have two distinctive processes accomplishing the same result. My work so far does not bear this out, and in this discussion I assume that the action in both cases is the same.

It is certainly of vital importance to the development of a theory, and, in turn, to the development of the flotation process, to know whether we are dealing with two actions or with one. I believe that we are dealing with identically the same action in either case. It is, of course, self-evident that in using such sensitive reagents as oils in the complicated mixtures that exist in a diluted ore-pulp a great many chemical changes are bound to take place. In tracing several reactions I am constantly consulting the literature on organic and inorganic chemistry, and while many reactions are totally unknown, there are enough facts available to illustrate the enormous possibilities. Another strong factor in flotation is the oxidation of the reagents, caused by the enormous volumes of air rapidly dispersed through the pulp. Without going into the possibilities or probabilities, I believe it can be stated that the majority of the soluble frothing-agents will be chemically changed on entering the flotation-cell, on account of the alkaline and acid mixtures, and in contact with the metallic oxides and other substances present. A certain amount of research-work in that direction has convinced me that it is possible to make, under the identical conditions present in the flotation process, oils from non-oils. A number of the so-called soluble frothing-agents become insoluble after a period of agitation. Furthermore, I have found that, when the soluble frothing-agents do produce results, with absolutely no oil used and with particular care being taken that no oil should be in the circuit, hydrophilic colloids were present when the frothing took place. On the other hand, no frothing can be obtained with strictly saturated hydrocarbons. In making this statement, I must point out that, in order to obtain saturated hydrocarbons, absolutely free from organic oils and acids, one must go to considerable trouble. All the commercial products contain sufficient organic substances to yield misleading results. In comparing thousands of tests, I find that, first of all, if I wish to float with oil, the oil must have a soluble portion.

I believe that under the proper treatment most organic oils used are almost entirely soluble. Whether they first go into solution and then re-precipitate as hydrophilic colloids is undetermined, but I believe that these oils are not present as droplets but as hydrophilic colloids. This seems to be borne out when flotation is accomplished by soluble substances. It has been manifest in all tests that I have made, that these soluble substances are precipitated as hydrophilic colloids. Seldom are conditions such that an oil may be formed from these soluble substances, although I think this is entirely possible, but it does not seem to be necessary. If it is not necessary then the flotation phenomena evidently exist without the presence of oil, but never without the presence of hydrophilic colloids. The suspicion must naturally arise then that the hydrophilic colloids are directly responsible for flotation. The proper form of these colloids, of course, has much to do with the coherence of the froth. They may be present in the proper chemical composition but not in the proper physical form, and an unstable froth or perhaps a completely negative result may follow. For instance, we all know that heat has a great deal of influence on flotation, and I have observed that finely disseminated colloids at, say, 30°C., will group themselves closely at 50°, and will re-assemble in drops at 80°, and in that condition they will be useless for flotation. Raising the temperature, however, to the boiling point, flotation may again be accomplished. These temperatures, of course, are relative to the substance used. The 'critical temperature' is always that temperature which causes the colloids to collect into droplets or into drops.

The second method of varying these results is by the addition of quite insoluble oils, that is, saturated hydrocarbons. The action due to the saturated hydrocarbons or mineral oils is almost identical with the action of heat, that is to say, these hydrocarbons have a faculty for collecting the colloids, thus ending their usefulness. This action is the underlying reason for the error of the assumption that more than 1% of oil will not give any result. Most commercial products contain a certain amount of mineral oil, and when used in excess introduce enough mineral oil to collect the hydrophilic colloids, and hence to eliminate their usefulness, while, perhaps, when a small amount of mineral oil is added to any suitable quantity of organic oil, the resulting formation of colloids will be such that the elastic limit of the groups is increased, and their capabilitity of surrounding air and forming air-bubbles and froth is therefore improved. The absorption of water by these colloids has a great bearing on the process because, in the concentration of 5 tons of ore into one ton of concentrate with the application of 10 lb. of oil per ton of ore, 50 lb. of oil is used to lift one ton of concentrate, assuming that the ordinary depth of a flotation-cell is 2 ft. This then exerts 4000 ft-lb. of energy, or is a condition under which such energy is rendered available. Without being hydrated, the oil would not have the necessary elasticity for surrounding sufficient air to exert that enormous power, but having absorbed so much water and having thereby enlarged its bulk many times, a comparatively small amount of original reagent used is sufficient to carry the mineral.

The above leaves open the question of the preferential affinity of these colloids to mineral. While I have adhered for years to the theory of a chemical reaction between the mineral and the flotation reagent, I consider it entirely possible nevertheless that we are dealing with a surface-tension pure and simple. The fairly successful operation of various surface-tension methods is illustrative of this possibility, and the lack of complete success may lie only in the fact that the surface-area exposed was too small, and the success of the froth-flotation method may lie only in the increase of surface-area. The fact that different minerals will froth with different reagents does not necessarily exclude that possibility. Different minerals unquestionably will require for their successful flotation a changed condition as to the surface-tension, and that may be influenced by the type of colloids produced. I am at present successfully floating lead from zinc, not making an imperfect separation, but as nearly complete as that of any other mineral floated from its gangue, and I have strong indications that the surface-tension was responsible for this effect. I have ventured, in the above, to give my views, not to convert others to my theory, but merely to furnish material for further discussion.

ALFRED SCHWARZ.

Webb City, Missouri, November 22.

IN precipitating copper from sulphate solution with SO_2, the solution is neutralized with lime and treated with the sulphur di-oxide until it has dissolved a proportion of the gas equal to that of the contained copper. The precipitation of metallic copper takes place instantly when this solution is heated to 160° C. under a pressure of 100 lb. per square inch.

PYRITE from domestic sources is quoted at 20c. per unit of contained sulphur, this being the highest price yet reached.

Displacement-Tanks

By WALTER S. WEEKS

The general principle of displacement-tanks for ieasuring the air-consumption of rock-drills is known) all mining men, but it takes a good deal of experi-ienting to construct them so that they will work prop-rly. It is hoped that these few suggestions will be elpful.

The elements of the displacement-tanks are two closed anks connected with a pipe at the base and equipped 'ith valves so that air may enter one while leaving the ther. The air in the two tanks is always separated by piston of water, which flows back and forth from one ank to the other. The change in level of the water in ne tank multiplied by the area of the tank is the olume of air that has flowed out.

It will occur at once to the reader that there are two ossible ways of measuring the volume of air displaced. n the first method we place on one tank a gauge-glass hat reaches from the top to the bottom. We can read he gauge-glass at the start and after a given elapsed ime, say, a minute, read it again. From these readings 'e can compute the flow of air per minute. The main rawback to this method is that the tanks must be per-ectly regular from top to bottom, or else the volume iust be calibrated in many places.

In the second method we place two small gauge-glasses a one tank, one at the top and one at the bottom, and nd the volume between two marks on the glasses. If e take the time for the water to rise from the lower ark to the higher with a stop-watch, the volume flowing er minute can be calculated. With this type of meter, anks of any shape can be used.

The air entering the tanks must do so at constant pres-re. A reducing-valve is needed. I have tried many nds and the most satisfactory one is comparatively in-pensive. It is made by the Westinghouse Air Brake). and is known as the S 4 locomotive governor. The ice is $21. It should be fitted with a hand-wheel so at it may be regulated easily. It should be arranged shown in Fig. 1 (*a*). (All references will be to Fig. 1 less otherwise specified.) The control-pipe *b* should t be connected to a tank because the valve continu-sly exhausts air to the atmosphere. A lubricator r ould be placed on the control air-line feeding very thin to the valve.

The air, after passing the reducing-valve, must enter st one tank and then the other and when it is entering e it must be leaving the other. These operations re-ire a four-way valve. The stock four-way valve is tly and generally leaks. An excellent four-way valve i be made with four single valves *c* arranged as shown, th levers, so that they can all be thrown together. The valves are made by the Westinghouse company and are known as 1-inch cut-out valves. They cost $1.80 each. In the pipe *d* connecting the tanks at the bottom should be placed a quick-closing gate-valve *e*. It is necessary to measure the volume of but one tank. Glancing at Fig. 1, place a mark on the gauge-glass *f* and fill with water to that point through the pipe *g*. Close the gate-valve. Draw off water through the pipe *g*, carefully measuring it until it comes in sight on the glass *h*. Now draw carefully until an even number of cubic feet is reached and mark point on the glass *h*. When the water in tank II rises from the lower mark to the higher, the measured volume of air has been displaced.

To obviate the necessity for watching the gauge-glasses I have arranged for a bell to ring at the two critical points. It is done by fitting up spark-plugs so that a contact is made between the tank and the sparking point of the plug by means of a float. The principle is shown in Fig. 2. An ordinary automobile-plug fitted with a float is tapped in at *i*. A plug built as shown in the drawing is placed at *k*. The plug is so constructed that the point of ringing can be controlled by screwing the centre point in or out. The plug *k* is placed on a line with the mark on the gauge-glass. A water-glass should be placed opposite the plug *i*, as shown.

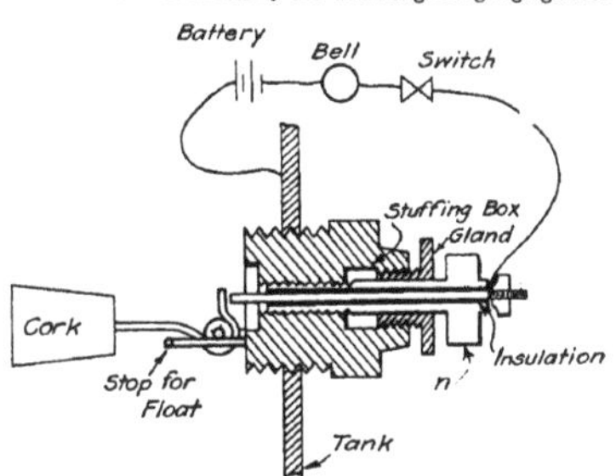

FIG. 2. CONTACT-PLUG

To put the apparatus in working order the water is run up in tank I until the bell rings. The gate-valve is closed and the water in tank II is brought to the lower mark. The gate-valve is then opened and the water is forced to the upper mark. The gate-valve is again closed and the spark-plug *k* adjusted by turning the nut *n* (Fig. 2) until the bell rings at this point. Thus it is seen that between two ringings of the hell the measured volume of air has flowed out. The time is taken with a stop-watch.

To test a drill, connect the hose at 1 and start the drill running. Run water into tank I until the bell rings. Throw the valves and while the water is falling to the level of the spark-plug *i* adjust the reducing-valve. Start the stop-watch when the bell stops ringing and stop it when the bell starts again by contact being made at *k*. To hold the pressure constant the reducing-valve must be changed slightly owing to the small change in hydraulic head on the air. A run of any duration may be made by throwing the valves at each ringing of the bell and keeping account of the number of throws.

The volume between marks on tank II should be about 16 or 18 cu. ft. for ordinary drill-testing. Means should be provided to exhaust the air in the tanks to the atmospheric pressure if necessary. In the tanks shown in the drawing it is accomplished by a three-way valve at *m*. Charts may be plotted for different gauge-pressures with time in seconds between bells, against free air per minute.

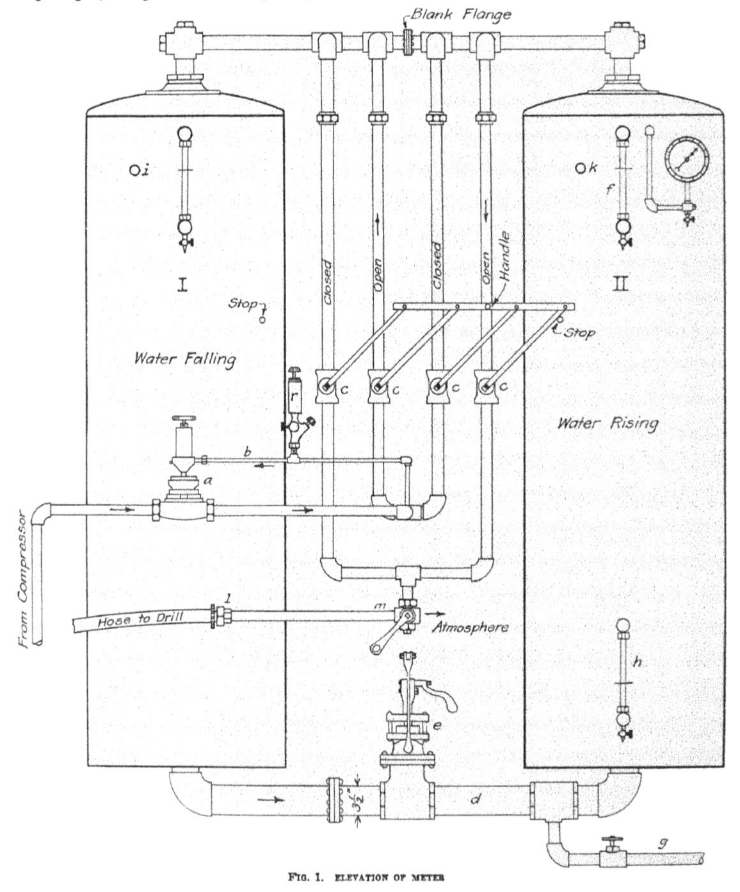

FIG. 1. ELEVATION OF METER

Microscopic Features in Silver-Deposition

By F. N. GUILD

*Since the epoch-making paper by Graton and Murdoch on the copper minerals,[1] the value of the study of polished surfaces of ores by means of the reflecting microscope has been fully appreciated. Following these researches the method has been found of equal interest and value in the investigation of many other classes of opaque minerals. Thus Tolman and Rogers, in applying these methods to the magmatic ores,[2] have arrived at important conclusions regarding the relationship of the ore-minerals to the rock-forming silicates. While investigating certain copper and silver specimens from the Silver King mine, near Prescott, Arizona, I became impressed with the need of more detailed knowledge of the relationships of the silver minerals. I therefore investigated, by means of the reflecting microscope, all the classes of silver ores available and published the results.[3] Some of the more important features are reproduced in this paper.

While it has been held that the rich silver minerals have been deposited later than the other sulphides, the microscopic details of the process have not been described, neither has it been shown how the lean silver minerals break down into those richer in silver that replace other minerals or migrate to other portions of the deposit, thus giving rise to a process of enrichment similar to that observed in most copper deposits. Frequently the copper minerals are also present, and the process of copper enrichment has then progressed side by side with that of silver enrichment. This is well illustrated in Fig. 7, where chalcopyrite has been replaced by bornite, and then has later been replaced by chalcocite and stromeyerite. The ore has been further enriched by the break-down of stromeyerite into native silver. In silver deposits pyrite is most often found to be the first sulphide deposited. It may be preceded by arsenopyrite when that is present. Sphalerite and tetrahedrite are the characteristic early sulphides to follow. Tetrahedrite usually shows rather indifferent relations to sphalerite, but is occasionally found definitely replacing it as shown in Fig. 4. Galena is later than tetrahedrite and in those deposits rich in tetrahedrite it is found actively replacing it. The general order of deposition of the various classes of silver ores will be outlined in the résumé.

*Abstract by author from paper in 'Economic Geology,' Vol. 12, page 297.

[1]'The Sulphide Ores of Copper; Some Results of Microscope Study,' Trans. A. I. M. E., Vol. 45, p. 26.

[2]'A Study of the Magmatic Sulphide Ores,' Leland Stanford Junior University Publication, 1916.

[3]'A Microscopic Study of the Silver Ores and Their Associated Minerals,' 'Econ. Geol.,' Vol. 12, p. 297.

GALENA. This mineral is believed to be one of the early sources of the richer silver minerals, and therefore it is a matter of great interest to determine the form in which silver is found in galena. There are at least three possibilities. First, the silver molecule, usually thought to be argentite, may be present as a solid solution or isomorphous constituent; second, it may occur as inclusions or microscopic particles of definite minerals; and, third, it may occur in the form of sub-microscopic particles. Nisson and Hoyt have conducted laboratory experiments with the hope of throwing light on the subject.[4] They found that lead sulphide when fused was capable of absorbing or holding in solid solution on cooling somewhat less than 0.2% silver sulphide. The excess separated out in the form of definite grains of argentite as shown by their photo-micrographs. These experiments were not performed under the same condition of temperature as that which existed in the vein when the deposition of the galena took place and therefore are not conclusive. It is possible that quite different relationships between the two sulphides exist at the temperature of deposition from aqueous solution. Nevertheless field observations have strongly suggested conclusions similar to those arrived at by Nisson and Hoyt. Thus, in the silver deposits of Bingham, Utah, Boutwell found those ores to assay highest in silver which contained the largest amount of some other silver mineral, in this case tetrahedrite (freibergite). Ranking next to these were the 'black sulphides' thought not to contain tellurides, and last of all the pure galena samples.[5] In the silver deposits near Lake City, Colorado, Irving and Bancroft report that "little of the argentiferous galena is rich in silver in any of these districts unless accompanied by tetrahedrite or some rich secondary silver mineral." Surfaces were polished, and when no admixtures were observed the assay-values indicated only 10 to 15 oz. per ton.[6]

A large number of specimens of argentiferous galena from various localities was investigated by the study of polished surfaces and by quantitative determinations, and many instructive features were observed. On etching the surface with nitric acid or hydrogen peroxide all galena grains showing over 0.10% silver revealed an appreciable number of spots of either argentite or tetrahedrite, and frequently both. These spots are frequently so minute that even when present in

[4]'Silver in Argentiferous Galena Ores,' 'Econ. Geol.,' Vol. 10, p. 172.

[5]'Economic Geology of the Bingham Mining District,' Prof. Paper, U. S. Geol. Surv., No. 38, p. 113.

[6]'Geology and Ore Deposits near Lake City, Colorado,' U. S. Geol. Surv. Bull. No. 478, p. 56.

great abundance, the percentage of silver was not raised above 0.39. Such a specimen from Rimini, Montana, is illustrated in Fig. 1. In this specimen the microscopic inclusions are of four distinct minerals, three of which could be identified. Tetrahedrite appeared harder than galena and could be recognized by its relief. Moreover, it could be compared with other grains in the specimen of sufficient size to permit of micro-chemical tests. Argentite appeared slightly softer than galena, and was practically invisible before etching. Ruby silver could also be distinguished by its bluish tint as compared with galena. The fourth mineral was either some of the sulpho-salt minerals of silver or some soft lead or copper mineral. Only when the spots were of considerable size, in fact large enough to be visible with the hand lens, did the percentage reach 0.50 silver. These, then, begin to show evidences of later addition of silver minerals, with the exception of those showing tetrahedrite. The mineral then has the characteristic appearance of residual grains left behind in the replacement of tetrahedrite by galena. Ores that owe their richness in silver to tetrahedrite have therefore not been enriched by subsequent additions. The later entrance of the rich silver minerals is illustrated in Fig. 2 and 3. In Fig. 2, representing a polished specimen from Tonopah, Nevada, the spots are proustite, and are associated with veinlets of the same mineral that do not show in the photograph. In Fig. 3, taken from a specimen obtained at the Reco mine, Sandon, B. C., pyrargyrite is seen following cleavage-directions in galena. The specimen, as assayed by methods described below, contained 0.54% silver.

On the other hand, specimens of argentiferous galena that show on etching only occasional spots of silver minerals, were found to contain less than 0.10% silver. Thus, galena from the Bunker Hill mine, in Idaho, only now and then showing one spot in the field of view (No. 5, Leitz objective) contained 0.08% silver. A specimen from the Hercules mine, with occasional spots, gave 0.108% silver. Areas in the same hand-specimen, with 25 to 30 spots in the field of view, contained 0.24% silver. A specimen of galena from the Old Yuma mine, Pima county, Arizona, showed no spots over large areas. This contained only 0.016% silver. These results show that silver may exist in galena to the extent of nearly 0.10% in the form of sub-microscopic particles or as solid solutions. Perfect isomorphous mixtures with the silver minerals are certainly not formed. Above the limit mentioned, spots of silver minerals begin to appear in constantly increasing numbers as the percentage of silver rises. The number given above, when converted into silver sulphide, is close to the figure found by Nisson and Hoyt for artificial mixtures.

The method of procedure in making the determinations described consisted in selecting from the polished surfaces as pure areas as possible, etching with nitric acid, and, after examining with the reflecting microscope, breaking with a sharp awl sufficient of the material for scorification and cupellation. From 100 to 500 milligrammes was found sufficient except in the case of the very lean specimens, when one gramme was taken when possible. The spots were found to be grouped unevenly, hence care had to be exercised not to cut too deeply, as that would decrease the possibility of representing different conditions from those recorded on the surface.

TETRAHEDRITE. Tetrahedrite, when argentiferous (freibergite), becomes an important source of silver. Analyses in Hintze's 'Handbuch der Mineralogie' show as high as 36.90% silver. This is usually present as an isomorphous mixture of some silver mineral. It may, however, be enriched by the later addition of ruby silver which appears in cracks and cavities. Microscopic investigations show tetrahedrite to be a more prolific source of the rich later silver minerals than is galena. Indeed, much of the galena in silver deposits has replaced tetrahedrite and its silver content is due to residual spots of silver-bearing tetrahedrite as well as, perhaps, argentite resulting from the break-down of the complex molecule.

Separate determinations of silver have been made on tetrahedrite and galena grains in specimens of typical rich silver ores, from 50 to 100 milligrammes being secured from the individual grains by means of a sharp awl. The tetrahedrite, even under high power, was seen to be perfectly homogeneous, showing the silver to be isomorphously mixed, while the galena as usual showed spots. In a specimen from the Nettie L. mine, Ferguson, B. C., tetrahedrite grains were found to contain 7.27% silver while the galena contained but 0.23%. A similar specimen from the Silver King mine, Arizona, showed 6.06% silver for the tetrahedrite grains, while the galena contained 1.24%. Determinations on specimens from many other localities show similar results.

The microscope shows that tetrahedrite easily gives way to attacking solutions, and it may thus form the starting point for a series of reactions by which minerals richer in silver are developed. These have been observed to still further break down into native silver. In Fig. 5, from the Silver King mine, Arizona, tetrahedrite is seen with a complex system of veinlets of stromeyerite which in places enlarge to areas of considerable size. This seems to be quite a characteristic mode of break-down of argentiferous tetrahedrite, another specimen from an unknown locality in Colorado being shown in Fig. 6. The alteration has set in around quartz grains and along cracks. In Fig. 6, the complicated structure of stromeyerite due to an excess of the chalcocite molecule is faintly seen.

Tetrahedrite as observed on polished surfaces varies considerably in hardness, a condition which causes some confusion in identification because of its great variability in composition. As shown by analyses given in Dana's 'System of Mineralogy,' the copper varies from 10.8 to 44.08%; silver from a trace to 31.29%; zinc from nothing to 7.25%; and iron from 0.64 to 8.24%. Polytelite is thought to belong here, being a lead-zinc variety with less than 1% of copper. In addition, the mineral grades into tennantite, the arsenical variety of fahlerz. Although it varies in hardness from 3 to 4,

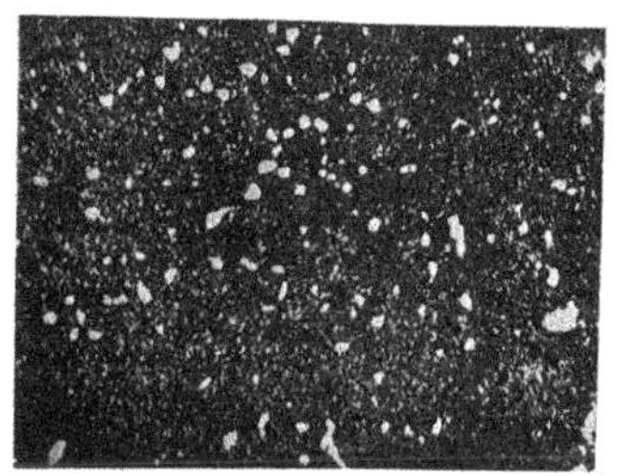

FIG. 1. RIMINI, MONTANA

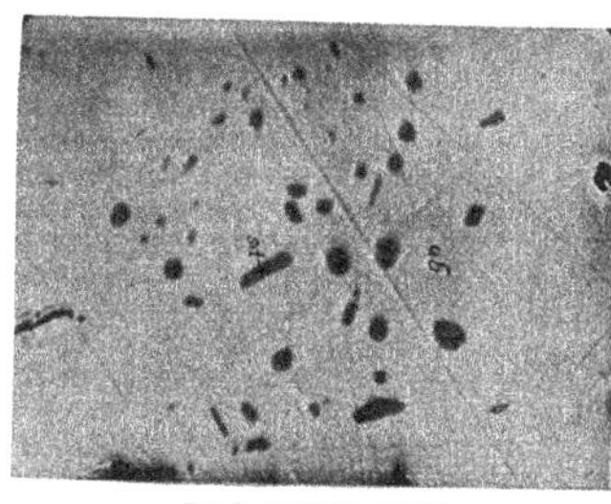

FIG. 2. TONOPAH, NEVADA

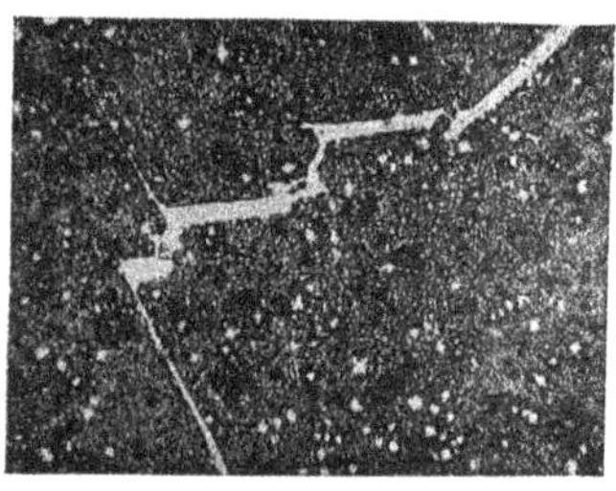

FIG. 3. SANDON, B. C.

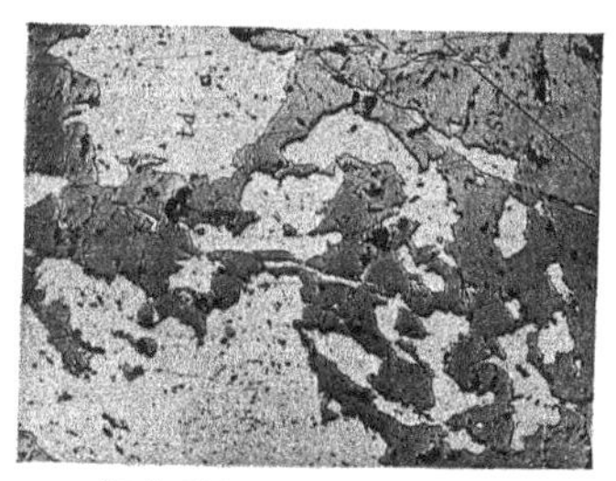

FIG. 4. SILVERSMITH MINE, SANDON, B. C.

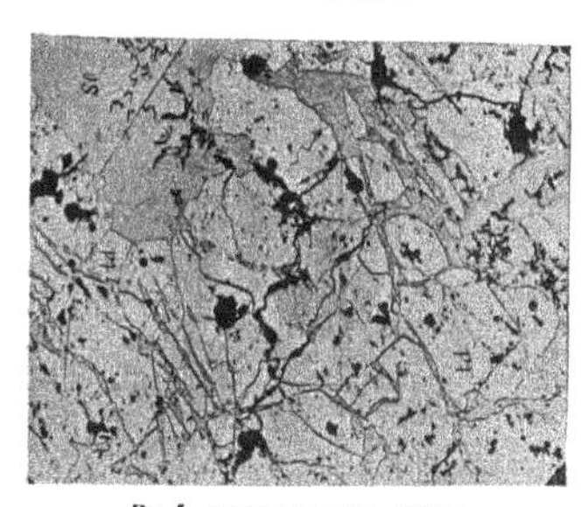

FIG. 5. SILVER KING MINE, ARIZONA

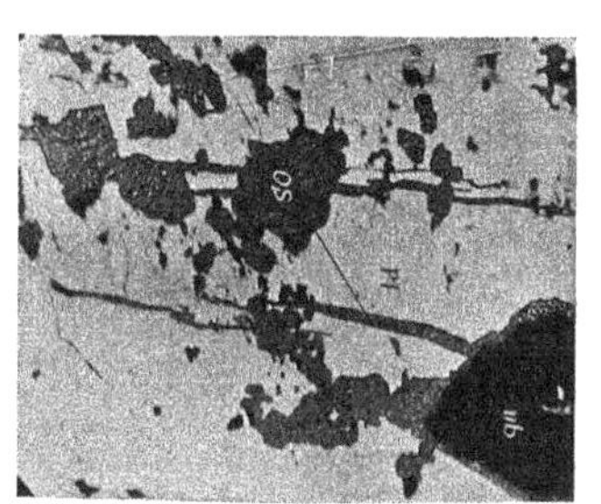

FIG. 6. COLORADO

all varieties show more relief than the silver minerals, a very important feature in identification.

Stromeyerite, $Cu_2S \cdot Ag_2S$, is one of the rarer ore-minerals of silver and but little regarding its characteristics or paragenesis is found in published descriptions of examination with the reflecting microscope. Murdoch describes it as showing a purple tint when viewed side by side with chalcocite and to have sometimes a smooth and sometimes a ragged surface.[7]

In the silver ores thus far studied stromeyerite has been found to present two types of occurrence. First, in veinlets and areas of considerable size, associated with tetrahedrite and doubtless derived from it. This type

[7]'The Microscopic Determination of the Opaque Minerals,' p. 117, 1916.

has already been mentioned under tetrahedrite, and is illustrated in Fig. 5 and 6. Second, as a replacement-product of bornite. In this type it appears to play much the same rôle as chalcocite in the copper ores. There are scattered residual grains of pyrite more or less completely replaced by chalcopyrite, which in its turn has altered to bornite, later to be replaced by chalcocite and stromeyerite. All of these transformations, with the exception of the first, is illustrated in Fig. 7, a photomicrograph of a polished specimen from the Silver King mine in Arizona. Pyrite though present in the specimen does not appear in the photograph. An incipient replacement of bornite in a specimen from the same locality is illustrated in Fig. 8. Replacements less often observed are galena by stromeyerite, and chalcopyrite by stromeyerite.

Microscopic investigation points toward stromeyerite being a definite molecule, probably $Cu_2S \cdot Ag_2S$, and not a mixture of varying amounts of copper and silver sulphides. Many samples of stromeyerite show a very complicated structure under the microscope, and this has been found to be due to an excess of the chalcocite molecule which is not isomorphously mixed with the other constituents. The structure, shown in Fig. 7, is brought out more plainly by etching with potassium cyanide solution. Stromeyerite, which appears smooth and homogeneous under the microscope, approaches the composition of the simple double salt, $Cu_2S \cdot Ag_2S$.

Pyrargyrite, $3Ag_2S \cdot Sb_2S_3$, and proustite, $3Ag_2S \cdot As_2S_3$ are the ruby silvers, and these rank among the most important of the ores of that metal. They are thought to have been deposited late in the history of ore deposition. Thus, Spurr describes pyrargyrite as coating crevices that cut the primary ore; it is also found in the oxidized material, and is thought to have been deposited by descending solutions.[8] In the same district argentite, polybasite, and stephanite are thought to be in part primary. Ransome holds that the ruby-silver minerals, together with stephanite and polybasite, are as characteristic of downward enrichment as chalcocite.[9] The early minerals of silver are confined mainly, if not entirely, to tetrahedrite and galena. They are therefore held to be the source of the later enriched products. Tetrahedrite seems to be the most prolific source, as shown by microscopic and chemical investigations, as well as from the fact that the rich silver minerals are most often compounds of arsenic and antimony. The results obtained from the study of polished surfaces are in accord with the earlier geological observations, at least in respect to the late deposition of these minerals. They are most frequently observed replacing sphalerite, tetrahedrite, and galena. Thus, in Fig. 2, proustite is seen in spots in argentiferous galena. In Fig. 3, pyrargyrite is observed as a microscopic veinlet following cleavage-lines in galena. Photographs might be added in which these minerals are found replacing sphalerite and tetrahedrite in a similar manner.

Microscopic evidence is in favor of considering proustite and pyrargyrite as not isomorphous. They are frequently found mingled with great complexity, but the contact between the two minerals has always been found to be sharp and definite. The fading of one into the other, which would indicate mutual solubility, has not been observed. Fillings of cavities have been studied from Tonopah, Nevada, in which the first layer to be deposited consisted of proustite, the central portions being later filled with pyrargyrite. Pyrargyrite is more easily affected by the electric arc than proustite, and therefore greater detail may be brought out for the purposes of photographing. The ruby silvers are frequently found with spots and veinlets of native silver, resulting from the break-down of the complex silver molecule. These minerals may be distinguished from each other by micro-chemical tests as well as by simple microscopic tests of small fragments broken from the polished surface by means of a sharp point. Proustite is light red in thin fragments, while pyrargyrite is deep cherry red on the thin edges, approaching opacity in the thicker centres. The streaks are also characteristic, proustite being cherry red, while pyrarygyrite is dark purplish red or brown. This test may be made on small fragments, the particles being crushed on glazed paper. These minerals may be distinguished from tetrahedrite by their inferior relief, and by their failure to respond to the micro-chemical test for copper.

Stephanite, $5Ag_2S \cdot Sb_2S_3$, and polybasite, $9Ag_2S (Cu_2S) \cdot Sb_2S_3$, are found replacing the earlier sulphides, galena and tetrahedrite, very much as in the case of the ruby silvers. They cannot be always distinguished by means of the reflecting microscope alone. Micro-chemical tests and tests on small fragments are necessary. Stephanite, like argentite, is opaque on the thinnest edges, but may be distinguished from it by its lack of sectility. Polybasite may be red on thin edges, similar to pyrarygyrite, but it reacts for copper.

Argentite, Ag_2S, from the numerous descriptions of its occurrence, would seem to be either a hypogene or supergene mineral. Great masses of it were found in the upper zones of the Comstock lode. Such bonanzas are usually associated with oxidized material, and the evidence seems to be conclusive that they were deposited by supergene solutions, yet in the Comstock lode, for example, argentite has been identified at a depth of 3000 ft. below the surface.[10] Other examples might be cited showing evidence of its hypogene origin. In the specimens studied during this investigation, argentite is rarely found actively replacing the earlier sulphides. It is most often in small particles filling cavities in gangue, and associated with galena and with the late silver minerals as pyrargyrite, proustite, and others. On the other hand, argentite seems to yield easily to the attack of vein-solutions, and the resulting replacements and products

[8]'Geology of the Tonopah Mining District,' U. S. G. S., Prof. Paper, No. 42.

[9]'Criteria of Downward Sulphide Enrichment,' 'Econ. Geol.,' Vol. 5, p. 211, 1910.

[10]Emmons, 'The Enrichment of Sulphide Ores,' U. S. G. S. Bull. 529, p. 120.

of break-down show interesting features. In almost every specimen the mineral is seen going over to native silver to some extent. This may appear in the form of thousands of spots of such minute size that a high power occur in the areas of argentite, and both argentite and stromeyerite are replaced by native silver and less often by a late generation of chalcopyrite. The sectility of argentite, as observed by scratching the polished surface

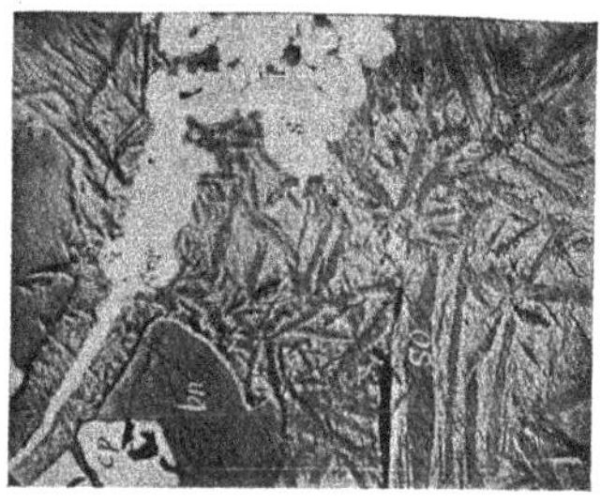

FIG. 7. SILVER KING MINE, ARIZONA

FIG. 8. SILVER KING MINE, ARIZONA

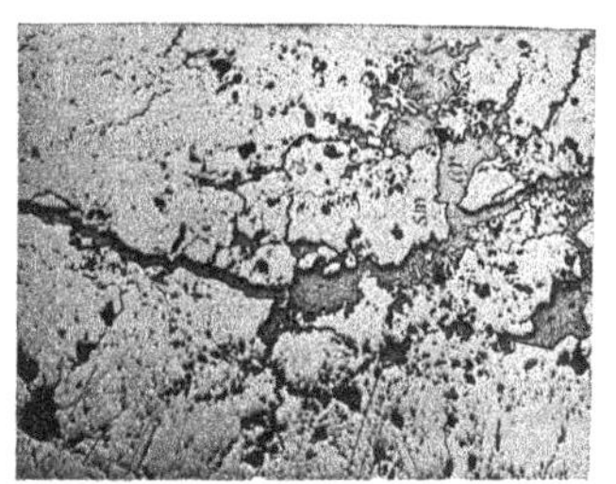

FIG. 9. COBALT, ONTARIO

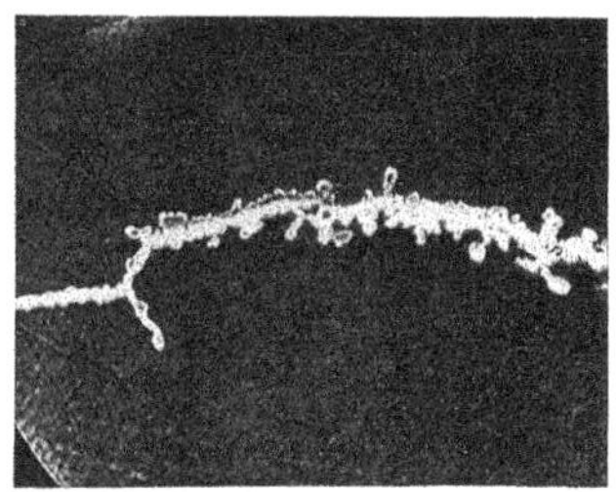

FIG. 10. BUTTE, MONTANA

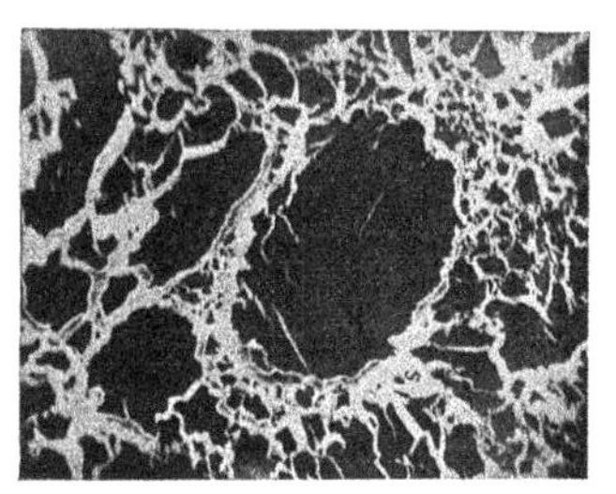

FIG. 11. SILVER KING MINE, ARIZONA

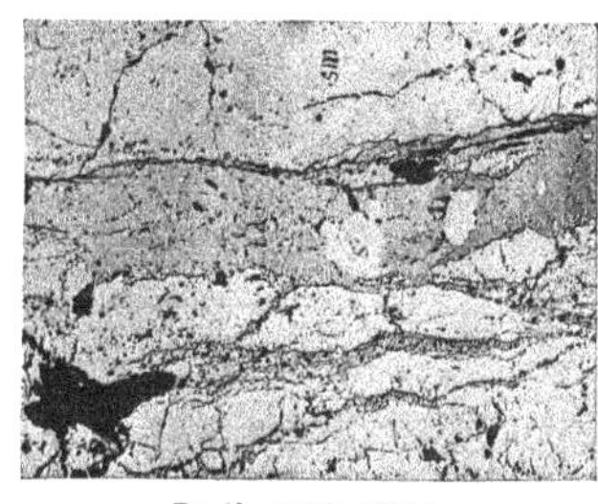

FIG. 12. COBALT, ONTARIO

is necessary to make them visible, or it may take place in the form of veinlets, or as crude segregated masses. Argentite replacing smaltite in the form of wandering veinlets is illustrated in Fig. 9, being a polished specimen from Cobalt, Ontario. Patches of stromeyerite with a sharp point, is sufficient for identification.

Native silver is believed to have been formed ordinarily by the alteration of the earlier silver minerals, the real source of which has been mainly tetrahedrite. It has frequently been described as coating the ruby silvers, polybasite, and others. It sometimes persists to great depths, as in the Aspen district of Colorado, where it is found crossing barite 900 ft. below the surface.[11] At Creede, Colorado, it is reported by Emmons to be found 1200 ft. below the surface. It has generally been assumed that silver is typically a supergene mineral, yet the compounds of silver are known to be weak, breaking down under a variety of conditions. As early as 1843 Bischof found that superheated steam was capable of reducing argentite to metallic silver, which then appeared in aborescent shapes much as in nature. Other minerals of silver treated in the same way yielded similar results.[12] These experiments have been repeated from time to time and modified, Moesta even finding that the reaction could take place at 100°C. From considerations of this kind, coupled with geological data, Kato thinks that native silver in certain ore deposits of Japan was deposited by hypogene solutions.[13] From Vogt's article on native silver at Kongsberg a similar conclusion concerning the paragenesis of some of the metal at that locality might be reached.[14]

In the investigation of the silver ores many cases have been observed where the complex silver minerals have broken down into native silver. The occurrence of native silver in the form of thousands of microscopic specks in stromeyerite and in argentite has been frequently observed. This is the first step in the break-down of the silver mineral, solution, transportation, and re-crystallization being necessary for the production of the delicate structure so frequently seen. Fig. 10 illustrates a veinlet of silver in an argentite specimen from Butte, Montana. A somewhat similar veinlet in stromeyerite is shown in Fig. 7. Here it expands into areas of considerable size. Fig. 11 illustrates the characteristic manner in which stromeyerite breaks down into silver and chalcocite. This is a common occurrence at the Silver King mine near Prescott, Arizona. The silver is arranged in a beautiful filiform structure, the branches of which envelop individual chalcocite grains, some of the finer filaments even extending into the cleavage-cracks of the chalcocite, thus completing the intricate design. Occasionally areas are found where the whole design is roughly oriented with reference to cleavage-directions of chalcocite. All of these features are well brought out by etching with potassium-cyanide solution, when the outline of the individual grains as well as their cleavage is made clearer. These areas grade into stromeyerite, when the native silver disappears altogether, or is confined to borders, veinlets, or clumps of rather rounded outline. The causes which have been responsible for the filiform structure now become clear. Stromeyerite has broken down to chalcocite and silver. The chalcocite has re-crystallized into definite grains of varying size. The silver in re-crystallizing has formed around these grains, extending everywhere into the minutest cracks. Frequently the silver in re-crystallizing has extended into the cracks and surrounded grains of the gangue-minerals. Thus well-formed hexagonal crystals of quartz are found with a mirror-like film of silver. Native silver resulting from the break-down of ruby silver has been observed in specimens from Durango, Mexico, where the silver appears in much less delicate forms. In Fig. 12, spots of native silver are seen in a veinlet of tetrahedrite crossing smaltite; this is from the Cobalt district, Ontario. In similar specimens from the same locality the tetrahedrite is seen breaking down and forming ruby silver, the latter further breaking down to metallic silver.

In Fig. 13, also from the Cobalt district, silver is seen following a veinlet of niccolite in smaltite. This association calls to mind the researches of Palmer, who found niccolite to be an especially good precipitant of silver from silver sulphate solutions.[15] Fig. 14 shows the replacement of smaltite by silver. The euhedral crystals of smaltite embedded in silver is a characteristic feature of this class of ores. Although native silver is the natural end-product in the transformation of the various silver minerals, it may itself be the object of the further attack of supergene solutions containing chlorides or hydrogen sulphide. Its alteration to cerargyrite has been observed in specimens from the Stonewall Jackson mine near Prescott, Arizona, where grains of native silver embedded in siderite show all gradations into the chloride.

Identification of silver minerals on polished surfaces of opaque minerals with the reflecting microscope was confidently expected to enable the ore-minerals to be determined with the same degree of accuracy as has long been the case with the transparent minerals. Failure to realize this has been the cause of a frequently repeated criticism of this method. It must be admitted reluctantly that, at the present time, the chief use of the reflecting microscope in the study of ores consists in determining the relationships of the minerals rather than their identification. A good summary of the work thus far accomplished in methods of determination has been given by Murdoch.[16] From this it will appear that very little detailed study of the silver minerals had been made. In Murdoch's tables they are grouped according to color, hardness, and reaction with various chemicals, such as KCN, HNO_3, and others. Argentite, polybasite, and stephanite are put down as grayish white; silver, dyscrasite, and huntilite as creamy white, and proustite and pyrargyrite as bluish white. His investigations on the action

[11] Lindgren, quoted by Emmons, 'The Enrichment of Sulphide Ores,' U. S. G. S. Bull. 529, p. 118.

[12] Bishof, 'Einige Bemerkungen über die Bildung der Gangmasses,' Pogg. Ann. 60, p. 285.

[13] Kato, 'The Ore Deposits in the Environs of Hanano-Yama, Province of Nagato, Japan,' Meiji College of Technology Jour., Vol. 1, No. 1, p. 32.

[14] Vogt, 'Ueber die Bildung des gediegenen Silbers, besonders des Kongsberger Silbers, durch secondärprocesse aus Silberglanz und anderen Silbererzen,' 'Zeit. prakt. Geol.,' Vol. 7, p. 113.

[15] 'Studies in Silver Enrichment,' 'Econ. Geol.,' Vol. 9, p. 664.

[16] 'Microscopic Determination of the Opaque Minerals,' pp. 4-16.

of reagents were confined mainly to the study of the tarnishing produced on the polished surfaces. I have not found this method very useful in distinguishing the silver minerals. He has employed mainly the following three methods: First, the observation of color, relief, and habit, as observed on polished surfaces; second, the study of minute fragments broken from the polished surface by a sharp point and transferred to a glass slip for examination with an ordinary microscope; and, third, by micro-chemical tests on small fragments secured in the same manner. Etching with nitric acid has been found useful in the study of argentiferous galena, the silver minerals in these being less easily attacked by this reagent; and also for bringing out structure for the purpose of photographing, especially in the case of some of the rarer silver minerals. Potassium-cyanide solution has been found especially valuable in developing the complicated structure of stromeyerite and in distinguishing it from chalcocite. Hardness and color are best compared with galena, since that mineral is the one most commonly associated with the silver minerals. In this respect the tables of Murdoch have been of great value. All of the minerals, with the exception of tetrahedrite, are softer than galena, and therefore they stand out in negative relief. Even tetrahedrite, as it becomes highly argentiferous, approaches galena in hardness. The normal varieties are considerably harder. Proustite is probably the only one that can be distinguished by color alone. It has a characteristic pale bluish tint, rarely mistaken for that of other minerals. Pyrargyrite, in my opinion, is not perceptibly blue, being practically of the same tint as tetrahedrite and polybasite. It can, however, readily be distinguished from tetrahedrite when in contact with it by its inferior hardness, and from polybasite by micro-chemical tests described below.

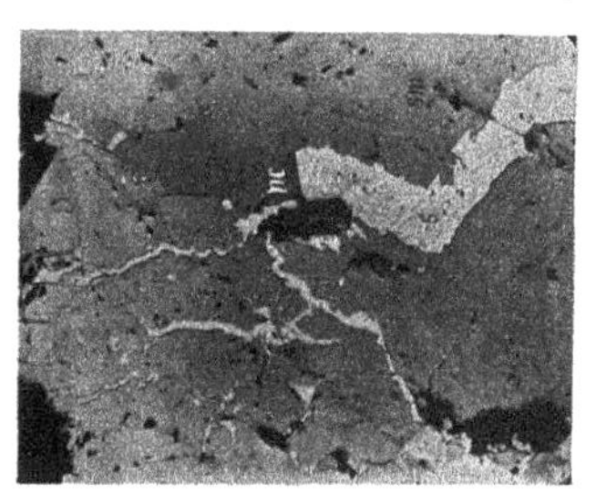

FIG. 13. COBALT, ONTARIO

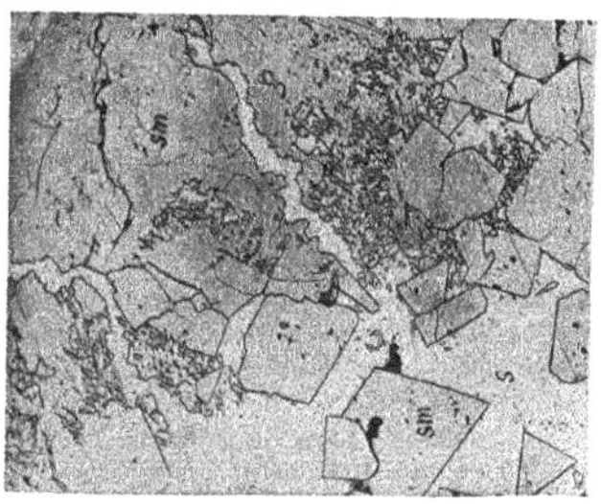

FIG. 14. COBALT, ONTARIO

Tests on fragments are made by breaking from the polished surface minute quantities of the mineral by means of a sharp awl, transferring to a glass slip, and examining with a high-power microscope. Argentite and stephanite are opaque even on the thinnest edges. Proustite is transparent even in the thicker portions, and is of a beautiful red color, shading into amber on the thinnest edges. Pyrargyrite is transparent and of a deep cherry red color only on thin edges, the thicker portions being practically opaque. Polybasite, and some specimens of tetrahedrite, react in the same way as pyrargyrite, but the presence of copper in these minerals is a distinguishing feature. In making these tests it is well to use as intense a light as possible, and care should be used in focusing, otherwise interference of light on sharp edges of black fragments may produce red tints. Proustite and pyrargyrite may be distinguished by their streak on white paper. In securing the material by means of a sharp awl, the sectility of argentite is a sufficient test for that mineral. Micro-chemical tests are made by securing a minute portion of the powder in the manner described above, transferring to a glass slip, dissolving in nitric acid by warming slightly, evaporating nearly to dryness, taking up in a drop of distilled water, and proceeding as outlined below. The presence of silver is learned by adding a drop of dilute chlorhydric acid. The characteristic curdy precipitate is satisfactory evidence. Comparative tests may be made on known minerals so as to be able to judge if the silver be present in large or small amounts. Copper is determined by adding potassium ferrocyanide to the slightly acid drop. The peculiar red precipitate is thoroughly characteristic of this element, and may sometimes be seen in the presence of iron if the drop of reagent be allowed to stand without rapid mixing. Some phenomenon of diffusion permits the two precipitates to be seen. These two tests are those most often required for distinguishing the silver minerals, and it has been found much more satisfactory to remove the minute amount of material required than to attempt to make the test directly on the surface. Moreover, by this method the specimen is much less damaged for further work. Having obtained the material in solution, tests may be made for other elements according to the methods outlined by Chamot in his 'Elementary Chemical Microscopy.' In the investigation of the cobalt-nickel-silver minerals the dimethylglyoxime test for nickel has been found especially useful in studying the associated minerals. Nickeliferous pyrrhotite, probably containing not more than 3% nickel, respond nicely to this test, even in the presence of the large amount of iron. The test is applied by making the drop alkaline with ammonia and adding a minute grain of the reagent. By using the solid reagent the bright red needles seem to concentrate about the solid crystal, thus making the test more delicate. The test has been found especially valuable in distinguishing fibrous iron disulphide (marcasite) from millerite.

Summarizing, it is found that stephanite and argentite are both opaque on the thinnest edges, but argentite may be distinguished by its sectility; pyrargyrite and polybasite are each red on thin edges, but polybasite reacts for copper; polybasite and tetrahedrite may each be red on thin edges, each react for copper, but tetrahedrite is easily distinguished by its greater hardness; proustite is distinguished by its blue tint under the reflecting microscope, and its transparency and bright red color in fragments; stromeyerite, when compared with chalcocite, may be recognized by its purplish tint. It also nearly always shows a perfectly characteristic complicated structure, due to excess of the chalcocite molecule.

A delicate micro-test for arsenic and antimony for distinguishing arsenical and antimonial varieties of tetrahedrite, is greatly needed for this class of work. Berg recommends for the sulpho-salt minerals the following test: Dissolve a fragment in potassium-hydroxide solution, then add chlorhydric acid; arsenic, if present, is thrown down as a lemon-yellow precipitate; the corresponding antimony minerals give the orange sulphide.[17] While this works well for pure compounds, it has been found to be of limited value when applied to the complex mixtures found among this class of minerals.

RÉSUMÉ. (1) Tetrahedrite and argentiferous galena are the early sources of the rich silver minerals, even including native silver. Much of the galena has replaced tetrahedrite, so the latter mineral becomes the more important of the two. Moreover chemical tests show this mineral to be the most important source.

(2) The minerals antedating the deposition of silver in the vein are, in the order of sequence, arsenopyrite, pyrite, and sphalerite.

(3) Galena which does not show abundant evidence of later addition or enrichment has not been observed to contain more than 0.35% silver, the average of 15 specimens from rich silver-lead deposits being 0.20%. This may be present in solid solution or as sub-microscopic particles up to nearly 0.10%. Above that amount it appears as spots of definite minerals identified as tetrahedrite, or argentite, or both. Specimens of galena, with more than about 0.35% silver, show evidence of later addition of ruby silver or other rich silver minerals in the form of veinlets and other secondary features.

[17]Berg, 'Mikroskopische untersuchung der Erzlagerstätten,' p. 46.

(4) The complicated structure of stromeyerite seen in some specimens is the result of a mixture of stromeyerite and chalcocite. Stromeyerite is believed to be a definite double salt of silver sulphide and copper sulphide. Chalcocite is probably able to hold some silver sulphide in solid solution, but the limit of solubility has not been worked out.

(5) Micro-chemical and other tests, applied on fragments secured from the polished surfaces, are considered more useful in identifying silver minerals than the etching and tarnishing methods.

(6) The order of deposition of minerals in silver deposits is outlined below. Important deviation from this order has not been observed.

1. Silver-lead-zinc series:
   (1) pyrite, (2) sphalerite, (3) tetrahedrite, (4) galena, (5) ruby silver, polybasite, stephanite, (6) silver.
2. Copper-silver series.
   (1) pyrite, (2) chalcopyrite, (3) bornite, (4) chalcocite, stromeyerite, and probably argentite, (5) silver. Galena when present is later than chalcopyrite.
3. Cobalt-silver series:
   (1) smaltite-chloanthite, (2) niccolite, breithauptite, (3) argentite, (4) silver and bismuth.

ZINC-BLENDE from the Joplin district contains a metal known as gallium, which is now being extracted in a small way and is being sold at $8 per dram, or about $240 an ounce, according to George W. Waring, of Webb City, Missouri. The discovery of gallium in the zinc ore was made by a watchman at the plant of the Bartlesville Zinc Co. He noted silvery globules on some lead residue that had emerged from the process of re-distillation. He scraped off some of the globules and took them to a chemist who found that it was gallium, a very rare metal. Mr. Waring recently received about an ounce of gallium from Washington for experimentation. He describes it as having some remarkable qualities. If a drop or two is spilled into a plate it instantly scatters over the surface, covering it completely and making a better mirror than quicksilver. A few drops of weak sulphuric acid dropped on the plate will cause the metal to re-assemble into a globule. C. E. Siebenthal, of the U. S. Geological Survey, had a queer experience with gallium. By accident he touched his spectacles to the metal. It immediately covered the glasses and the aluminum frame of the spectacles fell to pieces.

BAUXITE in 1916 showed an increase in production of 43% over the previous year. Producers of metallic aluminum consumed about 300,000 tons, chemical manufacturers 80,000, and makers of abrasives and refractories 45,000 tons. For the latter purpose it is fused in the electric furnace, broken, and sized, being sold under the names of alundum, aloxite, carborundum, exolon, and lionite. The artificial abrasives made from bauxite in 1916 amounted to 30,708 tons, valued at $2,139,230.

American Smelting v. Bunker Hill

ABSTRACT OF REPLY-AFFIDAVITS

By E. L. NEWHOUSE:

The Bunker Hill company knew, as Mr. Bradley's affidavit shows, that in 1905, when the contract was made, the plaintiff company did have a majority of the Coeur d'Alene ores. This is also indicated by the contract itself. This implied to them, and to all other mining men, that other smelters needing lead ore for fluxing might be compelled to bid for the smaller tonnage a smelting rate out of which they would expect to make little or no profit, or even an actual loss, in order to obtain the lead necessary as a collector in their own smelting operations. It was not contemplated that these 'fluxing' contracts, as we call them, were to determine the price of the Bunker Hill, but that the Bunker Hill rate was to be governed by a rate applicable to a majority of the tonnage of the Coeur d'Alene district, as fixed by plaintiff company, so long as it had a majority of the tonnage, and as fixed by other companies whenever the event arose that plaintiff did not receive a majority of that tonnage. This latter contingency has never arisen.

My understanding has always been that, prior to the year 1905, the Bunker Hill company did not ship any material proportion of its mine output to smelters below the grade of 30% in lead, or above the grade of 75% in lead.

My understanding at the time of the contract was that they shipped none of their product within these grades, except neligible quantities. We have not the figures of their actual shipments between those dates, and I notice that the Bunker Hill has not produced them. Schedule No. 2 of our complaint shows our information upon this subject. I believe that an actual production of the figures will show that, prior to 1905, the Bunker Hill company did not ship more than 10% of its entire shipping product in the form of a product below 30% and over 75% of lead, taking the aggregate of those two grades. Schedule No. 2 shows accurately what has been shipped since 1905, within the limitations of cut-offs. It is probable that the Bunker Hill books show ore in transit, but so far as average shipments were concerned, our schedule is correct since 1905.

I do not agree with Mr. Bradley's statement that the low-grade products, as such, are desired by near-by smelters.

The books of the Tacoma plant show, as indicated in Mr. Rust's affidavit, that the Bunker Hill company shipped to Tacoma, in the 10-year period from 1894 to 1904, substantially 1700 tons of a product of 30% lead or less, or an average of 170 tons a year. During the same period, according to our records, it did not ship to East Helena (which was the smelter at all times nearer the Bunker Hill mine than any smelter in the United States to which they shipped) any product containing 30% lead or less.

The clause in the first paragraph of the contract, providing that the average product delivered shall be of approximately the same yearly average analysis and lead assay as the shipments made from the mines of the Bunker Hill company during any of the 12 years immediately preceding, was put in entirely for our benefit, and not at all for the benefit of the Bunker Hill. The defendant company was protected under the immediately preceding clause, providing that the product should be of a lead assay between 30% and 75%, and therefore did not need the 'average product' clause in order to enable it to deliver ore containing any percentage of lead content between those two extremes. From our standpoint, however, it was very important that the product should be of the average analysis that had theretofore prevailed. It would be a decided disadvantage to a smelter to have the ores run with a very low average or with a very high average, even though it were between the limits of the two figures named.

While I have not been able to inform myself of the exact percentage which the defendant's product averaged during the 12 years referred to, my recollection is that during those years it averaged in the fifties, and the figures of 53% in the Tacoma differential rate existed in the Tacoma contract, and as I recall, the statement was made that that was supposed to represent approximately the average shipment of the defendant company. There is no fact connected either with the milling or the smelting business, that I am aware of, which would make the 'average clause' important to the mining and milling company, but that clause is of very great importance to a smelting company, because a substantial departure from such an average as I have indicated would very greatly increase the difficulties of the smelting operation.

I note that Mr. Bradley says that the idea of inserting the 'average clause' was in order that he might be able to mix ores carrying 30% and under with ores carrying 75% and over, so as to make an intermediate product. The 'average clause' would not enable him to mix a low-grade product under 30%, and a high-grade product over 75%, any more than he could do so without it. On the contrary, the clause restricts the mining company and favors the smelter, because it compels the delivery of a product having a like average as that which it had previously delivered. As a matter of fact, however, in the ordinary practice of milling, with reference to which we made this contract, there would be very little high-grade and very little low-grade. The high-grade would only normally occur when rich pockets in the mine were reached. This is seldom, and the tonnage is always small. No other mine in the Coeur d'Alene has ever shipped a product containing over 75% lead, except in insignificant amount.

The clause of paragraph one of the contract, that the average product delivered should be of approximately the same yearly analysis and lead assay as shipments made during any year in the 12 years preceding, like the last section of paragraph ten of the contract (which is to the effect that shipments will be made as far as practicable of approximately equal amounts during each month) was intended for the benefit of the smelting company, as it is obviously to the interest of such company to have deliveries of ores and concentrates and slimes to it made in substantially uniform quantities and of approximately uniform qualities. This is desirable in carrying on its business.

The clause upon which Mr. Bradley relies—"The mining company, in making shipments, may exercise its option as to lead content of ores, slimes and concentrates wherever the same may be shipped"—does not limit the above statement. It immediately follows the clause known as the 'Tacoma differential', which made the price of a certain amount of tonnage shipped to Tacoma vary either upward or downward from the regular price, according as the percentage of lead should be above or below 53%. The clause was put in with the intent of providing that, while we could insist upon our general average as stipulated in the first and tenth clauses of the contract, we could not control the lead contents of the individual shipments so long as the Bunker Hill was maintaining its contract average, this, because otherwise we might prevent the Bunker Hill from getting the benefit of that differential.

It must be borne in mind that the Tacoma differential provided for a different rate at Tacoma than at any other smelter, based upon the percentage of lead. On ore running less than 53% the Bunker Hill got a monetary advantage of 10 cents for every 1% of lead. Reciprocally, it lost 10 cents for every 1% of lead exceeding 53%. This differential did not apply to shipments to any of our other smelters, hence the necessity of the above clause. Otherwise, the plaintiff might have directed the Bunker Hill company to ship any individual lot, containing less than 53%, to East Helena, for example, thus depriving it of the Tacoma differential, or it might direct it to ship a lot, containing more than 53%, to Tacoma, thus imposing an additional cost upon defendant. The intent of the clause was, that the Bunker Hill although bound to produce the product called for by the other clauses of the contract, might so arrange its individual shipments as to take advantage of the Tacoma differential, and this necessitated their control of the lead content of individual shipments "wherever the same might be shipped," so long as the same were produced as the contract required. The clause above quoted, it will be noted, gives the mining company the option only in making *shipments*. It makes no reference to *producing*, which is fully covered by other clauses.

Referring to paragraph nine of the answer, I note that the tonnage of the Hercules is stated to be 1500 tons per month. This was true in the year 1905, but that mine rapidly increased its tonnage until in 1912 it was shipping approximately 5000 tons per month and in 1914 over 7000 tons per month.

In some of the affidavits of defendant reference is made to the purchase by plaintiff of lead ores from Missouri mines. The lead ores of Missouri contained practically nothing but lead, and no silver. The method of treating them is therefore entirely different from that of the silver-lead ores of the Coeur d'Alene district. They are never mixed in with dry ores containing gold and silver and cannot be used as a collector in the treatment of these ores for the reason that to mix them with these ores necessitates then the further refining of the entire product of the lead mine, which is a very material additional expense. It is true that we have increased the output of our Missouri mines to take care of lead sales demands, but this increase has reached its limit and if we should now lose the Bunker Hill ore we would be hampered in our lead sales department in the manner shown in our opening affidavits. Moreover, the production from the Missouri mines will rapidly fall off when lead goes below 7 cents a pound, since much of that production comes from ores which can only be mined and treated profitably with a very high-priced lead.

Referring to that portion of Mr. Bradley's affidavit in which he mentioned as other minerals in the Bunker Hill ore, "salts of silver, copper, zinc, antimony, iron, tin, nickel, cobalt, cadmimum, arsenic, titanium, phosphorus, aluminum, manganese, and magnesium, etc," I will say that while some of these minerals are found theoretically in the laboratory by the most minute chemical analysis, they are not found in what are known as commercial quantities and are not found in quantities now exceeding the quantities existing at the time of making the contract.

Referring to the lead smelters standing ready until June 1915 to buy ore when it was prepared for shipment, I know of no fact or cimcumstance whereby the situation on or since June 1915 is different in this regard to the situation prior to that date, excepting only that owing to the war risks above described these smelters now are asking higher rates than they did before.

Mr. Bradley mentions a number of smelting plants and states that plaintiff is now in control of all but two of said smelting plants. The two plants there listed not controlled by plaintiff are those of the Pennsylvania Smelting Co., at Carnegie, Pennsylvania, and the Ohio & Colorado Smelting & Refining Co., at Salida, Colorado. The other plants there listed were acquired by plaintiff at or before the time of the execution of the contract and there has been no change in that respect since that date, so far as concerns the control of those companies. Since the making of that contract additional lead smelters have been constructed by other companies, as already pointed out. All of the plants controlled by plaintiff have, during all said times, been managed with equal honesty and fair dealing, so far as I have ever been informed, without discrimination either as between themselves or as to mining companies.

Mr. Bradley intimates that the art of concentrating can be developed to the extent of producing a high concentrate and thus avoiding metal losses. It is not true that high concentration reduces losses. Mr. Larson carries this a step further and proposes by a hydro-metallurgical chemico-electrical process to abolish smelting altogether. The problems suggested by these contentions have not as yet arisen; such commercial methods have not as yet been developed, and they do not now constitute any point at controversy between these parties.

At the time the contract was made I by no means considered that defendant had less than five years ore in reserve, or that its active life would be limited by any such period. My knowledge of the Coeur d'Alene district and my investigations led me to believe that a 25-year contract for defendant's output would be valuable and that ore would be delivered under it for that full term—assuming, of course, that defendant should continue mining at the rate theretofore and since prevailing.

Referring to Mr. Bradley's statement as to the understanding reached in the negotiations early in March 1905, I do not deem it important to set out here the different conclusions reached at different times in those negotiations, I consider that the contract speaks for itself and that nothing would be gained by stating the conclusions or views of the parties other than the language finally embodied in it.

In connection with Mr. Bradley's reference to the purchase of the Everett smelter by the Federal Mining & Smelting Co., one of the subsidiary companies of plaintiff, it should be remembered that the plaintiff purchased the Everett smelter from the Federal company in 1903, and the ownership of that smelter by plaintiff was well known to defendant prior to and during the year 1905. The Federal company in 1903 made an arrangement with plaintiff similar to that which defendant made with plaintiff in 1905, namely, to sell its smelter and received from plaintiff at the same time an ore-purchase or smelting contract. I have already stated that the only effect of inserting in the contract a clause referring to the shipment of from 27 to 37 tons of metallic lead per day to Tacoma was to give defendant a special price and treatment charge on that quantity of product, there being corresponding provisions to that extent in the then existing contract between defendant and the Tacoma company. That full advantage has been received by the defendant from the making of the contract to the present day.

The suggestion as to the idea of providing for the same yearly average analysis and lead-assay as to ore previously shipped is not at all in accordance with my understanding. We have always acted on the understanding that this was for the protection of the plaintiff and was to assure it a continuation by the defendant of a product averaging as it had theretofore averaged and a product of substantially the same grade and character. So far as concerns Mr. Bradley's claim that a low-grade product might be shipped to Tacoma under that clause, I think it is sufficient to say that when the point arose as to the possibility of shipment of low-grade in 1906, we elected to purchase the entire low-grade product of defendant at the Tacoma rates and have ever since continued ready to do so.

Referring to the contracts of defendant outstanding in March 1905, I call attention to the terms of our contracts in relation thereto. The reference in Mr. Bradley's affidavit to his desire, expressed in the correspondence, to ship crude ore to Tacoma, and the position of the plaintiff with regard thereto, is misleading. Prior to the time that we exercised our option to take both the defendant's high and low-grade products, it is true that under the contract crude ore shipped to Tacoma would have to run at least 30% lead. After the exercise of our option above-mentioned there was no longer any question but that we were entitled to receive defendant's entire output.

Mr. Bradley recognizes that from 1906 to 1915 the Tacoma company met competitive bids on the high and low-grade products. The fact is we have met them up until not only 1915 but since and have been and are now willing to meet them. It should be borne in mind, however, that the usual operation of defendant's plant produces a product mostly within the normal grade containing from 30 to 75% lead and whenever and wherever that grade of product is produced, whether in its mining, milling, or smelting operations, the plaintiff is entitled to it at the rate prescribed for normal grade. The defendant has no right to re-treat the product of that character so that it will be removed out of the normal grade and put into high-grade. Plaintiff has never recognized the propriety of any such practice.

The examination of defendant's mine referred to by Mr. Bradley, made by Mr. Pope Yeatman, was in no way related to the dealings between the parties to this action.

Mr. Bradley says that he claimed that in 1910 plaintiff was not purchasing a majority of Coeur d'Alene ores for the reason that we should not count the ores to the Selby and Tacoma smelters. Mr. Bradley knew that we owned and we have at all times since owned and controlled all of the stock of both these smelters. He always dealt with these smelters as American Smelting & Refining plants and shipments were made to each of them as plaintiff directed, these plants, like all our other plants, being managed and operated by plaintiff.

Mr. Bradley says that they let contracts to their architects for their smelter as early as September 1915, while Mr. Brush did not write his letter notifying them of our position until November 1915. We did not know that they had let any kind of a contract at the time the letter was written. We still understood that they had the matter of constructing a smelter under advisement and it was not until February 1916 that they finally notified us that they had determined to construct a smelter. Actual construction was not begun until March or April 1916.

I note that Mr. Bradley states that in defendant's new smelter they are smelting lead ores, some of which carry 8% zinc. Not knowing of the percentage of the ores smelted carrying this percentage of zinc, no conclusions can be based on that statement except that I do know that the Bunker Hill ores do not average 8% zinc, while the Hecla, Sierra Nevada, and other ores (which the Bunker Hill has contracted to smelt) do run higher in zinc. Hence, I conclude that the defendant is diluting the other ores with the ores from the Bunker Hill mine. In my judgment, Kellogg is by no means a logical point for a lead-smelting plant. We ourselves examined the Coeur d'Alene for a point at which to erect a smelter and determined that a smelter at that location was not warranted because of the absence of dry-ore supply, increased cost of coke, and lack of limestone, and the Bunker Hill people themselves were in great doubt whether to locate their new smelter at Kellogg or at tidewater.

Mr. Bradley names a number of mines that, he says, are new mines. These are old mines, with the exception of the Interstate Callahan, which is principally a zinc mine. The Bunker Hill company itself owns practically all the stock of the Sierra Nevada, the stock of the Caledonia, and other companies either directly or through Mr. Bradley, Mr. Easton, or Mr. Folsom, and has a lease on the Ontario. The Marsh mine is not a profitable mine; the Green Hill Cleveland is a mere extension of the orebody of one of the Federal mines that has been operated for over 20 years.

In answer to further statements by Mr. Bradley, I would say that neither plaintiff nor any of its subsidiary companies has given to the Federal company, or to the Stewart Mining Co., or to any other mining company in the Coeur d'Alene any better rate or terms than those which are now given to defendant under the contract. For a certain kind of low-grade ore, we are giving a special rate to the Federal company, but we have offered the same rate to the defendant on the same character of product. The defendant attempted shipping that class of ore and decided that it was not as profitable as the method followed theretofore and hence has not availed itself since of the method thus offered and tried. I desire further to call attention to the fact that the freight-rate from Kellogg to Tacoma at the time of the making of the contract was $4 and that it has since remained at that rate.

By F. W. Bradley:

Replying to the reply affidavit of Edgar L. Newhouse: I have no accurate information as to the exact tonnage of Coeur d'Alene ores now being produced, but I do know that the tonnage varies from day to day and my best information is that the plaintiff herein has not been and is not now purchasing a majority of Coeur d'Alené ores of the general character referred to in said contract.

Defendant has never owned or leased the mining claims of the Sierra Nevada Co. For several years prior to 1910

is not true that the Bunker Hill company either directly or indirectly has ever held a lease on the Ontario.

Replying to the reply affidavit of W. R. Rust:

It is not true that the stockholders of the Bunker Hill company ever controlled the three Alaska companies described in the affidavit of said Rust or that they ever owned a majority of the stock of either of said companies. Said W. R. Rust in a letter to me dated February 1, 1905, reported "since May last we (Tacoma Smelting) have been earning at the rate of $553,334.40 per year."

The statement contained in said affidavit to the effect that a contract was drawn up between Baruch and the stockholders of the Tacoma Smelting Co. and signed by all individual stockholders of the Tacoma company prior to the meeting in San Francisco is not true. The above referred to contract was solicited in New York City without my presence and without my knowledge, and was drawn up in New York City on January 9, 1905, between B. M. Baruch and W. R. Rust; and it was made a condition of this Baruch deal that 25-year contracts be made for the entire output of the Douglas Island and Bunker Hill & Sullivan mines. On January 23, 1905, Mr. Rust presented his said contract with Mr. Baruch to William H. Crocker (director and treasurer of the defendant) and myself, in San Francisco, and asked us to consent to the deal on behalf of the Bunker Hill company. We refused to do so unless the deal be modified so as to guarantee the Bunker Hill its then existing freight and treatment rates, and in addition thereto all concessions and betterments that might thereafter be granted Coeur d'Alene mines by the smelters. The defendant's position in the matter was wired to New York City the same day, and on February 20, 1905, Mr. Rust again appeared in San Francisco, and this time with Mr. Baruch, to resume the negotiations. I never signed any Baruch deal contract on behalf of the defendant prior to signing the contract.

It is not true as stated by Mr. Rust that it was ever understood by the defendant Bunker Hill company, or by me, that said company should after the consummation of contract continue to treat its product substantially in the same manner as it had theretofore treated its products from the mine. It is true that this contention was made by Mr. Rust after the contract was entered into, to wit: About the first of the year 1906, and Judge Lindley was asked to give his opinion upon the matter and he agreed with me that we had the right to control the grade of our product wherever the same might be shipped. Rust consulted other counsel and he advised that they agreed with this construction. Subsequent to the resolution of the Tacoma company the said Rust and said Tacoma company repeatedly treated the same as having expired.

The so-called iron middling is concentrate containing less than 30% in lead; is produced in the concentrating plants of the company and is not essentially different from any other Bunker Hill concentrate containing an equal percentage in lead.

It is not true that it has been the customary practice for the defendant before shipping any of its product elsewhere to inform plaintiff that it had a bid therefor or the nature of such bid or to consult plaintiff with reference thereto.

The closing of the Tacoma lead smelter did materially injure the Bunker Hill company, because there was one less lead smelter in need of lead ore, and consequently there was more lead ore for other lead smelters. Poorer terms would naturally be offered by the smelting companies when there was less demand for lead ores.

It is not true, as stated by Mr. Rust, that after June 1, 1915, the Tacomá company demanded merely half of the price above $4.10 per hundred for lead. The Tacoma company also demanded 10% of the price up to $4.10 per hundred pounds on the theory that plaintiff herein could keep up the price of lead and it and its associated companies were entitled to be rewarded. The average price of lead had not been $4.10 per hundred pounds up to 1905; but on the contrary had been $5 per hundred pounds.

IN PERU the Cerro de Pasco Mining Co., an American organization, and the Backus & Johnston Mining Co., which is now controlled by the former, produced about 95% of the total output of fine copper during 1916, the total production being estimated at 41,625 long tons with a value of $25,928,712. Of this output the Cerro de Pasco Co. produced 31,250 tons, and the Backus & Johnston Co. 8638, a total of 39,888. It was shipped in the form of copper bars. The production of Peruvian copper in long tons during the last five years was as follows: In 1912, 27,400; in 1913, 27,500; in 1914, 25,070; in 1915, 31,890; and in 1916, 41,625. The indications are that the 1917 output will be considerably higher than in any previous year.

SANTA EULALIA, the famous silver district situated 14 miles east-southeast from the city of Chihuahua, Mexico, was known as early as 1591, but the first important discoveries were made in 1702. From 1705 to 1790 it produced ₱130,748,314, being at the rate of ₱1,538,100 per annum. From that time forward for 94 years the annual output averaged only ₱23,900. During the last 15 years the production of silver ore from the mines of Santa Eulalia has greatly increased through the efforts of American companies. Part of the ore is smelted by the American Smelting & Refining Co. at its Chihuahua plant, and part is shipped to the smelter at El Paso.

THE HOMESTAKE MINE in South Dakota produced from 1875 to the end of 1916 a total of $148,141,385, of which amount $37,826,116 was paid to the stockholders in dividends. During the last five years the mine has been worked at a greater profit than ever before. The assay-value of the ore last year was $4.08 per ton. This is by no means the lowest average value, however. In 1911 the ore milled contained only $3.57 per ton, and in that year the profits distributed were $1,310,400.

Licenses for Use of Explosives

Any person in the United States found with explosives in his possession after November 15, and who does not have a license issued by the Federal government showing the purpose for which the explosives are to be used, will be at once arrested and fined up to $5000 or sent to prison for one year. If the circumstances warrant, violators of this order may be fined $5000 and in addition given the one year in prison. This is the principal clause in a war ,measure passed by the last Congress which is now being put into effect by the Bureau of Mines. Francis S. Peabody, of Chicago, a well-known coal-operator familiar with the use of explosives, large amounts of which are used in the coal-mining industry, has been appointed by the Secretary of the Interior to act as assistant to the director of the Bureau of Mines in the enforcement of the law. The director is empowered to utilize the services of all United States officers and all police officers of the States, including the city-police forces, county sheriffs, deputies, constables, and all officers in any way charged with police duties. Persons apprehended in plots to blow up factories and bridges will be turned over to the authorities for prosecution under Federal or State laws. Most States have specially severe punishments for these crimes. New York has an extreme penalty of 25 years imprisonment for the placing of dynamite with intent to blow up property. The penalty provided in this Federal war measure is merely to cover the illegal possession of explosives. The law provides that everyone who handles explosives must have a license. The manufacturer, the importer, and the exporter must have licenses issued by the Bureau of Mines in Washington. The seller of explosives and the purchaser of explosves must also have licenses, these to be issued generally by county clerks, or other local officers who are authorized to administer oaths. There will be at least one licensing officer in each county, and more agents will be designated if the county is sufficiently large to warrant it. If a State has laws providing for a system of licensing persons manufacturing, storing, selling, or using explosives, the State officials authorized to issue such State licenses shall be designated as federal licensing-agents; also city officials qualified to issue city-explosives licenses will be given authority to issue federal licenses. A federal license will not relieve any person from securing licenses required under State laws and local ordinances. In each State there will be appointed a State explosive inspector, who will represent the Bureau of Mines in the administration of the law within the State. Only citizens of the United States or of countries friendly to the United States and the Allies may so obtain licenses. Contractors, mining companies, quarrymen, and others using large quantities of explosives, which are handled by employees, may issue explosives to their employees only through those employees holding a license, called a foreman's license. The purchaser of dynamite, in obtaining a license, must state definitely what the explosive is to be used for, and will be held accountable for its use as stated, and the return of any explosives that may be left. With the strict enforcement of this law, the Federal authorities hope to prevent explosives falling into the hands of evilly-disposed persons and to put a stop to further dynamite plots.

Active Nitrogen

*It has been known for a long time that near-vacuum tubes frequently show luminosity of the contained gas after discharge of electricity through the tube. In the case of air R. J. Strutt found that this is due to a phosphorescent combustion occurring between nitric oxide and ozone, both formed during the discharge. He also found that other phosphorescent combustions are observed in ozone, notably of sulphur, sulphuretted hydrogen, acetylene, and iodine. With moderate discharge of electricity it was at first supposed that pure nitrogen gave no afterglow, but when a jar-discharge was used with a spark-gap the glow was readily obtained. In order to examine the properties of the gas while showing this phenomenon, the vacuum tube through which the discharge was passed was connected with an observation vessel, and a current of gas was drawn through it by a powerful air-pump. One remarkable effect on the appearance of the glow is produced by a change of temperature. If a long tube, through which a stream of glowing nitrogen passes, is moderately heated the glow is locally extinguished, but the luminosity is recovered as the gas passes into the cooler parts of the tube. If, on the other hand, the gas be led through a tube immersed in liquid air, it glows with increased brilliancy where it approaches the liquid air, though the luminosity is finally extinguished when it reaches the coldest part of the tube.

The glowing nitrogen has remarkable chemical properties. It combines with common phosphorus, at the same time producing much red phosphorus. In this behavior it resembles the halogens, chlorine, bromine, and iodine. It also combines with sodium, with mercury, and some other metals, in each case developing the line-spectrum of the metal concerned. It attacks nitric oxide with the formation of nitrogen peroxide, a more highly oxidized substance. Active nitrogen also attacks mercury, forming a compound that explodes when moderately heated. It appears that the phosphorescent nitrogen does not owe its activity to a state of condensation corresponding with that of ozone, the molecule of which consists of three atoms, O_3, the instability of the molecule being due chiefly to the tendency to the production of the more stable molecule which contains only two atoms, O_2. Active nitrogen appears to consist of separate atoms of the element.

*W. A. Tilden, 'Chemical Discoveries and Inventions of the 20th Century.'

LORDSBURG, NEW MEXICO

THE 85.—LAWRENCE.—UNITED STATES COPPER.

It is estimated that the production of the Lordsburg mining district for 1917 will total $2,250,000. The bulk of this production is from the mines of the 85 Mining Co. and the Lawrence Mining Co., operators of the Bonney mine. Lessees have shipped the balance of the ore in the district, there being no other operating and shipping companies.——The 85 Extension Mining Co. was a failure; it has given up its option on the Atwood mines and is in a bad financial condition.——The Green King mine also has proved to be one of little merit. An Ingersoll-Rand air-compressor and equipment was erected at the mine, but the returns from ore shipped would not justify operations. The mine has been closed-down and the equipment will be sold to miners in the Burro Mountain district.——The 85 Mining Co. has had one of its most successful years and is rushing the construction of its new concentration plant. Part of the machinery is now on the ground and more is coming in daily. The excavation work is completed. The mill will have a daily capacity of 400 tons and will treat custom ore besides the 85 company output. The plant should be in operation by next March.——At the Bonney mine the heavy flow of water has been adequately handled and development work is progressing. The output of the Bonney mine this year has been about 1200 tons per month. The ore averages $23 per ton. Considerable development work has been done by the Lawrence Mining Co., the No. 2 and 3 shafts have been connected and the No. 2 shaft extended to the 400-ft. level.——The Last Chance mine is operated now by Fairly, Wells & Sholly under lease and option from the El Centro Mining Co. Two carloads of high-grade gold and silver ore has been shipped and mining continues on a larger scale.——The Nelly Bly is still shipping but may close down before many months.——At Lee's Peak the Octo Mining Co. has been getting out some lead ore but no work to speak of has been done during the year.——The Atwood mine has been taken over by a company to be known as the South Chino Copper Co., recently organized in Boston, Massachusetts, and incorporated in Arizona. Berger and Russell of Globe, Arizona, are the principal operators, with Boston interests assisting. The company is operating steadily on the Atwood shaft and taking out fair ore.——At Santa Rita it is reported the Ivanhoe will be sold to Eastern interests. Owens and Kiner now are leasing this mine from the Hermosa Copper Co. of San Francisco. The Lucky Bill mine near the Ivanhoe at Vanadium is shipping steadily and J. B. Gilchrist is developing some property in the neighborhood with a three-compartment shaft.——The United States Copper Co. at Hanover has been revived with the purchase of interests in the company by C. B. Manville of the Johns-Manville Co. It is said to be in good condition for operating.——At Steeplerock mining is quiet. The Carlisle Mines Co. has practically closed, and the men have left camp. It is announced that the company will be reorganized for the continuance of operations at the Carlisle. The company has spent a great deal of money on this property. The Jim Crow mine has also been turned back to the owners by the La Jara Mining Co. George Utter is continuing work on this property. Other operators in the Steeplerock region are going ahead with their development. The close-down of the two big companies does not seem to have had any effect upon the optimism of the other owners and operators.——Gage, in Luna county, is again revived and a large number of people are employed at the old Victorio mines by the Gage Mining Co. A company from Louisiana is operating the property and meeting with success.——In Lincoln county mill equipment and amalgamating machinery has been received at the Vera Cruz mine (Carrizozo). Preparations are being made for extensive operations of the property.——Oil excitement is prevailing in numerous parts of the State, especially in the Pecos valley at Dayton and Carlsbad.——Labor conditions are excellent and no trouble has occurred in any of the metal mines. There was some I. W. W. agitation in the coal mines in the northern part of the State but the metal mines have been most fortunate this year.

THE 85 MINING CAMP

MONTERREY, MEXICO

SETTLEMENT OF CLAIMS.—A. S. & R. CO.

As soon as President Carranza announces the appointment of the commission to pass upon the claims for losses and damages to property as well as for personal injuries and deaths incurred during the long revolutionary period, the work of adjudicating the thousands of claims that are ready for filing will be started. Under the decree issued by the President the commission will have three years in which to complete its work. Where the Mexican government is to get the money to pay the enormous claims that are held against it by Americans and other foreigners no one seems to know. It is roughly estimated that the bills of American interests alone aggregate upward of two hundred million dollars gold. This sum is made up chiefly of losses and damages to mines, smelters, manufacturing plants, railroads, and plantations. British interests have suffered about fifty million dollars losses and damages. Besides these foreign-owned properties, there are many millions of dollars of domestic-owned holdings, belonging to non-combatants which the Government will be asked to settle. According to authoritative information from the City of Mexico, the Carranza government will seek to evade being held financially responsible for losses and damages that were inflicted by opposing or preceding warring forces. In other words, the Government will refuse to consider claims that were incurred during the regimes of Madero, Huerta, Diaz, or Villa. One of the requirements of the recent decree of President Carranza is that certificates of citizenship must be attached to all claims of foreigners filed with the commission.

The smelters which were recently placed in operation are meeting with difficulty in securing shipments of ore from the different mining camps owing to the renewed activity of groups of revolutionists and bandits. The American Smelting & Refining Co. will be forced to again close-down its smelters at Monterrey, Aguas Calientes, and Matehuala unless conditions improve. This company has ceased the construction and repair work that was in progress upon its smelter at Chihuahua, and all of the American employees of that plant have been moved to the United States to avoid the possibilities of their falling into the hands of Villa.

PACKARD, NEVADA

ROCHESTER COMBINED.—NEVADA PACKARD.

The Rochester Combined Mines Co., at Packard, has closed-down after a short but spectacular existence. Last December the company bought and optioned about 90 claims lying between Packard and Rochester, upon which practically no development work had been done, and immediately began the erection of a 350-ton mill. The mill is now nearly completed at a cost of over $300,000, but practically no ore has been developed in the mine. M. R. Thurston, now in charge of the Kanrohat mine near Round Mountain, recently took charge of the Combined company, but resigned as soon as he recognized the true conditions. The crew of over 100 men was discharged and the mill left in charge of a watchman. There is little chance of the mill ever being completed. Some ore has been opened on the extension of the Packard orebody near the Happy Jack shaft, and intelligent work might develop enough to justify a 50-ton mill. The company also located a townsite, selling lots to its employees and to business men, many of whom erected substantial cottages in the belief that the town would be a permanent one. The stock was floated by Louis A. Friedman, president of the Rochester Mines Co. The latter company continues to produce about 200 tons per day, most of which comes from the 800 and 900-ft. levels. Excellent ore is reported to have been developed by a winze sunk to the 1100-ft. level.——The Nevada Packard mill is handling slightly more than 100 tons daily of ore running about $8 per ton. Grading for the mill addition is nearly completed and erection of vats will start within a few days. On the surface ore, where the silver was contained in cyrargyrite, the mill made the remarkable extraction of 94 to 95%, but when the milling of sulphide ore began the extraction dropped to 87.5%. Tests have shown that the reduced extraction was due to the period of agitation being too short. Two 35 by 15-ft. agitators were purchased from the dismantled Buckhorn Mines Co. mill and will be added to give an uninterrupted agitation with change of solution, followed by the present counter-current treatment and filtration. The output will probably be increased to 115 tons per day. Sinking from mill-level will commence as soon as the necessary equipment arrives. Some rich ore has been opened in a previously unprospected block of ground between the contact orebodies and the caved C adit workings. The open-cuts are still furnishing a large tonnage.——Several thousand tons of low-grade ore from the Nenzel Crown Point and other Rochester properties are awaiting shipment to the smelter at Kennett, California, but are delayed by the condition of the chronic shut-down afflicting the Nevada Short Line railway. The smelting company has done much to revive many of the smaller properties of Nevada that are in a position to supply silicious ores. It is understood that only a nominal treatment charge is made.——The old camps of Mill City and Imlay on the Southern Pacific railroad between Lovelock and Winnemucca are again active. At the former camp, the long-deserted mill is being remodeled by Wilson and Johnson. Several mines are making small shipments of tungsten ore to the mill at Toulon, which is running close to capacity. The Imlay mine, at Imlay, a silver producer of the early days, is again producing, after a shut-down of many years.

MAMMOTH, UTAH

BINGHAM.—ORE SHIPMENTS.—KNIGHT DRAINAGE TUNNEL.

Bingham Mines Co. and its subsidiaries, Eagle & Blue Bell and the Victoria Consolidated, earned $28,000 in October, against $13,385 in September. The poor showing made in the last two months was due to the smelter embargo on lead ores. To offset this situation shipments of dry ore carrying chiefly copper and silver are being increased. In 10 months the three companies earned $747,500 as compared with $310,075 in the entire year 1916. A new strike has been made on the 1875-ft. level at the Eagle & Blue Bell.——The assessment of 1½c. per share, delinquent December 27, sales day January 15, has been levied by the Bullion Hill Mining Co., at Chloride, Arizona, controlled by the Knight interests.——The Cedar Talisman, near Milford, now being worked under lease by H. S. Joseph, is producing ore of an average value of 14% lead and 14 oz. silver.——A depth of 1975 ft. has been attained in the No. 1 shaft of the Iron Blossom Mining Co. This shaft will be sunk to the 2100-ft. level, and from there prospecting drifts will be sent out.——The ore shipments from the Tintic district during the week ended November 30 totaled 183 cars. The following are the shippers:

	Cars		Cars
Dragon	37	Gold Chain	10
Mammoth	24	Tintic Standard	8
Iron Blossom	21	Chief Con.	5
Centennial Eureka	14	Empire Mines	5
Eagle & Blue Bell	14	Ridge & Valley	5
Colorado	13	Other mines	17
Grand Central	10		

The Knight Drain Co. is progressing favorably. The company has purchased hundreds of claims which will be cut by the tunnel. The tunnel will tap the big mines at the south end at a depth of not less than 2000 ft. If it makes it possible to operate these mines below the water-level the company will receive a royalty on all ores shipped.

PLATTEVILLE, WISCONSIN

THE ZINC AND LEAD SITUATION.—LUCKY FIVE—WISCONSIN ZINC. —FRONTIER.

Complete returns from the zinc-lead districts of Wisconsin for the week ended November 24 show good results. The set-backs previously experienced on account of sleet storms were entirely overcome and roads from outlying mines to railway track showed much improvement, making ore deliveries less difficult. The gross recovery of zinc concentrate from all mines for the week totaled 5251 tons, against an even 4200 tons reported for the week preceding. At the same time there was a correspondingly generous contribution of raw ore to zinc refineries in the field, 102 cars, totaling 4051 tons, going this way. Some of the refining-plants that had been out of commission resumed operations, running on current production so that the reserve in the field, conservatively estimated at several thousand tons, was in no wise diminished during the week. Shipments of mine-run to smelter direct was shown in heavier volume, mainly through the aggressive buying tactics of the Grasselli Chemical Co. In addition the chemical company cut in on high-grade ore from refineries making the best record on straight out and out buying of any firm represented in the field. Refiners improved their time as well, although all finished ore ready for prompt delivery was not permitted to reach track; 41 cars, or 1673 tons, clearing for the week. This with mine-run to smelters made the total net deliveries out of the field for the week 2886 tons, a fairly respectable showing, all adverse conditions of labor and operating costs considered. Prices for zinc ore remained unchanged, the base per ton standing at $62 for top grade and 60%, with the range down to $57 per ton, base, on second and medium-grade ores. Lower grades evidently were in good demand as shipments of ore from independents to refiners sold on open market, and quotations were much heavier than in several weeks.

Lead-ore producers held determinedly to their sworn purpose not to sell any ore, having followed this programme for six consecutive weeks. Offerings of $75 per ton for 80% ore held firm, but sellers declare not a pound of lead ore will be offered for sale until the ground lost, when lead ore declined from $80 to $60 per ton, has been fully recovered. Evidence has been furnished that the deed has been suited to the word for not a car of lead ore has cleared from the field upon which current market quotations apply for six weeks. Production shows marked improvement of late, many of the larger operating groups running into rich ground from which recoveries of ore have been uniform and heavier than usual. The reserve in the field after careful tabulation shows about 2000 tons of choice lead concentrate snugly tucked away in bins. Operators declare this reserve will be permitted to increase until it may be seen what action the Government will take with regard to both zinc and lead before making final decisions.——Prices on iron pyrites showed no change whatsoever and refiners displayed no enthusiasm in marketing ores, a scarcity of cars for prompt loading making the shipment of zinc ore preferable. Fine from separating plants cleared exclusively, shipments averaging about 500 tons per week.——The carbonate-zinc ore miners of the northern districts of the field found little in current quotations to tempt increased operations and shipments were confined to a single car. Usually at this time of the year scores of small operating concerns turn their entire attention to underground work, but quite the contrary is true this season, miners finding little encouragement this year in the offerings held out for this class of ore. Labor is very scarce and unless mining is found profitable, crews are disbanded, quickly scattering to other sections of the field where the wages paid are now the highest ever known in this field.

The week was crowded with events of importance at all points in the field. In the Dodgeville district the surface rig of the Lucky-Five Mining Co., valued at $25,000, was entirely destroyed by fire. Explosions, quickly followed by fire breaking out at several points in the concentrator, leave no doubt the fire was of incendiary origin. The plant was partly covered against loss and will be re-built without delay. In the Linden district three new milling plants will be constructed, one now being under way. The other two will be erected on the Vial mine and the Treloar-Kickapoo leases. At Platteville foundations were run for a new concentrator on the New Rose mine. A new power and mining plant is also being erected for the Old Mexico Mining Co. of Chicago. The Wisconsin Zinc Co. is well along with a new 200-ton mill on the Copeland mine, in the Shullsburg district. At New Diggings the Chicago & Northwestern railway has completed a branch from its main line to the very doors of the big Skinner Separating works, incidentally giving quick and easy access to several large zinc-ore producers upon the immediate right of way. This branch will prove one of the biggest boons offered to zinc-mining men in this field. The Frontier Mining Co. has developed a wonderful producer in the Grotkin mine where two distinct veins have been determined through underground development. Two low-grade producers, the Hird mines No. 3 and 4, have been shut-down until operating and marketing conditions offer better returns. The C. S. & H. Mining Co. of Cuba City, recently equipped and set in operation, is developing into a heavy producer, deliveries already going four cars weekly. The Connecting Link Mining Co., recently incorporated for $60,000, has two shafts bottomed in ore, and at a recent meeting of stockholders it was elected to provide a modern surface plant. A rich strike of zinc ore was made in this same district during the week on the Hercules Mining Co. lease, the drillers worming their way through 16 ft. of highly mineralized areas. The company has been incorporated for $100,000; a power and milling plant will be provided without delay.——Two new producers are expected to materialize in the Hazel Green district, one at the old Jefferson mine for the Vinegar Hill Zinc Co. and the other at the McMillan Zinc Co., fully equipped with a new power and concentrating establishment. Both are in full operating order and shipments are expected soon.

ANCHORAGE, ALASKA

PROGRESS IN THE WILLOW CREEK DISTRICT.

It may roughly be said that the results of the season of 1917 in the Willow Creek district were only about three-quarters of what they might have been had normal conditions prevailed.

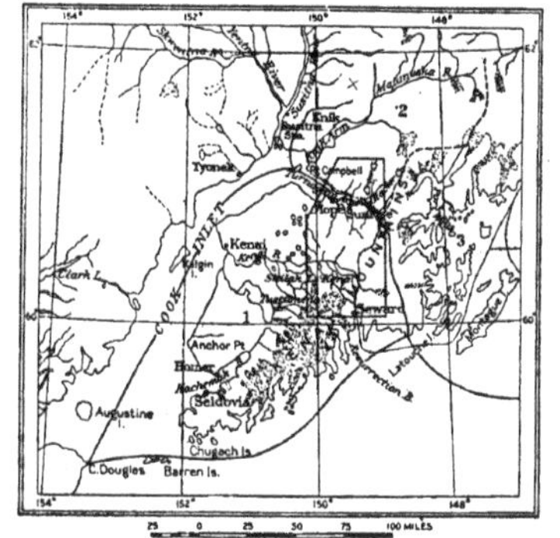

PART OF ALASKA

× Denotes position of Willow Creek

The district was greatly handicapped, as all gold mining districts have been, by the scarcity of labor and the high price of materials and supplies. Also, though an isolated district, Willow Creek was affected by the general unrest of labor, two of the mines having been forced to close for nearly a month on account of strikes. The fact that the steamship companies carrying freight to Alaska could switch their vessels to the Atlantic and secure a profitable war business has allowed these companies to hold their freight-rates excessively high, even though they handle an immense business. The high price of cyanide has caused some of the operators to pond their tailings to await more favorable conditions.

At the Martin mine the 10-ft. Lane slow-speed mill has operated for four and one-half months this season, having been closed nearly a month on account of labor troubles. The payroll has averaged about 20 men. Production will be $85,000 for the season, which is considered satisfactory. The ore assays $30 per ton, $24 being caught on the plates and $6 will be extracted by leaching when the price of cyanide is more favorable. During a large part of the summer the Lane mill was amalgamating $1.10 gold per minute, and for a 10-day period the clean-up amounted to $190 per hour. The bullion is high-grade, running about $19.50 per ounce. Feed for horses is excessive and this in turn has increased the freight on supplies from the Government railroad to the mines. Mr. Martin has introduced an innovation in the form of a gasoline-driven air-compressor and jack-hammer drills. This machine was brought into the district by Fred Laubner to be used in a contract for tunnel-driving at the Gold Cord mine, but was later removed to the Martin mine, and the results have been most gratifying. It is to be hoped that the other operators will follow suit and install labor-saving machinery. In a trench behind the Martin mill a new vein has been uncovered; this is believed to be the continuation of the Gold Cord. The Independence mine was not operated beyond doing annual assessment work. A new vein has been uncovered back of the present mill, and it is believed that this also may be a continuation of the Gold Cord.

The Milo Kelly group, north of the Independence, between it and the Gold Cord, has shown up well. The adit of the Gold Cord was advanced during the year along an altered crushed zone in the grano-diorite for a distance of about 180 ft. This zone varies from 2 to 8½ ft. in width, and is said to give an average assay-value, for over 100 ft. long, of $28 per ton. The gold, however, is not all free, being combined with pyrite. During the summer a tram was constructed from the mouth of the Gold Cord adit to the Independence mill, and the mill leased for a year. The owners of this property, Charles Horning, Charles Byron, and Frank Bartholf, and the Isaac brothers, expect to have tests made on the ore in order to determine the most advantageous flow-sheet and they expect to build a small mill next year. On a test-run of about 350 tons of ore from the adit, it was possible to recover only $12 per ton on the plates.

The Rae-Wallace company, which is developing the Trickster and the Rosenthal groups of claims on the east side of Fishhook creek has opened up the orebody by an adit and cross-cuts for about 175 ft. The vein is 3 ft. thick. Another vein of high-grade ore has been uncovered on the property in numerous cuts and pits, and it promises well. Contrary to current reports, it is not believed that the orebody thus uncovered can be a continuation of the Mabel vein. It is planned to push development work during the winter and to build a mill next summer. This company is controlled by Coeur d'Alene men and has excellent prospects. It is so situated that a jig-back aerial tram of about 1600 ft. will serve to transport the ore from the mine to the mill. A pan-amalgamation test gave an extraction of over 90% of the gold. An artificial lake is to be made half-way up the mountain by building a dam across the mouth of a small glacial cirque; this will impound sufficient water to supply a small mill during the season, when the run-off is slight, and will give sufficient head to furnish power. There is a small copper-bearing vein on the property, and a little more trenching may show that this vein also is worthy of development. The ore so far developed has been opened only on the mountain-tops at an elevation of 1000 to 1500 ft. above the valley, and presents opportunities for quick development.

On Archangel creek, the Fern & Goodell property and the Galkeme & Conroy properties were developed by assessment work, consisting largely of stripping the veins and driving short adits. These two properties have been taken under lease and option. Adjoining these properties to the south, the Talkeetna Mining Co., controlled by Mrs. D. M. Fulton, of Portland, Indiana, installed a 15-ton Denver quartz-mill, driven by a Pelton wheel during the summer. Near the close of the season 50 tons of ore was milled, returning $2000 worth of bullion. This company plans to install another tram and extend the power pipe-line farther up the mountain in order to catch a greater quantity of water before it has disappeared under the glacial wash. This extension will also give greater head on the Pelton wheel and the Denver quartz-mill simultaneously, which was heretofore impossible. It is also planned to block out and develop the orebody by lower adits during next summer.

George Anderson has had the Little Gem group under lease from Bob Hatcher. He has developed a small vein of high-grade ore and endeavored to use a hemp rope as an aerial tram to transport the ore from the mountain to the creek. Mr. Anderson drove 75 ft. of adit on a vein 14 in. wide and having a high-grade quartz stringer running through it from an inch to 2½ in. thick. He plans to float a company to work this property.

The Arch group property at the mouth of Sidney and Archangel creeks is being developed by open-cuts and pits. This group is owned by Colorado people. The Pilger group of claims, containing the northward extension of the Mabel vein, was represented during the year by trenching. It is planned to drive a cross-cut from near the bed of Archangel creek across the compact group. This should explore the Mabel vein from the lowest point at which it can be tapped by the shortest adit. The Mohawk Mining Co., on Sidney creek, bought two Gibson mills and two gas-engines, together with all necessary equipment. The Spokane capitalists who subscribed for a large block of the shares were disappointed later in the season because the ore failed to hold up in value on second examination. The Willow Creek Development Co., controlled by Anchorage business men, who have a lease and option on the claims of the Isaac brothers on Reed creek, has conducted discovery work in open-cuts and pits. About 500 lb. of ore was sacked to be shipped outside for test. On the results of this test depends the character of mill to be built next year. Major L. H. French has secured a lease and option on the Keystone group, owned by Messrs. Kelly, Laubner & Corlew, and on the Shough property on the Little Susitna river. He and his associates plan to do active development work during the coming year and hope to erect a mill.

The Mabel mine, owned by the Mabel Mining & Milling Co., of which William Bartholf is president and Fred Laubner is an important owner, made a five-months run during the summer. They employed an average of 16 men and milled 13 tons of $40 to $45 ore per day. Of the $45, only about $28 was plated in the mill, the rest having been ponded for further treatment at some future time. All of the mining thus far has been hand-work; consequently the cost has been high. During the year, the Denver quartz-mill and the penstock for the water-turbine were re-set.

On Craigie creek, the Gold Bullion mine operated for about five months, but was closed down for a short time by labor trouble. This property employed from 60 to 70 men during the year, of whom from 50 to 60 were in the mine. Power has been derived from water only; operations were hindered by shortage of water. This property is reported to have made a gross production of about $150,000. The operating company is buying the property from the owners under a contract calling for payments extending over a term of years, one payment of $160,000 out of a total of $450,000 becoming due this year. It is rumored, however, that an extension of time will be granted.

Charles Bartholf and Dave Miller report a discovery of 2 ft. of vein-matter carrying a high-grade streak 2½ inches thick, on Craigie creek, about three miles below the Gold Bullion. They plan to erect a small prospector's mill next season, so that they may be able to thoroughly test the property and, at the same time, pay expenses.

ALASKA

(Special Correspondence.)—Work at the Jualin Alaska mines has been suspended pending re-adjustments and preparations to start on a much larger scale. Work will be resumed before spring. It is intended to increase the present 10-stamp mill by an addition of a pebble-mill, bringing the capacity up to 150 tons per day. Work on the 7800-ft. adit, cross-cutting the entire property and draining the mine, will be resumed. The adit has already advanced 2000 ft. During the past 12 months the Jualin Alaska mines has recovered from its 10-stamp mill gold to the value of $197,296. The mine is controlled by Belgium and French capital with its head office in Paris, France.

Jualin, November 21.

(Special Correspondence.)—Labor conditions at Treadwell have greatly improved within the past few weeks and the scarcity of good men which prevailed during the summer has been almost relieved. At present 200 men are employed underground at the Ready Bullion mines, more than were working there previous to the cave-in of the other mines last April. Many of the old-time employees of the company have returned lately, some from the westward, where they went early in the year, and others from outside, where labor conditions are uncertain. The electric-steel furnace, recently completed for the Treadwell company, was requisitioned by the Government before shipment and sent to Watertown. A new furnace will be built and shipped as soon as possible. According to word just received, the Jualin mines will re-open before spring and will operate on larger scale. A Hardinge ball-mill is to be added to the present stamp-mill, bringing the capacity up to 150 tons per day. The drainage adit, which is already in 2000 ft., will probably be driven the full 5800 ft. necessary and will tap the orebody at a point 600 ft. below the present workings.

Treadwell, November 26.

ARIZONA

COCHISE COUNTY

(Special Correspondence.)—Negotiation between the Cananea Consolidated Copper Co. and the authorities at Mexico City have culminated in terms under which the company is able to resume operation. All American employees have been recalled to positions formerly held at Cananea. The terms of the agreements have not yet been made public.

Bisbee, December 1

MARICOPA COUNTY

(Special Correspondence.)—The Hatton property six miles north-west of Aguila is shipping one carload of manganese ore per week. Manganese properties to the south of Aguila are also shipping.

Phoenix, December 1.

MOHAVE COUNTY

(Special Correspondence.)—The old 63 mine in the Stockton Hills district has been purchased by G. E. Pierce of Kingman. A 400-ft. adit has been driven to cut a 30-in. vein of high-grade silver ore, running as high as 3000 oz. per ton. Several miles of road has been graded, and an air-compressor, gasoline engine, and other machinery erected.——The Walk Over mine, adjoining the Copper Giant, nine miles from Hackberry, is pushing development. The mine is equipped with a 25-hp. Fairbanks-Morse gasoline hoist, an 8 by 10 Ingersoll-Rand compressor, and jack-hammer drills.

Kingman, December 10.

(Special Correspondence.)—The main ore-shoot in the north drift of the 300-ft. level of the New Tennessee has been broken into. The ore is of good grade.——Recent developments on the 400-ft. level of the Gray Eagle shaft of the Tom Reed has opened up ore for 75 ft. which it is claimed averages $20 per ton. It is estimated that there is 500 ft. of backs in this orebody.

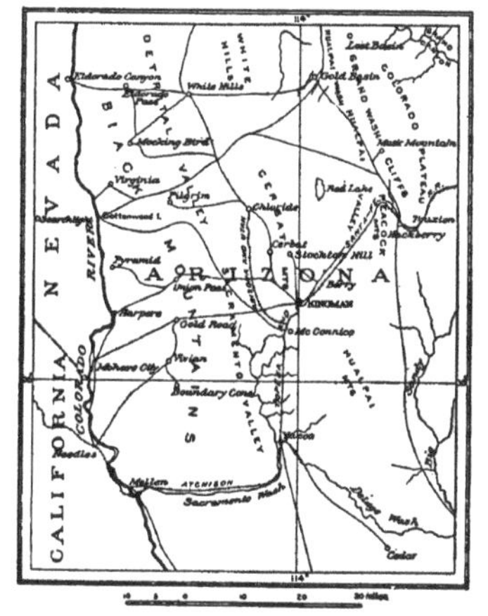

PART OF ARIZONA

Kingman, November 30.

PIMA COUNTY

(Special Correspondence.)—The Indiana Arizona silver-copper mine at Silverbell has been bonded to A. Knox, of Los Angeles, for $200,000.

Tucson, November 30.

PINAL COUNTY

(Special Correspondence.)—The first of four diamond-drill holes to test the orebody of the Magma Chief Copper Co. has been started. All drilling is under contract, the depth of the holes average 550 ft.——The last 300 ft. of the Magma Chief adit has been giving gold and silver in heavily iron-stained porphyry.

Florence, November 30.

YAVAPAI COUNTY

(Special Correspondence.)—A considerable flow of water has been cut in the Dundee Arizona shaft, which is down 500 ft., causing some delay. Twelve carloads of carbonate ore averaging 5% copper is being shipped per month.

Prescott, December 1.

The Loecy Pabst Gold Mining Co. has been organized to take over the Union mine, the property of the estate of the late John S. Jones. A 1200-ft. adit has been driven to facilitate the tramming of ore from the mine to the 10-stamp mill. In driving this two veins were cut, 8 ft. and 6 ft. wide, respectively. A chamber has been cut and an inclined shaft is being sunk on these veins, which are only 6 ft. apart. The average value of the gold and silver content is said to be $15 to $18 per ton. The mine is equipped with a 20-hp. gasoline hoist, a Sullivan air-compressor, four jack-hammer drills, a 10-stamp mill, and a Wilfley table.——Work is being continued at the Silver Belt mine and some high-grade silver-lead ore has been uncovered. The shaft is down 300 ft., and 400 ft. of levels has been driven.——Preliminary prospecting drifts and cross-cuts from the 140-ft. shaft on the American Eagle group of the Arizona Copper & Mining Co. have cut some good-grade chalcopyrite ore. The company proposes to do some extensive developing on a heavily-mineralized belt of schist ranging from 30 to 300 ft. wide.

YUMA COUNTY

(Special Correspondence.)—A number of producing mines and many prospects are being operated on the Harqua Hala mountains, 65 miles south-east of Parker. The Swansea and the Planet mines have both been producing steadily for over a year. The Empire-Arizona and the Billy March have shipped considerable high-grade ore.——The Mammoth Gold & Copper Co. is sinking a three-compartment shaft. Over 5000 ft. of development work has been done and ore running 7% copper and $7 in gold has been found.——At the Empire-Arizona, which has been worked intermittently for several decades and on which some 10,000 ft. of developing has been done, a 350-ft. level has been driven on a vein of oxidized copper ore, carrying $2 to $3 in gold.

Parker, December 1.

CALIFORNIA

INYO COUNTY

A large body of zinc ore has been uncovered by Utah miners nine miles from Darwin. Near the surface the orebody is said to have a width of 500 ft. Shipments are being made at the rate of 45 tons per day.——The Standard Tungsten Co. has increased the capacity of its plant and is handling a large tonnage.

SAN BERNARDINO COUNTY

(Special Correspondence.)—The Tom Reed Gold Mines Co. is sinking a vertical shaft in the hanging wall of a body of quartz on the Clipper mountains. The shaft is down 300 ft. and when a depth of 500 ft. is reached a station will be cut and the vein thoroughly explored.——The Tom Reed Gold Mines Co. of Arizona is developing a large body of quartz that is 50 ft. wide and in places stands up 100 ft. above the surrounding rock. The shaft, which is being sunk on the hanging wall, is down 300 ft., and will be continued another 200 ft., when a station will be cut and levels driven.——The Clipper Gold Mining Co. of Los Angeles is developing a claim adjoining the Tom Reed.——The Merger Mines Co., T. B. Bassett of New York, president, has taken over the Black Hawk mine, 36 miles east of Victorville. This property was located in 1887 and a 10-stamp mill erected. There is a 30-ft. vein which is said to carry $4 in gold.

San Bernardino, December 8.

(Special Correspondence.)—The Mohave United Mining & Milling Co. has uncovered a considerable body of ore, the extent of which has not yet been determined, running 20% copper and $18 in gold. The mine is situated at Crucero.

Crucero, December 8.

SIERRA COUNTY

The new 10-stamp mill at the Ireland mine, near Alleghany is in operation and running satisfactorily. Development at the mine is being pushed.

TUOLUMNE COUNTY

(Special Correspondence.)—Activity in chrome mining in the western part of the county is steadily increasing. It is estimated that shipments now amount to 1000 tons per month. This represents the output of eight properties that have been worked for only a few months. The operators who are making shipments regularly are Booker, Powell & Porter, Richards Bros., Pereira Bros., Eglin & Williams, Sell & Terry, Charles Gillis, Robert McCormick, and Charles Quigg. The properties are all situated in the serpentine belt west of the Mother Lode, and are traversed by the Sierra railroad. Several other prospects are being opened, among the most promising being those of John Bullock and Aaron Morgan.——With about 175 men on the payroll, the Shawmut mine is pushing development work, making changes in its surface plant and adding new machinery, including an electric hoist on the third level. The number of stamps in the mill is being reduced to 60, but a Hardinge ball-mill and a flotation plant are being put in. An adit 2800 ft. long connects the mill with the hoist, thus making possible the abandonment of the upper part of the shaft.——At the Dutch mine work is still confined chiefly to sinking and making underground improvements.——An electric-transmission line has been built to the Star King mine and operations have been resumed, after a suspension of several months.——Another good orebody is being opened at the Chaparral mine. The first Ellis ball-mill made was used at this mine.

Sonora, December 4.

COLORADO

BOULDER COUNTY

(Special Correspondence.)—George Chesebra and Robert Kermick are building a fluorspar mill. There is an abundance of medium-grade fluorspar in this county, especially in the Jamestown section, but it requires concentrating before it can be shipped at a profit. F. B. Cahill is the foreman. The U. S. Geological Survey has made a thorough examination of the county with a view to estimating available sources of the mineral. The St. Elmo Tungsten Mining Co. has been incorporated with J. Taylor Smith, president; George Pomeroy, secretary and treasurer; and U. A. Ashley, foreman. It has acquired the St. Elmo and Mountain Queen claims, one mile north-east of Nederland. Operations on the Mountain Queen have already been started. The Quay mine, near Nederland, leased by Snowden & Co. from the Tungsten company, has had a successful season.——The Wolf-Tongue mill at Nederland, C. E. Dewitt superintendent, is being worked at full capacity. The Vasca mine is running two mills, one at Stevens camp and one at Boulder, and is doing considerable development work at Stevens camp.

The output of the mines at Caribou this winter is far above that of any previous year and is evidence of the revival of silver mining. The ores average from 100 to 400 oz. per ton. The Potosi, under lease to Davis & Linstrom, is one of the most active in the district. A two-car shipment was sent out last week.——The Tungsten Metals Co., under the management of W. W. Charles and George Teal, has acquired the old Kekionga mine at Magnolia.——The Victoria mine, situated at Summerville, is producing high-grade ore, running from 55 to 300 oz. in silver. Frank J. Pickford is foreman; a large force of men is employed.

The Yellow Pine group, owned by Ralph Cotton, has had a successful season; the Michener being the heaviest producer.

The ore runs 10% copper, 18% lead, and 200 oz. in silver. Regular shipments are being made.——Charles Walstrom, who has a bond and lease on the old Fairfax property at Salina, has been adding new machinery, preparatory to more extensive work. The Logan mine, at Chrisman, is being worked by lessees and is making a good production.——Miner Smith, George Sale, Robert Smith, John Walgren, William Thompson, and James Francis are all shipping from their respective leases.——Around Eldora, work at the Huron and the Consolidated Leasing is going ahead and will be continued all winter.

Eldora, December 3.

LAKE COUNTY

(Special Correspondence.)—At a special meeting of the stockholders held at Denver, November 26 and 27, the Yak Mining, Milling & Tunnel Co., sold an undivided half interest in its property at Leadville to the American Smelting & Refining Co. With the sale the Yak company turned over to the A. S. & R. Co. the management of the property. The consideration for which the transfer was made has not been stated. The management of the Yak by the smelting company is expected to be a decided benefit not only to the property itself but also to the entire district. For almost a year operations at the Yak have been on a much smaller scale than has been customary. The big orebodies of the Cord and White Cap, which formed the greatest source of production, were exhausted several months ago, and, owing to continued trouble with the draining equipment in the Cord, White Cap, and Diamond, development has proceeded slowly and with difficulty. A number of new ore-shoots are reported to have been uncovered, but they are low grade and prospecting has not yet reached a stage where the value of these discoveries can be determined. The Yak has been one of the biggest dividend payers in the district, maintaining an unbroken record until this year. Expensive pumping equipment and driving development-drifts has consumed the return from shipments made during the past six months, and it is believed that this is the reason for turning the property over to the smelting company.

Leadville, December 3.

TELLER COUNTY

(Special Correspondence.)—The output of the Cripple Creek gold mining district for the month of November was 95,358 tons, with an average value of $10.95 per ton and gross bullion value of $1,044,697. The feature of the production was the heavy output of low-grade ores, namely, 57,600 tons, assaying below $2.20 per ton.——The dividends during the same period totalled $167,000, and were as follows: Cresson Consolidated G. M. & M. Co., 10c., $122,000; Golden Cycle M. & R. Co., 3c., $45,000.——The treatment at the several mills was:

Plant and location	Tons treated	Average value	Gross bullion value
Golden Cycle M. & R. Co., Colorado Springs	25,500	$23.00	$586,500
Portland G. M. Co., Colorado Springs	9,758	20.50	200,039
Smelters, Denver and Pueblo	2,500	55.00	137,500
Portland G. M. Co., Independence Mill	39,400	2.06	81,164
Portland G. M. Co., Victor Mill	18,200	2.17	39,494
Total	95,358	$10.95	$1,044,697

The Trail, Big Banta, Old Ironsides, and Lost Fraction lode mining claims, U. S. Mineral Survey No. 7812, containing 25,338 acres on Bull hill and Battle mountain, originally owned by the Requa Gold & Silver Mining Co., now expired by limitation, were sold at public auction on November 30, and were purchased by H. McGarry, of Colorado Springs, representing the United Gold Mines Co., owner of 70% of the capital stock of the defunct company, for $71,000. W. H. Hayes of Pasadena, and W. H. Wolfe of Parkersburg, submitted bids at the sale. Leases on the Trail and Big Banta lodes will continue, as the sale was made subject to existing written leases.——The Geraldine Mining Co., recently organized by Lamar investors, has secured sub-leases on blocks of the Victor mine, and a portion of the estate of the Anona Mining Co. adjacent, on the eastern slope of Bull Cliffs, and during the past week has laid track in the old Victor adit, from which the territory will be exploited. The Victor mine is owned by the Smith-Moffat Mines Co.——The Acacia Gold Mining Co. has opened a promising ore-shoot at the bottom or 1400-ft. level of the South Burns mine on Bull hill, beyond the junction of the Shurtloff-Eagles vein. The company and its lessees shipped nine cars of 300 tons of $30 grade during November.——The directors of the Isabella Mines Co. have appointed Frank Gunn of Cripple Creek superintendent and managing director to succeed Clarke G. Mitchell, who has voluntarily enlisted in the U. S. Aviation Signal Corps.

Cripple Creek, December 6.

IDAHO

SHOSHONE COUNTY

The Spokane Metals Recovery Co. has been capitalized at $100,000 and recorded at Wallace, Idaho, with principal place of business at Mullan. The company has a lease on a flat at Nine Mile, where 50,000 tons of tailing from the Rex mine has accumulated, and has also arranged to secure tailing from the Interstate Callahan, Success, and other mills. Samples show values running over $6 per ton. The company expects to make a profit of $1.50 per ton.

MAINE

HANCOCK COUNTY

The soundness of C. Vey Holman's judgment in re-opening the old Douglas mine in the summer of 1915 has been fully demonstrated. The American Smelting & Refining Co. took the property over and has 150 men on its payroll at the mine. It is meeting with satisfactory results. Mr. Holman, who was at one time State geologist, has long advocated the re-opening of some of the old mines of the State.

MONTANA

SILVER BOW COUNTY

(Special Correspondence.)—Benjamin B. Thayer, vice-president of the Anaconda Copper Mining Co., left Butte, after completing an inspection of all the properties of the company at Butte, Great Falls, Anaconda, and other parts of the State. From Butte he will go to Tooele, Utah, where he will inspect the International Mining & Smelting property. Mr. Thayer said the output of the mines was gradually getting back to normal, at the present time it equalled 88% of the production previous to last June, and that by the end of the year it would have recovered completely. Mr. Thayer found both mines and smelters in excellent condition.——The Anaconda Copper Mining Co. is now employing 9280 men and hoisting 13,772 tons of ore per day, 685 tons of which is zinc ore. In the last two days 169 miners have been given employment.——The Montana Institute of Mining Engineers held a banquet at the Silver Bow Club, November 16. Some interesting papers were read and discussed. Benjamin B. Thayer gave an account of what is being done by the Naval Board, of which he is a member. He was given a most cordial reception by his many old friends. C. W. Goodale read notes on the St. Louis meeting; Reno H. Sales discussed the manganese situation; E. M. Norris gave a splendid paper on the fire-proofing of the Butte shafts; C. L. Berrien explained the process of the hydraulic filling for extinguishing fires at the Leonard mines.——A gradual improvement is being shown in the labor situation in the Butte district. The Tramway mines, one of the largest producers in the district, resumed work on November 19. It had been idle since October 3, on account of repair-work. In the last two days 169 miners have been given employment.

Butte, December 3.

NEVADA

ELKO COUNTY

(Special Correspondence.)—At the Liberty Gold Mines Co., operating at Liberty, 64 miles north of Elko, a combination gravity-concentration and flotation plant will be erected early in 1918. The plant will have a capacity of 150 tons per day. On the 130-ft. level, the deepest point in the mine, the vein shows a width of 42 ft. with portions assaying $25 to $30 per ton in gold and copper. F. H. Bird, former superintendent of the Seven Troughs Coalition, is manager.——A promising deposit of cinnabar is being developed 90 miles north-west of Elko by the Ivanhoe Springs Co. This is the first discovery of an important quicksilver deposit in this part of Nevada. Development is confined largely to surface working; two retorts are operating on selected ore. W. C. Davis is manager. ——The old silver camp of Cornucopia is showing signs of a revival. Some promising strikes have been reported recently, and on the Peacock property a high-grade vein has been opened near the surface. It is a cross-vein and carries horn-silver and black sulphide. The camp is near Tuscarora.——Frank Middleton and Bruce Bros. are developing extensive deposits of shale in Burner basin, six miles east of Elko. Tests indicate the shale contains a high proportion of oil; it is planned to extract the gasoline and other products by the Crane method. Sinking of a 100-ft. shaft has begun.——L. R. Thatcher and associates have filed on several hundred acres of shale land near Pallisade, and started development. The shale is reported to test 80 to 90 gal. of petroleum per ton. Mr. Thatcher states a 100-ton plant will be erected, and also a custom smelter to treat silver-gold ores in this field. The smelter is to be operated by gas obtained from the shale beds.

Elko, December 9.

MINERAL COUNTY

(Special Correspondence.)—The United States Refining, Smelting & Mining Co. has arranged to treat a heavy tonnage of gold ore from the R. B. T. mine and silver-copper ore from the Luning-Idaho group. The gold ore, being highly silicious, will be shipped to the Mammoth smelter, near Kennett, and the silver-copper product to Utah plants. The mines are situated near Luning. The grade of the R. B. T. shipments will range from $20 to $40 per ton.——S. M. Summerfield of Mina, representing California capital, has purchased the Garnet group of tungsten claims, 18 miles east of Mina. The terms of purchase specify that the construction of a concentrating plant must be commenced within 90 days. It is reported the purchase involved $60,000, and that the Atkins-Kroll Co. is interested.——The group of tungsten properties at Sodaville have been re-opened by A. C. Beck and associates, and the mill is being repaired. Caterpillar tractors and self-dumping trailers have been ordered; mine development is progressing steadily.——The Montgomery Mountain Quicksilver Co. has arranged for the erection of a battery of D-shaped retorts. A large tonnage of profitable ore has been opened. The mine is near the Nevada-California line, five miles east of Montgomery Mountain station, and 40 miles south of Mina. F. C. Beedle is manager.——The furnace capacity of the plant of the Mina Quicksilver Co. is being increased from 14 to 28 retorts, which will enable a production of 10 flasks of quicksilver per day. On the 155-ft. level the vein is 11 ft. wide, and averages 7% mercury. The company is controlled by Manson & Humphrey of Reno.——An important strike is reported from the Aurora Consolidated, at Aurora. In the main adit an orebody of considerable size has been uncovered, and is reported to assay $25 to $30 per ton. The management is centring work at this point and sending a heavy tonnage to the 500-ton mill. Development is proceeding to determine extent of the lode. The Goldfield Consolidated holds 85% of the stock of the Aurora company.

Mina, December 2.

NYE COUNTY

Drift No. 19 at the MacNamara is progressing well. Stopes No. 1 and 2 on this level show an 8-ft. and 5½-ft. face, respectively, of good ore. Stoping on the Ohio vein on the 500-ft. level continues in good ore. The output last week was 507 tons.——The Jim Butler Tonopah Mining Co. during October milled 3500 tons of ore, resulting in a net profit of $28,242. At the Wandering Boy shaft raise No. 369 on the 300-ft. level continues on a 5-ft. face of medium-grade ore. The production the past week was 750 tons.

The Cash Boy commenced stoping on the 1700-ft. level, and is shipping the ore to the smelter at Keswick, California. The shipments are being sampled at Hazen, Nevada, by the Western Ore Purchasing Co. The drifts in the hanging wall and the foot-wall continue in excellent ore, and the raise in the hanging wall, which is being driven for a connection, shows a face of good ore.——The Rescue-Eula produced 66 tons of ore, the Cash Boy 50 tons, and miscellaneous 39 tons, making the week's production at Tonopah 10,125 tons with a gross value of $177,187.——The Tonopah Divide Mining Co. at Gold Mountain, which is situated six miles south of Tonopah on the Goldfield road, has opened a vein of silver ore. The property has previously shown only gold-bearing veins, so the present discovery was entirely unexpected. The vein has not been thoroughly prospected, but specimens from it are exceptionally rich. A large ore-bin is being erected. All the surrounding claims have been purchased by H. C. Brougher, who is president of the company.

NEW MEXICO

SOCORRO COUNTY

(Special Correspondence.)—The Socorro Mining & Milling Co. is making rapid progress in the re-building of its mill and mine plant. The local saw-mill has been secured, the company's own force being sent into the woods. All logs are in and sawed timber has been arriving in camp for the 10 days. All lumber needed for the re-building of the plant will have been delivered by the end of next week. Permanent buildings around the plant are starting; a 70-ft. steel head-frame will replace former structure and the plant will be more efficient throughout.——The Mogollon Mines Co. is running at capacity. All tonnage coming below the adit-level is now being handled through the new two-compartment shaft, which is equipped with two counter-balancing self-dumping skips.

The Oaks Co. is maintaining daily shipments from the Maud S. and Deep Down mines. The Central shaft is being re-timbered to water-level and sinking will soon be in progress. This company is also actively developing the Clifton, Eberle, and Pacific mines, and is shipping ore from all of them.

Mogollon, December 4.

TEXAS

TRAVIS COUNTY

(Special Correspondence.)—Development of the quicksilver deposits, the sulphur beds, and the silver and copper mines of the upper border region of Texas has been greatly stimulated by the existing high prices of these minerals, according to the statement of William B. Phillips, formerly director of the Bureau of Economic Geology and Technology of the University of Texas. Mr. Phillips has just returned from a trip through that region. The activities in the Terlingua quicksilver district are unusually brisk. Notwithstanding the fact that the mines are situated 90 to 100 miles from the nearest railroad point, the high price for which the metal is selling has caused the re-opening of old mines and of much prospecting work, which, it is to be hoped, will provide fruitful results.

The quicksilver mines in Texas are in the southern part of Brewster county, 400 miles west of San Antonio. The existence of cinnabar in that part of Texas has been known since 1894, but it was not until 1897-'98 that any efforts were made to utilize the discovery. Since then, however, the total production has been about 75,000 flasks.

Austin, December 7.

WASHINGTON

STEVENS COUNTY

An average of 180 tons of calcined magnesite is being maintained by the Northwest Magnesite Co., operating near Chewelah. A big reserve is being created while shipments go forward steadily. Meantime the big calcining plant is rapidly nearing completion. The third metal kiln has been put in operation and the six-mile tramway is virtually completed. The plant and tramway represent an investment of $250,000. Twenty carloads of calcined material has been shipped by the Valley Magnesite Co., operating in the same locality. As the cars contained about 40 tons, and each ton has an average value of $32.50, with an operating and hauling charge of $10, a net profit of $18,000 is shown on the shipment.

The United Copper Co. has passed its December dividend. Reasons assigned are that the smelter at Trail, B. C., is closed on account of a strike and it is also thought advisable to build up a little more surplus on account of the excess-profits war-tax. Shipments to the smelter at Tacoma are continued and it is expected that dividend payments will be resumed in January. The company has been disbursing $10,000 per month this year.

CANADA

BRITISH COLUMBIA

Andrew G. Larson, operating the Lucky Jim mines in the Kootenay district, B. C., in the capacity of receiver, has made an enviable record during the past two years. In that time he has satisfied a mortgage of $55,000 and the claims of unsecured creditors, aggregating $15,000, and established a reserve of $8000. In addition the mine has ore worth more than $8000 in transit and ore worth $5000 on hand. The total of these figures is $91,000. Production has been maintained at the rate of 50 tons per day.

ONTARIO

The following table, compiled by the Bureau of Mines, gives the mineral production of the Province of Ontario for the first nine months of 1917. The production during the same period of 1916 is given for comparison.

	Quantity		Value	
	1916	1917	1916	1917
Gold, oz.	363,955	343,490	$7,513,734	$6,754,535
Silver, oz.	16,203,091	15,236,002	9,750,040	12,001,875
Cobalt (metallic), lb.	172,055	205,866	146,467	438,732
Nickel (metallic), lb.	17,435	166,921	7,618	67,499
Nickel (oxide), lb.	54,152	10,831	6,381	3,025
Cobalt (oxide), lb.	378,732	276,769	231,947	323,162
Other cobalt and nickel compounds, lb.	57,026	276,217	22,890	30,025
Molybdenite, lb.	15,845	65,827	15,845	83,550
Copper ore, tons	1,715	2,658	21,685	33,419
Nickel in matte, tons	31,046	31,064	15,523,000	15,532,000
Copper in matte, tons	16,989	15,928	6,285,930	6,371,200
Iron ore (exported), tons	....	98,757	....	412,401
Pig-iron from domestic ore, tons	....	48,820	....	936,118
Lead, tons	....	540	....	136,948

1916 figures are not available for the last three items.

NICARAGUA

A new gold-silver-lead deposit has been discovered 12 miles south of Matagalpa, between the Leonesa and the Mina Verde gold mines. N. DeLaney, Gus Fraumberger, and W. H. De-Savigny are the principals, and claim to have 6 ft. of galena, running high in gold, silver, and lead. An engineer has been engaged who will leave for Nicaragua to investigate.

PERSONAL

Note: The Editor invites members of the profession to send particulars of their work and appointments. This information is interesting to our readers.

Hugh Rose is in London.

Frank W. Oldfield is at Mexico City.

R. S. Archibald is in the Black Hills region.

J. M. Callow was in San Francisco this week.

Charles Butters is due here from Washington.

John B. Wise has returned to Chuquicamata, Chile.

Harry D. Griffiths has left London to go to Burma.

J. G. Shannonhouse has returned from Mazatlan, Mexico.

Antony F. Lucas has returned to Washington from Texas.

John A. Rice has returned from a geologic inspection at Cananea, Mexico.

Philip Bradley, superintendent of the Alaska Juneau mine, visited San Francisco this week.

Walter H. Aldridge passed through San Francisco on his return from Arizona to New York.

W. F. Collins has returned from China and is helping to organize Chinese labor in France.

E. V. Daveler, superintendent of the mills of the Alaska Gold Mining Co., is in San Francisco.

J. B. Moore has resigned as mine superintendent for the Tigre Mining Co. and is now at San Antonio, Texas.

Harry J. Sheafe has completed his military training at Fort Leavenworth and is now at Chicago awaiting orders.

Frederic R. Weekes has moved his office from 71 to 42 Broadway, New York. He is returning from California to New York.

P. K. Lucke has resigned as consulting engineer to Cia. Minera de Peñoles in Mexico to resume private practice at San Antonio, Texas.

John J. Croston, lieutenant, has been transferred from the 23rd Engineers to the 27th Engineers, the Miners Regiment, at Camp Meade, Florida.

E. A. Holbrook, formerly professor in the mining department of the University of Illinois, has been appointed Supervising Mining Engineer and Metallurgist for the U. S. Bureau of Mines and will serve as superintendent of the station at Urbana, Illinois. W. B. Plank has been appointed mining engineer in the Mine Safety Section, with headquarters at the same station.

Obituary

Alfred Winter Evans, D.S.O., D.C.M., Lieutenant-Colonel, New Zealand Rifle Brigade, was killed in action on Octover 12, 1917. Evans was a mining engineer, Columbia '06. After graduation he worked in Utah and Arizona for a year or more before going to the Rand, South Africa, where he followed mining till 1911 when he went to New Zealand as general manager for the Consolidated Gold Fields of New Zealand Ltd., which position he was filling when war was declared. He had seen service in the South African war, was twice recommended for the Victoria Cross by the late Field Marshal Sir George White, and was awarded the D.C.M. with a commission. He was mentioned in dispatches after the Somme for gallantry in the field, and awarded the D.S.O. after the battle of Messines. Always a credit to his profession he possessed in a rare measure those traits of character that endeared him to his friends and commanded the respect even of those who differed from him. A friend testifies that he worked with Evans through many trying circumstances, especially in connection with mine-fires and the New Zealand miners strike of 1912, and never saw him weaken when tried, or knew him to send a man where he would not go himself.

Recent Publications

THORIUM MINERALS IN 1916. By W. F. Shaller. Mineral Resources of the United States 1916.

POTTERY IN 1916. By Jefferson Middleton. Mineral Resources of the United States 1916.

BARYTES AND BARIUM PRODUCTS IN 1916. By James H. Hill. Mineral Resources of the United States 1916.

SILICA IN 1916. By Frank J. Katz. Mineral Resources of the United States 1916.

ASPHALT, RELATED BITUMENS, AND BITUMINOUS ROCK IN 1916. By John D. Northrop. Mineral Resources of the United States, 1916.

CHROMITE IN 1916. By J. S. Diller. 17 pp., ill. U. S. Geological Survey.

BAUXITE AND ALUMINUM IN 1916. By James M. Hill. 11 pp. U. S. Geological Survey.

SILVER, COPPER, LEAD, AND ZINC IN THE CENTRAL STATES IN 1916. By J. P. Dunlop and B. S. Butler. 104 pp. U. S. Geological Survey.

FOURTEENTH ANNUAL REPORT OF THE DIRECTOR OF THE BUREAU OF SCIENCE, PHILIPPINE ISLANDS. By Alvin J. Cox. 70 pp. Manila Bureau of Printing. An account of the work accomplished by the Bureau in 1916.

THE ALASKAN MINING INDUSTRY IN 1916. By A. H. Brooks. 62 pp., map. Bulletin 662. U. S. Geological Survey.

OUR MINERAL SUPPLIES, FLUORSPAR. By Ernest F. Burchard. 7 Pp. Bulletin 666-CC. U. S. Geological Survey.

GYPSUM IN 1916. By Ralph W. Stone. 6 pp., map. U. S. Geological Survey.

HYDRAULIC CONVERSION TABLES AND CONVENIENT EQUIVALENTS. 23 pp. Water Supply Paper 425-C. U. S. Geological Survey.

WELL-DRILLING METHODS. By Isaiah Bowman. 138 pp., ill. Water-Supply Paper 257. U. S. Geological Survey.

THE PALESTINE SALT DOME, ANDERSON COUNTY, TEXAS. THE BRENHAM SALT DOME, WASHINGTON AND AUSTIN COUNTIES, TEXAS. By Oliver B. Hopkins. 27 pp., maps.. Bulletin 661-G. U. S. Geological Survey.

MANGANIFEROUS IRON ORES. By E. C. Harder. 13 pp. U. S. Geological Survey.

MINERAL RESOURCES OF THE KANTISHNA REGION, ALASKA. By Stephen Capps. 50 pp. map. U. S. Geological Survey.

PRODUCTION OF EXPLOSIVES IN THE UNITED STATES DURING 1916. By Albert H. Fay. 24 pp. Technical Paper 175, U. S. Geological Survey.

SPIRIT LEVELING IN NEVADA. By R. B. Marshall, Chief Geographer. 91 pp., Ill. Bulletin 654, U. S. Geological Survey.

MEN WHO RECEIVED BUREAU OF MINES CERTIFICATES OF MINE RESCUE TRAINING. Technical Paper 167. U. S. Bureau of Mines.

YEAR-BOOK OF THE BUREAU OF MINES, 1916. By Van H. Manning. 174 pp., ill. U. S. Bureau of Mines.

EFFECT OF LOW-TEMPERATURE OXIDATION ON THE HYDROGEN IN COAL AND THE CHANGE OF WEIGHT OF COAL ON DRYING. By S. H. Katz and H. C. Porter. 16 pp., ill. Technical Paper 98. U. S. Bureau of Mines.

COKE-OVEN ACCIDENTS IN THE UNITED STATES DURING 1916. By A. H. Fay. 22 pp. Technical Paper 173, U. S. Bureau of Mines.

CONTROL OF HOOKWORM INFECTION AT THE DEEP GOLD MINES OF THE MOTHER LODE, CALIFORNIA. 52 pp. Bulletin No. 139, U. S. Bureau of Mines.

LABORATORY DETERMINATION OF THE EXPLOSIBILITY OF COAL DUST AND AIR MIXTURES. By J. K. Clement and J. W. Lawrence. 35 pp. Ill. Technical Paper 141, U. S. Bureau of Mines.

DETERIORATION IN THE HEATING VALUE OF COAL DURING STORAGE. By Horace C. Potter and F. K. Ovitz. 38 pp. Ill. Bulletin 136, U. S. Bureau of Mines.

PREPAREDNESS CENSUS OF MINING ENGINEERS, METALLURGISTS, AND CHEMISTS. By A. H. Fay. 19 pp. Technical Paper 179, U. S. Bureau of Mines.

COKING OF ILLINOIS COALS. By F. K. Ovitz. 71 pp. Ill. Bulletin 138, U. S. Bureau of Mines.

ABSORPTION OF METHANE AND OTHER GASES BY COAL. By S. H. Katz. Technical Paper 147, U. S. Bureau of Mines, 22 pp. Ill.

EFFECT OF STORAGE UPON THE PROPERTIES OF COAL. By S. W. Parr. 44 pp. Ill. Bulletin No. 97, Engineering Experimental Station, University of Illinois, Urbana.

THE EFFECT OF MOUTHPIECES ON THE FLOW OF WATER THROUGH A SUBMERGED SHORT PIPE. By Fred B. Seely. 53 pp. Ill. University of Illinois Bulletin.

UNIVERSITY OF ARIZONA AND THE WAR. 22 pp.

ARIZONA STATE BUREAU OF MINES. Bulletin No. 75. 13 pp. University of Arizona.

SAMPLING MINERALIZED VEINS. By G. R. Fansett. 5 pp. Ill. Bulletin No. 66, University of Arizona.

OIL AND ITS GEOLOGY. By M. A. Allen. 34 pp. Ill. Bulletin No. 65, University of Arizona.

PROGRESS AND CONDITION OF THE ILLINOIS STATE MUSEUM OF NATURAL HISTORY. By A. R. Crook. 68 pp.

THE MINING OF THIN COAL-SEAMS AS APPLIED TO THE EASTERN COAL-FIELDS OF CANADA. By J. F. Kellock Brown. Bulletin No. 15. 135 pp. Ill. Maps. Mines Branch, Department of Mines, Canada.

TEST OF SOME CANADIAN SANDSTONES TO DETERMINE THEIR SUITABILITY AS PULPSTONES. By L. Heber Cole. 16 pp., ill. Mines Branch, Department of Mines, Canada.

MEASUREMENTS OF VARIOUS THERMAL AND ELECTRICAL EFFECTS, ESPECIALLY THE THOMPSON EFFECT, IN SOFT IRON. By Hall, Churchill, Campbell, and Serviss. 32 pp. Proceedings of the American Academy of Arts and Sciences.

THE RUSTLER SPRINGS SULPHUR DEPOSITS. By E. L. Porch, Jr., 71 pp. Ill. Bulletin 1722, University of Texas.

THE CLAYS OF PIEDMONT PROVINCE, VIRGINIA, Bulletin XIII. University of Virginia, 86 pp. Ill. Maps. An exhaustive study of the clays of Piedmont.

Publications by the U. S. Geological Survey:

PROFESSIONAL PAPER 108-F. A Fossil Flora from the Frontier Formation of South-Western Wyoming. By F. H. Knowlton. 37 pp., 23 plates.

BULLETIN 647. The Bull Mountain Coal Field, Musselshell and Yellowstone Counties, Montana. By L. H. Woolsey, R. W. Richards, and C. T. Lupton, compiled and edited by E. R. Lloyd. 218 pp., 36 plates, 2 text figures.

BULLETIN 660-B. Notes on the Greensand Deposits of the Eastern United States. By G. H. Ashley; Methods of Analysis of Greensand. By W. B. Hicks and R. K. Bailey. 34 pp., 1 plate, 1 text figure.

BULLETIN 661-B. Structure of the Northern Part of the Bristow Quadrangle, Creek County, Oklahoma., with Reference to Petroleum and Natural Gas. By A. E. Fath. 35 pp., 4 plates, 5 text figures.

BULLETIN 661-E. The Bowdoin Dome, Montana, a Possible Reservoir of Oil or Gas. By A. J. Collier. 17 pp., 1 text figure.

BULLETIN 666-Q. Our Mineral Supplies—Copper. By B. S. Butler. 4 pp., 1 plate.

BULLETIN 666-S. Our Mineral Supplies—Portland Cement. By E. F. Burchard. 5 pp., 1 text figure.

BULLETIN 666-V. Our Mineral Supplies—Iron. By E. F. Burchard. 12 pp., 1 text figure.

BULLETIN 666-BB. Our Mineral Supplies—Magnesite. By H. S. Gale. 3 pp.

MINERAL RESOURCES OF THE UNITED STATES, 1915. Part II, Non-metals. 1084 pp., 7 plates, 36 text figures, 2 inserts. Statistics of the production, importation, and exportation of non-metalliferous mineral substances in the United States in 1915—a consolidation of 35 advance chapters. Contains inserts showing coke produced in the United States, 1880-1915, and coal produced in the United States, 1807-1915.

THE METAL MARKET

METAL PRICES

San Francisco, December 11

Aluminum-dust (100-lb. lots), per pound	$1.00
Aluminum-dust (ton lots), per pound	$0.95
Antimony, cents per pound	16.00
Antimony (wholesale), cents per pound	15.25
Electrolytic copper, cents per pound	23.50
Pig-lead, cents per pound	6.50— 7.50
Platinum, soft and hard metal, respectively, per ounce	$105— 113
Quickselver, per flask of 75 lb	115
Spelter, cents per pound	9.50
Zinc-dust, cents per pound	20

ORE PRICES

San Francisco, December 11

Antimony, 45% metal, per unit	$1.00
Chrome, 34 to 40%, free SiO_2, limit 8%, f.o.b. California, per unit, according to grade	$0.60— 0.70
Chrome, 40% and over	$0.70— 0.85
Magnesite, crude, per ton	$8.00—10.00

The magnesite market remains practically the same; considerable calcined and raw magnesite has been offered, but no business to amount to anything has been reported.

Manganese: The Eastern manganese market continues fairly strong with $1 per unit Mn quoted on the basis of 48% material.

Tungsten, 60% WO_3, per unit	26.00
Molybdenite, per unit MoS_2	$40.00—45.00

EASTERN METAL MARKET

(By wire from New York)

December 11.—Copper is unchanged at 23.50c. all week. Lead is dull and firm at 6.50c. all week. Zinc is inactive and lower at 8 to 7.87c. Platinum is unchanged at $105 for soft metal and $113 for hard.

SILVER

Below are given the average New York quotations, in cents per ounce, of fine silver.

Date		Average week ending	
Dec. 5	85.62	Oct. 30	84.71
" 6	85.62	Nov. 6	89.25
" 7	85.87	" 13	86.18
" 8	85.87	" 20	85.62
" 9 Sunday		" 27	84.60
" 10	85.62	Dec. 4	84.70
" 11	85.62	" 11	85.70

Monthly Averages

	1915	1916	1917		1915	1916	1917
Jan.	48.85	56.76	75.14	July	47.52	63.06	78.92
Feb.	48.45	56.74	77.54	Aug.	47.11	66.07	85.40
Mch.	50.61	57.89	74.13	Sept.	48.77	68.51	100.73
Apr.	50.25	64.37	73.51	Oct.	49.40	67.86	87.38
May	49.87	74.27	74.61	Nov.	51.88	71.60	85.97
June	49.03	65.04	76.44	Dec.	55.34	75.70	....

Samuel Montagu & Co., of London, state: The net imports of silver into India during July 1917 were heavy. The total was 8,867,052 oz.; allowing for exports amounting to 2,460,227 oz., the net imports were 6,406,825 oz. The net imports of silver during April, May, June, and July amounted to the substantial weight of 33,702,629 oz. valued at about £5,314,000. Of this total the Indian government was responsible for no less than 30,552,-487 oz. The stock of silver at Shanghai on November 3 consisted of about 25,700,000 oz. of sycee-silver and $14,900,000, as compared with 25,200,000 oz. in sycee and $15,200,000 on October 27.

COPPER

Prices of electrolytic in New York, in cents per pound.

Date		Average week ending	
Dec. 5	23.50	Oct. 30	23.50
" 6	23.50	Nov. 6	23.50
" 7	23.50	" 13	23.50
" 8	23.50	" 20	23.50
" 9 Sunday		" 27	23.50
" 10	23.50	Dec. 4	23.50
" 11	23.50	" 11	23.50

Monthly Averages

	1915	1916	1917		1915	1916	1917
Jan.	13.60	24.30	29.53	July	19.09	25.66	29.67
Feb.	14.38	26.62	34.57	Aug.	17.27	27.03	27.42
Mch.	14.80	26.65	36.00	Sept.	17.69	28.28	25.11
Apr.	16.64	28.02	33.16	Oct.	17.90	28.50	23.50
May	18.71	29.02	31.69	Nov.	18.88	31.95	23.50
June	19.75	27.47	32.57	Dec.	20.67	32.89	....

Production reports for November are not as favorable as was at first expected. Anaconda's November production of 21,666,332 lb. of copper showed a decrease of 633,668 lb. from October. This was due to closing down of the Leonard mine for a week, occasioned by escaping gas from a fire on the 1300-ft. level. On December 1, 12,211 tons of copper ore was hoisted. Anaconda is now operating at 94% capacity and officials say, barring unforeseen setbacks, December production should reach between 25,000,000 and 26,000,000 lb. of copper. The Great Falls refinery is operating at 65% capacity. A preliminary statement of the Porphyries output for November shows a falling off of about 2,000,000 lb. as compared with October. Utah's output last month was 16,300,000 lb., whereas in October it was 18,100,000 lb. Miami, on the other hand, shows a satisfactory increase, its November production being given as 3,361,426 lb. as compared with 2,673,775 lb. in October.

ZINC

Zinc is quoted as spelter, standard Western brands, New York delivery, in cents per pound.

Date		Average week ending	
Dec. 5	8.00	Oct. 30	8.12
" 6	8.00	Nov. 6	7.42
" 7	7.87	" 13	7.81
" 8	7.87	" 20	7.98
" 9 Sunday		" 27	8.00
" 10	7.87	Dec. 4	8.00
" 11	7.87	" 11	7.91

Monthly Averages

	1915	1916	1917		1915	1916	1917
Jan.	6.30	18.21	9.75	July	20.54	9.90	8.98
Feb.	9.05	19.99	10.45	Aug.	14.17	9.03	8.58
Mch.	8.40	18.40	10.78	Sept.	14.14	9.18	8.33
Apr.	9.78	18.62	10.20	Oct.	14.05	9.92	8.32
May	17.03	16.01	9.41	Nov.	17.20	11.81	7.76
June	22.20	12.85	9.63	Dec.	16.75	11.26	....

LEAD

Lead is quoted in cents per pound, New York delivery.

Date		Average week ending	
Dec. 5	6.50	Oct. 30	5.66
" 6	6.50	Nov. 6	6.02
" 7	6.50	" 13	6.35
" 8	6.50	" 20	6.50
" 9 Sunday		" 27	6.50
" 10	6.50	Dec. 4	6.50
" 11	6.50	" 11	6.50

Monthly Averages

	1915	1916	1917		1915	1916	1917
Jan.	3.73	5.95	7.64	July	5.59	8.40	10.93
Feb.	3.83	6.23	9.01	Aug.	4.62	6.28	10.75
Mch.	4.04	7.26	10.07	Sept.	4.62	6.86	9.07
Apr.	4.21	7.70	9.38	Oct.	4.62	7.02	6.97
May	4.24	7.38	10.29	Nov.	5.15	7.07	6.38
June	5.75	6.88	11.74	Dec.	5.34	7.55	....

Statistics of the production of lead and zinc pigments in the United States, compiled by the U. S. Geological Survey, show that the marketed output of these pigments of domestic manufacture in 1916 reached a value of $60,376,319, as compared with a value of $43,336,939 reported for 1915, an increase of more than 39%.

The higher price of pig-lead in 1916 evidently curtailed the production of pigments made from it, and at the same time, in conjunction with other war conditions, largely increased the value. The decrease in quantity amounted to 11,901 tons, or about 6%; the value increased $6,447,583, or about 26%. It is not permissible to give separately the figures for sublimed lead, zinc oxide, and leaded zinc oxide, as to do so would disclose individual output. The combined production of these pigments in 1916 amounted to 135,606 short tons, valued at $23,515,803, as against 141,383 short tons, valued at $14,962,461, in 1915, a decrease of 4% in quantity and an increase of 57% in value. Manifestly war conditions were operative in this branch of the pigment industry also. As stated in the Survey's report on barium products, 51,291 short tons of lithopone of domestic manufacture, valued at $5,798,927, was marketed in the United States in 1916, as compared with 46,494 short tons, valued at $3,760,472, in 1915.

QUICKSILVER

The primary market for quicksilver is San Francisco, California being the largest producer. The price is fixed in the open market, according to quantity. Prices, in dollars per flask of 75 pounds:

Week ending

Date			
Nov. 13	100.00	Nov. 27	110.00
" 20	100.00	Dec. 4	115.00
		" 11	115.00

Monthly Averages

	15	1916	1917		1915	1916	1917
Jan.	.90	222.00	81.00	July	95.00	81.20	102.00
Feb.	.00	295.00	126.25	Aug.	93.75	74.50	115.00
Mch.	.00	219.00	113.75	Sept.	91.00	75.00	112.00
Apr.	.50	141.60	114.50	Oct.	92.90	78.20	102.00
May	19.00	90.00	104.00	Nov.	101.50	79.50	102.50
June	[illegible].00	74.70	85.50	Dec.	123.00	80.00	...

TIN

Prices in New York, in cents per pound.

Monthly Averages

	1915	1916	1917		1915	1916	1917
Jan.	34.40	41.76	44.10	July	37.38	38.37	62.60
Feb.	37.23	42.60	51.47	Aug.	34.37	38.88	62.53
Mch.	48.76	50.50	54.27	Sept.	33.12	36.66	61.54
Apr.	48.25	51.49	55.63	Oct.	33.00	41.10	62.24
May	39.28	49.10	63.21	Nov.	39.50	44.12	74.18
June	40.26	42.07	61.93	Dec.	38.71	42.55	....

ORES

Tungsten: The market has come to a standstill because of a railroad embargo on shipments leaving New York, causing buyers to abstain from the market. The embargo on imports has put a damper on further business because of expected delays in getting licenses. Also some tungsten-ore shippers in South America are of German nationality. Prices are largely nominal at $23 to $25 per unit for wolframite, and $26 to $27 for sheelite. One sale of wolframite, high grade, at $25 is noted. Ferrotungsten is unchanged at $2.25 per lb. of contained tungsten.

Molydbenum and antimony: There is but little change in these ores, with the market quiet. Molybdenite is quoted at $2.25 per lb. of MoS_2

Eastern Metal Market

New York, December 5.

Practically all the metal markets have come to a dead standstill. There is very little business outside of that occasioned by Government activity and by war-requirements.

Copper conditions are unchanged. It is a Government-controlled product.

Tin continues scarce and nominal at high prices. There is no activity.

Lead is in poor demand, but prices are firm at unchanged levels.

Zinc is still lifeless, and nominal at the same quotations as last week.

Antimony is more active and higher.

Aluminum continues dull.

In the steel-trade steel producers have been called upon for cost-sheets in the last week by the Federal Trade Commission. There is also expected renewed agitation of Government control and of price reductions from the same source, synonymous with the calling of the Pomerene bill in Congress. The outlook, in the opinion of 'The Iron Age,' is by no means encouraging in the face of production and distribution troubles. Steel and iron exports have been much reduced despite considerable demand from foreign sources. The Government is still in the market for large orders for plates and forging-steel for shells. There is little said, however, about shell-steel for the Allies. The November pig-iron output was unexpectedly larger than in October, or 106,859 tons per day as against 106,550 tons per day in the latter month.

COPPER

There is very little to say about the copper market in addition to what was said last week. There is no general market. The situation is absolutely under Government control. Distribution is evidently going on in an orderly and satisfactory manner. Very little is heard in the way of complaint. As stated last week, metal is sold only at the Government price of 23.50c. per lb. for delivery in the first quarter and also for prompt delivery to consumers having Government contracts. The settling of the jobbers' position has relieved the situation in this field, and orders of small lots are going at 5% above the Government price, and probably at a fair profit above cost in some cases. The November output of refined copper has been higher than that for October, but it is believed to be considerably less than the maximum in 1916. The statement is made that, after the needs of our own Government and those of the Allies are met, the supply for domestic consumption will be small, with the prospect that this state of affairs will continue for a number of months. Copper exports in 1916 were 784,000,000 lb., according to B. S. Butler of the U. S. Geological Survey. This compares with a 10-year average of 735,000,000 lb., and with a record exportation of 926,000,000 lb. in 1913. The 1916 domestic consumption is put at 1,430,000,000 lb. at an average price per pound of 24.6c., which compares with a 10-year average of 773,000,000 lb. at 16.3c. per lb. Our production of primary copper in 1916 is reported as 2,259,000,000 lb., of which 1,928,000,000 lb. was smelted from domestic ores. The 10-year average up to 1917 was 1,502,750,000 lb. produced, of which 1,201,400,000 lb. came from domestic ores.

TIN

The tin market has been very dull and quite neglected during the last week. Spot Straits continues unobtainable at the highest price ever recorded. It is therefore nominal at 80 to 81c., New York. Banca and Chinese tin are also scarce and are quoted nominal, at high prices. No. 1 Chinese was nominal yesterday at 75c., New York. These grades have been in better demand recently because of the shortage of Straits tin and therefore stocks have been drawn upon for sale and consumption until the supplies are less. The only new item of interest in the week is that the tin in warehouses, which was announced last week as commandeered by the Government, has been all released—except such as was appropriated for direct use by the Government. It is again in the possession of those who bought it, subject to distribution by the American Iron and Steel Institute, and under Government control. The market is now a completely regulated one, the same as copper. In general, demand is light, with little or no business reported. Buyers show no interest in any grade for any position. There is no prospect of immediate relief to the spot-situation. The entire market has come to a halt. Arrivals to December 4, inclusive, have been 325 tons, with the quantity afloat unreported. Spot Straits in the London market yesterday sold at £294 per ton, an advance of £10 per ton over last week.

LEAD

Conditions are entirely unchanged from those outlined in last week's letter. The market is dull, and there has been little or no business. Prices are firmly held at not less than 6.50c., New York, or 6.37½c., St. Louis, in the outside market. The American Smelting & Refining Co.'s quotation is unchanged at 6.25c., New York. There is no pressure to sell, neither is there any anxiety to buy, and the situation is therefore a stand-off. Lead in transit has sold during the week in carload lots and less at 6.75 cents.

ZINC

The market is still in the doldrums without interest or life. The trade seems to be waiting for the results of the new Zinc Committee's deliberation with the Government as to price-fixing and other matters, concerning which nothing yet has been made public. The market is bare of demand, and sales are few. Quotations, mainly nominal, are unchanged from last week, with prime Western for early delivery quoted at 7.75c., St. Louis, or 8c., per lb., New York. These prices are being shaded slightly in some cases, but not so as to greatly affect sentiment on the market. Spelter-exports in September, just announced, were large, amounting to 25,333 gross tons, of which 17,322 tons was made from foreign ores. This fact is regarded as explaining why the accumulation of stocks at smelters in the third quarter was not large. Zinc ore in the Joplin district has declined to $50 to $70 per ton as compared with $57.50 to $72.50 last week.

ANTIMONY

The effect of the President's proclamation of an embargo on imports has decidedly strengthened this market, and, as a result, as high as 16c. per lb., New York, duty paid, was bid yesterday for Chinese and Japanese grades. Sales are reported at 15.25c. per lb. It is expected that available supplies, especially for spot and early delivery, will be curtailed, and there has been a scramble to obtain those now available. It is too early to forecast the ultimate effect of the new-conditions on the market.

ALUMINUM

The market is stagnant and unchanged at 36 to 38c. per lb., New York, for No. 1 virgin metal, 98 to 99% pure.

CHRISTMAS greetings to our brave boys in the trenches and to the gallant lads that are preparing to go over the top in the coming year.

THIS Christmas is ironical. Jerusalem has been captured from the Moslem despite the German guns mounted in Bethlehem, and bombs have been thrown on motor-boats patroling the Jordan. The irony is that of the iron cross on which christendom is crucified.

REFERRING to developments at Packard, Nevada, we are informed that our local correspondent was misinformed concerning the true conditions, which are that a recent examination of the Rochester Combined property justifies the early completion of the mill and that the recent shut-down was due to temporary delays in financing the enterprise.

PRELIMINARY estimates indicate that the production of copper in the United States for 1917 will be about 150,000 tons less than in 1916, but about 150,000 tons more than in 1915. In 1916 the output was 2,311,000,000 pounds of copper refined in the United States; of this, 1,928,000,000 pounds was derived from domestic ores. The American increase was nearly 80% and the production represented 75% of the world's total output. The normal increase is between 6 and 7%. In 1913 the world's output was 1,100,000 tons. Now, despite all the hindrances to industry, it is about 30% more.

IN a recent issue of the 'South African Mining Journal' we note a letter from Mr. J. S. Curtis, known to American geologists as the author of the Eureka monograph, in which he corrects sundry careless statements crediting Mr. John Hays Hammond with being the originator of 'deep-level' mining on the Rand. As a matter of fact the South Reef was struck in the Village Main Reef shaft about the time when Mr. Hammond first arrived at Johannesburg. The Village Main Reef was a deep-level enterprise on the dip of the gold-bearing seams of banket that were being mined by the City & Suburban and Jubilee companies. The participation of Cecil Rhodes and others in deep-level projects came after this discovery—of 9-oz. ore, 12 inches wide—had been made on the initiative of Mr. Curtis in the Village Main Reef shaft.

RUMORS are current that the Department of Mines in British Columbia is considering the idea of discontinuing the publication of the annual report of the Provincial Mineralogist, Mr. William F. Robertson. We hope that this is untrue, for it would seem to us that any such step would be a blunder. The annual reports of the Provincial Mineralogist are recognized as the most reliable sources of information concerning the mining development of British Columbia and to them the public outside British Columbia is accustomed to turn for reliable guidance in such matters. We are surprised to learn that these reports are not fully appreciated at the place of their origin.

EDITORS have been honored by being joined recently by Mr. Taft, who has followed Mr. Roosevelt in his work as a publicist in print. It is greatly to the gain of journalism that men of such character and distinction should use it as a means of exerting influence. Among others of the fraternity, we note that Lieutenant J. L. Gallard of the 'Financial Times,' London, has been wounded at the front and that Captain Clem D. Webb, formerly of the 'South African Mining Journal,' has greatly distinguished himself. He and his son Lieutenant Stuart Webb, were decorated with the Military Cross on the same day. Captain Webb has four sons in the British Army, and other editors will join us in friendly envy of the record that he has made.

YESTERDAY evening, walking to the trans-bay ferry, the present writer saw a man spill some potatoes out of a paper bag, which had become torn by the excessive weight. The potatoes fell on the sidewalk; the owner, evidently annoyed by discovering that his paper bag was no good, allowed the remaining potatoes to fall out and did not take the trouble to collect them. For several minutes enough potatoes to make a meal for two families lay unclaimed and unheeded; even the newsboys seemed unaware of their opportunity; finally a man stooped to pick them up and we helped him to collect them. What would happen if such a thing occurred in a thoroughfare in Belgium, Poland, or northern France, or even in Germany or England? The eagerness to get hold of such excellent food would precipitate a riot, and the man that would not take the trouble to recover it would be deemed a crazy fool.

PLATINUM, a metal much needed in chemical laboratories, has become scarce, owing to the interruption of supply from the Ural mountains; therefore the shipment of 21,000 ounces that arrived from Vladivostok last week was most welcome. It is interesting to note that Mr. Norman C. Stines, a Californian, the manager of the Sissert copper mine, was responsible for collecting all this platinum, having been attached to the military staff of the American Embassy at Petrograd so as to obtain facilities for the purpose. Having collected the platinum by purchase at various mines, he handed it to Mr. F. W. Draper, of the Verk Isetz copper mine, and he brought it to the United States by way of the Siberian railway, Vladivostok, and San Francisco. Our readers will recall Mr. Draper's excellent article on the Kalata plant appearing in our issue of September 1, 1917.

ON another page we give the text of a patent obtained recently by Mr. Walter A. Scott, formerly of counsel for the Miami Copper Company and James M. Hyde in the litigation with Minerals Separation. It would appear that the learned lawyer has been so much in the atmosphere of flotation technology that he felt impelled to try his own hand at devising a method of operation. The result is patent No. 1,246,665; this reminds one of the so-called bubbles patent of Sulman & Picard, No. 793,808, in which "the pulp is submitted to the action of a current of air or other gas bubbles, the air or other gas being first suitably charged either with the vapor of a volatile oil, such as petroleum of low boiling-point, or with the spray of any other suitable volatile or non-volatile or fixed oil or the like [what verbiage! why not delete eight out of the last nine words?]. The oil may be sprayed or reduced to a state of such fine division that minute globules of the same can remain temporarily suspended in an air or other gas current by the use of any suitable spraying or atomizing device and the air-current introduced into the ore-pulp, preferably at the bottom, by means of a pipe or pipes provided with suitable perforations or by other suitable contrivance. The minute oil globules or the condensed vapors or volatile oils [what happens to the "non-volatile oils or the fixed oil or the like"?] attach themselves to the metalliferous particles in preference to the gangue." Mr. Scott uses a volatile oil—he dispenses with the non-volatile, the fixed, or the like—and appears therefore to know what he is about, for he says that the oil performs its useful function not in the ore-pulp as a whole but only in the film enclosing the bubbles of air. Thus he impregnates the air with the gasified oil and consequently all the oil is introduced as part of the gaseous volume that makes the bubble; accordingly none of the oil is dispersed through the pulp, and none of it can fail to perform its proper function, for in condensing it necessarily collects as a film around the bubble. The oil is heated by a steam-coil so that it is vaporized in the presence of a current of air under pressure, the resulting gaseous mixture of oil and air being forced through the porous bottom of the vessel containing the ore-pulp. As distinguished from the method of 793,808, Mr. Scott uses heat to vaporize his oil and so avoids the need for a "spraying or atomizing device." Successful experiments, we understand, have been made at Ray, by means of an injector into which hot oil and air under pressure are admitted, the injector being then placed on the bottom of the flotation-cell. In this way, as in Mr. Scott's patent, the intensive distribution of the oil required to bring it in contact with the mineral particles is performed without the aid of mechanical agitation and the oil is introduced to the mineral through the medium of the bubble-films in contact with which the flotation is effected.

Mine Taxation

The owners of mines, whether individuals, syndicates, or companies, have to face an increase of taxation incidental to war. In addition to the customary State, County, and Federal taxes, they will have to consider the payment of an extra income-tax and an excess-profit tax, in amount sufficient to modify their policy and mode of operation. The Act imposing the special war-tax is expressed in terms that are intricate and confusing. When the Treasury Department has decided a few doubtful cases, it will be known what is the intent of the enactment, but until then each man, assisted by those versed in official ways, must interpret for himself. In default of other guidance he will do well to study the pamphlet issued by the National City Company of Washington. In this digest of the war-tax it is stated that "the invested capital and net income for the taxable year and the average rate of income for the pre-war period (1911-1913 inclusive) must first be determined, according to the technical rules prescribed in the Act." Therefore first ascertain the rate of income earned on the invested capital, by dividing the net income by 1% of the capital. If the rate is less than 7%, there will be no excess-profit tax. If the rate earned is not over 9%, and is not greater than the average rate for the pre-war period, there will be no tax. If the rate earned is over 9%, or is greater than the average rate for the pre-war period, but does not exceed 15%, there may or may not be a tax, depending upon whether the excess over 9% or over the pre-war rate exceeds the specific exemptions allowed, namely, $3000 for corporations and $6000 for partnerships or individuals. If the rate earned does not exceed 15%, then income equivalent to the pre-war rate (between the limits of 7 and 9% as specified above) is exempted. If the pre-war rate was 7% or less, multiply the invested capital by 0.07 (7%); if the pre-war rate was 9% or more, multiply the capital by 0.09 (9%);

if the pre-war rate was between 7 and 9%, multiply the capital by the exact percentage of pre-war income. Deduct the exempted income, computed as above, from the net income for the year. From the remainder deduct the specific exemption of $3000 for a corporation or of $6000 for partnerships and individuals. After these deductions, the remaining income, if any, will be taxed at 20%. Pay it cheerfully. Next, where the rate of income exceeds 15%, the tax on this excess is additional to the tax on the first 15%, thus

Next 5% or less between 15 and 20%	is taxed	25%
" 5% " " " 20 " 25%	" "	35%
" 8% " " " 25 " 33%	" "	45%
Remaining income in excess of 33%	" "	60%

To determine the total tax, it is necessary to add the tax on the first 15% to the tax on each of the higher grades that the income reaches; but since the rates on the excess over 15% are all in fixed ratio to the income, without any deductions, it is possible to tabulate, for each enterprise, the aggregate tax on the excess over 15% expressed in percentages of the capital. To do this the first step is to define 'invested capital'. This may be (1) actual cash paid in; (2) the cash value of tangible property paid in (not exceeding in any event the par value of stock or shares issued therefor) at the time of payment, but if paid in prior to January 1, 1914, then the cash value on that date; (3) surplus, paid in or earned, and undivided profits, used or employed in the business, exclusive of undivided profit earned during the taxable year. Next, what is 'net income'? It is gross income less (1) expenses of maintenance and operation, including rentals, (2) losses sustained and charged off during the year, not compensated by insurance or otherwise, (3) depreciation, an allowance for wear and tear of plant and machinery, (4) interest paid on indebtedness, (5) taxes paid within the year, except income and excess-profit taxes, (6) depletion of resources. It would be a simple matter to make these subtractions if words meant the same thing to everybody. 'Depreciation' is interpreted variously; by improvident operators it is ignored; by conservative operators it is fully appraised annually. 'Depletion' will cover the exhaustion of ore-reserves, but here again individual practice differs widely and the tax-collector will not only fall back on awkward precedents but he may quote the Supreme Court's decision of last February whereby allowance for depletion was limited to 5%—a proportion ridiculously low in view of the short life of most metal mines. Unfortunately it is a fact that the book-keeping of mining companies is anything but scientific, that is, it fails to recognize sundry basic facts, of which two, at least, have been ignored persistently by promoters, who paid no attention whatever to the opportunity given to an unphilosophic tax-collector. (1) A mine is a wasting asset. (2) The dividend from a mine is not income, but a return of capital, until such time as the capital has been reduced. A mine has no goodwill; when exhausted it is a mere hole in the ground surrounded by dumps that disfigure the scenery. The general ignorance of the economics of mining, not only among miners themselves, but among the tax-collectors that now take advantage of this fact, is a serious obstacle to the correct incidence of taxation. It appears likely that reliable definitions of such fundamental terms as 'capital', 'income', 'depreciation', and 'depletion' will not be available until the Courts have handed down a large number of decisions. Much litigation may ensue, for the law is darkened by technical obscurities. Even a constitutional question could arise, because under the Act property may be taken without recompense, but no American cares to raise a constitutional question in time of war; the average citizen is willing to bear his part of the cost of the War, while disliking intensely to be mulcted for more than his proper share. The excess-profit tax is a piece of legislation that weighs inequitably upon many forms of mining; it was enacted in a hurry and will call for a great deal of patience and long-suffering on the part of the mining community. The one clear point in the law is the exemption of Senators and Congressmen from the incidence of the tax; that is a feature that calls for instant correction.

The Making of an American

In the interviews previously published, covering the careers of distinguished members of the profession, we have had to deal with the mining and metallurgy of gold, silver, copper, and lead. In this issue we give our readers the record of a conversation with an engineer famous for his success in exploring for oil. It is a curious fact that people will talk of the finding and winning of oil as if it were not 'mining'; in an interesting controversy between two distinguished engineers it was contended by one of them that the exploitation of an oil-field did not come under the definition of 'mining.' Presumably the liquid condition of mineral-oil and the fact that it is pumped, when it does not gush, suggests the notion that the art of extracting it from the earth is different from digging for ore. One might as well refuse to recognize hydraulicking and ground-sluicing as belonging to the ancient art, or, if that be stretching the analogy, then we may compare the winning of the liquid oil with the extraction of salt by solution in water, as a brine, or the raising of sulphur by melting it first with hot water. On the contrary, the finding of oil involves a more direct application of geology than ordinary metal mining, and the exploitation of an oil-bearing sand entails the application of engineering in its simplest and most direct forms, as we shall see in the interview with a notable exponent of this branch of technique. All those whose careers have been reviewed recently in these columns happen to be Americans of British descent, that is, they connect themselves immediately or remotely with English, Scottish, or Irish ancestors; and only one was not born in the United States. Our victim this time is Anthony F. Lucas, a Dalmatian by birth, a Montenegrin by origin, an Austrian by forced adoption, and an American by choice. Starting as a cadet in the Austrian service, he happened to visit this country and became interested in the mechanical engineering of a lumber camp. Soon after arrival he felt the pull of this great democracy and decided not to return to his Adriatic home, which he re-visited only once, on the occasion of his honeymoon. Whatever the peccadillo that caused him to leave the Austrian navy it is clear that he was not afraid to renew acquaintance with his former compatriots. He suggests that his immunity from arrest was due to the presence of his bride, and from what we know we can well believe that this charming and accomplished Texan lady completely disarmed any official that might have thought of making trouble. After some experience of metal mining in Colorado and elsewhere Captain Lucas went to Louisiana to explore for salt. He describes the chamber-and-pillar system of mining adopted by him and explains how it is necessary to use brine in drilling for salt, in order not to enlarge the bore, as would happen if fresh water were used. He found an enormous mass of salt for the celebrated actor, Joseph Jefferson, who took a philosophic or grandpaternal view of the discovery, deciding to leave it for the benefit of his descendants. The salt was 1750 feet thick. Here we may mention the fact that scientific men are not yet agreed as to the origin of the masses of salt found in Louisiana. A post-Cretaceous movement of a deep-seated magma produced results not fully understood, because the igneous rock did not reach the surface save perhaps at one point on the Gulf Coast. It is supposed that the heated solutions ascended through the disturbed strata, which contained salt and gypsum. The solutions dissolved the salt. The force of crystallization enlarged the openings and the accretion of the saline deposits produced a doming of the strata. The final salt deposit is not a horizontal bed, neither is it a simple lagoon sediment or precipitate. Captain Lucas published a paper on his exploratory work at Belle Isle in the September 1917 bulletin of the American Institute of Mining Engineers. Reference may also be made to an earlier paper by him on 'Rock Salt in Louisiana' in the transactions of the Institute for 1899. At Belle Isle, on the shore of the Gulf of Mexico, he discovered not salt only, but sulphur and oil as well. Indeed, the subsequent exploiters of this deposit found that the salt was so impregnated with oil as to be unmarketable. The association of these three mineral products—salt, sulphur, and oil—is most interesting to the geologist, but space will not permit us to linger on it. We must hasten to the big event. Captain Lucas had now been prospecting and geologizing along the Gulf Coast for five years; going inland, and exploring the coastal plain, he detected a mound at Beaumont in Texas. This mound indicated a dome, and the escape of sulphurous gas suggested that either sulphur or oil would be found underneath. So he leased the ground and started to drill, using a rotary tool, and therewith going safely through a layer of quicksand. At 575 feet he penetrated oil-sand, but almost immediately lost the well by collapse due to gas. He had saved a sample of the oil and with that sample set forth on a pilgrimage to obtain financial assistance. He tells us how he interested several well-known men and how others came to see him at Beaumont. He had spent all his money and was facing disaster at the moment when he thought he had justified his theory of dome structure. Experts of the Geological Survey and of the Standard Oil Company came to pat him on the back and dissuade him from so visionary an enterprise. He persisted, and at last obtained financial aid from Mr. J. M. Guffey. After spending $6000 more he struck oil and no mistake, a flood of oil that gushed forth with such violence as to hurl 1100 feet of casing and the head-blocks of the derrick out of sight. The oil gushed at the rate of 100,000 barrels per day; it made a 30-acre lake that overflowed to the sea, endangering both railroads and shipping. Thus the man that a few days before was facing bankruptcy now saw his wealth running to waste down the bayou. Then came the supreme test of engineering: to curb and control the flow of oil. Captain Lucas accomplished the feat by means of an ingenious device, which he describes, and thereby avoided the necessity for paying $10,000 to the woman that sent him a 90-cent collect telegram offering to do it by aid of absent treatment. It must have been more pleasant to receive the congratulatory message from the eminent oil expert, but that gentleman did not liquidate his obligation thereby, for the discouraging advices of sundry 'savants'—that is what the daily press calls them, we believe—caused him, as he says, to "sell his birthright for a mess of pottage." The Lucas, or Spindle Top, gusher has produced over 50 million barrels of oil without giving a proportionate reward to the discoverer; but he had the satisfaction of proving his geological hypothesis, and later he used his experience and reputation to get into some better-paying enterprises, and so, it is pleasant to record, he did not go unrecompensed. That recompense was not all in standard coin, for he acquired two things not measurable in common units; he gained fame and friends. His ideas and his work won him many friendships among geologists and engineers, who appreciate, among other things, the outstanding fact that he never succumbed to the importunities and blandishments of those that wanted to use his name for oil promotions. A big man physically, Captain Lucas has a big warm heart, spontaneously sincere and generous. Whole-hearted in all he does, he is enthusiastic, even impetuous. Just now he is so keenly interested in the War and in the success of the great cause for which the Allies are battling that he will endure no half-hearted patriotism or tepid sympathy for the wrongs of Belgium or Servia, for example. The young Dalmatian that rejected the Austrian service to become an American has been splendidly consistent, not only in becoming a supremely useful citizen of the United States, but in sealing his devotion to the flag of his adoption by giving a fine upstanding son to the American army now fighting under General Pershing in France. The younger Anthony sets the seal on his father's Americanization.

SPINDLE TOP, TEXAS, IN 1902

Anthony F. Lucas, and the Beaumont Gusher

AN INTERVIEW. BY T. A. RICKARD

Capt. Lucas, you are of European origin?

Yes, Sir, I was born in Dalmatia, Austria, in 1855. My forefathers, however, were of pure Montenegrin blood.

What was your father's occupation?

He was a ship-builder and ship-owner on the island of Lesina.

So you had an interest early in engineering. What was your education?

My family moved to Trieste, where I was educated in the high-school and after that in the Polytechnic of Gratz; at the age of 20 I was graduated as an engineer.

After graduating, what did you do?

I entered the Austrian navy as a midshipman and was promoted to second lieutenant. At that time an unpleasant incident made me very much dissatisfied with the rigor of the service, perhaps because of my Slav origin, so that I was glad to accept an invitation to pay a visit to an uncle of mine in this country. For that purpose I obtained a six months' leave of absence and came to the United States. That was in 1879, when I was 24 years old.

I have an idea that your father's name was not Lucas?

My father's name was Luchieh. The reason of the change was that when I came to America on a visit to my uncle, my father's brother, as I have stated, I found that my uncle had adopted the name of Lucas, owing to the difficulty that Americans had in spelling and pronouncing Luchieh. So, for the time, expecting to remain only three or four months, I permitted myself to be addressed as Lucas. When I decided to reside in the United States I retained this modification of the name.

What happened to make you remain?

An engineering problem was offered to me during my stay in Michigan. At that time Michigan was a lumber country, and Saginaw, where I resided, was its centre. A gang-saw was needing some improvement in design, and when asked if I could do it I agreed with pleasure, completing the design satisfactorily to the mill-men. Later on I was asked to supervise the erection of this gang-saw. Then, just as the end of my leave was approaching, I was offered a flattering engagement, which, after some consideration, I accepted, asking, however, for another six months' leave. At the end of that time I had made up my mind to become an American citizen; I made proper application and completed the change of allegiance by obtaining my final papers four years later at Norfolk, Virginia, on the 9th of May, 1885.

Did you ever return to Dalmatia?

Yes, in 1887, nearly ten years after, on my wedding

journey. My wife and I went to Trieste, Fiume, and Pola, and to the place where I was born, Spalatro, which, by the way, is an old Roman city erected by the emperor Diocletian. Although I had some fear of contact with the rigor of the Austrian law, I was not troubled. On the contrary, I was entertained at Pola by the officers of the Navy and ascribe my immunity to having good friends, but perhaps it was Mrs. Lucas' charm of manner or because she was unmistakably American. I was abroad one year.

On your return from your wedding journey, where did you settle?

I made Washington my home, and entered the profession of mechanical and mining engineering. First I went to the San Juan region in Colorado and prospected for gold, with some little success, but after two years I returned and began to look around for a good opening in the mining industry, and became employed as mining engineer at a salt mine in Louisiana, in 1893, at Petit Anse, where I practised the engineering and mining of salt for three years.

What was the nature of the exploitation?

I found the salt deposit only 20 ft. below the drift-soil, and the shaft 180 ft. deep. Mine and mill were in very bad condition owing to the fact that water had found its way into the mine and caused a large cave. The mill was antiquated. It required constant care to check the caving and water in the mine, and the ravages of the salt on the mill machinery.

Was there any other technical feature worth mentioning?

Yes, the system of mining. I opened long drifts in virgin ground, adopting an overhead method of mining, feasible only under such conditions. A gallery was started with a 7-ft. under-cut, 50 to 60 ft. wide, and from 200 to 300 ft. long. After clearing away this broken salt in tram-cars the second under-cut was started, 18 to 20 ft. in height, and when this was cleared the final mining was begun with the aid of threepod ladders upon which light hand-drills were placed, drilling batteries of holes 10 ft. deep. Thus six or eight holes shot down hundreds of tons of salt. This method proceeded until, by the time the height reached 50 ft., no more ladders were needed, owing to the increased volume of the salt. The roof, of course, was arched toward the 40-ft. pillars left standing. Suspiciously loose slabs of salt were pried down or shot down to make the roof safe, and by the time the men had finished there remained a mass of from 3000 to 5000 tons of mined salt ready to be trammed and hoisted to the mill. By this method not a board or stick of timber was used or needed.

When the mining in this chamber was completed another under-cut was started laterally and we thus had always in reserve one or more chambers of mined salt, say, 60 ft. wide, 60 ft. high, and 200 or 300 ft. long, each containing from 3000 to 5000 tons, at a cost of less than 14 cents per ton of salt mined. The salt was of unusual purity, 98.5 to 99% sodium chloride, the remainder being gypsum.

What did you do next?

During the three years that I was employed there I became acquainted with Joseph Jefferson, the famous actor, who had an island known by his name, a few miles off Petit Anse. Mr. Jefferson was anxious to bore for water, and had actually given a contract, but the contractor found difficulties in complying with the terms of his agreement, owing to the boulders and sand through which he had to drill. At this time Mr. Jefferson asked me if I could see my way to help the contractor. Although employed by the salt people, I accepted eagerly, and helped by introducing a method of driving the casing and succeeded in assisting the contractor to pass through the gravel bed. About 100 ft. deeper we struck what appeared to be solid rock, but upon analysis, it proved to be an enormous bed of salt. Then Mr. Jefferson asked me if I would continue the exploration in an advisory capacity, which I did. I purchased a diamond-drill from the Sullivan Machinery Co., of Chicago, and drilled to the depth of 2100 ft., still in salt, a total thickness of 1750 ft. of salt, without encountering any foreign substance.

What did you do with this discovery?

During my operations at Jefferson island rumors were spread by uncharitable persons that I was fooling Mr. Jefferson. I was asked, if I claimed I had so much salt, why did I haul carloads of salt from Petit Anse to Jefferson island. It is true I hauled several carloads of waste salt, but only to use as brine with which to bore, so that it would not enlarge the bore excessively. You will understand that if I used fresh water I would have dissolved the walls of the bore. Owing to this malicious gossip, Mr. Jefferson asked me if I had found enough salt, as he wanted to stop. I replied that I had salt enough to salt the earth; I was proceeding nicely, and was anxious to find the floor of the salt, when he stopped me at 2100 ft., thus balking a possible study in geology, for I wanted to learn on which geological formation this salt was resting.

That must be an enormous mass of salt, do you know of any larger?

The salt mines at Wielitzka, in Austrian Poland, are worked at over 3000 ft. in depth, and the potash salt mines in the Stassfurt district of Germany are worked to over 1500 ft. in depth, but the salt deposits of our Coastal Plain now surpass both of those in quantity and extent.

Probably you had an idea of mining from that floor upward?

No, because I found on subsequent borings that the salt mass was in one locality only 81 ft. from the surface, and to mine at a depth below 2100 ft. would have been costly and unnecessary when one could mine at a reasonable depth, say, 500 ft., and leave between 200 to 400 ft. of solid salt for a roof. Mr. Jefferson, however, decided to leave this deposit to his grandchildren, and I heard only recently that it has been sold.

So, Captain, you had to make another start?

Yes; in the meantime I had examined a small island known as Belle Isle, on the Gulf coast, one of a series of five islands now well-known for their salt deposits, and contracted with the owner to explore Belle Isle for minerals on my own account. I undertook to explore the land by perforations. Four wells were drilled for salt, discovering thereby a deposit not only of salt, but of sulphur and oil as well.

CRYSTAL OF SALT SHOWING INCLUSION OF OIL AND GAS

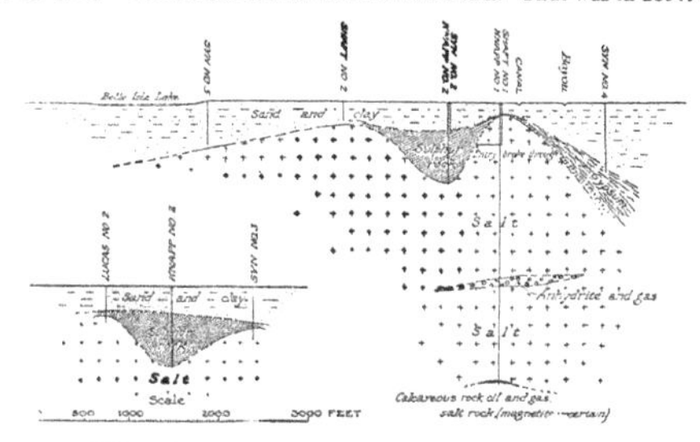

CROSS-SECTION OF RECENT WELLS, SHOWING THE POCKET OF SULPHUR, AT BELLE ISLE

How were these three minerals related in space?

The first well was a miss; the second well penetrated a 66-ft. bed of sulphur, and below that I discovered the matrix of a salt dome. By further boring I encountered oil-sand at a depth of 115 ft., and deeper down, at about 800 ft., I discovered a strong flow of petroleum gas. Resting there, I completed my contract, and acquired title to one-half of the mineral resources of the island.

You then proceeded to exploit the oil?

I did not have the money. The island was purchased later by the American Salt Co. and I received a consideration of $30,000 in bonds and $5000 in cash. This led me to study the accumulation of oil around salt masses, and I formed additional plans for prospecting other localities. Thus I began my investigations into the occurrence of oil on the Coastal Plain. That was in 1897. I fully expected that the American Salt Co. would put me in charge of their operations, but they did not make me the offer and I did not make any request. They sent a New York salt man and began large operations, involving over two million dollars.

Which of the three minerals did they exploit?

Salt only. Unfortunately, the method of mining salt in Wyoming county, New York, was not adapted to the deposits of the Coastal Plain. They started sinking a large working shaft where it showed nearest to the surface, at 115 ft., without having previously sounded the

ROOF OF CHAMBER IN THE WEEKS SALT MINE

deposit by boring to ascertain its conformation and purity. Salt is much in demand from the meat-packers of the interior, but this particular salt was unmarketable because it was impregnated with oil and gas. In bailing out the salt-cuttings and dumping them on the floor, they would explode like pop-corn. This was caused by the sudden liberation of gas coming in contact with the air. In my exploration of the various salt deposits of the Coastal Plain this was my only experience of the kind.

What did the contractors do?

They sank to a depth of 250 ft. and started driving toward the interior of the deposit, searching for purer salt. When I heard what they were doing, I telegraphed at once asking if the drift was in the right direction, and if they had sounded with a diamond-drill in that direction. Unfortunately, the next news was that they had passed through the salt and had penetrated quicksand. The mire of the marshes drove them out and they barely had time to save the men, thus losing the first shaft. It proved afterward that the salt in this locality formed a depression, 500 ft. westward, although it was connected with the main dome. They started another shaft westward, but made a series of woeful blunders. The salt did not occur so shallow there and they had to go down 276 ft. before they reached it. On top of this salt they found a layer of about 30 ft. of bad quicksand through which they could not pass. They then employed an expert shaft-sinker, named Sooy-Smith, who employed a freezing process. Another unfortunate incident was that Sooy-Smith never put the brine-pipes in the salt; he stopped in this quicksand, so that when the mass was frozen, preparatory to mining, they were just as badly off and could not pass the quicksand. They spent a large sum of money, finally abandoning the effort. They then proceeded to explore for oil, and after a series of efforts gave that up. The island was sold at public auction and is now the property of the New Orleans Mining Corporation.

That did not stop your prospecting, I feel sure. What did you do?

I explored, meantime, Weeks island, also in Louisiana, and there discovered a magnificent bed of salt now being worked with commercial success. Meanwhile further investigation led me to proceed. I selected a point known as Anse la Butte, six miles north of Lafayette, in Louisiana. In this locality I again discovered oil and salt, but not under favorable conditions. (Since that time, however, it has produced and is still producing large quantities of oil.) So I abandoned the discovery and went to Beaumont, Texas, about 70 miles west of Lafayette. There I was attracted by an elevation, then known locally as Big Hill, although this hill amounted merely to a mound rising only 12 ft. above the level of the prairie.

What led you to prospect there?

This mound attracted my attention on account of the contour, which indicated possibilities for an incipient dome below, and because at the apex of it there were exudations of sulphuretted hydrogen gas. This gas suggested to me, in the light of my experience at Belle Isle, that it might prove a source of either sulphur or oil, or both. I decided to test it therefore and leased all the ground that I could secure.

How many acres?

The hillock only covered 300 acres, of which I secured 220 acres; but I leased altogether about 27,000 acres in the vicinity in order to have ample scope for exploration, although this proved unnecessary, as no oil was ever found beyond the contour of the dome.

What was your first result?

This elevation had already been explored by three companies and none succeeded in penetrating below 250 ft. in depth.

Why?

Because a bed of quicksand was struck at about 200 ft. Knowing that they used cable-drilling apparatus, I decided that that must have been the reason for their failure, so I set to work with rotary-drilling tools. The rotary drill at that time was almost unknown and was only used for artesian-water wells of shallow depth on ranches and rice plantations. I penetrated the quicksand and soon realized that I was correct in my surmise of the reason why my predecessors had failed. I managed to pass the quicksand and bored to a depth of 575 ft., encountering an oil-sand but losing the well by gas collapse. I thought best, however, before proceeding with heavier rotary-drilling machinery to seek geological and financial aid, so I went to a number of capitalists and laid before them my plans and expectations; but they turned me down. I recall one instance, when a friend took me to see former Congressman Sibley, of Pennsylvania, to lay before him my project. He read me a lecture and stated that he could not participate in such a wild scheme; that unless I had a production of so many thousand barrels per day to count upon, he was obliged to decline; to which I replied, that if I had such a production I probably would not have come to see him.

You did not abandon your quest?

No, I went to others, for instance to H. C. Folger, Jr., of the Standard Oil Co., in February 1899. I had a bottle of the oil with me and it happened to be a bitter cold night in New York. Having my ideas on practical issues, I put the bottle out of my window in order to give it a cold test. The oil was 17° B. gravity; it did not seem to congeal at all, although I understood that the temperature outdoors was two or three degrees below zero. Happy in this test, I went to see Mr. Folger, and laid the matter before him, asking him to join me. I told him that I did not want money personally, only assistance in further prospecting and proving the field; but Mr. Folger, while he received me graciously, declined. However, he promised to send Call Paine, of Titusville, the then Standard Oil expert, to examine my scheme. Paine arrived a month later, with J. S. Cullinan, who later became President of the Texas Company. I showed them the location of my first shallow well and the heavy oil extracted therefrom. I also explained to them my 'nascent dome' theory. It ended, however, with Mr. Paine giving me a piece of well-meant advice, to wit, that there was no indication whatever to warrant the expectation of an oilfield on the prairies of south-eastern Texas, that he had been in Russia, Borneo, Sumatra, and Rumania, and in every oilfield of the United States, and that the indications I had shown him there had no analogy to any oilfield known to him, that I absolutely had no chance, as in fact there was not the slightest trace of even an oil-escape; in conclusion, he advised me to go back to my profession of mining engineering.

DAMON MOUND, TEXAS

That was a squelcher; did it crush you?

I retorted by showing him a demijohn of the heavy 17° B. oil, obtained at 575 ft. He characterized it as of no value and of no importance whatever, stating further that such heavy stuff could be found most anywhere. I am, however, convinced that Mr. Paine was sincere in his advice, which naturally shook my confidence.

Another incident happened during the fall of 1899 to shatter my faith in the venture. Charles W. Hayes, formerly Chief Geologist of the U. S. Geological Survey, accompanied by Edward W. Parker, formerly Chief Statistician of the Survey, dropped in on me to see what I was doing, and I explained to Mr. Hayes in detail my deductions of possible oil accumulations around great masses of salt, etc. He also discouraged me, saying that there were no precedents for expecting to find oil on the great unconsolidated sands and clays of the Coastal Plain, pointing to the great well drilled by the city of Galveston, Texas, over 3000 ft. in depth (a big undertaking at that time and one that cost the city nearly a million dollars), stating further that I had no seepages of oil, as leaders or indications, etc. I pointed to the great sulphur dome near Lake Charles, at that time in course of development by Herman Frasch, whereon a limited production of 1½ barrels of oil was obtained—a heavy oil it is true and a straw for me to grasp—but no encouragement could I get from Mr. Hayes, whom I of course knew as one of the best geologists in the country. Thereupon I began to seek co-operation and financial aid from some one better fortified than myself financially. This led me to enlist the aid of J. M. Guffey, of Pittsburgh, who took my proposition up, but to whom I was obliged to relinquish the larger part of my interest in the venture. In the first well that I drilled with Mr. Guffey's money, at a cost of less than $6000, at a depth of about 1100 ft., the oil came with a rush of over 100,000 bbl. per day. This was on the 10th day of January 1901. Next day I received a telegram from Mr. Paine extending to me his warmest congratulations on my success and saying that he was coming down to see the new wonder. He little knew, however, that he and Mr. Hayes were the indirect cause of my "selling my birthright for a mess of pottage."

SPINDLE TOP, TEXAS, EARLY IN 1902

Mr. Cullinan also appeared next day and became a large factor in the Spindle Top field, realizing millions thereby, and enabling him later to organize the Texas Company, a corporation now worth its hundred millions of dollars. Mr. Guffey, with Andrew Mellon, of Pittsburgh, organized the J. M. Guffey Petroleum Co., with a capital of $15,000. This capital was later largely increased, and later still the title of the company was changed to the Gulf Petroleum Company.

So Guffey became associated with you?

Yes, he was a noted petroleum operator in Pennsylvania. I entered into a contract with him that he should drill three wells, at least 1200 ft. deep, under my direction.

Did he undertake to supply the capital?

He undertook to stand the expense of three wells, and I was to superintend the operations. The first well, however, proved the Lucas gusher at a cost of less than $6000.

You proceeded to work?

I contracted with Al and Jim Hamill, of Corsicana, to drill three wells at $2 per foot, we to furnish the casings. I took charge of the drilling and chose a site where the now famous Spindle Top, or Lucas, gusher, was developed.

Kindly outline the events that led up to the escape of oil?

The first casing was 12 inches, reducing the diameter finally to 6 inches. When we reached the quicksand, at about 300 ft., we began to have trouble; our drill-pipes stuck. The 6-in. pipe was relieved by going over with an 8-in. pipe, which in turn became stuck, and I, knowing that the practicable limits had been reached, was in great distress to proceed with the work. I could not sleep that night, but toward morning the thought came to me that if a boiler having a 100-lb. pressure of water could be pumped full without any water coming out, there was a reason for it, and the reason was that it had a check-valve alongside. That proved an eye-opener to me. Much excited, I hastened to see my driller and explain what I wanted. Thereupon I designed a check-valve, made out of the boards of a pine box lying in my back yard, and with openings in the centre and a small rubber belt underneath I placed this check-valve between the coupling of the casings, and thus proceeded without any further trouble.

Did you patent your invention?

No, at the time I had it in mind to do so but was deluged with work and incipient possibilities. If I had done so I would have realized a large amount of money.

How did your drilling proceed by aid of the check-valve?

The drill-hole proceeded through various layers of quicksand to 800 ft., then struck lime and sulphur for about 250 ft., to a few feet less than 1100, and while we were lowering the casing, after having sharpened the bit, the casing (which was attached with a block and five strands of 2-in. cable) began to rise, increasing in momentum until the whole casing of about 1100 ft. was shot out to an unknown height, carrying the heavy block and the head blocks of the derrick with it into the air, followed by a gush of muddy water, our own drilling-water, then by rocks and fossils, and finally gas. It then settled down to a magnificent 6-in. stream of solid oil of 23° B., rising to a height of 200 ft. That was the celebrated gusher, which ran wild for 10 days, making a lake of nearly 30 acres, which was impounded by levees that I built to confine it.

Were you charged with damages?

I was served notice by the railroad company and by a marine insurance company at Port Arthur and Sabine Pass that ships were endangered, as a great deal of oil had found its way to these seaports, 16 miles below on the coast, and requesting me to remove the oil. The Southern Pacific railroad company also served notice on me to remove the oil because of the danger to the tracks, etc.

How did you control the flow of oil?

During the ten days that preceded the closing of the well, I had innumerable offers from irresponsible people to close it for me, because some newspaper had published a statement that I would give $10,000 to any one who would close the well. One of the offers was made to me by a woman in Illinois. She telegraphed that if I would put the $10,000 in a bank, subject to her order, she would use her occult power to discipline nature. However, I realized that it was up to me and my men to solve the problem. In 10 days we completed the construction of a steel-rail carriage to pass an 8-in. gate-valve over the 6-in. stream of oil. At 10 o'clock in the morning of the tenth day, with the aid of block and tackle, I started to drive and drag this valve, which was woven with the iron rails, to enter this powerful stream. At the first impact when this gate-valve, which was open, came in contact with the stream of oil, the derrick, badly shaken, but still holding, began to rock, and I remember picking up one of the fossils on the ground to throw at the horses so as to urge them to faster speed, until finally the valve passed the stream of oil, which was turned inside of it. Then we screwed the valve on the 8-in. casing, and that closed the gusher. This discovery of oil led to the development of a big oilfield, which became known as Spindle Top. It has produced over 50,000,000 bbl. of oil, and is still producing.

I hope, Captain, that you received a proper financial reward this time?

I did, but my chief reward was to have created a precedent in geology whereby the Gulf coast of the Coastal Plain has been and is now a beehive of production and industry. Owing to the fact that Mr. Guffey and the Mellon group had a lot of money and I had not, I accepted their offer and sold my interest to them for a satisfactory sum.

Did you stay there long?

I severed my connection with the Guffey Petroleum Co. about six months after the Beaumont discovery. In selling to them I retained the leases that I had acquired at High Island, near Galveston, an entirely different locality, 70 miles south-west of Spindle Top. I also secured land and drilled a well for the Guffey Petroleum Co. on another dome then known as Bryan Heights, on the Gulf, some 40 miles south-west of Galveston.

What success did you have?

In July, 1901, at 800 ft. deep, I struck a powerful force of sulphuretted hydrogen gas, which drove my men and everybody else off the location. Unfortunately, the company neglected that discovery and it became forfeited.

BURNING OF A MILLION BARRELS OF OIL AT SPINDLE TOP, TEXAS

in the next three years; subsequently it was taken up by speculators, who induced Eric L. Swenson, one of the directors of the National City Bank of New York, to proceed on the basis of positive knowledge of a sulphur bed existing there. Mr. Swenson organized the Freeport Sulphur Co. and is now exploiting it by aid of the method applied by Herman Frasch at a similar sulphur deposit, in Louisiana, by the process of melting the sulphur with hot water and forcing it to the surface by the pressure of hot air. This, together with the operations of the Union Sulphur Co., is one of the main sources of sulphur supply in the United States, and is proving most useful as an ingredient of explosives for warfare. Another well worth mentioning was located by me for the Guffey Petroleum Co., in 1901, on a tract of land known as the Damon Mound in Brazoria county, Texas, 30 to 40 miles from Houston.

What did you do there?

In order to obtain leases on this dome, I had to promise the owner, J. F. Herndon, that this well should be called by his name. The well, while it did not prove a failure, for the reason that it became clogged and ruined at 1600 ft., enabled us to ascertain the existence of a bed of sulphur and oil. This dome is now producing between 5000 and 10,000 bbl. per well per day.

So you severed your connection with the Guffey company?

CAPT. LUCAS DRILLING ON JEFFERSON ISLAND IN 1896. HE IS SHOWN SITTING, EXAMINING A CORE JUST REMOVED FROM THE DRILL.

Yes, I had no more interest in their affair. This promising field became abandoned until about two years ago when E. F. Sims, of New York, began to prospect the dome, which is over 100 ft. high and one of the largest on the Coastal Plain. The result was to bring in gushers that are now yielding 10,000 bbl. per well. On another part of the dome there was exploited a great sulphur deposit, and deeper down an enormous mass of rock-salt to unknown depth. I got nothing out of this, except that I had bought two tracts of land at the time I was drilling, and two additional tracts of land that were presented to me by the Cave heirs of Paducah, Kentucky. These two additional tracts of land of 10 acres were presented to me with the compliments of the heirs, but, not obtaining satisfactory results, I deeded them back. These tracts of land are now worth a considerable amount of money, as they are producing large quantities of oil, but I retain the two originally purchased tracts, covering 17 acres. The land is now very valuable.

How many years did you spend in this region?

About three years. After that I became connected with Sir Wheetman Pearson, now Lord Cowdray, who, as you know, has large oil interests in southern Mexico.

Please say something about your Mexican experience in oil?

In 1902 I went to Coatzacoalcos, now known as Puerto Mexico, where I located two fields, one known as the San Cristobal, and the other as Jaltipam, on the Tehuantepec railroad. The former is producing a beautiful light paraffine oil; the latter, a somewhat sluggish heavy oil; both, however, are attractive oilfields.

You were advisory engineer to Sir Wheetman?

Yes, and three years after my arrival in Mexico he made me a flattering offer to remain as managing engineer in Mexico, but, owing to other aims and interests, I declined, returning to Washington in 1905. Since then I have been practising the profession of consulting engineer. My work has taken me to various parts of the world, notably Algeria, North Africa, Russia, Rumania, Galicia, and Stassfurt, besides various oilfields in the United States.

Would you say something concerning the oil resources of the Tampico oilfield, which, I understand, today is so important a factor in supplying the Allies with fuel-oil?

In 1903, E. L. Doheny was operating in the neighborhood of Tampico, and his success induced Pearson to obtain large concessions there, but I had no part in it. I know that the production of the Tampico field is very large, and it contains large resources of petroleum, which is now supplied to ourselves and our allied fleet.

Have you been to any other Mexican oilfield?

Yes, through the instrumentality of Sir Wheetman, I went by way of Oaxaca in 1903 to the Pacific coast, where the son of President Diaz and other capitalists had been operating for oil, near Port Angel in the State of Vera Cruz. There were no safe ports there and they were obliged to beach barges of supplies towed from San Francisco, from which place they unloaded the machinery and pipes. I found them laboring under great difficulties, for the reason that they were drilling in pure syenite, which, as you know, is a crystalline rock in which engineers in general would not expect to find oil, although there were some oozes of neutral oil on the surface. This field has never produced oil. I discouraged the continuation of this work, but owing to the fact that they had already invested a large amount of money, they did not do it gracefully, and continued for another year and a half, until ultimately they had to give it up. Here I contracted dysentery, which kept me nearly a year confined to my home.

You have not been to any of the Californian oilfields?

No. The conditions there were not such as interested me, compared with the Coastal Plain in Louisiana and Texas.

By the way, Capt. Lucas, you have a son?

Yes, I have a son, born two years after my marriage. He is now a lieutenant with General Pershing's army in France.

"IN studying a map of Southern ore deposits with relation to the placing of a nitrate plant it became evident that pyrite was to be found in a stretch of the mountains running from northern Georgia to central Alabama, and just when this was found, there came into the office one of the most forceful of southern manufacturers, who entered with a statement that he was looking for a place—"not under the spotlight; I'm not a prima donna; just a man's job; something somebody else would shy at."

"Why not find the pyrite ore in your southern hills?" I asked.

"Never heard of the stuff, but if it's there, and you say we need it for the War I'll get it."

That was almost literally the conversation that has led to the opening of five mines yielding 400 tons per day, which it is promised before the winter is over will be increased to a thousand tons per day; and 30,000 tons per month is more than 15 good ships could bring from Spain to our coast if kept in a continuous circle.—Annual Report, Secretary of the Interior.

PLATINUM is the object of energetic prospecting, on account of its extraordinary price, and every district from which it had been reported in the past is again being combed. The public is also being invited to buy shares in numerous enterprises that have no larger foundation than the exuberant hopes and misleading promises of promoters. The U. S. Geological Survey has recently deemed it necessary to sound a warning against the schemes of wild-cat companies that propose to exploit the gravels of the Adirondack mountains in New York. The paltry gold content and the feeble traces of platinum in these sands have led to repeated disastrous attempts at placer-mining.

THE TOWN OF GOLD HILL, IN THE DEEP CREEK MINING DISTRICT

Mining in Utah

By BENJAMIN F. TIBBY

COAL. Operators report that there is a serious shortage of cars. Practically all the coal mines in the State are, at the present time, served by the Denver & Rio Grande system. The lack of cars to move coal is said to be due to the fact that the Rio Grande railroad is congested with through-freight and the physical condition of the railroad is such that unless this class of business is materially reduced the shortage of cars will continue. The independent coal-operators have made complaint to Washington regarding the situation and a member of the Federal Trade Commission was recently sent to Utah to investigate. It has been urged that the Government issue orders compelling the Rio Grande to divert freight received from the Western Pacific in favor of other railroads. At the present time, with one exception, the coal mines are working only two shifts per week. The exception is the Sunnyside, which is the only producer of coke in the State and for that reason is kept in continuous operation. Many of the metal mines are short of fuel; in several instances the railroads have commandeered coal consigned and already being transported to the mines. The coal-miners were recently granted an increase in wages that will give practically every employee an additional $1.40 per day. A census shows that in Salt Lake City last week there were 1500 families without coal and 3000 families with one ton or less. The same condition prevails in other parts of the State. All retail coal-dealers are without stock-piles and only the favorable weather has prevented extreme suffering. A campaign for the conservation of coal is being waged.

At the Peerless Coal Co.'s mines, near Castle Gate, in Carbon county, material is on the ground for the erection of a Wood rotary-dump tipple with electrically operated shake-screens. The mine is near the Denver & Rio Grande Western railroad and one mile from the Utah Coal railroad. The Utah Construction Co. and the Wasatch Grading Co. are grading for yard-facilities and for a 3700-ft. gravity-tramway from the main mine-entry to the railroad. The company owns 660 acres of land. There are five coal seams exposed, three of which are workable. Operations are to be confined to the No. 3 Castle Gate seam and the No. 4 seam. Each of these is 12 ft. thick. The coal seams dip at 7% but mining will be conducted in such a manner that there will be a 1% grade in favor of haulage. By January it is expected that the mine will be capable of producing 500 tons of coal per day.

OIL. Interest in oil is particularly keen at this time. It is reported that options are being sought on land at the north end of the Great Salt lake where oil-seepages with heavy asphaltum base have been known for some time. Conditions theoretically favorable for oil exist in the Henry mountains, and drilling is being started in this region. More attention, however, is being given to oil-shale. There are practically no oil-wells of commercial importance in the State, but there is a vast area of oil-bearing shale. The U. S. Geological Survey has estimated that Utah has 20,000,000,000 barrels of crude oil tied up in its shale, from which 2,000,000,000 barrels of gasoline may be extracted by ordinary methods of refining. It is said that there are 1400 square miles of oil-shale land in Utah. Numerous new oil-shale companies have sprung up and the laboratories at the University of Utah have been busy making distillation tests.

METAL MINES have been having a full measure of grief. The recent rapid decline in the price of lead has greatly curtailed output and profit from the lead mines. This, together with the smelter and car embargo, has produced a serious situation. A great many of the small operators have been forced to suspend work entirely while the big producers have cut their working forces materially. However, the mines throughout the State have had a period of exceptional prosperity due to the high metal-prices during the earlier months of the year. New development was greatly stimulated, as is shown by the unusually large number of small producers who continue to ship even under present market conditions. After the harvest a greater number of cars will be available for ore-shipments and except for the question of metal prices, the situation is rapidly righting itself.

The smelters have lifted the embargo and are accepting an increased tonnage from week to week. Shipments from Park City totalled 2930 tons for the week ended November 30 as compared with 2721 tons the week before. Shipments from Tintic for the past week aggregated 183 cars as compared with 230 cars for the week previous and 219 cars for the week before that.

Silver is essentially a by-product in the Utah mines, no strictly silver and but few gold-silver mines being in operation. For that reason the uncertainty of silver prices is not so vital, although in the case of the lead producers it is decidedly important, especially in view of the present lead quotations. A meeting has been called for December 4 by the Salt Lake chapter of the American Mining Congress. Representatives of the mining industry from all the States west of the Rocky Mountains, except California, are to attend. Measures are to be adopted for bringing before Secretary McAdoo the necessity for sustaining the silver market by urging Government regulation of price at some figure above the present quotation of 85 cents. The uncertain interpretation of the War Excess Profits Tax, as applied to mines, will also be considered.

There have been a number of important developments in the American Fork district. The Pacific mill is handling 200 tons of ore per day and is shipping a 30-ton car of concentrate every third day. Crude ore is also being mined and sent directly to the smelter. The Miller mine, under lease to George Ryan, has opened a body of galena ore 160 ft. below the stope from which the Tyng brothers shipped over $750,000 worth of ore several years ago. The new strike is just above the Lady Annie tunnel and occurs along a well-defined fissure striking NE-SW, which is characteristic of the mineralizing fissures of the district. The ore is galena carrying 25 to 47 oz. silver and 30 to 60% lead. The high-grade streak is from 18 to 24 in. wide. The lower-grade ore appears to be an impregnation of the blocky shale by the galena and occurs on both sides of the fissure. There is little oxidized material in the vein, indicating that below the limestone there has been but little alteration of the primary sulphides. The ore occurs about 250 ft. above the zone in the quartzite that has proved to be heavily mineralized in the Pacific and points to a continuation of the Miller fissures into the underlying quartzite. Work is being carried on at the Whirlwind, Beddmeyer & Borussia, Yankee, Bay State, Globe, Earl Eagle, Belorophan, Bog, Silver Lake, and South Park mines. Small strikes have been reported from several of these properties and nearly all are making irregular shipments. A late fall has helped materially in carrying on new work.

Encouraging reports continue to come in from the Deep Creek region and Tooele county. During the month of September 117 cars of ore were shipped over the Deep Creek railroad from the Gold Hill district. Of this amount the Western Utah Copper Co. shipped 112 cars, the Pole Star 3 cars, the Copperopolis 1 car, and the Garrison Monster 1 car. During the month of October 91 cars were shipped, 85 cars coming from the Western Utah Copper Co., 2 cars from the West Mountains Development Co., 1 car from the Monacca Mines Co., 1 car from the Seminole, 1 car from the Garrison Monster, and 1 car from the Pole Star. The reduced output for the month of October was due to the smelter embargo. At the Pole Star some high-grade copper ore has been uncovered in driving the No. 1 tunnel. The same company will sink its 150-ft. shaft to the 500-ft. level. At the 150-ft. level drifts are being extended both north and west. A strike of heavy sulphide copper ore has been made in the Western Utah extension. The Seminole company has recently made some shipments of ore containing rare metals.

The Emma Consolidated at Alta is shipping from 60 to 70 tons per day. October shipments averaged 65 tons per day, the ore netting $30 per ton after deducting smelter and railroad charges. The earnings for the month were approximately $60,000. Since shipments started in April about 8000 tons has been sent to the smelter. Snowfall at Alta is unusually light, being only a few inches against the normal amount of two to three feet. The Michigan Utah at Alta is shipping about 60 tons of ore daily. Shipments during October totalled 49 cars. Transportation difficulties in Little Cottonwood canyon prevented the moving of a larger tonnage. It is reported that the roads are in bad shape. The Michigan Utah has been a heavy shipper during the past month, 600 tons having been sent to the smelter in one week. The drift south on the Lavinia has been driven 100 ft. in ore and there is 5 ft. of ore in the breast extending both above and below the tunnel-level. The tramway has been put in condition for winter work. The Columbus-Rexall is shipping 35 tons of ore per day. The ore is being hauled to the sampler by team. Since the new orebody was cut three months ago there has been mined and shipped about 1900 tons of ore netting approximately $30,000. The adit of the Alta Tunnel & Transportation Co. in Silver Fork is now in 3000 ft. and work will be carried on throughout the winter. The Prince of Wales and Woodlawn properties in Honeycomb gulch are also laying in supplies so that work can be carried on through the winter. The Alta Consolidated is now hauling ore by team from Alta to the valley smelters. An ore-bin and chutes are about completed and a connection is to be made with the Michigan Utah tramway, avoiding the wagon haul to Tanner's Flat. The Alta railroad has been completed and additional equipment, including a Shay locomotive and 50 six-ton cars, has been ordered. The railroad is narrow-gauge with exceedingly heavy grades and sharp curves. The 30-ton cars originally provided were not found to be serviceable. Alta operators expect that this road will offer a substantial solution to the difficult ore-hauling problems from Little Cottonwood canyon.

The Tintic Standard at Eureka has about 80 men at work. Shipments average about $90 per ton running 70 oz. silver, 2½% copper, and 10% lead, with $6 in gold. Selected specimens show as much as 400 oz. silver and $35 in gold. Fifteen teams are hauling ore from the mine. Surveys have been made for a spur-track from the Denver & Rio Grande railroad. The Iron Blossom is sinking its No. 1 shaft below the 1900-ft. level. The only work below this level has been done with a diamond-drill. The drilling has demonstrated more favorable limestones than shown on the 1900. A number of lessees are working the older parts of the mine. On the 900-ft. level of the Eureka Mines Co.'s property the main drift has entered the east and west fissures where good ore has been found. Some additional work will be needed to determine the extent of this discovery, but the present indications are promising. Heavier shipments are now possible from the Gemini and the Ridge & Valley mines, railroad-cars being more plentiful. A large amount of ore is now available at the Gemini, and while it may be some time before the mine is permitted to move its regular output of about 500 tons per week, some of the accumulated ore can be shipped. Production from the Chief Consolidated for the quarter ended September 30 was derived from the 700, 800, 1600, and 1800-ft. levels. Shipments for the quarter were considerably curtailed on account of the smelter embargo. Men formerly employed on the ore have been put on the development work. Dump-stations are being made at the 600, 1200, and 1800-ft. levels. The Chief is rapidly completing the installation of the new electrically operated pumping-plant. Concrete reservoirs of 60,000 gal. capacity each have been constructed on the 600, 1200, and 1800-ft. levels, where the pumping units are to be placed. The water-level is at the 1800-ft. level. Important orebodies are known to extend below that point. The work which has been done indicates that no great difficulty will be experienced in handling the water. The ore from below the water-level is a sulphide product and different from anything that is found in other parts of the mine. Dur-

ENTRANCE OF THE ALTA ADIT, BIG COTTONWOOD DISTRICT

PART OF GOLD HILL, SHOWING THE OLD WOODMAN MILL AND REMAINS OF AN OLD SMELTER

ing October 2900 ft. of development was completed. A new unit to the compressor plant will be completed in a short time. The unit comprises a direct motor-driven compressor having a capacity of 2500 cu. ft. of free air per min. At the present time the Chief is using a Nordberg compressor of 1500 cu. ft. capacity and a Sullivan compressor of 750 cu. ft. capacity. The 400-ft. shaft on the Plutus, recently taken over by the Chief, has been repaired and hoisting machinery has been installed. Development work is planned on the 400-ft. level.

The Utah Copper Co. is installing an ore-crushing plant that will cost $250,000. The crusher, which will be the largest in the United States, is to be installed at the Arthur mill, and is of the gyratory type. An electrically operated Wood rotary car-dump will be used for unloading the railway-cars. The 38th quarterly report, covering the period ended September 30, has just been issued. The net profit for the quarter was $2,834,225 as compared with $10,563,541 for the second quarter. This large decrease in earnings is explained by the fact that the earnings for the third quarter were computed on the basis of 14.030c. for copper as compared with 27.977c. the previous quarter. During the third quarter there was removed a total of 1,230,677 cu. yd. of cap as compared with 1,129,198 cu. yd. for the second quarter. The Bingham and Garfield hauled an average of 32,019 tons of ore per day. During the quarter a total of 3,439,400 tons of ore was treated, being 141,000 tons more than for the preceding quarter. The average grade of the ore was 1.3059% copper as compared with 1.3881% copper for the second quarter. The average cost per pound of net copper recovered in the concentrate was 10.860c., which is an increase of 1.396c. as compared with the second quarter of 1917, when the average cost was 9.464c. This high cost was due partly to the heavy charge made against operations for increased State and Federal taxes, partly to the increase in wages, and partly to the low grade of the ore. During the month of June mining was resumed at the Utah Apex on a small scale and the working force gradually increased so that by the end of July operations were again on a normal basis. The company benefited by the high price of lead and silver during the quarter, but a sharp decline in the price of lead during the last month makes the outlook much less promising. The Utah Metal & Tunnel Co. has paid dividends to the amount of $548,140 for the year 1917. Owing to the recent slump in the lead market it has been necessary to reduce the working force.

The Spiro tunnel of the Silver King Consolidated Mining Co. at Park City is now in about 4600 ft. and is being driven at the rate of 10 ft. per day. The smelter embargoes have reduced the company's production from 50 tons to 40 tons per day. The mill is shipping a car of high-grade concentrate each week. A large amount of mill-ore is being taken from dumps accumulated in past years.

Considerable activity is reported in the Gold Strike district. Work is being done by the Baker, Gold Strike, Search Light, Gold Valley Gold, Bonanza, and others. The Baker is sinking a vertical shaft and is down 170 feet.

OZOKERITE. The American Chemical & Ozokerite Co. recently shipped ozokerite wax to the value of $12,000. The product sells for 75c. per pound at New York. The company is putting in a new crushing and milling plant, which will double the present output.

POTASH. The plant of the Minerals Products Co., at Alunite, near Marysvale, was destroyed by fire recently, entailing a loss of about $250,000. The fire was accidental, being caused by an explosion in the coal-dryer. The plant was new, having been in operation only about a year. This is the second fire within that time. About 500 men were employed and the plant was one of the largest of its kind in the United States, and has been producing about six carloads of high-grade potash each week. Re-building has started and it is announced that the plant will be re-constructed with double its former capacity.

GROWTH of chemical industry in the United States since 1914 has been phenomenal. The country now manufactures practically everything required in matters chemical. The increase in capital invested in the chemical industries was, in 1915, $65,565,000; in 1916, $99,244,000; and up to September 1917, $65,861,000 over the preceding year. New chemical industries are being started at an unprecedented rate, owing to war needs, under the energy of American chemists and physicists. Sulphuric acid, the chemical barometer, has doubled in production. In 1916, 6,250,000 tons of 50° Baumé acid was produced. The estimate for 1917 is much greater, and the production for 1918 will again greatly increase.

THE first discovery of gold in Montana, according to the U. S. Geological Survey, is accredited to a half-breed Indian, who found 'colors' in gravels near the mouth of Gold creek, a small stream that rises in the north-east corner of the Philipsburg quadrangle, in 1852, and the first workable placers discovered in Montana were found on this creek in 1862. A small quantity of gold was taken out by means of sluice-boxes near the present site of Pioneer. The deposit afterward exploited by the Hope mine was discovered in December 1864, and in 1865 Philipsburg was founded just south of the mine. Since that time the Philipsburg district has produced about $50,000,000 in gold and silver.

BY-PRODUCT coking plants doubled in capacity in the last three years, yet in 1918 the United States will make half her coke in beehive ovens. Light oil, which contains the benzene and toluene needed for explosives, jumped from 7,500,000 gal. in 1914 to 60,000,000 in 1917, and is again being largely increased. Production of ammonia has increased 100% in three years, and the visible supply is insufficient to meet the demands. Gasoline production has increased from 35,000,000 to 70,000,000 bbl. per annum since 1914.

A Flotation Patent

Herewith we give the text of Patent No. 1,246,665, dated November 13, 1917, granted to Walter A. Scott, of Chicago.

My invention is an improved flotation process. In the flotation process as previously practised, as is well known, certain reagents or modifying agents generally termed oils, although many of such substances are not oils, are added to a freely flowing ore pulp whereupon the introduction of air or some other gas into the pulp gives rise to the formation of hubles which collect the desired mineral, generally metalliferous sulfids, and reject the gangue.

The modifying agents used for this purpose comprise a great variety of oils, animal, vegetable and mineral oils, coal and wood tar and derivatives thereof, various alcohols, hydroxyl compounds and other substances, all of which I refer to herein, for the sake of brevity, as oils. Air, of course, is the most available gas and the one generally used, but in referring to air herein I include all gases which have a similar action.

Heretofore the air necessary in the operation of the process has been introduced into the ore pulp either by mechanical agitation which has the effect of beating in the air from the atmosphere, or by introducing air into the pulp through a porous medium forming part of the vessel in which the pulp is contained. Detailed explanation of the flotation process as heretofore practised is unnecessary in view of the fact that the process is widely known and described in many patents and publications, among them Patent No. 1,022,085 granted to J. M. Hyde April 2, 1912, Patent No. 1,104,755 granted to J. M. Callow July 29, 1914, and Patent No. 1,125,897 granted to J. M. Callow January 19, 1915.

In the operation described in the Hyde patent above referred to the air is beaten into the pulp by mechanical agitation, and in the operation described in the Callow patents above referred to the air is introduced under slight pressure through a porous medium, the pressure being just sufficient to over-balance the hydrostatic head of the pulp and to overcome the resistance of the porous medium. In the mechanical agitation method of introducing air the mechanical agitation imparted to the pulp for that purpose is sufficient to thoroughly disseminate the oil through the pulp, such distribution of the oil being necessary in order that it may accomplish its function. In the method of introducing air through a porous medium it is necessary that independent means for mixing the oil be adopted and such independent means have heretofore taken the form of mechanical agitation sufficient to disseminate the oil, but not for the purpose of introducing air into the pulp and not having that effect to any appreciable or useful extent, reliance being placed for the introduction of air upon forcing it through the porous medium after the mechanical mixing of the oil with the pulp.

I have ascertained that the oil performs its principal, if not its only, function in the film surrounding the air bubbles and my invention consists in introducing the oil in such manner that it is supplied directly to the bubble films without the necessity of disseminating the oil through the entire body of ore pulp. My invention consists in using a volatile oil, impregnating the air with the gasified volatile oil, and introducing the gaseous mixture of air and oil into the pulp whereupon sufficient of the oil collects in the bubble films to effect the desired

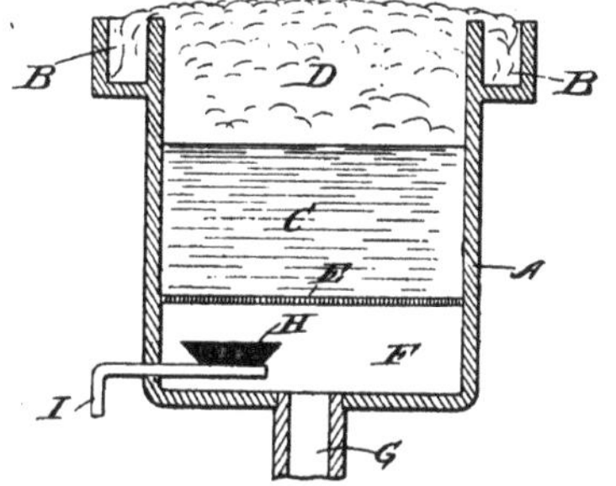

function. When operating according to my invention the necessity of mechanical agitation for the purpose of mixing the oil with the entire body of ore pulp is dispensed with and the amount of oil necessary for the purpose is greatly diminished in quantity by reason of the fact that the oil is applied directly to the bubble films where it is needed, without the necessity of distributing the oil throughout the body of the ore pulp in order to accomplish that purpose.

The drawing forming part of this application is a diagrammatic representation of means of applying my invention, and is a cross-sectional view of an apparatus designed for applying my invention by introducing the gaseous mixture of air and oil through a porous medium into the pulp.

In the figure A represents a containing vessel for the pulp, the same being provided with launders B into which the mineral bearing bubbles overflow. C represents the pulp contained in the vessel, A and D the column of mineral bearing bubbles represented as filling the upper part of the vessel A and overflowing into the launders B. E is a porous medium spaced some distance above the bottom of the vessel A, this porous medium being composed of several layers of canvas or other fabric or of some other porous medium such as unglazed porcelain, or other substance. Between the porous medium E and the bottom of the vessel A there is an air space F into which air under pressure is introduced through the pipe G. The air pressure in the space F is sufficient to support the head of pulp in the vessel A and to force the air through the pores of the medium E. H is a vessel containing some volatile oil and I represents a steam coil for heating the same, if necessary.

In operation the air supplied throught the inlet pipe G becomes impregnated with the gasified modifying agent which vaporizes from the vessel H and the mixture of air and gasified oil penetrates the permeable medium E and rises as bubbles through the pulp C. The effect of this is to form the bubble column D above the pulp, the bubbles forming the bubble column having collected the metalliferous mineral during their passage through the pulp and subsequently being overflowed with the mineral into the launders B. The bubbles collected in the launders B may be broken down by any suitable means, as by a spray of water and the precipitated mineral collected as concentrate. The concentrate so collected may, as customary, be retreated to further purify it, if necessary.

While I have shown certain specific means for effecting the mixture of ore with gasified oil, it will be obvious that my invention is not restricted to the specific means shown, nor is my invention restricted to the use of any particular volatile oil. A variety of suitable substances are volatile and many others such as coal tar, and the coal tar products contain volatile constituents.

What I claim is:

1. An ore concentration process comprising the operations of vaporizing a modifying agent and forming a mixture thereof with air, introducing said mixture through a fine texture porous medium into an ore pulp and forming bubbles to which certain mineral particles in the ore adhere as said bubbles rise through the pulp, causing said mineral-bearing bubbles to form a column of bubbles above the pulp and separating the mineral carried by the bubbles in the upper part of the bubble column from the remainder of the ore.

2. An ore concentration process comprising the operations of vaporizing a modifying agent and forming a mixture thereof with air free from modifying agent in the liquid state, introducing said mixture into an ore pulp in the form of bubbles to which certain mineral particles in the ore adhere as said bubbles rise through the pulp, causing said mineral-bearing bubbles to form a column of bubbles above the pulp and separating the mineral carried by the bubbles in the upper part of the bubble column from the remainder of the ore.

3. An ore concentration process comprising the operations of vaporizing a modifying agent and forming a mixture thereof with air free from modifying agent in the liquid state, introducing said mixture through a porous medium into an ore pulp and forming bubbles to which certain mineral particles in the ore adhere as said bubbles rise through the pulp, causing said mineral-bearing bubbles to form a column of bubbles above the pulp and separating the mineral carried by the bubbles in the upper part of the bubble column from the remainder of the ore.

4. An ore concentration process comprising the operations of vaporizing a modifying agent and forming a liquid free mixture thereof with air, introducing said mixture into an ore pulp under pressure not substantially greater than that necessary to sustain the hydrostatic head of the pulp, thus forming bubbles which are impelled upwardly through the pulp by their buoyancy only and to which certain mineral particles in the ore adhere, causing said mineral-bearing bubbles to form a column of bubbles above the pulp, and separating the mineral carried by the bubbles in the upper part of the bubble column from the remainder of the ore.

5. An ore concentration process comprising the operations of vaporizing a modifying agent and forming a mixture thereof with air, introducing said mixture through a fine texture porous medium into an ore pulp under pressure not substantially greater than that necessary to sustain the hydrostatic head of the pulp and to overcome the resistance of the porous medium to the passage of the gaseous mixture, thus forming bubbles which are impelled upwardly through the pulp by their buoyancy only and to which certain mineral particles in the ore adhere, causing said mineral-bearing bubbles to form a column of bubbles above the pulp, and separating the mineral carried by the bubbles in the upper part of the bubble column from the remainder of the ore.

6. An ore concentration process comprising the operations of exposing a body of liquid modifying agent having a relatively large exposed surface to a current of air thereby vaporizing part of said agent, introducing the mixture of air and vaporized modifying agent so formed free from any of such agent in liquid form into an ore pulp in the form of bubbles to which certain mineral particles in the ore adhere as said bubbles rise through the pulp, causing said mineral-bearing bubbles to form a column of bubbles above the pulp and separating the mineral carried by the bubbles in the upper part of the bubble column from the remainder of the ore.

7. An ore concentration process comprising the operations of exposing a relatively large surface of a liquid modifying agent containing a volatile constituent to a current of air thereby vaporizing part of said agent, introducing the mixture of air and vaporized modifying agent so formed free from any of such agent in liquid form into an ore pulp in the form of bubbles to which certain mineral particles in the ore adhere as said bubbles rise through the pulp, causing said mineral-bearing bubbles to form a column of bubbles above the pulp and separating the mineral carried by the bubbles in the upper part of the bubble column from the remainder of the ore.

8. An ore concentration process comprising the operations of exposing a relatively large surface of heated liquid modifying agent to a current of air thereby vapor-

izing part of said agent, introducing the mixture of air and vaporized modifying agent so formed free from any of such agent in liquid form into an ore pulp in the form of bubbles to which certain mineral particles in the ore adhere as said bubbles rise through the pulp, causing said mineral-bearing bubbles to form a column of bubbles above the pulp and separating the mineral carried by the bubbles in the upper part of the bubble column from the remainder of the ore.

9. An ore concentration process comprising the operations of exposing a body of heated liquid modifying agent having a relatively large exposed surface containing a volatile constituent to a current of air thereby vaporizing part of said agent, introducing the mixture of air and vaporized modifying agent so formed free from any of such agent in liquid form into an ore pulp in the form of bubbles to which certain mineral particles in the ore adhere as said bubbles rise through the pulp, causing said mineral-bearing bubbles to form a column of bubbles above the pulp and separating the mineral carried by the bubbles in the upper part of the bubble column from the remainder of the ore.

Sponge Iron in California

Many attempts have been made in recent years to manufacture sponge iron from ores and from the calcine left after burning pyrite for the manufacture of sulphuric acid. Rather elaborate experiments of this kind were made a few years ago by Frederick P. Laist, at Anaconda, his effort being to obtain a sponge iron suitable for use in precipitating copper from sulphate solutions. It did not, however, prove economical. In Spain similar attempts were made prior to the War, using large rotary furnaces fired with producer-gas. The results were so successful as to attract wide attention, and a company was formed in England under the leadership of Sir William McKenzie to exploit the process. The enterprise is reported to have advanced to the point of erecting experimental works at Birmingham. The great attraction that the manufacture of sponge iron holds for the world is the possibility of using liquid or gaseous fuel, apart from the merits of the product itself, which are very nearly those of wrought-iron, and to some extent those of mild steel. For the West it is a matter of great importance. California, for example, possesses large deposits of iron ore and an abundance of asphaltic oil, but no coal, nor is there cheap blast-furnace coke available from any source. Among the many that have experimented on the matter one of the latest is Herbert Lang, who has erected a small plant at Emeryville, near Oakland, Cal., where he has produced a good quality of sponge iron direct from Pit river magnetite, an ore containing 65% metallic iron. The ore is ground to 16 mesh, and is mixed into a paste with Kern river, 13° B., crude asphaltic base oil. The paste is molded into cylinders 8 in. diam. by 15 in. long. No flux was used in the earlier experiments, but the necessity for doing so is now seen, although the insoluble constituent in the Pit river magnetite is less than 2%. The cylinders are charged into the hearth of a reverberatory furnace and heated to the welding temperature of iron, that is, about 2000°F. The furnace is fired for eight hours with crude oil, maintaining a reducing atmosphere. The product holds together as separate masses, each representing an original cylindrical mass or briquette. These are removed from the furnace and squeezed in the same manner as blooms in the making of wrought-iron.

Phosphate Rock

Prior to 1914 about 40% of the phosphate rock produced in this country was exported. Most of the exported material is high-grade rock from Florida. Some think it unwise to allow more than 1,000,000 tons of high-grade phosphate to go from this country, as it did before the War, to enrich the soils of Europe. They believe American deposits of high-grade rock are exhaustible, and when the rock is sent abroad the phosphorus, on which the agricultural prosperity of a country partly depends, is irretrievably lost to us. When we use phosphate rock on our own soils the phosphorus in it can be removed in crops and in large part returned to the soil again in the by-products of the farm—green manure, farm manure, bone-meal, and tankage. Not only once, but over and over again, the phosphorus may play its part in producing crops. The economic loss in allowing the phosphate to leave the country is therefore manifold. The phosphate rock that is exported to gain a few million dollars annually would, if applied to our land, play its part in producing hundreds of millions of dollars in cotton and corn. Another and broader view, however, may be taken of the situation. The world's annual production of phosphate rock is about 6,000,000 tons. The phosphate reserves in the United States amount to about 6,000,000,000 tons. At the rate of mining during the last few years the high-grade deposits of the Eastern States cannot last long. In the Western States, however, there is 5,750,000,000 tons of high-grade material in the areas that have been examined and classified, no account being taken of rock carrying less than 30% tricalcium phosphate. These figures make the situation seem less discouraging. At the present rate of consumption in the entire world, the deposits in the western United States would supply the demand for about 900 years. The use of phosphate for agricultural and other purposes is bound to increase, but it will be a long time before it reaches 10,000,000 tons per year. Even if the world should demand a quantity so large as that, the Western States could supply it for more than 500 years. How long phosphate could be mined and shipped at prices that could be paid abroad is another question.—U. S. Geological Survey.

Before the War 40,000 tons of barite was imported from Germany for the manufacture of lithopone. Now five companies are producing this article from deposits in Tennessee, Kentucky, Virginia, and Missouri.

Searles Lake Potash

At the recent annual meeting, in London, of the Consolidated Gold Fields of South Africa, the chairman, Lord Harris, had this to say concerning the Californian potash enterprise in which his company is heavily interested:

"Now, as regards American Trona Corporation, you will, I feel sure, be anxious to hear the latest news of that venture. It has taken a longer time than we originally anticipated to reach the producing stage. We were unlucky in our first reduction plant, which failed to come up to expectations. I am happy to say that the second effort has been successful—I think I may say very successful. The producing stage was reached at the end of October last year, since when the plant at Searles Lake has been in continuous operation. Owing to the fact that the mixed salts produced are required in ever-increasing quantities for fertilizing purposes, and command a satisfactory price, it was decided not to press on with the completion of the refinery at San Pedro, which, consequently, will not be in operation till early in the coming year, by which time the second unit of plant at Searles Lake should be completed. All chemical difficulties have been overcome and the Grimwood process has proved its ability to solve the problem of separating all the constituent salts of the brine. Up to now, however, we have suffered considerable disappointment as regards the output of mixed salts, on which at present the Trona Corporation depends for the greater part of its income. For the seven months ended August 31, the output of these salts has amounted to 8348 tons, or an average of a little over 34 tons per day, as against the 70 tons which we were informed might be anticipated. This reduced output is in some measure due to the abnormal heat obtaining in the desert, where the bulk of the reduction plant is erected, from May to September inclusive, which retards the crystallization of the salts. With the return of cooler weather we may expect a considerably increased production, and that, after certain supplementary plant now on order has been installed, the daily output of the first unit will, it is estimated, amount to at least 70 tons. The most recent telegram as regards production shows that in October we had got up to 43½ tons, and had also sold 55 tons of high-grade potash for the chemical trade. You will be interested to know that the profits for the month of October amounted to $133,000; that was on an output of about a quarter of what we anticipate it will be when the second unit at Searles Lake and the refinery at San Pedro are complete, which we hope will be the case next April.

"The difficulty of obtaining and forwarding structural steel has greatly delayed the erection of the second unit, and has involved a serious increase of capital expenditure, which, however, is compensated for by the abnormal price of potash now ruling and likely to rule for some time to come. The near approaching completion of the second unit and of the refinery will result not only in a largely increased production of the mixed salts required for fertilizing purposes, but will also enable the corporation to produce a considerable amount of practically pure potash required by the chemical trade, as well as to segregate and market the borax, which is the most important of the corporation's by-products. The title question is still undecided. The hearing of the corporation's application for patent took place at Washington at the end of September, and we are informed that our representatives await the decision of the Commissioners with complete confidence. The corporation has done good work in helping to allay the potash famine in the United States which the War has caused, and we feel justified in believing that the Government of that allied country will not fail to show some recognition of this fact in its treatment of the corporation. Of course, we have to anticipate a falling off in the price of potash after the War, when its exportation from other parts of the world re-commences, but that will be compensated for by the production of by-products, which we believe has been satisfactorily proved at the San Pedro refinery.

Compensation Law in California

The legislature of 1917 amended the present State compensation laws in a number of particulars and the changes will become effective January 1, 1918. One of the important changes is that which requires every employer, except the State and all political sub-divisions or institutions thereof, to secure the payment of compensation in one of the following ways:

1. By insuring and keeping insured in an insurance company duly authorized to write compensation insurance in this State.

2. By securing from the Commission a certificate of consent to self-ensure, which may be given when satisfactory proof of the employer's ability to carry his own insurance is furnished.

The Commission will require such employer to deposit with the State Treasurer a surety bond or securities approved by the Commission, in an amount to be determined by the Commission. If an employer fails to secure the payment of compensation, he is subjected to an added liability, as his injured employee can not only claim compensation but may also bring an action at law for damages and shall be entitled in such action to the right to attach the property of the employer, to secure the payment of any judgment that may be obtained. In any such action the employer is denied the defences of tributory negligence, assumption of risk and negligence of fellow servant, and the negligence of the employer is presumed.

Those employers who wish to obtain a certificate of consent to carry their own insurance, or who may desire further information on any question relating to the compensation laws, should make application to the Industrial Accident Commission, 525 Market St., San Francisco.

Mining Companies and Excess Tax

The American Mining Congress, responding to invitation from the Commissioner of Internal Revenue, outlines a few respects in which the excess-profits tax law will work injustice and inequality of taxation to mining companies, taxing the conservative companies with small stock-issues at a relatively much larger amount than those with inflated capital.

The memorandum says:

A mine or an oil-well is a very different property from any other. Owners of mines and wells must seek, find, and extract their product from beneath the earth. This process generally requires an immense expenditure for development work, which expenditure is not carried to capital account. In the case of the great majority of smaller enterprises each mine is owned by a separate company, which cannot add to the purchase price of its properties expenditures for development work and losses sustained in proving other properties to be valueless, an advantage possessed by the large corporations. When the ore is mined out the capital is absolutely gone.

Now the War Revenue Act takes no account of the peculiar nature of mines and wells, giving them no benefit whatever from the discovery of new orebodies. So far as mines are concerned, the definition of 'invested capital' in section 207 of the Act sets a premium on all of the discreditable devices of the past 20 years intended to work up and inflate book-values as a basis for the issue of watered stocks.

Any technical limitation that prevents a company's fixing as its invested capital the real value of its property works an injustice to that company; and any technical provision that enables one company to fix its invested capital, and hence its deduction, at a larger amount than another company with exactly the same property and exactly the same income, works a great injustice to the latter.

We contend that section 207, unless it be construed as suggested in III below, will be particularly objectionable from either one of these standpoints, and that it will be particularly objectionable from the standpoint of mining companies of small capitalization. We therefore make three suggestions under the following heads:

I

The words "but in no case to exceed the par value of the original stock or shares specifically issued therefor" should be stricken out of section 207 (a) sub-division 2.

If it remains in the Act it will bring about exactly the opposite result from that intended by the framers of the law. Presumably their object was to prevent companies with inflated capital and watered stock, over-capitalized for stock-jobbing purposes, from taking advantage of their inflated capital to reduce their taxes.

Suppose a company on December 31, 1913, issues only $100,000 capital stock specifically for assets worth $1,000,000, setting up no surplus whatever. Under section 207, its 'invested capital' is only $100,000; whereas another company with the same assets and the same capital stock, organized on January 2, 1914, just two days later, will be allowed as 'invested capital' $1,000,000. So that the latter company will be entitled to a deduction of $93,000 as against $12,000 for the first company, unfortunately acquiring its assets in the year 1913.

Take the provision that invested capital as to property acquired prior to 1914 shall not exceed the par value of the stock issued therefor. Suppose two companies are incorporated in 1910 each with assets worth $1,000,000. One issues only $100,000 capital stock, while the other issues $10,000,000. And suppose that the property of each on January 1, 1914, is worth $10,000,000, and suppose the income of each for 1917 is $500,000. According to the strict provisions of section 207 (a) sub-division (2) the first company will probably be allowed only $100,000 as its capital invested, while the second will probably be allowed $10,000,000. If the percentage of deduction is 9%, the first company will be allowed to deduct from its $500,000 only 9% of $100,000 in addition to the $3000 fixed deduction, or $12,000 in all; while the second company will be allowed to deduct $903,000. In other words, the first company will have to pay an excess-profits tax on $488,000 income, while the second will pay no excess-profits tax whatever.

The injustice is particularly great to conservative mining companies, for many mining companies are largely over-capitalized, capitalized largely on their trust in the future. We simply ask that these companies should be compelled to pay exactly the same tax as the conservative ones owning properties of the same value and receiving the same income.

Par value of capital stock is in no sense a criterion of real value. In fact, the question may well be raised whether this provision which makes your excess-profits tax depend on the par value of your capital stock is not so artificial and unjust as to be unconstitutional.

II

The words "at the time of such payment" should be omitted from section 207 (a) Sub. 2 and (b) Sub. 2.

The attempt to limit 'invested capital' to the value of property at the time of its acquisition is peculiarly unjust in the case of the smaller mining company. This provision clearly discriminates against the mining company which only after the expenditure of large amounts of money and time finally succeeds in putting itself on a paying basis, which increases the values of its properties not by sitting and watching them grow, but by the expenditure of large amounts of money and labor and energy. As it has by these means increased the value of its properties, it should in all justice and fairness be permitted to use these values in determining its 'invested capital.'

Suppose a manufacturing plant worth $1,000,000 to be acquired for $1,000,000 stock in 1910, that it was worth $1,000,000 on January 1, 1914, but that by reason of wear and tear, falling off in business or other causes, it is today worth only $250,000. Under section 207 the 'invested capital' of this company is today $1,000,000, whereas its real assets are worth $250,000.

It is difficult enough to determine the value of a mining property as of today, but this provision makes it necessary today to attempt in some way to determine the value of all sorts of property as of four years ago, a task that will be productive of endless disputes and probably much litigation.

Finally, the Government itself, in its capital-stock tax-law, yearly makes a rough attempt to value the assets of a corporation by fixing the value of its capital stock. Surely it is not unreasonable to ask that the same principle be adopted when there is an effort to fix the ratio of income to invested capital or the amount of invested capital to be used in determining the amount to be deducted from income before applying the excess-profits tax.

III

The words "earned surplus" in section 297, Sub. 3, of the Act should be construed to include the values of ores and mineral deposits discovered, located, and developed in any mine since the organization of the company, and should be considered as 'invested capital' within the Act, to the extent that such developments add to the net values of the company in excess of its capital stock. Such values are 'earned' and constitute surplus.

Life of all mineral properties is limited, and in most, great fluctuation and differences in the value of the ore and mineral deposits occur over limited periods, due to exhaustion on the one hand of its ore by the ordinary process of mining, and the discovery of ore due to the expenditure of money and time in development work, on the other. For this reason the values of the mining properties are subject to constant change.

This is not generally so, however, in the case of manufacturing and mercantile corporations.

A mining corporation, therefore, differs from a manufacturing or mercantile business in that its source of supply (its ores) is a question of uncertainty; and the values of its assets dependent on the discovery of ore or mineral deposits through the expenditures of moneys and work, are subject to increase or decrease within limited periods of time.

The discovery of ore or mineral is not, however, in any sense to be considered unearned increment. Mining values discovered and so developed as to be readily valued within reasonable limits are values earned and produced as the direct results of (a) mining and engineering skill, (b) the expenditures of moneys and time in opening up the mine, and (c) the employment of labor. The fact that there is a hazard or chance incident to every such discovery does not make the discovery, when successful, any the less earned as the reward of and as the result of industry. There would be no discovery if there was no work done.

It would, therefore, follow that if through the discovery of mineral in place, such a substantial increase in the values of the company's property would occur that there would be a surplus of value over "the actual cash value of tangible property paid in other than cash, at the time of such payment" (section 207), that such value should be considered first as 'earned' for the reasons set forth above, and second, as 'surplus' or increase over the original capital invested, and should become within the definition of section 207 of the Act 'invested capital' and 'earned surplus.'

Unless a broad construction be given to the words "earned surplus" so as to include ore discoveries, as above, many mining companies will pay from 50% to 60% as a war tax, while companies in other lines will pay but 25% to 40% on the same value.

If, however, the value of this ore, so discovered and developed as the result of underground work, be considered 'invested capital' under the heading 'earned surplus,' a more uniform rate of taxation would be placed on the various mining industries, and the basis of taxation would become scientifically fixed at the present value of the mining property and not on a value fixed some time in the past nor based upon a 'value created' by the whim of the original incorporator of the company.

For the reasons given, it is submitted that ore so found and valued be considered as 'earned surplus' and 'invested capital.'

If the construction of the law above suggested be adopted, then it will not be necessary to amend the law as indicated in I and II above. It is therefore highly desirable for all interests that such a construction be adopted.

PRIOR to the War China had contributed largely to the supply of antimony, and the Chinese, considering the exceptional demand that would be created for this mineral for use in the manufacture of ammunition, proceeded to further development and production. For a time it was in prime demand, and the Chinese were able to dispose of their output at satisfactory profits. Large stocks were accumulated by speculators, but they failed to take into consideration the impetus that would be given to the antimony industry in other parts of the world through an increased demand. When large shipments from Bolivia and Spain began to arrive in the markets, prices began to fall, and continued steadily to decline during the year 1916. The result was that the speculators suffered heavy losses, and many Chinese mines and works had to discontinue operation.

NEW ZEALAND has offered a bonus of 8c. per pound for the first 100,000 lb. of retorted quicksilver that the mines in New Zealand produce under certain restrictions and conditions. There are said to be several rich deposits of quicksilver in the Dominion.

ANCHORAGE, ALASKA

Development at Cache Creek.

The Cache Creek Dredging Co., of which Bulkeley Wells is president, operated from May to September most satisfactorily. The season was cut short by floods about September 8. This marked the highest water in the district for 18 years. Thirty new buckets were taken up the Susitna river this year. The Government, through the Alaska Engineering Commission, offered great assistance to these placer operators by allowing them use of boats and crews to pull their freight up the river to McDougall. This company leases about 12 miles of the lower stretches of Cache creek from the Cache Creek Mining Co. This tract averages 600 ft. in width. It has a depth of about 7 ft. of gravel and a grade of 1½%. The 7-cu. ft. steam-driven dredge, which was installed at the mouth of Cache creek, has worked its way almost out of the large boulders that were so troublesome at first. At the time of starting work last spring the ground was found to be frozen almost three feet deep, which was unusual, as ordinarily it freezes only about six to eight inches. The altitude of these properties is about 2000 ft. and if the snow falls early in the fall before the ground freezes, and covers the ground like a blanket, the gravel can then be mined in the following spring without thawing by steam-points. During the season of 1917 it was practically impossible to secure any quantity of good

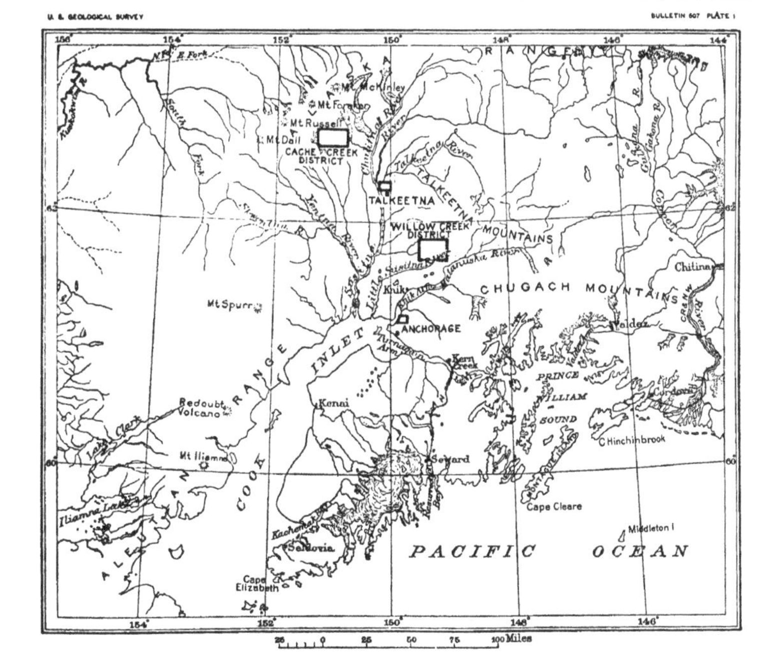

PART OF ALASKA, SHOWING POSITION OF CACHE CREEK

labor, although the men were paid $4 per day and board, amounting to $5 for common labor. The abundance of work all through Alaska and the States, and particularly the excessive wages paid in the base-metal mines, have made it difficult to secure men for work in isolated camps like the Cache Creek district. The men were restless, quitting without notice, and it required two weeks to make the trip to Anchorage to secure substitutes for those that had gone. The Cache Creek Dredging Co. is also operating properties 60 miles overland from McDougall. Cache creek is a tributary of the Kahiltna river, which flows into the Yentna and thence into the Susitna; the Susitna flowing into Cook's Inlet. This district is about 40 miles directly south from Mt. McKinley. The Government is building a wagon-road from Talkeetna, a station on the new railroad, into the district, and it is hoped that it will also erect a telephone line, which will mean a great deal to the operators in securing men and filling emergencies. The construction of this new wagon-road should effect a great saving for all the mines of the Cache Creek district, as at present supplies have to be transferred to small vessels, thence into boats, and then landed and hauled into the mines.

The Cache Creek plans to electrify the dredge next year, good for six months of the year, installing a hydro-electric plant on Cache creek. Up to this time, the dredge has been operated by steam generated by coal, which is mined from a five-foot bed of lignite outcropping about a quarter of a mile below the dredge. It has been necessary to employ ten men in the coal mine to get out the necessary supply, and ten men to handle coal between the mine and the dredge. The coal seams have a dip of only 18° and form part of the bedrock, together with sandstone and shale. It is believed by local miners that there is an enrichment of the placer over these coal croppings. Small quantities of scheelite and platinum occur in the gravels of this section and the U. S. Geological Survey is endeavoring to assist in every way in the development of these rare minerals.

T. D. Harris, manager for the Marysville Dredging Co. of California, with which the Cache Creek Dredging Co. is affiliated, has been looking over the plans for electrical installation. One of the great advantages of electrification is that the dredge will draw only three feet with electric motors, whereas with steam-boilers it drew four feet. It will also effect a great saving in labor. The plan is to dam the tailings in the fall of each year, closely behind the dredge and entirely across the river-bottom, so high that the dam will hold the water for a considerable distance ahead of the dredge and thus prevent the gravel from freezing. At the approach of summer the water from the pond will be liberated, and the gravel, having been protected, will require no thawing.

The Thunder Creek Mining Co., on which M. A. Ellis of Seattle is the principal owner, operated two hydraulic plants on Thunder creek and one on Windy creek. Both of these creeks are tributary to Cache creek. The season was good with the exception of the troubled period of the flood early in the fall, when pipe-lines and ditches were washed out. On the upper reaches of Thunder creek, Mr. Ellis has uncovered a large quantity of angular quartz, broken up and mixed with rounded pebbles. This condition on bedrock has been found to exist for 2000 ft. up and down the stream and extends from one rim-rock to the other, 600 ft. across. The owner is puzzled by this peculiar condition, and so far, has not been able to find a satisfactory explanation. [It may be glacial drift.—EDITOR.] He is opening up the deposit on one rim by trenching and by an adit. The angular quartz contains particles of free gold and occasionally a piece of quartz will be cracked into two parts held together by free gold. Mr. Ellis hopes to uncover the vein that is the source of this gold and then to sink on it. However, it is possible that the stream has eroded its bed down to a quartz vein that dips at nearly the same angle as the bed of the stream, so that the vein-quartz may spread over the stream bed. This company employed about 20 men. In 1918 the placer mines of Alaska will have to operate under the 8-hour law, as the quartz mines do now. It is feared that this may cause the abandonment of many small properties, particularly in outlying districts where labor is scarce, at best, and where the margin of profit is already small.

The Annabubb Mining Co., for which Louis Bubb of Susitna, is general manager, operated four hydraulic giants on Dollar creek, which is another tributary of Cache creek. At this property, as on Thunder creek, the lower 40 ft. of a 105-ft. gravel bank was found to be formed principally of angular quartz. The quartz here differs, however, in that it carries little, if any, gold, while the Thunder creek quartz is heavily mineralized. Floods spoiled an extraordinarily good season at this property. The face of the gravel-bank, which is worked for 100 ft. high, caved, covering part of the clean-up and some of the equipment of the company. Next year the owners hope to extend their ditch to Dutch creek, about 8 miles, on the opposite side of the watershed, whence they expect to secure a greater and more constant supply of water. All the mining operations are conducted at the head of Dollar creek, the tailing being dumped into Lower Dollar creek. This is not in accordance with the best practice, as the gravel at the lower end of Dollar creek has not yet been mined and the virgin ground there is being covered with worthless tailing. It is, however, the only feasible way to operate this creek until dredges can be set to work on the lower reaches of Dollar creek. The company employed 12 men during the season. It is probable that the placer miners of this district will ask $5 per day and board next season. The law provides that no miner shall work more than 8 hours during any 24, even though paid for such additional work.

MANHATTAN, NEVADA

WHITE CAPS.—UNION AMALGAMATED.—MANHATTAN CON.

The work at White Caps as reported for last week is as follows: On the fourth level raise 407 made 18 ft. for the week, and connection was made with the hanging-wall drift on the 300-ft. level on the shaft vein. This raise has been made for the entire distance between levels in good-grade ore. Stoping operations have commenced from this raise and the stope will be known as 407 stope. From the fifth level the 503 cross-cut has progressed for the week a distance of 23 ft., in shale for 190 ft.; this cross-cut is being extended south and west to reach the shaft orebody. In the mill several minor changes have been made with the object of reducing cost. These improvements produce a more uniform output with an increased tonnage. Conveyor belts from the roaster distribute the roasted product over the mill vats automatically. The mine has added to its accumulation of fuel-oil for the roaster; the local storage tanks have a capacity of 45,000 gal. and these will be filled to capacity in less than ten days. The Butters filter is being put in place and should be in running order in three days.

The week's work at the Union Amalgamated property is as follows: Cross-cut No. 601 advanced five feet in limestone. Drift 609 west in foot-wall advanced 12 ft., starting from 601 cross-cut and following the bedding planes of the formation. This drift is exposing oxidized quartz carrying free gold. The work of enlarging and timbering the Earl shaft for use as an operating shaft is progressing well. The shaft has been enlarged to two-compartment, straightened in the foot-wall, and re-timbered to the 256-ft. level. The new six-drill compressor is on its foundation and the motor is now being set in place for driving it. The building covering the compressor is completed. The company should be operating through the Earl shaft by January 1. Notice has been issued for a special stockholders meeting on December 29 to vote on re-organizing the company into an assessable corporation.

At the White Caps Extension property a sump is being sunk

in the shaft below the 400-ft. level. The timbers for the station at the 400-ft. level are in place and mining will be started east as soon as the sump is completed. The drift east will follow the formation bedding-planes and its objective will undoubtedly be the blue White Caps limestone belt, similar to the limestone that has carried the great orebodies developed by the White Caps company.

At the Manhattan Consolidated a station is being cut on the bottom level, preparatory to sinking a shaft. A sump back in the drift is being cut to hold the water, accumulating during the shaft-sinking. From this sump the station pump will handle the water direct to the surface. A 5 by 8 Gould triplex pump has been purchased by the company, which will handle 100 gal. of water per minute under a 500-ft. head. The miners have been taken out from the development of the east orebody from the 400-ft. level. The present dimensions of the east orebody shows a length of 125 ft.; the last 10 ft. the ore has been narrowing until the face of the drift shows only three feet of ore. A raise from the fourth to the third level on the hanging wall has been started for the purpose of providing better ventilation in the underground workings. The south-west drift from the third level on the free gold orebody has developed one good ore-shoot and three feet of vein material now shown in the face of the drift. The ore south from the shaft is found in pipes or chimneys so that as long as the development follows the ore-zone additional ore may come into the drift at any time.

TORONTO, ONTARIO

PETERSON LAKE TAILING.—KIRKLAND LAKE MINE.—BOSTON CREEK.

A decision of much interest to mining companies was rendered by Justice Middleton at Toronto on November 29 in an action brought by the Peterson Lake Mining Co. against the Dominion Reduction Co. over the ownership of a large quantity of tailing deposited by the Dominion Reduction Co. and its predecessor, the Nova Scotia Co., in Peterson lake. The tailing, which was formerly regarded as valueless, owing to the high price of silver and the feasibility of its recovery by flotation, may be worth now $1,000,000. The Peterson Lake Co. owns the bed of the lake, and the Nova Scotia Co., which erected a reduction mill in 1909, deposited its tailing there. When the property of the Nova Scotia passed to the Dominion Reduction Co. in 1912 this practice was continued with the consent of the Peterson Lake management. In 1915 the Dominion Reduction Co., thinking that the tailing might become valuable, obtained from the Peterson Lake permission to remove the deposits if ever they acquired value, provided the Peterson Lake had the right to direct the point of deposit. The claim of the Dominion Reduction to all tailing dumped subject to this agreement after July 2, 1915, was admitted, but the Nova Scotia claimed the ownership of all tailing deposited during the five years prior to that date. The judgment was in favor of the plaintiff company, the judge regarding the case as analogous to that of a land-owner dumping earth from an excavation on a neighbor's lot to fill up a hollow. As the practice of dumping tailing on adjacent properties has been a frequent one, several Cobalt companies, including the Coniagas, Beaver, Hudson Bay, Temiskaming, and McKinley-Darragh are adversely affected by this decision.

The mining companies have done their share in subscribing to the Canadian Victory Loan, the leading contributions from this source being as follows: Nipissing, $500,000; Mining Corporation of Canada, $500,000; Hollinger Consolidated, $500,000; Kerr Lake, $250,000; Temiskaming, $200,000; McIntyre, $100,000; Dome Mines, $50,000; Coniagas, $50,000; and McKinley-Darragh, $50,000.——At the Davidson the mill machinery is on the ground and construction is progressing rapidly.——The Newray, now under option to the Crown Reserve, has a 10-stamp test mill in operation, and exploration is being systematically undertaken.——The Beaver Consolidated has made its final payment on the purchase of the stock of the Kirkland Lake. On each payment of about $75,000 a block of stock passed to the purchasers, the first three payments being made in full. The last installment, however, was not fully met, owing to the heavy drains on the resources of the Beaver for development and the new mill so that a small amount of stock remains in outside hands.——The management of the Tough Oakes mine has been assumed by D. H. Angus, formerly of the Right-of-Way at Cobalt. Charles O'Connell, the late manager, has been placed in charge of the Patricia Syndicate at Boston Creek.——The Mondeau claims in the Boston Creek area are under option to the Kerr Lake of Cobalt. Sinking has been commenced and a mining plant will be erected as soon as possible.——The Lake Shore has completed its mill building, and machinery is being erected. Interests associated with the Lake Shore have taken an option on a group of claims in the new Lightning River gold camp near Lake Abitibi.——A winter road is being cut from Kirkland Lake to this field, about 30 miles distant. Many additional claims in Lightning river are being staked and a great rush is anticipated in the spring.

The Miller-Independence mine of Boston Creek is the first gold mine in Ontario to adopt flotation, which for the last two months has been in successful operation at its 35-ton mill. A Blake rock-crusher delivers into a 4 by 5-ft. ball-mill and the latter into a Grotch flotation machine.——The company is treating $8 ore and obtaining a 94% recovery. The amalgamating plates recover 30%, and the mercury traps a like proportion, the balance being in the form of a flotation concentrate that assays $200 per ton.

MAMMOTH, UTAH

BINGHAM DIVIDEND.—NORTH TINTIC MINES.

Just at the time when the Gemini and the Ridge mines were making an effort to clear up the accumulation of ore the smelting company came out with another embargo which limits the shipments from each of the properties to one car-load per day. Officers of the mining companies are now trying to line things up for another smelting contract, and it is thought that the new contract, when it becomes effective, will permit of heavier shipments even though it is not as desirable as the one which now exists and which expires this month. Considerable ore is now blocked out in the Gemini, Ridge, and Valley mines.——Last week the Tintic Standard mine shipped eight cars of ore. The ore is being sent to the surface through the new working shaft, new and old workings are now being properly connected, and everything is in shape for economically and rapidly handling the output.——Joseph S. Berry has suceeded John D. Dixon as a director in the Provo Mining Co. Mr. Berry has bought Mr. Dixon's stock of 50,000 shares, and also has acquired other large blocks of stock in the company.——A discovery of scheelite has been made in the raise above the main adit at the Frankie mine of the Woodman Mining Co.——The Bingham Mines Co. has declared a dividend of 50c. per share payable on January 1 to all stockholders of record December 20. Most of the Eagle & Blue Bell stock has now been exchanged for shares in the Bingham Mines. Six carloads, or about 300 tons of ore, left the Scranton mine in the North Tintic district last month. Work is carried on by lessees. R. J. Deighton of Salt Lake is manager.——At a meeting held at Provo on December 8 the stockholders of the Sioux Consolidated Mining Co. voted to sell the property to the Sioux Mines Co., recently organized with 1,000,000 shares of a par value of 10c. per share. Stockholders of the old company will receive one share of Sioux mines stock for two shares of Sioux Con.; 500,000 shares will remain in the treasury and will be used to liquidate the old company's debts, amounting to about $20,000, and to push development at the Sioux mine. The directors of the new company are Reed Smoot, C. E. Loose, P. G. Peterson, J. T. Farrer, and Reed L. Anderberg. With the exception of Mr. Peterson, who takes the place of T. R. Cutler, the board is the same as that of the old company.——A body of excellent galena has been opened at the Lehi Tintic and 900 sacks are awaiting shipment. The ore-body has been opened in three different places.——M. K. Nebo has a small force of men at work on the North Tintic Mining & Milling Co. ground, north of the Homansville holdings of the Chief Con., endeavoring to find the continuation of the Chief's manganese vein.——An inclined shaft has been sunk in the North Iron Blossom to a depth of 200 ft., where a small cave was discovered. This cave is well mineralized and assays show the presence of iron, silver, and gold.——The double-compartment shaft being sunk on the Eureka King property is down nearly 300 ft. Salt Lake people are in control of the Eureka King, which is one of the most promising in the North Tintic district.——When the donkey-engine is erected on the 500-ft. level, where a winze is being sunk, another shift will be put to work at the Zuma. This winze is down 35 ft. and has been in excellent ore the entire distance. ——Owing to an accident in the shaft the Chief Con. mine lost a couple of shifts work this week. One of the cars became loose while it was being hoisted to the surface and before the engines could bring the cage to a stop the wall plates in the shaft had been torn out for a distance of 50 ft. Cecil Fitch, superintendent, states that some of the development work, which has been going on at the Chief Con. mine, has been discontinued temporarily and that more attention will be given to the mining of the ore during the remainder of the year. He states that this new arrangement does not affect the Plutus, which is being developed through the Chief Con. mine.——Work has been discontinued at the Sioux-Alex adit, controlled by a Bingham corporation. During the past year the prospecting has been done by diamond-drills; seven holes have been put down. The deepest hole was sunk from the 800-ft. level, the drill penetrating the formation for a distance of 1600 ft.——The total shipments from the Tintic district amounts to 190 cars, twenty mines contributed to this amount, together with one car of bullion from the Tintic Milling Co. at Silver City.

SONORA, MEXICO

MOLYBDENUM AND TUNGSTEN ORES.

The St. Saba Mining Co., which is situated some 18 miles south of Douglas, Arizona, has resumed operations after being closed down for several years. Considerable development work has been done on this property and some high-grade lead-silver ore has been shipped in the past.——Several carloads of copper-silver ore are being shipped monthly from the Belen mine, which is situated in the Moctezuma district. S. M. Greenidge is in charge.——A large number of claims containing molybdenum have been denounced recently in the vicinity of Nacozari, Sonora. The molybdenum ore in the form of molybdenite is found in the mountains surrounding Nacozari, most of the mines being within a radius of 20 miles from that place. There is a great deal of disseminated ore in granite formation carrying one or two per cent molybdenum sulphide while numerous quartz stringers are found in the same formation carrying up to 30 or 40% of the sulphide ore. Options upon a number of claims are said to be held by Texas mining men who contemplate erecting a mill to treat the lower-grade ore. In the district of Sahuaripa, 200 miles south of the International line, George Fast and associates are working high-grade molybdenum and tungsten deposits. The ores occur as scheelite and molybdenite and considerable ore of both classes has been shipped during the past year. A carload of molybdenite which has just been shipped is said to have brought $38,000.——In the same district are the tungsten and molybdenum mines of J. S. Douglas and associates, from which a large tonnage of high-grade ore has been shipped in the past few years. This ore is packed overland on mule trains for a distance of 150 miles. The Cinco de Mayo group of mines, six miles north of El Tigre, is now being worked by the Mexican government, 50 men being employed. The property formerly belonged to the late Colonel Garcia but was confiscated by the Mexican government several years ago. It has produced several millions of dollars in high-grade silver ore.

The famous Chispas mine, in the Arispe district, is now being worked by its owner; high-grade silver ore is being shipped. This property was taken over and worked by the Mexican government for some time but several months ago was returned to its owner. It is reputed to be the richest silver mine in Sonora and has produced a large tonnage of high-grade ore.

PLATTEVILLE, WISCONSIN

LEAD, ZINC, AND PYRITE SHIPMENTS.—LINDEN DISTRICT.

Complete returns from the Wisconsin zinc-lead field for the month of November show it to be one of the best operating months of the year. While the labor situation had improved for several of the larger operating groups due to the closing down of a number of low-grade producers, weather and road conditions and lowering prices for ore militated against those isolated from immediate railway facilities. Published quotations on zinc ore remained stationary, namely $62 per ton, base, on premium and high-grade ore with the range down to $57 per ton on the medium and second grades.——Lead production through November showed decided improvement and the reserve in the field at the end of the month was estimated in excess of 2000 tons. Price offerings remained firm all through the month at $75 per ton, base, of 80% metal content, but little lead cleared.——Deliveries of ore from mines to refining plants in the field and from mines to smelter direct for the month were made as follows, by districts:

Districts	Zinc lb.	Lead lb.	Pyrites lb.
Benton	29,454,000	170,000	750,000
Mifflin	8,584,000	......	80,000
Galena	4,642,000	66,000	......
Linden	2,724,000	88,000	260,000
Shullsburg	1,444,000	90,000	......
Hazel Green	1,300,000	......	......
Highland	1,080,000	328,000	......
Platteville	986,000	......	......
Cuba	488,000	......	4,574,000
Dodgeville	346,000	......	......
Potosi	290,000	......	......
Mineral Point	114,000	......	......
Totals	51,452,000	742,000	5,664,000

High-grade blende from separating plants in the field was shipped in volume from the following works.

	Lb.
Mineral Point Zinc Co.	6,900,000
National Separators	5,350,000
Wisconsin Zinc Co.	4,294,000
Benton Roasters	260,000
Linden Zinc Co. (Linden)	104,000
	16,908,000

The gross recovery of crude concentrate from all mines for the month aggregated 25,728 tons; net deliveries to smelters, including refinery ore and ore sent from mines to smelters direct 14,598 tons.

Sales and distribution of zinc ore were made during the

month to representatives of the following buying agencies. Mineral Point Zinc Co., 8968 tons; Grasselli Chemical Co, 6717; National Separators (Cuba), 5346; Wisconsin Zinc Co., 4798; American Zinc Co. (Hillsboro, Ill.), 1287; M. & H. Zinc Co. (LaSalle, Ill.), 960; Illinois Zinc Co. (Peru, Ill.), 616; Benton Roasters, 711; Linden Zinc Co. (Cuba), 538; Lanyon Zinc Co., 385; Edgar Zinc Co., 324; American Metals Co. (Langeloth, Pa.), 182 tons.

Considerable activity was seen in the Linden district where Kletszch Bros,. of Milwaukee, purchased the Vial mine and the Wickes mine equipment. The latter was hastily dismantled and removed to the new mine site. On the Kickapoo lease the Milwaukee-Linden Development Co. drove a cross-cut to develop disseminated ore deposits. Rich borings on adjoining leaseholds was the main objective in this drive. Work of building a new mill on Optimo mine No. 4 has been suspended for the winter. This followed the shutting down of the Spring-Hill Mining Co., which became involved and was placed in the hands of a receiver, and it was stated that the Optimo company was negotiating for taking over the Spring-Hill mine and equipment. The Gilman mine, one of the best producers in the district, has nearly run its course, and the miners are pulling pillars. The mine is owned by the Milwaukee-Linden Development Co., formerly known as the Saxe-Pollard company.

A new industry spur about three miles long has been completed to the big zinc-ore refinery of the Wisconsin Zinc Co. at New Diggings, the cost of construction being borne by the main operating groups found in this section, namely, Wisconsin Zinc Co., New Jersey Zinc Co., Vinegar Hill Zinc Co., and the Fields Mining & Milling Co. of Chicago. Trains are in operation, connection having been made with the main line of the Chicago & Northwestern, three miles south of Benton. The new spur is giving great impetus to zinc mining in this section and many outlying producers are already engaged in building macadam roads to connect with the terminal of the new spur.

HOUGHTON, MICHIGAN

QUINCY.—MOHAWK.—ISLE ROYALE.—ALLOUEZ.—LA SALLE.

The one absorbing topic of conversation among the copper-mining interests of the Lake Superior district is the attitude of the Government in the regulation of prices for the output after January 1. While every effort is being made to maintain production up to normal conditions this is quite out of the question at the present time for two reasons. One is the shortage of men, 2000 having left to join the army; the other is the impossibility of securing good men by importation. The natural source of trammer supply, Europe, has been shut off for four years. Notwithstanding that the wage an average underground man can earn is at least 25% higher than at any previous time the fact remains that a shortage of labor continues. At Calumet the trammer that cannot average $100 per month now is a pretty slow worker. A man with a good physique can earn more money than the man who has the engineering ability to map out his work for him. Yet the shortage continues at most of the mines.——The costs of producing copper at all of the mines in the Lake district will be higher in 1917 when the books are made up than ever before, not even excepting the year of the strike 1913, in most of the mines. In addition to the top notch wages, to the 100% average increase in practically everything that the mining companies have to buy, to the shortage of labor, most of the mining companies during the year have laid out enormous sums for high-cost construction-work. While homes for employees and buildings for surface use cost more this year than ordinarily, the fact is that these improvements were absolutely necessary, especially the buildings for the higher-class workingmen. And the advance contracts made early in the year assured the corporations profit enough to pay for these necessary expenditures.

MINING COPPER IN MICHIGAN

MASS COPPER IN THE QUINCY

Never in the history of Quincy has there been such an unusual amount of mass copper secured as is now going directly to the smelter from this mine. It is coming in such large pieces that special methods for handling it, including a block and tackle at the shaft-house, are necessary. Chunks weighing eight tons are not unusual. The extent of this rich showing is not yet realized. It has been opened at 6500 ft.——Tonnage from the Kearsarges is increasing, the North Kearsarge contributing materially to the increase and the total from the Kearsarges and the old Osceola makes an average contribution in ore output equal to that which the Ahmeek secures, although, of course, the copper output is not as large. Last week the Kearsarges showed better than 80 cars per day, the grade of ore being a little better than average.——Mohawk is securing more men and increasing its output of ore; unless held up by snow storms it will make a better average in December. At present it is handling 70 cars of ore per day. Wolverine, the other producing Stanton property, is running 40 cars per day.——Production from the Isle Royale constantly increases. The three-stamp mill on the south side of Portage lake is running to full capacity. In addition to this the ore handled at the Point Mills plant, across the lake, has been increased from 12 to 20 carloads per day.——Ahmeek maintains production at 4700 tons per day.

Allouez has a substantial amount charged to plant improvements this year. Many new boilers have been added, and the whole surface plant gradually is being rehabilitated. An output of 1800 tons per day is being maintained.——Cherokee continues to find small mass and shot copper east of the 420-ft. level. East and west drifts on this level are in 70 ft.——The operations discontinued on account of lack of money and inability to interest capital. The property is owned by Duluth people. The land comprises 340 acres south-east of the Phoenix.——Stoping is being done only from the No. 1 shaft to the 15th level at the La Salle. Ore shipments run 500 tons per day, half from this shaft and half from No. 2. At No. 2 shaft driving is being done on the bottom on the 21st level. The La Salle shipped 15,600 tons in November.——The three mines of the Copper Range Co. are maintaining their production of copper. There may be a slight falling off for the November output but the total ore-tonnage for the month was the same as for October, both under the monthly average for 1916. Champion is now running better than 60,000 tons per month, Baltic 30,000, and Trimountain 20,000 tons. There was a slight increase in tonnage from Trimountain last month and a comparative falling off in Baltic. The general grade of ore was therefore not any better in November than in October. General underground conditions at the Champion are good. In the southerly shafts the showing has not improved of late but in the lower levels there are bunches of spectacular richness.

Production of ore from South Lake last month was 2345 tons.——The Lake output last month was 6000 tons.——Following the suspension of work at the Adventure the mine shipped 35 tons to the mill.——Ahmeek continues to break records. Its total output of ore now runs close to the combined output of the three Copper Range mines. Last month Ahmeek shipped 104,000 tons.——November showed a slight falling off in Franklin's output, which was 27,720 tons.——Mass is gradually increasing its output, now running 1000 tons per day; last month the output was 19,000 tons.——Superior shows a small increase for November. During November, Calumet & Hecla shipped 245,000; Isle Royale 13,000 to Point Mills plant and 54,000 to its own mill at Pilgrim; Osceola 97,500 tons.——At Seneca one boiler is working, one stack up and another ready to set up, and part of mining crew assembled. All machinery has been erected. Mining will start January 1.

The output of ore from Hancock Consolidated was 22,879 tons for November, a decline from the October one, due to shortage of labor. Indications point to better record for December.——There is no immediate intention of suspending operations at the Michigan, which now has its shaft down 120 ft.; a loading-station will soon be cut. Developments on the seventh level have been fairly good, particularly the drift and the cross-cut. This work is on the Ogimah lode.——While no Calumet & Hecla official is willing to admit that the White Pine, its Ontonagon county subsidiary, has little more than a fighting chance to become a big profit-maker, the assertion I made three years ago that the White Pine might some day take the place of the Calumet conglomerate is now more likely to be verified than at any time since this property was opened. The truth is that the Calumet & Hecla has got back the money it paid out for the opening of the White Pine.——There will be no further shipments of copper from this district by water this year. But rail shipments already are going forward for Government use.

TONOPAH, NEVADA

TONOPAH MINING.—TONOPAH EXTENSION.—TONOPAH BELMONT.—WEST END CONSOLIDATED.—MACNAMARA.—JIM BUTLER.

Last week the Tonopah Mining Co. milled 2650 tons of ore averaging $10.50 per ton. At the Silver Top 93 ft. of development has been done, 22 ft. at the Mizpah, and 90 ft. at the Sandgrass. At the Silver Top on the 340-ft. level the big stope on the Valley View vein has been filled with waste drawn from the surface, and high-grade ore is being taken from the hanging wall. Last week's production was 3050 tons.——The Tonopah Extension Mining Co. shipped 24 bars of bullion, valued at $47,000. This is the final November clean-up. At the No. 2 shaft 96 ft. of development has been done and 144 ft. at the Victor. On the 850-ft. level of the No. 2 shaft, raise No. 814 continued to advance on a 2-ft. face of ore. On the 1350-ft. level cross-cut No. 510 cut a vein showing 3 ft. of ore. East intermediate drift No. 505 continues on a 3-ft. face of ore. At the Victor on the 1440-ft. level cross-cut 608 cut a 2-ft. vein of ore. On the 1540-ft. level winze 1501 continued to advance on a full face of good ore. On the 1680-ft. level 1600 south cross-cut advanced 51 ft., while 1600 west drift advanced 24 ft., showing 5 ft. of ore. The production the past week was 2380 tons.——The Tonopah Belmont Development Co. declared the regular quarterly dividend of 12½c. per share. The plant at Millers is being dismantled and the machinery distributed between the Belmont mill at Tonopah, the Belmont-Wagner mill at Telluride, Colorado, and the Eagle Shawmut mill at Tuolumne, California. Considerable development is being done on the 800-ft. level. On the 1100-ft. level raise No. 91 on the Belmont vein holed through into the 1000-ft. level, exposing medium-grade ore. Raise No. 93 on the Western vein is unchanged. Development has been resumed on the 1200-ft. level, which is the deepest level. Raise No. 58 from the west cross-cut No. 6 is prospecting for the faulted segment of the Shaft vein. Last week's production was 2767 tons.——The West End Consolidated Mining Co. shipped 30 bars of bullion, valued at $49,845, which represents the final clean-up for November. At the Ohio shaft drift No. 526 made a connection with the 512 cross-cut. Drifts 535 and 536 continue in good ore. Winze 534, which has been sunk 35 ft. below the drift, shows excellent ore. Raise No. 814 on the 800-ft. level has advanced 60 ft. At the West End preparatory work on the 800-ft. level has been finished, and a cross-cut will be started to the Ohio. The output the past week was 1099 tons. At the Halifax Tonopah cross-cut 1708 on the 1700-ft. level has been started to avoid the heavy ground and large flow of water in the 1704 cross-cut. Cross-cuts 1018 and 1256 continue to make good progress.——The MacNamara Mining Co. shipped the final clean-up of bullion for November. The value of the shipment was $10,804. Cross-cut 700 on the 700-ft. level advanced 54 ft. in Montana breccia the past week. On the 725-ft. level drift No. 19 advanced on good ore, while raise No. 815 on the 800-ft. level continues in low-grade ore. The production for the past week was 541 tons.——The Jim Butler Tonopah Mining Co. is now shipping its ore to the Belmont mill for treatment, instead of shipping it to the plant at Millers. This arrangement will benefit both companies, as the Jim Butler saves the railroad haul to Millers, and the Belmont mill will more nearly reach its full capacity. Raise No. 369 on the 2nd level of the Wandering Boy continues on a 3-ft. face of fair ore. Raise No. 372 on the same vein is making good progress on a 2-ft. face of medium-grade ore. Raise No. 373, which was started to pick up the faulted segment of the vein, shows a full face of low-grade ore. The production the past week was 247 tons.——The Montana produced 122 tons, making the week's production at Tonopah 10,206 tons with a gross value of $178,605.

At the Cash Boy the west drift from the winze on the 1700-ft. level has been extended in the foot-wall and stoping is now being done in good ore. With the completion of the raise to the intermediate above, stoping has commenced on the east side of the winze as well. Lack of cars has prevented shipments of ore.——The North Star continues to raise from the 1050-ft. level on a good grade of ore. The production the past week was 56 tons.——The strike at Gold Mountain, on the Tonopah Divide Mining Co. estate, was made at a depth of 150 ft. It is reported that the vein has been cross-cut for 27 ft., showing good silver-sulphide ore. The orebody does not outcrop. The property is equipped with a compressor and jack-hammer drills. The Taylor-Govan claims on the west have been purchased by the Tonopah Divide Mining Company.

ARIZONA

COCHISE COUNTY

(Special Correspondence.)—It is reported that several thousand tons of good-grade vanadium and gold ore have been blocked out by the Dragon Mining & Development Co.——A serious break-down of the hoist at the boundary shaft of the Dragon copper lease will cause a shut-down of 30 days.

Bisbee, December 7.

(Special Correspondence.)—The Trinity Copper Co., which has been organized recently, is doing considerable development and construction work on its property in the northern part of the Swisshelm mountains. A 3-ton motor truck and hoist have been purchased and are on the property and a compressor and drill sharpener have been ordered. The ore carries silver, copper, lead, and zinc, while at one point high-grade bismuth ore has been exposed. About 20 men are employed under the superintendence of D. H. Crane.——Adjoining this property to the north is the Scheerer property, which has produced considerable lead-silver ore in the past. The property is now being worked under lease and bond and several carloads of lead-silver ore have been shipped during the past few months from the 200-ft. level. Some high-grade wulfenite has been mined and shipped from this level.——To the east of the Scheerer property is the Heney, from which is being taken high-grade silver-lead ore while high-grade zinc ore, in the form of calamine, has recently been discovered.——The Central Butte Mining Co., four miles south of Pearce, has seven men at work. It has widened the main shaft to the 200-ft. level and is now cutting a station. High-grade silver-copper ore is being taken from some surface workings.——The Warren Basin Copper Mining Co., in the Mule mountains, about five miles east of Warren, has started development work. Contractors have been engaged to sink a shaft to intersect a vein showing good copper-silver ore at the surface.——To the south of this property, Alfred Paul and associates of Douglas have shipped several carloads of manganese ore during the past few months.

Douglas, December 11.

GILA COUNTY

(Special Correspondence.)—The Old Dominion Copper Co. is working at full capacity in all departments. The smelter produced 3,000,000 lb. of copper for November. The cutting of the 1900-ft. station of the A shaft has been started, and as soon as completed sinking will be continued to the 2100-ft. level.——Lawton and Pollard have purchased the Burro pack outfit of the Ray Silver-Lead Mining Co. and have moved it to the McCallum claims from which five carloads of high-grade ore will be shipped per month.——The steel-work of the addition to the Hayden smelter will be completed in January; the brick-work of the furnace is almost finished. Excavation for the erection of the coal-grinding plant is under way; in future coal dust will be used as fuel throughout the smelter.

Globe, December 8.

MARICOPA COUNTY

(Special Correspondence.)—It is reported that the mine of the Consolidated Arizona Smelting Co. is ready for a larger production of ore, and the mill has been producing more concentrate than can be handled at the smelter. This excess, however, will be taken care of when the new roasting-plant is completed. At the Blue Bell mines the shaft is now down 1200 ft., the station cut, and development on that level about to commence.——The Noble Electric Steel Co. is working a large force of men on its manganese property, near Aguila, which was purchased last summer.——The Dolbeer interests are also pushing development on their property in the same district. ——It is rumored that a smelter will be erected for ferromanganese in the Aguila district.

Phoenix, December 7.

MOHAVE COUNTY

(Special Correspondence.)—Two different companies are operating on different parts of the Old McCracken mine. A road is being built from the mine to Yucca, the nearest railroad point.——The Hercules mine has commenced stoping on the 250-ft. level, and a good grade of ore is being hoisted.

Kingman, December 7.

(Special Correspondence.)—The discovery of ore made in the Gray Eagle claim of the Tom Reed Gold Mines Co. is regarded as the most important made in the district since the Aztec orebody was opened up last year. The Gray Eagle is directly south of the Big Jim claim of the United Eastern Mining Co. Oscar H. Hershey, who has made a study of the Tom Reed property, reported that in his opinion the Aztec-Big Jim vein was faulted and thrown to the south several hundred feet. To explore the faulted apex, the Gray Eagle shaft was sunk and a cross-cut run south to the vein. When first cut it gave little promise, being narrow and barren, but a drift was started toward the south and the vein was found to widen and at 25 ft. from the cross-cut to carry commercial ore. The drift is now in about 75 ft.; for the last 50 ft. the ore has averaged $20 per ton in gold. The vein runs in a north-westerly direction and the drift is still about 500 ft. from the end-line of the Red Lion Mining Co.'s property. The probability is that the ore-shoot extends farther than the property line, as a similar orebody on the lower portion of the vein extends for over 1000 ft. on Aztec and Big Jim ground. Since the discovery the stock of the company has more than doubled in value, and is now selling at 79c. per share.——The Telluride will start exploration work with drills very soon, and the Red Lion shaft, already down nearly 100 ft., will be sunk to the 400 or 500-ft. level within the next few months. Shaft-sinking is now under way at the Gold Ore, the Record Lode, and the Oatman Southern; cross-cutting is proceeding at the United Oatman, the Oatman United, the United Northern, and the old Gold Road property. Development work will be resumed on the Gold Road Bonanza and the Arizona Moss Back.——The United Eastern main shaft has reached its objective. As soon as a sump is sunk and a station prepared, cross-cutting will be started.

Oatman, December 10.

PINAL COUNTY

(Special Correspondence.)—It is reported that a high-grade gold strike has been made by the Queen Creek Copper Co. at Superior. A rich strike has also been made at the Fortuna Consolidated.

Florence, December 8.

SANTA CRUZ COUNTY

(Special Correspondence.)—It is reported that J. W. Bible of New Mexico has purchased the Australitz mine in the Oro Blanco district.

Nogales, December 6.

YAVAPAI COUNTY

(Special Correspondence.)—A motion has been made to throw the Hull Copper Co. into the hands of a receiver that will handle the affairs of the company during the litigation now on. The Blue Jacket mine has started to ship to Humboldt, and development work has been resumed.

Prescott, December 7.

YUMA COUNTY

(Special Correspondence.)—N. D. MacMyers of Prescott has taken a two-years bond and lease on the Dream Creek Copper Co., five miles west of Salome.

Yuma, December 8.

CALIFORNIA

ELDORADO COUNTY

(Special Correspondence.)—The Christiana Copper Co., which for the last two months has been unwatering and developing the old Seven Bells copper mine on the Charles Worthington place, has uncovered a fair-size body of high-grade copper ore and is erecting machinery. Frank C. Fox, manager, has purchased and will erect crushing and jig-concentrating machines.——The principal mines, mining properties, and ore deposits, in the county are being tabulated and indexed by Burr Evans, mining engineer of Placerville. Mr. Evans reports there is considerable inquiry and a good demand for properties containing gold, silver, copper, cinnabar, lead, zinc, cobalt, nickel, molybdenum, tungsten, manganese, chrome, strontium, barytes, bauxite, and magnesite.——As soon as the necessary equipment can be obtained a new double-compartment shaft will be sunk to the 600-ft. level, with three-shift crews sinking continuously, on the Shan Tsz gold-quartz mine, situated just four miles south-west of Placerville. H. DeC. Richards, of San Francisco, the supervising engineer, was on the property on November 25 to put things in readiness for starting the work. The Shan Tsz mine, which is owned by New York capitalists, is one of the best equipped gold quartz mines in Eldorado county. It contains a modern electric power and light plant and a 60-ton mill.

Placerville, December 7.

FRESNO COUNTY

(Special Correspondence.)—The Copper King Mining Co. has merged the Copper King mine with the Wabash, Copper King Extension, Kneiper, and Davis properties, giving the company control of 745 acres of land on the foothill copper belt of California. The work of developing the property is being pressed; several shipments to the Kennett and Tacoma smelters have been made and the property is expected to be shipping 50 tons of ore daily inside of 90 days. Ore shipped averages 7½% copper, 16% sulphur, 28% iron, with $1.80 in gold and one ounce of silver per ton. N. Treloar, of Salt Lake City, formerly manager for the Calaveras Consolidated, has been appointed general manager and engineer, having arrived at Copper King and taken charge of the property on December 12.

Fresno, December 12.

INYO COUNTY

(Special Correspondence.)—The Friday silver-lead property, 2½ miles from Queen station, has been purchased by Michael Yalovica and associates of Goldfield for a reputed price of $40,000. A cash payment has been made. A little work has been done near the surface, and a 30-in. shoot of $50 ore exposed. One shipment, averaging $140 per ton, has been made. Silver predominates, but some lead and a little gold occurs.

Bishop, December 4.

NEVADA COUNTY

The North Star Mines Co. has declared a dividend of 60c., payable December 22. During the current year the company has disbursed $250,000, bringing its total disbursement to $5,337,040.

TRINITY COUNTY

(Special Correspondence.)—The dredge of the Pacific Gold Dredging Co., which was moved four miles from the mouth of Morrison gulch to Carrville, has been running a week in its new location. The all-steel craft had to be torn to pieces to move it to the new site.

Carrville, December 11.

(Special Correspondence.)—The Bablew Mining Co. is equipped for steady production during the winter. One giant will be operated on the gravel bank, and a second monitor employed to clear away tailing. Extensive development work has been carried on under the new management and a large area of productive ground proved. The holdings are on the west end of the Paulsen ranch. V. P. Demens is manager.——The Jennings Gold Mining Co. is preparing for extensive hydraulicking along the Trinity river, five miles from Lewiston. Water is pumped into a reservoir 220 ft. above the mine, the pump being operated by a 200-hp. electric motor; 12 men are employed under Fred Dorman.——Dredging is quite active in this county. The Pacific Dredging Co. has re-built its steel dredge near Carrville. The deposits have been thoroughly tested and can be advantageously worked.——The Trinity Dredging Co. is operating near Lewiston with good results. ——Near Trinity Center the Estabrook Gold Dredging Co. is building a large boat. Unlike the newer dredges, the Estabrook dredge will have a wooden hull.——Prospecting of placer ground by dredging companies is active along the Trinity and tributary streams.

Lewiston, December 10.

COLORADO

CLEAR CREEK COUNTY

The Wasatch-Colorado Mining Co. at Silver Plume is working three shifts continuously and shipping 200 tons of lead-zinc ore per week.——The Scepter Mining Co. has completed its tramway from Democrat mountain to Georgetown, a distance of nearly two miles, and expects to start shipping shortly. J. C. Hershey is manager.——The Smuggler is making regular shipments to the Idaho Springs mill. J. Simpson is manager.

Grading for the new concentration-mill at Urad for the treatment of molybdenum ore has been completed and some of the foundations are in place. A 2000-ft. adit is being driven, and from the end a raise will be put up to connect with the surface and give ventilation. The ore occurs in a porphyry dike which seems to be impregnated through its whole width with molybdenite. The ore, after coarse crushing, will be ground finely in ball and tube-mills and then subjected to flotation.

LA PLATA COUNTY

(Special Correspondence.)—In the issue of November 17 it is stated that the American Smelting & Refining Co. had closed its plant at Durango on November 1. Mr. Gilbert's statement refers to the Silver Lake mill at Silverton, which has been run as a custom mill for the last few years, and not to the Durango smelter.

Colorado Springs, December 4.

TELLER COUNTY

(Special Correspondence.)—The production by lessees the El Paso Consolidated Gold Mining Co. during Novem totaled 300 tons of $18 to $20 grade.——The properties of Sedan Gold Mining Co., on Galena hill, were sold under a judgment issued from the District Court of El Paso county on December 8 to Hildreth Frost of Cripple Creek and Colorado

Springs for $20,257, including costs.——The shaft on the Rose Nicol property, on Battle mountain, was sunk 75 ft. during November by John Nichols, superintendent for the Camp Bird M. L. & P. Co., holding a 6-year lease on the property. It is a two-compartment shaft 4½ by 9 ft. in the clear.——The Victory Gold Mining Co., operating the Prince Albert group on Beacon hill under bond and lease, has commenced shipping from the Beacon claim.——The shaft on the Index property on Gold hill is being re-timbered and the mine machinery overhauled. The property has been leased to the El Paso Extension Co. Al Campbell is superintendent.——The Roosevelt tunnel of the Cripple Creek Deep Drainage & Tunnel Co. was advanced 115 ft. in November and in addition the Cresson drifts extended 43 ft., making a total of 158 ft. with only one shift working. The flow from the tunnel has decreased to 4150 gal. per minute.

Cripple Creek, December 12.

C. H. Hays of Parkersburg, acting for the El Paso Extension Co., has taken a five-year bond and lease on the Index mine, and expects to start development shortly. The Index was formerly operated by the Index Mining Co., and there is said to be a considerable body of low-grade ore in the mine.

WASHINGTON COUNTY

The Akron Oil & Gas Co. has struck a six-foot stratum of oil-sand at 1992 ft. that is giving a flow of oil, estimated at over 100 bbl. per day. It is the intention of the company to continue sinking in the hope of getting a more extensive flow.

IDAHO

BONNER COUNTY

The Armstead Mines, operating at Blacktail, on lake Pend Oreille, has broken another record for adit-driving. The main adit was advanced 489 ft. in the 28 working days of November. This performance beats the former record of 488 ft. in 29 working days of October.

LATAH COUNTY

A short course at the Idaho School of Mines, at Moscow, this winter is being planned with a view to preparing substitutes for the young professional and technical men who have gone to the army.

SHOSHONE COUNTY

The Hecla Mining Co. has declared a dividend of 5c. per share for December. The company has been paying 15c. per share but the low price of lead and uncertainty of the war-tax caused a reduction in the dividend rate for this month.

MONTANA

FERGUS COUNTY

(Special Correspondence.)—The plant at Hanover, being built by the Three Forks Portland Cement Co., is nearing completion.——The gypsum mill at Hanover has been operated regularly. The U. S. Gypsum Co. has done a considerable amount of development at its new Casofour plant, 10 miles east of Lewistown, and it will have a gypsum mill in operation next year.——At Kendall, North Moccasin mountains, the Barnes-King Development Co. is the chief operator. Work has been curtailed at the North Moccasin property on account of a labor shortage; nevertheless the company's output has averaged about $18,000 per month during the past six months. The Kendall mine, owned by the Barnes-King, has been worked successfully under lease. Development work has been done at the West Kendall, the North Kendall, and some other properties.——The Spotted Horse, Maginness, and Cumberland mines at Maiden, in the Judith mountains, have been operated by lessees. The Cumberland has made a successful record, the lessees having found a large body of high-grade ore. Development work has been done at the War Eagle and other properties near Maiden. A new cyanide mill has been built at Giltedge to treat low-grade ore and tailing from the old Gold Reef mill. N. J. Littlejohn, of Lewistown, has purchased the Mammoth property near Giltedge. It is equipped with a 50-ton cyanide plant. Development work this summer disclosed sufficient ore to justify operation of the mill next spring.——A good grade of copper ore has been found on the Sutter claims on Armell creek.——Silver-lead and zinc properties are active at Neihart in the Little Belt mountains. Production is also coming from Wolf creek and near Monarch. Several properties that have been idle since the early nineties have been purchased or leased and re-opened this summer.

Lewistown, December 6.

NEVADA

CLARK COUNTY

(Special Correspondence.)—From 75 to 100 tons of manganese ore is being shipped daily to Pittsburgh steel mills from a deposit recently opened 15 miles east of Las Vegas. The property is operated under lease by Connor, Gillin & McCoy, and a large tonnage of ore is exposed.——The main orebody has been recovered on the lowest level of the Yellow Pine Extension and is developing favorably. It ranges from six to seven feet wide and assays 35 to 40% zinc. A. J. Robbins is manager.——High-grade copper ore has been uncovered in the adit of the Red Streak. The ore carries copper glance, and is said to be among the richest ever opened in this district. Alex. Anderson is superintendent.——Copper ore has been found in the lower adit at the Platino, 300 ft. below surface. In the upper level development of high-grade ore continues.——Development of the main lead-zinc orebody on the 700-ft. level of the Yellow Pine group, north of the shaft, is progressing satisfactorily. The ore assays as high as on the upper levels. A four-inch vein of high-grade copper has been found on the 700-ft. level; the first appearance of copper in the deep workings of the mine.

Goodsprings, December 3.

LANDER COUNTY

(Special Correspondence.)—The Mercury Mining Co. has opened an extensive deposit of cinnabar near Ione. The furnaces are being repaired, more workmen engaged, and arrangements made for a steady output.—A promising cinnabar deposit, carrying a little gold, has been found near Phonolite by Joseph Beck.——Lessees are opening high-grade silver ore on the Kattenhorn group, near Maysville.——The Kenzevich-Govich lease has shipped a car of selected ore assaying 100 to 400 oz. of silver. The Kenzevich-Sly lease is developing a shoot assaying 200 oz. of silver.——Shipments of copper ore are going to the Kennett smelter from the Edison lease at Copper Basin, near Battle Mountain. The shaft is down 170 ft. and at 160 ft. the orebody is 22 ft. wide and averages 12% copper; 70 cars have been shipped.——The Copper Canyon Co. has erected a hoist.——The Goff & McDonald leases are shipping high-grade ore.——At Galena, Johnson & McGregor have erected a small concentrator to treat their silver-lead ore. A large vein of medium-grade ore has been developed and silver-lead concentrate is being shipped to custom plants. Several properties in this section are active, and some good discoveries have been reported.

Austin, December 1.

SOUTH DAKOTA

LAWRENCE COUNTY

(Special Correspondence.)—Announcement is made that the special dividend of $1 per share will be paid to Homestake stockholders with the regular payment during December. This will make $1.65 for the month and $8.80 per share for the year. The total distributed to stockholders during the year will reach $2,210,208. The second power plant on Spearfish creek has been completed and water will be turned into the flume about the middle of December.——The Trojan company has completed changes in its milling plant enabling it to handle 500 tons daily. The Two Johns mine has been added to the Trojan present holdings and regular shipments of ore are being made. In addition to this some custom ores are being handled. The Republic property in Black Tail gulch, which is leased by the company, is supplying ore to the Trojan plant. Motor-trucks are used in transporting the ore from the mine to the nearest railroad point.——At the Montezuma & Whizzers, which is under bond to J. T. Millikin, diamond-drilling continues. Mr. Millikin recently purchased the Oro Hondo property, which he has been developing for the past two years. The Golden Reward Co. has men at work at the Gilt Edge Maid in the Galena district and the property will be thoroughly developed.

PENNINGTON COUNTY

The Homelode Co. has suspended mining and milling temporarily and will prospect by means of diamond-drilling.——The Blue Lead copper property near Hill City will be worked by the new owners. One car of ore has been shipped to the smelter.——One car of copper ore has been shipped from the Black Hills copper property near Rochford.

WASHINGTON

FERRY COUNTY

(Special Correspondence.)—During the month of November the mines of Republic yielded 7150 tons of ore, as follows: Northport Smelting & Refining Co., from the Lone Pine mine, 4000 tons; Lone Pine-Surprise Consolidated Mining Co., from the Last Chance mine 1250 tons, Knob Hill mine 700 tons, lessees from the Quilp mine 1000 tons, Rathfon Reduction Works 200 tons.——The Last Chance mine is now in excellent working condition. A drift on the 500-ft. level has been driven 200 ft. on the vein. The vein is from 6 to 14 ft. wide. A raise has been carried up from the 500-ft. level 40 ft. to the old Harper workings; the pay-shoot is 180 ft. long, the ore averaging around $11 per ton. Stoping ground is now open from the 500-ft. level to the surface, and the ventilation is perfect throughout.

Republic, December 13.

STEVENS COUNTY

The first dividend to be declared by a magnesite company operating in this county is a 2% return to stockholders of the American Minerals Production Co., making a total of $30,000. The company is quarrying on a big deposit west of the town of Valley, and has been shipping both crude and calcine material for several months. It is headed by Thomas W. Cole of Chicago. Work on the railroad from Valley to the quarries is being rushed.——A disputed point over magnesite operations on account of industrial insurance applying has been ruled upon by the State Insurance Commission, with a decision that taking out magnesite is quarrying and not mining and the hauling of the product to kilns is incidental to the quarry work. However, it is expected that magnesite operations will be classified as mining in the public mind regardless of the decision by the commission.

WHITMAN COUNTY

The School of Mines of the State College of Washington announces that beginning January 7, 1918, a 12-weeks course in mining will be offered to prospectors and mining operators of the North-West. Instruction will be given in mining, metallurgy, geology, and ore deposits, assaying, chemistry, surveying, and ore testing. These courses will be given in the winter School of Mines under a faculty of seven professors from the School of Mines, and the Departments of Geology and Chemistry. Certificates will be issued to those satisfactorily completing any of the courses. A fee of $25 is charged. This is the only charge.

PERSONAL

Note: The Editor invites members of the profession to send particulars of their work and appointments. This information is interesting to our readers.

CARLOS W. VAN LAW is here.

ROBERT M. RAYMOND has returned to New York from Mexico.

C. F. BRANDT has returned to San Francisco from the Dutch East Indies.

JOHN BALLOT, chairman of Minerals Separation Ltd., is in San Francisco.

W. J. SHARWOOD, metallurgist at the Homestake, is here on a short holiday.

C. A. HOFFART has returned to Oroville from the Amur region of Siberia.

HARLOWE HARDINGE is First Lieutenant in the Signal Corps, A. E. F. in France.

W. A. PRICHARD has returned hither from New York, on his way back to Colombia.

RENSSELAER TOLL is at Secret Pass, near Kingman, Arizona, constructing a gold mill.

C. A. THOMAS, resident manager for the Yukon Gold Co., is now at Berkeley, California.

GELASIO CAETANI has been promoted to captain in the engineer corps of the Italian army.

E. L. S. WRAMPELMEIER, of Tucson, Arizona, has opened offices in the Hobart Bdg., San Francisco.

N. I. TSUSCHKOFF sailed on the 'Nippon Maru' for Yokohama on December 15, returning to Russia.

A. B. APPERSON, L. S. DINWOODY, and WILLIAM FOSTER, of Salt Lake City, were here during the past week.

DANIEL C. JACKLING has been selected by the Secretary of War to direct the construction of explosive-plants.

ANDREW NISBET has returned to California from the Federated Malay States, where he has been for the past three years.

C. J. HALL has returned from the Spassky copper mine, Siberia, and is now with the Garfield Smelting Co. at Garfield, Utah.

JACOB W. YOUNG has returner to Oakland from Alaska, having spent the summer on Slate creek, below the Chestochina glacier.

F. W. DRAPER has returned to San Francisco from the Verk Isetz copper mine, in the Ural region of Russia, and is now in New York.

P. G. BECKETT, who recently resigned as manager of the Dominion mine, has been appointed assistant to the president of the Phelps-Dodge Corporation.

N. TRELOAR, formerly manager of the Calaveras Con., has been appointed general manager and engineer of the Copper King mine in Fresno county, California.

AUGUSTUS MACDONALD has been appointed general superintendent of the mining and milling operations for the Primos Mining & Milling Co. at Lakewood, Colorado.

An unsigned letter from Washington, dated December 5, on 'Flotation Physics' has been received. The Editor desires to know the name of the writer.

JOHN H. BROWN died at Oakland on December 14. He was born in Pennsylvania 78 years ago and came to California across Panama in 1859, starting to mine at North San Juan in Nevada county and becoming superintendent of the American mine. In 1892 he went to Idaho and in 1893 he was appointed superintendent for the Leesburg Gold Mining Co. retaining this post for 10 years. In 1907 he went to Cobalt, Ontario, to take charge of stripping operations for the Nipissing company.

METAL PRICES

San Francisco, December 18

Aluminum-dust (100-lb. lots), per pound	$1.00
Aluminum-dust (ton lots), per pound	$0.95
Antimony, cents per pound	17.00
Antimony (wholesale), cents per pound	15.25
Electrolytic copper, cents per pound	23.50
Pig-lead, cents per pound	6.50— 7.50
Platinum, soft and hard metal, respectively, per ounce	$105— 113
Quicksilver, per flask of 75 lb.	115
Spelter, cents per pound	9.50
Zinc-dust, cents per pound	20

ORE PRICES

San Francisco, December 18

Antimony, 45% metal, per unit	$1.00
Chrome, 34 to 40%, free SiO_2, limit 8%, f.o.b. California, per unit, according to grade	$0.60— 0.70
Chrome, 40% and over	$0.70— 0.85
Magnesite, crude, per ton	$8.00—10.00

Manganese: The Eastern manganese market continues fairly strong with $1 per unit Mn quoted on the basis of 48% material.

Tungsten, 60% WO_3, per unit	26.00
Molybdenite, per unit MoS_2	$40.00—45.00

EASTERN METAL MARKET

(By wire from New York)

December 18.—Copper is unchanged at 23.50c. all week. Lead is dull and firm at 6.50c. all week. Zinc is inactive and lower at 7.75c. all week. Platinum is unchanged at $105 for soft metal and $113 for hard.

COPPER

Prices of electrolytic in New York, in cents per pound.

Date		Average week ending	
Dec. 12	23.50	Nov. 6	23.50
" 13	23.50	" 13	23.50
" 14	23.50	" 20	23.50
" 15	23.50	" 27	23.50
" 16 Sunday		Dec. 4	23.50
" 17	23.50	" 11	23.50
" 18	23.50	" 18	23.50

Monthly Averages

	1915	1916	1917		1915	1916	1917
Jan.	13.60	24.30	29.53	July	19.09	25.66	29.67
Feb.	14.38	26.62	34.57	Aug.	17.27	27.03	27.42
Mch.	14.80	26.65	36.00	Sept.	17.69	28.28	25.11
Apr.	16.64	28.02	33.16	Oct.	17.90	28.50	23.50
May	18.71	29.02	31.69	Nov.	18.88	31.95	23.50
June	19.75	27.47	32.57	Dec.	20.67	32.89	....

The production of copper at the Inspiration is only 2,500,000 lb. per month as compared with the nominal output of 10,500,000 to 11,000,000 lb. The shortage of labor is the cause. We understand that the company is engaging only the better type of American workmen.

SILVER

Below are given the average New York quotations, in cents per ounce, of fine silver.

Date		Average week ending	
Dec. 12	85.62	Nov. 6	89.25
" 13	85.62	" 13	86.18
" 14	86.12	" 20	85.62
" 15	85.87	" 27	84.60
" 16 Sunday		Dec. 4	84.70
" 17	85.87	" 11	85.70
" 18	85.87	" 18	85.38

Monthly Averages

	1915	1916	1917		1915	1916	1917
Jan.	48.55	56.76	75.14	July	47.52	63.06	78.92
Feb.	48.45	56.74	77.54	Aug.	47.11	66.07	85.40
Mch.	50.61	57.89	74.13	Sept.	48.77	68.51	100.73
Apr.	50.25	64.37	72.51	Oct.	49.40	67.86	87.38
May	49.87	74.27	74.61	Nov.	51.88	71.60	85.97
June	49.03	65.04	76.44	Dec.	55.34	75.70	....

Plans of the Guggenheims for the re-opening of the Chihuahua smelter in Mexico have been delayed owing to recent depredations of marauders in that section. At the operating points in Mexico which the American Smelting & Refining Co. has maintained none of the plants has been able to operate at capacity. With the demand for silver exceeding supply, enlarged operations by the Guggenheims in Mexico would result in some relief in this direction, although the greatest silver yield of Mexico comes from the southern part, practically inaccessible at present. Operations of the tin smelter of the American Smelting & Refining Co. in New Jersey have been showing a gradual increase and January output should be nearly up to capacity of 800 tons compared with 600 tons in October. Raw material comes from Bolivia in the shape of 60% product which the smelter—the only one in the United States—reduces to marketable form.

Samuel Montagu & Co. says: It has been reported from America, but not yet officially confirmed, that the British and United States governments have bought jointly 100,000,000 oz. of silver for delivery in 1918, of which 60,000,000 oz. is intended for India (at the rate of 5,000,000 oz. per month) and the remainder for the financing of troops in France and elsewhere. The rate indicated is 86c. for delivery in New York, and 85c. in San Francisco. The former figure works out at 40.11d. per standard ounce (based on 476%, the present U. S. exchange). This would compare favorably with 43.05d. per ounce standard, the minted Value of the rupee, calculating at 15 rupees to the sovereign. Should the news be correct, the assurance of so substantial a price for silver should stimulate individual mines to increase their output and it is quite possible that the total of 100,000,000 oz., large though it be, will not represent half of the world's output for next year.

ZINC

Zinc is quoted as spelter, standard Western brands, New York delivery, in cents per pound.

Date		Average week ending	
Dec. 12	7.75	Nov. 6	7.42
" 13	7.75	" 13	7.81
" 14	7.75	" 20	7.98
" 15	7.75	" 27	8.00
" 16 Sunday		Dec. 4	8.00
" 17	7.75	" 11	7.91
" 18	7.75	" 18	7.75

Monthly Averages

	1915	1916	1917		1915	1916	1917
Jan.	6.30	18.21	9.75	July	20.54	9.90	8.98
Feb.	9.05	19.99	10.45	Aug.	14.17	9.03	8.58
Mch.	8.40	18.40	10.78	Sept.	14.14	9.18	8.33
Apr.	9.78	18.62	10.20	Oct.	14.05	9.92	8.32
May	17.03	16.01	9.41	Nov.	17.20	11.81	7.76
June	22.20	12.85	9.63	Dec.	16.75	11.26	....

LEAD

Lead is quoted in cents per pound, New York delivery.

Date		Average week ending	
Dec. 12	6.50	Nov. 6	6.02
" 13	6.50	" 13	6.35
" 14	6.50	" 20	6.50
" 15	6.50	" 27	6.50
" 16 Sunday		Dec. 4	6.50
" 17	6.50	" 11	6.50
" 18	6.50	" 18	6.50

Monthly Averages

	1915	1916	1917		1915	1916	1917
Jan.	3.73	5.95	7.64	July	5.59	6.40	10.93
Feb.	3.83	6.23	9.01	Aug.	4.62	6.28	10.75
Mch.	4.04	7.26	10.07	Sept.	4.62	6.86	9.07
Apr.	4.21	7.70	9.38	Oct.	4.62	7.02	6.97
May	4.24	7.38	10.29	Nov.	5.15	7.07	6.38
June	5.75	6.88	11.74	Dec.	5.34	7.55	....

QUICKSILVER

The primary market for quicksilver is San Francisco, California being the largest producer. The price is fixed in the open market, according to quantity. Prices, in dollars per flask of 75 pounds:

Date		Week ending	
Nov. 20	100.00	Dec. 4	115.00
" 27	110.00	" 11	115.00
		" 18	115.00

Monthly Averages

	1915	1916	1917		1915	1916	1917
Jan.	51.90	222.00	81.00	July	95.00	81.20	102.00
Feb.	60.00	295.00	126.25	Aug.	93.75	74.50	115.00
Mch.	78.00	219.00	113.75	Sept.	91.00	75.00	112.00
Apr.	77.50	141.60	114.50	Oct.	92.90	78.20	102.00
May	75.00	90.00	104.00	Nov.	101.50	79.50	102.50
June	90.00	74.70	85.50	Dec.	123.00	80.00	....

TIN

Prices in New York, in cents per pound.

Monthly Averages

	1915	1916	1917		1915	1916	1917
Jan.	34.40	41.76	44.10	July	37.38	38.37	62.60
Feb.	37.23	42.60	51.47	Aug.	34.37	38.88	62.53
Mch.	48.76	50.50	54.27	Sept.	33.12	36.66	61.54
Apr.	48.25	51.49	55.63	Oct.	33.00	41.10	62.24
May	39.28	49.10	63.21	Nov.	39.50	44.12	74.18
June	40.26	42.07	61.93	Dec.	38.71	42.55	

Under arrangements being made by the Tin Committee of the American Iron and Steel Institute, moderate amounts of Banca tin at 72c., Pittsburgh, continue to be supplied to those who can show that the metal is urgently required for work on Government orders. It transpires that the amount of tin commandeered by the Navy Department in store, New York, was comparatively small, and only represented lots stored in warehouses under bond to the Government. It is believed the Navy requirements are being arranged for by the Tin Committee, and that tin not required from said commandeered lots will be released.

The situation is serious, and the relief must come from new arrivals, therefore what the trade is eagerly praying for is news of free granting of permits for the stocks they have coming to them from London. Large shipments are expected from London, as a result of the successful negotiations of the Tin Committee. There must be a great curtailment in the production of non-essentials into which tin enters, and manufacturers should see to it that this is continued until the situation is relieved.

Neither spot Straits nor No. 1 Chinese tin is offering. A Very small quantity of English Lamb & Flagg is on the market at 80c. per pound. Tin for shipment from London or Liverpool as soon as permit is obtained has been advanced sharply, 72c. now being asked. Tin for shipment from the East Indies, as soon as permit is obtained, is nominal at 71c., and February-March shipment at 70 cents.

Eastern Metal Market

New York, December 12.

Price-regulation is the principal topic in the metal markets. With copper and tin under Government control, lead is soon expected to follow, and spelter next. Activity is limited in all markets.

Copper is unchanged at the controlled prices.

Tin continues to be scarce and to advance nominally.

Lead is inactive, but firm, at unchanged prices.

Zinc is lifeless and lower.

Antimony is dull but steady.

Aluminum is quiet and unchanged.

The chief topic in the steel circles is the question of a further revision of prices, those now in effect having been fixed to January 1. A conference was held in Washington this week, but the result seems to indicate that few, if any changes, will be made, and that present levels will continue for a considerable time, despite the desire of some in authority for lower prices in basic products. The severe weather has interfered with traffic and production. Hundreds of thousands of tons of steel for Europe are held up at Atlantic ports or elsewhere by freight congestion.

COPPER

The market has settled down in such a routine of 'controlled activity' that no interest is taken in it. Practically it is bare of news. From indications the regulated Government distribution, as carried out by the Copper Producers' Committee, is working smoothly and to the entire satisfaction of consumers. Sales are being made rather freely to all consumers in large quantities at 23.50c. per lb., New York, for delivery in the first quarter, but very little is available for December except to those having Government contracts. The jobbing business is also proceeding smoothly at 24.67½c. per lb. or 5% above the Government price, and it is not unlikely that some sales are made at higher prices where the original cost was higher. The principal topic of interest now is the question of a revision of the fixed price of copper after January 31. In the original fixing of a price the end of January was put as the time limit for the application of the present 23.50c. price. A meeting is scheduled for Friday, December 14, to discuss the copper situation with the Government, including possibly the price question. The feeling in the trade is strong that if any re-adjustment is made it must be upward rather than downward, so as to stimulate production as much as possible. The sentiment, however, now seems to be that no change will be made at this time.

TIN

It develops that many consumers are short of tin, some of them being large ones. On the other hand, it is a fact that one large consumer has lately sold spot Banca tin to needy users, refusing any to all others. In general, the market is extremely dull, due to lack of offers, and quotations have continued to advance; but in all cases they have been and are nominal. On December 5, 6, and 7 spot Straits sold at 85c., New York, and on December 10 and yesterday at 86c., New York, the highest ever recorded. Business yesterday and Monday was very light. Cables are abominably late, greatly retarding business. The American Iron and Steel Institute took actual control of the market on Monday, December 10, without any official announcement. There are urgent advocates that import restrictions and licenses on tin be done away with as superfluous, leaving absolute control of the metal in the hands of the American Iron and Steel Institute. Without doubt there are serious delays and annoyances in this phase of the situation. There has been a moderate inquiry for prompt shipment from England, and a little business is reported at 71 to 72½c., New York, subject to license. Arrivals to date, inclusive, have been 460 tons, with 4840 tons reported afloat. The London market is again higher. Yesterday the quotation for spot Straits was £299 per ton, or £5 higher than last week.

LEAD

It is understood that the Government has closed for its requirements for the next 30 days, but the quantity and price are not announced. It is also expected that the price of lead will be fixed in a few days, but on what basis is still conjecture. The market in general has been quiet and without interest during the past week. It has been firm, with quotations unchanged at 6.50c., New York, or 6.37½c., St. Louis, in the outside market for early delivery. The 'Trust' price still stands at 6.25c., New York. Activity has been in spots, but the quantity sold has not been large. Producers are well sold up for six weeks to come.

ZINC

Dealers are decidedly pessimistic and call the market an absolutely lifeless one, with the prospects not of the most encouraging nature. Until there is a balance between supply and demand conditions are likely to remain as they are, with buying at a low ebb and with prices falling. Some think that, because of the general dropping out of smelters here and there, this balance has about been reached. It is estimated by some that production has fallen off 30%, while others put it at 50%. It is probable that an accurate survey, were this possible, would put the reduction in output at 40%, but there are still enough producers to meet the poor demand so that quotations have gradually declined again until yesterday prime Western was quoted at 7.50c., St. Louis, or 7.75c., New York, for early delivery, with forward positions ⅛ to ¼c. higher. Galvanizers and brass makers are not in the market nor has the Government's programme for shells had any effect. When this materializes in orders, its volume will be large, but just what effect it will have on the market is a matter of speculation. Price-fixing has made no progress publicly, neither is it known how much the Government will need of the various grades. The entire situation is uncertain and unsatisfactory.

ANTIMONY

The spurt in the market last week was short-lived. Attempts were evidently made to lift it on the strength of the effect of the embargo on imports, but this did not work out. Today the market is quiet at 15.25 to 15.50c. per lb., New York, duty paid, for the brands available.

ALUMINUM

The market continues dull and quiet at 36 to 38c. per lb., New York, for No. 1 virgin metal, 98 to 99% pure.

ORES

Tungsten: There is practically no change in prices, the range being from $26 per unit for scheelite down to $20 to $24 per unit for wolframite and other grades as 60% concentrates. Sales have been made at these prices, but there is a tendency to quietness as the end of the year approaches.

Molybdenum and antimony: Molybdenum is strong, with the demand active. Small sales are reported at from $2.20 to $2.30 per lb. of MoS_2 in 90% concentrates. There is nothing new in antimony ore.

A HAPPY NEW YEAR to the mining industry: good prices for metals and such use of them as shall enable the Allies to win a victorious peace!

TAXATION is a subject in which everybody is interested at this time. On another page we publish an analysis and a discussion of the excess-profit tax, with suggested modifications, by Mr. Philip Wiseman, the president of the United Eastern Mining Company, the most successful gold-mining enterprise in the Oatman district of Arizona. The fact that this memorandum was not intended originally for publication makes it only the more interesting and convincing. It seems to us to be a sane and reasonable protest, not by a man or by a group of men anxious to evade making their proportionate contribution to the national treasury at a time of crisis, but by citizens having a valid objection to inequitable taxation.

IN this issue we publish a description of the Government Railroad in Alaska, an enterprise of the greatest importance to the welfare and development of the North. This article is written by Mr. Theodore Pilger, a mining engineer from Butte, who spent the past summer in a tour of inspection in southern and south-eastern Alaska. The excellent map that goes with the text will indicate the route to be followed by the railway into the heart of the interior, or 'inside,' as the Alaskans call the region across the coast range, in distinction to the 'outside,' meaning the rest of the world. We join with our friends up north in hoping that this railroad will give aid to the exploration and development of a great mineral region. In the news columns of recent issues we have published notes, by Mr. Pilger, describing progress in the Willow Creek, Cache Creek, and Copper River districts.

THEORIES of delicately balanced conservation and utilization of mineral resources are giving way to practical economic development for the welfare of the nation. The plans brewing in Congress for a vast expansion of coal-mining and oil-production on previously withdrawn areas in the West, go so far as the proposal, made by congressmen who hold to the older standards, to recognize again the right of acquiring ownership in fee simple. This marks a return to the historic democratic principles of American development, which is founded on that sort of individual initiative and strength that actual possession encourages. The Government now seems disposed to take a true forward look toward the effective unfolding of the resources of the West.

FROM Cambridge comes the announcement of Mr. Theodore W. Richards, the president of the American Association for the Advancement of Science, requesting members not to attend the annual meeting at Pittsburgh "unless assured that they can render an important service to the country by doing so." This is said in order to lessen the difficulties of transportation, already acute. It is a wise move. Unnecessary travel is to be deprecated. Tourists are an absurdity in times like these. We read the other day, among a list of hotel arrivals, the name "A. Vandenberg, tourist," as having landed from an incoming Australian steamer. A man that starts to go on a tour at this juncture in the world's history ought to be placed in a lunatic asylum or tied in a sack and dropped in the nearest large body of salt water.

THE threat to place an embargo on freightage to gold mines has been disclaimed by Mr. R. S. Lovett, the chairman of the Priority Board. We are glad of that. The intimation that such freightage would be declared non-essential to the War came to us from such a source as to indicate that it was no false alarm, although now disavowed. To Senator Pittman of Nevada, we believe, is to be given the credit for putting the matter in a proper light before the authorities at Washington. We are amused to learn that sundry gentlemen in San Francisco suggested that the story was concocted to provide a Hindenburg retreat for the promoters of the two unfortunate gold-mining enterprises at Juneau. This is a fallacy, which is punctured further by the letter from Mr. J. Parke Channing dated December 3, on the subject, in our New York contemporary's issue of December 8. The luncheon at the Pacific Union Club given by Mr. F. W. Bradley in honor of Mr. Hennen Jennings and with the purpose of enabling Mr. Jennings to meet a number of representatives of the gold-mining industry in California took place on December 5. Mr. Jennings, in his capacity of consulting engineer to the U. S. Bureau of Mines, had come to San Francisco to obtain data on the status of gold mining in its relation to the War and to the taxation incident thereto. We take the opportunity of suggesting to managers of gold mines that they put themselves in communication with Mr. Jennings, so as to assist in the gathering of the data required by the Bureau of Mines.

BRET HARTE has said poetically that "the ways of a man with a maid be strange, yet simple and tame to the ways of a man with a mine when buying or selling that same." To the prospectus we owe as many humorous errors as to the school-boy's essays. The latest appears in the advertisement of the Gold Cross Mining Company, which offers stock for sale at St. Louis, to operate a mine near Prescott, Arizona. The general nature of the business includes "to bore, drill, prospect, and mine" for sundry metals, including "steel" and "brass." No wonder that one of our prospector friends writes to say that this is "something new" to him, and inquires if we know of "any books on the geology of brass." He adds that "one of the parties is a mining engineer with a great long handle to his name and knows all about brass." We expect he does. He must be an embodiment of it. As far as we know, no vein or other natural deposits of brass or even 'steal' have been discovered in Arizona or in any other of our rich and varied mineral regions, but both these products are made artificially for the purpose of facilitating the circulation of money, particularly from the pockets of the simple to those of the sophisticated.

DRAFTING of undergraduates in colleges and universities, especially young men dedicated to scientific pursuits, has appeared unwise to us and to many patriotic Americans. We have urged that Congress discriminate in favor of men already matriculated and in line of training for such scientific work as is of economic importance. It is pleasing now to record that the California State Council of Defense has directed attention to this subject, and that this body regards as an ineffectual solution of the problem the request sent from the War Department that students of science continue their studies until drafted, without previous enlistment. Indeed, it is ineffectual. Young men on the draft-list are kept in such uncertainty as to their future movements that it becomes quite impossible to concentrate the mind upon study in a way that can lead to good results. We know of cases where drafted men have been kept dangling for months, with orders to report at a training camp on a certain date, countermanded a few days later by orders to look forward to service next April, followed in a few days by notification to keep in readiness to report on call within 24 hours. Any man who has ever done serious study knows that it is impossible under such conditions of mental distraction as must result from these conflicting orders. Again we insist that our political leaders will ultimately have cause to repent their folly in having depleted the engineering and mining schools of the younger generation of specialists in case the War should continue a year or two longer. Trained men will be needed in greater numbers for the vitally important industrial ranks that make it possible for other men to fight for country and home on the battle-front. It is not exemption from military service that we advocate; it is the performance of national service in the technical branches where they are peculiarly and urgently required. It is a truism to assert that the scientific army is hardly less important than the army in the trenches. We would soon discover that fact calamitously if we had to recruit incompetents to do the work in our mines, our smelters, our steel works, and our chemical factories. A practical solution of the matter would be to detail training officers to the important universities, enlist the students in the army, and give them daily military training, but under regulations ensuring their continuance in college until graduation, with the outlook thereafter corresponding to their special aptitude and to the Government's greater needs at that particular time. We commend this to the serious consideration of Congress.

War Minerals

In previous issues we have drawn passing attention to the importance of the work being done by the War Minerals Committee, of which Mr. William Y. Westervelt is chairman, assisted by such representative economists and geologists as Messrs. Alfred G. White, David White, and W. O. Hotchkiss. The good work being done by these gentlemen is to be aided by a bill already drafted and submitted to the Mines Committees of both houses of Congress. This, therefore, is a good time to bespeak the interest of our readers. Even the man in the street is aware that this is a war in which metals are required in enormous amount; our readers, taking part directly or indirectly in the production of metals and minerals, are cognizant of the need for accelerating every effort in creating an adequate supply of the material used in the manufacture of munitions. The War found us unprepared to supply many important ingredients, which we had been accustomed to import, some of them from the enemy countries and others from the allied countries now unable to export to us on account of the scarcity of shipping. Domestic production has been hindered by labor problems, railway congestion, the incidence of new taxation, and other discouragements. In order to overcome these difficulties and to encourage domestic production it is necessary that authority be given to the President whereby he may have wide powers for this particular purpose. The bill to be submitted to Congress is modeled upon that which authorized the formation of the Food and Fuel administrations. The aim is to provide authority for stimulating the production of ores and minerals, and to conserve the supply of them, to assure equitable distribution, to improve and direct utilization, to prevent manipulation, to lessen our dependence upon importation, and to control prices.

In order to strengthen the Government's purchasing power it is necessary that the President shall be enabled to proclaim maximum prices. In order, however, that such action shall be taken wisely it is advisable that an advisory board be created capable of collecting information on which to base such control of production, involving interference with the normal operation of the laws of supply and demand. Mining cannot respond immediately to an increased demand, but it can be crippled suddenly by unwise restrictions; therefore any administrative control required for the War must be exercised judiciously. At first the Government endeavored to assist importation of war minerals, thereby discouraging the development of domestic deposits. The shortage of ships stopped that abruptly and it became evident to everybody that this country had to make intensive effort to find every war mineral within its own borders or in the adjacent friendly territory of Canada. In order to encourage private initiative in the exploration and exploitation of the necessary minerals, it is essential that the Government state its requirements, announce its policy in regard to importation, and guarantee a minimum, as well as a maximum, price over a definite period long enough to justify systematic search and development of mineral deposits. Capital will not be induced to take the risk inseparable from mining adventure unless the basic uncertainties of the market are removed or lessened. Any steps in this direction, in order to be taken intelligently, should be guided by an agency exercising control over the mineral industry of the whole country. It is true we have the Geological Survey and the Bureau of Mines, but neither of these lepartments has the machinery adequate for the purpose. Such machinery can be made out of the War Minerals Committee, which has already demonstrated ts energy and good sense. The Committee includes vorthy representatives of both the Geological Survey ınd the Bureau of Mines, and it can be assumed with onfidence that anything that the Committee does will e done with the willing aid of the two existing branches f administration. The head of the Committee, most ppropriately, is a mining engineer of wide experience. o a good start has been made. In order to proceed on roader lines and in a more systematic manner, it is exected that provision will be made by Congress for the unds needed to purchase surplus production during a efinite period; also that a clause in the bill will adjust e prices of imports after such prices have been estabshed, otherwise the liability of the Government may be nduly increased by its obligation to purchase at guarnteed minimum prices. In order to check profiteering d to protect the miners themselves, it will be necessary discipline the smelting and refining companies, which, return for a stable market for their products, will be revented from imposing unfair contracts on the minal producers. Any reserves of stock will be put under e control of the Government, in the national interest, d to check speculation or manipulation within the limition of the minimum and maximum prices. Another clause will protect the producer by giving the President the power to control the cost of supplies entering into production. Thus, it will be seen, a radical scheme of legislation has been devised to meet the abnormal conditions created by the War. We feel confident that it will be accepted in good spirit as a means to a great end, namely, the waging of war to a victorious peace. Interference with industry is annoying, if casual or unnecessary; regulation of industry in time of war is necessary; we are facing a crisis that justifies unusual measures and we feel confident that all those taking part in mining will accept loyally any enactment that will conduce to national efficiency.

Retrospect

The end of another year finds most of us serious, if not anxious. We are living in the midst of events that eclipse the importance of the individual experience. Our thoughts are projected continually beyond the narrow horizon of the individual life to that of the nation and of the world. We have been shaken out of our ordinary petty detachment into an awakened interest in the big things of existence. Many of our hopes and illusions have been shattered. The optimism that had led us to imagine that the jungle period of human evolution has passed forever, the belief that the ape and tiger in man had died with the reading of books has been falsified, the fond expectation that intellectual endowment would terminate savagery on land and sea has proved a delusion. We face a resurgence of brutishness, the appeal to physical strength, and the decision of an ungenerous and unchivalrous mode of fighting. In these winter days the outlook is obscure. The year 1917 had opened cheerfully; in the spring a new democracy came into being out of the ashes of a revolution; we hailed Russia as a comrade in the march of political progress. That hope has proved a Dead Sea apple. The destiny of Russia has passed into the hands of a group of crazy criminals; she has played the poltroon and the imbecile, yielding to cowardice and corruption her virility as a nation and her self-respect as a people. Russia has proved a quitter. If a people gets the kind of government it deserves, then Russia deserves her Germanized I. W. Ws to rule her henceforth. To us, who have known the Russians, their kindly character, their dogged courage, and the keen intelligence of their constitutional reformers, the debacle is a tragedy without mitigation. The Italian reverse is another lamentable event. Undoubtedly the sudden defeat on the Isonzo was due to treachery, to a refusal to fight on the part of those defending the extreme left flank. Poisoned by German propaganda they gave an opening through which the invader forced his way so promptly as to turn the whole battle-front and compel a disastrous retreat. That is now in the record of 1917; so also is the greater event of American participation, an event that goes far to compensate for the Russian collapse. But with the realization of the great service already done by the United States in heartening the cause of right there comes an appreciation of the fact that it was done so late as to endanger the result of the War and to prolong it by at least two years. The regret that our entry into the conflict should have been delayed is hardly softened by the conviction, which we share, that the United States took action as soon as her people were ready for it, and that earlier interference would have found the President at the head of a distracted people. As we entered late so the greater is the effort that we must now make to bring the War to a decision, the only decision that will allow the unhindered development of liberal ideas and the continued progress of representative government in our own country and in the world at large. We care greatly for the ideals of political freedom, but another feature of the War forces itself upon our attention at this season of supposed peace and goodwill. Increasingly we who are detached by distance from the scenes of carnage and destruction in Europe are coming to understand the blistering pain that afflicts those overwhelmed by the great calamity. This year has seen a splendid awakening of sympathy with the victims of the special frightfulness unloosed by the Huns and of the manifold horrors inseparable from any war. Not only has money in increasing amount gone forward for the alleviation of that suffering, but even those heretofore heedless of the misery across the Atlantic have begun to feel the pathos of it all and the pressing need for doing what is humanly possible to mitigate the fell consequence of invasion and conquest by the beast with the brains of an engineer. We have traveled a long way since our last year's end; as a people we have committed ourselves to the greatest enterprise in which this young republic has ever taken a hand; we have awakened to our real part in the comity of nations; we are about to prove our manhood. Therefore we face 1918 sadly, soberly, but firmly, strong in the purpose to see it through.

Perils of Inflation

Some years ago the Chinese revolutionary committee in San Francisco offered for sale an issue of paper money on behalf of the party seeking to supplant the dynasty of the yellow dragon with a democracy. At that particular moment there was no democratic Chinese government in existence; there was only the hope of one; so the committee offered the nominal equivalent of 10-dollar bills for five cents. The revolution succeeded and those who had gambled on the Chinese insurrection through the purchase of these bills reaped a handsome reward. That was fiat money, looked at from an end that is not often possible, and leading to a result that is even less frequently realized. No better example could be found to illuminate the fact that beneath fiat money rests nothing tangible. It is national pride that has made so many people in America at so many different times advocate currency inflation. The Government seems so big, so powerful, so representative of the opulence of a vast productive area, that the national credit appears illimitable. It is difficult for men whose financial experience has not exceeded the simple transactions of buying and selling commodities to comprehend the delicacy of the mechanism of credit in modern industrial life, and people with this simple experience constitute the great majority of the population. They cannot see why the Government might not issue billions of greenbacks to pay for war-supplies as well as to borrow treasury-bills from the people in exchange for another promise to pay in the form of a bond. Mr. Vanderlip has been forced to come to the rescue, and he is expounding the doctrines of sound finance in the Middle West in order to nip the growing heresy in the bud, when, as a matter of fact, his services are urgently needed in Washington for other purposes. Nevertheless the fiat-money craze is a disease so perilous that a prophylactic must be applied instantly; it is a disease as dangerous as typhus; it floats in the air and inoculates the passers-by, once it gets going; no more insidious and damaging germ than this could the Teutons spread throughout the country; it is worse than tetanus or anthrax, because our financial structure would crumble should the greenback fallacy get itself written into law, and the result would be universal distress and suffering. Gold would continue the actual gauge of value, and paper the measure of timid and declining faith. It would not take long, on that basis, before a five-dollar pair of shoes would sell for a hundred paper dollars. The movement of commerce would be blocked by insistence on contracts for payment in gold, while no bank would dare to let the gold escape from its vaults. The advantage of the bond is that the Government defers the date of redemption; it shifts the time for payment to a coming generation, and has to meet no more than current interest. The treasury-note, on the other hand, is a promise to pay on demand. What is it that the Government promises to pay? It is gold! Unless there is a reasonable proportion of gold in the vaults to redeem the issue on the demand of doubting Thomases it will not take long for the fiat-money enthusiasts to realize as keenly as the bankers that the Government cannot pay and that the paper is worthless for all present needs. The end of inflation is the explosion of credit. In the case of China the revolutionary paper grew in value because the revolutionists got possession of the solid metallic resources of the country in time to save the situation. In Mexico the grafters made off with the gold, while the people were forced at the muzzle of the rifle to accept the paper money in exchange for commodities and labor, and in the end, in order to extinguish the last trace of the obligation, the Government decreed that, on the payment of taxes, which were collectible in gold, an additional sum, of an equal face-value, must be paid in the fiat money. Mexico furnishes many instructive lessons in the things to avoid in politics and infinance, yet even Carranza, after sneaking from the duty of redeeming his convenient revolutionary currency, has planted the country squarely on a hard-money basis.

Physics of Flotation

The Editor:

Sir—On page 535, Trans. A. I. M. E., Vol. LV, R. J. Anderson, in summarizing the electrolytic and electrostatic phenomena of flotation, states that "bubbles in flotation are simply air-spaces, contained by a mantle of oil or of water and there is therefore nothing within to bear the charge. In case it could carry a charge, which would only be possible by the presence of contained ionized gases or water-vapor, the charge would be speedily dissipated by contact with the interfacial boundary." He then refers to Fahrenwald's article, 'The Electro-statics of Flotation,' M. & S. P., Vol. III, No. II, page 375 (March 11, 1917). He does not refer to James A. Block's article criticizing certain statements of Mr. Fahrenwald. This article is in the M. & S. P. of June 10, 1916.

O. C. Ralston, in the discussion of Mr. Anderson's paper, brought out some other points of interest, but also stated that "Fahrenwald's objections to this theory are sound." In the last paragraph of Mr. Ralston's discussion, on page 546, he makes this excellent remark: "Too many have been willing to contribute thoughts to the technical press and too few have busied themselves with experimental measurements to prove the theories proposed."

I therefore take great pleasure in enclosing a copy of some paragraphs from Sir J. J. Thomson's 'Discharge of Electricity Through Gases' (1898). He made a thorough investigation of the electrification of gases over twenty years ago. In all the literature that has been published on flotation, I have yet to see his researches quoted. L. S. Deitz Jr., now with the Bureau of Mines at Cornell, called my attention to these researches, soon after my articles in 1915 were written.

THOS. M. BAINS JR.

Minneapolis, Minn., November 2.

From J. J. Thomson's 'Discharge of Electricity Through Gases.'

COMMUNICATION OF A CHARGE OF ELECTRICITY TO A GAS

One of the most striking phenomena connected with the electrical properties of gases is the difficulty of directly communicating a charge of electricity to a gas in its normal condition. A very simple instance will suffice to show this: let us take the case of a charged metal plate which is insulated so well that there is no leakage of electricity across its supports; let this plate be in contact with the air or any other gas at a moderate temperature and let it be screened off from ultra-violet light and Röntgen rays, then the evidence of the best experiments we have proves that under these conditions the plate will suffer absolutely no loss of charge, provided the surface density of its electrification is less than a certain value. Thus, though myriads of molecules of the gas strike against the charged surface, they rebound from it without any electrification. To fully appreciate the significance of this result, we must remember the very large charges that can be carried by the gas under other conditions. The phenomena of electrolysis show that the charge on each unit surface of the plate could be carried thousands of millions of times over by a cubic centimetre of hydrogen at normal pressure and temperature. We must, I think, conclude that the inability of a gas (which when in a certain state has such an enormous capacity for carrying electricity) to take up when in its normal condition any of the charge of electricity from a body against which it strikes is very significant and suggestive.

Another fact which exhibits in perhaps even a more striking way the inability of the molecules of a gas to take up an electric charge is that the vapor arising from an electrified liquid is quite free from any charge of electricity. * * * In these experiments the molecules of the unelectrified vapor came from or through an electrified surface and the conditions here seem such that if a molecule could ever get electrified by contact, it would do so in these experiments. Thus we see that when an electrified liquid evaporates the electrified particles are left behind, just as the salt in a salt solution is left behind on evaporation.

ELECTRIFICATION OF GAS BY CHEMICAL MEANS

Chemical action is so frequently attended by electrical separation that we might expect that the most likely way to electrify a gas would be to make it one of the parties in the chemical reaction. Examples of electrification, apparently in a gas, produced by this method have been known for a long time, though it is only in recent years that experiments have been made to show that the electrification in these cases is not, in all cases, carried by dust which may either be present in the gas to begin with, or produced by the chemical reaction itself. One of the earliest-known cases of electrification produced in a gas by chemical action is that of the combustion of carbon. Pouillet found that when a carbon cylinder was being burnt, the cylinder was negatively electrified, while there was positive electrification in the gas over the cylinder. Lavoisier and Laplace showed that the same effect takes place with glowing coal. Reiss proved that there was positive electricity in the air near a glowing platinum spiral. Pouillet found that when a jet of hydrogen was burnt in the air there was a negative electrification in the unburnt hydrogen of the jet. Another case was discovered by Lavoisier and Laplace, who found that when hydrogen is rapidly liberated by the action of sulphuric acid on iron, a strong positive electrification comes off. This, with other cases of electrification by chemical means, was investigated by Mr. Enright. As a good deal of spray is produced by the bubbling of the hydrogen through the sulphuric acid, it has been suggested that the spray and not the gas may in this case be the carrier of the electrification. To test this point, Mr. Townsend recently made in the Cavendish laboratory a series of experiments on the electrification produced when a gas is liberated by chemical action. He found that when the hydrogen produced by the action of strong sulphuric acid on iron or zinc was passed through tubes fitted with plugs of tightly packed glass wool it retained after its passage through these plugs a strong positive electrification, thus showing that no ordinary spray could be the carrier of the electrification.

Using an open beaker, Mr. Townsend found that when the mixture of the sulphuric acid and iron was heated to about 94° C., there was at first strong positive electrification at the mouth of the beaker when the chemical action was very vigorous and the gas was coming off with great rapidity. But as the temperature fell and the rate of evolution of hydrogen diminished, the positive electrification diminished, and finally changed to negative. It was found, however, that the negative electrification, unlike the positive, was completely stopped not only by a plug of glass wool, but even by a layer of wire gauze. This seems to indicate that the negative electrification is carried by coarse spray, while the positive on the hydrogen, or at any rate on much smaller carriers. If there is positive electrification in the hydrogen, there must be an equal quantity of negative in the sulphuric acid; if this is dashed into spray, the spray will be negatively charged. The experiments indicate that when the gas is not coming off very vigorously, the negative electrification carried from the vessel exceeds the positive carried off by the gas; while when the gas comes off with great rapidity, the positive electrification carried by it far exceeds the negative carried by the spray. Mr. Townsend investigated several other cases by electrification produced when a gas is liberated by chemical means. He found that when chlorine was liberated by the action of hydrochloric acid on manganese di-oxide, the chlorine had a strong positive electrification, and that when potassium permanganate was heated, there was positive electrification in the oxygen evolved.

Electrification of Gases Liberated by Electrolysis

Mr. Townsend has also found that when a strong current is sent through a solution of sulphuric acid, so that there is copious liberation of hydrogen at one terminal and of oxygen at the other, there is positive electrification in the hydrogen, while the oxygen is either apparently unelectrified or has a small positive charge. It is perhaps to the point to mention that in this case the oxygen is liberated by a secondary process; the negative ion is SO_4 and the oxygen is liberated by the chemical action of this ion on the water. The positive electrification in the hydrogen is very much influenced by temperature. At the ordinary temperature of the laboratory there is very little electrification; when, however, the temperature is raised to 40 or 50° C., the electrification is very strong. When a solution of caustic potash is electrolyzed, there is, on the other hand, very little electrification in the hydrogen while the oxygen is negatively electrified, though the amount of this electrification is not nearly so large as that on the hydrogen in the preceding experiment. In this case the hydrogen is liberated by secondary chemical action and the great diminution in its electrification as compared with the previous case seems to show that a gas is more likely to be electrified from electrolysis when it forms one of the ions than when liberated by secondary chemical action. The amount of electrification in the oxygen rapidly increases with the temperature. If oil is added to the caustic potash solution, the sign of the electrification in the oxygen changes and the oxygen is positively instead of negatively electrified. The nature of the electrodes has a considerable influence on the amount of electrification which comes off with the gas.

Electrification by the Splashing of Liquids

One of the most effectual ways of charging a gas with electricity is by means of the splashing of liquids. It has been long known that there was something anomalous in the condition of atmospheric electricity at the feet of waterfalls, where the water fell upon rocks and broke into spray. Lenard investigated the subject with great thoroughness, and found that when a drop of water splashed against a metal plate [note that is what happens in the M. S. machines] a positive charge went to the water, while there was negative electrification in the surrounding air. This electrical separation is even more marked in the case of mercury than in that of water. A very simple way of showing the effect is to shake some mercury up vigorously in a bottle and then draw off the air. This will be found to have a negative charge; this charge is carried neither by dust nor by spray, for it will remain after the air has been sucked through glass wool or even through a coarse porous plate. [Note that this may be applied to the Callow process.] Lord Kelvin has shown that the reciprocal process of bubbling air through water also gives rise to electrification, the air which has bubbled through the water being negatively electrified.

To investigate the laws of electrification produced by splashing, it is often more convenient to measure the positive charge on the drop rather than the negative charge in the air. This can be done conveniently by the arrangement represented in Fig. 1.

A known quantity of the liquid to be investigated is owed to fall through the funnel A and falls on the tal plate and saucer C, which is carefully insulated l connected with one pair of quadrants of an elec-meter. The blower sends a strong current of air from neighborhood of the plate, so as to prevent the nega-e electrification in the air interfering with the indica-ns of the electrometer. The deflection of the electrom-r when a given quantity of liquid falls on the plate y be taken as an indication of the electrification pro-ced by the splashing of the liquid. By aid of an in-ument of this kind, we can investigate the circum-

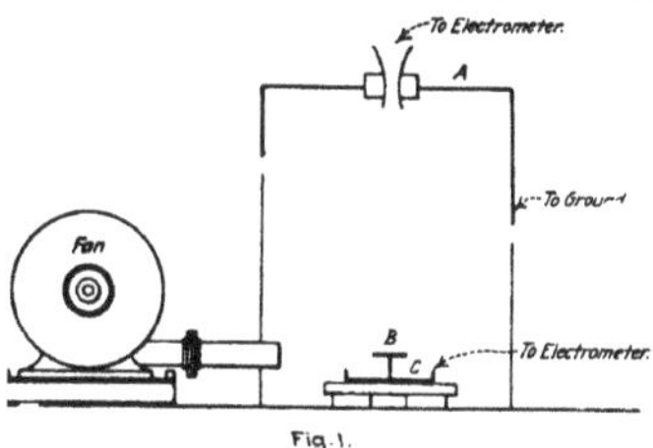

Fig. 1.

nces which affect the electrification. This electrifica-n is influenced to a remarkable extent by minute inges in the composition of the liquid. [Note this.] Lenard found that the electrification in the air in the ghborhood of the splashing plate, which was very rked with the exceptionally pure water of Heidelberg, s almost insensible with the less pure water of Bonn, ile the splashing of a weak salt solution electrified the in the neighborhood positively instead of negatively when water was pure. Thus, while the splashing of rain electrifies the air negatively, the breaking of res on the seashore electrifies it positively.

n some of the experiments which I made on this sub-: I found that the effects produced by exceedingly nute traces of some substances were most surprisingly te. Rosaniline, for example, is a substance which has great coloring power, so much so, indeed, that the r imparted to a large volume of water by a small ntity of rosaniline is sometimes given as an instance he extent to which matter can be subdivided; yet I d that the change produced in the electrification, to the splashing of drops, was appreciable in a solu-so weak as to show no trace of color. The effects of rescent solutions on the electrification of drops is cially great, but different kinds of solution act in rent ways. Thus rosaniline and methyl violet re-the effect, that is, they make the electrification on lrop negative in the air; while eosine and fluorescein, e other hand, increase the normal effect, that is, they the drop more strongly electrified positively than a of pure water, and produce a greater negative rification in the air.

e electrical effects of weak solutions are much greater than those of strong ones [Note this]; in fact, strong solutions of all substances tested gave little or no electrical effects by splashing. The effect produced by the addition of foreign substances to water may be represented by curves, in which the abscissae are proportional to the amount of the substance added to the water, and the ordinates to the electrification produced by splashing. We find that these curves are of three types, a, b, and c. Fig. 2.

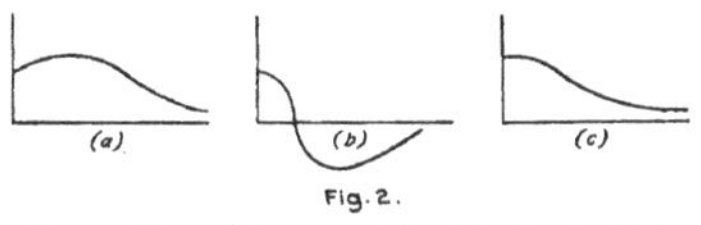

Fig. 2.

Curves of type (a) represent the behavior of solutions of phenol, eosine, fluorescein, where the addition of a small quantity of the substance increases the electrical effect.

Curves of the type (b) represent the behavior of solutions of potassium permanganate, chromium bioxide, hydrogen peroxide, rosaniline and methyl violet; here the addition of the substance begins by diminishing the electrification and finally reverses it.

Curves of the type (c) represent the behavior of solutions of zinc chloride, hydrochloric acid, and hydriodic acids, and in fact of most inorganic salts and acids; here the addition of the substance produces a diminution but not a reversal of the electrification.

The addition of strong oxidizing agents to the water seems, on the whole, to tend to reverse the normal effect, that is, it tends to make the electrification in the air positive, while the addition of the reducing agents seems to increase the normal effect.

We find that the electrification is produced by splashing of drops of substances as unlike as paraffin-oil [note], water, and mercury.

ELECTRIFICATION OF A GAS BY THE AID OF THE RÖNTGEN RAY

Of all the methods by which we can put a gas into a state in which it can receive a charge of electricity, none is more remarkable than that of the Röntgen rays. These rays, when passed through a gas, turn it into a conductor and enable it to receive a charge of electricity and the gas retains its conductivity and its power of being charged for some time after the rays have ceased to pass through it. * * * If the gas in its passage from the aluminum vessel to the tester was made to bubble through water, every trace of the conductivity seemed to disappear. The gas also lost its conductivity when forced through a plug of glass wool. * * * The properties of gases electrified when under the influence of the Röntgen ray differ somewhat from those of gases electrified by other means. Thus the electrification in a gas electrified under the Röntgen rays cannot pass through glass wool or bubble through water. [This nullifies Mr. Fahrenwald's experiments, in M. & S. P., Feb. 26, 1916.] It thus resembles the conducting power conferred on the gas by these rays. On the other hand, as we have seen when the electrification in the gas is by chemical means [or splashing of liquids], the electrification is able to pass through glass wool or *bubble through conducting liquids.* [Note this.]

Do We Need Gold?

The Editor:

Sir—Under the above heading, in your issue of December 1, Mr. Lester S. Grant asks "Why is gold mining?" From the context of his letter it appears that he, with many others, wants to know if gold mining, and probably other industries not essential to war, should be continued under the present world crisis.

Above all else, if this nation is to take part in financing the Allies, as well as in manning and financing an army of its own, it is manifestly essential that every dollar of new wealth from Nature's storehouse, should be brought into requisition and added to the financial strength of the nation, by the enrichment of its citizenry. The funds raised come from the citizens, from the profits of their industry, either through the purchase of bonds or through taxation. Again, a natural resource of energy lies dormant in every individual, which can be brought out and converted into new wealth only by its expenditure upon raw material. Are the reasons against the present mining of gold so obvious? The labor employed in the industry might or might not ultimately find its way into other channels; in the meantime, contemplate the loss of wealth to those at present so employed, the loss of purchasing power of the labor with the infinite reaction on production in innumerable lines, the proportionate loss of earning power along these lines, the lowered assessable valuation and the loss of profit, and their consequently lowered bond-purchasing power, to say nothing of the dormant, unprofitable, and untaxable condition of the natural material resources so suspended, and that cannot aid in purchasing bonds. As to the consumption of supplies, powder, machinery, foods—these operatives must live, must eat, must clothe themselves, whether mining gold, or doing something else. The increased wealth makes possible, through the expenditure of but a small percentage of itself, the provision for this additional consumption. Is not this relatively small item justified in the greater good the added wealth will engender? Every added dollar of new wealth lessens the burden on the dollar in circulation; every dollar's worth of new enterprise, of taxable values, lessens the burden of taxation on established industries, and every dollar of new wealth strengthens the arm of Uncle Sam in his fight for liberty and freedom. What of the investor in and the operator of gold mines? He must, in these times of high prices for labor, for explosives, for cyanide, for machinery, be a patriot. If not, he will not stay in the game. His ounce of gold brings not one cent more than it did in the time of peace. As a business proposition, it were wise to shut-down. Will he? But, looking farther, what can he do with his profits? Will not these compensate, in part at least, for the near-by loss in mining the gold? I believe, judiciously applied, the profits will compensate in full, while the patriotic side is not to be overlooked.

L. S. Ropes.

Helena, Montana, December 3.

The Threat to Gold Mining

The Editor:

Sir—Your timely editorial on this subject strikes me as particularly worthy of commendation. It certainly seems most short-sighted to do anything, by governmental action, that would decrease the output of gold.

The statement, frequently made in irresponsible publications, that the United States "has all the gold in the world" does not tend to convince the more careful-thinking ones.

In the financial column of the 'Chronicle' this morning it is stated that the public debts of the six leading belligerent nations on January 1, 1917, amounted to $71,372,000,000, and that the enormous total, on the coming New Year's day, with our United States debt added, "will probably pass the hundred-billion mark." While the argument is made in favor of the re-monetization of silver, it is surely a better argument in behalf of a more liberal treatment of the gold-mine operators. In fact, some governmental stimulation of the production of gold that could be brought about without interference with the production of the other metals would be the more rational course to pursue.

The argument was made, as long ago as the first of last April, that an ounce of gold was worth only $12 to $14 because of the decrease in its purchasing power. But, for that reason, should the hungry prospector and impecunious mine-owner be satisfied with an offer for his property, based on specious figures, thus made to the detriment of the yellow metal? Either one of two things: the value of an ounce of gold, above cited, is a case of mistaken judgment, on the part of those so asserting it, or they have an ulterior object in view.

Your question "are we to diminish the foundation of our financial security, at this time of world-crisis" is one that is quite likely to be answered in the near future, by a reversal of the idle talk of shutting-down the gold mines and a substitution of encouragement to the gold-mine operators, who have already suffered by the rise in cost of supplies and the increased scale of wages. If there is anything more fallacious than the attempt to prove that we cannot secure enough gold to pay the annual interest charge, I would like to know what it is.

Since 60% of the world's output of gold is produced under the British flag, and only 20% in the United States, it would seem a clear case of discrimination against ourselves, if now we were to enact any regulations which would still further restrict the production of gold in our country.

F. L. Sizer.

San Francisco, December 17.

ANCHORAGE, ALASKA

The Government Railroad of Alaska

By THEODORE PILGER

Since the mining business furnishes more than a half of the total freight-tonnage of American railroads, it is altogether fitting that the first railroad to be owned and operated by the Government should be built to tap what is hoped may prove to be one of the greatest undeveloped mining regions of the North American continent. The vast store of wealth that the territory of Alaska has contributed to the nation entitles it to substantial consideration, which can best be put in practical form by providing Alaska with a proper system of transportation.

During the years 1902 and 1903 the Alaska Central Railroad Co. built its line north from Seward. This company, being inadequately financed, was unable to carry out its purpose and soon went into bankruptcy. At later periods the bondholders and successors in interest constructed 71 miles of standard-gauge railroad from Seward north to Kern creek. The interests backing the company at this time were also practically in control of the Matanuska coalfields. In 1906 the Roosevelt order withdrawing the coalfields of Alaska from entry removed the principal source of freight anticipated by the railroad-builders and as they were again in financial straits the idea of building an Alaskan railroad by private capital was finally decided to be impracticable.

Shortly afterward came the Ballinger-Pinchot controversy which frightened capital to such an extent that it soon became apparent that the only capitalist who would build an Alaskan railroad to develop the territory was Uncle Sam. The Government could afford to build a railroad, develop the territory, and wait for the returns on its investment, while the ordinary capitalist would demand dividends from the beginning of operations on the road. However, it might have been that private capital would have entered into this work had special inducements been offered by the Government, in the way of land-grants, or a guarantee of bond-interest, but neither of these plans are in accord with the policy of the Administration.

It was at this time that the Guggenheims were planning to open their Kennecott mine in the upper Copper River basin by means of a railroad that would also tap the Bering River coalfield, where the Cunninghams had located large areas, and would have its terminal at Katalla bay, which is the nearest harbor to the coalfield. This plan was abandoned because of physical difficulties and also because of the withdrawal from entry of the Bering River coalfield. The Guggenheims then selected Cordova as their terminal port, built 191 miles of railroad from that town to the Kennecott copper mine, and called their railroad the Copper River & Northwestern Railroad.

The first serious agitation for a government-owned and government-operated railroad started in 1911, during the Taft administration, while Walter L. Fisher was Secretary of the Interior. He visited Alaska, and recommended that the coal-lands be leased, also that the Bering River and Matanuska Valley coals be tested by the Navy Department, and that a Government railroad be built from Seward to the Matanuska coalfield, should that coal prove satisfactory. In 1912, a commission, known as the Taft Commission, having examined all the feasible routes, recommended that the Government should build from the station of Chitina on the Copper River & Northwestern railroad to the town of Fairbanks in the interior.

With the Wilson administration came Franklin K.

Lane as Secretary of the Interior. Although he has never visited Alaska, he has taken a keen interest in the development of the territory. He personally pushed the Railroad Act, and the Coal Lands Leasing Act, both of which were passed in 1914. By the Railroad Act, the President of the United States is restricted to cause to be built not more than 1000 miles of railroad, and to expend not to exceed $35,000,000 thereon. President Wilson created the Alaska Engineering Commission, composed of W. C. Edes, chairman, Fredrick Mears, and Thomas Riggs, Jr., members. These are railroad men of wide experience and capability, having been engaged in engineering and construction work all their lives. They have fully justified the President's choice.*

This commission first made preliminary surveys of various proposed lines, and finally recommended the Seward-Broad Pass-Fairbanks route, which runs from Seward on the west shore of Resurrection bay, northward through the Kenai peninsula, thence westward along the north shore of Turnagain Arm, along the east shore of Knik Arm, across the Knik and Matanuska rivers, in a north-westerly direction through the Susitna valley, across Broad Pass, down the Nenana river, and up Gold stream to Fairbanks. It also recommended the route of a branch line, designated from a point two miles north from where the main line crosses the Matanuska river, running thence in an easterly direction into the Matanuska coalfield.

These recommendations having been adopted, the Commission then purchased the 71 miles of railroad from Seward to Kern creek for $1,500,000, although over $4,000,000 had been expended on this railroad by the original owners; and the 40 miles of narrow-gauge railroad known as the Tanana Valley Railroad was also purchased at a later date for the sum of $300,000. This bit of railroad goes north from Fairbanks and acts as a feeder to the main line. The reasons for the choice of the Seward-Fairbanks route by the Alaska Engineering Commission was that the railroad could be constructed with the least expense, as there were no great obstacles to overcome, and it was believed that this route would aid the development of the greatest amount of resources with benefit to the greatest number of people.

Broad Pass is a comparatively low and wide crossing, and there are no large streams or big glaciers to traverse. While the cuts and fills in many places are quite heavy, the greatest problems of construction are the securing and distribution of materials in a country that has no industrial development. The distribution of material and supplies for the most part has to be done over the snow in the winter, except where the work can be reached by waterways.

Beginning at Seward, which has an open harbor the year round, on Resurrection bay, the railroad passes near the placer mines on the Kenai river. There are several quartz mines at various points on the railroad side of the Kenai peninsula, and fairly well-developed placer and lode claims above Sunrise and Hope, two camps on the same peninsula. Passing near placers on Glacier creek, the railroad runs through the agricultural lands around the town of Anchorage that open out into the broad valleys of the Matanuska and Susitna rivers. On the upper Matanuska branch are the seams that, it is hoped, will develop into producers of Pacific Coast coal. The centre of this coal district is the new railroad town of Chickaloon.

Both the Matanuska and Susitna valleys have been eroded by glaciers, and the lower stretches of both are of the 'pot and kettle' topography common to glacially eroded valleys. This topography causes many lakes and swamps in the lowlands, and most of the profitable ranches in these valleys are situated on the benches above the flats. From these agricultural regions on both sides of the Susitna valley, reaching for about 200 miles, are many partly prospected mining districts, such as the Willow Creek district, Kashwitna Creek, Cache Creek, Montana Creek, Sunshine Creek, Yentna River, Talkeetna River, and the Broad Pass district. North of Broad Pass lies first the Nenana coalfield, which has great deposits of a high-grade lignite and farther north are the agricultural lands of the Tanana valley, which has been considerably developed to supply the markets of the interior. Many of the ranches in this region are well-equipped and have been farmed since 1905.

The main line of the railroad is 470 miles long; with the Matanuska branch it will cover a distance of a little more than 500 miles. Acting within the restrictions of the Alaska Railroad Bill, with regard to mileage, this will leave practically 500 miles of railroad that can be built in the territory without further legislation, provided that the total expenditure for the 1000 miles falls within the $35,000,000 appropriation to which the bill restricts the President. A reconnaisance survey has been made by the Commission, from the point where the main line turns north in the Susitna valley, following the Yentna river, crossing the main Alaskan range, down into the Kuskokwim valley. This valley is widely known for its agricultural and mining possibilities. The original estimate of the cost of the first 500 miles of the railroad was about $27,000,000, but the War has raised prices so greatly since the estimate was made that it will probably fall 25% short of the mark. This would run the cost of the authorized railroad now being built up to the total appropriation of $35,000,000, leaving no funds for the branch lines.

All the departments of the Government are working in close harmony with the Alaskan Engineering Commission in an attempt to render the work most fruitful in its results. The Bureau of Mines has an experimental station at Fairbanks and the Department of Agriculture its experimental station at Matanuska. The Federal Road Commission, which is a branch of the War Depart-

*Mr. Edes prior to his appointment had for many years been in the employ of the Southern Pacific Railway as location and construction engineer. Mr. Mears was formerly general superintendent and chief engineer of the Panama railroad under the supervision of Colonel Goethals, and Mr. Riggs was formerly in charge of the Alaska Boundary Survey.

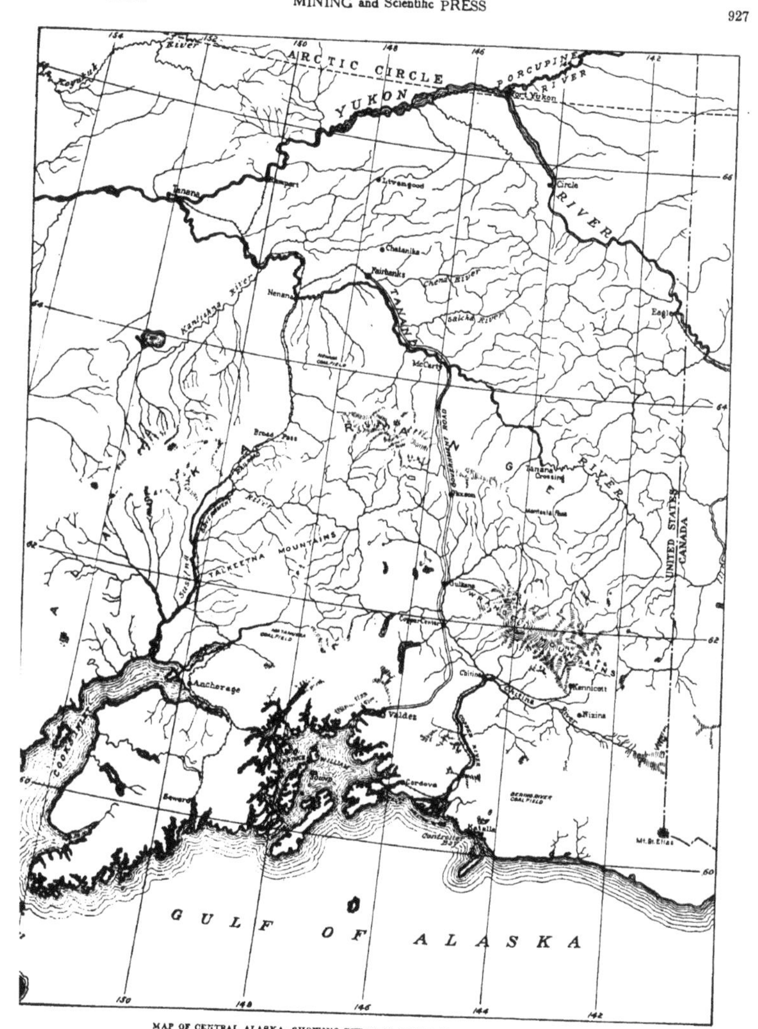

MAP OF CENTRAL ALASKA, SHOWING THE LINE OF THE GOVERNMENT RAILROAD

ment, is building good wagon-roads from the railroad at many points so that every possible inducement may be given to the development of tributary industries. The Geological Survey is pushing examinations and issuing reports on many new mining districts along the route. The Alaskan Engineering Commission has established a Telephone and Telegraph Department to handle the business along the line of the railroad. It appears that if some department of the Government would cover the territory lying away from the railroad with telephone lines so that dependable methods of communication could be had at all seasons it would do much for the development of the region.

The most difficult work has been in rock for about 30 miles along the north side of Turnagain Arm. Here the cost will be as much as $100,000 per mile for the grading and bridging alone. This should be completed by the spring of 1918, and the line completed through from Seward early in the summer. Some difficulty is being experienced in crossing the large terminal moraine below Spencer glacier on the Kenai peninsula, but it is not regarded as serious. The Commission has purchased a large amount of equipment, which it will have on its hands when this road is completed, and it is believed that it will be the policy of the Government to build railroad-feeders to the main line from both the east and the west to furnish sufficient tonnage to render the road profitable. No part of the railroad is expected to be endangered by snow-slides, although it is probable that there will be some trouble from the deep snows in the Susitna valley and between Turnagain Arm and Seward. Many snow-sheds will undoubtedly have to be built.

The railroad rates now in force are based on a doubling of the Western States Classification; thus passengers pay 6c. per mile, as against 12c. per mile over the Copper River & Northwestern. At the time that the railroad was before Congress, another bill was introduced to establish a steamship line from Seattle to Alaskan ports, to operate in conjunction with the railroad. This bill was rejected but it yet may be found that in order to develop Alaska rapidly, the Government will be obliged to go further than merely to build railroads. The freight-rates at present prevailing from Seattle to Anchorage on the average class of supplies are about $20 per ton, while the U. S. transport 'Crook,' which brings up supplies and handles passengers for the A. E. C., lands freight at Anchorage for from $8 to $9 per ton. Potatoes when sacked, 100 lb. per sack, will average 3 cu. ft. in volume. The base rate for potatoes is $14.50 per ton by volume. It being to the advantage of the steamship companies to charge by volume rather than by weight, the $14.50 rate would apply for 1335 lb. of potatoes instead of 2000 lb. When wharfage is added, the charges for potatoes laid on the dock is $30 per ton by weight. The first-class passenger fare on the steamship lines is $60, while the 'Crook' brings its people in for $30. If an employee of the Commission works three months for it, he will be taken outside for $15; and should he work two years for the Commission, he and his entire family will be taken outside and brought in again gratis. Some of the freight for the A. E. C. is brought from Seattle in barges towed by tugs, and but little business is given by the Commission to either the Alaska Steamship Co. or the Pacific Steamship Co., owned respectively by the American Smelting & Refining Co., and the D. C. Jackling interests. The Federal Shipping Board controls these freight-rates but it seems to be deaf to the appeals of Alaskans for relief from such exploitation.

Wages are good; every employee works an 8-hour day and receives time-and-a-half for over time. Ordinary crafts such as machinists, blacksmiths, and electricians receive $6 per day while their helpers receive $4.80. Train-men receive from $150 to $210 per month. At any place where board is furnished by the A. E. C. the charge is $1.25 per day. Civil service rules do not apply. Authority is passed from the three members of the Commission, who have charge of the three divisions, respectively, to the heads of the departments, thence to the lower divisions of the departments, and from these in turn to bosses and to the laborers. Each official handles the men under his control with absolute authority with regard to hiring, discharging, and ordering. There is no central point where men are hired or discharged, the men being directly responsible to their immediate superior. An employment office at Anchorage supplies men to the various outlying points, but this office acts more as a registration headquarters for laborers than as a hiring agency. The main portion of the grade is being constructed by small contractors called 'station-men', using hand-work and push-cars. The smaller portions of the road that are being constructed directly by the Commission are being graded by power-shovels and steam-trains. All bridges are constructed by day's pay directly under the authority of the Commission. In contrast to railroad-building in the States, it is noticeable that there are no large contractors, no large construction camps, and no really big piece of work going on at any one point. There is an absence of the general rush and driving work that marks railroad-building outside. This does not mean that efficient work is not being done, but that it is being done in an entirely different manner, and under entirely different circumstances, for an entirely different purpose, and with entirely different difficulties to overcome. The general Federal Compensation Act applies to all employees of the A. E. C., but so far no pension system has been established.

The entire system has been divided into the Anchorage, Fairbanks, and Seward divisions, which employ collectively about 4500 men. Of this number, 2900 are employed on the Anchorage division, about 900 on the Seward division, and about 700 on the Fairbanks division. Seward is considered the terminal of the railroad, and Anchorage the terminal of the Anchorage division, which is, however, the most important division of the three. At Anchorage the Government will attempt to solve the difficulties connected with the maintaining of

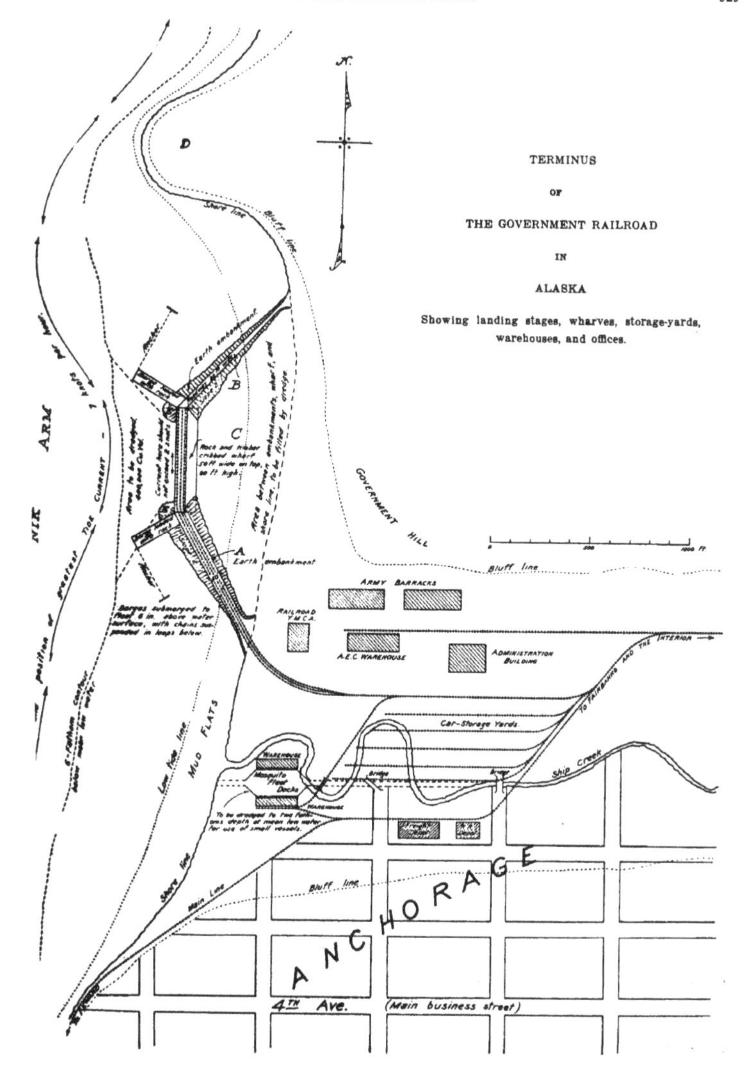

TERMINUS
OF
THE GOVERNMENT RAILROAD
IN
ALASKA
Showing landing stages, wharves, storage-yards, warehouses, and offices.
N.
D
Shore line
Bluff line
Earth embankment
B
C
A
Earth embankment
NIK ARM
GOVERNMENT HILL
Bluff line
ARMY BARRACKS
RAILROAD Y.M.C.A.
A.E.C. WAREHOUSE
ADMINISTRATION BUILDING
To Fairbanks and the Interior
Car-Storage Yards
Ship Creek
Bridge
MUD FLATS
Low tide line
Warehouse
Mosquito Fleet Docks
Shore line
Main Line
Bluff line
ANCHORAGE
4TH Ave.
(Main business street)

an open harbor on upper Cook's Inlet, by dredging and building a large cribbed stone wharf. Here will be built the bunkers to receive the Matanuska coal. The water of Knik Arm, at Anchorage is, of course, salt, and consequently does not freeze at the prevailing temperature, which has never exceeded -38°. The difficulty at Anchorage is the fact that large rivers and creeks bring with them blocks of ice, up to 40 ft. diam., and 4 ft. or more thick, which collect in Knik Arm above a constricted opening opposite Campbell Point. This ice settles on the mud flats at low tide, and freezes to the ground. On the return of the tide, the blocks of ice are lifted and carry with them considerable earth. This, repeated time after time, builds up combined ice and mud blocks that are a great menace to vessels lying at anchor. The tidal difference ranges from 36 to 42 ft., and this water at ebb and flow attains a velocity up to 7 knots per hour. It is the difficulty of withstanding the pressure of these blocks of mud-ice, traveling at such a velocity, that prevents uninterrupted shipping from Anchorage.

To overcome this the Commission has determined to build a wharf 80 ft. high, 1000 ft. long, which will give 6 fathoms at low water, and will allow for a 42-ft. rise in the tide. This wharf will be built in the re-entrant angle of a crescent shore-line, about ¾ mile north from the mouth of Ship creek. Referring to the diagram, *A* is a mole, or wing-dam, 50 ft. wide, built of earth, and riprapped on the Arm side. This will carry three standard-gauge railroad-tracks, to the coal-bunkers on the wharf proper. *B* is a second mole, 20 ft. wide at the top, riprapped on both sides, and extending to the other end of the 1000-ft. wharf *C*. Both of these moles, or shear-dams, are built with a 2:1 batter on both sides, and each will be about 1500 ft. long. The wharf will be built of cribbed timber, floated into position and sunk gradually at the proper places by being filled with rock. Openings will be left in the wharf to allow the rise and fall of the tide back of the wharf, as well as in front of it, so that the head of water will always be balanced on both sides of the moles until the space between the wharf and the shore-line has been completely filled. In front of the wharf about 400,000 cu. yd. of the ocean-bottom must be moved by dredges to give 6 fm. for vessels at low tide. This material, which is dredged from in front of the wharf, will be discharged back of it until sufficient area is filled to supply ground-storage for 100,000 tons of coal. It is planned to use one-half of the new dock for the coaling of vessels and the other half for commercial freight and passenger traffic. The southern embankment will carry the railroad-tracks that serve the dock and the northern embankment will be used as a highway for vehicles and pedestrians.

On the wharf, four cross-conveyors and electric cranes will load and unload vessels, no matter what the condition of the tide. It is believed that the current due to the ebb and flow of the tide at this sheltered point will not exceed two knots, and the ice running in the channel will be thrown by the natural point of land *D* away from the wharf so that it will not be troublesome. If, however, the blocks of ice should race up and down in front of the wharf, it is intended to build floating wings at points *E* and *E'*, which will throw the ice farther out and protect any vessel lying at anchor. These floating wings will be strong barges loaded heavily so that they will not rise much above water-level; and they will have chains suspended below them to catch any ice that might attempt to pass underneath. Upon the completion of this plan the Commission is certain that vessels will be able to enter Knik Arm on flow tide, discharge, load cargoes, and pass out on ebb tide on any day of the year. It is to be noted that whereas it is likely that most of the freight from the interior will be transferred to vessels at the wharf, which will save a 115-mile haul from Anchorage to Seward over two mountain passes, one 700 ft., the other 1100 ft. high, most of the passengers will probably go over the railroad from one of these points to the other, taking a four-hour trip by railroad in preference to a 20-hour trip by boat around the Kenai peninsula in stormy weather. In addition to the wharf built for use by ocean-going vessels, it is planned by the Commission to dredge for a small wharf nearer to the town of Anchorage, to accommodate the 'mosquito fleet' on Cook's Inlet. The new harbor improvements are expected to be sufficiently completed to handle traffic by the fall of 1918. William Gerig, who is in charge of this work at Anchorage, is a man of wide experience in dredging and harbor work; he was formerly in the employ of the Government on the Panama Canal and on the harbor of New York City.

The machine-shop and boiler-shop equipment together with the locomotive and a large part of the rolling stock came from the Panama Canal Zone. The passenger-service equipment at present in use is small and while it is sufficient for construction purposes it can hardly be considered first-class. It will be replaced with something better when the line is completed. The Anchorage-Seward connection should be finished by August 1918 and the entire road, as planned, should be ready by 1920, if Congress continues appropriations. To date, about $22,000,000 has been appropriated, and the $4,000,000 appropriation to complete the road from Seward into the coalfields is now pending.

The U. S. Bureau of Mines has established an experimental station at Fairbanks with Mr. Davis, formerly of the Arizona Experimental Station, in charge. This station will make assays and qualitative analyses at cost for prospectors and examine properties, as well as conduct tests on ores. The Bureau is endeavoring to discover platinum in the black sand of the placer mines by making tests on samples.

The Department of Agriculture has established an experimental farm near the town of Matanuska under the able charge of F. E. Rader. Mr. Rader has had much experience in Alaskan agriculture, having been formerly stationed at Fairbanks. The station is in the heart of the new agricultural district to be served by the railroad, and should be of great value to the farmers there, because

MAIN STREET OF ANCHORAGE IN WINTER

experiments will be carried out under working conditions. The other experimental farms of Alaska, at Kodiak, Sitka, Fairbanks, and Rampart, will work together in harmony with the Matanuska farm, but the conditions prevailing at each of the others is so different from the conditions along the railroad that experiments performed there would not be of much value in this district. Alaska has greater agricultural possibilities than are realized by persons unacquainted with the territory. Nowhere else are such delicious, large, and hardy vegetables grown, in the long hours of mellow sunlight during the summer months. Annual crops of barley, oats, and even wheat, are grown. Alaska is a mountainous country and only the wide valleys of the central interior and the gentler slopes of the hills are tillable to any extent. The soil is graded as 'poor' from a crop-producing standpoint, because of its shallowness. The ice cap remained in these northern latitudes for perhaps thousands of years after it had disappeared from the lands to the south, and since that time no great amount of soil has had time to develop by the growth and decay of vegetation, as has been the case in the States. The excessive moisture along the coastal region gives rise to a luxuriant growth of trees, bushes, and moss, but the trees having only a shallow depth for their roots outgrow their supply of nourishment and die, giving a characteristic appearance, since the green shades of the living trees and bushes contrasts everywhere with the gray streaks of dead tree trunks, and the rich reds, yellows, and browns of the moss.

This growth must be cleared away before the land can be placed under cultivation, and it takes much faith and unlimited hard work to develop a farm. Once developed, however, the soil is very productive, and good prices are received for the produce. Some of the Alaskan vegetables were sent outside this fall on the 'Crook' and from Seattle were shipped back to Valdez, a total distance of over 3000 miles. A well-organized farmers' society could have sold direct to Valdez merchants and saved over 2500 miles of water haul, and could also develop a local market for their products. More than this, such an organization could develop a market for choice vegetables in the States by judicious advertising, and an introduction of their sweet, tender, juicy vegetables. The crops of vegetables grown this year are more than sufficient for the local markets, and the Commission is endeavoring to assist the ranchers in stimulating a demand for Alaskan vegetables in the markets of the Pacific coast. The Army Commissary Department is testing Alaskan turnips for use in the cantonment at American Lake, Washington. The turnips fetch ½ to 1c. more per pound than turnips that are grown in the States, and Alaskan celery is always at a premium on the markets.

There is some loose talk of making either Anchorage or Seward the capital of Alaska, and there is great likelihood of the territory tributary to Anchorage being formed into another judicial district, which would make the fifth. The President has authority to either operate or lease the railroad. During the summer of 1917 the railroad was short between 500 and 1000 men, but in November there is an over-supply until work is resumed in April. Seward, the all-year-around port, is a city of about 2500 people, situated on the most beautiful bay in the whole territory. This harbor is land-locked, deep, and large enough to accommodate the vessels of the entire Navy without crowding. It has a small tide, is never troubled with ice, and is without doubt the finest harbor on the Pacific coast, except San Francisco. Seward is well built and has all the modern conveniences. It stands on an alluvial fan at the mouth of Lowell creek, a small glacial stream on the west side of Resurrection bay, and there is much room for expansion. The winter wind blows a gale out of the canyon of Resurrection river. Except for this the climate is excellent. The annual precipitation is 54 in., including a total snow-fall of 6 ft. The temperature extremes are −14°, and +80° F. The average temperature for the three summer months is +51° F., and for the three winter months +25° F.

Anchorage, on the level bench at the base of the Talkeetna mountains, and at the mouth of Ship creek, is a city with a population of between 6000 and 7000. It is only a little over two years old. One of its banks, which started with a capitalization of $25,000, now has deposits of over $750,000. It is a division point and the second ocean port for the Government railroad. It is certain to be the principal coal-freight terminal for nine months of the year, if not for twelve. The summers are mild and the winters are not severe; being farther back from the Pacific the average temperature is a few degrees lower, and the rain and snow-fall less than at Seward. The summer days never end. Only twilight intervenes between two warm mellow days, but the days of winter are correspondingly brief. It does not resemble any typical Alaskan town, and neither is it like a boom town of a new mining district in the States. The Government has a large and well-appointed hospital, and buildings for all the departments of the railroad's operations. There is a large post-office, a telephone, telegraph, and electric-light office, a fine Masonic temple, several churches, lodge quarters, Y. M. C. A., and club-rooms. It has the largest labor temple in Alaska, a good school, well graded streets, the thoroughfares having concrete and the side streets board walks. The sale of liquor is unlawful, but liberal allowances are made for private use, and it is a very wet 'dry' town. January 1918 will change this, however, for then all Alaska will come under prohibition. The town boasts a large number of pretty and substantial homes, including those built by the Government for its employees, having electric lights and modern sewerage. The surrounding country is remarkably beautiful. In the distance can be seen at all times the splendid stretch of the Alaskan range, and on clear days Mt. McKinley rears its snow-crowned head against the blue, 20,300 ft. high and 175 miles distant.

CANADIUM, a new metal, has been discovered in the Nelson mining district of British Columbia, and has been named for Canada. It is allied to the platinum group and occurs native in the form of grains and short crystalline rods and also as an alloy. Assays give three ounces or less per ton. Canadium has a brilliant lustre, and, like gold, silver, and platinum, does not oxidize when exposed to the air. It is softer than platinum, and its melting point is a great deal lower. The physical and chemical properties of the metal are to be studied at the chemical laboratory of the University of Glasgow.—'Daily Metal Reporter.'

TINNED copper is not superior in any way to tin-plate for roofing material and in view of its greater cost cannot compete with it. This has been shown by tests made by the Bureau of Mines. Tinned sheet copper is used for containing-vessels such as milk-cans and for fittings where corrosion is to be resisted. There seems to be no warrant for its use, and it is probable that such articles would be subject to pitting and corrosion of the same type as that shown in the tests if they were not worn out by abrasion before the corrosion takes place to any extent.

Foundations for Small Gas-Engines

To obtain the highest efficiency an engine should have a firm heavy foundation. This will reduce wear and tear from vibration, and will result in prolonged and better service. Rigidity and durability in the foundation are best obtained through the use of concrete. It should be mixed in the proportion of 1 part portland cement, 3 parts clean well-graded sand, and 5 parts crushed stone or gravel. The use of accurate measuring-boxes is advisable. The sand should pass a ¼-in. mesh sieve, and the crushed stone or gravel should pass readily through a 1½-in. opening. In no case should bank-run gravel be used without being screened. If the size of the engine warrants the use of a reinforced-concrete footing, a 1:2:4 mix should be used for the reinforced portion of the concrete. Sufficient water should be used to form a plastic mixture, but not enough to cause separation of the cement and aggregates when placed. After the position of the centre-line of the foundation has been determined, a pit two to four feet deep should be excavated, the same size as the footing. The concrete should be deposited to the depth determined in the plan. In order to key the engine foundation to the footing, stones three to four inches in diameter should be embedded in the portion of the footing under the engine so that they will protrude. A box-form eight inches larger in both length and width than the engine-base should be carefully set over the footing. The inside of the forms should be oiled to prevent the concrete from adhering. It is essential that the anchor-bolts for the engine be carefully spaced and so placed as to take care of any small variations in position. A templet should be used, also greased gas-pipes twice the diameter of the bolts. The purpose of the pipes is to provide for such slight adjustments of bolts as may be required. The anchor-bolts should be embedded in the concrete at least 18 in., and should be supplied with cast-iron washers at the lower ends. When the templet has been accurately set over the forms and the bolts so arranged that the tops are at a proper elevation, the concrete is deposited carefully and spaded into the form. The gas-pipes should be turned from time to time, thus preventing them from sticking to the concrete. Sticking may be prevented effectually by wrapping the pipe with oiled manila paper, taking care that it does not become torn while pouring the concrete. The concrete next to the form should be carefully spaded to prevent the formation of air-bubbles or pockets. Damp burlap should be placed over the form after the concrete is poured. This will ensure normal set. After 24 hours the form can be removed. The engine may be set and the bolts adjusted after 48 hours. Before the engine is set the gas-pipes should be removed, and when the engine is finally placed, the space around the bolts should be filled with 1:1 mortar. The engine should not be used until the concrete is at least two weeks old. If necessary to have an exhaust or drain-pipe, this may be set in the form before the concrete is deposited.

The War Excess-Profit Tax

By PHILIP WISEMAN

EXCESS PROFITS. It is evident from the discussions in Congress and in the press that the intent under Title II was to levy a tax upon abnormal profits, due to war conditions. While the text of Title II makes no further reference to 'excess profits,' are not the title and the evident intent sufficient to justify the Treasury department in ruling that its application should be only to cases where abnormal profits have been secured?

If the taxes imposed under Section 201 had not been intended to be applied to abnormal taxes and were to be additional sur-taxes, there would have been no occasion for Title II, as additional sur-taxes would more properly have been added to Title I of the Act.

With this view of the meaning and intent of Title II and its schedules, the first thing necessary is to determine whether or not a corporation has earned a profit in excess of what it would have earned under pre-war conditions, and in case it has done so, whether the additional profit is the result merely of enlarged operations or is due to abnormal prices received for the output.

If the 1917 profit is not in excess of the pre-war profits, the schedules in Title II would not be used. If an excess profit is indicated, the next point to determine is whether the increased, or excess profit, is due to abnormal prices for the products disposed of or merely to the use of additional capital upon which there has been only the pre-war rate of return. If the latter should prove to be the case, the schedules of Title II would not apply. The important thing to be determined is whether or not a profit in excess of pre-war figures has been obtained, and how, and the question of invested capital should be only secondary.

INVESTED CAPITAL. This term in some instances may be interpreted fairly by considering only the cash that may have been put into the enterprise, but probably in the great majority of cases other factors should be taken into consideration. There is a recognized relation between earning capacity and market-value and possibly in most cases the market-value may fairly be taken as the 'invested capital.'

That a corporation may not have desired to realize upon an increase in the value of its property over and above the original cash investment (which realizable value may be fairly indicated by the market-value of its shares), should not operate against it in the levying of the excess-profit tax by requiring the company to consider its invested capital to be a less amount than the property is reasonably worth. The purchaser of such a property would be taxed, under the provisions of Title II, upon the basis of the amount paid for the property, and this would result in the purchaser paying a lower tax than the selling company had been required to pay unless the latter had been permitted to consider the market-value of the property to be its invested capital. The mere transfer of the title of a property should not thus affect the taxes to be paid upon it.

For example: If one were to take up 320 acres of desert land and put water upon it at an expense of $25 per acre, the invested capital (omitting the small price paid the Government for the land) would be $8000. With water upon it, this land might be as fully productive as land for which $200 per acre would readily be paid. The invested capital of the one who had paid $200 per acre for 320 acres would be $64,000, while under strict and literal interpretation the one who had had the enterprise, ability, and judgment to develop hitherto worthless land would be allowed only $8000 as 'invested capital,' even though the improved land were readily salable for eight times that amount, and, assuming a net return of $50 per acre, the excess-profit tax he would have to pay would be nearly three and a half times that which the purchaser at $200 per acre would be required to pay. The injustice of such an interpretation of the term 'invested capital' in this or analogous cases seems obvious. The example cited applies with equal force to many industries, and particularly to oil and mining companies. Anyone developing the natural resources of the country is deserving of encouragement and should not be required to pay taxes at what amounts to a higher rate than would be paid by the one who purchases at its market-value property already producing or ready for production.

The Treasury department has recognized the fairness of taking the market-value of the property, rather than the actual cash investment, in the levying of the excise tax, and the same logic should apply, in equity, and the value of the property, as nearly as it can be determined, be considered a necessary factor in arriving at the amount of the invested capital.

The conduct of the business of a corporation may result in an increase of its assets. Its net assets are its invested capital, and the amount of its net assets is the principal factor in the determination by investors of the market-value of its shares.

The invested capital of a bank, or a similar institution, may be wholly cash, or cash and the amount it may have invested in bonds, securities, and loans. The invested capital would, of course, include its accumulated surplus. In other lines of business activity, the invested capital may be represented, in whole or in part, by the accumulation of some commodity, and its selling price determines the invested capital of the enterprise.

In the case of mining properties, the value of the ore-reserve would properly be considered as being a part of the invested capital. This would be the case even though the ore-reserves were regarded as surplus. 'Surplus' is usually thought of in terms of cash, but it may be in any form of property. Even in banking, where the exploitation of cash as a commodity is the principal purpose of the business, the surplus accumulated is not necessarily represented by cash, but may be represented entirely by the value of bonds and securities purchased and loans made. A great deal of the perplexity regarding the interpretation and possible application of Title II is due to the injection of the words 'invested capital.' As the intent of the Act, as was clearly indicated by the discussions in Congress, was to levy additional taxes upon abnormal profits realized, due to war conditions, in excess of pre-war profits, the amount of invested capital is in no sense a factor, excepting in cases where it is necessary to determine whether the excess profits realized are due to abnormal prices received for the output or to the use of additional capital upon which only the pre-war rate of profit was received.

PAR VALUE OF STOCKS. In Title II there are several references to the par value of stocks and no provision seems to have been made for valuing stocks at an amount in excess of par. In many instances the adherence to the view that stocks should not be considered as having a value in excess of par would cause great injustice in the levying of this tax, particularly upon oil and mining companies. With a bank, or similar institution, there is a quite direct relation between actual value and par value, for cash is the commodity dealt in, and, at the time of organization, bank stock is usually sold at par, though if it is desired to show a surplus, the stock may be sold above par. After a bank has been engaged in business for some time any further issue of stock it may make may be at a value considerably in excess of par, because the condition and extent of the bank's business justified it. On the other hand, the stock of a bank, for which par may have been paid, may have a value far below par, owing to unwise management. Therefore it has become an axiom with the intelligent investor that he must consider the actual value of the property behind the stock and not the par value ascribed to the stocks. It is the recognition of this fact that has caused many companies to organize in recent years without par value to their shares. While the par value of a stock is unimportant, it is important for the shareholders to know the number of shares issued, because the extent of his ownership in the property is thus defined. With conservative mining companies it has been the practice to capitalize at relatively low figures, and such conservatism has had the commendation of banking institutions and investors. The result is that with a satisfactory development a property may have a market-value far in excess of its capitalization. If a company is to be penalized for conservatism, it is equivalent to placing a premium on gross over-capitalization and the 'watering' of stock, practices that are frowned upon. If a stock is purchased at its market-price because of the known value of the property, and that price is in excess of the par value, an injustice is done to the holder of the stock if the company is not permitted to value at market-price any shares it may issue. The ultimate value of a mining property is rarely even approximately correctly estimated in the early stages of its development and only in exceptional cases is it deemed advisable to keep more than two or three years' supply of ore ready for extraction; consequently, any disturbance in the ratio between the increase of ore-reserves and their depletion affects the price of the stock.

DEPLETION. The net proceeds from the operation of a mining property do not represent the profit, though often so regarded. In the net proceeds of operation are included return of capital, interest on investment, and the profit. Just what percentage of the net proceeds each represents is difficult to determine. With the uncertainties attending further development, it is only equitable that a liberal deduction for depletion should be made from the net proceeds. Some mining properties have ore so low in grade that the success of the enterprise is dependent upon the assurance of an enormous tonnage prior to the installation of reduction plants. In such cases, it is possible to figure a much lower rate of depletion than where the ore-reserve is only a few years in advance of the production rate and the possibility of maintaining the ore-reserve is questionable.

GOLD-MINING COMPANIES. The interpretation and application of the provisions of Title II to gold-mining companies is a matter of much concern to such companies. These are required to pay an income tax, but they should not be obliged to pay an excess-profit tax under Title II, for the reason that they obviously cannot gain an 'excess profit.' Gold is the basis of all values. The Government, in establishing our financial system on a gold basis, fixed the price of gold at $20.67 per ounce, and thus it has become the measure of value for all other commodities. In the case of copper, zinc, lead, and other base metals there has been a great appreciation owing to war conditions, but gold-mining companies receive no more for their output than before the War, while, on the other hand, it has been necessary for them to pay greatly increased prices for labor, material, and supplies, so that in effect, gold has depreciated in value because its purchasing power is less, with the result that gold-mining companies have a less income under existing conditions than they would have had under pre-war conditions. The purchasers of stock in such companies pay the market-price for their stock, and the market-price is considered a fair gauge of the value of the property. If the schedules of Title II are levied against gold-mining companies, it would, for many of the shareholders, amount virtually to confiscation. The individual income-tax schedules under Title I very adequately take care of any large incomes received from gold mining as well as other sources.

Mining is justly regarded as a hazardous enterprise, and as the value of a mining property is reduced by each ton of ore removed from it, it is difficult to determine just what proportion of a year's production is return of capital, interest on investment, or profit. If the taxes to be imposed upon gold-mining companies practically amount to confiscation, it would seem there was considerable justification in the demand of stockholders that the gold properties cease production so long as the excess-profit tax schedules of Title II are levied.

Considering that gold is the basis of our financial structure, it would seem that the production of gold should be encouraged, rather than discouraged. Every dollar's worth of gold added to the country's supply permits a maximum banking credit of about 20 dollars. Only a few months ago, there was considerable alarm because a comparatively small amount of gold had to be shipped from the United States to other countries. There were suggestions recently that all gold coin should be withdrawn from circulation in order to strengthen our financial position.

The directors and stockholders of gold-mining companies have no desire to shirk the responsibility now resting upon every citizen, but it is desired to call attention to certain inequitable provisions of the War Revenue Act as it now stands, so that they may either be corrected by Congress, or interpreted equitably by the Treasury department, in case Congressional action is unnecessary.

On account of the many diverse businesses and industries in the country and the various manners in which they have originated and have been developed, it is exceedingly difficult to frame a law that will be sufficiently specific to cover equitably all the cases to come up for consideration. Therefore, any changes in the War Revenue Act should be with the idea of stating clearly in general terms the intent of the law, with such exceptions as a more thorough study of the Act would now suggest, and authorizing the Secretary of the Treasury to determine the equity in any cases submitted to him, rather than require him to adhere to a literal interpretation, which may be clearly inequitable. For example:

United Eastern Mining Co. This company was organized in 1913 with a capitalization of 1,500,000 shares of a par value of $1. Its orebodies were developed rapidly and the stock soon had a market-value far in excess of par. Production did not commence until January 1917. As the output is gold, with an almost negligible amount of silver, there has been no abnormal or excess profit, because the price for gold has remained at $20.67 per ounce, although in effect it is less, for its purchasing power has diminished owing to the rise in the price of labor and materials.

The net proceeds are actually less than they would have been under pre-war conditions. The purchase of another property was effected by the delivery of treasury stock. The market-value of this stock at the time was approximately five times par. It would have been perfectly feasible to have sold this stock for cash at the market-price and delivered the cash. The purchase for the number of shares delivered was only possible because the shares had a market-value, legitimately based upon proved ore, of approximately five times the par value. Is it equitable that it should militate against the company because shares were delivered instead of the cash for which those shares could have been sold? This company, as a gold-producer, should not be required to pay an excess-profit tax, and in no event should it be penalized for delivering its shares instead of the cash that could have been realized from the sale of the shares.

The depletion of the ore-reserve is proceeding at a relatively rapid rate; therefore, a high rate for depletion should be allowed.

Summary. The following suggestions are offered:

1. That the provisions under Title II be amended or interpreted in accordance with the discussions in Congress, so that it will clearly apply only to abnormal profits in excess of pre-war profits due to increased prices received for products.

2. That the provisions under Title II be amended by omitting 'invested capital' as a factor in determining excess profits, except in cases where the additional profit is due to the use of additional capital.

3. That if 'invested capital' be not omitted that it be interpreted to mean the fair market-value of a company's stock or its net assets.

4. That the provisions under Title II be amended so that the value of shares issued for property, etc., shall not be limited by the par value, but be taken at either the market-value of the shares or the market-value of the property for which the shares are exchanged.

5. That the provisions under Title II be amended to exempt gold-mining companies from its provisions until the Government authorizes a price in excess of $20.67 per ounce.

6. That Title II be amended to permit mining companies (or companies with similar wasting assets) to deduct an equitable amount for depletion in arriving at net profits.

7. That the War Revenue Act be amended so that the Secretary of the Treasury will be empowered to determine the merits of cases arising under its provisions and adjust them equitably.

An international monetary system based on silver as well as gold, to provide for the expanding credit of the world, due to war exigencies, is under consideration by the Treasury Department, in conjunction with allied and neutral governments. If the scheme is worked out along the line now being discussed, gold and silver would be put upon a ratio of coinage together forming the basis upon which the paper money of the United States and of the allied and neutral countries would be issued. In other words, bi-metallism would be restored in the money standard of the nation, but under such circumstances that the gold would not disappear from circulation, as would happen if the United States were to adopt it without such international agreement. France may view it kindly, but the attitude of Great Britain is doubtful.

Pig-Iron From Scrap-Steel

The electric furnace has made possible what may be regarded as almost a revolution in the steel industry. It is the conversion of scrap-steel or iron back into pig-iron. What may be called 'synthetic pig-iron' is now a commercial product; in other words, the original constituents of pig-iron are being made to re-unite in the condition originally assumed. This unusual achievement is another evidence of the adaptability of electrical energy to the production of results impossible by any other means. The new process is being applied commercially in a large electric-steel plant in the eastern part of the United States, high-grade or low-phosphorus pig-iron being made directly from ordinary scrap-steel. Not only is pig-iron being produced in large quantities, but wash-metal and iron and steel-castings are made in the same furnaces. The pig-iron sells in the open market as a competitor with regular low-phosphorus blast-furnace iron, the wash-metal goes to crucible-steel makers, and the iron-castings are sold to near-by users, or else are used by the company in its rolling mill.

The idea is not new, but this is the first record of its commercial exploitation in the United States. Late in 1912, Horace W. Lash, of the Garrett-Cromwell Engineering Co., of Cleveland, organized a corporation for the purpose of converting scrap-steel and iron into pig-iron in an electric furnace on the Pacific Coast. Pig-iron being expenisve in that part of the country, and the fact that it could be made by this process without the use of coke, charcoal, or coal, which are not cheaply available there, were the impelling reasons. It is not known that the company carried out its plans, but it is understood that Robert Trumbull, an engineer at one time identified with Heroult in his efforts to introduce his electric furnace in America, is now producing pig-iron from scrap in an electric furnace in Canada.

The Eastern concern that has been successfully operating, as stated above, is known as the Sweetser-Bainbridge Metal Alloy Corporation of Watervliet, New York. For several months it has been producing from a Ludlum electric furnace. About 50 tons per day is the present output from two 5-ton furnaces.

There is no secret about the process. It is made possible by the unusual metallurgical feats which the electric furnace can perform. The actual method used by the company is not revealed. In general, however, ordinary scrap-steel of any grade is introduced into the furnace cold. The process is basic. The material is melted and refined if necessary, depending on the quality of the scrap and the degree of refinement desired in the product. These considerations also influence the cost. After the desired refinement is reached, the slag is removed. The necessary amount of ferro-silicon and ferro-manganese is added to bring the iron to the composition desired and then the carbon is added in the form of fine coke and this is easily absorbed by the hot metal. Five or six 5-ton heats per day in each of two 5-ton furnaces are being regularly produced by the above or a similar process. There is no difficulty in its execution and no question as to the quality of the products. In respect to sulphur and phosphorus, they equal if they do not surpass any blast-furnace or cupola product, as shown by the analyses that follow. Regulation of the temperature or the composition is apparently simple, and a range from high-grade low-phosphorus iron and wash-metal down to the high-phosphorus foundry and other irons is claimed to be possible, and it probably is. The graphitic-carbon content is regulated by the introduction of varying percentages of silicon. The manganese can be made anything desired and the total carbon is not difficult to regulate. The amount of phosphorus and sulphur is readily placed at almost any required percentage, depending on the composition of the scrap charged and the degree of refinement before conversion into pig-iron or iron-castings.

The commercial possibilities of the new process are shown, so far as the composition of the products is concerned, by the following analyses made in ordinary operation:

Composition of Pig-Iron Made From Scrap-Steel

	Sample No. 1, %	Sample No. 2, %	Sample No. 3, %
Total carbon	3.60	3.73	3.38
Manganese	0.41	0.35	0.31
Silicon	1.77	1.52	0.15
Sulphur	0.007	0.008	0.028
Phosphorus	0.019	0.030	0.032

The average phosphorus-content of the iron so far made has averaged 0.023%. The graphitic carbon can be whatever is desired, or as high as 3.50%, depending on the silicon content.

Composition of Wash-Metal Made From Scrap-Steel

	Sample No. 1, %	Sample No. 2, %
Total carbon	3.55	3.83
Graphitic carbon	none	none
Manganese	0.37	0.32
Silicon	0.20	0.14
Sulphur	0.020	0.012
Phosphorus	0.014	0.018

Composition of Cast-Iron Casings Made From Scrap-Steel

	Sample No. 1, %	Sample No. 2, %
Total carbon	3.36	3.62
Graphitic carbon	....	3.08
Manganese	0.36	0.31
Silicon	1.30	1.15
Sulphur	0.019	0.014
Phosphorus	0.024	0.018

AN ELECTRIC FURNACE ABOUT TO BE CHARGED WITH SCRAP-STEEL. A LIFTING-MAGNET IS DELIVERING THE SCRAP

PILES OF PIG-IRON MADE FROM SCRAP-STEEL AND THE MOLDS IN WHICH THE METAL RUNS FROM THE ELECTRIC FURNACE

These analyses show how varied compositions can be controlled. The low-silicon content of the wash-metal permits the carbon to remain almost entirely in the combined state.

The pig-iron is reported to be very tough and strong. As to cast-iron in the form of castings, the experience of the Ludlum Steel Co. is cited. Cast-iron wobblers, for use in the company's rolling mill, which were purchased in the open market, were breaking at the rate of 10 to 12 per day. Wobblers from iron made in the electric furnace have been in use for two or three months without a case of breakage. The company also made the breaking-boxes on the mill of electric cast-iron. These should break easily when performing their normal function, but when made of this iron they were found to be entirely too tough, and their section had to be reduced considerably. It is believed that some explanation for these qualities may be found in the crystalline structure of the iron as well as in its greater purity compared with cupola-iron, this being contrary to cupola-practice, in which the sulphur rises with each re-melting in the electric process. In making steel castings the conversion loss is put at 5%. The cost of production is one of the vital points. So long as low-phosphorus pig-iron is selling at its present war price there can be no question as to the liberal profits obtainable. Whether in normal times such a method of making this grade or any other grade of pig-iron or castings is economically possible depends on several factors. First among these are the consumption of electricity and the grade of scrap used. If a relatively pure scrap-steel is the starting point, practically no refinement is necessary, and the kilowatt-hour consumption is at a minimum. The converse is true if poorer scrap, requiring more refinement, is used. The matter of other materials and the question of labor and electrode-charges are fairly constant in all cases. Owing to the fact that the Sweetser-Bainbridge Co. has been operating this process only for a short time and under rather imperfect conditions, actual data as to the cost of production are not available. Extensive experiments are being made to determine costs under various conditions and in various stages of the process, and these will be known in the near future. It is the estimate of the metallurgists of the company that the kilowatt-hour consumption of electricity per ton of pig-iron produced can be brought below 400 kw-hr. depending upon transformer-capacity, grade of scrap used, and the product desired. Electrode consumption and refractory renewals, it is said, are low, even under the prevailing abnormal costs of these materials. One melter and four men on each shift of eight hours are employed to operate the two furnaces, which, in 24 hours, are reported to produce 5 tons each, or 50 tons in all. With cheaper iron and steel-scrap in normal times it is claimed that the total cost would be lower.

The furnace in which the foregoing products are being made has come to be known as the Ludlum. Besides the two now making pig-iron, there are two 10-ton furnaces producing high-grade special and tool-steels in the new plant of the Ludlum Steel Co. This concern is a large producer of the new chromium permanent-magnet steel and it also makes chromium roller-bearing steels. There are two fields in which 'synthetic pig-iron' may play an important rôle. It is certainly practical and economical in districts where blast-furnace production is not possible as on the Pacific Coast. It is also likely that its purity and probably superior dynamic and static properties may make it a worthy competitor of charcoal-iron.

Corrosion of Barbed Wire

As a result of complaints of the short life of fence wire made by users, some 15 years ago, the U. S. Department of Agriculture made an investigation with a view to improving its resisting qualities. Notwithstanding this, there is continuous complaint of the short life of wire procurable today, as compared to some of the older brands of wire, much of which is still in service. A study of the analyses of a number of fence wires that have been in use for a number of years shows that those best resisting atmospheric influences generally contained greater or less quantities of copper, the rust-resisting quality of the wire increasing directly with the copper content. Much of the bessemer steel made 20 to 30 years ago contained copper, while that of more recent manufacture contains but a trace. The reason for this is that metallurgists have objected, and many still do object, to the presence of copper in structural steel, claiming that it makes the metal red-short, so, to meet this objection, care has been taken recently to eliminate copper from manufactured bessemer steel, hence its poor rust-resisting qualities.

The rate of corrosion of fence wire varies with the location. Not only do climatic conditions and the presence of industrial corrosive gases in the atmosphere influence the speed of corrosion but the position of the wire on the fence; whether it be the top or bottom strand, whether shaded, covered by bush, or next to a post or tree, plays a by no means inconsiderable part in the rate of its corrosion. The bottom wire on a fence generally will be found in a better state of preservation than a top wire, and wires shaded by trees and shrubbery, curiously, are less prone to rust than those in the open. All this makes it exceedingly difficult to obtain trustworthy data. The ideal condition for a corrosion test would seem to be in a strand of barbed wire, made by twisting together two wires of different composition, but even here there is the danger of an electrolytic action being set up which might act prejudicially on the wire that, under ordinary conditions, would be the most rust-resisting. Experiments are being conducted at a number of our universities with the view to finding an alloy, that, while not materially increasing the cost of the steel, will make it more resistant against atmospheric influences. Some success has already been attained and it is to be hoped that before long users of wire fences will be provided with a more satisfactory material.

CORDOVA, ALASKA

DEVELOPMENT IN THE COPPER RIVER DISTRICT.

The Kennecott Copper Co. overcame most of its labor troubles. The strike at Kennecott lasted about a month and at the present time the mine is operated full-handed. During the summer a new bunk-house was built; it has a capacity for 300 men, affords individual rooms, which are steam-heated, electric-lighted, clean, and well ventilated. A dining-room furnishing board at $1.25 per day is included in the same building. With copper at 23½c. as set by the Government the miners are paid $4.75 per day. The company has built a recreation hall, of frame construction, containing card-rooms, pool and billiard-rooms, reading-rooms, together with a moving-picture theatre which is open every Sunday night. The new Bonanza shaft, sunk on an incline of about 45° from the main-haulage adit, was completed to the 900-ft. level; a new electric hoist was installed and is now operating, using skips. The bunk-house at the mill has been enlarged to accommodate from 150 to 200 men, and a recreation hall, similar to the one at the mine, is available to the men. Two 250-hp. Erie City boilers, burning crude oil, and a 500-hp. steam-turbine will generate electric power for mine and mill.

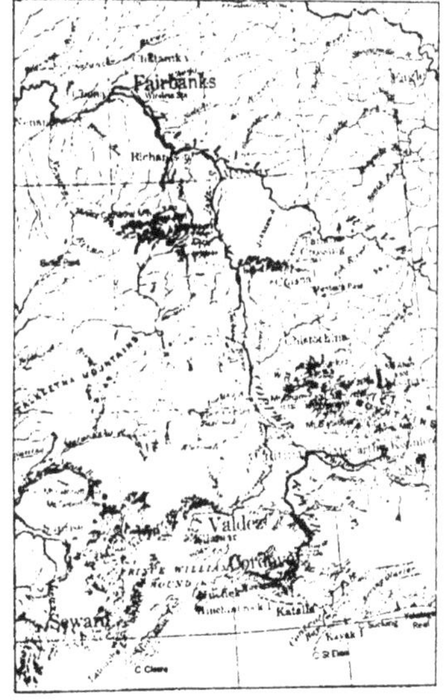

PART OF ALASKA, SHOWING POSITION OF COPPER RIVER

The Mother Lode Copper Mines Co., of which James J. Godfrey is president, has spent the summer in making improvements. The company built about 15 miles of wagon-road from McCarthy, on the C. R. & N. W. R. R., to the mill on McCarthy creek. This 15 miles of automobile road has 16 bridges and two tunnels; it was constructed without Federal or Territorial aid of any kind, without even co-operation with any other mine-owners; it is, therefore, natural that the Mother Lode company should exact a toll. Motor-trucks were used for hauling during the summer, and it is thought they will serve in winter. Two months in the fall and two in the spring will probably be the only periods wherein motor-truck hauling will be impracticable. The timber for the bridges was brought from Puget Sound. Although there is a great quantity of good timber near Cordova, it is found cheaper to pay the freight and steamship rates from Seattle to McCarthy. The combined rate from Seattle to McCarthy is $25 per M for 100,000-ft. lots, or over, and the freight-rate from Cordova to McCarthy is $30 to $35 per M. At Cordova, the lumber costs $20 per M and at Seattle $17. This gives a cost of from $50 to $55 per M for Alaskan lumber hauled 150 miles by rail, as contrasted with a cost of $42 for Puget Sound lumber hauled 1000 miles by water plus 150 miles by rail. As a matter of fact, it is merely boycotting Alaskan products and throwing a little freight-tariff into the treasury of the Alaska Steamship Co., which is owned by the benevolent gentlemen that control the C. R. & N. W. R. R., the Kennecott Copper mine, and the A. S. & R. Co. The railroad and steamship rates from McCarthy depend upon the grade of the ore, 60% plus ore bearing a rate of $32.50 per ton, between 50 and 60% a rate of $30 per ton, and under 20% a rate of $10 per ton. A flat smelter-rate, regardless of grade, of $2.50 is charged at Tacoma on any tonnage. Officials of the C. R. & N. W. R. R. claim that their present revenue is not now sufficient to pay the interest on the bonded indebtedness of the company, and so it is not likely that these freight-rates will be reduced.

At McCarthy station a 300-hp. oil-burning boiler and an Allis-Chalmers turbine generator were installed. From McCarthy to the lower camp, which is 1½ miles below the mine, a 15-mile transmission line was erected of No. 4 hard-drawn copper wire, carrying current at 16,000 volts. At the lower camp this current is transformed to 440 volts, and is then delivered to the mine. A telephone line has been constructed, and is now in use from the lower camp to the mine. A two-story building for the use of officials, with offices for the president and general manager, dining and living quarters, has been erected at the lower camp. The company expects to make a bull-jig product of 50% copper. They expect also to ship a screened product composed of pulverized glance and copper carbonate, which will assay 20% copper. The mine will send down 200 tons of ore per day over the 6000-ft. Leschen aerial tram. The concentrate will be shipped to the smelters and the tailing stacked to run through a leaching-plant at a later time. It is estimated that 24,000 tons of 20% ore is broken and lying in the stopes. The Pringle vertical shaft is to be sunk 600 ft. below the Pittsburg tunnel. For this purpose an electric hoist, manufactured by S. Flory & Co., of Bangor, Pennsylvania, has been placed underground. This is a second-motion herringbone-gear hoist having a capacity of 100 to 300 ft. per min. At the same time, the Rhodes tunnel is being driven 215 ft. below the Pittsburg tunnel to tap the Pringle shaft. This shaft will have two hoisting and one pump compartment and will be sunk in the orebody. The mine has been equipped with 25 piston and jack-hammer drills, and three electrically driven air-compressors of from 3 to 10-drill capacity. This company plans to build next year a 500-ton mill, installing a set of rolls, three Hancock jigs and Wilfley tables, together with a copper-leaching plant. It may be that a hydro-electric plant will be installed on McCarthy creek in 1919.

On the Kuskolina river the North Midas Copper Co., of which Ole Berg is manager, has developed a promising showing of gold and copper ore. The company has installed a Chicago Pneumatic Tool Co. compressor and will make regular shipments of gold ore 17 miles to Strelna, a station on the C. R. & N. W. R. R. The Chitina Kuskolina Mining Co., 12 miles from Strelna, installed a power-plant burning wood to drive its air-compressor. This company has an excellent copper prospect, but was disappointed during the past year by delayed shipments of machinery. It can do nothing of consequence until spring.

The Alaska Copper Corporation is operating the oldest property out from Strelna. This mine has been making periodic shipments for ten years. It has recently been taken over by Col. Stevenson of the Ladysmith Smelting Co., and will make regular shipments this winter. H. W. DuBois is manager. A considerable tonnage of moderate-grade copper ore has been developed by tunnels and a 300-ft. shaft. The property has been equipped with two Chicago Pneumatic Tool Co. air-compressors. The Alaska Westover Mining Co., of which C. C. Johnson is manager, has developed six feet of ore containing bornite and glance. This deposit occurs in the limestone and there are 3000 tons exposed by tunnels that could be mined during the winter. D. C. Bard made an examination of this property during the summer for the Ladysmith Smelter Co. interest.

SPOKANE, WASHINGTON

Northwest Magnesite.—United Copper.

The expenditures of the Northwest Magnesite Co. for the development of its quarries and the equipment of its plant at Chewelah will reach $450,000, according to R. S. Talbot, president, who has just returned from a trip to California. The first appropriation, made more than a year ago, was $250,000, but the operations and plans have since been expanded. Much of the additional equipment is being installed and will soon be ready for operation. Shipments are proceeding at the rate of six carloads daily; they are confined to the calcined material, which has about half the weight of the crude. H. K. Devereux, a mining engineer of Seattle, has been appointed manager for the company, with headquarters at Chewelah.——The United Copper Mining Co., operating at Chewelah, has purchased the Copper King mine, nearby, from H. S. Wales, who recently bought the property from the receiver for $75,000. The price paid by United Copper is not given out. The purchase will give the United company a mile and a half on the strike of its vein.

The magnesite industry, employing 400 and 600 men at Valley and Chewelah, has been awarded a decreased payroll premium under the State Industrial Insurance Administration. Until this ruling was made the mining and transportation of magnesite in wagons has all been classed as quarrying, one of the heavy-risk industrial activities, and 5% of the payroll has been collected to care for accidents. The Commission has segregated the employments in connection with magnesite mining, so that hauling the calcined product from the kilns to the railroad is put in a lower risk-class, and only 2½% of the payroll will be collected hereafter in this department. The remainder of operations are still regarded as in the quarrying class. The Commission has also put hauling of ore from metal mines, and lumber from mills, in the 2½% rate division. The new order complies with a request of the magnesite operators.——Mining men in the North-West are requested to send samples of ore to Spokane for exhibition at the annual North-West mining convention, February 11 to 17. Arrangements have been made for free transportation of these samples. The purpose in asking that ore be sent now is to give plenty of time for arrangement of the displays and also to avoid the likelihood of delayed freight trains which are more frequent later in the winter. Officers of the North-West Mining Association request that ore be sent from all promising prospects as well as from partly developed claims and mines. At previous conventions it has been noted that capital has frequently become interested in a prospect through the showing of ore.

MAYER, ARIZONA

Pocahontas.—Circle Peak.—Jerome Conger.—Verde Inspiration.

Half a dozen or more important strikes of copper and silver-lead ore have been made recently in the Mayer district. High-grade chalcopyrite has been found in the bottom of the 200-ft. shaft at the Arizona Copper mine, one and a half miles north-west of Mayer. The company has sufficient funds to sink the shaft 1000 ft. This discovery is especially interesting since it is close to the Butternut mine where some excellent ore has recently been found.——In the Copper Mountain district, the Monte de Cobre Mining Co., which recently took over the old Barbara mine, has gone through 40 ft. of ore in a new adit that has been run to tap a known body of copper ore. Indications point to this as one of the largest bodies of ore in the Copper Mountain district. The grade is not high but it is similar to the ore coming from adjoining properties in the schist.——South of the Barbara mine a strike of lead-silver ore has been made at the Pocahontas mine in a cross-cut at the 150-ft. level. The vein is about 12 ft. wide and averages $15 to $20 per ton. A carload of lumber has been purchased and a 50-ton flotation-mill will be erected as soon as the machinery can be assembled. This ore is similar to that of the Anderson-Birch mine, west of Humboldt, where flotation has been successful.——Two important strikes have been reported recently from near the old Hidden Treasure mine, south of Mayer. John Slak, of Mayer, has been developing a large group of claims called the Naco. The vein is about three feet with a rich centre, which will run from 60 to 70% copper. Two miles east of the Naco a group of claims is being developed by B. W. Gill. At one place a pit has been sunk 15 ft. on a strong vein carrying considerable glance and native copper. This vein outcrops for over 3000 ft. and is 20 ft. wide.——Between these two properties, J. D. Dillon and associates have opened the extension of the Naco vein with a 75-ft. open-cut which shows a large body of milling ore. These properties are near the railroad and are but a short distance from the Prescott-Phoenix highway.——A new camp will soon be established eight miles west of Mayer in the Pine Flat district by John A. Peacock, of Ft. Worth, Texas, who has organized the Circle Peak Mining Co. and taken over ten claims near the old Cumberlain mine and mill. About 2,000,000 ft. of pine timber covers the property. The ore carries copper, gold, and silver. The company also has an option on the Cumberlain property, which in the early days produced high-grade gold and silver ore. The wagon-road between these properties and Mayer is to be repaired.——Development work will be commenced in a few days on the Patton mine, which has been taken over by the recently organized Rio Tonto Copper Mines Co. P. J. McIntyre and W. D. Mahoney of Jerome are the principal owners. This is one of the best developed properties south of the Binghampton and Copper Queen mines.——The change in the officers and management that has been made at the annual meeting of the Big Reef Mining Co. will undoubtedly mean the early financing of the company and the resumption of the development work that, so far, has shown up some strong veins. It is understood that Eastern capital has been secured.——The Schist Belt Copper Co. has been organized by A. W. Davis, H. C. Storey, S. E. Wallace, Charles Hooker, and others. The property is north of the Silver Belt mine, west of Humboldt, and is on the strike of the Jesse and Union veins.——Indications are that there will be general activity in the Cherry Creek district during the coming year. Several new companies are being organized among the best mining men of the State.——J. H. Tribby, who is one of the oldest of the mine-owners in that district, has turned over four of his properties, the Evening Star, Climax, Three Metals, and the Shipper, to men residing at Pasadena, California, the deal having been consummated by W. A. Crawford of that city. The four properties are well developed, showing high-grade gold and copper ore.——In the same district, the Jerome Conger Mining Co. has acquired the old Conger mine. Henry Suder is president, Rudolph Baehr is vice-president, R. T. Belcher is secretary, and W. H. Hoover is treasurer. The company is to be financed by Ohio people. The ore carries gold and copper.——One of the best developed properties in the Cherry district is the Verde Inspiration, of which E. H. Meek is president and A. L. Hammond is secretary, both of Prescott. There are several parallel veins traversing the property; these have been extensively developd. At the Verde Inspiration mine several shafts have been sunk on different portions of the veins. The 20-ton stamp-mill has handled all the ore that has been mined up to now. The capacity of the mill is to be increased and concentration and cyanide equipment added. There is about 20,000 tons of oxidized gold ore blocked out and much ground has not been developed. The ore averages $15 per ton in gold.

DEADWOOD, SOUTH DAKOTA

HOMESTAKE.—ORO HONDO.—GILT EDGE-MAID.—TROJAN.

Richard Blackstone, superintendent for the Homestake Mining Co. since November 1914, has tendered his resignation to become effective January 1, 1918; he will be succeeded by B. C. Yates, who is promoted from the assistant superintendency. Mr. Blackstone has been with the Homestake for forty years, his first engagement being as civil engineer. He has had charge of nearly all of the construction done by the company, including the building of a railroad, later sold to the C. B. & Q., the development of water power for electric generation, the construction of the present water system, and numerous minor and incidental changes and improvements. Mr. Blackstone's energy and devotion to duty have been recognized by the board of directors in the substantial form of tendering him a handsome pension. Mr. Yates has been with the Homestake company for twenty years. Like Mr. Blackstone, his start was made in the engineering department. No assistant superintendent will be appointed at present.

John T. Milliken has purchased the Oro Hondo property, south of the Homestake, which he has been developing, under option, for the past three years. Deeds to the acreage were recently recorded in Lawrence county. This enterprise has been financed by Mr. Milliken in a search for the southerly extension of the Homestake orebodies. A shaft has been sunk to a depth of 2000 ft. On both the 1500 and 2000-ft. levels considerable exploratory work has been done, and on the latter extensive diamond-drilling. While Mr. Milliken has at all times refrained from announcing any of the results of this development, the fact that he has purchased the property and that he has further announced the purchase of a large hoist from the Alaska Treadwell, to be moved to the Oro Hondo and erected, causes the belief locally that he has found ore and that another large mining territory will be opened on the southerly trend of the Homestake ore-zone.——The encouraging prices of lead and silver have stimulated ore extraction in the Galena district, a section which was idle for years. Several properties have been yielding good ore, which, although it occurs in small bodies, forms the basis for payable enterprises.——Construction of a spur to the Gilt Edge-Maid mine, which has been taken over recently by local parties, is a preliminary step to the resumption of operations. The mine is known to contain a large body of low-grade gold-bearing porphyry, which, it is believed, can be profitably treated on a large scale. The ore can be mined economically, either by 'glory-holing' or by steam-shovels, and is in such quantity that large equipment and extensive operation are justified. Taken in connection

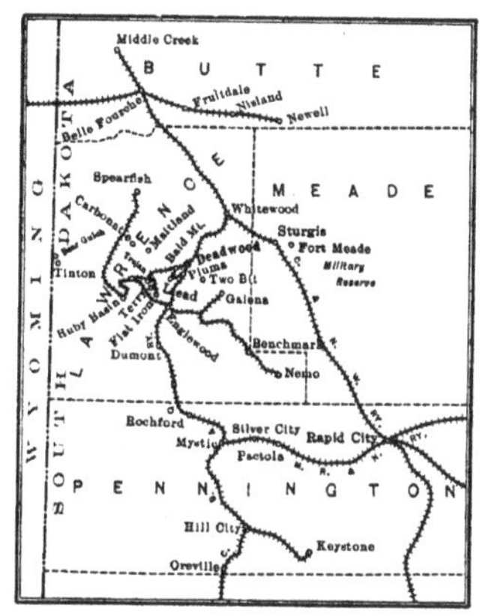

PART OF SOUTH DAKOTA

with adjoining and neighboring properties, it promises to develop into a large enterprise. In the past the mine produced a considerable tonnage of smelting grades and later a cyanide mill treated upward of 50,000 tons of ore containing more than $2 per ton.

Notwithstanding the high cost of materials and supplies and a decreased efficiency of labor, the Homestake during 1917 maintained a normal output and paid its usual monthly dividends regularly. Many of the skilled miners, attracted by higher wages paid in the copper camps and the Wyoming oil-fields, left the company, notwithstanding a wage increase of practically 50c. per day. The Homestake is paying practically $4 per day to miners, besides furnishing rubber clothes for workers in wet places. The company's tungsten output was large and commanded a good market price. Five stamps are in continuous operation on tungsten ore, preliminary to its concentration. The ore contains some free gold, which is amalgamated, and the tailing is sent to the cyanide plants for further treatment. The gold content of the ores is sufficient to pay all costs of mining and milling, leaving the tungsten concentrate as a clear gain from the operation. Completion of the second Spearfish hydro-electric plant is announced. This plant secures its water from Spearfish creek at a point above the first plant, and is returned to the stream for the benefit of that plant. This installation will give the company sufficient electric energy to drive all of its machinery in and about Lead, except the hoisting engines, which are steam-driven. This places the company in an enviable position regarding dependence on railroads and coal mines, and with the eventual electrification of hoisting plants coal will only be required for heating, and in emergency wood could be used for that purpose.

The adit being driven by the Mogul and Ofer companies to tap the North Lode, Mark Twain, and neighboring ore-shoots, is practically completed. From the portal of the adit to the Mogul mill, about half a mile, an aerial tram has been erected which will dispense with the necessity of railroad transportation of the ore from a large territory served by adit. For the past two years the Ofer company has been a steady shipper to the Mogul mill; it has a large tonnage of ore in sight.——Work is progressing rapidly on the construction of a loading-bin at Blacktail, where ore from the Republic, leased by the Trojan, will be delivered by team and motor truck, for transportation by the C. & N. W. railroad to the Trojan mill. A good supply of ore is being developed in the Republic; Trojan officials are well pleased with the showing so far made in this property.——Mill construction has been started at the Deadwood Zinc & Lead Co. property. Grading is completed and the stamps of the Lexington Hill mill are being removed to the new site. The company's operations are being financed by people of eastern South Dakota.

MAMMOTH, UTAH

UTAH ZINC.—TINTIC-DELAWARE.—EUREKA LILY.—GRAND CENTRAL.

The Majestic Mining Co., incorporated to take over and operate 13 claims in the Erickson district, will start operations shortly. Considerable development has been done. Two years ago Charles Stauffenberg and associates equipped the mine with machinery and operated on the 100 and 170-ft. levels, mining some good copper, gold, and silver ore.——The Utah Zinc Mining Co. has started to haul the ore from its property in the Erickson district. This company during the past year has been developing ground on which there is a fine showing of lead-silver ore. One carload has been shipped and there is enough ore sacked to load three railroad cars. This ore has to be hauled by team 30 miles, costing $7.50 per ton. The ore carries 40% lead and is sought by the smelters for fluxing on account of its high iron content.——The No. 1 shaft at the Iron Blossom is down 2000 ft. Frank Birch, the local manager of the Knight mines, says that it is the intention of the company to sink another 100 ft. and then send out drifts to cut the veins that have been developed on the upper levels.——The orebody in the Tintic-Delaware is increasing in size and richness. The two-foot lens of high-grade ore, running 80% lead, on the north side of the adit, has increased five feet, and the orebody, which showed a width of 20 ft. at the adit face, has opened to 25 ft. The road to the mine has been completed, but shipments to the smelter have been delayed because of cold weather. Attempts to make a shipment have been abandoned until sorting-sheds can be erected to protect the workmen.——The bad air condition in the Eureka Lily has been remedied temporarily by the use of a blower. The work in the mine has been handicapped for several weeks past by gases such as those found at the Tintic Standard immediately prior to the time the orebody was discovered. As a permanent relief from the gases, an incline is being run to connect with the old Lily shaft.——Recent development work in the Gold Chain resulted in the discovery of an orebody which some years ago a prospect drift had missed by only a few inches.——Most of the ore is now coming from the levels above 500 ft.; it carries gold, silver, and copper.——Another dividend of 4c. per share, or $24,000, has just been declared by the Grand Central Mining Co. at a meeting held recently at Provo. Checks were sent out on December 23 to shareholders of record on December 18. Reports from this mine indicate that one of the best orebodies that has been found in recent years is yielding a good portion of the present output. The ore was first discovered to the south of the shaft on the 2100-ft. level and since then it has been followed by means of a raise to the 2000-ft. level. The ore is low-grade, but the deposit is unusually large. A new piece of development work is to be started on the 2200-ft. level where a drift will be sent toward the north in an effort to find ore that has been developed on levels above.

TONOPAH, NEVADA

TONOPAH EXTENSION.—TONOPAH BELMONT.—WEST END CON.—JIM BUTLER.

The Tonopah Mining Co. milled 11,350 tons of ore during November, averaging $11 per ton, making a net profit of $38,000. At the Silver Top, 52 ft. of development has been done, 14 ft. in the Mizpah, and 88 ft. in the Sandgrass. At the Mizpah shaft the stope on the Mizpah vein on the 600-ft. level continues on a 3-ft. face. On the 1100-ft. level of the Sandgrass raise No. 1163 made a connection with the level above and stoping has commenced.——At the No. 2 shaft of the Tonopah Extension Mining Co. 160 ft. of development has been done, and 168 ft. in the Victor. On the 1350-ft. level of the No. 2 the west drift from the 510 cross-cut shows a 4-ft. face of ore, while the drift to the east has a 3-ft. face of ore. Raise No. 567 advanced 22 ft. on a 4-ft. face of ore. At the Victor the 1501 winze continues in a face of ore. On the 1680-ft. level the 1600 south cross-cut advanced 64 ft. toward the extension of the Murray vein. West drift 1600 advanced 26 ft. on the Merger vein. The entire face of the drift is in ore. The production during the past week was 2380 tons.——On the 700-ft. level of the Tonopah Belmont Development Co. east drift No. 721 on the Shoestring vein struck a fault, and a cross-cut is being driven south to cut the faulted segment. On the 800-ft. level east drift No. 8017 and west drift No. 8018 each show a full face of ore. Raise No. 73 on the 1000-ft. level continues on a 1½-ft. stringer of good ore on the North vein. Last week's production was 2582 tons.——At the Ohio shaft of the West End Consolidated Mining Co. drift No. 535 struck a small fault, but recovered the vein. Winze No. 534 continues on excellent ore. On the 555-ft. level drift No. 1 shows a full face of good ore. Raise No. 814 from the 800-ft. level has reached 100 ft. above the level. Drift No. 2 at the West End continues on a face of low-grade quartz. The output during the past week was 859 tons.——At the MacNamara Mining Co. the west cross-cut No. 700 advanced 45 ft. without change. The production the past week was 504 tons.——The Jim Butler Tonopah Mining Co. at the Desert Queen shaft exposed 4 ft. of medium-grade ore in raise No. 653, which was started from the intermediate below the 500-ft. level. At the Wandering Boy raise No. 372 on the 200-ft. level made a connection, proving a large body of good ore. Raise No. 373 at present shows a 3½-ft. face of medium-grade ore. The production during the past week was 936 tons.

The Montana produced 338 tons, the Halifax 55 tons, the Midway 15 tons, and miscellaneous 41 tons, making the week's production at Tonopah 7655 tons with a gross value of $133,962. As the Tonopah Mining Co., with an average of 2750 tons per week, gave no tonnage report for the week, the total production at Tonopah is not given in full.——The silver vein recently uncovered by the Tonopah Divide Mining Co. at Gold Mountain, shows considerable molybdenum minerals.

CALIFORNIA

ALAMEDA COUNTY

On December 12 Judge Henshaw, with Justice Shaw, Sloss, Melvin, and Angellotti of the Supreme Court of California concurring, upheld the decision of the lower court appointing a receiver to take charge of and operate the mines of the Western Magnesite Co. The receiver was originally appointed because of dissension among the directors of the company, preventing the operation of the magnesite mines. Now that the Supreme Court has confirmed the appointment of L. E. Boyle as receiver, operations at the mines will be resumed. The property of the Western Magnesite Co. is situated at Livermore and is said to contain an extensive deposit of excellent-grade magnesite.

CALAVERAS COUNTY

(Special Correspondence.)—The Stockton Ridge Consolidated Mining Co., which controls the Hexter gravel claims, has definitely shut-down. There is a 3800-ft. adit on the property, and it will take some time to remove the rails and dismantle the pumps. Everything is for sale, including the mill. When work ceased there was a large body of gravel in sight; the high cost of powder and labor was the cause of shutting-down. H. S. Chapman, the superintendent, connected at one time with the Greenwater boom, is leaving for Chile, South America, shortly. The company has spent in the neighborhood of $400,000.——The Buffalo gravel mine, situated in Chili gulch, has shut-down after a mill-test of several tons of gravel. The company erected a plant, including transformers, electric hoists, and Deister tables, before ascertaining what was in the ground.——A slight accident to two men in the Crazy Bird mine took place recently, otherwise the mine is running smoothly. The orebodies are proving themselves with depth. This is the only mine in the district offering work to miners.——The Fischer quartz mine has erected a head-frame, and is unwatering and re-timbering the shaft. San Francisco capital, headed by Benjamin F. Lyle, is interested.——A 30-ton mill-run has been made at the Buffalo mill from the Chili gulch claims owned by George Edwards and Frank Bernardi. The return gave a $12 recovery on the plate and one and a half tons of concentrate that assayed $1700 per ton.——The Maypole quartz mine, which is temporarily shut-down, has a 6-ft. vein at the face of the adit, which is in 365 ft. Samples have been taken, going as high as $68.72 per ton. The future of the district is in the quartz, and there are a lot of good prospects and partly developed mines awaiting the magic touch of capital to become steady producers.

Mokelumne Hill, December 17.

DEL NORTE COUNTY

(Special Correspondence.)—At the different mines within 30 miles of Crescent City, the water-shipping point, 5000 tons of high-grade chrome ore has been developed and 1000 tons mined and on the dump ready for delivery. The mines are situated 4½ to 13½ miles from the present wagon-road, but a good road is being constructed to connect them with Crescent City. These mines are owned and operated by George S. Barton of Grants Pass, Oregon.

Crescent City, December 18.

ELDORADO COUNTY

(Special Correspondence.)—Six miles south of Placerville, where Squaw Hollow creek empties into Martinez creek, just south of the Martinez gold mine, the Pacific Gold Mining Co., of South Dakota, is running a cross-cut west from Martinez creek 1000 ft. to cut the main Mother Lode of California at a vertical depth of 520 ft. The cross-cut is being driven at the rate of 200 ft. per month, by night and day shifts working with air-drills. It is now in more than 600 ft. and will be completed about the middle of February 1918. A 30-ft. vein of free-milling gold quartz, averaging $1.26 gold per ton, has already been cut. The work is in charge of Anthony Crafton.

Placerville, December 11.

SHASTA COUNTY

(Special Correspondence.)—The new 5-stamp mill of the Shasta Hills Mining Co. on the Sybil and Accident mines west of French gulch was started this week. E. Hillman is superintendent.——The Crescent Copper Co. has filed its articles of incorporation. The company has mines near Keswick. John E. Baker is general manager.

Redding, December 10.

(Special Correspondence.)—The Victor Power & Mining Co. at Harrison gulch has unwatered the Midas mine to the 600-ft. level.——The Shasta Hills Mining Co. has started up a new 5-stamp mill on the Accident & Sybil mine, four miles west of French Gulch.——Shipments of ore from the Bully Hill mine to the smelter at Kennett have been increased to three carloads per day.——The Yankee John mine, four miles west of Redding, is sinking its shaft 100 ft. deeper.——The Mammoth Copper Co. at Kennett, being unable to get prompt delivery of mining timbers by railroad from the mills up the canyon and in Oregon, is now having material hauled by auto-trucks from Shingle Shanty, eight miles from French Gulch. The haul is 41 miles.

Redding, December 16.

SIERRA COUNTY

(Special Correspondence.)—The Sierra-Alaska Mining Co., operating at Pike City, has arranged for a survey for a lower adit site on Oregon creek. The adit will be driven to cut the vein at a depth of over 500 ft. below the bottom of the shaft, and will be a mile long. Following its completion the shaft will be also deepened. The adit is designed to drain the property besides becoming the main working level. The vein in the shaft is stated to be yielding excellent ore.——Reese Bros. are developing the Mexican group, near Downieville, under a bond and purchase option. The 10 stamps formerly operated at the Hayes Consolidated have been moved to the Mexican mill, which at present contains five stamps; 15 men are employed and some underground work is proceeding, although the manager is centring work on surface improvements. The Mexican is owned by the Yorke Bros. of Downieville.——H. N. Yates of Pacific Grove and D. C. Smith of Meridian have completed an examination of the Irelan mine, near Alleghany, in which they are heavily interested. The new mill is running steadily on ore of good grade; recent developments have been encouraging.——The Hilda Mining Co. has completed a road from near Milton to its property, and is freighting lumber and supplies with two motor trucks. An excellent camp is being rushed to completion and, if the buildings can be finished before deep snow sets in, the property will be worked all the winter.——Rich gravel continues to be mined from the new deposit in the Hilo mine, above Downieville.——The Loftus Blue Lead Gravel Co. has everything in readiness for hydraulic mining along Slate creek. The company is also developing large areas of virgin territory in the vicinity of St. Louis, Howland Flat, and other old camps. The water supply for the giants has been greatly improved.

Downieville, December 16.

SISKIYOU COUNTY

(Special Correspondence.)—As a result of the closing of Klamath river the past weeks, prospectors had an opportunity to find the gold in the river bed and high-bed points. While no sensational strikes were reported, it is believed the miners were given a chance to find the gold deposits for wing-daming next year.——The placer-mine operators throughout the county are making special plans to save the platinum in gold-bearing gravel during the coming season.——H. J. Barton has leased his Beaver Creek placer mine to Owens, Baldwin & McAllister of Oak Bar, who are operating with a large crew.——The Mercury company recently purchased the Cowgill quicksilver mine, at Cinnabar Springs, and has completed a retorting plant operated by electricity.——Florin and Charles Le May are shipping ore, running 50% chromic oxide, from their Lime gulch mine.——Henry Muskgrave of Oak Bar is opening a large deposit of chrome in the Greenhorn district.——J. W. Dyer and Van Fleet of Yreka are developing the Dyer chrome prop rty below Hamburg on the Klamath river.——Fred Myers and Barney Nelson of Sawyers Bar in a recent clean-up on their Jackson Creek placer mine recovered a $200 gold nugget. ——The Grey Eagle Copper Mining Co., at Happy Camp, is operating with 50 men.——William Might, of Happy Camp, while at Yreka recently, exhibited $1000 worth of gold nuggets, which he cleaned up during the last three months in his Humbug Creek placer mine.——John Boyle, of Yreka, who is operating the R. S. Taylor quartz mine in that district, reports a rich strike of $1000 gold ore.

Hornbrook, December 7.

COLORADO

CLEER CREEK COUNTY

A snow-slide down the sheer side of Red mountain, near the Urad mines, carried away 20,000 ft. of timbers that had been framed for the new concentrator at the Molybdenum mines. The mill-site is safe from snow-slides, but the lumber had been unloaded near the site in a spot where slides often occur.

SAN MIGUEL COUNTY

Tonopah Belmont people have taken over the Alta mines in this county and have placed J. M. Fox in control as manager.

TELLER COUNTY

Production from the properties of the Granite Gold Mining Co., the Dillon, Granite, Dead Pine, Monument, and Gold Coin mines on Battle mountain, for November, amounted to 145 broad-gauge cars, or 4300 tons, which was the heaviest in the life of the company.——The ore averaged $21 per ton; it was shipped to the Golden Cycle M. & R. Co.'s plant at Colorado Springs. The company paid its regular bi-monthly dividend of 1c. amounting to $16,500, on December 10. The sum of $49,000 has been paid in dividends this year. The report of A. E. Carlton, president of the Cresson Consolidated Gold Mining & Milling Co., accompanying the December dividend checks, announced a new discovery of ore, carrying a gold content of 32.5 oz. or $650 per ton, in the raise between the 1600 and 1500-ft. levels. The ore shipped in November yielded a net return of $129,971 above operating cost and the company has cash in bank of $1,260,000, and 16 cars of ore in transit, of an estimated value of $16,000.——The November production from properties of the Elkton Consolidated Mining & Milling Co., on Raven hill, operated by lessees totaled 425 tons from the main shaft, with a bullion value of $8500. In addition, lessees on Tornado mine of the company shipped two cars of about $30 grade.——A five-year lease has been issued to the Patterson-Bradley Leasing Co., H. G. Gones, general manager, on the Specimen and Sacramento mines, Bull Hill, by Stratton's Cripple Creek Mining & Development Co. The new company is adding an electric hoist and compressor.—— The Clyde shaft, of the Millasler Mining Co., on Battle mountain, has reached a depth of 1300 ft. Charles Fish, the superintendent, is now cutting stations at the 1100 and 1300-ft. levels and will then continue sinking to the 1500-ft. point.—— A new surface discovery has been made on the New Boston, a Stratton Estate property on Womack hill, by Dworak and Green, lessees. The first shipment was sent out last week and is estimated at $20 grade.——The annual stockholders meeting of the Free Coinage Gold Mining Co. was held at South Altman, on Bull hill, on December 15, when the following directors and officers were re-elected: J. B. Neville, Denver, president and general manager; Charles Cavendar, Leadville, vice-president and treasurer; and A. S. Neville, Denver, secretary.

IDAHO

LATAH COUNTY

(Special Correspondence.)—A promising strike of copper ore has been made recently at the Mispah mine, the property of the Merger-Mining Co., of Palouse. The mine has been worked for a number of years and some 2000 ft. of development work has been accomplished; during the last two years, however, work has been confined to that required to secure the patent. Last spring Ernest Northrop, one of the principal owners, found some good float between the two adits the company had driven, and a little prospecting uncovered a body of carbonate ore. An adit was started, and has been driven for 50 ft. on a vein ranging from 5 to 15 ft. A second adit was started to cut the vein at a lower level, and this has cut a body of massive chalcopyrite and pyrrhotite in a gangue of micaceous schist. A trial car of the oxidized ore gave a return of 17% copper.

Moscow, December 22.

MONTANA

SILVER BOW COUNTY

The copper production of the Butte district is at normal or better for the first time since the beginning of the I. W. W. labor troubles in June last, according to a statement issued by the local office of the Anaconda Copper Mining Co., which showed a production of ore of 15,042 tons December 14, while the normal is estimated at 14,300 tons daily, without the ores of the North Butte company, which approximate 1700 tons per day. The production of zinc ore is below normal, but this is chiefly because the Anaconda has a five months stock of zinc concentrate at its electrolytic plant at Great Falls, awaiting reduction, in addition to a large stock of crude ore at the Washoe zinc concentrate, which it is calculated will keep the plant busy for some time. The spelter market is at a low ebb and under the prevailing mart the Anaconda has found it unprofitable to pay custom shippers 1½c. per pound for zinc and has reduced its settlement schedule to ½c. per pound. The Anaconda also has found it necessary to shut off receiving further zinc ores from the Coeur d'Alene and has effected a drastic curtailment in its own output of zinc ores from the Butte district.

NEVADA

MINERAL COUNTY

(Special Correspondence.)—Several of the largest mines at Candelaria, including the Chief of the Hills, Northern Belle, Mt. Diablo, and Holmes, have been taken under option by

Charles E. Knox, of the Montana-Tonopah Co., who is reported to be acting for Eastern capital. Large quantities of ore are in the old workings, and some rich shipments have been made recently by lessees. The prospective buyers intend to erect a modern milling plant.——On the 250-ft. level of the Nevada Rand, in the Rand district, No. 2 vein has been cut and shows a wide face of $130 ore; 16 tons is ready for shipment to the Hazen sampler. Ore shipped from the 180-ft. level of this vein carried $77 in gold and silver. Preparations are being made to open the vein on the 350-ft. level, and later to seek it on the 450. No. 1 vein has been cut on the 150, 180, and 250-ft. levels, and 35,000 tons, averaging $15 per ton, is blocked out. Charles Huber, of Tonopah, is president; W. V. Ruderow, Reno, secretary and representative of Eastern interests; S. E. Montgomery, Reno, consulting engineer.——Tests on ore from the Walker River copper group, near Mina, have proved satisfactory in the 20-ton experimental leaching-plant, and it is planned to erect a 100-ton unit in the spring. R. F. Banks is metallurgist, and Joseph Gelder manager. A new compressor has been added to the equipment, and development at the mine is proceeding steadily.——The Empire Nevada, under management of Joseph Gelder, is shipping 40 to 50 tons of first-grade ore daily. The bulk of this is from the Ross, Hart & Hair lease.

Mina, December 15.

STOREY COUNTY

(Special Correspondence.)—Two shipments of bullion, aggregating $40,000 to $45,000, have been made from the Mexican mill, representing a week's run on ore from the Union Consolidated. Splendid ore continues to develop on the 2300 and 2400-ft. levels, particularly in the southerly workings from the latter. New stopes are being opened and preparations made for prospecting new ground.——From the 2300-ft. level of the Mexican an easterly cross-cut has been started, to prospect territory adjacent to the rich area in the Union Consolidated.——The Sierra Nevada is repairing old workings on the 2500-ft. level and expects soon to reach a point where good ore was exposed 35 years ago.——Driving to the west from the 2700-ft. level of the Consolidated Virginia is progressing.

Silver City is showing activity, and several old properties are worked by lessees. In the Silver Hill, Robohm & Windisch is working several blocks of ground and keeping the Donovan mill supplied with ore running high in silver and gold.——Alex. Armstrong and L. S. Cain are opening a good ore-shoot in the Hayward, and Charles Silva and Harry Cardeu have taken a lease on the Lager Beer.——The Golden is being worked by Simon Conwell and associates, and Hickey & Hardwick are extracting good ore from the Oest group. The Mt. Bullion, Occidental, and other properties are active. With silver at present price much ore in old workings is being profitably extracted.

Virginia City, December 17.

OREGON

JOSEPHINE COUNTY

(Special Correspondence.) — Concentrating machinery is being erected at the Golconda chrome mine. The mine has been producing steadily during the summer.——The Oak Flat mine, leased by the California Chrome Co., has developed 2000 tons of ore, which will be ready for shipping next spring.——G. S. Barton has 400 tons of chrome ore ready for shipment at his mine 14 miles from Grants Pass.——In this county there are a number of deposits, isolated by lack of roads, which would be opened if chrome prices attained a higher level.

Grants Pass, December 17.

SOUTH DAKOTA

FALL RIVER COUNTY

Drilling for oil has commenced in the Ardmore field and the Hat Creek company has its well down over 500 ft. The Shiloh company and the DesMoines company are drilling also. It is expected that the oil-sands will be reached at a depth of about 2000 ft. and the presence of oil will be determined soon after the first of the year. At the Shiloh well the gas zone has been passed. Other lands have been leased in the same district and additional rigs will be placed in commission next year.

Lead, December 7.

LAWRENCE COUNTY

In the issue of December 22 it was stated in error that the Homestake had declared a special dividend of $1 per share, so that the total dividends paid by the company in 1917 amounted to $7.80 not $8.80, as stated, and the total distribution to stockholders to $1,959,048, not $2,210,208.

TEXAS

BROWN COUNTY

(Special Correspondence.)—A nine-foot orebody assaying 48% lead, 6% silver, and 4½% copper has been discovered upon the farms of Lee Baker, six miles west of Bangs. The ore is found at a depth of 90 ft. The property has been leased by Mr. Baker to the Knox County Copper Co., and a shaft is being sunk.

Bangs, December 17.

EL PASO COUNTY

(Special Correspondence.)—Charles Davis of El Paso has purchased at a receiver's sale the tin mine, situated in the Mt. Franklin range, 14 miles from El Paso, the consideration being $21,000. Mr. Davis was one of the original stockholders of the company that owned the property. He plans to resume development. This mine was opened in 1883 and more than $250,000 has been expended in the development of the orebody and the erection of a mill. The property comprises about 3500 acres. Shipments of tin were made from time to time, but on account of a disagreement among the stockholders the company was thrown into the hands of a receiver. An abundant supply of water for all purposes is available, there being three springs upon the property.

El Paso, December 15.

CANADA

BRITISH COLUMBIA

F. A. Starkey, a mining broker, announced on December 11 that one of the largest mining deals in British Columbia has practically been completed, the Utica Mining Co. negotiating with C. F. Caldwell for the purchase of the Sunset and Bell mines for $70,000 cash and $800,000 in stock in the new company.

Flotation will be used by the Canada Copper Co. in concentrating the ore of the Copper Mountain mine, for which it is building a concentrator of 3000 tons daily capacity. The concentrate will be nodulized in rotary kilns. Experiments have shown that the fine flotation concentrates can be made suitable for smelting in this way.

The Granby Consolidated Mining & Smelting Power Co. has ordered two large converters for its smelting plant at Anyox, which are expected to increase the capacity by 20%. The production of the Anyox smelter will be 30,000,000 lb. of copper this year. The additional equipment should increase this by 6,000,000 lb. and raise the total to 36,000,000 on the basis of the 1917 production. The production of 1917, however, is not quite representative, being 5,000,000 lb. less than in 1916. The rest of the plant has sufficient capacity to meet the demands of the new converters. It is believed that capacities in other departments are to be increased, as the cost of the converters would not equal the appropriations the Granby company is reported to have made for this smelter in 1917. The dimensions of the new converters are 20 by 20 ft., as compared with 12 by 12 ft., the size of three converters in use. The capacity of a new converter will be more than three times that of the three of less size. The erection of the first converter will be completed by February 1, it is believed, and the second soon after. One will be used continuously and the other will be held in reserve for emergencies. Anyox is receiving about 75,000 tons per month, of which 10,000 is from its lesser mines and its patrons and the remainder from Hidden Creek. Preparations are proceeding for the erection of a coking plant on Graves Point, Anyox. The ground has been graded for 30 by-product ovens. The coal will be removed by barges from the Vancouver island mine of the Granby company. It is estimated that the cost of the deposit and its development and of the coking plant and transportation equipment will be $2,500,000.

Mining men are noticing with much concern that there seems no likelihood of the Trail smelter being re-opened. The plant was closed a few weeks ago owing to a strike of the 1600 employees who demanded an all around eight-hour day. Only about 350 men would have been affected by the change. Apparently all negotiations for settlement have broken off. The fires in the furnaces have been drawn.

Production of zinc in British Columbia is steadily becoming more important. The value of zinc produced in 1909 was only $400,000. For 1910 it was $192,473, and for 1911, $129,092. Thereafter there was gradual improvement up to $346,125 for 1914. Then came a comparatively big increase, to $1,460,524 for 1915, and a still bigger advance in 1916, to $4,043,984. The total for zinc to the end of 1916 was $7,212,759.

The Consolidated Mining & Smelting Co. of Canada has declerad its usual quarterly dividend of 2½%. Payment will be made January 2 to stockholders of record on December 15. The aggregate of the disbursement will be $210,695. This will increase the total payments to $3,794,121. Payments this year aggregate $842,780. The declaration of this dividend at a time when the smelter at Trail, B. C., and some of the company's mines are idle created little surprise in mining circles in view of the stability and long standing of the corporation. There is enough lead on hand at Trail to supply the average demand for three months.

Yukon

The dredge of the Yukon Gold Co., which is a Guggenheim organization, has given employment to a large force of men for many months under the supervision of A. D. Hughes; it began elevating gravel on December 6. The try-out proved the big machine to be in perfect working order and it is now in steady operation, working two shifts of nine hours. The dredge was assembled a mile below Murray and was launched a month ago; it was used by the company in Alaska. It is of almost solid steel construction and has more the appearance of a battleship than a machine to separate gold from alluvial gravel. The dredge is headed down stream and after working a certain distance in that direction will be turned and worked up stream. The Yukon Gold Co. arranged with the Washington Water Power Co. to extend its transmission line to furnish electric power to drive the big machine. This necessitated the construction of 12 miles of new line, much of it over high mountains, across deep canyons, and through heavy timber. This line will prove of great benefit, as it makes electric power available for other mining operations along the route. The Yukon Gold Co. has acquired all the placer ground along Prichard creek, extending from the mouth of the creek to the source, a distance of 14 miles. Last year the ground was systematically prospected with Keystone drills for a considerable distance below and above Murray, and as a result the engineers of the company estimated the gold to be recovered from the ground prospected at approximately $2,000,000.

A rich strike in copper ore has been made on Williams creek, near Yukon Crossing, by Dawson men. A tunnel has been run into the hill 180 ft. and splendid ore has been found all the way. The strike is on one of the group of ten copper claims on Williams and Merrit creeks.

PERSONAL

Note: The Editor invites members of the profession to send particulars of their work and appointments. This information is interesting to our readers.

Morton Webber is at Butte.

B. F. Beckwith, of Anaconda, is here.

J. C. Hopper is at Silverton, Colorado.

C. F. Sherwood, of Salt Lake City, is here.

George A. Farish has returned from Salvador.

Rensselaer H. Toll is at Berkeley for the holidays.

H. L. Mead has returned to New York from Alaska.

J. P. Norris, of Rome, New York, is at Hamilton, Ontario.

J. K. Turner was here from Goldfield, Nevada, this week.

Robert H. Bedford has gone to Cuba on a mine examination.

Walter Karri-Davies has returned to Berkeley from New York.

D. C. Jackling has returned to San Francisco from Washington.

Nelson Dickerman has returned to San Francisco from Salt Lake City.

J. Roebling Jarvis, of Idaho City, Idaho, has gone to Belmar, New Jersey.

J. C. Kennedy, of Manhattan, Nevada, spent Christmas in San Francisco.

D. D. Muir is manager for mines for the U. S. Smelting Co. at Miami, Oklahoma.

William G. Reynolds is Captain of infantry (U. S. R.) at Fort Sheridan, Illinois.

W. H. Seamon Jr. has joined the 27th Engineers and is at Camp Meade, Maryland.

E. M. Rogers passed through San Francisco on his way from New York to Los Angeles.

E. J. Jennings, formerly foreman of the Chino leaching-plant, has joined the Navy.

Hugh F. Marriott is president elect of the Institution of Mining and Metallurgy, London.

Rush M. Hess is at St. Louis, on his way to Birmingham, Alabama, from Bouse, Arizona.

Roscoe H. Channing has obtained a commission as Captain in the Field Artillery, U. S. Army.

Richard Blackstone has tendered his resignation to the Homestake Mining Co., to become effective on January 1.

Frederick Lyon, vice-president of the U. S. Smelting, Refining & Mining Co., has been in California on a tour of inspection.

Victor C. Alderson, president of the Colorado School of Mines, was here before Christmas on his return from Los Angeles to Golden.

Brace C. Yates, assistant superintendent for the Homestake Mining Co., has been appointed superintendent in succession to Mr. Blackstone.

L. Douglas Anderson, superintendent of the Midvale smelter, has been appointed smelter manager for the United States Smelting Co. at Salt Lake City.

L. F. S. Holland, until recently field engineer for the Consolidated Arizona Smelting Co., has been appointed consulting engineer to the Arizona Mines & Reduction Co., with headquarters at Wickenburg, Arizona.

An unsigned letter from Washington, dated December 5, on 'Flotation Physics' has been received. The Editor desires to know the name of the writer.

Book Reviews

THE COMPOSITION OF TECHNICAL PAPERS. By Homer Andrew Watt, assistant professor of English in New York University. 16mo. Pp. 431. McGraw-Hill Book Co., Inc., New York. Price, $1.50 net.

This is an elementary book for students and engineers without university training. A little more than a third of the volume is devoted to examples of technical writing. Among the exemplars we recognize John Carty, Charles Going, and the greatest of all scientific expositors, Thomas Henry Huxley. Messrs. Carty and Going are to be felicitated on being in such company. Mr. Watt, like most writers on the subject, illustrates unintentionally the errors against which he preaches. For example, in the preface he says: "The division between the liberal and the professional aims is not, in the case of instruction in English composition, sharp, since any engineering student who receives one type of instruction will receive with it much of the value contained in the other type." This is true, but it is badly expressed. We suggest: "The division between the liberal and the professional aims of instruction in English composition is not sharp, since any engineering student that is helped by one kind of instruction will receive benefit from the other kind." It is well to place modifying words or phrases close to the words that they modify; it is also well to avoid the use of such words as 'case', 'instance', 'nature', and 'character'. See Quiller-Couch 'On Jargon' in 'The Art of Writing'. Immediately after the sentence quoted above we find the statement: "Every engineering student should certainly have a general course in English composition." Surely; and when he has done so he will place his adverb where it is most significant; he will write: "Certainly every engineering student should have a general course in English composition." A few lines lower we find: "But the bulk of whatever writing they will do *will* almost certainly *be* of a professional kind, and a specific course in the composition of technical papers *will*, therefore, also *be* of undoubted service." The disjointed portions of verbs are indicated by italic. Avoid the splitting of verbs. We suggest that the conjunction 'but' is continuative and therefore should not begin a sentence. 'Certainly' is an adverb that does not lend itself to qualification. 'Specific' has taken the place of 'special'. We venture to submit an alternative phrasing: "The larger part of their writing is likely to be done in the exercise of their profession, therefore a special course in the composition of technical papers is sure to be of service." So much for the preface. Chapter I begins with a quotation from "a well-known New York engineer," who is reported to have said: "An engineer who is inarticulate is quite as useless as one who is professionally incompetent." The assertion is nonsense. A competent engineer does not become incompetent because he is inarticulate; it is only his ability to transmit information that suffers. An incompetent engineer becomes pestiferous when he is articulate. Such statements are detrimental to interest in a subject having an importance that needs no exaggeration. As Mr. Watt himself says: "The man who has not the power to give his ideas clearly to others, the man whose thoughts are locked in his brain simply because he has not the ability to communicate them, suffers under a handicap which no amount of professional knowledge can possibly overcome." We believe that to be true and for that reason we rejoice to see the publication of books intended to arouse the engineering profession to an intelligent interest in technical writing. This book may serve that purpose.

T. A. R.

THE LIFE OF ROBERT HARE, AN AMERICAN CHEMIST. By Edgar Fahs Smith. Pp. 508. Ill. Index. Oct. de luxe. J. B. Lippincott Co., Philadelphia, 1917.

If one were to name the fathers of American chemistry, he would necessarily mention Hare and Silliman. In some way Silliman caught the public eye more than Robert Hare. It would be foolish to attempt to give precedence to either. They were bound to each other by ties of friendship such as seldom exist between men, so that in large measure the work of one was the work of the other, a building up through inter-communication of two of the most progressive scientific intellects of the age. Hare was a Philadelphian, and it is interesting to note that the man of native genius and the stimulus with which to develop him appeared together. Philadelphia was distinctly the centre of scientific culture and progress in the early days of the Republic. It was in Philadelphia that the oldest chemical society in the world, namely, the Chemical Society of Philadelphia, was founded in 1792. It was to Philadelphia that Franklin went to make his home, and where many of his notable scientific achievements were accomplished. Again, it was to Philadelphia that Joseph Priestly, fleeing from persecution in England, came to continue the work of his choice. Here Robert Hare was born in 1781. He was connected with a chemical industry, which, however, was not as acutely chemical in those days as it is in our own, his father being the owner of a brewery. It must be noted, however, that his father, an Englishman by birth, who came to America in 1773, was prominent in the public life of the day, and was no common artisan, as were many brewers of that period. He was a member of the Constitutional Convention of Pennsylvania, serving the State as senator, and he was also one of the trustees of the University of Pennsylvania. It was with this institution that his son's great life-work is associated in the public mind.

The achievements of Robert Hare are traced in this book with the appreciation of a biographer who is a master in the science that Hare did so much to establish, and he is, furthermore, a worthy successor in the chair which Hare occupied in the same university. The most sensational discovery which Hare made was that of the oxy-hydrogen blow-pipe, by which he was enabled to fuse platinum, which was then invested with a certain sanctity because of the inability of man to fuse it in the past. But the invention of the oxy-hydrogen blowpipe was of vastly more consequence in its demonstration of the real nature of combustion, a phenomenon that previously had not been understood. The account of the activities of Robert Hare from the period of this remarkable discovery in 1801 to his death in 1858 involves passing in review the higher development of physics and the great revelations of the chemical properties and relations of matter from the epoch of the misunderstandings of the alchemists, whose views of matter was just being supplanted at the time when Hare began his work. There is consequently no better way to trace the development of science through the first half of the nineteenth century than by a perusal of this work. Hare was equally distinguished in his directive influence upon the development of electrical knowledge. He devised the first electric furnace, which, however, he called a deflagrator, in which he converted charcoal into graphite, thus anticipating Acheson by about 70 years. Also, with this apparatus, he produced calcium carbide, elemental phosphorus, and calcium. It may be said that if Hare had known how to bring his remarkable discoveries to the attention of the world, or had cared to do so in the way that is now common among scientists, the advancement of practical science would have been much more rapid. Some of his most notable experiments were demonstrated before small groups of appreciative scientific associates, and then forgotten, so that their importance, from the standpoint of general advancement in industry, was overlooked. Ira Remsen, in 1915, called attention to the remarkable fact that work which he was pursuing and considered new at that period, he was astonished to find had been investigated by Robert Hare as early as 1821, and that similar conclusions had been reached at that time by this extraordinary pioneer in chemistry. The book is so full of valuable information that it will prove a great addition to any chemist's library.

NOTES ON MILITARY EXPLOSIVES. By Erasmus M. Weaver. 8vo. Pp. 382. Index. Fourth edition, revised and enlarged. John Wiley & Sons, Inc., New York, 1917. For sale by MINING AND SCIENTIFIC PRESS. Price, $3.25.

This authoritative treatise on a subject of pronounced importance and of great interest at this moment is enriched by an important discussion on 'The Rôle of Chemistry in the War', by Allerton S. Cushman, director of Industrial Research, Washington, D. C., in which is given an illuminating discussion of the basic chemical action of nitrogen, carbon, hydrogen, and oxygen in all explosives. It may be said, for the benefit of those who have not seen the book before, that General Weaver, who is chief of Coast Artillery in the U. S. Army, has given an introductory chapter on the principle of chemistry, which is so simple that any man of general education would be prepared for an undertaking of what follows, even though he had not devoted a large amount of attention to chemical subjects. He then discusses the various types of explosives, service-tests of explosives, and proper methods of storage and handling. As a convenient handbook, it is a most commendable treatise.

EXAMPLES IN BATTERY-ENGINEERING. By F. E. Austin. Pp. 90. Index. Published by the author. Hanover, N. H., 1917. For sale by MINING AND SCIENTIFIC PRESS. Price, $1.25.

This is another one of the practical little books on electricity for which Professor Austin has already acquired a wide reputation. He has the faculty of giving a clear statement to problems, so that they may be readily comprehended by students who are working their own way. Professor Austin points out that battery-engineering offers an opportunity to develop and to store energy where at present it is allowed to go to waste. For example, he calls attention to the need of gasoline, which is employed chiefly as a generator of power for automobiles. His suggestion is that storage batteries should be employed more generally, and that it would be possible to utilize the wasted energy of the winds by windmills for storing up electric power, which could be drawn upon as needed. He says "Two windmills each 25 ft. diam. properly arranged, would, in many parts of the world, furnish sufficient power for the needs of a fair-sized farm, and the necessary power for transporting farm products to market. * * * Another source of power may be the sun's rays acting upon a thermo-pile that may charge storage cells while the sun is shining. These could then supply low-pressure high-efficiency incandescent lamps during the hours of darkness." It is a suggestive and interesting little book.

MACHINE DRAWING. By Ralph W. Hills. Pp. 92. Ill. Index. McGraw-Hill Book Co., Inc., New York. For sale by MINING AND SCIENTIFIC PRESS. Price, $1.

This valuable little book was prepared for use in the Extension Division of the University of Wisconsin, and is the sort of thing that a student obliged to work his way alone will welcome. It covers the essentials of machine-design in its more simple aspects, and gives practical instruction in draughting and in the production of acceptable high-grade drawings and tracings.

REVIEW OF THE GEOLOGY OF TEXAS. By J. A. Udden, C. L. Baker, and Emil Böse. Pp. 164. Inserts and maps. Bulletin of the University of Texas No. 44.

The awakening interest in the mineral resources of Texas makes the re-appearance of this bulletin timely. It gives the geological history, with a special chapter on the economic minerals, aided by small maps as well as by a large colored map of the State, on which are valuable geologic columns.

Recent Decisions

MINER'S LIEN—'LABOR' DEFINED

'Labor' within a miner's lien statute means physical labor unequivocally performed on the property. Failure of a mining lien in foreclosure proceedings does not necessarily involve the validity of the defendant's indebtness to claimants.

Stuart *v.* Camp Carson Mining & Power Co. (Oregon), 165 Pacific, 359. June 6, 1917.

CO-TENANT OF MINERAL GROUND—INJUNCTION DENIED

The taking of minerals from land will not be enjoined at the suit of a party claiming only an undivided one-sixth interest in the land where the injunction might work great injustice upon defendant in its mining operations and there does not appear in the bill grounds for such a drastic proceeding. The remedy by injunction in favor of a co-tenant is to be sparingly exercised.

Wilmer *v.* Philadelphia & Reading Coal & Iron Co. (Maryland), 101 Atlantic, 538. June 26, 1917.

OIL AND GAS LEASE—FORFEITURE—CONSTRUCTION

An oil and gas lease which binds the lessor but is optional with respect to the lessee will be construed most strongly in favor of the party bound and against the party not bound. After a lessor has declared an oil lease forfeit and leased to another, the first lessee cannot come in and set up the failure of the second lessee to carry out his contract, in an attempt to re-establish his own right to possession. The lessor alone may avail himself of forfeiture clauses.

Bearman *v.* Dux Oil & Gas Co. (Oklahoma), 166 Pacific, 199. July 10, 1917.

MINING CONTRACT—HELD MERE LICENSE

Where, under the contract between the owner of a placer claim and another, that other was to take possession of the mine, furnish machinery and material to work it, and, at the end of the first summer's work, was to have the option of continuing operations on a royalty basis, or in lieu thereof, to be paid the difference between the cost of his plant and operations and the value of the gold obtained by him, it was held that the contract was neither a conveyance nor a lease, but a mere license.

In re Seward Dredging Co. (Alaska), 242 Federal, 225.

OIL WITHDRAWAL OF 1909—APPLICATION CONSTRUED

The right of oil locators to retain their locations after the petroleum land withdrawal of 1909 depends upon the bona fides of the location and the diligence with which discovery work is prosecuted upon each location taken up. It is not enough that a lessee of an entire section prosecuted discovery work on one-quarter thereof only, with a view to drilling on the remaining quarter sections only if petroleum was found in paying quantities on the first. The withdrawal was effective as to all those sections upon which work was not actually in progress. On the other hand, delay in the work caused by the necessity of first obtaining an adequate water supply for drilling purposes, which latter work was in progress, is excusable and does not work a forfeiture.

United States *v.* Brookshire Oil Co. (California), 242 Federal, 718.

United States *v.* North American Oil Consol. (California), 242 Federal, 723.

United States *v.* Thirty-two Oil Co. (California), 242 Federal, 730.

United States *v.* Record Oil Co. (California), 242 Federal, 746.

METAL PRICES

San Francisco, December 24

Aluminum-dust (100-lb. lots), per pound	$1.00
Aluminum-dust (ton lots), per pound	$0.95
Antimony, cents per pound	17.00
Antimony (wholesale), cents per pound	15.25
Electrolytic copper, cents per pound	23.50
Pig-lead, cents per pound	6.50— 7.50
Platinum, soft and hard metal, respectively, per ounce	$105— 113
Quicksilver, per flask of 75 lb	115
Spelter, cents per pound	9.50
Zinc-dust, cents per pound	20

The antimony market rules quiet. Business on the spot has been done at 18c, though no business of importance has been reported. Needle antimony is quoted at 9¼ to 9½c, while antimony ore is not offered at any price.

ORE PRICES

San Francisco, December 24

Antimony, 45% metal, per unit	$1.00
Chrome, 34 to 40%, free SiO_2, limit 8%, f.o.b. California, per unit, according to grade	$0.60— 0.70
Chrome, 40% and over	$0.70— 0.85
Magnesite, crude, per ton	$8.00—10.00

Manganese: The Eastern manganese market continues fairly strong with $1 per unit Mn quoted on the basis of 48% material.

Tungsten, 60% WO_3, per unit	26.00
Molybdenite, per unit MoS_2	$40.00—45.00

EASTERN METAL MARKET

(By wire from New York)

December 24.—Copper is unchanged at 23.50c. all week. Lead is quiet and steady at 6.40 to 6.50c. Zinc is inactive at 7.75c. all week. Platinum is unchanged at $105 for soft metal and $113 for hard.

COPPER

Prices of electrolytic in New York, in cents per pound.

Date		Average week ending	
Dec. 19	23.50	Nov. 13	23.50
" 20	23.50	" 20	23.50
" 21	23.50	" 27	23.50
" 22	23.50	Dec. 4	23.50
" 23 Sunday		" 11	23.50
" 24	23.50	" 18	23.50
" 25 Holiday		" 25	23.50

Monthly Averages

	1915	1916	1917		1915	1916	1917
Jan.	13.60	24.30	29.53	July	19.09	25.66	29.67
Feb.	14.38	26.62	34.57	Aug.	17.27	27.03	27.42
Mch.	14.80	26.65	36.00	Sept.	17.69	28.28	25.11
Apr.	16.64	28.02	33.16	Oct.	17.90	28.50	23.50
May	18.71	29.02	31.69	Nov.	18.88	31.95	23.50
June	19.75	27.47	32.57	Dec.	20.67	32.89	....

The Anaconda Copper Mining Co is estimated to have a net income of $30,000,000 for 1917 after all charges and taxes. This would be $13 per share. Such earnings compare with the net of $50,828,373, or $21.81 per share, in 1916. But last year was an exceptional one. There were no labor troubles. Operating costs were normal. Output was unhampered. The price of copper averaged more than 27c. per pound. This year Anaconda, like other mining companies, has had to face serious labor troubles since June. Through July and the greater part of August the properties were operated at about one-third of their capacity. The company's operation in Montana practically ceased August 25. Later the situation changed for the better, and the properties are now operating at 92% of capacity. Meanwhile abnormal costs of operation have prevailed and the peak has not yet been reached. The copper price of 23½c. fixed by the Government has obtained since October. It is estimated that Federal taxes will call for the payment of about $4,700,000 by Anaconda. Still Anaconda will in all probability return net income within $9 per share of the 1916 record-breaking profits of $21.81 per share.

The few hundred holders of Phelps-Dodge Corporation stock will have received $32 per share in dividends during 1917, comparing with $32.50 per share last year. Production was materially affected by the shut-down due to strike at the Detroit mine for over four months and the part curtailment of Copper Queen operations early last summer. Despite this interruption production as a whole will probably exceed that of the previous year.

Half of the Inspiration Copper Co.'s mill is at work. Ten of the 22 sections are now active. Others will follow as sufficient labor becomes available. The mine lacks underground men, a dearth of experienced miners being noted since the late strike; in the mill department the company has enough men. Scouts have been sent to various labor centres in search of men.

SILVER

Below are given the average New York quotations, in cents per ounce, of fine silver

Date		Average week ending	
Dec. 19	85.78	Nov. 13	86.18
" 20	86.37	" 20	85.62
" 21	86.62	" 27	84.60
" 22	86.62	Dec. 4	84.70
" 23 Sunday		" 11	85.70
" 24	86.62	" 18	85.38
" 25 Holiday		" 25	86.42

Monthly Averages

	1915	1916	1917		1915	1916	1917
Jan.	48.85	56.76	75.14	July	47.52	63.06	78.92
Feb.	48.45	56.74	77.54	Aug.	47.11	66.07	85.40
Mch.	50.61	57.89	74.13	Sept.	48.77	68.51	100.73
Apr.	50.25	64.37	72.51	Oct.	49.40	67.86	87.38
May	49.87	74.27	74.61	Nov.	51.88	71.60	85.97
June	49.03	65.04	76.44	Dec.	55.34	75.70	....

ZINC

Zinc is quoted as spelter, standard Western brands, New York delivery, in cents per pound.

Date		Average week ending	
Dec. 19	7.75	Nov. 13	7.81
" 20	7.75	" 20	7.98
" 21	7.75	" 27	8.00
" 22	7.75	Dec. 4	8.00
" 23 Sunday		" 11	7.91
" 24	7.75	" 18	7.75
" 25 Holiday		" 25	7.75

Monthly Averages

	1915	1916	1917		1915	1916	1917
Jan.	6.30	18.21	9.75	July	20.54	9.90	8.98
Feb.	9.05	19.99	10.45	Aug.	14.17	9.03	8.58
Mch.	8.40	18.40	10.78	Sept.	14.14	9.18	8.33
Apr.	9.78	18.62	10.20	Oct.	14.05	9.92	8.32
May	17.03	16.01	9.41	Nov.	17.20	11.81	7.76
June	22.20	12.85	9.63	Dec.	16.75	11.26	....

LEAD

Lead is quoted in cents per pound, New York delivery.

Date		Average week ending	
Dec. 19	6.40	Nov. 13	6.35
" 20	6.40	" 20	6.50
" 21	6.50	" 27	6.50
" 22	6.50	Dec. 4	6.50
" 23 Sunday		" 11	6.50
" 24	6.50	" 18	6.50
" 25 Holiday		" 25	6.46

Monthly Averages

	1915	1916	1917		1915	1916	1917
Jan.	3.73	5.95	7.64	July	5.59	6.40	10.93
Feb.	3.83	6.23	9.01	Aug.	4.62	6.28	10.75
Mch.	4.04	7.26	10.07	Sept.	4.62	6.86	9.07
Apr.	4.21	7.70	9.38	Oct.	4.62	7.02	6.97
May	4.24	7.38	10.29	Nov.	5.15	7.07	6.38
June	5.75	6.88	11.74	Dec.	5.34	7.55	...

The lead market shows a slight advance for spot delivery, as shipments are slow in arriving on account of the congested railroads. The trust price remains at 6.25c., but there are no sellers for spot lead under 7c., with buyers at 6.87½ cents.

QUICKSILVER

The primary market for quicksilver is San Francisco, California being the largest producer. The price is fixed in the open market, according to quantity. Prices, in dollars per flask of 75 pounds:

Week ending

Date			
Nov. 27	110.00	Dec. 11	115.00
Dec. 4	115.00	" 18	115.00
		" 25	115.00

Monthly Averages

	1915	1916	1917		1915	1916	1917
Jan.	51.90	222.00	81.00	July	95.00	81.20	102.00
Feb.	60.00	295.00	126.25	Aug.	93.75	74.50	115.00
Mch.	78	219.00	113.75	Sept.	91.00	75.00	112.00
Apr.	77	141.60	114.50	Oct.	92.90	78.20	102.00
May	75	90.00	104.00	Nov.	101.50	79.50	102.50
June	90	74.70	85.50	Dec.	123.00	80.00	...

TIN

Prices in New York, in cents per pound.

Monthly Averages

	1915	1916	1917		1915	1916	1917
Jan.	34.40	41.76	44.10	July	37.38	38.37	62.60
Feb.	37.23	42.60	51.47	Aug.	34.37	38.88	62.53
Mch.	48.76	50.50	54.27	Sept.	33.12	36.66	61.54
Apr.	48.25	51.49	55.63	Oct.	33.00	41.10	62.24
May	39.28	49.10	63.21	Nov.	39.50	44.12	74.18
June	40.26	42.07	61.93	Dec.	38.71	42.55	...

Tin continues scarce and strong. The quotation is only nominal as practically no Straits tin is to be had. For spot 90c. is quoted, with tin for delivery in a month's time offered at prices ranging from 72 to 75c. per pound.

ALUMINUM

There is no change in condition or demand. No. 1 virgin metal, 98 to 99% pure, is quoted at 36 to 38c. per lb., New York, for early delivery.

ORES

Tungsten: These ores are now to come under Government regulation so far as their importation is concerned, and the obtaining of licenses and their distribution is to be regulated by the American Iron and Steel Institute, the same as in the case of tin. Considerable business has been done in the past week at $22 to $26 per unit in 60% concentrates, depending on the grade of ore. Scheelite has again sold at $20 per unit. Ferro-tungsten is quoted at $2.40 to $2.50 per lb. of contained tungsten.

Molybdenum and antimony: There is more activity reported than in some time in molybdenite at $2.25 to $2.30 per lb. of MoS_2 in 90½ concentrates. No antimony ore has been offered.

Eastern Metal Market

New York, December 19.

Lack of activity in all quarters characterize the markets, and less business is reported than for a long time.

Copper is unchanged at the controlled prices at which a fair business is reported.

Tin was never so scarce and is nominal for nearly all grades for early delivery.

Lead is quiet, and a little easier, and price fixing has been officially abandoned.

Zinc continues lifeless and featureless but unchanged.

Antimony is quiet and a little lower.

In the steel markets production has fallen off considerably, due to lack of fuel, poor transportation facilities, and other causes, such as the severe storms. Work has stopped on Russian locomotives, and many of those already built have not been shipped and may be used on domestic roads. The most interesting general news is that arrangements have been made whereby the American Iron and Steel Institute, representing the War Industries Board, will regulate the issuing of import licenses for ferro-alloys and the ores from which they are made, including principally chrome ore, cobalt, ferro-manganese, manganese ore, molybdenum ore, titanium ore, tungsten ore, and vanadium ore. The object of this arrangement is to secure better co-operation in obtaining imports.

COPPER

It is generally believed now by the trade that there will be no change in the official price of copper for at least three months, although some seem to have desired an advance from 23.50c. per lb. to 25c. Official information regarding the market in general is hard to obtain, but the foregoing is generally accepted as a fact. Sales are being made to the general trade, even outside of Government work, at 23.50c. per lb., for any deliveries up to April 1, 1918. This is to large buyers. In the jobbing trade there has been considerable business done at the fixed price of 24.67½c. per lb. From one source it is estimated that production of refined copper in 1917 will exceed the total of 1916. For the first 10 months of this year the smelting output is placed at 1,645,187,271 lb., or at the rate of 164,518,727 lb. per month. At this rate it will amount to 1,975,000,000 lb. for the year, which compares with 1,927,850,548 lb. for 1916, according to the data of the U. S. Geological Survey. It seems probable that there will be sufficient metal for all consumers, as it is figured that war-needs do not require more than 75% of the output. Sales and distribution are being made under the direction of the Copper Producers' Committee and the War Trade Board to the apparent satisfaction of the entire trade. There has been no change in the London market in several weeks. Quotations there yesterday were £125 per ton for spot electrolytic, and £121 for futures.

TIN

In our letter last week the fact was referred to that urgent representation was being made that all import restrictions in the shape of licenses, and the like, be abandoned as superfluous. This movement has failed, and on December 12 the obtaining of import licenses became necessary in order to bring in any metal. Confusion since has been general, due to the fact that it has been difficult to obtain licenses, or else because they have been much delayed. This is illustrated by the fact that tin has commenced to burden the docks since the custom houses are not permitted to accept metal until the importer has obtained his licenses. It is hoped that this serious defect in the smooth operation of trade soon will be remedied. It is attended also with additional expenses to importers, as these cannot be saddled on to the buyer. There has been no trouble or delay with the arrangement in regard to guarantees as to purchases to be furnished by the American Iron and Steel Institute. The selling of Banca tin by a large consumer, referred to last week, has continued on a small scale to needy users. Inquiries for prompt shipment from England have continued, but there have been very few transactions because of the little metal available. As a whole, the market during the past week has been dull and uninteresting. There have been no offerings of spot Straits, which is quoted nominal at 85 to 86c. per lb., New York. Arrivals to December 16, inclusive, have been 1015 tons, with 4400 tons reported afloat. Spot Straits in London continues to advance, the quotation yesterday having been £6 higher than a week ago, or £305 per ton.

LEAD

For the present there will be no fixing of lead prices, according to the following statement given out on Tuesday after a meeting of lead producers and the War Industries Board: "The principal lead producers of this country at the invitation of the War Industries Board met with the Board this morning in Washington to discuss the advisability of putting restrictions upon the price and distribution of lead. After a protracted conference, and thorough consideration of the facts brought out in conference, the War Industries Board does not consider that regulatory action is at this time necessary or advisable." This decision comes as entirely unexpected, because a maximum price was regarded as probable. It is generally interpreted as meaning that there is ample lead for all needs and a probability that fair prices will continue. Domestic business the past week has been light, and the tone has been easier, with sales at 6.40c., New York, or 6.25c., St. Louis. The quotation of the leading producer is unchanged at 6.25c., New York. There has been an improvement in export inquiry. It is understood that negotiations are on as to a further supply for the Government.

ZINC

It is the same old story in the zinc market, with no prospect of any resuscitation. While there has been some buying recently by galvanizers, it has not been large because production of this commodity is at a low mark. Demand from this source, however, has been a little better recently. Nor have brass-makers been over-eager to purchase zinc of any grade. Quotations are low at 7.50c., St. Louis, or 7.75c., New York, for prime Western for early delivery, and the only fact that keeps them from going lower is that they are at or below the cost of production in many cases. Production is gradually lessening, but is still in excess of the demand. Considerable interest attaches to the opening of bids this week, Friday, by the Government on 1000 tons of grade 'C' zinc, which is only a little better than regular prime Western. Reports are persistent to the effect that railroad congestion is seriously interfering with the filling of contracts, especially for export. Price-fixing of zinc is not expected now because of the official decision not to regulate the lead market as announced on Tuesday, December 18.

ANTIMONY

Chinese and Japanese grades are quoted at 15c. per lb., duty paid, New York, for early deliveries, but demand is not large, and the market is quiet and featureless.